ASTRONOMICAL DATA

1 Light Year (L.Y.)	$= 9.461 \times 10^{15}$ m
1 Astronomical Unit (A.U.) (Earth-Sun distance)	$= 1.496 \times 10^{11}$ m
Earth-Moon distance	$= 3.844 \times 10^{8}$ m
Radius of Sun	$= 6.960 \times 10^{8}$ m
Radius of Earth	$= 6.378 \times 10^{6}$ m
Radius of Moon	$= 1.738 \times 10^{6}$ m
Mass of Sun	$= 1.989 \times 10^{30}$ kg
Mass of Earth	$= 5.974 \times 10^{24}$ kg
Mass of Moon	$= 7.35 \ \times 10^{22}$ kg

CONVERSION FACTORS

1 in.	= 2.54 cm (exactly)
1 yd	= 0.9144 m (exactly)
1 mi	= 1.6093 km
1 mi/h	= 0.4470 m/s
1 lb	= 0.45359237 kg (exactly)
1 lb (force)	= 4.44822 N
1 u	$= 1.6605 \times 10^{-27}$ kg
1 Cal	= 4186 J
1 hp	= 745.7 W

VECTOR RELATIONS

$$\mathbf{A} \cdot \mathbf{B} = |\mathbf{A}|\,|\mathbf{B}|\cos\theta_{AB}$$

$$|\mathbf{A} \times \mathbf{B}| = |\mathbf{A}|\,|\mathbf{B}|\sin\theta_{AB}$$

$$\mathbf{A} \times \mathbf{B} = \begin{vmatrix} \hat{\mathbf{i}} & \hat{\mathbf{j}} & \hat{\mathbf{k}} \\ A_x & A_y & A_z \\ B_x & B_y & B_z \end{vmatrix}$$

$$\mathbf{A} \times (\mathbf{B} \times \mathbf{C}) = (\mathbf{A} \cdot \mathbf{C})\mathbf{B} - (\mathbf{A} \cdot \mathbf{B})\mathbf{C}$$

$$(\mathbf{A} \times \mathbf{B}) \cdot \mathbf{C} = (\mathbf{C} \times \mathbf{A}) \cdot \mathbf{B} = (\mathbf{B} \times \mathbf{C}) \cdot \mathbf{A}$$

Parts 1 & 2 Combined

PHYSICS

For Science and Engineering

Jerry B. Marion
William F. Hornyak

University of Maryland

SAUNDERS GOLDEN SUNBURST SERIES

SAUNDERS COLLEGE PUBLISHING

Philadelphia   New York   Chicago
San Francisco   Montreal   Toronto
London   Sydney   Tokyo   Mexico City
Rio de Janeiro   Madrid

Address orders to:
383 Madison Avenue
New York, NY   10017

Address editorial correspondence to:
West Washington Square
Philadelphia, PA   19105

This book was set in Times Roman by York Graphic Services, Inc.
The editors were John Vondeling, Jay Freedman, and Janis Moore.
The art and design director was Richard L. Moore.
The text design was done by Phoenix Studio, Inc.
The cover design was done by Richard L. Moore.
The artwork was drawn by ANCO/Boston.
The production manager was Tom O'Connor.

Front cover: *Mathematical Models* by M.C. Escher. © BEELDRECHT,
          Amsterdam/VAGA, New York 1981. Collection Haags Gemeentemuseum.

PHYSICS FOR SCIENCE AND ENGINEERING, Parts 1 & 2 combined   ISBN 0-03-062831-8

234   032   987654321

CBS COLLEGE PUBLISHING
Saunders College Publishing
Holt, Rinehart and Winston
The Dryden Press

PREFACE

This is the combined version of a two-volume text for the standard introductory physics course for students of science and engineering. It is intended for courses requiring calculus either as a prerequisite or as a corequisite, and may be used in two- and three-semester sequences. Part 1 presents mechanics (Chapters 1–15), elasticity and fluids (Chapters 16–18), waves (Chapters 19 and 20), relativity (Chapter 21), and thermodynamics (Chapters 22–26). Part 2 covers electromagnetism (Chapters 27–39), optics (Chapters 40–45), and quantum physics (Chapters 46 and 47). These topics include more detail than can be conveniently covered in the time usually available. Such completeness does, however, allow the instructor many options in the design of a course. A number of these possibilities are listed in the accompanying Instructor's Manual.

We believe that theoretical generalizations are best understood by students after many simple special cases have been introduced. Our approach is therefore to present simple applications as we build toward the generalized formulations. The unity of the physics involved in many separate phenomena and summarized in such a generalization is then more forcefully demonstrated, and is accepted by the student almost as a necessary final step. The reverse procedure, of presenting the general case followed by special applications, has been found in our classroom experience to perplex many average students; if they fail to comprehend the general, they often find the particular mysterious.

Calculus Appendices. In our opinion and that of many of our colleagues, the greatest single obstacle in teaching introductory physics is the wide range of mathematical background and computational proficiency of the students. One semester of calculus as a prerequisite is helpful, but not always possible in the busy engineering and science curricula. Even with the prerequisite, the skills in simple differentiation and integration provided by the usual calculus course are seldom sufficient for solving problems in physics. Further, it may be desirable to explain phenomena in terms of techniques not presented in the first semester of calculus, including multiple integration (even if only masquerading as a series of single integrations), partial derivatives, line integrals, and elementary differential equations. Teachers often resort to handing out extra material covering these topics. Many students react with uninterest and suspicion: "If it's important that I know this material, why isn't it in the book we're using or covered in the math course we took?"

Here we offer a flexible approach that allows the instructor to address the problem of inadequate mathematical background. All of the usable calculus is presented in the form of seven special appendices. Each one appears in the text just before significant use of the techniques might be desired. To emphasize their supplementary nature, these appendices are shaded and set in a double-column format. The appendices DO NOT include any development of physics. Some or all of them may be assigned as regular reading or left simply as reference material.

We realize that, even with the aid of the mathematical appendices, the student's facility in using the calculus will be limited at best. Thus, the physics chapters introduce the use of calculus slowly. Particularly in the early chapters, special cases are first treated without calculus; it is then demonstrated that deeper insight requires the introduction of calculus. Experience has shown that such deliberate repetition provides a secure mental bridge for the student between the more familiar treatment and the new techniques. Finally, more general cases are presented that would be awkward or impossible to treat without calculus.

We have made a great effort to be precise and realistic in our presentation. The approximations necessary for a first-order treatment of any phenomenon are clearly spelled out. We have opted for an informal style that at the same time is concise and crisp and has a no-nonsense character. Precise terminology is introduced only where appropriate, anticipating its use in the student's later intermediate and advanced courses, but jargon is never used for its own sake. We have also sprinkled the text liberally with thought-provoking questions addressed to the student, which examine further implications of the current topic.

Special Topics. Since our approach in most cases follows the historical order of development, brief historical references are included where appropriate and useful. This ordering of topics is consistent with the fact that most chapters build in physical abstraction (and to some extent in mathematical complexity). The sequence also allows for sensible shortening of the required reading, depending on the instructor's emphasis and time constraints. With this in mind, we have included many "extended footnotes" (in smaller type and set off by triangle symbols) and Special Topic sections (generally at the end of a chapter, and distinguished by shading and double-column format). For the enjoyment and enlightenment of the student, a number of applications are given to topics of current interest. These include sound reproduction (hi-fi), rocketry, semiconductors, lasers and holography, and radiating systems (both electromagnetic and sound). Any or all of these may be omitted without loss of continuity in later chapters.

A number of the Special Topic sections involve practical and engineering applications. The intent is to bridge the gap between fundamental physics and detailed engineering application. A review of popular engineering texts has convinced us of the need to set the stage in this way; too often, such knowledge is blandly assumed to be part of the student's background. In addition, this may be the only exposure to these topics for many science majors. Under severe time constraints, of course, the Special Topics may be omitted.

Modern Physics. Although concepts from modern physics are freely introduced in the text, this is done only where truly necessary to avoid gross misconceptions. We do not interject examples from modern physics at every opportunity. For instance, conservation of energy and momentum are illustrated with objects that are at least laboratory-size. While it is true that low-energy elastic scattering at the subatomic level (e.g., proton-proton scattering) offers ideal examples, they raise too many extraneous questions about the particles themselves. Such questions are interesting but blur the focus of the example. Modern physics for its own sake is treated in separate chapters.

Examples and Problems. Strong emphasis is placed on problem-solving ability. Indeed, the ability to solve problems is viewed as the best proof of student understanding of the text material. To achieve this aim, a large number of worked-out examples are presented. The examples span all levels of difficulty, and in each instance situations are selected that emphasize the text presentation. An effort has been made to relate the examples to one another to reveal different aspects of a given physical system.

We have included many problems at the end of each chapter, grouped together by section number. Moderately difficult problems are marked with a dot (•) and more difficult ones with two dots (••). Usually there is also a set called Additional Problems, which draws together concepts developed in several sections and may include a few problems of greater difficulty. The answers to approximately half of the numerical problems appear in the back of the book. We use SI units (sometimes referred to as metric or MKSA units) throughout the text, and conversions to other familiar units are mentioned when appropriate.

Acknowledgements.   A preliminary edition of this text was class-tested in our introductory physics courses. We are grateful to the University of Maryland for the opportunity to try our ideas and presentation in the classroom, and to the students who provided many thoughtful suggestions for improvement.

We have also had the benefit of a great many reviews during the development and redrafting of the manuscript. Among those who have lent their expertise and encouragement are

Albert Altman
    University of Lowell
Angelo Bardasis
    University of Maryland
S. M. Bhagat
    University of Maryland
George Bowen
    Iowa State University
Alex Burr
    New Mexico State University
Thomas Cahill
    University of California, Davis
James Finley
    University of New Mexico
O. W. Greenberg
    University of Maryland
Thomas Griffy
    University of Texas, Austin
Donald Holcomb
    Cornell University
H. T. Hudson
    University of Houston
David King
    University of Tennessee, Knoxville
Carl Kocher
    Oregon State University
Victor Korenman
    University of Maryland
Andrew Kowalik
    Nassau Community College
Calman Levich
    Central Michigan University
Glenn Liming
    Mississippi State University

Eugene Loh
    University of Utah
David Markowitz
    University of Connecticut
John Marshall
    Louisiana State University
Col. Richard Minnix
    Virginia Military Institute
Francis Prosser
    University of Kansas
Arthur Quinton
    University of Massachusetts,
        Amherst
Andres Rodriguez
    College of the Pacific
Lance Rogers
    City College of San Francisco
Philip Roos
    University of Maryland
Stanley Shepherd
    Pennsylvania State University
Carl Tomizuka
    University of Arizona
Donald Treacy
    U.S. Naval Academy
William Van Wyngaarden
    California State Polytechnic
        University
Marcellus Wiedenbeck
    University of Michigan
Gary Williams
    University of California,
        Los Angeles

We would like especially to thank our colleague Angelo Bardasis, who not only reviewed the manuscript but also prepared the accompanying Study Guide. His close scrutiny of the manuscript and proofs has contributed greatly to the accuracy of the book.

We are particularly indebted to Jay Freedman for his many valuable suggestions and editorial assistance. Our typists, Theresa Bryant and Eleanor Fisher, were indispensable in preparing the manuscript.

Jerry B. Marion

William F. Hornyak

CONTENTS

*Sections marked as "Special Topics" can be omitted without loss of continuity.

       Problems   1134

Chapter 43   DIFFRACTION   1138

43–1   Fraunhofer Diffraction   1138
43–2   The Diffraction Grating   1147
43–3   X-Ray Diffraction   1153
43–4   SPECIAL TOPIC   Fresnel Diffraction   1158
       Problems   1160

Chapter 44   POLARIZED LIGHT   1163

44–1   States of Polarization   1163
44–2   Polarization by Reflection and Scattering   1166
44–3   Polarization by Selective Absorption—Dichroism   1169
44–4   The Law of Malus   1170
44–5   Birefringence   1173
       Problems   1179

Chapter 45   OPTICAL INSTRUMENTS   1182

45–1   Thick Lenses and Lens Combinations   1182
45–2   Simple Magnifiers   1189
45–3   Microscopes   1190
45–4   Telescopes   1196
45–5   Cameras   1199
45–6   SPECIAL TOPIC   The Eye and Vision   1204
       Problems   1208

Chapter 46   QUANTA, WAVES, AND PARTICLES   1212

46–1   Thermal Radiation   1212
46–2   The Photoelectric Effect   1220
46–3   The Compton Effect   1225
46–4   The Wave-Particle Duality   1228
46–5   The Uncertainty Principle   1234
46–6   SPECIAL TOPIC   Free-Particle Wave Functions   1239
       Problems   1243

Chapter 47   ATOMS   1247

47–1   Rutherford's Nuclear Atom   1247
47–2   The Bohr Model of the Hydrogen Atom   1249
47–3   SPECIAL TOPIC   Elliptical Orbits   1257
47–4   Model Refinements   1258
47–5   Multielectron Atoms   1263
47–6   Laser Radiation   1269
       Problems   1272

       Appendix A: Basic Algebra and Trigonometry   A.1
       Appendix B: Complex Numbers   B.1
       Answers to Selected Odd-Numbered Problems   C.1
       Index   I.1

INTRODUCTION

The word *physics* is derived from the Greek word φύσις, meaning "nature." Indeed, physics is a branch of natural science that deals with the properties and interactions of matter and radiation. Because physics is concerned with such fundamental ideas as these, it is generally considered to be the most basic of all the sciences.

The key to progress in the understanding of Nature is to base conclusions on the results of *observations* and *experiments,* analyzed by *logic* and *reason.* This basic approach to science is called the *scientific method.* This "method" is not a prescription for learning scientific truth; instead, the scientific method represents a philosophy of discovery that emphasizes the importance of *measurement* when dealing with problems of the real physical world. Theories are valuable (indeed, indispensable) in organizing the facts that have been gathered about the behavior of Nature. But at its roots, physics is an experimental science. The ultimate answer to any question concerning natural phenomena must be the result of experimental measurement.

In order to describe the natural universe, we utilize *concepts, theories, models,* and *laws.* Generally, a *theory* attempts to explain why Nature behaves as it does. Paradoxically, to construct a theory, we introduce certain *unexplained* fundamental abstractions or *concepts.* Thus, we consider the concepts of *energy, time, space,* and *electric charge* as "given," without offering an explanation for their existence. (Even so, we can still provide precise *definitions* for these concepts viewed as quantities.) The theory then asserts a connection between these concepts and some observed characteristics of interactions of matter. This connection is achieved by constructing a *model* to reflect the experimentally determined facts. Finally, the deductions from these models result in the *laws* of physics, which tell us *how* things behave in terms of the theory.

How are theories to be judged? If there are several contending theories relating to the same set of experimental facts, how do we decide which to adopt? There are three criteria for answering such questions—*predictive power, comprehensiveness,* and *simplicity.*

A theory, if it is to have merit, must be able to predict observable results of experiments as yet unperformed. When the experiments *are* performed the results must agree with the predictions of the theory within acceptable limits. Also, a greater degree of credence is associated with a theory that can relate in a comprehensive way to a wide variety of phenomena. Finally, we have a faith that Nature is inherently simple, so we are led to believe that a valid theory should be transparent, direct, and simple, with an economy of postulates and *ad hoc* assumptions. However, simplicity must not be equated with ease of comprehension. The theory of relativity is a model of simplicity and logical precision, based on only two postulates. But to comprehend fully the implications of this theory is a formidable task.

What is accomplished by constructing a theory? What does a theory really explain? The best answer is probably that our theories provide a point of view of Nature that permits us to assemble in a comprehensible form the essence of a variety of related observations concerning physical phenomena. Theories provide a means of reducing an enormous number of experimental results to a manageable number of precise statements. Progress consists of refining these theories and discovering links among them as new observational information becomes available and as new insights are developed. No theory is perfect, and none is all-encompassing. But we are confident in our expectation that our theories will provide ever better descriptions of Nature through continual evolution.

1-1 UNITS AND STANDARDS OF MEASURE

To describe physical quantities and processes in a precise and orderly way, it is essential that a system of measure first be established. The units for quantities such as length, time, and mass must be defined, and standards for these units must be provided so we can agree on the meaning of measurements. Having established a system of units, a measurement of some observable can then be quoted as a *numerical value* together with the appropriate *unit* for that quantity. We stress the necessity for developing the habit of *always* quoting physical quantities in terms of a numerical value and the *relevant unit of measure*. It makes no sense, for example, to quote a length as "60," for there is a considerable difference whether the units are centimeters or miles!

Most scientific measurements and most of the world's commerce are carried out in *metric* units. Engineering practice in this country uses the metric system more and more, and there is even a slow conversion to metric usage in the public sector. Within a matter of years the United States may join the rest of the industrialized world in the exclusive use of the metric system.

The metric unit of length is the *meter* (m), originally conceived as 10^{-7} of the

The conversion to metric units is taking place in stadiums around the country. (Courtesy of the Philadelphia Phillies. Photo by Paul H. Roedig.)

A krypton lamp used as a length standard at the U.S. National Bureau of Standards.

The krypton lamp is cumbersome in actual practice, so the operational standard for routine use is the less complicated stabilized iodine laser. (Both photos courtesy U.S. National Bureau of Standards.)

distance from the equator to the North Pole along a meridian passing through Paris. This ingenious but impractical standard was replaced, in 1889, by the distance between two finely drawn, parallel scribe marks on a bar of platinum-iridium. Copies of this *standard meter bar* were distributed from the central depository near Paris to standards laboratories throughout the world. Even in the classical Greco-Roman world, it was necessary to disperse stone tablets throughout the empire to standardize the length of the Roman foot.

To provide a truly universal standard for the meter, in 1961 an international agreement was reached to base the definition of the meter on an atomic radiation. Because all atoms of a particular type are identical, such a definition allows every laboratory to prepare a standard for the meter in exactly the same way. Accordingly, the meter is now defined to be 1,650,763.73 wavelengths of a particular orange radiation from krypton atoms (see the photographs on this page).

The range of distances encountered in the Universe is indicated in Fig. 1–1.

An atomic standard for *time* has also been established. Until recently, the *second* (s) was defined to be 1/86,400 of the mean solar day. The small but perceptible changes in the speed of the Earth's rotation render this definition of the second inadequate for precise experiments. In 1967, it was decided to define 1 s to be the time required for 9,192,631,770 complete vibrations of cesium atoms. (•What are some repetitive phenomena that are used in more common time-keeping devices?)

The range of time intervals found in the Universe is shown in Fig. 1–2.

The unit of mass in the metric system—the *kilogram* (kg)—was originally defined to be the mass of one liter (1 ℓ = 10^3 cm^3 = 10^{-3} m^3) of water. This definition has been refined by establishing a standard in the form of a platinum-iridium cylinder. The primary standard kilogram is maintained in the International Bureau of Weights and Measures, near Paris, and precision copies are available in most countries. It would, of course, be desirable to have an atomic standard for mass, just as we have for length and time. Although it is possible to compare the mass of one atomic

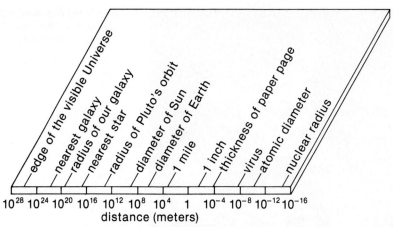

Fig. 1–1. Range of distances found in the Universe. Notice that the scale is *logarithmic,* with each scale division representing a factor of 10^4.

species with that of another to extremely high precision, we do not yet have a sufficiently precise way to determine the mass of the standard kilogram in terms of the mass of an atom. (The precision is, at best, a few parts per million.) When technology has progressed to the point that we can make such comparisons with high precision, an atomic standard for mass will probably be adopted.

The range of mass values in the Universe is indicated in Fig. 1–3.

The units of all mechanical quantities can be expressed in terms of the basic units of length, time, and mass by using various products of these units. That is, the units of such quantities as energy, momentum, and force can all be expressed by combinations of meters, seconds, and kilograms. We say that these units are *derived* from the basic units. For example, speed (or velocity) is expressed in terms of a derived unit, namely, meters per second, m/s.

In some cases, for convenience, a special name is given to the derived unit for a quantity. The metric unit of force is the kg $\cdot$ m/s^2 (Section 5–2), and we call this unit the *newton* (N). But this does not alter the fact that only three basic units—those of length, time, and mass—are all that are required to describe any mechanical quantity.

Fig. 1–2. Range of time intervals found in the Universe. The various periods indicated are representative only.

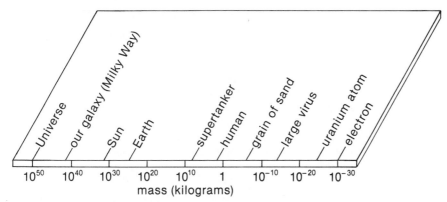

Fig. 1-3. Range of mass values found in the Universe.

Another type of derived unit is one that is constructed from some measured physical quantity and cast in a form that provides a useful and convenient unit. When describing astronomical objects and events, the relevant distances, expressed in meters, are very large numbers indeed. For example, the distance to Alpha Centauri (a member of the star grouping nearest the Earth) is 4.07×10^{16} m. To deal with such enormous distances, it is convenient to define a new length unit equal to the distance that light will travel (at a speed of 2.998×10^8 m/s) during 1 year (3.156×10^7 s). This unit is called the *light-year:* 1 L.Y. $= 9.46 \times 10^{15}$ m. In terms of this unit, the distance to Alpha Centauri is 4.3 L.Y. A similar unit, useful for distance measurements in the solar system, is the *astronomical unit* (A.U.), defined to be the distance from the Earth to the Sun, 1.50×10^{11} m.

Systems of Measure. The primary advantage of the metric system is that it is a *decimal* system. That is, the common multiples and subunits of the primary units are related by factors of 10.* Such units as the *centimeter* (10^{-2} m), the *kilometer* (10^3 m), and the *milligram* (10^{-3} g $= 10^{-6}$ kg) are often used. Sometimes a new term is employed instead of the standard prefix. For example, 10^3 kg is referred to as a *metric ton* instead of a megagram. The standard prefixes and the corresponding powers of 10 are listed in Table 1-1.

In this book we use the particular metric units and symbols that have been prescribed by the commission on "Le Système International d'Unités." We refer to these as *SI units.* A summary of the SI units used in this book is given inside the rear cover.

A few parts of the world (most notably, the United States) have not yet converted to the metric system, and still retain the British system of units. Sometimes it is necessary to convert from one of these systems to the other. In both systems, *time* is measured in seconds. The relationship connecting units of *length* can be expressed in several ways:

$$1 \text{ in.} = 2.54 \text{ cm (exactly)}$$
$$1 \text{ yd} = 0.9144 \text{ m (exactly)}$$
$$1 \text{ mi} = 1.609 \text{ km}$$

The situation with regard to the unit of *mass* is more confusing. To give a

---

*The only exception is for time intervals longer than a second. Entrenched habit requires us to use minutes, hours, days, etc., for such multiples. However, subsecond units are always decimal—we use milliseconds, microseconds, nanoseconds, etc.

TABLE 1–1. Prefixes and Powers of 10

MULTIPLE	PREFIX	SYMBOL
10^{12}	tera-	T†
10^{9}	giga-	G†
10^{6}	mega-	M
10^{3}	kilo-	k
10^{2}	hecto-	h
10	deka-	da
10^{-1}	deci-	d
10^{-2}	centi-	c
10^{-3}	milli-	m
10^{-6}	micro-	μ
10^{-9}	nano-	n
10^{-12}	pico-	p
10^{-15}	femto-	f
10^{-18}	atto-	a

†In the United States, 1 *billion* means 10^{9}, but in Europe it means 10^{12}; the meaning of T and G are unambiguous, so the term *billion* should be avoided.

complete discussion requires us to distinguish between "mass" and "weight," a topic considered in Chapter 5. Here we only point out that an object with a weight of 1 pound (lb) has a mass of exactly 0.45359237 kg. Note also that the *metric ton* (or *tonne,* 1000 kg) is equal to 2204.6 lb (weight), which is approximately the same as the British long ton, or 2240 lb.

Dealing with Units. We say that a physical quantity such as distance has *dimensions* of *length* and is measured in *units* of *meters;* that is, if x represents distance,* $[x] = L$. Every equation expresses an equality between quantities that have the same dimensions. Moreover, if any equation or formula involves a sum, each term in the sum must have the same dimensions. There are *dimensionless* quantities (numbers) such as π, $\sqrt{2}$, and sin 30° that also may appear in equations. Multiplying a quantity with dimensionality by a dimensionless number does not change the dimensionality.

Units and dimensions obey the familiar arithmetic of numbers. For example, suppose that we wish to write an equation that states, "distance is equal to speed multiplied by time." If distance is represented by x, speed by v, and time by t, we have $v \times t = x$. Dimensionally,

$$[v] \times [t] = [x]$$

Speed is measured in m/s, so the dimensionality of v is length per unit time, or LT^{-1}. Thus,

$$LT^{-1} \times T = L$$

Because $T^{-1} \times T = 1$, both sides of the equation have dimensions of length. When the quantities are expressed with units, we have

$$\left(20 \frac{m}{s}\right) \times (4\ s) = 80 \frac{m}{s} \times s = 80\ m$$

---

*A square bracket is used to denote the dimensionality of a quantity; thus, $[x] = L$ is read as "the dimensionality of x is length L."

The unit "s" appears both in the numerator and in the denominator of the result; cancellation leaves the final answer expressed in *meters,* as expected.

It is sometimes necessary to convert the value of a quantity expressed in one set of units to the equivalent value in a different set of units. To do this, we take any needed relationship of the form,

$$1 \text{ mi} = 5280 \text{ ft}$$

and write it as a unity ratio:

$$\frac{5280 \text{ ft}}{1 \text{ mi}} = 1$$

Then, we can multiply or divide any term in the expression to be converted by unity factors of this type. By properly choosing these factors, a value can be converted from one set of units to another. Notice that this procedure does not change the dimensionality of the quantity.

For example, to convert 40 mi/h to m/s, we write

$$40 \text{ mi/h} = \left(40 \frac{\text{mi}}{\text{h}}\right) \times \left(\frac{1.609 \text{ km}}{1 \text{ mi}}\right) \times \left(\frac{10^3 \text{ m}}{1 \text{ km}}\right) \times \left(\frac{1 \text{ h}}{3600 \text{ s}}\right)$$

$$= 17.88 \text{ m/s}$$

Notice how miles, hours, and kilometers cancel to leave the answer expressed in units of m/s.

In this book, when writing equations, we usually supply the units for each quantity. Thus, we write

$$x = vt = (30 \text{ m/s})(5 \text{ s}) = 150 \text{ m}$$

However, if the equation is lengthy, the explicit inclusion of the units may not be convenient. Then, we write

$$x = vt = (30)(5) = 150 \text{ m}$$

In these cases, it is implied that all values are expressed in the appropriate SI units. The units of the final answer are given explicitly.

1-2 DENSITY

One of the important physical properties that distinguishes one type of matter from another is *density*. Each sample of a particular substance, under identical physical conditions, has a mass that is directly proportional to its volume. Therefore, an important characteristic of a substance is the ratio of its mass to its volume—this is the density ρ:

$$\rho = \frac{M}{V} \tag{1-1}$$

In the metric system, density is measured in units of kg/m^3. However, densities are often expressed in units of g/cm^3. Some useful densities are listed in Table 1-2.

TABLE 1–2. Densities of Some Common Substances*

SUBSTANCE	DENSITY kg/m³	DENSITY g/cm³
Gold (Au)	1.93×10^4	19.3
Mercury (Hg)	1.36×10^4	13.6
Lead (Pb)	1.13×10^4	11.3
Copper (Cu)	8.93×10^3	8.93
Iron (Fe)	7.86×10^3	7.86
Aluminum (Al)	2.70×10^3	2.70
Water (H_2O)	1.00×10^3	1.00
Ice (H_2O)	0.92×10^3	0.92
Alcohol (C_2H_5OH)	8.06×10^2	0.806
Air ($N_2 + O_2$) ⎫ normal	1.293	1.293×10^{-3}
Helium (He) ⎬ pressure	0.1786	1.786×10^{-4}
Hydrogen (H_2) ⎭	0.08994	8.994×10^{-5}

*Near 0°C.

PROBLEMS

Section 1-1

1-1 In the discussion of unit conversion, a calculation was made to convert mi/h to m/s. Make this calculation again, using 1 in. = 2.54 cm (and other factors, as necessary) instead of 1 mi = 1.609 km.

1-2 When conversion to the metric system is complete in this country, what will the highway speed limit signs read?

1-3 Express one year in seconds.

1-4 What fraction of a mile is the "metric mile" of 1500 m?

1-5 Use the original definition of the meter and the fact that 1 mi = 1.609 km to find the diameter of the Earth in miles.

1-6 What is the factor that converts in.³ to cm³?

1-7 An athlete runs at constant speed and completes the 100-yd dash in 9.52 s. If he continues on at the same average speed, what will be his time for the 100-m distance?

1-8 Machinists sometimes work in units of *mils* (10^{-3} in.) or *microinches* (10^{-6} in.). Express both of these units in meters.

1-9 A *section* of land has an area of 1 mi² and contains 640 acres. How many m² are there in 1 acre?

1-10 Convert 3.65 yd² to cm².

1-11 In the *Gregorian calendar,* which we now use, every fourth year is a *leap year,* so the average length of a year is $365\frac{1}{4}$ days. The actual solar year (in 1980) was 365 d, 5 h, 48 min, 45.6 s. Assume the solar year is always this long (it varies slightly) and calculate how long it will be before our current calendar is out-of-step with the Sun by one day.

1-12 The distance x that an object moves during a time t is expressed as $x = v_0 t + \frac{1}{2}at^2$. Find the dimensions of a.

1-13 If an object is dropped from a height h, the speed v with which it strikes the ground depends only on h and on g, the acceleration due to gravity, together with a numerical (*i.e.,* dimensionless) factor. We know that $[h] = L$ and we state that $[g] = LT^{-2}$. Use only this information and obtain an expression for the dependence of v on h and g. (Your equation will, of course, be missing the numerical factor that would appear in the complete expression.)

Section 1–2

1–14 A part for an aircraft engine is manufactured from steel (same density as iron) and has a mass of 4.86 kg. If, instead, a magnesium-aluminum alloy ($\rho = 2.55$ g/cm^3) is used, what will be the mass of the part?

1–15 A flat circular plate of copper has a diameter of 48.6 cm and a mass of 62 kg. What is the thickness of the plate?

1–16 What is the mass of air (at normal conditions) in a room that has dimensions 8 m $\times$ 6 m $\times$ 3.2 m?

1–17 Find the density (the "weight density") of water in lb/ft^3.

1–18 What is the average density of (a) the Earth and (b) the Sun? [*Hint:* Use the data in the table inside the front cover.]

APPENDICES

In this book there are seven appendices (with two additional in Part II) that summarize various mathematical techniques. Appendix A, which is located in the traditional place at the rear of the book, gathers together the algebraic and trigonometric relationships that we use throughout the book. The remaining appendices deal with topics in the calculus and contain explanatory material, examples, and problems. To make access to these discussions easier, the appendix dealing with each new topic is placed so that it immediately precedes the chapter in which that technique is first used. Moreover, in the box at the beginning of each appendix, you will find an indication of where and how often that material is used in the text.

CALCULUS I

DIFFERENTIATION

The ideas summarized in this appendix are used throughout the text.

In the following chapters we discuss a variety of topics, many of which are rooted in everyday experience. We consider, for example, the motion of objects and the forces that affect such motions, the concepts of work and energy, the rotation of rigid bodies, and the orbital motion of the planets. All but the most trivial aspects of these considerations require a refinement of viewpoint and technique that can become quite demanding. Mathematical concepts and methods, some of which you may be meeting for the first time in this book, are needed to contend with these situations. The introduction of these tools broadens our ability to discuss physical phenomena in a meaningful way.

When you first encounter some of the new mathematical ideas, you may not appreciate fully that the intent is indeed to simplify the discussion of physical problems. In moments of such doubt, an historical reminder might prove reassuring. The starting point of all classical mechanics is based on the work of a towering genius, the extraordinary Englishman, Sir Isaac Newton (1642–1727). Yet, many of Newton's physical conjectures were not proven until he invented the methods of the calculus (which he called fluxions). Only then (1687) did he publish his monumental work entitled *Philosophiae Naturalis Principia Mathematica*. Although this important work is based on mathematics, the treatise did originate from Newton's contemplations of the *physical world*.

An important part of the rationale for including considerable "pure" mathematics in this book is to be able to present the mathematics from the physicist's point of view. Instead of introducing mathematical ideas for their own sake, we take the attitude (following Newton) that mathematics is a tool to assist us in understanding the physical world.

I-1 VARIABLES AND FUNCTIONS

A quantity to which a possibly unlimited number of values may be assigned is called a *variable*—for example, time, temperature, or the coordinate location of a particle. When two variables are related in such a way that giving the value of one variable determines uniquely the value of the other, we say that the second variable is a single-valued *function* of the first variable. For example, the water pressure exerted on the body of a skin-diver is a single-valued function of his depth below the water surface. However, there are cases in which more than one value for the second variable is possible for a given value of the first variable. In such cases, the second variable is a multiple-valued function of the first variable.

The first variable, to which any arbitrary value may be assigned, is called the *independent variable*. The second variable, whose value is in each instance determined by the functional relationship, is called the *dependent variable*. If we select the symbol x to represent the independent variable and the symbol y to represent the dependent variable, we write the connective relationship between the two quantities as $y = f(x)$ or $y = \phi(x)$ or $y = G(x)$, etc. The relationships $f(x)$, $\phi(x)$, and $G(x)$ are referred to as *functions* of x and are read "f of x," "ϕ of x," and "G of x," respectively. Depending on the nature of the physical situation, the independent variable may have assignable values only within some limited region called the *domain;* the corresponding values of the dependent variable define the *range* of the function.

The notation $f(x)$ stands for an operation on the variable x that is independent of the exact value assigned to x.

11

Thus, we might have $y = f(x) = \frac{1}{2}x^2 + 1$, and view $f(x)$ as an instruction requiring us to take the value of x (any of its allowed arbitrary values) and square it, multiply the result by $\frac{1}{2}$, and finally add 1 to that product. You may consider your electronic calculator to be a "function machine." When you enter a number and press the e^x key, the calculator performs various electronic steps that generate the desired exponential function, which is then displayed as a number.

A functional relationship such as $y = f(x)$ may be inverted to yield $x = g(y)$, thereby interchanging the roles of the dependent and independent variables.* For example, the relationship $y = f(x) = \frac{1}{2}x^2 + 1$ can be inverted to give $x = g(y) = \sqrt{2(y - 1)}$.

A particularly useful visual aid is the graphical plot of a function $y = f(x)$. To prepare a graph, first establish a coordinate frame with x- and y-axes at right angles, as in Fig. I–1. Then, points are located in the x-y plane thus defined that have coordinates corresponding to the functional relationship between x and y. The totality of such points $P(x, y)$ constitutes a curve, the graph of $f(x)$. Figure I–1a is the graph of $y = f(x) = \frac{1}{2}x^2 + 1$. Notice that whereas y is a single-valued function of x, the inverse function $x = g(y)$, shown in Fig. I–1b, is a *double-valued* function of y, corresponding to the assignment of $+$ or $-$ signs to the square-root function: $x = g(y) = \pm\sqrt{2(y - 1)}$. Each sign determines a particular *branch* of the function. (• Is there any significance to the region $y < 1$ in this case?)

I-2 LIMITS, INCREMENTS, AND INFINITESIMALS

Limits and Continuity. The precise mathematical definition of the *limit* is of basic importance both to the concept of *continuity* and in the fundamental operations of the calculus. We content ourselves here with a simple non-rigorous examination of the underlying principles of the limiting process.

We begin by considering the question of continuity. For a given function $f(x)$, imagine that the independent variable x assumes successive values ever closer to a fixed value $x = a$. If $f(x)$ approaches a definite value, say L, we write

$$\lim_{x \to a} f(x) = L$$

and we say that L is "the *limit* of $f(x)$ as x approaches a." The limit exists if and only if L is the same whether x approaches a from smaller values or from larger values. (The former is called the *left-hand limit* and the latter is called the *right-hand* limit. See Fig. I–2.)

A function $f(x)$ is said to be *continuous* at $x = a$ if the

Fig. I–1. (a) Graph of the single-valued function $y = f(x) = \frac{1}{2}x^2 + 1$. The point P is the unique point corresponding to $x = 2$, namely, $P(2, 3)$. (b) Graph of the double-valued function $x = g(y) = \pm\sqrt{2(y - 1)}$. The points Q and R both correspond to $y = 3$, namely, $Q(3, -2)$ and $R(3, +2)$. The point on the curve corresponding to $y = 1$ separates the upper and lower branches, and is called the *branch point*.

limit L exists and if the value assigned to the function at $x = a$ is such that $f(a) = L$. A function is continuous in an interval $b < x < c$ if it is continuous for every value of x within the interval.

Some examples will help to clarify these ideas. First, consider the function $y = x^2$ and its behavior in the vicinity of $x = 2$ (see Fig. I–3a). It is clear intuitively that, whether x approaches the value $x = 2$ from larger or from smaller values, we have $\lim_{(x \to 2)} f(x) = 4$. Also, the value of $f(x)$ at $x = 2$ is well defined:* $f(2) = 4$. This situ-

---

*The function $x = g(y)$ is referred to as the *inverse function* of $y = f(x)$, and is sometimes written $x = f^{-1}(y)$.

*For $f(x)$ at some specific value of x, say, $x = a$, we write $f(x = a) \equiv f(a)$. In the case here, $a = 2$.

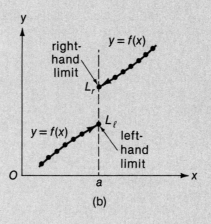

(a)

(b)

Fig. I–2. (a) The limit L of the function $f(x)$ exists at $x = a$ because the left-hand limit is equal to the right-hand limit. (b) This function has no limit at $x = a$.

(a)

(b)

Fig. I–3. Graphs of the functions
(a) $y = x^2$ and (b) $y = 1/x$.

ation is true for all values of x; therefore, $f(x) = x^2$ is continuous for all values of x.

Next, consider the function $y = 1/x$ (see Fig. I–3b). As x approaches $x = 0$ from negative values, it is easy to see that y acquires *negative* values of larger and larger magnitude without limit; in contrast, as x approaches $x = 0$ from positive values, y increases in *positive* magnitude without limit. Of course, the value of $y = 1/x$ at $x = 0$ is undefined, because division by zero must always be excluded. (● What is the limit of $y = 1/x$ as x becomes arbitrarily *large* in the positive sense?)

Example I–1

Show that (with θ expressed in radians)

$$\lim_{\theta \to 0} \frac{\sin \theta}{\theta} = 1$$

Solution:

A geometric proof of this limit is convenient. Refer to the diagram below. From the geometry, we can write an inequality connecting the arc $\overparen{CQC'}$, the chord $\overline{CPC'}$, and

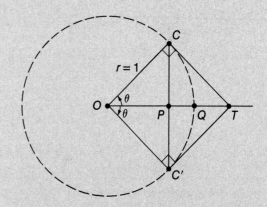

the two tangent lines, $\overline{CT}$ and $\overline{TC'}$:

$$\overline{CPC'} < \overarc{CQC'} < \overline{CT} + \overline{TC'}$$

In trigonometry, this corresponds to

$$2 \sin \theta < 2\theta < 2 \tan \theta$$

First, divide each term by $2 \sin \theta$; then, taking the reciprocal of each term reverses the inequality signs. The result is

$$1 > \frac{\sin \theta}{\theta} > \cos \theta$$

Thus, as θ approaches zero, $(\sin \theta)/\theta$ lies "sandwiched" between 1 and $\cos \theta$. But in the limit as $\theta \to 0$, we have $\cos \theta = 1$, and the proposition is proved. You should also consider the case in which θ approaches zero from negative values of θ, from which the proposition again follows. By defining the value of $(\sin \theta)/\theta$ to be 1 at $\theta = 0$, we render this function continuous at $\theta = 0$. (• Use your calculator to evaluate $\sin \theta/\theta$ for $\theta = 1, 0.1, 0.01, 0.001$, etc. Does it appear that a limit of 1 is being approached? Work in the *radian* mode!)

Example I–2

Show that

$$\lim_{x \to 0} (1 + x)^{1/x} = e = 2.718281828\cdots$$

Solution:

We can show numerically that this limit is being approached. (The rigorous proof of the limit is beyond the scope of this text.) The table gives values of $y = (1 + x)^{1/x}$ for $x \to 0$ from both negative and positive values. It is evident from the entries that the limit is approximately 2.718. (• To how many decimal places can you find the value of e with your calculator?)

x	0.1	0.01	0.001	0.0001
$y(-x)$	2.8680	2.7320	2.7196	2.7184
$y(+x)$	2.5937	2.7048	2.7169	2.7181

Some Useful Theorems.   Let $u(x)$ and $v(x)$ be functions such that

$$\lim_{x \to a} u(x) = L \quad \text{and} \quad \lim_{x \to a} v(x) = M$$

Then, we assert that

$$\lim_{x \to a} [ku(x)] = k \lim_{x \to a} u(x) = kL$$

(for all constants k)   **(I–1a)**

$$\lim_{x \to a} [u(x) + v(x)] = L + M \qquad \textbf{(I–1b)}$$

$$\lim_{x \to a} [u(x)v(x)] = LM \qquad \textbf{(I–1c)}$$

$$\lim_{x \to a} \left[\frac{u(x)}{v(x)} \right] = \frac{L}{M} \quad (M \neq 0) \qquad \textbf{(I–1d)}$$

$$\lim_{x \to a} [u(x)]^n = L^n \quad (n \geq 0, \text{ or } n < 0 \text{ if } L \neq 0) \quad \textbf{(I–1e)}$$

Increments and Infinitesimals.   Central to the differential calculus is the precise determination of the change in a function as the independent variable changes. The *increment* in any variable, corresponding to the change from one value to another, is the difference between the new and the old value. An increment in x is written Δx and is read "delta x." (Note that Δx does *not* indicate the product of Δ and x.) The increment in y is Δy, in θ it is $\Delta \theta$, and in $f(x)$ it is $\Delta f(x)$ or simply Δf.

Consider points in the neighborhood of $x = a$ that correspond to $x = a + \Delta x$. Then, the limiting process in which x approaches a ($x \to a$) can be viewed as one in which Δx approaches zero ($\Delta x \to 0$). A quantity such as Δx, which approaches zero as a limit, is referred to as an *infinitesimal* and is imagined ultimately to become (and remain) less than any pre-assigned positive number, however small.

Example I–3

Given the function $f(x) = 1/x$, determine the increment $\Delta f(x) = f(x + \Delta x) - f(x)$.

Solution:

Here, $f(x + \Delta x) = \dfrac{1}{x + \Delta x}$

Thus, $\Delta f = f(x + \Delta x) - f(x)$

$$= \frac{1}{x + \Delta x} - \frac{1}{x} = \frac{x - (x + \Delta x)}{x(x + \Delta x)}$$

$$= -\frac{\Delta x}{x^2 + x \Delta x}$$

Example I-4

Given the function $F(\theta) = \cos\theta$, determine the increment $\Delta F = F(\theta + \Delta\theta) - F(\theta)$.

Solution:

Here, $F(\theta + \Delta\theta) - F(\theta) = \cos(\theta + \Delta\theta) - \cos\theta$. Using Eq. A-29 in Appendix A, we have

$$\begin{aligned}\Delta F &= F(\theta + \Delta\theta) - F(\theta) \\ &= -2\sin(\theta + \tfrac{1}{2}\Delta\theta)\sin(\tfrac{1}{2}\Delta\theta)\end{aligned}$$

I-3 DIFFERENTIATION*

We begin by considering a function, $y = f(x)$, in which x has some fixed value, say, $x = a$. We now increment x by an amount Δx, thereby generating a new value for the function y, originally $y = f(a)$. Thus, we have

$$\Delta y = f(x + \Delta x) - f(x)$$

Now form the quotient,

$$\frac{\Delta y}{\Delta x} = \frac{f(x + \Delta x) - f(x)}{\Delta x}$$

The geometry of this expression is shown in Fig. I-4.

Fig. I-4. Illustrating the quotient $\Delta y/\Delta x$.

We define *the derivative* of $f(x)$, customarily denoted by the symbol dy/dx, to be†

---

*The results of this section are summarized in a table of useful derivatives inside the rear cover.

†We use the symbol $\equiv$ to mean "identically equal to" or "equal by definition."

$$\frac{dy}{dx} \equiv \lim_{\Delta x \to 0}\frac{\Delta y}{\Delta x} = \lim_{\Delta x \to 0}\frac{f(x + \Delta x) - f(x)}{\Delta x} \qquad \textbf{(I-2)}$$

To make explicit the fact that we are considering the derivative at the point corresponding to $x = a$, we use the obvious notation,*

$$\left.\frac{dy}{dx}\right|_{x=a} = \lim_{\Delta x \to 0}\left.\frac{f(x + \Delta x) - f(x)}{\Delta x}\right|_{x=a} \qquad \textbf{(I-3)}$$

When the limit in Eq. I-3 exists at $x = a$, the function is said to be *differentiable*† at $x = a$.

We may also view Eq. I-2 as a set of instructions for generating the derivative of the function $y = f(x)$. That is, we can write

$$\frac{dy}{dx} \equiv \frac{d}{dx}(y) \equiv \frac{d}{dx}f(x)$$

where we imply that d/dx is an operator that acts on the function $f(x)$ standing on the right, specifying that we perform the limiting operation indicated in Eq. I-2. To emphasize the operator character of d/dx, the equivalent symbol D_x is sometimes used; that is,

$$\frac{d}{dx}f(x) = D_x f(x)$$

This notation is particularly useful when discussing differential equations (see the appendix titled Calculus VII, in Vol. II).

The notation

$$\frac{df(x)}{dx} \equiv f'(x) \qquad \text{and} \qquad \left.\frac{df(x)}{dx}\right|_{x=a} \equiv f'(a)$$

is widely used in mathematics texts. We do not use this notation here because we frequently employ the prime to distinguish new variables—x and x', r and r', and so forth.

Derivative of the Product of a Constant and a Function. Let $v(x)$ be a differentiable function of x and let C be a constant. Then, for $y = Cv(x)$, we have

$$\Delta v = v(x + \Delta x) - v(x)$$

---

*Alternative notations appearing in mathematics texts are:

$$\left.\frac{dy}{dx}\right|_{x=a} = \lim_{x \to a}\frac{f(x) - f(a)}{x - a} = \lim_{\Delta x \to 0}\frac{f(a + \Delta x) - f(a)}{\Delta x}$$

†The existence of this limit at $x = a$ also insures continuity of $f(x)$ at $x = a$; however, it is easy to construct continuous functions that do not have unique limits for Eq. I-3 at some points and therefore do not have derivatives at those points. For example, such a function is $y = x^{2/3}$ at $x = 0$. See Example I-10.

and
$$\Delta y = C(v + \Delta v) - Cv = C \Delta v$$

so that
$$\frac{\Delta y}{\Delta x} = C \frac{\Delta v}{\Delta x}$$

Finally,
$$\frac{dy}{dx} = \lim_{\Delta x \to 0} C \frac{\Delta v}{\Delta x} = C \frac{dv(x)}{dx} \qquad \textbf{(I–4)}$$

The derivative of the product of a constant and a function is equal to the product of the constant and the derivative of the function.

Derivative of the Algebraic Sum of Functions.

Let $u(x)$ and $v(x)$ be two differentiable functions of x. Then, for $y = u(x) + v(x)$, we have

$$\Delta u = u(x + \Delta x) - u(x) \quad \text{and} \quad \Delta v = v(x + \Delta x) - v(x)$$

and
$$\Delta y = u(x + \Delta x) - u(x) + v(x + \Delta x) - v(x)$$
$$= \Delta u + \Delta v$$

so that
$$\frac{\Delta y}{\Delta x} = \frac{\Delta u}{\Delta x} + \frac{\Delta v}{\Delta x}$$

Finally,
$$\frac{dy}{dx} = \lim_{\Delta x \to 0} \left(\frac{\Delta u}{\Delta x} + \frac{\Delta v}{\Delta x} \right) = \frac{du(x)}{dx} + \frac{dv(x)}{dx} \qquad \textbf{(I–5)}$$

The derivative of the algebraic sum of functions is equal to the same algebraic sum of their derivatives.

Example I–5

Given the function $y = f(x) = 2x^2 + 1$, show by explicit calculation for $\Delta x = 1, 0.1, 0.01, 0.001, \ldots$, that

$$\lim_{\Delta x \to 0} \frac{\Delta y}{\Delta x} \bigg|_{x=3} = 12$$

Solution:

We have
$$\Delta x = 1: \quad \frac{\Delta y}{\Delta x} = \frac{f(3 + 1) - f(3)}{1}$$
$$= \frac{2(4)^2 + 1 - [2(3)^2 + 1]}{1} = 14$$

$$\Delta x = 0.1: \quad \frac{\Delta y}{\Delta x} = \frac{2(3.1)^2 + 1 - 19}{0.1}$$
$$= \frac{2(9.61) - 18}{0.1} = 12.2$$

$$\Delta x = 0.01: \quad \frac{\Delta y}{\Delta x} = \frac{2(3.01)^2 + 1 - 19}{0.01}$$
$$= \frac{18.1202 - 18}{0.01} = 12.02$$

Continuing, we find the values summarized in the table:

Δx	1	0.1	0.01	0.001	0.0001	0.00001
$\dfrac{\Delta y}{\Delta x}\bigg|_{x=3}$	14	12.2	12.02	12.002	12.0002	12.00002

Apparently, the limit of this process yields a value of 12.

Example I–6

Given the function $y = f(x) = 2x^2 + 1$, find the derivative at $x = 3$.

Solution:

We first form the quotient in Eq. I–2:

$$\frac{\Delta y}{\Delta x} = \frac{[2(x + \Delta x)^2 + 1] - [2x^2 + 1]}{\Delta x}$$

$$= \frac{2x^2 + 4x \Delta x + 2(\Delta x)^2 + 1 - 2x^2 - 1}{\Delta x}$$

$$= 4x + 2 \Delta x$$

Then,
$$\lim_{\Delta x \to 0} \frac{\Delta y}{\Delta x} = \frac{dy}{dx} = 4x$$

Thus,
$$\frac{dy}{dx} \bigg|_{x=3} = 4x \bigg|_{x=3} = 12$$

which is the result we found in Example I–5.

Example I–7

Find the general derivative of the function

$$y = f(x) = x^n$$

where n is a positive integer.

Solution:

After first obtaining $f(x + \Delta x)$, we eventually allow $\Delta x \to 0$. It is therefore appropriate to use an expansion of $(x + \Delta x)^n$ for which $\Delta x/x < 1$. Using the terminating series (Eq. A–13 in Appendix A), we have

$$(x + \Delta x)^n = x^n \left[1 + n\frac{\Delta x}{x} + \frac{n(n-1)}{2}\frac{(\Delta x)^2}{x^2} + \cdots \right.$$
$$\left. + \frac{(\Delta x)^n}{x^n} \right]$$

$$= x^n + nx^{n-1}(\Delta x)$$
$$+ \frac{n(n-1)}{2}x^{n-2}(\Delta x)^2 + \cdots + (\Delta x)^n$$

Thus,

$$f(x + \Delta x) - f(x) = (x + \Delta x)^n - x^n$$

$$= nx^{n-1}(\Delta x) + \frac{n(n-1)}{2}x^{n-2}(\Delta x)^2 + \cdots + (\Delta x)^n$$

Hence,

$$\frac{dy}{dx} = \lim_{\Delta x \to 0} \left[nx^{n-1} + \frac{n(n-1)}{2}x^{n-2}(\Delta x) + \cdots + (\Delta x)^{n-1} \right]$$

so that

$$\frac{dy}{dx} = \frac{d}{dx}(x^n) = nx^{n-1} \qquad \textbf{(I-6)}$$

This expression is in fact valid for any n, negative or positive, integer or non-integer; when $n = 0$, the result is *zero* (because $x^0 = 1$, and the derivative of any constant is identically zero). (• Can you show that the derivative of any constant is zero?)

Note that various powers of Δx appear in the expansion of $f(x + \Delta x) - f(x)$. As $\Delta x \to 0$ and Δx becomes an infinitesimal quantity, we refer to $(\Delta x)^2$ as an *infinitesimal of second order*, $(\Delta x)^3$ as an *infinitesimal of third order,* etc. It should be intuitively clear that as the limiting process continues toward zero without limit, quantities involving infinitesimals of higher order decrease in value faster as n increases; that is, a term involving $(\Delta x)^3$ becomes negligible sooner than a term involving $(\Delta x)^2$, which, in turn, effectively vanishes sooner than a term involving Δx.

Example I-8

Find the derivative of the function $y = f(\theta) = \sin \theta$.

Solution:

Using Eq. A-27 in Appendix A, with $x = \theta + \Delta\theta$ and $y = \theta$, we can write

$$\Delta y = \sin(\theta + \Delta\theta) - \sin\theta$$
$$= 2\cos(\theta + \tfrac{1}{2}\Delta\theta)\sin(\tfrac{1}{2}\Delta\theta)$$

Thus, $\quad \dfrac{d}{d\theta}(\sin\theta) = \lim_{\Delta\theta \to 0}\cos(\theta + \tfrac{1}{2}\Delta\theta)\dfrac{\sin(\tfrac{1}{2}\Delta\theta)}{(\tfrac{1}{2}\Delta\theta)}$

From the result in Example I-1, we can write

$$\lim_{\Delta\theta \to 0}\frac{\sin(\tfrac{1}{2}\Delta\theta)}{(\tfrac{1}{2}\Delta\theta)} = 1$$

Also, $\qquad \lim_{\Delta\theta \to 0}\cos(\theta + \tfrac{1}{2}\Delta\theta) = \cos\theta$

so that $\qquad \dfrac{dy}{d\theta} = \dfrac{d}{d\theta}(\sin\theta) = \cos\theta$

As in all such calculations, the angle θ must be expressed in radian measure.

Derivatives of Special Functions. The differentiation of certain functions requires more sophisticated treatment than indicated above. We now give a few examples to illustrate the use of these techniques.

Example I-9

Find the derivative of the function $y = f(x) = \ln x$.

Solution:

We have

$$\Delta y = \ln(x + \Delta x) - \ln x$$
$$= \ln\frac{x + \Delta x}{x} = \ln\left(1 + \frac{\Delta x}{x}\right)$$

Thus, $\qquad \dfrac{\Delta y}{\Delta x} = \dfrac{1}{\Delta x}\ln\left(1 + \dfrac{\Delta x}{x}\right)$

Multiplying the right-hand side of this equation by x/x, we obtain

$$\frac{\Delta y}{\Delta x} = \frac{1}{x}\cdot\frac{x}{(\Delta x)}\ln\left(1 + \frac{\Delta x}{x}\right) = \frac{1}{x}\ln\left(1 + \frac{\Delta x}{x}\right)^{x/\Delta x}$$

$$\frac{dy}{dx} = \lim_{\Delta x \to 0}\left\{ \frac{1}{x}\ln\left(1 + \frac{\Delta x}{x}\right)^{x/\Delta x} \right\}$$

$$= \frac{1}{x}\ln\left\{ \lim_{\Delta x \to 0}\left(1 + \frac{\Delta x}{x}\right)^{x/\Delta x} \right\}$$

Notice that we have interchanged the Lim and the ln operations; this is permissible if we are dealing with a continuous function, as is the case here. Now, using the result of Example I-2, we can write

$$\lim_{\Delta x \to 0}\left(1 + \frac{\Delta x}{x}\right)^{x/\Delta x} = e$$

We also have $\ln e = 1$, so

$$\frac{d}{dx}(\ln x) = \frac{1}{x}$$

Example I–10

Consider the function $y = x^{2/3}$, the graph of which is shown in diagram (a). Investigate the derivative of this function.

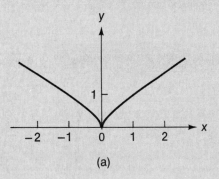

(a)

Solution:

According to Eq. I–6, we have

$$\frac{dy}{dx} = \frac{2}{3}x^{-1/3}$$

which is continuous everywhere except at $x = 0$, where dy/dx both increases and decreases without limit as $x \to 0$; the derivative at $x = 0$ is therefore not defined. The graph of dy/dx is shown in diagram (b).

(b)

Derivative of a Product of Functions.   Consider a pair of functions, $u(x)$ and $v(x)$, and form the product, $y = u(x)v(x)$. Then, if the variable x is incremented by Δx, we have

$$\Delta u = u(x + \Delta x) - u(x), \quad \text{and}$$
$$\Delta v = v(x + \Delta x) - v(x)$$

Also, $\Delta y = (u + \Delta u)(v + \Delta v) - uv$
$$= uv + u\,\Delta v + v\,\Delta u + \Delta u\,\Delta v - uv$$

Next, we form the quotient,

$$\frac{\Delta y}{\Delta x} = u\frac{\Delta v}{\Delta x} + v\frac{\Delta u}{\Delta x} + \frac{\Delta u\,\Delta v}{\Delta x}$$

As we proceed to the limit, $\Delta x \to 0$, the first two terms contain

$$\lim_{\Delta x \to 0}\frac{\Delta v}{\Delta x} = \frac{dv}{dx} \quad \text{and} \quad \lim_{\Delta x \to 0}\frac{\Delta u}{\Delta x} = \frac{du}{dx}$$

Using Eq. I–1c, the third term in $\Delta y/\Delta x$ vanishes:

$$\lim_{\Delta x \to 0}\frac{\Delta u\,\Delta v}{\Delta x} = \left[\lim_{\Delta x \to 0}\frac{\Delta u}{\Delta x}\right]\cdot\left[\lim_{\Delta x \to 0}\Delta v\right] = \frac{du}{dx}\cdot 0 = 0$$

Therefore,

$$\frac{d}{dx}(uv) = u\frac{dv}{dx} + v\frac{du}{dx} \qquad \textbf{(I–7)}$$

Note in particular that

$$\frac{d}{dx}(uv) \neq \left(\frac{du}{dx}\right)\left(\frac{dv}{dx}\right) \qquad (!)$$

Derivative of a Quotient of Functions.   Next, we consider the quotient of the functions $u(x)$ and $v(x)$, namely, $y = u(x)/v(x)$. Proceeding as in the case of a product of functions, we write

$$\Delta y = \frac{u + \Delta u}{v + \Delta v} - \frac{u}{v}$$
$$= \frac{v\,\Delta u - u\,\Delta v}{v(v + \Delta v)}$$

Then,

$$\frac{\Delta y}{\Delta x} = \frac{v\dfrac{\Delta u}{\Delta x} - u\dfrac{\Delta v}{\Delta x}}{v(v + \Delta v)}$$

In the limit, $\Delta x \to 0$ (which also implies $\Delta v \to 0$), and using Eq. I–1d we have

$$\frac{dy}{dx} = \frac{\displaystyle\lim_{\Delta x \to 0}\left(v\frac{\Delta u}{\Delta x} - u\frac{\Delta v}{\Delta x}\right)}{\displaystyle\lim_{\Delta v \to 0} v(v + \Delta v)}$$

Thus,

$$\frac{d}{dx}\left(\frac{u}{v}\right) = \frac{v\dfrac{du}{dx} - u\dfrac{dv}{dx}}{v^2}, \qquad v(x) \neq 0 \qquad \textbf{(I–8)}$$

Example I-11

Given the function $y = (x^2 - 2)(x + 2)$, find dy/dx.

Solution:

Write $(x^2 - 2) = u(x)$ and $(x + 2) = v(x)$. Then,

$$\frac{dy}{dx} = \frac{d}{dx}(u \cdot v) = u\frac{dv}{dx} + v\frac{du}{dx}$$

$$= (x^2 - 2)(1) + (x + 2)(2x)$$
$$= x^2 - 2 + 2x^2 + 4x$$
$$= 3x^2 + 4x - 2$$

Alternatively, we could write $y = (x^2 - 2)(x + 2) = x^3 + 2x^2 - 2x - 4$, obtaining $dy/dx = 3x^2 + 4x - 2$.

Example I-12

Given the function $y = 6x^3/(x^2 - 1)$, with $x^2 \neq 1$, find dy/dx.

Solution:

Write $6x^3 = u(x)$ and $(x^2 - 1) = v(x)$. Then,

$$\frac{dy}{dx} = \frac{d}{dx}\left(\frac{u}{v}\right) = \frac{v\dfrac{du}{dx} - u\dfrac{dv}{dx}}{v^2}$$

$$= \frac{(x^2 - 1)(18x^2) - (6x^3)(2x)}{(x^2 - 1)^2}$$

$$= \frac{6x^4 - 18x^2}{(x^2 - 1)^2} = \frac{6x^2(x^2 - 3)}{(x^2 - 1)^2}$$

($\bullet$ Does the derivative exist at $x = 1$? That is, does dy/dx, considered as a function of x, have the two limits equal to the value at $x = 1$?)

Chain Rule of Differentiation. It sometimes happens that we must deal with a function $y = f(u)$ in which the variable u depends, in turn, on some other variable, as in $u = g(x)$. We could make the substitution of $g(x)$ for u in $f(u)$, obtaining y as a function of x, namely, $y = f[g(x)]$. However, it is often more convenient to treat the functions in the following way. We write

$$\frac{\Delta y}{\Delta u} = \frac{f(u + \Delta u) - f(u)}{\Delta u}$$

and

$$\frac{\Delta u}{\Delta x} = \frac{g(x + \Delta x) - g(x)}{\Delta x}$$

Now,

$$\frac{\Delta y}{\Delta u} \cdot \frac{\Delta u}{\Delta x} = \frac{\Delta y}{\Delta x}$$

so that

$$\frac{\Delta y}{\Delta x} = \left[\frac{f(u + \Delta u) - f(u)}{\Delta u}\right] \cdot \left[\frac{g(x + \Delta x) - g(x)}{\Delta x}\right]$$

We take the limit, $\Delta x \to 0$ (which also means $\Delta u \to 0$), and use Eq. I-1c:

$$\frac{dy}{dx} = \operatorname*{Lim}_{\Delta u \to 0}\left[\frac{f(u + \Delta u) - f(u)}{\Delta u}\right] \cdot \operatorname*{Lim}_{\Delta x \to 0}\left[\frac{g(x + \Delta x) - g(x)}{\Delta x}\right]$$

Thus,

$$\frac{dy}{dx} = \frac{dy}{du} \cdot \frac{du}{dx} \qquad (I-9)$$

This is called the *chain rule of differentiation* and can be extended to any number of related variables; for example, if $y = f(u)$, $u = g(v)$, $v = h(w)$, and $w = i(x)$, then

$$\frac{dy}{dx} = \frac{dy}{du}\frac{du}{dv}\frac{dv}{dw}\frac{dw}{dx}$$

Example I-13

Given the functions $y = 2v^4 - v + 1$ and $v = x^2$, find dy/dx.

Solution:

$$\frac{dy}{dx} = \frac{dy}{dv} \cdot \frac{dv}{dx} = (8v^3 - 1)(2x)$$

$$= 2x(8x^6 - 1)$$

Successive Differentiation. Usually, the derivative of a function is, in turn, a differentiable function of the independent variable and can therefore itself be differentiated. (Reference to Example I-10 shows that sometimes the derivative cannot be differentiated at particular values of x.) The derivative of a derivative is referred to as the *second derivative* of the original function. The process may be continued by further differentiation, producing the third, or even higher, derivatives. We use the following special symbols to indicate the successive derivatives.

$$\frac{d^2y}{dx^2} \equiv \frac{d}{dx}\left(\frac{dy}{dx}\right), \qquad \frac{d^3y}{dx^3} \equiv \frac{d}{dx}\left(\frac{d^2y}{dx^2}\right)$$

and, in general,

$$\frac{d^ny}{dx^n} \equiv \frac{d}{dx}\left(\frac{d^{n-1}y}{dx^{n-1}}\right), \qquad n \geqslant 2$$

Example I–14

Given the function $y = 2x^3 + 7x^2 - 5$, find the derivatives of all orders.

For the first derivative, we have

$$\frac{dy}{dx} = 6x^2 + 14x$$

Clearly, this may be viewed as a new polynomial function of x, which in turn can be differentiated; thus,

$$\frac{d^2y}{dx^2} = \frac{d}{dx}\left(\frac{dy}{dx}\right) = \frac{d}{dx}(6x^2 + 14x) = 12x + 14$$

Further, $\frac{d^3y}{dx^3} = \frac{d}{dx}\left(\frac{d^2y}{dx^2}\right) = \frac{d}{dx}(12x + 14) = 12$

Finally, d^4y/dx^4 and all higher derivatives are *zero*.

I-4 APPLICATIONS OF THE DERIVATIVE

The Slope of a Function. Consider the function $y = f(x)$ and its graph shown in Fig. I–5. We show on this graph the terms that appear in the basic definition of the derivative (Eq. I–2):

$$\frac{dy}{dx} \equiv \lim_{\Delta x \to 0} \frac{\Delta y}{\Delta x} = \lim_{\Delta x \to 0} \frac{f(x + \Delta x) - f(x)}{\Delta x}$$

From Fig. I–5 you can see that $\Delta y/\Delta x$ is equal to the tangent of the angle θ:

$$\frac{\Delta y}{\Delta x} = \frac{f(a + \Delta x) - f(a)}{\Delta x} = \tan \theta$$

During the limiting process, $\Delta x \to 0$, the point Q moves ever closer to the point P. It seems intuitively clear that in this limiting process the secant line $\overline{PQ}$ approaches

Fig. I–5. The graph of $y = f(x)$, showing the geometric representation of the various terms appearing in Eq. I–2.

the tangent line $\overline{TT'}$ and that $\tan \theta$ approaches $\tan \alpha$. Indeed, this is rigorously true. The tangent of the angle θ is the *slope* of the secant line $\overline{PQ}$, and $\tan \alpha$ is the *slope* of the line $\overline{TT'}$ that is tangent to the curve at the point P.* That is,

$$\left.\frac{dy}{dx}\right|_{x=a} = \lim_{\Delta x \to 0} \left.\frac{f(x + \Delta x) - f(x)}{\Delta x}\right|_{x=a} = \tan \alpha$$

It was the study of this important "tangent problem" that led Gottfried Wilhelm Leibnitz (1646–1716) to discover the differentiation process. Leibnitz developed the calculus independently of the work of Newton. His discovery appeared in an essay published in Leipzig in 1684, a few years before Newton published his version.

Example I–15

Find the tangent to the curve $y = \frac{1}{3}x^2 - x$ at $x = 2$.

Solution:

We have

$$\frac{dy}{dx} = \frac{2}{3}x - 1$$

from which $\tan \alpha = \left.\dfrac{dy}{dx}\right|_{x=2} = \frac{2}{3}(2) - 1 = \frac{1}{3}$

or, $\alpha = 18.43°$

---

*By the *slope* of any line we mean the tangent of the angle of inclination of the line with the x-axis. If the equation of the line can be written as $y = mx + b$, the slope is m. If the equation of the line is given in *intercept form*, $(y/b) + (x/a) = 1$, the slope is $-(b/a)$.

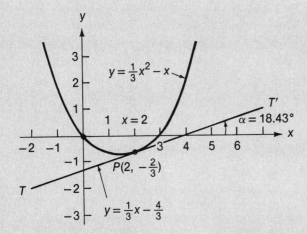

We can also find the equation of the tangent line $\overline{TT'}$ at the point corresponding to $x = 2$ on the curve $f(x)$. For the tangent line $\overline{TT'}$, write

$$y = mx + b$$

with

$$m = \tan \alpha = 1/3$$

Now, the point $P(x = 2, y = -2/3)$ is common to both the tangent line and the curve $y = f(x) = \frac{1}{3}x^2 - x$. Therefore,

$$-\tfrac{2}{3} = \tfrac{1}{3}(2) + b$$

or

$$b = -\tfrac{4}{3}$$

Thus, the equation of the tangent line $\overline{TT'}$ at $x = 2$ is

$$y = \tfrac{1}{3}x - \tfrac{4}{3}$$

The graph of $f(x)$ and the tangent line $\overline{TT'}$ at $x = 2$ are shown in the diagram.

Fig. I-6. The function $f(x)$ exhibits a local minimum at $x = a$, a local maximum at $x = c$, and a point of inflection at $x = b$.

Extremal Values. A function $y = f(x)$ has extremal values—local maxima or minima—at points for which the slope of the function vanishes, that is, $dy/dx = 0$. Figure I-6 illustrates this idea. At $x = a$ the slope is zero and $f(a)$ represents a local minimum. (The minimum is only "local" because $f(x)$ becomes very large negatively as $x \to \infty$.) Also, at $x = c$ the slope is zero and $f(c)$ represents a local maximum. At $x = b$ the slope is zero, but the function is neither maximum nor minimum; this point is called a *point of inflection*. We give a test later in this section to determine the nature of the extremal values at all such points.

Consider a particular case. Figure I-7a shows the function $y = f(x) = \frac{1}{2}x^3 - 2x^2 + 3$, and Fig. I-7b shows the first and second derivatives,

Fig. I-7. The graph of (a) the function $y = f(x) = \frac{1}{2}x^3 - 2x^2 + 3$ and (b) its first and second derivatives. The line $\overline{TT'}$ tangent to $f(x)$ at the point $P(1, 3/2)$ is also shown.

$$\frac{dy}{dx} = \tfrac{3}{2}x^2 - 4x \quad \text{and} \quad \frac{d^2y}{dx^2} = 3x - 4$$

First, we note that $f(x)$ is a cubic function with three real (distinct) roots, as indicated by the fact that there are three x-intercepts for the function (Fig. I–7a). For large values of $|x|$, the function becomes indefinitely large. Thus, $f(x)$ has one local (or relative) maximum and one local minimum.

The line tangent to $f(x)$ is horizontal (*i.e.*, $\tan \alpha = dy/dx = 0$) at two points: $x = 0$ (a local maximum) and $x = 8/3$ (a local minimum). If the derivative of a function $f(x)$ vanishes for certain values of x, there may be local extremal values for $f(x)$ at these points. (In the next paragraph we discuss the conditions under which $dy/dx = 0$ does not indicate an extremum.) How do we determine whether a particular extremum corresponds to a maximum or a minimum? If the slope dy/dx is positive for $x < a$, continually decreases as $x \to a$ from the left, reaches zero at $x = a$, and becomes increasingly more negative as x increases in the region $x > a$, then the function has a local maximum at $x = a$. Note that this is exactly the case for the function in Fig. I–7a at $x = 0$; the behavior of the slope is illustrated by the parabolic curve in Fig. I–7b. Conversely, if the slope is negative for $x < b$, zero at $x = b$, and positive for $x > b$, then there is a local minimum at $x = b$. In Fig. I–7a, this occurs at $x = 8/3$. Whether the slope is increasing or decreasing as x passes through an extremal point may, in most cases, be determined by noting the behavior of the *second* derivative, which is just the slope of the first derivative. In Fig. I–7b, $(d^2y/dx^2)|_{x=0} < 0$ (that is, dy/dx is *decreasing* through $x = 0$), whereas $(d^2y/dx^2)|_{x=8/3} > 0$ (that is, dy/dx is *increasing* through $x = 8/3$).

Before we state general rules for the location of extrema, consider the function $y = \tfrac{1}{4}x^3$, illustrated in Fig. I–8. In this case, both dy/dx and d^2y/dx^2 vanish at $x = 0$. Thus, the second-derivative test fails, and we must examine the behavior of the first derivative in the neighborhood of the candidate extremal point. We see that $dy/dx = \tfrac{3}{4}x^2$ remains positive on both sides of $x = 0$—that is, the slope does not change sign. Thus, $x = 0$ corresponds to *neither* a maximum *nor* a minimum. For the function $y = \tfrac{1}{4}x^3$, the vanishing of the slope at $x = 0$ signifies a point of inflection. Such a point separates a curve into arcs that have opposite directions of bending, as shown in Fig. I–8 at $x = 0$ (and in Fig. I–6 at $x = b$).

One final remark is in order. We can *invent* functions that are continuous and possess local extremal values at points where the derivative is infinite (undefined) or discontinuous. In Example I–10, we found that the function $f(x) = x^{2/3}$ has a minimum at the cusp point, $x = 0$, where dy/dx grows without limit. Also, the function $y = |a + bx|$ has a minimum at the break point, $x = -a/b$, where dy/dx is not defined. We call these func-

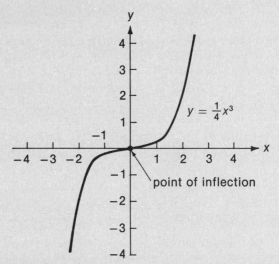

Fig. I–8. Graph of the function $y = \tfrac{1}{4}x^3$, which exhibits a point of inflection at $x = 0$. Both the first and second derivatives vanish at $x = 0$.

tions *pathological* because such ambiguous derivatives do not occur in the real physical world. Invented functions of this type can always be replaced by functions that have similar behavior but have continuous derivatives (see Problem I–24).

The rules for extrema are:

i) Extremal values for a function $y = f(x)$ may occur at those points, such as $x = a$, where $(dy/dx)|_{x=a} = 0$.
ii) If $(d^2y/dx^2)|_{x=a} > 0$, there is a local *minimum* at $x = a$.
iii) If $(d^2y/dx^2)|_{x=a} < 0$, there is a local *maximum* at $x = a$.
iv) If $(d^2y/dx^2)|_{x=a} = 0$, tests (ii) and (iii) fail; then, examine the sign of the first derivative near $x = a$. If, in passing through $x = a$ with x increasing, the derivative changes sign from positive to negative, there is a maximum; if the derivative changes sign from negative to positive, there is a minimum. If the derivative does not change sign at $x = a$, there is a point of inflection.

The graph of a function can often be sketched easily by noting the positions of the x- and y-intercepts, the points and the values of the extrema, and the behavior of the function for $x \to \pm\infty$.

Example I–16

Given a rectangle with sides l and m and a fixed perimeter p, find the values of l and m that give the largest enclosed area.

Solution:

We have $p = 2(l + m)$, or $m = \frac{1}{2}p - l$. Also, the area is $A = lm = l(\frac{1}{2}p - l)$. To convert to x, y notation, we identify A with y and l with x; then,

$$y = x(\tfrac{1}{2}p - x) = \tfrac{1}{2}px - x^2$$

A maximum value for y requires

$$\frac{dy}{dx} = \tfrac{1}{2}p - 2x = 0$$

Thus, $x = l = p/4$, from which $m = \frac{1}{2}p - l = p/4$, or $l = m$. Because $d^2y/dx^2 = -2$, we indeed have a maximum value for the area. This result proves the well-known fact that the square, among all possible rectangles with a given perimeter, encloses the greatest area.

Example I-17

Find the volume of the largest open box that can be made from a square piece of tin (with sides a) by cutting equal squares from each corner and turning up the sides, as indicated in the diagram.

Solution:

The volume of the open box is

$$V = x(a - 2x)^2$$

The extremal value of V results when $dV/dx = 0$; that is,

$$x \cdot 2(a - 2x)(-2) + (a - 2x)^2 = 0$$

Hence,

$$12x^2 - 8ax + a^2 = 0$$

or

$$(6x - a)(2x - a) = 0$$

Thus, we have extremal values for $x = \frac{1}{6}a$ and $x = \frac{1}{2}a$. From the figure we see that $x = \frac{1}{2}a$ gives cut-outs that leave no material; therefore, $V = 0$, which is clearly a mini-

imum. For $x = a/6$, we find $(d^2V/dx^2)|_{x=a/6} = 24x - 8a = 24(a/6) - 8a = -4a$; that is, $d^2V/dx^2 < 0$. Hence, $x = a/6$ yields the required maximum. Finally,

$$V = (a - \tfrac{1}{3}a)^2 \cdot \tfrac{1}{6}a = \tfrac{2}{27}a^3$$

I-5 DIFFERENTIALS

In arriving at the basic definition of the derivative (Eq. I-2), we took the limit, $\Delta x \to 0$, of the quotient $\Delta y/\Delta x$ to obtain

$$\frac{dy}{dx} = \frac{df(x)}{dx} = \operatorname*{Lim}_{\Delta x \to 0} \frac{\Delta y}{\Delta x}$$

Notice that we define in this way only dy/dx, not dy and dx separately. We now take the view that even dy/dx may be treated as the quotient of the quantities dy and dx, which we call *differentials*.

The quantity dy/dx is the slope of the line tangent to the curve $y = f(x)$ at a particular point. Thus, in Fig. I-9 at $x = a$, we see that $dy/dx|_{x=a} = \tan \alpha$. Using the definition of the tangent function as a guide, we can define the differential dy to be

$$dy \equiv \frac{df(x)}{dx}\bigg|_{x=a} \Delta x = \frac{df(x)}{dx}\bigg|_{x=a} dx \qquad \textbf{(I-10)}$$

where we define $\Delta x \equiv dx$.

When the variable x is incremented from $x = a$ to $x = a + \Delta x$, the change in the function $y = f(x)$ is

$$\Delta y = f(a + \Delta x) - f(a)$$

That is, Δy is the change in y *along the curve* $y = f(x)$ (from P to Q in Fig. I-9) that corresponds to the incremental change in x from a to $a + \Delta x$. In contrast, dy is the

Fig. I-9. The geometric representation of the increment Δy and the differential dy, evaluated at $x = a$, for the function $y = f(x)$. Here, $\Delta y = f(a + \Delta x) - f(a)$, and $dy = df(x)/dx|_{x=a} \cdot \Delta x$.

change in y along the tangent line $y = f(a) + (df/dx)|_{x=a}(\Delta x)$ (that is, from P to Q' in Fig. I-9) for the same incremental change in x. Moreover, Δy approaches dy in the limit $\Delta x \to 0$.

These results imply that in the immediate neighborhood of the point P, corresponding to $x = a$, small excursions Δx can be considered equally well to correspond to changes that follow the curve $f(x)$ or the tangent line $\overline{TT'}$. In this sense, the function $f(x)$ may be replaced locally (i.e., near $x = a$) by the equation of its tangent line erected at $x = a$. This leads to an important approximation for incremental changes in $f(x)$ when Δx is appropriately small, namely,

$$\Delta y \cong \frac{dy}{dx}\bigg|_{x=a} \cdot \Delta x \qquad \text{(I-11)}$$

A useful extension of Eq. I-11 is also possible. Because $f(x + \Delta x) = f(x) + \Delta y$, we have, also for suitably small Δx,

$$f(a + \Delta x) \cong f(a) + \frac{dy}{dx}\bigg|_{x=a} \cdot \Delta x \qquad \text{(I-12)}$$

Example I-18

Find an approximate expression for the area of a circular ring that has a radius r and width Δr.

Solution:

For a circle with a radius r, the area is $A = \pi r^2$. Therefore,

$$\frac{dA}{dr} = 2\pi r$$

so that

$$\Delta A \cong \left(\frac{dA}{dr}\right)\Delta r = 2\pi r \, \Delta r$$

(In the notation of Eq. I-10, we see this commonly used expression written as $dA = 2\pi r dr$.)

The *exact* expression for ΔA is

$$\Delta A = A(r + \Delta r) - A(r)$$
$$= \pi(r + \Delta r)^2 - \pi r^2$$

$$\Delta A = 2\pi r \Delta r + \pi(\Delta r)^2$$

$$= 2\pi r \, \Delta r + \pi(\Delta r)^2$$

In the figure, the shaded area on the left is the true area of the ring, whereas the approximate expression yields the area on the right, namely, the circumference of the circle $2\pi r$ multiplied by the strip width Δr. For small values of $\Delta r/r$, the error in the approximate expression, $\pi(\Delta r)^2$, is negligible compared with the approximate value of ΔA; that is,

$$\frac{\text{error}}{\Delta A} = \frac{\pi(\Delta r)^2}{2\pi r \, \Delta r} = \frac{(\Delta r)}{2r} \ll 1$$

Example I-19

Given $y = f(x) = x^2 \sin \pi x$, find an approximate value for $f(x)$ at $x = 5/8$ in terms of the more easily calculated exact value of $f(x)$ at $x = \frac{1}{2}$.

Solution:

Using Eq. I-12, we have

$$f(\tfrac{5}{8}) \cong f(\tfrac{1}{2}) + \frac{dy}{dx}\bigg|_{x=1/2} \cdot (\tfrac{5}{8} - \tfrac{1}{2})$$

Now, $\quad f(\tfrac{1}{2}) = (\tfrac{1}{2})^2 \sin \tfrac{1}{2}\pi = \tfrac{1}{4}$

and $\quad \dfrac{dy}{dx} = \pi x^2 \cos \pi x + 2x \sin \pi x$

from which $\quad \dfrac{dy}{dx}\bigg|_{x=1/2} = \pi(\tfrac{1}{2})^2 \cos \tfrac{1}{2}\pi$
$$+ 2(\tfrac{1}{2}) \sin \tfrac{1}{2}\pi = 1$$

Thus, $\quad f(\tfrac{5}{8}) \cong \tfrac{1}{4} + 1\,(\tfrac{1}{8}) = \tfrac{3}{8} = 0.375$

Actually, $\quad f(5/8) = \left(\dfrac{5}{8}\right)^2 \sin \dfrac{5\pi}{8} = 0.361$

The approximate value is about 4 percent larger than the true value.

Example I-20

Given that $\sqrt{25} = 5$, evaluate $\sqrt{26}$ using Eq. I-11.

Solution:

Writing $y = x^{1/2}$, we have $dy/dx = \tfrac{1}{2}x^{-1/2}$. Then,

$$y(26) \cong y(25) + \frac{dy}{dx}\bigg|_{x=25} \cdot (26 - 25)$$

Thus, $\sqrt{26} \cong \sqrt{25} + \dfrac{1}{2} \cdot \dfrac{1}{\sqrt{25}} \cdot 1 = 5 + \dfrac{1}{10} = 5.10$

The actual value is $\sqrt{26} = 5.0990195.\ldots$

PROBLEMS

Section I-1

I-1 Plot the graph of $G(v) = \frac{1}{3}v^2 + \sin(\frac{1}{2}\pi v)$ as a function of v in the domain $-3 < v < +3$. (Notice that the argument of the sine function is in radians. See Appendix A-10.)

I-2 Use an electronic calculator to find the positive solution of $\frac{1}{3}v^2 + \sin(\frac{1}{2}\pi v) = 3$. [Hint: Because $|\sin\theta| \leqslant 1$, we can use a method of successive approximations by writing the equation in the form $v = \sqrt{9 - 3\sin(\frac{1}{2}\pi v)}$. For the first approximation, use $v_1 = \sqrt{9} = 3$. For the second approximation, use $v_2 = \sqrt{9 - 3\sin(\frac{1}{2}\pi v_1)}$. Continue, using $v_{n+1} = \sqrt{9 - 3\sin(\frac{1}{2}\pi v_n)}$, until there is less than 0.1 percent difference between two successive calculated values of v. Be certain that your calculator is set to operate in the radian mode!]

I-3 The diagram shows the graph of the single-valued function $y = f(t) = v_0 t - \frac{1}{2}gt^2$. (a) What is the illustrated domain for this function? What is the corresponding range? (b) Construct the graph of the inverse function, $t = g(y)$. Is $g(y)$ a multiple-valued function? (c) What are the values of the domain and the range corresponding to the inverse of the original function?

I-4 Make a graph of the function $y = f(x) = \frac{1}{4}x^4 - x^2$ for $-\frac{5}{2} < x < +\frac{5}{2}$. Determine the inverse function $x = g(y)$. [Hint: Set $x^2 = z$ and use the quadratic formula, Eq. A-6 in Appendix A.] If, in $x = g(y)$, the variable x is to be real, what is the allowable domain for y? [Hint: Because $x = g(y)$ is a multiple-valued function of y, you must consider each branch separately.]

Section I-2*

I-5 In each of the following cases for $y = f(x)$, verify the stated increment of the function, $\Delta y = f(x + \Delta x) - f(x)$.
 (a) $f(x) = mx + b$ $\Delta y = m\,\Delta x$
 (b) $f(x) = e^x$ $\Delta y = e^x(e^{\Delta x} - 1)$
 (c) $f(x) = \ln x$ $\Delta y = \ln\left(1 + \dfrac{\Delta x}{x}\right)$

I-6 Calculate the value of e ($2.718281828\ldots$) using the power series (Eq. A-11 in Appendix A),

$$e^x = 1 + \frac{x}{1!} + \frac{x^2}{2!} + \frac{x^3}{3!} + \cdots + \frac{x^n}{n!} + \cdots$$

If an accuracy of ± 1 in the fifth decimal place is required, estimate the value of n to which the sum must be carried. Prepare a table of calculated values of e for the sum carried to the values $n = 1, 2, 3, \ldots$. [Hint: You will find it convenient to accumulate each term into the memory of your calculator and to recall the sums as needed.]

I-7 Prepare a table of values to determine $\mathrm{Lim}_{s\to\infty}\Big[1 + \frac{1}{2} + \frac{1}{3} + \frac{1}{4} + \cdots + \frac{1}{s} - \ln s\Big]$ for $s = 2, 4, 8, 16, 32$. (The memory accumulator function of your calculator will be helpful here.) Compare your limit with the known limit for this series, namely, $0.5772157\ldots$. This limit is known as *Euler's constant*. (It is evident that the error in the sum after n terms is approximately $1/n$; that is, the series converges very slowly.)

I-8 Prepare a table of values, carried to 5 digits, for $(10^x - 1)/x$, with $x = 1/2, 1/4, 1/8, 1/16, \ldots$, to a value of x as small as the capacity of your calculator permits while still yielding 5 digits for the quantity $10^x - 1$. Compare your results with the known value,

$$\mathrm{Lim}_{x\to 0}\frac{10^x - 1}{x} = \ln 10 = 2.302585093\ldots$$

I-9 Calculate the square root of a number N by an *iterative* process as follows:

 i) Guess a value for the square root; let this be A.
 ii) Compute a new candidate value B for the square root using $B = \frac{1}{2}[A + (N/A)]$.
 iii) Obtain yet a better value C using $C = \frac{1}{2}[B + (N/B)]$.
 iv) Repeat to the limit of your calculator capacity.

*The properties of logarithmic and exponential functions are reviewed in Appendix A.

(•Can you see why this procedure works? What is the value of C when $B \cong C$?)

The steps i) through iv) can be made a continuous operation by calculating* $A + (N \div A) = \div 2 = (\text{STO}) + (N \div \text{RCL}) = \div 2 = (\text{STO}) + (N \div \text{RCL}) = \div 2 = (\text{STO}) \ldots$ Try this process for $N = 10$ and at least two guesses for A, such as $A = 5$ or $A = 3$. Repeat for $N = 19$. Note how the rapidity of convergence depends on the shrewdness of your guess for A (i.e., how close A actually is to $\sqrt{N}$). Compare your final answers with those obtained by using the square-root function key on your calculator.

Section I-3

I-10 Determine the derivative of each of the following functions by explicit use of the definition (Eq. I-2). Compare your results with those obtained by using the table of derivatives inside the rear cover.

(a) $y = \dfrac{1}{x}$ [*Hint:* Refer to Example I-3.]

(b) $y = \cos \theta$ [*Hint:* Refer to Example I-4 and use the result of Example I-1.]

(c) $y = \tan \theta$ [*Hint:* Make use of Eq. A-24 in Appendix A, and the result of Example I-1.]

I-11 Using the table of derivatives inside the rear cover, verify the following:

(a) $\dfrac{d}{dx} (mx + b) = m$

(b) $\dfrac{d}{dt} (\tfrac{1}{2}at^2 + v_0 t + s_0) = at + v_0$

(c) $\dfrac{d}{dx} (\sqrt{a^2 - x^2}) = -\dfrac{x}{\sqrt{a^2 - x^2}}$

I-12 Verify the following:

(a) $\dfrac{d}{dx} (2ax)^{1/2} = \sqrt{\dfrac{a}{2x}}, \quad x \neq 0$

(b) $\dfrac{d}{dx} (\sin 2x) = 2 \cos 2x$

(c) $\dfrac{d}{d\theta} \left(\dfrac{\cos \theta}{\theta} \right) = -\dfrac{\theta \sin \theta + \cos \theta}{\theta^2}, \quad \theta \neq 0$

I-13 Find the values of the first and second derivatives of the following functions at the given value(s) of the indicated independent variable:

(a) $y = x \sin x$,   at $x = \pi/2$

(b) $s = \tfrac{1}{2}at^2 + v_0 t + s_0$,   at $t = 0$

(c) $u = e^{-\alpha t} \cos \omega t, \alpha > 0$,   at $t = 0$ and $t \to \infty$

I-14 Consider $y = x^n = u(x)v(x)$ with $u(x) = x^{n-m}$ and $v(x) = x^m$ $(n > m > 1)$. Show, by using the rule for the derivative of a product of functions (Eq. I-7), that one obtains the same result as in Example I-7, namely, $dy/dx = nx^{n-1}$.

I-15 According to Eq. A-9 in Appendix A, the power series expansion for $\ln (1 + x)$ is

$$\ln (1 + x) = x - \frac{x^2}{2} + \frac{x^3}{3} - \frac{x^4}{4}$$
$$+ \cdots + (-1)^{n-1} \frac{x^n}{n} + \cdots$$

Show by differentiating both sides of this equation that a result consistent with the binomial theorem is obtained.* Refer to Eq. A-15 in Appendix A.

I-16 According to Eq. A-11 in Appendix A, the power series expansion for e^x is

$$e^x = 1 + \frac{x}{1!} + \frac{x^2}{2!} + \frac{x^3}{3!} + \cdots + \frac{x^n}{n!} + \cdots$$

Show that differentiating both sides of this equation maintains the equality.*

Section I-4†

I-17 Differentiate each of the following functions and verify the value of the slope at the indicated point.

(a) $y = \sin x$: $\tan \alpha = \sqrt{2}/2$ at $x = \pi/4$
(b) $y = e^{-ax}$: $\tan \alpha = -a$ at $x = 0$
(c) $y = xe^{-x}$: $\tan \alpha = 0$ at $x = 1$

I-18 In each of the following cases, find the equation of the line tangent to the curve at the indicated value of x. Make a sketch of each case.

(a) $y = \tfrac{1}{2}x^2 + 1$,   at $x = 2$
(b) $y = (x - 1)^3$,   at $x = 0$ and at $x = 1$
(c) $y = e^{-x}$,   at $x = 1$

I-19 Determine the extremal values for the following functions, specifying whether each is a maximum, a minimum, or a point of inflection. Sketch a graph for each case. $(a, b > 0)$

(a) $y = \dfrac{ax}{x^2 + 1}$ (b) $y = (x - a)^3$ (c) $y = \sin \tfrac{1}{2}\pi x$

I-20 What point on the curve $y = \tfrac{1}{4}x^2$ is nearest the fixed
•   point $(0, 4)$? [*Hint:* The distance ℓ between two points (x_1, y_1) and (x_2, y_2) is given by $\ell^2 = (x_2 - x_1)^2 + (y_2 - y_1)^2$.]

---

*The symbol STO represents the "store in memory" key, and RCL represents the "recall from memory" key. Some calculators may have different symbols for these keys.

*In general, for such an equality to exist requires careful convergence tests for the derived series. All functions we consider satisfy the required conditions.

†Here, as elsewhere in this book, a solid circle (or circles) below a problem number denotes a more difficult problem.

I–21 What is the diameter of a thin cylindrical tin can with a fixed volume V that requires the least amount of tin (a) if the can is open at the top and (b) if the can is closed at both ends?

I–22 An open rain gutter made from thin sheet metal has sides and a horizontal bottom, all with equal length. The sides make equal angles with the bottom, as shown in the diagram. What should be the width of the opening across the top for maximum carrying capacity?

I–24 Contrast the behavior of the two functions,
• $y = |a + bx|$ and $y = \sqrt{(a + bx)^2 + \varepsilon^2}$, and their derivatives. Use the positive branch of the square root and take ε to be a constant with $\varepsilon^2 \ll 1$. Be sure to consider the cases in which $(a + bx)^2 \ll \varepsilon^2$ and $(a + bx)^2 \gg \varepsilon^2$. Do these functions have any extremal values? If so, for what values of x? Which of these two functions would you expect to correspond to a real physical situation? (These two functions are considered again in Problem 2–53.)

Section I–5

I–25 Evaluate the differential dy for each of the following functions:
(a) $y = 3x^3 + 7x^2 - 2x + 1$
(b) $y = \sin(2\pi x^2)$
(c) $y = e^{-ax} \sin(\frac{1}{2}\pi x)$

I–26 A circular band around the Earth at the Equator is to have its length increased by 1 m. What is the required increase in the radius of the band? Is there any difference between the exact solution to this problem and a solution based on the use of differentials? Would your answer be any different if we were talking about the Moon instead of the Earth? Explain.

I–27 Find an approximate expression for the volume of a thin spherical shell with a radius r and a shell thickness Δr. Can you give a geometric interpretation of your answer? [*Hint:* The volume of a solid sphere is $V = (4\pi/3)r^3$; use Eq. I–11.]

I–28 All six sides of a cubical box are 2 mm thick, and the volume of the box is $10^3 \, \text{cm}^3$. Use Eq. I–11 to determine the volume of material needed to construct the box. Give a geometric interpretation of your answer.

APPENDIX

CALCULUS II

INTEGRATION

The material in this appendix is first required for the discussion of the general equations of motion for a particle, Section 2–6. The techniques of integral calculus are used sparingly until Chapter 7; thereafter, they are used freely.

We are all familiar with a number of mutually inverse operations, such as addition and subtraction, multiplication and division, raising to a power and extracting a root, and so forth. We also recognize the following pairs as examples of inverse functional operations:

$$y = \tfrac{1}{2}x^2 - 1 \longleftrightarrow x = \pm\sqrt{2(y+1)}$$
$$y = \sin x \longleftrightarrow x = \sin^{-1} y$$
$$y = \ln x \longleftrightarrow x = e^y$$

In this appendix, we inquire into the nature of the operation that is inverse to differentiation.

II-1 INTEGRATION AS ANTIDIFFERENTIATION

In the preceding appendix, we studied the differentiation operation, in which a function $f(x)$ is given and we are asked to find the derivative $df(x)/dx$. That is, the derivative of a function $y = f(x)$ is a function $g(x)$ that is obtained by performing the operation

$$g(x) = \frac{dy}{dx} = \frac{df(x)}{dx}$$

Now, if we are given a function $g(x)$, which is the derivative of some unspecified function $f(x)$, how do we find $f(x)$? We imagine performing an inverse operation on $g(x)$ that yields $f(x)$. We represent this operation with a notation in which the derivative is viewed as the quotient of

two differentials; then, we write $dy/dx = g(x)$ in the differential form, $dy = g(x)\,dx$. The antiderivative or integration process is written symbolically as*

$$y = \int dy = \int g(x)\,dx \qquad \textbf{(II–1)}$$

Our task is to find $y = f(x)$ such that its derivative is $g(x)$. We can do this only to within an additive constant C (the *constant of integration*) because $g(x) = d[f(x)]/dx = d[f(x) + C]/dx$. We therefore refer to Eq. II–1 as an *indefinite integral*.

The differential calculus gives a precise prescription for obtaining the derivative of any function, $y = f(x)$, by using the rule given in Eq. I–2. No corresponding prescription exists in integral calculus for performing the inverse operation. Each integration must be considered individually; the integration process is a tentative one, requiring verification that the result, if differentiated, does in fact reproduce the original integrand. The task of evaluating integrals is made easier by the use of tables of standard integrals. To use such tables it is only necessary to reduce the integrand function of the desired integral to a form that is tabulated. You will find it worthwhile to become familiar with the use of integral tables.†

---

*The *integral sign* ∫ has its historical origin as a distorted symbol for the letter S (implying a sum, as we discuss in Calculus III). We refer to $\int g(x)\,dx$ as the integral of the function $g(x)$ (the *integrand*).

†A list of useful integrals appears inside the rear cover.

Exact Differentials. Sometimes we are required to evaluate an integral of the form

$$y = \int du(x)$$

where $du(x)$ is called an *exact differential*. Then,

$$y = u(x) + C$$

For example, suppose that $u(x) = x^2$, so that $du(x) = 2x\,dx$; thus,

$$y = \int 2x\,dx = x^2 + C$$

Or, suppose that $u(x) = \sin x$, so that $du(x) = \cos x\,dx$; thus,

$$y = \int \cos x\,dx = \sin x + C$$

Evaluation of an integral becomes trivial if the integrand can be converted into an exact differential.

Change of Variable. An unfamiliar integral can often be converted into a familiar one by making a clever substitution for the variable. For example, the integral

$$\int 2x \cos x^2\,dx$$

is not a familiar integral. However, let us make the substitution

$$u = x^2 \quad \text{and} \quad du = 2x\,dx$$

Then, we have

$$\int 2x \cos x^2\,dx = \int \cos u\,du = \sin u + C = \sin x^2 + C$$

We must often deal with integration variables that appear in combination with constants. Again, a simple variable change puts these integrals into standard form. For example, using $v = ax$ and $dv/a = dx$, we have

$$\int e^{ax}\,dx = \int e^v \frac{dv}{a} = \frac{1}{a}e^v + C = \frac{1}{a}e^{ax} + C$$

or, without making the substitution explicitly, we could write

$$\int e^{ax}\,dx = \frac{1}{a}\int e^{ax}\,d(ax) = \frac{1}{a}e^{ax} + C$$

There are no general rules that prescribe the best way to choose a variable substitution in every case. Facility in handling these situations is gained largely through experience.

Example II–1

Given the following functions $g(x)$, find $y = f(x)$ in each case such that $df(x)/dx = g(x)$:
(a) $g(x) = x^2$ (c) $g(x) = a$
(b) $g(x) = x^n,\ n > 0$ (d) $g(x) = \sin x$

Solution:

(a) We have

$$dy = x^2\,dx$$

Then, using Eq. II–1, we can write

$$y = \int x^2\,dx = \tfrac{1}{3}x^3 + C$$

This result is evidently correct, because

$$\frac{df(x)}{dx} = \frac{d}{dx}(\tfrac{1}{3}x^3 + C) = x^2$$

Similarly,

(b) $\quad y = \int x^n\,dx = \dfrac{1}{(n+1)}x^{n+1} + C,\ n \neq -1$

(c) $\quad y = \int a\,dx = ax + C$

(d) $\quad y = \int \sin x\,dx = -\cos x + C$

Differentiate the function $y = f(x)$ in each of the above cases to verify that the original function $g(x)$ is obtained.

Example II–2

Given the function $g(x) = 2x^2 - 6x - 3$, find $y = f(x)$ such that $df(x)/dx = g(x)$.

Solution:

We have

$$dy = (2x^2 - 6x - 3)\,dx$$

Thus,

$$y = \int (2x^2 - 6x - 3)\,dx = 2\int x^2\,dx - 6\int x\,dx - 3\int dx$$

$$= \tfrac{2}{3}x^3 - 3x^2 - 3x + C$$

II-2 INTEGRATION OF HIGHER DERIVATIVES

How do we proceed if we are given the second derivative of a function and are asked to find the first derivative? That is, given $d^2F(x)/dx^2$, how do we find $dF(x)/dx$? If we write $dF(x)/dx = f(x)$, then the first derivative of $f(x)$ is (by definition) the second derivative of $F(x)$:

$$\frac{df(x)}{dx} = \frac{d}{dx}\left(\frac{dF(x)}{dx}\right) = \frac{d^2F(x)}{dx^2}$$

Thus, the problem reduces to the evaluation of the integral of an exact differential that we have already solved, namely,

$$\int \frac{d^2F(x)}{dx^2}\,dx = \int \frac{df(x)}{dx}\,dx = f(x) + C = \frac{dF(x)}{dx} + C$$

In general, the integral of the nth derivative of a function is equal to the $(n-1)$th derivative of the function (to within an additive constant).

Example II-3

Given $d^2y/dx^2 = x^3 - 2x + 7$, find dy/dx and $y = f(x)$.

Solution

We have

$$\frac{d^2y}{dx^2} = \frac{d}{dx}\frac{dy}{dx} = x^3 - 2x + 7$$

so that

$$\frac{dy}{dx} = \int (x^3 - 2x + 7)\,dx$$

$$= \tfrac{1}{4}x^4 - x^2 + 7x + C_1$$

Then,

$$y = f(x) = \int (\tfrac{1}{4}x^4 - x^2 + 7x + C_1)\,dx$$

$$= \tfrac{1}{20}x^5 - \tfrac{1}{3}x^3 + \tfrac{7}{2}x^2 + C_1x + C_2$$

Boundary Conditions. If it is required to integrate a given function to obtain $f(x)$, the constants that appear with each integration can be evaluated only if more information is given. In fact, if we integrate $d^nf(x)/dx^n$ to obtain $f(x)$, there will be n constants of integration, which can

be evaluated if we are given the value of $f(x)$ at n different points. For example, if we integrate $d^2f(x)/dx^2$ twice to find $f(x)$, we might be given the value of $f(x)$ for two values of the independent variable, say, $x = a$ and $x = b$, from which the two integration constants, C_1 and C_2, could be determined. We could equally well use the value of $df(x)/dx$ at, say, $x = a$ to determine C_1, and the value of $f(x)$ at $x = b$ (or even at $x = a$) to determine C_2. (● Can you also evaluate C_1 and C_2 by using the value of $df(x)/dx$ at two different points, such as $x = a$ and at $x = b$?)

The additional quantities that must be specified in order to make definite the solution of an integration problem are called the *boundary conditions* on the solution.

Example II-4

Given $d^2f(x)/dx^2 = x^2 - 2x + 1$ and the boundary conditions $f(0) = 3$ and $f(-2) = -1$, find $f(x)$.

Solution:

With

$$\frac{d^2f(x)}{dx^2} = x^2 - 2x + 1$$

we have

$$\frac{df(x)}{dx} = \tfrac{1}{3}x^3 - x^2 + x + C_1$$

and

$$f(x) = \tfrac{1}{12}x^4 - \tfrac{1}{3}x^3 + \tfrac{1}{2}x^2 + C_1x + C_2$$

Substituting $x = 0$ into this solution and using $f(0) = 3$, we find $C_2 = 3$. Similarly, substituting $x = -2$ and using $f(-2) = -1$, we find $C_1 = 5$. Therefore,

$$f(x) = \tfrac{1}{12}x^4 - \tfrac{1}{3}x^3 + \tfrac{1}{2}x^2 + 5x + 3$$

Example II-5

Given $d^2f(x)/dx^2 = x - 3$ and the boundary conditions $df(x)/dx\big|_{x=1} = -2$ and $f(1) = +2$, find $f(x)$.

Solution:

Here,

$$\frac{d^2f(x)}{dx^2} = x - 3$$

from which

$$\frac{df(x)}{dx} = \tfrac{1}{2}x^2 - 3x + C_1$$

When $x = 1$, we have

$$-2 = \tfrac{1}{2} - 3 + C_1$$

or $\qquad C_1 = \tfrac{1}{2}$

Thus, $\qquad \dfrac{df(x)}{dx} = \tfrac{1}{2}x^2 - 3x + \tfrac{1}{2}$

Therefore, $\qquad f(x) = \tfrac{1}{6}x^3 - \tfrac{3}{2}x^2 + \tfrac{1}{2}x + C_2$

Now, $f(1) = +2$, so that

$$2 = \tfrac{1}{6} - \tfrac{3}{2} + \tfrac{1}{2} + C_2$$

from which $\qquad C_2 = \tfrac{17}{6}$

Finally, $\qquad f(x) = \tfrac{1}{6}x^3 - \tfrac{3}{2}x^2 + \tfrac{1}{2}x + \tfrac{17}{6}$

Example II-6

Assume that the rate at which knowledge is developed, dK/dt, is directly proportional to the knowledge K already present. If the time required for accumulated knowledge to double is 15 years, how much more knowledge will there be by the year 2000 than was present in the year 1900?

Solution:

We have $K = K(t)$, and we know that

$$\frac{dK}{dt} = \lambda K(t)$$

where λ is a proportionality constant. Rearranging and integrating both sides, we find

$$\int \frac{dK}{K} = \int \lambda \, dt$$

$$\ln K = \lambda t + \ln C$$

so that $\qquad \ln \dfrac{K}{C} = \lambda t$

Then, taking the exponential of both sides of this equation, we have

$$K(t) = Ce^{\lambda t}$$

Next, in view of the 15-year doubling time, we write

$$K(t + 15\text{ y}) = Ce^{\lambda(t+15\text{ y})} = 2K(t)$$

Dividing the second of these equations by the first gives

$$e^{\lambda(15\text{ y})} = 2$$

so that $\qquad \lambda = \dfrac{\ln 2}{15\text{ y}} = 0.0462 \text{ y}^{-1}$

Thus, $\qquad K(t) = Ce^{0.0462\,t}$

where t is measured in years. We now write

$$K(1900) = Ce^{(0.0462)(1900)}$$

and $\qquad K(2000) = Ce^{(0.0462)(2000)}$

Then, upon dividing, we find

$$\frac{K(2000)}{K(1900)} = e^{(0.0462)(100)} = e^{4.62} = 101.6$$

That is, knowledge increases by a little more than a factor of 100 each century.

PROBLEMS

Section II-1

II-1 Verify the following:

(a) $\displaystyle\int (2x^2 - 7)\, dx = \tfrac{2}{3}x^3 - 7x + C$

(b) $\displaystyle\int \cos\frac{2\pi}{\tau} t\, dt = \frac{\tau}{2\pi}\sin\frac{2\pi}{\tau} t + C$

(c) $\displaystyle\int e^{-at}\, dt = -\frac{1}{a}e^{-at} + C$

II-2 Given the function $g(x) = \sin x \cos x$, find $f(x)$ such that $df(x)/dx = g(x)$.

II-3 Determine the integral of $x\sqrt{2x + 1}\, dx$. [*Hint:* Make a change in variable.]

II-4 Evaluate $\displaystyle\int \frac{u^2}{u^3 + 1}\, du$. [*Hint:* Make a change of variable.]

Section II-2

II-5 Given $d^n f(x)/dx^n$, find $f(x)$ for the following cases:
(a) $d^2 f(x)/dx^2 = 2x + 1$
(b) $d^2 f(x)/dx^2 = \cos(\frac{1}{2}\pi x)$
(c) $d^3 f(x)/dx^3 = \frac{1}{2}(x + a)^2$

II-6 Given $df(x)/dx = ax + b$ and $f(0) = 0$, find $f(x)$.

II-7 Given $d^2 f(t)/dt^2 = a$, $df(t)/dt = v_0$ at $t = 0$, and $f(0) = s_0$, find $f(t)$.

II-8 Given $d^2 f(t)/dt^2 = a$, $f(0) = s_0$, and $f(1) = s_1$, find $f(t)$.

II-9 Given $d^2 f(t)/dt^2 = e^{-t}$, $df(t)/dt = 0$ as $t \to \infty$, and $f(0) = 0$, find $f(t)$.

II-10 Given $d^2 f(x)/dx^2 = \sin(\frac{1}{2}\pi x)$, $df(x)/dx = 0$ when $x = 1$, and $f(1) = (2/\pi)^2$, find $f(x)$.

II-11 Assume a model of the Earth's population growth in
• which the rate of population increase, dN/dt, is directly proportional to the population N. The world population in 1970 was 3.63×10^9. If the population doubles every 35 years, what will be the world population in the year 2020? The area of the Earth's inhabited continents is 1.36×10^8 km². How much area will each person have, on the average, in the year 2500 (assuming that this growth pattern is maintained)?

KINEMATICS OF LINEAR MOTION

We live in a Universe of continual motion. In every piece of matter, the atoms are in a state of unceasing agitation. We move around on the Earth's surface while the Earth moves in its orbit around the Sun. Although the stars in the sky seem to remain motionless as the Earth revolves beneath them, they too are in motion. Even the enormous collections of stars—the galaxies such as our own Milky Way—are moving through the vastness of space.

Every physical process involves motion of some sort. Because motion is such an important feature of the world around us, it is the logical subject with which to begin the study of physical phenomena. The ideas developed here are applied in all of the topics that follow—in describing planetary motion, in discussing fluids and electric current, and in studying the behavior of atoms. Motion is an essential common feature in all physical processes.

In this chapter we begin the study of *mechanics,* a subject that is developed for much of the remainder of this volume. It is appropriate that the discussion of physical topics start with mechanics because this was the first of the subdivisions of physics to be developed, and because it has become one of the cornerstones of modern physical science.

Mechanics can be logically divided into two parts, *kinematics* and *dynamics.* Kinematics is concerned with the geometrical *description* of motion; it does not address the question of the *cause* of motion. Dynamics, on the other hand, relates the motion of objects to their properties and to the forces that act on them.

We begin by treating the kinematics of simple straight-line motion. This introduction to the subject permits the unhurried development of the mathematical tools required for accurate descriptions of the way objects move.

Abstractions and Idealizations. To make the discussions easier, it is useful to introduce various abstractions and idealizations. In this chapter we begin dealing with the motion of objects. The real objects of everyday experience all have measurable size and recognizable physical structure. The actual motion of objects is in general quite complicated and involves rotations (changes in orientation) as well as translations (changes in spatial position). You can certainly appreciate this statement if you have ever witnessed the wobbly flight of a poorly thrown football!

The discussions can be simplified by introducing the concept of a *particle,* which is the physical counterpart of the idea of the mathematical point. A particle has no

33

physical size or structural features and is not capable of rotational motion. (However, we eventually endow these particles with certain properties such as mass and electric charge.)

In our model for particle motion we imagine a trajectory or path that is followed by the particle. In time, the particle occupies *every point* on this path in *continuous succession,* just as we suppose real objects do. This motion occurs whether we observe it or not, and we imagine that there is possible a mode of observation that in no way alters the motion. We therefore introduce the idealization of a "neutral observer." We take for granted the notion of a straight line and, indeed, all the propositions of Euclidean geometry.

Are these various abstractions realistic? Is the concept of a particle even plausible? First, we must appreciate that a *model* is developed to represent in some simple but approximate way an aspect of the real physical world. Models are used whenever we do not understand completely some physical process or if the correct theory is so cumbersome that it does not permit calculations to be made easily. A model is considered successful if it describes the way Nature behaves to the level of accuracy we demand. An unsophisticated view of mechanics is sufficient, for example, to design a toy cart; but a highly developed theory is necessary to account for the detailed motions of Earth satellites.

We should also note that apparently pointlike particles do exist in Nature. There are now theoretical speculations that *electrons,* particles found in all atoms, have a size that is no larger than about 10^{-18} m; in fact, with present experimental techniques, electrons have *no* measurable size. Thus, electrons can be considered to be ideal pointlike particles for all our purposes. So also for atoms and, usually, for grains of sand or pebbles. Indeed, the objects that we wish to represent as particles are required only to have a size that is small compared with the dimensions of the trajectory we are considering, or their sizes must be irrelevant to the motion studied. Thus, to a reasonable approximation, the Earth (and, indeed, any planet) may be considered to be a particle for purposes of discussing its orbital motion around the Sun. (The concept of the center of mass, to be introduced later, makes this approximation better than might be judged simply on the basis of the scale of sizes involved.) The Sun and stars, in turn, may be considered particles when galactic motions are involved.

It is extremely useful to model the behavior of Nature in terms of particles—if we keep in mind the limitations. The idea of particles is used in many of the discussions that follow.

2-1 POSITION AND DISPLACEMENT

Consider a straight-line scale along which the progress of a moving particle can be followed. On this *coordinate line* locate an origin (which we usually label O) and mark off distance intervals in a convenient unit of length, such as meters. Label the intervals on one side of the origin with increasing positive numbers and label those on the other side with increasing negative numbers. In this way, we construct a *directed* coordinate line. The motion of a particle along such a coordinate line can be followed by recording the position s occupied by the particle at the corresponding time t. These time-position pairs are written as (t_1, s_1), (t_2, s_2), and so on, and are marked along the directed coordinate line as shown in Fig. 2–1.

It is evident from Fig. 2–1 that the position of the particle is determined by its distance from the origin O. The choice of some other point as the origin would have

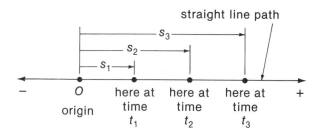

Fig. 2–1. The progress of a particle moving in a straight line is recorded in terms of the time-position pairs (t_1, s_1), (t_2, s_2), and so forth.

been equally acceptable. It seems clear that no essential feature of the description of a particle's motion should depend on how the origin is chosen.

It is usually convenient to pick arbitrarily a reference instant for time zero and to quote the times t_1, t_2, and so on, in terms of the *elapsed time* after the reference instant t_0. There is no requirement that the particle be at the coordinate origin O at time t_0 (although the location of the origin can be adjusted to make this correspondence).

The motion of a particle can be represented in graphical form as in Fig. 2–2a, where the time-coordinate pairs, (t_0, s_0), (t_1, s_1), (t_2, s_2), and so on, are plotted as open circles. We assume that there is an infinite density of additional pairs (t, s) that correspond to intermediate points of progress, some of which are shown as solid dots in Fig. 2–2a. In this way we arrive at the concept of s as a continuous function of t (see Section I–2), and write

$$s = f(t) \qquad \text{or} \qquad s = s(t)$$

The notation in each of these equations means "the position s as a function of the time t." The points in Fig. 2–2a can be connected with a smooth (continuous) curve to obtain the graphical representation of $s(t)$ shown in Fig. 2–2b. To indicate that the position function $s(t)$ is to be evaluated at some particular time—for example, at $t = 10\,\text{s}$—we can write with equivalent meaning,

$$s(t = 10\,\text{s}), \qquad s(10\,\text{s}), \qquad s(10)$$

Fig. 2–2. (a) A graph of the coordinate locations s at different times t for a moving particle, s_1 at t_1, s_2 at t_2, etc., shown by open circles, plus possible intermediate values shown with solid dots. (b) The abstraction of s as a function of t, that is, $s = s(t)$, based on the observed data points.

(a)   (b)

In the last instance we suppose it is clear from the context that the time value is given in seconds. This notation is used throughout the text.

To further facilitate the description of motion, we introduce a quantity called the *displacement,* the distance between two points actually occupied by the particle as it moves along the path. Thus, the (algebraic) distance $s_2 - s_1$ is the displacement of the particle during the elapsed time interval from t_1 to t_2.

Example 2–1

Suppose that the position of a particle traveling in a straight line is given by the expression

$$s = s(t) = bt + c$$

with $b = 5.0 \text{ m/s}$ and $c = -3.0 \text{ m}$; t is expressed in seconds and s in meters. What is the coordinate location of the particle at elapsed times $t = 0.5 \text{ s}$ and $t = 10 \text{ s}$?

Solution:

For $t = 0.5 \text{ s}$, we have

$$
\begin{aligned}
s(0.5 \text{ s}) &= (5.0 \text{ m/s})(0.5 \text{ s}) - (3.0 \text{ m}) \\
&= (2.5 \text{ m}) - (3.0 \text{ m}) \\
&= -0.5 \text{ m}
\end{aligned}
$$

The negative sign here indicates that the position of the particle at $t = 0.5 \text{ s}$ is to the *left* of the origin. For $t = 10 \text{ s}$, we have

$$
\begin{aligned}
s(10 \text{ s}) &= (5.0 \text{ m/s})(10 \text{ s}) - (3.0 \text{ m}) \\
&= 47.0 \text{ m}
\end{aligned}
$$

Where was the particle at $t = -1 \text{ s}$ (that is, at a time 1 s *earlier* than the arbitrarily chosen reference instant)?

$$
\begin{aligned}
s(-1 \text{ s}) &= (5.0 \text{ m/s})(-1 \text{ s}) - (3.0 \text{ m}) \\
&= (-5.0 \text{ m/s}) - (3.0 \text{ m}) \\
&= -8.0 \text{ m}
\end{aligned}
$$

which is a point 8.0 m to the left of the origin.

▶ *Physically Allowable Functions.* What kinds of mathematical functions $s(t)$ can be used to represent physically allowable motions? We have the intuitive feeling that only *continuous* functions are acceptable because a particle, in moving from one location to another, must occupy *all* intermediate positions in time sequence. For example, the position function $s(t) = a/t$ is not an acceptable position function near $t = 0$ because it is not defined at that point.

The position s and the time t are physical *observables;* that is, they can be measured with instruments such as meter sticks, clocks, and the like. Because a physical observable can be *observed,* we require that it correspond to a mathematically *real* quantity (as distinct from an imaginary quantity). If we have the position function

$$s(t) = \sqrt{at - b}$$

then physical reality can be ascribed only to times $t \geqslant b/a$, for which the radicand is non-negative. For $t < b/a$ we would say that $s(t)$ represents a "physically unrealizable" motion (because $s(t)$ is imaginary for this range of values for t). Furthermore, each *branch* of the function (that is, $s = +\sqrt{at - b}$ and $s = -\sqrt{at - b}$ corresponds to a separate history of the motion because the particle cannot be at two positions at the same time.

In Section 2–2 we point out that a physically realizable position function $s(t)$ not only must be continuous, but also it must have continuous derivatives. In fact, we identify the time derivative of $s(t)$ as the *velocity* and the second time derivative as the *acceleration*. Almost always, our approximations are such that we do not inquire about derivatives of order higher than the second. ◄

2-2 VELOCITY

Suppose that the position function of a particle is $s = s(t)$. We now inquire about the *rate* of the motion described by this function. At a certain instant t_a, the particle is at the position P, for which $s = s(t_a)$; at some elapsed time Δt later, it is at the position Q, for which $s = s(t_a + \Delta t)$, as shown in Fig. 2–3. During the time interval Δt, the particle has undergone a displacement $\Delta s = s(t_a + \Delta t) - s(t_a)$. The average time rate of the motion is called the particle's *average velocity** $\bar{v}$ and is defined to be

$$\bar{v} = \frac{s(t_a + \Delta t) - s(t_a)}{(t_a + \Delta t) - t_a} = \frac{\Delta s}{\Delta t} \tag{2-1}$$

According to Eq. 2–1, the dimensions of velocity are those of distance divided by time; that is, $[v] = LT^{-1}$. We usually measure velocity in units of *meters per second* (m/s or m · s⁻¹), although we may sometimes use cm/s, km/s, or km/h.

The quantity $\bar{v}$ gives the *average* velocity of a particle that undergoes a total displacement Δs during the time interval Δt. It should be noted that $\bar{v}$ depends on both t_a and Δt (see Fig. 2-3b). The information provided by the value of $\bar{v}$ concerning the progress of the motion is incomplete and may even be misleading. For

---

*An overbar is used on a symbol to represent an *average* value.

Fig. 2–3. (a) The positions of a particle at $t = t_a$ (P) and $t = t_a + \Delta t$ (Q) along a coordinate line. (b) The graph of the position function $s = s(t)$, showing the relevant quantities.

(a)

(b)

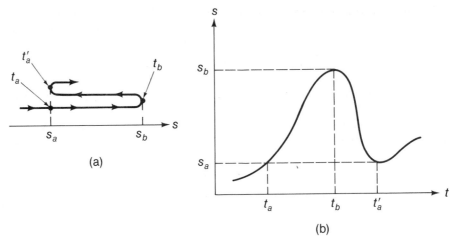

Fig. 2–4. (a) Motion of a particle, $s_a \rightarrow s_b \rightarrow s_a$, during the time interval $\Delta t = t_a' - t_a$. (b) A position-time graph of the motion for the same time interval, showing that the total displacement is $\Delta s = 0$, so the average velocity $\bar{v}$ is also zero. However, the average speed during the interval is *not* zero: $\Delta \ell / \Delta t = 2(s_b - s_a)/(t_a' - t_a)$.

example, Fig. 2–4 shows the motion of a particle that is at a position s_a at time t_a and moves to s_b at time t_b, finally returning to s_a at time t_a'. For the time interval $\Delta t = t_a' - t_a$, the total displacement is $\Delta s = 0$. Thus, $\bar{v} = 0$ for this time interval even though the particle undergoes considerable motion between times t_a and t_a'. To give a means for describing this motion, we introduce the idea of *average speed*, defined as the *total path length* $\Delta \ell$ traveled by the particle divided by the corresponding time interval Δt:

$$\text{average speed} = \frac{\Delta \ell}{\Delta t}$$

For motion along a straight line, the average velocity and the average speed are equal unless the particle reverses its direction of motion during the time interval considered. Then, $\Delta \ell / \Delta t$ is always numerically greater than $\bar{v}$ (as in Fig. 2–4, where $\bar{v} = 0$).

We can give more precise information concerning the motion of a particle by defining the *instantaneous velocity*. Imagine a limiting process (see Section I–2) in which the time interval Δt is allowed to decrease to smaller and smaller values. Hold t_a fixed, and at each of the successively smaller values of Δt evaluate the average velocity $\bar{v}$. As Δt is made smaller and smaller, $\bar{v}$ changes less and less. In fact, $\bar{v}$ approaches a limiting value as $\Delta t \rightarrow 0$. (We assume that a limiting value actually exists.) By allowing Δt to approach zero, the *instantaneous velocity* at t_a is determined. This velocity is denoted by v (without an overbar).

Now that *average velocity* and *instantaneous velocity* have been defined, we really do not need the qualifier "instantaneous." The term *velocity* is used to mean the instantaneous value. When we require the velocity to be averaged over a particular time or space interval, the term *average velocity* is used.

For motion along a straight line, the velocity v may be either positive or negative; that is, the particle may be moving either to the right or to the left. Notice that the instantaneous values of the velocity and the speed are equal in magnitude. Speed is a positive definite quantity, but velocity can be negative.

Figure 2–5a is a position-time graph that shows the average velocity,

$$\bar{v} = \frac{\Delta s}{\Delta t} = \frac{s(t_a + \Delta t) - s(t_a)}{\Delta t} = \tan \beta$$

That is, $\bar{v}$ is equal to the *slope* of the line $\overline{PQ}$ that connects the end-points of the interval considered.

Figure 2–5b shows the limiting process for obtaining the velocity at time t_a. As $\Delta t \to 0$, the line $\overline{PQ}$ becomes the line $\overline{TT'}$, which is tangent to the curve $s(t)$ at the point P (that is, for time t_a). Thus,

$$v(t_a) = \operatorname*{Lim}_{\Delta t \to 0} \frac{\Delta s}{\Delta t} = \operatorname*{Lim}_{\Delta t \to 0} \frac{s(t_a + \Delta t) - s(t_a)}{\Delta t} = \tan \alpha \qquad \textbf{(2-2)}$$

We recognize v to be the *general derivative* of $s(t)$ evaluated at t_a (see Section I–3). For any differentiable function $s(t)$, we have

$$v(t_a) = \operatorname*{Lim}_{\Delta t \to 0} \frac{s(t + \Delta t) - s(t)}{\Delta t} \Bigg|_{t=t_a}$$

In the usual calculus notation, we write

$$v(t) = \frac{ds}{dt} = \frac{d}{dt} s(t) \quad \text{and} \quad v(t_a) = \frac{ds}{dt} \Bigg|_{t=t_a} \qquad \textbf{(2-3)}$$

The velocity is the general time derivative of the position function. Note particularly, in contrast with the definition of $\bar{v}$, that the instantaneous velocity does not depend

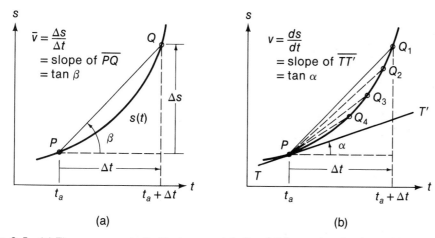

(a)   (b)

Fig. 2–5. (a) The average velocity $\bar{v}$ between points P and Q is equal to the slope of the line $\overline{PQ}$. (b) The velocity v at time t_a is obtained by progressively decreasing the time interval Δt, thus moving the point at the end of the interval, given by $s(t_a + \Delta t)$, ever closer to the point P, which corresponds to $s(t_a)$. During the limiting process, we have $Q_1 \to Q_2 \to Q_3 \to \cdots \to P$, and $\beta \to \alpha$, where $\tan \alpha$ is the slope of the tangent line $\overline{TT'}$ at P.

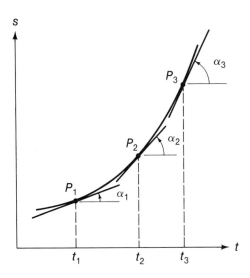

Fig. 2–6. The velocity of a particle at any time is equal to the slope of the distance-time curve at that time. Thus, $v(t_1) = \tan \alpha_1$.

on Δt. The velocity v of an automobile, for example, is the quantity indicated by the speedometer reading.

For most practical cases, Eq. 2–3 is evaluated at a specific time such as t_a. Figure 2–6 emphasizes the point that the velocity at a particular time is equal to the slope of the distance-time curve at that time. Also, note that neither $\bar{v}$ nor v can depend on the choice for the location of the origin of the coordinate line, because only the displacements Δs (that is, the differences in position coordinates) are involved in the calculations.

Example 2–2

Given the position function

$$s(t) = \tfrac{1}{4}gt^2 + 2bt$$

where $g = 1$ m/s² and $b = 1$ m/s, evaluate the average velocity $\bar{v}$ during elapsed time intervals starting at $t_a = 2$ s for which $\Delta t = 1$ s, 0.5 s, 0.1 s, 0.01 s, and 0.001 s.

Solution:

Evaluating $s(t_a) = s(2\text{ s})$, we find

$$s(2\text{ s}) = \tfrac{1}{4}(1 \text{ m/s}^2)(2 \text{ s})^2 + 2(1 \text{ m/s})(2 \text{ s}) = 5 \text{ m}$$

Then, using Eq. 2–1 for $\bar{v}$, we have

$$\bar{v} = \frac{s(2\text{ s} + \Delta t) - (5 \text{ m})}{\Delta t}$$

For $\Delta t = 1$ s,

$$\bar{v} = \frac{s(3\text{ s}) - (5 \text{ m})}{1 \text{ s}} = \frac{\tfrac{9}{4} + 6 - 5}{1} \text{ m/s} = 3.250 \text{ m/s}$$

Substituting the remaining values of Δt, we obtain the following results for $\bar{v}$:

Δt (s)	1	0.5	0.1	0.01	0.001
$\bar{v}$ (m/s)	3.250	3.125	3.025	3.0025	3.00025

From these values for $\bar{v}$ it appears that in the limit $\Delta t \to 0$, we have $v = 3$ m/s (exactly). We can see the dependence of $\bar{v}$ on Δt if we write

$$\Delta s = s(t_a + \Delta t) - s(t_a)$$
$$= [\tfrac{1}{4}g(t_a + \Delta t)^2 + 2b(t_a + \Delta t)] - [\tfrac{1}{4}gt_a^2 + 2bt_a]$$
$$= [\tfrac{1}{4}gt_a^2 + \tfrac{1}{2}gt_a \Delta t + \tfrac{1}{4}g(\Delta t)^2 + 2bt_a + 2b \Delta t] - [\tfrac{1}{4}gt_a^2 + 2bt_a]$$
$$= (\tfrac{1}{2}gt_a + 2b) \Delta t + \tfrac{1}{4}g(\Delta t)^2$$

Then,

$$\bar{v} = \frac{\Delta s}{\Delta t} = \tfrac{1}{2}gt_a + 2b + \tfrac{1}{4}g \Delta t$$

For $t_a = 2$ s, $g = 1$ m/s^2, and $b = 1$ m/s, we find

$$\bar{v}(2 \text{ s}) = (3 + \tfrac{1}{4}\Delta t) \text{ m/s}$$

and the dependence of $\bar{v}$ on Δt for this case is exhibited explicitly. It is easy to verify that this expression leads to the values of $\bar{v}$ listed in the table above. It is also clear that in the limit $\Delta t \to 0$ the velocity at $t = 2$ s is indeed exactly 3 m/s.

The instantaneous velocity at $t = 2$ s can also be obtained by differentiating $s(t)$ and evaluating the result at $t = 2$ s. We find

$$v(t) = \frac{ds}{dt} = \frac{d}{dt}s(t) = \tfrac{1}{2}gt + 2b$$

so that

$$v(2 \text{ s}) = \tfrac{1}{2}(1 \text{ m/s}^2)(2 \text{ s}) + 2(1 \text{ m/s}) = 3 \text{ m/s}$$

Example 2–3

Suppose that the position function of a particle is

$$s(t) = \frac{v_0}{\alpha}(1 - e^{-\alpha t})$$

where $\alpha > 0$ and v_0 are constants. What are the dimensions of α and v_0? What is the velocity of the particle for $t = 0$, $t = 1/\alpha$, and $t \to \infty$?

Solution:

Because the exponent of e must be dimensionless (•Why?), we have $[\alpha] = T^{-1}$. Also, the dimension of $s(t)$ is $[s] = L$; hence, $[v_0/\alpha] = L$, or $[v_0] = LT^{-1}$. That is, the dimensions of v_0 are those of *velocity*.

Differentiating $s(t)$ with respect to time, we obtain the velocity:

$$v(t) = \frac{d}{dt}s(t) = \frac{v_0}{\alpha}(-1)(-\alpha)e^{-\alpha t}$$

$$= v_0 e^{-\alpha t}$$

Notice that differentiating with respect to *time* leads to a derivative whose dimensions are those of the original function divided by time.

For the various values of t, we have

$$v(0) = v_0$$

$$v(1/\alpha) = v_0 e^{-1} \cong 0.368 \, v_0$$

$$\operatorname*{Lim}_{t \to \infty} v(t) = 0$$

Example 2–4

Two objects A and B are connected by a rigid rod that has a length L. The objects slide along perpendicular guide rails, as shown in the figure. If A slides to the left with a constant speed V, find the velocity of B when $\alpha = 60°$.

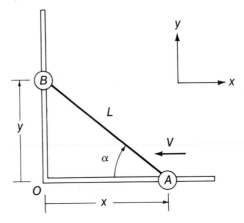

Solution:

We are given that $dx/dt = -V$, where the negative sign indicates that A moves toward the left. We are required to find dy/dt, the velocity of B. Because AOB is a right triangle, we can obtain a relationship connecting x and y by writing

$$x^2 + y^2 = L^2$$

from which we see that x and y are related functions of time. Differentiating this expression with respect to time, we find

$$2x \frac{dx}{dt} + 2y \frac{dy}{dt} = 0$$

or,

$$\frac{dy}{dt} = -\frac{x}{y} \frac{dx}{dt}$$

Therefore,

$$\frac{dy}{dt} = \frac{x}{y} V = V \operatorname{ctn} \alpha$$

When $\alpha = 60°$, $\operatorname{ctn} \alpha = 1/\sqrt{3}$, and we have

$$\frac{dy}{dt} = V/\sqrt{3}$$

(•What is the velocity of B when $\alpha = 0°$?)

Example 2–5

A particle's position function is $x(t) = 2at^2$, with $a = 1 \text{ m/s}^2$, for $t < 1$ s; and is $x(t) = bt + c$, with $b = 1 \text{ m/s}$ and $c = 1 \text{ m}$, for $t > 1$ s. By definition, $x(1 \text{ s}) = 2 \text{ m}$. Describe the way in which the velocity of the particle changes with time.

Solution:

The position function is continuous for all values of t, including $t = 1$ s (see the diagram below). However, the velocity is

$$v = \frac{dx}{dt} = 4at, \qquad \text{for } t < 1 \text{ s}$$

$$v = \frac{dx}{dt} = b, \qquad \text{for } t > 1 \text{ s}$$

The velocity is not defined at $t = 1$ s, as can be seen in the right-hand figure. Nevertheless, the *average velocity* is defined everywhere. For example, the average velocity during the interval from $t = 0$ to $t = 2$ s is

$$v = \frac{x(2 \text{ s}) - x(0)}{2 \text{ s}} = \frac{3 \text{ m} - 0}{2 \text{ s}} = 1.5 \text{ m/s}$$

A (nearly) discontinuous change in the velocity could be produced by striking the particle a sharp blow.

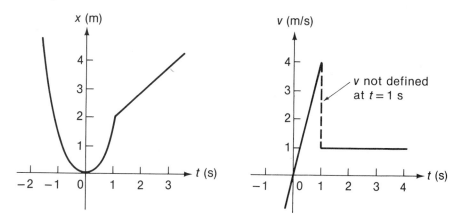

The difficulties that arise because a function is nondifferentiable at a finite number of points can always be avoided, at least formally, by introducing a truly physical but somewhat more complicated function. Thus, to describe the motion of the particle in this example, we could find a function that is curved at $t = 1$ (instead of having a sharp change) and therefore has a continuous derivative. We then take the attitude that the simple nondifferentiable function is a "caricature" of the truly physical function and is accurate everywhere except at the troublesome points in question. (See also Problems 2–52 and 2–53 at the end of this chapter.)

2-3 ACCELERATION

Both the velocity and the position of a particle may be functions of time. A particle that moves in such a way that it "speeds up" or "slows down" undergoes accelerated motion, and the time rate of change of the velocity is called the *acceleration* of the particle.

The *average acceleration* $\bar{a}$ of a particle during a time interval Δt, starting at t, is defined to be

$$\bar{a} = \frac{v(t + \Delta t) - v(t)}{\Delta t} = \frac{\Delta v}{\Delta t} \tag{2-4}$$

The *acceleration* a at the instant t is defined to be the limit of Eq. 2–4 as Δt approaches zero (see Fig. 2–7); thus,

$$a = \underset{\Delta t \to 0}{\text{Lim}} \frac{v(t + \Delta t) - v(t)}{\Delta t}$$

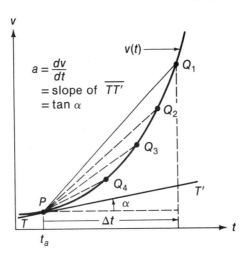

Fig. 2–7. The acceleration a at time t_a is obtained by progressively decreasing the time interval Δt. This moves the point Q at the end of the interval ever closer to P. In the limit, $\Delta v / \Delta t$ equals the slope of the tangent line $\overline{TT'}$ at the point P which corresponds to $v(t_a)$. Compare Fig. 2–5a for the case of velocity.

and we say that *the acceleration is the general time derivative of the velocity*. Because the velocity is the time derivative of the position function, the acceleration is in turn the second derivative of the position function; that is,

$$a(t) = \frac{dv}{dt} = \frac{d}{dt}\frac{ds}{dt} = \frac{d^2s}{dt^2} \tag{2-5}$$

The velocity at a particular instant is equal to the slope of the position-time graph at that instant (Fig. 2–8a). In the same way, the acceleration at a particular instant is equal to the slope of the velocity-time graph at that instant (Fig. 2–8b). It

Fig. 2–8. (a) Velocity is equal to the slope of the *s-t* graph. (b) Acceleration is equal to the slope of the *v-t* graph.

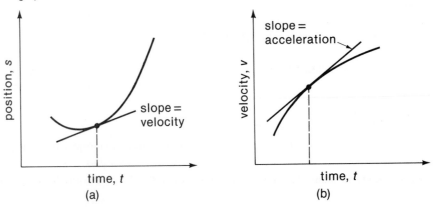

is important to understand that $v = 0$ does *not* imply $a = 0$. (•Can you think of examples from everyday experience in which $v = 0$ but $a \neq 0$?)

Note that the dimensions of acceleration are $(LT^{-1})/T = LT^{-2}$. Acceleration is usually quoted in units of *meters per second per second* or *meters per second squared,* which is written as m/s².

Example 2–6

The velocity of a particle increases quadratically with time according to the function

$$v(t) = \tfrac{1}{2}\beta t^2$$

where β is a constant with the appropriate dimensions.
(a) What is the average acceleration during a 2-s interval starting at $t = 1$ s?
(b) What is the instantaneous acceleration at $t = 1$ s and $t = 3$ s?

Solution:

(a) We have

$$v(t + \Delta t) = \tfrac{1}{2}\beta(t + \Delta t)^2 = \tfrac{1}{2}\beta t^2 + \beta t(\Delta t) + \tfrac{1}{2}\beta(\Delta t)^2$$

Therefore,

$$\bar{a} = \frac{v(t + \Delta t) - v(t)}{\Delta t} = \beta t + \tfrac{1}{2}\beta(\Delta t)$$

Thus, for $t = 1$ s and $\Delta t = 2$ s, we find

$$\bar{a} = \beta + \tfrac{1}{2}\beta(2 \text{ s}) = 2\beta \text{ m/s}^2$$

Alternatively, we have

$$v(t + \Delta t) = v(1 \text{ s} + 2 \text{ s}) = \tfrac{1}{2}\beta(3 \text{ s})^2 = 4.5\beta \text{ m/s}$$

$$v(t) = v(1 \text{ s}) = \tfrac{1}{2}\beta(1 \text{ s})^2 = 0.5\beta \text{ m/s}$$

and

$$\bar{a} = \frac{\Delta v}{\Delta t} = \frac{4.5\beta - 0.5\beta}{2 \text{ s}} = 2\beta \text{ m/s}^2$$

(b) For the acceleration, we have

$$a = \frac{dv}{dt} = \frac{d}{dt}(\tfrac{1}{2}\beta t^2) = \beta t \text{ m/s}^2$$

Therefore,

$$a(1 \text{ s}) = \beta \text{ m/s}^2$$

and

$$a(3 \text{ s}) = 3\beta \text{ m/s}^2$$

Example 2–7

The position function of a particle is

$$x(t) = A \sin \omega t$$

where A and ω are constants with appropriate dimensions. Find the velocity and the acceleration.

Solution:

We have

$$v(t) = \frac{dx}{dt} = A\omega \cos \omega t$$

and

$$a(t) = \frac{dv}{dt} = \frac{d^2x}{dt^2} = -A\omega^2 \sin \omega t$$

By comparing the position function $x(t)$ with the resulting acceleration function $a(t)$, we notice that

$$\frac{d^2x}{dt^2} = -\omega^2 x$$

This equation describes "simple harmonic motion"—for example, the up-and-down oscillations of a block attached to a spring or the back-and-forth motion of a pendulum in a clock. We discuss this type of motion in detail in Chapter 15.

2-4   MOTION WITH CONSTANT ACCELERATION

Motion that takes place at constant velocity along a straight line is called *uniform linear motion*. In this case, with $v = v_0 = $ constant, the acceleration is necessarily zero:

$$a = \frac{dv}{dt} = 0 \qquad \text{for uniform motion}$$

When the velocity is constant, we also have $\bar{v} = v_0$. Suppose that at time $t = 0$ the position of a particle is s_0. Then, after an elapsed time t its position is $s(t)$, and we have

$$\bar{v} = v_0 = \frac{s(t) - s_0}{t - 0}$$

or

$$s(t) = v_0 t + s_0 \tag{2-6}$$

Equation 2–6 is the position function of a particle that moves with constant velocity v_0 and passes through the position s_0 at $t = 0$. This equation includes the case $v_0 = 0$, that is, $s(t) = s_0$, in which the particle remains at a certain position and experiences no displacement with time.

Next, consider the motion that results when the acceleration is constant and not necessarily zero. In this case, the acceleration a is not a function of time, so $\bar{a} = a$. If the velocity at $t = 0$ is v_0 and the velocity at time t is $v(t)$, we have

$$\bar{a} = a = \frac{v(t) - v_0}{t - 0}$$

or

$$v(t) = at + v_0 \tag{2-7}$$

We can determine the position function for this case by noting that when the velocity is a linear function of time, the average velocity during any time interval (as we see in Fig. 2–9) is just one half the sum of the initial and final velocities at the ends of the interval.*

---

*Actually, it is not trivial to prove this proposition by algebraic means. In Section 2–6 we obtain the same results by integration, thereby avoiding any reference to average values.

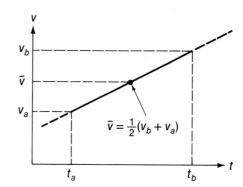

Fig. 2-9. The velocity of a particle varies linearly with time, with $v(t_a) = v_a$ and $v(t_b) = v_b$. The average velocity $\bar{v}$ for the interval $t_b - t_a$ is the mean between v_b and v_a; that is, $\bar{v} = \frac{1}{2}(v_b + v_a)$.

Now, the distance traveled (if always in the same direction) is always given by $\bar{v}t$. Therefore, we can write

$$s(t) = \bar{v}t + s_0 \qquad (2\text{-}8)$$

Then, for $\bar{v}$ we use the expression (see Fig. 2-9)

$$\bar{v} = \tfrac{1}{2}[v(t) + v_0]$$

Next, we substitute for $v(t)$ from Eq. 2-7 and write

$$\bar{v} = \tfrac{1}{2}[(at + v_0) + v_0] = \tfrac{1}{2}at + v_0$$

Then, using this expression for $\bar{v}$ in Eq. 2-8, we find

$$s(t) = \tfrac{1}{2}at^2 + v_0 t + s_0$$

This equation gives the position of a particle that moves with constant acceleration a, with velocity v_0 and position s_0 when $t = 0$. We summarize our results:

For	$a = \text{constant}$	
we have	$v(t) = at + v_0$	(2-9)
and	$s(t) = \tfrac{1}{2}at^2 + v_0 t + s_0$	(2-10)
with	$v = v_0 \quad \text{and} \quad s = s_0 \quad \text{at } t = 0$	

Note that these equations remain correct when $a = 0$; then, the equation for $s(t)$ reduces to Eq. 2-6.

It is reassuring to note that Eqs. 2-9 and 2-10 are consistent with our previous definitions. If the expression for $s(t)$ is used in the defining equations for v and a, we find

$$v(t) = \frac{ds}{dt} = \frac{d}{dt}(\tfrac{1}{2}at^2 + v_0 t + s_0)$$

$$= at + v_0$$

and

$$a(t) = \frac{dv}{dt} = \frac{d}{dt}(at + v_0) = a = \text{constant}$$

Figure 2–10 shows the position function $s(t)$ for a particle moving with constant acceleration (Eq. 2–10); the curve is a *parabola*. The position of the particle at $t = 0$ is $s = s_0$. The velocity at $t = 0$ is v_0 and is equal to the slope of the curve at this point. At any other time t, the velocity of the particle is equal to the slope of the curve at the corresponding point. (●Look again at Example 2–5. What is the acceleration for $t < 1$ s? For $t > 1$ s?)

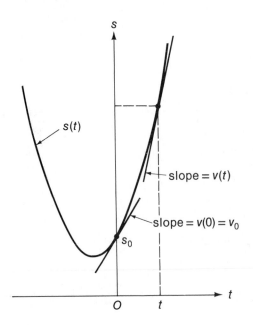

Fig. 2–10. The graph of the position function $s(t)$ for a particle moving with constant acceleration is a *parabola*. The position s_0 and the velocity v_0 at $t = 0$ are indicated.

It is sometimes useful to have an expression based on Eqs. 2–9 and 2–10 but which has no explicit reference to time. From Eq. 2–8 we have

$$s(t) = \bar{v}t + s_0$$

Then, using $\bar{v} = \frac{1}{2}(v + v_0)$, we find

$$s - s_0 = \frac{1}{2}(v + v_0)t$$

From Eq. 2–9 for $v(t)$, we have $t = (v - v_0)/a$. Thus,

$$s - s_0 = \frac{1}{2}(v + v_0)\frac{(v - v_0)}{a}$$

and the variable t has been eliminated. Then,

$$s - s_0 = \frac{v^2 - v_0^2}{2a}$$

or

$$v^2 - v_0^2 = 2a(s - s_0) \qquad \textbf{(2–11)}$$

Example 2–8

A particle is observed at time $t = 0$ to be at the coordinate position $x_0 = 5$ m and moving with a velocity $v_0 = 20$ m/s. The particle undergoes a constant deceleration (*i.e.*, an acceleration in the direction opposite to the velocity). If, 10 s after the initial observation, the particle has a velocity $v = 2$ m/s, what is the acceleration? What is the position function? How long a time will elapse before the particle returns to $x = 5$ m?

Solution:

In this problem,
$$v_0 = 20 \text{ m/s}$$
$$v(10 \text{ s}) = 2 \text{ m/s}$$
$$x_0 = 5 \text{ m}$$

with a and $x(t)$ to be determined.

Using $v = at + v_0$ from Eq. 2–9, we have

$$a = \frac{v - v_0}{t} = \frac{(2 \text{ m/s}) - (20 \text{ m/s})}{10 \text{ s}} = -1.80 \text{ m/s}^2$$

The position function, from Eq. 2–10, is

$$x(t) = (-0.9 \text{ m/s}^2)t^2 + (20 \text{ m/s})t + (5 \text{ m})$$

For the particle to return to $x = 5$ m, we have

$$5 \text{ m} = (-0.90 \text{ m/s}^2)t^2 + (20 \text{ m})t + (5 \text{ m})$$

or
$$t[(-0.90 \text{ m/s}^2)t + (20 \text{ m/s})] = 0$$

Hence,
$$t = 0 \quad \text{and} \quad t = \frac{20 \text{ m/s}}{0.90 \text{ m/s}^2} = 22.22 \text{ s}$$

The first solution, $t = 0$, is the initial condition, so the return to $x = 5$ m occurs at $t = 22.22$ s.

Example 2–9

In the American sport of drag racing, the performances are reported in British units instead of metric units. Thus, the length of the course is set at 1/4 mile (mi) and the speeds are measured in miles per hour (mi/h or mph).

A drag racer in the Top Fuel Class sometimes achieves a speed of 240 mi/h at the end of the 1/4-mi strip. What must be the acceleration (assumed constant) to reach this speed from a standing start ($v_0 = 0$)?

Solution:

We have
$$s - s_0 = 1/4 \text{ mi}$$
$$v_0 = 0$$
$$v = 240 \text{ mi/h}$$

with the acceleration a to be found.

Using Eq. 2–11 and solving for a, we have

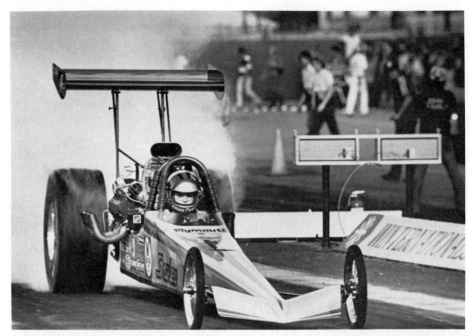

World Champion Shirley Muldowney reached a speed of 255.58 mi/h at the end of a quarter mile during the 1979 Winternationals. On three separate occasions she has driven that distance from a standing start in 5.77 seconds. (Courtesy of Atco Raceway and Chrysler Corporation.)

$$a = \frac{v^2 - v_0^2}{2(s - s_0)} = \frac{(240\ \text{mi/h})^2 - 0}{2(1/4\ \text{mi})}$$

$$= 115{,}200\ \text{mi/h}^2$$

Almost no one has a "feeling" for the magnitude of the acceleration expressed in these units. However, recall that acceleration is the change in the velocity per unit time. We can understand a velocity value expressed in miles per hour and a short time interval measured in seconds. Therefore, multiply the above acceleration value by $(1\ \text{h}/3600\ \text{s})$:

$$a = 115{,}200\ \frac{\text{mi}}{\text{h}} \times \frac{1}{3600}\ \frac{\text{h}}{\text{s}}$$

$$= 32\ (\text{mi/h})/\text{s}$$

which states that the velocity increases by 32 mi/h each second (or 51 km/h each second). Expressed in these mixed units, the value of the acceleration in this case is much easier to understand.

Example 2–10

A truck travels at a constant speed of 80 km/h and passes a more slowly moving car. At the instant the truck passes the car, the car begins to accelerate at a constant rate of 1.2 m/s² and passes the truck 0.5 km farther down the road. What was the speed of the car when the truck passed it?

Solution:

The elapsed travel times of the truck and the car between the two passing situations are the same. Using the truck's speed, we have

$$t = \frac{s}{v} = \frac{0.5 \text{ km}}{80 \text{ km/h}} = 6.25 \times 10^{-3} \text{ h} = 22.5 \text{ s}$$

Thus, for the car, we have

$$s - s_0 = 0.5 \text{ km} = 500 \text{ m}$$
$$a = 1.2 \text{ m/s}^2$$
$$t = 22.5 \text{ s}$$

with v_0 the unknown to be found. Using Eq. 2–10, we have

$$s - s_0 = \frac{1}{2}at^2 + v_0 t$$

or

$$(500 \text{ m}) = \frac{1}{2}(1.2 \text{ m/s}^2)(22.5 \text{ s})^2 + v_0(22.5 \text{ s})$$

Therefore,

$$v_0 = \frac{500 \text{ m} - 303.8 \text{ m}}{22.5 \text{ s}} = 8.72 \text{ m/s}$$

$$= (8.72 \text{ m/s})(3600 \text{ s/h})(10^{-3} \text{ km/m})$$

$$= 31.4 \text{ km/h}$$

2-5   FREE FALL

We now seek to apply the equations that have been developed to a familiar situation, namely, an object that falls freely near the surface of the Earth. The problem of falling bodies had been addressed in classical Greek times by Aristotle (384–322 B.C.), the most prominent of the early natural philosophers, whose teachings dominated scientific thought. Aristotle's approach to science was basically to set forth "philosophic truths" resulting from logical deductions, not "observational facts" obtained from experiment. Some of Aristotle's ideas about physical phenomena were useful, but many were terribly wrong. For example, he claimed as self-evident that, under similar circumstances, a heavy object would fall to the Earth more rapidly than would a lighter object. It is true that a feather and a rock will fall at different rates, but this is due to the effects of air resistance and does not represent the more fundamental effect of gravity. (•How would Aristotle compare the rates of fall of two rocks separately and when tied together?) Neither Aristotle nor his followers sought to perform the simple experiment that would have tested his basic premise (and proven it false). Nevertheless, so completely did Aristotelian doctrine influence succeeding generations that fundamental advances in physical science were stifled in Western Civilization for nearly two thousand years.

In the 16th century, a new attitude toward scientific thought was emerging. During this age of discovery, new ideas began to develop in science, and the Aristotelian philosophy was being overturned. Science began to be guided by conclusions based on experiment and observation, coupled with logic and reason—the *scientific method.*

By taking this new approach, Galileo Galilei (1564–1642), the great Italian physicist and astronomer, was able to establish mechanics as a mature science through his careful experiments and well-constructed logical arguments. Indeed, many of the concepts we have treated to this point had their origins in Galileo's

(Courtesy of Education Development Center, Newton, MA)

monumental work, *Discourses and Mathematical Demonstrations Concerning Two New Sciences Pertaining to Mechanics and Local Motion,* published in 1638. In contradiction to Aristotelian lore, Galileo put forward the hypothesis that all objects falling near the Earth's surface accelerate at the same constant rate. One conclusion that can be drawn from this hypothesis is that all objects released from the same height above the Earth will fall to the ground in equal times. Galileo tested this prediction and found it to be true.*

Consider now the ideal case of an object that falls freely without any interfering effects due to wind or air resistance; this is the situation that would prevail in a vacuum. (For the case of retarded motion, see Section 6–3.) We postulate, along with Galileo, that all bodies (particles) that fall freely near the Earth's surface experience a constant downward acceleration caused by the pull of the Earth's gravity. The value of this acceleration is determined by experiment and is given the symbol g:

$$g = 9.80 \text{ m/s}^2 \tag{2-12}$$

Fig. 2–11. The position s and the velocity v of an object released from rest.

In Section 5–3 we discuss the association of this quantity with the law of universal gravitation and the properties of the Earth.

It is convenient to have a set of equations representing the application of Eqs. 2–9 and 2–10 to a particle moving under the influence of the Earth's gravity. The constant acceleration g is always directed downward. Therefore, if we choose the upward direction as positive, the sign of the particle's acceleration is negative, $a = -g$. Then,

For	$a = -g$
we have	$v(t) = -gt + v_0$ (2-13)
and	$y(t) = -\frac{1}{2}gt^2 + v_0 t + y_0$ (2-14)
with	$v = v_0$ and $y = y_0$ at $t = 0$

Notice also that we can use Eq. 2–11 to obtain an expression for the downward velocity v of an object dropped (that is, $v_0 = 0$) through a height $h = s - s_0$:

$$v = \sqrt{2gh} \tag{2-15}$$

With Eqs. 2–13 and 2–14 we can easily calculate the velocity and the position of an object that is released from rest and allowed to fall freely. Figure 2–11 shows the results for the first few seconds after release. (The upward direction is positive.)

---

*But not, as legend would have it, by dropping a cannon ball and a musket ball from the Tower of Pisa. The demonstration of the effect by dropping objects with different masses from a tower was carried out in 1586 by a Dutch mathematician, Simon Stevinus (1548–1620), not by Galileo.

▶ *Variations in the Value of g.* Because of the Earth's rotation and its nonspherical shape, the gravitational acceleration at sea level varies with the latitude λ. The value of g at the Equator ($\lambda = 0°$) is 9.78 m/s², and the value steadily increases to 9.83 m/s² at the poles ($\lambda = \pm 90°$). The most convenient expression for computing the variation of g with latitude is

$$g = 9.7803185(1 + \alpha \sin^2 \lambda + \beta \sin^4 \lambda) \text{ m/s}^2 \qquad \text{(2-16)}$$

where $\qquad\qquad \alpha = 0.0052789; \; \beta = 0.00002346$

This expression is based on the internationally adopted constants of the *Geodetic Reference System 1967*.

In this book we use $g = 9.80$ m/s², unless the variation with latitude (or with altitude) is specifically called for. The value $g = (9.80 \pm 0.01)$ m/s² holds, at sea level, for all of the conterminous United States (as well as for Galileo's Tuscany). That is, $g = 9.80$ m/s² is accurate to 0.1 percent for most of our purposes. The value of g also has a variation with the height above sea level, decreasing with increasing altitude (see Chapter 7). When we discuss motion near the Earth's surface, we confine our attention to altitudes below about 4000 m, so the value of g falls within the quoted limits (± 0.01 m/s²) of 9.80 m/s². In British units the corresponding value of g is 32.2 ft/s². For some engineering purposes, "standard gravity" means $g = 9.80665$ m/s², a value appropriate for a latitude of approximately 45° at sea level. ◀

Example 2–11

A stone is thrown vertically upward from the ground with an initial velocity of 25 m/s. How long is required for the stone to reach its maximum height, and how high does it rise? With what velocity does the stone strike the ground on its return? How long does the entire round trip take?

Solution:

In this problem, with the upward direction positive, we have

$$a = -g = -9.80 \text{ m/s}^2$$

$$\text{with initial conditions } v_0 = +25 \text{ m/s} \quad \text{and} \quad y_0 = 0$$

The maximum height occurs at the time $t = T$ when the upward velocity becomes zero. (•Why?) Thus, from Eq. 2–13 for $v(t)$,

$$v(T) = 0 = -gT + v_0$$

or $\qquad\qquad T = \dfrac{v_0}{g} = \dfrac{25 \text{ m/s}}{9.80 \text{ m/s}^2} = 2.55 \text{ s}$

The maximum height* y_m can be calculated from Eq. 2–11, where $s_0 = 0$ and $s = y_m$ for $v = 0$:

$$-v_0^2 = 2ay_m$$

or $\qquad\qquad y_m = \dfrac{-v_0^2}{2a} = \dfrac{v_0^2}{2g} = \dfrac{(25 \text{ m/s})^2}{2(9.80 \text{ m/s}^2)} = 31.9 \text{ m}$

You should verify that the same result is obtained if $T = 2.55$ s is used in Eq. 2–14 for $y(t)$.

The velocity of the stone upon its return to ground level also can be determined from Eq. 2–11. Here, $s = s_0 = 0$, and we have

$$v^2 - v_0^2 = 0$$

*Throughout this book we use a subscript m to indicate the *maximum* value of a quantity.

so

$$v = \pm v_0$$

The desired value is $v = -v_0 = -25 \, \text{m/s}$ because the negative sign indicates *downward* motion. (•What is the significance of the other root, $v = +v_0$?)

To find the time of the return to ground level, we use Eq. 2–14 for $y(t)$ with $y = y_0 = 0$; then,

$$-\tfrac{1}{2}gt^2 + v_0 t = 0$$

or

$$t(v_0 - \tfrac{1}{2}gt) = 0$$

This equation has the solutions

$$t = 0 \qquad \text{and} \qquad t = \frac{2v_0}{g} = 2T$$

Thus, $t = 2T = 5.10 \, \text{s}$ is the duration of the round trip.

The diagrams below show the motion graphically in terms of $y(t)$ and $v(t)$. In particular, note the symmetric behavior of the motion about $t = T$.

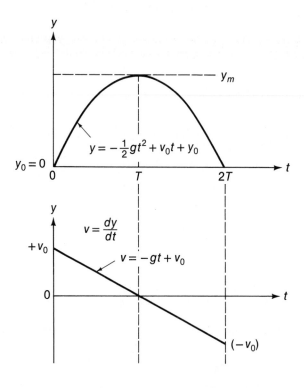

Example 2–12

A falling object requires 1.50 s to travel the last 30 m before hitting the ground. From what height above the ground did it fall?

Solution:

Let the velocity upon hitting the ground be v_2 and let the velocity be v_1 at the time $\Delta t = 1.50 \, \text{s}$ earlier. Then, with the *downward* direction positive (so that $a = +g$), we have

$$v_2 = v_1 + g\,\Delta t = v_1 + (9.80 \text{ m/s}^2)(1.50 \text{ s})$$
$$= v_1 + 14.70 \text{ m/s} \qquad\qquad\qquad (1)$$

We also know that the distance Δy fallen during the time Δt can be related to the *average* velocity by

$$\Delta y = \tfrac{1}{2}(v_1 + v_2)\,\Delta t$$

so
$$v_1 + v_2 = \frac{2(30 \text{ m})}{(1.50 \text{ s})} = 40 \text{ m/s} \qquad\qquad (2)$$

Adding Eqs. (1) and (2), we find

$$2v_2 = (40 + 14.70) \text{ m/s}$$

or
$$v_2 = \tfrac{1}{2}(54.70 \text{ m/s}) = 27.35 \text{ m/s}$$

Finally, solving Eq. 2–15 for h, we have

$$h = \frac{v_2^2}{2g} = \frac{(27.35 \text{ m/s})^2}{2(9.80 \text{ m/s}^2)} = 38.16 \text{ m}$$

2-6 GENERAL EQUATIONS OF MOTION

The basic equations that we have been using (Eqs. 2–9 and 2–10) are valid *only* for cases in which the acceleration is *constant* (including the case $a = 0$). When the acceleration itself is a function of time, we must return to the basic definitions for the acceleration, velocity, and position functions.

We now ask, "If the acceleration $a(t)$ is given, how do we find $v(t)$ and finally $s(t)$?" We have

$$a(t) = \frac{d}{dt}v(t)$$

We require the function $v(t)$ such that its time derivative is $a(t)$. The necessary inverse operation involves the *antiderivative* or the *integration* operation (Section II–1), and we write

$$\int dv(t) = v(t) = \int a(t)dt \qquad\qquad (2\text{–}17)$$

It must be remembered that the integration of $a(t)$ determines $v(t)$ only to within an arbitrary constant, the *constant of integration* (Section II–1). The evaluation of this constant requires that additional information be provided. This information is called the *boundary condition* for the problem (Section II–2).

Once we have determined $v(t)$ by integrating $a(t)$, we may proceed to find $s(t)$ by integrating $v(t)$; that is,

$$\int ds(t) = s(t) = \int v(t)dt \qquad\qquad (2\text{–}18)$$

This integration produces yet another (independent) constant of integration that requires another given boundary condition for its evaluation. (If the given boundary conditions are the values of $v(t)$ and $s(t)$ at $t = 0$, they are referred to as the *initial conditions*.) The final expression for the position function $s(t)$ is the *equation of motion* of the particle.

Example 2–13

Given that the acceleration of a particle is constant, $a(t) = a$, find the equation of motion if at $t = 0$ the velocity is v_0 and the position is s_0.

Solution:

Here,
$$v(t) = \int a\,dt = at + C$$

Now, $v(0) = v_0$, so that $C = v_0$. Thus,

$$v(t) = at + v_0$$

Then,
$$s(t) = \int (at + v_0)\,dt$$

or
$$s(t) = \tfrac{1}{2}at^2 + v_0 t + C'$$

We know that $s(0) = s_0$; hence, $C' = s_0$. Finally,

$$s(t) = \tfrac{1}{2}at^2 + v_0 t + s_0$$

This is precisely Eq. 2–10 for $s(t)$. You can see that the method used here, based on integral calculus, is much more direct than the earlier derivation. Also, note that it was unnecessary to deal at all with average quantities.

Example 2–14

Given that $a(t) = \beta t^2$, where β is a constant with appropriate dimensions, find $v(t)$ such that at $t = \tau$ the velocity is $v(\tau) = v_0$.

Solution:

We have
$$v(t) = \int a(t)\,dt = \int \beta t^2\,dt$$

Performing the integration, we find

$$v(t) = \tfrac{1}{3}\beta t^3 + C$$

where C is the constant of integration. We know that $v(\tau) = v_0$, so

$$v_0 = \tfrac{1}{3}\beta \tau^3 + C$$

or
$$C = v_0 - \tfrac{1}{3}\beta \tau^3$$

Finally,
$$v(t) = \tfrac{1}{3}\beta(t^3 - \tau^3) + v_0$$

(•Is the time derivative of this expression for $v(t)$ equal to $a(t)$?)

Example 2–15

Find the position function for the particle in Example 2–14 if at $t = \tau$ the particle is at the origin, that is, $s(\tau) = 0$.

Solution:

We have
$$v(t) = \tfrac{1}{3}\beta(t^3 - \tau^3) + v_0$$

Hence,
$$s(t) = \int [\tfrac{1}{3}\beta(t^3 - \tau^3) + v_0]\,dt$$

Performing the integration, we find

$$s(t) = \tfrac{1}{3}\beta(\tfrac{1}{4}t^4 - \tau^3 t) + v_0 t + C'$$

Now, when $t = \tau$, $s(\tau) = 0$; then, $C' = \tfrac{1}{4}\beta\tau^4 - v_0\tau$. Thus,

$$s(t) = \tfrac{1}{3}\beta(\tfrac{1}{4}t^4 - \tau^3 t + \tfrac{3}{4}\tau^4) + v_0(t - \tau)$$

(•Is the time derivative of this expression for $s(t)$ equal to $v(t)$?)

PROBLEMS

Section 2-1

2–1   Suppose that the straight-line motion of a particle is described by the expression $s(t) = \tfrac{1}{2}at^4 - bt^2$, where a and b are constants with appropriate dimensions. Make a graph of this function and describe the history of the particle's motion. For what range of values of t does $s(t)$ represent possible positions for the particle?

2–2   Consider the position function, $s(t) = \pm s_0\sqrt{1 + \alpha t}$, where $s_0 = 1$ m and $\alpha = \pm 1\,\text{s}^{-1}$. Make a graph that shows $s(t)$ for both possibilities for the sign of α. For what times t can the function $s(t)$ represent the position of a physical particle?

2–3   A particle moves according to the position function $s(t) = \alpha t^3 + \beta t$. Describe qualitatively the motion if $\beta > 0$. What feature or features of the motion change if $\beta = 0$?

Section 2-2

2–4   In the celebrated tortoise-and-hare race, the hare runs the first 100 m of a 300-m course in 40 s and then rests for 2 h. Upon awakening, the hare runs the remainder of the race in 1 min, only to have the tortoise just beat him at the finish line. (a) If the tortoise moved at constant speed throughout the race, what was this speed? (b) What were the average speeds of the hare during the first 40 s and during the last minute of the race?

2–5   A particle moves one half of the distance from A to B at a constant velocity of $v_1 = 10$ m/s and moves the remainder of the distance at a constant velocity of $v_2 = 40$ m/s. What was the average velocity $\bar{v}$ for the entire trip? Does $\bar{v}$ equal $\tfrac{1}{2}(v_1 + v_2)$? Explain.

2–6   An athlete can run a 1500-m race in 3 min and 50 s. (a) What is his average speed in meters per second?

(b) If his actual speed near the finish line is 20 percent greater than his average speed and if he wins over his nearest rival, who is running at almost the same speed, by an extra forward lean of 3.0 cm, what is the difference in time between first and second place? Does it make sense to measure the elapsed time of such races to an accuracy of a millisecond? Explain. [*Hint:* Near the finish line, small changes in position Δs are associated with small time intervals Δt such that $\Delta s / \Delta t \cong ds/dt = v$.]

2-7 The position function of a particle traveling a straight-line path is $s(t) = \frac{1}{2}at^2 + bt + c$, with $a = 1\ \text{m/s}^2$, $b = -1\ \text{m/s}$, and $c = 2\ \text{m}$. (a) Make a sketch of $s(t)$. (b) With what average velocity does the particle travel from $s = 3/2\ \text{m}$ to $s = 7/2\ \text{m}$? (c) With what velocity does it arrive at $s = 7/2\ \text{m}$? (d) What was the velocity at $s = 3/2\ \text{m}$? Does your answer have any particular significance? (e) Show on your sketch the lines whose slopes correspond to these velocities.

2-8 A particle moves according to the position function
• $s(t) = \alpha t^3 + \beta t^2 + \delta t$, where $\alpha = 1\ \text{m/s}^3$, $\beta = -9\ \text{m/s}^2$, and $\delta = 24\ \text{m/s}$.
(a) Make a sketch of $s(t)$.
(b) Determine $v(2\ \text{s})$, $v(4\ \text{s})$, and $v(4.5\ \text{s})$.
(c) Evaluate $\bar{v}$ and the average speed $\Delta \ell / \Delta t$ for the following time intervals: 0–2 s, 0–4 s, and 0–4.5 s.

2-9 A man whose height is 1.8 m walks directly away
•• from a street light that is 5.0 m above street level. If he walks with a constant velocity of 2.0 m/s, at what rate does his shadow lengthen? [*Hint:* Make appropriate use of the similar triangles present in the geometry of this problem.]

Section 2-3

2-10 A particle is moving with a velocity $v_0 = 60\ \text{m/s}$ at $t = 0$. Between $t = 0$ and $t = 15\ \text{s}$ the velocity decreases uniformly to zero. What was the average acceleration during this 15-s interval? What is the significance of the sign of your answer?

2-11 A particle travels with a velocity $v_0 = 10\ \text{m/s}$ during the interval from $t = 0$ to $t = 5\ \text{s}$. Then, it travels for an additional 10 s at a uniformly increasing velocity $v(t) = 3(t - t_0) + v_0$, where $t_0 = 5\ \text{s}$. What was the average acceleration for the total 15-s interval?

2-12 An automobile starts from rest and in 8 s linearly increases its speed to 100 km/h. (a) What was the average acceleration of the automobile? (b) What was its acceleration at $t = 4\ \text{s}$ and at $t = 8\ \text{s}$? Explain.

2-13 In Example 2-7, suppose that $A = 5\ \text{cm}$ and $\omega = 3\ \text{rad/s}$. What are the values of the maximum speed and the maximum acceleration? (See Section I-4.)

2-14 The position function of a particle is $s(t) = at^2 + bt + c$, with $a = 5\ \text{m/s}^2$, $b = 2\ \text{m/s}$, and $c = -1\ \text{m}$.
(a) What is the distance traveled by the particle during the time interval from $t = 1\ \text{s}$ to $t = 3\ \text{s}$?
(b) What is the average velocity $\bar{v}$ during this interval?
(c) What is the average acceleration during this interval? Is $\bar{a} = \bar{v}/\Delta t$? Explain.

2-15 The position function of a particle is $x(t) = At \sin \omega t$,
• with $A = 1\ \text{m/s}$ and $\omega = 2\pi\ \text{s}^{-1}$. What is the acceleration (a) at $t = 5/4\ \text{s}$ and (b) at $t = 2\ \text{s}$?

Section 2-4

2-16 A Cessna 150 aircraft has a lift-off speed of about 125 km/h. (a) What minimum constant acceleration does this require if the aircraft is to be airborne after a take-off run of 250 m? (b) What is the corresponding take-off time? (c) If the aircraft continues to accelerate at this rate, what speed will it reach 25 s after it begins to roll?

2-17 A rocket is launched from the surface of the Earth with an upward acceleration of 60 m/s² held constant by a governing mechanism. After 10 s of flight, what is the velocity of the rocket and to what height has it risen?

2-18 An automobile traveling initially at a speed of 60 m/s is accelerated uniformly to a speed of 85 m/s in 12 s. How far does the automobile travel during the 12-s interval?

2-19 A truck is moving with a speed of 60 m/s. The driver applies the brakes and slows the vehicle (uniformly) to a speed of 40 m/s while traveling a distance of 250 m. What was the acceleration of the truck?

2-20 The driver of a car traveling at 90 km/h observes a hazard on the road and applies the brakes, giving a constant deceleration of 2.5 m/s². (a) If the driver's reaction time is 0.2 s, how long does it take to stop after sighting the hazard? (b) What distance does the automobile travel before coming to rest?

2-21 An automobile is cruising at a speed of 100 km/h on
• a highway with a posted speed limit of 80 km/h. As the automobile passes a parked highway patrol car, the officer accelerates his car at a uniform rate, reaching 60 km/h in 10 s; he continues to accelerate at the

same rate until he catches the speeding car. (a) How long did the chase last? (b) How far from the parked position of the patrol car was the speeder overtaken? (c) What was the speed of the patrol car as it overtook the speeder?

Section 2-5

2-22 A balloon is rising at a constant rate of 10 m/s when a stone is dropped from the gondola by simply releasing it to fall freely. (a) If the balloon is 80 m above the ground at that time, how long does it take for the stone to hit the ground? (b) With what velocity does the stone strike the ground? [*Hint:* Consider carefully the proper initial velocity to ascribe to the stone on its release.]

2-23 An irate physics student, on learning of his failing exam grade, wants to throw a rock through the glass skylight in the lecture-hall ceiling 10 m overhead. If breaking the glass requires a projectile velocity of at least 7 m/s, with what minimum speed must he hurl the rock if its flight starts from 1 m above the floor?

2-24 A body falling from rest travels $\frac{1}{3}$ of the total distance of fall during the last second. Find (a) the height from which it was released and (b) the total time of fall.

2-25 A stone is dropped from a cliff 200 m high. If a sec-
• ond stone, thrown vertically downward from the cliff 1.50 s after the first stone is released, strikes the base of the cliff at the same instant as the first stone, with what velocity was the second stone thrown?

2-26 The Soviet ground-to-air missile known as the SA-10 can sustain an acceleration 100 times that of gravity for a distance of 1.4 km. What is the duration of the accelerated phase of the flight? What is the final velocity of the missile?

2-27 A ball is thrown directly downward with an initial velocity of 8 m/s from a height of 30 m. When does the ball strike the ground?

2-28 A startled student sees beer cans being thrown up-
•• ward past his dormitory window from a party below. The height of the window is 1.5 m from ledge to top. Using a stopwatch, the student determines that one of the cans is in view for a total of 0.6 s (upward plus downward flight). How high above the window ledge did the beer can rise before starting to fall? [*Hint:* What is the relationship between the time to rise through the distance of the window opening and the time to fall through the same distance?]

Section 2-6

2-29 Given that the acceleration of a particle is $a(t) = \beta t$, find the position function $x(t)$ if at $t = 0$ we have $x(0) = 0$, and if at $t = \tau$ we have $x(\tau) = \gamma$.

2-30 Suppose that the velocity of a particle is given by
•
$$v(t) = \alpha(1 - e^{-\beta t})$$

Find (a) the acceleration and (b) the position function $x(t)$ if at $t = 0$, $x(0) = 0$. (c) What is the limit of $v(t)$ as $t \to \infty$ if $\beta > 0$? (β is real.) (d) What is the value of $v(0)$? (e) What is the limit of the acceleration as $t \to \infty$?

2-31 A computer-controlled engine drives a vehicle with
• an acceleration that increases proportionately with time and reaches a velocity of 20 km/h after 20 s, starting from rest. What distance does the vehicle travel in 1 min after starting?

2-32 A particle moves according to the position function
•• $x(t) = at^3 + bt^2 + ct$, with a $= 2$ m/s^3, b $= -5$ m/s^2, and c $= 2$ m/s. Find the acceleration at those points for which (a) $x(t) = 0$ and (b) $v(t) = 0$.

2-33 A particle moves with an acceleration
• $a(t) = 3\alpha t + 2\beta$, with $\alpha = 1$ m/s^3 and $\beta = 1$ m/s^2. Determine $s(t)$ such that $v(1\,\text{s}) = 4$ m/s and $s(1\,\text{s}) = 3$ m.

Additional Problems

2-34 Consider the position function

$$s(t) = \pm A \sqrt{\frac{t}{t - t_0}}$$

with $A = 1$ m and $t_0 = 1$ s. Make a plot of this function. For what values of t can a physical particle follow this position function?

2-35 A car makes a 200-km trip at an average speed of 40 km/h. A second car starting 1 h later arrives at their mutual destination at the same time. What was the average speed of the second car?

2-36 A glider is approaching a landing, moving at a con-
• stant velocity along a path sloping 15° down from horizontal. The Sun is 45° above the horizon and directly behind the glider. How fast, in terms of its glide velocity, is the plane's shadow moving along the ground? [*Hint:* Use the law of sines.]

2-37 A tennis ball falling vertically strikes the ground with a velocity of 20 m/s and rebounds with a velocity of

14 m/s. If the total time of contact with the ground was 0.01 s, what was the average acceleration produced by the impact?

2-38 The position function of a particle is $s(t) = \alpha t^3 + \beta t$,
• with $\alpha = -1/10$ m/s^3 and $\beta = 6$ m/s. (a) What is the farthest positive excursion reached after starting ($t = 0$)? (b) With what velocity does the particle return to the origin? (c) What is the acceleration (in units of g) 10 s after starting?

2-39 In studies designed to investigate the physiological effects of large accelerations on human beings, Lt. Col. John L. Stapp rode a rocket-propelled sled that was brought to rest from an achieved speed of 285 m/s within 1.5 s (New Mexico, 1954). What was the value of the deceleration (assumed constant) experienced by Col. Stapp? Express the result in terms of g. (He survived with minor blood vessel damage.)

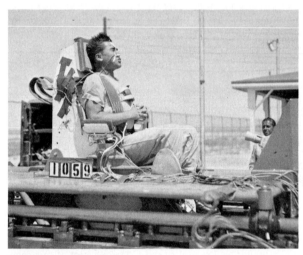

(Courtesy U.S. Air Force)

2-40 Five seconds after starting from rest, a car reaches a speed of 80 km/h. Assuming constant acceleration, find the distance covered during the last second.

2-41 A motorist is traveling along an interstate highway at the illegal speed of 120 km/h. (The posted speed limit is 90 km/h.) At kilometer-post 38 his radar detector buzzes, having been activated by the beam from a highway patrol car at kilometer-post 40. If the reliable range of the patrol car's radar speed meter is 800 m, at what rate must the motorist decelerate to avoid a speeding ticket?

2-42 A boy runs as fast as he can to "hitch a ride" on the
• back of a truck. When he is still 20 m from the truck, the truck begins to move forward with a constant acceleration of 0.8 m/s^2. If the boy just catches the truck, how fast was he running? [*Hint:* Sketch the two position functions and inquire about the special significance of the fact that the boy *just* reached the truck.]

2-43 A truck traveling at a constant speed of 80 km/h passes a more slowly moving car. The instant the truck passes the car, the car begins to accelerate at a constant rate of 1.2 m/s^2 and passes the truck 0.5 km farther down the road. What was the speed of the car when it was passed by the truck?

2-44 An ideally elastic ball (see Section 9–3) bounces repeatedly from a hard surface, rising 10 cm vertically after each bounce. How many times per minute does the ball strike the surface?

2-45 An athlete can throw a certain object vertically to a
• height of 15 m in Austin, Texas, where the latitude is $\lambda = 30° 15'$. How much does he fall short of this height in Stockholm, where $\lambda = 59° 15'$, if he throws the object with the same initial velocity in each case? [*Hint:* Use Eq. 2–16.]

2-46 A ball is thrown vertically upward with an initial
•• velocity of 12 m/s. One second later, a second ball is thrown upward directly in line with the first ball with a velocity of 16 m/s. (a) At what time and (b) at what height do the two balls collide? (c) Is the first ball rising or falling at the time of collision?

2-47 A toy rocket is equipped with an engine that provides a constant acceleration of 40 m/s^2. If the rocket is fired vertically and if the engine burns for 3 s, to what maximum height will the rocket rise? [*Hint:* After the fuel is exhausted, the rocket "coasts" upward.]

2-48 A ball is thrown directly upward from ground level
• with an initial velocity v_0. The ball rises to a height h and then lands on a roof at a height $\frac{1}{2}h$. If the entire motion required 3.5 s, find the velocity v_0 and the height h.

2-49 At a certain instant, a ball is thrown downward with a
•• velocity of 8 m/s from a height of 40 m. At the same instant, another ball is thrown upward from ground level directly in line with the first ball with a velocity of 12 m/s. (a) How long after release do the balls collide? (b) At what height does this collision occur? (c) In what direction is the second ball moving when the collision occurs?

2-50 Determine the position function $s(t)$ for a particle if
•• its acceleration is given by the following expressions.
Assume that at $t = 0$ in each case, $s(0) = s_0 = 0$ and
$v(0) = v_0 = 0$.
(a) $a(t) = \alpha t^3 + \beta t$; α and β constants
(b) $a(t) = A \sin (\omega t + \delta)$; A, ω, and δ constants
(c) $a(t) = At e^{-\mu t}$; A and μ constants
[*Hint:* For part (c), use $\int u \, dv = uv - \int v \, du$, with
$u = At$ and $dv = e^{-\mu t} \, dt$.]

2-51 Two trucks are approaching each other on a long,
•• straight stretch of highway in Arizona. Each proceeds
at a constant speed of 50 km/h. When they are 20 km
apart, a roadrunner (a strange bird that prefers to run
on the ground at high speeds rather than fly) just in
front of one truck runs at a speed of 70 km/h toward
the other truck. When it arrives just in front of that
truck, it turns and runs back toward the first truck. In
cartoon fashion, it runs back and forth between the
two trucks until the fateful final moment when it is
sandwiched between the colliding trucks. Although
the roadrunner makes an infinite number of trips
before the "end," it travels a finite distance in a finite
time. What total distance does the roadrunner cover?
Can you show that it makes an infinite number of
trips in doing so? [*Hint:* Use Eq. A–3 in Appendix A
for the sum of the geometric series.]

2-52 Consider the physically allowable position function
••

$$x(t) = \frac{2t^2 + (t + 1)e^{\alpha(t-1)}}{1 + e^{\alpha(t-1)}}$$

where x is measured in meters when t is given in
seconds. When α (measured in s^{-1}) is large compared
with unity, this function is closely represented by the
much simpler caricature function,

$$x(t) = \begin{cases} 2t^2, & t \leqslant 1 \\ t + 1, & t > 1 \end{cases}$$

(This function is not differentiable at $t = 1$.)

(a) Take $\alpha = 10 \, s^{-1}$ and, by making graphs, compare the two functions for $0 \leqslant t \leqslant 2 \, s$. Also,
graph the *difference* between the functions, using
a larger scale.
(b) Show that the velocity at $t = 1 \, s$ for the case of
the physical function is $v = 2.5 \, m/s$, independent of the value of α. Compare this value of v
with those obtained from the caricature function
as $t \to 1 \, s$ from the left and from the right.
(c) Show that the acceleration for the case of the
physical function is $a = (2 - 1.5\alpha) \, m/s^2$ at
$t = 1 \, s$.

2-53 Consider the physically acceptable position function
••

$$x(t) = + \sqrt{(t + 1)^2 + \varepsilon^2}, \qquad \varepsilon \ll 1$$

and its caricature function,

$$x(t) = |t + 1|$$

which is not differentiable at $t = -1$. In each case, x
is measured in meters when t is given in seconds.
(a) Contrast these two functions by using $\varepsilon = 1/10$.
Be sure to consider $(t + 1)^2 \ll \varepsilon^2$ as well as
$(t + 1)^2 \gg \varepsilon^2$.
(b) Show that, independent of the value of ε, the velocity for the case of the physical function is zero
at $t = -1 \, s$.
(c) Show that the acceleration for the physical function is

$$a(t) = + \frac{\varepsilon^2}{[(t + 1)^2 + \varepsilon^2]^{3/2}}$$

Make a sketch of this acceleration function in the
vicinity of $t = -1 \, s$ for $\varepsilon = 1/10$.
(d) The square root in the physical function suggests
another possible path, namely,

$$x'(t) = - \sqrt{(t + 1)^2 + \varepsilon^2}$$

Could a real particle ever "jump" from one
square-root branch to the other?

CHAPTER 3

VECTORS AND COORDINATE FRAMES

In the preceding chapter we discussed the motion of particles confined to straight-line paths. The study of this restricted case permitted the introduction of the basic kinematic concepts of position, velocity, and acceleration without the additional burden of considering the more complicated geometric factors of the three-dimensional world. We must eventually turn to the description of motion in its most general three-dimensional form, so we now introduce the geometric framework for that description.

3-1 COORDINATE FRAMES

We begin by asking how we specify the position of a particle. A particle that is constrained to move along a fixed curve can be located unambiguously with respect to some selected origin by specifying only one algebraic (positive or negative) real number, namely, the distance along the curve from some selected point. Similarly, a particle that always moves on a surface, flat or curved, can be located by specifying two numbers. A position in three-dimensional space, in turn, requires quoting three numbers.

To make matters specific in each case, we locate the position of a particle in terms of a *frame of reference* or *coordinate frame*. The familiar Cartesian* or rectangular coordinate frame is one possibility. In a Cartesian frame, such as that in Fig. 3–1, the position of a particle located at the point A would be given by quoting the three coordinates, x_A, y_A, z_A.

In physics, we are usually interested in *changes* that take place in the physical world. For example, we might be concerned with the time variation of some physical observable at point A of Fig. 3–1, or how this observable is related to the physical condition at some other point B. Or our interest might be the time-developed history of the motion (the *trajectory*) of a particle that passed through point A.

---

*Named for the famous French mathematician Rene Descartes (Latinized form: Renatus Cartesius) (1596–1650), who first made graphs of equations by plotting the curves on a rectangular coordinate system.

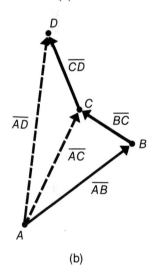

Fig. 3–1. A three-dimensional view of the location of a point A in a Cartesian or rectangular coordinate frame.

(a)

To help sort out various observables in space, we use an appropriate coordinate frame. We emphasize at the outset that such coordinate frames are products of our own invention, devised solely to facilitate the description of events that we study. Clearly, *any laws of Nature we discover cannot in any way depend on our special choice of a coordinate frame* (i.e., on the choice of the spatial orientation of the frame or on the location of its origin). The laws of Nature must be *invariant* to the choice of the coordinate frame used to describe the situation.*

3–2 DISPLACEMENT

When a particle moves from one position to another, we say that the particle undergoes a *displacement*. In Fig. 3–2a, a particle moves from A to B, and we represent the displacement $\overline{AB}$ by the straight-line segment connecting A to B. An arrowhead is placed at B, indicating that the displacement proceeds *from A to B*. In so doing, we imply nothing concerning the actual path followed by the particle in moving from A to B; for example, the motion could have been along the curved path shown by the dashes in Fig. 3–2a. In this description we are concerned only with the *end points A and B*.†

If the particle proceeds from B to a new position C, it undergoes a new displacement $\overline{BC}$, thereby producing a total displacement from A which is represented by the *single* displacement $\overline{AC}$, shown in Fig. 3–2b. Yet a further displacement from C to D, or $\overline{CD}$, leads to a total displacement from A represented by the single displacement $\overline{AD}$. The total displacement $\overline{AD}$, although it consists of three separate parts, depends only on the location of point D relative to point A. In the same way, the displacement $\overline{AB}$ in Fig. 3–2a may be viewed as the sum of a large number of small displacements that follow the curved path from A to B.

With Fig. 3–2b, we intuitively arrive at the geometric concept of displacement (arrow) addition. The sum of two displacements, $\overline{AB}$ and $\overline{BC}$, is the resultant displacement, $\overline{AC}$, obtained by drawing the arrow connecting the point A to the point C

(b)

Fig. 3–2. (a) The directed-line representation of the displacement from A to B, $\overline{AB}$. (b) The representation of continued displacements, A to B to C to D, illustrating the notion of a sum of displacements.

---

*We restrict attention to nonaccelerating coordinate frames; see Section 5–2.

†The initial point A is the tail end (or *origin*) and the end point B is the arrow tip (or *terminus*) of the displacement $\overline{AB}$.

(with the arrow tip at *C*). This type of sum is clearly more complicated than the simple addition of two numbers.

These ideas lead us to the concept of *vectors*. Many quantities in Nature possess the same essential properties as displacement arrows, namely, *magnitude* and *direction*. In general, vectors are quantities that have magnitude and direction and obey the same rules of combination as displacements. Physical quantities in this class include velocity, acceleration, force, momentum, and many others.

Besides vectors, there are quantities in Nature that require only a magnitude (and an appropriate scale unit) for a complete specification. These quantities—called *scalars*—do not have any associated direction. Physical quantities in this class include mass, energy, temperature, electric charge, and many others.

3-3   BASIC VECTOR CONCEPTS

Fig. 3–3. The three directed line segments are all parallel and have the same length. Thus, **A**, **A′**, and **A″** are different representations of the *same* vector.

In printed matter, vectors are usually represented by bold-face characters, such as **A**, **E**, **r**, and **ω**. In handwritten material the most convenient way to indicate vectors is by a small arrow above the symbol: $\vec{A}$, $\vec{E}$, $\vec{r}$, and $\vec{\omega}$. If we wish to refer only to the *magnitude* of the vector **A**, we use light-face type, *A*. When we want to call attention to the vector property of a quantity but need only its magnitude, we write $|\mathbf{A}|$; of course, we have, identically, $A \equiv |\mathbf{A}|$. Note that *A*, the magnitude of **A**, is always a non-negative real (positive definite) quantity: $A \geqslant 0$.

It is often useful to represent a vector in a graphic way by means of a directed line segment (arrow). The length of the line segment (on some convenient scale) corresponds to the magnitude of the vector, and the arrow indicates the direction of the vector. The origin of the line segment may be *any* point. Thus, we can represent the *same* vector by different parallel line segments of equal length, as in Fig. 3–3. Vectors are characterized by magnitude and direction, nothing more.

Multiplication of a Vector by a Scalar.   The product of a vector **A** and a dimensionless number *n* is defined to be a new vector *n***A** having the *same* direction as **A** but altered in magnitude: $|n\mathbf{A}| = n|\mathbf{A}| = nA$. When $n < 0$, the product *n***A** is defined to have a direction *opposite* to **A** and to have a magnitude $|n\mathbf{A}| = |n||\mathbf{A}| = |n|A$. In particular, if $n = -1$, the *negative* of the vector is generated. In the event that *n* represents a physical quantity with dimensions, the product *n***A** will correspond to a new physical quantity (see Section 3–7). Figure 3–4 illustrates several examples of vectors multiplied by scalars. In Sections 3–5 and 3–6 we discuss the multiplication of a vector by a vector.

Unit Vectors.   For many vector operations it proves convenient to use the idea of *unit vectors*. We may associate with any vector **A** a unit vector* $\hat{\mathbf{u}}_A$ that has the same direction as **A** but has unit (dimensionless) length, namely,

$$\hat{\mathbf{u}}_A \equiv \frac{\mathbf{A}}{|\mathbf{A}|} = \frac{\mathbf{A}}{A} \tag{3-1}$$

Fig. 3–4. The result of multiplying the vector **A** by positive and negative dimensionless numbers.

Then, we can write $\mathbf{A} = A\hat{\mathbf{u}}_A$.

Unit vectors may also be selected to designate particular directions in space even though they are not associated with a specific vector. The utility of such choices will become clear as we proceed.

---

*Unit vectors in this book are always indicated by a caret ⌃ over the symbol.

Addition and Subtraction of Vectors.   The vector sum of two vectors, **A** and **B**, is defined by the graphical construction (geometrical addition) shown in Fig. 3–5. To carry out this summation, first translate the vector **B** parallel to itself until its tail end (origin) is at the arrow tip (terminus) of **A**, as shown in Fig. 3–5b. Then, the vector resultant **R** of the addition, **R** = **A** + **B**, is obtained by drawing a vector connecting the tail end of **A** to the arrow tip of **B**. Alternatively, the vector **A** can be translated parallel to itself until its tail end is at the arrow tip of **B**, as shown in Fig. 3–5c. Then, **R′** = **B** + **A** is obtained by drawing a vector from the tail end of **B** to the arrow tip of **A**. Evidently, **R** = **R′**. This equality,

$$\mathbf{R} = \mathbf{A} + \mathbf{B} = \mathbf{B} + \mathbf{A} = \mathbf{R}' \qquad (3\text{–}2)$$

exhibits the *commutative* property of vector addition.

The so-called *parallelogram construction* for vector addition is illustrated in Fig. 3–5d; this method is *entirely equivalent* to the arrow method described above. The two vectors, **A** and **B**, are translated until their tail ends coincide. The resultant vector **R**, the included diagonal between **A** and **B**, is obtained by completing the construction of the parallelogram with the aid of the dashed lines, which are parallel to **A** and **B**.

The negative of a vector, −**B**, has already been defined, so we can now define the *subtraction* of vectors, **A** − **B**, to be the addition of **A** and −**B**, as indicated in Fig. 3–6.

Fig. 3–6. The graphical representation of **R** = **A** − **B** = **A** + (−**B**).

Vector addition is *associative* as well as commutative. The associative property is expressed by writing

$$\mathbf{R} = \mathbf{A} + \mathbf{B} + \mathbf{C} = (\mathbf{A} + \mathbf{B}) + \mathbf{C} = \mathbf{A} + (\mathbf{B} + \mathbf{C}) \qquad (3\text{–}3)$$

The first equality, **R** = **A** + **B** + **C**, is an extension of the addition law to three vectors. Refer to Fig. 3–7. First, **B** is translated so that its tail end matches the arrow tip of **A**. Then, **C** is translated so that its tail end matches the arrow tip of **B**. Finally,

Fig. 3–5. The graphical or geometrical addition of two vectors, **A** and **B**, also illustrating the commutative nature of vector addition.

Fig. 3–7. (a) The successive addition of vectors to produce the sum, **R** = **A** + **B** + **C**. (b) The grouped addition **R** = (**A** + **B**) + **C**. (c) The grouped addition **A** + (**B** + **C**).

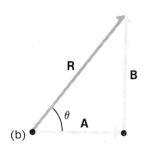

(a)

(b)

the tail end of **A** is connected to the arrow tip of **C** to yield **R**. The use of parentheses in the expression $(\mathbf{A} + \mathbf{B}) + \mathbf{C}$ means that we first add **A** and **B**, obtaining an intermediate sum, to which we add **C**. The two algebraic laws expressed by Eqs. 3–2 and 3–3 tell us that when vectors are added, the result is the same for any order of the addition and for any subgrouping of the added vectors.

Example 3–1

What is the resultant vector, $\mathbf{R} = \mathbf{A} + \mathbf{B}$, if $A = 3$ and $B = 4$, and if **A** is perpendicular to **B**? Give the direction of **R** relative to **A**.

Solution:

First translate **B** so that its tail end matches the arrow tip of **A**. Then, draw the resultant vector **R**, as shown in diagram (b). The Pythagorean theorem gives $R^2 = A^2 + B^2$ or $R = \sqrt{3^2 + 4^2} = 5$. Because **A** and **B** are at right angles, $\tan \theta = B/A = 4/3$, so that $\theta = 53.13°$.

Example 3–2

If the vectors **A** and **B** ($A = 3$, $B = 4$) have an included angle of $\alpha = 60°$ between them, what is the resultant, $\mathbf{R} = \mathbf{A} + \mathbf{B}$?

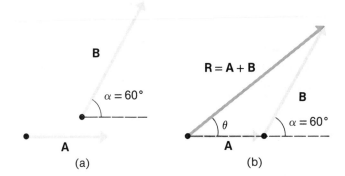

(a)　　　　(b)

Solution:

The resultant vector **R** is constructed as shown in diagram (b). Using the law of cosines (see Eq. A–37 in the Appendix), we have

$$R^2 = A^2 + B^2 + 2AB \cos \alpha$$
$$= 3^2 + 4^2 + 24 \cos 60° = 37$$

Therefore,
$$R = \sqrt{37} = 6.08$$

From the law of sines (see Eq. A–36 in the Appendix), we have

$$\frac{B}{\sin \theta} = \frac{R}{\sin (180° - \alpha)} = \frac{R}{\sin \alpha}$$

or
$$\sin \theta = \frac{B}{R} \sin \alpha = \frac{4}{\sqrt{37}} \sin 60° = 0.569$$

from which
$$\theta = 34.7°$$

3-4 VECTOR COMPONENTS

In some types of problems, there is a *preferred* direction in space—for example, the direction of motion or of a force or of an electric field. It is often required to determine the *component* of a vector **A** in the preferred direction (called the *projection* of **A** onto this axis). To do this, construct a unit vector $\hat{\mathbf{u}}$ at the origin of the vector **A** and in the preferred direction. Then, the component of **A** in the direction of $\hat{\mathbf{u}}$ is defined to be

$$A_u = A \cos \theta_u \tag{3-4}$$

where θ_u is the angle between $\hat{\mathbf{u}}$ and **A** measured in the plane defined by these vectors. We also define the *vector component* of **A** in the direction of $\hat{\mathbf{u}}$ to be

$$\mathbf{A}_u = (A \cos \theta_u)\hat{\mathbf{u}} \tag{3-5}$$

Note carefully that A_u has associated with it an algebraic sign; A_u is *not* simply the magnitude of $\mathbf{A}_u$. Figure 3–8 shows two cases illustrating this point. In Fig. 3–8a we see a vector **A** that has a vector component $\mathbf{A}_u$ in the direction of $\hat{\mathbf{u}}$; here, $A_u > 0$. In Fig. 3–8b we see a vector **B** that has a vector component $\mathbf{B}_u$ in the direction opposite to $\hat{\mathbf{u}}$; here $B_u < 0$. The proper sign of the component is always given by the term $\cos \theta_u$.

For a vector **A**, we have defined the following quantities:

> *magnitude* of **A**: $A = |\mathbf{A}|$, a non-negative real scalar
> *component* of **A**: A_u, a (signed) scalar
> *vector component* of **A**: $\mathbf{A}_u$, a vector in the direction of $\hat{\mathbf{u}}$

A particular vector is *zero* only if its component in every direction vanishes.

Vectors in Two Dimensions. To describe general two-dimensional vector operations, it is convenient (although not essential) to introduce a set of coordinate axes. We define dimensionless unit vectors, $\hat{\mathbf{i}}$ and $\hat{\mathbf{j}}$, that are directed parallel to the positive x- and y-axes, respectively, as indicated in Fig. 3–9. Then, a vector **A** has vector components in the x- and y-directions given by

$$\mathbf{A}_x = A_x\hat{\mathbf{i}} = (A \cos \theta_x)\hat{\mathbf{i}}$$
$$\mathbf{A}_y = A_y\hat{\mathbf{j}} = (A \cos \theta_y)\hat{\mathbf{j}}$$

where θ_x and θ_y are the angles between **A** and $\hat{\mathbf{i}}$ and between **A** and $\hat{\mathbf{j}}$, respectively (Fig. 3–9a).

Because we have chosen x- and y-axes that are mutually perpendicular in two dimensions, we have, in Fig. 3–9a, $\theta_x + \theta_y = \pi/2$. Then, $\cos \theta_y = \cos (\pi/2 - \theta_x) = \sin \theta_x$. Thus, we need to specify only the angle θ_x, which we write without a subscript. That is, the vector components are

$$\left. \begin{array}{l} \mathbf{A}_x = A_x\hat{\mathbf{i}} = (A \cos \theta)\hat{\mathbf{i}} \\ \mathbf{A}_y = A_y\hat{\mathbf{j}} = (A \sin \theta)\hat{\mathbf{j}} \end{array} \right\} \tag{3-6}$$

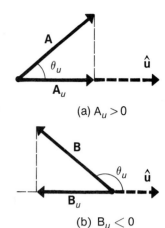

(a) $A_u > 0$

(b) $B_u < 0$

Fig. 3–8. The vector components in the direction of $\hat{\mathbf{u}}$, illustrated for two different vectors.

Fig. 3–9. Resolution of the vector **A** into rectangular vector components, $\mathbf{A}_x = A_x\hat{\mathbf{i}}$ and $\mathbf{A}_y = A_y\hat{\mathbf{j}}$. In (a), $A_x = A \cos \theta_x$ and $A_y = A \cos \theta_y$. In (b), $A_x = A \cos \theta$ and $A_y = A \sin \theta$. The angle θ is positive when measured in the counterclockwise sense, as shown. The unit vectors $\hat{\mathbf{i}}$ and $\hat{\mathbf{j}}$ indicate only the directions of the x- and y-axes. These vectors can be freely translated to any point where the tail end of **A** might be located; this in no way influences the calculation of the vector components.

(a)

(b)

It is evident from Fig. 3–9 that

$$\mathbf{A} = A_x\hat{\mathbf{i}} + A_y\hat{\mathbf{j}} \qquad (3\text{–}7)$$

Thus, Eqs. 3–6 and 3–7 are *inverse transformations,* one set giving the vector components of **A** and the other giving the vector **A** in terms of the components.

We also have, from Fig. 3–9,

$$A^2 = A_x^2 + A_y^2 \qquad (3\text{–}8)$$

and

$$\theta = \tan^{-1}\frac{A_y}{A_x} \qquad (3\text{–}9)$$

The signs of A_x and A_y determine the quadrant in which θ lies. (Notice that Eq. 3–8 also follows from Eqs. 3–6 when we use the trigonometric identity $\sin^2\theta + \cos^2\theta = 1$.)

By applying the above ideas, we arrive at a particularly simple method for computing analytically the sum of a number of given vectors. Consider Fig. 3–10, which shows the vector addition $\mathbf{R} = \mathbf{A} + \mathbf{B}$. Evidently,

$$\mathbf{R}_x = R_x\hat{\mathbf{i}} = (A_x + B_x)\hat{\mathbf{i}}$$
$$\mathbf{R}_y = R_y\hat{\mathbf{j}} = (A_y + B_y)\hat{\mathbf{j}} \qquad (3\text{–}10)$$

so that

$$\mathbf{R} = R_x\hat{\mathbf{i}} + R_y\hat{\mathbf{j}}$$
$$= (A_x + B_x)\hat{\mathbf{i}} + (A_y + B_y)\hat{\mathbf{j}} \qquad (3\text{–}11)$$

Also,

$$R^2 = (A_x + B_x)^2 + (A_y + B_y)^2 \qquad (3\text{–}12)$$

and

$$\theta = \tan^{-1}\frac{A_y + B_y}{A_x + B_x} \qquad (3\text{–}13)$$

Any number of coplanar vectors can be added by a straightforward extension of the above equations.

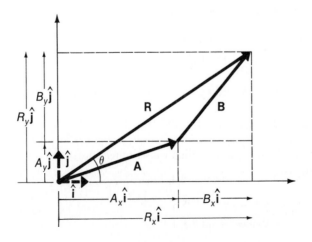

Fig. 3–10. A graphical representation of the vector addition $\mathbf{R} = \mathbf{A} + \mathbf{B}$, employing the components (A_x, A_y) and (B_x, B_y). We have $R_x = A_x + B_x$ and $R_y = A_y + B_y$.

Example 3-3

A ship travels a straight course (using land-based references) of 130 km directed 60° north of east. How far north and east of the starting point does the ship travel?

Solution:

Orient the x- and y-axes so that x corresponds to east and y to north. We then identify A_E with A_x and A_N with A_y. Thus,

$$A_E = A \cos 60° = (130 \text{ km}) \cos 60° = 65.0 \text{ km}$$
$$A_N = A \sin 60° = (130 \text{ km}) \sin 60° = 112.6 \text{ km}$$

Example 3-4

In flat, open country a hiker travels due east for 5.12 km and then turns due south, proceeding for another 3.87 km before resting. Locate this rest position relative to the starting point.

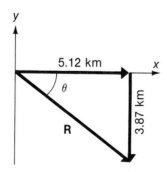

Solution:

We have $$R^2 = (5.12 \text{ km})^2 + (-3.87 \text{ km})^2$$

or $$R = \sqrt{(5.12)^2 + (3.87)^2} \text{ km} = 6.42 \text{ km}$$

$$\theta = \tan^{-1}\left(\frac{-3.87 \text{ km}}{5.12 \text{ km}}\right)$$

Because $A_x > 0$ and $A_y < 0$, the angle θ lies in the fourth quadrant. Thus,

$$\theta = -37.1°, \text{ south of east} \qquad (\text{or } +322.9°)$$

(•You might find it worthwhile to solve this problem graphically. Try a scale of 1 cm = 0.5 km. Try to estimate the accuracy of your results before comparing them with the analytic solution.)

Example 3–5

Next, we return to Example 3–2 and obtain the solution in terms of the components of the two vectors. It is convenient to select a two-dimensional rectangular coordinate system with the x-axis in the direction of **A**. (Vectors **A** and **B** are both in the x-y plane.) In problems of this sort, with no additional information given, we are free to select the most convenient coordinate frame.

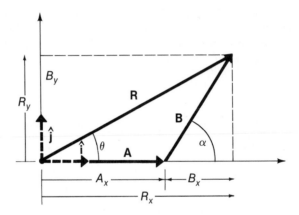

Solution:

According to the construction in the diagram, we have

$$R_x = A_x + B_x = A + B \cos \alpha = 3 + 4 \cos 60° = 5$$
$$R_y = A_y + B_y = 0 + B \sin \alpha = 4 \sin 60° = 3.46$$

Then, $$R = \sqrt{R_x^2 + R_y^2} = \sqrt{(5)^2 + (3.46)^2} = 6.08$$

Finally, $$\theta = \tan^{-1} \frac{R_y}{R_x} = \tan^{-1} \left(\frac{3.46}{5} \right)$$

Because $A_x > 0$ and $A_y > 0$, the angle θ lies in the first quadrant. Thus,

$$\theta = 34.7°$$

The results in the two examples agree, as expected. It is important to note that the introduction of the coordinate frame was *not* a necessity but *only* a convenience (particularly so if you did not remember the laws of cosines and sines!). This convenience would have become even more evident if we had the task of adding three or more vectors. (•What do you suppose would be the result of selecting the axes rotated by, say, 15° clockwise? See Problem 3–5.)

Vectors in Three Dimensions. The previous discussion of components and vector components is readily generalized to three dimensions. If we are required to add two vectors (no matter how they are oriented in space), we can translate one vector by parallel displacement so that its tail end matches the arrow tip of the other vector. These two vectors now define a plane, and a rectangular coordinate frame constructed in that plane permits us to use the methods we have just established for analyzing vectors in two dimensions. However, given three or more arbitrary vectors, the selection of a plane containing all the vectors is not always possible, and to add such vectors it is convenient to use a three-dimensional coordinate system.

We again make use of a rectangular or Cartesian coordinate frame and introduce the additional unit vector $\hat{\mathbf{k}}$ along the direction of the z-axis, as shown in Fig. 3–11. Then, the vector components of **A** are:

$$\left.\begin{aligned} \mathbf{A}_x &= A_x\hat{\mathbf{i}} = (A\cos\alpha)\hat{\mathbf{i}} \\ \mathbf{A}_y &= A_y\hat{\mathbf{j}} = (A\cos\beta)\hat{\mathbf{j}} \\ \mathbf{A}_z &= A_z\hat{\mathbf{k}} = (A\cos\gamma)\hat{\mathbf{k}} \end{aligned}\right\} \tag{3-14}$$

The angles, α, β, and γ are measured in planes containing **A** and the unit vector $\hat{\mathbf{i}}$ (for α), **A** and the unit vector $\hat{\mathbf{j}}$ (for β), and **A** and the unit vector $\hat{\mathbf{k}}$ (for γ). The quantities $\cos\alpha$, $\cos\beta$, and $\cos\gamma$ are the *direction cosines* of the vector **A**. These cosines are not all independent because

$$\cos^2\alpha + \cos^2\beta + \cos^2\gamma = 1 \tag{3-15}$$

(•Can you prove this using Eq. 3–14?)

When the components A_x, A_y, and A_z are given, **A** can be determined by calculating the magnitude of **A**, namely A, and the three direction cosines. These quantities are obtained from

$$A = \sqrt{A_x^2 + A_y^2 + A_z^2}$$

and

$$\cos\alpha = \frac{A_x}{A}, \qquad \cos\beta = \frac{A_y}{A}, \qquad \cos\gamma = \frac{A_z}{A} \tag{3-16}$$

Two vectors are equal if and only if the three components of one are individually equal to the corresponding components of the other. That is, if $A_x = B_x$, $A_y = B_y$, and $A_z = B_z$, then **A** = **B**. Therefore, vector equations such as **A** = **B**, **A** = n**B**, or **A** = 0 symbolically represent three separate equations involving the individual components.

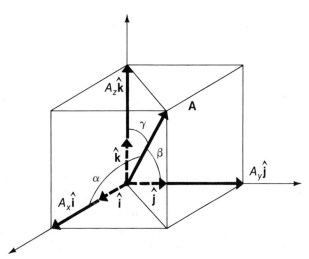

Fig. 3–11. A three-dimensional view of a (right-handed) rectangular coordinate frame showing the resolution of a vector **A** into its (Cartesian) vector components, $A_x\hat{\mathbf{i}}$, $A_y\hat{\mathbf{j}}$, and $A_z\hat{\mathbf{k}}$. We have $A_x = A\cos\alpha$, $A_y = A\cos\beta$, and $A_z = A\cos\gamma$.

Example 3–6

Determine the components of a vector **A** that makes equal angles with the x-, y-, and z-axes.

Solution:

It is given that $\alpha = \beta = \gamma = \theta$; therefore, using Eq. 3–15, we have

$$3 \cos^2 \theta = 1$$

or

$$\cos \theta = \frac{1}{\sqrt{3}}; \qquad \theta = 54.7°$$

Thus,

$$A_x = A_y = A_z = A \cos \theta = \frac{A}{\sqrt{3}} = 0.577\,A$$

Example 3–7

For a particular rectangular coordinate frame, three vectors are given as

$$\mathbf{A}_1 = 6\hat{\mathbf{i}} + 2\hat{\mathbf{j}} - \hat{\mathbf{k}}$$
$$\mathbf{A}_2 = -2\hat{\mathbf{i}} + 2\hat{\mathbf{k}}$$
$$\mathbf{A}_3 = 3\hat{\mathbf{i}} - 2\hat{\mathbf{j}}$$

What is the vector sum, $\mathbf{R} = \mathbf{A}_1 + \mathbf{A}_2 + \mathbf{A}_3$?

Solution:

We have $\qquad R_x = A_{1x} + A_{2x} + A_{3x} = +6 - 2 + 3 = 7$

Similarly, $\qquad R_y = +2 + 0 - 2 = 0$
$$R_z = -1 + 2 + 0 = 1$$

Thus, $\qquad\qquad \mathbf{R} = 7\hat{\mathbf{i}} + \hat{\mathbf{k}}$

and $\qquad\qquad R = \sqrt{7^2 + 1^2} = \sqrt{50} = 7.07$

$$\cos \alpha = \frac{7}{7.07}, \qquad \text{or} \qquad \alpha = 8.1°$$

$$\cos \beta = \frac{0}{7.07}, \qquad \text{or} \qquad \beta = 90°$$

$$\cos \gamma = \frac{1}{7.07}, \qquad \text{or} \qquad \gamma = 81.9°$$

Because $\beta = 90°$, we see that the resultant vector **R** lies entirely in a plane parallel to the x-z plane, and makes an angle of 8.1° with the x-axis.

3-5 THE SCALAR PRODUCT OF TWO VECTORS— THE DOT PRODUCT

When two vectors are to be added, they must represent the same physical quantity—that is, they must have the same dimensionality. Thus, a displacement vector can be added only to another displacement vector. However, when two vectors are multiplied, the product represents a new physical quantity with dimensionality different from that of the original vectors. Because vectors have direction as well as magnitude, it is clear that vector multiplication cannot follow the rules that apply for the multiplication of scalar quantities. In fact, it proves useful to define two different types of vector multiplication, one that yields a scalar and one that yields a vector. The first of these, which we discuss in this section, is called by several names—*scalar product, inner product,* and *dot product.* We symbolically define and write the dot product as

$$\mathbf{A} \cdot \mathbf{B} = |\mathbf{A}||\mathbf{B}| \cos \theta_{AB} = AB \cos \theta_{AB} \qquad \text{(3-17)}$$

where θ_{AB} is the angle between the two vectors $\mathbf{A}$ and $\mathbf{B}$ ($0° \leqslant \theta_{AB} \leqslant 180°$). This type of multiplication of two vectors yields a *scalar* result.

The fact that we can write $\mathbf{A} \cdot \mathbf{B} = A(B \cos \theta_{AB}) = B(A \cos \theta_{AB})$ makes it clear that the dot product is a symmetric function of $\mathbf{A}$ and $\mathbf{B}$; in other words,

$$\mathbf{A} \cdot \mathbf{B} = \mathbf{B} \cdot \mathbf{A}$$

This equation expresses the *commutative* property of the dot product.

Geometrically, we can consider the dot product as either the projection of $\mathbf{A}$ onto the vector direction of $\mathbf{B}$ multiplied by the magnitude of $\mathbf{B}$, or the projection of $\mathbf{B}$ onto the vector direction of $\mathbf{A}$ multiplied by the magnitude of $\mathbf{A}$. Figure 3–12 illustrates these two points of view.

Note that if we set $\mathbf{B}$ equal to a unit vector $\hat{\mathbf{u}}$, then

$$\mathbf{A} \cdot \hat{\mathbf{u}} = A \cos \theta_{Au} = A_u$$

which is equivalent to Eq. 3-4 in defining the component of $\mathbf{A}$ in the direction of $\hat{\mathbf{u}}$. We can also write

$$\left.\begin{aligned} \mathbf{A} \cdot \hat{\mathbf{i}} &= A_x \\ \mathbf{A} \cdot \hat{\mathbf{j}} &= A_y \\ \mathbf{A} \cdot \hat{\mathbf{k}} &= A_z \end{aligned}\right\} \qquad \text{(3-18)}$$

Consider the dot product $\mathbf{A} \cdot \mathbf{A}$, for which $\cos \theta_{AA} = \cos 0 = 1$. Hence, we have

$$\mathbf{A} \cdot \mathbf{A} = A^2 = A_x^2 + A_y^2 + A_z^2 \qquad \text{(3-19)}$$

Thus, the dot product of a vector by itself is equal to the square of its magnitude.

The dot product of two nonzero vectors is zero if and only if the vectors are mutually perpendicular (for then, $\cos \theta_{AB} = 0$).

Fig. 3–12. Two equivalent graphical representations of the dot product, $\mathbf{A} \cdot \mathbf{B}$. (a) $A(B \cos \theta_{AB})$. (b) $B(A \cos \theta_{AB})$.

(a)

(b)

The unit vectors $\hat{\mathbf{i}}$, $\hat{\mathbf{j}}$, and $\hat{\mathbf{k}}$ have the following dot products:

$$\left.\begin{array}{c} \hat{\mathbf{i}} \cdot \hat{\mathbf{j}} = \hat{\mathbf{j}} \cdot \hat{\mathbf{k}} = \hat{\mathbf{k}} \cdot \hat{\mathbf{i}} = 0 \\ \hat{\mathbf{i}} \cdot \hat{\mathbf{i}} = \hat{\mathbf{j}} \cdot \hat{\mathbf{j}} = \hat{\mathbf{k}} \cdot \hat{\mathbf{k}} = 1 \end{array}\right\} \qquad \textbf{(3–20)}$$

The dot product obeys the *distributive law;* thus,

$$\mathbf{A} \cdot (\mathbf{B} + \mathbf{C}) = \mathbf{A} \cdot \mathbf{B} + \mathbf{A} \cdot \mathbf{C} \qquad \textbf{(3–21)}$$

A useful application of the distributive law follows immediately by considering the product

$$\mathbf{A} \cdot \mathbf{B} = (A_x\hat{\mathbf{i}} + A_y\hat{\mathbf{j}} + A_z\hat{\mathbf{k}}) \cdot (B_x\hat{\mathbf{i}} + B_y\hat{\mathbf{j}} + B_z\hat{\mathbf{k}})$$

Of the nine terms that result in expanding this dot product, we see that six vanish by virtue of Eqs. 3–20. The remaining three terms yield

$$\mathbf{A} \cdot \mathbf{B} = A_xB_x + A_yB_y + A_zB_z \qquad \textbf{(3–22)}$$

The angle between two vectors can be obtained by combining Eqs. 3–17 and 3–22 and using Eq. 3–16 for the magnitudes:

$$\cos \theta_{AB} = \frac{\mathbf{A} \cdot \mathbf{B}}{AB} = \frac{A_xB_x + A_yB_y + A_zB_z}{\sqrt{(A_x^2 + A_y^2 + A_z^2)(B_x^2 + B_y^2 + B_z^2)}} \qquad \textbf{(3–23)}$$

Example 3–8

Determine the dot product of the two vectors $\mathbf{A} = 3\hat{\mathbf{i}} + 7\hat{\mathbf{k}}$ and $\mathbf{B} = -\hat{\mathbf{i}} + 2\hat{\mathbf{j}} + \hat{\mathbf{k}}$.

Solution:

Here,
$$\begin{array}{ccc} A_x = 3, & A_y = 0, & A_z = 7 \\ B_x = -1, & B_y = 2, & B_z = 1 \end{array}$$

Therefore, using Eq. 3–22, we find

$$\mathbf{A} \cdot \mathbf{B} = (3)(-1) + (0)(2) + (7)(1) = -3 + 0 + 7 = +4$$

Example 3–9

What is the angle between the two vectors of Example 3–8?

Solution:

We have
$$A^2 = (3)^2 + (7)^2 = 58$$
$$B^2 = (-1)^2 + (2)^2 + (1)^2 = 6$$

Therefore, using Eq. 3–23, the result is

$$\cos \theta_{AB} = \frac{\mathbf{A} \cdot \mathbf{B}}{AB} = \frac{+4}{\sqrt{58 \cdot 6}} = 0.214$$

or

$$\theta_{AB} = 77.6°$$

Example 3–10

If $\mathbf{C} = \mathbf{A} + \mathbf{B}$, give a geometric interpretation of $\mathbf{C} \cdot \mathbf{C}$.

Solution:

We have

$$\mathbf{C} \cdot \mathbf{C} = (\mathbf{A} + \mathbf{B}) \cdot (\mathbf{A} + \mathbf{B}) = \mathbf{A} \cdot \mathbf{A} + \mathbf{A} \cdot \mathbf{B} + \mathbf{B} \cdot \mathbf{A} + \mathbf{B} \cdot \mathbf{B}$$

Because

$$\mathbf{A} \cdot \mathbf{B} = \mathbf{B} \cdot \mathbf{A}$$

we find

$$C^2 = A^2 + B^2 + 2\mathbf{A} \cdot \mathbf{B}$$

or

$$C^2 = A^2 + B^2 + 2AB \cos \theta_{AB}$$

This result will be recognized as the law of cosines (a version of Eq. A–37 in Appendix A) for the triangle formed by the three vectors $\mathbf{A}$, $\mathbf{B}$, and $\mathbf{C}$. Notice that for a right triangle (that is, $\theta_{AB} = 90°$), we have the Pythagorean theorem, $C^2 = A^2 + B^2$.

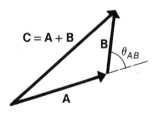

3-6   VECTOR PRODUCT OF TWO VECTORS— THE CROSS PRODUCT

In the preceding section we discussed a type of vector multiplication that produces a *scalar product*. We now define a way to combine two vectors to produce a *vector product*. In order to distinguish this new operation from the dot product, we call this product the *cross product,* the *vector product,* or the *outer product,* and write

$$\mathbf{C} = \mathbf{A} \times \mathbf{B}$$

The magnitude of $\mathbf{C}$ is defined to be

$$|\mathbf{C}| = |\mathbf{A}||\mathbf{B}| \sin \theta_{AB}$$

or

$$C = AB \sin \theta_{AB} \qquad\qquad \textbf{(3–24)}$$

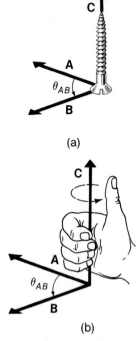

C

A

θ_{AB}

B

(a)

C

A

θ_{AB}

B

(b)

Fig. 3–13. The right-hand convention for the vector product (cross product) **C = A × B**.

where θ_{AB} is the angle between the vectors **A** and **B** ($0° \leqslant \theta_{AB} \leqslant 180°$). The direction of **C** is defined to be *perpendicular* to a plane formed by **A** and **B**.* Because the direction of the perpendicular to a plane is ambiguous, we specify the direction by a convention called the *right-hand rule*. The perpendicular to the plane containing **A** and **B** is taken in the sense of advance of a right-handed screw rotated from the first vector **A** to the second vector **B** through the smaller angle between the vector directions (i.e., the angle that is less than 180°), as shown in Fig. 3–13a. Alternatively, if the right hand is held as in Fig. 3–13b, with the fingers curling in the direction of rotation carrying **A** into **B** through the smaller angle θ_{AB}, the thumb then points in the direction of **C**. (•Is it really necessary to specify that the rotation is through the *smaller* angle θ_{AB}? Will you obtain the correct direction for **C** if you rotate from **A** to **B** through the larger angle? Remember the significance of a vector with a negative sign.)

The cross product is not a commutative operation because

$$\mathbf{A} \times \mathbf{B} = -\mathbf{B} \times \mathbf{A}$$

Notice also that $\mathbf{A} \times \mathbf{A} = 0$. (•Why is this so?)

A Cartesian coordinate system is said to be *right-handed* if the rotation of the *x*-axis toward the *y*-axis produces a right-hand-rule direction that corresponds to the direction of the *z*-axis. Then, in a right-handed system, the unit vectors obey the relationships†

$$\hat{\mathbf{i}} \times \hat{\mathbf{j}} = \hat{\mathbf{k}}, \quad \hat{\mathbf{j}} \times \hat{\mathbf{k}} = \hat{\mathbf{i}}, \quad \hat{\mathbf{k}} \times \hat{\mathbf{i}} = \hat{\mathbf{j}}$$
$$\hat{\mathbf{i}} \times \hat{\mathbf{i}} = \hat{\mathbf{j}} \times \hat{\mathbf{j}} = \hat{\mathbf{k}} \times \hat{\mathbf{k}} = 0$$

(3–25)

Cross products obey the distributive law of multiplication:

$$\mathbf{A} \times (\mathbf{B} + \mathbf{C}) = \mathbf{A} \times \mathbf{B} + \mathbf{A} \times \mathbf{C}$$

If we write **A** and **B** in Cartesian component form, we obtain

$$\mathbf{A} \times \mathbf{B} = (A_x\hat{\mathbf{i}} + A_y\hat{\mathbf{j}} + A_z\hat{\mathbf{k}}) \times (B_x\hat{\mathbf{i}} + B_y\hat{\mathbf{j}} + B_z\hat{\mathbf{k}})$$
$$= (A_yB_z - A_zB_y)\hat{\mathbf{i}} + (A_zB_x - A_xB_z)\hat{\mathbf{j}}$$
$$+ (A_xB_y - A_yB_x)\hat{\mathbf{k}}$$

(3–26)

This rather formidable expression does offer a simplification, in that the second and third vector components can be generated from the first by a process called *cyclic permutation*. Consider the permutation, $x \rightarrow y \rightarrow z \rightarrow x \rightarrow y$, etc., on all subscripts and on the unit vectors. Then, in the first vector component in Eq. 3–26 consider such a permutation: $A_y \rightarrow A_z$, $B_z \rightarrow B_x$, $A_z \rightarrow A_x$, $B_y \rightarrow B_z$, and $\hat{\mathbf{i}} \rightarrow \hat{\mathbf{j}}$. We observe that this generates the second vector component. Continuing, we generate the third from the second by the permutations $A_z \rightarrow A_x$, $B_x \rightarrow B_y$, and so forth.

Perhaps the easiest way to reproduce Eq. 3–26 is by formal evaluation of the determinant

$$\mathbf{C} = \mathbf{A} \times \mathbf{B} = \begin{vmatrix} \hat{\mathbf{i}} & \hat{\mathbf{j}} & \hat{\mathbf{k}} \\ A_x & A_y & A_z \\ B_x & B_y & B_z \end{vmatrix}$$

(3–27)

---

*If **A** and **B** are collinear, no plane is defined. But then $\theta_{AB} = 0$, so that $C = 0$ automatically.
†Much grief will be avoided if a right-handed coordinate frame is *always* used, for only then are Eqs. 3–25 valid.

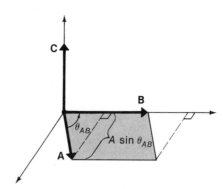

Fig. 3-14. The area of the parallelogram defined by the vectors **A** and **B** is represented by the vector **C** = **A** × **B**.

Developing this determinant by the first row results* in Eq. 3-26.

An interesting application of the cross product is in the representation of areas. Consider a parallelogram whose sides are formed by the vectors **A** and **B**, as in Fig. 3-14. We can represent the area of this parallelogram by the vector **C** = **A** × **B**, where $C = AB \sin \theta_{AB}$ is the magnitude of the area, and where the direction of the vector **C** is perpendicular to the plane of the parallelogram, as shown in the diagram. Although the association of a direction with an area may seem unusual, this concept is useful in many circumstances.

We state without proof a useful vector identity that involves the cross product:

$$\mathbf{A} \times (\mathbf{B} \times \mathbf{C}) = (\mathbf{A} \cdot \mathbf{C})\mathbf{B} - (\mathbf{A} \cdot \mathbf{B})\mathbf{C} \qquad (3\text{-}28)$$

Example 3-11

Evaluate the cross product **A** × **B** with $\mathbf{A} = 3\hat{\mathbf{i}} + 7\hat{\mathbf{j}} - \hat{\mathbf{k}}$ and $\mathbf{B} = \hat{\mathbf{i}} - \hat{\mathbf{j}}$.

Solution:

We have

$$\mathbf{C} = \mathbf{A} \times \mathbf{B} = \begin{vmatrix} \hat{\mathbf{i}} & \hat{\mathbf{j}} & \hat{\mathbf{k}} \\ +3 & +7 & -1 \\ +1 & -1 & 0 \end{vmatrix}$$

$$= \hat{\mathbf{i}}[(+7)(0) - (-1)(-1)] - \hat{\mathbf{j}}[(+3)(0) - (-1)(+1)] + \hat{\mathbf{k}}[(+3)(-1) - (+7)(+1)]$$

or

$$\mathbf{C} = -\hat{\mathbf{i}} - \hat{\mathbf{j}} - 10\hat{\mathbf{k}}$$

---

*Developing a determinant by its first row means

$$\begin{vmatrix} a & b & c \\ d & e & f \\ g & h & i \end{vmatrix} = a\begin{vmatrix} e & f \\ h & i \end{vmatrix} - b\begin{vmatrix} d & f \\ g & i \end{vmatrix} + c\begin{vmatrix} d & e \\ g & h \end{vmatrix}$$

Then,

$$\begin{vmatrix} e & f \\ h & i \end{vmatrix} = ei - hf, \quad \text{and so forth.}$$

Example 3–12

Consider the triangle formed by the vectors **A**, **B**, and **C** = **A** + **B**. Derive the law of sines using vector methods.

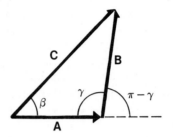

Solution:

We have

$$\mathbf{A} \times \mathbf{C} = \mathbf{A} \times (\mathbf{A} + \mathbf{B}) = 0 + \mathbf{A} \times \mathbf{B}$$

or

$$AC \sin \beta = AB \sin (\pi - \gamma) = AB \sin \gamma$$

Cancelling A and rearranging, we have

$$\frac{B}{C} = \frac{\sin \beta}{\sin \gamma}$$

and similarly for the other ratios. See Eq. A–36 in Appendix A.

3-7   VECTORS IN PHYSICS

The Dimensionality of Vectors.   In the preceding sections we have concentrated on the mathematical aspects of vectors. We now seek to use these quantities to describe situations and events in the real physical world.

An important difference between mathematical vectors and physical vectors is that the latter always have physical dimensions associated with them. In Chapter 2 we learned that velocity is the rate of change of the position function, $v = \Delta s/\Delta t$. All three quantities in this expression have dimensions: $[v] = LT^{-1}$, $[\Delta s] = L$, and $[\Delta t] = T$. We know that displacement, represented here by Δs, and the velocity are actually vector quantities. Thus, we can write the expression for velocity in vector notation as

$$\Delta \mathbf{s} = \mathbf{v} \, \Delta t$$

We write the equation in this form to show clearly that the multiplication of the vector **v** by the scalar Δt results in another vector, $\Delta \mathbf{s}$. This is the physical equivalent of the mathematical concept of multiplying a vector **A** by a number n. We emphasize the fact that the vectors, $\Delta \mathbf{s}$ and **v**, have *different* dimensions.

How do the rules of addition apply to physical vectors? Can we add the vector **v** to the vector $\Delta \mathbf{s}$? The answer is an emphatic *NO*. Physical vectors can be added only if they have the same dimensionality. One displacement vector can be added to another displacement vector, and two velocity vectors can be added, but *NEVER* can a displacement vector be added to a velocity vector. The addition of vectors with different dimensionality is physically meaningless.

The Equivalence of Physical Vectors. In a formal mathematical sense, the vector **C** is equivalent to the sum of the vectors **A** and **B** when $\mathbf{C} = \mathbf{A} + \mathbf{B}$. It is an important fact concerning the way Nature behaves that this same statement applies to physical vectors. For example, when two forces, $\mathbf{F}_A$ and $\mathbf{F}_B$, are simultaneously applied to a particle, the particle responds precisely as if the single force, $\mathbf{F}_C = \mathbf{F}_A + \mathbf{F}_B$, had been applied. This is an experimentally verified fact. This result is suggested, but not guaranteed, by the mathematics of vectors. We have no way of knowing whether a particular mathematical result or technique has any applicability to the physical world until it has been established by experimental test. (Albert Einstein once remarked that he did not understand why mathematics, which is purely a product of the mind, is so well suited to describing the real physical world.)

The Position Vector. One of the most important physical vectors is the *position vector* **r**. One major objective of mechanics is to describe the motions of particles under various influences. This may be accomplished by introducing a coordinate frame with an origin and oriented axes. The vector that describes the location of a particle with respect to this origin is the position vector **r**, obtained by placing the tail end at the origin and the arrow tip at the particle's coordinate $P(x, y, z)$, as shown in Fig. 3–15a. The vector **r** is therefore fixed by these two points, and, in a sense, is only a different way of quoting $P(x, y, z)$ relative to the chosen coordinate frame.

In Fig. 3–15b, we see two position vectors, $\mathbf{r}_1$ and $\mathbf{r}_1'$, which locate the point P in two arbitrarily chosen coordinate frames, S and S'. The magnitudes and directions of $\mathbf{r}_1$ and $\mathbf{r}_2$ are totally different from those of $\mathbf{r}_1'$ and $\mathbf{r}_2'$, but we can see in the figure that the displacement of the particle from the point P to the point Q is the same: $\overline{PQ} = \mathbf{r}_2 - \mathbf{r}_1 = \mathbf{r}_2' - \mathbf{r}_1'$. The magnitudes of the vectors are equal, that is,

Fig. 3–15. (a) The location of the instantaneous position $P(x, y, z)$ of a particle in coordinate frame S by means of the position vector **r**. (b) Although the individual position vectors, $\mathbf{r}_1$, $\mathbf{r}_1'$ and $\mathbf{r}_2$, $\mathbf{r}_2'$, are not the same in the two coordinate frames, S and S', the displacement vector $\overline{PQ} = \mathbf{r}_2 - \mathbf{r}_1 = \mathbf{r}_2' - \mathbf{r}_1'$ is the same.

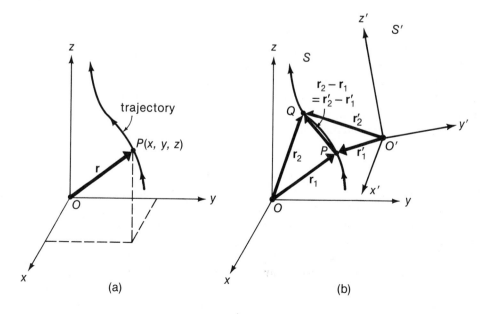

(a)

(b)

$|\mathbf{r}_2 - \mathbf{r}_1| = |\mathbf{r}_2' - \mathbf{r}_1'|$, and the spatial directions of $\mathbf{r}_2 - \mathbf{r}_1$ and $\mathbf{r}_2' - \mathbf{r}_1'$ are likewise the same. Thus, the displacement $\overline{PQ}$ is the same in all coordinate frames—it is an *invariant* quantity.

Vector Functions. It is often possible to describe physical phenomena by using vector quantities. Some of these vectors—such as velocity and force—do not have the dimensions of length. Nevertheless, the components of these vectors are specified by their values at particular points in ordinary coordinate space. Such vector quantities are called *vector functions* and are written

$$\mathbf{A}(\mathbf{r}) = A_x(\mathbf{r})\hat{\mathbf{i}} + A_y(\mathbf{r})\hat{\mathbf{j}} + A_z(\mathbf{r})\hat{\mathbf{k}}$$
$$= A_x(x, y, z)\hat{\mathbf{i}} + A_y(x, y, z)\hat{\mathbf{j}} + A_z(x, y, z)\hat{\mathbf{k}}$$

As this form implies, the components A_x, A_y, and A_z are functions of the spatial location (x, y, z). The vector quantity $\mathbf{A}$ need not have the dimensionality of length (for example, $\mathbf{A}$ may represent any vector quantity, such as velocity, force, or momentum). Operations such as sums, dot products, and cross products involving vectors of this type can be carried out in the ways specified in the preceding sections. These operations can *always* be defined exclusively in terms of the components, such as A_x, A_y, and A_z.

(a)

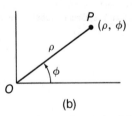

(b)

Fig. 3–16. (a) The point P is located with respect to the origin O by the coordinates (x, y). (b) The same point P is located in terms of the polar coordinates (ρ, ϕ).

Fig. 3–17. Unit polar vectors at a point P. Notice that $\hat{\mathbf{u}}_\rho$ and $\hat{\mathbf{u}}_\phi$ have different directions for different points P. However, $\hat{\mathbf{u}}_\rho$ and $\hat{\mathbf{u}}_\phi$ are always mutually perpendicular.

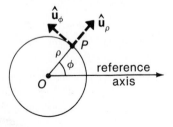

3-8 POLAR COORDINATES

To this point we have used only rectangular coordinate systems in our discussions. Because of the nature of a particular problem—for example, the description of circular motion—it may prove more convenient to use a different type of coordinate system, such as *plane polar coordinates*.

Figure 3–16 shows two equivalent ways to locate a point P with respect to a reference point (origin) O. In Fig. 3–16a the rectangular coordinates of the point are (x, y). Figure 3–16b shows the same point located in terms of the polar coordinates (ρ, ϕ), where the angle ϕ is measured counterclockwise from a reference line taken to coincide with the x-axis. The relationships connecting the various coordinates are:

$$\left. \begin{array}{ll} x = \rho \cos \phi & \rho = \sqrt{x^2 + y^2} \\ y = \rho \sin \phi & \phi = \tan^{-1}(y/x) \end{array} \right\} \qquad \textbf{(3–29)}$$

In polar coordinates, the direction of a vector is specified in terms of the unit vectors $\hat{\mathbf{u}}_\rho$ and $\hat{\mathbf{u}}_\phi$, as shown in Fig. 3–17. The unit vector $\hat{\mathbf{u}}_\rho$ is directed radially outward from the origin O. The unit vector $\hat{\mathbf{u}}_\phi$ is perpendicular to $\hat{\mathbf{u}}_\rho$ and is in the direction of increasing ϕ. An arbitrary vector $\mathbf{A}$ can always be written in terms of unit vectors. For rectangular and polar coordinates, we have

$$\left. \begin{array}{l} \mathbf{A} = A_x\hat{\mathbf{i}} + A_y\hat{\mathbf{j}} \\ \mathbf{A} = A_\rho\hat{\mathbf{u}}_\rho + A_\phi\hat{\mathbf{u}}_\phi \end{array} \right\} \qquad \textbf{(3–30)}$$

The quantity A_ρ is the *radial component* of $\mathbf{A}$, and A_ϕ is the *azimuthal component*. The squared magnitude of a vector (for example, the velocity vector) is

$$v^2 = \mathbf{v} \cdot \mathbf{v} = (v_\rho\hat{\mathbf{u}}_\rho + v_\phi\hat{\mathbf{u}}_\phi) \cdot (v_\rho\hat{\mathbf{u}}_\rho + v_\phi\hat{\mathbf{u}}_\phi)$$
$$= v_\rho^2 + v_\phi^2 \qquad \textbf{(3–31)}$$

where $\hat{\mathbf{u}}_\rho \cdot \hat{\mathbf{u}}_\phi = 0$ because the unit vectors are mutually perpendicular.

PROBLEMS

Section 3-3

3-1 A car is driven due west a distance of 30 km and then due south a distance of 50 km. Make an appropriate vector diagram, and use both graphic and analytic means to determine the car's displacement from the starting point. Compare the results.

3-2 The vector **S** has a magnitude of 2.4 units and is directed due north. The vector **T** has a magnitude of 1.6 units and is directed due west. Find the direction and magnitude of (a) the vector **S** + **T** and (b) the vector **S** − 3**T**.

Section 3-4

3-3 The vector **C** has a magnitude of $|C| = 3$ units and a direction with respect to the unit vector $\hat{\mathbf{u}}$ specified by $\theta_u = 227°$, measured counterclockwise from $\hat{\mathbf{u}}$ to **C**. What is the component C_u?

3-4 Given two vectors, $\mathbf{a} = 3\hat{\mathbf{i}} + 2\hat{\mathbf{j}}$ and $\mathbf{b} = -\hat{\mathbf{i}} + 7\hat{\mathbf{j}}$, find a third vector **c** such that $\mathbf{a} + \mathbf{b} + \mathbf{c} = 0$.

3-5 Repeat Example 3-5 but take the coordinate frame oriented as shown in the diagram. Is either the magnitude of the resultant or the angle between the resultant and the vector **A** different? Comment.

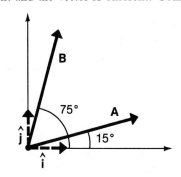

3-6 The vector **A** has a magnitude of 5 units and makes an angle of 20° with the x-axis. The vector **B** has a magnitude of 4 units and makes an angle of 50° with the x-axis. Find **C** = **A** + **B**.

3-7 The addition of two vectors, **a** and **b**, yields a resultant **s** such that $s = a + b$. What are the possible relationships of the original two vectors?

3-8 Given that $\mathbf{A} = 2\hat{\mathbf{i}} + 3\hat{\mathbf{j}}$ and $\mathbf{B} = 3\hat{\mathbf{i}} - 4\hat{\mathbf{j}}$, find the magnitudes and directions of (a) **C** = **A** + **B** and (b) **D** = 3**A** − 2**B**.

3-9 Two vectors, **A** and **B**, lie in the x-y plane and are perpendicular to each other. Show that their components are related by the expression $A_x B_x = -A_y B_y$.

3-10 Find the angle between the two vectors $\mathbf{A} = 5\hat{\mathbf{i}} + \hat{\mathbf{j}}$ and $\mathbf{B} = -2\hat{\mathbf{i}} + 4\hat{\mathbf{j}}$.

3-11 What is the component of the vector $\mathbf{A} = 2\hat{\mathbf{i}} + 3\hat{\mathbf{j}}$ in the direction of the unit vector $\hat{\mathbf{u}} = \frac{1}{\sqrt{17}}(4\hat{\mathbf{i}} + \hat{\mathbf{j}})$?

3-12 Find the vector sum of the three vectors

$$\mathbf{a} = -3\hat{\mathbf{i}} - 2\hat{\mathbf{j}} + 7\hat{\mathbf{k}}$$
$$\mathbf{b} = \hat{\mathbf{i}} + 3\hat{\mathbf{j}} + 3\hat{\mathbf{k}}$$
$$\mathbf{c} = -5\hat{\mathbf{k}}$$

3-13 A small boat sails out on the still waters of a deep lake. It leaves the landing and proceeds due north for 1000 m, then due east for 2000 m, whereupon the bottom is accidentally holed and the boat sinks straight down 300 m to the lake floor. Locate the sunken boat relative to the landing.

3-14 Given the vectors $\mathbf{A} = 3\hat{\mathbf{i}} - 2\hat{\mathbf{j}} + 2\hat{\mathbf{k}}$, $\mathbf{B} = 4\hat{\mathbf{i}} + 3\hat{\mathbf{j}} - 2\hat{\mathbf{k}}$, and $\mathbf{C} = 3\hat{\mathbf{i}} + 2\hat{\mathbf{j}} + 3\hat{\mathbf{k}}$, find the projection of $\mathbf{D} = \mathbf{A} - \mathbf{B} + \mathbf{C}$ onto the x-y plane.

3-15 Refer to the vectors **A**, **B**, and **C** in the preceding problem. Find the vector **E** such that $\mathbf{A} + \mathbf{B} + \mathbf{C} + \mathbf{E} = 0$.

Section 3-5

3-16 Find the dot product of the two vectors $\mathbf{B} = 6\hat{\mathbf{i}} + \hat{\mathbf{j}} - 2\hat{\mathbf{k}}$ and $\mathbf{D} = -\hat{\mathbf{i}} + 3\hat{\mathbf{j}}$. What is the angle between **B** and **D**?

3-17 Show that the two vectors $\mathbf{a} = 9\hat{\mathbf{i}} + \hat{\mathbf{j}} - 4\hat{\mathbf{k}}$ and $\mathbf{b} = 3\hat{\mathbf{i}} - 7\hat{\mathbf{j}} + 5\hat{\mathbf{k}}$ are mutually perpendicular.

3-18 Using vector methods, find the acute angle of an equilateral parallelogram if one diagonal is twice as long as the other.

Section 3-6

3-19 Find the cross product of the two vectors $\mathbf{C} = 2\hat{\mathbf{i}} - \hat{\mathbf{j}}$ and $\mathbf{D} = \hat{\mathbf{i}} + 2\hat{\mathbf{j}} - 3\hat{\mathbf{k}}$. What is the angle between these vectors?

3-20 Find the cross product of the two vectors $\mathbf{C} = 2\hat{\mathbf{i}} - 4\hat{\mathbf{j}} + 2\hat{\mathbf{k}}$ and $\mathbf{D} = -3\hat{\mathbf{i}} + 6\hat{\mathbf{j}} - 3\hat{\mathbf{k}}$. What special relationship exists between these two vectors?

3-21 Given $\mathbf{A} = 2\hat{\mathbf{i}} + 3\hat{\mathbf{j}} + \hat{\mathbf{k}}$ and $\mathbf{B} = \hat{\mathbf{i}} + 3\hat{\mathbf{j}} + 2\hat{\mathbf{k}}$.
• What is the angle between $\mathbf{A} \times \mathbf{B}$ and the z-axis?

3-22 If, in Eq. 3-27, we have
•

$$\frac{A_x}{B_x} = \frac{A_y}{B_y} = \frac{A_z}{B_z}$$

show that C vanishes identically. If the above postulated relationships hold, what can you say about the direction cosines of $\mathbf{A}$ and $\mathbf{B}$?

3-23 Find the area of the triangle determined by the two vectors $\mathbf{a} = 2\hat{\mathbf{i}} + 3\hat{\mathbf{j}}$ and $\mathbf{b} = \hat{\mathbf{i}} - 2\hat{\mathbf{j}} + \hat{\mathbf{k}}$. Find the two unit vectors that are normal (perpendicular) to the plane containing this triangle.

Section 3-8

3-24 Quito, Ecuador, and Kampala, Uganda, both lie
• approximately on the Equator. The longitude of Quito is 78°30′ W and that of Kampala is 32°30′ E. What is the distance from Quito to Kampala (a) along the shortest surface path and (b) along a direct (through-the-Earth) path? (The radius of the Earth is 6.38×10^6 m.)

Additional Problems

3-25 Two particles are located at $\mathbf{r}_1 = 3\hat{\mathbf{i}} + 7\hat{\mathbf{j}} - 2\hat{\mathbf{k}}$ and $\mathbf{r}_2 = -2\hat{\mathbf{i}} + 6\hat{\mathbf{k}}$, respectively. Find the vector distance from the particle labeled 1 to the particle labeled 2. Give both the magnitude of the vector and its orientation with respect to the x-, y-, and z-axes. [*Hint:* Find $\mathbf{r} = \mathbf{r}_1 - \mathbf{r}_2$.]

3-26 Two vectors with magnitudes of 5 and 6 units, respectively, have an included angle of 45° between their directions. A third vector, with a magnitude of 7 units, is perpendicular to the plane containing the first two. Determine the magnitude of the sum of the three vectors and its direction relative to the first two vectors. [*Hint:* Use component resolutions in a carefully chosen three-dimensional Cartesian frame.]

3-27 What is the geometric condition for three vectors with magnitudes of 7, 24, and 25 units to add to zero?

3-28 Given a rectangular room with dimensions
•• $4 \text{ m} \times 5 \text{ m} \times 6 \text{ m}$, as shown in the diagram. A bug starting at point A in the corner crawls to point B along a straight line path on the floor, then crawls up the wall from B to the corner point C, also along a straight line path. (a) What is the value of x such that

the total path A to C is a minimum? [*Hint:* You may use either the calculus or geometric considerations based on a "developed plan" of the room with the walls folded onto the plane of the floor.] (b) A second insect flies directly from A to C. How much shorter is its trip than the minimum for the crawling bug?

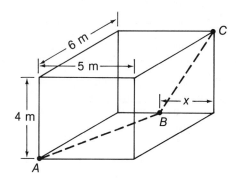

3-29 The coordinates (x, y, z) of two points are $(2, 1, 3)$ and $(4, -2, 2)$. Find the angle between the two lines formed by connecting these points to the origin.

3-30 Define two vectors with unit length to be
•

$$\mathbf{a} = \cos \alpha \, \hat{\mathbf{i}} + \sin \alpha \, \hat{\mathbf{j}}$$
$$\mathbf{b} = \cos \beta \, \hat{\mathbf{i}} + \sin \beta \, \hat{\mathbf{j}}$$

By forming the dot product $\mathbf{a} \cdot \mathbf{b}$ and the cross product $\mathbf{a} \times \mathbf{b}$, derive the addition laws for the sine and cosine.

3-31 Using vector methods, show that the diagonals of an
•• equilateral parallelogram are perpendicular.

3-32 In view of Eq. 3-14, two arbitrary unit vectors, $\hat{\mathbf{a}}$ and
• $\hat{\mathbf{b}}$, can be written in terms of their direction cosines as

$$\hat{\mathbf{a}} = \cos \alpha_a \, \hat{\mathbf{i}} + \cos \beta_a \, \hat{\mathbf{j}} + \cos \gamma_a \, \hat{\mathbf{k}}$$
$$\hat{\mathbf{b}} = \cos \alpha_b \, \hat{\mathbf{i}} + \cos \beta_b \, \hat{\mathbf{j}} + \cos \gamma_b \, \hat{\mathbf{k}}$$

Derive an expression for the cosine of the angle between $\hat{\mathbf{a}}$ and $\hat{\mathbf{b}}$.

3-33 Find the area of a parallelogram determined by the vectors $\mathbf{a} = \hat{\mathbf{i}} - 2\hat{\mathbf{j}} + 3\hat{\mathbf{k}}$ and $\mathbf{b} = 2\hat{\mathbf{i}} + \hat{\mathbf{j}} - \hat{\mathbf{k}}$. Find the unit vector that is normal to the plane containing $\mathbf{a}$ and $\mathbf{b}$, and having the direction of $\mathbf{a} \times \mathbf{b}$.

3-34 Show that $\mathbf{A} \cdot (\mathbf{B} \times \mathbf{C}) = \begin{vmatrix} A_x & A_y & A_z \\ B_x & B_y & B_z \\ C_x & C_y & C_z \end{vmatrix}$

3-35 Show explicitly by vector methods that if $\mathbf{C} = \mathbf{A} \times \mathbf{B}$, then $\mathbf{C}$ is perpendicular to both $\mathbf{A}$ and $\mathbf{B}$.

3-36 Show that the expression in Problem 3-34 may be
• interpreted as the volume of a parallelopiped with edges formed by the three vectors, as illustrated in the diagram below.

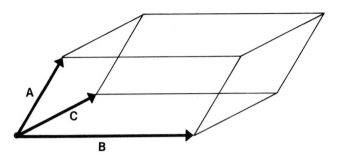

3-37 Find the volume of a parallelopiped with edges
•

$$\mathbf{a} = 3\widehat{\mathbf{i}} - 2\widehat{\mathbf{j}}, \qquad \mathbf{b} = \widehat{\mathbf{i}} + 2\widehat{\mathbf{j}} - \widehat{\mathbf{k}}, \qquad \mathbf{c} = 3\widehat{\mathbf{k}}$$

3-38 Show that three (nonvanishing) vectors **A**, **B**, and **C** are coplanar if and only if

$$\begin{vmatrix} A_x & A_y & A_z \\ B_x & B_y & B_z \\ C_x & C_y & C_z \end{vmatrix} = 0$$

KINEMATICS IN TWO AND THREE DIMENSIONS

In the discussion of kinematics in Chapter 2 we considered motion along a straight line. Now that vector methods have been added to our collection of mathematical tools, we are prepared to extend the discussion of kinematics to motion in two and three dimensions. We consider here the motion of (pointlike) particles, but the results are also applicable to rigid objects with extent—blocks, balls, and similar objects—if we restrict attention to *translational* motion and omit (for the moment) any consideration of *rotational* effects. Pure translation is illustrated in Fig. 4–1a. In this case, the orientation of the body remains fixed as the motion proceeds. That is, the two reference points, *P* and *R*, travel along parallel paths (the dashed and solid curves, respectively, in Fig. 4–1a). Knowledge of the motion of one point—for example, *R*—determines completely the motion of any other point, such as *P*, and thus determines the motion of the body as a whole. *In this sense,* the object moves as if it were a particle.

Fig. 4–1. Two-dimensional motion of a body (a) with only translational motion, and (b) with both translational and rotational motion. The inset shows how the rotational motion produces changes in the relative orientation of two points, *R* and *P*, on the body.

(a)

(b)

In Fig. 4–1b we show a case of motion with both translation and rotation. Here, the orientation of point P with respect to R changes continually because of the rotation. The inset shows this rotational aspect of the motion. The motion of point R alone no longer determines the motion of the body as a whole. This complication is treated in later chapters.

4-1 THE POSITION VECTOR

In order to describe the motion of a particle, we first adopt a convenient coordinate frame. At any particular instant the position of the particle is specified by the position vector $\mathbf{r}(t)$. Recall (Section 3–7) that we construct this vector with its tail end at the origin of the coordinate frame and its arrow tip at the instantaneous position of the particle. As the particle moves, the position vector (the tip) defines the path or *trajectory* of the particle with respect to the chosen coordinate origin. Figure 4–2 illustrates a particular trajectory in a Cartesian coordinate frame.

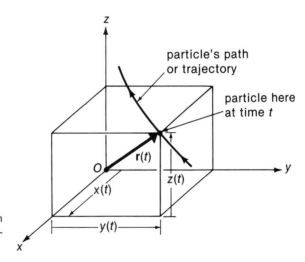

Fig. 4–2. The use of the position vector $\mathbf{r}(t)$ to describe the path followed by a particle.

Quoting the position vector (or position function) $\mathbf{r}(t)$ is equivalent to specifying the x, y, and z coordinates of the particle as functions of time, with

$$\mathbf{r}(t) = x(t)\hat{\mathbf{i}} + y(t)\hat{\mathbf{j}} + z(t)\hat{\mathbf{k}} \tag{4-1}$$

We are interested in following the changes in $\mathbf{r}(t)$ as the particle proceeds along its trajectory. For simplicity, consider a particle that moves in the x-y plane, as shown in Fig. 4–3. At a particular instant t_1, the particle is at the point P given by $\mathbf{r}(t_1)$; at a later time t_2, the particle is at point Q given by $\mathbf{r}(t_2)$. During the time interval $t_2 - t_1$, the particle undergoes a displacement corresponding to*

$$\begin{aligned} \mathbf{r}_{12} &= \mathbf{r}(t_2) - \mathbf{r}(t_1) \\ &= [x(t_2)\hat{\mathbf{i}} + y(t_2)\hat{\mathbf{j}}] - [x(t_1)\hat{\mathbf{i}} + y(t_1)\hat{\mathbf{j}}] \\ &= [x(t_2) - x(t_1)]\hat{\mathbf{i}} + [y(t_2) - y(t_1)]\hat{\mathbf{j}} \end{aligned} \tag{4-2}$$

---

*The double subscript on $\mathbf{r}$ is used to identify the initial and final points respectively, or the displacement, either through the time parameter, as here, or through the spatial designations, P and Q. Thus, we could equally well write $\mathbf{r}_{12}$ or $\mathbf{r}_{PQ}$.

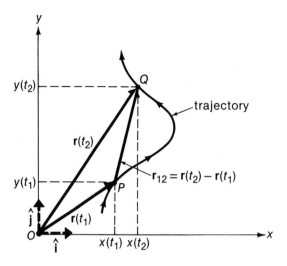

Fig. 4–3. Illustrating the displacement $\mathbf{r}_{12}$ of a particle executing planar motion during the time interval $t_2 - t_1$.

We can also imagine $t_2 - t_1$ to be a small incremental quantity Δt. The corresponding incremental displacement is then $\Delta\mathbf{r}_{12}$ (or, simply, $\Delta\mathbf{r}$). In the limit $\Delta t \to 0$, we have

$$d\mathbf{r} = dx\,\widehat{\mathbf{i}} + dy\,\widehat{\mathbf{j}} \tag{4–3}$$

This quantity is called the *differential displacement vector* (in two dimensions).

Example 4–1

Suppose that the trajectory of a particle is $\mathbf{r}(t) = x(t)\widehat{\mathbf{i}} + y(t)\widehat{\mathbf{j}}$ with $x(t) = at^2 + bt$ and $y(t) = ct + d$, where a, b, c, and d are constants that have appropriate dimensions. What displacement does the particle undergo between $t = 1$ s and $t = 3$ s?

Solution:

Using Eq. 4–2, we have

$$\Delta\mathbf{r} = [x(3) - x(1)]\widehat{\mathbf{i}} + [y(3) - y(1)]\widehat{\mathbf{j}}$$

Now,

$$x(3) = 9a + 3b \qquad y(3) = 3c + d$$
$$x(1) = a + b \qquad y(1) = c + d$$

Hence,

$$\Delta\mathbf{r} = (8a + 2b)\widehat{\mathbf{i}} + 2c\,\widehat{\mathbf{j}}$$

4-2 THE VELOCITY VECTOR

Suppose that the position vector of a particle is $\mathbf{r}(t_1)$ at time t_1 and is $\mathbf{r}(t_2)$ at time t_2. During the time interval $t_2 - t_1$, the particle undergoes a displacement $\mathbf{r}_{12} = \mathbf{r}(t_2) - \mathbf{r}(t_1)$. In analogy with the definition of average velocity for straight-line motion (Eq. 2–1), we now define the *average vector velocity* during this time interval to be

$$\bar{\mathbf{v}} = \frac{\mathbf{r}(t_2) - \mathbf{r}(t_1)}{t_2 - t_1} = \frac{\mathbf{r}_{12}}{t_2 - t_1} \tag{4-4}$$

Notice that $\bar{\mathbf{v}}$ is indeed a vector quantity because it involves the multiplication of a vector $\mathbf{r}_{12}$ by a scalar $1/(t_2 - t_1)$. The magnitude of $\bar{\mathbf{v}}$ is $|\mathbf{r}_{12}|/(t_2 - t_1)$ and the direction of $\bar{\mathbf{v}}$ is the same as that of $\mathbf{r}_{12}$. Of course, $\bar{\mathbf{v}}$ and $\mathbf{r}_{12}$ have different dimensions; nevertheless, we find it convenient on occasion to draw velocity vectors in the same figures that represent trajectories in coordinate space, as in Fig. 4–4.

We can rewrite Eq. 4–4 in the form

$$\bar{\mathbf{v}} = \frac{\mathbf{r}(t + \Delta t) - \mathbf{r}(t)}{\Delta t} = \frac{\Delta \mathbf{r}}{\Delta t} \tag{4-5}$$

Aside from shifting emphasis, this equation has precisely the same meaning as Eq. 4–4 if we identify t with t_1, Δt with $t_2 - t_1$, and $\Delta \mathbf{r}$ with $\mathbf{r}_{12}$. The (instantaneous) *vector velocity* is now defined as the general time derivative,

$$\mathbf{v} = \lim_{\Delta t \to 0} \frac{\mathbf{r}(t + \Delta t) - \mathbf{r}(t)}{\Delta t} = \frac{d\mathbf{r}}{dt} = \frac{d}{dt}\mathbf{r}(t) \tag{4-6}$$

Figure 4–4 illustrates Eqs. 4–5 and 4–6. We see that $\mathbf{v}$ is always tangent to the trajectory at the point that locates the particle position. It is also evident from this figure that $\mathbf{v}$ is, in general, a function of time. Thus, at the time t we express the position vector of the particle as $\mathbf{r}(t)$ and the velocity as $\mathbf{v}(t)$. When we are interested only in the magnitude of $\mathbf{v}(t)$, we write $|\mathbf{v}(t)|$ or $v(t)$, which we call the *speed* of the particle. (Note that $|\mathbf{v}(t)| \geq 0$.)

We wish to emphasize that $\mathbf{r}(t)$ can vary with time *both* in magnitude and in direction. The velocity is a measure of both of these possible changes. Thus, for example, a particle that travels in a circular path with a radius R and centered at the origin has $|\mathbf{r}(t)| = R$, which is a constant, yet the particle has a nonzero velocity. We return to a discussion of this interesting type of motion later in this chapter.

The velocity vector can always be given in component form. Thus, in three dimensions, we have

$$\mathbf{v}(t) = \frac{d}{dt}\mathbf{r}(t) = \frac{d}{dt}[x(t)\hat{\mathbf{i}} + y(t)\hat{\mathbf{j}} + z(t)\hat{\mathbf{k}}]$$

Now, for rectangular coordinates, the unit vectors $\hat{\mathbf{i}}$, $\hat{\mathbf{j}}$, and $\hat{\mathbf{k}}$ do not change with time; hence,

$$\begin{aligned} \mathbf{v}(t) &= \frac{dx}{dt}\hat{\mathbf{i}} + \frac{dy}{dt}\hat{\mathbf{j}} + \frac{dz}{dt}\hat{\mathbf{k}} \\ &= v_x\hat{\mathbf{i}} + v_y\hat{\mathbf{j}} + v_z\hat{\mathbf{k}} \end{aligned} \tag{4-7}$$

where the velocity components are

$$v_x = \frac{dx}{dt}, \qquad v_y = \frac{dy}{dt}, \qquad v_z = \frac{dz}{dt}$$

(a)

(b)

Fig. 4–4. (a) Illustrating the finite displacement, $\overline{PQ} = \Delta\mathbf{r}$, and the average velocity $\bar{\mathbf{v}}$ (dashed line). (b) As the limit in Eq. 4–6 is developed, the point Q approaches P without limit. The velocity $\mathbf{v}(t)$ is tangent to the trajectory at P.

The vector components of the velocity are

$$\mathbf{v}_x = \frac{dx}{dt}\hat{\mathbf{i}}, \qquad \mathbf{v}_y = \frac{dy}{dt}\hat{\mathbf{j}}, \qquad \mathbf{v}_z = \frac{dz}{dt}\hat{\mathbf{k}}$$

Also, for the speed v we have

$$v = |\mathbf{v}| = \sqrt{v_x^2 + v_y^2 + v_z^2} \qquad\qquad \textbf{(4–8)}$$

Example 4–2

Suppose that the position vector function for a particle is given as $\mathbf{r}(t) = x(t)\hat{\mathbf{i}} + y(t)\hat{\mathbf{j}}$, with $x(t) = at + b$ and $y(t) = ct^2 + d$, where $a = 1$ m/s, $b = 1$ m, $c = 1/8$ m/s^2, and $d = 1$ m. (a) Calculate the average velocity during the time interval from $t = 2$ s to $t = 4$ s. (b) Determine the velocity and the speed at $t = 2$ s.

Solution:

(a) For the average velocity, we have

$$\overline{\mathbf{v}} = \left(\frac{x(4) - x(2)}{4\,\text{s} - 2\,\text{s}}\right)\hat{\mathbf{i}} + \left(\frac{y(4) - y(2)}{4\,\text{s} - 2\,\text{s}}\right)\hat{\mathbf{j}}$$

$$= \left(\frac{5\,\text{m} - 3\,\text{m}}{2\,\text{s}}\right)\hat{\mathbf{i}} + \left(\frac{3\,\text{m} - 1.5\,\text{m}}{2\,\text{s}}\right)\hat{\mathbf{j}}$$

$$= (\hat{\mathbf{i}} + 0.75\hat{\mathbf{j}})\,\text{m/s}$$

(b) For the velocity components, we have

$$v_x = \frac{dx}{dt} = a = 1\,\text{m/s}$$

$$v_y = \frac{dy}{dt} = 2ct = (\tfrac{1}{4}\,\text{m/s}^2)t$$

Therefore,

$$\mathbf{v} = v_x\hat{\mathbf{i}} + v_y\hat{\mathbf{j}} = (1\,\text{m/s})\hat{\mathbf{i}} + (\tfrac{1}{4}\,\text{m/s}^2)t\hat{\mathbf{j}}$$

$$\mathbf{v}(2) = (1\,\text{m/s})\hat{\mathbf{i}} + (0.5\,\text{m/s})\hat{\mathbf{j}}$$

and the speed is

$$|\mathbf{v}|_{t=2\,\text{s}} = \sqrt{(1\,\text{m/s})^2 + (0.5\,\text{m/s})^2} = 1.118\,\text{m/s}$$

The position functions, $x(t)$ and $y(t)$, are shown in diagrams (a) and (b). We can eliminate the time parameter by solving for t in the expression given for $x(t)$, and substituting $t = (x - b)/a$ into the expression given for $y(t)$. Then, we obtain

$$y = y(x) = \tfrac{1}{8}(x - 1)^2 + 1$$

with x and y both given in meters. This is the equation of a parabola, which is illustrated in diagram (c).

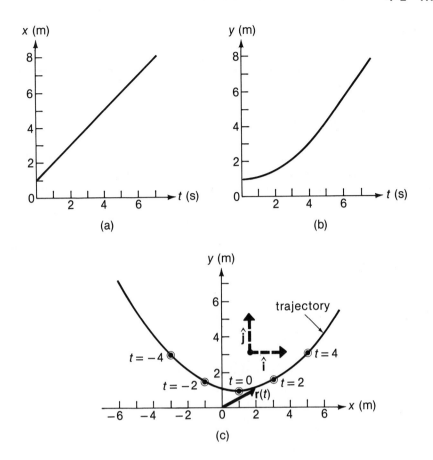

(a)

(b)

(c)

Example 4-3

A particle moves along a straight-line path in the direction of a constant unit vector $\hat{\mathbf{u}}$. The time-dependent position function is

$$\mathbf{r}(t) = \mathbf{r}_0 + (v_0 t + \tfrac{1}{2}a_0 t^2)\hat{\mathbf{u}}$$

where v_0 and a_0 are constant scalars, and where $\mathbf{r}_0$ is a constant vector in some coordinate frame. Find the vector velocity at time t.

Solution:

From the defining equation (Eq. 4-6), we have

$$\mathbf{v}(t) = \frac{d\mathbf{r}(t)}{dt} = \frac{d}{dt}[\mathbf{r}_0 + (v_0 t + \tfrac{1}{2}a_0 t^2)\hat{\mathbf{u}}]$$

$$= \frac{d\mathbf{r}_0}{dt} + (v_0 t + \tfrac{1}{2}a_0 t^2)\frac{d\hat{\mathbf{u}}}{dt} + \left[\frac{d}{dt}(v_0 t + \tfrac{1}{2}a_0 t^2)\right]\hat{\mathbf{u}}$$

Because $\mathbf{r}_0$ and $\hat{\mathbf{u}}$ are independent of time, only the last term survives. Upon differentiating, we find

$$\mathbf{v}(t) = (v_0 + a_0 t)\hat{\mathbf{u}}$$

Note that although $\mathbf{r}(t)$ is referred specifically to the chosen coordinate frame (through the vector $\mathbf{r}_0$), the velocity vector is independent of this choice.

4-3   THE ACCELERATION VECTOR

Suppose that the velocity of a particle at time t is $\mathbf{v}(t)$ and that a time interval Δt later it is $\mathbf{v}(t + \Delta t)$, as shown in Fig. 4-5. We define the *average vector acceleration* during this time interval to be

$$\bar{\mathbf{a}} = \frac{\mathbf{v}(t + \Delta t) - \mathbf{v}(t)}{\Delta t} = \frac{\Delta \mathbf{v}}{\Delta t} \tag{4-9}$$

The (instantaneous) *vector acceleration* is obtained by taking the limit of Eq. 4-9 as $\Delta t \to 0$; thus,

$$\mathbf{a} = \lim_{\Delta t \to 0} \frac{\mathbf{v}(t + \Delta t) - \mathbf{v}(t)}{\Delta t} = \frac{d\mathbf{v}}{dt} = \frac{d}{dt}\mathbf{v}(t) \tag{4-10}$$

We again emphasize that $\mathbf{v}(t)$ can change with time *both* in magnitude and in direction. Each of these changes is reflected in the resulting vector $\mathbf{a}$. For example, a particle traveling in a circular path has a velocity vector that continually changes direction (always being tangent to the circular path). Thus, the particle undergoes an acceleration even if the speed remains constant.

Finally, it is useful to introduce the vector components for the acceleration $\mathbf{a}$. If we write

$$\mathbf{v}(t) = v_x(t)\hat{\mathbf{i}} + v_y(t)\hat{\mathbf{j}} + v_z(t)\hat{\mathbf{k}}$$

then

$$\mathbf{a} = a_x\hat{\mathbf{i}} + a_y\hat{\mathbf{j}} + a_z\hat{\mathbf{k}}$$

$$= \frac{dv_x}{dt}\hat{\mathbf{i}} + \frac{dv_y}{dt}\hat{\mathbf{j}} + \frac{dv_z}{dt}\hat{\mathbf{k}} \tag{4-11}$$

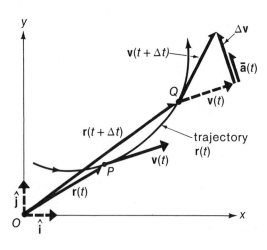

Fig. 4-5. The trajectory of a moving particle. The velocity increment is $\Delta\mathbf{v} = \mathbf{v}(t + \Delta t) - \mathbf{v}(t)$. The average acceleration $\bar{\mathbf{a}}$ is in the same direction as $\Delta\mathbf{v}$.

where the acceleration components are

$$a_x = \frac{dv_x}{dt} = \frac{d^2x}{dt^2}, \qquad a_y = \frac{dv_y}{dt} = \frac{d^2y}{dt^2}, \qquad a_z = \frac{dv_z}{dt} = \frac{d^2z}{dt^2}$$

Example 4-4

A particle has the velocity function $\mathbf{v}(t) = bt^2\hat{\mathbf{i}} + ct\hat{\mathbf{j}}$, with $b = 3$ m/s^3 and $c = -4$ m/s^2. (a) What is the velocity at $t = 1$ s and at $t = 3$ s? (b) What is the average acceleration during this time interval? (c) What is the acceleration at $t = 1.5$ s?

Solution:

(a) We have

$$\mathbf{v}(1) = (3\hat{\mathbf{i}} - 4\hat{\mathbf{j}}) \text{ m/s}$$
$$\mathbf{v}(3) = (27\hat{\mathbf{i}} - 12\hat{\mathbf{j}}) \text{ m/s}$$

(b) The average acceleration can be calculated using Eq. 4-9 if we identify $t = 1$ s and $\Delta t = 3$ s $- 1$ s $= 2$ s; that is,

$$\bar{\mathbf{a}} = \frac{\mathbf{v}(3) - \mathbf{v}(1)}{2 \text{ s}}$$

$$= \frac{(27 - 3)\hat{\mathbf{i}} \text{ m/s}}{2 \text{ s}} - \frac{(12 - 4)\hat{\mathbf{j}} \text{ m/s}}{2 \text{ s}}$$

$$= (12\hat{\mathbf{i}} - 4\hat{\mathbf{j}}) \text{ m/s}^2$$

(c) The acceleration is

$$\mathbf{a}(t) = \frac{d}{dt}\mathbf{v}(t) = 2bt\hat{\mathbf{i}} + c\hat{\mathbf{j}}$$

Then,
$$\mathbf{a}(1.5) = (9\hat{\mathbf{i}} - 4\hat{\mathbf{j}}) \text{ m/s}^2$$

Example 4-5

A particle travels at constant speed, but with no restriction on its direction of motion. Show that the acceleration at any instant is always perpendicular to the velocity.

Solution:

Because $|\mathbf{v}(t)|$ is a constant, $v^2(t) = \mathbf{v}(t) \cdot \mathbf{v}(t)$ is also a constant. Thus,

$$\frac{d}{dt}(\mathbf{v} \cdot \mathbf{v}) = 0$$

But
$$\frac{d}{dt}(\mathbf{v} \cdot \mathbf{v}) = \mathbf{v} \cdot \left(\frac{d\mathbf{v}}{dt}\right) + \left(\frac{d\mathbf{v}}{dt}\right) \cdot \mathbf{v} = 2\mathbf{v} \cdot \left(\frac{d\mathbf{v}}{dt}\right)$$

Therefore,
$$\mathbf{v} \cdot \frac{d\mathbf{v}}{dt} = \mathbf{v} \cdot \mathbf{a} = 0$$

which (for $\mathbf{a} \neq 0$) insures that we always have $\mathbf{v} \perp \mathbf{a}$.

To understand this result more fully, notice that any nonzero acceleration **a** that is not perpendicular to the velocity **v** must have a component in the direction of **v** or opposite to it. Such a component will have the effect of changing the speed. Thus, we must have either **a** = 0 or **v** $\perp$ **a**.

4-4 MOTION WITH CONSTANT ACCELERATION

In many practical cases we consider a particle moving with a constant acceleration $\mathbf{a}_0$. This means that in an arbitrarily chosen coordinate frame, the components, a_{0x}, a_{0y}, and a_{0z}, are all constants. By a straightforward extension of the discussions in Chapter 2, we can obtain the velocity functions by integration:

$$\left.\begin{array}{l} v_x(t) = a_{0x}t + v_{0x} \\ v_y(t) = a_{0y}t + v_{0y} \\ v_z(t) = a_{0z}t + v_{0z} \end{array}\right\} \tag{4-12}$$

where v_{0x}, v_{0y}, and v_{0z} are the constants of integration and correspond to the components of the velocity at $t = 0$, namely, $\mathbf{v}_0$. We can also write Eq. 4–12 in vector form as

$$\mathbf{v}(t) = \mathbf{a}_0 t + \mathbf{v}_0 \tag{4-13}$$

Integrating once more gives the position function components,

$$\left.\begin{array}{l} x(t) = \frac{1}{2}a_{0x}t^2 + v_{0x}t + x_0 \\ y(t) = \frac{1}{2}a_{0y}t^2 + v_{0y}t + y_0 \\ z(t) = \frac{1}{2}a_{0z}t^2 + v_{0z}t + z_0 \end{array}\right\} \tag{4-14}$$

or in vector form,

$$\mathbf{r}(t) = \frac{1}{2}\mathbf{a}_0 t^2 + \mathbf{v}_0 t + \mathbf{r}_0 \tag{4-15}$$

In these equations the additional constants of integration, x_0, y_0, and z_0, correspond to the components of the particle's coordinate position $\mathbf{r}_0$ at $t = 0$.

The above equations also hold, of course, for $\mathbf{a}_0 = 0$. In this case,

$$\mathbf{r}(t) = \mathbf{v}_0 t + \mathbf{r}_0$$

and the ensuing motion is simple uniform motion in the direction of the constant velocity $\mathbf{v}_0$.

Ballistic Trajectories.* In any case of constant acceleration, there will be a

_____

*The path followed by an unpowered and unguided projectile is called a *ballistic trajectory*. A ballistic flight is one in which the projectile is given an initial boost and is then allowed to coast to its target.

simplification of the equations of motion if we select one of the coordinate axes to be parallel to the direction of the acceleration vector. The motion of a projectile under the influence of gravitational acceleration is typical of this class of problems. Once launched, such objects as baseballs, golf balls, or bullets are subject only to the downward acceleration of gravity, if we neglect air resistance effects.

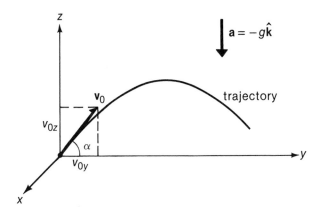

Fig. 4-6. The flight path of a struck golf ball. The initial position (the tee) is at the origin ($\mathbf{r}_0 = 0$); the initial velocity $\mathbf{v}_0$ and the ensuing motion are in the y-z plane. The angle between $\mathbf{v}_0$ and the horizontal plane is α.

Consider the flight path of a golf ball driven from a tee with an initial speed v_0. Orient a Cartesian coordinate frame so that the z-axis is taken positive in the upward direction. Thus, $\mathbf{a}_0 = -g\hat{\mathbf{k}}$, with $g = 9.80 \text{ m/s}^2$. The origin is most conveniently taken to be the tee, so that $\mathbf{r}_0 = 0$. Neglecting aerodynamic effects, the golf ball will travel in a vertical plane, which we take to be the y-z plane, as indicated in Fig. 4-6. In selecting the axes in this special way, the three-dimensional problem has been reduced to one in two dimensions. The equations of motion (Eqs. 4-14) become

$$\left. \begin{array}{l} x(t) = 0 \\ y(t) = v_{0y}t \\ z(t) = -\tfrac{1}{2}gt^2 + v_{0z}t \end{array} \right\} \tag{4-16}$$

The angle α specifies the angle between the horizontal plane and the direction with which the ball was initially lofted. Therefore, $v_{0y} = v_0 \cos \alpha$ and $v_{0z} = v_0 \sin \alpha$. We can eliminate t from Eqs. 4-16 by substituting $t = y/(v_0 \cos \alpha)$ into the expression for $z(t)$. This yields

$$z = -\left(\frac{g}{2v_0^2 \cos^2 \alpha} \right)y^2 + (\tan \alpha)y \tag{4-17}$$

which is the equation of a parabola. Figure 4-7 shows the position functions, $y(t)$ and $z(t)$, as well as the velocity vector at various points along the trajectory.

The time when the golf ball is at the level of the ground (assumed flat) is obtained by setting $z(t) = 0$ in Eqs. 4-16. The two solutions are

$$t = 0 \quad \text{and} \quad T = \frac{2v_{0z}}{g} = \frac{2v_0}{g} \sin \alpha$$

The first solution is the time the ball was driven; the second solution corresponds to the end of the flight. The range R, defined as $R = y(T)$, is obtained either by substituting the value of T into $y(t)$ given by Eq. 4-16 or by setting $z = 0$ in Eq. 4-17.

The rider and his motorcycle are in a ballistic trajectory. (Courtesy of *Cycle World*.)

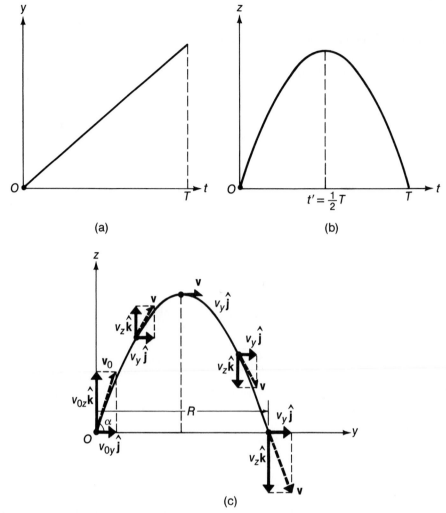

Fig. 4–7. (a) and (b) The position functions, $y(t)$ and $z(t)$, for the flight of a struck golf ball. The time T refers to the time at which the ball strikes the ground. (c) The trajectory z as a function of y, also showing the vector components of the velocity at various points. The distance R is the horizontal range.

(•Why?) The result is

$$R = y(T) = \frac{2v_0^2}{g} \sin \alpha \cos \alpha = \frac{v_0^2}{g} \sin 2\alpha \tag{4-18}$$

where we have used $\sin 2\alpha = 2 \sin \alpha \cos \alpha$ (Eq. A–34 in Appendix A).

For a given initial velocity v_0, what angle α will produce the longest drive (i.e., the largest R)? Evidently this occurs when $\sin 2\alpha = 1$, or $2\alpha = \pi/2$. The answer can also be obtained by setting $dR/d\alpha$ equal to zero (see Section I–4). Thus,

$$\frac{dR}{d\alpha} = \frac{2v_0^2}{g} \cos 2\alpha = 0, \qquad 0 \leqslant \alpha \leqslant \pi/2$$

or
$$2\alpha = \pi/2$$

Thus, $\alpha = \pi/4$ (or $45°$) gives the maximum range, namely,

$$R_m = \frac{v_0^2}{g} \qquad \textbf{(4-19)}$$

The correct description of the trajectories of projectiles (cannon balls) was first given by Galileo. He deduced that the maximum range would be achieved for $\alpha = 45°$ (neglecting air resistance effects).

For a given v_0 and angle α, at what time t' will the golf ball reach maximum height? To determine t', we set dz/dt equal to zero, obtaining

$$\frac{d}{dt}z(t) = -gt + v_{0z} = 0$$

or
$$t' = \frac{v_{0z}}{g} = \frac{v_0}{g}\sin\alpha \qquad \textbf{(4-20)}$$

Physically, this means that maximum height corresponds to $v_z = 0$. (•Why?) We see that

$$t' = \tfrac{1}{2}T$$

which emphasizes the symmetric nature of the quadratic function $z(t)$. The descending portion of $z(t)$ is simply a reversal of the motion that brought the ball to the top of its trajectory. The golf ball strikes the ground with exactly the same speed with which it was lofted, and at that point it makes a descending angle with the horizontal exactly equal to α. (•Can you verify these assertions?)

The maximum height reached by the ball is readily determined by substituting the time t' (Eq. 4–20) into the equation for the vertical motion, $z(t)$; thus,

$$z_m = -\tfrac{1}{2}gt'^2 + v_{0z}t' = -\tfrac{1}{2}g\left(\frac{v_{0z}}{g}\right)^2 + v_{0z}\left(\frac{v_{0z}}{g}\right)$$

Hence,
$$z_m = \frac{v_{0z}^2}{2g} = \frac{v_0^2}{2g}\sin^2\alpha = \tfrac{1}{4}R\tan\alpha \qquad \textbf{(4-21)}$$

where the last equality follows from Eq. 4–18.

Example 4-6

A student stands at the edge of a cliff and throws a stone horizontally over the edge with a speed of 18 m/s. The cliff is 50 m above a flat horizontal beach, as shown in the diagram. How long after being released does the stone strike the beach below the cliff? With what speed and angle of impact does it land?

Solution:

We orient the coordinate axes as shown in the diagram so that the trajectory of the stone lies in the x-z plane. We have from Eqs. 4–14

$$x(t) = v_0 t, \qquad z(t) = h - \tfrac{1}{2}gt^2$$

When the stone lands on the beach, we have $z = 0$, so that the corresponding time t' is

$$t' = \sqrt{\frac{2h}{g}} = \sqrt{\frac{2(50 \text{ m})}{9.80 \text{ m/s}^2}} = 3.19 \text{ s}$$

On impact,

$$v_z = \frac{dz}{dt} = -gt' = -(9.80 \text{ m/s}^2)(3.19 \text{ s}) = -31.26 \text{ m/s}$$

$$v_x = \frac{dx}{dt} = v_0 = 18 \text{ m/s}$$

Therefore, the speed is

$$v = \sqrt{v_x^2 + v_z^2} = \sqrt{(31.26 \text{ m/s})^2 + (18 \text{ m/s})^2} = 36.1 \text{ m/s}$$

Also,

$$\tan \beta = \frac{v_z}{v_x} = \frac{-31.26 \text{ m/s}}{18 \text{ m/s}}$$

so that

$$\beta = -60.1°$$

Example 4–7

An often-quoted problem involves a jungle native with a blow pipe, who is intent on hitting a falling tree-dwelling animal with his dart. He releases his dart at the precise instant when the animal lets go from a branch and starts to fall. The remarkable fact is that no matter what the height from which the animal falls and no matter what the speed of the released dart, the dart will find its mark provided only that the blow gun is aimed exactly at the animal at the start of its fall. Show that these statements are correct.

Solution:

We select the origin of a coordinate frame at the business end of the blow pipe, with the z-axis vertical. Then, the initial position of the dart is $\mathbf{r}_{d0} = 0$, so the vector equation of motion for the dart is

$$\mathbf{r}_d(t) = -\tfrac{1}{2}gt^2 \widehat{\mathbf{k}} + \mathbf{v}_{d0}t$$

At first we imply nothing about either the magnitude v_{d0} or the direction of $\mathbf{v}_{d0}$. (Thus, we do not assume that $\mathbf{v}_{d0}$ and $\mathbf{r}_{a0}$ are parallel; see the diagram.) The falling animal is initially at the position $\mathbf{r}_{a0}$ and has initial velocity zero; that is, $\mathbf{v}_{a0} = 0$. Thus, the equation of motion for the animal is

$$\mathbf{r}_a(t) = -\tfrac{1}{2}gt^2 \widehat{\mathbf{k}} + \mathbf{r}_{a0}$$

If the dart is to strike the animal at some later time t', we must have

$$\mathbf{r}_d(t') = \mathbf{r}_a(t')$$

or

$$-\tfrac{1}{2}gt'^2 \widehat{\mathbf{k}} + \mathbf{v}_{d0}t' = -\tfrac{1}{2}gt'^2 \widehat{\mathbf{k}} + \mathbf{r}_{a0}$$

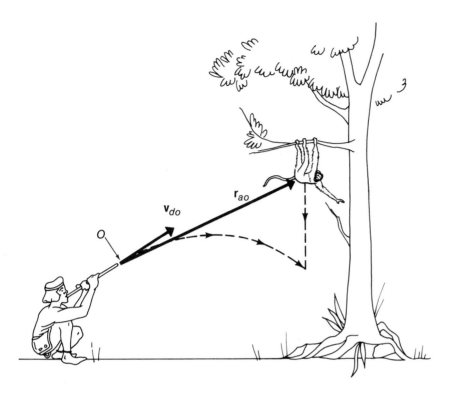

Therefore,

$$\mathbf{v}_{d0}t' = \mathbf{r}_{a0}$$

That is, the blow gun must be aimed *directly* at the animal as it begins to fall. The strike will occur at the time

$$t' = \frac{r_{a0}}{v_{d0}}$$

This time does, of course, depend on the initial distance and the velocity imparted to the dart. Notice that the solution is obtained by using vector methods without the necessity of specifying a particular coordinate frame.

(•How would you work out the solution if components were used instead of vectors? Remember, at first you *do not know* that $\mathbf{v}_{d0}$ is parallel to $\mathbf{r}_{a0}$—this must be proven.)

Example 4-8

Suppose that the position function of a particle is

$$\mathbf{r}(t) = \tfrac{1}{2}\mathbf{a}_0 t^2 + \mathbf{v}_0 t + \mathbf{r}_0$$

with $\mathbf{a}_0$, $\mathbf{v}_0$, and $\mathbf{r}_0$ suitable constant vectors. Show that

$$\mathbf{v} \cdot \mathbf{v} = 2\mathbf{a}_0 \cdot (\mathbf{r} - \mathbf{r}_0) + \mathbf{v}_0 \cdot \mathbf{v}_0$$

This expression is the vector generalization of Eq. 2-11.

Solution:

By differentiating the position function, we have

$$\frac{d}{dt}\mathbf{r}(t) = \mathbf{v} = \mathbf{a}_0 t + \mathbf{v}_0$$

Thus,

$$\mathbf{v} \cdot \mathbf{v} = (\mathbf{a}_0 \cdot \mathbf{a}_0)t^2 + 2(\mathbf{a}_0 \cdot \mathbf{v}_0)t + \mathbf{v}_0 \cdot \mathbf{v}_0$$
$$= \mathbf{a}_0 \cdot (\mathbf{a}_0 t^2 + 2\mathbf{v}_0 t) + \mathbf{v}_0 \cdot \mathbf{v}_0$$

But the position function can be rearranged to yield

$$2(\mathbf{r} - \mathbf{r}_0) = \mathbf{a}_0 t^2 + 2\mathbf{v}_0 t$$

Therefore,

$$\mathbf{v} \cdot \mathbf{v} = 2\mathbf{a}_0 \cdot (\mathbf{r} - \mathbf{r}_0) + \mathbf{v}_0 \cdot \mathbf{v}_0$$

Example 4–9

We conclude this section by considering a simple case in which the acceleration vector actually varies with time. This is a straightforward extension of the analysis presented in Section 2–6. Suppose that $\mathbf{a}(t) = (bt + c)\hat{\mathbf{i}}$, with $b = 6\ \text{m/s}^3$ and $c = 2\ \text{m/s}^2$. (a) What is the velocity at any time t, if at $t = 1\ \text{s}$ the velocity is $\mathbf{v}(1) = (3\ \text{m/s})\hat{\mathbf{i}} + (2\ \text{m/s})\hat{\mathbf{j}}$? (b) What is the position function at any time t, if at $t = 1\ \text{s}$ we have $\mathbf{r}(1) = 0$?

Solution:

(a) The velocity is found by integrating $\mathbf{a}(t)$; thus,

$$\mathbf{v}(t) = (\tfrac{1}{2}bt^2 + ct)\hat{\mathbf{i}} + \mathbf{C}$$

where $\mathbf{C} = C_x\hat{\mathbf{i}} + C_y\hat{\mathbf{j}} + C_z\hat{\mathbf{k}}$ is an arbitrary *vector* constant, because the expression for $\mathbf{v}(t)$ is actually the result of three separate component integrations:

$$v_x(t) = \int a_x(t)\,dt = \int (bt + c)\,dt = \tfrac{1}{2}bt^2 + ct + C_x$$

$$v_y(t) = \int a_y(t)\,dt = \int 0\,dt = 0 + C_y$$

$$v_z(t) = \int a_z(t)\,dt = \int 0\,dt = 0 + C_z$$

Now, we know that at $t = 1\ \text{s}$

$$\mathbf{v}(1) = (3\ \text{m/s})\hat{\mathbf{i}} + (2\ \text{m/s})\hat{\mathbf{j}}$$

Also, using the expressions for the velocity components, we find

$$\mathbf{v}(1) = (5\ \text{m/s} + C_x)\hat{\mathbf{i}} + C_y\hat{\mathbf{j}}$$

where $C_z = 0$ because $\mathbf{v}(1)$ does not have a z-component. Thus, we identify

$$\mathbf{C} = (-2 \text{ m/s})\hat{\mathbf{i}} + (2 \text{ m/s})\hat{\mathbf{j}}$$

so that

$$\mathbf{v}(t) = [(3t^2 + 2t - 2)\hat{\mathbf{i}} + 2\hat{\mathbf{j}}](\text{m/s})$$

(b) Integrating $\mathbf{v}(t)$, there results

$$\mathbf{r}(t) = [(t^3 + t^2 - 2t)\hat{\mathbf{i}} + 2t\hat{\mathbf{j}}](\text{m}) + \mathbf{C}'$$

When $t = 1$ s, we know that $\mathbf{r}(1) = 0$; hence,

$$\mathbf{r}(1) = 0 = [0\hat{\mathbf{i}} + 2\hat{\mathbf{j}}](\text{m}) + \mathbf{C}'$$

Thus, $\mathbf{C}' = (-2 \text{ m})\hat{\mathbf{j}}$, and we have

$$\mathbf{r}(t) = [(t^3 + t^2 - 2t)\hat{\mathbf{i}} + (2t - 2)\hat{\mathbf{j}}](\text{m})$$

4-5 CIRCULAR MOTION

The motion of a particle that is constrained to move in a circular path is of sufficient general interest to warrant special attention. Although such motion can be analyzed quite rigorously by using Cartesian coordinates, a polar coordinate frame (Fig. 3–17) is a more natural choice. Consider motion for which $\rho = R$ is a constant. Therefore, only the single variable ϕ is needed to describe the motion, and we write $\phi(t)$ to emphazise the time dependence. (In a Cartesian frame we would require both of the variables, x and y; they are $x = R \cos \phi$ and $y = R \sin \phi$.)

The instantaneous time rate of change of $\phi(t)$ is called the *angular frequency of rotation* and is designated by the symbol ω (Greek omega); thus,

$$\omega = \frac{d\phi}{dt} \tag{4-22}$$

It is often convenient to measure ϕ in *radians** and ω in *radians per second* (rad/s). The dimensionality of ω is $[\omega] = T^{-1}$, that is, *inverse time.*

Motion at Constant Speed. The simplest type of circular motion is that in which the particle moves with a constant speed. Then, the angle ϕ increases by equal amounts in equal time intervals, and, hence, ω is a constant. It follows that this type of motion—called *uniform circular motion*—is *periodic* or *cyclic*. The particle repeatedly arrives at any arbitrarily selected position at regular intervals of time. This repetition time is called the *period T.* The angle ϕ increases by 2π radians during each time interval T; therefore,

$$\omega = \frac{2\pi}{T} \tag{4-23}$$

---

*See Section A–10 of Appendix A for a discussion of radian measure.

Another useful concept is the *frequency* of the motion, the number of cycles or revolutions executed per unit time. Frequency is usually measured in *cycles per second,* which we abbreviate as *hertz* (Hz); sometimes the unit *revolution per minute* (rpm) is used. The frequency ν is related to the period T by

$$\nu = \frac{1}{T} = \frac{\omega}{2\pi} \qquad \text{(4-24)}$$

A particle undergoing uniform circular motion travels an arc length $2\pi R$ (the circumference of the circular path) in a time equal to the period. Therefore, the speed v of the particle is

$$v = \frac{2\pi R}{T} = \omega R \qquad \text{(4-25)}$$

The velocity vector of a particle is always tangent to the particle's path (recall Fig. 4-4). For the case of uniform circular motion, the velocity **v** is therefore always perpendicular to the particle's position vector **r**, as indicated in Fig. 4-8. Although the velocity is constant in magnitude, its direction changes continually. Such motion implies the existence of an *acceleration.*

Figure 4-8 shows the finite changes in **r** and **v** that result from the motion during the time increment Δt. However, we are really interested in the limiting situation in which $\Delta t \to 0$, and, consequently, $\Delta\phi \to 0$. In this limit, the velocity vectors $\mathbf{v}(t + \Delta t)$ and $\mathbf{v}(t)$ become more nearly parallel, and $\Delta\mathbf{v}$ becomes directed more nearly toward the center of motion O. The acceleration of the particle is

$$\mathbf{a} = \frac{d\mathbf{v}}{dt} = \lim_{\Delta t \to 0} \frac{\Delta\mathbf{v}}{\Delta t} = \lim_{\Delta\phi \to 0} \frac{\Delta\mathbf{v}}{\Delta t}$$

In this limit the acceleration is directed toward the center of the circle and is therefore called the *centripetal* (or *center-seeking*) acceleration. The magnitude of the acceleration is determined from the fact that the triangles involving **r** and **v** are similar isosceles triangles (see Fig. 4-8). (•Can you see that the included angles are equal?) Therefore, we can write

$$\frac{\Delta v}{v} = \frac{\Delta r}{R} \qquad \text{or} \qquad \frac{\Delta v}{\Delta r} = \frac{v}{R}$$

Then,

$$a = \lim_{\Delta t \to 0} \frac{\Delta v}{\Delta t} = \lim_{\Delta t \to 0} \frac{\Delta v}{\Delta r} \cdot \frac{\Delta r}{\Delta t} = \frac{v}{R} \lim_{\Delta t \to 0} \frac{\Delta r}{\Delta t}$$

or

$$a = \frac{v^2}{R} = R\omega^2 \qquad \text{(4-26)}$$

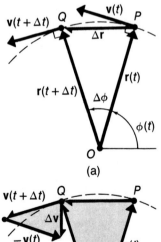

Fig. 4-8. Uniform circular motion. (a) The particle travels with a constant speed, so that $v(t) = v(t + \Delta t) = v$. (b) The velocity increment $\Delta\mathbf{v}$ is equal to $\mathbf{v}(t + \Delta t) - \mathbf{v}(t)$. The shaded triangles are similar because $-\mathbf{v}(t)$ is perpendicular to $\mathbf{r}(t)$ and $\mathbf{v}(t + \Delta t)$ is perpendicular to $\mathbf{r}(t + \Delta t)$.

Figure 4–9 shows the location of the particle at four different times, together with the corresponding vectors $\mathbf{r}(t)$, $\mathbf{v}(t)$, and $\mathbf{a}(t)$.

Example 4–10

A particle travels with uniform speed in a circular path with a radius $R = 2$ m and a period $T = 0.25$ s. Evaluate ω and ν. What is the speed? What is the centripetal acceleration?

Solution:

Using Eqs. 4–23 and 4–24, we have

$$\omega = \frac{2\pi}{T} = \frac{2\pi \text{ rad}}{0.25 \text{ s}} = 8\pi \text{ rad/s}$$

$$\nu = \frac{1}{T} = 4 \text{ cycles/s} = 4 \text{ Hz}$$

From Eq. 4–25, $\qquad v = \omega R = (8\pi \text{ rad/s})(2 \text{ m}) = 16\pi \text{ m/s}$

and from Eq. 4–26, we find

$$a = R\omega^2 = (2 \text{ m})(8\pi \text{ rad/s})^2 = 128\pi^2 \text{ m/s}^2$$

Fig. 4–9. The coordinate positions $\mathbf{r}(t)$ of a particle executing uniform circular motion at various selected times. Also shown superimposed are the vectors corresponding to the velocity $\mathbf{v}(t)$ and the acceleration $\mathbf{a}(t)$.

Example 4–11

The Moon circles the Earth at a distance $R_M = 3.84 \times 10^8$ m with a period $T_M = 27.3$ days. Compute the centripetal acceleration of the Moon in units of g. (Note that 1 day = 86,400 s.)

Solution:

We can rewrite Eq. 4–26 as

$$a = \frac{v^2}{R} = \left(\frac{2\pi R_M}{T_M}\right)^2 \frac{1}{R_M} = \left(\frac{2\pi}{T_M}\right)^2 R_M$$

$$= \frac{4\pi^2(3.84 \times 10^8 \text{ m})}{[(27.3 \text{ d})(8.64 \times 10^4 \text{ s/d})]^2} = 2.72 \times 10^{-3} \text{ m/s}^2$$

so that $\qquad a/g = \dfrac{2.72 \times 10^{-3} \text{ m/s}^2}{9.80 \text{ m/s}^2} = 2.78 \times 10^{-4}$

Fig. 4–10. The unit vectors $\hat{\mathbf{u}}_\rho$ and $\hat{\mathbf{u}}_\phi$ and the position vector $\mathbf{r}$.

Motion with Variable Speed. Consider now the case of circular motion with variable speed. This is most readily accomplished by using the polar unit vectors $\hat{\mathbf{u}}_\rho$ in the radial direction and $\hat{\mathbf{u}}_\phi$ in the azimuthal or ϕ-direction (see Fig. 4–10, which shows Fig. 3–17 redrawn for convenience).

The fact that the motion is restricted to a circular path is expressed by writing the position vector as

$$\mathbf{r} = \rho\hat{\mathbf{u}}_\rho = R\hat{\mathbf{u}}_\rho$$

(a)

(b)

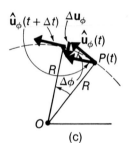

(c)

Fig. 4–11. Geometric representation of $\Delta\mathbf{u}_\rho$ and $\Delta\mathbf{u}_\phi$ corresponding to a time increment Δt.

Because $\rho = R = $ constant, the velocity of the particle is

$$\mathbf{v} = \frac{d\mathbf{r}}{dt} = R\frac{d\hat{\mathbf{u}}_\rho}{dt}$$

To evaluate $d\hat{\mathbf{u}}_\rho/dt$, refer to Fig. 4–11a, which shows the unit vectors $\hat{\mathbf{u}}_\rho$ and $\hat{\mathbf{u}}_\phi$ at times t and $t + \Delta t$, corresponding to the movement of the particle from position $P(t)$ to $P(t + \Delta t)$. During this same time increment Δt, the angle ϕ changes by $\Delta\phi$. In Fig. 4–11b the vector $\hat{\mathbf{u}}_\rho\,(t + \Delta t)$ has been translated to point $P(t)$ to facilitate calculating $\Delta\mathbf{u}_\rho = \hat{\mathbf{u}}_\rho(t + \Delta t) - \hat{\mathbf{u}}_\rho(t)$. Similarly, in Fig. 4–11c, $\hat{\mathbf{u}}_\phi(t + \Delta t)$ has been translated to the point $P(t)$ in order to calculate $\Delta\mathbf{u}_\phi$. The unit vectors $\hat{\mathbf{u}}_\rho$ and $\hat{\mathbf{u}}_\phi$ have unit magnitude, so the triangles in Figs. 4–11b and 4–11c that involve the (small) angle $\Delta\phi$ are both isosceles triangles. Consequently, $|\Delta\mathbf{u}_\rho| = (1)(\Delta\phi)$. It is also apparent that, in the limit as $\Delta t \to 0$ (and, hence, $\Delta\phi \to 0$), the direction of $\Delta\mathbf{u}_\rho$ is in the direction of increasing ϕ, i.e., in the direction of $+\hat{\mathbf{u}}_\phi$. Thus,

$$\Delta\mathbf{u}_\rho = (\Delta\phi)\hat{\mathbf{u}}_\phi$$

Then, we can write

$$\frac{d\hat{\mathbf{u}}_\rho}{dt} = \lim_{\Delta t \to 0}\left(\frac{\Delta\phi}{\Delta t}\right)\hat{\mathbf{u}}_\phi = \frac{d\phi}{dt}\hat{\mathbf{u}}_\phi = \omega\hat{\mathbf{u}}_\phi \qquad (4\text{–}27)$$

Hence,

$$\mathbf{v} = R\omega\hat{\mathbf{u}}_\phi \qquad (4\text{–}28)$$

For the case of circular motion, the speed v is always given by $v = R\omega$. Equation 4–28 gives the expected result, $\mathbf{v} = v\hat{\mathbf{u}}_\phi$, showing explicitly that $\mathbf{v}$ is tangent to the circular path.

The acceleration of the particle is

$$\mathbf{a} = \frac{d\mathbf{v}}{dt} = \frac{d}{dt}(R\omega\hat{\mathbf{u}}_\phi) = R\frac{d}{dt}(\omega\hat{\mathbf{u}}_\phi)$$

or

$$\mathbf{a} = R\omega\frac{d\hat{\mathbf{u}}_\phi}{dt} + R\frac{d\omega}{dt}\hat{\mathbf{u}}_\phi \qquad (4\text{–}29)$$

For the case of variable speed, ω is no longer a constant and, hence, $d\omega/dt$ is nonzero. For these cases, we find it convenient to define the *angular acceleration*, which we denote by the symbol α:

$$\alpha \equiv \frac{d\omega}{dt} = \frac{d^2\phi}{dt^2} \qquad (4\text{–}30)$$

To complete the evaluation of $\mathbf{a}$ in Eq. 4–29, we also need the expression for $d\hat{\mathbf{u}}_\phi/dt$. In Fig. 4–11c we show the geometry of $\Delta\mathbf{u}_\phi = \hat{\mathbf{u}}_\phi(t + \Delta t) - \hat{\mathbf{u}}_\phi(t)$. In this case, we have $|\Delta\mathbf{u}_\phi| = (1)(\Delta\phi)$, and $\Delta\mathbf{u}_\phi$ is directed toward the center of the path; thus,

$$\Delta\mathbf{u}_\phi = -(\Delta\phi)\hat{\mathbf{u}}_\rho$$

Then, we can write

$$\frac{d\hat{\mathbf{u}}_\phi}{dt} = \lim_{\Delta t \to 0} \left(-\frac{\Delta\phi}{\Delta t} \right) \hat{\mathbf{u}}_\rho = -\frac{d\phi}{dt} \hat{\mathbf{u}}_\rho = -\omega\hat{\mathbf{u}}_\rho \qquad (4\text{-}31)$$

Notice the difference in sign between the expressions for $d\hat{\mathbf{u}}_\rho/dt$ (Eq. 4-27) and $d\hat{\mathbf{u}}_\phi/dt$ (Eq. 4-31). (•Can you account for this difference by examining Fig. 4-11b, c?) For the acceleration **a**, we have from Eq. 4-29

$$\begin{aligned} \mathbf{a} &= a_\rho\hat{\mathbf{u}}_\rho + a_\phi\hat{\mathbf{u}}_\phi \\ &= -R\omega^2\hat{\mathbf{u}}_\rho + R\alpha\hat{\mathbf{u}}_\phi \end{aligned} \qquad (4\text{-}32)$$

In this equation, we recognize the first term as the centripetal acceleration, which now may be variable in magnitude because ω is a possible function of time. This term can also be written as $-(v^2/R)\hat{\mathbf{u}}_\rho$, where we understand that v has its instantaneous value. In addition, there is a new term in the expression for **a**, namely, the *tangential acceleration* $R\alpha\hat{\mathbf{u}}_\phi$, which has a magnitude proportional to α and which is directed tangent to the path of the particle.

The unit vectors, $\hat{\mathbf{u}}_\rho$ and $\hat{\mathbf{u}}_\phi$, are mutually perpendicular, so the magnitude of the acceleration is

$$|\mathbf{a}| = \sqrt{a_\rho^2 + a_\phi^2} = R\sqrt{\omega^4 + \alpha^2} \qquad (4\text{-}33)$$

The Equations of Motion by Integration. Consider the case of circular motion with a constant angular acceleration α_0. The equations of motion can be obtained by integrating the expression $d\omega/dt = \alpha_0$. The first integration yields

$$\omega = \int \alpha_0 \, dt = \alpha_0 t + C_1$$

If at $t = 0$ we have $\omega(0) = \omega_0$, then $C_1 = \omega_0$, so that

$$\omega(t) = \alpha_0 t + \omega_0 \qquad (4\text{-}34)$$

Integrating $\omega = d\phi/dt$, often called the *angular speed,* we have

$$\phi = \int \omega \, dt = \int (\alpha_0 t + \omega_0) \, dt = \tfrac{1}{2}\alpha_0 t^2 + \omega_0 t + C_2$$

If at $t = 0$ we have $\phi(0) = \phi_0$, then $C_2 = \phi_0$, so that

$$\phi(t) = \tfrac{1}{2}\alpha_0 t^2 + \omega_0 t + \phi_0 \qquad (4\text{-}35)$$

Note the formal similarity of the expressions for $\omega(t)$ and $\phi(t)$ to the results for $v(t)$ and $s(t)$ in Eqs. 2-9 and 2-10. (The equations are only analogous, however, because the dimensions in the two cases are different.)

At any time t, the velocity can be written in the form of Eq. 4-28:

$$\mathbf{v}(t) = R\omega(t)\hat{\mathbf{u}}_\phi = R(\alpha_0 t + \omega_0)\hat{\mathbf{u}}_\phi \qquad (4\text{-}36)$$

The corresponding expression for the acceleration, using Eq. 4–32, is

$$\mathbf{a}(t) = -R[\omega(t)]^2 \hat{\mathbf{u}}_\rho + R\alpha \hat{\mathbf{u}}_\phi$$

$$= -R(\alpha_0^2 t^2 + 2\alpha_0 \omega_0 t + \omega_0^2)\hat{\mathbf{u}}_\rho + R\alpha_0 \hat{\mathbf{u}}_\phi \tag{4-37}$$

Example 4–12

At a particular instant, a particle traveling in a circular path with a radius of 0.5 m has a speed $v = 1.6$ m/s and an angular acceleration $\alpha = 16$ rad/s^2. What is the magnitude of the acceleration? What angle does the acceleration vector make with a tangent to the path?

Solution:

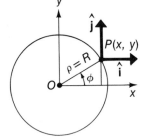

We have

$$\mathbf{a}_\rho = -\frac{v^2}{R}\hat{\mathbf{u}}_\rho = -\frac{(1.6\text{ m/s})^2}{0.5\text{ m}}\hat{\mathbf{u}}_\rho = (-5.12\text{ m/s}^2)\hat{\mathbf{u}}_\rho$$

Also,

$$\mathbf{a}_\phi = R\alpha\hat{\mathbf{u}}_\phi = (0.5\text{ m})(16\text{ rad/s}^2)\hat{\mathbf{u}}_\phi = (8\text{ m/s}^2)\hat{\mathbf{u}}_\phi$$

Hence,

$$|\mathbf{a}| = \sqrt{a_\rho^2 + a_\phi^2} = \sqrt{(-5.12)^2 + (8)^2} = 9.50\text{ m/s}^2$$

and

$$\tan\beta = \frac{a_\rho}{a_\phi} = \frac{-5.12}{8} = -0.64$$

or

$$\beta = -32.6°$$

Notice, in the diagram, the significance of the negative sign for β. If there were no angular acceleration ($\alpha = 0$), we would have $a_\phi = 0$ and $\beta = -90°$.

Example 4–13

Derive the results given by Eqs. 4–28 and 4–32 using a Cartesian coordinate frame.

Solution:

We have

$$x = R\cos\phi, \qquad y = R\sin\phi$$

and

$$\mathbf{r} = x\hat{\mathbf{i}} + y\hat{\mathbf{j}} = R\cos\phi\,\hat{\mathbf{i}} + R\sin\phi\,\hat{\mathbf{j}}$$

It follows that

$$\mathbf{v} = \frac{d\mathbf{r}}{dt} = -R\sin\phi\frac{d\phi}{dt}\hat{\mathbf{i}} + R\cos\phi\frac{d\phi}{dt}\hat{\mathbf{j}}$$

$$= R\omega(-\sin\phi\,\hat{\mathbf{i}} + \cos\phi\,\hat{\mathbf{j}})$$

(a)

where we have used the fact that the unit vectors $\hat{\mathbf{i}}$ and $\hat{\mathbf{j}}$ are constant in both magnitude and direction, and that $d\phi/dt = \omega$. We can also write

$$\mathbf{a} = \frac{d\mathbf{v}}{dt} = -R\left(\frac{d\omega}{dt}\sin\phi + \omega\cos\phi\frac{d\phi}{dt}\right)\hat{\mathbf{i}} + R\left(\frac{d\omega}{dt}\cos\phi - \omega\sin\phi\frac{d\phi}{dt}\right)\hat{\mathbf{j}}$$

$$= -R\omega^2(\cos\phi\,\hat{\mathbf{i}} + \sin\phi\,\hat{\mathbf{j}}) + R\alpha(-\sin\phi\,\hat{\mathbf{i}} + \cos\phi\,\hat{\mathbf{j}})$$

From diagram (b), we see that

$$\hat{\mathbf{u}}_\rho = \cos \phi \, \hat{\mathbf{i}} + \sin \phi \, \hat{\mathbf{j}}$$

$$\hat{\mathbf{u}}_\phi = -\sin \phi \, \hat{\mathbf{i}} + \cos \phi \, \hat{\mathbf{j}}$$

Hence,

$$\mathbf{v} = R\omega \hat{\mathbf{u}}_\phi$$

$$\mathbf{a} = -R\omega^2 \hat{\mathbf{u}}_\rho + R\alpha \hat{\mathbf{u}}_\phi$$

which are the same as the results we obtained previously.

Incidentally, from the above equations relating $\hat{\mathbf{u}}_\rho$ and $\hat{\mathbf{u}}_\phi$ to $\hat{\mathbf{i}}$ and $\hat{\mathbf{j}}$ we can see that

$$\frac{d\hat{\mathbf{u}}_\rho}{dt} = -\sin \phi \frac{d\phi}{dt} \hat{\mathbf{i}} + \cos \phi \frac{d\phi}{dt} \hat{\mathbf{j}}$$

$$= \omega(-\sin \phi \, \hat{\mathbf{i}} + \cos \phi \, \hat{\mathbf{j}}) = \omega \hat{\mathbf{u}}_\phi$$

and

$$\frac{d\hat{\mathbf{u}}_\phi}{dt} = -\cos \phi \frac{d\phi}{dt} \hat{\mathbf{i}} - \sin \phi \frac{d\phi}{dt} \hat{\mathbf{j}}$$

$$= -\omega(\cos \phi \, \hat{\mathbf{i}} + \sin \phi \, \hat{\mathbf{j}}) = -\omega \hat{\mathbf{u}}_\rho$$

which duplicate the results given in Eqs. 4–27 and 4–31.

unit circle

(b)

The Vector Representation of Circular Motion. In the discussions thus far, we have considered the angular frequency ω to be a scalar quantity. We may, in fact, define a *vector* counterpart $\boldsymbol{\omega}$ which is referred to as the *angular velocity*. The vector $\boldsymbol{\omega}$ is taken to have the magnitude of the angular frequency and a direction *perpendicular to the plane containing the circular path*. We use a right-hand rule to resolve the ambiguity of the two possible perpendicular directions. When the fingers of the right hand are curled in the direction of the (tangential) velocity, the thumb points in the direction of $\boldsymbol{\omega}$, as shown in Fig. 4–12. We usually draw the vector $\boldsymbol{\omega}$ along the axis of rotation. For circular motion, we can also define an *angular acceleration vector* $\boldsymbol{\alpha}$, which is equal in magnitude to the scalar α; the direction of $\boldsymbol{\alpha}$ is parallel to $\boldsymbol{\omega}$ if α is algebraically positive and is antiparallel to $\boldsymbol{\omega}$ if α carries a negative sign.

With these definitions we can write $\mathbf{v} = R\omega \hat{\mathbf{u}}_\phi$ (Eq. 4–28) in vector form:

$$\mathbf{v} = \boldsymbol{\omega} \times \mathbf{r} \qquad \text{(4–38)}$$

where we make use of the vector-product or cross-product multiplication of two vectors, as discussed in Section 3–6.

Figure 4–13 shows two cases in which Eq. 4–38 is applied. In Fig. 4–13a, the origin O is selected on the rotation axis and in the plane of rotation. In this case, $\boldsymbol{\omega}$ and $\mathbf{r}$ are perpendicular; $|\mathbf{r}| = R$, so $v = R\omega$, and the direction of $\mathbf{v}$ is perpendicular to the plane containing $\boldsymbol{\omega}$ (the axis) and $\mathbf{r}$. Consequently, $\mathbf{v}$ is tangential to the path, as required.

In Fig. 4–13b, the origin is on the axis but at an arbitrary point O'. In this case, $\boldsymbol{\omega}$ and $\mathbf{r}$ have an included angle γ; thus, $v = r\omega \sin \gamma$. But $r \sin \gamma = \rho = R$; therefore,

Fig. 4–12. Illustrating the right-hand-rule convention for determining the direction of the angular velocity vector ω with respect to the tangential velocity $\mathbf{v}$.

Fig. 4–13. The calculation of $\mathbf{v} = \boldsymbol{\omega} \times \mathbf{r}$ is the same, irrespective of the point on the rotation axis selected for the origin of the position vector $\mathbf{r}$. In (a), $v = R\omega$. In (b), $v = r\omega \sin \gamma = R\omega$.

(a)

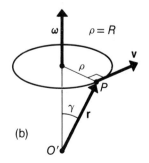

(b)

we again have $v = R\omega$. The direction of **v** is seen to be the same as in Fig. 4–13a. (•Why is this so?) These equalities suggest the invariant character of Eq. 4–38. In the next section we discuss the invariant nature of the acceleration **a**.

4-6   RELATIVE MOTION IN THREE DIMENSIONS

We now consider how two observers who have a constant relative velocity describe the motion of the same particle in terms of their own coordinate frames. We assume that the relative orientation of the two frames does not change with time, thereby ruling out any possible rotational motion of the two observers. Figure 4–14 illustrates the geometry for two coordinate frames, S and S'. The relationship between the two position vectors, **r** and **r'**, reckoned in the frames S and S', respectively, can be expressed as

$$\mathbf{r} = \mathbf{R} + \mathbf{r'} \tag{4–39}$$

Here, **R** is the vector locating the origin O' (of the S' frame) in the S frame, and $d\mathbf{R}/dt = \mathbf{u}$, the constant velocity of S' with respect to S. We assume the existence of a "universal clock" to which both observers may refer simultaneously; formally, this means $t' = t$. Differentiating **r** with respect to this universal time t gives

$$\frac{d\mathbf{r}}{dt} = \frac{d\mathbf{R}}{dt} + \frac{d\mathbf{r'}}{dt}$$

or

$$\mathbf{v} = \mathbf{u} + \mathbf{v'} \tag{4–40}$$

where **v** and **v'** are the velocities of the particle in the frames S and S', respectively.

If the particle has nonzero acceleration, a further differentiation of the expression for **v** gives

$$\frac{d\mathbf{v}}{dt} = \frac{d\mathbf{u}}{dt} + \frac{d\mathbf{v'}}{dt}$$

Because **u** is assumed constant, $d\mathbf{u}/dt = 0$, so that

$$\mathbf{a} = \mathbf{a'} \tag{4–41}$$

Fig. 4–14. Two frames S and S' used to describe the motion of a particle at point P. The velocity of S' relative to S is **u**, a constant.

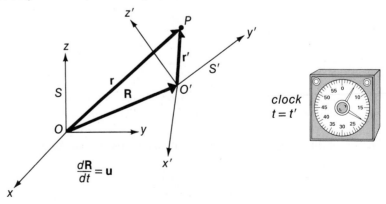

These equations for **r**, **v**, and **a** (together with $t' = t$) constitute a set of transformation equations—the *Galilean transformation equations*—connecting the two frames, S and S'.

We can draw an important conclusion from the result contained in Eq. 4–41. If an observer in a certain coordinate frame measures an acceleration **a** for a particle, then any other observer, whose coordinate frame moves *at constant velocity* with respect to the first frame, will measure the same acceleration vector for the particle. Acceleration (unlike velocity or position) is therefore an *invariant* quantity. Vectors such as **a** that are the same in all unaccelerated coordinate frames are called *free vectors*. We often use this important concept.

The derivation of the Galilean transformation equations suggests several questions. Does a "universal clock" actually exist? If each observer uses his own clock, how do they synchronize the clocks? How do the observers make "simultaneous" measurements of the particle's motion? These questions can be addressed properly only within the framework of the theory of relativity (Chapter 21).

Example 4–14

A boat crosses a river with a width $w = 160$ m in which the current flows with a uniform speed of 1.5 m/s. The steersman maintains a bearing (i.e., the direction in which his boat points) perpendicular to the river and a throttle setting to give a constant speed of 2 m/s with respect to the water. What is the velocity of the boat relative to a stationary shore observer? How far downstream from the initial position is the boat when it reaches the opposite shore?

Solution:

The key to solving problems of this type is to identify carefully the selection of the observers (and frames), S and S'. In this case, it is convenient to take the shore observer to be S and to consider a *hypothetical observer S'* drifting with the river, as shown in the diagram. The "particle" being observed is the boat.

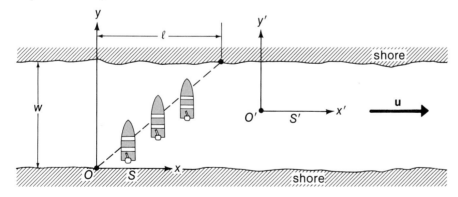

We are given that the river's velocity components are $u_x = 1.5$ m/s, $u_y = 0$, and those of the boat are $v'_y = 2$ m/s, $v'_x = 0$. Therefore, taking components of Eq. 4–40, we have

$$v_x = u_x + v'_x = 1.5 \text{ m/s}$$
$$v_y = u_y + v'_y = 2 \text{ m/s}$$

which give the velocity components as seen by the shore observer. The time to cross the river can be found using v_y:

$$t = \frac{w}{v_y} = \frac{160 \text{ m}}{2 \text{ m/s}} = 80 \text{ s}$$

Then we obtain for ℓ, the distance to the downstream landing point,

$$\ell = v_x t = (1.5 \text{ m/s})(80 \text{ s}) = 120 \text{ m}$$

Example 4–15

A child in danger of drowning in a river is being carried downstream by a current that flows uniformly with a speed of 2.5 km/h. The child is 0.6 km from shore and 0.8 km upstream of a boat landing when a rescue boat sets out. If the boat proceeds at its maximum speed of 20 km/h with respect to the water, what heading relative to the shore should the boatman take? What angle does the boat velocity **v** make with the shore? How long will it take the boat to reach the child?

(a)

(b)

Solution:

Identify the frame S' (origin O') with the child and frame S (origin O) with a stationary observer at the boat landing. Then, the child sees the boat approaching along a straight line with a speed v', as shown in diagram (a). (•Can you see why this is the case?)

At $t = 0$, the separation of the origins is $R_0 = \sqrt{h_0^2 + \ell_0^2}$. Therefore, the time required to make the rescue is

$$t = \frac{R_0}{v'} = \frac{\sqrt{(0.6 \text{ km})^2 + (0.8 \text{ km})^2}}{20 \text{ km/h}} \times 60 \text{ min/h} = 3 \text{ min}$$

Note that this time is independent of the speed of the river current. (•Why is this so?) The heading β' is determined from $\tan \beta' = h_0/\ell_0 = 0.75$ so that

$$\beta' = 36.9°$$

From the shore (frame S), the view is that shown in diagram (b). We have

$$v_x = v'_x - u = v' \cos \beta' - u = v' \frac{\ell_0}{R_0} - u$$

$$v_y = v'_y = v' \sin \beta' = v' \frac{h_0}{R_0}$$

Now, $R_0 = \sqrt{(0.6 \text{ km})^2 + (0.8 \text{ km})^2} = 1.00 \text{ km}$, so we obtain

$$v_x = \frac{0.8 \text{ km}}{1.0 \text{ km}} (20 \text{ km/h}) - 2.5 \text{ km/h} = 13.5 \text{ km/h}$$

$$v_y = \frac{0.6 \text{ km}}{1.0 \text{ km}} (20 \text{ km/h}) = 12.0 \text{ km/h}$$

The direction of **v** with respect to the shore line is given by

$$\tan \beta = \frac{v_y}{v_x} = \frac{12.0 \text{ km/h}}{13.5 \text{ km/h}} = 0.889$$

or

$$\beta = 41.6°$$

Note the similarity between this problem and Example 4–7.

Example 4–16

A science student is riding on a flat car of a train traveling along a straight horizontal track at a constant speed of 10 m/s. The student throws a ball into the air along a path that he judges to make an initial angle of 60° with the horizontal and to be in line with the track. The student's professor, who is standing on the ground nearby, observes the ball to rise vertically. How high does the ball rise?

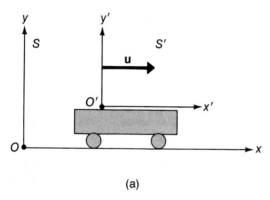

(a)

Solution:

Identify the student as the S' observer and the professor as the S observer. For the initial motion in S', we have

$$\frac{v_y'}{v_x'} = \tan 60° = \sqrt{3}$$

Then, because there is no x-motion in S, we can write

$$v_x = v_x' + u = 0$$

so that

$$v_x' = -u = -10 \text{ m/s}$$

Hence, the ball is thrown backwards in S'. Then,

$$v_y = v_y' = \sqrt{3} \, |v_x'| = 10 \sqrt{3} \text{ m/s}$$

Using $v_y^2 = 2gh$ (from the first equality in Eq. 4–21), we find

$$h = \frac{(10 \sqrt{3} \text{ m/s})^2}{2(9.80 \text{ m/s}^2)} = 15.3 \text{ m}$$

The motion of the ball as seen by the student in S' is shown in diagram (b). The view of the professor in S is shown in diagram (c).

(b)

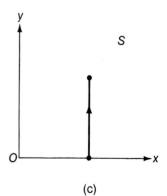

(c)

PROBLEMS

Section 4-1

4-1 The position vector of a particle is $\mathbf{r}(t) = \alpha t\,\hat{\mathbf{i}} + (\beta t^2 + \delta)\hat{\mathbf{j}}$, with $\alpha = 2$ m/s, $\beta = 8$ m/s², and $\delta = 1$ m. The y coordinate is a quadratic function of x. Determine this function.

Section 4-2

4-2 A particle moves in a circular path given by $x(t) = A \cos 2\omega t$, $y(t) = B \sin 2\omega t$, with $A = 6$ m, $B = 6$ m, and $\omega = 1\,\text{s}^{-1}$. Determine the vector $\Delta \mathbf{r}$ that describes the displacement from point P (at $t = 0$) to point Q (at $t = \pi/4$ s).

Section 4-2

4-3 A particle, initially at the position $\mathbf{r}_1 = (6\text{ m})\hat{\mathbf{i}} + (7\text{ m})\hat{\mathbf{j}}$, is at the position $\mathbf{r}_2 = (4\text{ m})\hat{\mathbf{i}} - (2\text{ m})\hat{\mathbf{j}}$ one second later. Find the average velocity and the average speed of the particle during this time interval.

4-4 The position function of a particle is $\mathbf{r}(t) = \alpha t^3\hat{\mathbf{i}} + \beta t^2\hat{\mathbf{j}}$. (a) What is the velocity at $t = 1$ s? (b) At $t = 1$ s the particle is $\sqrt{5}$ m from the origin and has a speed of 5 m/s. How far from the origin is the particle at $t = 2$ s?

4-5 The position function of a particle is $\mathbf{r}(t) = ct\hat{\mathbf{i}} + (bt - at^2)\hat{\mathbf{j}}$, where a, b, and c are positive constants with appropriate dimensions. At $t = 0$ the particle's trajectory makes a 45° angle with the x-axis. If the largest positive value attained by $y(t)$ is h, prove that the corresponding value of $x(t)$ is $2h$.

4-6 Suppose that the position function of a particle is $\mathbf{r}(t) = x(t)\hat{\mathbf{i}} + y(t)\hat{\mathbf{j}} + z(t)\hat{\mathbf{k}}$, with

$$x(t) = a \sin \omega t, \quad y(t) = a \cos \omega t, \quad z(t) = bt$$

where a, b, and ω are constants. Find the average velocity during the time interval from $t = 0$ to $t = 2\pi/\omega$. Find the velocity and the speed at time t. Give a geometrical description of the path followed by the particle.

Section 4-3

4-7 A particle moves in accordance with the functions $x(t) = (3\text{ m/s}^2)t^2 + (2\text{ m/s})t$ and $y(t) = (2\text{ m/s}^2)t^2 + (4\text{ m/s})t + (3\text{ m})$. Determine the velocity vector $\mathbf{v}(t)$ and the acceleration vector $\mathbf{a}(t)$. Find the magnitude and the direction of $\mathbf{a}$ at $t = 2$ s.

4-8 A particle has the position function

$$\mathbf{r}(t) = (t^3 + 1)\hat{\mathbf{i}} + (t^3 - 6t^2 + 12t)\hat{\mathbf{j}} + (3t - 6)\hat{\mathbf{k}}$$

where each term is expressed in meters when t is given in seconds. (a) Find the general expressions for the velocity and the acceleration. (b) At what time is the velocity in the y-direction a minimum? (c) Describe the behavior of $\mathbf{r}(t)$, $\mathbf{v}(t)$, and $\mathbf{a}(t)$ for times near the value found for part (b).

4-9 A particle has the cycloidal position function $\mathbf{r}(t) = x(t)\hat{\mathbf{i}} + y(t)\hat{\mathbf{j}}$, with
••
$$x(t) = R(\omega t - \sin \omega t) \quad \text{and} \quad y(t) = R(1 - \cos \omega t)$$

where R and ω are constants. Determine the velocity and the acceleration of the particle at any time t. For what values of the time is the particle momentarily at rest? What are the coordinate locations of these momentary rest positions and what are the corresponding accelerations? Does the magnitude of the acceleration depend on time? Show that the angle between the velocity vector and the acceleration vector is $\theta_{va} = \frac{1}{2}\omega t$. [*Hint:* Determine $\cos \theta_{va}$ by evaluating $\mathbf{v} \cdot \mathbf{a}$.]

Section 4-4

4-10 A boy throws a ball into the air as hard as he can and then runs as fast as he can under the ball in order to catch it. If his maximum speed in throwing the ball is 20 m/s and his best time for a 20-m dash is 3 s, how high does the ball rise?

4-11 A baseball is released from the thrower's hand 2 m
• above the level of a flat playing field. The initial velocity is 20 m/s in a direction making an angle of 30° with the horizontal. (a) What is the maximum height reached by the ball? (b) How far from the thrower does the ball strike the ground? (c) Find the time and the velocity at impact.

4-12 A driven golf ball just clears the top of a tree that is
• 15 m high and is 30 m from the tee, and then lands (with no roll or bounce) on the green, 180 m from the tee. What was the initial velocity imparted to the golf ball? [*Hint:* Use Eqs. 4-17 and 4-18 to find the angle α.]

4-13 A rifleman fires a bullet with a muzzle velocity of 500 m/s at a small stationary target 100 m away and at the same elevation as the rifle. At what angle of elevation must the rifle barrel be set? With this setting, how high above the target is the rifle aimed?

4-14 A ball on a floor is kicked, imparting to the ball
• an initial velocity $\mathbf{v}_0$. The ball strikes a wall that is a distance of 6 m from the original position, and then rebounds. The ball retraces its path and hits the floor at the exact spot from which it was kicked. What is

the minimum speed $\mathbf{v}_0$ that the ball must have to follow such a path? [*Hint:* At what angle must the ball strike the wall? What would have been the range of the ball if the wall had not been present?]

4-15 Two projectiles are launched with the same speed
 • but at angles $\frac{1}{4}\pi \pm \beta$ ($\beta < \frac{1}{4}\pi$) with the horizontal. Show that the projectiles have the same range. (This result was first deduced by Galileo.) Show also that the difference in flight times is

$$\Delta t_m = \frac{2\sqrt{2}\,v_0}{g}\sin\beta$$

Discuss the case $\beta = \frac{1}{4}\pi$.

4-16 A projectile is to be catapulted over a fortress wall
 • that is 20 m high by a siege *ballista* (a Roman catapult) capable of a launching speed of 22 m/s. The projectile is released at a height of 4 m above ground level. What is the farthest horizontal distance from the foot of the wall that the ballista may be set up? What elevation angle must be used at this distance? [*Hint:* Galileo would probably suggest that you first determine the elevation angle using Eq. 4–21.]

4-17 Water leaves the nozzle of a garden hose with a
 • speed of 10 m/s. At what angle above the horizontal is the nozzle pointed if the water falls on a flower bed 10 m away horizontally and 1 m lower than the nozzle? Assume the water stream behaves in the manner of projectiles. [*Hint:* Using $1 + \tan^2\alpha = 1/\cos^2\alpha$ in Eq. 4–17 reduces this equation to a quadratic in $\tan\alpha$.]

4-18 Photographs show that a competitive long jumper launches himself (i.e., his body center) at an angle of approximately 25° with respect to the horizontal. When a jumper lands in the pit, he doubles over to the extent that his body center is about 0.6 m lower than at take-off. What must be a jumper's take-off velocity in order to match the record jump of 8.90 m by Robert Beamon at the 1968 Olympics in Mexico City? Is this a reasonable velocity? Compare with that of a sprinter. [*Note:* In Mexico City, $g = 9.786$ m/s².]

Section 4-5

4-19 Find the tangential velocity and the centripetal acceleration of a particle glued to the tip of a fan blade which has a radius of 20 cm and rotates with an angular frequency of 1000 rpm.

4-20 A training simulator designed to expose astronauts to the accelerations encountered in rocket launches consists of a seat at the end of a 5-m boom that can rotate around a vertical axis. What rotational frequency (in rpm) is required to simulate a $7g$ acceleration? What is the tangential velocity at this acceleration? If the tangential acceleration is to be no larger than $g/30$, what is the minimum time to reach the required rotational frequency, starting from rest?

4-21 A disc with a radius $R = 0.1$ m starts from rest at $t = 0$ and is given a constant angular acceleration, $\alpha_0 = \pi$ rad/s. A particle glued to a point on the circumference has an initial (counterclockwise) angular location of $\phi_0 = \pi/2$ rad, measured from a fixed reference line. At $t = 1$ s, find the following quantities for the particle: (a) the angular position, (b) the angular velocity, (c) the tangential velocity, (d) the tangential and radial accelerations, and (e) the resultant acceleration.

4-22 Show that the results of Eqs. 4–34 and 4–35 for constant angular acceleration α_0 lead to an angular frequency ω at the end of a rotation through an angle $\phi - \phi_0$ given by

$$\omega^2 - \omega_0^2 = 2\alpha_0(\phi - \phi_0)$$

Compare Eq. 2–15.

4-23 A particle glued to the rim of a wheel with a radius of 20 cm has an angular position (measured counterclockwise) with respect to a fixed reference line given by $\phi(t) = \alpha t^3 + \beta t^2 + \pi/2$ radians, with $\alpha = 2$ s^{-3} and $\beta = -3$ s^{-2}. (a) Find the angular frequency and the acceleration of the particle. (b) At what time is the resultant acceleration entirely radial? What is its value at this instant? (c) At what time is the resultant acceleration entirely tangential? What is its value at this instant?

4-24 An object is located on the Earth's surface at a latitude λ. Calculate the centripetal acceleration that the object experiences by virtue of the rotation of the Earth about its axis. Express your answer in terms of g.

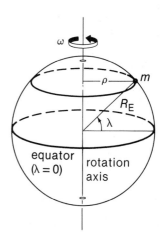

Section 4-6

4-25 A shopper in a department store can walk up a stationary (stalled) escalator in 30 s. If the normally functioning escalator can carry the standing shopper to the next floor in 20 s, how long would it take the shopper to walk up the moving escalator? Assume the same walking effort for the shopper whether the escalator is stalled or moving.

4-26 Two canoeists in identical canoes exert the same effort paddling in a river. One paddles directly upstream (and moves upstream), whereas the other paddles directly downstream. An observer on the riverbank reckons their speeds to be 1.2 m/s and 2.9 m/s. How fast is the river flowing?

4-27 A motor boat has two throttle positions on its engine: the high-speed position propels the boat at 10 km/h in still water and the low position gives half the maximum speed. The boat travels downstream on a river at half speed and returns to its dock at full speed. It takes 15 percent longer to make the upstream trip than it does to make the downstream trip. How fast is the river flowing?

4-28 A train is moving east on a straight track with a speed of 12 m/s. Raindrops falling straight down with respect to the ground make initial tracks on the train windows at an angle of 27° with respect to the vertical. What is the (vertical) speed of the raindrops with respect to the ground? What is the speed of the raindrops with respect to the train?

4-29 What compass bearing should a pilot set if he wishes to fly due north when there is a wind blowing from the west at 100 km/h and when the aircraft moves with an air speed of 350 km/h? What is the ground speed of the aircraft?

4-30 A boat requires 2 min to cross a river that is 150 m wide. The boat's speed relative to the water is 3 m/s and the river current flows at a speed of 2 m/s. At what possible upstream or downstream points does the boat reach the opposite shore?

Additional Problems

4-31 A particle follows a path $\mathbf{r}(t) = x(t)\hat{\mathbf{i}} + y(t)\hat{\mathbf{j}} + z(t)\hat{\mathbf{k}}$, with

$$x(t) = a \sin \omega t,$$
$$y(t) = b \cos \omega t,$$
$$z(t) = c \cos \omega t$$

and $a^2 = b^2 + c^2$. Determine the velocity, the speed, and the acceleration of the particle. Calculate the value of the dot product of velocity and acceleration.

Does your result agree with the conclusions of Example 4-5? Give a geometrical description of the path.

4-32 Show that two projectiles launched simultaneously
• near the surface of the Earth will collide if and only if they would have collided without benefit of the acceleration due to the Earth's gravity. Give an expression for the time of impact. Are these results in accord with Example 4-7? [*Hint:* Use vector notation, as in Example 4-7.]

4-33 The acceleration of a particle is $\mathbf{a}(t) = [(2 \text{ m/s}^3)t + (1 \text{ m/s}^2)]\hat{\mathbf{i}}$. If the velocity at $t = 1$ s is $\mathbf{v}(1) = (2 \text{ m/s})\hat{\mathbf{i}} + (7 \text{ m/s})\hat{\mathbf{k}}$, and if the position at $t = 0$ is $\mathbf{r}_0 = (3 \text{ m})\hat{\mathbf{j}}$, find the velocity function and the position function for the particle at any time t.

4-34 A football is kicked from a kicking tee with an initial velocity of 20 m/s at an angle of 45° with the horizontal. A receiver 60 m away in the direction of the kick starts to run toward the ball the same instant it is kicked. What is the slowest constant speed at which he can run in order to field the ball with a shoe-top catch?

4-35 A batter hits a pitched baseball 1 m above the ground, imparting to the ball a speed of 40 m/s. The resulting line drive is caught on the fly by the left fielder 60 m from home plate with his glove 1 m above the ground. If the shortstop, 45 m from home plate and in line with the drive, were to jump straight up to make the catch, instead of allowing the left fielder to make the play, how high above the ground would his glove have to be?

4-36 In attempting to kick a field goal from a point
• midway between the sidelines and 40 m from the goal post, the kicker imparts an initial velocity to the football of 20.8 m/s at an angle of 40° with the horizontal. (a) The crossbar of the goal post is 10 ft (3.05 m) above the ground. Was the kick successful? (b) Under the above conditions, for what range of angles of the initial velocity of the football would the ball clear the crossbar? [*Hint:* Note that $1/\cos^2\alpha = 1 + \tan^2\alpha$ used in Eq. 4-17 reduces this equation to a quadratic in $\tan \alpha$.]

4-37 A catapult hurls a stone with a velocity of 20 m/s
• at an angle of 40° above the horizontal from a wall that is 10 m above the level ground. What will be the increase in range of the projectile if the catapult is moved to a wall that is 20 m high? [*Hint:* Use Eqs. 4-16, but add a term z_0 to $z(t)$.]

4-38 A particle is traveling along a ballistic trajectory with
• initial velocity v_0 and elevation angle α. It is asserted that at the top of its trajectory the particle behaves momentarily as if it were traveling along a circular path with a speed equal to its actual speed. (a) What

is the radius R of such a circular path? [*Hint:* What must be the centripetal acceleration at the top of the trajectory?] (b) A mathematician claims that this radius is simply the *radius of curvature* for the actual path at the highest point, that is,

$$R^{-1} = \left| \frac{d^2z}{dy^2} \right|_{\text{apex}}$$

Use Eq. 4–17 and verify this assertion.

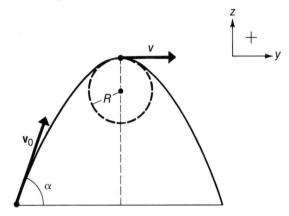

4-39 A particle originally at rest on top of a hemisphere with a radius R is given a horizontal velocity v_0. What is the *smallest* value of v_0 that will allow the particle to leave the surface immediately instead of starting to slide down the surface?

4-40 An LP record rotates at $33\frac{1}{3}$ rpm. What is the arc length along the record groove corresponding to one cycle of a tone with a frequency of 20,000 Hz (20 kHz) when the stylus is 10 cm from the record axis? (For comparison, the diameter of a human red blood cell is about 13×10^{-6} m.)

4-41 An athlete can throw a ball a maximum horizontal distance R. If air resistance effects are negligible, show that the athlete can throw the same ball to a maximum height $\frac{1}{2}R$.

4-42 An ultracentrifuge is designed to reach a radial acceleration of 500,000g. If the instrument has a radius of 1 cm, what angular frequency (in rpm) is required? Through how many complete rotations will the centrifuge turn to reach this final rotational frequency if

it undergoes a constant tangential acceleration of 10g and starts from rest?

4-43 Some science students are canoeing on a river. While heading upstream they accidentally drop overboard an empty beer can; then they continue paddling for 1 h, reaching a point 4 km further upstream. At this point they realize what has happened, and they turn around and head downstream. They catch up with and retrieve the floating beer can 10 km downstream from the turn-around point. (a) Assuming a constant paddling effort throughout, how fast is the river flowing? (b) What would be the canoe velocity in a still lake for the same paddling effort?

4-44 An elevator is moving upward at a constant velocity of 2.0 m/s. A bolt in the elevator ceiling 3 m above the elevator floor works loose and falls. (a) How long does it take for the bolt to fall to the elevator floor? (b) What is the velocity of the bolt just as it hits the elevator floor according to an observer in the elevator? (c) What would be the velocity according to an observer standing on one of the floor landings of the building? (d) What total distance do the two observers determine for the bolt's journey from ceiling to floor? [*Hint:* What is the initial velocity of the bolt?]

4-45 Two boats are moving along perpendicular paths on a still lake at night. One boat moves with a speed of 3 m/s, the other with a speed of 4 m/s. At $t = 0$, the boats are 300 m apart, and they manage to collide some time later. At what time did the collision occur? How far did each boat move between $t = 0$ and the time of collision?

4-46 An ocean liner on a calm sea is steaming due west with a speed of 20 km/h. Smoke from its funnel stretches out in a line 15° south of east. If the wind is blowing from the north, what is the speed of the wind? (Assume that the smoke particles take up the wind velocity as soon as they are emitted from the funnel.)

4-47 An airplane flies a straight-line course between two airports at a constant air speed. Show that the time for a round trip is always increased by the presence of a wind blowing with a constant velocity during the entire flight, independent of the direction of the wind.

THE PRINCIPLES OF DYNAMICS— NEWTON'S LAWS

We have by now examined in detail the methods for describing the motion of particles and objects that can be considered to behave as particles. In these discussions we have dealt with the time variations of the geometry of motion, the essential ingredient of the subject called *kinematics*. In this chapter we extend our view to include the question of *why* particles move as they do. The study of the way that other agencies or objects affect the motion of particles constitutes the subject called *dynamics*.

By the skillful combination of intuition and experimentation, Galileo began the modern development of the science of mechanics. Isaac Newton (who was born the same year that Galileo died, 1642) incorporated the findings of Galileo along with his own brilliant discoveries into a formulation of dynamics that has become known as *Newtonian mechanics*. Newton's theory is summarized in three laws that have wide validity and profound consequences. In the following pages we state and then discuss the implications of these laws.

Although Newton's laws have provided a remarkably simple and satisfying theory of mechanics, they nonetheless involve certain problems of logic. The particular line of development of the theory that we follow here is not the one taken by Newton. This discussion is neither more philosophically satisfying nor less subject to logical criticism than the original development. The difficulty that arises in any formulation of Newtonian theory is a result of the necessity to introduce simultaneously two concepts not considered in the subject of kinematics. These new concepts are *force* and *mass*. In stating his three laws of dynamics, Newton intermingled the definitions of these quantities with the rules of their behavior. No one has yet discovered an escape from this logical paradox. It is a tribute to Newton's insight that, in spite of this difficulty, he was able to formulate a set of laws that provides a correct description of the dynamics of bulk matter.

In the discussions here, an intuitive approach is taken in the introduction of the concept of force. By using an operational definition of this important quantity, we can avoid the force-mass dilemma. This procedure, although not completely free of criticism, offers a straightforward approach to Newtonian dynamics.

Newton's laws are not a perfect description of the way that Nature behaves in every domain. Within this century it has been discovered that Newtonian theory fails in the realm of atoms and nuclei and that it fails when extremely high velocities are involved. The investigation of these failures has led to the development of quantum theory and the theory of relativity. However, the durability of Newton's laws in satisfactorily accounting for observations in the everyday macroscopic world remains intact.

In this chapter we concentrate on developing and interpreting Newton's laws. In the next chapter, we direct our attention toward some interesting and important applications of Newtonian theory.

5-1 FORCE

The Basic Forces. We are all familiar with the everyday use of the word *force* to describe a push or a pull. Both animate and inanimate agencies can exert forces. Stretched springs and rubber bands or taut cables and ropes exert forces on objects attached to their ends. Liquids and gases exert forces on container walls, and they exert buoyant forces on immersed objects. Materials that are rubbed together exert frictional forces on each other. Machines can be used to exert forces that cut or press, and they can be used for locomotion by exerting forces against the ground. All of these examples involve *contact forces,* in which one object exerts a force on another object by coming into direct contact with it.

There is another type of interaction, exemplified by the *gravitational force,* in which a force acts between objects that need not be in physical contact with each other. Thus, the gravitational force is sometimes said to be an "action-at-a-distance" force. It is the gravitational force acting through the vacuum of space that maintains the Earth in its orbit around the Sun. It is also the force that keeps us "pinned" to the Earth's surface even as we move about, and it "holds" our atmosphere and oceans in captive attachment. The importance of the gravitational force in our lives is therefore self-evident, and it is the subject of a careful examination in this and in later chapters. Another important force that acts at a distance is the *electromagnetic force.* This is the basic force that acts between electrically charged objects. Much of modern life depends on devices that operate by utilizing the electromagnetic force. This too is a force that we examine in detail later.

The structure and behavior of the entire Universe can be ascribed ultimately to the action of only *four* fundamental forces. These are, in addition to the gravitational and electromagnetic forces, the strong (or hadronic) and the weak (or leptonic) forces that operate in the domain of nuclei and elementary particles. Even though all matter in the Universe is subject to the action of these four forces, they can really be compared only at the level of elementary particles. If we consider, for example, two protons, the relative strengths of the four forces when the particles are close together* are

Strong force	1
Electromagnetic force	10^{-2}
Weak force	10^{-13}
Gravitational force	10^{-38}

Notice how very weak the gravitational force is, even compared to the weak force.

---

*The comparison must be made at distances near 10^{-15} m because the strong and weak forces are not effective over greater distances.

Although the four basic forces have always been considered to be independent, there is emerging a new theoretical description that may succeed in unifying the electromagnetic, weak, and strong forces. Someday we may even have a successful description of Nature that is framed in terms of a *single* theoretical model encompassing all four basic forces.

The structures of atoms and molecules are governed almost exclusively by the electromagnetic interaction* (treated suitably within the context of quantum theory). When macroscopic objects are brought into contact, their constituent atomic systems partially penetrate one another, and the resulting interaction is manifest as a macroscopic force. This is the force that we have called a contact force. A detailed microscopic (atomic) treatment of the gross interaction, although theoretically possible, involves enormous complexity. Fortunately, it is not necessary to take this approach because quite accurate descriptions can be given in terms of macroscopic variables. Thus, contact forces, which are actually aggregate atomic effects involving mostly electromagnetic forces, can nonetheless be treated in terms of familiar macroscopic pushes and pulls.

Because the objects we consider in these early chapters are electrically neutral (that is, they possess as much negative as positive electric charge), macroscopic electromagnetic forces can be neglected. Nuclear effects, although they are strikingly evident in special large-scale applications such as nuclear explosions and in nuclear reactors, are not involved in everyday phenomena, due primarily to the extremely short range of the strong interaction. We are therefore justified, for the present, in neglecting these effects. The gravitational force, on the other hand, operates over large distances and is always an attractive force. Consequently, gravitational effects are cumulative, and when a large object—such as the Earth—is involved, the gravitational force becomes the dominant interaction.

Force—An Operational Definition.   We all know from experience that pulling at both ends of a strand of an elastic material or a spring produces an elongation of the object. It is also easy to discover that the greater the pulling effort, the greater is the elongation. This fact suggests how we might devise a means for exerting a definite and reproducible force on an object.

To be specific, suppose that we have constructed several identical, standard spring scales such as the one shown schematically in Fig. 5–1a. The spring is imagined to have a natural or relaxed length ℓ_0. If we pull on the two (hook) ends, we can stretch the spring to a total length ℓ; that is, $\Delta\ell = \ell - \ell_0$ represents the elongation of the spring. The force F necessary to produce this elongation is expected to be a function of $\Delta\ell$. We now describe an empirical way to discover the relationship connecting F and $\Delta\ell$.

Quite apart from the identical construction of the springs, we can verify their genuine equivalence by a simple experiment. Refer to Fig. 5–1b. The two springs, which have relaxed lengths ℓ_{10} and ℓ_{20}, are connected as shown. By adjusting the slide S, we automatically stretch both springs. If for every setting of S, we find that $\Delta\ell_1 = \ell_1 - \ell_{10}$ is equal to $\Delta\ell_2 = \ell_2 - \ell_{20}$, the two standard springs may be judged equivalent.† Imagine that several equivalent springs have been constructed and tested this way. We are now ready to determine the connection between F and $\Delta\ell$.

Refer next to Fig. 5–2, where one standard spring scale is shown balanced

---

*The weak and gravitational forces are ineffective, and the particles are too far apart for the strong force to play a role.

†Is this self-evident? Our intuition (based on experiences such as tug-of-war games) suggests that it is—and, indeed, that is correct. But there are sophisticated ideas involved in this simple statement. Think about this point again after studying Newton's third law.

(a)

(b)

Fig. 5-1. (a) Schematic of a spring scale. The length ℓ of the spring is determined by the position of the pointer P. (b) Arrangement for determining the equivalence of two standard spring scales.

against two other standard springs. Suppose that in the earlier arrangement of Fig. 5-1b the slide S is set so that there results a specific elongation, $\Delta \ell_s = \ell - \ell_0$. (Because we have established that the two springs are equivalent, this elongation refers to either spring.) The elongation $\Delta \ell_s$ corresponds to some definite but unknown spring force F_s. Then, in the arrangement of Fig. 5-2, we set the slide S so that $\Delta \ell_y = \Delta \ell_s$ and measure $\Delta \ell_x = \ell_x - \ell_0$. The elongation $\Delta \ell_x$ now corresponds to a force $2F_s$ (exerted by the left-hand spring). Next, we reset S in Fig. 5-2 such that $\Delta \ell_x = \Delta \ell_s$ and determine a new value of $\Delta \ell_y$. This defines a force equal to $\frac{1}{2}F_s$ (exerted by each of the right-hand springs). (•Can you generalize this procedure to define forces $3F_s$ and $\frac{1}{3}F_s$, and so forth?)

Fig. 5-2. Arrangement used to determine the relationship connecting the force with the elongation for standard springs.

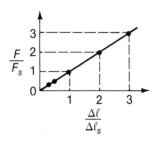

Fig. 5-3. Results of the hypothetical experiment performed using the arrangement of springs shown in Fig. 5-2. The graph is linear, so the springs obey Hooke's law.

From the spring measurements we can imagine obtaining data such as that shown in Fig. 5-3. This graph shows that F is directly proportional to $\Delta\ell$; that is,

$$F = \kappa\Delta\ell = \kappa(\ell - \ell_0) \tag{5-1}$$

The proportionality constant κ is the *force constant* for the particular set of standard springs. As long as they are not stretched too far, most springs obey Eq. 5-1.

The linear relationship connecting F with $\Delta\ell$ that is given by Eq. 5-1 is referred to as *Hooke's law.** By using a standard spring scale (being careful to confine the elongation to the range in which Hooke's law is valid), we have a simple way of applying any desired force. For example, if, in a second experiment, we wish to apply a force that is 2.46 times larger than that in a previous experiment, we simply set up conditions such that $\Delta\ell$ on the second occasion is 2.46 times $\Delta\ell$ in the first instance. We assert that our standard spring scale can be made to exert a force with a particular magnitude and with a direction along the spring axis (opposite to the direction of the spring extension).

In all of the discussion here, we have avoided making any statement as to what *force* actually is, except that it is a quantity with both magnitude and direction (a *vector* quantity). We have merely given an operational method for setting up forces that have any desired ratio, one to another. We now use these intuitive ideas concerning force in the discussion of Newton's laws of dynamics.

5-2 NEWTON'S LAWS OF MOTION

Frames of Reference.   Newton's laws of dynamics describe the motion of objects. In order to specify positions, velocities, and accelerations, it is necessary first to establish a *frame of reference*. That is, we require a background coordinate system so that precise meaning can be given to all of the kinematic quantities that enter the theory.

What type of reference frame shall we choose? Newton understood the important and profound point that a frame suitable for expressing the laws of dynamics must be a *nonaccelerating* reference frame. It is easy to see why. Have you ever walked on a moving merry-go-round (which undergoes continual centripetal acceleration) and noticed the sensation of strange forces acting on your body? And how would you describe the motion of a person walking on the ground in terms of a coordinate frame attached to the merry-go-round?

Newton realized that a reference frame attached to the Earth is not an unaccelerated frame. The rotational motion of the Earth results in a centripetal acceleration at the Equator (where the value is maximum) of $3.4 \times 10^{-2}\,\mathrm{m/s^2}$ or approximately 0.34 percent of g. Moreover, the Earth experiences a centripetal acceleration of $5.9 \times 10^{-3}\,\mathrm{m/s^2}$ due to its orbital motion around the Sun. If these accelerations can be tolerated or ignored in a particular situation, an Earth-based reference frame will be adequate for describing the dynamics of an object or system. Thus, for many purposes, an Earth-based frame is a satisfactory "unaccelerated" reference frame.

Newton sought to avoid the difficulties associated with a rotating Earth-based frame by choosing as his "primary" reference frame one that is based on the posi-

---

*Robert Hooke (1635–1703), an English physicist, stated the equivalent of this famous law in 1676 in the form of a Latin cryptogram, CEIIINOSSSTTUV, the solution of which is *ut tensio sic vis* ("as the extension, so the force").

tions of the stars. If the stars are in fixed positions in the sky, a reference frame based on these positions is well-defined and enduring. We now know that each star in the sky undergoes its own particular motion (Fig. 5–4); however, this motion is difficult or impossible to detect for stars that lie at great distances from the Earth. Thus, we might base a primary reference frame on these distant stars. This procedure will be adequate for most purposes, but it cannot be justified as a fundamental part of a precise theory. (We return to this point later and give a definition of a proper unaccelerated reference frame.)

Fig. 5–4. The current positions of the major stars in the constellation Scorpio and their expected positions as seen 100,000 years hence.

Once a primary unaccelerated reference frame has been established, it is possible to define any number of additional frames that, at most, move with constant velocity and without rotation with respect to the primary frame. All such frames are equally suitable for describing the dynamics of particles and systems.

Newton's First Law. According to Aristotelian doctrine, the "natural state" of an object is one of *rest,* and in order to sustain *any* motion, some outside agency must provide a propelling force. This must surely be true, it was argued, because all moving objects, of their own accord, eventually come to rest. In the 17th century, observations made by Galileo convinced him that this view of motion is incorrect. Instead, Galileo concluded that, in the absence of any external applied forces, the state of motion of a body is one of uniform motion—that is, the motion takes place with constant velocity, with zero velocity included as a possibility. Galileo reasoned that an object sliding over a rough horizontal surface eventually comes to rest because a retarding force—the force of friction—acts on the object. If the same object slides over a slippery surface, with the same initial velocity, it will move a greater distance before coming to rest. If this test is continued on smooth ice or an oiled surface, the stopping distance is increased. Further refinements are now possible. Standard modern laboratory equipment for performing nearly friction-free experiments includes the *air track* and the *air table.* These devices provide air cushions on which objects float, thereby permitting unimpeded horizontal motion. The remnant effect of friction in these experiments is several orders of magnitude smaller than if lubricants were used.

Students perform an experiment on a nearly frictionless air track. (Courtesy of Daedalon Corp.)

Using the results of his Earthbound experiments, Galileo correctly arrived at the conclusion that, in the absence of all friction, an object once set into motion will continue to move with constant velocity until acted upon by a force. Newton accepted Galileo's conclusion and incorporated it into his first axiom regarding the nature and causes of motion. We now refer to this statement as *Newton's first law of motion.* In Newton's words (translated from the original Latin version*):

I. Every body continues in its state of rest, or in uniform motion in a right line (straight line) *unless it is compelled to change that state by forces impressed upon it.*

In mathematical language Newton's first law becomes

$$I. \quad \mathbf{F} = 0 \quad \textit{implies} \quad \mathbf{v} = \text{constant} \qquad (5\text{–}2)$$

The first law does not really tell us anything about the nature of force. We have

---

*The theory of dynamics developed by Isaac Newton (1642–1727) is contained in his monumental work, *Philosophiae Naturalis Principia Mathematica,* published in 1686. (We usually refer to this treatise simply as the *Principia.*) The translated passages (except for the editorial parenthetical inserts) are from Andrew Motte, 1729.

here only a statement concerning the behavior of objects in the *absence* of any force. Even so, the first law is not an empty statement. Consider a body that experiences no outside forces.* This situation can exist only in the complete absence of other matter, for contact or action-at-a-distance forces (such as the gravitational force) always originate with other material bodies. However, we can approximate the situation by considering an object isolated in deep space. With the zero-force condition of the first law satisfied owing to the isolation, we know that the object moves with constant velocity in an unaccelerated reference frame, which we call an *inertial reference frame*. Instead of relying on the distant stars to define an appropriate unaccelerated reference frame, we can take the attitude that *any* frame in which any particle subject to no force moves with $\mathbf{v}$ = constant is, in fact, an inertial frame. That is, the first law can be considered to *define* the concept of an inertial reference frame.

Notice how fragile the argument is here. First we recognize that a reference frame is necessary in order to describe motion. We choose a primary frame defined by the distant stars or any frame not undergoing acceleration with respect to the primary frame. We then use these frames to formulate the first law of motion. Finally, we use the first law to define a proper reference frame (an inertial frame) for describing dynamical effects. Then, because all inertial frames are equally valid, we can attach no meaning to the concept of a "primary" reference frame! Nevertheless, in spite of the circular logic, Newton's laws provide us with the correct relationship connecting forces and accelerations.

Because the first law is valid in any inertial reference frame, there can be no physical distinction between a state of rest and one of constant velocity for an object. If an object is subject to no force, an observer who chooses to view the object from an inertial reference frame attached to the object will, of course, see the object at rest. But another observer who views the same object from a second frame that moves with constant velocity with respect to the first frame will see the object in uniform motion. Neither interpretation of the situation is more basic nor more correct than the other; both views are equally valid.

The state of motion of an isolated body will change only if it is acted upon by a force. This tendency to remain at rest or in uniform motion is due to a property of the body we call its *inertia*. Newton's first law is accordingly called the *law of inertia*, and we use the term *inertial reference frame* for any frame in which the law is valid.

The measure of the inertia of a body is its *mass*. In the second law we find how the motion of an object is determined by the force applied to it and by its mass.

Newton's Second Law.   Common experience suggests how an object responds to the application of a force. We observe that the stronger the push or pull applied to an object, the greater is its altered motion or *acceleration* (remember, $\mathbf{a} = d\mathbf{v}/dt$). We also observe that the same magnitude of push or pull produces different accelerations when applied to two different objects, such as a golf ball and a bowling ball.

In order to remove ourselves from various local extraneous influences, suppose that we conduct a series of experiments in deep space, completely isolated from all other objects. With a standard spring scale, we imagine applying a variety of controlled forces to some selected rigid test object, as indicated in Fig. 5–5. We observe the resulting acceleration of the object, measured in an inertial frame, as we apply a sustained force (by moving the disembodied hand shown in the figure). A definite

---

*Actually, it is not necessary to require that a body experience no outside force in order for the first law to be valid. It is required only that the *resultant* (or *net*) force be zero. But in following Newton's development of the theory there is, at this point, no clear definition of force. Therefore, one can safely consider only the case of no forces whatsoever.

Fig. 5-5. Experiment to determine the nature of the acceleration produced by an applied force, using a standard spring scale.

strength of sustained force is set by maintaining a fixed value of $\Delta \ell = \ell - \ell_0$. We observe that the acceleration produced is constant in magnitude and is directed along the spring axis (that is, in line with the direction of the applied force). If we increase the force by doubling $\Delta \ell$, we observe an acceleration that is also twice the magnitude of the original acceleration. By repeating such measurements with our test object, we conclude that the acceleration produced is directly proportional to the applied force. If we use the vector symbol $\mathbf{F}$ to represent the applied force and $\mathbf{a}$ for the resulting acceleration, we have as our discovered empirical law $\mathbf{F} \propto \mathbf{a}$, which we can write as

$$II. \quad \mathbf{F} = m\mathbf{a} \qquad\qquad (5\text{-}3)$$

where m is the proportionality constant (not yet identified).

Next, we use several standard springs, each exerting a constant force and acting simultaneously on our selected test object. We discover that the resulting acceleration is in the direction of the *resultant* force and has a magnitude again given by Eq. 5-3 if $\mathbf{F}$ now stands for the resultant or net force on the object. That is, $\mathbf{F}$ is the *vector sum* of all individual external forces acting on the object: $\mathbf{F} = \Sigma_i \mathbf{F}_i$.

The conclusion expressed in Eq. 5-3 is known as *Newton's second law* of dynamics. In his own words:

II. The change in motion (acceleration*) *is proportional to the motive force impressed; and is made in the direction of the right line* (straight line) *in which that force is impressed.*

The scalar quantity m appearing in Eq. 5-3 was introduced simply as a constant of proportionality. Suppose that we alter the shape of the test object (without altering the amount of material present) and then maintain its rigidity in the new form. We observe that the relationship between $\mathbf{F}$ and $\mathbf{a}$ is unchanged; that is, the constant m is the same as before. Next, we consider a second test object, which may even consist of a material with a completely different chemical composition. When this object is tested by itself using various forces, the values obtained for the accelerations indicate the same value of m as for our first test object. Now, we physically attach the second object to the first, and we find that the appropriate value of m for the combined unit is twice that for either object alone. This means that, for the same force, only half as large an acceleration results. Thus, the quantity m has an inertial quality, and we refer to m as the *inertial mass* of the object. Mass is an intrinsic property of matter, and, in a colloquial sense, we say that mass is a measure of the amount of matter in an object.

---

*We examine more closely the meaning of Newton's words when we discuss *momentum* in Chapter 9.

Because we have defined force in terms of the action of springs, Newton's second law permits a definition of mass in terms of the acceleration produced by a definite force. We can determine the mass of an object in terms of the mass of a standard (for example, the standard kilogram) by using a spring to apply the same force to each and then measuring the ratio of accelerations produced. If m_S represents the standard mass, the unknown mass m is

$$m = \frac{a_S}{a} m_S \qquad\qquad (5\text{-}4)$$

This definition of mass is essential in the formulation of the theory of dynamics.

In SI units, mass is measured in kilograms (kg) and acceleration is measured in meters per second per second (m/s²). Therefore, force is measured in units of kg · m/s², to which we give the special name *newton* (N):

$$1\,\text{N} = 1\,\text{kg} \cdot \text{m/s}^2$$

That is, a force of 1 N applied to an object with a mass of 1 kg will produce an acceleration of 1 m/s².

Example 5–1

(a) What is the mass of an object if a net applied force of 7 N produces an acceleration of 3 m/s²?

$F_2 = 5$ N

F_R

α

$F_1 = 7$ N

$m = \frac{7}{3}$ kg

Solution:

A direct application of Eq. 5–3 gives

$$m = \frac{F}{a} = \frac{7\,\text{N}}{3\,\text{m/s}^2} = \frac{7}{3}\,\text{kg}$$

(b) If an additional force of 5 N is applied to the same object, but at right angles to the 7-N force, what is the resultant acceleration?

Solution:

The resultant vector $\mathbf{F}_R$ has the magnitude

$$F_R = \sqrt{(7\,\text{N})^2 + (5\,\text{N})^2} = 8.60\,\text{N}$$

and makes an angle α with $\mathbf{F}_1$ given by

$$\tan \alpha = F_2/F_1 = 5/7$$

or $$\alpha = 35.5°$$

The resulting acceleration is in the direction of $\mathbf{F}_R$, with a magnitude

$$a = \frac{F_R}{m} = \frac{8.60\,\text{N}}{(7/3)\,\text{kg}} = 3.69\,\text{m/s}^2$$

Example 5–2

A stone with a mass of 100 g is tied to the end of a 2-m string and is whirled in a circular path at a constant speed of 3 m/s, as shown in the diagram. The other end of the string is fixed in an inertial frame. What is the force that the taut string exerts if this is the only force acting on the stone?

Solution:

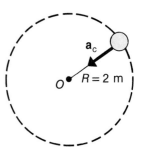

The stone travels in a circular path at constant speed, so this is the special case of $\rho = R = $ constant (Section 4–5). Thus, the stone undergoes a constant centripetal acceleration (directed toward the center O) with magnitude (Eq. 4–26).

$$a_c = \frac{v^2}{R} = \frac{(3 \text{ m/s})^2}{2 \text{ m}} = 4.5 \text{ m/s}^2$$

Therefore, the force exerted on the stone by the string is also directed toward the center O and has the constant magnitude,

$$ma_c = (0.1 \text{ kg})(4.5 \text{ m/s}^2) = 0.45 \text{ N}$$

Example 5–3

A flat disc with a mass of 2 kg slides across the frozen surface of a lake with an initial speed of 5 m/s. If the frictional force on the disc has a constant value of 4 N, in what distance will the disc come to rest?

Solution:

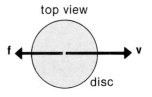

Because the disc has no vertical motion, we can ignore all vertical forces acting on the disc. (That is, whatever vertical forces *are* acting, they must give a zero resultant force.)

The given frictional force acts on the disc in a direction precisely opposite to its velocity, as indicated in the diagram. The acceleration produced by the frictional force is (taking the positive direction to be to the right)*

$$a = -\frac{f}{m} = -\frac{4 \text{ N}}{2 \text{ kg}} = -2 \text{ m/s}^2$$

We also know (Eq. 2–11) that $v^2 - v_0^2 = 2as$; because $v = 0$, we can write

$$s = -\frac{v_0^2}{2a} = -\frac{(5 \text{ m/s})^2}{2(-2 \text{ m/s}^2)} = 6.25 \text{ m}$$

Newton's Third Law.   On the basis of his observations, Newton presumed that all forces in Nature act in *pairs*. Consider the simple case of two objects, A and B, that have a mutual interaction but which are otherwise isolated. Newton reasoned that the interaction between these objects involves a pair of forces—the force exerted on A by B and that exerted on B by A. Let us call the first of these forces $\mathbf{F}_{AB}$ and the second $\mathbf{F}_{BA}$. Newton postulated that

---

*We frequently deal with equations that involve vector quantities. When vector components (Section 3–4) are involved, as is the case in this example, the scalar components carry an appropriate sign. The frictional force here is directed to the left, so in Newton's equation we write $-f$.

$$\text{III.} \quad \mathbf{F}_{AB} = -\mathbf{F}_{BA} \tag{5-5}$$

This relationship constitutes *Newton's third law,* and in his words,

III. To every action (force) *there is always opposed an equal reaction* (force); *or, the mutual actions* (forces) *of two bodies upon each other are always directed to contrary parts* (in opposite directions).

This law states that the forces in an action-reaction pair have the same magnitude but opposite directions. The symmetry of the situation (attaching no more importance to object A than B) permits us to call either $\mathbf{F}_{AB}$ or $\mathbf{F}_{BA}$ the "force" to which the other is then the "reaction force." Thus, Eq. 5-5 is the *law of action and reaction.*

Consider a test object A and a standard spring scale B (as in Fig. 5-5). The force exerted on the test object by the spring is the force $\mathbf{F}_{AB}$, and the force exerted on the spring by the test object is the reaction force $\mathbf{F}_{BA}$. Figure 5-6 shows these two forces, which are related by Eq. 5-5: $\mathbf{F}_{AB} = -\mathbf{F}_{BA}$.

Fig. 5-6. Action and reaction forces for the situation illustrated in Fig. 5-5.

The third law tells us that forces always occur in pairs. But notice carefully that the two forces in every action-reaction pair always act on *different* objects. In Fig. 5-6, the force $\mathbf{F}_{AB}$ acts on A, whereas the force $\mathbf{F}_{BA}$ acts on B. To determine the motion of an object, we need to know only the force acting *on* that object; it does not matter that the object exerts forces on other bodies.

Example 5–4

In deep space, some agency (a disembodied hand) applies a constant force of 2 N to a block A that is in contact with a block B, as shown in diagram (a). (a) Identify all mutual action-reaction force pairs. (b) If the mass of block A is $m_A = 1$ kg and that of block B is $m_B = 2$ kg, what is the resulting acceleration of the two blocks? What force acts on block B?

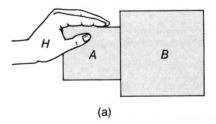

(a)

Solution:

Diagram (b) is an "exploded view" of the situation, isolating the three objects involved. (The hand and the two blocks remain in contact throughout the time during which we are considering the motion.)

(a) The force that the hand H exerts on block A is $\mathbf{F}_{AH}$; block A exerts the reaction force

F_{HA} on the hand. Block A also exerts a force F_{BA} on block B, and block B exerts the reaction force F_{AB} on block A. We observe that there are two separate pairs of mutual forces: one pair is $F_{AH} = -F_{HA}$, and the other pair is $F_{BA} = -F_{AB}$.

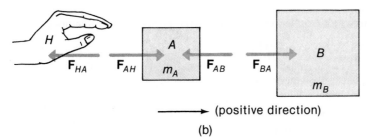

→ (positive direction)

(b)

(b) For the force on block B, using Eq. 5–3 and taking the direction to the right as positive, we have

$$F_{BA} = m_B a_B$$

Because the net force on block A is $F_{AH} - F_{AB}$, we again apply Eq. 5–3 with the result

$$F_{AH} - F_{AB} = m_A a_A$$

We add these two equations. Realizing that the blocks remain in contact so that $a_A = a_B = a$, we obtain

$$F_{AH} - F_{AB} + F_{BA} = (m_A + m_B)a$$

From the third law we know that F_{AB} and F_{BA} have the same magnitude; thus, we arrive at the result

$$F_{AH} = (m_A + m_B)a$$

Blocks A and B, because they are in contact, have the same motion. Consequently, we could consider them to be a single unit with a mass $m_A + m_B$ upon which there acts only the *external* force F_{AH}, resulting in a common acceleration a, as indicated in diagram (c). Then, we would obtain exactly the same expression for F_{AH}. In this view, the forces F_{BA} and F_{AB} are simply canceling internal forces that behave in the same way as all the internal molecular forces present in the blocks.

Finally, we have

$$a = \frac{F_{AH}}{(m_A + m_B)} = \frac{2\,\text{N}}{(1\,\text{kg} + 2\,\text{kg})} = \tfrac{2}{3}\,\text{m/s}^2$$

Also, from above, $F_{BA} = m_B a = (2\,\text{kg})(\tfrac{2}{3}\text{m/s}^2) = \tfrac{4}{3}\,\text{N}$

We learn from this example that when several objects participate in a common motion, we may conveniently group them together into a single unit provided that, in using Eq. 5–3, we consider *only* the forces that are externally impressed on the combined unit. Naturally, if we wish to learn the force that one of these objects exerts on another (as in the present case), we must consider the objects separately.

5-3 MASS AND WEIGHT

In this section we continue the discussion of the concept of mass and examine some of its important properties, particularly in the presence of the Earth's gravity.

The Gravitational Force.   Another of the important contributions of Isaac Newton was his formulation of the theory of gravitation. Newton postulated that every pair of particles exerts on one another a mutual attractive gravitational force that is directly proportional to the product of their masses and inversely proportional to the square of the distance between them. The magnitude of the gravitational force between two particles with masses m and M, separated by a distance r, can be expressed as

$$F_g = G\,\frac{mM}{r^2} \tag{5-6}$$

where G is the *universal gravitation constant* and has the value $6.673 \times 10^{-11}\,\text{N}\cdot\text{m}^2/\text{kg}^2$. In Section 14-2 Newton's formulation of the theory of gravitation is summarized and the force law is given in full vector notation (see Eq. 14-1). It is also shown, following Newton, that for solid spherical objects* the distance r is the distance between the *centers* of the objects.

▶ *The Determination of* G. The value of the gravitation constant G can be determined by measuring the force between a pair of objects with known masses that are separated by a known distance. The first experiment from which a reasonably accurate value for G could be obtained was carried out by an English chemist, Henry Cavendish (1731–1810), who reported his results to the Royal Society in 1798. For his measurement, Cavendish used an instrument (called a *torsion balance*) that was similar to the one invented independently by Charles Coulomb and used by him in studies of electric and magnetic forces. The type of torsion balance used by Cavendish (see Fig. 5–7) consists of a thin rod that carries two small balls with

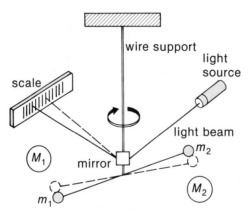

Fig. 5–7. Schematic diagram of the apparatus used by Henry Cavendish to determine the value of G.

masses m_1 and m_2. The rod is attached to the lower end of a wire that is suspended from a fixed support. The apparatus is allowed to come to rest with the large balls, M_1 and M_2, some distance away. This equilibrium position is measured by directing a light beam from a stationary source onto a mirror that is attached to the suspension wire; the point at which the reflected beam is seen on the scale is recorded.

---

*Precisely, for spherical objects whose density depends at most on the radial distance from the center.

Next, the large balls are moved close to the small balls; this corresponds to the situation shown in Fig. 5–7. Because M_1 attracts m_1 and M_2 attracts m_2, a torque* is exerted on the rod. The rod rotates through a small angle until the restoring torque due to the torsional force in the twisted wire just balances the gravitational torque. The angular displacement of the rod is determined by noting the new position of the light spot on the scale. The force necessary to twist the wire by a certain amount (the so-called torsional constant) is related to the period of oscillation of the system, which can be measured in a separate experiment. In this way Cavendish measured the gravitational force between known masses separated by a known distance. The value of G obtained from Cavendish's data is only 1 percent different from the value now accepted. ◀

Weighing the Earth.   The gravitational force exerted by the Earth (mass M_E) on an object (mass m) located on or near the Earth's surface is the *attractive* force,

$$F_g = G\frac{mM_E}{R_E^2}$$

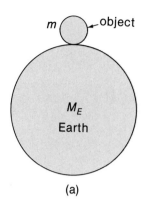

(a)

where R_E is the radius of the Earth. Now, in Section 2–5 we learned that near the Earth's surface any unsupported body falls downward with the constant acceleration g (assuming negligible air resistance). According to Newton's second law, accelerated motion requires the presence of a force; in this case, the force is the gravitational force,† and we write the second law as‡

$$F_g = mg$$

If we equate these two expressions for the gravitational force on the object, we see that the mass m can be eliminated; then, solving for M_E, we find

$$M_E = \frac{gR_E^2}{G} \tag{5-7}$$

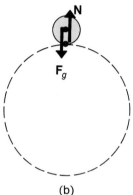

(b)

which expresses the mass M_E of the Earth in terms of the measurable quantities g, R_E, and G. Substituting the values of these quantities, we find $M_E = 5.98 \times 10^{24}$ kg. Thus, the Cavendish experiment, in which the value of G was determined, is sometimes called "weighing the Earth."

Force Pairs.   Consider an object with a mass m at rest on the Earth's surface. Figure 5–8 shows the situation, exaggerating the size of the object for illustrative purposes. In this condition there is no motion at all of the object relative to the Earth. As indicated in Fig. 5–8b, the object is acted upon by two forces—the Earth's attractive gravitational force $\mathbf{F}_g$ and the contact force $\mathbf{N}$ with the ground.§ Because the object is at rest with respect to the Earth, $\mathbf{F}_g + \mathbf{N} = 0$, so that

$$\mathbf{N} = -\mathbf{F}_g \tag{5-8}$$

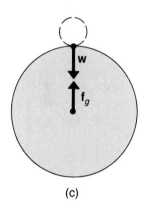

(c)

Fig. 5–8. (a) An object at rest on the Earth's surface. (b) The forces acting on the *object*. (c) The forces acting on the *Earth*. The forces shown together in (b) and (c) are *not* action-reaction pairs.

---

*The idea of torque is discussed in Chapter 10. A torque applied to an object tends to make the object rotate.

†The value of g varies slightly with geographic location and altitude. This variation is ascribed to the dependence of F_g on geometric features of the Earth and to kinematic effects due to the rotation of the Earth, *not* to any changes in m.

‡In writing this expression, we make the assumption that the mass m that appears in the gravitational force equation (Eq. 5–6) is the same as the inertial mass m that appears in Newton's second law. This point is discussed further on page 128.

§The contact force $\mathbf{N}$ is often referred to as the *normal* force because it is normal (i.e., perpendicular) to the surface of contact.

The Earth is also acted upon by two forces—the contact force **w** with the object and the attractive gravitational force $\mathbf{f}_g$ exerted by the object (Fig. 5–8c). Again, because there is no relative motion, $\mathbf{w} + \mathbf{f}_g = 0$, so that

$$\mathbf{w} = -\mathbf{f}_g \qquad (5\text{–}9)$$

Notice carefully that the forces appearing in Eq. 5–8 and those in Eq. 5–9 are *not* action-reaction pairs. The force pairs are

$$\mathbf{F}_g = -\mathbf{f}_g \qquad \text{and} \qquad \mathbf{N} = -\mathbf{w}$$

Comparing these equations with Eqs. 5–8 and 5–9, we conclude that all four forces—$\mathbf{F}_g$, $\mathbf{f}_g$, $\mathbf{N}$, and $\mathbf{w}$—have the same magnitude.

▶ *Inertial and Gravitational Mass.* In Section 5–2 we noted that the quantity m, the proportionality factor connecting the force applied to an object and its acceleration, is called the *inertial mass* of the object. The value of m determines the magnitude of the applied force necessary to produce a particular acceleration. An experiment designed to measure the inertial mass of an object could be carried out in deep space using a spring force. As shown in Fig. 5–9a, the spring force $\mathbf{F}_s$ applied to the object produces an acceleration $\mathbf{a}$. The gravitational force does not enter into such an experiment at all.

(a)

Fig. 5–9. (a) A deep-space experiment in which a spring force $\mathbf{F}_s$ applied to an object produces an acceleration $\mathbf{a}$. (b) The same object and spring scale are suspended near the Earth's surface. If the spring elongation is the same as in (a), the gravitational force $\mathbf{F}_g$ has the same magnitude as the spring force $\mathbf{F}_s$ in (a). If the object in (b) is now released, it will have an acceleration g equal to a only if the gravitational mass m' of the object is equal to its inertial mass m.

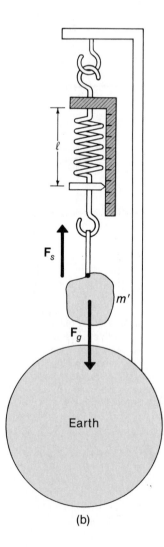

(b)

Next, imagine performing an experiment with the same spring scale and the same object at a point near the Earth's surface. The spring scale is attached to a post that is anchored to the Earth and the object is suspended from the lower end, as shown schematically in Fig. 5–9b. When the object hangs motionless, the downward gravitational force $\mathbf{F}_g$ is just balanced by the upward spring force $\mathbf{F}_s$. Let us suppose that we cleverly chose the conditions in the first case (Fig. 5–9a) so that the spring elongation is exactly the same in the two experiments. That is, the magnitude of the accelerating force in Fig. 5–9a is equal to the gravitational force in Fig. 5–9b.

Now, in the second experiment (Fig. 5–9b), there is no acceleration; hence, dynamical effects are not involved. The magnitude of the gravitational force depends on some property of the object, and we call this property the *gravitational mass* m'. This is the mass that appears in Newton's universal law of gravitation (Eq. 5–6).

We now ask a profoundly important question: Is the gravitational mass of an object equal to its inertial mass? We can answer this question by performing a third experiment. In Fig. 5–9b, we cut the string that attaches the object to the spring scale and measure the acceleration of the object as it falls toward the Earth. During the fall, the force acting on the object (namely, $\mathbf{F}_g$) has exactly the same magnitude as the spring force $\mathbf{F}_s$ that acted in the first experiment. Therefore, if indeed we have $m = m'$, the acceleration due to gravity g must exactly equal the acceleration a measured in the first experiment. If the accelerations are not equal, we would conclude that gravitational mass and inertial mass are not the same.

Clearly, the experiments we have been discussing would be very difficult to perform. However, there are equivalent experiments that *can* be performed with considerable precision to answer the same question. Newton was aware of this problem, and he conducted experiments with pendula to determine whether $m = m'$. He concluded that there was no difference between the two types of mass to within 1 part in 1000. In 1890, Eötvös* devised an ingenious method to test the equivalence of inertial and gravitational mass. Using two objects made from different materials, he compared the effect of the Earth's gravitational force with the inertial effect of the Earth's rotation. Eötvös was able to conclude that $m = m'$ to within about 5 parts in 10^9. In recent years, Robert Dicke of Princeton University and V. Braginskii at Moscow University have improved upon the Eötvös experiment and have established the equivalence of gravitational and inertial mass to within 1 part in 10^{12}.

There is nothing in Newtonian theory or in our experience that would lead us to expect gravitational and inertial mass to be the same. This is a remarkable result for which no one has a completely satisfactory explanation. However, the hypothesis of the *exact* equality of gravitational and inertial mass is an essential ingredient in the general theory of relativity. This important idea is called the *principle of equivalence*. ◄

Weight. Let us return to the discussion of the forces **w** and **N** that appear in Eqs. 5-8 and 5-9. The contact force **w** that an object exerts on whatever is supporting it is called the *weight* of the object. In the case illustrated in Fig. 5-8, the weight **w** of the object is the force it exerts on the Earth. Because there is no acceleration of the object in this instance, **w** is equal to the gravitational force $\mathbf{F}_g$ acting on the object. We stress, however, that $\mathbf{F}_g$ acts *on the object*, whereas **w** acts *on the surface of the Earth*. The magnitude of **w** is $w = mg$ (for this case of zero acceleration).

A different situation is shown in Fig. 5-10, where the mass m rests on the floor of an elevator that can be made to accelerate either up or down. Imagine that the elevator is accelerating upward with constant acceleration **a**. Further, let us imagine that, compared to the gravitational force of the Earth, the gravitational attraction between the object and the elevator is negligible. Then, focusing attention on m, and taking the upward direction to be positive, we have

$$N - F_g = ma$$

or

$$N = F_g + ma = mg + ma = m(g + a)$$

Here, **N** is the force the elevator floor exerts on the object and $\mathbf{F}_g$ is the gravitational force the Earth exerts on the object. The reaction force to **N** is **w**, the weight of the object, because this is the force the object exerts on the supporting elevator floor. These two forces have the same magnitude, so the weight of the object is

$$w = m(g + a) \qquad \textbf{(5-10a)}$$

The weight is therefore *increased* by the amount ma over the weight at rest. Note that the gravitational force on the object is still $F_g = mg$.

If the elevator accelerates downward with an acceleration a', we would have

$$w = m(g - a') \qquad \textbf{(5-10b)}$$

The weight is therefore *reduced* by the amount ma' from the weight at rest. Again, the gravitational force on the object remains $F_g = mg$.

If the cable that supports the elevator were to break, the elevator would be in free fall and would accelerate downward with $a' = g$. According to Eq. 5-10b, we

Fig. 5-10. (a) An object with a mass m is on the floor of an elevator that is accelerating upward. (b) Forces acting on the object. (c) Forces acting on the elevator.

---

*Roland von Eötvös (1848-1919), a Hungarian baron, who invented many scientific instruments, including a sensitive gravimeter used to determine the density of subsurface rock strata.

would have

$$w = 0 \qquad \text{(for } a' = g\text{)}$$

In this situation, we say that the object is *weightless,* although the gravitational force on the object remains equal to mg. An object is weightless if it does not exert a force on any supporting platform or suspending rope.*

Consider an artificial satellite that moves around the Earth in a circular orbit with a radius r (Fig. 5–11a). The satellite encounters no resistance, and even though

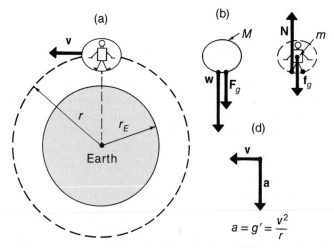

$$a = g' = \frac{v^2}{r}$$

Fig. 5–11. An astronaut in an orbiting artificial satellite is *weightless.*

the path is circular, the satellite undergoes free fall. At the position of the orbit, the acceleration due to gravity is g', which is equal to the centripetal acceleration of the satellite, v^2/r. Now, suppose that there is an astronaut in the satellite. The total force F_s acting on the satellite is the gravitational force F_g plus the weight w of the astronaut (Fig. 5–11b); this force is equal to the mass M of the satellite multiplied by its acceleration:

$$F_s = F_g + w = M\frac{v^2}{r} = Mg'$$

The total force f_a acting on the astronaut is the gravitational force f_g plus the (upward) contact force N (Fig. 5–11c); this force is equal to the mass m of the astronaut multiplied by his acceleration:

$$f_a = f_g - N = m\frac{v^2}{r} = mg'$$

Now, according to the usual definition, we have $F_g = Mg'$ and $f_g = mg'$. Substituting these expressions into the equations for F_s and f_a, we conclude that $w = 0$ and $N = 0$. Thus, the astronaut is weightless. This is the result we would expect: both the satellite and the astronaut are undergoing free fall, so the astronaut can exert no net

---

*In most other texts you will find *weight* defined as the gravitational force acting on an object: $\mathbf{w} = m\mathbf{g}$. Then, in the case of acceleration, mg is called the *true weight* of the object and the force exerted by the object on any support is called the *apparent weight.* This definition seems unnecessarily complicated (and it is almost never used in a consistent manner); moreover, it is also contrary to the usual sense of weight as the force exerted *by* the object.

force on the satellite. (The astronaut could push on the "ceiling" and the "floor" at the same time, but this would not result in the application of any net force and would not constitute *weight*.)

The definition of weight we have given here can be applied with a consistent interpretation in all situations. Even in complex circumstances involving accelerations **a** not in the direction of **g**—such as a block accelerating down an inclined plane or a person revolving in a fair-ground ride or an astronaut experiencing the "artificial gravity" of a rotating space station—the weight of the object or person is always equal to the (vector) force exerted *on* the supporting object or system. (See Problem 5-27.)

▶ *The Pound as a Unit of Force.* For many years engineers used almost exclusively the British system of units in which the *pound* is a unit of *force*. In this system, 1 pound (lb) is taken to be the weight of an unaccelerated object with a mass of exactly 0.45359237 kg at a location where the acceleration due to gravity has its "standard" value, $g = 9.80665 \text{ m/s}^2 = 32.1740 \text{ ft/s}^2$ (see Section 2-5). Thus, the relationship connecting the British and metric units of force is

$$w = mg$$
$$1 \text{ lb} = (0.45359237 \text{ kg})(9.80665 \text{ m/s}^2)$$
$$= 4.44822 \text{ N}$$

In the British system, the unit of mass, ingloriously called the *slug,* is defined as that mass which, when subjected to a net force of 1 lb, will undergo an acceleration of 1 ft/s².

A shopper in a grocery store must contend with an untidy and incorrect usage of the units of mass and weight regardless of the system that prevails. According to the *legal* definition, the pound is a unit of mass: 1 lb = 0.45359237 kg (exactly). However, in a store that operates under the British system, the shopper will find weight measured in *pounds;* if the metric system is used, she will find weight measured in *kilograms!*

We include mention of these points for reference when consulting older texts. In this book, we use only kilograms for mass and newtons for force. ◀

Example 5-5

The spaceship shown in the diagram is traveling in deep space and is undergoing an acceleration, relative to an inertial reference frame, which happens to be 9.80 m/s². What is the weight of a 2-kg object that rests on the "floor" of the spaceship cabin?

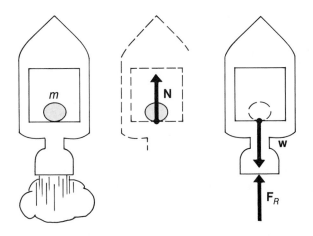

Solution:

According to an inertial-frame observer, the object ($m = 2 \text{ kg}$) is accelerating (along with the spaceship) at a rate g. Hence, the contact force N with the floor (which is the *only* force acting on the object) must be

$$\mathbf{N} = m\mathbf{a}$$

Because $a = g$, we have $N = mg$. Also, the weight w is the reaction force to N; therefore,

$$w = mg = (2 \text{ kg})(9.8 \text{ m/s}^2) = 19.6 \text{ N}$$

It would thus appear that, to an observer in the spaceship, the mass m rests on the floor with its natural (Earth-environment) weight. This result is consistent with the principle of equivalence. (•Suppose that an observer who is confined to the spaceship interior—with no windows—attempts to discover whether the ship is accelerating through space or is simply sitting on the Earth's surface. Can the observer, by performing experiments within the spaceship, distinguish between these two conditions? See Problem 5–14.)

5-4   SOME COMMENTS ON NEWTON'S LAWS

In setting down his three laws of dynamics, Newton relied heavily on intuitive ideas concerning the terms used in the statements of the laws, terms otherwise not defined in any rigorous way. For example, the second law gives the relationship between force and mass, but does not tell us what either force or mass really is. We might even paraphrase the statement of the second law to read: "An object will not accelerate until something makes it accelerate, and that something is defined as a force." Indeed, in one approach to Newton's laws, force is defined in terms of the acceleration it produces. Although this procedure has some merit, we have elected first to gain some intuitive appreciation for the concept of force by considering several simple experiments with springs. Additional experiments concerning the effect of force on the motion of objects then led us to the second law and the concept of mass. In this way we avoid the logical difficulties that confound us when Newton's laws are considered alone.

We have asserted that Newton's laws are valid in any inertial (nonaccelerating) reference frame. It is not difficult to see why this is so. In Section 4–6 we showed that two observers, S and S', in relative motion with a constant velocity will measure the same acceleration for a particle, $\mathbf{a} = \mathbf{a}'$. That is, acceleration is an invariant quantity, a free vector. If the frame of observer S is an inertial reference frame, he concludes that a force $\mathbf{F} = m\mathbf{a}$ is acting on the particle. Then, the frame of observer S' is also an inertial reference frame, and he concludes that a force $\mathbf{F}' = m\mathbf{a}'$ is acting on the particle. But $\mathbf{a} = \mathbf{a}'$, and we have

$$\mathbf{F} = m\mathbf{a} = m\mathbf{a}' = \mathbf{F}'$$

Thus, $\mathbf{F} = \mathbf{F}'$, and the second law has exactly the same form in each frame. We conclude that if $\mathbf{F} = m\mathbf{a}$ is valid in a frame S, it is also valid in any other frame S' that, at most, moves with a constant velocity and without rotation relative to S. Moreover, $\mathbf{F} = m\mathbf{a}$ is the invariant form of the second law in all inertial frames. *Force $\mathbf{F}$ and acceleration $\mathbf{a}$ are free vectors and have the same magnitude and direction regardless of the particular inertial reference frame used to describe them.*

Finally, we ask whether experiment shows the third law to be true in every situation. The law is readily verified for ordinary contact forces; however, problems may arise with action-at-a-distance forces when relative motion is involved. In these cases it is convenient to introduce the concept of a *mediating agency* or *field*.

We may simulate the effect of a field by considering the following case. Suppose in Example 5–4 that block A is *invisible* and has a mass $m_A \ll m_B$, as indicated in Fig. 5–12. In such a case, an observer would be inclined to ascribe the acceleration of

Fig. 5–12. The invisible block A plays the role of a field in transmitting the force from the hand H to the block B.

block B to an action-at-a-distance force between the hand H and the block B. However, the action-reaction forces that would be associated with a third-law relationship—namely, the force exerted by the hand and the reaction force exerted by the block B—are not quite equal in magnitude. (Compare the forces F_{AH} and F_{BA} in Example 5–4 when m_A becomes small.) The inequality is due to the inertial effect of the "mediating agency," the block A. The observer would conclude that the third law is not valid in this case, although Eq. 5–5 holds exactly for the action-reaction force pairs at the actual contact points, HA and AB, shown in Fig. 5–12.

When we are dealing with the genuine action-at-a-distance forces—namely, the gravitational and electromagnetic forces—we must be careful to reckon with the inertial effects of the fields. If one particle interacts with another particle through the intermediary of a field, an exact third-law relationship holds between each particle and the field. However, if the field inertia is ignored and each particle is assumed to act directly on the other, this can lead to an apparent violation of the third law. This effect will be manifest when accelerations are present and is especially important in many electromagnetic phenomena. (In the gravitational case the effect is so small that it can generally be ignored.)

We can see the effect of the field inertia by considering a situation in which a sudden change occurs. Suppose that the hand in Fig. 5–12 abruptly increases its force on block A (the "field"). The effect of such a change will be transmitted to block B, but this block will not respond immediately. Because the sudden push of the hand produces a series of elastic forces that act among the microscopic constituents of block A, a certain time interval is required for the impulse to propagate through the block. During this interval no valid third-law relationship exists between the force exerted by the hand and the reaction force exerted by block B. However, the equations $\mathbf{F}_{AH} = -\mathbf{F}_{HA}$ and $\mathbf{F}_{BA} = -\mathbf{F}_{AB}$ still hold exactly at the respective contact points. All mechanical impulses propagate with a finite speed that depends on the elastic properties of the medium. The electromagnetic and gravitational forces likewise propagate with a finite speed, the speed of light. Special care must therefore be exercised in dealing with the third law when fields are involved.

5-5   GENERAL ROTATION AND NEWTON'S LAWS IN ROTATING FRAMES

In Section 4–5 we considered in some detail the circular motion of a particle. We now extend this discussion to the most general type of motion in a plane by allowing both azimuthal and radial variations of position. This investigation leads directly to the question, "How does an observer stationed in a rotating (and, therefore, *non-inertial*) reference frame use Newton's laws to describe the motions of objects?" The answer to this question involves the introduction of new terms into Newton's equation.

General Motion in a Plane.   We begin by writing the position vector for a particle in polar coordinates in a suitable inertial frame (Fig. 5–13) as $\mathbf{r} = \rho\hat{\mathbf{u}}_\rho$. In Section

Fig. 5–13. Polar coordinates of a particle at a point P. Notice that the position vector $\mathbf{r}$ does not involve $\hat{\mathbf{u}}_\phi$.

4–5 we restricted attention to the case $\rho = $ constant; now, we allow ρ to vary with time. Thus, the velocity of the particle is expressed as

$$\mathbf{v} = \frac{d\mathbf{r}}{dt} = \frac{d}{dt}(\rho\hat{\mathbf{u}}_\rho) = \frac{d\rho}{dt}\hat{\mathbf{u}}_\rho + \rho\frac{d\hat{\mathbf{u}}_\rho}{dt}$$

The unit vectors, $\hat{\mathbf{u}}_\rho$ and $\hat{\mathbf{u}}_\phi$, move with the point P and therefore vary with time. From Eqs. 4–27 and 4–31, we have

$$\frac{d\hat{\mathbf{u}}_\rho}{dt} = \hat{\mathbf{u}}_\phi\frac{d\phi}{dt} = \omega\hat{\mathbf{u}}_\phi \qquad \textbf{(5–11a)}$$

$$\frac{d\hat{\mathbf{u}}_\phi}{dt} = -\hat{\mathbf{u}}_\rho\frac{d\phi}{dt} = -\omega\hat{\mathbf{u}}_\rho \qquad \textbf{(5–11b)}$$

where the angular speed is $\omega = d\phi/dt$. Then, we have

$$\mathbf{v} = \frac{d\rho}{dt}\hat{\mathbf{u}}_\rho + \rho\omega\hat{\mathbf{u}}_\phi = v_\rho\hat{\mathbf{u}}_\rho + v_\phi\hat{\mathbf{u}}_\phi \qquad \textbf{(5–12)}$$

Thus, the velocity vector $\mathbf{v}$ is separated into a *radial* component v_ρ and an *azimuthal* component v_ϕ, where

$$v_\rho = \frac{d\rho}{dt} \quad \text{and} \quad v_\phi = \rho\omega \qquad \textbf{(5–13)}$$

In Section 4–5 we considered only circular motion ($\rho = $ constant) and therefore had no counterpart of the radial velocity component v_ρ. The velocity $\mathbf{v}$ is always tangent to the path of the particle, so for the case of circular motion the azimuthal velocity v_ϕ is the same as the *tangential* velocity v_t. For the present case of general planar motion, the particle can move radially, so the path is not always in the direction of $\hat{\mathbf{u}}_\phi$. Then, the azimuthal velocity v_ϕ is no longer the same as the tangential velocity. In fact, $v_t = \sqrt{v_\rho^2 + v_\phi^2}$. (•Can you see why?)

The acceleration of the particle is

$$\mathbf{a} = \frac{d\mathbf{v}}{dt} = \frac{d}{dt}(v_\rho\hat{\mathbf{u}}_\rho + \rho\omega\hat{\mathbf{u}}_\phi)$$

The first term in this expression is

$$\frac{d}{dt}(v_\rho\hat{\mathbf{u}}_\rho) = \frac{dv_\rho}{dt}\hat{\mathbf{u}}_\rho + v_\rho\frac{d\hat{\mathbf{u}}_\rho}{dt}$$

$$= \frac{dv_\rho}{dt}\hat{\mathbf{u}}_\rho + v_\rho\omega\hat{\mathbf{u}}_\phi$$

Also,

$$\frac{d}{dt}(\rho\omega\hat{\mathbf{u}}_\phi) = \frac{d\rho}{dt}\omega\hat{\mathbf{u}}_\phi + \rho\frac{d\omega}{dt}\hat{\mathbf{u}}_\phi + \rho\omega\frac{d\hat{\mathbf{u}}_\phi}{dt}$$

$$= v_\rho\omega\hat{\mathbf{u}}_\phi + \rho\alpha\hat{\mathbf{u}}_\phi - \rho\omega^2\hat{\mathbf{u}}_\rho$$

where we have used Eqs. 5–11 and $\alpha = d\omega/dt$. Altogether, we have

$$\mathbf{a} = \left(\frac{dv_\rho}{dt} - \rho\omega^2\right)\hat{\mathbf{u}}_\rho + (\rho\alpha + 2v_\rho\omega)\hat{\mathbf{u}}_\phi \qquad \textbf{(5–14)}$$

Thus, the acceleration vector $\mathbf{a}$ is separated into a radial component a_ρ and an azimuthal component a_ϕ, where

$$a_\rho = \frac{dv_\rho}{dt} - \rho\omega^2 \quad \text{and} \quad a_\phi = \rho\alpha + 2v_\rho\omega \qquad \textbf{(5–15)}$$

The radial acceleration contains the usual centripetal acceleration term, $-\rho\omega^2$, together with the term dv_ρ/dt, not present for $\rho = $ constant, that arises from the radial motion. The azimuthal acceleration contains the term $\rho\alpha$, which is the same as the tangential acceleration for circular motion. In addition, there is a new term, $2v_\rho\omega$, which is called the *Coriolis acceleration*. Note that the expressions for a_ρ and a_ϕ reduce to Eqs. 4–32 when $\rho = R = $ constant.

Forces in a Rotating Reference Frame. Consider pure radial motion with respect to a plane that rotates with a constant angular speed ω about a fixed axis. This would be the situation, for example, if you were to walk along a radial line marked on a moving merry-go-round platform (Fig. 5–14). To the stationary (inertial) observer O, you are carried along by the platform because there is a horizontal force $\mathbf{F}$ exerted by the platform on your shoes (Fig. 5–14a). (This is a frictional force and could be measured by suitable strain gauges attached to the soles of your shoes.) This force can be expressed in terms of radial and azimuthal components by multiplying the general expression for the acceleration (Eq. 5–14, but with $\alpha = 0$) by your mass m; thus,

$$\mathbf{F} = m\left(\frac{dv_\rho}{dt} - \rho\omega^2\right)\hat{\mathbf{u}}_\rho + 2mv_\rho\omega\hat{\mathbf{u}}_\phi \qquad \textbf{(5–16)}$$

Another observer O' stationed at the center of the merry-go-round and rotating with it would see only your straight-line radial motion (Fig. 5–14b). If this observer has an unshakable faith in the correctness of Newton's laws in his reference frame, he will consider mdv_ρ/dt to be the important dynamical quantity and will write the second law as

$$m\frac{dv_\rho}{dt}\hat{\mathbf{u}}_\rho = \mathbf{F}'$$

Comparing this expression for $\mathbf{F}'$ with Eq. 5–16, we see that

$$\mathbf{F}' = \mathbf{F} + m\rho\omega^2\hat{\mathbf{u}}_\rho - 2mv_\rho\omega\hat{\mathbf{u}}_\phi$$
$$= \mathbf{F} + \mathbf{F}_\rho + \mathbf{F}_\phi \qquad \textbf{(5–17)}$$

Thus, in addition to the real force $\mathbf{F}$, the rotating observer must invoke the *pseudoforces*, $\mathbf{F}_\rho$ and $\mathbf{F}_\phi$, in order to account for the readings of the strain gauges on your shoes. These "forces" are due solely to the various accelerations

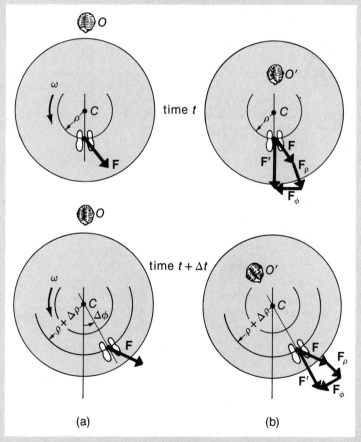

Fig. 5–14. (a) An inertial observer O views you walking along a radial line on a rotating merry-go-round platform. The actual external force that acts on your shoes is $\mathbf{F}$ (a frictional force). (b) According to observer O' who rotates with the platform, the motion is due to the radial force $\mathbf{F}'$ which is the sum of $\mathbf{F}$ (the only real force) and the pseudoforces $\mathbf{F}_\rho$ and $\mathbf{F}_\phi$, the centrifugal and Coriolis forces, respectively.

that are present in the rotating (noninertial) frame. The term $\mathbf{F}_\rho = m\rho\omega^2\hat{\mathbf{u}}_\rho$ is known as the *centrifugal* (outward directed) *force*, and the term $\mathbf{F}_\phi = -2mv_\rho\omega\hat{\mathbf{u}}_\phi$ is called the *Coriolis force*. Notice that it is *always correct* (and probably less confusing) to describe such situations by adopting the point of view of an inertial observer and using Eq. 5–16, properly accounting for the various accelerations.

The effect of the Coriolis force is also apparent for large-scale motions on or near the rotating Earth. For a complete description of such motions, we require the three-dimensional version of Eq. 5–16 or Eq. 5–17. However, consider the simple case of a projectile that is fired from the North Pole and moves along a ballistic trajectory. An observer in an inertial reference frame would see the projectile follow a curving path due to the gravitational attraction of the Earth but not deviating to the right or left (Fig. 5–15). The motion would be exactly described by Newton's laws for a single force, namely, the gravitational force exerted on the projectile by the Earth. This observer would also see the Earth rotating from west to east under the path of the projectile, but this rotation would not influence the motion of the projectile.

How does an Earth-bound observer view the situation? In his reference frame, the Earth observer sees the projectile deviating toward the west as it moves south. That is, because the Earth rotates under the path of the projectile, the observer sees it curve toward the right in the

Northern Hemisphere. (•What effect is observed in the Southern Hemisphere?) If the Earth observer insists that his reference frame is an inertial frame, he must invoke the Coriolis force to explain the motion of the projectile.

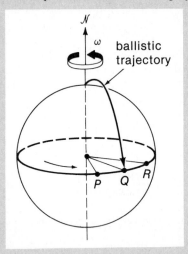

Fig. 5–15. An inertial observer's view of a projectile fired from the North Pole. The projectile is fired directly toward point Q, but during the flight of the projectile, the rotation of the Earth carries Q to R and P to Q. Consequently, the projectile strikes the Earth at P. To an Earth-observer, the projectile appears to have been deflected toward the west.

Example 5–6

In virtually all surveying methods, the "vertical" direction is determined by means of a *plumb line*, which consists of a stationary plumb bob that hangs from a string, the upper end of which is held fixed with respect to the Earth. What is the true orientation of the plumb line with respect to the geometric vertical (i.e., the radial line directed toward the center of the Earth)?

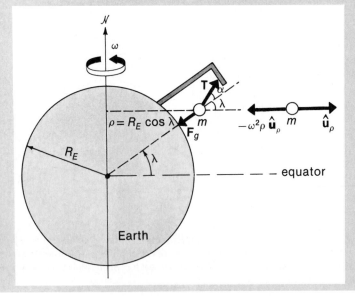

Solution:

As shown in the (exaggerated) diagram, the plumb line makes an angle α with the geometric vertical. The only "true" forces acting on the plumb bob m are the tension $\mathbf{T}$ in the string and the gravitational force $\mathbf{F}_g = m\mathbf{g}$ directed along the geometric vertical. The string lies in the plane that contains the Earth's polar axis and the support point of the string. The position of the bob is essentially at the Earth's surface.

The radial velocity of the bob is zero, so the only surviving term in Eq. 5-16 involves the centripetal acceleration. Then,

$$\mathbf{F} = \mathbf{T} + \mathbf{F}_g = -m\rho\omega^2\hat{\mathbf{u}}_\rho$$

where $\hat{\mathbf{u}}_\rho$ is in the plane of rotation of the bob, as shown in the diagram. Taking components perpendicular and parallel to $\hat{\mathbf{u}}_\rho$, we obtain

$$T\sin(\lambda + \alpha) - mg\sin\lambda = 0$$

$$T\cos(\lambda + \alpha) - mg\cos\lambda = -m\rho\omega^2 = -mR_E\omega^2\cos\lambda$$

Dividing these two equations and using the trigonometric identity for $\tan(\lambda + \alpha)$ (Eq. A-24 in the Appendix), we obtain

$$\tan(\lambda + \alpha) = \frac{\tan\lambda + \tan\alpha}{1 - \tan\lambda\tan\alpha} = \frac{g\tan\lambda}{g - R_E\omega^2}$$

Solving for $\tan\alpha$, and using the identity $1 + \tan^2\lambda = 1/\cos^2\lambda$, we find

$$\tan\alpha = \frac{R_E\omega^2\tan\lambda}{g(1 + \tan^2\lambda) - R_E\omega^2} = \frac{R_E\omega^2\sin\lambda\cos\lambda}{g - R_E\omega^2\cos^2\lambda}$$

Now, the rotation rate of the Earth is $\omega = 7.27 \times 10^{-5}$ rad/s, so $R_E\omega^2 = 3.4 \times 10^{-2}$ m/s². (•Can you verify these figures?) Because $g = 9.80$ m/s², it is evident that α is a very small angle, so we can write (using $\tan\alpha \cong \alpha$)

$$\alpha \cong \frac{R_E\omega^2\sin\lambda\cos\lambda}{g} \cong (3.4 \times 10^{-3})\sin\lambda\cos\lambda$$

For $\lambda = 45°$, the angle of deviation is 1.7×10^{-3} rad or about 1/10 of a degree.

From the result of Example 5-6, we see that the Earth's surface, in most practical cases, is a reasonably accurate inertial frame when laboratory-size dimensions are involved. However, even though the noninertial accelerations are small, if they are allowed to act over sufficiently long distances and times, important effects can result. The calculation of the trajectories of long-range ballistic projectiles, for example, must take these accelerations into account to avoid serious targeting errors.

PROBLEMS

Section 5-2

5-1 A 2-kg object originally judged to be at rest in an inertial frame is subjected to a constant force of 3 N for a period of 10 s. What final speed does it reach? What displacement does it undergo during the first 10 s? During the first 20 s?

5-2 The velocity of a particle with a mass of 4 kg that moves along the x-axis is determined at various instants. The results are: $v(t = 1\text{ s}) = 5$ m/s, $v(2) = 8$ m/s, $v(3) = 11$ m/s, and $v(4) = 14$ m/s. Write down the expression for $v(t)$ and give the acceleration of the particle. What force acts on the particle?

5-3 A 0.50-kg particle has an acceleration $\mathbf{a} = \alpha\hat{\mathbf{i}} + \beta\hat{\mathbf{j}}$, with $\alpha = 4$ m/s² and $\beta = -2$ m/s². What is the magnitude of the force that acts on the particle?

5-4 A rifle bullet with a mass of 12 g, traveling with a speed of 400 m/s, strikes a large wooden block, which it penetrates to a depth of 15 cm. Determine the magnitude of the frictional force (assumed constant) that acts on the bullet.

5-5 An object with a mass of 500 g is held by a string in a horizontal circular orbit with a diameter of 1 m. If the maximum tension the string can support before breaking is 100 N, with what linear speed will the object leave its orbit if it is spun ever faster until the string breaks?

5-6 What constant force is required to accelerate a 2000-kg automobile from rest to 90 km/h in 12.2 s?

5-7 What constant force must be applied to a 5000-kg space vehicle (in deep space) to accelerate it from rest to 10,000 km/h during a time interval of 1 y?

5-8 An automobile ($m = 2200$ kg) moves with a velocity $v = 32$ m/s on a level road. If a constant braking force of 6000 N is applied to the automobile wheels, what distance will the vehicle travel after the brakes are applied?

5-9   A particle with a mass $m = 4$ kg is known to be acted upon by two forces F_1 and F_2, shown in the diagram. The particle moves with the constant velocity shown. What third force must be acting on the particle?

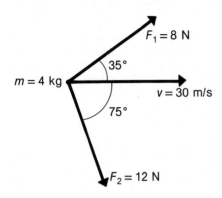

5-10 A freight train that consists of 100 equal-mass box
•   cars is pulled along a straight stretch of horizontal track by a diesel engine delivering a constant force of 10^6 N to the coupling of the leading car. Neglecting any frictional drag due to the tracks or air resistance, what is the tension in the front coupling of the leading car? In the front coupling of the last car? In the front coupling of the nth car, where n runs from 1 at the front to 100 at the end?

5-11 A rope with a mass $m = 0.8$ kg is attached to a block with a mass $M = 4$ kg. The rope-block combination is pulled across a horizontal frictionless surface by a 12-N force that is applied to the free end of the rope. What is the acceleration of the system? What force does the rope exert on the block?

5-12 A particle with a mass of 0.80 kg moves in a circular path with a radius of 2 m, completing each revolution in a time of 4 s. What is the magnitude of the force acting on the particle? If, at some instant, the velocity of the particle is $\mathbf{v} = v\hat{\mathbf{i}}$ and the motion is in the x-y plane what are the two possible vector expressions for the force? (Evaluate v.)

Section 5-3

5-13 An object with a mass of 2 kg is suspended from the ceiling of an elevator by a calibrated ideal (massless) spring scale, as shown in the diagram at right. (a) If the elevator is moving upward with a constant velocity v, what is the spring scale reading in newtons? (b) If the elevator is accelerating upward with a constant acceleration of 3 m/s², what is the spring scale reading? (c) If the elevator is accelerating downward with a constant acceleration of 2 m/s², what is the spring scale reading?

5-14 Suppose the experimenter in the accelerating spaceship of Example 5-5 throws a ball across the cabin from wall to wall (a distance of 5 m), giving the ball an initial velocity of 15 m/s parallel to the floor. If the spaceship is accelerating with $a = g$, how much closer to the floor is the impact point on the far wall than is the launching point? On the basis of your result, what answer would you give to the question asked at the end of Example 5-5? [*Hint:* Consider this problem from the point of view of an observer in an inertial frame. What motion does he ascribe to the ball? How long does it take for the ball to cross from wall to wall? How far did the laboratory itself move during this time?]

Problem 5-14

Problem 5-13

5-15 A vendor sets up shop in an elevator where he buys and sells radishes by the newton. (Formerly, he bought and sold by the pound.) The vendor buys a bundle of radishes when the elevator is accelerating downward at a rate of 3.5 m/s², and he sells the same bundle at the same price per newton when the elevator is accelerating upward at 2.5 m/s². What is the vendor's percentage profit?

SPECIAL TOPIC

Section 5-5

5-16 An observer rotates with a $33\frac{1}{3}$ rpm phonograph turntable while observing a bug with a mass $m = 0.5$ g crawling radially outward at a constant velocity of 5 cm/s, as shown in the diagram. When the bug is 15 cm from the center of the turntable, what are the horizontal forces acting on it? Use polar coordinates attached to the turntable. What actually exerts these forces on the bug?

5-17 A ballistic missile is fired from the North Pole and travels along a path that lies close to the Earth. What is the angular deviation for a 5000-km flight that lasts 30 min? By what distance will the missile miss the point at which it was directly aimed?

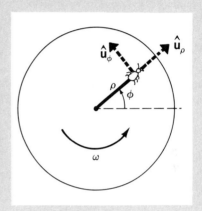

Additional Problems

5-18 An object with a mass of 3 kg is pulled upward by a string that has a breaking strength of 50 N. What is maximum acceleration the object can be given without breaking the string?

5-19 The largest caliber antiaircraft gun operated by the Luftwaffe during World War II was the *12.8-cm Flak 40.* This weapon fired a 25.8-kg shell with a muzzle velocity of 880 m/s. What propulsive force was necessary to attain the muzzle velocity within the 6.0-m barrel? (Assume constant acceleration and neglect the Earth's gravitational effect.)

5-20 A slingshot accelerates a 12-g stone to a velocity of 35 m/s within a distance of 5 cm. To what (constant) force is the stone subjected during the acceleration?

5-21 A flat disc with a mass of 500 g is sent sliding along a frozen pond with an initial speed of 10 m/s. If the sliding frictional force exerted by the ice surface on the disc is a constant 0.5 N, in what distance will the disc come to rest?

5-22 A block with a mass of 1 kg slides along a horizontal surface while constrained to a circular path of 50 cm radius by a string connected at the other end to a stationary frictionless pivot. Assume that the block experiences a frictional force with a magnitude of 2 N and in a direction always exactly opposed to the instantaneous velocity. If the block is given an initial tangential velocity of 5 m/s, find the string tension as a function of time thereafter.

5-23 A particle with a mass of 3 kg is at rest at $x = 3$ m, and then a force $F = (12$ N/m$)x$ is applied to it. What is the acceleration of the particle when it reaches $x = 5$ m? Determine $v(t)$ for the particle.

5-24 Two identical particle-like objects ($m = 15$ kg) are separated by a distance of 5 m in deep space. (a) What force does one object exert on the other? (b) What is the initial acceleration of either object? (c) Determine the time required for the separation of the objects to be reduced by 1 cm.

5-25 A particle ($m = 2.5$ kg) moves along the x-axis with
• a constant velocity ($v_x = 1.6$ m/s). In the region from $x = 5.0$ m to $x = 10.0$ m, the particle experiences a constant force $F_y = 8$ N that acts in the $+y$-direction. (a) If the particle was at $x = 0$ at $t = 0$, when will it escape the effects of the force? (b) What is the value of y when $x = 10.0$ m? (c) What is the velocity vector of the particle after it escapes the effects of the force?

5-26 What is the acceleration due to gravity at a height

above the Earth's surface equal to one Earth diameter?

5-27 (a) Refer to Fig. 5-11. Deduce numerical values
•   for all quantities shown if the mass of the capsule is 2×10^3 kg and the mass of the astronaut is 90 kg. Assume that the satellite is orbiting in a circular path 10^3 km above the Earth's surface, where $g' = 7.32$ m/s^2. (The Earth's radius is 6.38×10^3 km.) (b) Due to the brief firing of a retrorocket, the orbital speed of the satellite is reduced to 80 percent of the original value. Just after the firing, with the satellite still at its original distance above the Earth's surface, what is the weight of the astronaut?

5-28 A small bob with a mass of 250 g is suspended by a string from the ceiling of a truck that is accelerating at a constant rate of 2 m/s^2 as it moves along a flat, straight road, as shown in the diagram below. Assuming the bob to be suspended motionless with respect to the truck, what is the tension in the string? What angle does the string make with the vertical? (The force exerted on the bob by the string—the string tension—is directed along the string.)

5-29 A subway car travels at a speed of 30 km/h around an unbanked (flat) curve. The freely suspended hand straps are observed to swing out and hang at an angle of 10° from the vertical. What is the radius of the curve?

5-30 A grasshopper is observed to jump 2 m horizontally,
•   rising to a maximum height of 1 m along the way. During takeoff, the entire body mass of the grasshopper, considered to be concentrated at a single point, moves a distance of 1.5 cm while its hind feet are still in contact with the ground. What effective constant force, in terms of its weight, did the feet of the grasshopper exert on the ground? Could you do as well in terms of your body weight? [*Hint:* Use your knowledge of ballistic trajectories to determine the grasshopper launch speed v_0. This can be immediately related to the required (constant) acceleration

imparted by its legs. The grasshopper mass (a few grams) is not needed to make the calculation.]

5-31 Consider a platform that is rotating at a constant
•• angular speed $\omega = 2$ rad/s. The platform is located on the Earth's surface and its plane is perpendicular to **g**. A man ($m = 80$ kg) stands on and rotates with the platform. Make a diagram showing the forces that the man exerts on the platform when he stands a distance r from the axis of rotation. Identify the vector **w** that represents the man's weight. Calculate the weight w for $r = 1, 2, 4, 8, 12$ m. What direction does the man sense as "up"? Calculate the angle α that "up" makes with the vertical for the same values of r. Suppose that the platform is covered with soil and that seeds are planted uniformly throughout. If the platform continues to rotate while the plants grow, describe the appearance of the crop.

5-32 An object with a mass of 1 kg is at rest on the surface of the Earth at the Equator. What is the weight of the object? [*Hint:* Remember that the Earth rotates.]

5-33 A space station in the shape of a torus (a donut) and with an outer radius of 150 m revolves once every minute about its central axis. What is the weight of an 80-kg astronaut who is standing on the outer rim?

5-34 A pilot performing aerobatic maneuvers is in a vertical dive at a speed $v_0 = 600$ km/h. He pulls out of the dive along a circular arc while maintaining constant speed, eventually climbing vertically. (a) If the pilot is to experience an acceleration not to exceed $7g$, what is the minimum radius R of the circle? (b) How long after the start of the pullout at time t_0 will the aircraft be climbing vertically? (c) Construct a diagram of the forces acting on the aircraft when it has a heading 45° below the horizon.

APPLICATIONS OF NEWTON'S LAWS

In the preceding chapter we introduced and discussed Newton's laws of motion, and we presented a few examples to illustrate the principles involved. The range of application of Newtonian dynamics is surprisingly extensive. We now show the richness of these applications with further examples, some involving considerable detail. We emphasize that even complex problems can be solved by a systematic approach using Newton's laws. Of particular importance in this regard is the use of *free-body diagrams*.

6-1 FREE-BODY DIAGRAMS

In a number of instances in Chapter 5 we focused attention on a particular object among a collection of interacting objects. Then, we identified all of the forces acting on this singled-out object and, by applying Newton's laws, determined its motion. This procedure can be formalized by introducing the idea of the *free-body diagram*.

Consider the situation in Fig. 6–1a in which a block with a mass m is being pulled along a horizontal surface by a rope. At least five objects play a role in this example—the block itself, the rope, the hand, the horizontal surface, and the Earth. In Fig. 6–1b we isolate the block and show a free-body diagram with *all* of the forces (and *only* the forces) acting *on* the block. The state of motion of an isolated object—in this instance, the block—is completely determined by the sum of all the forces acting on it. A common error in analyzing situations of this type is the failure to distinguish which of the many forces present act *directly* on the isolated object and which do not. In this case, there is, first of all, the force exerted on the block by the rope, which we identify as the tension $\mathbf{T}$ in the rope. The gravitational force on the block due to the Earth is $\mathbf{F}_g$, with magnitude mg. The unknown resultant force $\mathbf{R}$ exerted on the block by the surface is conveniently represented by its two vector components, $\mathbf{N}$ (the component perpendicular to the surface) and $\mathbf{f}$ (the component parallel to the surface). The effect of $\mathbf{N}$ and $\mathbf{f}$ taken together is completely equivalent to the effect of $\mathbf{R}$ alone. (Resist the temptation to show $\mathbf{R}$ as well as $\mathbf{N}$ and $\mathbf{f}$ in the free-body diagram!) The component $\mathbf{f}$ is the *frictional force* acting on the block, and $\mathbf{N}$ is the *normal* (that is, the perpendicular) *constraining force*. (We consider these forces in detail later.) Finally, note that the hand does *not* exert a force directly on the block. The influence of the hand is transmitted through the rope; this is already accounted for in the tension $\mathbf{T}$.

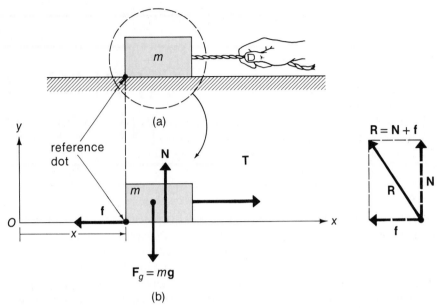

Fig. 6–1. Illustrating the concept of a *free-body diagram*. (a) The actual situation in which a block is pulled along a horizontal surface by a rope (and hand). (b) The block as the isolated object of interest, showing all the forces acting on the block *and* a convenient inertial coordinate frame of reference.

Next, we establish a convenient inertial coordinate frame of reference in order to describe the motion of the block. Because of the constraint introduced by the horizontal surface, the motion of the block is completely specified when we know the motion of any point on the block, such as the dot of paint shown in the lower left-hand corner of the block in Fig. 6–1. The motion of the block is described by giving at each instant the location of this dot with respect to the chosen coordinate frame.

Finally, we are ready to apply Newton's laws of motion to the isolated block. Recall that the second law, $\mathbf{F} = m\mathbf{a}$, is a *vector* equation. Usually, the easiest approach is to examine separately the various components of this equation.

Because no motion occurs in the vertical or y-direction, we have

$$N - mg = ma_y = 0$$

from which $$N = mg$$

In the horizontal or x-direction, we have

$$T - f = ma_x$$

Therefore, the motion in the x-direction is described by

$$a_x = \frac{d^2x}{dt^2} = \frac{1}{m}(T - f)$$

If T and f are constant, then a_x is constant, and this equation is easily integrated. We find (compare Eq. 2–10)

$$x(t) = x_0 + v_0 t + \tfrac{1}{2}\left[\frac{1}{m}(T - f)\right]t^2$$

Also, $$y(t) = 0$$

These equations completely describe the motion of the reference dot and, hence, the motion of the block.

Next, consider the more complicated case of two blocks with masses m_1 and m_2 that are rigidly coupled and are sliding across a horizontal plane, pulled by a force $\mathbf{F}$, as shown in Fig. 6-2a. Figure 6-2b is the free-body diagram for the block m_1 in terms of an inertial coordinate frame that has its x-axis along the horizontal surface and its y-axis perpendicular to this surface. Figure 6-2c is the free-body diagram for the block m_2, shown in a similar way. Notice that the tensions exerted on the two blocks by the linkage have opposite directions. That is, $\mathbf{T}_1 = -\mathbf{T}_2$, and we write for the x-components, $T_1 = T = -T_2$.

Fig. 6-2. (a) Two rigidly coupled blocks slide across a horizontal plane, pulled by a force **F**. (b) The free-body diagram for block m_1 alone. (c) The free-body diagram for block m_2 alone. (d) The free-body diagram for both blocks considered as a single unit.

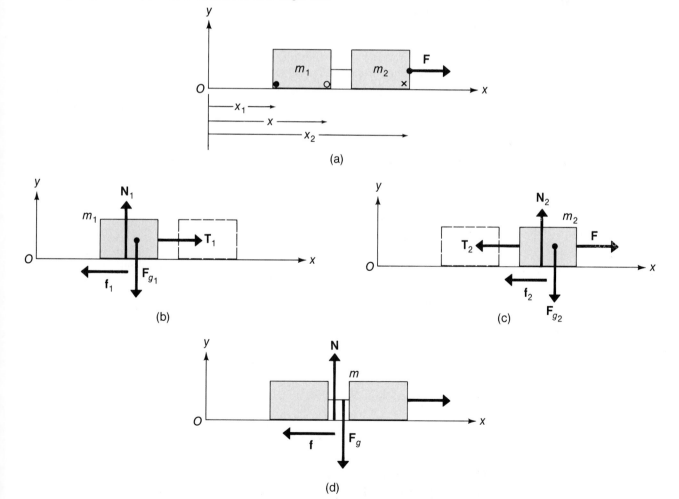

For the block m_1, Newton's second-law expressions for the forces in the x- and y-directions are

$$N_1 - m_1 g = 0$$

and
$$T - f_1 = m_1 a_x \qquad \textbf{(6-1)}$$

Similarly, for the block m_2, we have

$$N_2 - m_2 g = 0$$

and
$$F - T - f_2 = m_2 a_x \qquad \textbf{(6-2)}$$

When the two blocks together are considered as a unit (Fig. 6–2d), the tension T becomes an internal force and is *not* a force exerted on the isolated unit by any external agency. Thus, T does not appear in the free-body diagram for the coupled pair. Again taking x- and y-components, we find

$$N - mg = 0$$

and
$$F - f = m a_x \qquad \textbf{(6-3)}$$

where $N = N_1 + N_2$, $f = f_1 + f_2$, and $m = m_1 + m_2$.

A word is in order concerning the coordinate frames shown in Fig. 6–2b, c, and d. First, the frames are, in principle, independent of one another—they are required only to be inertial frames. At our option, we take them to be one and the same, as shown in Fig. 6–2a; this is clearly a convenient choice in the present case. The coordinate x_1 refers to the position of some selected identifying mark (a reference dot) on block m_1; similarly, the coordinate x_2 refers to some selected mark on block m_2. The coordinate x (without subscript) can refer to a completely separate mark (on either block or on the coupling). These coordinates are related at all times by

$$x_1 = x + \ell_1$$
$$x_2 = x + \ell_2$$

where ℓ_1 and ℓ_2 are constants. It then follows that $v_{1x} = dx_1/dt$, $v_{2x} = dx_2/dt$, and $v = dx/dt$ are all equal. Likewise, the accelerations are all equal:

$$a_{1x} = a_{2x} = a_x$$

It was in anticipation of this result that we placed no identifying subscripts on the acceleration terms in Eqs. 6–1, 6–2, and 6–3.

If we add Eqs. 6–1 and 6–2, we see that we obtain precisely Eq. 6–3. If we divide Eq. 6–1 by m_1 and divide Eq. 6–2 by m_2, then subtract one result from the other, we obtain

$$T\left(\frac{1}{m_1} + \frac{1}{m_2}\right) = \frac{F}{m_2} + \frac{f_1}{m_1} - \frac{f_2}{m_2}$$

or
$$T = \frac{m_1 F + m_2 f_1 - m_1 f_2}{m_1 + m_2}$$

What do we learn from this? If we are required only to find the acceleration a_x of the system, we should obviously consider the two blocks as a unit and construct directly the free-body diagram of Fig. 6–2d. In this case, there is no need to construct separately the free-body diagrams for m_1 and m_2. No error is committed by constructing the separate diagrams and arriving at Eqs. 6–1 and 6–2. However, using Fig. 6–2d is more direct and eliminates the need for determining the unknown tension T; at the very least, this method simplifies the required algebra. (In this respect the situation here is very similar to that discussed in Example 5–4.)

On the other hand, if we want to find the tension in the coupling, there is no recourse but to use the individual free-body diagrams. (•Can you see why it is permissible to use *any* two of the three diagrams?)

In summary, we give these general rules for solving problems in dynamics:
1) Draw carefully a diagram showing all the key features of the stated problem.
2) Depending on the questions raised, draw one or more *separate* free-body diagrams. In each instance show *all* the forces (and *only* the forces) acting *on* the isolated object through its interaction with external agencies. These, in general, will include contact forces and gravitational forces. (Do not show any *internal* forces.)
3) Select a set of inertial coordinate frames; you may select a different one for each free-body diagram. Of course, the constraints in the problem will relate these frames and the involved coordinates through various imposed conditions (such as those in the equations for x_1 and x_2 in the preceding example).
4) Apply Newton's laws of motion to *each* isolated object shown in its own free-body diagram, using component resolutions where appropriate.
5) An optional step, very useful in complicated situations, is to examine the behavior of the solutions for certain simplifying, extreme, or special conditions (for example, by allowing one of the masses or angles or velocities to become zero). This is particularly helpful for those cases in which you know or may be able to guess the result. At the very least, this step should provide additional insight into the significance of the solutions (and may reveal possible errors in the analysis).

Finally, a word of caution. Particularly when pressed for time, you may be tempted to forego or combine some of the above steps. This usually turns out to be a false economy, because there is the great risk of omitting relevant forces, of making sign errors, of incorrectly identifying the geometry, and so forth. It pays to be neat and orderly.

Ideal Strings. In many problems we are confronted with various objects that are connected by light strings (or cords or ropes). An important property of a string (one that is correct to a high degree of accuracy in ordinary cases) is that it can sustain only a force of tension. That is, a string can exert a *pull* but not a *push*. A string has zero internal resistance to bending and will therefore align itself *along* the direction of the applied tension force vector. Any attempt to make a string exert a compressional force will simply cause the string to collapse into an irregular shape.

Although strings obviously can stretch under tension, these changes in length are usually small compared to the other lengths or displacements that occur in problems. For our present purposes, we assume that every string has a constant length.

Strings obviously also possess mass. The effect of a string's mass is most easily appreciated by considering an example. In Fig. 6–3 we show two blocks with masses m_1 and m_2 that are tied together by a string with a length L and a total mass m_s. Through a second string, a disembodied hand applies a force **F** to the first block and thereby pulls the two blocks in straight-line motion in deep space, far from any possible interacting body. The figure also shows separate free-body diagrams for each of the two blocks *and* the connecting string.

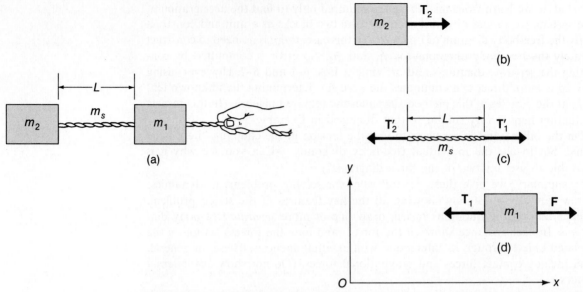

Fig. 6–3. (a) Two blocks that are tied together with a string are pulled along by a force **F**. The string has a length L and a mass m_s. (b), (c), and (d) Free-body diagrams for the block m_2, the connecting string m_s, and the block m_1, respectively.

We observe first that $\mathbf{T}_2$ and $\mathbf{T}'_2$ represent an action-reaction pair of forces, as do $\mathbf{T}_1$ and $\mathbf{T}'_1$. Then, taking components in the x-direction, we can write

$$F - T_1 = m_1 a_x$$
$$T_1 - T_2 = m_s a_x$$
$$T_2 = m_2 a_x$$

Adding all three equations gives

$$a_x = \frac{F}{m_1 + m_2 + m_s}$$

and using this result in the equation for $T_1 - T_2$ leads to

$$T_1 - T_2 = \frac{m_s}{m_1 + m_2 + m_s} F$$

For strings with small mass, $m_s \ll m_1$ and $m_s \ll m_2$, we note that the tensions T_1 and T_2 may be considered to be essentially equal; then, the expression for a_x becomes

$$a_x \cong \frac{F}{m_1 + m_2}$$

It is also instructive to divide the string shown in Fig. 6–3c into two unequal arbitrary parts with lengths ℓ and $L - \ell$. Figure 6–4 shows the separate free-body diagrams for the two parts (which are assumed to be homogeneous). Evidently,

$$T_\ell - T_2 = \frac{\ell}{L} m_s a_x$$

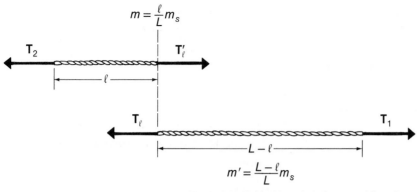

Fig. 6–4. The connecting string shown in Fig. 6–3 is divided into lengths ℓ and $L - \ell$.

and

$$T_1 - T_\ell = \frac{(L - \ell)}{L} m_s a_x$$

Substituting $a_x = F/(m_1 + m_2 + m_s)$ into the first of these equations, we obtain

$$T_\ell = T_2 + \frac{\ell}{L} \frac{m_s}{(m_1 + m_2 + m_s)} F \qquad\qquad \textbf{(6-4)}$$

We see that T_ℓ increases linearly with ℓ, from the value T_2 at $\ell = 0$ to the value T_1 at $\ell = L$. For the case $m_s \ll m_1$ and $m_s \ll m_2$, all the tensions are essentially equal: $T_\ell \cong T_1 \cong T_2$.

This example demonstrates how free-body diagrams for the divided sections of an object may be used to learn about the *internal* stresses in an object. (Note that for each of the divided sections, T_ℓ is an *external* force!)

An *ideal string* is defined to be a string that has negligible mass and has the same tension throughout its length; at each point the tension **T** is directed along the string.

Example 6–1

An elevator accelerates upward at a constant rate of $2\,\text{m/s}^2$. A uniform string with a length $L = 25\,\text{cm}$ and a mass $m = 2\,\text{g}$ supports a small block with a mass $M = 150\,\text{g}$ that hangs from the car ceiling. Find the tension along the string.

Solution:

The tension T_0 in the string at the ceiling support can be calculated by considering the free-body diagram for the entire string *plus* the block. If we take upward as the positive direction, we can write

$$T_0 - (m + M)g = (m + M)a$$

or
$$T_0 = (m + M)(g + a)$$
$$= (152 \times 10^{-3} \, \text{kg})(9.80 \, \text{m/s}^2 + 2.00 \, \text{m/s}^2)$$
$$= 1.794 \, \text{N}$$

The tension T at a point in the string a distance ℓ from the ceiling can be found by considering the free-body diagram for the upper portion of the string segment, which has a length ℓ. The mass of this segment is $m_\ell = (\ell/L)m$, so

$$T_0 - T - m_\ell g = m_\ell a$$

or
$$T = T_0 - m(g + a)\frac{\ell}{L}$$

$$= 1.794 \, \text{N} - (2 \times 10^{-3} \, \text{kg})(9.80 \, \text{m/s}^2 + 2.00 \, \text{m/s}^2)\frac{\ell}{0.25 \, \text{m}}$$

$$= (1.794 - 0.094\ell) \, \text{N}$$

with ℓ in meters.

We note that for $\ell = L$, that is, at the end attached to the block, we have

$$T = T_M = 1.770 \, \text{N}$$

As a check, consider the free-body diagram for the block alone. Then,

$$T_M - Mg = Ma$$

or
$$T_M = M(a + g) = (0.15 \, \text{kg})(11.80 \, \text{m/s}^2)$$
$$= 1.770 \, \text{N}$$

in agreement with the result above. (•What would be the tension in the string if its mass m were zero?)

Fig. 6–5. Free-body diagram for an ideal (massless) string that passes over an ideal (frictionless) peg. The magnitude of the contact force is $N = 2T \cos \alpha$.

Ideal Pegs, Pulleys, and Rods. In order to simplify the treatment of certain classes of problems, we often make use of several ideal components in addition to ideal strings. For example, the function of an ideal peg or pulley is simply to change the direction of the force exerted by a string. If the string is also ideal (massless), the tension in the string on either side of the peg has the same magnitude. The situation is illustrated in Fig. 6–5. The tension in each part of the string is T, and the net contact force N exerted on the string by the peg is directed along the bisector AA'. The resultant of these three forces acting on the string is *zero,* so the entire effect of the peg is to change the direction of the tension. (•How would the analysis change if the mass of the string were not zero?)

Figure 6-6 shows an arrangement of blocks with the connecting string passing over an *ideal pulley.* An ideal pulley has zero mass, and it is supported by and rotates freely on frictionless bearings. As the string is pulled over the pulley, the string causes the pulley to rotate in such a way that there is no slippage between them. (That is, the dots of paint shown in Fig. 6-6 remain adjacent during the motion.) Slippage does not occur because the string exerts a tangential force on the pulley rim due to the friction that exists between them. There is no conceptual difficulty in allowing friction between an ideal string and an ideal pulley because there are no energy losses associated with this nonslipping type of friction (see Section 6-2). (There would be losses if friction were present and slippage did occur.)

Finally, as far as the relevant forces are concerned, the ideal pulley and string behave just as the ideal peg and string shown in Fig. 6-5. The string tensions at both free ends are equal, and the force exerted by the pulley is also $N = 2T \cos \alpha$ and is likewise along the bisecting line AA'. In the following problems ideal pegs and pulleys are interchangeable elements—the main effect of either is simply to redirect the action line of the string tension force.

Another ideal component we sometimes use is the *ideal rod*. Such rods are absolutely rigid structural members and are employed as supporting devices or to maintain a fixed separation between parts of a system. An ideal rod cannot be extended, compressed, bent, or twisted. The rod may have mass or it may be massless. Forces are transmitted undiminished through ideal massless rods.

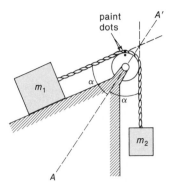

Fig. 6-6. The two blocks are connected by an ideal (massless) string that runs over an ideal pulley (massless and with frictionless bearings). The line AA' bisects the angle between the ends of the string.

Example 6-2

Consider the arrangement shown in Fig. 6-6, in which a block with a mass $m_1 = 2 \, \text{kg}$ slides on a frictionless, inclined plane with elevation angle $45°$; the mass of the hanging block is $m_2 = 3 \, \text{kg}$. An ideal connecting string runs over an ideal pulley. What is the linear acceleration of the blocks, and what forces are exerted on the pulley and on the inclined plane?

Solution:

In the free-body diagrams for the blocks we take the coordinate x_1 to lie along the inclined plane for m_1, and we take the coordinate x_2 to be vertically downward for m_2. Then, the equations for the force components in these directions are

$$T - m_1 g \sin 45° = m_1 a$$
$$m_2 g - T = m_2 a$$

Adding these equations gives

$$(m_2 - m_1 \sin 45°)g = (m_1 + m_2)a$$

from which
$$a = \frac{(m_2 - m_1 \sin 45°)}{(m_1 + m_2)}g = \frac{\left(3 - 2\frac{\sqrt{2}}{2}\right)}{(3 + 2)}(9.80 \, \text{m/s}^2)$$
$$= 3.11 \, \text{m/s}^2$$

$$\alpha = \tfrac{1}{2}(90° - 45°)$$

From the free-body diagram for block m_1, the force components perpendicular to the plane obey the relationship

$$N_1 = m_1 g \cos 45° = (2 \, \text{kg})(9.80 \, \text{m/s}^2)\frac{\sqrt{2}}{2}$$
$$= 13.9 \, \text{N}$$

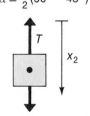

The force exerted on the inclined plane by the block is the reaction force to N_1.

The string tension T can be obtained from the free-body diagram for block m_2:

$$T = m_2(g - a) = (3 \text{ kg})(9.80 \text{ m/s}^2 - 3.11 \text{ m/s}^2)$$

$$= 20.1 \text{ N}$$

The string tension is needed to calculate N, the force exerted on the string by the pulley. From the free-body diagram for the string running over the pulley, we have

$$N = 2T \cos 22.5° = 2(20.1 \text{ N}) \cos 22.5°$$

$$= 37.1 \text{ N}$$

The force exerted on the pulley by the string is the reaction force to N.

Example 6–3

A block with a mass $m = 300$ kg is set into motion on a horizontal frictionless surface by using an ideal pulley-and-rope system, as shown in the diagram. What horizontal applied force F is required to produce an acceleration of 5 cm/s² for the block?

Solution:

The free-body diagram for the pulley is shown above. The force exerted on the pulley by the rope (this is equivalent to the reaction force to **N** in Fig. 6–5, with $\alpha = 0$) is simply the sum of the two forces F. Because the pulley is massless, we also have $T = 2F$. Moreover, $T = ma$, so that

$$F = \tfrac{1}{2}ma = \tfrac{1}{2}(300 \text{ kg})(0.05 \text{ m/s}^2)$$

$$= 7.5 \text{ N}$$

Notice that with this *block-and-tackle* arrangement, the force required to produce a particular acceleration is only half that required if the force were applied directly to the block. This block-and-tackle system is said to have a *mechanical advantage* of 2 (that is, $T/F = 2$).

Example 6–4

A ball with a mass $m = 0.20$ kg is whirled at the end of a 1.5-m string at an angular speed ω (in a gravity-free environment). The string will break under tension of 12 N. What is the maximum angular speed that can be tolerated by the string?

Solution:

The acceleration required to maintain the circular motion is the centripetal acceleration, $a_c = v^2/R = R\omega^2$, and is due to the tension in the string. We have

$$T = mR\omega^2$$

The angular speed at which the string will break is

$$\omega = \sqrt{\frac{T}{mR}} = \sqrt{\frac{12 \text{ N}}{(0.20 \text{ kg})(1.5 \text{ m})}} = 6.32 \text{ rad/s}$$

Example 6–5

A 0.50-kg stone is tied to the end of a string that has a length of 0.25 m and is rotated in a horizontal plane at a constant rate of 1.5 rev/s. The string makes an angle β with the vertical. Find the tension in the string and the angle β.

Solution:

The horizontal component of the tension, namely $T \sin \beta$, provides the centripetal acceleration, $a_c = v^2/R = R\omega^2$, required for the circular motion. Therefore,

$$T \sin \beta = mR\omega^2 = mL\omega^2 \sin \beta$$

from which
$$T = mL\omega^2$$
$$= (0.50 \text{ kg})(0.25 \text{ m})(2\pi \times 1.5 \text{ s}^{-1})^2$$
$$= 11.10 \text{ N}$$

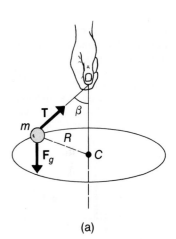

(a)

In the vertical direction there is no acceleration, so we have

$$T \cos \beta - F_g = 0$$

or,
$$\cos \beta = \frac{mg}{T} = \frac{(0.50 \text{ kg})(9.80 \text{ m/s}^2)}{11.10 \text{ N}} = 0.441$$

from which
$$\beta = 63.8°$$

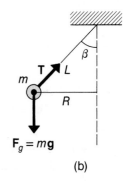

$F_g = mg$

(b)

6-2 FRICTION

When an effort is made to slide one object relative to another along the common surface of contact, the resistive force that acts along (tangent to) the interface is called *friction*.* We have referred to friction and its effects on several occasions; now we give specific attention to this subject.

Friction can be a serious practical problem or it can be quite useful. Friction in machinery wastes energy and causes wear. However, walking would be impossible without friction, and wheels would not roll across the ground. The details of the

---

*The first careful and systematic experimental study of friction was made by the French physicist Charles Augustin de Coulomb (1736–1806), who is more famous for his investigations of electric phenomena. Some of the earliest work on friction was by Leonardo da Vinci (1452–1519), whose clear understanding of the essentials of the subject is revealed by sketches and commentary in his famous *Notebooks*.

phenomenon of friction are complicated and not well understood in terms of any fundamental theory. Consequently, we give here only a qualitative overview of the subject.

Consider a book that is at rest on a flat tabletop. Suppose that you apply a horizontal force to the book by pushing on it with your finger in an effort to slide it along the tabletop. If the applied force is too small, the book does not move—it remains static. Evidently, there must be another horizontal force acting on the book, a force that just cancels your applied force. This force exerted on the book by the tabletop is referred to as the force of *static friction*. Note that this force is rather unusual, in that it is exactly equal and opposite to the force you apply, and it vanishes when you stop pushing! If you continue to increase your pushing effort, at some point the book will finally "break loose" and start to slide. During the sliding motion it is clear that a frictional force is still acting because, if you stop pushing, the book comes to rest (perhaps after first traveling a short distance). If you are careful, you may feel that the break-away force is slightly larger than the force you must exert to maintain the book sliding with a constant speed. The resisting horizontal force exerted on the sliding book by the tabletop is referred to as *sliding* (or *kinetic*) *friction*.

Although frictional processes are complex, simple empirical "laws" are found to apply in many situations. However, we should not expect these laws to be of the same basic character as Newton's laws of dynamics or the law of gravitation. The laws of friction are only approximate descriptions of the way that matter behaves under some (but not all) conditions. When we lack a fundamental theory for a process or find the theory too complicated to use in practical situations, we are forced to rely on approximate or empirical descriptions. We have already used one such empirical law, namely, Hooke's law, and later we introduce another famous example, Ohm's law of electric conductivity. None of these empirical laws is valid in all circumstances, but each is sufficiently accurate for many engineering and practical applications.

When an object is in contact with a surface, the component of the externally applied force that acts tangentially to the surface of contact is exactly balanced by the force of static friction between the object and the surface, up to a certain maximum value (Fig. 6-7a, b). Experimental evidence shows that this maximum force is approximately proportional to the loading (or normal) force pressing the two objects together. For example, an ordinary wood block in contact with a flat (dry) wood board may generate frictional forces that will balance applied tangential forces up to about 0.4 of the normal force that is exerted on the block by the board. If the applied tangential force exceeds this value, the block will slide relative to the board surface. It has been found that this maximum (static) force is independent of the area of contact between the objects. That is, the maximum force of static friction between a wood block and a board will be the same when the block rests on its largest side as when it rests on its smallest side. Thus, in ordinary cases we can write a relationship of direct proportionality between the maximum force of static friction, $f_{s,\max}$, and the normal force N:

Fig. 6-7. (a, b) For $f_s < f_{s,\max}$, the frictional force exactly balances the applied force; then, there is no acceleration. (c) When a force sufficient to cause motion is applied, the frictional force is equal to $\mu_k N$ and the acceleration is $(F - \mu_k N)/m$.

(a) $\mathbf{f}_1 = -\mathbf{F}_1$

(b) $\mathbf{f}_2 = -\mathbf{F}_2$

(c) $f_3 = \mu_k N$

$$f_{s,\max} = \mu_s N \qquad \text{(6-5)}$$

where the proportionality constant μ_s is called the *coefficient of static friction*.

If the applied tangential force exceeds $f_{s,\max}$, relative sliding occurs (Fig. 6-7c). Again, experimental evidence indicates that the force of sliding or kinetic friction is independent of the area of contact and is directly proportional to the normal force. Thus,

$$f_k = \mu_k N \qquad \textbf{(6-6)}$$

where μ_k is the *coefficient of kinetic friction*. The direction of the force $\mathbf{f}_k$ is always opposite to the direction of the relative velocity between the objects in contact.

Generally, $\mu_s > \mu_k$. (•Is this what you would expect?) Also, μ_k has some dependence on the relative velocity of the objects, the degree depending strongly on the specific nature of the objects, as we mention later. For some materials, the variation of μ_k with velocity is not pronounced, and it is often assumed (for ease of calculation!) that μ_k is constant.

The statements expressed by Eqs. 6-5 and 6-6, together with the observation that the frictional forces are independent of the area of contact, constitute the laws of friction. Notice carefully that the equation relating to static friction (Eq. 6-5) involves the maximum possible frictional force $\mu_s N$, whereas Eq. 6-6 expresses the force $\mu_k N$ that actually exists during sliding. It is important to distinguish clearly between these two situations.

An important effect associated with sliding friction is the conversion of some of the motional energy of the system into heat. For example, if you rub the palms of your hands together for a few seconds, the sensation of heat generation will be readily felt. We discuss the important topic of energy and its various forms in Chapter 7.

▶ *Dragster Tires.* An automobile (or any driven vehicle) is propelled forward by virtue of the frictional force that exists between the tires and the road surface. (Some details are given later in this section.) In order to achieve maximum acceleration, it is clearly necessary to maintain this frictional force as large as possible. We have argued that frictional forces, for particular materials in contact, depend only on the normal force. Why, then, do dragracers use tires that are much wider than normal tires? The extra width produces a greater area of contact between the tires and the road surface, but frictional forces do not depend on the contact area. When a dragracer accelerates his vehicle, starting from rest, the tires spin furiously. The best way to accelerate would be to maintain a drive power level that produces frictional forces just short of slipping (because $\mu_s > \mu_k$). But this is technically very difficult to achieve, so the dragracer simply applies maximum power and allows the tires to spin until the car's velocity increases to the point that slipping ceases. While the tires are slipping, considerable heat is generated. This causes actual melting of the tire rubber (and possibly also the road surface). The presence of liquid rubber (and asphalt) reduces the friction coefficient and the frictional force. By using dragster tires that are very wide, the increased contact area distributes the heat load and maintains the temperature sufficiently low that excessive melting does not occur. A greater frictional force and therefore a greater acceleration are thereby achieved by using wide tires instead of narrow ones. ◀

Example 6-6

A wooden crate with a mass of 100 kg rests on a flat, wooden floor. An effort is made to start the crate sliding by pulling on a rope that makes an angle of 30° with the horizontal, as shown in the diagram. If the coefficient of static friction between the crate and floor is 0.4, what is the minimum rope tension that will start a sliding motion?

Solution:

From the free-body diagram, we can write for the vertical force components,

$$N + F \sin 30° - mg = 0$$

so that
$$N = (100 \text{ kg})(9.80 \text{ m/s}^2) - F(0.5)$$
$$= (980 \text{ N}) - 0.5F$$

Just before sliding begins, we have $f = \mu_s N$. Then, for the horizontal force components, we have

$$F \cos 30° - f = 0$$

Thus,
$$F = \frac{\mu_s N}{\cos 30°} = \frac{0.4(980 \text{ N} - 0.5F)}{0.866}$$

Solving for F, we find
$$F = 368 \text{ N}$$

Example 6–7

A block of wood rests on an inclined plane, also of wood, with an adjustable angle of inclination α as indicated in the diagram. The angle α is slowly increased from zero, and the block is observed to start sliding when $\alpha = 23°$. What is the coefficient of static friction μ_s between the block and plane?

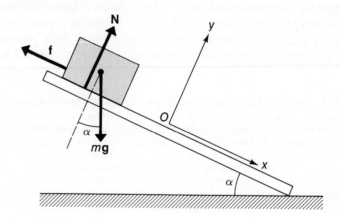

Solution:

In the direction normal to the plane (i.e., the y-direction), the force components are

$$N - mg \cos \alpha = 0$$

In the direction along the plane (i.e., the x-direction), we have

$$mg \sin \alpha - f = 0$$

Now, we form the ratio
$$\frac{f}{N} = \frac{mg \sin \alpha}{mg \cos \alpha} = \tan \alpha$$

Hence,
$$(f/N)_{max} = \mu_s = \tan 23° = 0.424$$

This maximum angle is sometimes called the *angle of repose*. Note that it is independent of the mass of the block.

Example 6-8

The crate in Example 6-6 is pulled along at a constant speed v. To maintain this motion, the required force applied at $30°$ to the horizontal is $F = 330$ N. What is the value of the coefficient of sliding friction at the speed v?

Solution:

The same free-body diagram applies, and we have, in the horizontal and vertical directions,

$$F \cos 30° = \mu_k N$$

and

$$N = (980 \text{ N}) - 0.5F$$

Thus,

$$\mu_k = \frac{(330 \text{ N})(0.866)}{(980 \text{ N}) - (0.5)(330 \text{ N})} = 0.351$$

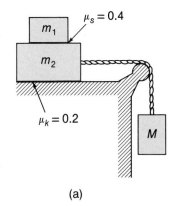

(a)

Example 6-9

Two blocks, $m_1 = 2$ kg and $m_2 = 4$ kg, are connected with a light string that runs over a frictionless peg to a hanging block with a mass M, as shown in diagram (a). The coefficient of sliding friction between block m_2 and the horizontal surface at the speeds involved is $\mu_k = 0.2$. The coefficient of *static friction* between the two blocks is $\mu_s = 0.4$. What is the maximum mass M for the hanging block if the block m_1 is *not* to slip on block m_2 while m_2 is sliding over the surface?

Solution:

This example involves both static and sliding friction. Assume that slipping of block m_1 is impending. The relevant free-body diagrams are shown in diagram (b). Using the diagram for the two-block system, we can write

$$N - (m_1 + m_2)g = 0 \qquad \text{(1)}$$

$$T - f = (m_1 + m_2)a \qquad \text{(2)}$$

and taking the downward direction to be positive, we have for the hanging block

$$Mg - T = Ma \qquad \text{(3)}$$

Adding Eqs. (2) and (3) gives

$$(M + m_1 + m_2)a = Mg - f$$

We also have, using Eq. (1), $f = \mu_k N = \mu_k(m_1 + m_2)g$

Thus,

$$a = \frac{M - \mu_k(m_1 + m_2)}{M + m_1 + m_2}g$$

From the free-body diagram for m_1, we have

$$N_1 - m_1 g = 0$$
$$f_1 = m_1 a$$

Notice that the only horizontal force acting on m_1 is the frictional force f_1; it is this force that

(b)

accelerates m_1 to the right. Just before slipping occurs, we find

$$\frac{f_1}{N_1} = \mu_s$$

or

$$\mu_s = \frac{a}{g} = \frac{M - \mu_k(m_1 + m_2)}{M + m_1 + m_2}$$

Solving for M, we have

$$M = \frac{(\mu_s + \mu_k)(m_1 + m_2)}{1 - \mu_s}$$

Hence,

$$M = \frac{(0.4 + 0.2)(2\ \text{kg} + 4\ \text{kg})}{(1 - 0.4)} = 6\ \text{kg}$$

If M were to exceed this value, the resulting greater acceleration would require $f_1 = m_1 a$ to exceed the maximum sustainable value of the quantity $\mu_s N_1 = \mu_s m_1 g$. Then, slipping between the blocks would result.

Example 6–10

A 5-kg block is initially at rest on an inclined plane, as shown in diagram (a). A force $F = 20\ \text{N}$ is applied to the block in a direction parallel to the plane. Determine the acceleration of the block if the coefficient of kinetic friction is $\mu_k = 0.42$.

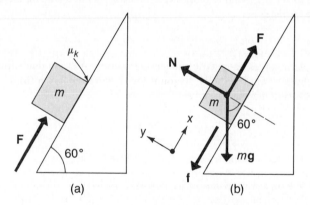

(a)   (b)

Solution:

Assume that the block will slide up the plane. Then, the free-body diagram is as shown in diagram (b), with the frictional force **f** directed down the plane. In the y-direction, we have

$$N = mg \cos 60° = (5\ \text{kg})(9.80\ \text{m/s}^2) \cos 60° = 24.5\ \text{N}$$

Hence,

$$f = \mu_k N = 0.42(24.5\ \text{N}) = 10.29\text{N}$$

Then, in the x-direction, we have

$$F - f - mg \sin 60° = ma$$

$$(20\ \text{N}) - (10.29\ \text{N}) - (5\ \text{kg})(9.80\ \text{m/s}^2) \sin 60° = (5\ \text{kg})a$$

from which

$$a = -6.55\ \text{m/s}^2$$

In this case it is *not* permissible to assume that the negative sign in the solution simply indicates that the acceleration is 6.55 m/s² *down* the inclined plane. The reason is that motion *down* the plane means that the frictional force **f** is directed *up* the plane, contrary to the free-body diagram that was used. Instead, the problem must be solved again, with **f** directed up the plane. Then, with the same coordinate frame, we find

$$F + f - mg \sin 60° = ma$$
$$(20 \text{ N}) + (10.29 \text{ N}) - (5 \text{ kg})(9.80 \text{ m/s}^2) \sin 60° = (5 \text{ kg})a$$

from which

$$a = -2.43 \text{ m/s}^2$$

Now the negative sign does indeed correctly indicate that the block accelerates down the plane.

When solving problems that involve directions of motion with friction present, careful attention must be given to insure that the frictional force is always directed opposite to the motion.

▶ *A Microscopic View of Friction.* What are the basic sources of friction? One of the obvious and important contributions to friction is the effect of adhesive molecular forces acting between the constituent molecules of the materials in contact. Another effect, which assumes importance when a slider with a hard surface moves across a softer surface, involves a plastic plowing resistance. In addition, one can identify various phenomena associated with wear, such as scoring, tearing, and microcutting, among others. Also, plastic waves or bulges may form and travel in the softer material in front of gross, high surface points of the harder material.

The molecular forces involved in friction are not only those between the bodies in contact, but they are also those due to any oxide layers or accumulations of dirt and other foreign substances coating the surfaces (occasionally, as with lubricants, deliberately introduced). In fact, in order for ordinary sliding motion to be possible, the presence of such foreign substances at the interface is essential. For example, consider two flat metallic surfaces that are cleaned, heated and polished in vacuum, and then brought together while still in vacuum. These ultraclean surfaces will tend to bond together, producing frictional forces ten times or more greater than would ordinarily act between the surfaces. In the cases we consider in this book, it is assumed that the surfaces are reasonably smooth, and that dirt, occluded air, and oxide layers are present. Under these conditions, simple sliding readily occurs.

At the atomic and molecular level, two surfaces in contact appear quite rough and have numerous gaps and voids, as illustrated in Fig. 6–8. Even metallic surfaces that are carefully polished will at best consist of a series of microscopic hills and valleys that range in cross section from about 100 to 1000 molecular diameters. Forces between molecules diminish rapidly with the distance of separation, essentially vanishing at distances greater than several molecular diameters. Consequently, the adhesion between the molecules at the touching, high points of the two surfaces provides the major contribution to the friction between the surfaces. During sliding, these bonds must be continually broken loose and then formed anew as different high points are brought into contact. Surface molecules in these interacting regions are set into rapid vibrational motion, generating heat (and occasionally sound). Also, pieces of one surface may be broken loose and then welded or bonded to the other surface. During this process, scratches and other forms of wear will appear.

In the previous discussion, the contact area we referred to is the gross macroscopic covering area A_c. The actual or real microscopic interaction area of contact, where the strong adhesive molecular forces are exerted, is the much smaller area A_r (Fig. 6–8). It is this area that supports the entire normal force load. Even under strong loading, $A_r \ll A_c$.

For a particular pair of surfaces, an increase in the normal force tends both to flatten the existing contact points by elastic and plastic deformations and to generate new contact points. Experiments show that, in ordinary cases, the increase in the area A_r is approximately proportional to the normal force, thereby leading to the observed relationship connecting the maximum sustainable static frictional force, $f_{s,\text{max}}$, and the normal force N (Eq. 6–5).

It is often stated that the friction coefficient μ_k depends on the degree of smoothness of the contacting surfaces, becoming smaller if the surfaces are smoother. In fact, smooth surfaces are often said to be "frictionless." This is simply *not so*. To a surprising extent, μ_k is independent of the roughness of the sliding surfaces—at least, for finishes commonly employed in engineering practice, and even for surfaces showing moderate wear. For extremely smooth and flat (and clean) surfaces, μ_k is actually *larger* than for the same

Fig. 6-8. A highly magnified view of the contacting surfaces of a book and a tabletop, showing that the actual or *real microscopic interaction* area, A_r, of the high points is only a small fraction of the gross, macroscopic, or *covering* contact area, A_c. The black dots in the right-hand diagram represent the individual microscopic areas of contact that sum to the area A_r. It is the area A_r that supports the normal load.

surfaces with rougher finish. For example, two pieces of glass have a larger value of μ_k if the surfaces are polished than if they are rough ground! Atomically smooth surfaces of crystals, produced by cleavage, also have very large friction coefficients.

However, surface cleanliness, the presence of oxide layers and adsorbed gases, and the introduction of lubricants *do* have considerable effect in determining values of μ_k. For example, Table 6–1 shows values of μ_k obtained for a hemispherical steel rider sliding on a hardened and polished flat steel surface. A comparable experiment, had it been performed in high vacuum with surfaces also cleaned in vacuum by heating and polishing, would have produced friction coefficients of about 5 or greater. Experiments in high vacuum with ultraclean copper surfaces in contact have yielded values of μ_k up to about 100. The exact values for the coefficients found in such experiments are extremely sensitive to the degree of cleanliness achieved.

TABLE 6-1.   Values of μ_k for Steel on Steel for Various Surface Conditions

SURFACE CONDITION	μ_k
Clean (ordinary chemical cleaning in air)	0.78
Sulfide layer	0.39
Lubricant film	0.32
Oxide layer	0.27
Oxide plus lubricant	0.19
Sulfide plus lubricant	0.16
Oleic acid film	0.11

Finally, what about the dependence of the coefficient of kinetic friction on the relative velocity between the surfaces in contact? The values of μ_k for all material pairs exhibit some velocity dependence, often to a marked degree. At very slow sliding speeds, μ_k increases with speed. At some point, a maximum is reached, and thereafter μ_k decreases with increasing speed; quite low values of μ_k are found for metals at high speeds (greater than several hundred m/s).

At very low velocities (in the range 10^{-8} m/s to 10^{-10} m/s), soft materials that have a tendency to creep*—for example, indium and lead—exhibit a strong dependence of μ_k on speed. Figure 6-9 shows the variation with speed of μ_k for a steel slider on unlubricated indium and lead. Notice the steep slopes of the curves, especially at very low speeds. (Note that a speed of 10^{-12} m/s corresponds to a displacement of only 0.01 mm in 4 months. In this extreme case, it is difficult to distinguish between "sliding" and "creep.")

For a pair of hard materials—such as titanium or steel—μ_k decreases with speed, as shown in Fig. 6-10. Notice that the dependence of μ_k on speed is much greater for steel than for titanium. The case of titanium is unusual—values of μ_k for most materials exhibit a more pronounced dependence on speed. ◄

Creep is the very suggestive name given to phenomena associated with a generally quite slow adjustment in displacement or deformation to the application of a sustained load.

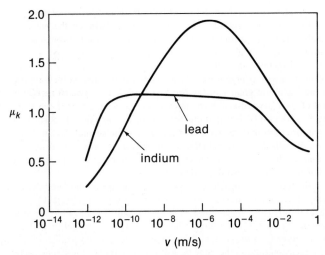

Fig. 6–9. The dependence on velocity of the coefficient of sliding friction μ_k for a steel slider on indium and lead, materials that exhibit a tendency to creep.

Fig. 6–10. The dependence of μ_k on speed for steel sliding on steel and titanium, both hard materials. The values were obtained without lubrication.

▶ *Wheels and Rolling Friction.* Consider an isolated, perfectly circular wheel that rolls across a flat surface. It is our experience that any rolling wheel will eventually come to rest, and we understand that the reason is the friction between the wheel and the surface. This *rolling friction* is rather different from static or sliding friction.

We could construct the force diagram for the rolling wheel as in Fig. 6–11. Here we see the forces acting on the wheel—the gravitational force $\mathbf{F}_g$, the normal force $\mathbf{N}$, and the frictional force $\mathbf{f}$. In the ideal situation, the wheel remains perfectly round and the surface remains perfectly flat, so there is a single point of contact at which both $\mathbf{f}$ and $\mathbf{N}$ act.

Now, look at Fig. 6–11 more closely. Although $\mathbf{f}$ is the rearward force that is required to slow the wheel, this force has a tendency* to make the wheel roll forward! That is, $\mathbf{f}$ tends to increase the rotational speed ω about the wheel axis. (Neither $\mathbf{N}$ nor $\mathbf{F}_g$ has this tendency.) Clearly, the surface cannot have such an effect on the wheel. We are forced to conclude that the frictional force is zero!

This is exactly correct. With the ideal conditions we have assumed, there is no instantaneous relative motion between the wheel and the surface at the point of contact (actually, a line), so there is no mechanism that can cause frictional losses. Thus, there is no dynamic friction, and the wheel will roll forward at constant velocity.

For a *real* case, the diagram shown in Fig. 6–11 is incorrect in several respects. First, both the wheel and the surface suffer deformations to an extent that depends on their particular elastic qualities. Figure

Fig. 6–11. The forces acting on a perfectly round wheel that rolls across a flat surface.

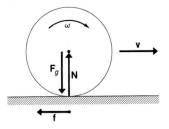

---

*This "tendency" is due to the *torque* exerted on the wheel by $\mathbf{f}$. Although we do not discuss this quantity in detail until Chapter 10, the basic concept is easy to understand.

Fig. 6–12. Schematic behavior of a real wheel rolling across a surface. Both the wheel and the surface are deformed. (The deformations are exaggerated in the figure.)

6–12 shows an exaggerated view of these deformations—the lower portion of the wheel is flattened and a shallow trench is formed in the surface. The line of contact between the wheel and the surface has expanded to a sizable area. The normal force is now distributed over the contact area. As the wheel rolls forward, the leading portion of the wheel and the surface are compressed while the trailing portion becomes relaxed. Consequently, the deformations are greater forward of the center of rotation. A large deformation is associated with a large force (Hooke's law); hence, the individual normal forces acting on various parts of the wheel are greater in the forward section, as indicated in Fig. 6–12.

The resultant of the rearward-acting frictional forces associated with the distributed normal force is shown as **f** in Fig. 6–12, and it is clear that this force tends to make the wheel roll forward, just as in Fig. 6–11. However, it is also clear that the net rotational effect of the normal force is in the opposite direction. The latter force counteracts the rotational effect of the frictional force and causes the observed rotational deceleration. Moreover, there is now a net horizontal force in the direction opposite to the motion, so the forward velocity of the wheel is also decreased.

The nature of the force of rolling friction is very complex. As the wheel and the surface deform during the rolling motion, some slippage does occur, so that sliding friction is present to a degree. The primary frictional effect over the area of contact, however, is one of static friction, which produces no energy losses. For many engineering purposes, the rolling friction that acts on an object such as a cylinder can be considered to exert a force f_r on the center of the cylinder and directed parallel to the supporting surface. This force of rolling friction follows the same type of law that applies for kinetic friction (Eq. 6–6), except that the factor that multiplies the normal force is inversely proportional to the cylinder radius R. Thus, we can write

$$f_r = \frac{\mu_r}{R} N \tag{6–7}$$

For most combinations of materials, μ_r lies in the range 10^{-5} m $\leq \mu_r \leq 10^{-3}$ m.

In most cases, the important source of frictional losses due to rolling is in the deformation process. No real material is perfectly elastic, so there are always energy losses when a material is compressed and then relaxed. This continual compression-relaxation cycle generates heat in the wheel and in the surface. It is for this reason that automobile tires become hot when run at high speeds.

What happens when the brakes are applied to a moving wheel? Figure 6–13 shows the situation in a schematic way. First, we must realize that normal drum-and-shoe (or disc) brakes cannot be effective on a single wheel or even on a pair of wheels connected by an axle; the braked wheel must be connected in some way to another wheel or other support in front or behind. (•Can you see why?) Therefore, we assume that Fig. 6–13 represents only one wheel of an automobile (or bicycle). When the brake shoes are pressed on the drum, the rotational motion of the wheel is retarded. This produces a forward force on the road, and the reaction force of the road on the tire is in the backward direction. This force greatly increases the force **f** shown in Fig. 6–13. The frictional force still attempts to increase the rotational motion of the wheel, but this is now opposed by both the normal force and the force exerted by the brakes.

The near-static conditions that exist at the tire-road interface can be upset if the braking action is severe. Then, skidding occurs. Because $\mu_k < \mu_s$, skidding results in a decrease of the frictional force tending to slow the vehicle, and the vehicle moves with dangerously high speed without directional control.

If the wheel is *driven,* the additional force exerted by the tire on the road is in the backward direction and the reaction force of the road on the tire is in the forward direction. The result is a forward acceleration. ◄

Fig. 6–13. The vector **N** here represents the effect of all the individual normal forces indicated in Fig. 6–12. Also, **f** here represents the effect of all the individual frictional forces that were not shown in Fig. 6–12. When the brakes are applied, the frictional force **f** is increased significantly and the vehicle is slowed.

Example 6–11

An automobile with a mass M starts from rest and accelerates at the maximum rate possible without slipping on a road with a coefficient of static friction $\mu_s = 0.5$. If only the rear wheels are driven and half the weight of the automobile is supported on these wheels, what amount of time is required to reach a speed of 100 km/h?

Solution:

The normal load on both rear wheels is $N = \frac{1}{2}Mg$, and the maximum sustainable frictional force is

$$f = \mu_s N = \tfrac{1}{2}\mu_s Mg$$

Thus, the acceleration of the automobile is

$$a = \frac{f}{M} = \tfrac{1}{2}\mu_s g$$

Therefore,

$$t = \frac{v}{a} = \frac{2v}{\mu_s g} = \frac{2(100 \text{ km/h})(10^3 \text{ m/km})}{0.5(9.80 \text{ m/s}^2)(3600 \text{ s/h})}$$

$$= 11.3 \text{ s}$$

Notice that the mass of the automobile is not required to obtain the solution.

Example 6-12

A driver is attempting to negotiate a flat (i.e., unbanked) highway curve that has a radius of curvature $R = 100$ m. What is the maximum speed v_m that the driver may use if his vehicle is not to skid? ($\mu_s = 0.50$.)

top view

end view

Solution:

The force equations are $\qquad N - mg = 0$

and

$$f = ma_c = m\frac{v^2}{R}$$

where we have used $a_c = v^2/R$ for the centripetal acceleration. It is evident that as the speed v is increased, the frictional force necessary to prevent skidding increases quadratically. The maximum value of f is $\mu_s N = \mu_s mg$, so we have

$$v_m = \sqrt{\mu_s Rg} = \sqrt{(0.50)(100 \text{ m})(9.80 \text{ m/s}^2)}$$

$$= 22.1 \text{ m/s} = 79.7 \text{ km/h}$$

Example 6-13

If the highway curve in Example 6–12 is banked instead of flat, what angle of inclination with respect to the horizontal is required to reduce the frictional force to zero for a speed of 22.1 m/s (80 km/h)?

Solution:

The force equation in the radial direction is

$$f \cos \alpha + N \sin \alpha = m \frac{v^2}{R}$$

and in the vertical direction we have

$$N \cos \alpha - f \sin \alpha - mg = 0$$

We multiply the first of these equations by $\cos \alpha$ and the second by $-\sin \alpha$, then add the results. Using $\sin^2 \alpha + \cos^2 \alpha = 1$, we find

$$f = m \left(\frac{v^2}{R} \cos \alpha - g \sin \alpha \right)$$

This is the frictional force required to maintain the circular motion (no skidding) for the inclination angle α. By properly selecting the angle α, the force can be made zero:

$$\tan \alpha = \frac{v^2}{gR} = \frac{(22.1 \text{ m/s})^2}{(9.80 \text{ m/s}^2)(100 \text{ m})} = 0.498$$

from which $\alpha = 26.5°$

SPECIAL TOPIC

6-3   VISCOUS FORCES

Galileo asserted that all objects falling freely near the surface of the Earth experience the same constant acceleration g. However, when an object moves through a fluid—even the air—it is observed that there is present a resisting force or drag. This retarding force is the result of friction between the moving object and the molecules of the liquid or gaseous medium. Galileo was aware of these resistive effects, and he knew that an object would fall "freely" only if they were negligible.

When a cannonball falls toward the Earth (the legendary experiment of Galileo at the Tower of Pisa), the effects of air resistance are small, and the acceleration of the cannonball is closely equal to g. However, there are many cases involving the motion of an object through a resisting (or *viscous*) medium in which the retarding force plays a crucial role.

Viscous drag is not a simple force; in fact, the magnitude of the force varies with the speed of the object in a complicated way. (These effects are discussed in Chapter 18.) At relatively low velocities, the flow of a medium past a smooth object produces a viscous force f_v that is (approximately) directly proportional to the velocity. The force

law that governs this case is called *Stokes' linear law of resistance*[*]:

$$f_v = -bv \qquad \textbf{(6-8)}$$

Figure 6–14 shows an object with a mass m falling toward the Earth. We assume that Stokes' law applies, so the object is subject to the gravitational force, $F_g = mg$, and the viscous force, $f_v = -bv$. Using Newton's second law, we can write

$$mg - bv = m \frac{dv}{dt} \qquad \textbf{(6-9)}$$

Even before solving Eq. 6–9 for $v(t)$, a great deal can be learned about the solution by examining the relationship itself. When the object is first released, we have $v = 0$, and hence the acceleration, $a = dv/dt$, is just equal to g. As a result of this acceleration during the first moments after release, the velocity v increases from zero to some small value. The presence of this velocity reduces the left-hand side of Eq. 6–9, thereby reducing the acceleration dv/dt.

---

[*]First formulated for the case of a sphere moving through a viscous medium by the British mathematician and physicist Sir George Stokes (1819–1903) in 1845.

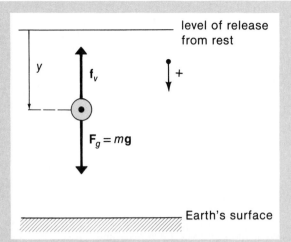

Fig. 6-14. An object falling to Earth with a retarding viscous force $\mathbf{f}_v$ directed opposite to its velocity. The distance from the position of release from rest is y, taken positive downward.

Thus, in the subsequent moments, v increases at a slower rate. This process continues and, eventually, v becomes sufficiently large that the left-hand side of Eq. 6-9 is reduced essentially to zero. (Of course, bv can never exceed mg.) In the limit $t \to \infty$, we have $mg - bv_\infty \to 0$, or

$$v_\infty = \frac{mg}{b} \quad \textbf{(6-10)}$$

This limiting velocity v_∞ is called the *terminal velocity*. Thus, we expect the solution to Eq. 6-9 to have the behavior illustrated in Fig. 6-15.

Returning to Eq. 6-9, we rewrite this expression in the form

$$\frac{dv}{v - \dfrac{mg}{b}} = -\frac{b}{m} dt$$

Fig. 6-15. The expected behavior of the solution $v(t)$ to Eq. 6-9.

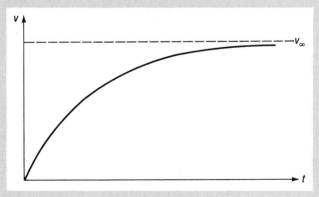

Integrating this equation, we have

$$\int \frac{dv}{v - mg/b} = -\frac{b}{m} \int dt$$

with the result

$$\ln(v - mg/b) = -\frac{b}{m} t + \ln C$$

where we have elected to write the constant of integration as $\ln C$. Taking the antilogarithm of this equation, we obtain

$$v - \frac{mg}{b} = Ce^{-(b/m)t}$$

When $t = 0$, we have $v = 0$; therefore, $C = -mg/b$. Then,

$$v(t) = \frac{mg}{b}(1 - e^{-(b/m)t}) = v_\infty(1 - e^{-(b/m)t}) \quad \textbf{(6-11)}$$

This result indeed conforms to the expected behavior shown in Fig. 6-15. We also see, as expected, that

$$\operatorname*{Lim}_{t \to \infty} v(t) = v_\infty = \frac{mg}{b}$$

We can integrate Eq. 6-11 to obtain the displacement y from the point of release as a function of time:

$$\int dy = \frac{mg}{b} \int (1 - e^{-(b/m)t}) dt$$

so that $\quad y(t) = \dfrac{mg}{b} t + \dfrac{m^2}{b^2} g e^{-(b/m)t} + C'$

When $t = 0$, $y = 0$; hence, $C' = -m^2 g/b^2$, and

$$y(t) = \frac{m^2}{b^2} g(e^{-(b/m)t} - 1) + \frac{mg}{b} t$$

$$= \frac{v_\infty^2}{g}(e^{-(b/m)t} - 1) + v_\infty t \quad \textbf{(6-12)}$$

For the extreme values of the time, the expression for $y(t)$ reduces to simple forms. When t is small compared with m/b ($t \ll m/b$), we can expand the exponential to obtain (see Eq. A-12 in Appendix A)

$$e^{-(b/m)t} = 1 - \frac{b}{m} t + \frac{1}{2} \frac{b^2}{m^2} t^2 - \frac{1}{6} \frac{b^3}{m^3} t^3 + \cdots$$

so that

$$y(t \ll m/b) = \frac{m^2}{b^2}g\left(1 - \frac{b}{m}t + \frac{1}{2}\frac{b^2}{m^2}t^2 + \cdots -1\right) + \frac{mg}{b}t$$

$$= -\frac{mg}{b}t + \frac{1}{2}gt^2 + \cdots + \frac{mg}{b}t$$

Finally, $y(t \ll m/b) \cong \tfrac{1}{2}gt^2$ **(6-13)**

Thus, the object falls in the normal way during the very short time interval before the viscous force becomes important.

For large values of t ($t \gg m/b$) Eq. 6-12 reduces to the approximate linear expression

$$y(t \gg m/b) = v_\infty t - v_\infty^2/g \qquad \textbf{(6-14)}$$

Example 6-14

For how long a time will a 90-kg skydiver fall before reaching 95 percent of his terminal velocity, $v_\infty = 65$ m/s?

Solution:

We find the coefficient b from Eq. 6-10:

$$b = \frac{mg}{v_\infty} = \frac{(90 \text{ kg})(9.80 \text{ m/s}^2)}{65 \text{ m/s}} = 13.6 \text{ kg/s}$$

Now, we have

$$\frac{v(t)}{v_\infty} = 0.950 = 1 - e^{-(b/m)t}$$

or

$$\frac{b}{m}t = \ln 20$$

Thus, $t = \dfrac{90 \text{ kg}}{13.6 \text{ kg/s}} \ln 20 = 19.8 \text{ s}$

Had the skydiver fallen *without* air resistance, the velocity acquired during this time would have been

$$v = gt = (9.80 \text{ m/s}^2)(19.8 \text{ s}) = 194 \text{ m/s}$$

which is considerably greater than the actual velocity,

$$0.950\, v_\infty = 61.8 \text{ m/s}$$

PROBLEMS

Section 6-1

6-1   Two blocks with masses $m_1 = 2$ kg and $m_2 = 3$ kg are suspended vertically by a connecting ideal string that runs over an ideal pulley, as shown in the diagram below. (a) Find the linear acceleration of the system. (b) What is the tension in the connecting string? (c) What is the tension in the string that supports the pulley?

6-2   Two blocks, $m_1 = 5$ kg and $m_2 = 3$ kg, are connected by an ideal string that runs over an ideal peg. The blocks are free to slide on frictionless inclined planes, as shown in the diagram. (a) Find the linear acceleration of the system. (b) What is the string tension? (c) What is the force exerted on the peg?

6-3   Two blocks, $m_1 = 6$ kg and $m_2 = 4$ kg, are connected by a homogeneous rope that has a mass of 1 kg, as shown in the diagram. A constant vertical force, $F = 150$ N, is applied to the upper block m_1. (a) What is the acceleration of the system? (b) What is the tension in the rope at the top end? (c) At the bottom end? (d) At the midpoint of the rope?

$F = 150$ N

m_1 | 6 kg

$m_s = 1$ kg

m_2 | 4 kg

6-4 The compound block-and-tackle arrangement shown in the diagram is used to raise a block with a mass of 300 kg. Calculate the force F necessary at the free end of the rope to impart an upward acceleration of 0.2 m/s² to the block. What is the mechanical advantage of this arrangement? (Assume that the block-and-tackle system and the rope are ideal and massless.)

F

M | 300 kg

6-5 Find the magnitude and direction of the force exerted on the patient's head by the cervical traction system illustrated, if $M = 1.8$ kg.

40°

M

6-6 A ball with a mass $m = 1.5$ kg is attached to a string that is anchored in a rigid ceiling, as shown in the diagram. The length of the string is $\ell = 2.2$ m. A horizontal force **F** draws the ball aside until the string makes an angle $\alpha = 20°$ with the vertical. In this position the ball is at rest. Determine F. If the force is removed, what will be the initial acceleration of the ball?

α | ℓ

F

m

6-7 A rope has a length of 12 m and a mass of 16 kg. The rope hangs from a rigid support. A man whose mass is 80 kg slides down the rope at a constant speed of 0.8 m/s. What is the tension in the rope at a point 6 m from the top when the man has slid to this point?

6-8 Three blocks ($m_1 = 6$ kg, $m_2 = 4$ kg, and $m_3 = 8$ kg) are connected as in the diagram on the next page. The surfaces and the pulleys are frictionless. Determine the motion of the system and find the tension in each string.

Problem 6–8

Section 6–2

6–9 A crate rests on the flat floor of a truck, which is traveling up a hill at an angle of 15°. If the coefficient of static friction between the crate and the truck floor is $\mu_s = 0.4$, what is the maximum acceleration the truck may have if the crate is not to slip?

6–10 Imagine a long stretch of smooth ice (such as a frozen river) over which a hockey puck can slide. If the initial velocity of the puck is v_0 and if the coefficient of kinetic friction between the puck and the ice has the constant value μ_k, derive an expression for the distance the puck will travel before coming to rest. Find the distance for the case in which $v_0 = 8\,\text{m/s}$ and $\mu_k = 0.12$.

6–11 Two blocks, $m_1 = 2\,\text{kg}$ and $m_2 = 5\,\text{kg}$, are connected by an ideal string that runs over an ideal pulley. The block m_1 is free to slide on an inclined plane as shown. The angle of inclination of the plane is 30°, and $\mu_k = 0.30$. (a) Calculate the linear acceleration of the system starting from rest. (b) Calculate the tension in the connecting string.

6–12 A block with a mass of 2 kg initially at rest is pushed
• by a horizontal force, $F = 20\,\text{N}$, as shown in the diagram. Assume that the coefficient of sliding friction is $\mu_k = 0.40$. In each of the following cases, it is necessary to know the direction of motion in order to state the direction of the frictional force. Consider this carefully. (a) Find the acceleration of the block if $\alpha = 10°$. (b) What is the acceleration if $\alpha = 70°$?

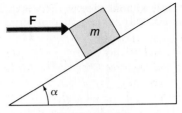

Problem 6–12

6–13 A crate is pulled along a horizontal floor with a
• rope that is maintained at a constant angle α above the horizontal. The coefficient of kinetic friction is μ_k. For a given constant force, show that the maximum acceleration along the floor results when $\tan \alpha = \mu_k$.

6–14 Refer to the diagram in Problem 6–12. Consider the
• situation in which $\alpha = 30°$, $m = 2\,\text{kg}$, and the coefficient of static friction is $\mu_s = 0.50$. Determine the maximum and minimum values of the force F for which the block will remain stationary. What will happen if $F > F_{\text{max}}$ or $F < F_{\text{min}}$?

6–15 A block with a mass m will be pinned by static friction against the wall of a cylindrical shell that rotates about a vertical axis if the angular speed of rotation exceeds a certain critical value ω_0. (a) Show that $\omega_0 = \sqrt{g/\mu_s R}$, where μ_s is the coefficient of static friction between the wall and the block, and where R is the radius of the cylinder. (b) A favorite application of this effect is a rather scary ride at some amusement parks in which the passengers are pinned against the wall of a large rotating cylinder. When the rotational speed exceeds ω_0, the floor on which the passengers were previously standing suddenly drops away! If $R = 5\,\text{m}$, and if the coefficient of static friction between clothing and the wall is $\mu_s = 0.20$, what is the critical angular speed in revolutions per minute (rpm) for this ride?

6–16 An automobile coasts in neutral and moves initially
• on a slick surface at a speed of 20 km/h while its

wheels roll without slipping. The total mass of the car, $m = 1000$ kg, is equally distributed on its four wheels. The relevant coefficients of friction between the tires and roadway are $\mu_s = 0.20$ and $\mu_k = 0.10$. (a) If a braking force is applied (rear wheels only) up to 80 percent of the maximum value supportable by static friction, what is the stopping distance and what is the stopping time? (b) In a panic application of the brakes, the rear wheels lock. Now, what is the stopping distance, and what is the stopping time? (c) If the force due to rolling friction for each of the four wheels is 2 percent of the force due to kinetic friction when a wheel is skidding, in what distance will the

car roll to a stop, and how long will it take? (Assume all other forms of resistance to the motion are negligible.) (d) In view of the answers to part (c), is it permissible to neglect rolling friction when solving parts (a) and (b)? Explain.

6-17 In deep space, a particle ($m = 0.6$ kg) slides in a circular path on the inner surface of a cylinder, the radius of which is $R = 1.8$ m. The coefficient of kinetic friction for this case is $\mu_k = 0.20$. If the particle has a speed $v_0 = 2$ m/s at $t = 0$, find the speed at any later time t. Comment on the behavior of $v(t)$.

SPECIAL TOPIC

Section 6-3

6-18 If a cotton ball has a terminal velocity of 0.8 m/s when falling through air, how long after release will it reach 99 percent of its terminal velocity?

6-19 An object is thrown downward with an initial velocity $\mathbf{v}_0$. (a) Show that
•
$$v(t) = v_\infty + (v_0 - v_\infty)e^{-(b/m)t}$$

(b) A skydiver, after descending for a while at a terminal velocity of 75 m/s, opens a parachute that re-

sults in a final terminal velocity of 10 m/s. How soon after opening his 'chute is the diver's speed reduced to 12 m/s? (Assume that Stokes' law is valid and also that the 'chute is fully deployed as soon as it is opened, not a very good assumption.)

6-20 Consider the general aerodynamic drag force on an
• object to be given by a combination of linear and quadratic terms, $\mathbf{f}_v = -(\alpha + \beta v)\mathbf{v}$, where $\alpha > 0$ and $\beta > 0$. Derive an expression for the terminal velocity of the object that is falling near the surface of the Earth.

Additional Problems

6-21 A painter whose mass is 80 kg is sitting in a light scaffold chair, as shown in the diagram at right. With what steady force must he pull on the rope to achieve an upward acceleration of 0.20 m/s²? Is there any limit to the acceleration he can achieve? (Neglect the mass of the chair, and take the rope and pulley to be ideal elements.)

6-22 Two blocks, $m_1 = 2$ kg and $m_2 = 5$ kg, are supported
• by two ideal pulleys and ideal connecting cords, as shown in the diagram at right. (a) Determine the acceleration of each block. (b) What is the tension in each of the cords shown? [Hint: Remember that the cord running around the two pulleys has a fixed length. From this, deduce the relationship between the accelerations of the two blocks.]

6-23 The three blocks shown in the diagram on p. 168 have
• masses $m_1 = 2$ kg, $m_2 = 3$ kg, and $m_3 = 5$ kg. The blocks are connected by ideal cords and pulleys, A and B. (a) Determine the acceleration of each block. (b) Determine the tension in each of the cords shown. [Hint: The cord that passes over pulley B has

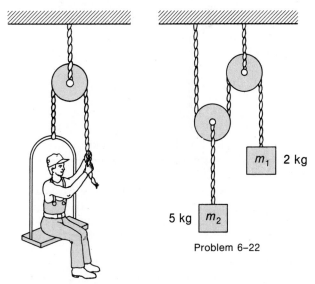

Problem 6-21

Problem 6-22

a fixed length; this means that the upward acceleration of m_1 relative to pulley B is equal to the downward acceleration of m_2 relative to pulley B. Newton's equations hold only in an inertial frame, so you must deduce the accelerations of m_1 and m_2 with respect to the ground in terms of the accelerations with respect to pulley B, using the acceleration of pulley B.]

6-24 Two blocks, $m_1 = 2$ kg and $m_2 = 5$ kg, are initially
• resting on the floor. They are connected by an ideal cord running over an ideal pulley, as shown in the diagram. Find the acceleration of each block and the pulley if the upward force applied to the pulley is: (a) 35 N, (b) 70 N, (c) 140 N. [*Hint:* Refer to the hints in the two previous problems.]

6-25 Two blocks are pulled by a 50-N force, as shown in the diagram. What is the minimum value of μ_s, the coefficient of static friction between the two blocks, that will prevent m_1 from slipping on m_2?

6-26 A block is free to slide down an inclined plane
•• that is equipped with wheels so that it can roll along a horizontal surface, as shown in the accompanying diagram. If the angle of inclination of the plane is 45°, and if the coefficient of static friction between the block and the plane is $\mu_s = 2/5$, show that the cart must have an acceleration to the right of at least $(3/7)g$ if the block is not to slide down the plane.

6-27 A daring skier slides freely down a 45° slope (without using his ski poles or angling his skis). The coefficient of sliding friction between his skis and the snow is $\mu_k = 0.080$. If he starts from rest, calculate his speed 100 m down the slope. How long does it take him to travel this distance?

6-28 A block slides down an inclined plane with an initial speed $v_0 = 3$ m/s. The plane has an angle of 20° with respect to the horizontal, and the coefficient of kinetic friction is $\mu_k = 2/3$. In what distance does the block come to rest?

6-29 A small twig slides down the roof of a house pitched
• at an angle of 30° with respect to the horizontal. The twig starts from rest at a point 1.5 m from the edge and strikes the ground 2 m from the house, which has its roof-line 4 m above the gound. What is the coefficient of kinetic friction between the twig and the roof? (Assume zero air resistance during the fall.)

6-30 Two blocks with masses $m_1 = 5$ kg and $m_2 = 3$ kg
• are tied together by an ideal string. The blocks slide

down an inclined plane that makes an angle of 30° with the horizontal, as shown in the diagram below. If the coefficient of kinetic friction between m_1 and the plane is 0.20, and that between m_2 and the plane is 0.30, find the acceleration of the system and the tension in the string. What would happen if the blocks were reversed in their positions, with m_2 leading the way down?

6–31 A block is pushed along a horizontal surface by a force F that makes an angle α with the horizontal, as shown in the diagram. Taking the coefficient of sliding friction to be μ_k, show that as α is increased, a critical angle will be reached at which forward motion is impossible, regardless of the magnitude of F. Find the expression for this critical angle.

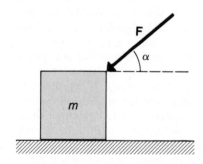

6–32 Two blocks, m_1 and m_2, are connected by an ideal
• string that runs over an ideal pulley. The block m_1 is free to slide on an inclined plane with an adjustable inclination angle α, as shown in the diagram below. It is observed that when α is less than 15°, the block m_1 slides up the incline, and when α is greater than 63°, it slides down the incline. (a) What is the coefficient of static friction between block m_1 and the inclined plane? (b) If $m_1 = 2$ kg, what is m_2?

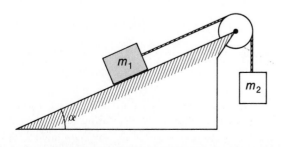

6–33 A stone with a mass m is tied to the end of a string
• and twirled in a vertical plane. Show that ω, the angular frequency of the motion, is related to the angle ϕ by the equation $\omega^2 = \omega_0^2 + (2g/R)(1 - \cos\phi)$, where ω_0 is the angular frequency corresponding to $\phi = 0$.

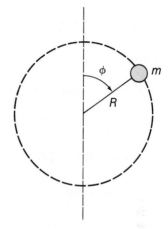

$$\left[Hint: \frac{dv_T}{dt} = R\frac{d\omega}{dt} = R\frac{d\omega}{d\phi}\frac{d\phi}{dt} = R\omega\frac{d\omega}{d\phi}, \text{ where } v_T \right.$$

is the tangential velocity. $\Big]$

6–34 A penny is placed on a record turntable that revolves at $33\frac{1}{3}$ rpm. If the coefficient of static friction between the penny and the turntable is $\mu_s = 0.25$, how far from the center of rotation may the penny be placed if it is to remain stationary on the turntable?

6–35 A jet airliner, traveling with a ground speed of 700 km/h, banks into a turn that has a radius $R = 8$ km. If passengers standing in the center aisle experience forces exerted on them by the floor that are entirely perpendicular to the floor, what is the banking angle α of the aircraft?

6–36 A lead ball and a piece of paper loosely crumpled
• into a spherical shape are dropped simultaneously from a tower that is 20 m high. Assume the terminal velocity for the lead ball to be 600 m/s, and that for the paper ball to be 60 m/s. How far above the ground is the paper ball when the lead ball strikes the ground? [*Hint:* Expand the exponential to order t^3.]

6-37 Equation 6-9 is actually a *differential equation.*
•• Sometimes equations of this type cannot be solved in terms of simple functions. In these cases, the solutions must be obtained by numerical methods. Use this approach to solve Eq. 6-9. First, write the equation in terms of finite increments:

$$\Delta v = \left(g - \frac{b}{m}v\right)\Delta t = g\left(1 - \frac{v}{v_\infty}\right)\Delta t$$

Use the notation

$$v_{i+1} = v_i + \Delta v_i \quad \text{and} \quad t_{i+1} = t_i + \Delta t_i$$

in which

$$\Delta v_i = g\left(1 - \frac{v_i}{v_\infty}\right)\Delta t_i$$

Consider a case for which $v_\infty = 100$ m/s. Take all intervals to be $\Delta t_i = 2$ s and let $v_0 = 0$ at $t_0 = 0$ (that is, for $i = 0$). From $v_0 = 0$, we find v_1 to be

$$v_1 = v_0 + \Delta v_0 \quad ; \quad \Delta v_0 = 2g\left(1 - \frac{0}{v_\infty}\right)$$

Then, $v_2 = v_1 + \Delta v_1 \quad ; \quad \Delta v_1 = 2g\left(1 - \frac{v_1}{v_\infty}\right)$

and $v_3 = v_2 + \Delta v_2 \quad ; \quad \Delta v_2 = 2g\left(1 - \frac{v_2}{v_\infty}\right)$

and so forth. Calculate numerical values of v_i by this process up to at least $i = 10$. Compare your results with the exact values calculated from the actual solution:

$$v_i = v_\infty(1 - e^{-(b/m)t_i})$$

Suggest a way for improving the above numerical method, which is accurate to only about 10 percent.

6-38 A steel block with a mass of 100 g is set to
•• sliding along a horizontal steel plate with an initial speed $v_0 = 300$ m/s. The coefficient of kinetic friction is

$$\mu_k(v) = 0.27\frac{(1 + 0.0044\,v)}{(1 + 0.064\,v)}$$

where v is in m/s. (This expression is commonly used for high-speed, steel-on-steel friction.) (a) Calculate the stopping time by using numerical integration. [*Suggestion:* Use a linear extrapolation, $v_f = v_i - a_i\,\Delta t$, with a_i evaluated at v_i by using $\mu_k(v_i)$. Then, v_f becomes v_i for the next interval Δt, etc. Start with $\Delta t = 100$ s, and switch to $\Delta t = 50$ s when $a_i\,\Delta t \cong \frac{1}{2}v_i$; switch again to $\Delta t = 25$ s when next this occurs, etc. Halt the process when $v_f \cong 2$ m/s.] (b) Determine an "average" value for μ_k by using $\bar{v} = \frac{1}{2}v_0$ in the expression for $\mu_k(v)$. From this, calculate the corresponding deceleration and determine the stopping time. Compare this result with that in (a) and with the exact answer, $T = 670$ s. [The exact analytic solution for the stopping time, given the initial speed v_0, is $T = 5.498\,v_0 - 1163.7\ln(1 + 0.0044\,v_0)$.] You should find that your result obtained by numerical integration is within 5 percent of the true answer, whereas the selected constant value of μ_k gives a result that is within 8 percent. You can therefore appreciate why there is such a strong temptation to assume that μ_k is independent of speed! If one selects some reasonable mean value for μ_k in the velocity range of interest, this approximation affords a useful first estimate of the solution. What would be the error in T if we used the low-speed limit for the friction coefficient, $\mu_k(v \to 0) = 0.27$?

CALCULUS III

DEFINITE INTEGRALS

Definite integrals are used extensively throughout the remainder of this book.

In the appendix, Calculus II, we discussed integration* simply as the operation that is inverse to differentiation. We now introduce a second general interpretation of the integral of a function, one that has great practical utility. In this new view, we find that integration can be looked upon as a summing operation. Then, it becomes possible to associate an *area* with an integral, thereby giving a geometric interpretation to the integration process.

III-1 DEFINITE INTEGRALS

The Basic Definition. First, recall (Section II-1) that the *indefinite integral* of a function $f(x)$ is expressed as

$$\int f(x)\, dx = F(x) + C \qquad \text{(III–1)}$$

We define the *definite integral* of $f(x)$ to be

$$\int_a^b f(x)\, dx = [F(x) + C]\Big|_a^b = [F(b) + C] - [F(a) + C]$$

---

*The integral calculus was invented independently by Isaac Newton and by Gottfried Leibnitz (1646–1716), a German philosopher and diplomat. Both men correctly perceived the important intimate connection between the antiderivative and the definite integral. The formulation of the calculus made by Leibnitz was somewhat simpler than Newton's and constitutes the basis of the subject as we know it today.

or, simply,

$$\int_a^b f(x)\, dx = F(b) - F(a) \qquad \text{(III–2)}$$

where the notation implies that we first substitute $x = b$ into $F(x) + C$, and then subtract the result of substituting $x = a$ into $F(x) + C$. We call b the *upper limit* of the integral and a the *lower limit* of the integral. At this point, Eq. III–2 should simply be viewed as a formal definition, but it has a fundamental importance in the integral calculus. Observe that Eq. III–2 results in a *numerical value,* not a function.

You should have no difficulty in proving that

$$\int_a^c f(x)\, dx = \int_a^b f(x)\, dx + \int_b^c f(x)\, dx \qquad \text{(III–3)}$$

and

$$\int_a^b f(x)\, dx = -\int_b^a f(x)\, dx \qquad \text{(III–4)}$$

Example III–1

Evaluate the definite integral of $f(x) = x^2$ (a) between the lower limit $a = 1$ and the upper limit $b = 3$, and (b) between $a = -1$ and $b = 3$.

Solution:

(a) $\displaystyle \int_a^b f(x)\, dx = \int_1^3 x^2\, dx = \frac{x^3}{3}\Big|_1^3 = \frac{(3)^3}{3} - \frac{(1)^3}{3} = \frac{26}{3}$

171

(b) $\int_{-1}^{3} x^2 \, dx = \frac{x^3}{3} \Big|_{-1}^{3} = \frac{(3)^3}{3} - \frac{(-1)^3}{3} = 9 + \frac{1}{3} = \frac{28}{3}$

Example III–2

Evaluate the integral of $\theta \cos \theta$ between $\theta = 0$ and $\theta = \pi/2$.

Solution:

Refer to the table of integrals inside the back cover.

$$\int_0^{\pi/2} \theta \cos \theta \, d\theta = (\theta \sin \theta + \cos \theta) \Big|_0^{\pi/2}$$

$$= \left(\frac{\pi}{2} \sin \frac{\pi}{2} + \cos \frac{\pi}{2} \right) - (0 \cdot \sin 0 + \cos 0)$$

$$= \frac{\pi}{2} - 1 = 0.5708$$

The Definite Integral as a Function of the Upper Limit.

The basic definition of the definite integral, Eq. III–2, involves an evaluation between two specific values of x, namely, $x = a$ and $x = b$. It is also permissible to consider the upper limit to be a free variable. In this event, we refer to the definite integral as a function of its upper limit, and write

$$\int_a^{x'} f(x) \, dx = F(x') - F(a) \tag{III–5}$$

Of course, there is no need to use the separate symbol x' for the new upper limit variable. Customarily, one simply uses x for the limit variable as well as for the running variable in $f(x)$ of the integrand. No confusion need arise owing to this usage.*

We also note that the constant of integration C appearing in Eq. III–1 can be associated with the value of the integral at one of the limits. If we write

$$\int_a^x f(x) \, dx = F(x) + C = F(x) - F(a)$$

---

*When emphasis is required, we distinguish the variable upper limit from the running variable in the integrand by replacing the latter with another symbol—say, u—and writing

$$\int_a^x f(u) \, du = F(x) - F(a)$$

It should be evident that the definite integral may also be considered to be a function of its lower limit. In view of Eq. III–4, nothing unusual is introduced by this exchange.

we see that $\qquad C = -F(a)$

Change of Variable. A useful technique for evaluating many integrals involves changing the variable (see Section II–1). When such a variable change is made, a definite integral can be evaluated using the new variable without returning to the original variable after the integration step. As the following examples show, new upper and lower limits corresponding to the original limits must be substituted before the integral can be evaluated.

Example III–3

Evaluate $\int_0^1 x \sqrt{x^2 + 1} \, dx$.

Solution:

Let $v = x^2 + 1$; then, $dv = 2x \, dx$. To determine the new upper and lower limits, we note that

$$\text{when } x = 0, \text{ we have } v = 1$$

and $\qquad$ when $x = 1$, we have $v = 2$

Therefore,

$$\int_0^1 x \sqrt{x^2 + 1} \, dx = \int_1^2 \frac{v^{1/2}}{2} \, dv = \frac{1}{3} v^{3/2} \Big|_1^2$$

$$= \tfrac{1}{3} (\sqrt{8} - 1) = 0.6095$$

Example III–4

Evaluate $\int_0^2 2x \, e^{x^2} \, dx$.

Solution:

Let $u = x^2$; then, $du = 2x \, dx$. The new limits are:

$$\text{when } x = 0, \text{ we have } u = 0$$

and $\qquad$ when $x = 2$, we have $u = 4$

Therefore,

$$\int_0^2 2x \, e^{x^2} \, dx = \int_0^4 e^u \, du = e^u \Big|_0^4 = e^4 - 1 = 53.60$$

Separable Forms. It sometimes happens that a problem involves a differential relationship of the form

$$f(y) \, dy = g(x) \, dx \tag{III–6}$$

where $f(y)$ is a function only of y and $g(x)$ is a function only of x. If we are given a boundary condition such as $y = k$ when $x = a$, we can incorporate these values as limits and write

$$\int_k^y f(y)\, dy = \int_a^x g(x)\, dx \qquad \textbf{(III–7)}$$

This is another way of handling the constant of integration in situations such as that in Example II–5.

Example III–5

Given $y^{1/2}\, dy = \sin\theta\, d\theta$ and that $y = 1$ when $\theta = \pi/4$, find the relationship between y and θ.

Solution:

Here, we have

$$\int_1^y y^{1/2}\, dy = \int_{\pi/4}^{\theta} \sin\theta\, d\theta$$

$$\frac{2}{3} y^{3/2}\Big|_1^y = -\cos\theta\,\Big|_{\pi/4}^{\theta}$$

$$\frac{2}{3}(y^{3/2} - 1) = -\cos\theta + \frac{\sqrt{2}}{2}$$

from which

$$y^{3/2} = -\frac{3}{2}\cos\theta + \left(1 + \frac{3\sqrt{2}}{4}\right)$$

or

$$y = (-\tfrac{3}{2}\cos\theta + 2.061)^{2/3}$$

Improper Integrals. It is sometimes the case that the variable in a problem must be integrated over an infinite interval. An integral with one or both limits equal to $+\infty$ or $-\infty$ is called an *improper integral*. For instance, when the upper limit is ∞, we mean

$$\int_a^{\infty} f(x)\, dx = \lim_{b\to\infty} \int_a^b f(x)\, dx \qquad \textbf{(III–8)}$$

Of course, the integral is meaningful only if this limit exists.

Example III–6

Evaluate the integral of $1/x^2$ between the limits $x = 1$ and $x = +\infty$.

Solution:

$$\int_1^{+\infty} \frac{dx}{x^2} = \lim_{b\to+\infty} \int_1^b \frac{dx}{x^2} = \lim_{b\to+\infty} \left(-\frac{1}{x}\right)\Big|_1^b$$

$$= \lim_{b\to+\infty} \left(-\frac{1}{b} + 1\right) = 1$$

III–2 THE DEFINITE INTEGRAL AS AN AREA

Areas of Simple Geometric Figures. We know that a rectangle with sides of lengths s and t has an area $A = st$. (•Can you think of any other way to define *area*?) We also know that the area of a triangle is one-half the product of the length l of its base and its altitude h, or $A = \frac{1}{2}lh$. The consistency of these two statements of area is demonstrated in Fig. III–1, where we see that the constructed (inscribing) rectangle has twice the area of the given triangle.

In a like manner, the area of any regular or irregular polygon can always be decomposed into the sum of the areas of triangles, as shown in Fig. III–2. This construction can be used to define the polygon area.

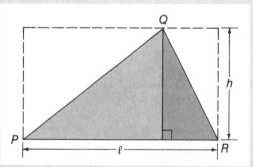

Fig. III–1. The area A of the triangle PQR is one-half the area of the constructed (inscribing) rectangle with sides l and h; that is, $A = \frac{1}{2}lh$.

Fig. III–2. The area of the irregular polygon $OPQRST$ represented as the sum of the areas of the triangles OPQ, OQR, ORS, and OST.

When we consider the area inside a "smooth" curve, such as a circle, we arrive at the concept of an area as the limit of a sum of areas derived from some geometric construction. Thus, following the ancient Greek mathematician and engineer Archimedes (287?–212 B.C.), who invented this technique, we can compute the area of a circle by inscribing within it a regular polygon with n sides. The area is then defined to be the limit of the area of the polygon, as the number of sides n increases indefinitely.

Example III–7

Show that the area of a circle with a radius R can be calculated from the limit of the area of a regular inscribed polygon with an ever greater number of sides.

$$\text{area} = \tfrac{1}{2}\ell h = \tfrac{1}{2}(2R\sin\alpha)(R\cos\alpha)$$
$$\alpha = \frac{2\pi}{2n} = \frac{\pi}{n}$$

Solution:

As indicated in the diagram, an inscribed n-sided regular polygon has an area n times that of the isosceles triangle with an apex half-angle $\alpha = \pi/n$; thus,

$$A_n = \tfrac{1}{2}n(2R\sin\alpha)(R\cos\alpha)$$

$$= nR^2 \sin\frac{\pi}{n}\cos\frac{\pi}{n}$$

Hence, the area of the circle is (using Eq. A–34 in Appendix A and the result of Example I–1)

$$A = R^2 \operatorname*{Lim}_{n\to\infty} n\sin\frac{\pi}{n}\cos\frac{\pi}{n} = R^2 \operatorname*{Lim}_{n\to\infty} n\left(\frac{1}{2}\sin\frac{2\pi}{n}\right)$$

$$= R^2 \operatorname*{Lim}_{n\to\infty} \pi \frac{\sin\dfrac{2\pi}{n}}{\dfrac{2\pi}{n}} = \pi R^2 \operatorname*{Lim}_{u\to0} \frac{\sin u}{u}$$

or
$$A = \pi R^2$$

(•Can you show that $A_n < A$, for any finite n?)

The Area under a Curve.   Consider next the area of a region enclosed by a curve $f(x)$, the x-axis, and ordi-

nates (vertical lines intercepting the curve) erected at $x = a$ and $x = b$. Such a region is shown shaded in Fig. III–3a. We refer to the area of this region as the area "under the curve." Taking a hint from the previous discussion, we try approximating the area by the sum of the areas of n inscribed rectangles. Each rectangle has an equal base length Δx constructed by dividing the interval, $b - a$, into n equal parts [that is, $\Delta x = (b - a)/n$], and each has a height $y_i = f(x_i)$ corresponding to the *minimum* ordinate within its subinterval Δx (see Fig. III–3b). Then,

$$A_n = \sum_{i=1}^{n} y_i \,\Delta x \leqslant A$$

so that A_n is a *lower bound* of A.

If we approximate the area by n circumscribing rectangles, each with base length Δx as before, but with height equal to the *maximum* ordinate within its subinterval Δx, we arrive at the situation shown in Fig. III–3c. Calling the sum of these rectangular areas $A'_n = \sum_{i=1}^{n} y_i \,\Delta x$, we now observe that $A'_n \geqslant A$, so that A'_n is an *upper bound* of A. We might expect (correctly) that the lower and upper bounds converge to the true area A as $n \to \infty$. Note that the interval defined by a and b remains fixed as we take the limit $n \to \infty$, thereby requiring Δx to become vanishingly small. Thus, for either inscribed or circumscribed rectangles, we have

$$A = \operatorname*{Lim}_{n\to\infty} \sum_{i=1}^{n} y_i \,\Delta x = \operatorname*{Lim}_{n\to\infty} \sum_{i=1}^{n} f(x_i) \,\Delta x \quad \textbf{(III–9)}$$

with
$$\Delta x = \frac{b - a}{n}$$

When the limit of the sum in Eq. III–9 exists as $n \to \infty$, the *Fundamental Theorem of Calculus* (which we give without proof) states that*

$$\int_a^b y\,dx = \int_a^b f(x)\,dx = \operatorname*{Lim}_{n\to\infty} \sum_{i=1}^{n} f(x_i)\,\Delta x \quad \textbf{(III–10)}$$

According to Eqs. III–9 and III–10, the definite integral of $f(x)$ is equal to the area under the curve defined by $f(x)$ between the ordinates erected at $x = a$ and $x = b$.

---

*The sums in Eqs. III–9 and III–10 need not involve equal base lengths Δx, and the ordinates y_i may have any value defined by $f(x)$ within a particular interval Δx_i. The sum so obtained,

$$\sum_{i=1}^{n} f(x_i)\,\Delta x_i,$$

is referred to as the *Riemann Sum* of $f(x)$ on the interval $a \leqslant x \leqslant b$. Properly, the Fundamental Theorem refers to this sum in the limit $n \to \infty$; Eq. III–10 is actually a special case of the general theorem.

Fig. III-3. The area associated with the function $y = f(x)$. (a) The shaded region enclosed by $f(x)$, the x-axis, and the ordinates at $x = a$ and $x = b$ has area A. (b) An approximation to the area A by the sum of n inscribed rectangular subdivisions. (c) An approximation to the area A by the sum of n circumscribing rectangular subdivisions.

Example III-8

Following the procedure used to obtain Eq. III-9, calculate the area under the straight line $y = \sigma x$, between ordinates at $x = 0$ and $x = \ell$, (a) using a set of inscribed rectangles, and (b) using a set of circumscribed rectangles. (c) Show that the results of (a) and (b) are consistent with the Fundamental Theorem (Eq. III-10).

Solution:

(a) The diagram below shows the interval $0 \leqslant x \leqslant \ell$, divided into n equal parts so that $\Delta x = \ell/n$. The minimum ordinates y_i are regularly spaced at values $x_i = (\ell/n)i$, with i running from 0 to $n - 1$; therefore, $y_i = \sigma(\ell/n)i$. Thus,

$$A_n = \sum_{i=0}^{n-1} y_i \, \Delta x = \sum_{i=0}^{n-1} \left(\sigma \frac{\ell}{n} i\right)\left(\frac{\ell}{n}\right) = \frac{\sigma \ell^2}{n^2} \sum_{i=0}^{n-1} i$$

From Eq. A-1 in Appendix A with $a = d = i$ (but with the

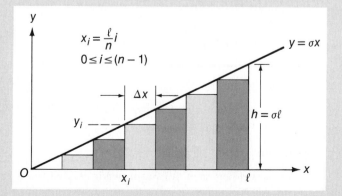

sum running from 0 to $n - 1$), we have

$$\sum_{i=0}^{n-1} i = \frac{(n - 1)n}{2}$$

Hence,

$$A_n = \frac{\sigma \ell^2}{n^2} \frac{(n - 1)n}{2} = \sigma \ell^2 \frac{(n - 1)}{2n}$$

Then,

$$A = \underset{n \to \infty}{\text{Lim}} A_n = \underset{n \to \infty}{\text{Lim}} \sigma \ell^2 \frac{(n - 1)}{2n}$$

$$= \tfrac{1}{2}\sigma \ell^2 = \tfrac{1}{2}(\sigma \ell)\ell = \tfrac{1}{2}h\ell$$

This agrees with the simple geometric definition of the area of a triangle with base length ℓ and altitude h.

(b) To obtain the set of circumscribing rectangles shown on page 176, we note that again the ordinates are equally spaced with $x_i = (\ell/n)i$ and $y_i = \sigma(\ell/n)i$; however, i now runs from 1 to n. Evidently,

$$A'_n = \sum_{i=1}^{n} y_i \, \Delta x = \sum_{i=1}^{n} \left(\sigma \frac{\ell}{n} i\right)\left(\frac{\ell}{n}\right) = \frac{\sigma \ell^2}{n^2} \sum_{i=1}^{n} i$$

Again, using Eq. A-1 in Appendix A (but now with the sum running from 1 to n), we have

$$\sum_{i=1}^{n} i = \frac{n(n + 1)}{2}$$

so that

$$A'_n = \frac{\sigma \ell^2 (n + 1)}{2n}$$

and
$$A = \lim_{n \to \infty} A'_n = \lim_{n \to \infty} \frac{\sigma \ell^2 (n+1)}{2n}$$

$$= \tfrac{1}{2}\sigma\ell^2 = \tfrac{1}{2}h\ell$$

In these calculations, we see that $A_n < A < A'_n$. In one case, A is approached as the limit from below, and in the other case it is approached from above.

(c) Finally, we express the area as

$$A = \int_0^\ell y \, dx = \int_0^\ell \sigma x \, dx = \sigma \cdot \tfrac{1}{2}x^2 \Big|_0^\ell = \tfrac{1}{2}\sigma\ell^2$$

which confirms our expectation.

"Negative" Areas.   When part or all of a function in the interval $a \leqslant x \leqslant b$ is negative, we associate a "negative" area with such portions. When both negative and positive contributions exist, we can define the total area "under the curve" to be the algebraic sum of the parts, as shown in Fig. III-4. The definite integral between the limits a and b automatically gives this sum for areas such as $A = A_1 + (-A_2)$ in Fig. III-4. (•Do you see why?)

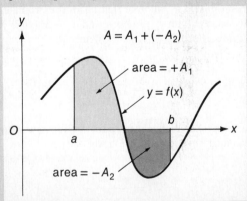

Fig. III-4. Determination of the area "under the curve" when $y = f(x)$ is negative for part of the region of interest. The total area A is defined to be $A_1 - A_2$. Alternative definitions are also possible.

Areas for Curves Given in Parametric Form.   It sometimes happens that the equation of a curve is given in parametric form. That is, a curve is specified by $x = f(t)$ and $y = g(t)$. The definite integral

$$A = \int_a^b y \, dx$$

can be readily evaluated when the functions $f(t)$ and $g(t)$ are continuous and single-valued. By direct integration, using $y = g(t)$ and $dx = (df/dt)dt$, we obtain

$$A = \int_{t_a}^{t_b} g(t) \, \frac{df(t)}{dt} \, dt \qquad \textbf{(III–11)}$$

Example III–9

Find the area under the hyperbola $xy = 2$, between the lines $x = 1$ and $x = 4$.

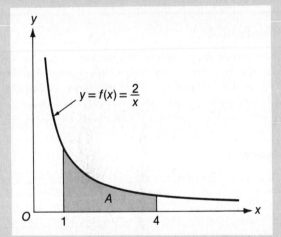

Solution:

Here,

$$A = \int_1^4 y \, dx = 2 \int_1^4 \frac{dx}{x} = 2(\ln x) \Big|_1^4$$

$$= 2 \ln 4 = 2.773$$

Example III–10

Find the area included between the line $y = 3x + 2$ and the parabola $y = 2x^2$.

Solution:

Because the bottom of the area shown in the diagram is not the x-axis, we must calculate the area a bit differently.

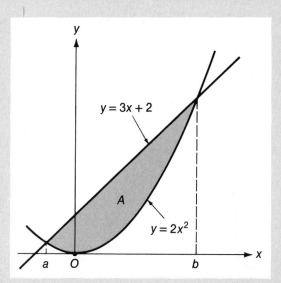

The shaded area can be considered in either of two exactly equivalent ways: (1) as the remainder when the area under the parabola is subtracted from the area under the straight line, or (2) as the limit of the sum of the areas of rectangles with width Δx and height $(3x + 2) - 2x^2$. In either case, the integral required by Eq. III-10 becomes

$$\int_a^b [(3x + 2) - 2x^2]\, dx$$

To determine the limits $x = a$ and $x = b$, we first find the points of intersection. At these points, we have $2x^2 = 3x + 2$, or $2x^2 - 3x - 2 = 0$; factoring gives

$$(2x + 1)(x - 2) = 0$$

so that $a = -\frac{1}{2}$, $b = +2$. Therefore,

$$A = \int_{-1/2}^2 [(3x + 2) - 2x^2]\, dx = \left(\frac{3x^2}{2} + 2x - \frac{2}{3}x^3\right)\Bigg|_{-1/2}^2$$

$$= \tfrac{14}{3} - (-\tfrac{13}{24}) = \tfrac{125}{24}$$

Example III-11

The parametric equations for an ellipse centered on the origin are

$$x = a \cos \phi, \qquad y = b \sin \phi$$

Find the area of the ellipse.

Solution:

We first calculate the area of the portion of the ellipse that lies in the first quadrant (the shaded region in the figure). Then, taking the total area to be 4 times this value, we have

$$A = 4 \int_0^a y\, dx$$

Now, $\qquad dx = -a \sin \phi\, d\phi$

and $\qquad \phi = \pi/2$ when $x = 0$
$\qquad\qquad \phi = 0$ when $x = a$

Hence,

$$A = 4 \int_{\pi/2}^0 (b \sin \phi)(-a \sin \phi\, d\phi)$$

$$= 4ab \int_0^{\pi/2} \sin^2 \phi\, d\phi = 2ab(\phi - \tfrac{1}{2}\sin 2\phi)\Bigg|_0^{\pi/2}$$

$$= \pi ab$$

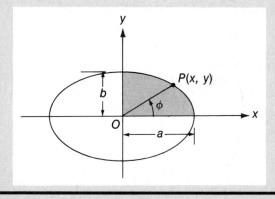

PROBLEMS

Section III-1

III-1   Verify the following definite integrals:

(a) $\displaystyle\int_1^2 (x^2 + 3)\, dx = \frac{16}{3}$ (b) $\displaystyle\int_{-1}^{+1} y^3\, dy = 0$

(c) $\displaystyle\int_0^\pi \sin \theta\, d\theta = 2$ (d) $\displaystyle\int_0^1 x e^x\, dx = 1$

III-2   Show that $y(t) = \displaystyle\int_1^t \frac{dt}{\sqrt{5 - t}} = 4 - 2\sqrt{5 - t}$

[*Hint:* Let $u = 5 - t$.]

III-3 Show that

(a) $\int_0^\theta \sin^2 \theta \, d\theta = \frac{1}{2}\theta - \frac{1}{4}\sin 2\theta$

(b) $\int_0^\theta \cos^2 \theta \, d\theta = \frac{1}{2}\theta + \frac{1}{4}\sin 2\theta$

[*Hint:* Use Eqs. A-32 and A-33 in Appendix A.]

III-4 Evaluate $\int_0^\pi \sin^3 \theta \, d\theta$. [*Hint:* Write $\sin^3 \theta = \sin^2 \theta \sin \theta = (1 - \cos^2 \theta) \sin \theta$ and let $x = \cos \theta$; make the corresponding changes in the limits.]

III-5 Evaluate $\int_0^a \dfrac{a^2 \, dx}{(x^2 + a^2)^{3/2}}$. [*Hint:* Consider the change in variable $\cos \beta = a/(x^2 + a^2)^{1/2}$; make the corresponding changes in the limits. Refer to the diagram for the geometry. This substitution is often useful.]

III-6 Given $\kappa x \, dx = mv \, dv$ (where κ and m are constants) and $x = 0$ when $v = v_0$. Find the relationship between x and v, using Eq. III-7.

III-7 Given $y^2 \, dy = e^{-at} \, dt$ and $t = 0$ when $y = 0$. Find the relationship between y and t.

III-8 Verify the following:

(a) $\int_1^{+\infty} \dfrac{dx}{x^2} = 1$ (b) $\int_0^{+\infty} ve^{-v^2} \, dv = \dfrac{1}{2}$

(c) $\int_{-\infty}^{+\infty} \dfrac{2x \, dx}{(1 + x^2)^2} = 0$

$\left[\textit{Hint:} \text{ Write } \int_{-\infty}^{+\infty} = \int_{-\infty}^{0} + \int_{0}^{+\infty} \right]$

Section III-2

III-9 In the following, use the appropriate definite integrals and determine the area under each of the given curves between ordinates erected at the stated values of $x = a$ and $x = b$:

(a) $y = x^2$; $a = 0$, $b = 1$

(b) $y = e^x$; $a = 0$, $b = 1$

(c) $y = \dfrac{4}{x}$; $a = 1$, $b = e$

(d) $y = (x + 9)^{-1/2}$; $a = 0$, $b = 16$

(e) $y = \sin x$; $x = 0$, $x = 3\pi/2$ (Note that a portion of this area is negative. You might find it instructive to sketch the graph of the integrand, indicating the limits.)

III-10 Prove the assertion that the area under the curve of a symmetric (even) function [one for which $f(-x) = f(+x)$], between the ordinates $x = -a$ and $x = +a$ is *twice* the area between ordinates $x = 0$ and $x = +a$, whereas for an antisymmetric (odd) function [one for which $f(-x) = -f(x)$], the area is *zero* for the same limits.

III-11 Find the area in the first quadrant enclosed (bounded) by the two curves $y = x/(1 + x^2)$ and $y = x/5$.

III-12 Find the area enclosed (bounded) by the intersecting curves $y^2 = 32x$ and $y = x^3$.

III-13 Find the total area under the curve known as the
• witch of Agnesi, $y = f(x) = 8a^3/(x^2 + 4a^2)$. [*Hint:* Use the same type of substitution suggested in Problem III-5. The integration is over the range $-\infty < x < +\infty$; what is the corresponding range of the angle β?]

III-14 Find the entire enclosed area of a circle with radius R given by the parametric equations $y = R \sin \phi$ and $x = R \cos \phi$, by calculating the area of the portion in the first quadrant as in Example III-11.

III-15 A student, given the task of finding the entire enclosed area of a circle of radius R, decides to base his answer on the area of the portion in the first quadrant for the function $y^2 + x^2 = R^2$. Faced with evaluating

$$A = 4 \int_0^R y \, dx = 4 \int_0^R (R^2 - x^2)^{1/2} \, dx$$

he decides on the variable substitution $x = R \cos \phi$. Contrast this approach with that called for in Problem III-14.

MECHANICAL ENERGY AND WORK

In previous chapters we applied Newton's laws of dynamics in a variety of situations involving the motion of particles. For the cases considered, solutions were obtained in which the particle's location, its velocity, and its acceleration were given as explicit functions of time. Thus, we were able to specify the motion of the particle through a set of equations for $\mathbf{r}(t)$, $\mathbf{v}(t)$, and $\mathbf{a}(t)$.

The concepts developed in this chapter permit us, in many cases, to express directly the relationship connecting position and velocity without obtaining a time-dependent solution, indeed, without any reference to time whatsoever. For example, suppose that a block slides down a ramp. We wish to know the velocity of the block at the bottom of the ramp but we are not really interested in *when* the block arrives there. The techniques considered here allow us to obtain solutions to such problems in a simple and direct way. Although the information obtained is less than that contained in the complete time-dependent solution, nonetheless many interesting characteristics of a system and its motion can be understood more easily. Moreover, we can treat in a straightforward way cases involving forces that are not constant with position, cases that would be difficult to treat by applying Newton's laws directly.

The procedure that we follow involves the concept of *energy*. The importance of this idea lies in the fact that although energy can exist in many different forms, the total energy content of an isolated system remains constant. This is the celebrated law of energy conservation. We begin the discussion by carefully refining our ideas about the everyday concepts of energy and the closely related quantity, work.

7-1 WORK DONE BY CONSTANT FORCES

Work Involved in Infinitesimal Displacements. We first consider in some detail the concept of *work*. Everyone is familiar with the colloquial use of this term in ordinary conversation. We usually think of work as involving physical activity and muscular effort or in terms of work done by machines. Indeed, work is done in all of these situations, but the definition needed here is more restricted and more precise. If a force acts on an object and causes it to undergo a displacement, we say that work was done by the force. To avoid the necessity of treating rotational effects, we again

Fig. 7–1. The block (a particle) is acted upon by a force **F** during a displacement $\Delta\mathbf{s}$. The displacement is imagined to be infinitesimally small so that, throughout the motion, the force **F** remains constant.

confine attention to particles and consider only the work done by forces during translational displacements.

Figure 7–1 shows a block (considered to be a particle) that is acted upon by a force **F**. During the time that this force is acting, the block undergoes an infinitesimal displacement $\Delta\mathbf{s}$. The infinitesimal mechanical work ΔW done by the agency responsible for the force **F** is defined to be the product $F\,\Delta s$ multiplied by the cosine of the angle between **F** and $\Delta\mathbf{s}$. (There may be other forces acting on the block, but we are interested here only in the work associated with **F**.) Even though it is generated by two vector quantities—force and displacement—work is a *scalar* quantity. Thus, we write

$$\Delta W = F\,\Delta s \cos\theta$$

We can write this definition in more general terms as

$$\Delta W = \mathbf{F} \cdot \Delta\mathbf{s} \qquad\qquad \textbf{(7–1)}$$

because the dot product means precisely $F\,\Delta s \cos\theta$.

Notice that $\Delta W = 0$ when $\cos\theta = 0$ or $\theta = \pi/2$. That is, when the force **F** and the displacement $\Delta\mathbf{s}$ are perpendicular, no work is done.

If the displacement $\Delta\mathbf{s}$ is zero, the work done is necessarily also zero. For example, consider a student holding a ball in his outstretched hand. Because the ball is stationary, the student does no work on the ball; nevertheless, he feels tired—he is convinced that he has done work. Indeed, the student *has* done work. However, this work is *internal work* that is done in the student's muscles. Work always involves (force) × (distance); in the case of internal muscular work, electrons and ions are moved by electric forces generated in the muscle fibers. Although work is done in the muscles, this work is not done on the ball and therefore does not manifest itself in any change in the position or physical characteristics of the ball.

The dimensionality of work is

$$[\Delta W] = [F][\Delta s] = (\mathrm{MLT^{-2}})(\mathrm{L}) = \mathrm{ML^2T^{-2}}$$

The SI unit of work is the $\mathrm{kg}\cdot\mathrm{m^2/s^2}$ or $\mathrm{N}\cdot\mathrm{m}$. We use the special name *joule* (J) for the unit of work:*

$$1\,\mathrm{J} = 1\,\mathrm{N}\cdot\mathrm{m} = 1\,\mathrm{kg}\cdot\mathrm{m^2/s^2}$$

Work Done by Several Forces Acting Simultaneously. When several forces

---

*This unit is named in honor of the English physicist James Prescott Joule (1818–1889), who contributed greatly to the understanding of the concepts of work and energy.

act on an object during a displacement, we can calculate the work done by each force individually. The total work done on the object is then the algebraic sum of the individual contributions:

$$\Delta W_{total} = \Delta W_a + \Delta W_b + \Delta W_c$$
$$= \mathbf{F}_a \cdot \Delta \mathbf{s} + \mathbf{F}_b \cdot \Delta \mathbf{s} + \mathbf{F}_c \cdot \Delta \mathbf{s}$$

Alternatively, we can obtain the result by first finding the resultant **F** of the individual forces:

$$\Delta W_{total} = (\mathbf{F}_a + \mathbf{F}_b + \mathbf{F}_c) \cdot \Delta \mathbf{s}$$
$$= \mathbf{F} \cdot \Delta \mathbf{s} \qquad (7\text{-}2)$$

Figure 7-2 shows the free-body diagram for a block that is being pulled at constant velocity across a horizontal surface by a force **F**. According to the definition, the amount of work done by the force **F** during a displacement Δs is $\Delta W_F = (F \cos \theta) \Delta s$. The normal force **N** and the gravitational force $\mathbf{F}_g$ act perpendicular to the surface, so $\theta = \pi/2$ for these forces, and $\Delta W = 0$ for each.

Fig. 7-2. Free-body diagram for a block that is pulled at constant velocity across a horizontal surface.

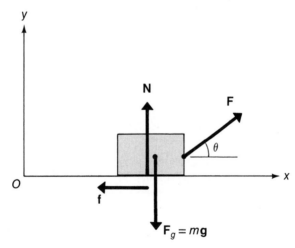

Negative Work.   What about the frictional force **f** in Fig. 7-2? This force acts on the block in a direction *opposite* to that of the displacement; that is, $\theta = \pi$ and $\cos \theta = -1$ for **f**. Consequently, $\Delta W_f = -f \Delta s$. The total work done *on the block* is the sum of the work done by all the forces. No work is done by either **N** or $\mathbf{F}_g$, so the total work is simply that done by **F** plus that done by **f**, namely,

$$\Delta W_{total} = \Delta W_F + \Delta W_f = F \Delta s \cos \theta - f \Delta s$$
$$= (F \cos \theta - f) \Delta s$$

Now, if the block moves with constant velocity, the horizontal forces that are acting must sum to zero; that is, $F \cos \theta - f = 0$. Thus, $\Delta W = 0$, and there is no net work done on the block.

How do we interpret this result? First, notice that the block's velocity is the same at the end of the displacement as at the beginning. Moreover, the block has not been deformed or altered in its composition. Thus, it seems reasonable that the net work done on the block is, in fact, zero. However, we know that the force **F** has done work on the block. Also, the force **f** has done work on the block, but this work is *negative*. This means that the *block* has actually done work. (We always interpret negative work in this way.) The block has done work in breaking the intermolecular bonds that extend between the block and the surface, bonds that are responsible for friction. In effect, then, the work done by the applied force **F** is done against friction. The result of this work is the generation of heat; this heat cannot be fully recovered and converted into mechanical work. (This point is discussed further on page 200).

Work Along a Finite Straight-Line Path. Later in this chapter, we discuss more complicated situations, but now we consider the simple case in which a *particular* force **F** that acts on an object is *constant* and the displacement of the object is constrained to a straight line. Other forces, in addition to **F**, may be present; for example, if **F** is not along **s**, other forces are required to maintain the straight-line motion. Then, using Eq. 7–2, the total work done by the constant force **F** in a sequence of N straight-line displacements $\Delta\mathbf{s}_n$ is*

$$W = \sum_{n=1}^{N} (\Delta W)_n = \sum_n \mathbf{F} \cdot \Delta\mathbf{s}_n$$

$$= \mathbf{F} \cdot \sum_n \Delta\mathbf{s}_n$$

If the total displacement is $\mathbf{s} = \Sigma\,\Delta\mathbf{s}_n$, we have

$$W = \mathbf{F} \cdot \mathbf{s} \qquad (\mathbf{F} = \text{constant}) \tag{7–2a}$$

If the end points of the displacement, A and B, are held fixed while the number of displacement increments $\Delta\mathbf{s}_n$ becomes larger and larger (and the size becomes correspondingly smaller and smaller), the sum becomes an *integral:*

$$W = \lim_{N \to \infty} \sum_{n=1}^{N} \mathbf{F} \cdot \Delta\mathbf{s}_n = \int_A^B \mathbf{F} \cdot d\mathbf{s} \tag{7–3}$$

Because **F** is constant, we again obtain

$$W = \mathbf{F} \cdot \int_A^B d\mathbf{s} = \mathbf{F} \cdot \mathbf{s}$$

In the equation for the work W, the term $\Delta\mathbf{s}_n$ is the actual vector displacement of the particle; consequently, the sign (and the magnitude) of $\mathbf{F} \cdot \Delta\mathbf{s}_n$ depends on the direction of motion of the particle relative to the force vector **F**. On the other hand $d\mathbf{s}$ is always taken to be in the positive direction of the selected coordinate frame. Thus, for linear motion along the x-axis, $d\mathbf{s}$ is always $+dx\,\hat{\mathbf{i}}$. The limits on the integral specify the direction of motion—for example, $\int_{x_1}^{x_2}$ for $x_1 \to x_2$ and $\int_{x_2}^{x_1}$ for $x_2 \to x_1$.

---

*When a sum is understood to run over a particular range, we use a more compact notation in which we abbreviate $\Sigma_{n=1}^{N} A_n$ to $\Sigma_n A_n$ or ΣA_n.

Example 7-1

A block that has a mass m is pushed across a horizontal surface at constant velocity. The coefficient of sliding friction is μ_k. What amount of work is done by the applied force **F** in moving the block through a distance s?

Solution:

The free-body diagram shown at right is essentially the same as that in Fig. 7-2. We must first express the force F in terms of the known quantities, m, μ_k, and θ. From the free-body diagram we see that

x-components:	$F \cos \theta - f = 0$
y-components:	$-F \sin \theta + N - F_g = 0$

Now,

$$f = \mu_k N = \mu_k (F \sin \theta + mg)$$

Then,

$$F \cos \theta - f = F(\cos \theta - \mu_k \sin \theta) - \mu_k mg = 0$$

from which

$$F = \frac{\mu_k mg}{\cos \theta - \mu_k \sin \theta}$$

Using $W = Fs \cos \theta$, we find

$$W = \frac{mgs\mu_k \cos \theta}{\cos \theta - \mu_k \sin \theta} = \frac{mgs}{(1/\mu_k) - \tan \theta}$$

This is the amount of work done by **F** against the retarding force of friction. (•Interpret the result when θ is increased to $\tan^{-1}(1/\mu_k)$.)

Example 7-2

Consider a block with a mass $m = 8$ kg that is being pushed up a frictionless ramp, as shown in the diagram. What amount of work is necessary to push the block at constant velocity from point A to point B, a distance of 65 cm, if the angle of inclination of the ramp is 32°?

Solution:

From the free-body diagram, we have

x-components:	$F - F_g \sin \theta = 0$
y-components:	$N - F_g \cos \theta = 0$

Now,

$$W = Fs = F_g s \sin \theta = mgs \sin \theta$$

$$= (8 \text{ kg})(9.80 \text{ m/s}^2)(0.65 \text{ m})(\sin 32°)$$

$$= 27.0 \text{ J}$$

Notice that by using $s \sin \theta = h$, the expression for W simplifies to

$$W = mgh$$

That is, the amount of work done in pushing the block up the ramp is exactly the same as that required to lift the block vertically through the same height. (•What amount of work is done on the block by the gravitational force?)

Work Involved in Two-Dimensional Motion.   An object moves in the *x-y* plane. Consider the work done on the object by a particular force **F** that is constant. Again, we allow the possibility of additional forces. We identify two points, *A* specified by (x_A, y_A) and *B* specified by (x_B, y_B), and ask how much work is done by **F** in displacing the object from *A* to *B*. We divide the path into a large number *N* of small displacements, $\Delta \mathbf{s}_n = \Delta x_n \hat{\mathbf{i}} + \Delta y_n \hat{\mathbf{j}}$. Because $\mathbf{F} = F_x \hat{\mathbf{i}} + F_y \hat{\mathbf{j}}$ is constant, we have

$$W = \sum_{n=1}^{N} \mathbf{F} \cdot \Delta \mathbf{s}_n = \sum_{n} (F_x \hat{\mathbf{i}} + F_y \hat{\mathbf{j}}) \cdot (\Delta x_n \hat{\mathbf{i}} + \Delta y_n \hat{\mathbf{j}})$$

$$= \sum_{n} F_x \Delta x_n + \sum_{n} F_y \Delta y_n$$

$$= F_x \sum_{n} \Delta x_n + F_y \sum_{n} \Delta y_n$$

or $$W = F_x(x_B - x_A) + F_y(y_B - y_A) \tag{7-4}$$

If we take the limit, $N \to \infty$, this result is equivalent to the integral (Eq. 7-3)

$$W = F_x \int_{x_A}^{x_B} dx + F_y \int_{y_A}^{y_B} dy = \int_{A}^{B} \mathbf{F} \cdot d\mathbf{s}$$

Thus, the work done consists of the sum of the *x*-component of the force multiplied by the *x*-component of the displacement and a similar term for the *y*-components. Figure 7-3 shows the situation.

As in the one-dimensional case, a sign convention applies in the evaluation of Eq. 7-4 and similar integrals. Even though the path followed may involve negative components for the displacement, we *always* write $d\mathbf{s} = dx\hat{\mathbf{i}} + dy\hat{\mathbf{j}}$ in $\mathbf{F} \cdot d\mathbf{s}$. By always taking the *lower limit* to correspond to the *initial position* and the *upper limit* to correspond to the *final position*, the sign associated with the displacement is correctly taken into account. The algebraic sign associated with **F**, which depends on the

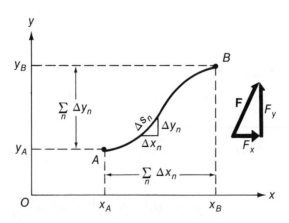

Fig. 7-3. Graphical representation of the work done by a constant force **F**, as a particle moves along the curved path from point *A* to point *B* in the *x-y* plane.

coordinate frame selection, must however *always* be properly included. Observe carefully in the following examples how this convention applies.

The three-dimensional generalization of the result expressed in Eq. 7–4 involves the additional term $F_z(z_B - z_A)$. The work done by a constant force $\mathbf{F}$ can then be expressed in vector notation as

$$W = \mathbf{F} \cdot (\mathbf{r}_B - \mathbf{r}_A) = \mathbf{F} \cdot \mathbf{s}$$

Example 7–3

A pendulum bob is displaced from its equilibrium position and is raised to a vertical height h, as shown in the diagram. If the bob is released from this height at rest, what amount of work has been done by the gravitational force when the bob arrives at the bottom of its swing?

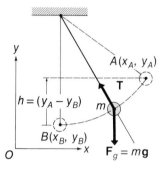

Solution:

Because $\mathbf{F}_g$ is constant, with $F_{gy} = -mg$ and $F_{gx} = 0$, we have from Eq. 7–4, noting that $y_B - y_A = -h$,

$$\begin{aligned} W_g &= F_{gx}(x_B - x_A) + F_{gy}(y_B - y_A) \\ &= 0(x_B - x_A) + (-mg)(-h) \\ &= mgh \end{aligned}$$

In this special case, we note that, although $\mathbf{T}$ is neither constant in direction nor constant in magnitude, $\mathbf{T}$ does no work on the bob because $\Delta\mathbf{s}$ and $\mathbf{T}$ are perpendicular at all times. Therefore, the total work done on the bob is just mgh.

7-2   WORK DONE BY VARIABLE FORCES DURING LINEAR MOTION

Next, we examine cases in which a particular force $\mathbf{F}$ that acts on a particle depends on the position of the particle. We postpone discussion of the general case and restrict the inquiry in this section to situations in which the particle is constrained to move along a straight-line path, say, the x-direction. If the force $\mathbf{F}$ acts only in the x-direction but varies with x, we have

$$\Delta W = F(x)\,\Delta x$$

What is the total work done by $F(x)$ during the displacement of the particle from the point A ($x = x_A$) to point B ($x = x_B$)? Dividing the interval into N increments Δx_n, we can write

$$W = \sum_{n=1}^{N} F(x_n)\,\Delta x_n$$

Although $F(x)$ is not constant, we can give a graphical or integral interpretation to this expression, as in Fig. 7–4, where $F(x)$ is shown as a function of x. A typical term

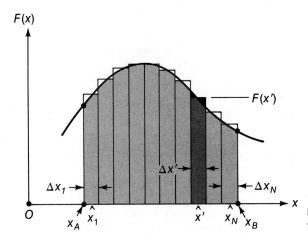

Fig. 7-4. The graphical representation of $\Sigma F(x)\,\Delta x$ and $\int F(x)\,dx$.

in the sum, such as $F(x')\,\Delta x' = \Delta W'$, is seen to be equivalent to the small, darkly shaded area. In the limit of an infinite number of vanishingly small infinitesimal steps, keeping x_A and x_B fixed, we have

$$W = \lim_{N\to\infty} \sum_{n=1}^{N} F(x_n)\,\Delta x_n = \int_{x_A}^{x_B} F(x)\,dx \qquad (7\text{-}5)$$

The small "error" areas (shown in black for $\Delta x'$) vanish in the limit, and the definite integral becomes precisely the shaded area (or the *area under the curve,* as explained in Section III–2) between the limits x_A and x_B.

If traveling the path from B to A involves the same force $F(x)$, the work done is

$$W(B \to A) = \int_{x_B}^{x_A} F(x)\,dx = -\int_{x_A}^{x_B} F(x)\,dx = -W(A \to B)$$

It is important to realize that this equation is not valid when friction is present because the frictional force depends on the direction of motion of the particle. When the path of a particle is reversed, the direction of the frictional force also reverses, so the work done against friction always has the same sign. Frictional effects are discussed later in connection with Eq. 7–10.

(•Suppose that **F** is *not* in the x-direction but instead has both magnitude and direction that vary with x. How would this affect the discussion leading to Eq. 7–5?)

Example 7–4

A spring with zero relaxed length and spring constant $\kappa = 50\ \text{N/m}$ moves a block by contracting from a stretched length of 30 cm to a length of 5 cm, as indicated in the diagram. The block slides on a horizontal frictionless surface. What is the amount of work done on the block by the spring?

Solution:

The net horizontal force on the block is

$$F = F_s = -\kappa x$$

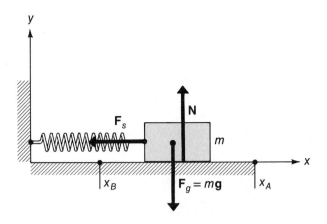

so the work done on the block by the spring is

$$W = \int_{x_A}^{x_B} F(x)\, dx = \int_{x_A}^{x_B} (-\kappa x)\, dx = -\tfrac{1}{2}\kappa x^2 \Big|_{x_A}^{x_B} = \tfrac{1}{2}\kappa(x_A^2 - x_B^2)$$

$$= \tfrac{1}{2}(50 \text{ N/m})[(0.30 \text{ m})^2 - (0.05 \text{ m})^2] = 2.188 \text{ J}$$

Example 7–5

An object with a mass m is released from rest at a distance R_0 from the center of the Earth and falls freely toward the Earth. What amount of work has been done on the object by the gravitational force when the object strikes the Earth's surface? (Neglect the motion of the Earth and the effects of the Earth's atmosphere.)

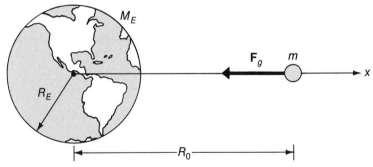

Solution:

The only force on the object is the gravitational force, which is directed toward the center of the Earth. The initial velocity of the object is zero, so the path followed is a straight line, which we take to be the x-axis.

Now, the gravitational force is (Eq. 5–6)

$$F_g = -G\frac{mM_E}{x^2}$$

where the negative sign indicates that $\mathbf{F}_g$ is in the direction of negative x. According to Eq. 5–7, $g = GM_E/R_E^2$, so we can rewrite F_g as

$$F_g = -mg\frac{R_E^2}{x^2}$$

Then, the work done by this force is

$$W = -mgR_E^2 \int_{R_0}^{R_E} \frac{dx}{x^2} = mgR_E^2 \cdot \frac{1}{x}\Big|_{R_0}^{R_E}$$

$$= mgR_E^2 \left(\frac{1}{R_E} - \frac{1}{R_0}\right)$$

If the object is released from a point close to the Earth at a height $h = R_0 - R_E$, then $R_0 \cong R_E$, and we can write

$$W(R_0 \cong R_E) = mgR_E^2 \frac{R_0 - R_E}{R_0 R_E} = mgh\frac{R_E}{R_0}$$

$$\cong mgh$$

This is the familiar result; compare Example 7–2.

7-3 THE WORK-ENERGY THEOREM

We continue to concentrate on motion constrained to a straight line. Suppose that the resultant of all forces acting on a particle is a force $F(x)$. The force may include contributions from position-dependent and frictional forces. If the motion is constrained to be along the x-axis, the force $\mathbf{F}$ has at most an x-component, so we write $F(x)$. Then, the Newtonian equation of motion is

$$F(x) = m\frac{d^2x}{dt^2} = m\frac{dv}{dt}$$

where we understand that the velocity has only an x-component. Multiplying this equation by dx gives

$$F(x)\,dx = dW = m\frac{dv}{dt}\,dx$$

$$= m\frac{dx}{dt}\,dv = mv\,dv$$

If the velocity is v_A at point A $(x = x_A)$ and is v_B at point B $(x = x_B)$, integrating the expression for dW yields

$$\int_A^B dW = m\int_{v_A}^{v_B} v\,dv = \tfrac{1}{2}m(v_B^2 - v_A^2)$$

or $$W(A \rightarrow B) = \tfrac{1}{2}mv_B^2 - \tfrac{1}{2}mv_A^2 = K_B - K_A \qquad (7\text{–}6)$$

Thus, the total work done by the *sum of all forces* acting on the particle during a displacement from A to B is equal to the change in $\tfrac{1}{2}mv^2$.

The quantity $\tfrac{1}{2}mv^2$ is the *energy* associated with a particle by virtue of its motion and is called *kinetic energy* (K). From Eq. 7–6, we see that kinetic energy has the same dimensions as work and is therefore also measured in joules. The kinetic energy acquired by the particle is derived ultimately from the energy possessed by the agencies responsible for the force **F**. Work is done in transferring the energy from these sources to the particle where it appears as kinetic energy. Thus, work is a transitory form of energy; work does not persist. However, the kinetic energy of a particle *does* persist as long as no forces act on the particle to change its speed.

The relationship (Eq. 7–6) that connects the work done to the change in kinetic energy is called the *work-energy theorem*: $W_{\text{total}} = K_B - K_A$. The power of this theorem lies in the fact that it holds for variable forces as well as constant forces, even in the presence of forces of friction.

Example 7–6

A block with a mass $m = 8$ kg is pushed up a frictionless inclined plane ($\theta = 32°$) by a force $F = 48$ N that acts parallel to the plane, as in Example 7–2. If the block starts from rest, how much work is done on the block and what is the speed of the block when it has been moved 0.65 m up the plane?

Solution:

The net force acting up the plane is

$$F_{\text{net}} = F - mg \sin \theta$$

Thus, the total work done on the block is

$$W = F_{\text{net}} s = (F - mg \sin \theta) s$$
$$= [(48 \text{ N}) - (8 \text{ kg})(9.80 \text{ m/s}^2)(\sin 32°)](0.65 \text{ m})$$
$$= 4.2 \text{ J}$$

Now, the initial speed is zero, so we have

$$W = \tfrac{1}{2}mv^2$$

or

$$v = \sqrt{\frac{2W}{m}} = \sqrt{\frac{2(4.2 \text{ J})}{8 \text{ kg}}} = 1.025 \text{ m/s}$$

Example 7–7

(a) A block with a mass of 0.5 kg is suspended from a spring that has a spring constant $\kappa = 50$ N/m and zero relaxed length. The block is pulled down a distance of 50 cm to a point x_A and released; at the same time there is imparted to the block an upward velocity of 2.0 m/s. What is the velocity v_B of the block when it first arrives at a position x_B that is 30 cm higher than the point of release?

Solution:

Choosing x positive downward, we have for the net force on the block

$$F = F_g - F_s = mg - \kappa x$$

To find the work done, we integrate $F \, dx$:

$$W(x_A \to x_B) = \int_{x_A}^{x_B} F \, dx = \int_{x_A}^{x_B} (mg - \kappa x) \, dx = (mgx - \tfrac{1}{2}\kappa x^2)\Big|_{x_A}^{x_B}$$
$$= mg(x_B - x_A) - \tfrac{1}{2}\kappa(x_B^2 - x_A^2)$$

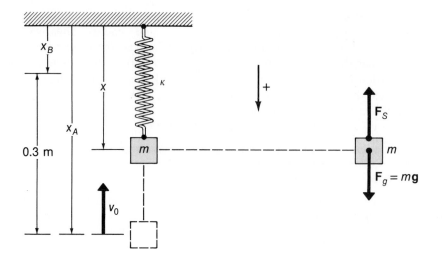

from which

$$W = (0.5 \text{ kg})(9.80 \text{ m/s}^2)(0.20 \text{ m} - 0.50 \text{ m}) - \tfrac{1}{2}(50 \text{ N/m})[(0.20 \text{ m})^2 - (0.50 \text{ m})^2]$$
$$= -1.47 \text{ J} + 5.25 \text{ J} = 3.78 \text{ J}$$

Using Eq. 7–6, we have

$$W = \tfrac{1}{2}mv_B^2 - \tfrac{1}{2}mv_A^2$$

or

$$v_B^2 = v_A^2 + \frac{2W}{m}$$

so

$$v_B = \sqrt{(-2.0 \text{ m/s})^2 + \frac{2(3.78 \text{ J})}{(0.5 \text{ kg})}}$$

$$= \pm 4.37 \text{ m/s}$$

When the block first arrives at x_B, it is traveling upward (i.e., in the direction of negative x); thus, the velocity is -4.37 m/s.

(b) What will happen if the initial velocity v_0 is *downward* at x_A ($v_0 = +2.0$ m/s)?

Solution:

In this case the block will first descend until the acceleration F/m (which is upward) brings it momentarily to rest; then, the block moves upward. When the block arrives back at x_A, the net work done by F during the round trip is zero, so the block now has an upward velocity $v_0 = -2.0$ m/s. The subsequent motion is therefore exactly the same as before!

Example 7–8

Imagine the situation in Example 7–4 to be modified by adding a frictional force between the block and the surface; let the friction coefficient be $\mu_k = 0.2$ (constant for all velocities achieved during the motion). Calculate the total work done by all of the forces acting on the block during its motion from $x_A = 30$ cm to $x_B = 5$ cm. If the block is released from rest at x_A, what is its velocity at x_B? (The mass of the block is $m = 0.5$ kg.)

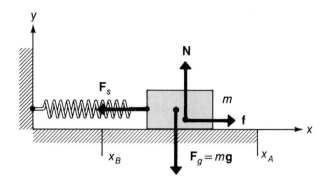

Solution:

The frictional force is

$$f = \mu_k N = \mu_k mg$$

and the work done by this force is

$$W_f = \mu_k mg \int_{x_A}^{x_B} dx = \mu_k mg(x_B - x_A)$$
$$= 0.20(0.5 \text{ kg})(9.80 \text{ m/s}^2)(0.05 \text{ m} - 0.30 \text{ m})$$
$$= -0.245 \text{ J}$$

The work done by the spring is the same as in Example 7–4:

$$W_s = \int_{x_A}^{x_B} (-\kappa x)\, dx = 2.188 \text{ J}$$

No work is done by the gravitational force or by the normal force. Therefore,

$$W = W_f + W_s = -0.245 \text{ J} + 2.188 \text{ J} = 1.943 \text{ J}$$

From Eq. 7–6 with $v_A = 0$, we have

$$v_B = \sqrt{\frac{2W}{m}} = \sqrt{\frac{2(1.943 \text{ J})}{0.5 \text{ kg}}}$$

$$= 2.788 \text{ m/s}$$

Example 7–9

Review the situation in Example 7–8 if the mass of the block is changed to 2.0 kg.

Solution:

Now, the work W_s is the same, but W_f increases to $2.0/0.5 = 4$ times the previous value:

$$W_s = 2.188 \text{ J} \quad \text{and} \quad W_f = -0.980 \text{ J}$$

so that

$$W = W_s + W_f = 1.208 \text{ J}$$

Then,
$$v_B = \sqrt{\frac{2W}{m}} = \sqrt{\frac{2(1.208\ \text{J})}{2.0\ \text{kg}}} = 1.099\ \text{m/s}$$

Notice that when the block arrives at $x_B = 5$ cm, the horizontal forces acting on the block are

$$F_s(x_B) = -\kappa x = -(50\ \text{N/m})(0.05\ \text{m})$$
$$= -2.50\ \text{N}$$
$$f = \mu_k mg = (0.20)(2.0\ \text{kg})(9.80\ \text{m/s}^2)$$
$$= 3.92\ \text{N}$$

That is, at x_B we have $|f| > |F_s|$ and the net force on the block at this point is directed toward the *right*. Thus, the block, which is moving toward the *left* at x_B, is experiencing a net retarding force that acts to decrease its speed. The block reaches x_B with a velocity (1.099 m/s) that is smaller than the velocity found for $m = 0.5$ kg (2.788 m/s). The block will always arrive at x_B if $W > 0$ for the path $x_B \rightarrow x_A$. (•Can you see why? What is the maximum value of m for which the block will reach x_B?)

▶ *Extension to Three Dimensions.*   The results we have just obtained can be generalized by allowing **F** to vary with (x, y, z). In vector notation, we have

$$\mathbf{F}(x, y, z) = m\mathbf{a} = m\frac{d\mathbf{v}}{dt}$$

Take the dot product of this equation with $d\mathbf{s}$, a differential displacement along the particle path. Then,

$$\mathbf{F} \cdot d\mathbf{s} = m\frac{d\mathbf{v}}{dt} \cdot d\mathbf{s} = m\frac{d\mathbf{s}}{dt} \cdot d\mathbf{v} = m\mathbf{v} \cdot d\mathbf{v}$$

The work done by **F** as the particle moves along its trajectory from point A (x_A, y_A, z_A) to point B (x_B, y_B, z_B) is given by the integral of $\mathbf{F} \cdot d\mathbf{s}$:*

$$\int_A^B \mathbf{F} \cdot d\mathbf{s} = m\int_{\mathbf{v}_A}^{\mathbf{v}_B} \mathbf{v} \cdot d\mathbf{v} = \tfrac{1}{2}m\int_{\mathbf{v}_A}^{\mathbf{v}_B} d(\mathbf{v} \cdot \mathbf{v})$$

$$= \tfrac{1}{2}m\int_{\mathbf{v}_A}^{\mathbf{v}_B} d(v^2) = \tfrac{1}{2}m(v_B^2 - v_A^2)$$

or
$$W(A \rightarrow B) = \tfrac{1}{2}mv_B^2 - \tfrac{1}{2}mv_A^2 = K_B - K_A \tag{7-7}$$

The path, $A \rightarrow B$, that is followed by the particle is not restricted in any way; thus, the path could be a very complicated three-dimensional trajectory. Moreover, there is no requirement that $\mathbf{v}_B$ be parallel to $\mathbf{v}_A$. In fact, Eq. 7-7 refers only to the *magnitudes* of $\mathbf{v}_B$ and $\mathbf{v}_A$, namely, the speeds v_B and v_A.

In Chapter 8 we discuss the general case of the evaluation of integrals of the form $\int_A^B \mathbf{F} \cdot d\mathbf{s}$ (called *path integrals* or *line integrals*). However, there are special cases in which the evaluation is particularly simple. For example, we might have a particle subject to a constant force plus a constraining force that acts always in a direction perpendicular to the direction of motion. We have already considered, in Example 7-3, a constraint force of this type. There, the string tension is a constraining force that restricts the pendulum bob to motion in a circular path. The tension acts always at right angles to the direction of motion of the bob, so $\mathbf{T} \cdot d\mathbf{s} = 0$. Thus, the tension does no work; only the gravitational force does work ($W = mgh$). Then, using Eq. 7-7 (with $v_A = 0$), we find the speed at the bottom of the swing, $v_B = \sqrt{2gh}$. Notice that this speed is the same as that acquired by an object sliding frictionlessly down a curved or

---

*If $w = \mathbf{u} \cdot \mathbf{v}$, then $dw = d(\mathbf{u} \cdot \mathbf{v}) = \mathbf{v} \cdot d\mathbf{u} + \mathbf{u} \cdot d\mathbf{v}$. Thus, with $\mathbf{u} = \mathbf{v}$, we can write $d(\mathbf{v} \cdot \mathbf{v}) = 2\,\mathbf{v} \cdot d\mathbf{v}$, which we use here.

wavy ramp (for example, a rollercoaster track), starting from rest and descending a total vertical distance h. In such a case, with friction absent, the force exerted by the ramp on the object is always perpendicular to the direction of motion.

In the cases considered here, it has been easy to evaluate the integral $\int_A^B \mathbf{F} \cdot d\mathbf{s}$ only because of the very special nature of the forces involved. However, these restrictions are not as "special" as they might seem. In fact, we encounter these cases sufficiently often to warrant further discussion in the form of additional examples. ◀

Example 7-10

(a) A particle with a mass m slides without friction down the complicated ramp shown in the diagram, never losing contact with the ramp. Let the heights (the values of y) at x_A, x_B, x_C, and x_D be, respectively, $h_A = 7$ m, $h_B = 4$ m, $h_C = 7.2$ m, and $h_D = -1$ m. The particle has an initial velocity, $v_A = 3$ m/s, directed downward and tangential to the ramp at x_A. What is the speed of the particle at x_B, x_C, and x_D?

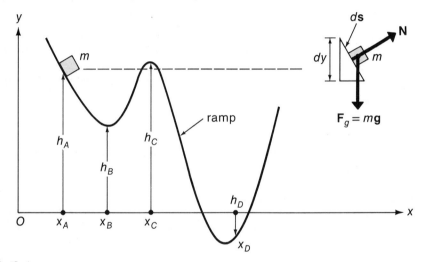

Solution:

It is clear from the diagram that $\mathbf{N} \cdot d\mathbf{s}$ is always zero, so work is done only by the gravitational force. To find $W(A \to B)$, we must evaluate

$$W(A \to B) = \int_A^B \mathbf{F}_g \cdot d\mathbf{s} = \int_A^B (-mg)\, dy$$

$$= mg(h_A - h_B)$$

Then, using Eq. 7-6, we find

$$v_B^2 = v_A^2 + \frac{2W}{m} = v_A^2 + 2g(h_A - h_B)$$

$$= (3 \text{ m/s})^2 + 2(9.80 \text{ m/s}^2)(7 \text{ m} - 4 \text{ m})$$

$$= 67.8 \ (\text{m/s})^2$$

Thus, $\qquad\qquad v_B = 8.23$ m/s

Also, $\qquad\qquad v_C^2 = (3 \text{ m/s})^2 + 2(9.80 \text{ m/s}^2)(7 \text{ m} - 7.2 \text{ m})$

$$= 5.08 \ (\text{m/s})^2$$

so that $\qquad\qquad v_C = 2.25 \text{ m/s}$

Finally, $\qquad\qquad v_D^2 = (3 \text{ m/s})^2 + 2(9.80 \text{ m/s}^2)(7 \text{ m} + 1 \text{ m})$

$$= 165.80 \ (\text{m/s})^2$$

with $\qquad\qquad v_D = 12.88 \text{ m/s}$

(b) How far up the ramp will the particle move after passing through x_D?

Solution:

The maximum height h_M will be reached at x_M where $v_M = 0$. That is,

$$0 = v_A^2 + 2g(h_A - h_M)$$

or $\qquad\qquad h_M = h_A + \dfrac{v_A^2}{2g}$

$$= 7 \text{ m} + \frac{(3 \text{ m/s})^2}{2(9.80 \text{ m/s}^2)}$$

$$= 7.46 \text{ m}$$

Example 7–11

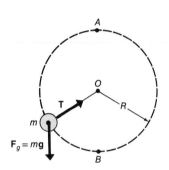

A rock with a mass m is tied to the end of a string and whirled in a circular path that lies in a vertical plane, as shown in the diagram. What is the minimum (tangential) speed v_0 that the rock must have at the bottom of the path if the rock is to be able to pass the top position of the circle while maintaining the string taut? (Notice that the angular speed of the rock is *not* constant.)

This problem is one of a class that requires an important technique for obtaining the solution. Energy considerations alone are not sufficient; Newton's law must also be applied at some particular point. Other similar cases are examined in Problems 7–22, 7–41, 7–42, and 7–44.

Solution:

At the top, the minimum speed occurs when the tension just vanishes. Then, the net downward force is $F_g = mg$. This force must give the centripetal acceleration required to maintain a circular path:

$$F_g = mg = m\frac{v_T^2}{R}$$

so $\qquad\qquad v_T^2 = gR$

Using Eq. 7–7, in which we identify the top position as A and the bottom position as B, we have

$$W(A \rightarrow B) = mgh = K_B - K_A$$

so that $\qquad\qquad mg(2R) = \tfrac{1}{2}mv_0^2 - \tfrac{1}{2}mv_T^2$

from which $\qquad\qquad v_0^2 = v_T^2 + 4gR = gR + 4gR = 5gR$

Finally, $\qquad\qquad v_0 = \sqrt{5gR}$

7-4 POTENTIAL ENERGY AND THE CONSERVATION OF ENERGY FOR LINEAR MOTION

There are many important situations in which the work done by a force on a body can be recovered at a later time. For example, if you do work in raising a box from the floor to a shelf, this amount of work can be recovered in the form of kinetic energy by allowing the box to fall to the floor. Because of this feature of reversibility, the box in its elevated position possesses stored energy that is called *potential energy*. (The box has the *potential* to do work by falling to the floor.)

Now consider raising an object against the force of gravity. Suppose that a student raises a box *very slowly* from the floor to a shelf at a height h (Fig. 7–5a). Raising the box "very slowly" means that at any instant the velocity of the box, and hence its kinetic energy, is essentially zero. The acceleration is then zero, so that the magnitude of the applied force $\mathbf{F}_a$ exerted on the box by the student is equal to that of the gravitational force, $F_g = mg$. Then, because the displacement is in the same direction as the applied force, $\theta = 0$ and $\cos \theta = 1$, so the work W_a done *by* the student is

$$W_a = \mathbf{F}_a \cdot \mathbf{s} = F_a h = mgh$$

During this process, the work W_g done by the gravitational force is

$$W_g = \mathbf{F}_g \cdot \mathbf{s} = -F_g h = -mgh = -W_a$$

where the negative sign arises because $\mathbf{F}_g$ is directed opposite to the displacement $\mathbf{s}$.

Fig. 7–5. (a) An object acquires a potential energy *mgh* by being raised to a height *h*. (b) Free-body diagram for the block as it is raised very slowly.

(a)

The energy possessed by the box on the shelf is a *positive* quantity relative to the level of the floor. We ascribe to the box a potential energy U_g equal to the negative of the work done by the gravitational force (which is itself a negative quantity). That is,

$$U_g = -W_g$$

This definition permits us to associate the potential energy of the box exclusively with the gravitational force and to eliminate any further reference to the role of the student.

Let us now generalize this idea, but with the continued restriction to one-dimensional cases involving motion along the x-axis. To accomplish this, it is convenient to use incremental displacements. Thus, for the work ΔW_n done on a body by the particular force $\mathbf{F}_n$ whose x-component is F_{nx}, we write

$$\Delta W_n = F_n \cos \theta_n \, \Delta x = F_{nx} \, \Delta x$$

As above, we introduce ΔU_n as the negative of ΔW_n:

$$\Delta U_n = -\Delta W_n$$

That is, ΔU_n is the negative of the incremental work done by the force $\mathbf{F}_n$ on the body.

First, consider forces $\mathbf{F}_n$ that depend *only* on the position of the body on which they act (that is, on x) and not, for example, on time or velocity (not even on the direction of velocity). Then,

$$\Delta U_n(x) = -F_{nx}(x) \, \Delta x$$

and
$$F_{nx}(x) = -\frac{\Delta U_n(x)}{\Delta x}$$

In the limit, $\Delta x \to 0$, we have the important relationship,

$$F_{nx}(x) = -\frac{dU_n(x)}{dx} \tag{7-8}$$

The integral of this expression yields the function $U_n(x)$. Notice that $U_n(x)$ is defined only to within a constant additive term, the usual constant of integration.

We refer to $U_n(x)$ as the *constrained potential-energy function*. The adjective "constrained" is used to call attention to the special one-dimensional constraint that has been imposed. In the presence of this constraint, the only requirement placed on $F_{nx}(x)$ is that it be integrable. (In Chapter 8 we see that this constraint has nontrivial consequences when we contrast the present case with the general three-dimensional case.)

We can express the resultant F_x of the various forces acting on a body as

$$F_x = \sum_{n=1}^{N} F_{nx}(x) + f \tag{7-9}$$

where we have included the dissipative frictional force f (which, because the motion is in the x-direction, acts only in the x-direction*) as well as the applied forces $F_{nx}(x)$ that depend only on x. Then, the resultant force has at most an x-component.

Using Eqs. 7–8 and 7–9, the work-energy theorem (Eq. 7–6) can be expressed as

$$\int_{x_A}^{x_B} F_x \, dx = \int_{x_A}^{x_B} \sum_n F_{nx}(x) \, dx + \int_{x_A}^{x_B} f \, dx = \tfrac{1}{2} m v_B^2 - \tfrac{1}{2} m v_A^2$$

Now, $\displaystyle \int_{x_A}^{x_B} \sum_n F_{nx}(x) \, dx = \sum_n \int_{x_A}^{x_B} F_{nx}(x) \, dx = -\sum_n [U_n(x_B) - U_n(x_A)]$

Hence, $\displaystyle \sum_n U_n(x_B) + \tfrac{1}{2} m v_B^2 = \sum_n U_n(x_A) + \tfrac{1}{2} m v_A^2 + \int_{x_A}^{x_B} f \, dx$

Because the various $U_n(x)$ are simply functions of x, we can write

$$U(x) = \sum_n U_n(x)$$

Finally,

$$U(x_B) + \tfrac{1}{2} m v_B^2 = U(x_A) + \tfrac{1}{2} m v_A^2 + \int_{x_A}^{x_B} f \, dx \qquad \textbf{(7-10)}$$

Because the frictional force **f** is always directed opposite to the displacement, the integral in Eq. 7–10 is to be evaluated with the following rule:

$$f < 0 \quad \text{if} \quad x_B > x > x_A$$
$$f > 0 \quad \text{if} \quad x_B < x < x_A$$

(•Can you see why this rule is necessary?)

Equation 7–10 expresses the law of *energy conservation* for a mechanical system. The quantity $E = U(x) + \tfrac{1}{2} m v^2$ (where v is the velocity at point x) is called the *mechanical energy* of the body. As before, we refer to $K = \tfrac{1}{2} m v^2$ as the *kinetic energy* (or motional energy). The total *constrained potential energy* $U(x)$, or any of its individual parts $U_n(x)$, is simply the work that must be done to bring the body to the position x from some arbitrary reference point where all of the $U_n(x)$ are defined to be zero.† But as Eq. 7–10 shows, the basic law always involves the *difference,*

---

*The frictional force f is not simply a function of position because it carries a sign that depends on the direction of motion at the point x. Technically, f is not a continuous function of x at a turnaround point, $x = x_T$, because $f(x_T)$ is not defined: $f(x \to x_T$, from the left$) = -f(x \to x_T$, from the right$)$.

†When we are interested in following the motion of a particular object, we usually refer to the potential energy (or the total energy) *of the body*. However, we should properly refer to the potential energy *of the system* because this energy is the result of the location of the body being studied relative to the body or bodies that produce the forces $\mathbf{F}_n$.

$U(x_B) - U(x_A)$ [or the corresponding differences, $U_n(x_B) - U_n(x_A)$]; consequently, the position of this arbitrary reference level is immaterial.

In the case of zero friction ($f = 0$), Eq. 7–10 asserts that the mechanical energy of a body is a constant, independent of position and, hence, independent of the time (because $x = x(t)$ describes the path followed by the body). Such a quantity is called a *constant of the motion*. Thus, in the absence of frictional forces, the mechanical energy of a one-dimensional system is a constant of the motion.

In Eq. 7–10, the integral of $f\,dx$ is always negative. Therefore, the content of this equation can be stated: *When a particle moves from a point x_A to a point x_B, the mechanical energy at x_B is equal to the mechanical energy at x_A less the dissipative work done by the particle against frictional forces.*

The various examples and problems in the last section demonstrate that care must be exercised in evaluating the work done by friction. For example, consider the coordinates $x_A < x_B < x_C$; then, for motion that takes a particle along the path $x_A \rightarrow x_C \rightarrow x_B$, account must be taken of the fact that the sign of f along the path $x_C \rightarrow x_B$ is different from that along the path $x_A \rightarrow x_C$.

There are many situations in which frictional forces can be neglected; in this approximation, we have

$$U(x) + \tfrac{1}{2}mv^2 = E = \text{constant} \quad \text{(no friction)} \tag{7–11}$$

where E is the constant total mechanical energy.*

Example 7–12

A particle with mass m moves across a horizontal surface under the influence of a potential $U(x) = \tfrac{1}{2}kx^2$. A frictional force acts on the particle, namely, $f(x) = \pm ax$, where the sign depends on the direction of motion at x. The particle is released from rest at $x = x_0$. What is the speed of the particle as it passes $x = 0$ for the first time?

Solution:

In this case, we have $x_A = x_0$ and $x_B = 0$. Then, $U(x_A) = \tfrac{1}{2}kx_0^2$ and $U(x_B) = 0$. Also, $v_A = 0$. Because $x_B < x_A$, we use $f(x) = +ax$. Thus, Eq. 7–10 becomes

$$\tfrac{1}{2}mv_B^2 = \tfrac{1}{2}kx_0^2 + \int_{x_0}^{0} ax\,dx$$

$$= \tfrac{1}{2}kx_0^2 - \tfrac{1}{2}ax_0^2$$

so that

$$v_B = \sqrt{\frac{k-a}{m}}\,x_0$$

The Potential Energy Curve.   It is instructive to examine a graphical representation of Eq. 7–11. Imagine that $U(x)$, for some particular physical situation, has the form shown in Fig. 7–6. The characteristics of the motion are determined by the

---

*Because $U(x)$ is defined only to within an additive constant, so is E.

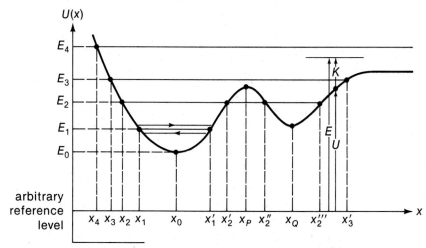

Fig. 7-6. The graphical representation of the one-dimensional constrained potential energy function $U(x)$ for an object. Also shown are various values of the total mechanical energy.

value of the total mechanical energy and by the shape of the potential energy function.

Solving Eq. 7-11 for v, we have

$$v = \pm \sqrt{\frac{2}{m}[E - U(x)]} \tag{7-12}$$

It is clear that v has a physically meaningful value only when $E \geqslant U(x)$; otherwise, Eq. 7-12 would have an imaginary solution. This means that the object cannot have a mechanical energy less than E_0, as shown in Fig. 7-6. When $E = E_0$, the velocity v is identically zero (for all values of the time).

Suppose that the energy is slightly greater than E_0, say, $E_0 + \varepsilon$. Then, as the particle moves away from x_0 in either direction, a force acts to return the particle to the point $x = x_0$. Points such as x_0 are referred to as points of *stable equilibrium*.

If $E = E_1$, the particle may be found anywhere in the interval $x_1 \leqslant x \leqslant x_1'$, with the velocity at any point given by Eq. 7-12. In Fig. 7-6, v_1 (corresponding to $E = E_1$) is maximum for $x = x_0$ (that is, for $E_1 - U(x)$ a maximum). As $x \to x_1$, the velocity vanishes, but $F = -dU(x)/dx|_{x=x_1}$ is greater than zero; thus, at $x = x_1$, there is a force (and, hence, an acceleration) to the right. The particle is in effect "reflected" back to the right. A similar reflection occurs at $x = x_1'$, and the particle oscillates back and forth in the *potential well* between $x = x_1$ and $x = x_1'$. Points such as x_1 and x_1' are referred to as *turning points*.

The case $E = E_2$ is very interesting. According to Eq. 7-12, the particle is allowed to be anywhere in the regions $x_2 < x < x_2'$ and $x_2'' < x < x_2'''$; however, these are *disconnected regions*. If the system is initially prepared with the particle somewhere in the region $x_2 < x < x_2'$, it can never enter the region $x_2'' < x < x_2'''$, for it could do so only by passing through the region $x_2' < x < x_2''$ where, from Eq. 7-12, v would have imaginary (i.e., unphysical) values. The region $x_2' < x < x_2''$ is referred to as a *potential barrier*. Barriers also exist for $x < x_2$ and $x > x_2'''$.

The case $E = E_3$ is similar to the case $E = E_1$, except that v now has a local minimum at $x = x_P$. [•Suppose that the particle is at $x = x_P$ (see Fig. 7-6) and has a total energy $E(x_P)$. What is the velocity of the particle? What is the character of the force that acts on the particle if it moves an incremental distance away from $x = x_P$? Compare with the character of the force in the vicinity of $x = x_0$. Can you see why points such as $x = x_P$ are called points of *unstable equilibrium*?]

(•Suppose that the particle is at a point $x > x_3'$ (see Fig. 7–6) and is moving to the left with total energy E_4. Describe the motion of the particle.)

Conservation of Energy. Equation 7–11 states that, in the absence of friction, the total mechanical energy of a system remains constant. This, in itself, is a useful result because friction can be neglected in many situations. However, if friction *is* important, we must use the energy equation in which the work done against friction is explicitly included (Eq. 7–10).

Does this mean that energy is no longer conserved because energy has been "lost" to friction? It does not—*if* we suitably redefine the system to include the places where the work done against friction has been absorbed.

It is a common experience that friction produces *heat*. In a later chapter we discuss in detail the fact that heat is a form of internal kinetic energy associated with the motions of the atoms and molecules that comprise a sample. The work done against friction is the energy that must be expended to break the molecular bonds that form between the sliding surfaces (see Section 6–2). As a result, the molecules at these points become highly agitated. Moreover, the extra molecular energy is shared with neighboring molecules, so the regions of the samples in the vicinity of the sliding surfaces become heated. The idea of energy conservation holds in this situation if we are careful to include in the tabulation of the system energy all of the energy associated with heat in the sliding surfaces. In fact, it was the realization, in the mid-19th century, that heat is a form of energy that led to the enunciation of the principle of energy conservation.*

As we examine a greater variety of situations involving different kinds of forces, we find that we must enlarge our view still further and allow for even more forms of energy. Indeed, we have come to adopt the attitude that if we do not find a balance of energy in a particular process, we invent a new form of energy that exactly makes up the difference! Following this attitude, we have invented, for example, chemical energy, electromagnetic energy, and nuclear energy. This is not really done to hide our ignorance about Nature, for once we have invented a new form of energy we must thereafter always include this energy in our calculations in the same way. If we have made a poor choice, we will rapidly come upon contradictions. Those forms that survive all tests become part of our concept of energy. Because of its enormous range of application, the energy conservation principle is one of the most important ideas in all of science.

Example 7–13

Consider a block (a *particle*) with a mass m that slides down a frictionless inclined plane, as shown in the diagram. Describe the motion of the block by using the idea of energy conservation.

---

*Credit for formulating the energy conservation principle is usually given to the German physiologist and physicist Hermann von Helmholtz (1821–1894) because of the detailed and forceful way in which he advanced the idea (1847). However, the concept had been announced earlier (1842) by the German physician and physicist Julius Robert Mayer (1814–1878), but he had considerable difficulty in finding a journal that would publish his paper because his concept was too novel. When it finally did appear, Mayer's paper aroused no particular interest. Finally, the papers by the English physicist James Prescott Joule (1818–1889), which appeared a few years later, firmly established the connection between heat and other forms of energy, thereby setting the stage for Helmholtz's pronouncements. The credit for the energy conservation principle should probably be divided among these three scientists.

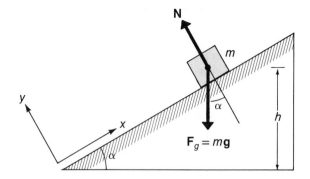

Solution:

The net force **F** acting on the block is directed down the plane:

$$F = F_x = -F_g \sin \alpha = -mg \sin \alpha = -\frac{dU(x)}{dx}$$

Then,
$$U(x) = mgx \sin \alpha + C$$

If we take $U(0) = 0$, we find $C = 0$. We know that

$$U(x) + \tfrac{1}{2}mv^2 = E = \text{constant}$$

so
$$mgx \sin \alpha + \tfrac{1}{2}mv^2 = E$$

Now, $x \sin \alpha = h$, so we can also write

$$mgh + \tfrac{1}{2}mv^2 = E$$

If the block slides, starting from rest at $x = x_0$ (with $x_0 \sin \alpha = h_0$), we have $E = mgh_0$ for the initial energy. Then,

$$mgh + \tfrac{1}{2}mv^2 = E = mgh_0$$

so that
$$\tfrac{1}{2}mv^2 = mg(h_0 - h)$$

The velocity at any height h is, therefore,

$$v = \sqrt{2g(h_0 - h)}$$

which is the same as the expression we obtained earlier for the case of free fall (see Eq. 2–15).

Notice that we can recover Newton's equation by differentiating the energy equation with respect to time. We have

$$\frac{d}{dt}(mgx \sin \alpha + \tfrac{1}{2}mv^2) = \frac{d}{dt}E$$

$$mg\frac{dx}{dt} \sin \alpha + mv\frac{dv}{dt} = 0$$

Canceling dx/dt with v and using $a = dv/dt$, we find

$$-mg \sin \alpha = ma$$

or
$$F = ma$$

as expected.

Example 7–14

Consider a thin rod with a length ℓ and a mass m that is in a vertical position, as shown in the diagram. What is the potential energy U of the rod with respect to the surface shown?

Solution:

The work dW done by gravity in raising the mass element $dm = (m/\ell)\, dx$ from the reference level to the position $h + x$ is $dW = -g(x + h)\, dm$; hence, $dU = -dW = (mg/\ell)$ $\cdot (x + h)\, dx$. Thus, the potential energy is

$$U = \int_0^\ell dU = \frac{mg}{\ell} \int_0^\ell (x + h)\, dx = \frac{mg}{\ell}(\tfrac{1}{2}x^2 + hx)\Big|_0^\ell$$

$$= \frac{mg}{\ell}(\tfrac{1}{2}\ell^2 + \ell h) = mg(h + \tfrac{1}{2}\ell)$$

The potential energy is the same as if we raised a total concentrated mass m from the reference level to a height $h + \tfrac{1}{2}\ell$.

Example 7–15

Consider a particle with a mass $m = 2$ kg for which the potential energy function is $U(x) = 3x - 5 - (x - 3)^3$, where U is measured in joules when x is in meters. Describe the motion for two values of the total energy, $E_1 = 3$ J and $E_2 = 8$ J. For example, what is the velocity of the particle at $x = 3$ m?

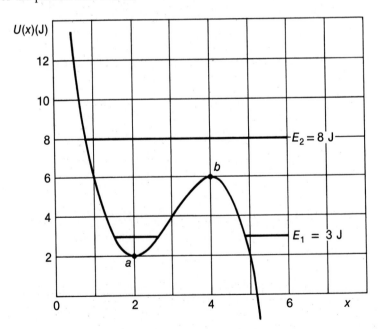

Solution:

Most of the important features of the motion are evident in the energy diagram. For example, the minimum allowed value of the total energy occurs at $x = 2$ m and is equal to 2.0 J. A particle, originally at $x = 2$ m with 2 J $< E < 6$ J, will undergo back-and-forth motion; if $E > 6.0$ J, the particle will *escape* from the potential well and will forever move off to the right.

At $x = 3$ m, there is no physically realizable motion for $E_1 = 3$ J. Now,

$$\tfrac{1}{2}mv^2 = E - U(x)$$

Thus, for $E_2 = 8$ J we have

$$\tfrac{1}{2}mv_2^2 = 8\,\text{J} - U(3\,\text{m}) = 8\,\text{J} - 4\,\text{J} = 4\,\text{J}$$

so that

$$v_2 = \sqrt{\frac{2(4\,\text{J})}{2\,\text{kg}}} = \pm 2\,\text{m/s} \qquad \text{at } x = 3\,\text{m}$$

7-5   POWER

Consider a force $\mathbf{F}(x, y, z)$ that is one of the forces acting on a particle. The work done by this force during a general displacement that requires a time Δt is

$$\Delta W = F_x\,\Delta x + F_y\,\Delta y + F_z\,\Delta z$$

To obtain the rate at which work is done, divide this equation by Δt and take the limit, $\Delta t \to 0$; this gives

$$\frac{dW}{dt} = \operatorname*{Lim}_{\Delta t \to 0} \frac{\Delta W}{\Delta t} = \operatorname*{Lim}_{\Delta t \to 0} \left(F_x\frac{\Delta x}{\Delta t} + F_y\frac{\Delta y}{\Delta t} + F_z\frac{\Delta z}{\Delta t} \right)$$

$$= F_x\frac{dx}{dt} + F_y\frac{dy}{dt} + F_z\frac{dz}{dt}$$

$$= F_x v_x + F_y v_y + F_z v_z$$

$$= \mathbf{F} \cdot \mathbf{v}$$

The rate at which work is done, dW/dt, is called the *power P*. Thus,

$$P = \frac{dW}{dt} = \mathbf{F} \cdot \mathbf{v} \tag{7-13}$$

The SI unit of power is the *joule per second* (J/s), to which we give the special name *watt* (W)*:

$$1\,\text{W} = 1\,\text{J/s}$$

The unit of power in the British system, namely, the *horsepower* (hp), is often used in engineering applications:

$$1\,\text{hp} = 745.7\,\text{W} \cong \tfrac{3}{4}\,\text{kW}$$

At what rate is energy expended against friction? As in Eq. 7-9, we separate the resultant F of the various forces acting on a body into the applied forces F_{nx} and the

---

*Named in honor of James Watt (1736–1819), the Scottish engineer who was responsible for the development of the first practical industrial steam engines.

frictional force f. In differential notation, we express the work done as

$$F_x \, dx = \sum_n F_{nx}(x) \, dx + f \, dx$$

We now use Newton's second law to write

$$F_x \, dx = m \frac{dv}{dt} \, dx = m \, dv \frac{dx}{dt} = mv \, dv$$

and we express Eq. 7–8 as

$$\sum_n F_{nx}(x) \, dx = -dU$$

so that the work done by friction is

$$f \, dx = dU + mv \, dv$$

Dividing by dt, we have $$f \frac{dx}{dt} = \frac{dU}{dt} + mv \frac{dv}{dt}$$

or $$fv = \frac{dU}{dt} + \frac{d}{dt}(\tfrac{1}{2}mv^2)$$

$$= \frac{d}{dt}(U + \tfrac{1}{2}mv^2)$$

The quantity $U + \tfrac{1}{2}mv^2$ is the *mechanical energy* E of the system (Eq. 7–11). Thus,

$$fv = \frac{dE}{dt} \tag{7–14}$$

Now, the frictional force $\mathbf{f}$ is always directed opposite to the velocity $\mathbf{v}$. Therefore, $\mathbf{f} \cdot \mathbf{v} = fv$ is always negative, and we conclude that $dE/dt < 0$. Thus, Eq. 7–14 states that the power dissipated in friction is equal to the rate at which mechanical energy is lost.

Example 7–16

The engine of an automobile develops 50 hp simply to overcome air and road resistance while maintaining a constant speed of 55 mi/h. What is the effective thrust (or force) developed by the engine (that is, the hypothetical force which, if pulling the automobile, would result in the same power expenditure)?

Solution:

First, convert the power and the speed to SI units:

$$P = 50 \text{ hp} \quad = (50 \text{ hp})(746 \text{ W/hp}) = 37{,}300 \text{ W}$$
$$v = 55 \text{ mi/h} = (55 \text{ mi/h})(1609 \text{ m/mi})/(3600 \text{ s/h}) = 24.6 \text{ m/s}$$

Then,
$$F = \frac{P}{v} = \frac{37,300 \text{ W}}{24.6 \text{ m/s}} = 1516 \text{ N}$$

Example 7-17

A block with a mass $m = 4 \text{ kg}$ slides, starting from rest, down an inclined plane ($\alpha = 35°$) from an initial height $h_0 = 6 \text{ m}$. The friction coefficient is $\mu_k = 0.25$. Find the kinetic energy of the block and the power being expended against friction when the block has reached the height $h = 2 \text{ m}$.

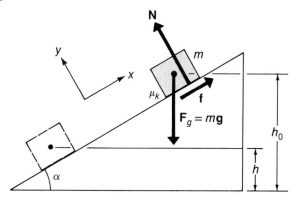

Solution:

The frictional force is

$$f = \mu_k N = \mu_k mg \cos \alpha$$
$$= (0.25)(4 \text{ kg})(9.80 \text{ m/s}^2)(\cos 35°)$$
$$= 8.03 \text{ N}$$

Using Eq. 7-10, we can write

$$mgh_0 = mgh + \tfrac{1}{2}mv^2 + f(x_0 - x)$$

where $(x_0 - x) \sin \alpha = h_0 - h$. Thus,

$$K = \tfrac{1}{2}mv^2 = mg(h_0 - h) - \mu_k mg \frac{\cos \alpha}{\sin \alpha}(h_0 - h)$$
$$= mg(1 - \mu_k \operatorname{ctn} \alpha)(h_0 - h)$$
$$= (4 \text{ kg})(9.8 \text{ m/s}^2)[1 - (0.25)(\operatorname{ctn} 35°)][(6 \text{ m}) - (2 \text{ m})]$$
$$= 100.8 \text{ J}$$

The power expended against friction at $h = 2 \text{ m}$ is

$$P = fv = f\sqrt{\frac{2K}{m}}$$
$$= (8.03 \text{ N})\sqrt{\frac{2(100.8 \text{ J})}{4 \text{ kg}}}$$
$$= 57.0 \text{ W}$$

(•Show, by explicitly differentiating the expression for K, that Newton's equation is recovered.)

PROBLEMS

Section 7-1

7-1 A student pulls a block along a frictionless horizontal ice surface, as shown in the diagram. (The student manages to walk across the frictionless surface by wearing spiked shoes.) If the angle α is maintained at a constant value of 30° and if the tension in the ideal massless rope is maintained constant at 3 N, how much work is done by the student in dragging the block a horizontal distance of 10 m? Assume the block slides on the ice at all times. Calculate separately the work done on the block by the rope tension, the work done by the block on the rope, the work done on the student by the rope tension, and the work done by the student on the rope. Do your answers give meaning to the idea that work is done on the block by the student, even though they are not in direct contact? (•What would happen if the rope had nonzero mass?)

7-2 Two balls with masses $m_1 = 10$ kg and $m_2 = 8$ kg hang from an ideal pulley, as shown in the diagram above right. What amount of work is done by the Earth's gravitational force on each ball separately during a downward vertical displacement of 50 cm by m_1? What is the *total* work done on each ball, including the work done by the string tension? Comment on any relationship you have discovered connecting these quantities.

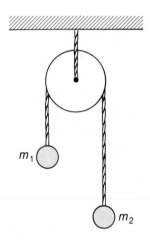

7-3 A block with a mass m slides frictionlessly down a complicated ramp, as shown in the diagram below. What amount of work is done by the gravitational force as the block slides down to the horizontal level, having started at a height h above this level? Are there any other forces acting on the block; if so, what amount of work is done by these forces?

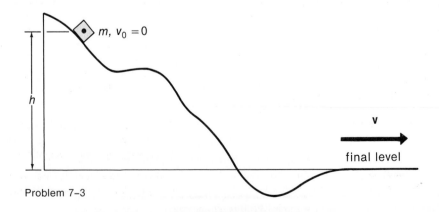

Problem 7-3

7-4 A pendulum bob with a mass m is released from position 1 shown in the accompanying diagram. Some time later, after the support cord has intercepted the two horizontal fixed pegs A and B, the bob is observed to be at position 2. How much work was done on the bob by gravity?

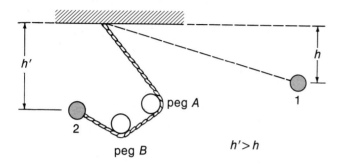

7-5 Two blocks with masses m_1 and m_2 are connected by a massless rope, as shown in the diagram. The ramp is frictionless. If m_1 rises in height by a distance h, how much work does gravity do on the two blocks during this motion? [*Hint:* Consider each block separately.]

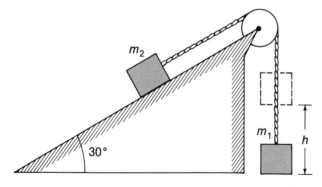

7-6 The excavation for a building is to be 40 m long,
● 15 m wide, and 10 m deep. The material to be removed has a density of 4 g/cm³. Each bucket of dirt is hoisted to a height of 3 m above the *original ground level* and dumped into a truck. Compute the minimum amount of work required for the complete excavation.

7-7 A strip of adhesive tape 15 cm long is stuck to a
●● flat surface. Assume that a fixed amount of work (4 J) must be done to separate the tape from the surface. Describe how to strip off the tape by applying a constant force with the smallest magnitude. Does your result agree with your experience?

Section 7-2

7-8 An object that is constrained to move along the x-axis is acted upon by a force $F(x) = ax + bx^2$, where

$a = 5$ N/m and $b = -2$ N/m². The object is observed to proceed directly from $x = 1$ m to $x = 3$ m. How much work was done on the object by the force? Does the process of integration take into account the fact that the force $F(x)$ changes sign in this interval?

7-9 An object that is constrained to move along the x-axis is observed to proceed from $x = 1$ m to $x = 5$ m and then to return to $x = 3$ m. What total amount of work was done on the object by the force $F(x) = (3$ N/m²$)x^2 - (2$ N/m$)x$? (This complicated motion results from the presence of other unspecified forces that also do work.)

7-10 The block shown in the diagram has a mass of 2 kg and is acted upon by a horizontal force $F(x) = a - bx$, where $a = 20$ N and $b = 10$ N/m. The coefficient of sliding friction between the block and the flat surface is $\mu_k = 0.20$. The block is initially at $x = 0$. What distance must the block travel if the net work done on the block by $F(x)$ and friction combined is to be exactly zero? At what point along the way does the net force on the block reach zero?

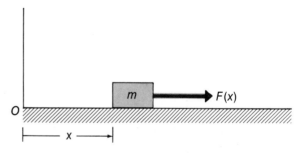

7-11 By adding a certain variable force, the block in Problem 7-10 is observed to proceed from $x = 0$ to $x = 7$ m and then to return to $x = 1$ m. What total amount of work was done in the combined motion by $F(x)$? How much work was done by the frictional force?

Section 7-3

7-12 A particle with a mass of 0.25 kg is dropped from rest and falls through a distance of 10 m. What amount of work is done on the particle by the gravitational force? What kinetic energy does the particle acquire? What is its velocity at the end of the 10-m fall? If the mass of the particle is changed to 2.5 kg, which of the values calculated will change and which will not change?

7-13 A particle with a mass m and an initial velocity v_0 begins to compress a spring. In what distance will the particle be brought to rest if the spring constant is κ?

7-14 A block with a mass m and an initial horizontal velocity v_0 slides across a horizontal surface with coefficient of friction μ_k. In what distance is the block brought to rest? (Use energy considerations here; compare Problem 6–10.)

7-15 A block with a mass $m = 10$ kg is lowered vertically by a rope, the other end of which is attached to a machine that exerts a tension $T = ky^2$, where $k = 2$ N/m² and where y is the distance by which the block has descended from its starting position. If the block starts from rest at $y = 0$, what velocity does it have after descending to $y = 10$ m?

7-16 In Problem 7–15, what is the farthest descent y_m of
•   the block? At what position will the speed of the block be maximum?

7-17 If the block in the diagram is released from rest at
•   position A, to what height h_B will it rise before coming to rest momentarily at position B?

the initial velocity at $x = 0$ is $\mathbf{v}_0 = +(1.5 \text{ m/s})\hat{\mathbf{i}}$, what is the velocity of the particle when it arrives at $x = 6$ m? Give a general expression for $\mathbf{v}$ as a function of x.

7-21 A particle with a mass of 2 kg is constrained to move
••   along the x-axis. The particle is subjected to a force $F(x) = (2 \text{ N/m})x - 4$ N. Consider the possibility of the particle starting with an initial velocity of $+1.5$ m/s and moving from $x = 0$ to $x = 6$ m. According to Eq. 7–6 the particle should arrive at $x = 6$ m with a velocity considerably greater than v_0. Yet the proposed trip is impossible. Why?

7-22 A small block with mass m slides down a frictionless ramp at the bottom of which is a loop-the-loop track with a radius R, as shown in the diagram. If the block is to stay in contact with the track at all times during

Problem 7-17

7-18 A block with a mass m slides down the rough surface of a plane that makes an angle α with the horizontal. If the block starts down the plane with a velocity v_0 and is observed to come to rest after traveling a distance ℓ (along the inclined plane), what is the coefficient of sliding friction?

7-19 A force F is applied to a block, originally at rest, as in
•   the diagram below. After the block has moved a distance of 6 m to the right, the direction of the horizontal component of the force F is reversed (but the magnitude remains the same). With what velocity does the block arrive at its starting point? [*Hint:* The block travels a total distance to the right greater than 6 m. Divide the motion into appropriate parts and apply Eq. 7–6 successively.]

7-20 A particle with a mass of 0.5 kg is acted on by a force
• $\mathbf{F} = (ax + b)\hat{\mathbf{i}}$, where $a = 2$ N/m and $b = 4$ N. If

the loop, what is the minimum height h from which the block can be released from rest?

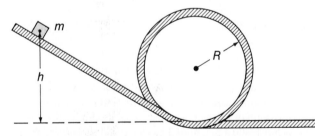

7-23 A soap-box derby cart has a length of 2 m and is
•   required to have a total mass (including the driver) of 100 kg. This particular cart and driver has a total mass of 70 kg, so it is required to carry an additional load of 30 kg. Where should the 30-kg block be placed to provide maximum velocity along the horizontal section at the bottom of the 20° accelerating ramp? First, assume that by simply placing the block in the *middle* of the cart, the velocity at the bottom of the ramp is 10 m/s. Then, determine the best (and the worst) that one can do by placing the block at other positions along the cart. (•If you were the driver of this cart, where would you plan to sit during the race?)

Section 7-4

7-24 A block with a mass of 3 kg starts at a height $h = 60$ cm on a plane with an inclination angle of 30°, as shown in the diagram below. Upon reaching the bottom of the ramp, the block slides along a horizontal surface. If, on both surfaces, the friction coefficient is $\mu_k = 0.20$, how far will the block slide on the horizontal surface before coming to rest? [*Hint:* Divide the path into two straight-line parts.]

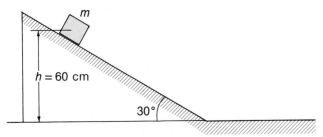

7-25 A 5-kg block is set into motion up an inclined plane with an initial velocity of 8 m/s. The block comes to rest after traveling 3 m along the plane, as shown in the diagram. The plane is inclined at an angle of 30° to the horizontal. (a) Determine the change in kinetic energy. (b) Determine the change in potential energy. (c) Determine the frictional force on the block (assumed to be constant). (d) What is the coefficient of kinetic friction?

7-26 In the diagram below, the two balls have masses $M = 2$ kg and $m = 0.5$ kg. Initially, the two balls are at
•• positions A and B, and both are at rest. If the distance L is 60 cm, with what velocity does M hit the vertical wall?

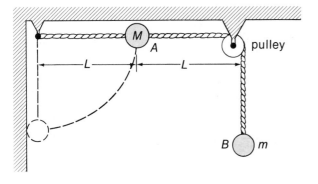

7-27 A limp rope with a mass of 0.2 kg and a length of 1 m
• is hung, initially at rest, on a frictionless peg that has a negligible radius, as shown in the diagram below. What vertical velocity of the rope is reached just as the end slides off the peg? [*Hint:* Review Example 7-14.]

7-28 A potential energy function often used to describe the interaction between two atoms in a diatomic molecule is the *Lennard-Jones potential,* $U(r) = U_0[(r_0/r)^{12} - 2(r_0/r)^6]$. (a) Sketch the function $U(r)$. (b) Show that $U(r)$ has a minimum value at $r = r_0$ and that this value is $-U_0$. (c) Find the values of r/r_0 for which $U(r) = -\frac{1}{4}U_0$.

Section 7-5

7-29 A crane is to lift 2000 kg of material to a height of 150 m in 1 min at a uniform rate. What electric power is required to drive the crane motor if 35 percent of the electric power is converted to mechanical power? (•Is it reasonable to neglect the kinetic energy in this process?)

7-30 A small automobile with a mass of 400 kg is powered by an engine delivering a maximum of 45 hp. At what maximum speed could it go up a 15° hill? (Neglect all frictional forces.)

7-31 A vehicle with a mass m is propelled in a straight
• horizontal line by an engine delivering a constant power P. If the vehicle starts from rest, show that the distance traveled in a time t is $s = \sqrt{8Pt^3/9m}$.

7-32 A block with a mass of 5 kg is pushed up a 20° inclined plane, starting from rest, by a constant *horizontal* force of 50 N. If the coefficient of sliding friction between the block and the ramp is $\mu_k = 0.20$, at what rate is heat generated by frictional losses when the block has been moved a distance of 0.5 m up the plane? At what rate is the block gaining kinetic energy at this point?

Additional Problems

7-33 A block with a mass of 2 kg is lowered vertically, starting from rest, by a student who maintains the acceleration of the block at $\frac{1}{5}g$. How much work is done by the student in lowering the block by 25 cm? What is the velocity of the block at this point?

7-34 A uniform plank with a mass of 50 kg and a length of 3 m rests on a frictionless horizontal surface, as shown in the diagram below. A student slowly raises the plank with the aid of the two ideal pulleys, so that it just dangles off the surface. What is the least amount of work the student must do?

7-35 A student slowly raises a block by pulling on the
•• end of a massless rope as he walks to the right in the way shown in the diagram. If the student moves a distance such that the angle α, initially $\alpha_0 = 30°$, is increased to $\alpha_f = 45°$, calculate how much work he does on the block by considering the motion of the block. Next, by direct calculation, find the amount of work done by the student on the rope. Assume an ideal pulley. Take $\ell = 10$ m and $m = 3$ kg. Compare the two answers.

7-36 A student, using a massless lever with a length ℓ, very slowly raises a large boulder with a mass M to a height h by applying a constant vertical force F, as indicated in the diagram. If the pivot point (or fulcrum) is $\frac{1}{4}\ell$ from the boulder, how much work was done by the student? Considering the lever and the

boulder together as a unit, show in a free-body diagram all the forces acting. Neglecting the kinetic energy of the lever and boulder, how much total work was done on this combined unit? Does the force at the pivot point do any work? Find the relationship between Mg and F.

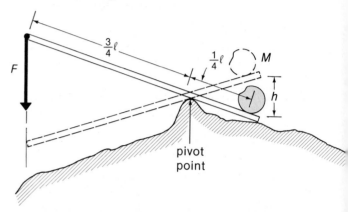

7-37 The world's records for throwing the shot put, the discus, and the javelin are approximately 22 m, 71 m, and 95 m, respectively. The corresponding masses of these missiles are 7.3, 2.0, and 0.80 kg. Assume that each throw is made at the optimum angle (at least, for the case of negligible air resistance) of 45°, and compute the initial kinetic energy in each case. Comment on the results. (•Does the efficiency of the muscles in throwing depend on the mass of the object thrown?)

7-38 A student whose mass is 80 kg stands in the middle
• of a long, uniform plank that has a mass of 40 kg. He raises the center of the plank (and himself as well) 1 cm above the initial horizontal position by pulling

on an ideal rope that runs over two ideal pulleys as shown in the diagram. The rope is attached to the plank at point B; assume that the contact point A remains fixed. Use the conservation of energy to calculate the tension in the rope. To what amount could the mass of the plank be increased before the student could no longer raise it off the ground?

7-39 A limp chain with a length of 1 m and a total mass of 0.5 kg rests on a table with 1/3 of its length hanging over the table edge. How much work is required to raise this part of the chain to the table top?

7-40 If the chain in Problem 7-39 is allowed to slide off the frictionless table, what is the velocity of the chain just as the trailing end leaves the table? (The height of the table top is >1 m.)

7-41 A particle with a mass m slides down a frictionless sphere with a radius R, starting (essentially) from rest at the top. Through what vertical distance y does the particle move before it loses contact with the sphere?

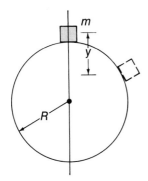

7-42 The diagram below shows a ball with a mass m attached to a string with a length l. The peg is located a distance d directly below the support point. If the ball is to swing completely around the peg, starting from the position shown, prove that d must be greater than $3l/5$. [Hint: Refer to Example 7-11.]

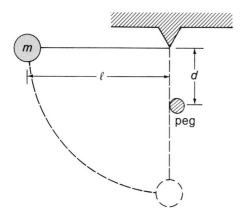

7-43 A thin rod with a mass m and a length l is pivoted
• at one end by a frictionless bearing. The rod is released from rest, standing erect on its pivoted end. What velocity does the free end have at the bottom of the swing? What is the tension in the bearing support at this instant? [Hint: Integrate over the length of the rod.]

7-44 A particle with a mass m is tied to the end of a string and is whirled in a circular orbit in a vertical plane. If the *total energy* of the mass is a constant, what is the string tension when the particle is at its lowest point compared with that when the particle is at its highest point?

7-45 A slingshot consists of a Y-shaped fork with the tips 12 cm apart. The elastic band has a relaxed length of 40 cm. A draw which just doubles the length of the band requires a force of 50 N. What speed would this draw impart to a 100-g rock?

7-46 A block starts from rest and slides along a plane with an inclination of 38°. After the block has traveled down the plane for a distance of 4 m, its velocity is 5 m/s. What is the coefficient of friction between the block and the plane?

7-47 A block of wood ($m = 1.6$ kg) is initially at the bottom of an inclined plane whose inclination angle is 25°. The coefficient of kinetic friction between the block and the plane is $\mu_k = 0.25$ and the coefficient of static friction is $\mu_s = 0.50$. The block is given an initial velocity of 2.5 m/s up the plane. How far along the plane will the block rise? Will the block then slide down the plane? Explain.

7-48 A block with a mass of 2 kg slides across a horizontal
• surface. The coefficient of sliding friction is $\mu_k = 0.35$. The block collides with a horizontal spring, which it compresses by 50 cm before coming momentarily to rest. If the spring constant is 30 N/m,

what was the velocity of the block at the instant of contact? How far back past this point does the recoiling block slide before coming to rest again?

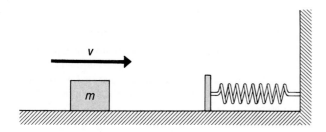

7–49 A particle with a mass $m = 0.1$ kg moves under the influence of a force and has a potential energy $U(x) = (\alpha + \beta x)x^2$, where $\alpha = 1$ J/m^2 and $\beta = -\frac{1}{2}$ J/m^3. Where are the equilibrium points of the motion [i.e., where is $F(x) = 0$]? Are these points of stable or unstable equilibrium? What minimum total energy must the particle have in order to escape from the potential well and move to $x \rightarrow +\infty$?

7–50 Consider the potential function $U(r) = U_0[(r_0/r)^5 - a(r_0/r)^3]$. Find the value of a that gives a minimum value of $U(r)$ at $r = r_0$. What is the value of $U(r_0)$? Find the values of r/r_0 for $U(r) = -\frac{1}{4}U_0$ (these are the turning points for $E = -\frac{1}{4}U_0$). [*Hint:* Use an iterative or successive approximation procedure. For the larger value of $r_0/r \equiv x_1$, write $x_1^5 = ax_1^3 - 1/4$, and for the smaller value of $r_0/r \equiv x_2$, write $ax_2^3 = x_2^5 + 1/4$. Try, as starting values, $x_1 = 10$ and $x_2 = 0.1$. (These values will give reasonable rates of convergence toward the true solutions.) Use these values in the right-hand sides of the above expressions to give new trial values for x_1^5 and ax_2^3; solve these expressions for the new values of x_1 and x_2. These values can, in turn, be substituted into the right-hand sides of the expressions to obtain better values of x_1^5 and ax_2^3. This procedure can be repeated until the desired accuracy is attained. Find values of x_1 and x_2 that are accurate to 5 percent.]

CALCULUS IV

PARTIAL DERIVATIVES

The idea of partial differentiation is used in Chapter 8 in connection with the discussion of conservative forces. The next extensive use of partial derivatives occurs in the study of wave motion (Chapter 19).

In preceding chapters we have been concerned with particle motions describable by means of functions of a single independent variable. These functions have been of the form $x(t)$ and $y(t)$, or, by eliminating t between such a pair of functions, they have been reduced to the form $y = f(x)$. There are many cases in which functions depend on more than a single independent variable. To deal with such functions, we must extend the mathematical techniques we have developed for dealing with functions of one variable. Accordingly, this appendix is concerned with the calculus of multi-variable functions.

IV-1 PARTIAL DERIVATIVES

Simple examples of functions of more than one independent variable can be found in elementary mathematical formulas. Thus, the area A of a rectangle with sides x and y is actually a function of two independent variables: $A = A(x, y) = xy$. Also, the volume V of a right circular cylinder is $V = V(r, h) = \pi r^2 h$; here, V is a function of r and h. In each of these cases, it is clear that the variables are truly independent: the choice of a value for one variable in no way influences the choice of a value for the other. We are therefore led naturally to considerations of general relationships of the type $u = f(x, y)$. In all of our discussions we assume that $f(x, y)$ is a continuous function of both x and y.

For functions of the type $u = f(x, y)$, it is of interest to inquire how u changes if we vary x alone while holding y fixed, and how u changes varying y alone while holding x fixed. In the first instance u in effect becomes a function of the single variable x, whereas, in the second instance, u becomes a function of the single variable y. Generalizing the usual definition of the derivative, we now define the *partial derivative* of u with respect to x (y held fixed) to be*

$$\frac{\partial u}{\partial x} \equiv \lim_{\Delta x \to 0} \frac{f(x + \Delta x, y) - f(x, y)}{\Delta x} \qquad \textbf{(IV–1)}$$

Similarly, we define the partial derivative of u with respect to y (x held fixed) to be

$$\frac{\partial u}{\partial y} \equiv \lim_{\Delta y \to 0} \frac{f(x, y + \Delta y) - f(x, y)}{\Delta y} \qquad \textbf{(IV–2)}$$

The idea of partial differentiation is readily generalized to functions of more than two variables. We can calculate the partial derivatives of a function such as $w = f(x, y, z)$ by, in turn, considering the variation of each variable alone while holding *both* of the other two variables fixed; these are $\partial w/\partial x$, $\partial w/\partial y$, and $\partial w/\partial z$.

If $u = f(x, y)$ is considered to be the equation of a surface in Cartesian coordinates, then Eqs. IV–1 and IV–2 are interpreted as the slopes of special lines that are tangent to this surface. In order to construct a three-dimensional image of the situation, we can associate the variables

---

*We use the cursive (or Byzantine) delta ∂ to imply partial differentiation. Sometimes we see the notation $\partial_x u$ or u_x for $\partial u/\partial x$.

x and y with the Cartesian coordinates x and y and the dependent variable u with the coordinate z. Consider the partial derivatives of $f(x, y)$ at the point P where $x = a$ and $y = b$. The shaded surfaces in Fig. IV–1 represent the same graph of $f(x, y)$, shown twice for convenience. In Fig. IV–1a the plane $y = b$ intersects the surface $u = z = f(x, y)$, giving the line of intersection RPQ or $z = f(x, y = b)$. In the plane $y = b$, $\partial z/\partial x$ is equivalent to the ordinary derivative dz/dx; therefore, $\partial z/\partial x$ is the slope of the tangent line TT' (which lies also in the plane $y = b$). That is,

$$\frac{\partial z}{\partial x} = \tan \alpha$$

Similarly, if we pass the plane $x = a$ through P, we obtain the line of intersection MPL or $z = f(x = a, y)$, as shown in Fig. IV–1b. Then, $\partial z/\partial y$ is equivalent to the derivative dz/dy in this plane. Hence, for the slope of the tangent line SS' through the point P and lying in the plane $x = a$, we have

$$\frac{\partial z}{\partial y} = \tan \beta$$

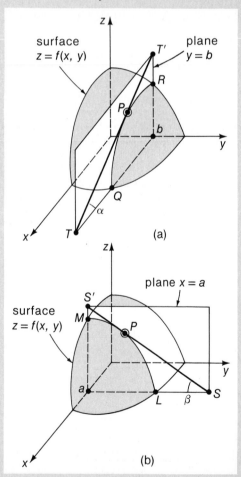

Fig. IV–1. Tangents to the surface at P taken along planes for which (a) $y = $ constant and (b) $x = $ constant.

Example IV–1

Given the function $z = \frac{1}{3}x^2 + y^2$, which describes a paraboloidal surface, find the slope of the curve of intersection between this surface and a plane (a) for the plane $y = 2$ and the point for which $x = 1$, and (b) for the plane $x = 1$ and the point for which $y = 2$.

Solution:

(a) Referring to Fig. IV–1a for guidance, we can write

$$\frac{\partial z}{\partial x} = \frac{2}{3}x$$

so that

$$\tan \alpha = \frac{\partial z}{\partial x}\bigg|_{x=1} = \frac{2}{3}$$

(b) Referring to Fig. IV–1b, we have

$$\frac{\partial z}{\partial y} = 2y$$

so that

$$\tan \beta = \frac{\partial z}{\partial y}\bigg|_{y=2} = 4$$

Notice that the specification of the plane of intersection, namely, $y = 2$ in (a) and $x = 1$ in (b), did not enter into the calculation of the slope. (●What does this mean? Can you give a geometric interpretation?)

Higher Partial Derivatives. The partial derivatives of $u = f(x, y)$ are, in general, functions of the same two variables, x and y. Consequently, these derivatives may, in turn, be differentiated partially with respect to either x or y. This leads to the concept of *second* partial derivatives (and, in fact, to partial derivatives of higher order). These are written as $\dfrac{\partial}{\partial x}\left(\dfrac{\partial u}{\partial x}\right) = \dfrac{\partial^2 u}{\partial x^2}$; $\dfrac{\partial}{\partial y}\left(\dfrac{\partial u}{\partial x}\right) = \dfrac{\partial^2 u}{\partial y \partial x}$; $\dfrac{\partial}{\partial x}\cdot\left(\dfrac{\partial^2 u}{\partial x \partial y}\right) = \dfrac{\partial^3 u}{\partial x^2 \partial y}$; and so forth.

We offer without proof a useful theorem for functions possessing continuous derivatives, namely: *The value of a partial derivative is independent of the order of differentiation.* That is,

$$\frac{\partial}{\partial x}\left(\frac{\partial u}{\partial y}\right) = \frac{\partial^2 u}{\partial x \partial y} = \frac{\partial^2 u}{\partial y \partial x} = \frac{\partial}{\partial y}\left(\frac{\partial u}{\partial x}\right)$$

This type of equality holds for derivatives of all orders.

Example IV–2

Verify the equality of the second partial derivatives taken in either order for the function $u = f(x, y) = x^3 - 4x^2y + 2xy^2 + 3y$.

Solution:

$$\frac{\partial^2 u}{\partial x \partial y} = \frac{\partial}{\partial x}\left(\frac{\partial u}{\partial y}\right) = \frac{\partial}{\partial x}(-4x^2 + 4xy + 3) = -8x + 4y$$

$$\frac{\partial^2 u}{\partial y \partial x} = \frac{\partial}{\partial y}\left(\frac{\partial u}{\partial x}\right) = \frac{\partial}{\partial y}(3x^2 - 8xy + 2y^2) = -8x + 4y$$

The Chain Rule. The chain rule for differentiating a function of a single variable states that if $y = f(v)$ and $v = g(x)$ are differentiable functions, then

$$\frac{dy}{dx} = \frac{dy}{dv}\frac{dv}{dx}$$

Suppose that $u = f(v)$ and v is a function of two variables, $v = g(x, y)$; then,

$$\frac{\partial u}{\partial x} = \frac{du}{dv}\frac{\partial v}{\partial x} \quad \text{and} \quad \frac{\partial u}{\partial y} = \frac{du}{dv}\frac{\partial v}{\partial y}$$

A more general form of the chain rule emerges when $u = f(v, w)$, where $v = g(x, y)$ and $w = h(x, y)$. This is a situation that arises when, for example, we change variables from v and w to x and y. Then, we have

$$\frac{\partial u}{\partial x} = \frac{\partial u}{\partial v}\frac{\partial v}{\partial x} + \frac{\partial u}{\partial w}\frac{\partial w}{\partial x} \quad \text{and} \quad \frac{\partial u}{\partial y} = \frac{\partial u}{\partial v}\frac{\partial v}{\partial y} + \frac{\partial u}{\partial w}\frac{\partial w}{\partial y}$$

Example IV–3

Given the functions $u = v^2$ and $v = x^2y + y^3$, find $\partial u/\partial x$ and $\partial u/\partial y$.

Solution:

$$\frac{\partial u}{\partial x} = \frac{\partial u}{\partial v}\frac{\partial v}{\partial x} = (2v)(2xy) = 2(x^2y + y^3)(2xy)$$

$$\frac{\partial u}{\partial y} = \frac{\partial u}{\partial v}\frac{\partial v}{\partial y} = (2v)(x^2 + 3y^2) = 2(x^2y + y^3)(x^2 + 3y^2)$$

IV–2 INCREMENTS, DIFFERENTIALS, AND TOTAL DERIVATIVES

The incremental change Δy in a function, $y = f(x)$, of a single independent variable x is

$$\Delta y = f(x + \Delta x) - f(x)$$

When Δx is small, the increment Δy is essentially the same as the differential dy (see Section I–5). Thus, for sufficiently small values of Δx, we have (see Eq. I–10 and Fig. I–9)

$$\Delta y \cong dy = \frac{df}{dx}dx = \frac{df}{dx}\Delta x$$

For a function $u(x, y)$ of two independent variables, the increment Δu is

$$\Delta u = f(x + \Delta x, y + \Delta y) - f(x, y)$$
$$= [f(x + \Delta x, y + \Delta y) - f(x, y + \Delta y)]$$
$$+ [f(x, y + \Delta y) - f(x, y)]$$
$$= \frac{\partial u}{\partial x}\Delta x + \frac{\partial u}{\partial y}\Delta y \qquad \textbf{(IV–3)}$$

As in the single-variable case (Fig. I–9), where $\Delta x = dx$,

Fig. IV–2. Geometric interpretation of Eq. IV–4 for the differential du. To calculate the incremental change du in the function $u(x, y)$ in proceeding from point P to point Q, first move along the arc PR for which $y = $ const. The change in u between P and R is $(\partial u/\partial x)\,dx$, as in Eq. I–10. Then, move along the arc RQ for which $x = $ const. The change in u between R and Q is $(\partial u/\partial y)\,dy$. Thus the total change in u is $du = (\partial u/\partial x)\,dx + (\partial u/\partial y)\,dy$, as in Eq. IV–4. Notice that in proceeding from P to Q for the surface illustrated, the quantities $\partial u/\partial x$, dx, $\partial u/\partial y$, and dy are all *negative*, so that $du > 0$, as is evident from the diagram.

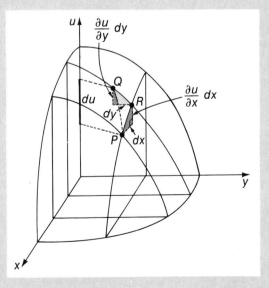

we now have $\Delta x = dx$ and $\Delta y = dy$. Thus, the *total differential* is

$$du = \frac{\partial u}{\partial x} dx + \frac{\partial u}{\partial y} dy \qquad \textbf{(IV-4)}$$

The geometric significance of this expression is illustrated in Fig. IV-2.

For a function $w(x, y, z)$ of three independent variables, the total differential is

$$dw = \frac{\partial w}{\partial x} dx + \frac{\partial w}{\partial y} dy + \frac{\partial w}{\partial z} dz$$

If the quantities x and y in Eq. IV-4 are functions of the single variable t, $x(t)$ and $y(t)$, the total derivative of u with respect to t is obtained simply by dividing Eq. IV-4 by dt:

$$\frac{du}{dt} = \frac{\partial u}{\partial x} \cdot \frac{dx}{dt} + \frac{\partial u}{\partial y} \cdot \frac{dy}{dt} \qquad \textbf{(IV-5)}$$

$\Delta x \, \Delta y$. When Δx and Δy are small increments—that is, when $\Delta x \ll (x$ or $y)$ and $\Delta y \ll (x$ or $y)$—the term $\Delta x \, \Delta y$ is of negligible importance compared with either $y \, \Delta x$ or $x \, \Delta y$. The terms Δx and Δy are called infinitesimals of *first order*; in comparison with these quantities, we can usually neglect terms such as $\Delta x \, \Delta y$, which are called infinitesimals of *second order*. In the diagram below, we see the relationship between $\Delta x \, \Delta y$ and the areas $y \, \Delta x$ and $x \, \Delta y$.

Example IV-4

Consider a rectangle with sides x and y. By how much does the area $A = xy$ change when x is increased by Δx and y is increased by Δy?

Solution:

The exact calculation gives

$$\Delta A = (x + \Delta x)(y + \Delta y) - A$$
$$= (xy + y \, \Delta x + x \, \Delta y + \Delta x \, \Delta y) - xy$$
$$= y \, \Delta x + x \, \Delta y + \Delta x \, \Delta y$$

On the other hand, using Eq. IV-3 yields the approximate result,

$$\Delta A \cong \frac{\partial A}{\partial x} \Delta x + \frac{\partial A}{\partial y} \Delta y$$

Now, for $A = xy$,

$$\frac{\partial A}{\partial x} = y \qquad \text{and} \qquad \frac{\partial A}{\partial y} = x$$

Therefore, $\qquad \Delta A \cong y \, \Delta x + x \, \Delta y$

Comparing the two expressions for ΔA shows that the error in the approximate form is in the neglect of the term

Example IV-5

The altitude h of a right circular cone is 20 cm at a certain instant and is decreasing at a rate of 3 cm/s. The radius r of the base is 10 cm and is simultaneously increasing at a rate of 2 cm/s. At what rate is the volume changing?

Solution:

The volume of the cone is $V = \frac{1}{3}\pi r^2 h$. Then,

$$\frac{\partial V}{\partial r} = \frac{2}{3}\pi rh \qquad \text{and} \qquad \frac{\partial V}{\partial h} = \frac{1}{3}\pi r^2$$

Using Eq. IV-5, we have

$$\frac{dV}{dt} = \frac{\partial V}{\partial r}\frac{dr}{dt} + \frac{\partial V}{\partial h}\frac{dh}{dt}$$

Substituting the quoted values, $r = 10$ cm, $h = 20$ cm, $dr/dt = +2$ cm/s, and $dh/dt = -3$ cm/s, yields

$$\frac{dV}{dt} = \frac{1}{3}\pi(10 \text{ cm})[2(20 \text{ cm})(2 \text{ cm/s}) + (10 \text{ cm})(-3 \text{ cm/s})]$$

$$= 523.6 \text{ cm}^3/\text{s}$$

This is the rate of *increase* of the volume.

PROBLEMS

Section IV-1

IV–1 Verify the following partial derivatives:

(a) $w = 6x^2y + 2x - 7y^2z$; $\quad \dfrac{\partial w}{\partial x} = 12xy + 2$

$$\dfrac{\partial w}{\partial y} = 6x^2 - 14yz$$

$$\dfrac{\partial w}{\partial z} = -7y^2$$

(b) $f(x, y) = \dfrac{y}{x - y}$, $x \neq y$; $\quad \dfrac{\partial f}{\partial x} = -\dfrac{y}{(x - y)^2}$

$$\dfrac{\partial f}{\partial y} = \dfrac{x}{(x - y)^2}$$

(c) $\rho = \sin \theta \cos \phi$; $\quad \dfrac{\partial \rho}{\partial \theta} = \cos \theta \cos \phi$

$$\dfrac{\partial \rho}{\partial \phi} = -\sin \theta \sin \phi$$

IV–2 If $f(x, y) = x^3y - xy^3$, show that $\dfrac{\partial^2 f}{\partial x^2} + \dfrac{\partial^2 f}{\partial y^2} = 0$.

IV–3 If $g(x, y, z) = 6x^3y^2 + \dfrac{z}{x} - 2y^4z$, show that

$$\dfrac{\partial^2 g}{\partial x \partial z} = \dfrac{\partial^2 g}{\partial z \partial x}.$$

IV–4 If $r = \sqrt{x^2 + y^2 + z^2}$, show that $\dfrac{\partial r}{\partial x} = \dfrac{x}{r}$, $\dfrac{\partial r}{\partial y} = \dfrac{y}{r}$, and $\dfrac{\partial r}{\partial z} = \dfrac{z}{r}$.

IV–5 If $r = \sqrt{x^2 + y^2 + z^2}$ and $u = \frac{1}{2}ar^2$, show by using the chain rule that

$$\dfrac{\partial u}{\partial x} = ax, \dfrac{\partial u}{\partial y} = ay, \text{ and } \dfrac{\partial u}{\partial z} = az.$$

Section IV-2

IV–6 Find the differential of each of the following functions:

(a) $u = f(x, y, z) = 6xy + y^2 + 2yz$
(b) $u = f(\rho, \phi) = e^{-\alpha\rho} \cos \phi$
(c) $w = f(\alpha, \beta) = \sin (\alpha - \beta)$

IV–7 The dimensions of a rectangular box are measured to be 1.21 m, 2.87 m, and 4.02 m. The accuracy of each of these measurements is ± 2 cm. What is the largest possible error in the computed volume? What is the corresponding percentage error?

IV–8 Find the approximate volume of aluminum in a cylindrical beverage can with an outside diameter of 7 cm and a height of 12 cm, if the wall has a uniform thickness of 0.5 mm.

FORCES AND POTENTIAL ENERGY

When the subject of work and energy was first discussed in Chapter 7, we limited consideration to cases of motion constrained to one-dimensional paths. We now remove this restriction and treat the general case of motion in three dimensions. By making this extension of the subject we are able to discuss the general characteristics of conservative forces.

8-1 FORCE FIELDS

If some agency exerts a force on a particle, the particle will follow one or another of a large number of possible trajectories, depending on the initial conditions of position and velocity. As the particle passes through a specific point (x, y, z) at a specific time t, the force on the particle has a magnitude and direction that we can express as $\mathbf{F}(x, y, z, t)$. We imagine that we are allowed an unlimited number of trials so that we can test experimentally the force on the particle at every conceivable point and time. In this way we are led to the idea that there exists throughout space an agency that *would exert* a force $\mathbf{F}(x, y, z, t)$ on a particle *if* the particle were actually at the specified point at the specified time. This is really what is meant by "the force $\mathbf{F}(x, y, z, t)$".

The idea that $\mathbf{F}(x, y, z, t)$ exists throughout space and, in particular, at points where no particle is located, is called the *force-field* concept, and we refer to $\mathbf{F}(x, y, z, t)$ simply as the *force field*. For a particular particle, the force field may be viewed as a property of the space (and time) in which the particle moves. The actual path or trajectory that is followed by the particle depends, in addition to the details of the force field, on the initial conditions and on any constraints on the motion that might be imposed. The specification of the force field in which a particle moves is one of the important pieces of information that must be provided in order to solve problems in mechanics. For all of the cases discussed in this volume, the force field is considered static, so that $\mathbf{F} = \mathbf{F}(x, y, z)$.

In the discussions in the preceding chapters we introduced various kinds of forces. Usually, these were *contact forces*—forces that result from the actual contact between two objects. For example, we have considered the push exerted by a person's hand and the pull exerted by a spring or a string. Suppose that the force exerted on a particle by some device has the form

$$\mathbf{F} = -\kappa\mathbf{r} \qquad (8\text{–}1)$$

where the origin ($\mathbf{r} = 0$) corresponds to a point of stable equilibrium. Such a spring-like force could be created by a two- or three-dimensional "spring machine," such as that discussed in Problem 8–11. At every point in space a particle that is attached to the machine will experience a force described by Eq. 8–1. We can illustrate this springlike force field by constructing a map of the force vectors, as in Fig. 8–1. Notice that the force vectors increase in magnitude with increasing distance from the origin, as prescribed by Eq. 8–1.

In addition to contact forces, there exist in Nature the so-called *action-at-a-distance forces*—forces that act between objects even though they are separated by empty space. The gravitational and the electromagnetic forces are action-at-a-distance forces.

Both the gravitational and the electric forces between pairs of particles at relative rest have the form of an inverse-square law (that is, $F \propto 1/r^2$). The basic force law in each of these cases is stated in terms of the force between *particles*, but in some situations it is possible to use exactly the same force equation for objects with extent (finite size). An important case involves the gravitational force exerted on objects by the Earth. In Chapter 14, we prove that if two spherically symmetric objects interact via the gravitational force, they act in the same way as do particles. Then, the gravitational force that the Earth (mass M_E) exerts on a particle (mass m) located a distance r from the center of the Earth (with $r > R_E$, the radius of the Earth) is given by

$$\mathbf{F}_g = -G\frac{mM_E}{r^2}\hat{\mathbf{u}}_r \qquad (8\text{–}2)$$

where $\hat{\mathbf{u}}_r$ is the unit radial vector that is directed outward from the center of the Earth (Fig. 8–2). According to this equation, the particle m experiences, at any point

Fig. 8–1. Map of a springlike force field, $\mathbf{F} = -\kappa\mathbf{r}$. Each arrow represents the force on a particle located at the base of the arrow.

Fig. 8-2. The gravitational force $\mathbf{F}_g$ on m is directed toward the center of the Earth.

in space, a force that has a direction toward the Earth's center and a magnitude inversely proportional to the square of the distance from the center. Figure 8-3 shows the *gravitational force field* that is experienced by a particle with a mass m attracted toward a spherical object with a mass M. Notice that the force vectors decrease in magnitude with increasing distance from the origin, as prescribed by Eq. 8-2. (Contrast Fig. 8-1.)

The abstraction of a force field in empty space gains an added elegance when applied to the gravitational and electric forces. In the gravitational case we introduce a vector quantity called the *gravitational field strength* $\mathbf{g}$ associated with an object of mass M that is the source of the field. Thus,

$$\mathbf{g} = \operatorname*{Lim}_{m \to 0} \frac{\mathbf{F}_g}{m} = -G\frac{M}{r^2}\hat{\mathbf{u}}_r \qquad (8\text{-}3)$$

The field represented by $\mathbf{g}$ is a property only of the source object M. Placing a particle with a mass m at any point in this field results in a force on m given by $\mathbf{F}_g = m\mathbf{g}$, where $\mathbf{g}$ is the field strength at the point.

The essential feature of every force field—whether due to a contact force or an action-at-a-distance force—is that associated with every point in space there exists a unique vector $\mathbf{F}(x, y, z)$ that specifies the force on a particle if placed at that point. This is the only aspect of a force field that concerns us at the moment. In particular, we now investigate the work and energy associated with a particle that moves in various force fields. In later chapters we return to the discussion of the properties of action-at-a-distance forces.

Fig. 8-3. Map of the gravitational force field surrounding a source object M. Each arrow represents the force on a particle at the position of the base of the arrow.

8-2 WORK AS A GENERAL LINE INTEGRAL

When we remove the restriction to motion along one-dimensional paths, several of the results obtained in Chapter 7 require important modifications. For example, suppose that a particle moves in a force field specified by

$$\mathbf{F} = \mathbf{F}(x, y, z) = F_x\widehat{\mathbf{i}} + F_y\widehat{\mathbf{j}} + F_z\widehat{\mathbf{k}}$$

where F_x, F_y, and F_z are all functions of x, y, and z. When the particle undergoes a displacement,

$$d\mathbf{s} = dx\widehat{\mathbf{i}} + dy\widehat{\mathbf{j}} + dz\widehat{\mathbf{k}}$$

the work done during this displacement is

$$dW = \mathbf{F} \cdot d\mathbf{s} = F_x\,dx + F_y\,dy + F_z\,dz$$

Again, as in Chapter 7, we point out that independent of the direction in which a path is traversed we *always* write* $d\mathbf{s} = dx\widehat{\mathbf{i}} + dy\widehat{\mathbf{j}} + dz\widehat{\mathbf{k}}$. The limits used in the integral for the total work correctly account for the direction of the motion; thus the total work done in moving from point $P(x_1, y_1, z_1)$ to point $Q(x_2, y_2, z_2)$ is given by

$$W = \int_P^Q dW = \int_P^Q F_x\,dx + \int_P^Q F_y\,dy + \int_P^Q F_z\,dz \qquad \textbf{(8-4)}$$

Each of these integrals is to be carried out over a single variable, but the integrand is a function of x, y, and z. How are we to interpret the indicated integration? The value of the integral, in general, depends on the exact path followed from P to Q, and is therefore ambiguous unless we specify the particular path that is followed by the particle. When the path is specified, the integral is called a *line integral* (sometimes, a *path integral* or *contour integral*).

A path in three-dimensional space can be described in parametric form by functions $x(t)$, $y(t)$, and $z(t)$, where the parametric variable t could (but need not) represent the time. For any value of t, the coordinates x, y, and z are uniquely determined. Then, as t varies over its range, (x, y, z) changes from (x_1, y_1, z_1) at P to (x_2, y_2, z_2) at Q along a unique path.

To emphasize that integrals such as those in Eq. 8-4 are to be evaluated along specified paths, we adopt the special symbol $\oint$ for these line integrals and write, for the work done along the path C,

$$W(C) = \oint_P^Q dW = \oint_P^Q \mathbf{F} \cdot d\mathbf{s}$$

$$= \oint_P^Q F_x\,dx + \oint_P^Q F_y\,dy + \oint_P^Q F_z\,dz \qquad \textbf{(8-5)}$$

When the parametric equations for x, y, and z are substituted into F_x, F_y, and F_z, the integrals reduce to simple definite integrals over the parameter t.

---

*We use $d\mathbf{s}$ to indicate a differential element of the actual path followed by a particle, whereas $d\mathbf{r}$ is an element of an arbitrary path (which may or may not be followed by the particle). In either case, the differential path element is expressed as $dx\widehat{\mathbf{i}} + dy\widehat{\mathbf{j}} + dz\widehat{\mathbf{k}}$.

Example 8-1

Consider a force field specified by

$$\mathbf{F} = 3Axy\hat{\mathbf{i}} + 3Ax^2\hat{\mathbf{j}} + \tfrac{1}{8}Axz\hat{\mathbf{k}}$$

with $A = 10/3\ \text{N/m}^2$. Evaluate the work done in moving a test particle from $P(0, 1, 0)$ to $Q(1, 3, 4)$ along the path C given by the parametric functions

$$x(t) = \tfrac{1}{2}at \qquad y(t) = a(t + 1) \qquad z(t) = at^2$$

where $a = 1$ m and where t is given in seconds. The test particle follows this path because other unspecified forces act on the particle in addition to the force $\mathbf{F}$ we are considering.

Solution:

First, we verify that the points P and Q are actually on the path specified by the parametric functions. Substituting $t_1 = 0$, we find the coordinates of P, and substituting $t_2 = 2$, we find the coordinates of Q. Then,

$$W(C) = \oint_P^Q dW = 3A \oint_P^Q xy\,dx + 3A \oint_P^Q x^2\,dy + \tfrac{1}{8}A \oint_P^Q xz\,dz$$

Substituting for x, y, and z, and noting that $dx = \tfrac{1}{2}a\,dt$, $dy = a\,dt$, and $dz = 2at\,dt$, we find

$$W(C) = 3a^3A \int_{t_1=0}^{t_2=2} (\tfrac{1}{2}t)(t + 1)(\tfrac{1}{2}dt) + 3a^3A \int_0^2 (\tfrac{1}{2}t)^2\,dt + \tfrac{1}{8}a^3A \int_0^2 (\tfrac{1}{2}t)(t^2)(2t\,dt)$$

$$= \tfrac{3}{4}a^3A \int_0^2 t(t + 1)dt + \tfrac{3}{4}a^3A \int_0^2 t^2\,dt + \tfrac{1}{8}a^3A \int_0^2 t^4\,dt$$

$$= \tfrac{3}{4}a^3A(\tfrac{1}{3}t^3 + \tfrac{1}{2}t^2)\big|_0^2 + \tfrac{1}{4}a^3At^3\big|_0^2 + \tfrac{1}{40}a^3At^5\big|_0^2$$

$$= \tfrac{63}{10}a^3A = \tfrac{63}{10}(1\text{ m})^3(\tfrac{10}{3}\text{ N/m}^2) = 21\text{ J}$$

Example 8-2

Next, consider a force field specified by

$$\mathbf{F} = -Ay\hat{\mathbf{i}} + Ax\hat{\mathbf{j}}$$

where A is measured in N/m. Calculate the amount of work done in moving a test particle from one fixed point $P\ (0, 1, 0)$ to another fixed point $Q\ (2, 2, 3)$ along three possible paths in this field:

(a) Path C: $x = f(t) = at \qquad y = g(t) = a(\tfrac{1}{2}t + 1) \qquad z = h(t) = \tfrac{3}{2}at$

(b) Path C': $x = f(t) = at \qquad y = g(t) = a(\tfrac{1}{4}t^2 + 1) \qquad z = h(t) = \tfrac{3}{8}at^3$

where $a = 1$ m and where t is given in seconds.

(c) Path C'': three straight-line segments:

1) $(0, 1, 0)$ to $(2, 1, 0)$, parallel to x-axis
2) $(2, 1, 0)$ to $(2, 2, 0)$, parallel to y-axis
3) $(2, 2, 0)$ to $(2, 2, 3)$, parallel to z-axis

Notice that the end points, P and Q, actually lie on all three paths. For paths C and C', point P corresponds to $t_1 = 0$ and point Q corresponds to $t_2 = 2$. These paths are shown in the diagram.

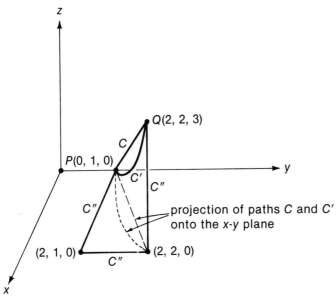

Solution:

(a) For path C, we have $dx = a \, dt$, $dy = \frac{1}{2}a \, dt$, and $dz = \frac{3}{2}a \, dt$. Although the function $z(t)$ is necessary in order to specify the path, neither F_x nor F_y contains z, and $F_z = 0$; consequently, the value of W does not depend on this function. Thus,

$$W(C) = -a^2 A \int_{t_1=0}^{t_2=2} (\tfrac{1}{2}t + 1)dt + a^2 A \int_0^2 \tfrac{1}{2}t \, dt$$

$$= -a^2 A(\tfrac{1}{4}t^2 + t)\big|_0^2 + a^2 A(\tfrac{1}{4}t^2)\big|_0^2$$

$$= -3a^2 A + a^2 A = -2a^2 A$$

(b) For path C', we have $dx = a \, dt$ and $dy = \frac{1}{2}at \, dt$. Again, W does not depend on $z(t)$. Thus,

$$W(C') = -a^2 A \int_0^2 (\tfrac{1}{4}t^2 + 1)dt + a^2 A \int_0^2 \tfrac{1}{2}t^2 \, dt$$

$$= -a^2 A(\tfrac{1}{12}t^3 + t)\big|_0^2 + a^2 A(\tfrac{1}{6}t^3)\big|_0^2$$

$$= -\tfrac{8}{3}a^2 A + \tfrac{4}{3}a^2 A = -\tfrac{4}{3}a^2 A$$

(c) For path C'', the line integrals must be evaluated by a different method. The work done is equal to the line integral of $\mathbf{F} \cdot d\mathbf{s}$. For the first part of the path the displacement is parallel to the x-axis, so $d\mathbf{s} = dx\hat{\mathbf{i}}$. Then, using the fact that $y = 1$ m along this path, we have

$$W(C'')_1 = \oint_{(0,1,0)}^{(2,1,0)} F_x \, dx = -a^2 A \int_0^2 dx = -2a^2 A$$

Next, along the second part of the path, we have $d\mathbf{s} = dy\hat{\mathbf{j}}$ and $x = 2$ m. Thus,

$$W(C'')_2 = \oint_{(2,1,0)}^{(2,2,0)} F_y \, dy = 2a^2 A \int_1^2 dy = 2a^2 A$$

Finally, $W(C'')_3 = 0$, so

$$W(C'') = W(C'')_1 + W(C'')_2 + W(C'')_3$$
$$= -2a^2A + 2a^2A + 0$$
$$= 0$$

We see that the work done by $\mathbf{F}$ in the displacement of the particle from P to Q is different along each of the three paths. Consequently, we draw the important conclusion that *there is no unique way to specify the potential energy of a particle in this field.*

Example 8–3

In the previous example we considered the field described by

$$\mathbf{F} = -Ay\hat{\mathbf{i}} + Ax\hat{\mathbf{j}}$$

We now change the sign of F_x in this expression and study the field

$$\mathbf{F}_+ = +Ay\hat{\mathbf{i}} + Ax\hat{\mathbf{j}}$$

Find the work done in displacing the particle between the same points, P and Q, along the same three paths as in Example 8–2.

Solution:

Because the only difference between the two fields is the sign of F_x, we can obtain the values of the integrals by referring to Example 8–2, making the appropriate sign changes. We find

$$W(C) = 3a^2A + a^2A = 4a^2A$$
$$W(C') = \tfrac{8}{3}a^2A + \tfrac{4}{3}a^2A = 4a^2A$$
$$W(C'') = 2a^2A + 2a^2A = 4a^2A$$

These results strongly suggest that the work done by the force $\mathbf{F}_+ = Ay\hat{\mathbf{i}} + Ax\hat{\mathbf{j}}$ in the displacement $P \to Q$ is *independent* of the path followed. (See also Problem 8–3.) This conjecture is, in fact, true. In the next section we examine the reason for the fundamental difference between these two apparently similar force fields.

Line Integrals in Polar Coordinates. In polar coordinates (Section 3–8), the force vector is expressed in terms of a radial component F_ρ and an azimuthal component F_ϕ (see Eq. 3–30):

$$\mathbf{F} = F_\rho \hat{\mathbf{u}}_\rho + F_\phi \hat{\mathbf{u}}_\phi$$

where F_ρ and F_ϕ can be functions of ρ and ϕ.

Figure 8–4 shows how we obtain the expression for an incremental displacement $\Delta\mathbf{s}$ in polar coordinates. For a displacement $\Delta\mathbf{s}'$ with the angle ϕ constant (Fig. 8–4a), the displacement vector $\Delta\mathbf{s}''$ is $\Delta\rho\hat{\mathbf{u}}_\rho$. If ρ is constant (Fig. 8–4b), the displacement vector is $\rho\,\Delta\phi\hat{\mathbf{u}}_\phi$. Thus, for an arbitrary displacement, we have $\Delta\mathbf{s} = \Delta\mathbf{s}' + \Delta\mathbf{s}''$, or

$$\Delta\mathbf{s} = \Delta\rho\hat{\mathbf{u}}_\rho + \rho\,\Delta\phi\hat{\mathbf{u}}_\phi$$

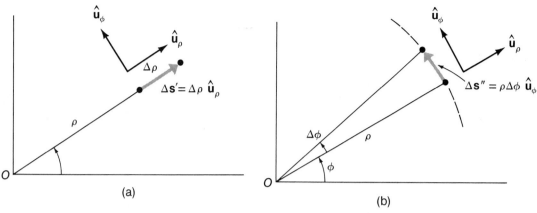

Fig. 8-4. Components of the displacement vector in polar coordinates. (a) The incremental displacement $\Delta\mathbf{s}'$ for ϕ = constant. (b) The incremental displacement $\Delta\mathbf{s}''$ for ρ = constant.

When the displacement $\Delta\mathbf{s}$ becomes vanishingly small, the increments are replaced by differentials:

$$d\mathbf{s} = d\rho\hat{\mathbf{u}}_\rho + \rho\,d\phi\hat{\mathbf{u}}_\phi \tag{8-6}$$

Then, the work done along some path from point P to point Q

$$W(C) = \int_P^Q (F_\rho\,d\rho + F_\phi\rho\,d\phi) \tag{8-7}$$

Example 8-4

A pendulum bob with a mass m is raised very slowly (with nearly zero acceleration) through a vertical height h from its equilibrium position P to the point Q by the application of a force $\mathbf{F}$ that acts always tangent to the circular arc followed by the bob. Calculate the amount of work done by $\mathbf{F}$.

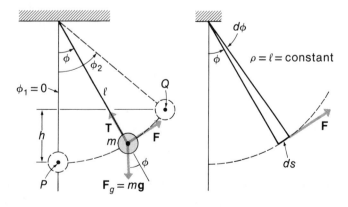

Solution:

We use polar coordinates, ρ and ϕ, with the origin at the point of suspension. Here, the radial coordinate ρ is constant in magnitude and equal to the length ℓ of the pendulum. Therefore, the differential displacement is $d\mathbf{s} = \rho\,d\phi\hat{\mathbf{u}}_\phi = \ell\,d\phi\hat{\mathbf{u}}_\phi$. The force $\mathbf{F}$ is in the direction $\hat{\mathbf{u}}_\phi$ or $\mathbf{F} = F(\phi)\hat{\mathbf{u}}_\phi$; then, because $\hat{\mathbf{u}}_\phi \cdot \hat{\mathbf{u}}_\phi = 1$, we have

$$W(C) = \oint_P^Q \mathbf{F} \cdot d\mathbf{s} = \int_{\phi_1=0}^{\phi_2} F(\phi)\ell\, d\phi$$

Next, we need the expression for $F(\phi)$. From the diagram, we see that

$$F = mg \sin \phi$$

Therefore, $$W(C) = mg\ell \int_0^{\phi_2} \sin \phi\, d\phi = -mg\ell \cos \phi \Big|_0^{\phi_2}$$

$$= mg\ell(1 - \cos \phi_2)$$

Now, $$\cos \phi_2 = \frac{\ell - h}{\ell} = 1 - \frac{h}{\ell}$$

so that $$W(C) = mgh$$

which is the expected result. Again, note that the tension $\mathbf{T}$ does no work. The gravitational force $\mathbf{F}_g$ does an amount of work that is the negative of the work done by $\mathbf{F}$. (•Can you see why?) Thus, the total amount of work done on the bob is zero.

Example 8-5

In Example 7–5 we calculated the work done by gravity in moving a particle along a radial line in the vicinity of the Earth. We now consider a similar case, but one in which we allow displacement along an arbitrary path. (For convenience, we limit the displacement to two dimensions along a plane that contains the Earth's center, but the result is valid for an arbitrary three-dimensional displacement.)

Solution:

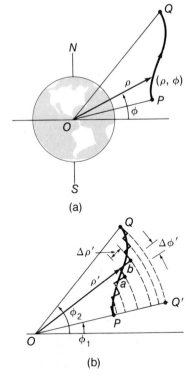

Diagram (a) shows the two-dimensional path $P \rightarrow Q$ that is followed by the particle. An arbitrary point on this path is identified by the angle ϕ (the latitude) and ρ (the radial distance from the center of the Earth). In diagram (b), this path has been subdivided into many small, alternately azimuthal and radial steps which, in the limit of an infinite number of such steps, correspond exactly to the actual curved path $P \rightarrow Q$. One such pair of steps, from point a to point b, is illustrated in diagram (b). The azimuthal part of the step is $\rho' \Delta\phi'$ and the radial part is $\Delta\rho'$.

The gravitational force $\mathbf{F}_g$ is given by Eq. 8–2:

$$\mathbf{F}_g = -G\frac{mM_E}{r^2}\hat{\mathbf{u}}_r$$

In proceeding from a to b, no work is done during the azimuthal part of the step because $\mathbf{F}_g$ is perpendicular to $\Delta\mathbf{s}$. During the radial part of the step, the work done is $F_g' \Delta\rho'$, where F_g' is the gravitational force at $r = \rho'$. The sum of all the azimuthal parts of the steps (along which no work is done) is equivalent to the arc $\overline{Q'Q}$; the sum of all the radial parts of the steps is equivalent to the radial displacement $\overline{PQ'}$. Thus, the total work done by $\mathbf{F}_g$ is the same for $P \rightarrow Q$ as for $P \rightarrow Q'$. Hence,

$$W(P \rightarrow Q) = W(P \rightarrow Q') = -GmM_E \int_{R_1}^{R_2} \frac{d\rho}{\rho^2}$$

$$= GmM_E \left(\frac{1}{R_2} - \frac{1}{R_1}\right)$$

(a)

(b)

$$\overline{OP} = R_1; \quad \overline{OQ} = \overline{OQ'} = R_2$$

Notice that $R_2 > R_1$, so $W < 0$ and the work done *by* the field was negative; this means that work was done *on* the field during the displacement $P \rightarrow Q$. Compare with the result obtained in Example 7–5.

Closed-Path Line Integrals. In many cases we are interested in calculating the total amount of work done in a displacement $P \rightarrow Q$ along one path followed by a displacement $Q \rightarrow P$ along a different path (Fig. 8–5). To denote such *closed-path* line integrals, we use the special symbol $\oint$. Usually, we are interested in paths that enclose nonzero areas.

Suppose that we evaluate the line integral of $\mathbf{F} \cdot d\mathbf{s}$ for the case in Example 8–2 by proceeding from P to Q along C' and then returning from Q to P along C. Then,

$$\oint \mathbf{F} \cdot d\mathbf{s} = \oint_P^Q \mathbf{F} \cdot d\mathbf{s} + \oint_Q^P \mathbf{F} \cdot d\mathbf{s}$$

Now, traversing the path in the reverse sense simply interchanges the limits on the integral. That is,

$$\oint_Q^P \mathbf{F} \cdot d\mathbf{s} = -\oint_P^Q \mathbf{F} \cdot d\mathbf{s}$$

Therefore, using the results in Example 8–2, we find

$$\oint \mathbf{F} \cdot d\mathbf{s} = -\tfrac{4}{3}a^2 A - (-2a^2 A) = \tfrac{2}{3}a^2 A$$

On the other hand, if we choose the path C'' for the $P \rightarrow Q$ displacement (with C again for $Q \rightarrow P$), we would have

$$\oint \mathbf{F} \cdot d\mathbf{s} = 0 - (-2a^2 A) = 2a^2 A$$

We see that the closed-path integral of $\mathbf{F} = -Ay\hat{\mathbf{i}} + Ax\hat{\mathbf{j}}$ depends on the path chosen. However, the $P \rightarrow Q$ line integral of the force $\mathbf{F}_+ = Ay\hat{\mathbf{i}} + Ax\hat{\mathbf{j}}$ was found in Example 8–3 to be independent of the path. Consequently, the closed-path integral for $\mathbf{F}_+$ is *zero* for any closed paths selected from amongst C, C', and C'' that include the points P and Q. However, as we see later, the vanishing of the closed-path integral for this force is quite generally true for whatever path is selected.

Of course, in the trivial case of a closed path associated with zero area, we always find $\oint \mathbf{F} \cdot d\mathbf{s} = 0$ for any force field that has no energy-loss mechanism (such as friction).

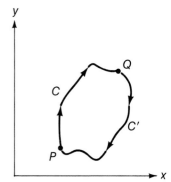

Fig. 8–5. A closed-path integral is performed by integrating from P to Q along C, followed by integrating from Q to P along C'.

SPECIAL TOPIC

8–3 CONSERVATIVE FORCES

In Section 7–1 we evaluated the work done by a *constant* force, $\mathbf{F} = F_x\hat{\mathbf{i}} + F_y\hat{\mathbf{j}} + F_z\hat{\mathbf{k}}$, along an arbitrary path between two points, $P(x_1, y_1, z_1)$ and $Q(x_2, y_2, z_2)$, with the

result

$$W(P \rightarrow Q) = F_x(x_2 - x_1) + F_y(y_2 - y_1) + F_z(z_2 - z_1)$$

That is, the work done was found to depend on the coordinates of the end points, P and Q, but not on the path connecting these points. In the notation introduced here, this conclusion can be expressed as

$$W(C) = \oint_P^Q \mathbf{F} \cdot d\mathbf{s} = \oint_P^Q F_x\, dx + \oint_P^Q F_y\, dy + \oint_P^Q F_z\, dz$$

$$= F_x \int_{x_1}^{x_2} dx + F_y \int_{y_1}^{y_2} dy + F_z \int_{z_1}^{z_2} dz$$

$$= F_x(x_2 - x_1) + F_y(y_2 - y_1) + F_z(z_2 - z_1) \quad \textbf{(8-8)}$$

If the path C is a *closed* path, then $x_2 = x_1$, $y_2 = y_1$, and $z_2 = z_1$, so that $W(C) = 0$. We conclude that the closed-path integral of $\mathbf{F} \cdot d\mathbf{s}$ is always zero for a constant force $\mathbf{F}$.

In the preceding section we obtained evidence that $\oint \mathbf{F} \cdot d\mathbf{s} = 0$ for the force $\mathbf{F}_+ = Ay\hat{\mathbf{i}} + Ax\hat{\mathbf{j}}$. Conversely, we know that there are other forces (for example, $\mathbf{F} = -Ay\hat{\mathbf{i}} + Ax\hat{\mathbf{j}}$) for which the closed-path integral is not zero. Must we always evaluate a number of line integrals to determine whether $\oint \mathbf{F} \cdot d\mathbf{s}$ is or is not zero for a particular force? Or is there some test we can apply directly to $\mathbf{F}$ in order to determine whether the closed-path integral vanishes?

In Chapter 7 we discovered that for the case of one-dimensional constrained motion it is useful to introduce the *potential energy function* $U(x)$ for which $F_x = -dU/dx$. We can extend this idea to the three-dimensional case and define a function $U(x, y, z)$ such that

$$F_x = -\frac{\partial U}{\partial x} \qquad F_y = -\frac{\partial U}{\partial y} \qquad F_z = -\frac{\partial U}{\partial z} \qquad \textbf{(8-9)}$$

where the derivatives are now *partial* derivatives (see Section IV-1).

The work done by the force $\mathbf{F}$ as a particle moves along a path C from $P(x_1, y_1, z_1)$ to $Q(x_2, y_2, z_2)$ is

$$W(C) = \oint_P^Q \mathbf{F} \cdot d\mathbf{s} = \oint_P^Q (F_x\, dx + F_y\, dy + F_z + dz)$$

Now, if Eqs. 8-9 are valid, this becomes

$$W(C) = -\oint_P^Q \left(\frac{\partial U}{\partial x}\, dx + \frac{\partial U}{\partial y}\, dy + \frac{\partial U}{\partial z}\, dz \right)$$

The integrand is just the total differential of $U(x, y, z)$ (see Eq. IV-4). Thus,

$$W(C) = -\int_P^Q dU = U(P) - U(Q) \qquad \textbf{(8-10)}$$

When point P coincides with point Q, we have $W(C) = 0$.

That is, when Eqs. 8-9 are satisfied, the closed-path integral vanishes for *any* closed path.

Any force for which $\oint \mathbf{F} \cdot d\mathbf{s} = 0$ is always true can be associated with a potential energy function $U(x, y, z)$ according to Eqs. 8-9. The work done by such a force in moving a particle from P to Q along some path can be recovered by moving the particle from Q back to P *along any path whatsoever*. For this reason, forces of this type are called *conservative forces*.

It is quite generally true, for the type of well-behaved functions we are considering, that the value of a second partial derivative does not depend on the order of differentiation. That is,

$$\frac{\partial}{\partial x}\left(\frac{\partial U}{\partial y} \right) = \frac{\partial^2 U}{\partial x \partial y} = \frac{\partial^2 U}{\partial y \partial x} = \frac{\partial}{\partial y}\left(\frac{\partial U}{\partial x} \right)$$

Now, if Eqs. 8-9 are satisfied, we can write

$$\frac{\partial^2 U}{\partial x \partial y} = \frac{\partial}{\partial x}\left(\frac{\partial U}{\partial y} \right) = -\frac{\partial}{\partial x} F_y$$

and

$$\frac{\partial^2 U}{\partial y \partial x} = \frac{\partial}{\partial y}\left(\frac{\partial U}{\partial x} \right) = -\frac{\partial}{\partial y} F_x$$

Then, it follows that

$$\frac{\partial F_x}{\partial y} = \frac{\partial F_y}{\partial x}$$

and also

$$\frac{\partial F_y}{\partial z} = \frac{\partial F_z}{\partial y} \qquad \text{and} \qquad \frac{\partial F_z}{\partial x} = \frac{\partial F_x}{\partial z}$$

$$\textbf{(8-11)}$$

This is the test we have been seeking. If Eqs. 8-11 are true, a potential energy function $U(x, y, z)$ exists and the force components can be obtained from U by using Eqs. 8-9. Then, the line integral reduces to the form of Eq. 8-10 and the closed-path integral vanishes, so $\mathbf{F}$ is conservative.

Example 8-6

Apply the test in Eqs. 8-11 to the force fields we have previously studied:
  (a) $\mathbf{F}_+ = Ay\hat{\mathbf{i}} + Ax\hat{\mathbf{j}} + 0\hat{\mathbf{k}}$
  (b) $\mathbf{F} = -Ay\hat{\mathbf{i}} + Ax\hat{\mathbf{j}} + 0\hat{\mathbf{k}}$

Solution:

(a) For $\mathbf{F}_+ = Ay\hat{\mathbf{i}} + Ax\hat{\mathbf{j}}$, we have

$$\frac{\partial F_x}{\partial y} = A, \qquad \frac{\partial F_y}{\partial x} = A$$

$$\frac{\partial F_y}{\partial z} = 0, \qquad \frac{\partial F_z}{\partial y} = 0$$

$$\frac{\partial F_z}{\partial x} = 0, \qquad \frac{\partial F_x}{\partial z} = 0$$

so that Eqs. 8–11 are satisfied. We can also see that the potential energy function is $U(x, y, z) = -Axy + \text{con-}$stant, for then

$$F_x = -\frac{\partial U}{\partial x} = Ay$$

$$F_y = -\frac{\partial U}{\partial y} = Ax$$

$$F_z = -\frac{\partial U}{\partial z} = 0$$

Thus, we have

$$W(C) = \oint_P^Q \mathbf{F_+} \cdot d\mathbf{s} = -\int_P^Q dU = U(P) - U(Q)$$

For $P = (0, 1, 0)$ and $Q = (2, 2, 3)$, as in Example 8–3, we find

$$W(C) = U(0, 1, 0) - U(2, 2, 3)$$
$$= 0 - (-4a^2)A$$
$$= 4a^2 A$$

which is the value we found repeatedly in Example 8–3 for the different paths. Thus, the force field specified by $\mathbf{F_+} = Ay\hat{\mathbf{i}} + Ax\hat{\mathbf{j}}$ is a conservative field.

(b) For $\mathbf{F} = -Ay\hat{\mathbf{i}} + Ax\hat{\mathbf{j}}$, we have

$$\frac{\partial F_x}{\partial y} = -A, \qquad \frac{\partial F_y}{\partial x} = A$$

so that Eqs. 8–11 are *not* satisfied and *no potential energy function exists for this field.* The force field described by $\mathbf{F} = -Ay\hat{\mathbf{i}} + Ax\hat{\mathbf{j}}$ is *not* a conservative field and the closed-path integral does not, in general, vanish.

▶ In Section 7–4 we introduced the idea of the *constrained* potential energy function for the case of motion constrained to a one-dimensional path. Let us now look more closely into the meaning of the qualifier "constrained."

A true potential energy function $U(x, y, z)$ cannot be associated with a nonconservative force. That is, for such forces, no function $U(x, y, z)$ can be found such that Eqs. 8–9 are generally satisfied. However, if a particle is allowed to move *only* along a particular curved path, a restricted potential energy function can be constructed. (The motion must, of course, be frictionless.) Such a

constrained potential energy function can give correctly only the force *tangential* to the path at any point on the path. This function cannot be used to determine the components of the restraining force that acts perpendicular to the path. A specific example will clarify these points. ◀

Example 8–7

Consider the force field described by $F_x = F_y = -Axy$ and $F_z = 0$, with $A = \frac{1}{5}\text{N/m}^2$. (•Can you show that this force field is *nonconservative*?) A bead with a mass of 40 g is constrained by some unspecified forces to move along a frictionless straight wire that coincides with the line $y = 2$ m, $z = 0$, as shown in the diagram. If the bead has a velocity $v_0 = 10$ m/s at $x = 0$, how far will the bead slide before it comes momentarily to rest?

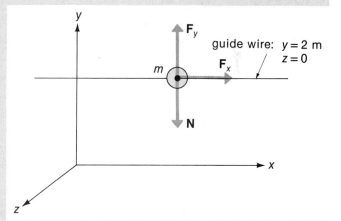

Solution:

Define a constrained potential energy function according to

$$U(x) = -\int F_x \, dx = +\int (Axy)_{y=2\,\text{m}} \, dx$$

$$= +\tfrac{1}{2}(2\,\text{m})Ax^2 + \text{const.}$$

For convenience, we take $U(0) = 0$, so the integration constant is zero. Then, using energy conservation, we can write

$$\tfrac{1}{2}mv_0^2 + U(0) = U(x)$$

or

$$\tfrac{1}{2}mv_0^2 = (1\,\text{m})Ax^2$$

from which

$$x = \sqrt{\frac{mv_0^2}{(2\,\text{m})A}} = \sqrt{\frac{(0.040\,\text{kg})(10\,\text{m/s})^2}{(2\,\text{m})(\tfrac{1}{5}\text{N/m}^2)}}$$

$$= +\sqrt{10}\,\text{m}$$

The constrained potential energy function is useful even though the force field is not conservative.

Example 8–8

Show that any (nondissipative) central force field is conservative.

Solution:

A central force is one that always acts in the direction toward (or away from) a specified origin. Thus, we can express such a force as $\mathbf{F} = f(r)r\hat{\mathbf{u}}_r$. (We can always factor r from $f(r)$ and display it explicitly; this simplifies the calculations we make.)

The chain rule for partial derivatives (Section IV–1) states that if $u = f(v)$ and $v = g(x, y, z)$, then

$$\frac{\partial u}{\partial x} = \frac{du}{dv}\frac{\partial v}{\partial x}$$

and similarly for $\partial u/\partial y$ and $\partial u/\partial z$. Consider the case in which $f(r) = f(\sqrt{x^2 + y^2 + z^2})$; thus,

$$\frac{\partial f(r)}{\partial x} = \frac{df(r)}{dr}\frac{\partial(\sqrt{x^2 + y^2 + z^2})}{\partial x} = \frac{x}{r}\frac{df(r)}{dr}$$

Also, $\dfrac{\partial f(r)}{\partial y} = \dfrac{y}{r}\dfrac{df(r)}{dr}$ and $\dfrac{\partial f(r)}{\partial z} = \dfrac{z}{r}\dfrac{df(r)}{dr}$

Now, the position vector $\mathbf{r}$ is

$$\mathbf{r} = x\hat{\mathbf{i}} + y\hat{\mathbf{j}} + z\hat{\mathbf{k}}$$

and the unit vector $\hat{\mathbf{u}}_r$ is

$$\hat{\mathbf{u}}_r = \frac{\mathbf{r}}{|\mathbf{r}|} = \frac{x}{r}\hat{\mathbf{i}} + \frac{y}{r}\hat{\mathbf{j}} + \frac{z}{r}\hat{\mathbf{k}}$$

Therefore, the components of $\mathbf{F}$ are

$$F_x = f(r)r\left(\frac{x}{r}\right) = xf(r), \qquad F_y = yf(r), \qquad F_z = zf(r)$$

Then,
$$\frac{\partial F_x}{\partial y} - \frac{\partial F_y}{\partial x} = x\frac{\partial f(r)}{\partial y} - y\frac{\partial f(r)}{\partial x}$$
$$= \frac{xy}{r}\frac{df(r)}{dr} - \frac{yx}{r}\frac{df(r)}{dr} = 0$$

and similarly for the other two combinations of partial derivatives. These equalities exist independent of the exact nature of $f(r)$. Thus, all central forces are conservative.* Examples are the spring force and the gravitational force, Eq. 8–2.

---

*The forces are further restricted not to depend on time or on the velocity of the particle.

PROBLEMS _____

Section 8–1

8–1 Sketch the force field for a springlike force, $\mathbf{F} = -\kappa(r - r_0)\hat{\mathbf{u}}_r$. Use an arrow diagram such as that in Fig. 8–1. Show the arrows on a plane that contains the origin.

Section 8–2

8–2 Refer to the force field in Example 8–1. Evaluate the line integral of $\mathbf{F} \cdot d\mathbf{s}$ along a path specified by the parametric functions

$$x(t) = \tfrac{1}{2}at, \qquad y(t) = a(t + 1), \qquad z(t) = 2at,$$

with $a = 1\,\text{m}$, between the same two points, $P(0, 1, 0)$ and $Q(1, 3, 4)$, used in the example. Is the closed-path integral of $\mathbf{F} \cdot d\mathbf{s}$ generally equal to zero for this force field?

8–3 Refer to Examples 8–2 and 8–3. Use the forces and the end points in these examples and evaluate the line integral of $\mathbf{F} \cdot d\mathbf{s}$ along the three straight-line segments, first from $(0, 1, 0)$ to $(0, 2, 0)$, then from $(0, 2, 0)$ to $(0, 2, 3)$, and finally from $(0, 2, 3)$ to $(2, 2, 3)$. Compare these results with those obtained in the examples, and comment.

8–4 Evaluate $\int_P^Q (F_x\,dx + F_y\,dy)$ for the functions $F_x(x, y) = Ay$ and $F_y(x, y) = Bx^2$, with $A = 1\,\text{N/m}$ and $B = 1\,\text{N/m}^2$, along the path C given by $x(t) = 2at$ and $y(t) = a(t^2 - 1)$, with $a = 1\,\text{m}$, for the terminal points $P(t = 0)$ and $Q(t = 1)$.

8–5 Evaluate $\int_P^Q (F_x\,dx + F_y\,dy)$ for the functions $F_x(x, y) = A(x + y)$ and $F_y(x, y) = Bx^2y$, with $A = \tfrac{1}{2}\,\text{N/m}$ and $B = 1\,\text{N/m}^3$, along a quadratic path, $y = ax^2 + b$, from the point $P = (1, 0)$ to the point $Q = (2, 2)$. [*Hint:* First, evaluate a and b by requiring P and Q to lie on the path.]

8-6 Evaluate the line integral of Problem 8–5 along a path consisting of two straight line segments $P \rightarrow R \rightarrow Q$, with $P = (1, 0)$, $R = (2, 0)$, and $Q = (2, 2)$. [*Hint:* Note that along the path $P \rightarrow R$ we have $y = 0$ and $dy = 0$, whereas along $R \rightarrow Q$ we have $x = 2$ m and $dx = 0$.]

8-7 Given three functions, $F_x(x, y, z) = A_1 xyz$,
• $F_y(x, y, z) = A_2(9x^2 - y^2)$, and $F_z(x, y, z) = A_3(3x - y - 3z)$, with $A_1 = \frac{5}{2}$ N/m³, $A_2 = 1$ N/m², and $A_3 = 1$ N/m, evaluate the line integral of $F_x \, dx + F_y \, dy + F_z \, dz$ along the curve C given by $x(t) = at$, $y(t) = a(3t - 2)$, and $z(t) = \frac{1}{2}at^2$, with $a = 1$ m, between the points $P(t = 0)$ and $Q(t = 2)$. What are the Cartesian coordinates of P and Q? Give a geometric description of the projection of the path C onto the three coordinate planes—the x-y plane, the x-z plane, and the y-z plane.

8-8 Consider the two-dimensional force field specified by $F_x = Axy$ and $F_y = A(x^2 - y^2)$. Evaluate the line integral of $\mathbf{F} \cdot d\mathbf{s}$ along the closed path of straight lines, $(0, 0) \rightarrow (1, 0) \rightarrow (1, 1) \rightarrow (0, 1) \rightarrow (0, 0)$.

8-9 A particle with a mass m slides slowly along a circular frictionless ramp from point P to point Q. During the descent, the particle is restrained by a tangential force that allows the motion to proceed slowly. Calculate explicitly the work done by this force.

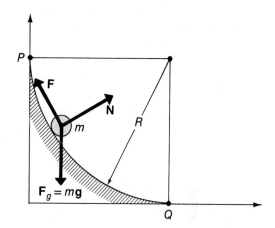

SPECIAL TOPIC

Section 8-3

8-10 In Example 8–1 we considered a force field specified by $\mathbf{F} = 3Axy\hat{\mathbf{i}} + 3Ax^2\hat{\mathbf{j}} + (A/8)xz\hat{\mathbf{k}}$. Determine whether this field is conservative.

8-11 Consider the two-dimensional force due to four
• identical ideal springs with force constant κ, each under a small extension at the equilibrium position. Each spring has one end in common with the other springs and one end that slides frictionlessly on one of four guide rods arranged in a square, as shown in the diagram. Show that the force on a particle attached to the common point is

$$\mathbf{F} = -2\kappa x\hat{\mathbf{i}} - 2\kappa y\hat{\mathbf{j}}$$

Is this force conservative? Show that the potential function is

$$U(x, y) = \kappa(x^2 + y^2) + \text{constant}$$

Express $\mathbf{F}$ and U in terms of ρ and ϕ. [*Hint:* $\hat{\mathbf{i}} = \cos\phi\hat{\mathbf{u}}_\rho - \sin\phi\hat{\mathbf{u}}_\phi$ and $\hat{\mathbf{j}} = \sin\phi\hat{\mathbf{u}}_\rho + \cos\phi\mathbf{u}_\phi$.]

8-12 A particle with a mass m is constrained to move in the x-y plane while in the presence of a force field described by $F_x = -\kappa x$, $F_y = -\kappa y$, and $F_z = -\gamma(x^2 + y^2)^{1/2}$. Obtain a constrained potential energy function $U(x, y)$ appropriate for the x-y plane. What is the normal force required to keep the particle in the x-y plane? If the particle is originally at the origin and is given an initial velocity $\mathbf{v}_0$ (directed in the x-y plane), what will be its farthest excursion from the origin?

8-13 If in Problem 8-12 the coefficient of sliding friction between the particle and the plane is μ_k, what is the particle's farthest excursion from the origin, given the same initial conditions?

8-14 By using the derived potential energy function $U(x, y) = Ax^2y^2$, with $A = \frac{1}{2}$ J/m⁴, show that for the force components $F_x(x, y) = Bxy^2$ and $F_y(x, y) = Byx^2$, with $B = -1$ N/m³, the line integral from $P = (0, 0)$ to $Q = (1, 3)$ is equal to $-9/2$ J. Show that this value is independent of the path by calculating the line integral along the following paths: (a) $x(t) = (t - 2)$m and $y(t) = (t^2 - 2t)$m, with t given in seconds; (b) $y = mx + b$; and (c) two straight line segments, $(0, 0) \rightarrow (1, 0) \rightarrow (1, 3)$.

Additional Problems

8-15 Refer to Example 8-7. Let the guide wire exert a
• frictional force on the bead with a kinetic friction coefficient $\mu_k = 2/5$. If the bead has a velocity $v_0 = 10$ m/s at $x = 0$, how far will the bead slide before coming to rest? How large must be the coefficient of static friction at this point in order to prevent the bead from beginning to move again?

8-16 A ferry barge that can be hauled across a river by
• winches experiences a drag force due to the current in the river. The velocity of flow in the river is zero at each shore and has a maximum value of v_0 at the center of the river. With the coordinate axes chosen as in the diagram at right, the flow velocity is $\mathbf{v} = -v_0[1 - (x^2/a^2)]\hat{\mathbf{j}}$, where $2a$ is the width of the river. The drag force on the barge is directly proportional to $\mathbf{v}$, that is, $\mathbf{F}_b = A\mathbf{v}$. The winches are used to haul the barge slowly around the closed rectangular path, $(0, 0) \rightarrow (0, L) \rightarrow (a, L) \rightarrow (a, 0) \rightarrow (0, 0)$. Calculate the work done by the winches. Is the force on the barge conservative?

8-17 According to the definition of a line of force,
•• the direction of the tangent to a force line at any point is the same as the direction of the force $\mathbf{F}$ at that point. Consider a two-dimensional force field.
(a) Show that $F_y/F_x = dy/dx$, where $y = f(x)$ is the equation for a line of force.
(b) Apply the result of (a) to the case

Problem 8-16

$\mathbf{F}_+ = Ay\hat{\mathbf{i}} + Ax\hat{\mathbf{j}}$ and show that $y = f(x)$ is a family of hyperbolas with $y = \pm x$ as asymptotes.
(c) Show that $\mathbf{F}_+$ can be derived from the potential function $U(x, y) = -Axy$. Also, show that the set of functions $U(x, y) = $ constant are hyperbolas that are everywhere perpendicular to $\mathbf{F}_+$.
(d) Apply the result of (a) to the case $\mathbf{F} = -Ay\hat{\mathbf{i}} + Ax\hat{\mathbf{j}}$ and show that the force lines consist of a family of concentric circles with centers at the origin.
(e) In view of the result of (d), can you explain why the test condition (Eqs. 8-11) fails?

LINEAR MOMENTUM

The previous discussions have been devoted primarily to considerations of the motions of single particles. In this chapter and in the next, we extend the treatment to particle pairs and then to systems of particles, allowing for more general relative motions. We begin here by considering simple two-particle systems that are isolated and are therefore free of all external forces. In this situation, we find that a powerful physical principle—the principle of momentum conservation—greatly simplifies the analysis. The importance of momentum conservation in treating dynamic situations lies in the fact that the principle is valid no matter how complicated are the interaction forces between the particles. In Chapter 10 we study the effect of external forces acting on systems of particles.

9-1 THE LINEAR MOMENTUM OF A PARTICLE

In Section 5–2, some liberty was taken in translating Newton's words appearing in the *Principia*. His use of the Latin words, *"mutationem motus"* or, literally, "change in the quantity of motion," was simply rendered as "acceleration" when we referred to the motion of a single particle. From the usage in the *Principia*, it is quite clear Newton had in mind that "quantity of motion" should be a kinetic property of mass and velocity taken conjointly, namely, the product $m\mathbf{v}$. Therefore, Newton's second law should express the idea that force is equal to the rate of change of the quantity $m\mathbf{v}$; that is,

$$\mathbf{F} = \frac{d}{dt}(m\mathbf{v}) \tag{9-1}$$

which is, in fact, the way that Newton expressed this law. Carrying out the time differentiation, we have

$$\mathbf{F} = \frac{d}{dt}(m\mathbf{v}) = m\frac{d\mathbf{v}}{dt} + \mathbf{v}\frac{dm}{dt}$$

For those cases in which the mass m of the object remains constant with time, we have $dm/dt = 0$, and the equation for the force reduces to

$$\mathbf{F} = m\frac{d\mathbf{v}}{dt} = m\mathbf{a} \qquad (9\text{-}2)$$

which is the way we previously expressed the second law. There are, however, situations in which the mass *is* a function of time. We are then required to use Eq. 9–1 instead of Eq. 9–2; we discuss some of these cases in Section 9–5.

There exists today a vast quantity of empirical evidence that the concept embodied in Eq. 9–1, rather than that in Eq. 9–2, is the fundamental law of Nature, even in the subatomic, quantum, and relativistic realms of physics. We assert that Eq. 9–1 is correct because such empirical evidence demands it.

A word of caution is in order concerning the application of Eq. 9–1. In many problems involving so-called "variable mass," only mass *exchange* between two parts of a system is actually involved. By viewing the system *as a whole,* the total mass is, in fact, conserved. We return to this point in Section 9–5.

It is useful to introduce a special term for the kinematic quantity $m\mathbf{v}$ that Newton called *quantity of motion;* we refer to $m\mathbf{v}$ as the *linear momentum* of a particle, and write

$$\mathbf{p} = m\mathbf{v} \qquad (9\text{-}3)$$

In terms of the linear momentum, Newton's second law is expressed as

$$\mathbf{F} = \frac{d\mathbf{p}}{dt} \qquad (9\text{-}4)$$

9-2 IMPULSE

There are situations in which the force applied to an object or particle lasts only for a time very short compared with the total time that we study the motion of the object. For example, when a golf ball is struck with a club or when a baseball is struck with a bat, the time during which the ball and the club or bat exert contact forces on each other is very short compared with the time the ball is in flight.

In most such cases the applied force is an unknown and complicated function of the time during the interval of contact or impact. The force is then best characterized by the total time integral of its effect. We rewrite Eq. 9–4, treating it as a relationship between differentials:

$$\mathbf{F}(t)dt = d\mathbf{p}$$

Then, if t_1 represents the sensible onset of the interaction and t_2 its effective termination, we have

$$\mathbf{J} = \int_{t_1}^{t_2} \mathbf{F}(t)dt = \int_{\mathbf{p}_1}^{\mathbf{p}_2} d\mathbf{p} = \mathbf{p}_2 - \mathbf{p}_1 \qquad (9\text{-}5)$$

The quantity $\mathbf{J}$ is the time integral of the force and is referred to as the *impulse*.

Equation 9–5 is a *vector equation*, so the relationship is valid for each of the three perpendicular components of $\mathbf{F}(t)$. For example, when a pitched baseball is driven over second base toward center field, we need to consider only two Cartesian components, x for the horizontal coordinate (along the line from home plate to second base) and y for the vertical coordinate. Figure 9–1 gives a graphical representation of the possible history of a baseball that is struck by a bat. In this particular case, the struck ball has a horizontal momentum component slightly greater than the initial momentum of the pitched ball. In addition, the ball is given a vertical momentum component and becomes a "fly ball." Figure 9–1b for the vertical motion shows the pitched ball falling under the action of gravity before being given a large impulse by the bat. After impact the baseball continues to accelerate downward under the influence of the gravitational force, $F_g = -mg$. During the impact period, $t_2 - t_1$, the vertical component of the force is

$$F_y(t) = F_{yb}(t) + F_g$$

Fig. 9–1. The graphical representation of the time dependence of the force acting on a batted baseball and the linear momentum of the baseball. The bat first comes into contact with the ball at t_1 and loses contact at time t_2. (a) The actions in the horizontal direction, with the positive direction taken to be from the batter toward the outfield. (b) The behavior in the vertical direction, with the positive direction taken to be upward. The downward force of gravity, $F_g = -mg$, is also shown.

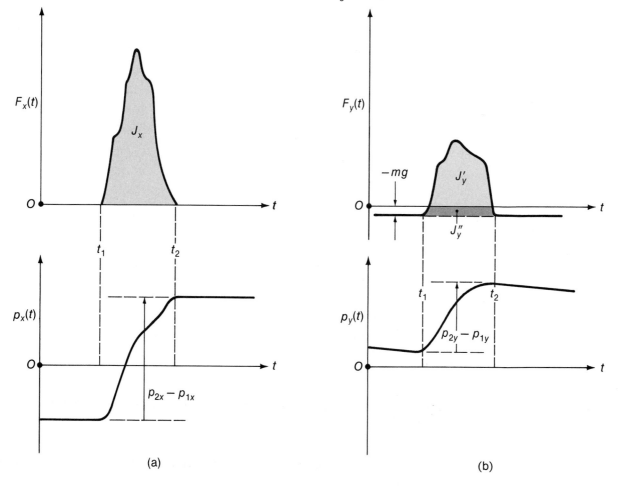

(a)   (b)

where the first term is the force supplied by the bat and the second is the force due to gravity. We have

$$J_y = \int_{t_1}^{t_2} F_y(t)dt = \int_{t_1}^{t_2} F_{yb}(t)dt - mg(t_2 - t_1)$$

$$= J_y' - J_y''$$

If $F_{yb} \gg F_g$ during the interval $t_2 - t_1$ (as indicated in the diagram), then $J_y' \gg J_y''$, and we have

$$J_y \cong \int_{t_1}^{t_2} F_{yb}(t)dt$$

In a situation involving the relatively strong but short-lived application of an impulsive or impact force, all other weaker forces may be ignored insofar as the momentum change during the impulse period is concerned. This is not to say that for time intervals before or after the application of the impulse these forces may be neglected. The struck baseball does, of course, follow a parabolic trajectory after the impact because of the influence of the Earth's gravity.

On occasion it is useful to consider the time average of the impulsive force. This is defined to be

$$\bar{\mathbf{F}} = \frac{1}{t_2 - t_1} \int_{t_1}^{t_2} \mathbf{F}(t)dt \qquad (9-6)$$

where, as before, t_1 is the time at which the force is applied and t_2 is the time at which it is removed. The importance of the impulse concept lies in the fact that *any* force $\mathbf{F}(t)$ that yields the same value for the integral of $\mathbf{F}(t)\,dt$ for the time interval $t_2 - t_1$ produces the *same* momentum change.

Finally, it is instructive to note the parallel structure of the two kinematic integrals we have thus far considered, namely, the change in momentum,

$$\mathbf{p}_2 - \mathbf{p}_1 = \int_{t_1}^{t_2} \mathbf{F}(t)dt$$

and the change in kinetic energy (Eq. 7–7),

$$K_2 - K_1 = \tfrac{1}{2}m(v_2^2 - v_1^2) = \int_A^B \mathbf{F}(x, y, z) \cdot d\mathbf{s}$$

The momentum change due to the action of the force $\mathbf{F}$ is obtained from an integral over *time,* whereas the change in kinetic energy due to the same force is obtained from an integral over *space.* These two relationships taken together completely specify the cumulative kinematic effects of the applied force $\mathbf{F}$.

Example 9–1

A small elastic ball with a mass of 100 g is dropped from a height of 1 m. It is observed to rebound from the horizontal floor to essentially its original height. If the contact with the floor lasts 1/50 s, what is the average impulsive force?

Solution:

From the conservation of energy we have, at the instant the ball makes contact with the floor,

$$\tfrac{1}{2}mv^2 = mgh$$

or

$$v = \sqrt{2gh} = \sqrt{2(9.80 \text{ m/s}^2)(1 \text{ m})}$$
$$= 4.43 \text{ m/s}$$

Because the ball rebounds to essentially its original height, the change in momentum is $2mv$. (•Can you see why this is the case?) Then, the impulse is

$$\bar{F}\,\Delta t = \bar{F}(t_2 - t_1) = 2mv$$

or

$$\bar{F} = \frac{2mv}{t_2 - t_1} = \frac{2(0.1 \text{ kg})(4.43 \text{ m/s})}{0.02 \text{ s}}$$
$$= 44.3 \text{ N}$$

Notice that this impulsive force is about 45 times greater than the gravitational force on the ball ($mg = 0.98$ N).

The time required for the ball to reach the floor from its original height can be obtained from

$$h = \tfrac{1}{2}gt^2$$

or

$$t = \sqrt{\frac{2h}{g}} = \sqrt{\frac{2(1 \text{ m})}{9.80 \text{ m/s}^2}} = 0.452 \text{ s}$$

This time is much greater than the time Δt during which the impulse is delivered:

$$\frac{t}{\Delta t} = \frac{0.452 \text{ s}}{0.02 \text{ s}} = 22.6$$

Therefore, we can reasonably treat the contact between the ball and the floor as an *impulse*, whereas the other portions of the trajectory cannot be considered in the same way.

9-3 THE TWO-PARTICLE INTERACTING SYSTEM

Consider two particles with masses m_1 and m_2 that interact with each other. For example, the particles could be coupled together by an ideal inertialess spring, but otherwise isolated in deep space. The particles and the spring are shown in Fig. 9-2.

The force exerted on m_1 by m_2 is $\mathbf{F}_{12}$ and that exerted on m_2 by m_1 is $\mathbf{F}_{21}$. We consider the ideal inertialess spring to be simply a schematic representation of the physical interaction between the particles. Then, from Newton's third law, we can write

$$\mathbf{F}_{12} = -\mathbf{F}_{21}$$

Also, from the second law applied to each particle separately, we have

$$\mathbf{F}_{12} = \frac{d\mathbf{p}_1}{dt} \quad \text{and} \quad \mathbf{F}_{21} = \frac{d\mathbf{p}_2}{dt}$$

Fig. 9-2. An isolated system of two particles, m_1 and m_2, interacting with each other through a force, shown here schematically as due to an ideal inertialess spring.

Adding these two expressions and using the third law, we find

$$\mathbf{F}_{12} + \mathbf{F}_{21} = \frac{d}{dt}(\mathbf{p}_1 + \mathbf{p}_2) = 0$$

or

$$\mathbf{p}_1 + \mathbf{p}_2 = \text{constant} \qquad \text{(9-7)}$$

Thus, we see that although the individual momenta, $\mathbf{p}_1$ and $\mathbf{p}_2$, might change with time due to the mutual interaction of the particles, the total momentum of the system,* $\mathbf{P} = \mathbf{p}_1 + \mathbf{p}_2$, is a constant of the motion; that is, the sum is independent of time. We refer to this result as the *conservation of linear momentum* of the isolated system.†

The nature of the force **F** did not enter into the considerations leading to Eq. 9-7. Therefore, it must be true that two otherwise isolated particles that happen to collide obeying some unknown and complicated force during impact will also obey the linear momentum conservation law.

We should remember that Eq. 9-7 is a vector equation involving all three perpendicular components. We frequently encounter cases in which the system is acted upon by a force that is due to an agency outside the system. If this external force is constant *in direction*, Eq. 9-7 is still valid for components in the plane perpendicular to the direction of this force (see Fig. 9-3.) It may also happen in cases of collisions that, during the impulse, external forces are either absent or act in a particular constant direction. Again, total or partial use of Eq. 9-7 may be appropriate.

We sometimes must deal with situations in which an object makes a collision with the surface of the Earth or with another object that is attached to the Earth. For example, suppose that a ball is thrown downward and strikes the ground at an angle of 45°, then rebounds at the same angle. It is easy to see that the horizontal component of momentum is conserved in this process because the horizontal velocity is the same after the collision as before. However, the vertical component of the ball's momentum changes from $-mv_y$ to mv_y, so that there is a net momentum change amounting to $2mv_y$. To an observer standing on the ground, it would appear that momentum is not conserved. But to an observer stationed in space (and equipped with very precise measuring apparatus), the Earth would be seen to approach the ball and to rebound from it, thereby conserving momentum. Because the Earth is so

Fig. 9-3. If the force acting on a moving particle has only an *x*-component, the *y*-component of the particle momentum will remain constant, but the *x*-component will change with time. (Of course, the total momentum **p** also changes with time.)

---

* We use **p** to indicate the momentum of a *particle* and **P** to indicate the momentum of a *system*.

† Notice that if we accept the validity of Newton's third law, then linear momentum conservation is indeed a *result*. It is possible to begin by considering momentum conservation as an experimental fact and then deduce the third law. This approach to dynamics is sometimes taken.

massive compared to everyday objects, we can reasonably neglect the exceedingly small recoil of the Earth.

Because force is a free vector, a particular force is represented by the same vector in any inertial reference frame. This fact led us to the conclusion, in Section 5-2, that Newton's second law is valid in any inertial frame. The same reasoning applies to momentum conservation, expressed by Eq. 9-7. The specific statement is: *An interacting isolated system of particles has a total system linear momentum that is conserved, i.e., is a constant of the motion, in any inertial frame of reference.** Of course, the precise value of the constant as measured in two inertial frames depends on the relative velocity of the frames. For example, if the frame S' moves with a velocity $\mathbf{u}$ relative to the frame S, we would have $\mathbf{P}' = \mathbf{P} - (m_1 + m_2)\mathbf{u}$. (•Can you see why?)

Example 9-2

Two particles with masses $m_1 = 2.0$ kg and $m_2 = 5.0$ kg are free to slide on a straight horizontal frictionless guide wire. The particle with the smaller mass is moving with a speed of 17.0 m/s and overtakes the larger particle, which is moving in the same direction with a speed of 3.0 m/s. The larger particle has an ideal inertialess spring ($\kappa = 4480$ N/m) attached to the side being approached by the smaller particle, as shown in the diagram. (a) What is the maximum compression of the spring as the two particles collide? (b) What are the final velocities of the particles?

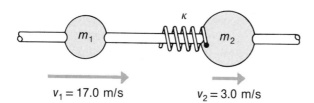

Solution:

(a) When the spring is under maximum compression, the two particles have zero *relative velocity* and, hence, each may be assigned the same guide-wire velocity, say v_0. Because no external forces are acting in the horizontal direction, we may use the conservation of linear momentum and write

$$m_1 v_1 + m_2 v_2 = (m_1 + m_2)v_0$$

Hence,
$$v_0 = \frac{(2.0 \text{ kg})(17.0 \text{ m/s}) + (5.0 \text{ kg})(3.0 \text{ m/s})}{2.0 \text{ kg} + 5.0 \text{ kg}}$$

$$= 7.0 \text{ m/s}$$

Initially, with no interaction between the particles, the total kinetic energy is

$$K = \tfrac{1}{2}m_1 v_1^2 + \tfrac{1}{2}m_2 v_2^2 = \tfrac{1}{2}(2.0)(17.0)^2 + \tfrac{1}{2}(5.0)(3.0)^2$$

$$= 311.5 \text{ J}$$

*Although we have considered only a two-body system so far, we show later that the concept extends to any number of particles acting as an isolated system.

At maximum compression of the spring, the total kinetic energy is

$$K_0 = \tfrac{1}{2}(m_1 + m_2)v_0^2 = \tfrac{1}{2}(2.0 + 5.0)(7.0)^2 = 171.5 \text{ J}$$

The difference in energy, $K - K_0 = 140.0$ J, must be stored in the spring. (•Why is this so?) Thus, the compression x of the spring is found from

$$\tfrac{1}{2}\kappa x^2 = K - K_0$$

or
$$x = \sqrt{\frac{2(140.0 \text{ J})}{4480 \text{ N/m}}} = 0.25 \text{ m}$$

(b) When the particles finally separate, both linear momentum and kinetic energy must be conserved (because the energy stored in the spring has been returned to the particles). Then,

$$m_1 v_1' + m_2 v_2' = m_1 v_1 + m_2 v_2 \tag{1}$$

and
$$\tfrac{1}{2} m_1 (v_1')^2 + \tfrac{1}{2} m_2 (v_2')^2 = \tfrac{1}{2} m_1 v_1^2 + \tfrac{1}{2} m_2 v_2^2 \tag{2}$$

In the momentum equation, Eq. (1), we substitute the numerical values of the masses and obtain

$$2v_1' + 5v_2' = (2.0)(17.0) + (5.0)(3.0) = 49 \text{ kg} \cdot \text{m/s} \tag{3}$$

We also substitute for the masses in the kinetic energy equation, Eq. (2), with the result,

$$2(v_1')^2 + 5(v_2')^2 = 623 \text{ J} \tag{4}$$

If we obtain v_2' in terms of v_1' from Eq. (3) and substitute this expression into Eq. (4), after some simplification, we find

$$(v_1')^2 - 14 v_1' - 51 = 0$$

Factoring, we obtain
$$(v_1' - 17)(v_1' + 3) = 0$$

Thus, we find a *pair* of solutions,

$$v_1' = +17 \text{ m/s} \quad \text{and} \quad v_2' = +3 \text{ m/s}$$
$$v_1' = -3 \text{ m/s} \quad \text{and} \quad v_2' = +11 \text{ m/s}$$

Of these, only the second set is physically realizable. (•Can you see why?) Verify that the kinetic energy K' after collision is equal to the kinetic energy K before collision.

Elastic and Inelastic Collisions. When one object collides with another object, deformations (generally, of both objects) occur during the time of contact. A striking example of deformation in a collision is shown in the high-speed photograph of a golf club colliding with a golf ball (page 241). In order to produce a deformation of a solid object, work must be done against the elastic forces within the body. In Example 9–2 this effect was represented by the work done on a spring. If the system is ideal, the deformed portions will recover completely and the stored energy will be returned to the colliding objects as kinetic energy. Then, there is no change in the

Fig. 9–4. Three types of collision processes.

(a) perfectly elastic

(b) inelastic

(c) completely inelastic

kinetic energy of the system caused by the collision, and we say that the collision is *perfectly elastic*. This is exactly the case illustrated in Example 9–2.

When we deal with real situations involving bulk matter, collisions are never perfectly elastic. A collision between two billiard balls or between two steel ball bearings is almost elastic, but there is always some loss of kinetic energy between the initial and final conditions. After all, we *hear* a click when two billiard balls collide, indicating that some energy has been converted into sound energy; additional energy is converted into nonacoustic forms. Any collision that is not elastic is called an *inelastic* collision. Any collision that results in a permanent deformation is necessarily inelastic.

In an inelastic collision, some of the initial kinetic energy is absorbed and converted into different forms of energy. If the amount of kinetic energy absorbed in a collision is the maximum that is allowed by momentum conservation, the collision is said to be *completely inelastic*. For example, if two railway cars collide and couple together, the collision is completely inelastic. (Any rebounding after collision will result in a larger value of the final kinetic energy; hence, coupling together provides for the maximum absorption of kinetic energy. See Problem 9–38.) In all collisions involving isolated objects, momentum conservation is valid, even if the collision is inelastic.

Figure 9–4 shows a simple example of each of three types of collision processes we have been discussing. The perfectly elastic case is an idealization.

(Courtesy of Dr. Harold E. Edgerton, MIT.)

Example 9–3

A block with a mass m_1 slides frictionlessly across a horizontal surface with a velocity $\mathbf{v}_1$ directly toward a stationary block with a mass m_2. What are the velocities, $\mathbf{v}_1'$ and $\mathbf{v}_2'$, of the blocks after an elastic collision?

Solution:

From momentum conservation, we have

$$m_1 v_1 = m_1 v_1' + m_2 v_2' \tag{1}$$

From energy conservation, we have (canceling the common factor $\frac{1}{2}$)

$$m_1 v_1^2 = m_1 (v_1')^2 + m_2 (v_2')^2 \tag{2}$$

Solve Eq. (1) for v_2' and substitute into Eq. (2). After some algebraic manipulation, the result is

$$(v_1')^2 - 2 \frac{m_1 v_1}{m_1 + m_2} v_1' + \frac{m_1 - m_2}{m_1 + m_2} v_1^2 = 0$$

This equation can be factored; thus,

$$(v_1' - v_1)\left(v_1' - \frac{m_1 - m_2}{m_1 + m_2} v_1\right) = 0$$

The two solutions are

$$v_1' = v_1$$

$$v_1' = \frac{m_1 - m_2}{m_1 + m_2} v_1$$

The first of these solutions corresponds to the uninteresting case in which the block m_1 is in front of the block m_2, so there is no collision! The second solution yields three interesting possibilities:

1) When $m_1 = m_2$, we have $v_1' = 0$. That is, the collision causes m_1 to come to rest, while m_2 is set into motion with $v_2' = v_1$. (•Can you see why?)
2) When $m_1 > m_2$, we have $v_1' > 0$, so that m_1 (and also m_2) moves in the same direction as before the collision.
3) When $m_1 < m_2$, we have $v_1' < 0$, so that the motion of m_1 is reversed while m_2 is set into motion with $v_2' > 0$. (•Can you see why?)

(•Verify that the relative velocity before collision, namely, $\mathbf{v}_1 - \mathbf{v}_2 = \mathbf{v}_1 - 0 = \mathbf{v}_1$, has the same magnitude as the relative velocity after collision, namely, $\mathbf{v}_1' - \mathbf{v}_2'$. This is a general characteristic of elastic collisions.)

Example 9–4

top view

m, K_0 m

initially

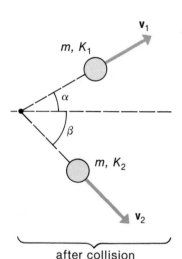

after collision

Two equal-mass hockey pucks are free to slide on a horizontal frictionless ice surface. Initially, one puck has a kinetic energy K_0 with a velocity $\mathbf{v}_0$ directed at the second puck, which is at rest. After colliding, the first puck is observed to move in a direction making an angle α with the original direction of motion, as shown in the diagram. The struck puck recoils in a direction defined by the angle β. (a) If the collision is perfectly elastic, what is the kinetic energy K_1 of the deflected puck? (b) What is the relationship between the angles α and β?

Solution:

(a) From momentum conservation, we have

$$m\mathbf{v}_0 = m\mathbf{v}_1 + m\mathbf{v}_2 \tag{1}$$

and for the components in the horizontal plane, we have

$$mv_0 = mv_1 \cos \alpha + mv_2 \cos \beta \tag{2}$$

$$0 = mv_1 \sin \alpha - mv_2 \sin \beta \tag{3}$$

From the conservation of energy, we have

$$K_0 = K_1 + K_2$$

$$\tfrac{1}{2}mv_0^2 = \tfrac{1}{2}mv_1^2 + \tfrac{1}{2}mv_2^2$$

or

$$v_0^2 = v_1^2 + v_2^2 \tag{4}$$

We may rearrange Eqs. (2) and (3) to give

$$v_2 \cos \beta = v_0 - v_1 \cos \alpha$$

$$v_2 \sin \beta = v_1 \sin \alpha$$

Squaring and adding yields

$$v_2^2 = v_0^2 - 2v_0 v_1 \cos \alpha + v_1^2$$

Eliminating v_2 between this equation and the energy equation, Eq. (4), we find

$$v_1 = v_0 \cos \alpha$$

from which we obtain the desired expression for K_1, namely,

$$K_1 = K_0 \cos^2 \alpha$$

(b) Writing the momentum Eq. (1) as $\mathbf{v}_0 = \mathbf{v}_1 + \mathbf{v}_2$ and squaring, we find

$$\mathbf{v}_0 \cdot \mathbf{v}_0 = v_0^2 = v_1^2 + v_2^2 + 2v_1v_2 \cos{(\alpha + \beta)}$$

Comparing this result with Eq. (4), we conclude that

$$\cos{(\alpha + \beta)} = 0$$

or, finally,
$$\alpha + \beta = \pi/2$$

We see that an elastic collision between two bodies with *equal masses* leads to scattering angles α and β that always add to $\pi/2$ (i.e., the directions of the final motions are at right angles). Notice that the angle α describing the motion of the deflected puck can never exceed 90°. Only if the struck object has a mass greater than that of the incident object will a backward deflection ($\alpha > 90°$) be possible. (The case $m_1 \neq m_2$ is treated in Problems 9–40 and 9–41.)

Example 9–5

A device known as a *ballistic pendulum* affords another good example of the application of the momentum principle. A ballistic pendulum can be used to measure the velocity of such projectiles as bullets, arrows, and darts. The device consists of a rather massive block of wood that is suspended by parallel cords and is initially hanging at rest, as shown in the diagram. The test projectile (for example, a bullet) is fired horizontally into the block, which is thick enough to bring the bullet to rest, embedded inside it. (Notice that the collision of the bullet with the block is completely inelastic.) The block and embedded bullet swing up to a maximum elevation h. From the known masses and h, find the initial velocity of the bullet.

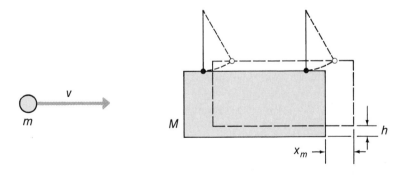

Solution:

Because the support cords are parallel, the pendulum swing of the block consists only of *translational motion*. Let m be the mass of the bullet and M be the mass of the block. During the very short time required to bring the bullet to rest within the block, the block moves only a negligible amount, and there are no external horizontal forces acting on the system of the block and bullet. (•Why is this so?) Therefore, we may use the conservation of linear momentum in the horizontal direction during this interval. Just before the bullet enters the block, the momentum of the system is

$$P = p_{\text{bullet}} = mv$$

When the bullet has come to rest and the block begins to move significantly, we have

$$P' = (m + M)V$$

where V is the horizontal velocity of the block and bullet combination. With $P = P'$, we find

$$V = \frac{m}{m + M}v$$

After the initial impact, the momentum of the system consisting of the block and embedded bullet is no longer conserved in the horizontal direction. (•What external horizontal force appears when $x > 0$? What happens to the momentum?)

The maximum height h of the block is reached when the velocity becomes zero. Using the conservation of energy, we can write

$$\tfrac{1}{2}(M + m)V^2 = (M + m)gh$$

Combining this with the previous result gives

$$v = \frac{m + M}{m}\sqrt{2gh}$$

Because h is generally small and difficult to measure, we can express this result in terms of x_m (the maximum horizontal displacement) and L (the length of the pendulum cord). We have

$$x_m^2 + (L - h)^2 = L^2$$

or

$$h = \frac{h^2 + x_m^2}{2L}$$

Now, when α is small,

$$h = L(1 - \cos\alpha) = \tfrac{1}{2}L\alpha^2 + \ldots$$

and

$$x_m = L\sin\alpha = L\alpha + \ldots$$

Therefore, $x_m^2 \gg h^2$, and the expression for h reduces to

$$h \cong \frac{x_m^2}{2L}$$

Thus,

$$v = \frac{(m + M)}{m}x_m\sqrt{g/L}$$

Note that the initial kinetic energy of the bullet, $\tfrac{1}{2}mv^2$, is in general much larger than the kinetic energy of the block (and bullet) at the instant the block begins to move. That is,

$$\tfrac{1}{2}mv^2/\tfrac{1}{2}(m + M)V^2 = \frac{m + M}{m} \gg 1$$

(•What happens to the lost kinetic energy?)

You should note carefully that, during the stopping time of the bullet, momentum is conserved but kinetic energy is not (this is a characteristic of an inelastic collision), whereas later, as the pendulum begins to swing, energy is conserved but the momentum of the block changes due to the unbalanced forces that then begin to act. (•Give these statements a careful interpretation.)

As a numerical example, consider a 10-g bullet fired into a 3-kg block which then moves a distance $x_m = 25$ cm with $L = 1$ m. Here,

$$h = \frac{x_m^2}{2L} = \frac{(0.25 \text{ m})^2}{2(1 \text{ m})} = 0.0312 \text{ m}$$

which verifies, for this example, that $x_m^2 \gg h^2$. Finally,

$$v = \frac{m + M}{m} \sqrt{2gh} = \frac{3.010 \text{ kg}}{0.010 \text{ kg}} \sqrt{2(9.80 \text{ m/s}^2)(0.0312 \text{ m})}$$

or $\qquad v = 235 \text{ m/s}$

Momentum conservation allows us to obtain a result for the velocity of the bullet even though the force exerted on the bullet by the block during the stopping time is extremely complicated (even unknowable). To obtain the result from the time-developed equations of motion using Newton's second law would be exceedingly difficult (even impossible).

9-4 CENTER OF MASS

When dealing with a system of particles or with a rigid body, we frequently inquire about the motion of the system "as a whole." In such situations it is useful to use the concept of the *center of mass*. The point that corresponds to the center of mass (C.M.) is the "effective mass center" of an object or system of particles. For a pair of equal-mass particles, we have the intuitive notion (which is quite correct) that the center of mass is located midway between the particles. Or, if one particle has a mass greater than that of the other, the C.M. is located closer to the more massive particle. The precise definition of the *center-of-mass vector* **R** of n particles is

$$\mathbf{R} = \frac{m_1\mathbf{r}_1 + m_2\mathbf{r}_2 + \cdots + m_n\mathbf{r}_n}{m_1 + m_2 + \cdots + m_n} = \frac{1}{M}\sum_{i=1}^{n} m_i\mathbf{r}_i \qquad (9\text{-}8)$$

with $\qquad M = \sum_i m_i$

It is important to realize that Eq. 9-8 is a *vector* equation and therefore represents three separate equations for the components of **R**, namely,

$$X = \frac{1}{M}\sum_i m_i x_i; \qquad Y = \frac{1}{M}\sum_i m_i y_i; \qquad Z = \frac{1}{M}\sum_i m_i z_i \qquad (9\text{-}9)$$

If the particles are in relative motion, these equations specify the C.M. coordinates at a particular instant.

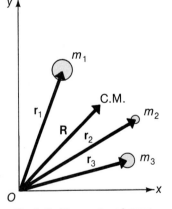

Fig. 9-5. The center-of-mass vector **R** for a collection of particles.

Example 9-6

Locate the C.M. of the set of four particles shown in the diagram on the next page. The particles are on the corners of a 2-m square.

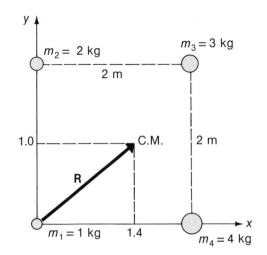

Solution:

This is a two-dimensional problem, so we are to find the coordinates (X, Y) of the C.M. We use Eqs. 9–9 with $M = 10$ kg; thus,

$$X = \frac{1}{10 \text{ kg}}[(1 \text{ kg})(0 \text{ m}) + (2 \text{ kg})(0 \text{ m}) + (3 \text{ kg})(2 \text{ m}) + (4 \text{ kg})(2 \text{ m})]$$

$$= \frac{14 \text{ kg-m}}{10 \text{ kg}} = 1.4 \text{ m}$$

$$Y = \tfrac{1}{10}[(1)(0) + (2)(2) + (3)(2) + (4)(0)]$$

$$= \tfrac{10}{10} = 1.0 \text{ m}$$

These coordinates for the C.M. are shown in the diagram. Notice that the C.M. is *not* located at the center of the square. (•But the *y*-coordinate is equal to one half the side of the square. Can you see why?)

Example 9–7

Prove that the C.M. of two particles is between the particles on a line that connects the two particles.

Solution:

It is an elegant application of vector analysis to consider this problem with an arbitrary location for the origin of the coordinate frame. We choose the arrangement shown in the diagram.

Let the relative coordinate be

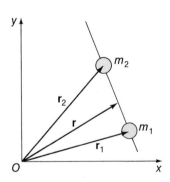

$$\boldsymbol{\rho} = \mathbf{r}_2 - \mathbf{r}_1$$

Then, Eq. 9–8 gives

$$\mathbf{R} = \frac{1}{M}(m_1\mathbf{r}_1 + m_2(\mathbf{r}_1 + \boldsymbol{\rho}))$$

$$= \frac{m_1 + m_2}{M}\mathbf{r}_1 + \frac{m_2}{M}\boldsymbol{\rho}$$

$$= \mathbf{r}_1 + \frac{m_2}{M}\boldsymbol{\rho}$$

However, the general vector equation for a line through the points defined by $\mathbf{r}_1$ and $\mathbf{r}_2$ is

$$\mathbf{r} = \mathbf{r}_1 + \lambda(\mathbf{r}_2 - \mathbf{r}_1)$$
$$= \mathbf{r}_1 + \lambda\boldsymbol{\rho}$$

where each value of λ $(-\infty < \lambda < \infty)$ corresponds to a particular point on the line. When $\lambda = m_2/M$, we have $\mathbf{r} = \mathbf{R}$, so the proposition is proved.

On occasion it proves convenient to separate a collection of particles into two (or more) groups, as in Fig. 9-6, and to rewrite Eq. 9-8 in the following way:

$$M\mathbf{R} = \sum_{i=1}^{\ell} m_i \mathbf{r}_i + \sum_{j=\ell+1}^{n} m_j \mathbf{r}_j$$

We can now write expressions for the center-of-mass coordinates of the ℓ particles (the *black* group) and for the $n - \ell$ particles (the *grey* group):

$$\mathbf{R}_B = \frac{1}{M_B} \sum_{i=1}^{\ell} m_i \mathbf{r}_i, \qquad M_B = \sum_{i=1}^{\ell} m_i \qquad \text{(9-10a)}$$

and

$$\mathbf{R}_G = \frac{1}{M_G} \sum_{j=\ell+1}^{n} m_j \mathbf{r}_j, \qquad M_G = \sum_{j=\ell+1}^{n} m_j \qquad \text{(9-10b)}$$

Combining these gives

$$M\mathbf{R} = M_B \mathbf{R}_B + M_G \mathbf{R}_G, \qquad M = M_B + M_G$$

Thus, a system of particles may be arbitrarily divided into separate subsystems that are non-overlapping (to avoid double counting), each with its own center of mass. The overall C.M. is then obtained by treating each subsystem as a particle with its mass concentrated at the subsystem center of mass.

The C.M. of an object with a finite size can be calculated by considering the object to consist of a large number of "particles." In these cases also we find application of the equations above. In the discussions of rigid-body motion in Chapter 11 we consider these matters in detail.

Fig. 9-6. A collection of particles may be separated into two arbitrary groups, here labeled *black* and *grey*. The C.M. vector for each group can be obtained, and then the C.M. vector for the entire collection can be related to these two vectors. According to the result of Example 9-7, the C.M. is located on the line connecting the points defined by $\mathbf{R}_B$ and $\mathbf{R}_G$.

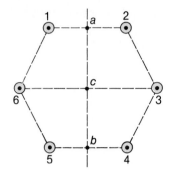

Example 9–8

Locate the C.M. of six particles with equal masses m, arranged to correspond in position to the six corners of a regular (planar) hexagon, as shown in the diagram.

Solution:

Divide the system into three subsystems, consisting of the pairs $1 + 2$, $3 + 6$, and $4 + 5$. The C.M. of subsystem $1 + 2$ is clearly (from Example 9–7) at point a, midway along a line connecting particles 1 and 2. Similarly, for subsystem $4 + 5$ the C.M. is at the midpoint b. These two subsystems, treated as particles each with a mass $2m$, have a resulting C.M. at point c, midway between a and b. The subsystem $3 + 6$ also has its C.M. at c. Hence, the C.M. of the entire group of six particles is also at point c, in the precise center of the hexagon. Evidently, the mass to be associated with this C.M. is $6m$. This corresponds to the intuitive guess that the C.M. of a symmetrically arranged set of equal-mass particles is at the geometric center of the system.

Momentum Conservation for a System of Particles.   Suppose that the masses of the n particles in Eq. 9–8 remain constant as the history of the system unfolds. Then, differentiating with respect to time gives

$$M\frac{d\mathbf{R}}{dt} = \sum_i m_i \frac{d\mathbf{r}_i}{dt}$$

Now, $d\mathbf{R}/dt = \mathbf{V}$ is the velocity of the C.M. of the system, and $m_i(d\mathbf{r}_i/dt) = \mathbf{p}_i$ is the momentum of particle i. Thus, we can write

$$MV = \sum_n \mathbf{p}_i = \mathbf{P} \tag{9–12}$$

That is, the linear momentum $\mathbf{P}$ of a system of particles is equivalent to that of a hypothetical particle with the entire mass M of the system moving with the velocity of the C.M.

Differentiating Eq. 9–12 with respect to time gives

$$\frac{d\mathbf{P}}{dt} = \sum_i \frac{d\mathbf{p}_i}{dt}$$

We suppose that there are two-body forces acting between each pair of the n particles; the force exerted on the particle labeled i by the particle labeled j is $\mathbf{F}_{ij}$. In addition there is an externally applied force $\mathbf{F}_i^e$ that acts on particle i. The total force on particle i is

$$\mathbf{F}_i = \mathbf{F}_i^e + \sum_{\substack{j=1 \\ j \neq i}}^{n} \mathbf{F}_{ij} = \frac{d\mathbf{p}_i}{dt}$$

where the index j runs from 1 to n but excluding $j = i$, thereby eliminating the unphysical self-interaction force $\mathbf{F}_{ii}$. Then, the total force acting on the system is

$$\mathbf{F} = \frac{d\mathbf{P}}{dt} = \sum_i \mathbf{F}_i = \sum_i \mathbf{F}_i^e + \sum_{i \neq j} \left(\sum_{j=1}^n \mathbf{F}_{ij} \right) = \sum_i \frac{d\mathbf{p}_i}{dt}$$

Now, according to Newton's third law, for every force $\mathbf{F}_{ij}$ that occurs in the last term in this equation, there is an equal and oppositely directed reaction force, $\mathbf{F}_{ji} = -\mathbf{F}_{ij}$. Therefore, the double sum vanishes by the cancellation of the individual forces in pairs. (•Can you verify this statement by writing out the sum for the case of three particles?) Then, writing $\mathbf{F}^e = \sum \mathbf{F}_i^e$ for the total external force, we have

$$\mathbf{F}^e = \frac{d\mathbf{P}}{dt} = M\frac{d^2\mathbf{R}}{dt^2} \qquad \text{(9-13)}$$

Thus, the C.M. of the system moves in exactly the same way as would a particle with the same mass and subjected to the same force. In the event that there are no external forces acting, we find

$$\mathbf{P} = \text{constant} \qquad \text{(no external forces)} \qquad \text{(9-14)}$$

That is, the total linear momentum $\mathbf{P}$ of an isolated system of particles is a constant of the motion. This implies that the C.M. of the system simply moves with a constant velocity $\mathbf{V}$. (Stated otherwise: a system that interacts only internally cannot accelerate itself.)

Example 9-9

Two large pucks, $m_1 = 0.5$ kg and $m_2 = 1.0$ kg, are free to slide on a horizontal frictionless surface. The pucks are connected by an ideal inertialess spring, as indicated in the diagram on the next page. An initial observation places the pucks at $\mathbf{r}_1 = 0$, $\mathbf{r}_2 = (0.3 \text{ m})\hat{\mathbf{i}} + (0.4 \text{ m})\hat{\mathbf{j}}$ with $\mathbf{v}_1 = (6 \text{ m/s})\hat{\mathbf{i}}$ and $\mathbf{v}_2 = 0$. Five seconds later, the puck m_1 is at coordinate $\mathbf{r}_1' = (10.6 \text{ m})\hat{\mathbf{i}}$ with a velocity $\mathbf{v}_1' = (3 \text{ m/s})\hat{\mathbf{i}} + (2 \text{ m/s})\hat{\mathbf{j}}$. Where is the other puck and what velocity does it have?

Solution:

We first determine the C.M. coordinate $\mathbf{R}$ and velocity $\mathbf{V}$ of the C.M. at time $t = 0$ (the initial observation). Using Eq. 9-8, we have

$$\mathbf{R} = \frac{(0.5 \text{ kg})(0 \text{ m}) + (1.0 \text{ kg})[(0.3 \text{ m})\hat{\mathbf{i}} + (0.4 \text{ m})\hat{\mathbf{j}}]}{(0.5 \text{ kg}) + (1.0 \text{ kg})}$$

or

$$\mathbf{R} = (0.2 \text{ m})\hat{\mathbf{i}} + (0.267 \text{ m})\hat{\mathbf{j}}$$

The time derivative of Eq. 9-8 can be expressed as

$$\mathbf{V} = \frac{(0.5 \text{ kg})(6 \text{ m/s})\hat{\mathbf{i}} + (1.0 \text{ kg})(0 \text{ m/s})\hat{\mathbf{j}}}{(0.5 \text{ kg}) + (1.0 \text{ kg})} = (2 \text{ m/s})\hat{\mathbf{i}}$$

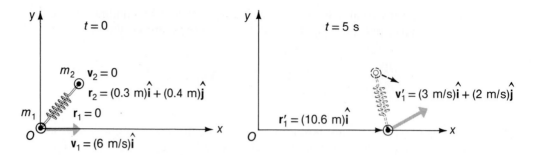

Because there are no external forces in the horizontal direction, the C.M. moves with a constant velocity **V**. Thus, at $t = 5$ s, the C.M. coordinate is

$$\mathbf{R}' = \mathbf{R} + \mathbf{V}t$$
$$= [(0.2 \text{ m})\hat{\mathbf{i}} + (0.267 \text{ m})\hat{\mathbf{j}}] + [(2 \text{ m/s})\hat{\mathbf{i}}](5 \text{ s})$$

or
$$\mathbf{R}' = (10.2 \text{ m})\hat{\mathbf{i}} + (0.267 \text{ m})\hat{\mathbf{j}}$$

Letting the unknown coordinates for m_2 be x and y, we have, using Eq. 9–8 at this later time,

$$M\mathbf{R}' = m_1\mathbf{r}_1' + m_2(x\hat{\mathbf{i}} + y\hat{\mathbf{j}})$$

$$(1.5 \text{ kg})[(10.2 \text{ m})\hat{\mathbf{i}} + (0.267 \text{ m})\hat{\mathbf{j}}] = (0.5 \text{ kg})(10.6 \text{ m})\hat{\mathbf{i}} + (1.0 \text{ kg})(x\hat{\mathbf{i}} + y\hat{\mathbf{j}})$$

Thus,

$$(15.3 \text{ kg} \cdot \text{m})\hat{\mathbf{i}} + (0.40 \text{ kg} \cdot \text{m})\hat{\mathbf{j}} = [(5.3 \text{ kg} \cdot \text{m}) + x]\hat{\mathbf{i}} + y\hat{\mathbf{j}}$$

from which $x = 10.0$ m and $y = 0.4$ m so that $\mathbf{r}_2' = (10.0 \text{ m})\hat{\mathbf{i}} + (0.40 \text{ m})\hat{\mathbf{j}}$

Momentum is conserved, so at $t = 5$ s we have

$$M\mathbf{V} = \mathbf{p}_1' + \mathbf{p}_2' = m_1\mathbf{v}_1' + m_2\mathbf{v}_2'$$

or
$$(1.5 \text{ kg})(2 \text{ m/s})\hat{\mathbf{i}} = (0.5 \text{ kg})[(3 \text{ m/s})\hat{\mathbf{i}} + (2 \text{ m/s})\hat{\mathbf{j}}] + (1.0 \text{ kg})\mathbf{v}_2'$$

Thus,

$$\mathbf{v}_2' = (1.5 \text{ m/s})\hat{\mathbf{i}} - (1 \text{ m/s})\hat{\mathbf{j}}$$

9-5 VARIABLE MASS

We now consider the behavior of objects and systems that have variable mass. We must emphasize, however, that in every case there is a larger system that has constant mass. Thus, when we say that a certain object has "variable mass," we mean that the object acquires mass from or loses mass to its surroundings. The object or system A that we study can be isolated from the global system of which it is a part by a bounding surface, as in Fig. 9–7. Then, mass (and momentum) can flow inward or outward through the boundary without affecting the mass of the global system.

Fig. 9–7. A system A is isolated from the surrounding global system by a bounding surface through which mass (momentum) can flow. The mass of the global system, of which A is a part, remains constant.

When mass flows inward, the time rate of change of the mass M of the system A is $dM/dt > 0$, and when mass flows outward, $dM/dt < 0$.

Mass Accretion. We first discuss the general case of mass accretion by a moving object. For example, when a raindrop falls through a region of saturated vapor in a cloud, the drop's mass will increase by vapor condensation on its surface.

Consider an object A that, at some time t, has a mass M and a velocity $\mathbf{v}$ in a certain inertial reference frame. The object encounters a mass Δm that has a velocity $\mathbf{u}$ in the same frame. During a time interval Δt, there occurs a complicated (or even unknown) interaction that binds together M and Δm. At a time $t + \Delta t$, the combined mass $M + \Delta m$ has a velocity $\mathbf{v} + \Delta\mathbf{v}$ in the inertial frame. The relevant global system, throughout the time interval Δt, consists of A and Δm, and has the constant mass $M + \Delta m$. The situation is illustrated in Fig. 9–8.

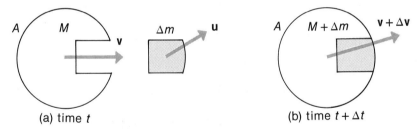

(a) time t　　　　　　　　　(b) time $t + \Delta t$

Fig. 9–8. An object A with a mass M accretes a mass element Δm in a time interval Δt. At time t, A has a velocity $\mathbf{v}$ and Δm has a velocity $\mathbf{u}$ in some inertial frame. At time $t + \Delta t$, A (now with mass $M + \Delta m$) has a velocity $\mathbf{v} + \Delta\mathbf{v}$ in the same frame. The mass of the global system, $M + \Delta m$, remains constant.

We suppose that an external force $\mathbf{F}^e$ also acts on the entire system. This force might include gravity, air resistance, and so forth. Note carefully that $\mathbf{F}^e$ does *not* include any force of interaction between A and Δm; such an internal force would not contribute to the momentum change of the system as a whole.

Using Newton's second law (Eq. 9–4), we have

$$\mathbf{F}^e = \frac{d\mathbf{P}}{dt} \cong \frac{\Delta\mathbf{P}}{\Delta t} = \frac{\mathbf{P}_f - \mathbf{P}_i}{\Delta t}$$

where $\mathbf{P}_f$ and $\mathbf{P}_i$ stand for the final and initial momenta, respectively, of the entire system. Thus,

$$\mathbf{P}_f = (M + \Delta m)(\mathbf{v} + \Delta\mathbf{v})$$
$$\mathbf{P}_i = M\mathbf{v} + \mathbf{u}\,\Delta m$$

Neglecting the second-order infinitesimal, $\Delta m \, \Delta \mathbf{v}$, and noting that $\Delta m = \Delta M$, we have

$$\mathbf{F}^e = M\frac{\Delta \mathbf{v}}{\Delta t} + \mathbf{v}\frac{\Delta M}{\Delta t} - \mathbf{u}\frac{\Delta M}{\Delta t}$$

Upon letting $\Delta t \to 0$, this equation can be written in the form

$$\left.\begin{aligned} M\frac{d\mathbf{v}}{dt} &= \mathbf{F}^e + (\mathbf{u} - \mathbf{v})\frac{dM}{dt} \\[2mm] &= \mathbf{F}^e + \mathbf{v}_r\frac{dM}{dt} \end{aligned}\right\} \qquad \textbf{(9--15)}$$

where $\mathbf{v}_r = \mathbf{u} - \mathbf{v}$ is the relative velocity of Δm with respect to A. Thus, the acceleration $d\mathbf{v}/dt$ of the object A depends not only on the external force $\mathbf{F}^e$ but on the rate at which mass is accreted (and on the velocity of the accreted mass). As we pass to the limit $\Delta m \to 0$, the force $\mathbf{F}^e$ becomes simply the external force acting on A alone.

In the first line of Eq. 9–15 we can combine the two terms involving time derivatives and write

$$\frac{d}{dt}(M\mathbf{v}) = \mathbf{F}^e + \mathbf{u}\frac{dM}{dt} \qquad \textbf{(9--16)}$$

This differs from the usual equation for Newton's second law (Eq. 9–1) by the additional term, $\mathbf{u}(dM/dt)$, which represents the accretion of momentum. This momentum is nonzero by virtue of the velocity $\mathbf{u}$. Indeed, in the event that $\mathbf{u} = 0$, Eq. 9–16 becomes identical to Eq. 9–1.

We must remember that Eqs. 9–15 and 9–16 are vector equations. In some situations we may wish to examine only one or two components of these vector equations.

Example 9–10

An empty freight flatcar, while traveling along a horizontal frictionless track, passes directly under a railroad hopper filled with sand, as shown in the diagram. At some point the hopper begins to release sand into the flatcar at a rate dm/dt. Find the force F that must be supplied by the locomotive to keep the flatcar moving at a uniform speed V_0.

Solution:

In this situation we are interested only in the component of Eq. 9–16 along the track. We note that although $\mathbf{u}$ is not zero, it has no horizontal component; to an observer on the ground (an inertial frame) the sand falls vertically downward. We identify $\mathbf{F}$ as the external force $\mathbf{F}^e$. Therefore, in terms of horizontal components, we have simply

$$F = \frac{d}{dt}(MV)$$

where M and V refer to the instantaneous mass and velocity of the flatcar with its sand load. Because V is maintained at the constant value V_0, we have

$$F = M\frac{dV}{dt} + V\frac{dM}{dt} = V_0\frac{dM}{dt}$$

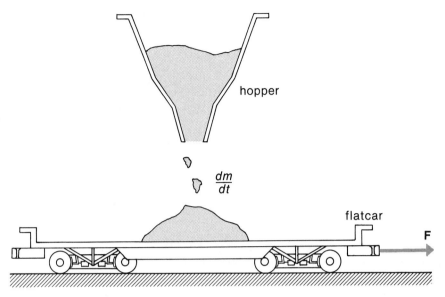

It is evident that $dM/dt = dm/dt$, the rate of sand accumulation by the flatcar, which may or may not be a constant. Thus,

$$F = V_0 \frac{dm}{dt}$$

It is interesting to inquire about the rate at which the locomotive does work. The locomotive power is equal to the product of the force and the velocity (Eq. 7–13):

$$P = \frac{dW}{dt} = FV = V_0^2 \frac{dM}{dt}$$

Because V is a constant, $V = V_0$, this is equivalent to

$$P = 2\frac{d}{dt}(\tfrac{1}{2}MV^2)$$

Thus, the locomotive does work at *twice* the rate at which the kinetic energy of the flatcar (with its sand load) is increasing. This apparent failure of the energy conservation principle is due to the fact that work must be done by the frictional forces between the arriving grains of sand and the flatcar surface in order to give these grains the horizontal velocity V_0, which they do not have as they leave the hopper. (•Can you relate the internal frictional force responsible for this result to that involved in the ballistic pendulum discussed in Example 9–5?)

Loss of Mass. There are physical situations in which an object continually ejects or otherwise loses mass during its motion. A familiar example of this effect is the motion of a rocket. We might suspect that in these cases the general solution would be similar to Eqs. 9–15 and 9–16. We now show that this expectation is, in fact, correct.

Consider an object A that, at some time t, has a mass M and a velocity **v** in a certain inertial reference frame. Some unknown interaction leads to the ejection from the object of a mass Δm during a time interval Δt. At the end of this interval, at time $t + \Delta t$, object A (now with mass $M - \Delta m$) has a velocity **v** + Δ**v**, and the ejected fragment Δm has a velocity **u**. Both of these new velocities are measured in the original inertial frame. The situation is illustrated in Fig. 9–9.

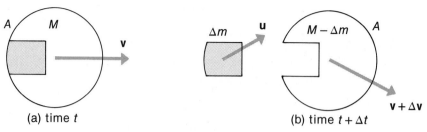

Fig. 9-9. An object A with a mass M ejects a fragment Δm during a time interval Δt. The initial and final velocities are shown relative to the same inertial frame.

In this case, we have $\Delta M = -\Delta m$ because the ejected mass increases with time, whereas the object A which is ejecting matter has a mass that decreases with time. We assume that an external force $\mathbf{F}^e$ also acts on the entire system. We can again write

$$\mathbf{F}^e = \frac{d\mathbf{P}}{dt} \cong \frac{\Delta \mathbf{P}}{\Delta t} = \frac{\mathbf{P}_f - \mathbf{P}_i}{\Delta t}$$

Now,

$$\mathbf{P}_f = (M - \Delta m)(\mathbf{v} + \Delta \mathbf{v}) + \mathbf{u}\,\Delta m$$

and

$$\mathbf{P}_i = M\mathbf{v}$$

Substituting these values into the equation for $\mathbf{F}^e$ and neglecting the second-order infinitesimal $\Delta m\,\Delta \mathbf{v}$, we have, in the limit $\Delta t \to 0$,

$$\left.\begin{aligned}
M\frac{d\mathbf{v}}{dt} &= \mathbf{F}^e + (\mathbf{u} - \mathbf{v})\frac{dM}{dt} \\[2mm]
&= \mathbf{F}^e + \mathbf{v}_r\frac{dM}{dt}
\end{aligned}\right\} \tag{9-17}$$

where, again, $\mathbf{v}_r = \mathbf{u} - \mathbf{v}$ is the relative velocity of Δm with respect to object A. Thus, we have exactly the same equation for the case of mass loss as for the case of mass accretion. However, we must remember that in Eq. 9-17, $dM/dt < 0$, whereas in Eq. 9-15, $dM/dt > 0$.

Rocket Propulsion.   A rocket engine ejects a stream of rapidly moving gas particles through nozzles located at its rear. These hot gases are the result of burning a fuel in the combustion chamber of the rocket, as indicated in Fig. 9-10. The mass of the rocket continually decreases as the fuel and oxidizer are consumed and ejected.

We identify m as the mass of burnt fuel and oxidizer that is ejected through the rocket nozzle. Initially, the rocket has a mass M (*including* the as-yet-unejected fuel and oxidizer) and a velocity $\mathbf{v}$ in some suitable inertial frame. Equation 9-17 then applies directly:

$$M\frac{d\mathbf{v}}{dt} = \mathbf{F}^e + \mathbf{v}_r\frac{dM}{dt} = \mathbf{F}^e - \mathbf{v}_r\frac{dm}{dt} \tag{9-18}$$

This equation is often called the *rocket equation*. The term $\mathbf{\Omega} = \mathbf{v}_r(dM/dt) = -\mathbf{v}_r(dm/dt)$ is referred to as the *thrust* of the rocket engine. Note that $\mathbf{v}_r$, the velocity of the ejected gases relative to the rocket, is to the *left* in Fig. 9-10; because $dM/dt < 0$, the thrust gives a contribution to $M(d\mathbf{v}/dt)$ that is to the *right*, thereby

Fig. 9-10. A schematic representation of a rocket and its key elements.

producing a positive acceleration. To achieve large accelerations, design engineers must deal with ejection speeds v_r that are very high and with combustion chambers and nozzles that are capable of large through-put.

Example 9-11

A spacecraft is stationary in deep space when its rocket engine is ignited for a 100-s "burn." Hot gases are ejected at a constant rate of 150 kg/s with a velocity relative to the spacecraft of 3000 m/s. The initial mass of the spacecraft (plus fuel and oxidizer) is 25,000 kg. Determine the thrust of the rocket engine and the initial acceleration in units of g. What is the final velocity of the spacecraft?

Solution:

We take the direction of motion of the spacecraft to be positive. Then, the thrust is

$$\Omega = -v_r \frac{dm}{dt} = -(-3000 \text{ m/s})(150 \text{ kg/s}) = 450{,}000 \text{ N}$$

From Eq. 9-18, the initial acceleration is

$$a = \frac{dv}{dt} = -\frac{v_r}{M}\frac{dm}{dt} = \frac{\Omega}{M} = \frac{450{,}000 \text{ N}}{25{,}000 \text{ kg}} \cdot \frac{g}{9.80 \text{ m/s}^2} = 1.84g$$

Next, we use Eq. 9-18 to write

$$M\frac{d\mathbf{v}}{dt} = \mathbf{v}_r \frac{dM}{dt}$$

or

$$\frac{dM}{M} = \frac{dv}{v_r}$$

Direct integration gives

$$\int_{M_i}^{M_f} \frac{dM}{M} = \frac{1}{v_r}\int_0^v dv$$

or

$$v = -v_r \ln(M_i/M_f)$$

Substituting numerical values for the 100-s burn, $M_i = 25{,}000$ kg, $M_f = M_i - (150 \text{ kg/s}) \cdot (100 \text{ s}) = 10{,}000$ kg, and $v_r = -3000$ m/s, we find

$$v = -(-3000 \text{ m/s}) \ln \frac{25{,}000}{10{,}000} = 2749 \text{ m/s} \quad (6150 \text{ mi/h})$$

Notice that this ratio of initial to final mass, $M_i/M_f = 2.5$, produces a final velocity v that is nearly equal to the exhaust velocity. To exceed the exhaust velocity by any substantial factor requires that the ratio M_i/M_f be much greater; that is, the payload is a small fraction of the entire rocket mass.

(a) time t

(b) time $t + \Delta t$

Fig. 9–11. (a) A jet engine at time t has a mass M (including fuel) and a velocity **v**. The amount of *still* air $\Delta \mu$ that *will* enter the air intake during the time interval Δt is also shown. (b) At the time $t + \Delta t$ the combustion gases, with mass $\Delta m + \Delta \mu$, are ejected with a velocity **u** in the inertial frame.

▶ *The Jet Engine.* Next, we consider the operation of air-breathing jet engines. Air enters at the front intake of these engines and, after compression, supports the combustion of the fuel; the exhaust gases are ejected with high velocities at the rear of the engine. Figure 9–11 shows the labeling of the masses and velocities involved.

To simplify matters, consider horizontal flight with no air resistance. That is, no external horizontal forces are acting, so linear momentum in the horizontal direction is conserved. The mass of air processed by the engine is $\Delta \mu$, the fuel burned is Δm, and the ejection velocity relative to the inertial frame is **u**. Then,

$$M\mathbf{v} + (0)(\Delta \mu) = (M - \Delta m)(\mathbf{v} + \Delta \mathbf{v}) + \mathbf{u}(\Delta m + \Delta \mu)$$

Rearranging, dividing by Δt, and proceeding to the limit $\Delta t \to 0$, we have

$$M\frac{d\mathbf{v}}{dt} = \mathbf{v}\frac{dm}{dt} - \mathbf{u}\left(\frac{dm}{dt} + \frac{d\mu}{dt}\right) \tag{9–19}$$

where the time derivatives, $d\mu/dt$ and dm/dt, are both *positive*. We again use $\mathbf{v}_r = \mathbf{u} - \mathbf{v}$ to represent the relative velocity of the exhaust gases with respect to the engine. Then, Eq. 9–19 becomes

$$M\frac{d\mathbf{v}}{dt} = -\mathbf{v}_r\frac{dm}{dt} - \mathbf{u}\frac{d\mu}{dt} \tag{9–20}$$

We now define the entire right-hand side of this expression to be the thrust Ω. The first term is the same as the thrust in the rocket equation (Eq. 9–18). Because $\mathbf{v}_r$ has a direction opposite to that of $\mathbf{v}$, this term represents a positive acceleration (i.e., to the right in Fig. 9–11). Next, look at the term $-\mathbf{u}\,d\mu/dt$, which involves the rate of through-put of the air. The relative exhaust speed v_r is usually at least several times greater than the forward speed of the engine with respect to the still air. Then, **u** is in the same direction as $\mathbf{v}_r$, so the term $-\mathbf{u}\,d\mu/dt$ represents a positive acceleration. In fact, even though $v_r > u$, the term $-\mathbf{u}\,d\mu/dt$ contributes the major part of the thrust because $d\mu/dt$ is usually 20 or so times greater than dm/dt (that is, the air-to-fuel ratio is 20 or more). ◀

Example 9–12

A jet aircraft is traveling at 500 mi/h (223 m/s) in horizontal flight. The engine takes in air at a rate of 80 kg/s and burns fuel at a rate of 3 kg/s. If the exhaust gases are ejected at 600 m/s relative to the aircraft, find the thrust of the jet engine and the delivered horsepower.

Solution:

We have (Eq. 9–20)

$$\Omega = -(-600 \text{ m/s})(3 \text{ kg/s}) - (-600 \text{ m/s} + 223 \text{ m/s})(80 \text{ kg/s})$$
$$= 1800 \text{ N} + 30{,}160 \text{ N}$$
$$= 31{,}960 \text{ N}$$

The delivered power is the force (or thrust) multiplied by the velocity:

$$P = \Omega v = (31{,}960 \text{ N})(223 \text{ m/s})$$
$$= 7.13 \times 16^6 \text{ W} = 9550 \text{ h.p.}$$

PROBLEMS

Section 9-1

9-1 A garden hose is held in a manner producing a right-angle curve near the nozzle as shown in the diagram. What force is necessary to hold the nozzle stationary if the discharge rate is 0.60 kg/s with a velocity of 25 m/s?

9-2 Sand is pouring onto a spring scale from a vertical height of 1 m at a rate of 0.5 kg/s. The maximum pile of sand the scale pan can support is 1 kg, with excess sand spilling over the side of the pan with negligible velocity, as shown in the diagram. When the equilibrium condition has been reached, what is the weight (force) reading of the scale?

Section 9-2

9-3 A ball with a mass of 0.15 kg is thrown directly against a wall with a velocity of 18 m/s, and it re-

bounds in the same direction with the same velocity. If the ball was in contact with the wall for 1.5 ms, what average force did the wall exert on the ball?

9-4 A standard 2-ounce tennis ball (mass = 57 g) acquires a speed of 40 m/s during a volley. The racket is in contact with the ball for 5 ms during a horizontal stroke. (a) What impulse is imparted to the ball? (b) What is the average force between ball and racket?

9-5
• A small ball is dropped from a height h onto a horizontal floor and is observed to rebound (on first bounce) to a maximum height ηh, with $0 \leqslant \eta \leqslant 1$. Show that the ratio of the total travel time from the instant of release back to the rebound height ηh, to the impulse time interval for an average impulsive force $\bar{F}$, is simply $\bar{F}/mg$.

Section 9-3

9-6 A block of wood with a mass of 0.6 kg is balanced on top of a vertical post 2 m high. A 10-g bullet is fired horizontally into the block. If the block and embedded bullet land at a point 4 m from the base of the post, find the initial velocity of the bullet.

9-7 Two blocks, with masses $m_1 = 2$ kg and $m_2 = 3$ kg, make a head-on collision while sliding on a frictionless horizontal surface. Initially their velocities are $v_1 = 5$ m/s and $v_2 = 4$ m/s, respectively. On collision they stick together and subsequently move as a single block. (a) What is the velocity of the combination after collision? (b) How much kinetic energy is lost in the collision? What happened to this energy?

9-8 A block with a mass of 2 kg rests on a horizontal frictionless surface and is attached to a wall with a horizontal ideal inertialess spring under no initial compression. (The spring constant is $\kappa = 250$ N/m.) A blob of putty having a mass of 0.5 kg and a hori-

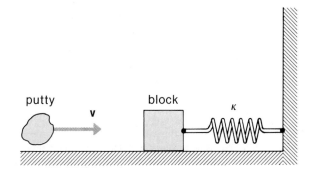

zontally directed velocity **v** collides with the block and sticks to it, as shown in the diagram. Determine v if the spring undergoes a maximum compression of 10 cm.

9-9 An automobile with a mass of 1500 kg is traveling due north at 90 km/h when it is struck by another automobile with a mass of 2000 kg and traveling due east at 60 km/h. If both cars lock together as a unit, find the resulting velocity (speed and direction) just after collision.

9-10 A ball with a mass m is attached to the end of a string
• with a length $L = 50$ cm and is released from a horizontal position, as shown in the diagram. At the bottom of its swing the ball strikes a block with a mass $M = 2m$ resting on a frictionless table. Assume that the collision is perfectly elastic. (a) To what height does the ball rebound? (b) What is the velocity of the struck block after impact?

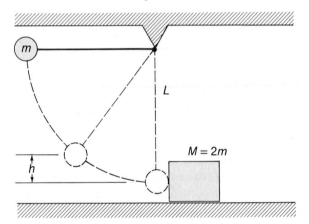

9-11 Two identical balls, each with mass m, undergo a
•• collision. Initially, one ball is stationary and the other has a kinetic energy of 8 J. The collision is inelastic, and 2 J of energy is converted to heat. What is the maximum deflection angle (α or β in the diagram) at which one of the balls may be observed?

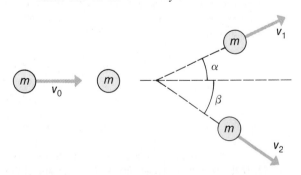

9-12 A 45-kg girl is standing on a plank that has a mass of 150 kg. The plank, originally at rest, is free to slide on a frozen lake, which is a flat, frictionless supporting

surface. The girl begins to walk along the plank at a constant velocity of 1.5 m/s *relative to the plank.* (a) What is her velocity relative to the ice surface? (b) What is the velocity of the plank relative to the ice surface?

9-13 In a collision process between two objects, the ratio of the relative velocity after collision to the relative velocity before collision is called the *coefficient of restitution ε.* A perfectly elastic collision has $\varepsilon = 1$ and a completely inelastic process has $\varepsilon = 0$; ordinary inelastic collisions have intermediate values of ε. If a ball is dropped from a height h onto a solid floor and rebounds to a height h', show that $h' = \varepsilon^2 h$.

9-14 A particle with a mass m collides elastically with a larger object whose mass is M. The particle rebounds along its original line of motion. If the massive object is initially at rest, what fraction of the particle's original kinetic energy is transferred to the object? Discuss the result (a) for the case of a golf ball incident on a bowling ball ($M/m \cong 500$) and (b) for a 100-kg block dropped from a height of 10 m to the Earth's surface.

Section 9-4

9-15 Find the position of the C.M. of five equal-mass particles located at the five corners of a square-based right pyramid with sides of length ℓ and altitude h.

9-16 The mass of the Sun is 329,390 Earth masses and the mean distance from the center of the Sun to the center of the Earth is 1.496×10^8 km. Treating the Earth and Sun as particles, with each mass concentrated at the respective geometric center, how far from the center of the Sun is the C.M. of the Earth-Sun system? Compare this distance with the mean radius of the Sun (6.960×10^5 km).

9-17 A bomb is at rest with respect to an inertial reference frame in deep space. The bomb explodes, breaking into three pieces. Two pieces with equal mass fly off with equal speeds of 50 m/s in directions perpendicular to one another. What is the velocity of the third piece relative to the other two if it has three times the mass of either of the other pieces?

9-18 Given two particles, $m_1 = 1$ kg and $m_2 = 2$ kg, the coordinates of which are known functions of time in a particular inertial frame: $\mathbf{r}_1 = 3t^2\hat{\mathbf{i}} + 2t\hat{\mathbf{j}}$ and $\mathbf{r}_2 = 3\hat{\mathbf{i}} + (2t^3 + 5t)\hat{\mathbf{j}} + 9\hat{\mathbf{k}}$, where the position vectors are given in meters for t in seconds. (a) What is the C.M. coordinate $\mathbf{R}$? (b) What is the C.M. velocity $d\mathbf{R}/dt$? (c) What is the C.M. acceleration $d^2\mathbf{R}/dt^2$?

9-19 A shell is fired with a uniform speed of 10 m/s at a

stationary target 200 m away, in deep space. Midway to the target the shell prematurely explodes into two equal fragments. One fragment strikes the target 8 s after the explosion. (a) Where is the second fragment at the moment the first fragment strikes the target? (b) Will the second fragment strike the target? If so, when?

Section 9-5

9-20 A railroad handcar has a mass, when empty, of
• 250 kg. The handcar, operatd by a 100-kg man, carries a small boulder ($m = 50$ kg) and coasts along a level, straight, frictionless track with a velocity of 2 m/s. The man throws the boulder out of the handcar in such a way that he observes the boulder to move at right angles to the track. (a) Use the straightforward application of the conservation of momentum to determine the velocity of the car after the boulder is thrown out. (b) Use Eq. 9-17 to obtain the answer to (a). (c) Repeat (a) and (b) for the case in which the boulder is seen by a stationary ground observer to leave the car at right angles to the track. (d) Does it matter in any of the above if the boulder's path is not horizontal? (e) Why is it unnecessary to give the *magnitude* of the boulder's velocity in the above?

9-21 A rocket with a total mass of 160,000 kg sits vertically on a launch pad. At what rate must gases be ejected on ignition if the thrust is to just overcome the gravitational force on the rocket? Assume a velocity for the ejected gases of 3000 m/s. What rate is required on ignition if an initial acceleration of $2g$ is to be achieved? What thrust does the rocket engine deliver in the above two cases?

9-22 A rocket with an initial total mass M_i is launched vertically from the Earth's surface. When the launch fuel has been completely burned, the rocket has reached an altitude small compared to the Earth's radius (so that the acceleration due to gravity may be considered constant during the burn). Show that the final velocity is

$$v = -v_r \ln (M_i/M_f) - gt_b$$

where the time of burn is

$$t_b = (M_i - M_f)(dm/dt)^{-1}$$

(M_f is the final total mass of the rocket, v_r is the exhaust gas velocity, and dm/dt is the constant rate of fuel consumption.)

9-23 A Titan-II rocket has a total mass of 145 metric tons, and the first-stage rocket engines deliver a total thrust

of 1.92×10^6 N. The fuel-oxidizer consists of kerosene and liquid oxygen (LOX) and gives an exhaust velocity of 2400 m/s. If the first-stage burn time is 150 s in a vertical ascent, calculate (a) the initial acceleration, (b) the rate at which gas is ejected, for constant thrust, and (c) the velocity reached at the end of first-stage burn (use the results of Problem 9-22).

9-24 A rocket for use in deep space is to have the capability of boosting a payload (plus the rocket frame and engine) of 3.0 metric tons to an achieved speed of 10,000 m/s with an engine and fuel designed to produce an exhaust velocity of 2000 m/s. (a) How much fuel and oxidizer is required? (b) If a different fuel and engine design could give an exhaust velocity of 5000 m/s, what amount of fuel and oxidizer would be required for the same task? Comment.

9-25 An aircraft uses a rocket-assisted takeoff unit (RATO) that is capable of discharging 50 kg of gas in 10 s with an exhaust velocity of 1200 m/s. The 3000-kg aircraft, on ordinary takeoff without RATO, requires a distance of 400 m to reach a lift-off speed of 200 km/h with a constant propeller thrust. (a) What is the thrust developed by the rocket? (b) With both the RATO and the normal propeller thrust providing a constant acceleration, what is the distance required for lift-off?

Additional Problems

9-26 A horizontal stream of water 10 cm in diameter strikes a stationary vertical wall and breaks into a low-velocity spray. If the discharge rate against the wall is 10^3 liters/min, what is the force of the stream against the wall?

9-27 A golf ball ($m = 46$ g) is struck a blow that makes an angle of 45° with the horizontal. The drive lands 200 m away on a flat fairway. If the golf club and ball are in contact for a time of 7 ms, what is the average force of impact? (Neglect air resistance effects.)

9-28 A hunter fires a shotgun that has a mass of 4 kg, discharging lead shot having a total mass of 35 g. The lightly held gun recoils with an initial speed of 2.9 m/s and is brought to rest against the hunter's shoulder in 0.07 s, a time long compared to the time required for the shot to leave the gun barrel. (a) What is the average force against the hunter's shoulder? (b) What is the muzzle velocity of the shot? (c) If the length of the gun barrel is 75 cm and if the shot travels down the barrel with constant acceleration, how long does it take the shot to clear the barrel? Is this time short compared to 0.07 s, as claimed?

9-29 A small tin can has a mass of 250 g and is suspended from a string 2 m long. The can is observed to swing through a maximum angle of 10° when a bullet with a mass of 10 g is shot through the can. If the bullet enters the can horizontally with a velocity of 350 m/s, calculate its velocity upon emerging from the can.

9-30 An attack helicopter is equipped with a 20-mm cannon that fires 130-g shells in the forward direction with a muzzle velocity of 800 m/s. The fully loaded helicopter has a mass of 4000 kg. A burst of 160 shells is fired in a 4-s interval. What is the resulting average force on the helicopter and by what amount is its forward speed reduced?

9-31 A loaded howitzer is rigidly bolted to a railroad flatcar. The mass of the combination is 20,000 kg. The flatcar is observed to be coasting along a level, straight (frictionless) track with a constant speed V. The howitzer is set to an elevation angle of 60° with the horizontal and then fired. After the howitzer is fired, the flatcar and attached howitzer are seen to recoil, moving in the opposite direction along the track with the same speed V. (a) If the muzzle velocity of the howitzer shell is 300 m/s, and if the shell has a mass of 50 kg, find V. (b) How high would this shell rise in the air?

9-32 A 75-kg man is standing in a 225-kg rowboat and is
• originally 10 m from a pier on a still lake. The man walks towards the pier a distance of 2 m relative to the boat, and then stops. Assuming that the boat can move through the water without resistance, how far is the man now from the pier?

9-33 A horizontal force of 0.80 N is required to pull a 5-kg block across a tabletop at constant speed. With the block initially at rest, a 20-g bullet fired horizontally into the block causes the block to slide 1.5 m before coming to rest again. Determine the speed v of the bullet. Assume the bullet to be embedded in the block.

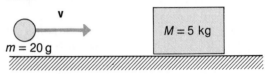

9-34 Two blocks rest on a frictionless surface with a compressed spring between them, as shown in the diagram. Block A has a mass of 1 kg and block B has a mass of 2 kg. The spring is initially compressed by 50 cm and has negligible mass. When both blocks are released simultaneously and the spring has dropped to the surface, block A is found to have a velocity of 2 m/s. (a) What is the velocity of block B? (b) How much energy was stored in the spring? (c) What fraction of the energy did each block receive? Why was it not evenly divided between the blocks? (d) What is the spring constant κ of the spring?

Problem 9-34

9-35 A puck with mass $M = 0.2$ kg is placed on a rough
• table at the midpoint between two identical springs with spring constant $\kappa = 200$ N/m. The springs are separated by a distance of 1 m. The coefficient of kinetic friction between the puck and the table is $\mu_k = 0.20$. The puck is set into motion by an impulse of 1 N·s in the direction of one of the springs. (a) Find the initial velocity and the initial kinetic energy of the puck. What frictional force acts on the puck? (b) What is the kinetic energy of the puck when it reaches the first spring? By how much is this spring compressed when the puck is momentarily at rest? (Ignore frictional forces over the distance of compression of the spring.) (c) With what kinetic energy does the puck arrive at the second spring? What will be the compression of this spring? (d) The puck runs back and forth between the two springs until it comes to rest as the result of the frictional

Problem 9-35

forces. What total distance does the puck move before coming finally to rest?

9-36 A block with a mass m slides down a frictionless wedge that has a mass M and an inclination angle α, as shown in the diagram. The wedge is free to slide on a frictionless horizontal flat surface. If the entire system is initially at rest with the leading edge of the block at an elevation h above the flat surface, show that, when the leading edge reaches the flat surface, the velocity of the wedge is

$$V = \left[\frac{2m^2 gh \cos^2 \alpha}{(m + M)^2 - (m + M)m \cos^2 \alpha} \right]^{1/2}$$

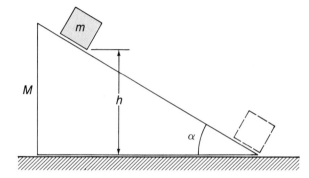

9-37 A ball with a mass m is attached to the end of a string of length L. The ball is released from a position such that the string makes an angle α_1 with the vertical. At the bottom of its swing the ball strikes a block with a mass M that is resting on a frictionless horizontal surface and the ball rebounds to an angle α_2, as indicated in the diagram. (a) Show that the velocity acquired by the block is

$$V = 4 \frac{m}{M} \sqrt{gL} \sin \left(\frac{\alpha_1 + \alpha_2}{4} \right) \cos \left(\frac{\alpha_1 - \alpha_2}{4} \right)$$

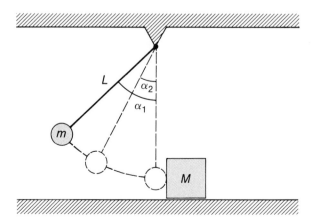

(b) What is the coefficient of restitution ε between the

ball and the block? (Refer to Problem 9–13 for the definition of ε.)

9-38 Two railway cars, m_1 and m_2, are moving along a track with velocities v_1 and v_2, respectively. The cars collide, and after the collision the velocities are v_1' and v_2'. Show that the change in kinetic energy, $K' - K$, will be maximum if the cars couple together. [*Hint:* Set $d(K' - K)/dv_1' = 0$ and show that $v_1' = v_2'$.]

9-39 A particle with a mass m has an initial velocity v_0 and collides head-on with a block that has a mass M and is initially at rest. The collision is inelastic and an amount of energy E is transferred from the particle to the block. Derive an expression in terms of m, M, and E, for the minimum value of v_0 for which the process is possible. [*Hint:* Derive an expression for the velocity of the block; this velocity must be *real*.]

9-40 A particle with a mass m_1 is incident with a velocity $\mathbf{v}_1$ on a stationary particle with a mass $m_2 = 4m_1$. The incident particle is scattered at an angle $\theta_1 = 65°$, and the struck particle recoils at an angle $\theta_2 = 55°$. Determine the velocity ratios, v_1'/v_2' and v_1'/v_1. Compare the initial kinetic energy with the final kinetic energy and determine whether the collision is elastic.

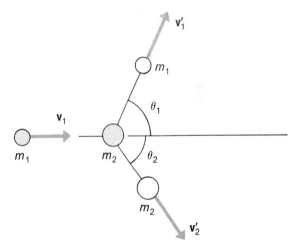

9-41 Refer to the diagram for Problem 9–40 and assume that $m_1 > m_2$. For an elastic collision, show that the maximum scattering angle θ_{1m} is obtained from $\cos^2 \theta_{1m} = 1 - m_2^2/m_1^2$.

9-42 A ball is dropped at $t = 0$ from a height $y_0 = 12$ m above a surface. At the instant of release a platform is at a height $h = 4$ m above the same surface and is moving upward with a constant velocity $v = 3$ m/s. The ball rebounds elastically from the moving platform. (a) To what height does the ball rise? (b) At what time does the ball strike the platform for the second time? Assume the platform to be very massive.

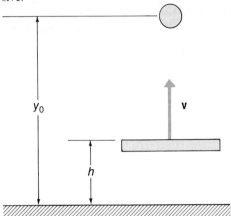

9-43 At $t = 0$, three particles are located and are moving as follows:

$m_1 = 2$ kg: (4, 2), $\mathbf{v}_1 = (3 \text{ m/s})\hat{\mathbf{j}}$
$m_2 = 3$ kg: (−3, 2), $\mathbf{v}_2 = (−2 \text{ m/s})\hat{\mathbf{i}}$
$m_3 = 5$ kg: (2, −2), $\mathbf{v}_3 = (−1 \text{ m/s})\hat{\mathbf{j}}$

Determine the position of the C.M. of this system at $t = 0$, 1, 2, and 3 s. Make a graph of the motion of the C.M. What force acts on the system?

9-44 A railroad tanker car has an empty mass of $1.5 \times$
• 10^4 kg and is carrying a 3×10^4 kg load of oil. The car is coasting along a level, straight, frictionless track at a speed of 2 m/s. A bottom drain accidentally opens, allowing all the oil to leak out at a uniform rate in 30 min. How far along the track did the car travel during the time required for all the oil to leak out for the following conditions? (a) The oil leaves the car drain straight downward relative to the car. (b) The oil leaves the car drain straight downward relative to the ground. [*Hint:* First, find the speed of the car as a function of time using Eq. 9–17.]

9-45 A limp, flexible chain with a mass m and a length ℓ
•• rests on a frictionless horizontal surface. One end of the chain is lifted vertically at a constant speed v, as indicated in the diagram. The required force F as a function of the length y already lifted is

$$F(y) = \frac{m}{\ell}(gy + v^2), \qquad y \leqslant \ell$$

(a) Derive this expression using Eq. 9–16. (b) Does the expression for $F(y)$ correspond to that obtained from the conservation of energy? Explain.

9-46 A flexible chain slides off a flat frictionless
•• table. The chain begins its slide lying in a straight line perpendicular to the table edge, over which a

small portion protrudes. The chain has a mass m and a length ℓ. The acceleration of the chain, in terms of the length y overhanging the edge, is given by

$$\ell \frac{d^2y}{dt^2} = gy$$

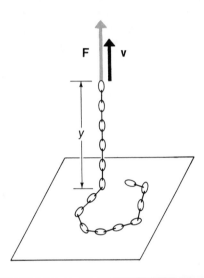

(a) Derive this expression using Eq. 9–15 and noting that the differential mass fragment $dm = (m/\ell)\, dy$ added to the overhang has $\mathbf{v}_r = 0$. [*Hint:* Do not forget to include the tension in the chain.] (b) Derive this expression from the conservation of energy. [*Hint:* Differentiate the expresssion for E.] (c) Why does conservation of energy apply in the present case but not in Problem 9–45?

9-47 A two-stage rocket has the following specifications:

Stage #2
   frame + engine
      + payload: $M_2 = 2000$ kg
   fuel + oxidizer: $m_2 = 3000$ kg

Stage #1
   frame + engine: $M_1 = 5000$ kg
   fuel + oxidizer: $m_1 = 20{,}000$ kg

gas velocity: $v_r = −2500$ m/s

The first stage of this rocket is fired while it is in deep space (initially at rest in some inertial frame). When this burn is completed and prior to igniting the second stage, the frame and engine of the first stage are jettisoned (at essentially zero separation velocity). (a) Show that at the end of the second-stage burn, the total achieved velocity v is

$$v = -v_r \ln \frac{(M_1 + M_2 + m_1 + m_2)(M_2 + m_2)}{(M_1 + M_2 + m_2)M_2}$$

(b) What is the numerical value of v? (c) What would be the achieved velocity of a rocket that has only a single stage with the same total mass, $M_1 + M_2 + m_1 + m_2$, and burns all the fuel, $m_1 + m_2$, in a single engine with the same gas velocity? (d) Recalculate (c) but subtract the 1000-kg mass of the redundant second-stage engine and accessories. Comment on the results of (b), (c), and (d).

CHAPTER 10

ANGULAR MOMENTUM AND THE DYNAMICS OF SYSTEMS

In this chapter we introduce a new dynamical quantity called *angular momentum,* a quantity that obeys a conservation principle similar to that for linear momentum. The angular momentum concept is useful in the discussion of the dynamics of a single particle and systems of particles. Of continuing interest is the case of the two-particle system. We treat this simple system in detail because there are many important cases that involve the interaction of just two objects. In Chapter 14 we use some of the results obtained here in the discussion of planetary and satellite motion.

10-1 ANGULAR MOMENTUM AND TORQUE FOR A PARTICLE

Angular Momentum. In Chapter 9 we found that *linear momentum* is an extremely useful concept because of the conservation principle associated with it. We now introduce a quantity called *angular momentum* (or sometimes the *moment of momentum*). The importance and usefulness of angular momentum lies in the fact that it, too, has an associated conservation principle. We begin by considering the angular momentum of a single particle; later, we generalize this concept for the case of two particles and then for systems of particles, and, finally (in Chapter 11) for rigid bodies.

For a single particle, we define the angular momentum ℓ about the origin O of a reference frame to be

$$\ell = \mathbf{r} \times \mathbf{p} = m\mathbf{r} \times \mathbf{v} \tag{10-1}$$

where $\mathbf{r}$ is the position vector of the particle with respect to O at some particular time and $\mathbf{p} = m\mathbf{v}$ is the simultaneous linear momentum. See Fig. 10-1.

The basic definition of the cross product, as prescribed in Section 3-6, requires

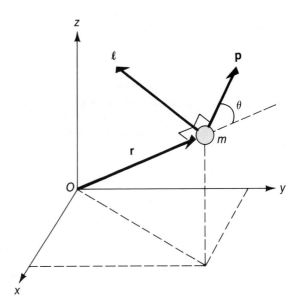

Fig. 10-1. Illustrating the definition of the angular momentum $\boldsymbol{\ell}$ of a particle about the origin O. With $\boldsymbol{\ell} = \mathbf{r} \times \mathbf{p} = m\mathbf{r} \times \mathbf{v}$, we see that the vector $\boldsymbol{\ell}$ is perpendicular to the plane containing $\mathbf{r}$ and $\mathbf{p}$. If θ is the angle between the positive directions of $\mathbf{r}$ and $\mathbf{p}$ (measured in the plane containing these vectors), we also have $|\boldsymbol{\ell}| = |\mathbf{r}||\mathbf{p}| \sin \theta$.

the vector $\boldsymbol{\ell}$ to be perpendicular to the plane containing $\mathbf{r}$ and $\mathbf{p}$ and to have a vector sense of direction given by the right-hand rule. Furthermore, the magnitude of $\boldsymbol{\ell}$ is*

$$|\boldsymbol{\ell}| = \ell = |\mathbf{r}||\mathbf{p}| \sin \theta = rp \sin \theta \tag{10-2}$$

where θ is the smaller angle between the positive directions of $\mathbf{r}$ and $\mathbf{p}$. We can also express the angular momentum in terms of a 3-by-3 determinant (see Eq. 3–27):

$$\boldsymbol{\ell} = \begin{vmatrix} \hat{\mathbf{i}} & \hat{\mathbf{j}} & \hat{\mathbf{k}} \\ x & y & z \\ p_x & p_y & p_z \end{vmatrix} = m \begin{vmatrix} \hat{\mathbf{i}} & \hat{\mathbf{j}} & \hat{\mathbf{k}} \\ x & y & z \\ v_x & v_y & v_z \end{vmatrix} \tag{10-3}$$

For many purposes, it is convenient to express the angular momentum of a particle in terms of its linear momentum components. In Fig. 10–2, $\mathbf{r}$ and $\mathbf{p}$ lie in the plane of the page, and we have

$$p_\perp = p \sin \theta$$

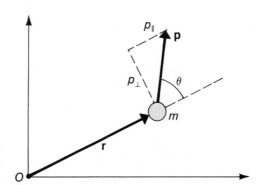

Fig. 10-2. Illustrating the components of the linear momentum $\mathbf{p}$: $p_\parallel$ is parallel to $\mathbf{r}$ and $p_\perp$ is perpendicular to $\mathbf{r}$. Both $\mathbf{r}$ and $\mathbf{p}$ lie in the plane of the page.

---

*Remember that, by definition, $0 \leqslant \theta \leqslant \pi$; therefore $\sin \theta \geqslant 0$.

Using Eq. 10–2, we can now write

$$\ell = rp_\perp$$

Thus, the angular momentum of the particle has a magnitude equal to the product of its distance r from the origin and the transverse or perpendicular component of its linear momentum. The parallel component $p_\parallel$ makes no contribution to the angular momentum. Notice that we can also write

$$\ell = mrv_\phi$$

where $v_\phi = v_\perp$ is the azimuthal velocity component (see Eq. 3–30).

It is evident that the dimensions of angular momentum are

$$[\ell] = [r][m][v] = L \cdot M \cdot (L/T) = L^2MT^{-1}$$

so the units can be expressed as joule-seconds (J · s).

In the important case of circular motion, with $|\mathbf{r}| = R = $ constant, the velocity $\mathbf{v}$ is perpendicular to the position vector $\mathbf{r}$. Then, the tangential velocity, $v_t = v$, is equal to the azimuthal velocity v_ϕ, so that $\ell = mRv$. But we also have $v = R\omega$; thus,

$$\ell = mR^2\omega \qquad \text{(circular motion)} \qquad \textbf{(10–4)}$$

Example 10–1

A particle with a mass of 500 g rotates in a circular orbit with radius $R = 22.4$ cm at a constant rate of 1.5 revolutions per second. What is the angular momentum with respect to the center of the orbit?

Solution:

As shown in the diagram, the angular momentum vector, $\boldsymbol{\ell} = \mathbf{r} \times \mathbf{p}$, is parallel to $\boldsymbol{\omega}$ (which lies along the rotation axis). The constant magnitude of $\boldsymbol{\ell}$ is $\ell = mR^2\omega$. With the motion repeating 1.5 times per second, we have

$$\omega = 2\pi(1.5 \text{ s}^{-1}) = 3\pi \text{ rad/s}$$

Thus,
$$\ell = (0.5 \text{ kg})(0.224 \text{ m})^2(3\pi \text{ s}^{-1})$$
$$= 0.236 \text{ J} \cdot \text{s}$$

Example 10–2

The conditions given in the previous example are exactly those of Example 6–5. In that example the stone was twirled at the end of a 25-cm-long string. To satisfy Newton's equations, the string was found to make an angle of 63.8° with the vertical. The radius of the circular orbit of the stone was, therefore, $R = L \sin \beta = (0.250 \text{ m}) \sin 63.8° = 0.224$ m, the value that we used in Example 10–1.

To illustrate the importance of the selection of the origin in determining the angular momentum, calculate the angular momentum with respect to the stationary or hand-held end of the string, shown in diagram (a).

(a)

(b)

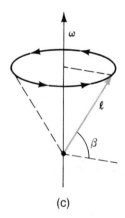

(c)

Solution:

We have $\mathbf{v} = \boldsymbol{\omega} \times \mathbf{r}$ (Eq. 4–38), so that $|\mathbf{v}| = v = \omega L \sin \beta = R\omega$, giving the same value as before. Now, however, for the angular momentum with respect to O (the hand), we have

$$\boldsymbol{\ell} = \mathbf{r} \times \mathbf{p} = m\mathbf{r} \times \mathbf{v}$$

which is *not* parallel to $\boldsymbol{\omega}$. In diagram (b), the vectors $\boldsymbol{\omega}$ (which corresponds to the axis of rotation) and $\mathbf{r}$ are instantaneously in the plane of the page. We see that $\boldsymbol{\ell}$ is inclined at an angle $\beta = 63.8°$ to the horizontal. The angular momentum has a constant magnitude ℓ, but its vector direction rotates or *precesses* about a vertical axis with the angular frequency ω, as shown in diagram (c).
We have

$$|\boldsymbol{\ell}| = \ell = mRL\omega$$
$$= (0.5 \text{ kg})(0.224 \text{ m})(0.250 \text{ m})(3\pi \text{ s}^{-1})$$
$$= 0.264 \text{ J} \cdot \text{s}$$

Note that the *component* of $\boldsymbol{\ell}$ parallel to $\boldsymbol{\omega}$ (the axis of rotation) has a constant value given by

$$\ell_z = \ell \sin \beta = (0.264 \text{ J} \cdot \text{s}) \sin 63.8°$$
$$= 0.236 \text{ J} \cdot \text{s}$$

This is exactly the value found in the previous example. We commonly refer to this component of $\boldsymbol{\ell}$ as the angular momentum *about the axis of rotation*. For a particle that rotates about the z-axis, this axial component of the angular momentum is simply the tangential momentum mv_t multiplied by the perpendicular distance R to the z-axis. In the case here,

$$\ell_z = \ell \sin \beta = mR(L \sin \beta)\omega = mR^2\omega$$
$$= mv_t R$$

---

*When a vector in a figure is perpendicular to the plane of the page, it is convenient to represent the vector by the symbol $\odot$ when it is pointing upward or out of the page and by the symbol $\oplus$ when it is pointing downward or into the page. These symbols are intended to remind you of the front view of the tip of an arrow and the rear view of the feather-end of the arrow.

Torque. It is a common experience that an object can be set into rotation by a force that is applied "off center." For example, a tangential push on the rim of a wheel will cause rotation, but a similar push along a line through the axle will not. The tendency for a force to produce rotation is proportional to the magnitude of the force and to the distance from the point of application to the center of rotation. The quantitative measure of this rotation tendency is the *torque* produced by the force.

Any vector can be expressed in terms of its component vectors. Therefore, a force **F** that is applied to a particle m, as in Fig. 10–3, can be expressed in terms of

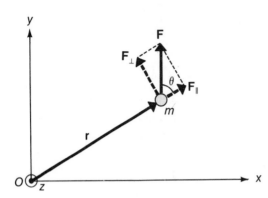

Fig. 10–3. Only the perpendicular component of **F** produces a torque about O.

a component $\mathbf{F}_\perp$ perpendicular to the position vector **r** and a component $\mathbf{F}_\parallel$ parallel to **r**. Only the component $\mathbf{F}_\perp$ can produce a torque about the point O. The magnitude of the torque produced by **F** about O is defined to be

$$\tau = rF_\perp = rF \sin \theta \qquad (10\text{–}5)$$

where θ is the smaller angle between the positive directions of **r** and **F**. Using vector notation, we can write, in analogy with the definition of angular momentum,

$$\boldsymbol{\tau} = \mathbf{r} \times \mathbf{F} \qquad (10\text{–}6)$$

Torque is also referred to as *moment of force*. Note that the vector $\boldsymbol{\tau}$, which results from the cross product of two vectors, is perpendicular to the plane containing **r** and **F**, with a direction sense given by the right-hand rule (Fig. 10–4).

Another way to calculate the torque produced by a force is illustrated in Fig. 10–5. The vector direction of **F** is extended as the dashed "line of action." The *lever*

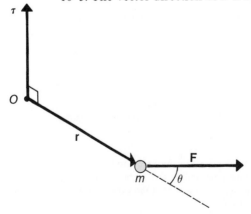

Fig. 10–4. The torque about O produced by the force **F** acting on a particle at the point defined by **r** is $\boldsymbol{\tau} = \mathbf{r} \times \mathbf{F}$.

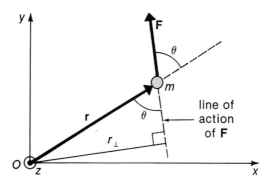

Fig. 10-5. The lever arm for the force **F** is $r_\perp$ and the torque about O is $\tau = r_\perp F = rF \sin \theta$.

arm of **F** (namely, $r_\perp$) is the perpendicular to the line of action. Now, $r_\perp = r \sin \theta$, so Eq. 10-5 can also be written as $\tau = r_\perp F$.

The torque vector $\boldsymbol\tau$ can also be expressed in determinant notation:

$$\boldsymbol\tau = \begin{vmatrix} \hat{\mathbf{i}} & \hat{\mathbf{j}} & \hat{\mathbf{k}} \\ x & y & z \\ F_x & F_y & F_z \end{vmatrix} \qquad \textbf{(10-7)}$$

The Connection Between Torque and Angular Momentum. Newton's second law can be expressed by stating that the net force acting on a particle is equal to the time derivative of the particle's linear momentum:

$$\mathbf{F} = \frac{d\mathbf{p}}{dt}$$

If we take the cross product of this expression with **r** (from the left), we have

$$\mathbf{r} \times \mathbf{F} = \mathbf{r} \times \frac{d\mathbf{p}}{dt} \qquad \textbf{(10-8)}$$

Now, we take the defining equation for angular momentum (Eq. 10-1) and differentiate with respect to time, obtaining

$$\frac{d\boldsymbol\ell}{dt} = \frac{d}{dt}(\mathbf{r} \times \mathbf{p}) = \frac{d\mathbf{r}}{dt} \times \mathbf{p} + \mathbf{r} \times \frac{d\mathbf{p}}{dt}$$

The first term on the right-hand side can be written as the cross product of a vector with itself, which therefore vanishes:

$$\frac{d\mathbf{r}}{dt} \times \mathbf{p} = m\frac{d\mathbf{r}}{dt} \times \mathbf{v} = m\mathbf{v} \times \mathbf{v} = 0$$

Then, using Eq. 10-8, we have

$$\frac{d\boldsymbol\ell}{dt} = \mathbf{r} \times \frac{d\mathbf{p}}{dt} = \mathbf{r} \times \mathbf{F}$$

But $\mathbf{r} \times \mathbf{F}$ is the torque, so we have

$$\boldsymbol{\tau} = \mathbf{r} \times \mathbf{F} = \frac{d\boldsymbol{\ell}}{dt} \tag{10-9}$$

Note carefully that *both* $\boldsymbol{\tau}$ and $\boldsymbol{\ell}$ are referred to the same origin that was selected to define the position vector $\mathbf{r}$.

The Conservation of Angular Momentum.   Equation 10–9 contains an important statement concerning the motion of a particle. The torque exerted on a particle is equal to the time rate of change of its angular momentum, $\boldsymbol{\tau} = d\boldsymbol{\ell}/dt$. Consequently, if there is *no torque* acting, the angular momentum must remain constant:

$$\boldsymbol{\tau} = 0 \quad \text{implies} \quad \boldsymbol{\ell} = \text{constant} \tag{10-10}$$

Thus, in the absence of an applied torque, *angular momentum is conserved.*

It should be pointed out that the principle of angular momentum conservation is not a new law of Nature. The essential ingredient of the principle is already contained within (linear) Newtonian dynamics. All solvable problems in mechanics can be solved, at least in principle, by using Newton's equations. In many instances, however, it proves convenient to use angular momentum conservation to treat a particular situation, but it is not required, as is demonstrated by some of the examples we give.

Example 10-3

It is instructive to consider again the familiar problem of an object falling vertically, now applying Eq. 10–6 to determine the equation of motion.

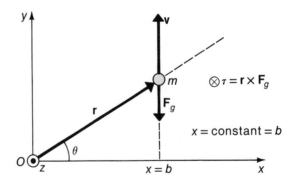

Solution:

We have
$$\boldsymbol{\tau} = \mathbf{r} \times \mathbf{F}_g = -mgr \sin\left(\theta + \frac{\pi}{2}\right)\hat{\mathbf{k}}$$
$$= -mgr \cos\theta\,\hat{\mathbf{k}}$$

or
$$\boldsymbol{\tau} = -mgb\,\hat{\mathbf{k}}$$

[We could also obtain this result by noting that $\mathbf{r} = b\hat{\mathbf{i}} + y\hat{\mathbf{j}}$ and $\mathbf{F}_g = -mg\hat{\mathbf{j}}$, whence,

$\tau = \mathbf{r} \times \mathbf{F}_g = (b\hat{\mathbf{i}} + y\hat{\mathbf{j}}) \times (-mg\hat{\mathbf{j}}) = -mgb\hat{\mathbf{i}} \times \hat{\mathbf{j}} = -mgb\hat{\mathbf{k}}.$] Now,

$$\frac{d\boldsymbol{\ell}}{dt} = m\frac{d}{dt}(\mathbf{r} \times \mathbf{v}) = m\frac{d}{dt}\left[vr\sin\left(\frac{\pi}{2} - \theta\right)\hat{\mathbf{k}}\right]$$

$$= m\frac{d}{dt}(vr\cos\theta)\hat{\mathbf{k}} = m\frac{d}{dt}(vb)\hat{\mathbf{k}}$$

or

$$\frac{d\boldsymbol{\ell}}{dt} = mb\frac{dv}{dt}\hat{\mathbf{k}}$$

Thus, with $\tau = d\boldsymbol{\ell}/dt = -mgb\hat{\mathbf{k}}$, we have

$$-mgb\hat{\mathbf{k}} = mb\frac{dv}{dt}\hat{\mathbf{k}}$$

or

$$\frac{dv}{dt} = -g$$

which is the expected result.

Example 10-4

It is also instructive to solve the problem posed in Example 6-5 by using Eq. 10-9. (Refer also to Examples 10-1 and 10-2.) Find the tension T and the angle β in terms of the given quantities, L, ω, and m.

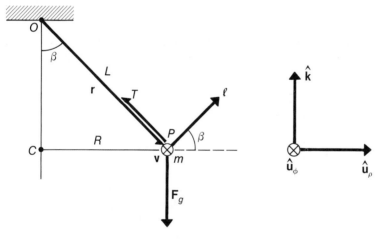

Solution:

The diagram locates the particle m relative to the origin O in terms of the vector $\mathbf{r} = \overline{OP}$. The polar unit vectors, $\hat{\mathbf{u}}_\rho$ and $\hat{\mathbf{u}}_\phi$, are used to describe the dynamics of the particle. In addition, we include the unit vector $\hat{\mathbf{k}}$ parallel to the rotation axis. The torque can be expressed as

$$\tau = (\mathbf{r} \times \mathbf{T}) + (\mathbf{r} \times \mathbf{F}_g) = 0 + Lmg\sin\beta\,\hat{\mathbf{u}}_\phi$$

The angular momentum, $\boldsymbol{\ell} = m\mathbf{r} \times \mathbf{v}$, can be written in terms of components in the directions $\hat{\mathbf{k}}$ and $\hat{\mathbf{u}}_\rho$. We note that

$$|\boldsymbol{\ell}| = \ell = mRL\omega$$

Then, $$\boldsymbol{\ell} = (mRL\omega \sin \beta)\widehat{\mathbf{k}} + (mRL\omega \cos \beta)\widehat{\mathbf{u}}_\rho$$

Now, $d\widehat{\mathbf{k}}/dt = 0$. (•Can you see why?) Thus,

$$\frac{d\boldsymbol{\ell}}{dt} = (mRL\omega \cos \beta) \frac{d\widehat{\mathbf{u}}_\rho}{dt}$$

But, from Eq. 4–27, we have $$\frac{d\widehat{\mathbf{u}}_\rho}{dt} = \omega\widehat{\mathbf{u}}_\phi$$

which confirms that $\boldsymbol{\ell}$ precesses about a vertical axis at the angular rate ω. Now, using $\boldsymbol{\tau} = d\boldsymbol{\ell}/dt$ gives

$$(Lmg \sin \beta)\widehat{\mathbf{u}}_\phi = (mRL\omega^2 \cos \beta)\widehat{\mathbf{u}}_\phi$$

or $$g \sin \beta = R\omega^2 \cos \beta \qquad (1)$$

Next, consider shifting the origin to C such that $\mathbf{r} = \overline{CP}$. Then,

$$\boldsymbol{\tau} = R(T \cos \beta - mg)\widehat{\mathbf{k}}$$

Also, we now have $$\boldsymbol{\ell} = mR^2\omega\widehat{\mathbf{k}}$$

so that $d\boldsymbol{\ell}/dt = 0$. Because $\boldsymbol{\tau} = d\boldsymbol{\ell}/dt = 0$, we have

$$\cos \beta = \frac{mg}{T} \qquad (2)$$

If we substitute from Eq. (2) for $\cos \beta$ in Eq. (1), there results

$$T = mL\omega^2 \qquad (3)$$

The two equations, (2) and (3), are exactly those found in Example 6–5, and therefore lead to the desired identical solution.

Example 10–5

A small disc with mass m is guided by a light string to execute circular motion on a frictionless tabletop, as indicated in the diagram. The string passes through a small hole in the table, and the lower end is held by a hand. We describe the motion in plane polar coordinates. An initial observation finds the disc at a radius r_0 and twirling with an angular frequency ω_0. By pulling on the string *slowly*, the radius is reduced to r. What is the new value of ω? How much total work was done by pulling the string? What force F must be exerted to accomplish this change in radius?

Solution:

Consider the disc to be a particle executing circular motion about an origin O at the position of the hole. The gravitational force $m\mathbf{g}$ that acts on the disc is just balanced by the normal force $\mathbf{N}$ that the tabletop exerts on the disc; that is, $\mathbf{N} = m\mathbf{g}$. These two forces have exactly canceling torques about O. The other force that acts on the disc is the tension in the string, which also generates no torque relative to O. Therefore, $\ell = mr^2\omega$ is a constant of the motion. Thus,

$$mr^2\omega = mr_0^2\omega_0$$

or $$\omega = \left(\frac{r_0^2}{r^2}\right)\omega_0$$

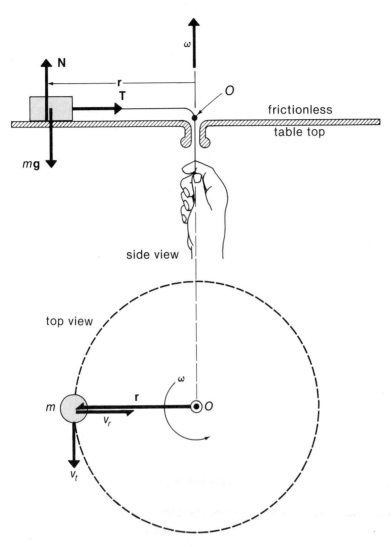

The kinetic energy of the disc at any instant is the sum of azimuthal and radial terms (see Eq. 3-31):

$$K = \tfrac{1}{2}m(v_\phi^2 + v_\rho^2)$$

The string is pulled *slowly*, so that $v_\rho \ll v_\phi$. Thus, the term involving v_ρ may be neglected, and the azimuthal velocity is essentially the tangential velocity (nearly circular path). That is,

$$K = \tfrac{1}{2}mv_t^2 = \tfrac{1}{2}mr^2\omega^2$$

The angular frequency ω varies as $1/r^2$, so the kinetic energy K is *not* a constant but increases as r decreases. This increase in energy is supplied by the work done in pulling the string. Thus,

$$W = \tfrac{1}{2}mr^2\omega^2 - \tfrac{1}{2}mr_0^2\omega_0^2$$

or

$$W = \tfrac{1}{2}mr_0^2\omega_0^2\left(\frac{r_0^2}{r^2} - 1\right)$$

Then, the force exerted is

$$F = -\frac{dW}{dr} = \frac{mr_0^4\omega_0^2}{r^3} = mr\omega^2 = \frac{mv_t^2}{r}$$

In the event that v_ρ is not negligibly small, we must provide an additional force to change the orbit radius. This force is

$$F' = -\frac{d}{dr}(\tfrac{1}{2}mv_\rho^2) = -\frac{dt}{dr}\frac{d}{dt}(\tfrac{1}{2}mv_\rho^2) = \frac{1}{v_\rho}\frac{d}{dt}(\tfrac{1}{2}mv_\rho^2)$$

where we have changed the sign to account for the fact that $v_\rho = dr/dt$ is negative. Then,

$$F' = m\frac{dv_\rho}{dt} = m\frac{d^2r}{dt^2}$$

which is a familiar result.

10-2 TWO-PARTICLE SYSTEMS

The idea of the center of mass, introduced in Chapter 9, is particularly useful in discussing the dynamics of systems of particles. By specifying the motions of the individual particles relative to the C.M. of the system, the equations of motion become much easier to interpret. We now consider the motion of particle systems, beginning with the simple two-particle case. We restrict attention to an *isolated* pair of particles (so that there are no external forces acting), and we specify that the particles interact only by central forces.

Reduction to the Equivalent Single-Particle Problem. Consider two particles with masses m_1 and m_2 located at positions identified by the vectors $\mathbf{r}_1$ and $\mathbf{r}_2$, respectively, in some inertial frame. A force acts between the particles, and we use $\mathbf{F}_{12}$ to represent the force *on* particle 1 exerted *by* particle 2. The corresponding reaction force is $\mathbf{F}_{21} = -\mathbf{F}_{12}$.

We can, in principle, find the equations of motion, $\mathbf{r}_1(t)$ and $\mathbf{r}_2(t)$, by solving the coupled equations that represent Newton's second law:

$$\mathbf{F}_{12} = m_1\frac{d^2\mathbf{r}_1}{dt^2} \tag{10-11}$$

$$\mathbf{F}_{21} = m_2\frac{d^2\mathbf{r}_2}{dt^2} \tag{10-12}$$

Although this procedure is always correct, an alternative approach is often more revealing of the physical nature of the motion and its relationship to the forces present. For this purpose we make use of the vector $\mathbf{R}$, the center-of-mass coordinate of m_1 and m_2, and we introduce the *relative coordinate* $\mathbf{r}$, as shown in Fig. 10-6. The expression that related $\mathbf{R}$ to $\mathbf{r}_1$ and $\mathbf{r}_2$ is obtained from Eq. 9-8:

$$\mathbf{R} = \frac{1}{M}\sum_i m_i\mathbf{r}_i = \frac{m_1\mathbf{r}_1 + m_2\mathbf{r}_2}{m_1 + m_2} \tag{10-13}$$

The relative coordinate $\mathbf{r}$ is

$$\mathbf{r} \equiv \mathbf{r}_{12} = \mathbf{r}_2 - \mathbf{r}_1 \tag{10-14}$$

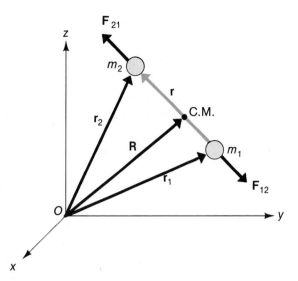

Fig. 10–6. A pair of interacting particles with masses m_1 and m_2. Two equivalent coordinate representations are shown—that consisting of the individual coordinates $\mathbf{r}_1$ and $\mathbf{r}_2$ in the inertial frame O, and that consisting of the relative coordinate $\mathbf{r} = \mathbf{r}_2 - \mathbf{r}_1$ and the center-of-mass coordinate $\mathbf{R}$. The central forces, $\mathbf{F}_{12}$ and $\mathbf{F}_{21}$, are shown as repulsive forces; they could equally well be attractive forces.

These equations for $\mathbf{R}$ and $\mathbf{r}$ can be inverted to obtain expressions for $\mathbf{r}_1$ and $\mathbf{r}_2$; after some algebraic manipulations, we find

$$\mathbf{r}_1 = \mathbf{R} - \frac{m_2}{M}\mathbf{r} \qquad \textbf{(10–15)}$$

$$\mathbf{r}_2 = \mathbf{R} + \frac{m_1}{M}\mathbf{r} \qquad \textbf{(10–16)}$$

where $M = m_1 + m_2$.

The equations of motion that represent the solution to the problem can be expressed equally well in terms of $\mathbf{r}_1(t)$ and $\mathbf{r}_2(t)$ or in terms of $\mathbf{R}(t)$ and $\mathbf{r}(t)$.

If we add Eqs. 10–11 and 10–12 and use the fact that $\mathbf{F}_{12} = -\mathbf{F}_{21}$, we obtain a particularly simple result:

$$M\frac{d^2\mathbf{R}}{dt^2} = 0 \qquad \textbf{(10–17)}$$

Thus, *the center-of-mass motion* $\mathbf{R}(t)$ *takes place with constant velocity and does not depend on the interaction between the particles.* (Compare the discussion leading to Eq. 9–14.)

If we multiply Eq. 10–11 by m_2 and Eq. 10–12 by m_1 and subtract the resulting expressions, we find, using Eqs. 10–15 and 10–16,

$$m_1 m_2 \frac{d^2\mathbf{r}}{dt^2} = M\mathbf{F}_{21}$$

We can rewrite this equation in terms of a quantity called the *reduced mass* μ, defined by

$$\mu \equiv \frac{m_1 m_2}{m_1 + m_2} = \frac{m_1 m_2}{M} \qquad \textbf{(10–18)}$$

We then obtain

$$\mu\frac{d^2\mathbf{r}}{dt^2} = \mathbf{F}_{21} \qquad \textbf{(10–19)}$$

Thus, *the relative coordinate* $\mathbf{r}(t)$ *of the two particles has exactly the same behavior as the coordinate of a single particle with mass* μ *moving in a force field* $\mathbf{F}_{21}$, *which is the mutual or internal force between the actual pair of particles.*

Kinetic Energy and Momentum. The total kinetic energy in the inertial frame O of the two particles, m_1 and m_2, can be expressed in terms of $\mathbf{V} = d\mathbf{R}/dt$, the velocity of the C.M., and $\mathbf{v} = d\mathbf{r}/dt$, the relative velocity of the particles. For the sum of the two individual kinetic energies, we have

$$K = \tfrac{1}{2}m_1 v_1^2 + \tfrac{1}{2}m_2 v_2^2$$

Now,

$$v_1^2 = \mathbf{v}_1 \cdot \mathbf{v}_1 = \left(\frac{d\mathbf{r}_1}{dt}\right) \cdot \left(\frac{d\mathbf{r}_1}{dt}\right)$$

and a similar expression holds for v_2^2. Substituting for $\mathbf{r}_1$ and $\mathbf{r}_2$ from Eqs. 10–15 and 10–16 and using Eq. 10–18, the result is

$$K = \tfrac{1}{2}M\left(\frac{d\mathbf{R}}{dt}\right) \cdot \left(\frac{d\mathbf{R}}{dt}\right) + \tfrac{1}{2}\mu\left(\frac{d\mathbf{r}}{dt}\right) \cdot \left(\frac{d\mathbf{r}}{dt}\right)$$

$$= \tfrac{1}{2}M\mathbf{V} \cdot \mathbf{V} + \tfrac{1}{2}\mu\mathbf{v} \cdot \mathbf{v}$$

or

$$K = \tfrac{1}{2}MV^2 + \tfrac{1}{2}\mu v^2 \tag{10-20}$$

Thus, the total kinetic energy of the system in frame O is the sum of the kinetic energy of a hypothetical particle with mass $M = m_1 + m_2$ moving with the velocity $\mathbf{V}$ of the C.M. (in frame O) *and* the kinetic energy of a hypothetical particle having a mass equal to the reduced mass, $\mu = m_1 m_2 / M$, moving with the relative velocity $\mathbf{v}$.

The linear momentum in frame O can also be expressed in terms of $\mathbf{V}$ and $\mathbf{v}$. According to Eq. 9–12, the momentum of the C.M. is

$$\mathbf{P} = \mathbf{p}_1 + \mathbf{p}_2 = M\mathbf{V} \tag{10-21}$$

Now, the *relative* linear momentum of the pair of particles is

$$\mu\mathbf{v} = \frac{m_1 m_2}{M}(\mathbf{v}_2 - \mathbf{v}_1)$$

By substituting the individual momenta, $\mathbf{p}_1 = m_1\mathbf{v}_1$ and $\mathbf{p}_2 = m_2\mathbf{v}_2$, we find

$$\mu\mathbf{v} = \frac{1}{M}(m_1\mathbf{p}_2 - m_2\mathbf{p}_1) \tag{10-22}$$

Note carefully that, whereas the total kinetic energy K requires *both terms* in Eq. 10-20 on the right-hand side, the total linear momentum is given by the *single term* $\mathbf{P} = M\mathbf{V}$ (*not* $\mathbf{P} + \mu\mathbf{v}$). Thus, when the particles (which interact through central forces) are viewed in the center-of-mass reference frame, the particles can possess kinetic energy but their total linear momentum is necessarily zero.

Constants of the Motion. If a pair of interacting particles is isolated and therefore subject to no external forces, various dynamical quantities remain constant in time. We have already seen, in Eq. 9-14, that the *total linear momentum* of an isolated system is constant:

$$\mathbf{P} = \text{constant} \qquad \text{no external forces} \qquad \textbf{(10-23)}$$

That the angular momentum is constant is an immediate consequence of the fact that we consider only pointlike particles with central forces acting between them. Under these circumstances, no torque is exerted on either particle, so the individual angular momenta, $\boldsymbol{\ell}_1$ and $\boldsymbol{\ell}_2$, are constant. Accordingly, the *total angular momentum,*[*] which is defined to be $\mathbf{L} = \boldsymbol{\ell}_1 + \boldsymbol{\ell}_2$, is also constant. (A proof is given in Section 10-3.) The value of the constant angular momentum of a pair of particles about their C.M. is equal to that of the equivalent single particle. Thus,

$$\mathbf{L} = \mu r^2 \omega = \text{constant} \qquad \textbf{(10-24)}$$

We assume that the force of interaction $\mathbf{F}$ between the particles is a central force that is derivable from a potential[†] $U(r)$. Then, the *total energy* of the two particles, another constant of the motion, is expressed as

$$E = \tfrac{1}{2}MV^2 + \tfrac{1}{2}\mu v^2 + U(r) = \text{constant} \qquad \textbf{(10-25)}$$

It is important to realize that the conservation of angular momentum and the conservation of energy are quite generally true for any two-particle system in which the particles interact via a *central force*. Notice also that when one of the particles has a mass much greater than that of the other, $m_2 \gg m_1$, the reduced mass becomes $\mu = m_1 m_2/(m_1 + m_2) \cong m_1$. Then, all of the expressions we have obtained for two-particle systems reduce to the single-particle counterparts. For example, Eq. 10-24 becomes $L = m_1 r^2 \omega = \text{constant}$.

Example 10-6

Two equal-mass cannon balls are tied together with a short chain (a favorite trick of medieval warriors) and are fired at an angle of 30° above the horizontal with a C.M. velocity of 120 m/s. Somewhere along the resulting trajectory, the chain breaks and one of the balls lands 9 s after the firing time, 1 km from the cannon and directly under its line of fire. If the chain has negligible mass compared with that of the cannon balls, where is the second ball at the instant the first ball lands?

---

[*] We use **L** to denote the angular momentum of a *system*; we reserve $\boldsymbol{\ell}$ for the angular momentum of an individual *particle*.

[†] We write $U(r)$ instead of $U(\mathbf{r})$ because **F** is a central force and so depends only on $|\mathbf{r}| = r$.

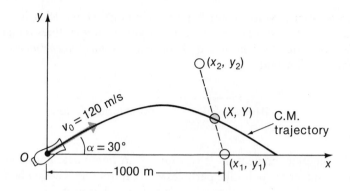

Solution:

The coordinates of the cannon ball that landed first are (x_1, y_1) with $x_1 = 1000$ m and $y_1 = 0$. The C.M. of the two balls follows the ballistic trajectory,

$$X = V_0 t \cos \alpha$$
$$Y = V_0 t \sin \alpha - \tfrac{1}{2} g t^2$$

At $t = 9$ s, $X = (120 \text{ m/s})(9 \text{ s}) \cos 30° = 935.3$ m

$$Y = (120 \text{ m/s})(9 \text{ s}) \sin 30° - \tfrac{1}{2}(9.80 \text{ m/s}^2)(9 \text{ s})^2 = 143.1 \text{ m}$$

With balls of equal mass, the C.M. coordinates (Eq. 9-9) reduce to

$$X = \tfrac{1}{2}(x_1 + x_2) \quad \text{and} \quad Y = \tfrac{1}{2}(y_1 + y_2)$$

Solving for the coordinates of ball 2 at $t = 9$ s, we find

$$x_2 = 2X - x_1 = 2(935.3 \text{ m}) - 1000 \text{ m} = 870.6 \text{ m}$$
$$y_2 = 2Y - y_1 = 2(143.1 \text{ m}) - 0 = 286.2 \text{ m}$$

Example 10-7

Two large pucks, with masses $m_1 = 0.5$ kg and $m_2 = 1.0$ kg, are free to slide on a horizontal, frictionless surface. They are connected by an ideal inertialess spring with a relaxed length $\ell_0 = 0.5$ m and a spring constant $\kappa = 10$ N/m. An initial observation at $t = 0$ places the pucks 2 m apart with puck 2 having a velocity with respect to the surface of 6 m/s directed away from puck 1, which is momentarily at rest on the surface.
(a) What is the velocity of the C.M.?
(b) What is the relative velocity v at $t = 0$?
(c) To what maximum length does the spring stretch?
(d) Will the pucks subsequently collide? If so, with what relative impact velocity?

Solution:

(a) No external forces act on the pucks, so the C.M. velocity is constant. We have

$$MV = m_1 v_1 + m_2 v_2$$

At $t = 0$, $\mathbf{v}_1 = 0$, so

$$V = \frac{m_2 v_2}{M} = \frac{(1.0 \text{ kg})(6 \text{ m/s})}{1.5 \text{ kg}} = 4 \text{ m/s, directed away from puck 1}$$

(b) The relative velocity is

$$\mathbf{v} = \mathbf{v}_2 - \mathbf{v}_1$$

At $t = 0$, $\mathbf{v}_1 = 0$, so $v = 6$ m/s.

(c) The reduced mass is

$$\mu = \frac{m_1 m_2}{M} = \frac{(0.5 \text{ kg})(1.0 \text{ kg})}{1.5 \text{ kg}} = \tfrac{1}{3} \text{ kg}$$

The constant value of the total energy can be evaluated at $t = 0$:

$$E = \tfrac{1}{2} M V^2 + \tfrac{1}{2} \mu v^2 + \tfrac{1}{2} \kappa (x_0 - \ell_0)^2$$
$$= \tfrac{1}{2}(1.5 \text{ kg})(4 \text{ m/s})^2 + \tfrac{1}{2}(\tfrac{1}{3} \text{ kg})(6 \text{ m/s})^2 + \tfrac{1}{2}(10 \text{ N/m})(2 \text{ m} - 0.5 \text{ m})^2$$
$$= 29.25 \text{ J}$$

When the spring has stretched to maximum length, the relative velocity will be zero. Then,

$$E = \tfrac{1}{2} M V^2 + \tfrac{1}{2} \kappa (x_m - \ell_0)^2 = 29.25 \text{ J}$$

from which

$$x_m = \sqrt{\frac{2}{\kappa}(E - \tfrac{1}{2} M V^2)} + \ell_0$$

$$= \sqrt{\frac{2(29.25 \text{ J}) - (1.5 \text{ kg})(4 \text{ m/s})^2}{10 \text{ N/m}}} + (0.5 \text{ m})$$

$$= 2.357 \text{ m}$$

(d) The particles will collide when $x = 0$:

$$E = \tfrac{1}{2} M V^2 + \tfrac{1}{2} \mu v_c^2 + \tfrac{1}{2} \kappa \ell_0^2 = 29.25 \text{ J}$$

Thus, the collision velocity is

$$v_c = \sqrt{\frac{2}{\mu}(E - \tfrac{1}{2} M V^2 - \tfrac{1}{2} \kappa \ell_0^2)}$$

$$= \sqrt{\frac{2(29.25 \text{ J}) - (1.5 \text{ kg})(4 \text{ m/s})^2 - (10 \text{ N/m})(0.5 \text{ m})^2}{1/3 \text{ kg}}}$$

$$= 9.798 \text{ m/s}$$

(•Would the values of x_m and v_c be affected if the problem were analyzed from the standpoint of an observer moving with the C.M.? What is the total energy of the system in the C.M. system?)

Example 10–8

Two isolated particles with masses m_1 and m_2 are connected by an ideal spring that has zero relaxed length and a spring constant κ. Describe how the particles move if it is observed that they rotate around their common C.M. in such a way that the separation remains constant.

Solution:

We are given that

$$\mathbf{F}_{21} = -\kappa \mathbf{r}$$

Because $|\mathbf{r}| = r_0 = $ constant, the force has a constant magnitude equal to the product of μ and

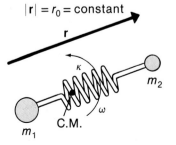

(a)

the centripetal acceleration:

$$\kappa r_0 = \mu r_0 \omega^2$$

or

$$\omega^2 = \kappa/\mu$$

Thus, the angular frequency ω is independent of the separation r_0. The value of r_0 depends on the angular momentum of the system. From Eq. 10–24,

$$r_0^2 = \frac{L}{\mu\omega} = \frac{L}{\sqrt{\mu\kappa}}$$

The paths followed by the particles are circles with radii equal to the magnitudes of the coordinate vectors $\mathbf{r}_1$ and $\mathbf{r}_2$. If we choose the origin O to coincide with the C.M., the expressions for r_1 and r_2 are given by Eqs. 10–15 and 10–16 with $\mathbf{R} = 0$:

$$r_1 = \frac{m_2}{M} r_0 \quad \text{and} \quad r_2 = \frac{m_1}{M} r_0$$

Suppose that we have $m_1 = 30$ kg, $m_2 = 10$ kg, $\kappa = 100$ N/m, and $L = 50$ kg $\cdot$ m^2/s. Then,

$$\mu = \frac{m_1 m_2}{m_1 + m_2} = \frac{(30 \text{ kg})(10 \text{ kg})}{40 \text{ kg}} = 7.50 \text{ kg}$$

$$\omega = \sqrt{\frac{\kappa}{\mu}} = \sqrt{\frac{100 \text{ N/m}}{7.50 \text{ kg}}} = 3.65 \text{ rad/s}$$

The separation r_0 is obtained from

$$r_0^2 = \frac{L}{\sqrt{\mu\kappa}} = \frac{50 \text{ kg} \cdot \text{m}^2/\text{s}}{\sqrt{(7.50 \text{ kg})(100 \text{ N/m})}} = 1.83 \text{ m}^2$$

or

$$r_0 = 1.35 \text{ m}$$

The individual radii are

$$r_1 = \frac{m_2}{M} r_0 = \tfrac{10}{40} \cdot (1.35 \text{ m}) = 0.34 \text{ m}$$

$$r_2 = \frac{m_1}{M} r_0 = \tfrac{30}{40} \cdot (1.35 \text{ m}) = 1.01 \text{ m}$$

The paths followed by the particles are shown in diagram (b).

The kinetic energy K is given by Eq. 10–20, in which $V = 0$ because the motion is described in the C.M. system. Because $v^2 = v_\rho^2 + v_\phi^2$ (Eq. 3–31) and $v_\rho = 0$ here, we have

$$K = \tfrac{1}{2}\mu v_\phi^2 = \tfrac{1}{2}\mu r_0^2 \left(\frac{L}{\mu r_0^2}\right)^2 = \frac{L^2}{2\mu r_0^2}$$

$$= \frac{(50 \text{ kg} \cdot \text{m}^2/\text{s})^2}{2(7.50 \text{ kg})(1.83 \text{ m}^2)} = 91.3 \text{ J}$$

The potential energy (referred to zero potential energy at zero relaxed length) is

$$U = \tfrac{1}{2}\kappa r_0^2 = \tfrac{1}{2}(100 \text{ N/m})(1.83 \text{ m}^2) = 91.3 \text{ J}$$

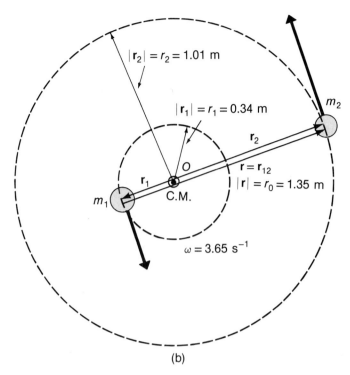

(b)

The fact that $K = U$ is a result of the *virial theorem* (see Problem 10–10).

10-3 MANY-PARTICLE SYSTEMS

The most convenient way to describe systems that consist of many particles is to identify the center-of-mass coordinate $\mathbf{R}$ and to specify the position of each particle with respect to the C.M. Thus, the particle labeled i has a position vector $\mathbf{r}_i$ in some inertial frame and a coordinate* $\mathbf{r}_{Ci}$ with respect to the C.M.; that is, $\mathbf{r}_i = \mathbf{R} + \mathbf{r}_{Ci}$, as shown in Fig. 10–7.

The Equations of Motion. In Section 9–4 we considered a system of particles subject to internal forces between each pair of particles and acted upon by forces

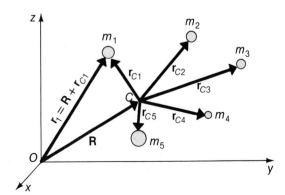

Fig. 10–7. Coordinates for describing a system of particles. The C.M. of the system is at C, which is located with respect to the origin O of an inertial reference frame by the vector $\mathbf{R}$. The individual particles are located with respect to C by the vectors $\mathbf{r}_{Ci}$. For m_1 the vector $\mathbf{r}_1 = \mathbf{R} + \mathbf{r}_{C1}$ is also shown.

*We use a subscript C to denote quantities that relate to the C.M. of the system.

applied from outside the system. We showed that when the total force on the system is calculated, the internal forces cancel in pairs. The resulting equation of motion is (see Eq. 9–13)

$$\mathbf{F}^e = \frac{d\mathbf{P}}{dt} = M\frac{d^2\mathbf{R}}{dt^2} \tag{10-26}$$

where $\mathbf{P}$ is the total linear momentum of the system (Eq. 9–12),

$$\mathbf{P} = \sum_i \mathbf{p}_i \tag{10-27}$$

Thus, the C.M. of the system moves in exactly the same way as would a particle with the same mass and subjected to the same applied force.

Kinetic Energy.   The total kinetic energy of the system referred to the inertial frame with origin O (Fig. 10–7) is

$$K = \tfrac{1}{2}\sum_i m_i\mathbf{v}_i^2 = \tfrac{1}{2}\sum_i m_i\left(\frac{d\mathbf{r}_i}{dt}\right)\cdot\left(\frac{d\mathbf{r}_i}{dt}\right)$$

$$= \tfrac{1}{2}\sum_i m_i\left(\frac{d\mathbf{R}}{dt} + \frac{d\mathbf{r}_{Ci}}{dt}\right)\cdot\left(\frac{d\mathbf{R}}{dt} + \frac{d\mathbf{r}_{Ci}}{dt}\right)$$

Using $\mathbf{V} = \dfrac{d\mathbf{R}}{dt}$ and $\mathbf{v}_{Ci} = \dfrac{d\mathbf{r}_{Ci}}{dt}$

we have $K = \tfrac{1}{2}\sum_i m_i(\mathbf{V}\cdot\mathbf{V} + 2\mathbf{V}\cdot\mathbf{v}_{Ci} + \mathbf{v}_{Ci}\cdot\mathbf{v}_{Ci})$

Now, $\tfrac{1}{2}\sum_i m_i\mathbf{V}\cdot\mathbf{V} = \tfrac{1}{2}\left(\sum_i m_i\right)V^2 = \tfrac{1}{2}MV^2$

and $\tfrac{1}{2}\sum_i m_i\mathbf{v}_{Ci}\cdot\mathbf{v}_{Ci} = \tfrac{1}{2}\sum_i m_i v_{Ci}^2$

Also, $\sum_i m_i\mathbf{V}\cdot\mathbf{v}_{Ci} = \mathbf{V}\cdot\left(\sum_i m_i\mathbf{v}_{Ci}\right) = \mathbf{V}\cdot\dfrac{d}{dt}\left(\sum_i m_i\mathbf{r}_{Ci}\right)$

The term $\sum m_i\mathbf{r}_{Ci}$ defines the position of the C.M. in the center-of-mass coordinate system and therefore vanishes. Consequently, the middle term in the expression for K vanishes, and we have

$$K = \tfrac{1}{2}MV^2 + \tfrac{1}{2}\sum_i m_i v_{Ci}^2 \tag{10-28}$$

That is, the total kinetic energy of the system is equal to the kinetic energy of the C.M. plus the sum of the kinetic energies of the individual particles relative to the C.M.

Angular Momentum and Torque. The angular momentum **L** of a system of particles is defined to be the sum of the individual angular momenta of the particles relative to the same origin; thus,

$$L = \sum_i \boldsymbol{\ell}_i = \sum_i \mathbf{r}_i \times \mathbf{p}_i$$

Using the particle coordinates, we can write

$$L = \sum_i m_i \left(\mathbf{r}_i \times \frac{d\mathbf{r}_i}{dt} \right)$$

$$= \sum_i m_i \left[(\mathbf{R} + \mathbf{r}_{Ci}) \times \frac{d}{dt} (\mathbf{R} + \mathbf{r}_{Ci}) \right]$$

$$= \sum_i m_i [(\mathbf{R} \times \mathbf{V}) + (\mathbf{R} \times \mathbf{v}_{Ci}) + (\mathbf{r}_{Ci} \times \mathbf{V}) + (\mathbf{r}_{Ci} \times \mathbf{v}_{Ci})]$$

The middle two terms are

$$\sum_i m_i (\mathbf{R} \times \mathbf{v}_{Ci}) = \mathbf{R} \times \left(\sum_i m_i \frac{d\mathbf{r}_{Ci}}{dt} \right) = \mathbf{R} \times \frac{d}{dt} \left(\sum_i m_i \mathbf{r}_{Ci} \right)$$

$$\sum_i m_i (\mathbf{r}_{Ci} \times \mathbf{V}) = \left(\sum_i m_i \mathbf{r}_{Ci} \right) \times \mathbf{V}$$

As we have just seen in the calculation of the kinetic energy, these two terms vanish. There remains

$$L = \sum_i m_i (\mathbf{R} \times \mathbf{V}) + \sum_i m_i (\mathbf{r}_{Ci} \times \mathbf{v}_{Ci})$$

$$= M(\mathbf{R} \times \mathbf{V}) + \sum_i m_i (\mathbf{r}_{Ci} \times \mathbf{v}_{Ci})$$

or

$$L = (\mathbf{R} \times \mathbf{P}) + \sum_i \mathbf{r}_{Ci} \times \mathbf{p}_{Ci} \tag{10-29}$$

Thus, the total angular momentum of a system of particles relative to the origin of some inertial frame is the vector sum of the angular momentum of the C.M. relative to the origin and the sum of the individual angular momenta of the particles relative to the C.M.

Notice that the total angular momentum **L** (Eq. 10–29) and the total kinetic energy K (Eq. 10–28) of a system each consist of *two* parts, whereas the total linear momentum **P** (Eq. 10–27) is expressed by a *single* term. The reason is that when a system is viewed from a reference frame at rest with respect to the C.M., the linear momentum is necessarily zero (as are the first terms in the expressions for **L** and K). However, the system can be rotating about some axis and would then have both kinetic energy and angular momentum.

If a torque acts on the particle i to change its angular momentum, $\boldsymbol{\tau}_i = d\boldsymbol{\ell}_i/dt$, we have

$$\frac{d\mathbf{L}}{dt} = \sum_i \frac{d\boldsymbol{\ell}_i}{dt} = \sum_i \boldsymbol{\tau}_i = \boldsymbol{\tau}$$

where we use $\boldsymbol{\tau}$ without a subscript to represent the *total* torque acting on the system relative to the same origin used to define **L**, $\boldsymbol{\ell}_i$, and $\boldsymbol{\tau}_i$. We have

$$\boldsymbol{\tau} = \sum_i \boldsymbol{\tau}_i = \sum_i \mathbf{r}_i \times \mathbf{F}_i \tag{10–30}$$

The total force $\mathbf{F}_i$ that acts on m_i consists of the external force $\mathbf{F}_i^e$ plus the sum of all the internal forces due to the other particles in the system. These internal forces are assumed to be *central forces*. When the torques due to these forces are summed over all particles, as specified in Eq. 10–30, they cancel in pairs.* The only forces that affect the total torque on the system are the external forces:

$$\boldsymbol{\tau} = \sum_i \mathbf{r}_i \times \mathbf{F}_i^e = \frac{d\mathbf{L}}{dt} \tag{10–31}$$

Conservation of Angular Momentum.   If we are dealing with an isolated system of particles so that all external forces are zero, and if the mutual pair forces are all central, we must have $\boldsymbol{\tau} = 0$; hence, **L** is a constant of the motion. This conclusion is valid in *any* inertial frame. (The frame chosen for the calculation of $\boldsymbol{\tau}$—Fig. 10–7—is an arbitrary inertial frame.) Thus, in solving a particular problem, any convenient frame may be chosen, and **L** will still be a constant of the motion. (The *value* of **L** will depend on the specific chosen frame.) For this case (namely, $\boldsymbol{\tau} = 0$), not only is the total angular momentum a constant of the motion, but the two parts of **L**, which are exhibited in Eq. 10–29, are *separately* constants of the motion (see Problem 10–23).

A further possible extension of the principle of angular momentum conservation must not be overlooked. External forces may act upon a system, but in such a way that they produce zero torque. (•Can you see how this might come about?) Then, the angular momentum remains constant, even though forces are acting. Furthermore, because Eq. 10–30 is a vector equation, it is also possible to have only one or perhaps two components of $\boldsymbol{\tau}$ vanish. In such a case, only the corresponding components of **L** are constants of the motion.

---

* •Can you show this? Write the expression for the torques due to the internal forces. Use Newton's third law to group the terms into pairs of the form $(\mathbf{r}_i - \mathbf{r}_j) \times \mathbf{F}_{ij}$. Argue that all such terms must vanish if the forces are central.

PROBLEMS

Section 10-1

10-1 A particle with a mass of 4 kg moves along the x-axis with a velocity $v = 15t$ m/s, where $t = 0$ is the instant that the particle is at the origin. (a) At $t = 2$ s, what is the angular momentum of the particle about a point P located on the $+y$-axis, 6 m from the origin? (b) What torque about P acts on the particle?

10-2 (a) For pure circular motion with the origin of the coordinate frame located at the center of the orbit, as in Example 10-1, show, using vector algebra, that

$$\boldsymbol{\ell} = m\mathbf{r} \times (\boldsymbol{\omega} \times \mathbf{r}) = mr^2\boldsymbol{\omega}$$

(b) For the same motion, but with the origin at an arbitrary point on the axis of rotation, as in Example 10-2, show that

$$\boldsymbol{\ell} = m\mathbf{r} \times (\boldsymbol{\omega} \times \mathbf{r}) = mr(r\boldsymbol{\omega} + \omega\mathbf{r}\cos\alpha)$$

Does this agree with the conclusions drawn in Example 10-2? [*Hint:* Use the vector identity for $\mathbf{A} \times (\mathbf{B} \times \mathbf{C})$ given in Eq. 3-28.]

10-3 What is the orbital angular momentum of the Earth relative to the center of the Sun? [Refer to the table of astronomical data inside the front cover.]

10-4 Two particles move in opposite directions along parallel straight-line paths that are separated by a distance d. If the two linear momenta are equal in magnitude, show that the angular momentum of this two-particle system is the same for any choice of the origin.

10-5 Consider a block with a mass m sliding down a frictionless plane set at angle α with the horizontal, as shown in the diagram. Find the acceleration of the block using Eq. 10-9. Is your result familiar? [*Hint:* Apply Eq. 10-9 twice, once with O and then with O' as the origin.]

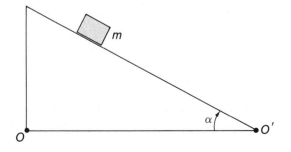

10-6 [All motion in this problem takes place in a vertical plane.] A ball with mass $m_1 = 4$ kg and an initial velocity $v_1 = 3$ m/s collides with another ball ($m_2 = 2$ kg), initially at rest, which is attached to a rigid massless rod of length 0.4 m that pivots freely about the opposite end. *Immediately after* the impact, the angular speed of m_2 about the pivot is ω, and m_1 moves with a speed $v_1' = 1.8$ m/s at an angle $\alpha = 35°$, as shown in the diagram. (a) Determine ω. (b) Find the maximum angle β that the rod will make with the vertical. (c) Is the collision elastic? If not, how much kinetic energy is transformed into heat? [*Hint:* After the collision, the linear momentum of m_2 is not conserved, whereas the angular momentum is conserved. (•Can you see why?)]

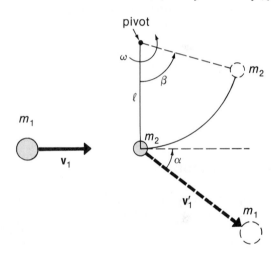

Section 10-2

10-7 Two 1-kg particles have trajectories $\mathbf{r}_1 = (3t^2 - 5t)\hat{\mathbf{i}} + 2t^3\hat{\mathbf{j}}$ and $\mathbf{r}_2 = 2t^3\hat{\mathbf{i}} + 2t^2\hat{\mathbf{j}}$, where $\mathbf{r}_1$ and $\mathbf{r}_2$ are in meters when t is given in seconds. (a) What is the total force (external plus mutual) acting on each of the two particles? (b) Determine the acceleration of the C.M., both by using the basic definition, Eq. 10-13, and by summing the forces found in (a) above, i.e., using Eq. 10-26. Do these results agree? (c) Calculate the relative velocity and acceleration of the two particles. (d) Is it possible, with the information given, to determine separately the mutual force $\mathbf{F}_{21}$?

10-8 A cannon fires an explosive shell for maximum range with a muzzle velocity of 500 m/s. (See Section 4-4.) At the top of its trajectory the shell detonates prematurely and explodes into two fragments with equal masses. One fragment is observed to drop *from rest* to the ground directly below the explosion. (a) Make a sketch of the paths followed by the fragments and by the C.M. (b) Where does the second fragment land? (c) How long after the explosion does each fragment strike the ground (assumed to be level)?

10–9 A container is dropped from a balloon that drifts
• toward the east with a uniform velocity of 50 m/s, holding a constant altitude of 2.5 km above the level ground. At some point in its descent, explosive bolts separate the container into two pieces, one with twice the mass of the other. Twenty seconds after the container was dropped, the lighter piece lands at a spot 200 m due south of a point directly under the balloon at the time that the piece impacts. Determine the position of the heavier piece when the lighter piece strikes the ground.

10–10 A two-particle system has a reduced mass μ. Show that in an inertial frame in which the C.M. is at rest, a potential of the form $U(r) = \kappa r^{n+1}$ has a solution $r = r_0 = \text{constant}$, such that the kinetic energy K and the potential energy $U(r_0)$ are related according to $K = \frac{1}{2}(n + 1)U(r_0)$. [This theorem is more generally true for the average values of K and $U(r)$ in all bounded cases, not just for $r = r_0 = \text{constant}$. The result is then known as the *virial theorem*.]

Additional Problems

10–11 A stone with a mass m is tied to the end of a string and twirled in a vertical plane. Show, using Eq. 10–9, that the angular frequency ω is related to the angle ϕ by the equation $\omega^2 = \omega_0^2 + (2g/R)(1 - \cos\phi)$, where ω_0 is the angular frequency at the top

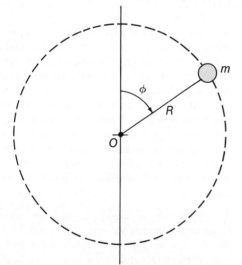

of the circular path ($\phi = 0$). [*Hint:* Use O as the origin and note that

$$\frac{d\omega}{dt} = \frac{d\omega}{d\phi}\frac{d\phi}{dt} = \omega\frac{d\omega}{d\phi}.]$$

Note that this result agrees with the statement given in Problem 6–33. (•Can you also derive the desired result using conservation of energy?)

10–12 Suppose, in Example 10–5, that the hand-held end
•• of the string is pulled down with a constant velocity u. (a) Is the angular momentum conserved? Obtain an expression for ω as a function of r. (b) Show that the angular acceleration $\alpha = d\omega/dt$ is given by

$$\alpha = -\frac{2r_0^2\omega_0 u}{r^3}$$

(c) Show that the spiral motion $r(\phi)$ is given by

$$r(\phi) = \frac{r_0}{1 - \beta\phi}$$

with $\beta = u/r_0\omega_0$. [*Hint:* Write $\omega = d\phi/dt = (d\phi/dr)/(dr/dt) = u(d\phi/dr)$ and integrate.] (d) Show that the required force is

$$F = mr_0\omega_0^2(1 - \beta\phi)^3$$

10–13 A cannon fires an explosive shell for some specific
• range (not necessarily the maximum range possible). The shell prematurely explodes at the top of its trajectory into two fragments with equal masses. One fragment lands directly under the burst and the other fragment takes twice as long to hit the ground after the burst. Where does this other fragment land? (Give your answer in terms of the range the unexploded shell would have had.)

10–14 Two particles that have equal mass are connected by a rod with negligible mass and rotate about an axis through the center of the rod and perpendicular to it. The length of the rod is increased by 13 parts in 10^6 due to a temperature rise. By what fraction does the angular velocity change?

10–15 Two particles with masses $m_1 = 3\text{ kg}$ and $m_2 = 2\text{ kg}$, which slide as a unit with a common velocity of 2 m/s on a level frictionless surface, have between them a compressed massless spring with $\kappa = 50\text{ N/m}$. The spring, originally compressed by 25 cm, is suddenly released, sending the two masses, which are not connected to the spring, flying apart from each other. The orientation of the spring with respect to the initial velocity is shown in the diagram.

(a) What is the relative velocity of separation after the particles lose contact? (b) What is the velocity of the C.M. after separation? (c) What are the speeds

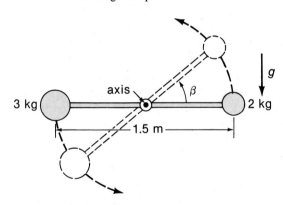

of m_1 and m_2 with respect to the frictionless surface after separation?

10-16 Two particles with masses of 3 kg and 2 kg are
•• connected by a massless stiff rod that has a length of 1.5 m. This device is free to rotate about a frictionless bearing located at the center of the rod with an axis perpendicular to the rod. The axis is horizontal and held fixed. Originally the device is horizontal and is released from rest, as shown in the diagram. Upon release the device begins to rotate about the axis with an angular speed ω.

(a) Determine the angular acceleration after the device has rotated through an angle β. (b) What is the corresponding angular momentum of the device? (c) What happens when an angle of $\beta = \pi/2$ is reached? (d) Prove that the total energy, kinetic energy plus gravitational potential energy, is conserved.

10-17 Two small flat discs, one with a mass of 3 kg and the
• other with a mass of 2 kg, are connected by a massless stiff rod 1.5 m long and are at rest on a frictionless horizontal surface. A third disc with a mass of 4 kg slides at a constant velocity of 3 m/s in a direction perpendicular to the rod, strikes the 2-kg mass, and sticks to it. (a) Determine the position and the

velocity of the C.M. of the entire system following the impact. (b) What is the angular frequency of rotation of the rod after impact? (c) Is energy conserved in this impact; if not, how much kinetic energy is lost?

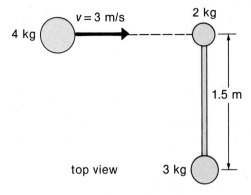

10-18 A hockey puck with a mass m slides on a friction-
•• less horizontal surface while tied by a massless string to a vertical post that has a radius R. Initially, the puck has a velocity v_0 perpendicular to the taut string of length s_0. As the motion proceeds, the string is wrapped around the post, shortening the distance from the puck to the center of the post, as shown in the diagram.

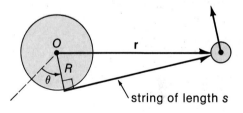

(a) Given the initial conditions, is it possible for the puck to move in any direction other than perpendicular to the string? (b) If the above suggestion is true, what work is done by the string tension on the puck? What do you then conclude about the kinetic energy and the velocity of the puck? (c) If we select the center of the post O as the origin, is the force on the puck a central force? What do you then conclude about the angular momentum of the puck relative to O? (d) Show that the angular frequency of rotation ω for the puck is $\omega = v_0 s/(s^2 + R^2)$, where s is the length of unwrapped string between the post and the puck, as in the diagram. (e) Show that $\ell = mv_0(s_0 - R\theta)$, where θ is the angle through which the string has wrapped around the post since the motion began. (f) Show that the elapsed time t to attain any length of string s is

$$t = \frac{R}{v_0}\left[\frac{s_0^2 - s^2}{2R^2} + \ln\frac{s_0}{s}\right]$$

10-19 We are given the trajectory of two particles, one with a mass $m_1 = 2$ kg with $\mathbf{r}_1(t) = (3t^2 + 7t)\hat{\mathbf{i}} - 2t\hat{\mathbf{j}}$, and the other with a mass $m_2 = 3$ kg with $\mathbf{r}_2(t) = 4t\hat{\mathbf{i}} - 3\hat{\mathbf{k}}$, with $\mathbf{r}_1$ and $\mathbf{r}_2$ expressed in meters when t is given in seconds. (a) Find the total angular momentum $\mathbf{L}$. (b) If you conclude that external forces are present, find their net torque relative to the origin. (c) Find the linear momentum of the system. (d) What is the net value of the external forces?

10-20 A small flat disc with a mass of 2 kg slides on a frictionless horizontal surface. The disc is held in a circular orbit by a 1.5-m string tied at one end to a pivot, as shown in the diagram. Initially, the disc has an orbit speed of 5 m/s. A 1-kg mass of putty is dropped onto the disc from directly above. If the putty sticks to the disc, what is the new period of rotation?

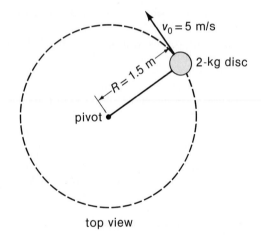

top view

10-21 Two particles with equal masses m are attached to a folding crossarm on a shaft that rotates in frictionless bearings, as shown in the diagram. Originally, both ends of the crossarm are fully extended to radii $r_1 = 1$ m, and the rotational frequency is $\omega_1 = 10$ rev/s. When two springs are triggered in

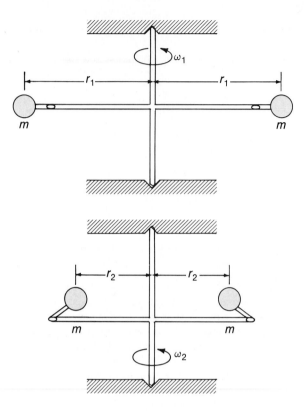

the crossarm, it folds so that both ends have radii $r_2 = 0.6$ m. What is the new angular velocity ω_2? Assume that the shaft and crossarm are massless.

10-22 Two particles in deep space, $m_1 = 2$ kg and $m_2 = 5$ kg, are connected by a massless stiff rod 50 cm long. This system is observed to rotate about the C.M. and perpendicular to the rod at an angular frequency of $\omega = 3$ rev/s. (a) What is the total angular momentum $\mathbf{L}$? (b) How large a torque must be applied to bring the system to rest in 8 s?

10-23 For a system that has constant angular momentum,
• show that the two terms in Eq. 10-29 are *separately conserved*. [*Hint:* Set $d\mathbf{L}/dt$ equal to zero.]

CALCULUS V

In Calculus III, concerning the evaluation of definite integrals, we introduced the idea that the integral of a function, $y = f(x)$, can be interpreted as the area under the curve defined by the function. We now extend this idea to a more general form of integration involving two and three variables. In virtually all cases we are concerned with objects that possess simple geometry so that the formal integrals can be evaluated by elementary means. Although the general formalism is introduced, we show how it is possible to obtain the desired results by evaluating only single integrals.

V-1 INTEGRALS IN TWO DIMENSIONS

Consider a function $f(x, y)$ of the two independent variables, x and y, that is continuous in a closed region S, including the boundaries. We use intersecting strips to partition the region S into small rectangles with widths Δx_i and Δy_j, as shown in Fig. V-1.

The double integral of $f(x, y)$ is formally defined to be the limit of the double series,

$$\iint_S f(x, y)dx\, dy = \lim_{\substack{m \to \infty \\ n \to \infty}} \sum_{j=1}^{n} \sum_{i=1}^{m} f(x_i, y_j)\, \Delta x_i\, \Delta y_j \quad \textbf{(V–1)}$$

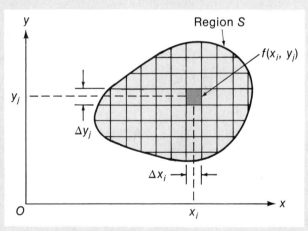

Fig. V–1. The region S is divided into small rectangles, $\Delta x_i\, \Delta y_j$.

Double Integrals as Areas. The simplest case of the double sum in Eq. V-1 is that for which $f(x, y) = 1$. Then, the addition of all the elements $\Delta x_i\, \Delta y_j$ within the closed region S yields the area of S. We therefore write the formal expression for the area of S as

$$A_S = \iint_S dA = \iint_S dx\, dy \quad \textbf{(V–2)}$$

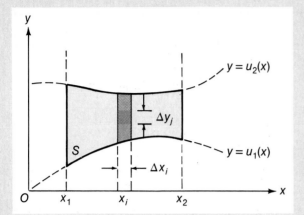

Fig. V–2. Geometry for the calculation of the area of the region S.

Figure V–2 shows two functions, $u_1(x)$ and $u_2(x)$. Consider the calculation of the area enclosed by these functions and the ordinates, x_1 and x_2, as indicated in the figure. To carry out the sums specified in Eq. V–1, we may sum first on Δx_i, then on Δy_j, or we may sum first on Δy_j, then on Δx_i. Consider summing first on Δy_j for some fixed x_i. This sum is required to extend only over the region S, so it generates the narrow, essentially rectangular strip shown in Fig. V–2 which has a width Δx_i and a height $u_2(x_i) - u_1(x_i)$. Thus, the area of the strip is $\Delta A_i = [u_2(x_i) - u_1(x_i)]\Delta x_i$. The area of the region S is given by

$$A_S = \underset{m \to \infty}{\text{Lim}} \sum_{i=1}^{m} \Delta A_i = \underset{m \to \infty}{\text{Lim}} \sum_{i=1}^{m} [u_2(x_i) - u_1(x_i)]\Delta x_i$$

In the limit, the area A_S is equal to the difference between two definite integrals; thus,

$$A_S = \iint_S dx\, dy = \int_{x_1}^{x_2} [u_2(x) - u_1(x)]\, dx$$

$$= \int_{x_1}^{x_2} u_2(x)\, dx - \int_{x_1}^{x_2} u_1(x)\, dx \qquad \textbf{(V–3)}$$

The first of these definite integrals is the area under the curve $u_2(x)$ between x_1 and x_2, whereas the second is the area under the curve $u_1(x)$ between the same limits. Thus, the difference is just the desired area of the region S.

Rectangular Coordinates. Figure V–3 shows two ways for the evaluation of an area in rectangular (Cartesian) coordinates. The region S is bounded by the coordinate axes and by the function $y = f(x)$ (Fig. V–3a) or by its inverse, the function $x = g(y)$ (Fig. V–3b).

In Fig. V–3a the sum is performed first on Δy so that $\Delta A = f(x)\,\Delta x$. Then, in the limit, the total area is given by

$$A_S = \int_0^a f(x)\, dx \qquad \textbf{(V–4a)}$$

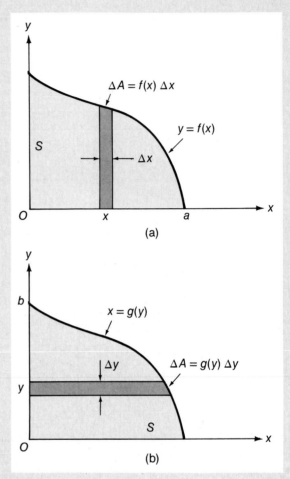

Fig. V–3. The area of the region S can be obtained by (a) summing first on Δy, then on Δx, or (b) summing first on Δx, then on Δy.

In Fig. V–3b the sum is performed first on Δx so that $\Delta A = g(y)\,\Delta y$. Then, in the limit, the total area is given by

$$A_S = \int_0^b g(y)\, dy \qquad \textbf{(V–4b)}$$

In practice, the choice between the two methods is made on the basis of simplicity.

Example V–1

Determine the area of a circle with radius R given by the equation $x^2 + y^2 = R^2$.

Solution:

The total area of S can be obtained by taking 4 times the area of the first quadrant, as indicated in the diagram. By performing the sum first on Δy, we generate the shaded strip, which has area $\Delta A = y\,\Delta x = \sqrt{R^2 - x^2}\,\Delta x$. Thus,

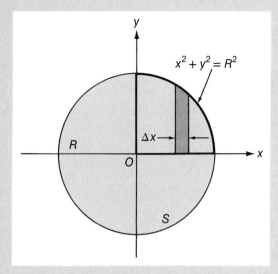

the total area is

$$A_s = 4 \int_0^R \sqrt{R^2 - x^2}\, dx$$

$$= 4[\tfrac{1}{2}x\sqrt{R^2 - x^2} + \tfrac{1}{2}R^2 \sin^{-1}(x/R)]_0^R$$

$$= \pi R^2$$

(•How would you obtain the area by summing first on Δx?)

Polar Coordinates. In some cases it is easier to evaluate the double sum over a planar region by using polar coordinates instead of rectangular coordinates. The element of area in polar coordinates, as shown in Fig. V–4, is $dA = \rho\, d\phi\, d\rho$. Thus, the area of a region S is expressed as

$$A_s = \iint_S \rho\, d\phi\, d\rho$$

The use of polar coordinates is indicated when the natural boundaries of the region are straight radial lines and a single-valued function, $\rho = \rho(\phi)$, as in Fig. V–5a, or circular arcs and a single-valued function, $\phi = \phi(\rho)$, as in Fig. V–5b.

Fig. V–4. The element of area in polar coordinates is $dA = \rho\, d\phi\, d\rho$.

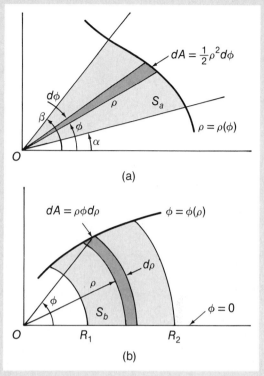

Fig. V–5. (a) The area of region S_a is obtained by integrating the triangular differential area from $\phi = \alpha$ to $\phi = \beta$. (b) The area of region S_b is obtained by integrating the annular differential area from $\rho = R_1$ to $\rho = R_2$.

In Fig. V–5a, the differential (triangular) area is $dA = \tfrac{1}{2}\rho \cdot (\rho\, d\phi) = \tfrac{1}{2}\rho^2\, d\phi$. Thus, the total area of the region S_a is

$$A_{S_a} = \tfrac{1}{2} \int_\alpha^\beta \rho^2(\phi)\, d\phi \qquad \textbf{(V–5a)}$$

In Fig. V–5b, the differential (annular) area is $dA = \rho\phi \cdot d\rho$. Thus, the total area of the region S_b is

$$A_{S_b} = \int_{R_1}^{R_2} \rho\phi(\rho)\, d\rho \qquad \textbf{(V–5b)}$$

Example V–2

Determine the area of a circle by integration using polar coordinates.

Solution:

Diagram (a) shows the differential area, $dA = \tfrac{1}{2}\rho^2\, d\phi$. The total area is (with $\rho = R$)

$$A_s = \tfrac{1}{2} \int_0^{2\pi} \rho^2\, d\phi = \tfrac{1}{2}R^2 \int_0^{2\pi} d\phi = \pi R^2$$

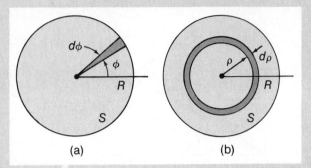

(a)   (b)

Diagram (b) shows the differential area, $dA = \rho\phi(\rho)\,d\rho$. The total area is (with $\phi(\rho) = 2\pi$)

$$A_S = \int_0^R \rho\phi(\rho)\,d\rho = 2\pi \int_0^R \rho\,d\rho = \pi R^2$$

Example V–3

Determine the area of the right-hand lobe of the *lemniscate*, $\rho^2 = a^2 \cos 2\phi$, shown in the diagram.

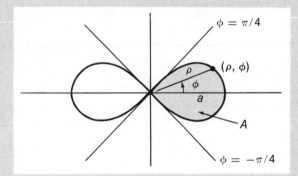

Solution:

The area of the lobe (shaded area) is

$$A = \tfrac{1}{2}\int_{-\pi/4}^{\pi/4} \rho^2\,d\phi = \tfrac{1}{2}\int_{-\pi/4}^{\pi/4} a^2 \cos 2\phi\,d\phi$$

$$= \tfrac{1}{4}a^2 \sin 2\phi\,|_{-\pi/4}^{\pi/4}$$

$$= \tfrac{1}{4}a^2[1 - (-1)] = \tfrac{1}{2}a^2$$

Moments. When the function $f(x, y)$ that appears in Eq. V–1 is equal to some integer power of x or y, the integral is called a *moment* of the particular region S of integration. For example,

$$M_x \equiv \iint_S y\,dx\,dy \qquad \textbf{(V–6a)}$$

is the first moment of the planar region S *with respect to the*

x-axis. The corresponding first moment with respect to the *y*-axis is

$$M_y \equiv \iint_S x\,dx\,dy \qquad \textbf{(V–6b)}$$

In a similar way, we define the *second moments* of S with respect to the x- and y-axes to be

$$I_x \equiv \iint_S y^2\,dx\,dy \quad \text{and} \quad I_y \equiv \iint_S x^2\,dx\,dy \qquad \textbf{(V–7)}$$

In Chapter 11 we use the idea of moments to discuss the properties of rigid bodies.

Notice the important point that when the function $f(x, y)$ is a function only of x (as in the expressions for M_y and I_y), the summation or integration is performed first on x for a fixed value of y. Similarly, when $f(x, y)$ involves only y, the summation is carried out first over y for a fixed value of x. (This is the case in Example V–4 which follows.) Be certain that you understand geometrically why this is so.

Example V–4

Determine the first moment M_x for the semicircular thin plate shown in the diagram.

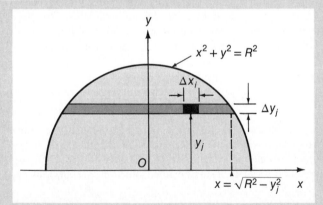

Solution:

Writing Eq. V–6a in the form of Eq. V–1, we have

$$M_x = \lim_{\substack{m\to\infty \\ n\to\infty}} \sum_{j=1}^{n} \sum_{i=1}^{m} y_j\,\Delta x_i\,\Delta y_j$$

We sum first on Δx_i for some fixed value of y_j. The result is the shaded horizontal strip with length $2\sqrt{R^2 - y_j^2}$ and width Δy_j, multiplied by y_j. Thus,

$$M_x = \lim_{n \to \infty} \sum_{j=1}^{n} 2y_j \sqrt{R^2 - y_j^2} \, \Delta y_j$$

This is simply the integral,

$$M_x = \int_0^R 2y \sqrt{R^2 - y^2} \, dy$$

$$= -\tfrac{2}{3}(R^2 - y^2)^{3/2} \big|_0^R = \tfrac{2}{3}R^3$$

V-2 INTEGRALS IN THREE DIMENSIONS

Equation V-1 can be extended in a simple way to three dimensions by writing

$$\iiint f(x, y, z) \, dx \, dy \, dz$$

$$= \lim_{\substack{m \to \infty \\ n \to \infty \\ p \to \infty}} \sum_{k=1}^{p} \sum_{j=1}^{n} \sum_{i=1}^{m} f(x_i, y_j, z_k) \, \Delta x_i \, \Delta y_j \, \Delta z_k \quad \textbf{(V–8)}$$

As in the case of two dimensions, the sums may be performed in any order.

Triple Integrals as Volumes. Again, consider the simple case in which $f(x, y, z) = 1$. Then, the addition of all the elements $\Delta x_i \, \Delta y_j \, \Delta z_k$ within the region S bounded by a simple closed surface, as in Fig. V-6, yields the volume of S. We therefore write the formal expression for the volume of S as

$$V_S = \iiint_S dV = \iiint_S dx \, dy \, dz \quad \textbf{(V–9)}$$

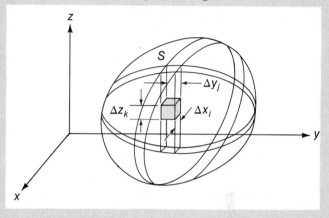

Fig. V-6. The sum (integral) of all the volume elements similar to the one illustrated here yields the volume of the region S.

Cylindrical and Spherical Coordinates. It sometimes happens that the volume of interest has some special symmetry. Then, it may be advantageous to express the

Fig. V–7. The important coordinate systems.

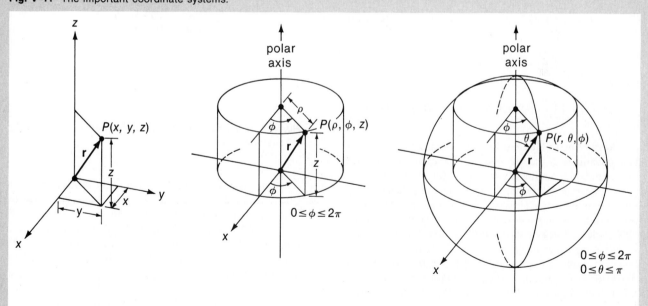

(a) Rectangular:

$x = \rho \cos \phi = r \sin \theta \cos \phi$
$y = \rho \sin \phi = r \sin \theta \sin \phi$
$z = \quad z \quad = r \cos \theta$

(b) Cylindrical:

$\rho = \sqrt{x^2 + y^2} = r \sin \theta$
$\phi = \tan^{-1} y/x = \phi$
$z = \quad z \quad = r \cos \theta$

(c) Spherical:

$r = \sqrt{x^2 + y^2 + z^2} = \rho \csc \phi$
$\theta = \tan^{-1} \sqrt{x^2 + y^2}/z = \tan^{-1} \rho/z$
$\phi = \tan^{-1} y/x = \phi$

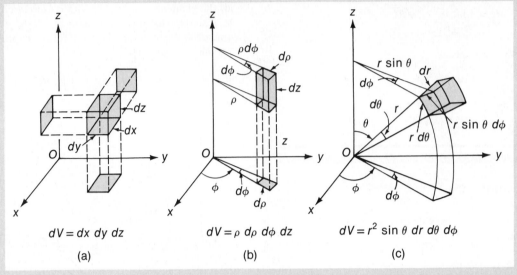

Fig. V-8. Differential volume elements for (a) rectangular coordinates, (b) cylindrical coordinates, and (c) spherical coordinates.

integral in Eq. V-9 in another coordinate system. The most commonly used coordinate systems, in addition to rectangular coordinates, are *cylindrical* coordinates and *spherical* coordinates. These three coordinate systems are illustrated in Fig. V-7 on the preceding page. The corresponding differential volume elements dV are shown in Fig. V-8. From these we see that the volume integrals are

$$V = \iiint_S dx\, dy\, dz \quad \text{(rectangular)} \qquad \textbf{(V-10a)}$$

$$V = \iiint_S \rho\, d\rho\, d\phi\, dz \quad \text{(cylindrical)} \qquad \textbf{(V-10b)}$$

$$V = \iiint_S r^2 \sin\theta\, dr\, d\theta\, d\phi \quad \text{(spherical)} \qquad \textbf{(V-10c)}$$

Reduction to Single Integrals. In many cases we can take advantage of the symmetry of the situation to reduce the formidable-looking expressions in Eqs. V-10 to simple single integrals. For example, consider the calculation of the volume of a right circular cone. As shown in Fig. V-9, the cone has a height h and a radius R (yielding a cone half-angle α). Because the cone is symmetric about its natural axis, we select a cylindrical coordinate frame to calculate the volume. We first perform the sums on ρ and ϕ for a fixed value of z, obtaining the area $\pi\rho^2$ of the circular segment shown in Fig. V-9. Now, $\rho = z \tan\alpha = zR/h$; therefore, we have

$$V = \iiint_S \rho\, d\rho\, d\phi\, dz = \int_0^h \pi\rho^2\, dz$$

$$= \int_0^h \pi \frac{R^2}{h^2} z^2\, dz = \tfrac{1}{3}\pi R^2 h$$

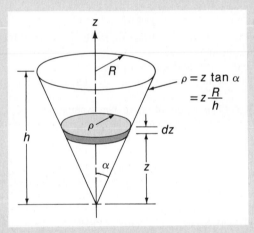

Fig. V-9. Cylindrical coordinates are used to calculate the volume of a cone.

Next, consider the calculation of the surface area of a sphere with radius R. An element of this area is defined by two cones, one with half-angle θ and the other with half-angle $\theta + d\theta$, as shown in Fig. V-10. The ring-shaped area thus defined has circumference $2\pi R \sin\theta$ and width $R\, d\theta$; hence, the area is $dA = 2\pi R^2 \sin\theta\, d\theta$. Thus, the surface area of the sphere is

$$A = \iint_S dA = 2\pi R^2 \int_0^\pi \sin\theta\, d\theta = 4\pi R^2$$

By using the expression we have just obtained for the surface area of a sphere, the volume of the sphere can be obtained by one further integration. Fig. V-11 shows a thin

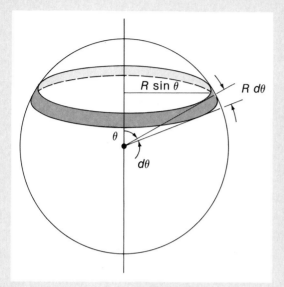

Fig. V-10. Geometry for calculating the surface area of a sphere.

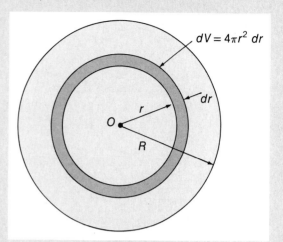

Fig. V-11. The volume of a sphere can be obtained by summing volume elements in the form of thin shells.

spherical shell within a sphere with radius R. The shell has a radius r and a thickness dr. The volume dV of the shell is equal to its area $4\pi r^2$ multiplied by its thickness dr; that is,

$dV = 4\pi r^2\, dr$. The volume of the entire sphere is found by summing these volume elements from $r = 0$ to $r = R$. Thus,

$$V = \int_0^R 4\pi r^2\, dr = \tfrac{4}{3}\pi R^3$$

PROBLEMS

Section V-1

V-1 Equation V-1 and Fig. V-1 (with $f(x_i, y_j) = 1$) suggest that one way to obtain the area of a region is to divide the entire region into rectangles with known area $\Delta x_i\, \Delta y_j$ and then to *count* the number of such rectangles. Use this procedure to determine the area of a circle. First, draw a circle with a radius of 4 cm. Section one quadrant into squares 4 mm × 4 mm. Count the number of *whole* squares that lie within the quadrant and multiply by $4 \cdot (16\ \text{mm}^2)$ to obtain an approximate area of the circle. Next, draw an identical circle and section one quadrant into squares 2 mm × 2 mm. Again, count the number of *whole* squares in the quadrant and determine an approximate area of the circle. Extrapolate this procedure to squares with size approaching zero to obtain your best estimate for the area of the circle. Compare your result with the true area.

V-2 Consider again the circle in Problem V-1 that has one quadrant sectioned into 4 mm × 4 mm squares. Count the number of whole squares and estimate the additional number represented by the partial squares. Make your estimate of the area of the circle and compare with the true value.

V-3 Determine the area between the functions $y = u_1(x)$ and $y = u_2(x)$ and between the ordinates $x = a$ and $x = b$ for the following cases:
a) $u_1(x) = 2x$; $u_2(x) = 3 + x^2$; $a = 1$; $b = 3$
b) $u_1(x) = 2x$; $u_2(x) = 5 \sin x$; $a = 0$; $b = \pi/2$

V-4 Determine the finite area between the intersecting curves of the functions $y^2 = 4x$ and $2x - y - 4 = 0$. (Sum first on Δx_i.)

V-5 Determine the area of the region bounded by the y-axis and $x = \pi/4$ that lies below $y = \cos x$ and above $y = \sin x$.

V-6 Show that the parabola $y^2 = 4x$ divides the circle $x^2 + y^2 = 12$ into two areas, the larger of which is 1.824 times the smaller. [*Hint:* Find the smaller area first and then use the known area of the circle.]

V-7 Find the area enclosed by the cardioid, $\rho = a(1 - \cos \phi)$, shown in the diagram on the next page.

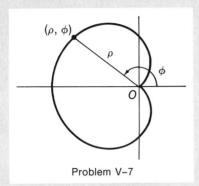

Problem V–7

ments, M_x and M_y, can be expressed as

$$M_x = \iint_S \rho^2 \sin \phi \, d\rho \, d\phi \quad \text{and} \quad M_y = \iint_S \rho^2 \cos \phi \, d\rho \, d\phi$$

V–11 Show that the second moment about a diameter of a circle with radius R is $\dfrac{\pi R^4}{4}$.

V–8 Determine the area outside the circle $\rho = R$ and inside the circle $\rho = 2R \cos \phi$.

V–9 Calculate the first and second moments of a rectangular area, $a \times b$, about one edge.

V–10 Use a polar coordinate system in which the x-axis corresponds to $\phi = 0$ and show that the first mo-

Section V-2

V–12 Calculate the surface area and the volume of a right circular cylinder with height h and radius R.

V–13 A right circular cone has a vertex half-angle β.
• The vertex coincides with the center of a sphere whose radius is R. What is the area of the portion of the spherical surface that lies within the cone?

ROTATION OF RIGID BODIES

In discussing the dynamics of translational motion we recognize that a solid object—if it is not too large compared with the scale of its trajectory—will behave as a pointlike particle. In our previous discussions we have been careful to exclude from consideration any rotational motion of the objects about internal axes. In this way we reduced all situations involving solid objects to the case of particle motion.

Objects in the real world, however, are not pointlike particles. A real object has a mass distribution associated with its size and shape, and the motion of a real object usually involves both translation and rotation. In this chapter, we study the rotational motion of an object about some point or axis located within or associated with the object. We limit our discussion here to the simple case of a rigid body rotating about an axis that is fixed in some inertial reference frame. In Chapter 12, we consider the more complicated case of simultaneous translational and rotational motion.

11-1 ANGULAR MOMENTUM

We now examine the dynamics of a rigid body that is constrained to rotate about an axis fixed in an inertial reference frame. A *rigid body* is one in which the relative coordinates connecting all of the constituent particles remain absolutely constant. That is, we allow the object to possess no elasticity, so that the constituent particles do not undergo any relative displacements when subjected to the accelerations of rotational motion.

Consider first the rotation of a particle about a fixed axis identified with the z-axis, as in Fig. 11-1. The particle moves in a circular path and its angular momentum *with respect to the origin O* is (Eq. 10-1)

$$\boldsymbol{\ell} = m\mathbf{r} \times \mathbf{v} \qquad (11\text{-}1)$$

The angular momentum vector $\boldsymbol{\ell}$ is perpendicular to both $\mathbf{r}$ and $\mathbf{v}$, and lies in the plane that contains the rotation axis and the position vector $\mathbf{r}$. As the particle rotates, $\boldsymbol{\ell}$ *precesses* (revolves) about the z-axis.

The vector $\mathbf{r}$ can be expressed as* (see Fig. 11-1)

$$\mathbf{r} = z\widehat{\mathbf{k}} + \rho\widehat{\mathbf{u}}_\rho$$

---

*To describe rigid-body motion, it is convenient to use *cylindrical* coordinates, which are just plane polar coordinates (Section 3-8) with the addition of the perpendicular z-axis.

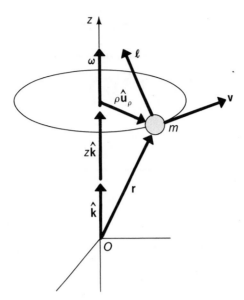

Fig. 11–1. The angular momentum of the particle is $\ell = m\mathbf{r} \times \mathbf{v}$ and the z-component is $\ell_z = m\rho^2\omega$.

where ρ is the perpendicular distance from the z-axis to the particle and where $\hat{\mathbf{u}}_\rho$ is a unit vector in the radial direction. Therefore, the angular momentum, $\ell = m(z\hat{\mathbf{k}} + \rho\hat{\mathbf{u}}_\rho) \times \mathbf{v}$, consists of two components. One component, $mz\hat{\mathbf{k}} \times \mathbf{v}$, is directed radially inward from the particle toward the rotation axis. The other component, $m\rho\hat{\mathbf{u}}_\rho \times \mathbf{v} = m\rho v\hat{\mathbf{k}}$, is directed along the z-axis; that is,

$$\ell_z = m\rho v\hat{\mathbf{k}} \tag{11–2}$$

But $v = \rho\omega$, so we can write

$$\ell_z = m\rho^2\omega \tag{11–3}$$

which is equivalent to Eq. 10–4. Notice that ℓ_z does not depend on z; hence, for the purpose of calculating *this* particular component of the angular momentum, *the location of the origin O along the rotation axis (the z-axis) is immaterial.*

We now extend the discussion to the case of a rigid body whose motion is restricted to rotation about a fixed axis. It is again convenient to choose the z-axis to coincide with the rotation axis and to use cylindrical coordinates. Figure 11–2 shows an irregularly shaped object rotating about an axis AA' which is the z-axis. The angular momentum of the differential mass element dm, just as the particle in Fig. 11–1, has a radial component and a z-component. Because we are dealing with a rigid body, only the z-component of the angular momentum influences the rotational motion of the object about the fixed axis. (The radial component is associated with the forces that the axis exerts on the support bearings.) Consequently, we single out for discussion the interesting z-component of the angular momentum.

For the differential mass element dm, we have in analogy with Eq. 11–3,

$$d\ell_z = \rho^2\omega \, dm \tag{11–4}$$

The z-component L_z of the angular momentum of the object is obtained by integrating the contributions of the individual mass elements. Thus,

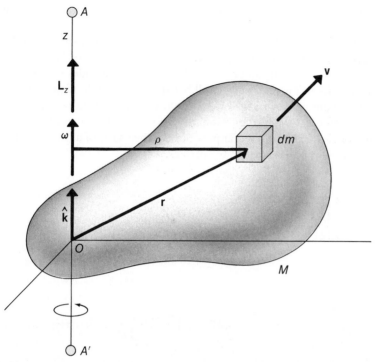

Fig. 11–2. The geometry of a rigid body undergoing rotation about a fixed axis AA' that coincides with the z-axis.

$$L_z = \int d\ell_z = \omega \int_M \rho^2 \, dm \qquad (11-5)$$

where ω can be removed from the integral because it is the same for every mass element.

The quantity L_z is referred to as the *angular momentum about the axis of rotation*. Just as for ℓ_z, the exact position of the origin O along AA' is immaterial for the purpose of calculating L_z.

The integral in Eq. 11–5 depends only on the geometrical properties of the object (and the position of the rotation axis) and is independent of any dynamic aspects of the motion. This integral is called the *rotational inertia* of the object and is denoted by the letter I:

$$I = \int_M \rho^2 \, dm \qquad (11-6)$$

We make clear that we have chosen the z-axis to correspond to the rotation axis by writing I_z for the rotational inertia. Then, the angular momentum about this axis is

$$L_z = I_z \omega \qquad (11-7)$$

The rotational inertia plays a role in rotational motion analogous to that of mass in translational motion.

It is important to understand that the value of I_z for a particular body depends on the distribution of the body's mass about the z-axis. Figure 11–3a shows a long, slender rod with a radius R_1 and a mass M whose axis coincides with the z-axis. The value of I_z for this rod is relatively small because R_1 is small. If this rod is squeezed into a stubby cylinder, as in Fig. 11–3b, the rotational inertia will increase because more of the mass is now farther from the z-axis. (The total mass M remains the same.) Finally, if the material is compressed into the shape of a thin disc (Fig. 11–3c), the value of I_z will be increased still further.

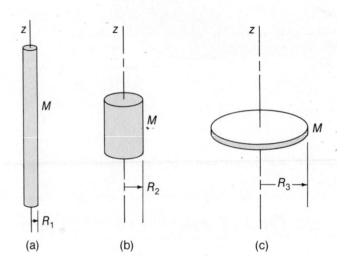

Fig. 11–3. Three different distributions about the z-axis of the same mass M of material. The rotational inertia values are $I_z(R_1) < I_z(R_2) < I_z(R_3)$.

In Section 11–4, calculations are carried out to obtain values of the rotational inertia for objects with several different shapes. Those results, together with some others, are summarized in Table 11–1. Note the dependence of I on the location of the rotation axis. Entries (c) and (d) and also (e) and (f) show how I depends on the axis for the *same* object.

11-2 ROTATIONAL DYNAMICS

The rotational motion of a rigid body about a fixed axis depends on the nature of the external forces applied to the body. For each such force we can calculate the component of the torque that is in the direction of the rotation axis. This *axial torque* is the only torque component that can influence the rotational motion. (Other components of the applied torque act only at the bearings of the rotation shaft.) We assume that the rotation axis is held in place by frictionless bearings. Such bearings can produce no axial torque and therefore cannot affect the rotational motion.

Following the same argument we made for ℓ_z in Section 11–1, we conclude that τ_z does not depend on z, and, hence, for the purpose of calculating the axial torque,

a) thin rod
$$I = \tfrac{1}{12} M\ell^2$$

b) rectangular plate
$$I = \tfrac{1}{12} M(a^2 + b^2)$$

c) solid cylinder
$$I = \tfrac{1}{2} MR^2$$

d) solid cylinder
$$I = \tfrac{1}{4} MR^2 + \tfrac{1}{12} M\ell^2$$

e) thin-walled cylinder or ring
$$I = MR^2$$

f) thin-walled cylinder or ring
$$I = \tfrac{1}{2} MR^2$$

g) solid sphere
$$I = \tfrac{2}{5} MR^2$$

h) hollow spherical shell
$$I = \tfrac{2}{3} MR^2$$

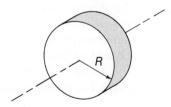

i) solid disc
$$I = \tfrac{1}{2} MR^2$$

j) annular disc or cylinder
$$I = \tfrac{1}{2} M(R_1^2 + R_2^2)$$

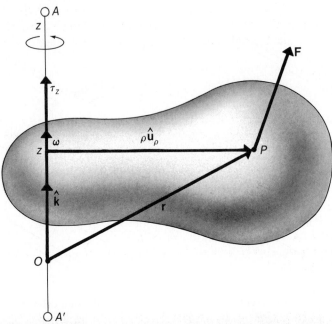

Fig. 11–4. The force **F** is applied to the body at a point P. The axial torque τ_z about O due to the force **F** does not depend on the location of O along the rotation axis AA'.

the location along the axis of the origin O is immaterial. (•Can you construct the argument for the torque?)

According to Eq. 11–7, $L_z = I_z\omega$. Differentiating with respect to time, we have (with I_z constant)

$$\frac{dL_z}{dt} = I_z\frac{d\omega}{dt} = I_z\alpha$$

where α is the angular acceleration. The time derivative of the angular momentum is equal to the applied torque (Eq. 10–9). Thus,

$$\tau_z = I_z\alpha \qquad\qquad \textbf{(11–8)}$$

If several forces are acting, τ_z is the sum of all z-components of the torques (including those at the bearings if bearing friction is present).

In the event that the applied torque has no net axial component (that is, $\tau_z = 0$), the axial component of the angular momentum remains constant (that is, $L_z =$ constant). This is simply a consequence of angular momentum conservation. Notice that we consider here only the *axial* components of torque and angular momentum. However, for an unconstrained object, we may have $\tau_z = 0$ but one or both of the other components of $\boldsymbol{\tau}$ may not be zero; then, $L_z =$ constant, but L_x and L_y can vary with time.

Radius of Gyration. Regardless of the shape of an object, we can always find a radial distance Γ from the axis of rotation at which the concentration of the entire mass M of the object would produce the same rotational inertia as that of the extended object itself. Then, the rotational inertia is equal to the value for a particle with a mass M at a distance Γ, namely, $I = M\Gamma^2$. We refer to this distance as the *radius of gyration* for the object about the particular axis; thus,

$$\Gamma^2 = \frac{I}{M} = \frac{1}{M}\int_M \rho^2 \, dm \qquad (11\text{-}9)$$

Example 11–1

A massless string is wrapped around a uniform solid cylinder that has a mass $M = 15$ kg and a radius $R = 6$ cm. The cylinder is free to rotate about its axis on frictionless bearings. One end of the string is attached to the cylinder and the free end is pulled tangentially by a force that maintains a constant tension of 2 N in the string.
(a) What is the angular acceleration α of the cylinder?
(b) What is the angular speed ω of the cylinder 2 s after the force is applied if the cylinder was originally at rest?

Solution:

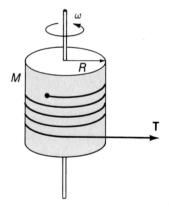

(a) The torque applied to the cylinder is constant and has the magnitude

$$\tau = RT = (0.06 \text{ m})(2 \text{ N}) = 0.12 \text{ N} \cdot \text{m}$$

The rotational inertia of the cylinder about its axis is

$$I = \tfrac{1}{2}MR^2 = \tfrac{1}{2}(15 \text{ kg})(0.06 \text{ m})^2$$
$$= 2.70 \times 10^{-2} \text{ kg} \cdot \text{m}^2$$

Thus, $\qquad \alpha = \dfrac{\tau}{I} = \dfrac{0.12 \text{ N} \cdot \text{m}}{2.70 \times 10^{-2} \text{ kg} \cdot \text{m}^2} = 4.44 \text{ rad/s}^2$

(b) The angular speed at $t = 2$ s is

$$\omega = \alpha t = (4.44 \text{ rad/s}^2)(2 \text{ s}) = 8.89 \text{ rad/s}$$

Example 11–2

(a) A grinding wheel in the shape of a thick solid disc with radius $R = 0.40$ m and mass $M = 100$ kg rotates about an axis through its center at an initial rate of 500 rpm (revolutions per minute). A tool to be sharpened is pressed against the rim and exerts a constant tangential frictional force of 25 N. In what time will the wheel be slowed to one half the initial rotation speed if no effort is made to maintain the speed? (Neglect bearing friction.)
(b) Suppose that the same grinding wheel, with the same initial rotation rate, requires 25 min to come to rest because of bearing friction. What is the (constant) magnitude of the axial torque produced by the bearing friction?

Solution:

(a) The axial torque produced by the tool is

$$\tau_z = -(0.40 \text{ m})(25 \text{ N}) = -10.0 \text{ N} \cdot \text{m}$$

Now,
$$I_z = \tfrac{1}{2}MR^2 = \tfrac{1}{2}(100 \text{ kg})(0.40 \text{ m})^2$$
$$= 8.0 \text{ kg} \cdot \text{m}^2$$

Thus, using Eq. 11–8, we have

$$\alpha = \frac{\tau_z}{I_z} = \frac{-10.0 \text{ N} \cdot \text{m}}{8.0 \text{ kg} \cdot \text{m}^2} = -1.25 \text{ rad/s}^2$$

By integrating $d\omega/dt = \alpha$, we obtain

$$\omega = \omega_0 + \alpha t$$

Here,
$$\omega_0 = 500 \text{ rpm} = (500 \text{ min}^{-1}) \times \frac{2\pi}{60 \text{ s/min}} = 52.36 \text{ rad/s}$$

Then,
$$t = \frac{\omega - \omega_0}{\alpha} = \frac{(-\tfrac{1}{2})(52.36 \text{ rad/s})}{-1.25 \text{ rad/s}^2} = 20.9 \text{ s}$$

(b) The final angular velocity is $\omega = 0$, so $\omega_0 + \alpha t = 0$. Thus, $\alpha = -\omega_0/t$, and

$$\tau_z = I_z\alpha = -\frac{I_z\omega_0}{t} = -\frac{(8.0 \text{ kg} \cdot \text{m}^2)(52.36 \text{ rad/s})}{25(60 \text{ s})}$$
$$= -0.279 \text{ N} \cdot \text{m}$$

Example 11–3

A massless string is wrapped around a pulley that has a radius $R = 10$ cm, a mass $M = 0.50$ kg, and a radius of gyration $\Gamma = 8$ cm about its axis. The free end of the string is attached to a hanging 2-kg block; the other end is fixed to a point on the rim of the pulley. Neglecting friction in the pulley bearing, find the linear acceleration of the hanging block and the force exerted by the bearing on the pulley.

Solution:

We assume that the string does not stretch, so the distance s that the block descends is equal to the arc length $R\phi$ through which the rim of the pulley moves; that is, $s = R\phi$. Then,

$$\frac{ds}{dt} = v = R\frac{d\phi}{dt} = R\omega \tag{1}$$

Also,
$$\frac{d^2s}{dt^2} = a = R\frac{d^2\phi}{dt^2} = R\alpha \tag{2}$$

Applying Newton's second law to the motion of the hanging block (with the positive direction taken downward), we have

$$mg - T = ma \qquad (3)$$

For calculating the torque, we take the clockwise direction to be positive. From Eq. 11–8, $\tau = I\alpha$; using (2) and $I = M\Gamma^2$, we find

$$\tau = TR = I\alpha = \frac{M\Gamma^2 a}{R} \qquad (4)$$

Eliminating T between (3) and (4), we obtain

$$mR^2 a + M\Gamma^2 a = mgR^2$$

from which

$$a = \frac{mgR^2}{mR^2 + M\Gamma^2} = \frac{g}{1 + \dfrac{M\Gamma^2}{mR^2}} \qquad (5)$$

$$= \frac{9.80 \text{ m/s}^2}{1 + \dfrac{0.5 \text{ kg}}{2 \text{ kg}}\left(\dfrac{8 \text{ cm}}{10 \text{ cm}}\right)^2}$$

$$= 8.45 \text{ m/s}^2$$

Then,

$$\alpha = \frac{a}{R} = \frac{8.45 \text{ m/s}^2}{0.10 \text{ m}} = 84.5 \text{ rad/s}^2$$

The force exerted by the bearing on the pulley is the normal force **N** shown in the free-body diagram for the pulley. The force equation for the vertical direction is

$$N - Mg - T = 0$$

We now know the acceleration a, so we can obtain the tension T from (3):

$$T = m(g - a)$$

Then, $N = T + Mg = m(g - a) + Mg$

$$= (2\text{ kg})(9.80\text{ m/s}^2 - 8.45\text{ m/s}^2) + (0.5\text{ kg})(9.80\text{ m/s}^2)$$

$$= 7.60\text{ N}$$

11-3 ROTATIONAL KINETIC ENERGY AND WORK

The kinetic energy of a rigid body rotating about a fixed axis can be conveniently given in terms of rotational quantities. For the situation illustrated in Fig. 11-5, the kinetic energy of the mass element dm is

$$dK = \tfrac{1}{2}v^2\,dm = \tfrac{1}{2}\rho^2\omega^2\,dm$$

and the kinetic energy for the entire body is

$$K = \tfrac{1}{2}\omega^2 \int_M \rho^2\,dm = \tfrac{1}{2}I\omega^2 \tag{11-10}$$

We sometimes refer to $\tfrac{1}{2}I\omega^2$ as the *rotational kinetic energy* of the body.

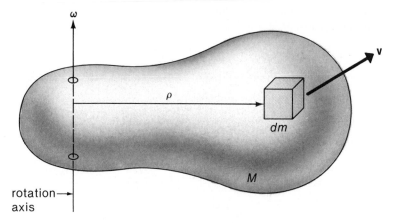

Fig. 11-5. The kinetic energy of the mass element dm is $\tfrac{1}{2}\rho^2\omega^2\,dm$.

We can also obtain the rotational counterpart of the work-energy theorem discussed in Section 7-3. Suppose that an applied force **F** acts on a rigid body at a point P and produces a displacement $\Delta\mathbf{s}$ in the plane perpendicular to the axis of rotation (Fig. 11-6). During the time Δt, the point P rotates through an angle $\Delta\phi$ and the displacement of P is $\Delta\mathbf{s} = \rho\,\Delta\phi\hat{\mathbf{u}}_\phi$. The work done by **F** during Δt is $\Delta W = \mathbf{F}\cdot\Delta\mathbf{s}$. Expressing **F** in cylindrical coordinates, we have

$$\Delta W = (F_\rho\hat{\mathbf{u}}_\rho + F_\phi\hat{\mathbf{u}}_\phi + F_z\hat{\mathbf{k}})\cdot(\rho\,\Delta\phi\hat{\mathbf{u}}_\phi)$$

$$= F_\phi\rho\,\Delta\phi$$

where $F_\phi\rho$ is just the torque about the rotation axis. Thus,

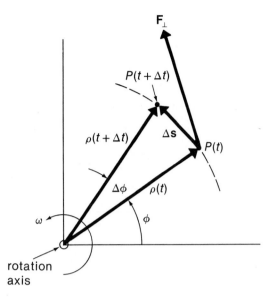

Fig. 11–6. A force **F** acts on a rigid body at point *P*. The component of **F** that lies in the plane perpendicular to the rotation axis is $\mathbf{F}_\perp = F_\rho \hat{\mathbf{u}}_\rho + F_\phi \hat{\mathbf{u}}_\phi$. During the time Δt, the force produces the displacement $\Delta \mathbf{s} = \rho \, \Delta\phi \, \mathbf{u}_\phi$, and does an amount of work equal to $F_\phi \rho \, \Delta\phi$.

rotation
axis

$$\Delta W = \tau \, \Delta\phi$$

so that

$$\frac{\Delta W}{\Delta t} = \tau \frac{\Delta\phi}{\Delta t}$$

In the limit $\Delta t \to 0$, we have

$$\frac{dW}{dt} = \tau \frac{d\phi}{dt} = \tau\omega \qquad\qquad \textbf{(11–11)}$$

(Recall that power P is dW/dt; consequently, this result can be expressed as $P = \tau\omega$.)

From Eq. 11–8, we can write

$$\tau = I\alpha = I\frac{d\omega}{dt}$$

so that

$$\frac{dW}{dt} = I\omega \frac{d\omega}{dt} = \frac{d}{dt}(\tfrac{1}{2}I\omega^2)$$

Thus,

$$dW = dK$$

and we again conclude that the work done by the applied force **F** is transformed completely into an increase in the rotational kinetic energy of the body (conservation of energy!).

We have now obtained expressions for all of the dynamical quantities involved in rotational motion. In Table 11–2 these quantities are compared with the corresponding ones that are used to describe rectilinear motion. Notice that these quantities are *analogous* only—for example, the first six entries in the table have different dimensions in the two cases and therefore represent completely different physical quantities.

TABLE 11-2.   Comparison of Translational and Rotational Quantities

TRANSLATION (LINEAR MOTION)		ROTATION (ABOUT A FIXED AXIS)	
Displacement*	$s = s_0 + v_0 t + \frac{1}{2}at^2$	Angular displacement*	$\phi = \phi_0 + \omega_0 t + \frac{1}{2}\alpha t^2$
Speed	$v = ds/dt$	Angular speed	$\omega = d\phi/dt$
*	$= v_0 + at$	*	$= \omega_0 + \alpha t$
Acceleration	$a = dv/dt$	Angular acceleration	$\alpha = d\omega/dt$
Mass	M	Rotational inertia	I
Force	$F = Ma$	Torque	$\tau = I\alpha$
Momentum	$P = mv$	Angular momentum	$L = I\omega$
Work	$W = \int F\, dx$	Work	$W = \int \tau\, d\phi$
Kinetic energy	$K = \frac{1}{2}Mv^2$	Kinetic energy	$K = \frac{1}{2}I\omega^2$
Power	$P = Fv$	Power	$P = \tau\omega$

*For constant acceleration (a or α).

Example 11–4

Refer again to the accelerated cylinder in Example 11–1. (a) What is the rotational kinetic energy of the cylinder at $t = 2$ s? (b) How much work was done by the applied force during the 2-s interval?

Solution:

(a) The rotational kinetic energy at $t = 2$ s is

$$K = \tfrac{1}{2}I\omega^2 = \tfrac{1}{2}(2.70 \times 10^{-2}\,\text{kg}\cdot\text{m}^2)(8.89\,\text{rad/s})^2$$

$$= 1.07\,\text{J}$$

(b) From the work-energy theorem, the work done is equal to the increase in kinetic energy, namely, 1.07 J. However, it is instructive to calculate the work done by using Eq. 11–11 and $\omega = \alpha t$:

$$dW = \tau\omega\, dt = \tau\alpha t\, dt$$

Integrating,

$$\int_0^W dW = \tau\alpha \int_0^t t\, dt$$

so that

$$W = \tfrac{1}{2}\tau\alpha t^2 = \tfrac{1}{2}(0.12\,\text{N}\cdot\text{m})(4.44\,\text{rad/s}^2)(2\,\text{s})^2$$

$$= 1.07\,\text{J}$$

as before.

Example 11–5

Refer to Example 11–3. Use energy conservation to obtain the acceleration of the block.

Solution:

Work is done on the system (block + pulley) corresponding to the decrease in the gravitational potential energy of the hanging block as it moves downward. Hence,

$$W = mgs$$

where s is the total distance of fall at time t.

The gain in kinetic energy of the system consists of two parts, $\frac{1}{2}I\omega^2$ for the pulley and $\frac{1}{2}mv^2$ for the block:

$$K = \tfrac{1}{2}I\omega^2 + \tfrac{1}{2}mv^2$$

where v and ω are the values at time t. We know that $I = M\Gamma^2$ and $v = R\omega$, so equating K to the work done gives

$$mgs = \tfrac{1}{2}\left(\frac{M\Gamma^2}{R^2} + m\right)v^2$$

Differentiating this equation with respect to time and using $ds/dt = v$, we find

$$mgv = m\left(\frac{M\Gamma^2}{mR^2} + 1\right)v\frac{dv}{dt}$$

Solving for $dv/dt = a$, we obtain

$$a = \frac{g}{1 + \dfrac{M\Gamma^2}{mR^2}}$$

which is just Eq. (5) in Example 11–3.

SPECIAL TOPIC

11-4 PROPERTIES OF RIGID BODIES

Center of Mass. Consider a rigid body in the form of a thin flat sheet or lamina S (Fig. 11–7). Because the sheet is thin, we can describe the mass distribution in terms of a *surface mass density* σ, measured in kg/m^2, which may be a function of position, $\sigma = \sigma(x, y)$.

The center of mass of the solid lamina can be obtained by a straightforward extension of the definition given in Section 9-4 for the case of n particles, each with mass m_i. We have (see Eq. 9–8)

$$\mathbf{R} = \frac{1}{M}\sum_{i=1}^{n} m_i\mathbf{r}_i, \qquad M = \sum_i m_i$$

Then, for the lamina,

$$M = \int_M dm = \iint_S \sigma\, dx\, dy \qquad \textbf{(11–12)}$$

$$\mathbf{R} = \frac{1}{M}\int_M \mathbf{r}\, dm = \frac{1}{M}\iint_S \sigma\mathbf{r}\, dx\, dy \qquad \textbf{(11–13)}$$

Equation 11–13 locates the C.M. in terms of the *vector* $\mathbf{R}$. If we let (X, Y) specify the coordinates of the C.M., we can write

$$\left.\begin{array}{l} X = \dfrac{1}{M}\displaystyle\iint_S \sigma x\, dx\, dy \\[2em] Y = \dfrac{1}{M}\displaystyle\iint_S \sigma y\, dx\, dy \end{array}\right\} \qquad \textbf{(11–14)}$$

(for S in the x-y plane)

The integrals in the expressions for X and Y are simply the *first moments* of the area S (see Section V–1).

These definitions are readily extended to the case of a three-dimensional object with arbitrary shape. For example, Eq. 11–13 becomes

$$\mathbf{R} = \frac{1}{M} \int_M \mathbf{r} \, dm = \frac{1}{M} \iiint_V \delta \mathbf{r} \, dx \, dy \, dz \quad \text{(11–15)}$$

where δ is the usual volume mass density.* In general, the calculation of $\mathbf{R}$ involves a triple integration over the volume V of the object; however, the integration can often be reduced to the evaluation of a single integral by taking advantage of the symmetry of the object (see Example 11–6). Equation 11–15 is a *vector* equation and represents three equations for X, Y, and Z similar to Eqs. 11–14.

The C.M. for any rigid body is a *fixed point* relative to the object, but the C.M. might not lie *within* the body. (This would be the case for the C.M. of a lamina in the shape of the letter *C*.)

Example 11–6

Determine the location of the C.M. of a solid right circular cone with an altitude h and a base radius R. The density is $\delta = $ constant.

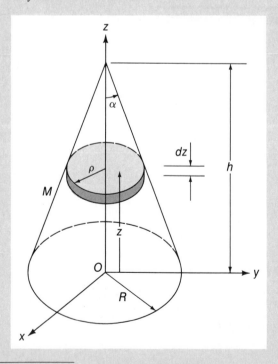

---

*In this chapter we use δ (instead of ρ) to represent the volume mass density in order to avoid confusion with the cylindrical coordinate ρ.

Solution:

As shown in the diagram, we imagine the cone to be subdivided into thin discs parallel to the base. One such disc, with radius ρ and thickness dz, is indicated in the diagram. Because of the symmetry, the C.M. of each disc is at its center, a point on the z-axis. In Section 9–4 we found that the C.M. of two objects (or groups of objects) lies on the line connecting the centers of mass of the objects. Therefore, the C.M. of the cone must also lie on the z-axis, and we need to determine only the z-coordinate. We use the z-coordinate of Eq. 11–15:

$$Z = \frac{1}{M} \int_M z \, dm = \frac{1}{M} \int_0^h z(\pi\rho^2 \delta \, dz)$$

Now, $\rho = (h - z)\tan\alpha = (h - z)(R/h)$, so

$$Z = \frac{\pi\delta R^2}{h^2 M} \int_0^h (h - z)^2 z \, dz = \frac{\pi\delta R^2}{h^2 M} \cdot h^4(\tfrac{1}{2} - \tfrac{2}{3} + \tfrac{1}{4})$$

$$= \frac{\pi\delta R^2 h^2}{12M}$$

The mass M of the cone is

$$M = \delta \int_0^h \pi\rho^2 \, dz = \frac{\pi\delta R^2}{h^2} \int_0^h (h - z)^2 \, dz$$

$$= \tfrac{1}{3}\pi\delta R^2 h$$

Then,

$$Z = \frac{\pi\delta R^2 h^2}{12(\tfrac{1}{3}\pi\delta R^2 h)} = \tfrac{1}{4}h$$

Rotational Inertia. For the case of a thin lamina that lies in the x-y plane (Fig. 11–3), Eq. 11–6 for the rotational inertia becomes

$$I_x = \int_M y^2 \, dm = \iint_S \sigma y^2 \, dx \, dy \quad \text{(11–16a)}$$

$$I_y = \int_M x^2 \, dm = \iint_S \sigma x^2 \, dx \, dy \quad \text{(11–16b)}$$

These integrals are simply the *second moments* of the mass distribution. The second moments are not a property solely of the object's geometry; they depend also on the location of the rotation axis.

Look again at the lamina in Fig. 11–7. What is the

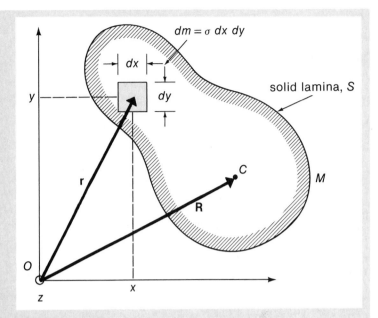

Fig. 11–7. A rigid solid lamina S lies in the x-y plane and has surface density σ. In diagrams such as this we denote the position of the center of mass by C.

rotational inertia for rotations about the z-axis? In this case, $\rho^2 = x^2 + y^2$, so

$$I_z = \int_M (x^2 + y^2)\, dm = I_x + I_y \qquad \textbf{(11–17)}$$

Caution: *This result is valid only for a thin lamina that lies in the x-y plane.*

Finally, consider a solid (three-dimensional) object.

For rotations about the z-axis (Fig. 11–8), the quantity ρ^2 is again equal to $x^2 + y^2$; therefore

$$I_z = \int_M (x^2 + y^2)\, dm = \iiint_V \delta(x^2 + y^2)\, dx\, dy\, dz \qquad \textbf{(11–18)}$$

with similar expressions for I_x and I_y. In this case, we do *not* have a simple relationship that connects I_x, I_y, and I_z.

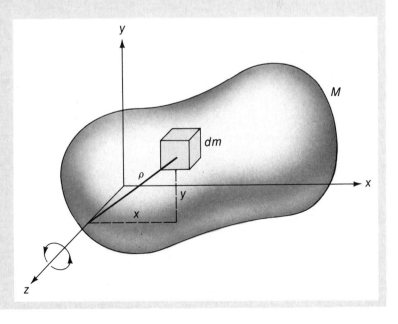

Fig. 11–8. To calculate I_z, we use $\rho^2 = x^2 + y^2$.

Example 11–7

Calculate the rotational inertia about the x-, y-, and z-axes for a uniform thin plate with dimensions $a \times b$ and with mass M if the y-axis coincides with an edge of the plate along the b dimension, as in the diagram.

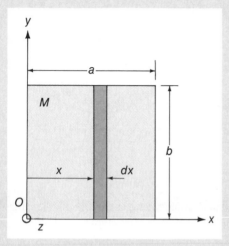

Solution:

The usual expression for I_y is

$$I_y = \int_M \rho^2 \, dm$$

Notice that every part of the shaded strip in the diagram is at the same distance x from the y-axis. Therefore, the rotational inertia of this strip is $dI_y = x^2 \, dm = x^2 \cdot \sigma b \, dx$, where $\sigma = M/ab$ is the surface density of the plate. Then,

$$I_y = \sigma b \int_0^a x^2 \, dx = Ma \cdot \tfrac{1}{3}a^3$$

$$= \tfrac{1}{3}Ma^2$$

From the diagram, it is clear that we also have

$$I_x = \tfrac{1}{3}Mb^2$$

Then, $\quad I_z = I_x + I_y = \tfrac{1}{3}M(a^2 + b^2).$

Example 11–8

Find the rotational inertia of a thin uniform circular disc with radius R, mass M, and surface density

$\sigma = M/\pi R^2$, about an axis through its center and perpendicular to the plane of the disc.

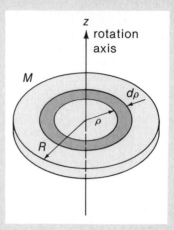

Solution:

Divide the disc into rings of width $d\rho$ and mass $dm = \sigma(2\pi\rho) \, d\rho$. Then,

$$I_z = \int_0^R \rho^2 \sigma(2\pi\rho) \, d\rho = 2\pi\sigma \int_0^R \rho^3 \, d\rho$$

$$= \tfrac{1}{2}\pi\sigma R^4 = \tfrac{1}{2}MR^2$$

which is the value given in Table 11–1(i).

Example 11–9

Determine the rotational inertia of a uniform solid sphere with radius R, mass M, and density $\delta = M/(4/3)\pi R^3$, about an axis through its center.

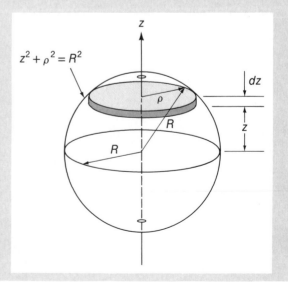

Solution:

By taking advantage of the cylindrical symmetry, the rotational inertia can be evaluated with a single integration. Divide the solid sphere into discs with differential thickness dz and radius ρ. According to the result of Example 11–8, the rotational inertia of a disc about its axis is $I_z = \frac{1}{2}MR^2$. Therefore, we can write

$$dI_z = \tfrac{1}{2}\rho^2\, dm$$

with

$$dm = \pi\rho^2\delta\, dz$$

Now, $z^2 + \rho^2 = R^2$, so we have

$$I_z = \int dI_z = \tfrac{1}{2}\pi\delta \int_{-R}^{+R} (R^2 - z^2)^2\, dz$$

$$= \tfrac{1}{2}\pi\delta[R^4 z - \tfrac{2}{3}R^2 z^3 + \tfrac{1}{5}z^5]_{-R}^{+R}$$

$$= \tfrac{8}{15}\pi\delta R^5 = \tfrac{2}{5}MR^2$$

which is the value given in Table 11–1(g).

The Parallel-Axis Theorem. We now prove a useful theorem that allows the computation of the rotational inertia of an object about any axis if the rotational inertia is known for a *parallel axis that passes through the center of mass* of the object.

Refer to Fig. 11–9. We wish to express the rotational inertia I_A about the axis AA' in terms of the rotational inertia I_C about the parallel axis BB' through the center of mass C of the object. The perpendicular position vector from the axis AA' to the mass element dm is $\rho_A \hat{\mathbf{u}}_A$, where $\hat{\mathbf{u}}_A$ is the unit vector $\hat{\mathbf{u}}_\rho$ referred to the axis AA'. The corresponding vector for the axis BB' is $\rho_B \hat{\mathbf{u}}_B$. The vector $\mathbf{h}$ extends from AA' to BB' and is perpendicular to both axes. The various vectors are shown in Fig. 11–9b.

We can express I_A as

$$I_A = \int_M \rho_A^2\, dm$$

Now, $\rho_A \hat{\mathbf{u}}_A = \mathbf{h} + \rho_B \hat{\mathbf{u}}_B$, so

$$\rho_A^2 = (\rho_A \hat{\mathbf{u}}_A) \cdot (\rho_A \hat{\mathbf{u}}_A) = (\mathbf{h} + \rho_B \hat{\mathbf{u}}_B) \cdot (\mathbf{h} + \rho_B \hat{\mathbf{u}}_B)$$
$$= h^2 + \rho_B^2 + 2\mathbf{h} \cdot (\rho_B \hat{\mathbf{u}}_B)$$

Therefore,

$$I_A = h^2 \int_M dm + \int_M \rho_B^2\, dm + 2\mathbf{h} \cdot \int_M \rho_B \hat{\mathbf{u}}_B\, dm$$

The integral in the last term defines the position of the

Fig. 11–9. The axis BB' passes through the C.M. of the object and is parallel to the axis AA'. The vectors $\rho_A \hat{\mathbf{u}}_A$, $\rho_B \hat{\mathbf{u}}_B$, and $\mathbf{h}$ are all in a plane perpendicular to both axes, as shown in (b). The point C is the center of mass.

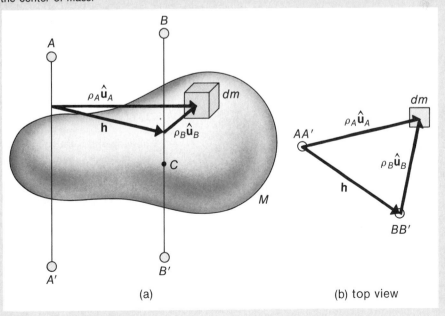

(a)   (b) top view

C.M. from an origin that is the C.M. itself (compare Eq. 11–15); consequently, this integral vanishes. Also,

$$\int_M dm = M \quad \text{and} \quad \int_M \rho_B^2 \, dm = I_C$$

Altogether, we have

$$I_A = I_C + Mh^2 \qquad \textbf{(11–19)}$$

which is the statement of the parallel-axis theorem. The utility of this theorem lies in the fact that it is almost always easier to calculate the rotational inertia about an axis through the C.M. than to make a direct calculation for some other axis.

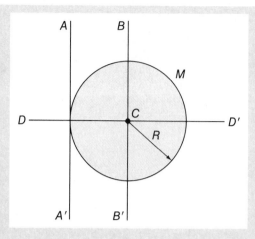

$$I_0 = I_B + I_D$$

There is nothing to distinguish the axis BB' from the axis DD'; therefore, $I_B = I_D \equiv I_C$, and we have

$$I_C = \tfrac{1}{2} I_0 = \tfrac{1}{4} MR^2$$

Example 11–10

Determine the rotational inertia of a thin uniform disc with radius R and mass M about an axis tangent to the edge of the disc (axis AA' in the diagram).

Solution:

From Example 11–8 (or Table 11–1[i]) we know that the rotational inertia of the disc about a perpendicular axis through the center C is $I_0 = \tfrac{1}{2} MR^2$. We can relate I_0 to the rotational inertia about the axes BB' and DD' by using Eq. 11–17:

Finally, we use the parallel-axis theorem to write

$$I_A = I_C + Mh^2$$
$$= \tfrac{1}{4} MR^2 + MR^2 = \tfrac{5}{4} MR^2$$

If the axis AA' remains fixed in space, any rotation of the disc about the BB' axis does not alter the value of I_A. (•Can you see why?)

PROBLEMS

Section 11-2

11-1   A solid uniform cylinder with a radius of 0.30 m and a mass of 12 kg is free to rotate about its axis, which is supported by frictionless bearings. A constant tangential force of 10 N is applied to the cylinder, originally at rest. (a) What is the torque applied to the cylinder? (b) What is the angular acceleration of the cylinder? (c) What is the angular velocity of the cylinder 3 s after the force is applied?

11-2   A uniform solid sphere with a radius of 20 cm and a mass of 200 kg is free to rotate about an axis that passes through the center of the sphere. A constant torque is applied to the sphere, and it reaches a rotation rate of 10 rpm in 5 s. (a) What is the angular

acceleration of the sphere? (b) What is the magnitude of the applied torque?

11-3   A phonograph turntable consists of a solid circular plate 30 cm in diameter that has a mass of 3.5 kg. The turntable is rotating at $33\tfrac{1}{3}$ rpm when the motor is switched off. A time of 105 s is required for the turntable to come to rest. (a) What is the (constant) angular acceleration? (b) How many revolutions does the turntable make after the motor is turned off? (c) What is the effective decelerating torque?

11-4   A massless string is wrapped around a pulley with a mass $M = 0.60$ kg, a radius $R = 8$ cm, and a radius of gyration $\Gamma = 4$ cm. The string slides (without friction) through a hook that supports a hanging block ($m = 2.4\,\text{kg}$) and the end is attached to a fixed point Q, as shown in the diagram. Determine the linear acceleration of the hanging block. [*Hint:* First, find the relationship connecting the linear acceleration of the hanging block and the angular acceleration of the pulley.]

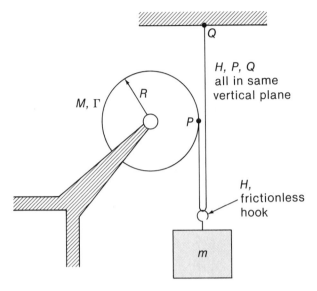

11-5 A thin rectangular plate with mass M and sides a and b rotates freely about a perpendicular axis through its center. The initial angular speed of the plate is ω_0. Suppose that the temperature increases so that all dimensions are enlarged by 1 part in 10^5. What is the fractional change in the angular speed? [*Hint:* Use partial differentiation.]

11-6 A block ($m_1 = 5\,\text{kg}$) slides without friction on a plane inclined at an angle of $30°$. This block is connected to a hanging block ($m_2 = 4\,\text{kg}$) by a massless string that runs over a pulley ($M = 0.50\,\text{kg}$) with a radius $R = 3\,\text{cm}$ and a radius of gyration $\Gamma = 2\,\text{cm}$. The string does not slip on the pulley. (a) Find the linear acceleration of the system. (b) Find the tension in the string attached to each of the blocks. Why are the two tensions not equal? (c) Find the

force exerted on the axle of the pulley by the bearings.

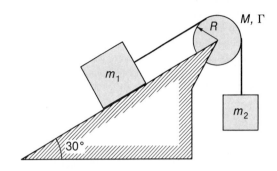

11-7 The diagram in Example 11–3 shows the string tension **T** producing a torque as if the string were acting at the rim point P. The string is actually attached to the rim at its end and is merely wrapped around the pulley to point P. Discuss the significance of this difference, making whatever assumption you wish regarding the nature of the static friction between the string and the pulley rim. [*Hint:* Consider dividing the strings into two parts at point P, taking the part wrapped around the pulley and the pulley itself as a *single* free body. Remember, the string is assumed to be massless.]

Section 11-3

11-8 An airplane engine delivers 3000 hp to a propeller that rotates at 2500 rpm. What is the torque exerted on the propeller shaft?

11-9 A solid uniform cylindrical shaft has a mass of 4.0 kg and a diameter of 2.0 cm, and is supported by frictionless bearings. A massless string with a length of 70 cm is wrapped around the shaft, which is set into rotation as the string is pulled with constant tension. The string unwinds, without slipping, in 2.0 s, and then pulls free. (a) What is the angular acceleration of the shaft? (b) What is the final angular speed? (c) Find the magnitude of the tension.

11-10 Assume that the Earth is a homogeneous sphere, and calculate the kinetic energy of rotation about its axis. (Use the table of astronomical data inside the front cover.) For how long could this energy be tapped to supply the world's energy needs (assumed constant at $2 \times 10^{20}\,\text{J/y}$) if, in the process, the length of the day is not to be increased by more than 1 min?

11-11 A girl stands at the center of a rotating platform. Initially, her arms are extended horizontally and she holds a 3-kg dumbbell in each hand; the platform rotates once every 3 s. If the girl drops her hands to her sides, what will be the new angular speed? Assume that the rotational inertia of the girl is constant at 5 kg · m²; the rotational inertia of the platform is 2 kg · m². The dumbbells (assumed to be particles) are first held 80 cm from the rotation axis and, when the girl drops her hands to her sides, they are 20 cm from the axis. (Neglect friction.) Calculate the initial and final values of the rotational kinetic energy. Account for the difference.

11-12 Two blocks, with masses M_1 and M_2, are connected by a massless string that runs over a pulley without slipping. The pulley is a uniform solid disc with a mass m and a radius R that rotates on frictionless bearings. Determine the linear acceleration of the blocks by considering energy conservation. [*Hint:* Refer to Example 11–5.]

SPECIAL TOPIC

Section 11-4

11-13 A solid rod 5 cm in diameter and 20 cm in length has a hole 3 cm in diameter drilled along the axis from one end to a depth of 10 cm. Locate the C.M. of the rod.

11-14 A straight piece of wire 20 cm in length is bent in two places through 90° to produce the shape indicated in the diagram. Determine the coordinates of the C.M.

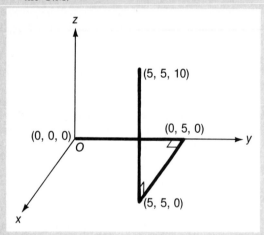

11-15 A thin uniform disc with a radius R has cut from it a circular hole with a radius a. The distance from the center of the disc O to the center of the hole is b ($a + b < R$), as indicated in the diagram. Find the position of the C.M. for the disc.

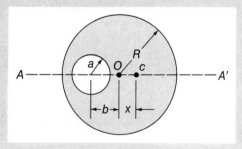

11-16 Determine the values of I_x, I_y, and I_z for a uniform thin plate with mass M and sides a and b, located in the x-y plane, as shown in the diagram.

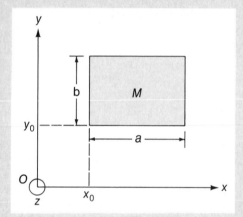

11-17 Locate the C.M. of a uniform lamina in the shape of an equilateral triangle with sides of length 20 cm.

11-18 Find the position of the center of mass of a uniform solid hemisphere with radius R and mass M. [*Hint:* Proceed as in Example 11–6.]

11-19 Show that the rotational inertia of a uniform thin rod, with a length ℓ and a mass M, about a perpendicular axis through the midpoint is $\frac{1}{12}M\ell^2$. Thus, verify the value given in Table 11–1(a).

11-20 A flat circular ring with a radius R is made from a thin wire that has a uniform linear mass density λ. Prove that the rotational inertia about an axis along any diameter is $\pi R^3\lambda$.

11-21 Show that the rotational inertia about the symmetry axis of a uniform annular cylinder with inner radius R_1, outer radius R_2, length ℓ, and mass M is $\frac{1}{2}M(R_1^2 + R_2^2)$. Thus, verify the value given in Table 11–1(j).

11-22 Determine the radius of gyration Γ about the symmetry axis of a right circular cone with altitude h

and base radius R. [*Hint:* Proceed as in Example 11–6 by using the rotational inertia for a thin disc obtained in Example 11–8.]

11-23 Determine the rotational inertia of a thin ring with a radius R and a mass M, about an axis that is tangent to the ring and lies in the plane of the ring by following the procedure of Example 11–10.

11-24 Verify the result of Example 11–10 by direct inte-
• gration. [*Hint:* The equation of a circle with origin O, as in the diagram at the right, is $r = 2R \cos \phi$. Refer to a table of integrals to evaluate the resulting integral over ϕ.]

11-25 Consider an annular cylinder with inner radius R_1,
• outer radius R_2, and length ℓ. Determine the rotational inertia of this cylinder about an axis that is perpendicular to the cylinder axis and coincides with a diameter of one end face of the cylinder. [*Hint:* Apply the parallel-axis theorem to the result for a thin annular ring.]

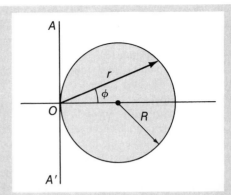

11-26 What is the rotational inertia of a cube with sides of length ℓ and a mass M, about an axis that coincides with one edge? [*Hint:* Start with the result for a rectangular plate—Table 11–1(b).]

11-27 Verify entry (d) in Table 11–1 by starting with
• entry (a).

Additional Problems

11-28 It is proposed to power a passenger bus with a massive rotating flywheel that is periodically brought up to its maximum speed (3500 rpm) by means of an electric motor. The flywheel has a mass of 1200 kg, a diameter of 1.8 m, and is in the shape of a solid cylinder. (•This is not an efficient shape for a flywheel that is designed to power a vehicle; can you see why?) (a) What is the maximum amount of kinetic energy that can be stored in the flywheel? (b) If the bus requires an average power of 30 hp, how long will the flywheel rotate?

11-29 The Sun rotates with a period of 24.7 days.* Suppose that the Sun were to collapse into a *neutron*

---
*The Sun is not a rigid body. This is the rotation period at the equator; the value increases with solar latitude.

star with a radius of 30 km. If the Sun is considered to be a rigid body in both conditions, what is the new rotation period? The radius of gyration of the Sun is approximately $0.3 R_S$. In the neutron-star state, assume that the density is uniform. Neglect the possibility of any mass ejection during the collapse. (This problem is intended only to establish an order-of-magnitude result, but neutron stars with solar mass have radii estimated to be as small as 10 km. One rotating neutron star—the *pulsar* in the Crab nebula—has a rotation period of 33 ms, the shortest period known.)

11-30 Two freely rotating components of a friction
• clutch, each with a mass of 10 kg, are initially separated, as shown in diagram (a) below. Component A

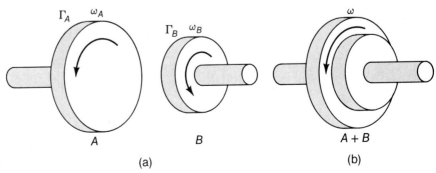

Problem 11–30   (a)   (b)

(with $\Gamma_A = 7$ cm) rotates at 200 rpm and component B (with $\Gamma_B = 5$ cm) rotates at 800 rpm in the same direcion as A. These components are brought together, as in diagram (b), and come to the same rotational speed in the time required for one disc to slide relative to the other by one-half turn. (a) What is the final angular speed of the pair? (b) How much kinetic energy is lost during the process? (c) Assume that all of the lost kinetic energy is dissipated in friction at the interface of the components. What is the average frictional torque that acts while the components are coming into synchronism? [*Hint:* Is angular momentum conserved?]

11-31 A pair of rocket motors AA are used to move a space station into deep space. The space station is a hollow torus, as shown in the diagram, similar to an automobile inner tube, with a connecting section that lies along a diameter. (a) It is desired to rotate the station about the BB' axis so that the astronauts walk about feeling a simulated gravitational acceleration. If they are to experience a centripetal acceleration equal to g at the radius R, what must be the angular speed of the rotation? (b) The rocket motors AA are used to produce the rotation about the BB' axis. The rocket motors produce a constant tangential thrust. From a stationary start the station acquires the required angular speed in 1000 seconds. What is the thrust (force) exerted by each rocket motor? (c) When the required angular velocity ω has been achieved, the motors are retracted to the center of the station at $A'A'$ for servicing and storage. The mass of each motor is $m = 255$ tonnes. What is the ratio of the original angular velocity ω to the final angular velocity ω'? (1 tonne $= 10^3$ kg.)

11-32 The lid of a box is a uniform rectangle with dimensions 0.5 m $\times$ 1.0 m; the hinge line is along one of the longer edges. The lid is allowed to fall, starting from a vertical position. What is the angular speed of the lid as it strikes the top of the box? [*Hint:* Use energy conservation.]

11-33 A string is wrapped around an unsupported solid cylinder that has a mass $M = 6$ kg and a radius $R = 5$ cm. A vertical force **F** is applied to the string such that the C.M. of the cylinder remains at the same height. (a) Show that the angular acceleration of the cylinder is $\alpha = 2g/R$. (b) What amount of work is done by the force during the 0.5-s interval after release? (c) Through what total distance does the free end of the string move during the same interval?

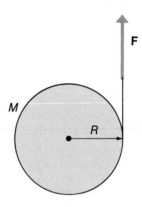

11-34 Two solid uniform discs are attached to a common axis which is free to rotate without friction (see the diagram on p. 319). A massless string is wrapped around each disc and supports a hanging block.

$R = 39.2$ m
$I_{BB'} = 7.84 \times 10^6$ tonne-m^2
(with engines at AA)

Problem 11–31

Disc A has a mass of 4 kg and a radius of 8 cm; the mass of the block hanging from this disc is 6 kg. Disc B has a mass of 6 kg and a radius of 10 cm; the mass of the block hanging from this disc is 5 kg. Find the linear acceleration of each block.

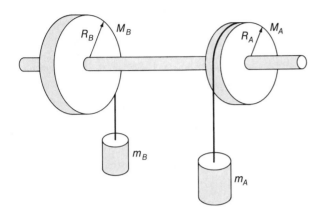

11-35 In 1784, George Atwood (1746–1807), a British
•• mathematician, devised a simple apparatus to "dilute" the effect of gravity so that accelerations could be accurately measured to determine the value of g. An *Atwood's machine* consists of a pulley with a mass $M_0 = 0.20$ kg, a radius $R = 7.5$ cm, and a radius of gyration $\Gamma = 2.5$ cm. Two blocks with nearly equal masses, $M_1 = 3.000$ kg and $M_2 = 2.950$ kg, are connected by a massless string that runs over the pulley, as shown in the diagram at right. The pulley bearings exert a constant frictional torque, $\tau = 1.5 \times 10^{-4}$ N·m. (a) Show that energy conservation can be expressed as

$$(M_1 - M_2)gs = \tfrac{1}{2}(M_1 + M_2)v^2 + \tfrac{1}{2}I\omega^2 + \tau\phi$$

where

s = vertical displacement of the blocks
v = linear velocity of the blocks
I = rotational inertia of the pulley
ω = angular speed of the pulley
ϕ = angular displacement of the pulley

(b) What is the linear acceleration of the blocks? What percentage of this acceleration is attributable to the pulley and what percentage is attributable to the bearing friction? (c) What time is required for M_1 to descend a distance of 50 cm, starting from rest? (d) Suppose that this Atwood's machine is used to determine a value for g. If the variation in the observer's reaction time (for starting and stopping a stopwatch, combined) is ± 0.08 s, what is the precision (expressed as a percentage) in the determination of g? (e) What appears to be the main limitation on the accuracy with which g can be determined? How could the experiment be improved?

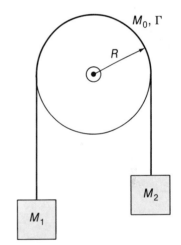

SPECIAL TOPIC

11-36 Show that the C.M. of the pie-shaped wedge of a uniform laminar circle illustrated in the diagram is a distance $x = (2R \sin \alpha)/3\alpha$ from the apex along the bisector. Examine X for $\alpha = \pi$ and $\alpha \to 0$. [*Hint:* Use polar coordinates.]

11-37 Determine the position of the C.M. of a figure cut from a uniform lamina bounded by the parabolic curve $x = ay^2$ and the ordinate $x = b$.

11-38 Determine the position of the C.M. of a thin circular disc with a radius R that has a nonuniform surface mass density given by $\sigma = \sigma_0 \cos(\phi/2)$ for

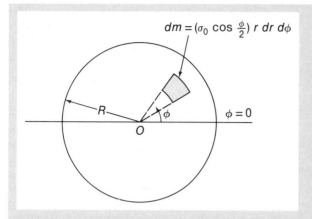

$$dm = (\sigma_0 \cos \tfrac{\phi}{2})\, r\, dr\, d\phi$$

$-\pi \leqslant \phi \leqslant \pi$, where ϕ is the azimuthal angle measured as shown in the diagram. [*Hint:* Use $\cos \phi = 1 - 2 \sin^2(\phi/2)$.]

11-39 A thin rod with a length ℓ and a mass M has a nonuniform linear mass distribution given by $\lambda = \lambda_0 x^2$, where x is measured from the center of the rod. Determine the rotational inertia about the perpendicular axis AA'.

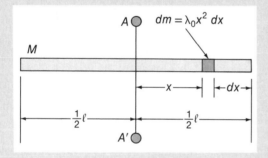

$$dm = \lambda_0 x^2\, dx$$

11-40 Refer again to the pie-shaped wedge in Problem 11-36. Show that the radius of gyration of the wedge about a perpendicular axis through the apex is $\Gamma = R/\sqrt{2}$.

11-41 Find the rotational inertia of a thin plate in the
 • shape of an equilateral triangle with a mass of 0.10 kg and sides of length 0.30 m, about a perpendicular axis through one corner. [*Hint:* Consider a series of strips perpendicular to the bisector line from the axis corner and use the parallel-axis theorem.]

11-42 Determine by direct integration the rotational
 • inertia of a thin ring with radius R and mass M about an axis tangent to the ring. [*Hint:* The equation of the ring with respect to an origin at the point of contact with the axis is $r = 2R \cos \phi$. The integral required is included in the list inside the back cover.]

11-43 Find the rotational inertia of a right circular
 •• cone which has an altitude h and a base radius R, about an axis along a diameter of the base. [*Hint:* Use the geometry shown in Example 11-6 and apply the parallel-axis theorem as in Example 11-10.]

11-44 The diagram below shows an assembly of four thin-shelled spheres and four slender connecting rods, all with the same mass, arranged in a square. Find the radius of gyration for this assembly about a perpendicular axis through the center of the square.

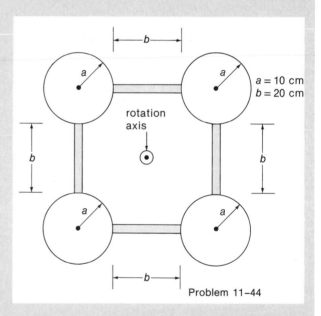

$a = 10$ cm
$b = 20$ cm

Problem 11-44

11-45 According to one model of the structure of the
 •• Sun, the density varies with the radial distance from the center of the Sun in the way shown in the table below. The radius of the Sun is R_S. Assume that the

r/R_S	0	0.05	0.10	0.15	0.20
(g/cm³)	148	125	86	56	36

r/R_S	0.30	0.40	0.60	0.80	1.00
(g/cm³)	12	4	0.5	0.1	0

Sun is a rigid body (this is not really a good assumption), and show that the radius of gyration about the solar rotation axis is approximately 0.3 R_S. [*Hint:* Consider the Sun to be composed of a series of shells and use Table 11–1(h) for each shell. Your result will depend to some extent on the precise way you choose to treat the contribution of each shell.]

11–46 A thin square plate with sides of 20 cm and a mass of 3 kg rotates about a perpendicular axis through its center. The plate is subjected to an axial torque, $\tau = \beta t$, where $\beta = 0.36 \text{ N} \cdot \text{m/s}$. (Neglect friction.) (a) What is the angular acceleration at the time $t = 2.5$ s after the torque is applied? (b) What is the corresponding angular speed if the plate was initially at rest? (c) Through what angle ϕ has the plate turned at time $t = 2.5$ s? (d) What is the angular momentum at the same time?

DYNAMICS OF RIGID BODIES

In this chapter we extend the treatment of rigid-body motion to more general situations than those discussed in the preceding chapter. We now remove the constraint permitting only rotation about a fixed axis, so that we may eventually study unrestricted motion. However, there is an intermediate stage of considerable importance. This is the class of problems in which the rotation axis of the body or system is not fixed but it is permitted to move only *parallel to itself* so that the orientation of the axis in space does not change. This type of combined translational and rotational motion occurs, for example, when a ball or a cylinder or a wheel rolls on a flat surface. Those cases in which torques act to *change* the orientation of the rotation axis are generally more complicated to treat than are the restricted cases we study first. At the end of this chapter we describe briefly some aspects of the general case.

12-1 THE RIGIDITY CONDITION AND MANY-PARTICLE SYSTEMS

We now undertake to apply the results obtained in Chapter 10 concerning many-particle systems to situations involving rigid bodies in the presence of external forces and the resulting torques. In contrast to the case of the two-particle problem, the general multi-particle problem cannot be solved exactly.* However, when the system of particles constitutes a rigid body, so that the object undergoes rotation and translation *as a whole,* the problem again is solvable.

We have seen that the general motion of a system of particles may be discussed conveniently in terms of the motion of the center of mass (C.M.) of the system together with the motions of the individual particles with respect to the C.M. The linear momentum **P** of a rigid body is defined by an obvious extension of our earlier definition for a system of particles (Eq. 9–12), namely,

$$\mathbf{P} = M\frac{d\mathbf{R}}{dt} = \int_M \frac{d\mathbf{r}}{dt}\,dm \tag{12-1}$$

---

*That is, general solutions cannot be obtained in terms of known mathematical functions except for very special situations. However, numerical calculations can be carried out to any desired accuracy, limited by the power of the computational facility and the patience of the investigator.

where $\mathbf{R}$ is the position vector of the C.M. and where $\mathbf{r}$ is the position vector of the differential mass element dm (see Fig. 12–1).

The translational motion of the C.M. depends only on the external applied forces $\mathbf{F}_n^e$:

$$\frac{d\mathbf{P}}{dt} = M\frac{d\mathbf{V}}{dt} = M\frac{d^2\mathbf{R}}{dt^2} = \sum_n \mathbf{F}_n^e \qquad \textbf{(12-2)}$$

The kinetic energy of the body is given by (compare Eq. 10–28)

$$K = \tfrac{1}{2}MV^2 + \tfrac{1}{2}\int_M v_C^2\, dm \qquad \textbf{(12-3)}$$

where $v_C^2 = \mathbf{v}_C \cdot \mathbf{v}_C$ and $\mathbf{v}_C = d\mathbf{r}_C/dt$. (We use a subscript C to denote quantities that relate to the C.M. of the body; see Fig. 12–1.) The results expressed by Eqs. 12–1, 12–2, and 12–3 are true *under all circumstances*.

The expression for the angular momentum of a rigid body can also be obtained by an extension of the previous results (refer to Eq. 10–29). For the angular momentum about the origin O, we write

$$\mathbf{L}_O = \mathbf{R} \times \mathbf{P} + \int_M (\mathbf{r}_C \times \mathbf{v}_C)\, dm \qquad \textbf{(12-4)}$$

where the angular momentum of the object with respect to its own C.M. is

$$\int_M (\mathbf{r}_C \times \mathbf{v}_C)\, dm = \mathbf{L}_C \qquad \textbf{(12-5)}$$

Fig. 12–1. Geometry for the discussion of the effects of forces acting on a rigid body. In diagrams such as this, we denote the position of the center of mass by C. The velocity of the C.M. is $d\mathbf{R}/dt = \mathbf{V}$.

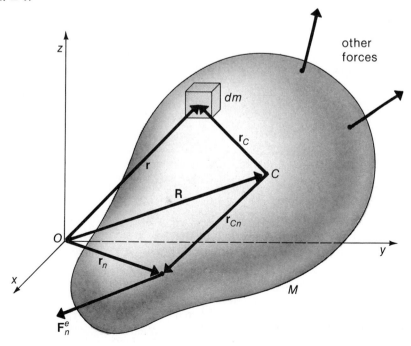

This quantity is sometimes called the *spin angular momentum*. The term $\mathbf{R} \times \mathbf{P} = M\mathbf{R} \times \mathbf{V}$ is the angular momentum of the C.M. with respect to the origin O (see Fig. 12–1); this is sometimes called the *orbital angular momentum*. Again, Eq. 12–4 is a *general result*.

The external forces $\mathbf{F}_n^e$ that are applied to a rigid body may give rise to torques, which are readily computed with respect to any desired point. We assume that we are dealing with a perfectly rigid body so that the internal forces do not influence the motion of the body as a whole. For the net torque τ_O about the origin O (Eq. 10–31), we have

$$\tau_O = \sum_n \mathbf{r}_n \times \mathbf{F}_n^e \tag{12–6}$$

where $\mathbf{r}_n$ locates the point of application of the force $\mathbf{F}_n^e$ with respect to the origin O (see Fig. 12–1). We also have $\mathbf{r}_n = \mathbf{R} + \mathbf{r}_{Cn}$, so that

$$\tau_O = \mathbf{R} \times \sum_n \mathbf{F}_n^e + \sum_n \mathbf{r}_{Cn} \times \mathbf{F}_n^e$$

The torque τ_O is equal to the time rate of change of the angular momentum $\mathbf{L}_O$; hence,

$$\tau_O = \mathbf{R} \times \sum_n \mathbf{F}_n^e + \sum_n \mathbf{r}_{Cn} \times \mathbf{F}_n^e = \frac{d}{dt}(\mathbf{R} \times \mathbf{P}) + \frac{d\mathbf{L}_C}{dt} \tag{12–7}$$

Now,

$$\frac{d}{dt}(\mathbf{R} \times \mathbf{P}) = \frac{d\mathbf{R}}{dt} \times \mathbf{P} + \mathbf{R} \times \frac{d\mathbf{P}}{dt}$$

$$= 0 + \mathbf{R} \times \sum_n \mathbf{F}_n^e$$

where the first term vanishes because it is equal to $M\mathbf{V} \times \mathbf{V}$. Then, Eq. 12–7 becomes

$$\sum_n \mathbf{r}_{Cn} \times \mathbf{F}_n^e = \frac{d\mathbf{L}_C}{dt} = \tau_C \tag{12–8}$$

This important general result states that the time rate of change of the angular momentum with respect to the C.M. is equal to the torque about the C.M. produced by the external forces. Notice that Eq. 12–8 is true *whatever* the nature of the C.M. motion. If the C.M. of an object is undergoing accelerated motion, any coordinate frame attached to the object is a noninertial frame; nevertheless, Eq. 12–8 is still valid.

We now specialize to the case of motion in which the angular momentum $\mathbf{L}_C$ maintains its orientation in space, although the magnitude may change due to the action of applied torques. Thus, the net externally applied torque τ_C must be parallel to $\mathbf{L}_C$. No restriction is otherwise placed on the applied forces and the torques they

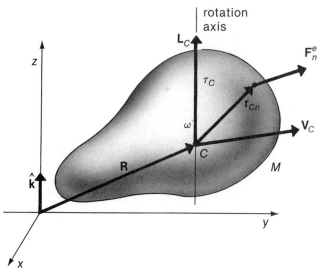

Fig. 12-2. The object moves in such a way that the rotation axis remains parallel to the z-axis.

produce. This situation corresponds to general translation of the C.M. of the object with rotation about the C.M. constrained to an axis that moves only parallel to itself. We take the z-axis to be parallel to the rotation axis, as shown in Fig. 12-2. In this case, we have (see Eqs. 11-5 through 11-8)

$$L_C = L_z = \omega \int_M \rho^2 \, dm = I_z \omega = I_C \omega \qquad \textbf{(12-9)}$$

and

$$\tau_C = \tau_z = \left(\sum_n \mathbf{r}_{Cn} \times \mathbf{F}_n^e \right)_z = I_C \frac{d\omega}{dt}$$

or

$$\tau_C = I_C \alpha \qquad \textbf{(12-10)}$$

where $\mathbf{r}_{Cn} \times \mathbf{F}_n^e$ has only a component that is parallel to the z-axis. The kinetic energy for this case is

$$K = \tfrac{1}{2} M V^2 + \tfrac{1}{2} I_C \omega^2 \qquad \textbf{(12-11)}$$

Example 12-1

Two massless ropes are wrapped around a uniform cylinder that has a mass M. The ropes are attached to the ceiling, as shown in the diagram on the next page. Initially, the unwrapped portions of the ropes are vertical and the cylinder is horizontal. Find the linear acceleration of the falling cylinder and the tension in the ropes.

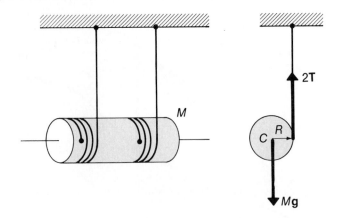

Solution:

Refer to the free-body diagram at the right. The tension in each rope is T, so the torque about the cylinder axis (which passes through the C.M.) is $2TR$. Also, $I_C = \frac{1}{2}MR^2$ (Table 11–1[c]); thus, $\tau_C = I_C\alpha$ becomes

$$2TR = \tfrac{1}{2}MR^2\alpha$$

If the downward direction is taken to be positive, the vertical acceleration a of the C.M. is obtained from

$$Mg - 2T = Ma$$

If the ropes do not slip on the cylinder, we have $a = R\alpha$. Then, eliminating T between the two equations above yields

$$a = \tfrac{2}{3}g$$

Finally, substituting this value of a into either equation containing T gives

$$T = \tfrac{1}{6}Mg$$

(•Why does the cylinder not swing to one side as it falls?)

Example 12–2

A flat uniform circular disc with a mass $M = 2\,\text{kg}$ and a radius $R = 10\,\text{cm}$ rests on a horizontal frictionless surface. A constant force $F = 5\,\text{N}$ is applied to the end of a string that is wrapped around the disc, thereby causing the disc to rotate about a vertical axis and to translate in the horizontal direction.
(a) Find the linear acceleration a of the C.M. of the disc and the angular acceleration α about the C.M.
(b) Find the acceleration a_s of the free end of the string.
(c) Show that the rate at which the applied force does work is equal to the rate at which the kinetic energy of the disc is increasing.

Solution:

(a) The linear acceleration of the C.M. is

$$a = \frac{F}{M} = \frac{5\,\text{N}}{2\,\text{kg}} = 2.5\,\text{m}/\text{s}^2 \tag{1}$$

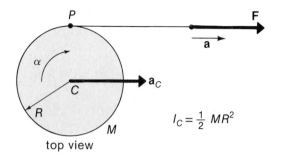

top view

$$I_C = \tfrac{1}{2} MR^2$$

Equating the torque about a vertical axis through the C.M. to $I_C\alpha$, we find

$$\alpha = \frac{\tau_C}{I_C} = \frac{FR}{\tfrac{1}{2}MR^2} = \frac{2F}{MR} = \frac{2(5\ \text{N})}{(2\ \text{kg})(0.10\ \text{m})} = 50\ \text{rad/s}^2 \tag{2}$$

(b) To obtain the acceleration of the free end of the string with respect to the horizontal surface, we note that the velocity v_0 of the string with respect to the surface at, say, point P is equal to the velocity of the point P with respect to the C.M. (namely, $v_T = R\omega$) *plus* the velocity of the C.M. with respect to the surface (namely, V):

$$v_0 = R\omega + V \tag{3}$$

Differentiating this expression with respect to time gives

$$a_s = \frac{dv_0}{dt} = R\alpha + a = \frac{3F}{M} = \frac{3(5\ \text{N})}{2\ \text{kg}} = 7.5\ \text{m/s}^2$$

(c) From conservation of energy (Eq. 12–11), we expect

$$\frac{dK}{dt} = Fv_0$$

or

$$\frac{d}{dt}(\tfrac{1}{2}MV^2 + \tfrac{1}{2}I_C\omega^2) = Fv_0 \tag{4}$$

from which

$$MVa + \tfrac{1}{2}MR^2\omega\alpha = Fv_0 \tag{5}$$

Substituting from (1) and (2) into (5) reproduces (3), thereby proving the assertion expressed in (4).

Example 12–3

Next, consider a system that consists of a student ($m = 70$ kg) and a plank. The plank is uniform and narrow, with a length $2b = 5$ m and a mass $M = 50$ kg. Initially, the plank is at rest on a frictionless horizontal surface. The student runs toward the plank in a direction at right angles to its length with a velocity $v = 3$ m/s, then jumps onto the end, as shown in diagram (a). Determine the position of the plank (with the student on its end) a time 1.2 s later.

Solution:

The system consists of a particle (the student) and a rigid body (the plank). In the final condition, the student-particle is attached rigidly to the end of the plank. There are no

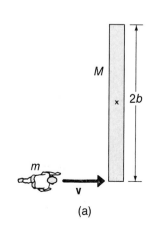

(a)

horizontal external forces acting on the system, so the horizontal linear momentum is conserved. In the initial condition the linear momentum P_i is the momentum of the student:

$$P_i = mv$$

and in the final condition,

$$P_f = (m + M)V$$

where V is the velocity of the C.M. of the system, as indicated in diagram (b). Requiring $P_i = P_f$ gives

$$V = \frac{mv}{m + M} = \frac{(70 \text{ kg})(3 \text{ m/s})}{70 \text{ kg} + 50 \text{ kg}} = 1.75 \text{ m/s}$$

The C.M. is located at a point that is a perpendicular distance d from the straight-line path of the running student. We can write

$$(m + M)d = Mb$$

(•Can you see why?) Then,

$$d = \frac{Mb}{m + M} = \frac{(50 \text{ kg})(2.5 \text{ m})}{70 \text{ kg} + 50 \text{ kg}} = 1.04 \text{ m}$$

There are no external torques acting on the system, so the angular momentum is conserved. Initially, the angular momentum about a vertical axis through the C.M. is

$$L_i = mvd$$

In the final condition,

$$L_f = I_C \omega = [md^2 + M\{\tfrac{1}{3}b^2 + (b - d)^2\}]\omega$$

(•Can you explain this equation?) Requiring $L_i = L_f$ gives

$$\omega = \frac{mvd}{I_C} = \frac{mvd}{md^2 + M\{\tfrac{1}{3}b^2 + (b - d)^2\}}$$

$$= \frac{(70)(3)(1.04)}{(70)(1.04)^2 + 50\{\tfrac{1}{3}(2.5)^2 + (2.5 - 1.04)^2\}}$$

$$= 0.762 \text{ rad/s}$$

Therefore, 1.2 s after the student lands on the plank, the C.M. will have moved a distance $Vt = (1.75 \text{ m/s})(1.2 \text{ s}) = 2.10 \text{ m}$ to the right, and the plank-plus-student will have rotated through an angle $\omega t = (0.762 \text{ rad/s})(1.2 \text{ s}) = 0.914 \text{ rad} = 52.4°$ about the C.M. Diagram (c) shows the initial and the final conditions.

The C.M. of the system moves in a straight line along the dotted path. Notice the initial *backward* motion of the end of the plank opposite the student.

12-2 ROLLING MOTION

We now consider the important case of cylinders and spheres rolling on flat surfaces. We must distinguish carefully between the case in which there is rolling *without* slipping and that in which slipping or sliding *does* occur. Consider first the case of a cylinder rolling without slipping on a horizontal surface, as illustrated in Fig. 12–3. The translational velocity of the center of the cylinder is **V** and the rotational speed about the center is ω. During the time interval Δt, the center moves to a position C', a distance $V \Delta t$ from the original position C. During the same interval, the point P on the rim of the cylinder, which was originally in contact with the surface, rotates to the position P', such that $\phi = \omega \Delta t$. A different point Q' on the rim is now in contact with the surface. We define rolling without slipping in the self-evident way, namely, that the length of the arc $\overparen{P'Q'}$ on the cylinder rim be equal to the distance of translational motion, $\overline{CC'}$. That is,

$$\overparen{P'Q'} = R\phi = R\omega\,\Delta t = \overline{CC'} = V\,\Delta t$$

so that

$$R\omega = V \qquad\qquad \textbf{(12–12)}$$

Rolling without slipping requires that Eq. 12–12 be satisfied *at all times*. Whenever $R\omega \neq V$, slipping occurs; we return later to the discussion of this case.

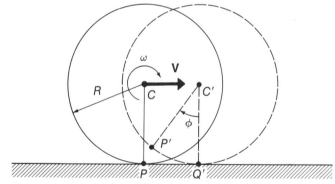

Fig. 12–3. Two positions of a rolling cylinder separated by a time interval Δt. The center of the cylinder moves with velocity **V** and the angular speed is ω. Rolling without slipping requires $V = R\omega$.

When an object rolls without slipping, it can be viewed as executing instantaneous pure rotation about the point or line of contact with the surface on which it is rolling. To see that this statement is indeed correct, refer to Fig. 12–4, where a cylinder is shown rolling without slipping. Consider a point Q that is a distance y from the contact point P along the line $\overline{PC}$ extended. The velocity $\mathbf{v}_Q$ of Q with respect to the surface at P can be found by adding the velocity $\mathbf{v}_{QC}$ of Q relative to C to the velocity **V** of C relative to P. Now, $\mathbf{v}_{QC} = (y - R)\omega\hat{\mathbf{i}}$, where $\hat{\mathbf{i}}$ is a unit vector in the horizontal direction. Then,

$$\mathbf{v}_Q = \mathbf{V} + (y - R)\omega\hat{\mathbf{i}}$$

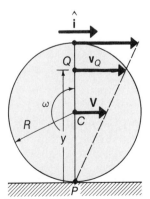

Fig. 12–4. The geometry for discussing rolling without slipping.

Because the velocities are all in the same direction, we can write

$$v_Q = V + (y - R)\omega$$
$$= R\omega + (y - R)\omega$$
$$= y\omega$$

The interpretation of this last equality is that point Q instantaneously rotates with the angular speed ω about the contact point P.

In this analysis the point Q can be anywhere along the extended line $\overline{PC}$. When $y = R$, for example, we have $v_Q = V$. For any other point, $v_Q = Vy/R$; hence, the velocity vector $\mathbf{v}_Q$ is perpendicular to the line $\overline{PC}$ with magnitude proportional to the distance from $\overline{PC}$ to the dotted line shown in Fig. 12–4.

▶ *The Velocity of a General Point.* The argument we have just made is strictly valid only for points along the line $\overline{PC}$ and its extension in Fig. 12–4. What about a general point in the cylinder, such as point Q in Fig. 12–5? Here, $\mathbf{r} = \mathbf{R} + \mathbf{r}_c$ locates Q with respect to P. Taking the vector product of this expression with ω from the left, we have

$$\omega \times \mathbf{r} = \omega \times \mathbf{R} + \omega \times \mathbf{r}_c$$

Because the wheel is assumed to roll without slipping, we have, from Eq. 12–12, $\omega \times \mathbf{R} = \mathbf{V}$. Also, $\omega \times \mathbf{r}_c = \mathbf{v}_{QC}$ is just the instantaneous velocity of Q with respect to C. Therefore,

$$\omega \times \mathbf{r} = \mathbf{V} + \mathbf{v}_{QC}$$

We identify the right-hand side of this equation as the velocity of Q with respect to the point P; hence,

$$\mathbf{v}_Q = \omega \times \mathbf{r}$$

thereby proving that *any* point Q may be viewed as undergoing instantaneous rotation about the point of contact P with the angular velocity ω. ◀

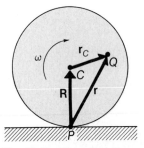

Fig. 12–5. The point Q undergoes instantaneous rotation about the contact point P.

Kinetic Energy of a Rolling Object. The kinetic energy of an axially symmetric object that rolls without slipping is

$$K = \tfrac{1}{2}MV^2 + \tfrac{1}{2}I_C\omega^2 \qquad \textbf{(12–13)}$$
$$= \tfrac{1}{2}(MR^2 + I_C)\omega^2$$

where V is the velocity of the C.M. and where I_C is the rotational inertia about the symmetry axis (which passes through the C.M.). The object may also be considered to be instantaneously rotating about the contact point P with angular speed ω, so

$$K = \tfrac{1}{2}I_P\omega^2$$

[Notice that this result is consistent with the parallel-axis theorem (Section 11–4). Using Eq. 11–19, we have

$$I_P = I_C + MR^2$$

Substituting this expression for I_P into $K = \tfrac{1}{2}I_P\omega^2$ reproduces Eq. 12–13. That is, for the purpose of calculating the kinetic energy, viewing the object as rotating instantaneously about the contact point is equivalent to using the parallel-axis theorem.]

Example 12-4

A uniform cylinder with a mass M rolls without slipping down an inclined plane. Find the linear acceleration of the center of the cylinder.

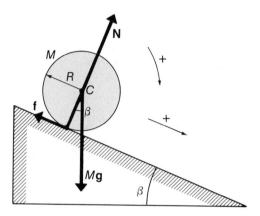

Solution:

In the free-body diagram we take the downward direction along the plane to be positive, so the force equation is

$$Mg \sin \beta - f = Ma$$

For the torques about the C.M. of the cylinder, we take clockwise rotation to be positive and write

$$\tau_C = fR = I_C\alpha = \tfrac{1}{2}MR^2\alpha$$

Rolling without slipping means

$$R\alpha = a$$

Solving these three equations for a by eliminating f, we find

$$a = \tfrac{2}{3}g \sin \beta$$

and for the frictional force we obtain

$$f = \tfrac{1}{3}Mg \sin \beta$$

The normal force N is $\qquad N = Mg \cos \beta$

Then, $\qquad\qquad\qquad \dfrac{f}{N} = \tfrac{1}{3}\tan \beta$

In order to prevent slipping, the coefficient of static friction must be equal to or larger than this value; thus,

$$\mu_s \geqslant \tfrac{1}{3}\tan \beta$$

(•Can you see why this is so? Refer to the discussion in Section 6-2.)

Next, let us examine the energy balance in this problem. We have, at any time,

$$K = \tfrac{1}{2}MV^2 + \tfrac{1}{2}I_C\omega^2$$
$$= \tfrac{1}{2}MV^2 + \tfrac{1}{4}MR^2\omega^2$$

Using $V = R\omega$, we find

$$K = \tfrac{3}{4}MV^2$$

Now, the increase in kinetic energy which we have just obtained is equal to the decrease in potential energy less the work done against friction W_f. If s represents the displacement along the plane, we can write

$$K = Mgs \sin\beta - W_f$$

or

$$\tfrac{3}{4}MV^2 = Mgs \sin\beta - W_f$$

Differentiating with respect to time, we obtain

$$\tfrac{3}{2}MVa = MgV \sin\beta - \frac{d}{dt}W_f$$

Using the previous result $a = \tfrac{2}{3}g \sin\beta$, to substitute for a, we find

$$MgV \sin\beta = MgV \sin\beta - \frac{d}{dt}W_f$$

(a) $t = 0$

from which we see that $dW_f/dt = 0$. Thus, as the motion proceeds, W_f does not change, and we conclude that $W_f = 0$. This means that *no work is done against the static frictional force*. The reason is that in rolling without slipping there is no relative motion between the cylinder and the plane at the contact point, so no work is done by or against the static frictional force. (If the motion involved slipping, work would be done against the kinetic frictional force. Be certain that you understand the difference between these two situations.)

In Section 6–2 we argued that ideal rolling without slipping along a horizontal surface involves no friction between the rolling object and the surface. In this example, however, we have ideal rolling without slipping that *does* involve a frictional force. Notice how the presence of friction is *demanded* here in order to produce rolling. (•If the cylinder were rolling *up* the plane, what would be the direction of **f**?) The magnitude of the frictional force varies with the sine of the inclination angle and so vanishes for horizontal rolling, as we argued previously.

(b) $0 < t < T$

Example 12–5

A uniform solid sphere with a mass $M = 2$ kg and a radius $R = 10$ cm is set into rotation with an angular speed $\omega_0 = 60$ rad/s. At $t = 0$ the sphere is dropped a short distance (without bouncing) onto a horizontal surface, as shown in diagram (a) at left. There is friction between the sphere and the surface.

(a) What is the angular speed of rotation when the sphere finally rolls without slipping at time $t = T$?

(b) What amount of kinetic energy is lost by the sphere between $t = 0$ and $t = T$?

(c) Show that the result of (b) is equal to the work done against the frictional force that acts to cause the sphere to roll without slipping.

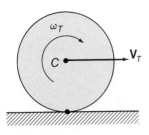

(c) $t = T$

Solution:

(a) Refer to the free-body diagram (b). We take the positive direction to be to the right.

The translational acceleration of the C.M. of the sphere is obtained from the force equation,

$$Ma = f \tag{1}$$

(•Can you see why the frictional force $\mathbf{f}$ that acts on the sphere is directed toward the *right?*) We can write $a = dV/dt$ and integrate Eq. (1) between $t = 0$ and $t = T$, obtaining

$$\int_{V_0=0}^{V_T} dV = \frac{1}{M} \int_0^T f \, dt \tag{2}$$

We see presently that the desired solution can be obtained without knowing the explicit time dependence of f.

We take the clockwise direction of rotation to correspond to positive torques, so

$$\tau_C = -fR = I_C \alpha \tag{3}$$

where $I_C = \frac{2}{5}MR^2$ is the rotational inertia of the sphere about an axis through the C.M. We can write $\alpha = d\omega/dt$ and integrate Eq. (3), obtaining

$$\int_{\omega_0}^{\omega_T} d\omega = \omega_T - \omega_0 = -\frac{5}{2R}\left(\frac{1}{M}\int_0^T f \, dt\right) \tag{4}$$

Eliminating the (unknown) integral of f between Eqs. (2) and (4), we find

$$\omega_T - \omega_0 = -\frac{5}{2}\frac{V_T}{R}$$

At $t = T$, we have pure rolling motion, so $\omega_T R = V_T$; then,

$$\omega_T = \tfrac{2}{7}\omega_0 = \tfrac{2}{7}\,(60 \text{ rad/s}) = 17.14 \text{ rad/s} \tag{5}$$

Notice that this result does *not* depend on the coefficient of friction between the sphere and the surface.

(b) The initial kinetic energy is

$$K_0 = \tfrac{1}{2}I_C\omega_0^2 = \tfrac{1}{2}\cdot\tfrac{2}{5}MR^2\cdot\omega_0^2 \tag{6}$$

$$= \tfrac{2}{10}(2 \text{ kg})(0.10 \text{ m})^2(60 \text{ rad/s})^2$$

$$= 14.4 \text{ J}$$

The kinetic energy of the sphere at $t = T$ is

$$K_T = \tfrac{1}{2}MV_T^2 + \tfrac{1}{2}I_C\omega_T^2 \tag{7}$$

Using $V_T = R\omega_T$, $I_C = \frac{2}{5}MR^2$, and $\omega_T = 2\omega_0/7$ from Eq. (5), we find

$$K_T = \tfrac{1}{2}I_C\omega_0^2 = \tfrac{2}{7}K_0$$

Therefore, the change in kinetic energy is

$$K_T - K_0 = -\tfrac{5}{7}K_0 = -\tfrac{5}{7}\,(14.4 \text{ J})$$

$$= -10.29 \text{ J}$$

where the negative sign indicates a *loss* of kinetic energy.

(c) To calculate directly the work done against the frictional force **f**, we must first determine the amount of slippage between the sphere and the surface. During a time interval dt (for $0 < t < T$), a point on the perimeter of the sphere rotates through an arc length $R\omega\, dt$, while the center moves a distance $V\, dt$. The amount of slippage or relative displacement ds is

$$ds = (R\omega - V)\, dt$$

To check this expression, notice that at $t = 0$ we have $V = 0$, so $ds = R\omega\, dt$, as expected. Also, at $t = T$, we have $V = V_T = R\omega$, so $ds = 0$; that is, when slipping ceases, the relative displacement becomes zero.

The work done against the frictional force is

$$dW_f = f\, ds = fR\omega\, dt - fV\, dt$$

For f in the first term we use Eq. (3), and for f in the second term we use Eq. (1); hence,

$$dW_f = -I_C\alpha\omega\, dt - MaV\, dt$$
$$= -I_C\omega\, d\omega - MV\, dV$$

where we have used $\alpha = d\omega/dt$ and $a = dV/dt$. Integrating between $t = 0$ and $t = T$, we find

$$W_f = -\tfrac{1}{2}I_C(\omega_T^2 - \omega_0^2) - \tfrac{1}{2}MV_T^2$$

Using Eqs. (6) and (7), this becomes

$$W_f = K_0 - K_T = 10.29\ \text{J}$$

which states that the work done against friction is equal to the decrease in kinetic energy.

SPECIAL TOPIC

12-3 GENERAL ROTATION

In this section we describe some of the features of the general case of the rotation of a rigid body. First, we write the relevant equations in a coordinate frame whose origin O_C is located at the C.M. of the object. We require Eqs. 12-5 and 12-8 (see Fig. 12-1):

$$\mathbf{L}_C = \int_M (\mathbf{r}_C \times \mathbf{v}_C)\, dm$$

$$\boldsymbol{\tau}_C = \frac{d\mathbf{L}_C}{dt}$$

Also, using Eq. 4-34, we can write

$$\mathbf{v}_C = \frac{d\mathbf{r}_C}{dt} = \boldsymbol{\omega} \times \mathbf{r}_C$$

where $\boldsymbol{\omega}$ is the angular velocity of the object (and of dm as well) with respect to the O_C coordinate frame. Substituting for $\mathbf{v}_C$ in the expression for $\mathbf{L}_C$ gives

$$\mathbf{L}_C = \int_M [\mathbf{r}_C \times (\boldsymbol{\omega} \times \mathbf{r}_C)]\, dm \qquad \textbf{(12–14)}$$

All of the equations above are completely general; in particular, they are valid in any convenient reference frame whose origin O_C is located at the C.M. of the object. In this frame the Cartesian unit vectors are $\hat{\mathbf{i}}_C$, $\hat{\mathbf{j}}_C$, and $\hat{\mathbf{k}}_C$, as indicated in Fig. 12-6.

Because the C.M. of the object can be in motion (even accelerating), we require an inertial reference frame to describe this motion in terms of Newton's laws. In Fig. 12-6 we also show this frame, with origin O and unit vectors $\hat{\mathbf{i}}$, $\hat{\mathbf{j}}$, and $\hat{\mathbf{k}}$. The vector $\mathbf{R}$ locates the C.M. of the object in this coordinate frame.

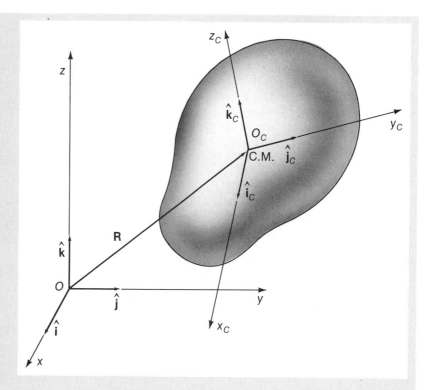

Fig. 12–6. The inertial frame O has unit vectors $\hat{\mathbf{i}}$, $\hat{\mathbf{j}}$, and $\hat{\mathbf{k}}$. The frame O_C has its origin at the C.M. of the object.

Consider now the orientation of the axes whose origin is fixed at the C.M. of the object. Bear in mind that we eventually wish to differentiate Eq. 12–14 with respect to time in order to relate it to the applied external torques. Then, two possibilities for the orientation suggest themselves. In the first case we select a fixed orientation of the O_C frame relative to the O frame. This has the virtue of making the time derivatives $\hat{\mathbf{i}}_C$, $\hat{\mathbf{j}}_C$, and $\hat{\mathbf{k}}_C$ all zero. However, we now encounter a substantial difficulty because, as the object rotates, it changes its orientation with respect to the O_C frame. This causes the terms involving $\mathbf{r}_C$ (which identifies a particular dm) to vary with time. To overcome this problem we choose a second set of coordinate axes with an origin O_B that also coincides with O_C but whose axes are *fixed* with respect to the object. This frame is called the *body-centered frame*. To call attention to this fact we label the unit vectors in this frame $\hat{\mathbf{i}}_B$, $\hat{\mathbf{j}}_B$, and $\hat{\mathbf{k}}_B$. The vector $\boldsymbol{\omega}$ represents both the angular velocity of the rigid body *and* the angular velocity of the body-centered coordinate frame.

The equations for the important dynamical quantities may now be expressed in terms of the body-centered unit vectors. For example, the angular momentum of the object referred to the rotating axes is

$$\mathbf{L}_B = \int_M [\mathbf{r}_B \times (\boldsymbol{\omega} \times \mathbf{r}_B)]\, dm \qquad \textbf{(12–15a)}$$

or

$$\mathbf{L}_B = L_x \hat{\mathbf{i}}_B + L_y \hat{\mathbf{j}}_B + L_z \hat{\mathbf{k}}_B \qquad \textbf{(12–15b)}$$

and the angular velocity is

$$\boldsymbol{\omega} = \omega_x \hat{\mathbf{i}}_B + \omega_y \hat{\mathbf{j}}_B + \omega_z \hat{\mathbf{k}}_B \qquad \textbf{(12–16)}$$

However, when we wish to relate the torques to the time rate of change of the angular momentum $\mathbf{L}_B$, we must take account of the fact that the unit vectors, $\hat{\mathbf{i}}_B$, $\hat{\mathbf{j}}_B$, and $\hat{\mathbf{k}}_B$, have a time variation when referred to the O axes.

The time derivative of $\mathbf{L}_B$ is given by

$$\frac{d\mathbf{L}_B}{dt} = \frac{dL_x}{dt}\hat{\mathbf{i}}_B + \frac{dL_y}{dt}\hat{\mathbf{j}}_B + \frac{dL_z}{dt}\hat{\mathbf{k}}_B$$

$$+ L_x \frac{d\hat{\mathbf{i}}_B}{dt} + L_y \frac{d\hat{\mathbf{j}}_B}{dt} + L_z \frac{d\hat{\mathbf{k}}_B}{dt}$$

We can write $d\mathbf{r}_B/dt = \boldsymbol{\omega} \times \mathbf{r}_B$, from which, as special cases, we obtain

$$\frac{d\hat{\mathbf{i}}_B}{dt} = \boldsymbol{\omega} \times \hat{\mathbf{i}}_B, \quad \frac{d\hat{\mathbf{j}}_B}{dt} = \boldsymbol{\omega} \times \hat{\mathbf{j}}_B, \quad \text{and} \quad \frac{d\hat{\mathbf{k}}_B}{dt} = \boldsymbol{\omega} \times \hat{\mathbf{k}}_B$$

Substituting these relationships into the expression for the time derivative of $\mathbf{L}_B$ gives the desired result:

$$\tau_B = \frac{d\mathbf{L}_B}{dt}$$

$$= \left(\frac{dL_x}{dt}\hat{\mathbf{i}}_B + \frac{dL_y}{dt}\hat{\mathbf{j}}_B + \frac{dL_z}{dt}\hat{\mathbf{k}}_B\right) + \boldsymbol{\omega} \times \mathbf{L}_B \qquad (12\text{--}17)$$

where the term in parentheses represents the time variation of the angular momentum *in the body-centered frame,* and where $\boldsymbol{\omega} \times \mathbf{L}_B$ represents the effect of the rotation of the body-centered frame with respect to the inertial frame O. This equation is known as *Euler's equation.*[*]

Dynamic Imbalance. Consider an eccentrically mounted dumbbell that rotates about a fixed axis (Fig. 12–7). The two balls have masses m and are connected by a rigid massless rod with a length $2l$. The center of the rod is attached at a fixed angle γ to a shaft AA' about which the dumbbell rotates with a constant angular velocity $\boldsymbol{\omega}$. For convenience, we select the z_B-axis to coincide with the connecting rod and the x_B-axis to lie in the plane formed by

Fig. 12–7. The dumbbell rotates at the fixed angle γ about the axis AA'. The unit vectors for the body-centered frame of the dumbbell are shown.

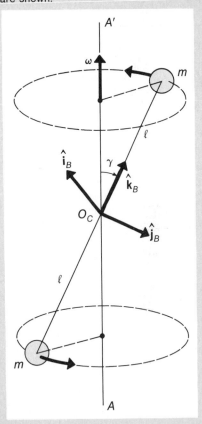

the rod and the rotation axis AA'; we require a right-handed coordinate frame, so the y_B-axis is determined by the choices for the other two axes.

The position vector for the upper ball is $\mathbf{r}_B = l\hat{\mathbf{k}}_B$ and for the lower ball it is $\mathbf{r}'_B = -l\hat{\mathbf{k}}_B$. The components of the angular velocity vector are

$$\omega_x = \omega \sin\gamma, \qquad \omega_y = 0, \qquad \omega_z = \omega \cos\gamma$$

We consider the dumbbell balls to be particles, so Eq. 12–15a becomes

$$\begin{aligned}\mathbf{L}_B &= m\mathbf{r}_B \times (\boldsymbol{\omega} \times \mathbf{r}_B) + m\mathbf{r}'_B \times (\boldsymbol{\omega} \times \mathbf{r}'_B) \\ &= 2ml^2\omega\{\hat{\mathbf{k}}_B \times [(\sin\gamma\,\hat{\mathbf{i}}_B + \cos\gamma\,\hat{\mathbf{k}}_B) \times \hat{\mathbf{k}}_B]\} \\ &= 2ml^2\omega\sin\gamma\,\hat{\mathbf{i}}_B \qquad (12\text{--}18)\end{aligned}$$

That is, the angular momentum vector is perpendicular to the connecting rod and is in the plane formed by the rod and the rotation axis AA'.

Because $\mathbf{L}_B$ is constant in the body-centered frame, in Eq. 12–17 the term in parentheses vanishes, and we have

$$\begin{aligned}\tau_B &= \boldsymbol{\omega} \times \mathbf{L}_B \\ &= 2ml^2\omega^2\sin\gamma\,(\sin\gamma\,\hat{\mathbf{i}}_B + \cos\gamma\,\hat{\mathbf{k}}_B) \times \hat{\mathbf{i}}_B \\ &= 2ml^2\omega^2\sin\gamma\cos\gamma\,\hat{\mathbf{j}}_B \qquad (12\text{--}19)\end{aligned}$$

That is, the torque τ_B is perpendicular to the connecting rod and also perpendicular to $\mathbf{L}_B$. This torque is supplied (in some unspecified way) to the rotation shaft AA' by the bearing supports.

The expressions we have obtained for $\boldsymbol{\omega}$, $\mathbf{L}_B$, and τ_B are valid at all times as the dumbbell rotates. In an inertial frame O_C fixed with respect to the axis AA', $\boldsymbol{\omega}$ is a constant vector but the angular momentum $\mathbf{L}_C$ and the torque τ_C are not constant vectors. In this frame, the magnitudes of $\mathbf{L}_C$ and τ_C are constant but their directions continually change, following the rotation of the x_B-axis and y_B-axis, respectively. Thus, the angular momentum vector $\mathbf{L}_C$ precesses about the rotation axis AA', as shown in Fig. 12–8.

The fact that τ_C precesses about the axis AA' means that the shaft bearings must provide the corresponding torque to the shaft. The presence of any mechanical slack in the system will cause a vibration at the frequency ω which may produce undesirable results. Such a condition for the system we are studying here is called *dynamic imbalance.* For irregular objects mounted on rotation shafts, dynamic imbalance can be detected only when the object is rotating. In order to correct the objectionable vibration due to dynamic imbalance of automobile tires, the tires must be "spin balanced" with weights added to align the angular momentum vector with the rotation axis. This type of imbalance cannot be corrected completely by "static balancing."

---

[*]First derived in 1758 by the Swiss mathematician Leonhard Euler (1707–1783).

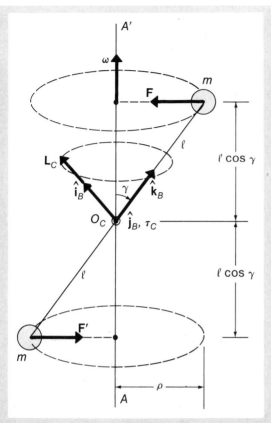

Fig. 12-8. The precession of the angular momentum vector $\mathbf{L}_C$ in the inertial frame O_C. The vector $\mathbf{L}_C$ is shown at the instant the two balls of the dumbbell are in the plane of the page. At this instant, the unit vector $\hat{\mathbf{j}}_B$ and the torque vector τ_C are directed out of the page, toward the reader.

There is a simple explanation for the torque that is supplied to the rotation axis (Eq. 12-19). The force $\mathbf{F}$ that must be exerted on the upper ball m in order to maintain its circular motion about the axis AA' and the associated centripetal acceleration is $F = m\rho\omega^2 = m\ell\omega^2 \sin\gamma$, directed as shown in Fig. 12-8. The force $\mathbf{F}'$ on the lower ball has the same magnitude but the opposite direction. Each force is a distance $\ell \cos\gamma$ from O_C, so the total torque about O_C produced by the two forces is $2m\ell^2\omega^2 \sin\gamma \cos\gamma$, just as we found in Eq. 12-19.

The Spinning Top. Finally, we consider some of the simple features of gyroscopic motion. The description of the general motion of a spinning top or a gyroscope supported at a fixed point is tedious and complicated. There is, however, a particular way of releasing a spinning top so that its subsequent motion is easy to understand. Figure 12-9a shows a spinning top that is free to rotate in the universal frictionless bearing at the support point O. The top has a mass $M = 0.30$ kg, which we assume is concentrated entirely in a thin ring with radius $R = 5$ cm. The

support point O is a distance $\ell = 8$ cm from the C.M. of the top, located at the center of the ring.

We describe the motion with respect to an inertial reference frame with origin at the point O. The simplest gyroscopic action of the top consists of a rapid spinning about its own axis with a spin angular velocity $\boldsymbol{\omega}_s$ together with a slow precession of the C.M. about the vertical axis with a

Fig. 12-9. (a) The motion of a simple gyroscope about the frictionless bearing at O. The vertical axis is the *precessional axis,* and the axis of the top is the *spin axis.* (b) The change of the angular momentum during a time interval Δt.

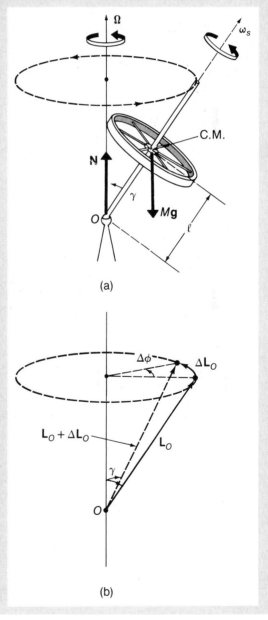

(a)

(b)

constant precessional angular velocity Ω. Throughout the motion, the inclination angle γ remains constant. How do we arrange for this type of motion?

Refer to Fig. 12–9a. We postulate motion with the C.M. always at the same height. Therefore, the bearing force N must be equal to Mg. Consequently, the torque is

$$\tau_0 = N\ell \sin \gamma = Mg\ell \sin \gamma \qquad \textbf{(12–20)}$$

The torque vector $\boldsymbol{\tau}_0$ is perpendicular to the plane defined by the precessional axis and to the spin axis, and is directed inward (away from the reader). Now, we have assumed that the precessional motion is slow, so $\omega_s \gg \Omega$. Then, we can neglect the angular momentum about the precessional axis and write

$$L_0 \cong I\omega_s = MR^2\omega_s \qquad \textbf{(12–21)}$$

where $I = MR^2$ is the rotational inertia of the top about its own axis.

Figure 12–9b shows the change in angular momentum $\Delta \mathbf{L}_0$ during a short time interval Δt. We also see in this figure that $\Delta L_0 = (L_0 \sin \gamma) \Delta \phi$. Then, we can write

$$\Omega = \frac{d\phi}{dt} = \frac{d\phi}{dL_0}\frac{dL_0}{dt} = \frac{1}{L_0 \sin \gamma}\frac{dL_0}{dt} = \frac{\tau_0}{L_0 \sin \gamma} \qquad \textbf{(12–22)}$$

or, in vector notation,

$$\boldsymbol{\tau}_0 = \boldsymbol{\Omega} \times \mathbf{L}_0$$

Assume that $\omega_s = 1200$ rpm and $\gamma = 20°$; then, we have

$$L_0 = MR^2\omega_s = (0.30 \text{ kg})(0.05 \text{ m})^2(20 \cdot 2\pi \text{ rad/s})$$
$$= 0.0942 \text{ kg} \cdot \text{m}^2/\text{s}$$

and

$$\tau_0 = Mg\ell \sin \gamma = (0.30 \text{ kg})(9.8 \text{ m/s}^2)(0.08 \text{ m})(\sin 20°)$$
$$= 0.0804 \text{ N} \cdot \text{m}$$

The precessional frequency is

$$\Omega = \frac{\tau_0}{L_0 \sin \gamma} = \frac{0.0804 \text{ N} \cdot \text{m}}{(0.0942 \text{ kg} \cdot \text{m}^2/\text{s})(\sin 20°)}$$
$$= 2.495 \text{ rad/s} = 23.8 \text{ rpm}$$

so that we indeed have $\omega_s \gg \Omega$.

The motion of the top will follow this description if it is released with the required precessional angular frequency, namely, $\Omega = 23.8$ rpm.

What would happen if all of the conditions were the same except that the top is released (spinning at the rate ω_s) with the C.M. at rest? The top would first begin to fall, and the resulting rotation about a horizontal axis would give rise to a torque that would produce a *counterclockwise* displacement, with the C.M. eventually rising to its original height. Thus, the C.M. would undergo a cusplike motion, as indicated in Fig. 12–10a. This type of motion is called *nutation*. If the C.M. were given a small horizontal velocity component upon release, the resulting motion would be that in either Fig. 12–10b or c, depending on the direction of the push.

The Moments of Inertia. The definition of the angular momentum about the C.M. given in the body-centered frame is (Eq. 12–15)

$$\mathbf{L}_B = \int_M [\mathbf{r}_B \times (\boldsymbol{\omega} \times \mathbf{r}_B)] \, dm$$

If we use the vector identity (Eq. 3–28), $\mathbf{A} \times (\mathbf{B} \times \mathbf{A}) =$

Fig. 12–10. Different nutational motions of a spinning top corresponding to different initial conditions. The curved paths represent the motion of the C.M.

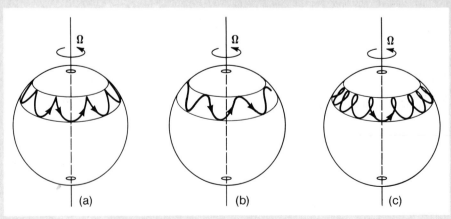

$A^2B - A(A \cdot B)$, the expression for L_B becomes

$$L_B = \int_M [r_B^2 \omega - r_B(r_B \cdot \omega)]\, dm \qquad (12\text{-}23)$$

where $r_B^2 = x_B^2 + y_B^2 + z_B^2$. Then, expanding all terms, we find

$$L_B = L_x \hat{i}_B + L_y \hat{j}_B + L_z \hat{k}_B \qquad (12\text{-}24)$$

with

$$\left.\begin{aligned}
L_x &= \omega_x I_{xx} - \omega_y I_{xy} - \omega_z I_{xz} \\
L_y &= -\omega_x I_{yx} + \omega_y I_{yy} - \omega_z I_{yz} \\
L_z &= -\omega_x I_{zx} - \omega_y I_{zy} + \omega_z I_{zz}
\end{aligned}\right\} \qquad (12\text{-}25)$$

and

$$\left.\begin{aligned}
I_{xx} &= \int_M (y_B^2 + z_B^2)\, dm; \quad I_{xy} = I_{yx} = \int_M x_B y_B\, dm \\[4pt]
I_{yy} &= \int_M (x_B^2 + z_B^2)\, dm; \quad I_{xz} = I_{zx} = \int_M x_B z_B\, dm \\[4pt]
I_{zz} &= \int_M (x_B^2 + y_B^2)\, dm; \quad I_{yz} = I_{zy} = \int_M y_B z_B\, dm
\end{aligned}\right\} \quad (12\text{-}26)$$

These quantities are the various *second moments* of the mass distribution referred to the body-centered coordinate frame. There are 9 of these quantities: those of the form I_{xx}, I_{yy}, and I_{zz} are called the *diagonal elements,* and those of the form I_{xy}, I_{xz}, I_{yz}, etc., are called the *off-diagonal elements* or the *products of inertia*. The diagonal element I_{xx} is just the rotational inertia about the x_B axis, and similarly for I_{yy} and I_{zz}. Notice that the products of inertia occur in pairs: $I_{xy} = I_{yx}$, etc. Thus, there are only 6 independent elements among the 9 second moments.

Notice that the second moments do not vary with time. This is, in fact, the reason for introducing the body-centered coordinates.

The angular momentum L_B is sometimes expressed by using the following notation:

$$L_B = \{I\} \cdot \omega \qquad (12\text{-}27)$$

where $\{I\}$ represents the collection of the 9 second moments and is called the *moment of inertia tensor*. The multiplication of $\{I\}$ and ω is carried out in the special way that reproduces Eqs. 12-24 and 12-25.

It is possible to orient the body-centered axes for any arbitrarily shaped object in such a way that all of the products of inertia vanish. Although it is not obvious, this statement is *always* true. Such axes are called the *principal axes* of the rigid body and are an intrinsic characteristic of the particular object. If the object possesses some regular or symmetric features, the principal axes are usually easy to identify. The principal axes of a uniform cylinder, for example, correspond to the cylinder axis and to any pair of mutually perpendicular axes that are perpendicular to the cylinder axis.

For the case in which the body-centered axes correspond to the principal axes, the expression for the angular momentum (Eqs. 12-24 and 12-25) assumes a particularly simple form:

$$L_B = I_{xx} \omega_x \hat{i}_B + I_{yy} \omega_y \hat{j}_B + I_{zz} \omega_z \hat{k}_B \qquad (12\text{-}28)$$

Because the principal moments are often different ($I_{xx} \neq I_{yy} \neq I_{zz}$), even in this case it is not possible to express the angular momentum as a scalar quantity I multiplying the angular velocity vector ω. This means that, in general, L_B is *not* parallel to ω, as we discovered in the discussion of dynamic imbalance. The quantity that connects L_B and ω (Eq. 12-27) is clearly *not* a scalar; in fact, as we have mentioned, $\{I\}$ is a more complicated quantity called a *tensor*.

The angular momentum vector L_B can be expressed as $L_B = I\omega$ only in the special case that L_B and ω are collinear. This occurs only when ω coincides with one of the principal axes.

PROBLEMS

Section 12-1

12-1 A meter stick, initially at rest on a horizontal frictionless surface, is given a horizontal impulsive blow at a right angle to its long dimension. Just after the blow is delivered, one end of the stick (the end marked 0 cm) has zero velocity relative to the surface. At what point was the blow delivered?

12-2 A stick with a length $b = 1.5$ m and a mass $M = 0.8$ kg is initially at rest on a flat frictionless surface. A hockey puck with a mass $m = 0.5$ kg strikes one
•

end of the stick at a right angle to its long dimension. The puck approaches the stick with a velocity $v_i = 8$ m/s and after the collision it continues in the same straight line with a velocity $v_f = 4$ m/s. (a) What is the velocity of the C.M. of the stick after impact? (b) What is the angular velocity of the stick after impact? (c) What fraction of the original kinetic energy is lost in the (inelastic) collision? [*Hint:* Consider a description of the situation in a *fixed* coordinate frame with origin at the *original* position of the C.M. of the stick.]

12–3 A solid uniform sphere with radius $R = 0.20$ m and mass $M = 50$ kg is at rest in an inertial reference frame in deep space. A bullet with mass $m = 20$ g and a velocity $v = 400$ m/s strikes the sphere along the line shown in the diagram and rapidly comes to rest within the sphere at point P. Determine the subsequent motion of the sphere and the imbedded bullet.

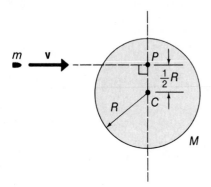

Section 12-2

12–4 A uniform cylinder with a radius $R = 15$ cm rolls without slipping on a horizontal surface with a linear speed of 2 m/s. To what maximum height h will the cylinder roll up the plane shown in the diagram?

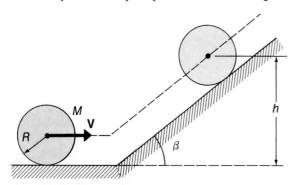

12–5 A spool has a mass of 2 kg, an inner radius $R_1 = 3$ cm, and an outer radius $R_2 = 5$ cm; the radius of gyration about the axis of the spool is $\Gamma = 4$ cm. A constant horizontal force of 5 N is applied to the free end of a massless thread that is

wrapped around the inner cylinder of the spool. If the spool rolls without slipping, calculate the linear acceleration along a horizontal surface. What is the minimum coefficient of static friction required to prevent slipping?

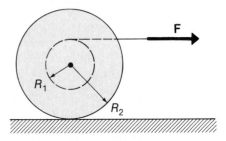

12–6 An object with cylindrical symmetry has an outer
•   radius R and a radius of gyration Γ about its axis. When placed on its flat end on a plane with variable angle of inclination, sliding begins when the angle reaches 15°. When the object is allowed to roll down the plane, the maximum angle at which rolling without slipping takes place is 30°. What is the ratio R/Γ? What is the coefficient of static friction? What possible type of simple object are we dealing with here?

12–7 Two identical uniform discs, each with radius
• $R = 8$ cm and mass $M = 8.0$ kg, are connected by a massless string that is wrapped around the discs in opposite directions, as indicated in the diagram. One disc is mounted on stationary frictionless bearings so that it can rotate freely about its axis. The other disc is initially held at the same height as the captive disc and then released to fall. During the fall, the string unwinds from both discs. Find the linear acceleration and the angular acceleration of the falling disc and the angular acceleration of the captive disc. Also, find the tension in the string.

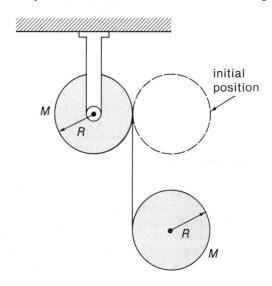

Section 12-3

12-8   A gyroscope is rotating at a rate of 300 rad/s about a horizontal axis. A string is tied to one end of the axis and is held vertically, as shown in the diagram. The mass of the flywheel is 2 kg and the radius of gyration is 5 cm. The mass of the entire gyroscope is 2.5 kg and its C.M. is 8 cm from the string. The gyroscope is released from a horizontal attitude and the C.M. begins to descend vertically with a linear acceleration of $\frac{1}{3}g$. What is the tension T in the string? What is the precessional angular velocity Ω?

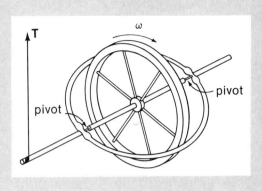

Additional Problems

12-9   A uniform thin rod with a mass $M = 0.60$ kg and a
•   length of 0.30 m stands on the edge of a frictionless table, as shown in the diagram. The rod is struck a horizontal impulsive blow, $J = 6$ N · s, at a point 0.20 m above the table top, driving the rod directly off the table. Determine the orientation of the rod and the position of its C.M. 1 s after the blow is struck.

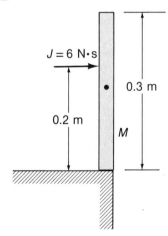

12-10 A 30-kg rectangular box with a uniform mass distribution has dimensions 0.5 m × 0.5 m × 1.0 m. The box slides on one of its small faces along a horizontal surface that has a coefficient of sliding friction $\mu_k = 0.40$. A light rope exerts a horizontal force F at a point 10 cm above the center of a 0.5 m × 1.0 m face, perpendicular to this face and in the direction of motion. (a) What is the maximum value of F that will cause the box to move without tipping over? (b) What is the acceleration that results from the force in (a)? (c) Suppose that the box is at rest and that the coefficient of static friction is 0.60. What is the maximum force F that

can be applied before motion occurs? Will this motion be sliding or tipping?

12-11 A rectangular crate with dimensions 0.6 m ×
•   0.6 m × 1.2 m is at rest on one of its small sides on the flat bed of a pickup truck. The crate has a uniform mass distribution and a total mass of 80 kg. The crate is oriented with its sides parallel to the sides of the truck. The coefficient of static friction between the crate and the truck bed is 0.60. The truck is moving at a speed of 60 km/h when the driver sights a hazard and applies a constant decelerating braking force. What is the least time in which the truck can be brought to a stop if the crate is neither to tip nor to slide on the truck bed?

12-12 A string is wrapped around the center portion of
••   a spool that has a uniform mass distribution with a total mass M and the shape shown in the diagram on the next page. A constant force $F = \frac{1}{2}Mg$ is applied to the free end of the string at an angle γ with the horizontal. The coefficient of static friction between the spool and the surface is 0.40. Show that the spool will both slide and roll unless $\gamma \geqslant 81.4°$ or $\gamma \leqslant 30.0°$. In what direction will the spool move in each case? [*Hint:* Note that $a \sin \psi + b \cos \psi = \sqrt{a^2 + b^2} \sin(\psi + \beta)$, where $\tan \beta = b/a$.]

12-13 A uniform thin meter stick ($M = 0.20$ kg) is
••   balanced vertically on one end that rests on a horizontal surface. The stick is given a (negligibly) small impulse that causes it to fall over. Calculate the velocity with which the C.M. of the stick strikes the surface (a) if the end on the surface does not slip, and (b) if the end on the surface slides along the surface without friction. (c) Show that the vertical acceleration of the C.M. upon striking the surface is $a = \frac{3}{4}g$ in both of the above cases. (d) Calculate the

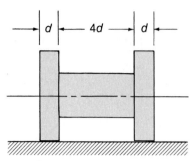

Problem 12–12

reaction forces on the meter stick in cases (a) and (b) just as the stick strikes the surface. [*Hint:* Express the energy conservation condition for an arbitrary angle β of the meter stick with the vertical and differentiate with respect to time.]

12–14 A marble with a radius $b = 1$ cm and a mass $M = 15$ g is released from rest at the top of a loop-the-loop where $h = 27$ cm, as shown in the diagram. The marble rolls without slipping completely around the track, which has an inner radius $R = 10$ cm. (a) Find the velocity of the marble at the top of the loop. (b) Find the reaction forces on the marble when at the top of the loop. (c) What is the minimum height h from which the marble can be released and still maintain contact with the track for the entire trip?

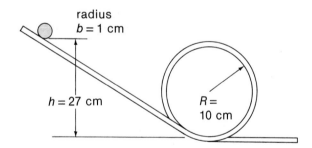

12–15 A light string is wrapped around a solid cylinder that has a radius $R = 5$ cm and a mass $M = 12$ kg. The cylinder is on a plane with an inclination angle of 32°. The free end of the string passes over an inertialess and frictionless pulley to a block with mass $m = 2$ kg as in the diagram at right. Find the linear acceleration of the cylinder and of the hanging block. [*Hint:* What is the relationship between the linear acceleration of the cylinder and that of the block?]

12–16 A cue stick strikes a cue ball and delivers a horizontal impulse J in such a way that the ball rolls without slipping as it starts to move. At what height above the ball's center (in terms of the radius of the ball) was the blow struck?

12–17 A billiard player gives a horizontal impulse to a cue
  • ball by striking the ball with the cue tip at a point $\frac{4}{5}$ of the ball radius above the ball's center. The ball begins to move, rolling and slipping, with a velocity V_0. Eventually, friction causes the ball to roll without slipping at a velocity V_1. Show that $V_1 = 9V_0/7$.

12–18 A solid sphere with a radius $R = 5$ cm and a mass
  •• $M = 4$ kg rolls down a plane that has an inclination angle $\gamma = 45°$. (a) Show in general that rolling without slipping requires the coefficient of static friction to be greater than $(2/7) \tan \gamma$. (b) In the case here, we actually have $\mu_s < (2/7) \tan \gamma$, so some slipping occurs. Then, use $\mu_k = 1/5$ for the coefficient of sliding friction and show that the linear acceleration a and the angular acceleration α are related by $a = (8/5)R\alpha$. Find both a and α. (c) What time T is required for the sphere to move 1 m down the plane, starting from rest at $t = 0$? (d) At $t = T$, find V and ω. (e) At $t = T$, find the kinetic energy K. (f) Of the 1 m of motion, how much was by rolling and how much was by sliding? What frictional force was acting and what amount of work W_f was done against friction? (g) Calculate the change in potential energy during the 1-m movement. Is this equal to $K + W_f$?

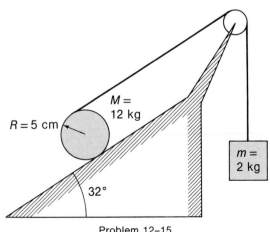

Problem 12–15

12–19 A plank with a mass $M = 6$ kg rides on top of
• two identical solid cylindrical rollers that have
$R = 5$ cm and $m = 2$ kg. The plank is pulled by a
constant horizontal force $F = 6$ N applied to the
end of the plank and perpendicular to the axes of
the cylinders (which are parallel). The cylinders roll
without slipping on a flat surface. There is also no
slipping between the cylinders and the plank. Find
the acceleration of the plank and of the rollers.
What frictional forces are acting?

FORCES IN EQUILIBRIUM

An important special case in the study of rigid-body motion is that in which the forces applied to a body do not produce either translational acceleration or angular acceleration. A body in such a situation is said to be in *dynamic equilibrium*. Note that this broad definition *does not* require the body to be at rest—it may, in fact, be in translational motion with constant linear velocity or in rotational motion with constant angular velocity or both simultaneously. Of course, the conditions of dynamic equilibrium permit the object to be at rest, and this is an important case with many applications.

13-1 CONDITIONS FOR DYNAMIC EQUILIBRIUM

A rigid body that undergoes neither translational acceleration of its C.M. nor rotational acceleration about its C.M. has $\mathbf{a} = 0$ and $\boldsymbol{\alpha} = 0$. Then, the equations of motion (Eqs. 12–2 and 12–8) reduce to

$$\frac{d\mathbf{P}}{dt} = 0 \qquad \text{and} \qquad \frac{d\mathbf{L}_C}{dt} = 0$$

where $\mathbf{P}$ is the linear momentum of the C.M. of the body and where $\mathbf{L}_C$ is the angular momentum relative to the C.M. These equations imply that the vector sums of all forces and all torques acting on the body must be zero:

$$\left. \begin{array}{c} \displaystyle\sum_n \mathbf{F}_n^e = 0 \\[2em] \displaystyle\sum_n \boldsymbol{\tau}_{Cn} = 0 \end{array} \right\} \tag{13-1}$$

The Torque Theorem. We have stated the equilibrium conditions in terms of torques calculated with respect to the C.M. of the object. This is, in fact, unnecessarily restrictive, as we now prove.

Suppose that the sum of all forces $\mathbf{F}_n^e$ acting on a body is zero. Consider calculating the net torque produced by these same forces with respect to some arbitrary point P. The situation is illustrated in Fig. 13–1, which shows only one of the forces

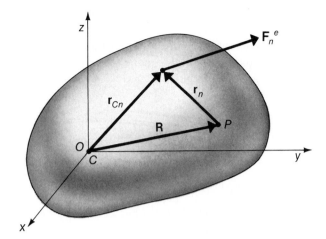

Fig. 13-1. The force $\mathbf{F}_n^e$ is *one* of the forces (whose vector sum is zero) acting on the object. The center of mass C is located at the origin and P is any arbitrary point.

$\mathbf{F}_n^e$. The position vector $\mathbf{r}_n$ from the arbitrary point P to the point of application of the force $\mathbf{F}_n^e$ can be expressed in terms of $\mathbf{R}$, the vector from the C.M. to the point P, and $\mathbf{r}_{Cn}$, the vector that locates the point of application of the force with respect to the C.M. Evidently, we have

$$\mathbf{r}_{Cn} = \mathbf{R} + \mathbf{r}_n$$

Then, the expression for the torque becomes

$$\sum_n \boldsymbol{\tau}_{Cn} = \sum_n \mathbf{r}_{Cn} \times \mathbf{F}_n^e$$

$$= \sum_n (\mathbf{R} + \mathbf{r}_n) \times \mathbf{F}_n^e$$

$$= \mathbf{R} \times \sum_n \mathbf{F}_n^e + \sum_n \mathbf{r}_n \times \mathbf{F}_n^e$$

The first term vanishes because the net applied force is zero. The second term is just the net torque calculated with respect to the point P. Therefore,

$$\sum_n \boldsymbol{\tau}_{Cn} = \sum_n \boldsymbol{\tau}_{Pn}$$

Thus, when the torque with respect to the C.M. vanishes, the torque with respect to *any arbitrary point* also vanishes for the case in which the net applied force vanishes. The equilibrium conditions in their most general form become

$$\sum_n \mathbf{F}_n^e = 0$$

$$\sum_n \boldsymbol{\tau}_n = 0$$

(13-2)

These equations specify the conditions of *dynamic equilibrium*. It is important to realize that these are *vector* equations and therefore represent *six* independent component equations. In many cases, the forces involved are confined to a plane. We can choose x- and y-axes in this plane, with the z-axis perpendicular to the plane. Then, the six equilibrium equations reduce to three:

$$\left.\begin{array}{l} \sum F_x^e = 0 \\[6pt] \sum F_y^e = 0 \\[6pt] \sum \tau_z = 0 \end{array}\right\} \tag{13-3}$$

where the summation index n has been suppressed, and where the torques can be taken about any convenient point.

Center of Gravity.   In many of the problems that we consider, an object is acted upon by a gravitational force. In Section 14–5 we show that the gravitational forces acting on the individual constituents of an extended object with mass M can be represented by a single force Mg acting at a point called the *center of gravity* of the object. If the acceleration due to gravity **g** is constant over the extent of the object (this is the case for ordinary objects near the Earth's surface), the center of gravity coincides with the center of mass of the object.

Stable and Unstable Equilibrium.   In the discussion of the potential energy function in Section 7–4, it was pointed out that a particle is in a condition of *stable equilibrium* when the potential energy U has a (local) *minimum*. In such a case, any small displacement of the particle away from the equilibrium point produces a force that tends to restore the particle to the equilibrium position. The same is true for extended objects. Consider the cone in Fig. 13–2a, which rests on its base on a flat surface. If the cone is given a small displacement by tipping, the C.M. (point C) tends to rise—that is, work must be done to effect the displacement and the potential energy increases. If the cone is released after a small displacement, it will return spontaneously to its original position. The resting condition in Fig. 13–2a is clearly one of stable equilibrium.

A particle—or an extended object—is in a condition of *unstable equilibrium* when the potential energy has a (local) *maximum*. Then, any small displacement, such as the tipping of the cone in Fig. 13–2b, will cause the C.M. to be lowered and the potential energy will decrease. When released, the object will proceed spontaneously to seek a condition of still lower potential energy—it will fall.

An object is said to be in a condition of *neutral equilibrium* when a small displacement causes no change in the potential energy. For example, a homogeneous sphere rolling on a horizontal surface is in a condition of neutral equilibrium.

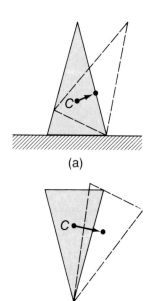

Fig. 13–2. (a) The cone is in a condition of stable equilibrium; a small tipping displacement *raises* the C.M. Upon release, the cone will return to its original position. (b) The cone is in a condition of unstable equilibrium; a small tipping displacement *lowers* the C.M. Upon release, the cone will fall.

13-2   APPLICATIONS OF THE EQUILIBRIUM CONDITIONS

We now give several examples of the application of Eqs. 13–3 to situations involving dynamic equilibrium. In these examples and the accompanying problems, we assume that the objects are all rigid bodies. The geometric configurations have been carefully contrived to produce mathematically solvable problems involving exactly as many unknowns as there are constraints. The mechanical structures that are met in engineering practice and elsewhere in the real world usually have many more constraints than the unknowns contained in the equilibrium equations (Eqs. 13–3). Any case in which the number of constraints exceeds the number of degrees of freedom for a perfectly rigid structure leads to a mathematical solution that is at least partially indeterminate. For example, a three-legged stool that rests on a level floor leads to a completely solvable problem in that the load supported by each leg can be calculated. Conversely, in the problem of an absolutely rigid four-legged chair that rests on a perfectly rigid level floor, no unique solution can be given for the load distribution. The difficulty disappears when we realize that the chair and the floor are not perfectly rigid structures and that they deform under applied forces. If we include the elastic properties of the objects, generally expressible in terms of *additional* equations, a unique mathematical solution becomes possible.

Example 13–1

A meter stick with a mass $m = 0.50$ kg is supported in a stationary horizontal position by vertical spring scales at each end. A block with a mass $M = 0.35$ kg is suspended from the meter stick by a massless string at the 25-cm mark. What are the force readings on the two scales?

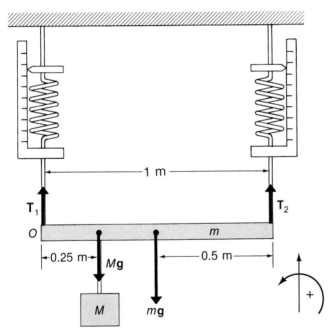

Solution:

There are no horizontal forces acting on the meter stick, so we have only one force equation:

$$\sum F_y^e = 0: \qquad\qquad T_1 + T_2 - Mg - mg = 0$$

from which
$$T_1 + T_2 = (0.35 \text{ kg} + 0.50 \text{ kg})(9.80 \text{ m/s}^2)$$
$$= 8.33 \text{ N}$$

Thus, the force equation yields the *sum* of the force readings on the scales, but not the individual values. We find these by applying the torque condition for the torques about the zero-cm end of the meter stick. With counterclockwise torques taken to be positive, we have

$$\sum \tau_0 = 0:$$

$$T_1 \times 0 - (0.35 \text{ kg})(9.80 \text{ m/s}^2)(0.25 \text{ m}) - (0.50 \text{ kg})(9.80 \text{ m/s}^2)(0.50 \text{ m}) + T_2(1.00 \text{ m}) = 0$$

from which $T_2 = 0.86 \text{ N} + 2.45 \text{ N} = 3.31 \text{ N}$

Then, $T_1 = 8.33 \text{ N} - 3.31 \text{ N} = 5.02 \text{ N}$

By selecting the point O as the reference point for calculating the torques, we eliminated any contribution from T_1 and directly obtained the value of T_2. Calculate the torques with respect to some other point, such as the center of the meter stick or a point 1 m to the *left* of O. The resulting equation will involve both T_1 and T_2. Show that when this equation is combined with the force equation, the same values for T_1 and T_2 result.

Notice that by writing *two* torque equations, one for each end of the meter stick, the values of T_1 and T_2 can be found without using the force equation at all. For any particular problem, a little thought should reveal the point or points about which the torques should be calculated to yield simple and useful relationships.

Example 13–2

A block with a mass $m = 20 \text{ kg}$ is suspended from one end of a 3-m plank (mass $M = 50 \text{ kg}$) that is supported by two piers, as shown in the diagram. What forces, F_1 and F_2, do the piers exert on the plank?

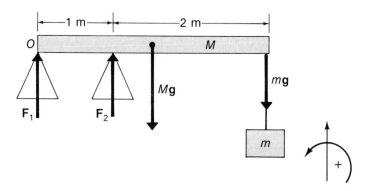

Solution:

There is only one force equation, that for the vertical direction:

$$\sum F_y^e = 0:$$ $$F_1 + F_2 - Mg - mg = 0$$

from which $$F_1 + F_2 = (50 \text{ kg} + 20 \text{ kg})(9.80 \text{ m/s}^2)$$
$$= 686 \text{ N}$$

We choose point O for calculating the torques and we let counterclockwise torques be positive. Then,

$$\sum \tau_0 = 0:$$ $$(1 \text{ m})F_2 - (1.5 \text{ m})Mg - (3 \text{ m})mg = 0$$

so that $$F_2 = (1.5 \text{ m})(50 \text{ kg})(9.8 \text{ m/s}^2) + (3 \text{ m})(20 \text{ kg})(9.8 \text{ m/s}^2)$$
$$= 1323 \text{ N}$$

Then,
$$F_1 = 686 \text{ N} - F_2 = 686 \text{ N} - 1323 \text{ N}$$
$$= -637 \text{ N}$$

We interpret the negative sign of F_1 in the following way. When we constructed the free-body diagram for the plank, we drew the vector $\mathbf{F}_1$ *upward*. The fact that the result for F_1 carries a negative sign means that the diagram is incorrect; $\mathbf{F}_1$ actually is directed *downward*. This is a general result: if the direction of a vector in a free-body diagram is chosen opposite to its actual direction, the result will contain a negative sign, thereby revealing the original incorrect choice.

Example 13-3

A block with a mass $M = 100$ kg hangs from one end of a uniform beam that has a length of 3 m and a mass $m = 20$ kg. The end opposite the hanging block is hinged to a vertical wall. A horizontal cable with negligible mass is attached to the beam at a point 2 m from the hinged end and holds the beam in equilibrium at an angle of 30° with respect to the horizontal, as shown in the diagram.

(a) Find the tension in the cable and the forces, F_x and F_y, exerted on the beam by the hinge.

(b) At what point Q on the wall should the cable be attached to yield the minimum cable tension for holding the beam in the position shown in the diagram?

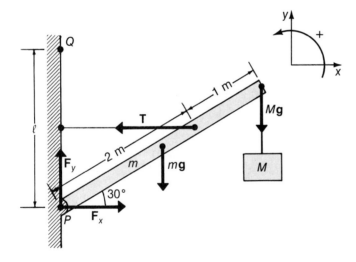

Solution:

(a) The three unknowns in this problem are T, F_x, and F_y. If we select P as the reference point for calculating the torques, we can obtain the value of T directly because both F_x and F_y have zero torque about P. Taking the counterclockwise direction to be positive, we have

$$\sum \tau_P = 0: \quad 2T \sin 30° - (1.5 \text{ m})(20 \text{ kg})(9.80 \text{ m/s}^2) \cos 30°$$
$$- (3 \text{ m})(100 \text{ kg})(9.80 \text{ m/s}^2) \cos 30° = 0$$

from which
$$T = 2801 \text{ N}$$

$$\sum F_x^e = 0: \quad F_x - T = 0$$

so that
$$F_x = 2801 \text{ N}$$

$$\sum F_y^e = 0: \quad F_y - mg - Mg = 0$$

from which
$$F_y = (100 \text{ kg} + 20 \text{ kg})(9.80 \text{ m/s}^2)$$
$$= 1176 \text{ N}$$

(b) Only the component of **T** that is perpendicular to the beam is effective in producing a torque about the point P. This component is maximum, and hence T is a minimum, when **T** itself is perpendicular to the beam. Then, the distance from P to Q is $\ell = (2 \text{ m}) \sec 60° = 4.00 \text{ m}$, and the tension T' is obtained from

$$2T' - (1.5 \text{ m})(20 \text{ kg})(9.80 \text{ m/s}^2) \cos 30°$$
$$- (3 \text{ m})(100 \text{ kg})(9.80 \text{ m/s}^2) \cos 30° = 0$$

which gives
$$T = 1400 \text{ N}$$

Example 13–4

A long uniform plank rests against a frictionless vertical wall. The coefficient of static friction between the plank and the floor is $\mu_s = 0.35$. What is the smallest angle γ at which the plank can remain stationary?

Solution:

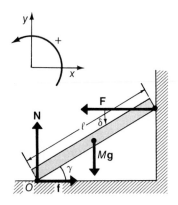

We have

$$\sum F_x^e = 0: \qquad\qquad f - F = 0 \qquad\qquad (1)$$

$$\sum F_y^e = 0: \qquad\qquad N - Mg = 0 \qquad\qquad (2)$$

For the torques about point O, we have

$$\sum \tau_0 = 0: \qquad \ell(F \sin \gamma) - \tfrac{1}{2}\ell(Mg \cos \gamma) = 0 \qquad (3)$$

from which
$$F = \tfrac{1}{2}Mg \text{ ctn } \gamma$$

Combining this result with (1) and (2) gives

$$\frac{f}{N} = \tfrac{1}{2}\text{ctn } \gamma$$

As γ is decreased, f/N increases and eventually reaches the value $f/N = \mu_s = 0.35$, at which point the plank will begin to slip. Thus,

$$\text{ctn } \gamma_{\text{min}} = 2\mu_s = 0.70$$

so that
$$\gamma_{\text{min}} = 55.0°$$

If friction exists between the top of the ladder and the wall, the system becomes indeterminate. (•Can you see why?)

Example 13–5

A uniform sphere with a mass $M = 5$ kg is pulled at constant velocity up an inclined plane by a string that is attached to its surface. The string applies a force parallel to the plane, whose angle of inclination is $33°$. If the coefficient of kinetic friction between the sphere and the plane is $\mu_k = 0.42$, find the string tension and the angle β that locates the point of attachment of the string, as shown in the diagram.

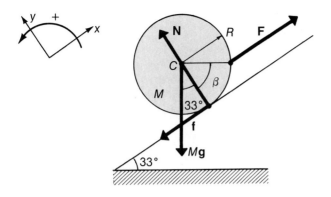

Solution:

The force equations are

$$\sum F_y^e = 0: \qquad\qquad N - Mg \cos 33° = 0$$

$$\sum F_x^e = 0: \qquad\qquad F - f - Mg \sin 33° = 0$$

Using $f = \mu_k N$, and solving for F, we have

$$F = Mg(\sin 33° + 0.42 \cos 33°)$$
$$= 43.95 \text{ N}$$

For the torques about the center C of the sphere,

$$\sum \tau_C = 0: \qquad\qquad R(F \cos \beta) - fR = 0$$

Substituting for f and using the above value for F, we find

$$\cos \beta = \frac{f}{F} = \frac{\mu_k Mg \cos 33°}{F}$$

$$= \frac{(0.42)(5 \text{ kg})(9.80 \text{ m/s}^2) \cos 33°}{43.95 \text{ N}} = 0.393$$

so that
$$\beta = 66.9°$$

Example 13-6

A dresser with a mass $M = 60$ kg is pulled at constant velocity across a level floor by a force **F**, applied as in the diagram on the next page. If the coefficient of kinetic friction between the feet of the dresser and the floor is $\mu_k = 0.20$, find the force F and the normal forces, N_F and N_R, on the front and rear legs, respectively.

Solution:

The force equations are (with N_R and N_F representing the normal force on *both* rear legs and *both* front legs, respectively)

$$\sum F_y^e = 0: \qquad N_R + N_F = Mg = (60 \text{ kg})(9.80 \text{ m/s}^2) = 588 \text{ N}$$

$$\sum F_x^e = 0: \qquad\qquad -f_R - f_F + F = 0$$

Using $f = \mu_k N$, we have

$$F = \mu_k(N_R + N_F) = 0.20(588 \text{ N})$$
$$= 117.6 \text{ N}$$

The torques about the line connecting the rear legs (point O) give

$$\sum \tau_0 = 0: \qquad -(0.4 \text{ m})Mg - (0.6 \text{ m})F + (0.8 \text{ m})N_F = 0$$

from which

$$N_F = \frac{(0.4 \text{ m})(60 \text{ kg})(9.80 \text{ m/s}^2) + (0.6 \text{ m})(117.6 \text{ N})}{0.8 \text{ m}}$$
$$= 382.2 \text{ N}$$

Then,

$$N_R = 588 \text{ N} - N_F$$
$$= 205.8 \text{ N}$$

Notice that $N_F > N_R$ because the force **F** tends to tip the dresser forward. (See also Problem 13–18.)

Example 13–7

One way that a small force applied to one end of a rope can balance a large force applied to the other end is to wrap the rope around a rough post. This principle is often used by sailors and mountaineers to manipulate heavy loads. Relate the forces at the ends of a rope to the length of rope wrapped around an intervening post if the coefficient of static friction is μ.

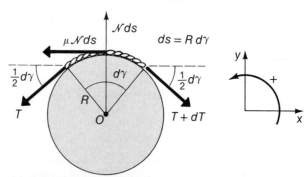

Solution:

The diagram shows a small section of the rope which subtends an angle $d\gamma$ at the center of the post. We are interested in the maximum force that can be sustained by the rope without slipping, so we assume that the static frictional force tangent to the rope is everywhere equal to the maximum value. Let the normal force on the rope be $\mathcal{N}$ per unit length at any point. Thus, for any segment of the rope with differential length ds, the force is $\mathcal{N}\, ds$.

The force equations are

$$\sum F_y^e = 0: \qquad \mathcal{N}\, ds - (T + dT)\sin\frac{d\gamma}{2} - T\sin\frac{d\gamma}{2} = 0 \qquad \textbf{(1)}$$

$$\sum F_x^e = 0: \qquad (T + dT)\cos\frac{d\gamma}{2} - T\cos\frac{d\gamma}{2} - \mu\mathcal{N}\, ds = 0 \qquad \textbf{(2)}$$

For very small angles $d\gamma$, we have $\cos(d\gamma/2) \cong 1$ and $\sin(d\gamma/2) \cong d\gamma/2$; then, also $dT \ll T$. Therefore,

$$T\frac{d\gamma}{ds} = \mathcal{N} \qquad \textbf{(1a)}$$

and

$$\frac{dT}{ds} = \mu\mathcal{N} \qquad \textbf{(2a)}$$

Note that Eq. (2a) is consistent with the torque equation taken with respect to the center O of the post:

$$\sum \tau_0 = 0: \qquad \mu\mathcal{N}R\, ds + TR - (T + dT)R = 0$$

which just yields Eq. (2a).

Dividing Eq. (2a) by Eq. (1a), we obtain

$$\frac{1}{T}\frac{dT}{d\gamma} = \mu$$

Integrating this equation from ($\gamma = 0$, $T = T_0$) to (γ, T), we write

$$\int_{T_0}^{T}\frac{dT}{T} = \mu\int_0^{\gamma}d\gamma$$

which gives

$$\ln\frac{T}{T_0} = \mu\gamma$$

and, finally,

$$T = T_0 e^{\mu\gamma}$$

where the length ℓ of rope wrapped around the post is $\ell = R\gamma$.

To give a specific example, suppose $\mu = 0.40$ and the rope goes once around the post so that $\gamma = 2\pi$. Then, the ratio of the two forces is

$$\frac{T}{T_0} = e^{\mu\gamma} = e^{(0.40)(2\pi)} = 12.3$$

and a 100-N force can sustain a force of 1230 N. (•Where does the additional force of 1130 N ultimately come from?)

PROBLEMS

13-1 A student manages to get his car stuck in a snow drift. Not at a loss, having studied physics, he attaches one end of a stout rope to the vehicle and the other end to the trunk of a nearby tree, allowing for a small amount of slack. The student then exerts a force **F** on the center of the rope in the direction perpendicular to the car-tree line, as shown in the diagram. If the rope is inextensible and if the magnitude of the applied force is 500 N, what is the force on the car? (Assume equilibrium conditions.)

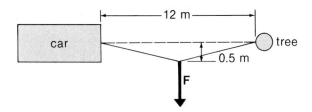

13-2 An automobile with a mass of 1500 kg has a wheel base (the distance between the axles) of 3.0 m. The C.M. of the automobile is on the center line at a point 1.2 m behind the front axle. Find the force exerted by the ground on each of the four wheels.

13-3 A block with a mass of 50 g is suspended from a meter stick at the 15-cm mark. The meter stick, which has a uniform density, will balance in a horizontal position if a pivot is placed at the 37-cm mark, as indicated in the diagram. What is the mass of the meter stick?

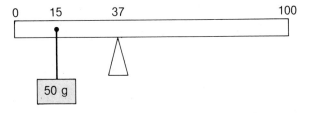

13-4 A trap door has a uniform density and a mass of 50 kg. The door is hinged along one side and is partially open, making an angle of 30° with the horizontal. A rope that is attached to the side opposite the hinge and is parallel to the floor holds the door in this position, as shown in the diagram (above right). (a) What is the tension in the rope? (b) What are the vertical and horizontal components of the force that the hinge exerts on the door?

13-5 A homogeneous door has a height of 2.2 m, a width of 1.0 m, and a mass of 25 kg. The door is hinged at two points, one 0.30 m from the top and the other

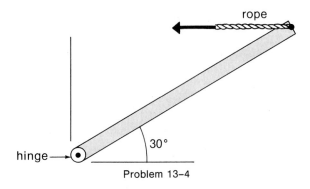

Problem 13-4

0.30 m from the bottom. Assume that each hinge supports half of the door's weight. Find the horizontal and vertical components of the forces that the hinges exert on the door. (•Can you see why it is necessary to stipulate the equality of the vertical components of the hinge forces?)

13-6 To a reasonable approximation, the human arm may be considered to be a simple set of lever components, hinges, and tension-producing muscles. Suppose that a 10-kg block is held in the hand with the forearm horizontal and perpendicular to the upper arm, as shown in diagram (a). Use the dimensions given in diagram (b) and find: (a) the tension T in the biceps, (b) both the compressional (vertical) and bending (horizontal) components of the force exerted on the humerus by the forearm at the elbow joint. Compare these results with the weight of the block. Can you see why it is tiring simply to hold a heavy block with the arm horizontal?!

(a)

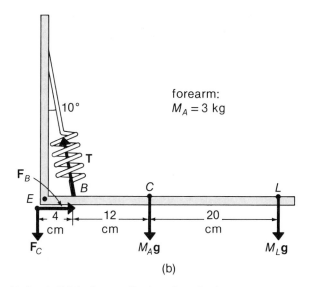

forearm:
$M_A = 3$ kg

(b)

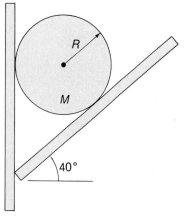

Problem 13–8

13-7 A 100-kg lawn roller is to be raised over a curb that has a height of 10 cm. The roller has a radius of 20 cm and essentially all of the mass is uniformly distributed within the drum. The handle of the roller is attached to the ends of the drum axis, as shown in the diagram. What is the least force F that will succeed in pulling the roller over the curb? What is the angle γ for this condition? [*Hint:* Treat the problem as one of static equilibrium. Find the force that just reduces to zero the normal force on the drum at the lower ground level.]

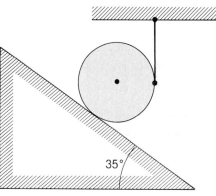

Problem 13–9

shown in the diagram below. What is the minimum value of the coefficient of static friction between the sphere and the plane that will allow this equilibrium situation?

13-8 A sphere with radius $R = 0.2$ m and mass $M = 12$ kg rests between two smooth planks, as shown in the diagram (above right). What force does each plank exert on the sphere?

13-9 A sphere is in static equilibrium as shown in the diagram (above right). The elevation angle of the inclined plane is 35°. What is the minimum coefficient of static friction necessary for the string to be vertical?

13-10 A sphere rests on a plane that has an inclination angle of 35° while tied with a horizontal string, as

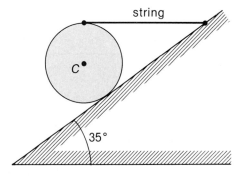

13-11 Two planks, each with a length of 2.50 m, are joined end-to-end by a hinge at point A, as shown in the diagram on the next page. The planks stand on a frictionless horizontal surface, forming an isosceles triangle. A cord with a length of 1.5 m is attached to the midpoints of the two planks. Each plank has a mass of 10 kg. (a) Find the tension in the rope. (b) Find the hinge forces.

Problem 13–11

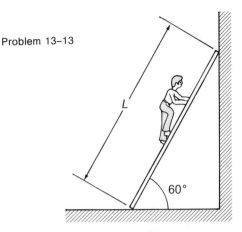

Problem 13–13

13–12 A 20-kg sign that has a uniform density hangs in front of an establishment. The sign is suspended by one hinge and a cable, as shown in the diagram. Find the tension in the cable and the components of the force that the hinge exerts on the sign.

13–13 A ladder with a uniform density and a mass m rests against a frictionless vertical wall at an angle of 60°. The lower end rests on a flat surface where the coefficient of static friction is $\mu_s = 0.40$. A student with a mass $M = 2m$ attempts to climb the ladder. What fraction of the length L of the ladder will the student have reached when the ladder begins to slip?

13–14 A post with a length L and a mass M rests with one
  • end on the ground. The post is held in a vertical position by a thin wire that makes an angle γ with the horizontal and by a horizontal force **F** applied at a point that is a distance ℓ above the ground. The coefficient of static friction between the post and the ground is μ_s. Show that there is a minimum height ℓ

for the point of application of the force **F** which allows a force of any magnitude to be applied without the bottom of the post slipping. Show that $\ell/L = 0.591$ for $\gamma = 60°$ and $\mu_s = 0.40$. This is an example of a *self-locking* system, a system that is particularly stable under the application of unbalancing forces. [*Hint:* Write the three equilibrium equations for the point O. Then, eliminate the tension T and solve for the ratio f/N. Finally take the limit, $F \rightarrow \infty$.]

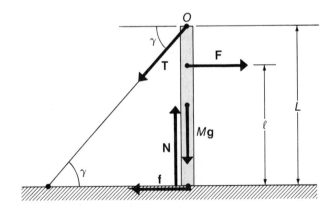

13–15 The boom of a crane has a length of 12 m and a
  •• mass that is small compared to all other masses in the structure. A 2000-kg load is suspended from the boom, as shown in diagram (a). The support cable is connected from the midpoint C of the boom to the winch at point Q which is 3 m directly above the lower end of the boom at P. (a) What is the cable tension when the boom angle is 45°? (b) What are the components of the force that the pivot exerts on the boom at P? (c) At what point on the boom should the cable be attached in order to produce the smallest cable tension and still hold the load in equilibrium? [*Hint:* The geometry will be simplified by finding the missing entries (r, s, and α) in diagram (b).]

Problem 13-15

(a)

(b)

13-16 A uniform flexible cable has a mass of 100 kg and is suspended between two fixed points, A and B, at the same level, as shown in the diagram. At the support points the cable makes angles of 30°. (a) Find the components of the force exerted on each support by the cable. (b) Find the tension in the cable at the lowest point.

13-17 A rod that has a length ℓ and negligible mass is
• placed against a vertical wall and is held in a horizontal position by a cord attached to the extended end. The cord makes an angle of 60° with the wall. A block with a mass m is suspended from the rod at a distance x from the wall. The coefficient of static friction between the end of the rod and the wall is $\mu_s = 0.20$. Show that the hanging block can have any mass whatsoever and the equilibrium will be maintained provided that $x \geqslant 5\ell/(5 + \sqrt{3})$. (Note the similarity of this situation to that in Problem 13-14.)

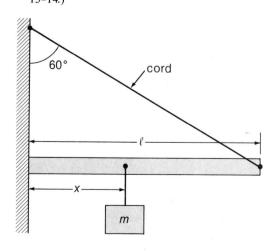

13-18 Refer to Example 13-6. How large can the coefficient of kinetic friction μ_k become before the dresser will actually tip forward?

13-19 A massless string with a length ℓ is attached to a
• fixed support at a point A and runs over a frictionless peg that is stationary at point B. The two points are at the same level and are separated by a distance $2b$. A block with a mass m is suspended from the free end of the string below the peg. Another identical block is attached to a pulley with negligible mass which rides at the center of the V formed by the string, as shown in the diagram. Use the energy principle and show that the system is in equilibrium when $y = \ell - 4b/\sqrt{3}$.

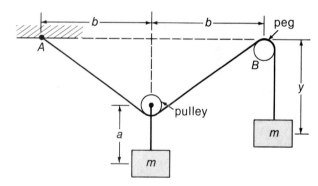

13-20 A hollow spherical shell is partially filled with an irregular mass of hardened cement and rests on a

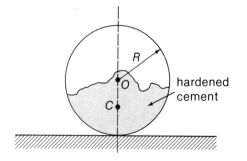

horizontal surface. Use the energy principle to show that the sphere will be in an equilibrium condition when the C.M. (point C) is directly below the center of the sphere (point O).

13-21 A solid homogeneous cube with sides of length b
•• rests balanced on the top of a cylindrical surface that has a radius R, as shown in the diagram. Show that the system is in a condition of stable equilibrium if the friction present is sufficient to prevent sliding and if $R > \frac{1}{2}b$. [*Hint:* Allow the cube to rotate without slipping through an angle α. Find the new height of C. Take α to be small and expand $\cos \alpha$ and $\sin \alpha$.]

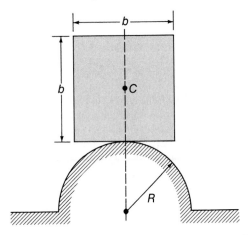

13-22 A rope with negligible mass hangs over a rough peg that has a diameter of 15 cm. A 15-kg block is attached to one end of the rope. The coefficient of kinetic friction between the rope and the peg is $\mu_k = 0.20$. What downward force **F** must be applied to the free end of the rope in order to lower the block at a constant velocity?

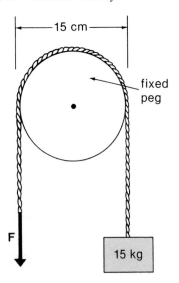

13-23 Determine the tension in the cable BC in the diagram. Also, determine the horizontal and vertical components of the force exerted on the massless strut AB at pin A.

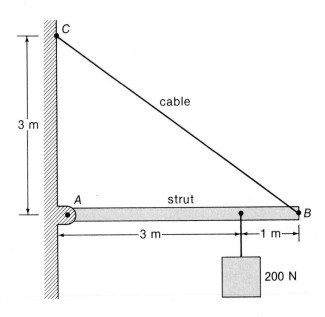

13-24 A uniform block with a height of 0.6 m and a width of 0.3 m is at rest on a plank, as shown in the diagram. The coefficient of static friction between the block and the plane is $\mu_s = 0.40$. (a) When one end of the plank is slowly raised, will the block eventually slide or tip over? At what angle θ will this occur? (b) Reconsider the situation for $\mu_s = 0.50$ and $\mu_s = 0.60$.

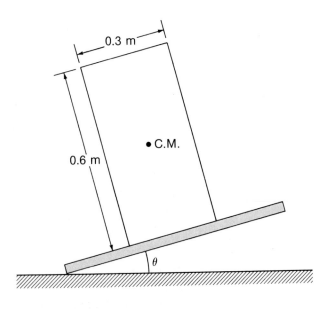

GRAVITATION AND PLANETARY MOTION

Newton's theory of gravitation, first published in 1687, united the wealth of astronomical observations that had been accumulated over the centuries with the body of knowledge that is the science of physics. With Newton's theory, it suddenly became possible to explain in a simple way all of the observational data that previously had only been speculated upon. Because of the enormous importance of this development, it is appropriate to preface the discussion of Newton's law of gravitation and its ramifications with a brief look at the historical background of astronomical observations and theories.

14-1 EARLY ASTRONOMY AND COSMOLOGY

The ancient astronomers in several regions of the world—Greece, Egypt, the Arabic countries, China, India, and Mexico—recognized the high degree of orderliness in the movements of the Sun and the planets against the background of the stars. They understood how these movements are correlated with the day-night cycle, the annual progression of the seasons, and the periodic return of the planets to particular regions of the sky. The instruments that were invented for naked-eye astronomy permitted remarkably precise observations. Accurate calendars were developed, detailed star maps were made, and the wanderings of the planets in the sky were recorded. These advances formed the basis of scientific astronomy.

Greek Astronomy. The concept of the Earth as a sphere floating in space can be traced to the Greek mathematician and philosopher Pythagoras (c582–c497 B.C.) and the members of his school. At the time, there was no empirical evidence that the Earth had a spherical shape; the speculation was based on the Greeks' infatuation with a belief in the symmetry and harmony of universal order. This preoccupation with the mathematical perfection of the sphere was elevated to a central cosmological principle by Plato (c427–c347 B.C.) and the philosophers of his school. Plato's model of the Universe was based on the belief that the heavenly bodies are perfect and must therefore move in exactly circular paths along precisely spherical transparent spheres.

Plato believed the Earth to be a stationary sphere, with the stars located in fixed positions on a spherical shell surrounding the Earth. This shell (the celestial sphere) rotated about the Earth's polar axis and carried the stars around the Earth once each day. The Sun, the Moon, and the planets revolved on spheres of their own with different individual periods designed to match the observed motions relative to the fixed stars. Plato placed all of these shells between the central Earth sphere and the shell of the fixed stars. The Sun's sphere, for example, shared the general daily rotation of the stars but had a superimposed motion of its own that progressed in the opposite direction with a period of one year. The planets Mars, Jupiter, and Saturn, with superimposed counter-rotational periods exceeding one year (namely, 1.88 y, 11.9 y, and 29.5 y, respectively), were placed in spheres between the Sun and the stars in order of increasing period. Notice that a superimposed period that is infinitely long would represent no motion at all relative to the fixed stars. Thus, the logic in placing the planet with the longest period farthest from the Earth and closest to the stars led to the correct ordering of the planets.

The planets Mercury and Venus were placed (again, correctly) between the Earth and the Sun because they have periods less than one year (88 days and 225 days, respectively). The Moon was placed inside the planetary and solar orbits. Beyond the stellar shell was the *prime mover* shell which, in Plato's model, was responsible for driving the entire planetary engine of concentric spheres. Each sphere drove the next inner sphere, and the proper amount of slippage was allowed to reproduce the correct planetary periods.

The cosmological model of Pythagoras and Plato, as modified by later Greek astronomers, still contained a serious flaw in that it could not account for one of the prominent features of the motion of the planets relative to the stars. The planets were observed to move relative to the stellar background in paths that are not simple. On occasion a planet's steady forward motion across the celestial sphere was interrupted by a puzzling backward or *retrograde* motion, as shown in Fig. 14–1 for Mars. Other problems with the simple model were that the apparent motion of the Sun and the planets did not proceed with exact uniform angular motion in the sky, and there also was a readily observable variation in the brightness of the planets over periods of weeks.

These various problems with the Pythagorean scheme were overcome by the invention of *epicycles*. In this modification of the model, the essential importance of uniform circular motion was not surrendered. But instead of moving on a single

Fig. 14–1. The retrograde motion of Mars through the Gemini constellation during late 1960 and early 1961. The position of the planet on the first of each month is shown. The path has the shape of a loop because the planes in which the Earth and Mars revolve do not exactly coincide.

sphere, each planet was imagined to move in a small circle—the epicycle—the center of which, in turn, revolved around the Earth in a larger circle called the *deferent* (Fig. 14–2).

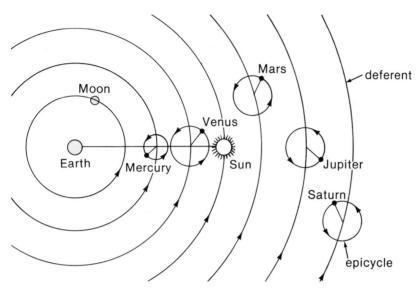

Fig. 14–2. The Ptolemaic system of the motion of planets (not to scale), put forward in about 150 A.D. Each planet moves on a circular *epicycle* and the center of each epicycle moves on a circular *deferent*. The centers of the epicycles of Mercury and Venus lie on a straight line connecting the Earth and the Sun.

The Ptolemaic System.   The epicycle theory was developed over four centuries by the Greek astronomers, and the whole grand scheme was finally summarized in about 150 A.D. by Ptolemy, in his famous work, *Almagest.** The model, usually called the *Ptolemaic system,* is shown in simplified form in Fig. 14–2. Notice that the centers of the epicycles of the inner planets, Mercury and Venus, are constrained to move always in line with the Sun. In this way the Ptolemaic system accounts for the fact that these two planets never appear at positions in the sky that are very far from the Sun. With revolving epicycles, Ptolemy could explain not only retrograde motion but also the changing brightness of the planets (they are brightest when nearest the Earth and dimmest when farthest away).

Today, the Ptolemaic system appears highly artificial, indeed, preposterous. But in ancient times it seemed quite reasonable to suppose that the Earth was the center of the solar system, even of the entire Universe. This belief, coupled with the conviction that the circle was the only "perfect" geometric form, led directly to the development of the Ptolemaic system. The refined system was a reasonably successful theory for interpreting the observations and predicting the future positions of the planets, if the times were not too long.

Copernicus and the Heliocentric Theory.   There was no challenge to Ptolemy's cosmological system for more than 1500 years. But during the period of the Renaissance, a new spirit of free inquiry characterized the revival of learning. Scien-

---

*Ptolemy (Latin name, Claudius Ptolemaeus) was a Greek (possibly an Egyptian) astronomer, mathematician, and geographer. Ptolemy's book *Megale mathematike syntaxis* ("Great mathematical composition") survived among the Arabs after the fall of the Roman Empire; they called it *Almagest* ("The Greatest"). The book was lost to Western civilization until 1175, when it was translated from Arabic into Latin and thereafter dominated astronomical thinking until the Renaissance.

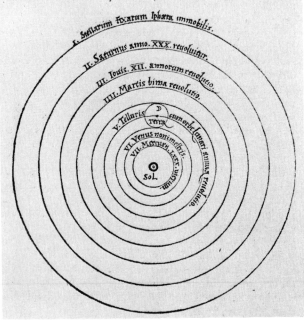

NICOLAI COPERNICI

net, in quo terram cum orbe lunari tanquam epicyclo contineri
diximus. Quinto loco Venus nono menfe reducitur. Sextum
deniq; locum Mercurius tenet, octuaginta dierum fpacio circū
currens, In medio uero omnium refidet Sol. Quis enim in hoc

Fig. 14-3. The Copernican heliocentric model of the solar system, as it appeared on a page in *De Revolutionibus*. (The epicycles on which the planets were presumed to move are not shown here.) (Courtesy of Chapin Library, Williams College.)

tific thought emerged from centuries of inactivity and began a vigorous development.

It was during this period that the Polish astronomer Nicolaus Copernicus (1473–1543) proposed a new model of the solar system. In the Copernican scheme, the Sun is located at the center of the solar system, and the Earth and the other planets revolve around the Sun in circular orbits. Actually, this idea was not new; a heliocentric theory of planetary motion had been put forward by the Greek astronomer Aristarchus (c320–c250 B.C.), based on an even earlier speculation by Heracleides (c388–c315 B.C.). But the suggestions of the Greek astronomers were only suggestions. Copernicus developed the idea fully and used the model to make detailed calculations of planetary positions.

Copernicus published his heliocentric theory in *De Revolutionibus Orbium Coelestium* ("On the Revolutions of Heavenly Spheres"), which appeared in 1543 as the author lay on his deathbed. Figure 14–3 shows the Copernican scheme with the Sun at the center and with the planets in order of increasing period—Mercury, Venus, Earth, Mars, Jupiter, and Saturn.* The largest circle in the diagram represents the stellar sphere. In order to maintain uniform circular motion (the Greek idea of the perfect circle was still strong), Copernicus found it necessary to introduce epicyclic motion. (The epicycles are not shown in Fig. 14–3.) In all, the Copernican scheme actually contained eight more epicycles than did the Ptolemaic system!

One of the advantages of the Copernican scheme was that it could account in a

---

*The outermost planets were discovered only after the invention of the telescope—Uranus (1781), Neptune (1846), and Pluto (1930).

natural way for retrograde planetary motion. In Fig. 14–4 we see the situation for an outer planet. The period of the Earth's motion around the Sun is shorter than that of the planet, so the Earth moves from A to G while the planet moves from a to g. As the Earth moves through the positions $A \rightarrow B \rightarrow C \ldots$, the planet moves through the corresponding positions $a \rightarrow b \rightarrow c \ldots$. To an observer on the Earth, the planet moves against the background of the distant stars along a track $a' \rightarrow b' \rightarrow c' \ldots$. The motion is generally from east to west, but as the Earth overtakes and passes the planet in its orbit, the track exhibits a retrograde (west-to-east) motion, $c' \rightarrow d' \rightarrow e'$. Notice also that because the distance between the planet and the Earth changes, there is a corresponding change in the apparent brightness of the planet. The brightness increases as the distance decreases ($Aa \rightarrow Bb \rightarrow Cc \rightarrow Dd$) and then decreases as the distance increases ($Dd \rightarrow Ee \rightarrow Ff \rightarrow Gg$). Thus, two of the troublesome features of the early Greek models are easily corrected in the heliocentric system.

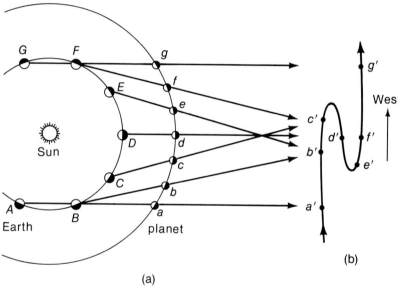

(a)

(b)

Fig. 14–4. The retrograde motion of planets is explained in a natural way with the heliocentric theory. (a) As the Earth moves $A \rightarrow B \rightarrow C \ldots$, an outer planet moves $a \rightarrow b \rightarrow c \ldots$. (b) The apparent track of the position of the planet in the sky as seen by an observer on the Earth.

Kepler and Elliptic Orbits. The next major advance in the theory of planetary motion was made by the German astronomer Johannes Kepler (1571–1630). The data used by Kepler were obtained by the Danish astronomer Tycho Brahe (1546–1601), who built observatories in Denmark and later in Prague for the purpose of making precise measurements of planetary positions. Kepler tried a variety of planetary models, including Brahe's curious system (see page 364), in an effort to interpret the data that had been accumulated, particularly those concerning Mars. All of these attempts failed until Kepler finally abandoned the idea of circular orbits. Trying other geometric forms, he decided that planetary motions are best described in terms of *ellipses.** Although he was mystically inclined toward the "perfect" circle, Kepler

---

*The orbit of Mars has the largest deviation from a circle (eccentricity) among all the planets Kepler could study. If he had concentrated on any other planet, he might never have noticed that the orbit is elliptical, not circular.

became the first to break with the ancient Greek idea of the uniform circular motion of planets and to propose a planetary model with elliptic orbits. The need for all manner of epicycles to describe planetary wanderings thus disappeared.

In 1609 Kepler published his famous work, *Astronomia Nova* ("New Astronomy"), which contained the first two of his three laws of planetary motion:

1. Every planet moves around the Sun in an orbit that is an ellipse, with the Sun located at one focus.

2. A straight line from the Sun to the planet sweeps out equal areas in equal time intervals.

Kepler's third law appeared in 1619 in *Harmonices Mundi* ("Harmony of the World"), buried amidst a mass of turgid mysticism that contained very little else of value:

3. The ratio of the square of the period of a planet to the cube of the length of the semimajor axis of its orbit is the same for all planets.

Kepler had a keen but rudimentary notion concerning gravity. In *Astronomia Nova*, he wrote that gravity is a "mutual corporeal tendency of kindred bodies to unite together." He also expressed the belief that the Sun was the source of the motive force responsible for the planetary motions. Kepler thereby anticipated the assertion proved by Newton more than 50 years later.

This allegorical scene published in 1651 shows planetary models being compared. Notice that the Copernican system at the left does not have as much "weight" as the curious Earth-centered system devised by Tycho Brahe in 1582. The Ptolemaic system, on the ground, seems to have been discarded. Nevertheless, the Ptolemaic system was still being taught at Harvard as late as 1650. (The New York Public Library.)

14-2 NEWTON'S GRAVITATION HYPOTHESIS

In his three laws of planetary motion, Kepler had given an accurate *description* of the way that planets orbit around the Sun. But by the time Newton turned his attention to the problem, nearly 50 years later, no one had provided a *fundamental theory* to support Kepler's laws. In 1665, Newton moved from Cambridge to his mother's farm in Woolsthorpe to escape the Black Plague that had begun to creep from London into the countryside. There, he studied quietly for two years, making important discoveries in optics, mathematics, and mechanics. During this period, Newton developed the theory of color, the differential calculus, and the first two laws of dynamics; in addition, he made a significant start on the theory of gravitation, which he finally completed and published in 1687.

Newton's law of gravitation is expressed by the equation*

$$\mathbf{F}_{12} = -G\frac{m_1 m_2}{r_{21}^2}\hat{\mathbf{u}}_r \qquad (14\text{--}1)$$

where the vector notation is illustrated in Fig. 14–5. The vector $\mathbf{F}_{12}$ is the force exerted *on m_1 by m_2*. The location of m_1 relative to m_2 is specified by the vector $\mathbf{r}_{21} = r_{21}\hat{\mathbf{u}}_r$, where $\hat{\mathbf{u}}_r$ is the unit vector in the direction from m_2 to m_1. Notice that $\mathbf{r}_{21} = -\mathbf{r}_{12}$, so that $\mathbf{F}_{21} = -\mathbf{F}_{12}$, in agreement with the requirement of Newton's third law.

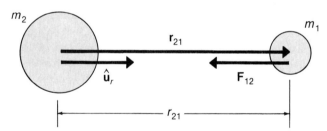

Fig. 14–5. The gravitational force on m_1 due to m_2 is $\mathbf{F}_{12}$.

The quantity G that appears in Eq. 14–1 is the *universal gravitation constant*, the best present value of which is

$$G = 6.673 \times 10^{-11}\,\text{N} \cdot \text{m}^2/\text{kg}^2$$

with an uncertainty of about 0.06 percent. (See Section 5–3 for a description of the Cavendish experiment which leads to a value for G.)

Newton originally hypothesized that the effect of gravity due to a particular body should diminish with the distance from that body in the same way that the intensity of light diminishes with the distance from a source, namely, as the inverse square of the distance.† By this reasoning, Newton introduced the $1/r^2$ factor into

---

*The gravitational force is the only force we discuss in this chapter, so we dispense with the usual subscript g on **F**.

†The inverse-square law for light intensity was not formulated, even in rudimentary form, until 1604 when Kepler deduced the correct relationship.

the equation for the gravitational force. He went on to provide a basis for this assertion by using Kepler's third law of planetary motion. For a planet that revolves around the Sun with a period T in a circular or nearly circular orbit with a radius r, the third law can be expressed as $T^2 \propto r^3$. The period is $T = 2\pi r/v$ and the force F necessary to maintain the circular motion is the mass m of the planet multiplied by the centripetal acceleration, v^2/r. Combining these three relationships, there results $F \propto m/r^2$. Newton understood this force to be the gravitational force exerted by the Sun on the planet. From Newton's third law, it is evident that the mass M of the Sun must enter the expression for the force in the same way that the planetary mass m enters. That is,

$$F = G\frac{mM}{r^2}$$

where G is the proportionality constant. This expression for F, deduced from Kepler's third law, is just the magnitude of the vectorial equation, Eq. 14-1.

Thus far we have made the tacit assumption that we are dealing with the gravitational interaction between *point* objects. Because such objects have no extent, there is no question about the meaning of the distance between objects. However, in calculating the gravitational force between the Earth and the Moon, are we to use the distance from surface to surface, or from center to center, or what? Newton reasoned that because the *entire* mass of the Earth attracts the *entire* mass of the Moon, the distance that enters the expression for the force law should be the distance between the *centers* of the objects. But Newton could not prove this assertion. With his brilliant theory incomplete, Newton chose not to make public the progress he had made. While devising the mathematical tools to solve the problem, Newton invented the integral calculus. Finally, he was able to make a proper calculation of the gravitational force between extended objects (see Section 14-5). The gravitational force law was then fully justified, and he included an exposition of the theory in the *Principia*, published in 1687.

The importance of Newton's gravitational force law is not just that it correctly accounts for the motion of the planets around the Sun, but that it describes the gravitational force between *any* pair of objects. Newton's law of gravitation is truly a *universal* law; it applies, as far as we know today, to every pair of objects in the Universe.

Example 14-1

The radius of the planet Mars is 3.40×10^6 m, and on the surface the acceleration due to gravity is 3.71 m/s². (Therefore, an astronaut on the Martian surface would have a weight about $\frac{1}{3}$ of his weight on the Earth's surface.) Use this information to calculate the mass M of Mars and its average density $\bar{\rho}$.

Solution:

In Section 5-3 we found that we can write two different equations for the gravitational force on an object with a mass m located near the Earth's surface, namely,

$$F = mg \quad \text{and} \quad F = G\frac{mM}{R^2}$$

We now interpret M and R as the values for Mars. Equating these expressions for F and solving for M, we find

$$M = \frac{gR^2}{G}$$

Substituting the known values of g, R, and G, we find

$$M = \frac{gR^2}{G} = \frac{(3.71 \text{ m/s}^2)(3.40 \times 10^6 \text{ m})^2}{6.67 \times 10^{-11} \text{ N} \cdot \text{m}^2/\text{kg}^2}$$
$$= 6.43 \times 10^{23} \text{ kg}$$

The volume of Mars is

$$V = \tfrac{4}{3}\pi R^3 = \tfrac{4}{3}\pi (3.40 \times 10^6 \text{ m})^3$$
$$= 1.646 \times 10^{20} \text{ m}^3$$

Then, the average density is

$$\bar{\rho} = \frac{M}{V} = \frac{6.43 \times 10^{23} \text{ kg}}{1.646 \times 10^{20} \text{ m}^3}$$
$$= 3.90 \times 10^3 \text{ kg/m}^3 = 3.90 \text{ g/cm}^3$$

which is slightly smaller than the average density of the Earth, namely, 5.50 g/cm³.

Example 14-2

Determine the gravitational force on a particle with mass m located a distance R from the center of a slender homogeneous rod with mass M and length ℓ, as shown in the diagram.

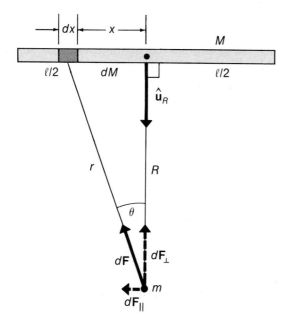

Solution:

The mass element $dM = (M/\ell)\, dx$ is a distance x from the center of the rod and a distance $r = \sqrt{x^2 + R^2}$ from the particle m. The force exerted on m by dM is $d\mathbf{F}$, the components of which are $d\mathbf{F}_\perp$ and $d\mathbf{F}_\parallel$. Another mass element at a distance x to the right of the midpoint of the rod also produces a perpendicular force component $d\mathbf{F}_\perp$; however, the parallel component is directed opposite to $d\mathbf{F}_\parallel$. Because the particle m is directly opposite the midpoint of the rod, the parallel force components all cancel. The total force $\mathbf{F}$ can then be obtained by integrating $d\mathbf{F}_\perp$ from $x = 0$ to $x = \ell/2$ and doubling the result. We have

$$d\mathbf{F}_\perp = -dF \cos\theta \,\hat{\mathbf{u}}_R = -\frac{Gm(M/\ell)\,dx}{x^2 + R^2} \cdot \frac{R}{(x^2 + R^2)^{1/2}} \hat{\mathbf{u}}_R$$

$$= -\frac{GmMR}{\ell} \cdot \frac{dx}{(x^2 + R^2)^{3/2}} \hat{\mathbf{u}}_R$$

Then,

$$\mathbf{F} = 2 \int d\mathbf{F}_\perp = -2\frac{GmMR}{\ell}\hat{\mathbf{u}}_R \int_0^{\ell/2} \frac{dx}{(x^2 + R^2)^{3/2}}$$

$$= -2\frac{GmMR}{\ell}\hat{\mathbf{u}}_R \cdot \frac{x}{R^2\sqrt{x^2 + R^2}}\Bigg|_0^{\ell/2}$$

$$= -2\frac{GmMR}{\ell}\hat{\mathbf{u}}_R \cdot \frac{\ell/2}{R^2\sqrt{(\ell/2)^2 + R^2}}$$

$$= -2\frac{GmM}{\ell R}\hat{\mathbf{u}}_R \cdot \frac{1}{\sqrt{1 + (2R/\ell)^2}}$$

Now, if the particle m is at a great distance from the rod, so that $R \gg \ell$, then in the denominator of the expression for $\mathbf{F}$, the factor unity can be neglected in comparison with $(2R/\ell)^2$. Thus, we have

$$\mathbf{F}(R \gg \ell) = -G\frac{mM}{R^2}\hat{\mathbf{u}}_R$$

which is the same as the result for a particle with mass M. This is to be expected because when $R \gg \ell$, the dimensions of the rod are no longer important and the rod is essentially pointlike.

14-3   GRAVITATIONAL POTENTIAL ENERGY

If a particle with a mass m is moved from a distance R_1 to a distance R_2 measured from another pointlike object with a mass M, the work done by the gravitational force is (see Example 8-5)

$$W = -GmM \int_{R_1}^{R_2} \frac{dr}{r^2} = GmM\left(\frac{1}{R_2} - \frac{1}{R_1}\right) \tag{14-2}$$

The associated change in potential energy is the negative of the work done (see Section 7-4); that is,

$$U_2 - U_1 = -GmM\left(\frac{1}{R_2} - \frac{1}{R_1}\right) \tag{14-3}$$

We define a convenient reference level for the potential energy by taking U to be zero for $R_1 \to \infty$. Then, for a separation $R_2 = r$, we have

$$U(r) = -G\frac{mM}{r} \tag{14-4}$$

Notice that we write $U(r)$ instead of $U(\mathbf{r})$ because the potential energy depends only on $r = |\mathbf{r}|$.

Suppose that the origin of a coordinate system is located at the position of the object M. The potential energy function for M and a particle m is defined for any point in space at which m might be located. At a particular point P, the force on m is related to the variation of U in the neighborhood of P. In general, the force will have x-, y-, and z-components that are related to the partial derivatives of U in the respective directions. However, in the event that M is a pointlike (or spherically symmetric) object, the situation is equivalent to a one-dimensional problem in that the potential energy function depends only on the distance r between m and M. Then, the force is equal to the negative of the derivative of $U(r)$ with respect to r (see Eq. 7–8). Using the notation defined in Fig. 14–6, we have*

$$\mathbf{F}(\mathbf{r}) = -\frac{dU(r)}{dr}\hat{\mathbf{u}}_r = -G\frac{mM}{r^2}\hat{\mathbf{u}}_r \qquad \textbf{(14–5)}$$

which is the expected result and demonstrates the consistency of our definitions of work, potential energy, and force.

$$U = -G\frac{mM}{r}$$

$$\mathbf{F} = -G\frac{mM}{r^2}\hat{\mathbf{u}}_r$$

Fig. 14–6. Potential energy and force for a particle with a mass m in the vicinity of a pointlike object with a mass M.

The Principle of Superposition. Throughout the discussions of force, we have assumed that the net effect of the application to an object of a number of forces, $\mathbf{F}_1$, $\mathbf{F}_2$, $\mathbf{F}_3$, ..., is simply the vector sum of the forces: $\mathbf{F}_{\text{net}} = \mathbf{F}_1 + \mathbf{F}_2 + \mathbf{F}_3 + \cdots$. This is actually one statement of a very important physical idea—the *principle of superposition*. It is a fact, established by experiment, that forces combine by ordinary (linear) vector addition. This result is not guaranteed by any theory: the introduction of a third particle into a two-particle system could conceivably alter the way in which the forces combine for two particles. However, according to experiment, when two or more particles exert gravitational (or other) forces on another particle, the net result is the vector sum of the individual forces, each calculated without regard for the presence of the other particle or particles. If the particles constitute an extended object, the summation procedure is replaced by an integration.

We assert that potential energy also combines by ordinary (linear) algebraic summation. For example, suppose that we assemble three gravitationally interacting particles into the configuration shown in Fig. 14–7. We imagine that the particles are originally dispersed infinitely far from one another. First, we bring M_1 into position; no work is involved in this process, and the potential energy of M_1 by itself is zero:

$$U_1 = 0$$

Next, we bring M_2 into position; the potential energy of the M_1-M_2 combination is

$$U_2 = -G\frac{M_1 M_2}{R}$$

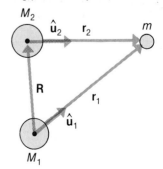

Fig. 14–7. The geometry of three gravitationally interacting pointlike objects.

---

*Notice that $\mathbf{F}(\mathbf{r})$ has the same meaning as $\mathbf{F}(x, y, z)$, a notation we used earlier.

Finally, we bring the third particle m into position. The potential energy that results from this process, we assert, is

$$U_3 = -G\frac{M_1 m}{r_1} - G\frac{M_2 m}{r_2}$$

That is, U_3 is calculated by considering the potential energies of the M_1-m and M_2-m combinations *separately and independently*. Then, by the same reasoning, the potential energy of the three-particle system is

$$U = U_1 + U_2 + U_3$$

$$= -G\frac{M_1 M_2}{R} - G\frac{M_1 m}{r_1} - G\frac{M_2 m}{r_2}$$

If M_1 and M_2 are held fixed (that is, if $\mathbf{R}$ is maintained constant), the force on m at any point is obtained from

$$\mathbf{F} = -\frac{\partial U}{\partial r_1}\hat{\mathbf{u}}_1 - \frac{\partial U}{\partial r_2}\hat{\mathbf{u}}_2$$

$$= -G\frac{M_1 m}{r_1^2}\hat{\mathbf{u}}_1 - G\frac{M_2 m}{r_2^2}\hat{\mathbf{u}}_2$$

Thus, the assumption of a linear algebra for the potential energy results in a linear force equation, as demanded by experiment. We conclude that potential energy also obeys the principle of superposition. (We can only infer this conclusion because forces are directly measurable whereas potential energies are not.)

Presumably, all of the basic forces in Nature, and their corresponding potential energies, obey the superposition principle—unless the forces involved are too large. It is difficult to give a precise meaning of "too large," but the strong force within nuclei and the gravitational force near extremely massive and dense objects (such as the curious astronomical objects called *black holes*), for example, seem to produce effects at variance with the principle of superposition. For all of the situations we consider, however, the principle is completely valid.

▶ *The Gravitational Field Strength.* Sometimes it is convenient to describe gravitational problems in terms of the *field* concept (see Section 8–1). An object with a mass M sets up in space surrounding it a condition to which another object (a *test object*) will respond by experiencing a force directed toward M. This "condition" is the *gravitational field* of M, and we say that M is the *source* of the field. Assume that M is spherically symmetric so that we can treat it as a pointlike object. Then, if the mass of the test object is m, we formally define the strength of the gravitational field to be

$$\mathbf{g}(\mathbf{r}) = \lim_{m \to 0}\frac{\mathbf{F}(\mathbf{r})}{m} = -G\frac{M}{r^2}\hat{\mathbf{u}}_r$$

where r and $\hat{\mathbf{u}}_r$ refer to the center of M.

This form for the definition emphasizes the fact that the quantity $\mathbf{g}$, a specific measure of the gravitational field, depends only on the source mass M. The quantity $\mathbf{g}$ also represents the acceleration that is experienced by a test object placed in the field of M. The force on a test object m in the field $\mathbf{g}$ is

$$\mathbf{F} = m\mathbf{g}$$

which is the vector form of the gravitational force equation we have used in earlier chapters. When the source mass is the Earth and $r = R_E$, the magnitude of $\mathbf{g}$ is the familiar constant, $g = 9.80 \text{ m/s}^2$.

We can also divide the expression for the potential energy (Eq. 14–4) by m to obtain a quantity that is characteristic of the source mass and not the test object. We call this quantity the *gravitational potential,* $\Phi(r)$:

$$\Phi(r) = -G\frac{M}{r}$$

In this chapter we make only a very limited use of the idea of field strength. However, when we turn to the discussion of electromagnetic forces, the field strength concept becomes important and we use it extensively. ◄

14-4 THE POTENTIAL ENERGY OF A SPHERICAL SHELL AND A PARTICLE

We now calculate the potential energy of a system that consists of a particle with a mass m at an arbitrary point P that is a distance r from the center of a thin spherical shell with mass M, thickness t, and radius R. We allow both $r > R$ and $r < R$. The geometry is shown in Fig. 14–8. We shall prove that the potential energy of m and the force acting on m due to the shell are the same as for a pointlike object with a mass M located at C. This interesting result also applies for the case of a solid sphere. (•Can you see why?)

We begin by considering a thin ring of material with density ρ (shown shaded in Fig. 14–8), every part of which is a distance s from the point P. The width of this ring is $R\,d\theta$, its thickness is t, and its radius is $R\sin\theta$. Consequently, the mass dM of the ring is

$$dM = R\,d\theta \cdot 2\pi R\sin\theta \cdot t \cdot \rho$$
$$= 2\pi R^2 t\rho \sin\theta\,d\theta \qquad \textbf{(14–6)}$$

Then, the potential energy of the ring and the particle m is

$$dU = -Gm\frac{dM}{s} = -\frac{2\pi GmR^2 t\rho \sin\theta\,d\theta}{s}$$

where we again take $U = 0$ for infinite separation (see Eq. 14–4).

To obtain the potential energy due to the entire shell, we must integrate this expression for dU. It is easier to integrate over s instead of over the angle θ, so we make a change of variable. Referring to Fig. 14–8, we see that the law of cosines can

Fig. 14-8. Geometry for the calculation of the potential energy of a thin spherical shell and a particle. The thickness of the shell is t.

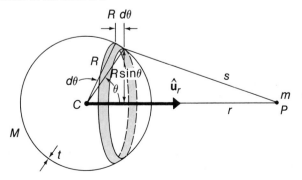

be used to write

$$s^2 = r^2 + R^2 - 2rR \cos \theta$$

Differentiating this expression with respect to θ for $R = $ constant and for a particular (fixed) value of r, we obtain

$$2s \frac{ds}{d\theta} = 2rR \sin \theta$$

so that
$$\frac{\sin \theta \, d\theta}{s} = \frac{ds}{rR}$$

Substituting this result into the equation for dU, we have

$$dU = -\frac{2\pi GmRt\rho}{r} ds \tag{14–7}$$

We now integrate over s to obtain the potential energy for the entire spherical shell. For $r > R$, the variable s ranges from $s = r - R$ (corresponding to $\theta = 0$) to $s = r + R$ (corresponding to $\theta = \pi$). Thus, we find

$$U(r > R) = \int dU = -\frac{2\pi GmRt\rho}{r} \int_{r-R}^{r+R} ds$$

$$= -\frac{2\pi GmRt\rho}{r} [(r + R) - (r - R)]$$

$$= -\frac{Gm}{r} \cdot (4\pi R^2 t\rho)$$

Now, $4\pi R^2 t$ is just the volume V of the thin shell, and the mass is $M = V\rho$. Therefore, the expression for U becomes

$$U(r > R) = -G\frac{mM}{r} \tag{14–8}$$

which is exactly the result we would obtain if the shell were a pointlike mass M located at the center (Eq. 14–4). Then, we obtain for the force on m,

$$\mathbf{F(r)} = -\frac{dU}{dr} \hat{\mathbf{u}}_r = -G\frac{mM}{r^2} \hat{\mathbf{u}}_r \qquad (r > R) \tag{14–9}$$

in agreement with Eq. 14–1.

Notice that a thick shell or even a solid sphere can be considered to be made up of a large number of thin shells. If each of these thin shells has uniform density, even though different shells may have different densities, the results for U and $\mathbf{F}$ still apply. That is, *the gravitational effect of a spherically symmetric distribution of matter on a particle is the same as that of a pointlike object with equal mass located at the center of the sphere.*

Next, consider the particle m to be located *inside* the shell ($r < R$). We again must integrate Eq. 14–7, but now the variable s ranges from $s = R - r$ (i.e., $\theta = 0$)

to $s = R + r$ (i.e., $\theta = \pi$). Thus,

$$U(r < R) = -\frac{2\pi GmRt\rho}{r} \int_{R-r}^{R+r} ds$$

$$= -\frac{2\pi GmRt\rho}{r} [(R + r) - (R - r)]$$

$$= -\frac{Gm}{R} \cdot (4\pi R^2 t\rho)$$

so that

$$U(r < R) = -G\frac{mM}{R} \qquad (14\text{-}10)$$

Because R is a constant, the force on m vanishes:

$$\mathbf{F(r)} = -\frac{dU}{dr}\hat{\mathbf{u}}_r = 0 \qquad (r < R) \qquad (14\text{-}11)$$

Thus, we have the remarkable result that for a uniform spherical shell the gravitational force on a particle inside the shell is *zero*. With the same argument that we used for $r > R$, we conclude that $\mathbf{F(r)} = 0$ within any spherically symmetric shell of matter.

Figure 14–9 illustrates the results we have obtained for the potential energy and the force. The graphs are shown for a thick shell with inner radius a and outer radius R. The case of a solid sphere corresponds to $a = 0$. For $a < r < R$, the exact forms of the functions depend on the radial variation of the density. If the density is constant, the force $F(r)$ is a linear function of r for $0 < r < R$ (see Problem 14–13).

Notice that in these calculations there are two independent sets of variables. First, there are the coordinates of the mass element within the object M over which the integration is carried out to obtain the potential energy. Second, there are the

Fig. 14–9. The gravitational potential energy and the force on a particle due to a spherical shell of matter.

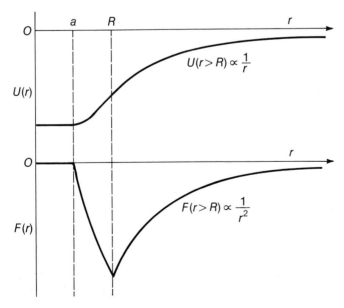

coordinates of the particle m which, although arbitrary, are nonetheless held fixed during the integration over M. Then, the force on m is obtained by differentiating the expression for the potential energy with respect to the coordinates of m (here given by r).

▶ Refer again to the situation in Example 14–2. Suppose that we approach the problem by first determining the potential energy function; then, we differentiate $U(R)$ to obtain the force on m. We have

$$dU = -\frac{Gm\,dM}{r} = -\frac{Gm(M/\ell)\,dx}{\sqrt{x^2 + R^2}}$$

We obtain $U(R)$ by integrating dU from $x = 0$ to $x = \ell/2$ and doubling the result. Thus,

$$U(R) = -2\frac{GmM}{\ell}\int_0^{\ell/2}\frac{dx}{\sqrt{x^2 + R^2}}$$

To evaluate this integral and then calculate $dU(R)/dR$ is tedious. However, we can simplify the procedure by taking advantage of the fact that the coordinates x and R are independent. Then, the differentiation of $U(R)$ with respect to R can be interchanged with the integration over x. That is,

$$\mathbf{F}(R) = -\frac{dU}{dR}\hat{\mathbf{u}}_R = 2\frac{GmM}{\ell}\hat{\mathbf{u}}_R\frac{d}{dR}\int_0^{\ell/2}\frac{dx}{\sqrt{x^2 + R^2}}$$

$$= 2\frac{GmM}{\ell}\hat{\mathbf{u}}_R\int_0^{\ell/2}\frac{d}{dR}\frac{1}{\sqrt{x^2 + R^2}}\,dx$$

$$= -2\frac{GmM}{\ell}\hat{\mathbf{u}}_R\int_0^{\ell/2}\frac{R\,dx}{(x^2 + R^2)^{3/2}}$$

The integral is now the same as that in Example 14–2, so we arrive at the same result,

$$\mathbf{F}(R) = -2\frac{GmM}{\ell R}\hat{\mathbf{u}}_R\cdot\frac{1}{\sqrt{1 + (2R/\ell)^2}} \qquad ◀$$

Example 14–3

If an object is propelled upward from the surface of the Earth with a sufficient velocity, it will overcome its gravitational bond to the Earth and will never fall back to its home planet. However, an object can escape from the Earth's gravity and still be bound within the solar system because of the enormous gravitational effect of the Sun. The minimum velocity required for an object to escape from the Earth (and yet remain within the solar system) is essentially the same as that required to remove the object to an infinite distance from the Earth, ignoring the effects of the Sun. (•Can you argue that this is reasonable?) Subject to this restriction, what is the value of the *escape velocity* v_E for the Earth?

Solution:

If an object has an initial kinetic energy $\frac{1}{2}mv_E^2$ at the Earth's surface, this energy will be completely expended in raising the initial gravitational potential energy, $-GmM_E/R_E$, to zero as $r \to \infty$. That is, the total energy of the object, $K + U$, becomes exactly zero as $r \to \infty$, and so must also be exactly zero at the Earth's surface. That is,

$$K + U = \tfrac{1}{2}mv_E^2 - G\frac{mM_E}{R_E} = 0$$

from which

$$v_E = \sqrt{2GM_E/R_E}$$
$$= \sqrt{\frac{2(6.67 \times 10^{-11}\,\mathrm{N \cdot m^2/kg^2})(5.98 \times 10^{24}\,\mathrm{kg})}{6.38 \times 10^6\,\mathrm{m}}}$$
$$= 1.12 \times 10^4\,\mathrm{m/s}$$
$$= 11.2\,\mathrm{km/s}$$

If an object is to be propelled to an infinite distance from the Earth, taking into account the effect of the Sun, the escape velocity increases to 43.5 km/s (see Problem 14–10). In fact, the gravitational potential energy of an object on the Earth's surface is about 93 percent *solar* and only about 7 percent *terrestrial*. (•Can you verify these numbers?)

14-5 GRAVITATIONAL FORCES ON EXTENDED OBJECTS

Many types of problems involve the gravitational force on an extended object. Any such object (which we consider to be perfectly rigid) consists of a large number of particles (atoms). When we use the phrase "the force on the object," we really mean the vector sum of the forces $\mathbf{f}_i$ acting on all the individual particles in the object. The gravitational force due to the Earth on a particle with a mass m_i is $\mathbf{f}_i = m_i\mathbf{g}$, where the acceleration, $\mathbf{g} = -g\hat{\mathbf{k}}$, is directed downward (toward the center of the Earth), as in Fig. 14–10. For a relatively small everyday object, the vector $\mathbf{g}$ is uniform over the dimensions of the object. Then, the total gravitational force on the object is

$$\mathbf{F} = \sum_i \mathbf{f}_i = \sum_i m_i\mathbf{g} = \left(\sum_i m_i\right)\mathbf{g} = M\mathbf{g}$$

where M is the total mass of the object.

Fig. 14–10. The individual particles m_i comprising an object are acted upon by the gravitational forces $m_i\mathbf{g}$. The sum of all these individual forces is equivalent to a single force that acts at C, the C.M. of the object.

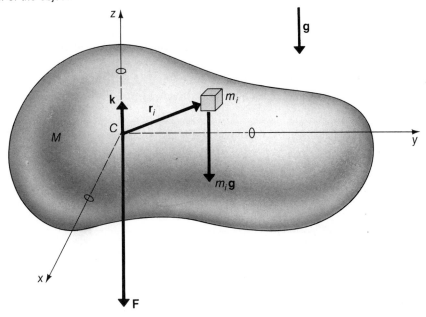

At what point on or in the object does this single effective force act? Intuition suggests that the point is the center of mass of the object. We can prove this in the following way. In Fig. 14–10, the point C is the C.M. of the object. If we calculate the total torque about C due to the gravitational force acting on each of the individual particles (Eq. 10–29) and sum the result, we find

$$\tau = \sum_i \mathbf{r}_i \times \mathbf{f}_i = \sum_i \mathbf{r}_i \times m_i\mathbf{g} = \left(\sum_i m_i\mathbf{r}_i\right) \times \mathbf{g}$$

The term in parentheses (see Eq. 9–8) is proportional to the position vector of the C.M. in the center-of-mass coordinate frame and therefore vanishes; hence, $\tau = 0$.

If a single force $\mathbf{F}$ is to represent the effect of all the individual forces $\mathbf{f}_i$, this force must also produce zero torque on the object. This will be the case only if the line of action of $\mathbf{F}$ passes through the C.M. of the object. Moreover, for any orientation of the object, $\mathbf{F}$ must still be directed through the center of mass. We conclude that $\mathbf{F}$ acts exactly *on* the C.M. of the object.

The point in a body at which the force of gravity can be considered to act is customarily referred to as the *center of gravity* of the object. The concept of the center of gravity is generally useful only in the event that the gravitational acceleration vector $\mathbf{g}$ is essentially uniform over the volume occupied by a body. In such a case, the center of gravity and the center of mass coincide. However, if an object is so large that $\mathbf{g}$ has some variation over the dimensions of the object, it is no longer possible to define a point identified as the center of gravity (except in the special case of spherical objects, as we show next). (•Can you see why this is the case?)

The Gravitational Force Between Spherically Symmetric Objects.   Next, consider the gravitational attraction between two spherically symmetric distributions of matter. If the objects are relatively close, as in Fig. 14–11, it is clear that the gravitational attraction due to one of the spheres is not uniform over the volume occupied by the other sphere. Nevertheless, we can argue that the vector $\mathbf{F}_{AB}$ representing the total gravitational force on sphere A due to sphere B acts on the center of mass of A (and vice versa).

In Fig. 14–11, we show a small mass element (particle) m in the sphere B. According to the results in the preceding section, the vector $\mathbf{f}_{mA}$ that represents the gravitational force on m due to A is directed toward the center of A. By Newton's third law, the force $\mathbf{f}_{Am}$ on A due to m is directed along this same line, as shown in Fig. 14–11. Because m is an arbitrary mass element in B, it must be true that the force on A due to *every* mass element in B is directed through the center of A. Also,

Fig. 14–11.   The gravitational force $\mathbf{F}_{AB}$ on sphere A due to sphere B acts on the C.M. of A.

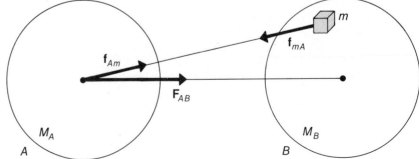

according to the previous results, the force due to A acting on m (and all of the similar elements in B) is the same as if the mass of A were located at its center point. Thus, we conclude that $\mathbf{F}_{AB}$, the total gravitational force on A due to B, acts exactly *on* the C.M. of sphere A. Moreover, because sphere B is spherically symmetric, the direction of $\mathbf{F}_{AB}$ is toward the center of B. By reversing the argument, we also conclude that $\mathbf{F}_{BA}$, the force on sphere B, acts exactly on the C.M. of B. The magnitude of each force is proportional to the product of the masses, $M_A M_B$, and depends inversely on the square of the distance between the centers.* Notice that this special result is obtained solely because we considered objects with spherically symmetric mass distributions. In fact, if object B has an irregular shape, the force $\mathbf{F}_{AB}$ still acts on the C.M. of A, but the force $\mathbf{F}_{BA}$ no longer acts on the C.M. of B (see Problem 14-15).

14-6 PLANETARY MOTION

One of the important applications of Newtonian gravitation theory is in the study of planetary motion. The Sun is by far the most massive object in the solar system. Consequently, the motion of every planet is dominated by the gravitational effect of the Sun and is influenced very little by the presence of the other planets. Thus, planetary motion in the solar system is the result of a collection of essentially two-body, planet-plus-Sun interactions.

As we learned in Section 10-2, the motion of a two-particle system is conveniently described in terms of the equivalent single particle. The two particles orbit about their center of mass O, as shown in Fig. 14-12a. The equivalent single particle, with mass $\mu = m_1 m_2 / M$, where $M = m_1 + m_2$, moves in an orbit around O' (Fig. 14-12b). The angle ϕ in the two descriptions of the motion is the same. According to Eqs. 10-14 through 16, we have

$$\left.\begin{array}{l} \mathbf{r} = \mathbf{r}_2 - \mathbf{r}_1 \\[2mm] \mathbf{r}_1 = -\dfrac{\mu}{m_1}\mathbf{r} \\[2mm] \mathbf{r}_2 = \dfrac{\mu}{m_2}\mathbf{r} \end{array}\right\} \qquad (14\text{-}12)$$

where the C.M. vector $\mathbf{R}$ is zero because we now refer $\mathbf{r}_1$ and $\mathbf{r}_2$ to the C.M. at O.

If m_1 and m_2 move in circular orbits around O, as in Example 10-8, the equivalent single particle also moves in a circular orbit around O'. (•Can you see why? Refer to the diagram in Example 10-8.) Likewise, if m_1 and m_2 move in elliptic orbits around O (Fig. 14-13a), μ moves in an elliptic orbit around O' (Fig. 14-13b). The points O and O' correspond to the *foci* of the respective ellipses.

In the case of planetary motion, the mass of the Sun (m_1) is much greater than the mass of the planet (m_2). Then, the C.M. of the system is close to the center of the Sun. That is, $\mu \cong m_2$, and the vector $\mathbf{r}$ is essentially the same as $\mathbf{r}_2$.

Conic Sections. An object that is subject to an inverse r^2 force will move along a path that is characterized as a *conic section*. This family of geometric curves consists of the circle, the ellipse, the parabola, and the hyperbola. A conic section (or

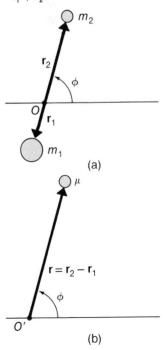

Fig. 14–12. (a) Two particles orbit around their common center of mass at O. (b) The situation in (a) is described in terms of the equivalent single particle with mass $\mu = m_1 m_2/(m_1 + m_2)$. The vector $\mathbf{r}$ is the *relative* coordinate of the particles, so, in magnitude, $r = r_1 + r_2$.

(a)

(b)

---

*Thus, the assertion made by Newton in his original formulation of the theory of gravitation is finally proven.

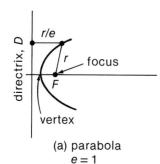

(a) parabola
$e = 1$

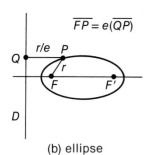

$\overline{FP} = e(\overline{QP})$

(b) ellipse
$0 < e < 1$

(c) hyperbola
$e > 1$

Fig. 14–14. The conic sections. (a) The parabola has $e = 1$. (b) The ellipse has $0 < e < 1$. (c) The hyperbola has $e > 1$. There is another branch of the hyperbola (not shown), which lies to the left of the branch illustrated and has the opposite curvature. The circle (not shown) has $e = 0$ and corresponds to an ellipse for which F and F' coincide.

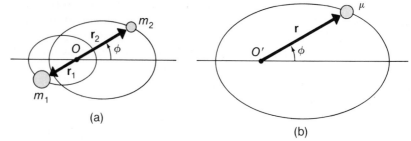

Fig. 14–13. (a) Elliptic motion of m_1 and m_2 around their C.M. at O. The point O is the right-hand focus of the smaller ellipse and the left-hand focus of the larger ellipse. (b) The corresponding elliptic motion of μ around O'.

conic) is a curve, every point of which has distances from a fixed point and a fixed line (not through the point) in a constant ratio to each other. The fixed point is called the *focus,* the fixed line is the *directrix,* and the constant ratio is the *eccentricity e.* Figure 14–14 shows the geometry for the noncircular conics. The definition is illustrated for the case of the ellipse (Fig. 14–14b), namely,

$$\overline{FP} = e(\overline{QP}) \tag{14–13}$$

The equation of a conic section can be obtained directly from the defining equation (Eq. 14–13). We select the origin of the coordinates at the focus F (Fig. 14–15). Then, we identify $\overline{FP} = r$ and $\overline{QP} = \delta + r \cos \phi$, where δ is the distance from the focus to the directrix D. Thus,

$$r = e(\delta + r \cos \phi)$$

from which

$$\frac{e\delta}{r} = 1 - e \cos \phi \tag{14–14}$$

This is the general equation for a conic; the various forms illustrated in Fig. 14–14 result for the values of the eccentricity indicated.

The Effective Potential Energy Function and the Characteristics of Planetary Orbits. The total energy of the planet-plus-Sun system is

$$E = K + U = \tfrac{1}{2}\mu v^2 + U(r)$$
$$= \tfrac{1}{2}\mu(v_r^2 + v_\phi^2) + U(r)$$

where the kinetic energy has been expressed in terms of the radial velocity v_r and the azimuthal velocity v_ϕ (refer to Example 10–5). Using $v_\phi = r\omega$ and $L = \mu r^2 \omega$ (Eq. 10–24), which is a constant of the motion, we can write

$$E = \tfrac{1}{2}\mu v_r^2 + \frac{L^2}{2\mu r^2} - G\frac{m_1 m_2}{r} \tag{14–15}$$

The term $L^2/2\mu r^2$ represents the kinetic energy of the system due to its angular motion.* It is often convenient to combine this term with $U(r)$ to form an *effective potential energy function* $V(r)$:

$$V(r) = \frac{L^2}{2\mu r^2} - \frac{k}{r} \qquad \textbf{(14-16)}$$

where $k = Gm_1 m_2 = G\mu M$. Thus,

$$E = \tfrac{1}{2}\mu v_r^2 + V(r) \qquad \textbf{(14-17)}$$

The effective potential energy function and its contributing parts are shown in Fig. 14-16. Notice how the two terms combine to produce a minimum. $V(r)$ approaches zero as $r \to \infty$.

Figure 14-17 shows more details of the motion in terms of $V(r)$. First, notice that the radial velocity v_r must be a real quantity; that is, $v_r^2 \geqslant 0$. Equation 14-17 shows that the total energy E has its minimum value when $v_r \equiv 0$ and $dV/dr = 0$. Differentiate Eq. 14-16 with respect to r and set the result equal to zero; after some algebraic manipulation this gives

$$E_{\min} = -\frac{\mu k^2}{2L^2} \qquad \textbf{(14-18)}$$

as indicated in Fig. 14-17. The fact that the radial velocity v_r is identically zero means that the value of r never changes; the radius remains fixed at $r = r_0$. Thus, the motion at minimum total energy follows a *circular path*.

Next, increase the total energy above the minimum value but subject to the restriction $E_{\min} < E < 0$. Then, v_r can be different from zero, and the radius changes with time between the values $r_{\min}$ and $r_{\max}$ (see Fig. 14-17). At an intermediate value of r, notice how K, V, and E are represented in the diagram. As r approaches either $r_{\min}$ or $r_{\max}$, the kinetic energy K approaches zero; that is, v_r decreases and finally equals zero at $r = r_{\min}$ and at $r = r_{\max}$. These positions are called the *turning points* of the motion, the points at which v_r changes sign. This type of

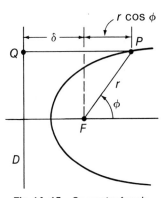

Fig. 14-15. Geometry for obtaining the general equation for a conic.

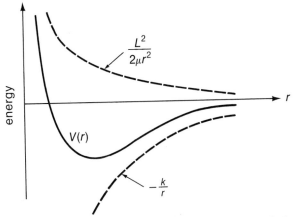

Fig. 14-16. The effective potential energy function $V(r)$ for the case of planetary motion. The ordinary potential energy function is $U(r) = -G\mu M/r = -k/r$.

---

*$L^2/2\mu r^2$ is sometimes called the *centrifugal energy*.

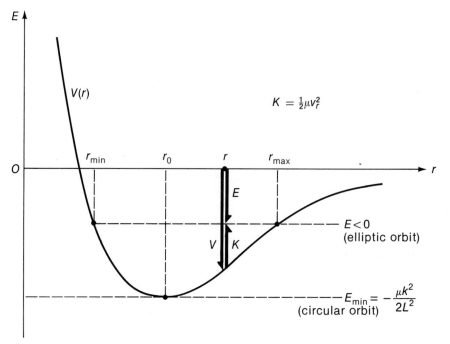

Fig. 14–17. The effective potential energy function for the case of planetary motion, showing radius values for two energies, corresponding to elliptic and circular orbits.

motion takes place along an *elliptic path*. (It is often said that the particle moves in the *potential energy well* $V(r)$ between the limits r_{min} and r_{max}.)

The values of r_{min} and r_{max} can be obtained from Eq. 14–15 by setting $v_r = 0$. Multiplying the resulting expression by r^2 and dividing by E, we obtain

$$r^2 + \frac{k}{E}r - \frac{L^2}{2\mu E} = 0 \qquad (14\text{–}19)$$

This is a quadratic equation for r, the solutions of which are

$$r = -\frac{k}{2E}\left[1 \pm \left(1 + \frac{2EL^2}{\mu k^2}\right)^{1/2}\right] \qquad (14\text{–}20)$$

These values of r, which we designate r_{min} and r_{max}, are both real (and positive) because E is negative and lies between $-\mu k^2/2L^2$ and zero. For planetary motion, the point r_{min} at which the planet is nearest to the Sun is called the *perihelion* and the point r_{max} of greatest separation is called the *aphelion*. The corresponding terms for motion around the Earth are *perigee* and *apogee*.

If the energy is made equal to zero or becomes positive, the motion is no longer bounded. In these cases, there is a minimum allowed value for r, but there is no maximum value. Thus, if the body moves toward O, the separation will decrease until the minimum value is reached. Then, v_r changes sign and the body thereafter recedes from O, never to return. (Compare the discussion of Fig. 7–6.) The motion for $E > 0$ is hyperbolic, and for $E = 0$ it is parabolic.

Additional Geometric Features of Elliptic Orbits. Figure 14–18 shows some of the important features of the ellipse that are useful for describing planetary motion. The distance a is called the *semimajor axis* of the ellipse; b is the *semiminor axis*.

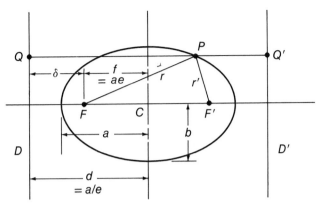

Fig. 14-18. Properties of the ellipse. The left-hand vertex is a point on the ellipse, and the defining equation for this point can be expressed as $(a - f) = e(d - a)$. Similarly, for the right-hand vertex, we have $(a + f) = e(a + d)$. Subtracting one of these equations from the other yields $f = ae$. Then, substitution of this result into either equation gives $d = a/e$. Notice also that $\delta + ae = a/e$, so that $e\delta = a(1 - e^2)$.

Notice that the distance from the center C to either focus is $f = ae$; also, the distance from the center C to either directrix is $d = a/e$.

In Fig. 14-18, notice that the ellipse may be represented in terms of the focus F and the directrix D or the focus F' and the directrix D'. For F and D, the defining equation (Eq. 14-13) for the point P is $r = e(\overline{QP})$, whereas for F' and D', the corresponding equation is $r' = e(\overline{Q'P})$. Adding these two equations and using the fact that $d = a/e$ (Fig. 14-18), we find

$$r + r' = e(\overline{QP} + \overline{Q'P}) = e \cdot 2d = 2a = \text{constant} \qquad \textbf{(14-21)}$$

Thus, an ellipse is the locus of points that have a constant sum of distances from two fixed points (the foci).

Some additional useful relationships for elliptic planetary motion can now be obtained. Refer to Fig. 14-19. The alternate definition of an ellipse expressed by Eq. 14-21 means that the distances $\overline{FA'F'}$ and $\overline{FBF'}$ are equal. Now, $\overline{FA'F'} = ae + a + (a - ae) = 2a$. Therefore, $\overline{FBF'} = 2a$, so that $\overline{FB} = \overline{BF'} = a$. Thus,

$$b = \sqrt{a^2 - (ae)^2} = a\sqrt{1 - e^2} \qquad \textbf{(14-22)}$$

Fig. 14-19. The distances $\overline{FB}$ and $\overline{BF'}$ are equal to the semimajor axis a. For the origin located at the left-hand focus F, the vertices A and A' correspond to r_{min} and r_{max}, respectively.

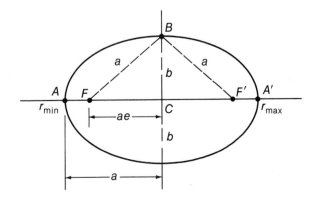

In Fig. 14–19, if the focus F is chosen as the origin, we see that

$$r_{min} = a(1 - e) \quad \text{and} \quad r_{max} = a(1 + e) \tag{14-23}$$

From the theory of quadratic equations (see Section A–3 in Appendix A), we know that the product of the roots, $r_{min}r_{max}$, is equal to the constant term in the quadratic (Eq. 14–19), namely, $-L^2/2\mu E$. Also, the sum of the roots, $r_{min} + r_{max}$, is equal to the negative of the coefficient of the linear term, namely, $-k/E$. Combining these two results and using Eqs. 14–23, we find

$$a(1 - e^2) = \frac{L^2}{\mu k} \tag{14-24}$$

which relates the angular momentum (L) to the size (a) and shape (e) of the orbit. The relationships we have obtained are often useful in analyzing the orbits of planets and space vehicles.

Kepler's Laws. In Section 14–1, we summarized the three laws of planetary motion that were discovered by Johannes Kepler in the early 17th century. Now that we have studied some of the details of planetary motion, we find that we must modify the laws as originally stated by Kepler.

Kepler believed that the planets move in elliptic orbits with the center of the Sun at one focus. We now see that the statement of this first law of Kepler must be changed to

I. A planet moves in an elliptic path with the focus at the position of the C.M. of the planet-Sun system.

For the Earth-Sun system the C.M. is located only 450 km from the center of the Sun, which represents a small correction to Kepler's law. (See also Problem 14–20.)

From our discussions, beginning in Section 10–1, we know that angular momentum is a conserved quantity for motion due to any central force—in particular, the gravitational force. Consider a particle that moves along the path PP' in Fig. 14–20. During the time Δt, the position line $r(t)$ sweeps out the triangular shaded region, the area of which is (approximately)

$$\Delta A = \tfrac{1}{2} r^2 \, \Delta \phi$$

Dividing by Δt and proceeding to the limit, $\Delta t \to 0$, we have

$$\operatorname*{Lim}_{\Delta t \to 0} \frac{\Delta A}{\Delta t} = \tfrac{1}{2} r^2 \operatorname*{Lim}_{\Delta t \to 0} \frac{\Delta \phi}{\Delta t}$$

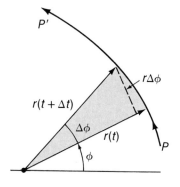

Fig. 14–20. Illustrating Kepler's second law.

or

$$\frac{dA}{dt} = \tfrac{1}{2} r^2 \omega = \frac{L}{2\mu} = \text{constant} \tag{14-25}$$

where we have used $L = \mu r^2 \omega = $ constant (Eq. 10–24). Thus, we see that Kepler's second law is valid if stated in the following way:

II. The position vector for a planet (measured from the C.M. of the planet-Sun system) sweeps out equal areas in equal time intervals; that is, dA/dt = constant.

Notice that Kepler's second law depends only on the fact that the force of interaction between the bodies is a *central force* so that angular momentum is conserved. In fact, the second law is valid for *any* central force.

Next, we write Eq. 14–25 in the form

$$dt = \frac{2\mu}{L}\, dA$$

If we integrate this expression over one complete period T, we find

$$T = \frac{2\mu}{L}A = \frac{2\mu}{L}(\pi ab) \qquad\qquad \textbf{(14–26)}$$

where A is the total area of the orbit; for an elliptic orbit the area is πab (see Example III–11). Now using Eqs. 14–22 and 14–24, and recalling that $k = G\mu M$, we find

$$T^2 = \frac{4\pi^2}{GM}a^3 \qquad\qquad \textbf{(14–27)}$$

Therefore, Kepler's third law, which states that $T^2/a^3 = $ constant for all planets, is not strictly true because the quantity M that appears in Eq. 14–26 is the combined mass of the Sun and the planet. Thus, M is slightly different for the different planets. That is,

III. The ratio of the square of the period (T^2) to the cube of the semimajor axis (a^3) is approximately the same for all planets.

Again, the required correction to Kepler's statement is small.

Planetary Data. Some of the important information concerning planetary motion and the properties of the planets is summarized in Table 14–1. Notice that the orbit semimajor axes are given in *astronomical units* (A.U.), equal to the length of the

TABLE 14–1.   Planetary Data*

PLANET	DIAMETER (EARTH = 1)	MASS (EARTH = 1)	SEMIMAJOR AXIS (A.U.)	SIDEREAL PERIOD (YEARS)	ECCENTRICITY OF ORBIT
Mercury	0.3824	0.0553	0.3871	0.24084	0.2056
Venus	0.9489	0.8150	0.7233	0.61515	0.0068
Earth	1	1	1	1.00004	0.0167
Mars	0.5326	0.1074	1.5237	1.8808	0.0934
Jupiter	11.194	317.89	5.2028	11.862	0.0483
Saturn	9.41	95.17	9.5388	29.456	0.0560
Uranus	4.4	14.56	19.1914	84.07	0.0461
Neptune	3.8	17.24	30.0611	164.81	0.0100
Pluto†	0.4	0.02	39.5294	248.53	0.2484

*The diameters and masses are those recommended by the International Astronomical Union in 1976 (except for more recent values for the diameter of Uranus and the mass of Pluto). The remaining data are 1974 values. (From J. M. Pasachoff, *Contemporary Astronomy*, 2nd ed.; Saunders College Publishing, Philadelphia, 1981.)

†The eccentricity of Pluto's orbit is so large that once each revolution it actually dips inside Neptune's orbit. This happened in 1979 and Pluto will remain closer to the Sun than Neptune until 1999 when once again it will become the "ninth" planet.

Earth's semimajor axis:

$$1 \text{ A.U.} = 1.496 \times 10^{11} \text{ m}$$

Also, we express diameters and masses in terms of Earth units:

$$1 \text{ Earth diameter}^* = 1.274 \times 10^7 \text{ m}$$
$$1 \text{ Earth mass} \quad = 5.974 \times 10^{24} \text{ kg}$$

The periods given are *sidereal periods,* that is, periods for motion with respect to the fixed stars. The units are *solar years* (approximately $365\frac{1}{4}$ mean solar days). The sidereal year is 20 min 24 s longer than the solar year because of the slow precession of the Earth's rotation axis with respect to the stars.

Example 14–4

An artificial Earth satellite is placed into an orbit described by the equation

$$\frac{8160 \text{ km}}{r} = 1 - 0.20 \cos \phi$$

(a) What are the lengths of the semimajor and semiminor axes? What is the altitude of the satellite at perigee and at apogee?
(b) What is the period of the motion?
(c) What is the satellite velocity at perigee and at apogee?

Solution:

Because the mass of the satellite is small compared with the Earth's mass, we have $\mu = m$ and the C.M. is located at the center of the Earth.

(a) Comparing the orbit equation with Eq. 14–14, we see that

$$e = 0.20$$

and $\qquad\qquad e\delta = 8160 \text{ km}$

Therefore, referring to Fig. 14–18, we can write

$$a = \frac{e\delta}{1 - e^2} = \frac{8160 \text{ km}}{1 - (0.20)^2} = 8500 \text{ km}$$

and from Eq. 14–22, we find

$$b = a\sqrt{1 - e^2} = (8500 \text{ km})\sqrt{1 - (0.20)^2} = 8328 \text{ km}$$

The Earth's center (the C.M. of the system and the focus of the orbit) is located at a distance ae from the center C of the orbit ellipse (see Fig. 14–18):

$$ae = (8500 \text{ km})(0.20) = 1700 \text{ km}$$

The various parameters of the orbit are shown in the diagram on the next page.

---

*This is an average value. "The radius" of the Earth usually means the equatorial radius, 6378 km.

The altitudes at perigee and apogee are

$$x = 8500 - (6380 + 1700) = 420 \text{ km at perigee}$$
$$y = 8500 - (6380 - 1700) = 3820 \text{ km at apogee}$$

(b) The period of the motion is obtained from Eq. 14-27, in which $M = M_E$:

$$T = \frac{2\pi}{\sqrt{GM_E}} a^{3/2}$$

$$= \frac{2\pi(8.50 \times 10^6 \text{ m})^{3/2}}{\sqrt{(6.67 \times 10^{-11} \text{ N} \cdot \text{m}^2/\text{kg}^2)(5.97 \times 10^{24} \text{ kg})}}$$

$$= 7.80 \times 10^3 \text{ s} = 130 \text{ min}$$

(c) At perigee and apogee, $v_r = 0$, so that the only component of the velocity is the azimuthal component. Using Eq. 14-26 and Eq. 14-22, we have

$$v_\phi = r\omega = \frac{L}{\mu r} = \frac{2\pi ab}{rT} = \frac{2\pi a^2 \sqrt{1 - e^2}}{rT}$$

For $v_\phi(\text{perigee})$, we need $r_{\min} = a(1 - e)$ (Eq. 14-23). Then,

$$v_\phi(\text{perigee}) = \frac{2\pi a}{T} \sqrt{\frac{1 + e}{1 - e}} = \frac{2\pi(8500 \text{ km})}{7.80 \times 10^3 \text{ s}} \sqrt{\frac{1 + 0.20}{1 - 0.20}}$$
$$= 8.39 \text{ km/s} = 3.02 \times 10^4 \text{ km/h}$$

For $v_\phi(\text{apogee})$, we use $r_{\max} = a(1 + e)$. Then,

$$v_\phi(\text{apogee}) = \frac{2\pi a}{T} \sqrt{\frac{1 - e}{1 + e}} = \frac{2\pi(8500 \text{ km})}{7.80 \times 10^3 \text{ s}} \sqrt{\frac{1 - 0.20}{1 + 0.20}}$$
$$= 5.59 \text{ km/s} = 2.01 \times 10^4 \text{ km/h}$$

The orbit parameters we have found here, using only a knowledge of the orbit equation, are typical of those for artificial Earth satellites.

Example 14–5

Determine the flight schedule for a space probe to be sent from the Earth to Mars.

Solution:

The vehicle is launched into an elliptic trajectory in such a way that the perihelion is at the launch point and the aphelion is at the point of rendezvous with Mars. Moreover, if the launch direction is the same as the direction of the Earth's motion around the Sun, the Earth's orbital velocity is added to the relative velocity of the probe with respect to the Earth when considering the vehicle's solar trajectory. These are the most favorable conditions for launch, and they produce the trajectory with the minimum energy expenditure. However, it is clear that the timing is critical and that the launch date must be carefully chosen so that the probe and Mars arrive simultaneously at the aphelion rendezvous point. The situation is illustrated in the diagram.

We see that the semimajor axis of the probe's orbit is easily obtained from the terrestrial and Martian orbits, which we assume here to be circular, namely,

$$a = \tfrac{1}{2}(1.0 \text{ A.U.} + 1.52 \text{ A.U.}) = 1.26 \text{ A.U.}$$

From Eq. 14–23, where the perihelion distance is $r_{\min} = 1.0$ A.U., we have

$$e = \frac{a - r_{\min}}{a} = \frac{1.26 \text{ A.U.} - 1.0 \text{ A.U.}}{1.26 \text{ A.U.}} = 0.206$$

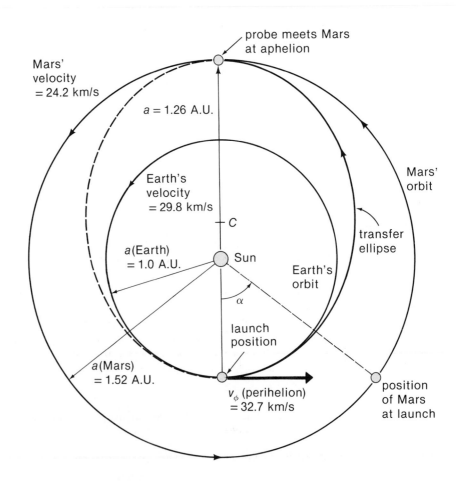

Now, using Eq. 14–24 and substituting $L = \mu r^2 \omega$, we find

$$\frac{L^2}{\mu k} = \frac{r^4 \omega^2}{GM} = a(1 - e^2)$$

where M is essentially the mass of the Sun. Then, at launch from the perihelion,

$$v_\phi(\text{perihelion}) = r_{\min}\omega = \sqrt{\frac{a(1 - e^2)GM}{r_{\min}^2}}$$

$$= \left[\frac{(1.26 \times 1.50 \times 10^{11})[1 - (0.206)^2](6.67 \times 10^{-11})(1.99 \times 10^{30})}{(1.50 \times 10^{11})^2}\right]^{1/2}$$

$$= 3.27 \times 10^4 \text{ m/s} = 32.7 \text{ km/s}$$

This is the required launch velocity in the Sun coordinate frame; relative to the Earth, this velocity is only $32.7 - 29.8 = 2.9$ km/s. However, we must realize that this is the probe velocity in solar orbit, calculated without regard to the gravitational presence of the Earth. If we consider the probe to be essentially free from the gravitational influence of the Earth when the gravitational force on the probe due to the Sun is 10 times that due to the Earth, we find that this occurs when the probe is at a distance of 130 Earth radii from the Earth. This distance, however, is only 5.5×10^{-3} A.U., so in our diagram the position at which the probe has a velocity of 32.7 km/s relative to the Sun is indistinguishable from the position of the Earth.

To calculate the launch velocity v_ℓ of the probe (at the Earth's surface) that is necessary to achieve a relative velocity $v_{\text{rel}} = 2.9$ km/s when the probe is free from the Earth's influence, we need to modify the calculation of the escape velocity v_E (Example 14–3). If the probe is to have an energy $\frac{1}{2}mv_{\text{rel}}^2$ when far from the Earth, then

$$\tfrac{1}{2}mv_\ell^2 = \frac{GmM_E}{R_E} + \tfrac{1}{2}mv_{\text{rel}}^2$$

from which

$$v_\ell = \sqrt{v_E^2 + v_{\text{rel}}^2}$$

Substituting $v_E = 11.2$ km/s and $v_{\text{rel}} = 2.9$ km/s, we find

$$v_\ell = 11.6 \text{ km/s}$$

Upon arrival at aphelion, the probe velocity relative to the Sun is (using angular momentum conservation)

$$v_\phi(\text{aphelion}) = \frac{(32.7 \text{ km/s})(1.0 \text{ A.U.})}{1.52 \text{ A.U.}} = 21.5 \text{ km/s}$$

The fact that the probe velocity at aphelion is less than the orbital velocity of Mars (24.2 km/s) means that Mars actually overtakes the probe to achieve the meeting. Again, we have not taken into account the gravitational effect of the planet. If the probe is allowed to collide with Mars without any braking by the firing of retrorockets, the impact velocity will be greater than $24.2 - 21.5 = 2.7$ km/s. (•Show that the impact velocity will be 5.8 km/s.)

The time for the Earth-to-Mars trip is just one half the period of the probe. Using Eq. 14–27, we find

$$t = \tfrac{1}{2}T = \frac{\pi a^{3/2}}{\sqrt{GM}} = \frac{\pi(1.26 \times 1.50 \times 10^{11})^{3/2}}{\sqrt{(6.67 \times 10^{-11})(1.99 \times 10^{30})}}$$

$$= 2.24 \times 10^7 \text{ s} = 0.709 \text{ y}$$

The Martian period is 1.881 y, so at launch the angle α shown in the diagram must be

$$\alpha = 180° \cdot \left[1 - \frac{0.709\ \text{y}}{\frac{1}{2}(1.881\ \text{y})}\right]$$

$$= 44.3°$$

Favorable launch conditions are not restricted precisely to the configuration specified by the angle α; in fact, launch can take place at any time during a period of a few weeks before or after the optimum angle is reached with essentially the same fuel expenditure. This favorable time period is called the *launch window*. For the case of Mars, such periods recur every 780 days.

PROBLEMS

Section 14-2

14-1 Two ocean liners, each with a mass of 40,000 metric tons, are moving on parallel courses, 100 m apart. What is the magnitude of the acceleration of one of the liners toward the other due to the mutual gravitational attraction?

14-2 In the experiment Cavendish performed (see Section 5-3), he used lead spheres with radii of 1 in. and 4 in. When the centers of two such spheres are separated by $r = 6$ in., the force of attraction is 2.76×10^{-8} N. Determine the value of G. (Refer to Table 1-2 for the density of lead.)

14-3 A length of wire with a mass M is bent into an arc of a circle with radius R, as shown in the diagram. A particle with mass m is located at the center of the circular arc. What gravitational force does the wire exert on the particle?

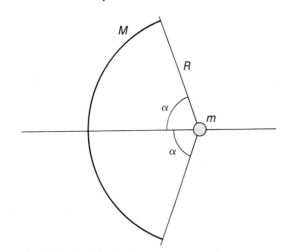

14-4 At what point along the line connecting the Earth and the Moon is there zero gravitational force on an object? (Ignore the presence of the Sun and the other planets.)

14-5 Calculate the fractional difference $\Delta g/g$ in the acceleration due to gravity at points on the Earth's surface nearest to and farthest from the Moon, taking into account the gravitational effect of the Moon. (This difference is responsible for the occurrence of the *lunar tides* on the Earth.)

14-6 What is the value of the acceleration due to gravity at the surface of (a) the Moon and (b) the Sun? Express the accelerations in units of g.

14-7 A straight thin rod has a mass M and a length L. Show that the gravitational force on a particle with a mass m located in line with the rod at a distance x from the near end is $F = GmM/x(x + L)$.

Section 14-3

14-8 What is the gravitational potential energy of a particle with a mass m located at the center of a thin homogeneous circular ring with a radius a and a mass M? What is the force on m? Explain physically the qualitative difference in the two results.

14-9 Show that the potential energy of a system consisting of four equal-mass particles M located at the corners of a square having sides of length d is $U = -(GM^2/d)(4 + \sqrt{2})$.

14-10 What minimum velocity must an object have at the surface of the Earth if it is to escape from the solar system (assumed to consist only of the Sun and the Earth) and proceed infinitely far away? Is the *direction* of the velocity important?

14-11 A particle is in a circular orbit around the Earth at a height h above the Earth's surface. If the amount of work required to raise the particle from the surface to the orbit height is equal to the kinetic energy of the particle in its orbit, show that $h = \frac{1}{2}R_E$, where R_E is the radius of the Earth.

14-12 Suppose that a vertical shaft is drilled through the Earth along a diameter. (a) If a particle is dropped into the shaft at the Earth's surface, with what speed will it pass through the center of the Earth? (b) If the particle is projected upward from the center of the Earth, what velocity is necessary to permit the particle to escape from the Earth?

Section 14-4

14-13 Construct a diagram similar to Fig. 14–9 for the case of a solid homogeneous sphere. Calculate the values of U and F for $r = 0$ and $r = R$. What is the functional form of the force for $0 < r < R$?

14-14 Consider a thin disc with a radius R and a mass M.
• A particle with a mass m is located on the disc axis a distance r from the center of the disc. (a) What is the gravitational potential energy of the particle? (b) What is the force on the particle? (c) Allow r to become large compared with R and show that both U and $\mathbf{F}$ reduce to the expressions for particles.

14-15 Consider the two gravitationally interacting ob-
•• jects shown in the diagram above right. The object A on the left is spherically symmetric and has a mass $M = 4$ kg. The object B on the right has the shape of a dumbbell with two spherical parts, B_1 and B_2; each part has a mass of 2 kg, and the rigid connecting section has negligible mass. (a) Show that the force exerted on A by B acts on the center of mass of A. (b) Show that the force exerted on B by A does not act on the center of mass of B. Find the position at which this force does act. (Simplify the calculations by expressing all results in units of G.) [*Hint:* Find the sum of the forces, $\mathbf{F}_1 + \mathbf{F}_2$, that act on A due to the two parts of B. Then, use Newton's third law. Check your result by calculating in two ways the torque on B about its C.M.]

Section 14-6

14-16 At what distance from the Earth's center will an artificial satellite in an equatorial circular orbit (that is, an orbit that lies entirely in the Earth's equatorial plane) have a period equal to 1 day? Such a *synchronous* (or *geostationary*) satellite moves in synchronism with the Earth, remaining in a fixed position above a point on the equator. Synchronous satellites are used extensively as fixed relay stations in the worldwide communications network.

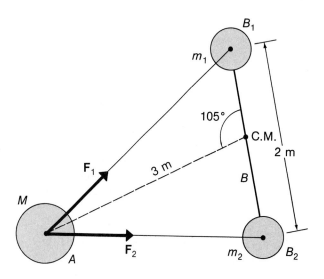

14-17 An artificial Earth satellite is "parked" in an equatorial circular orbit at an altitude of 10^3 km. What is the minimum additional velocity that must be imparted to the satellite if it is to escape from the Earth's gravitational attraction? How does this compare with the minimum escape velocity for leaving from the Earth's surface?

14-18 The largest moon of Jupiter, Ganymede, revolves around the planet in a nearly circular orbit with a radius of 1.07×10^6 km and a period of 7.16 days. Use this information and determine the mass of Jupiter. Express the result in units of the Earth's mass.

14-19 A satellite is in a circular orbit just above the surface of the Moon. (The radius of the moon is 1738 km.) (a) What is the acceleration of the satellite? (b) What is the speed of the satellite? (c) What is the period of the satellite orbit?

14-20 Determine the amount of error in Kepler's statement of his first law for the worst case in the solar system, namely, the planet Jupiter.

14-21 Show that the orbit equation (Eq. 14–14) for the case of elliptic motion can be expressed in rectangular coordinates (with origin at the center of the ellipse) as

$$\frac{x^2}{a^2} + \frac{y^2}{b^2} = 1$$

14-22 Find the lengths of the semimajor and semiminor
• axes, the eccentricity, and the locations of the foci, the vertices, and the directrices of the ellipse whose equation is $(x^2/25) + (y^2/16) = 1$.

14-23 A satellite with a mass of 500 kg is in a circular orbit at an altitude of 500 km above the Earth's surface. Because of air friction, the satellite eventually is brought to the Earth's surface, and it impacts with a velocity of 2 km/s. How much energy was absorbed by the atmosphere through friction?

14-24 Studies of the relationship of the Sun to the local galaxy—the Milky Way—have revealed that the Sun is located near the outer edge of the galactic disc, about 30,000 L.Y. from the center. Furthermore, it has been found that the Sun has an orbital velocity of approximately 250 km/s around the galactic center. (a) What is the period of the Sun's galactic motion? (b) What is the approximate mass of the Milky Way galaxy? Using the fact that the Sun is a typical star, estimate the number of stars in our local galaxy.

14-25 The general solution to the problem of three or more mutually gravitating bodies cannot be expressed in terms of any known mathematical functions. Solutions are possible, however, for certain special situations. For example, consider three identical particlelike objects that have equal masses M and are located at the vertices of an equilateral triangle with sides of length h. Show that when these particles move in the plane of the triangle, their relative positions will be maintained if they travel in circular orbits about the common C.M. with angular frequency $\omega = \sqrt{3GM/h^3}$.

14-26 The planet Mercury has an orbital period of 88.0
• days and a speed of 59.0 km/s at perihelion. (a) Use this information to determine the eccentricity of the orbit and the lengths of the semimajor and semiminor axes in A.U. Compare with the data in Table 14-1. (b) What is the difference between the orbital velocities at perihelion and at aphelion?

14-27 A satellite is placed into orbit around the Moon.
• The period of the motion is 124.2 min. The apolune (the lunar equivalent of *apogee*) is 310 km above the Moon's surface and the perilune is 20 km above the surface. The radius of the (spherical) Moon is 1738 km. (a) What is the eccentricity of the satellite's orbit? (b) Calculate the mass and the density of the Moon.

Additional Problems

14-28 Planet X has the same average density as the Earth but its mass is only one half that of the Earth. What is the value of g on the surface of planet X?

14-29 The acceleration due to gravity at a point varies inversely as the square of the distance of the point from the center of the Earth. (a) For heights h above the Earth's surface that are not too great, show that the fractional change in the acceleration is given approximately by $\Delta g/g \cong -2h/R_E$. (b) Use this expression to calculate the changes $\Delta g/g$ for $h = 10^3$ m and 10^6 m. Compare these values with those obtained from the exact equations.

14-30 Two identical rods with length l and mass M are
• placed in line with a separation x between the closer ends, as shown in the diagram.

(a) Show that the gravitational force between the rods is

$$F = G\frac{M^2}{l^2}\ln\left[\frac{(x+l)^2}{x(x+2l)}\right]$$

[*Hint:* Use the result of Problem 14-7.] (b) Show that this force reduces to

$$\lim_{x \gg l} F = G\frac{M^2}{x^2}$$

[*Hint:* Carry *all* expansions to at least order l^2/x^2.]

14-31 Assume the Earth to be a perfect homogeneous sphere. Imagine that a narrow tunnel is drilled completely through the Earth along a diameter. (a) Show that a particle with a mass m located in the tunnel a distance x from the center of the Earth experiences a gravitational force that obeys Hooke's law, $F = -\kappa x$, where

$$\kappa = G\frac{mM_E}{R_E^3} = g\frac{m}{R_E}$$

(b) In Section 15-2 we show that a particle subject to a Hooke's-law force has a period of motion equal to $T = 2\pi\sqrt{m/\kappa}$ (Eq. 15-12). Verify that the time required for the mass m to travel from one side of the Earth to the other is 42.2 min.

14-32 Assume that the Earth's crust (approximately 30 km
• deep) has a uniform density of 2.72 g/cm³. (a) What is the fractional change in the acceleration due to gravity $\Delta g/g$ upon descending from the surface to a depth of 15 km? (b) A spherical deposit of pure iron (density 7.86 g/cm³) has a diameter of 5 km and lies just beneath the Earth's surface. What is the fractional change $\Delta g/g$ measured on the surface immediately above the deposit? (c) A spherical cavity lies just under the Earth's surface and has a diameter of 5 km. What is the fractional change $\Delta g/g$ measured on the surface immediately above the cavity?

14-33 Two identical objects with mass M are initially separated by a very large distance. If each object is released from rest and allowed to gravitate toward the other, show that when the separation of the particles is R their relative velocity is $2\sqrt{GM/R}$.

14-34 At the time of a lunar eclipse, a 1-kg object is moved from a point on the Earth's surface farthest from the Sun to a point on the Moon's surface nearest the Sun. What is the total change in gravitational potential energy in this process? What velocity at the surface of the Earth would be necessary to accomplish this movement? Explain the significance of the fact that this velocity differs by only a small amount from the escape velocity, $v_E = 11.2$ km/s.

14-35 Find the gravitational potential energy for a system of eight stars, each with solar mass, located at the corners of a cube whose sides are 1 L.Y. in length. [*Hint:* For *n* objects, there are $n(n-1)/2$ independent pairs; thus, there are 28 pairs in this star system, many of which are equivalent.]

14-36 The period of Halley's comet is not exactly the same
• from one orbit to the next, due primarily to the perturbations caused by the massive planets Jupiter and Saturn. The variation is between 74 and 79 y; the average period is 76.5 y. The eccentricity of the orbit is 0.967. (a) Determine the lengths of the semimajor and semiminor axes in A.U. (b) Find the perihelion and aphelion distances in A.U. (c) Calculate the azimuthal velocities at perihelion and aphelion. Compare these with the Earth's average orbital velocity.

14-37 An equatorial synchronous satellite (see Problem
•• 14-16) has a mass of 10^3 kg. In order to bring the satellite to Earth at an equatorial point 180° away from its geostationary position, a retrorocket is fired to impart a pure decelerating azimuthal impulsive thrust. If the rocket discharges the exhaust gases

with a relative velocity of 2800 m/s, how much fuel and oxidizer must be burned? [*Hint:* Refer to Example 14-5. Also, note that the mass of the rocket is related to its velocity and the exhaust velocity in Problem 9-22, except that the term $-gt$ is not necessary in this problem.]

14-38 The two stars in a binary system are found to
• orbit around one another, each with a constant velocity of 40 km/s. The period of the motion is 20 days. Find the mass of each star and the separation of the stars. [*Hint:* Be careful to distinguish between the length of the semimajor axis for the equivalent single-particle orbit and that for the individual star orbits.]

14-39 One of the best studied binary star systems is
• that of *Sirius*. One member of the pair, Sirius A, is a bright star, whereas Sirius B is so faint that it was not observed until its existence had been inferred by studying the motion of Sirius A. The diagram shows the intertwining tracks of the two stars as they moved across the sky during the period from 1900 to 1970. The tracks A and B represent the star motions; track C represents the motion of the C.M. From this study, it has been found that the maximum separation of the stars is 32 A.U. and that the minimum separation is 8 A.U. Use the diagram to obtain estimates for the mass ratio m_A/m_B and for the period T. Find the following quantities: (a) the individual masses, m_A and m_B; (b) the orbit eccentricity e; and (c) the parameters of each orbit: a_A and b_A, a_B and b_B. (d) Sketch the two orbits. [*Hint:* Be careful to distinguish between the parameters of the equivalent single-particle orbit and those of the actual star orbits.]

CALCULUS VI

TAYLOR'S THEOREM

This useful theorem is needed for the derivation of Hooke's law in Section 15–1. In other applications, we usually achieve the desired results by using only the first two terms of the series (Eq. VI–1).

We now discuss a powerful mathematical theorem that finds wide application in many areas of physics. Consider a function $y = f(x)$. We inquire about the value of the function at points in the neighborhood of $x = a$. In Section I–5 we found that for suitably small values of Δx, we can write the approximation (Eq. I–12),

$$f(a + \Delta x) \cong f(a) + \frac{dy}{dx}\bigg|_{x=a} \cdot \Delta x \qquad \textbf{(VI–1)}$$

Now, the derivative, $dy/dx\,|_{x=a}$, is equal to the slope of the line that is tangent to $f(x)$ at $x = a$. Consequently, Eq. VI–1 for $f(a + \Delta x)$ is equivalent to evaluating $f(a + \Delta x)$ at the point Q' on the tangent line instead of at point Q on the actual curve (Fig. VI–1). As you can see in Fig. VI–1, the error QQ' involved in using the approximation in Eq. VI–1 will become smaller than any specified amount if Δx is made sufficiently small. Thus, there are situations in which the error introduced by the use of Eq. VI–1 is negligible.

Although the approximation given in Eq. VI–1 is simple and straightforward to apply, we can ask: How accurate is this approximation? Can we devise a procedure that gives a result even more accurate? We would, in fact, like to have a procedure that, in principle, can be extended to zero error.

Fig. VI–1. Geometric construction to illustrate the difference between the exact value of $f(a + \Delta x)$ at the point Q and the approximate value $f(a) + (dy/dx)|_{x=a} \cdot \Delta x$ at point Q'.

VI–1 TAYLOR'S SERIES

Consider a function $f(x)$ and allow the independent variable x to make an arbitrary excursion from a fixed value, $x = a$, to $x = a + \Delta x$. We restrict attention to functions that can be expanded in infinite power series in $\Delta x = x - a$. Assuming that $f(x)$ is such a function, we write

$$f(x) = b_0 + b_1(x - a) + b_2(x - a)^2 \\ + \cdots + b_s(x - a)^s + \cdots$$

for x in the vicinity of $x = a$ and with appropriate constants, b_0, b_1, and so forth. Taking successive derivatives of $f(x)$, we obtain

$$\frac{df(x)}{dx} = b_1 + 2b_2(x - a) + \cdots + sb_s(x - a)^{s-1} + \cdots$$

$$\frac{d^2f(x)}{dx^2} = 2b_2 + \cdots + s(s - 1)b_s(x - a)^{s-2} + \cdots$$

and so forth. Now, for $x = a$ (that is, $\Delta x = 0$), we have

$$f(a) = b_0$$

$$\left.\frac{df(x)}{dx}\right|_{x=a} = b_1$$

$$\left.\frac{d^2f(x)}{dx^2}\right|_{x=a} = 2b_2$$

$$\left.\frac{d^sf(x)}{dx^s}\right|_{x=a} = s!b_s$$

Substituting these values into the expression for $f(x)$, we find

$$f(x) = f(a) + \left.\frac{df(x)}{dx}\right|_{x=a} \cdot (x - a)$$

$$+ \frac{1}{2}\left.\frac{d^2f(x)}{dx^2}\right|_{x=a} \cdot (x - a)^2 + \cdots$$

$$+ \frac{1}{s!}\left.\frac{d^sf(x)}{dx^s}\right|_{x=a} \cdot (x - a)^s + \cdots \quad \textbf{(VI-2)}$$

We see that the previous expression (Eq. VI-1) is in fact only the first two terms of this expansion. The subsequent terms (if not zero) represent the additions needed to convert Eq. VI-1 into a precise expression.

A rigorous evaluation of the error made in using Eq. VI-2 when it is broken off at a particular term is beyond the scope of this discussion. However, in the event that Eq. VI-2 results in a convergent *alternating series* (i.e., a series in which successive terms alternate in algebraic sign), the error made in breaking off the series does not exceed numerically the absolute value of the first term discarded. For a nonalternating series that converges rapidly, the value of the first term of the discarded remainder series gives a rough indication of the error in most cases.*

Equation VI-2 is referred to as *Taylor's series* after its

---

*A rigorous proof that Eq. VI-2 is a valid convergent series and accurately represents the function $f(x)$ requires that

$$\underset{s\to\infty}{\text{Lim}} \left.\frac{d^sf(x)}{dx^s}\right|_{x=\xi} \cdot \frac{(x - a)^s}{s!} = 0 \quad \text{for } a < \xi < x$$

discoverer, Brook Taylor (1685–1731), an English mathematician. When $a = 0$, which leads to a power series expansion about $x = 0$, the series is referred to as *Maclaurin's series* (although it was actually first discovered by James Stirling). In this case the series takes the form

$$f(x) = f(0) + \left.\frac{df}{dx}\right|_{x=0} \cdot x + \frac{1}{2}\left.\frac{d^2f}{dx^2}\right|_{x=0} \cdot x^2$$

$$+ \cdots + \frac{1}{s!}\left.\frac{d^sf}{dx^s}\right|_{x=0} \cdot x^s + \cdots \quad \textbf{(VI-3)}$$

Taylor's theorem, represented by Eq. VI-2, is actually a rather startling proposition, for it asserts that knowing the value of a function and all its derivatives *at a point* uniquely determines the function *everywhere* (where the series is convergent).

Example VI-1

Suppose that a function, $y = f(x)$, has the value -2 at $x = 1$ and passes through this point with a slope -1 and a second derivative $+6$; all higher derivatives are zero. Using only these three values at the given point, find the expression for $f(x)$.

Solution:

From Eq. VI-1 we write

$$f(x) = f(1) + \left.\frac{df}{dx}\right|_{x=1} \cdot (x - 1)$$

$$+ \frac{1}{2}\left.\frac{d^2f}{dx^2}\right|_{x=1} \cdot (x - 1)^2 + 0$$

$$= -2 + (-1)(x - 1) + \tfrac{1}{2}(6)(x - 1)^2$$

$$= -2 - x + 1 + 3x^2 - 6x + 3$$

$$= 3x^2 - 7x + 2$$

and this expression for $f(x)$ is valid *everywhere*.

Example VI-2

In Example I-18 we gave an approximate expression for the area of a circular ring with a radius r and a width Δr, namely, $\Delta A \cong 2\pi r \, \Delta r$. By an exact calculation we found

$$\Delta A = 2\pi r \, \Delta r + \pi(\Delta r)^2$$

Compare these expressions with the corresponding Taylor's series.

Solution:

Using Eq. VI–2 we write

$$f(r + \Delta r) = A + \Delta A$$
$$= A + \frac{dA}{dr}\bigg|_{r=r} \cdot (\Delta r) + \frac{1}{2}\frac{d^2A}{dr^2}\bigg|_{r=r} \cdot (\Delta r)^2 + \cdots$$

Now, $f(r) = A = \pi r^2$

so $\frac{dA}{dr} = 2\pi r$ and $\frac{d^2A}{dr^2} = 2\pi$

with all higher derivatives identically equal to zero. Therefore, Eq. VI–2 terminates with three terms and we have

$$A + \Delta A = \pi r^2 + (2\pi r)(\Delta r) + \frac{1}{2}(2\pi)(\Delta r)^2$$

or $$\Delta A = 2\pi r \, \Delta r + \pi(\Delta r)^2$$

which agrees precisely with the exact calculation we previously made. Notice that the approximate expression for ΔA corresponds to breaking off the Taylor series after two terms.

Example VI–3

Using Eq. VI–2, derive the power series expansion,

$$(1 + \xi)^n = 1 + n\xi + \frac{n(n-1)}{2!}\xi^2$$
$$+ \frac{n(n-1)(n-2)}{3!}\xi^3 + \cdots$$

which holds for any n and $|\xi| < 1$.

Solution:

Here,

$$f(\xi) = (1 + \xi)^n$$

Hence, $$\frac{df}{d\xi} = n(1 + \xi)^{n-1}$$

$$\frac{d^2f}{dx^2} = n(n-1)(1 + \xi)^{n-2}$$

$$\frac{d^3f}{dx^3} = n(n-1)(n-2)(1 + \xi)^{n-3}$$

and so forth. Thus, for expansion about $\xi = 0$, we have

$$f(0) = 1$$

$$\frac{df}{d\xi}\bigg|_{x=0} = n$$

$$\frac{d^2f}{d\xi^2}\bigg|_{x=0} = n(n-1)$$

$$\frac{d^3f}{d\xi^3}\bigg|_{x=0} = n(n-1)(n-2)$$

and so forth. Finally, using Eq. VI–3 for Maclaurin's series, we have

$$(1 + \xi)^n = 1 + n\xi + \frac{n(n-1)}{2!}\xi^2$$
$$+ \frac{n(n-1)(n-2)}{3!}\xi^3 + \cdots$$

Notice that this is exactly the result of the binomial theorem, expressed in Eq. A–12 in Appendix A.

PROBLEMS

VI–1 Using the Maclaurin expansion, derive the series (Eq. A–11 in Appendix A),

$$e^x = 1 + x + \frac{x^2}{2!} + \frac{x^3}{3!} + \cdots + \frac{x^n}{n!} + \cdots$$

VI–2 Using the Maclaurin expansion, derive the series (Eq. A–41 in Appendix A),

$$\cos\theta = 1 - \frac{\theta^2}{2!} + \frac{\theta^4}{4!} - \frac{\theta^6}{6!} + \cdots$$

VI–3 Using the Maclaurin expansion, derive the series (Eq. A–9 in Appendix A),

$$\ln(1 + x) = x - \frac{x^2}{2} + \frac{x^3}{3} - \frac{x^4}{4} + \cdots$$

VI-4 Verify the series,

$$e^{-\theta} \sin \theta = \theta - \theta^2 + \frac{\theta^3}{3} - \frac{\theta^5}{30} + \cdots$$

VI-5 Use the result of Example VI-3 to obtain the expansion for $(1 + x)^{1/2}$, namely,

$$(1 + x)^{1/2} = 1 + \tfrac{1}{2}x - \tfrac{1}{8}x^2 + \tfrac{1}{16}x^3 - \cdots$$

With this expression, evaluate $\sqrt{10}$ correct to five decimal places, given that $\sqrt{9} = 3$ exactly. What is the error in the result if only the first two terms of the series are retained?

VI-6 A function $f(\theta)$ has the following properties: $f(0) = 0$; $df/d\theta|_{\theta=0} = 1$; $d^3f/d\theta^3|_{\theta=0} = -1$; $d^5f/d^5|_{\theta=0} = 1$; and so forth, with the odd derivatives evaluated at $\theta = 0$ alternating in sign. The even derivatives of $f(\theta)$ are all zero at $\theta = 0$. Determine $f(\theta)$. [*Hint:* If you do not recognize the series, refer to Appendix A.]

OSCILLATORY MOTION

There are many types of physical systems that undergo regular, repeating motions. Any motion that repeats itself at definite intervals is said to be *periodic,* such as the motion of the Earth around the Sun. Any periodic motion that carries a system back and forth between alternate extremes is said to be *oscillatory.* Some familiar examples of oscillatory motion are illustrated in Fig. 15–1.

The common feature of oscillatory systems is that the motion is initiated by displacing the system from a condition or position of *stable equilibrium.* When such a displacement is made, a *restoring force* acts to return the system to its equilibrium position (see Section 7–4). Upon release, the system moves toward the equilibrium position, but its inertia causes it to "overshoot" this point and produces a displacement in the opposite sense until the restoring force directs the motion back toward the equilibrium position. The result is a continuing oscillatory motion back and forth through the position of equilibrium.

It is a common observation that any nondriven oscillating system eventually comes to rest. Such motion is said to be *damped* and is due to energy-dissipative

Fig. 15–1. Three examples of oscillatory motion: (a) a block attached to a spring, (b) a rod anchored at one end, and (c) a simple pendulum consisting of a bob attached to a string.

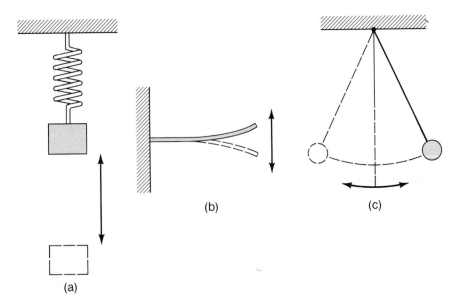

(b)

(c)

(a)

effects, such as friction. In order to maintain the oscillatory motion of a real system (such as the balance wheel in a watch), energy must be continually supplied from a compressed spring or a battery or some other source.

We first discuss oscillatory motion for the idealized case of no friction. Thus, no external energy source is required to sustain the motion undiminished. At the end of this chapter we briefly discuss damped motion. For the case of damped electrical oscillations, see Chapter 37.

15-1 RESTORING FORCES AND POTENTIAL ENERGY FUNCTIONS

Consider the simple, one-dimensional motion of a particle subject to a conservative force. Assume that there exists a point of stable equilibrium. At the coordinate location ξ, the particle has a potential energy $U(\xi)$, as indicated in Fig. 15–2. According to the discussion in Section 7–4, the force on the particle at ξ can be expressed as

$$F(\xi) = -\frac{dU(\xi)}{d\xi} \qquad (15-1)$$

Then, stable equilibrium occurs at the point (or points) for which $F(\xi) = 0$ or $dU(\xi)/d\xi = 0$, if, in addition, $d^2U(\xi)/d\xi^2 > 0$. (•Do you recall why this condition on the second derivative is necessary?) Evidently, the point ξ_0 in Fig. 15–2 is such a point of stable equilibrium.

Fig. 15–2. The potential energy function $U(\xi)$ for a particle subject to a one-dimensional force. The dashed line represents the "best quadratic function" fitted to $U(\xi)$ at $\xi = \xi_0$.

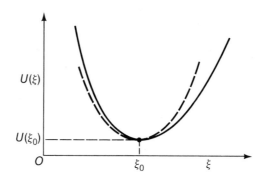

If we confine attention to small excursions of the particle from the point $\xi = \xi_0$, it seems reasonable that the potential energy function $U(\xi)$ can be represented as a constant term $U(\xi_0)$ plus a quadratic (parabolic) term that will duplicate the curvature at ξ_0. That is, we expect $U(\xi)$ to be of the form

$$U(\xi) = U(\xi_0) + \tfrac{1}{2}\kappa(\xi - \xi_0)^2 \qquad (15-2)$$

where, for future convenience, we have used $\tfrac{1}{2}\kappa$ for the coefficient of the quadratic term. Equation 15–2 is the simplest expression that exhibits the essential features of the general potential energy function illustrated in Fig. 15–2. (•Notice that $U(\xi)$ does not contain a linear term; can you see why?)

▶ *Taylor's Series Expansion of the Potential Energy Function.* We can obtain Eq. 15–2 for $U(\xi)$ in a rigorous way. Suppose that $U(\xi)$ is continuous and that all its derivatives are defined within the region we consider. Then, a fundamental theorem of mathematics states that this function can be expanded in a power series, called *Taylor's series* (see Section VI–1). Expanding $U(\xi)$ about the point ξ_0, we have

$$U(\xi) = U(\xi_0) + (\xi - \xi_0)\frac{dU(\xi)}{d\xi}\Big|_{\xi=\xi_0} + \tfrac{1}{2}(\xi - \xi_0)^2\frac{d^2U(\xi)}{d\xi^2}\Big|_{\xi=\xi_0}$$
$$+ \tfrac{1}{6}(\xi - \xi_0)^3\frac{d^3U(\xi)}{d\xi^3}\Big|_{\xi=\xi_0} + \cdots \tag{15–3}$$

where all of the derivatives are evaluated at $\xi = \xi_0$. Then, $U(\xi)$ gives the potential energy at the arbitrary point ξ. That is, the value of the function together with all its derivatives *at a single point* determine the value of the function *everywhere* within the region.

In Eq. 15–3, we have elected to expand $U(\xi)$ about the equilibrium point ξ_0 where we know that $dU(\xi)/d\xi|_{\xi_0} = 0$. Thus, the term involving $(\xi - \xi_0)$ vanishes. Moreover, we consider only small excursions, $\xi - \xi_0$, from the equilibrium position so that the term containing $(\xi - \xi_0)^3$ and all higher-order terms are small compared with the term containing $(\xi - \xi_0)^2$. Then, to this accuracy, we have

$$U_2(\xi) = U(\xi_0) + \tfrac{1}{2}(\xi - \xi_0)^2\frac{d^2U(\xi)}{d\xi^2}\Big|_{\xi=\xi_0}$$

where $U_2(\xi)$ represents the second-order or quadratic expansion of $U(\xi)$. Let us designate the value of the second derivative evaluated at ξ_0 by the symbol κ:

$$\kappa \equiv \frac{d^2U(\xi)}{d\xi^2}\Big|_{\xi=\xi_0} > 0$$

so that
$$U_2(\xi) = U(\xi_0) + \tfrac{1}{2}\kappa(\xi - \xi_0)^2$$

which is just Eq. 15–2.

The Taylor's series expansion for the potential energy function that we have just obtained consists of a constant plus a quadratic term. Consequently, we may consider this procedure equivalent to fitting the best parabola to the function $U(\xi)$ at $\xi = \xi_0$. This approximation is adequate for sufficiently small values of $\xi - \xi_0$. ◄

The force on the particle at the point ξ is found by differentiating $U(\xi)$; thus,

$$F(\xi) = -\frac{dU(\xi)}{d\xi} = -\kappa(\xi - \xi_0)$$

We use x to represent $\xi - \xi_0$, the (small) displacement from the equilibrium position. Then,

$$F(x) = -\kappa x \tag{15–4}$$

which we recognize as the simple linear Hooke's-law relationship we introduced in Section 5–1. From this analysis we conclude: *Any system that is displaced by a small amount from its equilibrium position will experience* (to first order) *a linear restoring force.* This fact accounts for the remarkably similar behavior of quite different systems (such as those illustrated in Fig. 15–1) when they are displaced slightly from their equilibrium positions.

If we apply Newton's second law to the motion of a particle with a mass m that is subject to the Hooke's-law force in Eq. 15–4, we have

$$F(x) = -\kappa x = ma$$

so that the acceleration a is given by

$$a = \frac{d^2x}{dt^2} = -\frac{\kappa}{m}x \qquad (15\text{-}5)$$

It is important to note that the acceleration is directly proportional to the displacement x measured from the equilibrium position and is *oppositely* directed (because $\kappa > 0$ and $m > 0$). Moreover, the acceleration is directly proportional to the strength of the restoring force, measured by κ, and is inversely proportional to the inertia of the system (which, for a particle, is the mass m). The description of *all* oscillatory motions involves these same ingredients.

Example 15–1

Consider the potential energy function, $U(\xi) = 3\xi - 5 - (\xi - 3)^3$, with U measured in joules when ξ is given in meters. Find the points of equilibrium and determine whether they are stable. Use the Taylor's series expansion to second order to approximate $U(\xi)$ for small displacements from the position of stable equilibrium and determine the strength parameter κ. Finally, calculate the error in using Eq. 15-2 for $\xi - \xi_0 = \frac{1}{10}\xi_0$.

Solution:

For the points of equilibrium we have

$$\frac{dU(\xi)}{d\xi} = 3 - 3(\xi - 3)^2 = 0$$

which gives

$$\xi = 2 \qquad \text{or} \qquad \xi = 4$$

Now,

$$\frac{d^2U(\xi)}{d\xi^2} = -6(\xi - 3)$$

which is equal to $+6$ at $\xi = 2$ and is equal to -6 at $\xi = 4$. Therefore, $\xi = 2$ is a local minimum and corresponds to a point of stable equilibrium. (•What can be said regarding the behavior of the function at $\xi = 4$?)

Because $U(2) = 2$ and $d^2U(\xi)/d\xi^2|_{\xi=2} = \kappa = 6$, the Taylor's series expansion to second order about the point $\xi = 2$ becomes (Eq. 15-2)

$$U_2(\xi) = 2 + 3(\xi - 2)^2$$

The diagram on the next page shows the actual function $U(\xi)$ and the Taylor's series $U_2(\xi)$.
When $\xi - \xi_0 = \frac{1}{10}\xi_0$, we have $\xi = 2 + \frac{2}{10} = 2.20$; then, the value of the actual function is $U(2.20) = 2.1120$, whereas the Taylor's series yields $U_2(2.20) = 2.1200$, which represents an error of only 0.4 percent in this case.

Incidentally, to verify the theorem stated in Eq. 15-3, note that for $U(\xi) = 3\xi - 5 - (\xi - 3)^3$ we have $d^3U(\xi)/d\xi^3|_{\xi=2} = -6$, and all higher derivatives are zero. Thus, Eq. 15-3 becomes (to all orders)

$$\begin{aligned}
U_\infty(\xi) &= 2 + \tfrac{6}{2}(\xi - 2)^2 - \tfrac{6}{6}(\xi - 2)^3 \\
&= 2 + (3\xi^2 - 12\xi + 12) - (\xi^3 - 6\xi^2 + 12\xi - 8) \\
&= -\xi^3 + 9\xi^2 - 24\xi + 22
\end{aligned}$$

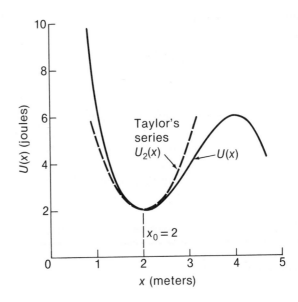

The actual function is

$$U(\xi) = 3\xi - 5 - (\xi - 3)^3$$
$$= -\xi^3 + 9\xi^2 - 24\xi + 22$$

which is identical to $U_\infty(\xi)$.

15-2 FREE HARMONIC OSCILLATIONS

We continue the discussion of oscillatory motion by examining the case of a block attached to a spring (Fig. 15-3). The block has a mass m and slides frictionlessly across a horizontal surface. The spring has a strength parameter κ and a relaxed length ℓ. The figure shows the block displaced by an amount x measured from the equilibrium position O where the spring has its relaxed length. If the spring is not stretched too far, the force exerted on the block can be represented approximately by the simple linear expression,*

$$F(x) = -\kappa x$$

From the discussion in Section 7-4, we know that a spring stretched by an amount x possesses stored energy $U(x)$ equal to the work done in stretching the spring:

$$U(x) = \int_0^x F_a(x)\, dx = \int_0^x \kappa x\, dx = \tfrac{1}{2}\kappa x^2$$

(Note that F_a is the force *applied* to the spring and is equal to $+\kappa x$; in Eq. 15-4, F

Fig. 15-3. A block with a mass m is attached to the free end of an ideal spring, the other end of which is fixed. The relaxed length of the spring is ℓ, corresponding to the block located at O $(x = 0)$.

---

*We could have begun our discussion by simply asserting that the systems we wish to study obey Hooke's law. But then we would have missed the point, which emerges from the analysis based on the Taylor's-series expansion, why so many systems behave in exactly this way.

is the force exerted on the particle *by* the spring and is equal to $-\kappa x$. F_a and F represent an action-reaction pair.)

At any instant the block has a kinetic energy equal to

$$K = \tfrac{1}{2}mv^2$$

Because friction is absent, the energy conservation principle can be easily applied. The total energy E of the system consists of the potential energy of the spring and the kinetic energy of the block, $E = U(x) + K$, and this total energy remains constant:

$$\tfrac{1}{2}\kappa x^2 + \tfrac{1}{2}mv^2 = E = \text{constant} \qquad \textbf{(15-6)}$$

The energy equation is shown graphically in Fig. 15-4. The total energy is E, and at the arbitrary displacement x from the equilibrium position O, the components of E are the potential energy $U(x)$ and the kinetic energy K.

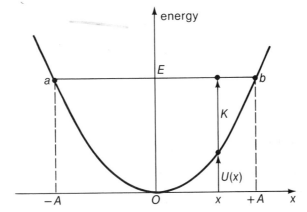

Fig. 15-4. Graph of the energy equation, Eq. 15-6. The total energy is E (a constant), and for any displacement x, we have $E = U(x) + K$. The maximum allowed excursion from the equilibrium position O is $\pm A$. The points a and b correspond to the turning points of the motion.

Notice in Fig. 15-4 that the kinetic energy K becomes zero when $x = \pm A$. Any further increase in the displacement would result in *negative* values for K. Because $K < 0$ represents an unphysical situation, we conclude that the maximum excursions from the equilibrium position occur for $x = \pm A$. The points on the energy curve labeled a and b are called the *turning points* of the motion (see Figs. 7-6 and 14-17 and the accompanying discussions). When, for example, the particle moves toward the right in Fig. 15-4, it must reverse its motion at the point $x = +A$ and then move toward the left. (The particle cannot stop and remain indefinitely at $x = +A$ because the spring exerts a force $F(A) = -\kappa A$ on the particle at this point.) Thus, the particle displacement is limited to $-A \leqslant x \leqslant +A$, and the particle moves back and forth within this interval.

Whenever the particle is at $x = \pm A$ it has maximum potential energy U_m and zero kinetic energy. Thus,

$$U_m = U(\pm A) = \tfrac{1}{2}\kappa A^2 = E, \qquad \text{at } x = \pm A \qquad \textbf{(15-7a)}$$

Whenever the particle passes through $x = 0$ it has maximum kinetic energy K_m and zero potential energy. Thus,

$$K_m = \tfrac{1}{2}mv_m^2 = E, \qquad \text{at } x = 0 \qquad \textbf{(15-7b)}$$

from which $\qquad v_m = \pm\sqrt{\dfrac{2E}{m}},\qquad$ at $x = 0$ $\qquad$ **(15–8)**

(•What is the significance of the $\pm$ sign in this expression?)

Simple Harmonic Motion. The discussion thus far suggests that some type of oscillatory motion is possible for the spring-block system, but we have not yet obtained any detailed description of the motion. We now turn our attention to finding this description. If we substitute E from Eq. 15–7a into Eq. 15–6 and solve for v, we find

$$\tfrac{1}{2}mv^2 + \tfrac{1}{2}\kappa x^2 = \tfrac{1}{2}\kappa A^2$$

or $\qquad v = \dfrac{dx}{dt} = \pm\sqrt{\dfrac{\kappa}{m}(A^2 - x^2)}$ $\qquad$ **(15–9)**

Integrating over time from 0 to t, for which the corresponding values of the displacement are x_0 and x, we have

$$\int_{x_0}^{x} \frac{dx}{(A^2 - x^2)^{1/2}} = \sqrt{\frac{\kappa}{m}}\int_0^t dt$$

where we select the positive value of the square root that appears in Eq. 15–9. The integral over x leads to an arcsine solution* (see the table of integrals inside the rear cover):

$$\left(\sin^{-1}\frac{x}{A}\right)\Bigg|_{x_0}^{x} = \sqrt{\frac{\kappa}{m}}\,t$$

or $\qquad \sin^{-1}\dfrac{x}{A} = \sqrt{\dfrac{\kappa}{m}}\,t + \sin^{-1}\dfrac{x_0}{A}$

It is now convenient to define the quantities

$$\phi_0 = \sin^{-1}\frac{x_0}{A} \qquad (0 \leqslant \phi_0 \leqslant 2\pi) \qquad \textbf{(15–10a)}$$

and $\qquad \omega_0 = \sqrt{\dfrac{\kappa}{m}}$ $\qquad$ **(15–10b)**

so that the solution can be expressed as

$$x(t) = A\sin(\omega_0 t + \phi_0) \qquad \textbf{(15–11)}$$

The argument of the sine function, $\phi(t) = \omega_0 t + \phi_0$, is referred to as the *phase angle*

---

*We consider here only this straightforward solution to the present problem. In the appendix Calculus VII (in Part 2), we discuss a general technique for solving simple differential equations that describe problems of this type.

of the motion. The constant ϕ_0 is called the *initial phase angle*, i.e., the value of $\phi(t)$ when $t = 0$. The coefficient A represents the magnitude of the maximum excursion from the equilibrium position and is called the *amplitude* of the motion. If only x_0 and A are given, the value of ϕ_0 is not uniquely determined. Of the two possible solutions for ϕ_0 obtained from Eq. 15–10a, we select the one consistent with the sign of the initial value of $v = dx/dt$.

We see that the solution, Eq. 15–11, is indeed periodic because the value of $x(t)$ (including its sign!) is repeated whenever $\phi(t)$ is increased by 2π. The time interval required for this increase is called the *period T* of the motion. Thus,

$$\phi(t + T) = \phi(t) + 2\pi$$

or

$$\omega_0(t + T) + \phi_0 = \omega_0 t + \phi_0 + 2\pi$$

from which $\omega_0 T = 2\pi$, so that $\omega_0 = 2\pi/T$, as we found in Eq. 4–23. Then,

$$T = \frac{2\pi}{\omega_0} = 2\pi \sqrt{\frac{m}{\kappa}} \qquad \textbf{(15–12)}$$

The motion of a system that takes place during a time interval T is referred to as one *cycle*. The number of such cycles that take place per unit of time is called the *frequency* of the motion, denoted by ν. Evidently,

$$\nu = \frac{1}{T} = \frac{1}{2\pi} \sqrt{\frac{\kappa}{m}} \qquad \textbf{(15–13)}$$

The metric unit of frequency is the *hertz** (Hz), which stands for one cycle per second. For example, if an oscillatory motion exactly repeats itself after each time interval $T = 0.020$ s, the frequency is

$$\nu = \frac{1}{T} = \frac{1}{0.020 \text{ s}} = 50 \text{ Hz}$$

The frequency at which a system undergoes free oscillations is called the *natural* or *characteristic* frequency of the system.

We now see that the constant ω_0 may be interpreted as the *angular frequency* of the oscillatory motion, with units of radians per second:

$$\omega_0 = 2\pi\nu = \sqrt{\frac{\kappa}{m}} \qquad \textbf{(15–14)}$$

---

*Named in honor of Heinrich Hertz (1857–1894), a German physicist who made great contributions to the theory of electric oscillations.

(In Section 15-3 we make the connection between the angular frequency ω_0 and the angular speed of circular motion. See also the discussion in Section 4-5.)

The velocity of the particle at any time t can be found by differentiating Eq. 15-11 with respect to time. We find

$$v(t) = \frac{dx}{dt} = A\omega_0 \cos(\omega_0 t + \phi_0) \qquad \textbf{(15-15)}$$

The acceleration is obtained, in turn, by a further differentiation:

$$a(t) = \frac{dv}{dt} = \frac{d^2x}{dt^2} = -A\omega_0^2 \sin(\omega_0 t + \phi_0) \qquad \textbf{(15-16)}$$

Notice that the basic equation of motion, $a = -(\kappa/m)x$ (Eq. 15-5), is satisfied by the expressions we have just obtained for $x(t)$ and $a(t)$, together with $\omega_0^2 = \kappa/m$. Moreover, we see that the behavior of the solution, $x(t)$, conforms to the general description of oscillatory motion given earlier in this section. Equation 15-11 shows that the maximum excursions of the displacement are $x = \pm A$, and these occur at times when $v = 0$ (Eq. 15-15). In addition, the velocity maxima occur when $x = 0$, as may be seen by comparing the expressions for $x(t)$ and $v(t)$ (Eqs. 15-11 and 15-15, respectively).

The time variations of the displacement, the velocity, and the acceleration are shown in Fig. 15-5 for the case $\phi_0 = 0$. Notice that the maxima (and the minima) of

Fig. 15-5. The variation with time (actually, $\omega_0 t$) of the displacement $x(t)$, the velocity $v(t)$, and the acceleration $a(t)$ for a simple harmonic oscillator. The curves are drawn for the special case $\phi_0 = 0$.

$x(t)$ and $v(t)$ are separated by intervals of $\pi/2$. (The sine and cosine functions are *out of phase* by $\pi/2$.)

Motion described by sines and cosines is referred to as *simple harmonic motion* (SHM) because the vibrations of musical instruments are described by these same functions. A system that undergoes SHM is called a *simple harmonic oscillator*.

Boundary Conditions. Equation 15–11 for the displacement $x(t)$ contains two adjustable parameters, namely, the amplitude A and the initial phase angle ϕ_0. To determine these parameters requires additional information in the form of *boundary conditions*. We can evaluate A and ϕ_0 if we are given any two values from amongst x, v, or a, except x and a at the same time. (•Can you see why?) Usually, x and v are given for $t = 0$, values which we designate x_0 and v_0. In this event the boundary conditions are referred to as the *initial conditions* because they involve the time $t = 0$. (See also Section II–2.) By virtue of the definitions, we have

$$\left.\begin{array}{l} x_0 = A \sin \phi_0 \\ v_0 = A\omega_0 \cos \phi_0 \end{array}\right\} \tag{15-17a}$$

Dividing the first of these equations by the second gives

$$\tan \phi_0 = \frac{\omega_0 x_0}{v_0} \tag{15-17b}$$

Squaring the equations and adding gives

$$A^2 = x_0^2 + \frac{v_0^2}{\omega_0^2} \tag{15-17c}$$

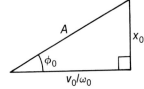

Fig. 15–6. Geometrical representation of the parameters involved in SHM. Compare Eqs. 15–17.

Equations 15–17 uniquely determine both ϕ_0 and A in terms of x_0, v_0, and ω_0. These two relationships are conveniently represented geometrically by the right triangle shown in Fig. 15–6.

Suppose, for example, that a simple harmonic oscillator is displaced to $x = C$ ($C > 0$) and then released from rest at $t = 0$. Then, we identify $x_0 = C$ and $v_0 = 0$. From Eqs. 15–17 we find $\phi_0 = \pi/2$ and $A = C$. The phase angle becomes $\phi(t) = \omega_0 t + \pi/2$, and the motion proceeds as indicated in Fig. 15–7 on page 406. This diagram shows one complete cycle of the motion, starting at $\omega_0 t = 0$ and with increments of $\pi/4$. The maximum velocity is $v_m = \omega_0 C$ and the maximum acceleration is $a_m = \omega_0^2 C$.

Example 15–2

A spring stretches by 3.0 cm from its relaxed length when a force of 7.5 N is applied. A particle with a mass of 0.50 kg is attached to the free end of the spring, which is then compressed horizontally by 5.0 cm from its relaxed length and released from rest at $t = 0$.
(a) What is the strength parameter (or *spring constant*) κ?
(b) Find the period, the frequency, and the angular frequency of the oscillations.
(c) Find the amplitude A and the initial phase angle ϕ_0 of the motion.
(d) Find the maximum velocity v_m and the maximum acceleration a_m.
(e) Find the total energy E of the system.
(f) Determine the displacement x, the velocity v, and the acceleration a when $\omega_0 t = \pi/8$.

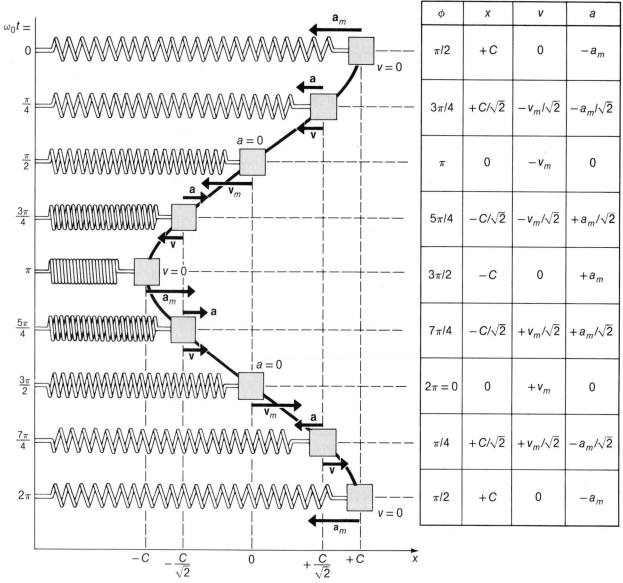

ϕ	x	v	a
$\pi/2$	$+C$	0	$-a_m$
$3\pi/4$	$+C/\sqrt{2}$	$-v_m/\sqrt{2}$	$-a_m/\sqrt{2}$
π	0	$-v_m$	0
$5\pi/4$	$-C/\sqrt{2}$	$-v_m/\sqrt{2}$	$+a_m/\sqrt{2}$
$3\pi/2$	$-C$	0	$+a_m$
$7\pi/4$	$-C/\sqrt{2}$	$+v_m/\sqrt{2}$	$+a_m/\sqrt{2}$
$2\pi = 0$	0	$+v_m$	0
$\pi/4$	$+C/\sqrt{2}$	$+v_m/\sqrt{2}$	$-a_m/\sqrt{2}$
$\pi/2$	$+C$	0	$-a_m$

Fig. 15–7. Graph of the simple harmonic motion, $x(t) = C \sin (\omega_0 t + \pi/2)$. Values of $\phi = \omega_0 t + \pi/2$ are shown from $\omega_0 t = 0$ through a complete cycle in increments of $\pi/4$. Also, $v_m = \omega_0 C$ and $a_m = \omega_0^2 C$.

Solution:

(a) The spring constant is

$$\kappa = -\frac{F}{x} = -\frac{(-7.5 \text{ N})}{0.03 \text{ m}} = 250 \text{ N/m}$$

(b) The angular frequency is

$$\omega_0 = \sqrt{\frac{\kappa}{m}} = \sqrt{\frac{250 \text{ N/m}}{0.50 \text{ kg}}} = 22.36 \text{ rad/s}$$

The frequency is $\qquad \nu = \dfrac{\omega_0}{2\pi} = \dfrac{22.36}{2\pi} = 3.559 \text{ Hz}$

The period is $\qquad T = \dfrac{1}{\nu} = \dfrac{1}{3.559 \text{ s}^{-1}} = 0.281 \text{ s}$

(c) The initial displacement is $x_0 = -5.0 \text{ cm}$; also, $v_0 = 0$, so the amplitude is (Eq. 15–17c)

$$A = \sqrt{x_0^2 + (v_0^2/\omega_0^2)} = |x_0| = 0.050 \text{ m}$$

The initial phase angle is given by Eq. 15–17b, taking account of the fact that $x_0 < 0$,

$$\phi_0 = \tan^{-1}(\omega_0 x_0/v_0) = \tan^{-1}(-\infty) = 3\pi/2$$

(•Can you see why $x_0 < 0$ in Eq. 15–17a implies $\pi \leqslant \phi_0 \leqslant 2\pi$?) Thus, we have

$$x(t) = (0.050 \text{ m}) \sin (22.36t + 3\pi/2)$$

(d) The maximum velocity is

$$v_m = A\omega_0 = (0.050 \text{ m})(22.36 \text{ s}^{-1}) = 1.118 \text{ m/s}$$

The maximum acceleration is

$$a_m = A\omega_0^2 = (0.050 \text{ m})(22.36 \text{ s}^{-1})^2 = 25.0 \text{ m/s}^2$$

(e) The total energy is (Eq. 15–7a)

$$E = \tfrac{1}{2}\kappa A^2 = \tfrac{1}{2}(250 \text{ N/m})(0.050 \text{ m})^2 = 0.3125 \text{ J}$$

(f) When $\omega_0 t = \pi/8$ we have

$$\phi(t) = \omega_0 t + \phi_0 = \pi/8 + 3\pi/2 = 13\pi/8 = 292.5°$$

Then, $\qquad x = (0.050 \text{ m}) \sin 292.5° = -0.0462 \text{ m}$

$$v = (1.118 \text{ m/s}) \cos 292.5° = +0.428 \text{ m/s}$$

$$a = -(25.0 \text{ m/s}^2) \sin 292.5° = +23.1 \text{ m/s}^2$$

Example 15–3

A horizontal spring-and-block system (Fig. 15–3) has an angular frequency $\omega_0 = 3\pi \text{ rad/s}$. Determine $x(t)$ if the block is displaced to $x_0 = 0.25 \text{ m}$ and given a velocity $v_0 = -1.5 \text{ m/s}$ at $t = 0$.

Solution:

The initial displacement is positive, so $0 \leqslant \phi_0 \leqslant \pi$ (Eq. 15–17a). Using Eq. 15–17b, we find

$$\phi_0 = \tan^{-1} \frac{\omega_0 x_0}{v_0} = \tan^{-1} \frac{(3\pi \text{ s}^{-1})(0.25 \text{ m})}{(-1.5 \text{ m/s})}$$

$$= \tan^{-1}(-1.571) = 2.138 \text{ rad} = 122.48°$$

From Eq. 15–7c, we obtain

$$A = \sqrt{(0.25\ \text{m})^2 + (-1.5\ \text{m/s})^2/(3\pi\ \text{s}^{-1})^2} = 0.296\ \text{m}$$

Therefore, the displacement is given by

$$x(t) = (0.296\ \text{m})\sin(3\pi t + 2.138)$$

Example 15–4

A real spring-and-block oscillator arranged horizontally as in Fig. 15–3 would be subject to the effects of sliding friction. These effects can be eliminated by hanging the spring and block vertically from a fixed support. Then, in addition to the spring force, we must take into account the gravitational force that acts on the block.

A spring with a relaxed length $\ell = 0.30\ \text{m}$ and spring constant $\kappa = 150\ \text{N/m}$ is suspended vertically, as shown in diagram (a) below. A block with a mass $m = 0.50\ \text{kg}$ is attached to the free end of the spring. The block is pulled down 5.0 cm below its equilibrium position and then released from rest.
(a) Find the angular frequency ω_0 of the oscillations.
(b) Determine the time-dependent behavior of the vertical motion of the block.

Solution:

Part (a) of the diagram shows the spring with relaxed length ℓ. In (b), attaching the block results in an equilibrium condition in which the spring is stretched by an amount d. In (c), the block is shown a distance y *below* its equilibrium position. In diagrams (b) and (c), the free-body diagrams are also shown.

The resultant force that acts on the block when it is displaced a distance y below the equilibrium position is

$$F = mg - \kappa(d + y)$$

where the upward direction is taken positive. At equilibrium, $y = 0$, the net force is zero, so

$$0 = mg - \kappa d$$

Combining these two equations, the resultant force on the block is given by

$$F = -\kappa y$$

Thus, the vertical spring-and-block system behaves in the same way as its horizontal friction-less counterpart, provided that the displacement y is measured—as logic demands—from the equilibrium position.

Therefore, in the present case,

$$\omega_0 = \sqrt{\kappa/m} = \sqrt{(150 \text{ N/m})/(0.50 \text{ kg})} = 17.32 \text{ rad/s}$$

At $t = 0$, we have $y_0 = -0.050$ m and $v_0 = 0$. Then, using Eqs. 15–17, we find

$$\phi_0 = 3\pi/2 \quad \text{and} \quad A = 0.050 \text{ m}$$

Hence, $\qquad\qquad y(t) = (0.050 \text{ m}) \sin (17.32t + 3\pi/2)$

It is instructive to consider the energy balance for this system. Even at equilibrium, the spring is stretched and possesses potential energy in the amount $U_0 = \frac{1}{2}\kappa d^2$. When the block is lowered to a point y below the equilibrium position, the total energy stored in the spring is

$$U_s(y) = \tfrac{1}{2}\kappa(d + y)^2$$

$$= \tfrac{1}{2}\kappa d^2 + \kappa yd + \tfrac{1}{2}\kappa y^2$$

There is an accompanying decrease in the gravitational potential energy, namely,

$$U_g(y) = -mgy = -\kappa yd$$

where we have used $mg = \kappa d$. Therefore, the total potential energy is

$$U(y) = U_s(y) + U_g(y) = \tfrac{1}{2}\kappa d^2 + \tfrac{1}{2}\kappa y^2$$
$$= U_0 + \tfrac{1}{2}\kappa y^2$$

Thus, the form of the potential energy function (shown in the diagram below) is exactly the same as that in Fig. 15–4. The only difference is the shift of the zero position by an amount U_0, corresponding to the potential energy of the spring at equilibrium. This is of no consequence, however, because only potential energy *differences* are meaningful, and the zero-point of the potential energy has no physical significance.

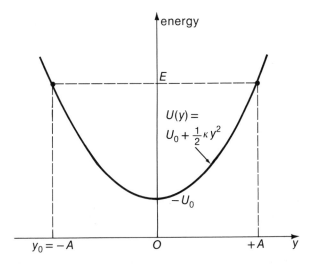

Example 15–5

Let us continue the analysis of the system in Example 15–4, now making the problem more realistic by including the effect of the mass of the spring, $m_s = 75$ g. Describe the motion of the system.

Solution:

At the instant when the length of the spring is L, its free end is moving with a velocity v, as shown in the diagram. Consider a small segment dx of the spring at a distance x from the fixed end. We assume that this mass element, $dm = m_s\, dx/L$, has a velocity* vx/L.

The kinetic energy of the mass element dm is

$$dK_s = \tfrac{1}{2}\left(\frac{vx}{L}\right)^2 dm = \tfrac{1}{2}v^2 x^2 \frac{m_s}{L^3}\, dx$$

and the total kinetic energy of the spring is

$$K_s = \tfrac{1}{2}v^2 \frac{m_s}{L^3}\int_0^L x^2\, dx = \tfrac{1}{2}v^2\left(\frac{m_s}{3}\right)$$

Thus, the kinetic energy of the spring is the same as that of a particle with $1/3$ the mass, located at the free end of the spring. The angular frequency of the motion is now

$$\omega_0 = \sqrt{\kappa/(m + \tfrac{1}{3}m_s)}$$
$$= \sqrt{(150 \text{ N/m})/(0.50 \text{ kg} + 0.025 \text{ kg})}$$
$$= 16.90 \text{ rad/s}$$

With the same boundary conditions as in Example 15–4, the displacement becomes

$$y(t) = (0.050 \text{ m}) \sin (16.90t + 3\pi/2)$$

The potential energy of the system is

$$U(y) = U_0' + \tfrac{1}{2}\kappa y^2$$

in which U_0' is the energy stored in the spring at the new equilibrium position, where the spring has been stretched beyond its *horizontal* relaxed length by the gravitational effect on the block and on its own mass.

(•Can you show that when the spring is held vertically it stretches by an amount $m_s g/2\kappa$ beyond its horizontal relaxed length?)

15-3 RELATIONSHIP OF SHM TO UNIFORM CIRCULAR MOTION

The idea of angular frequency ω was originally introduced in connection with the discussion of circular motion (see Section 4–5). This quantity appeared as the time derivative of the angular position of a particle moving along a circular path. For uniform circular motion we found that ω is related to the period T by $\omega = 2\pi/T$. In

---

*We assume that any disturbance propagates the length of the spring in a time that is very short compared with the period of the spring's harmonic motion.

the case of SHM we have a quantity ω_0 that also appears as the time derivative of an angle, namely, $\omega_0 = d\phi/dt$, where $\phi(t) = \omega_0 t + \phi_0$ is the phase angle of the motion.

There is, in fact, a high degree of similarity between simple harmonic motion and uniform circular motion. We can exploit this similarity to provide a simple means for visualizing the physical significance of the phase angle $\phi(t)$. Consider a particle that moves with a constant angular speed ω_0 around a circular path of radius A. Let the path lie in the x-y plane with the center at the origin O, as shown in Fig. 15-8. We refer to the instantaneous position Q of the particle as the *reference point* and to the circle in which the particle moves as the *reference circle*. The perpendicular projection of Q onto the y-axis is labeled P. Then, as the particle travels around the circle, the point P moves up and down along the y-axis. The line $\overline{OQ}$ makes an angle θ_0 with the x-axis at $t = 0$. At time t, this angle becomes $\omega t + \theta_0$, and the projection P of the reference point is described by the y-coordinate (Fig. 15-8a),

$$y(t) = A \sin(\omega t + \theta_0)$$

The (tangential) velocity of the particle is $v = \omega A$, and the y-component at time t is (Fig. 15-8b)

$$v_y(t) = \omega A \cos(\omega t + \theta_0)$$

The (centripetal) acceleration of the particle is $a = \omega^2 A$, and the y-component at time t is (Fig. 15-8c)

$$a_y(t) = -\omega^2 A \sin(\omega t + \theta_0)$$

If we identify ω with ω_0 and θ_0 with ϕ_0, these results for $y(t)$, $v_y(t)$, and $a_y(t)$ are identical with the corresponding equations (15-11, 15-15, and 15-16) for SHM. The analogy is therefore complete. *Simple harmonic motion can be described as the projection of uniform circular motion onto a fixed diameter of the circular path.* An observer who viewed the motion of the particle from the side, looking along the x-y plane from a distance $d \gg A$, would see the particle executing SHM.

15-4 EXAMPLES OF SIMPLE HARMONIC MOTION

We now give several additional examples of SHM. In each case we consider only motion in which the displacement from the equilibrium position has a small amplitude.

The Simple Pendulum. Consider a particle (the pendulum bob) that has a mass m and is suspended from a fixed support by a string of length ℓ. This system is in equilibrium when the bob hangs directly below the support point. A displacement away from this equilibrium position is conveniently measured in terms of the angle θ, as shown in Fig. 15-9a. The free-body diagram for the bob (Fig. 15-9b) shows the two forces that are acting. (We now use S for the tension, reserving T for the period.)

We now apply Newton's second law to the tangential component of the motion of the bob. The tangential acceleration is $a = d^2s/dt^2 = d^2(\ell\theta)/dt^2 = \ell(d^2\theta/dt^2)$, so we have

$$-mg \sin \theta = m\ell \frac{d^2\theta}{dt^2} \qquad (15\text{--}18)$$

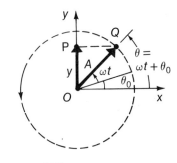

(a) $y = \overline{OP} = A \sin (\omega t + \theta_0)$

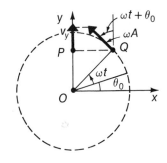

(b) $v_y = \omega A \cos (\omega t + \theta_0)$

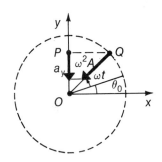

(c) $a_y = -\omega^2 A \sin (\omega t + \theta_0)$

Fig. 15-8. The representation of SHM as the projection of uniform circular motion onto a diameter corresponding to the y-axis:
(a) the coordinate $y = \overline{OP}$ of the reference point Q at time t,
(b) the y-component of the (tangential) velocity, $v = \omega A$, and
(c) the y-component of the (centripetal) acceleration, $a = \omega^2 A$.

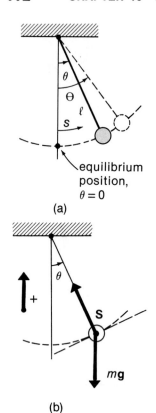

Fig. 15-9. (a) Geometry of a simple pendulum. The angular displacement θ is measured positive in a counterclockwise sense. (b) Free-body diagram for the pendulum bob.

The general solution of an equation of this type is complicated. However, if the motion of the pendulum is restricted to angles sufficiently small that $\sin \theta$ may be replaced* by θ, the solution is readily obtained. In this event, we have

$$\frac{d^2\theta}{dt^2} = -\frac{g}{\ell}\theta \qquad (15\text{-}19)$$

This equation has the same form as Eq. 15-5, so we have a case of SHM involving the angular coordinate θ. If Θ represents the amplitude of the angular motion, we can immediately write the solution as

$$\theta(t) = \Theta \sin(\omega_0 t + \phi_0) \qquad (15\text{-}20)$$

where

$$\omega_0 = \sqrt{\frac{g}{\ell}} \qquad (15\text{-}21)$$

The period of the motion is

$$T = \frac{2\pi}{\omega_0} = 2\pi \sqrt{\frac{\ell}{g}} \qquad (15\text{-}22)$$

Notice that the period of the pendulum depends only on its length and not on the mass of the bob. (•Argue on physical grounds whether the approximate expression for the period, obtained by using $\sin \theta \cong \theta$, yields values that are slightly too large or slightly too small.)

Be certain that you understand the distinction between the angular coordinate $\theta(t)$, which is confined to the relatively narrow range $-\Theta \leqslant \theta \leqslant +\Theta$, and the phase angle, $\phi(t) = \omega_0 t + \phi_0$, which increases without limit as t increases.

The solution to the pendulum problem contained in Eqs. 15-20 through 15-22 is only approximate; the expressions are useful only when the amplitude Θ is sufficiently small. Methods for obtaining the general solution are beyond the scope of this text, so we give without proof an accurate expression for the period T in terms of a power series in $\sin(\Theta/2)$:

$$T = 2\pi \sqrt{\frac{\ell}{g}} \left[1 + \tfrac{1}{4}\sin^2 \frac{\Theta}{2} + \tfrac{9}{64}\sin^4 \frac{\Theta}{2} + \cdots \right] \qquad (15\text{-}23)$$

Notice that the first term in this series is just Eq. 15-22. In Example 15-6 we see that the error in the period due to the use of the approximate expression (Eq. 15-22) is less than 0.5 percent for $\Theta = 15°$.

▶ *Historical Note on the Pendulum.* The observation that the period of a simple pendulum is independent of the amplitude (or nearly so, for small amplitudes) was first made by Galileo in 1593. Galileo watched the oscillations of a lamp that was suspended from the ceiling of a cathedral in Pisa and timed the oscillations by using his own pulse as a clock. From this crude measurement, Galileo correctly deduced the laws of pendulum motion. (The lamp oscillations were damped by friction, but, as we see later in this chapter, the period of a simple harmonic oscillator is insensitive to damping, if the damping is not too severe.)

About 1657, the Dutch physicist Christiaan Huygens (1629–1695) realized that a pendulum could be used to regulate the time-keeping of a clock. He also was the first to use a measurement of the period of a pendulum to determine the acceleration due to gravity g.

---

*For example, when $\theta = 15° = 0.2618$ rad, θ (in radians) exceeds $\sin \theta$ by only 1.15 percent.

Notice that the mass m of the pendulum bob occurs on both sides of the Newtonian equation of motion (Eq. 15–18) and therefore cancels. This results in equations for $\theta(t)$ and the period T that are independent of the mass of the bob. However, Newton realized that the quantity m on the left-hand side of Eq. 15–18 arises from the gravitational force and is therefore *gravitational mass,* whereas m on the right-hand side represents the *inertial mass* associated with acceleration and the second law. There was nothing in the theory then (nor is there now) to specify the equality of the two types of mass. Newton devised an ingenious method for comparing these two masses (an experiment we alluded to in Section 5–3). He constructed a hollow shell to use as a pendulum bob, and he placed in this shell different materials that he had determined by means of a balance to have the same weight (i.e., the same gravitational mass). In this way he was able to alter the composition of the bob while maintaining identical external conditions (such as air resistance). Newton measured the period of the pendulum for several different materials in the shell, always finding the same result. He concluded that to the precision of his measurements (about 1 part in 10^3), gravitational and inertial mass are equivalent. Modern versions of Newton's experiment have improved the precision of the equivalence to 1 part in 10^{12} (see Section 5–3). ◄

Example 15–6

A simple pendulum with a length of 80 cm is displaced by $15°$ from its equilibrium position and then released from rest. The mass of the bob is 0.30 kg.
(a) Use Eqs. 15–21 and 15–22 to calculate the angular frequency ω_0 and the period T to four significant figures.
(b) Determine Θ and ϕ_0, and give the resulting equation of motion, $\theta(t)$.
(c) Determine the maximum value of the angular speed, $\eta = d\theta/dt$.
(d) Find the maximum value of the tension S in the string. At what angle θ does this occur?
(e) Use Eq. 15–23 to calculate an accurate value for the period. Compare with the result in (a).

Solution:

(a) In order to calculate the angular frequency and period to four-place accuracy, we use $g = 9.806$ m/s^2, the value appropriate for sea level at latitude $45°$ (see Eq. 2–16). Then,

$$\omega_0 = \sqrt{\frac{g}{\ell}} = \sqrt{\frac{9.806 \text{ m/s}^2}{0.80 \text{ m}}} = 3.501 \text{ rad/s}$$

$$T = 2\pi\sqrt{\frac{\ell}{g}} = 2\pi\sqrt{\frac{0.80 \text{ m}}{9.806 \text{ m/s}^2}} = 1.795 \text{ s}$$

(b) The angular amplitude of the motion is, clearly, $\Theta = 15° = 0.262$ rad. Then, differentiating Eq. 15–20 and evaluating at $t = 0$, we have

$$\left.\frac{d\theta}{dt}\right|_{t=0} = \eta_0 = 0 = \omega_0\Theta\cos\phi_0$$

from which $\phi_0 = \pi/2$. Therefore, the equation of motion is

$$\theta(t) = 0.262\sin(3.50t + \pi/2)$$

(c) The angular speed, $\eta = d\theta/dt$, is maximum when $\phi(t) = 3.50t + \pi/2 = \pi, 2\pi, 3\pi,$ $\ldots$, that is, whenever $\theta = 0$. Hence,

$$\eta_m = \omega_0\Theta = (3.50 \text{ s}^{-1})(0.262) = 0.917 \text{ rad/s}$$

(d) Refer to Fig. 15–9b. The radial component of the acceleration of the bob is the centripetal acceleration, $a = v^2/\ell = (\ell\, d\theta/dt)^2/\ell = \ell\eta^2$. Then, Newton's second law for the radial motion can be expressed as

$$S - mg \cos \theta = m\ell\eta^2$$

Both $\cos \theta$ and η are maximum when $\theta = 0$, so

$$
\begin{aligned}
S_m &= m(g + \ell\eta_m^2) \\
&= (0.30\ \text{kg})[9.806\ \text{m/s}^2 + (0.80\ \text{m})(0.917\ \text{s}^{-1})^2] \\
&= 3.14\ \text{N}
\end{aligned}
$$

(e) Using Eq. 15-23, with $\Theta = 15°$, we find

$$
\begin{aligned}
T &= (1.795\ \text{s})[1 + \tfrac{1}{4}\sin^2 7.5° + \tfrac{9}{64}\sin^4 7.5° + \cdots] \\
&= (1.795\ \text{s})[1 + 0.00426 + 0.00004 + \cdots] \\
&= (1.795\ \text{s})[1.00430] = 1.803\ \text{s}
\end{aligned}
$$

We see that the error in using the approximate expression, $T = 2\pi\sqrt{\ell/g}$, amounts to 0.43 percent for $\Theta = 15°$.

The Torsional Pendulum. If a long fiber or wire or thin rod is twisted about its cylindrical axis, there is developed a restoring torque that follows a relationship similar to Hooke's law, namely,

$$\tau = -\Gamma\phi \qquad\qquad \textbf{(15–24)}$$

where ϕ is the angle of twist and where Γ is the *torsional constant* for the system, which depends on the elastic properties of the twisted element.

Consider two spherical particles, each with mass m and held a distance ℓ apart by a rigid rod with negligible mass. This dumbbell-shaped object is suspended by a wire that is attached to the midpoint of the rod with the other end clamped to a fixed support (Fig. 15–10). The rod is rotated by a small angle ϕ away from its equilibrium position, thereby twisting the wire. When the rod is released, the system will execute simple harmonic torsional oscillations.

We consider the dumbbell to be a free body, so $\tau = I_0\alpha$, where I_0 is the rotational inertia of the system about the suspension point, and where α is the angular

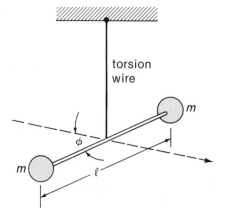

torsion
wire

Fig. 15–10. A torsional pendulum. The rigid rod connecting the two spheres has negligible mass. The rod is shown displaced from its equilibrium position by a small angle ϕ, and the support wire has been twisted by the same amount.

acceleration. Thus,

$$\tau = -\Gamma\phi = I_0\alpha = I_0\frac{d^2\phi}{dt^2}$$

from which

$$\frac{d^2\phi}{dt^2} = -\frac{\Gamma}{I_0}\phi$$

This equation has exactly the same form as Eq. 15–5, so the angular motion of the system is SHM. The period is, therefore,

$$T = 2\pi\sqrt{\frac{I_0}{\Gamma}} \qquad\qquad \textbf{(15–25)}$$

For the dumbbell system illustrated in Fig. 15–10, $I_0 = 2m(\ell/2)^2 = \frac{1}{2}m\ell^2$, so that $T = 2\pi\ell\sqrt{m/2\Gamma}$. A torsional system of this type was used by Henry Cavendish in his determination of the mass of the Earth (which also gave a value for the universal gravitation constant); see Section 5–3.

(•Suppose that you wish to determine the rotational inertia I_0 of an object with an axis of symmetry but with an otherwise complicated shape. Suppose also that you have a uniform cylinder with known dimensions and mass. How could you use torsional oscillations to determine I_0?)

The Physical Pendulum. We now examine the oscillatory motion of a system more realistic than the idealized simple pendulum we have been discussing. Consider the general rigid body shown in Fig. 15–11, which is suspended so that it can rotate on a horizontal axis, thereby becoming a *physical* (or *compound*) *pendulum*.

The torque about the point O is

$$\tau = -Mg\ell\sin\theta = I_0\alpha = I_0\frac{d^2\theta}{dt^2}$$

Fig. 15–11. A physical pendulum, consisting of a general rigid body that can rotate about a horizontal axis through the point O. Counterclockwise displacement angles are taken positive.

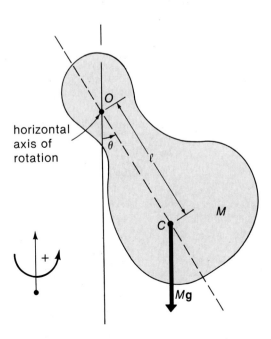

where I_0 is the rotational inertia about the point O. Again, we consider only small angular motions, so that $\sin\theta$ can be replaced by θ. Then, we can write

$$\frac{d^2\theta}{dt^2} = -\frac{Mg\ell}{I_0}\theta \tag{15-26}$$

This equation has the same form as Eq. 15-5 and therefore represents SHM with period

$$T = 2\pi\sqrt{\frac{I_0}{Mg\ell}} \tag{15-27}$$

Notice that this general result for the period of a physical pendulum reduces to that for the simple pendulum when $I_0 = M\ell^2$, as is appropriate for a particlelike bob and a massless string.

An interesting property of the physical pendulum was first realized by Henry Kater (1777–1835), an English physicist, and led to the development of an instrument, now known as *Kater's reversible pendulum*, used for precision measurements of the acceleration due to gravity g.

First, we note that by using the parallel-axis theorem (see Eq. 11-19), we can write

$$I_0 = I_C + M\ell^2$$

where I_C is the rotational inertia of the rigid body about an axis through its C.M. and parallel to the rotation axis through point O (see Fig. 15-11). Then, Eq. 15-27 can be expressed as

$$T = 2\pi\sqrt{\frac{(I_C/M\ell) + \ell}{g}} = 2\pi\sqrt{\frac{L}{g}}$$

where $L = (I_C/M\ell) + \ell$. Thus, the rigid body will have the same rotation period as a simple pendulum with a length equal to $(I_C/M\ell) + \ell$. Evidently, the rigid body will have the same oscillation period when suspended from any parallel rotation axis a distance ℓ from the C.M. Therefore, if we construct a circle with a radius ℓ about the C.M., any point on this circle can be selected for the point O. This circle is the outer dashed-line circle shown in Fig. 15-12. For any suspension point O, equilibrium results when the C.M. is directly below the chosen point, and small oscillations about all such points have the same period.

Next, we demonstrate that there exists another such circle, with radius ℓ' centered on the C.M., for which the period is also the same. (These two circles, with radii ℓ and ℓ', are referred to as *conjugate circles of oscillation*.) Select a point O' on the new circle, the inner dashed-line circle in Fig. 15-12, as the location of a parallel rotation axis. If the periods for the axes through O and O' are to be the same, we must have

$$\frac{I_C}{M\ell} + \ell = \frac{I_C}{M\ell'} + \ell'$$

or

$$I_C\ell' + M\ell^2\ell' - I_C\ell - M\ell\ell'^2 = 0$$

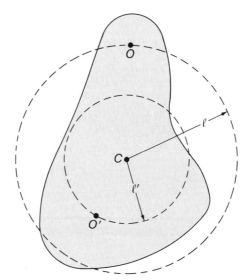

Fig. 15–12. A general rigid body, showing the conjugate circles on which any points O and O' at distances ℓ and ℓ' from the C.M. will yield the same oscillation period as a simple pendulum with a length $L = \ell + \ell'$.

from which
$$(\ell' - \ell)(I_C - M\ell\ell') = 0$$

One solution is the uninteresting case, $\ell = \ell'$; the other solution is

$$\ell' = \frac{I_C}{M\ell}$$

which we can use to write

$$\ell' + \ell = \frac{I_C}{M\ell} + \ell = L$$

If O, O', and the C.M. are collinear, with the C.M. between O and O', then $\ell + \ell'$ is equal to the separation of O and O' (refer to Fig. 15–12). This separation of the parallel rotation axes is the length $L = \ell + \ell'$ of the equivalent simple pendulum.

Figure 15–13 shows a schematic diagram of Kater's reversible pendulum. Two fixed (and inverted) knife-edged supports are separated by a precisely measured distance L. The two adjustable masses are moved along the shaft until the period T is the same for oscillations about each of the knife edges. Then, the local value of g is obtained from

$$g = \frac{4\pi^2 L}{T^2} \tag{15-28}$$

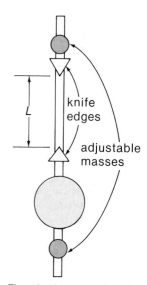

Fig. 15–13. Kater's reversible pendulum. When the adjustable masses are placed in positions that yield the same period for oscillations about either knife edge, the separation L of the knife edges is equal to the length of the equivalent simple pendulum.

Example 15–7

A uniform circular disc with a radius $R = 15$ cm is allowed to undergo small-amplitude oscillations about a point O located in its perimeter.
(a) Derive an expression for the period of the oscillations.
(b) What is the radius ℓ' of the conjugate circle on which the point O' can be chosen?
(c) If the periods for oscillations about O and O' are both $T = 0.9518$ s, what is the local value of g?

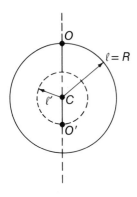

Solution:

(a) We have $L = (I_C/MR) + R$ and $I_C = \frac{1}{2}MR^2$; hence, $L = \frac{3}{2}R$. Then,

$$T = 2\pi \sqrt{\frac{L}{g}} = 2\pi \sqrt{\frac{3R}{2g}}$$

(b) Using $L = \ell + \ell'$ with $\ell = R$ and $L = \frac{3}{2}R$, we conclude that $\ell' = \frac{1}{2}R$.
(c) For the present case,

$$g = \frac{4\pi^2 L}{T^2} = \frac{4\pi^2(\frac{3}{2})(0.15 \text{ m})}{(0.9518 \text{ s})^2}$$
$$= 9.805 \text{ m/s}^2$$

SPECIAL TOPIC

15–5   OSCILLATIONS IN TWO DIMENSIONS

A variety of interesting patterns of oscillation are observed when two linear restoring forces act at right angles on a particle. The resulting motion of the particle is a superposition of SHM in each of the two directions. The most general type of such motion, in which there are different values for the two amplitudes, the two frequencies, and the two initial phase angles, is described by the parametric equations,

$$x(t) = A_x \sin(\omega_{0x}t + \phi_{0x})$$
$$y(t) = A_y \sin(\omega_{0y}t + \phi_{0y}) \qquad \textbf{(15–29)}$$

The path described by these equations is, in general, quite complicated. We can simplify matters considerably by requiring that the two frequencies be equal, $\omega_{0x} = \omega_{0y} = \omega_0$. We can also choose the instant that we designate $t = 0$ in order to make $\phi_{0x} = 0$. Then, with $\phi_{0y} = \phi_0$, we have

$$x(t) = A_x \sin \omega_0 t$$
$$y(t) = A_y \sin(\omega_0 t + \phi_0) \qquad \textbf{(15–30)}$$

The motion described by Eqs. 15–30 clearly is confined to a rectangle in the x-y plane with boundaries $\pm A_x$ and $\pm A_y$. The shape of the pattern that the particle describes while moving within this rectangle depends on the initial phase angle ϕ_0. For example, suppose $\phi_0 = \pi/2$. Then,

$$x(t) = A_x \sin \omega_0 t$$
$$y(t) = A_y \sin(\omega_0 t + \pi/2) = A_y \cos \omega_0 t$$

Squaring and adding these equations results in

$$\frac{x^2}{A_x^2} + \frac{y^2}{A_y^2} = 1$$

which is the equation of an ellipse (see Problem 14–21), and is illustrated in Fig. 15–14.

(•In Fig. 15–14, what is the direction of motion of the particle for $\phi_0 = \pi/2$? What would be the directions of motion for $\phi_0 = 0$, π, and $3\pi/2$?)

When $A_x = A_y$, the ellipse in Fig. 15–14 becomes a circle. But for initial phase angles other than 0, $\pi/2$, and π, and $3\pi/2$, the pattern again becomes an ellipse, with the axes at angles of 45° with respect to the x- and y-axes.

If the frequencies for the two components of the motion are different, the motion is repetitive only in the event that the ratio of the frequencies is equal to the ratio of two integers: $\omega_{0x}/\omega_{0y} = m/n$. Then, during the time that the

Fig. 15–14. The path of a two-dimensional simple harmonic oscillator with $\omega_{0x} = \omega_{0y}$ and $\phi_0 = \pi/2$ is an ellipse.

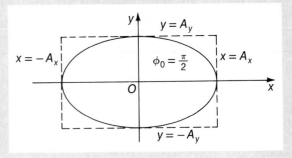

x-motion completes m full oscillations, the y-motion will complete exactly n full oscillations. The system is then in its original condition, ready to begin the cycle again. In such a case the motion describes a closed curve that is known as a *Lissajous' figure*.* A few of these patterns are shown in Fig. 15–15.

---

*Named after the French physicist, Jules Lissajous (1822–1880), who first analyzed these patterns.

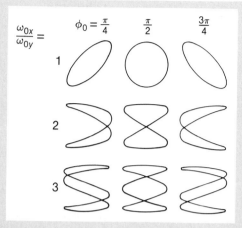

Fig. 15–15. Lissajous' figures for three different ratios of the frequencies and for three different initial phase angles. In each case, $A_x = A_y$.

SPECIAL TOPIC

15–6 DAMPED HARMONIC MOTION

In the preceding sections we have considered only cases that are free from any effects that cause energy loss. To analyze a real case, however, we must take account of the presence of frictional or viscous drag forces. It is often permissible to approximate the effects of such forces by including in the equation of motion a force term that is proportional to the velocity of the particle. (Refer to the discussion of viscous forces in Section 6–3.)

We now consider the consequences of including a retarding force of the form

$$F_v = -bv = -b\frac{dx}{dt}$$

Then, the equation of motion becomes

$$ma = -\kappa x + F_v$$

or

$$m\frac{d^2x}{dt^2} + b\frac{dx}{dt} + \kappa x = 0 \qquad \textbf{(15–31)}$$

This is a second-order linear differential equation with constant coefficients. We obtain the general solution to such equations by using the techniques discussed in the appendix, Calculus VII (Part II). For the present it is sufficient to note that the form of the solution to Eq. 15–31 depends on the relative magnitudes of the quantities $b^2/4m^2$ and κ/m. We are interested now in cases that in-

volve relatively small retarding forces. If the force coefficient b is sufficiently small that $b^2/4m^2 < \kappa/m$, the system undergoes *damped oscillatory motion*. The solution for this case is

$$x(t) = Ae^{-\beta t}\sin(\omega^0 t + \phi_0) \qquad \textbf{(15–32)}$$

where

$$\beta = b/2m > 0 \qquad \textbf{(15–33a)}$$

and*

$$\omega^0 = \sqrt{(\kappa/m) - (b/2m)^2} \qquad \textbf{(15–33b)}$$

Thus, the solution consists of an oscillatory term with a coefficient (the exponential) that continually *decreases* with time (because $\beta > 0$). Figure 15–16 shows a typical form of Eq. 15–32 in which $\phi_0 = \pi/2$. Notice that the new angular frequency ω^0 is shifted away from the undamped value, $\omega_0 = \sqrt{\kappa/m}$, by an amount that depends on the force coefficient b. When $b = 0$, which corresponds to the frictionless case, we have $\beta = 0$ and $\omega^0 = \omega_0$; thus, Eq. 15–32 becomes identical to Eq. 15–11, as it must.

The constants A and ϕ_0 are specified by the initial conditions:

$$x_0 = A\sin\phi_0 \qquad \textbf{(15–34a)}$$

$$v_0 = A(\omega^0\cos\phi_0 - \beta\sin\phi_0) \qquad \textbf{(15–34b)}$$

---

*Evidently, if ω^0 is to be *real*, the quantity within the radical sign must be positive. This leads to the stated restriction, $b^2/4m^2 < \kappa/m$. When $b^2/4m^2 \geq \kappa/m$, the solution does not represent oscillatory motion. We discuss these cases in Chapter 37.

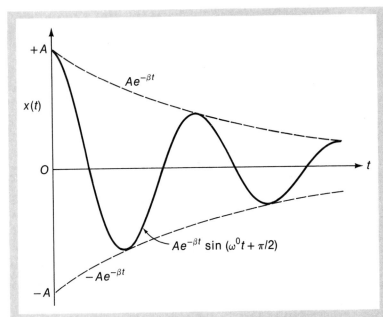

Fig. 15–16. The displacement $x(t)$ of a damped oscillator with initial phase angle $\phi_0 = \pi/2$. The dashed curves are the envelope functions, $\pm Ae^{-\beta t}$.

Example 15–8

Consider again the horizontal spring-block system illustrated in Fig. 15–3, with $m = 0.50$ kg and $\kappa = 250$ N/m. Now, the surface exerts a resistive force, $-bv$, on the block. The system is started at $t = 0$ with $\phi_0 = 3\pi/2$ and $A = 5.0$ cm. (Notice that this implies $v_0 \neq 0$.) At $t = 2$ s, the envelope of the oscillatory function has decreased to ± 1.0 cm.
(a) Find the value of the force coefficient b.
(b) Determine the angular frequency ω^0 and compare with the corresponding frictionless value ω_0.
(c) Find the values of the initial velocity v_0 and the initial acceleration a_0.

Solution:

(a) The decrease in the envelope between $t = 0$ and $t = 2$ s tells us that $e^{-2\beta} = \frac{1}{5}$. Solving for β,

$$2\beta = \ln 5$$

or

$$\beta = 0.805 \text{ s}^{-1}$$

Using $\beta = b/2m$, we find

$$b = 2m\beta = 2(0.50 \text{ kg})(0.805 \text{ s}^{-1})$$

$$= 0.805 \text{ kg/s} = 0.805 \text{ N} \cdot \text{s/m}$$

(b) The angular frequency is

$$\omega^0 = \sqrt{\frac{\kappa}{m} - \left(\frac{b}{2m}\right)^2} = \sqrt{\frac{\kappa}{m}} \sqrt{1 - \frac{b^2}{4m\kappa}}$$

$$= \omega_0 \sqrt{1 - \frac{(0.805 \text{ s}^{-1})^2}{4(0.50 \text{ kg})(250 \text{ N/m})}} = \omega_0 \sqrt{1 - 0.00130}$$

$$\cong \omega_0(1 - 0.00065)$$

Now, $\omega_0 = \sqrt{\dfrac{\kappa}{m}} = \sqrt{\dfrac{250 \text{ N/m}}{0.50 \text{ kg}}} = 22.361 \text{ rad/s}$

so $\omega^0 = (22.361 \text{ rad/s})(0.99935) = 22.361 \text{ rad/s}$

Thus, ω^0 differs from ω_0 by less than 0.1 percent even though the damping is sufficient to reduce the amplitude of the oscillations to 20 percent of the initial value after only 7 full oscillations (2 s).

(c) Using Eq. 15–34b,

$$v_0 = A(\omega^0 \cos \phi_0 - \beta \sin \phi_0)$$
$$= (0.050)[(22.346)\cos(3\pi/2) - (0.805)\sin(3\pi/2)]$$
$$= +0.0403 \text{ m/s}$$

SPECIAL TOPIC

15-7 FORCED OSCILLATIONS—RESONANCE

We now consider the behavior of an oscillatory system to which is coupled a driving force that varies sinusoidally with time at a frequency ω. We can again use as our model a horizontal spring-block combination, as shown in Fig. 15-17. Three forces act on the block—the restoring force exerted by the spring, the retarding force of friction, and the sinusoidal driving force. Newton's equation for the system is then the same as Eq. 15-31 with the addition of a term representing the driving force:

$$m\frac{d^2x}{dt^2} + b\frac{dx}{dt} + \kappa x = F_0 \cos \omega t \qquad \textbf{(15-35)}$$

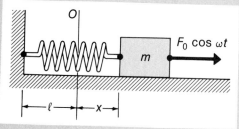

Fig. 15-17. An oscillatory system driven by a force $F_0 \cos \omega t$. The quantity x measures the displacement from O, the equilibrium position in the absence of the driving force.

The technique for solving equations such as Eq. 15-35 is discussed in the appendix Calculus VII (Part II). Here, we simply present the solution. When the driving force is first applied to the system, the motion that immediately follows is quite complicated. But after a time interval that is large compared with $1/\beta = 2m/b$ (see Eqs. 15-32 and 15-33a for the case $b^2/4m^2 < \kappa/m$), these *transient effects* subside due to the exponential factor $e^{-\beta t}$, and there remains the simple *steady-state* solution:

$$x(t) = A \sin(\omega t - \delta) \qquad \textbf{(15-36)}$$

Notice that the frequency of the system motion is ω, the frequency of the driving force. Also, the quantity δ represents the phase difference between the driving force and the system motion. Both A and δ are *completely determined* by the parameters of the system:

$$A = \frac{F_0}{\sqrt{(\kappa - m\omega^2)^2 + b^2\omega^2}} \qquad \textbf{(15-37a)}$$

$$\tan \delta = \frac{1}{b}\left(m\omega - \frac{\kappa}{\omega}\right), \qquad 0 \leqslant \delta \leqslant \pi \quad \textbf{(15-37b)}$$

If we use $\omega_0 = \sqrt{\kappa/m}$ and introduce the quantity $Q = m\omega_0/b$, we can express Eqs. 15-37 as

$$A = \frac{F_0/\kappa}{\sqrt{\dfrac{1}{Q^2}\dfrac{\omega^2}{\omega_0^2} + \left(\dfrac{\omega^2}{\omega_0^2} - 1\right)^2}} \qquad \textbf{(15-38a)}$$

$$\tan \delta = Q\left(\frac{\omega}{\omega_0} - \frac{\omega_0}{\omega}\right), \qquad 0 \leqslant \delta \leqslant \pi \quad \textbf{(15-38b)}$$

Because of the term $(\omega^2/\omega_0^2 - 1)^2$ in the denominator of the expression for A, the amplitude of the motion can become quite large. To find the angular frequency at which the amplitude is maximum, we calculate $dA/d\omega$ and set the result equal to zero, obtaining

$$\omega_R = \sqrt{\omega_0^2 - \frac{b^2}{2m^2}} \qquad \textbf{(15-39)}$$

This condition is called *amplitude resonance*. (•What is the smallest value of Q that permits an amplitude maximum?)

The three angular frequencies that have been defined can be expressed as follows:

$$\omega_0 = \sqrt{\kappa/m}$$

$$\omega^0 = \sqrt{(\kappa/m) - (b^2/4m^2)}$$

$$\omega_R = \sqrt{(\kappa/m) - (b^2/2m^2)}$$

so that $\omega_0 > \omega^0 > \omega_R$. In the event that the damping is small (i.e., $Q \gg 1$), the three frequencies are essentially the same. Thus, amplitude resonance occurs very close to $\omega = \omega_0$, the natural frequency of the undamped system.

The quantity Q measures the degree of damping in an oscillator and is called the *quality factor* or the "Q" of the system. If there is little damping, Q is very large and the amplitude at resonance becomes very large, namely, $A_m \cong Q(F_0/\kappa)$. On the other hand, if the damping is large and Q is very small, the resonance can be completely destroyed. Figure 15-18 shows the amplitude as a function of ω/ω_0 for several values of Q. (•Can you give a physical explanation why all of the curves approach unity for $\omega/\omega_0 \to 0$?)

Ordinary mechanical systems (e.g., loudspeakers) usually have Q's of a few to 100 or so. Quartz crystal oscillators may have values of 10^4 to 10^5; highly tuned electric circuits also have Q's in this range. Some laser systems have Q's as high as 10^{14}.

Fig. 15–18. The amplitude of a forced oscillation as a function of ω/ω_0, where ω is the driving frequency. The parameter Q is equal to $m\omega_0/b$, and $\omega_0 = \sqrt{\kappa/m}$.

Solution:

(a) When $\omega = \omega_0 = \sqrt{\kappa/m} = 22.36$ rad/s, we have

$$Q = \frac{m\omega_0}{b} = \frac{(0.50\text{ kg})(22.36\text{ s}^{-1})}{0.805\text{ N}\cdot\text{s/m}} = 13.89$$

and $A_m = Q\,\dfrac{F_0}{\kappa} = \dfrac{(13.89)(7.5\text{ N})}{250\text{ N/m}} = 0.417$ m

Also, $\omega_0 = 22.36\text{ s}^{-1}$

$$\omega^0 = 22.36(1 - 0.00065)\text{ s}^{-1} \cong \omega_0$$

$$\omega_R = 22.36(1 - 0.0013)\text{ s}^{-1} \cong \omega_0$$

so the amplitude resonance is very close to ω_0. From Eq. 15-38b, we see that $\tan\delta$ is very small, corresponding to δ essentially equal to zero. (•What is the physical significance of this result?)

(b) When $\omega = \frac{1}{2}\omega_0$, we find, using Eqs. 15-38,

$$A = \frac{(7.5\text{ N})/(250\text{ N/m})}{\sqrt{\dfrac{1}{(13.89)^2}\cdot\dfrac{1}{4} + \left(\dfrac{1}{4} - 1\right)^2}} = 0.0400\text{ m}$$

$$\delta = \tan^{-1}[13.89(\tfrac{1}{2} - 2)] = -1.523\text{ rad} = -87.3°$$

(c) When $\omega = 2\omega_0$,

$$A = \frac{(7.5\text{ N})/(250\text{ N/m})}{\sqrt{\dfrac{4}{(13.89)^2} + (4 - 1)^2}} = 0.0100\text{ m}$$

$$\delta = \tan^{-1}[13.89(2 - \tfrac{1}{2})] = 1.523\text{ rad} = +87.3°$$

Notice that the amplitude for $\omega = \omega_0$ is much larger than the values for the other two frequencies. Notice also that the amplitude for $\omega = \frac{1}{2}\omega_0$ is exactly 4 times greater than that for $\omega = 2\omega_0$. (•Can you prove that this is the case for any value of Q?) Both of these results are expected from the graph in Fig. 15-18.

Example 15–9

We continue to examine the horizontal spring-block system with $m = 0.50$ kg and $\kappa = 250$ N/m. As in Example 15–8, we have a damping force with $b = 0.805$ N $\cdot$ s/m. Now, we apply to the system a driving force, $F_0 \cos \omega t$, with $F_0 = 7.5$ N. Discuss the motion for (a) $\omega = \omega_0$, (b) $\omega = \frac{1}{2}\omega_0$, and (c) $\omega = 2\omega_0$.

PROBLEMS

Section 15-1

15-1   The potential energy function for a particle is $U(\xi) = a\xi^2 + b\xi + c$, with $a = 2$ J/m^2, $b = -4$ J/m, and $c = 1$ J. (a) Sketch the function $U(\xi)$. (b) Find any points of stable or unstable equilibrium. (c) Use the coordinate $x = \xi - \xi_0$ for the displacement from any point of stable equilibrium. Find the expression for the strength parameter κ that appears in Eq. 15-4. [*Hint:* Complete the square in the expression for $U(\xi)$.] (d) As an alternative to completing the square, as suggested in (c), use Eqs. 15-2 and 15-3 to obtain the same result.

15-2 Consider the gravitational potential energy of a pendulum bob (mass m) that is attached to a massless string with a length ℓ, as shown in the diagram. Use the angle θ as the coordinate variable, so that $\theta = 0$ represents the position of stable equilibrium. (a) Let the gravitational potential energy be zero for $\theta = 0$ and show that $U(\theta) = mg\ell(1 - \cos\theta)$. (b) Use Eq. 15-3 and obtain the expression for the potential energy correct to second order. (c) For this case, Eq. 15-4 becomes $F(\theta) = -\kappa\theta$. What is the expression for κ?

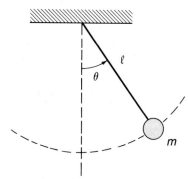

15-3 The *Lennard-Jones* potential function (first introduced in Problem 7-28) is often used to describe the interaction between two atoms in a diatomic molecule:

$$U(r) = U_0[(r_0/r)^{12} - 2(r_0/r)^6], \ r > 0 \text{ and } U_0 > 0$$

(a) Sketch the potential function $U(r)$. (b) Find any points of stable or unstable equilibrium. (c) Use Eq. 15-3 and derive an approximate expression for $U(r)$ that is correct to second order in the displacement from a position of stable equilibrium. (d) Determine the expression for κ (Eq. 15-4).

Section 15-2

15-4 A particle executes SHM with an amplitude of 10 cm and a frequency $\nu = 3$ Hz. Find the following: (a) the values of the maximum velocity and the maximum acceleration; (b) the velocity and the acceleration when the displacement is 8 cm; and (c) the minimum time for the particle to move from the equilibrium position to a displacement of 6 cm.

15-5 A block with a mass of 0.50 kg is attached to the free end of an ideal spring that provides a restoring force of 40 N per meter of extension. The block is free to slide over a frictionless horizontal surface. The block is set into motion by giving it an initial potential energy of 2 J and an initial kinetic energy

of 1.5 J. (a) Make a graph of the potential energy of the system for -0.5 m $\leqslant x \leqslant +0.5$ m. (b) Determine the amplitude of the oscillation—first, from the graph, and then by algebraic calculation. (c) What is the speed of the block as it passes through the equilibrium position? (d) At what displacements is the kinetic energy equal to the potential energy? (e) Find the angular frequency ω_0 and the period T of the motion. (f) If the initial displacement was $x_0 > 0$ and if the initial velocity was $v < 0$, determine the initial phase angle ϕ_0. (g) Write down the equation of motion $x(t)$.

15-6 Two blocks, with masses m_1 and m_2, are free to slide on a horizontal frictionless surface and are connected by an ideal spring with spring constant κ. Find the angular frequency of oscillatory motion for the system.

15-7 A particle that is attached to a vertical spring is pulled down a distance of 4.0 cm below its equilibrium position and is released from rest. The initial upward acceleration of the particle is 0.30 m/s². (a) What is the period T of the ensuing oscillations? (b) With what velocity does the particle pass through its equilibrium position? (c) What is the equation of motion for the particle? (Choose the upward direction to be positive.)

15-8 A particle that hangs from an ideal spring has an angular frequency for oscillations, $\omega_0 = 2.0$ rad/s. The spring is suspended from the ceiling of an elevator car and hangs motionless (relative to the elevator car) as the car descends at a constant velocity of 1.5 m/s. The car then stops suddenly. (a) With what amplitude will the particle oscillate? (b) What is the equation of motion for the particle? (Choose the upward direction to be positive.)

15-9 • A block with a mass m, when attached to a uniform ideal spring with strength parameter κ and relaxed length ℓ, executes SHM with angular frequency $\omega_0 = \sqrt{\kappa/m}$. The spring is then cut into two pieces, one with relaxed length $f\ell$ and the other with relaxed length $(1 - f)\ell$. The block is divided in the same fractions and the smaller part of the block is attached to the longer part of the spring; the remaining pieces are likewise joined. (a) What are the strength parameters for the two parts of the spring? (b) What are the angular frequencies of oscillation for the two systems?

15-10 A block with a mass of 0.20 kg, when hanging from a spring, is found to execute SHM with a period $T_1 = 0.550$ s. If the block is replaced by one with twice the mass, the period becomes $T_2 = 0.760$ s. (a) What is the mass of the spring? (b) What is the value of the spring constant κ of the spring?

Section 15-3

15-11 A particle is executing uniform circular motion with radius $R = 30$ cm in a vertical plane. The period of the motion is 3 s. A distant overhead light source produces a shadow of the particle on a horizontal plane. The velocity of the shadow, which executes SHM, is $v = -v_0 \sin \omega t$. Determine v_0 and ω. What is the expression for x, the coordinate of the shadow measured from its central position?

15-12 A particle executes SHM along a vertical line according to $y(t) = (60 \text{ m}) \sin (8t + \pi/8)$. (a) Determine the parameters for the corresponding particle that undergoes uniform circular motion in a vertical plane. Determine the position angle θ of the reference point when: (b) the particle is at $y = +40$ m and the velocity is directed downward, and (c) the particle has a downward acceleration equal to g and a velocity directed upward. (d) What are the shortest time intervals after $t = 0$ required for the particle to reach the positions referred to in (b) and (c)?

Section 15-4

15-13 A thin rod has a mass M and a length $L = 1.6$ m. One end of the rod is attached by a pivot to a fixed support and the hanging rod executes small oscillations about the pivot point. Find the frequency of these oscillations. If a particle with a mass M is added to the lower end of the rod, by what factor will the period change?

15-14 A small bob with a mass of 0.20 kg hangs at rest from a massless string with a length of 1.40 m. At $t = 0$ the bob is given a sharp horizontal blow that delivers an impulse, $J = \int F \, dt = 0.15$ N · s. (a) Find the amplitude Θ of the ensuing oscillations. (b) Determine the equation of motion of the bob, $\theta(t)$.

15-15 A sensitive balance for determining the masses of small objects consists of a thin uniform crossbar with a length of 10 cm and a mass of 100 mg which is attached at its midpoint to a fused quartz torsional fiber. The quartz fiber is stretched between the arms of a frame, as shown in the figure above right. When the crossbar is set into torsional oscillations, the period (for small oscillations) is found to be 8.0 s. (a) What is the torsional constant of the fiber? (b) If a small object with a mass of 10 μg is attached to the end of the crossbar at point A, what is the resulting vertical deflection at equilibrium?

15-16 A straight uniform wire with a length of 25 cm is
• bent in a right angle at its midpoint. The wire is

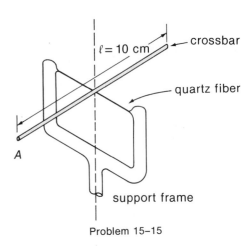

Problem 15–15

hung on a horizontal knife edge that acts as an axis, and it is set in oscillation with a small angular amplitude, as indicated in the diagram. Determine the period of the oscillations.

Problem 15–16

15-17 A thin uniform disc with a radius $R = 20$ cm swings
• about an axis that consists of a thin pin driven through the disc, as shown in the diagram. At what distance r from the center of the disc must the pin be located in order that the period of oscillation be a minimum? What is the period for this location of the pin?

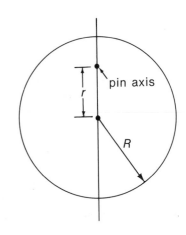

15-18 A physical pendulum consists of a bob with a radius $R = 5$ mm attached to a string whose mass is negligible in comparison with that of the bob. The distance from the point of support to the center of the bob is 25 cm. Find the period of this pendulum for oscillations with small amplitudes. Compare this period with that of the simple pendulum having the same length (but with a bob of negligible size).

SPECIAL TOPIC

Section 15-5

15-19 Make a sketch of the Lissajous' figure for the case $\omega_{0x}/\omega_{0y} = 3/2$ and $\phi_0 = \pi/2$.

SPECIAL TOPIC

Section 15-6

15-20 Show by direct substitution that $x(t)$ given by
• Eq. 15-32 is in fact a solution to the equation of motion, Eq. 15-31. [*Hint:* The equation that results from the substitution must be valid at any time t.]

15-21 A horizontal spring-block system (as in Example 15-8) has $m = 0.50$ kg and $\kappa = 250$ N/m. The frictional force, $-bv$, between the block and the surface produces a damped oscillatory motion with a period that is 1.25 times the period of the corresponding undamped motion. (a) What is the value of the force coefficient b? (b) After what time interval will the amplitude envelope function have decreased to $\frac{1}{3}$ the value at the beginning of the interval?

SPECIAL TOPIC

15-22 Derive an expression that gives the velocity as a function of time for a damped oscillator. Sketch $v(t)$ for the case $A = 1$ m, $\beta = 1\,s^{-1}$, and $\omega^0 = 2$ rad/s.

SPECIAL TOPIC

Section 15-7

15-23 Show that the amplitude A (Eq. 15-38a) is maxi-
• mum when $\omega^2 = \omega_R^2 = \omega_0^2(1 - 1/2Q^2)$. Show that this expression for ω_R is equivalent to Eq. 15-39. [*Hint:* The amplitude A is maximum when the square of the denominator in Eq. 15-38a is minimum.]

15-24 A horizontal spring-block system with $m = 0.50$ kg
•• and $\kappa = 250$ N/m experiences a frictional drag giving a Q of 1.50. A driving force, $(50$ N$)\cos \omega t$, is applied to the system. Find (a) the resonance angular frequency ω_R, (b) the amplitude and phase angle δ for $\omega = \omega_R$, (c) the amplitude for $\omega = \omega_0$, (d) the amplitude for $\omega = \frac{1}{3}\omega_0$, and (e) the amplitude for $\omega = 3\omega_0$.

Additional Problems

15-25 Consider the potential energy function for a particle, $U(\xi) = \xi^2(a\xi - b)$, with $a > 0$ and $b > 0$. (a) Sketch the function $U(\xi)$. (b) Find the points of stable and unstable equilibrium. (c) Expand the potential energy function about the minimum of $U(\xi)$ at $\xi = \xi_0$ and obtain the expansion correct to second order. (d) Introduce the coordinate $x = \xi - \xi_0$ as the displacement from stable equilibrium and show that the effective strength constant for the system is $\kappa = 2b$.

15-26 Consider the potential energy for a particle, $U(r) = U_0[(r_0/r)^5 - C(r_0/r)^2]$, with $r > 0$ and $U_0 > 0$. (a) Sketch the function $U(r)$. (b) Find the value of C for which stable equilibrium occurs at $r = r_0$. (c) Expand $U(r)$ about $r = r_0$ and obtain an expression correct to second order. (d) What is the effective strength constant κ for the system?

15-27 An interesting teeter-totter toy consists of a
•• pointed peg with two thin drooping arms having bobs of equal mass at the ends, as shown in the diagram. A properly designed toy will rock back and forth about the pivot point O without tipping over. (a) Show that the potential energy of the toy, when the peg is displaced by an angle β from the vertical, is $U(\beta) = 2mg \cos \beta(L - \ell \cos \alpha)$. Assume that the bobs are particles and choose the pivot point O as the zero reference for the potential energy. (b) Show that for stable equilibrium the

bobs must hang below the pivot point O when $\beta = 0$ (i.e., $\ell \cos \alpha > L$). (c) For the case $\ell = 3L$ and $\alpha = 60°$, show that the period for small oscillations is $T = 2\pi \sqrt{14L/g}$. Assume the plane containing the arms remains vertical.

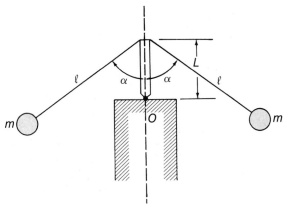

15-28 Show that Eq. 15-6 can be written in the form of an equation for an ellipse in the coordinate variable x and the momentum variable $p = mv$, namely,

$$\frac{p^2}{2Em} + \frac{x^2}{2E/\kappa} = 1$$

Sketch this ellipse in the x-p plane and determine the semimajor and semiminor axes.

Such representations of the history of oscillatory motions are called *phase diagrams*. [•Show that the area of the phase ellipse (see Example III-11) is $2\pi E/\omega_0$. As time advances, in what sense (clockwise or counterclockwise) does the history of the oscillatory motion unfold in the x-p plane?]

15-29 A flat plate P executes horizontal SHM by sliding across a frictionless surface with a frequency $\nu = 1.5$ Hz. A block B rests on the plate, as shown in the diagram, and the coefficient of static friction between the block and the plate is $\mu_s = 0.60$. What maximum amplitude of oscillation can the plate-block system have if the block is not to slip on the plate?

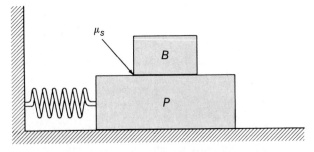

15-30 Consider a slender rod with mass $M = 4$ kg and length $\ell = 1.2$ m that is pivoted on a frictionless

horizontal bearing at a point $\ell/4$ from one end, as shown in the diagram. (a) Derive (from the definition) the expression for the moment of inertia of the rod about the pivot. (b) Obtain an equation that gives the angular acceleration α of the rod as a function of θ. (c) Determine the period of small amplitude oscillations about the equilibrium position.

frictionless horizontal
axis of rotation
$\perp$ to plane of page

15-31 A block rests on a flat plate that executes vertical SHM with a period of 1.2 s. What is the maximum amplitude of the motion for which the block will not separate from the plate?

15-32 A block with a mass of 2 kg hangs without vibrating at the end of a spring ($\kappa = 500$ N/m) that is attached to the ceiling of an elevator car. The car is rising with an upward acceleration of $\frac{1}{3}g$ when the acceleration suddenly ceases (at $t = 0$). (a) What is the angular frequency of oscillation of the block after the acceleration ceases? (b) By what amount is the spring stretched during the time that the elevator car is accelerating? (c) What is the amplitude of the oscillation and the initial phase angle observed by a rider in the car? Take the upward direction to be positive.

15-33 A block with a mass $M = 0.50$ kg is suspended at rest from a spring with $\kappa = 200$ N/m. A blob of putty ($m = 0.30$ kg) is dropped onto the block from a height of 10 cm; the putty sticks to the block. (a) What is the period of the ensuing oscillations? (b) Determine the equation of motion for the block-and-putty combination. (Take the upward direction to be positive.) (c) What is the total energy of the oscillating system?

15-34 A *seconds pendulum* is one for which each back-and-forth vibration requires exactly 1 s (that is, $T = 2$ s). When such a pendulum is carried aloft in a balloon, it is found to "tick" only 59.914 times per minute. What is the height of the balloon above sea level? [*Hint:* Use differentials.]

15-35 A simple pendulum ($\ell = 2$ m) oscillates with a *horizontal* amplitude of 20 cm. A 100-mg ladybug clings to the bottom of the pendulum bob. What is the minimum holding force that the bug must exert during the motion to prevent being detached from the bob?

15-36 Find the period of a simple pendulum whose bob is near the Earth's surface and whose suspension string is infinitely long. Show that this period is equal to that of an artificial Earth satellite in a low-altitude orbit.

15-37 A narrow tunnel is cut through the Earth along a chord. Show that the period of a particle that oscillates in this tunnel is the same as that of a particle oscillating in a tunnel along a diameter (84.5 min). [*Hint:* Refer to Problems 14-31 and 15-42.]

15-38 Two springs, with spring constants κ_1 and κ_2, are connected to a block with mass m in the three different ways indicated in the diagram. Determine for each case the equivalent spring constant κ that would produce the same force on the block.

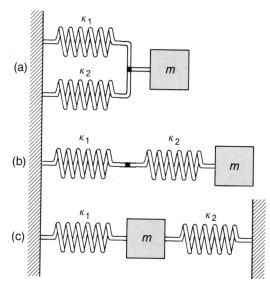

15-39 A thin rod with negligible mass supports at its lower end a particlelike object with a mass m. The upper end of the rod is free to rotate in the plane of the page about a horizontal axis through the point O in the diagram (above right). A horizontal spring is attached to a point on the rod, as shown. If the spring has its relaxed length when the rod is vertical, show that the period of small-amplitude oscillations is

$$T = 2\pi \left[\left(\frac{\kappa}{m} \right) \frac{\ell^2}{L^2} + \frac{g}{L} \right]^{-1/2}$$

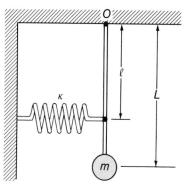

Problem 15-39

15-40 A horizontal spring-block system executes SHM with parameters A, $\phi_0 = 0$, and ω. At time $t = t_1$ (for which $\omega t_1 = \phi_1$) a blob of putty with a mass m is dropped straight down on the block, sticking to it. (a) Show that the new amplitude of oscillation is

$$A' = A \sqrt{\frac{M + m \sin^2 \phi_1}{M + m}}$$

(b) Show that the new angular frequency is $\omega' = \omega \sqrt{M/(M + m)}$. [*Hint:* Use conservation of horizontal linear momentum and re-evaluate the new total mechanical energy of the system.]

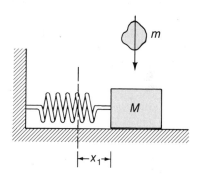

15-41 A wheel rotates uniformly at a rate of 300 rpm and drives a horizontal shaft by means of a Scotch yoke, as shown in the diagram on the next page. At $t = 0$ the drive pin is $\pi/4$ rad below the horizontal, as shown. (a) If the pin circle has a radius of 15 cm, write the equation of the point P on the shaft, using the midpoint of its motion as origin. (b) If the mass of the driven shaft is 5 kg, what maximum force must the drive pin exert on the yoke?

15-42 Show that a particle with a mass m moving in a narrow tunnel through the Earth along a diameter has acting on it a gravitational force with a Hooke's-law form, namely, $F = -\kappa x$, where x is the distance from the Earth's center and where $\kappa = gm/R_E$. (a) Verify that such a freely moving particle has a period of 84.5 min. (b) At what constant speed

Problem 15–41

would a low-altitude satellite have to travel so that
the perpendicular projection of its motion onto the
diameter of its circular orbit (made to correspond to
the tunnel) matches the motion of the oscillating
particle? (c) Is the satellite orbit referred to in (b) a
stable orbit?

15–43 A simple experiment shows that, when a standard
•• wire coat hanger is suspended in the positions
shown in the diagram, the periods for small oscillations about equilibrium are all the same. (a) Using
only this information and the dimensions in the diagram, determine the location of the C.M. of the coat
hanger. (b) What is the period of oscillation?
(c) Check your results experimentally.

15–44 A thin uniform rod executes small-amplitude oscillations about a perpendicular horizontal axis
through a point O that is a distance l above the
C.M. For what distance l will the oscillation period
be least?

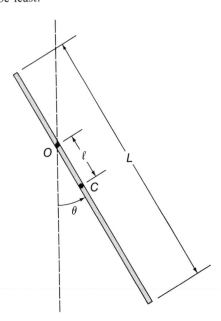

15–45 A 2-kg sphere, when falling through air, reaches a
• terminal velocity of 75 m/s (see Section 6–3). This
sphere, when suspended from an ideal lossless
spring, oscillates in air with a period $T = 0.30$ s.
(a) What is the spring constant κ? (b) What time
interval is required for the amplitude to decrease to
one half of the value at the beginning of the interval? (c) What fraction of the mechanical energy is
lost during the interval that the amplitude decreases
by half? (d) What is the angular frequency ω_R for
amplitude resonance of driven oscillations?
(e) What is the amplitude of driven oscillations
when $\omega \ll \omega_R$? (f) What is the Q of the system?

(g) What is the oscillation amplitude at resonance?
(h) What magnitude of constant force would produce the same displacement as the amplitude at resonance?

15–46 Consider a damped oscillator described by Eq. 15–
31. The total mechanical energy of the system is
$E = \frac{1}{2}mv^2 + \frac{1}{2}\kappa x^2$. Show that $dE/dt = -bv^2 = F_v v$,
which states that the rate of loss of energy is equal
to the power dissipated in the damping mechanism.

15–47 Consider a driven oscillator described by Eq. 15–35.
(a) Show that the instantaneous power input to the

oscillating system by the driving force is

$$P(t) = vF(t) = A\omega F_0(\cos^2 \omega t \cos \delta$$

$$+ \cos \omega t \sin \omega t \sin \delta)$$

(b) Show that the average input power is

$$\bar{P} = \frac{1}{T} \int_0^T P(t)\, dt = \tfrac{1}{2} v_m F_0 \cos \delta$$

where $v_m = A\omega$ is the maximum velocity of the block and where δ is the phase angle that appears in Eq. 15–36.

CHAPTER 16

DEFORMATIONS OF SOLIDS

To this point in our discussions we have assumed that solid objects (with the exception of springs and strings) retain a fixed shape under all conditions of applied forces. Actually, the application of a force to a real object causes it to undergo some deformation. Even a small load applied to a steel I-beam will deform the beam to an extent that can be measured with sensitive instruments.

The application of a force to an object produces complicated internal forces that depend on the composition of the object, its shape, and the way in which the force is applied. Every point within the object is then in a condition of *stress*. Any change in the shape or the volume of the object in response to these stresses is referred to as *strain*.

In this chapter we discuss the relationship of strain to stress in a variety of cases, including some practical engineering situations.

16-1 TENSION PRODUCED BY AXIAL FORCES

Consider first a test sample in the form of a uniform slender rod. The ends of this rod are clamped in the jaws, J and J', of a testing machine capable of applying any desired tensional force F to the sample (Fig. 16–1). The force applied in this way is *axial* and produces neither bending nor twisting of the sample.

Imagine a hypothetical transverse cut CC' through the rod near its middle. The portions of the rod bounded by the cut are in equilibrium, so the internal force at the cut also has the magnitude F and is collinear with the applied force, as shown in Fig. 16–1b. The *normal stress* at the cut is defined to be

$$s_n = \frac{F}{A} \tag{16-1}$$

where A is the cross-sectional area at CC'. The dimensionality of stress is force per unit area. The SI unit of stress is N/m², and this unit is given the special name *pascal* * (Pa):

$$1 \text{ Pa} = 1 \text{ N/m}^2$$

The *axial* (or *longitudinal*) *strain* experienced by the rod is defined to be the amount of elongation $\Delta \ell$ divided by the unstressed length ℓ (Fig. 16–1c):

$$e_\ell = \frac{\Delta \ell}{\ell} \tag{16–2}$$

Evidently, strain is a dimensionless quantity.

Figure 16–2 shows a typical stress-strain curve for a ductile material,† such as wrought iron, low-carbon steel, or copper. For small values of the stress, the strain is directly proportional to the stress. This linear relationship extends only to point B, called the *proportional limit*.

If the stress is increased beyond the proportional limit, a ductile material will exhibit a pronounced increase in strain for a small (even zero) increase in stress. This phenomenon is referred to as *plastic flow* and is represented by the region $\overline{CD}$ in Fig. 16–2. The stress at point C, where plastic flow begins, is called the *yield point* or the *elastic limit* for the sample. At this point the strain, for most ductile materials, is less than 1 percent. Beyond point D, changes in the microscopic structure of the material due to the effect of the strain (*strain hardening*) cause an increase in the resistance to further stretching.

Suppose that the sample is stressed to a point beyond the elastic limit—for example, point P in Fig. 16–2—and then the stress is slowly reduced to zero. In this process the sample follows a path $\overline{PR}$ in the stress-strain diagram, a path that is approximately parallel to the linear portion of the curve, $\overline{OB}$. The effect is to leave

Fig. 16–1. (a) A tensional force F is applied to a uniform slender rod in a test machine. (b) Free-body diagram for a portion of the sample formed by the hypothetical cut CC'. (c) The stress in the rod results in an axial (or longitudinal) strain equal to the elongation $\Delta \ell$ divided by the unstressed length ℓ.

Fig. 16–2. A typical stress-strain curve for a ductile material under axial tension.

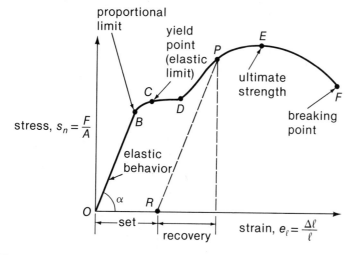

---

*Named for Blaise Pascal (1623–1662), a French mathematician and physicist famous for his experiments in hydrostatics.

†*Ductility* is the property that enables a metal to be extended appreciably without rupture, as in wire-drawing.

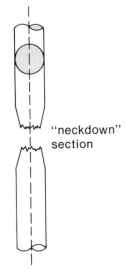

Fig. 16–3. When the ultimate strength of the material is exceeded, the sample "necks down" at a weak point and eventually ruptures there.

the sample at zero stress with a permanent elongation or *set* given by $\overline{OR}$ in Fig. 16–2.

The largest stress that the sample can sustain is reached at point E and is called the *ultimate strength* of the material (under tension). If the stress is increased beyond this point, the sample becomes "necked down" at some weak point (Fig. 16–3). Thereafter, the substance flows plastically until it finally ruptures at point F, the *breaking point*. Necking down occurs only for ductile materials. (From point E to point F, the stress-strain curve bends downward because the stress is measured in terms of the original cross-sectional area A. In the necked-down part of the sample, where the rupture finally occurs, the stress is actually increasing because the area is decreasing.)

The stress-strain curve for a brittle material, such as high-carbon steel or cast iron, is rather featureless, as shown in Fig. 16–4. (A brittle material is a *hard* material; it resists indentation and does not undergo plastic flow.) Again, we identify the proportional limit B. In contrast to the behavior of ductile materials, there is no well-defined elastic limit. Moreover, there is no neck-down region, so the stress-strain curve does not bend downward as in Fig. 16–2. Instead, there is a simple rupture at some maximum stress (point F).

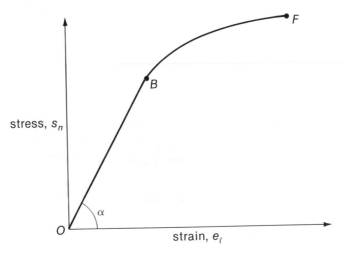

Fig. 16–4. A typical stress-strain curve for a brittle material under axial tension.

Values of the proportional limit and the ultimate strength for various types of steel are listed in Table 16–1.

▶ *Safety Factors.* The safe design value for the stress in a material under ordinary service conditions is called the *working stress* or *allowable stress*. Usually, the design stress value is arrived at by reducing the proportional limit, yield point, or ultimate strength by a certain factor called the *safety factor* for the material. For low-carbon structural steel (which contains approximately 0.30 percent C), the design stress for a member in simple tension is usually taken to be about 11.0×10^7 Pa, a value that is roughly one half the proportional limit for this type of steel (see Table 16–1); that is, the safety factor f is approximately 2. ◀

▶ *Stresses at Discontinuities.* The stress along a cross section of a test sample depends on the geometry of the sample and may not be constant. Consider an axially loaded rod that has an abrupt change in its cross section, as shown in Fig. 16–5a. The local stress at the corner of the junction is about twice that at the center of the rod. We define a *stress concentration factor k* such that the maximum stress at the discontinuity can be expressed as

$$s_m = k\frac{F}{A}$$

(16–3)

TABLE 16–1. Properties of Various Grades of Steel

COMPOSITION	TREATMENT	PROPORTIONAL LIMIT (10^8 Pa)[c]	ULTIMATE STRENGTH (10^8 Pa)
0.02% C[a]	hot rolled	1.11	2.92
0.20% C	cold drawn	3.80	5.57
0.30% C	annealed	2.41	4.82
0.49% C	tempered	4.67	6.68
0.93% C[b]	hardened	7.34	13.0
1.20% C	tempered	6.94	12.4
3.5% Ni	tempered	5.95	8.13
Chrome-vanadium	tempered	7.86	10.1

[a] Ingot iron
[b] Spring steel
[c] This notation means, for example, that the first entry in this column is 1.11×10^8 Pa. This convention is used throughout.

By smoothing the transition between the two sections of the rod with a fillet (Fig. 16–5b), the maximum stress can be considerably reduced:

$\frac{r}{D}$	0	1/16	1/8	1/4	1/2
k	2.00	1.75	1.50	1.20	1.10

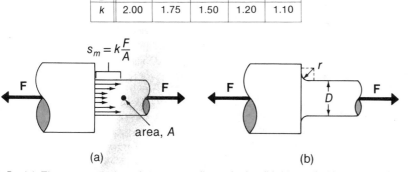

Fig. 16–5. (a) The concentration of stress at a discontinuity. (b) A rounded juncture reduces the maximum stress.

▶ *Compression.* Many structural members (e.g., columns) are required to sustain loads in *compression.* If the stress-strain curve for a ductile material (Fig. 16–2) is extended into the compressive region (that is, negative values for both stress and strain), it is found that the slope of the curve is the same as in the tensile region, although differences result for high-load conditions. Various composite materials, such as concrete, brick, and bone, have large ultimate strengths in compression but tend to fail easily in tension. The stress-strain relationships for these materials are linear in the compressive region but exhibit curvature in the tensile region.

A structural member that has a relatively small ratio of length to width tends to fail under extreme loads by *crushing.* However, a slender column will fail by *buckling.* Any irregularities in the shape of a column and the fact that a real load is never truly axial are important factors in determining the point at which buckling occurs. Consequently, safety factors used for compressive stresses in columns are usually larger than those used for tensile loads. ◀

Example 16–1

A structural rod consists of two coaxial cylindrical sections with diameters of 12 cm and 5 cm that are joined by a filleted surface, as shown in the diagram. The rod is made from cold-drawn 0.20-percent-carbon steel and the data in line 2 of Table 16–1 apply.

(a) Use a safety factor of 2.5 based on the proportional limit and determine the maximum safe axial load.

(b) What is the maximum stress in the rod when the maximum safe load is applied?

(c) What is the average stress in the larger diameter section when the maximum safe load is applied?

(d) Under what load would you expect failure?

Solution:

(a) If the concentration of stress at the discontinuity is well below the proportional limit, the maximum safe load can be determined from the average stress in the smaller diameter section. From Table 16–1, the proportional limit is 3.80×10^8 Pa; using Eq. 16–1 and introducing the safety factor $f = 2.5$, we have

$$F = \frac{s_n}{f} \cdot A = \frac{3.80 \times 10^8 \text{ Pa}}{2.5} \cdot \pi (0.025 \text{ m})^2$$

$$= 2.98 \times 10^5 \text{ N}$$

(b) The ratio r/D is $1/8$ for this rod, so the stress concentration factor is $k = 1.50$. Then, using Eq. 16–3, we find

$$s_m = k\frac{F}{A} = k\frac{s_n}{f} = \frac{1.5}{2.5} \times (3.80 \times 10^8 \text{ Pa})$$

$$= 2.28 \times 10^8 \text{ Pa}$$

This local concentrated stress is well below the proportional limit $(3.80 \times 10^8 \text{ Pa})$, so the omission of the stress concentration factor in (a) is justified.

(c) The average stress in the larger diameter section is

$$s_n = \frac{F}{A'} = \frac{2.98 \times 10^5 \text{ N}}{\pi (0.06 \text{ m})^2}$$

$$= 2.63 \times 10^7 \text{ Pa}$$

(d) According to Table 16–1, the ultimate strength of the steel from which the rod is made is 5.57×10^8 Pa. Failure will occur when the concentrated stress equals this value. Thus,

$$F = \frac{s_n}{k} \cdot A = \frac{(5.57 \times 10^8 \text{ Pa})}{1.50} \cdot \pi (0.025 \text{ m})^2$$

$$= 7.29 \times 10^5 \text{ N}$$

Fig. 16–6. The force pairs applied to the cube result in (a) tension and (b) shear. In the latter case the cube is not in equilibrium. (•What additional shear forces would produce an equilibrium condition?)

(a)

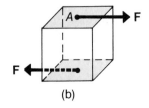

(b)

16-2 STRESS AND STRAIN

We now formulate the discussion of stress and strain in a way that permits us to analyze the *internal* forces that act within a stressed body. First, we define the types of stresses that enter the discussions. Figure 16-6a shows two equal forces applied perpendicular to opposite faces of a cube. These forces represent a pure *tension* and the corresponding normal stress is $s_n = F/A$. If the same forces were directed inward on the cube, they would represent pure *compression* and the corresponding stress would be $-F/A$. In Fig. 16-6b two equal forces are applied along opposite faces of a cube. These forces represent a pure *shear* and the corresponding tangential stress is $s_t = F/A$.

It sometimes happens that a force does not act exactly along or exactly perpendicular to an area. The area might be on the surface of an object supporting an applied force, or the area might be *within* an object, in which case the force would be an internal force. In either event, we can represent the force in terms of a component $\mathbf{F}_n$ that is normal to the area and a component $\mathbf{F}_t$ that is tangential to the area.

Figure 16–7a shows an irregular isolated object (which could be an internal section of a larger body). The orientation of the differential area ΔA on the surface of the object is specified by the outward-directed unit normal vector $\hat{\mathbf{n}}$ that is perpendicular to ΔA. A force $\Delta \mathbf{F}$ acts on the area ΔA, and Fig. 16–7b shows the normal component $\Delta \mathbf{F}_n$ and the tangential component $\Delta \mathbf{F}_t$ of this force.

The stress acting on the area ΔA likewise consists of two parts. The normal (or tensile) stress on ΔA is defined to be

$$s_n(\alpha) = \operatorname*{Lim}_{\Delta A \to 0} \frac{\Delta \mathbf{F} \cdot \hat{\mathbf{n}}}{\Delta A} \qquad (16\text{–}4)$$

and the tangential (or shear) stress acting on ΔA is defined to be

$$s_t(\alpha) = \operatorname*{Lim}_{\Delta A \to 0} \frac{|\Delta \mathbf{F} \times \hat{\mathbf{n}}|}{\Delta A} \qquad (16\text{–}5)$$

where the stresses depend on the angle α between $\hat{\mathbf{n}}$ and $\Delta \mathbf{F}$ (Fig. 16–7a).

Next, consider a rectangular rod that experiences a pure axial tension, as indicated in Fig. 16–8a. Construct an inclined section (shown as the shaded plane in the diagram) that makes an angle α with a perpendicular cross section of the rod. Figure 16–8b is the free-body diagram for the portion of the rod to the left of the inclined section. The force $\Delta \mathbf{F}$ that is exerted on the free body by the portion of the rod to the right of the section can be expressed in terms of components, $F_n = F \cos \alpha$ and $F_t = F \sin \alpha$.

The normal stress acting on the inclined area $A' = A/\cos \alpha$ is

$$s_n(\alpha) = \frac{F_n}{A'} = \frac{F}{A} \cos^2 \alpha = s_n \cos^2 \alpha \qquad (16\text{–}6)$$

and the tangential stress acting on A' is

$$s_t(\alpha) = \frac{|\mathbf{F}_t|}{A'} = \frac{F}{A} \sin \alpha \cos \alpha = \frac{1}{2} s_n \sin 2\alpha \qquad (16\text{–}7)$$

where $s_n = F/A$ is the stress on the exterior face.

(a)

(b)

Fig. 16–7. The force $\Delta \mathbf{F}$ acting on the area ΔA can be represented in terms of the normal component $\Delta \mathbf{F}_n$ and the tangential component $\Delta \mathbf{F}_t$. We sometimes refer to the *vector* differential area and write $\hat{\mathbf{n}} \, \Delta A$ or $\hat{\mathbf{n}} \, dA$.

Fig. 16–8. (a) A rectangular rod experiences pure axial tension due to equal forces applied to the ends of the rod. (b) A force F must act on the inclined section to hold the truncated free body in equilibrium.

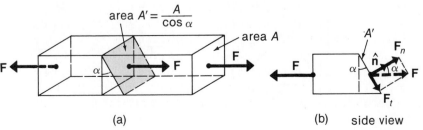

(a)

(b)   side view

Notice that the tensile stress $s_n(\alpha)$ is maximum when the shear stress $s_t(\alpha)$ is zero, namely, for $\alpha = 0$. Notice also that the shear stress is maximum for $\alpha = 45°$. (•Can you see how this follows?)

We have adopted the convention that the unit vector $\hat{\mathbf{n}}$ is directed *outward* from the isolated section (Fig. 16–7a). Then, for an outward directed force (i.e., tension), we have $s_n(\alpha) > 0$. Conversely, for an inward directed force (i.e., compression), we have $s_n(\alpha) < 0$. Thus, when an object is subjected to a fluid pressure, the normal stress is always negative. It is important to realize that the normal and the tangential forces that act on any surface within a stressed body depend on the *orientation* of that surface.

Figure 16–9 shows a solid cube under two different forms of loading that produce equilibrium. In each case, four forces with equal magnitudes are involved. In Fig. 16–9a the forces are applied in combined tension and compression. Now, imagine the cube to be sectioned by the plane PP'; then, the internal forces that act at this plane (shown in the first side view in Fig. 16–9a) are pure shear forces with magnitude $\sqrt{2}\,F$. (•Can you see why? What internal force must act on a section of the cube to produce equilibrium?)

Next, imagine the cube in Fig. 16–9a to be sectioned by the plane QQ'. In this case the internal forces that act at the plane are also pure shear forces, but they are directed in the opposite sense compared with those couples acting at the plane PP'. (•Do you recognize the action–reaction force pairs in Fig. 16–9?)

In Fig. 16–9b the externally applied forces place the cube in shear. The forces

Fig. 16–9. Two methods of applying four equal forces to a cube so that equilibrium results. The internal forces at the planes PP' and QQ' are shown for each case.

side views

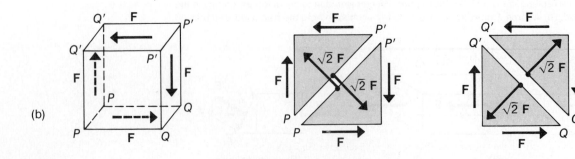

that act at the plane PP' are now pure tensile forces and those that act at the plane QQ' are pure compressional forces.

At either plane in Fig. 16–9b the normal stress is

$$s_n = \frac{\sqrt{2}\,F}{A'} = \frac{\sqrt{2}\,F}{\sqrt{2}\,A} = \frac{F}{A} = s_t$$

where s_t is the tangential stress on the exterior face. Similarly, the tangential stress at either plane in Fig. 16–9a is

$$s_t = \frac{\sqrt{2}\,F}{A'} = \frac{\sqrt{2}\,F}{\sqrt{2}\,A} = \frac{F}{A} = s_n$$

where s_n is the normal stress on the exterior face. Thus, the stresses on all the various faces and planes in this example have the same magnitude.

When a test rod is placed under axial tension, the resulting axial elongation is accompanied by a decrease in the rod's radius, as shown in Fig. 16–10. For stresses below the elastic limit, the decrease in radius is proportional to the tensile stress (for most materials).

The strain must now be described by *two* quantities. As before, we define the longitudinal (axial) strain to be

$$e_l = \frac{l' - l}{l} = \frac{\Delta l}{l} \qquad \text{(16–8)}$$

The transverse strain (perpendicular to the axis) is defined to be

$$e_{\text{tr}} = \frac{d' - d}{d} = \frac{\Delta d}{d} \qquad \text{(16–9)}$$

Fig. 16–10. A rod under axial tension stretches in length while decreasing in radius.

Notice that for pure tensile loads, $e_l > 0$ whereas $e_{\text{tr}} < 0$ (for pure compressive loads, the signs are reversed). For most materials the ratio e_{tr}/e_l is a constant, the negative of which is called Poisson's ratio*:

$$\sigma = -\frac{e_{\text{tr}}}{e_l} \qquad \text{(16–10)}$$

For most materials, σ lies between 0.2 and 0.4 (see Table 16–2).

The fractional change in the volume of the stressed material can be expressed as

$$\frac{\Delta V}{V} = \frac{(l + \Delta l)(d + \Delta d)^2 - ld^2}{ld^2}$$

$$= \left(1 + \frac{\Delta l}{l}\right)\left(1 + \frac{\Delta d}{d}\right)^2 - 1$$

---

*After Simeon Denis Poisson (1781–1840), French mathematician and physicist.

$$= \left(1 + \frac{\Delta\ell}{\ell}\right)\left(1 - \sigma\frac{\Delta\ell}{\ell}\right)^2 - 1$$

$$\cong \frac{\Delta\ell}{\ell}(1 - 2\sigma) = e_\ell(1 - 2\sigma) \tag{16-11}$$

where the terms containing $(\Delta\ell/\ell)^2$ and $(\Delta\ell/\ell)^3$ have been neglected. Note that a value $\sigma = 0.5$ would imply $\Delta V = 0$, that is, a volume-preserving deformation. No real material has a value of σ as large as 0.5. Equation 16–11, which does not depend on the geometry of the sample, is, in fact, a general result. Thus, the volume of any sample *increases* slightly under tension.

If the test rod were placed under compression, instead of tension, Eq. 16–10 would still be valid with, however, $\Delta\ell/\ell < 0$. Then, $\Delta V < 0$, which means that the volume *decreases* slightly under compression.

(a)   (b)

Fig. 16–11. Shear deformations of a solid cube. (a) Four equal forces produce an elongation of one diagonal and a shortening of the other. (b) Two equal forces applied to opposite faces produce a rotational strain; that is, every particle in the cube undergoes a small counterclockwise rotation.

Figure 16–11 shows two cases of a solid cube subjected to shear. In Fig. 16–11a the shear is the result of four equal forces applied tangentially to the faces of the cube. In Fig. 16–11b the shear is the result of two equal forces applied to opposite faces of the cube. In each case the shear strain is measured in terms of the angle γ, where

$$\gamma \cong \frac{\delta}{a} \tag{16-12}$$

16-3 ELASTIC MODULI

Young's Modulus. The direct proportionality between the normal stress and the strain in the region below the proportional limit (see Fig. 16–2) means that the material obeys Hooke's law. The constant of proportionality is called Young's modulus* Y, so that

$$s_n = Ye_\ell \tag{16-13}$$

---

*First used in this way by Thomas Young (1773–1829), an English physicist, known also for his pioneering experiments concerning the wave character of light.

For ductile materials, this relationship is valid for both tension and compression with the same value of Y. For composite materials, such as concrete and brick, the direct proportionality between s_n and e_ℓ holds only for compression. Remembering that $s_n = F_n/A$ and $e_\ell = \Delta\ell/\ell$, we can write

$$F_n = \frac{YA}{\ell}\Delta\ell = \kappa\,\Delta\ell \qquad (16\text{--}14)$$

We see that the Hooke's-law force constant κ is

$$\kappa = \frac{YA}{\ell} \qquad (16\text{--}15)$$

The Shear Modulus. Figure 16–11a shows the deformation of a cube due to a pure shear stress applied by four equal-magnitude forces. In Fig. 16–12 we show the same situation, with the addition of some construction lines and points. The shear angle γ is small, so the triangle CSQ' is approximately a 45°-45°-90° triangle. Then, $\overline{CQ'} = \sqrt{2}(\delta/4)$.

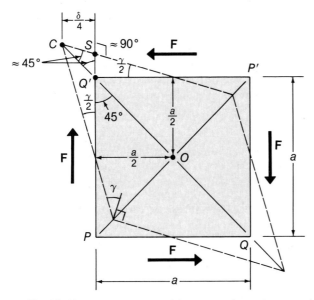

Fig. 16–12. A cube deformed by a pure shear stress.

The geometry of Fig. 16–12 is the same as that in Fig. 16–9b. Consequently, we again consider the applied shear forces to produce tensile forces that act along the direction of the diagonal $\overline{QQ'}$; the corresponding stress is $s_n = F/a^2$. Moreover, the compressional forces that act along the diagonal $\overline{PP'}$ may be considered to produce a compressional stress also with magnitude F/a^2. Both of these stresses produce an elongation along the diagonal $\overline{QQ'}$ (compare Fig. 16–11a). The longitudinal strain resulting from the tensile stress is $e_\ell = s_n/Y$. The compressional stress gives rise to a strain equal to σe_ℓ (see Eq. 16–10*). Thus,

$$\frac{\Delta\ell}{\ell} = \frac{\overline{CQ'}}{\overline{OQ'}} = \frac{s_n}{Y} + \sigma\frac{s_n}{Y}$$

---

*For the compressional stress, the elongation along $\overline{QQ'}$ is a *transverse* strain and is positive.

or
$$\frac{\sqrt{2}(\delta/4)}{\sqrt{2}(a/2)} = \frac{s_n(1 + \sigma)}{Y}$$

Then, we have

$$\gamma = \frac{\delta}{a} = \frac{2s_n(1 + \sigma)}{Y}$$

We define the *shear modulus* μ to be the proportionality constant connecting the shear stress s_t and the shear angle γ. In the discussion of the situation in Fig. 16–9b, which has the same geometry as the present case (Fig. 16–12), we argued that $s_t = s_n$. Therefore, we can now write

$$s_t = \mu\gamma \qquad (16\text{–}16)$$

with
$$\mu = \frac{Y}{2(1 + \sigma)} \qquad (16\text{–}17)$$

That is, Young's modulus Y and Poisson's ratio σ completely determine the shear modulus μ.* The shear modulus is sometimes called the *modulus of rigidity*.

   Compressibility and the Bulk Modulus.   Consider a cube with sides a that is subjected to equal normal compressive forces F_n on all six faces. The cube is then under a *pressure p*, where

$$p = \frac{\text{force}}{\text{area}} = -\frac{F_n}{a^2} = -s_n \qquad (16\text{–}18)$$

(Remember, a compressive stress is negative, so $s_n < 0$ and $p > 0$.) Considering one pair of forces acting on opposite faces, we can use Eqs. 16–11 and 16–13 to write

$$\frac{\Delta V}{V} = e_t(1 - 2\sigma) = \frac{s_n}{Y}(1 - 2\sigma)$$

The total effect of all six forces is three times larger:

$$\frac{\Delta V}{V} = \frac{3s_n}{Y}(1 - 2\sigma)$$

$$= -\frac{3p}{Y}(1 - 2\sigma) \qquad (16\text{–}19)$$

The quantity $3(1 - 2\sigma)/Y$ is called the *compressibility* λ. Thus

$$\frac{\Delta V}{V} = -\lambda p \qquad (16\text{–}20)$$

---

*Real materials do not obey Eq. 16–17 precisely. Consequently, you will find some discrepancies in Table 16–2 if you use two of the three quantities in Eq. 16–17 to calculate the third.

TABLE 16-2. Elastic Constants[a]

MATERIAL	YOUNG'S MODULUS Y (10^{10} Pa)	SHEAR MODULUS μ (10^{10} Pa)	BULK MODULUS β (10^{10} Pa)	POISSON'S RATIO σ[b]
Aluminum	7.0	2.6	7.8	0.35
Brass	9.0	3.5	6.8	0.28
Copper	12.4	4.5	13.1	0.36
Duraluminum[c]	6.9	2.8	4.6	0.25
Glass	5.5	2.3	3.1	0.20
Gold	7.9	2.8	16.6	0.41
Iron, wrought	19.0	7.6	12.7	0.25
Lead	1.6	0.54	5.2	0.45
Nickel	20.7	7.3	43.1	0.42
Platinum	16.7	6.4	13.9	0.30
Quartz	8.2	3.0	10.5	0.37
Steel (0.30% C)	20.0	8.1	12.3	0.23
Steel (0.60% C)	19.6	8.0	12.1	0.23
Tungsten	35.5	14.8	19.7	0.20

[a] These values are approximate; the exact values depend critically on the preparation of the particular sample.

[b] Average value.

[c] Aluminum alloy containing 4 percent Cu, 0.7 percent Mn, and 0.5 percent Mg.

The reciprocal of the compressibility is called the *bulk modulus* β:

$$\beta = \frac{1}{\lambda} = -p\frac{V}{\Delta V} = \frac{Y}{3(1 - 2\sigma)} \qquad (16\text{-}21)$$

An object under pressure will *decrease* in volume, so $\Delta V/V < 0$. Thus, in Eq. 16-21 we must have $\sigma < \frac{1}{2}$. The fact that work is required to compress an object also leads to $\sigma < \frac{1}{2}$. (•Can you see why?)

Values of the elastic moduli for some common materials are given in Table 16-2.

Example 16-2

A copper block has dimensions $a = 8$ cm and $b = 10$ cm, as indicated in the diagram on the next page. Two opposite square sides are subjected to equal compressive forces $F = 4 \times 10^4$ N.

(a) What is the fractional decrease in the distance between the stressed faces?
(b) By what fractional amount is the transverse area increased?
(c) What is the fractional change in volume?
(d) What are the normal and shear stresses on the (shaded) diagonal plane?

Solution:

(a) From Table 16-2 we find $Y(\text{Cu}) = 12.4 \times 10^{10}$ Pa. Then, using Eq. 16-14 (with $\ell = a$ and $A = b^2$), we have

$$\frac{\Delta a}{a} = \frac{-F}{Yb^2} = \frac{-4 \times 10^4 \text{ N}}{(1.24 \times 10^{11} \text{ Pa})(0.10 \text{ m})^2} = -3.23 \times 10^{-5}$$

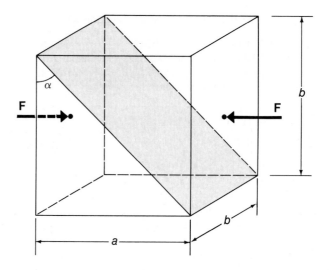

(b) The fractional increase in the transverse area is

$$\frac{\Delta A}{A} = \frac{(b + \Delta b)^2 - b^2}{b^2} \cong \frac{2 \, \Delta b}{b}$$

Now, $\sigma(\text{Cu}) = 0.36$, and from Eq. 16-10 we have

$$\frac{\Delta b}{b} = -\sigma \frac{\Delta a}{a}$$

so that

$$\frac{\Delta A}{A} = -2\sigma \frac{\Delta a}{a} = -2(0.36)(-3.23 \times 10^{-5})$$

$$= 2.32 \times 10^{-5}$$

(c)

$$\frac{\Delta V}{V} = \frac{(b + \Delta b)^2 (a + \Delta a) - ab^2}{ab^2}$$

$$\cong 2 \frac{\Delta b}{b} + \frac{\Delta a}{a} = \frac{\Delta a}{a}(1 - 2\sigma) \qquad \text{(see Eq. 16-11)}$$

$$= (-3.23 \times 10^{-5})(1 - 2 \cdot 0.36)$$

$$= -9.04 \times 10^{-6}$$

(d) The angle α is equal to $\tan^{-1}(a/b) = \tan^{-1} 0.8 = 38.66°$. Comparing with Fig. 16-8b and using Eqs. 16-6 and 16-7, we see that

$$s_n(\alpha) = \frac{-F}{b^2} \cos^2 \alpha = \frac{-4 \times 10^4 \, \text{N}}{(0.10 \, \text{m})^2} \cos^2 38.66°$$

$$= -2.44 \times 10^6 \, \text{Pa}$$

$$s_t(\alpha) = \frac{F}{2b^2} \sin 2\alpha = \frac{4 \times 10^4 \, \text{N}}{2(0.10 \, \text{m})^2} \sin 2(38.66°).$$

$$= 1.95 \times 10^6 \, \text{Pa}$$

SPECIAL TOPIC

16-4 APPLICATIONS OF ELASTICITY

We now consider two important applications of elasticity, namely, torsion in a rod and the bending of a beam. Both of these cases involve nonuniform (inhomogeneous) stresses.

Torsion. Consider a cylindrical rod with a radius R and a length ℓ, the upper end of which is clamped in a fixed position. A torque τ is applied to the lower end of the rod, as shown in Fig. 16–13a. This causes the longitudinal surface fiber $\overline{PQ}$ to twist through an angle ϕ with respect to the cylinder axis. The angle γ_R represents the shear strain at the outer surface ($r = R$). Figure 16–13b shows an internal portion of the rod between radii r and $r + dr$, with axial length $d\ell$. The distortion of the rectangular column clearly indicates that the rod is in shear.

At the outer surface of the rod, we have

$$\gamma_R = \frac{\overline{QQ'}}{\overline{PQ}} = \frac{\delta}{\ell} = \frac{R}{\ell}\phi$$

For a rod we usually have $R \ll \ell$, so ϕ can be large even though γ_R is small. Now, the shear angle is proportional to the radius (•Can you see why?), so we can write

$$\gamma(r) = \frac{r}{R}\gamma_R = \frac{r}{\ell}\phi$$

The shear stress s_t on the shaded areas in Fig. 16–13b is also a function of r. Using Eq. 16–16, we have

$$s_t(r) = \mu\gamma(r) = \mu\frac{r}{\ell}\phi$$

This stress on the area $r\,dr\,d\alpha$ corresponds to a tangential force (in this case, a differential of second order) given by

$$d^2F = s_t(r)r\,dr\,d\alpha$$

The total tangential force on the differential annular area is

$$dF = \int_0^{2\pi} d^2F = \int_0^{2\pi} [s_t(r)r\,dr]\,d\alpha$$

$$= 2\pi s_t(r)r\,dr = 2\pi\frac{\mu\phi}{\ell}r^2\,dr$$

Fig. 16–13. (a) A cylindrical rod is subjected to a torque τ. The longitudinal surface fiber $\overline{PQ}$ is twisted to the new position $\overline{PQ'}$, producing a shear angle γ_R. (b) An enlarged annular section of the cylinder, showing the shear of the region $(r\,d\alpha)(dr)(d\ell)$. (c) The stress s_t acting on the face area $r\,dr\,d\alpha$ associated with the tangential force $d^2\mathbf{F}$.

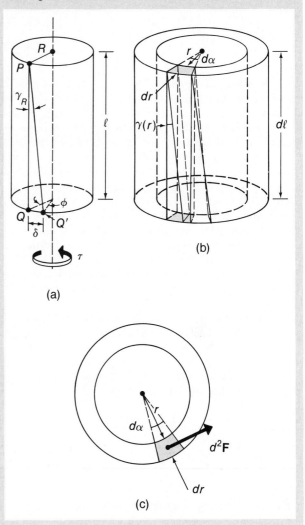

This force produces a torque about the cylinder axis given by

$$d\tau = r\,dF = 2\pi\frac{\mu\phi}{\ell}r^3\,dr$$

The twisted rod is in equilibrium, so the integral of this

internal torque must equal the applied torque. Thus,

$$\tau = 2\pi \frac{\mu\phi}{\ell} \int_0^R r^3 \, dr$$

from which

$$\tau = \frac{\pi\mu R^4}{2\ell}\phi \qquad \textbf{(16–22)}$$

The twist angle ϕ is proportional to the applied torque τ, and the torsional constant that was introduced in Eq. 15–24 is

$$\Gamma = \frac{\pi\mu R^4}{2\ell} \qquad \textbf{(16–23)}$$

Bending of Beams. Figure 16–14a shows a beam that is supported at its ends, S and S', and bears a concentrated load L at an arbitrary position. The beam deflects and reaches an equilibrium condition with some complicated shape. In doing so, the upper surface of the beam comes under compression and contracts, whereas the lower surface comes under tension and elongates. Between these regions of opposite response to the load there must be a surface NN', called the *neutral surface,* along which there is zero stress and zero strain. Although this surface has a complicated shape, we can approximate a differential section by a cylindrical surface with a radius of curvature R, as shown in Fig. 16–14a and enlarged in Fig. 16–14b. This latter view also shows the longitudinal stresses and how they change from compressive to tensile at the neutral surface.

Within the isolated section of the beam (shaded in Fig. 16–14a) consider a longitudinal strip located a distance ξ from the neutral surface and with a differential thickness $d\xi$ (Fig. 16–14b). The length of this strip is ℓ at the neutral surface and $\ell + d\ell$ at the distance ξ. From the geometry of Fig. 16–14b we see that

$$\frac{\ell + d\ell}{\ell} = \frac{R + \xi}{R}$$

or

$$\frac{d\ell}{\ell} = \frac{\xi}{R}$$

The corresponding longitudinal stress is (Eq. 16–13)

$$s_n = Ye_\ell = Y\frac{d\ell}{\ell} = Y\frac{\xi}{R} \qquad \textbf{(16–24)}$$

This linear dependence of the stress on the distance from the neutral surface is illustrated schematically in Fig. 16–14b by the sequences of arrows representing the stress.

The end faces of the longitudinal strip have area $w\,d\xi$, where w is the width of the strip at the distance ξ (see Fig.

Fig. 16–14. (a) A horizontal supported homogeneous beam is subjected to a load L. The neutral surface is NN'. (b) Enlarged view of the isolated shaded section in (a), showing the longitudinal stresses. (c) Enlarged end view of the same section.

16–14c).The force on each end face is

$$dF = s_n w \, d\xi = \frac{Yw}{R} \xi \, d\xi$$

This force produces a *bending moment* or torque $d\tau$ about an axis through the neutral surface given by (refer to Fig. 16–14c)

$$d\tau = \xi \, dF = \frac{Yw}{R} \xi^2 \, d\xi$$

so that the total bending moment is

$$\tau = \frac{Y}{R} \int w\xi^2 \, d\xi \qquad \textbf{(16–25)}$$

In general, the width w of the beam will be a function of ξ depending on the design shape (see Fig. 16–14c). The integral in Eq. 16–25 is recognized as the second moment of the transverse area of the beam about an axis through the neutral surface NN' (refer to Eqs. V–1). Equation 16–25 is the *flexure formula* for beams and represents the starting point for any calculation of beam deflection for a particular loading condition. (•Can you see from Eq. 16–25 why structural beams have the cross-sectional shape of an "I"?)

The evaluation of the integral in Eq. 16–25 requires the location of the neutral surface NN'. We can determine the position of NN' in the following way. In Fig. 16–14a consider the portion of the beam $ABCD$. Because an equilibrium condition exists, the total force on the face CD of the transverse section must be zero. Thus,

$$F = \int dF = \frac{Y}{R} \int w\xi \, d\xi = 0 \qquad \textbf{(16–26)}$$

That is, the first moment of the transverse area of the beam about an axis through the neutral surface is zero. Therefore, the neutral surface passes through the geometric center (or centroid) of the area.

Example 16–3

A beam with a length ℓ and with negligible mass is cantilevered (fixed at one end) and supports at its free end a concentrated load in the form of a block with a mass M, as shown in the diagram. When unloaded, the beam is horizontal. Determine the vertical deflection of the end of the loaded beam.

Solution:

The portion of the beam to the right of the point $P(x, y)$, including the load M, is shown in the free-body

(a)   (b)

diagram (b). The top fibers of the beam are in tension and the bottom fibers are in compression. The corresponding internal forces acting on the cross section of the beam at P produce a counterclockwise torque τ' which just balances the torque (also about P) due to the load, namely, $\tau' = (\ell - x)Mg$. The torque τ that acts on the cross section of the beam to the left of P is clockwise. Thus,

$$\tau = -\tau' = -(\ell - x)Mg$$

If the deflection of the beam is small, we can write for the radius of curvature,* $R \cong (d^2y/dx^2)^{-1}$. Then, using Eq. 16–25, we have

$$\tau = -(\ell - x)Mg = Y\frac{d^2y}{dx^2}\mathcal{I}$$

where $\qquad \mathcal{I} = \int w\xi^2 \, d\xi$

In the present case the center of curvature lies below the beam and $d^2y/dx^2 < 0$ throughout the length of the beam. Thus,

$$\frac{d^2y}{dx^2} = -\frac{Mg}{Y\mathcal{I}}(\ell - x)$$

from which $\quad \dfrac{dy}{dx} = -\dfrac{Mg}{Y\mathcal{I}}(\ell x - \tfrac{1}{2}x^2 + C_1)$

For the cantilevered beam, $dy/dx = 0$ at $x = 0$, so $C_1 = 0$. Then,

$$y(x) = -\frac{Mg}{Y\mathcal{I}}(\tfrac{1}{2}\ell x^2 - \tfrac{1}{6}x^3 + C_2)$$

At $x = 0$ we have $y = 0$, so $C_2 = 0$. Finally, for $x = \ell$, we have the required vertical deflection,

$$y(\ell) = -\frac{Mg\ell^3}{3Y\mathcal{I}}$$

---

*It is shown in calculus texts that the radius of curvature of the curve represented by the function $y = y(x)$ is $R = [1 + (dy/dx)^2]^{3/2}/(d^2y/dx^2)$. If the deflection is small, we have $dy/dx \ll 1$ and $R \cong (d^2y/dx^2)^{-1}$.

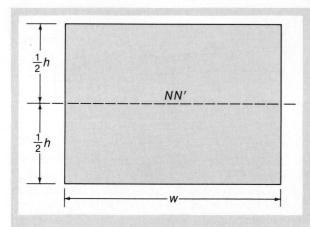

If the beam has a rectangular cross section, the neutral surface passes through the center of the beam as shown in the diagram. Thus,

$$g = w \int_{-h/2}^{+h/2} \xi^2 \, d\xi = \tfrac{1}{12} w h^3$$

For a brass bar with $\ell = 1\,\text{m}$, $w = 1\,\text{cm}$, and $h = 2\,\text{cm}$, and with a load $M = 2\,\text{kg}$, the deflection is

$$y(\ell) = \frac{4Mg\ell^3}{Ywh^3} = \frac{4(2\,\text{kg})(9.8\,\text{m/s}^2)(1\,\text{m})}{(9.0 \times 10^{10}\,\text{Pa})(0.01\,\text{m})(0.02\,\text{m})^3}$$
$$= 1.09\,\text{cm}$$

PROBLEMS

Section 16-1

16-1   What is the maximum axial tensile load that can be applied to a cold-drawn 0.20-percent-C steel rod that has a diameter of 8 mm?

16-2   An 800-kg load is being raised by a force applied to an annealed 0.30-percent-C steel rod with a diameter of 1.2 cm. What is the maximum upward acceleration that can be tolerated if a safety factor of 2.5 based on the proportional limit is used?

Section 16-3

16-3   A wire with a length of 3 m and a diameter of 1 mm is stretched 1.5 mm when a block with a mass of 5.0 kg is suspended from one end, the other end being attached to a rigid ceiling. What is Young's modulus for the material of the wire?

16-4   A simple pendulum consists of 5 m of No. 20 copper wire (diam. = 0.08118 cm) to which is attached a 1-kg bob. The period of the pendulum is measured. Next, a 10-kg bob is substituted for the 1-kg bob. By what amount does the period change? (Consider both bobs to be particles.)

16-5   Two 1-m lengths of wire, one of copper and one of tungsten, are joined end-to-end. The copper wire has a diameter of 0.50 mm. When a 100-kg block is suspended from one end, the combined length of wire stretches by 6.00 cm. What is the diameter of the tungsten wire?

16-6   A glass sphere 20 cm in diameter is subjected to a pressure of 4×10^7 Pa. By what amount does the radius of the sphere decrease?

16-7   A sandstone block in the shape of a cylinder with a height of 1 m and a diameter of 60 cm is subjected to a compressional load uniformly distributed over the top surface. The ultimate strength of the material is 4.6×10^7 Pa in compression and 1.16×10^7 Pa in shear. At what load will the block fail? [*Hint:* First, determine the orientation of the plane that will fail under the least stress.]

16-8   Two flat plates, 8 cm wide, are bolted together with two 1-cm diameter bolts, as shown in the diagram below. This joint is subjected to two longitudinal forces F. The plate material has an ultimate strength of 4.45×10^8 Pa in tension and the bolt material has an ultimate strength of 3.08×10^8 Pa in shear.

top view

side view

Determine the minimum thickness of the plates if the plates are not to fail in tension (across the plane TT') before the bolts fail in shear. Use a stress concentration factor $k = 2.0$ in the plates because of the presence of the bolt holes.

16-9 By how much will the density of an aluminum block change if it is subjected to a pressure of 2×10^8 Pa on all surfaces? (Refer to Table 1-2.)

Section 16-4

16-10 A steel rod has a length of 0.50 m, a diameter of 1 cm, and a shear modulus $\mu = 8.0 \times 10^{10}$ Pa. Calculate the twist angle ϕ and the strain angle γ_R for an applied torque of 100 N · m.

16-11 The integral that leads to the torque equation (Eq. 16-22) depends on r^3. This suggests that the rim portions of a rod provide the major contribution to the torque. (a) Show that a cylindrical hollow pipe with inner radius R_1 and outer radius R_2 results in a torque $\tau = \pi\mu(R_2^4 - R_1^4)\phi/2l$ (b) Given a hollow pipe and a solid rod made from the same material and both with the same mass per unit length, determine the ratio of the twist angles produced by the application of the same torque to each object. Do you agree with the engineering maxim, "Put the material where the stresses are!"?

16-12 A horizontal cantilevered steel beam has a
•   width $w = 5$ cm, a thickness $h = 2$ cm, and an un-

supported length $l = 80$ cm. The mass of the beam is negligible in comparison with the concentrated load of 300 kg that is placed on the free end. Young's modulus for the steel is 20×10^{10} Pa. (a) Calculate the deflection of the loaded end of the beam. (b) Where is the tensile stress in the beam a maximum? (c) Determine the magnitude of the maximum tensile stress in the beam.

16-13 A horizontally cantilevered beam carries no extra
••   load as in Example 16-3. Instead, the load is the mass M of the beam itself, which is distributed uniformly along the beam's length. (a) Show that the deflection of the unsupported end is $y(l) = -Mgl^3/8Yg$. [*Hint:* The mass of the beam to the right of the point $P(x, y)$ is $M(l - x)/l$.] (b) Determine the deflection of a steel beam with $l = 80$ cm, $w = 5$ cm, $h = 2$ cm, $\rho = 7.86$ g/cm³, and $Y = 20 \times 10^{10}$ Pa. (If you have solved Problem 16-12, compare the two deflections and determine whether the neglect of the mass of the beam in the previous problem was justified.)

Additional Problems

16-14 A steel rod for which $Y = 20.0 \times 10^{10}$ Pa rests on a horizontal surface. The diameter of the rod is 4.0 cm and the length is 6.0 m. By how much will the length of this rod decrease when it is stood on end? (The density of the steel is 7.90 g/cm³.)

16-15 A 2-m ideally rigid rod is suspended from a ceiling by two 1-m wires, as shown in the diagram. Wire A is a copper wire with a diameter of 1.6 mm; wire B is a brass wire with a diameter of 1.0 mm. Initially, the rod is precisely horizontal. (a) If an 80-kg block is hung from C, the midpoint of the rod, what angle will the rod make with the horizontal? (b) Locate the point D from which the block must be hung if the rod is to remain precisely horizontal.

16-16 A force of 4.0 N applied along the top face of a 5-cm cube of Jello that sits on a plate results in a

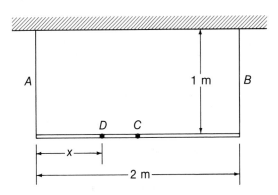

displacement $\delta = 5$ mm of a top edge, as shown in the diagram on the next page. (a) What is the shear modulus μ for Jello? (b) Jello consists mainly of water (which is essentially incompressible), so $\sigma \cong \frac{1}{2}$. What is Young's modulus for Jello?

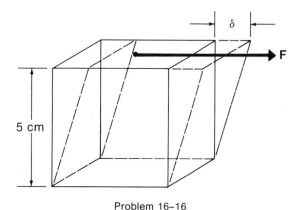

Problem 16–16

16–17 A block with a mass of 1.2 kg is attached to the lower end of an aluminum wire whose upper end is fixed. The length of the wire is 2 m and its diameter is 1 mm. (a) Determine the period of small-amplitude vertical oscillations of the block. (b) What is the greatest mass that the wire can support without exceeding the proportional limit of 1.2×10^8 Pa?

(c) What mass should be attached to the wire to give the maximum amplitude for SHM, noting that the wire cannot sustain compression? (•Do you see the reason for this restriction?) (d) Determine the maximum amplitude for SHM with the 1.2-kg block attached.

16–18 A cable made from annealed 0.30-percent-carbon steel is designed to support a loaded elevator with a mass of 2000 kg during a maximum upward acceleration of 1.6 m/s². Assume that the values of the various properties of steel listed in Tables 16–1 and 16–2 apply to the cable. (a) What is the minimum diameter of the cable? (Use a safety factor of 3, based on the proportional limit.) (b) At what acceleration will an overloaded cab with a mass of 10,000 kg cause the cable to break? (Assume that the cable has the diameter found in (a) and that the break will occur when the ultimate strength of the steel is reached.) (c) By what amount will a 50-m cable stretch if the cab (with a mass of 1600 kg) accelerates at 1.6 m/s²? (Assume the cable diameter found in (a).)

SPECIAL TOPIC

16–19 The main propeller shaft of a yacht has a length of 4.6 m and a diameter of 8 cm. The shaft is made from 0.30-percent-carbon steel. What is the twist angle ϕ of the shaft when it delivers a constant power of 500 kW (670 hp) at 200 rpm?

16–20 A flat disc with a mass of 100 g and a diameter of 10 cm is suspended horizontally by a 1-m quartz fiber that is attached to the center of the disc. The disc is observed to undergo torsional oscillations with a period of 300 s. Find the torsional constant and the diameter of the quartz fiber.

16–21 A horizontally cantilevered diving board has a length $\ell = 3$ m, a width $w = 35$ cm, and a thickness $h = 4$ cm. A student whose mass is $M = 65$ kg stands at the free end of the board and causes small-amplitude bending oscillations. Assume that Young's modulus for the board is $Y = 1.4 \times 10^{10}$ Pa. Neglect the mass of the board and show that the period of the oscillations is $T = 2\pi \sqrt{M\ell^3/3Y\mathcal{I}}$. Evaluate T.

16–22 A horizontally cantilevered I-beam with a length of 2 m supports a concentrated load of 2000 kg at the free end. The standard 10-cm I-beam has the following properties:

$t = 1.0$ cm
$w = 7.29$ cm
cross-sectional area = 19.68 cm²
$\mathcal{I}_{11} = 295$ cm⁴
$\mathcal{I}_{22} = 42$ cm⁴
mass = 15.7 kg/m
Assume that Young's modulus is $Y = 20 \times 10^{10}$ Pa.
(a) Which way would you orient the beam to

achieve the smallest vertical deflection? (b) In the case (a), what is the deflection? (c) Another cantilevered beam made from the same material and with the same length but with a square cross section is to be used to support the same load and give the same deflection. What is the width of this beam? (d) Compare the amount of steel required to make the two beams (i.e., compare the cross-sectional areas). Does this indicate that the I-beam is a useful engineering concept?

STATIC FLUIDS

Matter in the solid state generally offers considerable resistance to all changes in shape. Liquids and gases, in contrast, do not have rigid structure or form. These states of matter—which together we call *fluids*—are easily altered in shape. Liquids generally have low compressibility so they deform in shape without appreciable change in volume. Gases, on the other hand, readily change volume and expand to fill completely any container. Neither liquids nor gases can permanently sustain a shearing stress.

In this chapter and in the next, we examine the properties of fluids that flow readily, such as water and other liquids with low viscosity and including gases under ordinary pressures. We exclude highly viscous substances, such as pitch, tar, wax, and glass, all of which flow so slowly that they usually do not deform significantly during the time intervals of interest in the discussions here.

17-1 STRESSES IN A FLUID AT REST

A fluid is a substance that offers no resistance to slowly applied forces that change the shape of the substance while conserving its volume. Figure 17–1 shows a small parallelepiped of a fluid that is isolated as a free body within a larger volume of the fluid. The surrounding fluid exerts forces on the parallelepiped and slowly deforms it into the shape shown by the dashed line. The fluid offers no resistance to this volume-conserving deformation, so the forces do no work on the fluid. Therefore, the forces that act on the top and bottom surfaces of the parallelepiped (the shaded surfaces in Fig. 17–1) can have no tangential components. (Notice that this

fluid

volume-
conserving
deformation

Fig. 17–1. The deformation of an isolated parallelepiped of fluid within a large volume of fluid.

agrees with the statement that a fluid at rest can sustain no shearing stress.) Thus, the only forces that act on the shaded surfaces must be perpendicular to those surfaces. This conclusion is not altered by the orientation of the parallelepiped. It follows that *a stress in a fluid exerts only a normal force on any area located within the fluid*. The force per unit area is the *pressure, p = F/A* (see Eq. 16–18). Pressure is measured in N/m² or Pa.

In these discussions, we again find it convenient to use a vector representation of a differential area (compare Fig. 16–7). We write dA for the magnitude of the differential area, which is a part of a closed surface, as indicated in Fig. 17–2. The unit normal vector $\hat{\mathbf{n}}$ is perpendicular to the area, so we have

$$\text{Vector differential area} = \hat{\mathbf{n}}\, dA \qquad (17\text{–}1)$$

where we take the positive sense of $\hat{\mathbf{n}}$ to be outward from the surface. If the external pressure at the position of the differential area is p, the force exerted on the area dA is

$$d\mathbf{F} = -\hat{\mathbf{n}}\, p\, dA \qquad (17\text{–}2)$$

The pressure in a fluid may vary from point to point. If an isolated volume of fluid, such as that shown in Fig. 17–2, is at rest, the integral of $d\mathbf{F}$ over the entire surface must vanish.

Consider the triangular area ABC shown in Fig. 17–3. The unit vector $\hat{\mathbf{n}}$ that is normal to this area makes angles α, β, and γ with respect to the coordinate axes. The area of the projection of ABC onto the x-y plane (that is, the area of the triangle AOB) is

$$\Delta(AOB) = \Delta(ABC)\cos\gamma \qquad (17\text{–}3a)$$

(•Can you see why? Construct perpendicular lines from points O and C to the line $\overline{AB}$; these are the altitudes of the two triangles.) Similarly,

$$\Delta(AOC) = \Delta(ABC)\cos\beta \qquad (17\text{–}3b)$$

$$\Delta(BOC) = \Delta(ABC)\cos\alpha \qquad (17\text{–}3c)$$

Fig. 17–2. A differential area of the surface of an isolated fluid volume is represented by $\hat{\mathbf{n}}\, dA$, where $\hat{\mathbf{n}}$ is the outward directed unit vector normal to the plane of dA.

Fig. 17–3. The triangular area ABC has a unit normal vector $\hat{\mathbf{n}}$. The area of ABC projected onto the x-y plane is $\Delta(AOB) = \Delta(ABC)\cos\gamma$.

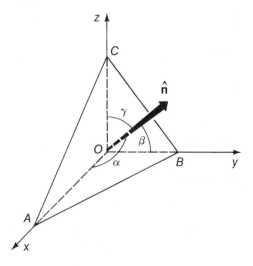

If the area ABC is located in a fluid at a position where the pressure is p, the force on the area is

$$\mathbf{F} = -p\,\Delta(ABC)\hat{\mathbf{n}} \tag{17-4}$$

The z-component of this force is $F_z = F\cos\gamma$, or

$$F_z = -p\,\Delta(ABC)\cos\gamma = -p\,\Delta(AOB) \tag{17-5}$$

where we have used Eq. 17–3a. Similar results follow for the components F_x and F_y. We can now draw an important conclusion. Suppose that we wish to find the component in a particular direction of the force produced by the pressure in a fluid acting on an arbitrary surface, such as the triangle $\Delta(ABC)$ in Fig. 17–3. This force component is equal to the pressure multiplied by the area of the projection of that surface onto the plane perpendicular to the force component.

The volume $AOBC$ in Fig. 17–3 can be considered to be a free body of fluid surrounded by fluid at a pressure p. Then, the requirement of equilibrium leads directly to Eq. 17–5 for F_z and to its counterparts for the x- and y-components. (•Can you see why?)

The result we have obtained is also valid for curved surfaces. For example, consider a hemisphere that is immersed in a fluid with a uniform pressure p, as shown in Fig. 17–4. The projected area of the hemisphere is πR^2, so the x-component of the fluid force on the curved surface is $F_x = -\pi R^2 p$. This result can be established by equilibrium considerations or by direct integration (see Problem 17–4). Because of the symmetry of the hemisphere about the x-axis as drawn in Fig. 17–4, the x-component of the force, $\mathbf{F}_x$, is the *total* force acting on the hemispherical surface. (•Can you see why?)

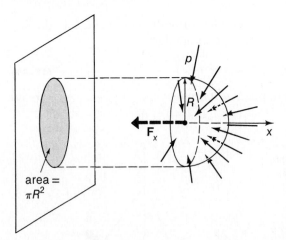

Fig. 17-4. The x-component of the fluid force on the hemispherical surface is $F_x = -\pi R^2 p$.

Example 17-1

The pressure on a small plane area $ABCD$ within a fluid is 500 N/m². The direction of the unit vector of the area is specified by the angles $\alpha = 90°$, $\beta = 60°$, $\gamma = 30°$, as shown in the diagram. What are the x-, y-, and z-components of the external fluid force exerted on $ABCD$? ($\overline{AB} = 0.1$ m and $\overline{AD} = 0.2$ m.)

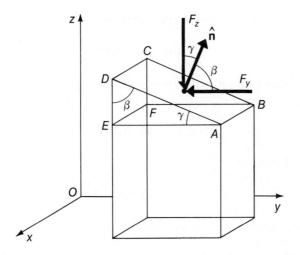

Solution:

The z-component of the force is equal to the pressure multiplied by the projected area $ABFE$:

$$F_z = -p(ABFE) = -p(ABCD)\cos \gamma$$
$$= -(500\ \mathrm{N/m^2})(0.1\ \mathrm{m} \times 0.2\ \mathrm{m})\cos 30°$$
$$= -8.66\ \mathrm{N}$$

Similarly, we find

$$F_y = -p(CDEF) = -p(ABCD)\cos \beta$$
$$= -(500\ \mathrm{N/m^2})(0.1\ \mathrm{m} \times 0.2\ \mathrm{m})\cos 60°$$
$$= -5.00\ \mathrm{N}$$

Because $\alpha = 90°$, we have $F_x = 0$.

Units of Pressure. The SI unit of pressure is the *pascal* (Pa):

$$1\ \mathrm{Pa} = 1\ \mathrm{N/m^2}$$

Also acceptable in the SI system are two other units. *Standard atmospheric pressure* (1 atm) is defined to be

$$1\ \mathrm{atm} = 101,325\ \mathrm{Pa}$$

which is the average pressure due to the atmosphere at sea level. The other unit is the *bar* (for "barometric pressure"):

$$1\ \mathrm{bar} = 10^5\ \mathrm{Pa}$$

from which 1 atm = 1.013 bar.

Another (non-SI) pressure unit that is often used is the *torr* (named for Evangelista Torricelli, who invented the mercury barometer), which is the pressure exerted by a column of mercury 1 mm in height:

$$1 \text{ atm} = 760 \text{ torr}$$

In some engineering applications pressure is still measured in *pounds per square inch* (lb/in.² or psi),

$$1 \text{ atm} = 14.70 \text{ lb/in.}^2$$

and in inches of mercury,

$$1 \text{ atm} = 29.92 \text{ in. Hg}$$

Meteorological reports in the United States often give atmospheric pressures in units of inches of mercury.

Compressibility of Liquids. When a pressure is applied to a confined liquid, the volume will actually decrease by a small amount. The *compressibility* λ of a substance is defined to be the fractional change in volume per unit of applied pressure. We have (compare Eq. 16–20)

$$\lambda = -\frac{1}{V}\frac{dV}{dp} \qquad (17\text{–}6)$$

Table 17–1 lists values of λ for various liquids. We leave for later chapters the discussion of the compressibility of gases.

TABLE 17–1.   Compressibility of Liquids (at 20°C)

LIQUID	$\lambda = -(1/V)\,dV/dp$	
	$(10^{-11} \text{ Pa}^{-1})$	$(10^{-6} \text{ atm}^{-1})$
Acetone	125	127
Benzene	93	94
Ether	187	189
Glycerine	21	21
Mercury	3.7	3.8
Water	46	47

Although liquids can be compressed, the fractional volume changes are relatively small even for pressures of several hundred atmospheres. Accordingly, when we discuss the properties of liquids, we usually consider them to be incompressible. Conversely, the compressible nature of gases is important in many situations.

Example 17–2

What is the fractional change in the density of sea water between the surface (where the pressure is 1 atm) and a depth of 5.2 km (where the pressure is 500 atm)?

Solution:

The density is $\rho = M/V$, so $d\rho = -M\,dV/V^2$; then dividing by $\rho = M/V$ gives

$$\frac{d\rho}{\rho} = -\frac{dV}{V}$$

and using Eq. 17–6 we have

$$\frac{\Delta\rho}{\rho} = \lambda\,\Delta p = [47 \times 10^{-6}\,(\text{atm})^{-1}] \times (500\,\text{atm} - 1\,\text{atm})$$

$$= 0.025 \text{ or } 2.5 \text{ percent}$$

Even at the tremendous pressure of 500 atm, the effect on the density is sufficiently small that it can be neglected for most purposes.

17-2 PRESSURE WITHIN A FLUID

We now consider the pressure within a fluid column due to the gravitational force of the Earth acting on the fluid. Two cases are of importance, namely, incompressible liquids and compressible gases.

Hydrostatic Pressure. Consider a tank that is filled with a liquid, as shown in Fig. 17–5. Select as a free body a thin horizontal slab of liquid at a depth z below the surface. The slab has a thickness dz and an area A. Let the pressure at depth z be p and the pressure at depth $z + dz$ be $p + dp$. The slab is in equilibrium, so the force equation in the vertical direction is (with downward positive)

$$-(p + dp)A + \rho g A\,dz + pA = 0$$

Simplifying, we obtain $\qquad dp = \rho g\,dz$

We assume that the fluid is incompressible so that ρ is constant (see Example 17–2). Then, upon integrating, we find

$$p(z) = p_0 + \rho g z \qquad\qquad \textbf{(17–7)}$$

where the integration constant is p_0, the pressure on the liquid surface ($z = 0$); if the surface is exposed to air, p_0 corresponds to atmospheric pressure. We see that the pressure within a liquid depends linearly on the depth.

If the tank is closed at the top by a piston that exerts a pressure p_a on the surface of the liquid, the constant p_0 in Eq. 17–7 is replaced by the applied pressure p_a. That

Fig. 17–5. A small horizontal slab of fluid within a tank of fluid is isolated as a free body.

Fig. 17–6. A small horizontal slab of air within the atmosphere is isolated as a free body.

is, the pressure at any depth z is ρgz plus the applied external pressure. This fact was first recognized by Blaise Pascal, and the statement of *Pascal's principle* is: *A pressure that is applied to the surface of a confined (incompressible) liquid is transmitted undiminished to every point within the liquid.*

Atmospheric Pressure. Next, we obtain the expression for the decrease in pressure with height in the atmosphere. Because air is compressible, we must consider the density to be a function of the altitude y above sea level; that is, $\rho = \rho(y)$. Consider the thin slab of air in Fig. 17–6. The force equation is (with the upward direction positive)

$$pA - (p + dp)A - \rho(y)gA\,dy = 0 \qquad \text{(17–8)}$$

so that

$$dp = -\rho(y)g\,dy$$

In order to obtain an expression for $p(y)$, we need to know the way in which the density depends on y. Let us assume that the atmosphere is at a constant temperature so that the density varies in direct proportion to the pressure;* that is,

$$\rho(y) = \rho_0\frac{p}{p_0} \qquad \text{(17–9)}$$

where $\rho(0) = \rho_0$ and $p(0) = p_0$. Inserting this relationship into Eq. 17–8 and integrating, we have

$$\int_{p_0}^{p} \frac{dp}{p} = -\frac{g\rho_0}{p_0}\int_{0}^{y} dy$$

from which

$$\ln\frac{p}{p_0} = -\frac{g\rho_0}{p_0}y$$

or, finally,

$$p(y) = p_0 e^{-\alpha y} \qquad \text{(17–10)}$$

where $\alpha = g\rho_0/p_0$. Thus, in this approximation, the atmospheric pressure decreases exponentially with altitude. Equation 17–10 is called the *law of atmospheres*.

Example 17–3

(a) The greatest depth in the oceans is approximately 11.0 km (in the Mariana Trench, western Pacific). What is the pressure at this depth? (The density of sea water is 1030 kg/m³.)

(b) What is the pressure at an altitude of 11.0 km? (The density of air at sea level is 1.29 kg/m³ for a temperature of 0° C; see Table 1–2.)

---

*See Section 23–2. The air temperature does, of course, vary with altitude; consequently, the result we obtain here is only approximately correct.

Solution:

(a) Using Eq. 17-7, we have

$$p(z) = p_0 + \rho g z$$

Then, $p(11.0 \text{ km}) = 1 \text{ atm} + \dfrac{(1030 \text{ kg/m}^3)(9.80 \text{ m/s}^2)(11.0 \times 10^3 \text{ m})}{1.013 \times 10^5 \text{ Pa/atm}}$

$= 1097 \text{ atm}$

(The actual pressure measured at this depth is about 2 percent greater. The difference is due to the compressibility of water, which we have neglected here.)

(b) We use Eq. 17-10 in which the value of α is

$$\alpha = \frac{g\rho_0}{p_0} = \frac{(9.80 \text{ m/s}^2)(1.29 \text{ kg/m}^3)}{1.013 \times 10^5 \text{ Pa}}$$

$$= 1.25 \times 10^{-4} \text{ m}^{-1}$$

Then, $$p(y) = p_0 e^{-\alpha y}$$

so that $$p(11.0 \text{ km}) = (1 \text{ atm})e^{-(1.25 \times 10^{-4} \text{ m}^{-1})(11.0 \times 10^3 \text{ m})}$$

$$= 0.253 \text{ atm}$$

(The "U.S. Standard Atmosphere" gives, for an altitude of 11.0 km, a pressure of 0.224 atm, about 10 percent smaller than the value we have obtained. The difference is due primarily to our neglect of the variation of temperature with altitude.)

Example 17-4

Find the torque exerted by the water behind a dam about the toe OO' of the dam shown in the diagram. The depth of water is $h = 20$ m and the width of the dam is $L = 100$ m. The upstream face of the dam is vertical.

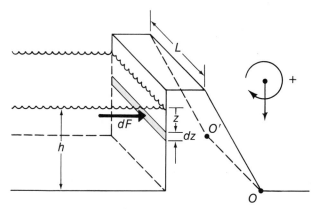

Solution:

The differential force dF that acts on a horizontal strip of height dz and width L on the upstream face of the dam at a depth z is

$$dF = p(z)L \, dz$$

From Eq. 17-7, omitting the atmospheric pressure p_0 (•Why?), we have $p(z) = \rho g z$. Hence, the differential torque produced by dF about OO' is

$$d\tau = \rho g L(h - z)z \, dz$$

Integrating, we find

$$\tau = \rho g L \int_0^h (h - z)z \, dz$$

$$= \tfrac{1}{6}\rho g L h^3$$

$$= \tfrac{1}{6}(10^3 \text{ kg/m}^3)(9.80 \text{ m/s}^2)(100 \text{ m})(20 \text{ m})^3$$

$$= 1.31 \times 10^9 \text{ N} \cdot \text{m}$$

17-3   THE MEASUREMENT OF PRESSURE

Manometers.   A convenient device for measuring pressure, particularly gas pressure, is a *manometer*. This instrument consists of a static liquid column that is used to balance the pressures applied to the two ends of the column. Figure 17–7 shows two types of manometers: (a) an *open* manometer and (b) a *differential* manometer.

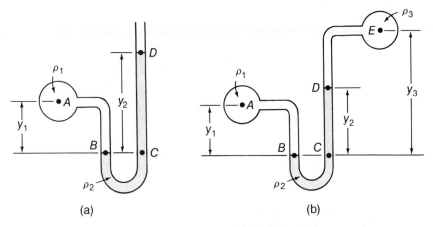

Fig. 17–7.   (a) An open manometer. (b) A differential manometer.

In an open manometer, one end of the liquid column is open to the atmosphere. Thus, in Fig. 17–7a, the pressure at point D is atmospheric pressure p_0. At points B and C the pressure is the same (•Explain.) and is equal to $p_0 + \rho_2 g y_2$, where y_2 is the height of the manometer liquid that is supported by the pressure at C. The pressure at point A in the vessel is

$$p_A = p_0 + \rho_2 g y_2 - \rho_1 g y_1 \qquad \text{(17–11)}$$

If the fluid in the vessel is a gas, the density ρ_1 generally is sufficiently small that the last term in Eq. 17–11 can be neglected. Then,

$$p_A(\text{gas}) = p_0 + \rho_2 g y_2 \qquad \text{(17–11a)}$$

Notice that the pressure in the vessel can be less than atmospheric pressure (even vacuum); then, point B will be above point D, and $y_2 < 0$.

The pressure p_A in Eq. 17–11a is called the *absolute pressure*. The difference between p_A and atmospheric pressure, $p_A - p_0$, is called *gauge pressure*. Any type of gauge that relates the pressure in a vessel to atmospheric pressure shows directly the gauge pressure. Thus, in Fig. 17–7a, the gauge pressure in A is indicated directly by the column height difference y_2.

In Fig. 17–7b, the difference in pressure between points A and E is

$$p_A - p_E = \rho_3 g(y_3 - y_2) + \rho_2 g y_2 - \rho_1 g y_1 \qquad \textbf{(17-12)}$$

Again, if the fluids in both vessels are gases, the terms involving ρ_1 and ρ_3 can be neglected, giving

$$p_A(\text{gas}) - p_E(\text{gas}) = \rho_2 g y_2 \qquad \textbf{(17-12a)}$$

The Barometer.   Instruments that are designed to measure atmospheric pressure are called *barometers*. The common mercury barometer is essentially an open tube manometer (Fig. 17–8a) in which the vessel is evacuated so that point B stands above point D. The usual construction of a mercury barometer, shown in Fig. 17–8, consists of a long tube closed at one end.* The tube is filled with mercury, then inverted, and the open end placed in a reservoir of mercury. A vacuum space forms above the mercury column.† Then, Eq. 17–11a becomes (with $p_A = 0$ and $-y_2 = h$)

$$p_0 = \rho g h \qquad \textbf{(17-13)}$$

where ρ is the density of mercury ($13.5955 \times 10^3 \text{ kg/m}^3$ at $0°$ C). The height of the mercury column at standard atmospheric pressure ($101{,}325$ Pa) and standard gravity (9.80665 m/s^2) is

$$h = \frac{p_0}{\rho g} = \frac{101{,}325 \text{ Pa}}{(13.5955 \times 10^3 \text{ kg/m}^3)(9.80665 \text{ m/s}^2)}$$
$$= 0.76000 \text{ m}$$

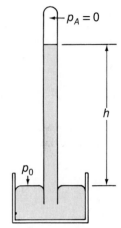

Fig. 17–8. A mercury barometer (or Torricelli tube).

Example 17–5

In the U-tube arrangement shown in the diagram both chambers, A and B, are initially open to the atmosphere. The right-hand section of the tube contains a column of oil with an initial height y_B; the density of the oil is $0.835 \times 10^3 \text{ kg/m}^3$. The left-hand section contains water which extends into the right-hand section, giving an effective height y_A for the water column. The density of water (at $20°$ C) is $0.998 \times 10^3 \text{ kg/m}^3$. Additional gas is now pumped into chamber B and the chamber is sealed. The new pressure in this chamber is p_B and the interface between the oil and water has been lowered by an amount $P_f - P_i = \delta = 1$ cm. What is the pressure in chamber B? (The diameter of the U-tube is much smaller than that of the chambers; consequently, when the interface is depressed by 1 cm, the positions of the liquid surfaces change by negligible amounts.)

*The mercury barometer was invented in 1643 by Evangelista Torricelli (1608–1647), an Italian physicist and mathematician who was a student of Galileo and succeeded his mentor at the Florentine academy.

†Actually, this space contains a small amount of mercury vapor, amounting to about 2×10^{-7} atm.

Solution:

If we use ρ_w for the density of water and ρ_{oil} for the density of oil, we can express the initial condition as

$$p_0 + \rho_w g y_A = p_0 + \rho_{oil} g y_B$$

After the pressure in chamber B has been increased to p_B, we can write

$$p_0 + \rho_w g(y_A + \delta) = p_B = \rho_{oil} g(y_B + \delta)$$

Subtracting the first equation from the second, we obtain

$$p_B - p_0 = (\rho_w - \rho_{oil})g\delta$$
$$= (998 \text{ kg/m}^3 - 835 \text{ kg/m}^3)(9.80 \text{ m/s}^2)(0.01 \text{ m})$$

so that $p_B = 1 \text{ atm} + 15.97 \text{ Pa}$
$$= 1.000158 \text{ atm}$$

17-4 BUOYANCY AND ARCHIMEDES' PRINCIPLE

Consider an irregularly shaped object at rest within a fluid, as shown in Fig. 17-9a. Because the fluid pressure increases with depth, the force exerted by the fluid on the surface of the object is greater for those portions more deeply immersed. The net effect on the entire body is an upward or lifting force, which is called the *buoyant force*.

The magnitude and line of action of the buoyant force can be determined in the following way. First, note that the fluid pressure on the surface of the immersed object does not depend on the material of which the object is composed. For example, suppose that we remove the body and replace it with an exactly equal volume of the fluid, as in Fig. 17-9b. This replaced fluid is in static equilibrium under the combination of the same pressure-developed surface force that acted on the object (the buoyant force) and the gravitational force that acts vertically downward through the center of gravity of the fluid. Thus, the gravitational force precisely balances the pressure-developed or buoyant force. We conclude that the buoyant force F_b that is exerted by a fluid with a density ρ on an object with a volume V is

$$F_b = \rho V g \qquad\qquad (17\text{-}14)$$

Fig. 17-9. (a) An irregularly shaped object immersed in a fluid. (b) The object replaced by an equal volume of fluid. The surface forces exerted by the surrounding fluid are the same in each case.

(a)   (b)

We can summarize the situation with the following statement: *A body that is entirely or partially submerged in a fluid is buoyed up by a force equal in magnitude to the weight of the displaced fluid and directed upward along a line that passes through the center of gravity of the displaced fluid.* This is the statement of *Archimedes' principle.* *

(•Can you see why the buoyancy principle is valid for a body partially immersed in a fluid as well as for a body that is completely immersed? •Can you see why the principle is valid for compressible as well as for incompressible fluids? •What is the buoyant force on an object immersed in a fluid that is undergoing free fall?)

Stability of Floating Objects.   Stability considerations for vessels floating in water have an obvious importance in marine engineering. Figure 17–10 shows a cross section of a floating ship both erect and with a small angular displacement or heel. The gravitational force $\mathbf{F}_g$ on the vessel is directed vertically downward through the center of gravity G and has the magnitude mg where m is the mass of the ship. In equilibrium, the upward buoyant force $\mathbf{F}_b$ must equal $\mathbf{F}_g$ in magnitude. Moreover, $\mathbf{F}_b$ acts through the center of gravity of the displaced water, point B, called the *center of buoyancy*. Points B and G do not coincide, but because the ship is in equilibrium the points must lie on the same vertical line.

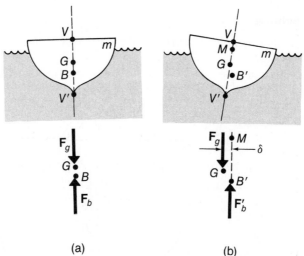

Fig. 17–10. (a) A cross section of the hull of a floating vessel in equilibrium. (b) The same vessel with a slight heel.

(a)

(b)

Suppose that the ship heels slightly, as shown in Fig. 17–10b, due to some external influence such as waves or wind but not because of shifting cargo; thus, point G remains fixed in the vessel. In this condition the shape and volume of the displaced water will be different from those in the erect position (unless the vessel has cylindrical symmetry about the axis of heel). Then, the center of buoyancy will shift to a new position B'. A vertical line through B' intersects the line $\overline{VV'}$ at M. The

*The basic idea that is central to the principle of buoyancy was first conceived by Archimedes (c287 B.C.–212 B.C.), one of the greatest scientists of ancient Greek times. According to the legend (which is probably true), Archimedes made the discovery while in his bath. He had been asked by the king of Syracuse, Hieron II, to determine the purity of a new gold crown without destroying it. As he stepped into his bath, Archimedes noticed that the water level rose by an amount equivalent to the volume of the immersed part of his body and that he was buoyed up in the process. He saw that by comparing in this way the volume of the crown with the volume of an equal weight of known pure gold, he could determine the purity of the crown. He was so pleased with this discovery that he is supposed to have leaped from his bath and to have run naked to the palace shouting "Eureka!"—"I've got it!" Subsequently, measurements showed the crown to have been adulterated with silver, and the goldsmith was executed.

point M is called the *metacenter* of the vessel, and for small angular displacements the position of M remains essentially unchanged. If M lies *above* G, the torque about G produced by the new buoyant force $\mathbf{F}'_b$ will tend to right the vessel; thus, the equilibrium is stable. Because $F'_b \cong F_b = F_g = mg$, the restoring torque is $\tau = mg\delta$, where δ is the horizontal displacement between $\mathbf{F}_g$ and $\mathbf{F}'_b$. Conversely, if M lies *below* G, the state of equilibrium is unstable, and the ship will tend to capsize. (•Can you construct the diagram for this case?)

The *metacentric height* $\overline{GM}$ is usually 20 to 60 cm for large ocean-going vessels. These ships exhibit slow steady roll with long periods. If the metacentric height is too large, the rolling motion is faster and not as smooth; this condition is less comfortable but it is safer than a long-period roll. (•Is there any risk in removing the cargo from a freighter without replacing it with ballast, or in standing up in a canoe?)

Example 17–6

What fraction of an iceberg lies beneath the surface of the sea? (Assume the density of sea water to be $\rho_w = 1.028 \times 10^3 \text{ kg/m}^3$ and that of sea ice to be $\rho_i = 0.917 \times 10^3 \text{ kg/m}^3$. Both values actually depend on the salinity of the water.)

Solution:

For equilibrium conditions, the gravitational force F_g on the iceberg equals the buoyant force F_b.

$$F_g = \rho_i(V_1 + V_2)g \quad \text{and} \quad F_b = \rho_w V_2 g$$

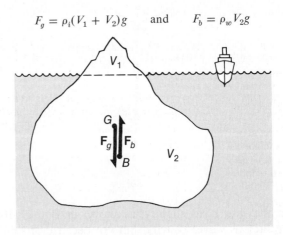

so that

$$\frac{V_2}{V_1 + V_2} = \frac{\rho_i}{\rho_w} = \frac{0.917 \times 10^3 \text{ kg/m}^3}{1.028 \times 10^3 \text{ kg/m}^3} = 0.892$$

That is, only about 11% of the iceberg is visible above the water.

SPECIAL TOPIC

17–5 SURFACE PHENOMENA

In liquids (and in solids) the average distance between molecules is about the same as a molecular diameter, so the molecules are essentially in contact with their nearest neighbors. For example, the diameter of a water molecule is approximately 3.1 Å (0.31 nm), and the spacing between adjacent molecules in water is about the same.

The molecules of a liquid experience strong attractive (*cohesive*) forces that resist attempts to separate the molecules. Work is therefore required to remove molecules from the surface of a liquid. Part of this work is done against atmospheric pressure, but the greater part is done against the liquid cohesive forces. To remove water molecules from liquid water to produce water vapor requires 2.45×10^6 J/kg (at atmospheric pressure and 20° C). This amount of work is called the *heat of vaporization* for water. The portion of this amount of work that is done against the liquid forces is $L = 2.32 \times 10^6$ J/kg.

The cohesive intermolecular forces on a liquid surface produce a number of interesting effects that we now discuss.

Surface Tension. Work is required to remove molecules from a liquid, so any volume of liquid possesses a certain amount of (negative) potential energy. Some of this potential energy is associated with the interior molecules and some is associated with the surface molecules. The work required to remove an interior molecule from a liquid is greater than that necessary to vaporize a surface molecule, because the interior molecule must first be brought to the surface by parting the surface molecules against their attractive forces. This means that a greater amount of negative potential energy is associated with each interior molecule compared with each surface molecule.

Intermolecular forces have short ranges. That is, the attractive force that one molecule exerts on another effectively vanishes for separations that exceed a few molecular diameters. Consequently, the molecules in a liquid interact only with their nearest neighbors, and the potential energy associated with a group of molecules is therefore proportional to the number of molecules. Thus, the interior potential energy is proportional to the volume of the liquid and the surface potential energy is proportional to the surface area.

Any physical system will tend spontaneously toward a condition of minimum potential energy. In a liquid, this means that the tendency is toward increased negative potential energy by maximizing the volume and minimizing the surface area. A fixed volume of free liquid will therefore assume the shape of a sphere because this shape has the least surface area for a given volume.*

Because the surface area of a liquid in equilibrium is minimum, work is required to cause any increase in the surface area. The amount of work required per unit of increased area is called the *surface tension* γ of the liquid,

---

*Falling raindrops are not exactly spherical because of air resistance effects. However, if a globule of oil is introduced into a liquid of the same density but with which it does not mix, the globule will be static and will assume a truly spherical shape. This observation was made in 1873 by the Belgian physicist Joseph Plateau. Perhaps you have also observed this effect in an oil-and-vinegar salad dressing. Spherical drops can also be formed in a free-fall environment, such as an orbiting space vehicle.

and we write

$$\frac{dW}{dS} = \gamma \qquad (17\text{–}15)$$

The units of γ are J/m^2 or, equivalently, N/m.

We can make an estimate of the magnitude of γ for the case of water by using the value for the liquid contribution to the heat of vaporization, namely, $L = 2.32 \times 10^6$ J/kg (at 20° C). In Fig. 17–11a we see a molecule with a diameter d being removed from a liquid and entering the vapor region above the surface. The attractive forces exerted on the molecule by the remaining surface molecules have short ranges. Consequently, when the molecule has been moved several molecular diameters away from the surface, the attractive intermolecular forces effectively vanish; let us assume this distance is about $3d$. During the movement through the distance $3d$, the molecule experiences an attractive average molecular force $\overline{F}$. Then, the work done in vaporizing the molecule from the surface layer is $3\overline{F}d$. If the mass of the molecule is m, the work done per kilogram of liquid is equal to L, the liquid contribution to the heat of vaporization:

$$L = \frac{3\overline{F}d}{m}$$

Figure 17–11b shows an interior molecule being forced into the surface region, thereby enlarging the surface area. The distance through which this molecule must

Fig. 17–11. (a) The path of a surface molecule entering the vapor above a liquid. (b) Forcing an interior molecule into the surface region of a liquid, thereby increasing the area.

be moved to reach the surface is about $\frac{1}{2}d$. We assume that the average force required to effect this movement is the same as that required to remove the surface molecule to a distance $3d$. Hence, the work done in bringing a molecule to the surface is $W = \frac{1}{2}\bar{F}d$. Each molecule added to the surface creates a square of new surface with area d^2 into which the molecule fits. The work done per unit area is equal to the surface tension:

$$\gamma = \frac{W}{A} = \frac{\frac{1}{2}\bar{F}d}{d^2} = \frac{\bar{F}}{2d}$$

Substituting for $\bar{F}$ from the expression for L, we find

$$\gamma = \frac{mL}{6d^2}$$

The mass of a water molecule is $m = 3.0 \times 10^{-26}$ kg. Then, substituting $L = 2.32 \times 10^6$ J/kg and $d = 3.1 \times 10^{-10}$ m, we find $\gamma \cong 0.12$ N/m. This result is actually about 50 percent greater than the measured value for water. Even so, it seems remarkable that such a simple atomic model is able to yield a value for a macroscopic quantity that is this close to the actual value.

Surface tension values for several liquids are given in Table 17–2.

The molecules in a static liquid are in equilibrium—at least on average, realizing that they are continually being jostled about due to thermal motions. Whether a molecule is well inside the interior region or on the surface of the liquid, this equilibrium results from a sum of a number of interactions with neighboring molecules. Such interactions will be repulsive when the adjacent molecule is momentarily interpenetrated, and they will be attractive when the two molecules have larger separations. Only when a surface molecule is moved outward, beyond its average equilibrium position, does a net surface-directed attractive

force result. (In many books, you will find illustrations or statements to the effect that a molecule at the surface experiences a net downward force. Such statements are simply incorrect.)

Pressure Differentials. Consider a small gas bubble with a radius R that is within a liquid. The surface area of the interface between the gas and the liquid is $S = 4\pi R^2$. Imagine a hypothetical (or *virtual*) increase in radius by an amount dR from the equilibrium value R. Then, the surface area would increase by an amount $dS = 8\pi R\,dR$. According to Eq. 17–15, the work required to increase the surface area is

$$dW = 8\pi\gamma R\,dR$$

We assume that the bubble is sufficiently small that the liquid pressure over the surface is constant; we take the value to be p_0. The gas and vapor pressure within the bubble is p. Then, a small area σ of the bubble surface experiences an outward pressure-developed force equal to $(p - p_0)\sigma$. The work that would be done by this force against the inward surface-tension force during an increase in radius dR is $(p - p_0)\sigma\,dR$ (see Fig. 17–12). Because all parts of the surface are equivalent, we can obtain the total work done by replacing σ with the total area $4\pi R^2$. Thus,

$$dW = (p - p_0)4\pi R^2\,dR$$

Then, equating the two expressions for dW, we find*

$$p - p_0 = \frac{2\gamma}{R} \quad \text{(gas bubble in a liquid)} \quad \textbf{(17–16)}$$

---

*Although a radial displacement dR was imagined in this derivation, no actual displacement was required to arrive at Eq. 17–16 and dR does not appear in this expression. For this reason, derivations of this type are said to involve the *principle of virtual work*.

TABLE 17–2. Surface Tension of Various Liquids

LIQUID	FORMULA	IN CONTACT WITH	TEMPERATURE (°C)	γ (10^{-3} N/m)
Acetone	C_3H_6O	air or vapor	0	26.2
Benzene	C_6H_6	air	20	28.8
Carbon tetrachloride	CCl_4	vapor	20	27.0
Ethyl alcohol	C_2H_6O	vapor	20	22.8
Glycerol	$C_3H_8O_3$	air	20	63.4
Mercury	Hg	air	15	487
		vacuum	20	472
Water	H_2O	air	0	75.6
			20	72.8
			40	69.6
			80	62.6
			100	58.6

Fig. 17-12. A spherical bubble of gas in a liquid. A small element σ of the surface is imagined to undergo a radial displacement dR.

Consider an arbitrary differential surface element σ of a liquid-air interface. Suppose that the principal radii of curvature of the surface in two perpendicular directions are R_1 and R_2, as shown in Fig. 17-13. If there exists across the surface a pressure differential, $p - p_0$, the equivalent of Eq. 17-16 is

$$p - p_0 = \gamma \left(\frac{1}{R_1} + \frac{1}{R_2} \right) \qquad \textbf{(17-17)}$$

In the event that $R_1 = R_2 = R$, this expression reduces to Eq. 17-16.

Linear Stress. An alternate view of surface tension is sometimes more appropriate than the basic definition

Fig. 17-13. An arbitrary surface element has a maximum and a minimum radius of curvature. These are the principal radii of curvature, R_1 and R_2, as shown here. Arcs on the surface with these radii are mutually perpendicular.

Fig. 17-14. The force **F** stretches the soap film. The surface area of the film is $2\ell x$.

given by Eq. 17-15. Suppose that a soap film is stretched over the area enclosed by a U-shaped frame and a crosspiece C that can slide smoothly along the frame, as shown in Fig. 17-14. A force **F** is applied to the crosspiece, which then moves a distance dx. This increases the surface area of the soap film (which has two sides) by an amount $dS = 2\ell\,dx$. The work done by the force is $F\,dx$. Then, using Eq. 17-15, we have

$$dW = F\,dx = 2\gamma\ell\,dx$$

from which $\qquad\qquad F = 2\gamma\ell \qquad \textbf{(17-18)}$

The edge of the two-sided film surface that adheres to the crosspiece has a length 2ℓ. Thus, the surface tension γ is just the *force per unit length* exerted by the film along its edge. This force lies in the fluid surface and is directed perpendicular to the edge.

We can use this definition of surface tension to obtain the pressure differential for the case of the gas bubble just discussed. Imagine the bubble to be cut in half along a diameter. Then, we ask what forces act on one of the hemispheres. According to the discussion in connection with Fig. 17-4, the pressure-developed force is $(p - p_0)\pi R^2$. This force is just balanced by the force due to the surface tension of the other hemisphere, which is $2\pi R\gamma$. Equating these two expressions for the forces reproduces Eq. 17-16.

Contact Angles at Liquid-Solid Interfaces. At the edge of a liquid, where it meets a solid, the liquid surface is curved. This effect is seen for a liquid in a container or for a liquid standing on a flat surface. For example, a drop of pure water on a clean horizontal glass surface is

Fig. 17–15. (a) A wetting liquid has $\beta < \pi/2$. (b) A nonwetting liquid has $\beta > \pi/2$. Notice that in each case β is measured from the surface to the tangent plane *through* the liquid.

curved downward at the edge, as indicated in Fig. 17–15a. Conversely, a drop of mercury on a glass plate has an edge that is curved upward, as shown in Fig. 17–15b. The angle between the solid surface and the tangent to the liquid surface at the extreme edge of the liquid is called the *contact angle*. We say that a liquid "wets" the surface when the contact angle β is less than $\pi/2$ (Fig. 17–15a); the liquid is "nonwetting" when $\beta > \pi/2$ (Fig. 17–15b).

In Fig. 17–16 we identify the interface forces that act in equilibrium at point P, the common point of the liquid, the gas, and the solid. The vectors **f** represent forces per unit length of the edge of the liquid and act along directions perpendicular to the edge at point P. The force $\mathbf{f}_{LG}$ is the liquid-gas interface force; this is the force we have previously associated with the surface tension γ for the case of a liquid in contact with only a gas. The force $\mathbf{f}_{SG}$ is the solid-gas interface force, and the force $\mathbf{f}_{SL}$ is the solid-liquid interface force. The force $\mathbf{f}_{\perp}$ is the *adhesive* force between the liquid and the solid and is perpendicular to the surface at P.

The equilibrium force equations are

$$f_{LG} \cos \beta + f_{SL} - f_{SG} = 0$$

Fig. 17–16. The interface forces that act along the edge of a liquid standing on a solid surface.

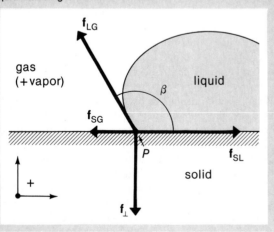

$$f_{LG} \sin \beta - f_{\perp} = 0$$

Then, the contact angle is given by

$$\cos \beta = \frac{f_{SG} - f_{SL}}{f_{LG}} \cong -\frac{f_{SL}}{f_{LG}} \qquad \textbf{(17–19)}$$

where we have neglected f_{SG} in comparison with f_{SL} because the solid-gas force is almost always relatively weak. Thus, $\cos \beta < 0$ and $\beta > \pi/2$. This corresponds to the case of a nonwetting liquid (Fig. 17–16). For a wetting liquid, we have $\beta < \pi/2$, so $\mathbf{f}_{SL}$ has the direction *opposite* to that shown in Fig. 17–16. That is, the solid exerts a force on a wetting liquid that tends to extend the edge away from the main body of the liquid (Fig. 17–15a).

Table 17–3 lists values of the contact angle for several situations.

TABLE 17–3.   Values of the Contact Angle β for Various Liquid-Solid Interfaces

MATERIALS	β
Water-glass[a]	0°
Water-silver	90°
Water-paraffin	106°
Kerosene-glass	26°
Mercury-copper	0°
Mercury-glass[b]	148°
Most organic liquids-glass	0°–10°

[a]For pure water; slight contaminations may increase the value to 20° or so.

[b]For pure mercury; slight contaminations may reduce the value to 140° or so.

Capillarity.   If a small-bore tube (a *capillary*) is partially immersed in a liquid, the height at which the liquid stands in the tube is determined by surface tension effects. Figure 17–17a shows the case of a wetting liquid, for which $\beta < \pi/2$. The interface force between the solid and the liquid is capable of supporting a column of liquid with a height h. If the liquid does not wet the tube, we have $\beta > \pi/2$, and the liquid in the tube is depressed, as shown in Fig. 17–17b. In each case, the curved surface inside the capillary is called the *meniscus*.

For the case of a wetting liquid (Fig. 17–17a), the vertical force per unit length along the interface edge exerted by the capillary wall on the liquid is f_{SL}. Because $f_{SG} \ll f_{SL}$, we have $f_{SL} \cong f_{LG} \cos \beta = \gamma \cos \beta$. Thus, the total upward surface-tension force developed along the perimeter of the solid-liquid interface within the capillary is $2\pi R\gamma \cos \beta$, where R is the capillary radius. This force supports the liquid column of height h, so we have at equilibrium

$$2\pi R\gamma \cos \beta = \rho g(\pi R^2 h)$$

Fig. 17-17. Capillary action for a tube in (a) a wetting liquid and (b) a nonwetting liquid. The radius of the tube is R and the radius of the meniscus is a.

from which*

$$h = \frac{2\gamma \cos \beta}{\rho g R} \qquad (17\text{-}20)$$

where ρ is the density of the liquid.

Equation 17-20 also holds for the case of a nonwetting liquid (Fig. 17-17b), where $\beta > \pi/2$ so that $h < 0$, indicating a depression of the meniscus below the level of the liquid outside the tube. (•Can you show this?)

Example 17-7

Obtain the expression for capillary rise, Eq. 17-20, from energy considerations.

Solution:

The column of liquid that has risen above the surface of the liquid reservoir is shown in the diagram. The energy of this liquid column consists of the potential energy U plus the surface energy associated with surface tension.

The potential energy is equal to the weight mg of the

---

*This expression is reasonably accurate if $h \gg R$; otherwise, the geometry of the meniscus surface must be more carefully taken into account. For β near zero, a more sophisticated analysis gives γ in terms of h and R: $\gamma \cong \frac{1}{2}\rho g R h[1 + (R/3h)]$.

liquid multiplied by the height $\frac{1}{2}h$ to which its C.M. has been raised:

$$U = \tfrac{1}{2}mgh = \tfrac{1}{2}(\rho \pi R^2 h)gh = \tfrac{1}{2}\pi \rho g R^2 h^2$$

To calculate the surface energy, we neglect the solid-gas contribution, as we did in writing Eq. 17-19. Now, $f_{LG} = \gamma_{LG} = \gamma$, the ordinary liquid-gas surface tension. In this case we also have $f_{SL} = \gamma_{SL}$. Then, the surface energy consists of two parts—that due to the meniscus and that due to the cylindrical surface. The meniscus energy does not depend on h,† so we write this energy simply as C, a constant. Thus, the surface energy is

$$E_s = C + \gamma_{SL}(2\pi R h)$$

---

†This is the case if $h \gg R$. See previous footnote.

From Eq. 17-19, we have $\gamma_{SL} = -\gamma \cos \beta$. Therefore,

$$E_s = C - 2\pi Rh\gamma \cos \beta$$

so that the total energy is

$$E = U + E_s = \tfrac{1}{2}\pi\rho g R^2 h^2 + C - 2\pi Rh\gamma \cos \beta$$

The liquid column will be in equilibrium when E is a minimum. The value of h for which this occurs is found from

$$\frac{dE}{dh} = 0 = \pi\rho g R^2 h - 2\pi R\gamma \cos \beta$$

Thus,

$$h = \frac{2\gamma \cos \beta}{\rho g R}$$

which is just Eq. 17-20.

We did not need to make the simplifying assumption, $\gamma_{SG} \cong 0$, in order to obtain this result. (•Can you see how we could have used the first form of Eq. 17-19 and included the solid-gas area of the capillary?)

Example 17-8

A mercury barometer (Fig. 17-8) is to be made from a glass tube with an inner radius R. If the capillarity correc-

tion on the barometric height y is to be less than 0.5 percent, what is the minimum value for R?

Solution:

For normal atmospheric pressure, we have $y = 0.760$ m, so the capillarity correction h must be

$$|h| < (0.005)(0.760 \text{ m}) = 3.8 \times 10^{-3} \text{ m}$$

Then, using Eq. 17-20,

$$R > \left| \frac{2\gamma \cos \beta}{\rho g h} \right|$$

$$= \left| \frac{2(472 \times 10^{-3} \text{ N/m})\cos 148°}{(13.6 \times 10^3 \text{ kg/m}^3)(9.80 \text{ m/s}^2)(3.8 \times 10^{-3} \text{ m})} \right|$$

so that $R > 1.58 \times 10^{-3} \text{ m} = 1.58$ mm.

In this case we have $R/3h = 1.58/(3 \times 3.8) = 0.14$. Consequently, we expect that meniscus effects such as those mentioned in the footnote to Eq. 17-20 will introduce an inaccuracy of the order of 10 percent into this calculation.

PROBLEMS

Section 17-1

17-1 If a 1-megaton nuclear weapon is exploded at ground level, the peak overpressure (that is, the pressure increase above normal atmospheric pressure) will be 0.2 atm at a distance of 6 km. What force will be exerted on the side of a house with dimensions 4.5 m × 22 m due to such an explosion?

17-2 Use the fact that normal atmospheric pressure is 1.013×10^5 N/m^2 and that the radius of the Earth is 6.38×10^6 m to calculate the mass of the Earth's atmosphere.

17-3 A plane triangular area that is immersed in a fluid intercepts the positive axes of a coordinate system at a distance of 0.1 m from the origin along each of the axes. The force **F** that the fluid exerts on the area has an x-component, $F_x = 20$ N. What is the fluid pressure (assumed uniform over the area)? Determine the vector expression for the force.

17-4 Refer to Fig. 17-4. Show by direct integration that the x-component of the fluid force on the hemispherical surface is $F_x = -\pi R^2 p$.

17-5 In 1654, Otto von Guericke of Magdeburg demonstrated the effect of air pressure by placing together two hemispherical steel shells with diameters of "nearly three-quarters of a Magdeburgian ell" (about 0.8 m) and pumping out the air from the enclosed volume. He then had two teams of eight horses pull in opposite directions on the hemispheres in an unsuccessful attempt to separate the shells as illustrated on the opposite page. With what force would each team of horses have had to pull to break apart the shells? [*Hint:* Refer to Fig. 17-4.]

Section 17-2

17-6 (a) What is the difference in atmospheric pressure between the base and the top of the World Trade

An engraving shows von Guericke's demonstration of air pressure (Problem 17–5). Diagrams I and II at the top illustrate the hemispheres and the gasket that provided an air-tight seal between them. (Reprinted by permission of The Bettmann Archive, Inc.)

Center building (412.4 m)? [*Hint:* Write Eq. 17–8 in increment form and use $\rho(y) \cong \rho_0$.] (b) What is the atmospheric pressure on top of Mt. Everest (8.85 km)?

17-7 What is the pressure on a diver at a depth of 40 m in sea water ($\rho = 1030$ kg/m³)?

17-8 The door to a room has dimensions 2.00 m × 0.70 m and is suspended on frictionless hinges. What minimum force is required to open this door against an atmospheric pressure difference of 1 percent? Where is this force applied?

17-9 A swimming pool has dimensions 30 m × 10 m and a flat bottom. When the pool is filled to a depth of 2 m with fresh water, what is the total force due to the water on the bottom? On each end? On each side?

17-10 What must be the contact area between a suction cup (completely exhausted) and a ceiling in order to support the weight of an 80-kg student?

17-11 Describe the operation of a hydraulic jack, shown in the diagram (above right). What are the functions of the various valves, V, V', and X? The load piston has a diameter of 30 cm and the pump piston has a diameter of 2 cm. What force f on the pump piston is required to raise a load of 5 tonnes (5,000 kg)? If

the hydraulic fluid has a compressibility $\lambda = 40 \times 10^{-11}$ (Pa)$^{-1}$, what is the fractional increase in density under this load?

17-12 The hydraulic press shown in the diagram is used to

work = 500 J

$h = 5.0$ mm

M

$D = 8.0$ cm

raise the mass M through a height of 5.0 mm by performing 500 J of work at the small piston. The diameter of the large piston is 8.0 cm. (a) What is the mass M? (b) What is the pressure in the press?

17-13 The dam gate shown in the diagram rotates about an axis through the hinge at A. If the width of the gate is 1.5 m, what counterclockwise torque must be applied about A to hold the gate closed?

Section 17-3

17-14 Normal atmospheric pressure is 1.013×10^5 Pa. The approach of a storm causes the height of a mercury barometer to drop by 20 mm from the normal height. What is the atmospheric pressure? (The density of mercury is 13.59 g/cm^3.)

17-15 Blaise Pascal duplicated Torricelli's barometer using (as a Frenchman would) a red Bordeaux wine as the working liquid. The density of the wine was 0.984×10^3 kg/m^3. What was the height h of the wine column for normal atmospheric pressure? (Refer to Fig. 17-7 and use $g = 9.80$ m/s^2.) Would you expect the vacuum above the column to be as good as for mercury?

17-16 The mercury in a simple barometer (Fig. 17-8) stands at a height of 76.05 cm. When a small amount of air is allowed to enter the space above the mercury column, the height drops to 73.42 cm. What is the absolute pressure in the space above the column expressed in torr? Expressed in pascals? What is the gauge pressure in torr?

17-17 A simple U-tube that is open at both ends is partially filled with water (above right). Kerosene ($\rho_k = 0.82 \times 10^3$ kg/m^3) is then poured into one arm of the tube, forming a column 6 cm in height, as shown in the diagram. What is the difference h in the heights of the two liquid surfaces?

17-18 The open manometer shown in the diagram (above right) is used to determine the pressure in a vessel containing oil with a density of 0.85×10^3 kg/m^3. If the mercury stands at the heights indicated in the figure, find the gauge pressure at point A.

Problem 17-17

Problem 17-18

17-19 Determine the difference in pressure between points A and B shown in the diagram below. (The density of kerosene is 0.82×10^3 kg/m^3.)

Section 17-4

17-20 A Styrofoam slab has a thickness of 10 cm and a density of 300 kg/m³. What is the area of the slab if it floats just awash in fresh water when a 75-kg swimmer is aboard?

17-21 A crown that is supposed to be gold has a mass of 2.30 kg. When the crown is immersed in water and weighed, the apparent mass is 2.18 kg. Could the crown be of pure gold? (The density of gold is 19.3×10^3 kg/m³.)

17-22 A frog in a hemispherical cockleshell finds that he just floats without sinking in a pea-green sea (density 1.35 g/cm³). If the cockleshell has a radius of 6 cm and has negligible mass, what is the mass of the frog?

17-23 A hollow brass tube (diam. = 4.0 cm) is sealed at one end and filled with lead shot to give a total mass of 0.20 kg. When the tube is floated in pure water, what will be the depth z of the bottom of the tube?

17-24 A pan of water is suspended from two spring scales, A and C, as shown in the diagram at right. A block of aluminum ($\rho = 2.70 = 10^3$ kg/m³) is suspended from a third scale, B. The scales are calibrated to read in mass units (for $g = 9.80$ m/s²): $A = C = 0.40$ kg and $B = 0.60$ kg. The string S that suspends the aluminum block is not lengthened until the block is immersed in the water, as shown by the dashed lines. What are the new readings of the three scales?

Problem 17-23

Problem 17-24

17-25 A beaker that is partly filled with water has a mass of 0.30 kg. The beaker rests on a platform scale. Next, a 250-g piece of copper ($\rho = 8.93 \times 10^3$ kg/m³) is suspended so that it is completely immersed in the water but does not rest on the bottom of the beaker. What is the reading on the scale?

SPECIAL TOPIC

Section 17-5

17-26 Derive the relationship in Eq. 17-20 using Eq. 17-16. Take $R \ll h$ and assume the surface of the meniscus to be spherical with a radius a, as in Fig. 17-17.

17-27 Show that the gauge pressure inside a thin-filmed soap bubble with radius R and surface tension γ is $4\gamma/R$. What is the gauge pressure (in atm) inside a bubble with a radius of 1.5 cm and with $\gamma = 30 \times 10^{-3}$ N/m?

17-28 A capillary tube with an inside diameter of 0.25 mm can support a 10-cm column of a liquid that has a density of 0.93×10^3 kg/m³. The observed contact angle is $\beta = 15°$. What is the surface tension of the liquid?

17-29 A 1-g globule of mercury is pressed between paral-
  • lel glass plates until the liquid covers an area of 30 cm². Use the information in Tables 17-2 and 17-3 to determine the force that must be exerted on the plates.

17-30 Two parallel plates are spaced apart by a distance d. If the plates are partially immersed in a liquid with density ρ, show that the liquid will rise to a height $h = 2\gamma \cos \beta / \rho g d$, where γ is the surface tension and β is the contact angle. (a) Obtain the result by balancing the forces on the liquid column. (b) Obtain the result by calculating the pressure difference across the liquid-air interface. [*Hint:* Use Eq. 17-17. What are R_1 and R_2?]

Additional Problems

17-31 Show that for small changes in altitude y above sea level, the expression for the atmospheric pressure given by Eq. 17-10 reduces to $p(y) = p_0 - \rho_0 g y$. Explain why this result appears similar to Eq. 17-7.

17-32 What would be the thickness of the Earth's atmosphere if the density were constant at the sea-level value (1.29 kg/m³) and if the sea-level pressure were 1 atm?

17-33 What is the altitude below which lies 90 percent of the Earth's atmosphere (by mass)?

17-34 A cylindrical diving bell (the bottom of which is open) has a height of 2 m. The bell is lowered into the sea until the water rises 0.80 m inside the bell. (a) At what depth is the top of the bell? (b) At this depth, what absolute air pressure inside the bell would prevent any water from entering the bell? [*Hint:* Assume that the pressure of a fixed quantity of gas at a constant temperature is inversely proportional to the confining volume. This assumption is the same as that expressed by Eq. 17-9. (•Can you see why?)]

17-35 The upstream face of a dam makes an angle α with
  • the vertical, as shown in the diagram. The width of the dam is L, and the depth of water behind the dam is h. (a) Show that the water exerts on the dam a horizontal force, $F_x = \frac{1}{2}\rho g L h$, with a line of action $\frac{1}{3}h$ above the heel of the dam at H. (b) Show that the water exerts on the dam a downward force, $F_y = \frac{1}{2}\rho g L h^2 \tan \alpha$, with a line of action $\frac{1}{3}h \tan \alpha$ to the right of the heel H.

17-36 The mercury column in a simple barometer stands at a height of 75.92 cm. An amount of pure water is introduced into the barometer column and forms a 1.20-cm layer on top of the mercury, as shown in the diagram. The mercury column now stands at a height of 74.08 cm. What is the pressure in the space above the water? (This is the *vapor pressure* of water at the particular temperature, namely, 20° C.)

17-37 An evacuated steel ball with an outside diameter of 10.0 cm floats in fresh water barely submerged. The density of the steel is 7.86×10^3 kg/m³. What is the wall thickness of the ball?

17-38 A *hydrometer,* shown in the diagram on the next page, is a device for measuring liquid density. The essential parts are a weighted volume V_0 and a uniform stem that floats partly out of the liquid. A particular hydrometer with a stem diameter $d = 8.5$ mm floats in fresh water at a depth measured by $x = 2.00$ cm. When floating in methyl alcohol, $x = 11.00$ cm. The density of methyl alcohol is 794 kg/m³ and that of water at the same temperature is 998 kg/m³. Determine the volume V_0. What is the total mass of the hydrometer?

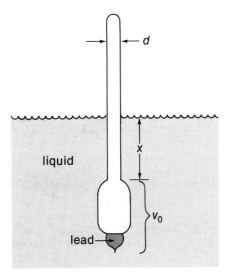

17-39 A beam balance is used to determine the mass of an aluminum object. The balance weights are made of brass. What percentage error in the measurement would be introduced by neglecting air buoyancy effects? (The relevant densities are 2.70×10^3 kg/m³ for aluminum, 8.41×10^3 kg/m³ for brass, and 1.29 kg/m³ for air.)

17-40 A helium-filled balloon supports a total load (including the mass of the balloon structure) of 1200 kg and is in equilibrium at an altitude where the atmospheric pressure is 0.58 atm. Assume an isothermal atmosphere at 0°C. (a) What is the altitude of the balloon? (b) What is the volume of the balloon? [Hint: The balloon material is not stretched, so the helium pressure is equal to atmospheric pressure. The gas densities are given in Table 1-2 for standard conditions. Recall that the gas densities vary in direct proportion to the pressure.] (c) If hydrogen is substituted for helium (equal volumes), what total load could the balloon support at the same altitude? [Hint: See Table 1-2.]

17-41 A cork is tied with a string to the bottom of a beaker
• and floats totally submerged when water is poured into the beaker. The beaker is placed at the edge of a large platform that rotates with an angular frequency ω about a vertical central axis. Show that buoyancy effects cause the line of the string holding the cork to tip inward, toward the center of rotation, by an angle $\alpha = \tan^{-1}(R\omega^2/g)$, where R is the equilibrium radius of the cork. (•What can you say about the shape of the water surface?)

17-42 Refer to Problem 17-23. The tube is depressed a small amount from its position of equilibrium and is then released from rest. Ignoring frictional effects, show that the tube undergoes vertical SHM. Determine the frequency of the motion.

17-43 Two concentric tubes are dipped vertically into an
• unknown liquid. The larger tube has an inside diameter of 6 mm, and the smaller tube has an outside diameter of 4 mm. The liquid rises to a height of 2.0 cm in the inner tube and to a height of 1.5 cm in the region between the tubes. What is the wall thickness of the smaller tube? The density of the liquid is 0.92×10^3 kg/m³ and the contact angle is $\beta = 20°$. What is the surface tension?

17-44 A nonwetting liquid is poured onto a flat hori-
•• zontal surface (e.g., mercury on glass) and forms a pool of a certain size. (The shape of the pool is arbitrary.) Show that the thickness of the pool is $h = \sqrt{2(1 - \cos\beta)\gamma/\rho g}$, where ρ is the density of the liquid, β is the contact angle, and γ is the surface tension. The expression for h is correct for the case in which $h \ll L$, the perimeter of the pool. For mercury on glass, show that $h \cong 3.6$ mm. [Hint: Use the principle of virtual work, as in the derivation of Eq. 17-16. Proceed as follows. First, calculate the change in surface energy of the pool for the liquid-air interface portion of the surface if h is decreased by an amount Δh. Next determine the work done by the surface tension along the perimeter of the liquid-solid contact edge. Then, the sum of these

changes in energy must balance the change in gravitational potential energy.]

17-45 Two large rectangular glass plates are in contact
• along one pair of vertical edges. The opposite edges are separated so that there is an angle α between the plates, as shown in the diagram on the next page. The plates are partially immersed vertically in a liquid. Show that the liquid edge between the plates forms an equilateral

hyperbola (that is, $xy = $ const). [*Hint:* Use Eq. 17–17.]

17–46 A thin aluminum disc with a diameter of 2 cm and a thickness of 1 mm floats on a water surface due to surface tension. The maximum mass M that can be placed on the disc without causing it to sink depresses the top surface of the disc 1.5 mm below the water surface. The angle of the surface intersection is 90°, as shown in the diagram above. Determine the mass M. (Assume $T = 20°$ C.)

FLUID DYNAMICS

In this chapter we consider fluids in motion. The analysis of the dynamics of continuous media such as fluids is much more complicated than the corresponding task for particle systems or rigid bodies. Consequently, we give here only an overview of the subject of fluid dynamics, including simplified discussions of some of the more interesting situations.

18-1 GENERAL CONSIDERATIONS

Fluid Particles and Path Lines. We often find it convenient to describe the dynamics of fluids by following the motion of a fluid "particle." Such a particle is defined to be a sample of the fluid with a size sufficiently small that the macroscopic properties of the fluid change in a continuous way from particle to particle. On the other hand, the size of a fluid particle must be sufficiently large that, on average, there are no net effects of the random thermal motions of its molecules.

The trajectory of a fluid particle, as it moves in accordance with Newton's laws, is called a *path line*. We can imagine that a specific particle is identified by means of a colored speck or with a tiny droplet of dye. Then, a time-exposure photograph would reveal the path line of the particle (Fig. 18-1).

Steady Flow and Velocity Fields. In the condition referred to as *steady flow*, every fluid particle that passes through a particular point generates the same path line. The constant, gentle flow of water in a quiet stream is steady flow, as is the smooth, slow, and constant flow of a fluid through a pipe. Steady flow is an important case of fluid motion, one that we consider later in detail.

The study of fluids by following the motion of specific fluid particles was introduced by Lagrange.* An alternative point of view was developed by Euler, in which the description is made in terms of a *velocity field* $\mathbf{v}(\mathbf{r}, t)$ of the fluid. In the Eulerian scheme, we consider the velocity at a particular instant of all the particles in the fluid. That is, we specify $\mathbf{v}(\mathbf{r})$ at all points $P(\mathbf{r})$ at the time t. At other times $\mathbf{v}(\mathbf{r})$ at $P(\mathbf{r})$ might be different, so, in general, we write $\mathbf{v}(\mathbf{r}, t)$. But steady flow requires that $\mathbf{v}$ be independent of time, so that every particle passes through $P(\mathbf{r})$ with the same velocity $\mathbf{v}(\mathbf{r})$.

Fig. 18-1. A *path line* is the line traced out by a single particle (here identified by coloring) as it moves with a flowing fluid.

---

*Joseph Louis Lagrange (1736–1813), French-Italian mathematician, physicist, and astronomer. Lagrange developed very general equations from which all problems in mechanics can be solved (at least, in principle).

One further consideration is necessary in the definition of steady flow. Suppose that a boat moves slowly with constant velocity across a calm lake. To a stationary shore observer, water at some point near the path of the boat is first at rest, then it moves when disturbed by the passage of the boat, and finally it returns to a condition of rest after the boat has passed. On the other hand, an observer stationed on the boat, who looks at a point that bears a fixed reference with respect to the boat, sees motion of the water that he judges to be a steady flow. That is, this observer sees every water particle pass through the point with the same velocity. Consequently, we amend the definition of steady flow to include cases in which the fluid appears to undergo steady flow with respect to *some* inertial reference frame. This permits us to use a steady-flow analysis when discussing the uniform motion through a fluid of various objects (airfoils, balls, and so forth).

In one type of nonsteady flow of a fluid, the flow conditions change so slowly that at a particular instant the flow appears to be steady. In principle it is no more difficult to analyze such motions than it is to treat true steady flow. However, we forego a discussion of slowly changing steady flow.

Nonsteady flow can also involve *turbulent flow*. In this case the fluid particles that pass through a particular point do not have the same history but, instead, follow randomly related path lines. For example, the smoothly flowing water in a stream becomes turbulent when rocks are encountered and "white water" rapids occur. Finally, there exists a type of nonsteady flow in which the flow pattern oscillates with time but does not become randomly turbulent in the process.

Density and Viscosity. The condition of a static fluid can be specified by giving the pressure $p(\mathbf{r})$ and the density $\rho(\mathbf{r})$ at every point within the fluid. In the event that the fluid is in motion, these quantities can have a time dependence, and then we write $p(\mathbf{r}, t)$ and $\rho(\mathbf{r}, t)$.

When considering a liquid, we usually assume incompressibility, so that $\rho = $ constant. We may also consider gases to have constant density if we exclude large-scale phenomena, such as meteorological problems, and if we restrict the flow velocities to be less than about half the speed of sound in the gas (that is, $v \lesssim 150$ m/s for air under normal conditions).

An important force that influences the motion of fluids is the frictional or drag force called *viscosity*. Viscosity results from the shear exerted by one layer of a fluid on an adjacent layer as they slide past each other. These shear stresses arise from the attractive forces that exist between the molecules in the two layers. Thus, viscosity is an inherent property of all real fluids. In some cases, however, when the effects of viscosity are relatively small, we neglect this complication in the interests of simplicity.

18-2 STREAMLINES AND CONTINUITY

Streamlines are used to describe the velocity field within a flowing fluid. At any point along a streamline, such as the points P, Q, and R in Fig. 18-2, the tangent to the line indicates the direction of the fluid velocity at that point. Notice that a *path line* refers to the trajectory of a single particle, whereas a *streamline* refers to the picture at a particular instant of the velocity directions of a number of particles. Along a streamline the velocity can vary (in magnitude and in direction). (•At what kinds of places will a streamline lead to a point where the velocity is *zero*?)

Any individual fluid particle must travel along a path whose tangent has the same direction as that of the fluid velocity at every point. An essential feature of

Fig. 18-2. The direction of the velocity of a fluid particle at a point, such as P, Q, or R, is the same as the direction of the tangent to the streamline through that point.

streamline

steady flow is that the streamlines are independent of time. Thus, the path line of a particle coincides with a streamline for the case of steady flow.

A method for making streamlines visible is shown schematically in Fig. 18-3. A distinctive dye is released into the flowing liquid through tiny tubes located at various places within the containing pipe. The dye does not immediately mix with the water and is carried along on a streamline, thereby making it visible. In the case of a flowing gas, as in a wind tunnel, the streamlines are often indicated by smoke streamers (see the photograph below).

Fig. 18–3. (a) Streamlines in a flowing liquid are made visible by releasing dye into the liquid.

(b) The flow pattern around an airfoil is revealed by smoke particles that move along the streamlines. The flow is from right to left, corresponding to motion of the airfoil through the gas from left to right. (Courtesy of NASA)

A concept related to the streamline is that of a *stream tube* or *tube of flow*. Imagine a tubelike surface generated by constructing all of the streamlines that pass through points on the perimeter C_1 of a small plane area ΔA_1, as shown in Fig. 18-4. This tube can be cut by another plane area ΔA_2 farther downstream, thereby creating a second perimeter C_2. The result is a volume defined by the tube wall and the two end areas, ΔA_1 and ΔA_2, whose perimeters are the curves, C_1 and C_2. Fluid particles can never enter or leave a particular tube of flow by passing through the tube wall. (•Can you see why?)

In Fig. 18-4 we suppose that the areas, ΔA_1 and ΔA_2, are sufficiently small that the flow velocity is essentially constant over each surface, being $\mathbf{v}_1$ at ΔA_1 and $\mathbf{v}_2$ at ΔA_2. Define two vector areas, $\hat{\mathbf{n}}_1 \Delta A_1$ and $\hat{\mathbf{n}}_2 \Delta A_2$, where $\hat{\mathbf{n}}_1$ and $\hat{\mathbf{n}}_2$ are unit normal vectors that are directed outward from the enclosed volume. The density of the fluid is ρ_1 at position 1 and is ρ_2 at position 2. Then, during a time Δt, a mass $\Delta m_2 = (\rho_2 \mathbf{v}_2 \cdot \hat{\mathbf{n}}_2 \Delta A_2) \Delta t$ passes through ΔA_2 (*leaving* the enclosed volume), and a mass $\Delta m_1 = -(\rho_1 \mathbf{v}_1 \cdot \hat{\mathbf{n}}_1 \Delta A_1) \Delta t$ passes through ΔA_1 (*entering* the enclosed volume).

Fig. 18–4. A tube of flow defined by streamlines passing through the small closed flat curve C_1. The area defined by the perimeter of C_1 is ΔA_1, the unit normal to which is $\hat{\mathbf{n}}_1$, directed *outward* from the enclosed volume. The average fluid velocity through the surface ΔA_1 is $\mathbf{v}_1$. Similar considerations apply for C_2 and ΔA_2.

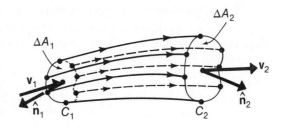

If there is no accumulation of mass within the volume, we have $\Delta m_1 = \Delta m_2$, or

$$\rho_1 \mathbf{v}_1 \cdot \hat{\mathbf{n}}_1 \Delta A_1 + \rho_2 \mathbf{v}_2 \cdot \hat{\mathbf{n}}_2 \Delta A_2 = 0 \qquad \text{(18-1)}$$

For incompressible fluids, we have $\rho_1 = \rho_2$, and Eq. 18-1 becomes

$$\mathbf{v}_1 \cdot \hat{\mathbf{n}}_1 \Delta A_1 + \mathbf{v}_2 \cdot \hat{\mathbf{n}}_2 \Delta A_2 = 0 \qquad \text{(incompressible fluids)} \qquad \text{(18-2)}$$

Equations 18-1 and 18-2 are the *equations of mass continuity* and express the *continuity condition*.

In order to satisfy the continuity condition, the flow velocity must increase when the area is decreased and the streamlines bunch together. This is the case in the narrow throat in the pipe shown in Fig. 18-3.

18-3   BERNOULLI'S EQUATION

We now derive the equation of motion for the steady flow of an ideal fluid—one that is incompressible and has no viscosity. Consider a fluid particle in the form of a narrow tube of flow that has length Δs along a central streamline, as shown in Fig. 18-5. The cross-sectional area of the tube is ΔA, and the mass of the enclosed volume of fluid is $\Delta m = \rho \, \Delta s \, \Delta A$.

Fig. 18-5. Free-body diagram for a fluid particle considered to be a short section of a narrow tube of flow that surrounds a central streamline. The position coordinate s is measured along the streamline.

The pressure on the upstream face of the tube is p, and the pressure on the downstream face is $p + (dp/ds) \, \Delta s$. The net force along the streamline produced by the variation in pressure (the *pressure gradient*) is

$$F_{\text{press}} = p \, \Delta A - \left(p + \frac{dp}{ds} \Delta s \right) \Delta A = - \frac{dp}{ds} \Delta s \, \Delta A$$

For the flow direction indicated in Fig. 18-5, we have $dp/ds < 0$.

The component of the gravitational force on the fluid particle that acts along the streamline is $-\rho g \, \Delta s \, \Delta A \cos \alpha$, where the upward direction is taken to be positive (see Fig. 18-5). Now, $\cos \alpha = dh/ds$, so the resultant force acting on the particle in the direction of the streamline is

$$F_{\text{net}} = - \left(\frac{dp}{ds} + \rho g \frac{dh}{ds} \right) \Delta s \, \Delta A \qquad \text{(18-3)}$$

The acceleration of the particle along the streamline is*

$$a = \frac{dv}{dt} = \frac{dv}{ds}\frac{ds}{dt} = v\frac{dv}{ds} \qquad (18\text{-}4)$$

Then, Newton's equation becomes

$$F_{\text{net}} = \Delta m\, a$$

$$-\left(\frac{dp}{ds} + \rho g\frac{dh}{ds}\right)\Delta s\,\Delta A = (\rho\,\Delta s\,\Delta A)\,v\frac{dv}{ds}$$

which can be written as

$$\frac{dp}{ds} + \rho g\frac{dh}{ds} + \rho v\frac{dv}{ds} = 0$$

or

$$dp + \rho g\, dh + \rho v\, dv = 0 \qquad (18\text{-}5)$$

We can integrate Eq. 18-5 *along a streamline* between two points labeled by the subscripts 1 and 2 with the result

$$(p_2 - p_1) + \rho g(h_2 - h_1) + \tfrac{1}{2}\rho(v_2^2 - v_1^2) = 0 \qquad (18\text{-}6)$$

where h_1 and h_2 are the heights of the two points above an arbitrary reference level. This equation is valid for any two points along a streamline, so we can write

$$p + \rho gh + \tfrac{1}{2}\rho v^2 = \text{constant} \qquad \text{steady flow of an ideal fluid} \qquad (18\text{-}7)$$

Equations 18-5, 18-6, and 18-7 are equivalent forms of *Bernoulli's equation*† for nonviscous steady flow (with the integrated forms restricted to the case of an incompressible fluid). The quantity $p + \rho gh$ is referred to as the *static pressure* and $\tfrac{1}{2}\rho v^2$ is called the *dynamic pressure*.

Notice the similarity between the derivation of Bernoulli's equation and that of the work-energy equations (Eqs. 7-6 and 7-10). Indeed, in Section 18-4 we find that Bernoulli's equation is just the expression of the energy conservation principle applied to the special case of ideal streamline flow.

In all of the examples and problems that follow in this section, we assume the ideal case of steady flow of a nonviscous, incompressible fluid so we can use Bernoulli's equation.

---

*We have limited our considerations here to the case of steady flow. In the general case, the velocity field of the fluid can be a function of both position and time. That is, the velocity of a fluid particle may change because of an explicit time dependence of the velocity and also because the flow along a streamline during a time Δt carries the particle to a position where the velocity is different. Then, we would have $v = v(s, t)$ and the acceleration would be expressed as $a = dv/dt = (\partial v/\partial s)(ds/dt) + \partial v/\partial t$. For the case of steady flow the explicit time dependence disappears; then, we have $\partial v/\partial t = 0$, and $\partial v/\partial s$ becomes dv/ds.

†Usually attributed to the Swiss mathematician Daniel Bernoulli (1700–1782), in his *Hydrodynamica* (1738); however, Bernoulli's discussion of the relationship between pressure and velocity in this book was obscure. The first to obtain the integrated equation (Eq. 18-7) was Leonhard Euler.

Example 18–1

A tank is filled to a depth h_1 with a liquid that has a density ρ. The liquid escapes into air through a small hole in the bottom of the tank. What is the escape velocity v_2?

Solution:

Bernoulli's equation can be applied only to points that lie on the same streamline. (In general, the constant that appears in Eq. 18–7 will be different for different streamlines.) However, in the situation here, we assume that the surface area A_1 is much larger than the orifice area A_2. Then, the fluid particles on the surface have essentially zero velocity. Because the values of p and h are the same across the surface, the constant in Eq. 18–7 is the same for the streamlines through all surface points, such as a, b, and c. Consequently, we can apply Bernoulli's equation between any point on the surface and a point in the orifice. With $v_1 = 0$, $h_2 = 0$, and $p_1 = p_2 = p_0$, Eq. 18–6 becomes

$$\rho g h_1 = \tfrac{1}{2}\rho v_2^2$$

from which

$$v_2 = \sqrt{2gh_1}$$

Thus, the velocity of the liquid emerging from the bottom of the tank is the same as that of an object dropped through a distance equal to the depth of the liquid. This result is known as *Torricelli's theorem*. Notice that the velocity of the emerging fluid is independent of direction; if the fluid were guided by a curved pipe, it would squirt to the original height h_1 (neglecting frictional effects).

When a jet of liquid escapes freely from an orifice, the streamlines, which converge toward the orifice, continue to converge for a short distance beyond the orifice. This effect is shown in the diagram below. The minimum cross section of the jet VV', where the streamlines are parallel, is called the *vena contracta*. The cross-sectional area at VV' is about 60 to 70 percent of that at the orifice OO'. (This contraction is not to be confused with the further contraction of the stream as it falls to lower elevations; see Problem 18–6.) The velocity profile across the *vena contracta* is much more uniform than in the plane of the orifice, as shown in the diagram.

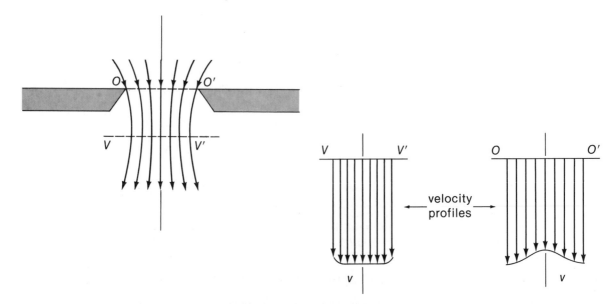

Example 18-2

Water flows over a rectangular weir or spillway in a dam. Assume that the head y_0 of water above the crest of the spillway is sufficiently small that the flow is steady. The geometry is shown in the diagram. Determine the volume flow rate of water passing over the spillway in terms of the head y_0.

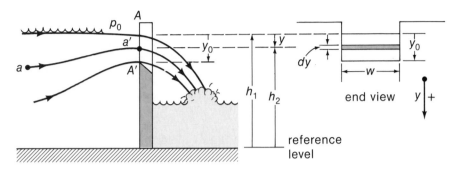

Solution:

As water flows over the spillway, the surface of the upstream basin gradually drops. Thus, any streamline such as aa' must eventually connect with a point on the surface. We assume that the basin is very large so that the fluid particles on the surface have essentially zero velocity—that is, $v_1 = 0$. We also assume that the water flowing over the spillway is vented underneath so that there is atmospheric pressure on all sides of the stream. Then, along AA' the water begins to fall freely, so there is no fluid pressure, and $p_2 = p_0$. (•Can you see why?) Using Eq. 18-6 between point a' and a point on the basin surface, we have

$$p_0 + \rho g h_2 + \tfrac{1}{2}\rho v_2^2 = p_0 + \rho g h_1$$

from which
$$v_2 = \sqrt{2g(h_1 - h_2)} = \sqrt{2gy}$$

The rate of flow of water over the spillway is

$$q = \int_0^{y_2} vw \, dy = w\sqrt{2g} \int_0^{y_0} y^{1/2} \, dy$$

$$= \tfrac{2}{3}w\sqrt{2g}\, y_0^{3/2}$$

Typically, the flow rate in practice is only about 60 percent of the value calculated from this expression, owing to frictional losses and to the fact that the actual height of the water above the crest of the spillway is less than the value y_0 used in the calculation (refer to the diagram).

18-4 THE ENERGY EQUATION

We now consider the energy balance in any device or machine through which a fluid flows. The device could be an ordinary pipe or nozzle, or it could be a pump, compressor, turbine, or other complicated system. Figure 18-6 shows schematically a general device, the essential features of which are in the volume enclosed by the dashed line.

We make no restriction on the properties of the fluid or on the nature of the flow

Fig. 18–6. Fluid flows through a generalized device (contained within the dashed line). The average entrance velocity of the fluid is $\bar{v}_1$ and the average exit velocity is $\bar{v}_2$. The device could be as simple as a straight pipe.

through the device. For example, the fluid could be viscous and the flow could be turbulent. Fluid with density ρ_1 and pressure p_1 enters the device with an average velocity $\bar{v}_1$ at a height h_1 above some convenient reference level. The corresponding quantities at exit are $\rho_2, p_2, \bar{v}_2$, and h_2. We do not need to include explicitly the variation in fluid pressure or gravitational potential energy across the diameter of the pipe. (See the remark following Eq. 18–9.)

Consider the flow of a mass m of fluid through the system. The work done *on* the fluid at the entrance to the device is equal to the product of the force $p_1 A_1$ and the distance moved, $\ell_1 = m/\rho_1 A_1$, that is, $W_1 = mp_1/\rho_1$. The work done *by* the fluid at the exit is $W_2 = mp_2/\rho_2$. There might also be work done on the fluid by a mechanical rotor (W_{in}, per unit mass) or by heat supplied (Q_{in}, per unit mass), as indicated in Fig. 18–6. (Note that W_{in} and Q_{in} could be negative. •What would this mean?) Then, the total energy input to the fluid from all these sources is

$$E_{in} = m\left(W_{in} + Q_{in} + \frac{p_1}{\rho_1} - \frac{p_2}{\rho_2}\right)$$

What net amount of energy is carried out of the system by the fluid? The net outflow of kinetic energy is $\frac{1}{2}m(\bar{v}_2^2 - \bar{v}_1^2)$. Also, there is a change in the potential energy of the fluid between entrance and exit amounting to $mg(h_2 - h_1)$. Finally, there may be a change in the internal energy of the fluid (the energy associated with thermal motions of the constituent molecules) due to heat gained or lost by the fluid. Let the internal energy per unit mass be u_1 at entrance and u_2 at exit. Then, the net amount of energy carried out of the system by the fluid is*

$$E_{out} = m(u_2 - u_1 + gh_2 - gh_1 + \tfrac{1}{2}\bar{v}_2^2 - \tfrac{1}{2}\bar{v}_1^2)$$

Conservation of energy requires $E_{in} = E_{out}$, so we have

$$W_{in} + Q_{in} + \frac{p_1}{\rho_1} - \frac{p_2}{\rho_2} = u_2 - u_1 + g(h_2 - h_1) + \tfrac{1}{2}(\bar{v}_2^2 - \bar{v}_1^2) \qquad \textbf{(18–8)}$$

Note that we have obtained this equation without assuming any special properties of the fluid and without any restriction on the nature of flow.

---

*We assume that no thermal energy leaves the fluid by the conduction of heat through the pipe walls, although we could explicitly include such a term in the energy equation.

Next, we specialize the energy equation (Eq. 18–8) to the case of steady, non-viscous, incompressible flow with a uniform velocity profile ($v = \bar{v}$). Moreover, we allow no mechanical or thermal energy input ($W_{\text{in}} = 0$ and $Q_{\text{in}} = 0$); then, the internal energy of the fluid cannot change, so $u_2 - u_1 = 0$. (•Can you see why?) The result is

$$p_1 + \rho g h_1 + \tfrac{1}{2}\rho v_1^2 = p_2 + \rho g h_2 + \tfrac{1}{2}\rho v_2^2 \qquad \textbf{(18–9)}$$

which is just Bernoulli's equation (Eq. 18–6).

In Section 17–2 we found that the fluid pressure at a depth z is $\rho g z$ (Eq. 17–7). This result means that the quantity $p + \rho g h$ appearing in Eq. 18–9 has the same value at any point on each cross section of the fluid, such as AA' in Fig. 18–7.

Fig. 18–7. Because $p + \rho g h$ is constant across any cross section of the pipe, the liquid rises to the same level regardless of the depth of the side-pipe if the flow velocity is the same at each point.

Bernoulli's equation applies for very special flow along a streamline of an ideal fluid. The energy equation, on the other hand, is valid under the most general conditions, including the nonsteady, nonstreamline flow of real fluids. When the energy equation is specialized in the way required to obtain Eq. 18–9, the fluid becomes ideal and there is no longer any distinction between the confining pipe and a tube of flow, so Bernoulli's equation results! Although by "artful" application Bernoulli's equation can sometimes be used for the case of real fluid flow, it is always correct to use the energy equation.

Example 18–3

A *Venturi meter** is a device for measuring the flow rate of fluids in pipes. The construction is shown in the diagram on the next page. A manometer tube, containing a liquid with a density ρ', is placed between the main part of the pipe with area A_1, and a constricted part of the pipe with area A_2. In practice, A_2 is $\tfrac{1}{3}$ to $\tfrac{1}{5}$ of A_1. The fluid has a density ρ. Assume that the height difference h in the manometer is a true measure of the static pressure difference between the points of attachment to the pipe. Assume further that the fluid is incompressible and experiences no appreciable frictional losses in flowing from position 1 to position 2. Obtain the expression for the rate of flow of the fluid through the pipe in terms of h.

Solution:

Apply Eq. 18–8 between positions 1 and 2 with the result,

$$\tfrac{1}{2}\bar{v}_1^2 + \frac{p_1}{\rho} = \tfrac{1}{2}\bar{v}_2^2 + \frac{p_2}{\rho}$$

*After G. B. Venturi (1746–1822), an Italian physicist, whose studies inspired the invention of this device.

Next, equate the static pressures at points a and b in the manometer:

$$p_1 + \rho g y_1 = p_2 + \rho g y_2 + \rho' g h$$

Using the fact that $y_1 - y_2 = h$, we have

$$\frac{p_1 - p_2}{\rho} = gh\left(\frac{\rho'}{\rho} - 1\right)$$

For an incompressible fluid, $\bar{v}_1 = \bar{v}_2 A_2 / A_1$. Combining these expressions, we find

$$q = \bar{v}_2 A_2 = A_2 \left[\frac{2gh(\rho'/\rho - 1)}{1 - (A_2/A_1)^2}\right]^{1/2}$$

Because this expression does not include frictional effects, the calculated value of q is always too large; nevertheless, the expression is accurate to better than 10 percent for a wide range of conditions.

Example 18–4

Water flows through a submerged orifice in a dam. The geometry is indicated in the diagram below. Determine the amount of energy per unit mass of flow that is dissipated by friction in the fluid behind the orifice.

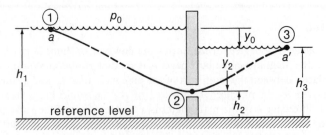

Solution:

We use the energy equation connecting the horizontal surface of the upstream basin (position 1), the orifice (position 2), and the horizontal surface of the downstream basin (position 3). The basins are sufficiently large that the particles on the surfaces have essentially zero velocity. Then, the portion of the energy equation (Eq. 18–8) that refers to position 1 is

$$① = p_0 + \rho g h_1 + \rho u_1$$

At position 2 the static pressure is $p_2 = p_0 + \rho g y_2$, and the average flow velocity is $\bar{v}_2$, so we

can write

$$\textcircled{2} = p_0 + \rho g y_2 + \rho g h_2 + \tfrac{1}{2}\rho \bar{v}_2^2 + \rho u_2$$

At position 3 we have $\bar{v}_3 = 0$; however, there is now an increase in the internal energy of the water due to the frictional dissipation of the kinetic energy of flow through the orifice. Thus,

$$\textcircled{3} = p_0 + \rho g h_3 + \rho u_3$$

Now, equating $\textcircled{1}$ and $\textcircled{2}$ we have

$$u_2 - u_1 = g(h_1 - h_2 - y_2) - \tfrac{1}{2}\bar{v}_2^2$$
$$= g y_0 - \tfrac{1}{2}\bar{v}_2^2$$

We may reasonably assume that streamline flow exists between stations $\textcircled{1}$ and $\textcircled{2}$; hence, $u_2 = u_1$, or

$$g y_0 = \tfrac{1}{2}\bar{v}_2^2$$

Equating $\textcircled{1}$ and $\textcircled{3}$, we have

$$u_3 - u_1 = g(h_1 - h_3) = g y_0$$

We conclude that the potential energy $g y_0$, which may be associated with the difference in surface level heights on the two sides of the orifice, first appears as flow kinetic energy at the orifice position 2 and then later is converted by friction *entirely* into internal energy of the downstream water. Notice that this is always the case when water passes from a calm basin at one level to another calm basin at a lower level. (•Do you see why?)

This situation cannot be analyzed by assuming nonviscous flow along a hypothetical streamline *aa'*, because such a streamline does not exist (•Can you see why it is not possible to write a consistent set of streamline equations connecting points 1, 2, and 3?)

Example 18–5

One method of determining fluid velocities (particularly gas velocities) involves the use of a Pitot tube,* the geometry of one type of which is shown in the diagram. Obtain the expression for v_1 in terms of the manometer height difference h.

---

*Named for Henri Pitot (1695–1771), a French astronomer and mathematician, who first used such a device in 1732 to measure the flow velocity in the River Seine at Paris.

Solution:

First, notice that at the entrance to the tube that faces the fluid flow, the flow divides, producing a *stagnation point* where $v_2 = 0$.

Apply Eq. 18–8 between positions 1 and 2 with the result,

$$p_1 + \tfrac{1}{2}\rho v_1^2 = p_2 \tag{1}$$

Next, equate the static pressures at points *a* and *b* in the manometer:

$$p_a = p_1 + \rho g y_1 + \rho' g h = p_b = p_2 + \rho g (y_1 + h) \tag{2}$$

Solving Eq. 2 for $p_2 - p_1$ and substituting from Eq. 1, we find

$$p_2 - p_1 = gh(\rho' - \rho) = \tfrac{1}{2}\rho v_1^2$$

from which

$$v_1 = \sqrt{2gh\left(\frac{\rho'}{\rho} - 1\right)}$$

(•Why is the static pressure port, position 1, placed upstream and far from the central tube?)

Example 18–6

A pump delivers oil with a density of 850 kg/m³ at a rate of 10 ℓ/s (10^{-2} m³/s). The input to the pump is from a pipe with a diameter of 7.5 cm at a suction gauge pressure of -0.25 atm. The discharge of the pump is at a gauge pressure of 3.0 atm into a pipe with a diameter of 3.0 cm. There is negligible heat flow and the oil temperature remains constant. Determine the mechanical power output of the pump.

Solution:

Here we have $Q_{in} = 0$, $u_2 - u_1 = 0$, and $h_2 - h_1 = 0$. The average velocity of the fluid at the entrance to the pump is

$$\bar{v}_1 = \frac{10^{-2} \text{ m}^3/\text{s}}{\pi(3.75 \times 10^{-2} \text{ m})^2} = 2.26 \text{ m/s}$$

The average discharge velocity is

$$\bar{v}_2 = \frac{10^{-2} \text{ m}^3/\text{s}}{\pi(1.5 \times 10^{-2} \text{ m})^2} = 14.15 \text{ m/s}$$

The pressure difference between the output of the pump and the input is (3.00 atm) − (−0.25 atm) = 3.25 atm = 3.28×10^5 Pa. Then, using Eq. 18–8, we have (per kilogram of oil)

$$W_{in} = \tfrac{1}{2}[(14.15 \text{ m/s})^2 - (2.26 \text{ m/s})^2] + \frac{3.28 \times 10^5 \text{ Pa}}{850 \text{ kg/m}^3}$$

$$= 483.4 \text{ J/kg}$$

The power requirement is

$$P = W_{in}q\rho = (483.4 \text{ J/kg})(10^{-2} \text{ m}^3/\text{s})(850 \text{ kg/m}^3)$$
$$= 4109 \text{ W} \quad (5.50 \text{ hp})$$

18-5   VISCOSITY

When one layer of a fluid slides over an adjacent layer, a drag force develops at the layer interface. This fluid friction is called *viscosity*. As in the case of friction in the rubbing action between solids, viscous forces generate heat and result in the loss of mechanical energy.*

The frictional forces that are present at the interface between two solids in motion result in very little transfer of matter from one surface to the other. In the case of sliding fluid layers, however, the viscous force exerted by one layer on the other does bring about the exchange of molecules between the layers. Some of the rapidly moving molecules from the higher velocity layer are transferred (diffuse) to the lower velocity layer, and vice versa. This amounts to a diffusion of momentum which results in a decrease in the relative velocity between the layers and a conversion of transport kinetic energy into thermal energy.

When a fluid undergoes turbulent flow, the diffusion is on a macroscopic scale, involving the transfer of fluid parcels that are very small in size but nevertheless contain very large numbers of molecules. The analysis of the diffusion of momentum in turbulent eddies and vortices is beyond the scope of this book.

In the case of steady flow, with viscosity present, the fluid layers slip past one another while maintaining their gross individual identity. This type of flow is called *laminar* (layered) *flow*.

The molecular description of laminar viscous flow for liquids is quite different from that for gases. In a gas the mean spacing between molecules is considerably larger than the range of molecular forces. Consequently, the diffusion of molecules from one layer to the next transfers momentum only when collisions occur between the intruding molecules and the local molecules in the layer. Raising the temperature of a substance increases the molecular velocities and therefore increases the collision rate. Thus, the viscous effects in a gas increase as the temperature and the general thermal motions increase. (See Chapter 24 for a discussion of the viscosity of gases based on the kinetic theory of gases.)

In a liquid the mean spacing between molecules is small, and strong intermolecular forces are always present. Viscosity in liquids results from a combination of molecular diffusion and strong cohesive shearing forces. Viscosity effects for liquids are generally several orders of magnitude greater than for gases and tend to decrease with increasing temperature.

Although viscosity is the result of complicated molecular interactions, viscous effects in the laminar flow of all fluids (both gases and liquids) can be described by a simple expression that involves a single empirical coefficient for each substance.

The Viscosity Coefficient.   In Section 16-2 we discussed the fact that when a shearing force is applied to a solid, there is developed an internal shear stress which, at equilibrium, just balances the applied force. In this condition the solid sustains a

---

*In the energy equation (18-8), notice that we distinguish between fluid energy u associated with internal thermal motion and transport kinetic energy $\frac{1}{2}v^2$, both per unit mass.

static shear strain; that is, it keeps the same deformed shape as long as the force is applied. If a shearing force is applied to a fluid, however, the resulting shear stress produces flow instead of developing a static strain. Indeed, we can define a fluid to be a substance that cannot maintain a permanent shear strain.

When a solid moves through a fluid or when a fluid flows over a solid, the layer of fluid in direct contact with the solid surface experiences no motion with respect to the surface—*the relative velocity between the solid surface and the contacting layer of fluid is zero.** We make frequent use of this fact in describing fluid flow.

Consider a viscous fluid that fills the space between a pair of parallel plates. Suppose that the lower plate is held fixed while the upper plate is set into motion with a velocity $\mathbf{v}_0$ that is parallel to the plate (Fig. 18-8). The velocity of the fluid in the layer in contact with the upper plate is $\mathbf{v}_0$, whereas the fluid layer in contact with the lower plate remains at rest. At equilibrium, there is a uniform shear between the plates, which produces a linear velocity profile from P to Q given by $v(y) = v_0(y/d)$, as shown in Fig. 18-8a. The block labeled A in Fig. 18-8b represents a small portion of the fluid at some instant. At subsequent times, this same quantity of fluid will have the shapes B, C, D, . . . , and so forth. Thus, the shear strain (Eq. 16-12) increases uniformly with time at the rate v_0/d.

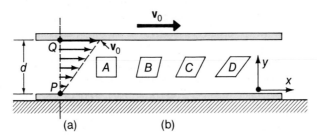

Fig. 18-8. (a) Velocity profile for laminar flow of a viscous fluid. (b) A portion of fluid represented by the block A undergoes a shear strain that increases uniformly with time.

In order to maintain the motion of the upper plate with velocity $\mathbf{v}_0$, a tangential (shearing) force $\mathbf{F}$ must be applied to the upper plate. (•What force must be applied to the lower plate to hold it stationary?) If the area of the upper plate is A, the shear stress applied to the fluid is $s_t = F/A$. The ratio of the shear stress s_t to the strain rate v_0/d is called the dynamic shear modulus (compare Eq. 16-16), or simply the *viscosity coefficient* η:

$$\eta = \frac{F/A}{v_0/d} \tag{18-10}$$

The quantity η is found to depend only on the molecular properties of the fluid (and on the temperature) and *not* on the geometry of the particular flow condition.

In a more general case of laminar viscous flow, the velocity profile and, consequently, the shear may not be uniform. Consider first the important case of viscous flow in a straight pipe or duct. A differential slab of fluid oriented parallel to the flow direction is shown in Fig. 18-9. The lower face of the slab moves with velocity v_x and the upper face moves with velocity $v_x + \Delta v_x$. The shear strain rate across the slab is $\Delta v_x/\Delta y$, where x is in the direction of the local streamlines and y is perpendicular to the streamlines. The shear stress is $s_t = \Delta F/\Delta A$, where ΔA is the area of the slab

---

*You can verify this proposition by attempting to blow away *all* of the dust on a smooth table top. Regardless of how hard you blow, there will remain a fine layer of dust. This indicates that the air layer immediately adjacent to the table surface remains at rest and is therefore unable to carry away the finest dust particles.

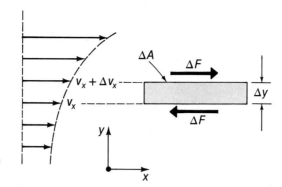

Fig. 18-9. The shearing of a fluid slab in general viscous flow.

surface. Then, the viscosity coefficient becomes

$$\eta = \frac{\Delta F/\Delta A}{\Delta v_x/\Delta y}$$

and in the limit $\Delta y \rightarrow 0$ and $\Delta A \rightarrow 0$, we have

$$s_t = \eta \frac{\partial v_x}{\partial y} \qquad \textbf{(18-11)}$$

It is evident from the expressions involving the viscosity coefficient η that the dimensions are those of presssure (F/A) multiplied by those of time (L/LT^{-1}). Thus, the SI units for the viscosity coefficient are Pa $\cdot$ s. Values of the viscosity coefficient are sometimes stated in *poise* (P) or its submultiples:

$$1 \text{ Pa} \cdot \text{s} = 10 \text{ P} = 10^3 \text{ cP} = 10^7 \,\mu\text{P}$$

The values of η for several substances are given in Table 18-1 and indicate the substantial variation of the viscosity coefficient with temperature. There is only a small dependence of η on pressure for liquids.

Poiseuille's Law. One of the important applications of the shear stress equation for parallel flow (Eq. 18-11) is concerned with laminar flow in a straight cylindrical pipe. Because of the cylindrical symmetry of the pipe, such flow consists of a series of concentric, telescoping layers, with the central portion flowing most rapidly and with the outermost layer stationary at the pipe wall.

TABLE 18-1. Viscosity Coefficient η for Several Substances

T (°C)	GLYCERINE (Pa $\cdot$ s)	MERCURY (10^{-3} Pa $\cdot$ s)	WATER (10^{-3} Pa $\cdot$ s)	BENZENE (10^{-3} Pa $\cdot$ s)	AIR (10^{-6} Pa $\cdot$ s)	HYDROGEN (10^{-6} Pa $\cdot$ s)
0	12.11	1.68	1.787	0.912	17.1	8.4
20	1.49	1.55	1.002	0.652	18.1	8.7
40	0.35	1.45	0.653	0.503	19.0	9.1
60	0.12	1.37	0.466	0.392	20.0	9.5
80	—	1.30	0.355	0.329	20.9	9.8
100	—	1.24	0.282	—	21.8	10.2
200	—	1.05	—	—	25.8	12.1

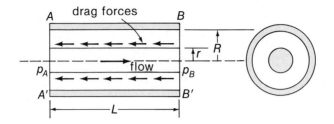

Fig. 18-10. A viscous fluid flows in a straight cylindrical pipe. The drag forces shown act on the cylinder with surface area $2\pi rL$.

Consider a cylinder of fluid with a radius r and a length L centered on the pipe axis, as shown in Fig. 18-10. The pressure difference, $p_A - p_B$, drives the fluid; this motion is opposed by the drag force, $-\eta \cdot 2\pi rL\,dv/dr$, that acts on the surface of the cylinder of radius r. At equilibrium we have

$$\pi r^2(p_A - p_B) = -2\pi rL\eta\,dv/dr$$

Integrating from r to R, the pipe radius, we have

$$\frac{p_A - p_B}{2\eta L}\int_r^R r\,dr = -\int_v^0 dv$$

so that

$$v(r) = \frac{p_A - p_B}{4\eta L}(R^2 - r^2) \tag{18-12}$$

from which we see that the velocity profile is parabolic.

Figure 18-11 shows the fluid entering the pipe through a flared input. Then, at the input plane AA', the velocity profile is nearly uniform except for a thin layer adjacent to the pipe wall. Somewhat farther along the pipe, at plane BB', the flow has begun to develop a parabolic velocity profile but there is a core of residual uniform flow. Finally, at plane CC', parabolic flow, following Eq. 18-12, is fully developed. The distance along the pipe required to develop parabolic flow depends on the pipe diameter and the flow velocity. A typical developmental length is 50 pipe diameters.

The volume flow rate through the pipe for fully developed parabolic laminar flow can be obtained by using Eq. 18-12 together with the continuity condition, which we express as $dq = v\,dA = v(r)\cdot 2\pi r\,dr$. Thus,

$$q = \frac{(p_A - p_B)\pi}{2\eta L}\int_0^R (R^2 - r^2)r\,dr$$

Fig. 18-11. Velocity profiles along a pipe for viscous laminar flow.

or

$$q = \frac{\pi R^4}{8\eta} \cdot \frac{p_A - p_B}{L} \qquad (18\text{--}13)$$

This result is known as *Poiseuille's equation.**

Poiseuille's result given in Eq. 18–13 is typical of several "flow equations" that we encounter in different areas of physics. For example, the heat conduction equation and Ohm's law of electricity are similar to Poiseuille's equation when the flow quantity is expressed in the form

$$q = \frac{\mathcal{F}}{\mathcal{R}} \qquad (18\text{--}14)$$

In the case of fluid flow, the driving force $\mathcal{F}$ is equal to the pressure difference multiplied by the cross-sectional area,

$$\mathcal{F} = \pi R^2 (p_A - p_B) \qquad (18\text{--}15a)$$

and the resistance $\mathcal{R}$ to the flow is

$$\mathcal{R} = \frac{8\eta L}{R^2} \qquad (18\text{--}15b)$$

(•Can you see why $\mathcal{R}$ is directly proportional to η and L and depends inversely on R^2?)

Example 18–7

A straight horizontal pipe with a diameter of 1 cm and a length of 50 m carries oil with a density of $\rho = 930 \text{ kg/m}^3$ and a viscosity coefficient of 0.12 Pa · s. At the entrance to the pipe, the oil temperature is 20°C. The discharge rate is 0.80 kg/s at atmospheric pressure.
  (a) Find the gauge pressure at the pipe input.
  (b) Determine the maximum stream velocity of the oil.
  (c) Show that the internal friction heating rate is

$$P = \frac{p_A - p_B}{\rho} \frac{dm}{dt}$$

where $dm/dt = q\rho$ is the rate of mass flow through the pipe. Determine the rate of heating in the 50-m length of pipe.

Solution:

  (a) The flow rate q is

$$q = \frac{1}{\rho} \frac{dm}{dt} = \frac{0.80 \text{ kg/s}}{930 \text{ kg/m}^3} = 8.60 \times 10^{-4} \text{ m}^3/\text{s}$$

---

*After Jean Louis Marie Poiseuille (1799–1869), a French physiologist noted for his study of blood flow. The result was derived and tested by Poiseuille in 1844.

Then, using Eq. 18–13, the pressure difference is

$$p_A - p_B = \frac{8\eta L q}{\pi R^4} = \frac{8(0.12 \text{ Pa} \cdot \text{s})(50 \text{ m})(8.60 \times 10^{-4} \text{ m}^3/\text{s})}{\pi(5 \times 10^{-3} \text{ m})^4}$$

$$= 2.10 \times 10^7 \text{ Pa} = 207 \text{ atm (gauge)}$$

(b) The maximum stream velocity occurs at the center of the pipe. Using Eq. 18–12 with $r = 0$, we have

$$v_m = \frac{(p_A - p_B)R^2}{4\eta L} = \frac{(2.10 \times 10^7 \text{ Pa})(5 \times 10^{-3} \text{ m})^2}{4(0.12 \text{ Pa} \cdot \text{s})(50 \text{ m})}$$

$$= 21.9 \text{ m/s}$$

(c) The difference between the input mechanical power and the output mechanical power represents the rate of heat generation by internal friction. According to Eq. 7–13, power is equal to the product of force and velocity. Here, the force is the pressure multiplied by the differential area $2\pi r\, dr$ of a cylindrical lamina. Thus, the rate of heat generation is

$$P = 2\pi(p_A - p_B) \int_0^R v(r)r\, dr$$

Substituting for $v(r)$ from Eq. 18–12, we have

$$P = \frac{\pi(p_A - p_B)^2}{2\eta L} \int_0^R (R^2 - r^2)r\, dr$$

and using Eq. 18–13 to identify q, we obtain

$$P = q(p_A - p_B) = \frac{p_A - p_B}{\rho} \frac{dm}{dt}$$

(•Can you see how this result follows directly from Eq. 18–8?)

If the pipe is insulated against heat loss, all of the friction-generated heat appears as increased internal energy. This amounts to

$$P = \frac{2.10 \times 10^7 \text{ Pa}}{930 \text{ kg/m}^3} \cdot (0.80 \text{ kg/s})$$

$$= 18.1 \text{ kW}$$

SPECIAL TOPIC

18-6   DIMENSIONAL ANALYSIS

Many physical problems, particularly those involving the mechanics of fluids, do not permit exact theoretical solutions. In such cases, it is often possible to gain some insight concerning the behavior of the system by using an orderly technique called *dimensional analysis*.

The broad purpose of dimensional analysis is to find, without actually solving the problem and by using only dimensional reasoning, basic relationships that connect the relevant variables. In engineering practice it is often necessary to test the behavior of a device by using a scale model instead of constructing a full-size prototype. New designs for aircraft are tested as models in wind tunnels; also, model testing in water tanks is used to determine the effects of changes in ship-hull designs. Any relationships connecting the variables of a system that have been discovered by dimensional analysis will indicate the proper way to scale the variables so that the model performance will be *dynamically similar* to that of the actual device.

To understand the meaning of dynamic similarity, consider the flow pattern of a fluid moving around a fixed sphere, as in Fig. 18–12a. To model this situation, it is not sufficient merely to reduce the size of the sphere. Figure 18–12b shows the flow around a sphere with half the diameter of the original sphere for the same fluid velocity v_0. The pattern of streamlines and vortices is considerably different in the two cases. However, if we double the fluid velocity to $2v_0$ for the half-size model, as in Fig. 18–12c, we find the same flow pattern as for the original sphere (Fig. 18–12a). By using a $\frac{1}{10}$-scale model, we would find dynamic similarity with the full-size case by using a fluid velocity of $10v_0$ (assuming incompressibility of the fluid). Thus, we find that dynamic similarity is achieved in this case if the fluid velocity and the size of the object are scaled in such a way that the product of these variables remains constant. The procedures of dimensional analysis show us how to predict that such a relationship exists.

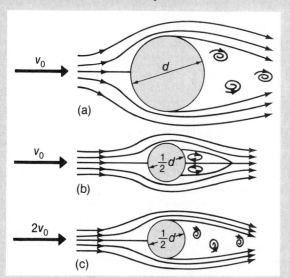

Fig. 18–12. (a) A fluid with a velocity v_0 flows around a sphere with a diameter d. (b) The pattern of streamlines and vortices is different for a sphere with a diameter $\frac{1}{2}d$. (c) If the velocity is doubled to $2v_0$, the flow is dynamically similar to that for the original sphere.

Procedure. Any equation involving physical variables must be *dimensionally homogeneous*. That is, every term in an equation must have exactly the same physical dimensions. For example, in the equation for the displacement in uniformly accelerated motion, $s(t) = \frac{1}{2}at^2 + v_0 t + s_0$, each term consists of quantities whose combined dimensions are length. We continue to consider length (L), time (T), and

mass (M) to be the fundamental physical dimensions of mechanics.

The first step in the dimensional analysis of a problem is to decide, using physical insight based on familiarity with related physical situations, what physical quantities could conceivably play important roles as variables. Suppose that there are ξ such quantities. Suppose further that there are ν fundamental dimensions involved in the problem.* Then, the desired equation connecting the variables can be expressed as a relationship among a minimum of $\xi - \nu$ *independent dimensionless groups of variables*. This theorem is referred to as *Buckingham's Pi theorem*, enunciated in 1914.

Consider a case involving 6 physical variables, r, s, t, u, v, and w; thus, $\xi = 6$. Suppose also that these variables involve all 3 of the fundamental dimensions of mechanics; thus, $\nu = 3$. Then, according to the Pi theorem, there are $\xi - \nu = 3$ dimensionless groups which, for this particular case, we might write as $r^2 s/uw^3$, v/u, and rt^3. Then, the desired equation is

$$f(r^2 s/uw^3, v/u, rt^3) = 0 \qquad \textbf{(18–16)}$$

where f is some function of the three new variables that are the dimensionless groups. The number of variables in the problem has been reduced from 6 to 3, a considerable simplification. Now, if we are interested in the dependence of one of the original variables, say s, on the others, we can rewrite Eq. 18–16 as

$$s = C\frac{uw^3}{r^2}\Phi(v/u, rt^3) \qquad \textbf{(18–17)}$$

Thus, we have explicitly determined the dependence of s on the variables r, u, and w. The dimensionless constant C is not determined, but it usually results from geometric considerations and most often is in the range $2\pi \geq C \geq \frac{1}{2}\pi$.

In Eq. 18–17, Φ is a new function of the variables v/u and rt^3. Although the function Φ is unknown, we *do* know that if the variables r, t, u, and v are changed in such a way that v/u and rt^3 remain constant, there can be no change in Φ. That is, Eq. 18–17 gives the explicit dependence of s on r, u, and w if v/u and rt^3 remain constant.

Example 18–8

Use dimensional analysis to determine an expression for the period P of a simple pendulum.

---

*Usually ν will be 3 or less, corresponding to which among L, T, and M are important. However, more complicated problems than those considered here might require such additional fundamental dimensions as temperature and electric charge.

Solution:

First, we must list the relevant variables on which the period might depend. These should include the acceleration g due to gravity and the length ℓ of the pendulum, and perhaps the amplitude A of the oscillations. We assume for the moment that the mass m of the bob is not important. We therefore have four variables, with $[P] = T$, $[\ell] = L$, $[g] = LT^{-2}$, and $[A] = L$, which require only two fundamental dimensions, L and T. Thus, we have $\xi = 4$ and $\nu = 2$, so we need to form $\xi - \nu = 2$ dimensionless groups.

To form these groups, we notice that the period P has dimensions only of T; the only other variable with dimensions including T is g, so one of the groups must contain both P and g. This group must also contain a variable with dimensions of length, either ℓ or A. From our experience with oscillatory systems, we expect that the size of the system (here, the length ℓ) is more closely associated with the period than is the amplitude. Therefore, we choose one dimensionless group to consist of P, ℓ, and g, and we write $[P^\alpha \ell^\beta g^\gamma] = 1$ or $T^\alpha L^\beta L^\gamma T^{-2\gamma} = 1$.

The second group must contain the remaining variable, A. This variable can only be coupled with ℓ to form a dimensionless group. Therefore, we have $[A^\delta \ell^\epsilon] = 1$ or $L^\delta L^\epsilon = 1$.

From the first dimensional equation, we have $\alpha - 2\gamma = 0$ and $\beta + \gamma = 0$, and from the second equation, $\delta + \epsilon = 0$. We have three equations with five unknowns, so we arbitrarily choose $\alpha = 1$ and $\delta = 1$. Then, $\gamma = \frac{1}{2}$, $\beta = -\frac{1}{2}$, and $\epsilon = -1$. We can now write the desired function as

$$f(P\ell^{-1/2}g^{1/2}, A/\ell) = 0$$

Solving for the period P, we find

$$P = C\sqrt{\frac{\ell}{g}}\,\Phi(A/\ell)$$

(•How significant is the arbitrary choice, $\alpha = 1$ and $\delta = 1$? Would the result have been different if we had chosen $\alpha = 2$?)

In Section 15–4, we found that the period of a simple pendulum is given by (Eq. 15–23)

$$P = 2\pi\sqrt{\frac{\ell}{g}}[1 + \tfrac{1}{4}\sin^2\frac{\Theta}{2} + \tfrac{9}{64}\sin^4\frac{\Theta}{2} + \cdots]$$

Thus, the dimensional analysis technique gives the correct dependence of the period on ℓ and g. Even the auxiliary function Φ has the correct form because $\Theta = \sin^{-1}(A/\ell)$. (•Suppose that we consider the "amplitude" of the oscillations to be the *angular* amplitude Θ instead of the *displacement* amplitude A. How would this influence the dimensional argument?)

Notice that the result for the period, obtained without the benefit of any theory of oscillations, correctly predicts that a pendulum on the Moon with a length $\ell_M/g_M = \ell_E/g_E$ and with an amplitude $A_M/\ell_M = A_E/\ell_E$ will have exactly the same period as a pendulum with length ℓ_E and amplitude A_E on Earth, even though the constant C and the function Φ are unknown!

Suppose that we had guessed (incorrectly) that the mass m of the bob also influences the period of the pendulum. This would have increased ξ to 5 and ν to 3, still leaving $\xi - \nu = 2$. However the groups are now formed, the exponent of m is always zero because m is the only variable involving the dimension M. Thus, dimensional analysis shows that the period of a simple pendulum is independent of the mass of the bob. (•Can you generalize this conclusion into a rule?)

In this example, there is an arbitrariness in the choice of the two dimensionless groups because ℓ and A have the same dimensions. In fact, there is often an arbitrariness when $\xi - \nu$ is 2 or greater. Thus, some skill is required in selecting the dimensionless groups. Experience with similar systems is usually the best guide.

The Reynolds Number. In many cases of viscous fluid flow, an important factor is the ratio of the inertial force to the viscous force. This dimensionless ratio is called the *Reynolds number,*[*] N_R.

The inertial force is proportional to (mass) × (acceleration) or (volume) × (density) × (velocity/time) = $(\ell^3)(\rho)(v/t) = \ell^3\rho v/(\ell/v) = \ell^2\rho v^2$, where we have eliminated the explicit appearance of time as a variable. Here, ρ is the density of the fluid and v is a characteristic velocity in the problem, such as the flow velocity in a pipe or the velocity of an object moving through air. The quantity ℓ is a characteristic length, such as the diameter of a pipe or the square root of the cross-sectional area of a moving object.

The viscous force (see Eqs. 18–10 and 18–11) is proportional to $s_t\ell^2$, with s_t proportional to $\eta v/\ell$. Thus, the viscous force is proportional to $\eta v\ell$. Finally, the Reynolds number is

$$N_R = \frac{\ell^2\rho v^2}{\eta v\ell} = \frac{\rho v\ell}{\eta} \qquad \textbf{(18–18)}$$

For example, suppose that a liquid with a density ρ and coefficient of viscosity η flowing in a pipe with a diameter ℓ is observed to break into unsteady turbulent flow when the flow velocity exceeds a critical value v_c. Then, because the equality of the Reynolds numbers means dynamically similar conditions, we expect that another liq-

---

[*] After Osborne Reynolds (1842–1912), English physicist.

uid, with ρ' and η', flowing in a pipe with diameter ℓ' will break into turbulent flow for velocities exceeding v_c', where

$$v_c' = \frac{\rho\ell/\eta}{\rho'\ell'/\eta'} v_c$$

When a subsonic test of a model airfoil is made in a wind tunnel, dynamic similarity with the real situation is achieved if the Reynolds number for the model test is made the same as that for the full-size airfoil in actual flight. Notice that the form of the Reynolds number N_R (namely, $N_R \propto v\ell$) justifies the comments made earlier in connection with the discussion of Fig. 18–12.

Other Dynamical Ratios.   The behavior of the free surface of a liquid is involved in problems concerning, for example, the propagation of surface waves or the flow of water in an open duct. In these situations, an important dimensionless quantity is the ratio of the inertial force to the gravitational force. The inertial force is again proportional to $\rho\ell^3 g$. Thus, the ratio of interest is $\ell^2\rho v^2/\rho\ell^3 g = v^2/\ell g$. The square root of this ratio (also dimensionless) is called the *Froude number, N_F*:

$$N_F = \frac{v}{\sqrt{\ell g}} \qquad (18\text{–}19)$$

When the compressibility of the air or other gas must be included in a problem, the elastic force involved is proportional to $\beta\ell^2$, where β is the bulk modulus of the gas (Eq. 16–21). The ratio of the inertial force to the elastic force is called the *Cauchy number, N_C*:

$$N_C = \frac{\ell^2\rho v^2}{\beta\ell^2} = \frac{\rho v^2}{\beta} \qquad (18\text{–}20)$$

The ratio of the pressure-developed force, $\Delta pA = \Delta p\ell^2$, to the inertial force is called the *Euler number, N_E*:

$$N_E = \frac{\Delta p\ell^2}{\ell^2\rho v^2} = \frac{\Delta p}{\rho v^2} \qquad (18\text{–}21)$$

Another dynamical ratio that is often used, particularly in aerodynamics, is the ratio of the velocity v to the speed of sound c in the medium. This ratio is called the *Mach number, N_M*:

$$N_M = \frac{v}{c} \qquad (18\text{–}22)$$

The Mach number is not independent of the Cauchy number because $c = \sqrt{\beta/\rho}$ (see Eq. 20–5), but it is used when the emphasis is on the relative velocity between an object and a gaseous medium (as in aircraft design).

Example 18–9

It is desired to determine the drag force on a high-speed airfoil by testing a scale model in a wind tunnel. The drag force F_D should depend on the following factors: the relative air velocity v, the air density ρ, the viscosity coefficient η for air, the bulk modulus β for air, and a characteristic length ℓ of the airfoil. Determine the dependence of the drag force F_D on the suggested relevant variables.

Solution:

We must express $f(F_D, v, \rho, \eta, \beta, \ell) = 0$ in terms of dimensionless variables. In this case, we have $\xi = 6$ and $\nu = 3$, so there are at least $\xi - \nu = 3$ dimensionless groups. One of these groups should be the ratio of the drag force to the inertial force, that is, $F_D/\ell^2\rho v^2$. We also expect that the Reynolds number and the Cauchy number are relevant. Thus, the desired equation is

$$f(F_D/\ell^2\rho v^2, N_R, N_C) = 0$$

Solving for the drag force, we have

$$F_D = (\tfrac{1}{2}\rho v^2)\ell^2 C_D \qquad (1)$$

where C_D is the *drag coefficient*,

$$C_D = \Phi(N_R, N_C)$$

In Eq. 1, we have inserted the customary factor $\tfrac{1}{2}$ because $\tfrac{1}{2}\rho v^2$ represents the dynamic pressure (as in Eq. 18–9); the remaining term, ℓ^2, is interpreted as the exposed area of the object.

Experiments show that the drag coefficient depends primarily on the Reynolds number; this is strictly true in the incompressible regime. Most objects, for a variety of high-speed conditions, have C_D values of approximately unity (see Fig. 18–13 on the next page for the case of a cylinder).

18-7 TURBULENCE

The character of the flow of an incompressible viscous fluid around a submerged object is determined almost exclusively by the Reynolds number N_R. The ratio of the drag force to the inertial force on an object is equal to the drag coefficient, $C_D = F_D/\frac{1}{2}\ell^2\rho v^2$, as we found in Example 18-8. In this general expression for C_D, we interpret ℓ^2 as the frontal area of the object exposed to the fluid.

Let us consider the case of a sphere with a radius R that moves through a fluid. The frontal area is just πR^2, so the drag coefficient is $C_D = F_D/\frac{1}{2}\pi R^2\rho v^2$. When the Reynolds number is less than about 0.5, experiment shows that $C_D = 24/N_R$. Then, the drag force becomes

$$F_D = \tfrac{1}{2}\pi R^2\rho v^2 \cdot C_D = \tfrac{1}{2}\pi R^2\rho v^2 \cdot \frac{24}{\rho v(2R)/\eta}$$

where we have substituted the diameter of the sphere ($2R$) for the characteristic length ℓ in the expression for the Reynolds number. Thus,

$$F_D = 6\pi\eta Rv \qquad (N_R \lesssim 0.5) \qquad \textbf{(18-23)}$$

This result is known as *Stokes' law*, which we alluded to in Section 6-3. In general, for low viscosities, the drag force on any object is proportional to the velocity.

The value of C_D for a sphere continues to decrease with increasing N_R until $N_R \cong 100$. Then, in the range from

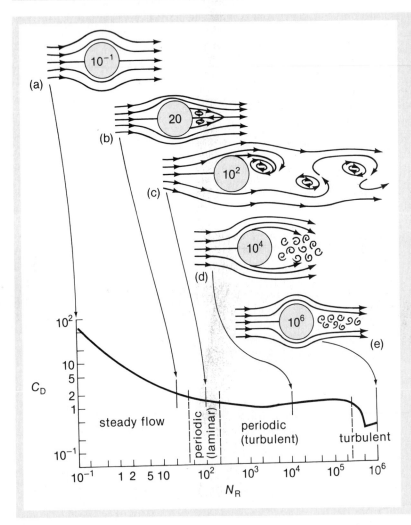

Fig. 18-13. Variation of the drag coefficient C_D with the Reynolds number N_R for a cylinder. The nature of the flow is also shown schematically for several values of N_R. (Adapted from R. P. Feynman, R. B. Leighton, and M. Sands, *The Feynman Lectures on Physics;* Addison-Wesley, Reading, Mass., 1964, Vol. II, pp. 44–48.)

about 10^2 to about 10^5, the drag coefficient is approximately constant at a value of $\frac{1}{2}$. In this regime the drag force is proportional to v^2:

$$F_D = \tfrac{1}{4}\pi R^2 \rho v^2 \qquad (10^3 \lesssim N_R \lesssim 10^5) \quad \textbf{(18-24)}$$

This result is known as *Newton's law of resistance*.

In the Stokes'-law region, the viscous force is dominant. In the Newton's-law region, however, the viscous force is relatively unimportant; indeed, Eq. 18-24 for the drag force in this region does not depend on the viscosity of the fluid. Instead, high-speed drag is due primarily to the force (the *pressure drag*) that results from pressure differences around the body arising from the separation of the streamlines from the object.

Figure 18-13 shows the variation with N_R of the drag coefficient for a cylinder oriented with its axis perpendicular to the flow direction. As for the case of the sphere, the inertial force is substantially weaker than the viscous force for small values of N_R. Then, the streamlines extend smoothly around the cylinder and into the downstream region (Fig. 18-13a). At the opposite extreme, $N_R \gtrsim 10^6$, the inertial force is dominant and the cylinder leaves a thoroughly turbulent wake (Fig. 18-13e).

At a flow rate corresponding to $N_R \cong 20$, the flow pattern changes more or less abruptly from the attached streamline flow associated with smaller values of N_R. As indicated in Fig. 18-13b, the flow separates from the cylinder along two lines on the downstream surface, and in the process two stationary (stable) vortices develop. In the region $20 \lesssim N_R \lesssim 40$, these vortices become unstable. At first, the vortices oscillate irregularly and then, at about $N_R = 40$, they break away alternately from each side. The detached vortices move downstream with the fluid, as shown in Fig. 18-13c and in the photograph at right. This stream of vortices is called the *Kármán vortex trail*.[*] This behavior persists for values of N_R up to about 200. For larger N_R, the detached vortices become increasingly turbulent, although they retain a quasi-periodic character (Fig. 18-13d). When N_R reaches a value of approximately 3×10^5, the drag coefficient decreases abruptly and a well-defined turbulent wake develops (Fig. 18-13e).

The variation of the drag coefficient with N_R is the result of competition between two types of drag—viscous or *skin-friction drag* and *pressure drag*. Skin-friction drag is due directly to the retarding effects of the tangential shear stresses caused by viscosity. Pressure drag depends on the distortion produced in the velocity field, which is also a result of viscous effects. Therefore, although the viscous force on an object is small compared to the pressure drag

---

[*] After Theodor von Kármán (1881–1963), noted Hungarian-American physicist and aeronautical engineer.

force for $N_R \gtrsim 10$, viscosity nevertheless plays the crucial role in determining the total drag on the object.

Pressure drag arises when the streamlines fail to pass completely around an object and instead separate from the rear surface, creating vortices (Fig. 18-13b). The pressure in the separated wake region is low compared with the pressure in the attached flow region. This pressure difference multiplied by the cross-sectional area of the separated wake yields the pressure drag force. This phenomenon sets in for $N_R \cong 10$, and the wake area increases slowly to equal the total frontal exposed area. For $10^3 \lesssim N_R \lesssim 10^5$, the drag coefficient remains approximately constant and the pressure drag force is proportional to v^2 because the dynamic pressure $\frac{1}{2}\rho v^2$ is the controlling factor. Throughout this region, the wake is turbulent.

When the Reynolds number for the flow past a cylinder reaches a value of about 3×10^5, the flow in the thin layer of fluid near the surface suddenly changes from laminar to turbulent. A turbulent layer permits the rapid exchange of momentum and this delays the onset of separated flow. The wake becomes thoroughly turbulent, but the cross-sectional area is smaller (Fig. 18-13e). Consequently, the pressure drag is decreased to about one-quarter of its former value.[*] For still larger values of N_R, the drag coefficient increases slowly.

---

[*] By roughening the leading part of the cylinder, the flow near the surface can be induced to become turbulent, with a resulting decrease in the drag coefficient, for N_R values substantially below 3×10^5, the value for the natural onset of turbulence. This effect can also be caused by turbulence in the fluid that approaches the object. The subject of turbulence is extremely complex and all predictive methods concerning turbulent behavior are based on empirical results. Indeed, von Kármán once said that "only God understands turbulence."

(From Prandtl, L., and Tietjens, O. G.: *Applied Hydro- and Aerodynamics.* McGraw-Hill Book Co., New York, 1934. By permission of the publisher.)

The streamlines for *nonviscous* flow around a cylinder closely resemble those shown in Fig. 18–13a. However, there is no drag whatsoever on the cylinder in this case, and the flow patterns of the types shown in Fig. 18–13b, c, d, and e never develop. The drag force is entirely a result of viscous effects.

The drag experienced by an object depends not only on the shape of the upstream surface but on the contour of the rear section as well. The photograph below shows the flow patterns in a wind tunnel for a cylinder and for a streamlined strut. Whereas the cylinder exhibits separation and wake turbulence, the more gradual contour of the strut permits a smooth flow over the entire surface so that separation does not develop. Consequently, the primary drag on a streamlined surface, such as an airfoil, is skin-friction drag, not pressure drag.

Lift. When a fluid flows past a stationary sphere or

Wind tunnel photograph of the flow past a cylinder (a) and a streamlined strut (b). The cylinder exhibits separation and wake turbulence. (From Prandtl, L., and Tietjens, O. G.: *Applied Hydro- and Aerodynamics*. McGraw-Hill Book Co., New York, 1934. By permission of the publisher.)

(a)

(b)

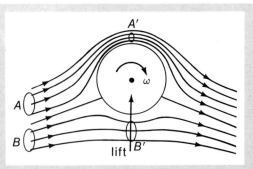

Fig. 18–14. Asymmetric viscous flow around a rotating sphere.

cylinder, the streamlines and the resulting pressure distribution are symmetric. Consequently, there is no net transverse force on the object. However, rotation of the sphere or cylinder produces an asymmetry in the streamlines because the fluid at the surface is dragged along by the rotation, as shown in Fig. 18–14. If we apply the continuity condition to the two tubes of flow, AA' and BB', we see that at A', where the cross-sectional area of the tube is small, the stream velocity must be greater than at B', where the area is large. Thus, in accordance with the energy equation, the pressure at A' is lower than that at B', thereby producing a net upward force or *lift*. This phenomenon is called the *Magnus effect*, after its discoverer,* and is responsible for the curved motion of spinning baseballs, tennis balls, Ping-Pong balls, and so forth.

The flow of a viscous fluid around a nonrotating but asymmetrically shaped object, such as an airfoil, also results in a lifting force (Fig. 18–15). The lift is caused by the bunching together of the streamlines above the upper surface of the airfoil, thereby causing the pressure to be lower than the pressure below the airfoil. For a particular airfoil shape, the lift (and the drag) depend on the *angle of attack,* measured between the geometric chord of the airfoil and the direction of the far upstream flow (Fig. 18–16). In anal-

---

*H. G. Magnus (1802–1870), German scientist, first demonstrated the presence of the transverse force in 1852.

Fig. 18–15. The flow of air around an asymmetric airfoil produces a bunching together of the streamlines above the airfoil. The pressure is therefore lower above the airfoil than below, so there is a net aerodynamic lift on the airfoil.

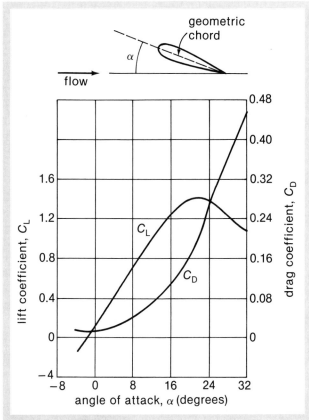

Fig. 18–16. Lift and drag coefficients for an airfoil at $N_R = 3.0 \times 10^6$.

ogy with the drag coefficient introduced in Example 18–8, $C_D = F_C/\frac{1}{2}\rho v^2 A$, we define a *lift coefficient*, $C_L = F_L/\frac{1}{2}\rho v^2 A$, where A is the frontal exposed area when $\alpha = 0$. Figure

18–16 gives C_D and C_L as functions of α for a particular airfoil* for $N_R = 3.0 \times 10^6$.

In Fig. 18–16, notice that the lift coefficient suddenly begins to decrease at an attack angle of about 20°. This loss of lift, or *stall*, is brought on by the separation of the flow from the airfoil, which results in a pressure above the upper surface greater than that caused by attached flow.

Flow Through Pipes. The channeling effect of the wall of a pipe permits smooth laminar flow with a parabolic velocity profile (Eq. 18–12) to take place for Reynolds numbers up to about 2000. For $2000 \le N_R \le 4000$ there is a critical transition region in which the flow is extremely sensitive to all sorts of small disturbances. Finally, for $N_R \ge 4000$, the flow is completely turbulent. In this region, experiments show that the flow rate through a straight pipe is given by another simple flow equation with the same form as Eq. 18–14, namely,

$$q = \frac{\mathcal{F}}{\mathcal{R}} \qquad \text{(18–25)}$$

where

$$\mathcal{F} = \pi R^2 (p_A - p_B) \qquad \text{(18–26a)}$$

$$\mathcal{R} = \frac{\eta L}{\pi R^2}\phi \qquad \text{(18–26b)}$$

$$\phi = 0.209(N_R)^{3/4} \qquad \text{(18–26c)}$$

The coefficient ϕ is based on evidence gathered by H. Blasius in 1913.

---

*N.A.C.A. airfoil 2418. N.A.C.A. (National Advisory Committee on Aeronautics) was the forerunner of N.A.S.A. (National Aeronautics and Space Administration).

PROBLEMS

Section 18–2

18–1   Water flows at a rate of 2.0×10^{-4} m³/s in a pipe with a diameter of 3 cm. A small crack (0.1 mm × 3 mm) develops, and the output flow rate decreases by 1 percent. (a) What is the average velocity of the water that squirts from the crack? (b) What is the average velocity of the water delivered by the pipe?

18–2   Rain falls vertically and accumulates at a rate of 10 cm/h. The rain is intercepted by a plane that has a surface area of 2.0 m² and is inclined to the hori-

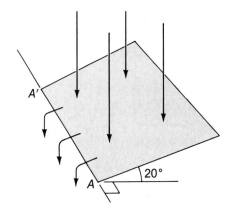

zontal by an angle of 20°. At what rate (in cm³/s) does the rain water flow across the edge AA'?

18-3 Air is forced at a rate of 150 g/s through a pipe with a diameter of 10 cm. What is the average flow velocity of the air? (The density of air is 1.293 kg/m³.) Can the flow be considered incompressible in this case?

Section 18-3

18-4 A large water tank has a 1-cm² hole in its side at a point 2 m below the surface. What is the mass discharge rate (in kg/s) through the hole?

18-5 A stream of water (initially horizontal) flows from a small hole in the side of a tank. The stream is 2.5 m from the side of the tank after falling 0.7 m. How far below the surface of the water in the tank is the hole?

18-6 The surface of water in a tank stands 2.5 m above the bottom where a small hole allows water to drain out in a vertical stream. Ignore the dynamic contraction of the stream immediately below the hole (i.e., ignore the formation of the *vena contracta*). At what distance below the bottom of the tank will the stream have a cross-sectional area equal to one half the hole area?

18-7 A siphon is used to drain water from a tank, as indicated in the diagram. The siphon has a uniform diameter d. Assume steady flow. (a) Derive an expression for the volume discharge rate at the end of the siphon. (Select the reference level at point 3.) (b) What is the limitation on the height of the top of the siphon above the water surface? Remember, liquids can sustain almost no tensile stress.

18-8 Suppose that the spillway in Example 18-2 is triangular in shape, as shown in the diagram. Obtain the

expression for the volume flow rate over the spillway. Assume steady flow.

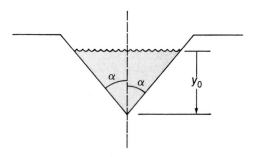

18-9 A stream flows smoothly down a 2° slope. At one • point the stream has a rectangular cross section with a depth of 15 cm and a width of 60 cm. At another point, 100 m farther down the slope, the width is 100 cm. If the average flow velocity is 4 m/s at the upper point, what is the depth of the stream at the lower point? Assume steady flow.

Section 18-4

18-10 The static pressure p in a tube filled with a flowing fluid can be measured with an open manometer, as shown in the diagram. Derive an expression for the static pressure at the tube wall in terms of h_1, h_2, ρ, ρ', and the atmospheric pressure p_0.

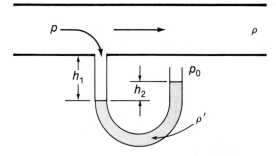

18-11 Water flows at a rate of 2 m³/min through a pipe with a diameter of 10 cm. (a) What is the average flow velocity? (b) What is the average flow velocity in a region of the pipe that has a constriction with a diameter of 6 cm? (c) What is the static pressure difference (in atm) between the main pipe and the constriction?

18-12 The nozzle system shown in the diagram is a variation of the Venturi meter (Example 18-3) for measuring the flow rate of a fluid. Unlike flow in the Venturi meter, however, the fluid emerging from the nozzle is allowed to expand freely in the pipe, resulting in considerable downstream turbulence. Show that the volume flow rate q is given by the same expression as for the Venturi meter. Assume that the right-hand tube of the manometer senses the static pressure at the nozzle—plus the usual fluid pressure. This is a good approximation in practice. (•Why is it that Bernoulli's equation cannot be used here?)

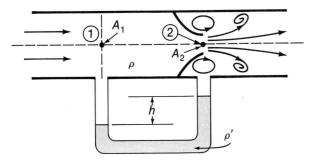

18-13 A pump that delivers 10 hp of mechanical power is
• used to raise water through a height of 30 m at a mass flow rate of 20 kg/s. The water flows through 200 m of rough-surfaced pipe that has an inside diameter of 5.0 cm. If the suction and output pressures are both 1 atm, what is the rate of frictional heating of the water (in W/m)? What fraction of the pump power is dissipated through friction?

18-14 The *Prandtl tube*, shown in the diagram, is a device
• similar to a Pitot tube (Example 18-5) for measuring fluid flow velocities. Show that the expression for the velocity v_1 is the same as that found for the

Pitot tube. What advantage does the Prandtl tube have compared with the Pitot tube?

Section 18-5

18-15 Two tubes that carry water have equal input pressure applied and both exhaust to air. Tube *A* has a diameter of 4 mm and a length of 10 m; tube *B* has a diameter of 6 mm and a length of 20 m. What is the ratio of the discharge rates q_B/q_A?

18-16 Water at 20° C is discharged into air at a rate of 2.0×10^{-2} m³/s from a 20-m length of garden hose that has a diameter of 1.90 cm. What input pressure (in atm) is required?

SPECIAL TOPIC

Section 18-6

18-17 Consider the flow of three different fluids through a pipe with a diameter of 10 cm. At what flow velocity would a Reynolds number of 100 be reached for (a) air ($\rho = 1.29$ kg/m³), (b) water ($\rho = 1.00 \times 10^3$ kg/m³), and (c) glycerine ($\rho = 1.26 \times 10^3$ kg/m³)? Use Table 18-1.

18-18 As a skydiver falls, his speed increases until the drag force (see Eq. 1 in Example 18-9) is equal to the gravitational force. Determine the *terminal velocity* of an 80-kg skydiver whose drag coefficient is $C_D = 0.5$ (a) when his arms and legs are spread out so that the exposed frontal area—corresponding to ℓ^2—is equal to 0.6 m², and (b) when he is drawn up essentially into a spherical ball. [*Hint:* The density of a person is approximately the same as that of

water. •Do you have some personal evidence that this is so?]

18-19 Consider a longitudinal compressional disturbance (a *sound wave*) that propagates through a long metal rod. In such a situation, the speed of sound *c* is expected to depend on Young's modulus *Y* and on the density ρ of the rod. Determine the dependence of *c* on *Y* and ρ. (The undetermined constant is just unity in this case, so the result of the dimensional argument is entirely correct.)

18-20 An oceangoing ship has a water-line length of 200 m and a cruising speed of 15 knots (0.517 m/s). A geometrically similar model of the hull is constructed to study the effects of generated wave resistance by towing the model in a water tank. If the model has a length of 2 m, what is the appropriate tow velocity? Neglect viscous effects.

18-21 In the design study of an automobile that is to cruise at 90 km/h, a 1/15-scale model is constructed for testing in water. If the important factors to be considered are the inertial force and the viscous force, at what speed should the model be tested in water? Assume a temperature of 20° C in both cases and use Table 18-1.

18-22 A study is to be made to predict the pressure distribution in flowing water under different gravity conditions (for example, in accelerated systems). A
  •

1:10 scale model is constructed and mercury (density = 13.6×10^3 kg/m³) is used. The system temperature is 40° C. The Froude, Euler, and Reynolds numbers are judged to be relevant, and each has the same value in the mercury case as in the water case. If the mercury experiment is carried out at sea level and the pressure difference between two points in the system is measured to be 0.4 atm, what is the corresponding pressure difference and what is the effective value of g in the full-scale water system? (Use Table 18-1.)

SPECIAL TOPIC

Section 18-7

18-23 Consider incompressible laminar viscous flow in a straight pipe. (a) Show that the space-averaged velocity is $\bar{v} = \frac{1}{2}v_0$, where v_0 is the flow velocity along the axis. (b) Determine the Reynolds number, $N_R = \rho\bar{v}\ell/\eta$, for the case in Example 18-7. (Consider ℓ to represent the diameter of the pipe.) Is the flow actually laminar?

18-24 Show that the terminal velocity (see Section 6-3) of a small sphere with density ρ and radius R that is falling near the Earth's surface through a viscous fluid with density ρ_0 is

$$v_\infty = \frac{2}{9}\frac{R^2 g}{\eta}(\rho - \rho_0)$$

where the effect of buoyancy has been included. (Assume Stokes' law.)

18-25 Use the result in Problem 18-24 in the following cases. (a) What is the radius of a fog droplet whose terminal velocity in air (at 20° C) corresponds to a Reynolds number of 0.5? (The density of air at 20° C is 1.20 kg/m³.) (b) What is the radius of a sphere of aluminum ($\rho = 2.70 \times 10^3$ kg/m³) sinking in glycerin ($\rho_0 = 1.26 \times 10^3$ kg/m³) at 20° C with $N_R = 0.5$? (c) Find the terminal velocity v_∞ in each of the above cases.

18-26 An N.A.C.A. 2418 airfoil (Fig. 18-16) presents a
  •• thickness of 25 cm to the airstream when $\alpha = 0$. (a) At what airspeed does $N_R = 3.0 \times 10^6$ when the airfoil moves through air at an altitude of 4200 m where the density is 0.80 kg/m³? (b) If the attack angle is $\alpha = 16°$, what is the drag force per meter of wingspan for the conditions in (a)? What is the lift force per meter of wingspan? (Use Fig. 18-16.) (c) If the wingspan is 10 m, what mass could the wing support? (For a finite wing, end effects are important. The vortices that form at the ends of the wing produce a "downwash" that reduces the effective angle of attack. A typical reduction is about 4°, so use $\alpha_{\text{eff}} = 12°$ to determine C_L.) (d) Assume that the airfoil has a uniform chord of 1.3 m. Estimate the difference in the average flow velocity along the upper and lower surfaces of the airfoil. (Use Eq. 18-8 and neglect the dissipation of energy by friction.)

18-27 A straight horizontal pipe with a diameter of 10 cm and a length of 20 m discharges water at a rate of 10 kg/s. The water temperature is 20° C. (a) What is the Reynolds number N_R? (b) What is the pressure differential between the ends of the pipe? (Assume that the Blasius formula is correct—Eqs. 18-25 and 18-26.)

Additional Problems

18-28 The free surface of water in a tank is a distance h above ground level. At what distance below the free surface should a small hole be drilled into the tank wall so that the emerging water jet will strike the ground the greatest distance from the tank?

18-29 A horizontal convergent duct has an entrance area of 6 m² and an outlet area of 1.5 m². Air at atmospheric pressure and with a uniform velocity of 20 m/s enters the duct. Assume that the air temperature remains at 20° C. (The density of air at 20° C is 1.20 kg/m³.) With what velocity does the air emerge from the duct? What is the air pressure at the outlet? [*Hint:* The density of air for constant temperature is proportional to the pressure. However, assume that the density remains constant and then justify this assumption. The very small change in pressure for a substantial increase in velocity is a characteristic of low-velocity airflow.]

18-30 An enclosed cylindrical tank with a radius R and a
• height h is half filled with water. The air in the top part of the tank is maintained at an absolute pressure p. A hole with a diameter d at the bottom of the tank is then opened, as indicated in the diagram. (a) Derive an expression for the upward thrust exerted by the escaping water. (b) If $R = 30$ cm, $h = 2$ m, and $d = 4$ cm, and if the tank mass is $m = 25$ kg, what is the minimum pressure p required to provide an upward acceleration of the tank? [*Hint:* Consult Section 9-5.]

18-31 What modification in the expression for q given in Example 18-3 is required if the Venturi tube is in-

clined at an angle α to the horizontal and if the axial distance between the manometer connections is L?

18-32 A suction pump located just above a water surface operates with an intake pressure of 1 atm and delivers 3.0 hp of mechanical power to rotating blades that raise water through a pipe with a uniform diameter. What is the maximum height above the surface to which 20 ℓ/s (2×10^{-2} m³/s) of water can be delivered? Neglect frictional effects.

18-33 The viscosity of a certain liquid is determined (at 40° C) by measuring the flow rate through a tube under a known pressure difference between the ends. The tube has a radius of 0.70 mm and a length of 1.50 m. When a pressure difference of $\frac{1}{20}$ atm is applied, a volume of 292 cm³ is collected in 10 min. What is the viscosity of the liquid? What is the liquid?

SPECIAL TOPIC

18-34 The analysis of wave propagation on the free sur-
• face of a liquid is rather complicated. One important characteristic of any wave motion is the *wavelength* λ, defined to be the distance between successive wave crests or troughs. Use dimensional analysis to deduce the dependence of the wave velocity c on the wavelength λ, and on other variables in the following cases: (a) For waves with ordinary wavelengths running over deep water, the wave velocity c might be expected to depend on the acceleration g due to gravity and on the wavelength λ. (b) For waves with very small wavelengths (ripples), the wave velocity might be expected to depend on the wavelength and also on the density ρ and the surface tension γ of the water. (c) For waves with very long wavelengths running in shallow water, the wave velocity depends only on g and on the depth

of the water. (d) The correct expression for the wave velocity when all relevant factors are considered is

$$c = \sqrt{\left(\frac{g\lambda}{2\pi} + \frac{2\pi\gamma}{\rho\lambda}\right)\tanh\frac{2\pi h}{\lambda}}$$

where the hyperbolic tangent is $\tanh x = (e^x - e^{-x})/(e^x + e^{-x})$. Do your results agree with the form of this expression in the appropriate limits? Discuss. [*Hint:* For large x, $\tanh x \cong 1$; for small x, $\tanh x \cong x$.]

18-35 A long cylinder with a diameter of 1 cm moves through water at a velocity of 1 cm/s in a direction perpendicular to its axis. The water temperature is 20° C. (a) Calculate the Reynolds number N_R. (b) What is the drag force on the cylinder per meter of length?

CHAPTER 19

MECHANICAL WAVES

One of the important and interesting types of motion that bulk matter can undergo is *wave motion*. In this distinctive type of motion, the particles that constitute the bulk matter participate in the motion in a highly organized manner. Wave motion in a material medium consists of a strain or other disturbance that is transmitted successively from one part of the medium to the next by interactions among the particles. Thus, wave motion in bulk matter is a propagating elastic disturbance. Because all matter is elastic to some degree, waves can propagate through any material substance—solid, liquid, or gas.

A familiar example of wave motion is the occurrence of waves on the surface of a body of water. If you observe a cork or some other bit of matter floating on the surface, you will notice that the cork undergoes an up-and-down bobbing motion as the waves move across the surface of the water. That is, the waves travel steadily in some direction without any corresponding transport of the water or floating matter. However, if a wave encounters a cork floating in still water, the cork will be set into motion as the result of receiving energy and momentum from the wave. The bulk matter through which a wave propagates acts as a medium for the transport of energy and momentum even though no part of the bulk matter participates in the long-range motion of the wave.

19-1 GENERAL CHARACTERISTICS OF WAVES

Wave propagation can take place in systems that are essentially one-dimensional, two-dimensional, or three-dimensional in character. A disturbance that propagates along a taut string or a spring constitutes a one-dimensional wave. Waves on the surface of a body of water are two-dimensional waves. And a pressure disturbance (sound wave) that propagates through air is a three-dimensional wave.

The impacts of raindrops on a pond produce surface waves that travel outward in ever-widening rings. An imaginary circle that connects the points on a particular outward-moving wave crest constitutes a circular *wavefront*. Or, we could define different wavefronts by connecting other obviously related parts of the wave, such as the points in a particular ring of depression. These various wavefronts all propagate radially outward from the central impact point.

A sharp disturbance of the air, such as a hand clap or an explosion, will produce pressure waves that expand in three dimensions from the source. The wavefronts in such cases are concentric spheres that propagate radially outward.

Raindrops falling on a pond produce circular surface waves that propagate radially outward from the impact points. (Photo by Jay Freedman.)

Waves can also be classified according to the type of motion that the medium undergoes. If the displacement of the medium is at right angles to the direction of propagation of the wave, the wave is said to be *transverse*. For example, if the end of a taut horizontal string is flipped up and down, a transverse disturbance will propagate along the string (Fig. 19–1a). If the displacement of the medium is back and forth along the direction of propagation of the wave, the wave is called a *longitudinal* or *compressional* wave. For example, if the end of a coiled spring is pushed and then pulled, a longitudinal disturbance (or compressional wave pulse) will propagate along the spring (Fig. 19–1b).

Fig. 19–1. (a) A transverse wave. (b) A longitudinal wave.

Some waves are combinations of transverse and longitudinal displacements. Both types of displacement are involved, for example, in water waves and in *torsional* waves which represent a propagating twisting motion of a rod or a spring.

One of the characteristics of a transverse wave is its *polarization*. If we select a coordinate frame with the x-axis along a taut string, the transverse displacements of the string will lie in the y-z plane. If all of these displacements, at every position along the string, are parallel, the wave is said to be *linearly polarized* (Fig. 19–2). In the event that these parallel displacements are in the y-direction, we describe the wave as a transverse wave propagating in the x-direction, linearly polarized in the y-z plane in the direction of the y-axis. If the end of the string is moved uniformly in a circular path, the resulting wave is *circularly polarized*. The idea of polarization is not applicable to longitudinal waves. (•Why?)

Fig. 19–2. A wave that is linearly polarized in the direction of the y-axis.

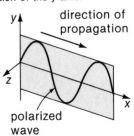

This chapter is concerned exclusively with waves in mechanical (material) systems. However, there are important types of waves that do not require any material medium for propagation. For example, electromagnetic waves—including radio waves, visible light, heat radiation, and X rays—consist of propagating electric and magnetic fields that travel through vacuum with the velocity of light. (These waves can also pass through matter, but in doing so they suffer absorption in varying degrees.)

Electromagnetic waves are produced by electric charge that undergoes acceleration, and it appears that *gravitational waves* are produced by mass that undergoes acceleration (for example, in cataclysmic astronomical events). These gravitational waves can also propagate through vacuum (and through matter), but they are extremely difficult to detect, so we do not yet have unambiguous experimental proof of their existence. The discovery that electromagnetic waves and (probably) gravitational waves can propagate through vacuum has altered in a fundamental way our concept of the nature of empty space. In Chapters 38 and 39 we discuss some aspects of electromagnetic waves and radiation.

19-2 WAVE FUNCTIONS

The Phase. A solitary wave pulse that propagates along a horizontal taut string can be generated by flipping the end of the string up and down (Fig. 19-1a). The mathematical expression that describes the wave disturbance and its propagation as a function of time is called the *wave function* of the pulse. Consider an infinitely long string and select a coordinate frame O with the x-axis along the string, as in Fig. 19-3. Suppose that a single linearly polarized pulse travels along the string in the direction of increasing values of x. We let y represent the transverse displacement of the string. Then, $y = f(x, t)$ is the wave function for the pulse and depends on the two independent variables, x and t.

Fig. 19-3. A single pulse travels along a taut string with a velocity v.

In the ideal case, the pulse travels along the string with constant velocity while maintaining its shape. Suppose that the pulse is viewed by an observer in a coordinate frame O' that moves in the x-direction with the same velocity v at which the pulse moves along the string (see Fig. 19-3). This observer would see a stationary pulse with a fixed shape described by some function $y' = f(x')$. The connection between the coordinates in the two frames is given by

$$x = x' + vt$$
$$y = y'$$

Consequently, the wave function in the frame O is

$$y = f(x - vt) \qquad \textbf{(19-1a)}$$

If the pulse were traveling in the opposite direction and had a shape described in O' by $y' = g(x')$, the wave function in O would be

$$y = g(x + vt) \qquad \text{(19–1b)}$$

Thus, the requirement that the pulse travel with a constant velocity while maintaining its shape demands that the wave function depend on x and t in the coupled form, $u = x + vt$ or $u = x - vt$. The quantity, $u = x \pm vt$, is called the *phase* of the wave.

▶ *The Phase Velocity.* It is instructive to examine the relationships contained in Eqs. 19–1 from a different point of view. Suppose that a wave pulse travels along a string in the positive x-direction and has the wave function $y = f(x - vt)$. Fix attention on some feature of the pulse, e.g., the peak, and follow its progress as the pulse moves. Let y_0 be the peak displacement which occurs for the specific value u_0 of the phase. At time t_1 the pulse peak is at coordinate x_1, and at a later time t_2 the peak is at coordinate x_2 (Fig. 19–4). To give the same phase u_0 at every instant, we must have

$$u_0 = x_1 - vt_1 = x_2 - vt_2$$

so that

$$v = \frac{x_2 - x_1}{t_2 - t_1}$$

This is just the result we would expect for a pulse (or a particle) traveling with the velocity v.

Fig. 19–4. A wave pulse travels to the right with a velocity v along a taut string. The location of the pulse is shown at times t_1 and t_2.

The equation we have just obtained for v resulted from the requirement that the phase u_0 of the pulse peak be a constant, independent of time. But this must be true for any selected feature of the pulse. Thus, for *all* phases u we must have $du/dt = 0$. Because u is a function of both x and t, we write*

$$\frac{du}{dt} = \frac{\partial u}{\partial x} \cdot \frac{dx}{dt} + \frac{\partial u}{\partial t}$$

Explicitly, $u = x - vt$, so we have $\partial u/\partial x = 1$ and $\partial u/\partial t = -v$. Then,

$$\frac{du}{dt} = 0 = 1 \cdot \frac{dx}{dt} - v$$

or

$$\frac{dx}{dt} = v$$

That is, any feature of the wave pulse has a coordinate location x that moves with a velocity $v = dx/dt$, which is called the *phase velocity* of the wave. ◀

Superposition. In a medium that supports wave propagation there may be several sources of waves. For example, the photograph on page 505 shows many different wave sources and the resulting complicated pattern of wavefronts crossing one another. When two waves cross or merge, the instantaneous displacement of the medium is simply the linear sum of the individual displacements that would have occurred for each wave alone. When such a linear addition law is valid for combin-

---

*See Section IV–1 for the definition of the partial derivative.

ing waves, the waves are said to be *linear* and the addition obeys the *principle of superposition.* We consider only systems that obey this principle.

If two separately generated waves with the same polarization are traveling in opposite directions on a string, the individual displacements can be expressed as $y_1 = f(x - vt)$ and $y_2 = g(x + vt)$. *At all times*—before, during, and after these waves overlap and combine—the total displacement of the string is the linear superposition of these waves; that is,

$$y = f(x - vt) + g(x + vt)$$

If the displacement y, at the time of overlap, is *less* than the displacement of either wave alone, the waves are said to exhibit *destructive interference* (Fig. 19–5a). If the waves combine to produce a displacement *greater* than that of either wave alone, the interference is *constructive* (Fig. 19–5b). In order to obtain *complete* destructive interference it is necessary that the two waves have exactly the same shape, but with opposite displacements. Even so, the complete cancellation exists only at the instant of overlap.

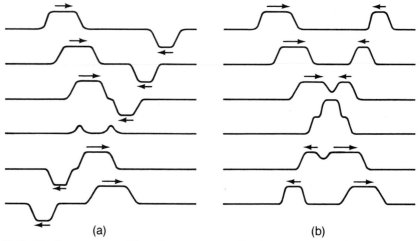

(a)   (b)

Fig. 19–5. The linear superposition of two waves traveling in opposite directions on a taut string. (a) Destructive interference. (b) Constructive interference.

The Wave Equation. Waves in a medium are described by a *wave equation* that gives the space and time dependence of disturbances in that medium. For waves that propagate in one dimension (corresponding to the x-direction), the wave equation is

$$\frac{\partial^2 \eta}{\partial x^2} = \frac{1}{v^2} \frac{\partial^2 \eta}{\partial t^2} \qquad \text{(19–2)}$$

where $\eta(x, t)$ represents a generalized displacement from equilibrium (e.g., the displacement of a string, a pressure excursion, etc.).

---

*The superposition principle is valid for mechanical waves if the displacements are "sufficiently small." We give only this vague criterion because the way in which nonlinear effects enter is quite complicated. For electromagnetic waves in vacuum, however, linear superposition *always* holds.

▶ *Derivation of the General Wave Equation.* Consider waves that are propagating along a one-dimensional system whose direction coincides with the x-axis. Suppose that two waves are present, described by $\eta_1 = f(x - vt) = f(u)$ and $\eta_2 = g(x + vt) = g(w)$. We wish to find the equation involving x and t that is satisfied by $\eta = \eta_1 + \eta_2$, *regardless* of the particular forms of the wave functions, $f(u)$ and $g(w)$. This equation should involve the space and time derivatives of η, so we calculate

$$\frac{\partial \eta}{\partial t} = \frac{\partial \eta_1}{\partial t} + \frac{\partial \eta_2}{\partial t}$$

$$\frac{\partial \eta_1}{\partial t} = \frac{df}{du}\frac{\partial u}{\partial t} = -v\frac{df}{du}$$

$$\frac{\partial \eta_2}{\partial t} = \frac{dg}{dw}\frac{\partial w}{\partial t} = +v\frac{dg}{dw}$$

Hence,

$$\frac{\partial \eta}{\partial t} = v\left(-\frac{df}{du} + \frac{dg}{dw}\right) \qquad \textbf{(19-3)}$$

Similarly, for the space derivatives we have

$$\frac{\partial \eta}{\partial x} = \frac{\partial \eta_1}{\partial x} + \frac{\partial \eta_2}{\partial x} = \frac{df}{du} + \frac{dg}{dw} \qquad \textbf{(19-4)}$$

Because there are present waves traveling in both directions along the system, the time derivatives of η_1 and η_2 differ by a sign. Consequently, no simple relationship exists between $\partial\eta/\partial t$ (Eq. 19–3) and $\partial\eta/\partial x$ (Eq. 19–4). We must proceed to the second derivatives in order to find the desired relationship connecting the derivatives:

$$\frac{\partial^2 \eta}{\partial t^2} = \frac{\partial^2 \eta_1}{\partial t^2} + \frac{\partial^2 \eta_2}{\partial t^2}$$

with

$$\frac{\partial^2 \eta_1}{\partial t^2} = \frac{\partial}{\partial t}\left(-v\frac{df}{du}\right) = -v\left(\frac{d}{du}\frac{df}{du}\right)\frac{\partial u}{\partial t} = v^2\frac{d^2 f}{du^2}$$

and

$$\frac{\partial^2 \eta_2}{\partial t^2} = \frac{\partial}{\partial t}\left(+v\frac{dg}{dw}\right) = v\left(\frac{d}{dw}\frac{dg}{dw}\right)\frac{\partial w}{\partial t} = v^2\frac{d^2 g}{dw^2}$$

Hence,

$$\frac{\partial^2 \eta}{\partial t^2} = v^2\left(\frac{d^2 f}{du^2} + \frac{d^2 g}{dw^2}\right) \qquad \textbf{(19-5)}$$

In a similar way we find

$$\frac{\partial^2 \eta}{\partial x^2} = \frac{d^2 f}{du^2} + \frac{d^2 g}{dw^2} \qquad \textbf{(19-6)}$$

Now, combining Eqs. 19–5 and 19–6 we have the desired relationship:

$$\frac{\partial^2 \eta}{\partial x^2} = \frac{1}{v^2}\frac{\partial^2 \eta}{\partial t^2}$$

which is the wave equation for the system (Eq. 19–2). Regardless of the particular forms of $f(u)$ and $g(w)$, these functions—separately or simultaneously—are solutions of the one-dimensional wave equation.

For a wave propagating in some direction in a three-dimensional medium, the generalized form of the wave equation is

$$\frac{\partial^2 \eta}{\partial x^2} + \frac{\partial^2 \eta}{\partial y^2} + \frac{\partial^2 \eta}{\partial x^2} = \frac{1}{v^2}\frac{\partial^2 \eta}{\partial t^2} \qquad \textbf{(19-7)}$$

Notice that η represents any scalar quantity that propagates as a wave (for example, the pressure in a gas for the case of a sound wave). Also, η can be any component of a vector that describes wave motion (for example, the displacement vector **r** or the electric field vector **E**). ◀

Example 19-1

Two waves are propagating along a taut string that coincides with the x-axis:
$y_1 = A\cos[k(x - vt)]$; $y_2 = A\cos[k(x + vt) + \phi_0]$.

(a) What value of ϕ_0 will produce constructive interference at $x = 0$? Destructive interference?

(b) For each of the values of ϕ_0 found in (a), write the total wave function, $y = y_1 + y_2$.

(c) Find the points along the string that are always stationary.

Solution:

The linear superposition of the two waves gives

$$y = y_1 + y_2 = A\{\cos[k(x - vt)] + \cos[k(x + vt) + \phi_0]\}$$

Using Eq. A–28 in Appendix A, this expression simplifies to

$$y(x, t) = 2A\cos(kx + \tfrac{1}{2}\phi_0)\cos(kvt + \tfrac{1}{2}\phi_0)$$

(a) At $x = 0$, we have

$$y(0, t) = 2A\cos\tfrac{1}{2}\phi_0\cos(kvt + \tfrac{1}{2}\phi_0)$$

For maximum displacement (either sign for y), $\cos\tfrac{1}{2}\phi_0$ must be ± 1. Thus, for constructive interference, we have

$$\phi_0(\text{constructive}) = 0,\ 2\pi,\ 4\pi,\ \ldots$$
$$= 2n\pi \quad (n = 0, 1, 2, \ldots)$$

For destructive interference ($\cos\tfrac{1}{2}\phi_0 = 0$), we have

$$\phi_0(\text{destructive}) = \pi,\ 3\pi,\ 5\pi,\ \ldots$$
$$= (2n + 1)\pi \quad (n = 0, 1, 2, \ldots)$$

(b) Substituting ϕ_0 for the case of constructive interference, we find

$$y_c(x, t) = 2A\cos kx\cos kvt$$

For the case of destructive interference, we have

$$y_d(x, t) = 2A\sin kx\sin kvt$$

(c) For both cases there are positions of zero displacement for all values of t. These occur at points for which $\cos kx = 0$ in the case of constructive interference and at points for which $\sin kx = 0$ in the case of destructive interference. That is,

$$kx = \tfrac{1}{2}(2n + 1)\pi \quad (\text{constructive})$$
$$kx = n\pi \quad (\text{destructive})$$
$$n = 0, 1, 2, \ldots$$

Such points of permanent zero displacement are called *nodes*.

(•Suppose that one of the waves in this example is polarized in the z-direction and that the other remains polarized in the y-direction. How would this affect the analysis? The situation is not simple!)

Example 19–2

We now inquire how a wave with a particular shape is caused to travel along a string. Suppose that we have an infinitely long taut string. We choose the x-axis of a coordinate frame to coincide with the string and we locate the origin at some convenient point. We wish to start a wave with the particular shape, $y_1 = f(x - vt)$, traveling in the positive x-direction by moving the string at the origin in some appropriate manner. How must we move the string and what will be the result?

Solution:

Suppose that the desired wave, $y_1 = f(x - vt)$, is a pulse that has the shape shown in the diagram. Let the imposed motion of the string at the origin required to produce this pulse be $y_0(t) = h(vt)$. One might have the initial thought that $y_0(t) = y_1(0, t)$, but this is not correct. When the string is pulled up and down at the origin, not only will there be a wave pulse that moves along the string in the $+x$-direction, but there will be a pulse with the same shape (actually, the mirror image) that moves in the $-x$-direction. Thus, in addition to the desired wave, $y_1 = f(x - vt)$, the wave $y_2 = g(x + vt)$ is necessarily generated by the imposed motion at the origin. According to the superposition principle, we have at $x = 0$,

$$y_0(t) = y_1(0, t) + y_2(0, t)$$

or

$$h(vt) = f(-vt) + g(vt)$$

This equation for $h(vt)$ must hold at *all times*. Moreover, symmetry along the (homogeneous) string requires corresponding parts of the two wave pulses to have the same displacements. That is,

$$g(vt) = f(-vt)$$

so that

$$h(vt) = 2f(-vt)$$

In order that $y_2(x, t)$ represent propagation of a wave pulse along the negative x-direction, the algebraic sign of x in the argument must be the same as that of vt. Thus, the complete solutions are

$$y_1(x, t) = f(x - vt)$$
$$y_2(x, t) = f(-x - vt)$$
$$y_0(t) = 2f(-vt)$$

(a)

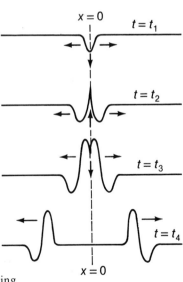

(b)

Part (a) of the diagram shows the imposed motion, $y_0(t)$, and part (b) shows the resulting development of the two wave pulses as the string is displaced at the origin.

19-3 PROPAGATION OF TRANSVERSE WAVES ON A TAUT STRING

Consider a transverse wave pulse that travels along a taut string, as shown in Fig. 19–6. We apply Newton's law to the segment of the displaced string that lies between x and $x + \Delta x$. There is a uniform string tension F, which, at every point, is directed tangent to the string. Neglecting gravitational effects (which are usually small), Newton's equation for the string segment is

$$F(\sin \alpha_2 - \sin \alpha_1) = \Delta m\, a_y$$

where Δm is the mass of the segment. When the displacement is small we can write

$$\sin \alpha_1 \cong \tan \alpha_1 = \left.\frac{\partial y}{\partial x}\right|_{x}$$

$$\sin \alpha_2 \cong \tan \alpha_2 = \left.\frac{\partial y}{\partial x}\right|_{x+\Delta x}$$

$$\Delta m \cong \mu\, \Delta x$$

where μ is the mass per unit length of the string (i.e., the linear mass density). Now, $a_y = \partial^2 y/\partial t^2$, so combining the equations above we have

$$F \operatorname*{Lim}_{\Delta x \to 0} \frac{\left(\left.\frac{\partial y}{\partial x}\right|_{x+\Delta x} - \left.\frac{\partial y}{\partial x}\right|_{x}\right)}{\Delta x} = \mu\, \frac{\partial^2 y}{\partial t^2}$$

The limit in this equation is just the definition of the second partial derivative. Thus, we find

$$\frac{\partial^2 y}{\partial x^2} = \frac{\mu}{F}\frac{\partial^2 y}{\partial t^2} \qquad\qquad \textbf{(19–8)}$$

Comparing Eq. 19–8 with the general one-dimensional wave equation (Eq. 19–2) shows that the wave velocity v is given by

$$v = \sqrt{\frac{F}{\mu}} \qquad\qquad \textbf{(19–9)}$$

Fig. 19–6. A taut string is given a small transverse displacement so that a wave pulse propagates to the right. A segment of the string is isolated for analysis.

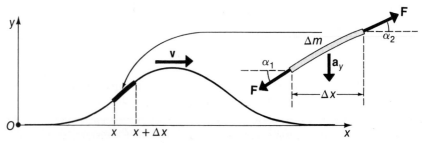

This form for the velocity exhibits a characteristic common to all mechanical waves. The velocity of a wave increases with increasing "strength" of the restoring force (here, the tension F), and the velocity decreases with increasing "inertia" of the system (here, the mass density μ).

Another important aspect of a traveling wave is the associated energy transport. Figure 19–7 shows a hypothetical cut in a displaced string at coordinate x. Assume that there is a wave pulse propagating to the right. At what rate does the string to the left of the cut do work on the part of the string to the right of the cut? This rate represents the *power* transmitted across the cut, and is equal to the product of the *perpendicular* component of the tension and the velocity of the string at x. (•Can you see why?) For small displacements, we have (see Fig. 19–7)

$$F_\perp = -F\sin\alpha \cong -F\tan\alpha = -F\frac{\partial y}{\partial x}$$

The power transmitted to the right is

$$P(x, t) = F_\perp v_\perp$$

Thus,

$$P(x, t) = -F\left(\frac{\partial y}{\partial x}\right)\left(\frac{\partial y}{\partial t}\right) \qquad\qquad \textbf{(19–10)}$$

For a cut at a point on the trailing edge of the pulse (as in Fig. 19–7), notice that $\partial y/\partial x > 0$ but $\partial y/\partial t < 0$, so that $P > 0$. If the cut is on the leading edge, we have $\partial y/\partial x < 0$ but $\partial y/\partial t > 0$, so again $P > 0$.

Fig. 19–7. The part of the string to the right of the hypothetical cut at x is driven by the force F which is due to the tension in the string to the left of the cut. At the cut, the velocity in the y-direction is $\partial y/\partial t$ and is downward.

Example 19–3

Verify that the velocity v at which the crest of a wave pulse travels along a taut string is given by Eq. 19–9.

Solution:

The diagram shows an isolated, symmetrically displaced segment of a string with length Δx and mass $\Delta m \cong \mu\,\Delta x$. The radius of curvature at the crest is R. (The displacement y is not the same as the radius R.)

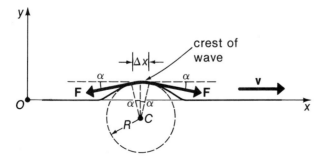

In the frame of reference that moves with the pulse (that is, moves to the right with velocity **v**), the string particles at the peak of the pulse move in a circular arc with tangential velocity $-\mathbf{v}$. Thus, the vertical force that provides the required centripetal acceleration of Δm

is given by

$$2F \sin \alpha \cong 2F\alpha = 2F\left(\frac{\frac{1}{2}\Delta x}{R}\right) = F\frac{\Delta x}{R}$$

Setting this expression equal to $\Delta m \, a_c = \Delta m \, v^2/R = \mu \, \Delta x \, v^2/R$, we obtain

$$F\frac{\Delta x}{R} = \frac{\mu \, \Delta x \, v^2}{R}$$

from which

$$v^2 = \frac{F}{\mu}$$

Example 19–4

The Gaussian pulse shown in the diagram,

$$y(x, t) = Ae^{-\sigma(x-vt)^2}$$

is traveling along a taut string in which the tension is F.
(a) Calculate the power transmitted as a function of time at a point x.
(b) Evaluate the total energy contained in the pulse.

Solution:

We have

$$\frac{\partial y}{\partial x} = -2A\sigma^2(x - vt)e^{-\sigma(x-vt)^2}$$

and

$$\frac{\partial y}{\partial t} = +2A\sigma v^2(x - vt)e^{-\sigma(x-vt)^2}$$

Substituting these expressions into Eq. 19–10, we find

$$P(x, t) = 4A^2\sigma^2 Fv(x - vt)^2 e^{-2\sigma(x-vt)^2}$$

This result shows that $P(x, t)$ also represents a wave pulse that travels in the positive x-direction with the velocity v of the displacement pulse.
(b) The power transmitted to the right through $x = 0$ as a function of time is

$$P(0, t) = 4A^2\sigma Fv^3 t^2 e^{-2\sigma v^2 t^2}$$

The total energy in the pulse is obtained by integrating this expression from $t = -\infty$ to $t = +\infty$. That is,

$$E = \int_{-\infty}^{+\infty} P(0, t) \, dt = 4A^2\sigma Fv^3 \int_{-\infty}^{+\infty} t^2 e^{-2\sigma v^2 t^2} \, dt$$

This integral has a form that is often encountered in physical problems. However, the integral cannot be evaluated by elementary means. In tables of integrals, you will find the value

$$\int_{-\infty}^{+\infty} u^2 e^{-\alpha u^2} \, du = \frac{1}{2}\sqrt{\frac{\pi}{\alpha^3}}$$

If we identify $\alpha = 2\sigma v^2$, we obtain

$$E = A^2 F \sqrt{\frac{\pi\sigma}{2}}$$

(•Can you verify this result?)

Example 19–5

A taut string has a linear mass density $\mu = 6 \times 10^{-3}$ kg/m and is under a tension of 400 N. A Gaussian-shaped pulse (see Example 19-4) with $A = 5$ cm and $\sigma = 20$ m^{-2} propagates along the string.

(a) What is the velocity of the pulse?

(b) What is the full width at half maximum displacement (FWHM) of the pulse? What time interval is required for this width to pass a given station along the string?

(c) What is the total energy in the pulse?

Solution:

(a) According to Eq. 19–9, the velocity of the pulse is

$$v = \sqrt{\frac{F}{\mu}} = \sqrt{\frac{400 \text{ N}}{6 \times 10^{-3} \text{ kg/m}}} = 258.2 \text{ m/s}$$

(b) At $t = 0$, for example, the displacement is

$$y(x, 0) = Ae^{-\sigma x^2}$$

When $y(x, 0) = \frac{1}{2}A$, we have $e^{-\sigma x^2} = \frac{1}{2}$, so that

$$x = \sqrt{\frac{\ln 2}{\sigma}}$$

Therefore, the FWHM is

$$\Delta x = 2\sqrt{\frac{\ln 2}{\sigma}} = 2\sqrt{\frac{\ln 2}{20 \text{ m}^{-2}}} = 0.372 \text{ m}$$

The wave velocity is $v = 258.2$ m/s, so one pulse width will pass a fixed station during a time Δt, where

$$\Delta t = \frac{\Delta x}{v} = \frac{0.372 \text{ m}}{258.2 \text{ m/s}} = 1.44 \times 10^{-3} \text{ s}$$

(c) Using the result for the total energy E in Example 19–4, we have

$$E = A^2 F \sqrt{\frac{\pi\sigma}{2}} = (0.05 \text{ m})^2 (400 \text{ N}) \sqrt{\frac{\pi(20 \text{ m}^{-2})}{2}}$$

$$= 5.60 \text{ J}$$

19-4 HARMONIC WAVES

Suppose that one end of a long taut string is driven in the transverse direction by a simple harmonic force. Then, a sinusoidal wave (or simple harmonic wave, SHW) will propagate along the string. Strictly, the wave will be a *pure* sinusoidal wave only if the string is infinitely long and has been driven harmonically since $t = -\infty$ so that a true steady-state condition exists. For a linearly polarized SHW propagating in the positive x-direction, the most general form of the wave function is

$$y(x, t) = A \cos[k(x - vt) + \phi_0] \qquad \textbf{(19-11)}$$

In this expression, the argument of the cosine term, $\phi = k(x - vt) + \phi_0$, is called the *phase angle* of the wave and ϕ_0 is the *initial phase angle* (see Fig. 19-8). The coefficient A is the *amplitude* of the wave, that is, the magnitude of the maximum transverse displacement; $-A \leqslant y \leqslant A$ for all t. The quantity k is called the *wave number* and has dimensions $[k] = L^{-1}$.

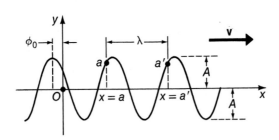

Fig. 19-8. A simple harmonic wave with amplitude A and wavelength λ shown at $t = 0$. The quantity ϕ_0 is the initial phase angle.

Another quantity that is useful for describing a simple harmonic wave is the *wavelength* λ, the repetition length of the wave. The wavelength is defined to be the x-distance between any two successive corresponding points on the wave (that is, two points that have the same displacement and direction of motion). (•Why is it necessary to specify "and direction of motion"?) In Fig. 19-8, the displacement and motion at $y(a, t)$ is first repeated at $y(a', t)$, so that $\lambda = a' - a$. Because the point $x = a$ may be anywhere on the wave, we can write $y(x + \lambda, t) = y(x, t)$. Between $x = a$ and $x = a'$, the phase angle increases by 2π, so

$$k[(x + \lambda) - vt] + \phi_0 = k[x - vt] + \phi_0 + 2\pi$$

from which $k\lambda = 2\pi$, or

$$k = \frac{2\pi}{\lambda} \qquad \textbf{(19-12)}$$

The time factor kv can be simplified by introducing the *period* T, which is the repetition time of the wave. The period is defined to be the time interval between successive identical displacements and motions of the wave at a particular point; that is, we can write $y(x, t - T) = y(x, t)$. In terms of the phase angle, we have

$$k[x - v(t - T)] + \phi_0 = k[x - vt] + \phi_0 + 2\pi$$

from which $kvT = 2\pi$, or

$$kv = \frac{2\pi}{T}$$

The *frequency* of the wave is $\nu = 1/T$ (see Eq. 15-13). Then,

$$kv = \frac{2\pi}{\lambda} \cdot v = \frac{2\pi}{T} = 2\pi\nu$$

or

$$v = \lambda\nu \qquad\qquad \textbf{(19-13)}$$

This equation expresses the fact that ν waves pass the (arbitrarily selected) origin each second, each wave having a length of λ meters. Thus, the feature that was at the origin at $t = 0$ is carried to a distance $\lambda\nu$ meters at $t = 1$ s. This corresponds to the same phase velocity that was defined in Section 19–2. Note also that the equation is of the form prescribed by dimensional analysis, namely, $[v] = LT^{-1} = [\lambda][\nu]$.

Equation 19–13 is valid for *all* types of simple harmonic waves. For each type of wave we must use the appropriate velocity v—for example, approximately 340 m/s for sound waves in air and 3×10^8 m/s for light waves in vacuum (or air).

The various quantities that describe the properties of waves can be used to express the wave function in several equivalent ways. If we include the idea of the *angular frequency,* $\omega = 2\pi\nu = kv$ (see Eq. 15–14), we have, for example,

$$v = \lambda\nu = \frac{2\pi}{k}\nu = \frac{\omega}{k} \qquad\qquad \textbf{(19-14)}$$

Then,
$$
\left.\begin{aligned}
y(x, t) &= A \cos\left[k(x - vt) + \phi_0\right] \\
&= A \cos\left[kx - \omega t + \phi_0\right] \\
&= A \cos\left[2\pi\left(\frac{x}{\lambda} - \frac{t}{T}\right) + \phi_0\right]
\end{aligned}\right\} \qquad \textbf{(19-15)}
$$

These expressions represent a SHW that propagates in the positive x-direction. (We assume that the end at $x = -\infty$ has been driven since $t = -\infty$.)

We can imagine that a harmonic wave is generated by driving the point $x = 0$ instead of one end. Then, two identical waves are set up simultaneously, one traveling in the $+x$-direction and the other in the $-x$-direction (as in Example 19–2). If we designate these waves y^+ and y^-, respectively, we have

$$
\begin{aligned}
y^+(x, t) &= A \cos(kx - \omega t + \phi_0), & x \geqslant 0 \\
y^-(x, t) &= A \cos(-kx - \omega t + \phi_0) & \\
&= A \cos(kx + \omega t - \phi_0), & x \leqslant 0
\end{aligned} \qquad \textbf{(19-16)}
$$

▶ There is a more compact way of writing these wave functions. Create a vector $\mathbf{k}$ with magnitude $k = 2\pi/\lambda$ that has a direction described by a unit vector in the direction of propagation. Then, using the vector coordinate $\mathbf{x} = x\hat{\mathbf{i}}$, we have

$$y(x, t) = A \cos(\mathbf{k}\cdot\mathbf{x} - \omega t + \phi_0)$$

For $x > 0$, the wave travels to the right, so $\mathbf{k} = k\hat{\mathbf{i}}$; for $x < 0$, the wave travels to the left, so $\mathbf{k} = -k\hat{\mathbf{i}}$.

Differentiating the phase angle $\phi = \mathbf{k}\cdot\mathbf{x} - \omega t + \phi_0$ with respect to time, we obtain

$$\mathbf{k}\cdot\frac{d\mathbf{x}}{dt} - \omega = 0$$

or
$$\mathbf{k} \cdot \mathbf{v} = \omega \qquad\qquad (19\text{-}17)$$

The quantity $\mathbf{k}$ is often called the *propagation vector.* ◄

Example 19-6

A simple harmonic wave propagates in the positive x-direction along a taut string that has a linear mass density of 5 g/m and is under a tension of 100 N. The amplitude of the wave is 5 cm and the wavelength is 75 cm.

(a) What is the phase velocity of the wave?

(b) What is the frequency of the wave? What is the period?

(c) At $t = 0$, the wave has its maximum positive transverse velocity at $x = 0$. Evaluate the initial phase angle ϕ_0 and write the complete wave function.

(d) What is the phase angle for $x = 0$ at $t = 1$ ms? What is the corresponding displacement?

(e) What is the magnitude of the transverse velocity?

(f) What is the average power transmitted along the string? (Average over one period.)

(g) What is the total energy per wavelength in the wave?

Solution:

(a) The phase velocity is (Eq. 19-9)

$$v = \sqrt{\frac{F}{\mu}} = \sqrt{\frac{100 \text{ N}}{5 \times 10^{-3} \text{ kg/m}}} = 141.4 \text{ m/s}$$

(b) Using Eq. 19-14, we have

$$\nu = \frac{v}{\lambda} = \frac{141.4 \text{ m/s}}{0.75 \text{ m}} = 188.5 \text{ Hz}$$

and
$$T = \frac{1}{\nu} = \frac{1}{188.5 \text{ Hz}} = 5.30 \times 10^{-3} \text{ s} = 5.30 \text{ ms}$$

(c) Write the wave function in the form

$$y(x, t) = A \cos (kx - \omega t + \phi_0)$$

Then,
$$v_{\perp} = \frac{\partial y}{\partial t} = A\omega \sin (kx - \omega t + \phi_0)$$

If $v_{\perp}$ is maximum at $x = 0$ for $t = 0$, the initial phase angle must be $\phi_0 = \pi/2$ (or an odd multiple thereof). Then,

$$y(x, t) = A \cos (kx - \omega t + \pi/2)$$

Notice that the magnitude of $y(x, t)$ is maximum, $|y(x, t)| = A$, when the velocity $v_{\perp}$ is zero, and that the magnitude of $v_{\perp}$ is maximum, $|v_{\perp}| = A\omega$, when $y(x, t)$ is zero.

(d) The phase angle for $x = 0$ and $t = 1$ ms is

$$\phi = kx - \omega t + \pi/2$$
$$= 2\pi \left(\frac{x}{\lambda} - \frac{t}{T} \right) + \frac{\pi}{2}$$

$$= 2\pi \left(0 - \frac{1.0 \times 10^{-3}\,\text{s}}{5.30 \times 10^{-3}\,\text{s}} \right) + \frac{\pi}{2}$$

$$= 0.385$$

Then, $\qquad y = A \cos \phi = (5\,\text{cm}) \cos 0.385 = 4.63\,\text{cm}$

(e) The magnitude of the transverse velocity is

$$|v_\perp| = A\omega = 2\pi A\nu$$
$$= 2\pi(0.05\,\text{m})(188.5\,\text{s}^{-1})$$
$$= 59.2\,\text{m/s}$$

(f) The power transmitted through the position x at time t is given by Eq. 19-10:

$$P(x, t) = -F\left(\frac{\partial y}{\partial x}\right)\left(\frac{\partial y}{\partial t}\right)$$

With $\phi_0 = \pi/2$, we have

$$\frac{\partial y}{\partial x} = -Ak \cos (kx - \omega t)$$

$$\frac{\partial y}{\partial t} = A\omega \cos (kx - \omega t)$$

Then, $\qquad P(x, t) = FA^2\omega k \cos^2 (kx - \omega t)$

Averaging over one period, we obtain

$$\bar{P} = \frac{1}{T} \int_t^{t+T} P(x, t)\, dt$$

$$= \frac{FA^2\omega k}{T} \int_t^{t+T} \cos^2 (kx - \omega t)\, dt$$

$$= \frac{FA^2\omega k}{T} \cdot \frac{T}{2} = \frac{1}{2}FA^2\omega k$$

(•Can you verify this integration by changing the variable to $u = kx - \omega t$? Would you obtain the same result by choosing $x = 0$ and integrating from 0 to T?)
Evaluating $\bar{P}$ using $k = 2\pi/\lambda$ and $\omega = 2\pi\nu$, we find

$$\bar{P} = \frac{2\pi^2 FA^2\nu}{\lambda} = \frac{2\pi^2(100\,\text{N})(0.05\,\text{m})^2(188.5\,\text{s}^{-1})}{0.75\,\text{m}}$$

$$= 1.24\,\text{kW}$$

(g) The energy per wavelength is

$$E = \bar{P}T = (1.24 \times 10^3\,\text{W})(5.30 \times 10^{-3}\,\text{s})$$
$$= 6.57\,\text{J}$$

(•Can you see why E is the energy per wavelength?)

19-5 LONGITUDINAL WAVES IN A SOLID ROD

We now consider a different type of one-dimensional wave motion, namely, compressional waves propagating in a solid rod. Suppose that a sharp axial blow is delivered to the end of a freely suspended rod. The material of the rod at the struck end undergoes a momentary axial compression. A compressional pulse then propagates through the rod as each cross section of the rod material moves back and forth parallel to the rod axis. Waves generated in this way are *longitudinal waves*. In the next chapter, we discuss the similar situation of longitudinal waves in a gas.

The equation that describes this type of wave motion is (see the derivation below)

$$\frac{\partial^2 \xi}{\partial x^2} = \frac{\rho}{Y} \frac{\partial^2 \xi}{\partial t^2} \tag{19–18}$$

where ξ represents the longitudinal displacement from equilibrium of the rod material, and where ρ and Y are the density and Young's modulus, respectively, of the rod material.

If we compare Eq. 19–18 with Eq. 19–2, we see that the phase velocity of the compressional wave is

$$v = \sqrt{\frac{Y}{\rho}} \tag{19–19}$$

This velocity, in which shear and torsional effects are neglected, is called the *extensional wave velocity*. Values for several materials are given in Table 19–1.

TABLE 19–1.
Extensional Wave Velocities
for Thin Rods of
Various Materials

MATERIAL	v (km/s)
Beryllium	12.87
Iron	5.20
Glass, Pyrex	5.17
Copper	3.75
Platinum	2.80
Silver	2.68
Gold	2.03
Lucite	1.84
Lead	1.21
Polyethylene	0.92

▶ *Derivation of the Wave Equation for Waves in a Solid Rod.* We consider a rod sufficiently thin that the microscopic motion of the material in any cross section is the same; that is, we neglect shear effects. We also assume that there are no frictional losses. The part of a long rod near the coordinate x is shown in Fig. 19–9.

The cross-sectional area of the rod is S. (In this chapter and in the next, we reserve A to represent amplitude.) Let the stress at x be s. Then, at $x - \frac{1}{2}\Delta x$, the stress is $s - (\partial s/\partial x)(\frac{1}{2}\Delta x)$ and at $x + \frac{1}{2}\Delta x$ it is $s + (\partial s/\partial x)(\frac{1}{2}\Delta x)$. The net force on the entire section of length Δx considered to be a free body is

$$F = S\left[-\left(s - \frac{\partial s}{\partial x}\frac{\Delta x}{2}\right) + \left(s + \frac{\partial s}{\partial x}\frac{\Delta x}{2}\right)\right]$$

$$= S\frac{\partial s}{\partial x}\Delta x$$

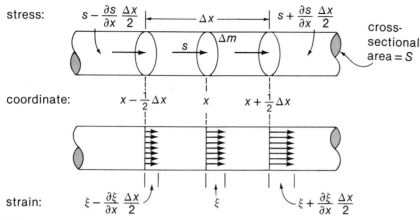

Fig. 19-9. Details of the stresses s and strains ξ in a section Δx of a long thin rod through which propagates a compressional wave. Note carefully the distinction between x, the equilibrium coordinate of matter in the rod, and ξ, the longitudinal displacement of this matter from equilibrium.

(Remember that the outward normal to the volume of length Δx is to the left at $x - \frac{1}{2}\Delta x$ and is to the right at $x + \frac{1}{2}\Delta x$.) The mass Δm of the section is $\rho S \, \Delta x$, where ρ is the volume mass density. We let ξ represent the displacement of matter from its equilibrium position (see Fig. 19–9). Thus, the acceleration at x is $\partial^2\xi/\partial t^2$, and Newton's second law is expressed as

$$F = S\frac{\partial s}{\partial x}\Delta x = (\rho S \, \Delta x)\frac{\partial^2\xi}{\partial t^2}$$

or

$$\frac{\partial s}{\partial x} = \rho\frac{\partial^2\xi}{\partial t^2} \qquad (19\text{--}20)$$

The displacement ξ results in a strain, $\partial\xi/\partial x$. Now, Young's modulus is (see Eq. 16–13)

$$Y = \frac{\text{stress}}{\text{strain}} = \frac{s}{\partial\xi/\partial x}$$

Then,

$$s = Y\frac{\partial\xi}{\partial x}$$

and

$$\frac{\partial s}{\partial x} = Y\frac{\partial^2\xi}{\partial x^2} \qquad (19\text{--}21)$$

Equating the two expressions for $\partial s/\partial x$ (Eqs. 19–20 and 19–21) gives

$$\frac{\partial^2\xi}{\partial x^2} = \frac{\rho}{Y}\frac{\partial^2\xi}{\partial t^2}$$

which is just Eq. 19–18 quoted above. ◀

19-6 THE REFLECTION AND TRANSMISSION OF WAVES AT A BOUNDARY

We have been considering waves propagating along strings (and rods) assumed to be infinite in length. We now study the effects that are introduced because a system is *bounded*. The boundaries may be in the form of terminations, with a fixed or a free end, or they may be the result of discontinuous changes in the density of the string. Consider first the latter possibility.

Suppose that two very long strings, with linear mass densities μ_1 and μ_2, are joined at $x = 0$, forming a single taut string system with tension F throughout.

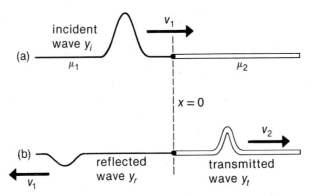

Fig. 19–10. Reflected and transmitted waves are generated when a wave encounters a discontinuity. The case illustrated corresponds to $\mu_2 = 4\mu_1$. The strings are under a uniform tension F.

Imagine that a train of pulses generated at $x = -\infty$ strikes the junction point from the left; one of these pulses is shown in Fig. 19–10a. After the initial pulse (the *incident wave*) has interacted with the junction, there are two additional possibilities for continuing disturbances of the string, namely, a wave that propagates from the junction toward the right (the *transmitted wave*) and a wave that propagates from the junction toward the left (the *reflected wave*), as indicated in Fig. 19–10b. Consequently, in the region $x < 0$, we must deal with the incident and reflected waves, whereas in the region $x > 0$, we have only the transmitted wave.

We can analyze the situation most readily (and with complete generality) by considering harmonic waves. For the incident wave, we write (Eq. 19–15)

$$y_i(x, t) = A_i \cos (k_1 x - \omega_1 t + \phi_0), \qquad x \leqslant 0 \qquad \text{(19–22a)}$$

Because the junction point acts as a source for the transmitted and reflected waves, we use the wave functions given in Eq. 19–16; thus,

$$y_r(x, t) = A_r \cos (k_1 x + \omega_1 t - \phi_0), \qquad x \leqslant 0 \qquad \text{(19–22b)}$$

$$y_t(x, t) = A_t \cos (k_2 x - \omega_2 t + \phi_0), \qquad x \geqslant 0 \qquad \text{(19–22c)}$$

According to the superposition principle, we have

$$y(x, t) = \begin{cases} y_i(x, t) + y_r(x, t), & x \leqslant 0 \\ y_t(x, t), & x \geqslant 0 \end{cases} \qquad \text{(19–23)}$$

Because the strings are connected, we have, at $x = 0$,

$$y_i(0, t) + y_r(0, t) = y_t(0, t) \qquad \text{(19–24)}$$

This equation must hold for all values of t. Notice that $y_r(0, t) = A_r \cos (\omega_1 t - \phi_0) = A_r \cos (-\omega_1 t + \phi_0)$. Then, $y_r(0, t)$ will have the same phase as $y_i(0, t)$ and $y_t(0, t)$ only if $\omega_2 = \omega_1$, so this must be true. (Henceforth, we use ω for the common value of the angular frequency.) Equation 19–24 also imposes a condition on the amplitudes, namely,

$$A_i + A_r = A_t \qquad \text{(19–25)}$$

The two string segments are considered to be ideal strings, so they can sustain only tension. Then, Newton's third law requires identical slopes for the two segments at $x = 0$. (•Can you explain why?) That is,

$$\frac{\partial}{\partial x}(y_i + y_r)\bigg|_{x=0} = \frac{\partial}{\partial x}(y_t)\bigg|_{x=0}$$

This condition yields

$$A_i k_1 \sin(-\omega t + \phi_0) + A_r k_1 \sin(\omega t - \phi_0) = A_t k_2 \sin(-\omega t + \phi_0)$$

from which

$$(A_i - A_r)k_1 = A_t k_2$$

Combining this result with Eq. 19-25 gives

$$\frac{A_r}{A_i} = \frac{k_1 - k_2}{k_1 + k_2} \quad \text{and} \quad \frac{A_t}{A_i} = \frac{2k_1}{k_1 + k_2} \tag{19-26}$$

These expressions give the amplitudes of the reflected and transmitted waves in terms of the amplitude of the incident wave and the wave numbers for the two parts of the system.

Equations 19-26 were obtained without any reference to the physical properties of the string or to the tension. The results are, in fact, correct for *any* type of SHW that meets a discontinuity in a medium.

The angular frequency ω and the tension F are the same throughout the string system. Then, noting that $k = \omega/v = \omega\sqrt{\mu/F}$, we can express the amplitude ratios as

$$\frac{A_r}{A_i} = \frac{\sqrt{\mu_1} - \sqrt{\mu_2}}{\sqrt{\mu_1} + \sqrt{\mu_2}} \quad \text{and} \quad \frac{A_t}{A_i} = \frac{2\sqrt{\mu_1}}{\sqrt{\mu_1} + \sqrt{\mu_2}} \tag{19-27}$$

We can obtain other useful forms for the amplitude ratios by expressing them in terms of the phase velocities, $v_1 = \omega/k_1$ and $v_2 = \omega/k_2$. Thus,

$$\frac{A_r}{A_i} = \frac{k_1 - k_2}{k_1 + k_2} = \frac{\dfrac{1}{v_1} - \dfrac{1}{v_2}}{\dfrac{1}{v_1} + \dfrac{1}{v_2}} = \frac{v_2 - v_1}{v_2 + v_1} \tag{19-27a}$$

$$\frac{A_t}{A_i} = \frac{2k_1}{k_1 + k_2} = \frac{\dfrac{2}{v_1}}{\dfrac{1}{v_1} + \dfrac{1}{v_2}} = \frac{2v_2}{v_2 + v_1} \tag{19-27b}$$

Because of the requirement at $x = 0$ expressed by Eq. 19-24, the shapes of the incident, reflected, and transmitted pulses must all be the same in the time variable. However, the phase velocity is different for the two string segments, so the spatial shapes of the pulses scale directly with the velocities. That is, if $v_2 = \frac{1}{2}v_1$, the transmitted pulse will have a spatial width equal to one half that of the incident pulse. On the other hand, the reflected pulse will have the same spatial width as the incident pulse. Figure 19-10 illustrates just this case, namely, $\mu_2 = 4\mu_1$ and $v_2 = \frac{1}{2}v_1$. The

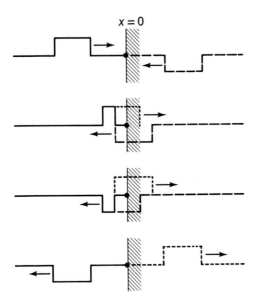

Fig. 19–11. Reflection of a pulse from a fixed support at $x = 0$. The reflected pulse (dashed line) is imagined to emerge from the wall as the incident pulse (dotted line) penetrates the wall. The superposition of these two pulses is the actual pulse (solid line).

amplitude ratios are $A_t/A_i = \frac{2}{3}$ and $A_r/A_i = -\frac{1}{3}$. Thus, the reflected pulse has a displacement opposite to that of the incident pulse, as shown in Fig. 19–10b.

Some special cases concerning Eqs. 19-26 and 19-27 are noteworthy. First, if $\mu_2 = \mu_1$, so that the discontinuity is removed, we find $A_r/A_i = 0$ and $A_t/A_i = 1$. That is, there is no reflected wave and the entire wave is transmitted unchanged, as we expect.

In the event that $\mu_2 > \mu_1$, the ratio A_r/A_i is negative and the displacement of the reflected pulse is opposite to that of the incident pulse. This means that the reflected pulse has a phase difference of π with respect to the incident pulse. (•Can you see why?) When $\mu_2 < \mu_1$, the ratio A_r/A_i is always positive, and the displacement of the reflected pulse is the same as that of the incident pulse.

Next, suppose that the string is fastened to a fixed support at $x = 0$. This is equivalent to letting μ_2 become indefinitely large. Then, $A_r/A_i = -1$, and the entire wave is reflected with a phase change of π. This situation is illustrated in Fig. 19–11, where we represent the reflection by imagining that the incident wave disappears into the wall as the reflected wave emerges from the wall. The actual wave is the superposition of these two waves for $x < 0$.

Finally, we consider the reflection of a pulse from a free end, that is, $\mu_2 = 0$. We can provide a tension F at the free end ($x = 0$) and yet allow the string vertical movement by attaching the end to a ring that slides frictionlessly on a shaft that is perpendicular to the string. In this case, $A_r/A_i = 1$, and the entire wave is reflected with no phase change. This situation is shown in Fig. 19–12, using the same scheme as in Fig. 19–11.

(We have not mentioned the amplitude ratio A_t/A_i for either of the cases $\mu_2 \rightarrow \infty$ or $\mu_2 = 0$, because the situation must properly be analyzed by considering the flow of energy in the system. In this way, one finds that there is no transmitted wave in either case.)

Example 19–7

One end of a thin aluminum rod (diam. = 1 mm) is welded to the end of a copper rod with the same diameter. Both rods are very long and are under a uniform tension of 1000 N.

Fig. 19-12. Reflection of a pulse from a free end. Tension at the free end ($x = 0$) is provided by attaching the end to a ring that can slide frictionlessly on a fixed shaft.

The pertinent data are

$$Y(\text{Cu}) = Y_1 = 11.2 \times 10^{10} \text{ Pa} \qquad Y(\text{Al}) = Y_2 = 6.9 \times 10^{10} \text{ Pa}$$
$$\rho(\text{Cu}) = \rho_1 = 8.93 \text{ g/cm}^3 \qquad \rho(\text{Al}) = \rho_2 = 2.70 \text{ g/cm}^3$$

(a) Calculate the velocities for transverse waves on the two rods.

(b) Calculate the velocities for longitudinal waves on the two rods.

(c) What is the reflection factor A_r/A_i for a longitudinal pulse approaching the junction along the copper rod? For a transverse pulse?

Solution:

(a) For transverse waves we have

$$v_1 = \sqrt{\frac{F}{\mu_1}} = \sqrt{\frac{F}{\rho_1 S}} = \sqrt{\frac{1000 \text{ N}}{(8.93 \times 10^3 \text{ kg/m}^3)\pi(0.5 \times 10^{-3} \text{ m})^2}} = 377.6 \text{ m/s}$$

$$v_2 = \sqrt{\frac{F}{\mu_2}} = \sqrt{\frac{F}{\rho_2 S}} = \sqrt{\frac{1000 \text{ N}}{(2.70 \times 10^3 \text{ kg/m}^3)\pi(0.5 \times 10^{-3} \text{ m})^2}} = 686.7 \text{ m/s}$$

(b) For longitudinal waves we have

$$v_1 = \sqrt{\frac{Y_1}{\rho_1}} = \sqrt{\frac{11.2 \times 10^{10} \text{ N/m}^2}{8.93 \times 10^3 \text{ kg/m}^3}} = 3542 \text{ m/s}$$

$$v_2 = \sqrt{\frac{Y_2}{\rho_2}} = \sqrt{\frac{6.9 \times 10^{10} \text{ N/m}^2}{2.70 \times 10^3 \text{ kg/m}^3}} = 5055 \text{ m/s}$$

(c) Using Eq. 19-27a, we find

$$\text{Transverse: } \frac{A_r}{A_i} = \frac{v_2 - v_1}{v_2 + v_1} = \frac{686.7 - 377.6}{686.7 + 377.6} = 0.290$$

$$\text{Longitudinal: } \frac{A_r}{A_i} = \frac{5055 - 3542}{5055 + 3542} = 0.176$$

Example 19–8

A taut string $(x \leqslant 0)$ with a linear mass density μ and subject to a tension F is attached to a machine at $x = 0$. A wave, $y_i(x, t) = A \cos(kx - \omega t)$, is incident on the machine. What input characteristic must the machine possess if there is to be no reflection of the wave at $x = 0$?

Solution:

The vertical force that the string transmits to the machine is

$$F_\perp = -F \frac{\partial y}{\partial x}\Big|_{x=0} = FAk \sin(-\omega t)$$

The vertical velocity is

$$v_\perp = \frac{\partial y}{\partial t}\Big|_{x=0} = A\omega \sin(-\omega t)$$

so we can write

$$F_\perp = \frac{Fk}{\omega} v_\perp$$

Using $v = \omega/k = \sqrt{F/\mu}$, we have

$$F_\perp = \sqrt{F\mu}\, v_\perp$$

The force $F_\perp'$ that is exerted by the machine on the string is the reaction force to $F_\perp$, that is, $F_\perp' = -\sqrt{F\mu}\, v_\perp$. Thus, the machine must exert on the string a force proportional to the vertical velocity and oppositely directed. This is just the characteristic of the frictional force provided by viscous damping. This damping force must absorb the entire input power carried by the wave. The reflection-free termination of any type of transmission line (mechanical or electrical) requires the same kind of resistance (or *impedance*) matching.

19-7   STANDING WAVES

We now consider the interesting and important case of wave motion on a taut string with finite length, that is, a string with *both* ends terminated in some way. A familiar example of systems in this class is the family of stringed musical instruments.

Consider first the case of a string that is stretched between two fixed supports a distance L apart. We choose a coordinate frame in which the x-axis coincides with the string in its rest position. Imagine that the string is set into motion by delivering to it a transverse blow at some point along its length. This disturbance of the string initiates *two* waves that travel in opposite directions along the string. When these waves arrive at the end terminations, they will be reflected and will then travel back along the string. If there are no energy losses, each reflection will be total and the waves will propagate back and forth along the string forever.

The wave propagation on the terminated string can be analyzed in a simple yet general way by assuming harmonic waves. Because waves travel in both directions along the string, we again use the wave functions of Eq. 19-16, namely,

$$y^+(x, t) = A^+ \cos (kx - \omega t + \phi_0)$$
$$y^-(x, t) = A^- \cos (kx + \omega t - \phi_0)$$
(19-28)

We can expand these wave functions by using Eqs. A-22 and A-23 in Appendix A, obtaining

$$y^+(x, t) = A^+ \cos kx \cos (\omega t - \phi_0) + A^+ \sin kx \sin (\omega t - \phi_0)$$
$$y^-(x, t) = A^- \cos kx \cos (\omega t - \phi_0) - A^- \sin kx \sin (\omega t - \phi_0)$$

According to the superposition principle, the total displacement of the string is

$$y(x, t) = y^+(x, t) + y^-(x, t)$$
$$= (A^+ + A^-) \cos kx \cos (\omega t - \phi_0) + (A^+ - A^-) \sin kx \sin (\omega t - \phi_0)$$
(19-29)

Take the origin $x = 0$ to be at the left-hand end support and take $x = L$ to be at the right-hand end support. Now, the displacement y must be zero at $x = 0$ for all values of t. This requires the first term in Eq. 19-29 to vanish for all t, and this is possible only if $A^+ = -A^- \equiv \frac{1}{2}A$. Then, the expression for $y(x, t)$ becomes

$$y(x, t) = A \sin kx \sin (\omega t - \phi_0)$$
(19-30)

We also require y to be zero at $x = L$ for all t. Excluding the uninteresting case, $A = 0$, we must have

$$\sin k_n L = 0$$

that is, $k_n L = n\pi$, with $n = 1, 2, 3, \ldots$. If we recall that $k = 2\pi/\lambda$ (Eq. 19-12), then $2\pi L/\lambda_n = n\pi$; that is,

$$\lambda_n = 2L/n \qquad n = 1, 2, 3, \ldots$$
(19-31)

For the string terminated by fixed supports at both ends, only those SHW solutions are possible for which the wavelength obeys Eq. 19-31. Thus, we can express the wave function as

$$y(x, t) = A \sin (n\pi x/L) \sin (\omega t - \phi_0) \qquad n = 1, 2, 3, \ldots$$
(19-32)

Notice that at those positions x for which $\sin (n\pi x/L) = 0$, the displacement y is zero for all t. Such positions along the string are called *nodes*. The ends of the string, $x = 0$ and $x = L$, are always nodes. There are other positions, called *antinodes*, for which $\sin (n\pi x/L) = 1$. At these positions, the displacement has its maximum excursion, from $-A$ to $+A$, as time progresses.

These waves have features—the nodes and antinodes—that occur at fixed positions. Accordingly, they are called *standing waves*. When strings vibrate in this manner, sound waves are excited in the surrounding air and we hear the tone generated by the string. (Thus, our assumption that there are no energy losses is not strictly correct!)

The standing wave that has $n = 1$ is called the *fundamental* (or *first harmonic*). The wave for $n = 2$ is the *second harmonic*, the wave for $n = 3$ is the *third harmonic*, and so forth. The wavelength λ_n is governed by the string length L; the frequency then depends on the phase velocity $v = \sqrt{F/\mu}$ (Eq. 19–9). Therefore, by altering the tension F, the string can be tuned to any particular frequency, $\nu_n = v/\lambda_n$. The first four harmonics that are possible for a string with a length L are shown in Fig. 19–13.

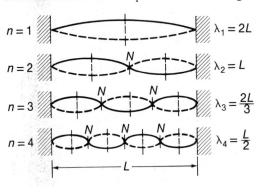

Fig. 19–13. Standing-wave patterns on a string with a length L. The fundamental has $n = 1$; the next three higher harmonics correspond to $n = 2$, 3, and 4. The nodal points are designated by N. The curves represent the envelopes of the oscillations of the strings (i.e., the greatest excursion reached by each point of the string).

Energy in a Standing Wave. The mechanical energy of a wave on a string is the sum of the kinetic energy and the potential energy contributions. For an element of string dx with mass $\mu\,dx$ and transverse velocity $v_\perp = \partial y/\partial t$, the kinetic energy is $dK = \frac{1}{2}\mu(\partial y/\partial t)^2\,dx$. The total kinetic energy of a string with length L is

$$K = \tfrac{1}{2}\mu \int_0^L \left(\frac{\partial y}{\partial t}\right)^2 dx \tag{19–33}$$

Fig. 19–14. A segment of string with arc length $d\ell$, horizontal length dx, and vertical length dy.

To obtain the corresponding expression for the potential energy, consider the segment of string illustrated in Fig. 19–14. For the triangle shown, we have $d\ell^2 = dx^2 + dy^2$, so

$$d\ell = dx \left[1 + \left(\frac{dy}{dx}\right)^2\right]^{1/2}$$

We assume, as throughout these discussions, that the displacement is sufficiently small that $dy/dx \ll 1$. Then, using the approximation in Eq. A–13 in Appendix A, we can write

$$d\ell \cong dx \left[1 + \tfrac{1}{2}\left(\frac{dy}{dx}\right)^2\right] \tag{19–34}$$

The string tension F is directed along $d\ell$, so the energy stored by virtue of the work done in stretching the original length dx to the extended length $d\ell$ is $dU = F(d\ell - dx)$. Then, using Eq. 19–34, we have $dU = \frac{1}{2}F(\partial y/\partial x)^2\,dx$, where we have written the partial derivative because y is actually a function of both x and t. The total potential energy of the string is, with $F = \mu v^2$,

$$U = \tfrac{1}{2}\mu v^2 \int_0^L \left(\frac{\partial y}{\partial x}\right)^2 dx \tag{19–35}$$

Standing waves on a stretched rubber tube. The waves are generated by vibrating the tube with a small amplitude at a point near a node. (From *PSSC Physics,* 2nd ed., 1965: D. C. Heath & Co. and Educational Development Center, Newton, MA.)

Then, the total mechanical energy becomes

$$E = \tfrac{1}{2}\mu \int_0^L \left[\left(\frac{\partial y}{\partial t}\right)^2 + v^2\left(\frac{\partial y}{\partial x}\right)^2\right] dx \qquad \textbf{(19–36)}$$

Let us apply this energy equation to the case $n = 1$ with $\phi_0 = 0$. We have

$$y(x,\ t) = A \sin kx \sin \omega t, \qquad k = \pi/L$$

Then,

$$\frac{\partial y}{\partial x} = Ak \cos kx \sin \omega t$$

$$\frac{\partial y}{\partial t} = A\omega \sin kx \cos \omega t$$

Thus,

$$E = \tfrac{1}{2}\mu A^2\omega^2 \int_0^L (\sin^2 kx \cos^2 \omega t + \cos^2 kx \sin^2 \omega t)\, dx$$

where we have used $\omega^2 = k^2v^2$. Now,

$$\int_0^L \sin^2 kx\, dx = \int_0^L \sin^2 \frac{\pi x}{L}\, dx = \frac{L}{\pi} \int_0^\pi \sin^2 u\, du = \tfrac{1}{2}L$$

and

$$\int_0^L \cos^2 kx\, dx = \tfrac{1}{2}L$$

so that
$$E = \tfrac{1}{4}\mu A^2 \omega^2 L (\cos^2 \omega t + \sin^2 \omega t) \qquad (19\text{–}37)$$

Before obtaining the final result by setting $\cos^2 \omega t + \sin^2 \omega t = 1$, it is interesting to note that the term involving $\cos^2 \omega t$ represents the kinetic energy, whereas the $\sin^2 \omega t$ term represents the potential energy. We see again a characteristic of all vibrating systems, namely, that K is maximum when U is minimum and conversely. Thus, in a standing wave, the energy oscillates between being completely kinetic and being completely potential; the time difference between each such excursion is one quarter period. Figure 19–15 illustrates this energy exchange over one complete cycle or period.

Finally, we have for the total mechanical energy of the string,

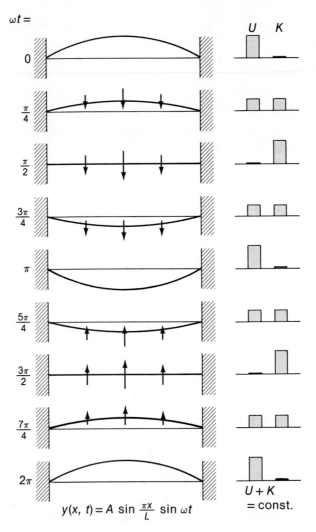

$\omega t =$

$y(x, t) = A \sin \frac{\pi x}{L} \sin \omega t$

$U + K$ = const.

Fig. 19–15. The exchange of energy between kinetic and potential forms during one period of the fundamental vibration of a stretched string.

$$E = \tfrac{1}{4}\mu A^2 \omega^2 L \qquad\qquad (19\text{-}38)$$

(Compare with the derivation leading to Eqs. 15–7a and b).

Example 19–9

Consider a string with a length L and a density μ that is under a tension F. One end of the string is fastened to a fixed support at $x = 0$, and the other end is attached to a ring that slides frictionlessly on a perpendicular shaft at $x = L$. Determine the possible standing-wave patterns.

Solution:

We start with the general wave functions, $y^+(x, t)$ and $y^-(x, t)$, of Eqs. 19–28. After applying the condition that a node exist at $x = 0$, we have (Eq. 19–30)

$$y(x, t) = A \sin kx \sin (\omega t - \phi_0)$$

At $x = L$, the ring slides frictionlessly, so the perpendicular component of the force is zero: $F_\perp = 0$. Now (see Fig. 19–7), $F_\perp = -F \sin \alpha \cong -F \tan \alpha = -F(\partial y/\partial x)$. Thus, the boundary condition at $x = L$ is

$$\left. \frac{\partial y}{\partial x} \right|_{x=L} = 0$$

This gives $\cos k_n L = 0$, from which $k_n L = \tfrac{1}{2} n\pi$, with $n = 1, 3, 5, 7, \ldots$. Then, using $k = 2\pi/\lambda$, the condition is

$$\lambda_n = 4L/n \qquad n = 1, 3, 5, 7, \ldots$$

Finally, the wave function for the standing wave is

$$y(x, t) = A \sin (n\pi x/2L) \sin (\omega t - \phi_0) \qquad n = 1, 3, 5, 7, \ldots$$

The diagram shows the fundamental standing-wave pattern and the next two lowest allowed harmonics. Notice the contrast with the standing waves shown in Fig. 19–13 for a string fixed at both ends. Only the *odd* harmonics are allowed in the present case.

SPECIAL TOPIC

19-8 COMPLEX WAVES

Fourier Analysis. In 1807 the French mathematician and military engineer Joseph Fourier (1768–1830) pointed out that any function can be represented as an infinite series of sine and cosine terms. If a function $f(x)$ is periodic with repetition length L, Fourier's theorem* states that

$$f(x) = \tfrac{1}{2}A_0 + \sum_{n=1}^{\infty} \left(A_n \cos \frac{2\pi n}{L}x + B_n \sin \frac{2\pi n}{L}x \right)$$

(19–39)

where the coefficients are given by

$$A_n = \frac{2}{L} \int_0^L f(x) \cos \frac{2\pi n}{L}x \, dx, \qquad n = 0, 1, 2, \ldots$$

(19–40a)

$$B_n = \frac{2}{L} \int_0^L f(x) \sin \frac{2\pi n}{L}x \, dx, \qquad n = 1, 2, 3, \ldots$$

(19–40b)

If the function $f(x)$ is relatively uncomplicated, the sum of only the first few terms in the series may give a satisfactory representation. For example, consider the rectangular step function with amplitude h and width $\frac{1}{2}L$, shown in Fig. 19–16. This function is periodic with repetition length L and therefore can be represented by a Fou-

---

*Fourier's 1807 paper was not rigorous; P. G. L. Dirichlet (1805–1859) first gave a rigorous proof of this important theorem in 1829.

Fig. 19–16. A rectangular step function.

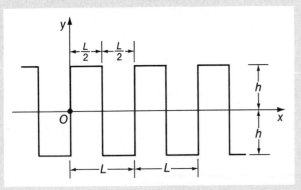

rier series. Notice that in the interval $0 < x < L$, the function starts from zero, exhibits first a positive segment, then a negative segment, and finally ends at zero. This is the general behavior of a sine function, so we expect that the term $\sin(2\pi/L)x$ will be important in the series. Notice also that the rectangular step function is an *odd* function about the origin, that is, $f(-x) = -f(x)$. The sine function is also an *odd* function, $\sin(-x) = -\sin x$, whereas the cosine function is an *even* function, $\cos(-x) = +\cos x$. Therefore, we do not expect the cosine terms to appear in the Fourier series. Moreover, the rectangular step function is also an odd function about $x = \frac{1}{2}L$. Consequently, the even sine terms do not appear in the series. Thus, the series reduces to

$$f(x) = \sum_{n \text{ odd}} B_n \sin \frac{2\pi n}{L}x$$

Evaluating the coefficients, we have

$$f(x) = \frac{4h}{\pi} \sum_{n \text{ odd}} \frac{1}{n} \sin \frac{2\pi n}{L}x \qquad \textbf{(19–41)}$$

Letting $\theta = 2\pi x/L$, we have

$$f(\theta)/h = 1.273 \sin\theta + 0.424 \sin 3\theta + 0.255 \sin 5\theta + \cdots$$

The first two terms of this series, their sum, and the rectangular step function are shown in Fig. 19–17. The two terms alone already give the general appearance of the step function.

Fig. 19–17. The first two terms, and their sum, of the Fourier series representation of a rectangular step function.

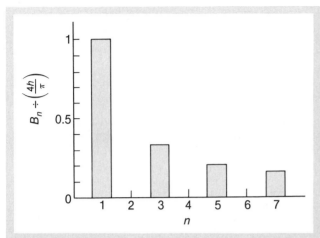

Fig. 19-18. Amplitude spectrum for the rectangular step function of Fig. 19-17.

The Fourier decomposition of a function is easy to visualize in an *amplitude spectrum,* a graphical display of the coefficients of the sinusoidal terms in the Fourier series. Figure 19-18 shows the amplitude spectrum for a rectangular step function. Only the odd terms are present and the amplitudes are proportional to $1/n$.

We can also suppose that the rectangular step function is actually a propagating wave form. That is, Fig. 19-16 and Eq. 19-41 show the wave and its Fourier representation at time $t = 0$. To recover the propagating aspect of the wave it is only necessary to replace x by $x - vt$; thus,

$$y(x, t) = \frac{4h}{\pi} \sum_{n \text{ odd}} \frac{1}{n} \sin\left(\frac{2\pi n}{L}(x - vt)\right) \quad \textbf{(19-42)}$$

which is a wave train of rectangular pulses propagating to the right.

Standing Waves. Consider again a taut string that is terminated by fixed supports at $x = 0$ and $x = L$. The only standing waves that can exist on such a string are those that have a node at each end of the string; these waves are described by *sine* functions. If the string is displaced in some way and released at time $t = 0$, we have

$$y(x, 0) = f(x) = \sum_n B_n \sin \frac{n\pi}{L} x \quad \textbf{(19-43)}$$

(•Notice that the argument of the sine function is now $n\pi/L$ instead of $2\pi n/L$; can you see why?) The time variation of the superposition of standing waves is (compare Eq. 19-32 with $\phi_0 = \pi/2$)

$$y(x, t) = \sum_n B_n \sin \frac{n\pi}{L} x \cos \omega_n t \quad \textbf{(19-44)}$$

where $\omega_n = k_n v = n\pi v/L$. By evaluating the coefficients B_n at $t = 0$, the behavior of the wave is determined for all subsequent times.

This type of analysis is valid for a *plucked* string—one that is released from rest. If instead the string is *struck,* the string is in motion at $t = 0$ and an additional initial condition, $\partial y / \partial t|_{t=0}$, must be imposed. The result still has the appearance of Eq. 19-44 except that $\sin \omega_n t$ terms are present.

Example 19-10

A taut string with a length L is displaced by raising the center point of the string a distance h so that the string has the shape of an equilateral triangle with base L and altitude h, as shown in the diagram. The string is then released from rest. Determine the Fourier expansion of the wave function $y(x, t)$ by using

$$B_n = \frac{2}{L} \int_0^L y(x, 0) \sin \frac{n\pi}{L} x \, dx$$

Solution:

The initial displacement of the string is given by

$$y(x, 0) = f(x) = \begin{cases} \dfrac{2h}{L} x, & 0 \leqslant x \leqslant \tfrac{1}{2} L \\[2ex] \dfrac{2h}{L}(L - x), & \tfrac{1}{2} L \leqslant x \leqslant L \end{cases}$$

Writing the expression for B_n in two parts, we have

$$B_n = \frac{2}{L} \int_0^{L/2} \frac{2h}{L} x \sin \frac{n\pi}{L} x \, dx$$

$$+ \frac{2}{L} \int_{L/2}^L \frac{2h}{L}(L - x) \sin \frac{n\pi}{L} x \, dx$$

$$= \frac{8h}{n^2 \pi^2} \sin \frac{n\pi}{2}$$

Then, the result is

$$y(x, 0) = \frac{8h}{\pi^2} \sum_{n=1}^{\infty} \frac{1}{n^2} \sin \frac{n\pi}{2} \sin \frac{n\pi}{L} x$$

and

$$y(x, t) = \frac{8h}{\pi^2} \left[\sin \frac{\pi}{L} x \cos \omega_1 t - \frac{1}{9} \sin \frac{3\pi}{L} x \cos 3\omega_1 t \right.$$
$$\left. + \tfrac{1}{25} \sin \frac{5\pi}{L} x \cos 5\omega_1 t - \cdots \right]$$

This expansion shows that the most prominent feature of the motion (the leading term) is the fundamental vibration of the string. The other terms represent the *odd* harmonics, whose amplitudes decrease as $1/n^2$. No even harmonics are present. (•Can you explain why?)

SPECIAL TOPIC

19-9   ATTENUATION AND DISPERSION

Any mechanical wave loses energy by some sort of frictional process in the medium through which it propagates. This results in the decrease of the wave amplitude with time, and the wave eventually disappears because all of the energy has been absorbed by the medium. The situation is similar to that of the damped harmonic oscillator, discussed in Section 15-6. For an idealized viscous medium in which the damping force is proportional to the velocity of the material, a term, $+\beta \, \partial \eta / \partial t$, must be added to the right-hand side of the general wave equation (Eq. 19-2 or 19-7). (Compare with Eq. 15-31, where a similar term is added to the oscillator equation.) This damping term introduces into the solution an exponential factor $e^{-\alpha t}$, where $\alpha = \tfrac{1}{2} v^2 \beta$. In this approximation, the damping is independent of the frequency. Figure 19-19a shows the exponential decay of a single pulse as it moves through a "lossy" medium. (Compare Fig. 15-16.)

The effect of viscous damping for standing waves on a taut string is to produce an exponential decay of the amplitude. In addition, the damping disrupts the simple relationship that connects the angular frequency and the phase velocity for an undamped wave, namely, $\omega_n = n\pi v/L$, and gives, instead, $\omega_n = (n\pi v/L)[1 - \beta^2 v^2/4n^2]^{1/2}$. If the damping is appreciable, a quality of dissonance is noticeable in the sound produced by vibrating strings. In many cases, the damping depends on the frequency because of the characteristics of the medium and because of the radiated loss of energy. This usually leads to an increase of the damping coefficient α with increasing frequency. Thus, the higher harmonics decay in amplitude more rapidly than the lower harmonics. This has the effect of "rounding off" the sharp features of a wave, eventually leading to a vibration that consists entirely of the fundamental.

For waves in certain media (for example, surface waves on water), the phase velocity v depends on the frequency (or wavelength). If a wave is analyzed into its Fourier components at an instant t, then farther along the line of propagation, at a time $t + \Delta t$, the various frequency components will have different spatial relationships to one another. The superimposed amplitudes then produce a wave with a shape different from that of the wave at time t. A group of pulses propagating in this situation experiences two important effects. First, each pulse broadens as it moves through the medium, as illustrated in Fig. 19-19b; this phenomenon is called *dispersion*. Second, the velocity of some distinctive feature of the group, such as the largest peak, is different from the phase velocity. The velocity with which the wave group moves is called the *group velocity*.

Fig. 19-19. The progress of a wave pulse as observed at time t_0 and at regular intervals thereafter. (a) With exponential damping due to viscous forces. (b) With dispersion of the wave resulting from a frequency dependence of the phase velocity.

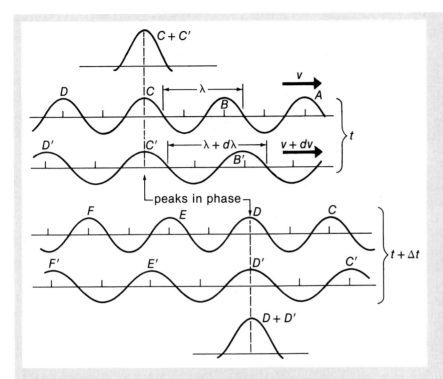

Fig. 19–20. Propagation of two wave components in the same medium, one with wavelength λ and phase velocity *v*, and the other with wavelength λ + *d*λ and phase velocity *v* + *dv*. The rate of propagation of the sum peaks, such as C + C' and D + D', determines the group velocity of the wave.

To obtain an expression for the group velocity, consider a wave that consists of many different frequency components and focus attention on the two components that make the greatest contribution to the wave system. Let the wavelengths of these waves in the medium be λ and λ + *d*λ, and let the corresponding phase velocities be *v* and *v* + *dv*. Figure 19–20 shows these two wave components at time *t* and at time *t* + Δ*t*. At *t* the wave crests C and C' are exactly in phase, and their sum produces a peak that is the most prominent local feature of the two waves. At *t* + Δ*t*, the crest D' of the wave component with wavelength λ + *d*λ, moving with the greater velocity, has overtaken the crest D of the wave λ. This coincidence of crests locates the new position of the peak D + D'. During the time Δ*t*, the crest D' moved a distance $(v + dv)\Delta t$, which is *d*λ greater than the distance $v\Delta t$ moved by the crest D. Therefore,

$$v\Delta t + d\lambda = (v + dv)\Delta t$$

or

$$\Delta t = \frac{d\lambda}{dv}$$

During this same time, the position of the sum peak moved a distance $v\Delta t - \lambda$. Consequently, the sum-peak velocity or *group velocity* v_g is

$$v_g = \frac{v\Delta t - \lambda}{\Delta t} = v - \frac{\lambda}{\Delta t}$$

or

$$v_g = v - \lambda \frac{dv}{d\lambda} \qquad (19\text{–}45)$$

If the phase velocity is independent of wavelength (or frequency), we have $dv/d\lambda = 0$. Then, the group velocity v_g is equal to the phase velocity *v*, as we expect.

PROBLEMS

Section 19-2

19-1 Suppose that you hear a thunder clap 16.2 s after seeing the associated lightning stroke. The speed of sound waves in air is 334 m/s and the speed of light in air is 3.0×10^8 m/s. How far are you from the lightning stroke?

19-2 A stone is dropped into a deep canyon and is heard to strike the bottom 10.2 s after release. The speed of sound waves in air is 334 m/s. How deep is the canyon? What would be the percentage error in the depth if the time required for the sound to reach the canyon rim were ignored?

19-3 Two rectangular wave pulses travel along a string, each with a speed of 10 m/s. At $t = 0$ the pulses are approaching each other, as shown in the diagram. Sketch the shape of the string at times $t = 6 \times 10^{-3}$ s to $t = 18 \times 10^{-3}$ s in intervals of 2×10^{-3} s.

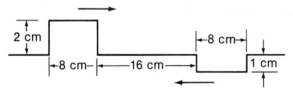

19-4 The wave function for a pulse is

$$y(x, t) = Ae^{-\sigma(x+vt)^2}$$

with $A = 3$ cm, $\sigma = 10$ m^{-2}, and $v = 10$ m/s. (This is called a *Gaussian* function.) (a) Sketch the shape of the pulse at $t = 0$ by determining $y(x, 0)$ for $x = 0$, ± 0.2 m, ± 0.4 m, and ± 0.6 m. What is the full width at half maximum displacement (FWHM) for the pulse? (b) Sketch the displacement of the string at $x = 0$ as a function of time, that is, $y(0, t)$, as the pulse passes. [*Hint:* Can you relabel the axes of the sketch in (a) to obtain the desired result? You should first determine the direction of propagation.]

19-5 The wave function for a pulse is

$$\eta(x, t) = \frac{0.1 \, C^3}{C^2 + (x - vt)^2}$$

with $C = 4$ cm. At some point x the displacement η is observed to decrease from its maximum value to one half the maximum value in 2×10^{-3} s. (a) What is the speed of the pulse? (b) What is the full width at half maximum displacement (FWHM) for the pulse? (c) Sketch the shape of the pulse at

$t = 0$ by determining $\eta(x, 0)$ for $x = 0$, ± 2 cm, ± 4 cm, ± 6 cm, ± 8 cm, and ± 10 cm.

19-6 Two transverse linearly polarized waves travel along
● a taut string in the negative x-direction. The wave functions are

$$y(x, t) = A \sin [k(x + vt) - \phi_0]$$
$$z(x, t) = A \sin [k(x + vt) - \phi_0]$$

(a) What is the plane of polarization of the combined wave? (b) Discuss in general terms the effect of changing the second wave to

$$z'(x, t) = A \sin [k(x + vt) - \phi_0 + \pi/2]$$

by studying the behavior of the string as a function of time at $x = 0$. Consider $kvt = \phi_0 + \alpha$, with $\alpha = n(\pi/4)$, $n = 0, 1, 2, \ldots, 8$. Is it appropriate to refer to such a wave as a *circularly polarized wave*?

Section 19-3

19-7 A copper wire with a length of 5 m has a mass of 0.060 kg and is under a tension of 750 N. With what velocity will transverse waves propagate along the wire?

19-8 An earthquake on the ocean floor in the Gulf of Alaska induces a *tsunami* (sometimes called a "tidal wave") that reaches Hilo, Hawaii, 4450 km distant, in a time of 9 h 30 min. Tsunamis have enormous wavelengths (100–200 km), and for such waves the propagation velocity is $v \cong \sqrt{g\bar{d}}$, where $\bar{d}$ is the average depth of the water. From the information given, find the average wave velocity and the average ocean depth between Alaska and Hawaii. (This method was used in 1856 to estimate the average depth of the Pacific Ocean long before soundings were made to give a direct determination.)

19-9 A string with a mass M and a length L hangs from a ceiling. Derive an expression for the velocity of transverse wave pulses on the string as a function of the distance x from the free end. [*Hint:* What is the tension in the string due to its own weight at a distance x from the bottom?]

19-10 In a gravity-free environment, a piece of string with
● its ends joined can be made to assume a circular shape and to spin about its center with a tangential speed v_0. A sharp blow produces a small distortion

in the string. Show that the distortion will travel around the string's path while maintaining a fixed position relative to the string. [*Hint:* First, determine the tension in the undisturbed string.]

19-11 A triangular wave pulse on a taut string travels
•• in the positive x-direction with a velocity of 50 m/s. The linear mass density of the string is $\mu = 0.080$ kg/m. The shape of the pulse at $t = 0$ is shown in the diagram below. (a) Using $u = x - vt$, verify that

$$y(x, t) = y(u)$$
$$= (h/L)(L + u), \qquad -L < u < 0$$
$$= (h/L)(L - u), \qquad 0 < u < L$$
$$= 0, \qquad u > L \text{ or } u < -L$$

(b) Show that

$$P(x, t) = P(u) = Fv(h^2/L^2), \qquad -L < u < L$$
$$= 0 \qquad u > L \text{ or } u < -L$$

(c) Show that the total energy in the pulse is $E = 2Fh^2/L$. Find the value of E.

Section 19-4

19-12 The wave function for a linearly polarized wave on a taut string is (in SI units)

$$y(x, t) = (0.35 \text{ m}) \sin(10\pi t - 3\pi x + \pi/4)$$

(a) What is the velocity of the wave? (Give the speed and the direction.) (b) What is the displacement at $t = 0$, $x = 0.1$ m? (c) What is the wavelength and what is the frequency of the wave? (d) What is the maximum magnitude of the transverse velocity of the string?

19-13 A transverse SHW has a period $T = 25$ ms and travels in the negative x-direction with a velocity of 30 m/s. At $t = 0$, a particle on the string at $x = 0$ has a displacement of 2.0 cm and a velocity $v_\perp = -2.0$ m/s. (a) What is the amplitude of the wave? (b) What is the initial phase angle? (c) What is the magnitude of the maximum transverse velocity? (d) Write the wave function for the wave.

19-14 A vibrating source located at $x = 0$ executes vertical
• SHM at a frequency of 100 Hz and excites transverse waves on an infinitely long taut string. At $t = 0$, the velocity of the source has one half of its maximum (positive) value of 30 m/s and is increasing. The string tension is 400 N and the linear mass density of the string is 50 g/m. (a) Write an equation for the SHM of the source that is compatible with Eqs. 19-16. What is the value of ϕ_0? (b) What is the phase velocity and what is the wavelength of the waves? (c) What is the amplitude of the waves? (d) Write the wave functions for the two waves that are generated (one for $x > 0$ and one for $x < 0$).

19-15 What is the average rate at which power is transmitted along the string in Problem 19-12 if $\mu = 75$ g/m? What is the total energy per wavelength in the wave?

19-16 A SHW on a taut string with $\mu = 0.10$ kg/m has a total energy of 4 J per wavelength. The maximum transverse velocity of the string is 15.0 m/s, which is one half of the propagation velocity of the wave. (a) What is the wavelength of the wave? [*Hint:* First, show that $E = \frac{1}{2}\mu\lambda v_{\perp,max}^2$.] (b) What is the tension in the string? (c) What is the frequency of the wave? (d) What is the amplitude of the wave?

Section 19-5

19-17 Use the value of Young's modulus for platinum given in Table 16-2 and the density value of 21.4 g/cm³, and calculate the velocity of a pure compressional wave in a thin platinum rod. Compare your result with the value given in Table 19-1.

19-18 Young's modulus for copper can be obtained from the known density of copper (8.96 g/cm³) and a measurement of the propagation velocity of a compressional pulse generated by a sharp blow to one end of a thin copper rod. Suppose that the blow has a duration of 120 µs. The precision desired in the measurement of the modulus is 1 percent, and the timing accuracy is equal to the pulse duration. What is the minimum length of the rod used for the measurements? (You will need an approximate value of v for copper in order to design the experiment. Use the value given in Table 19-1.)

19-19 A coiled spring (such as a Slinky toy) has a mass $M = 0.5$ kg, a length $\ell = 2$ m, and a spring constant $\kappa = 6$ N/m. Show that the extensional wave velocity is $v = \ell\sqrt{\kappa/M}$ and determine its value.

Section 19-6

19-20 Two strings, one with twice the linear mass density

of the other, are joined to make a single taut line. A SHW with an amplitude of 10 cm travels along the lighter string and approaches the junction. (a) What are the amplitude and phase (with respect to the incident wave) of the transmitted wave? (b) What are the same quantities for the reflected wave? (c) If the incident wave carries an average power of 100 W, what is the reflected average power? [*Hint:* Refer to Example 19-6 for an expression for $\bar{P}$.]

19-21 A long string with a linear mass density μ has a
• section of string with density 2μ inserted into its length, as shown in the diagram. A single pulse travels along the string toward the heavier section. Determine the magnitude and the shape of the first pulse to be transmitted past the heavy section. (•Why is it necessary to specify the "first" pulse?)

19-22 A 2-m length of string with linear mass density μ
•• is fastened at one end to a massive wall, and the other end is joined to a string with a density 3μ, as shown in the diagram. A single sharp pulse with peak amplitude A travels with a velocity of 20 m/s along the heavier string toward the wall. Sketch the distribution of pulses that are reflected back along the heavier string. (Show at least 4 pulses.) Show that eventually all the energy in the incident pulse is reflected. (An infinite number of reflections is required.) [*Hint:* Note that the energy carried in the pulse is proportional to the square of the amplitude; see Example 19-4. You will also need the result for the sum of an infinite number of terms in a geometric progression; see Eq. A-4 in Appendix A.]

Section 19-7

19-23 Two strings with equal lengths and densities are held fixed at their ends. What is the ratio of the tensions in the strings that is required to give the same vibration frequency when one string vibrates in a mode with two segments (that is, with one node between the ends) and the other string vibrates in three sections (that is, with two nodes between the ends)?

19-24 A violin string has a length of 35 cm and a mass of 0.41 g; the string tension is 90 N. What are the lowest three harmonic frequencies for the open string (i.e., with no fingering)?

19-25 A string on a cello has a length of 0.70 m and the tension is adjusted to sound concert A ($\nu_A = 440$ Hz) when played without fingering (this is the fundamental). (a) By what amount must the tension be increased if the fundamental tone is to be an octave higher ($\nu'_A = 2\nu_A$)? (b) By what amount must the length be shortened by fingering, with the original tension, to produce a fundamental tone that is an octave higher than middle C ($\nu_C = 261.6$ Hz)?

19-26 A thin rod, clamped at its midpoint, has a fundamental longitudinal vibration frequency of 1000 Hz. At what points could the clamp be placed to produce a fundamental frequency of 4000 Hz? [*Hint:* The position of the clamp is necessarily a node, and each end is an antinode.]

19-27 Two strings with different linear mass densities are joined together and stretched between two fixed supports. The string system is set into vibration with the standing-wave pattern shown in the diagram. Determine the length ratio L_2/L_1 if $\mu_2/\mu_1 = 5/4$.

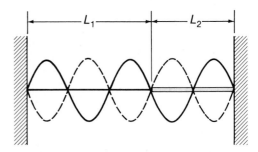

19-28 A 50-kg ball is attached to one end of a 1-m cord
• that has a mass of 0.10 kg. The other end of the cord is attached to a ring that can slide frictionlessly on a horizontal shaft, as shown in the diagram. A horizontal blow is delivered to the cord and excites the

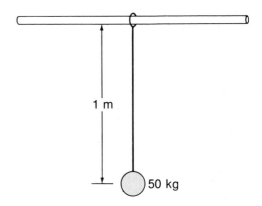

fundamental vibration with a maximum transverse velocity of 15 m/s. Assume that the ball remains essentially stationary as the cord vibrates. (a) What is the frequency of the fundamental vibration? (b) What is the amplitude of the motion? (c) If the blow is delivered as an impulse to a stationary cord, show that the wave function for the cord is

$y(x, t) = A \sin (\pi x/2L) \sin \omega t$, where the ball is located at $x = 0$. (d) Show that the energy in the system is given by Eq. 19-38 and calculate this energy. (e) Determine the period of the pendulum motion of the hanging ball and compare with the period of the vibrating string. Is it reasonable to assume that the hanging ball remains stationary?

Section 19-8

19-29 A taut string is given, at $t = 0$, the displacement
 • shown in the diagram. Show that the Fourier coefficients are given by $B_n = (32h/3n^2\pi^2) \sin (n\pi/4)$. Write out the first four terms in the expansion for $y(x, t)$. Discuss the fact that the expansion differs from that found in Example 19-10 only in a 6 percent reduction in the amplitudes of the odd harmonics and in the appearance of a significant second harmonic.

19-30 Consider the propagating triangular (or "saw-
 •• tooth") wave shown in the diagram. Show that the Fourier representation of this wave is

$$y(x, t) = \frac{8h}{\pi^2} \sum_{m=1}^{\infty} \frac{1}{m^2} \cos \frac{2\pi m}{L}(x - vt), \qquad m \text{ odd}$$

[*Hint:* The required integral is included in the list inside the rear cover.]

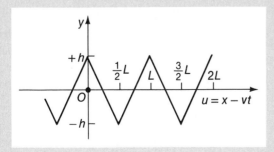

19-31 A string that is stretched between a pair of fixed
 • supports is displaced in the manner shown in the diagram, where $\epsilon \ll L$. The string is then released from rest at $t = 0$. Show that in the limit $\epsilon/L \to 0$, the Fourier expansion of the standing wave is

$$y(x, t) = \frac{4h}{\pi} \sum_{n=1}^{\infty} \frac{1}{n} \sin \frac{n\pi}{L} x \cos \frac{n\pi v}{L} t, \qquad n \text{ odd}$$

19-32 Use the expression for $y(x, 0)$ from Problem 19-31
 • to show that

$$4[1 - \tfrac{1}{3} + \tfrac{1}{5} - \tfrac{1}{7} + \tfrac{1}{9} - \tfrac{1}{11} + \ldots] = \pi$$

Evaluate the sum to the term $\frac{1}{31}$ and compare with the value of π. What can you say about the rate of convergence of the series? [*Hint:* Let $h = 1$ and $x = \frac{1}{2}L$.] Determine the value of n to which the sum must be carried to obtain an accuracy of at least 0.001 for π. [*Hint:* Take advantage of the alternating character of the series.]

Section 19-9

19-33 Transverse wave pulses travel with a phase velocity $v = 200$ m/s along a taut string. As pulses travel along this string, they decrease in size by one half in moving 2.08 m due to losses in the string. Assume that the damping is proportional to the velocity. Determine the damping coefficients, α and β.

19-34 Consider waves traveling on a water surface under conditions such that the phase velocity is $v = \sqrt{g\lambda/2\pi}$. (This corresponds to the case of waves with ordinary wavelengths running over deep water; see Problem 19-41.) A particular wave disturbance consists of many superimposed Fourier components with important contributions in the wavelength region near λ_0. Determine the group velocity of this wave group.

Additional Problems

19-35 A transverse wave pulse, $y(x, t) = Ae^{-\sigma(x-vt)^2}$,
• travels along a taut string, with $A = 3$ cm, $\sigma = 10$ m^{-2}, and $v = 10$ m/s. (a) Determine the radius of curvature of the wave at $x = 0$ for $t = 0$. [*Hint:* When $\partial y/\partial x \ll 1$, then $R \cong (\partial^2 y/\partial x^2)^{-1}$. Is this condition met at $x = 0$, $t = 0$?] (b) What is the centripetal acceleration (in units of g) of a small portion of the string at $x = 0$, $t = 0$? (c) What vertical force acts on a 10-g piece of the string at $x = 0$, $t = 0$? (d) The linear density of the string is $\mu = 0.5$ kg/m. Express the result of (c) in terms of the string tension.

19-36 The point $x = 0$ on an infinitely long taut string is displaced in the manner shown in the diagram. The wave velocity on the string is 12 m/s. Sketch the shape of the propagating waves at $t = 3 \times 10^{-3}$ s.

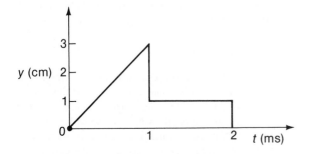

19-37 The elastic limit of a piece of steel wire is 2.7×10^9 Pa. What is the maximum velocity at which transverse wave pulses can propagate along this wire without exceeding this stress? (The density of the steel is 7.86 g/cm^3.)

19-38 Transverse pulses propagate with a velocity of 200 m/s along a taut copper wire that has a diameter of 1.5 mm. What is the tension in the wire? (The density of copper is given in Table 1-2.)

19-39 The wave function of a wave is (in SI units) $y(x, t) = 0.20 \sin(80\pi t - 4\pi x)$. Rewrite this wave function in the three forms given in Eq. 19-15. Determine the values of v, λ, ω, ν, k, T, and ϕ_0.

19-40 For longitudinal waves in a solid rod, show that the ratio of the particle velocity to the wave velocity at any point is equal to the ratio of the stress at that point to Young's modulus, or

$$\left| \frac{v_{particle}}{v_{wave}} \right| = \frac{|s|}{Y}$$

19-41 The velocity of surface waves on a body of water
• depends on the wavelength λ of the wave, the depth h of the water, the density ρ and surface tension γ of the water, and the acceleration g due to gravity. The expression is

$$v = \sqrt{\left(\frac{g\lambda}{2\pi} + \frac{2\pi\gamma}{\rho\lambda} \right) \tanh \frac{2\pi h}{\lambda}}$$

where the hyperbolic tangent is $\tanh x = (e^x - e^{-x})/(e^x + e^{-x})$. (a) For the case of waves in deep water $(h \gg \lambda$ and $\lambda \gg \sqrt{\gamma/\rho g})$, show that $v \cong \sqrt{g\lambda/2\pi}$. (b) For waves in shallow water $(h \ll \lambda$ and $\lambda \gg \sqrt{\gamma/\rho g})$, show that $v \cong \sqrt{gh}$. (c) For "surface tension waves" $(h \gg \lambda$ and $\lambda \ll \sqrt{\gamma/\rho g})$, show that $v = \sqrt{2\pi\gamma/\rho\lambda}$. [*Hint:* For large x, $\tanh x \cong 1$; for small x, $\tanh x \cong x$.]

19-42 Use the expression in Problem 19-41 and calculate the velocity of ripples ($\lambda = 3$ cm) on the surface of a deep lake. ($\gamma = 7.6 \times 10^{-2}$ N/m.)

19-43 Use the expression in Problem 19-41 and calculate the velocity of sea swells far from shore when $\lambda = 2$ m. What will be the velocity of these waves when they run into shallow water with $h = 0.5$ m? What will be the velocity when $h = 0.1$ m? What is the period of oscillation of a small floating object in each case? (•In view of these results, can you explain why ocean waves that are incident on a straight beach at any angle when far offshore eventually approach the beach parallel to the shoreline? This phenomenon—more often associated with light waves—is called *refraction*.)

19-44 In Example 19-6, we found that the power in a wave, averaged over one period, is $\bar{P} = \frac{1}{2}FA^2k\omega$, where A is the wave amplitude. Apply this result to Eqs. 19-27 and show that, averaged over one period, the power in the transmitted wave plus that in the reflected wave is equal to the power in the incident wave.

19-45 A steel wire in a piano has a length of 0.70 m and a mass of 4.3×10^{-3} kg. To what tension must this wire be stretched in order that the fundamental vibration correspond to concert middle C, $\nu_C = 261.6$ Hz?

19-46 A violin string has a length of 0.35 m and is tuned to concert G, $\nu_G = 392$ Hz. Where must the violinist place his finger to play concert A, $\nu_A = 440$ Hz? If this position is to remain correct to one half the width of a finger (i.e., to within 0.6 cm), by what fraction may the string tension be allowed to slip?

SPECIAL TOPIC

19-47 Show that the Fourier component of the asym-
•• metric "sawtooth" wave shown in the diagram is

$$\eta(x, t) = \frac{2h}{\pi} \sum_{n=1}^{\infty} \frac{(-1)^{n+1}}{n} \sin\left[\frac{2n\pi}{L}(x - vt)\right]$$

Sketch the sum of the first three terms and compare with the original wave. [*Hint:* The relevant integral is included in the list inside the rear cover.]

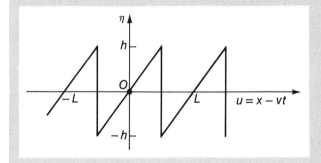

19-48 Consider surface waves on water with a depth
• $h \gg \lambda$. (Refer to the expression in Problem 19-41.) (a) Show that the minimum wave velocity is achieved for $\lambda_0 = 2\pi\sqrt{\gamma/\rho g}$. (b) Determine the value of λ_0. (c) What is the group velocity of a wave group whose most important Fourier components have wavelengths near λ_0? What is the corresponding dispersion?

SOUND

The propagation of pressure variations through a medium such as air is particularly interesting and important because pressure changes can be detected by the auditory apparatus of the ear. If the frequency of the pressure variations is in the range from about 15 Hz to about 20,000 Hz, the human ear produces the physiological sensation of *hearing*. Waves that produce such a response are called *sound waves*. Many species of animals and insects can detect pressure waves with frequencies much higher than those that can be heard by humans. Bats and dolphins, for example, can hear sounds with frequencies above 100,000 Hz. However, we restrict attention here to the human audible range from 15 Hz to 20,000 Hz.*

Sound waves with frequencies above 20,000 Hz are referred to as *ultrasonic* waves. By using special laboratory apparatus it is possible to produce sound waves with frequencies up to about 1 GHz = 10^9 Hz. Such waves have wavelengths in air as small as 300 nm, comparable with the wavelength of ultraviolet light. If the frequency of a sound wave is below 15 Hz, it is called an *infrasonic* wave. The motions of large scale matter, such as those produced by earthquakes and volcanic activity, can generate infrasonic waves.

20-1 THE SPEED OF SOUND WAVES

Compressional Waves. We consider first the propagation of sound waves for the one-dimensional case of a narrow tube that contains a gas and is closed at one end by a freely sliding piston (Fig. 20–1a). The result we obtain for the speed of sound in this special case is also valid for propagation in an unlimited (unbounded) gas.†

Suppose that the gas in the tube is at equilibrium with the ambient (outside) pressure p_0 and with the ambient density ρ_0. Consider what happens when the piston moves through a simple cycle. Starting from rest, the piston first moves to the right,

---

*These limits vary widely from person to person, depending on age and on the history of exposure to high-intensity sounds. Usually, a person's ability to hear high-frequency sounds deteriorates with age; by middle age, most individuals have a high-frequency limit of 12,000 to 14,000 Hz. People who have been exposed for long periods to high-intensity sounds (such as performers in rock bands and their fans, or people who work near aircraft) often have severely impaired high-frequency hearing.

†The reason why this is so for longitudinal waves in a gas or a nonviscous liquid, although it is *not* true for longitudinal waves in a solid, is not simple. We leave the discussion of this point to the end of this section.

stops instantaneously at the point of maximum excursion, and then returns to the original position. While the piston moves to the right, the gas in front of the piston is compressed, so the pressure and the density are raised above the equilibrium values. During the other half of the cycle, while the piston moves to the left, the pressure and density are lowered below the equilibrium values. The result of the piston movement is a pressure variation that propagates along the tube in the form of an overpressure pulse or *compression* followed by an underpressure pulse or *rarefaction*. If the piston continually repeats this motion as SHM, a simple harmonic wave consisting of alternating compressions and rarefactions in the gas will propagate steadily along the tube.

For a piston speed that is less than the wave speed, the pressure variation is in phase with the piston velocity. If the ratio of the wavelength λ of the wave to the tube diameter D is $\lambda/D \geq 2\pi$, the wave propagates along the tube as a plane wave; that is, the amplitude and phase depend only on the longitudinal position x and on the time t, and *not* on position within the tube on any cross section. Figure 20–1 represents this case.

▶ *The Wave Equation.* The derivation of the wave equation for longitudinal pressure disturbances in a gas-filled tube closely follows that for longitudinal waves in a thin rod (Section 19–5). Consider a section Δx of the gas in a tube, as shown in Fig. 20–2. Let ξ represent the displacement of the gas matter from equilibrium at the coordinate x. The excursion of the pressure from the equilibrium pressure p_0 is the *acoustic pressure* δp, and the total pressure at x is $p = p_0 + \delta p$. The equilibrium value of the volume of the gas in the region considered is $V = S\,\Delta x$, and the change in volume of the original quantity of gas due to the acoustic pressure is $\Delta V = S(\partial\xi/\partial x)\,\Delta x$. Using the definition of the bulk modulus (Eq. 16–21), we can write $\Delta V/V = -\delta p/\beta$, or

$$\frac{\partial\xi}{\partial x} = -\frac{\delta p}{\beta} \tag{20–1}$$

Fig. 20–1. (a) Longitudinal wave of alternating compressions C and rarefactions R produced by the oscillatory motion of a piston that executes SHM. The displacement of the piston relative to the wavelength λ is exaggerated. (b) The piston displacement x_P and velocity v_P and the gas pressure and density are shown as functions of time. The quantities $\delta p(t)$ and $\delta\rho(t)$ represent the deviations from the equilibrium values for the pressure and density, respectively. The deviations, δp and $\delta\rho$, compared with p_0 and ρ_0 are exaggerated. The quantity δp is called the *acoustic pressure.*

(a)   (b)

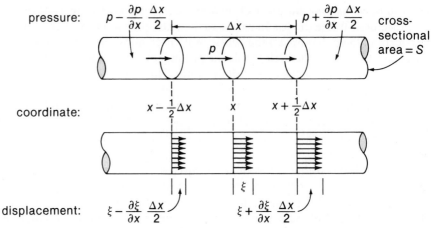

Fig. 20–2. Details of the pressure p and the matter displacement ξ in a section Δx of gas in a tube.

The net force acting on the quantity of gas is (recalling that pressure is a negative stress)

$$F = S\left[\left(p - \frac{\partial p}{\partial x}\frac{\Delta x}{2}\right) - \left(p + \frac{\partial p}{\partial x}\frac{\Delta x}{2}\right)\right]$$

$$= -S\frac{\partial p}{\partial x}\Delta x$$

The equilibrium density of the gas is ρ_0, so the mass in the section Δx is $\rho_0 S\,\Delta x$. Thus, Newton's equation of motion is

$$-S\frac{\partial p}{\partial x}\Delta x = \rho_0 S\,\Delta x\,\frac{\partial^2\xi}{\partial t^2}$$

Now,

$$\frac{\partial p}{\partial x} = \frac{\partial}{\partial x}(p_0 + \delta p) = \frac{\partial(\delta p)}{\partial x} = -\beta\frac{\partial^2\xi}{\partial x^2}$$

where we have used Eq. 20–1 for δp. Then, Newton's equation becomes

$$\frac{\partial^2\xi}{\partial x^2} = \frac{\rho_0}{\beta}\frac{\partial^2\xi}{\partial t^2} \qquad\qquad\qquad \textbf{(20–2)}$$

which is the desired wave equation. ◄

The Velocity of Compressional Waves. In Section 19–2 we learned that an equation of the form of Eq. 20–2 is a wave equation which is satisfied by any function $\xi(x, t) = \xi(u) = \xi(x \pm vt)$, where the wave velocity is (compare Eq. 19–21)

$$v = \sqrt{\frac{\beta}{\rho_0}} \qquad\qquad\qquad \textbf{(20–3)}$$

The compression of a gas produced by an applied pressure has the effect of raising the temperature of the gas unless the heat generated can be conducted away (see Section 23–2). For the situations we consider here, the heat conductivity of any gas is insufficient to transport the heat generated in the regions of compression to the cooler regions of rarefaction, at least for frequencies below about 10^5 Hz. Thus, we neglect heat flow between the various regions of a gas that supports an acoustic wave. That is, we assume conditions described as *adiabatic* (i.e., with no heat transfer), and we need to use in our equations the adiabatic bulk modulus β_a, which can be shown to be

$$\beta_a = \gamma p_0 \tag{20-4}$$

where γ is a number very close to 1.40 for all gases that consist of diatomic molecules, such as H_2, O_2, N_2, and air (see Section 25–3). Thus, the phase or wave velocity for sound in diatomic gases is

$$v = \sqrt{\frac{\beta_a}{\rho_0}} = \sqrt{\frac{1.40 \, p_0}{\rho_0}} \qquad \text{(for diatomic gases)} \tag{20-5}$$

For air at $0°$ C and normal atmospheric pressure, we have $p_0 = 1.013 \times 10^5$ Pa and $\rho_0 = 1.293 \text{ kg/m}^3$. Then, the value calculated from Eq. 20–5 is $v_{\text{calc}} = 331.2$ m/s, which is to be compared with the measured value, $v_{\text{meas}} = 331.5$ m/s. At a temperature T (in degrees C), the speed of sound $v(T)$ in an ideal gas* is

$$v(T) = v(0)\sqrt{1 + \frac{T}{273}} \qquad \text{(ideal gas)} \tag{20-6}$$

where $v(0)$ is the speed at $T = 0°$ C and where $T + 273$ is the absolute temperature of the gas.

Table 20–1 lists the speed of sound in various unbounded media at $0°$ C. For the solid media, the speeds given are for longitudinal waves.

▶ *Longitudinal Waves.* In Section 19–5 we obtained an expression for the velocity of longitudinal waves in a thin rod. Because the longitudinal stress-strain relationship for a thin rod involved only Young's modulus, we found

$$v_t = \sqrt{\frac{Y}{\rho}}$$

In Section 16–2 we learned that accompanying a longitudinal strain there is a lateral strain determined by the value of Poisson's ratio σ. However, longitudinal strains in an unbounded (isotropic and homogeneous) solid are *not* accompanied by lateral strains. The proper expression for the speed of longitudinal waves in such media is†

$$v'_t = \sqrt{\frac{\beta + \frac{4}{3}\mu}{\rho}} \tag{20-7}$$

where μ is the shear modulus (see Eq. 16–17). Gases and (nonviscous) liquids cannot sustain shear, so $\mu = 0$ for these substances; thus, the expression for v'_t reduces to Eq. 16–3.

Transverse waves cannot be set up in ideal fluids because such media cannot support shearing effects. In an unbounded solid medium, however, we do observe transverse waves. The speed of such waves is

$$v'_t = \sqrt{\frac{\mu}{\rho}} \tag{20-8}$$

Thus, $v'_t > v'_t$. Deep earthquakes excite both longitudinal and transverse seismic waves that travel through

---

*See Section 23–2 for the definition of an ideal gas.
†The derivation of this expression is beyond the scope of this text.

TABLE 20–1.
Speed of Sound (at 0° C)

MEDIUM	v (m/s)
Gases:*	
Air	331.5
Argon	319
Carbon dioxide	259
Chlorine	206
Helium	965
Hydrogen	1284
Liquids:	
Fresh water	1437
Sea water	1471
Mercury	1484
Solids:†	
Aluminum	6420
Gold	3240
Lead	1960
Glass, Pyrex	5640
Lucite	2680
Polyethylene	1950

*At normal atmospheric pressure.

†Notice that these values for longitudinal waves in unbounded media are different from the values listed in Table 19–1 for extensional waves in thin rods.

the Earth with different speeds. In the Earth's mantle, Poisson's ratio σ is about $\frac{1}{4}$; then, $\beta = 5\mu/3$ and, hence, $v'_t = \sqrt{3}v'_t$. The velocities in the mantle, for depths less than about 300 km, are $v'_t = 4.5$ km/s and $v'_t = 8.1$ km/s, a ratio of 1.8. When an earthquake occurs, distant seismograph stations will first detect the longitudinal wave, followed by the transverse wave. Generally, the speed of longitudinal seismic waves is about twice that of transverse waves (except in the molten core of the Earth, which cannot support transverse waves). The difference between the arrival times, when measured by several stations, permits an accurate determination of the location of the earthquake. ◀

Example 20–1

The wave function for a longitudinal sound wave in an air-filled tube is

$$\xi(x, t) = A \cos (kx - \omega t + \phi_0)$$

The amplitude of the motion of the driving piston is 2 mm and the frequency is 1000 Hz. The temperature of the air is 20° C and the equilibrium density is $\rho_0 = 1.20$ kg/m^3.
(a) Obtain expressions for the velocity of the gas matter, the acoustic pressure, and the density excursion.
(b) Obtain an expression for the power (averaged over a period) that is transported along the tube.
(c) Determine the initial phase angle ϕ_0 for the case in which the piston, whose average position is $x = 0$, undergoes excursion from this average position that constitutes SHM described by $y(t) = B \sin \omega t$.
(d) Show that the average power delivered to the piston is equal to the average power transported along the tube (b).

(e) Determine the speed of the wave and the wavelength.

(f) Obtain the amplitudes of the gas matter, the acoustic pressure, and the density excursion.

(g) Find the *intensity* of the wave, $I = \bar{P}/S$, which is the average power per unit cross-sectional area.

Solution:

(a) The velocity v_g of the gas matter is

$$v_g = \frac{\partial \xi}{\partial t} = A\omega \sin (kx - \omega t + \phi_0)$$

The acoustic pressure δp is (using Eq. 20–1)

$$\delta p = -\beta_a \frac{\partial \xi}{\partial x} = Ak\beta_a \sin (kx - \omega t + \phi_0)$$

where β_a is the adiabatic bulk modulus.

To obtain an expression for the density excursion, we note that $\rho = M/V$, so $\Delta\rho/\rho = -\Delta V/V$. Then, using $\Delta V/V = -\delta p/\beta_a$ (Eq. 16–21), we can write

$$\delta\rho = \frac{\rho_0}{\beta_a} \delta p = Ak\rho_0 \sin (kx - \omega t + \phi_0)$$

From these results we see that δp and $\delta\rho$ are in phase with v_g and that all three quantities differ in phase by $\pi/2$ compared with ξ.

(b) The power transported past a cross section of the tube at a point x at time t is (compare Eq. 19–10)

$$P(x, t) = \mathbf{F} \cdot \mathbf{v}_g = S \, \delta p v_g = -S\beta_a \left(\frac{\partial \xi}{\partial x}\right)\left(\frac{\partial \xi}{\partial t}\right)$$

$$= S\beta_a A^2 \omega k \sin^2 (kx - \omega t + \phi_0)$$

Now, the average of $\sin^2 u$ over a period is $\frac{1}{2}$ (see Example 19–6). Thus,

$$\bar{P} = \tfrac{1}{2}S\beta_a A^2 \omega k$$

(c) The motion of the piston is transferred directly to the adjacent air molecules; thus, $B = A$ and $dy/dt = v_g$. With $y = A \sin \omega t = B \sin \omega t$, we have

$$\frac{dy}{dt} = A\omega \cos \omega t = v_g(0, t) = A\omega \sin (-\omega t + \phi_0)$$

from which we conclude that $\phi_0 = \pi/2$ because $\sin (\pi/2 - \omega t) = \cos \omega t$.

(d) The instantaneous power delivered to the piston is (using $\phi_0 = \pi/2$)

$$P(t) = S \, \delta p \frac{dy}{dt} = SA^2 k\beta_a \omega \cos^2 \omega t$$

The average power delivered per cycle is

$$\bar{P} = \tfrac{1}{2}SA^2 k\beta_a \omega$$

which is the same as the average power transported along the tube (b), as it must be.

(e) For the speed of the sound wave, we use Eq. 20–6 and the value of $v(0)$ for air in

Table 20–1:

$$v(20°\,C) = (331.5 \text{ m/s}) \sqrt{1 + \tfrac{20}{273}} = 343.4 \text{ m/s}$$

The wavelength is
$$\lambda = \frac{v}{\nu} = \frac{343.4 \text{ m/s}}{1000 \text{ s}^{-1}} = 0.3434 \text{ m}$$

(f) For the various amplitudes we have*

$$v_{gm} = A\omega = (2 \times 10^{-3} \text{ m})(2\pi \times 10^3 \text{ s}^{-1}) = 12.57 \text{ m/s}$$

$$\delta p_m = Ak\beta_a = A\left(\frac{2\pi}{\lambda}\right)(\gamma p_0) = (2 \times 10^{-3} \text{ m})\left(\frac{2\pi}{0.3434 \text{ m}}\right)(1.40 \times 1.013 \times 10^5 \text{ Pa})$$
$$= 5.190 \times 10^3 \text{ Pa}$$

$$\delta\rho_m = Ak\rho_0 = A\left(\frac{2\pi}{\lambda}\right)\rho_0 = (2 \times 10^{-3} \text{ m})\left(\frac{2\pi}{0.3434 \text{ m}}\right)(1.20 \text{ kg/m}^3)$$
$$= 0.04391 \text{ kg/m}^3$$

Notice that the maximum acoustic pressure δp_m and the maximum density excursion $\delta\rho_m$ are small compared to the equilibrium values:

$$\frac{\delta p_m}{p_0} = \frac{5.190 \times 10^3 \text{ Pa}}{1.013 \times 10^5 \text{ Pa}} = 5.1 \text{ percent}$$

$$\frac{\delta\rho_m}{\rho_0} = \frac{0.04391 \text{ kg/m}^3}{1.20 \text{ kg/m}^3} = 3.7 \text{ percent}$$

(g) The intensity of the wave is $I = \bar{P}/S$, so that

$$I = \tfrac{1}{2}A^2 k\beta_a \omega \qquad\qquad \textbf{(20–9)}$$

For the present case, we have

$$I = \tfrac{1}{2}A^2\left(\frac{2\pi}{\lambda}\right)(\gamma p_0)\omega$$

$$= \tfrac{1}{2}(2 \times 10^{-3} \text{ m})^2\left(\frac{2\pi}{0.3434 \text{ m}}\right)(1.40 \times 1.013 \times 10^5 \text{ Pa})(2\pi \times 10^3 \text{ s}^{-1})$$

$$= 3.26 \times 10^4 \text{ W/m}^2$$

This is an extremely high sound intensity; it might be encountered in special industrial applications, but not in normal listening (see Table 20–2).

Example 20–2

A thrust with a very short duration Δt is imparted to the gas in a tube by a piston that moves with constant velocity u, as shown in the diagram. The equilibrium values of the pressure and density are p_0 and ρ_0, respectively. Show in impulse approximation that the velocity v of the resulting pulse is $\sqrt{\beta_a/\rho_0}$ (Eq. 20–3).

---

*As usual, a subscript m indicates *maximum value*, which, in these cases, is the amplitude.

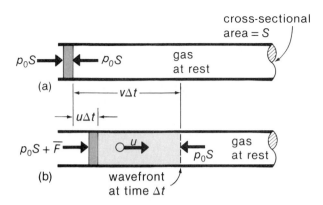

(a)

(b) wavefront
at time Δt

Solution:

Diagram (a) shows the situation immediately before the impulse is applied to the piston; diagram (b) represents the situation at time Δt later. During the time Δt, the wavefront progresses to a point $v\,\Delta t$ from the original position of the piston. The gas between the piston at time Δt and the position of the wavefront has a uniform velocity u imparted to it by the piston. The momentum of this mass of gas is $\rho_0 Svu\,\Delta t$. The average impulsive force $\overline{F}$ gives rise to an acoustic pressure, $\delta p = \overline{F}/S$, which produces the volume change, $\Delta V = -Su\,\Delta t$. The original volume of this mass of gas was $V = Sv\,\Delta t$. Now, from the definition of the bulk modulus (Eq. 16–21), we have

$$\frac{\delta p}{\beta_a} = -\frac{\Delta V}{V} = \frac{u}{v}$$

Then,
$$\delta p = \beta_a \frac{u}{v}$$

The impulse delivered to the gas is $\overline{F}\,\Delta t = S\,\delta p\,\Delta t$. Setting this equal to the momentum gained by the gas gives

$$S\,\delta p\,\Delta t = S\beta_a \frac{u}{v}\Delta t = \rho_0 Svu\,\Delta t$$

so that
$$v = \sqrt{\frac{\beta_a}{\rho_0}}$$

Example 20–3

A *Helmholtz resonator* is an acoustic device that consists of a rigid enclosure with a volume V coupled to the atmosphere through an open pipe with radius r and length ℓ, as indicated in the diagram. Derive an expression for the natural frequency ν_0 of oscillation of the system.

Solution:

The diagram shows a hypothetical displacement of the air mass in the pipe intruding a distance x into the main volume of the resonator. This intrusion compresses the air in the resonator by an amount $\Delta V = \pi r^2 x$. This compression, in turn, produces a pressure on the intruding mass given by

$$\delta p = -\beta_a \frac{\Delta V}{V} = -\beta_a \frac{\pi r^2 x}{V} = -\frac{v^2 \rho_0 \pi r^2}{V}x$$

The restoring force is

$$\delta F = \pi r^2 \, \delta p = -\frac{v^2 \rho_0 \pi^2 r^4}{V} x = -\kappa x$$

where the force constant κ for the volume of gas is

$$\kappa = v^2 \rho_0 \pi^2 r^4 / V$$

The mass of air oscillating in the pipe is $m = \pi r^2 \ell \rho_0$. The natural frequency is therefore (see Eq. 15–13)

$$\nu_0 = \frac{1}{2\pi} \sqrt{\frac{\kappa}{m}} = \frac{v}{2\pi} \sqrt{\frac{\pi r^2}{V\ell}}$$

(To obtain this expression we have made the implicit assumption that the wavelength, $\lambda_0 = v/\nu_0$, is large compared to all linear dimensions of the system. •Can you see why this assumption is necessary?)

When the system is excited, there is actually a small mass of air just beyond the open pipe that is also set into motion. Taking this into account increases the effective length of the pipe and gives a more accurate expression for the resonant frequency:

$$\nu_0 = \frac{v}{2\pi} \sqrt{\frac{\pi r^2}{V(\ell + 8r/3\pi)}}$$

Notice that even if $\ell = 0$, there is an effective mass, $8r^3\rho_0/3$, associated with the vent opening.

20-2 STANDING WAVES IN PIPES

We now consider the propagation of longitudinal waves in gas-filled pipes with finite length and investigate the effects of the pipe termination (whether open or closed) on these waves. We first examine the case of a pipe closed at $x = 0$, as shown in Fig. 20–3a. The wave function for the wave incident on the closed end from the left is $\xi_i(x, t)$; the reflected wave is described by $\xi_r(x, t)$. The gas matter immediately adjacent to the closed end has zero displacement. Thus, the required boundary condition on the waves, using superposition, is

$$\xi_i(0, t) + \xi_r(0, t) = 0 \qquad \text{(closed end)} \qquad \textbf{(20–10a)}$$

This condition is analogous to that discussed in Section 19–6 for the case of transverse waves on a taut string terminated by a rigid support.

Next, consider a pipe terminated at $x = 0$ with an open end, as shown in Fig. 20–3b. Now, we assume that the pressure is described by a continuous function and at $x = 0$ has a constant value equal to the ambient pressure p_0. (This assumption is valid only if the pipe diameter is small compared to the wavelength.) This requires the acoustic pressure δp to be zero at $x = 0$. Because $\delta p = -\beta_a \partial \xi / \partial x$, we can write

$$\left. \frac{\partial \xi_i(x, t)}{\partial x} \right|_{x=0} + \left. \frac{\partial \xi_r(x, t)}{\partial x} \right|_{x=0} = 0 \qquad \text{(open end)} \qquad \textbf{(20–10b)}$$

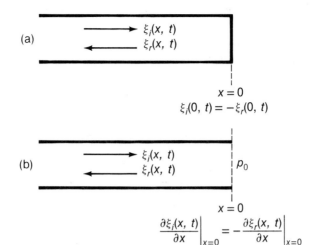

Fig. 20-3. Reflection of incident longitudinal waves $\xi_i(x, t)$ (a) from the closed end of a pipe and (b) from the open end of a pipe.

(a)

$\qquad \xi_i(x, t)$
$\qquad \xi_r(x, t)$

$x = 0$
$\xi_i(0, t) = -\xi_r(0, t)$

(b)

$\qquad \xi_i(x, t)$
$\qquad \xi_r(x, t)$

p_0

$x = 0$

$$\left.\frac{\partial \xi_i(x, t)}{\partial x}\right|_{x=0} = -\left.\frac{\partial \xi_r(x, t)}{\partial x}\right|_{x=0}$$

This condition is analogous to that discussed in Section 19–6 for the case of a transverse wave on a taut string terminated by a frictionless ring that slides on a fixed shaft.

The restriction that the pipe diameter be small compared to the wavelength is equivalent to specifying that there be no substantial sound radiation from the open end. Any such radiation would remove energy from the wave in the pipe, and consequently the reflected wave would have an amplitude smaller than that of the incident wave. Requiring the total absence of radiation from the end of the pipe is equivalent to requiring the total absence of friction for the case of the string terminated with a sliding ring. Moreover, the assumption that the sliding ring has zero mass is equivalent to ignoring the vibration of the mass of air just outside the open end of the pipe. These radiation and mass effects also introduce a phase difference between the incident and reflected waves.

Waves that propagate in terminated gas-filled pipes must obey the condition of Eq. 20–10a at every closed end and the condition of Eq. 20–10b at every open end. For particular wavelengths, standing waves are set up in pipes, just as they are on terminated strings. An interesting example is the case of organ pipes. Figure 20–4 shows the lowest frequency standing wave, the fundamental mode, of an organ pipe that is open at the nondriven end. Figure 20–4a indicates schematically the flow pattern that produces the turbulent excitation of the air in the pipe as air is blown against the lip edge and escapes through the slot at O. Although the opening in the

Fig. 20-4. An organ pipe open at the nondriven end. The graphs show the variation of ξ, v_g, δp, and $\delta \rho$ for the fundamental mode of vibration. The nodes are indicated by N. The wavelengths of the harmonics are given by $\lambda_n = 2L/n$, $n = 1, 2, 3, 4, \ldots$ (Compare Eq. 19–31.)

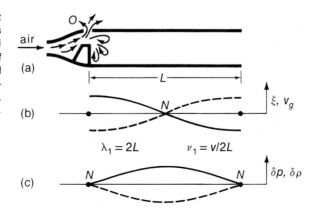

O

air

(a)

$|\leftarrow \qquad L \qquad \rightarrow|$

(b)

N

ξ, v_g

$\lambda_1 = 2L$ $\qquad \nu_1 = v/2L$

(c)

$N \qquad\qquad N$

$\delta p, \delta \rho$

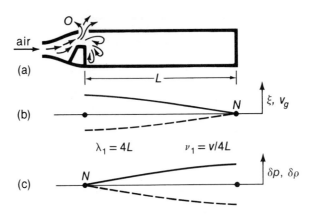

Fig. 20–5. An organ pipe closed at the nondriven end. The graphs show the variation of ξ, v_g, δp, and $\delta \rho$ for the fundamental mode of vibration. The nodes are indicated by N. The wavelengths of the harmonics are given by $\lambda_n = 4L/n$, $n = 1, 3, 5, 7, \ldots$. (Compare Example 19–9.)

pipe at O is not as large as the completely open end opposite, the pressure at the driven end is essentially constant and equal to the ambient pressure p_0. Also shown in Fig. 20–4 are the variations of ξ, v_g, δp, and $\delta \rho$. (•Can you explain the reasons for each shape?) Figure 20–5 illustrates the same quantities for an organ pipe that is closed at the nondriven end. (•Can you explain why only the *odd* harmonics are allowed in this case?)

One of the largest pipe organs in the world is played daily in the John Wanamaker department store in Philadelphia. Designed by George Audsley for the Louisiana Purchase Exposition in St. Louis, it was rebuilt in Wanamaker's store in 1911. Various additions brought the total to 30,067 pipes, the largest of which is a wooden gravissima 64 ft (19.5 m) in length. (Courtesy of John Wanamaker's, Philadelphia.)

Example 20–4

A third harmonic standing wave ($n = 3$) is set up in a 3-m organ pipe that is open at the nondriven end. The pipe diameter is 20 cm and the amplitude of the acoustic pressure is 0.5 percent of atmospheric pressure.
(a) Write expressions for $\xi(x, t)$, $v_g(x, t)$, $\delta p(x, t)$ and $\delta \rho(x, t)$.
(b) Determine the values of the amplitudes of the quantities in (a).
(c) What is the frequency of the standing wave?
(d) Determine the total wave energy in the pipe.

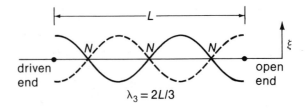

Solution:

The diagram shows the displacement curve for the $n = 3$ standing wave. Here, we have $\lambda_3 = 2L/3$.

(a) The displacement can be represented as

$$\xi(x, t) = A \cos \frac{3\pi}{L} x \sin \frac{3\pi v}{L} t$$

Then, with $v_g(x, t) = \partial \xi / \partial t$, we find

$$v_g(x, t) = \frac{3\pi v}{L} A \cos \frac{3\pi}{L} x \cos \frac{3\pi v}{L} t$$

For the acoustic pressure we have

$$\delta p(x, t) = -\beta_a \frac{\partial \xi}{\partial x} = -v^2 \rho_0 \frac{\partial \xi}{\partial x}$$

$$= \frac{3\pi v^2}{L} \rho_0 A \sin \frac{3\pi}{L} x \sin \frac{3\pi v}{L} t$$

Finally, the density excursion is (see Example 20–1)

$$\delta \rho(x, t) = \frac{\rho_0}{\beta_a} \delta p = \frac{1}{v^2} \delta p$$

$$= \frac{3\pi}{L} \rho_0 A \sin \frac{3\pi}{L} x \sin \frac{3\pi v}{L} t$$

(•Where are the nodes and antinodes of v_g, δp, and $\delta \rho$? Do your results agree with inferences based on Fig. 20–4?)

(b) Using $v = 331.5$ m/s, $p_0 = 1.013 \times 10^5$ Pa, and $\rho_0 = 1.293$ kg/m³, the various amplitudes are

$$\delta p_m = (0.5 \times 10^{-2})(1.013 \times 10^5 \text{ Pa})$$
$$= 506.5 \text{ Pa}$$

$$A = \delta p_m \frac{L}{3\pi v^2 \rho_0} = (506.5 \text{ Pa}) \frac{3 \text{ m}}{3\pi (331.5 \text{ m/s})^2 (1.293 \text{ kg/m}^3)}$$
$$= 1.135 \times 10^{-3} \text{ m}$$

$$v_{gm} = \frac{3\pi v}{L} A = \frac{3\pi (331.5 \text{ m/s})}{3 \text{ m}} (1.135 \times 10^{-3} \text{ m})$$
$$= 1.182 \text{ m/s}$$

$$\delta \rho_m = \frac{3\pi}{L} \rho_0 A = \frac{3\pi}{3 \text{ m}} (1.293 \text{ kg/m}^3)(1.135 \times 10^{-3} \text{ m})$$
$$= 4.610 \times 10^{-3} \text{ kg/m}^3$$

(c) The wavelength for the third harmonic wave is $\lambda_3 = 2L/3$, so the frequency is

$$\nu_3 = \frac{v}{\lambda_3} = \frac{3v}{2L} = \frac{3(331.5 \text{ m/s})}{2(3 \text{ m})} = 165.8 \text{ Hz}$$

(d) The kinetic energy at any time t is (compare Eq. 19–33)

$$K(t) = \tfrac{1}{2} \rho_0 S \int_0^L \left(\frac{\partial \xi}{\partial t} \right)^2 dx$$

where S is the cross-sectional area of the pipe; thus, $\rho_0 S$ represents the mass of the air per unit length in the pipe. Then,

$$K(t) = \tfrac{1}{2}\rho_0 S \frac{9\pi^2 v^2}{L^2} A^2 \cos^2 \frac{3\pi v}{L} t \int_0^L \cos^2 \frac{3\pi}{L} x \, dx$$

$$= \frac{9\pi^2 v^2}{4L} \rho_0 S A^2 \cos^2 \frac{3\pi v}{L} t$$

The potential energy is (compare Eq. 19–35)

$$U(t) = \tfrac{1}{2}\rho_0 S v^2 \int_0^L \left(\frac{\partial \xi}{\partial x}\right)^2 dx$$

$$= \tfrac{1}{2}\rho_0 S v^2 \frac{9\pi^2}{L^2} A^2 \sin^2 \frac{3\pi v}{L} t \int_0^L \sin^2 \frac{3\pi}{L} x \, dx$$

$$= \frac{9\pi^2 v^2}{4L} \rho_0 S A^2 \sin^2 \frac{3\pi v}{L} t$$

Thus, the total energy is

$$E = K(t) + U(t)$$

$$= \frac{9\pi^2 v^2}{4L} \rho_0 S A^2$$

$$= \frac{9\pi^2 (331.5 \text{ m/s})^2}{4(3 \text{ m})} (1.293 \text{ kg/m}^3)\pi(0.10 \text{ m})^2 (1.135 \times 10^{-3} \text{ m})^2$$

$$= 0.0426 \text{ J}$$

▶ Our discussion of standing waves in organ pipes has been based on a simplified model. In a real organ pipe, for example, the position of the lip is not actually a node of acoustic pressure, as indicated in Figs. 20–4 and 20–5, nor is the variation of pressure with position precisely sinusoidal. Figure 20–6 shows realistic standing wave patterns for the acoustic pressure in a flute. A flute is similar to an organ pipe in that it is excited near one end by blowing air. Note the variation in acoustic pressure near the driven end and the effective nodal positions in the vicinity of the open holes in the flute body. Notice also that the effective wavelengths of the harmonic waves are not related in a simple way as are the wavelengths in an ideal pipe. In the body of the pipe the waves have the normal appearance. However, near the ends (particularly the closed end) the pressure deviates substantially from that in an ideal pipe. Indeed, the ends of pipe instruments are sometimes altered to produce pressure deviations that result in more pleasant tones. ◀

Fig. 20–6. Acoustic pressure variations for three harmonic waves in a flute. The waves are excited by breath and lip variations at the embouchure hole. The dashed lines indicate the effective sinusoidal wave patterns.

20-3 WAVES IN THREE-DIMENSIONAL MEDIA

We now investigate the propagation of sound waves in an isotropic homogeneous gaseous medium. Consider the wave generated by the pulsations of a sphere that executes SHM by expanding and contracting about some mean radius. This wave is a harmonic *spherical wave* that propagates outward with the phase velocity v in all directions from the center of the pulsating sphere located at the origin.

Because of the spherical symmetry, the intensity of the wave (i.e., the average power transmitted per unit area) is uniform over any spherical surface centered on the origin. The wave traveling through a small volume surrounding some point Q on one of these spherical surfaces is essentially a *plane wave*, as indicated in Fig. 20-7.

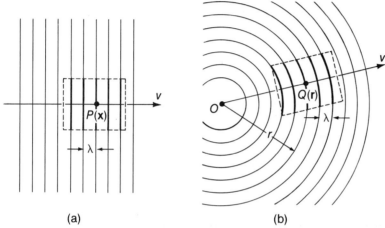

(a)   (b)

Fig. 20-7. Three-dimensional waves. (a) A plane wave, in which the wavefronts are parallel planes. (b) A spherical wave, in which the wavefronts are concentric spheres. In each case the wavefronts are separated by the wavelength λ of the wave. In a small volume surrounding point $Q(\mathbf{r})$ in (b), the wavefronts are nearly parallel, as in the vicinity of point $P(\mathbf{x})$ in (a). As the distance from O to Q increases, the wave more closely resembles a plane wave.

In Section 19-4 we learned that a wave source located at the origin and vibrating with SHM will generate a (one-dimensional) wave with a phase angle $\phi(x, t) = \mathbf{k} \cdot \mathbf{x} - \omega t + \phi_0$, where $\mathbf{x} = x\hat{\mathbf{i}}$ and where the propagation vector $\mathbf{k}$ has a magnitude $k = 2\pi/\lambda$ and a direction at coordinate $\mathbf{x}$ corresponding to that of the wave propagation. For the case of a three-dimensional wave we take the displacement $\xi(\mathbf{r}, t)$ to have a phase $\phi(\mathbf{r}, t) = \mathbf{k} \cdot \mathbf{r} - \omega t + \phi_0$ and to have an amplitude that depends, at most, on some power of $|\mathbf{r}| = r$. We therefore postulate that the wave function for the spherical wave has the form

$$\xi(\mathbf{r}, t) = Br^n \cos (kr - \omega t + \phi_0) \qquad \textbf{(20-11)}$$

where $\mathbf{k} \cdot \mathbf{r} = kr$ because both $\mathbf{k}$ and $\mathbf{r}$ in this case have the direction $\hat{\mathbf{u}}_r$.

According to Eq. 20-9, the intensity of a plane wave is $I = \frac{1}{2}A^2 k\beta_a \omega$. Using $\beta_a = \rho_0 v^2$ and $k = \omega/v$, this becomes

$$I = \frac{1}{2}A^2 v\rho_0 \omega^2 \qquad \textbf{(20-12)}$$

In these expressions for I we interpret the amplitude A as the combination Br^n in Eq. 20-11. Then, the total acoustic power per period that is transported by the wave through a spherical surface with radius R centered on the origin is

$$\overline{P} = 4\pi R^2 I = 4\pi R^2 \cdot \tfrac{1}{2}B^2 R^{2n} v \rho_0 \omega^2$$

If there are no losses of energy due to absorption, the total average power must be independent of the arbitrarily selected value of R; this requires $n = -1$. Thus, we conclude that the wave function for the spherical wave is

$$\xi(\mathbf{r}, t) = \frac{B}{r}\cos(kr - \omega t + \phi_0) \tag{20-13a}$$

with

$$B = \sqrt{\overline{P}/2\pi v \rho_0 \omega^2} \tag{20-13b}$$

where $\overline{P}$ is the average acoustic power per period radiated by the pulsating sphere.

The purely radial oscillation of a sphere is called the *breathing mode* of the sphere's vibrations. Other distinctive vibration modes are possible; for example, one hemisphere could have a decreasing radius while the other hemisphere has an increasing radius, and vice versa—this is a *dipole* oscillation.

Reflection. The familiar echoes of sounds from building or canyon walls is the result of the *reflection* of sound waves that strike these surfaces. The phenomenon of reflection is complicated by the fact that all ordinary materials absorb part of the incident wave energy. Nevertheless, we can make a useful analysis by assuming ideal, nonabsorptive reflection conditions. Then, a plane wave incident obliquely on a flat surface obeys a simple law of reflection. The wave at the surface may be viewed as the superposition of two waves—one that travels parallel to the surface and one that travels perpendicular to the surface. The parallel component is unaffected by the presence of the surface. However, because there can be no perpendicular displacement of the gas matter at the wall, Eq. 20–10a must be satisfied, and the perpendicular component is reflected completely. The result, illustrated in Fig. 20–8a, is a reflected wave with a wave front that makes the same angle with respect to the surface as does the incident wave front. That is, the angle of reflection is equal to the angle of incidence for the ideal reflection of a plane wave.

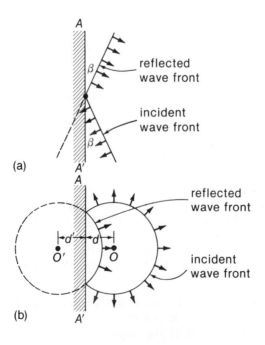

(a)

(b)

Fig. 20–8. The reflection of (a) a plane wave and (b) a spherical wave from an ideal reflecting surface AA'. For the plane wave, the angle of incidence β is equal to the angle of reflection β. (●Can you use the rule for the reflection of a plane wave to explain why a reflected spherical wave appears to originate behind the surface at a depth $d' = d$?)

Example 20-5

A rectangular room with dimensions $L_x \times L_y \times L_z$ has ideally reflecting walls. Obtain the wave function $\xi(x, y, z, t)$ that describes the standing waves in the room. Determine the allowed frequencies.

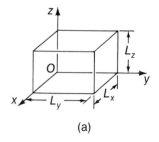

(a)

Solution:

Because the surfaces are perfectly reflecting, the acoustic pressure must be maximum at each wall of the room. Therefore, we expect the wave function for the acoustic pressure to have the form

$$\delta p(x, y, z) = \delta p_m \cos \frac{n_x \pi}{L_x} \cos \frac{n_y \pi}{L_y} \cos \frac{n_z \pi}{L_z} \sin(\omega t - \phi_0)$$

where n_x, n_y, and n_z can be any positive integer or zero: 0, 1, 2, 3, ... This wave function satisfies the boundary condition at each wall. If the expression also obeys the general wave equation (Eq. 19-7), it must be the desired solution. The general wave equation is

$$\frac{\partial^2 \eta}{\partial x^2} + \frac{\partial^2 \eta}{\partial y^2} + \frac{\partial^2 \eta}{\partial z^2} = \frac{1}{v^2} \frac{\partial^2 \eta}{\partial t^2}$$

If we identify η with the acoustic pressure and differentiate, we find

$$\frac{\partial^2 \eta}{\partial x^2} = \frac{\partial^2 (\delta p)}{\partial x^2} = -\left(\frac{n_x \pi}{L_x}\right)^2 \delta p$$

and similarly for y and z; also

$$\frac{\partial^2 \eta}{\partial t^2} = \frac{\partial^2 (\delta p)}{\partial t^2} = -\omega^2 \delta p$$

The wave equation will be satisfied if we require

$$\frac{\omega^2}{v^2} = \left(\frac{n_x \pi}{L_x}\right)^2 + \left(\frac{n_y \pi}{L_y}\right)^2 + \left(\frac{n_z \pi}{L_z}\right)^2$$

Then, the frequencies of the standing waves are

$$\nu_{n_x n_y n_z} = \tfrac{1}{2} v \left[\left(\frac{n_x}{L_x}\right)^2 + \left(\frac{n_y}{L_y}\right)^2 + \left(\frac{n_z}{L_z}\right)^2\right]^{1/2}$$

If L_z is the largest dimension of the room, the lowest allowed frequency is ν_{001}. (•Can you describe this standing wave?)

Diagram (b) shows measured values of the sound intensity level* β near the center of a living room of ordinary size due to the sound produced by a constant-level input to a 30-cm (12-inch) diameter bass reflex speaker. The speaker was placed against the center of the wall with dimension L_y. There is a close correspondence of the loudness peaks with the calculated frequencies of the room resonance modes. Notice also that there is an increase in average loudness with tone frequency that corresponds roughly with the increase in the density (close-

---

*Sound intensity level is discussed in Section 20-6. The scale of β is logarithmic and is given in decibels (dB). See Eq. 20-22.

(b)

ness in frequency) of room resonances. Thus, the room itself has an effect on the sound delivered by the loudspeaker. This effect can be corrected by the insertion of electric equalization circuits into the amplification system.

20-4   INTERFERENCE EFFECTS*

Beats. An interesting case of interference comes about when two waves with slightly different frequencies combine to produce *beats*. Figure 20–9 shows two such waves and their superposition. Because the amplitudes of the waves are equal, completely destructive interference occurs at regular intervals and the wave amplitude reaches zero.

Let the two displacement functions be

$$\xi_1 = A \cos(k_1 x - \omega_1 t)$$
$$\xi_2 = A \cos(k_2 x - \omega_2 t)$$

---

*The general subject of interference is discussed in detail in connection with light phenomena in Chapter 42.

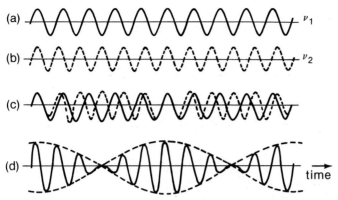

Fig. 20-9. (a, b) Two waves with the same amplitude but with slightly different frequencies. (c) The waves of (a) and (b) shown on the same axis. (d) The superposition of the two waves. The wavelength of the rapid oscillation is about the same as those of the original waves, but the amplitude is modulated on a longer time scale, as indicated by the dashed line.

The superposition at $x = 0$ yields

$$\xi = \xi_1 + \xi_2 = A(\cos \omega_1 t + \cos \omega_2 t)$$

because $\cos(-\alpha) = \cos \alpha$. Now we use the identity (see Eq. A-28 in Appendix A), $\cos \alpha + \cos \beta = 2 \cos \frac{1}{2}(\alpha + \beta) \cos \frac{1}{2}(\alpha - \beta)$, to write

$$\xi = 2A \cos \frac{\omega_1 + \omega_2}{2} t \cos \frac{\omega_1 - \omega_2}{2} t$$

The first cosine term involves the average frequency, $\bar{\omega} = \frac{1}{2}(\omega_1 + \omega_2)$, and the second term involves the difference frequency, $\Delta \omega = \omega_1 - \omega_2$. Thus,

$$\xi = (2A \cos \tfrac{1}{2} \Delta \omega t) \cos \bar{\omega} t \qquad \textbf{(20-14)}$$

Because $\Delta \omega \ll \bar{\omega}$, the term $\cos \bar{\omega} t$ varies rapidly compared with $\cos \frac{1}{2} \Delta \omega t$. We can therefore consider the term in parentheses in Eq. 20-14 to be a slowly varying amplitude, $A'(t) = 2A \cos \frac{1}{2} \Delta \omega t$, that *modulates* the *carrier* term $\cos \bar{\omega} t$.

When the quantity $|A'(t)|$ is large, the sound intensity is large, and when $A'(t) = 0$, the intensity is zero. Each successive recurrence of the maximum of $|A'(t)|$ is called a *beat*. Two such maxima occur during each cycle, so the beat frequency is equal to the frequency difference, $\nu_1 - \nu_2$. For example, if two nearly identical tuning forks have natural frequencies of 440 Hz and 441.8 Hz, the combination of these tones will produce a beat frequency of 1.8 Hz, or one beat every 0.556 s. Beat frequencies as large as about 7 Hz are discernible by the ear. For higher beat frequencies, the ear does not respond to the modulations and no tone variation is heard.

The occurrence of beats is used in the tuning of stringed musical instruments, such as pianos and violins. (•Can you explain how this is done?) Notice also that two strings with fundamental tones that differ by nearly one octave (i.e., a frequency ratio of 2 to 1) will produce beats between the string with the higher fundamental and the second harmonic of the other string. This effect can also be used in tuning instruments.

Interference of Two Point Sources. Consider two small identical pulsating

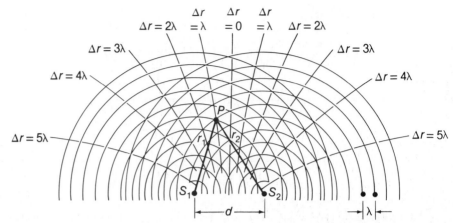

Fig. 20–10. The interference pattern produced by two small spheres pulsating in phase. Each circular arc represents a wavefront crest with a particular algebraic sign and is separated from the next arc by one wavelength λ. Whenever two such wavefronts intersect, there is constructive interference. Such a point is P, where $|r_1 - r_2| = \lambda$.

spherical sources, S_1 and S_2, that are separated by a distance d. The sources are driven in SHM exactly in phase and with the same amplitude. Harmonic spherical waves (described by Eq. 20–11) are radiated by each source. These spreading waves overlap and combine, producing a pattern of constructive and destructive interference effects, as indicated in Fig. 20–10. Points in space, such as the one represented by r_1 and r_2 in the diagram, that have $\Delta r = |r_1 - r_2|$ equal to an integer number of wavelengths, constitute a family of surfaces on which constructive interference occurs. Thus, the condition for constructive interference is

$$\Delta r = |r_1 - r_2| = n\lambda \qquad n = 0, 1, 2, 3, \ldots$$

In Fig. 20–10 the curves indicate the lines along which these surfaces intersect with a plane that passes through the sources, S_1 and S_2. Interlaced between these curves (actually, surfaces) are those along which destructive interference takes place. The condition for destructive interference is

$$\Delta r = |r_1 - r_2| = (n + \tfrac{1}{2})\lambda \qquad n = 0, 1, 2, 3, \ldots$$

Interference of water waves radiating from two in-phase sources. Note the similarity with Fig. 20–10. (Along the "lines" that diverge from the sources, the interference is destructive.) (From *PSSC Physics*, 2nd ed., 1965: D. C. Heath & Co. and Educational Center, Newton, MA.)

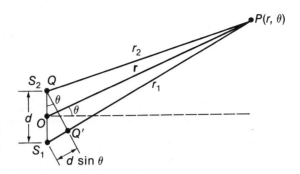

Fig. 20-11. Geometry for calculating interference effects at large distances from the sources: $r_1 \gg d$, $r_2 \gg d$. In this limit, $|r_1 - r_2| = d \sin \theta$.

At large distances from the sources, the interference curves asymptotically approach straight lines that radiate from a point midway between S_1 and S_2. (●What is the nature of the surface that corresponds to one of these straight interference lines?) The geometry of this situation is shown in Fig. 20-11, where the origin O is the midpoint between S_1 and S_2.

We construct a line QQ' that is perpendicular to **r** and approximately perpendicular to r_1 and r_2. Thus, $\Delta r = |r_1 - r_2| \cong d \sin \theta$. If the sources pulsate in phase with equal amplitudes, the interference conditions at point $P(r, \theta)$ are

$$\left. \begin{array}{l} \text{Constructive: } d \sin \theta = n\lambda \\ \text{Destructive: } \quad d \sin \theta = (n + \tfrac{1}{2})\lambda \end{array} \right\} n = 0, 1, 2, 3, \ldots \qquad \textbf{(20-15)}$$

Radiation from a Rigid Circular Piston. Consider the SHM of a simple model of a loudspeaker, namely, a rigid circular piston that is mounted in an infinitely large wall, as in Fig. 20-12. We can imagine the piston surface to be divided into a large number of small areas. Any two of these areas, as they vibrate, behave as do the sources S_1 and S_2 in Fig. 20-11. Thus, we can consider the surface of the piston to consist of many interfering sources.

For very low vibration frequencies (i.e., very long wavelengths), the piston acts essentially as a point source, and the radiation pattern is *isotropic* (equal intensities in all directions of the hemisphere). Now, suppose that the frequency is increased from a low initial value. Eventually, a frequency will be reached such that two points on opposite sides of the piston rim (points a and b in Fig. 20-11) will produce destructive interference for $\theta = \pi/2$, which corresponds to $R = \tfrac{1}{2}\lambda$. The radiation pattern begins to deviate significantly from isotropy at this frequency. At still higher frequencies, the pattern becomes more concentrated along the piston axis OO' (i.e., $\theta = 0°$). The actual behavior is a complicated function of θ and $kR = 2\pi R/\lambda$.

Fig. 20-12. A rigid circular piston is mounted in an infinite wall (or *infinite baffle*).

$kR = 1$ $DF = 2.4$

$kR = 5$ $DF = 25.7$

$kR = 10$ $DF = 100$

Fig. 20–13. Radiation patterns for a circular piston with radius R that vibrates to produce a wavelength $\lambda = 2\pi/k$. Each semicircular arc represents a factor of 10 in intensity. (Thus, in (c) the intensity at $\theta = 60°$ is about 0.004 of the intensity at $\theta = 0°$.) The *directivity factor* (DF) is the relative power that would be radiated into the entire hemisphere if the radiation were isotropic with the actual power level at $\theta = 0°$. (Adapted from L. L. Beranek, *Acoustics,* McGraw-Hill, New York, 1954; p. 102.)

The sound intensity as a function of θ for $r \gg R$ is shown in Fig. 20–13 for three values of kR. The complete radiation patterns correspond to the surfaces that are generated by revolving the curves shown about the $\theta = 0°$ axis. Notice that these patterns verify the assertion that the radiation is essentially isotropic for long wavelengths (small kR).*

Example 20–6

An infinitely long flat rigid ribbon piston of width w is mounted in an infinite wall and is driven harmonically. Determine the smallest angle θ for complete destructive interference, resulting in zero radiation intensity for that direction.

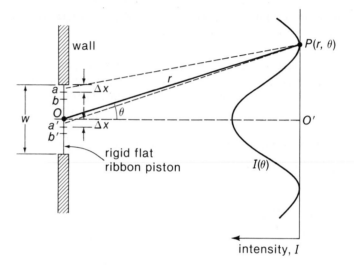

Solution:

Consider two narrow strips each of width Δx, one—labeled a—at the top edge of the ribbon piston, and the other—labeled a'—just below the midpoint O. The radiation intensity at $P(r, \theta)$ due to these two sources will be zero if $\frac{1}{2}w \sin \theta = \frac{1}{2}\lambda$. This is the minimum angle for complete destructive interference. (•Can you see why?) Assuming that $r \gg w$, a second pair of strips—b and b'—just below the first pair will obey the same condition. The same will be true for successive pairs of strips, extending completely over the ribbon. (We have, in effect, performed an integration over the width of the ribbon.) The result is that there will be zero intensity at the angle θ if $w \sin \theta = \lambda$. A similar analysis applies for the diffraction of light by a slit (Chapter 43).

20–5   THE DOPPLER EFFECT

It is a common experience that when an ambulance or other vehicle with a siren passes you on the street, you hear a dramatic decrease in the frequency of the siren at the moment the vehicle passes—the familiar *whee-oo* sound. This frequency change is a result of the *Doppler effect,* named for the Austrian physicist Christian Johann

---

*In Section 20–1 it was stated that a wave would propagate as a plane wave along a pipe with diameter D if $\lambda/D \geq 2\pi$. This condition is equivalent to $kR \leq \frac{1}{2}$ and corresponds to the isotropy condition here.

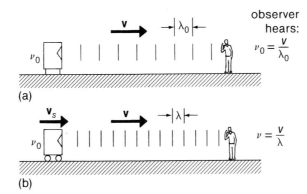

Fig. 20-14. (a) The stationary observer hears a tone with the same frequency ν_0 as that emitted by the stationary source. (b) The same observer hears a tone with a higher frequency when the source is approaching.

Doppler (1803–1853), who first made extensive studies of this phenomenon and in 1842 developed the theory we now discuss.

Suppose that a sound source at rest emits waves with a frequency ν_0 that travel through still air with a velocity v to a stationary observer some distance away, as indicated in Fig. 20-14a. The observer hears a tone with the frequency $\nu_0 = v/\lambda_0$. Now, suppose that the source moves toward the observer with a velocity v_s, as in Fig. 20-14b. The distance moved by the source during one period $T_0 = 1/\nu_0$ is $v_s T = v_s/\nu_0$. Therefore, when two successive wave crests are emitted, the distance between the crests will be less than λ_0 by the amount of movement v_s/ν_0. Thus, the wavelength perceived by the observer is

$$\lambda = \frac{v}{\nu} = \lambda_0 - \frac{v_s}{\nu_0} = \frac{v}{\nu_0} - \frac{v_s}{\nu_0}$$

and the tone heard by the observer has the frequency

$$\nu = \frac{\nu_0}{1 - (v_s/v)} \qquad (20\text{-}16)$$

When $v_s > 0$ (i.e., source approaching the observer), Eq. 20-16 shows that $\nu > \nu_0$, so that the observer hears a tone with a frequency higher than the natural frequency ν_0 he would hear if the source were at rest. On the other hand, if the source recedes from the observer, we have $v_s < 0$ and $\nu < \nu_0$. Consequently, when a source moves past an observer, the frequency of the tone drops suddenly.

▶ *The General Doppler Equation.* Suppose that a source of harmonic waves with frequency ν_0 moves with a constant subsonic velocity $\mathbf{v}_s$ relative to (still) air. At time $t = 0$ the source (at S_0) emits the crest of a particular wave. This wave travels through the air with the speed v as an expanding sphere centered on the position of the source at $t = 0$, namely, S_0. At time t the crest forms a circle with radius vt (the circle labeled A in Fig. 20-15 on the next page) and at that instant sweeps past a stationary observer at point P. The position of the source at time t is S_t, with $\overline{S_0 S_t} = \mathbf{v}_s t$. Because the speed of light is so much greater than v or v_s, the observer at P can visually determine the angle α_s' between $\mathbf{v}_s$ and $\overline{S_t P}$.

The next crest is emitted at time T, where $T = 1/\nu_0$ is the period of the source. During the interval between emitting the two crests, the source moves from S_0 to S_T, a distance $v_s T$. The second wavefront at time t is a circle with radius $v(t - T)$, the circle labeled B in Fig. 20-15. The arrival of these two crests at P defines for the observer the wavelength λ. Referring to Fig. 20-15, we see that

$$\lambda = vt - v(t - T) - v_s T \cos \alpha_s$$
$$= \frac{1}{\nu_0}(v - v_s \cos \alpha_s) \qquad (20\text{-}17)$$

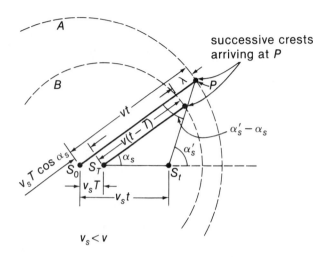

Then, the frequency perceived by the stationary observer is $\nu = v/\lambda$, or

$$\nu = \frac{\nu_0}{1 - (v_s/v) \cos \alpha_s} \tag{20–18}$$

Fig. 20–16. The component of the observer's velocity **u** in the direction of the advancing wave crests is $u \cos \alpha$.

Notice that Eq. 20–18 contains the angle α_s, whereas the readily observed angle is α_s'. If $v_s \ll v$, we have $\alpha_s' \cong \alpha_s$. Although this is often the case, the two angles are not generally equal. For example, think of the situation in which a jet aircraft flies past and you perceive the sound coming from a point considerably behind the aircraft. This effect is called the *aberration of sound* and occurs when $\alpha_s \neq \alpha_s'$. We can relate the two angles by using the law of sines. We find

$$\frac{v_s t}{\sin (\alpha_s' - \alpha_s)} = \frac{vt}{\sin (\pi - \alpha_s')}$$

from which

$$\sin (\alpha_s' - \alpha_s) = \frac{v_s}{v} \sin \alpha_s' \tag{20–19}$$

Next, in order to obtain a completely general expression for ν, we allow the observer to move at a constant subsonic velocity **u** relative to the still air. The progress of the wavefronts of successive crests and of the moving observer are shown in Fig. 20–16. The velocity of any particular wave crest with respect to the observer is $v + u \cos \alpha$. Therefore, the number of crests detected by the observer during a time Δt is

$$N = (v + u \cos \alpha)\frac{\Delta t}{\lambda}$$

The observer perceives the frequency of the sound to be $N/\Delta t$, or

$$\nu = \frac{N}{\Delta t} = \frac{1}{\lambda}(v + u \cos \alpha)$$

Fig. 20–17. Geometry of the Doppler effect.

Substituting for λ from Eq. 20–17, we find

$$\nu = \nu_0 \left(\frac{v + u \cos \alpha}{v - v_s \cos \alpha_s} \right) \tag{20–20}$$

It is important to recognize the positive sense in which the angles α_s and α are defined. Figure 20–17 shows the relevant geometry. When the source moves directly toward the observer, $\alpha_s = 0°$; when the source moves directly away from the observer, $\alpha_s = 180°$. The observer can determine the source velocity vector $\mathbf{v}_s$ and the angle α_s' by direct visual measurements. Then, he can deduce α_s by using Eq. 20–19. Finally, α is measured with respect to the line of advancing wave crests, $\overline{S_0 P}$. The aberration effect ($\alpha_s' \neq \alpha$) is significant when the source velocity v_s is not small compared to the wave velocity v (see Problem 20–42). ◀

▶ *Shock Waves and Sonic Booms.* In the discussion of the Doppler effect we considered only subsonic velocities for the source and the observer, that is, $v_s < v$ and $u < v$. However, some objects *can* move through the air with supersonic velocities—for example, jet aircraft, rifle bullets, and rockets.

Consider an object S that moves through air with a constant velocity v_s that exceeds v. The source object runs ahead of the outgoing waves, and the pileup of waves produces a cone-shaped *shock front,* as illustrated in Fig. 20-18. The faster the object moves, the greater is the degree to which the front is swept back and the smaller is the half-angle β of the cone. This angle β is given by

$$\sin \beta = \frac{v}{v_s} \qquad (20\text{-}21)$$

A ground observer who is under the flight path of a supersonic jet aircraft will experience a sharp pressure increase when the shock front arrives. This is followed by a decrease in the pressure to a value below the ambient pressure. Finally, when the shock front associated with the rear of the aircraft passes, there is another sharp increase in pressure back to the ambient value. Typically, the time difference between the arrival of the two shock waves is $\frac{1}{50}$ s or so. Consequently, the observer hears a single loud *crack* or *sonic boom.* The pressure difference at the shock front is sufficiently large that nearby buildings may experience some minor structural damage (usually limited to broken windows). For this reason supersonic aircraft (SST's and military jets) are permitted supersonic flight only over uninhabited regions.

The ratio of the speed of an object to the speed of sound for the same conditions is called the *Mach number,* $N_M = v_s/v$ (see Eq. 18-22). Thus, a moving object produces a shock front whenever $N_M \geqslant 1$. ◀

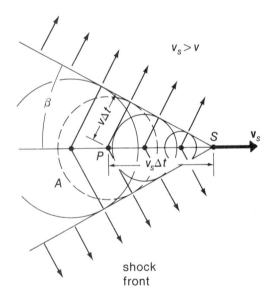

Fig. 20-18. When an object moves with a velocity $v_s > v$, a swept-back cone-shaped shock front is generated. An object at P at time t moves to S at time $t + \Delta t$; during this interval the disturbance at P has propagated outward as the spherical wavefront A.

Example 20-7

An automobile horn has a frequency $\nu_0 = 400$ Hz. The horn is sounded continuously as the automobile is driven in a straight line with a velocity of 90 km/h = 25 m/s past a stationary observer; the minimum distance between the automobile and the observer is 100 m. Determine the frequency of the tone heard by the observer as a function of the position of the automobile along its path.

Solution:

Because $v_s \ll v$, aberration effects are negligible and we have $\alpha'_s \cong \alpha_s$. The observer is stationary, so we use Eq. 20-18 to calculate ν as a function of α_s. Then, the distance x from the point of closest approach is $x = (-100 \text{ m}) \operatorname{ctn} \alpha_s$. For example, take $v = 343.4$ m/s (see

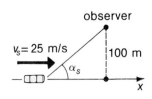

Example 20–1e) and choose $\alpha_s = 60°$:

$$\nu = \frac{\nu_0}{1 - (v_s/v)\cos\alpha_s} = \frac{400 \text{ Hz}}{1 - (25/343.4)\cos 60°}$$

$$= 415.1 \text{ Hz}$$

and
$$x = (-100 \text{ m})\,\text{ctn}\,60° = -57.7 \text{ m}$$

The diagram below shows the variation in ν calculated in this way.

Example 20–8

An observer rides with a moving sound source directly toward a large vertical wall. The vehicle has a constant speed of 30 km/h = 8.33 m/s and the source has a frequency of 100 Hz. What beat frequency is heard by the observer due to the combination of the direct and reflected sounds?

Solution:

First, consider an observer at the wall. For him, the velocity of the source is $v_s = u$, and his own velocity is zero. Therefore, with $\alpha_s = 0°$, Eq. 20-20 gives

$$\nu_w = \nu_0 \frac{v}{v - u}$$

for the frequency heard by the observer at the wall. The reflected sound now comes from a stationary source and is heard by the moving observer. Then, Eq. 20-20 gives

$$\nu = \nu_w \frac{v + u}{v}$$

Substituting for ν_w, we find
$$\nu = \nu_0 \frac{v + u}{v - u}$$

The beat frequency is

$$\nu - \nu_0 = \nu_0\left(\frac{v + u}{v - u} - 1\right) = \nu_0 \frac{2u}{v - u}$$

$$= (100 \text{ Hz})\frac{2(8.33 \text{ m/s})}{(343.4 \text{ m/s} - 8.33 \text{ m/s})}$$

$$= 4.97 \text{ Hz}$$

20-6 SOUND INTENSITY

In order to describe the enormous dynamic range of sound intensities that occur in Nature, it is customary to introduce a logarithmic scale. Two sounds that differ by a factor of 10 in intensity (average acoustic power per unit area) are said to differ by *1 bel* (1 B).* A more commonly used unit is the *decibel* (dB), equal to 0.1 B. Given a sound with an intensity I (W/m²) and a reference intensity I_0, the *sound intensity level* (in dB) of the sound is expressed as†

$$\beta = 10 \log (I/I_0) \quad \text{in dB} \qquad \textbf{(20–22)}$$

In acoustical work the reference intensity is usually taken to be

$$I_0 = 10^{-12} \text{ W/m}^2$$

which corresponds to the threshold of hearing in most individuals (see Table 20-2). At the other extreme, a rock band can produce a sound intensity level of about 115 dB, just below (?) the threshold of pain.

▶ According to Eq. 20-9, the intensity of a wave can be expressed as $I = \frac{1}{2}A^2 k\beta_a\omega$. By using the relations $\delta p_m = Ak\beta_a$, $v^2 = \beta_a/\rho_0$, and $\omega = kv$, we find that

$$I = \frac{1}{2}\,\delta p_m^2/v\rho_0 \qquad \textbf{(20–23)}$$

That is, the intensity of a sound wave expressed in terms of the acoustic pressure amplitude δp_m is *independent* of the frequency (or wavelength). Thus, we can compare the intensities of sound waves by examining the pressure ratio, even if the frequencies are different.

We define a reference-level root-mean-square (rms) acoustic pressure amplitude to be

$$\delta p_{\text{rms}} = \frac{1}{\sqrt{2}}\,\delta p_{m0} = \sqrt{I_0 v\rho_0}$$

$$= \sqrt{(10^{-12} \text{ W/m}^2)(331.5 \text{ m/s})(1.293 \text{ kg/m}^3)}$$

$$= 2.07 \times 10^{-5} \text{ Pa}$$

TABLE 20–2. Sound Intensity Levels

SOURCE	β (dB)	I (W/m²)
(Physical damage)	140	100
(Painful)	120	1
Very loud thunder	110	0.1
Subway train	100	10^{-2}
Noisy factory	90	10^{-3}
Heavy street traffic	80	10^{-4}
Average factory	70	10^{-5}
Conversation	60	10^{-6}
Quiet home	40	10^{-8}
Quiet whisper	20	10^{-10}
Rustling leaves	10	10^{-11}
Threshold of hearing	0	10^{-12}

---

*This unit is named in honor of Alexander Graham Bell (1847–1922), Scottish-American inventor of the telephone.

†Notice that $\log (I_1/I_0)$ gives the sound intensity level in *bels*.

This standard (often taken to be simply 2×10^{-5} Pa) corresponds to the intensity level 0 dBA (0 dB *absolute*). Then,

$$\beta = 20 \log \frac{\delta p_m}{\delta p_{\text{rms}}} \quad \text{(in dBA)} \qquad \textbf{(20-24)} \blacktriangleleft$$

20-7   THE EAR AND HEARING

The human ear is a marvelous piece of sound-detection apparatus; it has an enormous sensitivity range, good frequency response and direction-locating capability, and requires relatively little maintenance. Figure 20-19 shows schematically the three important parts of the ear—the outer, the middle, and the inner ear sections. The outer ear begins with the auditory canal, which transmits sound waves to the *tympanic membrane* (or *eardrum*). The canal structure resembles a closed-end acoustic pipe with a diameter of about 0.7 cm and a length of about 2.7 cm. The canal behaves approximately as a broadly tuned quarter-wavelength resonator that generates a pressure maximum at the eardrum (compare Fig. 20-5c). The central frequency for sound enhancement in the canal is about 3200 Hz, which accounts, in part, for the fact that the human auditory system is most sensitive for a range of frequencies near 3 kHz. The slightly conical eardrum, the closed end of the auditory canal, has an area of about 70 mm² and a thickness of approximately 0.1 mm.

The middle ear is an air-filled chamber with a volume of about 2 cm³ that contains three small bones, collectively called *ossicles* (bony levers). These bones, shown in Fig. 20-19, are named according to their shapes—the *malleus* (hammer), the *incus* (anvil), and the *stapes* (stirrup). The leverage system of the ossicles transmits the displacements of the eardrum to a membrane called the *oval window*. The middle ear is not an isolated chamber; it connects to the upper throat by a small tube called the *Eustachian tube,* which acts as a safety device to prevent changes in the ambient air pressure from distorting the eardrum. The diameter of the Eustachian tube is sufficiently large to allow the slow passage of air to adjust for changes in the ambient air pressure but is sufficiently small that the rapid pressure changes due to sound waves can still be transmitted through the eardrum. The tightness of the coupling of the incus and malleus to the eardrum is controlled by muscles that reflexively loosen to protect the middle and inner ear from any sudden dangerous increase in sound pressure.

The inner ear is the site of the loudness and frequency sensing apparatus that transmits electric nerve pulses to

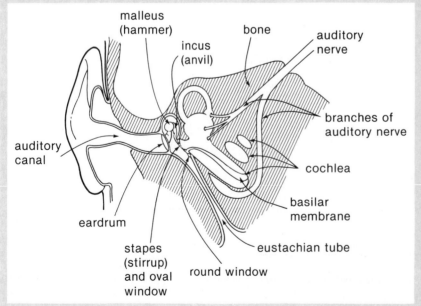

Fig. 20-19. Important parts of the human auditory system.

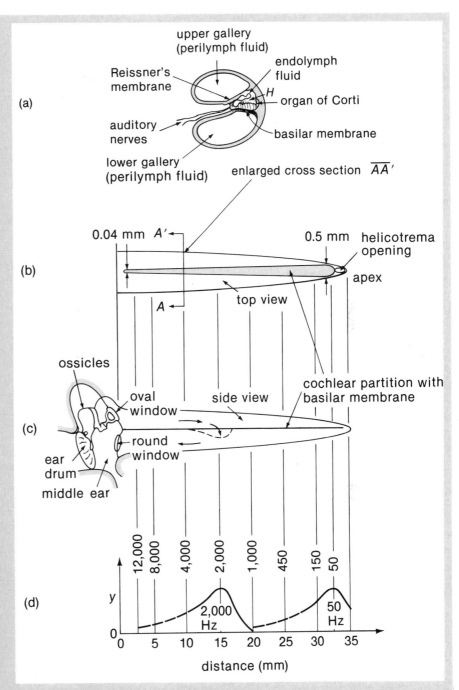

Fig. 20–20. (a) An enlarged cross section of the cochlea along *AA'* (see part *b*). (b) The uncoiled cochlea, top view and (c) side view. (d) Positions of the maximum nerve responses for various frequencies. The displacement of the basilar membrane is shown for tones of 2000 Hz and 50 Hz.

the brain and results in the sensation of hearing. This extraordinary sensing device, the *cochlea,* consists of a slightly tapered tube that is coiled in the form of a snail shell (see Fig. 20–19). Figure 20–20 shows an *uncoiled* view of the cochlear tube, which has a mean diameter of about 1.5 mm and a length of approximately 3.5 cm. The cochlea

is divided into two roughly equal, fluid-filled chambers called the upper and lower galleries.

In the important middle frequency range from about 500 Hz to 5000 Hz (which includes almost all speech tones), the resonance effect in the auditory canal produces a pressure gain at the eardrum of about a factor of 2. The ossicles couple the motion of the eardrum to the oval window; the leverage in this system increases the transmitted pressure by another factor of 2. Finally, the area of the oval window is about 3 mm², so that between the oval window and the eardrum (area about 70 mm²) there is a pressure gain of a factor of 20 or more. The overall pressure amplification between the incident sound wave and the cochlear fluid is nearly 100.

A small opening (the *helicotrema*) connects the upper gallery of the cochlea to the lower gallery at the small end or *apex* (see Fig. 20–20b). The lower gallery terminates with the *round window* back at the middle-ear end. The round window is essentially a pressure relief device for the fluid in the rigid bone-encased cochlear chambers, particularly for low frequency thumps.

The partition between the upper and lower galleries, which extends almost the entire length of the cochlea, consists of a number of membranes, the most important of which is the *basilar membrane*. This membrane tapers from a width of about 0.04 mm at the oval window end (where it is quite stiff) to about 0.5 mm at the apex (where it is rather flabby). The basilar membrane acts as a Fourier analyzer of the frequency components in the incident sound wave by producing maxima in the shearing strain in the sensory cells at appropriate positions along the membrane. The narrow taut end generates large strains for high frequencies (short wavelengths), whereas the wide slack end generates large strains for low frequencies (long wavelengths).

The ripple produced in the basilar membrane by a sound wave with a particular frequency is due to a combination of a standing wave along the entire length of the two connected galleries and some sort of hydraulic "short circuit," particularly effective at high frequencies, in which the inertia of the fluid in the upper gallery produces a downward deflection of the basilar membrane. Thus, the pressure increase is communicated directly to the lower gallery without the necessity for the pressure wave to pass through the helicotrema. Figure 20–20c illustrates this effect for a 2000-Hz signal.

Figure 20–20d shows the location of the maximum strains along the membrane for waves with various frequencies. Notice that over most of the length of the basilar membrane, a 5-mm change in position corresponds to a factor of 2 in frequency. The strains in the membrane cause a bending of sensory hair cells located in the *organ of Corti;* these hair cells are located at *H* in Fig. 20–20a. The flexing of the hair cells results in electric signals that are transmitted by the auditory nerve fibers to the brain for processing. The cochlear system analyzes the harmonic content of complex sound waves in terms of the locations and magnitudes of the strains in the basilar membrane.

Loudness. The term "loudness" is used to describe the sensation of hearing as perceived by the human ear. That is, *loudness* involves a subjective judgment of sound quality, whereas *intensity* is an objective quantity that can be measured with instruments. The intensity of a sound wave is independent of the frequency (Eq. 20–23), but the loudness varies markedly with frequency. For example, the faintest 1000-Hz sound that an "average" ear can detect

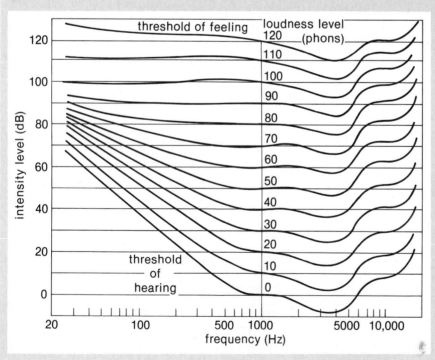

Fig. 20–21. Loudness curves for pure tones. The *loudness level* in *phons* is taken equal to the intensity level for a frequency of 1000 Hz. These curves are referred to as *Fletcher-Munson curves,* after H. Fletcher and M. Munson, who first constructed such curves in 1933.

corresponds to an intensity level of 0 dB (or 10^{-12} W/m² or 2×10^{-5} Pa). On the other hand, the minimum detectable intensity level for a 100-Hz sound is approximately 37 dB, or about 5000 times the intensity of the 1000-Hz sound.

The *threshold of hearing* curve in Fig. 20-21 gives the intensity levels for sounds of various frequencies that are barely audible; this curve corresponds to a *loudness level* of zero *phons*. Also shown in this figure are curves for different loudness levels; each loudness level in phons is set equal to the intensity level at 1000 Hz. For example, an individual will judge a 100-Hz sound at 71 dB to have the same loudness as a 1000-Hz sound at 60 dB. (Follow the 60-phon curve from 100 Hz to 1000 Hz.) Notice that the frequency dependence of the ear's sensitivity becomes less as the loudness level increases. Notice also that the ear is most sensitive to sounds with frequencies of 3000–4000 Hz, which corresponds to the resonant frequency range of the auditory canal.

Pitch. The psychological judgment of the frequency of a pure tone is called *pitch*. The unit of pitch is the *mel*. The subjective pitch sensation associated with a 1000-Hz tone at a loudness level of 40 phons is defined to be 1000 mels. (The loudness level must be specified because the perception of pitch is strongly dependent on loudness.) If another pure tone (at the same loudness level) is judged to have twice the pitch of the 1000-Hz tone, this tone is assigned a pitch of 2000 mels. Figure 20-22 shows a frequency-pitch curve for a constant loudness level of 40 phons. (Frequency-pitch perception varies from person to person, and such curves are constructed by averaging the results for many individuals.) Notice that the variation with frequency of pitch is quite similar to the variation with frequency of the position of maximum strain along the basilar membrane.

Timbre. When a single note is played on a musical instrument, the sound wave that is produced is usually rich in harmonic content. The degree to which the various har-

monics are present is different for different instruments. Thus, playing the note *A* (440 Hz) on a violin or on a piano will produce different intensities for the harmonics at 880 Hz, 1320 Hz, 1760 Hz, and so forth. Figure 20-23 shows the harmonic spectra for two different musical instruments.

Because the harmonics are produced with different intensities, we can easily distinguish a note played on a piano from the same note played on a violin or clarinet. Even the same type of instrument made in slightly different ways will have distinctive tonal qualities. Violins of the great makers—the Amati, Guarneri, and Stradivari families—are easily distinguished from ordinary fiddles. The combined harmonic quality of the produced sound is called *timbre*.

The harmonic content in notes sounded by musical instruments is further enhanced by the nonlinear coupling mechanism of the ossicles in the middle ear. A pure tone with frequency ν_1 will evoke a response that contains the harmonic frequencies, $2\nu_1$, $3\nu_1$, $4\nu_1$, and so forth. It is not surprising that the ear responds in a closely similar way to two notes whose frequency ratio is 2:1. Such a pair of notes are said to be an *octave* apart or to span a *musical interval* of one octave. This similarity in tone value for two notes an octave apart occurs even though the pitch of the higher note is not perceived to be twice that of the lower note.

When two frequencies, ν_1 and ν_2, are present simultaneously, the nonlinear response of the ear also produces *combination tones* with frequencies $\nu_2 - \nu_1$ and $\nu_2 + \nu_1$. Then, including the generation of harmonics, a host of additional tones are perceived: $2\nu_2 - \nu_1$, $\nu_2 + 2\nu_1$, $2(\nu_2 - \nu_1)$, and many others.

Musical Scales. When pairs of notes with fundamental frequency ratios that are small whole numbers are sounded together, the various tones perceived by the ear form a compact tonal progression. The notes together are judged smooth or *constant*. (See Problem 20-31.) The fa-

Fig. 20-22. The lower curve shows the average frequency-pitch relationship. The upper curve shows the strain position along the basilar membrane. The arrows indicate that a frequency of 1500 Hz corresponds to 1300 mels and to a strain position of 18 mm, measured from the oval window end.

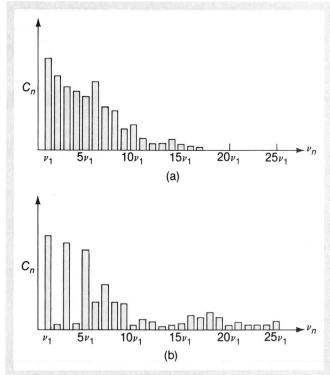

(a)

(b)

Fig. 20–23. (a) The harmonic spectrum for the note B ($\nu_1 = 493.88$ Hz) played on a violin. (b) Spectrum for the note B flat ($\nu_1 = 233.08$ Hz) played on a clarinet. In obtaining spectra such as these, it is customary to make the Fourier analysis in a series of terms of the form $C_n \cos(\omega_n t + \delta_n)$. Then, the coefficients C_n are related to the A_n and B_n in Eqs. 19–40 according to $C_n = \sqrt{A_n^2 + B_n^2}$. The phase information contained in the δ_n is discarded in presenting the harmonic spectra. (Spectra courtesy of Richard E. Berg.)

miliar scale, *do, re, mi, fa, so, la, ti, do,* with the second *do* twice the frequency of the first *do,* defines the eight notes and seven intervals that make up an octave. Musical notes are labeled by the letters A through G. If the note *do* corresponds to C, the octave becomes $CDEFGABC'$. The next octave is $C'D'E'F'G'A'B'C''$; and so forth. Each note in one octave has twice the frequency of the corresponding note in the next lower octave; thus, $\nu_{D''}:\nu_{D'}:\nu_D = 4:2:1$.

Early in the history of music it was discovered that certain combinations of three notes—called the *major triads*—are especially pleasing to the ear. These combinations are CEG, FAC', and GBD', whose frequencies are in the ratios $4:5:6$. These triads can be used to establish the *major diatonic just scale* in C. Some algebraic manipulation yields the frequency ratios in Table 20–3. (•Can you verify these results? See Problem 20–30).

Table 20–3 also shows, in the bottom row, the ratio of the frequency of each note to the next lower note. Notice that there are two short intervals with ratios of 16/15 and

five longer intervals with the nearly equal ratios of 9/8 or 10/9. This scale poses a problem if we wish to have *do* start on any note (*key* signature) and simultaneously have sensibly the same frequency relationship in the new triads. The problem is solved by inserting five new notes into the long intervals in each octave. These are the *sharp* (♯) or *flat* (♭) notes that correspond to the black keys on a piano. (The notes that are neither sharp nor flat are referred to as *natural*.) In addition, the frequencies of all the notes are altered slightly to produce *exactly* the same frequency ratio, namely, $2^{1/12} = 1.05946$, between any two adjacent notes. The scale arrived at in this way is called the *equally tempered scale*. Table 20–4 gives some of the frequencies on this scale near concert A (the fifth A from the bottom note on an 88-key piano), which is taken to be 440.00 Hz.*

Because the frequency ratios have been slightly com-

---

*In scientific work a frequency of 256.00 Hz is commonly used for middle C (given as 261.63 Hz in Table 20–4).

TABLE 20–3. Frequency Ratios in the Major Diatonic Just Scale

Note	C	D	E	F	G	A	B	C'
Frequency Ratio, Relative to C	1 1.000	9/8 1.125	5/4 1.250	4/3 1.333	3/2 1.500	5/3 1.667	15/8 1.875	2 2.000
Interval		9/8	10/9	16/15	9/8	10/9	9/8	16/15

TABLE 20–4. Some Frequencies of the Equally Tempered Scale

NOTE	ν (Hz)
C	261.63
C$^\sharp$ or D$^\flat$	277.18
D	293.66
D$^\sharp$ or E$^\flat$	311.13
E	329.63
F	349.23
F$^\sharp$ or G$^\flat$	369.99
G	392.00
G$^\sharp$ or A$^\flat$	415.30
A	440.00
A$^\sharp$ or B$^\flat$	466.16
B	493.88
C$'$	523.25
C$'^\sharp$ or D$'^\flat$	554.37
D$'$	587.33

promised to produce the equally tempered scale, these ratios are no longer exact ratios of small whole numbers. For example, $\nu_G/\nu_C = 1.4983$, instead of 3/2 as on the diatonic scale. It is interesting that musicians whose instruments have continuously adjustable frequencies for their note values (such as the stringed instruments) will accommodate to the fixed-pitch instruments when playing with an orchestra, yet they will automatically slide into a diatonic scale when playing unaccompanied.

Finally, the selection of 440.00 Hz for concert A, although widely accepted, has not always been the universal standard. Historically, the frequency has been as low as 374.2 Hz (for the organ of L'Hospice Comptesse, Lille, 1700) to as high as 567.3 Hz (the North German church pitch, 1619). Tuning to a higher pitch makes an orchestra or organ sound more *brilliant*; this tends to be popular, but vocalists sometimes have difficulty coping with these higher pitches.

PROBLEMS

Section 20-1

20-1 What is the ratio of the speed of sound in air on a hot summer day ($T = 38°$ C) to that on a very cold winter day ($T = -28°$ C)?

20-2 Calculate the speed of sound in helium at $0°$ C and normal atmospheric pressure. The density of helium for these conditions is 0.1786 kg/m^3; the value of γ for helium (a monatomic gas) is 1.67. Compare your result with that in Table 20–1.

20-3 A long thin copper rod is given a sharp compressional blow at one end. The sound of the blow, traveling through air at $0°$ C, reaches the opposite end of the rod 6.4 ms later than the sound transmitted through the rod. What is the length of the rod? (Refer to Table 19–1.)

20-4 Use the information in Table 16–2 and calculate the velocities of the three different types of wave propagation that are possible in copper.

20-5 A sound wave in air at $0°$ C has an intensity of 10^{-6} W/cm^2 (a *loud* sound). (a) What is the amplitude of the acoustic pressure? (b) If the wavelength of the wave is 0.80 m, what is the displacement amplitude?

20-6 An empty 2-liter jug is used as a Helmholtz resonator. The neck of the jug has a diameter of 5.0 m and a length of 4.0 cm. (a) What is the resonant frequency ν_0? (Use the expression that includes the length correction.) (b) What amount of liquid must be left in the jug if the resonant frequency is to be middle C ($\nu_C = 261.6$ Hz)?

Section 20-2

20-7 The longest pipe on an organ that has pedal stops is often 16 ft (4.88 m). What is the fundamental frequency (at $0°$ C) if the nondriven end of the pipe is (a) closed and (b) open? (c) What will be the frequencies at $20°$ C?

20-8 The fundamental of an open organ pipe corresponds to middle C (261.6 Hz). The third harmonic of a closed organ pipe has the same frequency. What are the lengths of the two pipes?

20-9 A tuning fork whose natural frequency is 440 Hz is placed immediately above a tube that contains water. The water is slowly drained from the tube while the vibrations of the tuning fork excite the air in the tube above the water level. It is found that the sound is enhanced first when the air column has a length of 58.0 cm and next when it has a length of 96.6 cm. Use this information to obtain a value for the speed of sound in air. What is the temperature of the air?

20–10 A clever engineering student sets out to design an organ pipe whose pitch (frequency) will be independent of temperature. Having read Section 22–3, he knows that most materials expand approximately linearly with increasing temperature so that the length L at a temperature T (in degrees C) is related to the length L_0 at $0°$ C according to $L = L_0(1 + \alpha T)$, where α is a constant. The student therefore decides to use a material for the pipe that will increase in length with temperature to compensate for the increase in the speed of sound with temperature. Determine the value that will permit this scheme to work. Most materials have values of α in the range from about 10^{-4} to about 10^{-6} per degree C. Is the scheme really practical?

20–11 A standing wave in a pipe with a length of 0.80 m is described by

$$\xi(x, t) = A \sin \frac{2\pi}{L} x \sin \left(\frac{2\pi v}{L} t + \frac{\pi}{4} \right)$$

(a) Locate the positions of all the nodes N and antinodes $\overline{N}$ in ξ, v_g, δp, and $\delta\rho$. (b) How is the pipe terminated? (c) What is the wavelength of the wave? What is the frequency? Is this the fundamental mode of vibration? If not, what is the fundamental frequency? (d) At $t = 0$ the acoustic pressure at $x = \frac{1}{2}L$ is 0.2 percent of atmospheric pressure. What is the value of the displacement amplitude A? (e) Determine the total energy in the wave if the cross-sectional area of the pipe is 25 cm².

Section 20–3

20–12 What is the average acoustic power radiated into air at $20°$ C by a sphere that pulsates with SHM between radii of 14.8 cm and 15.2 cm with a frequency of 1200 Hz? (The density of air at $20°$ C is 1.20 kg/m³.)

20–13 What must be the displacement amplitude (i.e., the radial excursion) of a sphere with a radius of 18 cm if the radiated acoustic power is to be twice that of a sphere with a radius of 10 cm and a displacement amplitude of 1.5 mm, both spheres having the same pulsation frequency?

20–14 A rectangular room has dimensions 8 m × 6 m ×

3 m. Determine the 12 lowest frequencies for standing waves in the room for an air temperature of $20°$ C.

Section 20–4

20–15 A tuning fork with unknown frequency beats against a standard tuning fork with frequency 440 Hz, producing a beat pulsation every 200 ms. What are the possible frequencies for the first tuning fork? A small amount of wax is placed on a prong of the first tuning fork, and the beat period increases. How do you choose between the possible values?

20–16 Two identical piano strings are intended to have middle C (261.6 Hz) as the fundamental. When played together the strings produce a beat frequency of 4.1 Hz. By what fractional amount must the tension in one string be changed so that the strings will vibrate with the same frequency?

20–17 Two small spheres that pulsate harmonically are separated by a distance of 0.5 m and are driven in phase with a frequency of 3000 Hz. (a) Determine the locus of points on a sphere (5.0 m radius with center midway between the vibrating spheres) that correspond to the first two interference maxima on either side of the equator. (The equator corresponds to $\theta = 0°$ in Fig. 20–11.) (b) Do the same for the first two minima. (c) If each source radiates an average power of 0.5 W, what is the sound intensity (W/m²) on the equator of the sphere? [*Hint:* Should you add amplitudes or intensities?]

20–18 Two probes that touch the surface of a pool of
• water vibrate in phase and set up expanding circular wavefronts. The waves produce an interference pattern as shown in Fig. 20–10. (a) If the two probes are separated by a distance $d = 4\lambda$, write an equation for the branch of the interference maxima that has $\Delta r = \lambda$. Identify the type of curve. (b) Does the asymptotic limit of this curve for large distances agree with Eq. 20–15?

20–19 A loudspeaker has a rectangular vibrating surface, 20 cm × 150 cm, and is mounted in the center of a stage facing an audience seated on a single horizontal floor. (Refer to Example 20–6.) (a) For the most efficient radiation of sound to the audience, should the long dimension of the loudspeaker be oriented vertically or horizontally? (b) For a tone with a frequency of 1000 Hz, determine the angle of the first intensity minimum in a vertical plane. (Assume the orientation found in (a).)

Section 20-5

20-20 A trumpet player traveling in an open convertible plays concert A (440 Hz) on his horn. A stationary observer directly in the path of the automobile hears the note $A^{\#}$ (466.16 Hz). What is the speed of the automobile?*

20-21 A sound source with $\nu_0 = 200$ Hz moves relative to still air with a speed of 15 m/s. Determine the wavelengths that would be measured by a stationary observer in the directions $\alpha_s = 0°$, $45°$, $90°$, $135°$, and $180°$. What are the corresponding frequencies? (Take $v = 331.5$ m/s)

20-22 The engine noise from a jet aircraft seen directly overhead appears to originate at a point $15°$ behind the aircraft. What is the speed of the aircraft?

20-23 A moving sound source S and an observer P are
• approaching a point Q along paths with an included angle of $30°$, as shown in the diagram above right. Let $v_s = 15$ m/s, $\nu_0 = 200$ Hz, and $u = 18$ m/s; take $v = 331.5$ m/s. (a) What frequency is heard by the observer when S and P are equidistant from Q? (Assume that $\alpha_s' = \alpha_s$.) (b) What is the actual value of $\alpha_s' - \alpha_s$? (c) Calculate the true value of the frequency ν and compare with the approximate result in (a).

20-24 A bullet fired from a rifle travels at Mach 1.38 (that

---

*When Doppler first tested his theory (1842), he used musicians playing on a moving railway flatcar, with trained observers on the ground identifying the notes played.

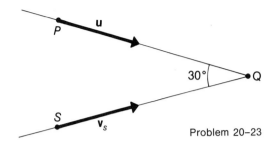

Problem 20-23

is, $N_M = 1.38$). What angle does the shock front make with the path of the bullet?

20-25 By altering the ingredients in the propellant charge used to fire a 12-g bullet, the shock front angle is decreased from $26°$ to $21°$. By how much was the kinetic energy of the bullet increased?

Section 20-6

20-26 An explosive charge is detonated at a height of several kilometers in the atmosphere. At a distance of 400 m from the explosion the acoustic pressure reaches a maximum of 10 W/m². Assuming that the atmosphere is homogeneous over the distances considered, what will be the sound intensity level (in dB) at 4 km from the explosion? (Sound waves in air are absorbed at a rate of approximately 7 dB/km.)

20-27 Determine the maximum excursion of air molecules (at $0°$ C) in a 1-kHz sound wave that has an intensity (a) of 0 dB and (b) of 120 dB.

Section 20-7

20-28 A sound wave with an acoustic pressure amplitude of 0.5 Pa is incident on a human ear. Assume an atmospheric temperature of $10°$ C and an air density of 1.25 kg/m³. (a) What is the acoustic intensity of the wave in W/m² and in dBA? (b) If the frequency is 3000 Hz, what is the loudness level in phons? (c) If the displacement of the eardrum is $\frac{1}{3}$ of the displacement amplitude of the wave, by how much does the eardrum move? (d) Where along the basilar membrane of the ear would the maximum strain develop?

20-29 The pitch of a pure tone at 40 phons is judged to be 1500 mels. (a) What is the frequency of the sound?

(b) At what position along the basilar membrane will the strain be maximum? (c) What frequency (at the same loudness) would be judged to have $\frac{1}{3}$ of the pitch of the original sound? What is the location of the maximum strain in this case? (d) Compare the ratio of the two frequencies and the ratio of the strain positions measured from the apex of the cochlea.

20-30 Require that the three major triads—CEG, FAC',
• and GBD'—all have frequency ratios $4:5:6$. Then, derive the frequency ratios to C for the major diatonic just scale and verify the entries in Table 20-3.

20-31 Two pure tones, with frequencies $\nu_1 = 200$ Hz and
• $\nu_2 = 300$ Hz, are sensed simultaneously by an ear.

Because of the nonlinear response of the ear, various combination and harmonic tones are generated. These tones have frequencies $\nu = m\nu_1 \pm n\nu_2$, with $m = 0, \pm1, \pm2, \pm3, \ldots$ and $n = 0, \pm1, \pm2, \pm3, \ldots$ (but $\nu > 0$). (a) List all of the tones that are heard with $\nu \leqslant 800$ Hz and $|m| \leqslant 4$, $|n| \leqslant 4$. (b) Suppose that ν_2 is changed to 282 Hz, no longer a simple-whole-numbers ratio with $\nu_1 = 200$ Hz. Again, list all of the tones heard below 800 Hz. Comment on the comparison.

Additional Problems

20-32 By what percentage does the speed of sound change when a cold front passes and the air temperature drops from 34° C to 11° C?

20-33 Suppose that the volume V of the resonator in Example 20-3 is a continuation of the cylinder with radius r and has an additional length ℓ'. Consider the case in which $\ell = \ell' = \frac{1}{2}L$, where L is the total length of the cylinder. Ignore the length correction, $\Delta \ell = 8r/3\pi$, and determine the oscillation frequency ν_0. Compare this result with the fundamental frequency for a pipe of length L with one end open. Could the small discrepancy be due to the fact that λ_0 is not sufficiently large compared with L?

20-34 At low frequencies, a loudspeaker that is mounted
• in a very large wall (an *infinite baffle*) acts as a simple mechanical oscillator with a natural frequency ν_0. Consider a speaker with a diameter of 20 cm which has an effective moving mass (speaker piston plus the voice coil) of 10 g and a spring constant (provided by the piston mounting) of 2000 N/m. (a) What is the natural frequency ν_0, neglecting any damping resistance? (b) The speaker is next converted into an *acoustic suspension* speaker by mounting it in one wall of a completely enclosed cabinet that has dimensions 0.5 m × 0.5 m × 0.2 m. What is the spring constant of the air enclosed in the cabinet for small displacements of the speaker piston? (c) The total effective spring constant for the acoustic suspension speaker is the sum of the two spring constants. (•Can you explain why it is appropriate to add the spring constants?) What is the natural frequency of the system?

20-35 A large-diameter loudspeaker (called a *woofer*) is to
• be mounted in a *bass reflex* enclosure in order to give optimum damping for the woofer's free-air resonance at $\nu_0 = 45$ Hz. When the enclosure is a Helmholtz resonator with the same natural frequency ν_0, the energy absorbed from the oscillating woofer by the air in the enclosure is maximum, which results in the required optimal damping. Find the dimensions of the enclosure cabinet if height:width:depth = 5:3:2 and if the port open-

ing is a circular hole with a diameter of 16 cm. (Neglect the conical volume of the woofer.)

20-36 An organ pipe sounds concert *A* (440 Hz). If the pipe were filled with carbon dioxide, what would be the new frequency of the tone? What is the (approximate) note that would be sounded? (Refer to Table 20-4.)

20-37 An organ pipe is tuned to middle *C* (261.63 Hz) at 0° C. At what temperature would the pitch rise to $C^\sharp$? (Neglect any change in length of the pipe. Refer to Table 20-4.)

20-38 The length of the flute shown in Fig. 20-6 is 30.5 cm from the center of the embouchure hole to the center of the fifth tone hole. Estimate the fundamental frequency of the flute, assuming that the sketch in Fig. 20-6 is to scale. To what note does this correspond? (Refer to Table 20-4.)

20-39 The two pulsating spheres shown in Fig. 20-11 are driven with a phase difference of π (exactly out of phase). Determine the conditions for interference maxima and minima at the point $P(r, \theta)$.

20-40 A stationary tuning fork that produces a pure tone with $\nu_0 = 250$ Hz is approached directly by an observer traveling with a speed of 10 m/s. A steady cross wind is blowing with a speed of 5 m/s. What is the frequency of the tone heard by the observer?

20-41 A supersonic aircraft is flying parallel to the ground.
• When the aircraft is directly overhead, an observer sees a rocket fired from the aircraft. Ten seconds

later the observer hears the sonic boom, followed 2.8 s later by the sound of the rocket engine. What is the Mach number of the aircraft?

20-42 Equation 20-18 for the Doppler-shifted frequency ν
• (for $u = 0$) is valid for $\alpha'_s \cong \alpha_s$. If $v_s \ll v$, show that the correction $\Delta\nu$ required to account for the actual difference between α'_s and α_s is $\Delta\nu \cong -\nu_0(\alpha'_s - \alpha_s)^2$.

20-43 A small sphere pulsates with SHM and radiates an average acoustic power of 10 W. (a) What is the intensity in dB (relative to 10^{-12} W/m²) at a distance of 20 m from the sphere? (b) What is the acoustic pressure amplitude at 20 m? (c) At what distance would the intensity be 3 dB lower than at 20 m?

SPECIAL RELATIVITY

In the preceding two chapters we discussed wave motion in mechanical systems and we learned how such waves depend upon the properties of the material media along or through which the waves propagate. Electromagnetic waves—in particular, *light* waves—have the unique feature that they can propagate through empty space without the benefit of any material medium. Prior to this century, scientists could not conceive how light could propagate through *vacuum*. There must be some medium, they argued, whose excursions would permit wave motion. There was no *apparent* medium in space to support the propagation of light, so an *invisible* medium—the *ether*—was invented to serve the purpose.

The ether was a strange stuff, indeed. Light and other electromagnetic radiations could propagate through the ether, yet it offered no perceptible drag on the motion of planets. In fact, the ether seemed to have *no* observable properties—its sole function was to transmit electromagnetic waves! Nevertheless, the ether theory maintained a tenuous existence. In 1905, however, a crucial new idea was contributed by Albert Einstein (1879–1955). In a single bold stroke, Einstein swept away the ether theory and replaced it with his elegant *relativity theory*.

Einstein's theory represents a new way of thinking about space and time. The theory predicts effects that are readily observable only when speeds approaching the speed of light are involved. Our intuition was not developed in such a fast-moving world, so some of the ideas of relativity seem strange and forced. Nevertheless, every prediction of the theory has been verified by numerous experiments. Relativity theory is the only theory capable of explaining phenomena that take place at high speeds. Moreover, when the speeds involved are small compared with the speed of light, the relativistic theory becomes identical to the Newtonian theory. Thus, Einstein's theory is truly universal and all-encompassing.

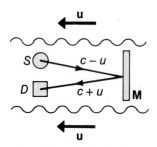

Fig. 21–1. The ether has a velocity **u** relative to the apparatus for a light experiment (a source S, a mirror M, and a detector D, which are in fixed positions with respect to one another).

21-1 THE ETHER EXPERIMENTS

In 1881, Albert Michelson* began a series of experiments designed to detect the presence of ether. To understand the magnitude of the effect he sought to measure, consider the situation in Fig. 21-1. The apparatus consists of a light source S, a mirror M, and a detector D, all fixed with respect to one another. If the entire

---

*Albert A. Michelson (1852–1931) was the first American to win a Nobel Prize in one of the sciences (1907).

apparatus moves through the ether to the *right,* this is the equivalent to the ether streaming past the apparatus with a velocity **u** directed toward the *left,* as shown in Fig. 21–1.

Assume that light propagates through the ether in the same way that sound propagates through air. Then, the velocity of a light signal from the source to the mirror is $c - u$, and the velocity from the mirror to the detector is $c + u$, where c is the velocity of light in the ether. If the distance from the source and detector to the mirror is ℓ, the transit time t for the signal to travel the path $S \to M \to D$ is

$$t = \frac{\ell}{c - u} + \frac{\ell}{c + u} = \frac{2\ell}{c}\left(\frac{1}{1 - u^2/c^2}\right)$$

$$= \frac{2\ell}{c}\left(1 + \frac{u^2}{c^2} + \frac{u^4}{c^4} + \cdots\right) \tag{21-1}$$

where we have used Eq. A–14 in Appendix A to expand the factor in parentheses. Thus, in principle it should be possible to determine the ether velocity u by measuring ℓ, c, and t. Now, the Earth moves around the Sun with an orbital speed of approximately 30 km/s, and the direction of motion changes during the year. Therefore, there must be some time during the year when the Earth moves relative to the ether with a speed of at least 30 km/s. Because $u \ll c$, the term u^4/c^4 and all higher-order terms in Eq. 21–1 are negligible. Then, the change in the transit time due to the ether velocity is $\Delta t \cong (2\ell/c)(u^2/c^2)$. The velocity of light is $c \cong 3 \times 10^5$ km/s, so $u^2/c^2 \cong 10^{-8}$. If $\ell = 30$ m, the time difference that must be detected is $\Delta t \cong 2 \times 10^{-15}$ s, a value that is too small to be measured by ordinary means.

If the transit time could be measured for only one leg of the path—for example, for $S \to M$—the result would be $t' = \ell/(c - u) \cong (\ell/c)(1 + u/c)$, which depends linearly on u/c. Then, $\Delta t' \cong 10^{-11}$ s, a time interval very much longer and within the realm of direct measurement. Unfortunately, there is no way to measure the velocity of light for a one-way transit from one point to another. (●Can you explain why?)

Michelson showed great ingenuity in his approach to the problem. Using mirrors, he set up split beams of light so that he could measure the time difference for two round-trip paths with closely equal lengths but at right angles to each other. Michelson's apparatus is shown schematically in Fig. 21–2. The two right-angle paths should yield different transit times because their motions through the ether involve different components of **u**. Michelson conducted the experiments throughout the year. To remove any effects due to the failure of the two path lengths, ℓ_1 and ℓ_2, to be exactly equal, in each measurement he interchanged the paths by rotating the apparatus through 90°. To Michelson's surprise (and disappointment), he found NO effect! In an improved experiment in 1887, in which he was joined by E. W. Morley, a null result was again found, even though the sensitivity was 40 times that required to detect a velocity with respect to the ether of 30 km/s.

The Michelson-Morley experiment has been repeated many times and in several variations. In every case a null result was obtained. One of the most sensitive experiments was performed with laser beams; an ether velocity of 1 *centimeter per second* could have been observed (!), but again no effect was found. The conclusion that must be drawn from these experiments is that the motion of a light source and a detector relative to the ether cannot be detected.

For a while following the announcement of the null result of the Michelson-Morley experiment, physicists continued to hold to the notion that the ether did exist. They argued that there was some unknown compensating effect that could

Fig. 21-2. Schematic representation of the Michelson-Morley experiment. The light beam from the source S is split by the half-silvered mirror M. One half of the beam is reflected toward M_1 and the other half is transmitted toward M_2. The returned beams enter the detector D, which can be used to determine the time difference for the two paths, $2l_1$ and $2l_2$. (The method involves observing the interference effects of the two beams.)

account for the null result. During the years 1895 to 1904, the Dutch physicist H. A. Lorentz (1853–1928) developed a theory in which all material objects in motion through the ether undergo a contraction in length of the dimension parallel to the relative velocity **u**. The magnitude of this contraction, which was imagined to be a *real* change of length, was exactly the correct amount to cancel the effect expected in the Michelson-Morley experiment. A more satisfactory explanation was provided, as we see in Section 21-5, by Einstein's relativity theory (1905); nevertheless, the

Albert Einstein (1879–1955) developed the special theory of relativity while employed as a junior official in the Patent Office in Berne, Switzerland. His first paper on the subject was published in 1905, the same year in which he published important papers on the photoelectric effect and on Brownian motion. Einstein was awarded the 1921 Nobel Prize in physics for his theory of the photoelectric effect (*not* for his theory of relativity!). His general theory of relativity was contained in a paper published in 1915. Einstein directed his own research institute at the Kaiser Wilhelm Institute in Berlin until 1933, when he decided to resign and go to the United States, away from the oppression of Hitler's Germany. He established himself at the Institute for Advanced Studies at Princeton University. Einstein's last decades were spent on two great undertakings—to find an all-encompassing geometrical description of physical phenomena and to find an end to the threat of nuclear warfare. Unfortunately, he was not successful in either mission. (Courtesy of Dr. Otto Nathan, Estate of Albert Einstein, and the American Institute of Physics Center for History of Physics.)

Lorentz theory was not experimentally refuted until the 1932 experiment of Kennedy and Thorndike. This experiment was similar to that of Michelson and Morley except that a difference in the path lengths, ℓ_1 and ℓ_2, was deliberately introduced. Also, the apparatus was maintained in a fixed position and carefully stabilized so that observations could be made continually over a period of months. The two paths differed in length by $2(\ell_2 - \ell_1)$ compared with the Michelson-Morley paths. Because of the change in direction of the Earth's motion during the observation period, the changing contraction of the path-length difference should have produced a corresponding change in transit time. However, no change in the time difference for the two paths could be found over the lengthy period of observation.

The problems associated with the various ether theories were finally resolved by Einstein in 1905 with the publication of his *special theory of relativity*. It is interesting that Einstein was led to the formulation of his theory by considerations of problems in electrodynamics, not by the results of the Michelson-Morley experiment. Indeed, it seems that Einstein was not even aware of the details of this experiment when he developed his theory.

21-2 EINSTEIN'S POSTULATES

Einstein based his special theory* on only two fundamental postulates:

I. All inertial reference frames are equivalent with respect to all the laws of physics.

II. The speed of light in vacuum is a constant c, *independent of the motion of either the source or the detector.*

The first postulate restates Newton's idea of the equivalence of inertial frames, but it extends the concept to *all* physical laws, not just the Newtonian laws of dynamics. This postulate denies the concepts of "absolute rest" and "absolute motion" because no inertial reference frame has a preferred status compared to any other inertial frame.

The second postulate explains the null result of the Michelson-Morley experiment with trivial simplicity because, in effect, the experiment involves the measurement of the velocity of light under different conditions. According to the first postulate, no inertial frame can have a preferred status. Then, the second postulate states that there is no significance in thinking about the velocity of light other than with respect to the observer who measures it. All observers in inertial frames measure the same velocity c for any light beam.

Einstein's postulates effectively demolished the ether theory. The postulates denied the possibility of detecting the ether, so the ether was rendered superfluous. According to current theory and belief, electromagnetic waves can travel freely through vacuum and require no medium, material or ethereal, for propagation. In vacuum, light and all other electromagnetic radiations travel with the velocity c, the best measured value for which is

$$c = 2.99792458 \times 10^8 \text{ m/s}$$

with an uncertainty of about one unit in the last decimal place. This corresponds to

---

*The theory is "special" because it deals with the special case of nonaccelerated reference frames. Accelerated systems and gravity are treated in Einstein's *general theory of relativity,* published in 1915.

the remarkable precision of 3 parts per billion! For most of our discussions, we use the approximate value,

$$c = 3.00 \times 10^8 \, \text{m/s}$$

21-3   THE RELATIVISTIC NATURE OF SIMULTANEITY

In the discussion of relative motion in Section 4–6 we assumed the existence of a "universal clock" that may be referred to by all observers (including those in motion with respect to the clock) to establish the "absolute time" of an event. We now make a careful examination of this proposition.

An *event* is an instantaneous occurrence at a definite (point) location. Two events are judged to be *simultaneous* if they are observed to occur at the same time. For a universal time to exist and to have meaning, it follows that if one inertial observer finds a pair of events to be simultaneous, all inertial observers (including those in motion with respect to the first observer) must also find the same events to be simultaneous. Events that are easy to analyze are those that emit light flashes while leaving marks to identify the locations of the events (for example, lightning strikes).

To draw conclusions regarding the time-ordering of events, we must arrange for signaling between events and observers and also between observers. This signaling is best carried out at the greatest possible speed. As we see later, no signal can convey information at a speed greater than the speed of light. Moreover, the speed of light is central to Einstein's theory. Therefore, we assume that all signaling is carried out with light beams.

The simultaneity of events is judged by the arrival sequence, at the observer's station, of light flashes from the events. For two such events at different locations that are actually simultaneous, an observer will see the light flash from the closer event before the arrival of the flash from the more distant event. In any case, because each event leaves a mark on the ground, the location of the events in the observer's frame can be determined by distance measurements made at leisure. Hence a correction for the difference in the times required for the light flashes to reach the observer can readily be made. The simultaneity of two events can then be deduced if the corrected times of arrival of the light signals agree. Thus, there is no conceptual difficulty in deducing simultaneity or the time-ordered sequence of observed events.

Relative Simultaneity.   Consider two observers who are at rest in their own inertial reference frames, which are in relative motion. If one observer judges two events at different locations to be simultaneous, will the other observer agree? To answer this question we discuss an example suggested by Einstein, illustrated in Fig. 21-3. A railway car that carries an observer O' moves past a stationary observer O with a constant velocity $\mathbf{u}$ toward the right.* A lightning bolt strikes the end of the car at A' and leaves burn marks on the car at A' and on the track at the immediately adjacent point A. Another bolt strikes B' and leaves burn marks on the car at B' and on the track at B. From later examinations of the marks A and B on the track, O deduces that his observing position was midway between these points. By making measurements in the car, O' establishes that his observing position was midway between the ends of the car, A' and B', and he also determines that the lightning bolts did indeed strike and leave marks at the ends of the car.

---

*Only the relative motion of the observers' frames is physically meaningful, but it is convenient to identify one observer as "stationary" and one as "moving."

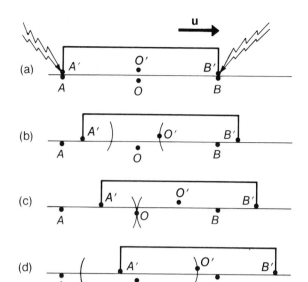

Fig. 21-3. Einstein's example illustrating the relativistic nature of simultaneity. The observers O and O' do not reach the same conclusion regarding the time sequence of events at AA' and BB'. Expanding spherical light pulses from AA' and BB' are shown at successively increasing times in (a), (b), (c), and (d), as reckoned by observer O.

When the lightning bolts strike, expanding spherical light pulses are emitted from AA' and from BB'. Let us examine the situation first from the viewpoint of observer O. The light flashes reach O at the same instant, as indicated in Fig. 21-3c. Because O later finds that he was midway between A and B, he concludes that the lightning strikes were simultaneous.

According to O, observer O' is moving to the right, so O argues that O' intercepts the light pulse from B' at an earlier time (Fig. 21-3b) than he intercepts the pulse from A' (Fig. 21-3d). Observer O knows that this is the true sequence of events because the light flashes arriving at O' are redirected toward O by the mirror that O' wears on his hat. If O sees the reflected flash from B' arrive before the reflected flash from A', he can correctly infer that this same sequence of flashes was also observed by O'. However, we must be careful *not* to assert that the time *interval* between the arrival of the light flashes at O' is judged to be the same by both of the observers. We return to this point in Example 21-4.

Thus far in the discussion we have come to no unusual conclusion nor have we invoked either postulate of relativity theory. We now ask the crucial question: How does O' interpret the fact that the flash from B' arrived at his position in the middle of the car before the arrival of the flash from A'? According to the second postulate, the velocity of the light flashes from the two sources is the universal constant c. Because the distances $A'O'$ and $B'O'$ are the same, the earlier arrival time of the B' flash must mean, so O' argues, that this flash originated at an earlier time than did the A' flash. Thus, O' concludes that the lightning strikes were *not* simultaneous. (•How would O judge the sequence of flashes from A and B if O' determined that the flashes from A' and B' are simultaneous?)

The observers O and O' do not agree on the time sequence of the lightning strikes because the observers are in relative motion and because the events take place at different locations. The observers do agree on the arrival sequence of the flashes at the position of O' because these events (the arrivals of the flashes) occur at the same location. In fact, all inertial observers, regardless of their relative motions, will always agree on the time sequence of events that take place at the same location. For example, suppose that the B' flash is a message warning O' that the A' flash is a potentially lethal burst of laser radiation. If the B' flash arrives before the A' flash, O' can move out of the way of the A' flash. But if the flashes arrive simultaneously, or if A' precedes B', O' will be killed by the laser radiation. After the events occur, all observers must agree that O' is either dead or alive. If one observer (for example, O) sees the B' flash arrive first and sees O' avoid the laser radiation, all observers must find O' alive and must agree that the B' flash preceded the A' flash.

The Synchronization of Clocks. We now consider a general scheme for determining the time sequence of events. Suppose that an observer, equipped with a standard clock, is located at a fixed point in an inertial reference frame. We attach a special significance to the time that the observer records for events *that occur at his location*. The observer's reckoning of the time of occurrence of such events is called *proper time* and the time intervals between such events are called *proper time intervals*. We allow the intuitive notion that two events are simultaneous if they take place at the same location and if the local clock indicates that the time of occurrence is the same for both events.

Next, consider two observers at rest at different points, A and B, in the same inertial frame. An event occurs at A which is also observed, by means of light signals, at location B. How do the two observers establish a consistent relationship between their clock readings for such events?

Let the observers at A and B be equipped with identical standard clocks that have been shown to run at the same rate when at rest. How do the two observers go about the coincident setting of their clocks to an agreed-upon zero, for example, 7:30 p.m.?

The observers agree to set their clocks by using light beams to signal between their known locations in the inertial frame. The clock at A is taken to be the master reference clock. This clock is set to $t_A = 0$ (or 7:30 p.m.) at the instant a light pulse is directed toward B. When the observer at B receives this light signal, he sets his clock to the time $t_B = L/c$, which is the time required for the light signal to travel from A to B. When clocks at rest are set in this manner at different locations in an inertial reference frame, the clocks are said to be *synchronized*. Evidently, we can imagine clocks synchronized in this way to be located at every point of interest within the inertial frame. (•When an observer looks around at all of these clocks, do they all indicate the same time?)

Suppose that two events occur at different locations within a synchronized frame of reference. Each event has an associated position $\mathbf{r}$ and a local time t. If the time of one event recorded by the associated local clock is the same as that of the other, the events are *simultaneous*.

How does an observer in an inertial frame S describe the motion of a particle in that frame? When the observer sees the particle at a point specified by the position vector $\mathbf{r}_S$, he also reads and records the *local time* t_S indicated on a synchronized clock *at* $\mathbf{r}_S$ (*not* the time on a clock at the observer's position). Because the observer reads the clock at the position of the particle (that is, at the position where he *sees* the particle), he always obtains the correct local time t_S for that position of the particle regardless of his distance from that point. The entire history of the particle's motion can then be described by a series of coordinates $\mathbf{r}_S$, t_S. Because the time that the particle is at a particular point is always specified by the *local* clock, all observers in S, regardless of their locations, will describe the motion of the particle in terms of the same set of coordinates $\mathbf{r}_S$, t_S. It is such a charting of the course of a particle in frame S, giving its location, velocity, and so on, that is meant when one says ". . . the observer in S sees . . . ." We now ask the interesting question, "How does an observer in the inertial frame S', which is in motion relative to S, describe the motion of the particle?"

To answer this question, we must first inquire how to synchronize the clocks in S' with those in S. We return to this point in Section 21–6.

21-4 THE RELATIVISTIC NATURE OF TIME INTERVALS

We now consider an experiment to measure time intervals in moving reference frames. Imagine a light pulse that is returned directly to its point of origin by a mirror placed a distance L from the source. The time required for the light pulse to make the round trip is the time interval of interest. This time interval can be measured by a local clock positioned at the light source. The clock records the round-trip time as the *proper time interval,* $\Delta t = 2L/c$. Any observer who performs this experiment in his own inertial frame with the same source-mirror distance L and using an identical standard clock will obtain the same result for Δt; this is guaranteed by the relativity postulates.

How do observers in relative motion compare time intervals? Figure 21-4 illustrates the geometry of the two inertial reference frames, S and S'. The observer O' in the S' frame has a light pulser and a mirror a distance L away, so he can establish a proper time interval $\Delta t' = 2L/c$ in his frame (Fig. 21-4a). The S' frame moves to the right with velocity $\mathbf{u}$ relative to the S frame, and the experiment performed by O' is observed by O in the S frame (Fig. 21-4b). The source and mirror M' are seen by O to move to the right through the positions x_0, x_1, and x_2, as indicated in the diagram. At x_0 the light pulse leaves the source, and the reflected pulse returns to the source at x_2. The S frame is equipped with its own synchronized clocks at x_0 and x_2. The difference in the clock readings at x_0 and x_2 when the source (and associated detector) passes through these points determines the time interval Δt in S that corresponds to the time interval $\Delta t'$ in S'. Evidently, $x_2 - x_0 = u\,\Delta t$; the point x_1 is midway between x_0 and x_2, so $x_1 - x_0 = \frac{1}{2}u\,\Delta t$. Now, the distance that the light pulse travels from x_0 to the mirror above x_1 is $\frac{1}{2}c\,\Delta t$. Then, from the right triangle in Fig. 21-4b we can write

$$(\tfrac{1}{2}c\,\Delta t)^2 = L^2 + (\tfrac{1}{2}u\,\Delta t)^2$$

which can be simplified to

$$\Delta t = \frac{2L}{c}\left(1 - \frac{u^2}{c^2}\right)^{-1/2}$$

Fig. 21-4. (a) The observer O' in S' establishes a time interval $\Delta t' = 2L/c$ by using a source of light pulses and a stationary mirror M'. (b) The S' frame moves to the right with a velocity $\mathbf{u}$ relative to the S frame, and the observer O in S views the light-pulse experiment performed by O'.

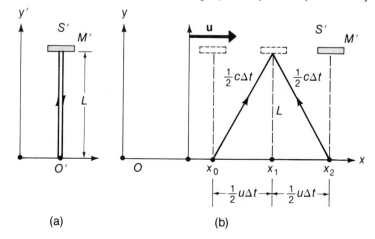

(a)   (b)

The factor $2L/c$ is just the time interval $\Delta t'$, as measured in the S' frame, that corresponds to the time interval Δt in S:

$$\Delta t = \Delta t' \left(1 - \frac{u^2}{c^2}\right)^{-1/2} \qquad \text{(21-2)}$$

It is customary in discussing relativity theory to use the following abbreviations:

$$\beta = u/c < 1 \qquad \text{(21-3a)}$$

$$\gamma = (1 - u^2/c^2)^{-1/2} = (1 - \beta^2)^{-1/2} > 1 \qquad \text{(21-3b)}$$

Then, Eq. 21-2 becomes

$$\Delta t = \gamma \, \Delta t' \qquad \text{(21-4)}$$

Notice the velocity restriction $\beta < 1$ or $u < c$; otherwise, γ would be infinite or imaginary, thereby yielding unphysical results for the time intervals.

How do we interpret the result expressed by Eq. 21-2 or 21-4? We can consider the travel of the light pulse from the source to the mirror and back to the source to represent one "tick" of a clock. Then, the duration Δt of this tick, as measured by O in S, is *longer* than the duration $\Delta t'$ as measured by O' in S'. That is, a moving clock always appears to run more slowly than an identical clock at rest with respect to the observer. This effect is called *time dilation*. (The time interval dilates, or *expands*, for a clock in motion.)

Notice that the interval $\Delta t'$ is a proper time interval in S', whereas the determination of the interval Δt requires clock readings at two different locations and is therefore *not* a proper time interval in S. Thus, the two measurements are not symmetric. According to observer O in S, the experiment performed by O' in S' required a longer time than did the same experiment performed by O in S; that is, $\Delta t > \Delta t'$. This is a general result, valid for any type or orientation of clocks used. The time interval between two events that is measured as a proper time interval in one reference frame will always appear longer when measured in a moving reference frame, in which the interval necessarily cannot be specified in terms of a proper time interval in that frame.

Example 21-1

An astronaut is traveling in a space vehicle that has a speed of $0.50\,c$ relative to the Earth. The astronaut measures his pulse rate to be 75 per minute. Signals generated by the astronaut's pulse are radioed to Earth when the vehicle is moving perpendicular to a line that connects the vehicle with an Earth observer. What pulse rate does the Earth observer measure? What would be the pulse rate if the speed of the space vehicle were increased to $0.99\,c$?

Solution:

The astronaut measures the interval between each of his pulses to be $\Delta t' = (60\ \text{s/min})/(75\ \text{min}^{-1}) = 0.80$ s. The relativistic factor γ is

$$\gamma = (1 - \beta^2)^{-1/2} = [1 - (0.50)^2]^{-1/2}$$
$$= 1.155$$

According to Eq. 21-4, the time interval between pulses as measured by the Earth observer is

$$\Delta t = \gamma \, \Delta t' = (1.155)(0.80 \text{ s})$$
$$= 0.924 \text{ s}$$

Thus, the Earth observer records a pulse rate of $(60 \text{ s/min})/(0.924 \text{ s}) = 64.9 \text{ min}^{-1}$.

At a relative speed $u = 0.99\, c$, the relativistic factor γ increases to 7.09 and the pulse rate recorded by the Earth observer decreases to 10.6 min^{-1}. That is, the lifespan of the astronaut (reckoned by the total number of his heartbeats) is much longer as measured by an Earth clock than by a clock aboard the space vehicle. (•Does the astronaut sense that he has a longer lifespan?)

Notice that the astronaut's beating heart is a kind of clock. Notice also that the radio transmissions generated by the heartbeats constitute a repetitive signal from a moving source. Consequently, the signal rate detected by the Earth observer is subject to both time dilation and the Doppler effect. Because we chose to calculate the rate at the instant when the space vehicle is moving perpendicular to the line connecting the vehicle and the Earth observer, the usual Doppler effect is absent. (That is, $\alpha_s = 90°$ in Eq. 20-18.) There remains, however, the time dilation effect, which is often called the *transverse Doppler shift* (see Eq. 21-24).

21-5 THE RELATIVISTIC NATURE OF LENGTH

Closely associated with the time dilation effect is the dependence of length measurements on the state of relative motion between the observer and the object measured. The *proper length* of an object is defined to be its length determined in a reference frame in which the object is at rest.

Suppose that a particular meter stick is passed around to observers in various inertial reference frames. With the meter stick at rest in his frame, each observer prepares an identical copy which then serves as the length standard in his frame. What results are obtained by two observers in relative motion who measure the length of a meter stick at rest with respect to one of the observers? Again, we imagine the experiment performed with light beams. As shown in Fig. 21-5, an inertial frame S' moves to the right with a velocity **u** relative to the frame S. A meter stick that lies parallel to the x'-axis is at rest in S'.

The observer in S' sets up a pulsed light source at one end (A') of the meter stick and a mirror M' at the other end (B'). The transit time for the round trip of a light pulse from the source to the mirror and back to the source is $\Delta t' = 2\ell'/c$.

As the apparatus in S' passes through the frame S, the experiment is also observed by the S observer. However, the situation is now more complicated. The light pulse is initiated at time t_0 at the end A_0 and arrives at the end B_1 after a time interval $\Delta t_1 = t_1 - t_0$. During this time the end B moves from B_0 to B_1, a distance $u \, \Delta t_1$. If ℓ is the length of the meter stick in the S frame (the value of which is still to be determined), we have, using the relativity postulates,

$$c \, \Delta t_1 = \ell + u \, \Delta t_1$$

or

$$\Delta t_1 = \frac{\ell}{c - u}$$

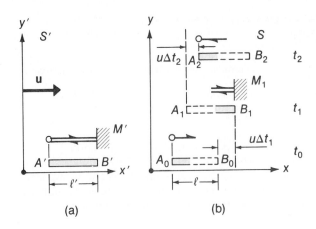

Fig. 21–5. (a) The length ℓ' of a meter stick at rest in S' is determined by using a light pulser and a standard clock. (b) The experiment is also observed in frame S. The positions of the pertinent end of the meter stick are shown at the times t_0 (when the pulse leaves the source), t_1 (when the pulse reaches the mirror), and t_2 (when the pulse returns to the source).

During the return flight of the pulse to the source, the end A moves by an amount $u \, \Delta t_2$, thereby shortening the path. Then,

$$c \, \Delta t_2 = \ell - u \, \Delta t_2$$

so that

$$\Delta t_2 = \frac{\ell}{c + u}$$

The total round-trip transit time is reckoned by the S observer to be

$$\Delta t = \Delta t_1 + \Delta t_2 = \frac{\ell}{c - u} + \frac{\ell}{c + u}$$

$$= \frac{2\ell}{c(1 - \beta^2)} = \frac{2\ell}{c} \gamma^2$$

Now, according to Eq. 21–4 for the time dilation effect, we have $\Delta t = \gamma \, \Delta t'$. Then, using $\Delta t' = 2\ell'/c$, we find

$$\ell = \ell'/\gamma \tag{21–5}$$

Because $\gamma \geqslant 1$, the moving meter stick (as measured by the S observer) is found to have a length ℓ that is *shorter* than the proper length ℓ'. If the meter stick were brought to rest in S, then $\gamma = 1$, and the observer would measure the length to be ℓ'. The shortening of length for moving objects is known as the *length contraction* effect. This contraction takes place only along the direction of motion, as we show in the following section. (See also Example 21–5.)

Are the time dilation and length contraction effects *real?* If we define a "real" effect to be one that can be measured, these relativistic effects, which have been confirmed by numerous experiments, are certainly real. The meaningful way to quote the length of an object or the duration of an event is in terms of the *proper length* or the *proper time interval*. The corresponding values that would be measured in a moving inertial frame can then be obtained for any particular case by using Eq. 21–4 or 21–5.

Example 21–2

An astronaut is traveling in a space vehicle that has a speed of 0.50 c relative to the Earth. The astronaut and an Earth observer compare the length of the standard meter stick that each has in his rest frame with the length of the other's meter stick. The measurements are made when each meter stick is held parallel to the relative velocity **u**. By what amount does each claim that the other's meter stick is shorter?

Solution:

With $u = 0.50\,c$, we have $\gamma = 1.155$ (see Example 21-1). Then, according to Eq. 21-5, each observer reckons the other's (moving) meter stick to be shorter by an amount

$$\Delta \ell = \ell' - \ell = \ell'(1 - \gamma^{-1})$$
$$= (1\ \text{m})[1 - (1.155)^{-1}]$$
$$= 0.134\ \text{m} = 13.4\ \text{cm}$$

At a relative speed $u = 0.99\,c$, we have $\gamma = 7.09$, so that $\Delta \ell = 85.9$ cm.

As L. Marder states (in *Time and the Space Traveler*), "Think of two witches on identical broomsticks. As they glide past each other, each notes with pride that her own status symbol is the longer!"

21–6   THE LORENTZ-EINSTEIN TRANSFORMATION EQUATIONS

Consider again an inertial frame S' that moves with a velocity **u** relative to another inertial frame S. Our task is to relate the four coordinates of an event, x', y', z', t', measured by the observer O' in S' to the four coordinates, x, y, z, t, of the *same* event measured by the observer O in S. We assume, without any loss of generality, that the x', y', z' axes are parallel to the x, y, z axes and that S' moves to the right relative to S along the x-x' axes. Assume further that at some instant the origins of the two frames coincide and that master clocks, one at the location of each origin, are set to zero, marking this coincident event.

In Section 4–6 we discussed the *Galilean transformation equations* that connect the coordinates and time in one reference frame to the coordinates and time in another reference frame in relative motion. For the geometry we have just specified, these transformation equations become

$$\left.\begin{aligned} x &= x' + ut \\ y &= y' \\ z &= z' \\ t &= t' \end{aligned}\right\} \tag{21-6}$$

These equations are based on the notion that there exists a single *universal time,* which can be read on either master clock by the observer in either frame and which is applicable for all locations in either frame. We have just learned that differences in the measurement of length and time in moving frames lead to the time dilation and length contraction effects. Thus, the Galilean transformation equations, which do not predict these effects, cannot be relativistically correct.

Space-Time Coordinate Transformations.   As the S' frame moves relative to the S frame, at some instant the origins coincide. At this instant the standard clock at the origin of each frame is set to zero. These clocks are then used to synchronize all other clocks in the respective frames.

A fixed space point x in frame S appears to move in frame S' with a velocity $-\mathbf{u}$. (This is the meaning of the statement that the two frames have a relative velocity u.) We have

$$x' = \kappa(x - ut) \tag{21-7a}$$

where the scale factor κ (to be determined) is included to allow for the possibility of a scale contraction of the type we discovered in Section 21-3. (The scale factor κ may depend on the magnitude of $\mathbf{u}$ but not on its direction.) According to the first postulate of relativity, we have the symmetric relationship,

$$x = \kappa(x' + ut') \tag{21-7b}$$

where the sign of u has been changed to reflect the point of view of S when observing a fixed space point x' in S'.

Suppose that a light flash is initiated at the instant $t = t' = 0$ at the position of the coincident origins of S and S'. (See Fig. 21-6.) Observers in S follow the progress of the spherical wavefront in the x-direction with $x = ct$. According to the second postulate, observers in S' follow the progress of the spherical wavefront in the x'-direction with $x' = ct'$. Therefore, solving Eqs. 21-7a and 21-7b for ct' and ct, we can write

$$ct' = \kappa(c - u)t$$
$$ct = \kappa(c + u)t'$$

Multiplying one of these equations by the other yields

$$c^2 tt' = \kappa^2(c^2 - u^2)tt'$$

from which

$$\kappa = (1 - u^2/c^2)^{-1/2} = (1 - \beta^2)^{-1/2} = \gamma$$

Thus, we find—with no surprise—that the scale factor κ is just equal to the relativistic factor γ. Substituting $\kappa = \gamma$ into Eqs. 21-7a and 21-7b yields the transformation $x \leftrightarrow x'$. Then, combining these results gives the transformation $t \leftrightarrow t'$. Altogether, we have

$$x = \gamma(x' + ut'); \qquad x' = \gamma(x - ut)$$
$$t = \gamma\left(t' + \frac{ux'}{c^2}\right); \qquad t' = \gamma\left(t - \frac{ux}{c^2}\right) \tag{21-8a}$$

The light flash initiated at $t = t' = 0$ at the coincident origins of the two frames advances as a spherical wavefront in each frame. In S, the wavefront is described by

$c^2t^2 = x^2 + y^2 + z^2$, and in S' by $c^2t'^2 = x'^2 + y'^2 + z'^2$. Thus,*

$$x^2 + y^2 + z^2 - c^2t^2 = x'^2 + y'^2 + z'^2 - c^2t'^2$$

Now, it can be shown (see Problem 21–8) that for the geometry of Fig. 21–6, we have

$$x^2 - c^2t^2 = x'^2 - c^2t'^2$$

Thus,

$$y^2 + z^2 = y'^2 + z'^2$$

Because there is complete symmetry in the plane perpendicular to the direction of **u**, we conclude that

$$y' = y; \qquad z' = z \tag{21-8b}$$

These six equations (Eqs. 21–8a,b) are referred to as the *Lorentz-Einstein transformation equations*.

Notice that $\gamma \cong 1$ for everyday conditions, namely, $u \ll c$. Then, Eqs. 21–8 reduce to the Galilean equations (Eqs. 21–6).

We emphasize again that Eqs. 21–8 relate the four coordinates x, y, z, t to the four coordinates x', y', z', t' for the *same* event observed in both of the inertial frames S and S'. The coordinate t refers to the local time at the space point $P(x, y, z)$ measured with a clock synchronized to the reference clock at the origin of the S frame. The coordinate t' refers to the local time at the space point $P'(x', y', z')$

---

*To be precise we should write

$$x^2 + y^2 + z^2 - c^2t^2 = K(x'^2 + y'^2 + z'^2 - c^2t'^2)$$

It can be argued that the homogeneity and isotropy of space require $K = +1$.

Fig. 21–6. Frame S' moves to the right with constant velocity **u** relative to frame S. A light flash is emitted from the common origin at the instant the origins coincide. Each observer (O and O') sees a spherical wavefront spreading out from the origin in his own frame. These wavefronts are shown at the times t and t' in the respective frames.

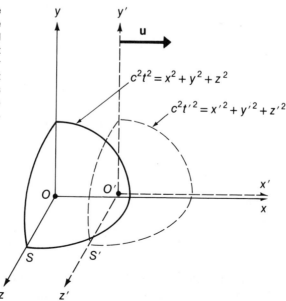

measured with a clock synchronized to the reference clock at the origin of the S' frame. (Recall that these two reference clocks were both set to zero at the instant the origins coincided.)

▶ *Simultaneity, Time Dilation, and Length Contraction According to the Lorentz-Einstein Equations.* Imagine a series of events in S that occur at various points along the x-axis. The synchronized local clocks in S reveal that these events are all simultaneous. The same events are observed in S' to occur at times $t' = \gamma(t - ux/c^2)$ according to synchronized local clocks in S'. Thus, in S', the events are *not* simultaneous. In fact, the events occur at earlier and earlier times in S' with increasing position x of the events in S. The Lorentz-Einstein transformation equations automatically include the relativistic nature of simultaneity that we expect.

Suppose that two events occur at the same space point in the S' frame, x', y', z', at proper times t'_1 and t'_2. Then, the proper time interval between the events is $\Delta t' = t'_2 - t'_1$. In the S frame, the same two events occur at *different space points*, namely, $x_1 = \gamma(x' + ut'_1)$ and $x_2 = \gamma(x' + ut'_2)$. Also, $t_1 = \gamma(t'_1 + ux'/c^2)$ and $t_2 = \gamma(t'_2 + ux'/c^2)$, so that $\Delta t = t_2 - t_1 = \gamma(t'_2 - t'_1) = \gamma \Delta t'$. This is just the time dilation result we found previously. (•Can you also show that a proper time interval Δt in S is observed to be dilated to $\gamma \Delta t$ in the S' frame?)

Next, consider a meter stick of length ℓ' that is at rest along the x'-axis of the S' frame with one end at x'_1 and the other end at x'_2. That is, $\ell' = x'_2 - x'_1$ is the proper length of the meter stick in the S' frame. These two space coordinates are related to those in S by

$$x'_1 = \gamma(x_1 - ut_1) \qquad \text{and} \qquad x'_2 = \gamma(x_2 - ut_2)$$

To determine the length of the (moving) meter stick in S, namely, $\ell = x_2 - x_1$, the observers in S must mark the locations of the ends of the meter stick at the same local time, that is, $t_1 = t_2$. Then, $x'_2 - x'_1 = \gamma(x_2 - x_1)$, or $\ell = \ell'/\gamma$. The length of the moving meter stick is therefore measured to be shorter than its proper length by the factor γ. This is the length contraction effect we obtained earlier.

There is a feature of length contraction contained in this analysis that was not mentioned in the earlier discussion. If the meter stick had been placed in frame S' parallel to either the y'- or z'-axes, no contraction would have been observed because $y'_2 - y'_1 = y_2 - y_1$ and $z'_2 - z'_1 = z_2 - z_1$. That is, *relativistic length contraction produces a foreshortening only in the direction of motion.* ◀

Velocity Additions. According to the Galilean transformation equations, velocities combine by simple algebraic addition. For example, suppose that a train (the frame S') moves past an observing position (the frame S) with a velocity u. If a passenger in the train throws a ball with a velocity v'_x relative to the train and along the direction of motion of the train, the S observer will measure the velocity to be $v_x = v'_x + u$.

The Galilean velocity addition rule (which can be used if either $v'_x \ll c$ or $u \ll c$) requires modification in relativity theory. The correct relationship (which we derive below) is

$$v_x = \frac{v'_x + u}{1 + \dfrac{v'_x u}{c^2}} \tag{21-9}$$

In the event that either v'_x or u is small compared with c, the term $v'_x u/c^2$ becomes negligible and the Galilean equation results.

An interesting feature of the relativistic velocity addition rule is that it makes clear the limitation on the velocity of any particle or system. Suppose that the relative velocity of the S and S' frames is $u = 0.9\,c$. Suppose also that a particle moves with a velocity $v'_x = 0.9\,c$ in the S' frame. According to the Galilean velocity addition rule, the S observer would measure the particle's velocity to be $v_x = v'_x + u = 0.9\,c + 0.9\,c = 1.8\,c$. However, the relativistic expression (Eq. 21-9) yields

$$v_x = \frac{v_x' + u}{1 + \dfrac{v_x' u}{c^2}} = \frac{0.9\,c + 0.9\,c}{1 + \dfrac{(0.9\,c)(0.9\,c)}{c^2}} = 0.995\,c$$

so that $v_x < c$. In fact, v_x will not exceed c even if $v_x' = c$ and $u = c$, for then,

$$v_x = \frac{c + c}{1 + \dfrac{(c)(c)}{c^2}} = c$$

That is, c "plus" $c = c$! We therefore reinforce our conclusion that the relative velocity of two objects (or two frames or an object and a frame) cannot exceed the velocity of light.

▶ *The Velocity Transformation Equations.* Consider a particle that moves with a constant velocity $\mathbf{v}'$ in frame S'. This velocity is equal to the quotient $\Delta\mathbf{r}'/\Delta t'$, which involves local time values from synchronized clocks at points defined by $\mathbf{r}'$ and $\mathbf{r}' + \Delta\mathbf{r}'$. (●An observer at an arbitrary location in S' records the value of the particle's arrival time that he *sees* on illuminated clocks at $\mathbf{r}'$ and $\mathbf{r}' + \Delta\mathbf{r}'$; is the difference in these times correctly $\Delta t'$?) The time and space increments have corresponding values in S, which we can obtain from Eqs. 21–8:

$$\Delta x = \gamma(\Delta x' + u\,\Delta t')$$

$$\Delta y = \Delta y'$$

$$\Delta z = \Delta z'$$

$$\Delta t = \gamma\left(\Delta t' + \frac{u\,\Delta x'}{c^2}\right)$$

Then,

$$\frac{\Delta x}{\Delta t} = \frac{\gamma(\Delta x' + u\,\Delta t')}{\gamma\left(\Delta t' + \dfrac{u\,\Delta x'}{c^2}\right)} = \frac{\dfrac{\Delta x'}{\Delta t'} + u}{1 + \dfrac{u}{c^2}\dfrac{\Delta x'}{\Delta t'}}$$

Taking the limit, $\Delta t \to 0$ and $\Delta t' \to 0$, we obtain

$$v_x = \frac{v_x' + u}{1 + \dfrac{uv_x'}{c^2}} \qquad\qquad \textbf{(21–10a)}$$

which is the result we asserted above. For the other components of $\mathbf{v}$, we have

$$\frac{\Delta y}{\Delta t} = \frac{\Delta y'}{\gamma\left(\Delta t' + \dfrac{u\,\Delta x'}{c^2}\right)} = \frac{\dfrac{\Delta y'}{\Delta t'}}{\gamma\left(1 + \dfrac{u}{c^2}\dfrac{\Delta x'}{\Delta t'}\right)}$$

so that

$$v_y = \frac{v_y'}{\gamma\left(1 + \dfrac{uv_x'}{c^2}\right)} \qquad\qquad \textbf{(21–10b)}$$

Similarly,

$$v_z = \frac{v_z'}{\gamma\left(1 + \dfrac{uv_x'}{c^2}\right)} \qquad\qquad \textbf{(21–10c)}$$

where γ involves the velocity u (not v).

For the inverse transformations, we find (replacing u with $-u$)

$$v'_x = \frac{v_x - u}{1 - \frac{uv_x}{c^2}} \tag{21-11a}$$

$$v'_y = \frac{v_y}{\gamma\left(1 - \frac{uv_x}{c^2}\right)} \tag{21-11b}$$

$$v'_z = \frac{v_z}{\gamma\left(1 - \frac{uv_x}{c^2}\right)} \tag{21-11c} \blacktriangleleft$$

Example 21-3

Two inertial frames, S and S', have a relative velocity $u = 0.50\,c$ along their common x-axes, as in the preceding discussions. Master clocks at both origins are set to zero when the origins coincide. Two simultaneous light flashes are observed at $(x_1, y_1, z_1, t_1) = (100$ m, 20 m, 20 m, 10^{-6} s$)$ and at $(x_2, y_2, z_2, t_2) = (200$ m, 30 m, 30 m, 10^{-6} s$)$. At what coordinates are these light flashes observed in S'?

Solution:

With $u = 0.50\,c$, we have $\gamma = 1.155$ (see Example 21-1). Then, using Eqs. 21-8a and 21-8b, we find, for the first flash,

$$
\begin{aligned}
x'_1 &= \gamma(x_1 - ut_1) \\
&= 1.155[(100 \text{ m}) - (0.50)(3.00 \times 10^8 \text{ m/s})(10^{-6} \text{ s})] \\
&= -57.75 \text{ m} \\
y'_1 &= y = 20 \text{ m} \\
z'_1 &= z = 20 \text{ m}
\end{aligned}
$$

$$
\begin{aligned}
t' &= \gamma\left(t_1 - \frac{ux_1}{c^2}\right) \\
&= 1.155[(10^{-6} \text{ s}) - (0.50)(100 \text{ m})/(3.00 \times 10^8 \text{ m/s})] \\
&= 0.962 \times 10^{-6} \text{ s} = 0.962 \text{ } \mu\text{s}
\end{aligned}
$$

Similarly, we find, for the second flash,

$$
\begin{aligned}
x'_2 &= +57.75 \text{ m} \\
y'_2 &= 30 \text{ m} \\
z'_2 &= 30 \text{ m} \\
t'_2 &= 0.770 \text{ } \mu\text{s}
\end{aligned}
$$

Thus, in frame S', the two events are judged not to be simultaneous, with the event at location (x'_2, y'_2, z'_2) observed to occur earlier by 0.192 μs. Notice that *both* time dilation and synchronization effects contribute to the lack of simultaneity in the S' frame.

Example 21-4

Analyze in detail the question of simultaneity in Einstein's train example (Fig. 21-3), first from the point of view of the ground observer O, then from the point of view of the moving observer O'.

Solution:

The observer O measures the distance $\overline{AB}$ along the tracks between the burn marks to be the proper length ℓ_0. This observer, having determined the lightning flashes to be simultaneous, claims that the length $\overline{A'B'}$ of the train is also ℓ_0; however, O measures only the contracted length. The proper length of the train, as measured by O', is $L_0 = \gamma \ell_0$.

The observer O argues, on the basis of synchronized clocks in frame S, that the light from A' reaches O' a time $t(A')$ after the lightning strike at A', where $\frac{1}{2}\ell_0 + u t(A') = c t(A')$, or $t(A') = \ell_0/2(c - u)$. Similarly, the time delay $t(B')$ in the flash that reaches O' from B' is $t(B') = \ell_0/2(c + u)$. Thus, according to observations made by O in frame S, the time interval between the arrivals of the flashes at O' is $\Delta t = t(A') - t(B') = \ell_0 u/c^2(1 - u^2/c^2) = \gamma^2 \beta \ell_0/c$. Now, O notes that the clocks in S' run more slowly than his own clocks, so he predicts that O' observes this time interval to be $\Delta t' = \Delta t/\gamma = \gamma \beta \ell_0/c = \beta L_0/c$.

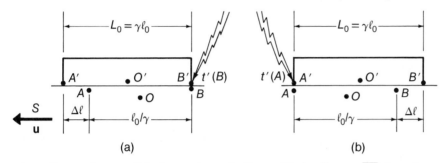

(a) (b)

According to observer O' in the moving train (frame S'), the distance $\overline{AB}$ along the tracks is the contracted length ℓ_0/γ, whereas the proper length of the train is L_0. According to synchronized clock determinations in frame S', the instant $t'(B)$ when B and B' were coincident corresponds to diagram (a); the instant $t'(A)$ when A and A' were coincident corresponds to diagram (b). During the time interval, $\Delta t' = t'(A) - t'(B)$, the ground (frame S) moves a distance $\Delta \ell = (\gamma - 1/\gamma)\ell_0$. Because the ground moves to the left with a velocity u, as seen by O', the time interval is $\Delta t' = \Delta \ell/u = (\ell_0/u)(\gamma - 1/\gamma) = (\gamma \ell_0/u)(1 - 1/\gamma^2) = \beta L_0/c$, which agrees with the conclusion reached by O.

The observer O', noting that the clocks in S run more slowly than his own clocks, judges that O measures the time interval between the lightning strikes to be $\Delta t_0 = \Delta t'/\gamma = \beta \ell_0/c$, according to the clocks in S. Because O has determined that the events at AA' and BB' were simultaneous, O' concludes that the S-frame clock at A must lag behind the clock at B by an amount $\beta \ell_0/c$.

Example 21–5

A rod with a proper length ℓ_0 lies at rest in the x'-y' plane of an inertial frame S' and makes an angle θ_0 with the x'-axis, as shown in the diagram. The frame S' moves with a velocity $\mathbf{u}$ relative to another frame S, with the x- and x'-axes parallel. Determine the length ℓ and the angle θ that the rod makes with the x-axis in the S frame.

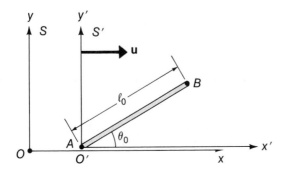

Solution:

In frame S', the coordinates of the ends of the rod, points A and B, are $x'_A = 0$, $y'_A = 0$, and $x'_B = \ell_0 \cos\theta_0$, $y'_B = \ell_0 \sin\theta_0$. The diagram shows the locations of points A and B at a time t as measured by synchronized clocks in S. Using Eqs. 21-8a at the instant t as determined in the S frame, we have

$$x'_A = 0 = \gamma(x_A - ut) \qquad\qquad y'_A = 0 = y_A$$
$$x'_B = \ell_0 \cos\theta_0 = \gamma(x_B - ut) \qquad y'_B = \ell_0 \sin\theta_0 = y_B$$

Hence, $x_B - x_A = (\ell_0/\gamma)\cos\theta_0$ and $y_B - y_A = \ell_0 \sin\theta_0$

Now,
$$\ell = \sqrt{(x_B - x_A)^2 + (y_B - y_A)^2}$$
$$= \ell_0 \left[\frac{\cos^2\theta_0}{\gamma^2} + \sin^2\theta_0\right]^{1/2} = \ell_0[(1 - \beta^2)\cos^2\theta_0 + \sin^2\theta_0]^{1/2}$$
$$= \ell_0[1 - \beta^2 \cos^2\theta_0]^{1/2}$$

The angle as measured in S is given by

$$\tan\theta = \frac{y_B - y_A}{x_B - x_A} = \gamma \tan\theta_0$$

Thus, the rod is both contracted in length and rotated in space. Note carefully that the length ℓ was determined from the two points (x_A, y_A) and (x_B, y_B), corresponding to the *same time t*. What would be "seen" by a single observer in frame S? (Refer to Problem 21-26.)

Example 21-6

The identical twins, Castor and Pollux, have identical synchronized clocks. Pollux boards a spaceship which accelerates briefly to a speed u and thereafter coasts at this speed to a star a distance L from the Earth. (Neglect any relative motion between the Earth and the star.) Upon reaching the star, Pollux rapidly reverses the velocity of his spaceship and heads back toward Earth, again coasting essentially the entire distance at the speed u. After a brief deceleration period at the end of his journey, Pollux lands on Earth and compares his clock reading and his biological age to those of Castor. They find that the elapsed clock time and the biological aging for Castor is $T_C = 2L/u$, whereas for Pollux these amount only to $T_P = 2L/\gamma u$. Why should Pollux be younger than Castor? From his spaceship, Pollux saw Castor (and the Earth) speed away from him, travel a certain distance into space at the speed u, then turn around and travel back at the same speed. Why, then, should not Castor be younger? This is the substance of the famous *twin paradox*.

Solution:

Castor's measurement of the proper time interval for the trip is $T_C = 2L/u$, but he knows that Pollux's clock runs more slowly and will record the lesser time $T_P = 2L/\gamma u$. Once his spaceship has reached coasting speed, Pollux determines that the distance from the Earth to the star has the contracted value L/γ. Thus, Pollux also finds that the duration of the trip is $T_P = 2L/\gamma u$.

The apparent paradox involves the explanation that Pollux gives for the larger time interval measured by Castor and for his greater aging. The crucial point is that Pollux is not in a single inertial frame throughout the trip, whereas Castor does remain in such a frame. During the first half of his trip, Pollux is in a frame S' that moves toward the star with a speed u relative to Castor's frame S. For the last half of the trip, Pollux is in a frame S'' that moves toward the Earth with the same relative speed u. During the initial acceleration, the turnaround at the star, and the deceleration preparatory to landing, Pollux recognizes the accelerations in terms of the forces he experiences. Castor, of course, experiences no such forces.

We have assumed that the star is at rest in Castor's Earth frame S. Therefore, we can imagine the existence of a clock on the star that has been synchronized with Castor's clock on Earth. When Pollux's spaceship, upon leaving Earth, has acquired its coasting speed u, Pollux claims that the star clock is not synchronized with the Earth clock but runs ahead of the Earth clock by an amount $uL/c^2 = \beta L/c$. (This is the same conclusion as that reached by O' in Example 21–4; see the last sentence in that example.)

Pollux arrives at the star, still moving with frame S', in a time $L/\gamma u$. According to Pollux, the clocks in S run slowly, so he concludes that the elapsed time measured in S is $L/\gamma^2 u$. Pollux's spaceship now decelerates rapidly and comes to rest with respect to the star. Pollux is now in frame S and he perceives that the Earth clock and the star clock are in synchronization. This can be the case, he reasons, only if, during his brief deceleration period, the Earth clock gained (and Castor aged) an additional time uL/c^2! This first half of the trip, as measured by Castor's Earth clock, required a time $uL/c^2 + L/\gamma^2 u = (L/u)(\beta^2 + 1 - \beta^2) = L/u$. By similar reasoning, the return trip in frame S'' also requires a time L/u. Thus, Pollux can account for the time interval for the entire trip as measured by Castor, namely, $T_C = 2L/u$.

(Notice that the accelerations that Pollux experiences in going from S to S' at the Earth and later from S to S'' at the star *do not* cause any loss of local synchronization because each change is imagined to take place essentially at a point.)

21-7 RELATIVISTIC MOMENTUM AND MASS

The first postulate of relativity theory is that all physical laws are the same in all inertial reference frames. Two observers set out to study one of these laws—momentum conservation—in a collision between two particles, A and B. One observer is in the inertial frame S and the other observer is in the frame S', which moves with the constant velocity **u** relative to S along their x-x' axes (see Fig. 21–7). The particles A and B are initially at rest in the frames S and S', respectively, in positions such that a grazing collision will take place when the frames pass one another. That is, after the

Fig. 21–7. An elastic collision between two identical particles, A and B. The particles are initially at rest in their respective frames, which have a relative velocity **u**. After the grazing collision, each particle has a z-component of velocity.

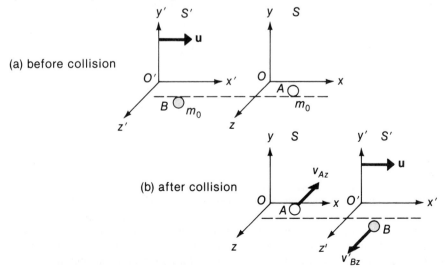

collision occurs, the particles have small *transverse* velocities (in the z-z′ direction), but the velocities in the x-x′ direction are essentially unchanged. Before the collision, neither particle has any momentum transverse to the direction of motion; all inertial observers agree on this point. Because momentum must be conserved, the total transverse momentum of A and B must be zero after the collision; again, all inertial observers will agree on this point.

Before performing this experiment, the two observers compare the masses of A and B when the particles are at rest and find them to be equal: $m_A = m_B \equiv m_0$. After the collision takes place, each observer measures the velocity of his particle (v_{Az} and v'_{Bz}) and reports a value for the momentum: the S observer finds $p_{Az} = m_0 v_{Az}$, and the S′ observer finds $p'_{Bz} = m_0 v'_{Bz}$. The observers are happy to discover that their values agree, $v_{Az} = -v'_{Bz}$ and $p_{Az} = -p'_{Bz}$, thus confirming momentum conservation for the transverse components. In a repeat of the experiment, S′ measures the transverse velocity of particle A (the particle originally at rest in S). He finds $v'_{Az} = v_{Az}/\gamma$, in accordance with Eq. 21–11c with $v_{Ax} = 0$. This velocity, which is smaller than v'_{Bz}, means that momentum conservation does not hold *unless* the mass m_A of particle A is greater than the value m_0 reported by the S observer. In fact, momentum conservation is restored (as demanded by the first postulate) if m_A varies with the velocity u so that the S′ observer finds $m'_A = m_0\gamma$. Then, the momentum of particle A as measured by the S′ observer is $p'_{Az} = m'_A v'_{Az} = (m_0\gamma)(v_{Az}/\gamma) = m_0 v_{Az} = -m_0 v'_{Bz} = -p'_{Bz}$. While the S′ observer is drawing this conclusion about particle A, the S observer is discovering the same result for particle B. Both observers agree that *the inertial mass of a particle in motion with respect to an observer is greater than the mass of an identical particle at rest with respect to the observer.* The mass of a particle at rest is called the *rest mass* or the *proper mass* and is given the symbol m_0. Then, the mass when moving with a velocity u is

$$m(u) = m_0\gamma = \frac{m_0}{\sqrt{1 - u^2/c^2}} \qquad (21\text{–}12)$$

This prediction of relativity theory has been verified to high precision in numerous experiments with various values of $\beta = u/c$. Some values of $m/m_0 = \gamma$ are given in Table 21–1, and Fig. 21–8 shows the results graphically. As $u \to c$, the mass becomes indefinitely large, but u can never equal c, so that m remains finite.

TABLE 21–1.   Relativistic Mass Increase with Velocity:
$m/m_0 = \gamma = (1 - \beta^2)^{-1/2}$

$\beta = u/c$	m/m_0	$\beta = u/c$	m/m_0
0.01	1.00005	0.90	2.294
0.10	1.005	0.95	3.203
0.50	1.155	0.99	7.089
0.80	1.667	0.999	22.37

Example 21–7

What is the velocity of an electron that has been accelerated in some device (for example, a *synchrotron*) to the point that its mass is 10 times its rest mass?

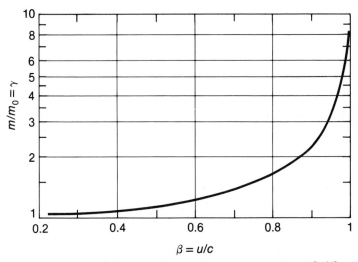

Fig. 21–8. The variation of inertial mass with velocity: $m/m_0 = \gamma = (1 - \beta^2)^{-1/2}$, with $\beta = u/c$.

Solution:

Using Eq. 21-12, we can write

$$10m_0 = m_0\gamma = \frac{m_0}{\sqrt{1 - \beta^2}}$$

Then,

$$\sqrt{1 - \beta^2} = \tfrac{1}{10}$$

so that

$$1 - \beta^2 = 0.01$$
$$\beta^2 = 0.99$$
$$\beta = \sqrt{0.99} = 0.995$$

Thus,

$$v = \beta c = 0.995 \times (3.00 \times 10^8 \text{ m/s})$$
$$= 2.985 \times 10^8 \text{ m/s}$$

Example 21–8

What is the fractional increase in mass for a jet aircraft that travels with a speed of 900 km/h (250 m/s)?

Solution:

Using Eq. 21-12, we can write

$$\Delta m = m - m_0 = m_0(\gamma - 1) = m_0[(1 - \beta^2)^{-1/2} - 1]$$

In this case, we have $v \ll c$ and $\beta \ll 1$, so we can expand the expression for γ by using Eq. A–14 in Appendix A:

$$\Delta m = m_0(1 + \tfrac{1}{2}\beta^2 + \tfrac{3}{8}\beta^4 + \cdots - 1) \cong \tfrac{1}{2}m_0\beta^2$$

Thus, the fractional mass increase is

$$\frac{\Delta m}{m_0} \cong \tfrac{1}{2}\beta^2 = \frac{1}{2}\frac{v^2}{c^2} = \frac{1}{2}\frac{(250 \text{ m/s})^2}{(3 \times 10^8 \text{ m/s})^2}$$

$$= 3.5 \times 10^{-13}$$

For an aircraft with a mass of 10^5 kg, Δm is approximately 35 μg. Thus, the mass is increased by an undetectably small amount.

21-8 RELATIVISTIC ENERGY

We can use the work-energy theorem of Newtonian mechanics (Section 7–3) to formulate a relativistic definition of energy. Consider a particle with a proper mass m_0 that is initially at rest in an inertial reference frame S. A force $\mathbf{F}$ in the x-direction acts on the particle and accelerates it through a distance ℓ. At the end of this accelerated motion, the particle has a velocity v. We can write*

$$F = \frac{dp}{dt} = \frac{dp}{dv}\frac{dv}{dx}\frac{dx}{dt} = v\frac{dp}{dv}\frac{dv}{dx}$$

so that

$$W = \int_0^\ell F\,dx = \int_0^v v\frac{dp}{dv}\,dv$$

From the results in the preceding section, we can write

$$\mathbf{p} = m\mathbf{v} = m_0\gamma\mathbf{v} \qquad (21\text{--}13)$$

The reference frame S' in which the particle has the proper mass m_0 is the frame in which the particle is at rest. After the period of acceleration, this frame moves with a velocity $u = v$ with respect to the frame S. Thus, we have

$$\frac{dp}{dv} = m_0\frac{d}{dv}\left[\frac{v}{(1 - v^2/c^2)^{1/2}}\right] = \frac{m_0}{(1 - v^2/c^2)^{3/2}}$$

Hence,

$$W = \int_0^v \frac{m_0 v\,dv}{(1 - v^2/c^2)^{3/2}} = \frac{m_0 c^2}{(1 - v^2/c^2)^{1/2}} - m_0 c^2$$

or

$$W = mc^2 - m_0 c^2 \qquad (21\text{--}14)$$

As in Newtonian dynamics, we associate the work W done on a particle with the increase in its kinetic energy K. The quantity $m_0 c^2$ in Eq. 21–14 does not depend on the velocity v and is called the *rest-mass energy* (or *rest energy*) of the particle. The

---

*If the equation $\mathbf{F} = d\mathbf{p}/dt$ is used in the most general sense, with $\mathbf{F}$ allowed to change in direction, the situation becomes quite complicated. In fact, in this case, the force $\mathbf{F}$ and the acceleration $\mathbf{a}$ are no longer in the same direction! Thus, if m is to be the quantity that connects the force $\mathbf{F}$ with the acceleration $\mathbf{a}$, it is clear that m cannot be a simple scalar quantity. Indeed, in such situations, the concept of inertial mass is of limited value.

total energy E of the particle is defined to be $E = mc^2 = m_0\gamma c^2$. Thus, Eq. 21-14 becomes

$$E = mc^2 = K + m_0c^2 \qquad \text{(21-15)}$$

When the velocity v is small compared with c, $v \ll c$, the relativistic factor γ can be expanded according to Eq. A-14 in Appendix A, with the result

$$K = m_0\gamma c^2 - m_0c^2 = m_0c^2[(1 - v^2/c^2)^{-1/2} - 1]$$
$$= m_0c^2\left(1 + \frac{1}{2}\frac{v^2}{c^2} + \frac{3}{8}\frac{v^4}{c^4} + \cdots - 1\right)$$
$$= \tfrac{1}{2}m_0v^2\left(1 + \frac{3}{4}\frac{v^2}{c^2} + \cdots\right)$$

so that
$$K \cong \tfrac{1}{2}m_0v^2 \qquad (v \ll c)$$

Thus, the relativistic expression for the kinetic energy reduces to the Newtonian result for sufficiently small velocities.

We can obtain another expression for the total energy E of a particle as follows. From Eq. 21-13 we can write.

$$p^2 = \mathbf{p}\cdot\mathbf{p} = m_0^2\gamma^2v^2$$

Also, $E^2 = m^2c^2 = m_0^2\gamma^2c^4$, so that

$$E^2 - p^2c^2 = m_0^2\gamma^2c^4 - m_0^2\gamma^2v^2c^2$$
$$= m_0^2c^4\gamma^2(1 - v^2/c^2) = m_0^2c^4$$

or

$$E^2 = p^2c^2 + m_0^2c^4 \qquad \text{(21-16)}$$

There are elementary physical entities—such as *photons* and *neutrinos*—that have zero proper mass. For such "particles," we have $|\mathbf{p}| = E/c$.

The Conservation of Mass-Energy. Let us rewrite Eq. 21-14 as

$$m - m_0 = \Delta m = \frac{W}{c^2} \qquad \text{(21-17)}$$

In this form, the equation suggests that an amount of work W done on a particle or system results in an increase in the mass of the system equal to $\Delta m = W/c^2$. There is no restriction on the source of the energy W that is added to the system, so we conclude that inertial mass is attributed to all forms of energy—radiant energy, potential energy, thermal energy, and so forth.

The idea of the interchangeability of mass and energy is one of the most important aspects of relativity theory. An immediate consequence of this idea is that two of the classical conservation laws—the conservation of mass and the conservation of energy—are merged into a single law. According to relativity theory, the quantity that is always conserved in any process is the *total energy*, $E = mc^2$, which includes the rest-mass energy, m_0c^2. This means that any change in the internal energy of an isolated system (for example, a change in the thermal energy or the potential energy) is accompanied by a corresponding change in the rest-mass energy of the particles that make up the system. (See Example 21–10 for a case in which kinetic energy is converted into rest-mass energy.)

The fact that c is a very large velocity means that an enormous amount of energy will result from the conversion of even a small amount of mass. This is exactly the case, for example, in the nuclear processes of fission and fusion. To give an idea of the scale involved, the conversion of $1 \text{ ng} = 10^{-9} \text{ g} = 10^{-12} \text{ kg}$ of matter would provide sufficient energy to raise a 10-ton truck to a height of 1 m.

Example 21–9

When a nuclear fusion reactor finally becomes operational, it will probably produce energy by a reaction in which nuclei of deuterium and tritium are fused together, yielding a neutron and a helium nucleus. The masses involved are

$$\begin{aligned}
\text{Deuterium:} \quad & m(D) = 2.01410 \text{ u} \\
\text{Tritium:} \quad & m(T) = 3.01605 \text{ u} \\
\text{Neutron:} \quad & m(n) = 1.00867 \text{ u} \\
\text{Helium:} \quad & m(He) = 4.00260 \text{ u}
\end{aligned}$$

where the mass unit is $1 \text{ u} = 1.6605 \times 10^{-27} \text{ kg}$.
(a) What amount of energy is released in each D + T fusion reaction?
(b) What amount of energy is released per kilogram of tritium consumed?
(c) The annual consumption of electric energy in the United States is approximately 2.5×10^{12} kWh. (The *total* U.S. energy consumption is about 10 times this value.) How many kilograms of tritium must be provided per year to yield this amount of electric energy if the efficiency of converting fusion energy to electric energy is 35 percent?

Solution:

(a) The initial reacting mass is $m(D) + m(T) = 5.03015$ u, and the final mass is $m(n) + m(He) = 5.01127$ u. Thus, the energy released is

$$E = \Delta mc^2 = (5.03015 \text{ u} - 5.01127 \text{ u})(1.66 \times 10^{-27} \text{ kg/u}) \cdot (3.00 \times 10^8 \text{ m/s})^2$$
$$= 2.82 \times 10^{-12} \text{ J per reaction}$$

The energy that is released in the reaction appears in the form of the kinetic energy of the resultant particles, the neutron and the helium nucleus.

(b) The energy released per kilogram of tritium consumed is

$$E = (2.82 \times 10^{-12} \text{ J}) \cdot \frac{1}{(3.016 \text{ u})(1.66 \times 10^{-27} \text{ kg/u})}$$
$$= 5.63 \times 10^{14} \text{ J/kg}$$

(c) The amount of tritium consumed annually would be

$$M = \frac{[2.5 \times 10^{15}(\text{J/s})\text{h}](3600 \text{ s/h})}{(0.35)(5.63 \times 10^{14} \text{ J/kg})} \text{ per year}$$

$$= 4.57 \times 10^4 \text{ kg/y}$$

or slightly more than 45 metric tons per year. (Note that the amount of mass (Δm) actually converted to energy is only 286 kg/y!) To generate the same amount of electric energy by using coal would require more than 900 *million* tons per year.

Example 21–10

A particle with a proper mass m_0 that moves with an initial velocity v' in some inertial reference frame collides with an identical particle that is at rest in the same frame. The collision is completely inelastic, and the two particles fuse together to yield a composite particle with a proper mass M_0.
(a) Determine the mass M_0.
(b) Determine the amount of energy that is converted into mass in the fusion process.

Solution:

In solving problems of this type, it is usually easier to proceed by working in a reference frame in which the total momentum of the system is zero. Such a frame defines the *center-of-momentum* system (which, for particle collision processes, is equivalent to the center-of-mass system). In this system, the two particles approach one another with equal speeds v, and after collision, the composite particle M_0 is at rest.

(a) Before collision, each particle has a mass $m = m_0\gamma$, so the total energy of the system is

before:

after:

M_0

$$E = E_1 + E_2 = m_0\gamma c^2 + m_0\gamma c^2$$

$$= 2m_0\gamma c^2$$

After collision, the composite particle has a proper mass M_0. Because this particle has no momentum, Eq. 21–16 reduces to

$$E = M_0 c^2$$

The total energy E is conserved in the collision process. Therefore, comparing the two expressions for E, we conclude that

$$M_0 = 2m_0\gamma$$

Thus, the final rest mass M_0 of the system is *greater* than the initial rest mass $2m_0$. The reason for this increase in mass is that the initial kinetic energy of the particles has been converted into mass.

(b) The change in proper mass of the system is

$$\Delta m = M_0 - 2m_0 = 2m_0\gamma - 2m_0 = 2m_0(\gamma - 1)$$

Thus, the amount of energy converted into mass is

$$\Delta m c^2 = 2m_0 c^2(\gamma - 1)$$

To obtain a Newtonian expression for $\Delta m c^2$, we assume that $v \ll c$, so we can expand γ to obtain

$$\Delta mc^2 = 2m_0c^2 \left(1 + \frac{1}{2}\frac{v^2}{c^2} + \cdots - 1\right)$$

$$\cong 2m_0c^2 \cdot \frac{1}{2}\frac{v^2}{c^2}$$

$$= 2 \cdot (\tfrac{1}{2}m_0v^2)$$

$$= 2K$$

where $K = \frac{1}{2}m_0v^2$ is the Newtonian kinetic energy of each particle before collision. Because no energy escapes from the system, we conclude that the initial kinetic energy is converted into heat and that this heat contributes to the mass M. Thus, the entire kinetic energy of the system before collision is converted into mass that resides in the composite particle, just as we guessed in part (a).

SPECIAL TOPIC

21-9   THE DOPPLER EFFECT FOR LIGHT

The Doppler effect for sound that we studied in Section 20-5 has a counterpart for the case of light. Because of the unique properties of light, we must now analyze the situation using relativistic considerations. The geometry we use is shown in Fig. 21-9. As S' moves past S, the clocks in both frames are set to zero at the instant the two origins coincide. Suppose that a light source L' is located at the origin of the S' frame and is at rest in that frame. The

Fig. 21–9. Geometry for discussing the Doppler effect for light. The observers are located at P,P' and the light source is perceived by the observers to be L and L'.

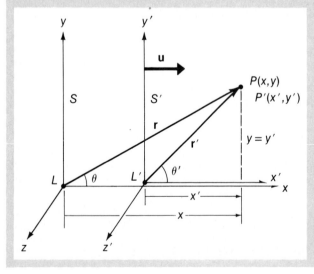

source emits light with the single frequency ν', as measured by the S' observer; this is the *proper frequency* of the source. In the S frame, the source moves with a velocity u.

Consider a labeled wave crest that leaves the coincident origins of the two frames at the time $t = t' = 0$. According to the S' observer, this wave crest travels from the origin L' to the point $P'(x', y')$ along the ray $\mathbf{r}'$ which makes an angle θ' with the x'-axis. The S' observer at $P'(x', y')$ detects this wave crest at the time t'. According to the S observer, this wave crest travels from the origin L to the point $P(x, y)$ along the ray $\mathbf{r}$ which makes an angle θ with the x-axis. The S observer at $P(x, y)$ detects this wave crest at the time t. The points $P'(x', y')$ and $P(x, y)$ coincide at the instant measured as t' in the S' frame and measured as t in the S frame. What is the wave frequency ν that is measured by the S observer?

An expanding spherical wave in the S' frame is described by (see Eq. 20–13a)

$$\psi' = \frac{A'}{r'}\cos\left(k'r' - \omega't'\right)$$

$$= \frac{A'}{r'}\cos 2\pi\nu'\left(\frac{r'}{c} - t'\right)$$

The corresponding expression for the S frame is

$$\psi = \frac{A}{r}\cos 2\pi\nu\left(\frac{r}{c} - t\right)$$

Now, the labeled wave crest left the coincident origins of the two frames at the instant of zero time in each frame.

This same crest was detected at the coincident points, $P'(x', y')$ and $P(x, y)$, at times measured to be t' by the S' observer and t by the S observer. Consequently, the phases of the wave functions, ψ' and ψ, must be equal. That is,

$$\nu'\left(\frac{r'}{c} - t'\right) = \nu\left(\frac{r}{c} - t\right) \qquad \textbf{(21-18)}$$

From the geometry in Fig. 21-9, we see that

$$r' = x' \cos \theta' + y' \sin \theta'$$
$$r = x \cos \theta + y \sin \theta$$

Substituting these expressions into Eq. 21-18 along with the Lorentz-Einstein transformations for x', y', and t' (Eqs. 21-8), we obtain

$$\nu'\gamma\left(t - \frac{ux}{c^2}\right) - \frac{\nu'}{c}\gamma(x - ut)\cos\theta' - \frac{\nu'}{c}y\sin\theta'$$

$$= \nu t - \frac{\nu}{c}x\cos\theta - \frac{\nu}{c}y\sin\theta \qquad \textbf{(21-19)}$$

If Eq. 21-19 is to be true for any possible observation point $P(x, y)$ and the corresponding time t, the coefficients of x, y, and t on the two sides of the equation must separately be equal. Equating the coefficients of t yields

$$\nu = \nu'\gamma(1 + \beta \cos \theta') \qquad \textbf{(21-20)}$$

This equation expresses the Doppler frequency shift for light.

By equating the coefficients of x and y, we can obtain expressions for $\cos \theta$ and $\sin \theta$ in terms of θ', or we can obtain the inverse relationships (all of which are equivalent); for example,

$$\left.\begin{array}{l} \sin \theta' = \dfrac{\sin \theta}{\gamma(1 - \beta \cos \theta)} \\[3mm] \cos \theta' = \dfrac{\cos \theta - \beta}{1 - \beta \cos \theta} \end{array}\right\} \qquad \textbf{(21-21)}$$

The fact that θ and θ' are not equal is called the *aberration of light*.

If we substitute the expression for $\cos \theta'$ into Eq. 21-20, we obtain a more useful expression for the Doppler frequency, namely,

$$\nu = \frac{\nu'}{\gamma(1 - \beta \cos \theta)} \qquad \textbf{(21-22)}$$

This equation is easier to use than Eq. 21-20 because it

expresses the frequency ν in terms of θ, the angle directly measured by the S observer.

Remember, the S observer is located at $P(x, y)$. When $0 \leqslant \theta \leqslant \pi/2$, the S observer perceives the source *approaching*. Then, Eq. 21-22 shows that $\nu > \nu'$ and the frequency measured by the S observer is higher than the proper frequency ν'. The maximum effect occurs when the source approaches directly; that is, when $\theta = 0$. Then, Eq. 21-22 gives

$$\nu = \nu'\sqrt{\frac{1 + \beta}{1 - \beta}} \qquad (\theta = 0) \qquad \textbf{(21-23a)}$$

Also, when $\pi/2 \leqslant \theta \leqslant \pi$, we find $\nu < \nu'$. In fact, for $\theta = \pi$, the source recedes directly from the S observer and the frequency is minimum:

$$\nu = \nu'\sqrt{\frac{1 - \beta}{1 + \beta}} \qquad (\theta = \pi) \qquad \textbf{(21-23b)}$$

Equations 21-23 give the frequencies for the *longitudinal* Doppler effect. (•Can you show that these expressions are the same as those for the case of sound in the limit $\beta \ll 1$?) There is, in addition, for light, an effect that does not occur for sound. This is the *transverse* Doppler shift, which results when $\theta = \pi/2$. Then, Eq. 21-22 gives

$$\nu = \frac{\nu'}{\gamma} \qquad (\theta = \pi/2) \qquad \textbf{(21-24)}$$

This result can be viewed as a manifestation of time dilation: moving clocks (such as the moving light source) run more slowly than stationary clocks, so the frequency measured by the S observer is less than the proper frequency ν'. (See Example 21-1.)

By expanding the radicals in the various formulas, it is easy to show that the longitudinal Doppler effect varies as β, whereas the transverse effect varies as β^2. This means that the transverse effect is smaller and more difficult to detect than is the longitudinal effect. Nevertheless, experiments have confirmed all aspects of the relativistic Doppler effect.

In Section 20-5, we discussed the Doppler effect for sound. We have now obtained rather different expressions for the equivalent effect for light. In the case of light, the relativity postulates place the observer and the light source on entirely equivalent footing in the sense that only their relative motion is important. In the case of sound, this symmetry is broken by the presence of a material medium (air) through which the wave propagates. Either the source or the observer (or both) may be in motion with respect to the medium, and each motion separately influences the total Doppler effect. The influence of this basic difference between the two cases also extends to the aberration effect.

Example 21-11

Astronomical studies of the light from distant galaxies have shown that these galaxies are all receding from us and, therefore, that the Universe is expanding. The recessional velocities have been determined from measurements of the Doppler shifts of distinctive features (*spectral lines*) in the light from these galaxies. It has been found that there is an approximately linear relationship between the recessional velocity u of a particular galaxy and the distance d to the galaxy: $u \cong Hd$, where $H \cong 2.3 \times 10^{-5}$ km/s $\cdot$ L.Y. is the *Hubble constant*, named for the American astronomer, Edwin Hubble, who discovered this important empirical relationship. (The quantity 1 light year = 1 L.Y. = 9.46×10^{15} m is the distance light will travel in vacuum in one year.)

In the light from a distant galaxy in the constellation *Boötes*, there is observed a prominent spectral feature at a wavelength of 4470 Å (due to light from calcium atoms). When the light from a stationary laboratory source is analyzed, this same feature has a wavelength of 3940 Å. What is the distance to the galaxy in Boötes?

Solution:

The recessional velocity can be calculated by using

Eq. 21-23b for the case $\theta = \pi$. We have

$$\frac{v}{v'} = \sqrt{\frac{1 - \beta}{1 + \beta}}$$

Now, the wavelength is $\lambda = c/v$, so that

$$\frac{\lambda'}{\lambda} = \sqrt{\frac{1 - \beta}{1 + \beta}} = \frac{3940 \text{ Å}}{4470 \text{ Å}} = 0.881$$

Solving for β, we find $\beta = 0.126$, so that

$$u = \beta c = (0.126)(3.00 \times 10^8 \text{ m/s}) = 3.78 \times 10^4 \text{ km/s}$$

Using the Hubble formula, we obtain

$$d = \frac{u}{H} = \frac{3.78 \times 10^4 \text{ km/s}}{2.3 \times 10^{-5} \text{ km/s} \cdot \text{L.Y.}} = 1.6 \times 10^9 \text{ L.Y.}$$

Galaxies are known that are receding from us far more rapidly than is this galaxy in Boötes. The galaxy known as 3C 123 has $\beta \cong 0.46$. One *quasi-stellar object* (or *quasar*) has a Doppler shift that corresponds to a recessional velocity that is 91 percent of the velocity of light!

PROBLEMS

Section 21-3

21-1   Apply the ether hypothesis to the example of the lightning strikes on the railway car. If the car moves through the stationary ether with a velocity u, show that O' (as well as O) will argue that the lightning strikes were simultaneous. Why, then, is it necessary to invoke the postulates of relativity? Is there anything curious about a "stationary" ether?

Sections 21-4, 21-5, 21-6

21-2   Suppose that the train in the Einstein example, Fig. 21-3 and Example 21-1, is a Japanese "bullet" train with a proper length $L_0 = 200$ m and with a maximum speed of 275 km/h. Let the ground observer O judge the lightning flashes at AA' amd BB' to be simultaneous. (a) Calculate the time interval $\Delta t'$ by which the moving observer O' judges the events to be nonsimultaneous. (b) Calculate the time interval

Δt_0 by which O' judges clocks at A and B to be out of synchronization. (c) From these results, can you see why these and similar effects escape our everyday observation?

21-3   Repeat Problem 21-2 with the speed of the train increased to $u = 0.6c$.

21-4   Two events are observed in an inertial frame S at
•   coordinates $(x_1, 0, 0, t_1)$ and $(x_2, 0, 0, t_2)$, with $x_1 > x_2$ and $t_1 > t_2$. A second inertial frame S' moves with a velocity $\mathbf{u}$ relative to S with the x-axes of the two systems coinciding. Master clocks at the origins of both systems are set to zero when the two origins coincide. (a) Show that the observers in S' will judge the two events to occur at the same point in space if $u = (x_1 - x_2)/(t_1 - t_2)$. (b) If u has the value given in part (a), show that the time interval $t'_1 - t'_2$ measured in S' is

$$t'_1 - t'_2 = \sqrt{(t - t_2)^2 - (x_1 - x_2)^2/c^2}$$

(c) Because we must have $u < c$, show from part (a) that no frame S' can be found in which the two events are spatially coincident if $(x_1 - x_2) > c(t_1 - t_2)$. (•Does this also agree with the result found in part (b)?) (d) If $(x_1 - x_2) > c(t_1 - t_2)$, a frame S' can be found in which the two events are simultaneous. Show that such a frame must have velocity u relative to the frame S given by $u = c^2(t_1 - t_2)/(x_1 - x_2)$.

21-5 A rod that has a proper length of 1 m makes an angle of 30° with the x-axis of an inertial frame S' in which it is at rest. S' moves with the constant velocity **u** with respect to a parallel frame S along the x-x' axes. Observers in S determine that the rod makes an angle of 60° with the x-axis. (a) What is the relative velocity u of the two frames? (b) What is the length of the rod as measured in the frame S?

21-6 An astronaut leaves the Earth in a spaceship that moves with $\beta = 24/25$ and travels to a star that is 10 L.Y. away. (1 L.Y. = 1 light year = 9.46 × 10^{15} m.) The astronaut spends one year on a planet of the star and then returns to Earth, again with $\beta = 24/25$. Assume that the star, the planet, and the Earth are all essentially at rest with respect to one another. (a) According to his own clock, how long was the astronaut away from the Earth? (b) According to an Earth clock, what was the duration of the astronaut's trip? (c) After returning to Earth, how much less has the astronaut aged than his backup man who remained on Earth?

21-7 A standard clock whose "tick" duration is 1 μs is carried by an Earth satellite that is in a circular orbit with a radius of two Earth radii. What time difference between an Earth clock and the satellite clock is accumulated each revolution? (Assume that the satellite undergoes straight-line motion. Is this a reasonable assumption? Explain.)

21-8 Show that the Lorentz-Einstein transformation leaves the quantity $x^2 - c^2t^2$ invariant, that is, $x^2 - c^2t^2 = x'^2 - c^2t'^2$. [Hint: Use Eqs. 21-8a to write x'^2 and t'^2; then, calculate $x'^2 - c^2t'^2$.]

21-9 Two particles are emitted in a nuclear reaction and travel in opposite directions, each with a speed $\frac{3}{5}c$ in the laboratory frame. What is the relative velocity of the two particles as determined by an observer in the rest frame of one of the particles? (A laboratory observer says that the relative velocity of the two particles is $\frac{6}{5}c$. Does this violate any principle of relativity?)

Sections 21-7, 21-8

21-10 What is the velocity of a particle that has a momen-

tum equal to m_0c? Express the kinetic energy K and the total energy E of this particle in units of m_0c^2.

21-11 The kinetic energy of a particle is to be expressed by the Newtonian formula, $\frac{1}{2}m_0v^2$, with an error of no more than 1 percent. What is the maximum velocity that the particle can have?

21-12 Refer to the statement of the situation in Example 21-10. Show that, in the reference frame in which one of the particles is at rest, the velocity of the composite particle M_0 is $V = \gamma v'/(\gamma + 1)$, where $\gamma = (1 - v'^2/c^2)^{-1/2}$.

21-13 The average amount of solar radiation received at the top of the Earth's atmosphere on a surface that is perpendicular to the Earth-Sun line is 1.37 kW/m^2. (a) At what total rate (in W) is energy being radiated by the Sun? (b) At what rate (in kg/s) does the mass of the Sun decrease? What is the fractional rate of decrease per year? (c) How long could the Sun maintain its present energy output before decreasing in mass by 10 percent?

21-14 In a typical nuclear fission event, a neutron is absorbed by a uranium nucleus which then undergoes fission, producing nuclei of krypton and barium together with two additional neutrons; that is,

$$U + n \rightarrow Kr + Ba + 2n$$

The relevant masses are

$m(U) = 235.04392$ u		$m(Kr) = 91.92569$ u
$m(n) = 1.00867$ u		$m(Ba) = 141.91647$ u

where 1 u = 1.6605 × 10^{-27} kg. (a) What amount of energy is released per fission event of this type? (b) How much energy is released by the fission of 1 kg of uranium? (c) The annual usage of electric energy in the United States amounts to approximately 2.5 × 10^{12} kWh. How much uranium would supply this energy? (Assume that the conversion efficiency from fission energy to electric energy is 32 percent, a typical value for present-day fission reactors.)

21-15 The Stanford Linear Accelerator (SLAC) is used to accelerate electrons to a speed close to the speed of light by doing an amount of work equal to 8.0 × 10^{-10} J on each electron. The proper mass of an electron is 9.11 × 10^{-31} kg. (a) What is the ratio of the electron kinetic energy to the electron rest-mass energy? (The electrons that emerge from SLAC are said to be *extreme relativistic* electrons.) (b) Determine the value of β for the SLAC electrons. (c) If a SLAC electron raced a burst of light to the Moon, by what distance would the light win? (d) What is the ratio of the mass of a SLAC electron to the electron proper mass?

21-16 In contrast to the previous problem, consider the

case of protons (proper mass, 1.67×10^{-27} kg) accelerated in the machine at the National Accelerator Laboratory (NAL) by imparting to each proton an amount of energy equal to 8.0×10^{-10} J. Repeat the calculations called for in Problem 21-15.

SPECIAL TOPIC

Section 21-9

21-17 Derive Eqs. 21-21. Proceed as follows. First, equate the coefficients of x and y in Eq. 21-19 and obtain expressions for $\cos \theta$ and $\sin \theta$ in terms of θ'. Then, combine these expressions and solve for $\sin \theta'$ and $\cos \theta'$ in terms of θ.

21-18 Obtain Eq. 21-22 by using Eqs. 21-20 and 21-21.

21-19 A prominent feature in the spectrum of atomic hydro-
•   gen is the so-called H_α line, which has a wavelength of 656.3 nm = 6563 Å. (a) What is the expected shift in the wavelength of this line in the light from a star that is approaching at a velocity of 50 km/s? (b) This particular line is observed in the light from the Sun. The difference between the wavelengths of the line measured for light from the receding and approaching limbs (the edges) of the solar disc is 0.085 Å. Assume that this difference is due entirely to the Doppler effect and determine the rotation period of the Sun.

21-20 A highway patrol radar unit operates with a wave-length of 10 cm. The radar beam is directed toward a car that is approaching with a speed of 100 km/h. The reflected signal from the car is mixed with a portion of the transmitted wave to produce beats that are a measurement of the speed of the car. What is the beat frequency in this case?

21-21 The star α *Centauri* is at a distance of 4.40 L.Y. and has a proper motion (motion perpendicular to the line of sight) of 3.68 arc second per year (3.68″ y⁻¹). The calcium spectral line that has a proper wave-length of 3968 Å is observed to be shifted by −0.29 Å in the light from α *Centauri*. Calculate the total velocity (magnitude and direction) of the star relative to the solar system.

Problem 21-21

21-22 An Earth satellite is in a circular orbit that has
•   a radius of two Earth radii and passes directly over a tracking station Q. The satellite carries a transmitter whose proper frequency is 2.0×10^7 Hz. What is the frequency shift of the signal measured at Q when the satellite is receding from the station at an angle of 45° above the horizon, as shown in the diagram?

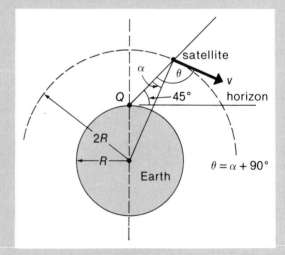

Problem 21-22

Additional Problems

21-23 In his 1905 paper on the special theory of relativity, Einstein states: "We conclude that a balance-clock at the equator must go more slowly, by a very small

amount, than a precisely similar clock situated at one of the poles under otherwise identical conditions." Assume that the equatorial clock undergoes straight-line motion and show that after a century the polar clock will lead the equatorial clock by approximately 0.0025 s.

21-24 Show that Eq. 21-5 for the length contraction effect can also be obtained if the *S* observer measures the time required for the meter stick to pass a fixed point in his system and then multiplies this time by *u*.

21-25 In an inertial frame *S'*, a light beam travels in the
• *x'*-direction with a speed *c/n* through a transparent medium. (The quantity *n* is called the *refractive index* of the medium; $n > 1$, typically, 1.3 to 1.6.) The frame *S'* moves with a constant velocity **u** with respect to a parallel frame *S* along the *x-x'* axes. Show that the speed of the light in the medium as measured in *S* is $v = c/n + u(1 - n^{-2})$, neglecting terms of order u^2/c^2 and higher. The quantity $(1 - n^{-2})$ is called the "drag coefficient" of the medium.

21-26 Consider an illuminated cube at rest in an inertial
•• frame *S'* with the sides of the cube parallel to the *x'y'z'*-axes. The frame *S'* moves with a constant velocity **u** with respect to a parallel frame *S* along the *x-x'* axes. An observer in *S*, a great distance from the cube, photographs the cube when the line of sight to the cube is parallel to the *y-y'* axes, as indicated in the diagram. Demonstrate that the photograph will show a cube with the same dimensions as measured in *S'* but apparently rotated through an angle $\theta = \cos^{-1}(1/\gamma)$. [*Hint:* A photograph of an object records light that arrives at the film simultaneously from all visible points on the object. Thus,

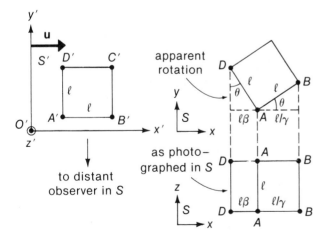

the light arriving from *D'* left that corner earlier than the light arriving from *A'*. Where was *D'* at the instant the light left relative to the position of corner *A'*?] Your analysis should show that, when an observer *sees* a rapidly moving three-dimensional object, he detects a rotation of the object instead of a simple foreshortening. For example, an illuminated sphere would still appear as a sphere! This interesting result of relativity theory went undiscovered until 1959 (J. Terrell).

21-27 Show that a relativistic particle with kinetic energy *K* and proper mass m_0 has a momentum *p* given by

$$cp = (K^2 + 2Km_0c^2)^{1/2}$$

Show that this expression reduces to the Newtonian result for $K \ll m_0c^2$. Determine *cp* in units of *K* for $K/m_0c^2 = 0.020, 0.20, 2.0,$ and 20.

21-28 Elementary particles called *muons* are created high in the atmosphere by energetic cosmic-ray particles interacting with atmospheric nuclei. Muons are unstable and undergo radioactive decay with a proper *half-life* of 1.52 µs. (The *half-life* of a sample is the time interval during which one half of the sample will decay.) If the muons are produced at an altitude of 9.0 km above sea level and one half reach sea level, what is the ratio of the muon kinetic energy to the rest-mass energy?

21-29 Two identical particles with proper mass m_0 ap-
•• proach each other in a head-on collision, each with a speed *v* in some inertial frame. (a) Show that the relativistic kinetic energy of one particle in the rest frame of the other is $K' = 2\gamma^2 m_0 v^2$, with $\gamma = (1 - v^2/c^2)^{-1/2}$. (b) What is the expression for *K'* in the Newtonian limit? (c) Suppose that a beam of particles is accelerated in some machine to $\beta = 0.80$ and that these particles, with kinetic energy K_1, are allowed to strike a target of identical particles at rest in the laboratory. This same machine (with appropriate auxiliary equipment) may be used to conduct a "colliding beam" experiment in which two particle beams, each with $\beta = 0.80$, are allowed to collide head-on. Operating in this colliding-beam mode, the machine is equivalent to one that must accelerate the particles to a much higher energy K_2 in order to yield the same collision kinematics for a stationary-target experiment. Calculate K_2/K_1.

PERIODIC TABLE OF THE ELEMENTS

Key:

26	Atomic number (Z)
Fe	Element symbol
55.847	Atomic mass of naturally occurring isotopic mixture; for radioactive elements, numbers in parentheses are mass numbers of most stable isotopes

§ The International Union for Pure and Applied Chemistry has not adopted official names or symbols for these elements.

Note: Atomic masses shown here are 1977 IUPAC values.

When the chemical elements are arranged in order of increasing atomic number (that is, the number of protons in the nucleus), many of their chemical and physical properties display a recurring or periodic character. Beginning with the independent work of Dimitri Mendeleev and Lothar Meyer in the 1870's, successive refinements and discoveries of unknown elements have brought the periodic table to the form shown here.

The table shows the symbol, atomic number, and atomic weight of each element. The atomic weight is an average of the number of protons and neutrons in the atoms of a natural sample of the element. This information will be needed in Chapters 23 and 24.

TEMPERATURE AND HEAT

This chapter begins a series of five chapters in which we discuss a number of important topics concerning the form of energy* called *heat*. This general subject, referred to as *thermodynamics,* deals with the thermal properties and the related behavior of macroscopic (bulk) matter as well as the underlying microscopic (molecular) basis for this behavior.

In this introductory chapter we give a quantitative description of temperature and heat for macroscopic systems. A later chapter is devoted to the molecular view contained in the *kinetic theory of matter*. At the beginning of the discussion of macroscopic thermodynamics, we discover that the three fundamental dimensions—length, time, and mass—are not sufficient for all purposes. It is necessary to introduce a new and important variable called *temperature*. Qualitatively, the temperature of an object indicates how "hot" or "cold" the object is.† Our first task, therefore, is to give a reliable, precise, and reproducible definition of temperature, one that permits the numerical specification of the degree of "hotness" or "coldness." In short, we must devise a useful *thermometer* and define a *temperature scale* by which it can be calibrated.

22-1 TEMPERATURE AND THERMAL EQUILIBRIUM

When a substance is heated and becomes "hotter"—that is, when the temperature of the substance is increased—several of its physical properties undergo measurable change. These property changes include the volume expansion of solids, liquids, and gases maintained at constant pressure, the increase in pressure of gases at constant volume, the increase in electric resistance of wires, the generation of thermoelectric currents in thermocouple circuits, and the change in color of glowing objects. These *thermometric properties* have all been used in the construction of thermometers. One selects as the *thermometric substance* some material that conveniently exhibits one of these properties. The addition of a scale or other numerical readout converts the material into a working thermometer.

---

*Review Chapter 7 for the important ideas relating to energy and energy conservation.

†In Chapter 24 we give a *microscopic* definition of temperature and relate it to the energy of the molecules in the sample.

Fig. 22-1. A common, colored-alcohol thermometer. The alcohol in the capillary tube stands at a height *x* above some reference zero and provides a measure of the temperature.

We could, for example, use the thermal volume expansion of alcohol (suitably colored with dye to make it easily visible) for the operation of a thermometer. A small amount of the alcohol is sealed in a glass tube that has a narrow bore and a reservoir bulb at one end (Fig. 22-1). The level at which the alcohol stands in the tube is a measure of the total volume of alcohol. When the temperature changes, so does the level of the alcohol. By calibrating the height of the alcohol above some reference zero in terms of the response to known temperatures, the device becomes a thermometer. We inquire in Section 22-2 how we accomplish this calibration.

Thermal Equilibrium. It is useful in thermodynamics to introduce the notion of a *system*. A system may be as simple as a small lump of matter or as complex as a locomotive engine. A system is in *thermal equilibrium* with its surroundings when it neither gains heat energy from nor loses heat energy to those surroundings. Equilibrium is a thermally quiescent state characterized by no discernible change with time of any thermometric property of the system. Two systems that are in thermal contact* are in thermal equilibrium with each other when neither system gains heat energy from the other.

We have the freedom to specify the boundaries of a system in any convenient way. For example, we could define as a system a certain volume of ice water that consists of solid ice and liquid water in intimate contact and in a state of thermal equilibrium. The important point here is not that the thermometric properties of the ice and the water are the same (they are not), but that neither set of properties changes with time. On the other hand, we could equally well consider the ice as one system and the water as a second system, with the two in thermal contact and in equilibrium. From this we can infer that for any system (simple or complex) to be in a state of equilibrium requires that there be no heat transfer between any of the various parts of the system. (•Can you see why this is so?)

There is a further generalization: Two systems that are in thermal contact and in equilibrium are considered to be still in equilibrium even if they are later separated and thermally isolated from each other (and from their surroundings).

Suppose that system *C* is in a state of equilibrium as determined by monitoring its thermometric properties. Another system *A*, also in equilibrium and monitored, is

---

* *Thermal contact* means that there is the possibility of heat flow, by some means, between the systems due simply to their proximity to each other. The systems need not be in physical contact, although this is the usual situation.

brought into thermal contact with system *C*. If none of the thermometric properties of either system change as a result of the contact, the two systems are in thermal equilibrium with each other. A similar test is subsequently performed with system *C* and another system *B*, indicating that *C* and *B* are also in equilibrium with each other. What can be said about the relative thermal states of systems *A* and *B*? Experiments have demonstrated that the thermal equilibrium relationship is a *transitive property*, so that *A* is also in equilibrium with *B*. That is, for the case of thermal equilibrium, $C \rightleftarrows A$ and $C \rightleftarrows B$ implies $A \rightleftarrows B$. This conclusion is referred to as the *zeroth law of thermodynamics**:

If two systems are each in thermal equilibrium with a third system, the two systems are also in thermal equilibrium with each other.

Suppose that the system *C* we have been discussing is a thermometer such as the one shown in Fig. 22–1. The alcohol height would be the same whether the thermometer were placed in thermal contact with system *A* or with system *B*. Thus, all three systems—*A*, *B*, and *C* (the thermometer)—would be at the same temperature. That is, the zeroth law also implies that systems in equilibrium with each other are at the same temperature. (If this were not the case, temperature would not be a useful concept!)

22-2 THE MEASUREMENT OF TEMPERATURE

We now return to the discussion of how a thermometer can be calibrated. Imagine that we use an alcohol thermometer such as that illustrated in Fig. 22–1 (or a similar thermometer filled with mercury). We select two easily reproducible and convenient physical systems and *arbitrarily* associate with each a numerical value for the temperature. Two such choices are the *ice point* and the *steam point* for water. At the ice point, ice and air-saturated pure water are in equilibrium at a pressure of 1.00 atm. The temperature of the ice point is assigned the value 0.00 degrees on the Celsius scale† (0.00° C). At the steam point (assigned a temperature of 100.00° C), steam and boiling water are in equilibrium at a pressure of 1.00 atm.

To calibrate the thermometer, it is first inserted into a mixture of ice and water. When the thermometer has reached equilibrium, as indicated by a steady height of the liquid column, this height is marked and labeled 0.00° C. Next, the thermometer is inserted into boiling water, and when the new equilibrium column height is reached, this height is marked and labeled 100.00° C.

The bore of the thermometer tube is assumed to be uniform and to have a volume that is negligible in comparison with that of the reservoir bulb. We also make the *assumption* that there exists a linear relationship between the liquid column height and the temperature.‡ Then, the distance between the marks that correspond

---

*So-called because of its importance in establishing the fundamental thermodynamic concept of temperature on a firm foundation. The other numbered laws of thermodynamics are discussed in later chapters.

†Originally devised in 1742 by the Swedish astronomer and inventor Anders Celsius (1701–1744). The name of this temperature scale was changed, by international agreement in 1948, from centigrade to Celsius.

‡We make the assumption of linearity here, but when we eventually adopt a theoretically superior definition for temperatures—namely, Lord Kelvin's definition based on the Carnot cycle (see Section 26–3)—linearity will be postulated for that scale. Then, in principle, a judgment of the linearity of all other scales can be based on this theoretical standard.

to 0° C and 100° C may be divided into 100 equal intervals, each representing a temperature interval of one degree Celsius (1 deg C). Thus, if the distance between the 0° C and 100° C marks is ℓ, then a column height x, referred to the 0° C mark, corresponds to a temperature

$$t_C = 100 \frac{x}{\ell} \, °C \qquad\qquad (22\text{-}1)$$

By extending the column-height scale to negative values of x (as suggested in Fig. 22–1), temperatures below 0° C can be measured.

The Constant-Volume Gas Thermometer.   It is found experimentally that when a fixed amount (mass) of gas is confined in a container with constant volume, the pressure of the gas increases as the temperature is raised. This thermometric property of a gas may be used as the basis of an accurate thermometer that possesses, in addition, considerable intrinsic thermodynamic interest.

The essential details of a constant-volume gas thermometer are illustrated schematically in Fig. 22–2. The thermometer bulb, shown immersed in a sample liquid at some temperature, is connected by a small-bore tube to a mercury manometer, which is used to measure the gas pressure. For each measurement, the reservoir column R is raised or lowered to place the mercury level in column A opposite the reference zero on the scale. In this way, the volume of the gas is maintained at a constant value. The difference in mercury levels in columns A and B is a direct measure of the gas pressure. Each millimeter of height difference corresponds to a pressure of 1 torr (where 760 torr = 1 atm). Notice that the manometer registers *absolute pressure* and is therefore independent of atmospheric pressure. (•Can you see why?)

Fig. 22–2.   A constant-volume gas thermometer. The difference in the heights of the mercury columns in A and B is a direct measure of the pressure of the gas.

A precise and readily reproducible temperature has been selected by international agreement to serve as the primary temperature standard. This is the temperature at which water, ice, and water vapor exist together in thermal equilibrium—the so-called *triple point of water*. (A triple-point cell of the type used in practical thermometry is shown in Fig. 22–3.) The three phases of water exist together *only* at a pressure of 4.58 torr. The corresponding temperature, which is also unique, is arbitrarily defined to be 273.16 degrees on a new temperature scale called the *Kelvin*

Fig. 22–3. A triple-point cell. The cell contains pure water and is sealed after all air has been removed. The temperature of the thermal bath is adjusted until it is visually observed that water, ice, and water vapor exist together in the cell.

vapor (no air)

ice

water

triple-point cell

thermal bath (water + ice)

gas thermometer bulb

scale (or *absolute scale*). The size of the degree on this scale (the kelvin, K) is the same as the Celsius degree. On the Kelvin scale, the temperature of the triple point of water is designated 273.16 K (read: 273.16 *kelvins;* formerly, 273.16 *degrees Kelvin*).

Consider performing the following experiments. The bulb of a constant-volume gas thermometer is filled with oxygen. The bulb is placed inside a triple-point cell and the pressure p_0 determined by the manometer reading is, say, 300 torr. Next, this gas thermometer is used to determine the steam-point temperature by observing the new manometer pressure p and using the assumed linear relationship,*

$$T = (273.16 \text{ K}) \cdot \frac{p}{p_0} \qquad \qquad \textbf{(22–2)}$$

The result for the steam point is $T_s = 373.35$ K. This temperature is indicated by point a in Fig. 22–4, where the temperature T_s of the steam point determined by using Eq. 22–2 is plotted as a function of the gas pressure p_0 at the triple point. If the initial pressure p_0 is changed to 150 torr by removing some of the oxygen from the bulb, the steam-point temperature decreases to 373.25 K (point b in Fig. 22–4). The results of many such measurements for different pressures p_0 at the triple point are represented by the straight line for oxygen (O) in Fig. 22–4. The temperature intercept ($p_0 = 0$) is $T_s = 373.15$ K. Also shown in the figure are the results obtained using air, nitrogen (N), helium (He), and hydrogen (H). The important result is that

---

*We use T to represent temperature on the Kelvin scale.

Fig. 22–4. Results of the determination of the temperature T_s of the steam point using different gases in a constant-volume gas thermometer. The pressure p_0 is the pressure of the gas at the triple point. For very dilute gases ($p_0 \to 0$), all results tend toward a value of 373.15 K.

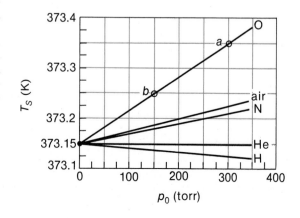

for any gas used in the thermometer, the temperature intercept is the same, namely, 373.15 K. That is, all dilute gases exhibit the same pressure–temperature behavior at constant volume. (This is one of the properties we ascribe to an imaginary substance called an *ideal gas*. We make extensive use of the concept of an ideal gas in later discussions.) A modification of Eq. 22–2 now gives a precise definition for temperature in terms of pressure measurements with a constant-volume gas thermometer with any gas as the thermometric substance:

$$T = (273.16 \text{ K}) \cdot \lim_{p_0 \to 0} \frac{p}{p_0} \tag{22–3}$$

Temperature Scales. By using a gas thermometer in the manner just described, the ice-point temperature is determined to be $T_i = 273.15$ K. Thus, the triple point for water is 0.01 K higher than the ice point. We also observe that $T_s - T_i = 100$ K, the same difference as for the Celsius scale. In fact, the coefficient, 273.16 K, in Eq. 22–3 was chosen to give exactly this result. The relationship between the two scales is

$$T = t_C + 273.15 \tag{22–4}$$

where T is measured in kelvins (K) and t_C is measured in °C.

Another temperature scale still used in some English-speaking countries is the *Fahrenheit scale** in which the temperature t_F is measured in *degrees Fahrenheit* (°F). The connection between the Fahrenheit and Celsius scales is given by

$$t_F = \tfrac{9}{5}t_C + 32.00° \tag{22–5}$$

On the Fahrenheit scale, the ice-point temperature is 32.00° F and the steam-point temperature is 212.00° F.

In engineering practice, the *Rankine temperature scale* is sometimes used. The size of the Rankine degree is the same as that of the Fahrenheit degree, and the zero point of the Rankine scale is the same as that of the Kelvin scale. Then, on the Rankine scale, the triple-point temperature is $\tfrac{9}{5} \cdot 273.16 = 491.69°$ R. In general, we have

$$t_R = \tfrac{9}{5}T = t_F + 459.68° \tag{22–6}$$

Practical Thermometry. Any particular type of thermometer has a limited range of usefulness. For example, there is a practical limit on the lowest temperature that can be measured with a gas thermometer because any gas will liquefy at a sufficiently low temperature. The gas best suited for low temperature measurements is helium, which has the lowest liquefaction temperature of any gas (4.2 K at normal atmospheric pressure).

To make measurements over a wide temperature range, it is necessary to use various thermometric devices, even some that lack a strict linear relationship connecting the thermometric property and the temperature. The use of such devices requires that a number of temperature reference points be established for purposes

---

*Named for Gabriel Fahrenheit (1686–1736), German-Dutch physicist, who devised the first precision thermometer, a mercury-filled instrument.

TABLE 22–1. Some Reference Temperatures on the International Practical Temperature Scale (IPTS)

EQUILIBRIUM STATE	T (K)	t_C (°C)
Triple point of water[a]	273.16	0.01
Triple point of hydrogen	13.81	−259.34
Boiling point of hydrogen[b]	20.28	−252.87
Boiling point of neon[b]	27.402	−246.048
Triple point of oxygen	54.361	−218.789
Boiling point of oxygen[b]	90.188	−182.962
Boiling point of water[b]	373.15	100.00
Freezing point of zinc[b]	692.73	419.58
Freezing point of antimony[b]	903.89	630.74
Freezing point of silver[b]	1235.08	961.93
Freezing point of gold[b]	1337.58	1064.43

[a] Primary standard.
[b] At normal atmospheric pressure.

of calibration. Some of the reference temperatures on the International Practical Temperature Scale (IPTS) are given in Table 22–1.

The electric resistance of a metallic conductor varies with temperature, so this thermometric property can be used to construct a thermometer. Platinum has excellent chemical and physical properties and is widely used in *resistance thermometers* that are used in the temperature range from about −260° C to about 600° C. Very precise and sensitive methods have been developed for measuring resistance values, and it is possible to detect temperature changes as small as 10^{-4} deg C by using this technique.

A circuit that consists of two wires of different metals and connected with two junctions will develop a voltage within the circuit if the junctions are at different temperatures. This is the operating principle of the *thermocouple*. Usually, the cold junction of the thermocouple is placed in a bath that is maintained at the ice point, whereas the hot junction is placed in contact with the sample whose temperature is to be determined (see Fig. 22–5). Thermocouples constructed from wires of platinum and (90%–10%) platinum-rhodium are often used for measurements in the temperature range from about 600° C to about 1000° C.

Very high temperatures—even above the melting points of most metals—can be determined by optical means that do not require contact with the sample. Devices for making such measurements are called *optical* (or *radiation*) *pyrometers*. An electrically heated filament in a pyrometer is viewed through a suitable filter simultaneously with the sample, which might be, for example, the interior of a furnace. The current through the comparison filament is adjusted so that its image vanishes when

Fig. 22–5. A thermocouple circuit. One junction (the cold junction) of two wires, *a* and *b*, made from different metals is placed in an ice-point bath; the other junction (the *hot* junction) is in thermal contact with the sample.

Fig. 22-6. Range of temperature found in the Universe.

superimposed on the image of the sample. The "color temperature" of the filament and sample are then the same, and the sample temperature is determined from the current calibration of the filament. Optical pyrometers are commonly used to measure temperatures above about 1000° C; they are useful for temperatures up to about 3600° C (the melting point of tungsten).

Absolute Zero. The form of Eq. 22-3 clearly suggests that there is a lower limit to the temperature scale, namely, 0 K. However, this temperature cannot be reached in practice and can only be judged to be the theoretical limit of the extrapolation of experimental data. This limit is called the *absolute zero* of temperature. The impossibility of achieving absolute zero is known as the *Nernst heat theorem*, which can be stated in the following way:

It is not possible by any procedure, no matter how idealized, to cool a system to an equilibrium state of absolute zero in a finite number of operations.

Various experimental techniques can be used to reach temperatures very close to absolute zero. For example, by pumping away the vapor above a sample of liquid helium, the temperature of the helium can be reduced below 1 K. Still lower temperatures can be reached, for very small samples, by cyclic magnetization procedures. By using the ionic magnetization of crystals, temperatures near 10^{-3} K can be reached, and nuclear magnetization methods permit temperatures as low as 10^{-6} K to be achieved!

Example 22-1

In a laboratory experiment, a student uses a simple constant-volume gas thermometer to determine the Celsius temperature of absolute zero. The student finds the gas pressure at the ice point to be 182 torr, and at the steam point he finds $p = 246$ torr. What value does the student calculate for the Celsius temperature of absolute zero?

Solution:

The two measurements are shown on the diagram. From the two similar triangles, we can write

$$\frac{|t_{0C}|}{182 \text{ torr}} = \frac{100 \text{ deg}}{(246 - 182) \text{ torr}}$$

or

$$|t_{0C}| = (100 \text{ deg}) \cdot \tfrac{182}{64} = 284 \text{ deg}$$

Hence, the student finds $t_{0C} = -284°\text{ C}$, compared with the actual value, $t_{0C} = -273.15°\text{ C}$. This is typical of the results obtained with the apparatus usually available in an elementary physics laboratory. Notice that in this experiment, the careful extrapolation to zero pressure (Fig. 22-4) was ignored.

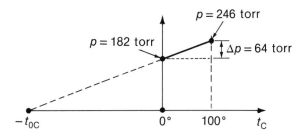

22-3   THERMAL EXPANSION

Most substances expand when heated at constant pressure. (A notable exception, as discussed later, is water in the temperature range from $0°$ C to $4°$ C.) We restrict attention here to liquids and solids; the behavior of gases requires special attention (see Chapter 23). We also omit discussion of cases that involve a change of phase of the substance—for example, cases in which melting or boiling occurs.

Consider a thin rod with a length ℓ_0 that is heated, causing its equilibrium temperature to change from T_0 to T. If the length at the temperature T is ℓ, a length change, $\Delta\ell = \ell - \ell_0$, has occurred due to the temperature change, $\Delta T = T - T_0$. Then, the *average coefficient of linear expansion* $\bar{\alpha}$ is defined to be

$$\bar{\alpha} = \frac{1}{\ell_0} \frac{\Delta\ell}{\Delta T} \qquad\qquad \textbf{(22-7)}$$

(We could equally well express $\bar{\alpha}$ in terms of t_C.) Because the linear expansion coefficient varies somewhat with temperature, it is more precise to define α for a particular temperature by writing the *coefficient of linear expansion* as

$$\alpha = \alpha(T) = \frac{1}{\ell} \frac{d\ell}{dT} \qquad\qquad \textbf{(22-7a)}$$

where it is understood that the derivative is evaluated at the temperature T. For example, the value of α for stainless steel is 0.81×10^{-5} deg^{-1} at $-100°$ C and is

$1.43 \times 10^{-5}\ \mathrm{deg}^{-1}$ at $700°$ C. The value of α for most substances depends only weakly on pressure.

We can also define expansion coefficients for area and for volume. For example, the coefficient of volume expansion is defined to be

$$\beta(T) = \frac{1}{V}\frac{dV}{dT} \tag{22-8}$$

For a particular substance, the volume expansion coefficient is just three times the linear expansion coefficient: $\beta = 3\alpha$. To see this, consider a parallelopiped with sides ℓ_1, ℓ_2, and ℓ_3, so that the volume is $V = \ell_1\ell_2\ell_3$. Then,

$$\beta = \frac{1}{V}\frac{dV}{dT} = \frac{1}{\ell_1\ell_2\ell_3}\left[\ell_2\ell_3\frac{d\ell_1}{dT} + \ell_1\ell_3\frac{d\ell_2}{dT} + \ell_1\ell_2\frac{d\ell_3}{dT}\right]$$

$$= \frac{1}{\ell_1}\frac{d\ell_1}{dT} + \frac{1}{\ell_2}\frac{d\ell_2}{dT} + \frac{1}{\ell_3}\frac{d\ell_3}{dT}$$

$$= \alpha + \alpha + \alpha = 3\alpha \tag{22-9}$$

This is actually a general result and does not depend on the geometrical shape of the object (see Problem 22-9).

Because α and β vary with temperature, the expansion due to a large temperature change cannot be calculated precisely with a single coefficient. It is then necessary to use an expression such as

$$\ell(T) = \ell_0[1 + a(T - T_0) + b(T - T_0)^2 + c(T - T_0)^3 + \ldots]$$

where $a = \alpha(T_0)$. Usually, the quadratic and higher-order coefficients are small; typically, the quadratic term represents a 0.1 percent correction.

Some typical values of α for solids and β for liquids are given in Table 22-2. In this discussion, we have assumed that the object being heated (or cooled)

TABLE 22-2. Thermal Expansion Coefficients for Some Substances at 20° C[a]

SUBSTANCE	α (deg^{-1})	SUBSTANCE	β (deg^{-1})
Aluminum	2.30×10^{-5}	Alcohol (ethyl)	1.12×10^{-3}
Brass, yellow	1.90×10^{-5}	Alcohol (methyl)	1.20×10^{-3}
Concrete	$1.2\ \ \times 10^{-5}$	Benzene	1.24×10^{-3}
Copper	1.67×10^{-5}	Carbon tetrachloride	1.24×10^{-3}
Glass, Pyrex	$3.2\ \ \times 10^{-6}$	Gasoline	$9.5\ \ \times 10^{-4}$
Gold	1.42×10^{-5}	Glycerine	$5.1\ \ \times 10^{-4}$
Invar	$1.2\ \ \times 10^{-6}$	Mercury	1.82×10^{-4}
Iron, wrought	1.12×10^{-5}	Olive oil	$7.2\ \ \times 10^{-4}$
Lead	2.87×10^{-5}	Turpentine	$9.7\ \ \times 10^{-4}$
Quartz, fused	$4.2\ \ \times 10^{-7}$	Water	2.07×10^{-4}
Silver	1.90×10^{-5}		
Steel, stainless	1.05×10^{-5}		
Tungsten	$4.5\ \ \times 10^{-6}$		

[a]Notice that the units of α are deg^{-1}; we do not need to specify whether this means *per degree Celsius* or *per absolute degree* (kelvin) because they are the same.

is unconfined and can freely change its dimensions. We also made the implicit assumption that all temperature changes take place slowly. When these conditions are not met, stresses with considerable magnitudes may develop in an object. In the construction of bridges, railroad tracks, concrete walls and roads, and other large structures that are exposed to wide temperature variations, thermal expansion must be allowed for to prevent severe damage from buckling.

In some applications, materials are required that have very small expansion coefficients. For example, chronometer parts, pendulum rods, balance arms of scales, and other precision measuring devices must have dimensions that are stable when the temperature changes. For these applications, fused quartz or special alloys such as Invar (36 percent nickel steel) are often used because they have expansion coefficients that are 1/10 or less of the coefficients for ordinary materials.

The Expansion of Water. When a quantity of water is cooled from the steam point, its volume decreases, as is the case for normal liquids, until a temperature of 4° C is reached. Further cooling, from 4° C to 0° C, causes the volume to *increase*. It is convenient to show this anomalous behavior of water in terms of the variation with temperature of the *specific volume,* or volume per unit mass (the inverse of *density*). Figure 22–7 shows the specific volume (in units of cm³/g) for water in the temperature range from 0° C to 100° C.

In winter, an open body of water, such as a pond or lake, cools primarily because of exposure to colder air. The cold surface water, with higher density, sinks to the bottom. But when the surface temperature falls below 4° C, the density of the surface water is *less* than that of the underlying layers, so the colder water no longer sinks. The water with temperature nearest to 0° C remains on the surface. Consequently, it is the *surface* of a body of water that freezes first. This permits the water near the bottom to remain liquid, thereby protecting aquatic life. If water behaved as a normal liquid, the coldest water would always be found at the bottom. Then, a

Fig. 22–7. The temperature variation of the *specific volume* (the inverse of *density*) for water. The behavior between 0° C and 4° C is anomalous. Specific volume is commonly expressed in cm³/g although these are not proper SI units. Sometimes the specific volume is expressed in ml/g; then, the value at 4° C is 1.0000 because 1 ml = 1.000027 cm³.

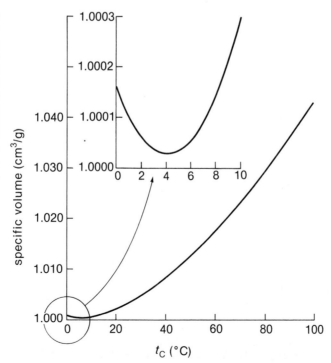

pond or a lake would freeze from the bottom up, with an increased likelihood that the entire volume would freeze solid.

Not only does water expand as it cools from 4° C to 0° C, but upon freezing, it expands still further. Again, this is contrary to the behavior of most substances. The expansion upon freezing is substantial, amounting to about 8.3 percent. The freezing of water trapped in cracks of rock or roadway materials can often exert sufficient stress to enlarge the existing fissures. In this way, the material suffers progressive deterioration with the freeze-thaw cycle.

Example 22-2

The measurement of the coefficient of volume expansion for liquids is complicated by the fact that the container also changes size with temperature. The diagram shows a simple means for overcoming this difficulty. With this apparatus, $\bar{\beta}$ for a liquid is determined by measurements of the column heights, h_0 and h_t, of the liquid columns in the U-tube. One arm of the tube is maintained at 0° C in a water-ice bath, and the other arm is maintained at the temperature t_C in a thermal bath. The connecting tube is horizontal. Derive the expression for $\bar{\beta}$ in terms of h_0 and h_t.

water-ice bath, 0°C

liquid sample

thermal bath, temperature t_C

Solution:

The pressure at the free surface of each liquid column is the same (atmospheric pressure). The system is static (no liquid flow), so the pressure at the bottom of each column is likewise the same. Hence, $\rho_t g h_t = \rho_0 g h_0$, or $\rho_0/\rho_t = h_t/h_0$. Now the volume V_0 of a unit mass of liquid at 0° C is $V_0 = 1/\rho_0$, and at $t°$ C it is $V_t = 1/\rho_t$. Then, we have for a unit mass,

$$\frac{\rho_0}{\rho_t} = \frac{V_t}{V_0} = 1 + \bar{\beta} t_C = \frac{h_t}{h_0}$$

Solving for $\bar{\beta}$, we find the desired relationship,

$$\bar{\beta} = \frac{h_t - h_0}{h_0 t_C}$$

$t_C = 100°C$

$D = 2.0$ cm

$\ell = $ 15.0 cm

Example 22-3

Two parallel surfaces of incompressible material are at a temperature of 100° C. A brass rod with a diameter of 2 cm and a length of 15 cm at a temperature of 20° C just fits between the parallel surfaces, as shown in the diagram. What is the compressional stress in the rod when its temperature has risen to the temperature (100° C) of the confining surfaces?

Solution:

If the rod had been free to expand, its length would have increased by an amount $\Delta l = \bar{\alpha} l_0 \Delta t_C = (1.90 \times 10^{-5} \deg^{-1})(0.15 \text{ m})(80 \deg) = 2.28 \times 10^{-4}$ m, where we have used for $\bar{\alpha}$ the value for $20°$ C given in Table 22–2. Because the length of the rod is actually held constant, there results a thermally induced compressive strain of $e_l = (2.28 \times 10^{-4} \text{ m})/(0.15 \text{ m}) = 1.52 \times 10^{-3}$. Young's modulus for brass is $Y = 9.0 \times 10^{10}$ Pa (see Table 16–2), so the stress is (Eq. 16–13)

$$s = Ye_l = (9.0 \times 10^{10} \text{ Pa})(1.52 \times 10^{-3})$$
$$= 1.37 \times 10^8 \text{ Pa}$$

This corresponds to a compressive force of

$$F = s(\tfrac{1}{4}\pi D^2) = (1.37 \times 10^8 \text{ Pa})(\tfrac{1}{4}\pi)(0.020 \text{ m})^2$$
$$= 4.30 \times 10^4 \text{ N}$$

(•Was the length of the rod required to make the calculation?)

22-4 HEAT

It was not until the 19th century that scientists accepted the view that *heat* is simply one form of energy. Indeed, this realization is crucial in the formulation of the energy conservation principle (see Section 7–4). In earlier times, however, the prevailing theory viewed heat as a massless and invisible substance called *caloric*. According to this theory, every body contains an amount of caloric that depends on the temperature of the body—a hot object contains more caloric than does a cold object. Also, caloric was assumed to repel itself. Therefore, when a hot object is placed in contact with a cold object, caloric should flow from the hot object to the cold object until a common temperature is reached; experiment shows that this is indeed the behavior of objects at different temperatures that are placed in contact.

The caloric theory could also explain a number of other observations concerning heat and temperature effects. For example, when a body is heated, caloric flows in and surrounds the atoms. Because caloric repels itself, as more caloric is added, the atoms are pushed farther apart and the body expands; again this prediction of the theory agrees with experiment. (•How could you use the caloric theory to account for the conversion of water to steam?)

The caloric theory appeared to be a comprehensive and useful description of heat phenomena. But the observations made in 1798 by Count Rumford (Benjamin Thompson, American-British physicist, 1753–1814) led eventually to the downfall of the theory. While overseeing the boring of cannon barrels in a Munich arsenal, Rumford was impressed by the amount of heat that was generated in the process. Every object was supposed to contain a definite amount of caloric, but Rumford observed that heat continued to be generated as long as the boring tool scraped away at the barrel. In fact, if the cannon had actually contained that much caloric at the beginning of the boring process, it should have melted! Rumford concluded that heat could not be a material substance. He believed that the scraping of the boring tool inside the barrel set into vibratory motion the particles comprising the cannon; this vibratory motion was then responsible for the sensation of heat. Thus, the generation of heat was due to the mechanical work done on the boring tool and continued as long as this work was being done.

Let us now consider carefully the concept of heat on the macroscopic level. We must first make a clear distinction between *heat* (energy) and *internal energy*. It was precisely the failure to appreciate this difference that was the basic fault of the caloric theory.

Heat is energy that is transferred from one body or system to another solely as the result of the temperature difference between them. Thus, a "quantity of heat" only has meaning in the sense that an energy flow exists between bodies at different temperatures. The direction of this heat (energy) flow is always from the body at the higher temperature to the body at the lower temperature.

A statement that a body or system *contains* a certain amount of heat is, in a strict sense, meaningless. *The energy content of a system in equilibrium is its internal energy.* Heat energy can be extracted from a system in a manner that reduces its internal energy, and, conversely, the internal energy can be increased by the addition of heat energy. However, the internal energy of a system can be changed by many means in addition to the flow of heat energy into or out of the system. For example, mechanical, acoustic, electromagnetic, gravitational, nuclear, or other forms of energy exchange can be used to increase or decrease the internal energy of a system. A very important feature of the internal energy of a system is that it depends *only on the equilibrium temperature** and *not at all* on the means of energy exchange by which that temperature was achieved.

It is possible for a system continually to supply heat energy to other systems at lower temperatures while the supplying system remains at constant temperature, provided there is energy input to the system by some means, not necessarily thermal in character. This was a feature lacking in the caloric theory, and precisely the point realized by Rumford in his observation of heat generation in the boring of cannon barrels.

Specific Heat and Heat Capacity. The specific heat c of a substance is defined as the amount of heat per unit mass that must be added to the substance to raise its temperature by one Celsius degree (1 deg C). To give a complete definition of specific heat it is also necessary to indicate the experimental conditions under which the heating process is carried out. The two most common methods involve heating at constant volume or at constant pressure. For most liquids and solids it is much easier to make the necessary measurements at constant pressure (usually atmospheric pressure). It is possible to relate the results of measurements made in this way to the results of measurements made at constant volume by a theoretical calculation. Usually, the differences in the two types of measurement are slight owing to the smallness of the volume expansion coefficients for liquids and solids. For gases, however, the two methods of measurement yield substantial differences that cannot be ignored. Therefore, we consider at first only liquids and solids; we come to the discussion of the specific heats of gases in Chapter 24.

The thermodynamic reference standard for quantity of heat is the heat required to raise the temperature of 1 kg of pure water from 14.5° C to 15.5° C at a pressure of 1 atm; this amount of heat is designated *one Calorie* (1 Cal). (A smaller unit, based on the heat required to raise the temperature of one *gram* of water by one degree, is called the *calorie*, with a lower case c; evidently, 10^3 cal = 1 Cal.) Thus, the specific heat of water at 15° is equal to 1.0000 Cal/kg · deg.

---

*The dependence of the internal energy exclusively on temperature is exactly true only for an ideal gas but is approximately true for real substances. (See Section 24–2.)

The specific heat of water actually varies slightly with temperature—but less than 1 percent between 0° C and 100° C. For example, to raise the temperature of 1 kg of water from 79.5° C to 80.5° C requires 1.00229 Cal. The variation with temperature of the specific heat of water is shown in Fig. 22–8. The fact that the specific heat of water is nearly constant is very useful in the determination of specific heat values for other substances in the temperature range from 0° C to 100° C, as we demonstrate in Section 22–5.

Fig. 22–8. The variation with temperature of the specific heat of water. On the scale of this diagram, the IT Calorie and the 15° Calorie are indistinguishable.

Because heat is a form of energy, quantity of heat should properly be expressed in SI energy units, namely, *joules*. In Section 22–5 we describe an experiment for determining the relationship between the Calorie and the joule. The Calorie we have just defined—called the 15° C Calorie—is found to equal 4186 J:

$$1 \, (15° \, \text{C}) \, \text{Cal} = 4186 \, \text{J}$$

There is also defined, by international agreement, a secondary or practical standard called the International Table Calorie (IT Cal):

$$1 \, (\text{IT}) \, \text{Cal} = \frac{1 \, \text{kWh}}{860} = \frac{3.60 \times 10^6 \, \text{J}}{860} = 4186 \, \text{J}$$

To the number of significant figures given, these two definitions of the Calorie are equivalent. There is yet another definition, the *thermochemical Calorie,* arbitrarily set exactly equal to 4184 J:

$$1 \, (\text{th}) \, \text{Cal} = 4184 \, \text{J}$$

Some engineering calculations require a conversion from *British thermal units* (B.T.U.) to calories:

$$1 \, \text{Cal} = 3.968 \, \text{B.T.U.}$$

In this book we use SI units for most purposes, and we usually express thermal energies in *joules*. However, heat measurements have traditionally been made in units of *calories*. We sometimes use as an alternate unit the *kilogram calorie* or *Calorie* because this unit involves the specification of mass in kilograms, the SI unit of mass. Table 22–3 lists specific heat values for various substances both in J/kg · deg and in Cal/kg · deg. (Again, it does not matter whether "deg" stands for "deg C" or "kelvin".)

TABLE 22–3.   **Specific Heat Values for Various Substances**
(at a pressure of 1 atm)

SUBSTANCE	TEMPERATURE RANGE (°C)	SPECIFIC HEAT c (OR $\bar{c}$)	
		(Cal/kg · deg)	(J/kg · deg)
Aluminum	0–100	0.216	904
Brass	20–100	0.0917	384
Carbon (amorphous)	25–80	0.168	703
Copper	15–100	0.0931	390
Eythyl alcohol	20	0.572	2394
Glass (crown)	10–50	0.161	674
Glass (flint)	10–50	0.117	490
Gold	0–100	0.0316	132
Ice	−10	0.530	2219
Iron	0–300	0.122	511
Lead	0–300	0.0326	136
Mercury	20–100	0.0330	138
Methyl alcohol	20	0.600	2512
Silver	20–100	0.0560	234
Tin	0–100	0.0556	233
Wood (typical)	—	0.42	1760

The specific heat values in Table 22–3 are average values $\bar{c}$ for the quoted temperature ranges. That is,

$$\bar{c} = \frac{1}{m}\frac{\Delta Q}{\Delta T} \qquad \textbf{(22–10)}$$

where ΔQ is the heat required to raise the temperature of the substance by ΔT. The specific heat at a particular temperature T is defined to be*

$$c = c(T) = \frac{1}{m}\frac{dQ}{dT} \qquad \textbf{(22–10a)}$$

The specific heat c is defined for a chemically pure or homogeneous substance. In the event that a system consists of a mixture of substances, we refer to a quantity H, called the *heat capacity* of the system (an unfortunate term!), where

---

*Usually, derivatives (and related differentials) refer to changes in various finite (non-infinitesimal) quantities. Thus, we would ordinarily write

$$\Delta Q = Q(T + \Delta T) - Q(T) \qquad \text{and} \qquad dQ = \frac{dQ}{dT}\,dT$$

and we would interpret $Q(T + \Delta T)$ to be the heat content at a temperature $T + \Delta T$ and $Q(T)$ to be the heat content at a temperature T. However, we have emphasized that it is incorrect to refer to the heat content of a body; instead, we use the concept of internal energy. Although dQ is indeed a differential quantity, it is so only in the sense that it takes the system from one thermodynamic state to another. In some texts this distinction is emphasized by using the symbol $đQ$ (meaning the "inexact differential" of Q). However, we use ΔQ and dQ, with the understanding that this special interpretation holds.

$$H = \sum_i m_i \bar{c}_i \qquad\qquad \textbf{(22-11)}$$

The quantity of heat Q that must be supplied to such a mixture in order to raise the temperature by an amount ΔT is

$$Q = \left(\sum_i m_i \bar{c}_i\right)\Delta T = H\,\Delta T \qquad\qquad \textbf{(22-12)}$$

Latent Heat. The familiar transitions of liquid water to steam and of ice to liquid water are called *phase transitions*. These phase changes for water are typical of the (gas) $\rightleftharpoons$ (liquid) $\rightleftharpoons$ (solid) transitions that occur suddenly at well-defined temperatures. These transitions are called *first-order phase transitions* and involve a striking change in the long-range order of the atoms and molecules in the substance. A change in the internal energy always accompanies a phase transition. In all cases of phase transitions, we imagine that heat is added or removed slowly so that the substance is always at a uniform temperature.

There are many forms of phase transitions in addition to the obvious first-order changes. For example, at some particular temperature a substance might undergo an alteration in its crystal structure, or a sudden change in the electric conductivity might occur, transforming an ordinary conductor into a superconductor. A further consideration is the fact that some substances—plastics, glasses, tars, and so forth—undergo (solid) $\rightleftharpoons$ (liquid) transitions, not at a well-defined temperature, but over a broad range of temperatures. Although these cases are very interesting, we confine our discussions here to simple first-order phase transitions.

Suppose that we have an isolated sample of ice at some temperature below the freezing point, say, $-25°$ C. The sample is maintained at atmospheric pressure and heat is supplied at a slow constant rate. The recorded temperature as a function of time is shown in Fig. 22-9. From the initial condition at a, the ice warms up to the melting point ($0°$ C) at b. The temperature remains constant from b to c as the ice melts. At any particular time between b and c there is a definite fraction of the sample in the form of ice—100 percent at b and 0 percent at c. At the intermediate point b', 80 percent of the sample is in the form of ice. If the heat input were arrested at this point, then, because the system is thermally isolated, an equilibrium state would persist, with the fraction of ice constant at 80 percent. The system would then be in a condition called *phase equilibrium*—that is, the coexistence of two different phases in equilibrium.

From c to d, the temperature of the sample, now all water, increases at about half the rate that occurred during the interval a to b. The temperature continues to

Fig. 22-9. A temperature-time curve for water heated at a constant rate at atmospheric pressure. The sample is thermally isolated from its surroundings (except for the heat source). The lines $\overline{ab}$, $\overline{cd}$, and $\overline{ef}$ have different slopes, reflecting the different specific heat values for ice, water, and steam.

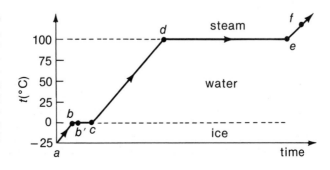

increase until the boiling temperature (100° C) is reached at *d*. From *d* to *e*, heat is added to boil the water while the temperature remains constant. Boiling takes place as the system proceeds through an infinite number of states that are in phase equilibrium between the liquid and gaseous (steam) phases. At *e* all of the water has been converted into steam, and any further heating causes a continual increase in the temperature of the steam.

If the path from *a* to *f* is reversed by cooling the system, the temperature will first decrease until the condensation point of steam is reached and liquid water begins to appear. This temperature is the same as the boiling temperature, namely, 100° C. To be precise, we should refer not to the boiling or condensation temperature, but to the temperature at which phase equilibrium exists between the gaseous and liquid phases. Further cooling of the system carries it to *c* where freezing begins. Again, the freezing and melting temperatures are the same (0° C), and, properly, this temperature should be referred to as the temperature for phase equilibrium between the solid and liquid phases.

The conversion of ice to water involves changes in the long-range order of the water molecules. These changes are associated with a change in the internal energy of the system. The amount of energy that is required to convert 1 kg of ice to water at 0° C is called the *latent heat of fusion*, L_f; thus,

$$L_f \text{ (water)} = 79.71 \text{ Cal/kg}$$
$$= 3.337 \times 10^5 \text{ J/kg}$$

This is the energy that must be *supplied* to melt ice completely at 0° C, and it is the energy that is *released* when water freezes completely at 0° C.

The energy involved in the complete phase change at 100° C is called the *latent heat of vaporization*, L_v:

$$L_v \text{ (water)} = 539.12 \text{ Cal/kg}$$
$$= 2.2567 \times 10^6 \text{ J/kg}$$

TABLE 22–4.   **Latent Heats of Transition (at 1 atm)**

SUBSTANCE	FUSION			VAPORIZATION		
	t_C (°C)	L_f		t_C (°C)	L_v	
		Cal/kg	J/kg		Cal/kg	J/kg
Nitrogen[a]	−210.00	6.13	2.567×10^4	−195.80	47.64	1.994×10^5
Ethyl alcohol	−144.4	24.9	1.042×10^5	78.3	204.	8.54×10^5
Mercury	−38.86	2.730	1.143×10^4	356.58	70.62	2.956×10^5
Water	0.00	79.71	3.337×10^5	100.00	539.12	2.2567×10^6
Sulfur[b]	115.18	12.74	5.33×10^4	444.60	68.5	2.869×10^5
Lead	327.3	5.533	2.316×10^4	1750	205.0	8.581×10^5
Zinc	419.58	26.98	1.1295×10^5	911	422.2	1.7675×10^6
Antimony	630.74	38.99	1.632×10^5	1440	134.0	5.61×10^5
Silver	961.93	21.07	8.82×10^4	2163	555.1	2.323×10^6
Gold	1064.43	15.8	6.61×10^4	2808	406.34	1.7009×10^6
Copper	1083.1	49.2	2.06×10^5	2566	1129	4.726×10^6

[a] At a pressure of 93.9 torr for fusion.
[b] For the fusion transition from monoclinic crystal to liquid.

This is the energy that must be *supplied* to convert water completely to steam at 100° C, and it is the energy that is *released* when steam condenses to water at 100° C.

Table 22-4 gives the values of the *latent heats of transition*, L_f and L_v, for several substances at atmospheric pressure.

Example 22-4

A block of ice with a mass $m = 10$ kg is moved back and forth over the flat horizontal top surface of a large block of ice. Both blocks are at 0° C, and the force that produces the back and forth motion acts only horizontally. The coefficient of kinetic friction (wet ice on ice) is $\mu_k = 0.060$. What is the total distance traveled by the upper block relative to the supporting block if 15.2 g of water is produced?

Solution:

The 15.2 g of water arises from the melting of an equivalent amount of ice (Δm) due to the friction between the blocks. The amount of heat required is $Q = L_f \Delta m$ and the amount of work done is $W = f \Delta s$, where f is the frictional force, $f = \mu_k N = \mu_k mg$. Setting Q equal to W and solving for Δs, we find

$$\Delta s = \frac{L_f \Delta m}{\mu_k mg} = \frac{(3.337 \times 10^5 \text{ J/kg})(15.2 \times 10^{-3} \text{ kg})}{0.060(10 \text{ kg})(9.80 \text{ m/s}^2)}$$

$$= 863 \text{ m}$$

22-5   CALORIMETRY

Figure 22-10 is a schematic drawing of a multipurpose device called a *calorimeter*. The inner vessel C is the calorimeter proper, and the outer jacket is a heat shield that reduces heat losses from the inner vessel. The water in the jacket is usually maintained at the average temperature prevailing in the calorimeter during an experiment. The calorimeter proper is filled with a measured amount of water, and the heat capacity for the container, stirrer, and thermometer combined is obtained from separate experiments. The device is used for the determination of the heat energy delivered to the calorimeter by any object immersed in the water by measuring the change in temperature of the water. Then, the value of an unknown quantity in Eq. 22-12 can be obtained in the manner described in Example 22-5.

A calorimeter of the type shown in Fig. 22-10 is used to measure temperature changes between initial and final equilibrium states. Another type of calorimeter—a *continuous-flow calorimeter*—is illustrated in Fig. 22-11. This device is particularly suited for measurements of the heat energy released by various substances during combustion. This energy is called the *heat of combustion, L_c*.

Under conditions of steady water flow through the calorimeter, a steady burning rate, and equilibrium temperature values T_1 and T_2 for the input and exhaust water, respectively, the rate of heat generation dQ/dt is

$$\frac{dQ}{dt} = \bar{c}(T_2 - T_1)\frac{dm}{dt} \qquad \textbf{(22-13)}$$

where $\bar{c}$ is the average specific heat of water for the temperature interval $T_2 - T_1$

Fig. 22-10. The essential features of a calorimeter.

Fig. 22-11. The essential features of a continuous-flow calorimeter used for the measurement of heats of combustion for gases.

TABLE 22-5. Heats of Combustion for Various Substances[a]

SUBSTANCE	PHASE[b]	L_c (Cal/kg)	SUBSTANCE	PHASE[b]	L_c (Cal/kg)
Acetylene	g	12,000	Hydrogen	g	33,900
Alcohol (ethyl)	l	7,180	Kerosene	l	11,100
Alcohol (methyl)	l	5,300	Oil (crude)	l	11,500[c]
Carbon (pure)	s	8,080	Oil (heavy)	l	10,200[c]
Coal (anthracite)	s	7,050[c]	Oil (light)	l	10,000[c]
Coal (bituminous)	s	7,300[c]	Propane	g	11,900
Coke	s	7,000[c]	Wood (beech)[d]	s	4,170
Gas (natural)	g	8,400[c]	Wood (oak)[d]	s	3,990
Gasoline	l	11,500	Wood (pine)[d]	s	4,420

[a] Combustion in air at an air temperature of 25° C and at a pressure of 1 atm.
[b] Phase: g (gas), l (liquid), s (solid).
[c] Typical value.
[d] Approximately 12 percent water content.

(that is, $\bar{c}$ is essentially 1.0 Cal/kg · deg), and where dm/dt is the mass flow rate of water through the calorimeter. (See Example 22-6 for an application of this type of calorimeter.)

Table 22-5 gives values for the heats of combustion for various substances burned in air.

Example 22-5

A calorimeter of the type shown in Fig. 22-10 contains 0.80 kg of water at a temperature of 15.0° C. The heat capacity of the container, stirrer, and thermometer combined is $H = 0.0102$ Cal/deg. A 0.40-kg sample of molten lead is poured into the calorimeter. The final (equilibrium) temperature of the system is 25.0° C. What was the initial temperature of the lead? (The specific heat of molten lead in the temperature range here is $\bar{c}' = 0.0375$ Cal/kg · deg.)

Solution:

In the temperature interval, 15° C to 25° C, the average specific heat of water is $\bar{c}_w = 0.999$ Cal/kg · deg (see Fig. 22-8). Therefore, the amount of heat Q_c absorbed by the calorimeter is

$$Q_c = (\bar{c}_w m_w + H) \Delta T_{cal}$$
$$= [(0.999 \text{ Cal/kg} \cdot \text{deg})(0.80 \text{ kg}) + (0.0102 \text{ Cal/deg})](10.0 \text{ deg})$$
$$= 8.094 \text{ Cal}$$

The molten lead cools from the unknown initial temperature t_C to its freezing temperature of 327° C (see Table 22-4), thereby releasing an amount of heat Q_1:

$$Q_1 = \bar{c}' m_{Pb}(t_C - 327° \text{ C})$$
$$= (0.0375 \text{ Cal/kg} \cdot \text{deg})(0.40 \text{ kg})(t_C - 327° \text{ C})$$
$$= (0.015 \text{ Cal})(t_C - 327° \text{ C})$$

The complete freezing of the lead releases an amount of energy Q_2:

$$Q_2 = L_f m_{Pb} = (5.533 \text{ Cal/kg})(0.40 \text{ kg})$$
$$= 2.213 \text{ Cal}$$

where we have used the value of L_f given in Table 22–4. Cooling the solid lead from 327° C to 25° C releases an amount of energy Q_3. For this calculation we use the value of $\bar{c}$ for lead given in Table 22–3. Thus,

$$Q_3 = \bar{c}m_{Pb}(327° \text{ C} - 25° \text{ C})$$
$$= (0.0326 \text{ Cal/kg} \cdot \text{deg})(0.40 \text{ kg})(302 \text{ deg})$$
$$= 3.938 \text{ Cal}$$

Then,

$$Q_c = Q_1 + Q_2 + Q_3$$
$$8.904 \text{ Cal} = (0.015 \text{ Cal})(t_C - 327° \text{ C}) + 2.213 \text{ Cal} + 3.938 \text{ Cal}$$

from which

$$t_C = 456.5° \text{ C}$$

Example 22–6

A continuous-flow calorimeter of the type shown in Fig. 22–11 is used to measure the heat of combustion of methane gas. The flow rate of water through the calorimeter is 0.972 kg/min. The input temperature is 10.0° C, whereas the output temperature is 75.0° C. The methane gas is supplied to be burned at a mass flow rate of 4.8×10^{-3} kg/min. What is the heat of combustion of methane?

Solution:

The average specific heat of water in the temperature integral, 10° C to 75° C, is essentially 1.000 Cal/kg · deg (see Fig. 22–8). Then, using Eq. 22–13, we find

$$\frac{dQ}{dt} = \bar{c}\Delta T \frac{dm}{dt} = (1.000 \text{ Cal/kg} \cdot \text{deg})(65 \text{ deg})(0.972 \text{ kg/min})$$
$$= 63.18 \text{ Cal/min}$$

This heat input to the water is derived from the burning methane supplied at a rate of 4.8×10^{-3} kg/min. Thus,

$$L_c = \frac{63.18 \text{ Cal/min}}{4.8 \times 10^{-3} \text{ kg/min}} = 13,160 \text{ Cal/kg}$$

The Mechanical Equivalent of Heat.   One of the important steps in the development of the concept of heat as a form of energy and the energy conservation principle was the determination by Joule* in 1849 of the ratio of heat energy units to mechanical energy units. Figure 22–12 shows schematically a modern version of the type of apparatus used by Joule (and others) for the determination of the mechanical equivalent of heat. The water-filled calorimeter contains a rotating set of paddles that are driven at constant speed by an external motor. The calorimeter is fitted with a set of interlacing stationary paddles to enhance the churning of the water. The input mechanical energy is transformed into heat through the effects of the viscosity of the water. The calorimeter container itself is restrained from rotating about the

---

*James Prescott Joule (1818–1889), an English physicist. Joule was the son of a wealthy Manchester brewer and had the advantage of a home laboratory supplied by his father. Here he conducted a long series of experiments that demonstrated conclusively the equivalence of heat energy and mechanical energy. Joule's 1849 value was 1 Cal = 4154 J, less than 1 percent from the present accepted value.

Fig. 22–12. A calorimeter used to measure the mechanical equivalent of heat.

drive shaft by the torque provided by two weights, each with mass m, that hang from two cords connected tangentially to the rim of the calorimeter. The amount of mechanical energy input after n rotations of the movable paddles is (see Section 11–3)

$$E = \tau \, \Delta\phi = (2mgR)(2\pi n) = 4\pi nmgR$$

where R is the radius of the calorimeter rim. After the value of E is corrected for the amount of heat absorbed by the calorimeter structure, the measured increase in temperature of the water provides a means of determining the mechanical equivalent of heat.

22-6   HEAT TRANSFER

There are three basic and distinguishable mechanisms by which heat can be transferred from a hotter to a colder body or from a hotter to a colder portion of a body. These heat transfer mechanisms are *convection, radiation,* and *conduction.* We discuss briefly the first two of these processes, and then devote most of our attention to the case of conduction.

Convection.   Convection takes place primarily in fluids (gases and liquids). In these media, heat is transferred by matter that moves as the result of either forced or naturally induced currents. Most substances expand when heated so that their densities decrease. Thus, the heated portion of a fluid is buoyed upward and rises, delivering heat to the cooler fluid above. If a favorable confining geometry exists, this buoyant action sets up a natural circulation called *free* (or *natural*) *convection.* Convection involves the actual macroscopic flow of matter, which can be readily detected. For example, convective effects are responsible, in part, for the movement of water in the oceans and air in the atmosphere. Moreover, after the Earth was formed it was cooled by convective currents in the molten interior. Heated matter can also be forced to move by using pumps or fans; this type of circulation is called *forced convection.* Figure 22–13 illustrates several convective circumstances in a house heated by a hot water system.

The mathematical analysis of convective processes is very complicated. Engineering design relies primarily on empirical information.

Radiation.   In the radiative transfer of heat, energy is carried by *electromagnetic radiation.* Every body, at any temperature above absolute zero, radiates electromagnetic waves from its surface. At each temperature this radiation is distributed over a characteristic wavelength spectrum. For temperatures below about $10^4{}^\circ$ C, the radiation consists primarily of visible light and radiation with longer wave-

Fig. 22–13. Convective circulation in a house heated by a hot water system. The water heated by the furnace is circulated through the radiator by forced convection; in the case here, this circulation is assisted by natural convection. The room air is heated primarily by natural convection. The products of combustion that result from the burning of the fuel in the furnace leave through a flue driven by a slight pressure difference produced by the injected gas, but primarily by a natural convective draft.

lengths.* Between temperatures of about 300° C and about 3000° C, the major portion of the radiation has wavelengths longer than visible light and is called *infrared* (IR) radiation. At the temperature of the Sun's surface—about 6000° C—the radiation is primarily in the visible portion of the spectrum. For temperatures above about $10^{4°}$ C, the dominant spectral component in the emitted radiation has a wavelength shorter than that of visible light; such radiation is called *ultraviolet* (UV) radiation.

Not only does the wavelength spectrum of emitted thermal radiation change with temperature, but so does the amount of radiated energy. In fact, the radiated power increases with the fourth power of the absolute temperature.

Suppose that two objects at different temperatures are placed, without touching, inside an evacuated container that has perfectly reflecting inner walls. The two objects will exchange energy by radiation (and only by this process). The colder object will emit less energy than it receives from the hotter object and vice versa. This process of energy transfer by radiation will continue until an equilibrium state is reached in which each object emits as much energy as it absorbs. Then, the objects will be at the same temperature. We defer a detailed discussion of radiation phenomena until we have developed the necessary background material (Section 46–1).

Conduction. Heat transfer by conduction usually involves substances in the solid phase. (Conduction does take place in fluids, but this process is very similar to *diffusion,* which we discuss later in connection with the kinetic theory of matter.) In the conduction process, heat is transferred between two objects or between two parts of an object through the intervening matter without any macroscopic (visible) motion of the object or its parts. The random thermal atomic motion that constitutes the internal energy of the substance is transferred from one region to another by atom-atom collisions. No long-range displacements of the individual atoms occur during this process.

Isotherms are the familiar curved lines on weather maps that connect ground stations reporting the same temperature. This concept may be extended to the three-dimensional case of solid matter. We define an *isothermal surface* to be a surface within a solid that connects points all at the same temperature. Heat will flow through the solid in the direction in which the rate of change of the temperature is greatest. These lines, which are perpendicular to the isothermal surfaces, are called the *temperature gradient lines.* They may be viewed as heat transfer flow lines or thermal current lines.

Figure 22–14 shows two isothermal surfaces at temperatures $T + \Delta T$ and T. Also shown is a differential volume element with facing surfaces $\Delta A_\perp$ that lie on the isothermal surfaces and are separated by a distance $\Delta \ell_\parallel$ measured parallel to $\hat{\mathbf{n}}$, the local normal to the isothermal surface. The rate of heat transfer between the two surfaces $\Delta A_\perp$ is shown by experiment to be proportional to $\Delta A_\perp$ and to $-dT/d\ell_\parallel$, which is equal to the limit of $\Delta T/\Delta \ell_\parallel$ as $\Delta \ell_\parallel \to 0$. (Notice that the flow direction $\hat{\mathbf{n}}$ is always taken positive in the direction opposite that of the temperature gradient. The designations $\perp$ and $\parallel$ are with respect to $\hat{\mathbf{n}}$.) The constant of proportionality, designated by κ, is called the *thermal conductivity* of the substance. Then, for the rate of heat flow through a portion $\Delta A_\perp$ of an isothermal surface with a temperature gradient $dT/d\ell_\parallel$ evaluated at the specific position of $\Delta A_\perp$, we have

$$\frac{dQ}{dt} = -\kappa \, \Delta A_\perp \frac{dT}{d\ell_\parallel} \tag{22–14}$$

---

*The visible spectrum of light extends from about 700 nm (7000 Å) at the red end to about 400 nm (4000 Å) at the violet end.

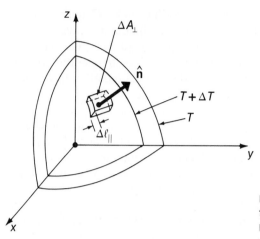

Fig. 22–14. Geometry for discussing the flow of heat through a solid in which a temperature gradient exists.

Notice that κ has units of Calories (or joules) per second per meter per degree: $Cal/s \cdot m \cdot deg$.

Table 22–6 lists some thermal conductivity values for various materials. The thermal conductivity for a substance usually varies considerably with temperature. The values given are average values for the temperature range from $0°$ C to $100°$ C. A "perfect thermal conductor" would have $\kappa = \infty$, whereas a "perfect thermal insulator" would have $\kappa = 0$.

TABLE 22–6.   Thermal Conductivities for Various Materials[a]

SUBSTANCE		SUBSTANCE	
Metals	κ **(Cal/s $\cdot$ m $\cdot$ deg)**	**Gases[b,e]**	κ **(Cal/s $\cdot$ m $\cdot$ deg)**
Aluminum	0.050	Air	6.2×10^{-6}
Brass	0.026	Hydrogen	44.2×10^{-6}
Copper	0.092	Oxygen	6.4×10^{-6}
Gold	0.070	Nitrogen	6.2×10^{-6}
Lead	0.0083	**Composite Materials**	
Silver	0.1000		
Tin	0.0150	Asbestos (fiber)	0.19×10^{-4}
		Brick (common red)	1.5×10^{-4}
Liquids[b]		Concrete	0.81×10^{-4}
		Cork	0.72×10^{-4}
Alcohol (ethyl)[c]	0.400×10^{-4}	Glass (window)	2.5×10^{-4}
Alcohol (methyl)[c]	0.483×10^{-4}	Ice	4.0×10^{-4}
Transformer oil[d]	0.424×10^{-4}	Paper	0.70×10^{-4}
Water $0°$ C	1.35×10^{-4}	Porcelain	2.5×10^{-4}
$20°$ C	1.43×10^{-4}	Rock wool	0.10×10^{-4}
$40°$ C	1.50×10^{-4}	Sawdust	0.12×10^{-4}
$60°$ C	1.56×10^{-4}	Snow (compact)	0.51×10^{-4}
$100°$ C	1.60×10^{-4}	Styrofoam	0.024×10^{-4}
		Wood (fir) $\parallel$ to grain	0.30×10^{-4}
		$\perp$ to grain	0.09×10^{-4}

[a] Average values for $0°$ C $\leqslant t_C \leqslant 100°$ C.
[b] Convection suppressed.
[c] $20°$ C.
[d] $70°$ C–$100°$ C.
[e] $27°$ C; essentially independent of temperature and pressure ($\leqslant 10$ atm).

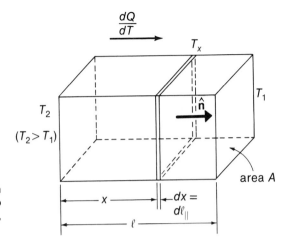

Fig. 22-15. Heat conduction through a rectangular bar that connects two surfaces at constant temperatures, T_2 and T_1.

Consider the flow of heat along a homogeneous rectangular bar (Fig. 22-15) that connects two surfaces at temperatures T_1 and T_2. Assume that the surfaces have infinite heat capacity and that, therefore, they remain at constant temperature throughout the flow process. Assume further that the bar has been in place for a time sufficient to bring about equilibrium conditions for the temperature distribution in the bar, and that there is negligible heat loss through the sides of the bar. In this case, the isothermal surfaces are parallel to the end faces of the bar. With $T_2 > T_1$, the heat flow is directed from the end at T_2 toward the end at T_1 along lines that are parallel to the length of the bar; that is, the flow is parallel to $\hat{\mathbf{n}}$. Then dQ/dt must be constant along the length of the bar. (•Can you see why?) For this simple situation, we can take $\Delta A_\perp$ to be the entire cross-sectional area A of the bar; also $d\ell_\parallel = dx$, so that

$$\frac{dT}{dx} = -\frac{1}{\kappa A}\frac{dQ}{dT} = a \text{ (constant)}$$

Integrating this expression gives $T_x = ax + b$. At $x = 0$, we have $T_x = T_2$; hence, $b = T_2$. At $x = \ell$, we have $T_x = T_1$; hence, $a = -(T_2 - T_1)/\ell$. Finally,

$$T_x = T_2 - \frac{(T_2 - T_1)}{\ell}x \qquad\qquad \textbf{(22-15)}$$

In addition, for the rate of heat flow, we have

$$\frac{dQ}{dt} = \kappa\frac{(T_2 - T_1)A}{\ell} \qquad\qquad \textbf{(22-16)}$$

Equation 22-15 shows that each isothermal surface, which is parallel to the end faces, has a temperature T_x that depends linearly on the distance x from the hotter end of the bar. (This situation corresponds to the curve for $t \to \infty$ shown in Fig. 22-16.)

If we imagine the temperature difference $\Delta T = T_2 - T_1 = \mathcal{F}$ to be the "force" that "drives" the heat flow, we can also consider the quantity $\mathcal{R} = \ell/\kappa A$ to be the "resistance" to the flow. Then, we can write the rate of heat flow as

$$\frac{dQ}{dt} = \frac{\mathcal{F}}{\mathcal{R}}$$

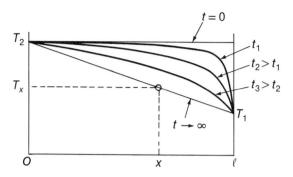

Fig. 22–16. Time dependence of the temperature distribution along a bar whose ends are maintained at constant temperatures, T_2 and T_1. The curves are for the case in which the bar has an initial temperature T_2 throughout and is inserted between the constant temperature surfaces at time $t = 0$.

Compare this equation with similar expressions in Section 18–5 for the case of fluid flow and in Section 31–3 for the case of electric current.

Figure 22–16 shows the approach to the equilibrium temperature distribution for the case in which the bar has initial uniform temperature T_2 and is isolated between the two surfaces at time $t = 0$.

Heat conduction problems require the simultaneous determination of the mathematical description of the isothermal surfaces and the rate of heat flow. This procedure often involves considerable mathematical difficulties. However, in situations that involve certain simple geometric arrangements, as in the case just discussed, it may be possible to guess the shapes of the relevant isothermal surfaces, thereby simplifying the problem significantly.

Example 22–7

A steam pipe with an outside diameter of 5.00 cm has a uniform temperature of 100° C. The pipe, which was installed many years ago, is insulated with a 6.00-cm layer of asbestos fiber. (Because of the environmental problems associated with asbestos, different materials are now being used.) If the room temperature is 20° C, what is the heat loss to the room per hour per meter of pipe length?

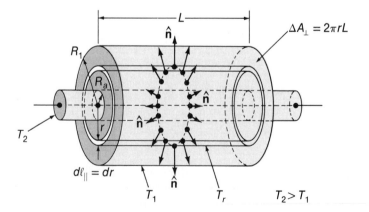

Solution:

The system has a cylindrical symmetry, so we conclude that the isothermal surfaces are concentric cylinders and that heat flows radially outward along lines parallel to the normal vectors $\hat{\mathbf{n}}$, as shown in the diagram. Let the temperature at radius r be T_r and set $d\ell_\| = dr$ and $\Delta A_\perp = 2\pi rL$. When equilibrium conditions have been attained, the total outward heat flow must be independent of r. (•Can you see why?) Using

$$\frac{dQ}{dt} = -\kappa \frac{dT}{dr} \cdot 2\pi r L$$

we can write

$$r\frac{dT}{dr} = -\frac{1}{2\pi L \kappa}\frac{dQ}{dt} = a \text{ (constant)} \qquad (1)$$

Next, we write

$$dT = a\frac{dr}{r}$$

Integrating, we find

$$T_r = a \ln r + b$$

At $r = R_2$, we have $T_r = T_2$, and at $r = R_1$, we have $T_r = T_1$. Thus,

$$T_2 = a \ln R_2 + b$$
$$T_1 = a \ln R_1 + b$$

Subtracting one of these equations from the other gives the expression for a, which can be substituted into Eq. (1) with the result

$$\frac{dQ}{dt} = \frac{2\pi L \kappa (T_2 - T_1)}{\ln (R_1/R_2)} \qquad (2)$$

We have the additional result that

$$T_r = T_2 - (T_2 - T_1)\frac{\ln (r/R_2)}{\ln (R_1/R_2)}$$

That is, the temperature of the isothermal surfaces varies logarithmically with the radial distance r.

Evaluating the heat flow, using the appropriate value of κ from Table 22–6, we find

$$\frac{dQ}{dt} = \frac{2\pi(1 \text{ m})(0.19 \times 10^{-4} \text{ Cal/s} \cdot \text{m} \cdot \text{deg})(80 \text{ deg})}{\ln (8.50 \text{ cm}/2.50 \text{ cm})}$$
$$= 7.80 \times 10^{-3} \text{ Cal/s} = 28.1 \text{ Cal/h}$$

for a 1-m length of the pipe. (•What is the temperature in the middle of the layer of insulation?)

PROBLEMS

Section 22-2

22–1 During a vacation in Budapest you become ill and the doctor reports that your temperature is 39.6° C. How much of a fever do you have? (Normal body temperature is 98.6° F.)

22–2 The highest recorded temperature on Earth was 136° F, at Azizia, Libya, in 1922. The lowest recorded temperature was −127° F, at Vostok Station, Antarctica, in 1960. Express these temperature extremes in degrees Celsius.

22-3 A helium gas thermometer indicates a pressure of 86.3 torr at the triple point of water and a pressure of 4.4 torr at the temperature of the sample. What is the sample temperature in degrees Celsius? (Assume that the pressures involved are sufficiently low that Eq. 22-3 applies.)

Section 22-3

(When necessary assume that α and β are independent of temperature, so that $\alpha = \bar{\alpha}$ and $\beta = \bar{\beta}$.)

22-4 A copper telephone wire is strung, with little sag, between two poles that are 35 m apart. How much longer is the wire on a summer day with $t_C = 35°$ C than on a winter day with $t_C = -20°$ C?

22-5 The New River Gorge bridge in West Virginia is a steel arch bridge 518 m in length. What total expansion must be allowed for between temperature extremes of $-20°$ C and $35°$ C?

22-6 A pendulum clock, made entirely of brass, is ad-
 • justed to keep perfect time at a room temperature of 18°C. During a power failure the air conditioner is out of service and the temperature rises to 27° C. By how much per day does the clock run slowly? [*Hint:* Refer to Section 15-4.]

22-7 A hole with a diameter of 2.00 cm is drilled through a stainless steel plate at room temperature (20° C). By what fraction ($\Delta r/r$) will the hole be larger when the plate is heated to 150° C?

22-8 A cylindrical stainless steel plug is inserted into a circular hole with a diameter of 2.60 cm in a brass plate. When the plug and the plate are at a temperature of 20° C, the diameter of the plug is 0.010 mm smaller than the diameter of the hole. At what temperature will the plug be squeezed tight?

22-9 Equation 22-9 was derived for the case of a parallelepiped. Show that the same result is obtained for the case of a sphere.

22-10 The mercury column in a barometer stands at a height of 76.02 cm (and therefore indicates a pressure of 760.2 torr) when the temperature is 0° C. What will be the indicated pressure for the same atmospheric pressure on a warm day when the temperature is 30° C?

22-11 A tungsten wire with a diameter of 0.20 mm is stretched until the tension is 50 N. The wire is then clamped to a stout aluminum bar when both are at a temperature of 20° C. What will be the tension in the wire when the wire and the bar are brought to a temperature of 150° C? (Young's modulus for tungsten is $Y = 34 \times 10^{10}$ Pa.)

Section 22-4

22-12 A metal block is made from a mixture of aluminum (2.4 kg), brass (1.6 kg), and copper (0.8 kg). What amount of heat (in Cal) is required to raise the temperature of this block from 20° C to 80° C?

22-13 A sheet of mill iron with an area of 1 m² and a thickness of 5 mm is cooled from 600° C to 250° C by sprayed water delivered at 25° C. Assume that the water is converted to steam at 100° C upon contact with the iron. How much water is required to cool the iron? (Take $\bar{c}$ for iron in this temperature range to be 0.145 Cal/kg · deg.)

22-14 A friction brake is used to stop a flywheel that has a mass of 10 kg and a radius of gyration of 0.40 m. The initial rotational speed of the flywheel is 2000 rpm. (a) How much heat (in Cal) is generated in the slowing process? (b) If the brake unit, with a mass of 6.0 kg and an initial temperature of 60° C, absorbs all of the heat generated, what temperature will it reach? (Take $\bar{c}$ for the brake to be 0.15 Cal/kg · deg.)

Section 22-5

22-15 A 0.450-kg sample of copper at a temperature of 87.2° C is immersed in 0.350 kg of water at 10.00° C contained in an aluminum calorimeter vessel with a mass of 30.2 g. The final equilibrium temperature of the system is 18.12° C. What is $\bar{c}$ for copper in this temperature range? (Neglect the heat capacity of the stirrer and the thermometer.)

22-16 A 300-g piece of tin is placed on a large block of ice, and this system is thermally isolated from its surroundings. As the tin cools to the temperature of the block (0° C), 18.2 g of ice melts. What was the original temperature of the piece of tin?

22-17 A heated iron rod ($m = 0.10$ kg) is immersed in a calorimeter that contains 0.20 kg of water at a temperature of 20.0° C. The heat capacity of the calorimeter (excluding the water) is 0.015 Cal/deg. If the final temperature of the system is 45.0° C, what was the original temperature of the iron rod?

22-18 A *bomb calorimeter* consists of a sealed chamber

Problem 22-11

that is immersed in the water of an ordinary calorimeter (Fig. 22–10). By burning a solid or liquid substance in the bomb and measuring the temperature rise of the water, the heat of combustion of the substance can be determined. (For gases, the continuous-flow method is used—Fig. 22–11.) The burning of 5.40 g of North Dakota lignite in a bomb calorimeter results in an increase of the water temperature from 20.2° C to 36.1° C. The calorimeter contains 0.90 kg of water, and the heat capacity of the remainder of the calorimeter is $H = 0.23$ Cal/deg. What is the heat of combustion for this type of coal? Compare with the values given in Table 22–5.

22–19 Water is heated from 10° C to 82° C in a residential hot water heater at a rate of 70 liters per minute (ℓ/min). Natural gas with a density of 1.12 kg/m³ is used in the heater, which has a transfer efficiency of 32 percent. What is the gas consumption rate in m³/h? (See Table 22–5.)

22–20 An automobile engine delivers a power of 60 hp while using 5.2 gallons of gasoline per hour. What is the overall efficiency of the engine? (1 U.S. gal = 3.79 ℓ; the density of gasoline is 740 kg/m³. See Table 22–5.)

Section 22-6

22–21 The wall of a factory, with dimensions 6.0×15.0 m, is made from common brick and has a thickness of 0.25 m. (a) If the outside temperature is a steady $-7.0°$ C, while the inside temperature is constant at 24° C, what amount of heat is lost by conduction through the wall per hour? (b) How many gallons (1 U.S. gal = 3.79 ℓ) of crude oil burned at 30 percent efficiency per 8-h working day does this heat loss represent? (The density of crude oil is 919 kg/m³. Use the information in Table 22–5.)

22–22 A closed cubical box with sides of length 0.50 m is constructed from wood that has a thickness of 2.0 cm. When the outside temperature is 20.0° C, an electric heater inside the box must supply 110 W of power to maintain a uniform air temperature of 85.0° C within the box. What is the average thermal conductivity of the wood used?

22–23 A compound slab with a cross-sectional area A is composed of a thickness ℓ_1 of a material with a thermal conductivity κ_1, and a thickness ℓ_2 of a material with κ_2. The temperatures of the surfaces are maintained at T_1 and T_2, as indicated in the diagram. (a) Show that the rate of heat flow through the compound slab can be expressed as $dQ/dt =$

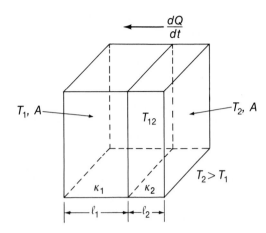

$(T_2 - T_1)/\Re$, where $\Re = \Re_1 + \Re_2 = (\ell_1/\kappa_1 A) + (\ell_2/\kappa_2 A)$. (b) Show that the temperature T_{12} of the interface is $T_{12} = T_1 + (T_2 - T_1)\Re_1/(\Re_1 + \Re_2)$.

22–24 Suppose that the wall in Problem 22–21 is insulated with 3.0 cm of rock wool. For the same temperature conditions, what is the ratio of heat loss rate with the insulation in place to the original heat loss rate?

22–25 A completely enclosed calorimeter made of copper with a wall thickness of 1.0 mm has a surface area of 335 cm². Mechanical work is delivered to the water at a rate of $\frac{1}{4}$ hp by means of stirring paddles. When equilibrium is achieved, the water in the calorimeter has a uniform temperature. What is the difference between the water temperature and the outside temperature?

22–26 One end of an aluminum rod that has a length of 35.0 cm and a diameter of 1.0 cm is immersed in boiling water at atmospheric pressure. The other end is in thermal contact with a large block of ice at 0° C. After the system has reached equilibrium (with no losses through the sides of the rod), what is the melting rate of the ice (in kg/h)?

Additional Problems

22–27 At what absolute temperature are the Celsius and Fahrenheit temperatures numerically equal? At what absolute temperature is the Fahrenheit temperature twice the Celsius temperature?

22–28 Express the temperature of the triple point of water in °F and in °R.

22–29 Imagine a steel belt that encircles the Earth at the Equator with a snug fit when the temperature of the belt is a uniform 20° C. What would be the uniform space between the belt and the surface of the Earth if the temperature of the belt were increased to 21° C? (Assume that the Equator is a perfect circle.)

22-30 (a) Show that the length ℓ of a rod can be expressed as

$$\ell = \ell_0 \left(1 + \int_{t_{0C}}^{t_C} \alpha(t_C)\, dt_C\right)$$

(b) In the temperature range from $0°$ C to $100°$ C, the thermal expansion coefficient for aluminum is given approximately by $\alpha(t_C) = (2.25 \times 10^{-5}\text{ deg}^{-1}) + (2.3 \times 10^{-8}\text{ deg}^{-2})t_C$. Find the change in length, $\Delta\ell = \ell - \ell_0$, when a 1-m aluminum wire is heated from $0°$ C to $100°$ C. (c) Determine $\alpha(t_C = 20° \text{ C})$ and also $\bar\alpha$ for the temperature range from $0°$ C to $100°$ C. Compare with the entry for aluminum in Table 22-2.

22-31 Steel rails for an interurban rapid transit system form a continuous track that is held rigidly in place in concrete. If the track was laid when the temperature was $0°$ C, what is the stress in the rails on a warm day when the temperature is $25°$ C? What fraction of the yield strength of 52.2×10^7 Pa does this stress represent?

22-32 A very thin spherical shell of silver has a radius of 5.00 cm at a temperature of $10°$ C. If the shell is heated until it has a uniform temperature of $30°$ C, find (a) the change in radius, (b) the change in surface area, and (c) the change in volume.

22-33 A glass container with a volume V is filled to the brim with a liquid when the temperature of both the container and the liquid is T. The system is now heated to a uniform temperature $T + \Delta T$. (a) Show that the volume of liquid that overflows is $\Delta V = V(\bar\beta_\ell - \bar\beta_g)\,\Delta T$, where $\bar\beta_\ell$ and $\bar\beta_g$ are the average volume expansion coefficients of the liquid and the glass, respectively. (b) A one-liter carafe is filled to the brim with white wine at the uniform (refrigerated) temperature of $5°$ C. The carafe is made of ordinary glass ($\bar\beta_g = 3.2 \times 10^{-5}\text{ deg}^{-1}$), and 6.8 cm^3 of wine overflows when the carafe is warmed to room temperature ($25°$ C). What is the average coefficient of volume expansion $\bar\beta_\ell$ for the wine?

22-34 A mercury thermometer, similar in design to the one shown in Fig. 22-1, has a reservoir bulb that is 5 mm in diameter and 2 cm in length, and a capillary tube that has a bore diameter of 0.20 mm. The thermometer is made of Pyrex glass, and only the bulb is subjected to temperature changes when in use. What is the total distance of movement of the upper surface of the mercury column in the tube when the bulb temperature changes from $0°$ C to $100°$ C?

22-35 A spherical shell of copper is completely filled, at a temperature of $10°$ C, with a liquid that has a bulk

modulus $B_\ell = 2.05 \times 10^9$ Pa and a volume thermal expansion coefficient $\bar\beta_\ell = 0.207 \times 10^{-3}\text{ deg}^{-1}$. (a) Show that the outward pressure Δp on the shell that results from a temperature increase ΔT is $\Delta p = B_\ell(\bar\beta_\ell - 3\bar\alpha_s)\,\Delta T$, where $\bar\alpha_s$ is the linear expansion coefficient for the shell material and where the bulk modulus for the shell material is $B_s \gg B_\ell$. (b) What is the pressure Δp when the temperature is increased to $100°$ C? What fraction of the yield strength of 34.0×10^7 Pa for copper does this stress represent? [*Note:* The hypothetical liquid in this problem resembles water. However, the properties of water are too variable with pressure and temperature to justify citing it as the liquid in question here.]

22-36 An approximate expression for the specific heat of water in the temperature range $0° \text{ C} \leqslant t_C \leqslant 100°$ C is $c = 1.00680 - (5.509 \times 10^{-4})t_C + (9.277 \times 10^{-6})t_C^2 - (3.768 \times 10^{-8})t_C^3$. The coefficients carry appropriate units so that when the temperature is expressed in $°$C, the specific heat is given in Cal/kg · deg. (a) Calculate the amount of heat required to raise the temperature of 1 kg of water from $0°$ C to $100°$ C. (b) What is the average value of c in this temperature interval? (c) Determine the temperature in this interval at which the specific heat is minimum. Compare your result with the graph shown in Fig. 22-8.

22-37 A calorimeter contains 80 g of ice and 300 g of water in equilibrium. The heat capacity of the remainder of the calorimeter is 0.0193 Cal/deg. If a 550-g block of aluminum at an initial temperature of $100°$ C is dropped into the calorimeter, what will be the final temperature of the system?

22-38 A 4-ounce (118.3-mℓ) whiskey drink (half ethyl alcohol and half water by mass), originally at a temperature of $25.0°$ C, is cooled by adding 20 g of ice at $0°$ C. If all the ice melts, what is the final temperature of the drink? (Assume that there are no heat losses and that the specific heat can be calculated as for a simple mixture. The density of the whiskey is 0.919 g/cm^3.)

22-39 The Trümmelbach waterfall in Switzerland has a height of approximately 400 m. Assume that all of the available gravitational potential energy of the water at the top of the falls is converted to heat energy in the water at the bottom of the falls. What should be the temperature increase of the water due to the 400-m drop? (Joule actually made such a measurement while on his honeymoon in 1847.)

22-40 A 45-g lead bullet is fired with a velocity of 250 m/s at a steel plate. The bullet is stopped completely by the plate, with 9/10 of the initial kinetic energy appearing as heat in the bullet. What is the temperature increase of the bullet?

22–41 A simplified version of the calorimeter shown in
• Fig. 22–12 is used in an elementary physics laboratory for the determination of the mechanical equivalent of heat. Each counterbalancing weight has a mass of 2.50 kg and the calorimeter rim has a radius of 8.00 cm. The calorimeter proper contains 1.200 kg of water, and the heat capacity of the remainder of the calorimeter is 0.072 Cal/deg. A total of 470 rotations of the paddle shaft increases the temperature of the water from 14.00° C to 16.21° C. Use this information to determine the mechanical equivalent of heat. (Why might this simple experiment yield too small a value?)

22–42 A long tungsten wire with a diameter of 0.0115 mm is encased in a porcelain sheath that has a diameter of 3.00 mm. The wire is heated electrically, and at equilibrium, the temperature of the wire is 1200° C and the temperature of the porcelain surface is 150° C. What power (in W/m) is being supplied to the tungsten wire by the electric current that flows through it?

22–43 Each of two panes of glass has a thickness of 3 mm.
• Compare the rate of heat flow through a double-pane window constructed by leaving a 3-mm air gap between the panes with the rate of heat flow through a single 3-mm pane for the same temperature difference. Are double pane windows good thermal insulators?

22–44 Suppose that the steam pipe in Example 22–7 is fit-
•• ted with a 2.00-cm-thick layer of Styrofoam surrounding the asbestos insulation. (a) What is the new rate of heat loss (in Cal/h) per meter of length? (b) What is the temperature at the interface between the asbestos and the Styrofoam?

22–45 The cold air outside a cabin has a temperature T_0.
• This air blows through a crack at an average speed of 2.5 m/s and enters the cabin while still at the temperature T_0. A strip of paper with a thickness of 0.18 mm is pasted over the crack. What is the ratio of heat loss by the cabin air without the patch and

with the patch? (The specific heat of air is 0.496 Cal/kg · deg.)

22–46 Consider a simplified model of a house heating
•• problem. The outside surface area of the house is 300 m², and the wall thickness is 10 cm, with an effective heat conductivity $\bar{\kappa} = 5.0 \times 10^{-6}$ Cal/s · m · deg. The interior contents of the house (including the occupants) have a combined mass of 2000 kg and an effective specific heat $\bar{c} = 0.20$ Cal/kg · deg. Take the outside nighttime temperature to be constant at 0° C. Assume that during the day, the house and its contents have been heated to a uniform interior temperature of 20° C. (a) If the furnace is turned off for the night, what interior temperature would the occupants find upon waking up 6 h later? (Assume that the house cools uniformly. Neglect the heat capacity of the walls.) (b) How much heat must the furnace supply to heat the interior to the former temperature of 20° C? (Assume that this heat is supplied in such a short time interval that none escapes to the outside.) (c) If the furnace had been left on for the 6-h period to maintain a constant interior temperature of 20° C, how much heat would the furnace have provided? (d) How much heat was saved by shutting off the furnace for the night?

22–47 On a cold winter day, the uppermost layer of water
•• in a lake is at a temperature of 0° C. Colder air blows in and maintains the surface at −15° C, causing the lake to freeze over. Consider the ice formation to proceed by heat loss due to conduction through the ice already formed. Show that the rate of formation of the ice layer of thickness x is $dx/dt = (\kappa_{ice}/x)(\Delta T/L_f)$. What is the thickness of the ice 2 h after freezing begins?

22–48 A radioactive source generates heat at a rate of
•• 296 W in a spherical capsule with a diameter of 2.0 cm. The capsule is embedded in the center of a brass sphere that has a diameter of 20.0 cm. Assume that all of the nuclear radiation is confined within the capsule. What is the temperature of the capsule if the surface of the brass sphere has a constant equilibrium temperature of 20° C?

CHAPTER 23

THE BEHAVIOR OF GASES

In this chapter, we are concerned primarily with the description of the *macroscopic* behavior of gases in various states of thermal equilibrium. We consider first the properties of an *ideal gas* and then the behavior of real substances. Although we concentrate on the gaseous phase of matter, we are also interested in the transitions to other phases.

It is our goal in Chapter 24 to give a description of the *microscopic* or molecular behavior of gases. Some of these ideas are introduced in this chapter, so we begin by presenting a brief résumé of some of the important features of the atomic theory of matter.

23-1 THE ATOMIC THEORY OF MATTER

According to the atomic theory, any sample of bulk matter consists of an enormous number of tiny bits of matter called *atoms*. Because there is obviously a great variety in the types of bulk matter, there must be at least some variety in the types of atoms. It is now known that atoms do indeed exist in a variety of types called *elements*, of which there are more than 100 chemically differentiable species. An atom is defined as *the smallest unit of matter that can be identified as a particular chemical element.*

Atomic and Molecular Sizes. A number of different types of measurements have demonstrated that atoms have sizes ("diameters") that range from about 0.15 nm (1.5 Å) to about 0.35 nm (3.5 Å), with the more massive atoms tending to have the larger diameters on the average. An atomic "size" refers to the extent of the (not well-defined) region around the atomic center (or *nucleus*) that is occupied by the atom's orbiting electrons. Most of the mass of an atom is concentrated, in the form of *protons* and *neutrons,* in the atomic nucleus, even though the size of the nucleus is much less than that of the atom. Nuclear diameters range from about 1.5 fm to about 15 fm. (Note that 0.15 nm $= 1.5 \times 10^{-10}$ m, whereas 1.5 fm $= 1.5 \times 10^{-15}$ m.) To better appreciate the scale of these sizes, imagine a baseball to be magnified to the size of the Earth, with the constituent atoms undergoing the same magnification. Then, an atom of medium size would become as large as a baseball. If this atom were further magnified to the size of the largest stadium (diameter about 250 m), the nucleus at this new magnification would be about as large as a small pea.

A combination of two or more atoms bound together is called a *molecule*. If the atoms are of different chemical elements, the molecule represents a chemical *compound*. The uniform properties of pure chemical substances, such as water or ordinary table salt, suggest a model for molecules in which a definite number of atoms of a particular type is associated with each molecular species. Thus, a molecule of water consists of two hydrogen atoms (symbol H) and one oxygen atom (symbol O), and we use the chemical notation H_2O to represent the water molecule (Fig. 23–1). Also, the basic unit of salt contains one sodium atom (symbol, Na) and one chlorine atom (symbol, Cl), and we write NaCl for sodium chloride. Sodium can be prepared in elemental (atomic) form, Na, but hydrogen, oxygen, and chlorine all occur as diatomic molecules, H_2, O_2, and Cl_2.

The force that acts between two atoms is a complicated function of the distance between the centers of the interacting atoms. The most meaningful way to discuss such interactions is in terms of the corresponding potential energy function. Figure 23–2 shows the potential energy of the diatomic hydrogen molecule H_2 as a function of the separation of the atomic centers. The force is related to the potential energy according to $F = -dU/dr$. In Fig. 23–2 we see that U is minimum and therefore that $F = 0$ at $r = 0.74$ Å, which is the equilibrium separation. For separations smaller than this value, the interatomic force is increasingly *repulsive*. For larger separations the force is *attractive* and begins to decrease rapidly with distance for $r \gtrsim 1$ Å. The force is effectively zero for separations greater than about 4 Å.

Fig. 23–1. Schematic representation of the chemical reaction that converts hydrogen and oxygen into water. A chemical reaction is simply the rearrangement of the constituent atoms into a new molecular form.

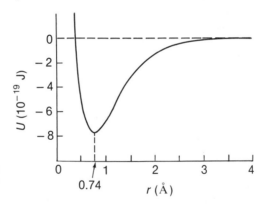

Fig. 23–2. The potential energy U of the H_2 molecule as a function of the distance between the atomic centers.

The equilibrium separation of the atoms in the hydrogen molecules is somewhat less than the "diameter" of the hydrogen atom (1.02 Å). This is the usual case because the atoms in a molecule slightly overlap at the equilibrium separation. Moreover, in solids and liquids, the molecules also "touch" (see Fig. 23–3). In contrast, the gas molecules in a dilute gas are, on the average, many molecular diameters apart.

Atomic and Molecular Masses. The chemical properties of a particular element are determined by the number of electrons in the atoms of that element. The number of electrons in an electrically neutral atom is exactly equal to the number Z of protons in the nucleus; Z is called the *atomic number* of the element.

The nuclei of a particular element may contain different numbers N of neutrons. Such nuclei, all with the same number of protons but with different numbers of neutrons, are called *isotopes* of that chemical element.

The total number of particles in a nucleus is called the *mass number A* of the particular isotope: $A = Z + N$. In order to distinguish one isotope of an element from another, it is customary to write the mass number as a superscript to the

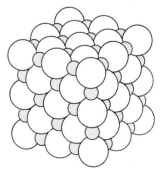

Fig. 23–3. This model of "touching" atoms was constructed by William Barlow in 1898 to represent the structure of the sodium chloride crystal. The small spheres are sodium atoms (actually, *ions*) and the large spheres are chlorine atoms (ions).

chemical symbol. Thus, the carbon ($Z = 6$) that is found in Nature consists of the isotopes ^{12}C (with $N = 6$), ^{13}C (with $N = 7$), and ^{14}C (with $N = 8$). The isotope ^{12}C (the "normal" isotope) has a natural abundance of 98.89 percent (that is, it constitutes 98.89 percent of the carbon atoms in a natural sample).

Most hydrogen (99.985 percent) occurs as ^{1}H; the nucleus of ^{1}H is a single proton (Fig. 23–4a). The remainder of natural hydrogen (0.015 percent) is "heavy hydrogen," ^{2}H, or *deuterium*. The deuterium nucleus contains one neutron in addition to the proton (Fig. 23–4b). A third isotope of hydrogen, ^{3}H, has two nuclear neutrons (Fig. 23–4c). This isotope, called tritium, is radioactive, and only traces are found in Nature.

(a) ^{1}H (b) ^{2}H (c) ^{3}H
$M = 1.007825$ u $M = 2.014102$ u $M = 3.016094$ u

Fig. 23–4. Schematic diagrams of the three isotopes of hydrogen: (a) ^{1}H, (b) ^{2}H, and (c) ^{3}H.

Using modern techniques, it is possible to measure the mass of one isotope relative to another with high precision. For example, the mass of a hydrogen atom (^{1}H) relative to the mass of a carbon atom (^{12}C) is known to 1 part in 10^{8}. Atomic masses are usually expressed in terms of a special unit called the *unified atomic mass unit* (symbol u), where 1 u is defined to be exactly $\frac{1}{12}$ of the mass of the isotope ^{12}C. In terms of u, the mass of every atom is approximately (but not exactly) equal to its mass number A. For example,

$$M(^{1}H) = 1.007825 \text{ u}$$
$$M(^{59}Co) = 58.9332 \text{ u}$$
$$M(^{197}Au) = 196.9666 \text{ u}$$

In terms of the kilogram, we have

$$1 \text{ u} = 1.66057 \times 10^{-27} \text{ kg}$$

with an uncertainty of about one in the last decimal. Atomic masses can usually be compared with a precision nearly 100 times greater than they can be given in kilograms.

The elements that occur in Nature consist of a mixture of isotopes. (Several elements—for example, fluorine, aluminum, cobalt, and gold—have only a single stable isotope.) Consequently, the quoted atomic mass of a particular element is actually a weighted sum of the masses of the various isotopes. The atomic masses shown on page 610 correspond to the naturally occurring isotopic mixtures.

When two or more atoms combine to form a molecule of a chemical compound, the *molecular mass* of the substance is simply the sum of the atomic masses of the constituents.

Example 23–1

There are two stable isotopes of copper with the isotopic masses and natural abundances given below. What is the atomic mass of natural copper?

$$^{63}\text{Cu: }62.9296 \text{ u} \qquad (69.09\%)$$
$$^{65}\text{Cu: }64.9278 \text{ u} \qquad (30.91\%)$$

Solution:

The weighted sum is

$$M(\text{Cu}) = (62.9296 \text{ u})(0.6909) + (64.9278 \text{ u})(0.3091)$$
$$= 63.5472 \text{ u}$$

This is the atomic mass of copper as found in Nature.

Example 23–2

Using the atomic masses shown on page 610, determine the molecular masses of (a) water, H_2O, (b) methyl alcohol, CH_3OH, and (c) uranium oxide, U_3O_8.

Solution:

We have

(a) $M(H_2O) = [2 \times (1.00797 \text{ u})] + [1 \times (15.9994 \text{ u})] = 18.0153 \text{ u} \cong 18 \text{ u}$

(b) $M(CH_3OH) = [4 \times (1.00797 \text{ u})] + [1 \times 12.01115 \text{ u})] + [1 \times (15.9994 \text{ u})]$
$$= 32.0424 \text{ u} \cong 32 \text{ u}$$

(c) $M(U_3O_8) = [3 \times (238.03 \text{ u})] + [8 \times (15.9994 \text{ u})] = 842.085 \text{ u} \cong 842 \text{ u}$

Avogadro's Number and the Mole. In molecular theory, particularly in the study of gases, it proves convenient to measure the quantity of a substance in terms of the number of molecules instead of the mass. For this purpose we employ a unit called the *mole* (abbreviation, *mol*). The term *molecule* refers to the basic unit of a substance, whether an element or a compound, that occurs naturally. Thus, a molecule of hydrogen is H_2, but a molecule of helium is He (an atom). With this meaning, we define one mole (1 mol) to be the amount of a particular substance that contains as many molecules as there are atoms in exactly 0.012 kg of ^{12}C. Thus, the mass of 1 mol of ^{12}C (molecular mass $M = 12$ u) is just 12 g. The mass in grams of 1 mol of any substance is equal to the molecular mass of that substance expressed in u.* That is, 1 mol of H_2O ($M \cong 18$ u) contains approximately 18 g and 1 mol of CH_3OH ($M \cong 32$ u) contains approximately 32 g (refer to Example 23-2).

It is clear that the number of molecules in a mole must be the same for all substances. This number is called *Avogadro's number*, in honor of the Italian physicist Amedeo Avogadro, Count of Quaregna (1776–1856).† The presently accepted value of Avogadro's number is

---

*Chemists refer to the mole as the *gram molecular weight*. (The definition actually refers to *mass*, not *weight*, but the distinction is not important here.)

†Avogadro, in 1811, put forward the hypothesis that equal volumes of all gases (at the same temperature and pressure) contain equal numbers of particles. (This brilliant idea was not generally accepted until 50 years later.) Avogadro was careful to state that the particles to which his hypothesis refers are *molecules* (as we have defined the term). One of the reasons that Avogadro's hypothesis was not quickly accepted was the vigorous opposition expressed by the prominent English chemist John Dalton (1766–1844), whose views were the result of the fact that he did not understand the distinction between atoms and molecules. Indeed, Avogadro coined the term *molecule* and gave it the modern interpretation.

$$N_0 = 6.02205 \times 10^{23} \text{ molecules/mol}$$

with an uncertainty of about 3 units in the last decimal. Notice particularly that Avogadro's hypothesis is not valid if stated in terms of *atoms* instead of *molecules*. For example, it is *not* true that there are N_0 atoms in 1 mol of water.

23-2   THE EQUATION OF STATE (IDEAL GASES)

Consider a quantity of a homogeneous, chemically pure substance that constitutes a simple thermodynamic system in a state of thermal equilibrium. The variables that define completely the *thermodynamic state* of the system are the temperature T, the pressure p, the volume V, and the mass m. The functional relationship that connects these variables,

$$F(T, p, V, m) = 0$$

is called the *equation of state* of the system. We may, of course, write this relationship in other forms, such as

$$T = f(p, V, m)$$

The function F (or f) must be determined by experiment.

One of the earliest relevant studies was carried out in 1662 by Robert Boyle (1627–1691), an Irish chemist, who investigated the pressure dependence of the volume of a confined gas (air) at constant temperature. Boyle discovered that, for a large range of values for p and V separately, the product pV remained constant (for constant temperature*). Thus, for a fixed quantity of gas, any two states, (1) and (2) with $T_1 = T_2$, have the pressure-volume relationship

$$p_1 V_1 = p_2 V_2$$

This equality expresses *Boyle's law*. Accordingly, we can combine the variables p and V and write the equation of state as $T = f(p, V, m) = f(pV, m)$; solving for pV, this relation can be expressed as

$$pV = g(T, m)$$

Avogadro's hypothesis states that equal volumes of all gases, under the same conditions of pressure and temperature, contain the same number of molecules. Using this idea, pV can be expressed as a function of the temperature as

$$pV = n\phi(T)$$

where n is the number of moles in the sample.

Two French physicists, Jacques Charles (1746–1823) and Joseph Louis Gay-Lussac (1778–1850), discovered independently that a fixed amount of any confined

---

*Boyle was rather lucky in his experiments because he took no precautions to maintain the temperature constant; however, his data indicate that there was little temperature variation of his apparatus. In fact, Boyle initiated the practice of carefully describing experiments and reporting data so that others could check the results.

gas at a constant pressure has a volume that depends linearly on the temperature. Thus, we can revise further the expression for the equation of state and write

$$pV = n(a + bT)$$

where the constants a and b depend on the scale used to measure the temperature. This relationship is known as the *law of Charles and Gay-Lussac*. In Section 22–2, we found that when the Kelvin temperature scale is used, the pressure of a gas becomes (extrapolates to) zero at zero temperature; thus, on this scale, the constant a is zero. Also, the constant b is traditionally written as R, so that the expression for the equation of state becomes

$$pV = nRT \qquad\qquad \textbf{(23–1)}$$

This equation expresses the *ideal gas law*.

The quantity R—called the *universal gas constant*—must be determined by experiment; the result is

$$R = 8.3144 \text{ J/mol} \cdot \text{K}$$

with an uncertainty of about 3 in the last decimal.

It is important to remember when applying the ideal gas law equation that p refers to the *absolute pressure* (not the gauge pressure*) and that T refers to the *kelvin* or *absolute temperature*.

How well do gases obey the ideal gas law? In Section 22–2, it was found that the temperature of the gas in a constant-volume gas thermometer is proportional to the pressure in the limit of low pressures (that is, for dilute gases). (See Fig. 22–4 on page 615.) Thus, real gases can be expected to obey the ideal gas law only in this limit. A hypothetical gas for which Eq. 23–1 is exactly correct under all conditions is called an *ideal gas*.

Dalton's Law of Partial Pressures. When a gas sample consists of a mixture of gases—n_1 moles of type 1, n_2 moles of type 2, and so forth—the ideal gas law equation becomes

$$pV = (n_1 + n_2 + n_3 + \cdots)RT$$

We may define the so-called *partial pressures* by the relations

$$p_1 V = n_1 RT, \qquad p_2 V = n_2 RT, \qquad \text{etc.}$$

Then, we have

$$p = p_1 + p_2 + p_3 + \cdots$$

This equation expresses *Dalton's law of partial pressures:* The total pressure of a mixture of gases is equal to the sum of the individual partial pressures, each obtained as if it were the only gas in the volume V.

---

*Recall that $p(\text{absolute}) = p(\text{gauge}) + p(\text{atmospheric})$.

Example 23–3

Determine the volume of 1 mol of an ideal gas at a temperature of $0°$ C and a pressure of 1 atm (standard temperature and pressure, commonly abbreviated as STP).

Solution:

The temperature that corresponds to $t_C = 0°$ C is $T = 273.15$ K; also, 1 atm $= 1.01325 \times 10^5$ Pa. Hence,

$$V = \frac{nRT}{p} = \frac{(1 \text{ mol})(8.3144 \text{ J/mol} \cdot \text{K})(273.15 \text{ K})}{1.01325 \times 10^5 \text{ Pa}}$$

$$= 2.2414 \times 10^{-2} \text{ m}^3$$

$$= 22.414 \text{ } l$$

This is a useful number to remember: 1 mol (at STP) $\to$ 22.4 l.

Example 23–4

The ideal gas law equation (Eq. 23–1) is valid for a real gas if the pressure is sufficiently low. Look at the upper line in Fig. 22–4, which describes the results of experiments performed with oxygen. If the initial (triple-point) pressure is 150 torr (point b), the temperature determined for the steam point is 373.25 K, only 0.10 K in error. This implies that oxygen at pressures near 150 torr and at temperatures between $0°$ C and $100°$ C behaves in a manner closely resembling an ideal gas. Use the ideal gas law equation to determine the density of oxygen at a pressure of 150 torr and at a temperature of 273.16 K.

Solution:

From the table of atomic masses shown in Fig. 47–12 we find $M(O_2) = 32.00$ u $= 32.00$ g/mol. Consider 1 mol of O_2 under the prescribed conditions. Then, using the ideal gas law equation, we find

$$\rho = \frac{m}{V} = \frac{m}{\dfrac{nRT}{p}} = \frac{(32.0 \times 10^{-3} \text{ kg})(150/760)(1.01325 \times 10^5 \text{ Pa})}{(1 \text{ mol})(8.3144 \text{ J/mol} \cdot \text{K})(273.16 \text{ K})}$$

$$= 0.282 \text{ kg/m}^3$$

Example 23–5

A sample of air with a mass of 100.00 g, collected at sea level, is analyzed and found to consist of the following gases:

$$\text{nitrogen (N}_2) = 75.52 \text{ g}$$
$$\text{oxygen (O}_2) = 23.15 \text{ g}$$
$$\text{argon (Ar)} = 1.28 \text{ g}$$
$$\text{carbon dioxide (CO}_2) = 0.05 \text{ g}$$

plus trace amounts of neon, helium, methane, and other gases.

(a) Calculate the partial pressure of each gas when the pressure is 1 atm $= 1.01325 \times 10^5$ Pa.

(b) Determine the volume occupied by the 100-g sample at a temperature of $15.00°$ C and a pressure of 1 atm. What is the density of the air for these conditions?

(c) What is the effective molecular mass of the air sample?

(d) What is the fractional molar composition of air?

Solution:

(a) Using the table on page 610, we find the molecular masses of the air components to be $M(N_2) = 28.01$ u, $M(O_2) = 32.00$ u, $M(Ar) = 39.95$ u, and $M(CO_2) = 44.01$ u. Thus, the number of moles of each gas in the sample is

$$n(N_2) = \frac{75.52 \text{ g}}{28.01 \text{ g/mol}} = 2.6962 \text{ mol}$$

$$n(O_2) = \frac{23.15 \text{ g}}{32.00 \text{ g/mol}} = 0.7234 \text{ mol}$$

$$n(Ar) = \frac{1.28 \text{ g}}{39.95 \text{ g/mol}} = 0.0320 \text{ mol}$$

$$n(CO_2) = \frac{0.05 \text{ g}}{44.01 \text{ g/mol}} = 0.0011 \text{ mol}$$

The total number of moles is

$$n_0 = \sum_i n_i = 3.4527 \text{ mol}$$

Then, the partial pressure of N_2 is

$$p(N_2) = \frac{2.6962 \text{ mol}}{3.4527 \text{ mol}} \cdot (1.01325 \times 10^5 \text{ Pa})$$

$$= 0.7912 \times 10^5 \text{ Pa}$$

Similarly, $p(O_2) = 0.2123 \times 10^5$ Pa

$p(Ar) = 0.0094 \times 10^5$ Pa

$p(CO_2) = 0.0032 \times 10^5$ Pa

(b) Solving the ideal gas law equation for V and using $T = 273.15 + 15.00 = 288.15$ K, we find

$$V = \frac{n_0 RT}{p} = \frac{(3.4527 \text{ mol})(8.3144 \text{ J/mol} \cdot \text{K})(288.15 \text{ K})}{1.01325 \times 10^5 \text{ Pa}}$$

$$= 8.164 \times 10^{-2} \text{ m}^3$$

Then, $$\rho = \frac{m}{V} = \frac{100 \times 10^{-3} \text{ kg}}{8.164 \times 10^{-2} \text{ m}^3} = 1.225 \text{ kg/m}^3$$

(c) The 100-g sample must have an appropriate molecular mass to yield n_0 moles of gas; that is,

$$M(\text{air}) = \frac{100 \text{ g}}{3.4527 \text{ mol}} = 28.96 \text{ u}$$

(d)   The fractional molar compositions are

$$f(N_2) = n(N_2)/n_0 = 2.6962/3.4527 = 0.7809$$
$$f(O_2) = n(O_2)/n_0 = 0.7234/3.4527 = 0.2095$$
$$f(Ar) = 0.0093$$
$$f(CO_2) = 0.0003$$

Example 23–6

A *McLeod gauge* is a simple device for measuring low pressures. As shown in part (a) of the diagram below, the gauge consists of a large bulb V and a capillary tube C that has a narrow bore. The relatively large volume V of the bulb (down to point A) is connected directly to the system whose pressure p is to be determined. The mercury reservoir R is now raised and the volume V is sealed off at point A. As the reservoir is raised further, the gas trapped in V is compressed into the smaller volume V' of the capillary tube, as indicated in part (b) of the diagram. Show that the pressure p is given by

$$p \cong \frac{V'}{V} \rho g h$$

where h is the difference in height of the mercury column (as shown in the diagram), and where ρ is the density of mercury. If $h = 1.2$ cm and $V/V' = 4 \times 10^4$, determine the pressure p.

(a)   (b)

Solution:

The difference in pressure between p' in the volume V' and the pressure p in the system is

$$p' = p + \rho g h$$

We assume that the temperature is uniform throughout the system, so we can use Boyle's law

to write

$$p'V' = pV$$

Combining these two equations yields

$$p = \frac{V'}{V - V'} \cdot \rho g h \cong \frac{V'}{V} \rho g h$$

where we make the approximation, $V \gg V'$. (•Can you see that this is equivalent to $p \ll \rho g h$?) Using the values given, we find

$$p = (4 \times 10^4)^{-1}(13.6 \times 10^3 \text{ kg/m}^3)(9.80 \text{ m/s}^2)(0.012 \text{ m})$$
$$= 0.0400 \text{ Pa} = 3.95 \times 10^{-7} \text{ atm} = 3.00 \times 10^{-4} \text{ torr}$$

23-3 THE EQUATION OF STATE (REAL GASES)

The equation of state that connects the thermodynamic variables, p, V, and T, for a fixed mass m of a substance defines a three-dimensional surface. We can imagine a Cartesian representation of such a surface by identifying V with the x-axis, T with the y-axis, and p with the z-axis. Before considering the complicated equation-of-state surfaces for real substances, we first examine the simpler situation for an ideal gas.

The Ideal Gas. The surface that represents the ideal gas law (Eq. 23-1), $pV = nRT$, is shown in Fig. 23-5.* Contour lines are indicated for equally spaced values of constant temperature (*isothermal lines*), such as *ab*, for equally spaced values of constant volume (*isochoric lines*), such as *cd*, and for equally spaced values of constant pressure (*isobaric lines*), such as *ef*. Also shown in the figure are the projections of the isothermal lines onto the *p-V* plane and the projections of the isochoric lines onto the *p-T* plane. The constant-volume lines in the *p-T* plane are straight lines that radiate from the point that represents absolute zero. The constant-

*A point that defines an equilibrium state of the gas lies *on* the surface.

Fig. 23–5. The equation-of-state (or *p-V-T*) surface for an ideal gas. Also shown are the projections of constant-temperature lines onto the *p-V* plane and of constant-volume lines onto the *p-T* plane.

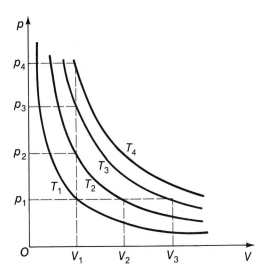

Fig. 23–6. The p-V diagram for an ideal gas, showing several isothermal lines.

temperature lines in the p-V plane are concentric rectangular hyperbolas. This latter representation of the isothermal lines is quite useful and is shown again in more detail in Fig. 23–6. The temperatures here—T_1, T_2, T_3, and T_4—are equally spaced, with $T_2 = 2T_1$, $T_3 = 3T_1$, and $T_4 = 4T_1$. According to the ideal gas law for a gas with a fixed volume V_1, the pressures that correspond to (V_1, T_1), (V_1, T_2), (V_1, T_3), and (V_1, T_4) are also equally spaced, so that $p_2 = 2p_1$, $p_3 = 3p_1$, and $p_4 = 4p_1$, as shown in Fig. 23–6. Similarly, the points (p_1, T_1), (p_1, T_2), and (p_1, T_3) correspond to the equally spaced volumes V_1, $V_2 = 2V_1$, and $V_3 = 3V_1$.

Real Substances. The p-V-T surface for a real substance is much more complicated than the simple surface that describes an ideal gas. For mixtures, solutions, and alloys, the situation is even more complex. Therefore, we restrict attention here to a typical chemically pure substance. Figure 23–7 illustrates the p-V-T surface for such a case. This diagram shows a number of contour lines such as gh, which represents a constant-temperature line (*isotherm*), and ij, which represents a constant-pressure line (*isobar*).

In Fig. 23–7 the region $ABCDE$ specifies the pure gaseous phase of the substance. We use this terminology because, if the temperature is greater than the *critical temperature* T_c (corresponding to the isotherm BCD), no liquefaction of the gas is possible, regardless of how much the pressure is increased. (That is, no liquid droplets would ever form to indicate the presence of a liquid-gas interface.) For temperatures substantially above T_c—such as the isotherm AE—the relationship of p, V, and T closely approximate that specified by the ideal gas law. The projection of this isotherm onto the p-V plane results in a line $A'E'$ that is nearly the same as the rectangular hyperbola $pV = nRT = $ constant.

For low pressures and below the critical temperature—at a point such as k—the substance is also in the gaseous phase. But as the pressure is increased (and simultaneously heat is removed to maintain constant temperature), a point e is reached at which liquefaction begins (liquid droplets appear) and a liquid-gas interface is formed. If heat removal is continued, even without a further increase in pressure, the fraction of liquid present continues to increase and the volume decreases because of the condensation. Finally, at point d, all of the substance has been converted to the liquid phase. If, at some point along the line ed, the removal of heat is stopped and the system is thermally isolated, the substance will remain in this state

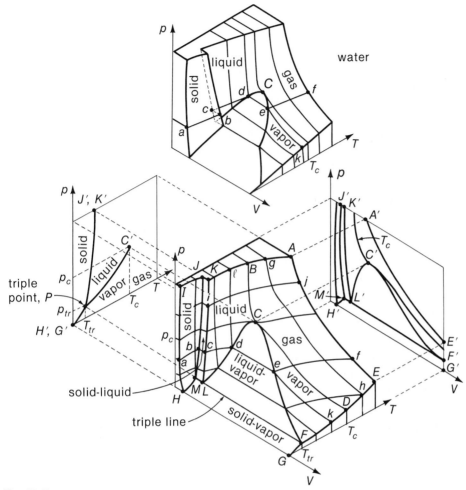

Fig. 23-7. The p-V-t surface for a typical real substance. The diagram for water (at the top) illustrates the case of a substance that expands upon freezing.

of two-phase equilibrium. At the point e, there is zero liquid and the substance is said to be a *saturated vapor*. At the point d, the substance is all liquid (and is called a *saturated liquid*).

The behavior of the substance in the gaseous region *CFD* is no different from that in the region *ABCDE*. However, because the substance can be liquefied at the temperatures that exist in region *CFD*, this region is called the *vapor phase* of the substance. The highest pressure p_c at which the liquid and vapor phases coexist (at the temperature T_c) is called the *critical presssure*. The point (p_c, T_c) is called the *critical point* (point *C* in Fig. 23-7). The critical point corresponds to the condition in which the densities of the liquid and vapor phases have become equal. Then, there is no longer any distinction between the two phases; the substance behaves essentially as a gas but has a density closer to that of a liquid. (The density of water at the critical point is 0.325 g/cm^3.) (•What is the value of the latent heat of vaporization at the critical point?)

If the pressure on a substance in the liquid phase is increased (at constant temperature), the state of the system will be described by a line such as $d\ell$. Eventually, a point will be reached (above ℓ and not shown in Fig. 23-7) at which solidification (or crystallization) begins. At first there is formed a two-phase, solid-liquid system; finally, when the heat of fusion has been removed, the substance is entirely solid.

Figure 22–9 (page 627) shows a constant-pressure heating curve for water. The points that correspond to the labeled points on the heating curve are indicated in the p-V-T diagram at the top of Fig. 23–7, namely, *abcdef*. The break in the temperature rise *bc* in Fig. 22–9 is identified with the solid-liquid phase line *bc* in Fig. 23–7. Similarly, the break *de* corresponds to the liquid-vapor line *de* in Fig. 23–7. Water does not behave in the normal way at the solid-liquid phase transition, as can be seen by comparing this region on the two p-V-T surfaces in Fig. 23–7. (•Does this mean that the heating curve for a normal substance will have an appearance different from that of Fig. 22–9?)

The critical point for water occurs for $T_c = 647.2$ K and $p_c = 218.3$ atm. Thus, if a heating curve were constructed for a very high pressure ($>p_c$), the break *de* would disappear. However, the break *bc* would become compound because several distinct phases of ice exist at high pressures (see Fig. 23–9).

If one starts with a substance in the gaseous phase at sufficiently low temperature, it is possible to convert the substance directly to the solid phase, bypassing liquefaction, by increasing the pressure and removing heat. This process (and its inverse) is called *sublimation*. In Fig. 23–7, *LMHGF* corresponds to the dual-phase, solid-vapor region. Notice that sublimation can occur only for temperatures below the triple-point temperature T_{tr}. The *triple line* on the p-V-T surface is the isotherm for the triple-point temperature along which the three phases of the substance can exist together. (This line corresponds to the single point P—the *triple point*—in the projection onto the p-T plane.)

In the projection of the p-V-T surface onto the p-T plane, the liquid-vapor phase boundary generates a line that appears to terminate abruptly at the critical point C'. However, it is clear in the view of the complete surface that there is no physical discontinuity at this point. Notice that each dual-phase boundary—solid-vapor, liquid-vapor, and solid-liquid—projects into a line in the p-T plane.

Information concerning the p-V-T surface of a substance is often presented in the form of projections onto the p-T and p-V planes. These important projections are shown again in more detail in Fig. 23–8.

Table 23–1 lists the critical-point temperature T_c and pressure p_c as well as the triple-point temperature T_{tr} and pressure p_{tr} for several pure substances.

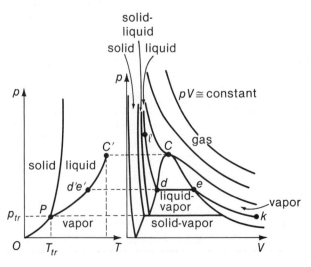

Fig. 23–8. Projections of the p-V-t surface for a typical pure substance onto the p-T and p-V planes. Compare Fig. 23–7.

TABLE 23-1. Critical-Point and Triple-Point Data

SUBSTANCE	T_c (K)	p_c (atm)	T_{tr} (K)	p_{tr} (torr)
Hydrogen	33.2	12.8	13.84	52.8
Nitrogen	126.	33.5	63.18	94.
Oxygen	154.6	50.1	54.35	1.14
Ammonia	405.6	112.5	195.40	45.57
Carbon dioxide	304.	72.9	216.55	3880.
Sulfur dioxide	430.9	77.7	197.68	1.256
Water	647.2	218.3	273.16	4.58

23-4 PHASE TRANSITION BEHAVIOR

Vapor Pressure. Suppose that an amount of water is introduced into a sealed container from which the air has been completely evacuated. If the amount of water is small, it will completely vaporize, producing a certain pressure of the vapor within the container. However, if a larger amount of water is introduced into the container, the vaporization will not be complete—a quantity of liquid water will remain on the bottom of the container. The space above the water surface can contain no additional vapor at that temperature. This is the condition of *saturation,* and the pressure of the vapor is called the *saturated vapor pressure* (or often, simply the *vapor pressure*) for that particular temperature. Saturation exists along a line such as *de* in the *p-V-T* diagram shown in the upper part of Fig. 23-7 (or the same line in Fig. 23-8 for a normal substance). The temperature is constant along this line, and the liquid and vapor phases coexist in equilibrium. Notice that, if the volume of the container is increased (with the temperature held constant by the addition of heat), a greater fraction of the water will be vaporized until finally, at *e*, no liquid water remains. Any further increase in the volume will result in the lowering of the pressure below the saturated value, along the curve *ek*.

As a function of temperature, the saturated vapor pressure follows the line *PC'* shown in the *p-T* diagrams in Figs. 23-7 and 23-8. For most substances, the saturated vapor pressure p_v is a rapidly increasing function of *T*. Some values of p_v for water are given in Table 23-2. Notice that p_v exceeds atmospheric pressure (760 torr) when the temperature is greater than 100° C.

Although we have phrased the discussion of vapor pressure in terms of the vaporization of a liquid into an evacuated space, this is actually an unnecessary restriction. If two or more different gases that do not interact chemically (for exam-

TABLE 23-2. Saturated Vapor Pressure and Vapor Density for Water at Various Temperatures

t_C (°C)	p_v (torr)	ρ (kg/m³)	t_C (°C)	p_v (torr)	ρ (kg/m³)
0	4.58	0.00485	70	233.53	0.1984
10	9.21	0.00941	80	355.1	0.2938
15	12.79	0.0128	90	525.8	0.4241
20	17.51	0.0173	100	760.0	0.598
25	23.76	0.0230	110	1074.5	0.827
30	31.71	0.03035	120	1488.9	1.122
40	55.13	0.0511	130	2025.6	1.498
50	92.30	0.0832	140	2709.5	1.968
60	149.19	0.1305	150	3568.7	2.550

ple, water vapor and air) are present in a container, their partial pressures contribute independently to the total pressure (Dalton's law of partial pressures). When sufficient water vaporizes to reach saturation in a vessel that contains air, it is often said that "the air is saturated with water vapor." However, the saturation condition does not depend at all on the presence of the air. Saturation is reached when the rate at which liquid molecules vaporize from the liquid surface is equal to the rate at which molecules strike the surface and are absorbed. This equilibrium (saturation) condition is independent of the existence of other gases in the vessel.

Boiling. At the *boiling point* of a liquid, the vapor pressure of the liquid is equal to the external pressure. Water has a vapor pressure equal to normal atmospheric pressure (760 torr) at a temperature of 100° C (Table 23–2), so this is the normal boiling-point temperature for water. However, if the ambient pressure is reduced (for example, on a mountaintop), the boiling-point temperature is correspondingly lowered. At a height of 6 km (20,000 ft), normal atmospheric pressure is approximately 355 torr, and the boiling-point temperature of water is about 80° C. (Cooking times for foods in boiling water are therefore somewhat longer at mountain altitudes than at sea level.)

If the temperature of a liquid is increased to the boiling-point temperature by the addition of heat, the continued heating of the liquid will not result in any further increase in temperature. Instead, the liquid will vaporize to completion at the boiling-point temperature. During boiling, vapor bubbles are produced throughout the liquid volume and cause the familiar agitation of the liquid that we recognize as active boiling.

The boiling-point temperature of a substance is affected not only by pressure but by the addition of a nonvolatile solute to the solvent liquid. Such an addition lowers the vapor pressure by a small amount and causes the boiling-point temperature to increase slightly. The addition of 1 mol of any (nonelectrolyte) substance such as glucose, sucrose, or ethylene glycol to 1 kg of water (thereby producing a *one molal* solution) will elevate the boiling-point temperature by 0.512 deg when the pressure is 1 atm. For example, ethylene glycol ($C_2H_6O_2$) has a molecular mass of 62.0 u = 62.0 g/mol. If 100 g of glycol is added to 1 kg of water, the solution will be $100/62.0 = 1.6\ m$ (that is, 1.6 molal). Then, the new boiling-point temperature will be $100.00° C + (1.6)(0.512\ deg) = 100.82° C$.

Melting. An increase in pressure always increases the boiling-point temperature of a substance. On the other hand, a pressure increase will lower the melting-point temperature of some substances and will raise it for others. If a substance expands in volume on melting, a pressure increase will cause the melting-point temperature to be raised. However, if a substance (such as water) contracts on melting, a pressure increase will cause the melting-point temperature to be lowered. This difference is due to the difference in slope, $\partial p/\partial T$, of the solid-liquid dual-phase lines for the two types of substances (refer to the two p-V-T surface diagrams in Fig. 23–7).

The change in melting-point temperature with pressure is a small effect for most substances. For water, the average rate of decrease of the melting-point temperature with increasing pressure (for $p \lesssim 1000$ atm) is approximately 9.1×10^{-3} deg/atm. The effect is complicated by the fact that there exist several different phases of ice (different crystal structures) in the high-pressure region (see Fig. 23–9). At a pressure of 2047 atm, the melting-point temperature is reduced to −22.0° C, which corresponds to the triple point for water, ice I (the common form of ice), and ice III. Any additional increase in pressure causes the melting-point temperature to increase. (In Fig. 23–9 notice that the slope of the p-T line between liquid water and ice changes sign at the triple point mentioned.) Notice that at a pressure of 25,000 atm, ice will still be present at a temperature of 100° C.

Fig. 23-9. The *p-T* diagram for water, emphasizing the high-pressure region. Notice the seven different phases of ice. The vapor-phase region is not apparent at this scale. [Adapted from G. C. Kennedy, in *American Institute of Physics Handbook,* D. E. Gray, Editor, McGraw-Hill, New York, 1963, 2nd ed., pp. 4–40.]

Because ice has a lower melting-point temperature when under pressure, an ice surface near 0° C becomes very slippery for an ice skater. The relatively high pressure beneath the skate blades causes the ice to melt and provides a thin layer of water on which the blades travel with little friction. After the skater passes, the pressure returns to atmospheric pressure and the water refreezes.

Just as the boiling-point temperature is raised by the addition of a nonvolatile solute to the liquid, so is the melting-point temperature lowered in the same way. Both effects are caused by a decrease in the vapor pressure due to the addition of the solute. (•Can you see why this results in the observed effects?) A 1-*m* solution of nonelectrolyte substance in water will depress the melting (freezing) point to a temperature of −1.86° C. Thus, a 5-*m* solution of glycol in water will result in a freezing-point temperature of −9.3° C. (Ethylene glycol is the primary ingredient of *antifreeze.*) For substances that are electrolytes (that is, ionizing substances)—such as NaCl and acids—the situation is somewhat more complicated. In general, these substances affect both the boiling-point and freezing-point temperatures to a greater extent than do nonelectrolyte substances (such as glycol and glucose). A practical application of the depression of the melting-point temperature by the addition of solutes is the spreading of calcium chloride ($CaCl_2$) on icy road surfaces; this lowers the freezing-point temperature and promotes the melting of the ice at temperatures below 0° C.

Humidity. The partial pressure of water vapor in air at ordinary temperatures is small. For example, the maximum partial pressure (that is, the saturated vapor pressure) at 20° C is only about 2 percent of atmospheric pressure (see Table 23-2). Nevertheless, water vapor in air has a great effect on climatic conditions and on personal comfort.

The ratio of the actual partial pressure of water vapor in air to the saturated vapor pressure at the same temperature is called the *relative humidity.* When the relative humidity is low, the air is *dry*. Then, water evaporates quickly because the rate at which water molecules are vaporized is much greater than the rate at which they return to the water surface and are absorbed. In low humidity conditions, body perspiration readily evaporates and we feel cooled by this process. (•Why?) However, if the relative humidity is high, the evaporation rate is low and one has a feeling of oppressive mugginess. If the relative humidity reaches 100 percent, there can be no net evaporation—wet clothing or damp skin will never dry under such a condition.

Example 23–7

When the air temperature is 30° C, a piece of metal is cooled until, at 20° C, water droplets appear on the surface. What is the relative humidity?

Solution:

The air in the immediate vicinity of the metal is cooled by conduction to the temperature of the metal. The appearance of water droplets means that the cooled air (at 20° C) is saturated at this temperature. Thus, the partial pressure at 30° C is equal to the saturated vapor pressure at 20° C, namely, 17.5 torr (see Table 23–2). Hence,

$$\text{relative humidity} = \frac{\text{actual partial pressure (30° C)}}{\text{saturated vapor pressure (30° C)}}$$

$$= \frac{17.5 \text{ torr}}{31.7 \text{ torr}} = 0.55 = 55\%$$

The temperature at which condensation first appears is called the *dew point,* and this technique for determining the relative humidity is called the *dew-point method.*

SPECIAL TOPIC

23-5 EMPIRICAL EQUATIONS OF STATE

The ideal gas law is remarkably accurate for very dilute gases. However, the law becomes progressively less accurate as the pressure is increased and the mean spacing between the molecules is reduced. This failure is due primarily to the increasing importance of intermolecular forces at the smaller average distances and to the finite size of the molecules themselves.

Figure 23–10 illustrates the behavior of air under a variety of temperature and pressure conditions. The curves show the quantity pV/nT which, for an ideal gas, should equal $R = 8.314 \text{ J/mol} \cdot \text{K}$. This value is attained for all of the temperatures shown for pressures less than about 0.3 atm. Thus, air closely resembles an ideal gas in this regime.

The smooth behavior of the curves in Fig. 23–10 suggests that we can make a power-series expansion of the form

$$\frac{pV}{n} = p\mathcal{V} = RT + B_1(T)p + B_2(T)p^2 + \cdots \quad \textbf{(23–2)}$$

where $\mathcal{V} = V/n$ is the volume per mole or the *molar volume* of the substance. The temperature-dependent coefficients—$B_1(T)$, $B_2(T)$, and so forth—are called the *first virial coefficient,* the *second virial coefficient,* and so forth. An equation of the form of Eq. 23–2 is called a *virial equation* or *virial expansion.*

The temperature at which the coefficient $B_1(T)$ vanishes is called the *Boyle temperature,* the temperature at which Boyle's law is valid for the widest range of pressure. This temperature separates those curves that tend upward in Fig. 23–10 from those that tend downward. Evidently,

Fig. 23–10. Pressure-temperature characteristics of air. A value of $pV/nT = 8.314\ \text{J/mol}\cdot\text{K}$ is expected for an ideal gas. At low pressures, air closely resembles an ideal gas. Note the logarithmic pressure scale.

for air, the Boyle temperature T_B is between 300 K and 400 K; in fact, $T_B = 346.8$ K. At this temperature, air obeys the ideal gas law to within a few parts per thousand for pressures up to 40 atm or so.

Van der Waals' Equation. In 1873, the Dutch physicist Johannes van der Waals (1837–1923) showed that real substances can be more accurately described if the ideal gas law equation is modified to the form

$$\left(p + \frac{a}{\mathcal{V}^2}\right)(\mathcal{V} - b) = RT \qquad \textbf{(23–3)}$$

where the parameters a and b are only slightly dependent on temperature and must be determined empirically for each substance. The term $a/\mathcal{V}^2$ that augments the externally imposed pressure p arises from the attractive nature of the intermolecular force. The dependence on $1/\mathcal{V}^2$ is the result of the short-range character of this force. We can see this in the following way. The number of neighboring molecules with which a particular molecule can interact is

$$N \cong \frac{4\pi}{3}\frac{r^3}{\mathcal{V}}N_0$$

where r is the range of the intermolecular force and where

N_0 is Avogadro's number. The number of interacting pairs is $N(N-1)/2 \cong \frac{1}{2}N^2$, so the increased pressure due to this intermolecular effect is proportional to $1/\mathcal{V}^2$.

The quantity b that modifies the molar volume $\mathcal{V}$ is approximately equal to N_0 multiplied by the volume of a single molecule. Thus, $\mathcal{V} - b$ represents the free volume available to each mole of the substance.

Van der Waals' equation gives a cubic dependence of the pressure p on the molar volume $\mathcal{V}$. Figure 23–11 shows several of these cubic curves for different constant temperatures. The isotherm $T = T_c$ that passes through the critical point C yields a single real root for $\mathcal{V}$ when $p = p_c$ (actually, three equal roots). For lower temperatures, there are three distinct real roots for $\mathcal{V}$ corresponding to any particular p, and for higher temperatures, there is only one real root.

It can be readily shown (see Problem 23–19) that the van der Waals' constants, a and b, can be expressed in terms of the critical pressure p_c, the critical temperature T_c, and the critical molar volume $\mathcal{V}_c$ as

$$a = \frac{27}{64}\frac{R^2T_c^2}{p_c}; \qquad b = \frac{\mathcal{V}_c}{3} = \frac{RT_c}{8p_c} \qquad \textbf{(23–4)}$$

According to the second of these equations, we have

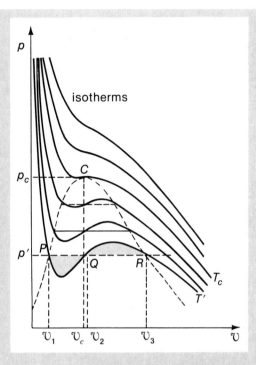

Fig. 23–11. Isotherms of a typical substance based on van der Waals' equation. The point C is the critical point. The three roots ($\mathcal{V}_1$, $\mathcal{V}_2$, and $\mathcal{V}_3$) of the cubic equation are shown for an isotherm, $T' < T_c$. In the region of the liquid-vapor phase (inside the dashed line), van der Waals' equation produces curved cubic lines instead of the straight isotherms that actually occur in this region (see Fig. 23–7). Maxwell showed that each straight isotherm, such as PQR, divides the curved cubic lines into two regions that have equal areas (the shaded regions in the diagram). This determines the positions of the roots.

$$\frac{p_c \mathcal{V}_c}{RT_c} = \tfrac{3}{8}$$

This relationship connecting the critical-point parameters should be valid for *any* substance. Actual values of $p_c\mathcal{V}_c/RT_c$ for common substances vary from 0.23 to 0.32; for example, nitrogen has a value of 0.292. It is not surprising that the prediction of this simple theory does not agree precisely with experiment; the more remarkable fact is that the values of $p_c\mathcal{V}_c/RT_c$ for many substances are so nearly constant. Notice that van der Waals' equation attempts to describe the liquid-vapor phase of the substance (the region enclosed by the dashed line in Fig. 23–11); such a description is beyond the capability of ideal gas theory.

It can also be shown (see Problem 23–21) that the first virial coefficient $B_1(T)$ that appears in the expansion (Eq. 23–2) of van der Waals' equation is

$$B_1(T) = b - \frac{a}{RT} \qquad \textbf{(23-5)}$$

Then, using Eqs. 23–4, we find the Boyle temperature,

$$T_B = \tfrac{27}{8} T_c$$

For nitrogen, this expression gives $T_B(N_2) = 425$ K, whereas the observed value is 327 K.

Equation 23–5 shows that $B_1(T)$ vanishes (and the ideal gas law equation becomes remarkably accurate) when $T = T_B$. That is, the unusual accuracy of the ideal gas law for temperatures near the Boyle temperature is *not* due to the *absence* of the effects of the short-range forces and the finite molecular size; instead, it is due to the near cancellation of these two important effects!

PROBLEMS

Section 23–1

23–1 What is the mass (in kg) of the light isotope of hydrogen, ^1H?

23–2 Natural lithium consists of two isotopes, ^6Li (7.42 percent abundance) and ^7Li (92.58 percent abun-

dance). The atomic mass of the isotope ^7Li is 7.016004 u. What is the atomic mass of ^6Li? [*Hint:* Use the information in the table shown in Fig. 47–12.]

23–3 The element chlorine has two stable isotopes, ^{35}Cl (34.96885 u) and ^{37}Cl (36.96590 u). What are the natural abundances of these isotopes? [*Hint:* Use the information in the table shown on page 610.]

23-4 How many grams of hydrogen are there in 1 kg of ethyl alcohol (C_2H_5OH)?

23-5 The density of butyl alcohol (C_4H_9OH) is 0.80567 g/cm^3. How many molecules of butyl alcohol are there in 1 cm^3?

23-6 Lord Rayleigh (1842–1919), an English physicist, was able to estimate the size of an oil molecule by allowing a droplet of the oil ($m = 8 \times 10^{-4}$ g, $\rho = 0.90$ g/cm^3) to spread out over a water surface. He found that the maximum area covered was 0.55 m^2. When he attempted to spread the oil over a greater area, the film tore apart. Rayleigh concluded that the area of 0.55 m^2 represented a monomolecular layer so that its thickness corresponded to the size of the oil molecule. What did Rayleigh find for the size of the molecule?

23-7 Estimate the "size" of a water molecule by assuming that each molecule in a sample occupies a cubic box that is in contact with all of the neighboring boxes.

Section 23-2

23-8 Gas is confined in a tank at a pressure of 10.0 atm and a temperature of 15° C. If one half of the gas is withdrawn and the temperature is raised to 65° C, what is the new pressure in the tank?

23-9 An automobile tire is inflated using air originally at 10° C and normal atmospheric pressure. During the process, the air is compressed to 28 percent of its original volume and the temperature is increased to 40° C. What is the tire pressure? After the car is driven at high speed, the tire air temperature rises to 85° C and the interior volume of the tire increases by 2 percent. What is the new tire pressure? Express each answer in Pa (absolute) and in lb/in.^{2} (gauge). [1 atm = 14.70 lb/in.^{2}]

23-10 A diving bell in the shape of a cylinder with a height of 2.50 m is closed at the upper end and open at the lower end. The bell is lowered from air into sea water ($\rho = 1.035$ g/cm^3). The air in the bell is initially at a temperature of 20° C. The bell is lowered to a depth (measured to the bottom of the bell) of 45.0 fathoms or 82.3 m. At this depth the water temperature is 4° C, and the bell is in thermal equilibrium with the water. (a) How high will sea water rise in the bell? (b) To what minimum pressure must the air in the bell be raised to expel the water that entered?

23-11 A 2.0-ℓ flask is filled with 1.50 g of helium gas (He) and sealed. The temperature of the flask is then reduced to $-150°$ C. What is the pressure of the helium gas?

23-12 The level of mercury in the simple barometer
   • shown in the diagram stands at a height $h = 75.9$ cm when the temperature is 15° C. The evacuated space in the tube above the mercury has a height of 8.0 cm. The inside diameter of the tube is 6.5 mm. When a small amount of nitrogen is injected into this space, the mercury level drops to a height $h' = 62.2$ cm. There is no temperature change during the process. Determine the mass m of the nitrogen that was injected.

23-13 A mixture of oxygen (O_2) and helium (He) at a tem-
   • perature of 15° C and at a pressure of 1 atm has a density of 0.980 kg/m^3. (a) What is the partial pressure of each gas? (b) What fraction (by mass) of the mixture is oxygen?

Section 23-4

23-14 A steel tank is partially filled with water and is sealed when the temperature is 20° C and the atmospheric pressure is 1 atm. What is the gauge pressure in the tank when the temperature is raised to 120° C? (Neglect any change in volume of the tank or of the water.)

23-15 At what temperature will water boil on the top of Mt. Whitney, California ($h = 4.42$ km)? [*Hint:* Use the law of atmospheres, $p = p_0 e^{-\alpha y}$, with $\alpha = 0.125$ km^{-1}; see Example 17-3.]

23-16 A 250-g sample of glucose ($C_6H_{12}O_6$) is dissolved in

3.0 kg of water. What is the boiling point of the solution?

23-17 What is the relative humidity when the air temperature is 25° C and the dew point is 15° C?

23-18 A warehouse has a volume of 1800 m³. At a temperature of 30° C, the relative humidity is 70 percent. How much water must be removed from the air to reduce the relative humidity to 40 percent?

SPECIAL TOPIC

Section 23-5

23-19 Show that when the three roots of van der Waals' equation are all equal (at the critical point), we have $\mathcal{V}_c = 3b$, $p_c = a/27b^2$, and $T_c = 8a/27Rb$, from which Eqs. 23-4 follow. [*Hint:* Expand $(\mathcal{V} - \mathcal{V}_c)^3 = 0$; write Eq. 23-3 in powers of $\mathcal{V}$ and compare like terms.]

23-20 Assume that the molecules of diatomic gases such as nitrogen and oxygen consist of touching, hard-sphere atoms. Use the fact that the critical molar volumes for nitrogen and oxygen are, respectively, 90.0 cm³/mol and 78.0 cm³/mol, and determine the radii of nitrogen and oxygen atoms.

23-21 (a) Show that van der Waals' equation can be written in a virial expansion of the form
•

$$pV = RT + A_1(T)/\mathcal{V} + A_2(T)/\mathcal{V}^2 + \cdots$$

with

$$A_1(T) = (RTb - a); \qquad A_2(T) = b^2RT;$$
$$A_3(T) = b^3RT$$

and so forth.

(b) Compare the expansion in (a) with Eq. 23-2 and show that $B_1(T) = b - a/RT$.

Additional Problems

23-22 A cylinder with a tight-fitting movable piston originally contains gas at atmospheric pressure and a temperature of 20° C. The piston is moved so that the gas is compressed to $\frac{1}{15}$ of its original volume, at which point the pressure is 22 atm (absolute). What is the final temperature of the gas?

23-23 A bubble of marsh gas rises from the bottom of a fresh-water lake at a depth of 4.2 m and a temperature of 5° C to the surface where the water temperature is 12° C. What is the ratio of the bubble diameter at the two locations? (Assume that the bubble gas is in thermal equilibrium with the water at each location.)

23-24 A classroom with dimensions 8 m × 6 m × 3.5 m contains air at a temperature of 18° C. What is the mass of air in the room? (The effective molecular mass of air is 28.96 u—see Example 23-5.)

23-25 What is the average distance between molecules in air at a temperature of 25° C and a pressure of 1 atm?

23-26 A 2.0-ℓ flask at a temperature of 15° C contains a
• mixture of argon (monatomic, Ar) and carbon dioxide (CO_2). The partial pressure of the argon is

350 torr and that of the carbon dioxide is 850 torr. (a) What is the mass of each gas in the flask? (b) What is the density of the mixture?

23-27 The variation of atmospheric pressure with altitude
•• was derived in Section 17-2 on the basis of an isothermal atmosphere. The temperature of the atmosphere actually decreases with altitude at a rate (called the *lapse rate*) $\gamma = -dT/dy$. Although γ does vary with meteorological conditions, the value is approximately constant—about 0.6° C decrease per 100 m increase in altitude. (a) If γ is constant, show that $T(y) = T_0 - \gamma y$, where T_0 is the sea-level temperature. (b) Show that the law of atmospheres (see Example 17-3) becomes

$$p(y) = p_0 \left(\frac{T_0 - \gamma y}{T_0} \right)^{Mg/R\gamma}$$

where $M = 28.96$ u is the effective molecular mass of air (see Example 23-5). (c) Show that

$$\operatorname*{Lim}_{\gamma \to 0} p(y) = p_0 e^{-(Mg/RT_0)y}$$

which is the usual law of atmospheres. [*Hint:* Use $\operatorname{Lim}\ (x \to 0)(1 + x)^{1/x} = e$; see Example I-2.] (d) Calculate the expected pressure and temperature at the top of Mt. Everest ($y = 8882$ m), when sea-level conditions are $15°$ C and 1 atm. Compare your results with the U.S. Standard Atmosphere values for this altitude, namely, $-43°$ C and 0.309 atm.

23-28 A simple mercury barometer, such as that in Problem 23–12, has an inside diameter of 6.5 mm, and the evacuated space has a height of 8.0 cm above the mercury surface. The vapor pressure of mercury at $15°$ C is 7.8×10^{-4} torr. (a) What error in the determination of the atmospheric pressure is introduced by ignoring the presence of the mercury vapor? (b) What is the mass of the mercury vapor in the space above the mercury column?

23-29 A hospital sterilizer contains saturated steam at a gauge pressure of 2.50 atm. What is the temperature inside the unit? [*Hint:* Interpolate in Table 23–2.]

23-30 A 0.50-kg sample of sucrose ($C_{12}H_{22}O_{11}$) is dissolved in 1.00 kg of water. (a) What is the expected boiling-point temperature of the sugar solution? (b) What is the expected freezing-point temperature of the solution?

23-31 On a winter day, a room with dimensions 6 m $\times$ 5 m $\times$ 3.5 m is heated to a temperature of $20°$ C with an accompanying reduction in relative humidity to 20 percent. How many liters of water must be vaporized to raise the relative humidity to 40 percent?

23-32 What is the mass of water vapor in a room with dimensions 8 m $\times$ 6 m $\times$ 2.8 m when the temperature is $20°$ C and the relative humidity is 45 percent?

23-33 A 2.5-m^3 sample of air at $40°$ C is sealed in a container. When the temperature is lowered to $10°$ C, 35 cm^3 of water collects in the bottom of the container. What was the original relative humidity of the air sample?

23-34 A room with a volume of 200 m^3 is at a temperature of $25°$ C and relative humidity of 60 percent. A bucket of water is placed in the room, which is then sealed. How many liters of water will eventually evaporate from the bucket?

23-35 A tropical air mass at a temperature of $30.0°$ C
•• passes over the Gulf of Mexico and becomes saturated with water vapor. The pressure is then 755.5 torr. The air mass moves northward and eventually reaches the Appalachians, where it abruptly rises; as a result, the temperature drops to $10°$ C and the pressure decreases to 710.0 torr. If the air remains saturated, how much water vapor (in g/m^3) was released by the air mass as rain?

SPECIAL TOPIC

23-36 In 1906, the French chemist Pierre Berthelot (1827–
•• 1907) made a modification in van der Waals' equation that significantly improved the accuracy. Berthelot's equation is

$$\left(p + \frac{a'}{T\mathcal{V}^2}\right)(\mathcal{V} - b') = RT$$

with $a' = (16/3)p_c\mathcal{V}_c^2 T_c$, $b' = \mathcal{V}_c/4$, and $p_c\mathcal{V}_c = (9/32)RT_c$. (These values do not yield three equal roots for $\mathcal{V}_c$ at (p_c, T_c); consequently, the region near the critical point is not well described.) (a) Show that $a' = (27/64)R^2T_c^3/p_c$ and $b' = (9/128)RT_c/p_c$. (b) Show that the virial equation now becomes $p\mathcal{V} = RT + (b' - a'/RT^2)p + \cdots$ (c) Show that the Boyle temperature is now $T_B = \sqrt{6}T_c$. (d) The Boyle temperature for nitrogen is 327.2 K. Use the information in Table 23–1 and calculate T_B for nitrogen, using van der Waals' equation and Berthelot's equation. Compare the various values. (e) At $T = 300$ K and $p = 70$ atm, the empirical value of $p\mathcal{V}/RT$ for hydrogen is 1.042. Use the information for hydrogen in Table 23–1 and calculate $p\mathcal{V}/RT$ from van der Waals' equation and from Berthelot's equation. Compare the various results.

KINETIC THEORY

In the preceding two chapters the behavior of gases was discussed from a *macroscopic* point of view. We now examine the underlying *microscopic* description of gases and establish the relationship between these two ways of viewing gas dynamics. When certain quantum effects are included, it is revealed that these two descriptions are completely equivalent—a remarkable triumph of modern physical theory. A complete discussion is, however, beyond the scope of this book, and we limit consideration here to the essential features and a few applications.

From the previous discussions we know that dilute gases are well described by the ideal gas law, at least for temperatures above the triple-point temperature. This suggests the possibility that real gases can be described in terms of an uncomplicated microscopic model. Accordingly, we begin by making some simplifying assumptions.

First, we accept the atomic hypothesis that a gas consists of a very large number of small particles (molecules) and that for a particular pure, stable chemical substance these molecules have a definite atomic structure and are identical in all respects. Molecules have sizes that are small compared with the mean separation of the molecules in a dilute gas. Consequently, we ignore (at first) the molecular size and treat molecules as ideal pointlike particles. Likewise, the shortness of the range of intermolecular forces (also small compared with the mean separation of the molecules in a dilute gas) permits us to ignore the effects of these forces on the average.

In view of these simplifying assumptions, we can consider a molecule in a gas to move in a straight-line path until it collides with another molecule (or with a wall of the confining vessel). During such encounters both elastic and inelastic collisions may occur. Inelastic collisions involve the conversion of some kinetic energy into internal atomic or molecular excitations. In other collisions the inverse process may occur in which atomic or molecular excitation energy is converted into kinetic energy. These processes that involve internal excitations can be important even at modest temperatures and for dilute gases, as we see later. However, we consider at first that only elastic collisions take place. These molecular encounters, even though elastic, can result in significant changes in the speeds and directions of motion of the molecules. Consequently, any particular molecule follows a zigzag path, with abrupt changes in its motion at short and random time intervals. We assume that Newton's laws of motion are obeyed in all of these processes.

If a gas is in a state of thermal equilibrium, the total kinetic energy of the gas is necessarily constant. Because the number of molecules present in any macroscopic gas sample is very large, the *fraction* of the molecules with a particular speed and direction of motion will remain constant. Thus, there is present an *average velocity distribution,* which we wish eventually to determine. Only when we are considering phenomena that involve a very small sample of molecules do fluctuations from this average distribution become important (as, for example, in the case of Brownian motion, the jiggling about of dust or similar visible particles due to molecular impacts).

24-1 KINETIC THEORY OF PRESSURE IN GASES

We first establish one of the important links between the macroscopic and microscopic views of gas behavior. The pressure exerted by a gas on the walls of a confining vessel may be imagined to result from the large number of collisions that the gas molecules make with the walls. We assume that these collisions are elastic, so only the component of the molecular velocity perpendicular to the wall is changed (from v_x to $-v_x$). The resulting change in linear momentum of the molecule is associated with the perpendicular force that is exerted on the molecule by the wall during the collision. The reaction to this force—the force exerted on the wall by the molecule—contributes to the total gas pressure on the wall.

Consider a small area A of a confining wall that we take to coincide with the *y-z* plane (Fig. 24–1). Let us examine the collisions between the wall and those molecules that have velocities with *x*-components between v_x and $v_x + \Delta v_x$, and with *any* velocity components in the *y-* and *z*-directions. For brevity, we refer to "the molecules with velocity components v_x."*

Fig. 24–1. Geometry for considering the collisions of gas molecules with a wall.

Construct a rectangular box with end areas A and sides with length $v_x \Delta t$ parallel to the *x*-axis. Let $\Delta \mathcal{N}(v_x)$ be the number of molecules per unit volume with velocity components v_x and any components in the *y-* and *z*-directions. The number of such molecules that strike the wall area A during the time Δt is just the number that occupy the volume $v_x \Delta t A$ shown in Fig. 24–1. This number is $v_x \Delta t A \, \Delta \mathcal{N}(v_x)$. For molecules that move exactly along the *x*-direction, this expression is clearly valid. However, molecules within the volume $v_x \Delta t A$ might have velocities with *y-* or *z*-components that will cause the molecules to leave the volume through the sides, top, or bottom before striking the area A. Even though we consider a differential volume, it is still necessary to deal with a very large number of molecules. For every molecule that leaves the volume there is another with the same velocity components, originally outside the volume, that will enter and strike A. The same reasoning applies to molecules within the volume and moving toward A that collide with other molecules and are scattered out of the volume without striking A. Just as many molecules are scattered *into* the volume and strike A as are scattered *out of* the volume.

When a molecule collides elastically with the wall, the *x*-component of the velocity changes from v_x to $-v_x$, and the momentum changes by an amount $2\mu v_x$, where μ is the mass of the molecule. The force exerted on the area A during the time interval Δt is ΔF; then, using Newton's second law in the form that equates $\Delta F \, \Delta t$ to the change in momentum, we have

$$\Delta F \, \Delta t = [2\mu v_x][v_x \, \Delta t A \, \Delta \mathcal{N}(v_x)]$$

---

*There is no molecule that has a velocity *precisely* equal to v_x (for example, 7.68310765 m/s); consequently, the only meaningful way to refer to the molecular velocities is to count the number of molecules whose velocities lie within a specified range, say, from v_x to $v_x + \Delta v_x$.

Thus, the pressure on A due to these collisions is

$$\Delta p = \frac{\Delta F}{A} = 2\mu v_x^2 \, \Delta \mathcal{N}(v_x) \tag{24-1}$$

Because of the complete lack of long-range molecular order,* a dilute gas has a velocity distribution that is *isotropic;* that is, the velocity distributions for the x-, y-, and z-components are exactly the same. Also, the distributions for the components $+v_x$ are the same as for the components $-v_x$, and similarly for the y- and z-directions. Moreover, if for some reason the initial conditions were such that isotropy did not prevail, collisions between molecules would rapidly bring about an equilibrium condition with an isotropic velocity distribution. Thus, we have

$$\Delta \mathcal{N}(v_x) = \Delta \mathcal{N}(-v_x) = \tfrac{1}{2} \Delta \mathcal{N}(|v_x|)$$

so that Eq. 24-1 becomes

$$\Delta p = \mu v_x^2 \, \Delta \mathcal{N}(|v_x|)$$

and the total pressure due to all velocity components v_x is

$$p = \mu \sum_{|v_x|=0}^{\infty} v_x^2 \, \Delta \mathcal{N}(|v_x|) \tag{24-2}$$

If we consider v_x^2 and calculate the average over the distribution of velocity components v_x, we have†

$$\langle v_x^2 \rangle = \frac{\Sigma v_x^2 \, \Delta \mathcal{N}(|v_x|)}{\Sigma \, \Delta \mathcal{N}(|v_x|)} \tag{24-3}$$

where the denominator is equal to the total number of molecules per unit volume:

$$\sum_{|v_x|} \Delta \mathcal{N}(|v_x|) = \mathcal{N} \tag{24-4}$$

Combining Eqs. 24-2, 24-3, and 24-4, there results

$$p = \mu \mathcal{N} \langle v_x^2 \rangle \tag{24-5}$$

The isotropy condition is expressed as

$$\langle v_x^2 \rangle = \langle v_y^2 \rangle = \langle v_z^2 \rangle$$

Moreover, $\qquad\qquad v^2 = \mathbf{v} \cdot \mathbf{v} = v_x^2 + v_y^2 + v_z^2$

Thus, we can write $\qquad \langle v_x^2 \rangle = \langle v_y^2 \rangle = \langle v_z^2 \rangle = \tfrac{1}{3} \langle v^2 \rangle$

---

*We assume that the confining vessel has ordinary laboratory dimensions so we rule out (for now) any gravitational effects.

†We use angled brackets to indicate an average over a *distribution;* an overbar continues to represent an average over time or space.

so that
$$p = \tfrac{1}{3}\mu\mathcal{N}\langle v^2\rangle \tag{24-6}$$

or, using $\mu\mathcal{N} = \rho$ for the mass density, we have*

$$p = \tfrac{1}{3}\rho\langle v^2\rangle \tag{24-7}$$

This equation expresses a remarkably simple relationship between a microscopic property, namely, the molecular velocity distribution (properly, the speed distribution) and a macroscopic concept, namely, the pressure. Note also that, ignoring gravitational effects, the pressure on the area A is the same as the pressure on any part of the interior surface of the confining vessel, regardless of the shape. (•Can you see why?)

It is important to realize that the velocities (speeds) that appear in the derivation of Eq. 24-7 are the speeds of the molecules *between* collisions. If a label could somehow be attached to an individual molecule, we would see it progress through the gas along a zigzag path, as mentioned earlier. The actual *transport speed* of the molecule would be only a small fraction of its average (or *root-mean-square*) speed, $v_{\mathrm{rms}} = \langle v^2\rangle^{1/2}$. At atmospheric pressure and near room temperature, v_{rms} for most common gases is in the range from about 400 m/s to about 1800 m/s; these speeds are about the same as the speed of sound in the same gases. Thus, if you open a bottle of perfume in one corner of a room by popping the cork, a considerably longer time will be required for a person across the room to detect the perfume by smell† than to hear the sound of the bottle being opened. The process by which individual molecules travel through a gas (or liquid or solid) is called *diffusion* (see Section 24-4).

Example 24-1

Dry air at 15° C and at 1 atm consists of nitrogen (N_2) with a partial pressure of 0.791×10^5 Pa and a mass fraction of air of 75.5 percent, oxygen (O_2) with a partial pressure of 0.212×10^5 Pa and a mass fraction of 23.2 percent, plus a number of other small molecular fractions (see Example 23-5). Calculate the root-mean-square speeds of the nitrogen and oxygen molecules. Take the density of dry air at the stated conditions to be $\rho_0 = 1.225$ kg/m³.

Solution:

The density of nitrogen is $0.755\rho_0 = (0.755)(1.225 \text{ kg/m}^3) = 0.925$ kg/m³. Therefore, solving Eq. 24-7 for $v_{\mathrm{rms}} = \langle v^2\rangle^{1/2}$, we find

$$v_{\mathrm{rms}}(N_2) = \left[\frac{3p}{\rho}\right]^{1/2} = \left[\frac{3(0.791 \times 10^5 \text{ Pa})}{0.925 \text{ kg/m}^3}\right]^{1/2} = 506 \text{ m/s}$$

Similarly, for oxygen, the density is $(0.232)(1.225 \text{ kg/m}^3) = 0.284$ kg/m³. Therefore,

---

*Note carefully the differences between such quantities as $\langle x\rangle^2$, $\langle |x|\rangle^2$, and $\langle x^2\rangle$. For example, suppose that the values of x are 1, -1, 3, and -3; then, we have

$$\langle x\rangle^2 = [\tfrac{1}{4}\{1 + (-1) + 3 + (-3)\}]^2 = 0$$
$$\langle |x|\rangle^2 = [\tfrac{1}{4}\{1 + 1 + 3 + 3\}]^2 = 4$$
$$\langle x^2\rangle = [\tfrac{1}{4}\{(1)^2 + (-1)^2 + (3)^2 + (-3)^2\}] = 5$$

†The human olfactory nerve is quite sensitive for some substances. As small a concentration as one molecule of methyl mercaptan (CH_3SH) in 10^{12} air molecules can be detected.

$$v_{\text{rms}}(O_2) = \left[\frac{3(0.212 \times 10^5 \text{ Pa})}{0.284 \text{ kg/m}^3)} \right]^{1/2} = 473 \text{ m/s}$$

Notice that the oxygen molecules, which are more massive than the nitrogen molecules, have a lower rms speed. We can write

$$\frac{v_{\text{rms}}(N_2)}{v_{\text{rms}}(O_2)} = \left[\frac{M(N_2)}{M(O_2)} \right]^{\alpha}$$

where M represents the molecular mass (in u). Determine the value of α. The reason for the simple result for α will become evident in the following section.

24-2   KINETIC INTERPRETATION OF TEMPERATURE

It is useful to continue the analysis beyond the result expressed in Eq. 24–6. If the volume of the gas in question is V and there are n moles of gas present, we have $\mathcal{N} = nN_0/V$, where N_0 is Avogadro's number. Thus,

$$\tfrac{1}{2}\mu\langle v^2 \rangle = \tfrac{3}{2}\frac{p}{\mathcal{N}} = \tfrac{3}{2}\frac{pV}{nN_0}$$

Now, for a sufficiently dilute gas, the ideal gas law equation is valid, so we substitute $pV = nRT$, and obtain

$$\tfrac{1}{2}\mu\langle v^2 \rangle = \tfrac{3}{2}\frac{R}{N_0}T$$

We define a new constant, the *Boltzmann constant** k, to be

$$k \equiv \frac{R}{N_0} = \frac{8.3144 \text{ J/mol} \cdot \text{K}}{6.02205 \times 10^{23} \text{ mol}^{-1}} = 1.38066 \times 10^{-23} \text{ J/K}$$

Thus, we arrive at the important relationship,

$$\tfrac{1}{2}\mu\langle v^2 \rangle = \tfrac{3}{2}kT \tag{24-8}$$

The kinetic energy of a single molecule is $K = \tfrac{1}{2}\mu v^2$, so we recognize $\tfrac{1}{2}\mu\langle v^2 \rangle = \langle K \rangle$ as the *average translational kinetic energy* per molecule.

Using Eq. 24–8, we can express the rms speed as

$$v_{\text{rms}} = \langle v^2 \rangle^{1/2} = \sqrt{\frac{3kT}{\mu}} \tag{24-9}$$

Remember that μ is measured in kg, *not* in u.

---

*Named in honor of Ludwig Boltzmann (1844–1906), Austrian physicist and pioneer developer of modern statistical thermodynamics.

Notice that in the derivation of Eq. 24–8 we made the implicit assumption that both the containing vessel and the center of mass of the gas are at rest in some inertial reference frame. Thus, any kinetic energy associated with the C.M. of the gas does not influence its thermodynamic behavior. (For example, placing the vessel on a platform that moves with constant velocity does not alter the temperature of the gas.)

The total translational kinetic energy of n moles of gas is $nN_0\langle K \rangle$, which we identify as the *internal energy* U of the gas:

$$U = nN_0\langle K \rangle = \tfrac{3}{2}nRT \qquad\qquad \textbf{(24–10)}$$

This expression is valid for an ideal gas that possesses only translational kinetic energy. If rotational and vibrational modes of internal energy are also present, we write a more general expression for the internal energy that includes the contribution of these modes (see Eq. 24–14).

Equation 24–10 represents the important conclusion that the internal energy of an ideal gas *depends only on the temperature*. Thus, to describe an equilibrium state of such a gas, it is necessary to know *only* its temperature; it is *not* necessary to know the history of how the state was formed, whether by adding or removing heat or by adding or removing energy in any other form. The internal energy is therefore called a *state function*.*

Quantum effects cannot be ignored completely in interpreting Eq. 24–10. For example, if the temperature of a substance approaches absolute zero, the internal energy does not reduce to zero, as this equation implies. According to quantum theory, there is a residual *zero-point energy* that remains even in the limit $T \to 0$ K.

Degrees of Freedom and the Equipartition of Energy. We have thus far considered only the translational motion of gas molecules, which involves the energy terms $\tfrac{1}{2}\mu v_x^2$, $\tfrac{1}{2}\mu v_y^2$, and $\tfrac{1}{2}\mu v_z^2$. But molecules may undergo motions in addition to translation. For example, a molecule might rotate about its C.M., thereby introducing energy terms such as $\tfrac{1}{2}I_x\omega_x^2$, $\tfrac{1}{2}I_y\omega_y^2$, and $\tfrac{1}{2}I_z\omega_z^2$. Or the component atoms of the molecule might execute vibrational motion in the molecular center-of-mass frame, giving energy terms of the form $E_{\text{vib}} = \tfrac{1}{2}\mu' v_{\text{vib}}^2 + \tfrac{1}{2}\kappa r^2$. (•What is the interpretation of μ' in this expression?) Each such independent mode of possible motion is referred to as a *degree of freedom* of the molecule.

It is a characteristic of the various modes of motion that each involves an energy term that depends on the square of the relevant dynamical quantity. In fact, we can express the dynamical energy of a molecule as

$$E = \sum_{i=1}^{\nu} \tfrac{1}{2}\lambda_i \xi_i^2 \qquad\qquad \textbf{(24–11)}$$

where λ_i is some (usually constant) molecular property, such as an effective mass, a rotational inertia, or an effective spring constant; and where ξ_i is either a velocity variable or a displacement variable of an excited mode. Each term in the sum is

---

*The internal energy of a real substance is also a state function. However, in addition to the dependence on temperature, the internal energy of a real substance may have some dependence on the other state variables (pressure and volume); see Problem 25–16.

called an *energy partition*. Translational motion of the molecule as a whole contributes three energy partitions (corresponding to the directions of the three coordinate axes) to Eq. 24–11. Each of the three rotational degrees of freedom, when excited, contributes an additional energy partition. However, each vibrational degree of freedom contributes *two* energy partitions—one involving the vibrational kinetic energy, $\frac{1}{2}\mu'v_{\text{vib}}^2$, and the other involving the Hooke's-law potential energy, $\frac{1}{2}\kappa r^2$. The quantity ν in Eq. 24–11 represents the total number of energy partitions for the system.

A system that consists of molecules with only the translational degrees of freedom excited has three energy partitions ($\nu = 3$), and Eq. 24–11 becomes

$$E = \sum_{i=1}^{3} \tfrac{1}{2}\mu v_i^2 = \tfrac{1}{2}\mu v_x^2 + \tfrac{1}{2}\mu v_y^2 + \tfrac{1}{2}\mu v_z^2 = \tfrac{1}{2}\mu v^2$$

Averaging over the molecules in the system and using Eq. 24–8, we can write

$$\langle E \rangle = \tfrac{1}{2}\mu\langle v_x^2 \rangle + \tfrac{1}{2}\mu\langle v_y^2 \rangle + \tfrac{1}{2}\mu\langle v_z^2 \rangle$$
$$= \tfrac{1}{2}\mu\langle v^2 \rangle = \tfrac{3}{2}kT$$

We have argued that in an isotropic system containing many molecules in thermal equilibrium the averaged energy terms are all equal; that is, $\frac{1}{2}\mu v_x^2 = \frac{1}{2}\mu v_y^2 = \frac{1}{2}\mu v_z^2$. It then follows that an average energy of $\frac{1}{2}kT$ is associated with each translational degree of freedom; that is,

$$\tfrac{1}{2}\mu\langle v_x^2 \rangle = \tfrac{1}{2}\mu\langle v_y^2 \rangle = \tfrac{1}{2}\mu\langle v_z^2 \rangle = \tfrac{1}{2}kT$$

This important conclusion regarding the average energy associated with translational motion can be extended, by using statistical arguments, to other forms of molecular motion as well. That is, in equilibrium, the average energy associated with *each* partition that is excited has the same value, namely, $\frac{1}{2}kT$. Thus, we can write

$$\langle E \rangle = \tfrac{1}{2}\nu kT \tag{24–12}$$

This general conclusion is referred to as the *equipartition of energy theorem.**

The determination of the number of degrees of freedom that a molecule possesses requires some quantum considerations. The simplest molecules are those of the noble gases—helium, neon, argon, krypton, xenon, and radon. These are monatomic elements, so the *molecules* are actually *atoms*. The electrons of these elements are arranged in closed shells of great stability.† As a result, the atoms of the noble gases have featureless structures—that is, they are *spherically symmetric*. One might imagine that such a spherical atom could undergo rotations about a central axis, as could a symmetric beach ball. However, for a quantity to be physically meaningful, there must be some way to *measure* it; this is a basic postulate of quantum theory. If

---

*First deduced by the great Scottish mathematical physicist, James Clerk Maxwell (1831–1879), who was also responsible for developing electromagnetic theory. Important contributions to the statistical interpretation of energy were made by Boltzmann and by J. Willard Gibbs (1839–1903), the first great American physicist.

†It is for this reason that the noble gases are "noble"—namely, there are no valence electrons to form chemical bonds with other atoms, so they remain aloof and do not participate in chemical reactions, nor do they form diatomic molecules after the fashion of H_2, N_2, O_2, and so forth.

an atom is spherically symmetric, there is no reference mark to use for a measurement of its rotation. We must conclude that such an atom *cannot rotate*.*

A diatomic molecule possesses cylindrical symmetry about an axis that passes through both atomic centers. Rotation about the symmetry axis is not measurable, but rotations about the two mutually perpendicular axes are measurable. Consequently, a diatomic molecule has two rotational degrees of freedom and two corresponding energy partitions. If the molecule is rigid, so that vibration is not possible, the total number of partitions is $\nu = 3 + 2 = 5$. If a diatomic molecule can undergo vibration, with the atoms moving back and forth along the line that connects their centers, this corresponds to an additional degree of freedom and contributes two additional partitions—one associated with the vibrational kinetic energy and one associated with the potential energy. Altogether, there are three translational partitions, two rotational partitions, and two vibrational partitions. Thus, $\nu = 7$, and $\langle E \rangle = \frac{7}{2}kT$.

For a triatomic molecule, such as H_2O, in which the three atoms are not collinear, there exists no axis of spatial symmetry so there are three rotational degrees of freedom. In addition, three independent modes of vibration are possible (see Fig. 24–2), and there are two energy partitions associated with each mode. Therefore, $\nu = 3 + 3 + (2 \times 3) = 12$. (If the molecule is rigid, $\nu = 3 + 3 = 6$.)

Characteristic Excitations.   Does a collection of molecules exhibit all of its possible motions at any temperature? To answer this question we must again invoke a quantum feature that is manifest in the atomic domain. The possible energies of the various modes of molecular motion can occur only in discrete units or *quanta*. The smallest excitation of a particular mode (corresponding to one quantum of energy) depends on the specific molecule. Some typical values of the minimum excitation energy for different modes are given in Table 24–1. Also given are the corresponding temperatures, $T_\theta = E_{\min}/k$. The first line of the table indicates the quantization of energy imposed on the translational motion by the finite size of the containing vessel. This limitation, for vessels of ordinary size, is negligible, as the values indicate. The last line in Table 24–1 is included to show the energy of a typical internal atomic excitation (in the optical region). Most molecules dissociate into their atomic constituents before such excitations take place; it was for this reason that the partitions associated with these excitations were not included in the previous discussion.

Suppose that a sample of n moles of a diatomic gas is heated, beginning at a temperature well below about 100 K. At such a low temperature the energy is not sufficient to excite either the rotational or vibrational modes (see Table 24–1). Thus, the sample consists almost exclusively of molecules executing only translational mo-

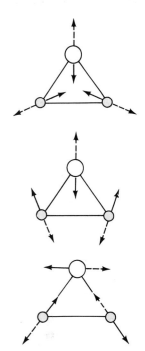

Fig. 24-2. The three independent modes of vibration of a triatomic molecule. In each mode, the C.M. of the molecule remains fixed (or moves with constant velocity).

TABLE 24–1. Typical Molecular Energies Imposed by Quantum Conditions

MODE	APPROXIMATE MINIMUM ENERGY (J)	CORRESPONDING TEMPERATURE T_θ (K)
Translation (within a box)	10^{-38}	10^{-15}
Rotation	10^{-21}	10^2
Vibration	10^{-20}	10^3
Internal (electronic)	10^{-19}	10^4

*Q. "But suppose that the atom really *does* rotate, even though we cannot detect the rotation. What then?"

A. "If there is no conceivable way to detect the rotation or its effects, it is meaningless to claim that rotation exists."

tion, with an internal energy $U = \frac{3}{2}nRT$. As the temperature is raised (but still below 100 K), rotational motion begins to be evident because some inelastic collisions involving the more rapidly moving molecules have sufficient C.M. energy to excite a rotational mode in one (or perhaps both) of the colliding molecules. The small population of such states is continually added to and depleted as a result of the collisions. At each temperature the molecules that possess rotational energy represent some fixed percentage of the total sample. At a temperature somewhat above 100 K, most of the molecules execute rotational motion; then, the full value of the partition associated with the rotational degree of freedom is developed. As a temperature of 1000 K is approached, the collisions become more energetic and more frequent, and now the population of vibrating molecules begins increasing. At a temperature somewhat above 1000 K, most of the molecules also participate in vibrational motion. Then, the full value of the partition for a diatomic molecule, $\nu = 7$, is obtained.

Specific Heats of Gases.   The equipartition principle has an important influence on the specific heats of gases. In order to facilitate comparisons, consider the *molar specific heat C*, which is the specific heat per mole of the gas sample. In particular, consider the specific heat for the case in which the gas is confined to a fixed volume. To distinguish this specific heat from that obtained under constant pressure conditions, we use a subscript v and write C_v for the molar specific heat at constant volume. With the volume constant, any flow of heat into or out of the gas directly affects the internal energy U (see Section 25–2). Then, we define C_v as (compare Eq. 22–10a)

$$C_v(T) = \frac{1}{n}\frac{dU}{dT}\bigg|_V \qquad \textbf{(24–13)}$$

where the notation emphasizes that the changes take place at constant volume.

Now, it is convenient to express the total internal energy of n moles of gas as (compare Eq. 24–10)

$$U(T) = \tfrac{1}{2}\nu(T\,;\,T_\theta)nRT \qquad \textbf{(24–14)}$$

where $\nu(T\,;\,T_\theta)$ is a slowly varying function of T that exhibits change primarily near the temperatures T_θ (see Table 24–1). Thus, $\partial\nu/\partial T \cong 0$ between different values of T_θ. Then, Eq. 24–13 becomes

$$C_v(T) \cong \tfrac{1}{2}\nu(T\,;\,T_\theta)R \qquad (T \text{ not near } T_\theta) \qquad \textbf{(24–15)}$$

Figure 24–3 shows $2C_v/R = \nu$ as a function of temperature for several gases. As expected, for the monatomic gas argon (Ar), we find $\nu = 3$. The curve for hydrogen (H_2) clearly exhibits the transition from $\nu = 3$ to $\nu = 5$. Similarly, the curves for nitrogen (N_2) and chlorine (Cl_2) show the transition from $\nu = 5$ to $\nu = 7$, and that for steam (H_2O) shows the beginning of the departure from $\nu = 6$. The empirical value for the characteristic temperature of the vibrational mode for chlorine is 820 K, whereas for nitrogen it is 3400 K and for hydrogen it is 6300 K. Thus, at a temperature of 1000 K, the curve for chlorine has almost reached its full equipartition value of $\nu = 7$ due to the addition of the full contribution from the vibrational mode, whereas the nitrogen curve is about halfway to its full value and the hydrogen curve has just begun to indicate the presence of a small population of molecules in the vibrational state.

Fig. 24-3. Values of $2C_v/R = \nu$ as a function of temperature for several gases. Note the logarithmic temperature scale.

Example 24-2

One mole of neon fills a 1-ℓ flask at 15° C. The neon found in Nature consists of three isotopes with the following proportions by mass:

$$^{20}\text{Ne}: 90.92 \text{ percent } (19.9924 \text{ u})$$
$$^{21}\text{Ne}: 0.26 \text{ percent } (20.9938 \text{ u})$$
$$^{22}\text{Ne}: 8.82 \text{ percent } (21.9914 \text{ u})$$

(a) What is the total internal energy of the gas?
(b) Determine $\langle v^2 \rangle$ for each isotope.
(c) What is the molecular mass of naturally occurring neon?
(d) Calculate the partial pressure of each isotopic component.

Solution:

(a) According to Eq. 24–10, the total internal energy of 1 mol of neon (a monatomic gas) at 15° C = 288.2 K is

$$U = \tfrac{3}{2}nRT = \tfrac{3}{2}(1 \text{ mol})(8.314 \text{ J/mol} \cdot \text{K})(288.2 \text{ K})$$
$$= 3594 \text{ J}$$

(b) From Eq. 24–8, we have

$$\langle v^2 \rangle = \frac{3kT}{\mu}$$

where μ is the mass of a single molecule (for neon, a single atom). Using $1 \text{ u} =$

1.6605×10^{-27} kg, we have

$$\langle v^2 \rangle_{^{20}\text{Ne}} = \frac{3(1.380 \times 10^{-23} \text{ J/K})(288.2 \text{ K})}{(19.9924 \text{ u})(1.6605 \times 10^{-27} \text{ kg/u})}$$

$$= 3.594 \times 10^5 \text{ m}^2/\text{s}^2$$

$$\langle v^2 \rangle_{^{21}\text{Ne}} = 3.423 \times 10^5 \text{ m}^2/\text{s}^2$$

$$\langle v^2 \rangle_{^{22}\text{Ne}} = 3.267 \times 10^5 \text{ m}^2/\text{s}^2$$

(c) The molecular mass of naturally occurring neon is

$$M(\text{Ne}) = [(0.9092)(19.9924 \text{ u})] + [(0.0026)(20.9938 \text{ u})] + [(0.0882)(21.9914 \text{ u})]$$

$$= 20.171 \text{ u}$$

(This value differs slightly from that listed in Figure 47–12 because of the uncertainty in the natural abundance percentages.)

(d) According to Eq. 24–7, we can write the partial pressure p_i of each isotopic component, $p_i = \frac{1}{3}\rho_i \langle v_i^2 \rangle$. Now, 1 mol of naturally occurring neon consists of

$$(0.9092)(19.9924 \text{ g/mol})(1 \text{ mol}) = 18.177 \text{ g of } {}^{20}\text{Ne}$$

$$(0.0026)(20.9938 \text{ g/mol})(1 \text{ mol}) = 0.054 \text{ g of } {}^{21}\text{Ne}$$

$$\underline{(0.0882)(21.9914 \text{ g/mol})(1 \text{ mol}) = 1.940 \text{ g of } {}^{22}\text{Ne}}$$

$$20.171 \text{ g of Ne}$$

For a volume of $1 \, \ell = 10^{-3} \text{ m}^3$, we have

$$p({}^{20}\text{Ne}) = \frac{1}{3}\left[\frac{18.177 \times 10^{-3} \text{ kg}}{10^{-3} \text{ m}^3}\right](3.594 \times 10^5 \text{ m}^2/\text{s}^2)$$

$$= 2.1776 \times 10^6 \text{ Pa}$$

$$p({}^{21}\text{Ne}) = 0.0062 \times 10^6 \text{ Pa}$$

$$\underline{p({}^{22}\text{Ne}) = 0.2112 \times 10^6 \text{ Pa}}$$

$$p(\text{Ne}) = 2.395 \times 10^6 \text{ Pa}$$

To check this result, use the equation of state for an ideal gas, $pV = nRT$; then,

$$p(\text{Ne}) = \frac{(1 \text{ mol})(8.314 \text{ J/mol} \cdot \text{K})(288.2 \text{ K})}{10^{-3} \text{ m}^3} = 2.396 \times 10^6 \text{ Pa}$$

which agrees with the pressure just calculated to within round-off error.

Example 24–3

An amount of heat $Q = 1.365$ Cal is required to raise the temperature of a 0.500-kg sample of iodine vapor (I_2) from $300°$ C to $400°$ C. It is also determined that C_v for iodine vapor is constant (to within 1 part in 10^3) from the boiling-point temperature ($185.2°$ C) to about $1500°$ C. From this information, determine the molecular mass of I_2.

Solution:

Because the specific heat is constant over such a large temperature range, we can reasonably suppose that the full equipartition value, $\nu = 7$, has been developed. Thus,

$$C_v = \tfrac{1}{2}\nu R = \tfrac{1}{2}(7)(8.314 \text{ J/mol} \cdot \text{K})$$
$$= 29.10 \text{ J/mol} \cdot \text{K}$$

(The experimental value is 29.01 J/mol · K.)

The heat required to raise the temperature by ΔT for a sample with mass m and molecular mass M is

$$Q = \frac{m}{M} C_v \, \Delta T$$

Thus,

$$M = \frac{m C_v \, \Delta T}{Q} = \frac{(0.500 \text{ kg})(29.10 \text{ J/mol} \cdot \text{K})(100 \text{ K})}{(1.365 \text{ Cal})(4186 \text{ J/Cal})}$$
$$= 254.6 \times 10^{-3} \text{ kg/mol}$$
$$= 254.6 \text{ u}$$

According to the table on page 610, the *atomic* mass of iodine is 126.9045 u. Hence, the molecular mass is $M = 253.81$ u, which is quite close to the value just calculated.

▶ *Specific Heats of Solids.* The theory of specific heats for solids is simple only for crystalline substances and for temperatures that do not involve phase transitions. The atoms in a crystalline solid are held in nearly fixed positions in a lattice whose structure varies from one substance to another. Thus, translational and rotational motions are suppressed; only vibrational thermal motion is allowed. Because vibrations can take place independently in three mutually perpendicular directions, the fully developed energy partition number is $\nu = 2 \times 3 = 6$.

The quantum nature of the vibrations of atoms in solids can be expressed, as in the case of gases, by a characteristic temperature called the *Debye temperature* T_D. The theory of solids developed by Debye* allows for the strong coupling of the vibrations of nearby atoms by treating the crystal as an elastic isotropic medium, incorporating the quantum feature of discrete energy excitations. This coupling of the vibrating atoms plays the same role as the molecular collisions in gases in producing an equilibrium thermodynamic state on a macroscopic scale.

In the Debye theory, the specific heat C_v is a function of temperature T and the Debye temperature T_D, with the additional feature that $C_v \to 0$ as $T \to 0$. Because $\nu = 6$, the expected value of C_v when $T \gg T_D$ is $C_v = \tfrac{1}{2}\nu R = 3R = 24.94$ J/mol · K. For many substances, $T_D < 250$ K, so that C_v should be close to 25 J/mol · K at room temperature. The observation that many molar specific heats are approximately equal to this value was first made by Dulong and Petit in 1819.

If the variation of the specific heat of a crystalline solid can be characterized by a single temperature—namely, T_D—it should be possible to construct a single curve to represent the temperature dependence of the specific heat of all such solids by plotting C_v versus T/T_D. Figure 24-4 shows the temperature variation of C_v for four substances with rather different values of T_D (see Table 24-2). At a temperature of 288 K (15° C), there is a wide spread in the values of C_v. However, when the data for different substances are plotted as a function of T/T_D (Fig. 24-5), there is a remarkable agreement.

At low temperatures ($T/T_D \lesssim 0.15$), the Debye theory predicts that C_v is proportional to

---

*Peter Debye (1884–1966), Dutch-American physicist and chemist, winner of the 1936 Nobel Prize in chemistry.

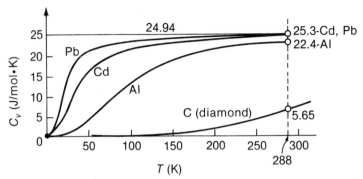

Fig. 24-4. The molar specific heats C_v for four solids that have different values of T_D (see Table 24-2). Values of C_v for solids are usually obtained by making small theoretical corrections to the experimentally determined values of C_p.

the *cube* of T/T_D. In fact,

$$C_v(\text{Debye}) = (1945 \text{ J/mol} \cdot \text{K})\left[\frac{T}{T_D}\right]^3 \tag{24-16}$$

The excellent agreement between this simple result and the experimental data is shown in Fig. 24-5.

Fig. 24-5. Values of C_v for different substances plotted as a function of T/T_D. At low temperatures ($T/T_D \lesssim 0.15$), the specific heat follows closely a T^3 law.

In substances that are good conductors of electricity, one or two electrons (occasionally more) become detached from each atom and move about more-or-less freely within the material. These *conduction electrons* constitute a kind of *electron gas* within the substance. However, this "gas" does not behave as a true gas; instead, the electrons obey a special quantum distribution that is appropriate for electrons. The electrons make a contribution to the specific heat of a substance, which is given approximately by

$$C_e = \gamma T \tag{24-17}$$

where, for most conductors, γ is not too different from 10^{-3} J/mol $\cdot$ K². At low temperatures, then, the molar specific heat at constant volume is

$$C_v = C_e + C_v(\text{Debye}) \tag{24-18}$$

The electronic correction to the specific heat is small at room temperature (about 0.3 percent for nickel), but it becomes increasingly important at very low temperatures. For example, at $T/T_D = 0.1$, the ratio $C_e/C_v(\text{Debye})$ for nickel is approximately 0.15 (see Problem 24-8).

Although C_v approaches zero as $T \rightarrow 0$, it should not be inferred that the internal energy also tends to zero at the absolute zero of temperature. As in the case of gases, the zero-point

TABLE 24-2.   The Debye
Temperature T_D for Several Solids

SUBSTANCE	T_D (K)
Lead (Pb)	88
Cadmium (Cd)	168
Silver (Ag)	215
Aluminum (Al)	398
Nickel (Ni)	413
Calcium fluoride (CaF$_2$)	474
Carbon (C, diamond)	1860

energy remains and the atoms of a solid continue to vibrate about the stable lattice points
even at $T = 0$. ◄

Example 24-4

A 150-g crystal of CaF_2 (fluorspar) is heated from 17 K to 50 K. What is the increase in
internal energy of the sample?

Solution:

The molecular mass of CaF_2 is

$$M = (40.08 \text{ u}) + 2(19.00 \text{ u}) = 78.08 \text{ u}$$

Hence, the number of moles in the sample is

$$n = \frac{m}{M} = \frac{150 \text{ g}}{78.08 \text{ g/mol}} = 1.92 \text{ mol}$$

From Table 24-2, the Debye temperature for CaF_2 is $T_D = 474$ K. Thus, using Eq. 24-16, we
find

$$U_2 - U_1 = n \int_{T_1}^{T_2} 1945 \left[\frac{T}{T_D}\right]^3 dT = \frac{1945 \, n}{4} \cdot \frac{T_2^4 - T_1^4}{T_D^3}$$

$$= \frac{(1945 \text{ J/mol} \cdot \text{K})(1.92 \text{ mol})}{4} \cdot \frac{(50 \text{ K})^4 - (17 \text{ K})^4}{(474 \text{ K})^3}$$

$$= 54.06 \text{ J}$$

24-3   MOLECULAR SPEEDS IN GASES

In the preceding discussions, we have made use of the fact that the molecules of
a gas are in a state of agitated random motion, with the molecular speeds described
by some distribution whose rms value depends only on the temperature. The expression for the molecular speed distribution was first obtained by James Clerk Maxwell
in 1860. Maxwell's derivation (which we present below) results in a function $dN(v)$
that gives the number of molecules in a gas sample with speeds between v and
$v + dv$. The original derivation was made rigorous by Boltzmann, and it is now

known as the *Maxwell-Boltzmann distribution:*

$$dN(v) = 4\pi N \left[\frac{\mu}{2\pi kT} \right]^{3/2} v^2 e^{-\mu v^2 / 2kT} \, dv \qquad (24\text{-}19)$$

where $N = \int dN(v)$ is the total number of molecules in the sample, where μ is the mass of a single molecule, and where T is the absolute temperature of the gas. Notice that for a particular molecular species, the distribution is a function only of the temperature. After we digress to derive the Maxwell-Boltzmann distribution, we examine some of the properties of this important function and discuss some of its consequences.

▶ *Derivation of the Maxwell-Boltzmann Distribution.* The crucial features of the derivation are best described in Maxwell's own words (with some slight changes of notation) from a paper presented to the British Association in 1859:

"Let N be the whole number of particles. Let v_x, v_y, v_z be the components of the velocity of each particle, and let the number of particles for which v_x lies between v_x and $v_x + dv_x$ be $Nf(v_x)dv_x$, where $f(v_x)$ is a function of v_x to be determined.

"The number of particles for which v_y lies between v_y and $v_y + dv_y$ will be $Nf(v_y)dv_y$, and the number for which v_z lies between v_z and $v_z + dv_z$ will be $Nf(v_z)dv_z$, where f always stands for the same function.

"Now the existence of the velocity v_x does not in any way affect that of the velocities in v_y or v_z, since these are all at right angles to each other and independent, so that the number of particles whose velocity lies between v_x and $v_x + dv_x$, and also between v_y and $v_y + dv_y$ and also between v_z and $v_z + dv_z$ is

$$Nf(v_x)f(v_y)f(v_z) \, dv_x \, dv_y \, dv_z$$

If we suppose the N particles to start from the origin at the same instant, then this will be the number in the element of volume $dv_x \, dv_y \, dv_z$ after unit of time, and the number referred to unit of volume will be

$$Nf(v_x)f(v_y)f(v_z)$$

"But the directions of the coordinates are perfectly arbitrary, and therefore this number must depend on the distance from the origin alone, that is

$$f(v_x)f(v_y)f(v_z) = \phi(v_x^2 + v_y^2 + v_z^2)$$

Solving this functional equation, we find

$$f(v_x) = B^{1/3}e^{-\beta v_x}, \qquad \phi(v_x^2 + v_y^2 + v_z^2) = Be^{-\beta(v_x^2 + v_y^2 + v_z^2)}\text{"}$$

where the constant in the exponential has been chosen to be $-\beta$ because the distribution must approach zero as $v \to \infty$. Thus, we have

$$dN(v) = Be^{-\beta(v_x^2 + v_y^2 + v_z^2)} \, dv_x \, dv_y \, dv_z \qquad (24\text{-}20)$$

This distribution gives the number of molecules that simultaneously have the x-component of velocity between v_x and $v_x + dv_x$, the y-component between v_y and $v_y + dv_y$, and the z-component between v_z and $v_z + dv_z$. Properly, we should now write $dN(\mathbf{v})$.

In order to convert Eq. 24–20 to a distribution that gives the number of molecules with speeds between v and $v + dv$, it is convenient to use spherical coordinates in *velocity space*. In

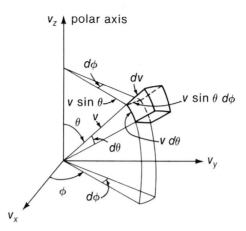

Fig. 24–6. Spherical coordinates for velocity space.

Fig. 24–6, we see that the rectangular volume element $dv_x\, dv_y\, dv_z$ becomes, in spherical coordinates,

$$dv_x\, dv_y\, dv_z = v^2 \sin\theta\, d\theta\, d\phi\, dv$$

Then,

$$dN(v) = Bv^2 e^{-\beta v^2} \sin\theta\, d\theta\, d\phi\, dv$$

This distribution gives the number of molecules with speeds between v and $v + dv$, moving in the direction between θ and $\theta + d\theta$ and between ϕ and $\phi + d\phi$.

To obtain the desired distribution, integrate over the angles θ and ϕ that specify the direction of motion. The integration over ϕ gives 2π; thus,

$$dN(v) = 2\pi Bv^2 e^{-\beta v^2}\, dv \int_0^\pi \sin\theta\, d\theta$$

$$= 4\pi Bv^2 e^{-\beta v^2}\, dv \qquad \textbf{(24–21)}$$

Finally, the two constants, B and β, must be evaluated. First, integrate over $dN(v)$ and set the result equal to N, the total number of molecules in the sample. That is,

$$\int dN(v) = 4\pi B \int_0^\infty v^2 e^{-\beta v^2}\, dv = N$$

This is a standard integral that can be found in integral tables:

$$\int_0^\infty x^2 e^{-\beta x^2}\, dx = \frac{1}{4\beta}\sqrt{\frac{\pi}{\beta}}$$

from which

$$B = N\left[\frac{\beta}{\pi}\right]^{3/2} \qquad \textbf{(24–22)}$$

Next, we use Eq. 24–9, which can be written as

$$\langle v^2 \rangle = \frac{3kT}{\mu}$$

The average value of v^2 is

$$\langle v^2 \rangle = \frac{\int v^2 N(v)dv}{\int N(v)dv}$$

Thus,

$$\frac{3kT}{\mu} = \frac{4\pi B}{N} \int_0^\infty v^4 e^{-\beta v^2} \, dv$$

This is another standard integral:

$$\int_0^\infty x^4 e^{-\beta x^2} \, dx = \frac{3}{8\beta^2} \sqrt{\frac{\pi}{\beta}}$$

so that

$$\frac{3kT}{\mu} = \frac{4\pi B}{N} \cdot \frac{3}{8\beta^2} \sqrt{\frac{\pi}{\beta}} \qquad (24\text{--}23)$$

Combining Eqs. 24-22 and 24-23, we find

$$B = N \left[\frac{\mu}{2\pi kT} \right]^{3/2}; \qquad \beta = \frac{\mu}{2kT} \qquad (24\text{--}24)$$

Altogether, we have

$$dN(v) = 4\pi N \left[\frac{\mu}{2\pi kT} \right]^{3/2} v^2 e^{-\mu v^2 / 2kT} \, dv \qquad (24\text{--}25)$$

which is the distribution we have been seeking—the Maxwell-Boltzmann distribution. ◀

The Maxwell-Boltzmann distribution has a number of interesting properties. The expression for v_{rms} (Eq. 24–9) has already been used in deriving Eq. 24–25, so the *root-mean-square speed* is known to be

$$v_{rms} = \langle v^2 \rangle^{1/2} = \sqrt{\frac{3kT}{\mu}} \qquad (24\text{--}26a)$$

We can also show (see Problem 24–9) that the *most probable speed* (that is, the speed for which $dN(v)/dv$ is maximum) is

$$v_m = \sqrt{\frac{2kT}{\mu}} \qquad (24\text{--}26b)$$

Finally, the *average speed* can be obtained by writing

$$\langle v \rangle = \frac{\int v \, dN(v)}{\int dN(v)}$$

$$= 4\pi \left[\frac{\mu}{2\pi kT} \right]^{3/2} \int_0^\infty v^3 e^{-\mu v^2 / 2kT} \, dv$$

Again, this is a standard integral:

$$\int_0^\infty x^3 e^{-\beta x^2} \, dx = \frac{1}{2\beta^2}$$

Using this result, we find

$$\langle v \rangle = \sqrt{\frac{8kT}{\pi\mu}} \qquad (24\text{--}26c)$$

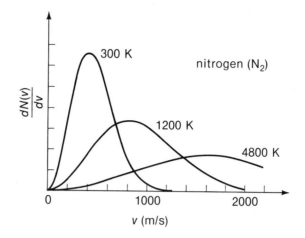

Fig. 24-7. The Maxwell-Boltzmann distribution for nitrogen molecules at three different temperatures.

These speeds are in the ratios

$$v_m : \langle v \rangle : v_{rms} = \sqrt{\frac{2}{3}} : \sqrt{\frac{8}{3\pi}} : 1$$

$$= 0.8165 : 0.9213 : 1$$

Figure 24-7 shows $dN(v)/dv$ as a function of v for nitrogen molecules at three different temperatures. As the temperature is increased, the distribution not only shifts toward higher speeds, but it also becomes more spread out.

A universal function to express the Maxwell-Boltzmann distribution for all cases can be obtained by writing the speed coordinate in relative units, $\xi = v/v_{rms}$, and by multiplying $dN(v)/dv$ by v_{rms}/N. Thus,

$$\frac{v_{rms}}{N} \cdot \frac{dN(v)}{dv} = 6 \sqrt{\frac{3}{2\pi}} \xi^2 e^{-3\xi^2/2} \qquad \textbf{(24-27)}$$

This function is shown in Fig. 24-8. Notice the locations of v_m, $\langle v \rangle$, and v_{rms}.

Fig. 24-8. The universal Maxwell-Boltzmann distribution function, Eq. 24-27.

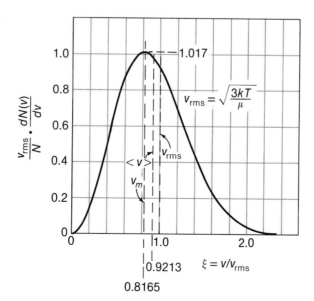

The Boltzmann Statistical Factor. The analytical methods of statistical thermodynamics that yield the equipartition of energy principle also lead to another general principle of great importance. This principle states that, for a collection of molecules in equilibrium, the number of molecules N_i with *total energy E_i* is proportional to the *Boltzmann statistical factor;* that is,

$$N_i = Ae^{-E_i/kT} \qquad (24\text{-}28)$$

Thus, the ratio of the number of molecules with energy E_i to the number with energy E_j is

$$\frac{N_i}{N_j} = \frac{e^{-E_i/kT}}{e^{-E_j/kT}} = e^{-(E_i - E_j)/kT}$$

If we consider only the translational motion of gas molecules in equilibrium at a temperature T, the number of molecules $dN(v)$ within the speed interval dv is proportional to the product of two statistical factors. The first is the volume of velocity space that corresponds to the speed interval dv, namely, $dv_x\, dv_y\, dv_z$ in rectangular coordinates or $4\pi v^2 dv$ in spherical coordinates. The second is the Boltzmann factor, which now becomes $e^{-\mu v^2/2kT}$. Thus, we have

$$dN(v) = 4\pi Av^2 e^{-\mu v^2/2kT}\, dv$$

When the constant A is evaluated by setting $\int dN(v) = N$, we obtain the Maxwell-Boltzmann distribution, Eq. 24-25! That is, the statistical approach of Boltzmann yields exactly the same result as Maxwell's (less rigorous) method.

The Boltzmann factor has broad applicability. A many-component system in thermal equilibrium has a certain number of degrees of freedom that contribute to the possible total energies E_i. The energy distribution for the components of the system is described by a function that contains the Boltzmann factor $e^{-E_i/kT}$ multiplied by the volume elements associated with the degrees of freedom. When only the translational degrees of freedom are considered the appropriate volume element is $dv_x\, dv_y\, dv_z$, as cited above. For other applications of the Boltzmann factor, see the example that follows and also Sections 30-5, 31-6, and 46-1.

Example 24-5

Consider a gas sample in a vessel of fixed volume and in equilibrium at a temperature T. Let the sample be in a gravitational field with a uniform downward acceleration **g**. Determine the pressure in the vessel as a function of height.

Solution:

Let the confining vessel have dimensions $a \times b \times c$, as shown in the diagram. Each molecule in the gas sample has a kinetic energy $\frac{1}{2}\mu v^2 = \frac{1}{2}\mu v_x^2 + \frac{1}{2}\mu v_y^2 + \frac{1}{2}\mu v_z^2$, and a potential energy, μgy, referred to the plane $y = 0$ indicated in the diagram. According to Eq. 24-28, the number of molecules within a spatial volume, $dx\, dy\, dz$, and simultaneously in a velocity volume, $dv_x\, dv_y\, dv_z$, is

$$dN(\mathbf{v}, \mathbf{r}) = Ae^{-[\mu(v_x^2 + v_y^2 + v_z^2)/2\, +\, \mu gy]/kT}\, dx\, dy\, dz\, dv_x\, dv_y\, dv_z$$
$$= A\{e^{-\mu(v_x^2 + v_y^2 + v_z^2)/2kT}\, dv_x\, dv_y\, dv_z\} \cdot \{e^{-\mu gy/kT}\, dx\, dy\, dz\}$$

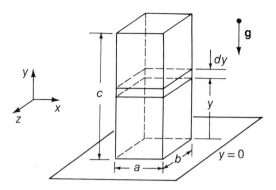

where the second equality emphasizes the fact that the six relevant coordinates are all *independent*. If we ask first for the speed distribution of the molecules in *any* part of the vessel, we must perform the appropriate integration over x, y, and z. This integration leaves unaltered the terms in the first pair of brackets (the velocity terms). Thus, the speed distribution does not depend on the space coordinates and is exactly the same as if g were zero. This is to be expected because we assumed the gas to be in equilibrium at a particular temperature (that is, the temperature is constant throughout the vessel).

If we ask next for the number of molecules within a volume $(ab)dy$ at a height y and with any velocity whatsoever, we must integrate over v_x, v_y, and v_z (in the range $-\infty$ to $+\infty$) and over x and z (in the ranges $0 \leqslant x \leqslant a$ and $0 \leqslant z \leqslant b$). The result in terms of a suitable constant $\mathcal{N}_0$ is

$$dN(y) = \mathcal{N}_0 ab e^{-\mu gy/kT}\, dy$$

Then, the number of molecules per unit volume is

$$\mathcal{N}(y) = \frac{dN(y)}{ab\, dy} = \mathcal{N}_0 e^{-\mu gy/kT}$$

so that $\mathcal{N}_0$ is seen to be the value of $\mathcal{N}(y)$ at $y = 0$.

The analysis we have made could be applied to a container of gas on an accelerating truck (•Can you see why?) or to the isothermal law of atmospheres discussed in Section 17–2. To convert the equation for $\mathcal{N}(y)$ into the form of Eq. 17–10, we write the ideal gas law equation as

$$p = \frac{n}{V}RT = \frac{N}{N_0 V}RT = \mathcal{N}kT$$

Using the above equation for $\mathcal{N}(y)$, we obtain

$$p(y) = p_0 e^{-\mu gy/kT}$$

Now, the density of the gas sample is $\rho = \mathcal{N}\mu$. Thus, the exponent becomes

$$\frac{\mu gy}{kT} = \frac{\rho_0}{\mathcal{N}_0} \frac{gy}{kT} = \frac{\rho_0}{p_0} gy = \alpha y$$

Therefore, we have $\qquad p(y) = p_0 e^{-\alpha y} \qquad (\alpha = \rho_0 g/p_0)$

which is just Eq. 17–10.

SPECIAL TOPIC

24-4   TRANSPORT PHENOMENA

Nonequilibrium conditions in a gas can lead directly to the transport of molecules from one region to another. An example of such a transport phenomenon is the *diffusion* of a gas due to a nonequilibrium density gradient; another example is the convective transfer of heat due to a temperature gradient. Also, nonequilibrium conditions are involved in the viscous flow of a gas as molecules move past one another with different velocities. Central to the explanation of this class of phenomena is the frequency with which the molecules collide with each other. Accordingly, we begin by introducing the important and related concept of the mean free path.

Mean Free Path.   The *mean free path* λ of a molecule is the average distance traveled between collisions with other molecules. A closely related quantity is the average time between collisions, or *mean free time* τ. If $\langle v \rangle$ is the average molecular speed relative to the C.M. of the gas, we have

$$\lambda = \langle v \rangle \tau \qquad \text{(24–29)}$$

Suppose that two molecules in a sample of a chemically pure gas approach each other with a relative velocity $\mathbf{v}_r$, as indicated in Fig. 24–9. Assume that all molecular collisions are elastic and that during these collisions the molecules behave as hard spheres with diameter d. The separation b between the centers of two molecules measured perpendicular to $\mathbf{v}_r$ (see Fig. 24–9) is called the *impact parameter*. When $b < d$, a collision will occur; when $b > d$, no collision will occur. (The case $b = d$ corresponds to a *grazing collision*.) Alternatively, a sphere with radius d, called the *sphere of action* (Fig. 24–9), can be associated with every molecule. As a molecule moves through the sample, its sphere of action sweeps out an area πd^2. If the

Fig. 24–9. A collision between the molecules will occur because the impact parameter b is less than the molecular diameter d. The relative velocity between the molecules is $\mathbf{v}_r$ (shown here in the frame in which the molecule at the right is at rest).

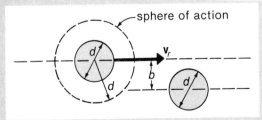
sphere of action

center of another molecule falls within this area, a collision will occur. The area πd^2 is called the *total scattering cross section* σ of the molecule. Thus,

$$\sigma = \pi d^2 \qquad \text{(24–30)}$$

Next, consider a molecule that moves through a collection of molecules. Let the average relative speed between molecules be $\langle v_r \rangle$ and let $\mathcal{N}$ be the number of molecules per unit volume. The average number of collisions that a particular molecule makes during a time interval Δt is equal to the cylindrical volume swept out by the molecule's sphere of action,* $\sigma \langle v_r \rangle \, \Delta t$, multiplied by the molecular density $\mathcal{N}$, or $\sigma \langle v_r \rangle \mathcal{N} \, \Delta t$. From the definition of the mean free time τ, the number of collisions during Δt is also equal to $\Delta t / \tau$. Equating these two expressions for the number of collisions, we have

$$\frac{\Delta t}{\tau} = \sigma \langle v_r \rangle \mathcal{N} \, \Delta t$$

or

$$\tau = \frac{1}{\sigma \langle v_r \rangle \mathcal{N}} \qquad \text{(24–31)}$$

Substituting this result into Eq. 24–29, we find

$$\lambda = \frac{\langle v \rangle}{\langle v_r \rangle} \cdot \frac{1}{\sigma \mathcal{N}} \qquad \text{(24–32)}$$

Next, we need an expression for the ratio $\langle v \rangle / \langle v_r \rangle$. There exist exact methods for calculating this quantity, but they are cumbersome and complicated. We obtain the same result by using a simpler method, but one that involves making a "reasonable assumption."

Consider two molecules whose velocities relative to the C.M. of the gas are $\mathbf{v}$ and $\mathbf{v}'$. A collision between these molecules involves the relative velocity, $\mathbf{v}_r = \mathbf{v} - \mathbf{v}'$. Thus,

$$v_r^2 = \mathbf{v}_r \cdot \mathbf{v}_r = v^2 + v'^2 - 2(\mathbf{v} \cdot \mathbf{v}')$$

Taking average values, we obtain

$$\langle v_r^2 \rangle = \langle v^2 \rangle + \langle v'^2 \rangle - 2\langle \mathbf{v} \cdot \mathbf{v}' \rangle$$

Now, the directions of $\mathbf{v}$ and $\mathbf{v}'$ are randomly distributed. Consequently, the angle between $\mathbf{v}$ and $\mathbf{v}'$ assumes all pos-

---

*With each collision, there is a change in the direction of motion of the selected molecule. The "cylindrical volume" is therefore a zigzag connection of short cylindrical segments.

sible values within the sample, so the average value is zero; that is, $\langle \mathbf{v} \cdot \mathbf{v}' \rangle = 0$. Moreover, because the colliding molecules are identical, we have $\langle v^2 \rangle = \langle v'^2 \rangle$. Thus, $\langle v_r^2 \rangle = 2\langle v^2 \rangle$. Taking square roots, we obtain

$$\langle v_r^2 \rangle^{1/2} = \sqrt{2}\langle v^2 \rangle^{1/2}$$

That is,

$$\frac{v_{\text{rms}}}{v_{r,\text{rms}}} = \frac{1}{\sqrt{2}}$$

We make the "reasonable assumption" that the same ratio applies for the *average values* of v and v_r, namely, $\langle v \rangle / \langle v_r \rangle = 1/\sqrt{2}$. Then, Eq. 24-32 becomes

$$\lambda = \frac{1}{\sqrt{2}\,\sigma \mathcal{N}} \qquad \textbf{(24-33)}$$

In this derivation, we have assumed that the colliding molecules behave as ideal hard spheres. However, the intermolecular force does not suddenly drop to zero for some separation d between the molecular centers. For low relative speeds, the effective value of d tends to be larger than for high relative speeds (because the longer interaction time allows a greater contribution from the weak, longer-range part of the force). Consequently, we expect λ to show a slight increase with temperature. For example, the values of λ for the noble gases are found to increase with temperature as $T^{1/6}$.

▶ The significance of the concept of the mean free path can be understood by asking for the probability $P(x)$ that a molecule will travel a distance x without suffering a collision. The probability that a molecule will collide with another molecule in traveling a distance dx is dx/λ. Thus, the probability that *no* collision occurs in dx is $(1 - dx/\lambda)$. The probability of surviving without a collision in the distance $x + dx$ is, therefore,

$$P(x + dx) = \left(1 - \frac{dx}{\lambda}\right)P(x)$$

which is just the probability of traveling a distance x multiplied by the probability of traveling an additional distance dx, each without a collision. Expanding $P(x + dx)$ in a Taylor's series (see Section VI-1 in Part 1) we have

$$P(x + dx) = P(x) + \frac{dP(x)}{dx} \cdot dx + \dots$$

Equating the two expressions for $P(x + dx)$, we find

$$P(x) + \frac{dP}{dx} \cdot dx = \left(1 - \frac{dx}{\lambda}\right)P(x)$$

from which

$$\frac{dP(x)}{P(x)} = -\frac{1}{\lambda}\,dx$$

We integrate this expression between $x = 0$ and x, noting that

$P(0) = 1$ (•Can you see why?). There results

$$P(x) = e^{-x/\lambda} \qquad \textbf{(24-34)}$$

Using this expression, we find that the probability that a molecule will travel a distance $x = 4\lambda$ without a collision is 0.018 (1.8 percent), whereas the probability of surviving a collisionless path of length $x = \frac{1}{4}\lambda$ is 0.78 (78 percent). ◀

We now use the idea of the mean free path and Eq. 24-33 to discuss the viscosity and the diffusion of gases.

Viscosity of Gases. In Section 18-5, the concept of *viscosity* was introduced as a measure of the dynamic shear effect in fluids. Consider a fluid that has a flow velocity parallel to the x-axis with a magnitude that depends in some way on the coordinate y; that is, $\mathbf{u} = u_x(y)\hat{\mathbf{i}}$. As the fluid layers slide past each other, shear stresses s_t are exerted across the planes that separate adjacent layers (Fig. 24-10). The viscosity coefficient η is equal to the ratio of the shear stress s_t to the rate of shear strain $\partial u_x/\partial y$. Thus,

$$s_t = \eta \frac{\partial u_x}{\partial y} \qquad \textbf{(24-35)}$$

We now examine the molecular description of viscosity and obtain an estimate of the magnitude of η for gases. As we proceed, we must keep in mind the distinction between the relatively slow flow speed of the fluid (that is, u_x) and the much larger thermal speed of the molecules (that is, $\langle v \rangle$).

Consider a fluid sample in which there are $\mathcal{N}$ molecules per unit volume and in which the average thermal speed of the molecules is $\langle v \rangle$. The random motion of the molecules is equivalent to 1/3 of them moving in each of the x-, y-, and z-directions, with equal numbers moving in the positive and negative directions. Thus, for example, at any particular instant, 1/6 of the molecules move in the $+y$-direction and 1/6 move in the $-y$-direction.

Fig. 24-10. Two layers of fluid move with different speeds. The more rapidly moving layer above the level $y = $ constant exerts a shear force F_t parallel to the area A on the layer beneath. The shear stress is $s_t = F_t/A$.

Fig. 24-11. During a time Δt, 1/6 of the molecules in the volume $A\langle v\rangle \Delta t$ will pass through the area A from above and an equal number will pass through from below.

Figure 24-11 shows an area A that separates two regions of a fluid column. The two slabs with thickness $\langle v\rangle \Delta t$ correspond to those shown in Fig. 24-10. During a time Δt, 1/6 of the molecules in a volume $A\langle v\rangle \Delta t$ will pass through the area A from above and an equal number will pass through from below. The molecules that reach the level y from above will, on the average, have experienced their most recent collisions at the level $y + \lambda$. We may therefore presume that these molecules share the flow velocity, $u_x(y + \lambda)$, appropriate for that level. Consequently, these molecules transport to the level y an x-directed momentum equal to $\frac{1}{6}\mathcal{N}\mu A\langle v\rangle u_x(y + \lambda)\Delta t$. The momentum transported downward per unit area and per unit time is

$$\frac{\Delta(\text{momentum})_{x+}}{A\,\Delta t} = \tfrac{1}{6}\mathcal{N}\mu\langle v\rangle u_x(y + \lambda)$$

The corresponding momentum transported upward to the level y is

$$\frac{\Delta(\text{momentum})_{x-}}{A\,\Delta t} = \tfrac{1}{6}\mathcal{N}\mu\langle v\rangle u_x(y - \lambda)$$

The difference between these expressions represents the force per unit area or the shear stress s_t at the level y. That is,

$$s_t = \tfrac{1}{6}\mathcal{N}\mu\langle v\rangle[u_x(y + \lambda) - u_x(y - \lambda)]$$

Now, we can write

$$u_x(y + \lambda) \cong u_x(y) + \lambda\frac{\partial u_x(y)}{\partial y}$$

$$u_x(y - \lambda) \cong u_x(y) - \lambda\frac{\partial u_x(y)}{\partial y}$$

Therefore, the expression for the shear stress becomes

$$s_t = \tfrac{1}{3}\mathcal{N}\mu\langle v\rangle\lambda\frac{\partial u_x}{\partial y}$$

Comparing this result with Eq. 24-35, we see that

$$\eta = \tfrac{1}{3}\mathcal{N}\mu\langle v\rangle\lambda \qquad \textbf{(24-36)}$$

We now specialize to the case of gases and substitute for $\langle v\rangle$ from Eq. 24-26c and for λ from Eq. 24-33. The result is*

$$\eta = \frac{2}{3\sigma\sqrt{\pi}}\sqrt{\mu kT} \qquad \textbf{(24-37)}$$

The surprising fact about Eq. 24-37 is that it shows η to be independent of the gas molecular density $\mathcal{N}$, and this also means (because $p = \mathcal{N}kT$) that η is independent of pressure. This conclusion is confirmed by experiment; for example, $\eta(\text{air}) = 1.81 \times 10^{-5}$ Pa·s at 20° C and 1 atm, whereas $\Delta\eta/\Delta p = 2.4 \times 10^{-8}$ Pa·s/atm for the same conditions. Equation 24-36 indicates that η is proportional to $\mathcal{N}$; however, an increase in $\mathcal{N}$ results in more frequent collisions so that λ is reduced in inverse proportion to $\mathcal{N}$. In Eq. 24-33, it is apparent that the product $\lambda\mathcal{N}$ is a constant.

When the pressure is very low or very high, the viscosity coefficient is no longer independent of pressure. For very low pressures, the mean free path may become comparable with the size of the confining vessel; this causes η to decrease with decreasing pressure. For sufficiently high pressures, the mean free path becomes comparable with molecular sizes. Then, the analysis changes to a description of the liquid phase and becomes considerably more complicated than for the dilute gases we have been discussing.

The viscosity coefficients for real gases are essentially independent of pressure. However, the temperature dependence of η in most cases is more pronounced than $T^{1/2}$, which is predicted by Eq. 24-37. This behavior is due primarily to the increase in λ (and the corresponding decrease in σ) with increasing temperature, as mentioned earlier.

Example 24-6

Argon gas at atmospheric pressure and at 20° C is confined in a vessel with a volume of 1 m³. The effective hard-sphere diameter of the argon atom is 3.10×10^{-10} m.

(a) Determine the mean free path λ.

---

*An exact calculation for the case of ideal hard-sphere molecules distributed in speed according to the Maxwell-Boltzmann function yields a coefficient $5\sqrt{\pi}/16 = 0.554$ instead of $2/3\sqrt{\pi} = 0.376$.

(b) Find the pressure (at 20° C) when $\lambda = 1$ m.

(c) Find the pressure (at 20° C) when $\lambda = 3.1 \times 10^{-10}$ m.

(d) Calculate the viscosity coefficient for the original conditions (1 atm and 20° C).

Solution:

Using $p = \mathcal{N}kT$ and $\sigma = \pi d^2$, Eq. 24–33 becomes

$$\lambda = \frac{kT}{\sqrt{2}\pi p d^2} \qquad (1)$$

(a) Using Eq. (1) for λ, we have

$$\lambda = \frac{(1.381 \times 10^{-23} \text{ J/K})(293 \text{ K})}{\sqrt{2}\,\pi(1.013 \times 10^5 \text{ Pa})(3.10 \times 10^{-10} \text{ m})^2}$$
$$= 9.36 \times 10^{-8} \text{ m} = 93.6 \text{ nm}$$

(b) Equation (1) shows that $p_1\lambda_1 = p_2\lambda_2$. Taking $p_1\lambda_1$ from (a) and with $\lambda_2 = 1$ m, we find

$$p_2 = \frac{(1 \text{ atm})(9.36 \times 10^{-8} \text{ m})}{1 \text{ m}} = 9.36 \times 10^{-8} \text{ atm}$$

(c) For $\lambda_2 = 3.1 \times 10^{-10}$ m, we have

$$p_2 = \frac{(1 \text{ atm})(9.36 \times 10^{-8} \text{ m})}{3.1 \times 10^{-10} \text{ m}} = 302 \text{ atm}$$

(d) Using Eq. 24–37 together with $\sigma = \pi d^2$, we find

$$\eta = \frac{2}{3\pi^{3/2}d^2}\sqrt{\mu kT}$$

$$= 0.120\frac{[(39.94 \times 1.66 \times 10^{-27} \text{ kg}) \cdot (1.38 \times 10^{-23} \text{ J/K})(293 \text{ K})]^{1/2}}{(3.10 \times 10^{-10} \text{ m})^2}$$
$$= 2.04 \times 10^{-5} \text{ Pa} \cdot \text{s}$$

For argon at 1 atm and 20° C, the measured value for the viscosity coefficient is 2.23×10^{-5} Pa · s, and the inferred mean free path is 94.3 nm. Such close agreement between experiment and the predictions of the simple theory is usually found only for the noble gases.

Diffusion. Consider a dilute gas of similar molecules that is confined in a vessel.* Let one component of

---

*We limit consideration to cases in which the molecules all have the same mass and the same hard-sphere diameter (or approximately so). The diffusion of helium through air, for example, involves more complications than we care to treat here.

the gas be labeled in some way. (For example, we might use radioactive atoms.) Suppose that the *concentration* or number density of these labeled molecules, $\mathcal{N}'$ (per unit volume), is not constant throughout the vessel but depends on the position coordinate y so that $\mathcal{N}' = \mathcal{N}'(y)$. We assume that the *total* number density $\mathcal{N}$, which includes $\mathcal{N}'$, is constant throughout the vessel; in this way, we avoid the complication of any net mass motion within the gas. The process by which the labeled molecules achieve equilibrium is called *diffusion*. The simplified case considered here, in which the molecules are of the same type, is known as *self-diffusion*.

Let the *flux* (that is, the number flow rate) of labeled molecules through a unit area parallel to the *x-z* plane at a position y be J_y. Thus, during a time interval Δt, the number of labeled molecules that pass through an area A parallel to the *x-z* plane is $J_y A \Delta t$. We expect J_y to be proportional to the number density gradient, $\partial \mathcal{N}'/\partial y$; indeed, this is confirmed by experiment. The *coefficient of self-diffusion* D is defined by the equation*

$$J_y = -D\frac{\partial \mathcal{N}'}{\partial y} \qquad (24\text{--}38)$$

We take D to be a positive quantity, so the negative sign in Eq. 24–38 is necessary to indicate that the flow of the labeled molecules is from the region with the high concentration to the region with the low concentration.

Because the molecules in the sample are essentially identical, they can be characterized by $\langle v \rangle$ and λ, quantities that depend on the total number density $\mathcal{N}$. The flow of the labeled molecules through an area A at the position y during a time interval Δt again consists of an upward flow and a downward flow (see Fig. 24–11). As in the analysis of viscosity, we can write

$$J_y A\,\Delta t = \tfrac{1}{6}\langle v \rangle A\,\Delta t[\mathcal{N}'(y - \lambda) - \mathcal{N}'(y + \lambda)]$$

The terms in the square brackets can be expressed as in the case of viscosity; then, we find

$$J_y = -\tfrac{1}{3}\langle v \rangle\lambda\frac{\partial \mathcal{N}'}{\partial y}$$

Comparing this result with Eq. 24–38, we conclude that

$$D = \tfrac{1}{3}\langle v \rangle\lambda \qquad (24\text{--}39)$$

Substituting for $\langle v \rangle$ and λ from Eqs. 24–26c and 24–33, respectively, we obtain

---

*Equation 24–38 was first formulated in 1855 by the German physiologist Adolf Fick (1829–1901) and is known as *Fick's law*.

$$D = \frac{2}{3\sigma\sqrt{\pi}\mathcal{N}}\sqrt{\frac{kT}{\mu}} \qquad (24\text{-}40)$$

If we divide Eq. 24-36 for η by Eq. 24-39 for D, we find the remarkably simple relationship,

$$\frac{\eta}{D} = \mathcal{N}\mu = \rho \qquad (24\text{-}41)$$

where ρ is just the mass density. For example, argon gas, at 1 atm and 20° C, has $\rho = 1.60\ \text{kg/m}^3$ and $\eta = 2.23 \times 10^{-5}\ \text{Pa}\cdot\text{s}$. According to Eq. 24-41, we expect the self-diffusion coefficient to be $D = \eta/\rho = 1.39 \times 10^{-5}\ \text{m}^2/\text{s}$; the experimental value is $1.76 \times 10^{-5}\ \text{m}^2/\text{s}$. The comparison seems reasonable in view of the simplicity of the theory.

Typical values for the kinetic properties of various dilute gases are listed in Table 24-3.

Example 24-7

Use the value of the self-diffusion coefficient D for argon given in Table 24-3 to determine the atomic diameter of argon. (Use $\sigma = \pi d^2$.)

Solution:

We can use $p = \mathcal{N}kT$ to express D (Eq. 24-40) in terms of the pressure:

$$D = \frac{2}{3\sigma\sqrt{\pi p}}\sqrt{\frac{k^3 T^3}{\mu}}$$

Then,

$$\sigma = \pi d^2 = \frac{2}{3\sqrt{\pi}Dp}\sqrt{\frac{k^3 T^3}{\mu}}$$

Substituting the values of the various quantities, we have

$$\pi d^2 = \frac{2}{3\sqrt{\pi}(1.76 \times 10^{-5}\ \text{m}^2/\text{s})(1.013 \times 10^5\ \text{Pa})} \cdot$$

$$\left[\frac{(1.38 \times 10^{-23}\ \text{J/K})^3(293\ \text{K})^3}{39.94 \times 1.66 \times 10^{-27}\ \text{kg}}\right]^{1/2}$$

$$= 2.11 \times 10^{-19}\ \text{m}^2$$

Thus,

$$d = \sqrt{\frac{2.11 \times 10^{-19}\ \text{m}^2}{\pi}} = 2.59 \times 10^{-10}\ \text{m}$$

This is to be compared with the value $(3.10 \times 10^{-10}\ \text{m})$ used in Example 24-6.

TABLE 24-3.   Properties of Dilute Gases (at 1 atm and 20° C)

GAS	DENSITY, ρ (kg/m³)	SELF-DIFFUSION COEFFICIENT, D ($10^{-5}\ \text{m}^2/\text{s}$)	VISCOSITY COEFFICIENT, η ($10^{-5}\ \text{Pa}\cdot\text{s}$)	AVERAGE SPEED, $\langle v\rangle$ (m/s)	MEAN FREE PATH, λ (nm)
Argon	1.604	1.76	2.229	395	94.3
Hydrogen	0.0809	14.3	0.887	1755	166.4
Krypton	3.366	0.90	2.496	272	
Methane	0.645	2.08	1.098	605	
Neon	0.810	4.73	3.138	557	
Nitrogen	1.125	2.00	1.757	471	88.7
Xenon	5.303	0.44	2.274	218	

PROBLEMS

Section 24-1

24-1   The speed distribution of 25 molecules is

$v(10^2\ \text{m/s})$	1	2	3	4	5	6	7	8	9	10	>10
$\Delta\mathcal{N}(v)$	1	2	4	5	4	3	3	2	0	1	0

(a) Plot the distribution function $\Delta\mathcal{N}(v)$ as a function of v. (b) Find the average speed, $\langle v\rangle$. (c) Find the root-mean-square speed, v_{rms}. (d) Determine the ratio, $v_{\text{rms}}/\langle v\rangle$, and compare with $\sqrt{3\pi/8}$, which is the value appropriate for an ideal gas.

24-2   A vessel contains 5.0 mg of nitrogen (N_2) at a pressure of 0.10 atm. An experiment shows that the rms speed is $v_{rms} = 300$ m/s. (a) Determine the volume of the container. (b) What is the average spacing between molecular centers? Compare your value with the diameter of a nitrogen atom (1.06×10^{-10} m).

24-3 At a temperature of $15.0°$ C and a pressure of 1 atm, 5.551 g of xenon (Xe, a monatomic noble gas) occupies 1.000 ℓ. (a) What is the atomic mass of xenon? (b) What is the rms speed of the xenon atoms in the container? (c) Assume that the confining vessel is a cube, 10 cm on a side. What is the difference in the gravitational potential energy of an atom at the top of the cube and one at the bottom of the cube? Compare this result with the average kinetic energy of an atom ($\frac{1}{2}\mu\langle v^2\rangle$). Comment.

Section 24-2

24-4 Determine the rms speeds for (a) hydrogen (H_2), helium (He), and carbon dioxide (CO_2) at $0°$ C; (b) oxygen (O_2), argon (Ar), and ammonia (NH_3) at $100°$ C.

24-5 Calculate the specific heat at constant volume c_v (in Cal/kg $\cdot$ deg) for hydrogen sulfide (H_2S) at $15°$ C. Base your calculation on Eq. 24-15 and assume that no substantial vibrational motion is present. (The experimental value is 0.178 Cal/kg $\cdot$ deg.)

24-6 According to the Debye theory, the molar specific heat of a crystalline solid is one half the Dulong-Petit value (that is, $C_v = 12.47$ J/mol $\cdot$ K) for $T/T_D = 0.250$. At what temperatures will diamond, aluminum, cadmium, and lead have this half value? Compare your results with Fig. 24-4.

24-7 The molar specific heat of iron pyrite (FeS_2) is 0.653 J/mol $\cdot$ K at $T = 45$ K. What is the Debye temperature for FeS_2? (The experimental value is $T_D = 645$ K.) [*Hint:* Use the Debye T^3 law.]

24-8 In nickel (Ni), the electrons make a substantial contribution to the specific heat at low temperatures. For nickel, $\gamma = 7.3 \times 10^{-3}$ J/mol $\cdot$ K^2. (a) Calculate and compare the values of C_e and C_v (Debye) for nickel at $T = 40$ K. (b) Assume that the expressions for C_e and C_v (Debye), Eqs. 24-16 and 24-17, are valid to 0 K. Determine the increase in internal energy per mole of nickel due to an increase in temperature from 0 K to 40 K.

Section 24-3

24-9 Show that the most probable speed for the Maxwell-Boltzmann distribution is $v_m = \sqrt{2kT/\mu}$.

24-10 A vessel contains 1000 molecules of helium in equilibrium at a temperature of 250 K. Find the number of molecules expected to have speeds in the ranges 0-250 m/s, 250-500 m/s, 500-750 m/s, ..., up to 2250-2500 m/s. [*Hint:* First determine v_{rms}, then transform the speed at the midpoint of each interval into the equivalent value of $\xi = v/v_{rms}$. Use Fig. 24-8 and find $\Delta N(v) = N\Delta v/v_{rms}$. Are your results consistent with $\Sigma \Delta N(v) = N$?]

24-11 An elevator with a height of 2 m is at rest. The air in the elevator is at normal atmospheric pressure and has a density of 1.29 kg/m^3. The elevator is now sealed (air tight) and accelerates upward at 3g. (a) Show that the center of mass of the air in the elevator is

$$Y = \frac{1}{\alpha} + h(1 - e^{\alpha h})^{-1}$$

where $\alpha = 4g\rho_0/p_0$ and where h is the height of the elevator. (The required integral can be found in the table inside the rear cover.) (b) Show that $\Delta Y = \frac{1}{2}h - Y \cong \frac{1}{12}\alpha h^2$, when $\alpha h \ll 1$. [*Hint:* Expand e^x and $(1 + x)^{-1}$.] (c) Evaluate ΔY.

SPECIAL TOPIC

Section 24-4

24-12 Show that for an ideal gas Eq. 24-33 for the mean free path can be expressed as

$$\lambda = \frac{1}{\sqrt{2}\pi d^2} \frac{kT}{p}$$

24-13 Calculate D using the values of η and ρ in Table 24-3 for each gas listed and compare with the value given in the table.

24-14 What value of the self-diffusion coefficient for xenon is expected at a pressure of 1/10 atm and a temperature of $100°$ C?

24-15 Use the values of D and η for krypton in Table 24-3, and from each determine the effective hard-sphere diameter of the krypton atom. Compare the results.

24-16 The viscosity coefficients for helium and xenon are
•

1.961×10^{-5} Pa·s and 2.274×10^{-5} Pa·s, respectively (at 1 atm and 20° C). Determine the ratio of the effective hard-sphere diameters for these atoms. Does the ratio of the atomic volumes correspond to the ratio of the atomic masses?

Additional Problems

24-17 Calculate the specific heat at constant volume for air at 15° C using Eq. 24-15. Assume air to consist of 76 percent (by mass) nitrogen N_2 and 24 percent oxygen O_2. Assume also that neither component molecule undergoes any significant vibrational motion. Compare your result with the observed value of 0.171 Cal/kg·deg.

24-18 Some measured values of the molar specific heat for
• aluminum at low temperatures are given in the accompanying table. Plot C_v/T as a function of T^2 and determine the Debye temperature T_D and the electronic coefficient γ for aluminum.

T(K)	4	6	8	10
$C_v(10^{-2}$ J/mol·K)	0.80	1.72	2.97	5.02

SPECIAL TOPIC

24-19 Use Eq. 24-34 for the probability that a molecule
• will travel a distance x before suffering a collision and show that the average distance $\langle x \rangle$ between collisions is just equal to the mean free path λ. [*Hint:* Use an expression for $\langle x \rangle$ equivalent to that given for $\langle v \rangle$ immediately following Eq. 24-26b. The required integral is listed in the table inside the rear cover.]

24-20 Determine the mean free time τ between molecular collisions in nitrogen (a) at 1 atm and 20° C and (b) at 1/20 atm and 100° C.

24-21 The molecular mass of hydrogen (H_2) is 2.016 u, and the effective hard-sphere diameter of the molecule is 2.35×10^{-10} m. Calculate the values of ρ, D, η, $\langle v \rangle$, and λ at 1 atm and 20° C. Compare your results with the values given in Table 24-3.

THERMODYNAMICS I

In this chapter we are concerned primarily with a comprehensive version of the energy conservation principle known as the *first law of thermodynamics*. At first, attention is focused on a simple system that consists of a confined gas sample. Heat energy is allowed to enter or leave the sample. Also, the confining volume is allowed to change so that mechanical work can be done on or by the gas. Any temperature changes that accompany these processes indicate changes in the internal energy of the sample. We seek to understand the relationship connecting these variables as the system undergoes change.

In these discussions we often refer to processes by which a system changes from one thermodynamic state to another. That such a change can occur means that the system is not in an equilibrium state. However, if the change is made sufficiently slowly, the system remains arbitrarily close to equilibrium at all stages in the process. A process of this type is said to be *quasi-static*.

If a system in equilibrium is disturbed, it will achieve a new equilibrium state in some characteristic time (called the *relaxation time*) after the disturbance is removed. For a process to be quasi-static, it must take place during a time that is long compared with the relaxation time. During such a slow process the system can be considered to pass through a large (strictly, infinite) number of equilibrium states. This allows a unique temperature to be defined for every point in the process. (To slow a process from its natural rate of progress requires deliberate intervention or control on the part of an external agent.)

A quasi-static process is an idealization of a real process that is valid (or nearly so) only for differential changes in equilibrium states. A finite change can never be truly quasi-static. We are forced to accept this lack of precision because it is impossible to maintain a system indefinitely in an equilibrium thermodynamic state. (This would require a perfect thermal insulator, which does not exist.)

25-1 MECHANICAL WORK DONE BY A GAS SYSTEM

Consider a system that consists of a gas confined in a cylinder closed with a frictionless, tight-fitting piston. The cylinder is immersed in a large constant-temperature heat bath or reservoir that constitutes the environment of the system (Fig. 25-1). Heat flows readily through the cylinder walls so that the gas is maintained

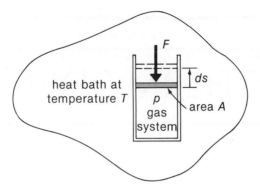

Fig. 25–1. A gas system is confined in a cylinder with a movable piston. The gas is maintained at the temperature T by the surrounding heat bath.

always at the temperature of the surrounding heat bath.* The gas (sufficiently dilute) obeys the ideal gas law.

Suppose that the gas is in thermal equilibrium at the temperature T and is in mechanical equilibrium by virtue of an external force F applied to the piston. The gas pressure is then $p = F/A$, where A is the surface area of the piston. Let the external force be slowly reduced so that the gas expands slightly and the piston moves upward by a differential displacement ds. During this process an amount of work $F\,ds = (pA)\,ds$ is done by the gas on the agency responsible for the force F. Now, $A\,ds = dV$ is the amount of volume expansion, so the work done by the gas is

$$dW = p\,dV \qquad\qquad \textbf{(25–1)}$$

Fig. 25–2. The slow isothermal expansion of a gas from an initial state (p_a, V_a, T). The work done in the differential expansion dV at the pressure p is $dW = p\,dV$. The total work done is the sum of all such contributions and corresponds to the shaded area under the p-V curve. The gas obeys the ideal gas law; therefore, once the initial state (p_a, V_a, T) is specified, only p_b or V_b (but not both) is required to define the final state reached after the isothermal expansion.

Initially the gas is in a thermal equilibrium state (p_a, V_a, T) and is in mechanical equilibrium with the force $F_a = p_a A$. By a large number of small, slow reductions in the external force, the system is allowed to proceed through a corresponding large number of quasi-static states, eventually arriving at the final equilibrium condition, (p_b, V_b, T) and $F_b = p_b A$. At each step in the process, heat is exchanged between the gas and the heat bath by whatever amount necessary to maintain the gas at the temperature T. (In this case heat is actually supplied to the gas by the heat bath.) This constant-temperature or *isothermal* process is shown in Fig. 25–2. The quantity of gas in the cylinder is fixed, so throughout the process we have $pV = nRT = $ constant. Thus, the total amount of work *done by the gas* in expanding from the volume V_a to V_b is

$$W_{ab} = \int_{V_a}^{V_b} p\,dV = nRT \int_{V_a}^{V_b} \frac{dV}{V}$$

from which

$$W_{ab} = nRT \ln \frac{V_b}{V_a} \qquad \text{isothermal process} \qquad \textbf{(25–2)}$$

*Ideal walls that are perfect heat conductors are called *diathermic walls*.

That is, the work done W_{ab} is equal to the definite integral that corresponds to the area under the p-V curve between V_a and V_b; this is the shaded area in Fig. 25-2.

▶ *The General Nature of* $dW = p\,dV$. Consider a gas that is confined within an irregular container that has a volume V and a surface area A (Fig. 25-3). Now, allow the container to undergo an arbitrary expansion by an amount dV. As shown in the figure, the differential area dA is displaced by an amount $d\mathbf{s}$ in a direction that makes an angle θ with the outward-directed normal $\hat{\mathbf{n}}$ to dA. The corresponding change in volume is a differential volume of a second order, which we write as* $d^2V = (dA\hat{\mathbf{n}})\cdot d\mathbf{s} = (dA)(ds)\cos\theta$. The work d^2W done during the displacement is* $d^2W = (dF)(ds)\cos\theta = p(dA)(ds)\cos\theta$, where we have made use of the fact that the force exerted by any static fluid on an area is perpendicular to that area. Thus,

$$d^2W = p\,d^2V$$

so that

$$dW = \int d^2W = \int p\,d^2V = p\int d^2V = p\,dV$$

Thus, Eq. 25-1 is generally valid, even for irregularly shaped volumes and arbitrary displacements. ◀

Fig. 25-3. The arbitrary expansion by an amount dV of a gas system with an initial volume V.

Because Eq. 25-1 is always valid, we can integrate this equation and write a general expression for the work done by a gas during an *arbitrary* (not necessarily isothermal) expansion that carries the system from a state (p_a, V_a, T_a) to a state (p_b, V_b, T_b) along a thermodynamic path specified by $p = p(V, T)$:

$$W_{ab} = \int_{V_a}^{V_b} p(V, T)\,dV \tag{25-3}$$

Figure 25-4 illustrates two possibilities for the work done by an expanding gas between the same two states. In each case the work done W_{ab} is equal to the area

---

*Notice that d^2V and d^2W are *second-order differentials* because they involve the product of two first-order differentials, dA and ds. A *double* integration is required to obtain V or W.

Fig. 25-4. The work done by an expanding gas depends on the path $p(V, T)$ that the system follows.

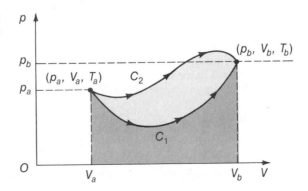

under the particular curve, $p = p(V, T)$. It is clear that $W_{ab}(C_2) > W_{ab}(C_1)$ and that, in general, the work done depends on the thermodynamic process involved (that is, on the path followed between the states).

When a thermodynamic path consists of a smooth progression of small changes specified by an infinite number of quasi-static intervening states, a reversal of the external conditions will return the system (and the environment) to its original state. Such a process is said to be *reversible*. In Fig. 25–4, if the imposed conditions carry the system from the state (p_a, V_a, T_a) to the state (p_b, V_b, T_b) along the path C_1, a reversal of the conditions will cause a compression that will carry the system along C_1 back to the state (p_a, V_a, T_a). When this process is completed, both the system and the environment will be in exactly the original condition. A perfectly reversible process is an idealization that cannot be achieved by a real system.

Suppose that the gas discussed above is confined in a cylinder with a tight-fitting movable piston. The initial position of the piston defines a volume V_a. The piston is suddenly pulled outward to a position that defines a volume V_b. In the natural (that is, highly probable) course of events, the gas will fill the newly created volume, $V_b - V_a$. Such an *uncontrolled* expansion is *irreversible*. Next, consider the unnatural (that is, highly improbable, though energy-conserving) process in which the gas occupying the volume $V_b - V_a$ suddenly and spontaneously collapses into the volume V_a, whereupon the piston is drawn inward to its original position defining the volume V_a. This hypothetical process is also irreversible. The *reversible (controlled)* path C_1 discussed above is any one of an infinite number of possible series of small steps that proceed through quasi-static states that connect the initial and final equilibrium states.

Suppose that the gas system in Fig. 25–4 is carried from the state (p_a, V_a, T_a) to the state (p_b, V_b, T_b) along the reversible path C_2 and then is returned (by a compression) along the reversible path C_1 (in a direction opposite to that shown in the figure). Such a combined process that returns a system to its original state is called a *thermodynamic cycle*. For the case illustrated in Fig. 25–4 the gas system does an amount of work represented by the shaded area enclosed by the two curves. (•Can you see why?) In the event that the cycle path encloses a nonzero area (as in the case here), the environment does not return to its original condition when the cycle is completed. (•How has the environment changed?) We consider in detail several interesting and practical thermodynamic cycles in the next chapter.

Example 25–1

A gas system that consists of n moles of an ideal gas is in an initial state (p_a, V_a, T_a). The system is carried along the straight-line path C_{ab} shown in the diagram from the initial state to the state (p_b, V_b, T_b) with $p_b = 2p_a$ and $V_b = 2V_a$. The pressure is then reduced (at constant volume) so that the system follows the path C_{bc} to the state (p_c, V_c, T_c). Finally, the system undergoes an isobaric compression along path C_{ca} back to the state (p_a, V_a, T_a).

(a) How much work was done by the gas along each of the paths, C_{ab}, C_{bc}, and C_{ca}?
(b) What was the net amount of work done by the gas during the complete cycle?
(c) What are the temperatures T_b and T_c?

Solution:

(a) The equation for the straight-line path C_{ab} is $p = p_a(V/V_a)$. Thus,

$$W_{ab} = \int_{V_a}^{V_b} p \, dV = \frac{p_a}{V_a} \int_{V_a}^{2V_a} V \, dV = \frac{1}{2} \frac{p_a}{V_a} V^2 \Big|_{V_a}^{2V_a} = \frac{3}{2} p_a V_a$$

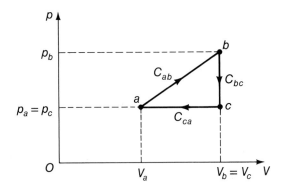

$W_{bc} = 0$ (•Can you see why?)

$$W_{ca} = \int_{V_c}^{V_a} p \, dV = p_a \int_{2V_a}^{V_a} dV = p_a V \Big|_{2V_a}^{V_a} = -p_a V_a$$

The fact that $W_{ca} < 0$ means that work was done *on* the gas during the compression. (Notice that the required negative sign appears automatically if the lower limit of the integral is taken to correspond to the initial state of the process and if the upper limit is taken to correspond to the final state. *This practice should always be followed.*)

(b) The net amount of work done by the gas was

$$W = W_{ab} + W_{bc} + W_{ca} = (\tfrac{3}{2} + 0 - 1)p_a V_a = \tfrac{1}{2} p_a V_a$$

This is just the area of the triangle enclosed by the paths C_{ab}, C_{bc}, and C_{ca}.

(c) Using the ideal gas law equation, we have

$$nR = \frac{p_a V_a}{T_a} = \frac{p_b V_b}{T_b} = \frac{(2p_a)(2V_a)}{T_b}$$

so that

$$T_b = 4T_a$$

Also,

$$\frac{p_a V_a}{T_a} = \frac{p_c V_c}{T_c} = \frac{p_a(2V_a)}{T_c}$$

so that

$$T_c = 2T_a$$

25-2 THE FIRST LAW OF THERMODYNAMICS

Simple Gas Systems. In the discussion of gas systems we have considered three forms of energy. *Heat* is the energy transferred between systems because of an existing temperature difference. *Internal energy* is the energy associated with the random thermal motions of the molecules in the system. *Mechanical work* is done as the result of the volume expansions or compressions that the system undergoes. For a differential change in the thermodynamic state of a gas system during a quasi-static process we have

> dQ: the heat *added to* the system
>
> $dW = p \, dV$: the mechanical work *done by* the system
>
> dU: the *increase* in the internal energy of the system

The energy conservation principle can now be written as

$$dQ = dW + dU = p\, dV + dU \tag{25-4}$$

This expression is the differential form of the *first law of thermodynamics* for a gas system. There is no new physical idea contained in this law—it is simply the energy conservation principle applied to a thermodynamic system. Integrating Eq. 25-4 we have

$$Q_{ab} = \oint_{V_a}^{V_b} p(V, T)\, dV + (U_b - U_a) \tag{25-5}$$

where the system changes from an initial state (p_a, V_a, T_a) to the final state (p_b, V_b, T_b) via the path C defined by $p = p(V, T)$. We call attention to the fact that the work done depends on the path followed by writing the integral sign as $\oint$. Whenever a process involves a volume expansion, the work done by the system is $W_{ab} > 0$, whereas $W_{ab} < 0$ for a volume contraction or compression.

In the previous discussions we have emphasized the fact that the internal energy U of an ideal system depends only on the equilibrium temperature. Thus, the internal energy is a function of the thermodynamic state; U is a *state function,* as are p, V, and T. Consequently, the change in internal energy during a process does *not* depend on the process path, but *only* on the difference in temperature between the initial and final states. This feature of the internal energy is indicated in Eq. 25-5, where we write $U_b - U_a$ for the integrated change in U.

By using Eq. 24-14 we can express the change in internal energy for a gas system as

$$U_b - U_a = \tfrac{1}{2}\nu nR(T_b - T_a) \tag{25-6}$$

where ν is the energy partition number, which is $\nu = 3$ for monatomic gases (He, Ne, Ar, and so forth) and is $\nu = 5$ for diatomic gases (H_2, N_2, O_2, and so forth) near room temperature (see Section 24-2). A temperature increase always means an increase in internal energy; a temperature decrease always means a decrease in internal energy.

Let us examine the idea of internal energy in a different way. First, we rewrite Eq. 25-5 in the form

$$U_{ab} = Q_{ab} - W_{ab} \tag{25-7}$$

Imagine that experiments are performed in which externally imposed conditions carry a system from an initial state (p_a, V_a, T_a) to the state (p_b, V_b, T_b) by various processes (Fig. 25-5). The paths C_1 and C_2 represent possible quasi-static processes, whereas I represents an *irreversible* process. (An irreversible process cannot properly be indicated on a p-V diagram because it does not take place via quasi-static steps.) The quantities Q_{ab} and W_{ab} are measured, and it is discovered that $Q_{ab} - W_{ab}$ has the same value for each process! This energy difference—which is independent of the path and even whether or not the process is reversible—therefore depends only

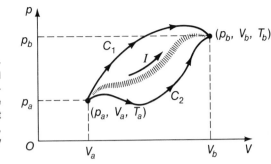

Fig. 25–5. Three possible paths connecting equilibrium states. Paths C_1 and C_2 represent reversible quasi-static processes. Path I represents an irreversible process. (An irreversible process is not properly representable on a p-V diagram, so we use a series of dashes to show such "paths.")

on the initial and final states. Thus, on the basis of experimental evidence, we can ascribe a quantity U to each state such that $U_b - U_a = U_{ab} = Q_{ab} - W_{ab}$. (Notice the similarity of this thermodynamic case to that of a conservative mechanical system for which a potential energy can be defined; see Section 8–3.) Because the work done W_{ab} depends on the path followed between the initial and final states whereas U_{ab} is independent of the path, we must conclude that Q_{ab} also depends on the path followed.

The General Behavior of Systems. Because of the universal nature of the energy conservation principle, much of the discussion of simple gas systems applies also to general systems that undergo any general process. The system isolated for consideration may consist of any combination of gaseous, liquid, or solid matter. In addition, many different forms of energy exchange between the system and its environment may be present. Work may be done in addition to the mechanical work $p\,dV$ associated with volume changes. For example, if the interface of the system and its environment involves a surface tension σ, a change dA of the boundary surface is accompanied by an energy change $\sigma\,dA$. Moreover, electric and magnetic work could be done, or a chemical reaction or phase change could involve energy changes. In spite of the variety of energy forms that can be involved, we can still express the first law for the system in the form

$$Q_{ab} = W_{ab} + U_{ab} \qquad (25\text{–}8)$$

where W_{ab} includes all of the possible ways that work can be done by the system. Depending on the circumstances, the quantities that appear in Eq. 25–8, including the separate terms that contribute to W_{ab}, can be either positive or negative. It is therefore advantageous to view Eq. 25–8 as an energy balance record: *The energy that enters a system (in the form of heat) is equal to the energy that leaves the system (in the form of work done by the system on the environment) plus the increase in the internal energy of the system.*

Notice that the processes connecting states of the system are not restricted to be quasi-static. It is required only that the initial and final states be equilibrium states so that each has a well-defined temperature. Only then is the internal energy a meaningful concept.

The first law of thermodynamics has an important limitation. The law states whether energy considerations permit a particular process to carry a system from one equilibrium state to another. But the law does *not* state whether this process will actually occur. That is, a certain process might be entirely consistent with the energy conservation principle, but even so it does not take place. Questions of this sort are addressed by the *second law of thermodynamics,* which we discuss in the next chapter.

Example 25–2

One mole of an ideal monatomic gas is carried by a quasi-static isothermal process (at 400 K) to twice its original volume.

(a) How much work W_{ab} was done by the gas?

(b) How much heat Q_{ab} was supplied to the gas?

(c) What is the pressure ratio, p_b/p_a?

(d) Suppose that a constant-volume process is used to reduce the original pressure p_a to the same final pressure p_b. Determine the new values for W'_{ab}, Q'_{ab}, and U'_{ab}.

Solution:

(a) We use Eq. 25–2 to find the work done in an isothermal expansion:

$$W_{ab} = nRT \ln \frac{V_b}{V_a} = (1 \text{ mol})(8.314 \text{ J/mol} \cdot \text{K})(400 \text{ K}) \ln 2 = 2305 \text{ J}$$

This is the work done by the gas during the expansion.

(b) For an isothermal process, $U_{ab} = 0$; therefore,

$$Q_{ab} = W_{ab} = 2305 \text{ J}$$

That is, energy enters the system in the form of heat, and exactly the same amount of energy is expended as work; no energy is added to or subtracted from the system's internal energy because the temperature does not change.

(c) Using the ideal gas law equation with $T_a = T_b$, we have

$$\frac{p_b}{p_a} = \frac{V_a}{V_b} = \tfrac{1}{2}$$

(d) For the constant-volume process, we have $W'_{ab} = 0$. The new final temperature is (using $V'_b = V_a$ and $p'_b = p_b = \tfrac{1}{2}p_a$)

$$T'_b = T_a \frac{p'_b}{p_a} = (400 \text{ K})(\tfrac{1}{2}) = 200 \text{ K}$$

Then, using Eq. 25–6 with $\nu = 3$, we have

$$U'_{ab} = \tfrac{1}{2}\nu nR(T'_b - T_a)$$
$$= \tfrac{1}{2}(3)(1 \text{ mol})(8.314 \text{ J/mol} \cdot \text{K})(200 \text{ K} - 400 \text{ K})$$
$$= -2494 \text{ J}$$

This represents a *decrease* in the internal energy of the gas. Finally, using the first law, we find

$$Q'_{ab} = W'_{ab} + U'_{ab} = -2494 \text{ J}$$

That is, during this process an amount of heat is rejected by the system to the environment, and the internal energy decreases by the same amount.

25-3   APPLICATIONS OF THE FIRST LAW
OF THERMODYNAMICS

Specific Heats of an Ideal Gas.   In Section 24–2 we discussed the specific heat of a gas for constant-volume processes. We can now understand the importance of this restriction. Because the volume remains constant, the mechanical work done is zero. Then, according to the first law, $U_{ab} = Q_{ab}$; that is, the heat added to the gas appears entirely in the form of internal energy. Thus, the heat dQ_v added at constant volume in order to raise the temperature from T to $T + dT$ is

$$dQ_v = nC_v\, dT = dU \qquad\qquad \textbf{(25–9)}$$

where C_v is the molar specific heat at constant volume (compare Eq. 24–13). The process that corresponds to Eq. 25–9 is shown in Fig. 25–6 as path C_1.

The constant-pressure path C_2 in Fig. 25–6 also connects the initial state with a state at the temperature $T + dT$ and therefore gives the same increase in the internal energy. But work is done along the path C_2, so we have

$$dQ_p = p\, dV + dU \qquad\qquad \textbf{(25–10)}$$

The molar specific heat at constant pressure is defined to be

$$C_p = \frac{1}{n}\frac{dQ_p}{dT} \qquad\qquad \textbf{(25–11)}$$

Combining Eqs. 25–9, 25–10, and 25–11, we find

$$nC_p\, dT = nC_v\, dT + p\, dV$$

Moreover, along path C_2 the ideal gas law can be expressed as

$$p\, dV = nR\, dT$$

Together, these last two equations (cancelling $n\, dT$ throughout) provide the important connection between the specific heats C_v and C_p, namely,

$$C_p - C_v = R \qquad\qquad \textbf{(25–12)}$$

Table 25–1 gives experimental values of the molar specific heats for various gases. It is apparent that Eq. 25–12—derived for the case of an ideal gas—is a correct description of the specific heats of real gases, to an accuracy of a few percent.

According to the discussion in Section 24–2, the value of ν that appears in Eq. 25–6 is expected to be $\nu = 3$ for a monatomic gas. For a diatomic gas the fully developed equipartition values are expected to be $\nu = 5$ (rigid) or $\nu = 7$ (nonrigid). For a triatomic gas, if the atoms in the molecule are not in line, we expect $\nu = 6$ (rigid) or $\nu = 12$ (nonrigid). Combining Eqs. 25–6 and 25–9, we see that (compare Eq. 24–15)

$$C_v = \tfrac{1}{2}\nu R \qquad\qquad \textbf{(25–13)}$$

Fig. 25–6.   Constant-volume (C_1) and constant-pressure (C_2) paths used to determine the relationship between the specific heats, C_V and C_p.

TABLE 25–1. Specific Heats for Various Gases (at 1 atm and 15° C)

	GAS	C_p (J/mol·K)	C_v (J/mol·K)	$C_p - C_v$ (J/mol·K)	$\gamma = C_p/C_v$
Monatomic gases	Ar	20.80	12.47	8.33	1.67
	He	20.80	12.47	8.33	1.67
	Ne	20.76	12.64	8.12	1.64
Diatomic gases	Air	29.01	20.68	8.33	1.403
	Cl_2	34.12	25.16	8.96	1.36
	H_2	28.76	20.43	8.33	1.41
	HCl	29.93	21.39	8.54	1.40
	N_2	29.05	20.72	8.33	1.40
	NO	29.26	20.93	8.33	1.40
	O_2	29.47	21.10	8.37	1.40
Polyatomic gases	CO_2	36.63	28.09	8.54	1.30
	H_2S	34.12	25.45	8.67	1.34
	NH_3	37.30	28.46	8.84	1.31
	CH_4	35.50	27.08	8.42	1.31
	C_2H_4	42.19	33.61	8.58	1.26

Thus, for monatomic gases, we expect $C_v = \frac{3}{2}R = 12.47$ J/mol·K and also

$$\gamma = \frac{C_p}{C_v} = \frac{\frac{3}{2}R + R}{\frac{3}{2}R} = \frac{5}{3} = 1.67$$

These predicted values for C_v and γ are in good agreement with the experimental results. (•Make similar comparisons for the diatomic and triatomic gases in Table 25–1.)

Adiabatic Expansion. In Section 25–1 we discussed the expansion of a simple gas system during an isothermal quasi-static process. Another important type of quasi-static expansion process is one that occurs with no exchange of heat. Such a process is called an *adiabatic* expansion.

Again, we consider a gas that is confined in a cylinder with a frictionless, tight-fitting piston (Fig. 25–7a). The cylinder walls and piston are imagined to be perfect heat insulators, so that there is no heat exchange between the gas system and the environment. The appropriate path equation for the adiabatic expansion involves $dQ = 0$, $dW = p\,dV$, and $dU = nC_v\,dT$. Thus, the first law becomes

$$0 = p\,dV + nC_v\,dT \qquad (25\text{–}14)$$

Expressing the ideal gas law equation, $pV = nRT$, in differential form, we have

$$p\,dV + V\,dp = nR\,dT \qquad (25\text{–}15)$$

Substituting $dT = -(p/nC_v)\,dV$ from Eq. 25–14 into Eq. 25–15, we obtain

$$C_v p\,dV + C_v V\,dp = -pR\,dV \qquad (25\text{–}16)$$

Using $C_p = C_v + R$ (Eq. 25–12) and $\gamma = C_p/C_v$, Eq. 25–16 reduces to

TABLE 25–2. Summary of Nomenclature

$dp = 0$	isobaric
$dT = 0$	isothermal
$dU = 0$	
$dQ = 0$	adiabatic

(a)   (b)

Fig. 25–7. (a) The insulated cylinder allows no heat exchange between the gas system and the environment. (b) As the gas expands adiabatically, it follows a path, $p = p(V, T)$, that cuts across the family of isotherms.

$$\gamma \frac{dV}{V} = -\frac{dp}{p}$$

Integrating, we find $\gamma \ln V = -\ln p + \text{constant}$

from which

$$p V^{\gamma} = \text{constant} \qquad\qquad \textbf{(25–17)}$$

This is the adiabatic path equation that we require. The value of the *constant* in Eq. 25–17 depends on the initial state (p_a, V_a, T_a) of the system.

Figure 25–8 shows several isothermal and several adiabatic paths constructed for a sample consisting of 1 mol of an ideal diatomic gas ($\gamma = 1.40$). These curves provide an accurate picture of the behavior of such a sample. In other illustrations, however, we give only schematic representations of isothermal and adiabatic paths so that the diagrams can be easily interpreted.

According to the ideal gas law, we can write $p_a V_a / T_a = p_b V_b / T_b$; and from Eq. 25–17, we have $p_a V_a^{\gamma} = p_b V_b^{\gamma}$. Therefore, once the initial state has been specified, we require one (and only one) of the quantities p_b, V_b, or T_b to identify uniquely the final state. Because $\gamma > 1$, the adiabatic paths in a p-V diagram decrease with increasing V more rapidly than do the isotherms; every adiabatic path therefore cuts across the family of isotherms (Fig. 25–8).

The work done by the gas during an adiabatic expansion can be expressed in terms of the temperature change, $T_b - T_a$. We have $Q_{ab} = 0$ for the adiabatic process, so the first law can be written as

$$W_{ab} = -U_{ab} = -nC_v(T_b - T_a) \qquad\qquad \textbf{(25–18)}$$

Because $T_b < T_a$ (see Fig. 25–7b), we see that $W_{ab} > 0$, as expected for an expansion process.

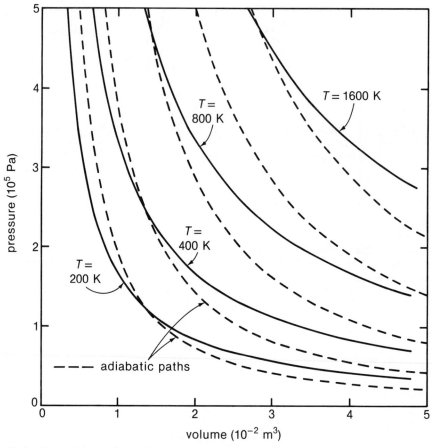

Fig. 25–8. The solid and dashed lines show, respectively, isothermal and adiabatic paths for a 1-mol sample of an ideal diatomic gas (for which $\gamma = 1.40$).

The work done can also be expressed in terms of p and V. From the ideal gas law equation, we can write $nT_a = p_a V_a / R$ and $nT_b = p_b V_b / R$. Substituting these expressions into Eq. 25–18, the result is

$$W_{ab} = \frac{C_v}{R}(p_a V_a - p_b V_b) \qquad \text{(25–19)}$$

Using $R = C_p - C_v$ (Eq. 25–12) and $\gamma = C_p / C_v$, Eq. 25–19 can be expressed as

$$W_{ab} = \frac{p_a V_a - p_b V_b}{\gamma - 1} \qquad \text{(25–20)}$$

This result can also be obtained by the direct integration of $p \, dV$, using $p = (\text{constant})/V^\gamma$ (see Problem 25–9).

Example 25–3

How much work is required to compress 5 mol of air at 20° C and 1 atm to 1/10 of the original volume by (a) an isothermal process and (b) an adiabatic process? (c) What are the final pressures for the two cases?

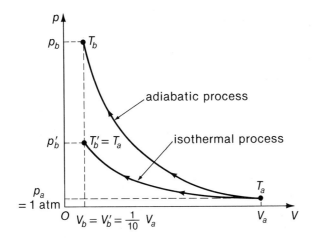

Solution:

(a) The work done on the gas is $-W_{ab}$. For the isothermal process, we use Eq. 25-2:

$$-W_{ab} = -nRT\ln\frac{V_b'}{V_a} = -(5\text{ mol})(8.314\text{ J/mol}\cdot\text{K})(293\text{ K})\ln\tfrac{1}{10}$$

$$= 2.80 \times 10^4\text{ J}$$

(b) For the adiabatic process, let us use Eq. 25–18. We must first determine the final temperature T_b. From Eq. 25–17, we have

$$p_b V_b^{\gamma} = p_a V_a^{\gamma}$$

and from the ideal gas law equation, we have

$$\frac{p_b V_b}{T_b} = \frac{p_a V_a}{T_a}$$

Dividing the first of these two equations by the second gives

$$T_b = T_a\left(\frac{V_a}{V_b}\right)^{\gamma-1} = (293\text{ K})(10)^{0.403} = 741\text{ K}$$

where we have used $\gamma(\text{air}) = 1.403$ (see Table 25–1). Thus, the work done on the gas is (Eq. 25–18)

$$-W_{ab} = nC_v(T_b - T_a) = (5\text{ mol})(20.68\text{ J/mol}\cdot\text{K})(741\text{ K} - 293\text{ K})$$

$$= 4.63 \times 10^4\text{ J}$$

where we have used the value of C_v for air given in Table 25–1.

(c) For the isothermal process, we have $p_b'V_b' = p_a V_a$; thus,

$$p_b' = p_a\left(\frac{V_a}{V_b'}\right) = (1\text{ atm})(10) = 10\text{ atm}$$

For the adiabatic process, we have $p_b V_b^{\gamma} = p_a V_a^{\gamma}$; thus,

$$p_b = p_a\left(\frac{V_a}{V_b}\right)^{\gamma} = (1\text{ atm})(10)^{1.403} = 25.3\text{ atm}$$

A Two-Phase System. A quantity of water is placed in a cylinder that is closed with a frictionless, tight-fitting piston. The cylinder has diathermic (perfectly conducting) walls. The entire assembly is immersed in a heat bath that has an adjustable temperature T. The external pressure on the piston is maintained at 1 atm. In the initial condition, the piston is in contact with the water. That is, the system under consideration consists entirely of the original mass of water.

The temperature of the heat bath is now increased to $100°$ C sufficiently slowly that the temperature of the water is always the same as that of the heat bath. During this process heat flows from the heat bath into the water. Next, the heat bath temperature is raised by an infinitesimal amount to $(100 + \varepsilon)°$ C and maintained at this value. Heat flows into the water and causes vaporization. The system now consists of liquid water plus steam. This situation is illustrated in Fig. 25–9.

Fig. 25–9. Heat flows through the diathermic walls of the cylinder from the heat bath into the water and causes the vaporization of the water. This produces a two-phase, water-plus-steam system.

The amount of heat Q required to vaporize a mass m of the water is

$$Q = mL_v$$

where L_v is the latent heat of vaporization for water (see Table 22–4). Due to the production of steam, the volume of the system increases and the piston is raised (see Fig. 25–9). At 1 atm and $100°$ C, the density of steam is ρ_s and the density of water is ρ_w. Then, the volume expansion of the system when a mass m of water is converted into steam at $100°$ C is

$$\Delta V = m\left(\frac{1}{\rho_s} - \frac{1}{\rho_w}\right)$$

The work done by the system at the constant pressure p_0 (1 atm) is

$$W = p_0\,\Delta V = mp_0\left(\frac{1}{\rho_s} - \frac{1}{\rho_w}\right)$$

Using the first law, the change in internal energy ΔU for the water is

$$\Delta U = Q - W = mL_v - mp_0\left(\frac{1}{\rho_s} - \frac{1}{\rho_w}\right)$$

Then, using $\rho_s = 0.598$ kg/m^3 and $\rho_w = 1040$ kg/m^3, we find

$$\frac{\Delta U}{m} = (2.257 \times 10^6 \text{ J/kg}) - (1.013 \times 10^5 \text{ Pa})\left(\frac{1}{0.598 \text{ kg/m}^3} - \frac{1}{1040 \text{ kg/m}^3}\right)$$

$$= 2.088 \times 10^6 \text{ J/kg}$$

Although an amount of heat equal to 2.257×10^6 J is required to convert 1 kg of water to steam at 1 atm and 100° C, only 2.088×10^6 J actually appears in the form of thermal motion of the molecules in the steam. The remainder of the energy, 0.169×10^6 J, is the work done by the steam in expanding at constant pressure. We may view the first amount of energy (2.088×10^6 J/kg) as the energy required to perform the microscopic work of separating the water molecules against the attractive intermolecular forces. The second amount of energy (0.169×10^6 J/kg) is the energy required to perform the macroscopic work of expansion. (This is the view we adopted in the discussion of the energetics of surface tension in Section 17-5.)

Free Expansion of a Gas.   In various discussions it has been implicit that the macroscopic definition of internal energy contained in the first law (Eq. 25-7) and the microscopic definition from kinetic theory (Eq. 25-6) refer to exactly the same quantity. An interesting test of this assumption was first devised and carried out by Gay-Lussac; the experiment was later repeated by Joule and subsequently refined still further by others.

The test consists basically of verifying that the internal energy of a gas defined by the first law does not depend on its volume. The experimental arrangement is shown schematically in Fig. 25-10. The gas is originally confined in the volume V_A, which is separated from the evacuated volume V_B by a closed valve. The entire system—defined to include *both* V_A and V_B—is immersed in a water bath. The valve is suddenly opened and the gas expands freely into the volume V_B. This expansion process (referred to as a *free expansion*) is uncontrollably rapid and does not proceed through a sequence of quasi-static states. A free expansion is an irreversible process.

The temperature of the water bath is measured carefully before and after the free expansion. The early measurements lacked precision and were inconclusive. However, refined versions of the experiment have demonstrated that the expansion produces essentially no change in the temperature of the water bath. From this result it follows that $\Delta Q = 0$ for the entire system, $V_A + V_B$. Because there was no change in the total volume, no work was done; that is, $W = 0$. Using the first law, we must conclude that $\Delta U = 0$. Thus, the internal energy of the gas does not depend on the volume in which it is confined or on the pressure.*

We may examine further the implications of the result of the free-expansion experiment by noting that according to the microscopic view, $\Delta U = 0$ means that the average molecular speed $\langle v^2 \rangle$ also remains constant. Suppose that $V_A = V_B$; then, after expansion, the gas density is smaller by a factor of two, but the pressure is also smaller by the same factor (because $p = \frac{1}{3}\rho\langle v^2 \rangle$, according to kinetic theory). Because the gas now occupies twice the volume, we have $p_f = \frac{1}{2}p_i$ and $V_f = 2V_i$. Thus, the relationship connecting the initial and final states is $p_i V_i = p_f V_f$. This, of course, is just Boyle's law!

thermometer

V_A V_B

valve

water

Fig. 25-10. Schematic representation of the free-expansion apparatus of Joule.

---

*At sufficiently high pressures and low temperatures, the internal energy of a real gas exhibits some dependence on pressure and volume.

PROBLEMS

Section 25-1

25-1 Show that the work done by a gas during an isobaric process is $W_{ab} = p(V_b - V_a)$.

25-2 A sample of ideal gas is expanded to twice its original volume of $1 \, m^3$ in a quasi-static process for which $p = \alpha V^2$, with $\alpha = 5.0 \, atm/m^6$, as shown in the diagram. How much work was done by the expanding gas?

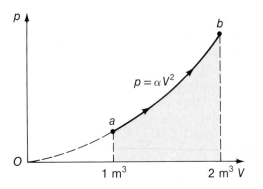

25-3 Argon gas ($m = 5.0 \, g$) undergoes an isothermal expansion at $20°$ C to twice its original volume. (a) How much work was done by the gas? (b) If the original pressure was 1 atm, what is the final pressure?

25-4 A 1-mol sample of an ideal gas is carried around the thermodynamic cycle shown in the diagram. The cycle consists of three parts—an isothermal expansion ($a \rightarrow b$), an isobaric compression ($b \rightarrow c$), and a constant-volume increase in pressure ($c \rightarrow a$). If $T = 300 \, K$, $p_a = 5 \, atm$, and $p_b = p_c = 1 \, atm$, determine the work done by the gas per cycle.

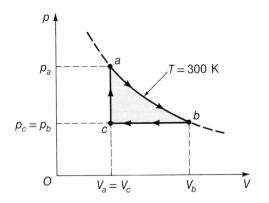

25-5 The *compressibility* of a gas is defined to be

$-(1/V)(dV/dp)$. Show that the isothermal compressibility of an ideal gas is $1/p$.

Section 25-2

25-6 Many of the thermodynamic ideas we have discussed apply to liquids and solids as well as to gases. A beaker contains 1.2 ℓ of mercury at $20°$ C. The beaker is open to the atmosphere. Heat is slowly added to the mercury and its temperature is increased to $100°$ C. During this process the pressure is constant at 1 atm. Neglect any thermal expansion of the beaker. Use the tabulated values for $\bar{\beta}$ (Table 22-2) and $\bar{c}$ (Table 22-3) for mercury. (Note that $\bar{c}$ is the specific heat at constant pressure.) (a) Calculate the work done by the expanding mercury. (b) Determine the amount of heat added to the mercury. (c) What is the increase in the internal energy of the mercury? (d) In view of these results, is there any substantial difference between the specific heat at constant pressure and the specific heat at constant volume for mercury near atmospheric pressure?

25-7 A quantity of an ideal gas ($n = 4 \, mol$) is confined within a cylinder that has a movable piston and diathermic walls. The cylinder is immersed in a heat bath that is maintained at a temperature of 300 K. An external force F_0 is required to hold the piston in mechanical equilibrium. This force is suddenly increased to $10 \, F_0$ and remains at this value until the volume has decreased to one half of the original value. (The force is then adjusted to maintain this volume.) After this irreversible process has been completed, a new equilibrium is established at the smaller volume. (a) How much work was done by the external force? (b) What happened to this energy?

25-8 One mole of argon is confined in a cylinder with a movable piston at a pressure of 1 atm and at a temperature of 300 K. The gas is heated slowly and isobarically to a temperature of 400 K. The measured value of the molar specific heat at constant pressure for argon in this temperature range is $C_p = 2.5043 \, R$, and the measured value of pV/nT is $0.99967 \, R$. Calculate in units of R, carried to two decimals, the following quantities: (a) the work done by the expanding gas; (b) the amount of heat added to the gas; (c) the increase in the internal energy of the gas (using the first law); and (d) the increase in the internal energy of the gas (using kinetic theory—Eq. 25-6). Compare the results of (c) and (d).

Section 25-3

25-9 Show by direct integration of $p\,dV$, using $p = $ (constant)$/V^{\gamma}$, that the work done during an adiabatic process is given by Eq. 25-20.

25-10 The following measured values are determined for
• argon at 1 atm and 15° C:

$$pV/T = 208.0 \text{ J/deg (for a 1-kg sample)}$$
$$c_p = 0.1253 \text{ Cal/kg} \cdot \text{deg}$$
$$\gamma = 1.668$$

Using these data alone (together with the assumption that argon is an ideal gas), determine the mechanical equivalent of heat and compare it with the accepted value, 4186 J/Cal. (This is essentially the method used by J. R. Mayer in 1842.)

25-11 The *compressibility* of a gas is defined to be $-(1/V)(dV/dp)$. Show that the adiabatic compressibility of an ideal gas is $1/\gamma p$. (Compare Problem 25-5.) [*Hint:* Differentiate the adiabatic path equation, Eq. 25-17.]

25-12 The speed of sound in a gas is $v = \sqrt{\gamma p/\rho}$ (see Eq. 20-5). Assume that air is an ideal gas and calculate the speed of sound in air at 15° C. (Refer to Example 23-5 for the effective molecular mass of air.)

Additional Problems

25-13 Nitrogen gas ($m = 1.00$ kg) is confined in a cylinder with a movable piston exposed to normal atmospheric pressure. A quantity of heat ($Q = 25.0$ Cal) is added to the gas in an isobaric process, and the internal energy of the gas increases by 8.0 Cal. (a) How much work was done by the gas? (b) What was the change in volume? (c) What was the change in temperature?

25-14 An ideal gas is carried around the cycle shown in the diagram above right, which consists of alternating constant-volume and constant-pressure paths. (a) How much work is done by the gas per cycle? (b) What amount of heat is supplied to the gas per cycle?

25-15 A gas obeys the van der Waals equation of state (Eq. 23-3),

$$\left(p + \frac{a}{\mho^2}\right)(\mho - b) = RT$$

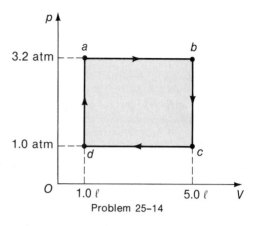

Problem 25-14

where $\mho = V/n$ is the molar volume. Derive an expression for the work done per mole of gas in an isothermal expansion.

25-16 For a gas that obeys the van der Waals equation of state (see Problem 25-15 or Eq. 23-3), it can be shown that

$$C_p - C_v \cong R\left(1 + \frac{9}{4}\frac{\mho_c T_c}{\mho T}\right)$$

where $\mho$ and T are the molar volume, $\mho = V/n$, and the temperature, respectively, and where $\mho_c$ and T_c are the corresponding values at the critical point. (Notice that the expression for $C_p - C_v$ implies a slight dependence of the internal energy on volume.) For argon, we have $T_c = 150.7$ K and $\mho_c = 75.3$ cm³/mol. Calculate the correction term in the expression for $C_p - C_v$ for argon at 0° C. [*Hint:* Use the fact that 1 mol of any gas at 0° C and 1 atm occupies 22.41 ℓ.]

25-17 Equation 23-2 is a *virial expansion* for the quantity
• pV/n. If only the first virial coefficient is nonzero, we have

$$\frac{pV}{n} = RT + B_1(T)p$$

(a) Derive an expression for the work done by the gas during an isothermal expansion from a volume V_a to a volume V_b.
  For air at 0° C, $B_1(T) = -13.5$ cm³/mol; also, take $V_a = 250$ cm³, $V_b = 1000$ cm³, and $n = 1$ mol. (b) Determine the work done by the air. (c) Calculate the work done during the expansion if air is considered to be an ideal gas. (d) Compare the results of (b) and (c). The initial pressure in this case is 89.6 atm. Examine the form of the virial expansion and comment on whether you expect a significant correction term $B_1(T)p$ for pressures near atmospheric pressure.

25-18 Argon gas ($n = 3$ mol), initially at a temperature of 20° C, occupies a volume of 10 ℓ. The gas undergoes a slow expansion at constant pressure to a volume of 25 ℓ; then, the gas expands adiabatically until it returns to its initial temperature. (a) Show the process on a p-V diagram. (b) What quantity of heat was supplied to the gas during the entire process? (c) What was the total change in internal energy of the gas? (d) What was the total amount of work done during the process? (e) What is the final volume of the gas?

25-19 An ideal monatomic gas is confined in a cylinder
•• with a cross-sectional area $A = 75$ cm² by means of a spring-loaded piston, as shown in the diagram. The relaxed length of the spring is ℓ_0 (no pressure difference across the piston); the force constant of the spring is $\kappa = 7500$ N/m. With the gas at its initial pressure, $p_a = 10^5$ N/m², the spring is compressed by an amount x_a, and the length of the gas-filled part of the cylinder is $\ell_a = 30$ cm. Heat is slowly added to the gas until it has expanded (and has compressed the spring) by an additional 10 cm; that is, $\ell_b = 40$ cm. At any point during the process, we have $\ell = \ell_a + (x - x_a)$.

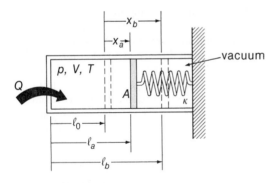

(a) Show that $p = p_a + (\kappa/A^2)(V - V_a)$. Plot this path from V_a to V_b in a p-V diagram. (b) Show that the work done by the expanding gas is

$$W_{ab} = \oint_{V_a}^{V_b} p \, dV = \tfrac{1}{2}(V_b - V_a)(p_b + p_a)$$

(c) Show that the result obtained in (b) is just equal to the work done on the spring, namely, $\tfrac{1}{2}\kappa(x_b^2 - x_a^2)$. (d) Evaluate W_{ab} by using the expressions in both (b) and (c). (e) Calculate the increase in internal energy U_{ab} of the gas. (f) Determine the amount of heat Q_{ab} supplied to the gas during the process.

25-20 A cylinder with a length of 1 m is divided into two
• compartments by a thin, tight-fitting, diathermic piston that is clamped in position, as shown in the

diagram. The left-hand compartment has a length of 0.4 m and contains 4 mol of neon at a pressure of 5 atm. The right-hand compartment contains helium at a pressure of 1 atm. The entire cylinder is surrounded by an adiabatic insulator. The initial temperature of each gas is 20° C. Now the piston is released and a new equilibrium condition is eventually reached. Locate the new position of the piston, which remains tight-fitting throughout.

25-21 A cylinder with adiabatic walls contains 0.40 ℓ of air
• at 20° C and 30 atm confined by a movable piston. A small quantity of gasoline (approx. 0.1 g) is introduced into the cylinder, and the mixture is ignited. The resulting constant-volume explosion yields a chemical energy of 1.2 Cal. After equilibrium is reached, the gas is expanded adiabatically to a volume of 2.0 ℓ. (a) How many moles of air did the cylinder originally contain? (b) Consider the gas in the cylinder to be air both before and after the ex-

plosion. What equilibrium temperature and pressure are reached after the explosion (but before the expansion)? (c) What is the temperature of the air after the adiabatic expansion? (d) Determine the amount of work done by the gas. (e) What was the efficiency of the conversion of chemical energy into mechanical work?

25-22 The top of a vertical cylindrical tank is closed by a tight-fitting piston with negligible mass. The column of ideal gas below the piston has a height h. A fluid with a density ρ is slowly poured onto the piston, and the piston descends a distance y before fluid spills over the rim of the cylinder. (a) Suppose that the system is perfectly insulated. Show that

$$h = \frac{y}{1 - \left(1 + \dfrac{\rho g}{p} y\right)^{-1/\gamma}}$$

(b) Suppose that the process is isothermal. Show that $h = y + p/\rho g$. (c) Suppose again that the process is isothermal. How much work is done in compressing the gas?

THERMODYNAMICS II

In the preceding chapter, the principle of energy conservation was applied to thermodynamic systems by using a generalized form of the *first law of thermodynamics*. This law permits us to determine whether or not a particular thermodynamic process—reversible or irreversible—can possibly take place. However, the first law alone is not sufficient to determine whether or not the process will actually occur. The fact that a process is consistent with energy conservation does not imply that the process will take place.

Natural processes—those that occur spontaneously, without any deliberate external intervention or control—are almost always irreversible. When such uncontrolled processes involve finite changes of state, they do not proceed through quasi-static equilibrium states (see Section 25–1). These uncontrolled processes have the important common feature that they take place in only one direction; the reverse process—even if it is energetically allowed—does not occur. For example, the spontaneous flow of heat is always from a hot object or region to a cooler object or region; heat does not flow spontaneously in the other direction. The free expansion of a gas will fill an evacuated volume; the gathering together of all the molecules into the original space is never observed. If two containers with different gases are connected, the gases will mix together; the separation of a mixture of gases into its molecular components will not occur spontaneously. All of these processes can be made to conform to the first law of thermodynamics, and yet they proceed only in one direction. The *second law of thermodynamics*—the principal subject of this chapter—is a precise statement of a broad generalization derived from observations of unidirectional processes. Moreover, the second law places severe restrictions even on processes that are thermodynamically reversible.

26-1 CYCLIC THERMODYNAMIC PROCESSES

A cyclic thermodynamic process is one in which a system acts through a series of steps and eventually returns to its original equilibrium state. The steps that constitute the cycle may be reversible or irreversible. If the system consists of a single homogeneous substance (the *working substance*) and if all of the steps are reversible, the cycle can be represented by a closed curve in a p-V diagram (Fig. 26–1).

The Carnot Cycle. The cycle that is most important to the theory of thermodynamics is the idealized reversible cycle of Carnot.* Figure 26–2 gives three different

Fig. 26–1. The p-V diagram for a general reversible thermodynamic cycle of a homogeneous system. The shaded area represents the net work done by the system per cycle.

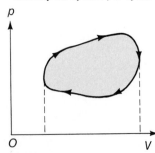

---

*Nicolas Léonard Sadi Carnot (1796–1832), French military engineer and physicist.

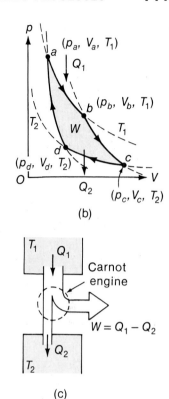

Fig. 26–2. The Carnot cycle. (a) Schematic diagram of the engine. The two heat reservoirs and the nonconducting stand are successively placed in contact with the diathermic base of the cylinder. (b) The *p-V* plot of the cycle. The paths *ab* and *cd* are along isotherms; *bc* and *da* are adiabatic paths. (c) The corresponding heat-flow diagram. The difference, $Q_1 - Q_2$, is equal to the work done by the system per cycle.

views of a *Carnot cycle*. Figure 26–2a is a schematic diagram of a Carnot engine; Fig. 26–2b shows the *p-V* plot for a Carnot cycle, and Fig. 26–2c is the corresponding heat-flow diagram.

Choose the beginning of the cycle at the equilibrium state *a*, characterized by (p_a, V_a, T_1), with the cylinder that contains the working substance in contact with a hot thermal reservoir at a temperature T_1 (Fig. 26–2a). Imagine that the base of the cylinder is perfectly conducting (diathermic) and that the walls and the piston are ideally nonconducting. The working substance undergoes a slow quasi-static isothermal expansion to state *b*, characterized by (p_b, V_b, T_1). This is accomplished by allowing heat, as needed, to flow from the hot reservoir through the diathermic base of the cylinder into the working substance. Let the total amount of heat delivered to the working substance during the expansion *ab* be Q_1, as indicated in Fig. 26–2b. During this part of the cycle the temperature and therefore also the internal energy of the working substance are constant. Consequently, the heat energy input is converted directly to the work done by the moving piston on whatever mechanical system is attached.

With the system at *b*, the cylinder is removed from contact with the hot reservoir and placed on the nonconducting stand. The system now undergoes a further expansion along the adiabatic path *bc* to the point *c*, characterized by (p_c, V_c, T_2). During this part of the cycle there is no heat exchange with the working substance; however, work is done at the expense of reducing the internal energy, and the temperature is thereby decreased to T_2.

Next, the cylinder is placed in contact with the cold thermal reservoir at temperature T_2. An external agency now slowly compresses the working substance isothermally to a predetermined volume V_d at point *d*, characterized by (p_d, V_d, T_2). During this part of the cycle the temperature and therefore also the internal energy of the working substance are constant. The work done on the working substance results in the rejection of an amount of heat Q_2 to the cold reservoir, as indicated in Fig. 26–2b.

Finally, the cylinder is removed from contact with the cold reservoir and again placed on the nonconducting stand. The volume V_d to which the system was carried by the isothermal compression cd was selected to allow an adiabatic compression to connect state d with the initial state a. This additional compression is also the result of work done on the working substance by an external agency and increases the system temperature from T_2 to T_1. This step completes the Carnot cycle and returns the system to its initial condition.

According to the first law, the amount of work W done by a system during a thermodynamic process is equal to the heat Q supplied to the system less the change in internal energy ΔU of the system (see Eq. 25-4). If a system is carried through a complete thermodynamic cycle, the initial and final temperatures are equal, so $\Delta U = 0$. For a system that undergoes a Carnot cycle, no heat is supplied to or rejected by the system during the adiabatic paths, bc and da (Fig. 26-2b). An amount of heat $Q_1 = Q_{ab} (Q_1 > 0)$ is supplied to the system during the isothermal expansion ab, and an amount $Q_2 = -Q_{cd} (Q_2 > 0)$ is rejected during the isothermal compression da. (The heat *supplied* to the system during this process is *negative;* that is, $Q_{cd} < 0$.) Thus, the first law can be written as

$$W = Q_1 - Q_2 \qquad \text{(heat engine)} \qquad \textbf{(26-1)}$$

Notice that Eq. 26-1 is applicable for a complete cycle of *any* heat engine (any device that operates in a cycle and converts heat energy into mechanical work) if we interpret Q_1 as the net amount of heat supplied to the engine (during whatever part of the cycle) and Q_2 as the net amount of heat rejected by the engine to some thermal reservoir.

The energy conservation statement contained in Eq. 26-1 is illustrated in the heat-flow diagram of Fig. 26-2c. The dashed circle is imagined to contain the Carnot engine. An amount of heat Q_1 is supplied to the engine by the hot reservoir at temperature T_1 and an amount of heat Q_2 is rejected to the cold reservoir at temperature T_2. The energy difference, $Q_1 - Q_2$, appears as work done by the system on an external mechanical agency.

Only a portion of the heat energy supplied to any engine is converted into mechanical work. The *thermal efficiency e* of an engine is defined to be the ratio of the work done (the *output*) to the heat supplied (the *input*). Thus,

$$e = \frac{W}{Q_1} = \frac{Q_1 - Q_2}{Q_1} = 1 - \frac{Q_2}{Q_1} \qquad \text{(heat engine)} \qquad \textbf{(26-2)}$$

Again this equation is applicable for any heat engine that operates in a cycle.

For a Carnot engine, the heat exchanges take place during the isothermal processes. Using Eq. 25-2, we have*

$$Q_1 = nRT_1 \ln (V_b/V_a)$$
$$Q_2 = nRT_2 \ln (V_c/V_d)$$

---

*Note that $Q_2 > 0$; we insure this by writing the logarithmic term as $\ln(V_c/V_d)$.

from which
$$\frac{Q_2}{Q_1} = \frac{T_2}{T_1} \frac{\ln(V_c/V_d)}{\ln(V_b/V_a)}$$ (26–3)

Along the isothermal paths we can write

$$p_a V_a = p_b V_b \quad \text{and} \quad p_c V_c = p_d V_d$$

and along the adiabatic paths we have

$$p_b V_b^\gamma = p_c V_c^\gamma \quad \text{and} \quad p_d V_d^\gamma = p_a V_a^\gamma$$

If we multiply together these four equations and cancel the common term, $p_a p_b p_c p_d$, we find

$$V_a V_b^\gamma V_c V_d^\gamma = V_a^\gamma V_b V_c^\gamma V_d$$
or
$$(V_a V_c)^{1-\gamma} = (V_b V_d)^{1-\gamma}$$

so that
$$\frac{V_c}{V_d} = \frac{V_b}{V_a}$$ (26–4)

Using this result in Eq. 26–3, we obtain

$$\frac{Q_2}{Q_1} = \frac{T_2}{T_1} \qquad \text{(Carnot cycle)}$$ (26–5)

Consequently, the efficiency of a Carnot engine can be expressed as

$$e = 1 - \frac{T_2}{T_1} = \frac{T_1 - T_2}{T_1} \qquad \text{(Carnot engine)}$$ (26–6)

This is the efficiency of an idealized Carnot engine (no heat losses through the walls, no losses due to friction). In Section 26–3 it is demonstrated that no heat engine of any type that operates between the temperatures T_1 and T_2 can have an efficiency greater than that specified by Eq. 26–6.

It is evident from Eq. 26–6 that the greater the temperature difference, $T_1 - T_2$, the greater will be the efficiency of a Carnot engine. Indeed, this is true of all thermodynamic cycles that convert heat into mechanical work (heat engines).

A heat engine can be imagined to run backwards, thereby operating as a *refrigerator*. Figure 26–3 shows the heat-flow diagram for a refrigerator. The input of mechanical work W is used by the working substance to transfer heat from a cold reservoir at temperature T_2 to a hot reservoir at temperature* T_1. The p-V plot for the Carnot refrigerator cycle is the same as that for the Carnot heat-engine cycle (Fig. 26–2b), except that the paths are traversed in the opposite directions. (Each step in a

---

*It is convenient to use T_1 for the temperature of the *hot* thermal reservoir, whether the device is a heat engine or a refrigerator.

$$Q_1 = Q_2 + W$$

T_1

$W \longrightarrow$

refrigerator

T_2 Q_2

Fig. 26–3. Heat-flow diagram for a refrigerator—a heat engine operating backwards.

Carnot cycle is *reversible,* so the entire cycle is reversible.) It is customary to express the first law for a refrigeration cycle in the form

$$Q_1 = Q_2 + W \qquad \text{(refrigerator)} \qquad \textbf{(26–7)}$$

where Q_1 is the amount of heat delivered to the hot reservoir and where Q_2 is the amount of heat removed from the cold reservoir by the input of the work W. Equation 26-7 appears to be a simple algebraic rearrangement of the heat-engine equation (26-1), but note that each of the quantities in Eq. 26-7 is the negative of the corresponding quantity in Eq. 26-1. We discuss some practical heat engines and refrigerators in Section 26-4.

Example 26–1

A Carnot engine uses as a working substance 0.15 mol of air (considered to be an ideal gas with $\gamma = 1.40$) and operates between the temperatures $T_1 = 550$ K and $T_2 = 300$ K. The maximum pressure in the cycle is $p_a = 20$ atm and the isothermal volume expansion ratio is $V_b/V_a = 5$ (see Fig. 26-2b).

(a) Determine the volume and the pressure at each of the points a, b, c, and d in the cycle (Fig. 26-2b).

(b) Calculate the amounts of heat, Q_1 and Q_2, exchanged per cycle.

(c) Find the net amount of work W done per cycle.

(d) Determine the thermal efficiency e of the engine.

Solution:

(a) The volume V_a can be obtained from the ideal gas law equation in which $T_a = T_1$:

$$V_a = \frac{nRT_1}{p_a} = \frac{(0.15 \text{ mol})(8.314 \text{ J/mol} \cdot \text{K})(550 \text{ K})}{20(1.013 \times 10^5 \text{ Pa})}$$

$$= 3.386 \times 10^{-4} \text{ m}^3 = 0.3386 \text{ } \ell$$

Then,
$$V_b = 5V_a = 5(0.3386 \text{ } \ell) = 1.693 \text{ } \ell$$

For the adiabatic path bc, we have $p_b V_b^\gamma = p_c V_c^\gamma$, which can be rewritten in the form $p_b V_b V_b^{\gamma-1} = p_c V_c V_c^{\gamma-1}$. Then, using the ideal gas law, this becomes

$$T_1 V_b^{\gamma-1} = T_2 V_c^{\gamma-1}$$

from which
$$V_c = V_b \left(\frac{T_1}{T_2} \right)^{1/(\gamma-1)}$$

$$= (1.693 \text{ } \ell) \left(\frac{550}{300} \right)^{1/0.40} = 7.705 \text{ } \ell$$

Solving Eq. 26–4 for V_d, we have

$$V_d = V_c \frac{V_a}{V_b} = (7.705 \text{ } \ell)(\tfrac{1}{5}) = 1.541 \text{ } \ell$$

Also,
$$p_b = \frac{p_a V_a}{V_b} = \tfrac{1}{5} p_a = \tfrac{1}{5}(20 \text{ atm}) = 4.00 \text{ atm}$$

$$p_c = p_b \left(\frac{V_b}{V_c} \right)^{\gamma} = (4.00 \text{ atm}) \left(\frac{1.693 \text{ } \ell}{7.705 \text{ } \ell} \right)^{1.40} = 0.479 \text{ atm}$$

$$p_d = \frac{p_c V_c}{V_d} = p_c \frac{V_b}{V_a} = 5 p_c$$
$$= 5(0.479 \text{ atm}) = 2.395 \text{ atm}$$

(b) Using Eq. 26–2, we have

$$Q_1 = nRT_1 \ln \frac{V_b}{V_a}$$
$$= (0.15 \text{ mol})(8.314 \text{ J/mol} \cdot \text{K})(550 \text{ K}) \ln 5$$
$$= 1104 \text{ J} \quad \text{(per cycle)}$$

From Eq. 26–5, we find

$$Q_2 = Q_1 \frac{T_2}{T_1} = (1104 \text{ J}) \left(\frac{300 \text{ K}}{550 \text{ K}} \right) = 602 \text{ J} \quad \text{(per cycle)}$$

(c) Using the results of (b), we obtain

$$W = Q_1 - Q_2 = (1104 \text{ J}) - (602 \text{ J}) = 502 \text{ J} \quad \text{(per cycle)}$$

(d) The efficiency of the engine is

$$e = \frac{T_1 - T_2}{T_1} = \frac{(550 \text{ K}) - (300 \text{ K})}{550 \text{ K}} = 0.455 = 45.5 \text{ percent}$$

or
$$e = \frac{W}{Q_1} = \frac{502 \text{ J}}{1104 \text{ J}} = 0.455$$

Example 26–2

A heat engine operates in a cycle that consists of an adiabatic expansion, an isothermal compression, and a constant-volume heating step, as indicated in the diagram on the next page. As in Example 26–1, the working substance is air (considered to be an ideal gas) and the operating temperatures are $T_1 = 550$ K and $T_2 = 300$ K. Determine the thermal efficiency of the engine.

Solution:

The efficiency e is given by Eq. 26–2,

$$e = 1 - \frac{Q_2}{Q_1}$$

so we need to calculate the amount of input heat Q_1 and the amount of rejected heat Q_2. The input occurs at constant volume, so $Q_1 = nC_v(T_1 - T_2)$; the rejection occurs during an isothermal process, so $Q_2 = nRT_2 \ln (V_b/V_a)$. According to the calculation of V_c in Example 26-1, we can write

$$V_b = V_a \left(\frac{T_1}{T_2}\right)^{1/(\gamma-1)}$$

Thus,
$$Q_2 = nRT_2 \ln \left(\frac{T_1}{T_2}\right)^{1/(\gamma-1)} = \frac{nRT_2}{\gamma - 1} \ln (T_1/T_2)$$

Therefore,
$$e = 1 - \frac{RT_2 \ln (T_1/T_2)}{C_v(\gamma - 1)(T_1 - T_2)}$$

Using $\gamma = C_p/C_v$, we have $C_v(\gamma - 1) = C_p - C_v = R$. Then,

$$e = 1 - \frac{T_2 \ln (T_1/T_2)}{T_1 - T_2}$$

(As in the case of the Carnot engine, Eq. 26-6, the efficiency here also increases as $T_1 - T_2$ is increased. •Can you see why?)

Substituting for T_1 and T_2, we find

$$e = 1 - \frac{(300 \text{ K}) \ln (550/300)}{(550 \text{ K}) - (300 \text{ K})} = 0.273 \text{ or } 27.3 \text{ percent}$$

The compression ratio in this cycle is

$$\frac{V_b}{V_a} = \left(\frac{T_1}{T_2}\right)^{1/(\gamma-1)} = \left(\frac{550 \text{ K}}{300 \text{ K}}\right)^{1/0.4} = 4.55$$

which is close to the value of 5 for the case in Example 26-1. Also, the values of T_1 and T_2 are the same. Nevertheless, the efficiency of the engine considered here is only *half* that of the Carnot engine. (For the reason, see the remark at the end of Example 26-3.)

26-2 THE SECOND LAW OF THERMODYNAMICS

It would clearly be advantageous to have a perfect heat engine, an engine capable of transforming heat into mechanical work with an efficiency of 100 percent. Figure 26-4 contrasts the heat-flow diagrams for an ordinary heat engine and a

Fig. 26-4. Heat-flow diagrams for an ordinary heat engine and a perfect heat engine. In the latter case, $Q_2 = 0$ and $W = Q_1$ so that $e = 1$.

(a) ordinary heat engine (b) perfect heat engine

perfect heat engine. In the latter case, no heat is rejected to the cold reservoir ($Q_2 = 0$), so $W = Q_1$; consequently, $e = 1$.

The counterpart to the perfect heat engine—the perfect refrigerator—would require no mechanical work to remove heat from a cold reservoir and deliver it to a hot reservoir. Figure 26-5 shows the heat-flow diagrams for an ordinary refrigerator and a perfect refrigerator. In the latter case, $W = 0$ and $Q_1 = Q_2$.

Fig. 26-5. Heat-flow diagrams for an ordinary refrigerator and a perfect refrigerator. In the latter case, $W = 0$ and $Q_1 = Q_2$.

(a) ordinary refrigerator (b) perfect refrigerator

Experience indicates that neither of these ideally perfect heat devices is possible. The statement that all such perfect devices are physically impossible is contained within the *second law of thermodynamics*. There are several equivalent statements of this important principle. The first formulation of the second law that we give was proposed by Kelvin (and by Planck):

It is impossible to remove heat from a system at a single temperature and have as the sole effect the conversion of this heat into mechanical work.

The Kelvin-Planck statement of the second law directly rules out the possibility of a perfect heat engine, as illustrated in Fig. 26-4b.

A second form of the law was proposed by Clausius*:

It is impossible to transfer heat continuously from a colder system to a warmer system as the sole result of any process.

The Clausius statement of the second law directly rules out the possibility of a perfect refrigerator, as illustrated in Fig. 26-5b.

It is interesting to note that the Kelvin and Clausius statements of the second law are equivalent in the sense that if either were false, the other would also be false. This proposition can be demonstrated in the following way. Suppose that the

---

*Rudolf Clausius (1822–1888), German physicist and one of the founders of thermodynamics. His statement of the second law was given in 1850.

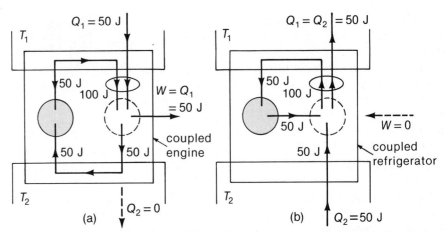

Fig. 26–6. (a) A postulated perfect refrigerator is coupled to an ordinary engine; the combination is still "perfect" and violates the Kelvin-Planck statement of the second law. (b) A postulated perfect engine is coupled to an ordinary refrigerator; the combination is still "perfect" and violates the Clausius statement of the second law.

Clausius statement is false. Imagine that an ordinary heat engine performs work with the rejection of heat to a cold reservoir (Fig. 26–6a). A perfect refrigerator, now possible, could transfer the rejected heat to the hot reservoir without the expenditure of any work. This heat would then be available to the engine to do more work. Using such a coupled device (Fig. 26–6a), all of the heat Q_1 removed from the hot reservoir would eventually appear as work; that is, $W = Q_1$ and $Q_2 = 0$. This would violate the Kelvin-Planck statement of the second law.

Now, suppose that the Kelvin-Planck formulation of the second law is false and that a perfect heat engine exists. Imagine that this engine is coupled to an ordinary refrigerator that removes heat from a cold reservoir and delivers to a hot reservoir an amount of heat greater by the amount of work done in the process (Fig. 26–6b). This excess heat is exactly the amount required to operate the perfect engine that drives the refrigerator. The net effect of the coupled engine and refrigerator is to transfer heat continuously from a cold reservoir to a hot reservoir without the need for any work provided by an external agency; that is, $Q_2 = Q_1$ and $W = 0$. This would violate the Clausius statement of the second law.

▶ *Perpetual-Motion Machines.* One often hears of claims concerning perpetual-motion machines. A perpetual-motion machine of the *first kind* is a device that would violate the first law by producing more work than the energy supplied to it. Claims for such machines are easy to refute by using an energy conservation argument. A device that would extract heat from the environment and do work without benefit of any temperature difference could be consistent with the first law. But such a machine—a perpetual-motion machine of the *second kind*—would violate the second law and, hence, is unrealizable. ◀

26-3 APPLICATIONS OF THE SECOND LAW OF THERMODYNAMICS

Carnot's Theorem. With remarkable insight for a time when the caloric theory was still in vogue, Sadi Carnot in 1824 stated a powerful thermodynamic theorem:

The efficiency of all reversible engines that absorb and reject heat at the same two temperatures is the same, and no irreversible engine working between the same two temperatures can have an efficiency as great.

Carnot's original proof of this theorem was based on caloric arguments and was therefore faulty. We now give a valid proof formulated by Clausius and Kelvin about 30 years later.

Consider two reversible engines, E and E', that operate between thermal reservoirs with temperatures T_1 and T_2 (with $T_1 > T_2$). Let the engine E' run backwards, thereby functioning as a refrigerator. Also, let the two engines be coupled so that the cycles have equal duration. (We can imagine that this is accomplished even though the working substances and pressures of the two engines may be different.) The devices are adjusted so that the work done by the engine is exactly that required to operate the refrigerator. This situation is illustrated schematically in Fig. 26-7.

Fig. 26-7. The work done by the engine E is exactly that required to operate the coupled refrigerator E'.

Suppose, at first, that the efficiency e of the engine E is greater than the efficiency e' of the engine E', which running backwards constitutes the coupled refrigerator; that is, $e > e'$. Thus, with the notation of heat flow quantities shown in Fig. 26-7, we can write

$$e = \frac{Q_1 - Q_2}{Q_1} > \frac{Q_1' - Q_2'}{Q_1'} = e'$$

Now, the work W done by E is equal to the work W' required by E'. Thus,

$$W = Q_1 - Q_2 = Q_1' - Q_2' = W' \qquad \textbf{(26-8)}$$

Using this result, the efficiency equation yields

$$\frac{1}{Q_1} > \frac{1}{Q_1'} \qquad \text{or} \qquad Q_1' > Q_1$$

Equation 26-8 can be expressed as $Q_1' - Q_1 = Q_2' - Q_2$; then, it follows that $Q_2' > Q_2$. The net result of the coupled process is that an amount of heat $Q_2' - Q_2 = Q_1' - Q_1$ is transferred from the cold reservoir (T_2) to the hot reservoir (T_1) without any work having been done by an external agency. But this violates the second law, so we must conclude that e *cannot be greater than* e'.

Suppose next that the operation of each (reversible) engine is reversed so that E' drives E, which now becomes a refrigerator. The same argument leads to the conclusion that e' cannot be greater than e. These two results imply that $e = e'$, so that the first part of Carnot's theorem is proved.

Finally, suppose that E' is a reversible engine whereas E is an irreversible engine. We can again show that e cannot be greater than e' (reversible), but the converse *cannot* be shown because E cannot be reversed. This leads to the second part of Carnot's theorem, namely, that for two engines operating between the same two temperatures,

$$e \text{ (irreversible)} < e \text{ (reversible)}$$

In the following section we discuss the idealized, reversible analogs of various types of practical heat engines, all of which are actually irreversible to some extent. Therefore, we realize from the outset that the calculated efficiencies, based on reversible cycles, represent only upper limits to the true efficiencies.

Example 26–3

Determine the thermal efficiency of an engine that operates on the reversible cycle shown in the diagram. The cycle consists of two isothermal volume changes alternating with two constant-volume heat exchanges.

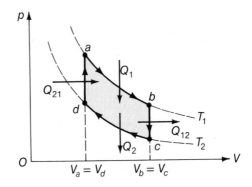

Solution:

The amount of heat added during the constant-volume step *da* is $Q_{21} = Q_{da} = nC_v \cdot (T_1 - T_2)$, and the amount rejected during the constant-volume step *bc* is $Q_{12} = -Q_{bc} = -nC_v(T_2 - T_1)$; thus, $Q_{21} = Q_{12}$. We take the attitude that the heat added Q_{21} is simply stored in the system until it is required to be returned as Q_{12}. Then, the only heat input is Q_1 and the net amount of heat added per cycle is $Q_1 - Q_2$. The first law (Eq. 26–1) therefore becomes $W = Q_1 - Q_2$. The thermal efficiency is

$$e = \frac{W}{Q_1} = \frac{Q_1 - Q_2}{Q_1} = 1 - \frac{Q_2}{Q_1}$$

For the isothermal step *ab* we have $Q_1 = nRT_1 \ln (V_b/V_a)$, and for *cd* we have $Q_2 = nRT_2 \ln (V_c/V_d)$. Also, $V_b = V_c$ and $V_d = V_a$, so that $Q_2/Q_1 = T_2/T_1$, and the efficiency becomes

$$e = 1 - \frac{T_2}{T_1}$$

which is the same as the efficiency of a Carnot engine operating between the same two temperatures (Eq. 26–6). Carnot's theorem applies to *any* reversible heat engine whose operation has the net effect of removing an amount of heat Q_1 from a hot reservoir at a temperature T_1, converting some of this energy into work W, and then rejecting the remainder, $Q_2 = Q_1 - W$, to a cold reservoir at a temperature T_2. The engine in Example 26–2, for example, does not operate between two fixed temperatures (because the heat input Q_1 takes place over the range of temperatures from T_2 to T_1), and the efficiency is correspondingly smaller than the Carnot efficiency.

The Thermodynamic Temperature Scale.   The statements of the first and second laws of thermodynamics and of Carnot's theorem involve only the energy quantities, Q_1, Q_2, and W. Although we recognized that one of the stable thermal reservoirs is "hot" and the other is "cold," the temperatures T_1 and T_2 were not specified or even defined, nor did we use the equation of state of the working substance.

The efficiency, $e = W/Q_1$, of all reversible engines that operate between two thermal reservoirs at different temperatures, *however defined,* is the same, independent of the working substance or the pressure and volume values around the cycle. That is, the efficiency depends *only* on the reservoir temperatures, which we now call θ_1 and θ_2. Because $e = 1 - (Q_1/Q_2)$, it follows that

$$\frac{Q_1}{Q_2} = f(\theta_1, \theta_2)$$

Kelvin realized that it is possible to define a new thermodynamic temperature scale (independent of the properties of any real substance) in terms of a Carnot engine operating in a cycle between the temperaures θ_1 and θ_2. Kelvin's definition of thermodynamic temperature is:

Any two temperatures, θ_1 and θ_2, are in the same ratio as the amounts of heat, Q_1 and Q_2, that are absorbed and rejected, respectively, by an engine executing any Carnot cycle between these two temperatures; that is, $Q_1/Q_2 = \theta_1/\theta_2$.

Thus, the ratio of the thermodynamic temperatures of two heat reservoirs can be specified in terms of the amounts of heat exchanged in a Carnot cycle. If one of these reservoirs is at the temperature of the triple point of water, arbitrarily taken to be 273.16 K, the Kelvin temperature is defined to be

$$\theta = (273.16 \text{ K})\frac{Q}{Q_{\text{tr}}} \qquad \textbf{(26-9)}$$

where θ is the temperature of the reservoir that supplies the amount of heat Q to the engine, and where Q_{tr} is the amount of heat rejected to the triple-point reservoir (see Fig. 26-8).

Only for an ideal gas is the temperature θ identical to the temperature T that appears in the ideal gas law equation. For a real gas we use the asymptotic definition of temperature (see Eq. 22-3),

$$T = (273.16 \text{ K}) \operatorname*{Lim}_{p_0 \to 0} \frac{p}{p_0} \qquad \textbf{(26-10)}$$

The determination of temperature by using Eq. 26-10 is not always feasible, particularly for temperatures below 1 K or so. However, there are many techniques for measuring amounts of heat exchanged, so that the use of Eq. 26-9 is possible. In these discussions we make no further distinction between the temperatures θ and T; we continue to use T to represent the Kelvin (absolute) temperature.

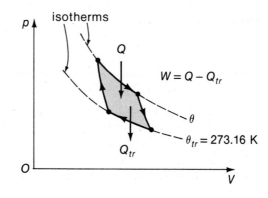

Fig. 26-8. A Carnot cycle used in the definition of the Kelvin temperature. The temperature θ may be either greater than or less than θ_{tr}.

26-4 PRACTICAL ENGINES AND REFRIGERATORS

The Gasoline Engine (Otto Cycle). The gasoline internal-combustion engine operates in a cycle that consists of six parts, only four of which, called *strokes,* involve motion of the piston. Although portions of the cycle are actually irreversible, and friction and heat losses are present, we may simulate the entire engine cycle by the idealized quasi-static cycle shown in Fig. 26–9. This cycle is known as the *air-standard Otto cycle.* Throughout this idealized cycle the working substance is treated as if it were air alone, ignoring the fuel vapor and the combustion products (see Problem 25–21).

Fig. 26-9. The air-standard Otto cycle for internal-combustion gasoline engines. This cycle describes the usual "four-cycle" (or four-stroke) automobile engine. (The description of the "two-cycle" engine used in motorboats and lawn mowers is slightly different.) The shaded area represents the work done per cycle: $W = Q_1 - Q_2$.

The six steps in the cycle are as follows:

(1) *Isobaric intake stroke, oa.* The gasoline-air mixture is drawn into the cylinder at atmospheric pressure p_0; the cylinder inlet valve remains open throughout this stroke. The volume increases from zero to V_a.

(2) *Adiabatic compression stroke, ab.* The inlet valve is closed and the gasoline-air mixture is compressed adiabatically from volume V_a to V_b.

(3) *Constant-volume ignition, bc.* The gasoline-air mixture (heated during the compression stroke) is ignited by an electric spark, thereby further increasing the temperature. The combustion process takes place in such a short time that there is essentially no motion of the piston; thus, $V_c = V_b$. This irreversible process is indicated in the diagram as an idealized constant-volume step; it does not constitute one of the strokes.

(4) *Adiabatic expansion or power stroke, cd.* The heated air and combustion products expand against and drive the piston; the volume increases from V_c to V_a. This is the stroke that delivers power to the crankshaft.

(5) *Constant-volume exhaust, da.* At the end of the power stroke, the exhaust valve is opened; this causes a sudden partial escape of the cylinder gases. There is essentially no piston motion during this part of the cycle, so we indicate *da* in the diagram as an idealized constant-volume step; thus, $V_a = V_d$.

(6) *Isobaric exhaust stroke, ao.* The exhaust valve remains open and the moving piston drives out the residual cylinder gases as the volume decreases from V_a to zero.

The heat input to the actual (changing) working substance of a real engine is the energy released by the combustion of the gasoline vapor in the cylinder during step *bc.* In a true engineering sense this is the available energy, but it is only partially utilized in doing work. Some of the available energy is lost during the valve exhaust step *da.* For the idealized air-standard Otto cycle using *n* equivalent moles of air, we have

$$Q_1 = nC_v(T_c - T_b)$$

$$Q_2 = nC_v(T_d - T_a)$$

and
$$W = Q_1 - Q_2$$

The thermal efficiency e of the engine is

$$e = \frac{W}{Q_1} = \frac{Q_1 - Q_2}{Q_1} = 1 - \frac{T_d - T_a}{T_c - T_b}$$

It is easy to show that $(T_d - T_a)/(T_c - T_b) = (V_b/V_a)^{\gamma-1}$ (see Problem 26–13). The ratio $r = V_a/V_b$ is called the *compression ratio* of the engine. Thus,

$$e = 1 - \frac{1}{r^{\gamma-1}} \qquad \text{(air-standard Otto cycle)} \qquad \textbf{(26–11)}$$

In practice, the compression ratio must be held below about 7, because if $r > 7$, the temperature T_b which is reached at the end of the compression stroke ab would be sufficiently high to ignite the fuel mixture before the proper moment specified by the occurrence of the spark. Such *preignition* causes "knocking" and loss of power. Using $r = 5$ and $\gamma(\text{air}) = 1.40$, we find

$$e = 1 - \frac{1}{(5)^{0.40}} = 0.475 \text{ or } 47.5 \text{ percent}$$

The actual or realized efficiency of a properly tuned gasoline engine is usually no more than about 25 or 30 percent due to heat losses, friction, finite acceleration times for engine parts, and the fact that the efficiency of an irreversible cycle falls short of that for a reversible cycle such as the idealized one we have assumed.

The Diesel Engine. An idealized Diesel cycle is illustrated in Fig. 26–10. This cycle also consists of four strokes. However, in contrast with the Otto cycle, only air is drawn into the cylinder during the isobaric intake stroke oa. The air is then adiabatically compressed, thereby raising the temperature above the ignition point of the fuel, which is injected into the cylinder at high pressure (point b in the cycle). The rate at which the fuel is injected is controlled so that the combustion, which takes place during the power stroke bc, occurs at essentially constant pressure (p_b). Power is also delivered during the adiabatic expansion stroke cd. The constant-volume exhaust through the exhaust valve in step da and the isobaric exhaust stroke ao are the same as in the Otto cycle.

Fig. 26-10. The air-standard Diesel cycle.

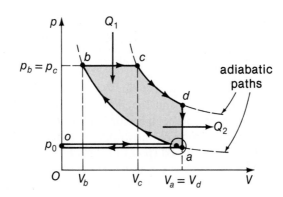

For the *air-standard Diesel cycle* using n equivalent moles of air, we have

$$Q_1 = nC_p(T_c - T_b)$$

$$Q_2 = nC_v(T_d - T_a)$$

and

$$W = Q_1 - Q_2$$

The thermal efficiency e of the engine is

$$e = \frac{W}{Q_1} = \frac{Q_1 - Q_2}{Q_1} = 1 - \frac{T_d - T_a}{\gamma(T_c - T_b)} \qquad \textbf{(26-12)}$$

In terms of the expansion ratio, $r_E = V_d/V_c = V_a/V_c$, and the compression ratio, $r_C = V_a/V_b$, the expression for the efficiency becomes (see Problem 26-14)

$$e = 1 - \frac{1}{\gamma} \cdot \frac{\left(\dfrac{1}{r_E}\right)^\gamma - \left(\dfrac{1}{r_C}\right)^\gamma}{\left(\dfrac{1}{r_E}\right) - \left(\dfrac{1}{r_C}\right)} \qquad \text{(air-standard Diesel cycle)} \qquad \textbf{(26-12a)}$$

There can be no preignition of the fuel in a diesel engine, so the compression ratio can be made much larger than in a gasoline engine. Typically, we might have $r_C = 15$ and $r_E = 5$; then using $\gamma = 1.40$, we find

$$e = 1 - \frac{1}{1.40} \cdot \frac{\left(\dfrac{1}{5}\right)^{1.40} - \left(\dfrac{1}{15}\right)^{1.40}}{\left(\dfrac{1}{5}\right) - \left(\dfrac{1}{15}\right)} = 0.558 \text{ or } 55.8 \text{ percent}$$

This efficiency is about $\frac{1}{6}$ greater than the efficiency of the gasoline engine just calculated. Diesel engines do, in fact, operate more efficiently than gasoline engines, with realizable efficiencies of about 30 or 35 percent.

The Steam Engine (Two-Phase Rankine Cycle). Consider now a steam-driven engine with a movable piston in which the working substance is water (present as liquid water and as steam). In previous discussions, the "engine" consisted entirely of the cylinder volume. We now enlarge the meaning of the term "engine" to include the boiler, the cylinder and piston, the condenser, and the feed pump, as shown schematically in Fig. 26-11. We ignore the losses of water that occur in real engines and follow a given mass of water through a complete cycle: boiler → cylinder → condenser → feed pump → boiler.

The idealized steam cycle (*Rankine cycle*) is shown on a p-V diagram in Fig. 26-12. The major part of the expansion occurs during the step ab as steam generated in the boiler is allowed to enter the cylinder. In this process, water is transformed from saturated liquid (point a) to saturated steam (point b). Along the path ab both the temperature and the pressure remain essentially constant. The steam intake valve is now closed, and a further adiabatic expansion bc takes place in the cylinder. At the completion of this stroke the cylinder contains a mixture of steam and a small percentage of liquid (point c). This mixture is exhausted from the cylinder into the condenser along the path cd. The condenser is a "heat exchanger" in which cooling water from the external cold reservoir absorbs the latent heat and reduces the nearly

Fig. 26–11. Essential features of a steam-driven piston engine.

saturated steam (point c) entirely to liquid (point d). The path cd represents an essentially isothermal-isobaric process. Next, the feed pump increases the pressure of the liquid water to match the boiler pressure p_1. This takes place along the nearly adiabatic path do and involves very little change in either volume or temperature; that is, point o is very close to $o'(p_1, T_2)$. Heat is supplied to the liquid water that enters the boiler, thereby raising its temperature to T_1 (point a). The process oa, which closes the cycle, involves a substantial temperature increase but only a small change in volume.

Except for the relatively minor deviations along paths bc and do, the Rankine cycle is very close to a Carnot cycle. The efficiency is somewhat less than the Carnot efficiency $(1 - T_2/T_1)$ because the heat added along the path oa takes place at temperatures ranging from slightly above T_2 to T_1, instead of entirely at T_1 as in a true Carnot cycle. For example, with $T_1 = 200°$ C and $T_2 = 35°$ C, the Carnot efficiency is 34.9 percent, whereas a careful calculation incorporating the measured properties of water gives an efficiency of 30.9 percent for a Rankine cycle. The realizable efficiency of such an engine might be 25 percent or slightly less.

The Rankine cycle with some modifications can be used to describe the operation of *steam turbines*. In these devices, steam passes through a set of fixed blades and moving blades that are attached to a rotating shaft. The moving blades experience both impulsive forces and jet or reaction forces, which provide the torque to drive the

Fig. 26–12. The Rankine steam-engine cycle.

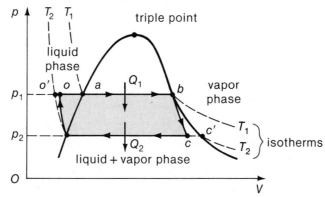

rotating shaft. The steam expands and cools as it moves through the turbine; the energy lost by the steam provides the energy to drive the shaft. Steam turbines have now replaced steam-driven piston engines in many situations because they can be operated at somewhat higher temperatures and therefore can achieve thermal efficiencies of 40 percent or more. This represents a vast improvement over the crude piston engines of the mid-18th century built by such pioneers as James Watt. These early devices had thermal efficiencies of only about 1 percent.

The Refrigerator (Two-Phase Cycle). The most common type of refrigerator operates in a two-phase cycle utilizing the liquid and vapor phases of the working substance. As in the case of the steam-engine cycle, we follow a given mass of the working substance (or *refrigerant*) through a complete cycle: compressor → condenser → liquid storage → throttle → evaporator → compressor. The usual refrigerant in ordinary household refrigerators is Freon-12 (CCl_2F_2); sulfur dioxide (SO_2) is often used in large commercial units. A schematic diagram of a two-phase refrigerator is shown in Fig. 26–13. In an ordinary household refrigerator only the evaporator is located inside the cold compartment. During each cycle the compressor does an amount of work W on the refrigerant using energy from an external source.

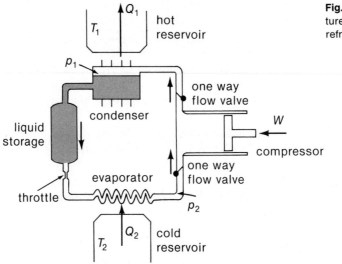

Fig. 26–13. Essential features of a common two-phase refrigerator.

The idealized cycle for a two-phase refrigerator is shown on a *p-V* diagram in Fig. 26–14. This cycle closely resembles the cycle for a Carnot refrigerator that operates between the temperatures T_1 and T_2. Starting at point *a* in the diagram, the saturated liquid in the storage tank at (p_1, T_1) is allowed to undergo an irreversible

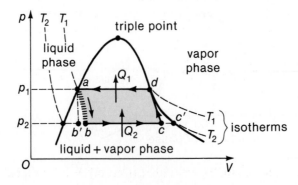

Fig. 26–14. A two-phase refrigerator cycle. The shaded area represents the work done on the working substance by the compressor. The irreversible throttling process *ab* could be replaced by an adiabatic, piston-controlled expansion *ab′*. However, this results in only a slight improvement in refrigerator performance and is therefore not economically worthwhile.

throttled expansion *ab* into the evaporator coils (see Fig. 26–13). During this process the pressure and temperature drop to (p_2, T_2), both of which are much lower than (p_1, T_1). The expansion causes vaporization of the liquid, which continues until point *c* is reached, close to vapor saturation. This evaporation process *bc* takes place at essentially constant pressure and constant temperature, and removes an amount of heat Q_2 from the cold reservoir (the cold compartment of the refrigerator). During the step *cd* the refrigerant, which is a nearly saturated vapor at *c*, is adiabatically compressed to saturated vapor at *d* where it is delivered at the higher pressure p_1 and higher temperature T_1 to the condenser coils. In passing through the coils (following the path *da*), the refrigerant gives up an amount of heat Q_1 to the hot reservoir (room air) at the temperature T_1. When this process is completed, the refrigerant is a saturated liquid at point *a*.

Most refrigerators are designed to maintain a temperature of a few degrees below $0°$ C in the cold compartment when the hot reservoir is at room temperature. It is therefore necessary to choose for the refrigerant a substance that will vaporize at about $-5°$ C at low pressure and will condense at about $20°$ C at high pressure. Many substances have been used as refrigerants, the most common of which is Freon-12.

What is the "efficiency" of a refrigerator? Because a refrigerator performs a function opposite that of an engine, it is not meaningful to define the efficiency in the same way. Instead, the performance of a refrigerator is specified by a figure of merit K_r, called the *coefficient of performance*:

$$K_r = \frac{Q_2}{W} = \frac{Q_2}{Q_1 - Q_2}$$

The quantity Q_2 is the amount of heat absorbed from the cold compartment at the temperature T_2. The quantity Q_1 is the amount of heat rejected into the room air; Q_1 is equal to the heat absorbed Q_2 plus the work W done on the refrigerant by the compressor. To the extent that the actual cycle may be represented by a reverse Carnot cycle, we have

$$K_r = \frac{T_2}{T_1 - T_2} \qquad \text{(Carnot cycle)} \qquad \textbf{(26–13)}$$

For a household refrigerator, typical temperatures are $T_1 = 300$ K $(80°F)$ and $T_2 = 260$ K $(10°F)$; then, the theoretical coefficient of performance is $K_r(\text{Carnot}) = 260/(300 - 260) = 6.5$. An actual refrigerator operating between the same two temperatures might have a coefficient of performance of 4 or 5.

Heat Pumps. Because a refrigerator rejects more heat to the room air than it absorbs from the cold reservoir $(Q_1 > Q_2)$, a refrigerator can be used to pump heat from the *outside* air (now used as the cold reservoir) to the *inside* air. Thus, a device operating on a refrigeration cycle can be used as an air conditioner during summer (pumping heat from inside to outside) and as a *heat pump* during winter (pumping heat from outside to inside). When the refrigerator is operating as a heat pump, the figure of merit for the device is

$$K_h = \frac{Q_1}{W} = \frac{T_1}{T_1 - T_2} \qquad \text{(Carnot cycle)} \qquad \textbf{(26–14)}$$

For $T_1 = 300$ K $(80°F)$ and $T_2 = 260$ K $(10°F)$, we find $K_h(\text{Carnot}) = 7.5$. In fact,

$K_h = K_r + 1$. (•Can you see why?) An actual heat pump might have $K_h = 5$ or 6 for these temperatures.

Example 26-4

A house loses heat through the exterior walls and roof at a rate of 5000 J/s = 5 kW when the interior temperature is 22° C and the outside temperature is $-5°$ C. Calculate the electric power required to maintain the interior temperature at 22° C for the following two cases:

(a) The electric power is used in electric resistance heaters (which convert all of the electricity supplied into heat).

(b) The electric power is used to drive an electric motor that operates the compressor of a heat pump (which has a coefficient of performance equal to 60 percent of the Carnot cycle value).

Solution:

(a) The amount of power required is just equal to the heat loss rate, namely, 5 kW.

(b) We have

$$K_h = 0.60 \, K_h(\text{Carnot}) = 0.60 \left(\frac{295}{295 - 268} \right) = 6.6$$

Thus, the power required is

$$P = \frac{W}{t} = \frac{Q_1/K_h}{t} = \frac{5 \text{ kW}}{6.6} = 0.758 \text{ kW}$$

That is, the heat pump supplies 6.6 times as much heat to the house as do the electric heaters for the same expenditure of electric power.

26-5 ENTROPY

We now return to the discussion of the Carnot cycle, making one important change in notation. In Fig. 26-2 and similar diagrams, we considered the heat rejected to the cold reservoir to be a positive quantity, $Q_2 > 0$; we recognized that this heat *leaves* the system by using a negative sign when expressing the work done as $W = Q_1 - Q_2$. We now adopt the convention that any Q represents heat *added* to the system; thus, rejected heat carries a negative sign, and we write the work done as $W = Q_1 + Q_2$. With this convention, Eq. 26-5 can be expressed as

$$\frac{Q_1}{T_1} + \frac{Q_2}{T_2} = 0$$

Figure 26-15 shows a complicated thermodynamic cycle, *abcdefgha,* that consists of four isothermal steps and four adiabatic steps. By constructing the continuation of the adiabatic steps along *bg'* and *c'f,* the shaded region is divided into three Carnot cycles. Because there are no heat exchanges during the adiabatic steps, the three cycles combined yield

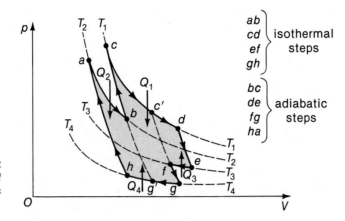

ab
cd } isothermal
ef } steps
gh

bc
de } adiabatic
fg } steps
ha

Fig. 26-15. A multipart cycle that consists of isothermal and adiabatic steps.

$$\frac{Q_1}{T_1} + \frac{Q_2}{T_2} + \frac{Q_3}{T_3} + \frac{Q_4}{T_4} = 0$$

where $Q_1 = Q_{cc'} + Q_{c'd}$ is the amount of heat added at the temperature T_1 during the isothermal step $cc' + c'd$; similarly, $Q_4 = Q_{gg'} + Q_{g'h}$. Altogether, we can write

$$\sum_{i=1}^{4} \frac{Q_i}{T_i} = 0 \qquad \textbf{(26-15)}$$

We can extend this idea of connected Carnot cycles to *any reversible cycle,* such as the one shown in Fig. 26-16. Imagine that this cycle is separated into a very large number of adjacent Carnot cycles, as indicated in the diagram. In the limit of an infinite number of cycles, the sum in Eq. 26-15 becomes an integral, and we have

Fig. 26-16. An arbitrary reversible cycle is separated into a large number of adjacent Carnot cycles.

$$\oint \frac{dQ}{T} = 0 \qquad \text{(reversible cycle)} \qquad \textbf{(26-16)}$$

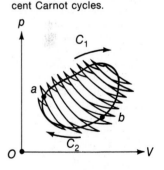

where the integration is to be carried out around the entire path. Select any two states of the system, such as those represented by the points a and b. Then,

$$\oint \frac{dQ}{T} = \oint_1 \int_a^b \frac{dQ}{T} + \oint_2 \int_b^a \frac{dQ}{T} = 0 \qquad \textbf{(26-17)}$$

The process is reversible, so we can write

$$\oint_2 \int_b^a \frac{dQ}{T} = - \oint_2 \int_a^b \frac{dQ}{T}$$

Using this result in Eq. 26-17 we find

$$\oint_1 \int_a^b \frac{dQ}{T} = \oint_2 \int_a^b \frac{dQ}{T} \qquad \textbf{(26-18)}$$

An infinite number of cycles may be drawn through the points that represent the definite states a and b, so we conclude that

$$\oint_a^b \frac{dQ}{T} = \text{constant } (a,b) \qquad\qquad \textbf{(26--19)}$$

This result is valid for *any reversible path* that connects the equilibrium states (p_a, V_a, T_a) and (p_b, V_b, T_b). The value of the constant depends only on the properties of these states.

We now assert that there exists a useful thermodynamic quantity, called the *entropy S*, which is a *state function* and therefore depends *only* on the nature of the equilibrium state with which it is associated. The *difference* in entropy between two states a and b is identified with the value of the integral of dQ/T between these states. Thus, we write Eq. 26–19 as

$$\oint_a^b \frac{dQ}{T} = S_b - S_a \qquad \text{(reversible path)} \qquad\qquad \textbf{(26--20)}$$

For a differential process, we have

$$\frac{dQ}{T} = dS \qquad\qquad \textbf{(26--20a)}$$

Notice that the units of entropy are J/K (or Cal deg^{-1}).

It is important to realize that the path C along which the integral in Eq. 26–20 is carried out can represent *any reversible process* that connects the two states. Thus, we may choose a path that makes easier any calculation that is required. Even if the actual path that carries the system from state a to state b is irreversible, the entropy difference, $S_b - S_a$, can still be evaluated by choosing a reversible path from a to b. This is possible because the entropy is a state function; the value of $S_b - S_a$ depends only on the properties of the states a and b, not on the path that connects them. In this regard, entropy is similar to the internal energy (see Section 25–2).

Entropy and the Second Law. The constraints that the entropy concept imposes on the second law of thermodynamics can be investigated by using Carnot's theorem (Section 26–3). Figure 26–17a shows a simple Carnot cycle; Fig. 26–17b shows another cycle that is identical except that the step ab is irreversible. The heat absorbed by the system during the reversible step ab is Q_1 and the heat absorbed during the irreversible step is Q_1'. The heat rejected during the step cd is the same for both cycles. Using the new convention regarding the sign of Q, the work done per cycle in Fig. 26–17a is $Q_1 + Q_2$, whereas the work done per cycle in Fig. 26–17b is $Q_1' + Q_2$. According to Carnot's theorem, the efficiency of the device that operates in the irreversible cycle is less than that of the device operating in the reversible cycle. That is,

$$e' = \frac{Q_1' + Q_2}{Q_1'} = 1 + \frac{Q_2}{Q_1'} < 1 + \frac{Q_2}{Q_1} = \frac{Q_1 + Q_2}{Q_1} = e$$

Now, $Q_2 < 0$, so we have

$$1 - \frac{|Q_2|}{Q_1'} < 1 - \frac{|Q_2|}{Q_1}$$

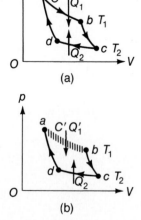

Fig. 26–17. (a) A simple Carnot cycle in which the isothermal step ab corresponds to the path C. (b) The same cycle with the step ab replaced by an irreversible process corresponding to the path C'. The rejected heat is $Q_2 < 0$.

from which we see that $Q_1' < Q_1$. For the reversible path C, we can write

$$S_b - S_a = \oint_a^b \frac{dQ}{T} = \frac{Q_1}{T_1} \qquad \text{(reversible path } C\text{)} \tag{26-21}$$

We also have

$$\oint_a^b \frac{dQ}{T} = \frac{Q_1'}{T_1} < \frac{Q_1}{T_1} = S_b - S_a$$

Hence,

$$S_b - S_a > \oint_a^b \frac{dQ}{T} \qquad \text{(irreversible path } C'\text{)} \tag{26-22}$$

We can combine Eqs. 26–21 and 26–22 and write

$$S_b - S_a \geqslant \oint_a^b \frac{dQ}{T} \tag{26-23}$$

where the equality pertains *only* to reversible processes.

An *isolated* (or *closed*) *system* is one that has no thermal contact with its surroundings, so that no heat exchanges can occur. Suppose that such a system undergoes a spontaneous transition from state a to state b. During this process, $dQ = 0$, so that $\int dQ/T = 0$. Then, Eq. 26–23 gives us the important statement that

$$S_b \geqslant S_a \qquad \text{(isolated system)} \tag{26-24}$$

No finite process in a real system is strictly reversible, so the inequality applies for all practical purposes.

Suppose that a system is prepared in some arbitrary initial state and is then allowed to undergo whatever spontaneous processes are available to it. As these processes occur, the system approaches ever more closely to a truly stable equilibrium state and the entropy of the system *increases* monotonically with time. Thus, the second law of thermodynamics leads to a correlation of natural or spontaneous processes with the direction of the evolution of time ("time's arrow"). In the examples and problems that follow, we find that the natural processes we expect to occur all involve increases in entropy. The entire Universe may be considered to be an isolated system, so it follows from the second law that the entropy of the Universe continually increases with time. Thus, the Universe is proceeding toward a state of uniform temperature in which heat engines can no longer do work because no temperature differences exist.

Free Expansion of a Gas. Consider again (as in Section 25–3) the free expansion of an ideal gas originally confined to a volume V_A and separated from an evacuated second volume V_B by a closed valve (Fig. 26–18a). Prior to the opening of the valve, the system (which is defined to consist of both volumes) is in an equilibrium state a. Opening the valve initiates a spontaneous irreversible process over which we have no further control. The system is isolated, so that $Q = 0$ and $W = 0$; it therefore follows from the first law that the change in the internal energy is zero and that the free expansion results in no temperature change. When both volumes

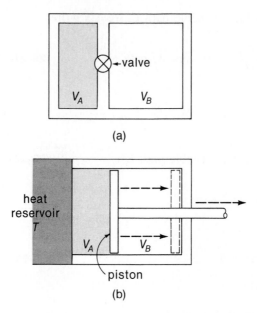

Fig. 26–18. (a) Container for a free expansion. (b) The slow movement of the piston achieves the same final state as the free expansion.

are filled and the flow of gas has ceased, the system is in a truly equilibrium final state *b*.

We cannot calculate the entropy change between states *a* and *b* by considering the free expansion process itself. (There is no well-defined path along which dQ/T can be integrated.) However, because the entropy change depends only on the initial and final states, we can make the calculation by choosing *any reversible path* that connects the same two states. For this purpose we choose an isothermal process in which the slow movement of a piston increases the volume of the gas from V_A to $V_A + V_B$ (Fig. 26–18b). We use the differential form of the first law, namely,

$$dQ = dU + p\,dV$$

But the temperature T is constant, so $dU = 0$; then, dividing by T and using $dQ/T = dS$, we have

$$dS = \frac{dQ}{T} = \frac{p}{T}dV = nR\frac{dV}{V}$$

where we have also made use of the ideal gas law. Thus,

$$S_b - S_a = \int_a^b \frac{dQ}{T} = nR\int_{V_A}^{V_A+V_B} \frac{dV}{V} = nR\ln\frac{V_A + V_B}{V_A} > 0 \qquad \textbf{(26–25)}$$

so that $S_b > S_a$. Thus, the irreversible spontaneous free expansion process results in an increase in entropy for the isolated system.

Both the free expansion and the reversible isothermal process carry the system from state *a* to state *b* with the same entropy change. What, then, distinguishes these two quite different processes? The answer is found by examining the entropy change in the *surroundings* of the system. In the free expansion, there is no heat exchange between the system and the surroundings, so the surroundings undergo no entropy change:

$$\text{Irreversible}\atop\text{free expansion:} \quad \begin{cases} \Delta S_{\text{system}} = +nR\ln\dfrac{V_A + V_B}{V_A} \\[2ex] \Delta S_{\text{surroundings}} = 0 \end{cases}$$

However, during the isothermal expansion, the surroundings of the system (that is, the heat reservoir at temperature T) delivers to the system an amount of heat Q equal to the work W done on the gas to accomplish the expansion, namely (see Eq. 25-2), $Q = W = nRT\ln[(V_A + V_B)/V_A]$. Because an amount of heat Q was *removed* from the surroundings, the entropy of the surroundings changes (decreases) by an amount $-Q/T = -W/T = -nR\ln[(V_A + V_B)/V_A]$. Thus,

$$\text{Reversible}\atop\text{isothermal expansion:} \quad \begin{cases} \Delta S_{\text{system}} = +nR\ln\dfrac{V_A + V_B}{V_A} \\[2ex] \Delta S_{\text{surroundings}} = -nR\ln\dfrac{V_A + V_B}{V_A} \end{cases}$$

We may consider the *Universe* to consist of the *system* we are discussing and its *surroundings*. Then, the entropy change of the Universe is*

$$\Delta S_{\text{Universe}} = \Delta S_{\text{system}} + \Delta S_{\text{surroundings}} \qquad\qquad \textbf{(26-26)}$$

For the irreversible free expansion, we find $\Delta S_{\text{Universe}} = +nR\ln[(V_A + V_B)/V_A]$; however, for the reversible isothermal expansion, we have $\Delta S_{\text{Universe}} = 0$. These are general results. For any irreversible process the entropy of the Universe increases, but for any reversible process the entropy of the Universe does not change. In a complex irreversible process, some part of the system may actually experience a decrease in entropy, but the net result of the process must be an increase in entropy for the system and for the Universe.

Temperature Equalization. Consider two objects, with object 1 at a higher initial temperature than object 2. The objects are thermally isolated from each other and from their surroundings. The objects are now brought into contact but are still isolated from the surroundings. Heat is transferred from the hotter to the colder object, and they eventually reach thermal equilibrium at some intermediate temperature. At all times during the equalization process, we have $T_1 > T_2$. Let an amount of heat ΔQ be transferred between the objects when one temperature is T_1 and the other is T_2. Then, the change in entropy ΔS of the system of two objects is

$$\Delta S = \Delta S_1 + \Delta S_2$$

where $\qquad\qquad \Delta S_1 = -\Delta Q/T_1 \qquad \text{and} \qquad \Delta S_2 = \Delta Q/T_2$

(The sign difference arises because heat is transferred from object 1 to object 2.) Thus,

---

*We simply assert this provable proposition, namely, that the entropy change of a system (here, the Universe) is equal to the sum of the entropy changes of its constituent parts.

$$\Delta S = \frac{T_1 - T_2}{T_1 T_2} \Delta Q > 0$$

This result is valid for each incremental transfer of heat, so we conclude that the entropy of the system increases continually throughout the process. Equilibrium is reached when the entropy change has achieved its maximum value.

Entropy of an Ideal Gas.   The differential form of the first law is

$$dQ = dU + p\, dV$$

Dividing by T and using $dS = dQ/T$ and the ideal gas law, we obtain

$$\frac{dQ}{T} = \frac{dU}{T} + nR\frac{dV}{V} = dS$$

Assume that the system is an ideal monotonic gas. Then, we have $U = \frac{3}{2}nRT$ and $dU = \frac{3}{2}nRdT$. Thus,

$$dS = nR\left(\frac{3}{2}\frac{dT}{T} + \frac{dV}{V}\right)$$

This expression for dS is an exact differential, which can be immediately integrated to give

$$S(T, V) = nR\left(\tfrac{3}{2}\ln T + \ln V\right) + S_0 \qquad\qquad \textbf{(26–27)}$$

where we indicate explicitly that $S = S(T, V)$ is a function of T and V. The constant of integration S_0 is called the *chemical constant* and depends only on the amount and the nature of the gas. According to the *Nernst heat theorem,* the absolute value of the entropy $S(T, V)$ of the system approaches zero as $T \rightarrow 0$. However, we deal here only with entropy *differences.*

Consider the free expansion of a fixed amount of gas from the volume V_A (state *a*) to the volume $V_A + V_B$ (state *b*) at a constant temperature. According to Eq. 26–27, we have

$$S_a = nR\left(\tfrac{3}{2}\ln T + \ln V_A\right) + S_0$$
$$S_b = nR\left[\tfrac{3}{2}\ln T + \ln(V_A + V_B)\right] + S_0$$

Then, $$\Delta S = S_b - S_a = nR\ln\left[(V_A + V_B)/V_A\right]$$

which is just the result we found in Eq. 26–25.

Entropy and the Availability of Work.   Consider a thermal reservoir (the *cold* reservoir) at a temperature T_r and an object at a higher temperature T that acts as a *finite* hot reservoir. Imagine that a reversible engine operating on an incremental Carnot cycle is connected to the hot object and to the cold reservoir. As the engine does work, the temperature of the object decreases continually, eventually reaching the reservoir temperature T_r. At this point no temperature difference exists and no further work can be done by the engine.

Let the total amount of heat delivered to the engine from the object be Q; let the total amount of work done by the engine be W; and let the amount of heat rejected by the engine to the cold reservoir be Q_r ($Q_r < 0$). During each incremental cycle we have

$$\Delta W = \Delta Q + \Delta Q_r$$

and also (from Eq. 26–5)

$$\frac{\Delta Q}{\Delta Q_r} = -\frac{T}{T_r}$$

where the negative sign is necessary because $\Delta Q_r < 0$. Thus, we find that

$$\Delta W = \Delta Q - \frac{\Delta Q}{T} T_r$$

Even though ΔQ and T are changing as the object is drained of its available internal energy, the quantity $-\Delta Q/T = \Delta S$ represents a continual entropy decrease of the object per cycle. Thus, with $\Delta S < 0$, we have

$$\Delta W = \Delta Q + T_r \Delta S$$

For the complete process, we have $\Sigma \Delta W = W$, and $\Sigma \Delta Q = Q$; also, $\Sigma \Delta S = S_f - S_i$, which is the difference in the entropy of the object between the final state and the initial state, where $S_f < S_i$. Thus, with $S_i - S_f > 0$, we have

$$W = Q - (S_i - S_f)T_r \qquad \textbf{(26–28)}$$

That is, the amount of heat Q that is extracted from the object is not entirely converted into work by the engine. The amount of work actually done is reduced by the entropy change of the (finite) hot reservoir. An even smaller amount of work would be realized if the engine were not perfectly reversible.

Temperature-Entropy Diagrams.   The area under a reversible thermodynamic path, $p = p(V, T)$, that is plotted in a p-V diagram represents the work done by the system during the corresponding process, namely, $W_{ab} = \int_a^b p\,dV$. If the entropy S is determined for every quasi-static state through which the system passes, the process can also be illustrated in a T-S diagram (Fig. 26–19a). The area under such a curve is equal to the heat added to the system during the process; that is, $Q_{ab} = \int_a^b T\,dS$. The T-S diagram for a Carnot cycle is rectangular (Fig. 26–19b). The horizontal parts of the cycle are the isothermal paths and the vertical parts are the adiabatic paths ($\Delta S = 0$).

(a)

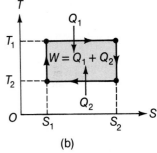

(b)

Fig. 26–19. (a) *T-S* diagram for an arbitrary reversible process. The area under the curve represents the heat added. (b) *T-S* diagram for a Carnot cycle.

Example 26–5

Consider an isolated system that consists initially of 0.50 kg of water in a beaker at a temperature of 3° C and a 10-g ice cube at 0° C (not in contact with the water). The ice cube is now placed in the water, and eventually the system reaches equilibrium with all of the ice melted. Calculate the change in entropy of the system and demonstrate that $\Delta S > 0$ for this process. (Neglect the heat capacity of the beaker.)

Solution:

The latent heat of fusion for water is $L_f = 79.7$ Cal/kg and the specific heat of water near $0°$ C is $c_w = 1.01$ Cal/kg · deg. The equilibrium temperature t_C is determined by equating the energy given up by the water to the energy gained by the ice. Thus,

$$m_i L_f + m_i c_w(t_C - 0°\,\text{C}) = m_w c_w(3°\,\text{C} - t_C)$$

$$(0.010\,\text{kg})(79.7\,\text{Cal/kg}) + (0.010\,\text{kg})(1.01\,\text{Cal/kg} \cdot \text{deg})t_C$$
$$= (0.50\,\text{kg})(1.01\,\text{Cal/kg} \cdot \text{deg})(3°\,\text{C} - t_C)$$

from which $t_C = \dfrac{718}{515.1} = 1.394°$ C

The change in entropy of the ice is

$$\Delta S_i = \left(\frac{\Delta Q}{T}\right)_{\text{melting}} + \left(\int \frac{dQ}{T}\right)_{\text{heating}}$$

where $dQ = m_i c_w\, dT$. Then, with $T_0 = 0°$ C $= 273.15$ K, we have

$$\Delta S_i = \frac{m_i L_f}{T_0} + m_i c_w \int_{T_0}^{T_0 + t_C} \frac{dT}{T}$$

$$= \frac{(0.010\,\text{kg})(79.7\,\text{Cal/kg})}{273.15\,\text{K}} + (0.010\,\text{kg})(1.01\,\text{Cal/kg} \cdot \text{deg}) \ln \frac{273.15 + 1.394}{273.15}$$

$$= 2.969 \times 10^{-3}\,\text{Cal deg}^{-1}$$

The change in entropy of the original amount of water is

$$\Delta S_w = m_w c_w \int_{T_0 + 3°\text{C}}^{T_0 + t_C} \frac{dT}{T}$$

$$= (0.50\,\text{kg})(1.01\,\text{Cal/kg} \cdot \text{deg}) \ln \frac{273.15 + 1.394}{273.15 + 3.00}$$

$$= -2.945 \times 10^{-3}\,\text{Cal deg}^{-1}$$

Hence, the total entropy change for the system is

$$\Delta S = \Delta S_i + \Delta S_w = 2.4 \times 10^{-5}\,\text{Cal deg}^{-1} \quad \text{or} \quad 0.100\,\text{J/K}$$

Thus, there is a net increase in the entropy of the system. The system is isolated, so there is also a net increase in the entropy of the Universe, as we expect for an irreversible process. Energy conservation (the first law) would not be violated if the process were to run backwards, thereby reconstituting the ice cube; however, such a process would involve an entropy *decrease* and would therefore violate the second law.

SPECIAL TOPIC

26-6   TIME'S ARROW

All everyday processes are irreversible.* What makes an irreversible process *irreversible?* Or, equivalently, why does the entropy of the Universe continually increase? Why does time always advance in the direction that conforms with entropy increase? Is a new fundamental law of physics required to account for the direction of "time's arrow" or can we understand these effects on the basis of a microscopic mechanical model (as in the case of kinetic theory)?

Irreversibility. When you dive from a diving board into a swimming pool, the process is clearly irreversible. You cannot gather energy from the water molecules, rise feet-first out of the water, and land on the diving board. A backwards-running movie of an event such as diving into a pool always provokes laughter because we appreciate that the situation is unreal. How can we characterize an irreversible process in a quantitative way?

Let us return to the case of the adiabatic free expansion of a gas and examine carefully the process that occurs. When the valve between the filled volume V_A and the evacuated volume V_B is opened, control of the system is lost and the expansion is free and irreversible. Let $V_A = V_B$ so that equilibrium is reached at twice the original volume and at one half the original pressure, but at the same temperature and internal energy. There would be no violation of the first law if the process were reversed and all of the gas molecules reassembled in the volume V_A. However, if we could photograph the molecules, a reverse running of the motion picture record of the expansion, showing all the molecules being drawn into V_A, would be recognized as unreal as the reversed diving situation.

Suppose that the volume V_A originally contains one molecule and that there are none in V_B. Now, when the valve is opened, if the molecule's velocity is directed toward the opening, it will pass freely into the volume V_B. After colliding a number of times with the walls in V_B, the molecule could return to V_A. Indeed, we can imagine this process repeated over and over, with the molecule shuttling back and forth between V_A and V_B. The duration of each visit will be different, but after many trips we would find the total dwell time in V_A to be closely equal to that in V_B. If we ran backwards the motion picture of this process, we would see nothing at all unusual! Even if there were two or three molecules originally in V_A, there is some chance that they will all migrate back into V_A from time to time. We would still be willing to accept as reasonable ei-

---

*How could you unburn a match or unscramble an egg?!

ther the forward-running movie or the backwards-running movie. The reversibility, of its own accord, of the free expansion of only a few molecules is clearly possible! Evidently, we recognize that the movie is running backwards only when we see that *many* molecules have collected exclusively in V_A. The explanation we are seeking is in some way connected with the laws of chance involving large numbers.

When equilibrium is reached in the free expansion process, the molecules move throughout the connected volume $V_A + V_B$ (with $V_A = V_B$) without interaction. We expect that on the average there will be just as many molecules in V_A as in V_B. The molecules are, of course, indistinguishable, but let us imagine that we can define a state of the system of N molecules by specifying whether each molecule is in V_A or in V_B. For example, with $N = 2$, there are four possible states, namely, AA, AB, BA, and BB, in obvious notation. That is, the probability that both molecules are in V_A is one in four, or $\frac{1}{4}$; the probability that there is one molecule in each volume (AB or BA) is $\frac{1}{2}$; and the probability that both molecules are in V_B is $\frac{1}{4}$. The probability that both molecules are *somewhere* is the sum of the individual probabilities, namely, *one*, as it must be. With $N = 4$, there are 16 different possible states, such as $AAAA$, $ABAA$, $BBAA$, and so forth. The probability is $\frac{1}{16}$ that all four molecules are in V_A; it is $\frac{4}{16} = \frac{1}{4}$ that any three molecules are in V_A; it is $\frac{6}{16} = \frac{3}{8}$ that any two molecules are in V_A; it is $\frac{4}{16} = \frac{1}{4}$ that only one molecule is in V_A; and the probability that no molecule is in V_A is $\frac{1}{16}$. (The sum of the probabilities is again *one*.) Even for such small numbers of molecules as $N = 2$ and $N = 4$, the most likely distribution of the molecules is clearly an equal division between V_A and V_B. For any N (taken for convenience to be an even number), the probability that all of the molecules are in V_A is

$$P(N) = \frac{1}{2^N}$$

The most probable distribution is that with $N/2$ molecules in V_A; this probability is

$$P(N/2) = \frac{N!}{\left[\left(\dfrac{N}{2}\right)!\right]^2 2^N}$$

The distribution of probabilities for states with m molecules in V_A is shown in Fig. 26-20.

A practical macroscopic sample of a gas contains of the order of Avogadro's number of molecules, that is, about $N_0 = 6 \times 10^{23}$ molecules. The average probability

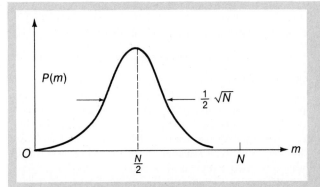

Fig. 26–20. The average probability of finding exactly m of the N molecules in V_A is $P(m) = \dfrac{N!}{m!(N-m)!}\dfrac{1}{2^N}$. The distribution has a "width" of $\frac{1}{2}\sqrt{N}$ about the most probable value, $m = N/2$. For large N, the distribution is essentially a narrow spike.

of finding all N_0 molecules in V_A is $P(N_0) = 1/2^{N_0}$, an exceedingly small number. The ratio of the probabilities for equal division of the molecules and for all in V_A is

$$\frac{P(N_0/2)}{P(N_0)} = \frac{N_0!}{\left[\left(\dfrac{N_0}{2}\right)!\right]^2} \cong 10^{2 \times 10^{23}}$$

which is an enormous number!*

In Fig. 26–20 we see that fluctuations of the number of molecules in V_A away from exactly $N/2$ are likely if the excursion is small. Fractionally, the fluctuations are effectively confined to values less than $(\frac{1}{2}\sqrt{N_0})/(\frac{1}{2}N_0) = 1/\sqrt{N_0} \cong 1.3 \times 10^{-12}$. Thus, the distribution is extremely narrow.

The reversibility of an "irreversible" process is not absolutely forbidden—it is simply very improbable. It is also the case that differential processes that carry a system only slightly away from equilibrium are readily reversible. "Time's arrow" is then simply an indicator of the direction of the most probable advance of a natural process.

Entropy. In the statistical theory of thermodynamics, the concept of entropy is associated with the ideas of *order* and *disorder*. A measure of the degree of disorder in a system is the number of microscopic ways in which the components of the system can be arranged without altering the macroscopic (thermodynamic) properties. This number is called the *thermodynamic probability*† Ω. The larger the value of Ω, the more disordered is the system. (Ω is sometimes called the *disorder number*.) In a detailed analysis of

---

*The number 10^{100} is so large that it has been given a special name—the *googol*. About how many googols are there in this probability ratio?

the dynamic mechanism of intermolecular collisions in a gas, Boltzmann found that the entropy S of a system is related to the thermodynamic probability Ω by‡

$$S = k \ln \Omega \qquad (26\text{–}29)$$

where k is Boltzmann's constant (see Section 24–2).

Consider two volumes, V_A and V_B, that contain, respectively, argon gas and helium gas at the same temperature and pressure. The partition between the two volumes is removed and the gases mix together (but no chemical reaction takes place). The original system had a degree of order in that only argon atoms could be found in V_A and only helium atoms could be found in V_B. When the gases are thoroughly mixed, this orderliness has disappeared. The value of Ω has therefore become larger and the entropy of the system has increased. Notice that this change in entropy is not accompanied by any change in the macroscopic properties of the system: The temperature and the pressure are exactly the same as in the original condition. The entropy increase is brought about solely as the result of the change in the distribution of the gas atoms in space.

Consider another system in the form of an isolated beaker of nutrient broth that contains some growing and dividing cells. Within a cell, various processes occur in which small molecules are organized into large molecules, such as the replication of DNA molecules from amino acids. Such processes clearly increase the degree of order within the cell and thereby *decrease* the entropy of the cell. However, the respiration of the cell *increases* the entropy of the broth. The net effect is an increase in the entropy of the system (broth plus cells).

The tendency for entropy to increase is entirely equivalent to the tendency for the most probable situation (largest Ω) to prevail. If we start with a system in a nonequilibrium state, the system will proceed spontaneously to states with ever higher degrees of disorder and therefore with ever larger values of Ω. At equilibrium Ω has its largest possible value. The equilibrium state of a system—the state toward which the system advances spontaneously—is

---

†Ω is actually a *relative probability* used to compare thermodynamic states. The value of Ω is not restricted to be between 0 and 1 (as would be the case for a true probability); Ω, in fact, is a very large number.

‡The logarithmic relationship between Ω and S comes about because entropy is additive and probability is multiplicative. Consider two systems in the states A and B; the entropy values are, respectively, $S_A = k \ln \Omega_A$ and $S_B = k \ln \Omega_B$. If these systems are parts of a larger system, the entropy of the combination is the sum, $S_A + S_B = k(\ln \Omega_A + \ln \Omega_B) = k \ln \Omega_A \Omega_B$. The product $\Omega_A \Omega_B$ represents the probability that one system is in state A and the other is in state B. For the larger system, we would write $S_{A+B} = k \ln \Omega_{A+B}$, where $S_{A+B} = S_A + S_B$ and $\Omega_{A+B} = \Omega_A \Omega_B$. This is exactly equivalent to the relationship for the conjoint system.

the state of maximum probability and maximum disorder. The increase in the entropy of the Universe—the steady procession from order to disorder—takes place because it is the most likely process to occur. Questions concerning the *origin* of this order—which are connected with questions concerning the origin of the Universe—remain unanswered.

Example 26-6

An ideal gas consisting of n moles, originally with a volume V_A, is expanded isothermally at the temperature T to a volume $V_A + V_B$. Show that the change in entropy calculated using Eq. 26-29 is equivalent to the change expressed by Eq. 26-25.

Solution:

Originally, the probability ω_i of finding the molecule labeled i in the volume V_A (and not elsewhere in the Universe) is proportional to V_A; that is, $\omega_i = \alpha V_A$. This is also the probability of finding each of the other molecules of the sample in V_A. Therefore, the total probability of finding all N molecules in V_A is the product of all the individual ω_i; that is,*

$$\Omega_A = \prod_{i=1}^{N} \omega_i = \alpha^N V_A^N$$

Thus,

$$S_A = k \ln \Omega_A = kN(\ln \alpha + \ln V_A)$$

For the final equilibrium state with volume $V_A + V_B$, we have

$$\Omega_{AB} = \alpha^N (V_A + V_B)^N$$

and

$$S_{AB} = kN[\ln \alpha + \ln(V_A + V_B)]$$

Finally, we obtain

$$S_{AB} - S_A = kN[\ln(V_A + V_B) - \ln V_A]$$
$$= nR \ln \frac{V_A + V_B}{V_A}$$

---

*The *product symbol* $\prod_{i=1}^{N} \omega_i$ means to take the product of the ω_i; that is, $\omega_1 \cdot \omega_2 \cdot \omega_3 \cdot \cdots \cdot \omega_N$. $\prod$ is therefore analogous to the symbol Σ for a sum.

where we have used $kN = kN_0 n = nR$. The result is the same as that found in the previous analysis (Eq. 26-25).

Example 26-7

Compare the thermodynamic probability of the equilibrium condition in Example 26-5 with the probability that 10 g of ice at 0° C and 0.50 kg of water at 3° C will be reformed.

Solution:

To return the equilibrium state to the original condition would require an entropy *decrease* of 0.100 J/K, just the value of the entropy *increase* found in Example 26-5. Let Ω_1 represent the thermodynamic probability of the equilibrium state and let Ω_2 represent the probability of the original ice-plus-water condition. Then, using Eq. 26-29, we have

$$S_1 = k \ln \Omega_1 \quad \text{and} \quad S_2 = k \ln \Omega_2$$

from which

$$S_1 - S_2 = k(\ln \Omega_1 - \ln \Omega_2)$$
$$= k \ln \frac{\Omega_1}{\Omega_2}$$

Thus,

$$\ln \frac{\Omega_1}{\Omega_2} = \frac{\Delta S}{k} = \frac{0.100 \text{ J/K}}{1.381 \times 10^{-23} \text{ J/K}} = 7.24 \times 10^{21}$$

which we can rewrite as

$$\log \frac{\Omega_1}{\Omega_2} = 0.434 \ln \frac{\Omega_1}{\Omega_2} = 3.14 \times 10^{21}$$

so that

$$\frac{\Omega_1}{\Omega_2} = 10^{3.14 \times 10^{21}}$$

Thus, the likelihood that the system will return to its original condition is so small that it is entirely negligible.

PROBLEMS

Section 26-1

26-1   A Carnot engine operates between the temperatures $T_1 = 100°$ C and $T_2 = 30°$ C. By what factor is the theoretical efficiency of the engine increased if the temperature of the hot reservoir is increased to $580°$ C (a boiler temperature common in power plants)?

26-2   A Carnot engine operates with a thermal efficiency of 40 percent, taking in 5 Cal of heat per cycle from a hot reservoir at a temperature of $250°$ C. (a) What is the temperature of the cold reservoir? (b) What amount of heat is rejected per cycle to the cold reservoir? (c) How much work is done by the engine per cycle? (d) If the working substance is 0.8 mol of an ideal (monatomic) gas, what is the expansion ratio during the isothermal expansion? (e) What is the expansion ratio during the adiabatic expansion?

26-3   An engine operates in a reversible cycle that consists of an isobaric expansion, an adiabatic expansion, and an isothermal compression, as shown in the diagram. Show that the efficiency of this engine is the same as that found in Example 26-2.

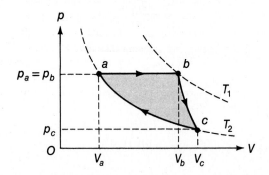

26-4   An engine operates in a reversible cycle that consists of an adiabatic expansion from a temperature $T_1 = 500°$ C to a temperature $T_2 = 200°$ C, followed by an isobaric compression and a constant-volume heating to close the cycle, as shown in the diagram above right. The working substance consists of 2.5 mol of a diatomic ideal gas. (a) Show that the efficiency of the engine is

$$e = 1 - \gamma \frac{T_2 - T_3}{T_1 - T_3}$$

(b) Derive an expression relating the three tempera-

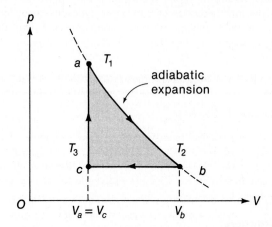

tures and determine the value of T_3. (c) Evaluate the efficiency e. (d) How much work is done by the engine per cycle?

26-5   A Carnot refrigerator is used to freeze 50 kg of water at $0°$ C by discharging heat into a room that acts as a hot reservoir with a temperature of $27°$ C. (a) What is the total amount of heat discharged into the room during the process? (b) How much energy must be supplied to operate the refrigerator? For how long must the 1-kW refrigerator motor run in order to accomplish the freezing of the water?

Section 26-2

26-6   Prove that two reversible adiabatic paths cannot intersect. [*Hint:* Assume that the paths do intersect and use them to construct a thermodynamic cycle that violates the second law.]

Section 26-3

26-7   An engine operates in a cycle between the temperatures $T_1 = 180°$ C and $T_2 = 100°$ C, and exhausts 4.8 Cal of heat per cycle while doing 1500 J of work per cycle. Compare the actual efficiency of the engine with that of a reversible engine operating between the same temperatures.

26-8   What is the minimum amount of work that must be done to extract 0.10 Cal of heat from a massive object at $0°$ C while rejecting heat to a hot reservoir at a temperature of $20°$ C?

26-9   A special low-temperature refrigerator that operates

on a Carnot cycle is found to deliver 0.50 Cal of heat to a hot reservoir at a Kelvin temperature of $\theta_1 = 5.21$ K while doing 1750 J of work. Determine the temperature θ_2 of the cold reservoir.

Section 26–4

26–10 A 50-hp crane is driven by an engine that operates in a cycle between a boiler temperature of 200° C and an exhaust temperature of 100° C. The cycle rate of the engine is 500 rpm. The realized efficiency of the engine is 40 percent of that of a reversible engine operating between the same temperatures. What is the rate of heat input required to operate the engine?

26–11 An electric generating plant produces 1250 MW of
• power by burning oil to drive steam turbines. Cooling water is supplied by a large river whose flow rate is 1800 m³/s. The electric generators have a combined efficiency of 95 percent; the thermal efficiency of the turbines is 40 percent; and the efficiency of transferring the fuel heat to the boiler is 80 percent. Assume that 1/10 of the river's flow is used for cooling, and assume that the heat of combustion of the fuel oil is 4.40×10^7 J/kg. (a) What is the total output power of the turbines? (b) What are the input and exhaust heat flow rates for the turbines? (c) If the steam enters the turbines at 580° C and leaves at 110° C, what is the efficiency of the turbines if they operate in a reversible cycle? Compare this with the actual overall efficiency. (d) How many barrels of fuel oil must be burned per hour to produce the output? (The density of the fuel oil is 920 kg/m³ and the volume of 1 bbl is 0.159 m³.) (e) What is the temperature rise of the cooling water that flows through the turbine condenser? If half of the heat delivered to the cooling water is dissipated into the air, what is the temperature rise of the river water when the cooling water has been thoroughly mixed?

26–12 A gasoline-powered automobile engine with a com-
• pression ratio $r = 6.8$ delivers 30 hp. The thermal efficiency of the engine is 45 percent of the theoretical value for an engine operating on the air-standard Otto cycle, and all of the chemical energy of the gasoline is realized as heat. What is the rate of gasoline consumption in ℓ/h? (The heat of combustion of the gasoline is 4.81×10^7 J/kg and the density is 740 kg/m³.) If the automobile travels at an average speed of 95 km/h, what "mileage" does the automobile achieve in km/ℓ? Express your answer also in mi/gal.

26–13 Show that for the air-standard Otto cycle, we have
• $1/r^{\gamma-1} = (T_d - T_a)/(T_c - T_b)$, with the notation of

Fig. 26–9. The quantity r is the compression ratio, V_a/V_b.

26–14 Show that the expression for the efficiency of a
• Diesel engine, Eq. 26–12, can be put in the form of Eq. 26–12a.

26–15 The diagram shows the *air-standard Sargent cycle,* which consists of two adiabatic steps, *ab* and *cd,* a constant-volume heat input step *bc,* and an isobaric compression *da.* Show that the efficiency of an engine operating in this cycle is

$$e = 1 - \gamma \frac{T_2 - T_3}{T_1 - T_4}$$

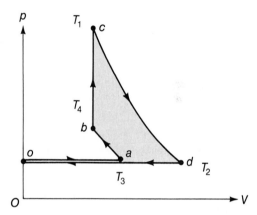

26–16 The heat discharged by an electrically driven refrigerator amounts to 250 Cal/h. The cold compartment is maintained at a temperature of $-10°$ C and the room air has a constant temperature of 27° C. The coefficient of performance for the refrigerator is 55 percent of that for a Carnot refrigerator. What is the average power consumption of the motor that drives the compressor?

Section 26–5

26–17 Determine the increase in entropy of 1 kg of water that is heated from a temperature of 0° C to 100° C.

26–18 A 3-mol sample of helium is carried by some unspecified process from an equilibrium state with a pressure of 1 atm and a temperature of 0° C to a state with a pressure of 10 atm and a temperature of 100° C. What is the change in entropy of the helium?

26–19 A 0.30 kg piece of aluminum at a temperature of 85° C is placed in 0.50 ℓ of water at a temperature of 25° C. The system is thermally isolated. Determine the increase in entropy of the system when it has reached equilibrium.

Section 26-6

26-20 By what common factor f must both the mass of the
• ice and the mass of the water be reduced in Examples 26-5 and 26-7 so that the reappearance of the initial state (ice cube in water) will have a probability of 10^{-6} of that for maintenance of the final equilibrium state? How many molecules now comprise the ice cube? Comment on your result. [*Hint:* Note that the entropy of an object is proportional to its mass.]

26-21 Determine the increase in entropy of a 1-mol sample of water when it is vaporized at 100° C and atmospheric pressure. Compare this value with the entropy increase for 1-mol samples of silver, mercury, and liquid nitrogen when these substances are also vaporized at the boiling-point temperature and at atmospheric pressure. (Refer to Table 22-4.) The close agreement of your results is an example of *the rule of Deprez and Trouton.* Comment on this rule in terms of the order-disorder aspect of entropy.

Additional Problems

26-22 A 2-mol sample of air (considered to be an ideal gas) is used in a Carnot engine that operates between temperatures of 450 K and 300 K. During the isothermal expansion step, the volume increases from 0.8 ℓ to 3.5 ℓ. (a) What amount of heat is absorbed from the hot reservoir per cycle? (b) What amount of heat is rejected to the cold reservoir per cycle? (c) What amount of work is done per cycle?

26-23 A house is heated by burning heating oil and thereby directly delivering an amount of heat Q_1 to the house. Kelvin pointed out that by using a heat engine and a heat pump in combination (as shown in the diagram) the amount of heat delivered to the house could be increased to

$$Q = Q_1 \cdot \frac{T_2}{T_1}\left(\frac{T_1 - T_3}{T_2 - T_3}\right)$$

Suppose that $T_1 = 200°$ C, $T_2 = 20°$ C, and $T_3 = 0°$ C. Show that $Q = 6.2\,Q_1$ so that by using Kelvin's scheme, 1 gal of heating oil gives the same heating effect as the direct heat from burning 6.2 gal of oil.

26-24 An average adult consumes about 90 Cal while running 1 mi. A cheeseburger from a typical fast-food service has an energy content of about 350 Cal. Is it more cost-effective to run a mile (powered by a cheeseburger) or to drive a mile in an automobile (powered by gasoline) that travels 12 miles per gallon? (Check current prices.)

26-25 In the tropics, the temperature of the ocean water at the surface averages about 30° C, whereas at a depth of 1 km the temperature is about 10° C. What is the maximum theoretical efficiency of an *ocean thermal gradient engine* that operates between these temperatures?

26-26 A solar power plant uses heat from a solar collector to make steam that drives turbine generators whose electric output is 1250 MW. The overall efficiency of the system (solar flux to electric output) is 30 percent. The amount of solar energy incident on a surface at ground level on a clear day in the southwestern part of the United States during the summer months is about 0.68 kW/m², averaged over the daylight hours. Calculate the area of the solar reflector required for the stated output (during daylight hours).

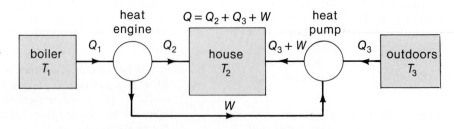

26-27 A small office building requires a heating rate of 10^5 Cal/h to maintain an inside temperature of 27° C during a winter day when the outside temperature is 5° C. The heat is provided by a heat pump that has a coefficient of performance K_h equal to 55 percent of that for an ideal Carnot system. What input power is required to drive the heat pump?

26-28 Determine the increase in entropy of the Universe when 1 kg of water slowly evaporates from an open beaker at a temperature of 20° C. (The latent heat of vaporization of water at 20° C is 585 Cal/kg.)

26-29 A Carnot engine that operates between temperatures of 250° C and 30° C delivers 2000 J of work per cycle. What is the change in entropy during one of the isothermal steps in the cycle? Sketch the cycle on a T-S diagram.

26-30 Consider two objects, one with mass m_1, specific
  • heat c_1, and temperature T_1, and the other with

mass m_2, specific heat c_2, and temperature T_2, with $T_2 < T_1$. The objects are brought into thermal contact inside an adiabatic container. The temperature of object 1 is decreased to T and the temperature of object 2 is increased to T' (consistent with energy conservation). (a) Show that the entropy of the system increases by an amount

$$\Delta S = m_1 c_1 \ln\frac{T}{T_1} + m_2 c_2 \ln\frac{T'}{T_2}$$

where

$$m_1 c_1 (T_1 - T) = m_2 c_2 (T' - T_2)$$

(b) Show that $d(\Delta S)/dT = 0$ requires $T = T'$ and therefore that this equilibrium condition is reached when ΔS is maximum. Discuss this result in terms of the probability concept of entropy. [*Hint:* Use $d(\Delta S)/dT = d(\Delta S)/dT' \cdot (dT'/dT)$ in the second term in the expression for ΔS; obtain dT'/dT from the energy condition.]

ELECTRIC CHARGE

Experimentation concerning *electromagnetic* phenomena dates from the time of ancient Greece. The accumulated wealth of information about electric and magnetic effects was finally summarized in a definitive theory of electromagnetism formulated during the 1860's by the great Scottish mathematical physicist, James Clerk Maxwell. These developments have led directly to the giant electric and electronic industries of today which furnish us with an ever-growing list of electromagnetic devices. Indeed, the application of electromagnetism to human requirements constitutes the foundation of our modern technologically oriented society. In the next several chapters we investigate the nature of electromagnetic phenomena and eventually summarize the governing principles and laws in an elegant and powerful set of equations developed by Maxwell.

27-1 EARLY HISTORY

Electrification. The first recorded observation of an electric effect was made in about 600 B.C. by a Greek mathematician and astronomer, Thales of Miletus (c640–c546 B.C.). He discovered that a piece of amber, having been rubbed, would attract small bits of straw and feathers. The word *electric* that we use to describe such phenomena is derived from the Greek word for amber (*elektron*). The next significant advance in the study of electric effects did not occur until the 16th century, when William Gilbert (1544–1603), physician to Queen Elizabeth I, investigated a variety of substances that exhibit the same electrification properties as amber. These substances he called *electrics,* materials that we now recognize to be *insulators.* Other materials, which are difficult to electrify by rubbing, he called *nonelectrics;* these materials are *conductors.*

In the early 18th century, Stephen Gray (1696–1736), an English electrical scientist, discovered that the state of electrification can be transferred from one object to another when the objects are connected by a piece of metal. Gray thereby demonstrated that the agency responsible for electrification can be caused to flow. Thus, electrification was shown to be a transitory, not an intrinsic, property of matter.

The discovery that two different types of electrification are possible was made by Charles François du Fay (1698–1739), a French chemist and physicist. Substances such as glass and mica were classified as *vitreous* materials, whereas amber, sulfur, hard rubber, and resins were classified as *resinous* materials. When rubbed with silk or cotton, the vitreous and resinous materials become oppositely electrified in the

745

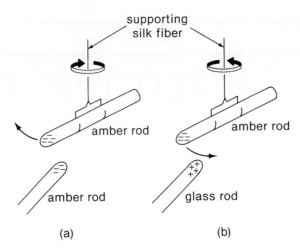

Fig. 27-1. (a) There is a repulsion between rods that have the same electrification. (b) There is attraction between rods that have different electrification.

sense that bodies with opposite electrification attract each other, but bodies with the same electrification repel each other. Figure 27-1 illustrates Du Fay's experiments. An electrified amber rod is affected oppositely by an electrified resinous material (amber) and an electrified vitreous material (glass).

The most able early American scientist was Benjamin Franklin,* whose numerous experiments and keen insight advanced the understanding of electric phenomena. The status of electrical science at the time (1748) of the publication of Franklin's *Experiments and Observations on Electricity* can be summarized as follows:

(a) All objects contain a substance called by Franklin an *electric fire* or *electricity*. The total amount of this substance in the world is constant.

(b) Rubbing together two objects, such as glass and silk, results in the transfer of electricity from one object to the other; one object gains a certain amount of electricity and the other loses an equal amount. Thus, glass and silk become oppositely electrified. In Franklin's convention, the glass becomes electrified in the positive (+) sense by *gaining* electricity, and the silk becomes equally electrified in the negative (−) sense by *losing* the electricity gained by the glass. However, when amber is rubbed with silk, the amber loses electricity and becomes electrified negatively, whereas the silk gains electricity and becomes electrified positively. According to Franklin, *positive* electricity is always transferred in such situations.

(c) Two objects electrified in the same sense, (+) and (+) or (−) and (−), exert repulsive forces on each other. Two objects electrified in the opposite sense, (+) and (−), exert attractive forces on each other.

(d) The mutual forces due to electrification are increased if the amount of electrification is increased or if the objects are brought closer to each other.

(e) Some materials (conductors), such as metals, readily conduct electricity, whereas others (nonconductors or insulators), such as wax and dry wood, do not. Most materials can be classified in one or the other of these categories. Even a metal can be electrified to some extent by rubbing with silk or cotton provided it is held apart from other objects by a nonconductor. A metal object electrified in this way will lose its electrification if any point of the object is touched by a conductor that is connected to the ground.

Franklin did not distinguish between the terms *electrified* and *charged*. We

---

*Benjamin Franklin (1706–1790), American scientist, patriot, and statesman. In the field of science he is best known for his experiments with lightning and for the invention of the lightning rod.

adopt the modern usage and henceforth refer to objects being *charged,* either positively or negatively.

Charging by Induction and Conduction. The process of charging an object by induction* is illustrated in Fig. 27–2. Two conducting metal spheres, *A* and *B*, are placed in contact and a negatively charged rod *C* is brought near to but not touching sphere *A* (Fig. 27–2a). According to Franklin's view, positive charge is attracted to the side of *A* nearer *C*, thereby leaving an equal negative charge on the opposite side of *B*. With *C* still present, *A* and *B* are separated. This leaves *A* with an excess positive charge and *B* with an equal negative charge (Fig. 27–2b). The charged rod *C* is now removed, and the charge states of the spheres remain the same (Fig. 27–2c).

The induction method illustrated in Fig. 27–2 produces charges on the spheres that are equal in magnitude and opposite in sign. How do we arrange for the two spheres to carry the *same* charge? Suppose that we have a positively charged rod; however, instead of *inducing* a charge on the sphere, suppose that the rod *C* touches sphere *A* and transfers charge to it by *conduction.* While spheres *A* and *B* (assumed to be identical) are still in contact, the rod is removed. The charge placed originally on *A* becomes equally distributed between the identical spheres. (•Can you see why?) When the spheres are separated, they carry equal charges of the same sign (+).

Induction can also be used to charge a single conducting object. A positively charged glass rod is brought near to but not touching an uncharged metal sphere. The Franklin positive charges on the sphere are repelled by the rod, leaving the side nearer the rod with an equal number of negative charges (Fig. 27–3a). A *ground wire*† is now brought into contact with the side of the sphere opposite the rod, thereby leading the positive charges to ground (Fig. 27–3b). With the charged rod still in place, the ground wire is disconnected (Fig. 27–3c). Finally, when the rod is removed, the sphere remains with a net negative charge which becomes uniformly distributed over the sphere (Fig. 27–3d).

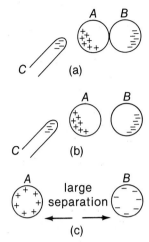

Fig. 27–2. The charging of two metal spheres by induction.

27-2 MATTER AND CHARGE

All ordinary matter is made up of atoms, and every atom, in turn, consists of a cloud of electrons that surrounds a tiny nucleus of protons and neutrons. Every atom, in its normal condition, contains exactly as many electrons as protons. Because the proton charge and the electron charge are equal in magnitude‡ but opposite in sign, and because the neutron carries no charge,§ all ordinary matter is therefore electrically neutral. The sign convention for electric charge (derived from Franklin's convention) is that a proton carries a *positive* charge ($+e$), whereas an electron carries a *negative* charge ($-e$).

The distribution of electric charge is almost always accomplished by the movement of *electrons;* the more massive positively charged nuclei remain essentially stationary in electric processes involving solids.‖ Thus, an object is given a negative charge by the addition of excess electrons or it is given a positive charge by the removal of electrons; the number of *atoms* is not changed in either case.

Fig. 27–3. The charging of a metal sphere by induction and the use of a ground wire.

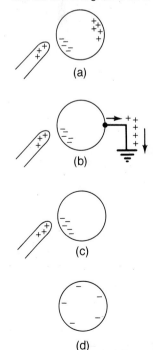

---

*Induction charging was discovered in 1753 by an English physicist, John Canton (1718–1772).

†A ground wire is a conducting metal wire, one end of which is in contact with the Earth, usually through a metal stake or pipe driven into the ground. The wire is then used to connect other objects electrically with the Earth, which acts as a virtually infinite reservoir to supply or absorb charge as needed.

‡Experiments have established this equality to an accuracy of better than 1 part in 10^{18}.

§The limit established for any neutron charge (of either sign) is 2×10^{-15} of the electron charge.

‖Positive charge often moves in electric processes in liquids and gases.

In certain types of materials (notably metals) a fraction of the atomic electrons are not bound to any particular atom but are free to move about in the material. These free electrons are able to conduct electricity, and these materials are called *conductors*. Even a weak external electric force applied to a conductor readily produces an electron flow (or *current*). (•Can you now see why the charges are distributed uniformly over the surfaces of the metal spheres in Figs. 27–2c and 27–3d?)

In a nonconducting material (an *insulator*) essentially all of the electrons are attached to atoms—that is, there are almost no free (conduction) electrons in insulators. (•Can you explain why the induction methods used to charge *conducting* objects, as shown in Figs. 27–2 and 27–3, cannot be used with *nonconducting* objects?)

When two different nonconducting materials, such as glass and silk, are rubbed together,* electrons are transferred from the surface of one material to the surface of the other. Thus, the material that gains electrons becomes negatively charged, and the material that loses electrons becomes positively charged by the same amount. The charged sites on the two insulators remain essentially fixed until the charge is neutralized or slowly leaks away (because the material is not a perfect insulator).

If you compare this qualitative summary of some of the features of modern electrical science with that for Franklin's time made in the preceding section, the similarity is clear. The primary exception is in regard to the sign of the charge that moves. Franklin chose as *positive* the charge that resides on glass when rubbed with silk, and he chose to consider charging effects to be the result of the movement of *positive* charge. It is now known that the sign of the charge on glass is the same as that of the proton charge. It is also known that charging effects are due to the movement of electrons (not protons). Nevertheless, the convention is to adhere to Franklin's choice for charge signs and for the direction of charge movement. Consequently, we consider electric current to flow in the direction in which positive charge *would* move, even though current is actually due to the motion of electrons in the opposite direction!

Charge Quantization.   In 1897, J. J. Thomson (1856–1940), an English physicist, demonstrated that there are particles (called electrons) that are common to all atoms and have a unique ratio of charge to mass. In 1909, R. A. Millikan† showed that all electrons, from whatever source, carry the same charge (which we designate $-e$). This is one example of the remarkable property of discreteness (or *quantization*) that exists in Nature. Thus, when an object becomes charged as the result of the addition or removal of some number of electrons, the object must carry a charge that is an integer multiple of e. In most situations, the number of electrons transferred is so large that a change of one electron, more or less, is undetectable. For this reason we often consider the variation of charge to be continuous.

It is now known that there exists in Nature, in addition to the electron and the proton, a host of additional particles. More than a hundred of these *elementary particles* are known. Without exception, the charge observed for each of these particles is either $+e$ or $-e$, or the particle carries zero charge. Thus, charge appears to obey a universal quantization rule, with only charges $+e$, $-e$, and 0 occurring in Nature.

One further point regarding charge quantization should be mentioned. An ex-

---

*Actually, the transfer of charge between two different nonconducting materials comes about because of a *contact potential difference*. Thus, charging occurs without any rubbing; the rubbing serves only to cause surface contact over a larger area. This effect is called *triboelectrification*.

†Robert A. Millikan (1868–1953), American physicist, was awarded the 1923 Nobel Prize in physics for his experiments with electrons. J. J. Thomson received the 1906 Prize for his pioneering work.

tremely successful model for the substructure of a large group of the known elementary particles involves entities called *quarks*. The various quarks that are required for a complete theory are postulated to carry charges of $\pm\frac{1}{3}e$ and $\pm\frac{2}{3}e$. The combinations in which quarks are present in the observed elementary particles are such that the net charge is always $+e$, $-e$, or 0. Although there is strong evidence that quarks exist *inside* elementary particles, a *free* quark has never been detected. It is even possible that a quark cannot exist except in close association with one or two other quarks.

Charge Conservation. An object can be given an electric charge by the addition or the removal of electrons. The charge that is *gained* by one object is *lost* by another. It is one of the fundamental conservation principles of Nature that *the total amount of charge in an isolated system remains constant.*

Under certain conditions and subject to certain restrictions, it is possible to *create* a pair of electric charges, one of which is negative ($-e$) and one of which is positive ($+e$). If the negative charge is an electron, the positive charge is a *positron* (a particle not found in ordinary matter, but which is identical to the electron except that it carries a positive charge*). Although such creation processes are possible, they do not violate the law of charge conservation because the total charge of the created pair is *zero*. Thus, the net charge of the system that contains the pair is not altered.

No positive or negative charge has ever been observed to be created or destroyed *by itself*. These processes *always* involve equal positive and negative charges so that the charge conservation principle is never violated.

27-3 COULOMB'S LAW

In 1785, the French physicist Charles Augustin de Coulomb (1736–1806) made the first direct measurements of the forces between charged objects. Coulomb performed these experiments with a torsion balance of his own design. The apparatus (Fig. 27–4) was similar to that used by Cavendish in his determination of the universal gravitation constant (see Section 5–3). Coulomb's balance was not as sophisticated as the delicate instrument used by Cavendish 12 years later because the forces measured by Coulomb were much larger than those in the gravitational experiments.

As shown in Fig. 27–4, *a* and *b* are small metal spheres that can be charged so that the mutual electrostatic force of attraction or repulsion can be measured for various separations and amounts of charge. Sphere *a* is connected to a counterweight by an insulating rod that is supported by a long fiber from an adjustable suspension head. The rod can pivot about the fiber axis, and the torsional twist of the fiber produces a restoring torque that opposes the torque due to the electrostatic force between *a* and *b*. Thus, an equilibrium condition is reached with the rod at some angle θ with respect to its neutral position (see the inset in Fig. 27–4). The torsional constant of the fiber is measured in a separate experiment involving the oscillation

---

*Electrons and positrons have opposite electric *and* magnetic properties.

suspension head

fiber

Fig. 27–4. Diagram of Coulomb's torsion balance as it appeared in his 1785 paper. (New York Public Library.) The inset shows the top view of the fiber-supported rod.

frequency of the system, as in the Cavendish experiment, so the electrostatic force can be computed in terms of the deflection angle θ (see Section 16–4).

The separation between the charged spheres can be changed by turning the suspension head or by moving the support rod of sphere b. Some changes in the amounts of charge on a and b can readily be made. For example, suppose that a measurement is made of the force between a and b for some arbitrary charge states of the spheres. Sphere b is now removed from the apparatus and brought into contact with an identical uncharged metal sphere. The two spheres share the charge equally, so when b is returned to the apparatus it carries exactly half the original charge. The effect on the force of reducing the charge on b by one half can now be determined.

On the basis of measurements of this type, Coulomb was able to formulate the law of electrostatic force, now called *Coulomb's law:*

The mutual electrostatic force between two point charges is directed along the line between the charges; the force is directly proportional to the product of the charges and is inversely proportional to the square of the distance between them. The force is repulsive if the charges have the same sign and is attractive if the charges have opposite signs.

In equation form, Coulomb's law is

$$\mathbf{F}_{12} = K \frac{q_1 q_2}{r_{21}^2} \hat{\mathbf{u}}_r \qquad\qquad \textbf{(27–1)}$$

Fig. 27–5. The geometry for Coulomb's law.

where $\mathbf{F}_{12}$ is the force exerted *on* charge q_1 *by* charge q_2, and where $\hat{\mathbf{u}}_r$ is a unit vector directed from q_2 toward q_1, as shown in Fig. 27–5. The proportionality factor K is the *electrostatic force constant.* The force exerted on q_2 by q_1 is directed opposite to $\mathbf{F}_{12}$; thus,

$$\mathbf{F}_{21} = -\mathbf{F}_{12}$$

which is consistent with Newton's third law.

The SI unit of charge is the *coulomb* (C), which is the amount of charge carried by 6.2414×10^{18} electrons.* That is, the electron charge has the magnitude

$$e = 1.6022 \times 10^{-19} \text{ C}$$

With the unit of charge defined, the electrostatic force constant K can be determined by experiment. When the charges are in vacuum (or air, for all practical purposes), we have

$$K = 8.98755 \times 10^9 \text{ N} \cdot \text{m}^2/\text{C}^2$$

A related constant is ϵ_0, called the *permittivity of free space:*

$$\epsilon_0 = \frac{1}{4\pi K} = 8.85419 \times 10^{-12} \text{ C}^2/\text{N} \cdot \text{m}^2$$

---

*The definition of the coulomb is properly made in terms of electric current; this definition is given in Section 34–4.

In terms of ϵ_0, Coulomb's law is written as

$$\mathbf{F}_{12} = \frac{1}{4\pi\epsilon_0} \frac{q_1 q_2}{r_{21}^2} \hat{\mathbf{u}}_r \qquad (27\text{-}2)$$

Apart from the substitution $K = 1/4\pi\epsilon_0$, the other quantities are the same as in Eq. 27-1.

It is verified by experiment that electrostatic forces have the same linear character as the forces we have already studied; that is, electrostatic forces obey the principle of linear superposition (see Section 14-3). Thus, when more than two point charges are present, the net force on any one charge is the linear vector sum of the electrostatic forces individually exerted by each of the other charges. If three charges are present, the force on q_1 is

$$\mathbf{F}_1 = \mathbf{F}_{12} + \mathbf{F}_{13} = K\frac{q_1 q_2}{r_{21}^2} \hat{\mathbf{u}}_2 + K\frac{q_1 q_3}{r_{31}^2} \hat{\mathbf{u}}_3$$

where $\hat{\mathbf{u}}_2 = \mathbf{r}_{21}/|\mathbf{r}_{21}|$ and $\hat{\mathbf{u}}_3 = \mathbf{r}_{31}/|\mathbf{r}_{31}|$.

Experiments have demonstrated that the electrostatic force on an object combines with forces of a nonelectric nature by simple vectorial addition.

Example 27-1

(a) Determine the magnitude of the electrostatic force F_e between two electrons separated by a distance of 10^{-10} m (1 Å), a typical atomic dimension. (b) Determine the ratio, F_e/F_g, of the electrostatic and gravitational forces that act between two stationary electrons. (The electron mass is 9.11×10^{-31} kg.)

Solution:

(a) Using Eq. 27-1, we have for the magnitude of the force

$$F_e = K\frac{e^2}{r^2} = \frac{(8.99 \times 10^9 \text{ N} \cdot \text{m}^2/\text{C}^2)(1.602 \times 10^{-19} \text{ C})^2}{(10^{-10} \text{ m})^2}$$

$$= 2.31 \times 10^{-8} \text{ N}$$

(b) The two forces are

$$F_e = K\frac{e^2}{r^2} \quad \text{and} \quad F_g = G\frac{m_e^2}{r^2}$$

Therefore,

$$\frac{F_e}{F_g} = \frac{K}{G}\left(\frac{e}{m_e}\right)^2 = \frac{(8.99 \times 10^9 \text{ N} \cdot \text{m}^2/\text{C}^2)}{(6.67 \times 10^{-11} \text{ N} \cdot \text{m}^2/\text{kg}^2)}\left(\frac{1.602 \times 10^{-19} \text{ C}}{9.11 \times 10^{-31} \text{ kg}}\right)^2$$

$$= 4.17 \times 10^{42}$$

Thus, the electrostatic force between electrons is enormously greater than the gravitational force, and the latter can be neglected in all calculations involving electrostatic forces between atomic constituents.

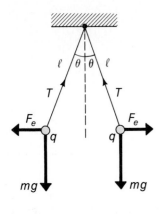

Example 27–2

Each of two identical small Styrofoam balls ($m = 5$ g) is charged uniformly to $+1.0\,\mu$C. The balls are suspended from a fixed point by nonconducting threads, each with a length $\ell = 1.0$ m and with negligible mass. At equilibrium, what is the angle 2θ between the threads?

Solution:

From the free-body diagram for either charge, we have

$$F_e = T \sin \theta \quad \text{and} \quad mg = T \cos \theta$$

Solving the second equation for T, substituting in the first equation, and using the result in Coulomb's law, we obtain

$$F_e = mg \tan \theta = K \frac{q^2}{(2\ell \sin \theta)^2}$$

from which

$$\tan \theta \sin^2 \theta = \frac{Kq^2}{4\,mg\,\ell^2}$$

$$= \frac{(8.99 \times 10^9 \text{ N} \cdot \text{m}^2/\text{C}^2)(1.0 \times 10^{-6} \text{ C})^2}{4(5 \times 10^{-3} \text{ kg})(9.80 \text{ m/s}^2)(1.0 \text{ m})^2}$$

$$= 0.0459$$

This equation for θ can be solved graphically or by a successive approximation procedure (see Problem 27–13). The result is $\theta = 0.358 = 20.5°$. Thus, the angle between the threads is $2\theta = 41.0°$.

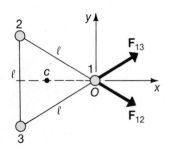

Example 27–3

Three point charges, each carrying a charge of $+1.5\,\mu$C, are located at the vertices of an equilateral triangle with sides $\ell = 30$ cm. Calculate the force exerted on one of the charges (for example, number 1) by the other two.

Solution:

Using the superposition principle, we can write for the force on charge 1,

$$\mathbf{F}_1 = \mathbf{F}_{12} + \mathbf{F}_{13} = (F_{12x} + F_{13x})\hat{\mathbf{i}} + (F_{12y} + F_{13y})\hat{\mathbf{j}}$$

From the geometry it is evident that

$$F_{12x} = +F_{13x} \quad \text{and} \quad F_{12y} = -F_{13y}$$

Hence, $$\mathbf{F}_1 = 2F_{12} \cos 30° \hat{\mathbf{i}} = \sqrt{3} F_{12}\hat{\mathbf{i}}$$

Now, $$F_{12} = K\frac{q^2}{\ell^2} = \frac{(8.99 \times 10^9 \text{ N} \cdot \text{m}^2/\text{C}^2)(1.5 \times 10^{-6} \text{ C})^2}{(0.30 \text{ m})^2}$$

$$= 0.225 \text{ N}$$

Thus, we have $\mathbf{F}_1 = 0.390\hat{\mathbf{i}}$ N

That is, the force on any charge is 0.390 N directed away from the center of the triangle at C.

PROBLEMS

27–1 Two pointlike objects carry charges of $+2.5\,\mu C$ and $-1.5\,\mu C$ and are separated by a distance of 20 cm. Determine the magnitude and the direction of the mutual electrostatic force between the objects. Suppose that each object is a conducting sphere with a diameter of 5 cm. What qualitative effect on the force does this change make if the separation distance (now measured between centers) remains the same? Place an upper limit on the new value of the force.

27–2 Repeat Problem 27–1 with the separation distance increased to 2.0 m. Comment on the importance of the size of the charged objects.

27–3 Two pointlike objects carry charges of $q_1 = +4.0\,\mu C$ and $q_2 = -3.0\,\mu C$ and are separated by a fixed distance of 50 cm. (a) Determine the point (other than infinity) at which a pointlike positive charge will experience zero net electrostatic force due to the two fixed charges. (b) Repeat (a) for a pointlike negative charge. (c) Comment on these results.

27–4 Repeat Problem 27–3 if q_2 is changed to $+3.0\,\mu C$.

27–5 Four pointlike charges, $q_1 = +8.0\,\mu C$, $q_2 = -4.0\,\mu C$, $q_3 = +4.0\,\mu C$, and $q_4 = -2.0\,\mu C$, are located in sequence at the corners of a square with sides 25 cm in length. Find the electrostatic force on q_2.

27–6 Four arbitrary pointlike charges are located at the corners of a rigid square plate. Prove that the net torque about the center of the plate is zero.

27–7 Three identical small Styrofoam balls ($m = 2$ g) are suspended from a fixed point by three nonconducting threads, each with a length of 50 cm and with negligible mass. At equilibrium the three balls form an equilateral triangle with sides of 30 cm. What is the common charge q that is carried by each ball?

27–8 In a simplified classical model of the hydrogen atom (the *Bohr model*) the electron orbits around the nuclear proton in a circular path with a radius of 0.53 Å (5.3×10^{-11} m). Consider the mass of the proton to be much greater than the mass of the electron and calculate the orbital period of the electron. Is a nonrelativistic calculation justified here?

27–9 Suppose that 1 g of hydrogen is separated into electrons and protons. Suppose also that the protons are placed at the Earth's north pole and that the electrons are placed at the south pole. What is the resulting compressional force on the Earth?

27–10 A particle with mass m and carrying a charge $+q$ is
• located midway between two fixed particles that carry equal charges $+q$ and are separated by a distance 2ℓ. Determine the frequency of oscillation of the middle charge when it is slightly displaced along the line joining the fixed charges. (The oscillating charge is restricted to move along the line joining the fixed charges.)

27–11 Two equal point charges separated by a distance 2ℓ
• are placed symmetrically on opposite sides of a large uniformly charged nonconducting plane, as shown in the diagram. Determine the locations on the plane where the net electrostatic force on a charge Q due to the two charges is maximum.

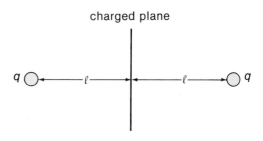

charged plane

27–12 A total charge q is divided equally among $2n$
• particles that are arranged with equal spacing around the perimeter of a circle with radius R. A point charge q' is placed a distance d from the plane of the circle along a perpendicular axis through the center, as shown in the diagram on the next page. We arrive at the idealization of a uniformly charged ring by allowing $n \to \infty$ while maintaining q fixed.

Show that the force on the charge q' is

$$F = K \frac{qq'}{(R^2 + d^2)^{3/2}}$$

regardless of whether the charges are discrete or constitute a uniformly charged ring.

2n charges, each
with charge $q/2n$

27-13 Refer to Example 27-2. Obtain the value for θ by using an approximation procedure to solve the equation, $\tan\theta \sin^2\theta = 0.0459$. Find a starting value θ_1 by assuming $\tan\theta \cong \sin\theta$, so that $\sin^3\theta_1 = 0.0459$. Then write

$$\sin\theta_{n+1} = (0.0459/\tan\theta_n)^{1/2}$$

Use the value of θ_1 for θ_n and find θ_2; use the value of θ_2 to find θ_3; and so forth. Continue listing each value of θ_n, until a value is found that yields the same value for θ_{n+1}. You must be careful when setting up a successive approximation procedure. Try to solve the equation by writing $\tan\theta_{n+1} = 0.0459/\sin^2\theta_n$. List the values of θ_n obtained using this procedure. What do you conclude? Finally, plot $\tan\theta$ and $0.0459/\sin^2\theta$ versus θ. Show that the intersection corresponds to the value of θ obtained by numerical calculation.

THE ELECTRIC FIELD

In Chapters 8 and 14 we discussed the field concept associated with the gravitational force. We now extend these considerations to the electromagnetic force and begin by examining the electrostatic case.

When an electric charge is placed at some point in space, this establishes everywhere a state of electric stress which is called an *electric field*. Any other charge that is placed at a certain point will interact with the field existing at that point. This interaction is identified with the electrostatic force that the first charge exerts on the second charge.

We restrict considerations of the electric field produced by a charge to a description in a coordinate frame at rest with respect to the charge. We ignore questions relating to the time required for the field to propagate to all points in space; indeed, this is the essential meaning of the electro*static* field.

28-1 ELECTRIC FIELD STRENGTH

The electric field produced by a stationary charge* Q can be investigated by measuring the force on a *test charge* q_0 placed at various points in space. (We imagine that the dimensions of q_0 are sufficiently small that the field is sampled by the test charge essentially at a point.) The *electric field intensity* or the *electric field strength* $\mathbf{E} = \mathbf{E}(\mathbf{r})$ at the point $\mathbf{r}$ is defined to be†

$$\mathbf{E} = \operatorname*{Lim}_{q_0 \to 0} \frac{\mathbf{F}_E}{q_0} \qquad (28\text{-}1)$$

where $\mathbf{F}_E = \mathbf{F}_E(\mathbf{r})$ is the force on the charge q_0 at the point $\mathbf{r}$. For convenience we construct a coordinate system whose origin coincides with the position of the charge Q. According to Coulomb's law (Eq. 27-2), the force on q_0 due to the charge Q is

$$\mathbf{F}_E = \frac{1}{4\pi\epsilon_0} \frac{Qq_0}{r^2} \hat{\mathbf{u}}_r \qquad (28\text{-}2)$$

---

*We have by now exhausted the supply of symbols to identify uniquely the various quantities we discuss. Consequently, in this and in the following chapters new meaning is given to some symbols already used. For example, the symbol Q (which represented *heat* in Chapters 22 through 26) is now used for *charge*. Whenever there is a chance for confusion, an explicit note is made.

†We take the limit $q_0 \to 0$ so that the test charge q_0 will not influence in any way the charges that are responsible for the field $\mathbf{E}(\mathbf{r})$.

where $\hat{\mathbf{u}}_r = \mathbf{r}/r$ (see Fig. 27–5). Thus, for the electric field strength at a point $\mathbf{r}$ due to a charge Q at the origin, we have, combining Eqs. 28–1 and 28–2,

$$\mathbf{E} = \frac{1}{4\pi\epsilon_0}\frac{Q}{r^2}\hat{\mathbf{u}}_r \qquad (28\text{-}3)$$

Note that if Q is positive, $\mathbf{E}$ is directed radially outward, whereas if Q is negative, $\mathbf{E}$ is directed radially inward. The direction of $\mathbf{E}$ at a particular point is always the same as the direction of the force on a *positive* test charge placed at that point. Because we have defined the field in the limit $q_0 \to 0$, our space consists only of the charge Q at the origin and the electric field that is everywhere defined by Eq. 28–3. If a charge q is now placed at the point $\mathbf{r}$, the force on q is

$$\mathbf{F}(\mathbf{r}) = q\mathbf{E}(\mathbf{r}) \qquad (28\text{-}4)$$

The direction of $\mathbf{F}(\mathbf{r})$ with respect to $\mathbf{E}(\mathbf{r})$ is determined by the sign of q.

It is evident from Eq. 28–4 (or Eq. 28–1) that the units of electric field strength are *newtons per coulomb*, N/C. In Section 29–1 we introduce other equivalent units for $\mathbf{E}$ (namely, *volts per meter,* V/m).

Because the superposition principle is valid for electrostatic forces (see Section 27–3), the electric field strength, $\mathbf{E} = \mathbf{E}(\mathbf{r})$, at the point $P(\mathbf{r})$ produced by the simultaneous presence of charges Q_i is (see Fig. 28–1)

$$\mathbf{E} = \mathbf{E}_1 + \mathbf{E}_2 + \cdots + \mathbf{E}_i + \cdots$$

$$= \frac{1}{4\pi\epsilon_0}\left[\frac{Q_1}{r_1^2}\hat{\mathbf{u}}_1 + \frac{Q_2}{r_2^2}\hat{\mathbf{u}}_2 + \cdots + \frac{Q_i}{r_i^2}\hat{\mathbf{u}}_i + \cdots\right]$$

or

$$\mathbf{E} = \sum_i \mathbf{E}_i = \frac{1}{4\pi\epsilon_0}\sum_i \frac{Q_i}{r_i^2}\hat{\mathbf{u}}_i \qquad (28\text{-}5)$$

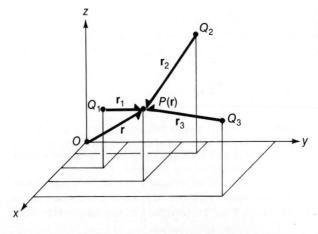

Fig. 28–1. The electric field at $P(\mathbf{r})$ is the superposition of the fields due to the individual charges Q_i.

where r_i is the distance from Q_i to the point $P(\mathbf{r})$, and where $\hat{\mathbf{u}}_i = \mathbf{r}_i/r_i$. (•Can you see how to construct the integral that yields $\mathbf{E}(\mathbf{r})$ for a continuous distribution of charge?)

Example 28–1

Two positive point charges, $Q_A = Q$ and $Q_B = 3Q$, are separated by a distance $\overline{AB} = 2\,\ell$. Determine the electric field strength, $\mathbf{E} = \mathbf{E}(Q_A) + \mathbf{E}(Q_B)$, along the bisector of the line $\overline{AB}$.

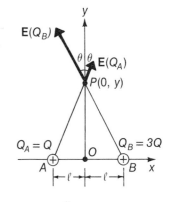

Solution:

Locate the coordinate axes as shown in the diagram so that the y-axis is the bisector of the line $\overline{AB}$. Then, the x-component of the field strength at the point $P(0, y)$ due to the charge Q_A is

$$E_x(Q_A) = \frac{1}{4\pi\epsilon_0} \frac{Q}{(\ell^2 + y^2)} \sin\theta = \frac{1}{4\pi\epsilon_0} \frac{Q\ell}{(\ell^2 + y^2)^{3/2}}$$

Similarly,

$$E_x(Q_B) = -\frac{3}{4\pi\epsilon_0} \frac{Q\ell}{(\ell^2 + y^2)^{3/2}}$$

Thus, the x-component of the total field strength is

$$E_x = E_x(Q_A) + E_x(Q_B) = -\frac{2}{4\pi\epsilon_0} \frac{Q\ell}{(\ell^2 + y^2)^{3/2}}$$

For the y-component, we find

$$E_y = E_y(Q_A) + E_y(Q_B)$$

$$= \frac{1}{4\pi\epsilon_0} \frac{Q}{(\ell^2 + y^2)} \cos\theta + \frac{3}{4\pi\epsilon_0} \frac{Q}{(\ell^2 + y^2)} \cos\theta$$

$$= \frac{4}{4\pi\epsilon_0} \frac{Qy}{(\ell^2 + y^2)^{3/2}}$$

Now, $\mathbf{E}(0, y) = E_x\hat{\mathbf{i}} + E_y\hat{\mathbf{j}}$, so that

$$\mathbf{E}(0, y) = \frac{1}{4\pi\epsilon_0} \frac{Q}{(\ell^2 + y^2)^{3/2}} (-2\ell\hat{\mathbf{i}} + 4y\hat{\mathbf{j}})$$

What is the behavior of $\mathbf{E}(0, y)$ when $y^2 \gg \ell^2$? Then, the x-component becomes negligible in comparison with the y-component; also, $(\ell^2 + y^2)^{3/2} \longrightarrow (y^2)^{3/2} = y^3$. Thus,

$$\mathbf{E}(0, y) = \frac{1}{4\pi\epsilon_0} \frac{4Q}{y^2} \hat{\mathbf{j}}, \qquad y^2 \gg \ell^2$$

This is just the result we would obtain for a *single* charge, $Q_A + Q_B = 4Q$, located at the origin.

In Example 28–1 we found that the field strength at a large distance from the charge system reduces to that for a single charge equal to the sum of the individual charges. This result is, in fact, an example of a broader generalization:

If a number of charges are located within a region of space, then at distances r *that are large compared with the dimensions of the region, the electric field strength* **E(r)** *is nearly equal to the distant solution,*

$$\frac{1}{4\pi\epsilon_0} \frac{\hat{\mathbf{u}}_r}{r^2} \sum_i Q_i$$

where **r** *is measured from any convenient origin within the region, and where the sum is an* algebraic *sum over all of the charges in the region.*

A similar result was obtained for the gravitational case in Section 14–5. Because gravitational forces are always attractive, the corresponding sum over the masses is never zero; however, there are important electric cases for which $\Sigma Q = 0$. When $\Sigma Q \neq 0$, the above expression for **E(r)** is referred to as the *monopole* (or *singlet*) approximation, and ΣQ is called the *monopole strength* of the charge distribution.

If a particular charge distribution has $\Sigma Q = 0$, its monopole strength vanishes, but the distant field strength is not necessarily zero. At large distances, the field strength can have no term with a $1/r^2$ dependence because such terms are always associated with ΣQ; however, there may be terms in the expression for the field strength that depend on $1/r^3$, $1/r^4$, and so forth. We next consider one such case, namely, the electric dipole.

The Electric Dipole. An important type of charge distribution—the *electric dipole*—consists of two charges, $+Q$ and $-Q$, with equal magnitudes and opposite signs (so that $\Sigma Q = 0$), separated by a fixed distance* 2ℓ. We are interested in the field strength due to such a dipole at distances $r \gg \ell$. Because a dipole has no monopole strength, the expression for the field strength contains no $1/r^2$ term. Therefore, in any expansion involving powers of ℓ/r, we must be careful to carry all of the important terms.

The geometry of a dipole is shown in Fig. 28–2. For the *x*-component of the field

---

*We write 2ℓ instead of ℓ to make easier some of the mathematical manipulations carried out here. Thus, we write the dipole strength (Eq. 28–7) as $p = 2\ell Q$. On other occasions we simply use ℓ for the charge separation and write $p = \ell Q$.

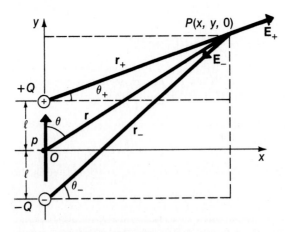

Fig. 28–2. Geometry for determining **E(r)** for a dipole with strength $p = 2\ell Q$. There is cylindrical symmetry about the axis of the dipole (the *y*-axis). Notice that $r_+^2 = x^2 + (y - \ell)^2$ and $r_-^2 = x^2 + (y + \ell)^2$.

strength at the point $P(x, y, 0)$, we have

$$E_x = \frac{Q}{4\pi\epsilon_0} \left(\frac{\cos\theta_+}{r_+^2} - \frac{\cos\theta_-}{r_-^2} \right)$$

For convenience, we write

$$\frac{4\pi\epsilon_0}{Q} E_x = \frac{x}{[x^2 + (y - \ell)^2]^{3/2}} - \frac{x}{[x^2 + (y + \ell)^2]^{3/2}}$$

Similarly, for the y-component, we obtain

$$E_y = \frac{Q}{4\pi\epsilon_0} \left(\frac{\sin\theta_+}{r_+^2} - \frac{\sin\theta_-}{r_-^2} \right)$$

and
$$\frac{4\pi\epsilon_0}{Q} E_y = \frac{y - \ell}{[x^2 + (y - \ell)^2]^{3/2}} - \frac{y + \ell}{[x^2 + (y + \ell)^2]^{3/2}}$$

Now, $r^2 = x^2 + y^2$, so we have

$$[x^2 + (y \pm \ell)^2]^{-3/2} = [x^2 + y^2 + \ell^2 \pm 2\ell y]^{-3/2}$$

$$= r^{-3} \left[1 + \frac{\ell^2 \pm 2\ell y}{r^2} \right]^{-3/2}$$

According to the table inside the rear cover, we can write

$$(1 + \xi)^{-3/2} = 1 - \tfrac{3}{2}\xi + \tfrac{15}{8}\xi^2 - \cdots$$

Then, expanding $4\pi\epsilon_0 E_x/Q$ and retaining terms with negative powers up to $1/r^5$, there results

$$\frac{4\pi\epsilon_0}{Q} E_x = \frac{x}{r^3} \left(1 + \frac{3y\ell}{r^2} - \frac{3\ell^2}{2r^2} + \cdots - 1 + \frac{3y\ell}{r^2} + \frac{3\ell^2}{2r^2} + \cdots \right)$$

$$= \frac{6xy\ell}{r^5} + \cdots$$

Similarly, for $4\pi\epsilon_0 E_y/Q$ we find, to order $1/r^5$,

$$\frac{4\pi\epsilon_0}{Q} E_y = \frac{y - \ell}{r^3} \left(1 + \frac{3y\ell}{r^2} - \frac{3\ell^2}{2r^2} + \cdots \right) - \frac{y + \ell}{r^3} \left(1 - \frac{3y\ell}{r^2} - \frac{3\ell^2}{2r^2} + \cdots \right)$$

$$= -\frac{2\ell}{r^3} + \frac{6y^2\ell}{r^5} + \frac{3\ell^3}{r^5} + \cdots$$

Therefore, the vector that describes the total field when $r^2 = x^2 + y^2 \gg \ell^2$ is

$$\mathbf{E} = \frac{Q}{4\pi\epsilon_0} \left(-\frac{2\ell}{r^3}\hat{\mathbf{j}} + \frac{6xy\ell}{r^5}\hat{\mathbf{i}} + \frac{6y^2\ell}{r^5}\hat{\mathbf{j}} + \frac{3\ell^3}{r^5}\hat{\mathbf{j}} + \cdots \right)$$

Fig. 28–3. Two different electric quadrupole configurations. In each case both the monopole strength ΣQ and the net dipole moment, $\mathbf{p} + (-\mathbf{p})$, vanish. In (b), the net dipole moment can be considered to be $\mathbf{p} + (-\mathbf{p})$ or $\mathbf{p'} + (-\mathbf{p'})$; in either case, the sum is zero. A charge distribution may have a quadrupole strength even though the monopole strength and the net dipole moment do not vanish.

Notice that the fourth term is always small compared with the first term, and so may be neglected. However, the second and third terms, although they also depend on $1/r^5$, may be comparable with the first term when y or x and y are large; these terms may not be neglected. Thus, the approximate expression for the field strength becomes

$$\mathbf{E} = \frac{Q}{4\pi\epsilon_0}\left(-\frac{2\ell}{r^3}\hat{\mathbf{j}} + \frac{6y\ell}{r^5}(x\hat{\mathbf{i}} + y\hat{\mathbf{j}})\right) \tag{28-6}$$

It is convenient to introduce the quantity $\mathbf{p}$, called the *electric dipole moment,* which is defined to be

$$\mathbf{p} = 2\ell Q\hat{\mathbf{j}} \tag{28-7}$$

when the y-axis has the direction from $-Q$ to $+Q$. Noting in Fig. 28-2 that $x\hat{\mathbf{i}} + y\hat{\mathbf{j}} = \mathbf{r}$ and that $6y\ell Q = 3pr\cos\theta = 3\mathbf{p}\cdot\mathbf{r}$, we can express the field strength in the compact form,

$$\mathbf{E}(\mathbf{r}) = \frac{1}{4\pi\epsilon_0}\left(-\frac{\mathbf{p}}{r^3} + \frac{3(\mathbf{p}\cdot\mathbf{r})\mathbf{r}}{r^5}\right), \qquad r \gg \ell \tag{28-8}$$

This expression, with $\mathbf{r} = x\hat{\mathbf{i}} + y\hat{\mathbf{j}} + z\hat{\mathbf{k}}$, is valid in three dimensions (because $\mathbf{p}$ is along the y-axis so there is symmetry about this axis).

Notice that, as expected, Eq. 28-8 contains no monopole ($1/r^2$) term. The surviving terms in the distant solution all are of order $1/r^3$.

It is possible to construct charge distributions for which both the monopole strength and the dipole strength vanish. Figure 28–3 shows two particularly simple *quadrupole distributions* that have this property. For either of these quadrupoles, $\mathbf{E}(\mathbf{r})$ varies as $1/r^4$ when $r \gg \ell$; see Problem 28–5.

28-2 LINES OF FORCE

A convenient way to visualize the electric field produced by a charge or a charge distribution is to construct a map of the *field lines* or *lines of force* of the field. We consider here only the electrostatic case (charges at rest). Lines of force are continuous curves, the tangent to which at any point gives the direction of $\mathbf{E}(\mathbf{r})$ at that point. When the field lines are drawn to a regular scale, the density of the lines in a particular region is proportional to the magnitude of $\mathbf{E}(\mathbf{r})$ in that region. Where the lines are closely spaced the field is strong; where the lines are sparse the field is weak.

Figure 28–4 shows the lines of force associated with a positive point charge. The lines radiate from the charge and spread out uniformly in three dimensions. Field lines originate on positive charges and terminate on negative charges. In the electrostatic case, a field line never closes on itself; every field line connects a positive charge with a negative charge. (Sometimes we imagine one of these charges to be at infinity.) Field lines never intersect, and they behave as if there were a tendency for the lines to repel one another.

The field lines for an electric dipole are illustrated in Fig. 28–5. Notice that all of

Fig. 28–4. The lines of force for a positive point charge. The lines radiate uniformly outward from the charge.

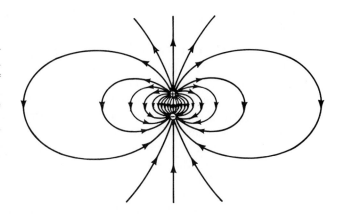

Fig. 28–5. The field lines for an electric dipole. For distances $r \gg l$, the pattern of field lines corresponds to Eq. 28–8. (See Fig. 29–6 for the field of a *point* dipole; in such a case, Eq. 28–8 is valid for all r.) The lines that leave the top of the diagram eventually curve back and join the appropriate lines that enter the bottom of the diagram.

the field lines originate on the positive charge and terminate on the negative charge. No lines proceed indefinitely far away without returning (this would imply a nonzero monopole strength).

Figure 28–6 illustrates the field produced by a positive charge $+3Q$ in combination with a negative charge $-Q$. The field is symmetric about the line connecting the charges, so the three-dimensional aspect of the field can be imagined by considering the field lines rotated about this symmetry axis. The individual field strengths, $\mathbf{E}_+$ and $\mathbf{E}_-$, due to the two charges are shown at the point P. The vector sum, $\mathbf{E} = \mathbf{E}_+ + \mathbf{E}_-$, is tangent to the field line that passes through P. The number of field lines that originate on the charge $+3Q$ is three times the number that terminate on the charge $-Q$. Thus, two-thirds of the lines originating on the charge $+3Q$ terminate at infinity where we imagine there exists an amount of charge $-2Q$ spread out uniformly; in this way we maintain an electrically neutral Universe. When viewed from a distance that is large compared with the separation of the charges, the outward-spreading lines become uniformly divergent; this tendency is evident in Fig. 28–6. (•Is this distant behavior consistent with the fact that the two charges have a monopole strength of $+2Q$?)

Fig. 28–6. Field lines for a charge distribution consisting of a positive charge $+3Q$ and a negative charge $-Q$. (See also Example 28–5.)

Fig. 28–7. Field lines for two equal positive point charges.

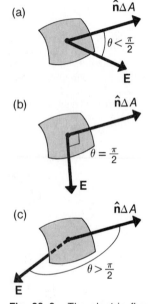

Fig. 28–8. The electric flux through an area ΔA. The angle between $\hat{\mathbf{n}}$ and $\mathbf{E}$ through ΔA is θ. (a) When $\theta < \pi/2$, the flux of $\mathbf{E}$ through ΔA is positive. (b) When $\theta = \pi/2$, the flux is zero. (c) When $\theta > \pi/2$, the flux is negative.

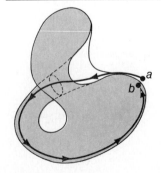

Figure 28–7 shows the lines of force associated with two equal positive point charges. In this case the tendency for the field lines to repel one another is quite evident. Note also that all of the lines diverge toward infinity and, at large distances, point directly away from the midpoint between the charges.

One of the important field configurations is the *uniform field*. In this situation the electric field strength is constant in magnitude and direction; for example, $\mathbf{E}(\mathbf{r}) = E_0\hat{\mathbf{i}}$, in which case the field lines are uniformly distributed parallel straight lines, all in the $+x$-direction. We can imagine such a field to exist between two parallel uniformly charged infinite planes, one carrying positive charge located at $x = -\infty$ and the other carrying negative charge located at $x = +\infty$. We often have occasion to discuss uniform fields and their effects.

Electric Flux. Let us now make quantitative the concept of the density of field lines as a measure of the field strength $\mathbf{E}$. Consider an incremental surface with area ΔA located in the field $\mathbf{E}$ (Fig. 28–8). We define an incremental scalar quantity,

$$\Delta\varphi = \mathbf{E} \cdot \hat{\mathbf{n}}\,\Delta A \qquad (28\text{–}9)$$
$$= E\,\Delta A \cos\theta$$

which is the *flux* of $\mathbf{E}$ through the surface area ΔA (Fig. 28–8). This element of area is represented by the vector $\hat{\mathbf{n}}\,\Delta A$, where $\hat{\mathbf{n}}$ is the unit vector perpendicular to the tangent plane that contains the area ΔA. When ΔA is part of a *closed* surface, $\hat{\mathbf{n}}$ is chosen to be directed *outward*.*

Suppose that a charge Q is located at the origin of some coordinate frame. Imagine that the charge is surrounded by a spherical surface with radius r and its center at the origin. Everywhere on this surface we have

$$\mathbf{E} = \frac{Q}{4\pi\epsilon_0 r^2}\hat{\mathbf{u}}_r \qquad (28\text{–}10)$$

The unit normal vector $\hat{\mathbf{n}}$ is directed outward from the surface and, in fact, for the spherical surface it is identical to the unit radial vector $\hat{\mathbf{u}}_r$. Therefore, with ΔA located on the surface, as indicated in Fig. 28–9, we can write

$$\hat{\mathbf{n}}\,\Delta A = \hat{\mathbf{u}}_r\,\Delta A$$

Then, the *total* flux of $\mathbf{E}$ through the entire spherical surface, $S_0 = 4\pi r^2$, is†

$$\varphi = \lim_{\Delta A_i \to 0} \sum_i \mathbf{E} \cdot \hat{\mathbf{n}}_i\,\Delta A_i = \oint_{S_0} \mathbf{E} \cdot \hat{\mathbf{n}}\,dA \qquad (28\text{–}11)$$

Note that Eq. 28–11 is applicable to any field originating within any closed surface. Using the expression for the field of a point charge, namely,

---

*We consider only closed surfaces that enclose a well-defined volume. We rule out such cases as the *Klein bottle*, a closed surface that is capable of holding a liquid but does not surround a definite volume. The path from a to b shown in the diagram lies entirely on the surface, yet it leads from one side of the surface to the other. The surface therefore has neither an "inside" nor an "outside."

†We use the integral symbol $\oint$ to indicate a surface integration over a *closed* surface or a line integration around a *closed* path.

$$\mathbf{E} \cdot \hat{\mathbf{n}} = \left(\frac{Q}{4\pi\epsilon_0 r^2}\hat{\mathbf{u}}_r\right) \cdot \hat{\mathbf{n}} = \frac{Q}{4\pi\epsilon_0 r^2}$$

we have for the spherical surface

$$\varphi = \frac{Q}{4\pi\epsilon_0 r^2} \oint_{S_0} dA = \frac{Q}{4\pi\epsilon_0 r^2} \cdot 4\pi r^2 = \frac{Q}{\epsilon_0} \qquad \textbf{(28-12)}$$

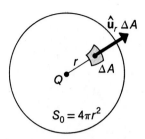

Fig. 28–9. The area ΔA lies on the spherical surface that surrounds the charge Q.

In a quantitative lines-of-force representation of the electric field we define a dimensional quantity N that measures the strength of the field. The number N of lines that radiate from a positive charge Q is equal to Q/ϵ_0; that is,

$$N = \frac{Q}{\epsilon_0} \qquad \textbf{(28-13)}$$

(•What are the dimensions of N?)

For a positive point charge, the lines of force radiate uniformly outward, as shown in Fig. 28–4 for those lines that lie in the plane of the page. The number of lines of force per unit area that pass through the surface of a sphere with radius r is

$$\frac{dN}{dA_\perp} = \frac{N}{4\pi r^2} = \frac{Q}{4\pi\epsilon_0 r^2}$$

so that

$$\frac{dN}{dA_\perp} = E \qquad \textbf{(28-14)}$$

According to the defining equation (Eq. 28–13), N is equal to the flux φ. Moreover, the electric field intensity $\mathbf{E}$ at any point has the direction of the line of force that passes through that point and, according to Eq. 28–14, is equal in magnitude to the number of lines per unit area that pass through a differential area oriented perpendicular to the line of force at that point. Thus, we may also refer to E as the *flux density* or *force-line density*.

Suppose that a point charge Q is surrounded by both a spherical surface S_0 centered on Q as before and an irregular surface S, as shown in Fig. 28–10a. Evidently, every one of the $N = Q/\epsilon_0$ lines that radiate from Q passes through S as well as through S_0. It follows that

Fig. 28–10. (a) Every line of force that passes through S_0 also passes through S. The net flux through each surface is Q/ϵ_0. (b) Every line of force that enters S' also leaves the surface. The net flux through S' is zero.

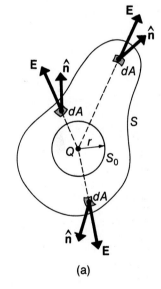

(a)

$$\varphi = N = \oint_S \mathbf{E} \cdot \hat{\mathbf{n}}\, dA = \frac{Q}{\epsilon_0} \qquad \textbf{(28-15)}$$

This result is *independent* of the shape of S; it is required only that S surround Q.

Next, suppose that the closed surface does *not* surround Q, as indicated in Fig. 28–10b. Then, every line of force that enters S' must also leave the surface. *Entering* the surface implies that the angle between $\mathbf{E}$ and $\hat{\mathbf{n}}$ is $\theta > \pi/2$ (as at P) so that $\mathbf{E} \cdot \hat{\mathbf{n}}\, dA < 0$. *Leaving* the surface implies that $\theta < \pi/2$ (as at P') so that $\mathbf{E} \cdot \hat{\mathbf{n}}\, dA > 0$. Thus, the *net flux* that passes through S' is *zero*. The generalization of these results leads to *Gauss's law*, which is discussed in Section 28–3.

Solid Angles. The geometric concept of the *solid angle* is very useful in describing electric flux and related quantities. Figure 28–11 shows an area S that has an arbitrary shape. The solid angle subtended by S at the origin O is defined to be S_r/r^2,

(b)

Fig. 28–11. Geometry for the definition of the solid angle of S subtended at O.

where S_r is the area of the radial projection of S onto a sphere with radius r centered on O.* We use the symbol Ω for solid angle, so that

$$\Omega = \frac{S_r}{r^2} \tag{28-16}$$

Solid angles are measured in *steradians* (sr). If the surface S, no matter how irregular, completely surrounds O (as in Fig. 28–10a), the projected area is $S_r = 4\pi r^2$ and the solid angle subtended by S at O is 4π sr.

If the surface S is a differential area dA with a normal $\hat{\mathbf{n}}$, as shown in Fig. 28–12, the corresponding differential solid angle $d\Omega$ is

$$d\Omega = \frac{dA\cos\theta}{r^2} = \frac{\hat{\mathbf{u}}_r \cdot \hat{\mathbf{n}}\, dA}{r^2} \tag{28-17}$$

Fig. 28–12. The differential solid angle subtended at O by the differential area dA is $d\Omega = \hat{\mathbf{u}}_r \cdot \hat{\mathbf{n}}\, dA/r^2$.

Consider a cone (of arbitrary cross section) that subtends a solid angle $d\Omega$ at the position of a positive point charge Q. The N lines of force from Q radiate outward uniformly in all directions. Therefore, the number of force lines that are contained within $d\Omega$ is

$$dN = \frac{N}{4\pi}\, d\Omega \tag{28-18}$$

Now, allow the cone to cut through a surface. When the charge Q is located *inside* a closed surface S (Fig. 28–13a), any differential solid angle cuts the surface an *odd* number of times, once more leaving than entering. Therefore, within the solid angle $d\Omega$ there is a net number dN of lines that leave the surface, and the total number is $(N/4\pi)\int_S d\Omega = N$.

The geometry for the case in which Q is located *outside* a closed surface S' is shown in Fig. 28–13b in more generality than in Fig. 28–10b. The net flux of lines through S' consists of dN entering in two places (that is, an outward flux of $-2\,dN$ lines) and dN leaving in two places (that is, an outward flux of $+2\,dN$ lines), so that the net flux is zero. With Q outside S, any differential solid angle cuts the surface an

---

*The definition is sometimes stated in terms of the radial projection onto a *unit sphere*, a sphere with unit radius. Then, the solid angle is numerically equal to the projected area.

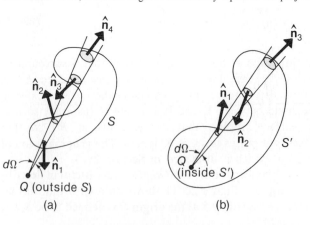

Fig. 28–13. Geometry of the solid angle subtended at a charge Q located (a) outside a closed surface and (b) inside a closed surface.

even number of times, as many times entering as leaving. Thus, the integral of the flux over the entire surface S' is zero. These results constitute a more general proof of the ideas presented in the discussion of Fig. 28–10.

Note carefully that the simple geometric results obtained here are a direct consequence of the fact that the electric field strength of a point charge depends on the inverse square of the distance from the charge. This, of course, is ultimately related to the $1/r^2$ dependence of the electric force between two point charges, as contained in Coulomb's law.

Example 28-2

Calculate the solid angle of a right circular cone with half-angle α subtended at its apex.

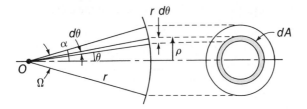

Solution:

From the geometry shown in the diagram, the differential solid angle $d\Omega$ of the annular ring with radius $\rho = r \sin \theta$ and area $dA = (2\pi\rho)(r\, d\theta)$ is

$$d\Omega = \frac{dA}{r^2} = \frac{(2\pi\rho)(r\, d\theta)}{r^2} = \frac{(2\pi r \sin\theta)(r\, d\theta)}{r^2} = 2\pi \sin\theta\, d\theta$$

Then,

$$\Omega = 2\pi \int_0^\alpha \sin\theta\, d\theta = 2\pi(1 - \cos\alpha)$$

Notice that when $\alpha = \pi$, the cone corresponds to the entire sphere and the expression for Ω appropriately yields $\Omega = 4\pi$. When $\alpha \ll 1$, we can write $\cos\alpha = 1 - \frac{1}{2}\alpha^2 + \ldots$, so that $\Omega = \pi\alpha^2$, or $\Omega = \pi(r\alpha)^2/r^2$. Now, $r\alpha$ is the radius and $\pi(r\alpha)^2$ is the area of the small circular segment defined by the intersection of the cone and the sphere of radius r. Thus, we see that the result for Ω is the same as writing $\Omega = S_r/r^2$ (Eq. 28–16).

28-3 GAUSS'S LAW

When Eq. 28–15 is generalized to include the idea of superposition of electric fields, the result is *Gauss's law** for the electric flux through a closed surface S that surrounds a distribution of charge:

$$\varphi_{\text{net}} = \oint_S \mathbf{E} \cdot \hat{\mathbf{n}}\, dA = \begin{cases} \dfrac{1}{\epsilon_0} \sum Q_{\text{in}} \\[2mm] \dfrac{1}{\epsilon_0} \displaystyle\int_V \rho\, dv \end{cases} \qquad \textbf{(28–19)}$$

*Karl Friedrich Gauss (1777–1855), German mathematician, astronomer, and physicist.

where ΣQ_{in} represents the algebraic sum of all the discrete charges *inside* the surface S. In the event that there is a continuous distribution of charge with a volume density ρ, the sum must be replaced by the integral of $\rho\, dv$ over the volume V occupied by the charge. (The volume V is surrounded by the surface S.) A closed surface that surrounds a charge distribution is called a *Gaussian surface*. Gauss's law is a powerful tool for analyzing the field strengths due to various charge distributions.

Example 28–3

An (unspecified) array of external charges produces a uniform electric field, $\mathbf{E} = E_x\hat{\mathbf{i}} + E_y\hat{\mathbf{j}}$, in a certain region of space. Show that there is no net flux through a cubical Gaussian surface located in this charge-free region, thereby verifying Gauss's law for this case. Choose the coordinate axes to lie along the edges of the cube, as shown in the diagram.

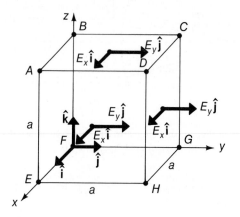

Solution:

The integral that must be evaluated, namely, $\oint_S \mathbf{E} \cdot \hat{\mathbf{n}}\, dA$, consists of the sum of contributions from all six faces of the cube. First, for the right face $DCGH$, we have (refer to the diagram)

$$\int_{DCGH} \mathbf{E} \cdot \hat{\mathbf{n}}\, dA = (E_x\hat{\mathbf{i}} + E_y\hat{\mathbf{j}}) \cdot \hat{\mathbf{j}} \int_{DCGH} dA = +E_y a^2$$

For the front face $ADHE$, we have

$$\int_{ADHE} \mathbf{E} \cdot \hat{\mathbf{n}}\, dA = (E_x\hat{\mathbf{i}} + E_y\hat{\mathbf{j}}) \cdot \hat{\mathbf{i}} \int_{ADHE} dA = +E_x a^2$$

For the top face $ABCD$, we have

$$\int_{ABCD} \mathbf{E} \cdot \hat{\mathbf{n}}\, dA = (E_x\hat{\mathbf{i}} + E_y\hat{\mathbf{j}}) \cdot \hat{\mathbf{k}} \int_{ABCD} dA = 0$$

For each corresponding parallel face the integral changes sign; thus,

$$\int_{ABFE} \mathbf{E} \cdot \hat{\mathbf{n}}\, dA = -E_y a^2$$

$$\int_{BCGF} \mathbf{E} \cdot \hat{\mathbf{n}}\, dA = -E_x a^2$$

$$\int_{EFGH} \mathbf{E} \cdot \hat{\mathbf{n}} \, dA = 0$$

The net flux is the sum of these six integrals, which vanishes:

$$\oint_S \mathbf{E} \cdot \hat{\mathbf{n}} \, dA = 0$$

Thus, Gauss's law is verified.

Example 28–4

Calculate the flux through a square surface with dimensions $2a$ by $2a$ due to a charge Q located a perpendicular distance a from the center of the square, as shown in diagram (a).

Solution:

The field strength due to the charge Q is

$$\mathbf{E} = \frac{Q}{4\pi\epsilon_0} \frac{\hat{\mathbf{u}}_r}{r^2}$$

Now $\hat{\mathbf{u}}_r = \mathbf{r}/r$, so we can write

$$\mathbf{E} = \frac{Q}{4\pi\epsilon_0} \frac{\mathbf{r}}{r^3} = \frac{Q}{4\pi\epsilon_0 r^3}(x\hat{\mathbf{i}} + y\hat{\mathbf{j}} + z\hat{\mathbf{k}})$$

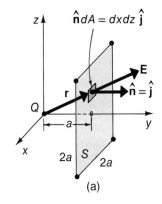

(a)

and $r^2 = x^2 + y^2 + z^2$.

The element of area on the surface S is $\hat{\mathbf{n}} \, dA = dx \, dz \, \hat{\mathbf{j}}$. Also, on S, we have $y = a$. Now, $\hat{\mathbf{i}} \cdot \hat{\mathbf{j}} = \hat{\mathbf{k}} \cdot \hat{\mathbf{j}} = 0$ and $\hat{\mathbf{j}} \cdot \hat{\mathbf{j}} = 1$; hence,

$$\mathbf{E} \cdot \hat{\mathbf{n}} \, dA = \frac{Q}{4\pi\epsilon_0} \frac{a}{r^3} \, dx \, dz$$

so that the flux through the open surface S is

$$\varphi = \int_S \mathbf{E} \cdot \hat{\mathbf{n}} \, dA$$

$$= \frac{Qa}{4\pi\epsilon_0} \int_{-a}^{a} \int_{-a}^{a} \frac{dx \, dz}{(x^2 + z^2 + a^2)^{3/2}}$$

$$= \frac{Qa}{\pi\epsilon_0} \int_{0}^{a} \int_{0}^{a} \frac{dx \, dz}{(x^2 + z^2 + a^2)^{3/2}}$$

This double integral can be evaluated by referring to the list of integrals inside the rear cover. The result is (see Problem 28–9)

$$\varphi = \frac{Qa}{\pi\epsilon_0} \cdot \frac{1}{a} \tan^{-1} \frac{z}{\sqrt{z^2 + 2a^2}} \bigg|_0^a = \frac{Q}{\pi\epsilon_0} \tan^{-1} \frac{1}{\sqrt{3}}$$

$$= \frac{Q}{6\epsilon_0}$$

Because of the particular geometry of this problem, the result for φ can be obtained

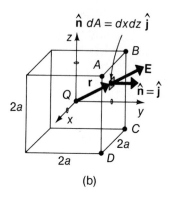

(b)

much more easily by making use of Gauss's law. Let the surface S be one face ($ABCD$) of a cube that surrounds the charge Q, as shown in diagram (b). The charge is at the *center* of the cube and is therefore located symmetrically with respect to each face. The net flux through the cubical surface is just Q/ϵ_0, so the flux through any one of the identical faces is $\frac{1}{6}$ as much. Thus,

$$\varphi = \frac{Q}{6\epsilon_0}$$

which is the result we obtained by the more laborious direct calculation.

▶ *Mapping Lines of Force.* Gauss's law combined with the superposition principle can be used to construct lines-of-force maps for various charge distributions. Consider the case of a positive charge Q_2 and a negative charge Q_1, as shown in Fig. 28–14. A particular line of force is shown originating on the charge Q_2 and terminating on the charge Q_1. Now, imagine the surface that is generated by rotating this line of force about the line that connects the two charges. Because force lines cannot cross, every line of force that originates on Q_2 within this surface is "trapped" within the surface and necessarily terminates on Q_1. Therefore, the *same* number of lines of force pierce any perpendicular cross-sectional plane that is constructed between Q_2 and Q_1. The flux through such a plane is equal to the algebraic sum of the fluxes due to each charge separately. For example, the net electric flux through the area dA on the perpendicular cross section shown in Fig. 28–14 is

$$d\varphi = \mathbf{E} \cdot \hat{\mathbf{n}} \, dA = \mathbf{E}_1 \cdot \hat{\mathbf{n}} \, dA + \mathbf{E}_2 \cdot \hat{\mathbf{n}} \, dA = d\varphi_1 + d\varphi_2$$

Then, the integral of $d\varphi$ over the entire cross-sectional area yields $\varphi_0 = \varphi_1 + \varphi_2$.

In the immediate vicinity of the charge Q_2 the total electric field strength (the flux density), $\mathbf{E}_1 + \mathbf{E}_2$, is completely dominated by the effect of Q_2 because of the inverse-square nature of the force law. Moreover, the lines of force leaving Q_2 are, in the immediate vicinity of Q_2, uniformly distributed. The total flux leaving Q_2 is Q_2/ϵ_0; the fraction of this flux within the cone defined by the tangent lines at angles α_2 is equal to Q_2/ϵ_0 multiplied by $d\Omega/4\pi$ (Eq. 28–18), where the solid angle of the cone is (see Example 28-2) $d\Omega = 2\pi(1 - \cos \alpha_2)$. All of this flux passes through every cross section, and so equals φ_0. That is,

$$\frac{Q_2}{\epsilon_0} \frac{2\pi(1 - \cos \alpha_2)}{4\pi} = \varphi_0 \tag{28–20a}$$

Using the same argument, the flux arriving at Q_1 is

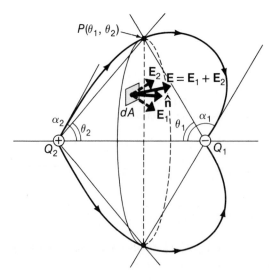

Fig. 28–14. Geometry for determining a line of force connecting Q_2 and Q_1.

$$-\frac{Q_1}{\epsilon_0} \frac{2\pi(1 - \cos\alpha_1)}{4\pi} = \varphi_0 \qquad \text{(28-20b)}$$

where the negative sign arises because $\hat{\mathbf{n}}$ is directed *inward,* that is, *toward* Q_1. (Remember, $Q_1 < 0$, so we still have $\varphi_0 > 0$.)

Every perpendicular cross section, such as the one shown in Fig. 28-14, defines two right circular cones, one with apex at Q_2 with half angle θ_2 and one with apex at Q_1 with half angle θ_1. The flux leaving Q_2 within the left-hand cone is equal to Q_2/ϵ_0 multiplied by the solid angle ratio; similarly for the flux arriving at Q_1 within the right-hand cone. The sum of these two fluxes is the total flux through the cross section, that is, φ_0. Thus,

$$\frac{Q_2}{\epsilon_0} \frac{2\pi(1 - \cos\theta_2)}{4\pi} - \frac{Q_1}{\epsilon_0} \frac{2\pi(1 - \cos\theta_1)}{4\pi} = \varphi_0 \qquad \text{(28-21)}$$

Equations 28-20 and 28-21 define completely the points $P(\theta_1, \theta_2)$ along any line of force that corresponds to a total enclosed flux φ_0. For a given value of φ_0, the solutions of Eq. 28-21 always give $\theta_1 \leqslant \alpha_1$ and $\theta_2 \leqslant \alpha_2$, as expected.

Example 28-5

Given two charges, $Q_2 = +3Q$ and $Q_1 = -Q$, plot the line of force that includes within its surface of revolution $\frac{3}{4}$ of all the flux arriving at Q_1.

Solution:

We are required to solve Eqs. 28-20 and 28-21 for the case $\varphi_0 = \frac{3}{4}(Q/\epsilon_0)$. From Eq. 28-20a, we have

$$\frac{3Q}{\epsilon_0} \frac{2\pi(1 - \cos\alpha_2)}{4\pi} = \frac{3}{4} \frac{Q}{\epsilon_0}$$

which yields

$$\cos\alpha_2 = \tfrac{1}{2}, \qquad \alpha_2 = 60°$$

From Eq. 28-20b, we have

$$-\frac{(-Q)}{\epsilon_0} \frac{2\pi(1 - \cos\alpha_1)}{4\pi} = \frac{3}{4} \frac{Q}{\epsilon_0}$$

which yields

$$\cos\alpha_1 = -\tfrac{1}{2}, \qquad \alpha_1 = 120°$$

Finally, Eq. 28-21 gives

$$\frac{3Q}{\epsilon_0} \frac{2\pi(1 - \cos\theta_2)}{4\pi} - \frac{(-Q)}{\epsilon_0} \frac{2\pi(1 - \cos\theta_1)}{4\pi} = \frac{3}{4} \frac{Q}{\epsilon_0}$$

from which we obtain

$$\cos\theta_2 = \tfrac{5}{6} - \tfrac{1}{3}\cos\theta_1$$

If we select values for θ_1 and solve this equation for θ_2, we determine the points $P(\theta_1, \theta_2)$. Some of these are:

θ_1	$\alpha_1 = 120°$	105°	90°	75°	60°	45°	30°	0°
θ_2	0ᵖ	23.1°	33.6°	41.7°	48.2°	53.3°	57°	$\theta_2 = 60°$

Plotting these points results in the line of force shown in the diagram on the next page. Notice that this line of force is just one of those shown in Fig. 28-6.

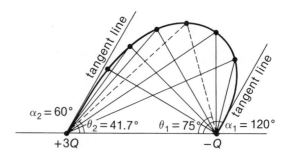

28-4 ELECTRIC FIELDS FOR CONTINUOUS DISTRIBUTIONS OF STATIC CHARGE

We now extend the discussion and consider the fields produced by continuous distributions of charge. We examine various cases in which the charges are held fixed by unspecified means. Later, we discuss the more realistic cases of the patterns of charge deposited on insulators and the patterns that are developed by charges on conductors.

The Field of a Uniformly Charged Infinite Filament. Consider an infinitely long filament that carries a uniform line distribution of charge (Fig. 28–15). Let the positive charge per unit length of the filament be λ. Then, the differential piece of filament dz carries the charge $dQ = \lambda\,dz$. At the point P, which is a perpendicular distance R from the filament, the electric field strength due to the charge on the segment dz at z is

$$d\mathbf{E}' = \frac{dQ}{4\pi\epsilon_0 r^2}\hat{\mathbf{u}}_r$$

where $r^2 = R^2 + z^2$. Evidently, there is a second piece of the filament, also with length dz, located at $-z$, which produces a field strength $d\mathbf{E}''$ at P that just cancels the z-component of $d\mathbf{E}'$ and gives an equal contribution in the radial direction. Thus,

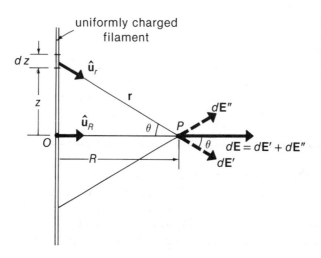

Fig. 28–15. An infinitely long filament carries a uniform charge λ per unit length. The amount of charge in the length dz is $dQ = \lambda\,dz$.

at P, we have

$$d\mathbf{E} = d\mathbf{E}' + d\mathbf{E}'' = \frac{2\lambda \, dz \cos\theta}{4\pi\epsilon_0 r^2} \hat{\mathbf{u}}_R$$

or

$$d\mathbf{E} = \frac{\lambda R \, dz}{2\pi\epsilon_0 (z^2 + R^2)^{3/2}} \hat{\mathbf{u}}_R$$

Integrating from $z = 0$ to $z = \infty$ we find (refer to the list of integrals inside the rear cover)

$$\mathbf{E}(R) = \frac{\lambda R \hat{\mathbf{u}}_R}{2\pi\epsilon_0} \int_0^\infty \frac{dz}{(z^2 + R^2)^{3/2}} = \frac{\lambda}{2\pi\epsilon_0} \cdot \frac{\hat{\mathbf{u}}_R}{R} \qquad \textbf{(28–22)}$$

The filament is infinitely long, so the field strength $\mathbf{E}$ has only a radial component and no z-component. Note that $\mathbf{E}$ depends inversely on the *first power* of the distance R.

This same result can be obtained by a straightforward application of Gauss's law. We deduce from the symmetry of the situation that $\mathbf{E}$ must be everywhere perpendicular to the filament. (•Can you see why this is the case?) Then, the dependence of $\mathbf{E}$ on R is easily found by using Gauss's law. As shown in Fig. 28–16, we surround a length ℓ of the filament with a cylindrical closed Gaussian surface whose axis coincides with the filament and which has circular end caps. Let the radius of the cylinder be R. The charge contained within the Gaussian surface is $q = \lambda \ell$. Because $\mathbf{E}$ is everywhere in the direction of $\hat{\mathbf{u}}_R$, there is no flux through either end cap. The outward flux through the cylinder wall is just $2\pi R\ell E(R)$; thus,

Fig. 28–16. A length ℓ of an infinite line charge is surrounded by a cylindrical Gaussian surface.

$$2\pi R\ell E(R) = \frac{q}{\epsilon_0} = \frac{\lambda\ell}{\epsilon_0}$$

or

$$E(R) = \frac{\lambda}{2\pi\epsilon_0} \cdot \frac{1}{R}$$

and vectorially, we have

$$\mathbf{E}(R) = \frac{\lambda}{2\pi\epsilon_0} \cdot \frac{\hat{\mathbf{u}}_R}{R}$$

Fig. 28–17. The geometry for calculating the electric field strength along the axis of a uniformly charged ring.

which is the same as the result obtained by explicit integration (Eq. 28–22).

Notice that Gauss's law can be applied to obtain the result easily in this case because the geometry is particularly simple so that the pattern of field lines can be readily deduced. Then, a Gaussian surface was constructed that permitted a straightforward integration. The usefulness of Gauss's law is generally limited to such situations.

The Field of a Uniformly Charged Circular Ring. Next, determine the electric field strength along the axis of a uniformly charged ring. The geometry is shown in Fig. 28–17. If the ring carries a total positive charge Q, the linear charge density along the ring is $\lambda = Q/2\pi a$, where a is the radius of the ring. In a small segment of the ring with arc length $ad\phi$ the amount of charge is $dQ = \lambda ad\phi$. This charge, combined with the charge on a similar segment of the ring at the other end of a common diameter (see Fig. 28–17), produces a contribution $d\mathbf{E}$ to the electric field that is

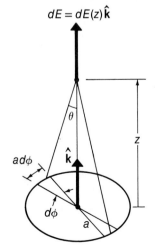

directed along the ring axis (the direction of $\hat{\mathbf{k}}$) because the transverse components cancel, as in the case of the infinite filament. Then, for the pair of ring segments, we have

$$d\mathbf{E} = \frac{2\lambda a \, d\phi \cos\theta}{4\pi\epsilon_0(z^2 + a^2)}\hat{\mathbf{k}}$$

and

$$\mathbf{E} = \frac{\lambda a z}{2\pi\epsilon_0(z^2 + a^2)^{3/2}}\hat{\mathbf{k}} \int_0^\pi d\phi$$

from which

$$\mathbf{E}(z) = \frac{Qz}{4\pi\epsilon_0(z^2 + a^2)^{3/2}}\hat{\mathbf{k}} \qquad (28\text{--}23)$$

The value of $E(z)$ as a function of z is shown in Fig. 28–18. The field strength reaches its maximum value at $z = a/\sqrt{2}$ where $E_m = Q/(6\sqrt{3}\pi\epsilon_0 a^2)$ (see Problem 28–13). In this case Gauss's law cannot be used because the situation does not have a sufficiently high degree of symmetry to permit the geometry of the field lines to be readily deduced.

Note that when $z^2 \gg a^2$, we have $(z^2 + a^2)^{3/2} \cong z^3$, so that

$$\mathbf{E}(z) \cong \frac{Q}{4\pi\epsilon_0 z^2}\hat{\mathbf{k}}, \qquad z^2 \gg a^2$$

which is just the distant-field (monopole) expression.

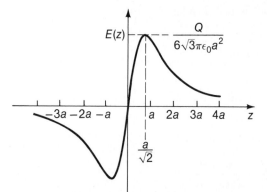

Fig. 28–18. The field strength E as a function of z along the axis of a uniformly charged ring.

Fig. 28–19. The geometry for calculating the electric field strength along the axis of a uniformly charged circular disc.

The Field of a Uniformly Charged Circular Disc. The expression for the electric field strength on the axis of a uniformly charged circular disc can be obtained by using the result for the charged ring (Eq. 28–23). If the disc carries a total positive charge Q that is distributed uniformly, the surface charge density is $\sigma = Q/\pi a^2$ (in C/m^2). Construct an annular segment of the disc with radius r and width dr, as shown in Fig. 28–19. The charge on this annulus is then $2\pi r\sigma \, dr$. The field strength $d\mathbf{E}$ at a point on the axis a distance z from the disc is given by Eq. 28–23 with the substitutions $\mathbf{E} \to d\mathbf{E}$, $a \to r$, and $Q \to 2\pi r\sigma \, dr$. Then, we have

$$d\mathbf{E} = \frac{z\sigma r \, dr}{2\epsilon_0(r^2 + z^2)^{3/2}}\hat{\mathbf{k}} \qquad (28\text{--}24)$$

The integral of this expression from $r = 0$ to $r = a$ gives the field strength due to the entire disc:

$$\mathbf{E} = \frac{z\sigma}{2\epsilon_0}\,\hat{\mathbf{k}} \int_0^a \frac{r\,dr}{(r^2 + z^2)^{3/2}}$$

This integral appears in the list inside the rear cover. The result is

$$\mathbf{E}(z) = \frac{\sigma}{2\epsilon_0}\left[1 - \frac{z}{(z^2 + a^2)^{1/2}}\right]\hat{\mathbf{k}} \qquad (28\text{-}25)$$

There are three limiting cases of this result that merit investigation. First, when $z^2 \gg a^2$, we expect the monopole term in Eq. 28-24 to prevail; indeed, we have

$$1 - \frac{z}{(z^2 + a^2)^{1/2}} = 1 - \frac{1}{(1 + a^2/z^2)^{1/2}}$$

$$= 1 - \left[1 - \frac{1}{2}\frac{a^2}{z^2} + \frac{3}{8}\frac{a^4}{z^4} - \cdots\right] \cong \frac{1}{2}\frac{a^2}{z^2}$$

Thus, we have the expected result,

$$\mathbf{E}(z) \cong \frac{\sigma a^2}{4\epsilon_0 z^2} = \frac{Q}{4\pi\epsilon_0 z^2}, \qquad z^2 \gg a^2$$

The other two limiting cases apply near the disc: $z^2 \ll a^2$ with a and σ constant, and $a^2 \gg z^2$ with z and σ constant. In the first of these cases we explore the region close to the center of a disc with a given radius a. In the second case we establish an arbitrary point that is a distance z from the disc; while holding σ constant, we allow the disc radius a and the total charge Q to increase without limit. In either case we have from Eq. 28-24, with σ constant,

$$\mathbf{E} = \left(\mathop{\text{Lim}}_{z \to 0} \text{ or } \mathop{\text{Lim}}_{a \to \infty}\right) \frac{\sigma}{2\epsilon_0}\left[1 - \frac{z}{(z^2 + a^2)^{1/2}}\right]\hat{\mathbf{k}}$$

or

$$\mathbf{E} = \frac{\sigma}{2\epsilon_0}\,\hat{\mathbf{k}} \qquad (28\text{-}26)$$

Thus, the field strength is essentially uniform in a region near the center of the disc.

The Field of a Uniformly Charged Infinite Flat Sheet. Consider an infinite flat sheet that carries a uniform positive surface charge density σ (Fig. 28-20). From the symmetry it is clear that $\mathbf{E}$ must be everywhere perpendicular to the sheet. That is, with z measured perpendicular to the sheet, we have $\mathbf{E} = E(z)\hat{\mathbf{k}}$ on the side $z > 0$ and $\mathbf{E} = -E(z)\hat{\mathbf{k}}$ on the side $z < 0$; however, the function $E(z)$ is not yet known.

Construct a Gaussian surface in the shape of a closed rectangular box that is perpendicular to the sheet and extends a distance z on either side. The ends each have area A. Because there is no flux through the sides of the Gaussian surface, Gauss's law is expressed as

$$2A\,E(z) = \frac{\sigma A}{\epsilon_0}$$

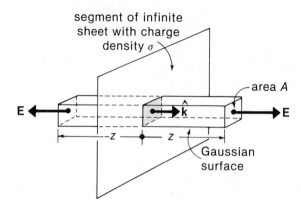

segment of infinite
sheet with charge
density σ

area A

$\hat{\mathbf{k}}$

E

E

$-z$ z

Gaussian
surface

Fig. 28–20. A Gaussian surface in the shape of a rectangular box is constructed perpendicular to an infinite flat sheet that carries a uniform charge density σ.

or

$$E(z) = \frac{\sigma}{2\epsilon_0}$$

which shows that the field strength is independent of z. Finally,

$$
\left.
\begin{aligned}
\mathbf{E} &= +\frac{\sigma}{2\epsilon_0}\hat{\mathbf{k}}, \qquad z > 0 \\[2ex]
\mathbf{E} &= -\frac{\sigma}{2\epsilon_0}\hat{\mathbf{k}}, \qquad z < 0
\end{aligned}
\right\}
\qquad \textbf{(28–27)}
$$

This is the same result obtained in the limiting case for the charged disc (Eq. 28–26). Thus, for the practical case of a uniformly charged flat sheet with large (but not infinite) dimensions, the field is uniform throughout the central region near the disc.

Fields of Uniform Spherical Charge Distributions.   Consider a thin spherical shell with a radius R that carries a total positive charge Q distributed uniformly over the surface (Fig. 28–21). Because the geometry and the charge distribution both have spherical symmetry, we expect the electric field strength to be everywhere radially outward, that is, $\mathbf{E} = E(r)\hat{\mathbf{u}}_r$. To determine the function $E(r)$, construct a spherical Gaussian surface with a radius r that is concentric with the charge distribution. Consider the possibilities $r > R$ and $r < R$.

For the case $r > R$, Gauss's law yields

$$4\pi r^2 E(r) = \frac{Q}{\epsilon_0}$$

Fig. 28–21. Application of Gauss's law to a spherical charge distribution.

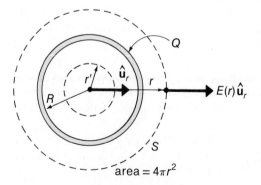

Q

r'

$\hat{\mathbf{u}}_r$

r

$E(r)\hat{\mathbf{u}}_r$

R

S

area $= 4\pi r^2$

so that

$$\mathbf{E}(r) = \frac{Q}{4\pi\epsilon_0}\frac{\hat{\mathbf{u}}_r}{r^2}, \qquad r > R \tag{28–28a}$$

Thus, the electric field strength in the region *outside* the charged sphere (that is, $r > R$) is the same as that produced by a point charge Q located at the center of the spherical shell.

For the case $r < R$, the Gaussian surface encloses no charge (see Fig. 28–21); hence,

$$\mathbf{E} = 0, \qquad r < R \tag{28–28b}$$

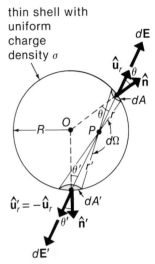

thin shell with uniform charge density σ

Fig. 28–22. Geometry for calculating the electric field strength at the arbitrary point P inside a uniform spherical distribution of charge.

The results expressed by Eqs. 28–28 are analogous to those obtained for the gravitational case (see Section 14–4). (•Why should the two cases give similar results?) Notice that the use of Gauss's law here gives the result immediately, whereas the evaluation of a complicated integral was required in the method used in Section 14–4.

▶ It is instructive to consider another proof that the field strength vanishes in the hollow region inside a spherical charge distribution. Refer to the geometry shown in Fig. 28–22. Select an arbitrary point P and construct two coaxial cones that subtend the same solid angle $d\Omega$ at P. The intersections of these cones with the spherical surface define the areas dA and dA'. The amounts of charge on these two areas are $dq = \sigma\,dA$ and $dq' = \sigma\,dA'$, respectively, where σ is the uniform density of charge on the sphere. The electric field strength at P due to these two charges is

$$d\mathbf{E} + d\mathbf{E}' = \frac{\sigma}{4\pi\epsilon_0}\left(\frac{dA}{r^2} - \frac{dA'}{r'^2}\right)\hat{\mathbf{u}}_r$$

where we have made use of the fact that $\hat{\mathbf{u}}_r' = -\hat{\mathbf{u}}_r$ (see Fig. 28–22). From the defining equation for solid angles (Eq. 28–17), we have

$$dA\cos\theta = r^2\,d\Omega$$
$$dA'\cos\theta' = r'^2\,d\Omega$$

Thus, the expression for the field strength becomes

$$d\mathbf{E} + d\mathbf{E}' = \frac{\sigma}{4\pi\epsilon_0}\left[\frac{r^2\,d\Omega/\cos\theta}{r^2} - \frac{r'^2\,d\Omega/\cos\theta'}{r'^2}\right]\hat{\mathbf{u}}_r$$

In Fig. 28–22 we see that $\theta = \theta'$ (because the triangle dA-O-dA' is isosceles). Consequently, the two terms in the brackets are equal, and we conclude that

$$d\mathbf{E} + d\mathbf{E}' = 0$$

The location of the point P is arbitrary. Therefore, when the integration over the complete solid angle is carried out, we have

$$\mathbf{E} = 0, \qquad r < R$$

which is the same result we obtained with Gauss's law. ◀

Example 28–6

Consider a spherical distribution of charge with a volume charge density $\rho(r)$, measured in C/m³, that depends at most on r. The charge is confined within a region with a radius R. Determine the electric field strength everywhere in space.

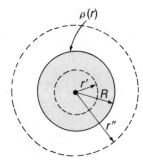

Solution:

The diagram shows the charge distribution and two Gaussian surfaces, one with radius $r' < R$ and one with $r'' > R$. Again, from the spherical symmetry we argue that $\mathbf{E}$ can depend only on r and not on the polar or azimuthal angles; $\mathbf{E}$ has a component only in the direction $\hat{\mathbf{u}}_r$. Thus, for $r' < R$, we have

$$4\pi r'^2 E(r') = \frac{Q'}{\epsilon_0}$$

Then, the electric field strength at $r = r'$ is

$$\mathbf{E}(r') = \frac{Q'}{4\pi\epsilon_0 r'^2}\hat{\mathbf{u}}_r$$

where Q' is the total charge within a sphere with radius r'; that is,

$$Q' = \int_0^{r'} 4\pi r^2 \rho(r)\, dr$$

For $r'' > R$, the Gaussian surface contains the entire charge Q; thus,

$$\mathbf{E}(r'') = \frac{Q}{4\pi\epsilon_0 r''^2}\hat{\mathbf{u}}_r$$

The field strength $E(r')$ can be evaluated only if the density function $\rho(r)$ is known. For the special case in which $\rho(r)$ is uniform, $\rho = Q/(4\pi R^3/3)$, we readily find that

$$Q' = \left(\frac{r'}{R}\right)^3 Q$$

Then,
$$\mathbf{E}(r') = \frac{r'Q}{4\pi\epsilon_0 R^3}\hat{\mathbf{u}}_r, \qquad r' < R$$

The magnitude of the field strength for this case is shown in the diagram. Again, it is instructive to compare this result with the gravitational case (Section 14–4).

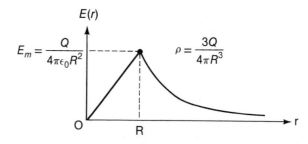

PROBLEMS

Section 28-1

28-1 A point charge $Q_1 = +3\,\mu C$ is placed at the origin and a point charge $Q_2 = -8\,\mu C$ is placed at $x = +0.20$ m. (a) Determine the electric field strength, $\mathbf{E}(x) = E(x)\hat{\mathbf{i}}$, at all points along the x-axis. (b) At what points, if any, apart from $|x| = \infty$, does $E(x) = 0$?

28-2 Show that Eq. 28-8 for the electric field strength of a dipole can be expressed in rectangular coordinates

as $E_x = \dfrac{3p}{4\pi\epsilon_0 r^3}\sin\theta\cos\theta$

and $E_y = \dfrac{p}{4\pi\epsilon_0 r^3}(3\cos^2\theta - 1)$

Also, show that

$|\mathbf{E}| = (E_x^2 + E_y^2)^{1/2} = (p/4\pi\epsilon_0 r^3)(3\cos^2\theta + 1)^{1/2}.$

28-3 Show that the spherical components of the electric field strength of a dipole are

$E_r = \dfrac{p\cos\theta}{2\pi\epsilon_0 r^3}, \qquad E_\theta = \dfrac{p\sin\theta}{4\pi\epsilon_0 r^3}, \qquad E_\phi = 0$

Also, show that $|\mathbf{E}| = (p/4\pi\epsilon_0 r^3)(3\cos^2\theta + 1)^{1/2}.$

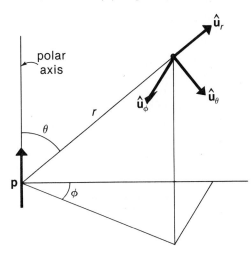

28-4 Consider a dipole in which each charge has $|Q| = 3\,\mu C$ and for which $2\ell = 2$ cm. (a) Determine the electric field strength at distances of 1 m and 10 m from the dipole along the dipole axis. (b) What would be the field strength at these dis-

tances for the positive charge alone? (c) What is the ratio of the monopole field strength to the dipole field strength at the two distances?

28-5 Calculate the electric field strength at points • along the x-axis ($|x| \gg \ell$) for the quadrupole charge distribution shown in the diagram.

Section 28-2

28-6 Construct a spherical surface S with radius R around an electric dipole. Show by explicit calculation that $\int_S \mathbf{E} \cdot \hat{\mathbf{n}}\, dA = 0$, where $\mathbf{E}$ represents the distant field (Eq. 28-8). [*Hint:* Use the results in Problem 28-3.]

28-7 A cube with sides of 0.20 m has its center at the origin of a rectangular coordinate frame and its faces perpendicular to the coordinate axes. The electric field strength is $\mathbf{E} = Kx\hat{\mathbf{i}}$, with $K = 100$ N/C · m. Determine the amount of charge within the cube.

28-8 Positive charge with a uniform linear density λ is • carried on four parallel infinite filaments that pass through the corners of a square, as shown in the diagram. (a) Sketch the lines of force in a plane $ABFG$ that is perpendicular to the filaments. (b) Determine the amount of electric flux that passes through the area $ABCD$.

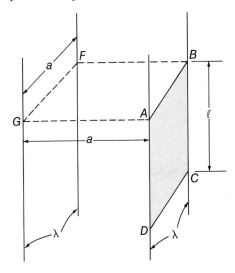

Section 28-3

28-9   Evaluate the double integral in Example 28-4 and obtain $\varphi = Q/6\epsilon_0$. For the integration over x you will need the integral of $dx/(x^2 + b^2)^{3/2}$, where $b^2 = z^2 + a^2$. This integral and the one for the resulting integration over z are contained in the list inside the rear cover.

28-10   Consider two infinite parallel sheets that are separated by a distance ℓ. Each sheet carries a uniform surface density σ of positive charge. Use (a) Gauss's law and (b) superposition to determine the electric field strength, both between and outside the sheets.

28-11   Rework Problem 28-10 with the condition that the sheets carry opposite charge densities, $+\sigma$ and $-\sigma$. Use superposition.

28-12   Plot the line of force $P(\theta_1, \theta_2)$ in Example 28-5
•• for the case $\varphi_0 = 4Q/3\epsilon_0$.

Section 28-4

28-13   Show that the maximum field strength E_m along the axis of a uniformly charged ring occurs at $z = a/\sqrt{2}$ (see Fig. 28-18) and has the value $Q/(6\sqrt{3}\pi\epsilon_0 a^2)$.

28-14   Two infinite parallel filaments carry opposite
• linear charge densities, $+\lambda$ and $-\lambda$, and are separated by a distance 2ℓ. Determine the electric field strength at any distant point $P(x, y)$, that is, any point for which $x^2 + y^2 \gg \ell^2$. Express the result vectorially in terms of the dipole moment per unit length, $\mathbf{p}_\ell = 2\ell\lambda\hat{\mathbf{j}}$.

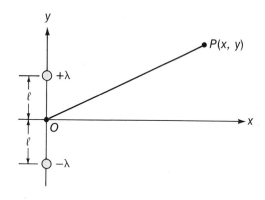

28-15   An infinitely long cylinder has a radius R and a uniform volume charge density ρ. Determine the electric field strength for $r < R$ and for $r > R$.

Additional Problems

28-16   Air will break down (lose its insulating quality) and sparking will result if the field strength is increased to about 3×10^6 N/C. (This field strength is also expressed as 3×10^6 *volts per meter*.) What acceleration will an electron experience in such a field? If the electron starts from rest, in what distance will it acquire a speed equal to 10 percent of the speed of light?

28-17   The electric field strength of a dipole at a distance of 10 m along the dipole axis is observed to be twice the value anticipated for a pair of charges, $+Q$ and $-Q$, separated by a distance of 2 cm. It is suspected that the reason for this discrepancy is that the positive charge is actually somewhat larger in magnitude than the negative charge. What fractional increase $\delta Q/Q$ of the positive charge would account for the observed field strength?

28-18   If a molecule AB is completely ionic in character, it consists of ions A$^+$ and B$^-$ separated by a certain internuclear distance 2ℓ. For singly charged ions the dipole moment would be $2\ell e$. (It is convenient to measure molecular dipole moments in units of eÅ, that is, the product of the electronic charge, 1.602×10^{-19} C, and the angstrom unit, 10^{-10} m.) Intermolecular separations can be determined by a variety of methods. Then, any deviation of the measured dipole moment below the value $2\ell e$ indicates the degree to which the molecular bond is less than completely ionic. That is, $100(p_{\text{meas}}/2\ell e) =$ percentage ionic character. (Most bonds are partly ionic and partly covalent.) The lithium fluoride (LiF) gas molecule has an internuclear separation of 1.52 Å and a dipole moment of 1.39 eÅ. On the other hand, the hydrogen iodide (HI) molecule has an internuclear separation of 1.62 Å but a dipole moment of only 0.080 eÅ. What is the degree of ionic bonding in these two molecules?

28-19   The water molecule has the structure shown on page 779, with O—H separations of 0.97 Å. The observed dipole moment of H_2O is 0.387 eÅ. What fraction of the O—H bond is ionic in character? (Refer to Problem 28-18.)

28-20   Refer to Fig. 28-6. Consider two lines of force that leave the charge $+3Q$ and diverge to infinity. If the angle between the lines near the charge is θ, show that the angle between the same lines at very large distances is $\frac{3}{2}\theta$.

28-21   A wire is bent into the shape of a square with sides $2a$ and carries a uniform linear charge density λ. Determine the electric field strength at a point that

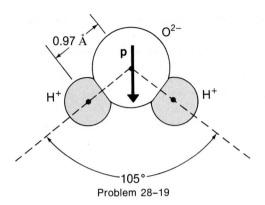

0.97 Å

O^{2-}

p

H^+

H^+

105°

Problem 28–19

is a perpendicular distance a from the center of the square. [*Hint:* Begin by determining the field strength along the perpendicular bisector of a length $2a$ of uniformly charged wire.]

28–22 A negative pointlike charge $-q$ with a mass m is
• constrained to move along the axis of a ring with a radius R that carries a uniformly distributed positive charge $+Q$. Locate the equilibrium position for the negative charge and show that the frequency for small oscillations about this position is $\omega = \sqrt{qQ/4\pi\epsilon_0 mR^3}$.

ELECTRIC POTENTIAL

In this chapter we study the interaction energy of various arrangements of charges. We emphasize the case in which the charges are at rest in the coordinate frame used for the description. When a test charge or a particle is allowed unrestrained motion, it is always assumed that the electric field to which the moving charged particle responds is produced by source charges that remain at rest in some coordinate frame. Moreover, this is the frame that we select for describing the motion of the particle. We make this choice instead of selecting a frame that moves with the particle because the latter would involve a moving (time-varying) electric field. The description of such fields involves complications that we do not wish to introduce at this point. Thus, for the present we restrict attention to the electrostatic case.

29-1 ELECTROSTATIC POTENTIAL ENERGY

Single Source Charge. Consider a point charge Q that is at rest at the origin of some coordinate frame. A second point charge q is observed to move along a curved path from point A to point B (Fig. 29-1). The actual path C that is followed by the charged particle is determined by the field in which it moves, that is, by the Coulomb force exerted on it by Q, as well as by other unspecified forces.

We now calculate the work done on q by the Coulomb force field of Q, ignoring all other energy exchanges that might take place. The fundamental definition of work, in differential form, is

$$dW = \mathbf{F} \cdot d\mathbf{r} \qquad (29\text{-}1)$$

where $\mathbf{F}$ here is the Coulomb force,

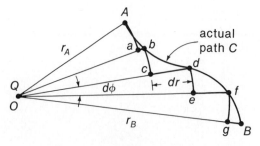

Fig. 29-1. The path C followed by a point charge q in the electrostatic field of the point charge Q is divided into differential radial and circular arc segments.

$$\mathbf{F} = \frac{qQ}{4\pi\epsilon_0 r^2}\hat{\mathbf{u}}_r \tag{29-2}$$

The total work done on q by this force as the particle moves from A to B is*

$$W(A \to B) \equiv W_{AB} = \frac{qQ}{4\pi\epsilon_0}\int_A^B \frac{\hat{\mathbf{u}}_r \cdot d\mathbf{r}}{r^2} \tag{29-3}$$

Although it might appear that the value of the line integral† in Eq. 29–3 depends on the path that connects the points A and B, we now show that the integral is, in fact, completely independent of the path. Imagine that the actual path $C = \widehat{AB}$ is divided into a large number of alternating differential radial and circular arc segments. In the limit of an infinite number of segments, the broken path becomes identical to the actual path. For clarity, Fig. 29–1 shows only a few such segments; thus, $\widehat{AB} = \widehat{Aa} + \overline{ab} + \widehat{bc} + \ldots + \widehat{gB}$. Along the circular arc segments, such as $\widehat{Aa}$ and $\widehat{bc}$, we have $d\mathbf{r} = r\,d\phi\hat{\mathbf{u}}_\phi$. Because $\hat{\mathbf{u}}_r \cdot \hat{\mathbf{u}}_\phi = 0$, these circular segments make no contribution to the line integral; that is, the force is perpendicular to the displacement, so the work done is zero. Along the radial segments, such as $\overline{ab}$ and $\overline{cd}$, we have $d\mathbf{r} = dr\hat{\mathbf{u}}_r$. Because $\hat{\mathbf{u}}_r \cdot \hat{\mathbf{u}}_r = 1$, the expression for W_{AB} reduces to

$$W_{AB} = \frac{qQ}{4\pi\epsilon_0}\int_{r_A}^{r_B} \frac{dr}{r^2} = -\frac{qQ}{4\pi\epsilon_0}\cdot\frac{1}{r}\bigg|_{r_A}^{r_B}$$

or

$$W_{AB} = \frac{qQ}{4\pi\epsilon_0}\left(\frac{1}{r_A} - \frac{1}{r_B}\right) \tag{29-4}$$

Thus, the work done W_{AB} depends only on r_A and r_B, the radial distances from Q to the points A and B, respectively; W_{AB} does not depend on the shape of the path that connects these points.

Suppose that the path C begins at A, follows some curve, and ends again at A; then, the points A and B coincide so that the line integral vanishes. That is,

$$W = \oint \mathbf{F} \cdot d\mathbf{r} = \frac{qQ}{4\pi\epsilon_0}\oint \frac{\hat{\mathbf{u}}_r \cdot d\mathbf{r}}{r^2} = 0 \tag{29-5}$$

In Section 8–3 it was pointed out that any force for which the closed-path line integral vanishes, as in Eq. 29–5, is a *conservative force*. For such forces the idea of *potential energy* is useful. Accordingly, we introduce the electrostatic potential energy $U(\mathbf{r})$ for the q-Q charge system and write

$$U(B) - U(A) = -W_{AB}$$

or

---

* We take this attitude throughout; when work is done *by* the field, $W > 0$, and when work is done *on* the field, $W < 0$.

† Line integrals are discussed in Section 8–2.

$$U(B) - U(A) = \frac{qQ}{4\pi\epsilon_0}\left(\frac{1}{r_B} - \frac{1}{r_A}\right) \tag{29-6}$$

Notice that W_{AB} is equal to the *negative* of the change in potential energy. (Compare the development in Section 7–4.) As in the case of mechanical systems, only *differences* in electric potential energy are physically meaningful. The potential energy of a system at a particular point is defined only to within an arbitrary additive constant. In the SI system, potential energy is measured in joules (J).

We may consider r_A to define the base reference level from which potential energy is measured. It is often convenient to let point A be at an infinite distance, $r_A = \infty$, so that $U(A) = U(\infty) = 0$. Then, with $r_B = r$ representing a general point, we have

$$U(r) = \frac{qQ}{4\pi\epsilon_0} \cdot \frac{1}{r} \tag{29-7}$$

$U(r)$ is the electrostatic potential energy of the charges q and Q when they are separated by a distance r.

Suppose that the point charge q is initially at a point defined by the position vector $\mathbf{r}_1$. If the particle is released from rest, it will move directly toward or away from Q (depending on the relative signs of the two charges) and will be accelerated by the Coulomb force due to Q. Upon arriving at the point $\mathbf{r}_2$, the particle will have acquired a kinetic energy K given by

$$K = \tfrac{1}{2}mv^2 = W_{12} = U(r_1) - U(r_2) = \frac{qQ}{4\pi\epsilon_0}\left(\frac{1}{r_1} - \frac{1}{r_2}\right)$$

That is, some of the original potential energy has been converted into kinetic energy. Notice that $U(r_2) < U(r_1)$ whether the particle moves toward or away from Q. (•Can you see why?) For the mechanical counterpart of this situation, refer to Example 7–5.

Several Source Charges. Next, consider the electrostatic potential energy associated with a group of point charges. First, examine the simple case of three charges, Q_1, Q_2, and q, shown in Fig. 29–2. To calculate the potential energy of this configuration, imagine that the charges are initially separated by infinite distances from each other and from the final position of the group. In accordance with the convention that $U(\infty) = 0$, we assign zero potential energy to the initial deployment of the charges.

The electrostatic force between each pair of charges obeys the superposition principle, so there is a potential energy associated with each pair in the final arrangement. The potential energy of the assembled charges is equal to the negative of the work done by the field in bringing the charges together into the final configuration (Fig. 29–2). First, bring the charge Q_1 to its position defined by the vector $\mathbf{r}_{01}$; no work is done during this process because the only field in the region is that of Q_1 itself. Next, charge Q_2 is brought to the position $\mathbf{r}_{02}$; the negative of the work done by the field during this process, namely, $-W = Q_1Q_2/4\pi\epsilon_0 R$, is equal to the potential energy contributed to the system. Finally, when q is brought to the position $\mathbf{r}$, there are two additional contributions to the total potential energy, namely, $qQ_1/4\pi\epsilon_0 r_1$ and $qQ_2/4\pi\epsilon_0 r_2$. Thus, the total potential energy of the system is

Fig. 29–2. A static configuration of the charges Q_1, Q_2, and q.

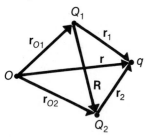

$$U(R, r_1, r_2) = \frac{q}{4\pi\epsilon_0}\left(\frac{Q_1}{r_1} + \frac{Q_2}{r_2}\right) + \frac{Q_1 Q_2}{4\pi\epsilon_0 R}$$

Suppose that Q_1 and Q_2 are held in fixed positions and q is allowed to move around in the electrostatic field generated by Q_1 and Q_2. Then, the Q_1-Q_2 contribution to the potential energy remains constant, and the reference level can be redefined so that

$$U(\mathbf{r}) = \frac{q}{4\pi\epsilon_0}\left(\frac{Q_1}{r_1} + \frac{Q_2}{r_2}\right) \qquad \textbf{(29-8)}$$

This expression is for the case in which only q is free to move, and represents the relevant potential energy when q is at the point defined by $\mathbf{r}$ (see Fig. 29-2).

If many point charges, $Q_1, Q_2, Q_3, \ldots$, are present in a fixed array in which only the point charge q is free to move, Eq. 29-8 can be generalized to

$$U(\mathbf{r}) = \frac{q}{4\pi\epsilon_0}\sum_i \frac{Q_i}{r_i} \qquad \textbf{(29-9a)}$$

where we again ignore the constant potential energy of the fixed array of charges Q_i. If the array becomes a continuous distribution with a charge density ρ, the sum in Eq. 29-9a is replaced by an integral:

$$U(\mathbf{r}) = \frac{q}{4\pi\epsilon_0}\int_Q \frac{dQ}{r_c} = \frac{q}{4\pi\epsilon_0}\int_V \frac{\rho\,dv}{r_c} \qquad \textbf{(29-9b)}$$

where r_c is the distance to q from the differential charge element $dQ = \rho\,dv$, and where the integration is over the entire distribution. Notice that in Eqs. 29-8 and 29-9 we have written $U(\mathbf{r})$, where $\mathbf{r}$ is the position vector of the charge q. In each case, the sum or integration involves the coordinates $r_i = |\mathbf{r} - \mathbf{r}_{Oi}|$ or $r_c = |\mathbf{r} - \mathbf{r}_{Oc}|$. Because the coordinates $\mathbf{r}_{Oi}$ and $\mathbf{r}_{Oc}$ are *fixed*, the result is a function of $\mathbf{r}$ alone. The examples at the end of this section clarify this point.

Because the superposition principle is valid, the electric force on a charge that is free to move in the electrostatic field of an array of fixed charges is necessarily a conservative force.

The Electrostatic Potential. In Section 28-1 the electric field strength $\mathbf{E}$ was defined to be the limit of $\mathbf{F}_E/q_0$ as $q_0 \to 0$. We now introduce a related quantity, the *electrostatic potential* Φ, by a similar definition, namely,

$$\Phi(\mathbf{r}) \equiv \lim_{q_0 \to 0} \frac{U(\mathbf{r})}{q_0} \qquad \textbf{(29-10)}$$

Combining this definition with Eqs. 29-9 gives

$$\Phi(\mathbf{r}) = \frac{1}{4\pi\epsilon_0}\sum_i \frac{Q_i}{r_i} \qquad \textbf{(29-11a)}$$

and

$$\Phi(\mathbf{r}) = \frac{1}{4\pi\epsilon_0} \int_Q \frac{dQ}{r_c} = \frac{1}{4\pi\epsilon_0} \int_V \frac{\rho \, dv}{r_c} \tag{29-11b}$$

We frequently refer to the *potential difference* between two points, A and B, in which case we write $\Phi(B) - \Phi(A)$ or $\Phi_B - \Phi_A$. As in the case of the electric potential energy we can choose the zero reference level for $\Phi(\mathbf{r})$ to be at infinity so that $\Phi(\infty) = 0$.

Potential and potential difference are measured in *joules per coulomb* (J/C). To this unit we give the special name *volt** (V); thus, $1 \text{ V} = 1 \text{ J/C}$. Because $1 \text{ J} = 1 \text{ N} \cdot \text{m}$, it follows that electric field strength, previously specified in N/C, can be expressed in the equivalent units of *volts per meter* (V/m).

When we wish to emphasize that the potential difference between two points is just the *voltage* between those points, we often use an equivalent notation, writing $V_{BA} \equiv \Phi_{BA} \equiv \Phi_B - \Phi_A$. If no confusion can arise, we simplify still further and write V for V_{BA}.

Example 29-1

Determine the electric potential due to two point charges, Q_1 and Q_2, along a perpendicular bisector of the line joining the charges.

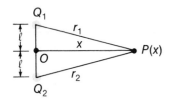

Solution:

The distances from Q_1 and Q_2 to $P(x)$ are $r_1 = r_2 = (x^2 + \ell^2)^{1/2}$, so that from Eq. 29-11a we have

$$\Phi(x) = \frac{1}{4\pi\epsilon_0}\left(\frac{Q_1}{(x^2 + \ell^2)^{1/2}} + \frac{Q_2}{(x^2 + \ell^2)^{1/2}}\right) = \frac{Q_1 + Q_2}{4\pi\epsilon_0} \cdot \frac{1}{(x^2 + \ell^2)^{1/2}}$$

Because ℓ is constant, the potential is a function only of x. When $x^2 \gg \ell^2$, the expression for the potential reduces to

$$\Phi(x) \cong \frac{Q}{4\pi\epsilon_0} \cdot \frac{1}{x}, \qquad x^2 \gg \ell^2$$

where $Q = Q_1 + Q_2$. Thus, $\Phi(x^2 \gg \ell^2)$ is the same as the potential of a single charge Q located at the origin O.

Example 29-2

Derive an expression for the electric potential at large distances from an electric dipole.

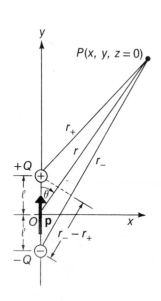

Solution:

We could proceed in a manner similar to that used in Section 28-1 to obtain the distant-solution expression for the electric field strength due to a dipole. However, because the potential is a scalar quantity, a simpler method is available. First, we write an *exact* expression for the potential at a point P, as shown in the diagram; thus,

---

* Named in honor of Count Alessandro Volta (1745–1827), Italian physicist, who constructed the first chemical battery or *voltaic cell*.

$$\Phi(\mathbf{r}) = \frac{Q}{4\pi\epsilon_0}\left(\frac{1}{r_+} - \frac{1}{r_-}\right) = \frac{Q}{4\pi\epsilon_0}\left(\frac{r_- - r_+}{r_+ r_-}\right)$$

Because the potential is required only at large distances (that is, $r \gg \ell$), we can use the approximate relationships (refer to the diagram),

$$r_- - r_+ \cong 2\ell\cos\theta \qquad \text{and} \qquad r_+ r_- \cong r^2$$

Then, the potential becomes

$$\Phi(\mathbf{r}) = \frac{Q}{4\pi\epsilon_0}\frac{2\ell\cos\theta}{r^2}$$

We again introduce the electric dipole moment, $\mathbf{p} = 2Q\ell\hat{\mathbf{j}}$, so that $2Q\ell\cos\theta = \mathbf{p}\cdot\mathbf{r}/r$. Thus, we have

$$\Phi(\mathbf{r}) = \frac{\mathbf{p}\cdot\mathbf{r}}{4\pi\epsilon_0 r^3}$$

The geometry of the dipole has cylindrical symmetry about the direction of $\mathbf{p}$; hence, the vector $\mathbf{r}$ in the expression for $\Phi(\mathbf{r})$ may be considered to be a general three-dimensional position vector.

Example 29-3

Calculate the electric potential along the axis of a uniformly charged ring with radius a.

Solution:

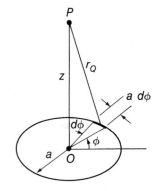

The distance from the ring segment to the point P is $r_c = (z^2 + a^2)^{1/2}$. The charge on this segment is $dQ = \lambda a\, d\phi$, where $\lambda = Q/2\pi a$ is the linear charge density on the ring. Using Eq. 29-11b, we have

$$\Phi(z) = \frac{1}{4\pi\epsilon_0}\int_Q \frac{dQ}{r_c} = \frac{a\lambda}{4\pi\epsilon_0(z^2+a^2)^{1/2}}\int_0^{2\pi} d\phi$$

$$= \frac{2\pi a\lambda}{4\pi\epsilon_0(z^2+a^2)^{1/2}} = \frac{Q}{4\pi\epsilon_0}\cdot\frac{1}{(z^2+a^2)^{1/2}}$$

Again, notice that when $z^2 \gg a^2$, the potential $\Phi(z)$ is the same as that for a point charge Q located at the origin O.

Example 29-4

Determine the electric potential along the axis of a uniformly charged disc with radius R.

Solution:

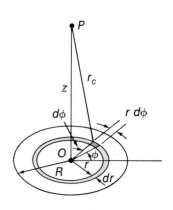

The surface charge density of the disc is $\sigma = Q/\pi R^2$. Thus, for the charge on the small annular ring shown in the diagram, $dQ = 2\pi r\sigma\, dr$. Using the result of Example 29-3 together with $r_c = (z^2 + r^2)^{1/2}$, we have

$$\Phi(z) = \frac{\sigma}{2\epsilon_0}\int_0^R \frac{r\, dr}{(z^2+r^2)^{1/2}}$$

$$= \frac{\sigma}{2\epsilon_0}(z^2 + r^2)^{1/2}\Big|_0^R = \frac{\sigma}{2\epsilon_0}[(z^2 + R^2)^{1/2} - z]$$

$$= \frac{Q}{2\pi\epsilon_0 R^2}[(z^2 + R^2)^{1/2} - z]$$

To investigate the behavior of the potential when $z^2 \gg R^2$, we expand the radical in the expression

$$\Phi(z) = \frac{Q}{2\pi\epsilon_0 R^2}[z(1 + R^2/z^2)^{1/2} - z]$$

Thus, we obtain

$$\Phi(z) = \frac{Q}{2\pi\epsilon_0 R^2}\left[z\left(1 + \frac{1}{2}\frac{R^2}{z^2} + \cdots\right) - z\right]$$

from which

$$\Phi(z) \cong \frac{Q}{4\pi\epsilon_0 z}, \qquad z^2 \gg R^2$$

which is the expected result.

29-2 FIELD STRENGTH-POTENTIAL RELATIONSHIPS

In the preceding section we obtained expressions for the direct determination of the electric potential for any static distribution of charge. The existence of an electric potential was assured in each case by the conservative nature of the force field associated with the static charges.

Suppose now that the electric field strength $\mathbf{E}(\mathbf{r})$ is given. If it is *also known* that $\mathbf{E}$ is produced solely by a static charge distribution, thereby guaranteeing that the field is conservative, we may write

$$U(B) - U(A) = -W_{AB} = -\oint_A^B \mathbf{F} \cdot d\mathbf{r}$$

or

$$U(B) = -q\oint_A^B \mathbf{E} \cdot d\mathbf{r} + U(A) \qquad (29\text{-}12)$$

where we have used $\mathbf{F} = q\mathbf{E}$. Now, the electric potential is defined by $\Phi(\mathbf{r}) = \lim_{q\to 0} U(\mathbf{r})/q$ (Eq. 29-10). Then, Eq. 29-12 becomes

$$\Phi(B) = -\oint_A^B \mathbf{E} \cdot d\mathbf{r} + \Phi(A) \qquad (29\text{-}13)$$

Finally, let point B be an arbitrary point in the field defined by the position vector $\mathbf{r}$, whereas the point A serves to establish the reference level for the potential. Then, we can write

$$\Phi(\mathbf{r}) = -\oint_A^{\mathbf{r}} \mathbf{E} \cdot d\mathbf{r} + \Phi(A) \qquad \textbf{(29–13a)}$$

(For economy we have not used a new symbol for the integration variable, but no confusion should arise.) Notice that the path C connecting point A and $\mathbf{r}$ can be *any* path, even a path selected solely for convenience. This is the case because the field $\mathbf{E}$ is produced by static charges only and is therefore a conservative field. The principle virtue of describing the situations in terms of the *potential* instead of the *potential energy* is that we are then able to deal with a quantity—the potential—that depends everywhere in space only on the source charges.

If the line integral of $\mathbf{E} \cdot d\mathbf{r}$ is calculated along a closed path (so that points A and B coincide), the integral vanishes; thus,

$$\oint \mathbf{E} \cdot d\mathbf{r} = 0 \qquad \textbf{(29–14)}$$

This is a general result for a static distribution of charge (a conservative field).

The Gradient. Suppose that $\mathbf{E}(\mathbf{r})$ is to be determined for a given potential $\Phi(\mathbf{r})$. Consider first the effect of a small displacement in the x-direction from a point $P(x, y, z)$ to a point $P(x + \Delta x, y, z)$, as shown in Fig. 29–3. From Eq. 29–13, we have

$$\Phi(x + \Delta x, y, z) - \Phi(x, y, z) = -\int_x^{x+\Delta x} E_x \, dx = -\bar{E}_x \Delta x$$

where $\bar{E}_x$ is the average value of E_x in the interval Δx. Thus,

$$\bar{E}_x = -\frac{\Phi(x + \Delta x, y, z) - \Phi(x, y, z)}{\Delta x}$$

Fig. 29–3. A change in the position of the point P results in an incremental change in the potential.

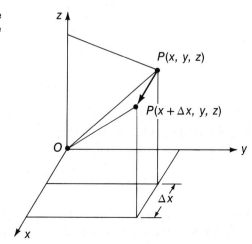

In the limit as $\Delta x \to 0$, $\bar{E}_x$ becomes $E_x(x, y, z)$ and the quantity on the right-hand side becomes the partial derivative* of Φ with respect to x; that is,

$$E_x(x, y, z) = -\frac{\partial \Phi(x, y, z)}{\partial x} \qquad \text{(29–15a)}$$

In a similar way we can show that

$$E_y(x, y, z) = -\frac{\partial \Phi(x, y, z)}{\partial y} \qquad \text{(29–15b)}$$

$$E_z(x, y, z) = -\frac{\partial \Phi(x, y, z)}{\partial z} \qquad \text{(29–15c)}$$

The result we are seeking is therefore

$$\mathbf{E}(x, y, z) = -\left[\frac{\partial \Phi}{\partial x}\hat{\mathbf{i}} + \frac{\partial \Phi}{\partial y}\hat{\mathbf{j}} + \frac{\partial \Phi}{\partial z}\hat{\mathbf{k}}\right] \qquad \text{(29–16)}$$

It is convenient to introduce a differential quantity called the *del* (or *nabla*) *operator,* which is defined as

$$\nabla \equiv \frac{\partial}{\partial x}\hat{\mathbf{i}} + \frac{\partial}{\partial y}\hat{\mathbf{j}} + \frac{\partial}{\partial z}\hat{\mathbf{k}} \qquad \text{(29–17)}$$

The application of this operator to a scalar quantity $f(x, y, z)$ that is differentiable in x, y, and z means

$$\nabla f(x, y, z) = \frac{\partial f}{\partial x}\hat{\mathbf{i}} + \frac{\partial f}{\partial y}\hat{\mathbf{j}} + \frac{\partial f}{\partial z}\hat{\mathbf{k}} \qquad \text{(29–18)}$$

Notice that when ∇ operates on a *scalar* quantity, the result is a *vector* quantity. Equation 29–16 can now be expressed in compact form as

$$\mathbf{E} = -\nabla\Phi \qquad \text{(29–19)}$$

Fig. 29–4. A family of equipotential curves, $\Phi(x, y) = $ constant.

In words, we say that the field strength $\mathbf{E}$ is the negative of the *gradient* of the potential Φ. Thus, to determine $\mathbf{E} = \mathbf{E}(x, y, z)$ at the point $P(x, y, z)$ we must know $\Phi(x, y, z)$ everywhere in the neighborhood of P so that the required differentiation can be carried out.

In Section 29–5 we discuss other applications of the del operator.

Equipotentials and the Gradient.   Consider a simple two-dimensional case in which the electric potential is $\Phi(x, y)$. The curves described by $\Phi(x, y) = $ constant are called *equipotential curves* (Fig. 29–4). In three dimensions, $\Phi(x, y, z) = $ constant describes *equipotential surfaces.*

---

*The partial derivative $\partial\Phi/\partial x$ is equal to the derivative of Φ with respect to x when the independent variables y and z are held constant. (See the appendix titled Calculus IV, in Vol. I.)

The electric potential at the point $P(x, y)$ is $\Phi(x, y)$. What is the change in potential $d\Phi$ between $P(x, y)$ and some neighboring point $P(x + dx, y + dy)$? Refer to Fig. 29-5a; we have

$$d\Phi = \Phi(x + dx, y + dy) - \Phi(x, y)$$

$$= [\Phi(x + dx, y + dy) - \Phi(x, y + dy)] + [\Phi(x, y + dy) - \Phi(x, y)]$$

where we have added and subtracted the term $\Phi(x, y + dy)$. Thus, we identify (see Eq. IV-4 in the appendix titled Calculus IV)

$$d\Phi = \frac{\partial\Phi}{\partial x}dx + \frac{\partial\Phi}{\partial y}dy \qquad (29\text{-}20)$$

Now, the vector displacement from $P(x, y)$ to $P(x + dx, y + dy)$ is

$$d\mathbf{r} = dx\,\hat{\mathbf{i}} + dy\,\hat{\mathbf{j}} \qquad (29\text{-}21)$$

Also, notice that

$$(\nabla\Phi) \cdot d\mathbf{r} = \left[\frac{\partial\Phi}{\partial x}\hat{\mathbf{i}} + \frac{\partial\Phi}{\partial y}\hat{\mathbf{j}}\right] \cdot (dx\,\hat{\mathbf{i}} + dy\,\hat{\mathbf{j}})$$

$$= \frac{\partial\Phi}{\partial x}dx + \frac{\partial\Phi}{\partial y}dy \qquad (29\text{-}22)$$

Comparing Eqs. 29-20 and 29-22 we see that

$$d\Phi = (\nabla\Phi) \cdot d\mathbf{r} \qquad (29\text{-}23)$$

Next, consider a displacement $d\mathbf{r}$ that is directed along the tangent to the equipotential curve through the point $P(x, y)$. This differential displacement is entirely along the equipotential and therefore results in no change of the potential; that is, $d\Phi = 0$. Consequently, we have

$$d\Phi = (\nabla\Phi) \cdot d\mathbf{r} = 0 \qquad (d\mathbf{r} \text{ along equipotential})$$

Because neither $\nabla\Phi$ nor $d\mathbf{r}$ is, in general, zero, we conclude that $\nabla\Phi$ is perpendicular to $d\mathbf{r}$; that is, $\nabla\Phi$ is perpendicular to the line tangent to the equipotential curve at point P (see Fig. 29-5b). For the three-dimensional case, $\nabla\Phi$ coincides with the vector that is normal to the equipotential surface $\Phi(x, y, z) = $ constant. Thus, the gradient of a potential function at a point is always normal to the equipotential through that point. Moreover, because $\mathbf{E} = -\nabla\Phi$, *the direction of* $\mathbf{E}$ *is also always normal to the equipotential through the point.*

Next, we note that $d\Phi$ has its maximum value when $d\mathbf{r}$ is parallel to $\nabla\Phi$ (Eq. 29-23). Then, we can write

$$\left(\frac{d\Phi}{dr}\right)_{\text{max}} = |\nabla\Phi| \qquad (29\text{-}24)$$

from which we conclude that *the gradient of* Φ *is always in the direction of the maxi-*

(a)

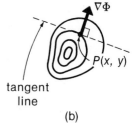

(b)

Fig. 29-5. (a) The displacement $d\mathbf{r}$ connecting points on two different equipotentials. (b) The gradient of Φ at the point $P(x, y)$ is perpendicular to the equipotential curve that passes through $P(x, y)$.

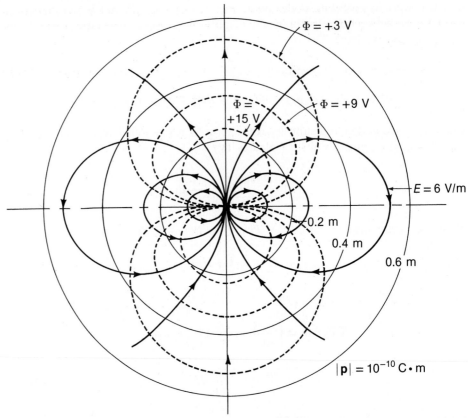

$\Phi = +15$ V

$\Phi = +9$ V

$E = 6$ V/m

0.2 m

0.4 m

0.6 m

$|\mathbf{p}| = 10^{-10}$ C · m

Fig. 29-7. (a) Three-dimensional map of a mountain showing various level curves. (b) Projecting the level curves onto the *x-y* plane results in a topographic map of the mountain.

Fig. 29-6. Field lines and equipotentials for a point dipole with a strength of 10^{-10} C · m. The dipole axis is directed vertically upward in the diagram. The full three-dimensional field is obtained by rotating the pattern of field lines about the dipole axis. Notice that the field lines and equipotentials are everywhere perpendicular.

(a)

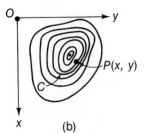

(b)

mum rate of change of Φ *with displacement*. In a map of equipotential curves (such as Fig. 29-4), the gradient has its largest values where the curves are closest together.

Figure 29-6 is a field diagram for a *point dipole* with strength $|\mathbf{p}| = 10^{-10}$ C · m. (A *point dipole* is produced by allowing $\ell \to 0$ in such a way that $2\ell Q$, and hence p, remains constant; the distant-field condition, $r \gg \ell$, is therefore met at all points in space.) Several of the field lines are shown (compare Fig. 29-5); notice the line with the indicated strength of 6 V/m. Also shown are the equipotential surfaces (corresponding to potentials of 3 V, 9 V, and 15 V). These equipotentials (and all others) are everywhere perpendicular to the field lines.

▶ A comparison of these results with topographic maps is instructive. Fig. 29-7 shows a three-dimensional map of a mountain on which are drawn lines that correspond to fixed elevations above sea level. If we take *x* and *y* as horizontal coordinates and *z* as the vertical coordinate, the constant-elevation lines (or *level curves*) have the mathematical form $z = h(x, y) = C$. A hiker who follows such a level curve around the mountain would be always at a fixed elevation. The level curves of Fig. 29-7a are shown projected onto the *x-y* plane in Fig. 29-7b. The point $P(x, y)$ is indicated in both of the representations.

The curves $h(x, y) = C$ are called *contour lines* in topography. You have probably seen such contour lines on topographic maps. By proceeding in a direction that is perpendicular to the contour line at any point, you make the locally steepest descent or ascent of the mountain. ◀

Example 29–5

Determine the electric potential at all points in space for a uniform spherical charge distribution that has a radius R.

Solution:

Let us calculate $\Phi(\mathbf{r})$ from $\mathbf{E}(\mathbf{r})$, which is readily obtained by using Gauss's law. From the results of Example 28–6 (page 775), we have

$$\mathbf{E}(\mathbf{r}) = \begin{cases} \dfrac{Q}{4\pi\epsilon_0 r^2}\,\hat{\mathbf{u}}_r, & r > R \\[2ex] \dfrac{Qr}{4\pi\epsilon_0 R^3}\,\hat{\mathbf{u}}_r, & r < R \end{cases}$$

We use Eq. 29–13a:
$$\Phi(\mathbf{r}) = -\oint_A^{\mathbf{r}} \mathbf{E}\cdot d\mathbf{r} + \Phi(A)$$

and select a reference level for Φ such that $\Phi(A) = 0$ at $r_A = \infty$. Because the electric field is due to static charges and is therefore conservative, we may integrate $\mathbf{E}\cdot d\mathbf{r}$ along any convenient path; we choose a radial path with $\hat{\mathbf{u}}_r\cdot d\mathbf{r} = dr$. Then, for $r \geqslant R$, we have

$$\Phi(r) = -\frac{Q}{4\pi\epsilon_0}\int_\infty^r \frac{dr}{r^2} = \frac{Q}{4\pi\epsilon_0}\cdot\frac{1}{r}\bigg|_\infty^r$$

$$= \frac{Q}{4\pi\epsilon_0}\cdot\frac{1}{r}, \qquad r \geqslant R \tag{1}$$

For $r < R$, we have

$$\Phi(r) = -\frac{Q}{4\pi\epsilon_0 R^3}\int_R^r r\,dr + \Phi(R)$$

where $\Phi(R)$ is obtained from Eq. (1). Thus,

$$\Phi(r) = -\frac{Q}{4\pi\epsilon_0 R^3}\cdot\frac{1}{2}r^2\bigg|_R^r + \frac{Q}{4\pi\epsilon_0 R}$$

$$= \frac{Q}{4\pi\epsilon_0 R}\left(\frac{3}{2} - \frac{r^2}{2R}\right), \qquad r < R \tag{2}$$

These results for $E(r)$ and $\Phi(r)$ are summarized in the diagram below.

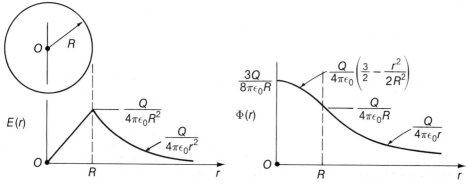

Example 29-6

The electric potential at large distances from a dipole **p** is (see Example 29-2)

$$\Phi(\mathbf{r}) = \frac{\mathbf{p} \cdot \mathbf{r}}{4\pi\epsilon_0 r^3}$$

Show that the gradient of this potential leads to Eq. 28-8, namely,

$$\mathbf{E}(\mathbf{r}) = \frac{1}{4\pi\epsilon_0}\left(-\frac{\mathbf{p}}{r^3} + \frac{3(\mathbf{p} \cdot \mathbf{r})\mathbf{r}}{r^5}\right)$$

Solution:

With $\mathbf{p} = 2\ell Q\hat{\mathbf{j}}$, and using the same geometry as in Fig. 28-2, we have

$$\Phi(x, y, z) = \frac{2\ell Q}{4\pi\epsilon_0 r^3}\hat{\mathbf{j}} \cdot \mathbf{r} = \frac{2\ell Q y}{4\pi\epsilon_0(x^2 + y^2 + z^2)^{3/2}}$$

The required partial derivatives are

$$-\frac{\partial\Phi}{\partial x} = \frac{2\ell Q}{4\pi\epsilon_0} \cdot \frac{3xy}{(x^2 + y^2 + z^2)^{5/2}} = \frac{2\ell Q}{4\pi\epsilon_0} \cdot \frac{3xy}{r^5}$$

$$-\frac{\partial\Phi}{\partial z} = \frac{2\ell Q}{4\pi\epsilon_0} \cdot \frac{3zy}{r^5}$$

and

$$-\frac{\partial\Phi}{\partial y} = \frac{2\ell Q}{4\pi\epsilon_0} \cdot \frac{3y^2(x^2 + y^2 + z^2)^{1/2} - (x^2 + y^2 + z^2)^{3/2}}{(x^2 + y^2 + z^2)^3}$$

$$= \frac{2\ell Q}{4\pi\epsilon_0}\left(\frac{3y^2}{r^5} - \frac{1}{r^3}\right)$$

Therefore,

$$\mathbf{E} = -\nabla\Phi = -\left[\frac{\partial\Phi}{\partial x}\hat{\mathbf{i}} + \frac{\partial\Phi}{\partial y}\hat{\mathbf{j}} + \frac{\partial\Phi}{\partial z}\hat{\mathbf{k}}\right]$$

$$= \frac{2\ell Q}{4\pi\epsilon_0}\left[-\frac{\hat{\mathbf{j}}}{r^3} + \frac{3y}{r^5}(x\hat{\mathbf{i}} + y\hat{\mathbf{j}} + z\hat{\mathbf{k}})\right]$$

$$= \frac{1}{4\pi\epsilon_0}\left[-\frac{\mathbf{p}}{r^3} + \frac{3(\mathbf{p} \cdot \mathbf{r})\mathbf{r}}{r^5}\right]$$

which is Eq. 28-8.

Example 29-7

Use $\mathbf{E} = -\nabla\Phi$ to determine the electric field strength on the axis of a uniformly charged disc.

Solution:

The expression for the potential along the disc axis that was obtained in Example 29-4 does not apply for a general point $P(x, y, z)$. Nevertheless, this expression can be used to find **E** on the axis. The reason is that the axial symmetry insures that **E** has vanishing components in directions perpendicular to the axis. If the disc axis corresponds to the z-axis, only E_z is nonvanishing; thus, we require only the z dependence of Φ to obtain the necessary partial derivative, $\partial\Phi/\partial z$. From the result of Example 29-4, we have

$$\Phi(z) = \frac{\sigma}{2\epsilon_0}[(z^2 + R^2)^{1/2} - z]$$

Then,

$$-\frac{\partial\Phi}{\partial z} = \frac{\sigma}{2\epsilon_0}\left[1 - \frac{z}{(z^2 + R^2)^{1/2}}\right]$$

from which

$$\mathbf{E} = -\frac{\partial\Phi}{\partial z}\hat{\mathbf{k}} = \frac{\sigma}{2\epsilon_0}\left[1 - \frac{z}{(z^2 + R^2)^{1/2}}\right]\hat{\mathbf{k}}$$

This is the same expression we obtained in Eq. 28-25.

29-3   CONDUCTORS

It has been pointed out that conducting materials (as distinct from nonconducting materials or insulators) contain large numbers of electrons—usually one or two per atom—that are free to move about within the material.* These free (conduction) electrons respond to applied electric forces, accounting for the conductive property of such materials. To charge a conductor negatively, one adds electrons; to charge a conductor positively, one removes some of the conduction electrons.

We now adopt the concept of an ideal, homogeneous, isotropic, perfectly conducting substance, which henceforth we simply call a *conductor*. We restrict attention to isolated conductors, either charged or neutral, possibly in the presence of externally generated electric fields. We further assume that only electric forces are important. In every case, we describe the situation only after the charges—placed on the conductors by some external agency—have achieved an equilibrium distribution. The electric field is then everywhere the result of the superposition of externally generated fields, if any, and the field produced by the conductor charge distributions themselves. In no case involving an equilibrium distribution of charge can there be a net electric field within the volume of a conductor.† If such a net interior field existed, it would exert a force on the conduction electrons that would produce a current of moving charges, in violation of the assumption of an equilibrium (static) charge distribution. Any static charge that is carried by a conductor must therefore reside entirely on the surface of the conductor. Moreover, the distribution of this charge must be such that the field it produces, when superposed with any external field, results in zero field within the conductor. When no external field is present, the surface charges are distributed in such a way that there is no internal electric field.

Figure 29-8 shows the result of placing an uncharged solid conducting sphere in an originally uniform electric field $\mathbf{E}_0$ (Fig. 29-8a). In the presence of the field, equal amounts of negative and positive charge are drawn toward the polar regions of the sphere ($\theta \cong 0$ and $\theta \cong \pi$). The equilibrium charge distribution that is induced on the surface of the sphere produces a uniform electric field $\mathbf{E}_C$ within the sphere (Fig.

---

*Not all materials can be classified as either conductors or insulators; an important exception is the group of *semi-conducting* materials, which are discussed in Section 31-6.

†We refer here only to the macroscopic field obtained by averaging over a sufficiently large region of the conductor that the effects of atomic structure have been removed.

Fig. 29-8. (a) The original uniform field $\mathbf{E}_0$. (b) The (dipole) field $\mathbf{E}_C$ produced by the charge distribution induced on the surface of the sphere. (c) The resultant field, $\mathbf{E} = \mathbf{E}_0 + \mathbf{E}_C$, with $\mathbf{E}$(inside) = 0. Notice that on the surface of the sphere neither the field $\mathbf{E}_0$ (a) nor the field $\mathbf{E}_C$ (b) is normal to the surface; only the *true resultant field* $\mathbf{E} = \mathbf{E}_0 + \mathbf{E}_C$ (c) must satisfy this requirement.

(a)

(b)

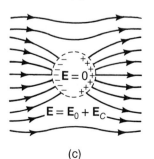

(c)

29–8b). The field $\mathbf{E}_C$ exactly cancels the external field $\mathbf{E}_0$ everywhere inside the sphere; thus, $\mathbf{E}$ (inside) $= \mathbf{E}_0 + \mathbf{E}_C = 0$ (Fig. 29–8c).

To obtain the solution (that is, the equation for $\mathbf{E}_C$) for even a simple problem such as that of a conducting sphere in a uniform field is a complicated affair, because the induced charge distribution and the final field configuration are both unknown. The method of solution must deal simultaneously in a self-consistent way with both of these related unknown quantities. Advanced techniques beyond the scope of this text are required to develop the solution for the general case of a conductor in an external field.

The basic characteristic we have discovered concerning the electrostatic case is that the electric field strength must be exactly zero everywhere within an isolated conductor. A number of important implications follow from this fact. For example, suppose that we construct a closed Gaussian surface that lies entirely within a conductor, as in Fig. 29–9. Then, because $\mathbf{E} = 0$ everywhere within the surface, we have

$$\oint_S \mathbf{E} \cdot \hat{\mathbf{n}} \, dA = 0 = \frac{Q}{\epsilon_0}$$

and the volume defined by the surface can contain no net charge; indeed, there can be no net charge anywhere inside the conductor.*

Next, consider the calculation of the integral of $\mathbf{E} \cdot d\mathbf{r}$ around a closed path, such as the path *abcda* shown in Fig. 29–10. The segment $\widehat{ab}$ of the path lies just outside and parallel to the surface of a conductor. The segment $\widehat{cd}$ lies just inside and parallel to the surface. The two end segments, with lengths Δx and $\Delta x'$, complete the closed path. Let the electric field strength just outside the surface be resolved into a tangential component $\mathbf{E}_t$ and a normal component $\mathbf{E}_n$, as indicated in Fig. 29–10. Orient the integration path so that $\widehat{ab}$ is parallel to $\mathbf{E}_t$. The closed-path line integral is (see Eq. 29–14)

$$\oint_{abcda} \mathbf{E} \cdot d\mathbf{r} = \widehat{ab}E_t - E_n\Delta x' + \widehat{cd} \cdot 0 + E_n\Delta x = 0$$

where we have explicitly indicated that the field strength inside the conductor (segment $\widehat{cd}$) is zero. In the limit $\Delta x \to 0$ and $\Delta x' \to 0$, and with $\widehat{ab}$ small but finite, we have

$$E_t = 0 \qquad \text{on the surface of a conductor} \qquad \textbf{(29–25)}$$

Finally, consider a small cylindrical Gaussian surface S that encloses a segment of a conducting surface, as shown in Fig. 29–11. The top cap of the cylinder, with area ΔA, lies just outside and parallel to the surface of the conductor. The bottom cap lies just below the surface, and the cylinder side wall of height Δx is perpendicular to the surface. Let the density of charge on the conductor surface be σ. Then, Gauss's law can be expressed as

$$\oint_S \mathbf{E} \cdot \hat{\mathbf{n}} \, dA = \frac{Q}{\epsilon_0}$$

Fig. 29–9. The electric field is zero inside the surface S, so no net charge is contained within the surface.

Fig. 29–10. Integration path to show that $E_t = 0$ at the surface of a conductor.

Fig. 29–11. The Gaussian surface used to evaluate E_n.

---

*We exclude the possibility that a charge exists in a hollow cavity in the conductor.

$$E_n \, \Delta A = \frac{\sigma \, \Delta A}{\epsilon_0}$$

where the only contribution to the integral arises from the top cap because the field is zero within the conductor and there is no tangential component to pierce the cylinder wall. Thus, we conclude that

$$E_n = \frac{\sigma}{\epsilon_0} \qquad \begin{array}{l} \text{on the surface} \\ \text{of a conductor} \end{array} \qquad \textbf{(29–26)}$$

The fact that the electric field vector on the surface of a conductor is everywhere normal to the surface means that *conductor surfaces are equipotentials.*

The basic properties of ideal isolated conductors under *electrostatic* conditions can be summarized by the following statements:

(1) *The field strength within a conductor is everywhere zero.*
(2) *The surfaces of conductors are equipotentials.*
(3) *On the surface of a conductor the field strength vector is everywhere normal to the surface:* $\mathbf{E} = \mathbf{E}_n = \dfrac{\sigma}{\epsilon_0}\hat{\mathbf{n}}.$

Charged Conducting Spherical Shell. Consider a thick conducting spherical shell with inner radius R_1 and outer radius R_2 that carries a net charge Q (see Fig. 29–12). The interior of the shell is empty and no additional fields due to external sources are present.

First, construct a spherical Gaussian surface with radius $r' < R_1$. Because there is no charge in the hollow region of the shell, it follows that

$$\mathbf{E} = 0, \qquad r < R_1$$

From this fact and the result stated in Eq. 29–26, we conclude that there can be no charge on the inner surface of the shell, $r = R_1$. The region of the shell itself is the interior of a conductor, so the field there is also zero:

$$\mathbf{E} = 0, \qquad R_1 \leqslant r < R_2$$

Finally, construct a spherical Gaussian surface with radius $r > R_2$. The net charge within this surface is equal to the charge Q on the shell. Thus,

$$4\pi r^2 E = \frac{Q}{\epsilon_0}$$

or

$$\mathbf{E} = \frac{Q}{4\pi\epsilon_0} \cdot \frac{\hat{\mathbf{u}}_r}{r^2}, \qquad r \geqslant R_2$$

When $r = R_2$, we have $E = E_n = \sigma/\epsilon_0$, where the surface charge density is $\sigma = Q/4\pi R_2^2$.

In Section 28–4 we used the fact that the electrostatic force depends on $1/r^2$ to prove by an explicit calculation that the electric field strength inside a uniform thin

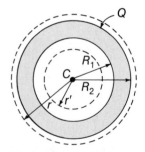

Fig. 29–12. A conducting spherical shell that carries a net charge Q.

spherical charge distribution is everywhere zero. To test experimentally whether $\mathbf{E} = 0$ in such a case requires that a suitable charge distribution be produced. In view of the results just obtained for the field of a charged conducting sphere, we can see how to produce the desired charge distribution. It is only necessary to place a charge on a conducting spherical shell—the charge will automatically arrange itself into the desired uniform distribution. (In fact, the charge on a conducting shell of arbitrary shape will distribute itself so that $\mathbf{E} = 0$ everywhere inside the shell.) The most precise way to test whether Coulomb's law involves exactly the *square* of the distance is to measure the field strength inside a charged conducting shell. Any deviation of the field strength from $\mathbf{E} = 0$ would indicate that the exponent n in $1/r^n$ is not exactly 2. The best experimental evidence indicates that $n = 2$ to within about 3 parts in 10^{16}.

It is often convenient to choose arbitrarily the zero reference level for electric potential and electric potential energy to be at infinity. When dealing with conductors, however, the Earth itself is sometimes chosen as the zero reference for potential. Although the Earth is not an ideal conductor, it represents a readily available and essentially inexhaustible reservoir for supplying (or accepting) electrons. An object can be connected by a wire conductor to a metal stake or pipe that is driven into the Earth. Such an object is said to be *grounded*. The object, the wire, the stake, and the Earth are then all at the same potential, which we choose to be $\Phi = 0$. A grounded object does not necessarily have zero surface charge; induction effects may cause electrons to be drawn from or discharged to the Earth reservoir.

Point Charge Above an Infinite Grounded Conducting Plane. Figure 29–13 shows the lines of force from a positive point charge Q in the region above an infinite flat conducting plane at zero potential. Because the plane is an equipotential surface, the lines of force are perpendicular to the plane at the surface. We know of another charge configuration that has a similar pattern of field lines, namely, a *dipole* (Fig. 29–6), if we identify the grounded plane with a plane midway between the dipole charges and oriented perpendicular to the line that connects the charges. This is an example of a general technique called *imaging* that, in certain cases, permits a straightforward solution of a difficult problem involving induced charges.

For the case of a point charge above a conducting plane (Fig. 29–13), we locate an image point charge, $Q' = -Q$, a distance $\ell' = \ell$ *below* the conducting plane and immediately beneath the actual point charge Q. In this substitute problem we calculate $\mathbf{E}$ (or Φ) for the actual charge plus the image charge as if the conducting plane were not present. Then, the part of the solution for $z \geqslant 0$ is the solution to the actual problem of the point charge above the conducting plane. This assertion arises from the fact that the electric field configuration for a particular charge distribution is

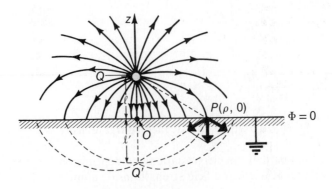

Fig. 29–13. Field lines for a positive point charge above an infinite conducting plane at zero potential. The method of images may be used to obtain this pattern. Notice the special symbol used to indicate *ground* or zero potential. As in Fig. 29–8, only the total field is normal to the surface at the surface; the component fields do not have this property.

unique (that is, has only one possible result). We simply assume the validity of this uniqueness proposition.

We use cylindrical coordinates with the z-axis perpendicular to the plane, as indicated in Fig. 29-13. As in Example 29-2, we have

$$\Phi(\rho, z) = \frac{Q}{4\pi\epsilon_0}\left(\frac{1}{r_+} - \frac{1}{r_-}\right)$$

with

$$r_{\pm} = [\rho^2 + (z \mp \ell)^2]^{1/2}$$

Notice that the expression for $\Phi(\rho, z)$ gives zero potential for $z = 0$, as required. The electric field strength at any point can be determined from $\Phi(\rho, z)$ in the usual way. (Convert the expression to $\Phi(x, y, z)$ and calculate the gradient.) However, to find the distribution of charge induced on the surface of the conducting plane we require only the field strength along the surface; this can be calculated directly, with the result (see Problem 29-14)

$$\mathbf{E}(\rho, 0) = -\frac{Q\ell}{2\pi\epsilon_0}(\rho^2 + \ell^2)^{-3/2}\,\hat{\mathbf{k}} \qquad (29\text{-}27)$$

Using this expression together with Eq. 29-26, we have for the density of the induced charge

$$\sigma(\rho) = -\frac{Q\ell}{2\pi}(\rho^2 + \ell^2)^{-3/2} \qquad (29\text{-}28)$$

This density function is shown in Fig. 29-14.

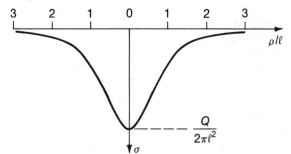

Fig. 29-14. Surface density of charge induced on a grounded conducting plane by a charge $+Q$ a distance ℓ above the plane.

Parallel Charged Conducting Planes. Next, consider the important case of two oppositely charged infinite parallel conducting planes separated by a distance ℓ (Fig. 29-15). When ℓ is sufficiently small, the results also apply for finite planes; this is the case for a *parallel-plate capacitor* (see Section 30-1).

The surface charge densities on the planes are uniform with values $+\sigma_0$ and $-\sigma_0$. The lower plane is grounded, so $\Phi(0) = 0$. From the geometry it is clear that $\mathbf{E}$ must everywhere be normal to the planes. Construct a Gaussian surface as shown in Fig. 29-15, with cap areas ΔA and with cylinder height Δy. The upper cap is within the upper conductor, where $\mathbf{E} = 0$. There is no flux through the cylinder wall, and the entire flux is through the lower cap. Thus, $E\,\Delta A = (\sigma_0/\epsilon_0)\Delta A$, so that

$$\mathbf{E} = -\frac{\sigma_0}{\epsilon_0}\hat{\mathbf{j}} \qquad (29\text{-}29)$$

Fig. 29-15. Equipotentials and lines of force for a pair of oppositely charged infinite parallel conducting planes.

The field strength is independent of y. Applying Eq. 29–13a, we have

$$\Phi(y) = -\int_0^y E\,dy + \Phi(0)$$

or

$$\Phi(y) = \frac{\sigma_0}{\epsilon_0}y \qquad (29\text{--}30)$$

Fig. 29-16. Field lines for a positive point charge in the presence of a grounded conducting sphere. The case illustrated corresponds to $R/\ell = m = 1/3$.

(a)

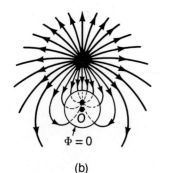

(b)

Thus, the equipotential surfaces are seen to be parallel to the conducting planes, with $\Phi(0) = 0$ at the grounded plane and $\Phi(\ell) = (\sigma_0/\epsilon_0)\ell$ at the upper plane. (This result is expected because the equipotential surfaces must everywhere be normal to the lines of force.) Let V represent the *potential difference* between the planes (measured in *volts*):

$$V = \Phi(\ell) - \Phi(0) = \frac{\sigma_0}{\epsilon_0}\ell$$

or

$$V = E\ell \qquad (29\text{--}31)$$

This simple result is useful in the discussion of capacitors (Section 30–1).

▶ *Point Charge and a Grounded Conducting Sphere.* Another case in which the simple application of imaging is successful is that of a point charge in the presence of a grounded conducting sphere. Suppose that a positive point charge Q is brought to a position a distance ℓ from the center of a conducting sphere with a radius R (Fig. 29-16). A grounding wire is used to connect the sphere to the Earth. The charge Q induces a charge on the sphere, and after equilibrium is reached (almost instantly!) the wire is disconnected, leaving a net charge Q' on the sphere. We now inquire about the potential of the system and the distribution of charge on the surface of the sphere.

The substitute image problem consists of replacing the sphere with a point charge Q' located a distance ℓ' from the center of the sphere. The symmetry of the situation demands that Q' be on the line connecting Q and the center of the sphere, as shown in Fig. 29-16. Also, because of the symmetry there can be no dependence of the potential on the azimuthal angle ϕ. Thus, the potential at the point $P(r, \theta)$ is

$$\Phi(\mathbf{r}) = \frac{Q}{4\pi\epsilon_0}\left(\frac{1}{r_+} - \frac{m}{r_-}\right)$$

where we let $Q' = -mQ$, and where

$$r_+ = [r^2 + \ell^2 - 2\ell r \cos \theta]^{1/2}$$
$$r_- = [r^2 + \ell'^2 - 2\ell' r \cos \theta]^{1/2}$$

Because the sphere is at ground potential, we require $\Phi(\mathbf{r}) = 0$ for all points on the surface of the sphere, namely, $r = R$. It follows immediately that $1/r_+ = m/r_-$ for $r = R$. Substituting $r = R$ in the expressions for r_+ and r_-, we find

$$m^2(R^2 + \ell^2 - 2\ell R \cos \theta) = R^2 + \ell'^2 - 2\ell' R \cos \theta$$

This equation must be valid for all values of θ, so the coefficients of the cosine terms and the angle-independent terms must separately be equal. (•Do you see why?) We obtain

$$m^2(R^2 + \ell^2) = R^2 + \ell'^2$$
$$m^2(2\ell R) = 2\ell' R$$

After some algebra, we find

$$\ell\ell' = R^2 \quad \text{and} \quad m = R/\ell$$

Figure 29-16 shows the lines of force for a point charge $+Q$ at a distance ℓ from the origin and an image point charge $Q' = -mQ$ at a distance ℓ' from the origin for the case $R/\ell = m = 1/3$. For the region $r \geqslant R$, these lines are exactly the same as those for the actual problem. (See also Fig. 28-6.)

Some interesting limiting cases for this solution should be noted. When $\ell \gg R$, m approaches zero and therefore the induced charge Q' also approaches zero. This corresponds to the case of an isolated charge. In the event that Q is close to the sphere, $\ell \cong R$, the value of m approaches unity and $Q' \cong -Q$. In this case, ℓ' also approaches R. If $\ell = R + \delta$, with $\delta \ll R$, we have

$$\ell' = \frac{R^2}{\ell} = \frac{R}{1 + \delta/R} \cong R - \delta$$

This result corresponds to the case of a point charge above a conducting plane. (•Why should this be so? The Earth is a conducting sphere. Does this mean that the "flat Earth" approximation may be used in laboratory-scale experiments?)

In general, we have $R < \ell$, so that $m < 1$ and $|Q'| < |Q|$. Thus, the induced image charge never exceeds the magnitude of the actual charge. ◄

29-4 THE MOTION OF CHARGES IN ELECTROSTATIC FIELDS

Consider now the general motion in an electrostatic field of an unconstrained particle with charge q and mass m. The field is produced by an array of fixed source charges, and the motion is described using a coordinate frame at rest with respect to these charges. We exclude the presence of conductors because the charges induced on conductors would influence the motion of the charged particle being studied. (However, see Problem 29-20.) Also excluded are any induction effects in insulators that might be present.

If the electrostatic field strength is $\mathbf{E}(x, y, z)$, the force on the charged particle is $\mathbf{F} = q\mathbf{E}$; hence, Newton's equation, $m\mathbf{a} = q\mathbf{E}$, yields

$$\mathbf{a}(x, y, z) = \frac{q}{m}\mathbf{E}(x, y, z) \tag{29-32}$$

The problem of finding the equation of motion for a particle in a given electrostatic field reduces to the mechanics problem in which $\mathbf{a}(\mathbf{r})$ and $\mathbf{v}_0$ are given and it is required to determine $\mathbf{r}(t)$. Notice that only the charge-to-mass ratio q/m of the particle enters in Eq. 29–32. Consequently, no study of the motion of the particle in an electrostatic field will reveal the value of either q or m separately.

The Electron Volt. Suppose that an electron is released from rest in a uniform electrostatic field. (We saw in Fig. 29–15 how such a field can be created.) Let the field be in the direction of the negative x-axis, so that $\mathbf{E} = -E_0\hat{\mathbf{i}}$. The electron $(q = -e)$ will be accelerated in the positive x-direction, with $a_x = (e/m)E_0$. The velocity of the electron after an elapsed time t will be $\mathbf{v}(t) = (e/m)E_0t\hat{\mathbf{i}}$ and the position will be $x(t) = \frac{1}{2}(e/m)E_0t^2 + x_a$, where x_a is the position at $t = 0$. In moving from the initial position x_a to a point x_b the electron will have acquired a kinetic energy equal to the work done by the field, namely,

$$\tfrac{1}{2}mv^2 = eE_0(x_b - x_a)$$

But $E_0(x_b - x_a)$ is just the potential difference, V_{ba}, between the two points; hence,

$$\tfrac{1}{2}mv^2 = e \times V_{ba}$$

If an electron (or any other particle with a charge e) is accelerated through a potential difference of 1 V, it will acquire an increase in kinetic energy equal to

$$e \times V = (1.602 \times 10^{-19}\ \text{C}) \times (1\ \text{V})$$

$$= 1.602 \times 10^{-19}\ \text{J}$$

This unit of work or energy is given the special name *electron volt*, abbreviated eV:

$$1\ \text{eV} = 1.602 \times 10^{-19}\ \text{J}$$

Energies associated with atomic phenomena are typically a few eV. For example, an energy of 5.1 eV is required to remove the least bound electron from an atom of sodium. X rays can have energies of a few thousand eV (keV). Nuclear reactions and transitions frequently involve energies of a few million eV (MeV). And elementary particle phenomena often involve energies of billions of eV (GeV). The electron volt is a general energy unit and can be used to specify the energy of any object, whether charged or not.

Electrostatic Deflection. The operation of devices such as television picture tubes depends on the deflection of an electron beam in an electric field. Consider a pencil-like electron beam that enters a uniform electrostatic field in a direction perpendicular to the field vector $\mathbf{E}$ (Fig. 29–17). The field is confined to the region $-\frac{1}{2}\ell \leqslant y \leqslant \frac{1}{2}\ell$ and $0 \leqslant x \leqslant d$; the extent in the z-direction is arbitrary. The field is produced by establishing two equipotentials, $\Phi = 0$ along $y = -\frac{1}{2}\ell$ and $\Phi = V_d > 0$ along $y = +\frac{1}{2}\ell$. As indicated in Fig. 29–17, the beam enters the field with each electron having a velocity $\mathbf{v}_0$ directed along the positive x-axis.

The velocity components are

$$\left. \begin{aligned} v_x &= v_0 = \text{constant} \\ v_y &= a_yt = (e/m)Et, \qquad t = x/v_0 \end{aligned} \right\} \tag{29–33}$$

At the point $P(d, y)$ where the beam emerges from the field, we have

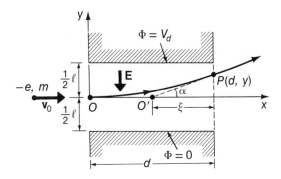

Fig. 29-17. An electron beam is deflected through an angle α by a uniform electrostatic field.

$$v_y = \frac{eE}{m}\frac{d}{v_0}$$

$$y = \frac{1}{2}a_y t^2 = \frac{1}{2}\frac{eE}{m}\frac{d^2}{v_0^2} \qquad (29\text{-}34)$$

The angle of deflection α is given by

$$\tan\alpha = \frac{v_y}{v_x} = \frac{eE}{m}\frac{d}{v_0^2} \qquad (29\text{-}35)$$

The point O' from which the deflected beam appears to emanate can be found by noting that $\tan\alpha = y/\xi$. Then, using Eqs. 29-34 and 29-35, we obtain $\xi = \frac{1}{2}d$. Thus, the apparent origin of the deflected beam is midway through the field in the x-direction.

It is usually the case in *electron guns* that electrons emitted by a heated cathode are accelerated to a high velocity v_0 by causing the electrons to pass through a potential difference V_a. Thus, $\frac{1}{2}mv_0^2 = eV_a$. The strength E of the deflecting field is related to the potential difference V_d by $E = V_d/\ell$ (Eq. 29-31). Substituting for v_0^2 and E in Eq. 29-35, we find

$$\tan\alpha = \frac{d}{2\ell}\frac{V_d}{V_a} \qquad (29\text{-}36)$$

Figure 29-18 is a schematic diagram of a *cathode-ray tube* (CRT), similar to

Fig. 29-18. Essential features of a cathode-ray tube. Plates are included for both horizontal and vertical deflection so that time-varying signals $V_d(t)$ can be displayed.

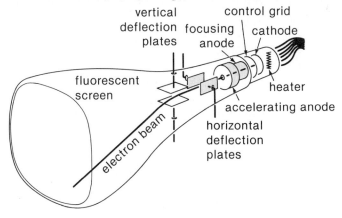

those found in oscilloscopes and in television receivers. The accelerating potential difference V_a in the electron gun is usually in the range from a few thousand volts to about 30,000 V in color television picture tubes.

Example 29–8

Determine the deflection angle α of an electron beam and the transit time through a uniform electrostatic deflecting field for the case in which $V_a = 2000$ V, $d = 4$ cm, $l = 1.5$ cm, and $V_d = 50$ V.

Solution:

Using Eq. 29–36, we have

$$\tan \alpha = \frac{d}{2l}\frac{V_d}{V_a} = \frac{(0.04\text{ m})(50\text{ V})}{2(0.015\text{ m})(2000\text{ V})} = \frac{1}{30}$$

so that

$$\alpha = 1.91°$$

To obtain the transit time from $T = d/v_0$ we need v_0:

$$v_0 = \sqrt{2\frac{e}{m}V_a}$$

For electrons, $e/m = 1.76 \times 10^{11}$ C/kg; thus,

$$v_0 = \sqrt{2(1.76 \times 10^{11}\text{ C/kg})(2000\text{ V})}$$
$$= 2.65 \times 10^7 \text{ m/s}$$

Then,

$$T = \frac{d}{v_0} = \frac{0.04\text{ m}}{2.65 \times 10^7 \text{ m/s}} = 1.51 \text{ ns}$$

Electric Dipole in a Uniform Electrostatic Field. Consider an electric dipole that consists of charges $+q$ and $-q$ separated by a distance $2l$, where l is the distance of each charge from the center of mass (C.M.) of the dipole, as indicated in Fig. 29–19. The dipole is placed in a uniform electrostatic field $\mathbf{E}$ with the dipole moment $\mathbf{p}$ at an angle θ with respect to $\mathbf{E}$. The force on each charge has the magnitude $F = qE$, with directions along and opposite to $\mathbf{E}$ as shown in Fig. 29–19. The net force on the dipole is zero, so there is no motion of the C.M.; however, there is a torque about the C.M. with magnitude

$$\tau_0 = 2lqE \sin \theta$$

which acts in the clockwise direction. This result can be expressed in compact vector notation as

$$\tau_0 = \mathbf{p} \times \mathbf{E} \qquad (29\text{–}37)$$

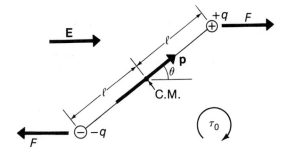

Fig. 29-19. An electric dipole with moment $p = 2\ell q$ in a uniform electro-static field.

where $p = 2\ell q$. The dipole is in a state of stable equilibrium when $\theta = 0$. (•What is the situation when $\theta = \pi$?)

When $\theta = \pi/2$, we let the potential energy of the dipole be U_0. (This corresponds to the work required to assemble the dipole in the absence of the field.) Then, the application of an external torque $\boldsymbol{\tau}_0'$ (equal to $-\boldsymbol{\tau}_0$) results in a rotation of the dipole to an angle θ and produces a change in the potential energy,

$$U - U_0 = -W(\theta) = -\int_{\pi/2}^{\theta} \tau_0'(\theta)\, d\theta = \int_{\pi/2}^{\theta} \tau_0(\theta)\, d\theta$$

$$= 2\ell qE \int_{\pi/2}^{\theta} \sin\theta\, d\theta = -pE\cos\theta$$

or, in vector notation,

$$U = -\mathbf{p} \cdot \mathbf{E} + U_0 \qquad\qquad \textbf{(29–38)}$$

29-5   MORE ABOUT ELECTROSTATIC POTENTIALS

We have already established some of the facts concerning electrostatic fields and potentials. When an electric field with strength $\mathbf{E}(\mathbf{r})$ is produced by an array of stationary charges, the field is conservative and the line integral $\int_A^B \mathbf{E} \cdot d\mathbf{r}$ is independent of the path C between the points A and B. If the path is closed so that the end point is the same as the starting point, the line integral vanishes (Eq. 29-14):

$$\oint \mathbf{E} \cdot d\mathbf{r} = 0$$

When this equation is satisfied for all possible closed paths in the field, we can define a potential function $\Phi(\mathbf{r})$, which is a scalar function with a unique value associated with every point in space. By selecting some point A as the reference point for the potential, we have (Eq. 29-13a)

$$\Phi(\mathbf{r}) = -\oint_A^r \mathbf{E}(\mathbf{r}) \cdot d\mathbf{r} + \Phi(A) \qquad \textbf{(29–39)}$$

If the potential $\Phi(\mathbf{r})$ is given, the field strength $\mathbf{E}(\mathbf{r})$ can be found by taking the negative of the gradient of $\Phi(\mathbf{r})$; that is,

$$\mathbf{E}(\mathbf{r}) = -\nabla\Phi(\mathbf{r}) \qquad \textbf{(29–40)}$$

where ∇ is the differential *del* operator, given in Eq. 29-17; thus,

$$\mathbf{E}(\mathbf{r}) = -\nabla\Phi = -\left(\frac{\partial\Phi}{\partial x}\hat{\mathbf{i}} + \frac{\partial\Phi}{\partial y}\hat{\mathbf{j}} + \frac{\partial\Phi}{\partial z}\hat{\mathbf{k}}\right) \qquad \textbf{(29–40a)}$$

We now seek to devise a test that can be applied to any field strength function $\mathbf{E}(\mathbf{r})$ to determine directly whether the field is conservative. Only in the event that the field is conservative can we define a potential function $\Phi(\mathbf{r})$.

First, let us understand more fully the relationship between $\mathbf{E}(\mathbf{r})$ and $\Phi(\mathbf{r})$ expressed by Eq. 29–39. Suppose that there exists a potential function obeying Eq. 29–40a. Then, we can write

$$-\oint_A^B \mathbf{E} \cdot d\mathbf{r} = \oint_A^B \left(\frac{\partial \Phi}{\partial x}\hat{\mathbf{i}} + \frac{\partial \Phi}{\partial y}\hat{\mathbf{j}} + \frac{\partial \Phi}{\partial z}\hat{\mathbf{k}}\right) \cdot$$

$$(dx\hat{\mathbf{i}} + dy\hat{\mathbf{j}} + dz\hat{\mathbf{k}})$$

$$= \oint_A^B \left(\frac{\partial \Phi}{\partial x}dx + \frac{\partial \Phi}{\partial y}dy + \frac{\partial \Phi}{\partial z}dz\right)$$

The integrand is just $d\Phi$ (see Eq. 29–20), so we have

$$-\oint_A^B \mathbf{E} \cdot d\mathbf{r} = \oint_A^B d\Phi = \Phi(B) - \Phi(A)$$

When B defines the point $\mathbf{r}$, this equation is identical to Eq. 29–39.

Now, Eq. 29–40a means that

$$E_z = -\frac{\partial \Phi}{\partial z} \quad \text{and} \quad E_y = -\frac{\partial \Phi}{\partial y}$$

If we differentiate the first of these expressions with respect to y and the second with respect to z, we have

$$\frac{\partial E_z}{\partial y} = -\frac{\partial^2 \Phi}{\partial y \partial z} \quad \text{and} \quad \frac{\partial E_y}{\partial z} = -\frac{\partial^2 \Phi}{\partial z \partial y}$$

Because the order of differentiation is immaterial, the two second partial derivatives are equal. Thus,

$$\frac{\partial E_z}{\partial y} - \frac{\partial E_y}{\partial z} = 0 \qquad \textbf{(29–41a)}$$

In the same way we find

$$\frac{\partial E_x}{\partial z} - \frac{\partial E_z}{\partial x} = 0 \qquad \textbf{(29–41b)}$$

$$\frac{\partial E_y}{\partial x} - \frac{\partial E_x}{\partial y} = 0 \qquad \textbf{(29–41c)}$$

The conditions expressed by Eqs. 29–41 constitute the test we have been seeking. If all three of these equations are satisfied, the function $\mathbf{E}(\mathbf{r})$ can indeed be derived from a potential function $\Phi(\mathbf{r})$. (Compare the discussion in Section 8–3.)

When three independent functions of x, y, and z are all zero, this is equivalent to stating that there exists a vector $\mathbf{G}(x, y, z)$ such that $G_x = G_y = G_z = 0$; that is, $\mathbf{G} = 0$.

If we identify Eq. 29–41a with G_x and the other two equations with G_y and G_z, we have

$$\mathbf{G} = \left(\frac{\partial E_z}{\partial y} - \frac{\partial E_y}{\partial z}\right)\hat{\mathbf{i}} + \left(\frac{\partial E_x}{\partial z} - \frac{\partial E_z}{\partial x}\right)\hat{\mathbf{j}}$$

$$+ \left(\frac{\partial E_y}{\partial x} - \frac{\partial E_x}{\partial y}\right)\hat{\mathbf{k}} = 0 \qquad \textbf{(29–42)}$$

We can express this same equation in determinant form by writing

$$\mathbf{G} = \begin{vmatrix} \hat{\mathbf{i}} & \hat{\mathbf{j}} & \hat{\mathbf{k}} \\ \dfrac{\partial}{\partial x} & \dfrac{\partial}{\partial y} & \dfrac{\partial}{\partial z} \\ E_x & E_y & E_z \end{vmatrix} = 0 \qquad \textbf{(29–42a)}$$

Given any field $\mathbf{E}(\mathbf{r})$, if Eq. 29–42 is valid, the field is conservative and $\mathbf{E}(\mathbf{r})$ is derivable from a potential function.

In Eq. 3–27 we expressed the vector (cross) product of two vectors in a determinant form similar to Eq. 29–42a; thus, if $\mathbf{C} = \mathbf{A} \times \mathbf{B}$, we have

$$\mathbf{C} = \begin{vmatrix} \hat{\mathbf{i}} & \hat{\mathbf{j}} & \hat{\mathbf{k}} \\ A_x & A_y & A_z \\ B_x & B_y & B_z \end{vmatrix}$$

If we identify the vector $\mathbf{B}$ here with $\mathbf{E}$ in Eq. 29–42a and the vector $\mathbf{A}$ with the del operator (so that $A_x = \partial/\partial x$, $A_y = \partial/\partial y$, and $A_z = \partial/\partial z$), we can express the vector $\mathbf{G}$ as

$$\mathbf{G} = \nabla \times \mathbf{E} = 0 \qquad \textbf{(29–42b)}$$

This vector, $\nabla \times \mathbf{E}$, is referred to as the *curl* of $\mathbf{E}$. Equations 29–42a and 29–42b may be viewed as simply short-hand ways of writing the differential relationships in Eq. 29–41.

In terms of this new vector function, we can state that if the curl of $\mathbf{E}$ vanishes, the field described by $\mathbf{E}$ is conservative. That is, the expressions

$$\nabla \times \mathbf{E} = 0 \quad \text{and} \quad \oint \mathbf{E} \cdot d\mathbf{r} = 0 \qquad \textbf{(29–43)}$$

are simply the differential and the integral formulations of the same fact concerning a conservative field, namely, that the field vector $\mathbf{E}$ can be derived from a unique potential function.

Stokes' Theorem. For some future discussions it is useful to answer the following question: If the curl of $\mathbf{E}$ does not vanish, and, hence, $\mathbf{E}$ does not represent a conservative field, what is the value of $\oint \mathbf{E} \cdot d\mathbf{r}$ for a particular closed path? In addition to the direct evaluation of the line integral, it is also possible to obtain a value, sometimes more conveniently, in terms of an integral involving $\nabla \times \mathbf{E}$. This constitutes *Stokes' theorem*.

Consider the line integral of $\mathbf{E} \cdot d\mathbf{r}$ along a closed path C, which for convenience is taken to lie in the x-y plane. The path is traversed in the counterclockwise sense, as indicated in Fig. 29–20a. Imagine the area enclosed by this path to be divided into any number of small, nonoverlapping regions. One such arrangement is shown in Fig. 29–20b. If $\mathbf{E} \cdot d\mathbf{r}$ is evaluated along each of the three paths in a counterclockwise sense, the algebraic sum, $\oint_{C'} + \oint_{C''} + \oint_{C'''}$, will equal the value of $\oint_C$ because each of the path segments $\overline{st}$ and $\overline{uv}$ is traversed twice in opposite directions and therefore makes no contribution. This leaves only the outer perimeter, which is traversed in the same sense as in the original integral $\oint_C$.

Figure 29–21 shows a path C with arbitrary shape. The

Fig. 29–20. The line integral of $\mathbf{E} \cdot d\mathbf{r}$ taken along the path C in (a) is equal to the sum of the integrals along paths C', C'', and C''' in (b).

(a)

(b)

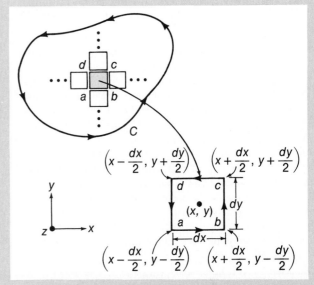

Fig. 29–21. The geometry of Stokes' theorem.

enclosed area is divided into a large number of nonoverlapping regions such as the square $abcd$. We evaluate the line integral of $\mathbf{E} \cdot d\mathbf{r}$ along the perimeter of this square, taking the segments ab, bc, cd, and da in turn. The result is

$$\oint_{abcda} \mathbf{E} \cdot (dx\hat{\mathbf{i}} + dy\hat{\mathbf{j}})$$

$$= \left(E_x - \frac{\partial E_x}{\partial y}\frac{dy}{2}\right)dx + \left(E_y + \frac{\partial E_y}{\partial x}dx\right)dy$$

$$- \left(E_x + \frac{\partial E_x}{\partial y}\frac{dy}{2}\right)dx - \left(E_y - \frac{\partial E_y}{\partial x}\frac{dx}{2}\right)dy$$

$$= \left(\frac{\partial E_y}{\partial x} - \frac{\partial E_x}{\partial y}\right)dx\,dy$$

We may also write this as

$$\oint_{abcda} \mathbf{E} \cdot d\mathbf{r} = (\nabla \times \mathbf{E})_z\,dx\,dy = (\nabla \times \mathbf{E})_z\,dA_z$$

When the direction of a closed path around an area dA is specified, we write $\hat{\mathbf{n}}\,dA$, where the direction of $\hat{\mathbf{n}}$ is given by a right-hand rule: Curl the fingers of the right hand in the direction in which the path is traversed; the extended thumb then points in the direction of $\hat{\mathbf{n}}$. With this convention, we can write

$$\oint_{abcda} \mathbf{E} \cdot d\mathbf{r} = (\nabla \times \mathbf{E}) \cdot \hat{\mathbf{n}}\,dA$$

When the results for all such differential quantities are summed, we have

$$\oint_C \mathbf{E} \cdot d\mathbf{r} = \int_S (\nabla \times \mathbf{E}) \cdot \hat{\mathbf{n}}\, dA \qquad \textbf{(29-44)}$$

where the path C is the perimeter of the surface S. This is the statement of Stokes' theorem.

The scope of this theorem is much greater than indicated here. The path C may be any irregular, nonplanar path, such as a closed wire loop that is bent in any way. The surface S is any surface for which C is the perimeter. To visualize this idea, suppose that a wire loop, bent into an arbitrary shape, is dipped into a soap solution. This produces a film with an area S and with a perimeter C. But this surface is only one of an infinite number of surfaces that have the same perimeter C. For example, you could produce some of these surfaces by gently blowing on the soap film to change its shape. All of these surfaces are equally appropriate for evaluating the surface integral in Eq. 29-44.

Example 29-9

Determine whether the field represented by $\mathbf{E} = \eta(-y\hat{\mathbf{i}} + x\hat{\mathbf{j}})$ is a conservative field. Evaluate the line integral of $\mathbf{E} \cdot d\mathbf{r}$ along the path shown in the diagram.

Solution:

We test for a conservative field by calculating the curl:

$$\nabla \times \mathbf{E} = \eta \begin{vmatrix} \hat{\mathbf{i}} & \hat{\mathbf{j}} & \hat{\mathbf{k}} \\ \dfrac{\partial}{\partial x} & \dfrac{\partial}{\partial y} & \dfrac{\partial}{\partial z} \\ -y & +x & 0 \end{vmatrix}$$

$$= \eta\left(0 - \frac{\partial x}{\partial z}\right)\hat{\mathbf{i}} - \eta\left(0 + \frac{\partial y}{\partial z}\right)\hat{\mathbf{j}} + \eta\left(\frac{\partial x}{\partial x} + \frac{\partial y}{\partial y}\right)\hat{\mathbf{k}}$$

$$= 2\eta\hat{\mathbf{k}} \neq 0$$

Thus, the field $\mathbf{E}$ is not conservative and no potential function can be defined.

The line integral for the path shown in the diagram is

$$\oint_C \mathbf{E} \cdot d\mathbf{r} = \int_0^a E_x(x, 0)\, dx + \int_0^b E_y(a, y)\, dy$$

$$+ \int_a^0 E_x(x, b)\, dx + \int_b^0 E_y(0, y)\, dy$$

$$= 0 + \eta ab + \eta ab + 0 = 2\eta ab$$

Using Stokes' theorem, Eq. 29-44, we have

$$\oint_C \mathbf{E} \cdot d\mathbf{r} = \int_S (\nabla \times \mathbf{E}) \cdot \hat{\mathbf{n}}\, dA$$

$$= 2\eta \int_S \hat{\mathbf{k}} \cdot \hat{\mathbf{k}}\, dx\, dy = 2\eta ab$$

and both methods yield the same result.

The name *curl* for the operation $\nabla \times \mathbf{E}$ suggests that it measures some rotational aspect of the field $\mathbf{E}$. In Example 29-9 a plot of the field lines would show them to be a set of concentric circles. If, instead of an electric field, $\mathbf{E}$ represented fluid streamlines, we would interpret the flow pattern as a circulation or a whirlpool. An electric field with a nonvanishing curl has a circulation, in some sense, although no matter actually flows. This is similar to calling $\mathbf{E} \cdot \hat{\mathbf{n}}\, dA$ the *flux* of $\mathbf{E}$, where there is again an analogy to fluid flow.

Gauss's Law Restated.   In Eq. 28-19 we stated Gauss's law in *integral* form:

$$\oint_S \mathbf{E} \cdot \hat{\mathbf{n}}\, dA = \frac{Q}{\epsilon_0} = \frac{1}{\epsilon_0} \int_V \rho\, dv \qquad \textbf{(29-45)}$$

where $\rho(x, y, z)$ is the volume charge density and where V is the volume enclosed by the surface S. The integrand $\mathbf{E} \cdot \hat{\mathbf{n}}\, dA$ is the electric flux that passes through the element dA of surface area. We now give a *differential* formulation of Gauss's law.

Suppose that a volume V, bounded by a surface S, is divided into a large number of cubes with sides dx, dy, and dz, as indicated in Fig. 29-22. For every interior cube, whatever flux *leaves* the cube through a particular face *enters* the adjacent cube through the corresponding face. Thus, the net outward flux, summed over all cubes, is equal to the flux that leaves the volume V through the surface S.

Now, calculate the flux through the six faces of a differential cell with volume $dv = dx\, dy\, dz$, as shown in Fig. 29-22. We have

$$\varphi = \oint_{\text{cell}} \mathbf{E} \cdot \hat{\mathbf{n}}\, dA = \frac{1}{\epsilon_0} \int_{\text{cell}} \rho(x, y, z)\, dv$$

Fig. 29–22. The geometry for obtaining a differential formulation of Gauss's law.

$$= \frac{1}{\epsilon_0}\rho(x, y, z)\, dx\, dy\, dz$$

For the top face of the cell, $\mathbf{E} \cdot \hat{\mathbf{n}}\, dA = E_n\, dx\, dy$ with $E_n = E_z + (\partial E_z/\partial z)(dz/2)$. For the bottom face, $E_n' = E_z - (\partial E_z/\partial z)(dz/2)$. The net outward flux through the upper and lower faces combined is therefore

$$\varphi_1 = (E_n - E_n')\, dx\, dy$$

$$= \frac{\partial E_z}{\partial z}\, dx\, dy\, dz$$

In a similar way we find the flux through the other two pairs of parallel faces:

$$\varphi_2 = \frac{\partial E_x}{\partial x}\, dx\, dy\, dz$$

$$\varphi_3 = \frac{\partial E_y}{\partial y}\, dx\, dy\, dz$$

The total flux leaving the cell is

$$\varphi = \oint_{\text{cell}} \mathbf{E} \cdot \hat{\mathbf{n}}\, dA = \left(\frac{\partial E_x}{\partial x} + \frac{\partial E_y}{\partial y} + \frac{\partial E_z}{\partial z}\right) dx\, dy\, dz$$

Comparing the two expressions for φ, we conclude that

$$\frac{\partial E_x}{\partial x} + \frac{\partial E_y}{\partial y} + \frac{\partial E_z}{\partial z} = \frac{\rho(x, y, z)}{\epsilon_0} \qquad \textbf{(29–46)}$$

This is the differential form of Gauss's law. It is possible to express this result in a compact way by using the del operator. We can write for the scalar (dot) product of ∇ and $\mathbf{E}$,

$$\nabla \cdot \mathbf{E} = \left(\frac{\partial}{\partial x}\hat{\mathbf{i}} + \frac{\partial}{\partial y}\hat{\mathbf{j}} + \frac{\partial}{\partial z}\hat{\mathbf{k}}\right) \cdot (E_x\hat{\mathbf{i}} + E_y\hat{\mathbf{j}} + E_z\hat{\mathbf{k}})$$

$$= \frac{\partial E_x}{\partial x} + \frac{\partial E_y}{\partial y} + \frac{\partial E_z}{\partial z}$$

This is just equal to the left-hand side of Eq. 29–46. Therefore, we have

$$\nabla \cdot \mathbf{E} = \frac{\rho}{\epsilon_0} \qquad \textbf{(29–47)}$$

We emphasize, as before, that this equation in shorthand notation has exactly the same meaning as Eq. 29–46. The quantity $\nabla \cdot \mathbf{E}$ is called the *divergence* of $\mathbf{E}$ and is a local measure of the strength of the source of the field; $\nabla \cdot \mathbf{E}$ gives the intensity of the flux that *diverges* from the source.

Table 29–1 at the top of the next page summarizes the relationships involving the del operator for the electrostatic case.

PROBLEMS*

Section 29–1

29–1 A small spherical object carries a charge of 8 nC. At what distance from the center of the object is the potential equal to 100 V? 50 V? 25 V? Is the spacing of the equipotentials proportional to the change in Φ?

---

*Assume a reference level for potential such that $\Phi(\infty) = 0$ unless the statement of the problem requires otherwise.

29–2 Two point charges, $Q_1 = +5$ nC and $Q_2 = -3$ nC, are separated by 35 cm. (a) What is the potential energy of the pair? What is the significance of the algebraic sign of your answer? (b) What is the electric potential at a point midway between the charges?

29–3 Four equal charges, $Q = 2\,\mu$C, are located at the corners of a square that is 0.50 m on a side. (a) What is the potential energy of the array?

TABLE 29–1.   Electrostatic Relationships Involving the Del Operator, $\nabla = \dfrac{\partial}{\partial x}\hat{\mathbf{i}} + \dfrac{\partial}{\partial y}\hat{\mathbf{j}} + \dfrac{\partial}{\partial z}\hat{\mathbf{k}}$

ELECTRIC POTENTIAL	STOKES' THEOREM	GAUSS'S LAW

$$\Phi(B) = -\oint_A^B \mathbf{E} \cdot d\mathbf{r} + \Phi(A)$$

$$\mathbf{E} = -\nabla\Phi$$

$$\nabla\Phi = \frac{\partial \Phi}{\partial x}\hat{\mathbf{i}} + \frac{\partial \Phi}{\partial y}\hat{\mathbf{j}} + \frac{\partial \Phi}{\partial z}\hat{\mathbf{k}}$$

$$\oint_C \mathbf{E} \cdot d\mathbf{r} = \int_S (\nabla \times \mathbf{E}) \cdot \hat{\mathbf{n}}\, dA = 0$$

$$\nabla \times \mathbf{E} = 0$$

$$\nabla \times \mathbf{E} = \left(\frac{\partial E_z}{\partial y} - \frac{\partial E_y}{\partial z}\right)\hat{\mathbf{i}}$$
$$+ \left(\frac{\partial E_x}{\partial z} - \frac{\partial E_z}{\partial x}\right)\hat{\mathbf{j}}$$
$$+ \left(\frac{\partial E_y}{\partial x} - \frac{\partial E_x}{\partial y}\right)\hat{\mathbf{k}}$$

$$\oint_S \mathbf{E} \cdot \hat{\mathbf{n}}\, dA = \frac{Q}{\epsilon_0}$$

$$= \frac{1}{\epsilon_0} \int_V \rho\, dv$$

$$\nabla \cdot \mathbf{E} = \frac{\rho}{\epsilon_0}$$

$$\nabla \cdot \mathbf{E} = \frac{\partial E_x}{\partial x} + \frac{\partial E_y}{\partial y} + \frac{\partial E_z}{\partial z}$$

(b) What is the electric potential at the center of the square? (c) What is the electric field strength at the center of the square?

29-4   Two point charges, $Q_1 = +5\,\mu C$ and $Q_2 = -2\,\mu C$, are separated by 50 cm. (a) Where, along a line that passes through both charges, is the potential zero (in addition to $R = \infty$)? (b) Determine the electric field strength at the point found in (a). If $\Phi = 0$ at this point, why is E not also equal to zero?

29-5   Determine the electric potential along a perpendicular bisector of a straight line charge with length L and uniform linear charge density λ. Show that when the distance becomes large compared with L, your result reduces to that for a point charge $Q = \lambda L$ located at the center of the line charge.

Section 29-2

29-6   The potential for a certain electric field is (in cylindrical coordinates) $\Phi(\rho, \phi, z) = A(\rho^2 \sin\phi \cos\phi + z^2 \tan\phi)$, where A has suitable dimensions. Obtain an expression for the field strength $\mathbf{E}$ in rectangular coordinates. [*Hint:* Use $x = \rho \cos\phi$, $y = \rho \sin\phi$, $z = z$.]

29-7   At a certain distance from a point charge the field intensity is 500 V/m and the potential is -3000 V.

(a) What is the distance to the charge? (b) What is the magnitude of the charge?

29-8   Use the expression obtained in Example 29-3 for the potential on the axis of a uniformly charged ring, and determine the field strength along the axis. (Why is it possible to do this?) Compare your answer with Eq. 28-23.

29-9   The potential for a certain two-dimensional electric field is $\Phi(x, y) = Axy$, where A has suitable dimensions. (a) Determine the field strength $\mathbf{E}$ everywhere in the x-y plane. (b) Determine the slope of the field line at a point (x, y) by calculating $\tan\beta = E_y/E_x$. (c) Determine the slope of the equipotential at a point (x, y) by calculating dy/dx for $\Phi = $ constant. (d) Are the field lines and the equipotentials mutually perpendicular? (e) Sketch some of the field lines and equipotentials.

29-10   Given the electric potential $\Phi(x, y) = \kappa(x^2 + y^2)$, where κ has suitable dimensions, show explicitly that the electric field strength $\mathbf{E}$ is everywhere perpendicular to the equipotential curves. Sketch the equipotentials and show $\mathbf{E}$ at some point.

29-11   (a) Use the expression for $\mathbf{E}$ obtained in Example 29-7 for the field strength on the axis of a charged disc, and derive an expression for the potential on the axis for the case in which the potential of the disc is taken to be zero. That is, determine $\Phi(z)$ with $\Phi(0) = 0$ instead of $\Phi(\infty) = 0$. (b) Determine the potential function for a charged infinite plane by allowing $R \rightarrow \infty$ while maintaining σ constant. (c) Explain why it was necessary to change the reference level for the potential before increasing R without limit.

Section 29-3

29-12 What charge Q must be placed on a sphere with a radius of 75 cm to produce a potential on the surface of 2×10^6 V?

29-13 A point charge Q is placed at the center of a spherical shell with inner radius R_1 and outer radius R_2 and which carries no net charge. Determine the field strength **E** everywhere. Sketch some of the lines of force.

29-14 Derive Eq. 29-27.

29-15 A ring with a radius of 6 cm carries a uniformly distributed total charge of 20 nC. The ring is placed 8 cm above and parallel to a grounded conducting plane. What is the surface charge density on the plane at a point directly beneath the center of the ring? [*Hint:* Refer to Eq. 28-23.]

29-16 A point charge $Q = 5 \, \mu C$ is located 15 cm from the center of a grounded conducting sphere that has a radius of 6 cm. What is the force exerted on Q by the induced charge on the sphere?

29-17 The conducting sphere shown in Fig. 29-16 is
•• grounded. However, suppose instead that the sphere is initially uncharged or is brought to some potential other than zero while in the presence of the point charge Q. The problem can then be reduced to one that involves the same image charge Q' plus an additional point charge Q'' located at the center of the sphere. (a) Why is the surface of the sphere still an equipotential after the addition of the charge Q''? (b) If the sphere is uncharged, what is the relationship among Q'', Q', and Q? (c) Determine the three charges if the sphere is uncharged and $Q = 2$ nC for $\ell = 3R = 60$ cm. (d) What is the potential for the sphere?

Section 29-4

29-18 A particle with a charge $q = 12$ nC and a mass $m = 2$ mg is released from rest at a point 5 cm from a charge $Q = 50$ nC that is stationary. (a) What is the initial acceleration of the particle in units of g? (b) What is the initial potential energy of the pair of charges? (c) What is the total amount of work done by the field in moving the particle q to infinity? (d) What is the speed of the particle q when it is a great distance from Q?

29-19 In Example 29-8, what is the magnitude of the transverse acceleration of the electrons in the beam as it passes between the deflecting plates? (Express the acceleration in units of g.)

29-20 Two parallel conducting plates, each 4 cm × 4 cm, are separated by 1.5 cm. (a) What magnitude of charge must be placed on each plate to produce a potential difference of 50 V? (b) An electron beam passes between the plates at a rate of 10^{-7} C/s. If each electron has a kinetic energy of 2.0 keV, calculate the amount of electronic charge that is between the plates at any instant. (c) Compare the two charges found in (a) and (b). Are image charge effects in the conducting plates likely to be of any significance in this case?

29-21 An electric dipole **p** is placed in a uniform electric field, $\mathbf{E} = E\hat{\mathbf{i}}$, that is confined between the planes $x = 0$ and $x = L$, as shown in the diagram. Calculate the work done by the field as the dipole is assembled at the indicated location, and show that the potential energy is $-\mathbf{p} \cdot \mathbf{E}$. (Bring the charges $\pm q$ separately from $x = \infty$.)

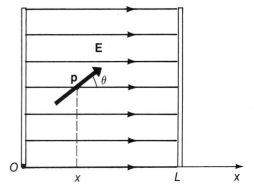

Section 29-5

29-22 Given that $\mathbf{E} = \eta(yz^2\hat{\mathbf{i}} + xz^2\hat{\mathbf{j}} + 2xyz\,\hat{\mathbf{k}})$, where
• η is a constant with suitable dimensions, calculate

$\nabla \times \mathbf{E}$ and determine whether the field is conservative. If the field is conservative, find the potential function $\Phi(x, y, z)$. [*Hint:* Use Eq. 29-39 and follow the path $(0, 0, 0) \rightarrow (x, 0, 0) \rightarrow (x, y, 0) \rightarrow (x, y, z)$.]

29-23 An electric field is given by $\mathbf{E} = \eta[(xy^2 + x^2y)\hat{\mathbf{i}}$
•• $+ x^2y\,\hat{\mathbf{j}}]$, where η is a constant with suitable dimensions. (a) Calculate the line integral of $\mathbf{E} \cdot d\mathbf{r}$ along the path C_1 $[(0, 0) \to (1, 0) \to (1, 2)]$, as indicated in the diagram. (b) Calculate the line integral of $\mathbf{E} \cdot d\mathbf{r}$ along the path C_2 ($y = 2x$) from $(0, 0)$ to $(1, 2)$. (c) Based on the results of (a) and (b), is the field $\mathbf{E}$ conservative? (d) Calculate the curl of $\mathbf{E}$. What can you conclude from this result? (e) Evaluate the closed-path line integral of $\mathbf{E} \cdot d\mathbf{r}$ along the path C_1 from $(0, 0)$ to $(1, 2)$, and then along the path C_2 from $(1, 2)$ to $(0, 0)$. Does this result confirm the conclusion reached in (d)? (f) Calculate the integral of $\nabla \times \mathbf{E} \cdot \hat{\mathbf{n}}\, dA$ over the surface S in the x-y plane bounded by C_1 and C_2. (That is, $\hat{\mathbf{n}}\, dA = dx\, dy\hat{\mathbf{k}}$.) Compare with the result of (e). [*Hint:* Along the path C_1, calculate

$$\oint_1 \mathbf{E} \cdot d\mathbf{r} = \int_0^1 E_x(y = 0)\, dx + \int_0^2 E_y(x = 1)\, dy$$

Along the path C_2, calculate

$$\oint_2 \mathbf{E} \cdot d\mathbf{r} = \int_0^1 E_x(y = 2x)\, dx + \int_0^2 E_y(x = \tfrac{1}{2}y)\, dy.]$$

29-24 Two parallel plates are located in the planes $x = 0$
• and $x = d$. Between the plates the potential is $\Phi(x) = V(x/d)^3$, where V is the potential difference placed across the plates. (a) Determine

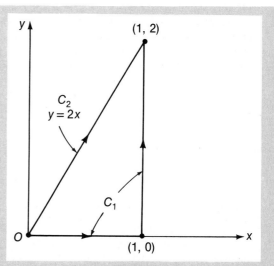

Problem 29-23

the field strength between the plates. (b) Calculate the surface charge density on the plates. (c) Determine the density of charge between the plates.

29-25 Gauss's law can be expressed as $\nabla \cdot \mathbf{E} = \rho/\epsilon_0$, and $\mathbf{E}$ can be derived from a potential Φ according to $\mathbf{E} = -\nabla\Phi$. These two expressions may be combined to yield Gauss's law in the form $\nabla \cdot (\nabla\Phi) = -\rho/\epsilon_0$. The operator $\nabla \cdot \nabla \equiv \nabla^2$ is called the *Laplacian operator*. Obtain the explicit expression of ∇^2 in rectangular coordinates and write out Gauss's law in this form.

Additional Problems

29-26 All of the corners, except one, of a 1-m cube are occupied by charges of $+1\,\mu\text{C}$. What is the electric potential at the empty corner?

29-27 Equal charges ($q = +2\,\mu\text{C}$) are placed at 30° intervals around the equator of a sphere with a radius of 1.2 m. What is the electrical potential (a) at the center of the sphere and (b) at the north pole of the sphere?

29-28 Three point charges are located as follows: $-Q$ at $x = -\ell$, $+2Q$ at $x = 0$, and $-Q$ at $x = +\ell$. Calculate the potential along the x-axis in the distant region ($|x| \gg \ell$).

29-29 (a) Determine the potential along the axis to the
• right of the uniform hemispherical charge distribution shown in the diagram. [*Hint:* Use $dQ = 2\pi\sigma R^2 \sin\theta\, d\theta$.] (b) Show that $\Phi(x) \cong Q/4\pi\epsilon_0 x$ for $x \gg R$, where $Q = 2\pi R^2\sigma$. (c) Calculate the field

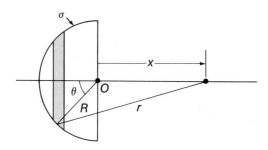

strength E at any point along the x-axis to the right of O.

29-30 A metal sphere is to be charged to a potential of 10^6 V. The field strength is not to exceed 3×10^6 V/m, the point at which the insulating quality of air breaks down and corona discharge or sparking takes place. (a) What is the minimum radius of the sphere? (b) For the minimum radius, what is the charge on the sphere?

29-31 Consider two infinitely long concentric cylinders. The inner cylinder has a radius R_1 and carries a linear charge density $+\lambda$. The outer cylinder has an inner radius R_2 and carries a linear charge density $-\lambda$. (a) Where on the outer cylinder does the charge reside? (b) Determine the electric field strength everywhere. (c) What is the potential difference, $V = \Phi(R_1) - \Phi(R_2)$? (d) Show that

$$\mathbf{E} = \frac{V}{\ln(R_1/R_2)} \cdot \frac{1}{\rho}\hat{\mathbf{u}}_\rho$$

where $\hat{\mathbf{u}}_\rho$ is the unit vector in the radial direction.

29-32 A point charge Q that is placed near a grounded conducting sphere, as in Fig. 29-16, produces an induced surface charge density distribution,

$$\sigma(\theta) = -\frac{mQ}{4\pi R^2}\frac{1 - m^2}{(1 + m^2 - 2m\cos\theta)^{3/2}}$$

where R is the radius of the sphere and where $m = R/\ell$. Construct a Gaussian surface in the shape of a sphere just outside the conducting sphere. Show by an explicit calculation that the total induced charge is $Q' = -mQ$. [*Hint:* Use Eq. 29-26 and $\hat{\mathbf{n}}\, dA = 2\pi R^2 \sin\theta\, d\theta\hat{\mathbf{u}}_r$.]

29-33 An electron is accelerated from rest through a potential difference of 2×10^6 V. (a) What is the final kinetic energy of the electron in eV? (b) What is the rest-mass energy mc^2 of the electron in eV? (c) Use the relativistic expression,

$$K = m_0 c^2[(1 - v^2/c^2)^{-1/2} - 1]$$

and calculate v/c. (d) What would be the value of v/c if an entirely classical (nonrelativistic) calculation were made?

29-34 A dipole consists of two particles, one with charge Q and mass m, and the other with charge $-Q$ and mass $2m$, separated by a distance L, as shown in the diagram. Show that the angular frequency of small oscillations of the dipole about its equilibrium position is $\omega = \sqrt{3QE/2mL}$ when placed in a uniform field E.

CHAPTER 30

CAPACITANCE AND DIELECTRICS

In this chapter we continue the investigation of the electrostatic fields produced by charged conductors, including the effects of induced charges, and adding the presence of insulators in the systems studied. When an insulating material is introduced into an electric field (of ordinary strength), the electrons in the material remain bound within their respective atomic or molecular domains. However, the field can alter the atomic charge distributions and produce *polarization* effects that cause substantial changes in the electric field at the macroscopic level. Insulators that exhibit polarization are referred to as *dielectrics*.

It is possible to determine the electrostatic field configuration for a system of insulators and charged conductors if the polarizability characteristics of the dielectric materials are known and if some geometric constants of the system can be determined. These geometric constants are associated with the idea of *capacitance*. Accordingly, we begin by establishing some of the features of capacitance in simple cases.

30-1 THE PARALLEL-PLATE CAPACITOR

A simple yet conceptually important device for storing electric charge is the *parallel-plate capacitor*. This device consists of two parallel conducting planes (the *capacitor plates*) that usually have the same shape and are spaced apart by a fixed distance s. One plate carries a charge $+Q$ and the other plate carries a charge $-Q$, as indicated in Fig. 30-1.

The charging of a capacitor can be accomplished by connecting the plates with a wire and then inducing a charge on one of the plates with a charged rod. An equal charge of the opposite sign will then reside on the other plate. When the wire is disconnected and the charged rod is removed, one plate will carry a charge $+Q$ and the other plate will carry a charge $-Q$.

If the dimensions of the plates are large compared with the spacing (that is, if $b \gg s$ in Fig. 30-1), the charges on the plates will be distributed over the inside surfaces in an essentially uniform manner. Thus, each plate, with area A, will have a surface charge density $\sigma = Q/A$. The resulting electric field between the plates will be practically the same as that for a pair of infinite parallel planes that carry the

Fig. 30-1. A parallel-plate capacitor with an arbitrary shape. The distance b represents a typical dimension of the capacitor.

same charge density σ (Fig. 29–15). If we ignore the edge effects associated with finite plates, this arrangement constitutes an *ideal parallel-plate capacitor.*

In practice, the usual way of charging a capacitor is by means of a battery. The composition and operation of a battery are discussed in Section 31–4; for our present purposes we need only to know that a battery produces a constant potential difference between its terminals. The standard symbol for a battery is

where terminal B (the *positive* terminal) is at a higher electric potential than is terminal A (the *negative* terminal); that is, $\Phi_B - \Phi_A = V_{BA} > 0$. When no confusion may arise, we simply write V for the potential difference V_{BA}. Figure 30–2 shows a parallel-plate capacitor connected to a battery so that a potential difference $\Phi_B - \Phi_A = V$ is established between the plates. Throughout this discussion the system that consists of the capacitor and the battery is assumed to possess no net charge and to be far removed from other conductors and charges. (Such a system is said to be "floating.") Later, we discuss briefly the situation when these conditions are not met.

Fig. 30–2. A battery is connected to a parallel-plate capacitor to establish a potential difference $\Phi_B - \Phi_A = V$ between the plates.

If the spacing s between the plates is small compared with the dimensions of the plates, the electric field strength is essentially uniform and we can write (Eq. 29–29)

$$\mathbf{E} = \frac{\sigma}{\epsilon_0}\hat{\mathbf{n}} = \frac{Q}{\epsilon_0 A}\hat{\mathbf{n}} \qquad (30\text{-}1)$$

where $\hat{\mathbf{n}}$ is a unit vector normal to the plates and in the direction from the positively charged plate toward the negatively charged plate. The potential difference between the plates is (see Eq. 29–31)

$$\Phi_B - \Phi_A = V = -\oint_A^B \mathbf{E} \cdot d\mathbf{r} = Es \qquad (30\text{-}2)$$

Combining Eqs. 30–1 and 30–2 we have

$$Q = \frac{\epsilon_0 A}{s} V \qquad (30\text{-}3)$$

Thus, the charge Q stored on either plate is related to the potential difference V between the plates by a geometry-dependent factor, $\epsilon_0 A/s$. In general, the proportionality constant that connects Q and V is called the *capacitance* C of the device or system; that is,

$$Q = CV \qquad (30\text{-}4)$$

For the case of the parallel-plate capacitor, we have

$$C = \frac{\epsilon_0 A}{s}, \qquad \text{parallel-plate capacitor} \tag{30-5}$$

From Eq. 30–4 we see that the units of capacitance are *coulombs per volt,* C/V, to which is given the special name *farad** (F); thus,

$$1\,\text{F} = 1\,\text{C/V}$$

The farad is a rather large unit, so we usually find capacitive circuit elements with capacitance values measured in μF (10^{-6} F) or pF (10^{-12} F).

▶ *Fringing Fields.* Because a real parallel-plate capacitor has a finite size, the electric field lines at the edges are not exactly straight and uniformly spaced as in the ideal case, nor is the field contained entirely between the plates. Figure 30–3a shows the field lines near the edge of a parallel-plate capacitor; this nonuniform portion of the field is referred to as the *fringing field.* Notice also the behavior of the equipotentials in this region. Figure 30–3b shows the idealized situation: the field terminates abruptly at the edge of the capacitor.

(b)

Fig. 30–3. (a) The fringing field near the edge of a parallel-plate capacitor. The charge density $\pm \sigma'$ on the outside of the plates is highly nonuniform. (b) The idealized field of a parallel-plate capacitor. All of the charge is on the inside surfaces.

Figure 30–3a suggests that the actual charge Q on a parallel-plate capacitor is slightly greater than the charge that results when the value of C given by Eq. 30–5 is used in Eq. 30–4. The charge Q on each plate includes *all* of the charge on *all* of the surfaces of each plate, most of which usually resides on the inside surface. In the "floating" condition that we have assumed, the total charge on each plate is the same in magnitude as that on the other plate, and opposite in sign. When a fringing field is present, as in Fig. 30–3a, we can write

$$C = \frac{\epsilon_0 A}{s} f \tag{30-5a}$$

where $f \geqslant 1$ is a correction factor that can be determined for a specific plate geometry. For a circular

---

*Named in honor of Michael Faraday (1791–1867), the great English physicist about whom we learn more in later chapters.

capacitor with radius R, $f = 1.167$ when $s/R = 0.10$ and $f = 1.023$ when $s/R = 0.01$; of course, in the limit, $s/R \to 0$, we have $f = 1$. ◄

Example 30-1

A circular parallel-plate capacitor with a spacing $s = 3$ mm is charged to produce an electric field strength of 3×10^6 V/m. What plate radius R is required if the stored charge Q is $1 \, \mu$C?

Solution:

The potential difference between the plates is

$$V = Es = (3 \times 10^6 \text{ V/m})(3 \times 10^{-3} \text{ m}) = 9 \times 10^3 \text{ V}$$

The capacitance value is

$$C = \frac{Q}{V} = \frac{1 \times 10^{-6} \text{ C}}{9 \times 10^3 \text{ V}} = 1.11 \times 10^{-10} \text{ F} = 111 \text{ pF}$$

From Eq. 30-5 we have $A = \pi R^2 = Cs/\epsilon_0$, so that*

$$R = \sqrt{\frac{Cs}{\pi\epsilon_0}} = \sqrt{\frac{(1.11 \times 10^{-10} \text{ F})(3 \times 10^{-3} \text{ m})}{\pi(8.85 \times 10^{-12} \text{ F/m})}} = 0.109 \text{ m} = 10.9 \text{ cm}$$

With $s/R = (3 \times 10^{-3} \text{ m})/(0.109 \text{ m}) = 2.75 \times 10^{-2}$, the fringing field correction factor can be calculated to be $f = 1.056$. Thus, the required area is 5.6 percent smaller and the required radius is 2.8 percent smaller than calculated with Eq. 30-5. The actual required radius is, therefore, $R' = 10.6$ cm.

30-2 THE SPHERICAL CAPACITOR

The fringing field effects associated with a parallel-plate capacitor are not present in a *spherical* capacitor because the field is totally confined within the region between the plates. Figure 30-4a shows a spherical capacitor that consists of an inner conducting sphere with a radius R_1 surrounded by a concentric conducting spherical shell with an inner radius R_2 and an outer radius R_3. We again assume an isolated floating system with no net charge.

Figure 30-4a indicates how the charging battery is connected to the spherical capacitor; Fig. 30-4b shows the potentials and the charges on the three surfaces. Because the outer shell of the capacitor is a solid conductor, we have $\Phi_2 = \Phi_3$ and $E = 0$ inside the shell. Applying Gauss's law to a spherical surface with radius between R_2 and R_3, we conclude that $Q_2 = -Q_1$. The system has no net charge, so we have $Q_3 = 0$. Next, applying Gauss's law to a spherical surface with radius greater than R_3, we conclude that $E = 0$ in the region outside the shell. If we take the potential at infinity to be $\Phi_\infty = 0$, it follows that $\Phi_2 = \Phi_3 = 0$ and $\Phi_1 = V$.

Next, surround the inner sphere with a spherical Gaussian surface of radius r

---

*Verify that the units of ϵ_0 (C^2/N · m^2) can be expressed as F/m.

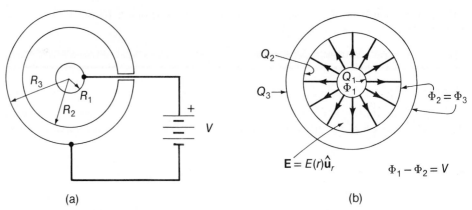

(a)   (b)

Fig. 30–4. (a) An isolated spherical capacitor. (b) Electric field lines and potential values for the spherical capacitor in (a).

$(R_1 \leqslant r \leqslant R_2)$; then, Gauss's law gives

$$\mathbf{E} = \frac{Q_1}{4\pi\epsilon_0} \cdot \frac{1}{r^2} \hat{\mathbf{u}}_r \tag{30–6}$$

Using Eq. 29–13a, the potential is

$$\Phi(r) = \Phi(R_1) - \frac{Q_1}{4\pi\epsilon_0} \int_{R_1}^{r} \frac{dr}{r^2} = V - \frac{Q_1}{4\pi\epsilon_0} \left(\frac{1}{R_1} - \frac{1}{r} \right)$$

When $r = R_2$, we have $\Phi(R_2) = 0$; hence,

$$Q_1 = 4\pi\epsilon_0 V \frac{R_1 R_2}{R_2 - R_1} \tag{30–7}$$

For the mode of charging illustrated in Fig. 30–4a, in which $Q_2 = -Q_1$ and $Q_3 = 0$, we can again define a stored charge Q with $Q = CV$. Notice that the inner conductor has only one surface (at $r = R_1$) and carries a charge Q_1. The outer conductor has two surfaces (at $r = R_2$ and $r = R_3$) and carries a total charge $Q_2 + Q_3 = Q_2 = -Q_1$ (because $Q_3 = 0$). Thus, the coefficient of V in Eq. 30–7 is identified as the capacitance of the spherical capacitor:*

$$C = 4\pi\epsilon_0 \frac{R_1 R_2}{R_2 - R_1} \tag{30–8}$$

In Fig. 30–4a the potential is zero on both surfaces of the shell of the spherical capacitor (that is, $\Phi_2 = \Phi_3 = 0$). Thus, we can define $\Phi(\infty) = 0$ in a practical way by imagining that R_2 and R_3 become indefinitely large. If we surround a piece of laboratory equipment with a large conducting cage that is grounded, we thereby create an equipotential that effectively corresponds to $\Phi(\infty) = 0$. In the case of the spheri-

---

*About a hundred years ago, the American physicist Henry Rowland (1848–1901) constructed a precision spherical capacitor and found agreement between the measured and the calculated capacitances to within 2 parts in 10^5.

cal capacitor, allowing $R_2 \to \infty$ yields $\mathrm{Lim}\,(R_2 \to \infty)\, C = 4\pi\epsilon_0 R_1$. Thus, in general, the capacitance of a sphere with radius R can be expressed as

$$C = 4\pi\epsilon_0 R \qquad\qquad \textbf{(30–9)}$$

▶ *Grounded Capacitors.* The manner in which a capacitor is grounded or brought near other conductors or charges can alter the charge distribution on the capacitor and can even influence the magnitude of the charge. Figure 30–5 shows two ways to attach a ground connection to a spherical capacitor. If the outer shell is grounded (Fig. 30–5a), this automatically gives $\Phi_3 = 0$; hence, this situation in all respects is the same as the case of the floating capacitor (Fig. 30–4). If the inner conductor is grounded (Fig. 30–5b), this gives $\Phi_1 = 0$ and $\Phi_2 = \Phi_3 = -V$, with the result that Q_1 and Q_2 are the same as in Fig. 30–5a. However, the charge Q_3 is no longer zero.

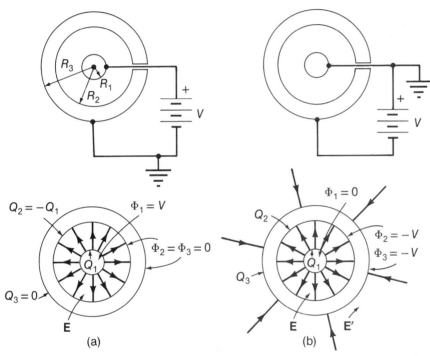

Fig. 30–5. (a) A ground connection is made to the outer conductor of a spherical capacitor. The resulting field is the same as that of a floating capacitor (Fig. 30–4). (b) The ground connection is made to the inner conductor of the capacitor, and the capacitor now carries a net charge.

The electric field in the region with $r \geqslant R_3$ is

$$\mathbf{E}' = \frac{Q_3}{4\pi\epsilon_0} \cdot \frac{1}{r^2}\hat{\mathbf{u}}_r \qquad\qquad \textbf{(30–10)}$$

Using Eq. 29–13a, the potential is

$$\Phi(r) = \Phi_\infty - \frac{Q_3}{4\pi\epsilon_0} \int_\infty^r \frac{dr}{r^2} = 0 + \frac{Q_3}{4\pi\epsilon_0} \cdot \frac{1}{r}$$

When $r = R_3$, we have $\Phi(R_3) = -V$; hence,

$$Q_3 = -4\pi\epsilon_0 R_3 V \qquad\qquad \textbf{(30–11)}$$

Thus, the total charge on the outer conductor, $Q_2 + Q_3$, is *not* equal to $-Q_1$ as in the previous case. Because the charges on the two conductors are not equal in magnitude, one cannot define a capacitance value according to the simple expression $C = Q/V$. A more complicated analysis is required to deal with these situations. ◀

Example 30–2

(a) Determine the capacitance of a spherical conducting space probe that has a diameter of 2.4 m.

(b) The probe passes through a region of space in which the density of ions is $\mathcal{N} = 8.0$ ions/cm³. The velocity of the probe with respect to a typical ion is 50 km/s. Assume that the probe sweeps up and retains all of the ions that it encounters and that each ion carries a charge $+e$. What is the potential of the probe (relative to $\Phi(\infty) = 0$) after moving through the ionic region for 20 minutes?

ion density $\mathcal{N}$

v

R

$|\!\leftarrow\! vt \!\rightarrow\!|$

volume of ions swept up in time t

Solution:

(a) Using Eq. 30–9, we have

$$C = 4\pi\epsilon_0 R = 4\pi(8.85 \times 10^{-12}\ \text{F/m})(1.2\ \text{m}) = 1.33 \times 10^{-10}\ \text{F}$$

(b) During the time interval t, the charge intercepted is (refer to the diagram)

$$Q = \mathcal{N}e(\pi R^2)(vt)$$
$$= (8.0 \times 10^6\ \text{m}^{-3})(1.60 \times 10^{-19}\ \text{C})\pi(1.2\ \text{m})^2\ (50 \times 10^3\ \text{m/s})(20 \times 60\ \text{s})$$
$$= 3.47 \times 10^{-4}\ \text{C}$$

The increase in potential, relative to $\Phi(\infty) = 0$, is

$$V = \frac{Q}{C} = \frac{3.47 \times 10^{-4}\ \text{C}}{1.33 \times 10^{-10}\ \text{F}} = 2.61 \times 10^6\ \text{V}$$

or more than *two million volts!* (•Why would you expect the charging action to cease after a while?)

Example 30–3

Determine the capacitance of the cylindrical capacitor shown in the diagram.

R_3

R_2

V

R_1

$|\!\leftarrow\! L \!\rightarrow\!|$

Solution:

We assume that the spacing between the cylinders is small; that is, $(R_2 - R_1)/R_2 \ll 1$ and $R_2/L \ll 1$. Then, the field is essentially radial and uniform in the region $R_1 \leqslant r \leqslant R_2$; thus, fringing effects can be ignored.

The electric field strength in the gap is easily determined by constructing a concentric cylindrical Gaussian surface with radius $R_1 < r < R_2$ and length L. We find

$$\mathbf{E} = \frac{Q_1}{2\pi L\epsilon_0} \cdot \frac{1}{r}\,\hat{\mathbf{u}}_r$$

where Q_1 is the total charge on the inner conductor. For $R_1 \leqslant r \leqslant R_2$, we have (using Eq. 29–13a)

$$\Phi(r) = \Phi(R_1) - \frac{Q_1}{2\pi L\epsilon_0}\int_{R_1}^r \frac{dr}{r} = V - \frac{Q_1}{2\pi L\epsilon_0}\ln\frac{r}{R_1}$$

Now, $\Phi_2 = \Phi_3 = 0$, as in the case of the spherical capacitor. Then $\Phi(R_2) = 0$, and we obtain

$$Q_1 = \frac{2\pi L \epsilon_0}{\ln (R_2/R_1)} V$$

Also, as in the case of the spherical capacitor, we have $Q_2 = -Q_1$ and $Q_3 = 0$. The capacitance of the cylindrical capacitor is again defined through the relationship $Q = CV$; thus,

$$C = \frac{2\pi L \epsilon_0}{\ln (R_2/R_1)}$$

Example 30–4

Two spherical conductors with radii R_1 and R_2 are separated by a sufficiently large distance that induction effects are negligible. The spheres are connected by a thin conducting wire and are brought to the same potential V relative to $\Phi(\infty) = 0$.
(a) Determine the capacitance C of the system, where $C = (Q_1 + Q_2)/V$.
(b) What is the charge ratio Q_1/Q_2?

Solution:

(a) If induction effects are negligible, the charges on the two spheres are distributed uniformly, although with different charge densities. According to Eq. 30–9, the common potential V of the spheres can be expressed as

$$V = \frac{Q_1}{4\pi\epsilon_0 R_1} = \frac{Q_2}{4\pi\epsilon_0 R_2}$$

Thus, with $Q = Q_1 + Q_2$, we have

$$Q = 4\pi\epsilon_0 (R_1 + R_2) V$$

and, hence,

$$C = \frac{Q_1 + Q_2}{V} = 4\pi\epsilon_0 (R_1 + R_2)$$

(b) The charge ratio is

$$\frac{Q_1}{Q_2} = \frac{R_1}{R_2}$$

Notice that the charge *density* is in the inverse ratio to the charge itself; that is,

$$\frac{\sigma_1}{\sigma_2} = \frac{Q_1/4\pi R_1^2}{Q_2/4\pi R_2^2} = \frac{R_2}{R_1}$$

From the result of Example 30–4 it is easy to understand that the charge density at any point on a conductor with arbitrary shape will be inversely proportional to the radius of curvature of the surface at that point. The charged conducting surface shown in Fig. 30–6 consists of a cone that has spherical end caps with quite different radii. The charge density and, hence, the electric field is more intense at the smaller end. If the radius of the tip is made very small (needlelike), the field strength for even a modest charge may be sufficiently high that corona discharge will take place. This, in fact, is the way that lightning arrestors work.

Fig. 30–6. The charge density is larger and, hence, the electric field is more intense at the end of the object that has the smaller radius of curvature.

30-3   ELECTROSTATIC FIELD ENERGY

Consider an uncharged two-conductor capacitor with a capacitance C. Imagine charging such a capacitor by moving small amounts of charge dq from one conductor to the other. Let q be the total amount of charge transferred before a particular instant, and let $V(q)$ be the potential difference between the conductors at that instant. The work that must be done to transport a charge dq from one conductor to the other when the potential difference is $V(q)$ is

$$dW = V(q)\,dq = \frac{q}{C}\,dq$$

The total amount of work required to move a total charge Q is therefore

$$W = \int dW = \int_0^Q \frac{q}{C}\,dq$$

so that

$$W = \frac{1}{2}\frac{Q^2}{C} \tag{30-12}$$

We take the attitude that this work appears in the form of potential energy U stored in the electric field of the capacitor system. Thus,*

$$U = \frac{1}{2}\frac{Q^2}{C} = \tfrac{1}{2}QV = \tfrac{1}{2}CV^2 \tag{30-13}$$

If the capacitor is discharged, for example, by connecting an electric motor between the conductors, current will flow through the motor, the shaft will turn, and mechanical work will be done. When the capacitor is completely discharged, the electric field will have disappeared and a total amount of mechanical work $\tfrac{1}{2}Q^2/C$ will have been done.

We have by now twice encountered the idea of potential energy associated with electrostatic fields. It is important to distinguish carefully between these two situations because they represent quite different ways of generating potential energy. Earlier, it was pointed out that the electric potential $\Phi(\mathbf{r})$ produced by a fixed array of charges determines the difference in the potential energy, $U(B) - U(A) = q[\Phi(\mathbf{r}_B) - \Phi(\mathbf{r}_A)]$, of a point charge q between the positions specified by $\mathbf{r}_A$ and $\mathbf{r}_B$. The potential energy U in Eq. 30-13, however, is the energy required to create the field of the charged array itself. (This is correct only in the macroscopic sense. We ignore any energy involved in the creation of the elementary charged particles that constitute the charges in the array. We consider only the energy required to move these charges in order to establish the desired field.)

Energy Density. Consider an ideal parallel-plate capacitor with capacitance $C = \epsilon_0 A/s$ that is charged to a potential difference V. According to Eq. 29-31, we have $V = Es$, where E is the magnitude of the field strength between the plates.

---

*Notice that W in Eq. 30-12 represents the work done *on* the field by an outside agency. The work done *by* the field is $W' = -W$, and the potential energy stored in the field is $U = -W' = W$.

Substituting these expressions for C and V into the last form of Eq. 30-13, we obtain

$$U = \tfrac{1}{2}CV^2 = \tfrac{1}{2}\epsilon_0 E^2 sA$$

The electric field between the plates is uniform, so we can define the *electrostatic energy density* to be $u = U/(\text{volume}) = U/sA$, or

$$u = \tfrac{1}{2}\epsilon_0 E^2 = \tfrac{1}{2}\epsilon_0 \mathbf{E} \cdot \mathbf{E} \qquad \text{(30-14)}$$

This result was obtained for the special case of a parallel-plate capacitor. However, the expression for u contains none of the parameters of the capacitor, so we expect the expression to be generally true. Indeed, Eq. 30-14 gives the correct electrostatic energy density for any electric field $\mathbf{E}$ in vacuum. (Compare Problem 30-24.)

The Force on a Capacitor Plate.   Suppose that the plates of a parallel-plate capacitor carry charges $\pm Q$. What force is required to maintain a separation s between the plates? First, calculate the work that would have to be done if the plates were moved apart by an additional distance ds.* Thus,

$$dW = F(s)\, ds$$

where $F(s)$ is the force we are seeking. The energy U stored in the electric field is

$$U = \frac{1}{2}\frac{Q^2}{C} = \frac{Q^2 s}{2\epsilon_0 A}$$

Then, the work dW would appear as an increase dU in the potential energy:

$$dU = \frac{Q^2}{2\epsilon_0 A}\, ds$$

Equating dW and dU, we find

$$F(s)\, ds = \frac{Q^2}{2\epsilon_0 A}\, ds$$

or

$$F(s) = \frac{Q^2}{2\epsilon_0 A} \qquad \text{(30-15)}$$

Example 30-5

Use Eq. 30-14 to make an explicit calculation of the energy stored in the field of a simple spherical capacitor. Show that $U = \tfrac{1}{2}Q^2/C$.

Solution:

According to Eq. 30-6, the electric field strength within the capacitor is

---

*The displacement is not actually made, so the work done is called the *virtual work*. This general technique for making calculations is referred to as the *principle of virtual work*. (See Section 17-5.)

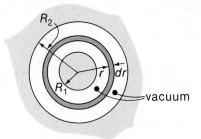

$$E(r) = \frac{Q}{4\pi\epsilon_0} \cdot \frac{1}{r^2}$$

Then, the energy stored in a spherical shell between r and $r + dr$ is

$$dU = u \, d(\text{volume}) = u(4\pi r^2 \, dr)$$

Using Eq. 30–14 and the expression for $E(r)$, we have

$$dU = \frac{Q^2}{8\pi\epsilon_0} \cdot \frac{dr}{r^2}$$

from which we obtain

$$U = \frac{Q^2}{8\pi\epsilon_0} \int_{R_1}^{R_2} \frac{dr}{r^2} = \frac{Q^2}{8\pi\epsilon_0} \cdot \frac{R_2 - R_1}{R_1 R_2}$$

Now, $C = 4\pi\epsilon_0 R_1 R_2 / (R_2 - R_1)$ for a spherical capacitor (Eq. 30–8); therefore,

$$U = \frac{1}{2} \frac{Q^2}{C}$$

which is the desired result.

Example 30–6

A circular parallel-plate capacitor has a radius of 30 cm and is charged to produce an electric field intensity of 3×10^6 V/m.

(a) What force is necessary to maintain the plates with a fixed separation?

(b) If the spacing between the plates is 3 mm, what amount of energy is stored in the capacitor field?

Solution:

(a) Using Eq. 30–15, we have

$$F(s) = \frac{Q^2}{2\epsilon_0 A} = \frac{C^2 V^2}{2\epsilon_0 A} = \frac{\epsilon_0 A}{2} \cdot \frac{V^2}{s^2} = \tfrac{1}{2}\epsilon_0 A E^2$$
$$= \tfrac{1}{2}(8.85 \times 10^{-12} \text{ C}^2/\text{N} \cdot \text{m}^2)\pi(0.30 \text{ m})^2(3 \times 10^6 \text{ V/m})^2$$
$$= 11.3 \text{ N}$$

(b) We have
$$U = (\tfrac{1}{2}\epsilon_0 E^2)(sA) = F(s)s$$
$$= (11.3 \text{ N})(3 \times 10^{-3} \text{ m})$$
$$= 3.39 \times 10^{-2} \text{ J}$$

By way of comparison, the energy stored in a typical 12-V automobile battery is about 2×10^6 J. A capacitor of enormous size would be necessary in order to store an equal amount of energy. In certain research and industrial applications, megajoule capacitors are actually used.

30-4 CAPACITOR NETWORKS

In problems that involve several capacitors, it is often convenient to introduce the idea of an *effective* or *equivalent capacitor*. That is, we seek a single capacitor which, when it replaces some network of capacitors, causes no change in the behavior of the remainder of the circuit to which it is attached. In diagrams we use the following descriptive symbol for a capacitor:

Series Connection. Consider two capacitors, C_1 and C_2, connected in *series* between the terminals A and D, as shown in Fig. 30-7a. When the switch S is closed, the capacitors, initially uncharged, become charged as the battery supplies an amount of charge Q. This charge accumulates on plate d so that $Q_d = +Q$. An induced charge, $Q_c = -Q_d = -Q$, appears on plate c. Because plates c and b (and the connecting wire) are isolated, we must have $Q_b = -Q_c = +Q$. Finally, the charge induced on plate a is $Q_a = -Q_b = -Q$. Thus, there are equal charges on the two capacitors, $Q_1 = Q_2 = Q$.

The middle pair of plates, b and c, have the same potential, $\Phi_b = \Phi_c$, so that the potential difference across the capacitor system can be expressed as $\Phi_D - \Phi_A = \Phi_d - \Phi_a = (\Phi_d - \Phi_c) + (\Phi_b - \Phi_a) = V$. Now, we have

$$\Phi_d - \Phi_c = \frac{Q_1}{C_1} \quad \text{and} \quad \Phi_b - \Phi_a = \frac{Q_2}{C_2}$$

Hence,

$$V = \frac{Q_1}{C_1} + \frac{Q_2}{C_2} = Q\left(\frac{1}{C_1} + \frac{1}{C_2}\right)$$

Fig. 30–7. (a) Two capacitors connected in series. (b) The capacitor C is equivalent to the series connection of C_1 and C_2.

If the equivalent capacitor C shown in Fig. 30-7b is to duplicate exactly the effect in the circuit of the C_1-C_2 combination, the same charge Q must appear on the plates when the switch is closed and the capacitor is charged to the voltage V. Thus,

$$V = \frac{Q}{C}$$

Comparing these two expressions for V, we see that

$$\frac{1}{C} = \frac{1}{C_1} + \frac{1}{C_2}$$

We can extend this argument to the case of N capacitors connected in series, with the

result

$$\frac{1}{C} = \frac{1}{C_1} + \frac{1}{C_2} + \frac{1}{C_3} + \cdots + \frac{1}{C_N} \qquad \text{series connection} \qquad \textbf{(30-16)}$$

Another way to view the series connection of two capacitors is to imagine the wire between them shrinking to zero length. The two common plates, which carry equal charges of opposite sign, combine into a single plate with zero effective thickness. Then, the separation between the two outer plates is $s_1 + s_2$, and

$$\frac{1}{C} = \frac{s_1 + s_2}{\epsilon_0 A} = \frac{s_1}{\epsilon_0 A} + \frac{s_2}{\epsilon_0 A} = \frac{1}{C_1} + \frac{1}{C_2}$$

Parallel Connection. Consider two capacitors, C_1 and C_2, connected in *parallel* between the terminals A and D, as shown in Fig. 30–8a. When the switch S is closed, the capacitors, initially uncharged, become charged as the battery supplies an amount of charge Q. This charge is shared by the two capacitors, so that $Q_1 + Q_2 = Q$. The potential differences, $\Phi_b - \Phi_a$ and $\Phi_d - \Phi_c$, are both equal to V; thus, we have

$$Q = Q_1 + Q_2 = C_1 V + C_2 V = (C_1 + C_2)V$$

The battery must supply the same charge Q to the equivalent capacitor C. That is,

$$Q = CV$$

Comparing these two expressions for Q, we see that

$$C = C_1 + C_2$$

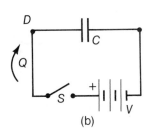

Fig. 30–8. (a) Two capacitors connected in parallel. (b) The capacitor C is equivalent to the parallel connection of C_1 and C_2.

We can extend this argument to the case of N capacitors connected in parallel, with the result

$$C = C_1 + C_2 + C_3 + \cdots + C_N \qquad \text{parallel connection} \qquad \textbf{(30-17)}$$

Again, there is an alternate way of viewing the parallel connection of two or more capacitors. Imagine moving the plates together until they are joined at the edges. Then, the area of each combined plate is $A_1 + A_2$, and

$$C = \frac{\epsilon_0(A_1 + A_2)}{s} = \frac{\epsilon_0 A_1}{s} + \frac{\epsilon_0 A_2}{s} = C_1 + C_2$$

Example 30–7

Show that the potential differences across the two capacitors in Fig. 30–7a are in inverse proportion to their capacitance values.

Solution:

For a series connection of two capacitors, we have

$$V = Q\left(\frac{1}{C_1} + \frac{1}{C_2}\right)$$

or

$$Q = \frac{C_1 C_2}{C_1 + C_2} V$$

Hence,

$$\Phi_d - \Phi_c = V_1 = \frac{Q_1}{C_1} = \frac{Q}{C_1} = \frac{C_2}{C_1 + C_2} V$$

and

$$\Phi_b - \Phi_a = V_2 = \frac{Q_2}{C_2} = \frac{Q}{C_2} = \frac{C_1}{C_1 + C_2} V$$

Therefore,

$$\frac{V_1}{V_2} = \frac{C_2}{C_1}$$

Example 30–8

(a) The diagram shows a network of capacitors between the terminals A and B. Reduce this network to a single equivalent capacitor.

(b) Determine the charge on the 4-μF and 8-μF capacitors when the capacitors are fully charged by a 12-V battery connected to the terminals.

(c) Determine the potential difference across each capacitor.

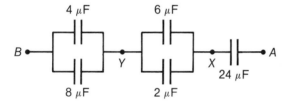

Solution:

(a) First, reduce the two parallel capacitors between B and Y to a single equivalent capacitor C_1:

$$C_1 = 4\,\mu\text{F} + 8\,\mu\text{F} = 12\,\mu\text{F}$$

Next, reduce the two parallel capacitors between Y and X to a single equivalent capacitor C_2:

$$C_2 = 6\,\mu\text{F} + 2\,\mu\text{F} = 8\,\mu\text{F}$$

The result of these two reductions is the equivalent network shown in the diagram below.

Finally, the three capacitors, C_1, C_2, and C_3, can be reduced to the single equivalent capacitor C, where

$$\frac{1}{C} = \frac{1}{12\,\mu\text{F}} + \frac{1}{8\,\mu\text{F}} + \frac{1}{24\,\mu\text{F}} = \frac{1}{4\,\mu\text{F}}$$

Thus, $$C = 4\,\mu\text{F}$$

(b) The total charge supplied by the battery to the capacitor network is $Q = CV = (4\,\mu\text{F})(12\,\text{V}) = 48\,\mu\text{C}$. The equivalent capacitors, C_1, C_2, and C_3, all have this same stored charge. The real capacitors that comprise C_1—the 4-μF and the 8-μF capacitors—share 48 μC of charge in direct proportion to their capacitance values. Therefore, the charge on the 4-μF capacitor is

$$Q_4 = \tfrac{4}{12} \times (48\,\mu\text{C}) = 16\,\mu\text{C}$$

and the charge on the 8-μF capacitor is

$$Q_8 = \tfrac{8}{12} \times (48\,\mu\text{C}) = 32\,\mu\text{C}$$

The charges on the 2-μF and 6-μF capacitors are shared in the same way. The 24-μF capacitor carries the full 48 μC of charge.

(c) The potential difference between points B and A is 12 V. That is,

$$\Phi_B - \Phi_A = (\Phi_B - \Phi_Y) + (\Phi_Y - \Phi_X) + (\Phi_X - \Phi_A)$$
$$= V_1 + V_2 + V_3 = V = 12\,\text{V}$$

Each of the capacitors, C_1, C_2, and C_3, carries the same charge Q. Thus, we have

$$V_1 = \frac{Q_1}{C_1} = \frac{Q}{C_1} = \frac{48\,\mu\text{C}}{12\,\mu\text{F}} = 4\,\text{V}$$

$$V_2 = \frac{Q_2}{C_2} = \frac{Q}{C_2} = \frac{48\,\mu\text{C}}{8\,\mu\text{F}} = 6\,\text{V}$$

$$V_3 = \frac{Q_3}{C_3} = \frac{Q}{C_3} = \frac{48\,\mu\text{C}}{24\,\mu\text{F}} = 2\,\text{V}$$

and $V_1 + V_2 + V_3 = 12$ V, as expected.

Although the reduction of the network in this example was straightforward, it should be noted that some types of two-terminal networks cannot be treated in this simple manner using only the series and parallel addition rules. For example, the network shown in the diagram below cannot be reduced using these techniques. (However, see Problem 30–32 for a readily solvable special case.)

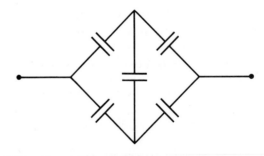

30-5 DIELECTRICS AND POLARIZATION

At about the time of Franklin's experiments with electricity it was discovered that the amount of charge that could be accumulated on a capacitor depends both on the geometry and on the insulating material between the conducting surfaces.* This effect can be demonstrated by inserting a slab of insulating material (a *dielectric*) between the plates of a parallel-plate capacitor. When a glass slab completely fills the space between the plates, it is found that a charge Q is accumulated on the plates for an applied potential difference only about one quarter of that required without the glass slab (Fig. 30–9). This implies that the capacitance value of the plates is increased by about a factor of 4 when the glass slab is used as a dielectric. The ratio of the capacitance C of a capacitor with a dielectric inserted to the capacitance C_0 with vacuum (or air) between the plates is called the *dielectric constant* κ of the material. That is,

$$\kappa = \frac{C}{C_0} \qquad (30\text{--}18)$$

Experiments show that $\kappa > 1$ for all dielectric materials. (Values of κ for some materials are given in Table 30–1 on page 830.)

Because a capacitor with a dielectric requires a potential difference of only $1/\kappa$ of that required to accumulate the same charge without the dielectric present, it follows that the electric field strength **E** in the dielectric is smaller by the same factor. That is,†

$$\mathbf{E} = \frac{1}{\kappa}\mathbf{E}_0 \qquad (30\text{--}19)$$

We now inquire how this reduction in field strength comes about.

Consider the effect on the molecules of a dielectric material when an external macroscopic electric field is applied. We distinguish two types of behavior, depending on whether the dielectric consists of a *polar* or a *nonpolar* material. In a polar material the electrons are distributed with respect to the positive nuclear charges in such a way that the individual molecules possess permanent dipole moments. In a nonpolar material, on the other hand, the charge distribution does not have this characteristic and the molecules do not possess permanent dipole moments. However, in the presence of an electric field, the positive and negative molecular charges in a nonpolar material experience oppositely directed forces and become displaced from each other by some average distance $\boldsymbol{\ell} = \ell\hat{\mathbf{u}}$, where $\hat{\mathbf{u}}$ is a unit vector in the direction of the applied field (Fig. 30–10b). The direction of $\boldsymbol{\ell}$ is taken to be from the negative charge center of the molecule to the positive charge center. Let the total individual molecular charge (of either sign) be q_0. Then, the quantity $q_0\ell$ is defined to be the induced molecular *dipole moment* **p** (compare Eq. 28–7):

$$\mathbf{p} = q_0\boldsymbol{\ell} \qquad (30\text{--}20)$$

---

*In Franklin's day electric charge was stored in a primitive type of capacitor called a *Leyden jar*. This device consists of a glass jar with inner and outer conducting surfaces of metal foil. A highly charged Leyden jar can be dangerous and must be handled with great care.

†We consider only homogeneous isotropic dielectric materials, and ignore the complications that may arise for crystalline substances. In the latter case, the dielectric "constant" is actually a tensor.

(a)

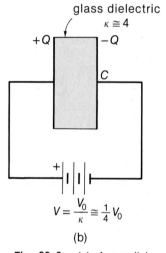

(b)

Fig. 30–9. (a) A parallel-plate capacitor with vacuum (or air) between the plates is charged to a potential difference V_0. (b) When a glass dielectric is inserted between the plates, a potential difference of only about $\frac{1}{4}V_0$ is required to place the same charge on the plates.

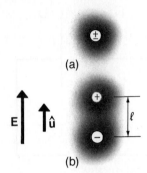

(a)

(b)

Fig. 30–10. (a) In the absence of an external electric field the positive charge center ($\oplus$) and the negative charge center ($\ominus$) coincide for a nonpolar molecule. (b) When a field is applied, the charge centers are displaced by an amount ℓ. (The displacement ℓ here corresponds to the quantity 2ℓ used in the earlier definition of the dipole moment—Eq. 28–7.)

Further, the dipole moment density or polarization **P** for a dielectric material is defined to be

$$\mathbf{P} = \mathcal{N}\mathbf{p}, \qquad \text{nonpolar dielectric} \tag{30–21}$$

where $\mathcal{N}$ is the number density of molecules (molecules per cubic meter). Experiments show that under ordinary circumstances the polarization **P** is proportional to the strength of the applied electric field.

In a polar dielectric material, each molecule possesses a permanent dipole moment $\mathbf{p}_0$ that results from its internal charge distribution. However, in the absence of an external electric field, the individual moments are randomly oriented so that the material as a whole has no dipole moment* (Fig. 30–11a). When an external field is applied, the molecular dipoles experience a torque that tends to align them with the field direction (see Section 29–4). Total alignment is not achieved because of the disorienting effects of thermal agitation (Fig. 30–11b). The resulting average alignment† $\langle \mathbf{p}_0 \rangle$ is in the direction $\hat{\mathbf{u}}$ of the field and, under ordinary circumstances, is proportional to the strength of the applied field. For a polar dielectric material the polarization **P** is expressed as

$$\mathbf{P} = \mathcal{N}\langle \mathbf{p}_0 \rangle, \qquad \text{polar dielectric} \tag{30–22}$$

The formal similarity of the expressions for **P** (Eqs. 30–21 and 30–22) does not mean that the behavior of polar and nonpolar molecules is the same. In fact, the dielectric constants for polar materials are about an order of magnitude larger than those for nonpolar materials.

Imagine that the space occupied by a dielectric is divided into a large number of identical cells (Fig. 30–12) with length L in the direction of the polarization vector **P** and with cross-sectional area d^2. Within each cell the actual molecular dipoles are replaced by fictitious charges $\pm q_P$ distributed on the front and rear faces of the cell, as shown in Fig. 30–12. These are *bound* charges that are not free to migrate through the dielectric. We consider only cases in which the polarization **P** in the medium is constant throughout. Then, for two adjacent cells, the common wall carries a charge $+q_P$ from one cell and $-q_P$ from the other. Thus, in the interior of the dielectric

---

*We ignore the small class of dielectric materials—the *ferroelectric* crystals such as Rochelle salt and barium titanate—that *do* possess macroscopic polarization even in the absence of an external electric field.

†The slanted brackets around **p** imply an average over the dipole moment distribution (see the footnote on page 666).

(a)

(b)

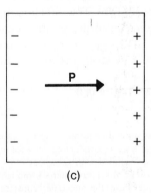

(c)

Fig. 30–11. (a) In the absence of an external electric field, the molecular dipoles are randomly oriented. (b) When a field is applied, the dipoles are partially aligned in the field direction $\hat{\mathbf{u}}$. (c) The net result is the appearance of bound surface charges on the ends of the material.

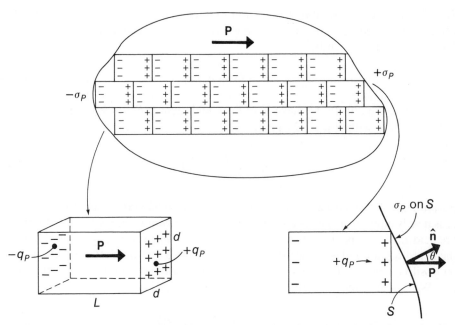

Fig. 30-12. A dielectric substance is divided into a large number of identical cells, each with a dipole moment density **P**. In the bulk material, the bound charges on the cell walls cancel throughout the interior, leaving a surface polarization charge density $\pm\sigma_P$.

material these charges cancel everywhere. However, bound charges remain on the macroscopic surface of the dielectric, as indicated in Fig. 30-12.

The polarization P is the dipole moment density. Therefore, the total dipole moment of the cell is equal to $P \times (Ld^2)$. Alternatively, the total dipole moment of the cell can be expressed as $q_P L$. From the two expressions for the total dipole moment we see that $P = |q_P|/d^2$. (The sign of the polarization charge $q_P \to \sigma_P$ is, of course, different at the two ends of the cell.)

In Fig. 30-12 the polarization vector **P** is perpendicular to the cell face with dimensions $d \times d$. The actual (slanted) surface area S makes an angle θ with the cell face such that $S \cos\theta = d^2$. The surface charge density that appears on the macroscopic surface of the dielectric is, therefore,

$$\sigma_P = \frac{q_P}{S} = \frac{q_P}{d^2}\cos\theta = P\cos\theta = \mathbf{P}\cdot\hat{\mathbf{n}} \qquad (30\text{-}23)$$

This effective polarization charge is still *bound* to neutral molecules in the dielectric. It is important to distinguish carefully between this bound charge and any *free* charge that might be deposited on the surface.

Figure 30-13 illustrates a case of uniform polarization in which a slab of a dielectric material is inserted between the plates of an ideal parallel-plate capacitor. Let the surface density of free charge on the plates be $\pm\sigma$, with a corresponding electric field $\mathbf{E}_0 = E_0\,\hat{\mathbf{u}}$ in the gaps between the plates and the dielectric slab (Fig. 30-13a). As before, we have $E_0 = \sigma/\epsilon_0$. The polarization **P** produces a polarization surface charge density, $\sigma_P = \mathbf{P}\cdot\hat{\mathbf{n}}$, on the two parallel faces of the dielectric. Figure 30-13b shows the construction of a Gaussian surface with faces that are parallel and perpendicular to $\hat{\mathbf{u}}$. The face within the lower conducting plate is in a region where the field is zero. The electric field strength inside the dielectric slab is $\mathbf{E} = E\hat{\mathbf{u}}$.

Fig. 30-13. (a) A dielectric slab is placed between the plates of a parallel-plate capacitor. The field strength in the gap is $E_0 = \sigma/\epsilon_0$. (b) A Gaussian surface is constructed to show that the field strength in the dielectric is $E = (\sigma - \sigma_P)/\epsilon_0 = \sigma/\epsilon_0\kappa$. (c) Field lines for the case $\kappa = 4$.

Therefore, there is a net flux EA through the face of the Gaussian surface that lies within the dielectric. The total charge inside the surface is $(\sigma - \sigma_P)A$, so that $EA = (\sigma - \sigma_P)A/\epsilon_0$. Thus,

$$E = (\sigma - \sigma_P)/\epsilon_0 \tag{30-24}$$

According to Eq. 30-19, we have $\mathbf{E} = \mathbf{E}_0/\kappa$ and, hence, $E = \sigma/\epsilon_0\kappa$. Combining this last expression with Eq. 30-24 gives

$$\sigma_P = \sigma\left(1 - \frac{1}{\kappa}\right) \tag{30-25}$$

Because $\kappa > 1$, we have $\sigma_P < \sigma$; however, for moderately large values of κ ($\kappa \gtrsim 10$), σ_P is nearly equal to σ. The presence of a large polarization charge greatly diminishes the field E within the dielectric, with the result $E \ll E_0 = \sigma/\epsilon_0$. Figure 30-13c shows the field lines for a glass dielectric with $\kappa = 4$. Notice that the field lines all originate on positive charges (free and polarization) and terminate on negative charges.

We can summarize these results with the expressions

$$
\begin{aligned}
\frac{1}{\epsilon_0}\mathbf{P} &= \mathbf{E}_0 - \mathbf{E} \\
&= \left(1 - \frac{1}{\kappa}\right)\mathbf{E}_0 \\
&= (\kappa - 1)\mathbf{E} \\
&= \chi\mathbf{E}
\end{aligned} \tag{30-26}
$$

where, in the last equality, the *electric susceptibility* χ of a dielectric material is

TABLE 30-1. Dielectric Properties of Various Substances

GASES[a]	$\chi \times 10^{6}$ [b]	SOLIDS[c]	κ
Helium, He	65	Paraffin	2.20
Hydrogen, H_2	254	Polyethylene	2.3
Air (dry, CO_2 free)	536	Polystyrene	2.56
Carbon dioxide, CO_2	922	Sulfur	3.44
LIQUIDS[d]	κ	"Ruby" mica	≈ 5.4
		Pyrex glass	≈ 5.6
Transformer oil	2.24	Glass (ordinary)	4 to 6
Benzene, C_6H_6	2.28	Sodium chloride	5.62
Chlorobenzene, C_6H_5Cl	5.71	Neoprene rubber	≈ 6.7
Ethyl alcohol, C_2H_5OH	24.3	High alumina porcelain	≈ 8.8
Methyl alcohol, CH_3OH	33.6	Titanium dioxide	173
Water, H_2O	80.4	Barium titanate	≈ 1500

[a] For 20° C, 1 atm.
[b] This notation means, for example, that the first entry is 65 × 10⁶.
[c] Near room temperature.
[d] For 20° C.

defined to be

$$\chi = \kappa - 1 \tag{30-27}$$

Values of the dielectric constant κ for various substances are listed in Table 30-1. Because κ is nearly unity for gases, it is more convenient to indicate the dielectric properties of gases in terms of the electric susceptibility χ.

For any dielectric material there is a maximum electric field strength that the material can withstand without breaking down and losing its insulating quality. This maximum tolerable field strength is called the *dielectric strength* of the material. Values of the dielectric strength for some materials are given in Table 30-2.

TABLE 30-2. Dielectric Strengths of Various Substances

SUBSTANCE	DIELECTRIC STRENGTH (10^6 V/m)
Air	3
Epoxy cast resin	16
Glass (ordinary)	9
Lucite	18–20
Paraffin	10
Polyethylene	18
Transformer oil	12

Example 30-9

A wafer of titanium dioxide has an area of 1 cm² and a thickness of 0.10 mm. Aluminum is evaporated on the parallel faces to form a parallel-plate capacitor.

(a) Calculate the capacitance.

(b) When the capacitor is charged with a 12-V battery, what is the magnitude of charge delivered to each plate?

(c) For the situation in (b), what are the free and polarization surface charge densities?

(d) What is the electric field strength E?

Solution:

(a) Combining Eqs. 30-5 and 30-18, we have

$$C = \kappa C_0 = \frac{\kappa \epsilon_0 A}{s} = \frac{(173)(8.85 \times 10^{-12} \text{ F/m})(10^{-4} \text{ m}^2)}{0.10 \times 10^{-3} \text{ m}} = 1.53 \times 10^{-9} \text{ F}$$

(b) The battery delivers the free charge Q:

$$Q = CV = (1.53 \times 10^{-9} \text{ F})(12 \text{ V}) = 18.4 \text{ nC}$$

(c) The surface density of free charge is

$$\sigma = \frac{Q}{A} = \frac{18.4 \times 10^{-9} \text{ C}}{10^{-4} \text{ m}^2} = 1.84 \times 10^{-4} \text{ C/m}^2$$

The surface density of polarization charge is, using Eq. 30–25,

$$\sigma_P = \sigma\left(1 - \frac{1}{\kappa}\right) = \sigma\left(1 - \frac{1}{173}\right) = 0.9942\,\sigma$$

(d) We have $E = E_0/\kappa$ and $E_0 = V/s$; hence,

$$E = \frac{V}{\kappa s} = \frac{12\text{ V}}{(173)(10^{-4}\text{ m})} = 694\text{ V/m}$$

▶ *Microscopic (Molecular) Considerations.* Whether we consider polar or nonpolar materials, the dipole moment of a molecule is developed in the presence of an average microscopic electric field at the position of the molecule. If, in a dielectric substance, there exists a polarization associated with an average macroscopic field strength **E**, what is the microscopic field strength $\mathbf{E}_e$ at the molecular site?

Figure 30–14 shows a portion of the interior of a dielectric in which we have constructed a spherical surface with a radius R centered on the position of the molecule under consideration. The value of R is sufficiently large that the effect of all the dipoles at distances greater than R can be represented in terms of a smoothly varying polarization charge σ_P on the spherical surface. All of the dipoles located within the spherical surface are treated individually. The electric field $\mathbf{E}_e$ at the molecular site is equal to the superposition of the macroscopic field **E** inside the dielectric, the field produced by the polarization charge on the surface of the sphere, and the field produced by the individual dipoles inside the sphere. (Notice that we exclude the field of the molecule at the site we are considering.)

In Fig. 30–14 we see that the density of polarization charge on the surface of the sphere is $\sigma_P = \mathbf{P} \cdot \hat{\mathbf{n}} = P \cos \theta$. Thus, the charge on the ring defined by θ and $\theta + d\theta$ is

$$dq_P = 2\pi R^2\sigma_P \sin\theta\,d\theta = 2\pi R^2 P \cos\theta \sin\theta\,d\theta$$

The symmetry insures that the corresponding electric field at the molecular site is in the direction $\hat{\mathbf{u}}$; hence, the polarization field is

$$d\mathbf{E}_P = \frac{dq_P}{4\pi\epsilon_0 R^2}\cos\theta\,\hat{\mathbf{u}} = \frac{P}{2\epsilon_0}\cos^2\theta \sin\theta\,d\theta\,\hat{\mathbf{u}} \qquad \textbf{(30–28)}$$

The integral of this expression over the range of θ is

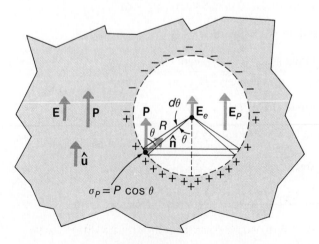

Fig. 30–14. Geometry for the calculation of the field strength $\mathbf{E}_e$ at the position of a molecule in a dielectric substance.

$$\mathbf{E}_P = \frac{\mathbf{P}}{2\epsilon_0} \int_0^\pi \cos^2 \theta \sin \theta \, d\theta$$

from which

$$\mathbf{E}_P = \frac{\mathbf{P}}{3\epsilon_0} \tag{30-29}$$

To obtain the electric field $\mathbf{E}_e$ at the molecular site, we must add to $\mathbf{E}_P$ the macroscopic field $\mathbf{E}$ and the field due to the individual molecules within the sphere surrounding the site. The calculation of this latter effect is complicated by the need to know the exact nature of any long-range order that exists in the arrangement of the molecules within the dielectric material. For the special (and important) case of a solid with a cubic crystalline lattice, it is easy to show (see Problem 30–36) that the nearest-neighbor molecules produce exactly zero net field at the site of the molecule. Moreover, for a number of other crystal geometries the net field is nearly zero. As a practical matter we confine considerations here to those cases in which we may take the field due to the nearby molecules to be zero. Then, the polarization-inducing field strength $\mathbf{E}_e$ at the molecular site is equal to $\mathbf{E} + \mathbf{E}_P$; that is,

$$\mathbf{E}_e = \mathbf{E} + \frac{\mathbf{P}}{3\epsilon_0} \tag{30-30}$$

This argument provides the justification for considering $\mathbf{E}_e$ to be the polarizing field that acts on a particular molecule. It should be emphasized that $\mathbf{E}_e$ does not include the field produced by the molecule itself. Indeed, the reason that E_e is larger than E is that the local field of the polarized molecule is directed opposite to $\mathbf{E}$.

The treatment of the induced dipole moment in nonpolar materials properly requires a calculation based on quantum theory. However, a simplified classical calculation provides at least a measure of the effect. Consider a simple model of an atom in which a pointlike nucleus with charge $+Ze$ is surrounded by a uniform distribution of electronic charge $-Ze$ out to an effective atomic radius R. The cloud of electronic charge is originally centered on the nucleus. But in the presence of a polarizing field $\mathbf{E}_e$, the center of the electronic cloud (still spherical) becomes displaced from the nucleus by an amount ℓ (see Fig. 30–15).

The magnitude of the electric field $\mathbf{E}_N$ produced by the displaced electronic cloud at the position of the nucleus can be determined by using Gauss's law with the surface shown in Fig. 30–15. We find

$$E_N = \frac{1}{4\pi \ell^2} \cdot \frac{Q}{\epsilon_0} = \frac{1}{4\pi \ell^2} \cdot \frac{Ze(\ell^3/R^3)}{\epsilon_0} = \frac{Ze}{4\pi\epsilon_0} \cdot \frac{\ell}{R^3}$$

The corresponding force on the nucleus, ZeE_N, is exactly balanced by the force, ZeE_e, due to the imposed field, thereby producing the equilibrium situation. Thus,

$$ZeE_e = \frac{(Ze)^2}{4\pi\epsilon_0} \cdot \frac{\ell}{R^3}$$

or

$$Ze\ell = 4\pi\epsilon_0 R^3 E_e$$

If we write the induced atomic dipole moment as $p = q_0\ell = Ze\ell$, we have

$$\mathbf{p} = \alpha \mathbf{E}_e \tag{30-31}$$

where α is the *atomic polarizability*. For this simple atomic model,

$$\alpha = 4\pi\epsilon_0 R^3 \qquad \text{nonpolar atom} \tag{30-32}$$

Atomic radii are of the order of 1 Å, so atomic polarizabilities are expected to be approx-

Fig. 30–15. In the presence of a polarizing field $\mathbf{E}_e$, the center of the electronic cloud of an atom is displaced from the nucleus by an amount ℓ. The nucleus is in equilibrium, with the force $\mathbf{F}_e$ due to the imposed field exactly balanced by the force $\mathbf{F}_N$ due to the displaced electron cloud.

**TABLE 30–3. Atomic Polarizabilities
of Some Elements**

ELEMENT	α (10^{-40} C · m²/V)
H	0.73
He	0.23
Li	13
C	1.7
Ne	0.44
Na	30
Ar	1.8
K	38

imately equal to

$$\alpha \cong 4\pi(8.85 \times 10^{-12}\,\text{C/V} \cdot \text{m})(10^{-10}\,\text{m})^3 \cong 10^{-40}\,\text{C} \cdot \text{m}^2/\text{V}$$

The entries in Table 30–3 are indeed close to this value. Notice the small values of α for the noble gases, He, Ne, and Ar, in which the electrons are tightly bound, and the larger values of α for Li, Na, and K, in which there is a loosely bound valence electron. In the latter case, the assumption of a displaced spherical electronic cloud is not really adequate.

Figure 30–16 pictures some polar molecules. These molecules have permanent dipole moments owing to their nonsymmetric structures. (Symmetric molecules such as H_2, N_2, CO_2, and C_2H_6 do not have dipole moments.) When a substance composed of polar molecules is placed in an electric field, the dipole moments $\mathbf{p}_0$ tend to align with the direction of the polarizing field $\mathbf{E}_e$. The alignment is not complete, however, because of the randomizing effects of thermal agitation. Consequently, the average value of the dipole moment for a polar molecule has a temperature dependence.

The energy of the molecular dipole in the polarizing field is $U = -\mathbf{p}_0 \cdot \mathbf{E}_e$ (see Eq. 29–38 with $U_0 = 0$). Let $d\eta$ represent the probability that the dipole moment $\mathbf{p}_0$ of a molecule makes an angle between θ and $\theta + d\theta$ with the direction of the field $\mathbf{E}_e$. In the absence of any

Fig. 30–16. Some typical polar molecules. The magnitudes of the permanent molecular dipole moments p_0 are given in units of 10^{-30} C·m.

long-range intermolecular forces, the equipartition theorem requires that $d\eta$ be proportional to the Boltzmann factor (see Eq. 24–28); thus,

$$d\eta = Ce^{-U/kT} d\Omega_\theta$$

where $d\Omega_\theta = 2\pi \sin\theta \, d\theta$ is the solid angle between θ and $\theta + d\theta$ and where C is a constant to be determined. Also, the projection of $\mathbf{p}_0$ along the direction of $\mathbf{E}_e$ is $p_0 \cos\theta$. Thus, the average dipole moment is

$$\langle p_0 \rangle = C \int_0^\pi e^{p_0 E_e \cos\theta/kT} (p_0 \cos\theta)(2\pi \sin\theta \, d\theta)$$

In ordinary circumstances the exponential factor is small: $p_0 E_e \cos\theta/kT \ll 1$. Then, the integral can be evaluated by expanding the exponential, with the result (see Problem 30–39) expressed in vector notation as

$$\langle \mathbf{p}_0 \rangle = \frac{p_0^2}{3kT} \mathbf{E}_e = \mathbf{p} \tag{30–33}$$

This result has the same form as Eq. 30–31 for the nonpolar case, namely, $\mathbf{p} = \alpha \mathbf{E}_e$, if we identify

$$\alpha = \frac{p_0^2}{3kT}, \qquad \text{polar molecule} \tag{30–34}$$

For both polar and nonpolar materials, we can express the macroscopic polarization as

$$\mathbf{P} = \mathcal{N}\mathbf{p} = \mathcal{N}\alpha\mathbf{E}_e \tag{30–35}$$

Using Eqs. 30–26, 30–27, and 30–30, we find, after some algebra,

$$\frac{\mathcal{N}\alpha}{3\epsilon_0} = \frac{\kappa - 1}{\kappa + 2} = \frac{\chi}{\chi + 3} \tag{30–36}$$

This result is known as the *Clausius-Mossotti relationship* and expresses the *microscopic* quantity $\mathcal{N}\alpha$ in terms of κ or χ, which represent *macroscopic* properties of the substance. The Clausius-Mossotti relationship applies strictly only for materials that have no long-range molecular order (with the exception of cubic crystals, for which the relationship does apply). Even liquid water exhibits a chainlike structure due to the attraction between the highly polarized molecules (see Fig. 30–16). ◀

Example 30–10

According to Table 30–3, the atomic polarizability of helium is $\alpha = 0.23 \times 10^{-40}$ C · m²/V.

(a) Calculate the electric susceptibility χ for helium gas at 20° C and 1 atm.

(b) Estimate the radius of a helium atom.

(c) Estimate the separation ℓ between the nucleus of a helium atom and the center of its electronic cloud when $E_e = 10^5$ V/m.

Solution:

(a) At 20° C and 1 atm, the number density of atoms in helium gas is $\mathcal{N} = 2.51 \times 10^{25}$ m⁻³. Using Eq. 30–36, we have

$$\frac{\kappa - 1}{\kappa + 2} = \frac{\mathcal{N}\alpha}{3\epsilon_0} = \frac{(2.51 \times 10^{25} \text{ m}^{-3})(0.23 \times 10^{-40} \text{ C} \cdot \text{m}^2/\text{V})}{3(8.85 \times 10^{-12} \text{ C/V} \cdot \text{m})} = 2.17 \times 10^{-5}$$

Thus, we see that κ is nearly equal to 1; hence, the denominator is $\kappa + 2 \cong 3$. Then,

$$\chi = \kappa - 1 = 3(2.17 \times 10^{-5}) = 65.1 \times 10^{-6}$$

which is in complete agreement with the value given in Table 30–1.

(b) We use Eq. 30–32 to estimate the atomic radius:

$$R = \left(\frac{\alpha}{4\pi\epsilon_0}\right)^{1/3} = \left(\frac{0.23 \times 10^{-40} \text{ C} \cdot \text{m}^2/\text{V}}{4\pi(8.85 \times 10^{-12} \text{ C/V} \cdot \text{m})}\right)^{1/3}$$
$$= 5.91 \times 10^{-11} \text{ m} = 0.59 \text{ Å}$$

(c) We have $Ze\ell = p = \alpha E_e$; thus,

$$\ell = \frac{\alpha E_e}{Ze} = \frac{(0.23 \times 10^{-40} \text{ C} \cdot \text{m}^2/\text{V})(10^5 \text{ V/m})}{2(1.60 \times 10^{-19} \text{ C})} = 7.2 \times 10^{-18} \text{ m}$$

which means that $\ell \ll R$ for ordinary field strengths.

Example 30–11

According to Fig. 30–16, the dipole moment of the water molecule is 6.2×10^{-30} C · m.
(a) Determine the polarizability α of the water molecule.
(b) Calculate the dielectric constant κ for saturated water vapor at 20° C.

Solution:

(a) Using Eq. 30–34, we have

$$\alpha = \frac{p_0^2}{3kT} = \frac{(6.2 \times 10^{-30} \text{ C} \cdot \text{m})^2}{3(1.38 \times 10^{-23} \text{ J/K})(293 \text{ K})} = 3.17 \times 10^{-39} \text{ C} \cdot \text{m}^2/\text{V}$$

(b) For saturated water vapor at 20° C, the vapor density is 17.3 g/m^3 (see Table 23–2). Thus, the number density is

$$\mathcal{N} = \frac{17.3 \text{ g/m}^3}{18.0 \text{ g/mol}} \times (6.02 \times 10^{23} \text{ molecules/mol})$$
$$= 5.79 \times 10^{23} \text{ molecules/m}^3$$

Using the Clausius-Mossotti relationship (Eq. 30–36), we find

$$\frac{\kappa - 1}{\kappa + 2} = \frac{\mathcal{N}\alpha}{3\epsilon_0} = \frac{(5.79 \times 10^{23} \text{ m}^{-3})(3.17 \times 10^{-39} \text{ C} \cdot \text{m}^2/\text{V})}{3(8.85 \times 10^{-12} \text{ C/V} \cdot \text{m})} = 6.91 \times 10^{-5}$$

from which $\chi = \kappa - 1 = 207 \times 10^{-6}$

This result is in close agreement with the experimental value.

30–6 THE DISPLACEMENT VECTOR

In the general case, the polarization **P** within a dielectric can be nonuniform. Then, within a volume V in the dielectric there can be a net polarization charge Q_P. Now, for the case of free charge, we have $\mathbf{E} \cdot \hat{\mathbf{n}} = \sigma/\epsilon_0$ and a corresponding expression for Gauss's law, namely, $\oint_S \mathbf{E} \cdot \hat{\mathbf{n}} \, dA = Q/\epsilon_0$. Thus, the polarization expression, $\mathbf{P} \cdot \hat{\mathbf{n}} = \sigma_P$, suggests that there exists a Gauss's law relationship involving the polarization **P**. We assert that this relationship is*

$$\oint_S \mathbf{P} \cdot \hat{\mathbf{n}} \, dA = -Q_P \qquad (30\text{–}37)$$

where the sign is opposite to that in the expression for Gauss's law for **E** because the polarization vector **P** is directed from negative to positive charges, whereas the electric field vector **E** is directed from positive to negative charges. Also, we have used the standard convention that $\hat{\mathbf{n}}$ is directed outward from the volume V. Notice that when **P** is uniform, the surface integral vanishes for any completely interior surface. (•Can you see why?) In this case we have $Q_P = 0$, in agreement with the discussion in Section 30–5.

Figure 30–17 shows a Gaussian surface S at the boundary of a dielectric material. If the surface density of polarization charge is σ_P, Gauss's law can be expressed as

$$\mathbf{P} \cdot \hat{\mathbf{n}}' \, \Delta A = -\sigma_P \, \Delta A$$

because the polarization is zero on the part of the Gaussian surface that lies outside the dielectric. Thus, we have

$$\mathbf{P} \cdot \hat{\mathbf{n}}' = -\sigma_P \qquad (30\text{–}38)$$

Notice that this result is the same as that expressed by Eq. 30–23 when we realize that $\hat{\mathbf{n}}$ in Eq. 30–23 is an outward normal at the dielectric surface whereas $\hat{\mathbf{n}}'$ in Eq. 30–38 is directed inward (see Fig. 30–17).

When there is present an effective polarization charge Q_P as well as free charge Q, we have

$$\oint_S \mathbf{E} \cdot \hat{\mathbf{n}} \, dA = \frac{1}{\epsilon_0}(Q + Q_P)$$

Shifting the term Q_P/ϵ_0 to the left-hand side of the equation and using Eq. 30–37, we obtain

$$\oint_S (\epsilon_0 \mathbf{E} + \mathbf{P}) \cdot \hat{\mathbf{n}} \, dA = Q \qquad (30\text{–}39)$$

This expression is often useful because the surface integral is expressed in terms of only the free charge Q, which can be easily specified.

The sum $\epsilon_0 \mathbf{E} + \mathbf{P}$ defines a new vector, called the *electric displacement* **D**:

$$\mathbf{D} = \epsilon_0 \mathbf{E} + \mathbf{P} \qquad (30\text{–}40)$$

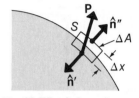

Fig. 30–17. A Gaussian surface S at the boundary of a dielectric material. The sides of the volume, with thickness Δx, are perpendicular to the boundary.

---

*In differential form we write $\nabla \cdot \mathbf{P} = -\rho_P$, where $\nabla \cdot \mathbf{P}$ is the divergence of **P**. (Compare Eq. 29–47.)

(a)

(b)

Fig. 30-18. (a) A Gaussian surface at a dielectric interface. (b) Path at the interface along which $\mathbf{E} \cdot d\mathbf{r}$ is evaluated.

Using this definition in Eq. 30-39, we have a new form of Gauss's law, namely,

$$\oint_{S} \mathbf{D} \cdot \hat{\mathbf{n}}\, dA = Q = \int_{V} \rho\, dv \tag{30-41}$$

The displacement $\mathbf{D}$ depends only on the *free charge;* even in the presence of matter, Q does not involve polarization charges.

According to Eq. 30-26, we have $\mathbf{P} = \epsilon_0(\kappa - 1)\mathbf{E}$; hence, we can also express the displacement vector $\mathbf{D}$ as

$$\mathbf{D} = \epsilon_0 \kappa \mathbf{E} = \epsilon \mathbf{E} \tag{30-42}$$

where $\epsilon = \epsilon_0 \kappa$ is called the *dielectric permittivity.*

The above equations are valid for dielectric materials that are homogeneous, isotropic, and linear (that is, $\mathbf{P} \propto \mathbf{E}$); such materials are called *Class A dielectrics.*

Boundary Conditions. The vectors $\mathbf{E}$ and $\mathbf{D}$ obey different boundary conditions at dielectric interface surfaces. Figure 30-18 shows how this comes about. To be completely general, assume that free charge with density σ exists on the interface surface. Applying Eq. 30-41 to the Gaussian surface S shown in Fig. 30-18a, we find

$$(D_{1\perp} - D_{2\perp})\, \Delta A = \sigma\, \Delta A$$

which can be written as

$$(\mathbf{D}_1 - \mathbf{D}_2) \cdot \hat{\mathbf{n}} = \sigma \tag{30-43}$$

That is, the component of the displacement $\mathbf{D}$ that is normal to the interface surface is increased by the presence of free charge on the surface. In the event that $\sigma = 0$, Eq. 30-43 becomes

$$D_{1\perp} = D_{2\perp} \qquad (\sigma = 0) \tag{30-44}$$

and the normal component of $\mathbf{D}$ is continuous across the surface if no free charge is present.

The electric field is conservative, so the line integral of $\mathbf{E} \cdot d\mathbf{r}$ around the path shown in Fig. 30-18b is equal to zero. Thus,

$$\oint \mathbf{E} \cdot d\mathbf{r} = (E_{1\parallel} - E_{2\parallel})\, \Delta x = 0$$

Hence,

$$E_{1\parallel} = E_{2\parallel} \tag{30-45}$$

and the tangential component of $\mathbf{E}$ is continuous across the surface.

The boundary conditions on the vectors $\mathbf{D}$ and $\mathbf{E}$ are illustrated in Fig. 30-19.

Figure 30-20 shows two cavities within a dielectric material—cavity (a) is long and slender, whereas cavity (b) is shallow and disc-shaped. In Fig. 30-20a the field

Fig. 30-19. At the boundary between two dielectric media the conditions on the vectors $\mathbf{D}$ and $\mathbf{E}$ are (a) the normal component of $\mathbf{D}$ ($D_{\perp}$) is continuous (for $\sigma = 0$) and (b) the tangential component of $\mathbf{E}$ ($E_{\parallel}$) is continuous.

(a)

(b)

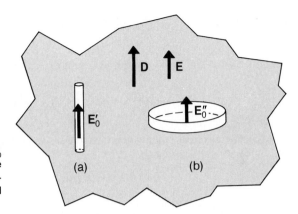

Fig. 30-20. The field strengths, $\mathbf{E}_0'$ and $\mathbf{E}_0''$, within the cavities can be determined by using the boundary-condition equations, Eqs. 30-44 and 30-45.

$\mathbf{E}_0'$ within the cavity is the same as the field $\mathbf{E}$ in the dielectric; that is, $\mathbf{E}_0' = \mathbf{E}$. In Fig. 30-20b the field $\mathbf{E}_0''$ within the cavity is $\mathbf{E}_0'' = \mathbf{D}/\epsilon_0$. (•Can you use the boundary conditions stated in Eqs. 30-44 and 30-45 to prove these assertions?)

The form of Eq. 30-41 suggests that we should view $\mathbf{D}$ as a vector that represents a field whose source is the free charge distribution ρ. Although this is true, the idea is useful only when the situation involves a geometry with a high degree of symmetry so that the directions of the lines of $\mathbf{D}$ are known beforehand. (This technique is demonstrated in the following examples.) If the situation is complicated and no obvious symmetry exists, more advanced methods are required to develop the solution.

Example 30-12

A conducting sphere with a radius R_1 carries a free charge Q and is surrounded by a concentric dielectric spherical shell with an outer radius R_2 and a dielectric constant κ. This isolated combination is in air ($\kappa \cong 1$).

(a) Determine the displacement $\mathbf{D}$ everywhere.
(b) Determine the electric field strength $\mathbf{E}$ everywhere.
(c) What is the polarization $\mathbf{P}$ within the dielectric?
(d) Obtain expressions for the effective polarization surface charges on the interfaces at radii R_1 and R_2.

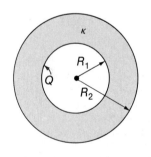

Solution:

(a) Inside the conducting sphere, $r < R_1$, the field is zero: $\mathbf{E} = 0$ and $\mathbf{D} = 0$. Using a Gaussian surface with $r > R_1$, Eq. 30-41 gives

$$4\pi r^2 \, D(r) = Q$$

or
$$\mathbf{D}(r) = \frac{Q}{4\pi r^2}\hat{\mathbf{u}}_r, \qquad r > R_1$$

(b) In the region $R_1 < r < R_2$, we have (Eq. 30-42) $\mathbf{E} = \mathbf{D}/\epsilon_0\kappa$; hence,

$$\mathbf{E}(r) = \frac{Q}{4\pi\epsilon_0\kappa r^2}\hat{\mathbf{u}}_r, \qquad R_1 < r < R_2$$

In the region $r > R_2$, we have $\mathbf{E} = \mathbf{D}/\epsilon_0$ because $\kappa(\text{air}) \cong 1$. Thus,

$$\mathbf{E}(r) = \frac{Q}{4\pi\epsilon_0 r^2}\hat{\mathbf{u}}_r, \qquad r > R_2$$

(c) Using Eqs. 30–40 and 30–42, we have

$$\mathbf{P}(r) = \mathbf{D}(r) - \epsilon_0\mathbf{E}(r) = \frac{Q(\kappa - 1)}{4\pi\kappa r^2}\hat{\mathbf{u}}_r, \qquad R_1 < r < R_2$$

(d) The effective polarization charge density on the interface at $r = R_1$ is

$$\sigma_P(R_1) = \mathbf{P}(r)\cdot\hat{\mathbf{u}}_r\Big|_{r=R_1} = -\frac{Q(\kappa - 1)}{4\pi\kappa R_1^2}$$

On the interface at $r = R_2$, we have

$$\sigma_P(R_2) = +\frac{Q(\kappa - 1)}{4\pi\kappa R_2^2}$$

Example 30–13

Determine the capacitance of a parallel-plate capacitor in which the region between the plates is partially filled with a dielectric slab, as shown in the diagram.

Solution:

The electric field strength in the air gaps is $E_0 = \sigma/\epsilon_0$, or

$$E_0 = \frac{Q}{A\epsilon_0}$$

Then, the electric displacement in the air gaps is $D_0 = \epsilon_0 E_0$. The vector $\mathbf{D}_0$ is perpendicular to the dielectric surface, and, according to Eq. 30–44, the normal component of the displacement is continuous across the surface; thus, $D = \epsilon_0\kappa E = D_0$. Hence, $\epsilon_0\kappa E = \epsilon_0 E_0$, so that

$$E = \frac{E_0}{\kappa}$$

(Notice that this result could have been obtained directly from Eq. 30–19.)

The potential difference between the plates can be expressed as

$$V = \Phi_b - \Phi_a = -\int_a^b \mathbf{E}\cdot d\mathbf{r}$$

Evaluating the integral along the dashed-line path shown in the diagram, we find

$$V = E_0\left(t_1 + \frac{d}{\kappa} + t_2\right) = \frac{Q}{A\epsilon_0}\left(t + \frac{d}{\kappa}\right)$$

where $t = t_1 + t_2$ is the total thickness of the air gap. The capacitance is defined to be $C = Q/V$, so that

$$C = \frac{A\epsilon_0}{t + d/\kappa}$$

Alternatively, we may view the capacitor as three capacitors in series, as shown at left.

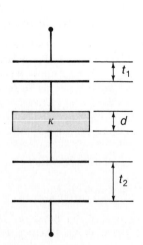

Using Eq. 30–16, we have

$$\frac{1}{C} = \frac{1}{C_1} + \frac{1}{C_2} + \frac{1}{C_3} = \frac{t_1}{A\epsilon_0} + \frac{d}{\kappa A\epsilon_0} + \frac{t_2}{A\epsilon_0} = \frac{1}{A\epsilon_0}\left(t + \frac{d}{\kappa}\right)$$

Finally,

$$C = \frac{A\epsilon_0}{t + d/\kappa}$$

as before.

PROBLEMS

Section 30-1

30–1   When a potential difference of 150 V is applied to the plates of a parallel-plate capacitor, the plates carry a surface charge density of 30 nC/cm². What is the spacing between the plates?

30–2   A small object with a mass of 350 mg carries a charge of 30 nC and is suspended by a thread between the vertical plates of a parallel-plate capacitor. The plates are separated by 4 cm. If the thread makes an angle of 15° with the vertical, what is the potential difference between the plates?

30–3   One plate of a parallel-plate capacitor (with area $A = 100$ cm²) is grounded. The ungrounded plate carries a fixed charge Q. When the ungrounded plate is moved 0.5 cm farther away from the grounded plate, the potential difference between the plates increases by 200 V. Determine the magnitude of the charge Q.

30–4   The ratio of the plate spacing to the radius for a circular parallel-plate capacitor is 1/10. Include fringing effects and determine the capacitance for a radius $R = 10$ cm.

Section 30-2

30–5   A spherical capacitor consists of a conducting ball with a diameter of 10 cm that is centered inside a grounded conducting spherical shell with an inner diameter of 12 cm. What capacitor charge is required to achieve a potential of 1000 V on the ball?

30–6   Show that a spherical capacitor for which the surface radii are nearly equal (that is, $R_2 - R_1 \ll R_1$) is equivalent to a parallel-plate capacitor with a spacing $s = R_2 - R_1$ and a plate area $A = 4\pi R_1^2$. If

this is so, the electric field strength in the gaps of the two equivalent capacitors should be the same for the same capacitor charge. Is this actually the case?

30–7   (a) A small spherical droplet of water has a radius of 2.5 mm. What is the charge on the droplet if its potential relative to $\Phi(\infty) = 0$ is 470 V? (b) Two such charged droplets unite to form a single larger spherical droplet without the loss of any charge. What is the potential on the larger droplet? (In each case, assume that the charge resides on the droplet surface.)

30–8   Compare the capacitance of a spherical capacitor with $R_1 = 7$ cm and $R_2 = 8$ cm with that of a cylindrical capacitor having the same radii and an axial length of 15 cm. (In each case the outer conductor is grounded.) Why are the capacitance values nearly equal?

30–9   Show that the expression found in Example 30–3 for the capacitance of a cylindrical capacitor reduces to that for a parallel-plate capacitor with dimensions $L \times 2\pi R_1$ when $s = R_2 - R_1$ and $s/R_1 \ll 1$. Argue that this result is reasonable.

30–10 Two conducting spheres with diameters of 0.40 m and 1.0 m are separated by a distance that is large compared with the diameters. The spheres are connected by a thin wire and are charged to 7 μC. (a) How is this total charge shared between the spheres? (Neglect any charge on the wire.) (b) What is the potential of the system of spheres relative to $\Phi(\infty) = 0$?

Section 30-3

30–11 Under standard conditions of temperature and pressure, dry air will break down when a field strength of approximately 3×10^6 V/m is reached. (a) What is the field energy density u under these conditions? (b) If this field strength were achieved in a parallel-plate capacitor with a plate area of 10 cm², what would be the magnitude of the force tending to pull the plates together?

30–12 Use $u = \frac{1}{2}\epsilon_0 E^2$ to calculate explicitly the electrostatic field energy stored in a cylindrical capacitor that carries charges $\pm Q$. Show that your expression leads to $U = Q^2/2C$.

30–13 A 4-μF capacitor is charged to a potential difference of 150 V. (a) What amount of energy is stored in the capacitor? (b) How much work would have to be done by an external agency to double the amount of charge on the capacitor?

Section 30-4

30–14 (a) Determine the equivalent capacitance for the capacitor network shown in the diagram. (b) If the network is connected to a 12-V battery, calculate the potential difference across each capacitor and the charge on each capacitor.

30–15 How should four 2-μF capacitors be connected to have a total capacitance of (a) 8 μF, (b) 2 μF, (c) 1.5 μF, and (d) 0.5 μF?

30–16 A conducting slab with a thickness d and area A is inserted into the space between the plates of a parallel-plate capacitor with spacing s and area A, as shown in the diagram. What is the value of the capacitance of the system?

30–17 A capacitor has square plates with sides of length ℓ. The plates are inclined at a small angle θ, as indicated in the diagram. Show that the capacitance of the plates is

$$C = \frac{\epsilon_0 \ell^2}{s_0}\left(1 - \frac{\ell\theta}{2s_0}\right)$$

[*Hint:* Assume that strips of width dx and length ℓ

constitute differential capacitors that are connected in parallel.]

Section 30-5

30–18 A coaxial cable consists of a central conductor with a diameter of 1 mm inside a thin outer conductor with a diameter of 7 mm. The dielectric material between the conductors is polystyrene. Calculate the capacitance per meter of length of the cable.

30–19 A rectangular parallel-plate capacitor has a dielectric slab that partially fills the space between the plates, as shown in the diagram. Show that the capacitance is

$$C = (\epsilon_0 b/s)[\kappa a - (\kappa - 1)x]$$

Does this expression reduce to the expected values when $x = 0$ and when $x = a$?

30–20 The capacitor of Problem 30–19 is charged to an amount Q and is then isolated. Show that the force **F** required to hold the dielectric slab in place is

$$\mathbf{F} = \frac{Q^2 s(\kappa - 1)}{2\epsilon_0 b[\kappa a - (\kappa - 1)x]^2}\hat{\mathbf{i}}$$

[*Hint:* Use the principle of virtual work (as in the derivation of Eq. 30–15). Be careful with signs!]

30–21 A parallel-plate capacitor has a spacing $s = 1.6$ mm. The capacitor, with air between the plates, is charged to a potential difference of 800 V and then isolated. Next, a slab of polystyrene is inserted, filling the space between the plates. (a) What is the electric field strength in the dielectric? (b) Determine the polarization P in the dielectric. (c) Calculate the surface densities of free charge (σ) and polarization charge (σ_P).

30–22 (a) Calculate the effective polarizing electric field strength E_e in helium gas at 20° C and 1 atm when $E = 10^5$ V/m. (b) Calculate E_e in water at

20° C for the same E. (c) In view of the results obtained for (a) and (b), would you expect the dipole-dipole interactions to be important in either case? Explain.

30-23 A parallel-plate capacitor can be charged to 3 C before breakdown occurs when air is between the plates. What maximum charge can the plates carry if the space between the plates is filled with (a) glass ($\kappa = 5$) or (b) polyethylene?

Section 30-6

30-24 Repeat the discussion leading to Eq. 30-14 for the case in which a dielectric material with dielectric constant κ completely fills the space between the plates of a parallel-plate capacitor. Show that $u = \frac{1}{2}\mathbf{E} \cdot \mathbf{D}$. (This result is valid only for *linear* dielectrics.)

30-25 Calculate explicitly the field energy stored in the capacitor of Example 30–13 by using the result of Problem 30–24, namely, $u = \frac{1}{2}\mathbf{E} \cdot \mathbf{D}$. Does the stored energy equal $Q^2/2C$, with C equal to the value given in Example 30–13?

30-26 One half of the space between the plates of a rectangular parallel-plate capacitor is filled with a dielectric material κ_1 and the other half is filled with a material κ_2, as indicated in the diagram. (a) Use the boundary conditions on the field vectors to show that $\sigma_1 = 2\kappa_1 Q/(\kappa_1 + \kappa_2)A$ and $\sigma_2 = 2\kappa_2 Q/(\kappa_1 + \kappa_2)A$, where $A = ab$. (b) Show that $V = \sigma_1 s/\epsilon_0\kappa_1 = \sigma_2 s/\epsilon_0\kappa_2$. (c) Show that $C = (\kappa_1 + \kappa_2)\epsilon_0 A/2s$.

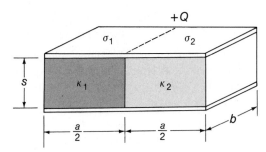

30-27 Derive the capacitance value given in Problem 30–26 (c) by considering the system to consist of two capacitors connected in parallel.

Additional Problems

30-28 A small object has a mass of 12 mg and carries a charge of 8 nC. When this object is released between the horizontal plates of a parallel-plate ca-

pacitor, it experiences a downward acceleration of $\frac{1}{3}g$. If the plate spacing of the capacitor is $s = 2.5$ cm, what is the potential difference between the plates?

30-29 Consider the Earth to be an isolated spherical conductor. (a) Calculate the capacitance of the Earth. (b) What excess charge on the Earth would produce a potential of 10^6 V relative to $\Phi(\infty) = 0$?

30-30 At one time it was suggested that the electron rest-mass energy $m_e c^2$ might be entirely electric field energy. Calculate the radius r_0 of a conducting sphere which, carrying a charge e, would have a field energy equal to $m_e c^2$. (A quantity $2r_0$ enters certain atomic calculations, and this quantity is referred to as the *classical electron radius*. The "radius" of the electron is now believed to be less than 10^{-18} m; experimentally, the electron is indistinguishable from a point object.)

30-31 The field energy of a uniform *volume* distribution of
• charge Q with radius R is greater than that for a uniform *surface* distribution of the same charge with the same radius. (•Can you see qualitatively why this is so?) Show that the field energy of the volume distribution is 6/5 times that of the surface distribution. [*Hint:* In each case, first calculate $E(r)$, then evaluate $\int 4\pi r^2 u(r) \, dr$.]

30-32 Five capacitors, each with $C = 2 \, \mu\text{F}$, are connected
• to form the network shown in the diagram. Determine the equivalent capacitance between terminals A and B. [*Hint:* From the symmetry, deduce the potential difference, $\Phi_a - \Phi_b$, when the network is connected to a battery.]

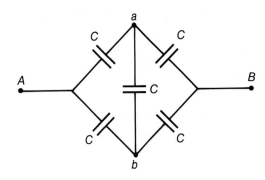

30-33 Show that the capacitance of the conducting sphere and dielectric shell combination of Example 30–12 is given by

$$C = \frac{4\pi\epsilon_0\kappa R_1 R_2}{R_2 + (\kappa - 1)R_1}$$

30-34 (a) Calculate explicitly the field energy stored in the dielectric shell of Example 30–12 ($R_1 < r < R_2$) by using the result of Problem 30–24, namely, $u = \frac{1}{2}\mathbf{E} \cdot \mathbf{D}$. (b) Calculate explicitly the field energy in the vacuum region ($r > R_2$). (c) Does the total stored energy equal $Q^2/2C$, with C equal to the value given in Problem 30–33?

30–35 The capacitor of Problem 30–19 is connected to a
• battery with terminal voltage V. This connection is maintained while the dielectric slab is held in place by the application of a force $\mathbf{F}$. Determine the required force $\mathbf{F}$. [*Hint:* Use the principle of virtual work (as in the derivation of Eq. 30–15). Here, you must equate the mechanical work $F\,dx$ to $dU' - dU''$, where dU' is the change in the energy stored in the field and $dU'' = V\,dQ$ is the additional work done by the battery.] Does your answer agree with the expression for $\mathbf{F}$ given in Problem 30–20?

30–36 In the discussion in Section 30–5 leading to Eq. 30–30, it was asserted that for the case of a cubic crystal lattice there is zero contribution to the field at the molecular site due to the nearby dipoles. Demonstrate explicitly that this is the case for the *nearest* neighbor dipoles of the molecule in question. For a cubic lattice these nearest neighbors (six in number) are distributed as in the diagram. Use the expres-

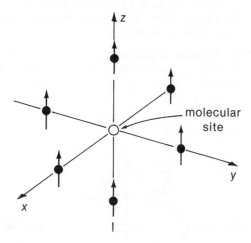

sions in Problem 28–3 to show that the net field due to these six dipoles at the molecular site is zero. [*Hint:* Notice that the four dipoles in the x-y plane all have $\theta = 90°$, whereas the two dipoles on the z-axis have $\theta = 0°$ and $\theta = 180°$.]

30–37 Use the values of $\chi(H_2)$ given in Table 30–1 to calculate the polarizability of hydrogen gas. If hydrogen gas is placed in an electric field with a strength of 10^6 V/m, what will be the effective displacement between the centers of positive and negative charge in a hydrogen molecule? Compare this with a nuclear radius. (The density of hydrogen gas is 0.0899 kg/m³.)

30–38 If the polarizability α of a substance remains constant, show that the Clausius-Mossotti relationship, when divided by the density ρ, is independent of the density. Thus, the term

$$\frac{1}{\rho} \frac{\kappa - 1}{\kappa + 2}$$

should be the same for both the liquid and gaseous phases of a substance. Test this conclusion for the case of hydrogen (H_2) by calculating κ(gas) using κ(liquid) = 1.2297 and the density values, ρ(gas) = 0.0899 kg/m³ and ρ(liquid) = 70 kg/m³. Compare the calculated value of κ(gas) with the measured value given in Table 30–1.

30–39 Obtain Eq. 30–33 by integrating the expression
• given. Because $p_0 E_e \cos\theta / kT \ll 1$, the exponential can be expanded: $e^x \cong 1 + x$. The integration is then simplified and the result is $\langle p_0 \rangle = 4\pi C p_0^2 E_e / 3kT$. The constant C can be evaluated by noting that the probability is unity for finding $\mathbf{p}$ at *some* angle between 0 and π. Show that $C = 1/4\pi$ and thereby verify Eq. 30–33.

30–40 Show that the boundary condition on the tangential component of $\mathbf{E}$ (Eq. 30–45) can be expressed as $(\mathbf{E}_1 - \mathbf{E}_2) \times \hat{\mathbf{n}} = 0$.

CURRENT AND RESISTANCE

Under equilibrium conditions—that is, when the charge distribution is static—the surface of a conductor is an equipotential and all of the field lines are perpendicular to the surface. Within the conductor the electric field is zero and no net charge is present. If this equilibrium condition is disturbed—for example, by placing a potential difference across the conductor—the electric field within the conductor is no longer zero. The existence of an internal field means that charges will move within the conductor; that is, an electric *current* will flow. Moreover, the conductor surface is no longer an equipotential.

In this chapter we study the special case in which there is a steady current within a conductor. This is the *direct current* (DC) case. In Chapter 37 we discuss the effects of nonsteady currents, in particular, the case of *alternating current* (AC).

31-1 ELECTRIC CURRENT

The instantaneous rate at which positive charge passes through a surface is called the *electric current* through the surface. Figure 31–1 shows a surface S that intersects a conductor through which charge is flowing. The current I through the surface S is equal to the rate dq/dt at which positive charge passes through the surface. That is,

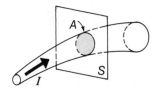

Fig. 31–1. A current $I = dq/dt$ flows along the conductor and through the surface S.

$$I = \frac{dq}{dt} \qquad \qquad (31-1)$$

In this chapter we are concerned with cases for which I is constant.

From Eq. 31-1 we see that the dimensions of current are charge per unit time. The units of current are *coulombs per second* (C/s), to which is given the special name *ampere**:

$$1 \text{ ampere (A)} = 1 \text{ C/s}$$

---

*Named in honor of the French physicist André-Marie Ampère (1775–1836), one of the founders of the subject of electrodynamics. In Section 34–4 we give a precise definition of the ampere.

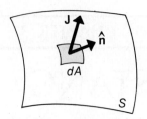

Fig. 31–2. The current dI through the differential element $\hat{\mathbf{n}}\,dA$ of the surface S is $dI = \mathbf{J}\cdot\hat{\mathbf{n}}\,dA$.

Current Density. In many situations it is convenient to refer to the *current density* **J** in a conductor. For the case illustrated in Fig. 31–1, the *average* current density in the conductor as it passes through the intersecting surface S is $J = I/A$ and the direction of **J** is in the direction of the current flow. In general, the current density is defined in terms of the flow through a differential element $\hat{\mathbf{n}}\,dA$ of a surface S, as indicated in Fig. 31–2. Here, we have

$$dI = \mathbf{J}\cdot\hat{\mathbf{n}}\,dA \qquad (31\text{–}2)$$

and the total current through any surface S is

$$I = \int_S \mathbf{J}\cdot\hat{\mathbf{n}}\,dA \qquad (31\text{–}3)$$

Fig. 31–3. Illustrating the effects of charge conservation for a steady current. (a) The same current I flows through S_1 as flows through S_2 or S_2'. (b) The current I_1 divides at the branch point into two parts such that $I_1 = I_2 + I_3$.

Usually, we are concerned with *open* surfaces. Current density is measured in A/m².

Figure 31–3 illustrates the effects imposed on the flow of current by charge conservation. If I represents a steady current in Fig. 31–3a, no charge can accumulate in the volume defined by the surfaces, S_1 and S_2, and the conductor wall. It follows that the same total current I flows through S_1 as flows through S_2. Moreover, the surface S_2 can be replaced by any surface such as S_2', whose perimeter is everywhere on the conductor wall, without affecting the value of the current.

In Fig. 31–3b the current I_1 divides into two parts, I_2 and I_3, that flow through separate strands of the conductor. If the current is steady, charge conservation requires $I_1 = I_2 + I_3$.

It should be apparent that electric current bears a strong similarity to fluid flow (see Chapter 18). The conditions we have discussed here—called the *continuity conditions*—are the result of steady current flow and charge conservation; the corresponding conditions for fluid flow are the result of incompressibility and matter conservation.

31-2 RESISTIVITY

On the microscopic level, electric current is due to the movement of the conduction electrons (charge, $-e$) or of ions (charge, one or more times $\pm e$). The individual movements of these charge carriers are governed by statistical thermodynamic factors, so we imagine that all macroscopic equations, such as Eq. 31–1, are obtained by appropriate averages over microscopic effects.

In a conducting wire, such as that illustrated in Fig. 31–4, the atoms remain in fixed positions, so the only charge carriers are the free electrons. (In Section 31–6 we consider conduction by positive charge carriers.) Let n be the number of free electrons per unit volume. Suppose that an electric field is imposed on the wire so that a conventional positive current flows to the right. In addition to their random thermal motions, the electrons experience a steady drift to the left in response to the electric field. We introduce the average free electron velocity $\langle \mathbf{v}\rangle$ and identify this with the drift velocity $\mathbf{v}_d = \langle \mathbf{v}\rangle$.* In the usual way we let $v_d = |\mathbf{v}_d|$ stand for the magnitude of

---

*The slanted brackets around **v** imply an average taken over the velocity distribution. See the footnote on page 666.

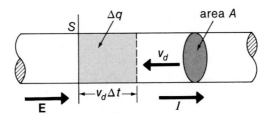

Fig. 31-4. The current I that flows to the right in the wire is the result of electrons drifting to the left with the average velocity $\langle \mathbf{v} \rangle = \mathbf{v}_d$.

$\mathbf{v}_d$. Then, in a time interval Δt, the amount of charge that passes through the surface S is the charge Δq contained within the volume $v_d A \Delta t$ (see Fig. 31-4); thus,

$$\Delta q = e n v_d A \, \Delta t \qquad (31\text{-}4)$$

The current I that flows in the wire is defined to be the net amount of positive charge that passes through the surface S per unit time. The *negative* electrons move toward the *left*, so the conventional current is directed toward the *right*. We have

$$I = \frac{\Delta q}{\Delta t} = e n v_d A \qquad (31\text{-}5)$$

A Classical Theory of Metals. To continue the discussion of charge movement in matter, we adopt a simple classical model for metallic conductors.* In this model it is supposed that some of the electrons are detached from their parent atoms and become free to move about within the material in much the same way as an electron gas. (Compare the discussion of the Debye theory of specific heats in Section 24-2.) When an electric field is applied, these free electrons drift with an average velocity $\mathbf{v}_d$ and give rise to a current expressed by Eq. 31-5. The current density in the material is

$$\mathbf{J} = -e n \mathbf{v}_d \qquad (31\text{-}6)$$

where the direction of $\mathbf{J}$ is *opposite* to the direction of the electron drift velocity $\mathbf{v}_d$.

In this classical electron-gas model of conductors, the electrons are imagined to move about within the material and to experience frequent collisions with the metal atoms. Between collisions, the electrons move in accordance with the force produced by the imposed electric field. Consider an electron that has a velocity $\mathbf{v}_0$ as the result of a particular collision. If the local field within the conductor is $\mathbf{E}$, the electron velocity at time t after collision will be

$$\mathbf{v} = \mathbf{v}_0 - (e/m_e)\mathbf{E}t \qquad (31\text{-}7)$$

where m_e is the electron mass. The drift velocity $\mathbf{v}_d$ that appears in Eq. 31-6 is just the average of Eq. 31-7 between collisions. The progress of an electron undergoing random collisions in the presence of an electric field is shown schematically in Fig. 31-5. Because $\mathbf{v}_0$ is distributed randomly in space, it follows that $\langle \mathbf{v}_0 \rangle = 0$. Denoting by τ the average time between collisions (the *mean free time*), the average of $\mathbf{v}$ in Eq. 31-7 becomes equal to $\mathbf{v}_d$,

---

*The consequences of this model were first worked out in 1900 by the German physicist Paul Drude (1863–1906).

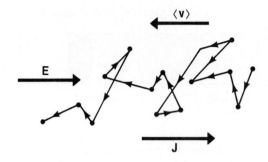

Fig. 31–5. Schematic representation of the path followed by a free electron in a conductor in which there is an electric field. Notice that the direction of the drift velocity $\mathbf{v}_d = \langle \mathbf{v} \rangle$ is opposite to the directions of $\mathbf{E}$ and $\mathbf{J}$.

$$\langle \mathbf{v} \rangle = -\frac{e\tau}{2m_e}\mathbf{E} = \mathbf{v}_d \tag{31-8}$$

(•Do you see how the factor $\frac{1}{2}$ arises when calculating $\langle \mathbf{v} \rangle$?) Substituting this expression for $\langle \mathbf{v} \rangle = \mathbf{v}_d$ into Eq. 31–6 yields

$$\mathbf{J} = \frac{e^2 n\tau}{2m_e}\mathbf{E}$$

which we can write as

$$\mathbf{J} = \frac{1}{\rho}\mathbf{E} = \sigma\mathbf{E} \tag{31-9}$$

with

$$\rho = \frac{1}{\sigma} = \frac{2m_e}{e^2 n\tau} \tag{31-10}$$

The quantity ρ is called the *resistivity* of the material and the reciprocal quantity σ is called the *conductivity*.* Many metals obey the relationship expressed in Eq. 31–9 that is predicted by the simple Drude model. These materials are referred to as *linear conductors* to emphasize the linear relationship between $\mathbf{J}$ and $\mathbf{E}$.

Resistivity ρ is measured in units of *ohm-meters* ($\Omega \cdot$ m), where $1\,\Omega = 1$ V/A (see Section 31–3). Conductivity σ is measured in the reciprocal units, namely, $(\Omega \cdot \text{m})^{-1}$, for which the SI term is *siemens per meter* (S/m).

Table 31–1 gives values of n (the free electron number density) for several metals. Also listed are the mean collision times τ calculated from Eq. 31–10 by using experimental values for ρ. The quantity ξ represents the number of free electrons per atom that was assumed to obtain the value of n.

Temperature Coefficients. Experiments show that the resistivity of a particular metal shows a nearly linear variation with temperature. Figure 31–6 shows this linear dependence for the case of copper. Because the number density n of free electrons in metals varies only slightly with temperature, Eq. 31–10 indicates that the mean collision time τ should be inversely proportional to the temperature. (The entries for τ in Table 31–1 reflect this variation.) The classical theory of metals (the Drude theory) is unable to predict correctly this relationship between τ and T. Calculations based on modern quantum theory, however, can account quite satisfactorily

---

*Although ρ and σ have previously been used as symbols for different quantities, the context readily identifies which usage is meant.

TABLE 31-1. Electron Number Densities n and Mean Collision Times τ for Various Metals

	METAL	ξ^a	n (10^{28} m^{-3})	τ (10^{-14} s) $T = 77$ K	$T = 273$ K	$T = 373$ K
Alkali Metals	Li	1	4.70	14.6	1.76	1.2
	Na	1	2.65	34	6.4	
	K	1	1.40	36	8.2	Molten
	Rb	1	1.15	28	5.6	
	Cs	1	0.91	17.2	4.2	
Noble Metals	Cu	1	8.47	42	5.4	3.8
	Ag	1	5.86	40	8.0	5.6
	Au	1	5.90	24	6.0	4.2
Group 2A Metals	Be	2	24.7	—	1.0	0.54
	Mg	2	8.61	13.4	2.2	1.5
	Ca	2	4.61	—	4.4	3.0
	Sr	2	3.55	2.8	0.88	—
	Ba	2	3.15	1.32	0.38	—
Other Metals	Al	3	18.1	13.0	1.6	1.1
	Pb	4	13.2	1.14	0.28	0.20
	Bi	5	14.1	0.144	0.046	0.032

^{a} Assumed number of free electrons per atom.

for the temperature dependence of τ (and, hence, also of ρ). (See Example 31–7.)

The rate at which the resistivity of a substance varies with temperature at a particular temperature can be expressed in terms of a *resistivity temperature coefficient* α:

$$\alpha = \frac{1}{\rho_0} \frac{d\rho}{dT}\bigg|_{T_0} \tag{31–11}$$

Fig. 31–6. Measured values of the resistivity for copper show the linear dependence on the temperature (except near absolute zero).

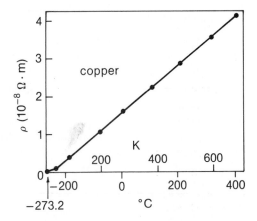

where $\rho_0 = \rho(T_0)$. If $\rho(T)$ were exactly linear, α would be a constant, and we would have

$$\rho(T) = \rho_0[1 + \alpha(T - T_0)] \qquad \text{(31–12)}$$

However, in general, α is not strictly constant, so we define an *average* temperature coefficient $\bar{\alpha}$ for a temperature range near T_A and write

$$\rho(T) = \rho_A[1 + \bar{\alpha}(T - T_A)] \qquad \text{(31–12a)}$$

where $\rho_A = \rho(T_A)$. Values of the resistivity and the temperature coefficient are given for some selected substances in Table 31–2. Notice the extremely wide variation in ρ for different types of materials:

pure metals	$\sim 10^{-8}\,\Omega \cdot m$
alloys	$\sim 10^{-6}\,\Omega \cdot m$
semiconductors	$\sim 10^{-5}\text{–}10^3\,\Omega \cdot m$
insulators	$\sim 10^8\text{–}10^{17}\,\Omega \cdot m$

Notice also that several of the values of α in Table 31–2 are *negative,* which means that the resistivity decreases with increasing temperature.

TABLE 31–2. Values of the Resistivity and the Temperature Coefficient for Various Substances

SUBSTANCE		$\rho\ (\Omega \cdot m)$	$\alpha\ (\text{deg}^{-1})$
Conductors:		ρ (at 0° C)	$\bar{\alpha}$ (0° C–100° C)
metals	Ag	1.47×10^{-8}	4.1×10^{-3}
	Cu	1.56×10^{-8}	4.3×10^{-3}
	Al	2.63×10^{-8}	4.4×10^{-3}
	Ni	6.38×10^{-8}	6.6×10^{-3}
	Fe	8.85×10^{-8}	6.5×10^{-3}
	Pt	9.83×10^{-8}	3.9×10^{-3}
alloys	Manganin	4.8×10^{-7}	2×10^{-5}
	Constantan	4.9×10^{-7}	-1×10^{-5}
	Nichrome	1.1×10^{-6}	4×10^{-4}
Semiconductors:		ρ (at 20° C)	α (at 20° C)
pure	C	3.5×10^{-5}	-5×10^{-3}
	Ge	0.45	-4.8×10^{-2}
	Si	640	-7.5×10^{-2}
Insulators:		ρ (at 20° C)	
	Amber	5×10^{14}	
	Glass	$10^{10}\text{–}10^{14}$	
	Lucite	$>10^{13}$	
	Quartz	7.5×10^{17}	
	Sulfur	10^{15}	
	Wood	$10^8\text{–}10^{11}$	

Example 31–1

A length of No. 8 copper wire, which has a diameter of 3.26 mm, carries a steady current of 20 A. (The maximum current in No. 8 copper wire with rubber insulation that is permitted by the National Electrical Code is 35 A.)

(a) What is the average current density J in the wire?

(b) Determine the average drift velocity v_d of electrons in the wire.

(c) Calculate the electric field strength E (assumed constant) along the wire.

Solution:

(a) The average current density is

$$J = \frac{I}{A} = \frac{I}{\pi R^2} = \frac{20 \text{ A}}{\pi(1.63 \times 10^{-3} \text{ m})^2} = 2.40 \times 10^6 \text{ A/m}^2$$

(b) In Table 31–1 we find $n(\text{Cu}) = 8.47 \times 10^{28} \text{ m}^{-3}$; then, using Eq. 31-6, we have

$$v_d = \frac{J}{en} = \frac{2.40 \times 10^6 \text{ A/m}^2}{(1.60 \times 10^{-19} \text{ C})(8.47 \times 10^{28} \text{ m}^{-3})} = 1.77 \times 10^{-4} \text{ m/s}$$

or less than 1/5 of a millimeter per second. This very small drift velocity is superimposed on the much larger random thermal speeds of the electrons (see Example 31–2).

(c) From Table 31–2, $\rho(\text{Cu}) = 1.56 \times 10^{-8} \, \Omega \cdot \text{m}$; then, using Eq. 31-9, we obtain

$$E = \rho J = (1.56 \times 10^{-8} \, \Omega \cdot \text{m})(2.40 \times 10^6 \text{ A/m}^2) = 0.037 \text{ V/m}$$

Example 31–2

(a) The classical mean free path for an electron moving through a collection of atoms with radius R is $\lambda_C = 1/\mathcal{N}\pi R^2$, where $\mathcal{N}$ is the number of atoms per unit volume (see Problem 31–4). Calculate λ_C for copper Cu^+, which has an ionic radius $R = 0.96$ Å.

(b) Assume that the free conduction electrons in a metal behave as a gas that follows the Maxwell-Boltzmann distribution law. Calculate the average thermal speed $\langle v \rangle$ at temperatures of 77 K, 273 K, and 373 K.

(c) Use the results obtained in (a) and (b) to determine $\tau(T) = \lambda_C/\langle v \rangle$ for the three temperatures.

(d) Show that the Drude model for resistivity leads to $\rho \propto \sqrt{T}$ instead of the observed behavior, $\rho \propto T$.

Solution:

(a) In the expression for λ_C, $\mathcal{N}$ is the number of atoms per unit volume. For the case of copper, we have $\xi = 1$, so that $\mathcal{N} = n$; in Table 31–1, we see that $n = 8.47 \times 10^{28} \text{ m}^{-3}$. Then, we find

$$\lambda_C = \frac{1}{n\pi R^2} = \frac{1}{(8.47 \times 10^{28} \text{ m}^{-3})\pi(0.96 \times 10^{-10} \text{ m})^2} = 4.08 \times 10^{-10} \text{ m}$$

This result is to be compared with the spacing of 2.55×10^{-10} m between nearest neighbor atoms in bulk copper. (See Problem 31–22.)

(b) For a Maxwell-Boltzmann distribution, the average velocity is given by the expression (Eq. 24–26c),

$$\langle v \rangle = \sqrt{\frac{8kT}{\pi m_e}}$$

where m_e is the electron mass. Thus, for the required temperatures, we find

$$\langle v \rangle(77\text{ K}) = \sqrt{\frac{8(1.38 \times 10^{-23}\text{ J/K})(77\text{ K})}{\pi(9.11 \times 10^{-31}\text{ kg})}} = 5.45 \times 10^4\text{ m/s}$$

Similarly, we obtain

$$\langle v \rangle(273\text{ K}) = 1.03 \times 10^5\text{ m/s}$$
$$\langle v \rangle(373\text{ K}) = 1.20 \times 10^5\text{ m/s}$$

(c) Using the above values for λ_C and $\langle v \rangle(T)$, we have

$$\tau(77\text{ K}) = \frac{\lambda_C}{\langle v \rangle(77\text{ K})} = \frac{4.08 \times 10^{-10}\text{ m}}{5.45 \times 10^4\text{ m/s}} = 0.75 \times 10^{-14}\text{ s}$$

$$\tau(273\text{ K}) = 0.40 \times 10^{-14}\text{ s}$$
$$\tau(373\text{ K}) = 0.34 \times 10^{-14}\text{ s}$$

These values, obtained from the simple Drude model, are 10 to 60 times smaller than those calculated from the measured resistivity values (see Table 31–1).

(d) According to Eq. 31–10, we have

$$\rho = \frac{2m_e}{e^2 n\tau}$$

Substituting $\tau = \lambda_C/\langle v \rangle$ and $\langle v \rangle \propto \sqrt{T}$, we find

$$\rho \propto \sqrt{T}$$

According to quantum theory, the average velocity for the conduction electrons in copper is approximately 1.2×10^6 m/s, nearly *independent* of temperature. Also, quantum theory predicts $\rho \propto T$, the correct dependence. See the discussion in Section 31–6 and Example 31–7.

▶ *Superconductors.* For materials that are good conductors, such as copper and silver, the resistivity decreases essentially linearly with temperature down to 10 K or so. As the temperature is decreased further, the resistivity assumes a constant value that it maintains for the lowest temperatures investigated (about 0.05 K). This behavior is shown in Fig. 31–7 for the case of sodium. (See also Fig. 31–6 for copper.)

In certain materials, called *superconductors,* there occurs at low temperatures a remarkable interaction among the electrons and the ionic lattice. As a result of this interaction, electrons are associated in bound pairs, called *Cooper pairs.* A Cooper pair is bound by an energy of the order of 10^{-3} eV even though the size of the pair may be as large as 10^{-6} m. The scattering of either member of the electron pair can occur only if the bond is broken. At very low temperatures the energy to break the pair bonds is unavailable. Thus, at least in principle, the mean free path of the pair becomes infinite and the resistivity becomes zero—not just "very small," but *zero!* The temperature that marks the transition between the normal phase and the superconducting phase of a material is called the *critical temperature* T_c. Figure 31–7 shows the transition of mercury into the superconducting phase at $T_c = 4.15$ K.* Table 31-3 lists the critical temperatures for several elements and compounds.

A current, once initiated in a superconducting material, will continue to flow (presumably) forever, even in the absence of any driving electric field. Such continuing currents in superconducting rings have been observed to persist for as long as $2\frac{1}{2}$ years (a time limited only by the patience of the investigator). These experiments have demonstrated that the resistivity of a superconductor is less than 10^{-25} $\Omega \cdot$ m. (Silver, the best normal conductor, has a resistivity of 1.47×10^{-8} $\Omega \cdot$ m; see Table 31-2.) Superconductors also have interesting and unusual magnetic and specific heat properties, but a discussion of these topics is beyond the scope of this text. ◄

Fig. 31–7. Resistivity versus temperature for the normal conductor sodium (Na) and for the superconductor mercury (Hg).

TABLE 31-3. Critical Temperatures for Various Superconductors

ELEMENT OR COMPOUND	T_c (K)
Nb_3Ge	23.2
Nb_3Sn	18.05
Nb	9.46
Pb	7.18
Hg	4.15
Sn	3.72
Al	1.19
Zn	0.88

31-3 OHM'S LAW

When a potential difference V is applied across a conducting object and causes a current I to flow, the ratio V/I is called the *resistance R* of the object:

$$R = \frac{V}{I}$$

(31–13)

The unit of resistance is the ohm (Ω), where $1\,\Omega = 1$ V/A.

The value of R for a particular conducting object (a *resistor*) depends on the material of which it is composed and on the geometry of its construction. However, for a large class of materials it has been found that R does *not* depend on the applied voltage (unless the voltage is very high). That is, for a resistor made from such a material, the ratio V/I is constant for all (ordinary) applied voltages V. Materials that follow this prescription are said to obey *Ohm's law*.†

We hasten to add that there are numerous materials and devices that do *not* obey Ohm's law. Some of these—for example, diodes, vacuum tubes, thermistors, and transistors—are specifically designed to have a nonlinear *V*-versus-*I* behavior.

---

*Superconductivity was discovered in 1911 by the Dutch physicist Heike Kamerlingh-Onnes (1853–1926), who observed that the resistivity of mercury dropped to zero at a temperature of approximately 4.2 K. For his discovery of this surprising new phase of matter, Kamerlingh-Onnes received the Nobel Prize for physics in 1913.

†After Georg Simon Ohm (1787–1854), the German physicist who discovered this effect in 1826.

Indeed, for a sufficiently high voltage, any material can be made to deviate from Ohm's law. Thus, Ohm's law is by no means a universal law of Nature; instead, it is an approximate relationship that is obeyed by many materials under restricted conditions of applied voltage and temperature.

Resistivity and Resistance. Consider the equipotential surface shown in Fig. 31–8. A small cubic volume has been constructed with sides $d\ell$ that are perpendicular to the equipotential. One face of the cube, with area ΔA, is tangent to the equipoten-

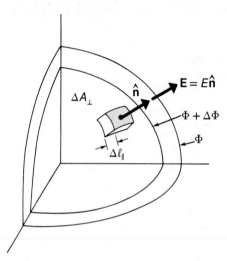

Fig. 31–8. Geometry for obtaining the current flow through the area $\triangle A_\perp$.

tial; the other parallel face is therefore tangent to an adjacent equipotential surface. (•Can you see why?) Suppose that there exists a current density **J** directed along the electric field vector **E** and, hence, also along the unit normal vector $\hat{\mathbf{n}}$, both of which are perpendicular to the equipotential surface. Then,

$$\mathbf{J} = \frac{\mathbf{E}}{\rho} = -\frac{1}{\rho}\frac{d\Phi}{d\ell_\parallel}\hat{\mathbf{n}}$$

where we have written $d\ell_\parallel$ for $d\ell$ to emphasize that the sides of the cube represent a displacement *parallel* to **J**. The (positive) current that flows through the area ΔA is

$$dI = \mathbf{J}\cdot\hat{\mathbf{n}}\,\Delta A = J\Delta A_\perp$$

where we have written $\Delta A_\perp$ for ΔA to emphasize that this area is oriented *perpendicular* to **J**. Combining these expressions, we find

$$dI = \frac{dq}{dt} = -\frac{\Delta A_\perp}{\rho}\frac{d\Phi}{d\ell_\parallel} \tag{31-14}$$

Notice the similarity between Eq. 31–14 and the expression for heat conduction (Eq. 22–14),

$$\frac{dQ}{dt} = -\kappa\Delta A_\perp\frac{dT}{d\ell_\parallel}$$

(•Can you see how the corresponding quantities in the two equations are analogous?)

Now, consider a rectangular bar with a length L that is constructed from a linear conducting material. The bar is placed between two parallel equipotential planes, as shown in Fig. 31-9. With $\Phi_a > \Phi_b$, both **E** and **J** are directed toward the right. The

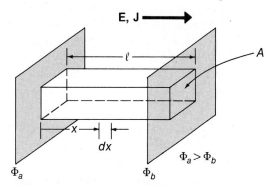

Fig. 31-9. The current through the bar is $I = (A/\rho\ell)(\Phi_a - \Phi_b)$.

continuity condition for this geometry requires **J** to be everywhere uniform and constant. (•Can you see why?) Applying Eq. 31-14, we have

$$I = -\frac{A}{\rho}\frac{d\Phi}{dx}$$

Then, upon integrating, we obtain

$$\Phi(x) = \Phi_a - \frac{\rho I}{A}x$$

Thus, the equipotential surfaces, as expected, are uniformly spaced between Φ_a and Φ_b. When $x = \ell$, we have $\Phi(\ell) = \Phi_b$. Therefore, solving for I gives

$$I = \frac{A}{\rho\ell}(\Phi_a - \Phi_b) \tag{31-15}$$

Equation 31-15 can be simplified by using the notation

$$V_{ab} = \Phi_a - \Phi_b \tag{31-16}$$

This notation means that $V_{ab} > 0$ when $\Phi_a > \Phi_b$; that is, V_{ab} represents the *potential drop* in going from point a to point b. (In Section 30-1 we used $V_{ba} = \Phi_b - \Phi_a$ to represent the *potential increase* in going from point a to point b; thus, $V_{ab} = -V_{ba}$.) We now write Eq. 31-15 as

$$I_{ab} = \frac{A}{\rho\ell}V_{ab} \tag{31-17}$$

where I_{ab} represents the (positive) current that flows from a to b. When no confusion can result, we drop the subscripts and write

$$I = \frac{A}{\rho\ell}V \tag{31-17a}$$

Comparing this expression with Eq. 31-13 shows that we have

$$R = \frac{\rho \ell}{A}, \qquad \text{straight bar} \qquad\qquad \textbf{(31–18)}$$

If the resistivity ρ does not depend on the voltage, the resistance R is constant and the resistor (the conducting bar) obeys Ohm's law.

Thus far, we have considered three different types of conduction problems, namely, fluid flow, heat flow, and now the flow of electric current. In each case there is a *flow rate, dQ/dt*, that is equal to the ratio of a *driving influence* $\mathfrak{F}$ to a *flow resistance* $\mathfrak{R}$; that is,

$$\frac{dQ}{dt} = \frac{\mathfrak{F}}{\mathfrak{R}}$$

(•Can you identify Q, $\mathfrak{F}$, and $\mathfrak{R}$ in each of the three situations?)

Example 31–3

(a) Calculate the resistance per meter of length for No. 20 copper wire at 20° C.

(b) Calculate the resistance at 0° C of a 2-m length of insulated manganin wire with a diameter of 1/10 mm that is wrapped in a tight coil to form a *wire-wound resistor*.

(c) Determine the change in the resistance of the manganin resistor in (b) if the temperature is increased to 100° C.

Solution:

(a) No. 20 copper wire (a common size of wire, often used as electronic hookup wire) has a diameter of 0.812 mm. Using the value in Table 31–2 for the resistivity ρ_0 of copper at 0° C, we find

$$\frac{R_0}{\ell} = \frac{\rho_0}{A} = \frac{1.56 \times 10^{-8}\, \Omega \cdot \text{m}}{\pi (0.406 \times 10^{-3}\, \text{m})^2} = 0.0301\ \Omega/\text{m}$$

and using Eq. 31–12a we can write

$$\frac{R_{20}}{\ell} = \frac{R_0}{\ell}(1 + \bar{\alpha}\,\Delta T)$$

$$= (0.0301\ \Omega/\text{m})[1 + (4.3 \times 10^{-3}\, \text{deg}^{-1})(20\ \text{deg})] = 0.0327\ \Omega/\text{m}$$

(b) For the manganin coil, again using the resistivity value in Table 31–2, we find

$$R_0 = \frac{\rho_0 \ell}{A} = \frac{(4.8 \times 10^{-7}\, \Omega \cdot \text{m})(2\ \text{m})}{\pi (0.5 \times 10^{-4}\, \text{m})^2} = 122.2\ \Omega$$

(c) At a temperature of 100° C we have

$$R_{100} = R_0(1 + \bar{\alpha}\,\Delta T)$$

so that
$$\Delta R = R_{100} - R_0 = R_0 \bar{\alpha}\,\Delta T$$

$$= (122.2\ \Omega)(2 \times 10^{-5}\, \text{deg}^{-1})(100\ \text{deg}) = 0.24\ \Omega$$

Manganin (84 percent Cu, 12 percent Mn, and 4 percent Ni) is a useful alloy for the construction of resistors when the application requires the resistance values to be stable under temperature changes.

Example 31-4

The annular gap between a coaxial pair of copper tubes is filled with pure silicon. The outer conductor is grounded ($\Phi_b = 0$), and the inner conductor is at a potential $\Phi_a = +2.0$ V. Calculate the total current for the dimensions shown in the diagram.

$r_1 = 0.8$ cm
$r_2 = 3.0$ cm
$L = 5.0$ cm

Solution:

Because the resistivity of copper is small compared with that of silicon, the copper surfaces may be considered to be equipotentials. Then, the electric field and the current density are everywhere radial; that is, $\mathbf{E} = E\hat{\mathbf{u}}_r$ and $\mathbf{J} = J\hat{\mathbf{u}}_r$. (•What are the shapes of the equipotential surfaces in the silicon?)

We use the differential form of Eq. 31–18, namely, $dR = \rho \, d\ell/A$, with $d\ell = dr$ and $A = 2\pi rL$. (•Why is it possible to use Eq. 31–18, which was obtained for the case of a straight bar?) Thus,

$$dR = \frac{\rho \, dr}{2\pi rL}$$

The current flows radially outward through successive cylindrical layers of material. Therefore, the total resistance is obtained by integrating dR from r_1 to r_2. We find

$$R = \int_{r_1}^{r_2} dR = \frac{\rho}{2\pi L} \int_{r_1}^{r_2} \frac{dr}{r} = \frac{\rho}{2\pi L} \ln(r_2/r_1)$$

Using $\rho(\text{Si})$ from Table 31–2 and the dimensions given in the diagram, we have

$$R = \frac{640 \, \Omega \cdot \text{m}}{2\pi(0.050 \text{ m})} \ln(3.0/0.8) = 2693 \, \Omega$$

Then, the current is $I = \dfrac{V}{R} = \dfrac{2.0 \text{ V}}{2693 \, \Omega} = 7.43 \times 10^{-4}$ A

31-4   EMF

We now consider methods of providing a constant potential difference between two points so that a steady current (DC) can flow between the points. One device capable of producing a constant potential difference is a *chemical cell* or *battery*. The

Fig. 31–10. Schematic diagram of a Daniell cell. The left-hand compartment contains zinc sulfate and a zinc electrode. The right-hand compartment contains copper sulfate and a copper electrode. The porous barrier allows the passage of sulfate ions but blocks the metallic ions. The EMF of the cell (no current flowing) is approximately 1.07 V.

essential common feature of all batteries is the immersion of two dissimilar metals in an electrolyte (ionic) solution or the sandwiching of an electrolytic paste between two plates of dissimilar metals. The first battery (Volta's "pile") consisted of a stack of zinc and copper discs separated by moistened pasteboard. The usual automobile battery consists of six lead–sulfuric-acid cells connected in series.

Many types of chemical cells have been devised, each slightly different in construction and mode of operation. To illustrate the basic operating principles of a chemical cell, we choose a simple device called a *Daniell cell*.* Figure 31–10 shows a schematic diagram of the cell, which consists of two compartments that are separated by a porous barrier. One compartment contains a dilute aqueous solution of zinc sulfate ($ZnSO_4$); a zinc rod (electrode) is inserted into this solution. The other compartment contains a saturated solution of copper sulfate ($CuSO_4$); a copper electrode is inserted into this solution. The two electrodes are connected externally by a wire and some resistive element (the *load*).

In solution, $ZnSO_4$ and $CuSO_4$ exist almost exclusively as ions—Zn^{2+}, Cu^{2+}, and SO_4^{2-}. The porous barrier that separates the two electrolyte solutions permits the passage of the SO_4^{2-} ions but effectively blocks the metallic ions. (If the Cu^{2+} ions were allowed to come into contact with the zinc electrode, a direct exchange of electrons would take place and no current would flow in the external circuit.)

Each zinc atom that leaves the zinc electrode and enters the solution as a Zn^{2+} ion leaves behind two electrons. In the completed circuit shown in Fig. 31–10, these two electrons travel through the connecting wire and the load to the copper electrode. A copper ion (Cu^{2+}) in the solution absorbs the electrons at the electrode surface; the neutral copper atom so formed becomes attached to (plated onto) the electrode. At the same time, an SO_4^{2-} ion passes through the barrier from the copper side to the zinc side, thereby maintaining a balance of charge. Electrically, the net effect of this process is the transfer of two electrons through the external circuit from the zinc electrode (the negative terminal or *cathode*) to the copper electrode (the positive terminal or *anode*). Chemically, the effect is a slow dissolving of the zinc electrode and the plating, at the same rate, of copper onto the copper electrode.

A chemical cell, when operating with a completed external circuit, converts chemical energy into electric energy. The processes that take place in the cell are *reversible*. That is, if a current is made to flow through the cell in the direction opposite to that of the cell's normal current, copper from the copper electrode will enter the solution as Cu^{2+}, and Zn^{2+} ions will be neutralized and plated onto the zinc electrode, thereby "charging" the cell. During this reactivation of the cell, electric

---

*After the English physicist and chemist John Frederick Daniell (1790–1845).

energy is converted into chemical energy. (Some energy is lost to heat in both the charging and the discharging of the cell.)

If the external circuit is disconnected, the chemical reactions within the cell will proceed until a certain potential difference (or *terminal voltage*) is developed between the electrodes. (•Can you see why the chemical reactions must eventually stop?) As long as there is no external connection, this potential difference will remain constant, with the anode at a potential higher than that of the cathode. This potential difference is called the EMF of the cell.* (For a Daniell cell the EMF is approximately 1.07 V.)

When an external circuit is connected to the cell, current flows through the cell and the voltage between the electrodes drops below the EMF of the cell because the cell has an effective internal resistance. (For the Daniell cell this internal resistance is usually between 0.75 Ω and 1.0 Ω.) Because of this effect, standard cells (that is, cells used as standards in the measurement of the EMF of other cells—see Section 32–3) are always used in circuits in such a way that no current flows through the cell; then, the voltage between the terminals is known to be the EMF of the cell.

Any device that converts nonelectric energy into electric energy is a source of EMF. Thus, chemical cells and batteries are sources of EMF, as are electric generators that convert mechanical energy into electric energy. Also, thermoelectric devices convert thermal energy into electric energy, and solar cells convert radiant energy into electric energy.

31–5 SINGLE-LOOP CIRCUITS

We now examine the potential-current behavior of a simple electric circuit that consists of a battery and an external resistive load. We represent the resistive load (a *resistor*) by the sawtooth symbol,

We imagine that all connections are made with hookup wire that has negligible resistance. The battery has a distributed internal resistance, due primarily to the resistivity of the electrolyte and the accumulation of polar ions on the electrodes; however, it is conventional to represent this distributed resistance as a single equivalent resistance r, as indicated in Fig. 31–11. For a fully charged low-voltage battery the internal resistance is usually less than 1 Ω, and the EMF (symbol, $\mathcal{E}$) varies between 0.75 V and 2.75 V per cell.

Consider the circuit shown in Fig. 31–12, where the internal resistance of the battery is indicated explicitly. The battery terminals are labeled a and b. The potential Φ_a at point a is, for convenience, set equal to zero by grounding this point. The continuity condition on the current means that the same current I flows through each part of the circuit.

Figure 31–13 shows the variation of the electric potential Φ around the circuit. Because the current is constant (steady-state condition), we have $\oint \mathbf{E} \cdot d\mathbf{r} = 0$ (Eq. 29–14). Evaluating this integral in steps, we have

Fig. 31–11. Schematic representation of a battery. The EMF of the battery is $\mathcal{E}$ and the internal resistance (actually a distributed resistance) is r.

---

*Historically, this potential difference was given the unfortunate name *electromotive force*. But this quantity is not a force and it does not carry the dimensions of force. We use here only the initials EMF as a name for this quantity and forego completely the use of the earlier, misleading term.

Fig. 31–12. A simple electric circuit that consists of a battery and an external load resistor R.

Fig. 31–13. The lower part of the diagram shows the circuit of Fig. 31–12 laid out in a straight line. The upper part of the diagram shows the variation of the potential Φ matched to the circuit. The net change in potential around the circuit (point a to point a) is zero.

$$\oint \mathbf{E} \cdot d\mathbf{r} = \int_a^{a'} E\,dr + \int_{a'}^{b} + \int_b^{c} + \int_c^{d} + \int_d^{a} E\,dr$$

$$(\Phi_{a'} - \Phi_a) + (\Phi_b - \Phi_{a'}) + (\Phi_c - \Phi_b) + (\Phi_d - \Phi_c) + (\Phi_a - \Phi_d) = 0$$

$$\mathcal{E} \qquad\qquad -V_{a'b} \qquad\qquad +0 \qquad\qquad -V_{cd} \qquad\qquad +0 \quad = 0$$

Using Eq. 31-13, the voltage drops can be expressed as $V_{a'b} = Ir$ and $V_{cd} = IR$. Thus, we have

$$\mathcal{E} - Ir = IR \qquad\qquad\qquad \textbf{(31–19)}$$

By writing the result in this form we emphasize the fact that the terminal voltage of the battery is $\mathcal{E} - Ir$ when a current I flows through the battery. Alternatively, we can write

$$\mathcal{E} = Ir + IR \qquad\qquad\qquad \textbf{(31–19a)}$$

This form of the result emphasizes the following rule.

In any circuit loop the sum of the rises in potential around the circuit equals the sum of the drops in potential.

(This is actually Kirchhoff's second circuit rule; see Section 32–2.)

Energy Relationships. When a charged particle that carries a charge $+q$ moves in a vacuum from a point a to a point b, thereby falling through a potential difference $V_{ab} = \Phi_a - \Phi_b > 0$, the particle acquires a kinetic energy equal to qV_{ab}. If, instead, this amount of charge passes through a resistor across which there is a potential difference $V_{ab} > 0$, the energy that would have appeared as kinetic energy is now dissipated in collisions within the resistor and appears as heat. Thus, the differential amount of this ohmic loss (also called resistive or Joule heating) within the resistor due to the passage of an amount of charge dq is

$$dW_R = V_{ab}\, dq$$

and using $V_{ab} = IR$ (Eq. 31–13), we have $dW_R = IR\, dq$. The rate at which heat is generated is the power, $P_R = dW_R/dt$; hence, $P_R = IR\, dq/dt$. Now, $dq/dt = I$, so the power dissipated in the resistor is

$$P_R = I^2R \tag{31-20}$$

Similarly, the power dissipated in the internal resistance r is

$$P_r = I^2r \tag{31-21}$$

Inside a battery, the charge dq is driven through a potential increase $\mathcal{E}$ by the chemical reactions in the cells. In this process the amount of chemical energy converted into electric energy is

$$dW_C = \mathcal{E}\, dq$$

Thus, the power delivered by the battery is $P_C = dW_C/dt = \mathcal{E}\, dq/dt$, or

$$P_C = \mathcal{E}I \tag{31-22}$$

The power dissipated in a resistor R when a current I flows is I^2R (Eq. 31–20). Using the Ohm's law relationship (Eq. 31–13), $V = IR$, we can express the power dissipation variously as

$$P = I^2R = IV = \frac{V^2}{R} \tag{31-23}$$

where V is the potential decrease across the resistor when the current I flows. Power is measured in *watts* (W), and from Eq. 31–23 we see that $1\text{ W} = 1\text{ A}^2\cdot\Omega = 1\text{ V}\cdot\text{A} = 1\text{ V}^2/\Omega$.

Conservation of Energy. If we multiply Eq. 31–19a by I, we have $\mathcal{E}I = I^2r + I^2R$. Then, substituting for each of these terms from Eqs. 31–20, 31–21, and 31–22, we find

$$P_C = P_r + P_R \tag{31-24}$$

This equation is the statement of energy conservation for the system, namely, the rate P_C at which chemical energy is converted to electric energy is equal to the rate, $P_r + P_R$, at which Joule heating losses occur.

Example 31-5

Two batteries and two external resistors are connected as shown in the diagram, using hookup wire with negligible resistance.
(a) Determine the current I.
(b) Calculate the power dissipated or generated in each of the circuit elements.

Solution:

Note that the batteries are connected so that they attempt to drive current in opposite directions around the circuit. Because $\mathcal{E}_1 > \mathcal{E}_2$, the current I flows counterclockwise as indicated in the diagram. Battery 1 *provides* power to the circuit, whereas battery 2 *absorbs* power. The excess power, $(\mathcal{E}_1 - \mathcal{E}_2)I$, appears as Joule heating.

(a) We have

$$\mathcal{E}_1 - \mathcal{E}_2 = I(r_1 + r_2 + R_3 + R_4)$$

so that

$$I = \frac{10.0\ \text{V} - 6.0\ \text{V}}{0.6\ \Omega + 0.2\ \Omega + 15\ \Omega + 8\ \Omega} = 0.1681\ \text{A}$$

(If we had guessed that the current flow is in the opposite direction, we would have found $I < 0$, thereby indicating an incorrect choice for the direction. •Can you see why this is the case?)

(b) Battery 1 is converting chemical energy into electric energy at a rate

$$P_{C1} = \mathcal{E}_1 I = (10.0\ \text{V})(0.1681\ \text{A}) = 1.681\ \text{W}$$

Battery 2 is converting electric energy into chemical energy at a rate

$$P_{C2} = \mathcal{E}_2 I = (6.0\ \text{V})(0.1681\ \text{A}) = 1.008\ \text{W}$$

Battery 1 is *discharging*, whereas battery 2 is being *charged*.
The power dissipated in the various resistive elements is

$$P_{r1} = I^2 r_1 = (0.1681\ \text{A})^2(0.60\ \Omega) = 0.017\ \text{W}$$
$$P_{r2} = I^2 r_2 = (0.1681\ \text{A})^2(0.20\ \Omega) = 0.006\ \text{W}$$
$$P_{R3} = I^2 R_3 = (0.1681\ \text{A})^2(15\ \Omega) = 0.424\ \text{W}$$
$$P_{R4} = I^2 R_4 = (0.1681\ \text{A})^2(8\ \Omega) = 0.226\ \text{W}$$

The energy balance equation is

$$\mathcal{E}_1 I = \mathcal{E}_2 I + I^2 r_1 + I^2 r_2 + I^2 R_3 + I^2 R_4$$
$$1.681 \text{ W} = 1.008 \text{ W} + 0.017 \text{ W} + 0.006 \text{ W} + 0.424 \text{ W} + 0.226 \text{ W}$$

which does indeed balance.

Example 31–6

A power supply with an EMF $\mathcal{E}$ and an internal resistance r provides current to a load resistor R. What must be the value of R for maximum power to be delivered to the load by the supply?

power
supply

Solution:

The current in the circuit is

$$I = \frac{\mathcal{E}}{r + R}$$

Then, the power delivered to the resistor R is

$$P = I^2 R = \frac{\mathcal{E}^2 R}{(r + R)^2}$$

Differentiating, we obtain

$$\frac{dP}{dR} = \mathcal{E}^2 \left(\frac{(r + R)^2 - 2R(r + R)}{(r + R)^4} \right)$$

Setting dP/dR equal to zero, we find that the power is maximum when $R = r$. That is, maximum power is delivered when the load is matched to the internal resistance. This is a particularly important result in AC circuits. (See, for instance, Example 37–8.)

SPECIAL TOPIC

31-6 THE MODERN ATOMIC VIEW OF ELECTRIC CONDUCTION

The Fermi-Dirac Distribution Law. In the classical Drude model (Section 31–2) we suppose that the thermal and electric properties of a substance can be calculated by considering the free electrons to constitute a "gas" that obeys the Maxwell-Boltzmann distribution law. The results obtained with this model are typically in error by a factor of 100 or so, although in some cases (for example, thermal conductivity) acceptable predictions result (as we now know) from fortuitous cancellations. During the early dec-

ades of this century the achievements of the Drude model were accompanied by considerable confusion.

The great success of the kinetic theory in explaining the properties of gases led to the attempt to apply the similar free electron gas model to the case of electric conductors. Kinetic theory is successful because the atoms and molecules in a gas (at ordinary temperatures) behave very much like Newtonian particles. Electrons, however, are definitely *not* Newtonian particles. During the 1920's it was finally realized that electrons are *quantum* entities and obey *quantum* rules. These rules severely restrict the number of electrons that can have momenta in a range between

p and $p + dp$ and simultaneously be in a volume dv. As a result, the gas of free electrons in a conductor does *not* obey the Maxwell-Boltzmann (M-B) distribution law; instead, these electrons are described by a drastically different function called the *Fermi-Dirac* (F-D) distribution law. In 1928 the German theorist Arnold Sommerfeld (1868–1951) modified the Drude theory of metals by substituting the F-D distribution law for the M-B distribution law. We now examine some of the results of Sommerfeld's theory of metals.

It is most convenient to discuss the free electron gas in terms of an *energy* distribution function. According to the F-D distribution law, the number of electrons per unit volume that have kinetic energies between ϵ and $\epsilon + d\epsilon$ is

$$dn(\epsilon) = \frac{8\pi}{h^3}(2\,m_e^3)^{1/2}\epsilon^{1/2}\,d\epsilon, \quad \epsilon < \epsilon_F \quad \textbf{(31–25)}$$

where m_e is the electron mass. The constant h, which appears often in quantum theory, is called *Planck's constant:*

$$h = 6.62618 \times 10^{-34}\,\text{J} \cdot \text{s}$$

with an uncertainty of about 4 in the last figure. The energy ϵ_F is the *Fermi energy* (see Problem 31–17):

$$\epsilon_F = \frac{1}{8}\left(\frac{3}{\pi}\right)^{2/3}\frac{h^2}{m_e}n^{2/3} = 3.65 \times 10^{-19}\,n^{2/3}\,\text{eV} \quad \textbf{(31–26)}$$

where n, the number density of free electrons, is measured in m^{-3}. Equation 31–25 is strictly true at 0 K, but it closely describes the distribution for all ordinary temperatures. The Fermi energy ϵ_F is the maximum electron energy at a temperature of 0 K.

Figure 31–14 compares the M-B distribution at $T = 4500$ K with the F-D distributions for copper at

$T = 0$ K and 4500 K. It is clear that the M-B and F-D distributions are quite different. (However, the two distributions approach one another in the limit of extremely high temperatures; see Problem 31–24.)

According to the M-B distribution law, the average energy of the electrons at a temperature T is (Eq. 24–8)

$$\langle \epsilon \rangle_M = \tfrac{1}{2}m_e v_{\text{rms},M}^2 = \tfrac{3}{2}kT \quad \textbf{(31–27a)}$$

whereas, according to the F-D distribution law, the average energy of the electrons is (see Problem 31–18)

$$\langle \epsilon \rangle_F = \tfrac{1}{2}m_e v_{\text{rms},F}^2 = \tfrac{3}{5}\epsilon_F = \tfrac{3}{5}kT_F \quad \textbf{(31–27b)}$$

where $v_{\text{rms},F}^2 = \langle v^2 \rangle_F$, and where T_F (the *Fermi temperature*) is approximately 81,600 K for copper and is more-or-less independent of the thermal temperature (for $T \lesssim 2000$ K). Thus, at room temperature (293 K), the ratio of the speeds for the two distributions is

$$\frac{v_{\text{rms},F}}{v_{\text{rms},M}} = \frac{\sqrt{\dfrac{3}{5}kT_F}}{\sqrt{\dfrac{3}{2}kT}} = \sqrt{\frac{2}{5}\frac{T_F}{T}} = \sqrt{\frac{2(81,600\,\text{K})}{5(293\,\text{K})}} = 10.6$$

The Work Function. We can understand the origin of the free electrons in a conductor by appealing to a simplified semi-classical picture. The electrons in an isolated, electrically neutral atom are bound to that atom by the electric attraction of the positive nuclear charge. Each electron in the atom has a discrete, well-defined energy. (This is an important result of the quantum description of atoms, an idea that is considered in detail in Chapter 47. For the present discussion it is only necessary to realize that atomic electrons occupy discrete energy states.) The bound electrons in an atom all have negative total energies

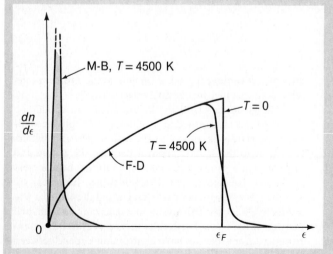

Fig. 31–14. The Maxwell-Boltzmann (M-B) and Fermi-Dirac (F-D) distributions for copper. The Fermi energy ϵ_F for copper is 7.04 eV. The area under each curve is the same. Notice that the F-D distribution is proportional to $\epsilon^{1/2}$ (Eq. 31–25) for energies nearly to ϵ_F. The peak of the M-B distribution is about 15 times higher than the maximum value shown.

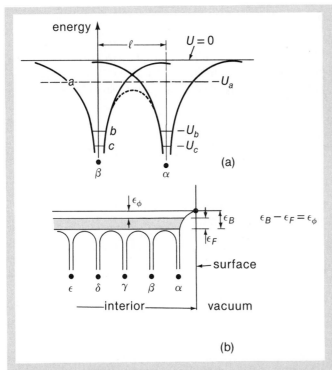

Fig. 31–15. (a) The solid curves represent the potential energy of an atomic electron in the electrostatic field of an isolated atom. The dashed curves represent the potential energy when two atoms, α and β, are separated by a distance ℓ. (b) The potential energy function for a linear array of atoms. The shaded region is the conduction band and contains the free electrons. ϵ_F is the Fermi energy, ϵ_B is the barrier energy, and $\epsilon_\phi = \epsilon_B - \epsilon_F$ is the work function.

with respect to zero energy for an infinite separation of an electron from the atom.

Each of the two solid curves in Fig. 31–15a represents the potential energy of an electron in the electrostatic field of an atom, α or β, each considered to be an isolated atom. The total energy of the least bound electron (one of the *valence* electrons) is $-U_a$. The energies $-U_b$, $-U_c$, and so forth, correspond to successively more tightly bound electrons. In the case of the copper atom, for example, we find $U_a = 7.72$ eV. This energy is referred to as the *first ionization potential;* it is the minimum energy that must be supplied to the atom to remove the least bound electron, thereby creating a Cu$^+$ ion.

The copper atoms in bulk matter have a separation between their nuclear centers of approximately 2.5 Å. When two atoms, α and β, have this separation, the sum of the two individual atomic potential energy functions becomes the dashed curve shown in Fig. 31–15a. The electrons labeled a, with total energy $-U_a$, are no longer bound to a particular atom but are shared by the two atoms, α and β. The electrons labeled b and c, however, remain bound to their respective atomic parents; these electrons are not shared.

Figure 31–15b shows the potential energy function for a linear array of atoms. The interatomic potential is depressed, as in Fig. 31–15a. However, for the surface atom, with no atom to its right, the potential at the surface remains at the normal (zero) level. This "potential barrier" acts to confine the electron gas within the conductor. These free or conduction electrons are distributed in energy as shown in Fig. 31–14, with the reference energy $\epsilon = 0$ corresponding approximately to the top of the rounded interatomic potential energy function. The shaded region in Fig. 31–15b is the *conduction band* (sometimes called the *Fermi sea* of electrons) and corresponds to the energy range $0 \leqslant \epsilon \leqslant \epsilon_F$. The energy ϵ_B measures the surface barrier height relative to the minimum energy in the conduction band. The *least* energy required to remove a conduction electron from the conduction band is ϵ_ϕ, called the *work function* of the material. Evidently,

$$\epsilon_\phi = \epsilon_B - \epsilon_F \qquad (31-28)$$

Values of ϵ_F, ϵ_ϕ, and ϵ_B, along with the first ionization potential ϵ_I, are given in Table 31–4 for various metals.

Contact Potential. When two dissimilar metals are placed in contact, there develops between them a *contact potential difference,* typically a volt or so. Thus, in Fig. 31–16a, which shows the metals 1 and 2 in contact, the contact potential difference is $\Phi_A - \Phi_B$. To see how this potential difference arises, consider the behavior of the Fermi sea of electrons at the junction point C. Figure 31–16b shows the energy states of the two metals before contact is made. In this condition, the potentials at the ends, C_1 and C_2, are the same. The work function $\epsilon_{\phi 1}$ of metal 1 is less than that of metal 2; that is, the electrons at the top of the Fermi sea in metal 1 have higher energies than the corresponding electrons in metal 2. Consequently, when

TABLE 31–4.   Fermi and Associated Energies for Various Metals

ELEMENT	ξ^a	FERMI ENERGY, ϵ_F (eV)	WORK FUNCTION, ϵ_ϕ (eV)	BARRIER ENERGY, ϵ_B (eV)	FIRST IONIZATION POTENTIAL, ϵ_I (eV)
Li	1	4.74	2.38	7.12	5.39
Na	1	3.24	2.35	5.59	5.14
K	1	2.12	2.22	4.34	4.34
Rb	1	1.85	2.16	4.01	4.18
Cs	1	1.59	1.81	3.40	3.89
Cu	1	7.04	4.46	11.46	7.72
Ag	1	5.49	4.64	10.13	7.57
Au	1	5.53	4.82	10.35	9.22
Be	2	14.30	3.92	18.22	9.32
Mg	2	7.08	3.64	10.72	7.64
Ca	2	4.69	2.80	7.49	6.11
Sr	2	3.93	2.35	6.28	5.69
Ba	2	3.64	2.49	6.13	5.21
Al	3	11.70	4.25	15.95	5.98
Pb	4	9.47	4.14	13.61	7.42
Bi	5	9.90	4.24	14.14	7.29

^{a}Assumed number of free electrons per atom.

Fig. 31–16. (a) Placing two dissimilar metals in contact results in a *contact potential difference*, $\Phi_A - \Phi_B$, of a volt or so. (b) The contact potential difference arises because electrons flow from the higher Fermi level to the lower Fermi level, leaving metal 1 positively charged and metal 2 negatively charged. The horizontal lines indicate schematically the electron energy states in the two metals.

the metals are brought into contact, electrons in metal 1 seek lower energy states by transferring to metal 2. This transfer lowers the Fermi level in metal 1 while raising it in metal 2. The number of electrons actually transferred in this process is exceedingly small compared with the total number of electrons present; consequently, the values of ϵ_ϕ and ϵ_F for each metal remain unchanged. When equilibrium is reached, with as many electrons per unit time crossing the junction in one direction as in the other, the Fermi levels coincide. This situation is illustrated schematically in Fig. 31–17. The transfer of electrons leaves metal 1 with a positive charge and metal 2 with a negative charge. The energy difference that an electron would have to overcome in order to move from point A to point B is $\epsilon_{\phi 2} - \epsilon_{\phi 1} = -e(\Phi_B - \Phi_A)$. Thus, the contact potential difference between points A and B is

$$V_{AB} = \Phi_A - \Phi_B = (\epsilon_{\phi 2} - \epsilon_{\phi 1})/e \qquad \textbf{(31–29)}$$

For example, if metal 1 is magnesium (Mg) and metal 2 is gold (Au), the values of ϵ_ϕ in Table 31–4 show that the contact potential difference is $V_{AB} = (4.82\,\text{eV} - 3.64\,\text{eV})/e = 1.18\,\text{V}$.

Thermionic Emission. At very high temperatures, the high energy tail of the F-D energy distribution (see Fig. 31–14) may contain a significant number of electrons with

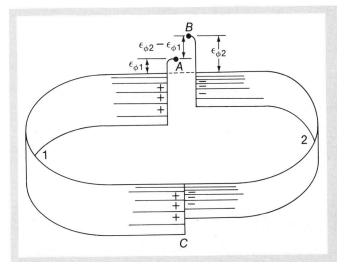

Fig. 31–17. The energy states within two dissimilar metals in contact. At equilibrium, the Fermi levels in the two metals are the same. Then, the contact potential difference between points A and B is $V_{AB} = \Phi_A - \Phi_B = (\epsilon_{\phi2} - \epsilon_{\phi1})/e$.

energies greater than the barrier energy, $\epsilon_B = \epsilon_F + \epsilon_\phi$. These *thermionic electrons,* shown in Fig. 31–18, are able to escape from the conductor surface. A nearby electrode that is maintained at a positive potential with respect to the conductor will collect the liberated electrons. The current so produced is called a *thermionic current.* This thermionic effect is used in electron guns, where the electrons are emitted from an electrically heated filament. To enhance the thermionic emission, these filaments are often coated with a metal that has a particularly low work function (for example, cesium).

Fig. 31–18. At high temperatures some of the electrons in the Fermi sea have energies greater than the barrier energy ϵ_B. These *thermionic electrons* can escape the conductor and can be collected by a nearby positive electrode.

Example 31–7

In a simplified version of the quantum theory of electric resistivity, the conduction electrons are imagined to undergo collisions only with the *imperfections* in an otherwise perfectly ordered crystalline lattice. (The effects of the ionic lattice centers themselves are already taken into ac-

count in determining the actual energy level scheme for the conduction electrons.) The imperfections may be chemical impurities, dislocations in the lattice, polycrystalline boundaries, or the thermal vibrations of the ionic centers about their 0 K equilibrium positions. Consider the case of copper with only thermal vibrations present. Only the *two* degrees of freedom in a plane perpendicular to the line of motion of the electron are relevant in determining the collision cross section. Then, we can use the equipartition theorem (see Section 24–2) to write, for isotropic vibrations,

$$\tfrac{1}{2}\eta \langle r_\perp^2 \rangle = kT$$

where η is the effective spring constant for lattice vibrations of the Cu^{2+} ions and where $\langle r_\perp^2 \rangle^{1/2}$ is the rms value of the vibrational amplitude. The collision mean free path is then

$$\lambda = \frac{1}{\mathcal{N}\pi \langle r_\perp^2 \rangle}$$

From specific heat data it is found that η for copper is 125 N/m.

(a) Evaluate λ for temperatures of 77 K, 273 K, and 373 K, and compare the results with the classical value λ_C determined in Example 31–2.

(b) Use the values of λ from (a) and the rms velocity $v_{rms,F}$ calculated from Eq. 31–27b to determine the mean collision time τ for temperatures of 77 K, 273 K, and 373 K.

(c) What is the temperature dependence of the resistivity ρ according to this model?

(d) The average gain in kinetic energy ΔK for a conduction electron between collisions is approximately $e\lambda E$, where E is the electric field strength. Determine ΔK for the case of copper at 273 K when the current density is 20 A/mm² (a typical value for a household circuit).

Solution:

(a) We have $\langle r_{\perp}^2 \rangle = 2kT/\eta$; thus, for the three temperatures, we have

$$\langle r_{\perp}^2 \rangle (77\,\text{K}) = \frac{2(1.38 \times 10^{-23}\,\text{J/K})(77\,\text{K})}{125\,\text{N/m}}$$

$$= 1.70 \times 10^{-23}\,\text{m}^2$$

$$\langle r_{\perp}^2 \rangle (273\,\text{K}) = 6.03 \times 10^{-23}\,\text{m}^2$$

$$\langle r_{\perp}^2 \rangle (373\,\text{K}) = 8.24 \times 10^{-23}\,\text{m}^2$$

The corresponding values for λ are

$$\lambda = \frac{1}{\mathcal{N}\pi\langle r_{\perp}^2 \rangle}$$

$$\lambda(77\,\text{K}) = \frac{1}{(8.47 \times 10^{28}\,\text{m}^{-3})\pi(1.70 \times 10^{-23}\,\text{m}^2)}$$

$$= 221\,\text{nm}$$

$$\lambda(273\,\text{K}) = 62.3\,\text{nm}$$

$$\lambda(373\,\text{K}) = 45.6\,\text{nm}$$

These values are to be compared with the classical result of $\lambda_C = 0.408$ nm, which is independent of temperature. (See Example 31–2.)

(b) According to Eq. 31–27b, the rms velocity is

$$v_{\text{rms},F} = \left(\frac{6}{5}\frac{kT_F}{m_e}\right)^{1/2}$$

$$= \left(\frac{6(1.38 \times 10^{-23}\,\text{J/K})(81{,}600\,\text{K})}{5(9.11 \times 10^{-31}\,\text{kg})}\right)^{1/2}$$

$$= 1.22 \times 10^6\,\text{m/s}$$

The average velocity for the F-D distribution is (see Problem 31–18)

$$\langle v \rangle_F = \sqrt{\frac{15}{16}}\,v_{\text{rms},F} = 1.18 \times 10^6\,\text{m/s}$$

Then, the mean collision times are given by $\tau = \lambda/\langle v \rangle_F$:

$$\tau(77\,\text{K}) = \frac{221 \times 10^{-9}\,\text{m}}{1.18 \times 10^6\,\text{m/s}} = 19 \times 10^{-14}\,\text{s}$$

$$\tau(273\,\text{K}) = 5.3 \times 10^{-14}\,\text{s}$$

$$\tau(373\,\text{K}) = 3.9 \times 10^{-14}\,\text{s}$$

At the higher temperatures (where η was determined), these calculated values are quite close to those listed in Table 31–1.

(c) From Eq. 31–10 we have $\rho = 2m_e/e^2 n\tau$, with $n = \mathcal{N}$. We also know that $\tau = \lambda/\langle v \rangle_F$. Then, using the above expressions for λ and $\langle r_{\perp}^2 \rangle$, we find

$$\rho = \frac{4\pi m_e k \langle v \rangle_F}{e^2 \eta}\,T = \text{constant} \times T$$

which is the observed temperature dependence. Thus, the Sommerfeld theory gives a much more accurate description of metals than does the earlier and completely classical Drude theory.

(d) The electric field strength is $E = \rho J$. Using the measured value for the resistivity of copper at 273 K, $\rho = 1.56 \times 10^{-8}\,\Omega \cdot \text{m}$ (see Table 31–2; the expression in (c) yields $1.59 \times 10^{-8}\,\Omega \cdot \text{m}$), we find

$$E = (1.56 \times 10^{-8}\,\Omega \cdot \text{m})(20 \times 10^6\,\text{A/m}^2) = 0.312\,\text{V/m}$$

so that

$$\Delta K \cong e\lambda E = (1.60 \times 10^{-19}\,\text{C})(62.3 \times 10^{-9}\,\text{m})(0.312\,\text{V/m})$$

$$= 3.11 \times 10^{-27}\,\text{J} = 1.94 \times 10^{-8}\,\text{eV}$$

Thus, the energy acquired by an electron between collisions is extremely small. Moreover, in these collisions the electrons are scattered in random directions. Consequently, after a collision, the probability is one half that an electron will have a net motion in the direction of the field **E**. Because the electron charge is $-e$, the effect of the field on such electrons is to decrease, not increase, their speeds. As a result of these frequent collisions and the attendant changes in the direction of motion, the electrons can never acquire kinetic energies greater than a tiny fraction of an eV.

Band Theory.   The Sommerfeld model was an important step in the development of the theory of metals, but it left a number of problems only partially resolved. The model was quite good for the alkali metals (Li, Na, K, Rb, and Cs—column 1A in the periodic table), but the accuracy suffered for the noble metals (Cu, Ag, and Au) and for the metals in column 2A (Be, Mg, Ca, Sr, and Ba). The most glaring deficiency of the model was that it failed to explain why some elements are good electric conductors and others are not. Why, for example, is aluminum a good conductor, whereas boron, which is one row higher in the same column of the periodic table, is an electric insulator?

The next important improvement in understanding the physical properties of solids was the development of *band theory*. This theory is based on a careful quantum analysis of the role of the atomic lattice centers in the solid. These centers are far more important than assumed in the previous theories, in which they served only to balance electric charge and to act as scattering sites for the free conduction electrons.

We begin by examining the energy level structure of a

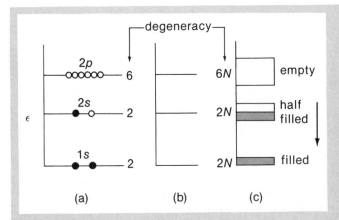

Fig. 31-19. (a) The lowest three energy states of an isolated lithium atom. The solid circles represent the occupied states and the open circles represent the possible but unoccupied spaces. In bulk matter, the atomic energy states (b) are enlarged into energy *bands* (c) whose occupancy follows the same scheme as in the atom (a).

simple metallic element such as lithium ($Z = 3$). Figure 31-19a shows the lowest three energy states of an isolated lithium atom. The solid circles represent the three atomic electrons of lithium in the lowest possible energy states. (The open circles represent additional but unoccupied spaces.) The number to the right of each state indicates the maximum number of electrons that can occupy that state. This number is called the *degeneracy* of the state. Notice that only two electrons can occupy the lowest energy state (the 1s state*) so that the third electron must occupy the second lowest energy state (the 2s state). That only a certain number of electrons are allowed to occupy each energy state is a result of a quantum rule called the *exclusion principle* (see Section 47-5).

When N atoms of lithium are collected together to form a bulk sample, the discrete energy states of an isolated atom (Fig. 31-19b) are broadened into *bands* (Fig. 31-19c). The degeneracy of each band is just N times the degeneracy of the corresponding atomic energy state. That is, each band actually consists of N states crowded into a narrow range of energies. Because N is a very large number for a sample of bulk matter (10^{23} or so), the spacing between the states is so small that the band represents an essentially continuous distribution of energies. In a bulk sample the electrons can possess only energies that correspond to the allowed energy bands; all other energies are forbidden. The highest occupied band is called the *valence band*.

What happens when an external electric field is applied to a bulk sample of lithium metal? In Example 31-6 we found that in a typical case the maximum kinetic energy that can be acquired by a conduction electron is of the order of 10^{-8} eV. This energy is completely negligible in comparison with the energy separation between the 1s and 2s bands in lithium (approximately 48 eV). Because the 1s band is completely filled, no energy state within the

band is available for the promotion (that is, movement to a higher energy state) of a lower energy electron. The *minimum* energy that a 1s electron can absorb is an energy that would lift it to the unfilled portion of the 2s band; this energy cannot be acquired in a collision with a conduction electron. Thus, the 1s electrons remain attached to their parent atoms and do not contribute to the current flow in the sample.

The 2s band, on the other hand, is only half filled. At low temperatures ($\lesssim 2000$ K) the 2s electrons occupy the lowest available energy states in the band with a Fermi-Dirac distribution as shown in Fig. 31-14. The spacing of the unoccupied energy states just above the Fermi energy ϵ_F is (see Problem 31-25)

$$\Delta\epsilon = \frac{2}{3}\frac{\epsilon_F}{n}$$

which gives $\Delta\epsilon = 6.8 \times 10^{-23}$ eV for a 1-cm³ sample of lithium. Therefore, 2s electrons can easily be promoted by collisions into the numerous vacant energy states above the Fermi energy. These electrons are the conduction electrons that participate in the current flow.

All good electric conductors have the feature that the highest occupied energy band is only partially filled. However, in a poor conductor (or insulator) the highest occupied energy band is completely filled and there is a substantial energy gap between the filled band and the nearest unfilled band. For example, Fig. 31-20a shows the band structure for carbon in the crystalline form of *diamond*. The lowest three bands are completely filled. The minimum energy required to promote an electron from a filled band into an unoccupied state in the lowest empty band is equal to the *gap energy* ϵ_g, which is 5.5 eV for diamond (Fig. 31-20a). Any possible applied electric field is totally inadequate to impart this amount of energy to an electron. Thus, electric conduction of the type we have been discussing does not occur.

Carbon can also exist in another crystalline form,

---

*This code for labeling the various atomic states is discussed in Section 47-5.

(a) diamond

(b)

Fig. 31–20. (a) The band structure for carbon in the crystalline form of diamond. The gap energy is substantial (5.5 eV), so diamond is not a conductor. (b) The band structure for carbon in the crystalline form of graphite. The gap energy is nearly zero, so graphite can be a conductor by virtue of thermally excited electrons that are raised into the empty conduction band.

namely, *graphite*, in which the atoms have geometric arrangement and spacing different from that in diamond. In graphite the gap energy is reduced almost to zero (Fig. 31–20b). Electrons can be promoted across the gap into the empty (conduction) band by *thermal* effects.

Figure 31–14 shows that the F-D distribution for a temperature above absolute zero contains a high-energy tail that extends above the Fermi energy ϵ_F. The extent of this tail is of the order of kT. At room temperature, $kT = 0.0252$ eV (see Problem 31–16). Thus, if the gap energy ϵ_g is small, some of the electrons in the highest filled band will be thermally excited into the empty conduction band. Those elements that have small gap energies ($\epsilon_g \lesssim 2$ eV) are called *intrinsic semiconductors*.

When an electron is promoted from the valence band into the conduction band, it leaves behind an unoccupied *hole* in the valence band. When another valence electron moves in to occupy this hole, it creates a hole at its previous site. Thus, a hole can "move" and act as a carrier of *positive*

charge. Both the holes and the electrons contribute to the current flow when an electric field is applied.

The conductivity of an intrinsic semiconductor depends on the density of thermally excited negative and positive charge carriers (electrons and holes). One of the results of the theory is that at an absolute temperature T the number density (per cm^3) of these carriers is given approximately by*

$$n_+ = n_- \cong (4.83 \times 10^{15} \text{ cm}^{-3})T^{3/2}e^{-(\epsilon_g/2kT)} \quad \textbf{(31–30)}$$

The gap energy for the intrinsic semiconductor germanium (Ge) is 0.67 eV (see Table 31–5). Thus, at room temperature, we find for germanium

$$n_+ = n_- \cong (4.83 \times 10^{15} \text{ cm}^{-3})(293)^{3/2}e^{-0.67/(2\times0.0252)}$$
$$\cong 4.1 \times 10^{13} \text{ cm}^{-3}$$

For copper, on the other hand, the number density of conduction electrons is $n = 8.5 \times 10^{22}$ cm^{-3} (see Table 31–1). When the different drift speeds for the two cases are taken into consideration, the ratio of the resistivity values at room temperature is found to be approximately 3×10^7 (compare Table 31–2).

The resistivity of a semiconductor depends strongly on the temperature. The usual increase in resistivity with temperature due to lattice vibrations is completely obscured by the increase in the number of charge carriers due to the rapidly changing exponential factor in Eq. 31–30. Thus, for semiconductors the resistivity decreases with temperature ($\alpha < 0$) instead of increasing, as is the case for conductors (see Table 31–2).

The number density of charge carriers at room temperature in diamond (no impurities) is, using Eq. 31–30 with $\epsilon_g = 5.5$ eV,

$$n_+ = n_- \cong 9.8 \times 10^{-29} \text{ cm}^{-3}$$

TABLE 31–5. Gap Energies ϵ_g for Selected Semiconductors

SUBSTANCE	ϵ_g (eV)
C (graphite)	0+
Sn (grey)	0.08
Te	0.35
PbS	0.37
Ge	0.67
Si	1.14
B	1.5

*The exponential term is just the *Boltzmann factor*, Eq. 24–28.

Fig. 31–21. (a) The diamondlike crystal structure of silicon. Each interatom line represents a covalent electron bond. (b) When one of the silicon atoms is replaced by an atom of arsenic the result is an As$^+$ ion in the lattice and a surplus electron that is easily excited into the conduction band.

Thus, an increase in the gap energy from 0.67 eV (germanium) to 5.5 eV (diamond) results in a decrease in the density of charge carriers by a factor of about 10^{42}. Diamond is indeed a very good insulator! In substances such as diamond, conduction actually takes place by the diffusion of ions through the solid.

Thermal excitation is not the only way to enhance the conductivity of a semiconductor material. The number of charge carriers can be permanently increased by the introduction into the material of a minute quantity of an appropriate impurity. For example, silicon has four valence electrons (as does germanium) and forms a diamondlike crystalline lattice in which each atom is connected to four other atoms by covalent bonds (Fig. 31–21a). Now, suppose that an arsenic (As) atom, which has five valence electrons, is substituted for one of the silicon atoms in the lattice, as indicated in Fig. 31–21b. (This process is called *doping*, and the arsenic is referred to as a *dopant*.) Four of the arsenic electrons are used to duplicate the silicon bonds, and there is one surplus electron. Thus, the crystal contains an As$^+$ ion, and the extra electron is so loosely bound (by about 0.05 eV) to the arsenic ion that it can be readily released by thermal excitation. In effect, the arsenic impurity atom has donated a negative charge carrier (an electron) to the material. When introduced into silicon (or germanium), arsenic is a *donor* atom, and the result is referred to as an *n-type* (*negatively doped*) material.

If an element with three valence electrons, such as gallium (Ga), is introduced into a silicon crystal, each dopant atom lacks one electron to complete the bonds to the four neighboring silicon atoms. To compensate for this deficiency, a gallium atom "steals" an electron from a silicon atom, thereby becoming a Ga$^-$ ion. This leaves an Si$^+$ ion or hole, which then acts as a positive charge carrier. The gallium impurity is called an *acceptor* atom, and the material is a *p-type* (*positively doped*) semiconductor.

Figure 31–22 shows the energy relationships of the donor and acceptor atoms to the valence and conduction bands of silicon. Because there are such small energy differences between the donor states and the conduction band and between the valence band and the acceptor states, electron and hole promotion is readily induced by thermal excitation.

Fig. 31–22. (a) Donor energy states for arsenic atoms in silicon (*n*-type). (b) Acceptor energy states for gallium atoms in silicon (*p*-type).

PROBLEMS

Section 31–1

31–1 A copper bus bar has a cross section of 5 cm × 15 cm and carries a current with a density of 2000 A/cm^2. (a) What is the total current in the bus bar? (b) What amount of charge passes a given point in the bar per hour?

31–2 A small sphere that carries a charge of 8 nC is whirled in a circle at the end of an insulating string. The rotation frequency is 100π rad/s. What average current does this rotating charge represent?

31–3 A coaxial conductor with a length of 20 m consists of an inner cylinder with a radius of 3.0 mm and a concentric outer cylindrical tube with an inside radius of 9.0 mm. A uniformly distributed leakage current of 10 μA flows between the two conductors. Determine the leakage current density (in A/m²) through a cylindrical surface (concentric with the conductors) that has a radius of 6.0 mm.

Section 31-2

31-4   A sample consists of $\mathcal{N}$ atoms per unit volume, each of which has an effective radius R. For a pointlike particle (an electron) that moves through the sample, show that the classical mean free path is $\lambda_c = 1/\mathcal{N}\pi R^2$.

31-5   The density of copper is 8.96 g/cm³ and the atomic mass is 63.55 u. Verify the free electron number density for copper quoted in Table 31-1.

31-6   Use the data in Table 31-1 to determine $\bar{\alpha}$ for copper in the temperature range from 273 K to 373 K. (Neglect any possible variation of n with temperature.) Compare your result with the value listed in Table 31-2.

31-7   An aluminum wire with a diameter of 0.1 mm has a uniform electric field of 0.2 V/m imposed along its entire length. The temperature of the wire is 50° C. (a) Use the information in Table 31-2 and determine the resistivity. (b) What is the current density in the wire? (c) What is the total current in the wire? (d) What is the drift speed of the conduction electrons? (e) What potential difference must exist between the ends of a 2-m length of the wire to produce the stated electric field strength?

31-8   The ionic radius of Ag⁺ is 1.26 Å. (a) Calculate the classical mean free path λ_c for electrons in silver. (Refer to Problem 31-4.) (b) Assume that the electrons behave like the molecules in an ideal classical gas and calculate the average speed $\langle v \rangle$ at 0° C. (c) Use the values obtained in (a) and (b) to determine the mean collision time at 0° C. Compare your result with the value listed in Table 31-1.

Section 31-3

31-9   Determine the resistance at 50° C of a 1.5-m length of platinum wire that has a diameter of 0.10 mm.

31-10 The diagram shows the annulus of silicon we considered in Example 31-4 ($r_1 = 0.8$ cm, $r_2 = 3.0$ cm, and $L = 5.0$ cm). In this case the silicon is placed between two thick copper plates with a potential difference of 2.0 V between them. What is the cur-

rent in the circuit? (The battery has negligible internal resistance.) Compare your result with that obtained in Example 31-4. (•Can you account qualitatively for the difference in the currents?)

31-11 An amber rod has a diameter of 1 cm and a length of 10 cm. A conductor is attached to one end of the rod and the other end is grounded. (a) Determine the axial resistance of the rod. (This resistance will be realized only in a perfectly dry environment. Any surface moisture will lower the value considerably.) (b) If the conductor is at a potential of 10⁴ V with respect to ground, what current flows in the rod? (c) At the rate determined in (b), how long would be required for a charge of 50 nC to leak from the conductor to ground?

Section 31-4

31-12 A Daniell cell (Fig. 31-10) delivers a current of 0.1 μA for one hour. During this time, how much zinc entered solution and how much copper was plated onto the anode?

Section 31-5

31-13 Two batteries are connected in series with a resistor R, as shown in the diagram on the next page: $\mathcal{E}_1 = 3.7$ V, $r_1 = 0.25 \Omega$; $\mathcal{E}_2 = 6.1$ V, $r_2 = 0.50 \Omega$. (a) Determine the current I if $R = 25 \Omega$. (b) What are the values of the EMF $\mathcal{E}'$ and the internal resistance r' of the equivalent single battery that would deliver to any resistance R the same current as would the actual two batteries?

31-14 A rotating mechanical DC generator with an internal (armature) resistance of 0.65Ω produces an EMF of 250 V. A load resistance of 10Ω is connected to the generator terminals. (a) What current flows through the load resistance? (b) What power is delivered to the load? (c) What amount of power is lost in the internal resistance? (d) If the frictional losses (including the *windage loss* due to the rotation

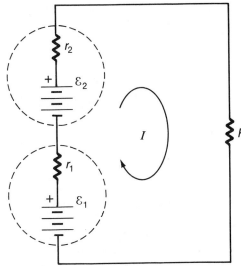

Problem 31-13

of the armature) amount to 600 W, what input mechanical power is required to produce the stated conditions? (e) With what efficiency is the input mechanical power delivered to the load resistor? (f) The generator has a magnetic field that is produced by an external source that requires 300 W of power. If this power is also counted as a loss, what is the overall efficiency?

31-15 In Example 31-5 an ammeter is inserted in series with the other elements in the circuit in order to measure the current. The internal resistance of the ammeter is 0.085 Ω. What is the fractional change in the current, $\Delta I/I$, that results from the presence of the ammeter?

SPECIAL TOPIC

Section 31-6

31-16 In terms of the speed v, the Fermi-Dirac distribu-
• tion law is

$$dn(v) = \frac{8\pi m_e^3}{h_3} \frac{v^2 \, dv}{1 + e^{(\frac{1}{2}m_e v^2 - \epsilon_F)/kT}} \quad \textbf{(A)}$$

where $dn(v)$ is the number of electrons per unit volume with speeds between v and $v + dv$; $h = 6.63 \times 10^{-34}$ J · s is Planck's constant, and ϵ_F is the Fermi energy, a constant that must be determined for each substance. For temperatures below about 1000 K, the exponent of e is a very large negative number, unless the electron kinetic energy $\frac{1}{2}m_e v^2$ is close to the Fermi energy ϵ_F. Then, we have

$$dn(v) \cong \frac{8\pi m_e^3}{h^3} v^2 \, dv \quad \textbf{(B)}$$

(a) Verify that at room temperature (293 K), $kT = 0.0252$ eV. (b) How close to ϵ_F can the electron kinetic energy be and still have the exponential factor $\leqslant 0.1$ at room temperature? (c) In view of the result obtained in (b), comment on the validity of the approximation (B) in the range $0 \leqslant \frac{1}{2}m_e v^2 \leqslant \epsilon_F$. (d) Show that (B) can be converted into the energy distribution law expressed by Eq. 31-25.

31-17 (a) The integral of the distribution (B) in Problem
• 31-16 gives n, the total number of free electrons per unit volume. Show that setting

$$n = \int_0^{v_F} dn(v)$$

where $v_F = \sqrt{2\epsilon_F/m_e}$ is the *Fermi velocity*, yields Eq. 31-26 for the Fermi energy, namely,

$$\epsilon_F = \frac{1}{8}\left(\frac{3}{\pi}\right)^{2/3} \frac{h^2}{m_e} n^{2/3}$$

(b) Use the values of n for lithium and copper listed in Table 31-1 and calculate ϵ_F for these elements. Compare with the values given in Table 31-4.

31-18 Use the distribution (B) in Problem 31-16.
• (a) Show that $\frac{1}{2}m_e\langle v^2\rangle_F = \frac{3}{5}\epsilon_F$.
(b) Show that $\langle v\rangle_F = \sqrt{15/16}\,v_{\text{rms},F}$, where $v_{\text{rms},F} = \langle v^2\rangle_F^{1/2}$.

[*Hint:* $\langle v^2\rangle_F = \int_0^{v_F} v^2 \, dn(v) \Big/ \int_0^{v_F} dn(v)$

$$\langle v\rangle_F = \int_0^{v_F} v \, dn(v) \Big/ \int_0^{v_F} dn(v)$$

where $v_F = \sqrt{2\epsilon_F/m_e}$.]

31-19 Use the values of ϵ_F for lithium and copper listed in Table 31-4 and show that the Fermi temperatures (Eq. 31-27b) for these elements are approximately 55,000 K and 81,600 K, respectively.

31-20 Determine the number density of negative (and positive) charge carriers, $n_- = n_+$, in intrinsic silicon at room temperature. (Refer to Table 31-5.)

Additional Problems

31-21 In the simple Bohr model of the hydrogen atom, a single electron is imagined to move around the nuclear proton in a circular orbit with a radius of 0.53×10^{-10} m. What average current does this motion represent? [*Hint:* Calculate the electron velocity by setting the Coulomb force equal to the centripetal acceleration times the electron mass.]

31-22 All of the metals listed in Tables 31-1 and 31-2 (except iron and zinc) have *face-centered cubic* crystalline structures. The diagram shows the elementary cell for such crystals. Eight atoms (open circles) form a cube with one atom (solid circle) centered in each of the six faces of the cube. (a) Determine the cube dimension *a* for copper. (Refer to Problem 31-5.) (b) Also for the case of copper, determine the distance between nearest-neighbor atoms (that is, between any corner atom and the nearby face-centered atom). [*Hint:* There are as many cubic cells as there are corners labeled *O*. Each face-centered atom is shared by two adjacent cells. Hence, there are $1 + (6/2) = 4$ atoms associated with each cell.]

31-23 An isolated parallel-plate capacitor has a plate area of 50 cm² and a plate separation of 5 mm. The gap

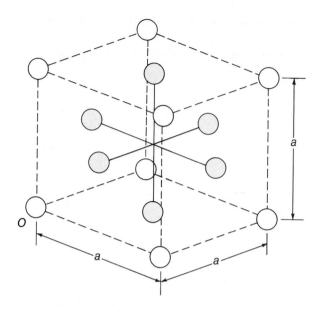

is filled with a solid slab of glass that has a dielectric constant $\kappa = 5.2$ and a resistivity $\rho = 2.0 \times 10^{13}\ \Omega \cdot$ m. The capacitor is charged to a surface charge density of 15 μC/cm². What is the initial current that flows between the plates?

SPECIAL TOPIC

31-24 For very high temperatures, the F-D distribution function (see Problem 31-16) is no longer given by expression A. In this situation the quantity ϵ_F is replaced by a new quantity μ (called the *chemical potential*) such that $e^{-\mu/kT} \gg 1$. Show that the F-D distribution function then becomes

$$dn(v) = Cv^2 e^{-\frac{1}{2}m_e v^2/kT}$$

which is the same as the M-B distribution.

31-25 For a conductor, show that the spacing of the unoccupied energy states just above the Fermi energy ϵ_F is given by

$$\Delta\epsilon = \frac{2}{3}\frac{\epsilon_F}{n}$$

31-26 In Example 31-6 it was mentioned that specific heat data for copper had been used to obtain the value $\eta = 125$ N/m for the effective spring constant for lattice vibrations of Cu^{2+} ions. Use this value to calculate Young's modulus for a single crystal of copper. (Take the spacing of copper atoms in the crystal to be 3.61 Å.) Compare your result with the value given in Table 16-2. (Note that the elastic moduli for single crystals are always much larger than those for an ordinary polycrystalline bulk sample.) [*Hint:* Convert the given *microscopic* value of η into a *macroscopic* value by multiplying by nV, where the volume V is equal to some area A times the spacing of the copper atoms.]

DC CIRCUITS

In Section 31–5 we considered single-loop circuits involving resistors and sources of EMF. Such circuits are easy to analyze because the same current flows through each of the circuit elements. Most circuits met in practice are more complicated and may involve several interconnecting loops. (A multiloop system within a circuit is called a *network*.) We begin by examining methods for reducing complicated networks to simpler ones. Then, general techniques for solving circuit problems are discussed. Attention here is restricted to direct-current (DC) situations in which all currents are constant in time. Alternating-current (AC) circuits are discussed in Chapter 37.

32-1 RESISTOR NETWORKS

We seek a method whereby a network of resistors can be replaced by a single effective or equivalent resistor in such a way that the remainder of the circuit is undisturbed. That is, if the network has terminals* a and b, the current between a and b and the potential difference V_{ab} must be the same with the equivalent resistor in place as for the original network.

First, consider a network of resistors connected in parallel. Figure 32–1 shows three resistors, R_1, R_2, and R_3, connected in parallel between terminals a and b. The potential difference (the voltage drop) V_{ab} is common to all three resistors. Moreover, the current I that enters the network is divided among the three resistors such that $I_1 + I_2 + I_3 = I$ (continuity of current). Then, using $I_1 = V_{ab}/R_1$ and similarly for

---

*We can imagine complicated networks that are connected to circuits through any number of terminals. In this section we consider only two-terminal networks; in Section 32–4 we discuss two simple three-terminal networks. The treatment of networks with four or more terminals is beyond the scope of this text.

Fig. 32–1. The parallel connection of three resistors (a) can be replaced by the single equivalent resistor R (b).

(a)

(b)

875

the other two resistors, we have

$$I = V_{ab}\left(\frac{1}{R_1} + \frac{1}{R_2} + \frac{1}{R_3}\right)$$

(32–1)

When the equivalent resistor is connected between the terminals a and b, the same current I must result for the same voltage drop V_{ab}. Thus,

$$I = \frac{V_{ab}}{R}$$

(32–2)

Comparing Eqs. 32–1 and 32–2, we see that

$$\frac{1}{R} = \frac{1}{R_1} + \frac{1}{R_2} + \frac{1}{R_3} \qquad \text{parallel connection}$$

(32–3)

Note also that the power dissipated in the equivalent resistor, $P = V_{ab}^2/R$, must equal the sum of the values representing the power dissipated in the individual resistors of the original network, namely, $V_{ab}^2/R_1 + V_{ab}^2/R_2 + V_{ab}^2/R_3$. Equating these two expressions for the power P, we have

$$\frac{V_{ab}^2}{R} = \frac{V_{ab}^2}{R_1} + \frac{V_{ab}^2}{R_2} + \frac{V_{ab}^2}{R_3}$$

which leads to the same result for R (Eq. 32–3). Clearly, the expression for the equivalent resistance R can be generalized to any number of parallel resistors.

Next, consider three resistors connected in series, as in Fig. 32–2a. Now, the same current I flows through each resistor. We write the voltage drop V_{ab} between the terminals a and b as the sum of the voltage drops across the individual resistors, namely, $V_{aA} = IR_1$ and similarly for the other two resistors. Thus,

$$V_{ab} = V_{aA} + V_{AB} + V_{Bb}$$
$$= I(R_1 + R_2 + R_3)$$

(32–4)

For the equivalent resistor, we have

$$V_{ab} = IR$$

(32–5)

Comparing Eqs. 32–4 and 32–5, we see that

$$R = R_1 + R_2 + R_3 \qquad \text{series connection}$$

(32–6)

Fig. 32–2. The series connection of three resistors (a) can be replaced by the single equivalent resistor R (b).

(•Can you obtain the same result by considering the power dissipation in the network and in the equivalent resistor?) Notice carefully that resistors and capacitors do *not* follow the same rules for series and parallel combinations; compare Eqs. 30–16 and 30–17 with Eqs. 32–3 and 32–6.

Example 32–1

Determine the currents I_1, I_2, and I_3 in the circuit illustrated in diagram (a).

Solution:

First, replace the two parallel resistors, R_2 and R_3, with the equivalent resistor R:

$$\frac{1}{R} = \frac{1}{5\,\Omega} + \frac{1}{10\,\Omega} = \frac{3}{10\,\Omega}, \quad \text{or} \quad R = \tfrac{10}{3}\,\Omega$$

This reduction is shown in diagram (b).

Now, the three resistors, R_1, R, and r, are in series, so the overall equivalent resistance is

$$R' = R_1 + R + r$$
$$= 3\,\Omega + \tfrac{10}{3}\,\Omega + 1\,\Omega = \tfrac{22}{3}\,\Omega$$

This reduction is shown in diagram (c).

The current I_1 can now be calculated:

$$I_1 = \frac{\mathcal{E}}{R'} = \frac{6\,\text{V}}{\tfrac{22}{3}\,\Omega} = \tfrac{9}{11}\,\text{A}$$

The currents through R_2 and R_3 are obtained by noting that the voltage drop across each of these resistors is $V = I_1 R$. Thus,

$$I_2 = \frac{V}{R_2} = \frac{I_1 R}{R_2} = \frac{(\tfrac{9}{11}\,\text{A})(\tfrac{10}{3}\,\Omega)}{10\,\Omega} = \tfrac{3}{11}\,\text{A}$$

$$I_3 = \frac{V}{R_3} = \frac{I_1 R}{R_3} = \frac{(\tfrac{9}{11}\,\text{A})(\tfrac{10}{3}\,\Omega)}{5\,\Omega} = \tfrac{6}{11}\,\text{A}$$

Also, $I_2 + I_3 = I_1$, as required.

(a)

(b)

(c)

32–2 KIRCHHOFF'S RULES

When an electric circuit consists of a number of interconnected loops with several sources of EMF, a simple reduction of the resistances is not sufficient to obtain a solution. For example, consider the circuit shown in Fig. 32–3. Suppose that values are given for both EMFs and for all of the resistances. It is required to determine the currents, I_1, I_2, and I_3. *Kirchhoff's rules** provide a systematic procedure for obtaining a sufficient number of linear equations to solve such problems.

---

*After the German physicist Gustav Kirchhoff (1824–1887).

Fig. 32–3. What are the currents I_1, I_2, and I_3?

Rule 1: The sum of the currents flowing away from any point in a circuit is equal to the sum of the currents flowing toward that point.

Rule 2: In traversing any closed path in a circuit, the algebraic sum of the EMFs is equal to the algebraic sum of the voltage (IR) drops.

The first rule is simply the current continuity condition; in Fig. 32–3 this rule applied at point *b* (or point *e*) means $I_1 + I_3 = I_2$.

The second rule is just the statement of $\oint \mathbf{E} \cdot d\mathbf{r} = 0$. Because the direction of the actual current flow may not be evident at the outset, how do we determine the proper algebraic signs to use in applying the second rule? (Notice that a selected path may be traversed in either of two directions.) To avoid the use of cumbersome notation such as V_{ab} and I_{cb}, we give a third rule:

Rule 3: The algebraic signs of the EMFs and the voltage (IR) drops to be used in applying Rule 2 are specified as follows:

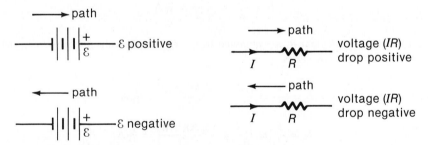

If we traverse the path *abefa* in Fig. 32–3, we find, using Rule 2 and the sign conventions in Rule 3,

$$\mathcal{E}_1 = (r_1 + R_1)I_1 - R_3 I_3$$

On the other hand, if we traverse the path in the opposite direction (that is, *afeba*), we find

$$-\mathcal{E}_1 = R_3 I_3 - (r_1 + R_1)I_1$$

which is just the negative of the first equation.

For the circuit in Fig. 32–3 there are three possible choices for paths, namely, *abcdefa*, *abefa*, and *bcdeb*. Along how many of these paths must the second rule be applied in order to solve the problem? ("Solving the problem" means finding the unknown *branch currents*, I_1, I_2, and I_3.) To attack such problems, first reduce to a minimum the number of unknown currents by repeated use of the first rule. Then, the number of remaining unknown currents is equal to the number of times the second rule must be applied. For the circuit in Fig. 32-3 the first rule gives one relationship, namely, $I_2 = I_1 + I_3$. Consider I_1 and I_3 to be the unknown currents; then, applying the second rule along *any two* of the three possible paths will suffice to solve the problem. (The third path would yield a result that is some linear combination of the first two and hence would give no new information.) It is always the case that the number of independent equations needed is equal to the number of unknowns.

Example 32–2

Determine the currents I_1, I_2, and I_3 in the circuit shown in the diagram (which is the same as that in Example 32–1) by using Kirchhoff's rules.

Solution:

At point a Kirchhoff's first rule gives

$$I_1 = I_2 + I_3 \tag{1}$$

For application of the second rule, select the two paths P (*eacde*) and Q (*abdca*). Then, using the sign conventions in Rule 3, we find

for path P: $\qquad (1\,\Omega + 3\,\Omega)I_1 + (5\,\Omega)I_3 = 6\ \text{V}$ $\qquad$ **(2)**

for path Q: $\qquad (10\,\Omega)I_2 - (5\,\Omega)I_3 = 0$ $\qquad$ **(3)**

Substituting for I_2 in Eq. (3) gives (with the units suppressed)

for path P: $\qquad 4I_1 + 5I_3 = 6$

for path Q: $\qquad 10I_1 - 15I_3 = 0$

Solving these equations yields

$$I_1 = \tfrac{9}{11}\ \text{A} \qquad \text{and} \qquad I_3 = \tfrac{6}{11}\ \text{A}$$

Finally, using Eq. (1), we have

$$I_2 = I_1 - I_3 = \tfrac{3}{11}\ \text{A}$$

These results are, of course, the same as those obtained in Example 32–1.

Example 32–3

Determine the currents, I_1, I_2, and I_3, in the circuit of Fig. 32–3 for the EMF and resistance values shown in the diagram on the next page.

Solution:

At point b, Kirchhoff's first rule gives

$$I_2 = I_1 + I_3 \tag{1}$$

To apply the second rule, we select two paths: path P, *abefa*, and path Q, *bcdeb*. Then, using

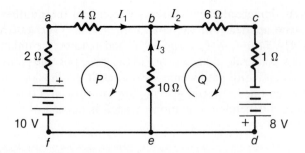

the sign conventions in Rule 3, we have

$$P: \quad (4\,\Omega + 2\,\Omega)I_1 - (10\,\Omega)I_3 = 10 \text{ V} \tag{2}$$

$$Q: \quad (6\,\Omega + 1\,\Omega)I_2 + (10\,\Omega)I_3 = 8 \text{ V} \tag{3}$$

Using Eq. (1) to substitute for I_2 in Eq. (3), we find (suppressing the units)

$$P: \quad 6I_1 - 10I_3 = 10 \tag{2'}$$

$$Q: \quad 7I_1 + 17I_3 = 8 \tag{3'}$$

Solving these two equations, we obtain

$$I_1 = \tfrac{125}{86} \text{ A} \quad \text{and} \quad I_3 = -\tfrac{11}{86} \text{ A}$$

so that
$$I_2 = I_1 + I_3 = \tfrac{57}{43} \text{ A}$$

Notice that the result for I_3 carries a negative sign. This simply means that I_3 actually flows in the direction *opposite* to that indicated in the diagram. Thus, it is not necessary to know beforehand the directions of all the currents in a circuit. Select the directions arbitrarily and use them consistently throughout the calculation. Then, any current for which the direction was chosen incorrectly will emerge from the analysis with a negative sign.

___

Example 32–4

Not all two-terminal resistance networks are reducible to a single equivalent resistor by using only the combination laws discussed in Section 32–1 for parallel and series connections. Kirchhoff's rules can be used to determine the single equivalent resistor for the network illustrated in the diagram.

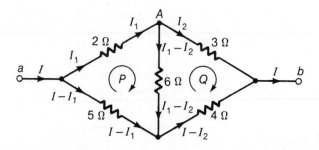

Solution:

First, use Kirchhoff's Rule 1 to reduce to a minimum the number of unknown currents. This is already done in the diagram; we have selected I_1 and I_2 as the unknown currents.

(Because we seek the equivalent resistance of the network, the current I serves merely as a reference current.) Next, choose the loops P and Q for the application of the second rule. We find (suppressing the units)

$$P: \quad 2I_1 + 6(I_1 - I_2) - 5(I - I_1) = 0$$

$$Q: \quad 3I_2 - 4(I - I_2) - 6(I_1 - I_2) = 0$$

These equations can be rewritten as

$$P: \quad 13I_1 - 6I_2 = 5I$$

$$Q: \quad -6I_1 + 13I_2 = 4I$$

Solving for I_1 and I_2, we obtain

$$I_1 = \tfrac{89}{133}I \quad \text{and} \quad I_2 = \tfrac{82}{133}I$$

Now, we have
$$V_{ab} = V_{aA} + V_{Ab}$$
$$= 2I_1 + 3I_2 = \tfrac{424}{133}I$$

Finally, the equivalent resistance is identified as

$$R = \tfrac{424}{133}\,\Omega$$

32-3   ELECTRIC MEASUREMENTS

Ammeters and Voltmeters. Devices that are used to measure electric current and voltage are called *ammeters* and *voltmeters*, respectively. Instruments of this type that use a mechanical movement (with a dial and pointer readout) are all based on a sensitive electromechanical device called a *galvanometer*.* A schematic diagram of a galvanometer meter movement is shown in Fig. 32–4a. At the center, mounted between the poles of a permanent magnet, is a coil that pivots on a jeweled bearing. A current passing through the coil produces a torque† that causes the coil to rotate against the restoring torque due to a spring. The angular deflection of the coil at equilibrium is proportional to the current through the coil. The amount of deflection is indicated by the position of the pointer on the scale (Fig. 32–4a). The sensitivity of

---

*Named for the Italian scientist Luigi Galvani (1737–1798), famous for his theory of animal electricity (galvanism).

†The reason for this torque is discussed in Section 33–4.

Fig. 32–4. (a) Schematic diagram of a galvanometer meter movement. When a current passes through the coil, a torque is produced that causes the coil to rotate until it is in equilibrium with the torque due to the spring. (b) A galvanometer meter movement used as an ammeter. The low-resistance shunt R_s bypasses most of the current I around the meter movement. (c) When the galvanometer is used as a voltmeter, a high resistance R_v is placed in series with the meter movement to limit the current through the meter. (d) Placement of an ammeter A and a voltmeter V in a circuit to measure the current through and the voltage across a resistor R. The circles marked V and A represent the complete subcircuits of (b) and (c), respectively.

(a) scale pointer spring N pole S pole coil of many turns pivot   (b)   (c)   (d) a galvanometer meter movement is typically 10^{-6} to 10^{-2} A for a full-scale deflection. The resistance of the many turns of fine wire used to form the coil is typically a few hundred ohms. Meters of this type must be calibrated by using known currents and marking the scales accordingly.

When a galvanometer is used as an ammeter to measure currents that are larger than the full-scale deflection current of the movement, the connection is made as indicated in Fig. 32–4b. The delicate meter movement is protected by diverting most of the current I (the current to be measured) through the low resistance *shunt* R_s.

Example 32–5

A galvanometer meter movement with a sensitivity of 1 mA for full-scale deflection and with an internal resistance $R_g = 300 \, \Omega$ is to be used in the construction of an ammeter and a voltmeter.

(a) If the movement is to be used in an ammeter that produces a full-scale deflection for a current of 1 A, what must be the value of the shunt resistance R_s?

(b) What series resistance R_v is required to convert the movement into a voltmeter with a full-scale deflection corresponding to 10 V?

(c) An ammeter and a voltmeter with the properties just described are used in the circuit of Fig. 32–4d with $\mathcal{E} = 6$ V and $R = 20 \, \Omega$. What percentage changes in the original current and voltage are introduced by inserting the meters into the circuit?

Solution:

(a) Refer to Fig. 32–4b. When $I = 1$ A, we wish to have $I_g = 10^{-3}$ A, corresponding to full-scale deflection. Thus, $V_g = I_g R_g = I_s R_s$ and $I = I_g + I_s$. From these expressions we find

$$\frac{R_g}{R_s} = \frac{I_s}{I_g} = \frac{I}{I_g} - 1 = 10^3 - 1 \cong 1000$$

so that

$$R_s = \frac{R_g}{10^3} = \frac{300 \, \Omega}{10^3} = 0.30 \, \Omega$$

(b) Refer to Fig. 32–4c. When $V = 10$ V, we again wish to have $I_g = 10^{-3}$ A, corresponding to full-scale deflection. Thus,

$$V = I_g(R_v + R_g)$$

from which

$$R_v = \frac{V}{I_g} - R_g = \frac{10 \text{ V}}{10^{-3} \text{ A}} - 300 \, \Omega = 9700 \, \Omega$$

(c) Refer to Fig. 32-4d. With the meters connected, the total resistance in the circuit can be reduced to an equivalent resistance R', where

$$R' \cong R_s + \frac{R R_v}{R + R_v}$$

Without the meters, the circuit resistance is R. The currents with and without the meters are, respectively,

$$I' = \frac{\mathcal{E}}{R'} \quad \text{and} \quad I_0 = \frac{\mathcal{E}}{R}$$

Then, after some algebra, the fractional change in the current is found to be

$$\frac{\Delta I}{I_0} = \frac{I' - I_0}{I_0} \cong \frac{R}{R_v} - \frac{R_s}{R} = \frac{20\ \Omega}{9700\ \Omega} - \frac{0.30\ \Omega}{20\ \Omega} = -0.013$$

which indicates a decrease in the current of 1.3 percent.

The fractional change in the voltage is found (after some algebra) to be

$$\frac{\Delta V}{\mathcal{E}} = \frac{V - \mathcal{E}}{\mathcal{E}} \cong -\frac{R_s}{R} = -\frac{0.30\ \Omega}{20\ \Omega} = -0.015$$

which indicates a decrease of 1.5 percent in the voltage across R.

The Wheatstone Bridge. Precision measurements of unknown resistances can be made by using the arrangement shown in Fig. 32–5, known as a *Wheatstone bridge.*[*] The fixed resistors, R_2 and R_3, are used in conjunction with the variable resistor R_1 to produce a condition that gives zero current through the galvanometer G when the switch S is closed.

Cutaway view of a helical potentiometer with a 10-turn capability. (From *Principles of Electronic Instrumentation,* 2nd ed., by A. J. Diefenderfer. Saunders College Publishing, 1979.)

Fig. 32–5. A Wheatstone bridge for making a precision determination of the unknown resistance R_x.

When $I_g = 0$, the bridge is *balanced,* and the potential difference V_{ab} between the points a and b is zero. This implies that

$$I_a R_x = I_b R_2$$
$$I_a R_1 = I_b R_3$$

Taking the ratio and solving for R_x, we find

$$R_x = \frac{R_2}{R_3} R_1 \qquad\qquad (32\text{–}7)$$

The variable resistor R_1 is usually either a helical slide-wire or a stepped resistor that allows settings to 4 or 5 figures. If the ratio R_2/R_3 is known, the measurement of R_x can then be made with high precision.

Potentiometer Circuits. All cells and batteries possess an internal resistance that makes the terminal voltage dependent on the current drawn from the cell. The EMF of a cell is equal to the terminal voltage only when there is no current through the cell. The measurement of the true EMF of a cell must therefore be made under

---

[*]Devised in 1843 by an English physicist, Sir Charles Wheatstone (1802–1875).

zero-current conditions. (A cell that serves as an EMF standard—such as a *Weston cell*—can be damaged if a current as small as a few microamperes is drawn for more than a few seconds.)

Figure 32–6 shows a *potentiometer circuit** for the precision comparison of an unknown EMF $\mathcal{E}_x$ with a standard EMF $\mathcal{E}_s$ under zero-current conditions. The cell (or battery) with EMF $\mathcal{E}_w$ is a rugged power source, called the *working cell*, that causes a current I_0 to flow through the potentiometer wire. We are not concerned about the EMF or the internal resistance of the working cell because its only function is to provide a constant current I_0 through R.

Fig. 32–6. A potentiometer circuit for comparing the unknown EMF $\mathcal{E}_x$ with the standard EMF $\mathcal{E}_s$.

To make the measurement, the potentiometer slide wire is adjusted until there is no current through the galvanometer G when the switch S is tapped momentarily closed in position 2. Let this setting of the potentiometer be R_{x2}. The process is repeated for position 1, yielding a value R_{x1}. The two measurements give

$$\mathcal{E}_x = I_0 R_{x1}$$
$$\mathcal{E}_s = I_0 R_{x2}$$

Taking the ratio and solving for $\mathcal{E}_x$, we find

$$\mathcal{E}_x = \frac{R_{x1}}{R_{x2}}\mathcal{E}_s \tag{32-8}$$

---

*The potentiometer itself is simply a long resistor wire with a sliding contact that can be positioned at any point between the ends of the wire. In Fig. 32–6 notice that the potentiometer R may be considered to be a three-terminal network.

SPECIAL TOPIC

32-4   NETWORK THEOREMS

A number of related theorems have been developed to facilitate the analysis of complicated circuits. By way of introduction to the subject we give here a few examples of these alternative methods for analyzing circuits.

Thévenin's Theorem and Norton's Theorem.
Thévenin's theorem states that any network of resistors and EMFs that is connected to a load resistor R_L through a pair of terminals, *a* and *b*, may be replaced by a single equivalent EMF $\mathcal{E}'$ and a series resistance r' (see Fig. 32–7a). The magnitude and polarity of $\mathcal{E}'$ is found by removing the load

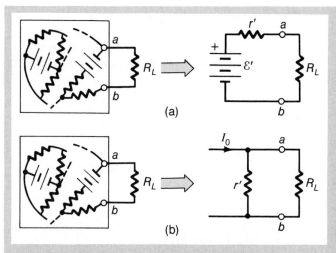

(a)

(b)

Fig. 32-7. (a) Illustrating Thévenin's theorem. (b) Illustrating Norton's theorem for the same circuit.

resistor R_L and measuring the potential difference between the terminals a and b. The series resistance is $r' = \mathcal{E}'/I_0$, where I_0 is the current that flows through a wire without resistance (that is, a "dead short") connected between the terminals a and b.

Norton's theorem states that the network may be replaced by a shunt resistance r' and a generator that provides a constant current I_0 (see Fig. 32-7b). The values of I_0 and r' for a particular network are the same in the two theorems. (•Can you see why?)

These two theorems are based on the fact that to determine the two quantities, $\mathcal{E}'$ and r' or I_0 and r', only *two* linearly independent measurements are needed. It is particularly convenient to make these measurements for infinite load resistance (open circuit between terminals a and b) and zero resistance (dead short between terminals a and b). The behavior of the network for other load resistors is completely characterized by these two measurements and the resulting values of $\mathcal{E}'$ and r' or I_0 and r'.

Maxwell Loop Currents and Superposition. Kirchhoff's rules for circuit analysis (Section 32-2) are expressed in terms of branch currents, in which each such current is the *total* current in the particular branch. Other procedures are also possible. For example, *Maxwell loop currents* are independent closed current loops that thread through a network. These loops may be internal or external to the network or may be combinations. Every branch of the network must have at least one loop current flowing through it; this establishes the number of loop currents required to solve the problem and automatically satisfies Kirchhoff's first rule. Any number of loop currents may flow through a branch; the branch current is then the algebraic sum of all loop currents in that particular branch.

A second technique involves the *superposition principle*. Any particular branch current is obtained by summing algebraically the branch currents produced by each individual EMF taken one at a time.* The following example illustrates the application of these techniques.

Example 32-6

Consider again the circuit discussed in Example 32-3. Determine the current in the 6-Ω resistor by (a) Thévenin's theorem, (b) loop currents, and (c) superposition.

(a)

Solution:

(a) Use Thévenin's theorem to replace the left-hand battery, its internal resistance, and the 4-Ω and 10-Ω resistors between terminals b and e (diagram (b)). The potential difference between b and e is

$$\mathcal{E}' = \left(\frac{10 \text{ V}}{2\,\Omega + 4\,\Omega + 10\,\Omega} \right) \times (10\,\Omega) = \tfrac{25}{4} \text{ V}$$

---

*All of the internal resistances associated with the sources of EMF remain in the circuit; only the EMFs themselves are considered one at a time.

(b)

With a dead short between b and e, the current is

$$I_0 = \frac{10 \text{ V}}{2\,\Omega + 4\,\Omega} = \tfrac{5}{3} \text{ A}$$

so that $\quad r' = \dfrac{\mathcal{E}'}{I_0} = \dfrac{25/4 \text{ V}}{5/3 \text{ A}} = \tfrac{15}{4}\,\Omega$

and the network becomes that shown in diagram (c).

(c)

Thus, we see that

$$I = \frac{8 \text{ V} + \frac{25}{4} \text{ V}}{\frac{15}{4}\,\Omega + 6\,\Omega + 1\,\Omega} = \tfrac{57}{43} \text{ A}$$

which is just the value obtained for I_2 in Example 32-3.

(d)

(b) To use the Maxwell loop current method, select internal loop currents that correspond to the voltage loops, P and Q, that were used in Example 32–2 (diagram (d)). Now apply Kirchhoff's second rule to the paths P and Q. We find

$$(2\,\Omega + 4\,\Omega + 10\,\Omega)I_1 - (10\,\Omega)I_2 = 10 \text{ V}$$
$$-(10\,\Omega)I_1 + (10\,\Omega + 6\,\Omega + 1\,\Omega)I_2 = 8 \text{ V}$$

Realizing that in Example 32–3 we have $I_2 = I_1 + I_3$, these equations are exactly the same as Eqs. (2) and (3) in that example. Hence, we find the same value for I_2, namely, $I_2 = 57/43$ A.

(c) To use the superposition method, first remove the 8-V EMF from the circuit (but leave the 1-Ω internal resistance), as indicated in diagram (e).

(e)

We now have a simple series-parallel circuit in which the equivalent resistance of the combination of the 10-Ω, the 6-Ω, and the 1-Ω resistors on the right is 70/17 Ω. Then, we have

$$I_2' = \left(\frac{10\,\Omega}{17\,\Omega}\right) \times \left(\frac{10 \text{ V}}{2\,\Omega + 4\,\Omega + 70/17\,\Omega}\right) = \tfrac{25}{43} \text{ A}$$

Next, remove the 10-V EMF from the original circuit (but leave the 2-Ω internal resistance), as indicated in diagram (f).

(f)

The equivalent resistance of the 2-Ω, the 4-Ω, and the 10-Ω resistors on the left is 60/16 Ω. Then, we have

$$I_2'' = \frac{8\text{ V}}{6\,\Omega + 1\,\Omega + 60/16\,\Omega} = \tfrac{32}{43}\text{ A}$$

Finally, we have

$$I_2 = I_2' + I_2'' = \tfrac{25}{43}\text{ A} + \tfrac{32}{43}\text{ A} = \tfrac{57}{43}\text{ A}$$

so that all of the methods yield the same result for I_2, as expected.

Transformations of Delta-Star Resistance Networks.

We now examine an interesting transformation property for a pair of *three-terminal* resistance networks. When three resistors are connected to the three terminals as in Fig. 32–8a, the arrangement is called a *delta* (sometimes a π) network. If the resistors are connected as in Fig. 32–8b, the arrangement is called a *star* (sometimes a T or a Y) network.

(a)   (b)

Fig. 32–8. (a) A *delta* network of resistors. (b) A *star* network of resistors.

The delta-star transformation theorem states that either of these three-terminal networks can be substituted for the other in a circuit (terminal for terminal*) without affecting the currents in the external circuit. The transformation equations are (see Problem 32–24)

$$\left.\begin{aligned}R_1 &= R_{12}R_{31}/G \\ R_2 &= R_{12}R_{23}/G \\ R_3 &= R_{23}R_{31}/G\end{aligned}\right\} \quad\text{and}\quad \left.\begin{aligned}R_{12} &= H/R_3 \\ R_{23} &= H/R_1 \\ R_{31} &= H/R_2\end{aligned}\right\} \quad \textbf{(32–9a)}$$

where

$$\left.\begin{aligned}G &= R_{12} + R_{23} + R_{31} \\ H &= R_1 + R_2 + R_3\end{aligned}\right\} \quad \textbf{(32–9b)}$$

*Connection to the center point of the star network is excluded.

Example 32–7

Use the delta-star transformation equations to determine the equivalent resistor for the network discussed in Example 32-4.

Solution:

In the first part of the diagram below, the right-hand section of the network is identified as a delta connection and the three terminals are labeled.

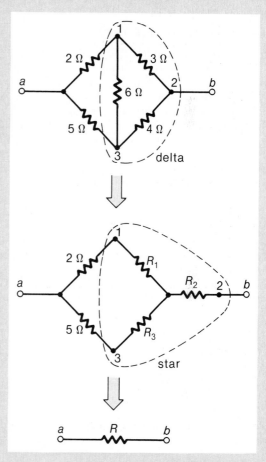

With the notation of Fig. 32–8 and Eqs. 32–9, we have

$$R_{12} = 3\,\Omega, \quad R_{23} = 4\,\Omega, \quad R_{31} = 6\,\Omega, \quad G = 13\,\Omega$$

Then, the resistances in the equivalent star network are

$$R_1 = R_{12}R_{31}/G = (3\,\Omega)(6\,\Omega)/(13\,\Omega) = \tfrac{18}{13}\,\Omega$$

$$R_2 = R_{12}R_{23}/G = (3\,\Omega)(4\,\Omega)/(13\,\Omega) = \tfrac{12}{13}\,\Omega$$

$$R_3 = R_{23}R_{31}/G = (4\,\Omega)(6\,\Omega)/(13\,\Omega) = \tfrac{24}{13}\,\Omega$$

The second part of the diagram above shows that the series combination $2\,\Omega + R_1$ is in parallel with the series combination $5\,\Omega + R_3$, and that this parallel network is in series with R_2. Thus, for the equivalent resistor R, we have (suppressing the units)

$$R = R_2 + \frac{(2 + R_1)(5 + R_3)}{2 + 5 + R_1 + R_3}$$

$$= \frac{12}{13} + \frac{(2 + \frac{18}{13})(5 + \frac{24}{13})}{2 + 5 + \frac{18}{13} + \frac{24}{13}} = \frac{424}{133}\,\Omega$$

which is just the result we obtained in Example 32–4.

32–5   THE SOLID-STATE DIODE

Electric power is usually generated and delivered to the user in the form of alternating current (AC). Because many applications require direct current (DC), it is necessary to have a simple means of converting (*rectifying*) AC to DC. This can be accomplished with a device, called a *diode*, that freely allows the passage of current in one direction but offers a high resistance to current flow in the opposite direction. We now describe briefly the operation of the most common type of diode in present use, a solid-state device composed of *p*-type and *n*-type semiconductor materials in conjunction.

Consider a piece of intrinsic semiconductor material (such as silicon or germanium) whose opposite sides have been converted, by appropriate doping, into *p*-type and *n*-type semiconductors. The interface boundary region is

Fig. 32–9. Electron energy band diagram for an unbiased *p-n* junction. The width of the junction region in commercial diodes is typically a few tens of nm.

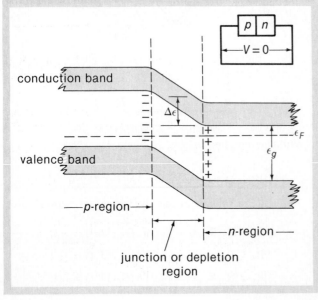

called a *p-n junction*. Suppose that the two sides of this diode are connected by using an external conductor, as indicated in the inset in Fig. 32–9. There is no potential difference between the *p*-side and the *n*-side; in this condition the diode is said to be *unbiased*. At equilibrium the Fermi energy ϵ_F is the same throughout the diode, and the electron energy bands take up the relationship shown in Fig. 32–9. An energy barrier $\Delta\epsilon$ is presented to any electron that attempts to move from the *n*-region to the *p*-region. (Notice the similarity of this situation to that of the contact potential discussed in Section 31–6.)

Let us consider how equilibrium conditions are established and maintained in a *p-n* junction. In the *n*-region there are many free electrons in the conduction band; in this region the electrons are the *majority charge carriers*. The flow of electrons from the *n*-region to the *p*-region is enhanced by the large density of conduction electrons but is diminished by the existence of the energy barrier $\Delta\epsilon$. Thermal effects impart to some of the electrons sufficient energy to overcome the barrier, and the net result is a weak electron flow from the *n*-region to the *p*-region. In the *p*-region the majority charge carriers are the *holes*, and the small number of electrons that exist in the conduction band are the *minority charge carriers*. Although the energy condition at the junction is favorable, there is only a weak flow of electrons from the *p*-region to the *n*-region because the electron density is so small. This flow rate is maintained by thermal effects that continually replenish the electrons in the *p*-region. At equilibrium the two oppositely directed electron currents just balance one another. A similar argument applies for the positive charge carriers (the holes).

In order to establish the equilibrium conditions, there was initially a flow of electrons from the *n*-region to the *p*-region and a flow of holes in the opposite direction. These effects deplete the junction region (or *depletion region*) of both types of charge carriers and create an excess electron concentration on the *p*-side of the junction and a corresponding excess hole concentration on the *n*-side. The junction region then possesses the relatively high resistivity of the intrinsic semiconductor material. The layers of excess charge on the boundaries of the junction region constitute a kind of capacitor and may be viewed as the source of the energy barrier $\Delta\epsilon$ (see Fig. 32–9).

How is the electron flow in a diode affected by the application of a voltage to the device? Figure 32–10 illustrates one way to apply an external voltage (called a *bias voltage*) to a solid-state diode. When the *p*-side is given a positive potential with respect to the *n*-side, the barrier energy $\Delta\epsilon$ is decreased. Then, because the density of electrons in the *n*-region is large, the flow rate of electrons into the *p*-region increases by a very large factor. This $n \rightarrow p$ flow of electrons completely overwhelms the thermally excited and already saturated $p \rightarrow n$ flow. Thus, the conventional current I (which flows from the *p*-region to the *n*-region) increases rapidly as the *forward bias voltage* is increased (see Fig. 32–10).

When the *p*-side of the diode is given a negative potential with respect to the *n*-side (this is the case of *reverse bias*), the increased barrier height essentially cuts off the $n \rightarrow p$ electron flow, but the $p \rightarrow n$ flow, which is already saturated, does not increase to any appreciable extent. The result is a very small $n \rightarrow p$ conventional current.

The behavior of the current I as a function of the bias voltage V for a solid-state diode is shown in Fig. 32–10. For an absolute temperature T, the current I is given by the expression

$$I = I_0[e^{eV/kT} - 1] \qquad (32\text{–}10)$$

where I_0 is typically about 10^{-5} A. If the reverse bias is increased to 15 V or so, electrons move across the junction region so rapidly that they can dislodge nonconduction

Fig. 32–10. The current through a solid-state diode as a function of the bias voltage V. The circuit shown allows the *p*-side of the diode to be given a positive or a negative potential with respect to the *n*-side.

electrons, which are then also accelerated across the junction. This cascading contribution to increased electron flow is called *avalanche breakdown* (see Fig. 32–10).

PROBLEMS

Section 32–1

32–1 Three resistors (10 Ω, 20 Ω, and 30 Ω) are connected in parallel. The total current through this network is 5 A. (a) What is the voltage drop across the network? (b) What is the current through each resistor?

32–2 What additional resistance must be placed in parallel with a parallel combination of 75-Ω and 50-Ω resistors to yield a resistance of 25 Ω for the entire network?

32–3 Determine the equivalent resistance between the terminals *a* and *b* for the network illustrated at left.

32–4 Three 100-Ω resistors are connected as shown in the diagram. The maximum power that can be dissipated in any one of the resistors is 25 W. (a) What is

Problem 32–3

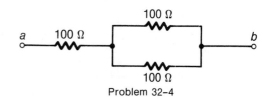

Problem 32–4

the maximum voltage that can be applied to the terminals a and b? (b) For the voltage determined in (a), what is the power dissipation in each resistor? What is the total power dissipation?

Section 32-2

32-5 Determine the current in each of the branches of the circuit shown in the diagram.

32-6 The two branch currents in the circuit shown in the diagram are $I_1 = \frac{1}{3}$ A and $I_2 = \frac{1}{2}$ A. Determine the EMFs, $\mathcal{E}_1$ and $\mathcal{E}_2$.

32-7 What is the EMF $\mathcal{E}_1$ of the battery in the circuit shown in the diagram?

$I_1 = 2$ A

Section 32-3

32-8 A galvanometer G with an internal resistance, $R_g = 50\,\Omega$, and a sensitivity of 1 mA for full-scale deflection is to be used as an ammeter with three different ranges. The currents for full-scale deflection are indicated in the diagram. Determine the required shunt resistors, R_1, R_2, and R_3, to an accuracy of 1 percent.

32-9 A galvanometer G with an internal resistance, $R_g = 50\,\Omega$, and a sensitivity of 1 mA for full-scale deflection is to be used as a voltmeter with three different ranges. The voltages corresponding to full-scale deflection are indicated in the diagram. Determine the required series resistors, R_1, R_2, and R_3, to an accuracy of 1 percent.

32-10 A galvanometer G with an internal resistance, $R_g = 50\,\Omega$, and a sensitivity of 1 mA for full-scale deflection is to be used as an ohmmeter (that is, to determine resistance values). The instrument incorporates an internal battery and a variable series resistor R_0, as shown in the diagram. Suppose that $\mathcal{E} = 3.0$ V and $r = 1\,\Omega$. R_0 is adjusted to give full-scale deflection when $R_x = 0$ (that is, when the terminals a and b are shorted together). Thus, full-scale deflection of the ohmmeter corresponds to $0\,\Omega$ and no deflection corresponds to $\infty\,\Omega$ (open terminals). (a) What is the required resistance R_0?

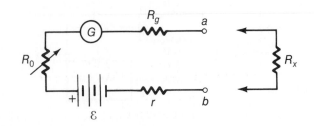

(b) What values of R_x correspond to deflections of 1/4, 1/2, and 3/4 of full scale? (c) What value of R_0 will be required when the battery voltage drops to 2.8 V?

32-11 Show that if the positions of the battery and the galvanometer in a Wheatstone bridge (Fig. 32-5) are interchanged, the same balance condition (Eq. 32-7) results.

SPECIAL TOPIC

Section 32-4

32-12 Determine the Thévenin and Norton circuit equivalents for the two circuits shown in the diagrams.

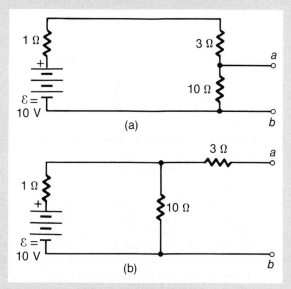

32-14 Use a delta-star transformation to find the equivalent resistance between the terminals a and b for the network shown in the diagram.

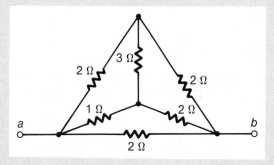

32-15 Determine the current I in the circuit shown in the
•   diagram, where each resistor has a resistance of 3 Ω.

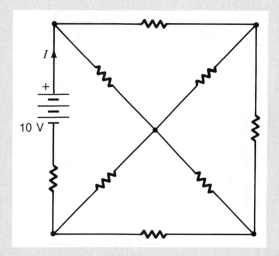

32-13 Use the superposition technique to determine the current in the 4-Ω resistor in the circuit shown in the diagram.

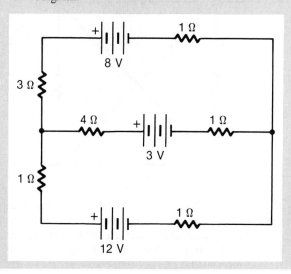

Section 32-5

32-16 A solid-state diode has $I_0 = 1.1 \times 10^{-5}$ A. Calculate the diode current I at room temperature for bias voltages of -0.5 V, -0.2 V, -0.1 V, 0 V, $+0.01$ V, 0.02 V, 0.05 V, 0.075 V, and 0.10 V. Make a graph of your results and compare with Fig. 32-10.

32-17 A solid-state diode is operated with a forward cur-

rent of 10 mA. What is the temperature coefficient, dI/dT, for this diode at room temperature? (Assume $I_0 \ll I$.)

32-18 A solid-state diode has $I_0 = 1.0 \times 10^{-5}$ A. Determine the differential resistance, dV/dI, for a forward bias of 0.10 V and for a reverse bias of -0.10 V. Assume room temperature.

Additional Problems

32-19 What must be the value of R_3 to make the equivalent resistance between the terminals a and b equal to R_1?

32-20 Determine the potential difference, $V_{ab} = \Phi_a - \Phi_b$, for the circuit shown in the diagram.

32-21 Show that the resistance between the terminals a and
•• b (the *input* resistance) for the infinite array of identical resistors shown in the diagram above right is $R = (1 + \sqrt{3})r$. [*Hint:* The resistor chain is infinite, so removing the first three resistors and moving the terminals to the dashed line does not alter the input resistance.]

Problem 32–21

32-22 The network shown in the diagram consists of eight
•• 1-Ω resistors. Determine the equivalent resistance between the terminals a and b. [*Hint:* If the symmetry of a network is such that two points are perceived to be at the same potential, these points may be connected by a resistanceless wire without altering the current distribution anywhere in the network.]

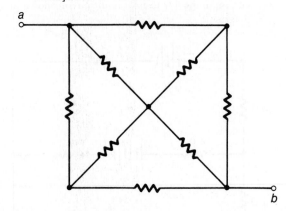

32-23 Twelve 1-Ω resistors are connected to form a cube,
•• as shown in the diagram on the next page. Show that the equivalent resistance between the terminals a and b is $\frac{5}{6}\Omega$. [*Hint:* Refer to the hint given in Problem 32–22.]

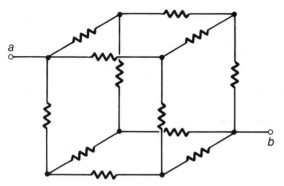

Problem 32–23

SPECIAL TOPIC

32-24 Derive the delta-star transformation equations
•• (Eqs. 32–9). [*Hint:* If a delta network is equivalent
to a star network, the resistance between any pair of
terminals in the delta network, with the third terminal unconnected, must be the same as the resistance
between the same pair of terminals in the star network, also with the third terminal unconnected.
This provides three equations which can be solved
for R_1, R_2, and R_3; these results can, in turn, be
inverted to obtain R_{12}, R_{23}, and R_{31}.]

32-25 Determine the value of R in the circuit shown in the
diagram such that maximum power is delivered to

R. [*Hint:* The easiest way to approach this problem
is to use Thévenin's theorem.]

32-26 Determine the current that flows through each of
the batteries in the circuit shown in the diagram.
Use any of the methods discussed in the text.

SPECIAL TOPIC

32-27 A solid-state diode with $I_0 = 1.0 \times 10^{-5}$ A is con-
•• nected at room temperature in the circuit shown in
the diagram. What current I flows in the circuit?
[*Hint:* Using $V = \mathcal{E} - IR$, obtain an expression of
the form $I = \alpha \ln(\beta/I)$. Assume a reasonable value
I_1 for I on the right-hand side and solve for
$I_2 = \alpha \ln(\beta/I_1)$. Substitute I_2 for I_1 and find
$I_3 = \alpha \ln(\beta/I_2)$. Continue this iterative procedure
until three-place accuracy is obtained.]

CHAPTER 33

THE MAGNETIC FIELD

In the preceding chapters electric fields were discussed for the simplifying time-independent (or electrostatic) condition. We now inquire into the behavior of magnetic fields by again considering the time-independent condition. For the magnetic case this condition is achieved in situations involving *steady currents*. In this chapter, we begin the discussion of the connection between electric and magnetic effects—this is the subject of *electromagnetism*.

33-1 EARLY HISTORY

In the 5th century B.C. the Greeks were aware that certain natural rocks had the curious property that they attracted each other and also attracted bits of iron. Because these rocks were plentiful in the district called Magnesia, in the western part of what is now Turkey, they became known as *magnets* (and the active mineral in the rocks is now called *magnetite,* Fe_3O_4).

As early as 1100 A.D. the Chinese seem to have constructed a primitive but functional magnetic compass. By the late 12th century European navigators were using a compass consisting of a long piece of magnetite floated on a splinter of wood in a basin of water. (Navigators called these rocks *lodestones*—"leading stones.")

The earliest scientific inquiry into the nature of magnetism was carried out in the 13th century by a French soldier-scientist, Pierre de Maricourt, commonly called Peregrinus. He fashioned a spherical lodestone and sketched on its surface a set of lines indicating the directions taken up by an iron needle placed at various points. The lines encircled the lodestone as do the meridian lines on an Earth globe, with intersections at two points. These points, which Peregrinus called the *poles* of the lodestone, are the points with the strongest power to attract bits of iron. He demonstrated that each of the fragments of a divided magnet has two poles, and that like magnetic poles repel whereas unlike magnetic poles attract one another. In this regard, magnetic poles are similar to electric charges; but, in sharp contrast with the electric case, magnetic poles always occur in closely associated pairs—isolated magnetic poles *do not* exist.*

More than 300 years elapsed before the next significant contribution to the understanding of magnetism was made by William Gilbert, whom we met in Section

---

*There have been theoretical conjectures that isolated magnetic poles (magnetic monopoles) should exist, but all attempts to observe these monopoles have been unsuccessful.

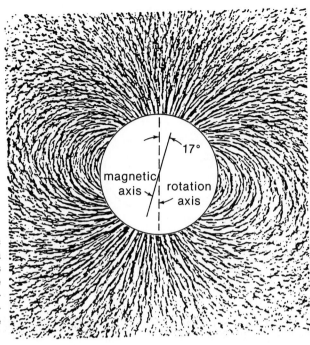

Fig. 33-1. The Earth's magnetic field closely resembles that of a magnetized sphere. The Earth's magnetic axis is inclined to the rotation axis at an angle of approximately 17°. Thus, the geomagnetic poles do not coincide with the geographic poles.

27-1 in connection with the discussion of electric charge. In his famous book *De Magnete* ("On the Loadstone and Magnetic Bodies, and on the Great Magnet, the Earth," published in 1600), Gilbert showed that the Earth has two magnetic poles and acts as a giant magnet. Figure 33-1 shows a simulation of the Earth's magnetism (its *magnetic field*) produced by a magnetized sphere. Iron filings, sprinkled on a flat piece of cardboard, are induced to become tiny compass magnets that align themselves with the magnetic field. Notice the striking similarity of this field to that of an electric dipole (compare Fig. 28-5). Notice also that a compass magnet that is free to rotate about a horizontal axis will take up an equilibrium position inclined to the local vertical. Measurements of this *dip angle* around the Earth are in agreement with those expected from the dipole model shown in Fig. 33-1.

The end of a compass magnet that points toward the geomagnetic north pole is designated the N pole of the magnet; the other end is the S pole. Because opposite magnetic poles attract, the Earth's geomagnetic north pole is actually an S pole. The Earth's magnetic axis is inclined with respect to its rotation axis; thus, the geographic and geomagnetic poles do not coincide (Fig. 33-1). At present, the geomagnetic north pole lies in the vicinity of Bathurst Island, Northwest Territories, Canada (approximately 73°N and 100°W), about 1900 km from the geographic north pole.* Because the two poles do not coincide, a compass will generally point slightly east or west of true north. The difference between the directions of geographic north and geomagnetic north is called the *declination* of the particular location. Figure 33-2 illustrates an easterly declination θ_E and a westerly declination θ_W.

The *direction* of magnetic field lines is taken to be the direction in which the N

Fig. 33-2. Because the Earth's magnetic north pole does not coincide with the geographic north pole, compass needles at the positions shown here exhibit an easterly declination θ_E and a westerly declination θ_W. The declination at a particular point varies slowly with time (by as much as 0.1° per year).

geographic (true) north

magnetic north

θ_E

θ_W (S pole)

meridians

*The exact locations of the geomagnetic poles are difficult to determine. Moreover, the positions of the poles wander slowly with time. The first determination of the position of the geomagnetic north pole was made in 1835 by Gauss, who used dip-angle and declination measurements from 84 stations around the world to place the pole at 78°N and 69°W.

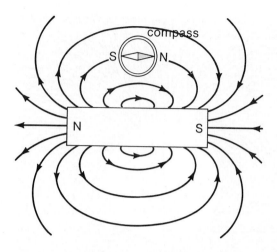

Fig. 33-3. Magnetic field lines for a simple bar magnet.

pole of a compass magnet points. Thus, in the region around a bar magnet the field lines run from the N pole to the S pole, as shown in Fig. 33-3.

For long slender magnets, the N and S poles can be considered effectively isolated from each other. Then, measurements can be made of the forces between individual like or unlike poles. Such an experiment was first carried out by the Rev. John Michell (1724–1793), an English geologist. Michell used a torsion balance of his own invention and discovered that the magnetic force between "isolated" magnetic poles follows an inverse square law with distance, just as in the case of electric charge.

A crucial discovery linking electricity and magnetism was made in 1819 by Hans Christian Oersted (1777–1851), a Danish physicist. Oersted found that a compass magnet is sensitive to the effect of current flowing in a nearby wire. That is, a

Fig. 33-4. The magnetic field lines in the vicinity of a current-carrying wire are circles concentric with the wire axis. (From *PSSC Physics,* 2nd ed., 1965: D. C. Heath & Co. and Educational Development Center, Newton, MA)

current-carrying wire is surrounded by a magnetic field. A current of a few amperes produces, within a few centimeters of the wire, a magnetic field with a strength comparable with that of the Earth's field. The iron-filing technique can be used to visualize the field in the vicinity of a current-carrying wire. Figure 33–4 shows that the field lines are circles concentric with the wire axis. Figure 33–5 shows the direction of these field lines as indicated by a compass magnet. Also illustrated is the right-hand rule for determining this direction: *Grasp the wire with the right hand, with the extended thumb in the direction of the current in the wire; then, the fingers encircle the wire in the same direction as the field lines.*

Soon after the connection between electric current and magnetism had been established by Oersted's discovery, other physicists rapidly expanded the investigation and added new details. By 1825 the mathematical theory of electromagnetism had been developed to such a degree by André-Marie Ampère (1775–1836) that his monograph on the subject is one of the most important in the history of physical science. Later, the experiments of Michael Faraday (1791–1867) and Joseph Henry (1797–1878), particularly relating to time-dependent effects, provided the foundation on which James Clerk Maxwell (1831–1879) was finally able to integrate the various discoveries into a universal theory of electrodynamics. We pursue these developments in Chapter 35.

Fig. 33–5. The direction of the field lines encircling a current-carrying wire is determined by a right-hand rule.

33-2 THE MAGNETIC INDUCTION VECTOR

The electric field strength vector **E** has been defined in terms of the force exerted on a positive test charge q_0 placed in the field (see Eq. 28–1). We now give a parallel definition of the corresponding vector for the magnetic field, namely, the *magnetic induction vector* **B**.

We select a region of space in which there is known to be a static magnetic field. Then, we place a test charge q_0 *at rest* at various points in the region and discover that the particle experiences no force at any point.* Unlike gravitational and electric fields, a static magnetic field exerts no force on a stationary charged particle. Next, we observe the test charge as it is moved with an arbitrary velocity **v** through various points in the region. We discover the following:

(a) The particle experiences a force $\mathbf{F}_M$ that is always perpendicular to **v** regardless of the instantaneous direction of the velocity.

(b) The magnitude of the force is proportional to each of the factors q_0, v, and $\sin \theta$, where θ is the angle between **v** and the direction of the field line at the position of the particle.

That is, the behavior of the force on the test charge is described by an expression of the form

$$\mathbf{F}_M = q_0 \mathbf{v} \times \mathbf{B} \qquad (33\text{–}1)$$

where $\mathbf{B} = \mathbf{B}(\mathbf{r})$ is a vector whose direction is the same as that of the magnetic field line that passes through the point **r**. We view Eq. 33–1 as the definition of $\mathbf{B}(\mathbf{r})$, the *magnetic induction* at the point **r**. The relative orientation of **v**, **B**, and $\mathbf{F}_M$ for a positive test charge is illustrated in Fig. 33–6.

In the SI system, the unit of magnetic induction is the tesla (T), where

Fig. 33–6. A positive test charge q_0 with velocity **v** experiences a force $\mathbf{F}_M$ due to the presence of the magnetic field **B**.

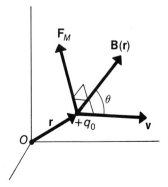

---

*What additional measurements would be necessary in order to conclude that the region is free of gravitational and electric fields?

$$1\,\mathrm{T} = 1\,\mathrm{N}\cdot\mathrm{s/C}\cdot\mathrm{m} = 1\,\mathrm{N/A}\cdot\mathrm{m}$$

In Section 35-2, we introduce the concept of magnetic flux and learn that **B** is also a measure of *magnetic flux density* (indeed, this is an alternate name for **B**). The SI unit for magnetic flux is the *weber* (Wb), and the unit for flux density (flux lines per unit area) is

$$1\,\mathrm{Wb/m^2} = 1\,\mathrm{T}$$

A unit for magnetic induction used in an earlier version of the metric system and still used extensively is the *gauss* (G):

$$1\,\mathrm{T} = 10^4\,\mathrm{G}$$

Geophysicists and planetary physicists frequently specify magnetic fields in *gammas* (γ), where

$$1\,\gamma = 10^{-5}\,\mathrm{G} = 10^{-9}\,\mathrm{T}$$

The magnitude of the Earth's magnetic field varies from about 30,000 γ at the equator to about 60,000 γ at the poles. The time variation of the Earth's field at a particular location can be as large as 150 γ per year. The magnitude of the Sun's magnetic field at the Earth's orbit is about 3 γ.

The Lorentz Force.   The force on a charged particle q in an electric field is (Eq. 28-4) $\mathbf{F} = q\mathbf{E}$. In the event that both electric and magnetic fields are present, the total force on the particle is the vector sum of the forces due to the electric and magnetic effects. This net electromagnetic force, or *Lorentz force,* is

$$\mathbf{F} = q(\mathbf{E} + \mathbf{v} \times \mathbf{B}) \tag{33-2}$$

The rate at which the combined electric and magnetic fields (each assumed to be time-independent) do work on a moving charged particle is $P = dW/dt = \mathbf{F}\cdot\mathbf{v}$, or

$$P = q(\mathbf{E} + \mathbf{v} \times \mathbf{B})\cdot\mathbf{v}$$

Now, $\mathbf{v} \times \mathbf{B}$ is a vector that is perpendicular to both **v** and **B**; consequently, $(\mathbf{v} \times \mathbf{B})\cdot\mathbf{v}$ is identically zero. That is, a time-independent magnetic field (of any shape) *can do no work* on a moving charged particle. The work done on the particle in such a field is due to the electric force alone.

Example 33-1

The Earth's magnetic field at Washington, D.C., has a magnitude of 56,900 γ with a dip angle of 71° and a declination of 6° W.

(a) Describe this field in local rectangular coordinates that correspond to geographic directions.

(b) Determine the force on an electron that moves horizontally and due south in this field with a velocity of 10^7 m/s.

(c) What is the acceleration of the electron in units of g?

(d) What is the kinetic energy of the electron and how does it change?

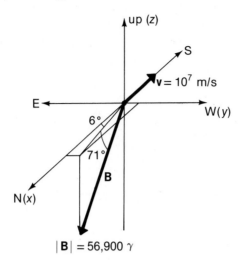

$|\mathbf{B}| = 56{,}900 \; \gamma$

Solution:

(a) The diagram shows the field vector $\mathbf{B}$ in local rectangular coordinates with geographic north along the x-axis, west along the y-axis, and upward along the z-axis. Then,

$$B_x = (56{,}900 \; \gamma) \cos(-71°) \cos 6° = 18{,}400 \; \gamma$$
$$B_y = (56{,}900 \; \gamma) \cos(-71°) \sin 6° = 1{,}940 \; \gamma$$
$$B_z = (56{,}900 \; \gamma) \sin(-71°) \qquad = -53{,}800 \; \gamma$$

(b) The charge of the electron is $q = -e = -1.60 \times 10^{-19}$ C and its velocity is $\mathbf{v} = -(10^7 \hat{\mathbf{i}})$ m/s. Thus, the force on the electron is $\mathbf{F} = -e\mathbf{v} \times \mathbf{B}$, or

$$\mathbf{F} = (-1.60 \times 10^{-19} \; \text{C})(10^7 \; \text{m/s}) \times \begin{vmatrix} \hat{\mathbf{i}} & \hat{\mathbf{j}} & \hat{\mathbf{k}} \\ -1 & 0 & 0 \\ 18.4 & 1.94 & -53.8 \end{vmatrix} \times (10^{-6} \; \text{T})$$

where the components of $\mathbf{B}$ have been expressed in tesla. Then, we have

$$\mathbf{F} = (0\hat{\mathbf{i}} + 86.1\hat{\mathbf{j}} + 3.10\hat{\mathbf{k}}) \times 10^{-18} \; \text{N}$$

(c) The mass of the electron is $m_e = 9.11 \times 10^{-31}$ kg, so its acceleration is (in units of $g = 9.80$ m/s²)

$$\mathbf{a} = \frac{\mathbf{F}}{m_e} = (0\hat{\mathbf{i}} + 96.4\hat{\mathbf{j}} + 3.47\hat{\mathbf{k}}) \times 10^{11} g$$

(d) We have $\beta = v/c = (10^7 \; \text{m/s})/(3 \times 10^8 \; \text{m/s}) = 0.033$, so the electron is "nonrelativistic." Hence, the kinetic energy is

$$K = \tfrac{1}{2}m_e v^2 = \frac{(9.11 \times 10^{-31} \; \text{kg})(10^7 \; \text{m/s})^2}{2(1.60 \times 10^{-19} \; \text{J/eV})} = 285 \; \text{eV}$$

Because no electric field is present and the magnetic field can do no work on the electron, the electron moves with this constant value of the kinetic energy.

Example 33-2

An electron is traveling through the Sun's chromosphere with a speed of $1/10$ the speed of light and is in a locally uniform magnetic field of 0.30 T produced by a nearby sunspot. The velocity vector of the electron makes an angle of $30°$ with respect to the direction of the magnetic field, as shown in diagram (a). Describe the motion of the electron.

(a)

Solution:

When the instantaneous velocity of the electron is $\mathbf{v}$, the acceleration is

$$\mathbf{a} = \frac{\mathbf{F}}{m_e} = -\frac{e}{m_e}\mathbf{v} \times \mathbf{B} = -\frac{e}{m_e}(v_x\hat{\mathbf{i}} + v_y\hat{\mathbf{j}} + v_z\hat{\mathbf{k}}) \times (B\hat{\mathbf{k}}) = -\frac{e}{m_e}B(v_y\hat{\mathbf{i}} - v_x\hat{\mathbf{j}})$$

The acceleration has no component in the z-direction, so the electron moves with a constant velocity component v_{0z} in this direction.

The component of the electron velocity that is parallel to the x-y plane is $\mathbf{v}_{xy} = v_x\hat{\mathbf{i}} + v_y\hat{\mathbf{j}}$, which is seen to be perpendicular to the acceleration $\mathbf{a}$. Moreover, the field does no work on the electron, so its kinetic energy remains constant; hence, the value of v^2 at any instant is $v_x^2 + v_y^2 + v_z^2 = v_0^2 = $ constant. Now, $v_z = v_{0z}$; thus, at all times, we have

$$v_{xy}^2 = v_x^2 + v_y^2 = v^2 - v_z^2 = v_0^2 - v_{0z}^2 = \text{constant}$$

Thus, we see that $\mathbf{v}_{xy}$ is perpendicular to $\mathbf{a}$ and has a constant magnitude. This means that the electron moves in a path whose projection onto the x-y plane is a circle. The motion also progresses with constant velocity in the z-direction, so the path is a uniform circular helix with radius R, as shown in diagram (b).

(b)

Because $\beta = v/c = 1/10$, the electron can be treated as a nonrelativistic particle. The centripetal acceleration experienced by the electron is constant and is easily evaluated at the instant when $v_x = 0$, as in diagram (a); we find

$$|\mathbf{a}_c| = \frac{v_{0y}^2}{R} = \frac{e}{m_e} B v_{0y}$$

Solving for the helix radius R, we find

$$R = \frac{m_e v_{0y}}{eB} = \frac{(9.11 \times 10^{-31} \text{ kg})(\frac{1}{10} \times 3.0 \times 10^8 \text{ m/s})\sin 30°}{(1.60 \times 10^{-19} \text{ C})(0.30 \text{ T})}$$

$$= 2.85 \times 10^{-4} \text{ m}$$

It is instructive to calculate the period T of the motion (the time required for one turn of the helix). Thus,

$$T = \frac{2\pi R}{v_{0y}} = \frac{2\pi m_e}{eB}$$

which is seen to be independent of the velocity v_0.* The value of T in this case is

$$T = \frac{2\pi(9.11 \times 10^{-31} \text{ kg})}{(1.60 \times 10^{-19} \text{ C})(0.30 \text{ T})} = 1.19 \times 10^{-10} \text{ s}$$

The time required for the electron to travel a distance $s = 1$ m in the z-direction is

$$t = \frac{s}{v_0 \cos 30°} = \frac{1 \text{ m}}{(\frac{1}{10} \times 3.0 \times 10^8 \text{ m/s})\cos 30°} = 3.85 \times 10^{-8} \text{ s}$$

Therefore, for each meter of travel in the z-direction, the electron executes $(3.85 \times 10^{-8} \text{ s})/(1.19 \times 10^{-10} \text{ s}) = 323$ helical turns.

---

*The corresponding angular frequency, $\omega_c = 2\pi/T = eB/m$, is called the *cyclotron frequency* because this is the frequency with which a particle revolves while being accelerated in a cyclotron.

The giant solar prominence of June 4, 1946. (Courtesy of the Harvard College Observatory.)

If the field is truly uniform, the electron will spiral about a particular field line. Because the helix radius here is so small, this is still the case even if the field lines are curved. Then, the electron will follow a particular field line as it curves through space. This effect is clearly evident in the photograph on page 901, which shows the giant looping arches of a solar prominence. Each arch represents a bundle of field lines about which electrons (and other charged particles) spiral.

33-3 MAGNETIC FORCES ON CURRENT CARRIERS

Consider a differential element $d\ell$ of a conducting wire that carries a steady current I (Fig. 33-7). When this wire is placed in a magnetic field, each of the moving charge carriers experiences a force due to the presence of the magnetic field. The sum of all the forces associated with the random thermal motion is zero. However, the average drift velocities of the electrons and any positive charge carriers that might be present are directed along the wire. Thus, the forces associated with the motions of the charge carriers are directed perpendicular to the wire ($\mathbf{F} \perp \mathbf{v}_d$). The cumulative effect of these individual forces is a net sideways force $d\mathbf{F}$ on the wire element. Because the positive and negative charge carriers drift in opposite directions, the magnetic forces on both types of carriers are in the *same* direction, as indicated in Fig. 33-7.

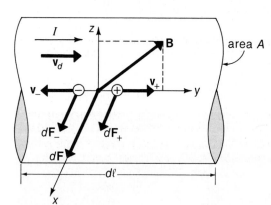

Fig. 33-7. The forces, $d\mathbf{F}_+$ and $d\mathbf{F}_-$, on the two types of charge carriers are in the same direction, namely, perpendicular to the wire. Thus, the force $d\mathbf{F}$ on the wire element is also perpendicular to the wire. For the drift velocities of the positive and negative charge carriers we use the simplified notation $\mathbf{v}_+$ and $\mathbf{v}_-$ instead of $\mathbf{v}_{d+}$ and $\mathbf{v}_{d-}$.

In most situations involving solid conductors, the charge carriers are electrons (but positive charge carriers might be present along with the electrons). In either case, we can identify an equivalent amount of net positive charge dq that moves with an average drift velocity $\mathbf{v}_d$ within the element of wire. Then, the force on the element $d\ell$ of the wire is

$$d\mathbf{F} = dq\,\mathbf{v}_d \times \mathbf{B}$$

Using $I = dq/dt$, we can write

$$d\mathbf{F} = I\,dt\,\mathbf{v}_d \times \mathbf{B} \tag{33-3}$$

The quantity $\mathbf{v}_d\,dt$ represents the average distance of drift $d\ell$ experienced by the electrons during the interval dt. Then the combination, $I\mathbf{v}_d\,dt = I\,d\ell$, is called a current element, where $d\boldsymbol{\ell}$ is a length element in the direction of the current.* Then, Eq. 33-3 becomes

$$d\mathbf{F} = I\, d\boldsymbol{\ell} \times \mathbf{B} \qquad \text{(33-4)}$$

This result is known as *Ampère's law of magnetic force.*

If the conductor is not a thin wire (essentially a one-dimensional conductor) and has a current density $\mathbf{J(r)}$, the force on a volume element dV is

$$d\mathbf{F} = \mathbf{J} \times \mathbf{B}\, dV \qquad \text{(33-5)}$$

Of course, isolated current elements do not exist. A real current element is always associated with a closed circuit loop (and a source of EMF). If a wire loop has sufficient rigidity to resist any deformation by the magnetic forces, or has assumed some equilibrium shape, the total magnetic force on the loop is

$$\mathbf{F} = I \oint d\boldsymbol{\ell} \times \mathbf{B} \qquad \text{(33-6)}$$

where the line integral is to be taken completely around the current loop. When the magnetic field is *uniform over the entire loop,* the integral vanishes identically (see Problem 33-9). Even though there is no net force on a current loop, the field can still exert a torque on the loop; we discuss this effect in Section 33-4.

Example 33-3

A current loop of eight turns is suspended from one pan of a beam balance into a uniform magnetic field that is perpendicular to the loop, as shown in the diagram at the top of the next page. The beam is balanced by placing weights in the left-hand pan. It is found that an additional 6.85 g must be placed in the left-hand pan when a current of 0.200 A flows through the loop in the direction indicated. Determine the magnitude B of the field.

Solution:

If we apply Eq. 33-6 to the three straight-line segments of the loop that lie within the field, we see that the forces cancel for the two vertical segments, so there remains only the downward force on the horizontal segment of length ℓ. The magnitude of this downward force is $F = I\ell B$ on each of the N turns in the loop. This force is balanced by the additional weight Δmg in the left-hand pan. Thus,

$$\Delta m\, g = IN\ell B$$

so that
$$B = \frac{\Delta m\, g}{IN\ell} = \frac{(6.85 \times 10^{-3}\,\text{kg})(9.80\,\text{m/s}^2)}{(0.200\,\text{A})(8)(0.075\,\text{m})}$$
$$= 0.559\,\text{T} \quad \text{or} \quad 5590\,\text{G}$$

---

*It is important to distinguish between $d\mathbf{r}$, which is the differential element of a path or displacement, and $d\boldsymbol{\ell}$, which is the differential element of a wire or circuit that carries a current. The direction of $I\, d\boldsymbol{\ell}$ is always the same as the direction of current flow; the direction of $d\mathbf{r}$ is arbitrary and must be specified in each situation.

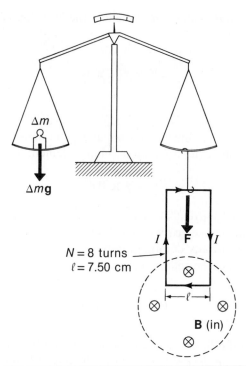

A refined version of this procedure was once a standard technique for measuring magnetic fields. Electronic methods (involving nuclear magnetic resonance) are much simpler and more accurate, and are now routinely used for such measurements.

The Hall Effect. When a current flows through a conductor that is in a magnetic field, the magnetic force should deflect the charge carriers and cause them to collect on one side of the conductor. Reasoning in this manner, the American physicist E. H. Hall (1855–1938) designed a series of experiments which, in 1879, demonstrated that the majority of the charge carriers in most metals bear a negative charge. (The negatively charged particles in matter—*electrons*—were discovered by J. J. Thomson somewhat later, in 1897.)

The *Hall effect* is easily understood in terms of the simple free-electron (or charge-carrier) model of Drude (Section 31–2). Figure 33–8 shows a portion of a long conducting strip that carries a current I along its length (taken to be the y-direction) in response to the externally applied electric field $\mathbf{E}_y$. For simplicity, consider the externally applied magnetic field $\mathbf{B}$ to be in the z-direction. The quantities $\mathbf{v}_+$ and $\mathbf{v}_-$ represent the drift velocities of the charge carriers. The two types of charge carriers have opposite signs and move in opposite directions. Consequently, the

Fig. 33–8. The Hall effect in a conducting strip. Charge carriers of either sign are deflected in the +x-direction. The accumulation of charge on the edge *A* results in a transverse electric field that is in the +x-direction if the majority charge carriers are negative.

magnetic force on each type is in the $+x$-direction (see Fig. 33–8). If the majority charge carriers are negative (electrons), the edge A of the strip will become negatively charged; the opposite face will experience a net loss of negative charge and will become positively charged. This separation of charge produces a transverse electric field $\mathbf{E}_x$ which, for *negative* majority charge carriers, is in the *positive x-*direction. For *positive* majority charge carriers, positive charge collects on edge A and $\mathbf{E}_x$ is in the negative x-direction. The potential difference between the opposite edges of the strip is called the *Hall EMF*. Thus, by establishing the direction of the transverse electric field (that is, the sign of the Hall EMF), the sign of the majority charge carriers can be determined. For most metals this sign is negative.

When equilibrium conditions in the strip have been attained, with the fully developed value of E_x and a steady current I, the x-component of the net force on a moving charge carrier must vanish and the net movement of the carriers must be in the y-direction. Thus,

$$F_x = q(E_x + v_y B_z) = 0$$

from which

$$E_x = -v_y B_z \qquad \text{(33–7)}$$

When the majority charge carriers are electrons, the current density is

$$\mathbf{J} = -en_- \mathbf{v}_- \qquad \text{(33–8)}$$

where n_- is the effective number of free electrons per unit volume. We have $v_y = v_- > 0$, so that

$$J_y = -en_- v_y \qquad \text{(33–9)}$$

Using Eq. 33–7 and substituting $v_y = -J_y/en_-$, we have

$$E_x = \frac{J_y B_z}{en_-}, \qquad \text{for electron carriers} \qquad \text{(33–10)}$$

It is customary to define the *Hall coefficient* R_H for a material to be

$$R_H = -\frac{E_x}{J_y B_z} = -\frac{1}{en_-}, \qquad \text{for electron carriers} \qquad \text{(33–11)}$$

where $E_x > 0$ (see Fig. 33–8). Evidently, for positive charge carriers, we have

$$R_H = -\frac{E_x}{J_y B_z} = \frac{1}{en_+}, \qquad \text{for positive carriers} \qquad \text{(33–12)}$$

where $E_x < 0$ (see Fig. 33–8).

Measurements of E_x, J_y, and B_z in a particular case not only will establish the sign of the charge carriers but will also yield the effective carrier density (if the carriers have a single sign). These effective carrier densities do not always agree with those based on the valence of the material (Table 31–1); see Table 33–1 and Problem 33–10. (When carriers with both signs are present, the equations are more complicated.)

TABLE 33-1.   Charge Carrier Densities for
Various Metals

METAL	n_- (10^{28} m^{-3}) Based on Valence[a]	n_- (10^{28} m^{-3}) Based on Hall Effect
Li	4.70	3.8
Na	2.65	3.2
K	1.40	1.5
Rb	1.15	1.2
Cs	0.91	0.9
Cu	8.47	12.7
Ag	5.86	7.5
Au	5.90	8.9

[a] See Table 31-1.

Notice that according to the simple Drude free-electron model presented here, the Hall coefficient (Eq. 33–11) depends only on the carrier density and not on any thermal or electric property of the conductor. Particularly important is the apparent lack of dependence of R_H on the mean collision time τ. Measurements show that R_H does in fact vary somewhat with temperature and with the strength of the magnetic field. However, for low temperatures and strong fields, the Hall coefficient for a carefully prepared sample of a pure metal does approach a constant value. Under these conditions, deduced values of n_- are approximately equal to those given in Table 31–1 for the alkali metals (Li, Na, K, Rb, and Cs); see Table 33–1. For other one-electron atoms (for example, the noble metals, Cu, Ag, and Au), the agreement is not nearly as good. For metals with valence >1 (for example, Be, Mg, In, and Al), the limiting values of R_H give the *opposite sign* and yield magnitudes for n that do not correspond to the actual valences. The explanation of these differences requires a quantum treatment of the effect.

Example 33–4

A flat ribbon of silver with a thickness $s = 0.20$ mm is used for a Hall-effect measurement of a uniform magnetic field that is perpendicular to the ribbon, as shown in the diagram. The Hall coefficient for silver is $R_H = -0.84 \times 10^{-10}$ m^3/C.

(a) What is the effective number density of charge carriers in silver?

(b) If a current $I = 20$ A produces a Hall EMF $\mathcal{E}_{ab} = 15\ \mu$V, what is the magnitude of the induction vector **B**?

Solution:

(a) Using Eq. 33–11, we have

$$n_- = -\frac{1}{eR_H} = -\frac{1}{(1.60 \times 10^{-19}\,\text{C})(-0.84 \times 10^{-10}\,\text{m}^3/\text{C})}$$
$$= 7.44 \times 10^{28}\,\text{m}^{-3}$$

On the basis of a valence of *one* for silver, we expect a charge carrier density of $5.86 \times 10^{28}\,\text{m}^{-3}$ (Table 33–1). Thus, in the Hall effect the effective number of electrons per atom of silver is $(7.44 \times 10^{28}\,\text{m}^{-3})/(5.86 \times 10^{28}\,\text{m}^{-3}) = 1.27$.

(b) From Eq. 33–11, we have

$$B_z = -\frac{E_x}{J_y R_H} = -\frac{(\mathcal{E}_{ab}/w)}{(I/sw)R_H} = -\frac{\mathcal{E}_{ab}s}{IR_H}$$

$$= -\frac{(15 \times 10^{-6}\,\text{V})(0.20 \times 10^{-3}\,\text{m})}{(20\,\text{A})(-0.84 \times 10^{-10}\,\text{m}^3/\text{C})}$$

$$= 1.79\,\text{T} = 17{,}900\,\text{G}$$

33–4 TORQUE ON A CURRENT LOOP

Consider a rectangular loop with dimensions $w \times \ell$ whose plane makes an angle θ with the x-y plane, as indicated in Fig. 33–9. A steady current I flows in the loop in the sense shown. The entire loop is within a uniform magnetic field **B** in the z-direction. We can use Eq. 33–4 to determine the forces that act on various parts of the loop.

Fig. 33–9. Geometry for calculating the torque on a current loop in a magnetic field.

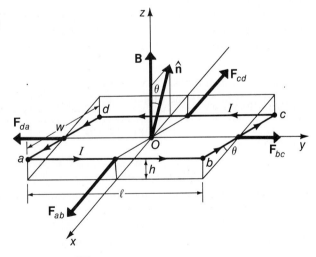

The force $\mathbf{F}_{ab}$ that acts on the side $\overline{ab}$ of the loop is in the x-direction and is given by

$$\mathbf{F}_{ab} = I\ell B\hat{\mathbf{i}}$$

The force $\mathbf{F}_{cd}$ that acts on the side $\overline{cd}$ has the same magnitude but is in the opposite direction:

$$\mathbf{F}_{cd} = -\mathbf{F}_{ab} = -I\ell B\hat{\mathbf{i}}$$

The forces on the sides $\overline{bc}$ and $\overline{da}$ are

$$\mathbf{F}_{bc} = -\mathbf{F}_{da} = IwB\hat{\mathbf{j}}$$

Thus, the net force that acts on the loop is zero,

$$\mathbf{F} = \mathbf{F}_{ab} + \mathbf{F}_{bc} + \mathbf{F}_{cd} + \mathbf{F}_{da} = 0$$

Moreover, $\mathbf{F}$ vanishes for any orientation of the loop.

Notice, however, that the loop experiences a net *torque* which tends to produce a rotation of the loop about the y-axis. Calculating this torque with respect to the origin O at the center of the loop, we find

$$\boldsymbol{\tau}_O = 2F_{ab}h\hat{\mathbf{j}}$$

where $h = \frac{1}{2}w \sin\theta$ is the height of the side $\overline{ab}$ above the x-y plane. Thus, we can write

$$\boldsymbol{\tau}_O = I\ell Bw \sin\theta\,\hat{\mathbf{j}} \qquad \text{(33–13)}$$

We now introduce a useful vector quantity associated with current-carrying loops. For a planar loop, such as that in Fig. 33–9, the *magnetic moment* **m** is defined to be a vector with a magnitude equal to the product of the loop area A and the current I, and with a direction along the normal $\hat{\mathbf{n}}$ to the loop:

$$\mathbf{m} = IA\hat{\mathbf{n}} \qquad \text{(33–14)}$$

(The direction of the positive sense of the normal $\hat{\mathbf{n}}$ is determined by the right-hand convention applied to the direction of the current.) For the loop shown in Fig. 33–9, we have

$$\mathbf{m} = I\ell w\hat{\mathbf{n}}$$

where

$$\hat{\mathbf{n}} = -\sin\theta\,\hat{\mathbf{i}} + \cos\theta\,\hat{\mathbf{k}} \qquad \left.\right\} \qquad \text{(33–15)}$$

Using these equations to express the magnetic moment, we can write the torque as

$$\boldsymbol{\tau}_O = \mathbf{m} \times \mathbf{B} \qquad \text{(33–16)}$$

(•Verify explicitly that Eq. 33–16 is equivalent to Eq. 33–13.)

Equation 33–16 is a general result for a planar loop with *any* shape in a (locally) uniform magnetic field. The case of nonplanar loops is discussed in Example 33–6.

Equation 33–16 is similar to the expression (Eq. 29–37) for the torque that acts on an electric dipole, namely, $\boldsymbol{\tau}_O = \mathbf{p} \times \mathbf{E}$. This suggests that a potential energy can be associated with the relative orientation of **m** and **B** corresponding to that found for the electric dipole, namely, $U = -\mathbf{p} \cdot \mathbf{E} + U_0$ (Eq. 29–38). Indeed, the relationship for the magnetic case is (see Problem 33–11)

$$U_M = -\mathbf{m} \cdot \mathbf{B} + U_0 \qquad (33\text{--}17)$$

Example 33–5

Derive an expression for the torque on a circular loop of radius a that lies in the x-y plane and carries a current I. The loop is within a uniform magnetic field, $\mathbf{B} = B_x\hat{\mathbf{i}} + B_y\hat{\mathbf{j}} + B_z\hat{\mathbf{k}}$.

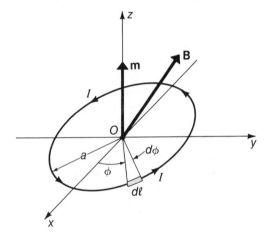

Solution:

Consider a differential segment $d\ell$ of the loop, as shown in the diagram. The location of the segment is given by

$$\mathbf{r} = a(\cos\phi\,\hat{\mathbf{i}} + \sin\phi\,\hat{\mathbf{j}})$$

Also,

$$d\ell = a(-\sin\phi\,\hat{\mathbf{i}} + \cos\phi\,\hat{\mathbf{j}})\,d\phi$$

The torque τ_0 about the origin O at the center of the loop is

$$\tau_0 = \oint \mathbf{r} \times d\mathbf{F} = \oint \mathbf{r} \times (I d\ell \times \mathbf{B})$$

Using the vector identity (see the table inside the front cover),

$$\mathbf{A} \times (\mathbf{B} \times \mathbf{C}) = (\mathbf{A} \cdot \mathbf{C})\mathbf{B} - (\mathbf{A} \cdot \mathbf{B})\mathbf{C}$$

we can write

$$\tau_0 = \oint (\mathbf{r} \cdot \mathbf{B})I d\ell - \oint (\mathbf{r} \cdot I d\ell)\mathbf{B}$$

Now, **r** is always perpendicular to $I d\ell$, so the second integral vanishes. We also have

$$\mathbf{r} \cdot \mathbf{B} = a(B_x \cos\phi + B_y \sin\phi)$$

Hence,

$$\tau_0 = a^2 I \int_0^{2\pi} (B_x \cos\phi + B_y \sin\phi)(-\sin\phi\,\hat{\mathbf{i}} + \cos\phi\,\hat{\mathbf{j}})\,d\phi$$

Of the four integrals in this expression, the two that involve $\sin\phi\cos\phi$ vanish. There remains

$$\tau_0 = a^2 I \int_0^{2\pi} (-B_y \sin^2\phi\,\hat{\mathbf{i}}\,d\phi + B_x \cos^2\phi\,\hat{\mathbf{j}}\,d\phi)$$
$$= \pi a^2 I(-B_y\,\hat{\mathbf{i}} + B_x\,\hat{\mathbf{j}})$$

The magnetic moment of the loop is $\mathbf{m} = \pi a^2 I\hat{\mathbf{k}}$, so we again have

$$\tau_0 = \mathbf{m} \times \mathbf{B}$$

This reinforces the assertion that Eq. 33–16 is completely general for planar loops.

Example 33–6

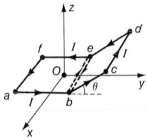

Determine the magnetic moment of the bent (nonplanar) rectangular rigid loop shown in the diagram. The straight-line segments $\overline{ab}$, $\overline{ef}$, and $\overline{fa}$ are in a plane that is parallel to the x-y plane; the segments $\overline{bc}$, $\overline{cd}$, and $\overline{de}$ are in a plane that is inclined by an angle θ to the x-y plane. The dimensions of the loop are $\overline{ab} = \overline{ef} = 1.0$ m, $\overline{bc} = \overline{de} = 1.5$ m, and $\overline{cd} = \overline{fa} = 0.80$ m; $\theta = 40°$, and $I = 3.5$ A in the sense shown in the diagram.

Solution:

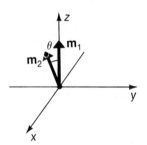

Imagine that we convert the bent loop into two planar loops by breaking the loop at b and e and inserting the legs $\overline{be}$ and $\overline{eb}$ to complete the smaller loops. With the same current in each loop, the currents in the legs $\overline{be}$ and $\overline{eb}$ flow in opposite directions and their magnetic effects cancel. Thus, the magnetic effect of the pair of loops is exactly the same as that of the original loop. The loop $abef$ has a magnetic moment $\mathbf{m}_1$; the loop $bcde$ has a magnetic moment $\mathbf{m}_2$. With the loops in a uniform magnetic field $\mathbf{B}$, the torque on the system is

$$\tau_0 = \mathbf{m}_1 \times \mathbf{B} + \mathbf{m}_2 \times \mathbf{B} = \mathbf{m} \times \mathbf{B}$$

where $\mathbf{m} = \mathbf{m}_1 + \mathbf{m}_2$ is the magnetic moment of the bent loop. The individual moments are

$$\mathbf{m}_1 = I(\overline{ab})(\overline{fa})\hat{\mathbf{k}} = (3.5 \text{ A})(1.0 \text{ m})(0.80 \text{ m})\hat{\mathbf{k}} = (2.80\hat{\mathbf{k}}) \text{ A} \cdot \text{m}^2$$
$$\mathbf{m}_2 = I(\overline{bc})(\overline{cd})(-\sin\theta\,\hat{\mathbf{i}} + \cos\theta\,\hat{\mathbf{k}})$$
$$= (3.5 \text{ A})(1.5 \text{ m})(0.80 \text{ m})(-\sin 40°\,\hat{\mathbf{i}} + \cos 40°\,\hat{\mathbf{k}})$$
$$= (-2.70\hat{\mathbf{i}} + 3.22\hat{\mathbf{k}}) \text{ A} \cdot \text{m}^2$$

Finally, we have

$$\mathbf{m} = \mathbf{m}_1 + \mathbf{m}_2 = (-2.70\hat{\mathbf{i}} + 6.02\hat{\mathbf{k}}) \text{ A} \cdot \text{m}^2$$

By using a similar technique to convert an irregular nonplanar loop into a set of differential planar loops, it can be shown that the magnetic moment becomes

$$\mathbf{m} = I \int_S \hat{\mathbf{n}}\, dA$$

where the integral is to be evaluated over any open surface S whose boundary is the loop.

The Galvanometer. In Section 32–3, we discussed the use of a galvanometer as the meter movement of ammeters and voltmeters. The explanation for the torque acting on the movable coil was necessarily left vague; we now return to this point. Figure 33–10 again shows the meter movement. The coil usually consists of a few hundred turns of fine wire; because τ_O acts equally on each turn, the meter's sensitivity is directly proportional to the number of turns. The coil is wound on an iron core that guides and concentrates the field lines, and the coil assembly with the attached pointer is pivoted about an axis that is perpendicular to the plane of the page. With no current through the coil, the spring is adjusted so that the pointer lines up with the zero point of the scale.

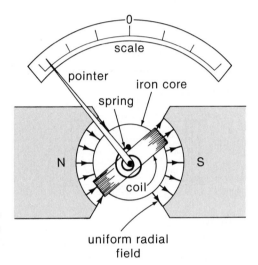

Fig. 33–10. Schematic diagram of a galvanometer movement. The concave pole faces provide a uniformly radial magnetic field in which the coil moves.

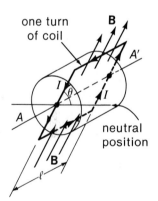

Fig. 33–11. When a steady current I flows through the coil of a galvanometer, a torque causes the coil to rotate about the axis AA'.

When a steady current I flows through the coil, a torque acts on each turn of wire, thereby producing a rotation of the coil assembly about the axis AA', as shown in Fig. 33–11. Depending on the direction of the current, the pointer will be deflected to the right or to the left of the zero point. The magnetic torque is opposed by the restoring torque due to the spring. A coiled spring can be made to produce a restoring torque that is proportional to the angular deflection. When such a spring is used in conjunction with a coil in a uniformly radial magnetic field (Fig. 33–10), the deflection of the pointer will be a linear function of the current in the coil. (•Can you see why a uniform radial field is required for a linear response?)

Let the coil consist of N turns of wire around a core that has a length ℓ and a radius R. The field at the coil is B and the torsional constant of the spring is Γ (see Eq. 15–24). Then, the equilibrium deflection angle θ for a current I is obtained from

$$\tau_{AA'} = 2NI\ell RB = \Gamma\theta \qquad (33\text{–}18)$$

Fig. 33–12. Schematic diagram of a DC electric motor.

DC Motors. The fact that a current-carrying coil experiences a torque in a magnetic field is basic to the design of electric motors. The torque will aid the rotation until the plane of the coil is perpendicular to the field direction. Then, $\sin\theta = 0$ and the torque vanishes. If provision is made for reversing the direction of the current in the coil whenever $\theta = 90°$, the rotation can be made continuous and the device becomes a motor. Figure 33–12 shows how this reversal can be accomplished in a simple way by delivering the current to the coil through a *split ring commutator*. The connection from the source of EMF to the ring is made through a pair of wire or graphite brushes. Because the split ring produces a reversal of the current in the coil every half cycle, the rotation is continuous. This is the principle of the *DC motor*.*

There exist many designs for practical electric motors. Usually, a number of multiturn coils are wound on the iron core; this assembly is called the *armature*. Also, an electromagnet is substituted for the permanent magnet shown in Fig. 33–12. One such design is illustrated in Fig. 33–13. The current I is delivered to the armature through the brushes, B_1 and B_2. The coil is wound in such a way that as the brushes slip to the next contact, the current always flows *into* the plane of the page in the coil sections on the N-pole side of the armature and *out of* the plane of the page in the coil sections on the S-pole side. This type of winding maintains a nearly constant torque on the armature and smooths the rotation.

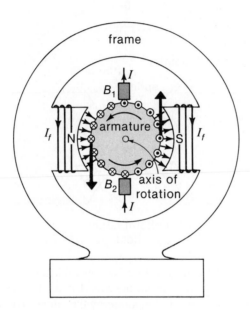

Fig. 33–13. A practical DC motor. The magnetic field is produced by the *field current* I_f in the windings that surround the pole pieces. Notice that the symbol $\odot$ (suggesting the tip of an arrow) is used to indicate that the direction of the current is *out* of the plane of the page; the symbol $\otimes$ (suggesting the feather end of an arrow) is used to indicate that the direction of the current is *into* the plane of the page. We often use this method to indicate direction.

33–5 ADDITIONAL APPLICATIONS OF THE LORENTZ FORCE

The Velocity Selector.   Suppose that a narrow beam of identical charged particles (for example, electrons) enters a region in which there exists a uniform electric field **E** at right angles to a uniform magnetic field **B**. The particles in the beam have a spectrum of velocities, but they all enter the field region perpendicular to both field vectors, as indicated in Fig. 33–14. The electric field produces an upward force on the electrons in the beam, whereas the magnetic field produces a downward force. There is a unique velocity v_0 for which the electric and magnetic forces exactly cancel. The value of v_0 is obtained by setting the Lorentz force (Eq. 33–2) equal to zero. Because $\mathbf{v}_0 \perp \mathbf{B}$, we find

$$v_0 = E/B \tag{33–19}$$

Electrons that have velocities less than v_0 are deflected upward and strike the exit wall at points such as *a* and *b*. Electrons that have velocities greater than v_0 are

---

*The electric motor was invented in 1831 by the American physicist Joseph Henry (1797–1878).

Fig. 33–14. Crossed electric and magnetic fields act as a *velocity selector*. Only electrons with $v_0 = E/B$ pass through the field region undeflected.

deflected downward and strike the wall at points such as c. Electrons that have the velocity v_0 (and only these electrons) are undeflected and pass through the exit slit S. The arrangement of crossed electric and magnetic fields is called a *velocity selector*.

The Mass Spectrograph. In 1913, J. J. Thomson demonstrated that charged particles with different masses can be separated by using combined electric and magnetic fields. Thomson's crude device was improved upon by F. W. Aston and by A. J. Dempster. By 1919 atomic masses were being measured with a precision of 1 part in 10^3; the best modern instruments are capable of precisions of about 1 part in 10^7.

Figure 33–15 shows a schematic diagram of an early *mass spectrograph*. Low-energy positively charged ions are produced in the source chamber a and are accelerated through a potential difference V. The ions enter a region with a uniform magnetic field **B** (which, in Fig. 33–15, is directed outward from the plane of the page) and are deflected in circular arcs. The ions are recorded on a photographic plate so that the distance x from the entrance slit b (the diameter of the orbit) can be measured.

Fig. 33–15. Basic features of an early Dempster mass spectrograph. Positive ions are produced in the source a. The entire system is maintained under vacuum to prevent scattering of the ions by gas molecules.

Consider a singly ionized ion ($q = +e$) with a mass M that enters the magnetic field with a velocity v obtained from

$$\tfrac{1}{2}Mv^2 = eV$$

Within the magnetic field, the force on the ion is always perpendicular to its velocity vector, which has a constant magnitude. Therefore, the ion follows a circular path with radius $R = \tfrac{1}{2}x$. The centripetal acceleration is provided by the magnetic field. Thus,

$$a_c = \frac{v^2}{R} = \frac{v^2}{\tfrac{1}{2}x} = \frac{F}{M} = \frac{evB}{M}$$

Expressing v in terms of V and solving for M gives

$$M = \frac{eB^2}{8V}x^2 \qquad (33\text{–}20)$$

Thus, the mass of the ion is obtained from measurements of B, V, and x. The relative masses of two ions accelerated at the same time involve only measurements of x and so can be obtained with high precision.

Figure 33–16 shows the photographic record obtained by Dempster for the element ytterbium (Yb). Not all Yb atoms have the same mass (because the nuclei contain different numbers of neutrons). Thus, the mass spectrograph places each of these *isotopes* at different distances x along the plate. Dempster's results show Yb isotopes with mass numbers between 168 and 176. Notice that mass numbers 169 and 175 are absent; these isotopes are radioactive and do not occur in Nature.

Fig. 33–16. Photographic record of ytterbium (Yb) isotopes obtained by Dempster in a mass spectrograph. (From *Introduction to Atomic Physics* by Henry Semat. Farrar & Rinehart, Inc., New York, 1939. By permission of the author, and courtesy of Prof. A. J. Dempster.)

Magnetic Confinement. If a charged particle is injected into a uniform magnetic field at some angle with respect to the field lines, the particle will spiral around a field line and will drift with a constant velocity along the field line (see Example 33–2). However, it is possible to produce a *nonuniform* field in which a particle is forced to move back and forth along a field line. The particle is therefore confined (or trapped) within the field.

Figure 33–17 shows a field that constitutes a *magnetic bottle*. The field is strong at the ends, a and b, and is weak in the central region c. Consider a positive ion at point f. If the ion has a velocity component directed toward the right (an *axial* velocity component), it will follow the spiral path shown in the diagram. In the region between a and c, the radial velocity component of the particle interacts with the magnetic field **B** (the lines of which are *diverging* in this region) to produce an acceleration of the particle toward the right. (•Verify this by examining **v** × **B**.) When the particle passes into the region between c and b, however, the field lines are *converging*. Then, the particle experiences an acceleration toward the left. This acceleration slows and eventually reverses the particle's axial drift toward the right. Point

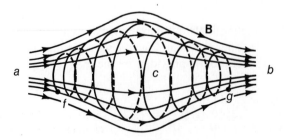

Fig. 33–17. A charged particle that is magnetically confined by the nonuniform field spirals back and forth between the points f and g.

g represents the extreme of the rightward motion. The particle then drifts to the left until it reaches the leftward extreme at point *f*. Thus, the particle spirals back and forth between the reflection (or *mirror*) points, *f* and *g*. (The sense of the rotational motion is always the same.)

Magnetic confinement devices are used in the study of high-temperature ionized gases (plasmas) to prevent the ions from striking and losing energy to the walls of the system. Unfortunately, particles tend to "leak" from these bottles for a variety of reasons, the major one having to do with ion-ion collisions at high plasma densities. These techniques may, however, eventually be perfected for use in nuclear fusion reactors that require plasmas at extremely high temperatures (10^6 K or so).

The Earth's Radiation Belts. The Earth's dipole magnetic field (see Fig. 33–1) is nonuniform and resembles a magnetic bottle field (Fig. 33–17) that is wrapped around the globe. In 1958, during the Explorer I satellite mission, James A. Van Allen discovered that the Earth's field contains trapped charged particles. These particles—electrons and protons—are confined in broad, somewhat overlapping shells or radiation belts, as indicated in Fig. 33–18. The inner region (labeled *P*) primarily contains protons and is centered about 3000 km above the Earth's surface. The outer region (labeled *E*) primarily contains electrons and is centered about 15,000 km above the Earth's surface. The particle densities in the radiation belts range from about 10^4 to about 10^5 per cm^3.

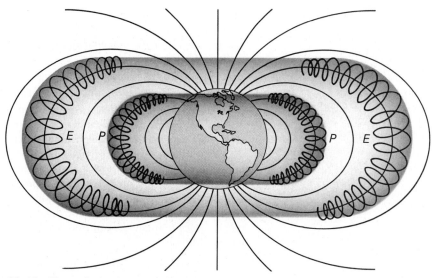

Fig. 33–18. The radiation belts consist of protons (*P*) and electrons (*E*) that are trapped in the Earth's nonuniform magnetic field.

PROBLEMS

Section 33-2

33-1 A proton moves in a uniform magnetic field **B** with a speed of 10^7 m/s and experiences an acceleration of 2×10^{13} m/s^2 in the $+x$-direction when its velocity is in the $+z$-direction. Determine the magnitude and direction of the field.

33-2 In a certain Cartesian reference frame the velocity of an electron is $\mathbf{v} = (2.5\hat{\mathbf{k}}) \times 10^7$ m/s and it experiences a force $\mathbf{F} = (4.0\hat{\mathbf{i}} + 3.0\hat{\mathbf{j}}) \times 10^{-13}$ N due to the presence of a uniform magnetic field. When the velocity is redirected and becomes $\mathbf{v} = (1.5\hat{\mathbf{i}} - 2.0\hat{\mathbf{j}}) \times 10^7$ m/s, the magnetic force on the electron is zero. Determine the components of the magnetic induction vector **B**. Why are both of the observations necessary to determine the field components?

33-3 A beam of protons (all with velocity **v**) emerges from a particle accelerator and is deflected in a circular arc with a radius of 0.45 m by a transverse uniform magnetic field of 0.80 T. (a) Determine the speed v of the protons in the beam. (b) What time is required for the deflection of a particular proton through an angle of 90°? (c) What is the energy of the particles in the beam?

33-4 A cosmic-ray proton in interstellar space has an energy of 10 MeV and executes a circular orbit with a radius equal to that of Mercury's orbit around the Sun (5.8×10^{10} m). What is the galactic magnetic field in that region of space?

33-5 The accelerating voltage that is applied to an electron gun is 15 kV, and the horizontal distance from the gun to a viewing screen is 35 cm. What is the deflection caused by the vertical component of the Earth's magnetic field (40,000 γ), assuming that any change in the horizontal component of the beam velocity is negligible?

33-6 A metal ball with a mass of 100 g is swung with an angular speed of 40π rad/s in a horizontal circle at the end of a taut string in the presence of a uniform vertical magnetic field of 100 G. (a) What amount of positive charge Q must be placed on the ball to double the tension in the string? (b) What is the direction of **B** relative to $\boldsymbol{\omega}$? (c) Would it be feasible to place the charge Q on the ball?

Section 33-3

33-7 A wire carries a steady current of 2.4 A. A straight section of the wire, with a length of 0.75 m along the x-axis, lies within a uniform magnetic field, $\mathbf{B} = (1.6\hat{\mathbf{k}})$T. If the current flows in the $+x$-direction, what is the magnetic force on the section of wire?

33-8 A wire carries a steady current of 3.0 A. A portion of the wire is located within a uniform magnetic field, $\mathbf{B} = (0.65\hat{\mathbf{i}})$T. This portion of the wire is composed of two straight-line segments, $\overline{AB}$ and $\overline{BC}$, that lie in the y-z plane. The three points are: $y_A = 2.0$ m and $z_A = 1.0$ m; $y_B = 6.0$ m and $z_B = 1.0$ m; $y_C = 6.0$ m and $z_C = 7.0$ m. (a) What is the magnetic force on each segment of the wire? (b) What is the

net magnetic force on the wire? (c) Consider the situation in which the wire segment in the field extends in a straight line from point A to point C. Demonstrate that the net magnetic force on the wire is the same as that found in (b).

33-9 Prove that the line integral in Eq. 33-6 vanishes identically if the loop lies entirely within a uniform magnetic field. [*Hint:* The constant vector **B** can be removed from the integral.]

33-10 A flat copper ribbon with a thickness of $\frac{1}{3}$ mm carries a steady current of 50 A and is located within a uniform magnetic field of 1.3 T that is directed perpendicular to the plane of the ribbon. If a Hall EMF of 9.6 μV is measured across the ribbon, what is the effective number density of free electrons in the copper ribbon? This result indicates what effective number of free electrons per atom? Compare with Table 33-1.

Section 33-4

33-11 Begin with Eq. 33-16 for the torque on a current-carrying loop in a uniform magnetic field, and derive Eq. 33-17 for the potential energy U_M relative to the condition $\theta = \pi/2$.

33-12 A coil of 10 turns of fine wire wound on a rectangular frame 8 cm × 10 cm carries a current of 10 A and is located within a uniform magnetic field of 300 G. (a) How many different geometric arrangements of coil and field produce maximum torque on the coil? Sketch one such arrangement and determine the torque τ_0. (b) How many different geometric arrangements of coil and field produce a condition of stable equilibrium? Unstable equilibrium? (c) How much work must be done by an external agency to rotate the coil from stable to unstable equilibrium?

33-13 A square loop of wire, 10 cm × 10 cm, has a mass

of 50 g and pivots about an axis AA' that corresponds to a horizontal side of the square, as shown in the diagram. The loop is in equilibrium in a uniform magnetic field of 500 G directed vertically downward when $\theta = 20°$. (a) What current flows in the loop and in what direction? (b) Determine the magnitude and direction of the force exerted on the axis by the bearings.

33–14 The coil of a galvanometer movement has an area of 1.0 cm² and consists of 100 turns of fine wire. The radial magnetic field at the position of the coil is 0.15 T, and the torsional constant of the spring is $\Gamma = 1.5 \times 10^{-7}$ N·m/deg. What is the angular deflection of the coil for a current of 1 mA?

Section 33–5

33–15 A crossed-field velocity selector has a magnetic field of 10^{-2} T. What electric field strength is required if 10-keV electrons are to pass through undeflected?

33–16 A Dempster mass spectrograph (Fig. 33–15) is used to determine the mass of deuterium (²H) relative to the known mass of ordinary hydrogen (¹H; M = 1.00783 u). The line due to singly ionized hydrogen molecules (¹H₂⁺) is observed on the photographic plate at $x = 16.20$ cm. A line corresponding to ionized atomic deuterium (²H⁺) is observed at a *smaller* value of x with a line separation of $\Delta x = 0.06$ mm. Determine the atomic mass of deuterium and compare with the value given in Fig. 23–4.

33–17 One type of mass spectrometer (developed in 1951) is based on the fact that the angular frequency of an ion in a uniform magnetic field about an axis parallel to the field direction is independent of the ion velocity. (This is the cyclotron frequency; see Example 33–2.) Low-velocity ions ($K \cong 10$ eV) are emitted from a source in short-duration bursts and are injected into a uniform magnetic field of 450 G. A measurement is made of the time required for a burst of ions to traverse 10 helical turns. How accurately must the transit time be measured in order to determine the mass of the ions to within 10^{-3} u?

33–18 A 5.0-keV proton is trapped in a magnetic bottle. At a certain instant the proton is moving in a circular path in a plane that is perpendicular to the axis of the magnetic bottle, as indicated in the diagram. The magnetic field has a magnitude of 600 G, and the field line at the position of the proton makes an angle $\theta = 15°$ with respect to the bottle axis as shown. (a) What is the velocity of the proton? (b) What is the radius of the path? (c) What is the radial acceleration a_r of the proton? (d) Show that

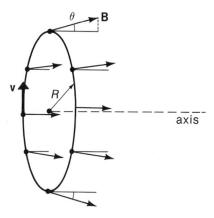

the ratio of the axial acceleration a_z to the radial acceleration is $a_z/a_r = -\tan \theta$. (e) What is the axial acceleration?

Additional Problems

33–19 A relativistic charged particle (charge e) that moves in a plane perpendicular to a uniform magnetic field B follows a circular path with radius $R = mv/eB$, where $m = m_0/(1 - v^2/c^2)^{1/2}$. (a) For an extreme relativistic particle ($K \gg m_0c^2$), show that $R \cong K/eBc$. (b) What is the orbit radius for a cosmic-ray proton with $K = 10^{12}$ eV in the spiral arm of our galaxy where the average field is 10^{-10} T? (Express your result in A.U., *astronomical units;* 1 A.U. = Earth–Sun distance, 1.50×10^{11} m.) (c) Calculate the cyclotron frequency for the particle.

33–20 One way to determine the ratio e/m for electrons (a method developed by H. Busch in 1922 and often used in laboratory demonstrations) involves producing monoenergetic electrons at a point source located in a uniform magnetic field. The field is adjusted until a sharp image of the source is produced at some distance L for those electrons that leave the source in directions making small angles with the field, as shown in the diagram. (a) The electrons in the source are accelerated through a voltage V, and the lowest focusing field is B_0. Show that $e/m = 8\pi^2 V/L^2B_0^2$. (b) Typical values for a laboratory exercise are $L = 15.0$ cm, $V = 1000$ V, and $B_0 = 45$ G. Determine the value of e/m from these data and compare with the actual value.

33-21 A length of copper wire with a square cross section, 1 mm $\times$ 1 mm, is formed into a circular ring with a radius of 8.0 cm. The ring carries a steady current of 20 A and is in static equilibrium in a uniform magnetic field of 0.5 T directed parallel to the ring axis. (a) Determine the tension in the wire. (b) What increase in circumference of the ring is caused by the presence of the field? (Young's modulus for copper is 12.4×10^{10} Pa.)

33-22 An artificial satellite circles the Earth in the equatorial plane at an altitude of 100 km, traveling from east to west. Along the satellite orbit the Earth's magnetic field is approximately 25,000 γ and is perpendicular to the equatorial plane. If the satellite has acquired a positive charge of 0.1 C, what force does the Earth's magnetic field exert on the satellite? Compare your result with the gravitational force on the satellite, the mass of which is 10^3 kg.

33-23 A small circular coil with a mass of 20 g consists of 19 turns of fine wire and carries a current of 1.0 A. The coil is located in a uniform magnetic field of 0.10 T with the coil axis originally parallel to the field direction. (a) What is the period of small-angle oscillations of the coil axis around its equilibrium position, assuming no friction and no mechanical restoring force? (b) If the coil is rotated through an angle of $10°$ from its equilibrium position and then released, what will be its angular speed when it passes through the equilibrium position?

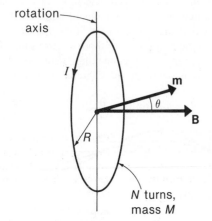

33-24 A DC motor operates from a 220-V DC power line. The motor's electromagnet is excited by a field coil whose resistance is 200 Ω. The field coil is connected in parallel with the motor armature, which has a resistance of 1.5 Ω (see the diagram). The armature current is 9.4 A. When the motor is delivering its rated output, there are frictional and windage losses amounting to 120 W. (a) What is the power input to the armature? (b) What is the total power loss in the armature, including Joule heating, friction, and windage? (c) What is the mechanical output power? (Express your result in watts and in horsepower.) (d) What is the total input current I? (e) What is the total input power? (f) What is the overall efficiency of the motor?

33-25 The diagram shows a modification of the Dempster mass spectrograph (Fig. 33-15) that was introduced by K. T. Bainbridge. In this device a velocity selector is placed after the entrance slit b and uses the same field **B** as in the main region of the spectrograph. Show that the selected ion mass is $M = (eB^2/2E)x$, where E is the electric field strength in the velocity selector.

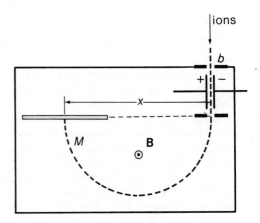

33-26 In a Dempster mass spectrograph (Fig. 33-15), the accelerating voltage is 20,000 V, and the ytterbium group of lines is centered at $\bar{x} = 25.0$ cm. (a) What magnetic field is required? (b) What is the range Δx for the isotopic lines from mass number 168 to 176?

SOURCES OF MAGNETIC FIELDS

In the preceding chapter the discussion centered on the interaction of steady magnetic fields with individual moving charged particles and with steady currents produced by the collective motion of charge carriers in conducting media. In this chapter we examine in detail the way in which currents give rise to magnetic fields, again restricting considerations to steady currents (DC). The effects of time-dependent fields are discussed in the following chapters.

34-1 THE BIOT-SAVART LAW

It is again convenient to use the concept of a differential vector current element $I d\ell$ and to ascribe to it an associated differential magnetic induction $d\mathbf{B}$ at every point in space. The connection between these quantities is known as the *Biot-Savart law*. In differential form the statement of the law is*

$$d\mathbf{B} = K_M \frac{I d\ell \times \hat{\mathbf{u}}_r}{r^2} \qquad \text{(34-1)}$$

In this expression, $d\mathbf{B}$ is the contribution to the field made by the current element $I d\ell$ at a point defined by the vector $\mathbf{r}$ from the current element, as shown in Fig. 34-1. The constant K_M is the magnetic counterpart of the constant K that appears in Coulomb's law (Eq. 27-1) and has the exact value

$$K_M = 10^{-7}\,\text{T} \cdot \text{m/A} = 10^{-7}\,\text{N/A}^2$$

(Refer to Section 34-4 for additional remarks concerning K_M and the definition of the ampere.)

In the discussion of electrostatics (Section 27-3), we found it convenient to introduce the permittivity of free space, $\epsilon_0 = 1/4\pi K$, and to use this quantity in

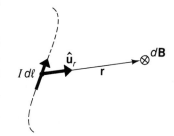

Fig. 34-1. Geometry for the Biot-Savart law. The unit vector is $\hat{\mathbf{u}}_r = \mathbf{r}/|\mathbf{r}|$. The direction of $d\mathbf{B}$ is *into* the plane of the page.

---

*Jean-Baptiste Biot (1774–1862) and Félix Savart (1791–1841) expressed the results of their investigations in terms of the forces experienced by current-carrying wires (1820). The differential formulation of the law is due to Pierre Simon Laplace (1749–1827).

Coulomb's law. We now introduce a quantity called the *permeability of free space,*

$$\mu_0 = 4\pi K_M = 4\pi \times 10^{-7} \text{ N/A}^2$$
$$= 4\pi \times 10^{-7} \text{ T} \cdot \text{m/A}$$

which we use in Eq. 34–1 to write the Biot-Savart law in the way it is usually stated, namely,

$$d\mathbf{B} = \frac{\mu_0 I}{4\pi} \frac{d\boldsymbol{\ell} \times \hat{\mathbf{u}}_r}{r^2} \tag{34–2}$$

A steady current implies the existence of a closed current loop. To determine the field **B** at some point P in space requires the addition (by linear superposition) of all the differential field components $d\mathbf{B}$ (Eq. 34–2) produced by elements around the entire loop. That is,

$$\mathbf{B} = \frac{\mu_0 I}{4\pi} \oint_{\substack{\text{current} \\ \text{loop}}} \frac{d\boldsymbol{\ell} \times \hat{\mathbf{u}}_r}{r^2} \tag{34–3}$$

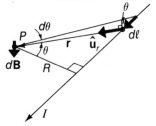

Fig. 34–2. Geometry for calculating the magnetic field produced by a steady current in a long straight wire.

Equation 34–3 can be used to calculate the magnetic field produced by a steady current in a long straight wire. Suppose that the perpendicular distance R from the point of interest P to the wire is small compared with the straight section of the wire that we consider. Thus, the wire is, in effect, infinitely long and we can ignore any contribution to the field at P from the part of the loop that connects the two ends by a return path at a great distance from P.

From the geometry in Fig. 34–2, we have

$$|d\boldsymbol{\ell} \times \hat{\mathbf{u}}_r| = d\ell \sin\left(\frac{\pi}{2} - \theta\right) = d\ell \cos\theta$$

Fig. 34–3. Field lines surrounding a wire that carries a steady current. (Compare Figs. 33–5 and 34–2.) The direction of the field lines is specified by the unit azimuthal vector $\hat{\mathbf{u}}_\phi$.

Also, $d\ell \cos\theta = rd\theta$ and $r \cos\theta = R$

Using these relationships in Eq. 34–3, we find for the magnitude of **B**,

$$B = \frac{\mu_0 I}{4\pi R} \int_{-\pi/2}^{\pi/2} \cos\theta \, d\theta = \frac{\mu_0 I}{2\pi R} \tag{34–4}$$

The direction of **B** is specified by the unit azimuthal vector $\hat{\mathbf{u}}_\phi$, as shown in Fig. 34–3. Then, we can write

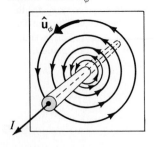

$$\mathbf{B} = \frac{\mu_0 I}{2\pi R} \hat{\mathbf{u}}_\phi \tag{34–4a}$$

The direction of the field at the point P is perpendicular to the plane that contains the wire and the point. Moreover, **B** has a constant magnitude for a given value of R. Thus, the field lines are circles concentric with the wire. This is the same conclusion that was reached in Section 33–1 (see Fig. 33–4). The direction of the field is specified by the right-hand rule that was given in connection with the discussion of Fig. 33–5. These properties of the field are summarized in Fig. 34–3.

According to Eq. 34–4, the magnitude of the magnetic induction decreases inversely with the perpendicular distance R from the wire. At a distance of 10 cm from a wire in which the current is 30 A, we have

$$B = \frac{(4\pi \times 10^{-7}\ \text{T} \cdot \text{m/A})(30\ \text{A})}{2\pi(0.10\ \text{m})}$$
$$= 6.0 \times 10^{-5}\ \text{T} = 0.60\ \text{G}$$

which is approximately the magnitude of the Earth's magnetic field near the surface.

Example 34–1

Determine the magnetic induction along the axis of a circular loop that carries a current I.

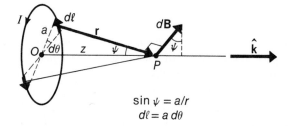

$$\sin \psi = a/r$$
$$d\ell = a\, d\theta$$

Solution:

At the point P a distance z along the loop axis, the field $d\mathbf{B}$ produced by the current element $Id\ell$ can be resolved into an axial component, $dB_{\parallel} = dB \sin \psi$, and a radial component, $dB_{\perp} = dB \cos \psi$. For each such current element there is a matching element directly opposite. The field components $dB_{\perp}$ exactly cancel, whereas the components $dB_{\parallel}$ add together. Therefore, using Eq. 34–3, we find

$$\mathbf{B} = B_{\parallel}\hat{\mathbf{k}} = \frac{\mu_0 I}{4\pi}\hat{\mathbf{k}} \oint \frac{a\, d\theta \sin \psi}{r^2}$$

With a, ψ, and r constant, and with $\oint d\theta = 2\pi$ and $\sin \psi = a/r = a/(z^2 + a^2)^{1/2}$, we have

$$\mathbf{B} = \frac{\mu_0 I a^2}{2(z^2 + a^2)^{3/2}}\hat{\mathbf{k}}$$

The Dipole Field. In Example 34–1, notice that when $z \gg a$, the result can be expressed as

$$\mathbf{B} = \frac{\mu_0}{2\pi z^3}\mathbf{m}$$

where $\mathbf{m} = \pi a^2 I\hat{\mathbf{k}}$ is the magnetic moment of the loop (see Eq. 33–14). In Section

28–1 we found that the electric field at large distances from an electric dipole is (Eq. 28–8)

$$\mathbf{E} = \frac{1}{4\pi\epsilon_0}\left(-\frac{\mathbf{p}}{r^3} + \frac{3(\mathbf{p}\cdot\mathbf{r})\mathbf{r}}{r^5}\right) \tag{34–5}$$

Taking $\mathbf{r}$ along the z-axis (the direction of $\mathbf{p}$), we have $\mathbf{r} = z\hat{\mathbf{k}}$ and $\mathbf{p} = p\hat{\mathbf{k}}$. Using these expressions in the equation for $\mathbf{E}$, we find

$$\mathbf{E} = \frac{1}{2\pi\epsilon_0 z^3}\mathbf{p}$$

(•Verify this result.) Thus, except for the different roles played by ϵ_0 and μ_0, the magnetic field along the axis of a current loop has the same behavior as the electric field along the axis of an electric dipole. Indeed, the distant field of a current loop has exactly the form of a dipole field, namely,

$$\mathbf{B} = \frac{\mu_0}{4\pi}\left(-\frac{\mathbf{m}}{r^3} + \frac{3(\mathbf{m}\cdot\mathbf{r})\mathbf{r}}{r^5}\right) \tag{34–6}$$

That is, a current loop is a *magnetic dipole*.

Equations 34–5 and 34–6 are the *distant-field* expressions for an electric dipole and a magnetic dipole, respectively. In each case the *near field* (where r is no longer much larger than the size of the dipole) is more complicated. Compare the field lines for the finite electric dipole (Fig. 28–5) with those for the point dipole (Fig. 29–6), for which the distant-field expression is valid for all r. Figure 34–4 shows the near field of a magnetic dipole in a plane that contains the magnetic moment vector $\mathbf{m}$. Notice that the field lines, although not circular, close on themselves just as in the case of a long straight wire (see Fig. 34–3). Do the field lines *always* close in this way? We answer this important question in Section 34–3.

Finally, we remark again that the Earth's magnetic field closely resembles the field of a magnetic dipole, at least for distances out to several Earth radii (see Fig. 33–1). The value of the Earth's magnetic moment is

$$m_E = 8.1 \times 10^{22}\,\text{A}\cdot\text{m}^2$$

(For distances greater than a few Earth radii, the magnetic effect of the stream of charged particles emitted by the Sun—the *solar wind*—severely distorts the dipole field of the Earth.)

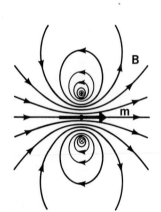

Fig. 34–4. Near field for a magnetic dipole. (•Can you define an N pole and an S pole for the dipole?)

Example 34–2

A *solenoid* is a helical coil that is constructed by tightly winding a single strand of fine wire around a straight, nonmagnetic cylindrical shell. Usually the diameter of the wire is so small compared with the diameter of the helix that a closely spaced winding produces a helix pitch that is negligible. Determine the magnetic induction B along the axis of a solenoid, with a length ℓ and a radius R, that has n turns per meter of wire carrying a current I. Refer to diagram (a).

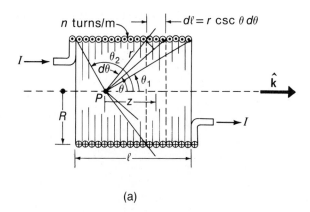

(a)

Solution:

Consider the magnetic field at a point P on the axis of the solenoid, as indicated in diagram (a). We use the result of Example 34–1 for the contribution $d\mathbf{B}$ of the turns that lie in the differential ring intercepted by two circular cones with half angles θ and $\theta + d\theta$, as shown. The axial length of this ring is $d\ell = r \csc \theta\, d\theta$, so the number of turns in the ring is $nd\ell = nr \csc \theta\, d\theta$. Thus, at point P, using $(z^2 + a^2)^{3/2} = r^3$, we have

$$d\mathbf{B} = \frac{\mu_0 nr \csc \theta \; IR^2\, d\theta}{2r^3}\,\hat{\mathbf{k}}$$

Using $r = R \csc \theta$, the expression for $d\mathbf{B}$ reduces to

$$d\mathbf{B} = \tfrac{1}{2}\mu_0 nI \sin \theta\, d\theta\,\hat{\mathbf{k}}$$

The field $\mathbf{B}$ is obtained by integrating between $\theta = \theta_1$ and $\theta = \theta_2$, as shown in diagram (a). The result is

$$\mathbf{B} = \tfrac{1}{2}\mu_0 nI(\cos \theta_1 - \cos \theta_2)\hat{\mathbf{k}}$$

This expression for the field on the solenoid axis is valid for all axial points P, whether inside or outside the solenoid. Diagram (b) shows the axial field B in units of $\mu_0 nI$ as a function of ξ, the position of P measured from the center of the solenoid in units of R, for the case $\ell = 4R$. Diagram (c) shows the magnetic field for this case. For this geometry the field along the axis, $|\xi| < 1$, is nearly constant with the approximate value $0.89\mu_0 nI$. In the event that the solenoid is very long ($\ell \gg R$), we have $\cos \theta_1 \to 1$ and $\cos \theta_2 \to -1$. Then, the axial field is

$$\mathbf{B} = \mu_0 nI\hat{\mathbf{k}}$$

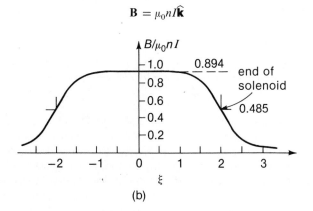

(b)

In Example 34–6 we find that for an infinite solenoid the field is constant with the value $\mu_0 nI$ everywhere within the solenoid.

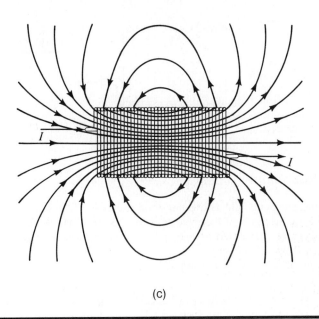

(c)

Example 34–3

Determine the magnetic induction produced by two long, straight, parallel wires that are a distance $2a$ apart, each carrying a current I in the same direction.

Solution:

The total field is the vectorial sum of the fields produced by the individual wires (principle of superposition). Diagram (a) illustrates the calculation of $\mathbf{B} = \mathbf{B}_1 + \mathbf{B}_2$ along the x- and y-axes. Along the x-axis the individual fields are entirely in the $\pm y$-direction. Consequently, the resultant field has only a y-component, $B_y(x, 0)$. Using Eq. 34–4 for each wire, we have, in the region $|x| > a$,

(a)

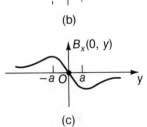

$$B_y(x, 0) = \frac{\mu_0 I}{2\pi} \left(\frac{1}{x - a} + \frac{1}{x + a} \right)$$

$$= \frac{\mu_0 I}{\pi} \frac{x}{x^2 - a^2}, \qquad |x| > a$$

For $|x| < a$, we have

$$B_y(x, 0) = \frac{\mu_0 I}{2\pi} \left(\frac{1}{a + x} - \frac{1}{a - x} \right)$$

$$= -\frac{\mu_0 I}{\pi} \frac{x}{a^2 - x^2}, \qquad |x| < a$$

These expressions for $B_y(x, 0)$ are shown in diagram (b). Notice that midway between the wires, we have $B_y(0, 0) = 0$. Also, at large distances ($|x| \gg a$) the field asymptotically approaches the field produced by a single wire along the z-axis that carries a current $2I$.

Along the y-axis the y-components of the two individual fields cancel, whereas the x-components add as shown in diagram (a). Thus, for $y > 0$, we have

$$B_x(0, y) = -2 \cdot \frac{\mu_0 I}{2\pi} \frac{\cos\theta}{(a^2 + y^2)^{1/2}}$$

$$= -\frac{\mu_0 I}{\pi} \frac{y}{a^2 + y^2}$$

This result is shown in diagram (c). (•Can you verify that this expression is also valid for $y < 0$?) Again, notice that at large distances ($|y| \gg a$) the field asymptotically approaches the field produced by a single wire along the z-axis that carries a current $2I$.

34-2 AMPERE'S CIRCUITAL LAW

In the discussion of the electric field produced by a static charge distribution (Section 29–2), we learned that the line integral of $\mathbf{E} \cdot d\mathbf{r}$ around any closed path is zero (Eq. 29–14). We now consider the static magnetic case and inquire about the line integral of $\mathbf{B} \cdot d\mathbf{r}$ around a closed path.

Figure 34–5 illustrates the case of a long straight wire that carries a steady current I directed out of the plane of the page. We wish to evaluate $\oint \mathbf{B} \cdot d\mathbf{r}$ for two closed paths—a path labeled ① that surrounds the wire and a path labeled ② that does not.* We can view $\mathbf{B} \cdot d\mathbf{r}$ as the product of B and the projection of $d\mathbf{r}$ onto $\mathbf{B}$. Because the field $\mathbf{B}$ around the straight wire has only an azimuthal component, the projection of $d\mathbf{r}$ onto $\mathbf{B}$ is $r \, d\phi$ for either path (see Fig. 34–5). Therefore, we have

$$\oint \mathbf{B} \cdot d\mathbf{r} = \oint B(r) r \, d\phi$$

Using Eq. 34–4 to write $B(r) = \mu_0 I / 2\pi r$, the line integral becomes

$$\oint \mathbf{B} \cdot d\mathbf{r} = \frac{\mu_0 I}{2\pi} \oint d\phi$$

For path ① the value of the integral is 2π, whereas for path ② the value is zero. Thus, we can express the result as

---

*Recall that $d\mathbf{r}$ represents an arbitrary displacement, whereas $d\boldsymbol{\ell}$, which appears only in combination with I, is always directed along a current filament. Refer to the footnote on page 221.

Fig. 34–5. Geometry for deriving Ampère's circuital law for a long straight wire that carries a steady current I. The line integral of $\mathbf{B} \cdot d\mathbf{r}$ does not depend on the shape of the integration path, even a path such as ① that loops back on itself. (•Can you see why?)

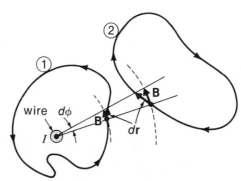

$$\oint \mathbf{B} \cdot d\mathbf{r} = \mu_0 I_p \qquad \text{(34-7)}$$

where I_p is the current enclosed by the integration path.

The result we have obtained in Eq. 34-7 is based on the special case of a steady current in a long straight wire. It was Ampère's brilliant guess that the equation is generally true for steady currents. Indeed, this result can be derived for any field produced by steady currents by using the general form of the Biot-Savart law, and experimental tests have shown the equation to be correct.

Suppose that we construct an open surface S with the perimeter coinciding with the integration path in Eq. 34-7. The total current I_p that is enclosed by the path must all flow through the surface S, as indicated in Fig. 34-6.* The current I_p can be expressed as the integral over S of the current density $\mathbf{J}$ (see Eq. 31-3); thus,

$$I_p = \int_S \mathbf{J} \cdot \hat{\mathbf{n}} \, dA$$

Then, we have

$$\oint \mathbf{B} \cdot d\mathbf{r} = \mu_0 \int_S \mathbf{J} \cdot \hat{\mathbf{n}} \, dA \qquad \text{(34-8)}$$

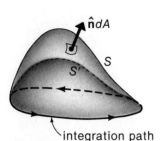

Fig. 34-6. Two arbitrary open surfaces, S and S', each with a perimeter corresponding to the integration path in Eq. 34-7 or 34-8. When a steady current flows and no charge accumulates (or disappears) between the two surfaces, it follows that the integrals of $\mathbf{J} \cdot \hat{\mathbf{n}} \, dA$ over both of the surfaces are the same. (See, however, the discussion in Section 35-4.)

The positive direction of $\hat{\mathbf{n}} \, dA$ is given by a right-hand rule: If the fingers of the right hand are curled in the direction of the integration path, the extended thumb points in the positive sense of $\hat{\mathbf{n}}$. Equation 34-8 is the most general statement of *Ampère's circuital law*.

As a practical matter, Ampère's circuital law is most useful for determining unknown magnetic fields for situations in which the geometry possesses a high degree of symmetry. When such symmetry is present, it is often possible to deduce the shape of the field lines, thereby simplifying the evaluation of the line integral.† (The integral is easy to evaluate when $\mathbf{B}$ lies along or perpendicular to the path of integration.) The examples at the end of this section illustrate a few applications of this type.

▶ Equation 34-8 is the *integral* formulation of Ampère's circuital law. We can obtain a *differential* formulation by using Stokes' theorem, which gives the connection between the line integral of a vector quantity and the surface integral of the curl of the vector. Applying Eq. 29-44 to the left-hand side of Eq.

---

*We must exclude such exceptional surfaces as the Möbius strip, which has but one edge and one side, and for which an outward directed normal $\hat{\mathbf{n}}$ is not defined! To construct a Möbius strip, take a strip of paper, give it a half twist, then glue the ends together. Can you identify more than one edge or one side?

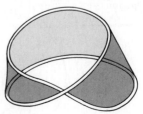

†This is analogous to the situation for Gauss's law; see the similar remark in Section 28-4.

34-8 gives

$$\oint \mathbf{B} \cdot d\mathbf{r} = \int_S (\mathbf{\nabla} \times \mathbf{B}) \cdot \hat{\mathbf{n}} \, dA \qquad \textbf{(34-9)}$$

where the surface integral is evaluated over any surface S whose boundary coincides with the path of the line integral, as in Fig. 34-6. Because Eqs. 34-8 and 34-9 are true for any path, it follows that the integrands on the right-hand sides are equal; that is,

$$\mathbf{\nabla} \times \mathbf{B} = \mu_0 \mathbf{J} \qquad \textbf{(34-10)}$$

which is the desired differential form of Ampère's law. Contrast this result with the electrostatic case, for which (Eq. 29-43)

$$\oint \mathbf{E} \cdot d\mathbf{r} = 0 \quad \text{and} \quad \mathbf{\nabla} \times \mathbf{E} = 0 \qquad \blacktriangleleft$$

Example 34-4

A long straight wire with a radius R carries a steady current I that is distributed uniformly over the cross section of the wire. Determine the magnetic induction everywhere.

Solution:

The current density in the wire is $J = I/A = I/\pi R^2$. To calculate the magnetic induction within the wire, we use a circular Ampèrian path (labeled ① in diagram (a)) that is centered on the wire axis; this path lies in a plane that is perpendicular to the wire axis. Because of the cylindrical symmetry, we expect that $\mathbf{B}$ is entirely azimuthal (no radial component) and is constant in magnitude along the path ①. The magnitude and radial dependence of $\mathbf{B}$ can be determined by using Eq. 34-8. The simplest surface for which ① is the perimeter is the circular disc S_1. Then,

$$\oint_① \mathbf{B} \cdot d\mathbf{r} = \mu_0 \int_{S_1} \mathbf{J} \cdot \hat{\mathbf{n}} \, dA$$

$$2\pi r B = \mu_0 \cdot \pi r^2 \cdot \frac{I}{\pi R^2}$$

from which we have

$$\mathbf{B} = \frac{\mu_0 I}{2\pi R^2} r \hat{\mathbf{u}}_\phi, \qquad r \le R$$

(a)

To calculate the field outside the wire, we use the circular path ② and split the integral over the circular disc S_2 into two parts—one integral over $S_2 - S_0$ and the other over $S_0 = \pi R^2$, the cross section of the wire. We find

$$\oint_② \mathbf{B} \cdot d\mathbf{r} = \mu_0 \int_{S_2 - S_0} \mathbf{J} \cdot \hat{\mathbf{n}} \, dA + \mu_0 \int_{S_0} \mathbf{J} \cdot \hat{\mathbf{n}} \, dA$$

$$2\pi r B = 0 + \mu_0 I$$

so that

$$\mathbf{B} = \frac{\mu_0 I}{2\pi r} \hat{\mathbf{u}}_\phi, \qquad r \ge R$$

(•Why is the surface integral over $S_2 - S_0$ identically zero?) Notice that we could have obtained the same results in this case by using Eq. 34–7. The variation of B with r is shown in diagram (b).

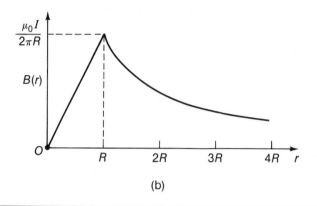

(b)

Example 34–5

Verify Ampère's circuital law by examining the axial magnetic field of a circular current loop.

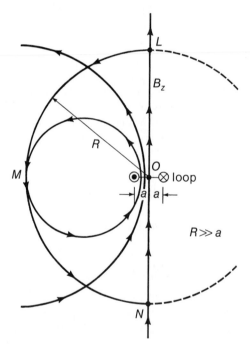

Solution:

Consider the closed Ampèrian path $NOLMN$ shown in the diagram. The path consists of a segment NOL along the axis of the current loop and a semicircular segment LMN that has a radius R. We need to calculate the Ampèrian path integral for the case in which $R \rightarrow \infty$. The field along the portion of the path segment NOL near O decreases relatively slowly with the distance from O. Thus, there is a significant contribution to the Ampèrian path integral from the segment NOL even for $R \rightarrow \infty$. However, in this limit the entire segment LMN is at a large distance from O ($R \gg a$, where a is the loop radius), and the field is a dipole field. Consequently, along the path LMN we find $B \propto 1/R^3$, whereas the path length is proportional to R. Thus, the integrand $\mathbf{B} \cdot d\mathbf{r}$ varies inversely with R^2, and in the limit, $R \rightarrow \infty$, the line integral along *LMN* vanishes. Thus, we have

$$\oint \mathbf{B} \cdot d\mathbf{r} = \int_{-\infty}^{\infty} B_z \, dz$$

where B_z is the field component along the loop axis, which was obtained in Example 34-1. Using this result, we find

$$\int_{-\infty}^{\infty} B_z \, dz = \tfrac{1}{2}\mu_0 I a^2 \int_{-\infty}^{\infty} \frac{dz}{(z^2 + a^2)^{3/2}}$$

$$= \tfrac{1}{2}\mu_0 I a^2 \left[\frac{z}{a^2(z^2 + a^2)^{1/2}} \right]_{-\infty}^{\infty}$$

so that

$$\oint \mathbf{B} \cdot d\mathbf{r} = \mu_0 I$$

Thus, Ampère's circuital law is verified for this case.

Example 34–6

Determine the magnetic induction produced by an infinitely long solenoid that consists of *n* turns per meter of wire carrying a current *I*.

Solution:

The diagram shows a portion of the solenoid cut by a plane that contains the axis of the coil. Consider the closed rectangular Ampèrian path *abcda* that lies in this plane.

To evaluate the line integral of $\mathbf{B} \cdot d\mathbf{r}$ we need to know the interior field $\mathbf{B}_1$ and the exterior field $\mathbf{B}_2$. What do we expect for the fields in this case? First, because the solenoid has an infinite length, the fields must be axial everywhere.* The interior field $\mathbf{B}_1$ is directed toward the right; the exterior field $\mathbf{B}_2$ (if it exists) is directed toward the left. (Refer to the diagrams in Examples 34–1 and 34–2.)

Using Eq. 34–7, we have for the path *abcda*,

$$B_1 \ell + B_2 \ell = \mu_0 n \ell I$$

Now, consider r_1 fixed but allow $r_2 \to \infty$. Then, the equation above requires B_2 to be constant. The field cannot have a constant nonzero value extending to infinity, so we conclude

---

*Because there is a net current *I* that flows axially along the solenoid, there is an azimuthal field, $B_\phi = \mu_0 I/2\pi r$, outside the solenoid. For reasonable values of *R* and *n*, we have $B_\phi \ll B_1$. (•Can you verify this?) In any event, $\mathbf{B}_\phi$ is perpendicular to *dr* along the integration path, so B_ϕ does not contribute to the integral of $\mathbf{B} \cdot d\mathbf{r}$. (•Can you show that $B_\phi = 0$ inside the solenoid?)

that the constant value of B_2 must be zero. Therefore, we have

$$B_1 = \mu_0 n I$$

and the interior field is constant, independent of r_1. (Compare Example 34–2.)

Example 34–7

Determine the magnetic induction inside a toroidal coil that has a mean radius R, as shown in the diagram. The coil has a total of N turns of wire carrying a current I.

Solution:

Choose as an integration path a circle with radius r that lies within the toroid, as indicated in the diagram. The symmetry demands that the field inside the toroid be everywhere azimuthal. Then, the integral of $\mathbf{B} \cdot d\mathbf{r}$ around the circular path yields $2\pi r B$. The total current linked by the integration path is $I_p = NI$. Then, using Eq. 34–7, we find

$$B = B_\phi = \frac{\mu_0 N I}{2\pi r}$$

If the mean radius R of the toroid is much larger than the cross-sectional dimensions of the ring, we have $r \cong R$. Then, with $n = N/2\pi R$ equal to the number of turns per unit length, we have

$$B = B_\phi = \mu_0 n I$$

(•Can you explain why this result is the same as that for the infinite solenoid obtained in Example 34–6? What is the nature of the field outside the toroid? It is not zero. Refer to the footnote for Example 34–6.)

34-3   GAUSS'S LAW FOR MAGNETIC FIELDS

The quantization of electric charge is a firmly established fundamental property of matter. Electric charge is always observed to occur in units of $\pm e$. Is there a corresponding elementary unit of "magnetic charge," a magnetic *monopole*? Although there have been theoretical speculations concerning magnetic monopoles, no convincing experimental evidence of their existence has ever been obtained. Therefore, the laws of electromagnetism are formulated on the basis that *there exists no fundamental isolated magnetic charge.*

If there are no magnetic monopoles, the lowest order of elementary magnetic induction is the *dipole*. On a macroscopic scale, magnetic dipole fields are set up by current loops of the type we have been discussing. On a microscopic (atomic) scale, a quantum treatment is necessary; nevertheless, a convenient model of atomic magnetism attributes the field to the effective current produced by electrons circulating around the nucleus. In addition to this orbital effect, elementary particles possess an *intrinsic* angular momentum and an associated magnetic moment. For example, the intrinsic angular momentum (or *spin*) of the electron is $s = \frac{1}{2}(h/2\pi) = 5.28 \times 10^{-35}$ J·s and the intrinsic magnetic dipole moment of the electron is $m_B = 9.27 \times 10^{-24}$ A·m², a quantity called the *Bohr magneton*.

What is the effect of the absence of magnetic monopoles on the shapes of magnetic field lines? Recall that electric field lines due to static charges always originate on positive charges and terminate on negative charges. Because there are no magnetic charges to originate or terminate magnetic field lines, these lines necessarily close upon themselves. It is for this reason that in the various cases we have studied the magnetic field lines have always been shown as nonintersecting closed loops. Although the *distant* electric and magnetic dipole fields obey the same mathematical description (see Eqs. 34–5 and 34–6), the *near* fields are quite different. Figure 34–7 shows the pattern of field lines for an electric dipole and for a magnetic dipole. Notice how the fields begin to resemble one another at large distances.

Gauss's Law. Figure 34–7 also shows a Gaussian surface S in each of the dipole fields. For the case of the electric dipole field (Fig. 34–7a), the surface encloses the positive charge $+Q$ of the dipole. According to Eq. 28–19 we have

$$\oint_S \mathbf{E} \cdot \hat{\mathbf{n}} \, dA = \frac{1}{\epsilon_0} Q$$

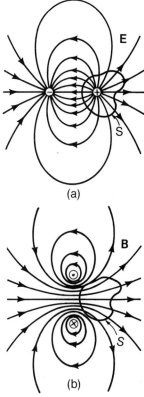

Fig. 34–7. (a) The electric field intensity due to an electric dipole. (b) The magnetic induction due to a magnetic dipole (a current loop). A Gaussian surface S is shown in each case.

The fact that this integral is *positive* means that there is a net *outward* flux of field lines piercing the surface, as we expect from the geometry (Fig. 34–7a).

For the case of the magnetic dipole field (Fig. 34–7b), every field line that *enters* the volume enclosed by the surface S also *leaves* the volume. Consequently, the net magnetic flux through the surface must be zero. Because magnetic monopoles do not exist, this is a general result: the net magnetic flux through *any* Gaussian surface in *any* magnetic field must be zero. That is, Gauss's law for magnetic fields is

$$\oint_S \mathbf{B} \cdot \hat{\mathbf{n}} \, dA = 0 \qquad \text{always} \qquad \textbf{(34–11)}$$

where S is a *closed* surface. In analogy with the electric case (see Eq. 28–11), the integral of $\mathbf{B} \cdot \hat{\mathbf{n}} \, dA$ is called the *magnetic flux* φ_M. The value of φ_M is always zero for a closed surface, but the flux can be nonzero for an open surface. (Be certain that you understand the distinction.)

▶ *The Differential Form of Gauss's Law.* In Section 29–5 we proved the *divergence theorem* as applied to an electrostatic field, namely,

$$\oint_S \mathbf{E} \cdot \hat{\mathbf{n}} \, dA = \int_V (\mathbf{\nabla} \cdot \mathbf{E}) \, dv = \frac{1}{\epsilon_0} \int_V \rho \, dv$$

where V is the volume enclosed by the surface S, and where ρ is the volume charge density in V. Because the second equality must be valid for an arbitrary volume V, we conclude that the integrands must be equal; that is,

$$\nabla \cdot \mathbf{E} = \frac{\rho}{\epsilon_0}$$

For the magnetic case, we have

$$\oint_S \mathbf{B} \cdot \hat{\mathbf{n}}\, dA = \int_V (\nabla \cdot \mathbf{B})\, dv = 0$$

Again, this result is valid for an arbitrary volume V; thus,

$$\nabla \cdot \mathbf{B} = 0, \qquad \text{everywhere} \tag{34–12a}$$

which means

$$\frac{\partial B_x}{\partial x} + \frac{\partial B_y}{\partial y} + \frac{\partial B_z}{\partial z} = 0 \tag{34–12b}$$

These equivalent expressions represent the differential formulation of Gauss's law for magnetic fields. ◄

34-4 FORCES BETWEEN CURRENT-CARRYING WIRES

Consider two infinitely long, straight, parallel wires separated by a distance a, carrying currents I_1 and I_2 in the same direction (Fig. 34–8). What force per unit length acts between the wires? Let $\mathbf{F}_{12}$ be the force that acts on a short length ℓ of wire 1 due to the field produced by the current in wire 2. This force can be calculated by noting that the magnetic field at the position of wire 1 produced by the entire (infinite) length of wire 2 is (see Eq. 34–4)

$$B_2 = \frac{\mu_0 I_2}{2\pi a}$$

The direction of $\mathbf{B}_2$ is shown in Fig. 34–8. Now, according to Eq. 33–4, the force on a current element $I\,d\boldsymbol{\ell}$ due to a field $\mathbf{B}$ is $d\mathbf{F} = I\,d\boldsymbol{\ell} \times \mathbf{B}$. Thus, for a length ℓ of wire 1, we have $F_{12} = I_1 \ell B_2$, or

$$F_{12}/\ell = I_1 B_2 = \frac{\mu_0 I_1 I_2}{2\pi a} \tag{34–13}$$

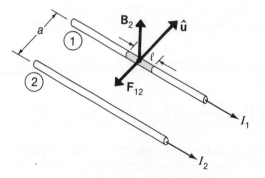

Fig. 34–8. $\mathbf{F}_{12}$ is the force on a length ℓ of wire 1 due to the field $\mathbf{B}_2$ produced by the current in wire 2.

Also, it is important to note that $\mathbf{F}_{12}$ is directed *toward* wire 2. That is, when the currents in the two wires flow in the same direction, the mutual force is *attractive;* when the currents flow in opposite directions, the mutual force is *repulsive.*

Definition of the Ampere.   The definition of the ampere in use at present is: *An ampere is that steady current which, when flowing in each of two long, straight, parallel wires separated by a distance of 1 m, causes each wire to exert on the other wire a force of 2×10^{-7} N for each meter of length of the wire.* With this definition and the expression for F_{12}/ℓ in Eq. 34-13, we see that $\mu_0/4\pi = K_M = 10^{-7}$ N/A^2, exactly, as was indicated in Section 34-1.

In Section 27-3, we gave a tentative definition of the *coulomb* in terms of the electron charge. Now that the ampere has been defined, we can give a proper definition of the coulomb: *A coulomb is the amount of charge that passes a particular point in a wire during a time of 1 s when a steady current of 1 A flows in the wire.* That is, $1\,C = (1\,A) \times (1\,s)$. This definition of the coulomb allows the precise evaluation of the constant K that appears in Coulomb's law. The result, as we stated in Section 27-3, is $K = 8.98755 \times 10^9$ N · m^2/C^2.

We now obtain an interesting result by calculating the ratio of the electric force constant K to the magnetic force constant K_M, namely,

$$\frac{K}{K_M} = \frac{1}{\epsilon_0 \mu_0} = \frac{8.98755 \times 10^9 \text{ N} \cdot \text{m}^2/\text{C}^2}{10^{-7} \text{ N}/(\text{C/s})^2} = (2.99792 \times 10^8 \text{ m/s})^2 = c^2 \quad \textbf{(34-14)}$$

The quantity on the right-hand side of this expression is c^2, the square of the speed of light! This striking equality can hardly be an accident; there must be a connection between electricity and magnetism, on the one hand, and light, on the other. The importance of the fact that the experimental value of K/K_M is close to the measured value of c^2 was first recognized by Gustav Kirchhoff. The great Scottish mathematical physicist James Clerk Maxwell (1831–1879) incorporated this idea into his theory of electromagnetism (1864) and asserted the equivalence of electromagnetic radiation and light waves. We discuss this aspect of Maxwell's theory in Chapter 38.

▶ *Newton's Third Law for Current Loops.*   We now prove that Newton's third law is obeyed in the interaction between any pair of current-carrying circuits. This is true in spite of the fact that the third law is not valid when applied to the forces between two *differential* current elements.

Consider the two differential current elements shown in Fig. 34-9, one in each loop. Let $d^2\mathbf{F}_{12}$ represent the force exerted *on* the current element $I_1 d\ell_1$. Then, we use Eq. 33-4 to write*

$$d^2\mathbf{F}_{12} = I_1 d\ell_1 \times d\mathbf{B}_2 \quad \textbf{(34-15)}$$

Fig. 34-9. The force $d^2\mathbf{F}_{12}$ is the force on the current element $I_1 d\ell_1$ due to the magnetic field produced by $I_2 d\ell_2$.

where $d\mathbf{B}_2$ is the magnetic field at the position of the current element $I_1 d\ell_1$, produced by $I_2 d\ell_2$. Using the Biot-Savart law (Eq. 34-2), we have

$$d\mathbf{B}_2 = \frac{\mu_0}{4\pi} \frac{I_2 d\ell_2 \times \mathbf{r}_{21}}{r_{21}^3}$$

where $\mathbf{r}_{21}/r_{21}$ has been substituted for the unit vector $\hat{\mathbf{u}}_r$. Using this expression for $d\mathbf{B}_2$ in Eq. 34-1, we find

$$d^2\mathbf{F}_{12} = \frac{\mu_0}{4\pi} \frac{I_1 I_2}{r_{21}^3} d\ell_1 \times (d\ell_2 \times \mathbf{r}_{21}) \quad \textbf{(34-16)}$$

---

*Notice that the differential force $d^2\mathbf{F}_{12}$ is a *second-order* differential because it involves the product of two first-order differentials, $d\ell_1$ and $d\mathbf{B}_2$.

It is easy to show that $d^2\mathbf{F}_{21}$, the force on the current element $I_2 d\boldsymbol{\ell}_2$ due to the magnetic field produced by the current element $I_1 d\boldsymbol{\ell}_1$, is *not* equal to $-d^2\mathbf{F}_{12}$! (See Problem 34-16.) Does this mean that Newton's third law is not valid in this situation? We must remember that isolated current elements do not exist. Therefore, to make a proper test of the third law, we should calculate the mutual forces between current loops, not simply between differential current elements.

Consider now the complete (stationary) loops of which $I_1 d\boldsymbol{\ell}_1$ and $I_2 d\boldsymbol{\ell}_2$ are elements. The total force $\mathbf{F}_{12}$ that acts on loop 1 due to the magnetic field produced by loop 2 is obtained by calculating the double integral of $d^2\mathbf{F}_{12}$ (Eq. 34-16) over the two loops. Thus,

$$\mathbf{F}_{12} = \frac{\mu_0 I_1 I_2}{4\pi} \oint_2 \oint_1 \frac{d\boldsymbol{\ell}_1 \times (d\boldsymbol{\ell}_2 \times \mathbf{r}_{21})}{r_{21}^3}$$

We can use the identity $\mathbf{A} \times (\mathbf{B} \times \mathbf{C}) = (\mathbf{A} \cdot \mathbf{C})\mathbf{B} - (\mathbf{A} \cdot \mathbf{B})\mathbf{C}$ to write the integrand as $[(d\boldsymbol{\ell}_1 \cdot \mathbf{r}_{21})d\boldsymbol{\ell}_2 - (d\boldsymbol{\ell}_1 \cdot d\boldsymbol{\ell}_2)\mathbf{r}_{21}]/r_{21}^3$. The first term gives rise to the integral

$$\oint_2 \oint_1 \frac{(d\boldsymbol{\ell}_1 \cdot \mathbf{r}_{21})d\boldsymbol{\ell}_2}{r_{21}^3} = \oint_2 d\boldsymbol{\ell}_2 \left(\oint_1 \frac{\mathbf{r}_{21} \cdot d\boldsymbol{\ell}_1}{r_{21}^3} \right)$$

The integral over loop 1 (the integral within the brackets) has the form of a closed-line integral of a time-independent inverse-square-law field, namely, $\mathbf{r}_{21}/r_{21}^3 = \hat{\mathbf{u}}_r/r_{21}^2$. We know that any such field is conservative and that the closed-line integral vanishes (see Eq. 29-5). Thus, the expression for $\mathbf{F}_{12}$ becomes

$$\mathbf{F}_{12} = -\frac{\mu_0 I_1 I_2}{4\pi} \oint_2 \oint_1 \frac{\mathbf{r}_{21}}{r_{21}^3}(d\boldsymbol{\ell}_1 \cdot d\boldsymbol{\ell}_2) \qquad \text{(34-17)}$$

This is the result we have been seeking—it gives the loop-loop interaction without reference to any specific magnetic field. Because $\mathbf{r}_{12} = -\mathbf{r}_{21}$, it is evident that $\mathbf{F}_{21} = -\mathbf{F}_{12}$. That is, Newton's third law is in fact valid for the physically realizable condition of complete current loops.

Example 34-8

Use Eq. 34-17 to calculate the force per unit length F_{12}/ℓ for the situation shown in Fig. 34-8.

Solution:

Consider the force on a length ℓ of wire 1 due to the entire (infinite) length of wire 2, as shown in Fig. 34-8. Using Eq. 34-17, we can write

$$\mathbf{F}_{12} = -\frac{\mu_0 I_1 I_2 \ell}{4\pi} \oint_1 \frac{\mathbf{r}_{21}}{r_{21}^3} d\boldsymbol{\ell}_1$$

The integral here is the same as that encountered in the calculation of the electric field of a uniformly charged infinite filament (see Section 28-4), namely,

$$\oint_1 \frac{\mathbf{r}_{21}}{r_{21}^3} d\boldsymbol{\ell}_1 = 2\frac{\hat{\mathbf{u}}}{a}$$

where $\hat{\mathbf{u}}$ is the unit vector shown in Fig. 34-8. Then, the force per unit length that acts on wire

l is given by

$$\mathbf{F}_{12}/\ell = -\frac{\mu_0 I_1 I_2}{2\pi a}\hat{\mathbf{u}}$$

which is the vector form of the result we obtained in Eq. 34–13. ◄

PROBLEMS

Section 34-1

34-1 Determine the magnetic induction at a point P that is a distance x from the corner of an infinitely long wire that is bent at a right angle, as shown in the diagram. The wire carries a steady current I.

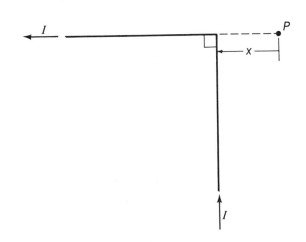

34-2 A thin, flat, infinitely long ribbon of width w carries a uniform current I. Determine the magnetic induction at a point P that is in the plane of the ribbon and is a distance x from one edge, as shown in the diagram. Test your result in the limit $w \to 0$.

34-3 Two long straight wires are connected by a circular section that has a radius R, as shown in the diagram. All three segments lie in the same plane and carry a steady current I. Determine the magnetic induction at the center O of the circular segment.

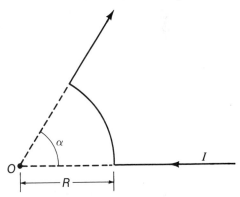

34-4 Assume that the Earth's magnetic field is that of a dipole with moment $m_E = 8.1 \times 10^{22}\,\text{A} \cdot \text{m}^2$. Calculate the magnetic induction at a point on the Earth's surface at the magnetic equator.

34-5 Determine the magnetic induction at a point P on the
•• axis (the z-axis) of a square current loop with sides $2a$ that carries a current I, as shown in the diagram.

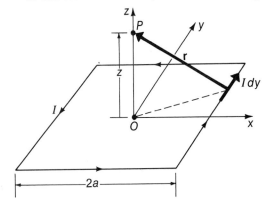

[*Hint:* First, show that the vector **r** is equal to $-x\hat{\mathbf{i}} - y\hat{\mathbf{j}} + z\hat{\mathbf{k}}$, where (x, y) locates the current element $Id\boldsymbol{\ell}$ and where $z\hat{\mathbf{k}}$ locates the point P. Then, use the symmetry of the situation to argue that only the z-component of **B** at P is nonvanishing. Finally, express $(Id\boldsymbol{\ell} \times \mathbf{r}) \cdot \hat{\mathbf{k}}$ in a form that permits evaluation of the integral.] Find the field at O by setting $z = 0$; compare with the field at the center of a circular loop with radius a (use the result of Example 34–1 with $z = 0$). Is the comparison reasonable? Explain.

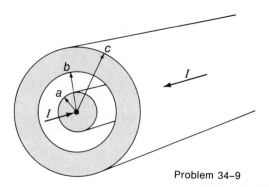

Problem 34–9

Section 34-2

34-6 A short solenoid, with a length of 10 cm and a radius of 5 cm, consists of 200 turns of fine wire that carries a current of 15 A. What is the magnetic induction at the center of the solenoid? For the same number of turns per unit length, what value of B would result for $\ell \to \infty$?

34-7 Use the results of Example 34–2 to determine as a function of length the magnetic induction B in the center of a solenoid with finite length and fixed radius. Make a graph of $B/\mu_0 nI$ as a function of $\eta = \ell/R$ for $0.5 \leqslant \eta \leqslant 20$.

34-8 Two infinite parallel conducting sheets carry equal surface currents of density **j** (A/m) in collinear but opposite directions, as shown in the diagram. (a) Show that the magnetic field between the sheets is uniform and that the field vanishes everywhere else. (b) Show that the magnetic induction between the sheets is $B = \mu_0 j$.

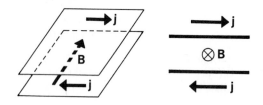

34-9 The long coaxial cable shown in the diagram above •• right carries a uniform current I in the center conductor, and this current is returned in the outer cylindrical shell. Determine the magnetic induction everywhere. Sketch a graph of your results.

34-10 Consider a toroidal coil of the type discussed in Example 34–7 that consists of 500 turns with a mean radius $R = 12$ cm and a square cross section, 4 cm × 4 cm. Determine the maximum and minimum values of the magnetic induction within the coil when the current is 25 A.

Section 34-3

34-11 Calculate the magnetic flux, $\varphi_M = \int_S \mathbf{B} \cdot \hat{\mathbf{n}} \, dA$, through the flat rectangular surface S, shown in the diagram, whose plane contains an infinitely long wire that carries a current I. (Notice that S is an *open* surface.)

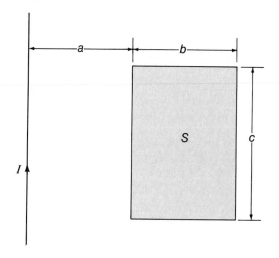

34-12 Determine by explicit calculation the magnetic flux, • $\varphi_M = \oint_S \mathbf{B} \cdot \hat{\mathbf{n}} \, dA$, through the *closed* surface S shown in the diagram on the next page. Use the results obtained in Example 34–3. Assume $R \gg a$ and $\epsilon \ll a$. [*Hint:* Note carefully the behavior of the magnetic induction **B** when $R \gg a$ and when $\epsilon \ll a$.]

34-13 A magnetic field is described by $\mathbf{B} = 2xyb\hat{\mathbf{i}} + y^2 a\hat{\mathbf{j}}$. •• (a) What is the relationship between the constants a and b? (b) Determine the steady current density **J** that produces this field.

Section 34-4

34-14 Calculate the force on the rectangular loop shown in the diagram, which carries a current $I_2 = 10$ A,

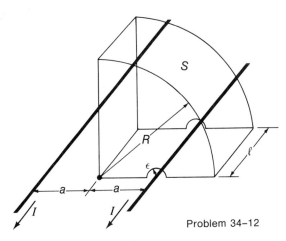

Problem 34–12

exerted by the field due to a long coplanar wire that carries a current $I_1 = 30$ A. (Take $a = 20$ cm, $b = 40$ cm, and $c = 50$ cm.)

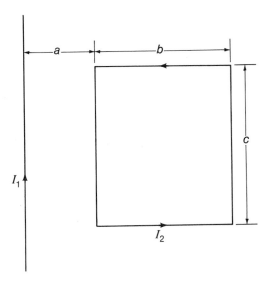

34–15 A 10-cm length of wire with a mass of 20 g is attached frictionlessly to the vertical segments of a wire in which flows a current I. The surrounding uniform horizontal field has $B = 10^4$ G and the direction shown in the diagram. What must be the current I to maintain the 10-cm wire in an equilibrium position?

34–16 Refer to the current elements in Fig. 34–9. Show in
• general that $d\boldsymbol{\ell}_1 \times (d\boldsymbol{\ell}_2 \times \mathbf{r}_{21})$ is not equal to $-d\boldsymbol{\ell}_2 \times (d\boldsymbol{\ell}_1 \times \mathbf{r}_{12})$ and, hence, that $d^2\mathbf{F}_{12} \neq -d^2\mathbf{F}_{21}$. [*Hint:* Use the vector identity for $\mathbf{A} \times (\mathbf{B} \times \mathbf{C})$.]

Additional Problems

34–17 Determine the magnetic induction at the center O of the current loop shown in the diagram.

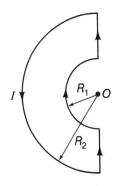

34–18 In an historic experiment conducted in 1878, the American physicist Henry A. Rowland demonstrated that a magnetic field is produced by a rotating charged disc. Consider a uniformly charged nonconducting disc, with total charge Q and radius R, that rotates with an angular speed ω. Derive an expression for the magnetic induction at the center of the disc. Determine the value of B for $Q = 0.1$ μC, $R = 10$ cm, and $\omega = 120\pi$ s^{-1}. Compare this field with that of the Earth. Does this comparison suggest that Rowland's experiment was easy to perform?

34–19 Show that the interaction energy U_{12} of two magnetic dipoles, $\mathbf{m}_1$ and $\mathbf{m}_2$, separated by a distance $\mathbf{r}$ is

$$U_{12} = \frac{\mu_0}{4\pi}\left[\frac{3(\mathbf{m}_1 \cdot \mathbf{r})(\mathbf{m}_2 \cdot \mathbf{r})}{r^5} - \frac{\mathbf{m}_1 \cdot \mathbf{m}_2}{r^3}\right]$$

34–20 Refer to Example 34–2. Compute values of $B/\mu_0 nI$ for $\xi = 0$, $\pm\frac{1}{4}$, $\pm\frac{1}{2}$, ± 1, ± 2, ± 4 for the case $\ell/R = 1$. Make a graph of the results and compare with the graph in Example 34–2, which is for the case $\ell/R = 4$.

34–21 An infinitely long cylindrical conductor contains
• a cylindrical cavity whose center line is a distance $\mathbf{a}$ from the center line of the conductor, as shown in the diagram. The conductor carries a uniform current density $\mathbf{J} = J\hat{\mathbf{k}}$, where $\hat{\mathbf{k}}$ is along the center line of the conductor. Show that the magnetic field in

the cavity is uniform and is described by $\mathbf{B} = \frac{1}{2}\mu_0 J\hat{\mathbf{k}} \times \mathbf{a}$. [*Hint:* Calculate the field at an arbitrary point P in the cavity by superimposing the interior field of a *solid* cylindrical conductor with a radius R_1 and a current density $\mathbf{J}$ onto the interior field of a *solid* cylindrical conductor with a radius $R_2 < R_1$ and a current density $-\mathbf{J}$. Note also that the appropriate azimuthal unit vectors can be expressed as $\hat{\mathbf{u}}_{\phi 1} = \hat{\mathbf{k}} \times \mathbf{r}_1$ and $\hat{\mathbf{u}}_{\phi 2} = \hat{\mathbf{k}} \times \mathbf{r}_2$.]

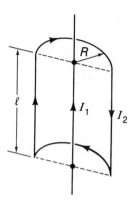

34–22   An infinitely long straight wire that carries a current $I_1 = 50$ A is partially surrounded by the loop shown in the diagram. The loop has a length $\ell = 35$ cm

and a radius $R = 15$ cm, and carries a current $I_2 = 20$ A. The axis of the loop coincides with the infinite wire. Calculate the force that acts on the loop.

34–23   A magnetic field is described by $\mathbf{B} = B_x(x, y)\hat{\mathbf{i}} + B_y(x, y)\hat{\mathbf{j}}$. (a) Show that

$$\frac{\partial B_x}{\partial x} = -\frac{\partial B_y}{\partial y}$$

$$\mathbf{J} = \frac{1}{\mu_0}\left[\frac{\partial B_y}{\partial x} - \frac{\partial B_x}{\partial y}\right]\hat{\mathbf{k}}$$

(b) If $\mathbf{J} = 0$ in the region of interest and if $B_x(x, y) = axy$, determine the most general expression for $B_y(x, y)$.

ELECTROMAGNETIC INDUCTION

In the preceding chapters the discussions were restricted to steady-state conditions; that is, we considered only electric fields produced by static charges and magnetic fields produced by steady currents. We now remove this restriction and allow the possibility of time-varying electric and magnetic fields. Consideration of this general situation leads to *Maxwell's equations*, which together with the Lorentz force equation yield a complete and elegant formulation of all of the laws of electromagnetism.

35-1 THE DISCOVERY OF ELECTROMAGNETIC INDUCTION

An electric field imposed on a conductor can cause a current to flow and this current can, in turn, produce a magnetic field. Does there exist a reciprocal effect whereby a magnetic field can produce an electric field? During the 1820's Michael Faraday* searched unsuccessfully for such an effect in steady-state situations. Finally, in 1831, he discovered that an electric field actually could be produced by a *time-varying* magnetic field. This is the phenomenon of *electromagnetic induction.*†

Figure 35–1a illustrates one of Faraday's early experiments. The apparatus consists of a bar magnet and a wire loop L that is part of a circuit that contains a galvanometer G to indicate any current flow. If the bar magnet (whose field is permanent and constant) is stationary with respect to the loop L, the galvanometer indicates no current in the circuit. However, if the magnet is moved along the axis of the loop, as indicated in Fig. 35–1a, a current does flow in the circuit. Figure 35–1b shows that the current flows in one direction as the magnet approaches the loop, and flows in the opposite direction after the magnet has passed through the loop, still moving along the axis.

Fig. 35–1. (a) When the magnet is in motion, a current is induced in the loop. (b) The current flows in one direction as the magnet approaches the loop and flows in the opposite direction after the magnet has passed through the loop.

---

*Michael Faraday (1791–1867), English chemist and physicist, was probably the most gifted experimental scientist of all time. His many accomplishments included the discovery of benzene, the formulation of the laws of electrolysis, the discovery of electromagnetic induction, and the construction of the first electric generator.

†Electromagnetic induction seems actually to have been discovered first by the American physicist Joseph Henry in 1830; however, his results were not published until after Faraday's announcement in 1831, so Faraday is accorded the credit for the discovery.

939

Fig. 35–2. When the switch P is closed and a current builds up in circuit ①, a current is induced to flow in circuit ②.

Faraday also discovered that a current can be induced in a circuit by a changing current in a nearby circuit, without any motion of either circuit. When the circuit ① shown in Fig. 35–2 is completed by closing the switch P, a surge of current is observed in circuit ②.* When a steady current is eventually reached in the primary circuit, the current in the secondary circuit becomes zero. If the switch P is now opened, a new current surge is observed in the secondary circuit (but in the opposite direction). Again, an EMF is induced in the loop of the secondary circuit only while the magnetic field through the loop is *changing*. This effect is called *mutual induction*. At about the same time as Faraday's work, Henry discovered that the changing current in a loop induces an EMF in the circuit itself; this effect is called *self induction*.

35-2 FARADAY'S LAW OF INDUCTION

Perhaps Faraday's greatest achievement in electrical science was the reduction of all the experimental evidence concerning electromagnetic induction to a simple quantitative law. *Faraday's law of induction* relates the instantaneous induced EMF $\mathcal{E}$ in a closed circuit to the time rate of change of the magnetic flux that links the circuit. This nonlocalized EMF produces the observed current in the circuit and is equal to the line integral around the circuit of the induced electric field intensity $\mathbf{E}$ at every point in the circuit. That is,

$$\mathcal{E} = \oint \mathbf{E} \cdot d\boldsymbol{\ell} \tag{35-1}$$

where $d\boldsymbol{\ell}$ is the differential element of path length at a point along the circuit where the induced electric field is $\mathbf{E} = \mathbf{E}(\mathbf{r})$. (Later we examine the field that is induced even when there is no circuit or conductor present.)

According to Eq. 29–14, the integral that appears in Eq. 35–1 vanishes for the electrostatic case. However, this result is no longer true when a changing magnetic field is present. Nor can we associate equipotential surfaces with such an induced electric field.

Consider the simple case of a stationary closed planar loop, as shown in Fig. 35–3. We select the direction indicated for the evaluation of the line integral around the loop. This direction, using the right-hand rule, defines the positive sense for the unit normal vector $\hat{\mathbf{n}}$ that is perpendicular to the plane surface S whose boundary is the loop. The law of induction states that

$$\mathcal{E} = \oint \mathbf{E} \cdot d\boldsymbol{\ell} = -\frac{d}{dt} \int_S \mathbf{B} \cdot \hat{\mathbf{n}} \, dA \tag{35-2}$$

where the magnetic field can be a function of position as well as time, $\mathbf{B} = \mathbf{B}(\mathbf{r}, t)$. The magnetic flux φ_M that links (or pierces) the open surface S is (see Section 34–3)

$$\varphi_M = \int_S \mathbf{B} \cdot \hat{\mathbf{n}} \, dA \tag{35-3}$$

---

*It is customary to refer to the initiating circuit (here, circuit ①) as the *primary* circuit and the responding or coupled circuit (here, circuit ②) as the *secondary* circuit.

Thus, the law of induction can be expressed in compact form as

$$\mathcal{E} = -\frac{d\varphi_M}{dt} \qquad (35-4)$$

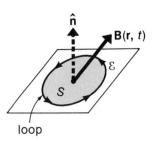

loop

Fig. 35-3. An EMF $\mathcal{E}$ is induced in the loop when the field **B** that links the loop changes with time.

Magnetic flux is measured in *webers* (Wb) and EMF is measured in *volts* (V).

The negative sign that appears in Eqs. 35-2 and 35-4 expresses *Lenz's law,* discovered in 1834 by Heinrich Lenz (1804-1865) and independently by Faraday. This law states that

The direction of an induced EMF is such that it opposes the change that produced it.

For example, if the field **B** in Fig. 35-3 were decreasing with time, $d\varphi_M/dt$ would be negative and, hence, $\mathcal{E}$ would be in the direction shown in the figure. An EMF in this sense produces a current in the loop that generates a magnetic field in a direction that opposes the decrease in the external field. On the other hand, if **B** were increasing with time, $\mathcal{E}$ and the resulting loop current would have a sense *opposite* to that indicated in the figure, and a magnetic field would be generated which would tend to oppose the increase in the external field.

According to Eq. 34-11, we have $\oint_S \mathbf{B} \cdot \hat{\mathbf{n}} \, dA = 0$ for *any* closed surface. Consider the closed surface shown in Fig. 35-4 which consists of two parts, S_1 and S_2, whose common boundary is a closed circuit loop. The unit normal vectors, $\hat{\mathbf{n}}_1'$ and $\hat{\mathbf{n}}_2$, are directed outward from the surface and are therefore in the positive sense as shown. We can write

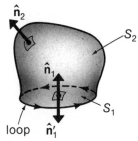

loop $\hat{\mathbf{n}}_1'$

Fig. 35-4. Geometry for evaluating the flux that links a circuit loop.

$$\oint_{S_1+S_2} \mathbf{B} \cdot \hat{\mathbf{n}} \, dA = \int_{S_1} \mathbf{B} \cdot \hat{\mathbf{n}}_1' \, dA + \int_{S_2} \mathbf{B} \cdot \hat{\mathbf{n}}_2 \, dA = 0$$

Now, consider the calculation of the flux φ_M that pierces the surface S_1 for the case in which the line integral around the boundary loop is taken in the sense shown in Fig. 35-4. Then, $\hat{\mathbf{n}}_1$ is positive, and because $\hat{\mathbf{n}}_1' = -\hat{\mathbf{n}}_1$, we have

$$\varphi_M = \int_{S_1} \mathbf{B} \cdot \hat{\mathbf{n}}_1 \, dA = -\int_{S_1} \mathbf{B} \cdot \hat{\mathbf{n}}_1' \, dA$$

Consequently,

$$\varphi_M = \int_{S_1} \mathbf{B} \cdot \hat{\mathbf{n}}_1 \, dA = \int_{S_2} \mathbf{B} \cdot \hat{\mathbf{n}}_2 \, dA$$

so that the flux φ_M that links a particular closed circuit can be calculated by evaluating the integral of $\mathbf{B} \cdot \hat{\mathbf{n}} \, dA$ over *any open surface* whose boundary is the circuit loop.

It is often the case (especially in practical applications) that a circuit loop consists of a number N of wire turns. If the coils are tightly wound (Fig. 35-5a) they are equivalent to separate loops, each bounding a surface (Fig. 35-5b). Then, the magnetic flux φ_M that links the entire coil is just N times the flux φ_M' that links a single turn. For the coil in Fig. 35-5 we have $\varphi_M = 2BA$, and, in general, the flux is

$$\varphi_M = N\varphi_M' \qquad (35-5)$$

Fig. 35-5. For calculating magnetic flux linkage, a tightly wound coil of N turns (a) may be considered to consist of the same number of individual loops (b).

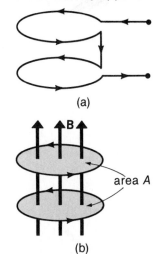

(a)

area A

(b)

The magnetic flux that links a circuit loop can vary with time for several reasons. The variation may result from a changing magnetic field, from relative motion between the circuit and other circuits or magnets that produce the linking flux, or from changes in the shape of the circuit itself. Moreover, any change in the loop current itself produces a time variation in the field that links the loop (self induction). The magnetic field **B** and the magnetic flux φ_M that appear in the different expressions of Faraday's law must include *all* sources of magnetic fields, and the time derivative that appears in Eq. 35–4 must be the *total* time derivative. These points are amplified in later discussions.

In this section we consider only the EMF induced by external magnetic fields generated by circuits *at rest* with respect to a rigid circuit loop selected for study. In the next section we discuss cases in which relative motion is involved. Also, for the present, self-induction effects are omitted. Because the superposition principle is valid, we may consider separately the self-induced EMF and the externally induced EMF, and obtain the total induced EMF by an algebraic sum of the two when this is required. (See Chapter 36.)

There is one additional restriction. We consider only cases in which the time variation of the field is *slow*. That is, the magnetic field (and the current) must not change appreciably during a time interval equal to the size of the circuit divided by the speed of light. Then, we can safely neglect effects associated with the finite propagation time of the field (which propagates with the speed of light).

Example 35–1

Calculate the EMF in a rigid planar loop with area A induced by a uniform time-varying external magnetic field directed at an angle θ with respect to the loop normal $\hat{\mathbf{n}}$, as shown in the diagram. The time dependence of B is given by $B_0 \sin \omega t$, where ω and B_0 are constants. Ignore self-induction effects. Assume $B_0 = 200$ G, $A = 100$ cm^2, $\theta = 30°$, and $\omega = 200\pi$ s^{-1}.

Solution:

The flux linkage is

$$\varphi_M = \mathbf{B} \cdot \hat{\mathbf{n}} A = B_0 A \cos \theta \sin \omega t$$

Using Eq. 35–4, we have

$$\mathcal{E} = -\frac{d\varphi_M}{dt} = -B_0 A \omega \cos \theta \cos \omega t$$

$$= -(200 \times 10^{-4}\text{ T})(10^{-2}\text{ m}^2)(200\pi\text{ s}^{-1}) \cos 30° \cos (200\pi\text{ s}^{-1})t$$

$$= -(0.109\text{ V}) \cos (200\pi\text{ s}^{-1})t$$

The negative sign means that the sense of $\mathcal{E}$ is *opposite* to the direction of the arrows indicated for the path in the diagram, when the cosine term > 0. (•What is the situation when the cosine term < 0?)

Example 35–2

A circular loop with a radius R consists of N tight turns of wire and is linked by an external magnetic field directed perpendicular to the plane of the loop. The magnitude of the field in the plane of the loop is $B = B_0(1 - r/2R) \cos \omega t$, where R is the radius of the loop,

circuit loop area $= A$

B

θ

$\hat{\mathbf{n}}$

$B = B_0\left(1 - \dfrac{r}{2R}\right)$

r

R

N turns

and where r is measured from the center of the loop, as shown in the diagram. Determine the EMF induced by B alone.

Solution:

The flux linkage is

$$\varphi_M = N\varphi_M' = NB_0 \cos \omega t \int_0^R \left(1 - \frac{r}{2R}\right) 2\pi r \, dr$$

$$= \frac{2\pi}{3} R^2 NB_0 \cos \omega t$$

Then, using Eq. 35-4, we have

$$\mathcal{E} = -\frac{d\varphi_M}{dt} = \frac{2\pi}{3} \omega R^2 NB_0 \sin \omega t$$

Example 35-3

A long straight wire carries a current $I = I_0 \sin (\omega t + \delta)$, and lies in the plane of a rectangular loop of N turns of wire, as shown in the diagram. The quantities I_0, ω, and δ are all constants. Determine the EMF induced in the loop by the magnetic field due to the current in the straight wire. Assume $I_0 = 50$ A, $\omega = 200\pi$ s^{-1}, $N = 100$, $a = b = 5$ cm, and $\ell = 20$ cm.

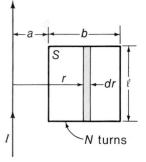

Solution:

The loop is the boundary of the plane surface S. The magnetic field produced by the current in the straight wire is perpendicular to S at all points on the surface. The magnitude of the field is (see Example 34-4)

$$B = \frac{\mu_0 I}{2\pi r}$$

Thus, the flux linkage is

$$\varphi_M = N\varphi_M' = \frac{\mu_0 NI\ell}{2\pi} \int_a^{a+b} \frac{dr}{r}$$

$$= \frac{\mu_0 NI_0\ell}{2\pi} \ln \left(\frac{a+b}{a}\right) \sin (\omega t + \delta)$$

Finally, the induced EMF is

$$\mathcal{E} = -\frac{d\varphi_M}{dt} = -\frac{\mu_0 NI_0\ell\omega}{2\pi} \ln \left(\frac{a+b}{a}\right) \cos (\omega t + \delta)$$

$$= -(2 \times 10^{-7} \text{ T} \cdot \text{m/A})(100)(50 \text{ A})(0.20 \text{ m})(200\pi \text{ s}^{-1}) \ln \left(\frac{5+5}{5}\right) \cos (\omega t + \delta)$$

$$= -(87.1 \text{ mV}) \cos (\omega t + \delta)$$

The term $\sin (\omega t + \delta)$ in the expression for the current in the straight wire does not change appreciably when ωt changes by 0.1 rad or less. Thus, the current does not change appreciably during a time interval $\tau < 0.1/(200\pi \text{ s}^{-1}) = 1.6 \times 10^{-4}$ s. We define a critical length, $c\tau = (3 \times 10^8 \text{ m/s})(1.6 \times 10^{-4} \text{ s}) = 4.8 \times 10^4$ m, equal to the distance to which field changes could be propagated during an interval of 1.6×10^{-4} s. This length is so much larger than any dimension of the loop or its distance from the wire that, although we consider the straight wire to be infinitely long, we can also safely ignore the field propagation effects in the vicinity of the loop. Moreover, the phase angle can be considered to be constant along the wire in the vicinity of the loop.

If the frequency ω were much larger, say, $200\pi \times 10^5 \text{ s}^{-1}$, the corresponding critical length would be only 48 cm. In this situation propagation effects would be important and the above expression for $\mathcal{E}$ would require modification. As a "rule of thumb" we can consider field propagation effects for circuits of laboratory size to be negligible for frequencies, $\nu = \omega/2\pi$, that are less than about 10^6 Hz.

35-3 MOTIONAL EMF

We wish to consider next the case of a conductor or circuit loop that is seen by an observer O to be in motion in the presence of an externally produced magnetic field. We restrict all speeds to be small compared with the speed of light in order to avoid relativistic effects. We begin by first examining the behavior of a single charge in the presence of a magnetic field. Suppose that a charge q is seen by an observer O to be moving with a velocity $\mathbf{u}$ in some inertial frame. If there is present a magnetic field $\mathbf{B}$ (but no electric field), the Lorentz force on the charge is (see Eq. 33–2)

$$\mathbf{F} = q\mathbf{u} \times \mathbf{B}$$

Suppose that a second observer O' is in an inertial frame that has a velocity $\mathbf{u}$ with respect to the frame of observer O. The charge q is seen by O' to be at rest and therefore subject to no magnetic force due to any static magnetic field that might be present. Nevertheless, O' must agree with O that a force $\mathbf{F}$ acts on the charge. (Recall that force is a free vector and is therefore the same in all inertial frames; see Section 5–4.) Because O' sees q at rest, he concludes that an electric field $\mathbf{E}'$ must be present in order to produce the force $\mathbf{F}$, namely,

$$\mathbf{F} = q\mathbf{E}' = q\mathbf{u} \times \mathbf{B}$$

Thus, we have

$$\mathbf{E}' = \mathbf{u} \times \mathbf{B} \tag{35-6}$$

Fig. 35–6. The differential element $d\ell$ of a conductor is in motion with a velocity $\mathbf{u}$ in the presence of a magnetic field $\mathbf{B}$.

Equation 35–6 is generally valid for all velocities u such that $u \ll c$. We conclude that if an inertial observer O determines that a magnetic field is present, another inertial observer O' in motion with respect to O will detect the additional presence of an electric field. We return to investigate the general case of field transformations in Section 35–5; for our present purposes this special case is sufficient.

Suppose that a differential element $d\ell$ of a conductor is moving with a velocity $\mathbf{u}$ in the presence of a magnetic field $\mathbf{B}$ in an inertial reference frame, as indicated in Fig. 35–6. To an observer moving with the conductor, the *only* forces that act on the charges in the conductor arise from the electric field $\mathbf{E}'$. This field produces an electric potential difference between the ends of the conductor element $d\ell$ given by

$$d\mathcal{E} = \mathbf{E}' \cdot d\ell = (\mathbf{u} \times \mathbf{B}) \cdot d\ell$$

If the element $d\ell$ is part of a closed conducting loop, the total induced EMF is

$$\mathcal{E} = \oint (\mathbf{u} \times \mathbf{B}) \cdot d\ell \qquad (35\text{-}7)$$

The quantity $\mathcal{E}$ is referred to as a *motional EMF*.

Let us apply Eq. 35-7 to the situation illustrated in Fig. 35-7. A rigid conducting loop lies partially within a uniform magnetic field $\mathbf{B}$ and is pulled with a constant velocity $\mathbf{u}$ by an external force $\mathbf{F}_e$. The line integral in Eq. 35-7 is to be evaluated around the closed path *abcdefa*. The segments *bc*, *cd*, and *de* lie outside the field and therefore give no contribution. There is likewise no contribution from the segments *ab* and *ef* because $\mathbf{E}' = \mathbf{u} \times \mathbf{B}$ is perpendicular to $d\ell$ along these segments. The remaining segment *fa* gives

$$\mathcal{E} = B\ell u \qquad (35\text{-}8)$$

with the sense shown by the arrows along the loop in Fig. 35-7.

Fig. 35-7. A conducting loop moves through a magnetic field $\mathbf{B}$. An EMF is induced in the loop with the sense indicated by the arrows. The unit normal $\hat{\mathbf{n}}$ for the surface $S = S_1 + S_2$ and the magnetic field $\mathbf{B}$ are both directed into the plane of the page.

Equation 35-8 represents the conclusion reached by the observer O' who moves with the loop. How is the situation analyzed by the stationary observer O who sees the loop in motion with a velocity $\mathbf{u}$? Observer O notes that the magnetic flux linking the loop is decreasing. Using Lenz's law, observer O determines that the EMF induced in the loop has the same sense as that found by O'. (•Can you see why?)

We apply Faraday's law (Eq. 35-2) for the case in which $\mathbf{B}$ is uniform and time-independent, and in which the surface is planar; we then have

$$\mathcal{E} = -\frac{d}{dt} \int_S \mathbf{B} \cdot \hat{\mathbf{n}} \, dA = -\mathbf{B} \cdot \hat{\mathbf{n}} \frac{d}{dt} \int_S dA \qquad (35\text{-}9)$$

The surface S consists of two parts, $S = S_1 + S_2$, and $\mathbf{B} = 0$ in S_2 (see Fig. 35-7). Thus, the integral is simply the area of S_1, namely, ℓx. Then,

$$\mathcal{E} = -\mathbf{B} \cdot \hat{\mathbf{n}} \frac{d}{dt}(\ell x) = -B\ell \frac{dx}{dt}$$

Now, $dx/dt = -u$, so we have

$$\mathcal{E} = B\ell u$$

which agrees precisely with the result obtained by O'. (The fact that the observers agree on the value of the EMF is the reason we wrote $\mathcal{E}$ instead of $\mathcal{E}'$ in Eq. 35–7, even though this result was obtained by observer O'.)

Next, we inquire about the energy balance as seen by the two observers. As long as some portion of the loop is within the magnetic field and moves with the constant velocity $\mathbf{u}$, the induced EMF remains constant. This situation is equivalent to a DC circuit in which there is a battery with an EMF $\mathcal{E}$. If the resistance of the entire loop is R, the constant current I that flows in the loop is

$$I = \frac{\mathcal{E}}{R} = \frac{B\ell u}{R}$$

The induced EMF therefore provides a power input to the loop given by

$$P = I\mathcal{E} = \frac{(B\ell u)^2}{R} \tag{35–10}$$

Both observers agree that this is the power input to the loop. According to observer O', this power is the result of the presence of the field $\mathbf{B}$ and the velocity $\mathbf{u}$ and is derived from the agency responsible for establishing the field $\mathbf{E}$, namely, the external force $\mathbf{F}_e$.

Observer O immediately attributes the power input to $\mathbf{F}_e$, thereby agreeing with O' as to the source of the power. Because there is no acceleration of the loop, $\mathbf{F}_e$ must be balanced by the net magnetic force on the loop; that is, according to O,

$$\mathbf{F}_e = -\mathbf{F}_L = -\oint I d\boldsymbol{\ell} \times \mathbf{B} = IB\ell \frac{\mathbf{u}}{u}$$

(•Can you show this?) The power delivered by the external force is

$$P = \mathbf{F}_e \cdot \mathbf{u} = IB\ell u = I\mathcal{E} = \frac{(B\ell u)^2}{R}$$

which is the same as the result obtained in Eq. 35–10. The power P supplied to the loop is dissipated in Joule heating.

▶ The current in the loop remains constant as long as some portion of the loop is within the field and the loop moves with constant velocity. Then, the magnetic field that is produced by the loop current is constant and does not give rise to any self-induced EMF. This is why we were able to ignore the field of the loop and why we obtained the simple energy relationship.

If the loop is set into motion while entirely within the field, we find $\mathcal{E} = 0$ and $I = 0$. (•Can you see why?) Then, as the leading edge emerges from the field, a current begins to flow and builds up to its final value as the self-induced field of the loop is created by virtue of work performed by the force $\mathbf{F}_e$. In the discussion above we considered the steady-state condition that ensues, with I constant and with the full self-induced field established. Finally, when the trailing edge of the loop leaves the field, the current decreases to zero as the self-induced field decreases to zero. The energy that was stored in the self-induced field is eventually dissipated in Joule heating. ◀

Example 35–4

A rectangular loop of wire rotates about an axis that is perpendicular to a uniform time-independent magnetic field $\mathbf{B}$, as shown in the diagram. Determine the induced EMF $\mathcal{E}$ at the slip-ring terminals.

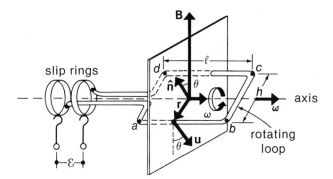

Solution:

First, obtain $\mathcal{E}$ as a motional EMF by using Eq. 35–7. Inspection of the diagram reveals that for the segments bc and da we have $\mathbf{u} \times \mathbf{B}$ perpendicular to $d\boldsymbol{\ell}$, so that these segments give zero contribution to the line integral. (We make no significant error by neglecting the small gap in the segment da. •Can you see why?) For the segments ab and cd we have $\mathbf{u} = \boldsymbol{\omega} \times \mathbf{r}$ and $\mathbf{u} \times \mathbf{B} = \frac{1}{2}h\omega B \sin\theta$. Then, performing the line integral around the loop, the result is

$$\mathcal{E} = h\ell\omega B \sin\theta$$

The loop area is $A = h\ell$; then, with $\theta = \omega t$, we have

$$\mathcal{E} = A\omega B \sin \omega t$$

The arrangement shown in the diagram is a simple alternating current (AC) generator in which the terminal voltage varies sinusoidally with time. Notice that in the case illustrated the circuit is not completed and no current flows. Consequently, there is no self-induced field and no corresponding contribution to the EMF. In this case we call $\mathcal{E}$ the *open-circuit EMF*. (If a load is connected to the terminals, time-varying currents will flow and the *back EMF* due to the self-induced field must be considered.)

We can also use Faraday's law to obtain the open-circuit EMF. The flux that links the loop is

$$\varphi_M = AB \cos\theta = AB \cos \omega t$$

Then, using Eq. 35–4, we have

$$\mathcal{E} = -\frac{d\varphi_M}{dt} = A\omega B \sin \omega t$$

Thus, the two methods give identical results. (•Can you verify that the two methods also give the same sense of the EMF?)

Example 35–5

A conducting disc with a radius R rotates about its own axis which is parallel to a uniform magnetic field $\mathbf{B}$, as shown in diagram (a). Two brushes, one contacting the axle and one contacting the rim of the disc, are connected to terminals across which the induced EMF $\mathcal{E}$ is measured. This arrangement is called a *Faraday disc dynamo* or a *homopolar generator*. Determine the open-circuit EMF $\mathcal{E}$.

(a)

Solution:

Again, first obtain $\mathcal{E}$ as a motional EMF. Consider a thin wedge-shaped section of the disc that subtends an angle $d\theta$ at the center of the disc. Select a length dr at a radius r as the differential conductor element within this wedge. (This element is shaded in diagram (a).) In the laboratory reference frame this element has a velocity $\mathbf{u}$, with $u = r\omega$. We have*

$$\mathbf{E}' = \mathbf{u} \times \mathbf{B} = r\omega B\hat{\mathbf{u}}_r$$

where $\hat{\mathbf{u}}_r$ is a unit vector in the radial direction; then,

$$d\mathcal{E} = \mathbf{E}' \cdot d\mathbf{r} = r\omega B \, dr$$

Integrating $d\mathcal{E}$ along the path $abcde$, only the segment cd gives a nonzero contribution. Thus,

$$\mathcal{E} = \int_0^R r\omega B \, dr = \tfrac{1}{2}\omega B R^2$$

(b)

Notice that *any* radial path along the disc would have given the same result. Even though an EMF is generated in all such wedge-shaped segments, they are all effectively in a parallel connection and so give the same total EMF.

The expression for $\mathcal{E}$ can also be obtained by using Faraday's law. Replace the integration path $abcde$ by the two paths, $abce$ and $cdec$, in which the segments ce and ec are traversed in opposite directions so their contributions to the integral cancel. No flux links the area defined by the path $abce$;† however, the flux through the area defined by the path $cdec$ is

$$\varphi_M = -\pi R^2 B \cdot \frac{\theta}{2\pi} = -\tfrac{1}{2}R^2 B\omega t$$

---

*We continue to use $\mathbf{u}$ for the velocity (because we reserve $\mathbf{v}$ for use later as another velocity); do not confuse $\mathbf{u}$ with the symbol $\hat{\mathbf{u}}$ for a unit vector.

†Notice that, when calculating the flux linkage, we can tolerate the open segment in the defining path because there is zero flux through the area.

The negative sign arises because $\hat{\mathbf{n}}$, the unit normal vector for the surface, is directed opposite to $\mathbf{B}$. Applying Eq. 35–4, there results

$$\mathcal{E} = -\frac{d\varphi_M}{dt} = \tfrac{1}{2}\omega BR^2$$

in agreement with the previous result.

Notice that we have calculated here the equilibrium open-circuit terminal voltage. As the disc begins to rotate, a charge separation occurs, with excess positive charge collecting around the rim of the disc and negative charge collecting at the center. The resulting electrostatic field just cancels the induced field, and at equilibrium there is no current flow. If a load is connected across the terminals of the dynamo, the current density $\mathbf{J}$ in the disc will have a complicated distribution, as indicated in diagram (b).

Fig. 35–8. The rotation of the conducting disc through the region of magnetic field $\mathbf{B}$ produces an eddy current in the disc.

Eddy Currents. Because the disc in Example 35–5 lies entirely within the magnetic field, the equilibrium condition is established in which no current flows. However, if the field were confined to a region of the disc, as indicated by the shaded area in Fig. 35–8, a distribution of current—called an *eddy current*—would be set up in the disc. Figure 35–8 shows that $\mathbf{J}$ in the region of the field is almost unidirectional (nearly radial) and is in a sense to produce a torque that opposes the rotational motion of the disc. (Recall that the force per unit volume is equal to $\mathbf{J} \times \mathbf{B}$—see Eq. 33–5.)

It is clear from energy considerations that in all cases involving rotation the eddy currents produce torques that *oppose* the rotation. Through Joule heating effects, eddy currents dissipate mechanical energy, the ultimate source of which is the external agency that produces the rotation. If there is no coupling to such an external agency, the energy is provided from the kinetic energy of the rotating object. Thus, the result of eddy currents is always to produce a braking action.

Fig. 35–9. (a) When the switch P is closed, what current flows between the plates of the capacitor C? (b) The current $I(t)$ flows into the surface S and the displacement current flows out of the surface.

35–4   THE MAXWELL DISPLACEMENT CURRENT

Consider the process of charging a capacitor by connecting it to a source of EMF (for example, a battery), as shown in Fig. 35–9a. When the switch P is closed, positive charge flows onto the top plate of the capacitor. The rate at which the charge flows is maximum at the instant the switch is closed and gradually decreases to zero as the capacitor charge approaches its full value, $Q_m = CV.$*

The upper plate of the capacitor may be completely surrounded by a closed surface S, as indicated in Fig. 35–9b. Conservation of charge requires that the rate at which charge accumulates on the upper plate, inside the surface S, be equal to the current that flows into the surface S. That is,

$$I(t) = \frac{dQ(t)}{dt} \qquad \textbf{(35–11)}$$

(a)

We now apply Gauss's law to the surface S, using the form expressed in Eq. 30–41.

---

*The exact mathematical nature of $Q(t)$ is not important for the present discussion. Such matters are discussed in Chapter 37.

Thus,

$$\oint_S \mathbf{D}(t) \cdot \hat{\mathbf{n}} \, dA = Q(t)$$

which is valid even if the region between the plates is filled with a dielectric. Taking the time derivative of this equation and combining with Eq. 35–11, we obtain

$$I(t) = \frac{d}{dt} \oint_S \mathbf{D}(t) \cdot \hat{\mathbf{n}} \, dA \qquad \text{(35–12a)}$$

In the event that the region between the plates is vacuum, we have $\mathbf{D} = \epsilon_0 \mathbf{E}$, so we can write

$$I(t) = \frac{d}{dt} \oint_S \epsilon_0 \mathbf{E}(t) \cdot \hat{\mathbf{n}} \, dA, \qquad \text{vacuum} \qquad \text{(35–12b)}$$

It is useful to consider the right-hand side of Eq. 35–12a or 35–12b to be a type of "current" that flows across the gap between the capacitor plates even though no actual transport of charge is involved. This permits I to be viewed as a current that flows continuously around the entire circuit of Fig. 35–9a. The new current term is known as *Maxwell's displacement current*.

The immediate consequence of this view is to modify Ampère's circuital law. Consider the magnetic field produced by a current $I(t)$ that flows in a circuit that contains a capacitor, as indicated in Fig. 35–10. Ampère's circuital law connects the line integral of **B** around a closed path with the total current that flows through *any open surface* whose boundary is the integration path. Thus, for the integration path shown in Fig. 35–10, either of the surfaces, S_1 or S_2, can be used in Ampère's law. Whereas ordinary current flows through surface S_1, no such current flows through surface S_2. This clearly indicates the necessity of some modification to the original formulation of Ampère's circuital law (Eq. 34–8). Taking guidance from Eqs. 35–12, which introduce the idea of displacement current, Maxwell modified Ampère's circuital law to read

$$\oint \mathbf{B} \cdot d\mathbf{r} = \mu_0 \int_S \mathbf{J} \cdot \hat{\mathbf{n}} \, dA + \mu_0 \frac{d}{dt} \int_S \epsilon_0 \mathbf{E} \cdot \hat{\mathbf{n}} \, dA$$

Fig. 35–10. For the integration path shown, either surface S_1 or surface S_2 may be used in Ampère's circuital law.

or

$$\oint \mathbf{B} \cdot d\mathbf{r} = \mu_0 \int_S \left(\mathbf{J} + \epsilon_0 \frac{d\mathbf{E}}{dt} \right) \cdot \hat{\mathbf{n}} \, dA, \qquad \text{vacuum} \qquad \text{(35–13)}$$

where the open surface S can be any surface whose boundary is the integration path around which $\mathbf{B} \cdot d\mathbf{r}$ is calculated, such as S_1 or S_2 in Fig. 35–10.

If the surface S_1 in Fig. 35–10 is chosen to include some of the fringing field of the capacitor, Eq. 35–13 remains correct if the contribution of the displacement current is included in the surface integral. (•Can you see how this follows by considering the closed surface S in Fig. 35–9b to consist of the two open surfaces, $S_1 + S_2$, in Fig. 35–10? Remember the sign convention for the unit normal vectors.)

In such an elegant theory as Maxwell's theory of electromagnetism, in which electricity and magnetism are treated on an equal basis, one somehow expects a symmetry in the behavior of these field quantities. Faraday's law expresses the fact that a time-varying magnetic field is accompanied by an electric field. By symmetry, then, a time-varying electric field should be accompanied by a magnetic field; this is exactly the effect of the displacement-current term in Maxwell's version of Ampère's law (Eq. 35–13).

The ultimate justification for Maxwell's modification of Ampère's law by the addition of the displacement-current term is in the experimental verification. It is difficult to observe directly the magnetic effects of the displacement current. However, in experiments that used very high frequencies to enhance the effect, the magnetic field in a capacitor has actually been detected. Even more important is the verification afforded by the observation of many predicted indirect effects, such as electromagnetic radiation (see Chapters 38 and 39).

Example 35–6

An alternating current, $I(t) = I_0 \cos \omega t$, is applied to a parallel-plate capacitor that consists of circular plates with radius R separated by a distance ℓ. Determine $\mathbf{E}(\mathbf{r}, t)$ and $\mathbf{B}(\mathbf{r}, t)$ everywhere (neglecting the fringing electric field). Assume $\omega R / 2\pi c \ll 1$, where c is the speed of light.

Solution:

The condition, $\omega R / 2\pi c \ll 1$, means that the fields are slowly varying; then, we can neglect field propagation effects (see Section 35–2) and can safely assume that $\mathbf{E}$ is uniform everywhere between the plates. The charge on the capacitor is $Q(t)$ and the current is $I(t) = dQ/dt$. If $Q(0) = 0$, we have

$$Q(t) = \frac{I_0}{\omega} \sin \omega t$$

Using the Gaussian surface S shown in diagram (a), we have

$$\pi R^2 E = \frac{Q}{\epsilon_0}$$

so that

$$\mathbf{E} = \frac{I_0}{\omega \pi R^2 \epsilon_0} \sin \omega t \,\hat{\mathbf{k}}, \qquad r \leqslant R$$

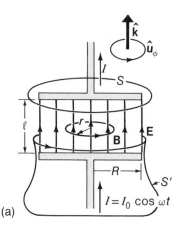

(a)

From the cylindrical symmetry, we see that the magnetic field is everywhere entirely azimuthal. In the region between the capacitor plates ($r \leqslant R$), we use Eq. 35-13 for the circular path with radius r shown in diagram (a) to obtain

$$2\pi r B_\phi = \frac{\pi r^2}{c^2}\frac{dE}{dt}$$

Then,
$$\mathbf{B} = \frac{\mu_0 I_0 r}{2\pi R^2}\cos\omega t\,\hat{\mathbf{u}}_\phi, \qquad r \leqslant R$$

Outside the plates ($r > R$), we find

$$2\pi r B_\phi = \frac{\pi R^2}{c^2}\frac{dE}{dt}$$

so that
$$\mathbf{B} = \frac{\mu_0 I_0}{2\pi r}\cos\omega t\,\hat{\mathbf{u}}_\phi, \qquad r > R$$

This is just the result expected from the Biot–Savart law (Eq. 34–4) for the surface S' that intercepts the external current I.

The approximations we have made are accompanied by an unavoidable inconsistency. Because **B** varies with time, the electric field **E** cannot be completely uniform between the plates. To illustrate this point, consider the line integral of **E** around the path (with sides perpendicular to the plates) that is shown in diagram (b). According to Faraday's law, the EMF $\mathcal{E}$ around this path is (Eq. 35–2)

$$\mathcal{E} = -\frac{d}{dt}\int_0^r B_\phi \ell\, dr = \frac{\omega\mu_0 I_0 \ell r^2}{4\pi R^2}\sin\omega t$$

However, if the electric field is completely uniform, we have

$$\mathcal{E} = \oint \mathbf{E}\cdot d\mathbf{r} = 0$$

To judge the magnitude of this inconsistency, let us attribute the Faraday EMF to a field variation ΔE, such that $\ell\,\Delta E = \mathcal{E}$ (Faraday). Then,

$$\frac{\Delta E}{E} = \frac{(\omega\mu_0 I r^2/4\pi R^2)\sin\omega t}{(I_0/\omega\pi R^2\epsilon_0)\sin\omega t} = \left(\frac{\omega r}{2c}\right)^2$$

which, according to the original assumption of slowly varying fields, is a very small quantity (because $r < R$). Thus, the field modification necessary to produce agreement with Faraday's law is negligible in this case.

Even for situations that possess simple geometry, completely consistent solutions $\mathbf{E}(\mathbf{r}, t)$ and $\mathbf{B}(\mathbf{r}, t)$ can be obtained only by using advanced mathematical techniques. In the present case, **E** is found to decrease slightly with increasing r so that the line integral of **E** for the path shown in diagram (b) yields exactly the value required by Faraday's law.

(b)

SPECIAL TOPIC

35–5 FIELD TRANSFORMATIONS

The Lorentz force on a charge is the same in all inertial frames. In the preceding section we used this fact to show that a moving observer attributes an electric field, $\mathbf{E'} = \mathbf{u} \times \mathbf{B}$, to a region of space in which an observer at rest with respect to the source of the magnetic field determines that only a magnetic field $\mathbf{B}$ exists. We continue this analysis and now inquire about general transformations of electromagnetic fields. Suppose that an observer O in an inertial frame determines by experiments with both moving and stationary charges that an electric field $\mathbf{E}$ and a magnetic field $\mathbf{B}$ are present. Then, what fields $\mathbf{E'}$ and $\mathbf{B'}$ will be measured in the same region by an observer O' in an inertial frame that moves with a velocity $\mathbf{u}$ with respect to the O frame?

Figure 35–11 shows two inertial frames, O and O', that have a constant relative velocity $\mathbf{u} = d\mathbf{R}/dt$. The velocity of the charge q is $\mathbf{v} = d\mathbf{r}/dt$ and $\mathbf{v'} = d\mathbf{r'}/dt$ in the frames O and O', respectively. Consider first the nonrelativistic case in which the speeds $\mathbf{u}$, $\mathbf{v}$, and $\mathbf{v'}$ are all small compared with the speed of light c.

First, suppose that observer O uses moving and stationary charges to determine that only a magnetic field $\mathbf{B}$ is present in the frame O. (We ignore all except electromagnetic forces.) In this frame a charge q that moves with a velocity $\mathbf{v}$ experiences a Lorentz force,

$$\mathbf{F} = q\mathbf{v} \times \mathbf{B}$$

Fig. 35–11. Geometry of the two inertial frames, O and O'. The relative velocity of O' with respect to O is $\mathbf{u} = d\mathbf{R}/dt = $ constant.

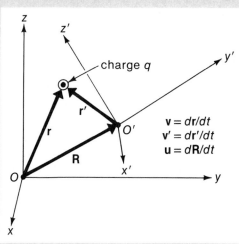

In frame O' the observer measures some combination of electric and magnetic fields, $\mathbf{E'}$ and $\mathbf{B'}$. The O' observer agrees that the force on q is $\mathbf{F}$, but he attributes this force to the action of both $\mathbf{E'}$ and $\mathbf{B'}$. Therefore, O' writes*

$$\mathbf{F} = q\mathbf{v'} \times \mathbf{B'} + q\mathbf{E'}$$

Equating the two expressions for $\mathbf{F}$ gives

$$\mathbf{v} \times \mathbf{B} = \mathbf{v'} \times \mathbf{B'} + \mathbf{E'} \qquad \text{(35–14)}$$

The electric field $\mathbf{E'}$ does not depend on the velocity $\mathbf{v}$ and therefore can be evaluated by considering the special case, $\mathbf{v} = \mathbf{u}$. Then, we have $\mathbf{v'} = \mathbf{v} - \mathbf{u} = 0$, and the term $\mathbf{v'} \times \mathbf{B'}$ vanishes. Hence,

$$\mathbf{E'} = \mathbf{u} \times \mathbf{B} \qquad \text{(35–15)}$$

This is just the result we obtained in Eq. 35–6.

Next, we inquire about the value of $\mathbf{B'}$. Substituting Eq. 35–15 for $\mathbf{E'}$ into Eq. 35–14, we obtain

$$(\mathbf{v} - \mathbf{u})\mathbf{B} = \mathbf{v'} \times \mathbf{B'}$$

But $\mathbf{v'} = \mathbf{v} - \mathbf{u}$, so we conclude that

$$\mathbf{B'} = \mathbf{B} \qquad \text{(35–16)}$$

That is, the O' observer finds the same magnetic field as does the O observer, even though he measures an additional electric field $\mathbf{E'}$.

What is the situation when observer O determines that only an electric field is present? In this case, it is instructive to shift attention to the sources of the fields. Suppose that in the O system there is an infinitely long straight filament that carries a uniform static charge with a linear density λ, as indicated in Fig. 35–12. The electric field $\mathbf{E}$ is entirely radial and is given by Eq. 28–22, namely,

$$\mathbf{E} = \frac{\lambda}{2\pi\epsilon_0} \frac{\hat{\mathbf{u}}_r}{r} \qquad \text{(35–17)}$$

The O observer measures no magnetic field in his reference frame.

The O' observer moves parallel to the filament with a

---

*Notice that we assume the charge q to be an invariant quantity. This is equivalent to asserting the validity of Gauss's law whether the charge inside the surface is stationary or in motion. Experiments show this proposition to be true.

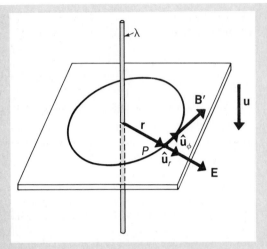

Fig. 35–12. Geometry of an infinite charged filament that is viewed by observers who are in relative motion.

velocity **u**. This observer measures the same charge density on the filament and hence he finds the same electric field strength at every point as does the O observer. In addition, the O' observer sees the charge moving and measures a current $I' = \lambda u$. (The direction of the current is *opposite* to the direction of **u**.) Therefore, the O' observer finds that there is in his frame an additional magnetic field which is the same as that due to an infinitely long current-carrying wire, as given by Eq. 34-4a, namely,

$$\mathbf{B}' = \frac{\mu_0 I'}{2\pi r}\,\hat{\mathbf{u}}_\phi \qquad (35\text{-}18)$$

where $\hat{\mathbf{u}}_\phi$ is the azimuthal unit vector, as shown in Fig. 35-12. Notice that $\mathbf{u} \times \mathbf{E}$ is in the direction of $-\hat{\mathbf{u}}_\phi$; then, combining Eqs. 35-17 and 35-18 gives

$$\mathbf{B}' = -\mu_0\epsilon_0 \mathbf{u} \times \mathbf{E} = -\frac{1}{c^2}\mathbf{u} \times \mathbf{E} \quad (35\text{-}19a)$$

and we also have

$$\mathbf{E}' = \mathbf{E} \qquad (35\text{-}19b)$$

In Eq. 35-19a we have used $\mu_0\epsilon_0 = 1/c^2$ (see Eq. 34-14). Although the transformation equations (Eqs. 35-19) were obtained for the special case of a charged filament, the features of the particular problem, such as the charge density λ, do not appear in the equations that relate the electric and magnetic fields. This suggests that the transformation equations are generally true, and indeed, this is the case.

These results can be summarized as follows. If the observer O detects electric and magnetic fields **E** and **B**, an observer O' moving with a velocity **u** relative to O will measure the fields

$$\mathbf{E}' = \mathbf{E} + \mathbf{u} \times \mathbf{B}$$
$$\mathbf{B}' = \mathbf{B} - \frac{1}{c^2}\mathbf{u} \times \mathbf{E} \qquad (35\text{-}20)$$

These are the nonrelativistic (or *Galilean*) transformation equations for the field vectors.

Both observers agree that the Lorentz force on a moving charge is

$$\mathbf{F} = q(\mathbf{v} \times \mathbf{B} + \mathbf{E}) = q(\mathbf{v}' \times \mathbf{B}' + \mathbf{E}') \qquad (35\text{-}21)$$

if terms of order uv/c^2 are neglected (this is the nonrelativistic approximation—see Problem 35-11).

Example 35–7

The fields associated with a long straight wire that carries a constant current $I = 50$ A are measured by an observer O at rest with respect to the wire and by an observer O' moving parallel to the wire (in the direction of the current) with a speed of 30 m/s. Determine the field quantities according to each observer at a distance of 5 cm from the wire.

Solution:

Consider first the description of the situation according to observer O. Independent of the polarity or density of the charge carriers in the wire, any segment of the wire always has zero net charge, for as many carriers of either sign enter the segment per unit time as leave it. Thus, there is no electric field outside the wire. The magnetic field measured by observer O is (Eq. 34-4a)

$$\mathbf{B} = \frac{\mu_0 I}{2\pi r}\,\hat{\mathbf{u}}_\phi$$

At $r = 5$ cm, we have

$$B = \frac{(2 \times 10^{-7}\text{ T}\cdot\text{m/A})(50\text{ A})}{0.05\text{ m}} = 2.0 \times 10^{-4}\text{ T} = 2.0\text{ G}$$

The observer O' measures the same magnetic field as O because $B' = B$ (Eq. 35-16 or 35-20). However, O' also measures an electric field,

$$\mathbf{E}' = \mathbf{u} \times \mathbf{B} = Bu\hat{\mathbf{u}}_r$$

or, at $r = 5$ cm,

$$|\mathbf{E}'| = (2.0 \times 10^{-4}\,\text{T})(30\,\text{m/s}) = 6.0 \times 10^{-3}\,\text{V/m}$$

This electric field strength is produced by a linear charge density λ',

$$|\mathbf{E}'| = \frac{\lambda'}{2\pi\epsilon_0 r}$$

so that $\lambda' = 2\pi\epsilon_0|\mathbf{E}'|r = 2\pi\epsilon_0 Bur = \epsilon_0\mu_0 Iu = \dfrac{1}{c^2}Iu$

$$= \frac{(50\,\text{A})(30\,\text{m/s})}{(3 \times 10^8\,\text{m/s})^2} = 1.67 \times 10^{-14}\,\text{C/m}$$

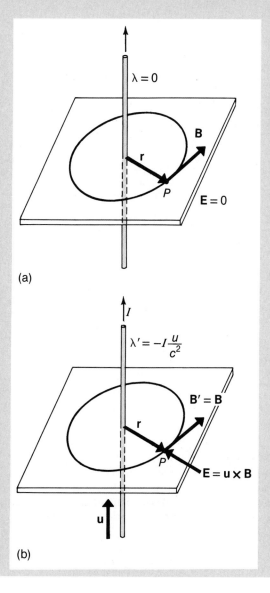

(a)

(b)

These results are illustrated in diagrams (a) and (b). (•Can you see why it is not correct to superimpose the observer's velocity $\mathbf{u}$ on the electron drift velocity $\mathbf{v}_d$ and to claim that O' sees a new effective drift velocity, $\mathbf{v}_d' = \mathbf{v}_d - \mathbf{u}$, and a corresponding new magnetic field $\mathbf{B}'$ ($\neq\mathbf{B}$)? *Hint:* What is the difference between a current-carrying wire and a filament that carries charge having a single polarity?)

Example 35–8

Determine in the nonrelativistic approximation the electric and magnetic fields associated with a point charge q that moves with a constant velocity $\mathbf{v}$ in an inertial reference frame O.

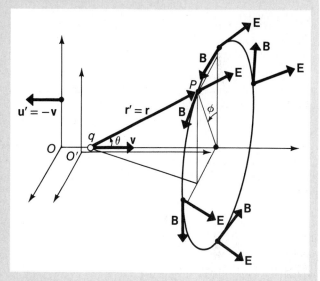

Solution:

Let the charged particle be at rest in an inertial frame O' and be moving with a velocity $\mathbf{v}$ in the inertial frame O, as indicated in the diagram. The Galilean transformations (Eqs. 35–20) are assumed to hold. We locate the point P at which we wish to determine the fields by the vector $\mathbf{r}' = \mathbf{r}$, measured from the position of the charged particle. Each observer finds that the fields are stationary with respect to the particle: the fields are static in the O' frame and they move with the particle in the O frame. We know the static field $\mathbf{E}'$ in the O' frame (and $\mathbf{B}' = 0$), and we wish to determine $\mathbf{E}$ and $\mathbf{B}$ in the O frame. We can do this by rewriting Eqs. 35–20 with the primed and unprimed quantities interchanged. Then, $\mathbf{u}'$ becomes the velocity of the O frame with respect to the O' frame; that is, $\mathbf{u}' = -\mathbf{v}$. Because $B' = 0$ and

$$\mathbf{E}' = \frac{1}{4\pi\epsilon_0}\frac{q}{r'^2}\hat{\mathbf{u}}_{r'}$$

we have

$$\mathbf{E} = \frac{1}{4\pi\epsilon_0} \frac{q}{r^2} \hat{\mathbf{u}}_r$$

Also,

$$\mathbf{B} = -\frac{q}{c^2}\mathbf{u} \times \mathbf{E}' = \frac{\mu_0 q}{4\pi}\frac{\mathbf{v} \times \hat{\mathbf{u}}_r}{r^2}$$

$$= \frac{\mu_0 q v}{4\pi}\frac{\sin\theta}{r^2}\hat{\mathbf{u}}_\phi$$

(•Can you see how to obtain the Biot-Savart law, Eq. 34–2, by replacing $q\mathbf{v}$ by $I\,d\boldsymbol{\ell}$?)

▶ *Relativistic Transformations.* The fact that the Lorentz force relationship (Eq. 35–21) is correct only if terms of order uv/c^2 are neglected suggests that the Galilean (nonrelativistic) field transformation equations are also approximations of the same order. For example, in the situation illustrated in Fig. 35–12 it is evident that $\mathbf{E}$ and $\mathbf{E}'$ cannot be exactly equal because the moving observer O' must measure a larger charge density, $\lambda' = \lambda(1 - u^2/c^2)^{-1/2}$, as a result of length contraction (see Eq. 21–5). And we also know that the velocity addition rule is different in the relativistic case (see Eq.

21–9). To obtain the relativistic transformation equations all such factors must be taken into account. It proves convenient to express the field transformations in terms of vector components parallel to and perpendicular to the direction of the velocity $\mathbf{u}$ that specifies the motion of O' with respect to O. The resulting equations are

$$\left.\begin{array}{ll} \mathbf{E}'_\perp = \gamma(\mathbf{E}_\perp + \mathbf{u} \times \mathbf{B}_\perp); & \mathbf{E}'_\parallel = \mathbf{E}_\parallel \\[2mm] \mathbf{B}'_\perp = \gamma(\mathbf{B}_\perp - \frac{1}{c^2}\mathbf{u} \times \mathbf{E}_\perp); & \mathbf{B}'_\parallel = \mathbf{B}_\parallel \end{array}\right\} \quad (35\text{--}22)$$

where $\gamma = (1 - u^2/c^2)^{-1/2}$.

Using Eqs. 35–22, the Lorentz force expressions are *exactly*

$$\mathbf{F} = q(\mathbf{v} \times \mathbf{B} + \mathbf{E}) = q(\mathbf{v}' \times \mathbf{B}' + \mathbf{E}')$$

where v, v', and u are related by the relativistic transformation equations of Section 21–6. When $u^2/c^2 \ll 1$, Eqs. 35–22 reduce to Eqs. 35–20 for the nonrelativistic case.

When a particle moves with a speed comparable with the speed of light, the relativistic factor γ becomes significantly greater than unity. The effect on the electric and magnetic fields associated with a rapidly moving charged particle are quite striking. Figure 35–13 shows how the electric field lines bunch together as $\beta = v/c$ increases. In the limit $\beta \to 1$, the field is concentrated in a thin sheet that is perpendicular to the direction of motion. ◀

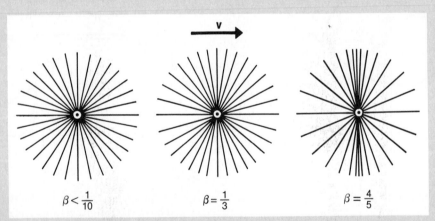

$$\beta < \tfrac{1}{10} \qquad\qquad \beta = \tfrac{1}{3} \qquad\qquad \beta = \tfrac{4}{5}$$

Fig. 35–13. The electric field lines for a moving charged particle for three different constant velocities, $\beta = v/c$.

SPECIAL TOPIC

35-6   MAXWELL'S FIELD EQUATIONS

In this section we summarize the equations that describe electric and magnetic fields in vacuum (or, for all practical purposes, in air). (The general relationships for fields in material media are given in Section 36–7.) The charges that are the sources of these fields are restricted to

move uniformly in some inertial reference frame. We imagine that this frame is populated with as many observers as are required to make simultaneous measurements of the electric and magnetic field vectors at all points of interest by using stationary and moving test charges. That is, $\mathbf{E}(\mathbf{r}, t)$ and $\mathbf{B}(\mathbf{r}, t)$ are considered to be experimentally determined quantities in the inertial reference frame.

Gauss's Laws. Gauss's law for the electric case (Eq. 28–19) is a completely general law:

$$\oint_S \mathbf{E} \cdot \hat{\mathbf{n}} \, dA = \frac{Q}{\epsilon_0} = \frac{1}{\epsilon_0} \int_V \rho \, dv \qquad (35\text{–}23)$$

where the integration is carried out at a particular instant and where Q is the instantaneous value of the net charge contained in the volume V that is bounded by the closed surface S. (As usual, the positive direction of $\hat{\mathbf{n}}$ is *outward*.)

In a situation that possesses a high degree of symmetry, the shape of the electric field can usually be deduced. Then, Eq. 35–23 may be used with a judiciously selected Gaussian surface to determine completely the electric field $\mathbf{E}$. Many examples are created with the required symmetry, but in the general case such a simplification is lacking. Although Gauss's law is valid in these situations, it does not provide a means for readily determining $\mathbf{E}$. Problems of this type are always attacked by starting with *differential* relationships. Gauss's law in differential form (Eq. 29–47) represents the first of the set of four equations called *Maxwell's field equations:*

I. $$\nabla \cdot \mathbf{E} = \frac{\rho}{\epsilon_0} \qquad (35\text{–}24)$$

Because magnetic monopoles do not exist, the magnetic quantities that correspond to Q and ρ are zero. Thus, Gauss's law for magnetic fields is (Eq. 34–11)

$$\oint_S \mathbf{B} \cdot \hat{\mathbf{n}} \, dA = 0 \qquad (35\text{–}25)$$

where S is any closed surface. In differential form this law becomes the second Maxwell equation (Eq. 34–12a), namely,

II. $$\nabla \cdot \mathbf{B} = 0 \qquad (35\text{–}26)$$

Ampère's Law. The integral form of Ampère's law, as modified by Maxwell to include displacement currents, is (Eq. 35–22)

$$\oint \mathbf{B} \cdot d\mathbf{r} = \mu_0 \int_S \left(\mathbf{J} + \epsilon_0 \frac{d\mathbf{E}}{dt} \right) \cdot \hat{\mathbf{n}} \, dA \qquad (35\text{–}27)$$

where the line integral can be taken around any closed path, and where S is any fixed surface that is bounded by the integration path. The positive direction of $\hat{\mathbf{n}}$ is determined by the usual right-hand rule. Equation 35–27 can be converted to differential form by using Stokes' theorem (as in obtaining Eq. 34–10), with the result

III. $$\nabla \times \mathbf{B} = \mu_0 \left(\mathbf{J} + \epsilon_0 \frac{\partial \mathbf{E}}{\partial t} \right) \qquad (35\text{–}28)$$

where we have written the partial derivative of $\mathbf{E}$ to take account of the possibility that $\mathbf{E}$ may depend on position as well as on time.

Faraday's Law. Faraday's law can be used to determine the induced EMF in a circuit, whether that EMF is due to a time variation of the magnetic field or to the motion of a circuit through the field. We now assert that the law is completely general and does not require that the path of the line integral correspond to a physical wire or circuit. That is, we write Eq. 35–2 as

$$\oint \mathbf{E} \cdot d\mathbf{r} = -\frac{d}{dt} \int_S \mathbf{B} \cdot \hat{\mathbf{n}} \, dA \qquad (35\text{–}29)$$

where the line integral can be taken around any closed path, and where S is any stationary surface that is bounded by the integration path. Notice that we write $d\mathbf{r}$ (indicating a *general* displacement) instead of $d\boldsymbol{\ell}$ (indicating a segment of a physical circuit). Equation 35–29 can be converted to differential form by using Stokes' theorem (see Eq. 29–44), with the result

IV. $$\nabla \times \mathbf{E} = -\frac{\partial \mathbf{B}}{\partial t} \qquad (35\text{–}30)$$

where we have written the partial derivative of $\mathbf{B}$ to take account of the possibility that $\mathbf{B}$ may depend on position as well as on time.

Example 35–9

In a certain region of space the electric field is entirely azimuthal around the z-axis of a reference frame and is given by $\mathbf{E} = E_\phi \hat{\mathbf{u}}_\phi = -E_0(r/a)\hat{\mathbf{u}}_\phi$. Describe this field and explain how it could be produced.

Solution:

Along any circular path around the z-axis, such as the dashed-line path with radius r shown in the diagram on the next page, there is a constant azimuthal field component E_ϕ. Consequently, the line integral of $\mathbf{E}$ around such a path is nonvanishing, which means that the field $\mathbf{E}$ cannot be derived from a potential function. Thus, the field is time-independent, yet nonconservative.

How is it possible to produce an entirely azimuthal electric field? Because the field lines do not originate and terminate on charges, the field is not due to any sort of static charge distribution. The only other possibility is that

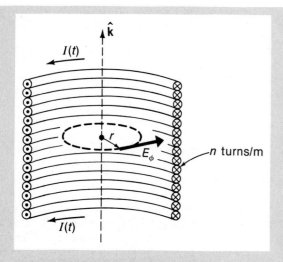

the electric field is due to a time-varying magnetic field. Maxwell's fourth equation (Eq. 35–30) is

$$\nabla \times \mathbf{E} = -\frac{\partial \mathbf{B}}{\partial t}$$

For the present case this equation becomes

$$\nabla \times \left(E_0 \frac{r}{a}\hat{\mathbf{u}}_\phi\right) = \frac{d\mathbf{B}}{dt}$$

Carrying out the curl operation (see Problem 35–14), there results

$$\nabla \times \left(E_0 \frac{r}{a}\hat{\mathbf{u}}_\phi\right) = 2\frac{E_0}{a}\hat{\mathbf{k}} = \frac{d\mathbf{B}}{dt}$$

Integrating with respect to time, we obtain

$$\mathbf{B} = 2\frac{E_0}{a}t\hat{\mathbf{k}}$$

(We could add a constant field $\mathbf{B}_0(\mathbf{r})$ to $\mathbf{B}$ without affecting the value of $\nabla \times \mathbf{E}$, but for simplicity we assume that no such extraneous field is present.) The field $\mathbf{B}$ could be produced, for example, by a current $I(t) = I_0 t/\tau$ flowing in a long solenoid, as indicated in the diagram. If the solenoid axis coincides with the z-axis of the coordinate frame used to describe $\mathbf{E}$, the cylindrical symmetry will yield the desired azimuthal electric field. That is, E_ϕ will have a constant magnitude along any circular path centered on the z-axis $\hat{\mathbf{k}}$.

The magnetic field within a long solenoid is $B = \mu_0 n I$ (see Example 34–6). Hence, the prescribed electric field could be produced by a solenoidal current

$$I(t) = \frac{2E_0}{\mu_0 na}t$$

where n is the number of turns per meter in the solenoid.

As a check on the result for B, we can use Faraday's law (Eq. 35–2 or 35–29) and integrate $\mathbf{E} \cdot d\mathbf{r}$ around the circular path shown in the diagram:

$$\oint \mathbf{E} \cdot d\mathbf{r} = -\frac{d}{dt}\int_S \mathbf{B} \cdot \hat{\mathbf{n}}\, dA$$

The surface S can be taken to be the plane circular area bounded by the integration path. Thus, we find

$$E_\phi \cdot 2\pi r = -\frac{dB}{dt} \cdot \pi r^2 = -2\frac{E_0}{a} \cdot \pi r^2$$

from which

$$E_\phi = -E_0 \frac{r}{a}$$

PROBLEMS

Section 35–2*

35–1 A rectangular single-turn loop, with dimensions $a = 10$ cm and $b = 15$ cm, lies in the x-y plane as indicated in the diagram on the next page. Within the loop there is a time-dependent magnetic field

---

*In these problems consider the effect *only* of the given current; that is, neglect the effect of any self-induced EMF. If desired, we could include these effects by using superposition.

$B = B_0(x \sin \omega t\hat{\mathbf{j}} + y \cos \omega t\hat{\mathbf{k}})$, with $B_0 = 120$ G/m and $\omega = 10^3 \text{ s}^{-1}$. Determine the EMF induced in the loop due to the changing magnetic field.

35–2 A toroid with a rectangular cross section ($a = 2$ cm by $b = 3$ cm) consists of 500 turns of wire that carries a current $I = I_0 \sin \omega t$, with $I_0 = 50$ A and a frequency $\nu = \omega/2\pi = 60$ Hz. A loop that consists of 20 turns of wire links the toroid, as shown in the

Problem 35-1

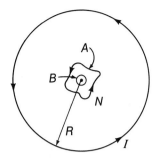

Problem 35-4

diagram. Determine the EMF induced in the loop by the changing current I.

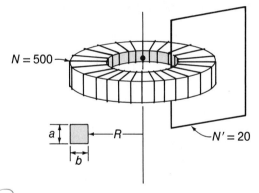

35-3 A long solenoid has n turns per meter and carries a current $I = I_0(1 - e^{-\alpha t})$, with $I_0 = 30$ A and $\alpha = 1.6$ s^{-1}. Inside the solenoid and coaxial with it is a loop that has a radius $R = 6$ cm and consists of a total of N turns of fine wire. What EMF is induced in the loop by the changing current? (Take $n = 400$ turns/m and $N = 250$ turns.)

35-4 A small irregular loop with area A consists of N turns of fine wire. This loop lies at the center and in the plane of a large single-turn loop with radius R that carries a current $I = I_0(1 + \sin \omega t)$. Obtain the expression for the EMF induced in the small loop by the changing current in the large loop.

Section 35-3

35-5 A conducting U-shaped frame has a conducting cross piece that moves with a velocity $v = 15$ m/s as shown in the diagram. The frame is in a uniform time-independent magnetic field **B** whose direction is perpendicular to the plane of the frame and whose magnitude is 800 G. Determine the induced EMF in the loop (a) by using the idea of motional EMF and (b) by using Faraday's law.

35-6 An automobile travels due north (magnetic north) at a speed of 120 km/h in a locality where the Earth's magnetic field has a magnitude of 56,000 γ and a dip angle of 70°. Calculate the magnitude of the EMF induced between the ends of a 1.3-m axle. What is the polarity of the western end of the axle?

35-7 A conducting rod with a length of 50 cm is rotated about one end with a constant angular velocity ω ($\omega = 360$ s^{-1}) in the presence of a uniform magnetic field **B** ($B = 200$ G) that is parallel to ω. What potential difference is induced between the ends of the rod? What is the polarity of the rotating end?

35-8 A disc with radius R and resistivity ρ is located in a
•   time-varying magnetic field, $B(t) = B_0 t/\tau$, that is directed parallel to the disc axis. Calculate the density $\mathbf{J}(r)$ of the eddy current induced in the disc. Assume $R = 20$ cm, $\rho = 6 \times 10^{-8} \Omega \cdot$m, $B_0 = 2.2$ T, and $\tau = 18$ s; assume also that Ohm's law, $\mathbf{J} = \mathbf{E}/\rho$, is valid. (•Why is it not necessary to be concerned with self-induced fields in this situation?)

Section 35-4

35-9 The electric field strength in a region of space is given by $\mathbf{E} = E_0(1 - r/a) \sin \omega t \hat{\mathbf{k}}$ for $r < a$ and $\mathbf{E} = 0$ for $r > a$. Determine the magnetic induction $\mathbf{B}$ everywhere within the region.

35-10 The electric field strength in a region of space is
•• given by $\mathbf{E} = E_0(t/\tau)\hat{\mathbf{k}}$. Determine the total displacement current through a ring with radius R whose axis coincides with the polar axis, as shown in the diagram. (a) First, use the bounding surface S, which is a flat disc whose perimeter is the ring. (b) Next, make an explicit calculation using the bounding surface S', which is a portion of a sphere with radius $2R$. [*Hint:* Make use of the geometry shown in the diagram at right.]

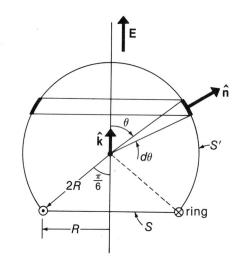

Section 35-5

35-11 Two inertial observers, O and O', who are in relative motion measure fields $\mathbf{E}$, $\mathbf{B}$ and $\mathbf{E}'$, $\mathbf{B}'$, respectively. Use the Galilean transformation equations (Eqs. 35-20) to show that the Lorentz force on a charge q determined by one observer is the same as that determined by the other, if terms of order uv/c^2 are negligible.

35-12 As viewed by an observer O, a rectangular conduct-
• ing slab moves with a velocity $\mathbf{v}$ in a uniform field $\mathbf{B}$, as shown in the diagram. A second observer O' moves with the slab. (a) Show that in the moving frame O' we have $\mathbf{E}$ (outside the slab) $= \mathbf{u} \times \mathbf{B}$ and $\mathbf{E}$ (inside) $= 0$. Show also that the surface densities σ of induced charge are σ ($-y$ face) $= +\epsilon_0|\mathbf{u} \times \mathbf{B}|$ and σ ($+y$ face) $= -\epsilon_0|\mathbf{u} \times \mathbf{B}|$. (b) Show that the surface charge densities are the same in both reference frames. Show also that $\mathbf{E}$ (outside) $= 0$ and $\mathbf{E}$ (inside) $= (\sigma/\epsilon_0)\hat{\mathbf{j}}$.

35-13 In 1832 Faraday proposed that electric current
• could be generated by the flowing water in the Thames with the apparatus illustrated in the diagram. Two conducting plates with lengths a and widths b are placed facing one another on opposite sides of the river, a distance w apart. The flow velocity in the river is $\mathbf{v}$ and the vertical component of the Earth's magnetic field is B. (a) Show that the current in the load resistor R is

$$I = \frac{abvB}{\rho + abR/w}$$

Problem 35–12

where ρ is the resistivity of the river water. (b) Calculate the short-circuit current (that is, take $R = 0$) if $a = 100$ m, $b = 5$ m, $v = 3$ m/s, $B = 50,000 \, \gamma$, and $\rho = 100 \, \Omega \cdot$ m.

SPECIAL TOPIC

Section 35-6

35-14 Show that the electric field $\mathbf{E} = -E_0(r/a)\hat{\mathbf{u}}_\phi$, which
• was studied in Example 35-9, can be expressed in
Cartesian coordinates as

$$\mathbf{E} = -E_0(x\hat{\mathbf{j}} - y\hat{\mathbf{i}})/a$$

Show that $\nabla \times \mathbf{E} = -(2E_0/a)\hat{\mathbf{k}}$.

35-15 Consider an electric field that has components
••

$$E_x = E_0(x^2/a^2) \quad \text{and} \quad E_z = -2E_0(x + y)z/a^2$$

(a) Assume that E_y is a function only of y and that
$\rho = 0$ everywhere. Determine E_y. (b) What magnetic induction $\mathbf{B}(t)$ is required to yield the curl of
the specified electric field? (c) Determine the cur-
rent density $\mathbf{J}$ necessary to produce $\mathbf{B}(t)$. (d) Verify
that the divergence of $\mathbf{B}(t)$ is zero.

Additional Problems*

35-16 A time-dependent magnetic dipole, specified by
• $\mathbf{m} = m_0 \sin \omega t\, \hat{\mathbf{k}}$, is located on the axis of a single-
turn circular loop of radius a. The distance from the
dipole to the center of the loop is b, as indicated in
the diagram. Determine the EMF induced in the
loop by the dipole variations. [*Hint:* To determine
the flux linkage with the loop, use the bounding
surface S, which is a segment of a sphere with center
at the position of the dipole and with radius
$R = (a^2 + b^2)^{1/2}$. The spherical components of the
dipole field are (compare Problem 28-3 for the electric case)

$$B_r = \frac{\mu_0 m \cos \theta}{2\pi r^3}, \qquad B_\theta = \frac{\mu_0 m \sin \theta}{4\pi r^3}, \qquad B_\phi = 0.]$$

location of the loop is 20,000 γ. Calculate the maximum EMF induced in the coil by the Earth's field.

35-18 A long solenoid, whose axis coincides with the x-
axis, consists of 200 turns per meter of wire that
carries a steady current of 15 A. A coil is formed by
wrapping 30 turns of thin wire around a frame that
has a radius of 8 cm. The coil is placed inside the
solenoid and mounted on an axis that is a diameter
of the coil and coincides with the y-axis. The coil is
then rotated with an angular speed of 4π radians per
second. (The plane of the coil is in the y-z plane at
$t = 0$.) Determine the EMF developed in the coil.

35-19 In Problem 35-18, replace the field of the solenoid
by a time-varying field, $B = (4 \times 10^{-3}\ \text{T}) \sin \omega t$,
where $\omega = 4\pi\ \text{s}^{-1}$. Determine the EMF developed
in the coil.

35-17 A square coil (20 cm × 20 cm) that consists of 100
turns of wire rotates about a vertical axis at
1500 rpm, as indicated in the diagram. The horizontal component of the Earth's magnetic field at the

*In these problems consider the effects *only* of the given currents; that is, neglect the effects of any self-induced EMFs.

35-20 A helicopter is hovering with its 4.6-m rotor blades revolving at a rate of 180 rpm. At the location of the helicopter the magnitude of the Earth's magnetic field is 50,000 γ and the dip angle is 65°. What EMF is developed between the rotor hub and the tip of a blade?

35-21 A 1-m conducting rod moves with its length parallel to the x-axis and with a velocity $\mathbf{v} = v_0\hat{\mathbf{j}}$, where $v_0 = 100$ m/s. There is present a magnetic field described by $\mathbf{B} = B_0(3\hat{\mathbf{i}} + 4\hat{\mathbf{j}} + 5\hat{\mathbf{k}})$, with $B_0 = 45$ G. What is the EMF induced between the ends of the rod?

35-22 N turns of wire are wound on a rigid rectangular
•• frame with dimensions $a \times b$. The coil moves across a horizontal surface (the x-y plane) with a velocity $\mathbf{v} = v_0\hat{\mathbf{j}}$, as shown in the diagram. There is present a uniform but time-varying magnetic field, $\mathbf{B} = B_0\hat{\mathbf{k}} \sin(\pi y/a) \sin \omega t$. Derive an expression for the EMF induced in the coil by the field B. [*Hint:* Consider the *total* time variation of the flux linkage by using

$$\frac{d\varphi_M}{dt} = \frac{dx}{dt}\frac{\partial\varphi_M}{\partial x} + \frac{dy}{dt}\frac{\partial\varphi_M}{\partial y} + \frac{dz}{dt}\frac{\partial\varphi_M}{\partial z} + \frac{\partial\varphi_M}{\partial t}.]$$

35-23 A single-turn wire loop is in the shape of an
• ellipse with semi-axes a and b. The loop is set into

Problem 35-22

oscillation such that $a = r(1 + \eta \sin \omega t)$, with $\eta \ll 1$ and with constant perimeter. The plane of the loop is perpendicular to a uniform magnetic field B. Show that, to lowest order in η, the EMF induced in the loop is $\mathcal{E} = 2\pi B\eta^2 r^2 \sin 2\omega t$. [*Hints:* The area of an ellipse is πab. For small η, the perimeter of the ellipse is approximately $\pi \sqrt{2(a^2 + b^2)}$.]

35-24 In Example 35-9 it was pointed out that a linearly
•• increasing magnetic field, $\mathbf{B} = 2(E_0/a)t\hat{\mathbf{k}}$, in a long solenoid produces an azimuthal electric field, $E_\phi = -E(r/a)$. (a) Calculate the rate of change of the magnetic flux through a circle with radius a centered on and perpendicular to the z-axis. (b) Calculate the rate of change of the flux through a circle with the same radius and in the same plane if the center is at $r = a$. Show that the two results are equal. [*Hints:* The equation of the path in (b) is $r = 2a \cos \phi$. It will be easier to perform the integration in (b) over an angle measured around the center of the circular path. Relate this angle to ϕ.]

INDUCTANCE AND MAGNETIC MATERIALS

In the preceding chapters, vacuum (or air) conditions were assumed for the discussions of magnetic fields. We now enlarge the scope to include the effects that magnetic materials have on the fields produced by electric currents and by changing electric fields. We also examine here the connection between these macroscopic magnetic effects and the microscopic behavior of the atoms that constitute the magnetic materials.

In Chapter 30 the idea of capacitance was used to facilitate the study of dielectric materials. We follow a parallel development here and first introduce the concept of *inductance*. Then, simple *inductors* are used to discuss measurements of the properties of magnetic materials.

36-1 MUTUAL INDUCTANCE

In Chapter 35 we considered the EMF induced in a circuit (the secondary) by a changing current in a nearby circuit (the primary). It was pointed out that Faraday's law of induction involves the time rate of change of the *entire* magnetic flux that links a circuit, including the flux due to any self-induced fields. However, to avoid discussing at that point the self-induced EMFs, we either omitted reference to the currents in the secondary circuits (remarking that they could be included by using superposition) or treated cases in which the secondary current was constant in time and therefore did not contribute any induced EMF of its own. In this section we proceed in the same way; in Section 36-2 we include self-induced EMFs.

Let us begin by considering a case in which no magnetic materials are present. Two coils are shown in Fig. 36-1. A (time-dependent) current I_1 flows in the primary coil ①, which consists of N_1 turns of wire, and produces a magnetic induction $\mathbf{B}_1$. The flux linkage with the secondary coil ②, which consists of N_2 turns of wire, is expressed as* $N_2 \varphi_{21}$, where $\varphi_{21} = \int_{S_2} \mathbf{B}_1 \cdot \hat{\mathbf{n}}_2 \, dA$. Using Faraday's law, the resulting induced EMF in the secondary coil is

$$\mathcal{E}_{21} = -\frac{d}{dt}(N_2 \varphi_{21}) \qquad (36\text{-}1)$$

We restrict attention to coils and loops in *fixed* geometry. Moreover, as in Chap-

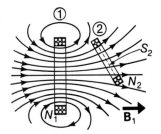

Fig. 36-1. The changing current in the primary coil ① induces an EMF in the secondary coil ②.

---

*In Eq. 35–5 we wrote this term as $N\varphi'_M$.

ter 35, only slowly varying currents are allowed. Then, because $\mathbf{B}_1$ depends linearly on I_1, the EMF $\mathcal{E}_{21}$ that is induced in coil ② is directly proportional to dI_1/dt; thus,

$$\mathcal{E}_{21} = -M_{21}\frac{dI_1}{dt}, \qquad \text{fixed geometry} \qquad \textbf{(36–2)}$$

where M_{21} is a strictly geometric factor that is called the *mutual inductance* of the pair of coils. Comparing Eqs. 36–1 and 36–2, we see that

$$M_{21} = \frac{N_2 \varphi_{21}}{I_1} \qquad \textbf{(36–3)}$$

That is, we can interpret M_{21} as the flux linkage with coil ② per unit amount of current in coil ①.

Suppose that we reverse the situation shown in Fig. 36–1, holding the geometry fixed. Then, a time-varying current I_2 flowing in coil ② (now the primary) induces an EMF $\mathcal{E}_{12}$ in coil ① (now the secondary). Thus,

$$\mathcal{E}_{12} = -M_{12}\frac{dI_2}{dt}$$

By using techniques more advanced than those at our disposal, it can be shown that

$$M_{12} = M_{21} = M$$

where we can now drop the subscripts and write M for the mutual inductance. For simple geometries, the equality of M_{12} and M_{21} can be demonstrated by explicit calculation—see Problems 36–2 and 36–5.

Inductance is measured in units of volts per ampere per second, abbreviated to the *henry* (H), named in honor of Joseph Henry. Thus,

$$1\ \text{H} = 1\ \text{V}/(\text{A}/\text{s}) = 1\ \text{V} \cdot \text{s}/\text{A} = 1\ \text{Wb}/\text{A}$$

The henry is a rather large unit; in practical cases, the millihenry (mH) or the microhenry (μH) is a more appropriate unit.

Example 36–1

Calculate the mutual inductance M between a toroid with $N_1 = 200$ turns and a linking coil with $N_2 = 10$ turns. The toroid has an inner radius $R = 6$ cm and a rectangular cross section as shown in the diagram, with $a = 3$ cm and $b = 4$ cm.

Solution:

Using Ampère's circuital law (Eq. 34–7) for a circuit at a radius r, we obtain

$$B_1 = \frac{\mu_0 N_1 I_1}{2\pi r}$$

The flux φ_{21} that links the coil is

$$\varphi_{21} = \int_{S_2} \mathbf{B}_1 \cdot \hat{\mathbf{n}}_2 \, dA_2$$

Because $\mathbf{B}_1$ is entirely azimuthal and exists only within the toroid, we have, with $dA_2 = b \, dr$,

$$\varphi_{21} = \frac{\mu_0 N_1 I_1 b}{2\pi} \int_R^{R+a} \frac{dr}{r} = \frac{\mu_0 N_1 b}{2\pi} I_1 \ln\left(\frac{R+a}{R}\right)$$

Therefore, we identify

$$M = M_{21} = \frac{N_2 \varphi_{21}}{I_1} = \frac{\mu_0 N_1 N_2 b}{2\pi} \ln\left(\frac{R+a}{R}\right)$$

Substituting the values of the various quantities, we find

$$M = (2 \times 10^{-7} \text{ H/m})(200)(10)(0.04 \text{ m}) \ln\left(\frac{0.09 \text{ m}}{0.06 \text{ m}}\right)$$

$$= 6.49 \times 10^{-6} \text{ H} = 6.49 \, \mu\text{H}$$

Notice that we have expressed μ_0 as $4\pi \times 10^{-7}$ H/m. (•Can you verify these units?) Notice also how much more difficult it would have been to calculate M_{12}.

36-2 SELF-INDUCTANCE

Consider an isolated coil, loop, or circuit with fixed geometry in which a current I flows. This current produces a magnetic field that links the circuit itself. If the current varies with time, so does the flux linkage. Then, according to Faraday's law, there results a self-induced EMF in the circuit. This EMF is sometimes referred to as a *back EMF* because its direction is such that it opposes any change in the driving EMF (Lenz's law).

For a coil with N turns and a linking flux $\varphi_M = N\varphi_M'$, the self-induced EMF is (see Eq. 35-4)

$$\mathcal{E} = -\frac{d}{dt}(N\varphi_M') \qquad (36\text{-}4)$$

If there are no magnetic materials present, the magnetic field and, hence, the flux linkage depends linearly on the current I in the coil, and the self-induced EMF is directly proportional to dI/dt; thus,

$$\mathcal{E} = -L\frac{dI}{dt} \qquad (36\text{-}5)$$

where L is a strictly geometric factor that is called the *self-inductance* (or, simply, the *inductance*) of the coil. Comparing Eqs. 36–4 and 36–5, we see that

$$L = \frac{N\varphi'_M}{I}$$

(36–6)

An *inductor* is a compact circuit element, such as a tightly wound coil of wire, that produces electromagnetic induction effects. When electric circuits are drawn, an inductor is indicated by the symbol

Example 36–2

Determine the self-inductance of a long solenoid, with a length ℓ and a radius R, that consists of N turns of wire as shown in the diagram. (Neglect end effects.)

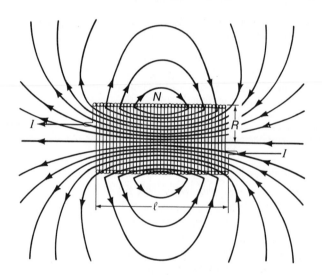

Solution:

The magnetic field inside a long solenoid is uniform and axial: $B = \mu_0(N/\ell)I$ (see Example 34–6). Therefore, the flux linking the N turns of the solenoid is

$$N\varphi'_M = N \cdot \pi R^2 \cdot \mu_0 \frac{N}{\ell} I$$

Referring to Eq. 36–6, we identify the inductance as

$$L = \frac{\mu_0 \pi R^2 N^2}{\ell}$$

To neglect end effects is to ignore the failure of some of the flux lines near the middle of the solenoid to link with the turns near the ends (refer to the diagram). Thus, the inductance of a solenoid with a finite length is less than the value calculated from the expression above. Recall the parallel situation with regard to the fringing field of a capacitor (Eq. 30–5a). In the

same way, we write

$$L = \frac{\mu_0 \pi R^2 N^2}{\ell} f$$

where $f \leqslant 1$. For a solenoid with $\ell/R = 1.0$ we find $f = 0.53$, and when $\ell/R = 5$ we find $f = 0.85$; of course, in the limit $\ell/R \to \infty$, we have $f = 1$.

If the coil has $R = 4$ cm and $\ell = 20$ cm, and consists of 250 turns, we find

$$L = \frac{(4\pi \times 10^{-7} \text{ H/m})\pi(0.04 \text{ m})^2(250)^2(0.85)}{0.20 \text{ m}}$$

$$= 1.68 \times 10^{-3} \text{ H} = 1.68 \text{ mH}$$

Example 36–3

Determine the self-inductance of the toroid shown in the diagram.

Solution:

According to the result obtained in Example 34–7, the magnetic induction inside an air-core toroid with mean radius b is

N = 250 turns

b = 8 cm

area A = 2 cm²

$$B = \mu_0 n I = \frac{\mu_0 N I}{2\pi b}$$

Then, the flux linking the N turns is

$$N\varphi'_M = N \cdot A \cdot \frac{\mu_0 N I}{2\pi b}$$

Referring to Eq. 36–6, we identify the inductance as

$$L = \frac{\mu_0 N^2 A}{2\pi b}$$

Substituting the given values, we obtain

$$L = \frac{(4\pi \times 10^{-7} \text{ H/m})(250)^2(2 \times 10^{-4} \text{ m}^2)}{2\pi(0.08 \text{ m})}$$

$$= 3.125 \times 10^{-5} \text{ H} = 31.25 \ \mu\text{H}$$

Coupled Inductors. Consider two coils that are magnetically linked—coil ① with N_1 turns and coil ② with N_2 turns. Suppose that the coils carry currents* i_1 and i_2, respectively, due to independent externally applied EMFs, $\mathcal{E}_{10}$ and $\mathcal{E}_{20}$, as indicated in Fig. 36–2. If i_1 is a function of time, the magnetic field due to i_1 will induce an EMF in coil ② that is proportional to the mutual inductance $M_{21} = M$ and also an EMF in coil ① that is proportional to the self-inductance L_1. We refer to these induced EMFs as $\mathcal{E}_{21}$ and $\mathcal{E}_{11}$ respectively. If i_2 is also a function of time, then

---

*We use i to indicate the instantaneous value of the current.

Fig. 36–2. Coupled inductors with external sources of EMF. The instantaneous currents in the two circuits are i_1 and i_2.

corresponding EMFs, $\mathcal{E}_{12}$ and $\mathcal{E}_{22}$, are generated. The total induced EMFs in the two coils are

$$\mathcal{E}_1 = \mathcal{E}_{10} + \mathcal{E}_{11} + \mathcal{E}_{12} = \mathcal{E}_{10} + \left(-L_1 \frac{di_1}{dt} \pm M \frac{di_2}{dt} \right)$$

$$\mathcal{E}_2 = \mathcal{E}_{20} + \mathcal{E}_{21} + \mathcal{E}_{22} = \mathcal{E}_{20} + \left(\pm M \frac{di_1}{dt} - L_2 \frac{di_2}{dt} \right)$$

(36–7)

The signs of the mutual inductance terms can be determined only when the sense of each coil winding is known. That is, the two fields that are produced by the coupling can either reinforce or oppose one another.

36-3   MAGNETIC FIELD ENERGY

Consider an inductor L that is connected to an external source of EMF. The resulting current $i(t)$ increases in some unspecified manner from zero to a final value I. While $i(t)$ is changing, there is a self-induced EMF (a *back* EMF), $\mathcal{E} = -L\,di/dt$, which causes energy to be absorbed by the inductor from the source of the external EMF at a rate

$$P = Li \frac{di}{dt}$$

The total amount of electrical work done by the source is

$$W = \int P\,dt = L \int_0^I i\,di$$

so that
$$W = \tfrac{1}{2}LI^2$$
(36–8)

Notice that the work done does not depend on the way in which $i(t)$ varies to achieve its final value I. Notice also that when a steady-state condition is reached, we have $di/dt = 0$, and power is no longer expended by the source.

What happens to the energy that is absorbed by the inductor? It proves convenient to associate this energy with the magnetic field produced by the current flow in the inductor.* (Compare the discussion of the case of electric field energy in Section 30–3.) Therefore, the electric energy W delivered to the inductor can be equated to

---

* We again restrict the current variations to be slow. Rapid variations lead to the radiation of energy away from the circuit.

the energy U_M stored in the magnetic field; that is,

$$U_M = \tfrac{1}{2}Li^2 \qquad \textbf{(36-9)}$$

Energy Density. Inside a long solenoid the magnetic field is uniform and has the magnitude (see Example 36-2) $B = \mu_0 NI/\ell$, neglecting end effects. Thus, the energy in the field is

$$U_M = \tfrac{1}{2}LI^2 = \frac{1}{2}\left(\frac{\mu_0 \pi R^2 N^2}{\ell}\right)\left(\frac{B\ell}{\mu_0 N}\right)^2$$

$$= \frac{B^2}{2\mu_0}(\pi R^2 \ell) = \frac{B^2 V}{2\mu_0}$$

where $\pi R^2 \ell$ is the volume V of the solenoid. We define the magnetic field *energy density* to be $u_M = U_M/V$; thus,

$$u_M = \frac{B^2}{2\mu_0} = \frac{1}{2}\frac{\mathbf{B}\cdot\mathbf{B}}{\mu_0}, \qquad \text{vacuum (or air) conditions} \qquad \textbf{(36-10)}$$

Although this expression was obtained for the special case of a solenoid, it contains none of the parameters of the solenoid and is, in fact, quite generally true for magnetic fields in vacuum.

Example 36-4

Determine the self-inductance of the coaxial inductor shown in the diagram.

Solution:

Assume that the inductor is sufficiently long ($\ell \gg b$) or the gap is sufficiently small ($b - a \ll b$) that end effects may be neglected. Then, the magnetic induction is entirely azimuthal, and there is no field for $r > c$. Ampère's circuital law can be used to find the magnetic induction in the three regions:

$$\mathbf{B}_1 = \frac{\mu_0 Ir}{2\pi a^2}\hat{\mathbf{u}}_\phi, \qquad r < a$$

$$\mathbf{B}_2 = \frac{\mu_0 I}{2\pi r}\hat{\mathbf{u}}_\phi, \qquad a < r < b$$

$$\mathbf{B}_3 = \frac{\mu_0 I}{2\pi r}\left(\frac{c^2 - r^2}{c^2 - b^2}\right)\hat{\mathbf{u}}_\phi, \qquad b < r < c$$

Next, Eq. 36–10 is used to calculate the field energy in the same three regions; the results are

$$U_1 = \frac{1}{2\mu_0}\int_0^a \left(\frac{\mu_0 I r}{2\pi a^2}\right)^2 2\pi r \ell \, dr = \frac{\mu_0 I^2 \ell}{16\pi}$$

$$U_2 = \frac{1}{2\mu_0}\int_a^b \left(\frac{\mu_0 I}{2\pi r}\right)^2 2\pi r \ell \, dr = \frac{\mu_0 I^2 \ell}{4\pi}\ln\frac{b}{a}$$

In a similar way we find

$$U_3 = \frac{\mu_0 I^2 \ell}{4\pi}\left(\frac{c^4}{(c^2 - b^2)^2}\ln\frac{c}{b} - \frac{3c^2 - b^2}{4(c^2 - b^2)}\right)$$

The total field energy is $U_M = U_1 + U_2 + U_3$. Then, we use Eq. 36–9 to identify the self-inductance per unit length; thus,

$$\frac{L}{\ell} = \frac{\mu_0}{8\pi}\left\{1 + 4\ln\frac{b}{a} + \left[\frac{4c^4}{(c^2 - b^2)^2}\ln\frac{c}{b} - \frac{3c^2 - b^2}{c^2 - b^2}\right]\right\}$$

Typical values for a coaxial cable are $a = 0.1$ mm, $b = 3$ mm, and $c = 3.2$ mm. In such a case we find

$$\frac{L}{\ell} = \frac{\mu_0}{8\pi}\{1 + 13.605 + [17.605 - 17.516]\} = \frac{\mu_0}{8\pi}(14.694)$$

Notice that the largest contribution is from U_2, corresponding to the field energy in the region between the two conductors. Finally,

$$\frac{L}{\ell} = \frac{1}{8\pi}(4\pi \times 10^{-7} \text{ H/m})(14.694) = 0.735 \; \mu\text{H/m}$$

(See also Problem 36–10.)

The Energy Stored by Coupled Inductors. Consider again the coupled inductors discussed in connection with Eqs. 36–7. If we multiply the first of these equations by i_1 and the second by i_2, we have the power absorbed by each of the coupled inductors; that is,

$$P_1 = L_1 i_1\frac{di_1}{dt} \pm M i_1\frac{di_2}{dt}$$

$$P_2 = \pm M i_2\frac{di_1}{dt} + L_2 i_2\frac{di_2}{dt}$$

Thus, the total power absorbed by both inductors is

$$P = L_1 i_1\frac{di_1}{dt} + L_2 i_2\frac{di_2}{dt} \pm M\left(i_1\frac{di_2}{dt} + i_2\frac{di_1}{dt}\right)$$

The last term in this equation can be expressed as

$$M\left(i_1\frac{di_2}{dt} + i_2\frac{di_1}{dt}\right) = M\frac{d}{dt}(i_1 i_2)$$

If both of the currents, i_1 and i_2, start from zero at the same instant and reach values of I_1 and I_2, respectively, also at the same instant, the total energy absorbed is

$$W = \int P\,dt = L_1\int_0^{I_1} i_1\,di_1 + L_2\int_0^{I_2} i_2\,di_2 \pm M\int_0^{I_1 I_2} d(i_1 i_2)$$

This energy is stored in the field, so $U_M = W$; that is,

$$U_M = \tfrac{1}{2}L_1 I_1^2 + \tfrac{1}{2}L_2 I_2^2 \pm M I_1 I_2 \qquad \text{(36–11)}$$

Inductors in Series. The existence of mutual inductance between two nearby inductors complicates the addition law for combinations of inductors. For example, if two coupled inductors are connected in series so that $I_1 = I_2 = I$, Eq. 36–11 becomes

$$U_M = \tfrac{1}{2}(L_1 + L_2 \pm 2M)I^2$$

In order to have the same stored energy, a single equivalent inductor must have an inductance

$$L = L_1 + L_2 \pm 2M, \qquad \text{series connection} \qquad \text{(36–12)}$$

The Force Between Coupled Inductors. Consider two long solenoids, one of which is a distance s inside the other, as indicated in Fig. 36–3. To determine the force between the solenoids, we calculate the work that would be done if the distance s were increased by an amount ds while holding constant the two currents, I_1 and I_2.*

Fig. 36-3. The force between the two solenoids can be calculated by using the method of virtual work.

It is a general principle that $F = -dU/ds$. In the case here, $U = U_M$ is given by Eq. 36–11, with L_1, L_2, I_1, and I_2 constant. As s is increased to $s + ds$, only the mutual inductance M changes. Thus,

$$F = \pm I_1 I_2 \frac{dM}{ds} \qquad \text{(36–13)}$$

The mutual inductance M can be obtained by calculating the magnetic flux of the inner solenoid that links the outer solenoid, namely,

---

*This is the method of *virtual work* that we have used previously—see Section 30–3.

$$N_2\varphi_{21} = \left(\frac{\mu_0 N_1 I_1}{\ell_1}\right) \cdot (\pi R_1^2) \cdot \left(\frac{s}{\ell_2} N_2\right)$$

Then, from Eq. 36–3, we see that

$$M = M_{21} = \frac{N_2\varphi_{21}}{I_1} = \frac{\mu_0 \pi R_1^2 N_1 N_2}{\ell_1 \ell_2} s$$

($\bullet$Can you deduce the same result by considering M_{12}?)

For the case illustrated in Fig. 36–3, the windings and currents are such that the inner coil is pulled into the outer coil. Thus, we require the positive sign in the expression for the force; that is, M *increases* as s increases. Then,

$$F = +\frac{\mu_0 \pi R_1^2 N_1 N_2 I_1 I_2}{\ell_1 \ell_2} \tag{36-14}$$

For example, with $N_1 = N_2 = 250$, $\ell_1 = \ell_2 = 20$ cm, $R_1 = 4$ cm, and $I_1 = I_2 = 25$ A, we find $F = 6.17$ N.

36-4   MAGNETIC MATERIALS

We now relax the restriction to vacuum conditions and inquire about magnetic fields within matter. A practical and conceptually convenient device for studying the magnetic properties of materials is the *Rowland ring*. The material to be investigated is formed into a toroid on which there is wound a tightly spaced coil of N turns, as indicated in Fig. 36–4. A secondary coil that consists of N' turns is looped around the toroid and is used to measure the EMF induced when the current in the primary coil is varied. The induction effects for a specimen material are compared with those for a toroid with an air (vacuum) core for otherwise identical conditions.

Fig. 36-4. A *Rowland ring* consists of a toroidal core of the specimen material tightly wrapped with a coil of N turns. We consider a toroid with $R \gg a$.

With a vacuum core, we find that the induction effects in the secondary coil are attributable to a toroidal magnetic field,

$$B_0 = \mu_0 \frac{NI}{2\pi R}$$

where $R \gg a$. With a core fabricated from the specimen material, the induction effects indicate a magnetic field

$$B = \kappa_M \mu_0 \frac{NI}{2\pi R}$$

The quantity
$$\kappa_M = B/B_0 \qquad \text{(36-15a)}$$

is called the *relative permeability* (that is, the permeability of the material relative to vacuum). We also use a quantity μ, called the *permeability,* where

$$\kappa_M = \frac{\mu}{\mu_0} \qquad \text{(36-15b)}$$

Finally, because κ_M is nearly unity for most materials, it is useful to introduce the magnetic susceptibility χ_M, where

$$\chi_M = \kappa_M - 1 \qquad \text{(36-16)}$$

Of course, we also have

$$\mu = (1 + \chi_M)\mu_0 \qquad \text{(36-17)}$$

Notice the similarity of the new magnetic parameters to those introduced in the discussion of dielectric materials (Section 30-5). The dielectric constant κ is always greater than one. However, the magnetic counterpart κ_M can be greater or less than one, and in the case of ferromagnetic materials κ_M is much greater than one. We therefore distinguish three major categories of behavior for magnetic materials, and we discuss these in turn.

At the atomic level, there are two fundamental sources for the magnetic properties of matter, although in most cases they are interconnected in complicated ways. One source is the circulation of electrons about their parent nuclei. This orbital motion is equivalent to an atomic current and gives rise to an atomic magnetic moment. These moments are governed by quantum rules and can take on only certain discrete values. Atomic magnetic moments are *quantized* in units of the *Bohr magneton,* $m_B = 9.27 \times 10^{-24} \text{ A} \cdot \text{m}^2$.

The second fundamental source is the magnetic moment associated with the intrinsic spin angular momentum of the individual electrons.* This magnetic moment is an intrinsic property of the electron, in the same way that charge and mass are intrinsic properties. The electron magnetic moment has a magnitude m_B.

Diamagnetism. Faraday discovered that a freely suspended needlelike piece of bismuth tends to align itself *perpendicular* to a uniform magnetic field, instead of taking up a parallel alignment as would a slender iron needle. Materials that exhibit this type of behavior are said to be *diamagnetic.* When an external magnetic field is imposed on a diamagnetic material, there is induced an atomic EMF. The accompanying current circulates in such a way that the resulting magnetic field opposes the external field (Lenz's law). That is, the *induced magnetization* of the material is directed opposite to the imposed field. Thus, the magnetic field within a diamagnetic material is smaller than the imposed field would be in vacuum. Consequently, diamagnetic materials have $\kappa_M < 1$. All matter is basically diamagnetic, but the effect is weak and it is usually overwhelmed by either of the other effects when one is present.

Paramagnetism. An atomic system that has a permanent magnetic moment will tend to align this moment with an external magnetic field to the extent permitted

---

*There is in addition a fundamental magnetic moment associated with the nuclear spin. In some special circumstances the nuclear magnetic moment produces important effects, but its existence need not be considered in describing the bulk magnetic properties of matter under ordinary conditions.

by thermal agitation. This behavior is similar to that of polar molecules whose permanent electric dipole moments tend to align with an external electric field. In the event that the permanent magnetic moment of a system is either very weak or non-existent, the diamagnetic behavior dominates. For example, the noble gases, with their completely symmetric closed electron shells, have zero permanent magnetic moments; these elements exhibit pure diamagnetism.

If the permanent magnetic moment of an atomic system is sufficiently large to overcome the inherent diamagnetism, the net alignment of the moments with the external field *increases* the magnetic induction within the sample. That is, the magnetization of the material is directed parallel to the imposed field. Such materials are said to be *paramagnetic* and have $\kappa_M > 1$. Liquid oxygen is an example of a strong paramagnetic substance; its large permanent magnetic moment is due to the combined effects of electron spin and orbital motion.

Because paramagnetism depends on the alignment of permanent magnetic moments with an applied field, the magnitude of the effect is limited by the randomization due to thermal agitation. Thus, we expect that κ_M for paramagnetic materials will decrease with increasing temperature. (•Should diamagnetism also be temperature-dependent?)

Ferromagnetism.   Certain materials exhibit a spontaneous ordering of their permanent magnetic moments within microscopic regions (called *domains*) even if no external magnetic field is applied. Because this phenomenon was first recognized for iron, it is referred to as *ferromagnetism*. This important effect is an entirely quantum effect involving the electron spins, and is restricted to a few elements. Iron, cobalt, and nickel comprise one group of ferromagnetic elements, and the other group consists of the rare-earth elements gadolinium and dysprosium. Ferromagnetism is also exhibited by various compounds and alloys of nonferromagnetic elements (for example, $CrBr_3$, Cu_2MnAl, and EuO).

The magnetic domain structure of Permalloy (22 percent iron, 78 percent nickel) as revealed by a microscope. The arrows indicate the magnetic axes of the individual domains. (After *Physics of Magnetism* by S. Chikazumi. John Wiley & Sons, Inc., New York, 1964.)

0.3 mm

When the net magnetic moments of the individual microscopic domains are oriented at random with respect to one another, the magnetic induction of the sample as a whole is zero—the sample is unmagnetized. However, even a relatively weak external magnetic field can induce changes in the size and orientation of the domains so that a very large magnetic induction results. The value of the relative permeability κ_M can reach 10^5 or even higher. (See Fig. 36–13.) It is this property of ferromagnetic substances that makes these materials valuable as permanent magnets and as cores for electromagnets. As in the case of paramagnetism, the effects of thermal agitation limit the achievable magnetization of a ferromagnetic sample. The ferromagnetism is maximum at 0 K and decreases with increasing temperature until it becomes zero at a characteristic critical temperature called the *Curie temperature* of the material. For higher temperatures the sample is paramagnetic.

Magnetization. When a sample of bulk matter is placed in a magnetic field, there develops, because of the presence of the field, an effective magnetic moment **m** per atom (or molecule). For the case of diamagnetic materials **m** is the induced moment, and for paramagnetic and ferromagnetic materials **m** includes an appropriate average over the distribution of directions of the permanent moments. Figure 36–5a shows a number of individual moments **m** in a sample of material with a volume $A\ell$. If there are $\mathcal{N}$ atoms (or molecules) per unit volume, the cumulative effect of the individual moments can be represented by the macroscopic magnetization vector **M** (Fig. 36–5b), where

$$\mathbf{M} = \mathcal{N}\mathbf{m} \tag{36–18}$$

The quantity **M** is referred to as the magnetization or the *magnetic moment density*.

Magnetic moments have dimensions of *current* $\times$ *area,* so the dimensionality of **M** is *current per unit length*. Because the magnetic moment of a coil depends on the number of loops that the current makes, it is customary to express magnetization in units of ampere-turns per meter, which we abbreviate as A · turns/m.

According to Ampère's view, magnetism and magnetic effects are due to elementary current loops. Thus, the magnetic moments shown in Fig. 36–5a are considered to result from the average currents of the circulating electrons in the constituent atoms (Fig. 36–6). It is evident in Fig. 36–6 that adjacent Ampèrian current segments are oppositely directed and therefore cancel throughout the interior of the sample. On the surface, however, there is no cancellation, so there is a net *effective magnetization current* I_M that circulates around the sample (Fig. 36–6). Notice that I_M is distributed continuously along the length ℓ of the sample. The total magnetic moment of the sample is $I_M A$. Now, the sample volume is $A\ell$, so the magnetic moment per unit volume is

$$M = \frac{I_M A}{A\ell} = \frac{I_M}{\ell} = j_M \tag{36–19}$$

where j_M is the surface magnetization current per unit length (A/m). For the special geometry of Fig. 36–5, M is numerically equal to j_M; in the general case, it is the tangential component of **M** at the surface that is equal to j_M.

Now let the sample we have been considering be a segment of a toroid with radii R and a ($R \gg a$) that is the core of a Rowland ring (Fig. 36–7). The toroid is wrapped with N turns of wire carrying a current I. If the ring core were vacuum, the current I would produce a uniform magnetic induction, $B_0 = \mu_0 NI/2\pi R$, within the core region; the magnetization would be zero. With the material core in place the induction is B, and application of Ampère's circuital law (Eq. 34–8) along the center line with radius R yields

$$2\pi RB = \mu_0(NI + 2\pi Rj_M)$$

or

$$B - \mu_0 j_M = \frac{\mu_0 NI}{2\pi R}$$

from which

$$B - \mu_0 M = B_0 \tag{36–20}$$

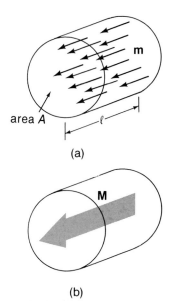

(a)

(b)

Fig. 36–5. (a) A sample of material contains atoms (or molecules) that have effective magnetic moments **m** in the presence of a magnetic field. (b) The net effect of these moments is represented by the magnetization **M**.

Fig. 36–6. The Ampèrian currents associated with the individual magnetic moments **m** (Fig. 36–5a) cancel throughout the interior of the sample, leaving an effective surface magnetization current I_M.

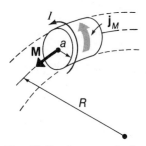

Fig. 36-7. A segment of a Rowland ring. The ring coil of N turns carries a current I. The surface magnetization current density $\mathbf{j}_M$ is associated with the magnetization $\mathbf{M}$.

Using $B = \kappa_M B_0$ and $\chi_M = \kappa_M - 1$, Eq. 36–20 can be rewritten in vector form as

$$\mu_0 \mathbf{M} = \mathbf{B} - \mathbf{B}_0 \tag{36–21a}$$

$$= (\kappa_M - 1)\mathbf{B}_0 \tag{36–21b}$$

$$= \chi_M \mathbf{B}_0 \tag{36–21c}$$

These expressions are valid for all homogeneous, isotropic magnetic media and are analogous to the expressions for uniformly polarized dielectrics (Eqs. 30–26). However, remember that in the magnetic case the susceptibility χ_M is negative for diamagnetic substances and is positive for paramagnetic and ferromagnetic substances. Values of χ_M are given in Table 36–1 for some typical diamagnetic and paramagnetic substances. The more complicated case of ferromagnetic materials is discussed later in this section and in Section 36–6.

TABLE 36–1. Magnetic Susceptibilities of Various Substances[a]

DIAMAGNETIC MATERIALS	χ_M
Bismuth	-1.7×10^{-4}
Mercury	-2.8×10^{-5}
Silver	-2.4×10^{-5}
Lead	-1.6×10^{-5}
Copper	-9.7×10^{-6}
Water	-9.1×10^{-6}
Hydrogen[b]	-2.2×10^{-9}
Nitrogen[b]	-6.7×10^{-9}

PARAMAGNETIC MATERIALS	χ_M
Oxygen[b]	$+1.9 \times 10^{-6}$
Magnesium	$+1.2 \times 10^{-5}$
Aluminum	$+2.0 \times 10^{-5}$
Tungsten	$+7.8 \times 10^{-5}$
Chromium	$+3.1 \times 10^{-4}$
Cerium	$+1.5 \times 10^{-3}$
Oxygen[c]	$+3.5 \times 10^{-3}$

[a] At 293 K.
[b] At 1 atm.
[c] Liquid, at 90 K.

Example 36–5

A solid toroid of bismuth has a core radius $a = 1$ cm and a toroid radius $R = 10$ cm. A winding of $N = 800$ turns around the toroid carries a steady current of 10 A. Determine the following quantities:

(a) the magnetic induction B_0;

(b) the magnetization M;

(c) the total magnetic induction B; and

(d) the total magnetization current I_M.

(e) The mass number of bismuth is 209 and its density is 9.8 g/cm³. What is the effective atomic magnetic moment of bismuth expressed in Bohr magnetons?

Solution:

(a) We have $R \gg a$, so the field within the toroid is essentially uniform and can be obtained from Ampère's circuital law; thus,

$$B_0 = \frac{\mu_0 NI}{2\pi R} = \frac{(2 \times 10^{-7} \text{ T} \cdot \text{m/A})(800)(10 \text{ A})}{0.10 \text{ m}}$$

$$= 0.016 \text{ T} = 160 \text{ G}$$

(b) Bismuth is diamagnetic, so the magnetization vector is directed opposite to the field $\mathbf{B}_0$. Using Eq. 36-21c and the value of χ_M from Table 36-1, the magnitude of the magnetization is

$$M = \frac{|\chi_M| B_0}{\mu_0} = \frac{(1.7 \times 10^{-4})(0.016 \text{ T})}{4\pi \times 10^{-7} \text{ T} \cdot \text{m/A}}$$

$$= 2.16 \text{ A} \cdot \text{turns/m}$$

(c) Using Eqs. 36-21a and 36-21c, we have

$$B = (\chi_M + 1)B_0 = (1 - 0.00017)B_0$$

$$= 0.99983 \, B_0$$

The susceptibility is so small that B is hardly distinguishable from B_0.

(d) The total magnetization current is

$$I_M = 2\pi R j_M = 2\pi R M = 2\pi (0.10 \text{ m})(2.16 \text{ A} \cdot \text{turns/m})$$

$$= 1.36 \text{ A} \cdot \text{turns}$$

which is to be compared with the driving ampere-turns, $NI = 8000 \text{ A} \cdot \text{turns}$.

(e) The effective atomic magnetic moment of bismuth is

$$\frac{m_{\text{eff}}}{m_B} = \frac{M}{\mathcal{N} \, m_B} = \frac{(2.16 \text{ A} \cdot \text{turns/m})(209 \times 10^{-3} \text{ kg/mol})}{(6.02 \times 10^{23} \text{ mol}^{-1})(9.8 \times 10^3 \text{ kg/m}^3)(9.27 \times 10^{-24} \text{ A} \cdot \text{m}^2)}$$

$$= 8.3 \times 10^{-6}$$

where we have used $\mathcal{N} = N_0 \, \rho/A$.

▶ *Atomic (Molecular) Considerations.* A detailed treatment of magnetization at the atomic level requires quantum theory. Because such a development is beyond the scope of this text, we limit the discussion here to a semi-classical model that at least indicates the trends of the effects and provides order-of-magnitude results for the various magnetic quantities.

We first discuss the simplified theory of diamagnetic susceptibility put forward in the early 1900's by Sir J. J. Larmor. This theory applies to atoms (and ions) that possess neither net orbital angular momentum nor net spin angular momentum, and hence have no permanent magnetic moment. In this category we have the noble gases (He, Ne, Ar, Kr, and Xe), the ions of the alkali metals (Li⁺, Na⁺, K⁺, Rb⁺, and Cs⁺), and the ions of the halide gases

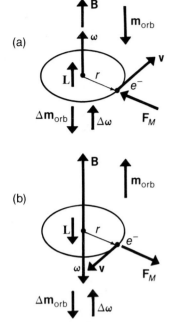

Fig. 36–8. A pair of electrons that occupy the same orbit are shown displaced for clarity. (a) The increasing field **B** causes the speed of one electron and its orbital magnetic moment to increase. (b) The changing field causes the speed and the magnetic moment of the other electron to decrease. However, $\Delta\mathbf{m}_{\text{orb}}$ is in the same direction for both electrons.

(F^-, Cl^-, Br^-, and I^-).* Each closed shell of these atoms and ions consists of pairs of electrons that circulate in opposite directions, as shown schematically in Fig. 36–8. Notice that the individual angular momenta **L** and magnetic moments $\mathbf{m}_{\text{orb}}$ of the electrons in these pairs are equal in magnitude and are oppositely directed. Consequently, these electrons contribute no net magnetic moment to the atom as a whole.†

Now, imagine that an external magnetic field is applied, rising from zero at $t = 0$ to **B** at time t. We consider the orbital planes of the atomic electrons to be perpendicular to **B**. Later, we take account of the possible random orientation of the orbital planes. Assume that the orbiting electrons are in mechanical equilibrium at $t = 0$ and at time t. Each circulating electron is equivalent to a current, $I = -e\omega/2\pi$. This current multiplied by the area πr^2 of the orbit yields the orbital magnetic moment (Eq. 33–15); that is,

$$\mathbf{m}_{\text{orb}} = (-e\omega/2\pi)(\pi r^2) = -\tfrac{1}{2}er^2\boldsymbol{\omega} \qquad \textbf{(36–22)}$$

When the field **B** is applied, each electron in Fig. 36–8 experiences a central magnetic force $\mathbf{F}_M = -e\mathbf{v} \times \mathbf{B}$. This force produces a change $\Delta\mathbf{m}_{\text{orb}}$ in each of the individual magnetic moments in a direction to oppose the applied field **B**, in accordance with Lenz's law. (•Can you verify this by referring to Fig. 36–8?) Thus, whereas the orbital moments $\mathbf{m}_{\text{orb}}$ exactly cancel, the induced moments $\Delta\mathbf{m}_{\text{orb}}$ add together!

Consider the motion of the electron in the circular orbit shown in Fig. 36–8a. For a central attractive but otherwise unspecified Coulomb force $F(r)$ due to the nucleus and modified by the presence of the other atomic electrons, we have

$$m_e r_0 \omega_0^2 = F(r_0) \qquad \text{(at } t = 0\text{)} \qquad \textbf{(36–23a)}$$

$$m_e r \omega^2 = F(r) + eBr\omega \qquad \text{(at time } t\text{)} \qquad \textbf{(36–23b)}$$

where m_e is the electron mass and where we have used $v = r\omega$.

The time t during which the magnetic field changes from zero to B is enormously longer than the period of the orbiting electrons.‡ Consequently, any change that might possibly occur in the electron orbit during one revolution must be extremely small. In particular, the electron orbit can be considered to remain a closed circular path. The average over a period of the torque τ that is applied to an orbiting object is equal to the change in the angular momentum ΔL per period Δt; that is, $\Delta L/\Delta t = \bar{\tau}$ (Eq. 10–9). For each of the atomic electrons in the pair this expression becomes

$$\frac{\Delta(m_e r^2 \omega)}{\Delta t} = \bar{\tau} = e\overline{Er} \qquad \textbf{(36–24)}$$

where $\overline{Er}$ is the average over a period of the product of the orbit radius r and the tangential electric field that acts on the electron. As the magnetic field increases from zero to B, the induced tangential electric field produces the torque $\bar{\tau}$ on the electron and causes the angular speed to increase from ω_0 to $\omega_0 + \Delta\omega$. Applying Faraday's law (Eq. 35–2) for one complete revolution of the electron in Fig. 36–8a, we have

$$\oint_{1 \text{ period}} \mathbf{E} \cdot d\mathbf{r} = -\frac{\Delta\varphi_M}{\Delta t}$$

---

*Notice that the alkali ions (each having lost an electron) and the halide ions (each having gained an electron) have closed electron shells; thus, the properties of these ions mimic those of the noble gases.

†In Larmor's theory no account is taken of the intrinsic magnetic moments of the electrons. In any event, according to quantum theory, the moments of a pair of electrons in the same orbit are oppositely directed and therefore also cancel.

‡A typical atomic electron moves in an orbit with a radius of about 0.5×10^{-10} m and with a speed of about 2×10^6 m/s. Thus, the period of the motion is about 10^{-6} s.

Thus,
$$2\pi \overline{rE} = \frac{\Delta(\pi r^2 B)}{\Delta t} \qquad (36\text{-}25)$$

Combining Eqs. 36-24 and 36-25, and proceeding to the limit in which the incremental quantities become differential quantities, we find

$$\frac{d}{dt}(r^2\omega) = \frac{e}{2m_e}\frac{d}{dt}(r^2 B)$$

Integrating from $t = 0$ to time t, the result is

$$r^2\omega - r_0^2\omega_0 = \frac{e}{2m_e}r^2 B \qquad (36\text{-}26)$$

A general theorem that was derived by Larmor in 1897 can be applied to the case under consideration. The theorem states that, independent of the nature of the central force $F(r)$, the application of the magnetic field B produces a new motion of the electron that is the same as the motion without the field present if the electron is viewed in a frame that rotates at an angular speed $\Delta\omega = \omega - \omega_0$. Thus, the orbit radius remains unchanged, $r = r_0$, and Eq. 36-26 yields

$$\Delta\omega = \omega - \omega_0 = \frac{eB}{2m_e} \qquad (36\text{-}27)$$

We now demonstrate that Larmor's theorem applied to Eqs. 36-23 leads to a result consistent with Eq. 36-27. For all practical cases, $\Delta\omega/\omega_0$ is a small quantity (see Problem 36-14), so $(\Delta\omega)^2$ can be neglected in comparison with ω_0^2. Therefore, we can write

$$\omega^2 = (\omega_0 + \Delta\omega)^2 \cong \omega_0^2 + 2\omega_0\,\Delta\omega$$

Next, substitute this expression for ω^2 into Eq. 36-23b and set $r = r_0$. Then, subtracting Eq. 36-23a gives

$$\Delta\omega = \frac{eB}{2m_e}(1 + \Delta\omega/\omega_0)$$

Because $\Delta\omega/\omega_0$ is a small quantity, the result for $\Delta\omega$ is the same as that in Eq. 36-27, thereby justifying the use of Larmor's theorem.

The change in the angular speed $\Delta\omega$ causes a change in the orbital magnetic moment $\mathbf{m}_{orb}$ of each electron in the pair given by (see Eq. 36-22)

$$\Delta\mathbf{m}_{orb} = -\tfrac{1}{2}er_0^2\,\Delta\omega = -\frac{e^2 r_0^2}{4m_e}\mathbf{B} \qquad (36\text{-}28)$$

where we have indicated explicitly that the radius r remains constant at the value r_0. Each pair of electrons in the atom contributes an amount $2\,\Delta\mathbf{m}_{orb}$ to the induced atomic moment. Averaging over the random orientations that the atom can have relative to the direction of the field, a multiplicative factor of 2/3 results. Thus, altogether we have for the total induced atomic magnetic moment, which is to be identified with $\mathbf{m}$ that appears in Eq. 36-18,

$$\mathbf{m} = -\frac{e^2}{6m_e}\mathbf{B}\sum_{i=1}^{Z} r_{0i}^2$$

$$\qquad (36\text{-}29)$$

$$= -\frac{Ze^2}{6m_e}\mathbf{B}\langle r_0^2\rangle$$

where Z is the total number of electrons in the atom or ion, and where $\langle r_0^2 \rangle$ is the mean square radius of the electron orbits. Thus, the induced atomic magnetic susceptibility can be expressed as (compare Eq. 36–21c)

$$\chi_M = \frac{\mu_0 m}{B} = -\frac{\mu_0 Z e^2}{6 m_e} \langle r_0^2 \rangle, \qquad \text{diamagnetic} \qquad (36\text{–}30)$$

This result, which we have obtained on the basis of a semi-classical model, is nevertheless the same as that obtained using quantum theory. The Li^+ ion of lithium (atomic number $= 3$) has two electrons in an orbit with rms radius $\langle r_0^2 \rangle^{1/2}$ of approximately 0.34 Å.* Thus,

$$\chi_M(Li^+) = -\frac{(4\pi \times 10^{-7}\ \text{N/A}^2)(2)(1.60 \times 10^{-19}\ \text{C})^2 (0.34 \times 10^{-10}\ \text{m})^2}{6(9.11 \times 10^{-31}\ \text{kg})}$$

$$= -1.36 \times 10^{-35} \text{ per ion}$$

The measured value of the magnetic susceptibility for lithium ions is approximately -1.5×10^{-35} per ion. Thus, the calculated value is in excellent agreement with the experimental value. Indeed, comparable agreement is found for all of the atoms and ions in the list given at the beginning of this discussion, if the values of $\langle r_0^2 \rangle$ are obtained from proper quantum calculations. In fact, the situation is often reversed, and susceptibility measurements are used to obtain values for $\langle r_0^2 \rangle$. In any event, the theory indicates that the values of magnetic susceptibilities should be near 10^{-35} to 10^{-34} per atom or ion.

To obtain the susceptibility for the bulk material,† it is necessary to multiply the atomic value by the number of atoms per cubic meter, typically a few times 10^{25} for gases. Thus, we expect susceptibility values near 10^{-9} for the noble gases (see Problem 36–15).

Although the simple Larmor theory does not apply to solid substances such as mercury, silver, and copper, all atoms are roughly the same size, so we expect the basic diamagnetic effect per atom to be of the same order of magnitude for all substances. For solids, the number density is usually 10^{28} to $10^{29}\ \text{m}^{-3}$; then, the magnitudes of the susceptibilities should be in the range from 10^{-7} to 10^{-5}. Inspection of Table 36–1 shows that this is indeed the case.

Lithium ions exhibit diamagnetism because the two electrons occupy symmetric orbits and therefore produce no net magnetic moment. Lithium atoms, on the other hand, have a third electron which occupies a different orbit. This electron does produce a net magnetic moment, so bulk lithium is *paramagnetic,* with a susceptibility of $+1.4 \times 10^{-5}$. The weak diamagnetic effect that is present in all atoms is overwhelmed by paramagnetic or ferromagnetic effects when one is present.

Paramagnetic effects, which are opposite to those of diamagnetism, also contribute only weakly to the magnetism of a sample. Paramagnetism results from the partial alignment with an applied magnetic field of the permanent magnetic moments of the atoms within the sample. To see why paramagnetism is a weak effect, compare the magnetic moment alignment energy, $\mathbf{m}_0 \cdot \mathbf{B}$ (see Eq. 33–17), with the atomic thermal kinetic energy, $\frac{3}{2}kT$. At room temperature (293 K), we have

$$\tfrac{3}{2}kT = \tfrac{3}{2}(1.38 \times 10^{-23}\ \text{J/K})(293\ \text{K}) = 6.1 \times 10^{-21}\ \text{J}$$

whereas for a substance with a permanent magnetic moment equal to one Bohr magneton m_B in a strong magnetic field of 2 T (20,000 G), we have

---

*This value for $\langle r_0^2 \rangle^{1/2}$ is the result of a quantum calculation that takes into account the shielding effect of the two electrons.

†In proceeding from the semi-classical result obtained for a single atom (which is the same as the quantum expression) to the susceptibility for a bulk sample, there is a difficulty in dealing with the boundary of the sample or of the field. This flaw in the semi-classical theory is overcome in the quantum calculation.

$$m_B B = (9.27 \times 10^{-24} \, \text{A} \cdot \text{m}^2)(2 \, \text{T}) = 1.8 \times 10^{-23} \, \text{J}$$

The fact that $\frac{3}{2}kT \gg m_B B$ means that the thermal agitation of the atoms prevents the complete alignment of the permanent magnetic moments. Thus, the situation is similar to that for the case of polar electric dipole moments discussed in Section 30–5. In fact, a calculation parallel to that leading to Eq. 30–33, namely,

$$\mathbf{p} = \langle \mathbf{p}_0 \rangle = \frac{p_0^2}{3kT} \mathbf{E} \tag{36-31}$$

yields for the magnetic case

$$\mathbf{m} = \langle \mathbf{m}_0 \rangle = \frac{m_0^2}{3kT} \mathbf{B} \tag{36-32}$$

where m_0 is the permanent atomic (or molecular) magnetic moment. This result is known as *Curie's law*, after Pierre Curie (1859–1906), who discovered the inverse temperature effect in 1895. The bulk magnetic susceptibility for paramagnetic substances is therefore

$$\chi_M = \frac{\mu_0 m_0^2}{3kT} \mathcal{N}, \qquad \text{paramagnetic} \tag{36-33}$$

Figure 36–9 shows the magnetization M of the salt chromium potassium alum in units of the maximum value M_m (which is realized only at 0 K) as a function of B/T for several temperatures near absolute zero. The data agree with Curie's law at small values of B/T. For a field as large as 2 T, temperatures above 10 K give $B/T < 0.2$, which is well within the range of validity of Curie's law. Of course, as B/T becomes large (at very low temperatures), the magnetization cannot exceed the limiting value, $M_m = m_0 \mathcal{N}$, which corresponds to the complete alignment of all the magnetic moments. The general behavior of M/M_m as a function of B/T is given by a quantum calculation and is shown in Fig. 36–9.

Fig. 36–9. The magnetization M of chromium potassium alum in units of the maximum possible magnetization, $M_m = m_0 \mathcal{N}$, as a function of B/T. The region in which Curie's law is valid, $B/T \le 0.3 \, T/K$, includes virtually all experimental conditions for which $T \ge 10 \, \text{K}$. The curve represents the result of a quantum calculation. (After *Physics*, 3rd Ed., by David Halliday and Robert Resnick. John Wiley & Sons, Inc., New York, 1978, p. 822. From Measurements by W. E. Henry.)

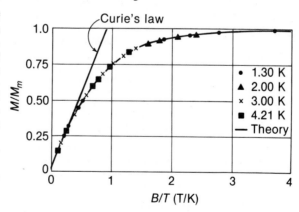

The atomic magnetic moment m_0 for a paramagnetic substance is typically about one Bohr magneton, m_B. Then, for $T = 293 \, \text{K}$ and $\mathcal{N} = 10^{28} \, \text{m}^{-3}$, the expected susceptibility is

$$\chi_M \approx \frac{(4\pi \times 10^{-7} \, \text{N/A}^2)(9.27 \times 10^{-24} \, \text{A} \cdot \text{m}^2)^2(10^{28} \, \text{m}^{-3})}{3(1.38 \times 10^{-23} \, \text{J/K})(293 \, \text{K})}$$

$$\approx 8.9 \times 10^{-5}$$

which is in the middle of the range of the values for paramagnetic substances that are listed in Table 36–1.

Because the ferromagnetic elements, Fe, Ni, and Co, have an unusual ordering of elec-

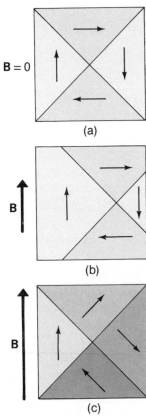

Fig. 36–10. (a) A region within an unmagnetized sample. (b) Magnetization by growth of the domain whose moment is parallel to **B**. (c) Magnetization by domain rotation.

trons in their orbital and spin states, the angular momenta of these atoms have abnormally large electron spin components. These spin components produce a strong interaction between neighboring atoms that leads to maximum stability when the atomic magnetic moments are aligned. This spontaneous tendency to align produces microscopic regions (domains) in which the moments are all aligned. In these domains, which have volumes of about 10^{-12} to 10^{-8} m^3 and contain 10^{17} to 10^{21} atoms, the local magnetization has its maximum value, $m_0 \mathcal{N}$.

Suppose that an increasing magnetic field is applied to an unmagnetized sample of iron. Those domains that happen to have moments aligned with the field increase in size at the expense of those not favorably aligned. At first, when the applied field is relatively weak, this process is reversible. That is, if the external field is reduced to zero, the domain boundaries return to their original positions and the sample is again unmagnetized. For a stronger applied field, however, the domain boundaries do not recover completely and the sample acquires a permanent magnetism that remains after the external field is removed.

A strong applied field can also produce permanent magnetization by causing unaligned domains to rotate and become more nearly parallel to the field direction. The shifts of the domain boundaries and the domain rotations both occur in discrete steps. In fact, it is possible to detect the electric impulses due to the induction effect as these realignments occur— this is called the *Barkhausen effect*. Figure 36–10 shows schematically how a region becomes magnetized by domain growth and by domain rotation.

When a ferromagnetic material is heated, the increased thermal agitation works against the spontaneous domain magnetization. At a critical temperature, called the *Curie temperature*, the ferromagnetic property of the material abruptly disappears. For temperatures above the Curie temperature, a ferromagnetic sample becomes simply paramagnetic. The transition from ferromagnetism to paramagnetism is an example of a phase change not unlike the situation in which the long-range atomic order of a crystal is suddenly destroyed when melting occurs. Table 36–2 lists the Curie temperatures for the ferromagnetic elements. Notice that the two rare-earth elements are not ferromagnetic at room temperature.

The magnetization of an iron sample by an applied external field is a complicated, nonlinear process that depends on the particular way that the sample was formed (cast, hot-rolled, alloyed, etc.). We consider some of the peculiarities of the magnetization of iron in Section 36–6.

TABLE 36–2. Curie Temperatures

ELEMENT	T_C (K)
Cobalt, Co	1395
Iron, Fe	1043
Nickel, Ni	631
Gadolinium, Gd	289
Dysprosium, Dy	85

Example 36–6

A solid toroid of iron has a core radius $a = 1$ cm and a toroid radius $R = 10$ cm. A winding of $N = 800$ turns around the toroid carries a steady current of 10 A. (Compare Example 36–5, where the toroid consists of bismuth.) Assume that the maximum magnetization of iron arises from the spin alignment of two electrons in each atom with $m_0 = 2m_B$. The induction in the iron core is measured to be 1.85 T. Determine the following quantities:

(a) the magnetization M;
(b) the saturation value M_m, with all domains aligned;
(c) the magnetization current I_M; and
(d) the values of κ_M, μ, and χ_M.

Solution:

(a) The value of the induction B_0 for a vacuum core was obtained in Example 36–5: $B_0 = 0.016$ T. Then, using Eq. 36–21a, we have

$$M = \frac{B - B_0}{\mu_0} = \frac{1.85 \text{ T} - 0.016 \text{ T}}{4\pi \times 10^{-7} \text{ T} \cdot \text{m/A}}$$
$$= 1.46 \times 10^6 \text{ A} \cdot \text{turns/m}$$

(b) The maximum magnetization is

$$M_m = m_0 \mathcal{N} = 2m_B \mathcal{N}$$

Now, $\mathcal{N} = N_0 \rho/A$, where the density of iron is $\rho = 7860$ kg/m³ and where the atomic mass is $A = 55.9$ g/mol $= 0.0559$ kg/mol. Then,

$$M_m = \frac{2m_B N_0 \rho}{A}$$

$$= \frac{2(9.27 \times 10^{-24} \text{ A} \cdot \text{m}^2)(6.02 \times 10^{23} \text{ mol}^{-1})(7860 \text{ kg/m}^3)}{0.0559 \text{ kg/mol}}$$

$$= 1.57 \times 10^6 \text{ A} \cdot \text{turns/m}$$

Thus,
$$\frac{M}{M_m} = \frac{1.46 \times 10^6 \text{ A} \cdot \text{turns/m}}{1.57 \times 10^6 \text{ A} \cdot \text{turns/m}} = 0.93$$

so that the magnetization is 93 percent of the full saturation value.

(c) The magnetization current is (see Eq. 36–19 with $\ell = 2\pi R$)

$$I_M = 2\pi R M$$
$$= 2\pi(0.10 \text{ m})(1.46 \times 10^6 \text{ A} \cdot \text{turns/m})$$
$$= 9.17 \times 10^5 \text{ A} \cdot \text{turns}$$

The direction of this magnetization current is the same as that of the coil current, the magnitude of which is $NI = (800 \text{ turns})(10 \text{ A}) = 8 \times 10^3 \text{ A} \cdot \text{turns}$.

(d) From Eq. 36–15, we have

$$\kappa_M = \frac{B}{B_0} = \frac{1.85 \text{ T}}{0.016 \text{ T}} = 116$$

From Eq. 36–16, we have

$$\mu = \kappa_M \mu_0 = (116)(4\pi \times 10^{-7} \text{ T} \cdot \text{m/A})$$
$$= 1.46 \times 10^{-4} \text{ T} \cdot \text{m/A}$$

Finally from Eq. 36–17, we have

$$\chi_M = \kappa_M - 1 = 115$$

Notice that we have $B/B_0 = 116$ in this case of a ferromagnetic material, whereas for bismuth, one of the strongest diamagnetic substances, we found $B/B_0 = 0.99983$ (Example 36–5). ◄

36-5 THE MAGNETIC INTENSITY VECTOR

In the preceding section a modification of Ampère's circuital law was introduced in order to relate the magnetic induction **B** in a magnetic material to the actual current in an encircling coil and the state of magnetization in the sample. We then wrote (see Eq. 36–1a)

$$\mu_0 \mathbf{M} = \mathbf{B} - \mathbf{B}_0$$

where $\mathbf{B}_0$ is the induction directly produced by the (real) coil current.* This relationship is valid for all homogeneous, isotropic magnetic media. Ampère's circuital law can now be expressed in a more general form as

$$\oint \left(\frac{\mathbf{B}_0}{\mu_0} \right) \cdot d\mathbf{r} = \oint \left(\frac{\mathbf{B}}{\mu_0} - \mathbf{M} \right) \cdot d\mathbf{r} = NI \qquad \textbf{(36–34)}$$

where NI is the total (real) current through any open surface whose boundary corresponds to the path of the line integration. It is convenient to introduce a new vector quantity, called the *magnetic intensity* **H**, defined as

$$\mathbf{H} = \frac{\mathbf{B}_0}{\mu_0} = \frac{\mathbf{B}}{\mu_0} - \mathbf{M} \qquad \textbf{(36–35)}$$

so that Ampère's circuital law becomes

$$\oint \mathbf{H} \cdot d\mathbf{r} = NI, \qquad \text{real currents} \qquad \textbf{(36–36)}$$

We can also write (using Eqs. 36–15 and 36–16)

$$\mathbf{B} = \mu_0 \kappa_M \mathbf{H} = \mu \mathbf{H} \qquad \textbf{(36–37)}$$

The field intensity **H** has the same dimensions as **M** and is therefore measured in units of $A \cdot turns/m$ (or, simply, A/m).

Boundary Conditions. In Section 30–6 we found that the electric field quantities, **E** and **D**, obey different boundary conditions at the interface between two different media. We now show that the magnetic field quantities, **B** and **H**, likewise obey different boundary conditions.

Figure 36–11a shows a small Gaussian pillbox surface S that has been constructed at the interface between two magnetic materials, ① and ②. Now, Eq. 34–11, which states that the magnetic flux φ_M through any closed surface is zero, is

---

*We distinguish between the effective surface magnetization currents and the externally applied coil currents by calling the latter "real" currents. However, the physical sources responsible for the magnetization currents are also real and produce observable effects.

still valid when magnetic materials are present. Thus, applying Gauss's law for magnetism to the situation in Fig. 36–11a with $\Delta y \to 0$, we find

$$(B_{1\perp} - B_{2\perp}) \, \Delta A = 0 \qquad \text{(36–38a)}$$

or

$$B_{1\perp} = B_{2\perp}$$

which we can express vectorially as

$$(\mathbf{B}_1 - \mathbf{B}_2) \cdot \hat{\mathbf{n}} = 0 \qquad \text{(36–38b)}$$

That is, *the normal component of* **B** *is continuous across the boundary between two media.*

To be completely general, consider a *real* surface current with a linear current density (in A/m), $\mathbf{j} = j\hat{\mathbf{t}}$, to flow at the interface between two media. As indicated in Fig. 36–11b, $\mathbf{j}$ and the tangential unit vector $\hat{\mathbf{t}}$ are directed into the plane of the page. Also shown is an integration path in a plane perpendicular to $\mathbf{j}$. Using Eq. 36–36, we obtain the result

$$(H_{1\parallel} - H_{2\parallel}) \, \Delta x = j \, \Delta x \qquad \text{(36–39)}$$

In the event that $\mathbf{j} = 0$, we have

$$H_{1\parallel} = H_{2\parallel} \qquad \text{(36–40)}$$

That is, *in the absence of any real surface currents, the tangential component of* **H** *is continuous across the boundary between two media.*

Magnetic Pole Density. We can now follow the line of development used in discussing polar dielectric materials (Section 30–5) and introduce the concept of a *fictitious* or effective induced surface *magnetic pole density* $\sigma_M = \mathbf{M} \cdot \hat{\mathbf{n}}$, which generates the field intensity **H**.

Figure 36–12 shows the field quantities **M**, **H**, and **B** for a permanently magnetized cylinder. The magnetization **M** is uniform and axial and is confined entirely to the interior of the cylinder (Fig. 36–12a). The magnetic intensity **H** (Fig. 36–12b) has an appearance similar to that of the electrostatic field between two uniformly charged circular discs that correspond to the end faces of the magnetized cylinder. There is considerable fringing field because the spacing between the discs is large compared to the diameter. The effective surface charge density, $\sigma_P = \mathbf{P} \cdot \hat{\mathbf{n}}$, for the electrostatic case corresponds to the effective surface magnetic pole density, $\sigma_M = \mathbf{M} \cdot \hat{\mathbf{n}}$.

The magnetic induction **B** (Fig. 36–12c) is exactly the same as the field that would be produced by a solenoid of finite length with a current density $j_M = M$ ampere-turns per unit length. Notice that the lines of **B** are continuous closed curves, whereas the lines of **H** are not. The latter originate on the face with pole density $+\sigma_M$ and terminate on the face with density $-\sigma_M$. However, notice that the lines of **H** are not perpendicular to these faces at the points of intersection.

(b)

Fig. 36–11. (a) A Gaussian surface used to obtain the boundary condition on **B**. (b) A line integration path used to obtain the boundary conditions on **H**. The (real) current density $\mathbf{j}$ is directed into the plane of the page. The tangential unit vector $\hat{\mathbf{t}}$ is in the direction of $\mathbf{j}$.

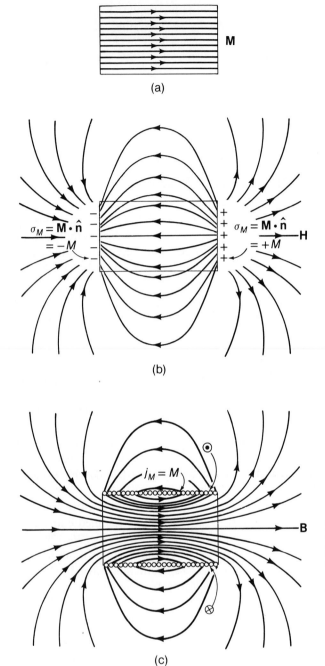

Fig. 36–12. (a) The magnetization **M**, (b) the magnetic intensity **H**, and (c) the magnetic induction **B** for a permanently magnetized cylinder. The effective surface magnetic pole density that corresponds to the magnetic intensity is $\sigma_M = \mathbf{M} \cdot \hat{\mathbf{n}}$. The magnetic induction is the same as that produced by a surface current density $j_M = M$.

36-6 PRACTICAL MAGNETIC MATERIALS

B-H **Curves.** Ferromagnetic materials, because they can achieve very high states of magnetization, find many important applications in magnetic devices of various sorts. Usually, these devices are fabricated from iron or from special iron alloys. As mentioned earlier, these materials have complicated nonlinear magnetization properties that depend strongly on the fabrication process and on the magnetic history of a sample.

Fig. 36–13. Magnetization curves for various types and alloys of iron. Note the logarithmic scales. (a) Supermalloy. (b) Mumetal. (c) 45 Permalloy. (d) USS motor steel, 26 gauge. (e) Pure iron sheet. (f) Hot-rolled iron sheet. The intensity *H* depends on the real current applied, and the induction *B* is the resulting field in the sample. Thus, "*H* is what you pay for, and *B* is what you get."

A convenient way to present the magnetic characteristics of a ferromagnetic material is in terms of a *B-H curve*, a plot of the magnetic induction *B* versus the magnetic intensity *H*. Figure 36–13 shows such curves for several ferromagnetic materials. Notice that a log-log graph is used to cover the wide range of values. Also shown for convenience are dashed lines along which the relative permeability κ_M is constant.

Example 36–7

A long solenoid with a radius $R = 4$ cm carries a current of 1 A in a coil that has a wrapping density of 200 turns per meter. Along the axis of the solenoid there is a cylinder with a radius $a = 2$ cm that is made from hot-rolled iron sheet. Assume that the *B-H* curve of Fig. 36–13f applies.

(a) Determine the field vectors, **H**, **B** and **M**, in the three regions indicated in the diagram.

(b) Compare the magnetic flux φ_M in the iron core with that in region ②.

Solution:

(a) We assume that the solenoid is infinitely long, so all of the field quantities are zero in region ① outside the solenoid. The magnetic intensity **H** in regions ② and ③ can be determined by applying Eq. 36–36 around the paths a and b, respectively. The results for the two paths are the same because the solenoid current I is the only real current enclosed. Thus,

$$\mathbf{H}_2 = \mathbf{H}_3 = \frac{NI}{\ell}\hat{\mathbf{k}} = (200 \text{ A} \cdot \text{turns/m})\hat{\mathbf{k}}$$

Notice that this result automatically gives agreement with the boundary condition on the tangential component of H at the core boundary (Eq. 36–40).

With $H_3 = 200$ A · turns/m, Fig. 36–13f gives $B_3 = 0.20$ T, so that $\mathbf{B}_3 = (2000 \text{ G})\hat{\mathbf{k}}$. On the other hand,

$$\begin{aligned} \mathbf{B}_2 = \mu_0\mathbf{H}_2 &= (4\pi \times 10^{-7} \text{ T} \cdot \text{m/A})(200 \text{ A} \cdot \text{turns/m})\hat{\mathbf{k}} \\ &= (2.51 \times 10^{-4} \text{ T})\hat{\mathbf{k}} \\ &= (2.51 \text{ G})\hat{\mathbf{k}} \end{aligned}$$

Thus, the induction in the iron core is much greater than the induction in the air region; that is, $B_3 \gg B_2$.

The value of the induction in region ② is just the value that we would have in region ③ if the iron core were not present. Therefore, in the core, we have

$$\kappa_M = \frac{B_3}{B_2} = \frac{0.20 \text{ T}}{2.51 \times 10^{-4} \text{ T}} \cong 800$$

which is in agreement with Fig. 36–13f.

The magnetization **M** is given by Eq. 36–35, namely,

$$\mathbf{M} = \frac{\mathbf{B}}{\mu_0} - \mathbf{H}$$

In region ② we have $\mathbf{B}_2 = \mu_0\mathbf{H}_2$, so that $\mathbf{M}_2 = 0$. In region ③ we have

$$M_3 = \frac{0.20 \text{ T}}{4\pi \times 10^{-7} \text{ T} \cdot \text{m/A}} - (200 \text{ A} \cdot \text{turns/m})$$

so that

$$\mathbf{M}_3 = (1.590 \times 10^5 \text{ A} \cdot \text{turns/m})\hat{\mathbf{k}}$$

(b) The magnetic flux in the iron core is

$$\begin{aligned} \varphi_3 = \pi a^2 B_3 &= \pi(0.02 \text{ m})^2(0.20 \text{ T}) \\ &= 2.51 \times 10^{-4} \text{ Wb} \end{aligned}$$

whereas the flux in the air region ② is

$$\begin{aligned} \varphi_2 = \pi(R^2 - a^2)B_2 &= \pi[(0.04 \text{ m})^2 - (0.02 \text{ m})^2](2.51 \times 10^{-4} \text{ T}) \\ &= 9.46 \times 10^{-7} \text{ Wb} \end{aligned}$$

Thus, the air region carries only 0.38 percent of the total magnetic flux. This illustrates the general feature that the magnetic flux in a system tends to be concentrated in any ferromagnetic material that is present.

Hysteresis. When a ferromagnetic material is magnetized, some of the domains are driven beyond the point at which the changes are reversible. Then, when the imposed field intensity H is decreased to zero, the material retains some degree of magnetization (that is, $B \neq 0$ even though $H = 0$). If the coil current that produces the imposed field is slowly cycled between maximum values in either direction, the corresponding values of H and B trace out a closed curve called a *hysteresis loop*[*] (Fig. 36–14). Starting from an initial unmagnetized state O, the magnetization of the sample follows the path $\overline{Oab}$, arriving at a maximum (saturated) induction B_m when the imposed field intensity is H_m. When H is reduced to zero, the corresponding value of B (point c) is called the *remanence* B_r. The value of the reverse field intensity that is necessary to reduce the induction to zero at point d is called the *coercive force* H_c (although H_c is not actually a "force"). The part of the curve between points c and d is referred to as the *demagnetization curve*. If the coil current is varied slowly, the curve efb is a rotated image of $bcde$.

In order for a material to be suitable for fabricating a good permanent magnet, the remanence must be large; that is, B_r must be a large fraction of the saturated induction B_m. The material should also have a large coercive force H_c so that relatively weak stray fields will not substantially affect the state of magnetization. A variety of special alloys are used in preparing permanent magnets.

Electromagnets. To perform various tests and experiments it is often necessary to have a strong (and variable) magnetic field within a certain working volume in air. This can be accomplished by bending a bar or rod of ferromagnetic material into some shape that brings the ends close together, forming a small air gap, as shown in Fig. 36–15a. A winding of N turns is wrapped around the core material and carries a current I. Most of the magnetic flux generated by the magnetizing current is guided by the core to the air gap. The stray or *leakage* flux, indicated by the field lines a, b, c, and d, usually represents a negligible part of the core flux φ_M. If the air gap is relatively small (that is, $\ell_g < \sqrt{A}$), the amount of fringing field around the gap is also small and may be neglected (see Fig. 36–15b). Then, the boundary condition on the normal component of **B** (Eq. 36–38a) gives $B_g \cong B_F$.

---

[*] The term *hysteresis* is used generally to describe a system whose state depends on its previous history. Hysteresis also occurs in dielectric and elastic materials.

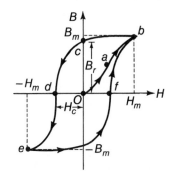

Fig. 36–14. The hysteresis loop for a typical ferromagnetic material. The various quantities indicated are defined in the text. The permeability μ of the material is not constant and is equal to the slope of the B-H curve at any point.

Fig. 36–15. (a) An electromagnet with a small air gap ℓ_g. (b) Detail of the magnetic induction in the vicinity of the air gap. The normal component of **B** is continuous, so $B_g \cong B_F$.

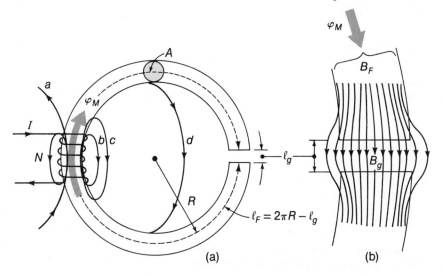

(a)   (b)

Ampère's circuital law (Eq. 36–36) can be applied along the centerline of the core with the result

$$NI = \ell_F H_F + \ell_g H_g$$

Now, we have $H_F = B_F/\mu_F$ and $H_g = B_g/\mu_0$; also $\varphi_M = B_F A \cong B_g A$. Then, the expression for Ampère's law becomes

$$NI = \frac{\varphi_M}{A}\left(\frac{\ell_F}{\mu_F} + \frac{\ell_g}{\mu_0}\right)$$

$$= \frac{\varphi_M}{A\mu_0}\left(\frac{\ell_F}{\kappa_M} + \ell_g\right) \qquad \text{(36–43)}$$

Example 36–8

It is required to produce a magnetic flux of 1.5×10^{-4} Wb in a cylindrical air gap that has a cross-sectional area of 3 cm² and a gap width of 3 mm. Available is an electromagnet that consists of a pure iron core with the appropriate cross section and a radius $R = 8$ cm. What current I must flow through the magnet's 100-turn coil to produce the desired flux in the gap?

Solution:

The magnetic induction is

$$B_g \cong B_F = \frac{\varphi_M}{A} = \frac{1.5 \times 10^{-4} \text{ Wb}}{3 \times 10^{-4} \text{ m}^2} = 0.50 \text{ T}$$

From Fig. 36–13e for pure iron sheet, we find that an induction of 0.50 T corresponds to $\kappa_M = 2400$. Thus, with $\ell_g = 0.003$ m and $\ell_F = 2\pi R - \ell_g = 0.500$ m, we use Eq. 36–41 to give

$$I = \frac{\varphi_M}{A\mu_0 N}\left(\frac{\ell_F}{\kappa_M} + \ell_g\right)$$

$$= \frac{1.5 \times 10^{-4} \text{ Wb}}{(3 \times 10^{-4} \text{ m}^2)(4\pi \times 10^{-7} \text{ T} \cdot \text{m/A})(100)}\left(\frac{0.500 \text{ m}}{2400} + (0.003 \text{ m})\right)$$

$$= 12.8 \text{ A}$$

In a practical situation, a somewhat larger current (by, perhaps, 10 percent) would be required because of the flux losses due to leakage and the fringing field.

36-7 MAXWELL'S EQUATIONS IN MATTER

In Section 35–6 Maxwell's field equations were given in both integral and differential form for the case of a vacuum medium. Table 36–3 gives the corresponding equations for the case of a material medium.

For a linear homogeneous isotropic medium, we have the following auxiliary equations:

TABLE 36–3. Maxwell's Equations in Matter

LAW	INTEGRAL FORM*	DIFFERENTIAL FORM
Faraday's law	$\oint \mathbf{E} \cdot d\mathbf{r} = -\dfrac{\partial}{\partial t} \int_S \mathbf{B} \cdot \hat{\mathbf{n}} \, dA$	$\nabla \times \mathbf{E} = -\dfrac{\partial \mathbf{B}}{\partial t}$
Gauss's law (electric)	$\oint_S \mathbf{D} \cdot \hat{\mathbf{n}} \, dA = \int \rho \, dV$	$\nabla \cdot \mathbf{D} = \rho$
Ampère's law	$\oint \mathbf{H} \cdot d\mathbf{r} = \int_S \left(\mathbf{J} + \dfrac{\partial \mathbf{D}}{\partial t} \right) \cdot \hat{\mathbf{n}} \, dA$	$\nabla \times \mathbf{H} = \mathbf{J} + \dfrac{\partial \mathbf{D}}{\partial t}$
Gauss's law (magnetic)	$\oint_S \mathbf{B} \cdot \hat{\mathbf{n}} \, dA = 0$	$\nabla \cdot \mathbf{B} = 0$

*For stationary surfaces S.

$$\mathbf{D} = \kappa \epsilon_0 \mathbf{E} = \epsilon \mathbf{E}$$
$$\mathbf{B} = \kappa_M \mu_0 \mathbf{H} = \mu \mathbf{H} \qquad \text{(36–42)}$$
$$\mathbf{J} = \sigma \mathbf{E}$$

For these materials, the electromagnetic energy density can be expressed as

$$u = \tfrac{1}{2}(\mathbf{E} \cdot \mathbf{D} + \mathbf{B} \cdot \mathbf{H}) \qquad \text{(36–43)}$$

(For the case of nonlinear magnetic materials, see Problem 36–22.)

PROBLEMS

Section 36-1

36-1 Two coils in fixed positions have a mutual inductance of 100 μH. What peak voltage is induced in one coil if a current, $I = I_0 \sin \omega t$, flows in the other coil? (Assume $I_0 = 10$ A and $\omega = 1000$ rad/s.)

36-2 A long solenoid consists of N_1 turns with a radius R_1. A second long solenoid (N_2 turns and radius R_2), with the same length ℓ, lies entirely within the first solenoid with the axes coinciding. Determine the mutual inductance of the pair of solenoids by first assuming that a current I flows in solenoid 1, and then by assuming that the same current flows in solenoid 2. Do you find $M_{12} = M_{21}$?

36-3 Derive an expression for the mutual inductance of a long straight wire and rectangular coil of N turns

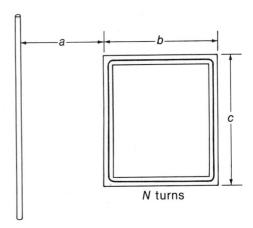

with dimensions as shown in the diagram. The wire is in the plane of the coil.

36-4　Two single-turn circular loops with radii R and $r \ll R$ have the same center and lie in the same plane. Show that the mutual inductance of the pair is $M = \mu_0 \pi r^2 / 2R$. What is the value of M for radii $r = 2$ cm and $R = 20$ cm?

36-5　A long straight wire coincides with the axis of a toroid of N turns that has an inner radius R and a rectangular cross section, $a \times b$, as shown in the diagram. (a) Determine the mutual inductance by considering the flux linkage produced by a current I in the long wire. (b) Determine the mutual inductance by considering the flux linkage produced by a current I in the toroidal coil, thereby demonstrating that $M = M_{12} = M_{21} = (\mu_0 Nb/2\pi) \ln[(R + a)/a]$.

Section 36-2

36-6　What amount of self-flux links an inductor coil of 10 mH when a current of 2 A flows through the coil?

36-7　A current $I = I_0 \sin \omega t$, with $I_0 = 5$ A and $\omega/2\pi = 60$ Hz, flows through an inductor whose inductance is 10 mH. What is the back EMF as a function of time?

36-8　An air-core Rowland ring has an inner radius $R = 10$ cm and consists of 500 turns of fine wire. The ring has a square cross section, 4 cm × 4 cm, as shown in the diagram. Determine the self-inductance of the ring.

36-9　A *twin-lead* cable consists of two parallel wires with radii a separated by a distance $2b$. If the two wires carry equal currents in opposite directions, show that the self-inductance per length ℓ of the twin-lead is

$$\frac{L}{\ell} = \frac{\mu_0}{\pi} \ln[(2b - a)/a]$$

(Neglect the fields within the wires.)

36-10　A long straight conductor with radius R carries a
•　steady uniform current I. Calculate the energy stored *inside* a length ℓ of the conductor. From this result determine the contribution from the internal flux to the inductance of the conductor. Verify this conclusion by calculating the inductance contribution determined from the internal flux linkage. [*Hint:* For the second part of the problem, the differential flux linkage is $d\varphi_M = (r^2/R^2)B(r)\,\ell\,dr$, because only a fraction, $\pi r^2/\pi R^2$, of the total current is linked by the flux. Refer to the diagram.]

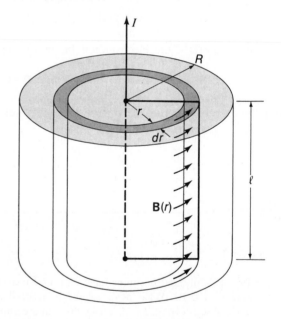

Section 36-3

36-11　It proves convenient to define a *coefficient of coupling* ξ for a pair of coupled inductors such that $\xi = M/\sqrt{L_1 L_2}$, with $\xi \leqslant 1$. Two coupled inductors, $L_1 = 50$ mH and $L_2 = 20$ mH, with $\xi = \frac{1}{3}$, are connected in series with all induced EMF's adding. (a) What is the effective inductance of the combination? (b) What amount of energy is stored in the field when a current of 2 A flows through the combination?

36-12　An air-core solenoid with a radius of 4 cm and a

length of 65 cm is tightly and uniformly wound with No. 18 wire that has a diameter of 1.2 mm, including insulation. What amount of energy is stored in the field when the steady current equals the maximum rated value of 6 A?

36-13 An external source of EMF maintains a constant current I in an inductor as it is moved from a field-free region to a point at which a steady flux φ_M links the inductor. Show that an amount of work $I\varphi_M$ was done by the external source during the change in position. Calculate the work done if $\varphi_M = 1.6 \times 10^{-3}$ Wb and $I = 10$ A.

Section 36-4

36-14 In the discussion of the Larmor theory it was mentioned that $\Delta\omega/\omega_0 = eB/2m_e\omega_0$ is a small quantity. Verify this by calculating $\Delta\omega/\omega_0$ for the typical values $\omega_0 = 10^{16}$ s^{-1} and $B = 8000$ G.

36-15 Calculate the Larmor diamagnetic susceptibility for the helium atom, using $\langle r_0^2 \rangle^{1/2} = 0.54$ Å. Then, determine χ_M for helium gas at 0°C and 1 atm. Compare your result with the measured value, -1.1×10^{-9}.

Section 36-5

36-16 A long thin needle of iron ($\kappa_M = 150$) is placed in a uniform magnetic field, $B_0 = 120$ G (in air), and is oriented parallel to $\mathbf{B}_0$. Neglect end effects and determine the three magnetic vectors $\mathbf{B}$, $\mathbf{H}$, and $\mathbf{M}$, inside the needle.

36-17 A disc of iron ($\kappa_M = 150$), with a diameter large compared with its thickness, is placed in a uniform magnetic field $B_0 = 120$ G (in air), and is oriented with the plane of the disc perpendicular to $\mathbf{B}$. Neglect end effects and determine the three magnetic vectors, $\mathbf{B}$, $\mathbf{H}$, and $\mathbf{M}$, inside the disc.

36-18 Modify the discussion leading to Eq. 36-10 to describe the situation in which the interior of the long solenoid is completely filled with a magnetic material whose relative permeability is κ_M. Show that $u_M = \frac{1}{2}\mathbf{B} \cdot \mathbf{H}$.

36-19 Show that Eq. 36-39 can be expressed in vectorial form as

$$[(\mathbf{H}_1 - \mathbf{H}_2) \times \hat{\mathbf{n}}] \cdot \hat{\mathbf{t}} = 0$$

and

$$(\mathbf{H}_1 - \mathbf{H}_2) \cdot (\hat{\mathbf{t}} \times \hat{\mathbf{n}}) = j$$

Why are *two* equations necessary to give the conditions on the tangential components of $\mathbf{H}$?

Section 36-6

36-20 A C-shaped yoke fabricated from pure iron has a
•• 500-turn magnetizing coil, as shown in the diagram. This electromagnet is used to convert a rectangular rod of Alnico-5 into a permanent magnet. During assembly of the yoke and rod, two 0.1-mm air gaps were inadvertently left at the ends of the rod. For the dimensions shown in the diagram and for a desired magnetic induction of 10,000 G in the Alnico rod (which occurs with $\kappa_M = 15$), determine the required current I in the magnetizing coil. (Use Fig. 36-13.) Assume that the median path (shown dashed in the diagram) can be used for the application of Ampère's law with the mean values of the fields. Assume also that fringing-field effects can be ignored. Is it worthwhile to attempt to reduce the air gaps or to improve the magnetic characteristics or geometry of the yoke?

36-21 A common geometry for an electromgnet is shown in
•• the diagram. The yoke is fabricated from pure iron. If a magnetic induction, $B = 0.5$ T, is required in the air gap, determine the current I that must flow in each of the identical magnetizing coils. Use Fig. 36-13. Neglect fringing-field effects and flux leakage. [*Hint:* At points a and b the main (gap) flux divides equally. Thus, Ampère's law may be applied either to path *abcda* or to path *abefa*.]

36-22 An iron-core Rowland ring has a central radius R, and
•• a cross-sectional area A, and is wrapped uniformly with N turns of wire (see Fig. 36-4). The input of electric energy to the coil is $dW = i(t)\,e(t)\,dt$, where $i(t)$ and $e(t)$ are the instantaneous values of the coil current and the back EMF, respectively. (a) Use Faraday's law to show that $dW = NAi(t)\,dB$. (b) Use Ampère's law to show that $dW = (2\pi RA)H\,dB$. (c) Show that the input electric energy W per cubic meter is equal to the area in a B-H diagram between the B-axis and the magnetization curve, $W/V = \int_0^B H\,dB$, where $V = 2\pi RA$. (d) Assume that the iron core is a linear magnetic material (that is, $B = \mu_0\kappa_M H$, with κ_M not dependent on H). Does the result of (c) agree with that of Problem 36-18, namely, that $\mu_M = \frac{1}{2}\mathbf{B}\cdot\mathbf{H}$? (e) Show that the energy loss per cubic meter in carrying the iron sample through one complete hysteresis loop is equal to the area of the loop in a B-H diagram.

Section 36-7

36-23 The gap of a circular parallel-plate capacitor is
• filled with a nonconducting material that has a dielectric constant $\kappa = 4$ and a relative permeability $\kappa_M = 30$. The (real) surface charge on each plate has the time dependence $\sigma = \pm\sigma_0\,t/\tau$, where $\sigma_0 = 10^{-6}$ C/m² and $\tau = 100$ ns. Determine the values of D, E, P, H, B, and M at time $t = 20$ ms for a point in the middle of the gap and 5 cm from the axis of symmetry.

Additional Problems

36-24 Two long solenoids consist of interlaced windings of N_1 turns and N_2 turns on the same nonmagnetic cylinder. Each solenoid has a length ℓ and a radius R. (a) Calculate L_1, L_2, and M. Neglect end effects. (b) What is the value of the coefficient of coupling, $\xi = M/\sqrt{L_1 L_2}$? (See Problem 36-11.) Is there any way to change the value of ξ for solenoids wound in this way?

36-25 Two small circular current loops with radii a and b are separated by a distance $\mathbf{r}$, as shown in the diagram above right. Consider the field of each loop to be a dipole field and derive an expression for the mutual inductance.

36-26 Derive an expression for the self-inductance of
•• the coaxial inductor discussed in Example 36-4 by using the defining equation (Eq. 36-6). [*Hint:* Find the flux linkage in the three cylindrical sections separately. This linkage in each region must be inter-

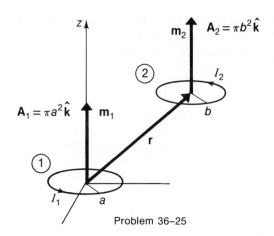

Problem 36–25

preted as the linkage with only the included current. That is, $d\varphi_M = (r^2/a^2)B(r)\ell\,dr$ for $r < a$; $B(r)\ell\,dr$ for $a < r < b$; $[1 - (r^2 - b^2)/(c^2 - b^2)]B(r)\ell\,dr$ for $b < r < c$; and zero for $r > c$. See also the hint for Problem 36–10.]

36-27 A thin spherical shell with radius R and a uniformly distributed surface charge Q spins with an angular velocity ω about an axis that coincides with a diameter of the sphere. Show that the magnetic moment is $\mathbf{m} = \frac{1}{3}QR^2\omega$. Calculate ω if $Q = e$, $R = 0.53$ Å (the radius of the hydrogen atom), and m is equal to one Bohr magneton. Compare your result with the angular speed of the electron in a hydrogen atom, $\omega_H = 4.13 \times 10^{16}$ rad/s.

36-28 A saturated magnetic induction of 10,000 G is developed in a homogeneous sample of iron with $\kappa_M = 2000$. (a) What is the energy density of the field within the sample? (b) What electric field strength E would be required to develop the same energy density in a homogeneous dielectric with $\kappa = 4$?

36-29 A long needlelike cavity (filled with air) is present
• in a uniformly magnetized piece of hot-rolled sheet iron. The needle axis is parallel to the magnetization $\mathbf{M}$ in the sample; $M = 1.5 \times 10^4$ A·turns/m. (a) Determine H and B in the sample. (b) Determine H and B in the cavity. (Neglect end effects.) [*Hint:* A trial and error method will be necessary. Use Fig. 36–13.]

36-30 When they are magnetized, magnetic materials undergo dimension changes. For example, iron increases in length, nickel decreases, and cobalt exhibits almost no change. The fractional change in length, $\lambda = \Delta\ell/\ell$, for (ordinary) polycrystalline iron when driven to saturation is $\lambda = 4 \times 10^{-6}$. What is the internal stress in such a sample of iron if the ends are held fixed? (The relevant elastic constants are given in Table 16–2.)

CALCULUS VII

SIMPLE DIFFERENTIAL EQUATIONS

We discuss briefly a technique for solving ordinary differential equations with constant coefficients. These results are needed to understand the origin of the equations that describe damped systems, such as the electric circuits in Section 37–3, and the discussion of dispersion that is given in Section 40–6.

Many of the problems encountered in physics and engineering involve relationships that connect the rates of change of various quantities. The statements of such problems naturally take the form of *differential equations*. Newton's equation, $F = ma$, is actually a differential equation, $d^2x/dt^2 - F(x)/m = 0$. Indeed, we solved a differential equation as early as Chapter 2 when we considered the case of constant acceleration and wrote $d^2x/dt^2 = a$. We then determined the solution of this equation to be $x(t) = \frac{1}{2}at^2 + v_0t + x_0$.

When a differential equation involves only a single independent variable, the equation is called an *ordinary* differential equation. We consider here only equations of this type. (If there are two or more independent variables, the equation involves partial derivatives and is called a *partial* differential equation. The wave equation, discussed in Section 19–2, is a partial differential equation because it involves the independent variables x and t.)

The most common type of differential equation in physics and engineering—an equation used to describe dynamical motion, vibrations of mechanical systems, AC electric circuits, and many other situations—is the linear second-order differential equation with constant coefficient. ("Second-order" means that the highest-order derivatives present are second derivatives.) Such equations are encountered in the analysis of time-dependent electric circuits (Sections 37–3 and 37–4). Accordingly, the discussion here is limited to one technique for solving these equations.

LINEAR SECOND-ORDER DIFFERENTIAL EQUATIONS WITH CONSTANT COEFFICIENTS

The most common type of second-order differential equation can be expressed as

$$\frac{d^2y}{dx^2} + a_1\frac{dy}{dx} + a_2y = E(x) \qquad \textbf{(VII–1)}$$

where a_1 and a_2 are constant coefficients. The equation is "linear" because any single term contains only y or dy/dx or d^2y/dx^2 but not any powers or products of these quantities. The function $E(x)$ is referred to as the *input* (or *driving*) function, and the solution $y(x)$ is referred to as the *output* function.

The notation is simplified by using the D operator, defined by

$$Dy \equiv \frac{dy}{dx}; \qquad D^2y = D(Dy) \equiv \frac{d^2y}{dx^2}$$

The differential equation may now be expressed as

$$(D^2 + a_1D + a_2)y = E(x) \qquad \textbf{(VII–2)}$$

The Case $E(x) = 0$. Consider first the case in which the input function is zero. When $E(x) = 0$, Eq. VII–2 is called a *homogeneous* equation. It can be factored into the

995

form

$$(D - \lambda_1)(D - \lambda_2)y = 0 \qquad \textbf{(VII–3)}$$

where the numbers λ_1 and λ_2 are the roots of the polynomial

$$\lambda^2 + a_1\lambda + a_2 = 0 \qquad \textbf{(VII–4)}$$

which is called the *auxiliary equation*.

From Eq. VII-3 it can be seen that the solutions to $(D - \lambda_1)y = 0$ and $(D - \lambda_2)y = 0$ are both solutions to the entire differential equation. Now, the equation

$$(D - \lambda_1)y = \frac{dy}{dx} - \lambda_1 y = 0$$

can be written as

$$\frac{dy}{y} = \lambda_1\, dx$$

so the solution is found by integrating both sides and taking the exponential:

$$y_1(x) = c_1 e^{\lambda_1 x}$$

where c_1 is the arbitrary constant of integration. Similarly, for $(D - \lambda_2)y = 0$, we obtain

$$y_2(x) = c_2 e^{\lambda_2 x}$$

Then, the most general solution to Eq. VII-3 is the *sum* of the individual solutions; thus,

$$y(x) = y_1(x) + y_2(x) = c_1 e^{\lambda_1 x} + c_2 e^{\lambda_2 x} \qquad \textbf{(VII–5)}$$

The fact that the sum of two individual solutions is also a solution follows from the linear character of the differential equation. (•Can you show this?) The general solution contains a number of arbitrary constants equal to the order of the equation (two).

There are two special cases of interest.

(a) *Complex roots.** If the differential equation VII-1 describes a physical process, the coefficients a_1 and a_2 are necessarily *real* quantities, and the solution $y(x)$ (Eq. VII-5) is also a *real* quantity. If there are any complex roots of the auxiliary equation (Eq. VII-4), these roots must appear in pairs that are complex conjugates. (•Can you see why? Both the sum of the roots and the product of the roots must

---

*The properties of complex numbers are summarized in Appendix B.

be real quantities.) For example, suppose that we find the roots $\lambda_1 = \alpha + i\beta$ and $\lambda_2 = \alpha - i\beta$, with α and β real. Then, Eq. VII-5 for the solution becomes

$$y(x) = y_1(x) + y_2(x) = c_1 e^{(\alpha+i\beta)x} + c_2 e^{(\alpha-i\beta)x}$$
$$= e^{\alpha x}(c_1 e^{i\beta x} + c_2 e^{-i\beta x}) \qquad \textbf{(VII–6)}$$

Now, this equation can be cast in different forms. For example, to make $y(x)$ a real quantity, the complex constants, c_1 and c_2, must be complex conjugates. We write $c_1 = \frac{1}{2}(A + iB)$ and $c_2 = c_1{}^* = \frac{1}{2}(A - iB)$, where A and B are real. Using the Euler relationship (Eq. B-6 in Appendix B), we find

$$y(x) = e^{\alpha x}(A \cos \beta x - B \sin \beta x) \qquad \textbf{(VII–7)}$$

Moreover, such an equation can always be expressed in the form

$$y(x) = a e^{\alpha x} \cos (\beta x - \delta) \qquad \textbf{(VII–8)}$$

where the amplitude a and the phase δ are to be determined by the boundary conditions.

(b) *Repeated roots.* If the solution to Eq. VII-4 yields a double root $\bar{\lambda}$, the general solution becomes

$$y(x) = (c_1 + c_2 x)e^{\bar{\lambda} x} \qquad \textbf{(VII–9)}$$

where c_1 and c_2 are arbitrary constants.

Example VII–1

Determine the general solution to the equation

$$\frac{d^2 y}{dx^2} - 4y = 0$$

Solution:

The auxiliary equation is $\lambda^2 - 4 = 0$, which can be factored to $(\lambda - 2)(\lambda + 2) = 0$. Thus, the roots are $\lambda_1 = 2$ and $\lambda_2 = -2$, and the general solution is

$$y(x) = c_1 e^{2x} + c_2 e^{-2x}$$

Example VII–2

Determine the general solution to the equation,

$$(D^2 - 4D + 5)y = 0$$

Solution:

The auxiliary equation is

$$\lambda^2 - 4\lambda + 5 = 0$$

The roots are complex:

$$\lambda_1 = 2 + i \quad \text{and} \quad \lambda_2 = 2 - i$$

Thus, the general solution is

$$y(x) = c_1 e^{(2+i)x} + c_2 e^{(2-i)x}$$
$$= e^{2x}(A \cos x + B \sin x)$$

where A and B are arbitrary real constants. (• Can you supply the intermediate steps?)

Example VII–3

Determine the general solution to the equation

$$(D^2 + 4D + 4)y = 0$$

Solution:

The auxiliary equation, $\lambda^2 + 4\lambda + 4 = (\lambda + 2) \cdot (\lambda + 2) = 0$, has the repeated root $\bar{\lambda} = -2$. Therefore, according to Eq. VII-9, the general solution is

$$y(x) = (c_1 + c_2 x)e^{-2x}$$

(• Can you demonstrate explicitly that $y(x)$ satisfies the original differential equation?)

The Case $E(x) \neq 0$. The solution to the differential equation VII-2 (called an *inhomogeneous* equation when $E(x) \neq 0$) consists of the sum

$$y(x) = y_c(x) + y_p(x) \qquad \textbf{(VII–10)}$$

The function $y_c(x)$, called the *complementary solution,* is found by setting $E(x)$ equal to zero. The function $y_p(x)$, called the *particular solution,* is specifically required to reproduce the input function $E(x)$ when substituted into the left-hand side of the differential equation. The complementary solution, which contains two arbitrary constants, is exactly the general solution just discussed for the case $E(x) = 0$. The particular solution contains *no* arbitrary constants and is to be found by any means whatsoever, limited only by the ingenuity of the investigator.

In physics and engineering problems, the function $E(x)$ usually involves only quantities from which a finite number of different terms can be generated by repeated differentiation. (By "terms" we mean the functional forms. Thus, $A \sin \beta x$ and $B \sin \beta x$, with any A and B, are not different terms; however, $M \sin \beta x$ and $Nx \sin \beta x$, with any M and N, *are* different terms.) The trial particular solution to an inhomogeneous equation is written as a linear combination of all the different generated terms, each with an arbitrary coefficient. These coefficients are then determined by substituting the trial solution into the original differential equation.

Example VII–4

Determine the general solution to the equation

$$\frac{d^2 y}{dx^2} - 4y = 2 \sin \Omega x$$

Solution:

To determine the complementary solution, the right-hand side of the equation is set equal to zero. Then, we have exactly the equation discussed in Example VII-1. Thus, the complementary solution is

$$y_c(x) = c_1 e^{2x} + c_2 e^{-2x}$$

Repeated differentiation of the input function, $E(x) = 2 \sin \Omega x$, gives only two terms, namely, $\sin \Omega x$ and $\cos \Omega x$. Accordingly, we try the particular solution

$$y_p(x) = A \sin \Omega x + B \cos \Omega x$$

Upon substituting $y_p(x)$ into the original differential equation, there results

$$-(\Omega^2 + 4)A \sin \Omega x - (\Omega^2 + 4)B \cos \Omega x = 2 \sin \Omega x$$

This equation must be valid for all x; consequently, the coefficients A and B are

$$A = -\frac{2}{\Omega^2 + 4} \quad \text{and} \quad B = 0$$

so that the general solution is

$$y(x) = c_1 e^{2x} + c_2 e^{-2x} - \frac{2}{\Omega^2 + 4} \sin \Omega x$$

PROBLEMS

VII-1 Determine the solutions to the following linear homogeneous differential equations.
(a) $(D^2 - 5D + 3)y = 0$
(b) $d^2y/dx^2 - 3dy/dx = 0$
(c) $(D^2 + 4)y = 0$
(d) $(D^2 - 3D + 2)y = 0$

VII-2 In Section 15-2 we found that the displacement of a simple harmonic oscillator with mass m and spring constant κ is given by $x(t) = A \sin(\omega_0 t + \phi_0)$, where $\omega_0^2 = \kappa/m$. The differential equation that describes the motion is $d^2x/dt^2 = -(\kappa/m)x$.
(a) Show that the general solution of this linear equation is

$$x(t) = c_1 \cos \omega_0 t + c_2 \sin \omega_0 t.$$

(b) Express A and ϕ_0 in terms of c_1 and c_2. [*Hint:*

Use the identity $\sin(\alpha + \beta) = \sin \alpha \cos \beta + \cos \alpha \sin \beta$.]

VII-3 Determine the general solutions to the following inhomogeneous equations:
(a) $(D^2 - a^2)y = a^2 x^2$
(b) $(D + \beta)y = Ae^{-\alpha x}$

VII-4 The harmonic oscillator of Problem VII-2 is subjected to a driving function $E(t) = E_0 e^{-\alpha t}$ for $t \geqslant 0$. Thus, the equation that describes the motion becomes

$$\frac{d^2x}{dt^2} + \frac{\kappa}{m}x = E_0 e^{-\alpha t}$$

Determine the motion of the oscillator for $t \geqslant 0$ if $x = 0$ and $dx/dt = 0$ at $t = 0$.

CIRCUITS WITH TIME-VARYING CURRENTS

In the preceding chapters we were concerned primarily with time-invariant currents (DC). We now consider the general case and let the currents vary with time.

It is usually possible to associate a characteristic time τ with a time-varying current. For example, the characteristic time associated with a sinusoidally varying current, $I_0 \sin \omega t$, is just the period $2\pi/\omega$. In order to avoid the complications that result when substantial amounts of electromagnetic radiation are emitted by a system, we require again (see Example 35–3) that all lengths ℓ associated with the system or circuit be small compared with a characteristic length $c\tau$. This is the distance to which electromagnetic effects could propagate (at the speed of light c) during a time equal to the characteristic time τ of the circuit. Then, with $\ell \ll c\tau$, the amount of radiation emitted is generally quite small. When this condition is not met, phase differences throughout the circuit will result, and substantial amounts of electromagnetic energy may be radiated away. Except in some simple cases (see Chapter 38), the general analysis then requires techniques beyond the scope of this text. Generally, these effects are not important for laboratory-scale systems driven by oscillating currents with frequencies below about 10^6 Hz, for then the characteristic length is 300 m or more. However, for current pulses with rise times of the order of 0.01 μs, we have $c\tau \approx 3$ m. Then, small circuit components and careful layout are required; also, the use of shielded cable may be necessary to suppress radiation effects.

When analyzing circuits we assume that all resistance, capacitance, and inductance occur only as *lumped circuit elements*. Thus, we consider that all substantial resistance in a circuit either occurs in localized resistor elements or is associated with well-defined isolated sections of the circuit. Similarly, we assume that all substantial capacitance and inductance occur in specific capacitors and inductors, respectively.

In a real circuit the resistors have resistance, of course, but they may also have some inductance and capacitance (especially if they are wire-wound resistors). Similarly, an inductor coil will have a small resistance and a capacitor will have a small conductivity. If the nature of the calculation requires taking into account these secondary effects, we replace each real circuit element with its approximate equivalent in terms of R, L, and C. For example, a "leaky" capacitor can be considered to be a capacitor and a (large) resistor in a parallel combination (Fig. 37–1a). An inductor

Fig. 37-1. (a) An ideal capacitor has only capacitance, but a real capacitor can be represented as a capacitance C and a resistance R in a parallel combination. (The resistivity ρ of a dielectric material is usually very large.) (b) The resistance R of a real inductor coil appears in series with the inductance L. (The resistivity ρ of an inductor wire is usually very small.)

coil with some resistance can be considered to be an inductor and a (small) resistor in a series combination (Fig. 37–1b).

The required condition, $\ell \gg c\tau$, means that the currents in the various branches of a circuit can be assumed to obey Kirchhoff's rules at any instant of time. (•Can you see why this is so?)

37-1 *RC* CIRCUITS

Charging a Capacitor. Consider first the process of charging a capacitor in the circuit shown in Fig. 37–2. We begin with zero charge on the capacitor. At $t = 0$ the switch S is closed to position a and a time-dependent current $I(t)$ begins to flow. The charge Q on the capacitor increases in accordance with $I = dQ/dt$. We apply Kirchhoff's second rule (see Section 32–2), proceeding clockwise around the circuit, to obtain

$$V_R + V_C = \mathcal{E}$$

or

$$IR + \frac{Q}{C} = \mathcal{E}$$

Fig. 37-2. Circuit for charging (and discharging) a capacitor C.

This equation can be written as

$$R\frac{dQ}{dt} + \frac{Q}{C} = \mathcal{E} \qquad (37\text{-}1)$$

Rearranging, we obtain

$$\frac{dQ}{Q - \mathcal{E}C} = -\frac{1}{RC}\,dt$$

Integration gives

$$\ln(Q - \mathcal{E}C) = -t/RC + \ln c_1$$

or

$$Q(t) = \mathcal{E}C + c_1 e^{-t/RC} \qquad (37\text{-}2)$$

where c_1 is the arbitrary constant of integration that is determined by the initial condition, namely, $Q = 0$ at $t = 0$. We find $c_1 = -\mathcal{E}C$, so that

$$Q(t) = \mathcal{E}C(1 - e^{-t/RC}) \qquad (37\text{-}3)$$

By differentiating $Q(t)$ we obtain the expression for the time dependence of the current. Thus,

$$I(t) = \frac{dQ}{dt} = \frac{\mathcal{E}}{R}e^{-t/RC} \qquad (37\text{-}4)$$

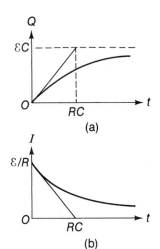

Fig. 37-3. (a) Time dependence of the charge Q on a capacitor C that is being charged through a resistance R. If the initial rate of charging were maintained, the capacitor would be fully charged after a time $\tau = RC$. (b) The current I corresponding to the rate of change of the charge in (a). (•Can you see the significance of the time RC in this diagram?)

The expressions for $Q(t)$ and $I(t)$ are illustrated graphically in Fig. 37-3.

The product RC has the dimensions of time and is referred to as the *time constant* τ of the circuit. For typical values, $R = 100\ \Omega$ and $C = 10^{-9}$ F, we find $\tau = RC = 10^{-7}$ s. The characteristic length associated with this circuit is $c\tau = 30$ m. Thus, a circuit with ordinary size (say, 1 m or less) will satisfy the condition that permits the neglect of radiation and phase-difference effects.

Figure 37-3a or Eq. 37-4 shows that the initial rate of charging of the capacitor is $dQ/dt = \mathcal{E}/R$ and that the rate decreases with time as the capacitor approaches the fully charged condition, $Q_\infty = \mathcal{E}C$. This charge is reached strictly only after an infinite time. However, for all practical purposes, the capacitor may be considered to be fully charged after a time of approximately 5τ, for then $Q/\mathcal{E}C = 1 - e^{-5} = 0.993$. Similarly, after a time of 5τ, the charging current is essentially zero.

Discharging a Capacitor. After the capacitor in Fig. 37-2 has been fully charged, the switch S is moved to position b, and the capacitor begins to discharge through the resistor R. The differential equation for the discharging condition can be obtained by simply setting $\mathcal{E} = 0$ in Eq. 37-1. The direction for positive current flow and the charge polarity are as shown in Fig. 37-2. Then, for Kirchhoff's rule around a clockwise path, we have

$$IR + \frac{Q}{C} = 0$$

or

$$I = \frac{dQ}{dt} = -\frac{Q}{RC} \qquad (37\text{-}5)$$

As expected, $I < 0$, which means that the current actually flows in the counterclockwise direction. The solution to this equation is obtained in a manner similar to that used in deriving Eq. 37-3. For $Q = Q_0$ at $t = 0$, we find

$$Q(t) = Q_0 e^{-t/RC} \qquad (37\text{-}6)$$

and the discharge current is

$$I(t) = \frac{dQ}{dt} = -\frac{Q_0}{RC} e^{-t/RC} \qquad (37\text{-}7)$$

Fig. 37-4. The discharge of a capacitor through a resistance R. (a) The charge on the capacitor as a function of time. (b) The discharge current as a function of time.

The expressions for $Q(t)$ and $I(t)$ are illustrated graphically in Fig. 37-4.

Sign Conventions.   It is useful to summarize the important rules concerning sign conventions for Q and I, and for V_C treated as a voltage drop. When we apply Kirchhoff's second rule and proceed along a path that crosses a capacitor from the positively charged plate to the negatively charged plate, we have $V_C = +Q/C$; for a path in the opposite direction, $V_C = -Q/C$. If the direction selected for positive current corresponds to that for increasing Q, we have $I = +dQ/dt$; for a direction that corresponds to decreasing Q, $I = -dQ/dt$.

In the discussions of the charging and the discharging of a capacitor, we had, for a clockwise path in each case, $V_C = +Q/C$ and $I = +dQ/dt$. For the discharging capacitor we could have chosen the positive direction for current flow to be counterclockwise, leaving the charge polarity the same (Fig. 37-5). Then, for a counterclockwise path, we would have $I = -dQ/dt$ (decreasing Q) and $V_C = -Q/C$ (negative to positive plate); also, $V_R = +IR$. Hence,

$$IR - \frac{Q}{C} = 0$$

Fig. 37-5. The positive direction for current flow for a discharging capacitor can be chosen opposite to that selected in Fig. 37-2.

so that

$$R\frac{dQ}{dt} + \frac{Q}{C} = 0$$

as before. Thus, $Q(t)$ is the same as Eq. 37-6, but $I(t)$ is now a positive quantity, as required by the choice in Fig. 37-5.

Energy Balance.   The Kirchhoff equation for a charging capacitor is $Q/C + IR = \mathcal{E}$. If we multiply this equation by $dQ/dt = I$, we obtain

$$\frac{Q}{C}\frac{dQ}{dt} + I^2 R = \mathcal{E}I$$

or

$$d\left(\frac{1}{2}\frac{Q^2}{C}\right) + I^2R\,dt = \mathcal{E}I\,dt \qquad (37\text{-}8)$$

The interpretation of this equation is that during a time interval dt an amount of energy $\mathcal{E}I\,dt$ supplied by the battery is expended partly in Joule heating, $I^2R\,dt$, and partly in increasing the stored energy in the electric field of the capacitor, $d(\frac{1}{2}Q^2/C)$. Thus, Eq. 37-8 expresses energy conservation for the system in which chemical energy is converted partly into heat and partly into field energy. (If electromagnetic radiation were emitted, Eq. 37-8 would no longer be adequate.)

The Kirchhoff equation for a discharging capacitor, $Q/C + IR = 0$, becomes

$$-d\left(\frac{1}{2}\frac{Q^2}{C}\right) = I^2R\,dt \qquad (37\text{-}9)$$

which states that the energy removed from the electric field of the capacitor appears in the form of Joule heat in the resistor.

Example 37-1

In the circuit of Fig. 37-2, $C = 0.1\,\mu\text{F}$, $\mathcal{E} = 3$ V, and $R = 1000\,\Omega$. (The internal resistance of the battery is small compared with R.) The capacitor is uncharged when the switch is closed to position a.

(a) Determine the initial current.

(b) Determine the charge Q_∞ on the capacitor when it is fully charged.

(c) Find the time required for the charge on the capacitor to reach 99 percent of Q_∞.

(d) Find the time required for the field energy in the capacitor to reach 99 percent of its final value.

Solution:

(a) The initial current is

$$I_0 = \frac{\mathcal{E}}{R} = \frac{3.0\text{ V}}{10^3\,\Omega} = 3\text{ mA}$$

(b) The final charge is

$$Q_\infty = \mathcal{E}C = (3.0\text{ V})(0.1 \times 10^{-6}\text{ F}) = 3 \times 10^{-7}\text{ C}$$

(c) For $Q = 0.99\,Q_\infty$, we have, using Eq. 37-3,

$$\frac{Q}{Q_\infty} = \frac{Q}{\mathcal{E}C} = 1 - e^{-t/RC} = 0.99$$

Thus,
$$e^{-t/RC} = 0.01$$

so that
$$t = -RC \ln 0.01 = -(10^3 \, \Omega)(10^{-7} \, \text{F})(\ln 0.01)$$
$$= 4.61 \times 10^{-4} \, \text{s}$$

(d) The stored energy is $U = \frac{1}{2}Q^2/C$. Thus

$$\frac{U}{U_\infty} = \frac{Q^2}{Q_\infty^2} = 0.99$$

so that
$$\frac{Q}{Q_\infty} = 0.995 = 1 - e^{-t/RC}$$

from which
$$e^{-t/RC} = 0.005$$

Finally,
$$t = -RC \ln 0.005 = -(10^3 \, \Omega)(10^{-7} \, \text{F})(\ln 0.005)$$
$$= 5.30 \times 10^{-4} \, \text{s}$$

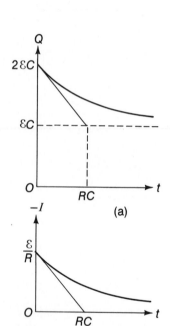

(a)

(b)

Example 37–2

Refer again to the RC circuit in Fig. 37–2. The capacitor initially carries a charge $2\mathcal{E}C$, and at $t = 0$ the switch is closed to position a, thereby making connection with the battery (EMF $\mathcal{E}$). Obtain expressions for $Q(t)$ and $I(t)$.

Solution:

The general solution for $Q(t)$ is given by Eq. 37–2. Substituting $Q = 2\mathcal{E}C$ at $t = 0$, we find $c_1 = \mathcal{E}C$. Therefore,

$$Q(t) = \mathcal{E}C(1 + e^{-t/RC})$$

For the charge polarity and current direction shown in Fig. 37–2, we have

$$I(t) = \frac{dQ}{dt} = -\frac{\mathcal{E}}{R}e^{-t/RC}$$

We see from the expression for $Q(t)$ and from diagram (a) below that the capacitor actually discharges from an initial charge $2\mathcal{E}C$ to a final charge $\mathcal{E}C$ (as $t \to \infty$). Also, the expression for $I(t)$ and diagram (b) shows that $I < 0$ for all t; that is, the current actually flows in the direction opposite to that indicated in Fig. 37–2.

Example 37–3

A 25-μF capacitor is left isolated after having been charged. Three hours later it is discovered that the capacitor has lost one half of its charge. What is the leakage resistance in the dielectric material of the capacitor?

Solution:

The equivalent circuit of the leaky capacitor is shown in Fig. 37–1a, and we see that the time constant is $\tau = RC$, where R is the dielectric resistance. Thus,

$$\frac{Q}{Q_0} = e^{-t/RC}$$

from which
$$\ln \tfrac{1}{2} = -t/RC$$

Then,

$$R = \frac{-3(3600 \text{ s})}{(25 \times 10^{-6} \text{ F})(\ln \frac{1}{2})}$$

$$= 6.23 \times 10^8 \ \Omega = 623 \text{ M}\Omega$$

Example 37–4

A capacitor C_1 that carries an initial charge Q_0 is connected to an initially uncharged capacitor C_2 by closing the switch S, as indicated in the diagram. Compare the energy stored in the capacitors after equilibrium is reached with the initial stored energy.

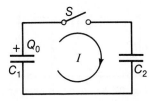

Solution:

After the switch is closed a current $I(t)$ flows, delivering charge from C_1 to C_2 until, at equilibrium, there is a charge Q_1 on C_1 and a charge Q_2 on C_2, where $Q_1 + Q_2 = Q_0$. In this condition, the two capacitor voltages are equal; that is, $V_1 = Q_1/C_1 = V_2 = Q_2/C_2$. These two conditions lead to

$$Q_1 = \frac{C_1}{C_1 + C_2} Q_0 \quad \text{and} \quad Q_2 = \frac{C_2}{C_1 + C_2} Q_0$$

The energy originally stored in C_1 was

$$U_0 = \frac{1}{2} \frac{Q_0^2}{C_1}$$

The energy stored in the final condition is

$$U_f = U_1 + U_2 = \frac{1}{2} \frac{Q_1^2}{C_1} + \frac{1}{2} \frac{Q_2^2}{C_2}$$

$$= \frac{1}{2} \frac{Q_0^2}{C_1 + C_2}$$

Thus, $U_f < U_0$, in apparent violation of the principle of energy conservation.

In any real circuit, there would be a certain amount of resistance R in the connecting wires. Then, using Kirchhoff's second rule, we would have

$$IR - \frac{Q_1}{C_1} + \frac{Q_2}{C_2} = 0$$

Multiplying by $I = -dQ_1/dt = dQ_2/dt$ gives

$$I^2R + \frac{Q_1}{C_1}\frac{dQ_1}{dt} + \frac{Q_2}{C_2}\frac{dQ_2}{dt} = 0$$

With $U = \frac{1}{2}Q^2/C$, we can write

$$I^2R + \frac{dU_1}{dt} + \frac{dU_2}{dt} = 0$$

Integrating, we find

$$\int_0^\infty I^2R \, dt = -\int_{U_0}^{U_f} d(U_1 + U_2) = U_0 - U_f$$

Thus, the difference in stored energy between the initial and final conditions, $U_0 - U_f > 0$, is accounted for by the Joule heating in the resistance R.

Suppose that the resistance R is made very small. Then, the characteristic time τ of the circuit also becomes very small. That is, $\tau = RC \to 0$, where $C = C_1C_2/(C_1 + C_2)$ is the effective capacitance of the series connection. Thus, when $R \to 0$, less energy is dissipated as heat, but more energy is lost as radiation, so that we still have $U_f < U_0$.

Fig. 37–6. A differentiating circuit. The output voltage $V(t)$ is the time derivative of the input voltage $\mathcal{E}(t)$.

The *RC* Differentiating Circuit. An interesting application of the simple RC circuit is an often used arrangement called a *differentiating circuit*, shown in Fig. 37–6. A time-dependent input voltage $\mathcal{E}(t)$ is applied to the RC circuit as indicated, and the output voltage $V(t)$ developed across the resistor is detected by a device such as an oscilloscope.* Using Kirchhoff's rule, we find

$$\frac{Q}{C} + IR = \mathcal{E}(t)$$

or

$$Q + IRC = C\mathcal{E}(t) \qquad \textbf{(37–10)}$$

In an appropriately designed circuit, RC is made sufficiently small that the term IRC in Eq. 37–10 can be safely neglected. Then, $Q \cong C\mathcal{E}(t)$, and we have

$$V(t) = IR = R\frac{dQ}{dt} \cong RC\frac{d\mathcal{E}(t)}{dt} \qquad \textbf{(37–11)}$$

In the approximation that RC is small, the output voltage $V(t)$ is proportional to the derivative of the input voltage $\mathcal{E}(t)$. Notice that we have here an example of a frequently encountered design dilemma, namely, that the approximation cannot be improved by making RC arbitrarily small because this would decrease the magnitude of the output voltage below a useful limit.

Figure 37–7 shows the response of a differentiating circuit to a square-wave input. If the input were a perfect square wave and if the circuit were ideal ($RC \to 0$),

---

*An oscilloscope draws negligible current from a circuit (as do certain other kinds of voltage-sensing devices) and therefore can be ignored when applying Kirchhoff's rules to the circuit.

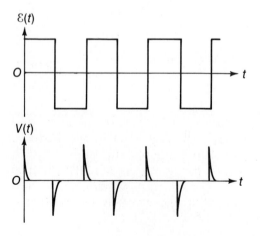

Fig. 37–7. The response $V(t)$ of a differentiating circuit is a series of spikes when the input voltage $\mathcal{E}(t)$ is a square wave.

the output voltage $V(t)$ would consist of sharp spikes with essentially zero width. The departure of the conditions from ideal results in voltage spikes with "tails." (For a discussion of an *integrating circuit,* see Problem 37–26.)

37-2 *RL* CIRCUITS

Energizing an Inductor. Consider the process of applying current to an inductor in the circuit shown in Fig. 37–8. At $t = 0$ the switch S is closed to position a and

Fig. 37–8. Circuit for energizing (and de-energizing) an inductor L.

a time-dependent (increasing) current $I(t)$ begins to flow. Because of this changing current in the inductor, there is produced a back EMF, $L dI/dt$, that opposes the driving EMF $\mathcal{E}$. In the application of Kirchhoff's rule, this back EMF appears as a voltage drop V_L. Thus, for a clockwise path in Fig. 37–8, we have

$$V_L + V_R = \mathcal{E}$$

or

$$L\frac{dI}{dt} + IR = \mathcal{E} \qquad \textbf{(37-12)}$$

This equation has the same form as Eq. 37–1, so the solution is

$$I(t) = \frac{\mathcal{E}}{R} + c_1 e^{-(R/L)t} \qquad \textbf{(37-13)}$$

where c_1 is the arbitrary constant of integration that is determined by the initial condition, namely, $I = 0$ at $t = 0$. We find $c_1 = -\mathcal{E}/R$, so that

$$I(t) = \frac{\mathcal{E}}{R}(1 - e^{-(R/L)t}) \qquad \textbf{(37-14)}$$

This expression for $I(t)$ is illustrated graphically in Fig. 37–9a. (•Do you understand the physical reason why the current is zero at $t = 0$, in contrast to the case illustrated in Fig. 37–3b? Consider the back EMF produced in the inductor.)

The quotient L/R is the time constant of the circuit. When $t \gg L/R$, the current approaches the asymptotic value, $I_\infty = \mathcal{E}/R$. For all practical purposes, we may consider this current to have been reached (and the inductor fully energized) after a time of approximately 5τ, for then $I = 0.993\,\mathcal{E}/R$.

Fig. 37–9. (a) Time dependence of the current in an *RL* circuit as the inductor is energized. (b) Time dependence of the current in the same *RL* circuit as the inductor is de-energized. (•Can you see the relationship between the time constant L/R and the slope of the curve in each case? Compare Fig. 37–3.)

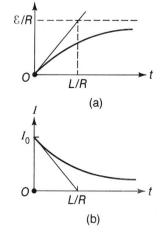

De-Energizing an Inductor. After the inductor in Fig. 37–8 has been fully energized, the switch S is moved to position b.* The current decreases from its initial value I_0 in accordance with Eq. 37–12, in which we now set $\mathcal{E} = 0$. Then, $c_1 = I_0$, so we have

$$I(t) = I_0 e^{-(R/L)t} \tag{37-15}$$

The time dependence of the de-energizing current is shown in Fig. 37–9b.

Energy Balance. Consider first the process of energizing the inductor. Multiplying Eq. 37–12 by I, we obtain

$$LI\frac{dI}{dt} + I^2 R = \mathcal{E}I$$

or

$$d(\tfrac{1}{2}LI^2) + I^2R\,dt = \mathcal{E}I\,dt \tag{37-16}$$

The interpretation of this equation is that during a time interval dt an amount of energy $\mathcal{E}I\,dt$ supplied by the battery is expended partly in Joule heating, $I^2R\,dt$, and partly in increasing the stored energy in the magnetic field of the inductor, $d(\tfrac{1}{2}LI^2)$.

The differential equation that described the de-energizing process is just Eq. 37–12 with $\mathcal{E} = 0$. Multiplying this equation by I leads to

$$-d(\tfrac{1}{2}LI^2) = I^2R\,dt \tag{37-17}$$

which states that the energy removed from the magnetic field of the inductor appears in the form of Joule heat in the resistor.

$N = 250$ turns

$b = 8$ cm

area $A = 2$ cm^2

Example 37–5

The inductor L in Fig. 37–8 is the toroid shown in the diagram, whose core consists of hot-rolled sheet iron. The resistance in the circuit is $R = 20\,\Omega$ and the EMF of the battery is $\mathcal{E} = 6$ V. Assume a linear magnetization curve for the iron, with $B = \kappa_M \mu_0 H$ and $\kappa_M = 300$ (a constant; see Fig. 36–13f). Neglect the battery internal resistance and the coil resistance of the inductor.

(a) What amount of energy is stored in the magnetic field of the inductor 1.0 ms after the switch is closed?

(b) What amount of energy has been supplied by the battery at $t = 1.0$ ms?

(c) What amount of energy has been expended as Joule heat in the resistor at $t = 1.0$ ms?

---

*The switch S is required to be of the "make before break" type; that is, contact with terminal b is made before contact with terminal a is broken. (•Can you see why this is necessary?)

Solution:

The self-inductance L of the iron-core toroid is*

$$L = \frac{\mu_0 \kappa_M N^2 A}{2\pi b} = 9.38 \times 10^{-3} \text{ H}$$

The magnetic induction in the iron is

$$B = \frac{\mu_0 \kappa_M N I}{2\pi b} = \frac{LI}{NA}$$

Using $I = I_\infty = \mathcal{E}/R = (6 \text{ V})/(20 \text{ }\Omega) = 0.3 \text{ A}$, we have

$$B = \frac{LI}{NA} = \frac{(9.38 \times 10^{-3} \text{ H})(0.3 \text{ A})}{(250)(2 \times 10^{-4} \text{ m}^2)} = 5.63 \times 10^{-2} \text{ T} = 563 \text{ G}$$

For such a low value of the maximum induction, the assumption that $B \propto H$ is quite good (see Fig. 36–13f). Thus, κ_M is indeed constant.

(a) The time constant for the circuit is $\tau = L/R = 0.47$ ms. Then, the current at $t = 1.0$ ms is

$$I = \frac{\mathcal{E}}{R}(1 - e^{-t/\tau}) = \frac{6 \text{ V}}{20 \text{ }\Omega}(1 - e^{-(1.0/0.47)}) = 0.264 \text{ A}$$

Hence, the amount of energy U_M stored in the magnetic field is

$$U_M = \tfrac{1}{2}LI^2 = \tfrac{1}{2}(9.38 \times 10^{-3} \text{ H})(0.264 \text{ A})^2 = 3.3 \times 10^{-4} \text{ J}$$

(b) The amount of energy U supplied by the battery from $t = 0$ to time t is found by using Eq. 37–14 for $I(t)$; thus,

$$U = \mathcal{E} \int_0^t I(t) \, dt = \frac{\mathcal{E}^2}{R} \int_0^t (1 - e^{-(R/L)t}) \, dt$$

$$= \frac{\mathcal{E}^2}{R}[t - \tau(1 - e^{-t/\tau})]$$

When $t = 1.0$ ms, we have

$$U = \frac{(6 \text{ V})^2}{20 \text{ }\Omega}[(10^{-3} \text{ s}) - (0.47 \times 10^{-3} \text{ s})(1 - e^{-(1.0/0.47)})]$$

$$= 1.05 \times 10^{-3} \text{ J}$$

(c) The amount of energy U_R expended as Joule heat in the resistor is

$$U_R = U - U_M = (1.05 \times 10^{-3} \text{ J}) - (3.3 \times 10^{-4} \text{ J})$$

$$= 7.2 \times 10^{-4} \text{ J}$$

(•Can you verify this result by integrating $RI^2(t) \, dt$?)

---

*See Example 36–3 and supply the factor κ_M to account for the presence of the iron core.

Example 37–6

Obtain an expression for the current $I(t)$ through the inductor L in the circuit shown in the diagram when the switch S is closed.

Solution:

Let the branch current in the inductor L be I, and let the branch current in the resistor R_2 be I'. Then, the battery current is $I + I'$ (Kirchhoff's first rule). Applying the second rule to the two loops, we find

$$(I + I')R_1 + L\frac{dI}{dt} = \mathcal{E}$$

$$(I + I')R_1 + I'R_2 = \mathcal{E}$$

By eliminating I' between these two equations there results

$$I\frac{R_1 R_2}{R_1 + R_2} + L\frac{dI}{dt} = \frac{R_2}{R_1 + R_2}\mathcal{E}$$

or, with obvious abbreviations,

$$IR' + L\frac{dI}{dt} = \mathcal{E}'$$

This equation has the form of Eq. 37–12, so the solution is (Eq. 37–14)

$$I(t) = \frac{\mathcal{E}'}{R'}(1 - e^{-(R'/L)t}) = \frac{\mathcal{E}}{R_1}(1 - e^{-(R'/L)t})$$

(•Can you obtain the same result by using Thévenin's theorem, discussed in Section 32–4?)

37–3   *LRC* CIRCUITS

Fig. 37–10. An *LRC* circuit. The capacitor carries an initial charge Q_0, and the switch is closed at $t = 0$.

Suppose that a capacitor C with an initial charge Q_0 is connected at time $t = 0$ to a resistor R and an inductor L in series, as shown in Fig. 37–10. Applying Kirchhoff's second rule gives

$$L\frac{dI}{dt} + RI - \frac{Q}{C} = 0$$

Now, $I = -dQ/dt$, so the Kirchhoff equation becomes

$$L\frac{d^2Q}{dt^2} + R\frac{dQ}{dt} + \frac{Q}{C} = 0 \tag{37-18}$$

The form of the solution to Eq. 37–18 depends on the relative values of two parameters defined to be $\beta = R/2L$ and $\omega_0^2 = 1/LC$. We discuss in turn the three cases, $\beta^2 > \omega_0^2$, $\beta^2 = \omega_0^2$, and $\beta^2 < \omega_0^2$. Finally, we compare the results and show graphically the forms of $Q(t)$ and $I(t)$ for the different cases.

Overdamping ($\beta^2 > \omega_0^2$). The general solution to Eq. 37–18 when $\beta^2 > \omega_0^2$ is found to be*

$$Q(t) = e^{-\beta t}(c_1 e^{\gamma t} + c_2 e^{-\gamma t}) \tag{37–19}$$

where

$$\beta = \frac{R}{2L}, \qquad \omega_0^2 = \frac{1}{LC}, \qquad \gamma = \sqrt{\beta^2 - \omega_0^2} \tag{37–20}$$

Equation 37–18 is a differential equation of second order (that is, the highest-order derivative present is the second derivative), so the solution contains two arbitrary constants, c_1 and c_2, that can be determined by imposing the boundary (initial) conditions. For the case here, at $t = 0$ we have $Q = Q_0$ and also $I = 0$. Applying these conditions to Eq. 37–19 and its time derivative, we find

$$c_1 + c_2 = Q_0$$

and

$$-\beta(c_1 + c_2) + \gamma(c_1 - c_2) = 0$$

Solving these equations for c_1 and c_2, and inserting the expressions into the general solution, we obtain

$$Q(t) = \frac{Q_0}{2\gamma} e^{-\beta t}[(\beta + \gamma)e^{\gamma t} - (\beta - \gamma)e^{-\gamma t}] \tag{37–21}$$

The current in the circuit is

$$I(t) = -\frac{dQ}{dt} = \frac{Q_0 \omega_0^2}{2\gamma} e^{-\beta t}(e^{\gamma t} - e^{-\gamma t}) \tag{37–22}$$

Critical Damping ($\beta^2 = \omega_0^2$). In this case, we have $\gamma = 0$, and the general solution takes the form

$$Q(t) = (c_1 + c_2 t)e^{-\beta t} \tag{37–23}$$

Imposing the same initial conditions, we find

$$c_1 = Q_0 \qquad \text{and} \qquad -\beta c_1 + c_2 = 0$$

Then, we have

$$Q(t) = Q_0(1 + \beta t)e^{-\beta t} \tag{37–24}$$

The current in the circuit is

$$I(t) = Q_0 \beta^2 t e^{-\beta t} \tag{37–25}$$

---

*The details are given in the section on page 1013.

Underdamping ($\beta^2 < \omega_0^2$). In this case, $\gamma = \sqrt{\beta^2 - \omega_0^2}$ is an imaginary quantity. The situation is most easily handled by introducing a *real* quantity, ω^0, defined by

$$\omega^0 = \sqrt{\omega_0^2 - \beta^2} \tag{37-26}$$

Then, we explicitly exhibit the imaginary character of the exponents by writing the solution (Eq. 37-19) as

$$Q(t) = e^{-\beta t}(c_1 e^{i\omega^0 t} + c_2 e^{-i\omega^0 t}) \tag{37-27}$$

In general, the constants c_1 and c_2 are complex. By using the Euler relation (Eq. B-6 in Appendix B), Eq. 37-27 can be expressed as

$$Q(t) = Ae^{-\beta t} \cos(\omega^0 t - \phi_0)$$

where A and ϕ_0 are real quantities that can be evaluated by imposing the initial conditions. The condition $Q = Q_0$ at $t = 0$ gives

$$A = \frac{Q_0}{\cos\phi_0}$$

and the condition $I = -dQ/dt = 0$ at $t = 0$ gives

$$\phi_0 = \tan^{-1}(\beta/\omega^0)$$

We define the phase angle ϕ_0 according to the right-triangle relationship shown in Fig. 37-11. Then, we see that $\cos\phi_0 = \omega^0/\omega_0$; thus,

$$A = \frac{Q_0\omega_0}{\omega^0}$$

Finally, we have

Fig. 37-11. The phase angle ϕ_0 is defined by the right-triangle relationship illustrated here.

$$Q(t) = \frac{Q_0\omega_0}{\omega^0}e^{-\beta t}\cos(\omega^0 t - \phi_0) \tag{37-28}$$

The current in the circuit is

$$I(t) = -Q_0\frac{\omega_0^2}{\omega^0}e^{-\beta t}\sin\omega^0 t \tag{37-29}$$

(•Can you obtain this result by differentiating Eq. 37-28 and using Fig. 37-11 and the identity $\sin(A + B) = \sin A \cos B + \cos A \sin B$?)

Comparison of the Solutions. Figure 37-12a shows the time dependence of the charge $Q(t)$ for the three cases we have just analyzed. The factor $e^{-\beta t}$ causes $Q(t)$ to decrease in all of the cases, but only for underdamping (case ③) is the general decrease accompanied by oscillation. The case of critical damping is of particular interest because it represents the most rapid approach to zero charge. That is, when the circuit is critically damped, the charge reaches (and remains smaller than) any

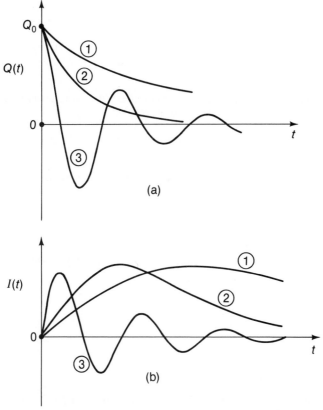

Fig. 37–12. (a) Time dependence of the charge $Q(t)$ for the three cases: ① overdamping, ② critical damping, and ③ underdamping. (b) Time dependence of the current $I(t)$ for the same three cases.

prescribed value in a time shorter than that required for over- or underdamped circuits with the same value of ω_0. When a circuit (or a mechanical system) is required to recover in the least time from some disturbance away from equilibrium, the conditions are adjusted to produce critical damping.

Figure 37–12b shows the time dependence of the current $I(t)$ for the three cases. The peak currents have been arbitrarily made equal for ease of comparison. (The degree of distortion can be judged from the fact that the integral of $I(t)$ over the range $0 \leq t \leq \infty$ is just Q_0. Thus, the area under each of the $I(t)$ curves must be the same.)

▶ *Solving the Differential Equation for an LRC Circuit.* The differential equation for an *LRC* circuit (Eq. 37–18) can be expressed as

$$\frac{d^2Q}{dt^2} + \frac{R}{L}\frac{dQ}{dt} + \frac{Q}{LC} = 0$$

This is a linear homogeneous equation of second order and can be solved by using the procedure described in the appendix Calculus VII. We construct the auxiliary equation,

$$\lambda^2 + \frac{R}{L}\lambda + \frac{1}{LC} = 0$$

and obtain two roots,

$$\lambda = -\frac{R}{2L} \pm \sqrt{\frac{R^2}{4L^2} - \frac{1}{LC}}$$

With the parameters defined in Eqs. 37–20, the expression for the roots becomes

$$\lambda = -\beta \pm \gamma$$

We distinguish three cases. First is the case of overdamping ($\beta^2 > \omega_0^2$), for which the roots, λ_1 and λ_2, are real and distinct. The general solution is (see Eq. VII–5)

$$Q(t) = c_1 e^{\lambda_1 t} + c_2 e^{\lambda_2 t} = e^{-\beta t}(c_1 e^{\gamma t} + c_2 e^{-\gamma t})$$

which is just the result quoted in Eq. 37–19.

Second is the case of critical damping ($\beta^2 = \omega_0^2$), for which $\gamma = 0$ and the two roots are equal. The general solution is (see Eq. VII–9)

$$Q(t) = (c_1 + c_2 t)e^{-\beta t}$$

which is just the result quoted in Eq. 37–23.

The final case is that of underdamping ($\beta^2 < \omega_0^2$), for which γ is an imaginary quantity. We use $\omega^0 = \sqrt{\omega_0^2 - \beta^2}$ to write

$$\lambda = -\beta \pm i\omega^0$$

Thus, the general solution is (see Eq. VII–6)

$$Q(t) = e^{-\beta t}(c_1 e^{i\omega^0 t} + c_2 e^{-i\omega^0 t})$$

which is just the result quoted in Eq. 37–27. ◄

Sinusoidally Driven *LRC* Circuits. Next, consider an *LRC* circuit that is driven by a source of repetitive voltage $\mathcal{E}(t)$. Now, any repetitive voltage, of whatever form, can be decomposed by Fourier analysis into a sum of sinusoidal terms (see Section 19–8). Consequently, if we obtain the response of the circuit to a general sinusoidal driving voltage, $\mathcal{E}(t) = \mathcal{E}_0 \cos \omega t$, the response to any repetitive driving voltage can be generated by the appropriate superposition of responses with different values of ω and $\mathcal{E}_0$.

Consider the circuit shown in Fig. 37–13. The relevant Kirchhoff equation is

$$L\frac{dI}{dt} + RI + \frac{Q}{C} = \mathcal{E}_0 \cos \omega t \qquad \textbf{(37–30)}$$

The general solution to this equation consists of the sum of the appropriate transient solution that we have just discussed and the particular solution constructed specifically to reproduce the driving term on the right-hand side of the equation. In the analysis here we restrict attention to the particular or steady-state solution.

Instead of proceeding with the analysis of Eq. 37–30 according to the methods described in Calculus VII, we introduce an elegant technique widely used in engi-

Fig. 37–13. An *LRC* circuit is driven by a sinusoidal voltage.

neering practice. To do this, we represent $Q(t)$, $I(t)$, and $\mathcal{E}(t)$ as *complex* quantities. For example, we write $\mathcal{E}(t) = \mathcal{E}_0 \cos \omega t$ as

$$\mathcal{E}(t) \rightarrow \mathcal{E}_0 e^{i\omega t} \tag{37–31}$$

Now, $Q(t)$, $I(t)$, and $\mathcal{E}(t)$ are physical quantities, so they are, of course, *real*. We understand that after having performed all required mathematical operations involving these complex quantities, we recover the final forms of the corresponding physical quantities by projecting the results onto the real axis.* Thus, we have, for example,†

$$Q(\text{physical}) \equiv \textbf{Re } \widehat{Q}(\text{complex})$$

Also, $$\textbf{Re } \widehat{\mathcal{E}} = \textbf{Re}(\mathcal{E}_0 e^{i\omega t}) = \mathcal{E}_0 \cos \omega t$$

The time differentiation and projection operations may be performed in either order; thus,

$$I = \textbf{Re } \widehat{I} = \textbf{Re } \frac{d\widehat{Q}}{dt} = \frac{d}{dt} \textbf{Re } \widehat{Q} = \frac{dQ}{dt}$$

Therefore, we can express Eq. 37–30 as

$$L\frac{d\widehat{I}}{dt} + R\widehat{I} + \frac{\widehat{Q}}{C} = \widehat{\mathcal{E}} \tag{37–32}$$

Fig. 37–14. The complex representation of the driving voltage, $\widehat{\mathcal{E}} = \mathcal{E}_0 e^{i\omega t}$. The physical voltage, $\mathcal{E}_0 \cos \omega t$, is given by the projection of $\mathcal{E}_0 e^{i\omega t}$ onto the real axis.

Before proceeding with the solution to Eq. 37–32, consider the behavior of the phasor (or *rotator*) $e^{i\omega t}$. The Argand diagram in Fig. 37–14 shows that a quantity that involves $e^{i\omega t}$, such as the voltage $\mathcal{E}_0 e^{i\omega t}$, corresponds to a complex number with constant magnitude that rotates in the counterclockwise direction with the constant angular speed ω. The projection of $\mathcal{E}_0 e^{i\omega t}$ onto the real axis, which gives the physical value, is just the originally specified voltage,‡ $\mathcal{E}_0 \cos \omega t$.

In practice we almost always measure the current in a circuit, not the charge on a capacitor. Accordingly, we convert Eq. 37–32 into an equation involving only the current by differentiating each term with respect to time. The result is

$$L\frac{d^2\widehat{I}}{dt^2} + R\frac{d\widehat{I}}{dt} + \frac{\widehat{I}}{C} = i\omega\mathcal{E}_0 e^{i\omega t} \tag{37–33}$$

We expect the response, that is, $\widehat{I}(t)$, to have the same time variation as the driving voltage.§ Therefore, we assume a steady-state solution of the form $\widehat{I}(t) = \widehat{I}_0 e^{i\omega t}$. Any possible phase difference between the current and the voltage is incorporated in the

---

*We can equally well take the projection onto the imaginary axis if we retain this convention throughout the calculation. (This is possible because the imaginary part β of a complex quantity, $\alpha + i\beta$, is also *real*.)

†A caret ⌢ over a symbol indicates a complex representation of the quantity.

‡Note the similarity between the present situation and that for the reference circle discussed in connection with simple harmonic motion (Section 15–3).

§Repeated differentiation of the time-dependent right-hand side of Eq. 37–33 yields a quantity always proportional to $e^{i\omega t}$.

complex constant $\widehat{I}_0$. Then, we obtain

$$\left(-\omega^2 L\widehat{I}_0 + i\omega R\widehat{I}_0 + \frac{\widehat{I}_0}{C}\right)e^{i\omega t} = i\omega\mathcal{E}_0 e^{i\omega t} \tag{37-34}$$

Cancelling the rotators $e^{i\omega t}$ and solving for $\widehat{I}_0$, we find

$$\widehat{I}_0 = \frac{\mathcal{E}_0}{R + i(\omega L - 1/\omega C)} = \frac{\mathcal{E}_0}{R + i(X_L - X_C)} \tag{37-35}$$

where

$$X_L = \omega L \tag{37-36a}$$

is the *inductive reactance,* and where

$$X_C = \frac{1}{\omega C} \tag{37-36b}$$

is the *capacitive reactance.*

   The fact that cancelling the factor $e^{i\omega t}$ in Eq. 37-34 leaves the expression for $\widehat{I}_0$ (Eq. 37-35) with no time dependence is consistent with the assumption that $\widehat{I}(t) = \widehat{I}_0 e^{i\omega t}$.

   Equation 37-35 can be written in a form that resembles Ohm's law if we introduce a quantity called the *complex impedance* $\widehat{Z}$, which in the present case is

$$\widehat{Z} = Ze^{i\delta} = R + iX \tag{37-37}$$

where X is the *total reactance,*

$$X = X_L - X_C = \omega L - \frac{1}{\omega C} \tag{37-38a}$$

and where

$$\delta = \tan^{-1}(X/R) \tag{37-38b}$$

The magnitude of $\widehat{Z}$ is (see Eq. B-19 in Appendix B)

$$|\widehat{Z}| = Z = \sqrt{R^2 + X^2} \tag{37-39}$$

Fig. 37-15. The complex impedance, $\widehat{Z} = R + i(X_L - X_C)$, for a series *LRC* circuit driven at an angular frequency ω.

$\widehat{Z} = Ze^{i\delta}$
$\widehat{Z} = |\widehat{Z}| = \sqrt{R^2 + X^2}$
$\delta = \tan^{-1}\dfrac{X_L - X_C}{R}$
$X_L = \omega L;\; X_C = \dfrac{1}{\omega C}$
$X = X_L - X_C$

   Figure 37-15 shows $\widehat{Z}$ and its components in an Argand diagram. Using the complex impedance in Eq. 37-35, we have the Ohm's law form

$$\widehat{I}_0 = \frac{\mathcal{E}_0}{\widehat{Z}} = \frac{\mathcal{E}_0}{Z}e^{-i\delta} = I_0 e^{-i\delta} \tag{37-40}$$

where

$$I_0 = \frac{\mathcal{E}_0}{Z} \tag{37-41}$$

Then, we also have

$$\widehat{I}(t) = \widehat{I}_0 e^{i\omega t} = I_0 e^{i(\omega t - \delta)} \tag{37-42}$$

Finally, the physical solution is

$$I(t) = \mathbf{Re}\,\widehat{I}(t) = I_0 \cos(\omega t - \delta)$$

or

$$I(t) = \frac{\mathcal{E}_0}{(R^2 + X^2)^{1/2}} \cos(\omega t - \delta) \qquad \textbf{(37-43)}$$

In this equation, δ is the phase difference between the driving voltage, $\mathcal{E}_0 \cos \omega t$, and the response current $I(t)$. This phase difference is introduced by the presence of the capacitive and inductive reactances; in a circuit that contains only purely resistive elements, the current is in phase with the driving voltage. The more common situation is that X_C and X_L are both nonzero; δ can still be zero, however, if both reactances have the same magnitude (Eq. 37–38a). This is discussed further in the section on resonance.

Let us rewrite Eq. 37–34 by dividing each term by $i\omega$; the result is

$$i\omega L\widehat{I} + R\widehat{I} + \frac{\widehat{I}}{i\omega C} = \widehat{\mathcal{E}}$$

We can use Eqs. B–9 in Appendix B to write $i = e^{i(\pi/2)}$ and $i^{-1} = e^{-i(\pi/2)}$. Then, we have

$$X_L\widehat{I}e^{i(\pi/2)} + R\widehat{I} + X_C\widehat{I}e^{-i(\pi/2)} = \widehat{\mathcal{E}} \qquad \textbf{(37-44)}$$

Now, define the complex voltage drops to be

$$\widehat{V}_L = X_L\widehat{I}e^{i(\pi/2)}, \qquad \widehat{V}_R = R\widehat{I}, \qquad \widehat{V}_C = X_C\widehat{I}e^{-i(\pi/2)} \qquad \textbf{(37-45)}$$

Hence, Eq. 37–44 becomes

$$\widehat{V}_L + \widehat{V}_R + \widehat{V}_C = \widehat{\mathcal{E}} \qquad \textbf{(37-46)}$$

Figure 37–16 shows Eqs. 37–44 and 37–46 represented in an Argand diagram. Here, we see graphically the various phase relationships among the applied EMF, the current, and the three voltage drops.

Power. The instantaneous power dissipation in a circuit is*

$$P(t) = \mathcal{E}(t)I(t) = [\mathbf{Re}\,\widehat{\mathcal{E}}(t)][\mathbf{Re}\,\widehat{I}(t)] \qquad \textbf{(37-47)}$$

Using $\widehat{\mathcal{E}}(t) = \mathcal{E}_0 e^{i\omega t}$ and $\widehat{I}(t) = I_0 e^{i(\omega t - \delta)}$, we have

$$P(t) = \mathcal{E}_0 I_0 \cos \omega t \cos(\omega t - \delta)$$

$$= \mathcal{E}_0 I_0 (\cos^2 \omega t \cos \delta + \cos \omega t \sin \omega t \sin \delta)$$

---

*The right-hand side of this equation is *not* the same as $\mathbf{Re}[\widehat{\mathcal{E}}(t)\widehat{I}(t)]$. (•Can you demonstrate the correctness of this statement?)

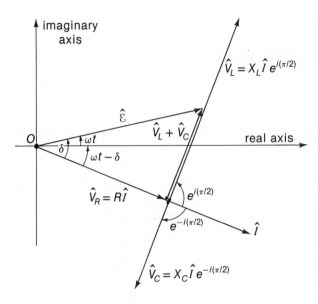

Fig. 37–16. The Argand diagram (complex representation) of Eqs. 37–44 and 37–46 for a series *LRC* circuit driven at an angular frequency ω. Notice that multiplying a complex quantity by $e^{i\theta}$ rotates that quantity counterclockwise through an angle θ in the Argand diagram. Thus, for example, $\hat{V}_L = X_L \hat{I} e^{i(\pi/2)}$ is directed perpendicular to $\hat{I}$, as shown.

With the identity $2 \cos\theta \sin\theta = \sin 2\theta$, the expression for $P(t)$ can be written as

$$P(t) = (\mathcal{E}_0 I_0 \cos \delta) \cos^2 \omega t + (\tfrac{1}{2}\mathcal{E}_0 I_0 \sin \delta) \sin 2\omega t \qquad \textbf{(37–48)}$$

The first term in this equation is always positive and represents the power dissipation in the resistive part of the circuit. The second term is oscillatory (with frequency 2ω) and represents the surging of energy back and forth between the electromagnetic fields associated with the capacitive and inductive parts of the circuit, and also the energy absorbed from and returned to the source.

The average power dissipation can be obtained by averaging Eq. 37–48 for $P(t)$ over one complete period, $2\pi/\omega$. The average of $\sin 2\omega t$ is zero, and the average of $\cos^2 \omega t$ is $\tfrac{1}{2}$. Therefore,

$$\overline{P} = \tfrac{1}{2}\mathcal{E}_0 I_0 \cos \delta \qquad \textbf{(37–49)}$$

The root-mean-square (rms) value calculated over a complete period for a sinusoidally varying quantity, such as $\xi(t) = \xi_0 \cos \omega t$ or $\xi_0 \sin \omega t$, is

$$\xi_{\mathrm{rms}} = \left[\frac{\omega}{2\pi} \int_0^{\omega/2\pi} \xi_0^2 \cos^2 \omega t \, dt \right]^{1/2} = \frac{\xi_0}{\sqrt{2}} \qquad \textbf{(37–50)}$$

In engineering practice it is customary to quote rms values for currents and voltages in AC circuits. Thus, we can express the power dissipation as

$$\overline{P} = \mathcal{E}_{\mathrm{rms}} I_{\mathrm{rms}} \cos \delta \qquad \textbf{(37–51)}$$

where $\mathcal{E}_{\mathrm{rms}} = \dfrac{\mathcal{E}_0}{\sqrt{2}}$ and $I_{\mathrm{rms}} = \dfrac{I_0}{\sqrt{2}}$ **(37–52)**

The peak voltage $\mathcal{E}_0$ for most household circuits is maintained at approximately 170 V, so that $\mathcal{E}_{rms} = (170 \text{ V})/\sqrt{2} = 120$ V. For this reason, we refer to "120-V" power lines.*

The term $\cos \delta$ in the expressions for $\overline{P}$ is referred to as the *power factor* for the circuit. For example, consider the case of a factory that is equipped with large pieces of rotating electrical machinery. Such machines represent very inductive loads; then, with X_L large compared with X_C and R, δ approaches $\pi/2$ and $\cos \delta$ becomes small. Then, the current I_0 that is necessary to operate the machines is much greater than would be the case for a power factor near unity. That is, the required current is large even though the average power remains modest. Because of the large value of I_0, the power loss in the transmission lines is increased. Under such conditions, the factory must correct the poor power factor or pay a surcharge to the power company to compensate for the line losses. One way to correct the power factor in this case would be to connect a low-loss capacitive load across the line.

Resonance. The maximum value of the current amplitude I_0 in a series *LRC* circuit occurs when $X = X_L - X_C = 0$. This is easy to see in the expression for $\widehat{I}_0$ (Eq. 37–35) or that for I (Eq. 37–43). This condition is called *resonance,* and the resonance frequency is designated by ω_0. At resonance, we have $X_L = \omega_0 L = 1/\omega_0 C = X_C$. That is,

$$\omega_0 = \frac{1}{\sqrt{LC}} \tag{37-53}$$

Resonance also yields $\delta = \tan^{-1}(X/R) = 0$ and $\cos \delta = 1$. Thus, the power factor is unity at resonance, so this is also the condition of maximum absorbed power.

We define a quantity called the *quality factor* (or, simply, the $\mathcal{Q}$) of the circuit according to

$$\mathcal{Q} = \frac{\omega_0 L}{R} = \frac{1}{R}\sqrt{\frac{L}{C}} \tag{37-54}$$

Then, we have

$$\tan \delta = \frac{X_L - X_C}{R} = \frac{\omega L - \dfrac{1}{\omega C}}{R} = \frac{\omega}{\omega_0}\mathcal{Q}\left(1 - \frac{\omega_0^2}{\omega^2}\right) \tag{37-55}$$

and also
$$I_0 = \frac{\mathcal{E}_0}{Z} = \frac{\mathcal{E}_0}{[R^2 + (X_L - X_C)^2]^{1/2}}$$

so that
$$I_0(\omega) = \frac{\mathcal{E}_0/R}{\left[1 + \dfrac{\omega^2}{\omega_0^2}\mathcal{Q}^2\left(1 - \dfrac{\omega_0^2}{\omega^2}\right)^2\right]^{1/2}} \tag{37-56}$$

It is apparent that the denominator of the expression in Eq. 37–56 has its minimum

---

*The line voltage is sometimes decreased by as much as 5 percent in peak-load situations to reduce power consumption (thereby producing a *brownout*). If the rms voltage is reduced below about 108 V, induction motors (of the type used in refrigerators and air conditioners) may be seriously damaged.

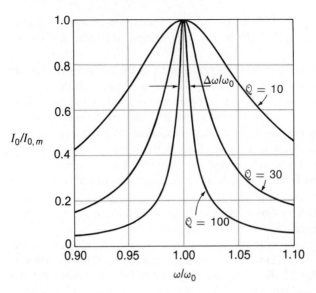

Fig. 37–17. The current amplitude for a series *LRC* circuit in the vicinity of the resonance frequency ω_0 for three different values of the Q of the circuit. The maximum current amplitude is $I_{0,m} = \mathcal{E}_0/R$. The full width at the one-half power points occurs for $I_0 = I_{0,m}/\sqrt{2}$ and is equal to $\Delta\omega/\omega_0 = 1/Q$.

value (namely, unity) when $\omega = \omega_0$. At this frequency, the current amplitude has its maximum value, $I_{0,m} = \mathcal{E}_0/R$. The resonance behavior of $I_0(\omega)$ is illustrated in Fig. 37–17 for three different values of Q. Although the current amplitude achieves its maximum value at $\omega = \omega_0$, the charge $Q_0(\omega) = I_0(\omega)/\omega$ achieves its maximum value at a slightly lower frequency (see Problem 37–9).

Figure 37–18 shows the behavior of the phase angle δ (Eq. 37–55) in the vicinity of the resonance frequency ω_0. Notice that below resonance ($\omega < \omega_0$) the capacitive reactance dominates, which produces $-\pi/2 < \delta < 0$, whereas above resonance ($\omega > \omega_0$) the inductive reactance dominates and $0 < \delta < \pi/2$.

In Fig. 37–15 we see that $\cos \delta = R/Z$. Therefore, the average power delivered to the resistor is (Eq. 37–49)

$$\bar{P}(\omega) = \tfrac{1}{2}\mathcal{E}_0 I_0 \cos \delta = \tfrac{1}{2}\mathcal{E}_0 \left(\frac{\mathcal{E}_0}{Z}\right)\left(\frac{R}{Z}\right)$$

$$= \frac{\mathcal{E}_0^2 R}{2Z^2} = \frac{\tfrac{1}{2}\mathcal{E}_0^2 R}{R^2 + (\omega L - 1/\omega C)^2} \tag{37–57}$$

Fig. 37–18. The phase angle δ in the vicinity of the resonance frequency ω_0 for the same three values of the Q of the circuit as used in Fig. 37–17. Notice that the one-half power points correspond to $\delta = \pm\pi/4$.

At resonance we have $Z = R$, so that

$$\overline{P}_m = \frac{\mathcal{E}_0^2}{2R} \qquad \text{(37–58)}$$

The two frequencies, ω_1 and ω_2, at which the power absorbed from the source is one half the maximum power, $\overline{P} = \frac{1}{2}\overline{P}_m$, are called the *one-half power points* and are such that (see Problem 37–8)

$$\omega_2 - \omega_1 = \Delta\omega = \frac{\omega_0}{Q} \qquad \text{(37–59)}$$

Thus, the larger the value of Q, the narrower is the resonance curve; this feature is evident in Fig. 37–17.

Example 37–7

The *LRC* circuit shown in the diagram has $L = 10\,\text{mH}$, $C = 1\,\mu\text{F}$, $R = 3.3\,\Omega$, and $\mathcal{E}_0 = 1\,\text{V}$.

(a) Determine the resonance frequency ω_0, the Q of the circuit, and the frequency interval between the one-half power points.

(b) Calculate the current amplitude at resonance, and determine the average power dissipated per period at resonance.

(c) Calculate the current amplitude and the average power dissipated per period at a frequency that is 5 percent lower than the resonance frequency.

Solution:

(a) The resonance frequency is

$$\omega_0 = \frac{1}{\sqrt{LC}} = \frac{1}{\sqrt{(10^{-2}\,\text{H})(10^{-6}\,\text{F})}} = 10^4\,\text{rads/s}$$

The Q of the circuit is

$$Q = \frac{\omega_0 L}{R} = \frac{(10^4\,\text{s}^{-1})(10^{-2}\,\text{H})}{3.3\,\Omega} = 30$$

The half-power interval is

$$\Delta\omega = \frac{\omega_0}{Q} = \frac{10^4\,\text{s}^{-1}}{30} = 333\,\text{s}^{-1}$$

This situation is represented by the middle curve in Fig. 37–17.

(b) At resonance, we have

$$I_{0,m} = \frac{\mathcal{E}_0}{R} = \frac{1 \text{ V}}{3.3 \text{ }\Omega} = 0.30 \text{ A}$$

$$\bar{P}_m = \frac{\mathcal{E}_0^2}{2R} = \frac{(1 \text{ V})^2}{2(3.3 \text{ }\Omega)} = 0.15 \text{ W}$$

(c) When $\omega = 0.95\omega_0 = 9500 \text{ s}^{-1}$, we have

$$X = X_L - X_C = \omega L - \frac{1}{\omega C}$$

$$= (9500 \text{ s}^{-1})(10^{-2} \text{ H}) - \frac{1}{(9500 \text{ s}^{-1})(10^{-6} \text{ F})}$$

$$= -10.3 \text{ }\Omega$$

Then, the impedance is

$$Z = \sqrt{R^2 + X^2} = \sqrt{(3.3 \text{ }\Omega)^2 + (-10.3 \text{ }\Omega)^2} = 10.8 \text{ }\Omega$$

Thus, the current amplitude is

$$I_0 = \frac{\mathcal{E}_0}{Z} = \frac{1 \text{ V}}{10.8 \text{ }\Omega} = 0.093 \text{ A}$$

which is approximately $\frac{1}{3}$ of the maximum current $I_{0,m}$.
 The phase angle is

$$\delta = \tan^{-1}\frac{X}{R} = \tan^{-1}\frac{(-10.3 \text{ }\Omega)}{(3.3 \text{ }\Omega)} = -1.26 \text{ rad} \qquad (-72.4°)$$

Then, the average power is

$$\bar{P} = \tfrac{1}{2}\mathcal{E}_0 I_0 \cos\delta = \tfrac{1}{2}(1 \text{ V})(0.093 \text{ A}) \cos 72.4° = 0.014 \text{ W}$$

which is about 10 percent of the maximum power $\bar{P}_m$.

Example 37–8

An electric generator operates at a frequency $\nu = 60$ Hz and develops a peak EMF of $120\sqrt{2}$ V. The generator has an internal resistance $r = 5 \text{ }\Omega$ and an effective series inductive reactance $X_L = 15 \text{ }\Omega$.

(a) What should be the load impedance for maximum power transfer to the load, which consists of resistive and capacitive elements in series, as shown in the diagram?

(b) Under the conditions determined in (a), what is the average power delivered to the load?

(c) What would be the power transfer if all capacitance were removed from the load?

Solution:

(a) For maximum power delivered to the entire circuit, we require a unity power factor, which corresponds to resonance for both the power and the current; this occurs for $X_C = X_L$. We also require $R = r$ for maximum power transfer to the load, as we found earlier for DC circuits (see Example 31–6). Thus,

$$X_C = X_L = 15\ \Omega$$

so that
$$C = \frac{1}{\omega X_C} = \frac{1}{2\pi(60\ \text{Hz})(15\ \Omega)} = 177\ \mu\text{F}$$

Also,
$$R = r = 5\ \Omega$$

In this condition, we have *impedance matching* between the generator and the load. The required capacitance (177 μF) is actually quite large. (Calculate the plate area required for a spacing of 1 mm and $\kappa = 3$.) This is a common problem that occurs in power transmission at the relatively low frequency of 60 Hz.

(b) The average power delivered to the entire circuit at resonance is

$$\overline{P}_m = \frac{\mathcal{E}_0^2}{2(r + R)} = \frac{(120\sqrt{2}\ \text{V})^2}{2(10\ \Omega)} = 1.44\ \text{kW}$$

This is the total power dissipated in the resistive circuit elements. Because $R = r$, one half of the power is delivered to the load; that is,

$$\overline{P}_m(\text{load}) = 0.72\ \text{kW}$$

(c) Removing the capacitance of the load but leaving the resistance, we have

$$X = X_L = 15\ \Omega, \qquad R + r = 10\ \Omega$$

Then, using Eq. 37–57, we have

$$\overline{P} = \frac{\mathcal{E}_0^2(R + r)}{2\,Z^2} = \frac{\mathcal{E}_0^2(R + r)}{2[(R + r)^2 + X_L^2]}$$

$$= \frac{(120\sqrt{2}\ \text{V})^2(10\ \Omega)}{2[(10\ \Omega)^2 + (15\ \Omega)^2]} = 0.44\ \text{kW}$$

and one half of this power is delivered to the load; thus,

$$\overline{P}(\text{load}) = 0.22\ \text{kW}$$

That is, the addition of the 177-μF capacitor to the purely resistive load increases the power delivered to the load by a factor of more than 3.

37-4   MAGNETICALLY COUPLED CIRCUITS

In some circuits there is magnetic coupling between a pair of inductors, as shown schematically in Fig. 37–19. Applying Kirchhoff's rule in terms of the two

Fig. 37–19. Magnetic coupling between inductors in an AC circuit.

loop currents, $\widehat{I}_1(t) = \widehat{I}_{01}e^{i\omega t}$ and $\widehat{I}_2(t) = \widehat{I}_{02}e^{i\omega t}$, results in a pair of coupled equations:

$$\left. \begin{array}{c} (R_1 + i\omega L_1)\widehat{I}_{01} + i\omega M\widehat{I}_{02} = \mathcal{E}_0 \\ i\omega M\widehat{I}_{01} + (R_2 + i\omega L_2)\widehat{I}_{02} = 0 \end{array} \right\} \tag{37–60}$$

where the rotator $e^{i\omega t}$ has been cancelled throughout. Standard techniques of complex algebra can be used to solve these coupled equations for the (complex) currents.

Example 37–9

The elements in the circuit shown in Fig. 37–19 have the values $R_1 = 2\,\Omega$, $R_2 = 5\,\Omega$, $\omega L_1 = 3\,\Omega$, and $\omega L_2 = 4\,\Omega$. The reactance associated with the mutual inductance (maximum coupling—see Problem 36–11) is $\omega M = \omega \sqrt{L_1 L_2} = \sqrt{3 \cdot 4}\,\Omega = 3.46\,\Omega$. The amplitude of the EMF is $\mathcal{E}_0 = 120\sqrt{2}$ V. Determine $I_1(t)$ and $I_2(t)$.

Solution:

The coupled equations (Eqs. 37–60) are

$$\left. \begin{array}{c} (2 + i3)\widehat{I}_{01} + i3.46\widehat{I}_{02} = 120\sqrt{2} \\ i3.46\widehat{I}_{01} + (5 + i4)\widehat{I}_{02} = 0 \end{array} \right\}$$

Solving for $\widehat{I}_{01}$ and $\widehat{I}_{02}$, we find

$$\widehat{I}_{01} = \frac{120\sqrt{2}(5 + i4)}{(2 + i3)(5 + i4) + 12} = (42.5\ \text{A})e^{-i0.411}$$

$$\widehat{I}_{02} = \frac{-i120\sqrt{2}(3.46)\cdot}{(2 + i3)(5 + i4) + 12} = (23.4\ \text{A})e^{i3.551}$$

We can also express the voltage and the currents as

$$\mathcal{E}(t) = (120\sqrt{2}\ \text{V})\cos \omega t$$
$$I_1(t) = (42.5\ \text{A})\cos(\omega t - 0.411)$$
$$I_2(t) = (23.4\ \text{A})\cos(\omega t + 3.551)$$

Transformers. An important application of magnetically coupled circuits is in the common device called a *transformer*. The purpose of a transformer is to increase or decrease the input or AC line voltage to the value required by the circuit receiving power. If the voltage supplied by the transformer is greater than the input voltage, the device is called a *step-up* transformer. Conversely, if the voltage is lowered, the device is a *step-down* transformer.

Figure 37–20 shows a schematic diagram of a step-up transformer. Two coils of wire are wound on an iron core. The *primary* coil (which is connected to an AC generator or power line) consists of N_1 turns; the *secondary* coil (which can be connected to the load R) consists of N_2 turns ($N_2 > N_1$).

Fig. 37–20. Schematic of an iron-core transformer. The voltage $V_1(t)$ across the primary is just equal to the output voltage $\mathcal{E}(t)$ of the generator.

We begin with the switch S open, so that no current flows in the secondary circuit. Moreover, we assume ideal conditions in which the primary circuit is without resistance or capacitance, so that the reactance is due entirely to the inductance of the primary coil. According to Eq. 37–38, we have $\delta = \tan^{-1}(X/R)$; with $R = 0$, we see that $\delta = \pi/2$. This means that the voltage $V_1(t)$ across the primary coil and the current $I_1(t)$ in the primary are 90° out of phase. Then, the power factor is $\cos \delta = \cos(\pi/2) = 0$; that is, there is no power expenditure when the secondary is an open circuit.

The (small) magnetizing current that flows in the primary coil induces a substantial magnetic flux φ_M in the iron core. We assume that there is no flux leakage from the core, so that the entire flux φ_M links the secondary coil. According to Faraday's law (Eq. 35–4), the EMF per turn is the same for the primary winding as for the secondary winding. Thus, the voltages across the primary and secondary coils are related by

$$\frac{V_1(t)}{N_1} = \frac{V_2(t)}{N_2} \quad \text{or} \quad \frac{V_{1,\text{rms}}}{N_1} = \frac{V_{2,\text{rms}}}{N_2} \qquad \textbf{(37–61)}$$

Now, suppose that the switch S is closed, thereby connecting the resistive load R to the secondary coil. A current $I_2(t)$ flows in the secondary circuit and power is delivered to the load (see Eq. 37–51):

$$\bar{P}_2 = V_{2,\text{rms}} I_{2,\text{rms}} \cos \delta$$

Because the flux in the iron core is large, the inductance of each coil is also large (see Eq. 36–6). Then, in general, we have $X_L = \omega L_2 \gg R$. This means that the phase angle in the secondary circuit is essentially zero and that the power factor, $\cos \delta$, is nearly unity.

The time-varying secondary current, $I_2(t)$, induces a back EMF that opposes the primary voltage, $V_1(t)$. However, because conditions are ideal, with no resistance or capacitance in the primary circuit, the voltage $V_1(t)$ must always equal the applied EMF, $\mathcal{E}(t)$. Thus, a new current, $I_1'(t)$, must flow in the primary circuit with the proper magnitude and phase to cancel the back EMF generated in the primary coil by the secondary current $I_2(t)$. The current $I_1(t)$ in the primary circuit now consists of the magnetizing current (small in magnitude and out of phase with V_1) and the induced current $I_1'(t)$ (larger in magnitude and in phase with V_1). That is, $I_1(t) \cong I_1'(t)$, and the power factor is essentially unity. (•Can you sketch the phasor diagram that shows $I_1'(t_1)$, $I_1(t_1)$, and the magnetizing current, also at $t = t_1$? Can you place $I_2(t_1)$ on this diagram?)

Energy conservation requires the generator to deliver power equal to that dissipated in the load R. Thus, we have, with $\delta = 0$ in both circuits,

$$\bar{P}_1 = V_{1,\text{rms}}I_{1,\text{rms}} = \mathcal{E}_{\text{rms}}I_{1,\text{rms}}$$
$$= V_{2,\text{rms}}I_{2,\text{rms}} = \bar{P}_2 \qquad \textbf{(37–62)}$$

Using Eq. 37–61, we can write

$$V_{2,\text{rms}} = \frac{N_2}{N_1}\mathcal{E}_{\text{rms}} \qquad \textbf{(37–63a)}$$

$$I_{2,\text{rms}} = \frac{N_1}{N_2}I_{1,\text{rms}} \qquad \textbf{(37–63b)}$$

SPECIAL TOPIC

37-5 SIMPLE AC NETWORKS

In the series LRC circuit discussed in Section 37–3, a separate complex impedance may be assigned to each of the circuit elements. Thus, we can write

$$\hat{Z}_R = R, \qquad \hat{Z}_L = i\omega L, \qquad \hat{Z}_C = -i/\omega C \qquad \textbf{(37–64a)}$$

Then, from the definition of impedance (Eq. 37–37), we have

$$Z = Z_R + Z_L + Z_C \qquad \textbf{(37–64b)}$$

We may, in fact, extend this view to any number of series impedances; thus,

$$\hat{Z} = \hat{Z}_1 + \hat{Z}_2 + \hat{Z}_3 + \cdots, \qquad \text{series} \qquad \textbf{(37–65)}$$
$$\text{connection}$$

The idea of complex impedance can also be used to simplify the analysis of situations that involve networks more complicated than a simple series connection. We assert without proof that *any* combination of individual impedances can be reduced to a single equivalent impedance using exactly the same rules that we applied in reducing DC resistance networks, *provided* that the different imped-

ances are treated throughout as complex quantities.* The result of such a reduction will, in general, depend on the frequency. Thus, for a number of impedances connected in parallel, we have†

$$\frac{1}{\hat{Z}} = \frac{1}{\hat{Z}_1} + \frac{1}{\hat{Z}_2} + \frac{1}{\hat{Z}_3} + \cdots, \qquad \text{parallel} \qquad \textbf{(37–66)}$$
$$\text{connection}$$

Parallel Resonance. Consider the parallel circuit arrangement shown in Fig. 37–21. The impedances of the two branches are

$$\hat{Z}_1 = -i/\omega C \qquad \text{and} \qquad \hat{Z}_2 = R + i\omega L$$

Then, after some complex-number algebra, we find (see Problem 37–14)

$$\hat{Z} = \frac{\hat{Z}_1 \hat{Z}_2}{\hat{Z}_1 + \hat{Z}_2} = \frac{R + i(\omega L - \omega R^2 C - \omega^3 L^2 C)}{\omega^2 C^2 [R^2 + (\omega L - 1/\omega C)^2]} \qquad \textbf{(37–67)}$$

When $\omega L = 1/\omega C$, the denominator of the expression

---

*If there exists any magnetic coupling between inductors, the mutual inductance effects must be included.
†The quantity $1/\hat{Z} \equiv \hat{Y}$ is called the *admittance*. Evidently, for a parallel connection, we have $\hat{Y} = \hat{Y}_1 + \hat{Y}_2 + \hat{Y}_3 + \cdots$.

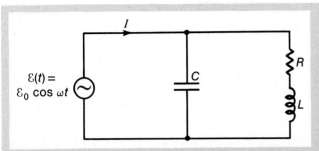

Fig. 37-21. A parallel *LRC* circuit is driven by a sinusoidal voltage.

for $\widehat{Z}$ becomes minimum, indicating the presence of a resonance. In fact, for parallel circuits, we can define several types of resonances—maximum impedance, maximum power dissipation, and unity power factor—all of which

occur at different frequencies. This situation is in contrast with that for series circuits, in which the resonance frequency is the same for minimum impedance, maximum power, and unity power factor.

For unity power factor, the imaginary part of $\widehat{Z}$ must vanish; this occurs for the frequency ω_R, which is obtained from

$$L^2 C \omega_R^2 = L - R^2 C$$

so that

$$\omega_R = \omega_0 \sqrt{1 - \mathbb{Q}^{-2}} \qquad (37\text{-}68)$$

where $\omega_0 = 1/\sqrt{LC}$ and $\mathbb{Q} = \omega_0 L/R$, as usual. If $\mathbb{Q}$ is large, we have $\omega_R \cong \omega_0$. In fact, for large $\mathbb{Q}$ all resonance frequencies are approximately equal to ω_0.

SPECIAL TOPIC

37-6 ELECTRO-DYNAMICAL ANALOGS

Figure 37–22a shows again the driven series *LRC* circuit discussed in Section 37–3. The differential equation for the capacitor charge Q is

$$L\frac{d^2Q}{dt^2} + R\frac{dQ}{dt} + \frac{Q}{C} = \mathcal{E}(t) \qquad (37\text{-}69a)$$

where we have used

$$I = \frac{dQ}{dt} \qquad (37\text{-}69b)$$

Fig. 37-22. (a) The basic driven series *LRC* circuit. (b, c, d) Mechanical analogs of the *LRC* circuit.

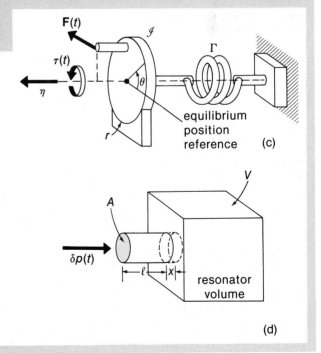

A number of important mechanical and electromechanical systems are described by differential equations that are analogous to the equation for the driven series *LRC* circuit.* It is therefore often convenient to analyze such systems in terms of the electrical equivalents.

Rectilinear Mechanical System. Figure 37–22b shows a mechanical system that consists of a block with mass m and a spring with force constant κ; the frictional resistance coefficient is b. The driving force $F(t)$ produces a time-dependent, straight-line displacement from equilibrium $x(t)$. Such a rectilinear system was discussed in Section 15–7, where we showed that the equation of motion is

$$m\frac{d^2x}{dt^2} + b\frac{dx}{dt} + \kappa x = F(t) \qquad \textbf{(37–70a)}$$

where the velocity is

$$v = \frac{dx}{dt} \qquad \textbf{(37–70b)}$$

Comparison of Eqs. 37–69 and 37–70 suggests that we can identify analogous quantities as in Table 37–1.

You should examine the correspondences between the results for mechanical systems given in Section 15–7 and those obtained for the *LRC* circuit in Section 37–3.

Rotational Mechanical System. Figure 37–22c shows a simple rotational mechanical system that has a rotational inertia $\mathscr{I}$ about the system axis, a frictional resistance coefficient r, and a torque constant Γ ($1/\Gamma$ is the *rotational compliance*). The driving torque $\tau(t)$ produces a time-dependent angular displacement from equilibrium $\theta(t)$. The equation of motion is

$$I\frac{d^2\theta}{dt^2} + r\frac{d\theta}{dt} + \Gamma\theta = \tau(t) \qquad \textbf{(37–71a)}$$

---

*The description of a mechanical system by Eq. 37–69a is expected to be good if the amplitude of oscillation is not too large. Any mechanical system that is stressed too severely will deviate from the conditions of a harmonic oscillator and will exhibit nonlinear effects. Because we assume sinusoidal driving voltages and resistive elements that obey Ohm's law, these electric circuits are always harmonic and linear.

where the angular speed is

$$\eta = \frac{d\theta}{dt} \qquad \textbf{(37–71b)}$$

(•Can you construct a table similar to Table 37–1 for this case?)

Acoustic System. Figure 37–22d shows a practical acoustic system in the form of a damped Helmholtz resonator (refer to Example 20–3). A time-dependent driving pressure $\delta p(t)$, measured with respect to the ambient pressure p_0, is applied to an open tube with length ℓ and cross-sectional area A that is connected in the manner shown to an otherwise closed volume V.* At the instant for which Fig. 37–22 applies, a column of air with length x that was originally in the tube has intruded into the resonator. The resulting compression of the air in the enclosed volume produces a back pressure that generates a force characterized by an effective acoustic spring constant κ_a. The moving acoustic mass m_a is the mass of the air in the tube. Altogether, we can write the equation of motion as†

$$m_a\frac{d^2x}{dt^2} + b_a\frac{dx}{dt} + \kappa_a x = A\delta p(t) \qquad \textbf{(37–72)}$$

where $\quad m_a = \rho_0 \ell A$

$\qquad \kappa_a = \dfrac{u^2\rho_0 A^2}{V} \quad$ (see Example 20–3)

in which we have used ρ_0 for the density of air and u for the speed of sound in air. We have also introduced into the equation of motion a frictional resistance term, $b_a\, dx/dt$, which results from the turbulence created when the air in the tube intrudes into the resonator volume.

---

*We assume that the characteristic time τ associated with the acoustic vibrations of the system is such that $u\tau \gg \ell$ or $V^{1/3}$, where u is the speed of sound in air. If this condition is not met, sound wave propagation effects must be included, thereby complicating the analysis to a considerable extent.

†Acoustic engineers replace the displacement x in this equation with the *moving air volume displacement* $\xi = Ax$. The coefficients are then also altered and have dimensions different from those in Eq. 37–72.

TABLE 37–1.   Comparison of Mechanical and Electrical Quantities

MECHANICAL QUANTITY	SYMBOL	ELECTRICAL QUANTITY	SYMBOL
mass	m	inductance	L
mechanical resistance	b	electric resistance	R
mechanical compliance (reciprocal spring constant)	$1/\kappa$	capacitance	C
displacement	x	charge	Q
velocity	v	current	I
driving force	F	driving voltage	$\mathcal{E}$

The Loudspeaker. In many types of systems a mechanical element is set into motion by passing a current through a coil that is attached to the element and is also in a magnetic field. Devices of this type include sensitive relays, solenoid-operated switches, actuators, locks, and so forth, and sound-reproducing equipment.

As an example of an electromechanical system, consider the operation of a loudspeaker, such as the one shown schematically in Fig. 37–23. A permanent magnet produces a radial magnetic field in which there is positioned a short solenoidal voice coil that is attached to a stiff conical piston. The outer rim of the piston is secured to a rigid frame by a surrounding flexible element. This suspension centers the piston and attached voice coil, yet it permits relatively free axial excursions when varying currents supplied by a suitable amplifier are passed through the coil. A speaker with a moving voice coil is called a *dynamic loudspeaker*.

The voice coil has N turns and a radius a, and it is located in a constant magnetic field B. We imagine that a sinusoidal current $\widehat{I}$ passes through the voice coil. Then, the axial force on the coil is

$$\widehat{F} = 2\pi a N B \widehat{I}$$

The resulting axial motion of the coil and piston may be described by the (complex) velocity $\widehat{v}$. The back EMF produced in the voice coil is

$$\widehat{\mathcal{E}} = 2\pi a N B \widehat{v}$$

Fig. 37–23. Schematic diagram of a loudspeaker.

Combining the expressions for $\widehat{F}$ and $\widehat{\mathcal{E}}$ yields

$$\frac{\widehat{\mathcal{E}}}{\widehat{I}} = \frac{D^2}{\widehat{F}/\widehat{v}}$$

where the electromechanical coupling constant is

$$D^2 = (2\pi a N B)^2 \qquad \textbf{(37–73)}$$

We define the motional electric impedance in the usual way to be $\widehat{Z}_{me} = \widehat{\mathcal{E}}/\widehat{I}$ and the motional mechanical impedance to be $\widehat{Z}_{mm} = \widehat{F}/\widehat{v}$. Then,

$$\widehat{Z}_{me} = \frac{D^2}{\widehat{Z}_{mm}} \qquad \textbf{(37–74)}$$

Example 37–10

Describe the behavior of a 12-in. (approx. 30-cm) low-frequency loudspeaker (a *woofer*) for frequencies near its natural resonance frequency when it is mounted in an enclosure with a very large volume (an *infinite baffle*).

Solution:

A typical loudspeaker would have characteristics similar to the following:

Voice coil: mass $m_c = 5$ g; radius $a = 1.5$ cm; $N = 120$ turns

Piston: mass $m_p = 5$ g; radius $r = 15$ cm

Magnetic field: $B = 8500$ G at position of coil

Flexible surround: effective spring constant $\kappa = 10^3$ N/m; flexing (frictional) resistance $b_m = 1.50$ kg/s

In the low-frequency range near the speaker resonance, the acoustic loading of the piston is described by an acoustic mass m_a that is nearly independent of frequency. This effective mass corresponds roughly to the mass of air in a cylinder with the radius r of the piston and with an axial length $8r/3\pi$ (see Example 20–3). Thus, $m_a \approx (8r/3\pi) \cdot (\pi r^2 \rho_0) = (8/3)(15 \text{ cm})^3(0.0012 \text{ g/cm}^3) = 10.8$ g. The sound energy that is radiated results in an effective resistive loading b_a that depends strongly on frequency; this resistance combines with that due to the surround, giving a resultant loading $b_m + b_a$.

The mechanical impedance of the loudspeaker is

$$\widehat{Z}_m = (b_m + b_a) + i[\omega(m_c + m_p + m_a) - \kappa/\omega]$$

with
$$b_m + b_a = 1.50 \text{ kg/s} + b_a$$
$$m_c + m_p + m_a = 5 \text{ g} + 5 \text{ g} + 10.8 \text{ g}$$
$$= 20.8 \text{ g} = 0.0208 \text{ kg}$$
$$\kappa = 10^3 \text{ N/m}$$

The loudspeaker mechanical system is therefore a damped series oscillator whose resonance frequency is

$$\omega_0 = \sqrt{\frac{\kappa}{m_c + m_p + m_a}} = \sqrt{\frac{10^3 \text{ N/m}}{0.0208 \text{ kg}}} = 219 \text{ s}^{-1}$$

or

$$\nu_0 = \frac{\omega_0}{2\pi} = \frac{219 \text{ s}^{-1}}{2\pi} = 34.9 \text{ Hz}$$

A piston with a diameter of 30 cm oscillating at a frequency of 35 Hz would radiate acoustic energy at a rate corresponding to an effective resistance $b_a \approx 0.13$ kg/s. The mechanical impedance at resonance is the purely resistive combination, $b_m + b_a \approx 1.63$ kg/s. Incidentally, the mechanical efficiency e of the loudspeaker at resonance is*

$$e = \frac{1}{2} \cdot \frac{b_a}{b_m + b_a} = \frac{1}{2} \cdot \frac{0.13 \text{ kg/s}}{1.63 \text{ kg/s}} = 0.040 = 4.0 \text{ percent}$$

The electromechanical coupling constant is given by Eq. 37–73:

$$D^2 = (2\pi aNB)^2 = [2\pi(0.015 \text{ m})(120)(0.85 \text{ T})]^2$$
$$= 92.4 \text{ N} \cdot \text{V} \cdot \text{s/A} \cdot \text{m}$$

Equation 37–74 shows that $\widehat{Z}_{\text{me}}$ and $\widehat{Z}_{\text{mm}}$ are reciprocally related. Thus, when the *mechanical* impedance Z_{mm} is *minimum* at the resonance frequency ω_0, the *electric* impedance Z_{me} is *maximum*. The mechanical system is equivalent to a *series LRC* circuit, whereas the electrical system is equivalent to a *parallel LRC* circuit. In the present case, the electric impedance at resonance is purely resistive:

$$Z_{\text{me}} = \frac{D^2}{Z_{\text{mm}}} = \frac{92.4}{1.63} = 56.7 \text{ }\Omega$$

The total electric input impedance to the voice coil consists of $\widehat{Z}_{\text{me}}$ plus the *blocked impedance* $\widehat{Z}_{0\text{e}}$ of the coil. This latter impedance is measured at the voice-coil terminals when the coil is blocked and prevented from moving

so that the motional impedance is not developed. The blocked impedance consists of the coil's DC resistance R_0 and its inductive reactance ωL_0; that is,

$$\widehat{Z} = \widehat{Z}_{\text{me}} + \widehat{Z}_{0\text{e}} = \widehat{Z}_{\text{me}} + R_0 + i\omega L_0$$

A typical speaker might have $R_0 = 4.0 \text{ }\Omega$ and $L_0 = 0.40$ mH. Then, at resonance, we find

$$\widehat{Z} = (56.7 \text{ }\Omega) + (4.0 \text{ }\Omega) + i(219 \text{ s}^{-1})(0.40 \times 10^{-3} \text{ H})$$
$$= (60.7 + i0.088) \text{ }\Omega$$

We see in this calculation a general feature of dynamic loudspeakers, namely, that the inductive part of the impedance is negligible at low frequencies, although it has undesirable effects in high-frequency sound drivers. The diagram shows Z as a function of frequency. At resonance, the impedance is 60.7 Ω; then, the impedance falls to the DC resistance value of 4 Ω before rising at the higher frequencies due to the inductive component.

(•At resonance, what fraction of the input electric power is transformed into mechanical power?)

*The dissipative damping for an infinite baffle includes the effects of the sound energy radiated into both the forward and the backward hemispheres. Only the forward radiation represents useful sound energy; hence, the factor ½ in the efficiency equation.

PROBLEMS

Section 37-1

37-1 A 6.0-V battery with negligible internal resistance is used to charge a 0.5-μF capacitor that is in series with a 100-Ω resistor. (a) How much energy is eventually stored in the capacitor? (b) How much energy was expended in heating the resistor? (c) How much energy was supplied by the battery? (d) By what factor(s) would the above results be altered if R were halved to 50 Ω?

37-2 A capacitor C is initially charged to a value $Q_0 = \mathcal{E}C$ with the polarity indicated in the diagram. Obtain an expression for the charge $Q(t)$ as a function of time t after the switch S is closed. Determine the instant at which $Q(t) = 0$. [*Hint*: Be careful with sign conventions.]

37-3 A capacitor C is charged in the circuit shown in Fig. 37-2. (a) After how many time constants will the energy stored in the capacitor equal 80 percent of the final value? (b) At the instant found in (a), what is the current expressed as a fraction of the initial current?

Section 37-2

37-4 When an inductor L is energized in the circuit shown in Fig. 37-8, the current reaches 90 percent of its asymptotic value 3.0 ms after the switch is closed. If $R = 10 \, \Omega$, what is the inductance L?

37-5 In the energizing circuit of Fig. 37-8, we have $L = 350$ mH, $R = 5 \, \Omega$, and $\mathcal{E} = 6.0$ V. Determine the following quantities 0.1 s after the switch is closed: (a) the rate at which energy is being stored in the magnetic field of the inductor; (b) the instantaneous power delivered to the resistor; and (c) the instantaneous rate at which energy is being expended by the battery. (d) Comment on the energy balance in the system at the instant in question.

37-6 To avoid any possible damage to the current-carrying coil of an electromagnet if the circuit is broken, the usual practice is to place a diode (often in the form of a neon bulb) in parallel with the coil, as indicated in the diagram. If the diode is connected as shown, it draws very little current when the switch S is closed because the back resistance R is very large. Now, if the circuit is broken (or the switch is opened), the magnet current is diminished slowly as it passes through the forward resistance r of the diode. (•What would happen if the diode were not present?) Suppose that $\mathcal{E} = 12$ V, $R_0 = 2 \, \Omega$, and $L = 5.0$ H. What should be the value of r if the voltage across the magnet is not to exceed 1500 V? At what rate must the diode be able to dissipate heat?

Section 37-3

37-7 Consider a series LRC circuit. Use the fact that $d\widehat{Q}/dt = \widehat{I}(t) = I_0 e^{i(\omega t - \delta)}$ to show that

$$\widehat{Q}(t) = \frac{\mathcal{E}_0}{i\omega Z} e^{i(\omega t - \delta)} = \frac{\mathcal{E}_0}{\omega Z} e^{i(\omega t - \delta - \pi/2)}$$

Then, show that the physical charge is

$$Q(t) = \frac{\mathcal{E}_0}{\omega Z} \sin(\omega t - \delta)$$

37-8 Use Eq. 37-57 to show that the frequencies ω_1 and ω_2 for which $\bar{P}(\omega) = \frac{1}{2}\bar{P}_m$ are given by

$$\omega_{1,2} = \pm \frac{R}{2L} + \sqrt{\frac{R^2}{4L^2} + \frac{1}{LC}}$$

and, hence, that

$$\Delta\omega = \omega_2 - \omega_1 = \frac{R}{L} = \frac{\omega_0}{Q}$$

37-9 Express the complex charge $\widehat{Q}(t)$ and the physical
• charge $Q(t)$ as in Problem 37-7. Show that the charge amplitude $Q_0 = \mathcal{E}_0/\omega Z$ is maximum at the frequency $\omega_Q = \sqrt{\omega_0^2 - R^2/2L^2}$. Verify that

$\omega_Q \cong \omega_0$ when the Q of the circuit is large. [*Hint:* Set $dQ_0^2/d\omega$ equal to zero.]

37–10 A 30-Ω resistor, an inductor with a reactance of 40 Ω at 60 Hz, and a capacitor with a reactance of 50 Ω at 60 Hz are connected in series to a 120-V, 60-Hz power line (peak voltage $\mathcal{E}_0 = 120\sqrt{2}$ V). (a) What is the impedance of the circuit? (b) What is the peak current in the circuit? (c) What is the phase difference between the current and the applied voltage? (d) What average power per cycle is absorbed by the circuit?

37–11 A piece of rotating machinery is connected to a
 • 240-V, 60-Hz power line (peak voltage $\mathcal{E}_0 = 240\sqrt{2}$ V). The load impedance of the unit is $(5 + i8)\,\Omega$. What value of capacitance must be placed in parallel with the machine to produce unity power factor? What amount of power is absorbed by the machine? If the machine delivers shaft power with an efficiency of 82 percent, what is the output mechanical power (in horsepower)?

Section 37–4

37–12 A transformer has $N_1 = 350$ turns and $N_2 = 2000$ turns. If the input voltage is $\mathcal{E}(t) = (170\text{ V}) \cos \omega t$,

what rms voltage is developed across the secondary coil?

37–13 The generator in the diagram has an internal resist-
 • ance $r = 0.5\,\Omega$ and produces a voltage $\mathcal{E}(t) = (170\text{ V}) \cos \omega t$, with $\omega/2\pi = 60$ Hz. (a) If a load resistor, $R = 20\,\Omega$, is connected to the generator terminals, what average power will be delivered to the load? (b) In Example 37–8 we found that maximum power is delivered to a load when the load impedance (here, the resistance R) is equal to the generator impedance (here, the resistance r). In the present case, we have $R \neq r$, but the load and generator impedances can be matched by connecting a transformer between the terminals AA' and BB'. What must be the turns ratio, N_1/N_2, of the (ideal) transformer that will maximize the transfer of power to the load R?

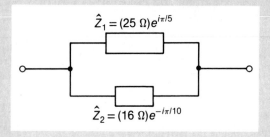

SPECIAL TOPIC

Section 37–5

37–14 Derive Eq. 37–67.

37–15 The impedance of a series circuit is $\hat{Z}_1 =$
 • $(25\,\Omega)e^{-i\pi/4}$ at $\omega_1 = 1000\text{ s}^{-1}$ and is $\hat{Z}_2 = (35.4\,\Omega)e^{i\pi/3}$ at $\omega_2 = 2000\text{ s}^{-1}$. Determine L, R, and C for the circuit. What is the resonance frequency ω_0? What is the Q of the circuit?

37–16 A 50-Ω resistor, an 80-mH inductor, and a 30-μF
 • capacitor are connected in series to a 120-V, 60-Hz power line (peak voltage $\mathcal{E}_0 = 120\sqrt{2}$ V). (a) Determine the individual reactances of the circuit elements and the total impedance. (b) What capacitance must be placed in parallel with the inductor to achieve series resonance for the circuit? (c) What is the peak current in the circuit with and without the added capacitance?

37–17 Determine the impedance at 60 Hz of the network shown in the diagram, above right.

37–18 Determine the impedance at 60 Hz of the network shown in the diagram below.

Section 37-6

37-19 A loudspeaker voice coil has a mass m_c, a density δ,
• a resistivity ρ, and a DC resistance R_0. Show that the
electromechanical coupling constant D^2 (Eq. 37-73)
for this coil in a magnetic field B is $D^2 =
B^2 R_0 m_c / \rho \delta$. Thus, D^2 does not depend on the cross-
sectional area of the wire or on the number of turns
in the coil. For the same resistance R_0 and the same
mass m_c, what conductor material would give the
best coupling (i.e., the *largest* value of D^2)? (Try
copper, aluminum, and silver.)

37-20 Estimate the Q value for the loudspeaker discussed
in Example 37-10. (Neglect any variation of b_a with
frequency.) What is the approximate half-power
frequency width $\Delta \nu$ for the electrical input impedance?

37-21 Suppose that the loudspeaker discussed in Example
• 37-10 is driven at a frequency $\nu = 200$ Hz. The
30-cm piston has an effective radiation resistance
$b_a = 4.43$ kg/s at 200 Hz; the value of m_a is essentially the same as at the resonance frequency
(35 Hz). Calculate (a) $\hat{Z}_{mm}$, (b) $\hat{Z}_{me}$, and
(c) $\hat{Z} = \hat{Z}_{me} + \hat{Z}_{0e}$.

Additional Problems

37-22 Verify by direct substitution that Eq. 37-3 for $Q(t)$
satisfies the differential equation, Eq. 37-1.

37-23 Use the fact that $e^{i(A+B)} = e^{iA}e^{iB}$ to show that

$$\cos(A + B) = \cos A \cos B - \sin A \sin B$$

and

$$\sin(A + B) = \sin A \cos B + \cos A \sin B$$

37-24 The parallel-plate capacitor shown in the diagram
• has a dielectric with constant κ and a capacitance
κC. At $t = 0$, the dielectric is suddenly removed.
(a) How much work was required to remove the
dielectric? (b) Obtain expressions for the charge
$Q(t)$ on the capacitor and for the current $I(t)$ in the
circuit. Make sketches of $Q(t)$ and $I(t)$. (c) How
much energy was dissipated as heat in the resistor?
(d) How much work was done on the battery?
(e) Account for the total energy balance.

37-25 The differentiating circuit shown in Fig. 37-6 is
• also referred to as a *high-pass filter*. To appreciate
this terminology, assume that $\mathcal{E}(t) = \mathcal{E}_0 \cos \omega t$ and
calculate the peak voltage ratio, $\xi(\omega) = V_0/\mathcal{E}_0$, for
values of ω equal to $\frac{1}{16}, \frac{1}{8}, \frac{1}{4}, \frac{1}{2}, 1, 2, 4,$ and 8 times
$\omega' = 1/RC$. Make a log-log plot of your results.
Verify that for frequencies $\omega \lesssim \omega'$ the ratio $\xi(\omega)$
decreases and asymptotically approaches a fall-off

rate of 6 dB per octave. (The dB measure of a voltage ratio is dB $= 20 \log_{10} \xi$; an *octave* means a frequency ratio of 2.)

37-26 The diagram shows an *integrating circuit* in which
the time constant, $\tau = L/R$, is large compared with
the time interval during which $\mathcal{E}(t)$ exhibits any substantial change. (a) Show that

$$V(t) = IR \cong \frac{R}{L} \int \mathcal{E}(t)\, dt$$

(b) Sketch $V(t)$ for the case in which $\mathcal{E}(t)$ is a
square-wave voltage (see Fig. 37-7).

37-27 The circuit in the diagram on the next page
•• represents a simple yet very effective *low-pass* filter.
Assume that the input voltage is $\mathcal{E}(t) = \mathcal{E}_0 \cos \omega t$.
Show that the peak voltage ratio is

$$\xi(\omega) = \frac{V_0}{\mathcal{E}_0} = \frac{1}{\left| 1 - \dfrac{\omega^2}{\omega_0^2} + i\dfrac{\omega}{\omega_0} Q \right|}$$

where, as usual, $\omega_0 = 1/\sqrt{LC}$ and $Q = \omega_0 L/R$. Assume $Q = 1$. Make a log-log graph of $\xi(\omega)$ as a func-
tion of ω/ω_0 in the range $0.1 \leqslant \omega/\omega_0 \leqslant 10$. The dB
measure of the voltage ratio is defined to be
dB $= 20 \log_{10} \xi$. Show that the slight peaking of ξ
for ω/ω_0 slightly less than 1 represents a rise of less
than 3 dB. Show also that for ω/ω_0 greater than

about 1.5, ξ decreases at a rate of 12 dB per octave. (An *octave* means a frequency ratio of 2.)

37–28 A resistor draws a peak current of 1.0 A when connected to a 120-V, 60-Hz power line. (The peak AC voltage is $\mathcal{E}_0 = 120\sqrt{2}$ V.) (a) What capacitance C must be placed in series with the resistor to reduce the peak current to 0.5 A? (b) Under the conditions determined in (a), what is the phase difference between the current and the applied voltage?

37–29 A small AC motor has an inductance of 40 mH and an effective series resistance of 25 Ω. The motor is connected to a 120-V, 60-Hz power line (peak voltage $\mathcal{E}_0 = 120\sqrt{2}$ V). (a) What is the inductive reactance X_L of the motor? (b) Determine the peak current I_0 and the phase angle δ_L between the current and the applied voltage.

37–30 A peak current of 12 A flows through an inductor
• coil when it is connected to a 120-V, 60-Hz power line (peak voltage $\mathcal{E}_0 = 120\sqrt{2}$ V). If the coil is placed in series with a 10-Ω resistor and again connected to the power line, the peak current is reduced to 8 A. Calculate the resistance and the inductance of the coil.

37–31 A series *LRC* circuit (see Fig. 37–10) has $L = 25$ mH, $R = 0$, and $C = 0.1$ μF. (a) What is the resonance frequency for this undamped (zero resistance) circuit? (b) What must be the value of R to shift the resonance frequency ω^0 by 0.1 percent? (c) What value of R is required for a 1 percent shift?

37–32 Determine the values of R' and C' in the equivalent series combination that represents the parallel *RC* connection (the "leaky" capacitor) shown in the diagram. Assume $R \gg X_C$.

37–33 The diagram shows the circuit for a *self-inductance bridge*. The driving voltage is a pure sine wave, usually with a frequency of 1000 Hz. The unknown element is the series r-L combination shown in the dashed box. The resistors R_1, R_2, R_3, and R_4 are adjusted to achieve a minimum sound level in the earphones. Show that when the bridge is balanced in this way, we have

$$r = \frac{R_1 R_3}{R_2} - R_4 \qquad \text{and} \qquad L = L_1 \frac{R_3}{R_2}$$

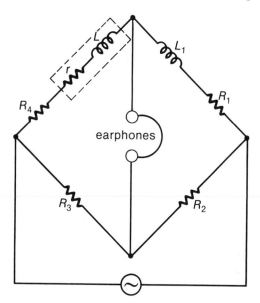

37–34 A 30-Ω resistor, a 100-mH inductor, and a 30-μF
• capacitor are connected in series to a 120-V, 60-Hz power line (peak voltage $\mathcal{E}_0 = 120\sqrt{2}$ V). Consider an instant when the line voltage is zero. (a) What amount of energy is stored in the capacitor? (b) What amount of energy is stored in the inductor? (c) What is the power delivered to the resistor? [*Hint:* $(\mathbf{Re}\,\widehat{Q})^2$ is *not* equal to $\mathbf{Re}\,\widehat{Q}^2$; similarly for $\widehat{I}$.]

ELECTROMAGNETIC WAVES

In previous chapters the discussions of electric and magnetic fields were limited primarily to fields produced by static charges or steady currents. When field variations were considered, we required the changes to be slow in order to avoid the complications associated with radiation effects.

In this chapter we enlarge the discussion to include situations in which a system produces time-varying electric and magnetic fields that propagate through space. This combined field constitutes *electromagnetic radiation* and carries energy away from the source. We examine here the simple case of a monochromatic (single-frequency) plane electromagnetic wave that travels through a medium in which there are no charges or currents.

(Before proceeding, you may find it useful to review the main features of mechanical waves in Chapter 19 and of sound in Chapter 20.)

38-1 ELECTROMAGNETIC PLANE WAVES

Basic Considerations. Many of the important characteristics of electromagnetic waves are revealed by a study of the simplest type of propagating electromagnetic disturbance, namely, a *plane wave*. We assume that the observer of the wave is sufficiently far from the source that the field vectors, $E(r, t)$ and $H(r, t)$, that describe the electromagnetic field are, at any instant, uniform over a plane that is perpendicular to the line that connects the source and the observer (Fig. 38-1). The medium through which the wave propagates has a uniform dielectric constant κ and a uniform permeability κ_M. The source, the medium, and the observer are all at rest in the same inertial reference frame. We consider a monochromatic wave in which the field variations are described by a single frequency ω.* Although all electromagnetic waves have the same basic character regardless of frequency, it is convenient to assign different names to the waves in different frequency regimes. Some typical frequencies and wavelengths are indicated in Fig. 38-2. We give special attention to

---

*Any arbitrary wave form can be decomposed by Fourier analysis into a sum of sinusoidal terms with different frequencies and amplitudes (see Section 19-8). Therefore, we need to consider here only the properties of a wave with a single frequency. All other wave forms can be obtained by superposition.

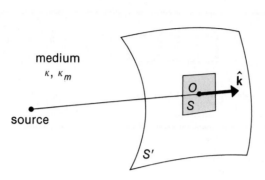

Fig. 38–1. An electromagnetic field disturbance radiates in all directions from a source. The surface S is a small portion of the spherical surface S' whose center is the source. If O is sufficiently far from the source, S can be considered to be planar, and an observer will find that the field vectors, **E** and **H**, are essentially uniform over S. Thus, the electromagnetic wave in the vicinity of the observer appears to be a *plane wave*. The direction of propagation of the wave is parallel to the line connecting the source and O; we take this line to be the z-axis ($\hat{\mathbf{k}}$) of the coordinate frame used to describe the wave.

the visible part of the electromagnetic spectrum when the subject of light is discussed in Chapter 40.

To describe the wave we use a coordinate frame in which the z-axis is in the direction of propagation of the plane wave, and we select the x-axis to coincide with the direction of the electric vector **E**. An electromagnetic wave is said to be *polarized* in the direction of its *electric* vector. Thus, the wave we are considering here is polarized in the x-direction. We have

$$\mathbf{E}(\mathbf{r}, t) = E_0 \cos(kz - \omega t)\hat{\mathbf{i}} \tag{38–1}$$

where $k = 2\pi/\lambda$ is the magnitude of the *propagation vector*,* $\mathbf{k} = k\hat{\mathbf{k}}$, whose direction is the direction of propagation of the wave. The quantity k is called the *wave number* and has dimensions $[k] = \mathrm{L}^{-1}$ (see Section 19–4). The propagation velocity (or phase velocity) of the wave is

$$v = \nu\lambda = (\omega/2\pi)\lambda = \omega/k \tag{38–2}$$

---

* Do not be confused by this expression! It means simply that the vector **k** has magnitude k and is in the z-direction ($\hat{\mathbf{k}}$).

Fig. 38–2. Frequencies and wavelengths for some of the bands in the electromagnetic spectrum. For most of the bands the end points are not well-defined.

We assert that a plane electromagnetic wave whose electric field is described by Eq. 38–1 has a magnetic field strength described by*

$$\mathbf{H}(\mathbf{r}, t) = H_0 \cos(kz - \omega t)\hat{\mathbf{j}} \qquad (38\text{--}3)$$

Thus, **H** is perpendicular both to **E** and **k** and is in phase with **E**. We proceed to show that the expressions given here for **E** and **H** satisfy Maxwell's equations, thereby justifying the *ad hoc* manner in which they have been introduced.† In this process we discover that the propagation velocity v for electromagnetic waves in vacuum is the same as the velocity of light c.

Figure 38–3a shows a portion of the wavefront of **E** and **H**. Over the wavefront, **E** is everywhere in the x-direction and **H** is everywhere in the y-direction. Figure 38–3b shows, at a particular instant, the variation of **E** and **H** along any line parallel to the direction of propagation (here, the z-direction). Notice that **E** has a changing orientation in the $\pm x$-direction and that **H** has a similar variation in the $\pm y$-direction.

(a)   (b)

Fig. 38–3. (a) A wavefront of **E** and **H** at a particular instant. Both **E** and **H** are transverse to the direction of propagation ($\hat{\mathbf{k}}$) of the wave. (b) The variation of **E** and **H** along the z-axis at a particular instant. The wave propagates in the $+z$-direction.

Faraday's Law. Consider two fixed rectangular plane surfaces with sides a and b, parallel to the x-y plane, and separated by a distance Δz, as shown in Fig. 38–4. As the plane wave that we have been considering propagates along the z-direction, the electric and magnetic field strengths at some instant take on the values indicated in the diagram. Faraday's law must be valid, so we equate the line integral of $\mathbf{E} \cdot d\mathbf{r}$ around the indicated path to the rate of change of magnetic flux through the area S that is bounded by the integration path. Thus,‡

$$\oint \mathbf{E} \cdot d\mathbf{r} = -\frac{\partial \varphi_M}{\partial t} = -\mu_0 \kappa_M \frac{\partial \varphi_H}{\partial t} \qquad (38\text{--}4)$$

where $\varphi_H = a \, \Delta z H$ is the flux of **H** through the area S (compare Eq. 35–2), and where we have used $\mathbf{B} = \mu_0 \kappa_M \mathbf{H}$ (Eq. 36–37). Traversing the path in the counter-

Fig. 38–4. Geometry for applying Faraday's law to the case of a plane electromagnetic wave.

---

*For the applications we study it turns out to be more convenient to write the electromagnetic wave relationships in terms of **H** instead of **B**.

†The derivations of the expressions for $\mathbf{E}(\mathbf{r}, t)$ and $\mathbf{H}(\mathbf{r}, t)$ are given in Section 39–4.

‡Because the field and, hence, φ_M and φ_H are functions of both position and time, the time derivative that appears in Eq. 35–2 must be written here as a *partial* derivative. Moreover, the space and time derivates of E and H are also partial derivatives, as in Eq. 38–5.

clockwise sense, as shown in Fig. 38–4, yields

$$\left(E + \frac{\Delta E}{2}\right)a + 0 - \left(E - \frac{\Delta E}{2}\right)a + 0 = -\mu_0\kappa_M a\,\Delta z\,\frac{\partial H}{\partial t}$$

from which

$$\frac{\Delta E}{\Delta z} = -\mu_0\kappa_M\frac{\partial H}{\partial t}$$

In the limit $\Delta z \to 0$, we have

$$\frac{\partial E}{\partial z} = -\mu_0\kappa_M\frac{\partial H}{\partial t} \tag{38–5}$$

Substituting into this equation the expressions for $E(\mathbf{r}, t)$ and $H(\mathbf{r}, t)$ (Eqs. 38–1 and 38–3) gives

$$-kE_0\sin(kz - \omega t) = -\mu_0\kappa_M\omega H_0\sin(kz - \omega t) \tag{38–6}$$

We originally assumed, by writing Eqs. 38–1 and 38–3 with the same factor, $kz - \omega t$, that **E** and **H** are *in phase*. Suppose that a phase difference δ is introduced by writing $\mathbf{H} = H_0\cos(kz - \omega t + \delta)\hat{\mathbf{j}}$. Then, we obtain an equation corresponding to Eq. 38–6 that contains $\sin(kz - \omega t + \delta)$ on the right-hand side and $\sin(kz - \omega t)$ on the left-hand side. But we require this equation to be valid for any time t. For this to be true, δ must be zero and the electric and magnetic field strengths are indeed in phase. Figure 38–3b has been drawn in accordance with this requirement.

From Eq. 38–6 we see that

$$\frac{E_0}{H_0} = \frac{\mu_0\kappa_M\omega}{k} = \mu_0\kappa_M\upsilon \tag{38–7}$$

If this condition is met, the expressions for $E(\mathbf{r}, t)$ and $H(\mathbf{r}, t)$ are consistent with Faraday's law.

Ampère's Law.   Figure 38–5 shows the same situation we have just discussed, except that a different integration path and surface S are now indicated. Using $\mathbf{D} = \epsilon_0\kappa\mathbf{E}$ (Eq. 30–42), Ampère's law (see Table 36–3) becomes

$$\oint \mathbf{H}\cdot d\mathbf{r} = \epsilon_0\kappa\frac{\partial\varphi}{\partial t} \tag{38–8}$$

Fig. 38–5. Geometry for applying Ampère's circuital law to the case of a plane electromagnetic wave.

The integration path shown in Fig. 38–5 is traversed in the clockwise sense, and φ is the electric flux through the area S. Thus,

$$\left(H + \frac{\Delta H}{2}\right)b + 0 - \left(H - \frac{\Delta H}{2}\right)b + 0 = -\epsilon_0\kappa b\,\Delta z\,\frac{\partial E}{\partial t}$$

The negative sign on the right-hand side of this equation results from the fact that the normal to the surface S is in the $-x$-direction ($\hat{\mathbf{n}} = -\hat{\mathbf{i}}$). Simplifying, we have

$$\frac{\Delta H}{\Delta z} = -\epsilon_0\kappa\frac{\partial E}{\partial t}$$

In the limit $\Delta z \to 0$, there results

$$\frac{\partial H}{\partial z} = -\epsilon_0 \kappa \frac{\partial E}{\partial t} \tag{38-9}$$

Substituting into this equation the expressions for $E(\mathbf{r}, t)$ and $H(\mathbf{r}, t)$ gives

$$-kH_0 \sin(kz - \omega t) = -\epsilon_0 \kappa \omega E_0 \sin(kz - \omega t) \tag{38-10}$$

(If a phase difference were introduced between E and H, we would again find $\delta = 0$, indicating that the field vectors are in phase.)

From Eq. 38-10 we see that

$$\frac{E_0}{H_0} = \frac{k}{\epsilon_0 \kappa \omega} = \frac{1}{\epsilon_0 \kappa v} \tag{38-11}$$

If this condition is met, the expressions for $\mathbf{E}(\mathbf{r}, t)$ and $\mathbf{H}(\mathbf{r}, t)$ are consistent with Ampère's law.

The Gauss Laws. Because there is no free charge in the medium, Gauss's law for the electric field (Eq. 28-19) becomes

$$\oint_S \mathbf{E} \cdot \hat{\mathbf{n}} \, dA = 0 \tag{38-12}$$

Fig. 38-6. Geometry for applying Gauss's law to the case of a plane electromagnetic wave.

Consider the Gaussian surface S shown in Fig. 38-6. The electric field vector $\mathbf{E}$ has only an x-component, so there is flux only through the sides of S that are parallel to the y-z plane. However, $\mathbf{E}$ has the same instantaneous value on any plane that is parallel to the x-y plane. Consequently, just as much electric flux emerges from the upper surface of S as enters the lower surface. Thus, the net flux that penetrates S is zero, so the expression for $\mathbf{E}(\mathbf{r}, t)$ is consistent with Gauss's law for the electric field.

The same argument can be made for the magnetic field. (•Can you see how to do this?) Thus, the expression for $\mathbf{H}(\mathbf{r}, t)$ is consistent with Gauss's law for the magnetic field.

The Propagation Velocity. Comparison of Eqs. 38-7 and 38-11 shows that Maxwell's equations require that

$$\frac{E_0}{H_0} = \mu_0 \kappa_M v = \frac{1}{\epsilon_0 \kappa v}$$

Solving for v gives

$$v = \frac{1}{\sqrt{\epsilon_0 \mu_0 \kappa \kappa_M}} \tag{38-13}$$

In vacuum, we have $\kappa = 1$ and $\kappa_M = 1$. Therefore, substituting the values of ϵ_0 and μ_0 we find (compare Eq. 34-14)

$$v = \frac{1}{\sqrt{(8.85419 \times 10^{-12} \text{ F/m})(4\pi \times 10^{-7} \text{ T} \cdot \text{m/A})}}$$
$$= 2.99792 \times 10^8 \text{ m/s}$$

That is,

$$v = \frac{1}{\sqrt{\epsilon_0 \mu_0}} = c \qquad \text{in vacuum} \qquad \text{(38-14)}$$

Thus, the propagation velocity of an electromagnetic wave in vacuum is just c, the velocity of light in vacuum, a result we anticipated in the remarks following Eq. 34–14. No restriction has been placed on the value of ω, so the result applies to *all* forms of electromagnetic radiation. Experiments have verified that, in vacuum, radio waves, microwaves, infrared radiation, visible light, and X rays all travel with the velocity c.

From Eqs. 38–11 and 38–13 we have

$$\frac{E_0}{H_0} = \sqrt{\frac{\mu_0 \kappa_M}{\epsilon_0 \kappa}} \qquad \text{(38-15)}$$

Notice that E_0 has dimensions of V/m, whereas H_0 has dimensions of A/m. Thus, E_0/H_0 has dimensions of V/A or *ohms*.

For vacuum conditions, we have $\kappa = 1$ and $\kappa_M = 1$; then,

$$\frac{E_0}{H_0} = \sqrt{\frac{\mu_0}{\epsilon_0}} \equiv Z_0 \qquad \text{vacuum} \qquad \text{(38-16)}$$

Substituting the values for μ_0 and ϵ_0, we find $E_0/H_0 = Z_0 = 377\ \Omega$. The quantity Z_0 is called the *impedance of free space* and is an important concept in radio-wave propagation and antenna theory.

For electromagnetic waves propagating through a material medium characterized by κ and κ_M, the velocity is

$$v = \frac{c}{\sqrt{\kappa \kappa_M}} = \frac{c}{n} \qquad \text{(38-17)}$$

The quantity $n = \sqrt{\kappa \kappa_M}$ is referred to as the *index of refraction* of the medium. In Chapter 40 we see how n plays a prominent role in the description of the propagation of light through transparent media. Both κ and κ_M (and hence n) have some dependence on frequency.

Example 38–1

A plane sinusoidal electromagnetic wave from a distant radio station is incident normally on a dielectric wall that has $\kappa = 1.25$ and $\kappa_M = 1.00$ for the particular frequency of the wave. The amplitude in air of the electric field of the incident wave (the *signal strength*) is $E_{0i} = 15\ \mu\text{V/m}$. The amplitude of the transmitted electric field E_{0t} (that is, the field that emerges from the other side of the wall) is given by* $E_{0t}/E_{0i} = 2/(n + 1)$.

---

*This relationship for the ratio E_{0t}/E_{0i} is developed in Chapter 40. For the purposes here the derivation is not important.

(a) Determine the amplitude H_{0i} of the associated incident magnetic field.
(b) What is the amplitude E_{0t} of the transmitted wave?
(c) What is the electromagnetic impedance Z_w of the wall?
(d) Determine the amplitude H_{0t} of the transmitted magnetic field.

Solution:

(a) The dielectric properties of air are essentially the same as those of vacuum, so we take the impedance of air to be $Z_0 = 377\,\Omega$ and use Eq. 38-16:

$$H_{0i} = E_{0i}/Z_0 = (15 \times 10^{-6}\,\text{V/m})/(377\,\Omega) = 3.98 \times 10^{-8}\,\text{A/m}$$

(b) We have $n = \sqrt{\kappa\kappa_M} = \sqrt{\kappa} = \sqrt{1.25} = 1.12$; then,

$$E_{0t} = 2\,E_{0i}/(n+1) = 2(15 \times 10^{-6}\,\text{V/m})/2.12 = 14.2\,\mu\text{V/m}$$

(c) The wall impedance is found from Eq. 38-15:

$$Z_w = E_{0t}/H_{0t} = Z_0/\sqrt{\kappa} = (377\,\Omega)/1.12 = 337\,\Omega$$

(d) Thus, the magnetic field amplitude is

$$H_{0t} = E_{0t}/Z_w = (14.2\,\mu\text{V/m})/(337\,\Omega) = 4.21 \times 10^{-8}\,\text{A/m}$$

38-2 ENERGY AND MOMENTUM IN ELECTROMAGNETIC WAVES

Energy Density. According to Eq. 36-43, the energy density for combined electric and magnetic fields in a linear medium is

$$u = \tfrac{1}{2}\mathbf{E} \cdot \mathbf{D} + \tfrac{1}{2}\mathbf{B} \cdot \mathbf{H} = \tfrac{1}{2}\epsilon_0 \kappa E^2 + \tfrac{1}{2}\mu_0 \kappa_M H^2$$

Squaring and rearranging Eq. 38-15, we can write

$$\tfrac{1}{2}\mu_0 \kappa_M H^2 = \tfrac{1}{2}\epsilon_0 \kappa E^2$$

We see that the energy density in an electromagnetic wave consists of equal contributions from the electric and magnetic fields. For a plane wave, the combined energy density is itself a propagating wave,[*]

$$u(z, t) = \epsilon_0 \kappa E^2 = \epsilon_0 \kappa E_0^2 \cos^2(kz - \omega t) \tag{38-18}$$

The time average of $u(z, t)$ over one complete period, $T = 2\pi/\omega$, at any point z is

$$\bar{u} = \frac{1}{T} \cdot \epsilon_0 \kappa E_0^2 \int_t^{t+T} \cos^2(kz - \omega t)\, dt$$

so that

$$\bar{u} = \tfrac{1}{2}\epsilon_0 \kappa E_0^2 \tag{38-19}$$

---

[*] Recall that any function of the form $f(kz - \omega t)$ represents a propagating wave; here, the function f is the square of the cosine function.

Fig. 38–7. Geometry for discussing the flow of energy through the fixed area A. In vacuum, $v = c$. (•What is the direction of N when E is in the $-x$-direction and H is in the $-y$-direction?)

This expression gives the average energy density in a monochromatic plane wave.

The Poynting Vector. The fact that the energy density in an electromagnetic wave is itself a propagating wave (Eq. 38–18) means that the field energy of a wave is carried along as the wave propagates through the medium. We define a quantity N which is the *energy flux* of a wave, the amount of energy that flows past a fixed point per unit time and per unit area (perpendicular to the flow direction). Accordingly, the units of N are W/m^2.

Consider the geometry shown in Fig. 38–7. During a time interval Δt a wavefront advances a distance $\Delta z = v\Delta t$. Thus, the field energy within the volume $A\Delta z$ passes through the fixed area A (shown shaded) during the interval Δt. Therefore, we can write

$$NA = \frac{uA\,\Delta z}{\Delta t}$$

from which we have

$$N = uv = \epsilon_0 \kappa v E_0^2 \cos^2(kz - \omega t)$$

Using Eq. 38–11, this expression becomes

$$N = E_0 H_0 \cos^2(kz - \omega t) = EH \tag{38–20}$$

We can represent the vector character of N by writing

$$\mathbf{N} = \mathbf{E} \times \mathbf{H} \tag{38–21}$$

which shows explicitly that the direction of N is the direction of propagation of the wave. (Check the direction in Fig. 38–3.) The vector N is called the *Poynting vector*.* In vacuum we have $v = c$, so that

$$\mathbf{N} = uc\hat{\mathbf{k}} \tag{38–21a}$$

Fig. 38–8. Energy changes within the volume V are described by the integral of the Poynting vector N over the bounding surface S.

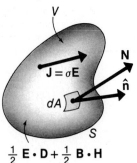

Although the Poynting vector was introduced in the context of energy flow in an electromagnetic plane wave, this vector has a more general and far greater significance. Maxwell's equations may be used to obtain an energy conservation equation for a closed region of space with a volume V bounded by a surface S, namely,

$$\oint_S \mathbf{N} \cdot \hat{\mathbf{n}}\, dA + \frac{\partial}{\partial t} \int_V (\tfrac{1}{2}\mathbf{E} \cdot \mathbf{D} + \tfrac{1}{2}\mathbf{B} \cdot \mathbf{H})\, dv + \int_V \mathbf{E} \cdot \mathbf{J}\, dv = 0 \tag{38–22}$$

In this equation, the surface integral of the Poynting vector N represents the outward flow of electromagnetic energy through the surface S (Fig. 38–8). The second term in the equation gives the rate of increase of electromagnetic energy within the volume V. (When there is an *outward* flow of energy the second term is *negative*.) The last term represents the Joule heating within V. In this way we can view *all* transfers of electromagnetic energy to take place via fields instead of current-carrying conduc-

---

*After the English physicist John Henry Poynting (1852–1914).

tors. Wires and conductors are necessary, not for the purpose of transmitting electromagnetic energy, but simply to give to the fields the shapes that are required. Then, the Poynting vector describes the energy transfer across the bounding surface of any volume in which electromagnetic energy is either stored or dissipated. Moreover, as will be shown in Example 38-5, the energy equation (Eq. 38-22) may be applied in cases that involve steady currents (and, therefore, no electromagnetic waves at all).

▶ *Interactions with Matter.* We have been considering the propagation of waves in material media whose electromagnetic properties are characterized by κ and κ_M. The resulting equations based on these quantities have an obvious simplicity. However, the application of these equations to the interactions of electromagnetic waves with the atoms and molecules in a material medium is actually far from simple. In fact, many of the various types of interactions that can take place—particularly for short-wavelength radiations—require quantum theory for a correct description.

Even for those cases in which the interaction of a wave with the medium can be characterized adequately by bulk properties that are treated according to the laws of classical physics, the analysis is nonetheless still complicated. One of the primary reasons is that the bulk electric and magnetic properties of real substances cause the phase velocity of the wave and the absorption of the wave energy to vary with frequency. This is an important consideration because every real wave actually consists of a superposition of waves with different frequencies. (A monochromatic wave is an idealization that is not achievable in practice. If a wave is truly to have a single, pure frequency, it must have identical properties throughout space—that is, the wave must have infinite extent and must have existed for an infinite time.)

A multi-frequency wave that propagates through a medium whose properties are frequency dependent will suffer both *attenuation* and *dispersion* (see Section 19-9). The presence of dispersion requires the introduction of the concept of the *group velocity* v_g in addition to the phase velocity or wave velocity v. In the X-ray and ultraviolet regions of the electromagnetic spectrum (see Fig. 38-2), most substances have a refractive index $n = c/v$ that is slightly less than unity. The fact that $v > c$ in these cases does not violate the principles of relativity theory because signals are transmitted and energy is transported with the group velocity, for which we always have $v_g < c$.

The index of refraction n for nonconducting media asymptotically approaches the value $\sqrt{\kappa\kappa_M}$ as the wavelength of the radiation becomes relatively long (radio waves). Then, we find $\kappa_M \cong 1$ and $n \cong \sqrt{\kappa}$, where κ is the DC value (that is, the value for $\nu \to 0$) listed in Table 30-1. ◀

Electromagnetic Momentum and Radiation Pressure. It is natural to expect a wave that transports energy to carry momentum as well. Indeed, a treatment of electromagnetic wave phenomena more detailed than we give here demonstrates that there exists a simple relationship connecting the Poynting vector **N** for a propagating wave in vacuum and the linear momentum density **G** of the wave, namely,

$$\mathbf{G} = \frac{1}{c^2}\mathbf{N} \qquad (38\text{-}23)$$

where **G** is momentum per unit volume and is measured in units of $J \cdot s/m^4$. Because $v = c$ for a wave in vacuum, we have $N = uc$; thus, the momentum density for a wave that propagates in the z-direction can also be expressed as

$$\mathbf{G} = \frac{1}{c}u\hat{\mathbf{k}} \qquad (38\text{-}24)$$

If a wave traveling in vacuum is incident on the surface of a substance that totally absorbs the wave, a pressure will be exerted on the surface. As indicated in Fig. 38-9, the total wave momentum contained within a volume $Ac\,\Delta t$ will be absorbed by the surface (which is perpendicular to the direction of **G**) during a time interval Δt. Thus, a force F is exerted by the wave on an area A of the surface, where

Fig. 38-9. By transferring momentum to a surface, an electromagnetic wave exerts a *radiation pressure*.

$$\text{Impulse} = (\text{Momentum density}) \cdot (\text{Volume})$$

$$F\Delta t = \frac{1}{c}u \cdot Ac\Delta t$$

The force per unit area is the *radiation pressure* p_{rad},

$$p_{rad} = \frac{F}{A} = u \tag{38-25}$$

As usual, pressure is measured in N/m² or Pa. The result expressed by Eq. 38-25 is independent of the nature of the absorption process.

▶ *Relativistic Considerations.* The relativistic relationship that connects the total energy U, the momentum p, and the proper mass M_0 for a particle is (see Eq. 21–16)

$$U^2 = p^2c^2 + M_0^2c^4$$

If we apply this equation to a radiation pulse, we take the proper mass M_0 to be zero. (●Because the speed of light is c in all inertial frames, what possible meaning could "rest mass" have for a radiation pulse?) Then, we have $U = pc$, and in terms of densities we write $u = Gc$, which is just the result expressed in Eq. 38–24. ◀

Example 38–2

A pulsed source of microwaves produces bursts of 20-GHz radiation, with each burst having a duration of 1 ns. A parabolic reflector ($R = 6$ cm) is used to produce a parallel beam of radiation, as indicated in the diagram. The average output power of the source during each pulse is 25 kW.

(a) What is the wavelength of the radiation?

(b) What is the total energy U in each pulse?

(c) Determine the average energy density $\bar{u}$ and the average magnitude of the Poynting vector, $\bar{N}$, during each pulse.

(d) Determine the average momentum density $\bar{G}$ and the average radiation pressure $\bar{p}$ on a totally absorptive surface that is perpendicular to the direction of propagation of the pulse.

(e) Calculate the values of E_0 and H_0 in the wave.

Solution:

(a) The wavelength of the radiation is

$$\lambda = \frac{c}{\nu} = \frac{3.00 \times 10^8 \text{ m/s}}{2 \times 10^{10} \text{ s}^{-1}} = 0.015 \text{ m}$$

which is in the microwave part of the electromagnetic spectrum (see Fig. 38–2).

(b) The total energy in the pulse is

$$U = \bar{P} \Delta t = (2.5 \times 10^4 \text{ J/s})(10^{-9} \text{ s})$$
$$= 2.5 \times 10^{-5} \text{ J}$$

(c) The length of the pulse is

$$\ell = c \Delta t = (3.00 \times 10^8 \text{ m/s})(10^{-9} \text{ s})$$
$$= 0.30 \text{ m}$$

so that the average energy density is

$$\bar{u} = \frac{U}{\pi R^2 \ell} = \frac{2.5 \times 10^{-5} \text{ J}}{\pi (0.06 \text{ m})^2 (0.30 \text{ m})}$$
$$= 7.37 \times 10^{-3} \text{ J/m}^3$$

Then, we have

$$\bar{N} = \bar{u}c = (7.37 \times 10^{-3} \text{ J/m}^3)(3.00 \times 10^8 \text{ m/s})$$
$$= 2.21 \times 10^6 \text{ W/m}^2$$

(d) The average momentum density is

$$\bar{G} = \frac{1}{c}\bar{u} = \frac{7.37 \times 10^{-3} \text{ J/m}^3}{3.00 \times 10^8 \text{ m/s}}$$
$$= 2.46 \times 10^{-11} \text{ J} \cdot \text{s/m}^4$$

and the average radiation pressure is

$$\bar{P}_{\text{rad}} = \bar{u} = 7.37 \times 10^{-3} \text{ Pa}$$

(e) The value of E_0 is obtained from Eq. 38–19 with $\kappa = 1$:

$$E_0 = \sqrt{\frac{2\bar{u}}{\epsilon_0}} = \sqrt{\frac{2(7.37 \times 10^{-3} \text{ J/m}^3)}{8.85 \times 10^{-12} \text{ C/V} \cdot \text{m}}}$$
$$= 4.08 \times 10^4 \text{ V/m}$$

Then, using Eq. 38–16, we have

$$H_0 = \frac{E_0}{Z_0} = \frac{4.08 \times 10^4 \text{ V/m}}{377 \ \Omega}$$
$$= 1.08 \times 10^2 \text{ A/m}$$

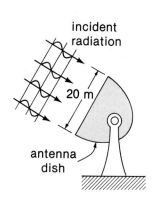

Example 38–3

A dish antenna with a diameter of 20 m is used to receive (at normal incidence) a radio signal from a distant source, as shown in the diagram. The signal is a continuous sinusoidal wave with amplitude $E_0 = 0.2 \ \mu\text{V/m}$. (This is a relatively low signal strength; most radio receivers require signal strengths of a few $\mu\text{V/m}$ for acceptable signal-to-noise ratios.) Assume that the antenna absorbs all of the radiation that is incident on its geometric cross section.

(a) Determine the value of H_0 for the incident wave.

(b) What is the flux of incident energy (that is, what is the average magnitude of the Poynting vector)?

(c) At what average power level is energy delivered to the antenna?

(d) What is the force exerted on the antenna by the incident radiation?

Solution:

(a) Both κ and κ_M are essentially unity for the medium (air) through which the wave propagates. Then, using Eq. 38–16, we have

$$H_0 = \frac{E_0}{Z_0} = \frac{2 \times 10^{-7} \text{ V/m}}{377 \, \Omega} = 5.31 \times 10^{-10} \text{ A/m}$$

(b) The average magnitude of the Poynting vector can be obtained from $N = EH$ (Eq. 38–20); thus,

$$\bar{N} = \left(\frac{E_0}{\sqrt{2}}\right)\left(\frac{H_0}{\sqrt{2}}\right) = \tfrac{1}{2}E_0H_0$$

$$= \tfrac{1}{2}(2 \times 10^{-7} \text{ V/m})(5.31 \times 10^{-10} \text{ A/m})$$

$$= 5.31 \times 10^{-17} \text{ W/m}^2$$

(c) The average power delivered by the wave to the antenna is

$$\bar{P} = \pi R^2 \bar{N} = \pi(10 \text{ m})^2(5.31 \times 10^{-17} \text{ W/m}^2)$$

$$= 1.67 \times 10^{-14} \text{ W}$$

▶ A major source of noise in receiver circuits is the effect of thermal agitation in resistors. The average noise power generated in this way is $\bar{P}_N = 4kT\,\Delta\nu$, where k is Boltzmann's constant, where T is the absolute temperature, and where $\Delta\nu$ is the *band width* of the receiver, that is, the frequency interval over which the receiver circuit is tuned to respond. For a wide-band receiver operating at UHF frequencies, we might have $\Delta\nu = 10^6$ Hz. Then, with $T = 300$ K, we find $\bar{P}_N = 4(1.37 \times 10^{-23} \text{ J/K})(300 \text{ K})(10^6 \text{ s}^{-1}) = 1.64 \times 10^{-14}$ W. This noise power is about equal to the power delivered by the wave to the antenna in this example. Then, the signal-to-noise ratio is $\bar{P}/\bar{P}_N \cong 1$, and special techniques are necessary to extract meaningful signals from the background noise generated within the receiving equipment itself. ◀

(d) The average radiation pressure exerted on the antenna by the wave is (using Eqs. 38–21a and 38–25) $\bar{p}_{\text{rad}} = \bar{u} = \bar{N}/c$. Then, the average force due to the incident radiation is

$$\bar{F} = \pi R^2 \cdot \bar{p}_{\text{rad}} = \frac{\pi R^2 \bar{N}}{c}$$

$$= \frac{\pi(10 \text{ m})^2(5.31 \times 10^{-17} \text{ W/m}^2)}{3.00 \times 10^8 \text{ m/s}}$$

$$= 5.56 \times 10^{-23} \text{ N}$$

which is a very small force indeed! ($\bar{F}$ corresponds to the weight of about 3400 hydrogen atoms.)

Example 38–4

A dust particle has a radius $r = 0.01$ mm and an average density $\bar{\rho} = 1.0 \text{ g/cm}^3$. This particle is in interplanetary space, where it is simultaneously attracted toward the Sun by the gravitational force and repelled by the radiation pressure of the intercepted solar radiation. The total radiant flux of the Sun is $\varphi_{\text{rad}} = 3.82 \times 10^{26}$ W. Assume that the radiation incident on the particle is totally absorbed.

(a) What is the ratio of the gravitational force to the radiation force acting on the particle?

(b) What is the smallest radius that such a particle can have if it is not to be blown out of the solar system?

Solution:

(a) The mass of the dust particle is $m = (4\pi/3)\bar{\rho}r^3$; therefore, when it is at a distance R from the Sun, the gravitational force on the particle is

$$F_g = G\frac{Mm}{R^2} = \frac{4\pi}{3}G\frac{M\bar{\rho}r^3}{R^2}$$

where M is the mass of the Sun. At the distance R, the magnitude of the Poynting vector associated with the solar flux is $N = \varphi_{\text{rad}}/4\pi R^2$. Then, the radiation force is

$$F_r = p_{\text{rad}} \cdot \pi r^2 = \frac{N}{c}\pi r^2 = \frac{1}{4}\frac{\varphi_{\text{rad}}r^2}{cR^2}$$

so that

$$\frac{F_g}{F_r} = \frac{\dfrac{4\pi}{3}G\dfrac{M\bar{\rho}r^3}{R^2}}{\dfrac{1}{4}\dfrac{\varphi_{\text{rad}}r^2}{cR^2}} = \frac{16\pi}{3}\frac{GM\bar{\rho}rc}{\varphi_{\text{rad}}}$$

$$= \frac{16\pi}{3}\frac{(6.67 \times 10^{-11})(1.99 \times 10^{30})(10^3)(10^{-5})(3.00 \times 10^8)}{3.82 \times 10^{26}}$$

where the value of each term is expressed in SI units. Finally,

$$\frac{F_g}{F_r} = 17.5$$

Notice that this ratio does not depend on the distance R from the Sun.

(b) If the radius r of the dust particle is sufficiently small that $F_g/F_r < 1$, the radiation force is greater than the gravitational force and the particle will be blown out of the solar system. The minimum radius for which this will not happen is

$$r_{\text{min}} = \frac{r}{17.5} = \frac{10^{-5}\text{ m}}{17.5} = 5.71 \times 10^{-7}\text{ m}$$

This dimension (which is 5710 Å) is about the same as the wavelength of light corresponding to the middle of the spectrum of visible light (yellow light).

In addition to the radiant flux from the Sun, there is also the outflow of high-speed charged particles (electrons and protons) that constitutes the *solar wind*. The impact of the solar wind particles on the dust particles contributes to the outward-directed force and significantly increases the value of r_{min}.

Example 38-5

Consider a long straight wire with a radius a that carries a uniform steady current I. Show that the integral of the Poynting vector over the surface of the wire gives a rate of input of field energy through the surface just equal to the Joule heating within the wire.

Solution:

Using Ampère's law, we find the magnetic field strength at the surface of the wire to be

$$\mathbf{H} = \frac{I}{2\pi a}\hat{\mathbf{u}}_\phi$$

as shown in the diagram. Whatever the resistance characteristics of the wire, homogeneity alone insures that the electric field, $\mathbf{E} = E\hat{\mathbf{k}}$, is constant everywhere inside and on the surface of the wire. Therefore, the Poynting vector on the surface of the wire is

$$\mathbf{N} = \mathbf{E} \times \mathbf{H} = -\frac{EI}{2\pi a}\hat{\mathbf{u}}_r$$

where we have used $\hat{\mathbf{k}} \times \hat{\mathbf{u}}_\phi = -\hat{\mathbf{u}}_r$. We integrate $\mathbf{N} \cdot \hat{\mathbf{u}}_r\, dA$ over the surface of a section of the wire with length ℓ, obtaining

$$\oint_S \mathbf{N} \cdot \hat{\mathbf{u}}_r\, dA = -(2\pi a \ell)\left(\frac{EI}{2\pi a}\right) = -EI\ell$$

The negative sign means that the energy flow is *inward*. The decrease in electric potential over the length ℓ of wire is $V = E\ell$. Thus, we see that $EI\ell = VI$ is just the Joule heating in the wire. If the material of the wire obeys Ohm's law, we have $E = J/\sigma = I/\pi a^2\sigma$, where σ is the conductivity; then,

$$EI\ell = \left(\frac{\ell}{\pi a^2\sigma}\right)I^2 = I^2 R$$

where R is the resistance of the length ℓ of the wire.

Notice that the result obtained here agrees with the energy conservation equation (Eq. 38–22) for the steady-state conditions of this problem.

PROBLEMS

Section 38-1

38-1   A monochromatic plane electromagnetic wave in vacuum has a propagation vector $\mathbf{k} = k\hat{\mathbf{k}}$. At time t, we have $\mathbf{E}(\mathbf{r}, t) = (200 \text{ V/m}) \cos(kz - \omega t)\hat{\mathbf{i}}$. Determine the magnetic vector $\mathbf{H}(\mathbf{r}, t)$.

38-2   The magnetic vector for a monochromatic plane electromagnetic wave in vacuum has only an x-component, $H_x = H_0 \cos(ky - \omega t)$, with $H_0 = 1.6 \times 10^{-6}$ T. What is the direction of the propagation vector $\mathbf{k}$? Determine the electric vector $\mathbf{E}(\mathbf{r}, t)$.

38-3   What is the propagation velocity of an electromagnetic wave in a medium for which $\kappa = 3.1$ and $\kappa_M = 200$?

38-4   Start with Eqs. 38–5 and 38–9 and show that E and H satisfy the wave equations,

$$\frac{\partial^2 E}{\partial z^2} = \epsilon_0 \mu_0 \kappa \kappa_M \frac{\partial^2 E}{\partial t^2}$$

$$\frac{\partial^2 H}{\partial z^2} = \epsilon_0 \mu_0 \kappa \kappa_M \frac{\partial^2 H}{\partial t^2}$$

[*Hint:* Consider differentiating Eqs. 38–5 and 38–9 with respect to z and t.]

38-5   Consider the superposition of two monochromatic
•   plane waves,

$$\mathbf{E} = E_0 \cos(kz - \omega t)\hat{\mathbf{i}}$$

$$\mathbf{E}' = 2E_0 \cos(kz - \omega t + \pi/4)\hat{\mathbf{j}}$$

Sketch $\mathbf{E}$, $\mathbf{E}'$, and $\mathbf{E} + \mathbf{E}'$ in the x-y plane for $kz - \omega t = \pi$, $3\pi/4$, $\pi/2$, $\pi/4$, and 0. Use these sketches to trace out the complete behavior of $\mathbf{E} + \mathbf{E}'$. This type of wave is said to be *elliptically polarized*.

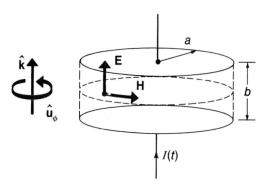

Section 38-2

38-6 A monochromatic plane electromagnetic wave in vacuum has $E_0 = 15.0$ V/m. What is the energy flux of the wave?

38-7 Monochromatic electromagnetic radiation is emitted isotropically at a rate of 40 W from a source whose size is small compared with the wavelength of the radiation ($\lambda = 1.0$ mm). At a distance $r = 2.0$ m, the waves in a small area are essentially plane waves. Evaluate the wave amplitudes, E_0 and H_0, at this distance. Compare each of the values with those for ordinary static fields; comment on the comparison.

38-8 A sinusoidal plane wave with wavelength $\lambda = 10$ m travels in vacuum parallel to the y-axis. The wave is polarized in the x-direction, and the average energy transport is 0.20 W/m². (a) Determine E_0 and H_0. (b) Write the equations for $\mathbf{E}$ and $\mathbf{H}$.

38-9 A radio wave propagates through a dielectric medium with dielectric constant κ and $\kappa_M = 1$. The field vectors, $\mathbf{E}$ and $\mathbf{H}$, for the plane wave are given by Eqs. 38-1 and 38-3, in which $E_0 = 1$ V/m, $H_0 = 3.2 \times 10^{-3}$ A/m, and $\nu = \omega/2\pi = 10^6$ Hz. (a) What is the dielectric constant of the medium? (b) What is the phase velocity of the wave? (c) What is the wavelength of the wave? (d) What is the average energy density in the wave? (e) What is the average energy transport (in W/m²)?

38-10 A 150-μs burst of 1-MW laser radiation is directed horizontally at a 10-mg bob that hangs at the end of a thin fiber whose length is 50 cm. Assume that the radiation is completely absorbed by the bob (and that there is negligible vaporization of the bob!). What is the maximum angle of deflection of the fiber?

38-11 The circular parallel-plate capacitor shown in the
•• diagram is slowly charged by a current $I(t)$. (a) Use Ampère's circuital law and the idea of displacement current to show that the vectors $\mathbf{E} = E\hat{\mathbf{k}}$ and $\mathbf{H} = H\hat{\mathbf{u}}_\phi$ at some point $r = a$ are related by $H(a) = \frac{1}{2}\epsilon_0 a\, \partial E/\partial t$. (Assume vacuum conditions. Assume also that $I(t)$ varies sufficiently slowly that $\mathbf{E}$ is essentially uniform in the region between the

plates.) (b) Evaluate the surface integral of the Poynting vector $\mathbf{N}$ over the outer cylindrical surface of the capacitor gap at $r = a$. Show that there is an *inward* flux,

$$-\oint_S \mathbf{N} \cdot \hat{\mathbf{n}}\, dA = \frac{\partial}{\partial t} \int_V \tfrac{1}{2}\epsilon_0 E^2\, dv$$

Interpret this result. (c) If the capacitor is charged by a battery through a series resistor R, giving a time constant $\tau = RC$, then the field E, instead of being constant, will have a radial variation. In such a case, it can be shown that $E = E(r) \cong E_0 \cos(r/\sqrt{2}c\tau)$, for $r \ll c\tau$. Assume $R = 1$ MΩ, $a = 5$ cm, and $b = 0.5$ cm, and estimate the ratio $[E(a) - E_0]/E_0$. Are radiation effects negligible? Is the situation changed if $R = 10$ Ω?

Additional Problems

38-12 At the top of the atmosphere the radiant flux from the Sun is $\bar{N} = 1.35 \times 10^3$ W/m². Although this radiation consists of a spectrum of frequencies, many of the interesting properties do not depend on frequency and can therefore be calculated by using the methods described for monochromatic waves. (a) What is the average energy density in the solar radiation at the top of the atmosphere? (b) What is the average momentum density? (c) What average force would the radiation exert on a completely absorbing surface with an area of 1 m² oriented perpendicular to the Earth-Sun line? (d) What are the average values of E_0 and H_0 in the wave?

38-13 A sinusoidal plane-polarized radio wave has a frequency $\nu = 500$ kHz and transports energy through air at a rate of 1 W/m². (a) Determine E_0 and H_0. (b) What is the energy density u in the wave? (c) What amount of field energy is contained within a cube with sides equal to the wavelength λ of the wave?

38-14 A sinusoidal plane-polarized wave has a wavelength $\lambda = 0.30$ m and transports energy through air at a rate of 2 W/m². The direction of propagation of the wave is in the x-y plane at an angle of 30° to the x-axis; the direction of the polarization is parallel to the z-axis. Write the equations that describe **E** and **H**.

38–15 Extend the discussion in Example 38–5 to the interior of the conductor by considering what happens on a cylindrical surface with $r < R$.

38–16 A long solenoid consists of many tightly wound turns
 • of wire. A slowly increasing current $I(t)$ flows through the wire, as indicated in the diagram. Assume vacuum conditions. Assume also that the variation of $I(t)$ is sufficiently slow that **H** may be considered to be uniform inside the solenoid. (a) Use Faraday's law of induction to show that $\mathbf{E} = -E\hat{\mathbf{u}}_\phi$ and $\mathbf{H} = H\hat{\mathbf{k}}$ at some circumferential point, $r = a$, are related by $E(a) = \frac{1}{2}\mu_0 a\, \partial H/\partial t$. (b) Evaluate the

integral of $\mathbf{N} \cdot \hat{\mathbf{n}}\, dA$ over the outer cylindrical surface of the solenoid at $r = a$ and show that the inward flux is

$$-\oint_S \mathbf{N} \cdot \hat{\mathbf{n}}\, dA = \frac{\partial}{\partial t} \int_V \tfrac{1}{2}\mu_0 H^2\, dv$$

Interpret this result.

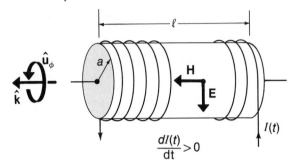

RADIATING SYSTEMS

We now consider methods for generating electromagnetic waves and for guiding these waves with various devices.

39-1 RADIATION FROM AN ACCELERATING CHARGE

Electromagnetic fields and electromagnetic waves originate from distributions of electric charge. In fact, we now demonstrate that electromagnetic radiation is produced by charges that undergo *acceleration*. By restricting attention to radiations with relatively long wavelengths (for example, radio waves), a satisfactory description can be made in terms of a classical model that involves alternating currents in conductors, called *antennas*. The electrons in these antennas are accelerated back and forth, thereby generating electromagnetic waves.

Suppose that a small positive charge q in vacuum is at rest in some inertial reference frame. At time $t = 0$ the charge is accelerated at a constant rate **a** for a brief interval τ and thereafter moves at whatever velocity it acquired as a result of the acceleration. Consider the change in the electric field of the charge due to this acceleration process. While the charge is at rest, the field lines **E** are straight radial lines emanating from q. The field lines for a positive charge that moves at constant velocity are also straight lines directed away from the charge (see Section 35–4). Consequently, when a charge is accelerated for a brief interval, a "kink" must appear in the field lines (which are continuous).

Figure 39–1 shows a particular electric field line, with segments corresponding to $t < 0$, $0 < t < \tau$, and $t > \tau$. We consider the case in which $\tau \ll t$, so that $\overline{PP'} \ll \overline{PQ} \cong vt$,

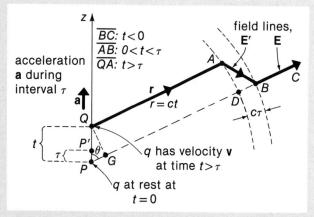

Fig. 39–1. Geometry of an electric field line for a charge that undergoes a uniform acceleration **a** for a time interval $\tau \ll t$.

where $v = a\tau$. From the diagram we see that

$$\overline{AD} = \overline{QG} = \overline{PQ}\sin\theta = a\tau t \sin\theta$$

The kink **E**′ propagates outward with the velocity c, so we also have

$$\overline{DB} = c\tau$$

The field line **E**′ has a tangential component E'_t and a radial component E'_r. The ratio of these components is $\overline{AD}/\overline{DB}$ (see Fig. 39–1); thus,

$$\frac{E'_t}{E'_r} = \frac{\overline{AD}}{\overline{DB}} = \frac{at}{c}\sin\theta$$

The radial component of the field is

$$E'_r = \frac{q}{4\pi\epsilon_0 r^2}$$

It follows that

$$E'_t = \frac{qat\sin\theta}{4\pi\epsilon_0 cr^2}$$

Using $r = ct$, we obtain the desired expression for the tangential component of the field, namely,

$$E'_t = \frac{qa\sin\theta}{4\pi\epsilon_0 c^2 r} \qquad (39\text{-}1)$$

The outward propagating disturbance that results from the acceleration is just the tangential component E'_t of the kink. Notice that $E'_t \propto 1/r$, **not*** $1/r^2$. In Fig. 39-2 the situation in Fig. 39-1 is redrawn to show only E'_t (now designated simply $\mathbf{E}$) and the accompanying vectors, $\mathbf{H}$ and $\mathbf{N}$. The electric vector $\mathbf{E}$ is in the direction $\hat{\mathbf{u}}_\theta$, and $\mathbf{N}$ is directed outward ($\hat{\mathbf{u}}_r$). Thus, $\mathbf{H}$ is in the direction $\hat{\mathbf{u}}_\phi$, as indicated in Fig. 39-2. The magnitude of $\mathbf{H}$ is obtained from $|\mathbf{H}|/|\mathbf{E}| = \sqrt{\epsilon_0/\mu_0}$ (see Eq. 38-16); thus,

$$\mathbf{E} = \frac{qa}{4\pi\epsilon_0 c^2}\frac{\sin\theta}{r}\hat{\mathbf{u}}_\theta \qquad (39\text{-}2a)$$

$$\mathbf{H} = \frac{qa}{4\pi c}\frac{\sin\theta}{r}\hat{\mathbf{u}}_\phi \qquad (39\text{-}2b)$$

$$\mathbf{N} = \mathbf{E} \times \mathbf{H} = \frac{q^2 a^2}{16\pi^2\epsilon_0 c^3}\frac{\sin^2\theta}{r^2}\hat{\mathbf{u}}_r \qquad (39\text{-}2c)$$

---

*Near an accelerating charge, a portion of the field disturbance does have a dependence on $1/r^2$. At large distances, as we consider here, this component is negligible compared with $E'_t \propto 1/r$.

Fig. 39-2. The field vectors, $\mathbf{E}$ and $\mathbf{H}$, and the Poynting vector $\mathbf{N}$ for the outward propagating disturbance due to an accelerating charge.

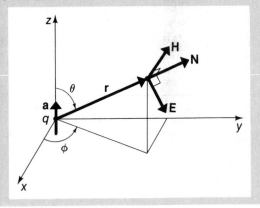

The method by which we have obtained these expressions is not rigorous; however, a precise calculation based directly on Maxwell's equations yields exactly the same results.

A polar diagram of $|\mathbf{N}|$ as a function of θ is called the *radiation pattern* for the particular system. Figure 39-3 shows the pattern for a uniformly accelerated charge, $|\mathbf{N}| \propto \sin^2\theta$.

Fig. 39-3. Radiation pattern for a charge undergoing uniform acceleration: $|\mathbf{N}| \propto \sin^2\theta$. Notice that no radiation is emitted along the line of acceleration.

Notice that $\mathbf{N} \propto \hat{\mathbf{u}}_r/r^2$. This dependence insures that the total energy flux through a spherical surface of radius r is the same for all distances r. (•Can you see why this is so?)

Example 39-1

A charge q undergoes a uniform acceleration $\mathbf{a}$. What is the total power radiated?

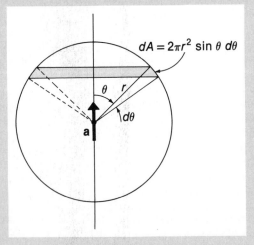

Solution:

The polar axis of a spherical coordinate frame is chosen to coincide with the direction of the acceleration vector $\mathbf{a}$, as shown in the diagram. The Poynting vector $\mathbf{N}$ is given by Eq. 39-2c. Thus, we have

$$P = \oint_S \mathbf{N} \cdot \hat{\mathbf{u}}_r\, dA$$

$$= \frac{q^2 a^2}{16\pi^2 \epsilon_0 c^3} \int_0^{2\pi} d\phi \int_0^{\pi} \sin^2\theta \cdot \sin\theta \, d\theta$$

$$= \frac{q^2 a^2}{8\pi \epsilon_0 c^3} \int_0^{\pi} \sin^3\theta \, d\theta$$

The integral is 4/3 (see the table inside the rear cover). Thus,

$$P = \frac{q^2 a^2}{6\pi \epsilon_0 c^3}$$

This is the *Larmor formula*, first derived by Sir J. J. Larmor in 1897.

▶ *Čerenkov Radiation.* A remarkable effect occurs when a charged particle, such as an electron, moves through a medium with a speed greater than the phase velocity of light in the medium. (This situation does not violate the principles of relativity theory because the theory requires only that the velocity of a particle be less than c, not less than the phase velocity in the medium, $v = c/n$.) A charged particle that moves with such a great speed sets up in the medium an electromagnetic shock wave that is similar to the pressure shock wave produced when an object travels faster than the speed of sound in a medium (see Section 20–5). The light emitted in the shock wave was discovered in 1934 by the Russian physicist Pauel Čerenkov and is called *Čerenkov radiation.*

When in motion, a charged particle may be viewed as continually creating along its path a set of outgoing electromagnetic waves. For $v < c/n$, these superimposed waves interfere destructively so that there remain no ordinary waves to carry away energy. If it were not for other forms of energy loss (for example, by ionization), the particle would move through the medium with constant speed (and constant energy). However, when $v > c/n$, the superposition of the waves produces constructive interference that results in a conical shock wave with vertex at the position of the particle, as shown in Fig. 39–4 (compare Fig. 20–18.) The shock wave consists of radiation with all frequencies ω for which $c/n(\omega) < v$. For each frequency ω the cone of radiation has a half angle $\theta_C(\omega)$ given by (compare Eq. 20–21)

$$\sin\theta_C(\omega) = \frac{c}{n(\omega)v} = \frac{1}{\beta n(\omega)} \qquad \textbf{(39–3)}$$

where $\beta = v/c$.

A theoretical analysis of Čerenkov radiation (first carried out by Ilya Frank and Igor Tamm) shows that the energy loss by the particle per unit path length, dW_C/dx, is proportional to $\int [1 - 1/\beta^2 n^2(\omega)]\omega \, d\omega$ for all ω such that $\beta n(\omega) > 1$. The weighting factor ω that appears in the integrand together with the fact that $n(\omega)$ is an increasing function of ω means that the emitted radiation is concentrated at the higher frequencies. Thus, when fast electrons pass through water, a characteristic bluish glow results. A variety of research instruments have been designed to determine the velocity of charged particles by measuring the Čerenkov radiation angle θ_C. ◀

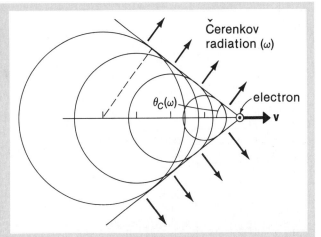

Fig. 39–4. The conical electromagnetic shock wave produced when a charged particle moves through a medium with a speed $v > c/n$. The shock wave consists of Čerenkov radiation.

39–2   ELECTRIC DIPOLE ANTENNA

A practical device for producing electromagnetic radiation consists of an oscillator and a dipole antenna, as illustrated in Fig. 39–5. To simplify the analysis we assume that the length 2ℓ of the antenna is short compared with the wavelength λ of the emitted radiation,* that is, $2\ell \ll \lambda$. An antenna of this sort is called a *Hertzian dipole.*† For such a dipole, we can consider the phase of the current in each antenna element to be the same at any instant along the entire length ℓ.

When the current I is directed as indicated in Fig. 39–5, positive charge $+q$ accumulates at the upper tip of the antenna while negative charge $-q$ accumulates at the lower tip. We can deduce the resulting radiation pattern by using the current I to determine $\mathbf{H}$ or by using the charge q to determine $\mathbf{E}$. For distances $r \gg \ell$, the distant-field analysis is valid and we again have plane waves with $|\mathbf{H}|/|\mathbf{E}| = \sqrt{\epsilon_0/\mu_0}$, for vacuum (see Eq. 38–16).

Suppose that the driving oscillator produces an antenna current $I(t) = I_0 \sin \omega t$. Because $2\ell \ll \lambda$, the current at any instant has nearly the same magnitude at every point on the antenna. The current causes charges $\pm q_0 \cos \omega t$ to accumulate at the tips of the antenna, where $q_0 = I_0/\omega$. Thus, we have in effect an oscillating electric dipole that consists of a pair of time-varying charges separated by a

---

*This requirement is called the *dipole approximation* and is used in the analysis of ideal antennas. However, in practical situations the requirement can be relaxed to $2\ell \lesssim \lambda/4$ without seriously degrading the results.

†After the German physicist Heinrich Hertz (1857–1894), who extensively studied the propagation of electromagnetic radiation in the 1880's.

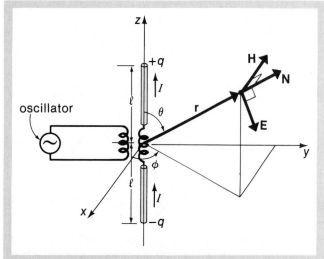

Fig. 39–5. Geometry of a Hertzian dipole antenna and the field vectors at a particular instant. Notice that the oscillator is inductively coupled to the antenna; this is common practice.

fixed distance 2ℓ. This dipole produces the same distant field as does a dipole that consists of a pair of fixed charges with magnitude $q_0 = I_0/\omega$ separated by a time-varying distance, $z = 2\ell \cos \omega t$. Then, the acceleration of each charge has the magnitude $a_z = 2\ell\omega^2 \cos \omega t$. Look again at Eqs. 39–2; we expect the product qa that appears in these equations now to be replaced by $q_0 \times 2\ell\omega^2 \cos \omega t$. To obtain $\mathbf{E}(\mathbf{r}, t)$ and $\mathbf{H}(\mathbf{r}, t)$, we must modify this expression for qa to take account of the fact that there is a finite propagation time between the source and the field point $\mathbf{r}$. This is the same effect we discussed in Section 20–3 for spherical sound waves. Thus, the term $\cos \omega t$ must be replaced by $\cos(kr - \omega t)$ where $k = 2\pi/\lambda$. (Compare Eq. 20–13a.) Altogether, then, we have for a Hertzian dipole antenna,

$$\mathbf{E}(\mathbf{r}, t) = \frac{I_0\ell\omega}{2\pi\epsilon_0 c^2} \frac{\cos(kr - \omega t)}{r} \sin \theta \, \hat{\mathbf{u}}_\theta \quad \textbf{(39–4a)}$$

$$\mathbf{H}(\mathbf{r}, t) = \frac{I_0\ell\omega}{2\pi c} \frac{\cos(kr - \omega t)}{r} \sin \theta \, \hat{\mathbf{u}}_\phi \quad \textbf{(39–4b)}$$

These expressions are valid for $2\ell \ll \lambda$ and $r \gg \lambda$. As before, the field vectors are mutually perpendicular, and the Poynting vector, $\mathbf{N} = \mathbf{E} \times \mathbf{H}$, is directed radially outward. Thus, the radiation intensity in a direction specified by the angle θ is proportional to $\sin^2 \theta$ (a *dipole radiation pattern*).

A precise calculation using Maxwell's equations produces expressions for the field vectors in the distant field that are the same as Eqs. 39–4. In addition, however, such a treatment reveals the presence of complicated near fields in which power surges back and forth between the field and the antenna.

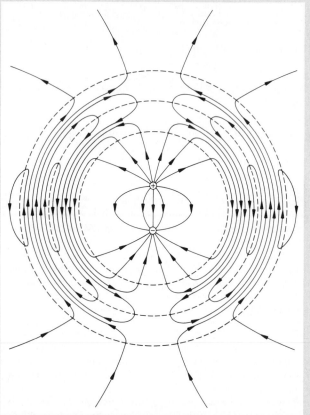

Fig. 39–6. The electric field for a sinusoidally driven Hertzian dipole after oscillating through one complete cycle and returning to the static condition. The near field is a static dipole field, but at large distance the field pattern approaches that for a plane wave and propagates outward with the speed c. (From *Principles of Physics* by J. P. Hurley and C. Garrod. Houghton Mifflin Co., Boston, 1978.)

Consider a Hertzian dipole that undergoes sinusoidal oscillations. Figure 39–6 shows the electric field lines for the radiation from such a dipole that has completed one cycle of oscillation and thereafter is static. The complete electric field is obtained by rotating this field pattern about a vertical axis through the dipole. Notice that the field lines near the dipole are the same as those for a static dipole (see Fig. 28–5). For the largest distances shown, a segment of the field begins to resemble a plane wave. The disturbance propagates outward with the speed c.

Example 39–2

A Hertzian dipole has $2\ell = 5$ cm and is driven at a frequency $\nu = 150$ MHz. The peak current in the antenna is $I_0 = 30$ A. Determine the average radiated power and the effective resistance (the *radiation resistance*) of the antenna.

Solution:

We could obtain the desired result by determining the Poynting vector from Eqs. 39–4 for **E** and **H**, then integrating $\mathbf{N} \cdot \hat{\mathbf{n}} \, dA$ over a convenient surface. Instead, it is easier to take the result of Example 39–1 for an accelerating charge, namely, $P = q^2 a^2 / 6\pi\epsilon_0 c^3$, and make the substitution of $2I_0\ell\omega$ for the peak value of qa. Then, the peak power is

$$P_0 = \frac{2I_0^2\ell^2\omega^2}{3\pi\epsilon_0 c^3}$$

Now, we use $\omega = 2\pi\nu = 2\pi c/\lambda$ to obtain the average power, $\bar{P} = \frac{1}{2}P_0$; thus,

$$\bar{P} = \frac{4\pi}{3}\frac{I_0^2}{\epsilon_0 c}\left(\frac{\ell}{\lambda}\right)^2$$

The wavelength of the radiation is $\lambda = c/\nu = 2.00$ m. Then, we find

$$\bar{P} = \frac{4\pi(30 \text{ A})^2}{3(8.85 \times 10^{-12} \text{ F/m})(3.00 \times 10^8 \text{ m/s})}\left(\frac{0.025 \text{ m}}{2 \text{ m}}\right)^2$$
$$= 221.9 \text{ W}$$

The radiation resistance R_{rad} of the antenna is defined by the relationship

$$\bar{P} = \frac{1}{2}I_0^2 R_{\text{rad}}$$

Using the expression above for $\bar{P}$, we have

$$R_{\text{rad}} = \frac{8\pi}{3\epsilon_0 c}\left(\frac{\ell}{\lambda}\right)^2$$

For vacuum, we have $c = 1/\sqrt{\epsilon_0\mu_0}$. Thus,

$$R_{\text{rad}} = \frac{8\pi}{3}\sqrt{\frac{\mu_0}{\epsilon_0}}\left(\frac{\ell}{\lambda}\right)^2 = \frac{8\pi}{3}Z_0(\ell/\lambda)^2$$

For the present case, in which $\lambda = 80\ell$, we find

$$R_{\text{rad}} = 0.493 \ \Omega$$

The Ohmic resistance of the antenna might be several ohms. Then, more power is expended in heat than is radiated into space. Consequently, a Hertzian dipole is not an efficient radiator. For efficient radiation, the antenna length must be comparable with the wavelength. But then the dipole approximation ($2\ell \ll \lambda$) is no longer valid and a more complicated analysis is required.

Example 39–3

Two identical Hertzian dipole antennas, separated by a distance d, are mounted parallel, as shown in diagram (a), where the axes of the dipoles are perpendicular to the plane of the page. The current in each antenna has the same magnitude, but the current in antenna B has a phase lag α compared with the current in antenna A. Determine the field strength E for the combination as a function of the azimuthal angle ϕ in the equatorial plane, $\theta = \pi/2$ (the plane of the page in the diagram).

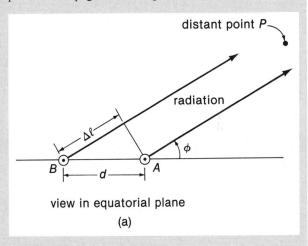

view in equatorial plane

(a)

Solution:

To arrive at a distant point P, the radiation emitted by antenna B must travel a distance different from that traveled by the radiation emitted by antenna A. This difference in path length, $\Delta\ell = d\cos\phi$, introduces a phase difference, $\delta = (2\pi d/\lambda)\cos\phi$, between the waves superimposed at P. In addition, the wave from antenna B has a phase lag α. Thus, the electric field strengths of the two waves are

$$E_A = E_0\cos(kr - \omega t)$$
$$E_B = E_0\cos\left(kr + \frac{2\pi d}{\lambda}\cos\phi - \omega t + \alpha\right)$$

Adding, we find

$$E = E_A + E_B$$
$$= 2E_0\cos\left(\frac{\pi d}{\lambda}\cos\phi + \frac{\alpha}{2}\right)\cos\left(kr - \omega t + \frac{\pi d}{\lambda}\cos\phi + \frac{\alpha}{2}\right)$$

(•Can you verify this result by using the various trigonometric identities in Appendix A of Part I?) Taking the ratio of the peak values of E and either E_A or E_B, we find

$$\left.\frac{E}{E_A}\right|_{\text{peak}} = 2\cos\left(\frac{\pi d}{\lambda}\cos\phi + \frac{\alpha}{2}\right)$$

(b)

(c)

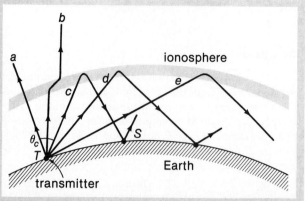

Fig. 39–7. Radio waves are incident on the ionosphere at various angles. Rays a and b have normal or near-normal incidence and pass through the layer into space. Ray c is the critical ray and corresponds to θ_c, the smallest angle for which total reflection occurs. The distance $\overline{TS}$ (the *skip distance*) defines the *blind region* that receives no reflected waves (*sky waves*).

For the case $d = \lambda/4$ and $\alpha = \pi/2$, this ratio function becomes a *cardioid*, as shown in diagram (b). Other values of d and α lead to the patterns shown in diagram (c).

39-3 GUIDED WAVES

Ionospheric Reflection. It is well known that radio signals can be received at great distances from the source, far beyond the range of line-of-sight transmission. This phenomenon is due to the guiding of the waves by reflection from a conducting layer of ionized air high in the Earth's atmosphere (see Fig. 39–7). The reflecting layer, called the *ionosphere*, actually consists of several bands at

altitudes of 100 to 400 km. The ionization is produced primarily by the ultraviolet component (the high-frequency component) of the incident solar radiation. The density of the atmosphere at altitudes of a few hundred kilometers is very low, and the electrons that are released in ionization processes have mean free paths of many meters. The great length of the mean free paths limits the rate of recombination of ions and electrons, with the result that the electron density is large—10^4 to 10^5 electrons/cm³. The electric field of an incident wave causes these electrons to oscillate, thereby absorbing energy and then reradiating it in the form of a reflected wave. Long-distance transmissions are possible by means of rays such as d and e in Fig. 39–7. If a wave is reflected by the Earth's surface (as at S), a secondary ionospheric reflection can take place, thereby increasing the transmission distance.

Waves Guided by Parallel Conducting Planes. Figure 39–8 shows a plane wave that progresses in the $+y$-direction by making repeated reflections from two parallel conducting planes separated by a distance a. In constructing this diagram we have selected a particular wave, one with the electric vector parallel to the x-axis and the propagation vector at an angle θ with respect to the z-axis. The electric vector of the incident wave (left-hand panel) shown in Fig. 39–8 is* (compare Section 20–3)

$$\mathbf{E}_i(\mathbf{r}, t) = -E_0 \cos (\mathbf{k}_i \cdot \mathbf{r} - \omega t)\hat{\mathbf{i}}$$
$$= -E_0 \cos (ky \sin \theta + kz \cos \theta - \omega t)\hat{\mathbf{i}}$$

---

*The propagation vector for the incident wave is $\mathbf{k}_i = k(\hat{\mathbf{j}} \sin \theta + \hat{\mathbf{k}} \cos \theta)$. The position vector is $\mathbf{r} = x\hat{\mathbf{i}} + y\hat{\mathbf{j}} + z\hat{\mathbf{k}}$. Hence, $\mathbf{k}_i \cdot \mathbf{r} = k(y \sin \theta + z \cos \theta)$. The propagation vector for the reflected wave is $\mathbf{k}_r = k(\hat{\mathbf{j}} \sin \theta - \hat{\mathbf{k}} \cos \theta)$, and $\mathbf{k}_r \cdot \mathbf{r} = k(y \sin \theta - z \cos \theta)$.

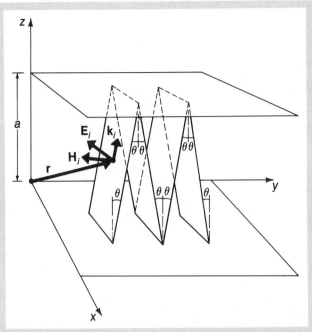

Fig. 39-8. A plane wave propagates in the y-direction by repeated reflections from a pair of parallel conducting planes.

For the reflected wave, we have

$$\mathbf{E}_r(\mathbf{r}, t) = +E_0 \cos{(ky \sin\theta - kz \cos\theta - \omega t)}\hat{\mathbf{i}}$$

The total electric field between the planes is equal to the superimposed sum,

$$\mathbf{E}(\mathbf{r}, t) = \mathbf{E}_i(\mathbf{r}, t) + \mathbf{E}_r(\mathbf{r}, t)$$
$$= 2E_0 \sin{(kz \cos\theta)} \sin{(ky \sin\theta - \omega t)}\hat{\mathbf{i}} \quad \textbf{(39-5)}$$

(•Verify this result by using the identities for $\cos{(A + B)}$ and $\cos{(A - B)}$.)

[The corresponding expression for the magnetic field strength $\mathbf{H}$ is somewhat more complicated because $\mathbf{H}$ has both y- and z-components. See Problem 39-6.]

Examining Eq. 39-5, we see that $\mathbf{E}(\mathbf{r}, t)$ is a wave that travels in the $+y$-direction with a wave number $k' = k \sin\theta = (2\pi/\lambda_0)\sin\theta$, where λ_0 is the free-space wavelength (the wavelength for propagation in vacuum). The phase velocity of the wave is (see Eq. 19-14)

$$v = \frac{\omega}{k'} = \frac{v\lambda_0}{\sin\theta} = \frac{c}{\sin\theta} \quad \textbf{(39-6)}$$

Although $v > c$, this does not violate the relativity principle, as we discuss later.

We assume that the parallel planes are perfect conductors; consequently, the electric field $\mathbf{E}$ must be zero for

$z = 0$ and for $z = a$. The factor $\sin{(kz \cos\theta)}$ automatically makes $\mathbf{E}$ vanish for $z = 0$. In order to have $\mathbf{E} = 0$ for $z = a$, we require $ka \cos\theta = n\pi$, for any integer n. Because $k = 2\pi/\lambda_0$, the wavelength λ_0 determines the permitted values of θ according to

$$\cos\theta = n\frac{\lambda_0}{2a}, \qquad n = 1, 2, 3, \cdots \quad \textbf{(39-7)}$$

These waves are referred to as *transverse electric modes of order n* or TE_n-mode transmissions. This designation arises from the fact that $\mathbf{E}$ is perpendicular (transverse) to the *plane of incidence,* a plane that contains both $\mathbf{k}_i$ and $\mathbf{k}_r$ (the y-z plane in Fig. 39-8). (There also exist independent modes of transmission—the TM_n modes—in which $\mathbf{H}$ is transverse to the plane of incidence.)

Notice that the longest (free-space) wavelength λ_0 that can be propagated between the planes occurs for $n = 1$ and $\cos\theta = 1$: thus,

$$\lambda_{0,m} = 2a \quad \textbf{(39-8)}$$

This value is referred to as the *cutoff wavelength.*

The phase velocity can be written in terms of λ_0 and a by noting that

$$\sin\theta = \sqrt{1 - \cos^2\theta} = \sqrt{1 - n^2(\lambda_0/2a)^2}$$

Using this expression for $\sin\theta$ in Eq. 39-6, there results

$$v = \frac{c}{\sqrt{1 - n^2(\lambda_0/2a)^2}}, \qquad n = 1, 2, 3, \cdots \quad \textbf{(39-9)}$$

The quantity $2\pi/k'$ is called the *guided wavelength* λ_g and is the wavelength associated with transmission along the y-direction. Evidently,

$$\lambda_g = \frac{\lambda_0}{\sin\theta}$$
$$= \frac{\lambda_0}{\sqrt{1 - n^2(\lambda_0/2a)^2}}, \qquad n = 1, 2, 3, \cdots \quad \textbf{(39-10)}$$

The velocity with which a signal can be propagated by the guided wave is called the *group velocity* v_g (see Section 19-9) and is equal to the component of the free-space velocity in the y-direction; thus,

$$v_g = c \sin\theta \quad \textbf{(39-11)}$$

This same result can be obtained by using Eq. 19-45, namely, $v_g = v - \lambda_g \, dv/d\lambda_g$; see Problem 39-7.

Combining Eqs. 39-6 and 39-11, we see that

$$vv_g = c^2 \quad \textbf{(39-12)}$$

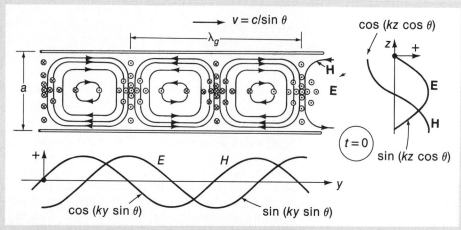

Fig. 39–9. The variation with y and z of the electric and magnetic fields between parallel conducting planes for the instant $t = 0$. As time increases, the fields propagate to the right with the velocity $v_g = c \sin \theta$.

Whereas the phase velocity v is always greater than c, the group velocity v_g is always less than c. A calculation of the time-averaged Poynting vector for the fields between the plates shows that energy flows in the y-direction with the group velocity v_g (see Problem 39–6). Thus, it is v_g that must obey the relativistic restriction ($v_g < c$).

Figure 39–9 shows the patterns of the electric and magnetic fields between the parallel conducting planes for the instant $t = 0$. The electric field is specified by Eq. 39–5, and the corresponding expression for the magnetic field is the subject of Problem 39–6. Notice that, as required, the electric and magnetic field lines are everywhere mutually perpendicular. Notice also that $\mathbf{E}$ and $\mathbf{H}$ have a phase difference of 90°.

TE and TM waves can also be propagated through hollow conducting pipes (called *waveguides*) that have rectangular or circular cross sections. These cases are treated by using techniques very similar to those emplyed here, although the details are more complicated. Many of the central features of the analysis—such as $v > c$, $v_g < c$, $v v_g = c^2$, and the existence of a cutoff wavelength—are also present for waveguides.

Example 39–4

Two parallel, perfectly conducting planes are separated by a distance of 2 cm.

(a) Calculate the cutoff wavelength $\lambda_{0,m}$ for the TE_1 mode.

(b) For a free-space wavelength $\lambda_0 = \lambda_{0,m}/\sqrt{5}$, determine θ, λ_g, v, and v_g for the TE_1 mode.

Solution:

(a) Using Eq. 39–8, we have $\lambda_{0,m} = 2a = 4$ cm.

(b) We use $n = 1$ and $\lambda_0 = \lambda_{0,m}/\sqrt{5} = 4/\sqrt{5}$ cm. Then, from Eq. 39–7, we find

$$\cos \theta = n \frac{\lambda_0}{2a} = \frac{1}{\sqrt{5}}; \qquad \theta = 63.4°$$

Eq. 39–10: $\lambda_g = \dfrac{\lambda_0}{\sin \theta} = \dfrac{4/\sqrt{5} \text{ cm}}{2/\sqrt{5}} = 2.00$ cm

Eq. 39–6: $v = \dfrac{c}{\sin \theta} = \dfrac{c}{2/\sqrt{5}} = 1.118c$

Eq. 39–11: $v_g = c \sin \theta = c(2/\sqrt{5}) = 0.894c$

Parallel-Wire Transmission Line.

In the types of problems considered in earlier chapters, oscillatory electric sources were connected to various circuit elements (R, L, and C) that determined the current and energy flow in the circuits. The presence of electric and magnetic fields was a secondary consideration, the effects of the fields being replaced by lumped (localized) capacitance and inductance. We now discuss a different approach in which the fields are shaped by the *distributed* capacitive and inductive properties of conductors. By way of example we consider a *balanced parallel-wire transmission line*.

Figure 39–10a shows two identical parallel conductors with radii a whose centers are separated by a distance b. The conductors are assumed to have negligible resistance. This transmission line is driven by an oscillating voltage source and terminated by a load resistance R_L. Figure 39–10b shows a segment of the line with length ℓ and an incremental element Δx of that segment. The current oscillates

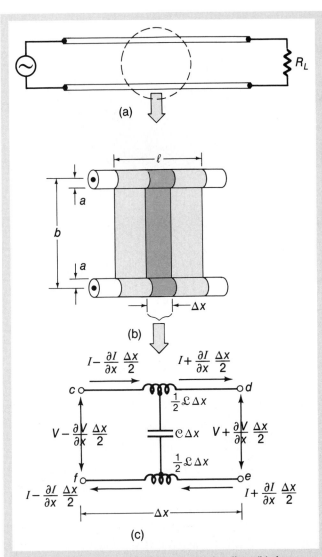

Fig. 39-10. (a) A parallel-wire transmission line. (b) A segment of the line with length ℓ. (c) Schematic circuit of the incremental element Δx, showing the distributed capacitance and inductance. The magnitude of the current at the center of the inductances is I and the voltage across the capacitance is V. Both I and V vary with x.

back and forth in the conductors, producing time-varying electric and magnetic fields. In discussing the currents in the line, we may think in terms of the effective distributed capacitance and self-inductance per unit length of line, $\mathcal{C} = C_l/\ell$ and $\mathcal{L} = L_l/\ell$, respectively, where C_l is the total capacitance of the length ℓ and similarly for L_l. We assume an inductance $\frac{1}{2}\mathcal{L}\,\Delta x$ to exist in each leg of the element Δx. The capacitance $\mathcal{C}\,\Delta x$ acts between the centers of the legs of the element. (See Fig. 39-10c.)

We apply Kirchhoff's second rule to the path $cdefc$ around the line element Δx (Fig. 39-10c); thus,

$$\frac{\mathcal{L}}{2}\Delta x\frac{\partial I}{\partial t} + \left(V + \frac{\partial V}{\partial x}\frac{\Delta x}{2}\right) + \frac{\mathcal{L}}{2}\Delta x\frac{\partial I}{\partial t} - \left(V - \frac{\partial V}{\partial x}\frac{\Delta x}{2}\right)$$

$$= 0$$

from which

$$\frac{\partial V}{\partial x} = -\mathcal{L}\frac{\partial I}{\partial t} \tag{39-13}$$

Applying the charge continuity condition, $\Delta I = -\partial Q/\partial t$, to the element Δx, we obtain

$$\left(I + \frac{\partial I}{\partial x}\frac{\Delta x}{2}\right) - \left(I - \frac{\partial I}{\partial x}\frac{\Delta x}{2}\right) = -\frac{\partial Q}{\partial t} - \frac{\partial}{\partial t}(V\mathcal{C}\,\Delta x)$$

from which

$$\frac{\partial I}{\partial x} = -\mathcal{C}\frac{\partial V}{\partial t} \tag{39-14}$$

Next, we differentiate Eq. 39-13 with respect to x and Eq. 39-14 with respect to t; thus,

$$\frac{\partial^2 V}{\partial x^2} = -\mathcal{L}\frac{\partial^2 I}{\partial x \partial t} \quad \text{and} \quad \frac{\partial^2 I}{\partial t \partial x} = -\mathcal{C}\frac{\partial^2 V}{\partial t^2}$$

Combining these equations,* we find

$$\frac{\partial^2 V}{\partial x^2} = \mathcal{L}\mathcal{C}\frac{\partial^2 V}{\partial t^2} \tag{39-15}$$

Also, differentiating Eq. 39-13 with respect to t and Eq. 39-14 with respect to x and combining the results leads to

$$\frac{\partial^2 I}{\partial x^2} = \mathcal{L}\mathcal{C}\frac{\partial^2 I}{\partial t^2} \tag{39-16}$$

We recognize Eqs. 39-15 and 39-16 to be wave equations, with $V(x, t)$ and $I(x, t)$ each propagating with the phase velocity

$$v = \frac{1}{\sqrt{\mathcal{L}\mathcal{C}}} \tag{39-17}$$

According to the results of Example 39-5 and Problem 39-8, we have

$$\mathcal{C} = \pi\epsilon_0/\ln(b/a) \tag{39-18a}$$

and

$$\mathcal{L} = (\mu_0/\pi)\ln(b/a) \tag{39-18b}$$

Thus,

$$v = \frac{1}{\sqrt{\mathcal{L}\mathcal{C}}} = \frac{1}{\sqrt{\epsilon_0\mu_0}} = c$$

*The two second partial derivatives of I are equal (see the appendix, Calculus IV, in Part I).

Fig. 39–11. The field patterns at a particular instant for a parallel-wire transmission line. The charge and current distributions are also shown in (a).

That is, the phase velocity v of a wave on a lossless transmission line (that is, a line with zero resistance) is just equal to the speed of light (in vacuum).

Assume that the solution of the wave equation for $V(x, t)$ has the form

$$V(x, t) = V_0 \cos (kx - \omega t) \qquad \text{(39–19)}$$

with $k = 2\pi/\lambda$, $\omega = 2\pi\nu$, and $\lambda\nu = v = \omega/k$. Then, from Eq. 39–14, we have

$$\frac{\partial I}{\partial x} = -\mathcal{C} V_0 \omega \sin (kx - \omega t)$$

so that

$$I(x, t) = \frac{\mathcal{C} V_0 \omega}{k} \cos (kx - \omega t) \qquad \text{(39–20)}$$

Comparing Eqs. 39–19 and 39–20, we see that $V(x, t)$ and $I(x, t)$, and hence $E(x, t)$ and $H(x, t)$, are all *in phase*. Figure 39–11 shows the behavior of these quantities at a particular instant. (•What is the direction of the Poynting vector for this case?) Notice that the fields E and H are in

phase, in marked contrast with the phase relationship we obtained for the case of propagation between parallel planes (Fig. 39–9).

The characteristic *impedance* of the transmission line is defined to be $Z_t = V(x, t)/I(x, t)$. Using Eqs. 39–19 and 39–20, we find

$$Z_t = \frac{V_0 k}{\mathcal{C} V_0 \omega} = \frac{1}{\mathcal{C} v} = \sqrt{\frac{\mathcal{L}}{\mathcal{C}}}$$

Then, using Eqs. 39–18 for $\mathcal{L}$ and $\mathcal{C}$, we obtain

$$Z_t = \frac{1}{\pi} \sqrt{\frac{\mu_0}{\epsilon_0}} \ln (b/a) = \frac{Z_0}{\pi} \ln (b/a) \qquad \text{(39–21)}$$

If the load resistance R_L that terminates the line (see Fig. 39–10a) is set equal to Z_t, all of the energy transmitted down the line will be absorbed by R_L and dissipated as heat. When the impedance is matched in this way, there is no reflected wave.

Example 39–5

Determine the capacitance per unit length of the parallel-wire transmission line shown in Fig. 39–10.

Solution:

Consider a short segment of the line with length ℓ that carries a charge $+Q$ on the upper conductor and a charge $-Q$ on the lower conductor. We can calculate the electric field due to the upper conductor by constructing the Gaussian surface shown in the diagram. We find, using Gauss's law,

$$2\pi r \ell E' = Q/\epsilon_0$$

If $a \ll b$, the potential difference V' between the wires that results from the presence of the field E' is

$$V' = \int_a^b E' \, dr = \frac{Q}{2\pi\epsilon_0 \ell} \ln (b/a)$$

The charge $-Q$ on the lower conductor contributes equally to the total potential difference. Hence,

$$V = 2V' = \frac{Q}{\pi\epsilon_0 \ell} \ln (b/a)$$

The capacitance per unit length is $\mathcal{C} = Q/V\ell$; thus,

$$\mathcal{C} = \pi\epsilon_0 / \ln (b/a)$$

which is the result quoted in Eq. 39-18a.

Example 39-6

The connection between a TV antenna and the receiver is often made by a 300-Ω balanced parallel-wire transmission line (the familiar plastic-covered "twin lead"). If the spacing between the wires is $b = 0.80$ cm, what wire diameter $2a$ is required? (Neglect the dielectric properties of the plastic separator.)

Solution:

Using Eq. 39-21, we have

$$Z = 300 \ \Omega = \frac{Z_0}{\pi} \ln (b/a)$$

or

$$\frac{b}{a} = e^{300\pi/377} = 12.2$$

so that the required wire diameter is

$$2a = \frac{2b}{12.2} = \frac{2(0.80 \text{ cm})}{12.2} = 0.13 \text{ cm}$$

39-4 THE WAVE EQUATION

We now use Maxwell's equations in differential form to give a derivation of the wave equation that describes electromagnetic radiation. It is assumed that the medium is linear, homogeneous, and isotropic, so that $\mathbf{D} = \epsilon_0\kappa\mathbf{E}$ and $\mathbf{B} = \mu_0\kappa_M\mathbf{H}$. It is also assumed that the region of interest is free of real charges and currents, so that $\rho = 0$ and $\mathbf{J} = 0$.

In Section 38-1 we showed, for these same conditions, that the postulated solutions for $\mathbf{E}(\mathbf{r}, t)$ and $\mathbf{H}(\mathbf{r}, t)$ (Eqs. 38-1 and 38-3, respectively) do in fact satisfy Maxwell's equations expressed in integral form. We proceeded in this manner because the general wave equation (which is a *differential* equation) cannot be obtained from the *integral* equations (at least, not without the *ad hoc* introduction of key aspects of the solution). To give such a derivation requires the use of the differential formulation of Maxwell's equations.

For the stipulated conditions of the medium, Maxwell's equations can be written in differential form as (see Table 36-3)

$$\nabla \cdot \mathbf{E} = 0 \quad \text{and} \quad \nabla \times \mathbf{E} = -\mu_0\kappa_M \frac{\partial \mathbf{H}}{\partial t}$$
$$\nabla \cdot \mathbf{H} = 0 \quad \text{and} \quad \nabla \times \mathbf{H} = +\epsilon_0\kappa \frac{\partial \mathbf{E}}{\partial t} \tag{39-22}$$

We seek a plane-wave solution, $\mathbf{E}(z, t)$, that is a function only of z and t. Then, it follows that $\partial E_x/\partial x = 0$ and $\partial E_y/\partial y = 0$; because the divergence of $\mathbf{E}$ is zero, we conclude that $\partial E_z/\partial z = 0$ also. Thus, E_z cannot be a function of z (nor of x or y). The easiest way to satisfy this requirement is to take $E_z = 0$.* We orient the coordinate axes so that $\mathbf{E}$ is parallel to the x-axis. (That is, $\mathbf{E}$ is polarized in the

---

* We could, instead, take $E_z = $ constant, but such a static field would play no essential role in wave propagation.

x-direction.) Thus,

$$\mathbf{E} = E_x(z, t)\hat{\mathbf{i}} \qquad (39\text{-}23)$$

The curl equation for **E** gives*

$$\frac{\partial E_x}{\partial z}\hat{\mathbf{j}} = -\mu_0\kappa_M\left(\frac{\partial H_x}{\partial t}\hat{\mathbf{i}} + \frac{\partial H_y}{\partial t}\hat{\mathbf{j}} + \frac{\partial H_z}{\partial t}\hat{\mathbf{k}}\right)$$

Equating vector components, we see that $\partial H_x/\partial t = 0$ and $\partial H_z/\partial t = 0$. Then, we have

$$\frac{\partial E_x}{\partial z} = -\mu_0\kappa_M\frac{\partial H_y}{\partial t} \qquad (39\text{-}24)$$

Although these conditions permit static magnetic fields in the *x*- and *z*-directions, the simplest case is that for which $H_x = 0$ and $H_z = 0$, so that **H** has only a *y*-component.

The curl equation for **H** gives†

$$-\frac{\partial H_y}{\partial z}\hat{\mathbf{i}} + \frac{\partial H_y}{\partial x}\hat{\mathbf{k}} = \epsilon_0\kappa\frac{\partial E_x}{\partial t}\hat{\mathbf{i}}$$

Equating vector components, we see that $\partial H_y/\partial x = 0$. Then, we have

$$-\frac{\partial H_y}{\partial z} = \epsilon_0\kappa\frac{\partial E_x}{\partial t} \qquad (39\text{-}25)$$

Now, the divergence of **H** is zero, and because we have $H_x = 0$ and $H_z = 0$, it follows that $\partial H_y/\partial y = 0$. Thus, **H** has only a *y*-component and H_y is a function only of *z* and *t*; that is,

$$\mathbf{H} = H_y(z, t)\hat{\mathbf{j}} \qquad (39\text{-}26)$$

Differentiating Eq. 39-24 with respect to *z* and Eq. 39-25 with respect to *t* gives

---

*The Maxwell equation for the curl of **E** is

$$\nabla \times \mathbf{E} = \begin{vmatrix} \hat{\mathbf{i}} & \hat{\mathbf{j}} & \hat{\mathbf{k}} \\ \dfrac{\partial}{\partial x} & \dfrac{\partial}{\partial y} & \dfrac{\partial}{\partial z} \\ E_x & 0 & 0 \end{vmatrix} = \frac{\partial E_x}{\partial z}\hat{\mathbf{j}} = -\mu_0\kappa_M\frac{\partial \mathbf{H}}{\partial t}$$

†The Maxwell equation for the curl of **H** is

$$\nabla \times \mathbf{H} = \begin{vmatrix} \hat{\mathbf{i}} & \hat{\mathbf{j}} & \hat{\mathbf{k}} \\ \dfrac{\partial}{\partial x} & \dfrac{\partial}{\partial y} & \dfrac{\partial}{\partial z} \\ 0 & H_y & 0 \end{vmatrix} = -\frac{\partial H_y}{\partial z}\hat{\mathbf{i}} + \frac{\partial H_y}{\partial x}\hat{\mathbf{k}} = \epsilon_0\kappa\frac{\partial \mathbf{E}}{\partial t}$$

$$\frac{\partial^2 E_x}{\partial z^2} = -\mu_0\kappa_M\frac{\partial^2 H_y}{\partial z\,\partial t}$$

and

$$-\frac{\partial^2 H_y}{\partial t\,\partial z} = \epsilon_0\kappa\frac{\partial^2 E_x}{\partial t^2}$$

Combining these equations, we find

$$\frac{\partial^2 E_x}{\partial z^2} = \epsilon_0\mu_0\kappa\kappa_M\frac{\partial^2 E_x}{\partial t^2} \qquad (39\text{-}27)$$

Also, differentiating Eq. 39-24 with respect to *t* and Eq. 39-25 with respect to *z* and combining the results leads to

$$\frac{\partial^2 H_y}{\partial z^2} = \epsilon_0\mu_0\kappa\kappa_M\frac{\partial^2 H_y}{\partial t^2} \qquad (39\text{-}28)$$

Equations 39-27 and 39-28 are the wave equations we have been seeking. In each case we see that the phase velocity is

$$v = \frac{1}{\sqrt{\epsilon_0\mu_0\kappa\kappa_M}} \qquad (39\text{-}29)$$

For vacuum conditions we have $\kappa = 1$ and $\kappa_M = 1$, so that $v = 1/\sqrt{\epsilon_0\mu_0} = c$.

The general solutions to the wave equations for **E** and **H** are (see Section 19-2)

$$\mathbf{E}(z, t) = E_0 f(z \pm vt)\hat{\mathbf{i}}$$

and

$$\mathbf{H}(z, t) = H_0 f(z \pm vt)\hat{\mathbf{j}}$$

For a sinusoidal wave traveling in the +*z*-direction with $k = 2\pi/\lambda$ and $\omega = 2\pi\nu$, the solutions can be expressed as

$$\begin{aligned} \mathbf{E}(z, t) &= E_0\cos(kz - \omega t)\hat{\mathbf{i}} \\ \mathbf{H}(z, t) &= H_0\cos(kz - \omega t)\hat{\mathbf{j}} \end{aligned} \qquad (39\text{-}30)$$

which are just the wave solutions that we postulated in Section 38-1.

We can find the relationship connecting E_0 and H_0 by substituting Eqs. 39-30 into either Eq. 39-24 or Eq. 39-25. Choosing the latter, we have

$$kH_0\sin(kz - \omega t) = \epsilon_0\kappa\omega E_0\sin(kz - \omega t) \qquad (39\text{-}31)$$

from which

$$\frac{E_0}{H_0} = \frac{k}{\epsilon_0\kappa\omega} = \sqrt{\frac{\mu_0\kappa_M}{\epsilon_0\kappa}}$$

which is just the result expressed by Eq. 38-15.

Note carefully the difference between the present treatment leading to the key equations (Eqs. 39–24 and 39–25) and the procedure followed in Section 38–1 that resulted in the *same* equations (Eqs. 38–5 and 38–9). In the discussion here we assumed only that **E** is a function of z

and t; all other field characteristics were then derived. In contrast, in Section 38–1 we assumed that both fields depend only on z and t, and in addition we assumed that **H** is perpendicular to **E** and also perpendicular to the propagation vector **k** (Eq. 38–3).

PROBLEMS

Section 39–1

39–1 An electron is released from rest in an electric field of strength 10^3 V/m. Calculate the rate at which the electron radiates energy as it accelerates.

39–2 An electron produced in a radioactive β-decay process travels through the water moderator of a "swimming pool" reactor. The kinetic energy of the electron is 1.25 MeV. The index of refraction of water for visible light is $n = 1.33$. The proper mass of the electron is $m_e = (0.511 \text{ MeV})/c^2$. (a) Determine the ray angle θ_C for the emitted Čerenkov radiation. (b) What is the minimum electron kinetic energy for exciting Čerenkov radiation in water?

Section 39–2

39–3 An airborne radio transmitter utilizes a Hertzian dipole antenna with length $2\ell = \lambda/20$. A receiver located at ground level 10 km directly below the transmitter requires a peak signal strength of $20 \, \mu$V/m for the desired signal-to-noise ratio. (a) What peak current must flow in the antenna? (b) What is the required average power input to the antenna? (c) What is the radiation resistance of the antenna?

39–4 Another antenna arrangement, an extension of that in
•• Example 39–3, involves the use of four parallel, equally spaced Hertzian dipoles, as shown in the diagram. With respect to antenna A, let the phase lags for antennas B, C, and D be α, 2α, and 3α, respectively. (a) Show that, for a distant point P, the ratio of the peak electric field for the array to that for any one antenna alone is

$$\frac{\sin\left[(4\pi d/\lambda)\cos\phi + 2\alpha\right]}{\sin\left[(\pi d/\lambda)\cos\phi + \alpha/2\right]}$$

(b) Sketch the radiation pattern for $d = \lambda/4$ and $\alpha = \pi/2$. (Be certain to include in your calculation the angles $\phi = 0$, $\pi/3$, $\pi/2$, $2\pi/3$, and π.) Because the radiation pattern has a sharp maximum along the line of the antennas, this arrangement is called an *end-fire array*. [Hint: $\sum_{n=0}^{n=3} \cos(A + nB) = \cos(A + \frac{3}{2}B)\sin(2B)/\sin(B/2)$. Exercise care in evaluating the result for $\phi = \pi$; expand both numerator and denominator for small arguments.]

Section 39–3

39–5 A radio receiver is located 200 km from a transmitter. A signal travels to the receiver via a direct line-of-sight path and also via a wave reflected from an ionospheric layer (the so-called *E layer*) at an altitude of 100 km. Make the approximation of a flat Earth and calculate the time difference between the arrivals of the two signals.

39–6 Determine the magnetic field intensity **H** that ac-
•• companies the electric field **E** specified by Eq. 39–5 for the case of a TE-mode wave guided between parallel conducting planes. The incident and reflected magnetic vectors are shown in the diagram on the next page. (a) Show that

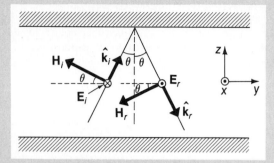

$$\mathbf{H}_i = H_0 \sin\theta \cos(ky\sin\theta + kz\cos\theta - \omega t)\hat{\mathbf{k}}$$
$$- H_0 \cos\theta \cos(ky\sin\theta + kz\cos\theta - \omega t)\hat{\mathbf{j}}$$
$$\mathbf{H}_r = -H_0 \sin\theta \cos(ky\sin\theta - kz\cos\theta - \omega t)\hat{\mathbf{k}}$$
$$- H_0 \cos\theta \cos(ky\sin\theta - kz\cos\theta - \omega t)\hat{\mathbf{j}}$$

(b) Add the expressions in (a) and show that

$$\mathbf{H} = \mathbf{H}_i + \mathbf{H}_r =$$
$$- 2H_0 \sin\theta \sin(kz\cos\theta) \sin(ky\sin\theta - \omega t)\hat{\mathbf{k}}$$
$$- 2H_0 \cos\theta \cos(kz\cos\theta) \cos(ky\sin\theta - \omega t)\hat{\mathbf{j}}$$

(c) Show that the time-averaged Poynting vector for the wave is

$$\bar{\mathbf{N}} = 2\epsilon_0 E_0^2 \sin^2(kz\cos\theta)(c\sin\theta)\hat{\mathbf{j}}$$
$$= uc\sin\theta\hat{\mathbf{j}}$$

where u is the energy density of the combined incident and reflected fields.

39-7   For guided waves between parallel conducting planes, show the following:

(a) $v = c\sqrt{1 + n^2(\lambda_g/2a)^2}$

(b) $\lambda_g\, dv/d\lambda_g = -cn^2(\lambda_g/2a)^2/\sqrt{1 + n^2(\lambda_g/2a)^2}$

(c) With $v_g = v - \lambda_g\, dv/d\lambda_g$, it follows that $v = c\sin\theta$.

39-8   Show that the inductance per unit length of the parallel-wire transmission line shown in Fig. 38-18 is $\mathcal{L} = (\mu_0/\pi)\ln(b/a)$, for $a \ll b$.

39-9   A sinusoidal wave with a peak current of 0.20 A travels along a balanced parallel-wire transmission line that consists of wires with radii $a = 1.0$ mm and spacing $b = 5.0$ cm. (a) What terminating load resistance R_L must be used so that there is no reflection from the load? (b) What is the peak voltage of the wave? (c) What average power is delivered to the load?

Section 39-4

39-10   For the plane-polarized wave, $\mathbf{E} = \mathbf{E}_0 \cos$
● $(\mathbf{k}\cdot\mathbf{r} - \omega t)$, show that the condition $\nabla\cdot\mathbf{E} = 0$ re-

quires that $\mathbf{E}_0$ be perpendicular to $\mathbf{k}$. Show that a similar result follows for $\mathbf{H} = \mathbf{H}_0 \cos(\mathbf{k}\cdot\mathbf{r} - \omega t)$.

39-11   (a) For the plane-polarized wave described by
● $\mathbf{E} = \mathbf{E}_0 \cos(\mathbf{k}\cdot\mathbf{r} - \omega t)$, show that

$$\nabla\times\mathbf{E} = -\mathbf{E}_0\times\mathbf{k}\sin(\mathbf{k}\cdot\mathbf{r} - \omega t)$$

(b) Use $\nabla\times\mathbf{E} = -\mu_0\kappa_M\,\partial\mathbf{H}/\partial t$ to show that

$$\mathbf{H} = [(\mathbf{k}\times\mathbf{E}_0)/k]\sqrt{\epsilon_0\kappa/\mu_0\kappa_M}\cos(\mathbf{k}\cdot\mathbf{r} - \omega t)$$

This proves that $\mathbf{E}$ and $\mathbf{H}$ are perpendicular and in phase. The result also demonstrates the correctness of Eq. 39-15.

Additional Problems

39-12   A Hertzian dipole can be used as a receiving antenna
●●   as well as a transmitting antenna. For an antenna with a length 2ℓ, an EMF of $2\ell E_0 \cos\omega t$ is developed when the antenna intercepts a sinusoidal plane wave with peak amplitude E_0 and with $\mathbf{E}$ parallel to the antenna. This antenna geometry and the equivalent circuit are shown on page 1065. Maximum power is absorbed from the wave when the load resistance R_L is equal to the radiation resistance R_{rad} of the antenna. (If the antenna has any reactance, this must be balanced by adding the complex-conjugate reactance to the load; that is, the antenna must be *tuned*.) Then, one half of the power in the wave is absorbed by the load and the other half is dissipated in the radiation resistance by being reradiated into space. (The incident wave induces currents in the antenna, causing it to act as a transmitter just as if the currents were induced by an attached generator.) Thus, the average power delivered to the load is

$$\bar{P} = \frac{(2\ell E_0)^2}{2}\cdot\frac{1}{4R_{rad}}$$

where $R_{rad} = (8\pi/3)Z_0(\ell/\lambda)^2$, as we found in Example 39-2 for a Hertzian dipole ($2\ell \ll \lambda$). Show that, independent of ℓ, the antenna presents an effective area A to the incident wave given by $A = (3/8\pi)\lambda^2$. [*Hint: $\bar{P} = \bar{N}A$.*]

39-13   An omnidirectional antenna aboard a satellite radiates an average power of 10 W at a frequency of 30 MHz. The satellite signal is received at a distance of 100 km by an antenna that has an effective cross-sectional area of $(3/8\pi)\lambda^2$. (Refer to Problem 39-12.) (a) Determine the average rate at which energy is intercepted by the antenna. (b) What is the peak value E_0 of the electric field of the intercepted wave?

Problem 39-12

39-14 An electromagnetic wave is traveling under
• conditions such that the phase velocity v is some
function of the wavelength $\lambda = v/\nu$. The function is
expressed in dimensionless form as $v = cf(\xi)$, with
$\xi = \lambda/b$, where b is an appropriate constant.
(a) Show that $\lambda \, dv/d\lambda = c\xi \, df/d\xi$. (b) According to
Eq. 19-46, the group velocity is $v_g = v - \lambda \, dv/d\lambda$.
Show that $vv_g = c^2(f^2 - f\xi \, df/d\xi)$. (c) In order to
meet the condition $vv_g = c^2$, show that we require
$v = c\sqrt{1 + (\lambda/b)^2}$. (d) The free-space wavelength
is $\lambda_0 = c/\nu$. Show that $v = c/\sqrt{1 - (\lambda_0/b)^2}$.

These conditions are met in the important cases
of wave propagation that we have studied, namely,
waves propagating through the ionosphere, waves
traveling between conducting planes, and waves
propagating inside hollow conducting pipes
(waveguides).

39-15 A *coaxial cable* (the cross section of which is
• shown in the diagram) has wave propagation characteristics that are similar to those of the parallel-
wire transmission line discussed in Section 39-3.

Consider the cable illustrated, which has an inner
conductor with a radius a, an outer conducting
jacket with a radius b, and a dielectric with constant
κ (and $\kappa_M = 1$) in the space between the conductors.
We can use the expression for the capacitance per
unit length for a coaxial cable obtained in Example
30-3, modified to account for the presence of the
dielectric, namely,

$$\mathcal{C} = 2\pi\epsilon_0\kappa/\ln{(b/a)}$$

The inductance per unit length is obtained from
Example 36-4 (where only the term for $a < r < b$ is
important); thus,

$$\mathcal{L} = (\mu_0/2\pi) \ln{(b/a)}$$

(a) The cable is driven at one end by a generator
connected between the inner and outer conductors.
The potential difference between the conductors is
$V(x, t) = V_0 \cos{(kx - \omega t)}$. Determine the expression for $I(x, t)$. (b) What is the characteristic impedance Z_0 of the cable? Evaluate the expression for Z_0
for $a = 0.4$ mm, $b = 5.0$ mm, and $\kappa = 4.1$.

LIGHT

In this chapter we single out for study that segment of the electromagnetic spectrum (see Fig. 38–2) to which the human eye is sensitive. This portion of the spectrum is termed *visible light* and corresponds approximately to the wavelength region from 4000 Å (violet) to 7000 Å (red).* The various colors of light in this range are identified in Table 40–1.

TABLE 40–1. Approximate Wavelengths for the Various Colors in the Visible Spectrum

COLOR	WAVELENGTH RANGE (Å)
Violet	4000–4500
Blue	4500–5000
Green	5000–5500
Yellow	5500–6000
Orange	6000–6500
Red	6500–7000

The Sun emits electromagnetic radiation with a much wider range of wavelengths than can be detected by the eye. However, the maximum in the solar intensity spectrum corresponds closely to the region of visible light (see Fig. 40–1). Presumably, genetic adaptation during the course of evolutionary history has been responsible for adjusting the sensitivity of the eye to match the solar emission maximum.

There are sources of light, such as the Sun, fire, and incandescent lamps, that emit radiation with a continuous spectrum of wavelengths. There are also sources, such as discharge tubes and arc lamps, that emit light with discrete wavelengths. For example, a hydrogen discharge tube produces intense lines† at wavelengths $\lambda = 656.3$ nm, 486.1 nm, 434.0 nm, 410.1 nm, and so forth. Also, in the following we refer frequently to the yellow light (the intense D lines) from luminous sodium vapor: $\lambda_{D_1} = 589.6$ nm and $\lambda_{D_2} = 589.0$ nm. (These lines are so close in wavelength that they appear as a single line in many instruments.)

---

*Wavelengths of radiation in or near the visible region are expressed variously in angstroms (10^{-10} m), micrometers (μm), and nanometers (nm). Thus, 4000 Å = 400 nm = 0.400 μm. You should be familiar with all of these units.

†The term "line" is used to denote light with a discrete wavelength because, when a spectrograph is used to observe the light, each wavelength corresponds to a line image on the recording film.

Fig. 40-1. Comparison of the solar radiation intensity spectrum with the sensitivity of the human eye. Notice that the solar spectrum at sea level contains numerous dips due to the absorption of radiation by various molecular species in the atmosphere.

The physical processes that are responsible for continuous spectra and for line spectra are quite different. Line spectra arise from a type of resonance phenomenon in atoms and molecules, whereas the origin of continuous spectra is more akin to noise generation.

Light plays such an important role in human existence that optical phenomena have been studied since the earliest times. The ancient Greeks knew that when a light ray is reflected from a mirror, the angle of incidence is equal to the angle of reflection. They were also aware of the bending (or refraction) of light as it passes from one transparent medium to another. The refraction of light was studied extensively in the 11th century by the Arabian mathematician Alhazen. In about 1280 spectacle lenses were invented and were adapted to correct faulty vision. In the 17th century, Galileo (1564–1642) devised a practical telescope and van Leeuwenhoek (1632–1723) developed the first microscope. These inventions paved the way for the detailed investigation of the large-scale and small-scale aspects of the Universe.

A casual observation reveals that an object placed in the light from a pointlike or nearly pointlike source, such as the Sun, produces a sharp shadow. That is, light does not appear to bend around (or be *diffracted* by) such an object, in contrast to the obvious diffraction of water waves when an obstacle is encountered. Thus, light seems to exhibit rectilinear propagation, and this apparent behavior led to the early notion that light consists of corpuscular matter. Although Isaac Newton did not commit himself to either a wave or a corpuscular description of light, he developed a theory of refraction that successfully accounted for the bending of light rays at the interface between two media in terms of an attractive interface force on light corpuscles. However, this theory predicted that the velocity of the corpuscles would be greater in a medium such as glass than in air; later measurements showed that this prediction is contrary to the actual case (see Problem 40-27).

A competing theory that treated light as a wave phenomenon was devised in 1678 (published 1690) by the Dutch physicist Christiaan Huygens (1629–1695), a contemporary of Newton. This theory did not gain much support because it failed to account for polarization. (Huygens had assumed light to be in the form of *longitudinal* waves, similar to sound waves.) The wave nature of light was finally convincingly demonstrated in 1801 by the interference experiments of Thomas Young (1773–1829). Additional evidence in support of the wave theory was supplied soon thereafter, particularly in a series of papers by Augustin Fresnel (1788–1827). By 1826 the corpuscular theory was dead. The theoretical analysis of Maxwell (1864) established a firm foundation for the wave theory of light based on the principles of electrodynamics.

The wave theory was unchallenged until the beginning of the 20th century, when it was found that a radical new idea was required to explain various puzzling phenomena involving the interaction of light with atomic systems. To account for the experimental results, Max Planck (1858–1947) and Albert Einstein (1879–1955) found it necessary to suppose that light consists ultimately of elementary *quanta* (or *photons*). We can think of a photon as a bundle of electromagnetic radiation that oscillates with a definite frequency and travels through space with the speed of light. Individual quanta carry energy and momentum, so light has some particlelike properties. But when the number of identical photons is very large, they exhibit the properties of a continuous wave with the same definite frequency and propagation speed as the quanta. Deviations from the predictions of the classical wave theory occur only when the photons interact individually with atoms. In 1913 the Danish physicist Niels Bohr (1885–1962) devised an atomic model for the emission and absorption of light quanta. This led eventually, in the 1920's, to the merger of the quantum theory of light with the emerging general quantum theory of matter.

In this chapter and in the next several chapters, we examine the nonquantum aspects of light. We begin by studying *geometric* (or *ray*) *optics,* in which the behavior of light is governed by the laws of reflection and refraction. This enables us to discuss the properties of lenses and mirrors. Then, we proceed to the topic of *physical* (or *wave*) *optics,* in which interference, diffraction, and polarization effects are important. Finally, we describe the properties of several types of optical instruments. The concluding two chapters of this volume are devoted to the exploration of some of the quantum aspects of light and matter.

40-1   MEASUREMENTS OF THE SPEED OF LIGHT

The speed of light was of interest to and was speculated on by the ancient Greeks. It was clear that the speed of light must be much larger than the speed of sound (perhaps infinite?) because there was a time delay between seeing a lightning flash and hearing the accompanying thunder. However, no direct measurements were attempted for many centuries.

The first recorded attempt to measure the speed of light was made by Galileo in about 1600. One night he and a companion took positions, each on a hilltop, with a separation of about 3 km. Each was equipped with a lantern and a cover. The experiment was conducted in the following way. First, Galileo removed the cover from his lantern. Upon seeing the light from Galileo's lantern, the companion on the other hilltop removed the cover from his lantern. Galileo reasoned that by measuring the time between the moment he uncovered his lantern and the moment he saw the light from his companion's lantern, he would be able to determine the speed of light. Although the principle of the measurement is correct, Galileo quickly realized that he was measuring, not the time of transit of light between the hilltops, but the combined reaction times of himself and his companion. The speed of light is far too great to measure in this crude way. (See Problem 40–1.)

The earliest reliable determination of the speed of light was made in 1676 by the Danish astronomer Olaus Roemer (1644–1710) in connection with his study of Jupiter's moons. Jupiter's innermost moon, Io, revolves around the planet with a period of 42.5 hours. The instant when Io passes behind the planet is easy to determine because the Sun's reflected light is suddenly cut off as the moon is eclipsed by Jupiter

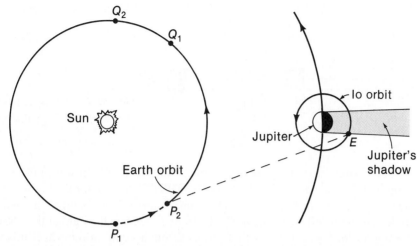

Fig. 40–2. Roemer observed that the instant when the moon Io moved into Jupiter's shadow (point E) came earlier than expected as the Earth moved from P_1 to P_2 and came later than expected during the movement from Q_1 to Q_2. The measurements led Roemer to conclude that the time required for light to traverse a distance equal to the Earth's orbit is approximately 22 min. (The value is actually 17 min.) The orbits are not drawn to scale.

(point E in Fig. 40–2). Roemer found systematic variations in the times of onset of the eclipses. When the Earth was moving *toward* Jupiter (from P_1 to P_2 in Fig. 40–2), the onset of the eclipse of Io came earlier and earlier than expected. However, when the Earth was moving *away from* Jupiter (from Q_1 to Q_2), the onset came later and later than expected. Roemer correctly concluded that the discrepancies were due to the time required for light to travel the changing distance from Jupiter to the Earth. He could not obtain a reliable value for the time difference between successive eclipses (see Problem 40–2), but he could measure the cumulative delay between the positions of the Earth when nearest Jupiter and when farthest from the planet. Thus, Roemer deduced that it required approximately 22 minutes for light to travel a distance equal to the diameter of the Earth's orbit around the Sun. If we use a value of 3.0×10^{11} m for the diameter of the Earth's orbit, the speed of light based on Roemer's data is 2.3×10^8 m/s. The modern value for the delay is approximately 17 min, giving $c = 3.0 \times 10^8$ m/s.

A second astronomical determination of the speed of light was reported in 1729 by the English astronomer James Bradley (1693–1762). He found that if a telescope is pointed in a certain direction to observe a particular star on one night, then six months later the telescope must be pointed in a slightly different direction to observe the same star. Figure 40–3 shows the situation for a star located on a line from the Sun that is perpendicular to the Earth's orbital plane. Because of the Earth's motion with velocity **v**, the telescope must be tilted away from the true direction to the star and toward the direction of **v**. (This effect is called the *aberration of light;* see Section 21–9.) The tilt angle α is given by* $\tan \alpha \cong v/c$. Bradley measured the difference in sighting angles six months apart, obtaining $2\alpha = 40.4''$ (arc seconds). Combined with $v = 30$ km/s, this value for 2α gives

$$c \cong \frac{v}{\tan \alpha} \cong \frac{v}{\alpha} = \frac{3.0 \times 10^4 \text{ m/s}}{20.2 \times \pi/(3600 \times 180)}$$
$$= 3.06 \times 10^8 \text{ m/s}$$

Fig. 40–3. The change in the apparent direction of a star between observations made six months apart was used by Brádley to determine the speed of light.

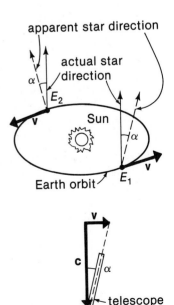

*The relativistic expression is $\tan \alpha = \beta(1 - \beta^2)^{-1/2}$, where $\beta = v/c$. When $\beta \ll 1$, this expression reduces, as in the present case, to $\tan \alpha \cong \beta$.

Fig. 40-4. Fizeau's toothed-wheel apparatus for measuring the speed of light. The flight path for the light was $\ell = 8.63$ km.

The first entirely terrestrial measurement of the speed of light was made by the French physicist Armand Fizeau (1819–1896) in 1849. The apparatus that was used in shown schematically in Fig. 40-4. Light from a source S passes through a lens L_1 and is reflected to the left by a mirror M_1. The lens L_1 focuses the light at the position of a rotating toothed wheel. Light that passes through a gap in the wheel travels a distance ℓ to a mirror M_2, where it is reflected and returns along the same path to the toothed wheel. If it is not blocked by a tooth of the wheel, some of the light is transmitted through the mirror M_1 and reaches the observer O. (M_1 is "half-silvered" and so partially reflects and partially transmits incident light.) The rotation speed of the wheel is adjusted until the returning light is entirely blocked by a tooth of the wheel and the observer sees no light. That is, the light travels the distance 2ℓ during the time required for the wheel to rotate from a gap to the next toothed section. In this way Fizeau obtained the value $c = 3.13 \times 10^8$ m/s (see Problem 40-3).

An improvement in Fizeau's method was made by J. B. L. Foucault (1819–1868), who substituted a rotating many-sided mirror for the toothed wheel. Foucault's value for the speed of light, obtained in 1862, was $c = 2.98 \times 10^8$ m/s, a value within 0.6 percent of the best modern determination. Also, by inserting a water-filled tube into the light path, Foucault demonstrated that the speed of light in water is less than that in air.

The rotating-mirror technique was carried to its height by Michelson,* who mounted an eight-sided mirror at the observatory on Mount Wilson in southern California. The return mirror was located on Mount San Antonio, 35 km away. The measurements made during the period 1923–1927 produced the value $c = 2.99798 \times 10^8$ m/s, with an uncertainty of 4 in the last decimal.

The most precise modern methods for determining the speed of light involve the simultaneous measurement of wavelength λ and frequency ν for a radiation by direct comparison with the atomic standards for length and time (see Section 1–2); then, the value of c is obtained from $c = \lambda\nu$. These techniques are capable of precisions of a few parts per *billion*. The currently accepted value is

$$c = 2.99792458 \times 10^8 \text{ m/s}$$

with an uncertainty of only 1 in the last decimal. In this book we usually use approximate values,

$$c \cong 2.998 \times 10^8 \text{ m/s} \quad \text{or} \quad c \cong 3.00 \times 10^8 \text{ m/s}$$

*Albert A. Michelson (1852–1931), one of the most renowned American scientists of his era and the first American to be awarded a Nobel Prize in science (1907). Michelson is also known for the famous ether-drift experiments performed with E. W. Morley (see Section 21–1).

40-2 WAVEFRONTS AND RAYS

Consider a monochromatic plane wave that propagates in a direction defined by the wave vector **k** with a time-dependent phase, $\phi(t) = \mathbf{k} \cdot \mathbf{r} - \omega t + \phi_0$ (see Section 19-4). At a particular time t, a plane of constant phase (a *wavefront*) is defined by all points $P(\mathbf{r})$ for which $\phi(t)$ has a particular fixed value. Figure 40-5a illustrates the case of a wave propagating in the z-direction so that $\mathbf{k} = k\hat{\mathbf{k}}$, where $k = \omega/c$. Values of **r** for which the phase is constant at a particular time t must have a constant value for $\mathbf{k} \cdot \mathbf{r} = kr \cos\theta = kz$. That is, the wavefronts are planes that are parallel to the x-y plane. In diagrams, a series of such planes are usually shown for wavefronts that differ in phase by 2π and are therefore spaced apart by one wavelength.

Rays are lines that are drawn perpendicular to the wavefronts and in the direction of the wave propagation (Fig. 40-5b). It is often convenient to describe optical effects in terms of rays instead of wavefronts. For this purpose we sometimes isolate a single ray, the idealization of a narrow beam of light.

Figure 40-6 shows the case of spherical waves expanding from a point source. At any point $P(\mathbf{r})$, the vectors **k** and **r** are parallel. Therefore, $\phi(t) = kr - \omega t + \phi_0$, and for a constant phase we must have $r = $ constant. Thus, the wavefronts are concentric spheres and the rays are radial lines.

The Huygens-Kirchhoff Principle. About 1680 Christiaan Huygens devised a clever scheme for determining the way in which a wavefront advances. Although intended to represent the propagation of mechanical waves in material media, Huygens' idea is also valid for light waves. According to *Huygens' construction,* the way in which a wavefront of arbitrary shape will move forward can be determined by considering every point on a given wavefront at any instant t to be the source of a secondary, outward propagating spherical wave (or *wavelet*). By drawing spheres with radii $c\,\Delta t$ about every point on the original wavefront and then constructing the *forward-going* envelope (which is tangent to the forward edge of each wavelet), the new position of the wavefront at time $t + \Delta t$ is determined (Fig. 40-7).

Huygens' construction is an incomplete concept because it does not address the question of why there are no backward-going wavelets. Kirchhoff eliminated this defect and converted Huygens' idea into a rigorous principle in 1882. The Huygens-Kirchhoff principle and its variations are used regularly in discussions of diffraction.

40-3 THE LAWS OF REFLECTION AND REFRACTION

A nonluminous object can be observed by virtue of the redirected (and perhaps altered) light that originated elsewhere. This process involves a variety of interactions between light and matter. Simple *diffuse reflection* occurs when light is incident on a rough irregular surface and is scattered in many directions, with some of the light scattered in the direction of the observer. For a white surface, diffuse reflection occurs equally for all wavelengths, so the reflected light has the same color or colors as the incident light. If a surface appears to have a particular color when irradiated with white light, this means that selective absorption or scattering has taken place. More complicated effects such as fluorescence and luminescence may also occur, in which the atoms of the irradiated material are induced to emit radiation with a characteristic color.

Specular Reflection. The simplest form of reflection is *specular reflection*. This type of reflection occurs when light is incident on a polished surface such as a mirror

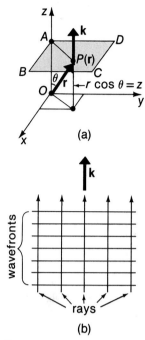

Fig. 40-5. (a) A *wavefront* of a plane wave consists of all points $P(\mathbf{r})$ for which the phase, $\phi(t) = \mathbf{k} \cdot \mathbf{r} - \omega t + \phi_0$, is constant. For the case illustrated, $\mathbf{k} = k\hat{\mathbf{k}}$, and the wavefronts, such as *ABCD*, are parallel to the x-y plane. (b) A *ray* is a line drawn perpendicular to the wavefronts and in the direction of the wave propagation vector $\hat{\mathbf{k}}$.

Fig. 40-6. The wavefronts expanding from a point source are concentric spheres, and the rays are radial lines.

Fig. 40–8. Time sequence of the reflection of three wavefronts incident on a plane mirror MM'. The light rays are perpendicular to the wavefronts. The vectors **k** and **k**′ are the propagation vectors for the incident wave and the reflected wave, respectively.

Fig. 40–7. Huygens' construction for determining the advance of a wavefront of arbitrary shape. Every point on the wavefront AB at time t is considered to be the source of a spherical wavelet. By drawing spheres with radii $c \, \Delta t$ and then constructing the *forward-going* envelope, the wavefront $A'B'$ at time $t + \Delta t$ is determined.

or a metal plate. A light ray that undergoes specular reflection will leave the surface in a well-defined direction. (In practice, all real surfaces exhibit both diffuse and specular reflection in varying degrees depending on the quality of the surface. We consider here only the idealized case of pure specular reflection.)

Figure 40–8 shows a time sequence of portions of three successive wavefronts, separated by a distance λ, as they are incident upon and are reflected by a flat mirror MM'. Notice that the *angle of incidence* θ_1 may be defined either as the angle that an incident wavefront makes with the mirror or the angle that an incident ray makes with the normal to the mirror. The *angle of reflection* is θ'_1, and for specular reflection we have $\theta_1 = \theta'_1$. This equality is demonstrated in Fig. 40–9, which shows in detail the reflection of an incident wavefront $\overline{AB}$ and its emergence a time Δt later as the reflected wavefront $\overline{DC}$.

The interval Δt is the time required for a Huygens wavelet emitted at point B on the incident wavefront to reach point C on the mirror. The segment $\overline{BC}$ corresponds to an incident ray. During this same interval Δt, the wavelet emitted at point A travels to point D. Thus, $\overline{DC}$ corresponds to the reflected wavefront.

A Huygens wavelet that originates at some intermediate point on the incident wavefront, such as point a, will first strike the mirror at point b. This creates another

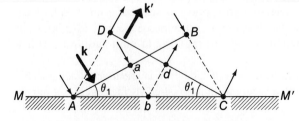

Fig. 40–9. Detail of the way in which the wavefront $\overline{AB}$ is transformed into the wavefront $\overline{DC}$ by reflection. The triangles $\triangle ABC$ and $\triangle ADC$ are congruent, so that $\theta_1 = \theta'_1$.

wavelet centered on b which in turn propagates to point d on the reflected wavefront, also with a total elapsed time Δt. (•Can you show that the elapsed time is Δt?)

Each of the ray segments $\overline{BC}$ and $\overline{AD}$ has length $v \Delta t$, where v is the wave velocity in the medium above the surface of the mirror. The angles $\angle ADC$ and $\angle ABC$ are both right angles. Therefore, the triangles $\triangle ABC$ and $\triangle ADC$, which have the common side $\overline{AC}$, are congruent. Thus, we conclude that the angle of incidence θ_1 is equal to the angle of reflection θ_1'; that is,

$$\theta_1 = \theta_1' \qquad \text{specular reflection} \qquad \textbf{(40-1)}$$

This is the *law of reflection.*

The wavelength λ of the radiation is not altered by the reflection process. (•Can you see why this statement is true?)

Refraction. When a light ray is incident on the interface between two transparent media, as in Fig. 40-10, it is observed that part of the light is reflected in the usual way, with $\theta_1 = \theta_1'$, whereas some of the light is transmitted into the second medium with $\theta_2 \neq \theta_1$. If medium ① is air and medium ② is a substance such as water or glass, it is observed that $\theta_2 < \theta_1$. This bending of light at an interface is called *refraction,* and θ_2 is the *angle of refraction.* Thus, when a light ray enters water from air, the ray is refracted *toward* the normal to the surface (Fig. 40-10).

Figure 40-11 shows a time sequence of portions of three successive wavefronts as they are incident upon and are refracted at the flat interface II' between two transparent media. (For simplicity, the reflected wavefronts have been omitted—they are as shown in Fig. 40-8.) The relationship connecting θ_1 and θ_2 is obtained by referring to Fig. 40-12, which shows in detail the refraction of an incident wavefront $\overline{AB}$ and its complete emergence in the second medium a time Δt later as the refracted wavefront $\overline{DC}$.

The analysis of refraction using Huygens' construction proceeds in a manner parallel to that for reflection. The Huygens wavelet emitted at point B (Fig. 40-12) arrives at point C on the interface after time $\Delta t = \overline{BC}/v_1$, where v_1 is the wave velocity in medium ①. During this same interval Δt, the wavelet emitted at point A travels to point D. The new (refracted) wavefront $\overline{DC}$ makes an angle θ_2 with the interface. In Fig. 40-12 we see that

$$\sin \theta_1 = \frac{\overline{BC}}{\overline{AC}}$$

$$\sin \theta_2 = \frac{\overline{AD}}{\overline{AC}}$$

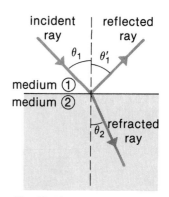

Fig. 40-10. When a light ray is incident on the interface between two transparent media, a reflected ray and a transmitted or refracted ray are observed.

Fig. 40-11. Time sequence of the refraction of three wavefronts incident on the flat interface II' between two transparent media.

Fig. 40-12. Detail of the way in which the wavefront $\overline{AB}$ is transformed into the wavefront $\overline{DC}$ by refraction at the interface II'. The angles θ_1 and θ_2 are related by Snell's law.

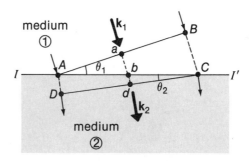

Dividing one of these equations by the other yields

$$\overline{AD} \sin \theta_1 = \overline{BC} \sin \theta_2 \tag{40-2}$$

Now, $\overline{BC} = v_1 \Delta t$ and $\overline{AD} = v_2 \Delta t$, where v_2 is the wave velocity in medium ②. Using these expressions in Eq. 40-2 and multiplying both sides by c, we have

$$\frac{c}{v_1} \sin \theta_1 = \frac{c}{v_2} \sin \theta_2$$

The ratio of the velocity of light in vacuum to the velocity in a medium is usually designated by the letter n. Thus, the *law of refraction* can be expressed as

$$n_1 \sin \theta_1 = n_2 \sin \theta_2 \tag{40-3}$$

This result is referred to as *Snell's law.** Because the ratio

$$n = \frac{c}{v} \tag{40-4}$$

appears in the law of refraction, the quantity n is called the *index of refraction* or the *refractive index* of the medium.

The value of the refractive index for a material depends somewhat on the wavelength of the incident light. (This effect gives rise to the phenomenon of *dispersion*—see Section 40-5.) When an index is quoted without specification of the wavelength, the value is, by custom, the value appropriate for yellow light (in particular, the yellow light emitted by sodium with $\lambda = 5893$ Å). Table 40-2 lists values of n for materials of various types.

Notice in Fig. 40-11 that a light wave (in fact, any type of wave) undergoes a change in wavelength upon refraction, whereas no such change occurs upon reflection. To insure a time-independent match of the boundary conditions of the electromagnetic field components, the frequency of a wave must be the same on either side of a boundary. However, the wavelength will be different on the two sides if the wave velocity is different (see Section 19-6). Thus,

$$\nu_1 = \frac{v_1}{\lambda_1} = \nu_2 = \frac{v_2}{\lambda_2}$$

so that

$$\lambda_2 = \frac{v_2}{v_1} \lambda_1 = \frac{n_1}{n_2} \lambda_1 \tag{40-5}$$

---

*After the Dutch mathematician Willebrord Snell (1591–1626), who discovered the result experimentally in about 1621.

TABLE 40-2. Refractive Indexes for Various Materials (for sodium light, $\lambda = 5893$ Å)

GASES (at 0° C and 1 atm)

Gas	n
Hydrogen (H_2)	1.000132
Air	1.000293
Ammonia (NH_3)	1.000376
Methane (CH_4)	1.000444
Carbon dioxide (CO_2)	1.000451
Sulfur dioxide (SO_2)	1.000686
Chlorine (Cl_2)	1.000773

LIQUIDS (at 20° C)

Liquid	n
Methyl alcohol (CH_3OH)	1.329
Water (H_2O)	1.333
Ethyl alcohol (C_2H_5OH)	1.362
Benzene (C_6H_6)	1.501
Carbon disulfide (CS_2)	1.625
Methylene iodide (CH_2I_2)	1.756

SOLIDS (at 20° C)

Solid	n
Ice (H_2O)	1.310
Fluorite (CaF_2)	1.434
Fused quartz (SiO_2)	1.458
Plastic (Lucite)	1.491
Glass (ordinary crown)	1.517
Rock salt (NaCl)	1.544
Plastic (polystyrene)	1.595
Glass (dense flint)	1.655
Diamond (C)	2.417

Example 40-1

A narrow beam of sodium yellow light is incident from air on a smooth surface of water at an angle $\theta_1 = 35°$. Determine the angle of refraction θ_2 and the wavelength of the light in water. Refer to Fig. 40-10.

Solution:

For all practical purposes, the index of refraction of air is 1 (see Table 40-2), so $n_1 = 1$. Also from Table 40-2, we have n_2(water) $= 1.333$. Then, using Snell's law (Eq. 40-3), we have

$$1 \sin \theta_1 = 1.333 \sin \theta_2$$

so that

$$\sin \theta_2 = \frac{1}{1.333} \sin 35°$$

from which

$$\theta_2 = 25.5°$$

The average wavelength of sodium yellow light in vacuum (or air) is $\lambda_1 = 5893$ Å. Therefore, using Eq. 40–5, we find the wavelength in water to be

$$\lambda_2 = \frac{n_1}{n_2}\lambda_1 = \frac{1}{1.333}(5893 \text{ Å}) = 4421 \text{ Å}$$

Example 40–2

A narrow beam of light passes through a plate of glass with a thickness $s = 4$ mm and a refractive index $n = 1.52$. The beam enters from air at an angle $\theta_1 = 30°$. Calculate the deviation d of the ray, as indicated in the diagram.

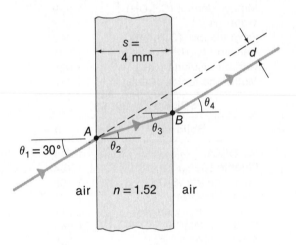

Solution:

Because the surfaces of the glass plate are parallel, the interior angles, θ_2 and θ_3, are equal. Then, applying Snell's law at A and B, we obtain

$$\sin \theta_1 = n \sin \theta_2 = n \sin \theta_3 = \sin \theta_4$$

Thus, $\theta_4 = \theta_1$, and the emergent ray is parallel to the incident ray. There is a deviation d of the ray but no angular displacement. (A glass plate therefore has no lens-like properties.)
The deviation d is

$$d = \overline{AB} \sin (\theta_1 - \theta_2)$$

We also have

$$s = \overline{AB} \cos \theta_2$$

Dividing one of these equations by the other gives

$$d = s\frac{\sin (\theta_1 - \theta_2)}{\cos \theta_2}$$

From Snell's law we have

$$\sin \theta_2 = \frac{1}{n} \sin \theta_1 = \frac{1}{1.52} \sin 30°$$

from which $\theta_2 = 19.2°$ and $\theta_1 - \theta_2 = 10.8°$. Thus,

$$d = (4 \text{ mm}) \frac{\sin 10.8°}{\cos 19.2°} = 0.794 \text{ mm}$$

Total Internal Reflection. Figure 40–10 shows that when the refractive index of the incident medium is less than that of the second medium (that is, $n_1 < n_2$), the ray is refracted *toward* the normal. In this circumstance, the ray will penetrate into the second medium for any incident angle θ_1. However, the situation is different when $n_2 < n_1$, the case illustrated in Fig. 40–13. Ray a is incident normally on the interface and passes through undeflected, whereas ray b is deflected in accordance with Snell's law, giving $\theta_2 > \theta_1$. As θ_1 is increased, there is eventually reached a critical angle θ_c, corresponding to ray c in Fig. 40–13, for which $\theta_2 = 90°$. Then, $\sin \theta_2 = (n_1/n_2) \sin \theta_c = 1$, whence

$$\theta_c = \sin^{-1} (n_2/n_1) \tag{40-6}$$

Fig. 40–13. When $n_2 < n_1$, total internal reflection occurs for all incident angles greater than the critical angle θ_c. (The reflected parts of rays a, b, and c are not shown.)

If θ_1 is made greater than θ_c, Snell's law shows that $\sin \theta_2 > 1$, an unphysical situation. Thus, when $\theta_1 > \theta_c$, no refraction occurs and the ray is completely reflected at the interface (ray d in Fig. 40–13). This phenomenon is called *total internal reflection*.

▶ *Fiber Optics.* When light is projected into the end of a glass or plastic rod with a small diameter as in Fig. 40–14, the angles with which the various rays strike the surface are greater than the critical angle. Such rays are "piped" through the rod by successive internal reflections, even along a curved path, and eventually emerge from the opposite end. Bundles of tiny rods (called *light pipes* or *optical fibers*) are used in a variety of applications that do not permit normal straight-line propagation. (For example, optical fibers allow a physician to see inside a patient's stomach.) Also, special fibers with low-loss characteristics are beginning to be used in the communications industry. A bundle of fibers can carry hundreds of times more telephone calls using light signals than can a bundle of copper wires of the same size using ordinary electrical signals. ◀

Fig. 40–14. Light is transmitted through a curved plastic rod (a *light pipe* or *optical fiber*) by successive internal reflections.

Example 40–3

As shown in the diagram, a light ray is incident normally on one face of a 30°–60°–90° block of dense flint glass (a *prism*) that is immersed in water.

(a) Determine the exit angle θ_4 of the ray.

(b) A substance is dissolved in the water to increase the index of refraction. At what value of n_2 will total internal reflection cease at point P?

Solution:

(a) From the geometry of the prism we see that $\theta_1 = 60°$ and $\theta_3 = 30°$. The refractive indexes are (see Table 40–2) $n_1 = 1.655$ and $n_2 = 1.333$. Then, using Snell's law, we find

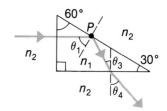

$$\sin \theta_4 = \frac{n_1}{n_2} \sin \theta_3 = \frac{1.655}{1.333} \sin 30° = 0.621$$

so that

$$\theta_4 = 38.4°$$

(b) Total internal reflection occurs when $(n_1/n_2) \sin \theta_1 > 1$. (For the initial situation here, $(n_1/n_2) \sin \theta_1 = 1.075$.) Total internal reflection will cease when $(n_1/n_2) \sin \theta_1 = 1$. Thus,

$$n_2 = n_1 \sin \theta_1 = (1.655) \sin 60° = 1.433$$

Fermat's Principle.   A powerful principle that specifies the paths followed by light rays was formulated in 1657 by the French mathematician Pierre de Fermat (1601–1665). *Fermat's principle* states that a light ray that connects two points within a medium follows a path for which the transit time of the light is minimum.

Figure 40–15 shows various conceivable paths that light might follow from point P to point Q. The time required for light to travel between the points along the particular path C is

$$t = \oint_P^Q \frac{ds}{v}$$

If the medium through which the light passes has a variable index of refraction, we can use $v = c/n$ to express the integral as

$$t = \frac{1}{c} \oint_P^Q n \, ds \qquad \text{(40–7)}$$

According to Fermat's *principle of least time,* the integral $\oint_P^Q n \, ds$, called the *optical path length,* must be minimum.

Bundles of optical fibers. (Courtesy of Edmund Scientific Company.)

Figure 40–16 shows the application of Fermat's principle to the refraction of light at a plane surface. Consider the passage of light from point A in a uniform medium with index n_1 to point B in a second uniform medium with index n_2, through an arbitrary interface point P. From A to P the least-time path is clearly a straight line whose optical path length is $n_1 \ell_1$. Similarly, the least-time path from P to B is a straight line with optical path length $n_2 \ell_2$. It remains to determine the location of point P, specified by the coordinate x. The total optical path length is

$$L = \oint_A^B n \, ds = n_1 \ell_1 + n_2 \ell_2$$
$$= n_1 \sqrt{a^2 + x^2} + n_2 \sqrt{b^2 + (d - x)^2}$$

Fig. 40–15.   According to Fermat's principle, light travels from P to Q along the path for which the transit time is least.

The minimum value of L is obtained by setting dL/dx equal to zero. Thus,

$$\frac{n_1 x}{\sqrt{a^2 + x^2}} = \frac{n_2(d - x)}{\sqrt{b^2 + (d - x)^2}}$$

or

$$n_1 \sin \theta_1 = n_2 \sin \theta_2$$

which we recognize to be Snell's law.

The law of reflection (Eq. 40–1) is also easy to obtain from Fermat's principle (see Problem 40–11).

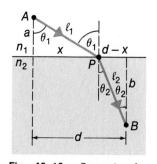

Fig. 40–16. Geometry for deriving Snell's law from Fermat's principle.

40-4 REFLECTION AND TRANSMISSION

When light is incident on the interface between two media, the fraction of the wave intensity that is reflected and the fraction that is transmitted depend on the refractive indexes of the media. Expressions for these fractions (called the *reflectance R* and the *transmittance T*) can be derived by applying the boundary conditions on the field vectors at the interface.

Normal Incidence. Consider light in the form of a plane electromagnetic wave that is incident normally on the flat interface between two media, represented by the *x-y* plane in Fig. 40–17. The propagation vector $\mathbf{k}_1$ of the incident wave has the direction of $\mathbf{E}_i \times \mathbf{H}_i$ (see Fig. 38–7). Moreover, the propagation vector for the reflected wave is $-k_1$, corresponding to the direction of $\mathbf{E}_r \times \mathbf{H}_r$. Because an electromagnetic wave always propagates in the direction of $\mathbf{E} \times \mathbf{H}$, it is clear (see Fig. 40–17) that there must be a change of orientation of the electric and magnetic vectors upon reflection. Figure 40–17 shows a *change of phase* for the magnetic vector but not for the electric vector; this is the case if $n_1 > n_2$. (For $n_1 < n_2$, the electric vector undergoes a phase change but the magnetic vector does not.)

Fig. 40–17. The electric and magnetic vectors of a plane electromagnetic wave at the boundary between two dielectric media for a particular instant. (The vector sets have been displaced from $z = 0$ for clarity.) The media are nonconducting ($\sigma = 0$), so the tangential components of both the electric and the magnetic vectors are continuous across the boundary. The diagram represents the case $n_1 > n_2$, and the subscripts represent the incident wave (i), the reflected wave (r), and the transmitted wave (t). The magnetic vector undergoes a phase change upon reflection but the electric vector does not.

Figure 40–17 represents the field vectors exactly at the interface for a particular instant. (The vector sets are displaced along the *z*-axis for clarity.) The media are perfect dielectrics ($\sigma = 0$), so there are no surface currents. Then, the boundary conditions on **E** and **H** (see Eqs. 30–45 and 36–40) specify that the tangential components of these vectors are continuous across the boundary between the two media. For the case of normal incidence, the field vectors have only tangential components. The orientation of the vector $\mathbf{E}_i$ is arbitrary, but for any orientation $\mathbf{E}_i$ is still tangential to the interface. We choose $\mathbf{E}_i$ to be polarized in the *x*-direction, with $\mathbf{k}_1$ parallel to the *z*-axis. Then, the boundary conditions on the electric and magnetic vectors can be expressed as

$$\left. \begin{array}{c} \mathbf{E}_i + \mathbf{E}_r = \mathbf{E}_t \\ \mathbf{H}_i + \mathbf{H}_r = \mathbf{H}_t \end{array} \right\} \tag{40-8}$$

or, in terms of the amplitudes, we have (refer to Fig. 40–17)

$$E_i^0 + E_r^0 = E_t^0 \tag{40-9}$$

and

$$H_i^0 - H_r^0 = H_t^0 \tag{40-10}$$

Now, according to Eq. 38-15, for a plane wave, we have

$$H^0 = \sqrt{\frac{\epsilon_0 \kappa}{\mu_0 \kappa_M}} E^0$$

Also, from Eq. 38-13, $v = 1/\sqrt{\epsilon_0 \mu_0 \kappa \kappa_M}$, so that $c = 1/\sqrt{\epsilon_0 \mu_0}$ (see Eq. 38-14). The media are nonferromagnetic, so $\kappa_M \cong 1$. Thus, $c/v = \sqrt{\kappa} = n$, and we have for all three waves

$$H^0 = \sqrt{\frac{\epsilon_0}{\mu_0}} \sqrt{\kappa}\, E^0 = \sqrt{\frac{\epsilon_0}{\mu_0}} n\, E^0 \tag{40-11}$$

Then, Eq. 40-10 becomes

$$n_1 E_i^0 - n_1 E_r^0 = n_2 E_t^0 \tag{40-12}$$

Solving Eqs. 40-9 and 40-12 for E_r^0 and E_t^0 in terms of E_i^0, the result is

$$E_r^0/E_i^0 = \frac{n_1 - n_2}{n_1 + n_2} \tag{40-13a}$$

$$E_t^0/E_i^0 = \frac{2n_1}{n_1 + n_2} \tag{40-13b}$$

(Compare these expressions with those obtained for the case of waves on a string, Eqs. 19-26.)

The quantity E_t^0/E_i^0 is called the *amplitude transmission coefficient*. Equation 40-13b shows that $E_t^0/E_i^0 > 0$ for all values of the refractive indexes. Thus, at the interface, the vectors $\mathbf{E}_i$ and $\mathbf{E}_t$ always have the same direction. That is, the incident and transmitted waves are in phase for all n_1 and n_2. On the other hand, Eq. 40-13a shows that the *amplitude reflection coefficient* E_r^0/E_i^0 is positive only when $n_1 > n_2$ (the case illustrated in Fig. 40-17). In this case, $\mathbf{E}_r$ and $\mathbf{E}_i$ have the same direction at the interface and are therefore *in phase*. When $n_1 < n_2$ (for example, when light is incident on glass or plastic from air or water), we have $E_r^0/E_i^0 < 0$. Then, $\mathbf{E}_r$ and $\mathbf{E}_i$ have opposite directions at the interface, and the incident and reflected waves *differ in phase* by π. In summary,

$$n_1 > n_2: \begin{cases} \mathbf{E}_r \text{ and } \mathbf{E}_i \text{ in phase, } E_r^0/E_i^0 > 0 \\ \mathbf{E}_t \text{ and } \mathbf{E}_i \text{ in phase, } E_t^0/E_i^0 > 0 \end{cases}$$

$$n_1 < n_2: \begin{cases} \mathbf{E}_r \text{ and } \mathbf{E}_i \text{ differ in phase by } \pi, E_r^0/E_i^0 < 0 \\ \mathbf{E}_t \text{ and } \mathbf{E}_i \text{ in phase, } E_t^0/E_i^0 > 0 \end{cases}$$

The reflectance R is defined to be the fraction of the incident energy flux that is reflected at the boundary. This energy flux is represented by the Poynting vector,

$\mathbf{N} = \mathbf{E} \times \mathbf{H}$ (Eq. 38–21), the time average of which is proportional to $E^0 H^0$. Then, using Eq. 40–11, we have

$$R = \frac{\overline{N}_r}{\overline{N}_i} = \frac{E_r^0 H_r^0}{E_i^0 H_i^0} = \frac{(E_r^0)^2}{(E_i^0)^2}$$

so that

$$R = \left(\frac{n_1 - n_2}{n_1 + n_2}\right)^2 \qquad \textbf{(40–14a)}$$

The transmittance T is defined in a similar way:

$$T = \frac{\overline{N}_t}{\overline{N}_i} = \frac{E_t^0 H_t^0}{E_i^0 H_i^0} = \frac{n_2 (E_t^0)^2}{n_1 (E_i^0)^2}$$

so that

$$T = \frac{n_2}{n_1} \left(\frac{2n_1}{n_1 + n_2}\right)^2 \qquad \textbf{(40–14b)}$$

Because energy is conserved, we have

$$R + T = 1 \qquad \textbf{(40–15)}$$

(•Can you verify this conservation condition by using Eqs. 40–14?)

Notice that R and T remain the same if n_1 and n_2 are interchanged. That is, the reflectance and transmittance values for a particular interface do not depend on the direction in which the light is incident on the interface.

Example 40–4

Ordinary crown glass has a refractive index of 1.517 (Table 40–2). When light is incident normally from air on a plate of crown glass, what fraction of the light is reflected and what fraction is transmitted?

Solution:

We have $n_1 = 1.000$ and $n_2 = 1.517$. Then, using Eqs. 40–14, we find

$$R = \left(\frac{n_1 - n_2}{n_1 + n_2}\right)^2 = \left(\frac{1.000 - 1.517}{1.000 + 1.517}\right)^2 = 0.0422$$

$$T = \frac{n_2}{n_1} \left(\frac{2n_1}{n_1 + n_2}\right)^2 = \frac{1.517}{1.000} \left(\frac{2 \times 1.000}{1.000 + 1.517}\right)^2 = 0.9578$$

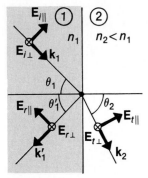

Fig. 40–18. A light wave is incident on a surface at an oblique angle. Only the electric vectors are shown, and each has been resolved into components that are perpendicular and parallel to the plane of incidence (the plane that contains the three propagation vectors). The perpendicular component $\mathbf{E}_{i\perp}$ has been chosen to be directed into the plane of the page; then, $\mathbf{E}_{r\perp}$ and $\mathbf{E}_{t\perp}$ have the same orientation. (•Why?) The diagram represents the various vectors at the boundary between two media for a particular instant. When $\theta_1 = 0$, this case becomes identical to that illustrated in Fig. 40–17.

from which we see that $R + T = 1$, as required. The small value of R agrees with the observation that ordinary glass readily transmits light in normal incidence.

Oblique Incidence. When light is incident normally on a surface, the electric and magnetic vectors of the wave are always tangential to the surface. However, when light is incident at an oblique angle the situation is more complicated. In this case it is convenient to define a *plane of incidence*, which is the plane that contains the propagation vectors $\mathbf{k}_1$, $\mathbf{k}_1'$, and $\mathbf{k}_2$ (see Fig. 40–18). The electric vector $\mathbf{E}_i$ of the incident wave can have any polarization around the direction defined by $\mathbf{k}_1$, but $\mathbf{E}_i$ can always be resolved into components, $\mathbf{E}_{i\perp}$ and $\mathbf{E}_{i\parallel}$, that are perpendicular and parallel, respectively, to the plane of incidence. These two vectors, together with the corresponding vectors for the reflected and transmitted waves, are shown in Fig. 40–18 for the case $n_1 > n_2$. For normal incidence, this case reduces to that illustrated in Fig. 40–17.

The perpendicular and parallel vector components can be treated separately by applying the appropriate boundary conditions. The resulting expressions for the amplitude reflection and transmission coefficients were first obtained by Fresnel in 1823. We show here, in Fig. 40–19, only the graphical results for a case of particular interest, namely, light incident from air ($n_1 = 1$) on glass ($n_2 = 1.5$). Figure 40–19a

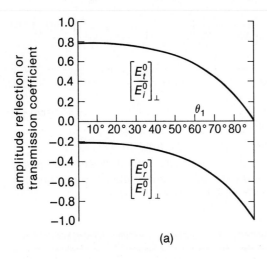

Fig. 40–19. Amplitude reflection and transmission coefficients for (a) the perpendicular component and (b) the parallel component of a wave incident from air ($n_1 = 1$) on glass ($n_2 = 1.5$). The reflection coefficient $(E_r^0/E_i^0)_\parallel$ becomes zero at Brewster's angle.

(a)

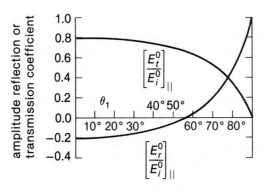

(b)

shows that the amplitude reflection coefficient for the perpendicular component is negative for all θ_1. This means that there is a phase change of π upon reflection of the perpendicular component for $n_2 > n_1$, in agreement with the conclusion reached for normal incidence (see Eq. 40–13a). There is no phase change upon transmission.

Figure 40–19b shows the peculiar behavior of the amplitude reflection coefficient for the parallel component. For incident angles θ_1 near 0°, the reflected wave exhibits a phase change, as does the perpendicular component. However, at an incident angle of about 56° in this case the reflection coefficient E_r^0 / E_i^0 changes sign. The angle, $\theta_B = \tan^{-1}(n_2/n_1)$, at which this occurs is called *Brewster's angle,** and corresponds to *zero* amplitude reflection coefficient for the parallel component of the incident beam. Figure 40–20 shows how the reflected field vector $\mathbf{E}_r$ changes with incident angle θ_1. It is really appropriate to describe the phase behavior of two vectors only when the vectors are collinear. Thus, for $\theta_1 \approx 0°$ we can say that there is a phase difference of π between $\mathbf{E}_i$ and $\mathbf{E}_r$. For almost all of the applications of reflection, we will be concerned with the case of near-normal incidence; the idea of phase change is useful in such situations.

Fig. 40–20. Variation of the reflected vector (parallel component) with the angle of incidence θ_1. When $\theta_1 = \theta_B$, we have $E_r^0 = 0$, and the reflection coefficient changes sign. The transmitted rays are not shown.

If light containing both perpendicular and parallel components is incident at Brewster's angle, only the perpendicular component is reflected. Thus, the reflected beam is completely polarized in the direction perpendicular to the plane of incidence. For this reason Brewster's angle is sometimes called the *polarizing angle*. The complete transmission of the parallel component at Brewster's angle occurs for both $n_2 > n_1$ and $n_2 < n_1$. In each case the angle between the transmitted ray and the direction of specular reflection is 90° (see Problem 40–15).

▶ When a light beam is incident on the interface between two media, the microscopic interaction between the beam and the atomic constituents of the media is quite complicated. However, from a study of geometric optics, we know the *results* of this interaction, namely, that there are reflected and transmitted wavefronts in particular directions.

The situation can be described in terms of an effective layer of atomic dipoles in the plane of the interface. These dipoles are driven by the fields in such a way that the resulting radiation is along the directions of the reflected and transmitted beams. In Section 39–2 we found that the radiation from an oscillating dipole (a dipole antenna) is proportional to $\sin^2 \alpha$, where α is measured from a line that coincides with the dipole axis. Thus, no radiation is emitted by a dipole along the direction corresponding to its axis.

Figure 40–21 shows a light beam incident on the interface between two media. The electric vectors are parallel to the plane of incidence. The dipoles in the interface layer are driven in the direction that produces maximum radiation along the direction of the transmitted beam. Thus, the oscillations of the dipoles are in a direction perpendicular to the transmitted ray (that is, $\alpha = 90°$), as shown in the diagram. When the incident angle is equal to Brewster's angle θ_B, the transmitted rays are perpendicular to the

Fig. 40–21. Geometry of the radiation produced by the layer of dipoles in the interface between two media when light is incident at Brewster's angle θ_B. Only the parallel component of the beam is shown.

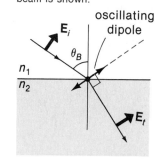

*The relationship connecting the refractive indexes and θ_B was discovered experimentally in 1811 by a Scottish physicist, Sir David Brewster (1781–1868).

direction in which specular reflection would be expected, the dashed line in Fig. 40–21 (see Problem 40–15). Thus, the direction of specular reflection would correspond to $\alpha = 0°$, so *no radiation is emitted in this direction*. The reflection coefficient of the parallel component of the incident beam is zero at Brewster's angle. (●Can you see why this dipole model predicts no similar effect for the perpendicular component?) ◄

40-5 DISPERSION

The index of refraction of most substances varies somewhat with the wavelength of the incident radiation. Figure 40–22 shows this variation in the visible region for three important types of glass and for fused quartz. When white light, which consists of light with all colors (wavelengths), enters a substance, the variation of the refractive index with wavelength causes the various colors to be spatially separated or *dispersed*.

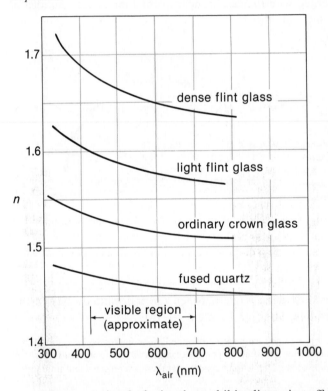

Fig. 40–22. The variation with wavelength (measured in air) of the index of refraction for three important types of glass and for fused quartz. Wavelengths near 600 nm = 6000 Å correspond to yellow light.

Fig. 40–23. A light ray is deviated by an amount δ upon passing through a prism. If the prism has the shape of an isosceles triangle, at minimum deviation the ray PQ is parallel to the base.

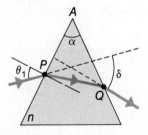

Prisms. A simple device that exhibits dispersion effectively is a triangular piece of glass, called a *prism*. Figure 40–23 illustrates the passage of a narrow monochromatic light beam or ray through a prism with a refractive angle α. The angle δ measures the *deviation* of the ray. As the incident angle θ_1 is varied, δ reaches a minimum value when $\overline{AP} = \overline{AQ}$. Under this condition of minimum deviation, we have (see Problem 40–17)

$$n \sin \frac{\alpha}{2} = \sin \frac{\alpha + \delta_{\min}}{2}$$ **(40–16)**

The measurement of $\delta_{\min}$ for incident light with different wavelengths is one method for obtaining curves of n versus λ such as those in Fig. 40–22.

glass prism

white
light

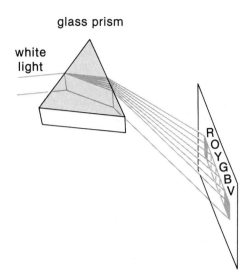

R
O
Y
G
B
V

Fig. 40–24. Dispersion of white light by a prism. The order of colors in the spectrum is red, orange, yellow, green, blue, and violet.

When white light is incident on a prism, the deviation angle δ is different for each color. Then, the white light is dispersed into a rainbow spectrum of colors, as indicated in Fig. 40–24. Because n(red) $< n$(violet) for most substances (see Fig. 40–22), red light is deviated less than violet light in passing through the prism.

The Rainbow. In the *rainbow* Nature has provided a spectacular effect that combines refraction, internal reflection, and dispersion. A rainbow can be observed when the Sun is relatively low in the sky (and is not obscured by the clouds) and is behind the observer. If the air contains droplets of rain that are sufficiently small that they are accurately spherical,* it is possible to see a set (or perhaps two sets) of circular arcs, each containing the spectrum of colors from red to violet.

The geometry of minimum deviation is important in describing the formation of a rainbow. Figure 40–25a shows three horizontal rays of sunlight that are refracted by a spherical droplet. Minimum deviation occurs for the ray bb'. (Notice that because of the reflection at R, the deviation is an obtuse angle, $\delta = 180° - \gamma$.) Other rays are spread out in various directions. However, only the rays at or near minimum deviation emerge practically parallel to one another and therefore reach the observer's eye with the greatest intensity. Calculation shows that for red light, $\gamma_R = 42.3°$, whereas for violet light, $\gamma_V = 40.4°$.

In Fig. 40–26 the observer is located at Q, and the x-axis is directed toward the Sun. When the observer looks away from the Sun, he sees a red arc with a cone half-angle $\gamma_R = 42.3°$ and a violet arc with a cone half-angle $\gamma_V = 40.4°$; between these two extreme arcs is the remainder of the rainbow spectrum.

In addition to the primary rainbow that is formed in the way shown in Fig. 40–25a, there is a secondary rainbow that is formed by two internal reflections (Fig. 40–25b), resulting in $\gamma_R = 50.6°$ and $\gamma_V = 53.9°$. Figure 40–26 indicates how the observer sees the secondary rainbow outside the primary rainbow. Notice the *opposite* ordering of colors in the two rainbows.

If the observer is standing on a flat plain and if the Sun is on the horizon, the observer will see semicircular rainbows. (This is the case illustrated in Fig. 40–26, in which the plain is the x-y plane.) The size of the rainbows is diminished when the

Fig. 40–25. (a) Formation of the primary rainbow with one internal reflection. (b) Formation of the secondary rainbow with two internal reflections.

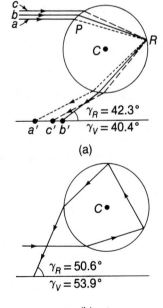

$\gamma_R = 42.3°$
$\gamma_V = 40.4°$

(a)

$\gamma_R = 50.6°$
$\gamma_V = 53.9°$

(b)

_____
*If the droplets are too small (for example, fog or mist particles), diffraction effects become important and the appearance of the rainbow is significantly altered.

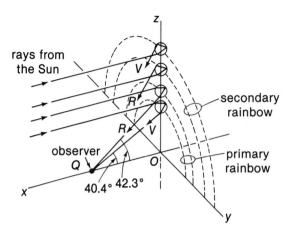

Fig. 40–26. Primary and secondary rainbows.

Sun is above the horizon, and no rainbow is visible at all if the Sun angle exceeds 42.3° above the horizon. However, an observer in an airplane may see a complete circular rainbow.

SPECIAL TOPIC

40-6   DISPERSION—AN ATOMIC DESCRIPTION

A revealing description of the dielectric constant and the related index of refraction for a medium can be made on the basis of a simple classical atomic model. In this theory it is assumed that the mutual interactions of the atoms may be ignored; therefore, the theory applies for the case of a dilute gas, but it also serves as a starting point for understanding the more complicated cases of liquids and solids.

First, assume that the electrons in an atom are bound by a Hooke's-law type force, so that for small excursions x from equilibrium, the restoring force can be expressed as $-m_e\omega_0^2 x$. The quantity ω_0 is the natural oscillation frequency for a particular electronic state. Any real substance has many natural frequencies. However, for the discussions here we select one such frequency, labeled ω_0, and consider the behavior of the substance when there is incident an electromagnetic wave with a frequency ω that is near ω_0. We expect some sort of resonance effect to occur when $\omega = \omega_0$.

In all of the situations that we consider, the incident electromagnetic wave does not impart a large velocity to the atomic electrons. Then, the magnetic force, which is proportional to v, can be neglected in comparison with the electric force.

Let the electric vector of the incident wave be polarized along the x-axis, so that, in complex notation, $\mathbf{E} = \hat{\mathbf{i}}\widehat{E} = \hat{\mathbf{i}}E_0 e^{-i\omega t}$. Then, the particular electron selected for study is subject to an electric force $-eE_0 e^{-i\omega t}$ (compare the development in Section 37–3).

Finally, account must be taken of the fact that the electron dissipates energy through collision and radiation processes. These effects can be approximated by including a resistive or damping force that is proportional to the electron velocity, namely, $-m_e\gamma\, dx/dt$.

Newton's equation for the electron, which we write in complex form, contains the three forces we have identified; thus,

$$m_e\left(\frac{d^2\widehat{x}}{dt^2} + \gamma\frac{d\widehat{x}}{dt} + \omega_0^2\widehat{x}\right) = -eE_0 e^{-i\omega t} \qquad \textbf{(40–17)}$$

This differential equation has the same form as the equation used to describe a driven LRC circuit (Eq. 37–33), so the solution may be written as (see Eq. 37–35)

$$\widehat{x}(t) = \widehat{x}_0\, e^{-i\omega t} = -\frac{eE_0 e^{-i\omega t}}{m_e[(\omega_0^2 - \omega^2) - i\gamma\omega]} \qquad \textbf{(40–18)}$$

The dipole moment acquired by the gas atom owing to the displacement of the electron is

$$\widehat{p} = -e\widehat{x} = \widehat{\alpha}\widehat{E} = \widehat{\alpha}E_0 e^{-i\omega t} \qquad (40\text{-}19)$$

where $\widehat{\alpha}$ is the polarizability of the atom (a complex quantity). (Compare Eq. 30-31.) Thus,

$$\widehat{\alpha} = -\frac{e}{\widehat{E}}\widehat{x} = \frac{e^2}{m_e[(\omega_0^2 - \omega^2) - i\gamma\omega]} \qquad (40\text{-}20)$$

We can now take the various quantities and expressions we obtained concerning polarization effects in dielectric media (Section 30-5) and modify them by the inclusion of the complex form of $\widehat{\alpha}$. Thus, the Clausius-Mossotti equation (Eq. 30-36) is written as

$$\frac{\widehat{\kappa} - 1}{\widehat{\kappa} + 2} = \frac{N_0\widehat{\alpha}}{3\epsilon_0} \qquad (40\text{-}21)$$

where the (complex) dielectric constant $\widehat{\kappa}$ has been introduced.

For most materials of interest, $\kappa_M = 1$; therefore, we may relate $\widehat{\kappa}$ to a complex index of refraction $\widehat{n}$, obtaining

$$\sqrt{\widehat{\kappa}} = \widehat{n} = n_1 + in_2 \qquad (40\text{-}22)$$

Gases have $\widehat{\kappa} \cong 1$; consequently, $|N_0\widehat{\alpha}/3\epsilon_0| \ll 1$, and after some algebra (see Problem 40-22b) we have, for ω near ω_0,

$$n_1 = 1 + \frac{N_0 e^2(\omega_0 - \omega)}{\omega_0\epsilon_0 m_e[4(\omega_0 - \omega)^2 + \gamma^2]} \qquad (40\text{-}23a)$$

$$n_2 = \frac{N_0 e^2\gamma}{2\omega_0\epsilon_0 m_e[4(\omega_0 - \omega)^2 + \gamma^2]} \qquad (40\text{-}23b)$$

What is the meaning of the fact that the index of refraction $\widehat{n}$ is a complex quantity? If $\widehat{n}$ is complex, so is the phase velocity, $\widehat{v} = c/\widehat{n}$. Now, an electromagnetic wave propagating in the z-direction through a medium is described by

$$\widehat{E} = E_0 e^{i(kz - \omega t)}$$

Then, if the medium has a refractive index $\widehat{n}$, we use $k = \omega/v$, $\widehat{v} = c/\widehat{n}$, and $\widehat{n} = n_1 + in_2$ to write

$$\widehat{E} = E_0 e^{-(\omega/c)n_2 z} e^{i(\omega n_1 z/c - \omega t)}$$

This expression represents a wave that propagates with a phase velocity c/n_1 and is attenuated by the factor $e^{-\beta z}$, where the attenuation coefficient is $\beta = \omega n_2/c$. Notice that β (through n_2) depends strongly on γ, which is the measure of the damping force that is responsible for the absorption and dissipation of wave energy. Because $\widehat{\kappa} \cong 1$, we have

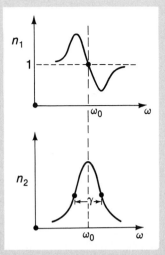

Fig. 40-27. The behavior of the real (n_1) and imaginary (n_2) parts of the complex index of refraction $\widehat{n}$ in the vicinity of the characteristic frequency ω_0. The index n_1 increases with ω except near $\omega = \omega_0$ (the region of anomalous dispersion).

$n_1 \cong 1$ and $n_2 \ll n_1$; thus, n_1 is essentially the same as the refractive index for the ideal nonabsorptive case.

Figure 40-27 shows how n_1 and n_2 depend on the frequency ω in the vicinity of ω_0. Notice that n_2 (and, hence, the absorption of wave energy by the medium) exhibits resonance behavior near $\omega = \omega_0$. In the frequency region for which n_2 is largest, called an *absorption band*, (the real part of) the refractive index *decreases* with frequency, contrary to the usual *increase* with frequency (see Fig. 40-22). This effect is called *anomalous dispersion*, and it always occurs in a frequency region for which the absorption is large. Indeed, the medium may be essentially opaque to the incident radiation in a region of anomalous dispersion.

A real substance exhibits a number of absorption bands. (Figure 40-28 shows the situation for quartz.) These bands have their origin in the various electronic excitation modes of the particular substance, and they may be correlated with the excitation modes discussed in connection with thermal properties (see Section 24-2). For low frequencies (generally, $\nu \lesssim 10^6$ Hz), absorption is usually unimportant in dielectric media, and n_1 is close to the ordinary static (DC) value of $\sqrt{\kappa}$, or more accurately, $\sqrt{\kappa\kappa_M}$. Substances that are transparent do not have resonances (absorption bands) in the visible part of the spectrum. However, the resonances that lie outside the visible region still influence the optical properties of the substance. These resonances are responsible for the general increase of n_1 with ω so that $n_{blue} > n_{red}$ (see Fig. 40-22).

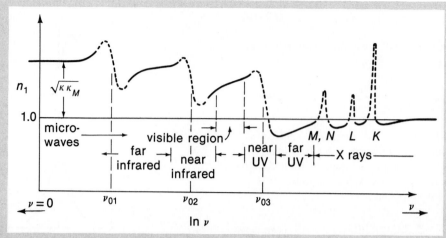

Fig. 40-28. The behavior of the real part of the refractive index n_1 of quartz as a function of the transmitted electromagnetic wave frequency ν. For fused quartz: $\nu \lesssim 10^8$ Hz, $\kappa = 3.75$ and $n_1 = \sqrt{3.75} = 1.94$; the resonances are $\nu_{01} = 1.45 \times 10^{13}$ Hz ($\lambda_{air} = 20.7\ \mu m$), $\nu_{02} = 3.39 \times 10^{13}$ Hz ($\lambda_{air} = 8.84\ \mu m$), and $\nu_{03} = 2.91 \times 10^{15}$ Ha ($\lambda_{air} = 103$ nm). The curve of n_1 is shown dashed in the regions of high absorption because the ordinary concept of wave propagation does not apply in such regions.

PROBLEMS

Section 40-1

40-1   In Galileo's attempt to determine the speed of light, he and his companion were located on hilltops 3 km apart. What time interval between uncovering his lantern and seeing the light from his companion's lantern would Galileo have had to measure in order to obtain a value for the speed of light?

40-2   Refer to Fig. 40-2. The onset of the eclipse of Io is observed when the Earth is at P_1. The next eclipse is observed when the Earth is at P_2. The true period of Io is 42.5 h. Compared with the time expected on the basis of the P_1 observation, how early was the P_2 observation? (Assume that Jupiter does not move appreciably in its orbit during the time interval and that P_1, P_2, and E are in a straight line.)

40-3   In his measurement of the speed of light, Fizeau used a wheel with 720 teeth and a wheel-to-mirror distance of 8633 m. The slowest angular speed at which the light was blocked by a tooth and did not reach the observer was found to be 756 rpm. What value did Fizeau obtain for c?

40-4   In 1932 Michelson and his collaborators determined the speed of light by using a 32-sided rotating mir-

ror to flash bursts of light down a path with a total length of 15 km. At what speed must the mirror rotate if the flash produced by light shining on one mirror face is reflected from the distant stationary mirror and returns to strike the adjacent face when it has moved to the exact position of the first face?

Sections 40-2 and 40-3

40-5   Prove that two mirrors with perpendicular faces, as shown in the diagram, will reflect an incident ray that lies in a mutually perpendicular third plane in such a way that the (twice) reflected ray is parallel to the incident ray for all angles θ. (•Can you extend the argument to three dimensions? Consider three mutually perpendicular mirrors. *Any* incident ray will be reflected in a direction precisely parallel to the incident ray. Such an arrangement of mirrors is called a *corner reflector*. These devices have many practical applications. For example, corner reflectors are sometimes used to line unlighted aircraft runways. •Can you see the purpose? An array of corner reflectors was left on the Moon by Apollo astronauts so that pulsed lasers could be used to

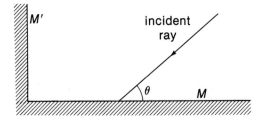

determine the precise Earth-Moon distance by radar ranging.)

40-6 Light is incident normally on a 1-cm layer of water that lies on top of a flat Lucite plate whose thickness is 0.5 cm. How much longer is required for light to pass through this double layer than is required to traverse the same distance in air?

40-7 A light ray is incident from air on the surface of a clear soft drink ($n_2 = 1.330$) that stands in a glass ($n_3 = 1.520$), as shown in the diagram. The angle of incidence is $\theta_1 = 50°$. Find the angles θ_2 and θ_3.

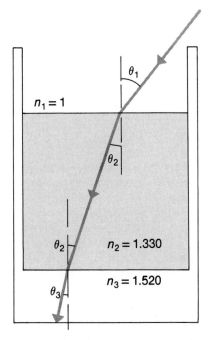

40-8 A cylindrical tank with an open top has a diameter of 3 m and is completely filled with water. When the setting Sun reaches an angle of 28° above the horizon, sunlight ceases to illuminate the bottom of the tank. How deep is the tank?

40-9 Light is incident from below on the interface between glass with index $n_1 = 1.520$ and a drop of liquid with index n_2, as shown in the diagram above right. When the angle of incidence θ_1 is increased to 74.0°, total internal reflection occurs (that is, $\theta_2 = 90°$). Determine the refractive index of the liquid.

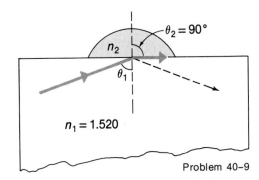

Problem 40-9

40-10 A microscope is focused on a small scratch on the top surface of a glass slide. When a cover slip with a thickness of 500 μm is placed on the slide, the microscope must be raised by 123 μm to bring the scratch again into focus. What is the index of refraction of the cover slip? (Assume that all rays entering the microscope make small angles with the normal to the slide.)

40-11 Use Fermat's principle to derive the law of reflection, $\theta_1 = \theta_1'$. Refer to the diagram.

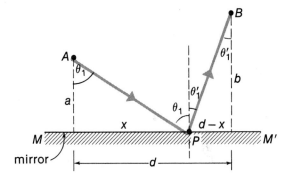

Section 40-4

40-12 When light is incident normally from air on a certain type of clear plastic, 94 percent of the light is transmitted and 6 percent is reflected. What is the refractive index of the plastic?

40-13 A beam of light is incident normally on a polished face of a diamond. (a) What is the ratio of the amplitude of the reflected electric field to that of the incident electric field? (b) What fraction of the incident (time-averaged) energy flux is reflected?

40-14 Use Fig. 40-19 to construct a graph of the dependence on θ_1 of the reflectances, $R_\perp$ and $R_\parallel$, for the two polarization components of a light wave incident from air ($n_1 = 1$) on glass ($n_2 = 1.5$). Construct also a graph of the reflectance R for *unpolarized* light, where $R = \frac{1}{2}(R_\perp + R_\parallel)$.

40-15 Show that when θ_1 is equal to Brewster's angle θ_B, there is an angle of 90° between the direction of the transmitted ray and the direction of the reflected ray. (For $\theta_1 = \theta_B$, reflection occurs only for the perpendicular component of the incident beam.)

Section 40-5

40-16 A prism made from fused quartz ($n = 1.4585$) has an apex angle $\alpha = 30°$. (a) What is the incident angle θ_1 for minimum deviation of sodium yellow light ($\lambda = 5893$ Å)? (b) Determine the minimum deviation δ.

40-17 Derive Eq. 40-16, which relates the minimum devi-
•• ation $\delta_{\min}$ to the refractive index n and the apex angle α of the prism. [*Hint:* Refer to the diagram. It is required that $d\delta/d\theta_1 = 0$. Use the relationship connecting δ with θ_1 and θ_2 to show that this condition implies $d\theta_2/d\theta_1 = -1$. Then, write $d\theta_2/d\theta_1 = (d\theta_2/d\gamma) \cdot (d\gamma/d\beta) \cdot (d\beta/d\theta_1) = -1$. Use Snell's law at points A and B to obtain the required relationships.]

40-18 A prism made from fused quartz has an apex angle of
• 30°. A beam of white light is incident on the prism with $\theta = 22.178°$, which yields minimum deviation for yellow light ($\lambda = 5893$ Å). The refractive in-

Problem 40–17

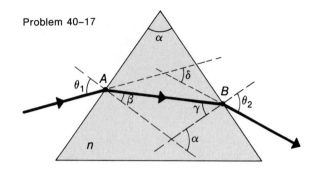

dexes for quartz for other wavelengths are $n(7000$ Å$) = 1.4553$ and $n(4000$ Å$) = 1.4702$. Determine the deviation δ for $\lambda = 4000$ Å and $\lambda = 7000$ Å. What is $\Delta\delta$ between the violet and red ends of the spectrum? [*Hint:* Refer to the diagram for Problem 40–17. Find β, γ, and θ_2, in that order, for each wavelength. Calculate to five significant figures.]

40-19 A hiker stands on a mountain peak near sunset and observes a (primary) rainbow caused by water droplets in the air about 8 km away. The valley is 2 km below the mountain peak and is entirely flat. What fraction of the complete circular arc of the rainbow is visible to the hiker?

SPECIAL TOPIC

Section 40-6

40-20 Derive the approximate relationships $\kappa_1 \cong n_1^2 - n_2^2$ and $\kappa_2 \cong 2n_1 n_2$, where $\hat{\kappa} = \kappa_1 + i\kappa_2$.

40-21 Use Eq. 40-20 for $\hat{\alpha}$. (a) Show that
•
$$\hat{\alpha} = \frac{e^2(\omega_0^2 - \omega^2)}{m_e[(\omega_0^2 - \omega^2)^2 + \gamma^2\omega^2]}$$
$$+ i\frac{\gamma e^2\omega}{m_e[(\omega_0^2 - \omega^2)^2 + \gamma^2\omega^2]}$$

(b) For $|N_0\hat{\alpha}/3\epsilon_0| \ll 1$, show that

$$n_1 = 1 + \frac{N_0 e^2(\omega_0^2 - \omega^2)}{2\epsilon_0 m_e[(\omega_0^2 - \omega^2)^2 + \gamma^2\omega^2]}$$

$$n_2 = \frac{N_0 e^2\gamma\omega}{2\epsilon_0 m_e[(\omega_0^2 - \omega^2)^2 + \gamma^2\omega^2]}$$

[*Hint:* Use the results in Problem 40–20.]

40-22 Use the results in Problem 40–21b. (a) For ω well
•

above the resonance frequency ω_0, show that

$$n_1 \cong 1 - \frac{N_0 e^2}{2\epsilon_0 m_e\omega^2} = 1 - \frac{N_0 e^2\mu_0}{8\pi^2 m_e}\lambda^2$$

(b) Show that the expressions for n_1 and n_2 given in Eqs. 40–23 are obtained when ω is near ω_0. [*Hint:* ω may be replaced by ω_0 everywhere except in $(\omega - \omega_0)$; thus, for example, $\omega_0^2 - \omega^2 = (\omega_0 + \omega)(\omega_0 - \omega) \cong 2\omega_0(\omega_0 - \omega)$.]

40-23 Use the result for $n_1 = n$ given in Problem 40–21b.
•• Show that

$$\frac{dn}{d\lambda} = -\frac{N_0 e^2\mu_0\lambda_0^4}{4\pi^2 m_e\lambda^3}$$

where $\lambda = 2\pi c/\omega$ and $\lambda_0 = 2\pi c/\omega_0$. Assume that this expression for $dn/d\lambda$ is valid for fused quartz in the violet and blue part of the spectrum, where the refractive index is influenced primarily by an absorption band centered at $\lambda_0 = 103$ nm. Use the values $n(400 \text{ nm}) = 1.47021$ and $n(500 \text{ nm}) = 1.46239$ to estimate the derivative at $\bar{\lambda} = 0.45 \, \mu\text{m}$ according to $dn/d\lambda \cong \Delta n/\Delta \lambda$; calculate the corresponding value of N_0.

Use the density of quartz ($\rho = 2.65 \text{ g/cm}^3$) and

its molecular mass ($M = 60.1 \text{ u}$) to determine the average number of electrons per SiO_2 molecule that are associated with the absorption resonance at $\lambda_0 = 103 \text{ nm}$.

Additional Problems

40–24 When the Sun is directly overhead, a narrow shaft of light enters a Mithran temple through a small hole in the ceiling and forms a spot on the floor 10.0 m below. (a) At what speed (in cm/min) does the spot move across the (flat) floor? (b) If a mirror is placed on the floor to intercept the light, at what speed will the reflected spot move across the ceiling?

40–25 A beam of sodium yellow light ($\lambda = 5893 \text{ Å}$) is incident normally on an optically flat fluorite plate that has a thickness of $20.00 \, \mu\text{m}$. (a) What is the optical path length through the fluorite plate? (b) What is the wavelength of the light in the plate? (c) What phase change does the light wave undergo in passing through the plate compared with a wave that travels the same distance in air?

40–26 The speed of sound in air is 332 m/s, and in Lucite it is 2680 m/s. Assume that there exists an acoustical analog of Snell's law with refractive indexes η. (a) What is the ratio $\eta(\text{Lucite})/\eta(\text{air})$? (b) With sound incident on Lucite from air, what is the maximum angle of incidence for which refraction occurs?

40–27 According to Newton's corpuscular theory of light, a "light particle" that passes into a medium from vacuum (air) experiences an attractive force at the interface due to the medium. The work W done on the light particle as it passes through the interface results in a negative potential energy, $U = -W$, relative to vacuum. Refer to the diagram and show that (a) $v_2 = \sqrt{v_1^2 - 2U/m}$ and (b) $n = (\sin \theta_1)/(\sin \theta_2) = v_2/v_1 = \sqrt{1 - 2U/mv_1^2}$ where m is the mass of a light corpuscle. Although the result for n gives $n > 1$ in agreement with experiment, the theory attributes this to $v_2 > v_1$ or $v_2 > c$, instead of $v_2 < c$, as given by the wave theory. [*Hint:* The attractive force that acts on the corpuscle can have no component tangent to the interface. (•Can you see why?)]

40–28 A beam of light is incident from air on a glass surface at an angle $\theta_1 = 57.8°$. The reflected and re-

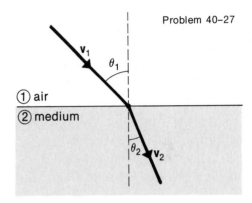

Problem 40–27

fracted rays are observed to be mutually perpendicular. (a) Determine the index of refraction n of the glass. (b) Show that $n = \tan \theta_1$, where $\theta_1 = \theta_B$ (Brewster's angle).

40–29 A light ray is incident on the interface between a diamond and air from within the diamond. What is the critical angle for total internal reflection? Use Table 40–2. (The smallness of θ_c for diamond means that light is easily "trapped" within a diamond and eventually emerges from the many cut faces; this makes a diamond more "brilliant" than stones with smaller n and larger θ_c.)

40–30 A light beam that is polarized parallel to the plane of incidence is incident at Brewster's angle on a glass plate with parallel surfaces, as indicated in the diagram. There is no reflection at the air-glass interface. Show that there is also no reflection at the glass-air interface, so that there is 100 percent transmission through the plate. (A glass plate oriented at Brewster's angle with respect to an incident beam constitutes a *Brewster window* for the parallel component of the beam.)

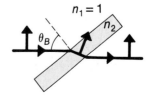

40-31 Radiations with very short wavelengths can pene-
•   trate into a metal (even though the absorption is
large). Thus, we can define an index of refraction
for X rays incident on silver. For $\lambda = 1.28$ Å, we
find $n = 1 - \delta$, with $\delta = 2.15 \times 10^{-5}$. (a) Show
that if the incident X rays make very small angles ψ_1
with the silver surface (this is called *grazing incidence*), there is no refracted ray; thus, the surface
acts as a perfect mirror if $\psi_1 < \sqrt{2\delta}$. (b) Use the
result given in Problem 40-22a, namely,
$n_1 \cong 1 - N_0 e^2 / 2\epsilon_0 m_e \omega^2$, to determine N_0 for silver.
(c) A model of refraction by a conductor considers
all of the atomic electrons to contribute to N_0. Compare this assumption with the result of (b).

40-32 Fused quartz exhibits the following dispersion:
•

$$n_1(500\,\text{nm}) = 1.46239$$
$$n_2(550\,\text{nm}) = 1.45997$$
$$n_3(600\,\text{nm}) = 1.45810$$

The group velocity v_g of a wave in a dispersive medium is related to the phase velocity v according to
(see Eq. 19-45)

$$v_g = v - \lambda(dv/d\lambda)$$

where λ is the wavelength in the medium. (a) Use
the refractive indexes n_1 and n_3 to estimate the derivative $dv/d\lambda$ at the median wavelength
($\bar{\lambda} = 550$ nm) and show that

$$\bar{\lambda}\frac{dv}{d\lambda} \cong \frac{550\,c}{n_2}\frac{\dfrac{1}{n_3} - \dfrac{1}{n_1}}{\dfrac{600}{n_3} - \dfrac{500}{n_1}}$$

(b) Determine the percentage difference between
the phase velocity and the group velocity for $\bar{\lambda}$.

GEOMETRICAL OPTICS

A variety of optical devices have been developed to aid the eye and extend its capabilities. These devices have the common feature that they employ curved surfaces to form images by reflection or refraction. The study of such image formation is called *geometrical optics* or *ray optics* and is the subject of this chapter. In Chapter 45 we apply these ideas to some specific optical instruments.

In the following chapters we discuss the modifications of the results obtained here that are required by diffraction effects due to the wave character of light. These effects are usually small owing to the smallness of the wavelength of light compared with the size of everyday objects. However, in some situations, particularly those involving the fine details of images, diffraction sets the ultimate limitation on the quality of the images.

41-1 PLANE MIRRORS

We introduce the idea of image formation by examining the (specular) reflection by a plane mirror of light from a point source. Figure 41–1a shows two of the rays that emanate from the point source O, the *object*. These rays are reflected by the mirror MM'. The normal ray OA is reflected back toward O. The oblique ray OB is reflected with equal angles of incidence and reflection θ. The backward extension of the reflected ray BC intersects the extension of ray OA at the point I. Notice that the intersection point I is independent of the angle θ. Thus, all of the rays from O that are reflected by the mirror appear to have originated at the point I, the *image point* of O. Because the triangle $\triangle OBI$ is isosceles, the image point I is the same distance behind the mirror that the object point O is in front of the mirror; that is,

$$s' = -s \qquad (41-1)$$

where we use the convention that distances are reckoned from the mirror surface, and are positive if they are measured to a point in front of the mirror and are negative if they are measured to a point behind the mirror.

Figure 41–1b shows the observer's eye intercepting several reflected rays from O. The observer perceives the source of the rays to be I, the image of the object O. Because the rays do not actually originate at the point I but only appear to do so, the image is said to be *virtual*. A virtual image can be seen by the eye (Fig. 41–1b) but it cannot be projected onto a screen or piece of film. If the eye is moved to a different position, the image point I remains fixed. (•Can you see why this is the case?)

Fig. 41–1. (a) The reflection of two rays, OA and OB, from an object O define the image point I. (b) An observer sees the reflected rays from O diverging from I, the (virtual) image of O.

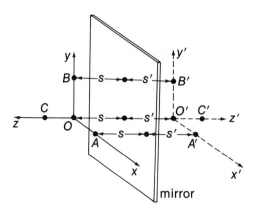

Fig. 41-2. The image of the right-handed coordinate frame *OABC* is the reversed or left-handed frame *O'A'B'C'*. The mirror is parallel to the *x-y* plane.

An extended object may be considered to be a collection of point sources of light. A mirror will form an image of each of these sources, thereby producing the image of the entire object. For example, consider the four point sources *O*, *A*, *B*, and *C*, located at the origin *O* and along the three axes of the right-handed coordinate frame in Fig. 41-2. The images of these sources are located at *O'*, *A'*, *B'*, and *C'*, respectively, and define a left-handed coordinate frame.* Thus, the mirror forms a reversed image of the object, in agreement with everyday experience.

Example 41-1

What is the minimum height of a mirror in which a 5′ 10″ (1.77-m) student can observe her full image?

Solution:

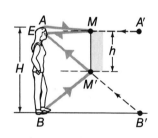

The top of the mirror at *M* needs to be no higher than the mid-height between *E* and *A*. The bottom of the mirror at *M'* needs to be no lower than the mid-height between *E* and *B*. Therefore, the height *h* of the mirror needs to be no more than half the height *H* of the student, or $h = \frac{1}{2}H = 2' \, 11'' \, (0.885 \, \text{m})$.

Example 41-2

Discuss the multiple images of a point source of light formed by a pair of mirrors.

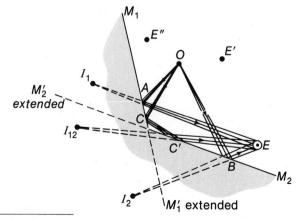

---

*A right-handed coordinate frame is one for which the unit vectors obey the relationship $\hat{\mathbf{i}} \times \hat{\mathbf{j}} = \hat{\mathbf{k}}$; a left-handed frame is one for which $\hat{\mathbf{i}} \times \hat{\mathbf{j}} = -\hat{\mathbf{k}}$.

Solution:

The point source is located at O, and the observer's eye is located at E. An image is formed in the usual way by each of the mirrors individually. The system of rays OAE determines the image point I_1 for mirror M_1 alone, and the system of rays OBE determines the image point I_2 for mirror M_2 alone. In addition to these images that involve only single reflections, there is a third image I_{12} formed by the system of rays $OCC'E$ that undergo reflection from both mirrors. The geometric construction in the diagram shows that I_{12} corresponds to the image point of I_1 in mirror M_2 extended. (•Can you provide the proof?) On the other hand, there is no image that corresponds to the image point of I_2 in mirror M_1 extended.

A simple rule governs the formation of multiple images:

(a) Determine the positions of all images I_n formed by single reflections from the mirrors M_n.

(b) A multiple image I_{nm} of I_n in mirror M_m will exist only if I_n is in *front* of mirror M_m extended.

In the situation illustrated in the diagram, image I_1 is in front of mirror M_2 extended, whereas image I_2 is behind mirror M_1 extended. Thus, the eye sees the image I_{12}, but there is no image I_{21} of the object.

Although rays exist that appear to originate in all image points determined by this rule, not all may reach the observer's eye at an arbitrary location. (•What images of O can be seen at the eye points E' and E''?) (•Can you extend these ideas to a system of three mirrors?)

41-2 SPHERICAL MIRRORS

Spherical reflecting surfaces—both concave and convex—possess interesting focusing properties. If the rays from a point source of light are brought together at a common point by reflection (or refraction) at a certain surface, that point is the *image* of the object point. Unlike a plane mirror, which forms an exact image of an object, a spherical mirror suffers from an intrinsic imperfection called *spherical aberration*. Figure 41–3a shows that a wide bundle of rays from a point source are not all reflected through the same point by a spherical mirror, so no well-defined image point results. However, if the bundle of rays emitted by a source is confined to a small region

Fig. 41–3. (a) Spherical aberration. The rays in a wide bundle from a point source are not all reflected through the same point. The image of the object O is therefore blurred. (b) A narrow bundle of rays near the axis does produce a sharp image. For a point source, the *axis* of the mirror is the line defined by the center C of the mirror and the object point O.

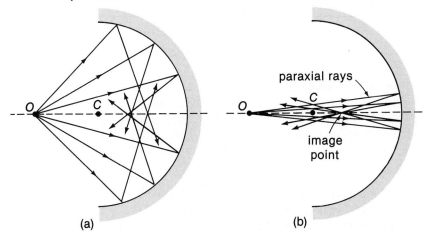

(a)   (b)

around the axis, the blurring of the image can be made suitably small (Fig. 41–3b). The rays that are incident along or near the axis of an optical system are called *paraxial rays.** We consider here only such rays, although in diagrams we often exaggerate the beam size in the interests of clarity. (See, however, Problems 41–25, 26, and 32.)

Concave Spherical Mirrors. Figure 41–4 shows a concave spherical mirror with radius of curvature R. The line joining the point source at O and the center of the mirror at C determines the mirror axis, which intersects the mirror at the *vertex* V. The location of the source O is specified by its distance s from the vertex. The ray OP is reflected from the mirror surface at P exactly as it would be from a flat mirror oriented tangent to the spherical surface at P. Thus, the incident and reflected rays at P make equal angles θ with the normal N to the spherical surface. All other rays from O for which the angle α is small (α ≪ 1) also intersect the axis at I (or very close to I). Thus, I is the image point of O. The distance to I from the vertex V is s′.

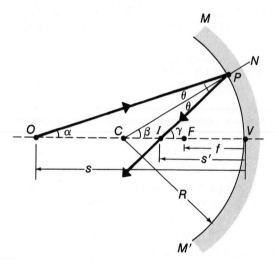

Fig. 41–4. Geometry for locating the focal point F of a concave spherical mirror. The ray OP is a paraxial ray (α ≪ 1); the angle α has been exaggerated for clarity.

Notice that the angle β is an exterior angle of the triangle △OPC and that the angle γ is an exterior angle of the triangle △CPI. Thus, β = α + θ and γ = β + θ; hence,

$$2\beta = \alpha + \gamma \tag{41–2}$$

Let the arc length $\overset{\frown}{PV}$ be ℓ; then, because $\overline{CV} = R$, we have $\beta = \ell/R$. If, in addition, $\alpha \ll 1$ and $\gamma \ll 1$, we have $\alpha \cong \ell/s$ and $\gamma \cong \ell/s'$. When these expressions for α and γ are substituted into Eq. 41–2, the result is

$$\frac{1}{s} + \frac{1}{s'} = \frac{2}{R} \tag{41–3}$$

The point at which an optical system causes incident parallel rays to converge is called the *focal point* of the system. When the source O in Fig. 41–4 is moved far away from the mirror (s → ∞), the incident rays become parallel and the image distance s′ approaches the limiting value R/2. Thus, the focal point F of the spherical mirror is located a distance f = R/2 from the vertex V. Equation 41–3 may now be

---

*The first systematic and reasonably complete description of image formation by paraxial rays was given by Gauss (1778–1855).

expressed as

$$\frac{1}{s} + \frac{1}{s'} = \frac{1}{f} \qquad \text{concave mirror} \qquad \textbf{(41–4)}$$

with

$$f = \tfrac{1}{2}R \qquad \text{concave mirror} \qquad \textbf{(41–5)}$$

It is important to note that the object and image points can be interchanged (that is, $s \to s'$ and $s' \to s$) without affecting Eq. 41-4. This is a property of any focusing system and is referred to as the *principle of reversibility*. Thus, the directions of the rays in Fig. 41-4 can be reversed and the diagram will still represent a realizable situation.

It is instructive to consider the behavior of s' as a function of s for a concave spherical mirror with radius of curvature R. The graph of this function is shown in Fig. 41-5. When s is very large, the value of s', as we have just seen, is asymptotically

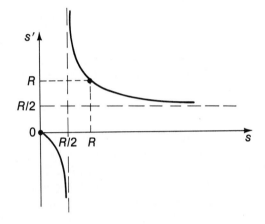

Fig. 41–5. The behavior of s' as a function of s for a concave spherical mirror with radius of curvature R.

equal to the focal length $f = R/2$. As s is decreased, s' increases and an image is formed with converging rays at I. Note that rays actually pass through the point I. Such an image is a *real* image, and it can actually be projected onto a screen or piece of film located at I. When s becomes equal to R, we have the symmetric situation, $s = R$ and $s' = R$, in which rays are emitted from and return to the center of the spherical surface. As s is decreased beyond this point, s' increases rapidly. When the object approaches the focal point F (that is, as $s \to f$), we have $s' \to \infty$. The ray geometry is reversible, so $s \to \infty$ means $s' \to f$, and $s \to f$ means $s' \to \infty$. Thus, a point source located at the focal point of a concave spherical mirror results in an outgoing beam of light parallel to the mirror axis.

When s is reduced still further (that is, when the object is placed between the focal point F and the vertex V), negative values of s' result. The meaning of $s' < 0$ is illustrated in Fig. 41-6, where the ray geometries for $s' > 0$ and $s' < 0$ are compared. Figure 41-6a shows that $s' > 0$ corresponds to the case in which I is located in front of the mirror and the rays converge to form a real image. Figure 41-6b shows that $s' < 0$ corresponds to the case in which I is located behind the mirror and the rays diverge from a virtual image. Finally, notice that for a plane mirror, we have $R = \infty$ and, hence, $f = \infty$. Then, for any object position, we find $s' = -s$. This is just the sign convention that was introduced in writing Eq. 41-1.

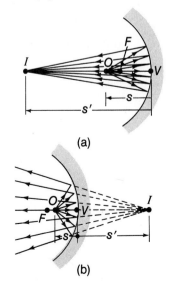

Fig. 41–6. (a) A concave spherical mirror produces a real image when the object O lies outside the focal point F (that is, when $s > f$ and $s' > 0$). (b) A virtual image is formed when O lies inside F (that is, when $s < f$ and $s' < 0$).

(a)

(b)

Convex Spherical Mirrors. Consider the convex spherical mirror with radius of curvature R shown in Fig. 41–7. The ray OP is reflected from the mirror surface at P with equal angles θ of incidence and reflection. Regardless of the position of the object O, the reflected rays appear to diverge from an image located behind the mirror. Thus, for any s, we have $s' < 0$, and the image formed by a convex spherical mirror is always virtual.

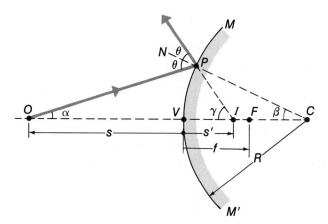

Fig. 41–7. Geometry for locating the focal point F of a convex spherical mirror with radius of curvature R.

In Fig. 41–7, θ is an external angle of the triangle $\triangle OPC$ and 2θ is an external angle of the triangle $\triangle OPI$. Thus, $\theta = \alpha + \beta$ and $2\theta = \alpha + \gamma$; hence,

$$-2\beta = \alpha - \gamma \tag{41-6}$$

We now extend the sign convention we have been using. The points C, F, and I are all located *behind* the mirror,* so the distances to these points as measured from the vertex V are taken to be *negative*. Thus, the focal length is $f = \overline{VF} < 0$, and the image distance is $s' = \overline{VI} < 0$. We also make a distinction between the *radius of curvature R* of a surface, which is always *positive*, and the distance $r = \overline{VC}$, which locates the center of curvature and is called the *surface radius*. For the case illustrated in Fig. 41–7, we have $\overline{VC} = r = -R$. (For the case of a concave mirror, we would have $r = +R$.) Further, the angles β and γ are both reckoned to be negative. As before, $\overparen{PV} = \ell$, so we have $\beta = -\ell/r$; then, with $\alpha \ll 1$ and $\gamma \ll 1$, we have $\alpha \cong \ell/s$ and $\gamma \cong -\ell/s'$. When these expressions for α and γ are substituted into Eq. 41–6, the result is

$$\frac{1}{s} + \frac{1}{s'} = \frac{2}{r} = -\frac{2}{R}$$

or

$$\frac{1}{s} + \frac{1}{s'} = \frac{1}{f} \qquad \text{convex mirror} \tag{41-7}$$

with

$$f = -\tfrac{1}{2}R \qquad \text{convex mirror} \tag{41-8}$$

---

*"In front of" and "behind" are judged by looking along the direction of the incident light.

Equations 41–4 and 41–7 are identical, so we can write for *all* spherical mirrors the so-called *Descartes formula,*

$$\frac{1}{s} + \frac{1}{s'} = \frac{1}{f} \qquad \textbf{(41–9)}$$

with

$$f = \tfrac{1}{2}r = \begin{cases} +\tfrac{1}{2}R & \text{concave mirror} \\ -\tfrac{1}{2}R & \text{convex mirror} \end{cases} \qquad \textbf{(41–10)}$$

The distances s, s', and f are positive when measured in front of the mirror and are negative when measured behind the mirror. Images formed in front of the mirror ($s' > 0$) are real, and images formed behind the mirror ($s' < 0$) are virtual. A convex mirror can form only virtual images of real objects.

The sign conventions for mirrors are summarized in Fig. 41–8.

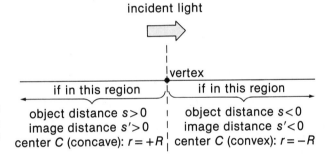

Fig. 41–8. Sign conventions for mirrors. The object distance s for a real object is always positive.

incident light

vertex

if in this region | if in this region

object distance $s > 0$ | object distance $s < 0$
image distance $s' > 0$ | image distance $s' < 0$
center C (concave): $r = +R$ | center C (convex): $r = -R$

Example 41–3

(a) A point source of light is located 20 cm in front of a concave spherical mirror with $R = 30$ cm. Determine the position and character of the image point.

(b) If the concave mirror is replaced by a convex mirror with $R = 30$ cm, determine the position and character of the new image point.

Solution:

(a) For the concave mirror, we have $f = \tfrac{1}{2}R = \tfrac{1}{2}(30\text{ cm}) = 15$ cm and $s = +20$ cm. Then, Eq. 41–9 gives

$$\frac{1}{s'} = \frac{1}{f} - \frac{1}{s} = \frac{1}{15\text{ cm}} - \frac{1}{20\text{ cm}} = \frac{1}{60\text{ cm}}$$

so that $s' = 60$ cm. The image point is located in front of the mirror and is *real*.

(b) For the convex mirror, we have $f = -\tfrac{1}{2}R = -\tfrac{1}{2}(30\text{ cm}) = -15$ cm and $s = +20$ cm. Then, Eq. 41–9 gives

$$\frac{1}{s'} = \frac{1}{f} - \frac{1}{s} = -\frac{1}{15 \text{ cm}} - \frac{1}{20 \text{ cm}} = -\frac{7}{60 \text{ cm}}$$

so that $s' = -60/7$ cm. The image point is located behind the mirror and is *virtual*.

Example 41–4

The inner surface of a hollow sphere with radius R is coated with a reflecting substance. A point source of light is located at a point O_1, which is $3R/7$ from the center C of the sphere. Determine the position and character of the image formed by rays that lie near the axis $\overline{O_1 C}$ and are reflected first from the surface nearest O_1.

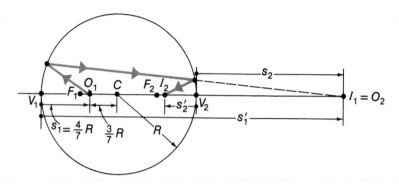

Solution:

The source point O_1 is a distance $s_1 = R - 3R/7 = 4R/7$ from the vertex V_1. If the right-hand hemisphere were missing, an image would be formed at I_1, a distance s_1' from V_1. The points O_1 and I_1 are to the right of the vertex V_1, but they are in *front* of the mirror according to our convention. Thus, s_1, s_1', r, and f are all positive. Using Eq. 41–9, we find

$$\frac{1}{s_1'} = \frac{1}{f} - \frac{1}{s_1} = \frac{2}{R} - \frac{7}{4R} = \frac{1}{4R}$$

so that $s_1' = 4R$

Because the right-hand hemisphere is present, the image at I_1 is not actually formed. However, the rays inside the sphere that converge toward I_1 define a *virtual object* O_2 for the second reflection. The corresponding object distance is *negative*: $s_2 = -(4R - 2R) = -2R$. For the right-hand mirror surface, $f = R/2$. Using Eq. 41–9 again, we obtain

$$\frac{1}{s_2'} = \frac{1}{f} - \frac{1}{s_2} = \frac{2}{R} - \frac{1}{-2R} = \frac{5}{2R}$$

so that $s_2' = \frac{2}{5}R$

Thus, the final image at I_2 is *real* and lies in front of the right-hand mirror surface.
(•Where would the image be formed if the source point were moved to F_1?)

Example 41–5

A narrow parallel beam of rays from a point source at infinity is incident on a concave spherical mirror with radius R. The rays make an angle α with the (arbitrarily defined) mirror axis, where $\alpha \ll 1$. Analyze the image formation for such rays.

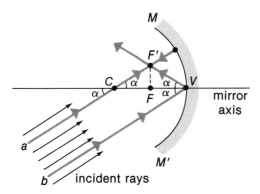

Solution:

From the bundle of incident rays, select ray *a* which passes through the mirror center *C* and ray *b* which is incident on the vertex *V*. Upon striking the mirror, ray *a* is reflected back toward *C*; ray *b* is reflected at the angle α, as shown in the diagram. The intersection of these rays defines the image point *F'*. (All of the other rays in the bundle also pass through *F'*. •Can you see why?)

The triangle $\triangle CF'V$ is isosceles, so *F'* lies directly above the focal point *F*, which is a distance $f = R/2$ in front of the vertex *V*. This result is independent of the angle α (if $\alpha \ll 1$), so the rays from all very distant objects are brought to focus in a plane that is perpendicular to the mirror axis. This plane is called the *focal plane* of the mirror, and lies a perpendicular distance $R/2$ in front of the vertex *V*.

Objects with Finite Size. The image formed by a mirror of an object with extent can be determined by considering the special geometry of certain easily constructed rays called *principal rays*. Figure 41–9 shows these rays emanating from the tip *X* of an object in the form of a luminous arrow. Ray *a* leaves *X* parallel to the axis and upon reflection passes through the focal point *F*. Ray *v* is directed toward the vertex *V* and is reflected at an angle α with respect to the axis equal to the angle of incidence. Ray *b* passes through the focal point *F* and upon reflection emerges parallel to the axis. Finally, ray *c* passes through the mirror center *C* and is reflected back along its original path. Any two of these four principal rays can be used to locate graphically the image point *X'* of the arrow tip. When solving any type of image problem, you should always make a sketch similar to Fig. 41–9. (This recommendation applies also for problems involving *lenses*.)

Fig. 41–9. Any two of the four principal rays (*a*, *v*, *b*, and *c*) may be used to locate the image of an off-axis object point.

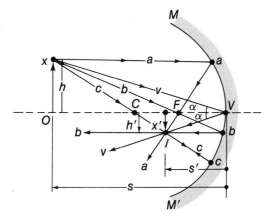

The height of the object arrow measured from the axis is h and the height of the image arrow is h'. Then, using the triangles formed by the incident and reflected portions of the vertex ray v, we have

$$\tan \alpha = \frac{h}{s} = -\frac{h'}{s'}$$

The negative sign is included because we choose arbitrarily to measure distances above the axis as *positive* and those below the axis as *negative*. In Fig. 41–9, we have $h > 0$ and $h' < 0$. When $h'/h < 0$, the image is *inverted* with respect to the object. We define the *transverse* or *lateral magnification* m to be

$$m = \frac{h'}{h} = -\frac{s'}{s} \qquad\qquad \textbf{(41–11)}$$

When $m < 0$, the image is inverted.

For a small object we can also define an *axial* or *longitudinal magnification* μ. Consider an object that has a length Δs along the mirror axis and is a mean distance s from the vertex; the image has a mean position s' and a corresponding axial length $\Delta s'$. The general relationship is (Eq. 41–9)

$$\frac{1}{s} + \frac{1}{s'} = \frac{1}{f}$$

Taking differentials, with f fixed, we find

$$-\frac{ds}{s^2} - \frac{ds'}{s'^2} = 0$$

We approximate ds by Δs and ds' by $\Delta s'$; then, the axial magnification is defined by the relationship

$$\mu = \frac{\Delta s'}{\Delta s} = -\left(\frac{s'}{s}\right)^2 = -m^2 \qquad\qquad \textbf{(41–12)}$$

Notice that m may be either positive or negative, depending on the relative signs of s and s'; however, $\mu < 0$ always. (•Can you deduce this from Fig. 41–5?) Equations 41–11 and 41–12 are completely general and are valid for all mirror and object geometries.

Figure 41–9 shows that when an object is located outside the focal point of a concave mirror (so that $s > f$), the image formed is real, inverted, and diminished in size. (•Can you apply the reversibility principle to Fig. 41–9 to deduce how the magnification depends on whether the object is outside the center C or is between C and F?) When the object is moved inside the focal point (so that $s < f$), the image becomes virtual, erect (that is, oriented the same as the object), and magnified, as shown in Fig. 41–10a. A concave mirror is therefore a *magnifying mirror* for objects located inside the focal point. (Your dentist uses such a mirror to examine the inner surfaces of your teeth.)

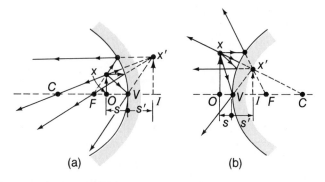

Fig. 41-10. (a) When an object is located inside the focal point of a concave mirror, the image is virtual, erect, and magnified. (b) A convex mirror always forms a virtual, erect, and diminished image, regardless of the location of the object.

(a)   (b)

Figure 41–10b illustrates the case for a convex mirror. Regardless of the position of the object, the image is always virtual, erect, and diminished.

Example 41-6

A small luminous object is located in front of a concave mirror whose radius of curvature is 30 cm. The image projected by the mirror onto a screen is three times as large as the object. Determine the location of the object and the screen.

Solution:

In view of the discussion concerning Fig. 41–9, a real image with $|m| > 1$ requires $s' > s > 0$ and an object position between C and F. It follows that $m = -3$, and using Eq. 41–11, we have $s' = 3s$. With $R = 30$ cm, Eq. 41–10 gives

$$f = \tfrac{1}{2}R = 15 \text{ cm}$$

Finally, using Eq. 41–9 we have

$$\frac{1}{s} + \frac{1}{3s} = \frac{1}{15 \text{ cm}}$$

from which

$$s = 20 \text{ cm}$$

and

$$s' = 3s = 60 \text{ cm}$$

Because $m < 0$, the image is inverted.

41-3   REFRACTION AT SPHERICAL SURFACES

One of the most useful optical elements is the *lens*. A lens is usually a thin, circular piece of glass (or other transparent substance) with spherical surfaces. To understand the image-forming properties of lenses, it is necessary first to examine, in this section, the nature of the refraction process at a spherical interface. In Section 41–4 we discuss the properties of different types of lenses.

Figure 41–11 shows a convex spherical interface with radius of curvature R that separates two media, one with index of refraction n and one with index n'. A point source of light is located at O. The line that connects O and the center C of the

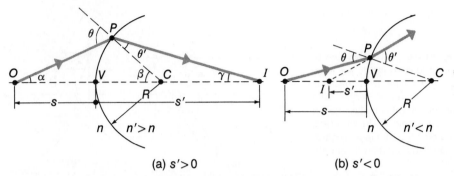

Fig. 41–11. Image formation by a convex spherical interface between two media. (a) $n' > n$, and an object position such that a real image is formed behind the interface: $s' > 0$. (b) $n' < n$, and a virtual image is formed in front of the interface: $s' < 0$.

surface defines the axis of the system and the vertex V. It is again convenient to measure distances from the vertex, so the object distance is $\overline{OV} = s$ and the image distance is $s' = \overline{VI}$.

For the case $n' > n$ (Fig. 41–11a) an incident ray OP may be sufficiently refracted toward the normal at P that the ray will intersect the axis at the image point I. Thus, a *real* image is formed *behind* the refracting surface. On the other hand, if $n' < n$ (Fig. 41–11b), an incident ray will be refracted away from the normal and will always appear to diverge from the image point I. Thus, a *virtual* image is formed in *front* of the refracting surface. The situation is therefore the opposite of that for image formation by mirrors, in which real images appear in front of mirrors and virtual images appear behind. For refracting systems, the "front" surface is the surface on which the light is first incident.

We can preserve the sign convention adopted for mirrors by making the interpretation that the image distance s' is positive for real images and negative for virtual images (Fig. 41–12). Moreover, the surface radius is $r = +R$ when the center C is located behind the vertex, as in Fig. 41–11. (If the surface were concave,* we would have $r = -R$.)

With these conventions, we can derive the equivalent of the Descartes formula (Eq. 41–9) for a refracting surface. Referring to Fig. 41–11a, we see that $\theta = \alpha + \beta$ and $\beta = \theta' + \gamma$. For paraxial rays, we can write $\sin \theta \cong \theta$ and $\sin \theta' \cong \theta'$; then, Snell's law becomes

$$n\theta \cong n'\theta' \tag{41–13}$$

---

* "Concave" and "convex" refer to the shape of the surface when looking along the direction of the incident light.

incident light

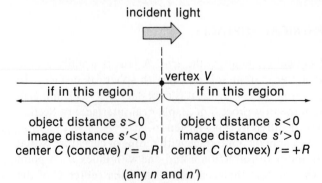

(any n and n')

Fig. 41–12. Sign conventions for spherical interface between two media. The object distance s for a real object is always positive.

Using the above expressions for θ and θ', Eq. 41–13 becomes

$$n\alpha + n'\gamma = (n' - n)\beta \qquad \textbf{(41–14)}$$

We also have $\widehat{PV} = \ell$, $\alpha \cong \ell/s$, $\gamma \cong \ell/s'$, and $\beta = \ell/r = \ell/R$ because $r = +R$ and all the angles are positive. Then, Eq. 41–14 can be expressed as

$$\frac{n}{s} + \frac{n'}{s'} = \frac{n' - n}{R} \qquad \text{convex surface} \qquad \textbf{(41–15a)}$$

For a concave surface, the derivation proceeds in a similar manner, except that $r = -R$ and the angles β and γ are negative. The resulting equation is (see Problem 41–13)

$$\frac{n}{s} + \frac{n'}{s'} = -\frac{n' - n}{R} \qquad \text{concave surface} \qquad \textbf{(14–15b)}$$

The equation for both the convex and concave cases can be combined in the form

$$\frac{n}{s} + \frac{n'}{s'} = \frac{n' - n}{r} \qquad \textbf{(41–16)}$$

with

$$\begin{array}{ll} r = +R & \text{convex surface} \\ r = -R & \text{concave surface} \end{array} \Bigg\} \qquad \textbf{(41–17)}$$

For either a convex or a concave surface, $s' > 0$ corresponds to a real image located behind the surface and $s' < 0$ corresponds to a virtual image located in front of the surface.

The behavior of s' as a function of s for a spherical refracting surface depends on the relative magnitudes of n and n'. Figure 41–13 shows s' versus s for the case of a convex surface with $r = +R$, (a) for $n' > n$ (compare Fig. 41–5) and (b) for $n' < n$. (•Can you explain the appearance of the latter curve?)

Fig. 41–13. The behavior of s' as a function of s for a convex spherical refracting surface with radius $r = +R$. (a) $n' > n$; (b) $n' < n$, for which the image is virtual ($s' < 0$) for all s.

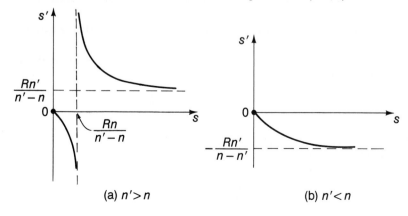

(a) $n' > n$ (b) $n' < n$

Objects with Finite Size. As is the case for mirrors, the image formed by a refracting surface of an object with extent can be determined by considering the geometry of the principal rays. Figure 41–14 shows the three principal rays emanating from the tip of an object arrow. (•Why is there no fourth principal ray as in the case of a mirror—Figs. 41–9 and 41–10?) In Fig. 41–13a a real image is formed behind the lens. When the object is moved close to the same refracting surface, as in Fig. 41–14b, a virtual image is formed in front of the surface.

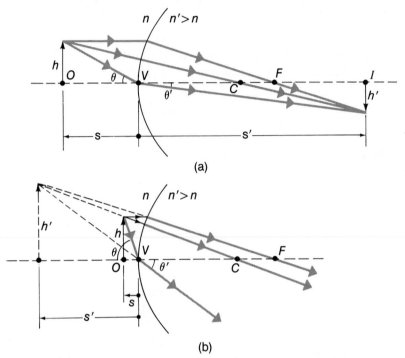

Fig. 41–14. Image formation by a convex refracting surface. The three principal rays are shown. (Compare Figs. 41–9 and 41–10.) (a) The image is real, inverted, and diminished. (b) The object is moved close to the surface, and the image is virtual, erect, and magnified. Notice that the location of the focal point *F* is determined by the incident ray that is parallel to the axis *OVC*.

Because the rays are paraxial in Fig. 41–14, we see that $h \cong s\theta$ and $h' \cong s'\theta'$. Then, using Eq. 41–13, the lateral magnification *m* can be expressed as

$$m = \frac{h'}{h} = \frac{s'\theta'}{s\theta} = -\frac{ns'}{n's} \qquad \textbf{(41–18)}$$

where the negative sign again means that the image is inverted when $s > 0$ and $s' > 0$.

The axial magnification μ is obtained by taking differentials of Eq. 41–16 with *n* and *n'* fixed; we obtain

$$-\frac{n\,ds}{s^2} - \frac{n'ds'}{s'^2} = 0$$

Thus, we find

$$\mu = \frac{\Delta s'}{\Delta s} = \frac{ds'}{ds} = -\frac{n'}{n}m^2 \qquad \textbf{(41–19)}$$

Example 41-7

A diver wears thin plastic goggles that, when underwater, assume a spherical shape with radius of curvature $R = 25$ cm due to the water pressure. A minnow with a length $h = 5$ cm swims directly in front of the diver's goggles at a distance $s = 75$ cm. What are the apparent location and size of the minnow as perceived by the diver? (Neglect any optical effect due to the composition of the goggles.)

Solution:

Using Eq. 41-16, we have, with $r = -R$,

$$\frac{1}{s'} = \frac{1}{n'}\left(\frac{n' - n}{r} - \frac{n}{s}\right) = \frac{1 - 1.333}{-25 \text{ cm}} - \frac{1.333}{75 \text{ cm}}$$

$$= -\frac{0.111}{25 \text{ cm}} = -\frac{1}{225 \text{ cm}}$$

or $\qquad\qquad s' = -225$ cm

The lateral magnification, from Eq. 41-18, is

$$m = -\frac{ns'}{n's} = \frac{(1.333)(-225 \text{ cm})}{(1)(75 \text{ cm})} = +4$$

Therefore, the apparent length of the minnow is

$$h' = mh = (4)(5 \text{ cm}) = 20 \text{ cm}$$

Example 41-8

A point source of light (in air) is placed at O_1, a distance $2R$ from the center of a glass sphere with radius of curvature R and refractive index $n = 1.50$. Find the image point of the source formed by paraxial rays.

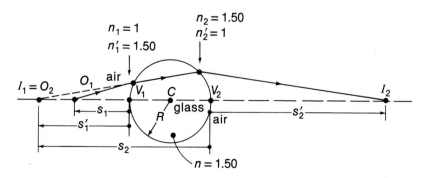

Solution:

The rays from the source at O_1 first strike the convex face of the sphere. Thus, in using Eq. 41–16, we have $r = +R$ for the surface radius. Also, $n_1 = 1$, $n_1' = 1.50$, and $s_1 = R$; we have

$$\frac{1}{s_1'} = \frac{1}{1.50}\left(\frac{1.50 - 1}{R} - \frac{1}{R}\right) = -\frac{1}{3R}$$

so that $\qquad\qquad s_1' = -3R$

Hence, the image formed by the first refraction is virtual and is located at I_1, a distance $3R$ in front of the vertex V_1.

The refracted ray within the sphere appears to have originated at the point I_1; thus, this point serves as the *virtual object** for the second refraction. The ray strikes the concave surface of the interface with vertex V_2. Thus, in using Eq. 41–16, we have $r = -R$ for the surface radius. Also, $n_2 = 1.50$, $n_2' = 1$, and $s_2 = 3R + 2R = 5R$. Equation 41–16 now gives

$$\frac{1}{s_2'} = \frac{1 - 1.50}{-R} - \frac{1.50}{5R} = \frac{1}{5R}$$

so that $\qquad\qquad s_2' = 5R$

Thus, the final image is located a distance $5R$ behind the rear surface of the sphere, and the image is real.

41-4 THIN LENSES

The laws that govern refraction at spherical interface surfaces can be used to examine the behavior of lenses of various types. In order to simplify the discussion, we restrict attention to thin lenses and paraxial rays. The study of these simplified cases exhibits all of the basic features of image formation by lenses. However, modern lens design is not limited by these restrictions because the time-consuming and tedious calculations that are necessary to analyze real lenses without approximations are now made with fast electronic computers. Not only is the design of lenses now carried out by computers, but so is the control of the manufacturing process. One result of this modern approach to lens production is that camera lenses today have much better optical quality and yet are less expensive than were those of twenty or thirty years ago.

A *thin lens* is one whose thickness along any line parallel to the lens axis is small compared with the radii of curvature of the lens surfaces. For example, the lens shown in Fig. 41–15 is *thin* because $t \ll R_1$ and $t \ll R_2$. Thus, it is permissible to consider the vertexes, V_1 and V_2, to coincide with the center of the lens at C. Lenses are commonly used in air, so we take $n_1 = 1$ and set $n_2 = n$, the refractive index of the lens glass.

Figure 41–15 shows the ray O_1P refracted by the interface surface at P. The object distance is $s_1 = \overline{O_1V_1} \cong \overline{O_1C}$, and the surface radius is $r_1 = +R_1$. The image that would be formed by this surface alone, if the medium n were extended, is real and its

---

*Rays actually intersect at the position of a *real* object or a *real* image; otherwise, we have a *virtual* object or a *virtual* image.

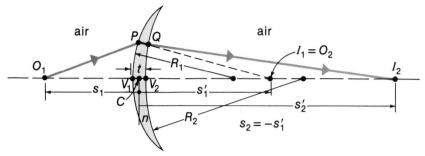

Fig. 41–15. Geometry for determining the image point formed by a thin lens of a point source located at O_1 on the lens axis. O_2 is a *virtual object* for the refraction at Q, and $s_2 < 0$.

position is given by Eq. 41–16; thus,

$$\frac{1}{s_1} = \frac{n-1}{R_1} - \frac{n}{s_1'} \qquad (41\text{-}20)$$

A second refraction of the ray occurs at Q. The (virtual) object O_2 for this refraction is the image I_1 that results from the first refraction. The interface surface is again convex, so the surface radius is $r_2 = +R_2$. Equation 41–16 gives

$$\frac{n}{s_2} = \frac{1-n}{R_2} - \frac{1}{s_2'} \qquad (41\text{-}21)$$

Now, $|s_1'| = \overline{V_1 I_1} \cong \overline{C I_1}$ and $|s_2| = \overline{V_2 O_2} \cong \overline{C I_1}$. According to the sign convention, $s_1' > 0$ and $s_2 < 0$; therefore, $s_2 = -s_1'$. (It is important to realize that this condition is always true. For example, if the image point I_1 is in front of V_1, we would have $s_1' < 0$; but, then, O_2 would be in front of V_2 so that $s_2 > 0$.) Substituting $s_2 = -s_1'$ into Eq. 41–21, we have

$$-\frac{n}{s_1'} = \frac{1-n}{R_2} - \frac{1}{s_2'}$$

and substituting this result into Eq. 41–20 gives

$$\frac{1}{s_1} + \frac{1}{s_2'} = (n-1)\left(\frac{1}{R_1} - \frac{1}{R_2}\right)$$

This equation can be rewritten to refer to the lens as a whole by dropping the subscripts on s_1 and s_2'. Moreover, the equation applies for either convex or concave surfaces if we replace R_1 and R_2 by r_1 and r_2. Thus,

$$\frac{1}{s} + \frac{1}{s'} = (n-1)\left(\frac{1}{r_1} - \frac{1}{r_2}\right) \qquad (41\text{-}22)$$

where s and s' are measured from the center C of the lens and, along with r_1 and r_2, follow the sign convention previously established. If either refracting surface of the lens is concave, the surface radius r for that surface is a negative quantity.

(a) $r_1 < 0$ $r_2 < 0$ (b) $r_1 = \infty$ $r_2 < 0$ (c) $r_1 > 0$ $r_2 < 0$ (d) $r_1 > 0$ $r_2 = \infty$ (e) $r_1 > 0$ $r_2 > 0$

(f) $r_1 > 0$ $r_2 > 0$ (g) $r_1 = \infty$ $r_2 > 0$ (h) $r_1 < 0$ $r_2 > 0$ (i) $r_1 < 0$ $r_2 = \infty$ (j) $r_1 < 0$ $r_2 < 0$

Fig. 41–16. Various lens surface combinations. Lenses (a) through (e) are *converging* (or *positive*) lenses with $|r_1| > |r_2|$ for (a) and $|r_2| > |r_1|$ for (e). Lenses (f) through (j) are *diverging* (or *negative*) lenses with $|r_1| > |r_2|$ for (f) and $|r_2| > |r_1|$ for (j).

Several different lens surface combinations are illustrated in Fig. 41–16. The lenses (a) through (e) are called *converging lenses* because they make a beam of parallel rays converge toward a focal point F' located behind the lens (Fig. 41–17a). The lenses (f) through (j) are *diverging lenses* in that they cause a beam of parallel rays to diverge from a focal point F' located in front of the lens (Fig. 41–17b). Notice in both cases that the point toward which the incident rays converge or away from which they diverge is labeled F'. This point is located behind a converging lens (Fig. 41–17a) and in front of a diverging lens (Fig. 41–17b). If the rays were incident from the right, the focal points would be the points labeled F. For a particular thin lens, the focal points, F and F', are equidistant from the lens.

Fig. 41–17. (a) A parallel beam that is incident on a converging lens is brought to a focus at F'. (b) A parallel beam that is incident on a diverging lens appears to diverge from a focal point F'.

As is the case for spherical mirrors, it is useful to define the focal length f of a lens as the distance s' when $s \rightarrow \infty$ (or vice versa). Thus, from Eq. 41–22, we have

$$\frac{1}{f} = (n - 1)\left(\frac{1}{r_1} - \frac{1}{r_2}\right) \qquad \textbf{(41–23)}$$

which is known as the *lens maker's formula*. Notice that for all of the lenses (a) through (e) in Fig. 41–16, Eq. 41–23 shows that $f > 0$; for this reason, such lenses are referred to as *positive lenses*. Similarly, all of the lenses (f) through (j) have $f < 0$ and are called *negative lenses*.

The lens equation (Eq. 41–22) now becomes

$$\frac{1}{s} + \frac{1}{s'} = \frac{1}{f} \qquad \textbf{(41–24)}$$

Fig. 41–18. Geometry for the Newtonian form of the lens equation, $xx' = f^2$.

This result is known as the *Gaussian form* of the lens equation. Notice that Eq. 41–24 is identical to the mirror equation (Eq. 41–9).

In the Gaussian form of the lens equation, the distances s and s' are measured from the center C of the lens. It is sometimes convenient to use a lens equation in which the object and image distances are measured from the focal points. We define (see Fig. 41–18)

$$x = s - f \quad \text{and} \quad x' = s' - f \tag{41-25}$$

Then, in terms of x and x', Eq. 41–24 becomes

$$xx' = f^2 \tag{41-26}$$

This result is known as the *Newtonian form* of the lens equation. The previous sign conventions apply, with the proper signs for x and x' given by Eqs. 41–25.

Objects with Finite Size. The image-forming properties of a thin lens for an object with extent are determined by considering the behavior of the principal rays, just as for the case of a single interface (Fig. 41–14). Figure 41–19 shows how the three principal rays from the tip of an object are used to determine the position of the image. Notice that the ray c passes essentially *undeviated* through the center C of the lens; this is the case only if the lens is *thin* (see Example 40–2).

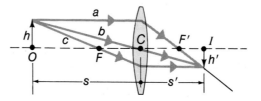

Fig. 41–19. The three principal rays, a, b, and c, are used to determine the position of the image I.

In the ray construction shown in Fig. 41–19, notice that the straight central ray c forms two similar triangles with the lens axis and the heights, h and h'. Thus, the lateral magnification can be expressed as

$$m = \frac{h'}{h} = -\frac{s'}{s} \tag{41-27}$$

where the negative sign again indicates an inverted image when $s > 0$ and $s' > 0$.

The location and the character of an image formed by a single converging lens are determined by the position of the object with respect to the focal point. Figure 41–20 illustrates the image formation by such a lens. If the object is located outside the focal point F (that is, farther from the lens than F), the image is real and inverted (Fig. 41–20a). (•How does the magnification change with object position? See Prob-

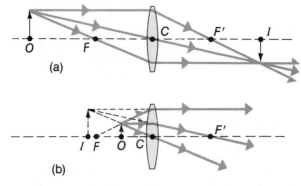

Fig. 41-20. Image formation by a single converging lens. (a) When $s > f$, we have $s' > f$; the image is real and inverted. (b) When $s < f$, we have $s' < 0$ and $|s'| > f$; the image is virtual and erect.

lem 41-35.) However, if the object is located inside the focal point, as in Fig. 41-20b, the image is virtual, erect, and magnified. Thus, a simple converging lens acts as a *magnifying glass* when a small object is placed close to and inside the focal point. Notice that a single converging lens does not form images between its focal points.

Figure 41-21 shows that a single diverging lens produces an erect virtual image inside the focal point F' regardless of the position of the object.

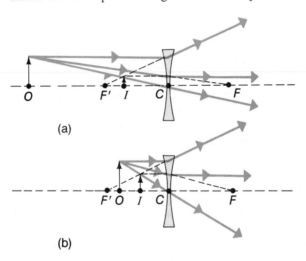

Fig. 41-21. Image formation by a single diverging lens. The image is virtual, erect, and diminished for both $s > |f|$ and $s < |f|$.

Example 41-9

The thin lens shown in the diagram is made from glass ($n = 1.520$) and has radii of curvature $R_1 = 10$ cm and $R_2 = 4$ cm.

(a) Determine the focal length f of the lens.

(b) A luminous object with a height $h = 2$ cm is located 30 cm in front of the lens. Determine the position, the character, and the size of the image.

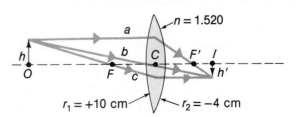

Solution:

(a) Using the lens maker's formula (Eq. 41–23), we find

$$\frac{1}{f} = (n - 1)\left(\frac{1}{r_1} - \frac{1}{r_2}\right) = (1.520 - 1)\left(\frac{1}{10\ \text{cm}} - \frac{1}{(-4\ \text{cm})}\right)$$

from which
$$f = +5.49\ \text{cm}$$

(b) Using Eq. 41–24 with $s = 30$ cm, we obtain

$$\frac{1}{s'} = \frac{1}{f} - \frac{1}{s} = \frac{1}{5.49\ \text{cm}} - \frac{1}{30\ \text{cm}}$$

from which
$$s' = +6.72\ \text{cm}$$

Alternatively, we can determine the value of s' by using Eq. 41–26. Thus,

$$x' = \frac{f^2}{x} = \frac{f^2}{s - f} = \frac{(5.49\ \text{cm})^2}{(30\ \text{cm}) - (5.49\ \text{cm})}$$
$$= +1.23\ \text{cm}$$

Then,
$$s' = x' + f = (1.23\ \text{cm}) + (5.49\ \text{cm})$$
$$= +6.72\ \text{cm}$$

as before.

The magnification is

$$m = -\frac{s'}{s} = -\frac{6.72\ \text{cm}}{30\ \text{cm}} = -0.224$$

so that the image is real and inverted, with a height

$$h' = mh = (-0.224)(2\ \text{cm}) = -0.448\ \text{cm}$$

where the negative sign means that h' is measured from the axis in the direction opposite that for h.

Example 41–10

Rework Example 41–9 in the event that the second surface of the lens is convex; that is, $r_2 = 4$ cm instead of -4 cm.

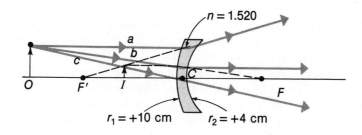

$r_1 = +10$ cm $r_2 = +4$ cm

Solution:

(a) The focal length is now

$$\frac{1}{f} = (1.520 - 1)\left(\frac{1}{10 \text{ cm}} - \frac{1}{4 \text{ cm}}\right)$$

from which

$$f = -12.82 \text{ cm}$$

Thus, the focal point F' is located 12.82 cm in front of the lens, as indicated in the diagram.

(b) Using Eq. 41–24 with $s = 30$ cm, we obtain

$$\frac{1}{s'} = \frac{1}{(-12.82 \text{ cm})} - \frac{1}{30 \text{ cm}}$$

from which

$$s' = -8.98 \text{ cm}$$

($\bullet$What do you find for s' by using Eq. 41–26?)

The magnification is

$$m = -\frac{s'}{s} = -\frac{(-8.98 \text{ cm})}{30 \text{ cm}} = +0.299$$

so that the image is virtual and erect with a height

$$h' = mh = (0.299)(2 \text{ cm}) = 0.598 \text{ cm}$$

Notice that, in each example, ray a leaves the object parallel to the axis and emerges from the lens directed toward or away from F'. Similarly, ray b leaves the object directed toward F and emerges from the lens parallel to the axis.

41-5 FERMAT'S PRINCIPLE AND GEOMETRIC ABERRATIONS

At the beginning of Section 41–2 it was pointed out that a spherical mirror is subject to an intrinsic defect called spherical aberration, which is the inability to bring all rays from an object point to the same focus. Only paraxial rays, which make relatively small angles of divergence with the axis, form a sharp image. As more divergent rays are permitted to participate, the aberration becomes increasingly noticeable.

What type of reflecting surface would be completely free from spherical aberration? That is, what type of surface is required such that *all* of the rays emanating from an object O are brought precisely to focus at an image point I? According to Fermat's principle (Section 40–3), the optical path length connecting points O and I, namely, $L = \int_O^I n \, ds$, must be a minimum in comparison with all neighboring paths of the same general character (that is, reflected from the same surface). Now, if *all* rays from O are to reach I, it follows that the optical path length must be the same for each ray. Moreover, if the medium in which the rays travel is optically homogeneous (that is, if n is everywhere the same), each ray must have the same geometric path length. Thus, if P is a point on the surface, the path length $\overline{OP} + \overline{PI}$ must be the same for all points P. The geometric figure that satisfies this requirement is an *ellipsoid* formed by the rotation about the axis OI of an ellipse with foci at O and I (see Fig. 41–22). All rays that reach I from O by reflection travel the same optical (geometric) path length; point I is a perfect image of point O.* The focusing is precise only for point O; rays emitted from a point near O will form a blurred image near I.

An ellipse is one of the family of geometric curves called *conic sections*. These curves are characterized by a parameter called the *eccentricity e*. For an ellipse, the range of e is $0 < e < 1$ (see Section 14–6). When the eccentricity is zero ($e = 0$), the two foci of the ellipse coincide and the figure is a *circle*. As the foci move farther apart, the eccentricity increases. Finally, when one focus (point O) has receded to infinity, we have $e = 1$ and the ellipse is no longer a closed curve. In fact, $e = 1$ characterizes a *parabola*. The rays from point O—now parallel rays from infinity—are still focused precisely at I. Point I now coincides with the focal point F of the mirror, which has become a *paraboloidal* surface (Fig. 41–23).

Fermat's principle of least optical path length also applies for refraction. Consider the spherical air-glass interface shown in Fig. 41–24. It proves convenient to specify the optical path length of the ray in terms of the parameter δ defined in Fig. 41–24. The axial ray OVI that connects the object point O with the image point I has the optical path length

$$L_O = \oint_O^I n \, ds = s + ns'$$

The optical path length of the ray OPI is

$$L_P = \sqrt{(s + \delta)^2 + h^2} + n\sqrt{(s' - \delta)^2 + h^2} \qquad \textbf{(41–28)}$$

We also have $h^2 + (R - \delta)^2 = R^2$, or $h^2 = 2R\delta - \delta^2$. Using this expression for h^2 in Eq. 41–28 gives

$$L_P = s[1 + 2\delta(R + s)/s^2]^{1/2} + ns'[1 + 2\delta(R - s')/s'^2]^{1/2}$$

---

*The same focusing effect occurs for sound waves. In elliptical "whispering galleries," a soft sound uttered at O will be easily heard at I but at no other point in the gallery. This effect, observed in the dome of St. Paul's Cathedral, London, was first explained by Lord Rayleigh (1842-1919).

Fig. 41–22. An ellipsoid has the property that $\overline{OP} + \overline{PI}$ is the same for all points P. The points O and I are the foci and determine the axis about which the ellipse is rotated to generate an ellipsoid.

Fig. 41–23. Rays that are parallel to the axis are brought to a precise focus at $I = F$ by a paraboloidal mirror surface. Parallel rays that make an angle with respect to the axis are brought to an approximate (but not precise) focus at a point on the focal plane (see Example 41–4).

Fig. 41–24. Fermat's principle can be used to demonstrate that only paraxial rays from the object point O are brought to a sharp focus at I.

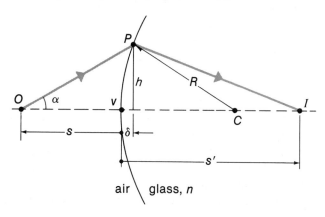

This expression for L_P is exact; however, by expanding in powers of δ and retaining only the linear term, we can show that L_P is approximately equal to L_0. Thus, to first order in δ, we have

$$L_P \cong s + ns' + \delta R\left(\frac{1}{s} + \frac{n}{s'}\right) + \delta(1 - n) \qquad \textbf{(41-29)}$$

Now, Eq. 41-16 for the present case, with $r = +R$, is

$$\frac{1}{s} + \frac{n}{s'} = \frac{n-1}{R}$$

Consequently, the last two terms in Eq. 41-29 sum to zero. There remains

$$L_P \cong s + ns'$$

Thus, $L_P \cong L_0$, as we set out to show. To the extent that spherical aberration can be ignored (or corrected), it follows that the wavefronts associated with *all* rays that leave the object at a given instant arrive at the image *in phase*. (This is an important point in the study of diffraction; see Fig. 43-1.)

This result means that only paraxial rays (for which δ is small) are focused at point I. Rays that make larger angles with the axis suffer spherical aberration and contribute to blurring of the image. Spherical abberation is only one of the various types of aberrations that affect refracting devices. (Another is *chromatic aberration*; see Problem 41-37.)

PROBLEMS

Section 41-1

41-1   A plane mirror is displaced a distance L perpendicular to its surface. Show that the image point for any fixed object point moves a distance $2L$.

41-2   A plane mirror is rotated through an angle α about an axis that lies in the plane of the mirror. Show that the reflected ray for any particular incident ray that is perpendicular to the axis rotates through an angle 2α.

41-3   Two plane mirrors intersect at an angle of $90°$. The line of intersection is the x-axis of a coordinate frame, and the y- and z-axes lie in the planes of the two mirrors. An object is at the point $(3, 4, 5)$ cm. Determine the positions of the image points.

41-4   Two plane mirrors intersect at an angle of $110°$.
•   (Compare Example 41-2.) An object is located on the bisector of this angle, a distance of 10 cm from the line of intersection. (a) Locate all of the image

points. (b) Determine the planes that divide the space in front of the mirrors into regions in which an observer may see various numbers of the possible images. Identify the images that are visible in each region.

Section 41-2

41-5   A concave spherical mirror has a radius of curvature $R = 50$ cm. A small luminous object is placed on the axis at various distances s from the mirror vertex. Determine the position, the size, and the character of the image for s equal to (a) 100 cm, (b) 35 cm, (c) 15 cm, (d) 2 cm, and (e) $s \rightarrow 0$. Comment on the latter result.

41-6   An object with a height of 3 cm is located on and is perpendicular to the axis of a mirror. A real image, 2 cm in height, is formed 100 cm from the object. What is the focal length of the mirror? Is the mirror concave or convex?

41-7 A concave mirror has a radius of curvature $R = 36$ cm. Find the two positions of the object for which the lateral size of the image is twice that of the object. What is the character of each of these images?

41-8 Construct a diagram analogous to Fig. 41-5 for a convex mirror with radius of curvature R.

41-9 A luminous cube whose sides have a length of 2 cm is placed in front of a convex mirror that has a radius of curvature of 18 cm. The center of the cube is on the axis. The face nearest the mirror is at a distance of 35 cm from the vertex and is perpendicular to the axis. Determine the position, the size, and the character of the image. [*Hint:* Consider differences measured from the center of the cube.]

41-10 An object with a height of 2 cm is placed 37.5 cm in front of a concave spherical mirror that has a focal length of 25 cm. A plane mirror, whose surface is perpendicular to the axis, faces the spherical mirror at a distance of 50 cm from the vertex. Determine the position, the size, and the character of the image that is formed by one reflection in each mirror.

Section 41-3

41-11 A transparent sphere of unknown composition is observed to form an image of the Sun on the surface of the sphere opposite the Sun. What is the refractive index of the sphere material?

41-12 A hemispherical piece of glass with radius $R = 10$ cm and index $n = 1.520$ has its plane face in contact with oil of the same index. Where in the oil should a luminous object be placed if the image is to be twice the size of the object? Is the image real or virtual?

41-13 The interface between two media with indexes n and n' is in the shape of a concave spherical surface with radius R, as shown in the diagram. Show that, for paraxial rays,

$$\frac{n}{s} + \frac{n'}{s'} = \frac{n' - n}{r}$$

where $r = -R$.

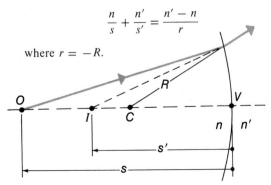

41-14 Construct a diagram analogous to Fig. 41-13 for a concave interface with surface radius $r = -R$. Consider both $n' > n$ and $n' < n$.

41-15 A plastic sphere with radius $R = 6$ cm and index $n = 1.360$ is suspended in air. A point source of light is placed 20 cm from the center of the sphere. Determine the position of the image point.

41-16 A glass sphere with radius $R = 10$ cm and index
 • $n = 1.520$ is coated with a reflecting layer over one hemisphere. An object with a height $h = 1.0$ cm is placed 15 cm in front of the clear surface of the sphere. Determine the position, the size, and the character of the final image. Sketch the principal rays.

Section 41-4

41-17 What is the focal length of a thin lens of the type shown in Fig. 41-16j if $r_1 = -10$ cm, $r_2 = -30$ cm, and $n = 1.520$?

41-18 An object with a height of 0.50 cm is placed 100 cm in front of a converging lens with a focal length $f = +30$ cm. Determine the position, the size, and the character of the image. Repeat the calculation for s reduced to 20 cm. Sketch the principal rays for each case.

41-19 A thin lens with refractive index n is immersed in a liquid with index n'. Show that the focal length f of the lens is given by

$$\frac{1}{f} = \left(\frac{n}{n'} - 1\right)\left(\frac{1}{r_1} - \frac{1}{r_2}\right)$$

41-20 Use the Newtonian form of the lens equation (Eq. 41-26) to show that the lateral magnification can be expressed as $m = -f/x = -x'/f$. [*Hint:* Construct the principal rays.]

41-21 An object has a height of 2 cm. Its image is formed by a converging lens on a screen that is 20 cm from the lens. When the lens is moved 10 cm closer to the screen (with the object fixed), another image is formed on the screen. Determine the focal length of the lens, the distance from the screen to the object, and the sizes of the two images. [*Hint:* Use the principle of reversibility.]

41-22 A spherical object is placed a certain distance from a screen. A lens located between the object and the screen produces an image with a radius a on the screen. The lens is moved to a second position (with the object fixed) such that an image with a radius b is formed on the screen. Show that the radius of the object sphere is $r = \sqrt{ab}$. [*Hint:* Use the principle of reversibility.]

41-23 A thin converging lens with a focal length of 20 cm
• is placed midway between an object and a plane mirror whose surface is perpendicular to the lens axis. The final image produced by the lens-mirror combination is erect and 1/7 the size of the object. Determine the distance between the object and the mirror. [*Hint:* First, use the Newtonian formula for the magnification given in Problem 41-20 to show that the overall magnification of the lens-mirror combination is $m = m_1 m_2 = f^2/x_1 x_2$, where x_1 is the distance from the lens focal point to the object and where x_2 is the distance from the image formed by the mirror to the other focal point. Then, show that $2y^2 - 4fy + f^2(1 - m^{-1}) = 0$, where $2y$ is the object-mirror distance.]

Section 41-5

41-24 A *hyperbola* may be defined as the curve composed
• of all the points that have the same difference of distances from two fixed points. As shown in the diagram, the two fixed points are the foci, F and F'; then, for the typical point P, we have $\overline{PF} - \overline{PF'} = $ constant. A *hyperboloid* is generated by rotating the hyperbola about the F-F' axis. Show that rays converging toward F are reflected by the first intercepted sheet of the hyperboloid and are brought to an exact focus at F'. Classify the figures of rotation of the conic sections—hyperboloid, paraboloid, ellipsoid, and sphere—according to the geometry of rays that are focused exactly.

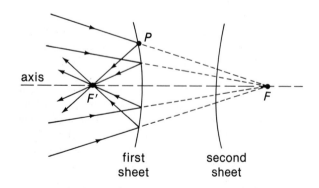

first
sheet

second
sheet

41-25 The angle α shown in the diagram specifies the posi-
• tion P at which a ray in a parallel bundle strikes a concave mirror. (a) Show that $x = R \sin \alpha / \sin 2\alpha$. (b) A measure of the spherical aberration of the mirror is $\delta = x - \frac{1}{2}R$. Show that

$$\delta = \frac{1}{4}R(\alpha^2 + \frac{5}{12}\alpha^4 + \cdots)$$

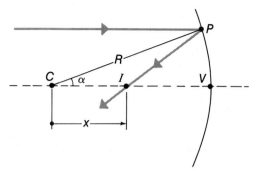

(c) For what angle α is the aberration δ equal to one percent of the focal length? [*Hint:* For (b), the expansions

$$\sin \theta = \theta - \theta^3/6 + \theta^5/120 - \cdots$$

$$\csc \theta = \theta^{-1} + \theta/6 + 7\theta^3/360 + \cdots$$

are useful.]

41-26 A light ray OP from an object in air is
• refracted by a convex spherical glass surface as shown in the diagram. Use an exact method of calculation to show the following:

(a) $\sin \theta = \dfrac{s + R}{R} \sin \alpha$

(b) $\sin \theta' = \dfrac{s + R}{nR} \sin \alpha$

(c) $s' = R\left(1 + \dfrac{\sin \theta'}{\sin \gamma}\right)$, where $\gamma = (\theta - \theta') - \alpha$

(d) Assume $s = 40$ cm, $R = 10$ cm, and $n = 1.520$. Calculate to five significant figures the location of the image for $\alpha = 0.5°$, $1°$, $2°$, $5°$, and $10°$. [*Hint:* Calculate the values of θ, θ', γ, and finally s' in that order.] (e) Estimate the angle α for which the exact value of s' deviates by 2 percent from that calculated by using Eq. 41-16.

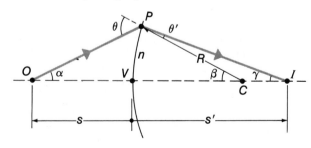

Additional Problems

41-27 Two plane mirrors are parallel and spaced 27 cm apart. A point source of light is located between them, a distance of 12 cm from one mirror. Multiple images of the source are formed behind each mirror. Find the position of the first two images behind each of the mirrors.

41-28 The Moon's diameter subtends an angle of 9.04×10^{-3} rad at the position of the Earth. The mirror of a telescope is a concave spherical surface with $R = 80$ m. What is the diameter of the Moon's image formed on the focal plane?

41-29 A small object is located 3 m below the surface of a pool of water. How far below the surface will the object appear to be when viewed from directly above? (a) Use Snell's law, assuming the angles to be small. (b) Show that the same result is obtained by using Eq. 41-16.

41-30 A glass rod has a diameter of 10 cm, a length of 30 cm, and a refractive index $n = 1.520$. One end of the rod is flat and the other end has a hemispherical convex surface (with $R = 5$ cm). An object with a height of 1 mm is placed on the rod axis, 40 cm in front of the vertex of the hemispherical surface. Determine the position, the size, and the character of the image. Sketch the principal rays. (As usual, the image formed by the first refraction serves as the object for the second refraction.)

41-31 A parallel beam of light enters a glass hemisphere perpendicular to the flat face, as shown in the diagram. The radius is $R = 6$ cm and the index of refraction is $n = 1.560$. Determine the point at which the beam is focused. (Assume paraxial rays.)

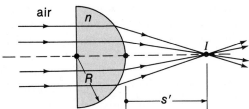

41-32 *Geometrical construction for refraction at a spheri-*
•• *cal surface.* The path of a ray upon crossing a spherical interface between two media can be obtained by the following geometrical construction. The spherical surface has a radius R. Construct two auxiliary concentric spherical surfaces with radii $(n'/n)R$ and $(n/n')R$, as shown in the diagram. The incident ray intersects the actual surface at P. Extend the incident ray until it intersects the outer sphere at A. Connect A with the common center C, thereby generating the intersection point B on the inner sphere. Then, the ray within the medium n' follows the path PB and intersects the optical axis at F. Show that this construction is consistent with Snell's law at point P. [*Hint:* Show that the triangles $\triangle CPA$ and $\triangle CPB$ are similar, and use the law of sines.] Use this procedure to construct the paths of three differ-

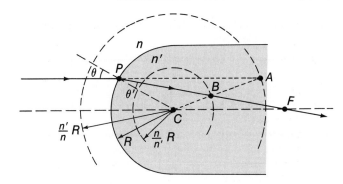

ent parallel incident rays and demonstrate the occurrence of spherical aberration. (Chose $n = 1$ and $n' = 1.5$.)

41-33 A thin lens of the type shown in Fig. 41-16d is made from glass with $n = 1.560$. When an object is placed 36 cm in front of the lens, a real image is formed 48 cm behind the lens. (a) What is the focal length of the lens? (b) What is the radius r_1 of the first surface?

41-34 An object is located in a fixed position in front of a screen. A thin lens, placed between the object and the screen, produces a sharp image on the screen when it is in either of two positions that are 10 cm apart. The sizes of images in the two situations are in the ratio 3:2. (a) What is the focal length of the lens? (b) What is the distance from the screen to the object? [*Hint:* Use the principle of reversibility.]

41-35 Show that, for a converging lens, unit magnification ($m = -1$) results when $s = 2f$. Under what condition is $m = +1$ approached?

41-36 Construct a graph of magnification m as a function of s for (a) a converging lens and (b) a diverging lens.

41-37 Because the index of refraction for all substances depends on the wavelength of the incident light, the focal length of a lens is different for different colors of light. This effect, called *chromatic aberration*, causes blurring of the colors in an image. What is the fractional difference in focal length $\Delta f/f$ between yellow and red light for a lens made from dense flint glass? Between yellow and violet light? Use λ (red) = 750 nm, λ (yellow) = 600 nm, and λ (violet) = 450 nm; refer to Fig. 40-22.

41-38 A converging lens is formed by creating an air cavity within a large block of glass with $n = 1.520$. The air cavity has the shape of an ordinary lens with two spherical surfaces. These surfaces have equal radii of curvature, and the "air lens" has a focal length $f = 40$ cm. Determine the radii. Sketch the lens and show the behavior of parallel rays that pass through the lens.

CHAPTER 42

INTERFERENCE

In Chapter 41 the image-forming properties of mirrors and lenses were discussed in terms of ray optics, without regard for all the wave aspects of light. This simplified approach is valid when the optical elements in a system have dimensions that are large compared with the wavelength of light and when ultimate detail of the image is not required. In this chapter (and in the next) we discuss situations in which the wave nature of light produces additional observable, even striking, effects of considerable interest and importance.

Maxwell's equations, which describe the behavior of all electromagnetic phenomena, are *linear* equations. This property permits the exact application of the superposition principle in the analysis of electromagnetic effects. Thus, we may view a complex situation, involving the simultaneous presence at a point of several different electromagnetic wavefronts, in terms of the vector sum of the separate field components. The overlapping waves that are combined might be the different Huygens' wavelets imagined to constitute a single propagating wavefront, or they might be waves that originate in separate and independent sources. At any particular instant the vector addition of the field components will give rise to constructive and/or destructive interference effects.*

Phenomena that involve the superposition of waves are usually classified either as *diffraction* effects or as *interference* effects, although there is no completely satisfactory way to make the distinction. If a wavefront initiated by a single source is allowed to propagate unimpeded through a homogeneous medium, the superposition of the Huygens' secondary wavelets results in straight-line propagation of the wave. However, if such a wave is partially blocked by an obstacle or is passed through an aperture, some of the Huygens' secondary sources are eliminated. This results in the spreading of the wave around the obstacle or from the aperture, and can also produce constructive and/or destructive interference effects within the transmitted beam. We choose to refer to such phenomena as *diffraction* effects.

When the superposition involves two or more isolated portions of a wavefront, or waves from independent sources, we refer to the resulting variation in the combined field amplitude as an *interference* effect. It should be emphasized, however, that diffraction and interference effects both result from the same basic process of superposition and are therefore not completely distinct effects. In this chapter we examine "interference" effects; Chapter 43 is concerned with "diffraction" effects.

---

*Some aspects of interference phenomena have already been treated in connection with sound (Section 20-4) and radiation from antennas (Example 39-3).

42-1 CONDITIONS FOR INTERFERENCE

In order for interference effects to be observed, it is necessary that several conditions be met. The most important of these is that the actual or effective sources must have the same frequency and a constant phase difference; this is the condition of *coherence*. It is relatively easy to establish coherence for electronically generated electromagnetic waves. For example, special electronic circuits are used to "frequency lock" and "phase lock" the oscillators that drive a pair of radio antennas. Light, on the other hand, consists of radiation from individual atoms. Each atomic transition produces a train of wave oscillations that has a finite length. In ordinary sources,* these atomic transitions occur randomly from atom to atom. Consequently, the resulting wave trains have random phases. If two such wave trains overlap, interference appropriate for the particular phase difference of the two waves will occur. However, light from a real source, averaged over the very large number of wave trains present, each with a random phase, produces no observable interference effect. Thus, *interference effects involving light ultimately require that the overlapping waves be generated by the same source.* This requirement for a single source can be met by using slits, prisms, and lenses to split the individual wave trains from a primary source into two or more secondary wave trains that are coherent (see Section 42-2).

If the pattern formed by constructive interference (intensity maxima) and destructive interference (intensity minima) is to be easily observed, additional (approximate) conditions should be satisfied. For the maximum effect produced by destructive interference, the vector sum of the two overlapping field components should be zero. The sum of two vectors can be zero only if the vectors are antiparallel and have the same magnitude. Hence, the interfering waves should have wavefronts that are parallel, or nearly so, and the field vectors should have the same polarization.

Finally, two light waves will interfere only if the wave trains overlap in space. Figure 42-1 is a schematic representation of a wave train for which the electric vector is polarized in a plane that is perpendicular to the *x*-axis.† Within the nominal length *L*, the wave train is essentially sinusoidal. In order for an oscillation to be a pure monochromatic wave, the frequency must be the same wherever and whenever it is measured; such a wave must have an infinite length. Any wave with *finite* length, such as the wave train in Fig. 42-1, consists of the superposition of waves with slightly different frequencies distributed in such a way that there is constructive interference in the region of length *L* and destructive interference everywhere else. Such waves are said to be *quasi-monochromatic* ("almost monochromatic"). The

---

* An exception is the class of devices that exhibit *laser* action; see Section 47-6.

† Because the electrons in matter acquire only small velocities through interactions with light waves, magnetic forces are almost always negligible in comparison with electric forces. Therefore, in describing interference and related effects, we concentrate on the behavior of the electric vectors.

Fig. 42-1. Schematic representation of a wave train (a plane wave) with finite length. This wave train consists of only a few oscillations, but an actual wave train would typically have a few million oscillations within its length *L*.

$$E = E_0 \cos(kx - \omega t)$$

wave trains from ordinary light sources have lengths that are typically a fraction of a meter to several meters.

The physical effects that are produced by incident light depend on the energy flux delivered to a surface. The Poynting vector **N** (W/m²) was introduced in Section 38–2 as a measure of energy flux. In the following discussions, we often refer to the *time-averaged* energy flux, a quantity called the *wave intensity* (symbol, I). For the wave train in Fig. 42–1, with $E = E_0 \cos(kx - \omega t)$, the wave intensity at a particular plane, $x = x_0$, is

$$I = \overline{N} = \overline{EH}$$

Now, $H = H_0 \cos(kx - \omega t)$ and $H_0 = \sqrt{\epsilon_0/\mu_0}\, E_0 = E_0/(377\ \Omega)$ (Eq. 38–16) for a wave propagating in vacuum (or air). Thus, $H = E/Z_0$; then,

$$I = \overline{E^2}/Z_0 \tag{42–1}$$

so that
$$I = (E_0^2/Z_0)\,\overline{\cos^2(kx - \omega t)} \tag{42–1a}$$
$$= \tfrac{1}{2}E_0^2/Z_0$$

where the units of I are W/m² and those of E_0 are V/m.

Example 42–1

Two coherent, equal-amplitude electromagnetic plane waves are propagating in the x-direction with parallel polarizations. The waves arrive at a plane, $x = x_0$, with their wavefronts essentially parallel. The phase difference of the two waves at that plane is δ. What is the wave intensity on the plane $x = x_0$? (For convenience, consider the waves to be infinite in extent, *i.e.*, purely monochromatic.)

$E_1 = E_0 \cos(kx_0 - \omega t)$

$E_2 = E_0 \cos(kx_0 - \omega t - \delta)$

Solution:

The diagram shows the time variation of the electric vectors, E_1 and E_2, for the two waves on the plane $x = x_0$. The superposition of the two waves gives

$$E = E_1 + E_2 = E_0[\cos(kx_0 - \omega t) + \cos(kx_0 - \omega t - \delta)]$$

from which
$$E = 2E_0 \cos\left(kx_0 - \omega t - \frac{\delta}{2}\right)\cos\frac{\delta}{2}$$

The (time-averaged) wave intensity is given by Eq. 42–1 as $I = \overline{E^2}/Z_0$. Now, in this case,

$$\overline{E^2} = 4E_0^2 \cos^2\frac{\delta}{2}\,\overline{\cos^2\left(kx_0 - \omega t - \frac{\delta}{2}\right)}$$

$$= 2E_0^2 \cos^2\frac{\delta}{2}$$

It is customary to let I_0 represent the intensity of one wave alone. Thus, $I_0 = \tfrac{1}{2}(E_0^2/Z_0)$, so that

$$I = 4I_0 \cos^2\frac{\delta}{2}$$

Note the important point that the intensity of two equal-amplitude superimposed waves that are in phase ($\delta = 0$) is *four times* the intensity of either wave alone, not simply twice the individual intensity.

42-2   YOUNG'S DOUBLE-SLIT EXPERIMENT

In an important experiment performed by Thomas Young in 1801, interference effects produced by light were clearly demonstrated. This was the first of several crucial experiments that eventually established the wave theory of light. From this experiment, Young was also able to determine approximately the wavelength of light.

The (nearly) monochromatic light from a source such as glowing sodium vapor is collimated (that is, formed into a narrow beam) by a narrow slit S in a barrier as shown in Fig. 42-2a; this slit acts as the primary light source.* Diverging waves radiate from this slit and impinge on the pair of narrow, parallel slits, S_1 and S_2, that are separated by a distance d and are equidistant from S. The result is that S_1 and S_2 become sources of coherent secondary waves of equal amplitude, which radiate outward and interfere in the region to the right of the slits, as indicated in Fig. 42-2a. The light that passes through the slits falls on a screen at a distance D from the slits (Fig. 42-2b). At a point $P(y)$ on the screen, the path length ℓ_2 from S_2 is longer than the path length ℓ_1 from S_1 by an amount $\Delta \ell \cong d \sin \theta$, when θ is small. (•Can you see how to obtain this expression for $\Delta\ell$?) Because the waves are in phase when they leave the slits S_1 and S_2, the wave from S_2 arrives at the point $P(y)$ with some phase lag δ relative to the wave from S_1. To convert $\Delta\ell$ into a phase difference, first divide by λ to measure the path-length difference in wavelengths and then multiply by 2π to convert to radians. Thus,

$$\delta = \frac{2\pi d}{\lambda} \sin \theta \qquad\qquad \textbf{(42-2)}$$

When θ is a small angle, the wavefronts of the interfering waves are nearly parallel.

---

*Young actually used sunlight passed through pinholes, but the essential features of the two arrangements are the same.

Fig. 42-2. (a) Interference of the coherent waves from the slits S_1 and S_2. Wavefronts corresponding to amplitude maxima of a particular sign (and therefore spaced apart by one wavelength λ) are shown for a particular instant. To the right of the slits, the dots correspond to instantaneous intensity maxima produced by constructive interference. (Compare Fig. 20–10 for the case of sound waves.) The lines ab, ef, and gh are *lines of antinodes*. (b) Geometry of Young's double-slit experiment.

Fig. 42-3. Intensity pattern along the screen in Young's double-slit experiment. The intensity maximum (bright fringe) labeled $m = 0$ is the central or zero-order bright fringe. The average intensity across the screen is $2I_0$. (Photo from *Atlas of Optical Phenomena* by M. Cagnet, M. Francon, and J. C. Thrierr. Springer-Verlag, 1962.)

Then, we have the conditions of Example 42–1, so the intensity at the point $P(y) = P(\theta)$ is

$$I(\theta) = 4I_0 \cos^2\left(\frac{\pi d}{\lambda}\sin\theta\right) \qquad \text{(42–3)}$$

For $\theta = 0$, there is an intensity maximum that corresponds to the intersection of the central line of antinodes $\overline{ef}$ with the screen (Fig. 42–2a). The next intensity maxima, on either side of the central maximum, have $\sin\theta = \pm\lambda/d$ and correspond to lines of antinodes $\overline{ab}$ and $\overline{gh}$. In fact, constructive interference and, hence, intensity maxima occur on the screen when

$$\sin\theta = \pm\frac{\lambda}{d}m, \qquad m = 0, 1, 2, 3, \ldots \qquad \text{(maxima)} \qquad \text{(42–4a)}$$

Destructive interference and, hence, intensity minima (*zero* intensity) occur when

$$\sin\theta = \pm\frac{\lambda}{d}(m' + \tfrac{1}{2}), \qquad m' = 0, 1, 2, 3, \ldots \qquad \text{(minima)} \quad \text{(42–4b)}$$

The result of this interference is to produce on the screen a pattern that consists of a series of alternating bright and dark bands or *interference fringes*. The central bright fringe ($m = 0$) is called the *zero-order fringe*; the next pair of bright fringes ($m = \pm 1$) are the *first-order fringes*, and so forth. (See Fig. 42–3.)

The intensity at each maximum is $4I_0$. Because the intensity curve has a cosine-squared shape (Eq. 42–3), the average intensity over the screen is $2I_0$, which is the same as the intensity would be if the sources were incoherent.

As a practical matter, with simple arrangements involving ordinary sources of (nearly) monochromatic light and ordinary slits, only the first 5 to 10 bright fringes on either side of the central maximum will be clearly visible. With intense, highly monochromatic laser sources and very narrow slits, many more fringes are visible. If sunlight is used, as in Young's original experiment, all fringes with $m \geq 1$ exhibit color structure and only the first few fringes are discernible. (•Can you give the reason for this limitation?)

The analysis we have made is based on the implicit assumption that the slits are sufficiently narrow that they represent line sources of light. Actually, the slits have finite size and this leads to diffraction effects that are superimposed upon the interference effects. The required modifications of the results are discussed in Section 43–2.

Example 42–2

In a double-slit experiment, the slits are separated by an amount $d = 0.4$ mm and the distance to the screen is $D = 2.0$ m. Determine the spacing between adjacent bright fringes if yellow sodium light ($\lambda = 5.89 \times 10^{-7}$ m) is used.

Solution:

The spacing is

$$\Delta y = D(\tan \theta_{m+1} - \tan \theta_m)$$

Now, θ is a small angle, so $\tan \theta \cong \sin \theta$; then,

$$\Delta y = D(\sin \theta_{m+1} - \sin \theta_m)$$

Using Eq. 42–4a, we obtain

$$\Delta y = D \cdot \frac{\lambda}{d}[(m + 1) - m]$$

$$= D \cdot \frac{\lambda}{d} = \frac{(2.0 \text{ m})(5.89 \times 10^{-7} \text{ m})}{0.4 \times 10^{-3} \text{ m}} = 2.94 \text{ mm}$$

The spacing between the dark fringes is the same.

Modifications of Young's Experiment. There are several modified forms of Young's double-slit experiment that are also very effective in exhibiting interference effects and demonstrating the wave character of light. The first of these experiments was performed by Fresnel in about 1815 and uses two flat mirrors, M_1 and M_2, whose planes make a small angle α with respect to one another (Fig. 42–4). Light from a monochromatic source S is reflected from the two mirror surfaces, thereby forming two images of the source that are slightly displaced. These images, S_1 and S_2, act effectively as a pair of (virtual) coherent sources that produce interference effects in the region where the two reflected wave systems overlap, as indicated in Fig. 42–4. Consequently, a pattern of interference of fringes will be observed on a nearby screen.

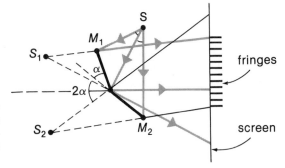

Fig. 42–4. Fresnel's double-mirror interference experiment. The angle α is actually small, but it has been exaggerated here to show the geometry more clearly.

Another method of producing coherent sources of light was devised by Billet and is illustrated in Fig. 42–5. Billet's experiment involves a split lens, with the two halves slightly displaced. The lens halves form images of the primary source S at S_1 and S_2. For an ideal lens (no spherical aberration), all light rays from S that pass through the upper lens half require the same time interval to reach S_1 (Fermat's principle, Section 41–5). Thus, the rays leave S_1 with the same initial phase. Because of the symmetry of the lens halves, the rays from S_1 and S_2 all have the same phase.

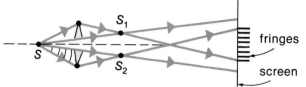

Fig. 42–5 Billet's split-lens arrangement. *S* is a source of monochromatic light. The space between the lens halves is filled with an opaque material to block the direct light.

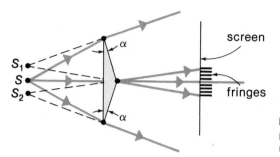

Fig. 42–6. Fresnel's double-prism experiment. *S* is a source of monochromatic light.

Consequently, S_1 and S_2 are secondary sources of coherent radiation, and a system of interference fringes is produced on the observing screen, as indicated in Fig. 42–5.

Another modification of Young's experiment was also invented by Fresnel. This arrangement involves a symmetric pair of prisms with very small refractive angles α. As shown in Fig. 42–6, the refraction by the prisms of the light from the primary source *S* produces two virtual images, S_1 and S_2, that act as sources of coherent light. Interference fringes are observed on the screen in the region where the secondary waves overlap. Further details of this experiment are given in Example 42–3.

Example 42–3

Determine the characteristics of the interference produced by a Fresnel double prism (Fig. 42–6).

Solution:

It is necessary to demonstrate first that a prism with a very small refractive angle α produces a displaced virtual image of the source *S*. Consider the prism geometry shown in diagram (a), in which the refractive angle α has been grossly exaggerated for clarity. The prism is in air and has a refractive index *n*. From triangle $\triangle ACD$, we have

$$\alpha = \beta + \gamma$$

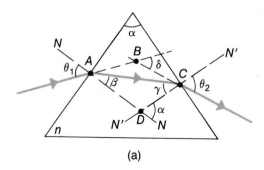

(a)

and from triangle $\triangle ABC$, we have

$$\delta = (\theta_1 - \beta) + (\theta_2 - \gamma)$$
$$= \theta_1 + \theta_2 - \alpha$$

For small angles, Snell's law can be expressed as

$$\theta_1 = n\beta \quad \text{and} \quad \theta_2 = n\gamma$$

Hence,

$$\delta = n(\beta + \gamma) - \alpha$$
$$= (n - 1)\alpha \tag{1}$$

Compare this result with Eq. 40–16 for the case in which α is small. Equation (1) shows that the deviation δ is independent of the incident angle θ_1. Consequently, when α is a very small angle, all of the rays that diverge from the source S undergo the same deviation δ. Then, for small angles θ, we can see from diagram (b) that a constant deviation of the rays from S results in a virtual image of the source at S_1.

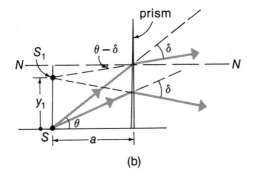

(b)

The position y_1 of S_1 is

$$y_1 = a[\tan\theta - \tan(\theta - \delta)] \cong a[\theta - (\theta - \delta)] = a\delta$$

which is independent of θ. Thus, the two virtual images S_1 and S_2 act as secondary sources of coherent light. The separation of the sources is

$$d = y_1 + y_2 = 2a\delta = 2a(n - 1)\alpha$$

The prism angles α (and the resulting deviations δ) must be of the order of minutes of arc in order to produce well-separated interference fringes. (See Problem 42–6.)

42-3   INTERFERENCE PRODUCED BY THIN FILMS

It is a common experience to see colored bands and rings in soap bubbles and in thin films of oil floating on water. These effects result from interference of light that is reflected into the viewer's eye from the two surfaces of a thin film.

Consider the reflection of light from a thin wedge-shaped film, as shown in Fig. 42–7. Each wave train of light from a source S is partially reflected and partially transmitted at points such as a, b, and c. All portions of each original wave train that eventually reach the observer's eye are coherent and will interfere.

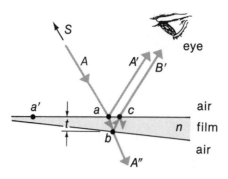

Fig. 42–7. Interference of light reflected from the surfaces of a thin, transparent film in air.

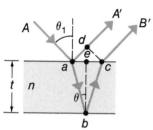

Fig. 42–8. Details of the reflection of light from a thin film near the point of incidence a.

Figure 42–8 shows the geometry of the reflection in more detail. The ray AB' clearly follows a longer optical path than does the ray AA'. For normal incidence, $\theta_1 = 0$, the path difference is $2nt$. For the general case of oblique incidence, we have* $\Delta\ell = n(\overline{ab} + \overline{bc}) - \overline{ad} = 2nt\cos\theta$. This path difference produces a phase difference $\delta = (2\pi/\lambda_a)\Delta\ell = (2\pi/\lambda_a) \cdot 2nt\cos\theta$, where λ_a is the wavelength of the light in air. There is an additional phase difference that is introduced because the reflection at a involves a phase change of π, whereas no phase change occurs for the reflection at b nor for the transmissions at a and c. (Refer to Section 40–4 for a discussion of the behavior of the phase upon reflection. When the incident medium has the smaller optical density and the incident angle is near $0°$, there is a phase change of π. When the incident medium has the greater optical density, there is no phase change.) Thus, the phase difference between rays AA' and AB' is

$$\delta = \frac{2\pi}{\lambda_a}\Delta\ell + \pi = \frac{2\pi}{\lambda_a}(2nt\cos\theta) + \pi \tag{42–5}$$

For near-normal incidence, the amplitudes of the two rays differ by only a few percent.

When the eye is focused on the plane of the film, interference fringes are observed. Each fringe of a particular color (or wavelength λ_a) corresponds to a band along which $nt\cos\theta$ is constant and, hence, the phase difference δ has a constant value. As the point of focus is moved along the film surface from a to a' (see Fig. 42–7), both the thickness t and the value of $\cos\theta$ decrease. (We assume that n is constant.) Consequently, the phase difference δ will remain constant only for light with decreasing wavelength. That is, the interference fringe varies in color from the red end of the spectrum at a toward the blue end at a'. For the next band, the additional path length difference is one wavelength λ_a. These variations produce the colored bands and rings that are observed in thin films.

The process just described will produce strong (high visibility) fringe patterns with extended light sources, such as the Sun or a flame, only for films whose thickness is not much more than a few wavelengths of visible light. Fringes are observable under these conditions because only light within a narrow range of incident angles can enter the pupil of the eye. For example, if an eye with a pupil diameter of 4 mm is located 1 m from a film, the angular diameter of the pupil is only $0.2°$. Then, the value of $\cos\theta$ across any pupil diameter is nearly constant. On the other hand, if the

---

*We can see this as follows: First, $\overline{ab} = \overline{bc}$, so that $\Delta\ell = 2n\,\overline{ab} - \overline{ad}$. Now, $\overline{ad} = 2\overline{ae}\cos(\pi/2 - \theta_1) = 2\overline{ae}\sin\theta_1 = 2n\,\overline{ae}\sin\theta$, where we have used Snell's law at point a. Then, $\overline{ad} = 2n\,\overline{ab}\sin^2\theta$, and $\Delta\ell = 2n\,\overline{ab} - 2n\,\overline{ab}\sin^2\theta = 2n\,\overline{ab}(1 - \sin^2\theta) = 2n\,\overline{ab}\cos^2\theta = 2nt\cos\theta$, where $t = \overline{ab}\cos\theta$. At normal incidence, $\Delta\ell = 2nt$, as required.

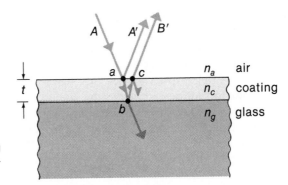

Fig. 42–9. An antireflective coating with refractive index $n_c \cong \sqrt{n_g}$ is applied to a glass surface.

film is thick, even small variations in $\cos\theta$ produce large changes in the phase difference δ. Then, the fringes become too closely spaced to be observed with the unaided eye. Consequently, interference fringes are seen clearly only for very thin films. (The aid of a telescope in near-normal geometry is required for the observation of fringes for very large path-length differences.)

The interference effects that are produced by thin films can be put to practical use. The reflectance of light incident normally on glass is about 0.04 (see Example 40–4). By application of a suitable thin coating to the surface, the reflectance can be reduced to about 0.01. The essential details are shown in Fig. 42–9. If the coating material is chosen such that

$$n_c \cong \sqrt{n_g} \qquad\qquad (42\text{–}6)$$

the relevant reflection and transmission coefficients for normal incidence combine to give nearly equal amplitudes for the first reflected ray AA' and the sum of the subsequent reflected waves such as AB' (see Problem 42–23). In this case we have $n_a < n_c < n_g$, so that both rays AA' and AB' undergo a phase change of π at the points a and c, respectively. (•Can you verify this statement?) Thus, the phase difference between rays AA' and AB' is due entirely to the path length of AB' within the coating. For normal incidence, this path length difference is $\Delta\ell = 2n_c t$. If the thickness of the coating is such that $\Delta\ell = \frac{1}{2}\lambda_a$, the combining out-of-phase amplitudes (which are almost the same in magnitude) result in nearly complete destructive interference. However, the cancellation can be complete only for one wavelength. Antireflective coatings are usually applied with thicknesses to cancel yellow-green light with $\lambda_a = 550$ nm, in the middle of the visible spectrum. Then, the reduction of the reflectance for red and blue light is substantial but not complete. It is for this reason that surfaces with antireflective coatings (for example, camera lenses) appear slightly purple—a combination of reflected red and blue light.

An effective coating material must satisfy Eq. 42–6 and must also be mechanically and chemically stable, moisture-resistant, and so forth. The most commonly used antireflective material is magnesium fluoride, MgF_2, which has $n_c = 1.38$. When applied to dense flint glass ($n_g = 1.656$ for $\lambda_a = 550$ nm—see Fig. 40–22), this substance is quite effective because $\sqrt{n_g} = 1.29$, which is close to n_c. A coating thickness of

$$t = \frac{\lambda_a}{4n_c} = \frac{550 \times 10^{-9} \text{ m}}{4 \times 1.38} = 100 \text{ nm}$$

which is called a *quarter-wavelength coating*, reduces the overall reflectance from about 4 percent to about 1 percent.

It is possible to construct a highly *reflective* surface by using a stack of quarter-wavelength films of alternating high refractive index and low refractive index. (•Can you see why this arrangement produces a large reflectance?) Such a film stack has the advantage, not possessed by a metallic surface, that little light is lost by absorption. It is possible to construct a stack that has $R = 0.995$ and $T = 0.003$.

Newton's Rings. It is possible to observe interference effects due to an air gap between two glass surfaces. This air gap may be a wedge-shaped space between two flat plates (see Example 42-4), or it may involve a curved surface. An example of the latter is the space between a flat plate and the spherical surface of a plano-convex lens (Fig. 42-10). Isaac Newton (as well as contemporaries Robert Boyle and Robert

Fig. 42-10. The geometry of Newton's rings. The lens has one flat surface and is called a *plano-convex* lens. (Photo courtesy of Bausch & Lomb.)

Hooke) noted that colored rings (called *Newton's rings*) are observed when looking straight down on such a lens that is illuminated by a nearby source. The interference effects are pronounced if monochromatic light is used and if the source and observer are close to the normal at the point of contact of the lens, as in Fig. 42-10. (Newton's rings are also observable in the transmitted light.)

If the radius R of the spherical lens surface is very large, the thickness of the air gap near the point of contact is of the order of the wavelength of light, and the analysis for the case of thin films applies. Consider a light ray at near-normal incidence. This ray will be partially reflected at points a and b (see Fig. 42-10), producing two interfering rays with a path-length difference of $2t$. The reflection at b involves a phase change of π, whereas no phase change results from the reflection at a. Then, interference maxima occur at radii r for which $4\pi t/\lambda_a = (2m + 1)\pi$, with $m = 0, 1, 2, \ldots$. Now, we also have $R^2 = (R - t)^2 + r^2$; for $t \ll R$, this expression gives $t \cong r^2/2R$. Thus, interference maxima produce bright rings with radii

$$r_m = \sqrt{(m + \tfrac{1}{2})R\lambda_a}, \qquad m = 0, 1, 2, \ldots \qquad \text{(42-7)}$$

At the point of contact ($r = 0$), the medium is effectively continuous, so there is no reflection. Consequently, the center of the ring system is dark, as shown in the photograph above.

Example 42–4

Two flat glass plates are separated at one edge by a thin spacer with a thickness h. For the case of near-normal incidence, locate the interference maxima with respect to the contact edge.

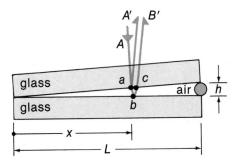

Solution:

The thickness of the glass plates is large compared with the wavelength of the incident light, so the observable interference effects are due entirely to the wedge-shaped air gap. The ray AB' undergoes a phase change of π upon reflection at point b. The thickness of the air gap at this point, $t = (x/L)h$, introduces an additional phase difference $\delta = 4\pi xh/L\lambda_a$. Hence, interference maxima occur at positions for which

$$\frac{4\pi xh}{L\lambda_a} + \pi = 2m\pi, \qquad m = 1, 2, 3, \ldots$$

At the contact edge, the medium is continuous, so there is no reflection and this edge is dark. The spacing Δx between successive bright fringes is given by

$$\frac{4\pi(x + \Delta x)h}{L\lambda_a} - \frac{4\pi xh}{L\lambda_a} = 2(m + 1)\pi - 2m\pi$$

from which

$$\Delta x = \frac{L}{2h}\lambda_a$$

For $L = 20$ cm, $h = 0.1$ mm, and $\lambda_a = 589$ nm (sodium yellow light), we find

$$\Delta x = \frac{20 \text{ cm}}{2(0.01 \text{ cm})}(5.89 \times 10^{-7} \text{ m}) = 5.89 \times 10^{-4} \text{ m}$$

$$= 0.589 \text{ mm}$$

42–4 THE MICHELSON INTERFEROMETER

In 1881, Albert A. Michelson used the idea of light interference to devise an instrument capable of making distance measurements with extremely high precision (within a fraction of a wavelength of light). The essential features of a *Michelson interferometer* are shown in Fig. 42–11. Light from an extended monochromatic

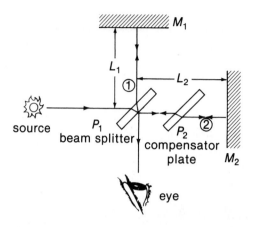

Fig. 42–11. Schematic diagram of the arrangement of plates and mirrors in a Michelson interferometer.

source is incident on a half-silvered glass plate P_1 (the "beam splitter"), where the original beam is partially reflected and partially transmitted. The reflected portion, labeled ① in the diagram, travels a distance L_1 to mirror M_1, which reflects the beam back toward P_1 where it is partially transmitted and arrives finally at the eye of the observer. The transmitted portion of the original beam, labeled ②, travels a distance L_2 to mirror M_2, which reflects the beam back toward P_1 where it is partially reflected and also arrives finally at the eye of the observer. The plate P_2 (the "compensator plate") is cut from the same piece of glass as P_1 and is introduced into beam ② to equalize the path lengths of the two beams. The two wave trains, having traveled these two different paths, are coherent when they arrive at the eye and therefore produce observable interference effects. The mirrors, M_1 and M_2, are closely but *not exactly* perpendicular; the small angular difference results in a slight inclination, at the eye, of one image of the extended source with respect to the other. The observer then sees a series of parallel interference fringes. (The effect is similar to that produced by the air gap of varying thicknesses discussed in Example 42–4.)

In a practical Michelson interferometer, the mirror M_1 can be translated along the direction perpendicular to its surface by means of a precision screw. As the length L_1 is changed, the pattern of interference fringes is observed to shift. By using a telescope with a cross-hair eyepiece, it is easy to see the fringes shift as L_1 is changed. A shift of the pattern by an amount corresponding to the distance between two successive bright fringes is observed for a change in L_1 equal to $\frac{1}{2}\lambda_a$. The distance that M_1 is moved (in units of the wavelength of the source light) is determined by counting the number of fringes that pass the cross-hair as the movement is made. A careful observer using a well-constructed instrument can detect a shift as small as 1/40 of a fringe spacing. This means that distances can be measured to about 75 Å.

It was pointed out in Section 42–1 that a wave train of light has a length that is typically of the order of a meter (see Fig. 42–1). Thus, when a particular wave train enters a Michelson interferometer and is divided by the "beam splitter" into two portions, these portions will not eventually overlap and interfere if the path-length difference, $L_1 - L_2$, is greater than the length of the wave train. This path-length difference in which some observable degree of interference can occur is called the *coherence length* of the radiation.

SPECIAL TOPIC

42–5 HOLOGRAPHY

In ordinary photography, a piece of film is used to record the intensity distribution of light from a source object. In a special type of photography, called holography,* the wavefronts themselves (both the phase and the intensity) are recorded by means of interference effects. A holograph therefore contains much more information than does a photograph. When a holographic image is reproduced, it presents a remarkable three-dimensional appearance. The image stands in space and may be viewed from different angles and distances. The effect is the same as viewing the object itself with all of its three-dimensional aspects.

Figure 42–12a illustrates the process of recording the holographic information on film. The incident light is a monochromatic plane wave produced by a laser (see Section 47–6). One portion of the incident wave train is directed at the film and constitutes the *reference plane wave*. The other portion illuminates the objects studied. At each

---

*From the Greek *holographos*, "written in full."

point on these objects, such as point *A* shown in Fig. 42–12a, a reflected *spherical object wave* is initiated, and a portion of this wave reaches the film. Laser radiation is highly monochromatic and has a coherence length of several meters. Consequently, if the scene that is viewed has dimensions of a few meters or less, the reference wave and the object waves will overlap and interfere at the film. For example, point *P* on the film is one point on a circular arc centered on O_A along which the intensity is maximum for object point *A*. The intensity distribution that is produced by object point *A* is determined both by the brightness of the reflected wave from *A* and by the phase relationship between the object wave and the reference wave at the film. The total pattern recorded on the film is the linear superposition of the contributions from all object points such as *A* and *B*. The transparent positive print of the developed film is called a *holograph* or *hologram* (see the photograph).

The reconstruction or viewing process is illustrated in Fig. 42–12b. A plane wave of monochromatic laser light with the *same wavelength* as that of the light used to expose the film is incident normally on the hologram. The trans-

Fig. 42–12. (a) Preparation of a hologram by the interference of the reference wave and the reflected object wave at the film. (b) Reconstruction of the image by irradiating the hologram with a plane wave of laser light.

Hologram of a barracuda. (Courtesy of Metrologic Instruments, Inc.)

mitted radiation can be analyzed in terms of three components. In the particular version of holography depicted in Fig. 42–12b, these components are:

(1) a reduced-intensity undeviated continuation of the incident plane wave (which is slightly diffracted);

(2) a diverging spherical wave (formed by diffraction) that appears to originate in a virtual image that corresponds in location and brightness to the object point (for example, point A); and

(3) a spherical wave (also formed by diffraction) that converges to form a real image of the object (and thereafter diverges); this image is reversed and is referred to as a *false image*.

In the process just described, we considered the use of

a thin flat film to obtain an essentially two-dimensional pattern on the film. A hologram prepared in this way can be viewed only with monochromatic light. If the image is to be the same size as the object, the reference and reconstruction waves must both be plane waves* with the same wavelength. Thus, if the hologram is prepared using red laser light, the full-size image will appear in red light.† However, if a film with a thick emulsion is used, the standing-wave pattern produced throughout the emulsion volume by interference can be recorded. Full-color holography can then be achieved by exposing the film using three lasers that emit red, green, and blue light (the colors used for color television), and then viewing the hologram with white light.‡

To prepare a hologram with a continuously operating laser, it is necessary that the objects remain absolutely stationary. A movement of a fraction of a wavelength will produce a blurred holographic image. However, three-dimensional motion-picture holography may be made feasible by using *pulsed* lasers.§ When recording, a laser burst with a duration of about 5 ns for each motion-picture frame should arrest sufficiently any ordinary stage motion. On reconstruction, each frame could be viewed with a longer burst duration in order to reduce flicker. (See Problem 42–17.)

---

*In methods that use spherical waves, magnified (or diminished) images are possible, and the perspective of the images may also be altered.

†If the hologram is irradiated with white light, each wavelength produces an image of a different size and no apparent reconstruction results.

‡This method is a variation of the color photography process invented in 1891 by the French physicist Gabriel Lippmann (1845–1921), for which he received the 1908 Nobel Prize in physics.

§Such motion pictures were first made in 1976 by NIFKI, the Soviet Cine and Photo Research Institute.

PROBLEMS

Section 42–1

42–1 Reconsider Example 42–1 for the case of two waves with unequal amplitudes, so that

$$E = E_1 \cos(kx_0 - \omega t) + E_2 \cos(kx_0 - \omega t - \delta)$$

Show that the time-averaged value of E^2 is

$$\overline{E^2} = \tfrac{1}{2}E_1^2 + \tfrac{1}{2}E_2^2 + E_1 E_2 \cos \delta$$

Show that this expression reduces to that obtained in Example 42–1 for the case $E_1 = E_2 = E_0$.

Section 42–2

42–2 Young's experiment is performed by passing monochromatic light through two narrow parallel slits that are separated by 0.30 mm. The bright fringes on a screen 1.5 m away are observed to have a spacing of 2.99 mm. What is the wavelength of the incident light?

42–3 Two narrow parallel slits are separated by 0.85 mm and are illuminated by light with $\lambda = 6000$ Å. What is the phase difference between the two interfering waves on a screen (2.8 m away) at a point 2.50 mm from the central bright fringe? What is the ratio of the intensity at this point to the intensity at the center of a bright fringe?

42-4 Young's experiment is performed by using white light. The two narrow slits are separated by 0.40 mm. The screen is at a distance of 3.0 m. (a) Locate the first-order maximum for yellow-green light ($\lambda = 550$ nm). (b) What is the separation between the first-order maxima for red light ($\lambda = 700$ nm) and violet light ($\lambda = 400$ nm)? (c) At what point nearest the central maximum will a maximum for yellow light ($\lambda = 600$ nm) coincide with a maximum for violet light ($\lambda = 400$ nm)? Identify the order of each maximum.

42-5 Consider a Fresnel double mirror with the geometry shown in the diagram. (a) Show that the distance between successive bright fringes on the screen is given by $\Delta y = \lambda D/2a\alpha$. (b) Locate the first-order bright fringe if $a = 25$ cm, $D = 1.25$ m, $\lambda = 5 \times 10^{-7}$ m, and $\alpha = 5'$ (5 arc minutes).

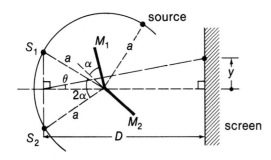

42-6 A Fresnel double prism has $\alpha = 10'$ and $n = 1.50$. (a) If the distance from the source to the prism is 25 cm, what is the distance between the two images, S_1 and S_2? (Refer to Fig. 42-6.) (b) What is the difference between successive interference maxima on a screen 1.5 m away when sodium vapor light ($\lambda = 589$ nm) is used?

Section 42-3

42-7 A soap film ($n = 1.45$) is contained within a rectangular wire frame. The frame is held vertically so that the film drains downward due to gravity and becomes thicker at the bottom than at the top, where the thickness is essentially zero. The film is viewed in white light with near-normal incidence, and the first violet ($\lambda = 420$ nm) interference band is observed 3 cm from the top edge of the film. (a) Locate the first red ($\lambda = 680$ nm) interference band. (b) Determine the film thickness at the positions of the violet and red bands. (c) What is the wedge angle of the film?

42-8 A soap film such as that in Problem 42-7 is viewed in yellow light ($\lambda = 589$ nm) with near-normal incidence. Interference fringes with a uniform spacing of 4.5 mm are observed. What is the variation in thickness of the film per centimeter of vertical distance?

42-9 A coating of magnesium fluoride is to be used on a glass plate to reduce the reflection of yellow light ($\lambda = 580$ nm) that is incident at an angle of 30°. What thickness should the coating be?

42-10 A thin film of glass ($n = 1.52$) with a thickness of 0.42 μm is viewed in white light at near-normal incidence. Visible light with what wavelength is most strongly reflected by the film?

42-11 The glass plates of Example 42-4 are immersed in water. What is the new spacing between bright fringes?

42-12 The air in the gap between the glass plates of Example 42-4 is replaced by methyl alcohol ($n = 1.329$). When sodium yellow light ($\lambda = 589$ nm) is normally incident, fringes with a spacing of 0.45 mm are observed. What is the wedge angle between the plates?

42-13 A plano-convex lens with a radius of curvature $R = 3.0$ m is in contact with a flat plate of glass. A light source and the observer's eye are located close to the normal, as shown in Fig. 42-10. The radius of the 50th of the bright Newton's rings is found to be 9.8 mm. What is the wavelength of the light produced by the source?

Section 42-4

42-14 A Michelson interferometer (Fig. 42-11) is illuminated with green light ($\lambda = 5300$ Å). How many fringes will be seen to pass the cross-hair in the viewing telescope as the mirror M_1 is moved from one end of a 1.0-cm gauge block to the other end? A new light source with a different color is substituted and the observer counts 36,200 fringes as the mirror is returned to its original position. What is the wavelength of the light from the new source?

42-15 An evacuated glass tube with parallel glass end plates and with an inside length of 10.0 cm is placed into one arm of a Michelson interferometer that uses sodium vapor light (Fig. 42-11). A gas is slowly admitted into the tube and the shift of the fringes is observed. When a gas pressure of 1.0 atm is reached, 99.5 fringes have passed the cross-hair of the viewing telescope. What is the refractive index of the gas? Identify the gas. (Use Table 40-2; the temperature is $0°$ C.)

42-16 The interference pattern observed in a Michelson interferometer that uses sodium yellow light ($\lambda = 589$ nm) has a fringe spacing of 0.60 mm. What amount of deviation from an angle of $90°$ between the mirrors, M_1 and M_2, could produce this fringe spacing?

SPECIAL TOPIC

Section 42-5

42-17 A holographic motion picture is to be made of a scene in which all motion takes place with a speed less than 10 m/s. The illuminating laser produces light with $\lambda = 5000$ Å. The amount of motion dur- ing the recording of any frame of the film cannot exceed $\lambda/10$. (a) What is the maximum allowable duration of the laser pulse? (b) What is the length of the corresponding wave train? Is this adequate for recording a scene with typical dimensions?

Additional Problems

42-18 A *Rayleigh refractometer*, shown schematically in the diagram, is a device used in precision measure- ments of refractive indexes of gases. Two identical glass cells, A and B, are placed behind a pair of narrow slits that are illuminated by a plane wave of blue-green light ($\lambda = 550$ nm). Each cell has an in- side length $\ell = 2.500$ cm, and $\ell \ll D$. When the cells are evacuated, the optical path lengths are identical and the central bright fringe is located at P. Gas is slowly admitted into cell A and the central fringe is observed to move upward. At a pressure of 1.00 atm (and a temperature of $0°C$) the central fringe is at Q, a distance equal to 20.5 fringe spac- ings from P. Determine the refractive index of the gas. Identify the gas. (Refer to Table 40-2.)

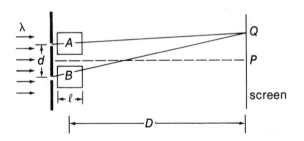

42-19 Young's double-slit experiment is performed with sodium yellow light ($\lambda = 5890$ Å) and with a slits- to-screen distance of 2.0 m. The tenth interference minimum (dark fringe) is observed to be 7.26 mm from the central maximum. Determine the spacing of the slits.

42-20 A Billet split lens (Fig. 42-5) is used with light of wavelength 475 nm. The focal length of the thin lens is $f = +25$ cm, and the lens halves are sepa- rated by an opaque spacer with a thickness of 0.30 mm. The primary source S is 50 cm behind the lens, and the screen is 1.5 m in front of the lens. (a) Locate the secondary sources, S_1 and S_2, and determine their separation. (b) What is the spacing of the fringes on the screen?

42-21 A 0.10-μm layer of carbon disulfide ($n = 1.625$) covers a glass plate ($n = 1.520$). What is the longest wavelength of light for which complete destructive interference will result when a normally incident beam is reflected?

42-22 A thin layer of liquid methylene iodide ($n = 1.756$) is sandwiched between two flat parallel plates of glass. What must be the thickness of the liquid layer if normally incident light with $\lambda = 600$ nm is to be strongly reflected?

42-23 Consider light at normal incidence on a glass plate
• coated with an antireflective material (Fig. 42-9). Show that equating the amplitude reflection coeffi- cients at the air-coating interface and the coating- glass interface leads to Eq. 42-6. (This result is only approximate because it ignores the contribution from multiple reflections and assumes that the re- flection coefficients are very much smaller than the transmission coefficients.)

42-24 Two glass plates with $L = 10$ cm are in contact along one edge and are separated by a 0.10-mm thick spacer at the opposite edge, as in Example 42-4. When monochromatic light is incident normally, fringes with a spacing of 0.34 mm are observed. What is the wavelength of the light?

42-25 Monochromatic light with continuously variable wavelength is incident normally on a thin film of oil ($n = 1.30$) that covers a glass plate. As the wavelength of the light is varied from 450 nm to 750 nm, complete destructive interference is observed for $\lambda = 500$ nm and for $\lambda = 700$ nm. Determine the thickness of the oil film. (Note that $n_{\text{oil}} < n_{\text{glass}}$.)

42-26 Sodium yellow light ($\lambda = 589$ nm) is used to produce Newton's rings in the arrangement shown in Fig. 42–10. The radius of the 30th bright ring is found to be 8.0 mm. What is the thickness of the air film at the position of this ring? What is the radius of curvature of the lens?

42-27 Sodium vapor light has a nominal wavelength
• of 5893 Å, but the light actually consists of two components of nearly equal intensity separated in wavelength by an amount $\Delta\lambda$. When a sodium vapor source is used in conjunction with a Michelson interferometer, one observes that as the mirror M_1 is moved, the fringes become sharp and clear, then soft and fuzzy, then sharp and clear, and so forth. The distance that M_1 must be moved between successive appearances of sharp fringes is 0.290 mm. What is the wavelength separation $\Delta\lambda$ of the two components of the light?

DIFFRACTION

Diffraction effects occur when incident waves interact with obstacles or apertures with finite size. Then, the waves bend around the obstacles or spread out from the apertures in the familiar manner of water waves. Although many effects involving light can be analyzed in terms of rectilinear propagation, careful investigation reveals that light waves also diffract into regions that would not be illuminated if only straight-line propagation were possible.*

It is convenient to separate the phenomena associated with the diffraction of light into two categories. *Fresnel diffraction* refers to cases in which the viewing screen or the observer's eye is relatively close to the diffracting obstacle or aperture. Then, the light is converging when it reaches the eye or screen (see Fig. 43–1a).

Somewhat simpler to analyze than Fresnel diffraction is the case in which both the source and the eye or screen are at very large distances from the diffracting obstacle or aperture. Then, the incident radiation is in the form of plane waves, as is the diffracted radiation that is directed toward an observing point. This situation is referred to as *Fraunhofer diffraction*.† It is usually arranged by using lenses to produce and to focus the parallel rays, thereby removing the source and the observation point to optical infinity (Fig. 43–1b). The discussion in this chapter is concerned primarily with Fraunhofer diffraction, although some aspects of Fresnel diffraction are described in Section 43–4.

43-1 FRAUNHOFER DIFFRACTION

Consider the Fraunhofer diffraction produced by a single slit with a finite width, as shown in Fig. 43–1b. A primary point source S of monochromatic light is located at the focal point of a converging lens L_1, which directs a beam of plane waves toward the slit A. The light that passes through the slit is focused by a second converging lens L_2 onto a screen that coincides with the focal plane of the lens. The details of this process are shown in Fig. 43–2.

A plane wavefront that arrives at the slit from lens L_1 may be replaced by an array of secondary (Huygens') line sources in the plane of the slit, each radiating into

---

*The first detailed observations of diffraction were made by an Italian physicist, Francesco Grimaldi (1618–1663), who noted the broadening of a beam of light upon passing through a narrow slit. Even earlier, Leonardo da Vinci (1452–1519) had recorded the observation of diffraction effects in his famous *Notebooks*.

† After the German optician and physicist Joseph von Fraunhofer (1787–1826).

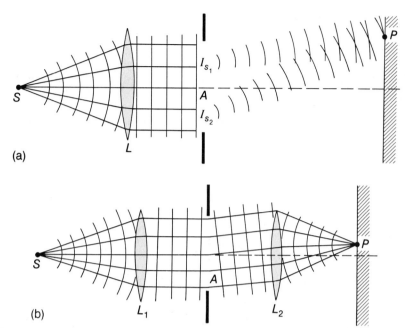

(a)

(b)

Fig. 43–1. (a) Fresnel diffraction. Light diverging from the source S is rendered parallel by lens L so that the radiation incident on the aperture A consists of plane waves. Radiation from the Huygens' secondary sources, such as s_1 and s_2, that are distributed over A combine at the observation point P. (b) Fraunhofer diffraction. As in (a), the radiation incident on the aperture A consists of plane waves. The diffracted beam of parallel rays is brought to a focus at P by lens L_2. Because plane waves are incident on A, the phase of the radiation is the same over the full extent of the aperture plane.

the region to the right of the slit. One such line source, a strip with a width dx parallel to the slit sides, is shown in Fig. 43–2. A narrow bundle of rays pq from this secondary source is focused by lens L_2 onto the viewing screen at point P. Additional rays from other similar secondary sources are also focused at P, and the light intensity there is obtained by superimposing all such rays.

All of the secondary Huygens' sources in the plane of the slit are coherent and radiate in phase; this phase condition is guaranteed by Fermat's principle (see Section 41–5). Choose the central ray rsu as the reference ray for determining the phase differences of the other rays as they arrive at P. The optical path length $\overline{pqP}$ is the

Fig. 43–2. The geometry for describing the Fraunhofer diffraction produced by a single slit.

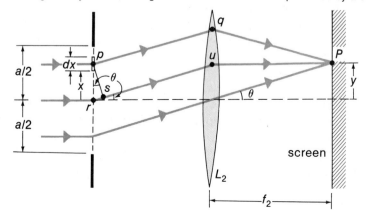

same as $\overline{suP}$. (•Can you see why?) Thus, the rays from the source at dx arrive at P with a phase lead, compared with the reference ray, corresponding to the optical path difference $rs \cong x \sin \theta$.

The slit is illuminated uniformly by the light from lens L_1. Consequently, the electric field amplitude dE_m at point P due to the secondary line source dx may be expressed as

$$dE_m = \xi \, dx$$

where ξ is a constant with appropriate dimensions. The reference ray is considered to have zero phase angle. If the angle θ is very small, all contributions to the electric field intensity on the screen may be considered to be parallel and to lie in the plane of the screen. Thus, the time-dependent electric field component at P due to the segment dx is

$$dE = \xi \, dx \cos(\omega t + \frac{2\pi}{\lambda} x \sin \theta)$$

In terms of a convenient new parameter, the phase angle α,*

$$\alpha = \frac{\pi a}{\lambda} \sin \theta, \tag{43-1}$$

the superposition of all contributions due to the secondary sources from $x = -a/2$ to $x = a/2$ is

$$E = \xi \int_{-a/2}^{a/2} \cos\left(\omega t + \frac{2\alpha}{a} x\right) dx$$

$$= \frac{\xi a}{2\alpha} [\sin(\omega t + \alpha) - \sin(\omega t - \alpha)]$$

$$= \xi a \left(\frac{\sin \alpha}{\alpha}\right) \cos \omega t$$

The intensity I is equal to $\overline{E^2}/Z_0$ (Eq. 42–1), so we can write

$$I = I_0 \left(\frac{\sin \alpha}{\alpha}\right)^2 \tag{43-2}$$

where $\frac{1}{2}\xi^2 a^2/Z_0$ has been set equal to I_0. (Compare Eq. 42–1a.)

Figure 43–3 shows the intensity I that is observed on the viewing screen. The central position ($\alpha = 0$) corresponds to a maximum because

$$\lim_{\alpha \to 0} \frac{\sin \alpha}{\alpha} = 1$$

In addition, there are secondary maxima that are given closely (but not exactly) by the condition $\sin \alpha = \pm 1$; that is,

$$\alpha = \frac{\pi a}{\lambda} \sin \theta \cong \pm\frac{1}{2}(2m + 1)\pi, \qquad m = 1, 2, 3, \ldots; \text{ approx. maxima} \tag{43-3a}$$

---

*Be careful to distinguish between the phase angle α and the geometrical angle θ.

Fig. 43-3. Intensity pattern on a viewing screen due to the Fraunhofer diffraction produced by a single slit. The parameter α is defined by Eq. 43-1. The intensity minima occur at points for which $(\sin \alpha)/\alpha = 0$, but there is no minimum for $\alpha = 0$. Thus, the angular width (from minimum to minimum) of the central peak is twice that of the secondary peaks. (Photo from *Atlas of Optical Phenomena* by M. Cagnet, M. Francon, and J. C. Thierr. Springer-Verlag, 1962.)

(•Can you see why this relationship is only approximate?) The intensity minima occur when $\sin \alpha = 0$ (excluding $\alpha = 0$); that is,

$$\alpha = \frac{\pi a}{\lambda} \sin \theta = \pm m\pi, \qquad m = 1, 2, 3, \ldots; \text{ minima} \qquad \textbf{(43-3b)}$$

It is customary to define the "half-width" $\Delta\alpha$ of the central maximum to be twice the value of α for which $I = \frac{1}{2}I_0$. Thus (see Problem 43-3),

$$\Delta\alpha = 2.783 \text{ rad} \qquad \textbf{(43-4)}$$

If the infinite slit we have been considering becomes a rectangular aperture, the diffraction pattern will contain contributions from the width a and the length b. In the direction perpendicular to the length of the rectangle, the intensity maxima are given by Eq. 43-3a. To find the maxima in the direction perpendicular to the width, the width a in Eq. 43-3a must be replaced by the length $b > a$, which means that the spacing between maxima in this direction (determined by the values of θ) will be *less* than in the other direction. This effect is shown clearly in the photograph on page 1143.

The condition for the location of the intensity minima in a Fraunhofer pattern (Eq. 43-3b) can be obtained without reference to the expression for the intensity given in Eq. 43-2. In Fig. 43-4 the slit aperture is divided into a large number of adjoining Huygens' equivalent slits. There are $2N$ such slits, labeled 1 through N in the upper half of the aperture and $1'$ through N' $(=N)$ in the lower half. The contributions from 1 and $1'$ will cancel at a point on a distant screen if

$$\frac{a}{2} \sin \theta = \frac{\lambda}{2}$$

This same condition applies for the cancellation of the contributions from the pairs 2-2', 3-3', . . . , N-N', and hence from the entire aperture. The condition we have obtained is just Eq. 43-3b with $m = +1$. (•Can you deduce the condition for $m = \pm 2, \pm 3, \ldots$?)

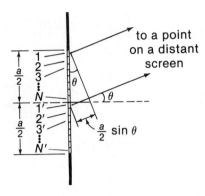

Fig. 43-4. Geometry for locating the intensity minima in a Fraunhofer diffraction pattern.

The diffraction pattern produced by a circular aperture was first calculated in precise form in 1835 by Sir George Airy (1801–1892), the English Astronomer Royal. The result is similar to that for the infinite slit, but the expression (Eq. 43-2) contains a complicated function instead of the simple sine function. The secondary intensity maxima occur for approximately the same values of θ, but the magnitudes for the case of the circular aperture are smaller than those for the case of the infinite slit (see Table 43-1).*

TABLE 43-1.   Positions of Diffraction Maxima and Minima

SINGLE SLIT		
	α	I/I_0
1st max.	0	1
1st min.	$\pi = 3.14$	0
2nd max.	$1.43\pi = 4.49$	0.0472
2nd min.	$2\pi = 6.28$	0
3rd max.	$2.46\pi = 7.72$	0.0169
3rd min.	$3\pi = 9.42$	0
4th max.	$3.47\pi = 10.90$	0.0083
CIRCULAR APERTURE		
	α	I/I_0
1st max.	0	1
1st min.	$1.22\pi = 3.83$	0
2nd max.	$1.64\pi = 5.14$	0.0175
2nd min.	$2.23\pi = 7.02$	0
3rd max.	$2.68\pi = 8.42$	0.0042
3rd min.	$3.24\pi = 10.17$	0
4th max.	$3.67\pi = 11.62$	0.0016

---

*Solutions for problems involving axial symmetry (as for circular apertures) are most readily expressed in terms of a special function, called the *Bessel function*, instead of sines and cosines. The first zero of this function corresponds to $\alpha = (\pi a/\lambda) \sin \theta = 1.220\pi$.

Diffraction pattern from a rectangular slit. (From *Atlas of Optical Phenomena* by M. Cagnet, M. Francon, and J. C. Thrierr. Springer-Verlag, 1962.)

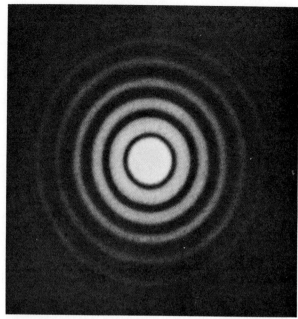

Diffraction pattern from a circular aperture. (From *Atlas of Optical Phenomena* by M. Cagnet, M. Francon, and J. C. Thrierr. Springer-Verlag, 1962.)

Example 43-1

A slit with a width of 0.030 mm is illuminated with plane waves of monochromatic light ($\lambda = 589$ nm) in the geometry shown in Fig. 43-3. The focal length of lens L_2 is $f_2 = 60$ cm.

(a) Determine the locations on the screen of the first minima.

(b) Determine the approximate locations on the screen of the first secondary maxima.

(c) Find the approximate ratio of the intensity of the first secondary maxima to the intensity of the central maximum.

(d) What is the width of the central maximum at half maximum intensity?

Solution:

(a) From Eq. 43-3b for $m = 1$, we find for the first minima,

$$\frac{\pi a}{\lambda} \sin \theta' = \pm \pi$$

or
$$\theta' = \pm \sin^{-1} \left(\frac{5.89 \times 10^{-7} \text{ m}}{3.0 \times 10^{-5} \text{ m}} \right) = \pm 0.0196 \text{ rad}$$

The position on the screen, measured from the central maximum, is obtained by referring to the ray that passes undeviated through the center of lens L_2 (see Fig. 43-2):

$$y' = \pm f_2 \tan \theta' \cong \pm (60 \text{ cm})(0.0196)$$
$$= \pm 1.18 \text{ cm}$$

(b) The approximate locations of the first secondary maxima can be found by using Eq. 43-3a with $m = 1$; thus,

$$\theta'' \cong \pm\sin^{-1}\frac{3\lambda}{2a} = \pm\sin^{-1}\frac{3(5.89 \times 10^{-7} \text{ m})}{2(3.0 \times 10^{-5} \text{ m})}$$

$$= \pm 0.0295 \text{ rad}$$

from which $$y'' = \pm f_2 \tan\theta'' \cong \pm 1.77 \text{ cm}$$

(c) The approximate intensity ratio is found by using Eq. 43–2,

$$\frac{I''}{I_0} = \left(\frac{\sin\alpha''}{\alpha''}\right)^2$$

together with Eq. 43–3a,

$$\alpha'' = \frac{\pi a}{\lambda}\sin\theta'' \cong \frac{3}{2}\pi$$

Thus, $$\frac{I''}{I_0} \cong \left(\frac{2}{3\pi}\sin(3\pi/2)\right)^2 = \frac{4}{9\pi^2} = 0.045$$

Compare Table 43–1.

(d) All relevant values of θ are small ($\theta \ll 1$), so we can write

$$\alpha = \frac{\pi a}{\lambda}\sin\theta \cong \frac{\pi a}{\lambda}$$

and $$\Delta\alpha = \frac{\pi a}{\lambda}\Delta\theta$$

Also, $$\Delta y \cong f_2\Delta\theta = \frac{f_2\lambda}{\pi a}\Delta\alpha$$

Then, using $\Delta\alpha = 2.783$ (Eq. 43–4), we find

$$\Delta y \cong \frac{(60 \text{ cm})(5.89 \times 10^{-7} \text{ m})}{\pi(3.0 \times 10^{-5} \text{ m})} \times (2.783) = 1.04 \text{ cm}$$

Note carefully that although the various values of θ in this example are all very small (this is the usual situation), the corresponding values of α are *not small*—indeed, we often have $\alpha \approx 1$. Therefore, caution must be exercised when making small-angle approximations.

Angular Resolution. Suppose that a narrow slit is illuminated with light from two incoherent sources, S_1 and S_2, that have an angular separation $\delta\theta$, as indicated in Fig. 43–5. The source-to-slit and slit-to-screen distances are sufficiently great that the conditions for Fraunhofer diffraction apply. Separate and independent contributions to the intensity, in the form of single-slit Fraunhofer patterns P_1 and P_2, are observed on the screen. If $\delta\theta$ is made smaller and smaller, the degree of overlap of the patterns will eventually reach a point at which it will no longer be possible to determine that two sources are present. That is, it will no longer be possible to *resolve* the two sources.

Figure 43–6 shows the intensity patterns for three different values of the angular separation of the two sources. What is the smallest angular separation $\delta\theta_{\text{min}}$ for which two incoherent sources can just be resolved? The *Rayleigh criterion* (an arbi-

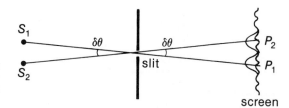

Fig. 43-5. Two incoherent sources, S_1 and S_2, produce separate and independent single-slit Fraunhofer patterns on the screen.

Fig. 43-6. Intensity distribution patterns on the screen of Fig. 43-5 for three different values for the angular separation of the two equal-intensity sources: $\delta\theta_a = 1.25\ (\lambda/a)$, $\delta\theta_b = \lambda/a$, and $\delta\theta_c = 0.75\ (\lambda/a)$. For (c), the images are not resolved.

trary rule) states that two point sources of monochromatic radiation with the same wavelength are just resolved if the position of the first minimum of one diffraction image coincides with the position of the central maximum of the other diffraction image. This is the situation illustrated in Fig. 43-6b, which shows a slight dip between the two maxima. If $\delta\theta$ is reduced by a small amount (Fig. 43-6c), the dip disappears and it is no longer obvious that two maxima are present. Referring to Fig. 43-3, we can see that the Rayleigh criterion for a single slit means $\delta\alpha_{min} \cong \pi$, or

$$\delta\theta_{min} \cong \frac{\lambda}{\pi a}\delta\alpha_{min} \cong \frac{\lambda}{a} \qquad \text{single slit} \qquad \text{(43-5a)}$$

The value of $\delta\theta_{min}$ is called the *angular resolution* of the slit. In Fig. 43-6, the intensity patterns correspond to $\delta\theta_a$, equal to 1.25, 1.00, and 0.75 times $\delta\theta_{min} = \lambda/a$. Thus, the Rayleigh criterion is seen to be a reasonable rule for specifying the angular separation of two just-resolved sources.

The position of the first minimum of a diffraction pattern relative to the position of the central maximum is slightly different for a circular aperture than for a single slit (see Table 43-1). When the Rayleigh criterion is applied, the angular resolution of a circular aperture with diameter d is given by

$$\delta\theta_{min} = \frac{1.22\lambda}{d} \qquad \text{circular aperture} \qquad \text{(43-5b)}$$

Diffraction effects can place a limit on the usefulness of certain high-resolution telescopes (see Problem 43-4).

Notice from Eqs. 43-5 that the angular resolution improves (that is, $\delta\theta_{min}$ becomes smaller) as the wavelength of the radiation is made smaller in comparison

Diffraction patterns corresponding to the distributions in Fig. 43-6. (From *Atlas of Optical Phenomena* by M. Cagnet, M. Francon, and J. C. Thrierr. Springer-Verlag, 1962.)

  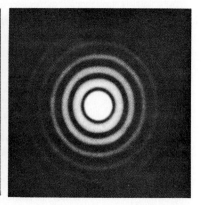

with the size of the aperture. For a particular aperture, two sources can be resolved better in blue light than in red light.

Example 43–2

A spy satellite circles the Earth at an altitude of 150 km and carries out surveillance with a special high-resolution telescopic camera having a lens diameter of 35 cm. If the angular resolution of this camera is limited by diffraction, estimate the separation of two small objects on the Earth's surface that are just resolved in yellow-green light ($\lambda = 5500\ \text{Å}$).

Solution:

The minimum angular separation is (Eq. 43–5b)

$$\delta\theta_{min} \cong \frac{1.22\,\lambda}{d} = \frac{1.22(5.50 \times 10^{-7}\ \text{m})}{0.35\ \text{m}} = 1.92 \times 10^{-6}\ \text{rad}$$

Hence, the minimum linear separation between the objects is

$$\delta x_{min} = h\delta\theta_{min} \cong (150 \times 10^{3}\ \text{m})(1.92 \times 10^{-6}\ \text{rad})$$
$$\cong 29\ \text{cm}$$

Factors other than diffraction usually place the limits on the definition that can be achieved by cameras. Nevertheless, the angular resolution of these special surveillance cameras is exceptionally good.

Example 43–3

At what maximum distance can the headlights of an approaching automobile be resolved by the eye, if the limitation is set by diffraction effects? Assume that the separation of the headlights is $\ell = 1.42\ \text{m}$ and that the diameter of the pupil of the human eye (in low-light conditions) is 4.0 mm. The mean wavelength of the light is 600 nm.

Solution:

Using Eq. 43–5b, we find

$$\delta\theta_{min} = \frac{1.22\,\lambda}{d} = \frac{1.22(6.0 \times 10^{-7}\ \text{m})}{4.0 \times 10^{-3}\ \text{m}} = 1.83 \times 10^{-4}\ \text{rad}$$

Hence, the maximum distance at which the headlights can be resolved is

$$y_m = \frac{\ell}{\delta\theta_{min}} = \frac{1.42\ \text{m}}{1.83 \times 10^{-4}} = 7.76 \times 10^{3}\ \text{m}$$
$$= 7.76\ \text{km}$$

Again, factors other than diffraction will set a smaller limit in practice.

43-2 THE DIFFRACTION GRATING

The wavelength of light can be determined with high precision by using a device called a *diffraction grating*. Such gratings are usually prepared by ruling a large number of narrow, closely spaced parallel grooves on a polished glass plate. Diffraction occurs for both reflected and transmitted light.*

Consider the Fraunhofer diffraction of light with wavelength λ by a transmission grating that consists of N equally spaced narrow slits (Fig. 43–7).† The *grating spacing* (the distance between successive slits) is b, and the direction to the point P of interest on the distant viewing screen is specified by the angle θ. For convenience, the ray that passes through the first slit, labeled ① in Fig. 43–7, is taken to be the reference ray with zero phase angle. The ray labeled ② has a path length that is shorter by an amount $b \sin \theta$ and therefore has a phase lead, compared with the reference ray, of $(2\pi b/\lambda)\sin \theta$. The ray labeled ③ has a phase lead of twice this amount, and so forth. If the contribution to the electric field E at point P due to the first slit is $E_0 \cos \omega t$, the superposition of the contributions from all N slits gives

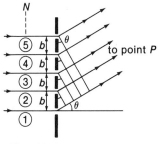

Fig. 43–7. Fraunhofer diffraction by a diffraction grating consisting of N equally spaced slits.

$$E = E_0 \sum_{n=0}^{N-1} \cos\left(\omega t + \frac{2\pi}{\lambda} nb \sin \theta\right) \tag{43-6}$$

The sum in Eq. 43-6 can be evaluated in a simple way by making use of the properties of complex numbers. In complex notation, the expression for E becomes

$$E = E_0 \mathbf{Re} \sum_{n=0}^{N-1} e^{i(\omega t + 2n\beta)}$$

$$= E_0 \mathbf{Re} e^{i\omega t} \sum_{n=0}^{N-1} e^{i 2n\beta} \tag{43-7}$$

where

$$\beta = (\pi b/\lambda) \sin \theta \tag{43-8}$$

The summation in Eq. 43-7 is a simple geometric progression, the sum of which is (see Eq. A-3 in Appendix A of Part 1)

$$a + ar + ar^2 + ar^3 + \ldots + ar^{N-1} = \frac{a(r^N - 1)}{r - 1}$$

where, in the present case, $a = 1$ and $r = e^{i2\beta}$. Thus,

$$E = E_0 \, \mathbf{Re} \, e^{i\omega t} \cdot \frac{e^{i2N\beta} - 1}{e^{i2\beta} - 1} \tag{43-9}$$

---

*The gratings are referred to as *reflection gratings* and *transmission gratings,* respectively. In each case, the set of undisturbed strips of surface between the grooves constitutes the array of line sources. Reflection gratings are often prepared by ruling lines on metallic mirrors. The first gratings were made by Fraunhofer (and consisted of parallel fine wires), but the first really successful ruled gratings were prepared by Henry Rowland in the 1870s. Rowland's ruling engine could score almost 6000 grooves per centimeter.

†For the moment we consider only the diffraction effects produced by the array of N slits, ignoring the single-slit diffraction due to each individual slit. The complete diffraction pattern, formed by the superposition of all effects, is discussed later in this section.

This equation can be put in the form (see Problem 43–5),

$$E = E_0 \, \mathbf{Re} \, e^{i[\omega t + (N-1)\beta]} \cdot \frac{\sin N\beta}{\sin \beta} \qquad \textbf{(43–10)}$$

from which

$$E = E_0 \frac{\sin N\beta}{\sin \beta} \cos[\omega t + (N-1)\beta] \qquad \textbf{(43–11)}$$

The intensity at P is proportional to the time average of E^2; thus,

$$I = \frac{I_0}{N^2} \left(\frac{\sin N\beta}{\sin \beta} \right)^2 \qquad \textbf{(43–12)}$$

where I_0 is the intensity for $\beta = 0$ (that is, $\theta = 0$).

▶ *Phasor Calculation of the Electric Field.* An instructive geometric interpretation of Eq. 43–7 can be made by using an Argand diagram. Figure 43–8 shows the vector addition of the complex electric vectors or phasors $\widehat{E}_0$ due to each slit of a six-slit array. (We consider here only the time-independent portions of the field vectors.) The phase of each phasor is advanced by an amount 2β compared with the preceding phasor. The sum of all six phasors is the electric field $\widehat{E}$ for the direction in space specified by β (that is, by the angle θ shown in Fig. 43–7). In the geometrical construction in Fig. 43–8, we see that $\angle OCB = 2\beta$. The point A is the midpoint of $\overline{OB} = E_0$, so that

$$\overline{OC} = \frac{E_0}{2 \sin \beta}$$

Also, the point D is the midpoint of $\overline{OG} = E$; thus, with $\angle OCD = N\beta$, it follows that

$$E = 2\overline{OD} = 2\overline{OC} \sin N\beta$$

from which

$$E = E_0 \frac{\sin N\beta}{\sin \beta}$$

This result is just the time-independent portion of Eq. 43–11.

The phase angle δ between E and the reference ray can also be obtained from Fig. 43–8. Notice that

Fig. 43–8. Argand diagram for the summation of the electric vectors due to the individual slits of a six-slit grating. For a regular polygon, each side subtends a central angle equal to the exterior angle between adjacent sides. Thus, in the case shown here, with the center of the polygon at C, we have $\angle OCB = 2\beta$.

$$\beta + \left(\frac{\pi}{2} - N\beta\right) + \delta = \frac{\pi}{2}$$

so that

$$\delta = (N - 1)\beta$$

in agreement with the phase angle that appears in Eq. 43–11. ◄

We continue by examining the result given by Eq. 43–12. The *principal maxima* of the intensity I occur when $\sin \beta = 0$; thus,

$$\beta = \frac{\pi b}{\lambda} \sin \theta = \pm m\pi \qquad m = 0, 1, 2, \ldots ; \text{ principal maxima} \qquad \textbf{(43–13)}$$

To calculate the value of I at the principal maxima, we use (see Problem 43–21)

$$\lim_{\beta \to \pm m\pi} \left(\frac{\sin N\beta}{\sin \beta}\right)^2 = N^2 \qquad \textbf{(43–14)}$$

Then, from Eq. 43–12, we have

$$I = I_0 \qquad \text{principal maxima} \qquad \textbf{(43–15)}$$

In Fig. 43–7 we can see how these maxima arise. For the constructive superposition of all of the rays, the difference in path length between successive rays must be an integer number of wavelengths; that is, the condition for the principal maxima is $b \sin \theta = \pm m\lambda$, in agreement with Eq. 43–13.

The value $m = 0$ corresponds to the central maximum. For $m = 1$, the two resulting maxima are referred to as the *first-order* maxima; for $m = 2$, the maxima are *second order;* and so forth. Between each pair of principal maxima are $N - 2$ additional maxima that correspond to $\sin N\beta \approx 1$; because $\sin \beta$ is also approximately equal to unity, the intensity of these secondary maxima, which does have a regular variation, is nevertheless of the order of I_0/N^2 (see Eq. 43–12). Between each pair of secondary maxima is a minimum for which $\sin N\beta = 0$ (but $\beta \neq \pm m\pi$). Figure 43–9 shows the principal and secondary maxima for the case $N = 9$.

As N is increased, the intensity at a principal maximum increases as N^2. This effect is due to two factors. First, the total power that reaches the viewing screen is proportional to N, the number of slits that are open to the incident plane wave. Second, the width of each principal maximum, which can be defined as the distance between the minima on either side, is proportional to $1/N$ (see Fig. 43–9). Thus, the power density (the intensity) at a principal maximum increases as N^2. Because the

Fig. 43–9. Intensity pattern for a grating with $N = 9$.

intensity of the secondary maxima decreases as $1/N^2$, increasing the number of slits in the grating both narrows the principal maxima and heightens the contrast with the secondary maxima. This is a point of the greatest importance for applications of diffraction gratings.

Resolving Power.   The quality of a diffraction grating can be judged by its resolving power, which is a measure of the ability of the grating to produce a principal maximum for radiation with wavelength λ that is separated from one for radiation with a nearby wavelength $\lambda' = \lambda + \Delta\lambda$. The resolving power of a grating is defined to be $R_g = \lambda/\Delta\lambda$, where $\Delta\lambda$ is the smallest difference between resolvable wavelengths. The angular separation $\Delta\beta$ between a principal maximum and the adjacent minimum is $\Delta\beta = \pi/N$ (see Eq. 43–12 or Fig. 43–9). Then, using the definition of β (Eq. 43–8) we have

$$\Delta\beta = \frac{\pi b}{\lambda}\cos\theta\,\Delta\theta = \frac{\pi}{N} \tag{43-16}$$

from which the angular change $\Delta\theta$ to descend from a principal maximum to the adjacent zero of intensity may be calculated. Suppose that there is a nearby principal maximum due to the presence in the source of radiation with wavelength $\lambda' = \lambda + \Delta\lambda$. According to the Rayleigh criterion, the two maxima are just resolved (in the same order m) when the separation is $\Delta\beta = \pi/N$. Differentiating Eq. 43–13 gives

$$\Delta\lambda = \frac{b}{m}\cos\theta\,\Delta\theta \tag{43-17}$$

Combining Eqs. 43–16 and 43–17, there results

$$R_g = \frac{\lambda}{\Delta\lambda} = mN \tag{43-18}$$

That is, the *resolving power* of a grating in any order m is proportional to the total number of slits in the grating. Practical considerations usually limit the size of simple gratings of the type discussed here to about 15 cm in length with a maximum of about 400,000 ruled grooves. Special gratings of different design may have first-order resolving powers as large as 10^6.

The Effect of Finite Slit Width.   To this point we have considered the slits in the diffraction grating to have negligible width. When the finite width of the slits is included in the calculation, the intensity given by Eq. 43–12 is modulated by the intensity function for diffraction by a single slit (Eq. 43–2). The complete expression for the intensity is just the product of these equations, namely,

$$I = \frac{I_0}{N^2}\left(\frac{\sin N\beta}{\sin \beta}\right)^2\left(\frac{\sin \alpha}{\alpha}\right)^2 \tag{43-19}$$

where $\alpha = (\pi a/\lambda)\sin\theta$ and where a is the slit width.

For the case of a double slit ($N = 2$), Eq. 43–19 can be reduced to (see Problem 43–22)

$$I = I_0 \cos^2\beta\left(\frac{\sin \alpha}{\alpha}\right)^2 \qquad \text{double slit} \tag{43-19a}$$

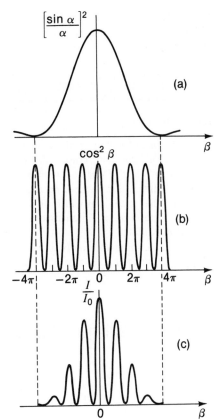

Fig. 43–10. Intensity variation in the Fraunhofer pattern due to a double slit with $b = 4a$. (a) The diffraction term, $(1/\alpha^2) \sin^2 \alpha$. (b) The interference term, $\cos^2 \beta$. (c) The product of (a) and (b) produces the observed intensity pattern.

Figure 43–10 shows the individual terms, $\cos^2 \beta$ and $(1/\alpha^2) \sin^2 \alpha$, together with their product, for the case $b = 4a$.

If the slit spacing is large compared with the slit width ($b \gg a$), the width of the central maximum of the diffraction term (the term $(1/\alpha^2) \sin^2 \alpha$ in Eq. 43–19a) is large compared with the distance between successive zeroes of the interference term ($\cos^2 \beta$). Then, for small angles θ, the diffraction term is essentially constant so that the variation of the intensity is due almost entirely to the interference term. This is the case that was discussed in Section 42–2 (see Fig. 42–3).

When $N > 2$, the individual secondary maxima, corresponding to those in Fig. 43–10c, become sharper and better defined. Figure 43–11 shows the intensity pattern

Fig. 43–11. Intensity pattern for a grating with $N = 9$ and a ratio of slit spacing to slit width of $b/a = 2.25$. The dashed curve represents the modulation of the grating pattern by the diffraction due to the finite width of the slits. (Photo from *Atlas of Optical Phenomena* by M. Cagnet, M. Francon, and J. C Thrierr. Springer-Verlag, 1962.)

for the case $N = 9$ and $b/a = 2.25$. Notice the important point that whereas the resolving power is greater in the higher orders ($R \propto m$, Eq. 43–18), the intensity at the principal maxima in these orders may be substantially diminished.

Example 43–4

A 2-cm diffraction grating contains a total of 10,000 equally spaced grooves, each of width $a = 0.8\,\mu$m. The grating is used in the Fraunhofer geometry with light that has a wavelength $\lambda = 550$ nm and with a focusing lens L_2 that has a focal length $f_2 = 50$ cm.

(a) What is the position on the viewing screen of the first-order principal maximum?

(b) What is the full width of the principal maxima?

(c) Determine the intensity at the first-order principal maximum relative to that at the central maximum.

(d) Determine the resolving power of the grating in first order.

(e) Principal maxima of how many orders are visible on the viewing screen?

Solution:

(a) The grating spacing b is

$$b = \frac{2 \times 10^{-2}\,\text{m}}{10,000} = 2 \times 10^{-6}\,\text{m}$$

The first-order principal maximum corresponds to (Eq. 43–13)

$$\frac{b}{\lambda}\sin\theta_1 = 1$$

so that

$$\theta_1 = \sin^{-1}(\lambda/b) = \sin^{-1}\left(\frac{5.50 \times 10^{-7}\,\text{m}}{2 \times 10^{-6}\,\text{m}}\right) = 16.0°$$

Referring to the diagram, we see that

$$y_1 = f_2 \tan\theta_1 = (50\,\text{cm})\tan 16.0° = 14.3\,\text{cm}$$

measured from the position of the central maximum. Conversely, the wavelength of the incident light can be determined by measuring the position on the screen of any principal maximum. However, wavelength determinations are usually made by measuring the separation between a maximum of the unknown radiation and a maximum (perhaps in a different order) due to radiation with a precisely known wavelength.

(b) The "full width" of a principal maximum means the distance between the zero intensity points on either side of the maximum. Using Eq. 43–16, we have for the first-order maximum,

$$2\,\Delta\theta = \frac{2\lambda}{Nb\cos\theta_1} = \frac{2(5.50\times10^{-7}\text{ m})}{10^4(2\times10^{-6}\text{ m})\cos 16.0°}$$

$$= 5.72\times10^{-5}\text{ rad}$$

Thus,
$$\Delta y = f_2\frac{d(\tan\theta)}{d\theta}\times(2\,\Delta\theta)$$

$$= \frac{f_2\times 2\,\Delta\theta}{\cos^2\theta_1} = \frac{(50\text{ cm})(5.72\times10^{-5})}{\cos^2 16.0°}$$

$$= 3.10\times10^{-3}\text{ cm}$$

(c) The slit width is $a = 0.8\ \mu\text{m}$, so the value of α at the first-order principal maximum is

$$\alpha = \frac{\pi a}{\lambda}\sin\theta_1 = \frac{\pi(8\times10^{-7}\text{ m})\sin 16.0°}{5.5\times10^{-7}\text{ m}}$$

$$= 1.26\text{ rad}$$

Thus, the intensity ratio is

$$\frac{I}{I_0} = \left(\frac{\sin\alpha}{\alpha}\right)^2 = \left(\frac{\sin 1.26}{1.26}\right)^2 = 0.571$$

(d) According to Eq. 43–18,

$$R_g = mN = 1\times 10{,}000 = 10{,}000$$

Hence, the principal maxima due to radiation with wavelength 550.055 nm or 549.945 nm could just be resolved from those of the given radiation ($\lambda = 550$ nm).

(e) According to Eq. 43–13,

$$m = \frac{b}{\lambda}\sin\theta$$

The largest conceivable value of θ is 90°, which corresponds to the ends of an infinite viewing screen. Then, $\sin\theta = 1$, so that the order m cannot exceed the value

$$m = \frac{b}{\lambda} = \frac{2\times10^{-6}\text{ m}}{5.5\times10^{-7}\text{ m}} = 3.6$$

Thus, principal maxima of only the first three orders appear on the viewing screen.

43–3 X-RAY DIFFRACTION

In 1895, Roentgen* discovered that when cathode rays (later called electrons) are rapidly stopped in a piece of metal, a penetrating radiation is produced. It was soon established that this radiation, called *X rays* by Roentgen, is a type of electromagnetic radiation with very short wavelength. X rays are usually produced by

---

*Wilhelm Konrad Roentgen (1845–1923), German physicist, who received the Nobel Prize for physics in 1901 for his pioneering studies of X rays. Roentgen made the first X-ray photograph (of his own hand) in 1895.

accelerating electrons through a potential difference of 10 to 300 kV, then allowing them to strike a metallic target, such as tungsten or molybdenum. The large deceleration of the electrons in the target results in the production of electromagnetic radiation in the manner discussed in Section 39–1. The X rays so produced have a continuous distribution of wavelengths with values in the fractional angstrom range.[*]

In order to observe the effects of diffraction, while avoiding the difficulties associated with using very small angles of incidence (*grazing* incidence), the grating spacing b must be of the order of the wavelength of the radiation analyzed. The regular array of atoms in a crystal forms a natural diffraction grating with a spacing that is typically a few angstroms. The scattering of X rays from the atoms in a cystalline lattice gives rise to diffraction effects very similar to those observed with visible light incident on ordinary gratings.

The suggestion that crystals could be used to diffract X rays was first made by von Laue.[†] The experiment, carried out in 1912, confirmed that a collimated beam of X rays incident on a single crystal is strongly scattered only in certain very sharply defined directions. A photographic film exposed to the scattered radiation reveals a pattern of spots that correspond to the directions of preferential scattering (Fig. 43–12). These patterns of spots, called *Laue patterns,* are related in complicated ways to the geometric properties of the target crystals.

Fig. 43–12. A monochromatic beam of X rays incident on a single crystal is diffracted strongly in certain directions determined by the geometry of the crystalline lattice.

Shortly after the experiment suggested by von Laue was successfully performed, W. L. Bragg[‡] used a different experimental arrangement to study X-ray diffraction. A single crystal cleaved along one of its lattice planes was used to *reflect* the X-ray beam (instead of *transmit* the beam, as in the von Laue experiment). This type of experiment is much easier to analyze in terms of the crystal geometry than is the transmission experiment, so we now consider this case.

The experimental arrangement used by Bragg is shown schematically in Fig. 43–13. Notice that the incident angle θ and the diffraction angle ϕ are measured from the crystal surface. For simplicity, consider the scattering of X rays from the atoms arranged in the symmetry planes of a simple cubic crystal, as indicated in Fig. 43–14. For constructive interference of the rays scattered from adjacent atoms in the horizontal plane, such as atoms A and B, the condition is

$$d \cos \theta - d \cos \phi = n\lambda, \qquad n = 0, 1, 2, \ldots$$

---

[*] Such radiation is referred to as *bremsstrahlung* (German, "braking radiation"). The interaction of the energetic electrons with individual target atoms may also produce X rays with well-defined characteristic wavelengths (see Section 46–2 and Fig. 47–16).

[†] Max von Laue (1879–1960), German physicist and winner of the 1914 Nobel Prize in physics for his work in X-ray analysis.

[‡] William L. Bragg (1890–1971), British physicist, who shared the 1915 Nobel Prize in physics with his father, William H. Bragg (1862–1942), for their research on X rays and crystal structure.

Neutron Laue photograph of NaCl. (From *Modern Physics and Quantum Mechanics* by Elmer E. Anderson. Saunders College Publishing, 1971. Courtesy of the author and Dr. E. O. Wollan.)

Fig. 43–13. The experimental arrangement used by Bragg to study X-ray diffraction by reflection from a single crystal.

Fig. 43–14. The Bragg scattering of X rays from the atomic planes of a cubic crystal.

For adjacent atoms in the vertical plane, such as atoms B and C, the condition is

$$d \sin \theta + d \sin \phi = m\lambda, \qquad m = 1, 2, 3, \ldots$$

The condition for simultaneous constructive interference from *all* atomic centers is obtained by eliminating the angle ϕ between the two equations above. This gives

$$\lambda_{nm} = \frac{2d}{n^2 + m^2}(n \cos \theta + m \sin \theta), \quad \begin{cases} n = 0, 1, 2, \ldots \\ m = 1, 2, 3, \ldots \end{cases} \qquad \textbf{(43-20)}$$

This relationship is known as the *Bragg equation*.

When $n = 0$, Eq. 43-20 becomes

$$\lambda_{0m} = \frac{2d}{m} \sin \theta, \qquad m = 1, 2, 3, \ldots \qquad \textbf{(43-21)}$$

The condition, $n = 0$, makes ϕ equal to θ and yields the equivalent of "specular reflection" from each horizontal plane. The value of m is referred to as the *order* of Bragg diffraction.

▶ When n is not zero, the angle of incidence θ and the angle of strong reflection ϕ are not equal. Then, introduce the angle ψ defined in Fig. 43-15a so that Eq. 43-20 may be rewritten as

$$\lambda_{nm} = \frac{2d}{\sqrt{n^2 + m^2}}(\sin \psi \cos \theta + \cos \psi \sin \theta)$$

$$= \frac{2d}{\sqrt{n^2 + m^2}} \sin(\theta + \psi) \qquad \textbf{(43-22)}$$

Now, define

$$d' = d \sin \psi = \frac{nd}{\sqrt{n^2 + m^2}} \qquad \textbf{(43-23a)}$$

and

$$\theta' = \theta + \psi \qquad \textbf{(43-23b)}$$

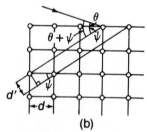

Then, Eq. 43-22 becomes

$$\lambda_{nm} = \frac{2d'}{n} \sin \theta' \qquad \textbf{(43-24)}$$

Fig. 43-15. Geometry for Bragg diffraction with $n \neq 0$.

Figure 43-15b shows the geometric interpretation of Eqs. 43-23 and 43-24 in terms of the beam incident at an angle $\theta + \psi = \theta'$ on an oblique crystal plane that is rich in atoms. This is equivalent to rotating the crystal with respect to the beam by an angle ψ to bring a new crystal plane into the horizontal direction. Finally, it may be shown (see Problem 43-25) that in terms of the angle ϕ we have

$$\lambda_{nm} = \frac{2d}{\sqrt{n^2 + m^2}} \sin (\phi - \psi) \qquad \textbf{(43-25)}$$

The same rotation that carries θ into $\theta' = \theta + \psi$ also carries ϕ into $\phi' = \phi - \psi$. Comparison of Eqs. 43-24 and 43-25 shows that $\theta' = \phi'$, which is the condition for specular reflection from the new crystal plane. Thus, we conclude that even $n \neq 0$ corresponds to a secondary specular reflection from some crystal plane that is rich in atoms. ◀

Example 43-5

The experimental arrangement shown in Fig. 43-13 is used as a *monochromator* to select a particular wavelength from the continuous distribution of wavelengths incident on the crystal. The X-ray source and the detector are placed so that $\theta = \phi = 10°$. The crystal is rock salt, which has a cubic structure with $d = 2.82$ Å. What is the wavelength of the strongly reflected radiation in first order?

Solution:

Specular reflection, $\phi = \theta$, requires $n = 0$. Then, using Eq. 43–21 with $m = 1$, we find

$$\lambda_{01} = 2d \sin \theta = 2(2.82 \text{ Å})\sin 10°$$
$$= 0.979 \text{ Å}$$

Example 43–6

When monochromatic X rays are incident on a sample that consists of a microcrystalline powder (a randomly oriented collection of microscopically small crystals), each diffraction spot for a single crystal of the same material becomes a diffraction *ring*, as indicated in the diagram. (•Can you see why rings are formed? Each single-crystal diffraction spot corresponds to a direction that makes the same angle with the direction of the incident beam, as shown in the diagram.) X rays with $\lambda = 0.75$ Å are incident on a sample that is 15 cm from the plane of the photographic film. The crystal structure of the sample is cubic with an atomic spacing $d = 2.75$ Å. Determine the radii of the rings that correspond to first- and second-order specular reflection.

Solution:

Specular reflection requires $n = 0$, so we use Eq. 43–21 with $m = 1$ and $m = 2$. Thus,

$$\lambda_{01} = 2d \sin \theta_1 \quad \text{and} \quad \lambda_{02} = d \sin \theta_2$$

But $\lambda_{01} = \lambda_{02} = 0.75$ Å, so that

$$\theta_1 = \sin^{-1}\left(\frac{0.75 \text{ Å}}{2(2.75 \text{ Å})}\right) = 7.84°$$

$$\theta_2 = \sin^{-1}\left(\frac{0.75 \text{ Å}}{2.75 \text{ Å}}\right) = 15.8°$$

Now, the total scattering angle in each case is 2θ (see the diagram). Therefore, the radii of the diffraction rings are

$$r_1 = (15 \text{ cm})\tan 2\theta_1 = (15 \text{ cm})\tan 15.68° = 4.21 \text{ cm}$$
$$r_2 = (15 \text{ cm})\tan 2\theta_2 = (15 \text{ cm})\tan 31.6° = 9.23 \text{ cm}$$

SPECIAL TOPIC

43-4   FRESNEL DIFFRACTION

Fresnel diffraction is the case in which the diffracting obstacle or aperture is relatively close to the viewing screen. Because the complete mathematical treatment of Fresnel diffraction is complicated and tedious, we consider here only simple cases that nevertheless illustrate the essential features of this type of diffraction.

Consider a plane wave of infinite extent that is directed towards an observation point P. It was Fresnel's idea to divide a wavefront S into a series of annular rings (called *Fresnel zones* or *half-period zones*), as indicated in Fig. 43–16. The distances r_n from P to the boundary circles

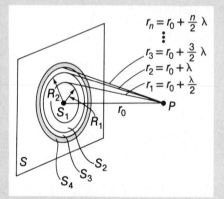

Fig. 43–16. The Fresnel zones have nearly equal areas S_n.

of the rings are such that the corresponding electric field strengths e_n at P differ successively in phase by π. This means that the distance from P to successive zone boundaries differs by one half wavelength (see Fig. 43–16); that is,

$$r_n = r_0 + \frac{n}{2}\lambda \qquad (43\text{–}26)$$

Further, the average phase difference between successive zones is also π.

The area S_n of the Fresnel zone bounded by the radii r_n and r_{n-1} is

$$S_n = \pi(R_n^2 - R_{n-1}^2)$$
$$= \pi[(r_n^2 - r_0^2) - (r_{n-1}^2 - r_0^2)]$$
$$= \pi(r_n^2 - r_{n-1}^2)$$

Using Eq. 43–26, this expression for S_n becomes

$$S_n = \pi[r_0\lambda + \tfrac{1}{4}(2n - 1)\lambda^2]$$

In a practical case, we would have $\lambda \ll r_0$, so the term involving λ^2 is negligible in comparison with the first term. Thus,

$$S_n \cong \pi r_0 \lambda \qquad (43\text{–}27)$$

It follows that the areas of the zones are all nearly equal and that the outer radius of each zone is proportional to the square root of the zone number:

$$R_n = \sqrt{n} R_1 \qquad (43\text{–}28)$$

where $R_1 = \sqrt{r_0\lambda}$.

The Fresnel zones, with nearly equal areas, represent coherent sources of radiation with nearly equal amplitudes. Let e_n be the contribution to the electric field at P due to the nth zone. The magnitude of e_n decreases with n, but only slowly, due to the increase in r_n and the increase in the angle of the line that joins S_n with P. Because the (average) phase of the radiation at P due to successive zones differs by π, the total field strength at P can be written as

$$E(P) = e_1 - e_2 + e_3 - e_4 + \cdots e_n + \cdots \qquad (43\text{–}29)$$

Now, an alternating infinite series of terms that decrease slowly has the property that the sum is equal to one half of the leading term;* that is,

$$E(P) = \tfrac{1}{2}e_1 \qquad (43\text{–}30)$$

Thus, *the electric field at P from **all** of the Fresnel zones on a plane wavefront has an amplitude equal to one half the amplitude due to the first zone and has the same phase as the contribution from that zone.*

Fresnel Diffraction from a Straight Edge. A Fresnel-zone construction can be used to describe the diffraction of a plane wave that encounters an obstacle with a straight edge. Figure 43–17 shows a plane wave with wavelength λ incident on a straight edge. We wish to determine the intensity of light on the viewing screen located a distance r_0 from the straight edge. At each position y on the screen, the intensity can be obtained by constructing a sys-

---

*Rewrite Eq. 43–29 as

$$E(P) = \frac{e_1}{2} + \left(\frac{e_1}{2} - e_2 + \frac{e_3}{2}\right) + \left(\frac{e_3}{2} - e_4 + \frac{e_5}{2}\right) + \cdots$$

To a good approximation, $e_2 = \frac{1}{2}(e_1 + e_3)$ and so forth, so that each term in brackets separately vanishes. This leads to Eq. 43–30.

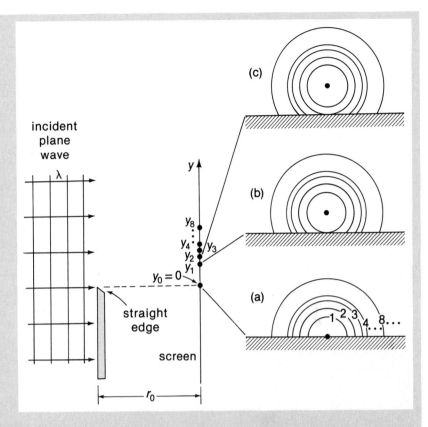

Fig. 43-17. Geometry for understanding the diffraction of light by an obstacle with a straight edge by using a Fresnel-zone construction.

tem of Fresnel zones in a plane that contains the straight edge and is parallel to the incident plane wave and the viewing screen. In each case, the center of the zone pattern is located on a line drawn perpendicular to the screen through the viewing point y (as in Fig. 43-16).

Consider first the point $y_0 = 0$ that is directly opposite the straight edge. From this position exactly one half of the Fresnel zone pattern is seen; the other half is obscured by the obstacle (Fig. 43-17a). Thus, the electric field at y_0 is just one half the field due to the total zone system, so the intensity is one quarter of that without the obstacle; that is, $I(y_0) = \frac{1}{4}I_0$.

As the observation point is moved up the screen, the zone system is also moved up. Then, more and more of the first zone contributes to the intensity. Finally, when the point $y_1 = R_1 = \sqrt{r_0\lambda}$ is reached, the entire first zone is effective (Fig. 43-17b). A further movement of the observation point up the screen allows the second zone to increase its contribution more rapidly than the other zones. Because the phase of this zone differs by π from that of the first zone, the intensity decreases between the points y_1 and y_2. In fact, the intensity is *maximum* at y_1 and *minimum* at y_2. Moreover, the intensity continues to oscillate (with decreasing amplitude) as y increases. For very large y, almost the entire zone pattern contributes, so that $\text{Lim}_{n \to \infty} I(y_n) = I_0$, and the screen is fully illuminated far

from the straight edge. Within the normal shadow region ($y < 0$), the intensity decreases monotonically (that is, without oscillations) from $\frac{1}{4}I_0$ at y_0, and the region rapidly becomes dark. This behavior of the intensity as a function of distance along the screen is illustrated in Fig. 43-18, where the independent variable is the dimensionless parameter $u = y\sqrt{2/r_0\lambda} = \sqrt{2}\,y/R_1$.

The Fresnel-zone method for analyzing diffraction effects is only approximate. (For example, we have assumed that the phase of the radiation from each zone is either 0 or π, whereas in fact the phase changes continuously.) Hence, we should not expect to obtain precise results for calculated quantities, such as the positions of the diffraction maxima and minima. Consider a case for which $\lambda = 550\,\text{nm}$ and $r_0 = 50\,\text{cm}$; then, $R_1 = \sqrt{r_0\lambda} = 0.524\,\text{mm}$. The values of the first two maxima and the intervening minimum are:

First maximum:
$$y_1 = R_1 \qquad\qquad = 0.524\,\text{mm} \ (0.452\,\text{mm})$$
First minimum:
$$y_2 = R_2 = \sqrt{2}R_1 = 0.741\,\text{mm} \ (0.697\,\text{mm})$$
Second maximum:
$$y_3 = R_3 = \sqrt{3}R_1 = 0.908\,\text{mm} \ (0.874\,\text{mm})$$

Fig. 43–18. The intensity of light diffracted by a straight edge. The normal shadow region ($u < 0$) is weakly illuminated near the straight edge.

(From *Atlas of Optical Phenomena* by M. Cagnet, M. Francon, and J. C. Thrierr. Springer-Verlag, 1962.)

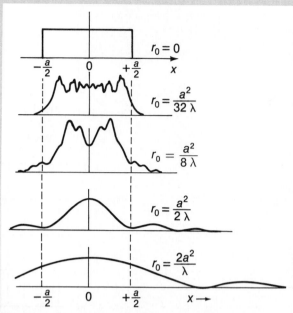

Fig. 43–19. Diffraction of a plane wave by a slit with width a for various slit-to-screen distances r_0.

The approximate values become closer to the actual values (given in parentheses) as n increases.

Figure 43–19 shows the diffraction pattern produced by a plane wave incident on a slit with width a and infinite length. The various patterns correspond to increasing slit-to-screen distances. Notice that as the ratio of the screen distance r_0 to the slit width a increases, the Fresnel pattern is transformed into a Fraunhofer pattern.

PROBLEMS*

Section 43–1

43–1   A plane wave with wavelength $\lambda = 500$ nm illuminates a narrow slit, and the diffracted light is focused on a screen by a lens with focal length $f_2 = 1.5$ m, as in Fig. 43–2. The first secondary maximum is located 4 mm from the central maximum. (a) What is the width of the slit? (b) Locate the approximate positions of the maxima with $m = 2$ and $m = 3$.

43–2   A plane wave with wavelength $\lambda = 650$ nm is incident on a narrow slit. The diffracted light has its first minimum at $\theta = \pi/2$ and thus illuminates the entire space beyond the slit. (a) Determine the width of the slit. (b) At what angle θ is the intensity equal to one half the maximum intensity?

43–3   Prove the result stated in Eq. 43–4 by solving the
•   equation $\sin \alpha = \alpha/\sqrt{2}$. Use an iterative procedure on an electronic calculator to obtain $\alpha = 1.39156$ and $\Delta\alpha = 2\alpha = 2.783$ rad.

43–4   What is the smallest angular separation between two stars that can be resolved by the Hale telescope

----

*In all grating problems, assume a Fraunhofer geometry (Fig. 43–7).

at Mount Palomar (diam. = 5.08 m), if the limitation is set by diffraction effects? Assume that the observation is made in yellow-green light ($\lambda = 550$ nm), and give your answer in seconds of arc.

Section 43–2

43–5 Supply the steps necessary to convert Eq. 43–9 into Eq. 43–10. [*Hint:* Use Eq. B–12 in Appendix B.]

43–6 A 5-cm diffraction grating has a total of 15,000 grooved lines, each with a width of 1.0 μm. The grating is illuminated with light that has a wavelength of 480 nm. (a) Determine the angles θ_m that correspond to the principal maxima of the first three orders. (b) How many complete orders are present in the entire space beyond the grating? (c) How many complete orders are present in the range $-\pi \leqslant \alpha \leqslant \pi$?

43–7 Young's interference experiment is performed (using Fraunhofer geometry) with two slits of width 0.20 mm, whose centers are separated by 0.80 mm. How many maxima (including the central maximum) lie in the region $|\alpha| \leqslant \pi$? Explain why these are the lines of high visibility.

43–8 A certain grating produces a first-order maximum
• with an intensity equal to one half that of the central maximum. Determine the intensities of the maxima of orders 2, 3, 4, and 5.

43–9 A 3-cm diffraction grating with 250 grooves per millimeter is used in third order. Monochromatic light produces a strong maximum at $\theta = 18.0°$. (a) What is the wavelength of the light? (b) A sec-

ond source of monochromatic light is now added. What is the difference in wavelength between the two radiations if they are just resolved in third order?

43–10 A diffraction grating with 100 grooved lines per millimeter is used to observe the yellow doublet emitted by sodium vapor ($\lambda_1 = 5895.93$ Å and $\lambda_2 = 5889.96$ Å). What is the minimum length of the grating if these spectral lines are to be resolved in first order?

Section 43–3

43–11 The highest-quality diffraction gratings have about 25,000 grooves/cm. Suppose that such a grating were used in the standard transmission mode (Fig. 43–7) to analyze X rays with a wavelength of 0.50 Å. At what angle θ would the first-order maximum appear? Comment on your result. (Neglect the practical question of whether a suitably thin grating could be constructed to transmit the X rays.)

43–12 X rays with $\lambda = 1.35$ Å are diffracted by a crystal with cubic lattice structure and exhibit a strong second-order maximum ($n = 0$, $m = 2$) at $\theta = 35.0°$. (a) What is the crystal lattice spacing d? (b) Determine the angles θ_m for all of the Bragg orders that are possible (for $n = 0$).

43–13 An X-ray diffraction experiment is performed in the geometry shown in Example 43–6, with the sample-to-film distance equal to 50 cm. The lattice spacing of the sample crystalline material is 2.20 Å. The third-order ring ($n = 0$, $m = 3$) has a radius of 8.0 cm. What is the wavelength of the X radiation?

SPECIAL TOPIC

Section 43–4

43–14 Light in the form of plane waves from a sodium vapor source is incident normally on a small circular aperture in an opaque barrier. At a point 1.0 m from the aperture and on its axis, exactly the first two zones of the Fresnel zone system are visible. What is the radius of the aperture? What is the intensity of the radiation at this observing point?

43–15 Monochromatic plane waves are incident normally on a circular aperture that has a diameter of 3.0 mm. As the viewing screen is moved away from the aperture, the central spot is observed to alter-

nate between being bright and being dark. When an aperture-to-screen distance of 2.0 m is reached, the central spot is dark. As the screen is moved farther from the aperture, the central spot becomes and remains bright. What is the wavelength of the light?

43–16 Figure 43–19 suggests that for distances r_0 that are greater than $a^2/2\lambda$ the central intensity of the diffraction pattern for a slit of width a remains a dominant maximum. Calculate the radii of the first two Fresnel zones for $r_0 = a^2/2\lambda$. Sketch the zone situation for $r_0 = a^2/2\lambda$ and for $r_0 \gg a^2/2\lambda$. Does this verify the suggested conclusion?

Additional Problems

43–17 Plane waves of light with $\lambda = 600$ nm illuminate a single slit. The diffraction pattern is projected onto a screen by means of a lens with focal length $f_2 = 1.50$ m, as in Fig. 43–3. The separation on the screen between adjacent minima is 4.0 mm. Determine the width of the slit.

43–18 Calculate the angular width $\Delta\theta$ at half maximum intensity for the central maximum of a Fraunhofer single-slit pattern for the cases $a = \lambda$, $a = 2\lambda$, and $a = 5\lambda$.

43–19 What is the smallest diameter of a telescope lens that would permit resolving two pointlike objects separated by 1 km on the central visible surface of the Moon, assuming that the limitation is set by diffraction? The objects are viewed in yellow-green light ($\lambda = 550$ nm). (The Earth-Moon distance is 3.84×10^5 km.)

43–20 According to Eq. 43–3a, the first secondary maximum
 • ($m = 1$) occurs for $\alpha = 3\pi/2$. Use Eq. 43–2 to show that this maximum actually occurs for $\alpha = 0.954 \times 3\pi/2$. (Compare Table 43–1.) [*Hint:* Set $dI/d\alpha = 0$. Make a sketch of the resulting expression to determine a reasonable starting point for a successive approximation calculation.]

43–21 Prove Eq. 43–14 (a) by using l'Hôpital's rule of
 •

calculus and (b) by making a series expansion of $\sin N\beta$.

43–22 Show that Eq. 43–19 for the case of Young's double-slit experiment ($N = 2$) can be expressed as (Eq. 43–19a)

$$I = I_0 \cos^2\beta \left(\frac{\sin \alpha}{\alpha}\right)^2$$

43–23 Consider a single slit of width a to be equivalent to a diffraction grating with N grooves and a grating spacing $b = a/N$. Show that the expression for the intensity due to a diffraction grating (Eq. 43–12) reduces to that for a single slit (Eq. 43–2) in the limit $N \to \infty$.

43–24 Consider a diffraction grating with $N = 9$ (see Fig. 43–11). What value of b/a will cause the fourth-order secondary maximum to occur at the position of the first-order secondary maximum of the diffraction term? (This change will enhance the intensity of the $m = 4$ maximum compared with that shown in Fig. 43–11.)

43–25 Derive the expression (Eq. 43–25) for λ_{nm} in terms of the angles ϕ and ψ.

43–26 Monochromatic X rays with $\lambda = 1.50$ Å are incident
 • on a crystal of rock salt ($d = 2.82$ Å). What are the angles θ and ϕ that correspond to the strong reflection in the mode with $n = 1$ and $m = 2$?

SPECIAL TOPIC

43–27 When a central hole and a sequence of annular rings that correspond to the odd-numbered Fresnel zones (for a particular r_0 and λ) are cut out of an opaque barrier, the result is called a *Fresnel zone plate*. Radiation with wavelength λ that is incident on such a zone plate is strongly focused at a point that is a distance r_0 from the plate. (•Can you explain this effect?) Microwaves with $\lambda = 5$ cm are to be focused at a point 50 cm from a zone plate. Calculate the radii R_n of the first six zones of the required plate.

43–28 Light with wavelength λ and intensity I_0 is incident normally on a circular aperture in an opaque barrier. The radius of the aperture is $\sqrt{r_0\lambda}$, where r_0 is the distance from the center of the aperture to the observing point P. What is the intensity of the light at P? The radius of the aperture is increased to $\sqrt{2r_0\lambda}$. What is the new intensity at P?

POLARIZED LIGHT

A number of interesting and important phenomena involving the transmission and reflection of light are sensitive to the state of polarization of the incident beam. Almost all ordinary light sources—such as the Sun, flames, and incandescent bulbs—produce unpolarized light. However, when such light is incident on certain substances, the randomly oriented polarization components in the incident beam interact with the medium in characteristically different ways. In this chapter we first discuss carefully the meaning of the various states of polarization; then, we treat a number of phenomena that involve polarized light.

44-1 STATES OF POLARIZATION

We have previously defined the state of polarization of an electromagnetic wave in terms of the direction along which the electric field vector oscillates (see Section 38–1). An electromagnetic wave is said to be *linearly polarized* if its electric vector $\mathbf{E}(\mathbf{r},t)$ always oscillates parallel to a fixed direction in space.* For example, a wave whose electric vector is expressed as

$$\mathbf{E}(\mathbf{r},t) = E_0 \cos (kz - \omega t)\hat{\mathbf{i}} \qquad \textbf{(44–1a)}$$

is a plane wave that is linearly polarized in the x-direction. Figure 44–1a shows a portion of this wave at the instant $t = 0$ (compare Fig. 38–3). The electric vector

---

*We use here the term *linearly polarized* wave, but elsewhere you may find these waves called plane-polarized waves; the terms are synonymous.

Fig. 44–1. (a) A linearly polarized electromagnetic wave. (b) The direction of linear polarization is indicated by a line segment with double arrowheads.

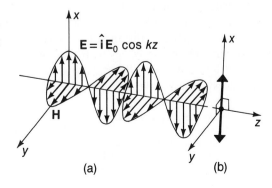

(a)   (b)

vibrates parallel to the x-axis with values between $+E_0$ and $-E_0$. It is customary to indicate this vibrational character in diagrams by means of a short line segment with double arrowheads as in Fig. 44–1b. The direction of the line segment corresponds to the direction of linear polarization.

The wave described by Eq. 44–1a is polarized in the x-direction and propagates in the z-direction. Because the direction of propagation corresponds to the direction of $\mathbf{E} \times \mathbf{H}$, it follows that the magnetic field vector is in the y-direction:

$$\mathbf{H}(\mathbf{r},t) = H_0 \cos(kz - \omega t)\widehat{\mathbf{j}} \qquad \text{(44–1b)}$$

Because the magnetic vector of a plane wave can always be deduced from a knowledge of the electric vector $\mathbf{E}(\mathbf{r},t)$ and the propagation vector $\mathbf{k}$, we usually do not specifically exhibit $\mathbf{H}(\mathbf{r},t)$. In any event, the interaction of light with matter is governed almost exclusively by the electric field of the wave. (See the comment in the footnote on page 1121.)

Suppose that a plane wave has an electric field

$$\mathbf{E} = E_x\widehat{\mathbf{i}} + E_y\widehat{\mathbf{j}} \qquad \text{(44–2)}$$

with components
$$\left. \begin{array}{l} E_x = E_{0x} \cos(kz - \omega t) \\ E_y = E_{0y} \cos(kz - \omega t + \alpha) \end{array} \right\} \qquad \text{(44–3)}$$

The nature of the polarization is determined by the amplitude ratio E_{0y}/E_{0x} and by the relative phase angle α. If $\alpha = 0$, Eqs. 44–3 describe a linearly polarized wave with the electric vector in the x-y plane and making an angle θ with respect to the x-axis, where

$$\theta = \tan^{-1}(E_{0y}/E_{0x}) \qquad \text{(44–4)}$$

Of particular interest is the case in which the components have equal amplitudes, $E_{0x} = E_{0y} = E_0$, and the phase is $\alpha = +\pi/2$ or $-\pi/2$. Figure 44–2 illustrates the propagation of $\mathbf{E}(\mathbf{r},t)$ for $\alpha = -\pi/2$. As the wave travels along the z-axis, $\mathbf{E}$ rotates at a uniform rate around the z-axis. To an observer looking toward the source of the wave, the direction of rotation is counterclockwise. As shown in Fig. 44–2, the tip of the electric field vector $\mathbf{E}$ moves on the surface of a circular cylinder and has the constant magnitude $(E_{0x}^2 + E_{0y}^2)^{1/2}$. Such a wave is said to be *circularly polarized*.*

In general, if α is not an integer multiple of $\pi/2$, then the magnitude of $\mathbf{E}$ varies smoothly between a maximum and a minimum value twice during each rotation; this results in *elliptical polarization*.

If an electric dipole antenna is driven by a sinusoidal current, the resulting radiation at a large distance in a certain direction from the antenna consists of approximately plane waves with a definite polarization (see Section 39–2). For the case of light, however, the radiating sources are atoms, and each atomic transition produces a single wave train with finite length. Thus, an ordinary light beam consists

---

*This counterclockwise rotation of the electric vector characterizes a wave with *left circular polarization*. With $\alpha = +\pi/2$, the sense of rotation is reversed and *right circular polarization* results. These rotations are rotations of the electric vector in space as the wave advances. Note, however, that at a fixed location, the time development of $\mathbf{E}(\mathbf{r},\ t)$ for $\alpha = -\pi/2$ (the case illustrated in Fig. 44–2) results in a *clockwise* rotation of the electric vector.

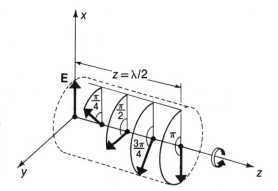

Fig. 44-2. A wave is circularly polarized if the electric vector has constant magnitude and rotates at a uniform rate around the propagation axis. The wave illustrated has *left* circular polarization.

of a very large number of uncorrelated wave trains, each arising from an individual atomic transition. Each of these wave trains has a specific linear polarization, but the polarization directions are randomly distributed in a plane that is perpendicular to the direction of propagation of the beam. Light that consists of wave trains with random polarizations is said to be *unpolarized* light. We now examine some of the characteristics of such light.

Figure 44-3 shows three different representations of a number of approaching wave trains that are randomly polarized. In (a) we see, at a particular instant and at a particular point in space, the randomly distributed electric vectors whose instantaneous sum is the resultant vector $\mathbf{E}_R$. Figure 44-3b shows the motion of the tip of $\mathbf{E}_R$ as a function of time at the same point in space. As a practical matter, the detection of light at this point involves an instrument (an electronic detector or the eye) that has some finite response time τ. That is, the detector is not capable of responding to the instantaneous value of $\mathbf{E}_R$ and gives a signal that is proportional to the average of $\mathbf{E}_R$ taken over a time τ. The period of a light wave is $\sim 2 \times 10^{-15}$ s, and the best detectors have response times $\sim 10^{-12}$ s. In most practical situations the output signal of the detector will be an average over very many more vector excursions than shown in Fig. 44-3b.

The energy deposited in the detector is proportional to

$$\overline{E_R^2} = \overline{E_{Rx}^2} + \overline{E_{Ry}^2} \tag{44-5}$$

where E_{Rx} and E_{Ry} are the instantaneous components of $\mathbf{E}_R$, and where the averages are taken over the time τ. The *average behavior* of E_{Rx} is indistinguishable from that of E_{Ry} or from that of a component of $\mathbf{E}_R$ in any other direction in the x-y plane. (•Can you see why this is the case?) Thus, if $\overline{E_R^2}$ corresponds to an energy flux I, then $\overline{E_{Rx}^2} = \overline{E_{Ry}^2}$ each correspond to an energy flux $\frac{1}{2}I$.

It is convenient to represent the average x- and y-vibrations as in Fig. 44-3c. However, these average vibrations are *not* coherent components of a single electric field vector.

It is important to note that for an ideal plane wave (a wave train of infinite length) whose direction of polarization makes equal angles of 45° with respect to the x- and y-axes of the chosen coordinate frame, we also have $E_x^2 = E_y^2 = \frac{1}{2}E^2$. Hence, each component of the wave transports an energy flux equal to $\frac{1}{2}I$. In this case, however, the two vibrations are components of a single vector and are therefore coherent. The time averages that are required to obtain the energy fluxes are taken over one period of the wave oscillation; hence, the values that result are independent of the detector response time τ. (•If the x- and y-axes were rotated about the z-axis,

Fig. 44-3. (a) The random distribution of electric vectors of a number of uncorrelated wave trains. The resultant vector is $\mathbf{E}_R$. (b) The time variation of the tip of $\mathbf{E}_R$. (c) Average vibrations corresponding to (b) resolved into x- and y-components.

(a)

(b)

(c)

Fig. 44-4. Instantaneous values of the electric vector at a particular location for a linearly polarized beam that consists of many uncorrelated, overlapping wave trains. Any detector will average over a time τ that includes many of these rapid variations and will produce a steady output signal.

would the energy flux carried by each component remain equal to $\frac{1}{2}I$? Contrast this conclusion with that for the unpolarized case.)

A polarized beam of light can be produced from a natural (unpolarized) beam in several ways, as discussed in the following sections. An incident unpolarized beam might be spatially separated into two beams, each possessing one of the vibrational polarizations shown in Fig. 44–3c. Or, one polarization component might be completely absorbed in a substance, thereby producing a single beam with a polarization in the perpendicular direction. In each case, the linearly polarized beam or beams still consist of a large number of uncorrelated and overlapping wave trains, so that the electric vector has a time variation such as that shown in Fig. 44–4.

If a polarizing device removes unequal amounts of energy flux from the two polarization components, the beam becomes *partially polarized*. For example, the partial polarization in the y-direction (which means that $I_y > I_x$) is expressed as

$$(PP)_y = \frac{I_y - I_x}{I_y + I_x} \tag{44-6}$$

Notice that this expression reduces to that expected for the limiting cases, $I_x = 0$ and $I_x = I_y$.

When an unpolarized beam is separated into two polarized beams, the individual wave trains have true coherent components in the two beams. Nevertheless, the two beams cannot produce interference effects because the electric vectors are mutually perpendicular. This is in contrast to the situation for longitudinal waves. If a longitudinal wave is separated by any means into two beams, the two beams will produce interference effects when recombined. The discovery that two separated polarized beams cannot be made to interfere constituted the first direct experimental proof of Thomas Young's assertion that light is a transverse wave phenomenon.*

44-2 POLARIZATION BY REFLECTION AND SCATTERING

Practical means for producing polarized light include a variety of techniques that involve
    (a) reflection and scattering,
    (b) absorption, and
    (c) double refraction.
In this section we consider polarization by the first of these methods.

Reflection. In Section 40–4 we learned that when a light wave is incident on the interface between two media at an angle equal to Brewster's angle θ_B, the amplitude reflection coefficient for the component with polarization parallel to the plane of incidence is zero, so

$$\tan \theta_B = \frac{n_2}{n_1} \qquad \text{Brewster's angle} \tag{44-7}$$

*The experiments were carried out jointly by Fresnel and François Arago (1786–1853) in 1817. Fresnel understood that the results implied strong support for the transverse wave theory of light (which, at the time, was controversial). Arago did not wish to take such a bold step, so the interpretation of the experimental results was published in Fresnel's name alone.

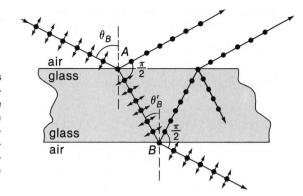

Fig. 44–5. An unpolarized wave is incident on a glass plate at Brewster's angle θ_B. The reflected wave is linearly polarized in the direction perpendicular to the plane of incidence. Notice that the double arrowheads indicate the parallel component and that the dots indicate the perpendicular component.

where the light is incident on medium 2 from medium 1. When light is incident on glass ($n_2 = 1.50$) from air ($n_1 = 1$), Brewster's angle is 56.3°. When light is incident on an air-glass interface from the glass side, we find $\theta'_B = 33.7°$.

The existence of Brewster's angle provides a practical means for producing polarized light. Figure 44–5 shows the geometry of the incident, reflected, and transmitted waves when the angle of incidence is equal to Brewster's angle. Because none of the parallel component is reflected at point A, the reflected wave is completely polarized in the perpendicular direction and the transmitted wave is partially polarized in the parallel direction.* If the glass plate has parallel surfaces, an incident angle equal to Brewster's angle θ_B at the air-glass interface guarantees that the transmitted wave is incident on the glass-air interface at Brewster's angle θ'_B appropriate for that interface (see Problem 40–30). Consequently, only the perpendicular component is reflected at B, and the transmitted wave becomes further depleted in the perpendicular component. By using a stack of glass plates, as in Fig. 44–6, this effect is enhanced so that the transmitted wave is almost completely polarized in the parallel direction. The reflected wave is completely polarized in the perpendicular direction. Even if the incident angle is not equal to Brewster's angle, each beam will have a substantial partial polarization if the incident angle is not close to 0° or 90°.

---

*This effect for the reflected beam was discovered in 1808 by the French engineer and physicist Étienne Louis Malus (1775–1812). The efficiency of polarization by this method is not high because only about 15 percent of the perpendicular component in the incident beam is actually reflected. (See Fig. 40–19 and Problem 44–16.)

Fig. 44–6. An unpolarized incident wave can be separated into two polarized waves by a stack of glass plates. The reflected wave is completely polarized in the perpendicular direction. Each reflection depletes the perpendicular component of the transmitted wave so that the emergent wave is almost completely polarized in the parallel direction.

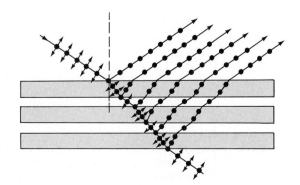

Example 44–1

Sunlight is reflected from a pool of water standing in a roadway. An observer finds that the reflected sunlight is completely polarized. What is the angle of the Sun above the horizon?

Solution:

The refractive index for water is $n_2 = 1.33$, so Brewster's angle is

$$\theta_B = \tan^{-1}\frac{n_2}{n_1} = \tan^{-1} 1.33 = 53.1°$$

Then, the angle of the Sun above the horizon is

$$\theta_S = 90° - 53.1° = 36.9°$$

The observer would need some type of polarization-sensitive device to determine that the reflected light is polarized because human eyes (except for those of some exceptional persons) do not distinguish between polarized and unpolarized light. It has been demonstrated, however, that some types of ants and beetles, and particularly the honeybee, can sense polarization.

Scattering. If a narrow beam of natural light is passed through air, water, or some other convenient transparent medium that contains a suspension of ultramicroscopic particles, the scattered light is found to be partially polarized, with the degree of polarization depending on the angle of scattering. Figure 44–7 shows an unpolarized beam that is directed along the z-axis. The atomic electrons in the scat-

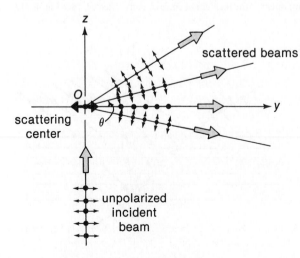

Fig. 44–7. Unpolarized incident light that is scattered through 90° is linearly polarized in the direction perpendicular to the plane of incidence.

tering center (particle) located at O are set into oscillation in response to the electric field of the incident wave. Because the incident beam is unpolarized, the electrons oscillate with a uniform distribution in the x-y plane. Now, according to the result obtained in Section 39–2, an oscillating dipole does not radiate along its direction of motion. Thus, the radiation emitted along the y-axis in Fig. 44–7 is due entirely to electron oscillations parallel to the x-axis and is therefore linearly polarized in the x-direction.* In fact, all of the radiation emitted in the x-y plane is polarized perpendicular to the plane of incidence associated with the particular direction of radiation. In other directions, the scattered light is partially polarized in accordance with the $\sin^2\theta$ distribution of an oscillating electric dipole.

Polarization by scattering occurs also for sunlight scattered by air molecules. The maximum effect is observed on a clear day, when there is little water vapor or dust in the air, and when the Sun is near the horizon (Fig. 44–8). Then, the light reaching an observer on the ground from directly overhead can be polarized to the extent of 70 or 80 percent.

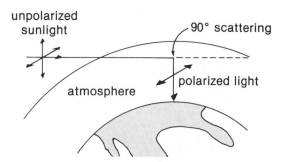

Fig. 44–8. Unpolarized sunlight becomes almost completely polarized upon scattering through an angle of 90° by air molecules.

44-3 POLARIZATION BY SELECTIVE ABSORPTION—DICHROISM

In 1815 Biot discovered that certain mineral crystals, such as *tourmaline*, exhibit selective absorption of light depending on its state of polarization. These crystals (called *dichroic crystals*) are anisotropic; they strongly absorb light that is polarized in a direction parallel to a particular plane in the crystal, but freely transmit light that is polarized in the perpendicular direction. Thus, when an incident wave of unpolarized light is passed through a dichroic crystal, one of the polarization components is absorbed and the emergent wave is highly polarized in the perpendicular direction (Fig. 44–9).

The most common type of dichroic polarizer in use is Polaroid H-sheet, invented by E. H. Land† in 1938. A plastic sheet of long-chain molecules of polyvinyl alcohol is heated and then stretched to many times its original length. During this process, the molecules become aligned along the direction of stretching. The sheet is then cemented to a rigid piece of plastic to stabilize its size, and it is next dyed with iodine. The iodine atoms attach themselves to the polyvinyl alcohol molecules and in effect become long parallel conducting chains. When unpolarized light is incident on a piece of H-sheet, the polarization component that is parallel to the chains of iodine atoms induces currents in the conducting chains and is therefore strongly absorbed.

---

*This is the same reasoning that we used in Section 40–4 to explain the polarizing effect at Brewster's angle.

†Edwin H. Land (1909–   ), who formed the Polaroid Corporation to manufacture and distribute polarizing materials, later developed the popular "picture-in-a-minute" system of photography.

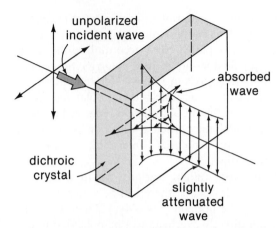

Fig. 44-9. Polarization by selective absorption of one polarization component in a dichroic crystal.

Consequently, the transmitted beam is polarized in the direction perpendicular to the direction of the molecules. The direction of polarization of the transmitted beam is called the *polarization axis* or *transmission axis* of the material.

The selective absorption of the different polarization components of a wave occurs for all types of electromagnetic radiation. Figure 44-10 shows an experiment

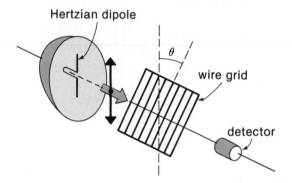

Fig. 44-10. A microwave experiment to demonstrate the selective absorption of polarized radiation.

to demonstrate selective absorption in the microwave region of the spectrum. A Hertzian electric dipole antenna (see Section 39-2) is placed at the focal point of a parabolic reflector so that it generates an essentially plane wave of microwave radiation, which is polarized in the direction parallel to the antenna axis. A conducting grid composed of parallel fine wires, spaced apart by a quarter of a wavelength or so, is placed between the antenna and a detector that is insensitive to polarization. The detector output signal is found to depend on the orientation of the wire grid with respect to the antenna axis. With the grid parallel to the antenna axis, $\theta = 0°$, the transmitted wave is almost completely eliminated, whereas with the grid perpendicular to the antenna axis, $\theta = 90°$, the wave is transmitted with only slight attenuation. The wire grid acts as a receiving antenna. The electric currents that are induced to flow in the wires result in the absorption of wave energy by Joule heating. The energy loss is very small for $\theta = 90°$ and is quite large for $\theta = 0°$. Thus, the transmission axis of the wire grid is perpendicular to the direction of the wires.

44-4 THE LAW OF MALUS

Suppose that a beam of unpolarized light with intensity I_0 is passed through a piece of Polaroid H-sheet (called a *polarizer*). The emerging beam is then linearly polarized in the direction of the polarizing axis of the H-sheet and carries an energy flux $I_1 = \frac{1}{2}I_0$, as indicated in Fig. 44-11. The beam next passes through a second

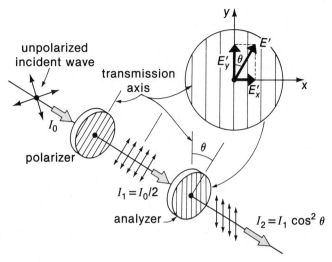

Fig. 44-11. Illustrating the law of Malus.

piece of H-sheet (called an *analyzer*) whose axis is at an angle θ with respect to the polarization direction of the entering beam. What is the intensity of the light transmitted through the analyzer? The answer to this question is readily deduced. The inset in Fig. 44–11 shows the electric vector $\mathbf{E}'$ of one of the wave trains in the beam entering the analyzer. The component of $\mathbf{E}'$ that is perpendicular to the transmission axis, namely, $E'_x = E' \sin \theta$, is extinguished, whereas the parallel component, $E'_y = E' \cos \theta$, is transmitted with only a slight attenuation. The associated energy flux is proportional to E'^2_y, so the total transmitted flux is given by the average over all such wave trains. The result is

$$I_2 = I_1 \cos^2 \theta \qquad \text{(44–8)}$$

This is the *law of Malus*, discovered in 1809.

(•Can you see how Polaroid sunglasses reduce glare? Remember, reflected sunlight is partially polarized. What should be the direction of the transmission axis in the sunglasses?)

▶ Ideally, the transmittance T_d for the *desired* component should be unity and the transmittance T_u for the *undesired* component should be zero. For H-sheet neither of these ideal values is realized. In fact, both T_d and T_u vary somewhat with wavelength. Averaged over the visible spectrum, the transmittance values for one type of H-sheet (not including reflective losses at the surfaces) are $\bar{T}_d \cong 0.83$ and $\bar{T}_u \cong 0.001$. By adding iodine and thereby reducing $\bar{T}_d$ to about 0.5, the value of $\bar{T}_u$ can be reduced to about 10^{-6}. Thus, the transmitted beam is reduced to about half the ideal intensity but the polarization is essentially complete. Some double-refracting crystals such as calcite (see Section 44–5) produce beams with $\bar{T}_u \lesssim 10^{-5}$. ◀

Example 44–2

A partially polarized beam of light has intensities I_x and I_y for the x- and y-components of the polarization, with $I_y > I_x$. The beam is passed through a piece of H-sheet, and the transmitted intensity $I(\theta)$ is measured as a function of the angle θ between the y-axis and the transmission axis. Show that Eq. 44–6 for the partial polarization in the y-direction is equivalent to the expression,

$$(PP)_y = \frac{I_{max} - I_{min}}{I_{max} + I_{min}}$$

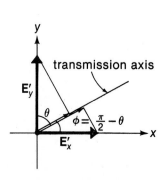

Solution:

The diagram shows the electric vectors, $\mathbf{E}'_x$ and $\mathbf{E}'_y$, corresponding to two randomly selected wave trains in the beam, one of which is polarized in the x-direction and the other in the y-direction. Each has a projection along the transmission axis that leads to an average intensity in accordance with the law of Malus. That is,

$$I(\theta) = I_y \cos^2 \theta + I_x \cos^2 \left(\frac{\pi}{2} - \theta \right) \tag{1}$$

$$= I_y \cos^2 \theta + I_x \sin^2 \theta$$

Because $I_y > I_x$, it follows that making the substitution $I_x \to I_y$ in Eq. (1) gives the inequality

$$I(\theta) \leq I_y \cdot (\cos^2 \theta + \sin^2 \theta) = I_y$$

Similarly, making the substitution $I_y \to I_x$ gives the inequality

$$I(\theta) \geq I_x$$

Thus, we conclude that $I_{max} = I_y$ and $I_{min} = I_x$, so that Eq. 44-6 becomes

$$(PP)_y = \frac{I_{max} - I_{min}}{I_{max} + I_{min}}$$

which is the desired result.

Example 44–3

An electromagnetic plane wave propagates in the z-direction. In a particular coordinate frame, the wave has $E_{x,\text{rms}} = (\overline{E_x^2})^{1/2} = 2$ V/m and $E_{y,\text{rms}} = (\overline{E_y^2})^{1/2} = 3$ V/m. This wave is passed through an H-sheet analyzer. The transmission axis of the analyzer makes an angle ϕ with respect to the x-axis.
Assume that the wave is linearly polarized.
(a) What is the energy flux in the incident wave?
(b) If the analyzer is rotated around the z-axis, what is the maximum transmitted intensity and for what angle ϕ does this occur?
Assume that the wave is only partially polarized.
(c) What is the energy flux in the incident wave?
(d) What is the degree of partial polarization?
(e) If the analyzer is rotated around the z-axis, what is the maximum transmitted intensity and for what angle ϕ does this occur?

Solution:

(a) For a linearly polarized plane wave, $E_{x,\text{rms}}$ and $E_{y,\text{rms}}$ are the true coherent components of the electric vector $\mathbf{E}_{\text{rms}}$, with $E_{\text{rms}} = (\overline{E_x^2} + \overline{E_y^2})^{1/2} = (2^2 + 3^2)^{1/2} = \sqrt{13}$ V/m. The energy flux in the incident wave is

$$I_0 = E_{\text{rms}} H_{\text{rms}} = \frac{E_{\text{rms}}^2}{Z_0}$$

$$= \frac{13 (\text{V/m})^2}{377 \, \Omega} = 34.5 \, \text{mW/m}^2$$

(b) The vibration axis of $\mathbf{E}_{\text{rms}}$ makes an angle ψ with respect to the x-axis, where

$$\psi = \tan^{-1}\frac{(\overline{E_y^2})^{1/2}}{(\overline{E_x^2})^{1/2}} = \tan^{-1}\tfrac{3}{2} = 56.3°$$

Maximum transmission results when $\phi = \psi$ (or when $\phi = \psi + \pi$). Then, the transmitted energy flux is equal to the incident flux, namely, $34.5\,\mathrm{mW/m^2}$.

(c) For a partially polarized plane wave, the intensities of the x- and the y-components must be added incoherently. Thus, the energy flux in the incident wave is

$$I_0' = \frac{E_{x,\mathrm{rms}}^2}{Z_0} + \frac{E_{y,\mathrm{rms}}^2}{Z_0} = \frac{4(\mathrm{V/m})^2 + 9(\mathrm{V/m})^2}{377\,\Omega}$$
$$= 34.5\,\mathrm{mW/m^2}$$

which is the same as the result obtained in (a): $I_0' = I_0$.

(d) The degree of partial polarization is given by Eq. 44–6; thus,

$$(PP)_y = \frac{E_{y,\mathrm{rms}}^2 - E_{x,\mathrm{rms}}^2}{E_{y,\mathrm{rms}}^2 + E_{x,\mathrm{rms}}^2} = \frac{9 - 4}{9 + 4} = 0.385$$

(e) As the analyzer angle ϕ is changed, each of the components separately obeys the law of Malus; thus (compare Example 44–2),

$$I'(\phi) = I_y'\cos^2\theta + I_x'\cos^2\phi$$
$$= I_y'\sin^2\phi + I_x'\cos^2\phi$$

Because $I_y' > I_x'$, the maximum intensity results when $\phi = \pi/2$. Then, the transmission axis is aligned with the polarization axis (the y-axis) of the stronger component of the wave, thereby allowing this component to pass through unattenuated. Hence,

$$I_{\mathrm{max}}' = \frac{E_{y,\mathrm{rms}}^2}{Z_0} = \frac{3^2\,(\mathrm{V/m})^2}{377\,\Omega} = 23.9\,\mathrm{mW/m^2}$$

44–5 BIREFRINGENCE

In substances such as air, water, and glass (which are *isotropic* materials), light waves propagate at the same speed in all directions regardless of the state of polarization. There are, however, a number of crystalline substances in which light, traveling in a particular direction, will propagate with one speed for one polarization component and with a different speed for the perpendicular component. The best known crystal of this type is *calcite*; the effect also occurs in quartz, tourmaline, and dolomite (see Table 44–1). Because light propagates with two different speeds in these crystals, they possess two distinct refractive indexes. When a light beam is incident on such a crystal, two refracted beams result. This phenomenon is called *double refraction* or *birefringence*. When an object is viewed through a birefringent crystal, two images are observed, one slightly displaced from the other,* as illustrated in the photograph on the next page.

The description of the anisotropy of birefringent crystals is assisted by distinguishing a unique direction in the crystal structures called the *optic axis*. Light

---

*This observation of double refraction was first made in 1669 by a Danish scientist, Erasmus Bartholinus (1625–1698). The phenomenon was also studied by Huygens and Newton.

TABLE 44–1.   Refractive Indexes for Several Uniaxial Crystalline
Substances ($\lambda = 589$ nm)

CRYSTAL	COMPOSITION	n_o	n_e
Ice	H_2O	1.309	1.313
Quartz	SiO_2	1.5442	1.5534
Zircon	$ZrO_2 \cdot SiO_2$	1.923	1.968
Wurtzite	ZnS	2.356	2.378
Calcite	$CaCO_3$	1.658	1.486
Tourmaline	oxides of Na, Fe, Al, B, Si	1.669	1.638
Dolomite	$CaMg(CO_3)_2$	1.681	1.500
Siderite	$FeCO_3$	1.875	1.635

waves that are polarized in a direction perpendicular to the optic axis propagate with the same speed in all directions. Such waves are called *ordinary waves* (or *o-rays*). At the crystal interfaces, these rays obey Snell's law with a refractive index labeled n_0. For example, calcite has $n_0 = 1.658$ (for sodium yellow light).

The component of a wave that is not polarized perpendicular to the optic axis has a speed that depends on the direction of propagation in the crystal. In calcite, for example, this speed is always greater than c/n_0. Such waves do not, in general, obey the ordinary laws of refraction; in fact, the wavefronts need not be perpendicular to the ray directions, as we show later. These waves are called *extraordinary waves* (or *e-rays*). The extraordinary wave with the greatest speed is the wave with polarization parallel to the optic axis. For such rays, Snell's law is valid, and the wave speed is c/n_e. For example, calcite has $n_e = 1.486$ (for sodium yellow light).

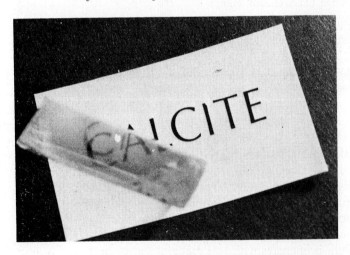

Calcite, a birefringent crystal. (From *Physics with Modern Applications* by Leonard H. Greenberg. Saunders College Publishing, 1978. Courtesy of the author and M. Velvick.)

Crystals such as calcite and tourmaline have a single optic axis and are referred to as *uniaxial* crystals. There is another type of crystal (*feldspar* and *mica* are examples) for which two optic axes are identifiable. These are referred to as *biaxial* crystals. There are no ordinary rays in biaxial crystals.

▶ Crystals that are optically isotropic possess highly symmetric crystal structures. An example of such a crystal structure is the cubic lattice of rock salt (NaCl) (see Fig. 23–3). In such crystals the electrons that are responsible for the optical properties of the substances are elastically bound to their equilibrium positions and have natural oscillation frequencies that are independent of the direction of displacement. Then, the speed of an electromagnetic wave in the crystal is independent of the direction of propagation (see the discussion in Section 40–5).

In the uniaxial crystal of calcite ($CaCO_3$), the lattice arrangement of the Ca atoms and the CO_3 groups forms a *rhombohedral* crystal, a kind of distorted rectangular structure, as shown in Fig. 44–12. The optic axis makes equal angles β with the crystal edges at the obtuse corners, such as A or A'. The natural frequency of electronic oscillation in the direction of the optic axis is greater than that in directions in a plane perpendicular to the optic axis. Thus, electromagnetic waves that excite oscillations in the direction of the optic axis (that is, waves that are polarized along the optic axis) have the higher wave speed associated with *e*-rays. ◄

Figure 44–13 shows two linearly polarized waves incident normally on a uniaxial birefringent crystal whose optic axis is in the plane of the page, as shown in the diagram. Notice that the wavefronts associated with the *o*-ray, which obeys Snell's law, propagate parallel to the crystal surface. The Huygens' wavelets (shown dashed in Fig. 44–13) are spherical with centers at the selected points a, b, and c on the interface. Notice also that the wavefronts associated with the *e*-ray propagate parallel to the crystal surface but that the *e*-ray itself is inclined to the surface normal. In this case, the Huygens' wavelets are ellipsoids of revolution with centers at the selected points a', b', and c' on the interface, and with the principal rotation axis of each ellipsoid in a direction parallel to the optic axis. The wave speed for an *e*-ray that travels in a direction making an angle α with the optic axis is given by (see Fig. 44–14)

$$u = c(n_0^2 \cos^2\alpha + n_e^2 \sin^2\alpha)^{-1/2} \qquad \textit{e}\text{-ray} \qquad \textbf{(44–9)}$$

Fig. 44–12. The rhombohedral structure of calcite. The optic axis makes equal angles, $\beta = 63.8°$, with the crystal edges at the obtuse corners where the edges make angles of 102°.

Fig. 44–13. Propagation of linearly polarized waves incident normally on a uniaxial birefringent crystal. The optic axis is in the plane of the page. The diagram corresponds to the case $n_e < n_o$.

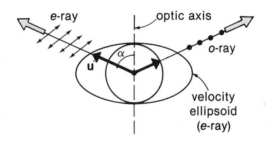

Fig. 44–14. The velocity of the *o*-ray (indicated by the circle) is independent of direction. The velocity of the *e*-ray, however, depends on direction and is greatest at right angles to the optic axis. The velocity ellipsoid for the *e*-ray is drawn for the case $n_e < n_o$.

The situation illustrated in Fig. 44–13 suggests a simple way to separate a light beam into linearly polarized components. Figure 44–15 shows how this is accomplished by using a birefringent crystal. The unusual refraction of the *e*-ray upon entering and leaving the crystal produces a displacement of this ray with respect to the straight-through *o*-ray.

Values of the refractive indexes, n_o and n_e, are given in Table 44–1 for some birefringent substances. If $n_e > n_o$, the substance is said to have *positive* birefringence; if $n_e < n_o$, the birefringence is *negative*.

Fig. 44–15. An unpolarized incident beam is separated into two beams that are linearly polarized in perpendicular directions upon passing through a birefringent crystal.

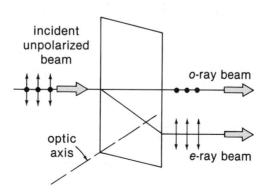

Example 44–4

One method for producing beams of polarizing light from an unpolarized beam makes use of a *Wollaston prism*, as shown in the diagram. The prism consists of two 30° pieces of calcite cemented together with the optic axes mutually perpendicular. Determine the angular separation, $\delta = \theta_{o2} + \theta_{e2}$, between the emergent beams for incident light with $\lambda = 589$ nm.

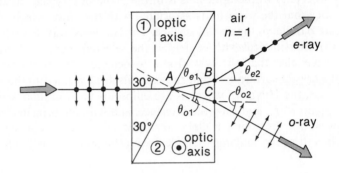

Solution:

In the diagram, the designations o-ray and e-ray refer to segment 2 of the prism. Notice that the polarization of the o-ray is such that in segment ① of the prism it is parallel to the optic axis; therefore, this ray travels with the speed c/n_e in segment ①. Similarly, the speed of the e-ray in segment ① is c/n_o. Then, using Snell's law for the e-ray at point A and the refractive indexes in Table 44–1, we have

$$n_o \sin 30° = n_e \sin \theta_{e1}$$

or

$$\theta_{e1} = \sin^{-1}\left(\frac{n_o \sin 30°}{n_e}\right) = \sin^{-1}\left(\frac{1.658 \sin 30°}{1.486}\right) = 33.91°$$

For the o-ray, we find $\quad n_e \sin 30° = n_o \sin \theta_{o1}$

or

$$\theta_{o1} = \sin^{-1}\left(\frac{n_e \sin 30°}{n_o}\right) = \sin^{-1}\left(\frac{1.486 \sin 30°}{1.658}\right) = 26.62°$$

At point B, the angle of incidence for the e-ray is $33.91° - 30° = 3.91°$. Thus,

$$n_e \sin 3.91° = n \sin \theta_{e2}$$

or, because $n = 1$,

$$\theta_{e2} = \sin^{-1}(n_e \sin 3.91°) = \sin^{-1}(1.486 \sin 3.91°) = 5.82°$$

The incident angle for the o-ray is $30° - 26.62° = 3.38°$. Thus,

$$n_o \sin 3.38° = n \sin \theta_{o2}$$

or

$$\theta_{o2} = \sin^{-1}(n_o \sin 3.38°) = \sin^{-1}(1.658 \sin 3.38°) = 5.61°$$

Hence, the angle between the two emergent beams is

$$\delta = \theta_{o2} + \theta_{e2} = 5.61° + 5.82° = 11.43°$$

Optical Phase Shifting. Suppose that a beam of linearly polarized light described by the vector **E** is passed through a thin piece of a negative birefringent crystal, with the crystal orientation shown in Fig. 44–16. The component of the wave

Fig. 44–16. A birefringent crystal can be used to convert a linearly polarized wave into a circularly polarized wave.

with polarization parallel to the optic axis (the *e*-wave) emerges from the crystal with a phase lead compared with the *o*-wave because the latter traverses a longer optical path by the amount $d(n_o - n_e)$. This path-length difference corresponds to a phase difference given by

$$\alpha = \frac{2\pi d}{\lambda}(n_o - n_e) \tag{44–10}$$

The emerging wave has components

$$E_x = E_0 \sin \theta \cos (kz - \omega t)$$
$$E_y = E_0 \cos \theta \cos (kz - \omega t - \alpha) \tag{44–11}$$

Of particular interest is the special case in which $\theta = \pi/4$ and $\alpha = \pi/2$. Then, a simple linearly polarized incident wave is transformed into a wave with left circular polarization (compare Eqs. 44–3). The crystal that produces this effect is called a *quarter-wave plate* (because $d(n_o - n_e) = \lambda/4$). (•Can another quarter-wave plate be placed in line with the first to reconstruct the original linearly polarized wave?) When the thickness of the crystal is such that $\alpha = \pi$, it is called a *half-wave plate*.

Example 44–5

What minimum thickness of quartz is required to produce circularly polarized light from a beam of linearly polarized light from a sodium vapor source?

Solution:

In this case, the crystal has *positive* birefringence, so the emergent light in the geometry of Fig. 44–16 has right circular polarization. (•Can you explain why?) Using Eq. 44–10, we have

$$\alpha = -\frac{\pi}{2} = \frac{2\pi d}{\lambda}(n_o - n_e)$$

Table 44–1 lists the values, $n_o = 1.5442$ and $n_e = 1.5534$; thus,

$$d = \frac{\lambda}{4(n_e - n_o)} = \frac{5.89 \times 10^{-7} \text{ m}}{4(1.5534 - 1.5442)}$$
$$= 1.60 \times 10^{-5} \text{ m} = 16.0 \; \mu\text{m}$$

Example 44-6

The birefringent crystal in Fig. 44–16 has a thickness appropriate to be a half-wave plate for the incident plane wave shown. What is the state of polarization of the transmitted wave?

Solution:

After the wave passes through the plate, the phase angle (Eq. 44–11) becomes $\alpha = \pi$. This change in phase represents a retardation of E_x by $\frac{1}{2}\lambda$, thereby bringing this component of the wave into phase with E_y. In effect, the x-component is reversed, and the linear polarization is shifted in angle by 2θ in a clockwise sense when the approaching wave is viewed as indicated in the diagram.

Photoelasticity. Certain substances, such as glass and some plastics, are optically isotropic under ordinary conditions but exhibit birefringence when stressed. These materials usually consist of long-chain molecules that, in the normal unstressed state, have random orientations. The lack of any order in the molecular arrangement means that the medium is isotropic, at least over lengths that are of the order of the wavelength of light. When stressed, however, the molecules tend to become aligned along the directions of tension. Such partial alignment of the molecules results in birefringent behavior.

Studies of the stress distribution in structures and structural members can be carried out by using models constructed from elastically birefringent (or *photoelastic*) materials. First, a polarizer and an analyzer are positioned with their transmission axes at right angles so that no light will pass through the combination. Then, the photoelastic model is placed between the polarizer and the analyzer and is subjected to the desired external stresses. The induced birefringence in the model usually causes the incident light to become elliptically polarized, so that some light now will pass through the analyzer. The degree of induced birefringence, and hence the amount of optical phase shift in the beam, depends on the magnitude of the internal stresses that are produced. The result is visible in the transmitted light. A bright-dark band structure is observed in which wide bands indicate low stress and narrow bands indicate high stress. An example of this type of photoelastic analysis is illustrated by the photographs.

(a)

The stress-birefringence pattern for a beam resting on two supports (a) when lightly loaded at the center and (b) when heavily loaded at the center. (From *Atlas of Optical Phenomena* by M. Cagnet, M. Francon, and J. C. Thrierr. Springer-Verlag, 1962.)

(b)

PROBLEMS

Section 44-1

44-1 Describe the state of polarization of the following
• waves:

   (a) $\mathbf{E} = E_0 e^{-i(z-ct)/l}\,\hat{\mathbf{i}}$

   (b) $\mathbf{E} = E_0 \cos(kz - \omega t)\hat{\mathbf{i}} \pm E_0 \cos(kz - \omega t)\hat{\mathbf{j}}$

   (c) $\mathbf{E} = E_0 \cos(kz - \omega t)\hat{\mathbf{i}} \pm E_0 \sin(kz - \omega t)\hat{\mathbf{j}}$

   (d) $\mathbf{E} = E_0 \cos(kz - \omega t)\hat{\mathbf{i}} \pm 2E_0 \sin(kz - \omega t)\hat{\mathbf{j}}$

   In each case, sketch $\mathbf{E}_x$, $\mathbf{E}_y$, and $\mathbf{E}$ for $t = 0$ and $0 \leq kz \leq \pi$ with intervals of $\pi/4$.

44-2 Describe the state of polarization of the following waves:

   (a) $\mathbf{E} = E_0 \cos(kz - \omega t)\hat{\mathbf{i}} \pm E_0 \cos(kz - \omega t - \pi/4)\hat{\mathbf{j}}$

   (b) $\mathbf{E} = E_0 \cos(kz - \omega t)\hat{\mathbf{i}} \pm E_0 \cos(kz - \omega t + 3\pi/4)\hat{\mathbf{j}}$

   In each case, sketch $\mathbf{E}_x$, $\mathbf{E}_y$, and $\mathbf{E}$ for $t = 0$ and $0 \leq kz \leq \pi$ with intervals of $\pi/4$.

44-3 Write an expression for $\mathbf{E}$ that describes a wave linearly polarized in a direction that makes an angle of $120°$ with respect to the x-axis.

Section 44-2

44-4 An unpolarized beam of light is reflected from a glass surface. It is found that the reflected beam is linearly polarized when the light is incident from air at an angle of $58.6°$. What is the refractive index of the glass?

44-5 A certain substance has a critical angle for internal reflection equal to θ_c and a Brewster's angle equal to θ_B. Show that $\sin \theta_c \tan \theta_B = 1$.

44-6 A glass plate ($n_g = 1.52$) is submerged in water ($n_w = 1.33$). What is the polarizing angle for light incident on the glass plate?

Section 44-3

44-7 Refer to Fig. 44-10. The component of the incident wave that induces currents in the wire grid is completely absorbed. Show that the transmitted intensity is $I(\theta) = I_0 \sin^2 \theta$, where I_0 is the intensity of the incident wave.

Section 44-4

44-8 A beam of unpolarized light is incident normally on two pieces of H-sheet that act as a polarizer-analyzer combination. What fraction of the incident energy flux is transmitted through the pair for angles 0, $\pi/6$, $\pi/4$, $\pi/3$, and $\pi/2$ between the two transmission axes? Assume that the transmittance for the desired component is $\overline{T}_d = 0.83$.

44-9 What angle between the transmission axes of a polarizer-analyzer pair of H-sheet should be chosen so that the emergent energy flux is 0.25 of the incident flux? Assume that the transmittance for the desired component is $\overline{T}_d = 0.83$.

44-10 Two pieces of H-sheet are positioned with their transmission axes at right angles, so there is no light transmitted through the pair. A third piece of H-sheet is placed between the original two pieces. Derive an expression for the intensity of light transmitted through the three-piece combination as a function of the angle θ between the transmission axes of the polarizer and the middle sheet. The incident light is unpolarized and has intensity I_0. Assume that the H-sheet is ideal and transmits 100 percent of the desired component of the incident beam.

Section 44-5

44-11 Light from a sodium vapor source ($\lambda = 589$ nm) is incident on a piece of calcite. What are the wavelengths in the crystal of the ordinary and extraordinary waves?

44-12 A possible design for a polarizer that is constructed from glass (n_g) and calcite (n_o, n_e) is shown in the diagram. The glass and the cement are chosen so that both have a refractive index equal to n_o for calcite. What is the largest prism angle θ for which the e-ray is totally reflected at the interface? (The transmitted beam then is the completely polarized o-ray.)

44-13 Left circularly polarized light ($\lambda = 589$ nm) is passed through a quarter-wave plate of calcite, as shown in the diagram. (a) Obtain an expression for the electric vector of the transmitted wave. (b) What is the thickness of the plate? (c) Describe the state of polarization of the transmitted beam.

44-14 What minimum thickness d of quartz is required to rotate the direction of linear polarization of an incident wave ($\lambda = 589$ nm) by $\pi/2$? (Refer to the diagram.)

Additional Problems

44-15 The beam that is transmitted through a multiplate polarizer (Fig. 44-6) contains some light with perpendicular polarization (due to multiple reflections, not shown in the figure). When this admixture is taken into account, the transmitted wave has a partial polarization in the parallel direction given by

$$(PP)_\| = \frac{m}{m + \left(\dfrac{2n}{n^2 - 1}\right)^2}$$

where m is the number of plates. How many plates of glass ($n = 1.50$) must be used to obtain a partial polarization of 75 percent?

44-16 Unpolarized light is incident on glass ($n = 1.50$) at Brewster's angle. Then, the amplitude reflection and transmission coefficients for the two polarization states are (see Fig. 40-19)

$$(E_r^0/E_i^0)_\perp = -0.384; \qquad (E_r^0/E_i^0)_\| = 0$$
$$(E_t^0/E_i^0)_\perp = 0.615; \qquad (E_t^0/E_i^0)_\| = 0.667$$

Consider only the first interface shown in Fig. 44–5. The intensity of the incident unpolarized wave is I_0. (a) Calculate the intensity of the completely polarized reflected wave. (b) Calculate the intensity of the parallel component of the transmitted wave. (c) Calculate the intensity of the perpendicular component of the transmitted wave. [*Hint:*

$$R = (E_r^0/E_i^0)^2 \quad \text{and} \quad T = \frac{n_2 \cos \theta_t}{n_1 \cos \theta_i} \left(\frac{E_t^0}{E_i^0}\right)^2$$

are the expressions for the reflectance and the transmittance.]

44–17 The direction of polarization of a linearly polarized wave is to be rotated through 90° by using two pieces of H-sheet. Describe how this can be accomplished while obtaining the maximum overall transmitted intensity.

44–18 Two pieces of H-sheet are positioned with their transmission axes parallel. A third piece of H-sheet is placed between the original two pieces. Derive an expression for the intensity of light transmitted through the three-piece combination as a function of the angle θ between the transmission axes of the polarizer-analyzer combination and the middle sheet. The incident light is unpolarized and has intensity I_0. Assume that the H-sheet is ideal and transmits 100 percent of the desired component of the incident beam. (Compare Problem 44–10.)

44–19 The electric field vector **E** of a monochromatic
• plane wave that propagates along the z-axis is

$$\mathbf{E} = E_0 \cos (kz - \omega t)\hat{\mathbf{i}} + E_0 \sin (kz - \omega t)\hat{\mathbf{j}}$$

(a) What is the state of polarization of the wave? (b) The wave is passed through a piece of H-sheet whose transmission axis makes an angle θ with the x-axis. Determine the electric field vector **E'** for the transmitted wave. What is the state of polarization of the transmitted wave? (c) What is the transmitted intensity as a function of θ? Assume that the H-sheet is ideal and transmits 100 percent of the desired component of the incident wave.

44–20 Calculate the wave velocity u/c for the e-ray in calcite if the wave ($\lambda = 589$ nm) travels in a direction that makes an angle $\alpha = 60°$ with the optic axis.

44–21 The diagram shows how an incident unpolarized beam is separated into two polarized beams upon passing through a *Rochon prism* of birefringent material. What are the states of polarization of the beams A and B? Determine the angular separation of the emergent beams if the prism is constructed from calcite. (Compare Example 44–4.)

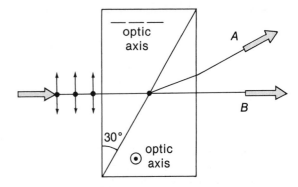

44–22 Repeat Problem 44–21 for a Rochon prism constructed from zircon.

44–23 A linearly polarized plane wave is passed through a half-wave plate of calcite. The optic axis of the crystal makes an angle θ with the direction of polarization of the incident wave. What is the direction of polarization of the emergent wave?

44–24 A tiny source of sodium vapor light is embedded in a calcite crystal and is pulsed on for an interval $< 10^{-12}$ s. Determine the shapes and the dimensions of the wavefronts corresponding to the o-wave and the e-wave at an instant 2×10^{-10} s after the source is pulsed.

44–25 Two pieces of H-sheet are positioned with their
• transmission axes at right angles so that there is no transmission through the pair. A half-wave plate is placed between the pieces of H-sheet with its optic axis at an angle θ with respect to the transmission axis of the first piece of H-sheet (the polarizer). The incident beam is unpolarized and has an intensity I_0. Determine the intensity of the light transmitted through the entire system. Assume that the H-sheet is ideal and transmits 100 percent of the desired component of the incident beam. (Compare Problem 44–10.)

CHAPTER 45

OPTICAL INSTRUMENTS

A large number of optical devices have been invented to assist the eye in studying very small and very distant objects and to correct impaired vision. The operation of these lens and mirror systems can be understood in terms of the laws of geometrical optics that have already been treated (Chapter 41). In this chapter we discuss a few examples of these devices, beginning with a single lens and continuing to multi-element systems.

45-1 THICK LENSES AND LENS COMBINATIONS

Principal Planes and Principal Points. The discussion in Section 41–4 was concerned with *thin* lenses, idealized lenses whose thickness along the optic axis is negligibly small. We now turn attention to cases in which the lens thickness is sufficiently large that it must be explicitly included in any calculation. The treatment of thick lenses is also applicable to systems of separated thin lenses, which are found in many optical instruments. Although the situation becomes more complicated, it is still possible to make a simple analysis of thick-lens systems by introducing some new geometric features of the lenses.

Figure 45–1 shows a thick lens that brings an incident parallel ray to the focal point F' in the region called the *image space* of the lens. The actual ray is refracted at each surface of the lens and follows the path $OABF'$. We can simplify the description of this ray within the lens by representing the transit in terms of a single refraction (at point B'). To do this, we introduce the *principal plane H'B'* and the *principal point* H' of the image space. As indicated in Fig. 45–1, the principal plane in the image space is located by the intersection of the incident parallel ray with the emergent ray that passes through the focal point F', when both rays are extended within the lens (dashed lines in Fig. 45–1). The principal plane and principal point in the object space (not shown in Fig. 45–1; see Fig. 45–2) are located by a similar construction with a ray incident from the right.

Figure 45–2 shows the geometry of a thick lens and the construction that locates the image I of the object O. For a particular lens, when the locations of the principal planes are specified and the focal length f is given, the positions of the focal points F and F' are then determined (see Fig. 45–2). These parameters are intrinsic properties of a particular lens and determine the image-forming properties of the lens.

We now consider the formation of an image by a thick lens. It is convenient to measure all distances from the focal planes or the principal planes. Then, both the

Fig. 45–1. The intersection of the extended rays, *OAB'* and *B'BF'*, locates the principal plane *B'H'* in the image space of a thick lens.

Fig. 45–2. Geometric construction to locate the image of an object produced by a thick lens. The focal planes are indicated by F and F'.

Gaussian and Newtonian forms of the lens equation apply (see Eqs. 41–24 and 41–26):

$$\frac{1}{s} + \frac{1}{s'} = \frac{1}{f} \qquad \text{Gaussian formula} \qquad \textbf{(45–1a)}$$

$$xx' = f^2 \qquad \text{Newtonian formula} \qquad \textbf{(45–1b)}$$

The principal rays that are used to locate the image of the object O are shown in Fig. 45–2. Notice that the "central" ray $OHH'I$ makes equal angles α with the lens axis; that is, OH and $H'I$ are parallel but offset rays. The hypothetical rays within the lens are again shown dashed. Notice also that in the limit of zero lens thickness, the principal planes coincide with the central plane of the resulting thin lens.

A Simple Two-Element Lens. Consider a simple lens system that consists of two separated thin lenses, as shown in Fig. 45–3. The lenses are separated by a distance t and have focal lengths f_1 and f_2. What is the focal length f for the lens combination? The treatment of the properties of a thick lens does not explicitly involve the fact that the lens has only two air-glass interfaces. Consequently we may discuss in the same way any system of lenses arranged in fixed geometry. Thus, the principal and focal planes are located just as for a thick lens. The parallel ray OF' originates in an object at an infinite distance so that $s_1 = \infty$. The lens L_1, if alone, would form an image at G; thus, G becomes the virtual object for lens L_2. Then, as indicated in Fig. 45–3, we have* $s_2 = -(f_1 - t)$. Now, the lens equation for the second lens is $1/s_2 + 1/s'_2 = 1/f_2$, which becomes

$$\frac{1}{s'_2} = \frac{1}{f_2} + \frac{1}{f_1 - t} \qquad \textbf{(45–2)}$$

---

*According to the convention on signs (see Section 41–4), s_2 is negative because G is to the right of lens L_2.

Fig. 45–3. Geometry for the calculation of the focal length f of a two-lens system.

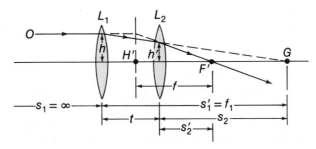

Also, from the similar triangles, we have

$$\frac{h}{s_1'} = \frac{h'}{-s_2} \qquad (45\text{-}3\text{a})$$

and

$$\frac{h}{f} = \frac{h'}{s_2'} \qquad (45\text{-}3\text{b})$$

where f is measured from the principal point H'. Dividing Eq. 45-3b by Eq. 45-3a, we find

$$\frac{1}{f} = -\left(\frac{1}{s_1'}\right)\left(\frac{s_2}{s_2'}\right) \qquad (45\text{-}4)$$

Then, using Eq. 45-2 together with $s_1' = f_1$ and $s_2 = -(f_1 - t)$, Eq. 45-4 gives

$$\frac{1}{f} = \frac{1}{f_1} + \frac{1}{f_2} - \frac{t}{f_1 f_2} \qquad (45\text{-}5)$$

Because the location of the focal point F' is determined by s_2' (which is known from Eq. 45-2), the position of the principal point H' is specified by the value of f calculated from Eq. 45-5.

Example 45-1

The lens shown in the diagram has a refractive index $n = 1.520$ and a thickness $t = 3.5$ cm, measured between the vertexes V_1 and V_2. The surface radii are $r_1 = +12.0$ cm and $r_2 = -16.0$ cm. Determine the focal length f and the locations of the focal point F' and the principal point H'.

Solution:

Consider an object O at infinity so that $s_1 = \infty$. Equation 41-16 for a spherical interface states that

$$\frac{n}{s} + \frac{n'}{s'} = \frac{n' - n}{r} \qquad (1)$$

Applying this equation to the first interface of the thick lens yields

$$\frac{1}{\infty} + \frac{1.520}{s_1'} = \frac{1.520 - 1}{12.0 \text{ cm}}$$

from which $\qquad\qquad\qquad\qquad\qquad s'_1 = +35.1 \text{ cm}$

The real image I_1 that would be formed by the first interface alone is the virtual object O_2 for the second interface. Thus,

$$s_2 = -(35.1 \text{ cm} - 3.5 \text{ cm}) = -31.6 \text{ cm}$$

Using Eq. (1) for this interface gives

$$\frac{1.520}{-31.6 \text{ cm}} + \frac{1}{s'_2} = \frac{1 - 1.520}{-16.0 \text{ cm}}$$

from which $\qquad\qquad\qquad\qquad\qquad s'_2 = +12.4 \text{ cm}$

That is, the focal point F' is 12.4 cm to the right of vertex V_2.

The focal length f can be obtained from Eq. 45-4, which is valid for thick lenses; thus,

$$\frac{1}{f} = -\left(\frac{1}{35.1 \text{ cm}}\right)\left(\frac{-31.6 \text{ cm}}{12.4 \text{ cm}}\right)$$

from which $\qquad\qquad\qquad\qquad\qquad f = 13.8 \text{ cm}$

These results for s'_2 and f determine the position of the principal point H', which is 13.8 cm − 12.4 cm = 1.4 cm to the left of vertex V_2.

The focal point F and the principal point H in the object space can be located by a similar calculation for a parallel ray incident from the right. The results are that F is located 12.7 cm to the left of vertex V_1 and that the principal point H is 13.8 cm − 12.7 cm = 1.1 cm to the right of vertex V_1.

Example 45-2

A telephoto lens for a camera should have a long focal length in order to produce a large image on the film surface. (Recall that the lateral magnification m is proportional to the focal length, $m = -f/x$; see Problem 41-20.) A second requirement is that the image plane F' should be close to the lens system so that the overall length of the telephoto lens plus camera body will not be inconveniently large. These seemingly contradictory design aims can be achieved by a lens combination whose principal point H' (from which the focal length is measured) lies on the *object* side of the lens system. Determine the focal length and the locations of the focal points and principal points for the two-element telephoto lens shown in the diagram.

Solution:

According to Eq. 45-5, the focal length f of the lens combination is given by

$$\frac{1}{f} = \frac{1}{10 \text{ cm}} + \frac{1}{-3 \text{ cm}} - \frac{8 \text{ cm}}{(10 \text{ cm})(-3 \text{ cm})}$$

from which $\qquad\qquad\qquad\qquad f = +30 \text{ cm}$

Consider an object at an infinite distance to the left of lens L_1 so that $s_1 = \infty$; then, $s_1' = f_1 = 10 \text{ cm}$. This image position is the position of the virtual object for lens L_2; hence, $s_2 = -(10 \text{ cm} - 8 \text{ cm}) = -2 \text{ cm}$, measured from L_2. The lens equation then gives

$$\frac{1}{s_2'} = \frac{1}{-3 \text{ cm}} - \frac{1}{-2 \text{ cm}}$$

from which $\qquad\qquad\qquad\qquad s_2' = 6 \text{ cm}$

Thus, the focal point F' is 6 cm to the right of lens L_2. With $f = +30 \text{ cm}$, the principal point H' is therefore $30 \text{ cm} - (8 \text{ cm} + 6 \text{ cm}) = 16 \text{ cm}$ to the *left* of lens L_1, as indicated in the diagram.

A similar calculation using a parallel ray incident from the right shows that the focal point F is 110 cm to the left of L_1 and that the principal point H is 80 cm to the left of L_1.

Notice that the length of the telephoto system, from the first lens L_1 to the focal plane, is 14 cm. If a single lens with the same focal length were used instead, the lens-to-focal-plane distance would be more than twice as large (30 cm).

▶ *Color Correction of Lenses.* Figure 40–22 shows the variation with wavelength of the refractive index for several types of glass used in lens making. As a result of this variation, the focal length of a lens is slightly different for different colors of light, so that a color-blurred image is formed. This effect is called *chromatic aberration*. A considerable improvement in color performance can be achieved by using a *compound lens* that consists of two closely spaced lenses made from different types of glass. One such arrangement, called an *achromatic doublet*, is shown in Fig. 45–4. For this case, in which two thin lenses are cemented together, we take the thickness t to be zero so that Eq. 45–5 becomes

$$\frac{1}{f} = \frac{1}{f_1} + \frac{1}{f_2} \qquad t = 0 \qquad\qquad \textbf{(45–6)}$$

Fig. 45–4. Positive and negative thin lenses, made from different types of glass, are cemented together to produce an *achromatic doublet*.

An achromatic doublet can be designed to produce equal focal lengths for any two wavelengths. In effect, the chromatic aberration of one lens cancels the chromatic aberration of the other lens at these wavelengths. Often, these wavelengths are chosen to be* λ_α(red) = 656 nm and λ_β(blue) = 486 nm; the focal length of the combination is then quoted, as usual, for λ_D(yellow) = 589 nm.

The lens maker's formula (Eq. 41–23) for a thin lens with surface radius r for the front spherical surface and r' for the rear surface is

$$\frac{1}{f} = (n - 1)\left(\frac{1}{r} - \frac{1}{r'}\right) \qquad\qquad \textbf{(45–7)}$$

where we use the sign convention of Fig. 41–16. For either lens, $f(n - 1) = \text{constant}$; thus, for lens L_1 at the wavelengths λ_α and λ_D, we can write

$$\frac{1}{f_{1\alpha}} = \frac{n_{1\alpha} - 1}{n_{1D} - 1} \cdot \frac{1}{f_{1D}} \qquad\qquad \textbf{(45–8a)}$$

and at the wavelengths λ_β and λ_D, we have

---

*λ_α and λ_β are the wavelengths of the strong H_α and H_β lines, respectively, in the spectrum of hydrogen; λ_D is the average wavelength of the yellow doublet of lines in the spectrum of sodium.

$$\frac{1}{f_{1\beta}} = \frac{n_{1\beta} - 1}{n_{1D} - 1} \cdot \frac{1}{f_{1D}} \tag{45–8b}$$

with a similar pair of equations for lens L_2. Also, Eq. 45–6 at the two wavelengths λ_α and λ_β can be expressed as

$$\frac{1}{f_\alpha} = \frac{1}{f_{1\alpha}} + \frac{1}{f_{2\alpha}} \tag{45–9a}$$

$$\frac{1}{f_\beta} = \frac{1}{f_{1\beta}} + \frac{1}{f_{2\beta}} \tag{45–9b}$$

We require $f_\alpha = f_\beta$; then, the substitution of Eqs. 45–8 and the counterparts for lens L_2 into Eqs. 45–9 leads to (after some algebra)

$$\frac{n_{1\beta} - n_{1\alpha}}{n_{1D} - 1} \cdot \frac{1}{f_{1D}} = -\frac{n_{2\beta} - n_{2\alpha}}{n_{2D} - 1} \cdot \frac{1}{f_{2D}} \tag{45–10}$$

Because $n_\beta > n_\alpha$ for all types of glass (see Fig. 41–22), Eq. 45–10 shows that $f_{1D}/f_{2D} < 0$; that is, the compound lens must consist of a positive lens and a negative lens (Fig. 45–4). We also have, at the wavelength λ_D,

$$\frac{1}{f_D} = \frac{1}{f_{1D}} + \frac{1}{f_{2D}} \tag{45–11}$$

Given the various refractive indexes and the desired value of f_D, Eqs. 45–10 and 45–11 can be solved simultaneously to yield the required values of f_{1D} and f_{2D}. Then, the four surface radii, r_1, r_1', r_2, and r_2', must be chosen to give these two focal lengths. If the two lenses are cemented together, a third condition is that $r_2 = r_1'$. To establish the four radii and complete the design of the lens, a fourth requirement may be set. Usually, this condition is utilized to minimize spherical aberration or some other geometrical aberration.

It is possible to achromatize a doublet lens system even if the two lenses are made from the same type of glass if the lenses are separated ($t \neq 0$). *Separated doublets* are sometimes used for eyepieces in optical systems of various sorts. (See Problem 45–23.)

With three lenses in combination (a lens *triplet*), the focal length can be made equal at three different wavelengths. Such lenses are called *apochromats* and are often used in high quality optical devices. (See diagram (b) in Example 45–3.)

Example 45–3

Diagram (a) shows a plano-convex doublet for which $r_1 = \infty$. The lens elements are cemented together; thus, $r_2 = r_1'$. Lens L_1 is made from crown glass that has $n_{1\alpha} = 1.5146$, $n_{1D} = 1.5170$, and $n_{1\beta} = 1.5226$; lens L_2 is made from flint glass that has $n_{2\alpha} = 1.5755$, $n_{2D} = 1.5795$, and $n_{2\beta} = 1.5897$ (compare Fig. 40–22). Determine the lens radii required to achromatize the doublet for a focal length $f_D = 50$ cm.

crown glass   flint glass

object space   image space

$r_1 = \infty$
$r_1' < 0$

$r_2' < 0$
$r_2 = R_1' < 0$

L_1 L_2 (a)

Solution:

Using Eq. 45–10, we find

$$\frac{1.5226 - 1.5146}{1.5170 - 1} \cdot \frac{1}{f_{1D}} = -\frac{1.5897 - 1.5755}{1.5795 - 1} \cdot \frac{1}{f_{2D}}$$

from which

$$\frac{1}{f_{1D}} = -1.5836 \cdot \frac{1}{f_{2D}} \qquad \text{(1)}$$

Equation 45–11 is

$$\frac{1}{50 \text{ cm}} = \frac{1}{f_{1D}} + \frac{1}{f_{2D}} \qquad \text{(2)}$$

Combining Eqs. (1) and (2) yields

$$\frac{1}{f_{1D}} = +0.05427 \text{ cm}^{-1} \qquad \text{and} \qquad \frac{1}{f_{2D}} = -0.03427 \text{ cm}^{-1}$$

Applying Eq. 45–7 to lens L_1 gives

$$0.05427 = (1.5170 - 1)\left(\frac{1}{\infty} - \frac{1}{r_1'}\right)$$

from which

$$r_1' = -9.526 \text{ cm} = r_2$$

Applying Eq. 45–7 to lens L_2 gives

$$-0.03427 = (1.5795 - 1)\left(\frac{1}{-9.526 \text{ cm}} - \frac{1}{r_2'}\right)$$

from which

$$r_2' = -21.82 \text{ cm}$$

The required lens surfaces have the curvatures shown in diagram (a). The focal lengths at wavelengths λ_α and λ_β for this lens are $f_\alpha = f_\beta = 50.05$ cm, a difference of only 0.5 mm from the focal length for yellow light ($\lambda_D = 50$ cm).

Diagram (b) shows the variation with wavelength of the focal length for the lens just designed, together with that for a single-element lens made from crown glass. The flatter the focal length curve, the better is the performance of the lens. It is apparent that a substantial

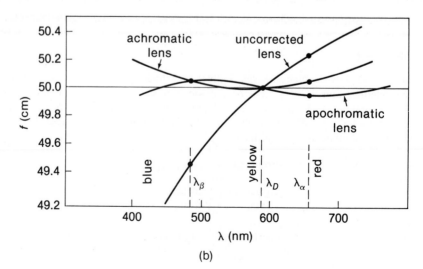

(b)

improvement results from achromatizing the lens. The diagram also shows the effect of adding a third element to the lens system to produce an apochromatic triplet. This lens has the same focal length for three wavelengths. The focal length curve is then even flatter than that for the achromatic lens. An even higher degree of color correction could be achieved by using more lens elements; however, the improvement is usually not worth the expense of constructing a more complicated lens system. ◄

45–2 SIMPLE MAGNIFIERS

To see an object in greatest detail, the largest possible image must be formed on the retina of the eye. This is easily accomplished by placing the object close to the eye. However, a normal human eye cannot focus properly on an object that is closer than about 25 cm. (An object at this distance is said to be at the *near point* of the eye.) Figure 45–5a shows an object being examined by an unaided eye; with the object at the near point, the subtended angle is θ. Figures 45–5b and 45–5c illustrate two ways to view the same object with a simple magnifier (or *magnifying glass*), which consists of a single converging lens. The object in Fig. 45–5b is positioned just inside the focal point of the lens to produce a virtual image at the near point of the eye. This method of viewing can produce eyestrain because the eye muscles must work to provide the necessary focusing effect. A more satisfactory arrangement (Fig. 45–5c) is to position the object at the focal point so that the virtual image appears at infinity; then, the retinal image is formed with the eye muscles completely relaxed.

Fig. 45–5. (a) An object at the near point is viewed by an unaided eye. (b) A simple magnifier produces a virtual image at the near point. (c) The object is positioned to produce a virtual image at infinity, which can be viewed by a relaxed eye.

By convention, the magnifying power of a device that produces a virtual image is defined in terms of the *angular magnification* γ. We define the angular magnification of an optical device to be the ratio of the tangents of the angles θ_{lens} and θ_{eye}, where θ_{lens} is the angle subtended by the image when viewed through the lens system, and where θ_{eye} is the angle subtended by the objected itself when placed at the near point of the eye. Thus,*

$$\gamma = \frac{\tan \theta_{lens}}{\tan \theta_{eye}} \tag{45–12}$$

Referring to Fig. 45–5b and c, we see that there are two possibilities for γ depending on the use made of the lens; that is,

$$\gamma = \frac{\tan \theta_b}{\tan \theta} \quad \text{and} \quad \gamma_\infty = \frac{\tan \theta_c}{\tan \theta}$$

---

*It is also common practice to define the angular magnification as the ratio of the angles themselves; that is, $\gamma = \theta_{lens}/\theta_{eye}$. For small angles, $\tan \theta \cong \theta$, and the definitions are approximately equivalent.

where θ is the subtended angle for the case of the unaided eye (Fig. 45–5a); that is,

$$\tan \theta = \frac{h}{25 \text{ cm}} \tag{45–13}$$

where we follow the usual practice and express the near-point distance in cm. We assume that the position of the eye coincides with that of the lens. Then, for the case illustrated in Fig. 45–5b, we have $s_b' = -25$ cm; hence,

$$\frac{1}{s_b} = \frac{1}{f} + \frac{1}{25 \text{ cm}}$$

Thus,

$$\tan \theta_b = \frac{h}{s_b} = h\left(\frac{1}{f} + \frac{1}{25 \text{ cm}}\right) \tag{45–14}$$

For the case illustrated in Fig. 45–5c, we have

$$\tan \theta_c = \frac{h}{f} \tag{45–15}$$

It follows that the angular magnifications (Eqs. 45–12) for these two cases can be expressed as

$$\gamma = \frac{25 \text{ cm}}{f} + 1 \tag{45–16a}$$

$$\gamma_\infty = \frac{25 \text{ cm}}{f} \tag{45–16b}$$

The distance from the object to the lens can be varied between s_b and s_c and the eye can still produce a retinal image in sharp focus. Typical values of γ are 2 or 3 for simple lenses, because aberrations usually preclude the use of lenses with shorter focal lengths and larger magnifications. For these low values of γ, the change in magnification as the object distance is changed between s_b and s_c is readily observable. Specially shaped single lenses can have values of γ as large as 10 or so.

45-3 MICROSCOPES

To obtain magnifications greater than those possible with simple magnifiers, it is necessary to use some form of *compound microscope*. The simplest such instrument consists of two lens elements, an *objective* and an *ocular* (or *eyepiece*), as shown in Fig. 45–6. We treat the objective and ocular as simple lenses, although in a practical microscope they would be multielement lenses (as shown in Fig. 45–7). The drawing on page 1192 shows a cutaway view of a modern compound microscope.

The angular magnification Γ of a compound microscope is defined to be the

Fig. 45-6. A simple two-lens microscope. The intermediate real image, formed at the focal point F of the ocular, serves as the object for the ocular, which in turn forms a virtual image at infinity.

ratio $\Gamma = \tan\theta_e / \tan\theta$, where θ_e is the angle subtended by the final image (see Fig. 45-6), and where θ is the angle that the object height h would subtend at the eye if placed at the near point (see Fig. 45-5a). Now,

$$\tan\theta = \frac{h}{25 \text{ cm}}$$

and

$$\tan\theta_e = \frac{h'}{f_e}$$

where f_e is the focal length of the eyepiece (Fig. 45-6). Thus,

$$\Gamma = \frac{\tan\theta_e}{\tan\theta} = \frac{h'}{h}\frac{25 \text{ cm}}{f_e}$$

The ratio h'/h is just the lateral magnification m of the objective (Eq. 41-27), and $(25 \text{ cm})/f_e$ is the angular magnification γ_∞ of the ocular (Eq. 45-16b); therefore, we have for the overall angular magnification of the microscope,*

$$\Gamma = m\gamma_\infty \qquad \text{microscope} \qquad \textbf{(45-17)}$$

for viewing by a relaxed eye (final image at infinity). For a final image at the near point, the angular magnification is

$$\Gamma = m\gamma \qquad \textbf{(45-17a)}$$

It is customary to refer to $|m|$ as the *magnifying power* of the objective and to γ_∞ as the magnifying power of the ocular. The values of $|m|$ and γ_∞ are both expressed as $\times 5$, $\times 10$, $\times 40$, and so forth.

Microscope Objectives. Modern microscope objectives are carefully designed multielement lens systems that have a high degree of correction for chromatic and spherical aberrations. Figure 45-8 shows cutaway sections of three objectives with different magnifying powers. Notice that, as the magnifying power is increased, more lens elements are required to minimize the aberrations.

Diffraction effects, primarily in the objective system, determine the ultimate resolution of a microscope and set the limit on the maximum usable magnification. For discussing microscope objectives, it is more convenient to cast in a different form the expression for the resolution that was developed in Section 43-1. The points O and O' shown in Fig. 45-9a are two object points that are just resolved according to the Rayleigh criterion (Section 43-1). The separation distance ξ is then the *linear*

Fig. 45-7. A compound microscope with multielement lenses. The "stop" is an aperture that limits the extreme off-axis rays. The lenses are positioned to produce a virtual image at infinity for viewing by a relaxed eye.

---

*For the situation shown in Fig. 45-6, we have $h'/h < 0$; thus, $\Gamma < 0$, and the final image is inverted.

RETINAL IMAGE

PROJECTION DISTANCE (250MM)

EYEPOINT

EYEPIECE

VIRTUAL IMAGE

INCLINED VIEWING HEAD

RELAY SYSTEM (1 × MAGNIFICATION)

FIRST SURFACE MIRROR

OBJECTIVES

NOSEPIECE

OBJECT

FOCUSING STAGE

DISC DIAPHRAGM

SUBSTAGE CONDENSER

COARSE FOCUS KNOB

OPTILUME ILLUMINATOR

FINE FOCUS KNOB

BASE

Sectional view of a modern compound microscope. (Courtesy of Bausch & Lomb Optical Co.)

× 8   × 45   × 90

Fig. 45–8. Sectional views of three modern microscope objectives. (Courtesy of Bausch & Lomb Optical Co.)

resolution of the lens. The central maximum of the image *I* is located at the position of the first diffraction minimum of the image *I'*. For a circular lens (neglecting all geometric aberrations), the resolution criterion means that the difference in path length between the extreme diffracted rays *O'BI* and *O'AI* is 1.22λ (see Eq. 43–5b). As shown in Fig. 45–9b, *O'B* is longer than *OB* by an amount ξ sin α, and *O'A* is shorter than *OA* (which is equal to *OB*) by the same amount. Thus, the difference in path length between *O'A* and *O'B* is 2ξ sin α. Because *AI* and *BI* are equal in length, *O'BI* is longer than *O'AI* by 2ξ sin α. Setting this equal to 1.22λ, we find

$$\xi = \frac{1.22\lambda}{2 \sin \alpha} \tag{45–18}$$

Fig. 45-9. Geometry for discussing the resolving power of a microscope objective.

Note carefully the distinction between the angle θ (the angle to which Eq. 43–5b applies) and the angle α (the half angle of the lens subtended at the object).

Equation 45–18 was obtained for the case in which the medium between OO' and the lens is air. The value of ξ can be decreased (that is, the resolution can be improved) by filling the space between object and lens with a transparent fluid, usually oil but sometimes water. Then, Eq. 45–18 must be modified to read (see Problem 45–8)

$$\xi = \frac{1.22\lambda}{2n \sin \alpha} \qquad \text{(45–18a)}$$

where n is the refractive index of the immersion fluid. This expression for the linear resolution is valid when O and O' are point sources of incoherent light. However, most microscopes use a complicated illumination system in which the light from O and O' is partially coherent. A detailed analysis by Ernst Abbe (1840–1905) reached the conclusion that Eq. 45–18a is a good estimate of the actual linear resolution if the factor 1.22 is replaced by unity.

An object that is viewed with a microscope must be placed at a definite distance from the objective, a distance determined by the focal length of the lens. Thus, the angle α in Eq. 45–18a has a fixed value for particular lens. In describing a lens, it is customary to specify the value of $n \sin \alpha$, which is called the numerical aperture (N.A.) of the lens:

$$\text{N.A.} = n \sin \alpha \qquad \text{(45–19)}$$

The numerical aperture is a measure of the effective angular opening of the lens for gathering light from the object. The linear resolution of the lens can now be expressed as*

$$\xi = \frac{\lambda}{2 \text{ N.A.}} \qquad \text{microscope} \qquad \text{(45–20)}$$

A microscope objective with a large numerical aperture is desirable, both to reduce the effects of diffraction and to increase the brightness of the image. With an air medium, values of N.A. as large as 0.95 have been obtained; with oil immersion, N.A. values up to about 1.6 are possible. For visible light with $\lambda = 550$ nm and with

---

*The reciprocal of ξ is sometimes called the *resolving power* of the lens: R.P. = 2 N.A./λ. When λ is expressed in cm, the resolving power corresponds to the number of lines per cm that can just be resolved.

Fig. 45–10. Dark-field illumination can be used to detect the light scattered by very small particles.

N.A. = 1.6, we find (using Eq. 45–20) $\xi = 170$ nm for the minimum separation between resolved objects.

Equation 45–18a refers to limitations imposed on image formation and the ability to recognize specific detail in the image. It is sometimes useful simply to be able to determine the presence or location of some very small particle (without requiring information about its shape). Figure 45–10 shows one arrangement, called *dark-field illumination*, for making such observations. The particles scatter the incident light and appear as luminous points in an otherwise dark field. Particles as small as 5 nm in diameter (about 1/100 of the wavelength of visible light) can be detected in this manner.

Microscope Oculars.   The oculars used in most microscopes consist of two plano-convex lenses mounted with a fixed separation. The first lens (nearer the objective) is called the *field lens* and the second is called the *eye lens*. With four refracting surfaces available, it is possible to choose the radii to minimize geometric aberrations as well as to correct for chromatic aberration. The corrections that can be made in this way are often sufficient so that compound lens elements are not required. (See Problem 45–23.)

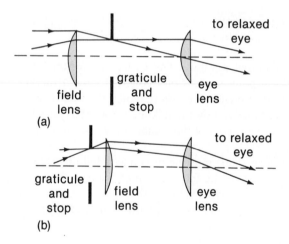

Fig. 45–11. Common oculars used in optical instruments. (a) Huygens ocular. (b) Ramsden ocular. In each case the stop excludes the extreme rays and limits the field of view.

Figure 45–11 illustrates two types of widely used oculars, the *Huygens* ocular and the *Ramsden* ocular. These oculars are used with all types of optical systems designed for visual observations (instead of photographic recording), such as microscopes, telescopes, and spectroscopes.

The diagram for the Huygens ocular (Fig. 45–11a) shows that the field lens focuses the already-converging rays to form a real intermediate image that is viewed by the eye lens. The diagram for the Ramsden ocular (Fig. 45–11b) shows that the converging rays form a real intermediate image in front of the ocular, and this image is then viewed by the two-lens system. In the Huygens ocular, the focal length of the field lens is usually two or three times greater than that of the eye lens; in the Ramsden ocular, the focal lengths of the two lenses are equal. The Huygens ocular is often used in ordinary microscopes because the placement of the stop between the lenses is particularly convenient when different oculars are substituted, as is often done in microscopy. (See the drawing on page 1192.) However, the Ramsden ocular possesses greatly reduced geometric aberrations compared with the Huygens ocular. Further improvement in the color performance of an ocular is often made by using a cemented doublet for one or both of the lenses.

An ocular may be provided with a cross-hair or a graticule and a field stop, as indicated in Figs. 45–7 and 45–11. These are located at the position of the intermediate image so that their sharp images are superimposed on the image of the object being studied. The cross-hair and graticule provide tracking and scale information; the stop excludes the extreme, poorly focused rays and outlines the observed field with a sharp boundary instead of a distracting fuzzy edge that would otherwise result.

Example 45–4

A normal human eye can just resolve two point sources of light that have an angular separation of approximately 3.4×10^{-4} rad, which corresponds to a linear separation of about 85 μm at a distance of 25 cm, the near-point distance (see Problem 45–18).

An oil-immersion objective with N.A. $= 1.5$ and magnification $\times 90$ is used in a microscope focused to produce a virtual image at the near point. What should be the magnification of the ocular if the resolving power of the microscope is to match that of the eye? (Assume that light with $\lambda = 550$ nm is used.)

Solution:

The linear resolution of the microscope is (Eq. 45–20)

$$\xi = \frac{\lambda}{2 \text{ N.A.}} = \frac{550 \text{ nm}}{2(1.5)} = 183 \text{ nm}$$

To match the linear resolution of the microscope to that of the eye, the overall angular magnification should be

$$\Gamma = \frac{85 \times 10^{-6} \text{ m}}{183 \times 10^{-9} \text{ m}} \cong 464$$

Thus, the required power of the ocular is $\gamma = \Gamma/m = 464/90 = \times 5.2$. Because observing in this manner would be tiring to the eye, the usual practice is to employ a higher-power ocular to increase the magnification by about 3 to 5 times. Thus, a more suitable ocular would have a power of $\times 15$, giving an overall magnification of $\Gamma = 90 \times 15 = 1350$. This value corresponds approximately to the maximum usable magnification for visual observation with an optical microscope.

Example 45–5

A compound microscope consists of a thin lens objective with focal length $f_1 = 1$ cm and diameter $d_1 = 1.5$ cm, together with a thin lens ocular with focal length $f_2 = 5$ cm. The lenses are mounted at opposite ends of a tube with length $\ell = 15$ cm. (Refer to Fig. 45–6.)

(a) At what distance from the objective should a specimen be placed to produce a final image at the near point of the eye?

(b) For an image at the near point, what is the overall angular magnification of the microscope?

(c) What is the numerical aperture of the objective? Is the overall resolution limited by the microscope or the eye?

Solution:

(a) With the final image at the near point, we have $s_2' = -25$ cm; hence,

$$\frac{1}{s_2} = \frac{1}{f_2} - \frac{1}{s_2'} = \frac{1}{5 \text{ cm}} + \frac{1}{25 \text{ cm}}$$

which yields
$$s_2 = 4.17 \text{ cm}$$

Then,
$$s_1' = \ell - s_2 = 15 \text{ cm} - 4.17 \text{ cm} = 10.83 \text{ cm}$$

Finally,
$$\frac{1}{s_1} = \frac{1}{f_1} - \frac{1}{s_1'} = \frac{1}{1 \text{ cm}} - \frac{1}{10.83 \text{ cm}}$$

from which
$$s_1 = 1.10 \text{ cm}$$

This is the required specimen-to-objective distance.

(b) Using Eq. 45–17a, we have (see Problem 45–7)

$$\Gamma = m\gamma = -\frac{s_1'}{s_1}\left(\frac{25 \text{ cm}}{f_2} + 1\right)$$

$$= -\frac{10.83 \text{ cm}}{1.10 \text{ cm}}\left(\frac{25 \text{ cm}}{5 \text{ cm}} + 1\right) = -59.1$$

(c) We have
$$\alpha = \tan^{-1}\frac{d_1}{2s_1} = \tan^{-1}\frac{1.5 \text{ cm}}{2(1.10 \text{ cm})} = 34.3°$$

Thus,
$$\text{N.A.} = n \sin \alpha = 1 \times \sin 34.3° = 0.564$$

The linear resolution of the microscope objective for light with $\lambda = 550$ nm is (Eq. 45–20)

$$\xi_m = \frac{550 \text{ nm}}{2(0.564)} = 488 \text{ nm}$$

The smallest detail in the specimen that is resolvable by the eye is (see Example 45–4)

$$\xi_e = \frac{85 \text{ μm}}{|\Gamma|} = \frac{85 \times 10^{-6} \text{ m}}{59.1} = 1440 \text{ nm}$$

Hence, the resolution is limited by the eye.

45–4 TELESCOPES

Telescopes perform two important functions. One is to produce an enlarged image of a distant object, and the other is to gather more light from the object than is possible with the unaided eye. A large useful magnification requires an objective with a large diameter in order to reduce diffraction effects and improve the resolution of the instrument. A large light-gathering power also requires a large-diameter objective lens (or mirror in the case of reflecting telescopes).

Figure 45–12 shows two types of simple telescopes that use lenses (*refracting telescopes*). Diagram (a) illustrates an *astronomical telescope*, which produces an inverted final image, and diagram (b) shows a *Galilean telescope*, which produces an erect final image. Both telescopes are shown giving images at infinity for a distant object (parallel incident rays); this is the condition appropriate for viewing by a relaxed eye. The erect final image that is produced by a Galilean telescope has an obvious advantage for terrestrial observations. For viewing astronomical objects, an inverted image presents no disadvantage. An astronomical telescope produces a real

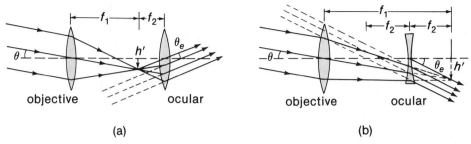

Fig. 45-12. (a) An astronomical telescope produces an inverted image of a distant object. (b) A Galilean telescope produces an erect image of a distant object.

intermediate image, so a Ramsden ocular with its desirable features can be used.

The angular magnification Γ of a telescope is defined in the same way as for a microscope, namely, $\Gamma = \tan \theta_e / \tan \theta$, where θ is the angle that the (distant) object subtends at the unaided eye, and where θ_e is the angle subtended by the final image. In Fig. 45-12 we see that

$$\tan \theta = h'/f_1 \qquad \text{and} \qquad \tan \theta_e = -h'/f_2$$

Hence,*

$$\Gamma = \frac{\tan \theta_e}{\tan \theta} = -\frac{f_1}{f_2} \qquad \text{telescope} \qquad \textbf{(45-21)}$$

Newton recognized the fact that a mirror does not suffer from chromatic aberration and therefore enjoys an advantage over a lens as the objective of a telescope. Moreover, the reflectance of a mirror is generally superior to the transmittance of a lens, thereby yielding a brighter final image. Also, spherical aberration can be considerably reduced (indeed, eliminated entirely for rays parallel to the optic axis) by

---

* Notice that the ocular of the Galilean telescope is a negative lens; thus $f_2 < 0$, which gives a positive value for Γ, indicating an erect final image.

Prism binoculars. The large distance between the objective and the ocular makes a telescope inconvenient to hold by hand. In a prism binocular, the overall length is shorter because the light path between objective and ocular is folding back upon itself by means of two 45°-45°-90° prisms, as shown here. The prisms also serve to produce a final image that is erect. Notice that the ocular here is a multielement lens; the objective is a cemented doublet. (Courtesy of Bausch & Lomb Optical Co.)

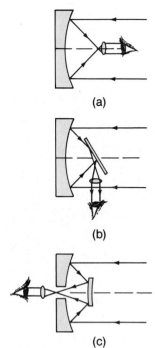

(a)

(b)

(c)

Fig. 45–13. Different methods for viewing an object with a reflecting telescope. (a) Observing at the primary focus. (This method is limited to very large telescopes, such as the Palomar Observatory instrument. See the photograph on page 1199.) (b) The Newtonian method of observation. (c) The Cassegrainian method. All reflecting telescopes have long focal lengths; the diagrams here show telescopes with focal lengths that are falsely short in order to conserve space.

using a paraboloidal mirror surface. Newton constructed the first *reflecting telescope* in 1667. Several methods have been devised for viewing an object with a reflecting telescope; three such arrangements are shown in Fig. 45–13.

Example 45–6

The Hale telescope (Fig. 45–14) has a mirror diameter of 5.08 m and a primary focal length of 17 m. (By using secondary mirrors, the focal length can be increased to 150 m.) For viewing at the primary focus (Fig. 45–13a), an observation cage with a diameter of 1.8 m is used to hold the observer. (See the photograph on page 1199.) Suppose that the ocular used has a focal length of 4 cm.

(a) What is the ratio of the light-gathering power of the telescope to that of the unaided eye? (Assume a pupil diameter of 2 mm.)

(b) What is the angular magnification for the two focal lengths given above?

(c) What is the angular resolution of the Hale telescope, if diffraction-limited? Compare with that of the unaided eye.

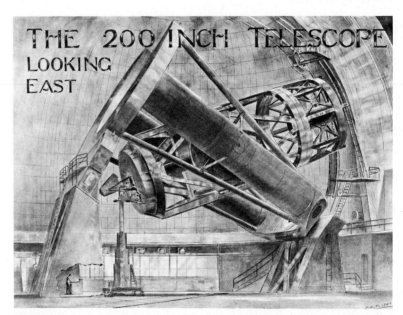

Fig. 45–14. This drawing of the 200-inch (5-m) diameter Palomar Observatory telescope, designed by the astronomer George E. Hale, was made by Russell W. Porter during the design phase of the project. The instrument was finally mounted in 1948 after 15 years of construction. This telescope is exceeded in size only by the 6-m Soviet telescope. (Courtesy of Palomar Observatory, California Institute of Technology)

Solution:

(a) Allowing for the light blocked by the observer's cage, the ratio of the light-gathering power of the telescope to that of the unaided eye is equal to the ratio of the collecting areas:

$$\frac{L_t}{L_e} = \frac{\pi[(5.08 \text{ m})^2 - (1.8)^2]}{\pi(2 \times 10^{-3} \text{ m})^2} = 5.6 \times 10^6$$

(b) Equation 45–21 applies for both mirror and lens objectives. (•Can you see why?) For the two mirror focal lengths given, we have

$$M_{\text{primary}} = -\frac{f_1}{f_2} = -\frac{17 \text{ m}}{0.04 \text{ m}} = -425$$

$$M_{\text{secondary}} = -\frac{150 \text{ m}}{0.04 \text{ m}} = -3750$$

(c) Using Eq. 43–5b with $\lambda = 550$ nm, we have for the Hale telescope

$$\delta\theta_{\text{min,H}} = \frac{1.22\lambda}{d} = \frac{1.22(550 \times 10^{-9} \text{ m})}{5.08 \text{ m}}$$

$$= 1.32 \times 10^{-7} \text{ rad} \qquad \text{or} \qquad 0.0272 \text{ arc sec}$$

For the unaided eye,

$$\delta\theta_{\text{min,}e} = \frac{1.22\lambda}{d} = \frac{1.22(550 \times 10^{-9} \text{ m})}{2 \times 10^{-3} \text{ m}}$$

$$= 3.36 \times 10^{-4} \text{ rad}$$

Thus,

$$\frac{\delta\theta_{\text{min,H}}}{\delta\theta_{\text{min,}e}} = \frac{1.32 \times 10^{-7} \text{ rad}}{3.36 \times 10^{-4} \text{ rad}} = 3.7 \times 10^{-4}$$

The actual resolution of the Hale telescope is limited by factors other than diffraction (for example, atmospheric turbulence), so the achieved value is about 5×10^{-6} rad or 1 arc sec. The main advantage of a large-diameter telescope is not that the diffraction effects are small but that the light-gathering power is large.

An astronomer rides in the observer's cage at the primary focus of the Hale telescope at the Palomar Observatory. (Courtesy of Palomar Observatory, California Institute of Technology.)

45-5 CAMERAS

Figure 45–15 shows a schematic view of a simple camera in which the essential optical component is a single converging lens that forms a real image on the film. For a camera of ordinary size, the focal length of the lens must be relatively short. The lens must also have a large angle of view. In a "35–mm" camera, the *field stop* (Fig. 45–15) defines a rectangular frame on the film that is 24 mm × 36 mm. A typical standard lens for such a camera has a focal length of 50 mm with an angle of 45°. To achieve a large light-gathering power, the lens would have a diameter of about 35 mm. To meet these geometrical requirements, while maintaining the smallest possible aberrations, places a severe burden on the lens design. Consequently, modern camera lenses necessarily consist of many individual lens elements. One type of camera lens that has been popular for many years is the Zeiss *Tessar* lens, shown in Fig. 45–16.

Fig. 45–16. Diagram of a Zeiss *Tessar* lens. Rudolph of Jena, in 1902, devised the Tessar-formula lens, which gives sharp images for angles of view up to 60° and for *f*-numbers as small as $\varphi = 3.5$. Such lenses helped make miniature 35-mm photography practical.

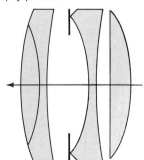

Fig. 45–15. Schematic diagram of a simple camera. The aperture stop limits the amount of light that enters the camera, and the field stop limits the exposed area of the film.

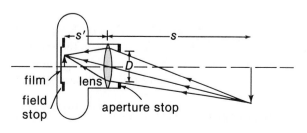

Lenses with angles of view larger and smaller than that of a standard 50 mm (focal length) lens are readily available. A wide-angle lens with a focal length of 28 mm has an angle of view of approximately 75°. A telephoto lens (see Example 45–2) with a focal length of 200 mm has an angle of view of about 12°.

(Courtesy of Olympus Camera Corp.)

Aperture Stops and Depth of Field. The amount of light that enters a camera is controlled by an *aperture stop* with variable diameter D. The maximum setting of the aperture stop is the lens diameter D_0. (For the Zeiss *Tessar* lens, Fig. 45–16, the aperture stop is located between the two lens pairs.) If the stop diameter is set to the maximum value D_0 and if the source is at a distance s, the amount of light that enters the camera is proportional to D_0^2/s^2. Now, the ratio of the amount of light incident on the film per unit area of the image to the amount of light radiated toward the camera per unit area of the object is just the inverse square of the lateral magnification m, namely, $1/m^2 = s^2/s'^2$. Thus, the brightness* of the image is proportional to

$$\frac{D_0^2}{s^2} \cdot \frac{s^2}{s'^2} = \frac{D_0^2}{s'^2}$$

Usually the distance s to an object being photographed is much larger than the focal length f of the camera lens. Then, the image distance s' is approximately equal to the focal length, so that the image brightness is proportional to D_0^2/f^2. The ratio f/D_0 is called the *f-number* (or the "speed") of the lens. We write†

---

*In technical literature, this quantity is referred to as the *illuminance* of the image.

†The *f*-number is often written as *f*/2.8 or *f*:3.5. To avoid confusion with the symbol for focal length, we write $\varphi = 2.8$ or $\varphi = 3.5$.

$$\varphi_0 = \frac{f}{D_0} \qquad \textit{f-number of lens} \qquad \textbf{(45–22a)}$$

The amount of light that passes through the lens at full aperture is therefore proportional to the inverse square of the *f*-number.

When the aperture stop is set to a diameter $D < D_0$ to reduce the amount of light arriving at the film, the setting is referred to as the *f-stop*,

$$\varphi = \frac{f}{D} \qquad \textit{f-stop of aperture} \qquad \textbf{(45–22b)}$$

Most cameras have click-stop aperture settings that correspond to successive ratios of approximately $\sqrt{2}$. Thus, we find $\varphi = 1.4, 2, 2.8, 4, 5.6, 8, 11, 16,$ and 22. A change of one click-stop of $\varphi = 4$ to $\varphi = 2.8$ (thereby increasing D) means that the amount of light admitted to the camera is doubled. Many cameras also have intermediate click-stops at $\varphi = 1.8, 2.4, 3.5,$ and possibly other values.

An important consideration in photography is the range of object distances for which a single setting of the camera focus will result in images with satisfactory sharpness. This range of distances, $\pm \Delta x$, is called the *depth of field* of the lens. Figure 45–17 shows converging rays from three object points at distances $x - \Delta x$, x, and

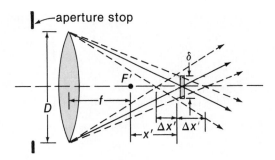

Fig. 45–17. Geometry for calculating depth of field.

$x + \Delta x$ (measured from the focal point F to the left of the lens—not shown). If the focus is set correctly for the distance x, rays that originate at the nearer and farther object points each produce blurred circles with diameter δ at the film. Because $\Delta x' \ll f$, we have

$$\frac{\delta}{D} \cong \frac{\Delta x'}{f + x'} \qquad \textbf{(45–23)}$$

Differentiating $xx' = f^2$ gives

$$\Delta x' = -\frac{f^2}{x^2} \Delta x \qquad \textbf{(45–24)}$$

Combining Eqs. 45–23 and 45–24, we obtain

$$\Delta x = -\frac{\delta(f + x')x^2}{Df^2}$$

Now, $x' \ll f$, so that neglecting x' in comparison with f, the resulting depth of

field is*

$$\text{D.F.} = \pm\Delta x = \pm\delta \cdot \frac{\varphi s^2}{f^2} \qquad \text{depth of field} \qquad \textbf{(45–25)}$$

In 35-mm photography, it is generally accepted that if a point source forms an image with a diameter δ no larger than 0.025 mm, the image is considered "sharp." With δ fixed, Eq. 45-25 shows that D.F. $\propto 1/f^2$. Thus, the depth of field for a wide-angle lens with $f = 28$ mm is substantially greater than that for a standard lens with $f = 50$ mm. Conversely, the depth of field for a telephoto lens is rather shallow.

Notice also that Eq. 45-25 predicts that the depth of field is directly proportional to the aperture stop φ. Lenses for 35-mm cameras, as well as cameras of other types, have engraved scales that show the depth of field for each combination of f-stop and focused subject distance. This scale permits a practice known as *field focusing*. The aperture is set to allow proper exposure for the ambient light level; then the infinity mark on the focusing ring is set opposite that f-stop on the depth-of-field scale (see the photograph on this page). Under these conditions, any subject between infinity and the near end of the depth of field will be in focus, and the camera may be used for rapid shooting without adjustments. Indeed, simple cameras that have no focusing mechanism rely entirely on depth of field to keep subjects in focus. For this reason they are designed with lenses having short focal lengths and small apertures.

The photograph on the left shows the scales on a typical camera lens. The indicator dot at the center is used with both the focusing and aperture scales. In this case, the lens is set for $\varphi = 11$ and a subject distance of about 12 m. The depth-of-field scale on each side of the indicator dot shows that subjects between about 9 m and 15 m will be in focus.

The photograph on the right shows the same lens set for field focusing. The aperture has been set to $\varphi = 16$, and the infinity mark of the focusing scale has been placed opposite the left-hand "16" on the depth-of-field scale. The right-hand "16" shows that the nearest subject that will be in focus is at 15 m; *any* subject between 15 m and infinity will be in focus. (Photographs courtesy of J. Freedman.)

▶ *Slide Projectors.* The basic operation of a slide projector is shown in Fig. 45–18. The condensing lens system, which consists of a pair of large plano-convex lenses, gathers light from the source and illuminates the slide, for example, at point *P*. The projection lens (usually an achromatic lens) then forms a sharp

---

*Note that the Gaussian and Newtonian formulas give $\pm\Delta x = \pm\Delta s$. (•Can you see why?)

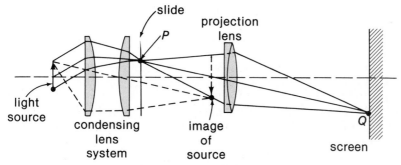

Fig. 45–18. Schematic diagram of a slide projector.

image of point P on the screen at point Q. The focal length and the placement of the projection lens are chosen to produce a large image of the slide on the screen (much larger than suggested in Fig. 45–18). The image of the source is actually at the position of the projection lens, but is shown displaced for clarity. This image should fill the projection lens. (•Can you see why?) The function of the condensing lens system is to direct a substantial portion of the light from the source toward the projection lens and to ensure that the slide is evenly illuminated from edge to edge. ◄

Example 45–7

A 35-mm camera with a 50-mm (focal length) lens is used to photograph an object at a distance of 5 m. An f-stop of $\varphi = 4$ is used.
(a) What is the distance x' of the image from the focal point of the lens?
(b) What is the depth of field at $s = 5$ m?
(c) What is the diameter of the first diffraction minimum of the image on the film of a point object when $\varphi = 22$?

Solution:

(a) We have $s = f + x$, so that

$$x = s - f = 5 \text{ m} - 0.05 \text{ m} = 4.95 \text{ m}$$

The distance x' is found by using the Newtonian formula, $xx' = f^2$; thus,

$$x' = \frac{f^2}{x} = \frac{(0.05 \text{ m})^2}{4.95 \text{ m}} = 5.05 \times 10^{-4} \text{ m}$$

or $x' = 0.51$ mm.
(b) The result obtained in (a) shows that $x \gg f$ and $x' \ll f$. Hence, Eq. 45–25 is valid. Using $\delta = 0.025$ mm, we find

$$\text{D.F.} = \pm\delta \cdot \frac{\varphi s^2}{f^2} = \pm(2.5 \times 10^{-5} \text{ m}) \cdot \frac{(4)(5 \text{ m})^2}{(0.05 \text{ m})^2}$$

$$= \pm 1.0 \text{ m}$$

Thus, all objects between $s = 4$ m and $s = 6$ m will be in satisfactory focus when the camera is focused to 5 m.
(c) Using Eq. 43–5b, $\delta\theta_{\min} = 1.22\,\lambda/D$, the diameter of the diffraction minimum is

$$d = 2f(\delta\theta_{\min}) = 2f \cdot \frac{1.22\,\lambda}{D} = 2.44\,\lambda\varphi$$

The average wavelength of white light is approximately 550 nm. Then, using $\varphi = 22$, we find

$$d = 2.44(550 \times 10^{-9} \text{ m})(22) = 2.95 \times 10^{-5} \text{ m} = 0.0295 \text{ mm}$$

Thus, the value of δ used to obtain the depth of field is approximately equal to the diameter of the central spot of the diffraction pattern for the largest f-stop ordinarily used in 35-mm photography.

The diffraction-limited angular resolution of the eye (2 mm diameter) is approximately 3.4×10^{-4} rad. Consequently, the diameter of the smallest resolvable detail on a photograph held at the near point of the eye is

$$d_e = 2(0.25 \text{ m})(3.4 \times 10^{-4}) = 1.7 \times 10^{-4} \text{ m} = 0.17 \text{ mm}$$

Thus, the accepted value of δ (namely, 0.025 mm) on a 35-mm negative would permit an enlargement of $d_e/\delta = (0.17 \text{ mm})/(0.025 \text{ mm}) = 6.8$ without any blurring detectable by the eye.

SPECIAL TOPIC

45-6 THE EYE AND VISION

The human eye is a marvelous optical instrument. It has a variable focal length that permits the formation of sharp images for a range of object distances from about 25 cm to infinity; it can function over an extremely wide range of light intensity; it produces color sensations; and it requires very little maintenance.

The eye of an adult human is a nearly spherical globe with a diameter of approximately 23 mm. Figure 45–19 shows the important features of the eye. An outer protec-

Fig. 45–19. Schematic diagram of the human eye.

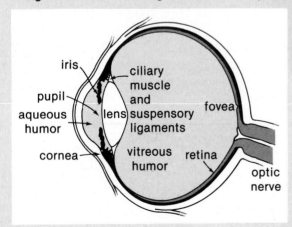

iris, ciliary muscle and suspensory ligaments, pupil, lens, fovea, aqueous humor, cornea, vitreous humor, retina, optic nerve

tive layer that consists of a tough elastic membrane called the *sclera* encloses the eye. At the front of the eye, this membrane leads into the *cornea*, which is a transparent material that admits light into the eye. The light then passes through a region filled with a clear liquid called the *aqueous humor*. The light that enters the *lens* is limited by the pupillary aperture in the *iris*, the colored part of the eye. The iris acts as a variable lens stop and allows the eye to adjust to differences in the intensity of the incident light. The cavity behind the lens is filled with a jellylike substance called the *vitreous humor*. When the eye views an object, the cornea and the lens form an image on the light-sensitive inner surface of the eye called the *retina*. Delicate nerve fibers leading from the retina unite to form the *optic nerve*, which carries the light-induced stimuli to the brain.

The refractive index of both the aqueous humor and the vitreous humor is approximately 1.336, almost the same as that of water. The lens is not a homogeneous structure; instead, the density, the hardness, and the refractive index increase toward the center. The outer region of the lens has an index of 1.405, whereas the inner region has an index of 1.454. This variation results in a partial correction for spherical aberration. Because the indexes for the various parts of the eye are not very different, most of the refraction of rays that enter the eye takes place at the air-cornea interface.

A normal relaxed eye will form a sharp retinal image for an object placed at any distance from about 6 m to infinity. For objects that are closer than 6 m to the eye, the focal length of the lens must be decreased to produce a sharp image on the retina. To achieve this change in focal length, the *ciliary muscle* contracts against the *suspensory ligaments* that support the lens. This reduces the radial tension on the lens and thereby increases its curvature. The ability to make this type of adjustment is called *accommodation*. The accommodation of the eye is limited, however, and objects closer than the *near point* cannot be brought into sharp focus. Typically, the near-point distance varies with age. For a 10-year-old child, the distance can be as small as 7 cm, increasing to 10 to 12 cm at age 20 years, reaching about 25 cm by age 40 years, and then rapidly increasing to 40 cm and greater after age 55 years.

The light-sensitive cells of the retina are of two types, the *rods* and the *cones*, as indicated in Fig. 45–20. The cones are color sensitive, but they require good lighting to be effective. The rods are more sensitive in low-light conditions, but they have little color response. The reduction in color perception under dim lighting conditions is quite apparent.

The rods and cones are not distributed uniformly over the retinal surface. The cones are concentrated in the central part of the retina, particularly in a region with a diameter of about 1/4 mm called the *fovea* (see Fig. 45–19), which contains only cones. The most distinct vision results when the image is formed on this dense collection of cones, corresponding to an angle of view of less than 1°. (For comparison, the angular diameter of the Moon as seen from the Earth is approximately 0.5°.)

Notice in Fig. 45–20 that the light sensor cells (the rods and cones) are not directly connected to the nerve fibers of the optic nerve. Instead, there is an interconnection by a different type of cells whose function is to analyze some of the visual information before stimuli are sent to the brain. Thus, a partial decoding of visual impressions takes place in the eye itself. The cone cells in the fovea, however, have a direct line to the brain.

At the point where the optic nerve passes through the retina, rods and cones are completely absent. Hence, an image formed at this point (called the *blind spot*) cannot be seen. This effect is easily verified. With your left eye closed, look steadily with your right eye at the cross in Fig. 45–21. Now move the book slowly away from your eye. At a distance of about 35 cm the dot will disappear, although it is visible at both nearer and farther distances. The blind spot corresponds to a viewing angle of approximately 17°.

An important feature of the response of the eye to light is the *persistence of vision*. When bright light flashes of very short duration are repeated at a rate of at least one flash every 1/10 or 1/20 second and are viewed by the central part of the retina, the source appears to be turned on continuously. This effect is sometimes called *flicker fusion*. (If the source is viewed peripherally, especially if the eye is fatigued, a higher flash rate is required to produce the persistence effect.) It is this characteristic of vision that makes practical the forms of motion pictures and television currently in use. The rapid projection of individual picture frames gives the appearance of smooth, nonflickering motion.

The human eye is sensitive over an intensity range of nearly 10^{10}. The upper limit corresponds to an intensity that causes the eye to experience discomfort, as might result from a momentary glance at the Sun. At the lower limit of intensity, as few as 10 or 20 photons with wavelengths in the middle part of the visible spectrum can be

Fig. 45–20. Scheme of the retina of the human eye. Light is incident from below. (Adapted from *The Anatomical Basis of Medical Practice* by Becker, Wilson, and Geweiler. Williams & Wilkins, Baltimore, 1971.)

Fig. 45–21. Geometric design to demonstrate the presence of the eye's blind spot. With your left eye closed, place the cross directly in front of the right eye and vary the distance from the page to your eye until the dot disappears.

detected when absorbed by the light receptors in the retina.*

When the light intensity varies, the eye adapts in part by changing the diameter of the pupil to admit more or less light. The range in diameter of the pupil is from about 2 mm to about 8 mm. The corresponding range in aperture area is 16:1, far less than the sensitivity range of the eye. The major part of the adaptation process is accomplished by chemical means in the retina. Such chemical changes are relatively slow, which explains why some time is required to adjust to low-light conditions upon coming indoors from a brightly lit outdoors.

The best angular resolution achieved by the eye results when the images are formed in the densely packed region of cones in the fovea. The ability of the eye to resolve two point sources of light is called *visual acuity*, defined to be the reciprocal of the angular resolution measured in minutes of arc. An eye that is able to resolve two point sources with an angular separation of 2.9×10^{-4} rad (1 arc min) is said to have a visual acuity of 1; if the sources are resolved only at a separation of 2 arc min, the visual acuity is 0.5; and so forth. The maximum acuity of the human eye is about 1. Experiments show that this value is achieved for a pupil diameter of about 2 mm and remains at this value as the pupil diameter is increased to full dilation. The improvement in acuity that should result from increased pupil diameter is offset by the increased aberrations due to the entry of nonparaxial rays at the larger diameters. The diffraction-limited angular resolution corresponding to a pupil diameter of 2 mm (see Eq. 43–5b) is 3.4×10^{-4} rad for $\lambda = 550$ nm. This corresponds to a visual acuity of 0.85. Thus, the eye is actually capable of slightly better angular resolution than that specified by the Rayleigh criterion. However, for the various calculations in the Examples and Problems, we have chosen to use the diffraction-limited value for a diameter of 2 mm. (The connection between angular resolution and cone spacing on the retina is the subject of Problem 45–18.)

Vision Defects.   The most common type of vision defect is the blurred image that results when the lens system of the eye is unable to focus the rays sharply on the retinal surface. Figure 45–22 illustrates the common defects. If the focal length of the relaxed eye is too short (due to an eyeball that is too long or to a cornea with too great a curvature), rays from a distant object converge in front of the retina (Fig. 45–22b). This defect is called *myopia* ("short-sightedness"). An unaided myopic eye can still see objects that are close by; the near-point distance of such an eye is less than for a normal eye.

---

*Only about 10 percent of the photons that enter the eye are eventually absorbed by the rods and cones; the remainder are reflected or are absorbed in other parts of the eye. However, a dark-adapted eye can detect, as a light flash, the passage through the eye of a single energetic atomic particle. Astronauts have confirmed seeing the flashes due to individual cosmic-ray particles.

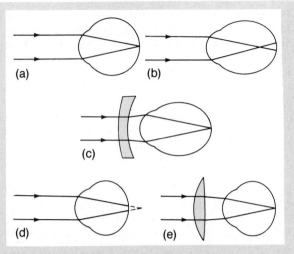

Fig. 45–22.   (a) Focusing of parallel rays for a normal eye. (b) Focusing for a myopic eye. (c) Correction of a myopic eye with a diverging lens. (d) Focusing for a hypermetropic (or presbyopic) eye. (e) Correction of a hypermetropic (or presbyopic) eye with a converging lens.

At the opposite extreme is *hypermetropia* ("long sightedness"), in which the focal length of the relaxed eye is too long (due to an eyeball that is too short or to a cornea with too small a curvature). In a hypermetropic eye the rays from a distant object converge toward a point beyond the retina (Fig. 45–22d). Such an eye has a near-point distance that is greater than normal, making close work difficult. Distant objects are still focused sharply.

As a person ages, the ability of the eye to accommodate lessens, and it becomes more difficult to focus on nearby objects. This effect, called *presbyopia* ("old sight"), is due to the loss of flexibility in the lens, which makes difficult the change of curvature required for near vision. Presbyopia is similar to hypermetropia, but distant vision remains unimpaired to closer distances.

Another type of defect results when the eye does not have symmetry about the optic axis. Then, the focal properties are different for horizontal and vertical line objects. This defect is called *astigmatism*.

Correction of Vision Defects.   The vision defects of myopia, hypermetropia, and presbyopia can be corrected by using simple spherical lenses, worn as *spectacles*, as indicated in Fig. 45–22c and e. (The correction of astigmatism requires lenses with cylindrical curvature.) Notice that the correction of myopia requires a diverging or negative lens ($f < 0$), whereas hypermetropia and presbyopia require a converging or positive lens ($f > 0$).

The focal lengths of lenses used for correcting vision defects are usually given in terms of *diopters*. The *power* of a lens with a focal length of f meters is

$$\text{power (in diopters)} = \frac{1}{f} \qquad \textbf{(45–26)}$$

For a normal eye, the effective lens-to-retina distance is $s' = 23$ mm. Thus, the diopter power of the relaxed eye (focused on an object at $s = \infty$) is given by

$$\frac{1}{f} = \frac{1}{s'} = \frac{1}{0.023 \text{ m}} = +43.5 \text{ diopters}$$

For an eye focused on an object at the near point, $s = 25$ cm, the diopter power is given by

$$\frac{1}{f} = \frac{1}{s'} + \frac{1}{s} = \frac{1}{0.023 \text{ m}} + \frac{1}{0.25 \text{ m}} = +47.5 \text{ diopters}$$

Thus, the accommodation of a normal eye amounts to 4 diopters.

Notice that when two thin lenses are used in contact with their optic axes coincident, the diopter power of the combination is simply the sum of the diopter powers of the two individual lenses. (See Eq. 45–6.)

The corrections that can be achieved by spectacle lenses can be duplicated by lenses that are fitted directly to the cornea; such lenses are called *contact lenses*. The direct contact of these lenses with the cornea gives them a special advantage over spectacle lenses. Most of the refraction of rays that enter the eye takes place at the air-cornea interface. When a contact lens is in place, tear fluid fills the narrow gap between the lens and the cornea. This fluid has an index of refraction that is nearly equal to that of the cornea. Consequently, the anterior corneal surface no longer contributes any significant amount of refraction, and essentially the entire refraction of the entering rays occurs at the front surface of the contact lens. A bulging cornea causes such severe abnormality of vision that ordinary spectacle lenses cannot achieve a satisfactory correction. With contact lenses, however, the abnormal corneal curvature can be negated and replaced by the proper curvature of the lenses. An additional advantage of contact lenses is that they have almost no effect on the apparent size of an object viewed through them. Ordinary spectacle lenses, although they can correct faulty focusing of the eyes, do affect the size of the image.

Example 45–8

A certain myopic eye has a near point of 11 cm and can focus sharply on an object only for distances out to 15 cm (the *far-point distance*).

(a) What is the diopter strength of the lens that is required to correct the distant vision?

(b) What is the near-point distance when the person uses the lens found in (a)?

(c) What is the diopter strength of the lens that is required to yield a near-point distance of 25 cm?

Solution:

(a) It is desired to move the far point to infinity. The corrective lens (assumed to be placed immediately in front of the eye) must form an image of an infinitely distant object at the far point of the eye. Thus, $s' = 0.15$ m; then,

$$\frac{1}{f} = \frac{1}{\infty} + \frac{1}{-0.15 \text{ m}} = -6.67 \text{ diopters}$$

Lenses are shaped to the nearest 1/4 diopter, so the lens power in this case would be given as −6.75 diopters. This is a rather strong correction.

(b) The near-point distance with the eye aided by the lens in (a) is given by

$$\frac{1}{s} = \frac{1}{f} - \frac{1}{s'} = -6.75 \text{ m}^{-1} - \frac{1}{-0.11 \text{ m}} = 2.34 \text{ m}^{-1}$$

from which

$$s = 0.43 \text{ m} = 43 \text{ cm}$$

Thus, correcting the far point gives a near point that is unsuited for close work or reading.

(c) Proper reading glasses would allow $s = 25$ cm; thus, with $s' = -11$ cm, the lens power is

$$\frac{1}{f} = \frac{1}{0.25 \text{ m}} + \frac{1}{-0.11 \text{ m}} = -5.09 \text{ diopters}$$

so that the lens would be ground to yield a strength of −5.0 diopters. A pair of bifocal eyeglasses would have an upper part consisting of lenses with a power of −6.75 diopters and a lower part consisting of lenses that are 1.75 diopters weaker.

PROBLEMS

Section 45-1

45-1 A symmetric thick lens has a focal length of 5 cm and principal planes that are separated by 2 cm. Locate the image of an object that is placed 4 cm from the center of the lens. Is the image real or virtual? Erect or inverted?

45-2 Two thin lenses, one with $f_1 = 50$ cm and the other with $f_2 = 30$ cm, are placed in contact with their optic axes coincident. Determine the focal length of the combination. Locate both focal points and both principal points.

45-3 The two lenses in Problem 45-2 are separated by a distance of 10 cm. Determine the focal length and locate the focal points and the principal points.

Section 45-2

45-4 A magnifying glass with a focal length of 10 cm is used by a person with normal vision. (a) Determine the angular magnifications γ and γ_∞. (b) For what range of object distances, s_b to s_c, will the image be in sharp focus? (Refer to Fig. 45-5.)

Section 45-3

45-5 The microscope described in Example 45-5 is refocused to permit viewing with a relaxed eye. (a) What is the new distance from the specimen to the objective? (b) What is the new magnification?

45-6 A compound microscope (Fig. 45-6) consists of an objective with a focal length of 6 mm and an eyepiece with a focal length of 20 mm. The tube length (the distance between the two lenses) is 15 cm. (a) What is the magnification of the microscope when adjusted for viewing by a relaxed eye? (b) For the condition in (a), what is the specimen-to-objective distance? (c) If the diameter of the objective is 3 mm, what is the numerical aperture (specimen in air) for $\lambda = 550$ nm? What is the linear resolution ξ?

45-7 A compound microscope is focused so that the final image appears at the near point of the viewer's eye (25 cm). Show that the angular magnification is

$$\Gamma = -\frac{s_1'}{s_1}\left(\frac{25 \text{ cm}}{f} + 1\right)$$

45-8 Justify the appearance of the refractive index n in Eq. 45-18a.

Section 45-4

45-9 The distance between the objective and the ocular of a simple astronomical telescope (Fig. 45-12a) is 40 cm, and the ocular has a magnification $\times 10$. What is the angular magnification of the telescope?

45-10 The objective lens of an astronomical telescope (Fig. 45-12a) has a focal length of 45 cm and a diameter of 5 cm; the ocular has a focal length of 1.5 cm. The telescope is used to view stars that emit light with average wavelength $\lambda = 550$ nm. (a) What is the angular magnification of the telescope? (b) What is the angular resolution (expressed in seconds of arc)? (c) What should be the distance between the lenses for optimum viewing of stars? What lens separation should be used when viewing an object at a distance of 15 m? (In both cases, assume viewing by a relaxed eye.)

45-11 A small Galilean telescope (Fig. 45-12b) has an objective with focal length $f_1 = 15$ cm and an ocular with focal length $f_2 = -3$ cm. A distant object is viewed by a relaxed eye. (a) What is the angular magnification of the telescope? (b) What is the distance between the lenses?

45-12 A reflecting telescope (Fig. 45-13a) consists of a spherical mirror with a radius of curvature of 1.5 m and a diameter of 10 cm together with an ocular with magnification $\times 15$. The telescope is used to view a light source with $\lambda = 550$ nm. (a) Determine the angular magnification of the telescope. (b) What is the angular resolution (expressed in seconds of arc)?

Section 45-5

45-13 A 35-mm camera has a lens with a focal length of 50 mm. By what amount must the lens-to-film distance be changed if the focus is to be shifted from a very distant object (essentially, $s = \infty$) to one at a distance of 1 m from the lens?

45-14 A 35-mm camera is equipped with a telephoto lens with focal length $f = 135$ mm. The object to be photographed is at a distance of 10 m. (a) What is the depth of field when the *f*-stop is $\varphi = 5.6$? (b) If the object is a point source of light with $\lambda = 550$ nm, what is the diameter of the first diffraction minimum on the film? (c) The *f*-stop is reset to admit four times as much light. What is the new depth of field? What is the new diameter of the diffraction minimum?

45-15 A slide projector for 35-mm transparencies (Fig. 45-18) is to project an image with size 1.2 m × 1.8 m on a screen at a distance of 8 m from the projection lens. What is the focal length of the required projection lens? What must be the distance from the slide to the projection lens?

SPECIAL TOPIC

Section 45-6

45-16 Assume that your eye is a sphere with a diameter of 23 mm. Determine the size of the retinal image when you focus on the following objects: (a) a printed dot on this page (diameter 0.3 mm) located at the near point of your (normal) eye; (b) a tennis ball (diameter 6.5 cm) at a distance of 10 m; (c) an eagle (wing span 1.5 m) at a distance of 200 m; and (d) the full Moon (angular diameter 31 arc min).

45-17 Assume that your eye is a sphere with a diameter of 23 mm. Determine the distance on your retina between the fovea and the optic nerve by measuring the distance from the cross to your eye in the blind-spot test using Fig. 45-21. What is the blind-spot angle in this test (the angle between the forward direction and the direction from the eye to the object not seen)?

45-18 (a) Determine the diffraction-limited angular resolution of the eye according to the Rayleigh criterion for a pupil diameter of 2 mm and for yellow-green light ($\lambda = 550$ nm). (b) What is the linear resolution of the eye at the near point (25 cm)? (c) Assume that the eye is a sphere with a diameter of 23 mm. Determine the diameter of the first diffraction minimum of the retinal image of a point source observed with a pupil diameter of 2 mm. (d) The cone cells in the fovea are closely packed and have diameters ranging from 0.0015 mm to 0.005 mm. In view of these sizes, comment on the result obtained in (c).

45-19 A person can accommodate to see clearly an object no closer than 1 m. What power lens would be required for this person to see the object clearly at 25 cm?

Additional Problems

45-20 The dimensions of a thick plano-convex lens are
• shown in the diagram. (a) Locate the principal points and the focal points. (b) An object is placed 10 cm to the left of the vertex of the convex surface. Locate the image.

45-21 A thin lens *L* with focal length $f = 8$ cm is
•• located 3.0 cm in front of a glass plate with a thickness of 0.5 cm, as shown in the diagram. Determine the focal length of the system. Locate the principal points and the focal points.

45-22 Redesign the achromatic doublet in Example 45-3,
• retaining $f_D = 50$ cm, but changing the radii so that a doubly convex lens results with $r_2' = -r_1 < 0$ and $r_2 = r_1'$.

45-23 A Huygens eyepiece, consisting of two plano-convex
•• lenses (Fig. 45–11a), is to be designed in such a way
that it has a magnification ×5 (when viewed by a
relaxed eye) and is achromatized for the wave-
lengths of the H_α and H_β lines ($\lambda_\alpha = 656$ nm and
$\lambda_\beta = 486$ nm, respectively). Both lenses are constructed from crown glass with $n_\alpha = 1.5146$,
$n_D = 1.5170$, and $n_\beta = 1.5226$, as in Example 45–3.
Let the distance t between the lenses be twice the
focal length of the eye lens. Find the lens radii that
give ×5 magnification for sodium yellow light (λ_D)
and equal focal lengths, $f_\alpha = f_\beta$, for wavelengths λ_α
and λ_β. Determine $f_\alpha = f_\beta$. (It is necessary to carry
the calculations to four significant figures.)

45-24 A magnifying glass is used by a person with normal
vision. When the glass is held close to the eye (the
situation assumed in obtaining Eqs. 45–16), the angular magnification is $\gamma = 4$. The glass is next
moved a distance of 1.5 cm from the eye. If the
image is located at the near point of the eye, what is
the new value of γ?

45-25 The first lens element of an oil-immersion microscope objective is a hemispherical plano-convex
lens, as shown in the diagram. The specimen to be
viewed is placed at the point P, which is a distance
$a = R/n$ from the lens. (R is the radius of curvature
of the lens and n is the refractive index of both the
immersion oil and the lens.) Show that a virtual
image is produced by the first element at point Q, a
distance $a' = nR$ from the lens. Show that this

image is completely free from spherical aberration
(*i.e.*, that Snell's law is valid for all angles). [*Hint:*
First, show that $\triangle CAQ$ is similar to $\triangle CAP$; then,
use the law of sines to show that Snell's law is exactly correct at point A. See also Problem 41–32.]

45-26 A microscope slide has a thickness of 2.0 mm. A
microscope that is focused on a scratch on the lower
surface of the slide must be raised by 1.32 mm to
focus on a scratch on the upper surface. Determine
the refractive index of the slide glass.

45-27 A telescope is focused to give a final image at the
• near point of a normal eye. Show that the angular
magnification is

$$\Gamma = -f_1\left(\frac{1}{f_2} + \frac{1}{25 \text{ cm}}\right)$$

45-28 Use the result of Problem 45–27 to calculate the
angular magnification of the Galilean telescope
described in Problem 45–11 when the object viewed
is at a distance of 15 m and the final image is at the
near point of the eye.

45-29 The 2.1-m diameter reflecting telescope at the Kitt
Peak National Observatory in Arizona has an *f*-
number $\varphi_0 = 2.63$. If a photograph of the Moon
(angular diameter 31 arc min) is made at the prime
focus, what is the diameter of the image? What is
the size of the smallest feature on the Moon's surface that can be seen if the angular resolution is
limited by diffraction? (Assume $\lambda = 550$ nm.)

45-30 A 35-mm camera that has a lens with a focal length
of 50 mm is used to photograph the Moon. The lens
opening is set to $\varphi = 3.5$. The angular diameter of
the Moon is 31 minutes of arc and its diameter is
3480 km. (a) What is the diameter on the film of the
Moon's image? (b) Calculate the diameter of the
diffraction spot for a point source of light on the
Moon if the wavelength of the light is 550 nm. (c) If
the lens aberrations and the grain size of the film
emulsion impose a resolution limit of 0.015 mm at
the film, is the image resolution limited by diffraction or by aberrations and grain size? (d) What is
the diameter of the smallest detail on the Moon that
can be resolved in the photograph?

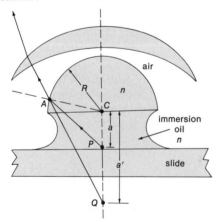

45-31 The near-point distance of a far-sighted (hypermetropic) person is 50 cm. (a) What is the diopter
strength of the lens that will correct the near-point

distance to 25 cm? (b) Assume that the eye is a
sphere with a diameter of 23 mm and that the pupil
diameter is 2 mm. What is the size of the blurred

retinal image of a point object located 25 cm from the unaided eye, which is focused to 50 cm? (c) Repeat part (b), assuming that squinting reduces the effective diameter of the pupil to 0.5 mm. Does squinting improve the sharpness of the image?

45-32 An optometrist finds that a patient has a near-point distance of 0.5 m and that she likes to read at a distance of 0.25 m. What is the accommodative power of the patient's eyes (which have a diameter of 23 mm)? What strength reading glasses should the optometrist prescribe?

45-33 A near-sighted (myopic) person has a far-point distance of 45 cm (refer to Example 45-8). (a) What is the diopter strength of the lens that will correct the far-point distance to infinity? (b) The eyes of the person have a diameter of 23 mm. If the diameter of the pupils is 3 mm, what is the size of the blurred retinal image formed by the relaxed eyes when unaided?

45-34 A magnifying glass has a focal length of 8 cm. (a) What is the angular magnification γ of the glass for a normal eye? (b) What angle θ_b (see Fig. 45-5b) would be subtended by the image of an object with a height of 1 mm when the image is at a distance of 25 cm? (c) The same glass is used by a near-sighted person (near-point distance = 11 cm) to view the same object. What angle θ_b is subtended by the image when it is at the near point of the myopic eyes? (d) Who makes better use of the lens, the *emmetrope* (the person with normal eyes) or the *myope* (the near-sighted person)?

QUANTA, WAVES, AND PARTICLES

In the final years of the nineteenth century, physics appeared to be a closed subject. A logically consistent, all-embracing set of basic physical laws was at hand to account in detail for all physical phenomena that had been observed in the macroscopic domain. Newtonian dynamics, Maxwellian electrodynamics, statistical thermodynamics, and optical theory (the main topics of the area called *classical physics*) were all well established and seemed capable of explaining every type of physical effect. What else of *real* importance could be discovered?

This state of complacency was soon jolted by a series of new discoveries in the microscopic domain of atoms and molecules. New and far-reaching questions were asked about the nature of radiation and matter, and about the properties and structure of atoms. During the early years of the twentieth century, an entirely new outlook concerning the microworld was being developed. Eventually, these views merged into modern *quantum theory*. In this chapter and the next, we give an introduction to these ideas following a general historical sequence.

46-1 THERMAL RADIATION

Every body, at whatever temperature above absolute zero, spontaneously emits electromagnetic radiation from its surface. This *thermal radiation* has a characteristic spectral distribution (amount of energy emitted per unit wavelength interval) that depends on temperature. Any solid, when heated to a temperature above about 900 K, emits some of its radiation in the visible part of the spectrum. For example, as the temperature of a body is raised through this region, it first glows dull red, then cherry red, and orange, finally becoming "white hot." At lower temperatures, the intensity of the visible radiation is too small to be observed easily. On the other hand, at the temperature of the Sun's surface (about 5800 K), the peak of the spectral distribution corresponds approximately to the center of the visible part of the spectrum (see Fig. 40–1).

It has been determined experimentally that the total amount R of electromagnetic radiation emitted by a unit area of surface is

$$R = e\sigma T^4 \quad (\text{W/m}^2) \tag{46-1}$$

where σ is a universal constant (called *Stefan's constant*):

$$\sigma = 5.6703 \times 10^{-8} \text{ W/m}^2 \cdot \text{K}^4$$

The quantity e (called the *emissivity*) is a dimensionless number, $0 < e < 1$. The emissivity is a characteristic of the particular substance and is a measure of the efficiency with which the body emits thermal radiation; an ideal radiator would have $e = 1$. The emissivity of a substance depends somewhat on temperature, tending to increase with increasing temperature. For temperatures in the range $250 \text{ K} \lesssim T \lesssim 1000 \text{ K}$, the emissivity of a bright shiny surface can be as small as 0.01. For a black oxidized metal surface, $0.6 \lesssim e \lesssim 0.8$, and for carbon black (soot), $e \cong 0.95$. Equation 46-1 is called the *Stefan-Boltzmann radiation law*.

When the radiation emitted from the interior of a cavity in a heated object is examined, it is found that the emissivity of the cavity is very nearly unity, independent of the composition of the object. The photograph shows a tungsten wire into which a small hole has been drilled and which has been heated to incandescence. Note that the radiation from the cavity appears more intense (brighter) than the radiation from the surface. Because a blackened body (for example, soot) has an emissivity very close to unity, thermal cavity radiation is sometimes referred to as *blackbody radiation*.

Figure 46-1 shows a comparison of the emission spectra for thermal cavity radiation at five different temperatures. It is found experimentally that the shape of the spectral curve for a particular temperature depends only on the temperature and not on the size or shape of the cavity nor on the material used for its walls.

Figure 46-2 shows schematically a thermal cavity radiator that consists of a cavity with arbitrary shape within a heated block of material; a small hole allows the escape of some of the radiation from the cavity. Because the probability of escape is small, radiation will bounce around inside and will interact many times with the cavity walls before emerging from the small hole. Consequently, it is reasonable to assume that the absorption and emission processes at the cavity walls are in thermal equilibrium.

Thermal cavity radiation from a hole drilled into a tungsten wire, which was then heated to incandescence. (From *Fundamentals of Physics*, 2nd ed., by D. Halliday and R. Resnick. John Wiley & Sons, New York, 1981.)

Fig. 46-1. Emission spectra for an ideal thermal cavity radiator at several different temperatures. The curves show the relative amounts of energy emitted per unit wavelength interval as functions of the wavelength.

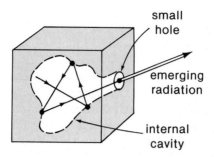

Fig. 46–2. Schematic diagram of a thermal cavity radiator.

It proves convenient, for the purpose of describing the thermal energy spectrum, to shift attention from the wavelength λ to the frequency ν. Thus, we denote by $u(\nu)$ the energy density per unit frequency interval of radiation with frequency ν inside the cavity. That is, the energy density in the frequency range from ν to $\nu + d\nu$ is $u(\nu)d\nu$. It can be shown (see Problem 46–5) that the amount of radiation that escapes through a small hole into the external hemisphere is*

$$R(\nu) = \tfrac{1}{4}cu(\nu) \qquad (\text{W} \cdot \text{s/m}^2) \qquad \textbf{(46–2)}$$

where $R(\nu)$ is the rate of radiation per unit area of the hole per unit frequency interval, and where c is the speed of light. Thus, the rate of radiation per unit area in the frequency range from ν to $\nu + d\nu$ is $R(\nu)d\nu$. The total power radiated per unit area is

$$R = \int_0^\infty R(\nu)d\nu = e\sigma T^4 \qquad (\text{W/m}^2) \qquad \textbf{(46–3)}$$

as required by Eq. 46–1.

The Rayleigh-Jeans Law. Rayleigh in 1900 and Jeans† in 1905 proposed a theoretical explanation for the shape of the thermal cavity radiation spectrum. The energy density function $u(\nu)$ is found to be independent of the shape of the cavity and its wall material. Therefore, we may use a simple geometry in which the cavity is a cube with sides a and the walls are made from an ideal conductor. The presumed thermal equilibrium in the cavity means that any radiant energy absorbed by the wall is reradiated, so the result is the same as if all of the radiation were simply reflected at the cavity wall. This requires that the field within the cavity be in the form of standing electromagnetic waves with $\mathbf{E} = 0$ at the walls. The situation is analogous to that of standing sound waves in a room (Example 20–5). Indeed, we can use here the result of that example, namely,

$$n_x^2 + n_y^2 + n_z^2 = (2\nu a/c)^2 \qquad \textbf{(46–4)}$$

which gives the allowed frequencies ν when n_x, n_y, and n_z take on all possible combinations of positive integers (including zero). Thus, each set of values (n_x, n_y, n_z) defines a possible standing-wave mode within the cavity.

How many standing-wave modes are there with frequencies less than or equal to a specified frequency ν? Equation 46–4 is just the equation for a sphere with radius

---

*The dimensions of $R(\nu)$ are power per unit area per unit frequency interval, $(\text{W/m}^2)/\text{s}^{-1} = \text{W} \cdot \text{s/m}^2$.
†Sir James Jeans (1877–1946), English mathematical physicist.

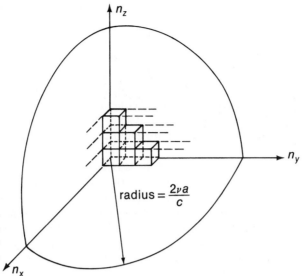

Fig. 46-3. Each possible standing-wave mode in the cavity corresponds to a unit cube in the three-dimensional space of n_x, n_y, and n_z.

$2\nu a/c$ in a Cartesian space with coordinates (n_x, n_y, n_z), as shown in Fig. 46-3. Thus, the number of possible standing-wave modes is equal to the number of unit cubes that can be placed inside an octant* of a sphere with radius $2\nu a/c$; thus

$$N(\nu) = \frac{1}{8} \cdot \frac{4\pi}{3} \left(\frac{2\nu a}{c}\right)^3 = \frac{4\pi}{3} \frac{a^3 \nu^3}{c^3} \qquad (46\text{--}5)$$

Differentiating $N(\nu)$, we obtain the number of modes per unit frequency interval,

$$\frac{dN(\nu)}{d\nu} = 4\pi a^3 \nu^2 / c^3 \qquad (46\text{--}6)$$

Now, there are two possible states of polarization for each standing wave, so the number of modes is doubled. Also, a^3 is just the cavity volume, so the number of independent standing waves per unit volume per unit frequency interval is

$$n(\nu) = 8\pi \nu^2 / c^3 \qquad (46\text{--}7)$$

If the average energy per standing wave is $\langle E \rangle$, the energy density per unit frequency interval is

$$u(\nu) = \langle E \rangle n(\nu) \qquad (46\text{--}8)$$

It was pointed out in Section 24-2 that for each quadratic term that appears in the expression for the total energy, there corresponds one degree of freedom. In Section 38-2 we learned that the energy density in an electromagnetic field in free space is proportional to $\epsilon_0 E^2 + \mu_0 H^2$ (here, E represents the electric field intensity). Thus, the number of degrees of freedom associated with each frequency mode of the electromagnetic field is 2. According to the equipartition of energy theorem, each

---

*An *octant* because n_x, n_y, and n_z are all positive integers (or zero).

Fig. 46-4. The Rayleigh-Jeans law (Eq. 46-9) can account only for the low-frequency part of the observed cavity radiation spectrum (solid curve).

degree of freedom has an average energy $\frac{1}{2}kT$ (Eq. 24-12). Thus, $\langle E \rangle = 2 \times \frac{1}{2}kT$, whence

$$u(v) = (8\pi v^2/c^3)kT \qquad (46\text{-}9)$$

This result, known as the *Rayleigh-Jeans law*, accounts quite well for the shape of the radiation spectrum for low frequencies, but it is seriously in error for high frequencies (see Fig. 46-4). Equation 46-9 predicts that $u(v)$ increases with frequency without bound. Thus, the cavity could contain an infinite amount of energy—clearly, an unphysical situation.

Planck's Law. In 1900, Max Planck* was also studying the shape of the thermal cavity radiation spectrum. He was able to account for the spectral shape for all frequencies, but only by introducing a strange and unorthodox idea. This idea was the first clue in a rapidly expanding body of evidence that led eventually to the development of modern quantum theory. To gain some appreciation for Planck's important first step, we now give some details of the statistical treatment of cavity radiation.

In the Rayleigh-Jeans model, each frequency mode of the radiation field can have *any* energy; it is required only that the *average* energy be equal to the equipartition value, $\langle E \rangle = kT$. Because the wall oscillators and the radiation field exchange energy under conditions of thermodynamic equilibrium, it should be possible to treat the problem by emphasizing the role of the wall oscillators. The wall oscillators also have two degrees of freedom (one associated with vibrations and one with potential energy—see the discussion preceding Eq. 24-11), so that the equipartition theorem again requires $\langle E \rangle = kT$.

Instead of considering average energies, Planck investigated the possibility that the energies of the wall oscillators are restricted to integer multiples of some particular amount of energy E_0. Thus, the lowest energy for a particular oscillator mode with frequency v would be zero, the next higher mode would have energy E_0, the next $2E_0$, then $3E_0$, and so forth.† Next, Planck assumed that the probability for exciting a particular energy of a mode is proportional to the Boltzmann statistical factor (Section 24-3). Then, the values of the relative probability $n(E)$ for the different oscillator energy states are:

E	$n(E)$
0	A
E_0	$Ae^{-E_0/kT}$
$2E_0$	$Ae^{-2E_0/kT}$
$3E_0$	$Ae^{-3E_0/kT}$
$\vdots$	$\vdots$
nE_0	$Ae^{-nE_0/kT}$

The total number of oscillators with frequency v is‡

$$N = A[1 + e^{-E_0/kT} + e^{-2E_0/kT} + \cdots]$$
$$= A[1 - e^{-E_0/kT}]^{-1} \qquad (46\text{-}10)$$

---

*Max K. E. L. Planck (1858-1947), German physicist, was awarded the 1918 Nobel Prize in physics for his development of the quantum concept.

†Planck's simple guess about the energies of the wall oscillators was quite fortuitous. The scheme that was assumed turns out to represent exactly the quantum theory result for the harmonic oscillator. (See Example 47-2.)

‡We use the expansion $(1 - x)^{-1} = 1 + x + x^2 + x^3 + \cdots$

The total energy associated with these oscillators is*

$$
\begin{aligned}
E_{\text{tot}} &= 0 \cdot A + E_0 A e^{-E_0/kT} + 2E_0 e^{-2E_0/kT} + \cdots \\
&= E_0 A e^{-E_0/kT}[1 + 2e^{-E_0/kT} + 3e^{-2E_0/kT} + \cdots] \\
&= E_0 A e^{-E_0/kT}[1 - e^{-E_0/kT}]^{-2}
\end{aligned}
\tag{46-11}
$$

Hence, the average oscillator energy $\langle E \rangle$ is

$$
\langle E \rangle = \frac{E_{\text{tot}}}{N} = \frac{E_0 e^{-E_0/kT}}{1 - e^{-E_0/kT}} = \frac{E_0}{e^{E_0/kT} - 1}
\tag{46-12}
$$

Using Eqs. 46-7 and 46-12 in Eq. 46-8, there results

$$
u(\nu) = \frac{8\pi\nu^2}{c^3} \frac{E_0}{e^{E_0/kT} - 1}
\tag{46-13}
$$

Planck realized that by substituting

$$
E_0 = h\nu
\tag{46-14}
$$

the thermal cavity radiation spectrum could be duplicated for all frequencies. The resulting expression for the energy density per unit frequency interval is

$$
u(\nu) = \frac{8\pi h\nu^3}{c^3} \frac{1}{e^{h\nu/kT} - 1} \qquad (\text{J/m}^3)/\text{s}^{-1}
\tag{46-15}
$$

This theoretical frequency distribution function is known as *Planck's law*.† The quantity h that represents the proportionality constant between the energy E_0 and the frequency ν (Eq. 46-14) is a new universal constant (called *Planck's constant*) that must be determined by experiment; the best present value is

$$
h = 6.62618 \times 10^{-34} \text{ J} \cdot \text{s}
$$

with an uncertainty of 4 in the last decimal.

Let us examine Eq. 46-15 in the range of low frequencies. Expanding the exponential, we find (see Eq. A-11 of Appendix A, in Volume I)

$$
\begin{aligned}
u(\nu) &= \frac{8\pi h\nu^3}{c^3} \frac{1}{\left(1 + \dfrac{h\nu}{kT} + \cdots\right) - 1} \\
&= \frac{8\pi\nu^2}{c^3} \frac{h\nu}{\dfrac{h\nu}{kT} + \cdots} \\
&\cong \frac{8\pi\nu^2}{c^3} kT
\end{aligned}
$$

---

*The expansion is $(1 - x)^{-2} = 1 + 2x + 3x^2 + 4x^3 + \cdots$

†Planck's formula was first presented in October 1900 simply as an empirical relationship that fits the blackbody data. Two months later, Planck gave a derivation of the formula based on the idea of quanta.

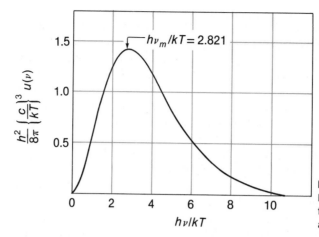

Fig. 46–5. Plotted in this way, the Planck radiation formula is valid for all frequencies and all temperatures.

The result shows that Planck's equation (which agrees with experiment for all frequencies) takes the form of the Rayleigh-Jeans law if only low frequencies are considered (see Fig. 46–4). Notice that we would obtain the same result for $u(\nu)$ by taking the limit $h \to 0$. This has the effect of removing the restriction to discrete energies. If all energies are allowed, we return to the conditions of the (incorrect) Rayleigh-Jeans model.

A universal graph of Planck's law (Eq. 46–15) can be obtained by plotting $(h^2/8\pi)(c/kT)^3 u(\nu)$ versus $h\nu/kT$, as shown in Fig. 46–5. The maximum of the distribution $u(\nu)$, for a particular temperature T, occurs for (see Problem 46–25)

$$h\nu_m/kT = 2.821 \qquad (46\text{–}16)$$

The wavelength distribution function that corresponds to Eq. 46–15 is expressed as energy density per unit wavelength interval:

$$u(\lambda) = \frac{8\pi hc}{\lambda^5} \frac{1}{e^{hc/\lambda kT} - 1} \qquad (J/m^3)/m \qquad (46\text{–}17)$$

The maximum value of $u(\lambda)$, for a particular temperature T, occurs for (see Problem 46–26)

$$\frac{hc}{\lambda_m kT} = 4.965 \qquad (46\text{–}18)$$

When this expression is written in the form $\lambda_m T = \text{constant}$, it is known as the *Wien displacement law.**

Planck used his theoretical expressions together with data from radiation experiments to determine values for the constants h and k. (See Problem 46–22.) A value for Stefan's constant σ could also be determined by using Eq. 46–15 for $u(\nu)$ in Eq. 46–2 and substituting the resulting expression for $R(\nu)$ into Eq. 46–3. The value that is obtained in this way agrees with the experimental value. (See Problem 46–23.)

Planck's radiation law was found to give an excellent representation of the vast

---

*Discovered in 1896 by Wilhelm Wien (1864–1928), German physicist, who was awarded the 1911 Nobel Prize in physics for his research on blackbody radiation. Wien had also shown, from thermodynamic considerations, that $u(\nu)$ had to be of the form $u(\nu) = \nu^3 f(\nu/T)$, where $f(\nu/T)$ is some differentiable function of the ratio ν/T. Referring to Eq. 46–13, we see that this conclusion requires that $E_0 = \text{const.} \times \nu$.

body of experimental data that had been obtained concerning thermal cavity radiation. The astonishing conclusion drawn by Planck was that it is *not* possible for the cavity oscillators to absorb and reradiate energy in a continuous manner; instead, the oscillators must gain and lose energy in discrete amounts, which Planck called energy *quanta*.

Planck did not go so far as to suggest that all electromagnetic radiation is quantized. (It remained for Einstein to make this crucial step in his theory of the photoelectric effect—Section 46–2.) However, if the cavity wall and the radiation field within the cavity are in thermodynamic equilibrium, the quantized absorption and reradiation of energy by the wall oscillators strongly suggests that the standing waves in the cavity also have quantized energies, $E = nE_0$ ($n = 0, 1, 2, \cdots$), with $E_0 = h\nu$. This is indeed the case.

When it was first presented, Planck's method for reproducing the thermal cavity radiation spectrum was regarded by most physicists as an interesting trick of no fundamental importance. In 1900, it was generally believed that all physical processes are *continuous*, so there was great reluctance to accept the idea of energy quantization proposed by Planck. However, with the derivation of the Planck radiation law, based on the quantum concept, the development of quantum theory had begun.

Example 46–1

A tungsten tube into which has been drilled a small hole is heated to a temperature of 2000 K. Radiation from both the surface and the hole is observed. The emissivity of tungsten at 2000 K is $e = 0.28$.

(a) At what frequency is the cavity radiation function $R(\nu)$ maximum?
(b) At what wavelength is $R(\lambda)$ maximum?
(c) Determine $R(\lambda)$ at λ_m for both the surface and the cavity radiations.
(d) Determine R for both the surface and the cavity radiations.

Solution:

(a) Using Eq. 46–16, we have

$$\nu_m = 2.821 \frac{kT}{h} = \frac{2.821(1.38 \times 10^{-23} \text{ J/K})(2000 \text{ K})}{6.63 \times 10^{-34} \text{ J} \cdot \text{s}}$$

$$= 1.17 \times 10^{14} \text{ Hz}$$

(b) Using Eq. 46–18, we find

$$\lambda_m = \frac{hc}{4.965 \, kT} = \frac{(6.63 \times 10^{-34} \text{ J} \cdot \text{s})(3.00 \times 10^8 \text{ m/s})}{4.965(1.38 \times 10^{-23} \text{ J/K})(2000 \text{ K})}$$

$$= 1.45 \times 10^{-6} \text{ m}$$

The values of ν_m and λ_m correspond to radiations in the infrared part of the spectrum. Notice that $\lambda_m \nu_m \neq c$. (•Do you understand why this is the case?)

(c) Equation 46–2 is also valid when R and u are expressed as functions of wavelength, $R(\lambda) = \frac{1}{4}cu(\lambda)$. Then, using Eq. 46–17, we have

$$R(\lambda) = \frac{2\pi hc^2}{\lambda^5} \frac{1}{e^{hc/\lambda kT} - 1}$$

For $\lambda = \lambda_m$, Eq. 46-18 gives $hc/\lambda kT = 4.965$; hence, for the cavity radiation,

$$R(\lambda_m) = \frac{2\pi(6.63 \times 10^{-34} \text{ J} \cdot \text{s})(3.00 \times 10^8 \text{ m/s})^2}{(1.45 \times 10^{-6} \text{ m})^5(e^{4.965} - 1)}$$

$$= 4.11 \times 10^{11} \text{ (W/m}^2)/\text{m}$$

For the radiation from the surface,

$$R_s(\lambda_m) = eR(\lambda_m) = 0.28(4.11 \times 10^{11} \text{ (W/m}^2)/\text{m})$$

$$= 1.15 \times 10^{11} \text{ (W/m}^2)/\text{m}$$

(d) Using Eq. 46-1, we have, for the cavity ($e = 1$),

$$R = \sigma T^4 = (5.67 \times 10^{-8} \text{ (W/m}^2)/\text{K}^4)(2000 \text{ K})^4$$

$$= 9.07 \times 10^5 \text{ W/m}^2$$

For the radiation from the surface, we have

$$R_s = eR = (0.28)(9.07 \times 10^5 \text{ W/m}^2)$$

$$= 2.54 \times 10^5 \text{ W/m}^2$$

46-2   THE PHOTOELECTRIC EFFECT

By the year 1890 it was well known that an isolated polished metal surface becomes positively charged when illuminated with ultraviolet light. Presumably, the UV light somehow removes negative charge from the surface, leaving it with a positive charge. This phenomenon is called the *photoelectric effect*. In 1900, Lenard proved that the ejected negative charge consists of *electrons* (which had been discovered by J. J. Thomson in 1897).

The photoelectric effect can be described by referring to the schematic experimental arrangement in Fig. 46-6. Ultraviolet light enters the evacuated glass tube and strikes a metallic cathode that has been polished to remove any oxide layer. The electrons that are ejected from the cathode pass through a wire grid and are absorbed by the collector cup. The collector current I is measured as a function of the grid potential V.

Suppose that monochromatic UV radiation with intensity L is incident on the

Fig. 46-6. Schematic arrangement for studying the photoelectric effect. The UV window is usually made from quartz, which passes UV radiation (glass does not). The potentiometer is used to adjust the potential of the grid relative to the cathode.

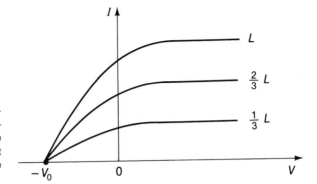

Fig. 46–7. Collector current I versus grid potential V for the photoelectric apparatus of Fig. 46–6. The cut-off potential $-V_0$ does not depend on the intensity L of the incident UV radiation.

cathode. As the grid potential V is varied, the collector current is found to change in the manner shown by the upper curve in Fig. 46–7. For all positive potentials, the current is essentially constant (saturated). However, when the potential is made negative, the current is reduced and becomes zero at the *cut-off potential* $-V_0$. This means that none of the ejected photoelectrons can overcome the retarding potential $-V_0$ and reach the collector. If the intensity of the UV light is changed, there is a proportional change in the saturated current but no change in the value of the cut-off potential (Fig. 46–7). Thus, the maximum kinetic energy of the electrons is

$$K_m = (-e)(-V_0) = eV_0 \qquad\qquad \textbf{(46–19)}$$

and K_m does not depend on the intensity of the incident radiation. (Later in this section we see why V_0 corresponds to the *maximum* kinetic energy.)

Another important feature of the photoelectric effect is that the maximum kinetic energy of the photoelectrons from a particular material depends only on the *frequency* of the incident radiation. Moreover, for each material there exists a characteristic threshold frequency v_0 for the emission of photoelectrons. Radiation with frequency lower than v_0 will not result in photoemission regardless of the intensity of the radiation.

Einstein's Theory of the Photoelectric Effect. At the time the photoelectric effect was first being investigated, it was known that electromagnetic waves could impart energy and momentum to electrons within matter. But classical electromagnetic theory was unable to account satisfactorily for the various experimental results that had been obtained.

In 1905, Albert Einstein published a beautifully simple theory of the photoelectric effect.* Einstein extended Planck's idea of the existence of quanta and boldly postulated that *all* electromagnetic radiation is quantized, with the energy of a quantum of radiation with frequency v given by $E = hv$. Quanta of electromagnetic radiation are called *photons*. Thus, in the photoelectric effect, the anode is showered by a flux of ultraviolet photons each with energy hv. In the anode, a single photon may interact with a single electron and cause it to be ejected from the cathode material. Einstein imagined this as a single-step, instantaneous process and rejected the possibility that two or more photons might act together to eject a particular electron. The photon disappears in the process, its entire energy having been consumed in freeing the electron from the material and giving the electron some kinetic energy. Let W_0 be

---

*Remarkably, it was nearly a decade after Einstein advanced his theory before there was available any high-quality experimental evidence to support his views. Einstein was awarded the 1921 Nobel Prize in physics for his theory of the photoelectric effect (not for his relativity theory, also proposed in 1905).

the least energy required to free an electron from the material. Then, the maximum kinetic energy of the ejected electron is

$$K_m = h\nu - W_0 \qquad\qquad (46\text{–}20)$$

Notice that the electron kinetic energy depends only on the photon energy $h\nu$ (and the particular value of W_0), and not on the rate at which photons strike the surface. The number of electrons released (called *photoelectrons*) is directly proportional to the number of incident photons (the UV intensity). Hence, the collector current in the experimental arrangement shown in Fig. 46–6, which consists of the ejected electrons, has a saturation value that depends on the intensity of the incident radiation. The (negative) retarding potential necessary to turn back the electrons so that they do not reach the collector to contribute to the current I is $eV_0 = K_m$. Thus,

$$eV_0 = h\nu - W_0 \qquad\qquad (46\text{–}21)$$

which is the photoelectric equation of Einstein.

In 1916, Millikan* published the results of a series of experiments in which he carefully measured the cut-off potential V_0 for several metals by observing the collector current as a function of photon frequency ν. According to Eq. 46–21, the graph of V_0 versus ν should be a straight line with slope equal to h/e and with intercept (the threshold frequency) ν_0, where $W_0 = h\nu_0$. Figure 46–8 shows the data from one of Millikan's experiments. Not only do the data follow a straight line, but the slope agrees with the ratio h/e calculated with known values. Thus, Einstein's theory was confirmed in dramatic fashion.

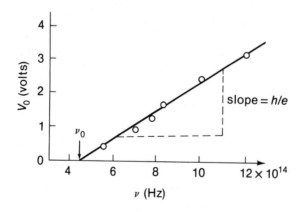

Fig. 46–8. In this experiment the cut-off frequency ν_0 was determined by Millikan to be 4.39×10^{14} Hz, corresponding to a work function of 1.82 eV.

▶ In Section 31–6, we discussed the energy distribution of free electrons in metals. The least binding energy of electrons in the Fermi sea is the work function ϵ_ϕ (see Fig. 46–9). Hence, we may identify ϵ_ϕ with the energy W_0 in Eq. 46–20. The least energetic photon that is capable of ejecting an electron from a metal therefore has an energy

$$h\nu_0 = W_0 = \epsilon_\phi$$

Thus, a determination of the threshold frequency ν_0 yields a value for ϵ_ϕ. Electrons ejected from below the top of the Fermi sea will have kinetic energies K that are less than $K_m = h\nu - \epsilon_\phi$ (Fig. 46–9a). ◀

---

*Robert A. Millikan (1868–1953), American physicist, won the 1923 Nobel Prize in physics for his work on the photoelectric effect and his measurements of the electron charge.

Fig. 46-9. (a) A photon with energy $h\nu$ ejects an electron with kinetic energy $K = h\nu - \epsilon$. (b) If the ejected electron comes from the top of the Fermi sea, the kinetic energy is $K_m = h\nu - \epsilon_\phi$, which is the maximum possible electron energy.

X-Ray Short-Wavelength Limit. In Section 43–3, it was mentioned that X rays produced by electron bombardment of metallic targets exhibit a broad continuous energy distribution (brehmsstrahlung) on which may be superimposed sharp spectral lines. Figure 46–10 shows X-ray spectral distributions from targets of two materials, tungsten and molybdenum, bombarded by electrons accelerated through the same voltage. (We defer discussion of the sharp spectral lines marked $K\alpha$ and $K\beta$ until Section 47–5.) Notice that both distributions exhibit a sharp cut-off at the same short-wavelength limit λ_{min}. In fact, the value of λ_{min} is the same for all target materials and varies inversely with the electron-accelerating voltage. Moreover, λ_{min} does not depend on the magnitude of the current represented by the stream of electrons that strikes the target.

The explanation of the behavior of λ_{min} can be given in terms of an "inverse" photoelectric effect. Electrons are accelerated through a potential V and acquire a kinetic energy $K = eV$. Upon striking the target, the electrons undergo a decelerating process in which X radiation is produced. Energy conservation requires that the most energetic photon that can be emitted has an energy $h\nu_m = K = eV$. Thus, the short-wavelength cut-off is given by

$$\lambda_{min} = \frac{c}{\nu_m} = \frac{ch}{eV} \tag{46–22}$$

Values of h determined by measuring V and λ_{min} are in agreement with those obtained by other methods.

Fig. 46-10. X-ray intensity distributions (spectra) obtained by the bombardment of tungsten (W) and molybdenum (Mo) targets with electrons accelerated through a potential of 35 kV. The cut-off wavelength λ_{min} is the same for both target materials.

Example 46–2

A mercury arc lamp provides 0.10 W of UV radiation at a wavelength of $\lambda = 2537$ Å (all other wavelengths having been absorbed by filters). The cathode of a photoelectric device (a *phototube*) consists of potassium and has an effective area of 4 cm². The anode is located at a distance of 1 m from the radiation source. The work function for potassium is $W_0 = \epsilon_\phi = 2.22$ eV (see Table 31–4).

(a) According to classical theory, the radiation from the arc spreads out uniformly in space as a spherical wave. What time of exposure to the radiation should be required for a potassium atom (radius 2 Å) in the anode to accumulate sufficient energy to eject a photo-electron?

(b) What is the energy of a single photon from the source?

(c) What is the flux of photons (number per second) at the cathode? To what saturation current does this flux correspond if the photoconversion efficiency is 5 percent (that is, if each photon has a probability of 0.05 of ejecting an electron)?

(d) What is the cut-off potential V_0?

Solution:

(a) The UV energy flux at a distance of 1 m is $0.1/4\pi$ W/m². The cross-sectional area of the atom is $\pi r^2 = \pi(2 \times 10^{-10} \text{ m})^2 = 4\pi \times 10^{-20}$ m². The energy required to eject a photo-electron is $E_0 = (2.22 \text{ eV})(1.60 \times 10^{-19} \text{ J/eV}) = 3.55 \times 10^{-19}$ J. Consequently, if energy could be accumulated by the atom from continuous radiation, the exposure time required to eject a photoelectron would be

$$t = \frac{3.55 \times 10^{-19} \text{ J}}{(0.1/4\pi \text{ W/m}^2)(4\pi \times 10^{-20} \text{ m}^2)} = 355 \text{ s (!)}$$

Experiments have demonstrated that the photoelectron current begins less than 10^{-12} s after the anode surface is first illuminated. This result is further evidence that the energy is absorbed in lumps or quanta (photons) and does not follow the classical prescription.

(b) The photon energy is

$$h\nu = \frac{hc}{\lambda} = \frac{(6.63 \times 10^{-34} \text{ J} \cdot \text{s})(3.00 \times 10^8 \text{ m/s})}{2537 \times 10^{-10} \text{ m}}$$

$$= 7.84 \times 10^{-19} \text{ J} \times \frac{1 \text{ eV}}{1.60 \times 10^{-19} \text{ J}}$$

$$= 4.90 \text{ eV}$$

A useful equation for converting photon wavelength into energy is

$$E(\text{eV}) = \frac{12,400}{\lambda(\text{Å})} \tag{46-23}$$

where the energy is given in eV when the wavelength is expressed in Å. (This formula is more precise than the calculation above, and gives $E = 4.89$ eV for the case at hand.)

(c) At the cathode (area 4×10^{-4} m²), the photon flux is

$$N = \left(\frac{0.1}{4\pi} \text{ W/m}^2\right)\frac{4 \times 10^{-4} \text{ m}^2}{7.84 \times 10^{-19} \text{ J}} = 4.06 \times 10^{12} \text{ photons/s}$$

With an efficiency of 5 percent, the photocurrent is

$$I = 0.05 \, Ne = (0.05)(4.06 \times 10^{12} \text{ s}^{-1})(1.60 \times 10^{-19} \text{ C})$$

$$= 3.25 \times 10^{-8} \text{ A} = 32.5 \text{ nA}$$

(d) Using Eq. 46-21, we have

$$V_0 = \frac{h\nu - W_0}{e} = \frac{4.86 \text{ eV} - 2.22 \text{ eV}}{e} = 2.64 \text{ V}$$

Example 46-3

Figure 46-10 shows the results of an experiment in which the cut-off wavelength was determined to be $\lambda_{min} = (0.353 \pm 0.002)$ Å. Use this result together with $e = 1.602 \times 10^{-19}$ C and $c = 2.998 \times 10^8$ m/s to calculate a value for h.

Solution:

Using Eq. 46-22, we find

$$h = \frac{\lambda_{min}\, eV}{c}$$

$$= \frac{[(0.353 \pm 0.002) \times 10^{-10}\ \text{m}](1.602 \times 10^{-19}\ \text{C})(35,000\ \text{V})}{2.998 \times 10^8\ \text{m/s}}$$

$$= (6.60 \pm 0.04) \times 10^{-34}\ \text{J} \cdot \text{s}$$

which is in agreement with the known value, $h = 6.626 \times 10^{-34}$ J $\cdot$ s.

46-3 THE COMPTON EFFECT

Studies of thermal cavity radiation and the photoelectric effect clearly indicate that light interacts with matter as a bundle of energy (a photon). Further evidence of the corpuscular aspect of radiation is obtained from an entirely different type of experiment.

According to classical (Maxwellian) theory (Section 38-2), an electromagnetic wave has an associated momentum density given by (Eq. 38-24)

$$G = u/c$$

This implies that a photon with energy $E = h\nu$ has an associated momentum,

$$p = \frac{E}{c} = \frac{h\nu}{c} = \frac{h}{\lambda} \tag{46-24}$$

As pointed out in the small-print paragraph on page 1044, the same conclusion can be drawn from Einstein's relationship (Eq. 21-16), $E^2 = p^2c^2 + M_0^2c^4$, by setting the photon proper mass M_0 equal to zero.

In 1922, A. H. Compton* carried out an experiment that verified the corpuscular character of the photon. X rays were produced by allowing energetic electrons to strike a molybdenum target located at T in the arrangement shown schematically in Fig. 46-11. The X-ray beam was scattered by a graphite scatterer S. The energy spectrum of the radiation scattered at an angle θ was analyzed by means of a Bragg reflection crystal. Compton found that the scattering of the strong $\lambda = 0.711$ Å radiation from molybdenum (the $K\alpha$ line in Fig. 46-10) produced a long-wavelength satellite line λ', as indicated in Fig. 46-12. He also showed that the wavelength difference $\lambda' - \lambda$ increases steadily with scattering angle.

---

*Arthur H. Compton (1892–1962), American physicist, received a share of the 1927 Nobel Prize in physics for the discovery of the effect that bears his name.

Fig. 46–12. Compton's data for the scattering of the Mo $K\alpha$ radiation (wavelength λ) from a carbon scatterer S at different angles θ. The wavelength λ' of the satellite line increases with θ. The primary line that is seen (with reduced intensity) at each angle is due to X rays each of which is scattered from an atom as a whole. Because an atom is much more massive than an electron, no wavelength shift results for the primary line. (Adapted from A. H. Compton.)

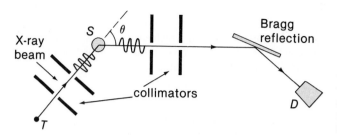

Fig. 46–11. Geometry of the Compton experiment. X rays produced by electron bombardment of the molybdenum target T are scattered through an angle θ at S. The scattered radiation is analyzed by Bragg reflection into the detector D.

Compton explained the presence of the satellite line (also called the modified line) and its behavior as a function of the angle θ in terms of a straightforward collision process between the incident X-ray photon and one of the atomic electrons in the scattering sample, as indicated in Fig. 46–13. The electron is treated as a free particle, because it turns out that the electron is bound in the sample with an energy that is small compared with the energy imparted to it in the collision process.*

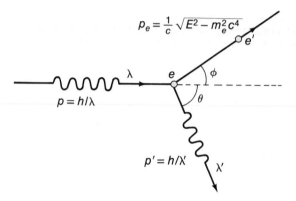

Fig. 46–13. The Compton scattering process. The electron e is treated as a free particle and is considered to be stationary prior to the collision.

The analysis of the collision between the photon and the electron is carried out by writing the equations that express momentum conservation in directions parallel and perpendicular to the direction of the incident photon, and the equation that expresses energy conservation. Because the X rays used in the experiment give to each electron a large velocity in the collision process, the relativistic expressions for the electron momentum and the electron energy must be used. The equations are (see Fig. 46–13):

$$\frac{h}{\lambda} = \frac{h}{\lambda'}\cos\theta + \frac{m_e v}{\sqrt{1-\beta^2}}\cos\phi \quad \left.\begin{array}{c} \\ \\ \end{array}\right\} \text{momentum} \quad \textbf{(46–25a)}$$

$$0 = \frac{h}{\lambda'}\sin\theta - \frac{m_e v}{\sqrt{1-\beta^2}}\sin\phi \quad \left.\begin{array}{c} \\ \\ \end{array}\right\} \text{conservation} \quad \textbf{(46–25b)}$$

$$\frac{hc}{\lambda} - \frac{hc}{\lambda'} = \frac{m_e c^2}{\sqrt{1-\beta^2}} - m_e c^2 \quad \left.\begin{array}{c} \\ \\ \end{array}\right\} \begin{array}{c}\text{energy} \\ \text{conservation}\end{array} \quad \textbf{(46–25c)}$$

where m_e is the electron rest mass, and where $\beta = v/c$. After some algebraic manipulations (see Problem 46–13), these equations yield the desired result:

---

*The energy of the Mo $K\alpha$ X ray is about 60 times greater than the binding energy of the most tightly bound electron in carbon.

$$\lambda' - \lambda = \frac{h}{m_e c}(1 - \cos \theta) \qquad \textbf{(46-26)}$$

This expression for the wavelength difference was found by Compton to fit the experimental data with great accuracy.

The quantity $h/m_e c$, which appears in the right-hand side of Eq. 46-26, is known as the *Compton wavelength of the electron*, and has the value

$$h/m_e c = 0.0243 \text{ Å}$$

Thus, when the scattering angle is 90°, the wavelength shift is just $\lambda' - \lambda = 0.0243$ Å.

The Compton experiment, probably more than any other, convinced scientists of the reality of the corpuscular character of radiation and the validity of the quantum concept.

Example 46-4

An X-ray photon with $\lambda = 0.711$ Å is Compton-scattered from an electron through an angle of 180°.
  (a) What is the energy of the incident X ray?
  (b) What is the wavelength shift of the X ray in the scattering process?
  (c) Determine the energy of the backscattered X ray.
  (d) Find the kinetic energy of the recoiling electron.

Solution:

  (a) Using Eq. 46-23, we have

$$E(\text{eV}) = \frac{12{,}400}{\lambda(\text{Å})} = \frac{12{,}400}{0.711} = 1.744 \times 10^4$$

or
$$E = 17.44 \text{ keV}$$

  (b) The wavelength shift is given by Eq. 46-26:

$$\lambda' - \lambda = (0.0243 \text{ Å})(1 - \cos 180°)$$
$$= 0.0486 \text{ Å}$$

  (c) and (d) The change in the energy of the X ray is equal to the kinetic energy acquired by the electron. The wavelength of the backscattered X ray is $\lambda' = \lambda + 0.0486$ Å $= 0.7596$ Å, so that the X-ray energy is

$$E' = \frac{12{,}400}{0.7596} \text{ eV} = 16.32 \text{ keV}$$

Hence, the electron energy is

$$K_e = E - E' = 17.44 \text{ keV} - 16.32 \text{ keV} = 1.12 \text{ keV}$$

46-4 THE WAVE-PARTICLE DUALITY

Photons. According to the discussions in the preceding chapters concerning optics experiments (such as those dealing with interference and diffraction effects), it is evident that in some situations electromagnetic radiation behaves as a wave. However, in other situations (such as photoelectric and Compton experiments), the radiation behaves as a corpuscular entity or particle. How do we account for this dual character? Niels Bohr* pointed out in stating his *principle of complementarity* that both wave and particle aspects are required for the complete description of both radiation and matter. Both aspects are always present and either may be revealed by an experiment. An experiment that involves a localized interaction of a photon will reveal its particle character. But an experiment that allows the unobserved interaction of (nonlocalized) radiation with, for example, a diffraction slit will reveal its wave character. The way in which the measurements are made determines whether a wave or a particle description would be most appropriate. (•Can you see the operation of this idea in the Compton effect?)

Suppose that a single atomic event results in the generation of an outgoing electromagnetic wave train corresponding to a single photon. If it is equally likely that the radiation is traveling in any direction, it is proper to describe the situation in terms of a spherical wave train expanding at the speed c, as indicated in Fig. 46–14.

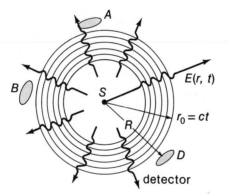

Fig. 46–14. An expanding spherical wave train of radiation from an atomic source S. The radiation is detected by the detector D.

Now, suppose that a radiation detector D (perhaps one that employs the photoelectric effect) subsequently signals the detection of the radiation at some specific point a distance R from the atomic source. *All* of the momentum and energy associated with the photon character of the radiation has now been localized at the detector site. How were the energy and momentum fluxes at regions such as A and B in Fig. 46–14 suddenly gathered in at the detector position D? There is no satisfactory way to answer such a question; instead, it is preferable to view the expanding wave as a *probability wave*, for which the energy density u at a point is proportional to the probability of detecting the photon at that point.† The Poynting vector **N** is now identified as a *probability flux*. For a spherical wave, the amplitude of the Poynting vector varies as $1/r^2$. This insures that the total radiant energy in the expanding wave is constant, and prior to detection (that is, $r_0 = ct < R$) it implies that the total probability of finding the photon *somewhere* on the wavefront is unity. If the photon

---

*Niels Bohr (1885–1962), Danish physicist, was one of the originators of modern quantum theory. More of Bohr's work is discussed in the following chapter.

†The probabilistic interpretation of quantum phenomena was originated by the German physicist Max Born (1882–1970), who received a share of the 1954 Nobel Prize in physics for his work on the fundamental concepts of quantum theory.

is detected by D or some other detector, the description of the state of the radiation abruptly changes because the probability for finding the photon at the detector site is now unity.

Note that even if the electromagnetic energy is not absorbed by the detector D, an event that would signal the arrival of a photon, the detector—an absorbing obstacle—still alters the progress of the wave because the associated probability wave is diffracted around the obstacle. The propagation of the photon is affected by the presence of the detector even if the detector does not absorb the photon. This conclusion is the same as that drawn from the conventional electromagnetic theory of diffraction.

De Broglie Waves. Compton's experiments in 1922 had clearly demonstrated the particle-like character of electromagnetic radiation. In 1924, Louis de Broglie* proposed in his doctoral thesis that, in view of the particle-like behavior of waves, it seems plausible that there should also be an identifiable wavelike behavior of material particles. He suggested that the wavelength associated with a moving particle is given by the same expression as that for a photon, namely (Eq. 46-24),

$$\lambda = \frac{h}{p} \qquad \text{de Broglie wavelength} \qquad \textbf{(46-27)}$$

where p is the (relativistic) momentum of the particle.

The de Broglie wavelength for ordinary objects is far too small to produce observable wave effects such as diffraction (see Problem 46-15). However, within three years after de Broglie's ingenious proposal, the wave properties of electrons had been demonstrated. The description of particles in terms of their wave properties is often called *wave mechanics*.

The Davisson-Germer Experiment. In 1927, the diffraction of electrons was observed by C. J. Davisson† and L. H. Germer working at the Bell Telephone Laboratories. The apparatus used is shown schematically in Fig. 46-15. A narrow beam of accelerated electrons was allowed to strike one face of a single crystal of nickel, and

---

*Louis Victor de Broglie (1892–   ), French physicist, was awarded the 1929 Nobel Prize in physics for his theory of the wave properties of matter.

†Clinton J. Davisson (1881–1958) shared the 1937 Nobel Prize in physics with the British physicist George P. Thomson (1892–1975) for their discoveries of electron diffraction by different methods.

Fig. 46–15. Schematic diagram of the apparatus used in the Davisson-Germer experiment.

Fig. 46–16. Polar diagrams of the intensity $I(\phi)$ of the scattered electrons at different angles ϕ. Each diagram corresponds to a particular electron-accelerating voltage.

the angular distribution of the scattered electrons was measured. The results are shown in Fig. 46–16. The middle graph shows that strong scattering (called Bragg reflection) occurs at $\phi = 50°$ when the electrons are accelerated through a potential of 54 V.

As shown in Fig. 46–17, the scattering angle $\phi = 50°$ corresponds to Bragg reflection from parallel crystal planes that are rich in atoms (compare Fig. 43–15).

Fig. 46–17. Geometry of the Bragg reflection in the Davisson-Germer experiment.

The lattice spacing for nickel is known from X-ray diffraction experiments to be $d = 2.15$ Å. In Fig. 46–17, we see that the spacing between the diffracting planes is $d' = d \sin \theta$. The Bragg law for first-order diffraction (Eq. 43–24 with $n = 1$) from the oblique plane gives

$$\lambda_B = 2d' \sin \left(\frac{\pi}{2} - \theta \right) = 2d \sin \theta \cos \theta = d \sin 2\theta$$

$$= d \sin \phi$$

This leads to

$$\lambda_B = (2.15 \text{ Å}) \sin 50° = 1.65 \text{ Å}$$

The de Broglie wavelength for 54-eV electrons is

$$\lambda = \frac{h}{p} = \frac{h}{\sqrt{2m_e E}} = \frac{6.63 \times 10^{-34} \text{ J} \cdot \text{s}}{\sqrt{2(9.11 \times 10^{-31} \text{ kg})(54 \text{ eV})(1.60 \times 10^{-19} \text{ J/eV})}}$$

$$= 1.67 \times 10^{-10} \text{ m} = 1.67 \text{ Å}$$

Thus, the calculated de Broglie wavelength agrees with the inferred Bragg wavelength.

Soon after the Davisson-Germer and Thomson experiments had demonstrated the diffraction of electrons, Stern and Estermann in 1930 succeeded in observing the diffraction of hydrogen and helium atoms. This experiment proved that bits of matter much more massive than electrons could also exhibit diffraction effects, and diffraction is one of the key phenomena associated with waves. Thus, the complete wave-particle symmetry of Nature imagined by de Broglie had been verified. Both radiation and matter have wavelike and particlelike characteristics. Depending on the experimental circumstances, one or the other of these attributes will be revealed.

▶ *The Electron Microscope.* The fact that energetic electrons have extremely short de Broglie wavelengths has been put to practical use in *electron microscopes.* In these devices, electric and magnetic fields are used to focus electrons by means of the electromagnetic forces that are exerted on moving charges, just as we studied in earlier chapters. The resulting deflections of the electron beams are similar to the refraction effects produced by glass lenses used to focus light in optical microscopes. By using electrons (with wavelengths of 0.05 Å or less) instead of visible light (with wavelengths near 5000 Å), the limitations on resolution imposed by diffraction effects can be largely overcome. Accelerating potentials from 30 kV to several MV have been used. Such high voltages give extremely short wavelengths and also give the electrons sufficient energy to penetrate specimens of reasonable thickness. The resolution that is achievable by an electron microscope is limited by lens aberrations and by scattering in the specimen, so it is somewhat poorer than predicted on the basis of diffraction effects. Nevertheless, a linear resolution of 5 to 10 Å is possible with a 50-kV instrument, and 2 Å is possible with special, high-voltage microscopes. (The best linear resolution for an optical microscope is about 2000 Å.)

The lenses of an electron microscope are actually electric and magnetic fields that achieve focusing because of special field configurations. Figure 46-18a shows paraxial trajectories of electrons that leave a point source O and pass through an electrostatic lens. The lens consists of two cylindrical conductors between which is maintained a constant potential difference $V_2 - V_1$. The bending of the electron paths in the field causes them to converge toward the point of focus I. Figure 46-18b shows the optical equivalent of this electrostatic lens.

Focusing can also be achieved with shaped static magnetic fields. Magnetic focusing requires strong fields with axial components confined to a relatively small space. Figure 46-19a shows how a magnetic field is concentrated in the vicinity of a small gap in the iron casing that surrounds a many-turn, current-carrying coil. Such a magnetic field will bring electrons from the source point O to a focus at I, but at the same time the electrons will be

Fig. 46-18. (a) Schematic of an electrostatic lens for focusing electrons. The conducting cylinder ② is at a higher potential than cylinder ①. The bulging equipotential lines indicate the field configuration. (b) The optical analog of the electrostatic lens consists of a converging and a diverging lens.

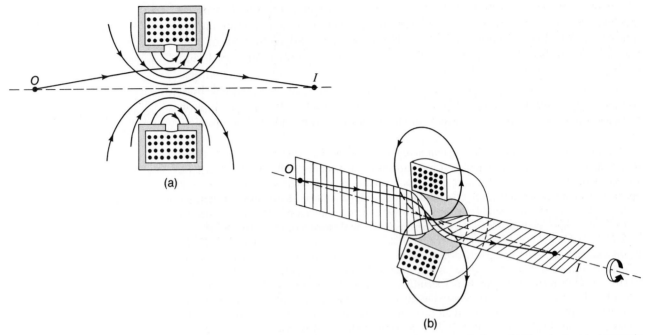

Fig. 46–19. (a) The focusing of an electron beam by a magnetic lens. (b) The electron trajectory is rotated about the lens axis.

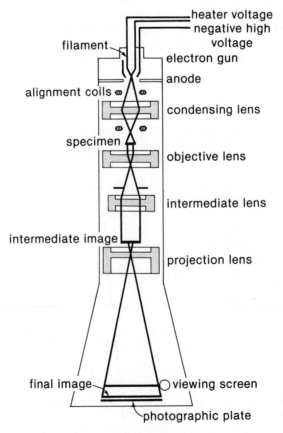

heater voltage
negative high
voltage
filament
electron gun
anode
alignment coils
condensing lens
specimen
objective lens
intermediate lens
intermediate image
projection lens

final image
viewing screen
photographic plate

Fig. 46–20. Schematic of the lens system in the RCA Model EMU-3 electron microscope. The electron gun involves an electrostatic lens, but all of the following lenses are magnetic.

rotated about the axis of the field. Figure 46–19b shows the plane that contains the electron trajectory and the field axis being rotated through an angle of 90°.*

Figure 46–20 shows the lens system of an early electron microscope (circa 1941). The magnetic condensing lens concentrates the beam from the electron gun onto the specimen. Electrons are scattered out of the beam by the thicker parts of the specimen. The transmitted beam therefore has spatial differences in density that correspond to the features of the specimen. The objective and intermediate lenses produce a real intermediate image, and the projection lens forms the final image, which can be viewed by means of a fluorescent screen or photographed on special film. Usually, specimens are viewed with magnifications from about 10^3 to 10^5 and with linear resolutions from about 5 Å to 50 Å or so.

A remarkable three-dimensional quality of the image can be achieved in modern *scanning electron microscopes*, in which the beam is scanned over the specimen in the same way that an electron beam is scanned over the face of a television picture tube. Some electron micrographs obtained with a scanning electron microscope are shown on the next page.

Example 46–5

An electron passes through an electrostatic lens of the type illustrated in Fig. 46–18a. Inside the first section of the lens, the electron kinetic energy is 20 keV. The potential difference between the lens sections is $V_2 - V_1 = 5$ kV.

(a) What is the electron wavelength in section 1 of the lens?

(b) What is the electron wavelength in section 2 of the lens?

Solution:

(a) The total energy $E = K + U$ is constant, as indicated in the diagram. Then, in section 1 we have

$$\lambda_1 = \frac{h}{p_1} = \frac{h}{\sqrt{2m_e K_1}} = \frac{6.63 \times 10^{-34} \text{ J} \cdot \text{s}}{\sqrt{2(9.11 \times 10^{-31} \text{ kg})(20 \times 10^3 \text{ eV})(1.60 \times 10^{-19} \text{ J/eV})}}$$

$$= 8.68 \times 10^{-12} \text{ m} = 0.0868 \text{ Å}$$

(b) In section 2, the kinetic energy is 5 keV greater, so

$$\lambda_2 = \frac{6.63 \times 10^{-34} \text{ J} \cdot \text{s}}{\sqrt{2(9.11 \times 10^{-31} \text{ kg})(25 \times 10^3 \text{ eV})(1.60 \times 10^{-19} \text{ J/eV})}}$$

$$= 0.0777 \text{ Å} \blacktriangleleft$$

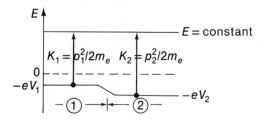

*Even for very strong (but thin) magnetic lenses, the rotation angle is less than 120°. Focusing in a uniform field (a very thick lens) requires a 360° rotation (see Problem 33–20).

Four photographs taken with scanning electron microscopes. (a) A pair of pincers, called a pedicellaria, with which the starfish *Stylasterias forreri* catches tiny fish. Note the scale (1 μ = 10^{-6} m). (From F. S. Chia and H. Amerongen, Canadian Journal of Zoology, *53*:748–755, 1975. Reproduced by permission of the National Research Council of Canada.) (b) The face of an ordinary house fly, magnified 50 times. (c) The foot of a tsetse fly, magnified 300 times. (d) Crystals of sodium benzoate, magnified 1500 times. (Parts (b), (c) and (d) courtesy of the Industrial Laboratory, Eastman Kodak Company.)

46-5 THE UNCERTAINTY PRINCIPLE

The fact that a wave can be associated with a material particle leads directly to certain profound and surprising conclusions regarding the observable behavior of the particle. An interesting point to consider is the spatial relationship between the corpuscular and wavelike attributes of a particle. We address this important topic before going on to discuss the relationships connecting the wavelength, the frequency, the energy, and the momentum of a particle, along with the question of the correspondence between the new wave description of matter and Newtonian mechanics.

In 1927 Werner Heisenberg* introduced an important new idea in the form of the *uncertainty principle*.† One statement of this principle is that it is impossible to locate the position of a particle and simultaneously to determine its momentum, each with unlimited experimental precision. The intrusion of the experimenter's apparatus in the attempt to measure either the position or the momentum with ever greater precision has the effect of decreasing the precision with which the other quantity can be determined.

Suppose that we prepare a collimated beam of rapidly moving particles. In an effort to localize one of these particles in space, we direct the beam toward an aperture that is closed with a shutter. The shutter is opened for a brief time and then is closed. We can make the intensity of the beam sufficiently low that a single particle passes through the aperture while the shutter is open. If the shutter is open for a very short time, the de Broglie wave train of the particle will be very short and well localized in space. Implicit in this statement is the idea that the de Broglie waves are associated with the probability of finding the particle at some particular place at some particular time (just as in the case of photons).

Figure 46–21a shows, at some instant, the truncated de Broglie wave train of a particle traveling in the $+x$-direction. This finite wave train (called a *wave packet*) consists of N oscillations of a sinusoidal wave with local wavelength $\lambda_0 = h/p_0$ that exists only in the region $-\frac{1}{2}\Delta x < x < \frac{1}{2}\Delta x$, where Δx is the total length of the wave train. In Section 19–8 we discussed the Fourier analysis of an arbitrary periodic function $f(x)$ with repetition length L in terms of a superposition of harmonic waves with different frequencies. In the present case, we might identify Δx as L; however, the function that describes the de Broglie wave train does not endlessly repeat, but is zero outside the range $-\frac{1}{2}\Delta x < x < \frac{1}{2}\Delta x$. This difference in the behavior of $f(x)$ requires that the usual Fourier sum (for a symmetric function), namely,

$$f(x) = \sum_{n=1}^{\infty} A_n \cos(k_n x) \qquad k_n = 2\pi n/L$$

---

*Werner Karl Heisenberg (1901–1976), German theoretical physicist, was awarded the 1932 Nobel Prize in physics for his formulation of the uncertainty principle.

†Sometimes referred to as the *principle of indeterminancy*.

Fig. 46–21. (a) The truncated de Broglie wave train (wave packet) for a particle that has passed through a briefly unshuttered aperture. (b) The wave components with wavelengths λ_1 and λ_2 are out of phase and cancel at the extremes of the wave packet.

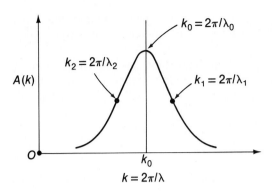

Fig. 46–22. The wave-number spectral function $A(k)$ for the wave packet in Fig. 46–21(a). The wavelengths λ_1 and λ_2 refer to the waves shown in Fig. 46–21(b).

be replaced by an integral, yielding

$$f(x) = \int_0^\infty A(k)\cos(kx)dk \qquad k = 2\pi/\lambda$$

The function $A(k)$ is the *wave-number spectrum*; for the wave packet of Fig. 46–21a, $A(k)$ is distributed about the peak at $k_0 = 2\pi/\lambda_0$ (Fig. 46–22). We need not be concerned here with the details other than to note that the wavelengths λ_1 and λ_2, which define the extent of the significant components of $A(k)$, must be exactly out of phase at the extremes of the wave packet so that they interfere destructively in the regions near $x = \pm\frac{1}{2}\Delta x$ (Fig. 46–21b*).

We write $\lambda_2 - \lambda_1 = \Delta\lambda$ and note that λ_1 and λ_2 are nearly equal to λ_0. Then, we can express the phase difference across half the region as

$$\tfrac{1}{2}N\,\Delta\lambda \cong \tfrac{1}{2}\lambda_0$$

Substituting $N \cong \Delta x/\lambda_0$, we find

$$\Delta x\Delta\lambda \approx \lambda_0^2 \qquad\qquad \textbf{(46–28)}$$

Using the de Broglie relationship, $\lambda = h/p$, the wavelength interval $\Delta\lambda$ corresponding to a momentum interval Δp is given (in magnitude) by

$$\frac{\Delta\lambda}{\lambda_0} = \frac{\Delta p}{p_0}$$

Substituting this relationship into Eq. 46–28 and noting that $\lambda_0 p_0 = h$ gives

$$\Delta x\Delta p \approx h \qquad\qquad \textbf{(46–29)}$$

This result is one form of the uncertainty principle.†

Note that when the particle is better localized (that is, when Δx is decreased), the momentum uncertainty Δp (and also $\Delta\lambda$) is increased. Thus, when the particle's

---

*Although not indicated in Fig. 46–21b, the Fourier component waves actually extend from $x = -\infty$ to $x = \infty$. The distribution of wavelengths in the function $A(k)$ is such that $f(x)$ is identically zero outside the range $-\frac{1}{2}\Delta x < x < \frac{1}{2}\Delta x$.

†This expression represents the result for optimum experimental conditions. When the usual imperfections in the apparatus and in technique are included, the possible precision is lessened, whence the result is often written as $\Delta x\,\Delta p \gtrsim h$.

position is determined more precisely, information concerning the particle's momentum is lost. If the momentum is to be determined very precisely, Δx (and, hence, N) necessarily increases. In fact, in the limit $\Delta p \rightarrow 0$, we have $N \rightarrow \infty$ (along with $\Delta\lambda \rightarrow 0$). This means that the de Broglie wave train becomes infinite in length and that the Fourier spectral function $A(k)$ is a single line corresponding to the wavelength λ_0. Thus, the wave train becomes monochromatic. In this limiting process the shutter has been effectively eliminated and any possibility of localizing the particle has been surrendered.

Physically, the result that we have obtained means that the operation of the shutter—its opening and closing—interferes with the passing particle. The shorter the duration of the open shutter, the greater is the perturbation of the momentum of the particle.

Next, we examine the possibility of localizing the particle in the direction *perpendicular* to its direction of motion. We can attempt to do this by passing the particle through a narrow slit with width Δy, as shown in Fig. 46–23. When a wave is

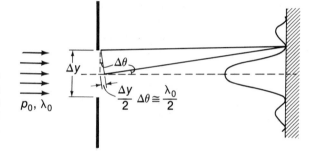

Fig. 46–23. The de Broglie wave of a particle is diffracted upon passing through a narrow slit. The first minimum in the diffraction pattern is located by the angle $\Delta\theta$. (Compare Fig. 43–4.)

incident on a narrow slit, diffraction inevitably results. The de Broglie wave of the particle is therefore subject to the same type of diffraction effect. Then, the first minimum in the diffraction pattern results when (see Fig. 46–23)

$$\Delta y \, \Delta\theta \cong \lambda_0 \qquad (46\text{–}30)$$

In order for a particle to stray away from its original direction of motion, it would need to acquire a momentum component in the transverse or y direction. To correspond to the half-width of the central maximum of the diffraction pattern, which is approximately equal to $\Delta\theta$, the transverse momentum would have a distribution with a half-width

$$\frac{\Delta p_y}{p_0} \cong \Delta\theta$$

Combining this relationship with Eq. 46–30 gives

$$\Delta y \, \Delta p_y \approx p_0 \lambda_0$$

so that

$$\Delta y \, \Delta p_y \approx h \qquad (46\text{–}31)$$

Thus, we arrive at the same form of the uncertainty principle for the transverse direction as for the direction of propagation (Eq. 46–29). In this case also we attribute the acquired transverse momentum to the interaction of the passing particle with the edges of the slit. This interpretation is consistent with the content of Eq. 46–31, because reducing Δp_y to zero requires $\Delta y \to \infty$, which simply means removing the slit together.

We are led by this line of reasoning to the conclusion that the Heisenberg uncertainty principle is a natural result of the wave-particle duality of matter. In three dimensions the principle is summarized by the statements*

$$\Delta x \Delta p_x \approx h$$
$$\Delta y \Delta p_y \approx h \qquad \text{(46–32)}$$
$$\Delta z \Delta p_z \approx h$$

The discussion to this point has centered on the localization of a *moving* particle. What is the situation if we attempt to determine the position of a *stationary* particle? According to the discussion in Section 45–3, the linear resolution of a microscope improves when the wavelength of the observing radiation is decreased (Eq. 45–20). In principle, then, it should be possible to improve the resolution as desired by using radiations with shorter wavelengths—UV light, X rays, or even γ rays.

Figure 46–24 shows the objective of an "X-ray microscope" that is used to "see" a stationary particle and thereby to localize its position in the x-direction. The particle is illuminated with a parallel stream of monochromatic X rays, as indicated in the figure. The angular aperture of the objective is θ, and the corresponding linear resolution is (Eq. 45–20)

$$\Delta x \cong \frac{\lambda}{2 \sin \theta} \qquad \text{(46–33)}$$

in which ξ has been replaced by Δx.

In order to "see" the particle, at least one X-ray photon must be deflected from its original path by an interaction with the particle. Figure 46–24 shows such a photon that is Compton-scattered through an angle α, acquiring in the process a transverse or x-directed momentum, $p_x \cong (h/\lambda) \sin \alpha$. The struck particle recoils in the opposite direction with the same momentum. The scattered photon will be focused by the microscope for any scattering angle in the range $-\theta \leqslant \alpha \leqslant \theta$. Consequently, the uncertainty in the particle momentum is

$$\Delta p_x \cong 2\frac{h}{\lambda} \sin \theta \qquad \text{(46–34)}$$

Multiplying Eqs. 46–33 and 46–34 yields

$$\Delta x \, \Delta p_x \approx h$$

and the uncertainty principle relationship again results. Any experiment, however

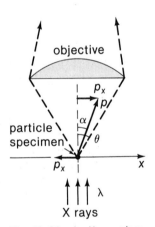

Fig. 46-24. An X-ray microscope is used to determine the position of a stationary particle.

---

*Using an argument more rigorous than that presented here, Heisenberg showed that $\Delta x \, \Delta p_x \geqslant h/4\pi$. This is the absolute minimum value for the product of the uncertainties, and occurs when the spectral function $A(k)$ (see Fig. 46–22) is a Gaussian function.

ingenious, designed to measure the location of a particle inevitably causes a loss of information concerning the particle's momentum, owing to interactions of the particle with the experimental apparatus.

Example 46-6

In the Davisson-Germer experiment, discussed in Section 46-4, the electrons in the beam had an energy of 54 eV and de Broglie wavelength $\lambda_0 = 1.67$ Å. Suppose that the beam has a momentum spread of ± 2 percent with an average momentum p_0. What is the minimum length of the de Broglie wave packet associated with an electron in the beam? How many oscillations N does the wave packet contain?

Solution:

The average momentum of an electron is

$$p_0 = \frac{h}{\lambda_0} = \frac{6.63 \times 10^{-34} \text{ J} \cdot \text{s}}{1.67 \times 10^{-10} \text{ m}} = 3.97 \times 10^{-24} \text{ kg} \cdot \text{m/s}$$

Thus, the momentum uncertainty (± 2 percent) is

$$\Delta p = 0.04 \, p_0 = 1.59 \times 10^{-25} \text{ kg} \cdot \text{m/s}$$

The corresponding uncertainty in position is

$$\Delta x \approx \frac{h}{\Delta p} = \frac{6.63 \times 10^{-34} \text{ J} \cdot \text{s}}{1.59 \times 10^{-25} \text{ kg} \cdot \text{m/s}}$$
$$= 4.17 \times 10^{-9} \text{ m} = 41.7 \text{ Å}$$

Hence, the number of oscillations in the wave packet is

$$N = \frac{\Delta x}{\lambda_0} = \frac{41.7 \text{ Å}}{1.67 \text{ Å}} = 25$$

SPECIAL TOPIC

46-6   FREE-PARTICLE WAVE FUNCTIONS

The preceding discussions were concerned with the de Broglie *wavelength* $\lambda = h/p$. In analogy with the situation for light photons, de Broglie postulated that the proper value of the *frequency* to be associated with a particle is given by

$$h\nu = E \qquad \text{(46-35a)}$$

where the total energy E is given by (see Eq. 21-16)

$$(E - U)^2 = p^2 c^2 + M_0^2 c^4 \qquad \text{(46-35b)}$$

or alternatively as

$$E = K + U + M_0 c^2 \qquad \text{(46-35c)}$$

with K the kinetic energy, U the potential energy, and $M_0 c^2$

(a)

(b)

(c)

Computer simulation of the growth of a two-slit interference pattern for electrons. (a) After the passage through the slit of 28 electrons. (b) After about 1000 electrons. (c) After about 10,000 electrons. There is no way to predict where a particular electron will strike the screen—only the probability can be given. After a very large number of electrons have passed through the slits, the pattern becomes the same as that for a beam of light photons (compare Fig. 42–3). (From *Physics I* by Elisha R. Huggins. W. A. Benjamin, New York, 1968.)

the rest-mass energy. (With both λ and ν defined, it remains for us to discover the expressions for the phase and group velocities of the de Broglie wave.)

We now use a heuristic argument to obtain an expression for the wave function for a particle traveling in a field-free region of space (that is, $U = 0$). We expect the wave function to have the form of a propagating plane wave; that is,

$$\psi = A \cos(\mathbf{k} \cdot \mathbf{r} - \omega t + \delta)$$

or perhaps

$$\psi = Ae^{i(\mathbf{k} \cdot \mathbf{r} - \omega t)} \qquad \textbf{(46–36)}$$

where $k = |\mathbf{k}| = 2\pi/\lambda$ and $\omega = 2\pi\nu$.

It develops that the appropriate expression for the wave function is the complex function given in Eq. 46–36. We have previously argued (see Section 46–4) that the probability of finding a photon at a particular point is pro-

portional to the value, at that point, of the energy density. The energy density, in turn, is proportional to the square of the electric field strength associated with the photon. Thus, we expect the equivalent probability for a particle to be proportional to the square of its wave function. Because the wave function is a complex quantity, we write

$$\text{Probability} \propto |\psi|^2$$

The wave function (or probability amplitude) ψ cannot be directly measured—only $|\psi|^2$ is a measurable physical quantity. The quantity ψ can be considered to be simply a mathematical function that is useful in computations.

Suppose that a free particle and its associated wave are traveling in the $+x$-direction. Then, with $E = K + M_0 c^2$ and $p = h/\lambda$, we have

$$\psi = Ae^{i(px - Kt)/\hbar}\, e^{-iM_0 c^2 t/\hbar}$$

where $\hbar = h/2\pi$. The factor $M_0 c^2/\hbar$ is seen to appear simply as a constant phase, which, in nonrelativistic situations, may always be omitted.* Thus, for a nonrelativistic free particle we expect that the appropriate wave function is

$$\psi = Ae^{i(px - Kt)/\hbar} \qquad \begin{array}{l}\text{nonrelativistic}\\ \text{free particle}\end{array} \qquad \textbf{(46–37)}$$

Next, we obtain the expressions for the phase velocity v_φ and the group velocity v_g for a de Broglie wave. For any sinusoidal wave the phase velocity is defined to be

$$v_\varphi = \lambda\nu = \omega/k \qquad \textbf{(46–38)}$$

so we use the same definition for a de Broglie particle wave. Then, with $U = 0$, and dropping the subscript from M_0, there results

$$v_\varphi = \lambda\nu = \lambda\frac{K}{h} = \frac{\lambda}{h}\frac{p^2}{2M} = \frac{\lambda}{h}\frac{(h/\lambda)^2}{2M}$$

$$= \frac{h}{2M\lambda} \qquad \textbf{(46–39)}$$

That is, the phase velocity depends on the wavelength. Thus, the de Broglie wave for a particle in empty space behaves as if it were traveling in a dispersive medium! In

---

*This is equivalent to shifting the reference level with respect to which the energy is measured. For a relativistic particle we should use Eq. 46–35b. However, the point is moot because a relativistic particle requires a more sophisticated mathematical treatment that involves methods beyond those at our disposal. Hence, we restrict consideration to nonrelativistic particles. (See, however, Problem 46–20.)

Section 19–9 we found that a wave described in this way has a group velocity given by (Eq. 19–45)

$$v_g = v_\varphi - \lambda \frac{dv_\varphi}{d\lambda} \qquad (46\text{–}40)$$

For the present case, we substitute for v_φ from Eq. 46–39 and obtain

$$v_g = \frac{h}{2M\lambda} + \frac{h}{2M\lambda} = \frac{h}{M\lambda} = \frac{p}{M}$$

Now, $p/M = v$ is just the particle velocity, so we are led to conclude that

$$v_g = v \qquad (46\text{–}41)$$

In Section 39–3 we found that the energy associated with a guided electromagnetic wave travels with the group velocity. It is therefore satisfying to find that the particle, which carries the energy flux, and the group aspect of the associated de Broglie wave travel together with the same velocity, $v_g = v$.

▶ *Wave Packet Dispersion.* Thus far we have been considering the case of a completely unlocalized free particle described by an infinite (complex) plane wave. Because the de Broglie wave for a freely propagating particle exhibits dispersion—that is, v_φ is a function of λ—we must reexamine the situation for the truncated wave train that is associated with a localized particle. At time $t = 0$ the length of the particle's wave packet is $(\Delta x)_0$ and the corresponding spread in momentum is $(\Delta p)_0$. Then, we expect the length $(\Delta x)_t$ of the packet at time t to be given by

$$(\Delta x)_t^2 = (\Delta x)_0^2 + t^2 (\Delta v)_0^2 = (\Delta x)_0^2 + t^2 \frac{(\Delta p)_0^2}{M^2}$$

where we assume that quantities describing uncertainties combine in the same quadratic manner that governs the combination of statistical probabilities (or experimental errors). Then, we have

$$(\Delta x)_t^2 = (\Delta x)_0^2 + \frac{t^2 h^2}{M^2} \frac{1}{(\Delta x)_0^2} \qquad (46\text{–}42)$$

Physically, this expression means that when the particle is located within the range $(\Delta x)_0$ at $t = 0$, the act of measurement causes an uncertainty of $(\Delta p)_0$ in the particle momentum. Then, the particle can move (in an unspecified direction) with a velocity in the range $(\Delta p)_0/M$, and during the time interval t it can move a distance as large as $[(\Delta p)_0/M] \times t = [h/M(\Delta x_0)] \times t$. Thus, at the time t, a new measurement of the particle position will locate the particle only within the range $(\Delta x)_t$.

Example 46–7

To see the significance of dispersion in a particle wave, determine, in different situations, the time interval required for the uncertainty in position $(\Delta x)_t$ to become double the initial value $(\Delta x)_0$.

Solution:

Solving Eq. 46–42 for t, we obtain

$$t = \frac{M(\Delta x)_0^2}{h} \sqrt{\frac{(\Delta x)_t^2}{(\Delta x)_0^2} - 1}$$

For the case at hand, we have $(\Delta x)_t = 2(\Delta x)_0$, so that

$$t = \frac{\sqrt{3} M(\Delta x)_0^2}{h}$$

(a) First, consider a person with a mass $M = 70$ kg whose center of mass can be determined to within $(\Delta x)_0 = 1$ mm. Then,

$$t = \frac{\sqrt{3}(70 \text{ kg})(10^{-3} \text{ m})^2}{6.63 \times 10^{-34} \text{ J} \cdot \text{s}} = 1.83 \times 10^{29} \text{ s} = 5.77 \times 10^{21} \text{ y}$$

Thus, the spreading of the wave packet associated with an object as large as a person is completely negligible. Wave mechanical effects are not important on the macroscopic scale.

(b) Because the doubling time t is proportional to $M(\Delta x)_0^2$, the effect of dispersion becomes important for sufficiently small M and $(\Delta x)_0^2$. Consider a dust particle with a diameter of 10 μm that has a mass $M = 10^{-12}$ kg. With a microscope, the position of such a particle can be established to within about 2000 Å (see Section 45–3); that is, $\Delta x_0 = 2 \times 10^{-7}$ m. Then,

$$t = \frac{\sqrt{3}(10^{-12} \text{ kg})(2 \times 10^{-7} \text{ m})^2}{6.63 \times 10^{-34} \text{ J} \cdot \text{s}} = 1.04 \times 10^8 \text{ s} = 3.30 \text{ y}$$

Even for an object as small as a dust particle, the position uncertainty does not depend significantly on time.

(c) Finally, consider an electron ($M = m_e = 9.11 \times 10^{-31}$ kg) whose position has been established to within 2 Å (the diameter of a typical atom); that is, $(\Delta x)_0 = 2 \times 10^{-10}$ m. Then,

$$t = \frac{\sqrt{3}(9.11 \times 10^{-31} \text{ kg})(2 \times 10^{-10} \text{ m})^2}{6.63 \times 10^{-34} \text{ J} \cdot \text{s}} = 9.52 \times 10^{-17} \text{ s}$$

Thus, the act of localizing an electron in such a small space causes a rapid spreading of the wave packet. Clearly, the behavior of particles in the atomic domain is distinctly different from that of macroscopic objects. In Chapter 47 we explore some of the quantum effects associated with atomic particles. ◀

The Schrödinger Equation.

What is the nature of the differential equation obeyed by the wave function for a nonrelativistic particle that is described by Eq. 46-37? Again, we give only a plausibility argument, not a rigorous derivation. (In fact, no rigorous derivation is possible. The equation must be postulated and then tested against experiments.) To be more general, suppose that the particle is in a region of space where it possesses a constant potential energy U. In view of Eqs. 46-35, the wave function (Eq. 46-37) must now be written as

$$\psi = Ae^{i(px-Et)/\hbar} \qquad \textbf{(46-43)}$$

with $E = K + U$. Taking the second partial derivative of ψ with respect to x gives

$$\frac{\partial^2 \psi}{\partial x^2} = -\frac{p^2}{\hbar^2}\psi = -\frac{2MK}{\hbar^2}\psi$$

from which

$$-\frac{\hbar^2}{2M}\frac{\partial^2 \psi}{\partial x^2} + U\psi = E\psi \qquad \textbf{(46-44)}$$

Taking the partial derivative of ψ with respect to t gives

$$\frac{\partial \psi}{\partial t} = -\frac{i}{\hbar}E\psi$$

or

$$-\frac{\hbar}{i}\frac{\partial \psi}{\partial t} = E\psi \qquad \textbf{(46-45)}$$

Combining Eqs. 46-44 and 46-45, we have

$$-\frac{\hbar^2}{2M}\frac{\partial^2 \psi}{\partial x^2} + U\psi = -\frac{\hbar}{i}\frac{\partial \psi}{\partial t} \qquad \textbf{(46-46)}$$

Although we obtained Eq. 46-46 by assuming that U is constant, the result is in fact the same if U is a function of position, $U = U(\mathbf{r})$. For the general case in three dimensions, we have

$$-\frac{\hbar^2}{2M}\nabla^2 \psi + U(\mathbf{r})\psi = -\frac{\hbar}{i}\frac{\partial \psi}{\partial t} \qquad \textbf{(46-47)}$$

where $\psi = \psi(\mathbf{r}, t)$, and where ∇^2 is the *Laplacian operator*; $\nabla^2 \psi$ stands for

$$\nabla^2 \psi = \frac{\partial^2 \psi}{\partial x^2} + \frac{\partial^2 \psi}{\partial y^2} + \frac{\partial^2 \psi}{\partial z^2}$$

Equation 46-47 is the famous *Schrödinger wave equation*,* the fundamental equation of wave mechanics and the basis for every calculation of a (nonrelativistic) quantum effect. This equation has been supremely successful in describing atomic effects of all sorts.

When the system is known to be in a steady-state (time-independent) condition, Eq. 46-45 can be used to eliminate the time dependence. The result is the *time-independent Schrödinger equation*:

$$-\frac{\hbar^2}{2M}\nabla^2 \psi + U(\mathbf{r})\psi = E\psi \qquad \textbf{(46-47a)}$$

Experimental Predictions. The wave function ψ represents as complete a statement as it is possible to make concerning the physical state of a system. For this reason, ψ is often referred to as the *state function* for the system.

Suppose that a particle is moving in a potential $U(\neq 0)$ and has an associated state function ψ that satisfies the Schrödinger equation. The function ψ is probabilistic in nature, so how can it be used to predict the results of experimental measurements that we might make? It is a fundamental postulate of quantum theory that the result of a measurement of a physical quantity is restricted to a particular set of allowed values, and that one cannot predict which of the allowed values will be obtained as a result of the experiment. However, if a large number of identical experiments is performed, there will be a distribution of measured values that corresponds to the probability of finding each allowed value in a particular experiment. In each such case there is a most likely result, which is referred to as the *expectation value* for the quantity measured and which is the statistical average of the distribution of allowed values. For a particle undergoing one-dimensional motion, the most likely value for the position is written as $\langle x \rangle$, and the most likely value for the momentum is written as $\langle p_x \rangle$.

Every set of measured values will exhibit a spread about the most likely value. According to the (classical) statistical theory of measurements, the amount of spreading (or the *statistical variance*) can be specified by the quantity $(\Delta x)^2 = \langle x^2 \rangle - \langle x \rangle^2$. The quantity Δx that is obtained from this expression is exactly the quantity Δx that appears (along with Δp_x obtained in a similar way) in the uncertainty principle relationship.

Newtonian dynamics is a successful theory of the behavior of macroscopic objects. Quantum theory is necessary for the description of the behavior of microscopic objects. However, quantum theory is completely general and

---

*This equation was put forward in 1926 by Erwin Schrödinger (1887-1961), Austrian physicist, who was awarded a share of the 1933 Nobel Prize in physics for his contributions to the foundations of quantum theory.

therefore applies to large-scale as well as small-scale systems. Does there exist a reconciliation between the Newtonian and quantum descriptions of macroscopic objects? Indeed there does. As is shown in advanced texts, when the expectation values of the quantum variables are calculated for any system, large or small, Newton's equation results:

$$\frac{d}{dt}\langle\mathbf{p}\rangle = -\langle\nabla U\rangle = \langle\mathbf{F}\rangle$$

However, for the case of a macroscopic object, such as a baseball, the de Broglie wavelength is exceedingly small, so diffraction effects are completely negligible. Thus, the variance (uncertainty) in the variables $\mathbf{r}$ and $\mathbf{p}$ is well below any possibility of direct measurement. This is one example of an important concept that is known as the *Bohr correspondence principle*:

In the classical limit (of objects with large size) *the predictions of wave mechanics agree with those of Newtonian theory.*

PROBLEMS

Section 46-1

46-1   What is the energy in eV of a photon of yellow-green light ($\lambda = 550$ nm)?

46-2   A vibrating electron in the wall of a cavity radiator emits a photon with $\lambda = 550$ nm. What is the effective spring constant of the electronic oscillator?

46-3   The Sun ($R = 6.96 \times 10^8$ m) radiates a luminous flux of 3.90×10^{26} J/s. (a) What is the effective blackbody temperature of the Sun? (b) According to Wein's law (Eq. 46-18), what is the wavelength λ_m that corresponds to the maximum radiation intensity?

46-4   The tungsten filament in an incandescent light bulb is operated at a temperature of 2800 K. The emissivity of tungsten at this temperature is 0.31. The surface area of the filament is 0.45 cm². (a) What input electric power is required to maintain the temperature of the filament? (b) At what frequency ν_m is the radiation intensity $R(\nu)$ maximum? (c) What is the ratio of $R(\nu_m)$ to $R(\nu)$ corresponding to light with $\lambda = 550$ nm? [*Hint:* Figure 46-5 will be helpful.]

46-5   The energy density inside a cavity radiator is $u(\nu)$.

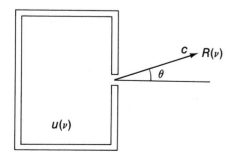

Show that the radiation energy flux $R(\nu)$ that escapes into the open hemisphere through a small hole is $R(\nu) = \frac{1}{4}cu(\nu)$. [*Hint:* The radiation inside the cavity travels in all directions, so at any instant only half is directed toward the wall with the opening. Per unit area of the opening, only a fraction $\langle\cos\theta\rangle$ of this flux escapes.]

46-6   According to the Rayleigh-Jeans analysis of thermal cavity radiation, how many standing-wave modes are there that have wavelengths greater than 5 μm in a cavity volume 1 mm on a side?

Section 46-2

46-7   What is the maximum wavelength of light that is capable of ejecting photoelectrons from a copper surface? From a cesium surface? (Refer to Table 31-4.)

46-8   Monochromatic light with wavelength 380 nm is incident on a metallic surface. A potential of 1.32 V is required to cut off the flow of photoelectrons. What is the work function of the metal? What is the maximum wavelength of light that is capable of ejecting photoelectrons from this surface?

46-9   What is the limiting wavelength λ_{min} for the continuous X-ray spectrum from an X-ray tube that operates with an anode-cathode potential difference of 45 kV? 25 kV?

Section 46-3

46-10 Calculate the momentum of each of the following: (a) an X-ray photon with $\lambda = 0.5$ Å, (b) a 54-eV electron, (c) a 5-MeV proton, and (d) a red blood cell ($M \approx 1.3 \times 10^{-12}$ kg) moving through an artery with a speed of 0.25 m/s.

46–11 An X-ray photon with $\lambda = 0.050$ Å is scattered through an angle of 90° by an electron that was initially at rest. What is the wavelength λ' of the scattered X ray?

46–12 An electron initially at rest backscatters ($\theta = 180°$)
 • a 2-MeV γ ray. What is the kinetic energy of the recoiling electron? First, show that

$$K_m = h\nu(1 + m_e c^2/2h\nu)^{-1}$$

46–13 Obtain Eq. 46–26 from Eqs. 46–25.
 •

Section 46-4

46–14 (a) The flux of solar radiation incident on the top of the Earth's atmosphere is 0.137 J/cm² · s. If the average wavelength of the solar radiation is 600 nm, what is the value of this flux in photons/m² · s? (b) The Mt. Palomar telescope (diam. = 5.08 m) is used to observe a star similar to the Sun at a distance of 50,000 light years (L.Y.). If the absorption of light in the Earth's atmosphere above the telescope is 20 percent, how many photons from the star are intercepted by the telescope per second? (1 L.Y. = 9.46×10^{15} m and Earth-Sun distance = 1.50×10^{11} m.)

46–15 Calculate the de Broglie wavelength for each of the following: (a) a 10-keV electron, (b) a 10-MeV proton, (c) a baseball ($M = 0.140$ kg) traveling at a speed of 45 m/s, (d) an "18-wheeler" truck ($M = 20$ metric tons) traveling at a speed of 120 km/h, and (e) the Earth in its orbit around the Sun.

46–16 Show that the de Broglie wavelength for a nonrela-

tivistic electron accelerated through a potential V volts is $\lambda = 12.3\, V^{-1/2}$ Å.

46–17 De Broglie postulated that the relationship $\lambda = h/p$ is valid for relativistic particles. What is the de Broglie wavelength for a (relativistic) electron whose kinetic energy is 3 MeV?

Section 46-5

46–18 The uncertainty in the time associated with the motion of a particle with velocity u is $\Delta t = \Delta x/u$, where Δx is the uncertainty in the position. The uncertainty ΔE in the particle's energy, $E = p^2/2M$, is associated with the uncertainty Δp in the momentum. Then, show that $\Delta E\, \Delta t \approx h$. This is another statement of the uncertainty principle.

46–19 The energy of a simple harmonic oscillator with
 • natural frequency ω_0 can be expressed as

$$E = \frac{p^2}{2M} + \tfrac{1}{2}M\omega_0^2 x^2$$

The least energy the oscillator may have results when x^2 is no larger than $(\Delta x)^2$ and p^2 is no larger than the corresponding value, $(\Delta p)^2 = h^2/(\Delta x)^2$. (a) Show that these requirements lead to $E \geqslant h\omega_0$. [*Hint:* Find the minimum value of $y = ax^{-2} + bx^2$ by setting dy/dx equal to zero.] (The exact solution from quantum theory is $E_{min} = h\omega_0/4\pi$.) (b) Estimate the minimum energy of a macroscopic harmonic oscillator for which $\omega_0 = 10^3$ rad/s. Compare your result with the minimum energy of a microscopic harmonic oscillator with $M = 10^{-15}$ kg which has a maximum speed of 1 cm/h. (c) The electronic oscillators in the wall of a cavity radiator may have frequencies ω_0 as large as 10^{15} rad/s. What is the corresponding minimum oscillator energy?

Section 46-6

46–20 For a relativistic free particle, de Broglie pos-
 • tulated that the frequency is given by

$$h\nu = E = \sqrt{p^2 c^2 + M_0^2 c^4}$$

(a) Assume that $\lambda = h/p$ is valid relativistically, and

show that the phase velocity v_φ is given by the dispersive relationship

$$v_\varphi = c\sqrt{1 + M_0^2 c^2 \lambda^2/h^2}$$

(b) Use Eq. 46–40 to show that the group velocity v_g is equal to the particle velocity v, just as in the nonrelativistic case (Eq. 46–41). (c) Show that $v_\varphi v_g = c^2$. (Compare Eq. 39–12.)

Additional Problems

46-21 Suppose that a human body with a surface area of 1.8 m² and a skin temperature of 34° C is in a room with walls at a temperature of 27° C. What is the net amount of radiant power emitted by the body (and supplied by the body's metabolic processes)? (For temperatures near 300 K, the radiation is in the infrared region, and the emissivity values for the skin and the room walls are approximately unity. Note that the emissivity of a surface is equal to the efficiency with which the body *absorbs* radiation as well as the efficiency with which it *emits* radiation.)

46-22 In his paper describing the new approach to the
• cavity radiation problem, Planck used his theoretical expressions to calculate values for h and k. He used the experimental results of Rubens and Kurlbaum, which gave

$$\mathcal{U} = \int_0^\infty u(\nu)d\nu = (7.6 \times 10^{-16} \text{ J/m}^3 \cdot \text{K}^4)T^4$$

and

$$\lambda_m T = 2.9 \times 10^{-3} \text{ m} \cdot k$$

where $\mathcal{U}$, equal to the integral of Eq. 46-15, is the total energy density in the cavity, and where $\lambda_m T$ is given by Eq. 46-18. Determine h and k from these data. [*Hint:* $\int_0^\infty x^3 (e^x - 1)^{-1} dx = \pi^4/15$.]

46-23 Use Eq. 46-15 for $u(\nu)$ in Eq. 46-2 and substitute
• the resulting expression for $R(\nu)$ into Eq. 46-3. Integrate using the hint in Problem 46-22, and evaluate Stefan's constant σ.

46-24 According to Planck's hypothesis, what is the average energy $\langle E \rangle$ of a set of oscillators in thermal equilibrium if the frequency is $h\nu = kT$? Express your answer in units of kT.

46-25 Derive Eq. 46-16 for ν_m which corresponds to the
• maximum value of $u(\nu)$ for a cavity radiator. (The numerical factor can be obtained by an iterative procedure.)

46-26 Derive Eq. 46-18 for λ_m which corresponds to the
• maximum value of $u(\lambda)$ for a cavity radiator. (The numerical factor can be obtained by an iterative procedure.)

46-27 A photoelectric experiment is performed using a clean lithium surface. The following values of the cut-off potential V_0 were obtained for the various listed monochromatic wavelengths of light.

λ (nm)	254	283	313	337	365	395
V_0 (V)	2.52	2.00	1.59	1.29	1.04	0.77

(a) Make a suitable plot of the data and determine a value for h/e. Compare with the accepted value.
(b) Obtain a value for the work function of lithium. Compare with the value quoted in Table 31-4.

46-28 Show that the energy of a photon that has been backscattered ($\theta = 180°$) in a Compton process has an energy that approaches $h\nu' = \frac{1}{2}m_e c^2 = 255$ keV in the limit of very-high-energy incident photons ($h\nu \to \infty$).

46-29 The average velocity of an atom with mass M that emerges from an oven at temperature T is $\langle v \rangle = \sqrt{kT/2\pi M}$. What is the average de Broglie wavelength of helium atoms that emerge from an oven at 300 K? Such beams were used by Stern and Estermann to demonstrate diffraction effects with atoms.

46-30 Determine the diffraction-limited resolution of an electron microscope that operates at 25 kV. Assume that the expression (Eq. 45-20) for an optical microscope applies with N.A. = 1.

46-31 The focal length of a magnetic lens of the type used in electron microscopes is given by

$$\frac{1}{f} = \frac{e}{8m_e V} \int_0^\ell B_z \, dz \cong \frac{e \overline{B_z^2} \ell}{8m_e V}$$

where B_z is the axial component of the magnetic field, and where ℓ is the length of the lens gap (see Fig. 46-19). If the magnification of the microscope is to be large, the focal length of the objective must be small. Typical values are $V = 25$ kV, $\ell = 3$ mm, and $\overline{B_z^2} = 1$ T². Determine f for this case.

46-32 De Broglie assumed that for a particle at rest in
•   an inertial frame S_0, the proper wave frequency ν_0
to associate with the particle is given by $h\nu_0 = M_0c^2$
and that the wave function is

$$\psi_0 = Ae^{-i2\pi\nu_0 t_0}$$

(a) Show that in a frame S that moves with a velocity v with respect to S_0, the particle wave function is

$$\psi = Ae^{-i2\pi\nu(t-vx/c^2)}$$

where $\nu = \nu_0(1 - \beta^2)^{-1/2}$. [*Hint:* Use the Lorentz
transformation, $t_0 = (t - vx/c^2)(1 - \beta^2)^{-1/2}$ (Eq.
21-8a).] (b) In frame S the particle appears to move
with velocity v. If ψ is to be associated with this
moving particle, is the phase velocity given by
$v_\varphi = c^2/v$? (c) If $v_\varphi = c^2/v$ and $\lambda = v/\nu$, is the relationship $\lambda = h/p$ valid in frame S? (d) Is $h\nu = Mc^2$
true in frame S?

ATOMS

By the beginning of the 20th century the existence of the electron as an elementary bit of matter had been firmly established by J. J. Thomson. Reasonable values for the electron charge $-e$ and for the ratio e/m_e had been determined. From the inferred value of m_e it was clear that atoms are much more massive than electrons. Atoms, under ordinary conditions, are electrically neutral, so they must contain equal amounts of negative charge (contributed by electrons) and positive charge. In 1904 Thomson proposed the first model of atomic structure. In this model the atom was considered to be a sphere of uniform positive charge (possessing most of the mass of the atom) in which were embedded electrons, uniformly distributed and sufficient in number to render the atom electrically neutral. Thomson's atomic scheme was called the *plum-pudding model* because the electrons were like the raisins in an English plum pudding.

Thomson's ingenious model survived only until 1911, when Ernest (later Lord) Rutherford* proposed his radically different *nuclear model* of the atom. The correctness of Rutherford's model has been established by numerous experiments and it is the model we still use to describe atomic phenomena of all sorts.

47-1 RUTHERFORD'S NUCLEAR ATOM

Soon after his move to Manchester University in 1907, Rutherford, assisted by Geiger and Marsden, began a study of the scattering of α particles† by thin foils of various metals. The apparatus used is shown schematically in Fig. 47–1a. The foils were of gold, platinum, and other metals, and had thicknesses of approximately 1 μm, corresponding to several thousand atomic diameters. It was observed that most of the particles passed through the foils with very little deflection from the original direction of motion. Because the foils consisted of so many layers of atoms, it seemed evident that there were no gaps between atoms for the α particles to pass through. Thus, the atoms themselves must be nearly transparent to the rapidly moving α particles.

---

*Ernest Rutherford (1871–1937), born in New Zealand, was one of the most influential physicists of the twentieth century. Most of the important early discoveries in nuclear physics are linked with his name. Rutherford received the 1908 Nobel Prize in chemistry for his discoveries in the field of radioactivity.

†Rutherford chose these particles as projectiles because they are emitted with high speeds ($v \approx 1.5 \times 10^7$ m/s) from various naturally radioactive substances such as radium. Rutherford had earlier (1908) shown that an α particle is the nucleus of a helium atom.

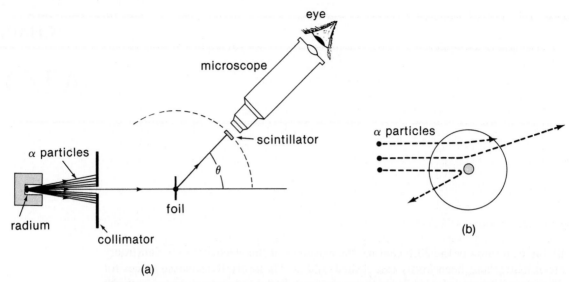

Fig. 47–1. (a) Schematic diagram of the apparatus used in the Rutherford scattering experiments. The source of α particles is a small amount of radium in a lead container. The collimated beam of α particles impinges on a thin metal foil. Particles scattered through an angle θ are detected by a scintillator that consists of a thin zinc sulfide screen. When a fast α particle strikes zinc sulfide, a flash of light is emitted, which the experimenter observes by means of a low-power microscope. The source, foil, and scintillator are enclosed in an evacuated chamber to eliminate the spurious scattering of α particles by air molecules. (b) Deflection of α particles according to Rutherford's nuclear model of the atom. Notice that the closer the encounter, the greater is the angle of deflection. (•Can you see why?)

Geiger and Marsden also discovered that occasionally an α particle would be scattered into the backward hemisphere—that is, the scattering angle θ was $>90°$ (Fig. 47–1b). A collision between an α particle and an electron cannot produce such a backscattering (•Can you see why? Refer to Section 9–3), nor can the accumulation of many small-angle scattering events. Thus, the atomic scattering centers were required to be more massive than the α particles and much smaller in size than the atom itself. Thomson's plum-pudding model was inadequate to explain the scattering results.

Rutherford's explanation of the α-particle scattering experiments was published in 1911. He had found that the scattering data could be accurately accounted for by assuming that an atom consists of a tiny nuclear core containing most of the mass of the atom and all of its positive charge, surrounded by the appropriate number of electrons. The diameter of the nucleus was determined to be no more than about 10^{-14} m (see Problem 47–1), compared with a diameter of about 10^{-10} m for a typical atom. In developing his theory, Rutherford had boldly asserted that Coulomb's law of electric force is valid at distances as small as 10^{-14} m.*

Rutherford's model of the atom resembled the solar system, with the nucleus as the Sun and the orbiting electrons as the planets. In the following sections we see how the nuclear model of the atom was refined and led eventually to the modern quantum theory of atoms.

---

*Thomson did not believe this assertion, but the Manchester mathematician C. G. Darwin promptly showed that no other force law nor any serious distortion could explain the results of the scattering experiments.

47-2 THE BOHR MODEL OF THE HYDROGEN ATOM

The science of optical spectroscopy had its beginning in 1802 when William Wollaston (1766–1828), English chemist and physicist, discovered a number of sharp dark lines (absorption lines) superimposed on the continuous band of color in the solar spectrum. Several years later these lines were studied in detail by the German optician Joseph von Fraunhofer (1787–1826), who cataloged and measured the wavelengths of nearly 600 lines in the solar spectrum. Subsequent laboratory experiments revealed that each element exhibits a distinct set of sharp emission lines. The wavelengths of the more intense lines in the spectra of hydrogen, calcium, and iron were found to correspond precisely to those of some of Fraunhofer's solar lines, thereby establishing the existence of these elements in the Sun. Later, experiments revealed sharp absorption and emission lines in all parts of the electromagnetic spectrum, including the X-ray, ultraviolet, and infrared regions. The two strong X-ray lines in the molybdenum spectrum shown in Fig. 46–10 are examples of such lines. It seemed that the sharp spectral lines were somehow due to atoms. However, the wavelengths did not follow any simple harmonic scheme (as is the case for vibrating strings and air columns), so the origin of the lines remained a mystery.

The first bit of progress was made in 1885 when Johann Balmer (1825–1898), a Swiss music teacher with an interest in numbers, discovered that the wavelengths of the lines in the optical spectrum of hydrogen (Fig. 47–2) could be represented by a simple formula, $\lambda = Am^2/(m^2 - 4)$, where A is a constant (3645.6 Å) and where $m = 3, 4, 5, \ldots$. Balmer obtained exceptionally good agreement between the wavelengths calculated from his formula and those that had been measured with high precision by the Swedish spectroscopist Anders Ångström (1814–1874).

Fig. 47–2. The Balmer series of lines in the spectrum of hydrogen. The series limit (3646 Å) corresponds to the substitution $n_i = \infty$ and $n_f = 2$ in Eq. 47–1.

A more useful form of the Balmer formula was given in 1890 by another Swedish spectroscopist, Johannes Rydberg (1854–1919), who expressed the relationship as

$$\frac{1}{\lambda} = \Re\left(\frac{1}{n_f^2} - \frac{1}{n_i^2}\right) \qquad n_i > n_f \qquad \textbf{(47–1)}$$

where $n_f = 2$ for the Balmer series of hydrogen lines. A similar formula was found to describe the spectral series of other elements. For hydrogen, the constant $\Re$ (the *Rydberg constant*) was determined experimentally to be

$$\Re_H = (1.0968 \pm 0.0005) \times 10^7 \text{ m}^{-1}$$

The value of $\Re$ for other elements was found to be essentially the same. Although the Balmer-Rydberg formula accurately represented the experimental data, the reason was completely unknown. The crucial step was supplied by Niels Bohr, who came as a visitor to Rutherford's laboratory in 1912. There he learned about Rutherford's model of the nuclear atom and accepted this new idea completely.

According to classical electromagnetic theory, an electric charge that is accelerated will produce electromagnetic radiation. In Rutherford's nuclear model of the atom, the "planetary" electrons undergo centripetal acceleration as they orbit around the nuclear "sun." A classical calculation shows that such an electron should radiate away its energy in about 10^{-6} s while spiraling in toward the nucleus. How, then, can an atom be a stable structure?

In a series of five papers published between 1913 and 1915, Bohr enunciated a bold new theory in which he incorporated the photon quantum hypothesis of Planck and Einstein into the Rutherford model of the atom. Bohr developed his theory on the basis of the following postulates* (which are given here in slightly modified and modernized form):

(1) Electrons move in circular orbits around the nucleus in accordance with the Coulomb law of electric force.
(2) Of the infinite number of orbital states that are possible according to classical theory, only those states are allowed for which the atomic system possesses certain discrete energies. (These states are sometimes called *stationary states*.)
(3) An atom can interact with radiation only by making a sudden and discontinuous transition from one of these stationary states to another, by emitting or by absorbing an electromagnetic photon (energy quantum) whose energy is exactly equal to the difference in energy between the two states.
(4) There exists a state of lowest energy of the atom (called the *ground state*) from which no further photon emission is allowed, even though in this state charges are undergoing accelerated motions.

To examine how these postulates apply, consider a hydrogen atom or a hydrogenlike ion that consists of a (pointlike) nucleus with mass M and charge Ze and a *single* orbital (pointlike) electron with mass m_e and charge $-e$. At first we consider only circular orbits for the electron and the nucleus about the center of mass of the system. The geometry is shown in Fig. 47–3, where ω is the angular speed of each particle about the C.M. The definition of C.M. means that $MR = m_e r$; hence,

$$r = \frac{M}{M + m_e}\rho \quad \text{and} \quad R = \frac{m_e}{M + m_e}\rho \qquad (47\text{–}2)$$

The kinetic energy of the atomic system is

$$K = \tfrac{1}{2}I\omega^2 = \tfrac{1}{2}(MR^2 + m_e r^2)\omega^2$$

Using the reduced mass, $\mu = m_e M/(M + m_e)$, and Eqs. 47–2, we have

$$K = \tfrac{1}{2}\mu\rho^2\omega^2 \qquad (47\text{–}3)$$

The potential energy of the system is

---

*Following the time-honored procedure of theorists, Bohr finally stated his postulates *after* he had developed his theory and worked out the main conclusions! This made explanations of his theory hard to follow at first, which led de Broglie to remark that Bohr had a predilection for "obscure clarity."

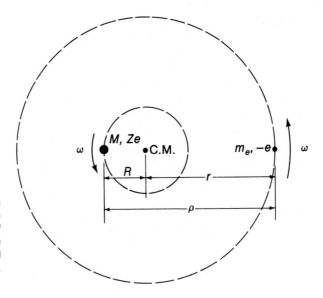

Fig. 47-3. The electron (mass m_e) and the nucleus (mass M) both orbit around the center of mass with angular speed ω. The distance R is much exaggerated here; in the hydrogen atom, $r/R = 1836$.

$$U = -\frac{1}{4\pi\epsilon_0}\frac{Ze^2}{\rho} \qquad (47\text{-}4)$$

The mass times the centripetal acceleration for either particle is equal to the Coulomb force; that is,

$$m_e r\omega^2 = MR\omega^2 = \mu\rho\omega^2 = \frac{1}{4\pi\epsilon_0}\frac{Ze^2}{\rho^2} \qquad (47\text{-}5)$$

Combining this expression with Eq. 47-4 gives

$$U = -\mu\rho^2\omega^2 \qquad (47\text{-}6)$$

Thus, the total energy of the system is, using Eqs. 47-3 and 47-6,

$$E = K + U = -\tfrac{1}{2}\mu\rho^2\omega^2 \qquad (47\text{-}7)$$

To this point, the calculation has been strictly classical in character. Bohr next made an inspired guess by introducing the idea that *angular momentum is quantized.* He saw that he could obtain the desired result of quantized energy states by assuming that the angular momentum L of the system can have only values that are integer multiples of a basic unit, $h/2\pi = \hbar$. Noting that $L = I\omega = \mu\rho^2\omega$, Bohr wrote

$$L = \mu\rho^2\omega = n\hbar, \qquad n = 1, 2, 3, \ldots \qquad (47\text{-}8)$$

Multiplying Eq. 47-5 by $\mu\rho^3$ and using the angular momentum quantization rule (Eq. 47-8), there results

$$\frac{Ze^2\rho\mu}{4\pi\epsilon_0} = \mu^2\rho^4\omega^2 = n^2\hbar^2$$

from which

$$\rho_n = \frac{4\pi\epsilon_0 \hbar^2}{Ze^2\mu}n^2 \qquad n = 1, 2, 3, \ldots \tag{47-9}$$

where the subscript n denotes the value of ρ for the particular value of n on the right-hand side of the equation.

Equations 47-8 and 47-9 may be used to substitute for the combination $\rho^2\omega^2$ that appears in the expression for the total energy (Eq. 47-7). The result is

$$E_n = -\frac{Z^2e^4\mu}{32\pi^2\epsilon_0^2\hbar^2} \cdot \frac{1}{n^2} \qquad n = 1, 2, 3, \ldots \tag{47-10}$$

which gives the allowed energy states for a hydrogen atom ($Z = 1$) or a hydrogenlike ion. The value $n = 1$ gives the lowest possible energy, corresponding to the ground state. The potential energy of the system is chosen to be zero for infinite separation of the electron and the nucleus (see Eq. 47-4). Then, the bound states have *negative* values of the total energy (Eq. 47-10). The system may also have positive energies, which correspond to unbound (scattering) states. The energy of such states is not quantized.

Suppose that an atom is initially in an excited energy state with $n = n_i > 1$. According to Bohr's third postulate, this atom may spontaneously make a transition to a final state of lower energy, which is characterized by $n = n_f < n_i$. In this process there is emitted a photon with energy

$$h\nu = E_{n_i} - E_{n_f} \qquad n_i > n_f$$

That is,

$$\frac{1}{\lambda} = \frac{1}{hc}(E_{n_i} - E_{n_f}) = \Re\left(\frac{1}{n_f^2} - \frac{1}{n_i^2}\right) \tag{47-11}$$

where

$$\Re = \frac{Z^2e^4\mu}{8\epsilon_0^2ch^3} = \frac{Z^2\mu}{m_e}\Re_\infty \tag{47-12a}$$

with

$$\Re_\infty = \frac{e^4m_e}{8\epsilon_0^2ch^3} \tag{47-12b}$$

The quantity $\Re_\infty$ is the Rydberg constant that would apply for a hydrogenlike ion with infinite mass. Because the ratio of the mass of the hydrogen nucleus (the *proton*) to the mass of the electron is $M_p/m_e = 1836.15 \pm 0.01$, there is only a small difference between $\Re_\infty$ and $\Re$ for the hydrogen atom. When the best values for the various constants are substituted into Eq. 47–12b, the result is

$$\mathcal{R}_\infty = 1.09737314 \times 10^7 \text{ m}^{-1}$$

with an uncertainty of about 1 in the last figure. This value is in complete agreement with that deduced from spectroscopic experiments. Using $\mu/m_e = M_p/(M_p + m_e) = 0.9994557$ and $Z = 1$ for the hydrogen atom, Eq. 47–12a gives $\mathcal{R}_H = 1.0967758$, in agreement with Rydberg's value quoted earlier (see Eq. 47–1).

The ground-state energy for the hydrogen atom is the value of E_n given by Eq. 47–10 with $n = 1$; that is,

$$E_1 = -13.60 \text{ eV}$$

To *ionize* an atom means to give sufficient energy to the atom to cause an electron to become unbound. Thus, the least energy that is required to ionize a hydrogen atom in its ground state is just 13.60 eV—this is the *ground-state binding energy* of the hydrogen atom.

A convenient way to display the energy states of the hydrogen atom is in the form of an *energy-level diagram* (Fig. 47–4). The energy of each state is given by

$$E_n = (-13.60 \text{ eV}) \cdot \frac{1}{n^2} \tag{47–13}$$

The internal excitation from the ground state ($n = 1$) that requires the least energy involves the promotion of the atom to the first excited state ($n = 2$); the required energy is

$$\Delta E_{12} = E_2 - E_1 = (-13.60 \text{ eV})\left(\frac{1}{2^2} - \frac{1}{1^2}\right) = 10.20 \text{ eV}$$

The energy of any allowed transition (involving the emission or absorption of radiation) is calculated in a similar way.

A number of allowed transitions and their wavelengths are indicated in Fig. 47–4. All of the (emission) transitions that terminate on a particular level are termed a *series*. Thus, each transition in the Balmer series has $n_f = 2$; these transitions produce photons of visible and near-ultraviolet light. The Lyman series of lines involves transitions to the $n = 1$ state; the emitted photons are more energetic than those in the Balmer series, and they occur in the UV part of the spectrum. The Paschen, Brackett, and Pfund series involve infrared transitions.*

The constants that appear in Eq. 47–9 can be grouped in a convenient form:

$$\rho_n = \frac{4\pi\epsilon_0 \hbar^2}{m_e e^2} \frac{m_e}{Z\mu} n^2 = a_0 \frac{m_e}{Z\mu} n^2$$

Because the electron mass m_e is so much smaller than any atomic mass, the ratio m_e/μ is approximately unity. Then, the radius ρ_n of the orbit of the reduced-mass particle is very closely equal to the radius r_n of the *electron* orbit about the C.M. (see

---

*The tenuous atmospheres of stars make better sources of atomic hydrogen radiation than laboratory devices. (•Can you imagine why this is the case?) As many as 50 lines in the Balmer series have been observed in stellar spectra.

Fig. 47-4. Energy-level diagram for the hydrogen atom. Some of the transitions in the first five series of lines are shown. The wavelengths are given in Å, except for the Brackett and Pfund series where the units are μm. The cross-hatched region above $n = \infty$ corresponds to the continuum of (unquantized) positive energy states.

Fig. 47–3); thus, very nearly,

$$r_n = \frac{a_0}{Z} n^2 \qquad\qquad (47\text{–}14)$$

The quantity $a_0 = 0.529$ Å is called the *Bohr radius* of the hydrogen atom (because $r_1 = a_0$ for $n = 1$ and $Z = 1$).

The various hydrogen-atom electron orbits and some transitions are shown in Fig. 47–5, where the orbits are shown to scale.

Fig. 47-5. Electron orbits and transitions in the hydrogen atom. The orbit radii are $r_n = (0.529 \text{ Å})n^2$.

The Correspondence Principle. Bohr's correspondence principle, mentioned in Section 46–6, can be stated in another way:

In the limit characterized by large quantum numbers and transitions involving small changes in quantum numbers, classical theory and quantum theory lead to the same predictions.

Let us see how the correspondence principle applies to the quantum number n introduced above. According to classical theory, an electron moving in a circular orbit will emit radiation with the frequency of its motion. In Bohr's theory the frequencies of the emitted photons are distinct from the orbital frequencies of the electron. However, consider a radiative transition from a hydrogen-atom state n to a state $n - \Delta n$, where $n \gg 1$ and $\Delta n = 1, 2, 3, \ldots$ The frequency of the motion in the initial Bohr orbit is (see Problem 47–5)

$$\nu_{orb} = \frac{\omega}{2\pi} = \frac{2\Re c}{n^3}$$

Now, Eq. 47–10 for the energy of the state n can be expressed as $E_n = -\Re ch/n^2$, and differentiating gives the incremental approximation

$$\Delta E = \frac{2\Re ch}{n^3} \Delta n$$

which is valid for sufficiently large n and small Δn. Then, the frequency of the radiated photon is

$$\nu = \frac{\Delta E}{h} = \frac{2\Re c}{n^3} \Delta n$$

Thus, $$\nu = \nu_{orb} \Delta n \qquad \Delta n = 1, 2, 3, \ldots$$

That is, the frequencies ν of the radiated photons are integer multiples of the orbital frequency and correspond to the fundamental ($\Delta n = 1$) and the harmonics ($\Delta n = 2, 3, \ldots$) associated with classical oscillators such as strings and pipes.

The correspondence principle (stated by Bohr in 1923) was a useful guide during the development of quantum theory, and it is still useful today in testing certain types of theories.

Example 47–1

An atom of deuterium (or "heavy hydrogen") has a nucleus that consists of a proton and a neutron and therefore has an atomic mass of approximately 2 u. Deuterium was discovered in 1932 by the American chemist H. C. Urey and his coworkers, who observed that each line in the Balmer series has a weak satellite. These close doublets are due to the fact that all natural samples of hydrogen gas contain a small fraction of the heavy hydrogen isotope. The two isotopes have slightly different values for the reduced mass μ, so transitions between states with the same values of n have slightly different wavelengths.

Use the values of λ_α, λ_β, and λ_γ given in Fig. 47–4 for the H_α, H_β, and H_γ lines. Then, calculate the doublet spacings, $\Delta\lambda_\alpha$, $\Delta\lambda_\beta$, and $\Delta\lambda_\gamma$, where $\Delta\lambda_\alpha = \lambda_\alpha(1) - \lambda_\alpha(2)$ and similarly for the other spacings. Assume that the masses of the proton, the deuterium nucleus (the *deuteron*), and the electron are, respectively, $m_p = 1.00728$ u, $m_d = 2.01355$ u, and $m_e = 5.486 \times 10^{-4}$ u.

Solution:

The wavelengths are

$$\lambda_\alpha(1) = 6562.8 \text{ Å}$$
$$\lambda_\beta(1) = 4861.3 \text{ Å}$$
$$\lambda_\gamma(1) = 4340.5 \text{ Å}$$

Using Eq. 47-11, we have for $\lambda_\alpha(1)$

$$\frac{1}{\lambda_\alpha(1)} = \mathcal{R}Q_\alpha = \mathcal{R}_\infty \frac{\mu(1)}{m_e} Q_\alpha$$

where

$$Q_\alpha = \left(\frac{1}{2^2} - \frac{1}{3^2}\right)$$

Then,

$$\frac{1}{\lambda_\alpha(2)} - \frac{1}{\lambda_\alpha(1)} = \mathcal{R}_\infty Q_\alpha \left(\frac{\mu(2) - \mu(1)}{m_e}\right) = \mathcal{R}_\infty Q_\alpha \frac{\mu(1)}{m_e}\left(\frac{\mu(2)}{\mu(1)} - 1\right)$$

$$\frac{\lambda_\alpha(1) - \lambda_\alpha(2)}{\lambda_\alpha(1)\lambda_\alpha(2)} \cong \frac{\Delta\lambda_\alpha}{[\lambda_\alpha(1)]^2} = \frac{1}{\lambda_\alpha(1)}\left(\frac{\mu(2)}{\mu(1)} - 1\right)$$

where we have used $\lambda_\alpha(2) \cong \lambda_\alpha(1)$. Thus,

$$\Delta\lambda_\alpha = \lambda_\alpha(1)\left(\frac{\mu(2)}{\mu(1)} - 1\right)$$

and similarly for the other lines. The term in brackets is

$$\frac{\mu(2)}{\mu(1)} - 1 = \frac{m_d}{m_p} \cdot \frac{m_e + m_p}{m_e + m_d} - 1$$

$$= \frac{2.01355}{1.00728} \cdot \frac{1.00783}{2.01410} - 1$$

$$= 2.728 \times 10^{-4}$$

Thus,

$$\Delta\lambda_\alpha = (6562.8 \text{ Å})(2.728 \times 10^{-4}) = 1.790 \text{ Å}$$
$$\Delta\lambda_\beta = (4861.3 \text{ Å})(2.728 \times 10^{-4}) = 1.326 \text{ Å}$$
$$\Delta\lambda_\gamma = (4340.5 \text{ Å})(2.728 \times 10^{-4}) = 1.184 \text{ Å}$$

The experimental values obtained in 1932 by Urey, Brickwedde, and Murphy were $\Delta\lambda_\alpha = 1.791$ Å, $\Delta\lambda_\beta = 1.313$ Å, and $\Delta\lambda_\gamma = 1.176$ Å, in good agreement with the results of calculations using modern values for the wavelengths.

SPECIAL TOPIC

47-3 ELLIPTICAL ORBITS

In his original paper on the hydrogen atom, Bohr remarked that elliptical as well as circular orbits should be possible. He did not pursue the matter because he was able to obtain the Rydberg wavelength equation using only circular orbits. However, within a few years, the German theorist Arnold Sommerfeld (1868–1951) had developed a detailed theory of elliptical orbits. One result of this investigation (and a similar study by H. A. Wilson) was a more general way to express the quantization condition. The *Wilson-Sommerfeld quantization rule* can be stated in the following way:

A physical system that executes periodic motion in some coordinate $q(t)$ possesses only quantized states for which the integral over a complete period of the associated momentum is an integer multiple of h; that is,

$$\oint p_q \, dq = n_q h \qquad \textbf{(47-15)}$$

In the Bohr model, the momentum associated with the angular coordinate ϕ is the angular momentum, $p_\phi = L$. Thus, the rule states that

$$\oint p_\phi \, d\phi = \int_0^{2\pi} L \, d\phi = n_\phi h \qquad \textbf{(47-16)}$$

Because L is constant, this condition yields

$$L = n_\phi \hbar \qquad n_\phi = 1, 2, 3, \dots \qquad \textbf{(47-17)}$$

which is just the Bohr angular momentum quantization rule.*

For an elliptical orbit (Fig. 47-6), there is, in addition to Eq. 47-16, an independent condition on the radial momentum, $p_\rho = \mu \, d\rho/dt$; thus,

$$\oint p_\rho \, d\rho = n_\rho h \qquad n_\rho = 0, 1, 2, \dots \qquad \textbf{(47-18)}$$

The integer n_ϕ is referred to as the *azimuthal quantum number*, and n_ρ is referred to as the *radial quantum number*.

---

*The value $n = 0$, which would otherwise be allowed according to the Wilson-Sommerfeld rule, is excluded on the grounds that such motion, with zero angular momentum, would correspond to the back-and-forth passage of the electron directly through the nucleus.

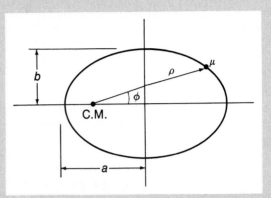

Fig. 47-6. Geometry of an elliptical orbit. According to the Sommerfeld theory, the semimajor and semiminor axes are $a = (a_0/Z)n$ and $b = (n_\phi/n)a$; the eccentricity of the orbit is $\epsilon = \sqrt{1 - n_\phi^2/n^2}$.

The sum, $n = n_\phi + n_\rho$, is called the *principal quantum number* because the energy of a state depends only on n; that is,

$$E_n = -\frac{Z^2 e^4 \mu}{32\pi^2 \epsilon_0^2 \hbar^2} \cdot \frac{1}{n^2} \qquad n = 1, 2, 3, \dots \qquad \textbf{(47-19)}$$

This result is an example of a quantum phenomenon called *degeneracy* in which several distinct states, in this case characterized by different values of n_ϕ, all have the same total energy.

Figure 47-7 shows, to scale, the three orbits that are allowed for $n = 3$. Notice that exactly *three* orbits are possible because n_ϕ can take on all integer values from one to n; that is, $n_\phi = 1, 2, 3, \dots, (n-1), n$. The semimajor

Fig. 47-7. The three allowed hydrogen-atom orbits for $n = 3$. The semimajor axis of each orbit is $n^2 a_0 = 9a_0$. The focus O corresponds to the C.M. of the system.

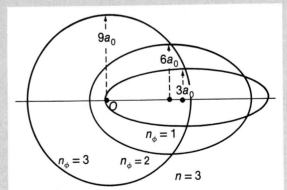

axes a (Fig. 47–6) all have the same length, equal to the Bohr orbit radius appropriate for the particular value of n.

Although the Wilson-Sommerfeld model gave a more general picture of atomic structure than did the original Bohr model, it proved inadequate to explain the spectra of systems with two or more electrons. In spite of the great ingenuity of the theorists, they achieved only minimal success in accounting for the properties of complex atoms. The classical-quantum model had been pushed to the limit. New ideas were needed.

Example 47–2

Use the Wilson-Sommerfeld quantization rule (Eq. 47–15) to obtain the energies of the allowed states of a simple harmonic oscillator.

Solution:

Consider a simple harmonic oscillator with mass m and spring constant k. The natural frequency of the oscillator is $\omega_0 = \sqrt{k/m}$. The (constant) total energy of the system is

$$E = K + U = \frac{p^2}{2m} + \tfrac{1}{2}kx^2$$

This expression for the energy can be rewritten in the form

$$\frac{p^2}{2mE} + \frac{x^2}{2E/m\omega_0^2} = 1$$

which is the equation of an ellipse in the variables p and x with semimajor axis $a = \sqrt{2E/m\omega_0^2}$ and semiminor axis $b = \sqrt{2mE}$.

The quantization rule involves the evaluation of $\oint p\,dx$ over one complete period or cycle. Diagram (a) shows that this integral is just the area of the p-x ellipse. Now, the area of an ellipse is πab, so

$$\oint p\,dx = \pi ab = nh$$

from which

$$2\pi E/\omega_0 = nh$$

Finally, using $\omega_0 = 2\pi\nu_0$, we have the desired result,

$$E = nh\nu_0 \qquad n = 1, 2, 3, \ldots$$

This result is illustrated in diagram (b), which shows ellipses that correspond to successive allowed states. Each ellipse has an area $h\nu_0$ larger than the next smaller ellipse of lower energy.

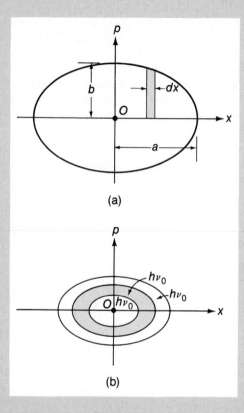

(a)

(b)

(Actually, the quantized energies of the harmonic oscillator expressed in terms of the values of the integral $\oint p\,dx$ were first obtained by Planck in connection with his investigation of thermal cavity radiation. Planck's result was then generalized by Wilson in 1915 and Sommerfeld in 1916 in the form of the Wilson-Sommerfeld quantization rule.)

47–4   MODEL REFINEMENTS

One of the first successes of de Broglie's new theory of particles and waves was to point out the underlying reason for the regularity of the discrete orbits in Bohr's model of the hydrogen atom. The de Broglie wavelength for the reduced-mass particle in the nth orbit is

$$\lambda_n = h/p_n = h/\mu\rho_n\omega$$

Bohr's angular momentum quantization condition (Eq. 47-8) can be expressed as

$$2\pi\mu\rho_n^2\omega = nh$$

Combining these two equations, there results

$$2\pi\rho_n = n\lambda_n \qquad n = 1, 2, 3, \dots \qquad \text{(47–20)}$$

which states that the circumference of the nth Bohr orbit $(2\pi\rho_n)$ is equal to n de Broglie wavelengths of the particle in that orbit. Thus, the de Broglie waves are *standing waves* and persist by virtue of their self-reinforcement (Fig. 47-8). If the orbit circumference were not *exactly equal* to an integer number of de Broglie waves, the waves would be out of phase and, instead of joining smoothly and reinforcing, would rapidly dissipate because of destructive interference.

De Broglie's idea of wave-particle duality, proposed in 1924, marked the demise of the "old" quantum theory based on classical mechanics and the Wilson-Sommerfeld quantization rule, and the beginning of the "new" quantum theory based on wave mechanics. Primarily from the work of Schrödinger, Heisenberg, and Born there emerged during the next two years a new view of atomic processes. The earlier ideas of electron orbits were abandoned and the descriptions were instead given in terms of waves and probabilities.

In 1926 Erwin Schrödinger published a paper in which he described the properties of the hydrogen atom in terms of the wave equation introduced in Section 46-6, namely,

$$-\frac{\hbar^2}{2\mu}\nabla^2\psi + U(x, y, z)\psi = E\psi \qquad \text{(47–21)}$$

where the potential energy U is just the Coulomb potential energy. Schrödinger showed that the solution of Eq. 47–21 led naturally to quantized energy states, and that the energies of these states are given by precisely the same expression as was found by Bohr (Eq. 47–10):

$$E_n = -\frac{Z^2e^4\mu}{32\pi^2\epsilon_0^2\hbar^2}\frac{1}{n^2} \qquad n = 1, 2, 3, \dots$$

The index n that appears in Schrödinger's expression for the energy does not indicate the angular momentum (as in Eq. 47-8), so it does not have the same interpretation as in the Bohr theory. In fact, the angular momentum is specified by a separate quantum number ℓ such that

$$L = \sqrt{\ell(\ell + 1)}\,\hbar \qquad \ell = 0, 1, 2, \dots, n - 1 \qquad \text{(47–22)}$$

For every value of n there are possible n different values of ℓ, all of which describe states that have the same energy (in the nonrelativistic approximation). Thus, to this

(a)

(b)

Fig. 47-8. (a) The pattern of standing de Broglie waves for the $n = 7$ Bohr orbit. In general, the circumference of the nth orbit is exactly equal to n de Broglie wavelengths of the particle in that orbit. (b) If n wavelengths do not match the orbit circumference, destructive interference results and standing waves cannot be supported.

point, the degree of degeneracy for the nth state is n, the same as that in the elliptical orbit model of Sommerfeld. However, in the Schrödinger treatment additional quantization phenomena appear, thereby increasing the degeneracy associated with n.

Not only is there quantization of the magnitude of the angular momentum vector **L**, but there is also a restriction on the allowed orientations of this vector in space. Suppose that a particular direction in space is specified as the polar axis. This direction could be that of an externally applied electric or magnetic field, or it might correspond to an axis of a piece of experimental apparatus. In any event, regardless of the way in which this preferred direction in space is chosen, a measurement of the projection of **L** in this direction (the component L_z) can yield only the values

$$L_z = m\hbar \qquad m = 0, \pm 1, \pm 2, \ldots, \pm \ell \qquad \text{(47–23)}$$

Because m can be determined from the spectra of atoms in a magnetic field, it is called the *magnetic quantum number,* and the associated atomic states are called the *magnetic substates.* There are $2\ell + 1$ of these substates for a particular ℓ, and there are n angular momentum states for a particular n. Finally, the total number of states with the same energy (that is, the total degeneracy) is n^2, according to the simple version of the Schrödinger theory that we have been discussing. (See Problem 47–7.)

Figure 47–9 shows the angular momentum vector **L** and its allowed projections L_z for the case $\ell = 2$. The vector **L** can be imagined as precessing about the polar axis in one of five possible orientations.

In quantum theory, the wave function ψ has a probabilistic interpretation. Thus, the quantity $|\psi(x, y, z)|^2 \, dv$ is proportional to the probability of finding the electron* within the volume element dv at the coordinate location (x, y, z). A quantity of particular interest is the probability of finding the electron between radii r and $r + dr$. This probability depends on both of the quantum numbers n and ℓ, and is written as $P_{n\ell}(r)$. Figure 47–10 shows these radial probability functions for all of the states with $n = 1, 2,$ and 3. The abscissae are expressed in units of $r/(a_0/Z)$. The states with the largest value of ℓ for a particular value of n correspond to the circular orbits in the Sommerfeld model (compare Fig. 47–7). The probability function for each of these states—P_{10}, P_{21}, and P_{32}—has a single maximum that occurs exactly at the position of the radius of the corresponding Bohr orbit. These points are indicated by the arrows above the curves in Fig. 47–10 at $r/(a_0/Z) = 1, 4,$ and 9.

The fact that the maxima of the probability functions for maximum ℓ occur exactly at the Bohr orbit radii is another manifestation of the correspondence principle. The circular Bohr orbits (the orbits of maximum n_ϕ in the Sommerfeld model where $n_\phi = \ell + 1$) represent the classical limit of the quantum theory. Notice that the quantum theory predicts that there is a substantial probability for finding the electron in regions that are forbidden according to the earlier (partly quantum, partly classical) theory.

Spin Angular Momentum. In 1925, S. A. Goudsmit and G. Uhlenbeck in Holland pointed out that the occurrence of closely spaced doublet lines in the spectra of the alkali elements could be explained if it is assumed that the electron possesses an intrinsic angular momentum. Classically, such an angular momentum would result from the electron spinning about its axis. For this reason the term *spin angular*

Fig. 47–9. The angular momentum vector **L** precesses about the z-axis in one of five possible orientations for $\ell = 2$.

---

*Strictly, ψ is the wave function for the *system,* not simply the *electron.* Therefore, $|\psi|^2 \, dv$ is proportional to finding the *reduced-mass particle* in dv. The distinction is not important for the purposes here.

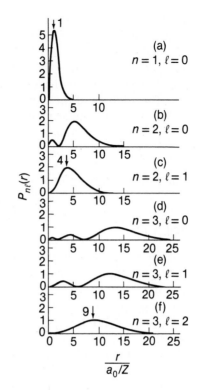

Fig. 47–10. Radial probability functions for the hydrogen atom or hydrogenlike ions for $n = 1$, 2, and 3. The maxima of the functions for the largest value of ℓ for each n are indicated by the arrows above the curves P_{10}, P_{21}, and P_{32}. These points occur exactly at the positions of the Bohr orbit radii, $r/(a_0/Z) = 1$, 4, and 9.

momentum or, simply, *spin* is used to describe this property. In nonrelativistic quantum theory there is no theoretical explanation for the origin of the mechanical spinning motion of elementary particles such as electrons, protons, and neutrons. We treat spin angular momentum as just another intrinsic property of the electron along with its mass and charge. The magnitude of the electron spin angular momentum $\mathbf{S}$ is equal to

$$S = \sqrt{s(s + 1)}\hbar \qquad s = \tfrac{1}{2} \qquad\qquad \textbf{(47–24)}$$

The measurement of the component of $\mathbf{S}$ in any preferred direction in space will produce the result

$$S_z = m_s \hbar \qquad m_s = \pm\tfrac{1}{2} \qquad\qquad \textbf{(47–25)}$$

By virtue of its spin, the electron has two internal quantum states corresponding to $S_z = +\tfrac{1}{2}\hbar$ and $S_z = -\tfrac{1}{2}\hbar$.

In 1928, Paul Dirac* produced a relativistic theory of the hydrogen atom by incorporating Einstein's special relativity theory into quantum theory. Dirac showed that it was a necessary consequence of the resulting relativistic quantum theory that the electron possess an intrinsic angular momentum. Thus, Dirac's theory is in complete accord with the earlier (pre-quantum-theory) discovery of Goudsmit and Uhlenbeck.

The inclusion in the theory of the relativistic motion of the electron and its spin leads to an energy-level scheme for the hydrogen atom in which the energies of the

---

*Paul A. M. Dirac (1902–   ), English theoretical physicist, shared the 1933 Nobel Prize in physics with Erwin Schrödinger for his work on quantum theory.

states are slightly different from those that result from the Bohr model.* In addition, some of the energy degeneracy is removed. Thus, the energy of the ground state ($n = 1$) is lowered by 1.81×10^{-4} eV and the $n = 2$ states with $\ell = 0$ and $\ell = 1$ are split apart by 0.91×10^{-4} eV. More recent refinements (quantum electrodynamics, abbreviated QED) have introduced additional energy-level corrections.

The addition of the angular momentum **L** and the spin angular momentum **S** yields the total angular momentum vector **J**; that is,†

$$\mathbf{J} = \mathbf{L} + \mathbf{S} \tag{47-26}$$

The rules for this sum are such that the allowed values of $J = |\mathbf{J}|$ are

$$J = \sqrt{j(j + 1)}\,\hbar \qquad j = |\ell \pm \tfrac{1}{2}| = \tfrac{1}{2}, \tfrac{3}{2}, \tfrac{5}{2}, \ldots \tag{47-27}$$

and

$$J_z = m_j \hbar \qquad m_j = \pm\tfrac{1}{2}, \pm\tfrac{3}{2}, \pm\tfrac{5}{2}, \ldots, \pm j \tag{47-28}$$

For each value of ℓ there are two values of j, except that only $j = \tfrac{1}{2}$ is allowed for $\ell = 0$.

The requirements for J and J_z can be visualized by referring to Fig. 47–11, which shows both **L** and **S** precessing about the vector $\mathbf{J} = \mathbf{L} + \mathbf{S}$; **J** itself precesses about the polar axis while maintaining a constant value for J_z. The concept of total angular momentum is useful in describing atomic states, as we find in Section 47–5.

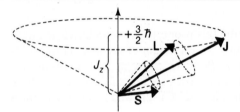

Fig. 47–11. The vectors **L** and **S** precess about **J** while **J** itself precesses about the z-axis. The projection J_z remains constant. The diagram corresponds to the state with $\ell = 1$ and $m_j = \tfrac{3}{2}$.

▶ *Wave Packets and the Hydrogen Atom.* In Example 46–7 we learned that a free particle that is localized at time $t = 0$ to within a dimension $(\Delta x)_0$ will spread its localization with time so that $(\Delta x)_t$ continually increases. Suppose, on the other hand, that the particle is confined to a box with impenetrable walls (corresponding to an infinite binding energy). Then, a wave packet localized to within $(\Delta x)_0$ at $t = 0$ will oscillate between the confining walls, alternately spreading with time and narrowing in such a way that the packet reforms to within $(\Delta x)_0$ at the completion of each cycle. The process repeats indefinitely because no probability flux can escape from the box.

The electron in a hydrogen atom experiences the confining potential due to the positive nuclear charge, but this potential is not infinitely binding. If an electron at $t = 0$ is represented by a wave packet that is localized within a volume $(\Delta x)_0 (\Delta y)_0 (\Delta z)_0$ at $\mathbf{r}_0$, the packet will rapidly spread with time to occupy the entire atomic volume. Furthermore, because the potential is finite, some of the probability flux will escape to infinite distance. That is, the solution of the Schrödinger equation for the hydrogen atom allows *unbound* as well as *bound* solutions. Thus, if a measurement shows that an electron is in the vicinity of a hydrogen nucleus, this does *not* guarantee that the electron is in a bound state; the electron could be in a bound state or in an unbound (scattering) state. The smaller the value of $|\mathbf{r}_0|$, the greater is the probability of a bound state.

Suppose that an experiment is to be carried out to determine the separation distance r between the electron and the proton in a hydrogen atom that is in one of its stationary states. What are the predicted results of such measurements, based on the Schrödinger treatment of the problem? The theory tells us to expect a spread in the measurements of r, with an average value equal to the expectation value $\langle r \rangle$ (see

---

*The relativistic correction for the motion of the electron in the hydrogen atom is not completely negligible. For the ground-state Bohr orbit, $\beta = v/c = 1/137$. (See Problem 47–9.) Therefore, we expect corrections of order $\beta^2 \approx 10^{-4}$ to result.

†Nuclear spin is also present, but its effect on the electron states is usually negligible.

Section 46–6), namely,

$$\langle r \rangle = \int_{\text{all space}} \psi^*(\mathbf{r}) r \psi(\mathbf{r}) \, dv$$

where $\psi(\mathbf{r})$ is the state function for the atom and where dv is the element of volume.

Measurements of the separation distance r provide a striking example of the Bohr correspondence principle. When the angular momentum quantum number ℓ has its maximum value, $\ell = n - 1$, the atomic state corresponds to a simple Bohr-model circular orbit. In this situation, we find (see Problem 47–21)

$$\langle r \rangle_{n, \ell = n-1} = n^2 a_0 \left(1 + \frac{1}{2n} \right)$$

where a_0 is the Bohr radius introduced in Eq. 47–14. This expression for $\langle r \rangle$ reduces to the appropriate circular orbit radius, $r_n = n^2 a_0$, in the limit of large n, in accordance with the correspondence principle. Moreover, the predicted spread in the measured values of r is given by the variance,

$$(\Delta r)^2 = \langle r^2 \rangle - \langle r \rangle^2$$

which, for $\ell = n - 1$, yields

$$\frac{\Delta r}{n^2 a_0} = \frac{\sqrt{2n + 1}}{2n} \qquad \ell = n - 1$$

In the limit of large n, this ratio approaches zero, indicating that the electron is in a well-defined circular orbit. Thus, the description of the hydrogen atom orbits becomes closer and closer to the classical theory as n becomes very large, again in accordance with the correspondence principle. For further details, see Problem 47–20. ◄

47-5 MULTIELECTRON ATOMS

The Pauli Exclusion Principle. The ground state of a normal (electrically neutral) atom with atomic number Z is the atomic state in which each of the Z electrons is in the state of lowest energy that is available to it. The strongest influences on the energies of these states are associated with effects that depend on the quantum numbers n and ℓ. It is customary to label the angular momentum states with a letter code; thus,

$$\ell = 0: \quad s\text{-state}$$
$$\ell = 1: \quad p\text{-state}$$
$$\ell = 2: \quad d\text{-state}$$
$$\ell = 3: \quad f\text{-state}$$
$$\ell = 4: \quad g\text{-state}$$
$$\ell = 5: \quad h\text{-state}$$

with higher ℓ-values following in alphabetical order. Then, an electron state labeled $3p$ means $n = 3$, $\ell = 1$; if two electrons are in the $3p$ state, this would be indicated by writing $3p^2$. Atomic states are formed by combining, according to quantum rules, all of the electron states. These atomic states are labeled by capital letters that correspond to the electron scheme; thus, an $L = 1$ state is a P state.*

---

*Care should be exercised to distinguish these designations, L, S, and J, of *atomic* states from the same symbols standing for the magnitudes of the angular momentum vectors for individual *electronic* states, as in Eqs. 47–21, 23, 25, and 26.

Key: 26 — Atomic number (Z)
Fe — Element symbol
55.847 — Atomic mass of naturally occurring isotopic mixture; for radioactive elements, numbers in parentheses are mass numbers of most stable isotopes

Fig. 47–12. Periodic table of the elements.

§ The International Union for Pure and Applied Chemistry has not adopted official names or symbols for these elements.

Note: Atomic masses shown here are 1977 IUPAC values.

In addition to the quantum numbers n and ℓ, the existence of spin and the coupling of the spin to the orbital angular momentum introduces the quantum numbers $j = \ell \pm \frac{1}{2}$ and m_j. An electron state is uniquely specified by the four quantum numbers* n, ℓ, j, and m_j. The ranges of values for these quantum numbers are

$$n = 1, 2, 3, \ldots$$
$$\ell = 0, 1, 2, \ldots, n - \ell$$
$$j = \ell - \tfrac{1}{2}, \ell + \tfrac{1}{2} \quad (\text{except } j = \tfrac{1}{2} \text{ for } \ell = 0)$$
$$m_j = -j, -j + 1, \ldots, j - 1, j$$

In the nineteenth century, chemists had realized that many of the physical and chemical properties of the elements could be organized in a systematic way and presented in the now-familiar *periodic table of the elements* (Fig. 47–12). In this table the elements are arranged according to *groups* and *periods*, with the members of each group (labeled I, II, . . . , VII, O) having similar properties. For example, the Group I elements (the alkali elements), Li, Na, K, Rb, Cs, and Fr, are all similar to hydrogen in that they participate in chemical reactions as if they have a single effective electron (called a *valence* electron). The Group O elements are the chemically inert noble gases, He, Ne, Ar, Kr, Xe, and Rn. The organization of the periodic table emphasizes the cyclic behavior of many chemical and physical properties. Each of

_____

*Alternatively, we could give n, ℓ, m, and m_s. (•Can you see why?)

these cycles (or *periods*) ends with a noble gas. Thus, the various periods terminate at the atomic numbers 2, 10, 18, 36, 54, and 86, suggesting that the atomic electrons exist in layers or *shells*. The significance of this electron shell structure and the meaning of the shell-closure numbers remained a mystery until a simple and elegant solution was proposed by Wolfgang Pauli.*

The existence of closed electron shells in atoms implies that an arbitrarily large number of electrons cannot be placed in a given shell. The reason for this restriction was expressed by Pauli in 1925 in the form of a principle now known as the *Pauli exclusion principle:*

No two electrons in an atom may occupy states that have the same set of quantum numbers n, ℓ, j, and m_j.

It seems quite remarkable that the details of atomic structure follow from a principle so simply stated.

"Building" the Elements. We have found that quantum number n is most important in determining the energy for the lowest electron states in the hydrogen atom. Consequently, if we imagine "building" the elements by adding electrons, we would expect that the electron states with lowest n are filled first, just as in the original Bohr model. As electrons are added (in the available states with lowest energy), the outermost electrons find the nuclear charge Ze electrically shielded by the accumulated "cloud" of inner electrons.† For a particular value of n, an electron in a state with low angular momentum has a greater probability for being found near the nucleus than does an electron in a state with high angular momentum.‡ Consequently, a high-ℓ electron experiences a greater shielding effect than does a low-ℓ electron and therefore has a smaller binding energy. This tendency of the states with high angular momentum to lie at higher energies (smaller binding energies) is shown in Fig. 47–13.

Figure 47–13 also shows, in a schematic way, the energy gaps (indicated by the horizontal dashed lines) that develop when all of the energy states in a major shell are occupied. These shells are tightly bound and stable structures that have spherically symmetric electron probability distributions with zero net angular momentum ($J = 0$). The number of electron states in each shell is governed by the fact that there are $2(2\ell + 1)$ states§ for each value of ℓ. Thus, there are two s-states, six p-states, ten d-states, and so forth, for each value of n (subject to the restriction that $\ell_{\max} = n - 1$).

The closed-shell elements are the noble gases ($Z = 2, 10, 18, 36, 54$). All of the other elements (with the exception of hydrogen) consist of an inner core that corresponds to a noble-gas closed shell plus outer electrons in the appropriate angular momentum states (see Table 47–2). For example, Fig. 47–14 shows the probability $P(r)$ for finding one of the three electrons in a normal lithium atom ($Z = 3$) at a distance r from the nucleus, regardless of the positions of the other two electrons. The large peak at small r corresponds to the two $1s$ electrons that constitute the

---

*Wolfgang Pauli (1900–1958), an Austrian-Swiss physicist, received the 1945 Nobel Prize in physics for his formulation of the exclusion principle.

†The electron probability function is larger near the nucleus for small values of n than for large values of n (see Fig. 47–10). Therefore, it is proper to speak of electrons with small n as "inner" electrons and those with large n as "outer" electrons.

‡This effect is easy to see in the classical Sommerfeld model with elliptical orbits. In Fig. 47–7 the orbit with the least angular momentum ($n_\phi = \ell + 1$) dips in closest to the nucleus.

§For each ℓ there are $2\ell + 1$ substates (see Eq. 47–23), and each of these can accommodate two electrons ($m_s = \pm\frac{1}{2}$).

Fig. 47–13. Angular momentum states in the various atomic electron shells. Each box can accommodate one electron. Shielding effects cause the raising in energy of the states with high angular momentum, so that the 3d states are in the fourth shell and similarly for the 4d, 5d, and 4f states.

helium core; the small peak corresponds to the 2s electron (the valence electron) in the second shell.

Figure 47–13 shows that the first shell closes when all of the $n = 1$ states are filled, and that the second shell closes when all of the $n = 2$ states are filled. However, the $n = 3$ high angular momentum states (the 3d states) occur at such a high energy because of the shielding effect that an energy gap develops, and the third shell closes at $Z = 18$. The 4s states are at a lower energy than are the 3d states, so the 4s states are filled first at the beginning of the fourth shell. This skipping over groups of states due to shielding occurs in all of the higher shells. There are exceptions even to this modified rule for the order of filling of states; notice the anomalous order of the states that are filled near Cr ($Z = 24$) and Cu ($Z = 29$), as indicated in Table 47–1.

Table 47–1 lists, for each element with $Z \leqslant 36$, the ground-state electron configuration as well as the net orbital angular momentum and total angular momentum for the atom as a whole. The code for the atomic states is L_J; thus, $P_{3/2}$ means $L = 1$

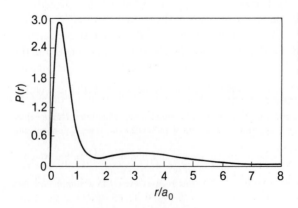

Fig. 47–14. The electron radial probability distribution function for the normal lithium atom. The two peaks correspond to electrons in two different shells.

**TABLE 47–1.   Ground-State Configurations
for Elements with $Z \leqslant 36$**

ATOMIC NUMBER Z	ELEMENT SYMBOL	ELECTRON STATES	ATOMIC STATE	IONIZATION POTENTIAL (eV)
1	H	$1s$	$S_{1/2}$	13.599
2	He	$1s^2$	S_0	24.588
3	Li	$[\text{He}]2s$	$S_{1/2}$	5.392
4	Be	——$2s^2$	S_0	9.323
5	B	——$2s^2 2p$	$P_{1/2}$	8.298
6	C	——$2s^2 2p^2$	P_0	11.260
7	N	——$2s^2 2p^3$	$S_{3/2}$	14.53
8	O	——$2s^2 2p^4$	P_2	13.618
9	F	——$2s^2 2p^5$	$P_{3/2}$	17.423
10	Ne	——$2s^2 2p^6$	S_0	21.565
11	Na	$[\text{Ne}]3s$	$S_{1/2}$	5.139
12	Mg	——$3s^2$	S_0	7.646
13	Al	——$3s^2 3p$	$P_{1/2}$	5.986
14	Si	——$3s^2 3p^2$	P_0	8.152
15	P	——$3s^2 3p^3$	$S_{3/2}$	10.487
16	S	——$3s^2 3p^4$	P_2	10.360
17	Cl	——$3s^2 3p^5$	$P_{3/2}$	12.967
18	Ar	——$3s^2 3p^6$	S_0	15.760
19	K	$[\text{Ar}]4s$	$S_{1/2}$	4.341
20	Ca	——$4s^2$	S_0	6.113
21	Sc	——$3d4s^2$	$D_{3/2}$	6.54
22	Ti	——$3d^2 4s^2$	F_2	6.82
23	V	——$3d^3 4s^2$	$F_{3/2}$	6.74
24	Cr	——$3d^5 4s$	S_3	6.765
25	Mn	——$3d^5 4s^2$	$S_{5/2}$	7.435
26	Fe	——$3d^6 4s^2$	D_4	7.87
27	Co	——$3d^7 4s^2$	$F_{9/2}$	7.864
28	Ni	——$3d^8 4s^2$	F_4	7.635
29	Cu	——$3d^{10} 4s$	$S_{1/2}$	7.726
30	Zn	——$3d^{10} 4s^2$	S_0	9.394
31	Ga	——$3d^{10} 4s^2 4p$	$P_{1/2}$	5.999
32	Ge	——$3d^{10} 4s^2 4p^2$	P_0	7.900
33	As	——$3d^{10} 4s^2 4p^3$	$S_{3/2}$	9.81
34	Se	——$3d^{10} 4s^2 4p^4$	P_2	9.75
35	Br	——$3d^{10} 4s^2 4p^5$	$P_{3/2}$	11.814
36	Kr	——$3d^{10} 4s^2 4p^6$	S_0	14.000

and $J = 3/2$. Also shown in Table 47–1 is the *ionization potential* of each element (compare Table 31–4). This is the energy that must be supplied to release the least bound electron from the neutral atom. Notice the characteristic manner in which this energy varies. It is small when the ejected electron is the only electron outside a closed shell; it is large when the electron must be removed from a closed shell. The variation of the ionization potential with Z is well described in terms of the energies of the electron states in the different shells.

The chemical and optical properties of an element depend most strongly on the outermost electrons in the atom (the valence electrons). For example, sodium (Na, $Z = 11$) has one $3s$ electron outside the closed neon shell. When sodium combines with chlorine to form NaCl (common table salt), the 3s electron is removed from the sodium atom (creating a Na$^+$ ion) and is absorbed by the chlorine atom, thereby completing the set of 3p states and closing the third shell of chlorine. The Cl$^-$ ion that is formed and the Na$^+$ ion both have stable closed electron shells. The ions are electrically attracted to one another and the pair is held together by *ionic bonding*. Other chemical compounds, such as LiF, KBr, and CsI, are formed in the same way.

The 3s electron is also involved in the transitions that are responsible for the prominent features of the optical spectrum of sodium. Because the inner neon core is inert and has zero angular momentum, the ground state of the sodium atom has $L = 0$ and $J = \frac{1}{2}$; this state therefore carries the designation $S_{1/2}$ (see Table 47–1). The least excited state available to the electron is the 3p state (see Fig. 47–13). When the electron is promoted into this state, the atomic state is either a $P_{1/2}$ or a $P_{3/2}$ state. (Because of the interaction between the spin angular momentum and the orbital angular momentum, the two P states are split apart by a small amount—see Problem 47–10.) The photons that are emitted when electrons make transitions from these states to the ground state constitute the close doublet of yellow lines in the sodium spectrum (the D lines at 5889.96 Å and 5895.93 Å). Figure 47–15 shows the D lines as well as a number of other prominent lines in the spectrum of sodium.

X-Ray Line Spectra. Those excitations of an atom that involve the tightly bound inner electrons give rise to X-ray line spectra. In discussing X-ray spectra it is customary to use the letter K to indicate $n = 1$ states, L for $n = 2$, M for $n = 3$, and so forth.

Consider the production of X rays by accelerating electrons through a potential of 35 kV and stopping them in a molybdenum (Mo) target. The accelerated electrons

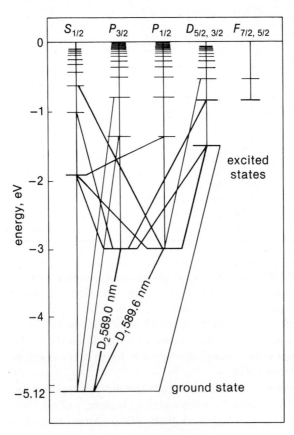

Fig. 47–15. Energy levels and prominent transitions in the sodium atom. The most intense lines in the visible part of the spectrum correspond to the $P_{3/2} \rightarrow S_{1/2}$ and $P_{1/2} \rightarrow S_{1/2}$ transitions (the yellow D lines).

have sufficient energy to eject *K*-shell electrons from the target atoms, because these *K* electrons are bound by 20,003 eV (see Fig. 47–16). An atom from which a *K*-shell electron has been removed is in an excited and unstable state. The most probable mode of deexcitation of such an atom is for one of the *L*-shell electrons to make a transition to fill the *K*-shell vacancy. This transition results in the emission of an X-ray photon. There are quantum *selection rules* that permit only transitions from the 2*p* states (not the 2*s* state) to the 1*s* ground state. Consequently, there results a closely spaced doublet of X-ray lines, the so-called *K*α lines. Somewhat less probable are transitions from the *M*-shell to the *K*-shell; these transitions also produce a doublet of lines, the *K*β lines. Figure 46–10 shows two strong X-ray lines superimposed on the continuous brehmsstrahlung spectrum. The line with the shorter wavelength is actually the resolved doublet of *K*β lines, and the line with the longer wavelength is actually the unresolved doublet of *K*α lines.

Fig. 47–16. X-ray energy-level diagram for molybdenum. The $n = 4$ and $n = 5$ levels, which are not shown, are crowded between the $n = 3$ levels and $E = 0$.

Example 47–3

Determine the wavelengths of the two *K*α lines from molybdenum.

Solution:

According to Fig. 47–16, the energies of the two lines (called $K\alpha_1$ and $K\alpha_2$) are

$$K\alpha_1: \quad E_1 = 20{,}003 \text{ eV} - 2525 \text{ eV} = 17{,}478 \text{ eV}$$
$$K\alpha_2: \quad E_2 = 20{,}003 \text{ eV} - 2630 \text{ eV} = 17{,}373 \text{ eV}$$

Then, using Eq. 46–23, we find

$$K\alpha_1: \quad \lambda_1 = \frac{12{,}400}{E_1} = \frac{12{,}400}{17{,}478} = 0.7095 \text{ Å}$$

$$K\alpha_2: \quad \lambda_2 = \frac{12{,}400}{E_2} = \frac{12{,}400}{17{,}373} = 0.7138 \text{ Å}$$

Thus, the doublet separation is

$$\Delta\lambda = 0.7138 \text{ Å} - 0.7095 \text{ Å} = 4.30 \times 10^{-3} \text{ Å}$$

Evidently, the apparatus used to obtain the spectrum shown in Fig. 46–10 did not have sufficient resolving power to separate the close doublet. (•How would the existence of the doublet affect the results of the Compton scattering experiment, Fig. 46–12?)

47-6 LASER RADIATION

Ordinary light sources emit photons with random time sequence and with random phases with respect to one another. The light from such a source is *incoherent*. It is possible to construct a light source which emits *coherent* light. Such a device is called a *laser*.*

---

*_Laser_ is an acronym for *l*ight *a*mplification by *s*timulated *e*mission of *r*adiation.

When an isolated atom is in an excited energy state, it will exist, on the average, for a characteristic time τ (called the *mean lifetime*) before spontaneously making a transition to a lower energy state by emitting a photon. If the energy of the initial state is E_1 and that of the final state (for example, the ground state) is E_0, the emitted photon will have an energy $h\nu_{10} = E_1 - E_0$. Ordinarily, a collection of atoms in the excited state E_1 will make spontaneous transitions to the state E_0 in a random or uncorrelated manner, with photons emitted in random directions and bearing no definite phase relationship.

If, on the other hand, an excited atom is in the presence of an electromagnetic field that oscillates exactly with the frequency ν_{10} (which could be produced by a passing photon wave train), the field will stimulate the atom to deexcite by emitting a photon *in phase* with the external field. The emitted photon will travel in the same direction as the passing photon and will oscillate exactly in phase with it. The two photons are therefore coherent. Evidently, the possibility exists for continuing this stimulation of radiation and building up an avalanche of coherent photon wave trains, a process called *laser action*.

When the passing photon encounters an atom in the state E_1 it will stimulate the emission of a photon in a deexcitation process. On the other hand, when the passing photon encounters an atom in the state E_0 it will be absorbed in an excitation process. The probabilities for these two processes are equal (except for statistical factors of order unity). Under ordinary conditions of thermal equilibrium, the ratio of the number N_1 of atoms in a sample that are in the state E_1 to the number N_0 in the state $E_0 < E_1$ is given by the Boltzmann factor:

$$\frac{N_1}{N_0} = e^{-(E_1 - E_0)/kT} < 1$$

Thus, more of the passing photons will be absorbed than will stimulate emission. For laser action to occur, the number of atoms in state E_1 must be made larger than the number in state E_0. That is, a *population inversion* must be created.

If the state from which laser radiation is desired has a mean lifetime that is typical of most atomic excited states (10^{-7} to 10^{-5} s), the population of atoms in this state cannot be made sufficiently large due to rapid spontaneous deexcitation. Then, stimulated emission cannot become effective. To achieve laser action, the radiating state must have a mean lifetime that is much longer than usual (10^{-3} to 10^{-1} s). Such a state is called a *metastable state*.

One way to accomplish a population inversion is to *pump* the atoms through a third state,* as indicated in Fig. 47–17a. (A device that operates on this principle is called a *three-level laser*.) The pumping radiation excites the atoms to the level E_2, which then deexcites (with some probability) to level E_1, the metastable state. Laser action involving the transition $E_1 \rightarrow E_0$ will then be initiated by one of the spontaneous photons $h\nu_{10}$.

One of the first substances that was found to exhibit laser action is ruby (a variety of corundum, Al_2O_3, that contains chromium as an impurity). The laser light is concentrated into a beam by using a ruby crystal in the form of a cylinder with the end surfaces polished accurately parallel, as shown in Fig. 47–18. One end is silvered to produce a mirror, and the other end is given only a partial coating of silver so that

---

*The level E_1 cannot be pumped directly because a long lifetime for level E_1 means that the transition $E_1 \rightarrow E_0$ is inhibited and, hence, the pumping transition $E_0 \rightarrow E_1$ is also inhibited.

Fig. 47–17. (a) Pumping of the state E_1 through state E_2. If state E_2 is a sharp (nondegenerate) atomic level, radiation with the precise frequency $\nu_{20} = (E_2 - E_0)/h$ is required to induce the upward transition $E_0 \rightarrow E_2$. In practice, the state E_2 is actually a band of states. Then, a range of frequencies will be effective in pumping, and a source of white light can be used. (b) A passing photon with energy $h\nu_{10}$ stimulates the deexcitation of the metastable state E_1. The two outgoing photons are coherent.

about half of the light incident on this end will escape. The laser state is pumped by means of a high-intensity source of white light that is pulsed on at regular intervals. One of the spontaneous photons triggers the laser action. Those photons that travel parallel to the cylinder axis are reflected from the ends and again traverse the crystal, stimulating the emission of additional photons. The process of reflection and further stimulation of emission is repeated many times and results in an intense beam of radiation in the crystal. Some of the photons that are incident on the partially silvered end are transmitted and constitute the external laser beam.

The light that emerges from a laser is a narrow, highly monochromatic, coherent beam. The divergence of a laser beam is determined by diffraction effects, so the divergence is approximately λ/d, where d is the initial diameter of the beam. Thus, a 1-mm-diameter beam of red light ($\lambda = 693$ nm) from a ruby laser will have a divergence of only about 7×10^{-4} rad (or $0.04°$).

Lasers have been used in such diverse applications as surveying, the drilling of microholes in hard materials, the "gluing" back of a detached retina in a human eye, and holography (Section 42–5).

Fig. 47–18. Schematic diagram of a ruby laser. The cylinder ends are polished accurately parallel so that the photons are reflected back and forth through the crystal, stimulating additional emissions and enhancing the beam.

Physicists use a laser (foreground) to check the alignment of 426 reflectors on the LAGEOS (Laser Geodynamic Satellite). From orbit, the satellite will reflect pulsed laser beams with which scientists can determine motions of the Earth's crust to within a few centimeters. Such data will be used in geophysical models that may eventually predict earthquakes. (Courtesy of NASA.)

PROBLEMS

Section 47-1

47-1 The α particles that are emitted by radium (those used in the Rutherford scattering experiment) have an energy of approximately 5 MeV. If such an α particle is incident on a nucleus of gold ($Z = 79$) and is scattered through an angle of $180°$, what is the distance of closest approach between the α particle and the nucleus? (Consider the gold nucleus to be a fixed scattering object and assume that only the Coulomb force is effective.) Such a calculation enabled Rutherford to place an upper limit on the nuclear size.

Section 47-2

47-2 The values of the Rydberg constant for atomic hydrogen and for singly ionized helium are
•

$$\Re_H = 1.096776 \times 10^7 \text{ m}^{-1}$$

$$\Re_{He} = 1.097223 \times 10^7 \text{ m}^{-1}$$

The ratio of the mass of the α particle to that of the proton is 3.9726. From these data determine the ratio of the proton mass to the electron mass.

47-3 What is the least amount of energy required to remove the remaining electron from a helium ion in its ground state? Use the Bohr model. The experimental value is 54.4 eV.

47-4 Consider a hydrogen atom. (a) Calculate the frequency ν of the $n = 2 \rightarrow n = 1$ transition and compare with the frequency ν_{orb} of the electron orbital motion in the $n = 2$ state. (b) Make the same calculation for the $n = 10{,}000 \rightarrow n = 9999$ transition. Comment on the results.

47-5 Show that $\omega/2\pi = 2\Re c/n^3$ for the circular Bohr orbits of the hydrogen atom, where ω is the orbital angular frequency.

SPECIAL TOPIC

Section 47-3

47-6 Apply the Wilson-Sommerfeld quantization rule to a rigid rotator with rotational inertia I. Show that the quantized energies are

$$E_l = l^2 h^2 / 2I \qquad l = 0, 1, 2, \ldots$$

Assume that the hydrogen molecule (H_2) is a rigid dumbbell-shaped rotator with a separation of 0.74 Å between the atomic centers. Calculate the energies of the first few excited states. What is the wavelength of the $l = 2$ to $l = 1$ transition?

Section 47-4

47-7 Prove that the degeneracy associated with the quantum number n is n^2. Assume that the electron has no spin, and enumerate the possible values of the quantum numbers l and m according to Eqs. 47-22 and 47-23.

47-8 Solution of the Schrödinger wave equation for the • ground state of the hydrogen atom shows that the probability $P_0(r)$ for finding the electron between the radii r and $r + dr$ is

$$P_0(r) = \frac{4r^2}{a_0^3} e^{-2r/a_0}$$

(a) Show that the maximum value of $P_0(r)$ occurs for $r = a_0$, the radius of the first Bohr orbit. (b) What is the probability for finding the electron outside the first Bohr orbit (that is, $r > a_0$)? (The integral is listed in the table inside the rear cover.)

47-9 Show that the velocity of the electron in the first (ground-state) Bohr orbit of the hydrogen atom is given by $\beta = v/c = 1/137$.

Section 47-5

47-10 The wavelengths of the photons emitted in the transitions from the $P_{1/2}$ and $P_{3/2}$ states to the

ground state of the neutral sodium atom are 5889.96 Å and 5895.93 Å. Calculate the energy splitting of this P-state doublet.

47-11 Use the information in the accompanying X-ray energy-level diagram for sodium to calculate the wavelength of the $K\alpha$ line (actually a doublet) corresponding to the $2p$-$1s$ transition.

47-12 According to the diagram in Problem 47-11, a K-shell electron in the neutral sodium atom has an energy of -1072 eV. Use this value to calculate the effective nuclear charge Z_{eff} which, if substituted into the Bohr expression for a hydrogenlike system, would give the same energy. Compare your result with the actual nuclear charge $Z = 11$. Compare also with the result of a theoretical calculation which shows that the effective nuclear charge experienced by an electron at the average radius of the $n = 1$ electrons is $Z - 2$, owing to the shielding effect of *all* atomic electrons.

Additional Problems

47-13 If the cross-hatched band that represents the unbound hydrogen atom states shown in Fig. 47-4 corresponds to thermal energies up to the average en-

ergy at room temperature (20° C), what is the width of the band?

47-14 The *positron* is a particle with the same mass as an electron but with a charge $+e$. It is possible to form a short-lived system, called *positronium,* that consists of an electron and a positron in orbits around their common center of mass. (a) What is the separation of the particles in the ground state? (b) Compare the wavelengths of the positronium transitions with the corresponding hydrogen atom transitions.

47-15 The *muon* is a particle with a mass $m_\mu = 206.8\, m_e$ that carries a charge of either $+e$ or $-e$. It is possible to form *muonic atoms* in which one of the atomic electrons is replaced by a negative muon. (a) Calculate the radius of the first Bohr orbit radius for a lead muonic atom, $Z(\text{Pb}) = 82$. (b) Calculate the difference in energy between the $n = 1$ and $n = 2$ states in a lead muonic atom. The observed value is 10.48 MeV. [*Hint:* In each case assume that the system consists only of a lead nucleus and a muon. (•Why is this reasonable?)]

47-16 Assume that the Earth's orbital motion around the Sun corresponds to a quantized angular momentum state (it does!). What quantum number l should be associated with this motion?

47-17 Consider an "atom" that consists of a neutron and an electron held together by the mutual gravitational force. What would be the ground-state binding energy of such an atom? What would be the radius of the first "Bohr orbit"? (Neutron mass = 1.008665 u and electron mass = 5.486×10^{-4} u.)

47-18 Make a graph of the ionization potential (Table 47-1) versus atomic number. Use Fig. 47-13 to explain any abrupt changes that the graph reveals.

SPECIAL TOPIC

47-19 Suppose that a particle with mass m moves (in one dimension) between two impenetrable walls separated by a distance d. (a) Apply the Wilson-Sommerfeld quantization rule and show that the allowed energies of the particle are

$$E_n = (h^2/8md^2)n^2 \qquad n = 1, 2, 3, \ldots$$

(b) Show that a set of states for which the de Broglie waves have a node at each wall give the same energies. (c) Calculate the energy difference between the first excited state and the ground state if the particle is an electron and $d = 1$ Å.

47-20 Determine ρ_min and ρ_max for the Sommerfeld hydrogen-atom orbit corresponding to $n = 3$, $n_\phi = 2$ (see Fig. 47-7). Compare these values with the locations of the two peaks in the $P_{31}(r)$ probability function (Fig. 47-10). Notice that $n_\phi = l + 1$.

47-21 In advanced texts it is shown that the average
•• values $\langle r \rangle$ and $\langle r^2 \rangle$ for a hydrogen atom in a state with quantum numbers n and l are

$$\langle r \rangle_{nl} = n^2 a_0 \left(1 + \frac{1}{2}\left[1 - \frac{l(l+1)}{n^2} \right] \right)$$

and

$$\langle r^2 \rangle_{nl} = (n^2 a_0)^2 \left(1 + \frac{3}{2}\left[1 - \frac{l(l+1)-1/3}{n^2} \right] \right)$$

(a) For the case of maximum $l\ (= n - 1)$, which corresponds to the simple circular orbits of the Bohr model, show that

$$\langle r \rangle_{n, l=n-1} = n^2 a_0 \left(1 + \frac{1}{2n} \right)$$

$$\langle r^2 \rangle_{n, l=n-1} = (n^2 a_0)^2 \left(1 + \frac{3}{2n} + \frac{1}{2n^2} \right)$$

(b) For a state with n and *any* allowed value of l, show that the variance, $(\Delta r)^2 = \langle r^2 \rangle - \langle r \rangle^2$, is

$$(\Delta r)_{nl}^2 = (n^2 a_0)^2 \left(\frac{1}{4} + \frac{1}{2n^2} - \frac{l^2(l+1)^2}{4n^2} \right)$$

(c) For the case $l = n - 1$, show that

$$(\Delta r)_{n, l=n-1} = \frac{n^2 a_0 \sqrt{2n+1}}{2n}$$

(d) The case $l = n - 1$ corresponds to the circular orbits of the Bohr model, and the ratio $(\Delta r)/n^2 a_0$ approaches zero for large n. However, in this same limit for other l, $(\Delta r)/n^2 a_0$ does not approach zero. Explain why this is so by using the Sommerfeld picture of elliptical orbits.

BASIC ALGEBRA AND TRIGONOMETRY

A-1 SERIES. The *arithmetic progression* or *series* S_n^A, to n terms, can be summed, with the result

$$S_n^A = a + (a + d) + (a + 2d) + (a + 3d)$$
$$+ \cdots + [a + (n - 1)d]$$
$$= na + \tfrac{1}{2}n(n - 1)d$$
$$= \tfrac{1}{2}n[(\text{first term}) + (\text{last term})] \qquad \textbf{(A-1)}$$

A *polynomial* is a series whose n terms are in ascending positive integer powers of a quantity, say r, which we can write in the form

$$S_n = a_0 + a_1 r + a_2 r^2 + a_3 r^3$$
$$+ \cdots + a_{n-1} r^{n-1} \qquad \textbf{(A-2)}$$

where the *coefficients, $a_0, a_1, a_2, \ldots, a_{n-1}$,* are *real* numbers. When the number of terms is allowed to increase without limit ($n \rightarrow \infty$: "when n becomes indefinitely large"), the resulting *infinite series* is referred to as a *power series*.

The simplest polynomial, to n terms, occurs when all the coefficients, $a_0, a_1, a_2, \ldots, a_{n-1}$, are equal to a number, say a. This defines the *geometric progression* or *series* S_n^G, which can also be summed, with the result

$$S_n^G = a + ar + ar^2 + ar^3 + \cdots + ar^{n-1}$$
$$= \frac{a(1 - r^n)}{1 - r} \qquad \textbf{(A-3)}$$

If $r^2 < 1$, the powers of r decrease in numerical value as n increases, and in the limit $n \rightarrow \infty$, r^n vanishes. Then

$$S_n^G(n \rightarrow \infty) = \frac{a}{1 - r} \qquad \textbf{(A-4)}$$

A-2 SIGMA NOTATION. It is customary to represent the sum of a series of quantities by means of the so-called *sigma notation*. Then, Eq. A–2 is expressed as

$$S_n = \sum_{i=0}^{n-1} a_i r^i = a_0 + a_1 r + a_2 r^2$$
$$+ a_3 r^3 + \cdots + a_{n-1} r^{n-1} \qquad \textbf{(A-5)}$$

That is, the Greek capital sigma (Σ) means "sum all of the quantities of the form that follows, over the range of the index specified below and above this symbol." In this case, the *index* is the letter i (but, of course, the index can be any letter or symbol because it does not appear in the result). Sometimes we economize by writing

$$\sum_i p_i \qquad \text{or} \qquad \sum p_i$$

where it is understood that we sum over the entire allowed range of i.

A-3 QUADRATIC EQUATIONS. A *quadratic equation* has the form

$$Ax^2 + Bx + C = 0$$

which we can express in terms of its *roots, x_1* and x_2, as

$$(x - x_1)(x - x_2) = 0$$

Then,

$$x_{1,2} = \frac{-B \pm \sqrt{B^2 - 4AC}}{2A} \qquad \textbf{(A-6)}$$

If $B^2 - 4AC > 0$, the roots are real and unequal; if $B^2 - 4AC = 0$, the roots are real and equal; and if $B^2 - 4AC < 0$, the roots are complex.

Comparing the two forms of the quadratic equation, we see that the sum of the roots and the product of the roots can be expressed as

$$x_1 + x_2 = -\frac{B}{A} \quad \text{and} \quad x_1 x_2 = \frac{C}{A} \quad \textbf{(A-7)}$$

A-4 LOGARITHMS.

The logarithm to the base m of a number a is the power to which m must be raised to produce a. That is,

$$\text{if} \quad m^y = a, \quad \text{then} \quad y = \log_m a$$

For example, $10^3 = 1000$, so we say that the logarithm to the base 10 of 1000 is 3.

For the real numbers a, b, and n, logarithms obey the following rules:

$$\left.\begin{array}{c} \log_m ab = \log_m a + \log_m b \\[4pt] \log_m \dfrac{a}{b} = \log_m a - \log_m b \\[4pt] \log_m (a^n) = n \log_m a \\[4pt] \log_m 1 = 0 \end{array}\right\} \quad \textbf{(A-8)}$$

Common logarithms have base 10, and we write $\log_{10} a$ or, simply, $\log a$. *Natural* or *Naperian logarithms* have base e ($e = 2.71828\ldots$), and we write $\ln a$. The above rules apply for logarithms with any base. The conversion between common and natural logarithms can be made by using

$$\ln a = \log_{10} a / \log_{10} e = 2.3026 \log_{10} a$$

We also note that

$$\log 10 = 1 \quad \text{and} \quad \ln e = 1$$

A useful infinite power series expansion for $\ln(1 + x)$ with $x^2 \leqslant 1$ is

$$\ln(1 + x) = x - \frac{x^2}{2} + \frac{x^3}{3} - \frac{x^4}{4}$$

$$+ \cdots + (-1)^{n-1} \frac{x^n}{n} + \cdots \quad \textbf{(A-9)}$$

where we show the alternation of sign by the device of using $(-1)^{n-1}$.

A-5 EXPONENTIAL FUNCTIONS.

For the real numbers a, n, and m, exponentials obey the following rules:

$$(a^n)(a^m) = a^{n+m}$$
$$(a^n)^m = a^{nm}$$
$$a^0 = 1$$

We also have

$$\ln e^x = x \ln e = x \quad \text{and} \quad e^{\ln x} = x \quad \textbf{(A-10)}$$

A useful infinite power series expansion for e^x with $x^2 < \infty$ is

$$e^x = 1 + \frac{x}{1!} + \frac{x^2}{2!} + \frac{x^3}{3!}$$

$$+ \cdots + \frac{x^n}{n!} + \cdots \quad \textbf{(A-11)}$$

Sometimes the notation $\exp x$ is used for e^x. The exclamation mark denotes the *factorial*, where $n! = 1 \cdot 2 \cdot 3 \cdot \cdots \cdot (n - 1) \cdot n$; by definition, $0! = 1$.

A-6 BINOMIAL THEOREM.

The general rule for calculating the nth power of the sum of two algebraic quantities, a and b, is given by the *binomial theorem*. If we let $b/a = \xi$ (with $\xi^2 \leqslant 1$), we can express the binomial theorem in the form

$$(a + b)^n = a^n \left(1 + \frac{b}{a}\right)^n = a^n (1 + \xi)^n$$

$$= a^n \left[1 + n\xi + \frac{n(n - 1)}{2!} \xi^2 + \frac{n(n - 1)(n - 2)}{3!} \xi^3 + \cdots \right.$$

$$\left. + \frac{n(n - 1)(n - 2) \cdots (n - r + 1)}{r!} \xi^r + \cdots \right]$$

$$\textbf{(A-12)}$$

When n is a positive integer, the series terminates and the rth term can be written as $\dfrac{n!}{(n - r)! r!} \xi^r$; then, the final term in the series is ξ^n. When $\xi = 1$, Eq. A-12 is valid only for $n > 0$; otherwise, the relationship is valid for any n. For example, if $n = \frac{1}{2}$ and $\xi^2 \leqslant 1$, we have

$$(1 \pm \xi)^{1/2} = 1 \pm \tfrac{1}{2}\xi + \frac{\frac{1}{2}(\frac{1}{2} - 1)}{2} \xi^2$$

$$\pm \frac{\frac{1}{2}(\frac{1}{2} - 1)(\frac{1}{2} - 2)}{3 \cdot 2} \xi^3 + \cdots$$

$$= 1 \pm \tfrac{1}{2}\xi - \tfrac{1}{8}\xi^2 \pm \tfrac{1}{16}\xi^3 - \cdots \quad \textbf{(A-13)}$$

The case $n = -\frac{1}{2}$ and $\xi^2 < 1$ is often encountered:

$$(1 \pm \xi)^{-1/2} = 1 \mp \tfrac{1}{2}\xi + \tfrac{3}{8}\xi^2 \mp \tfrac{5}{16}\xi^3 + \cdots \quad \textbf{(A-14)}$$

Other useful expansions are

$$(1 \pm \xi)^{-1} = 1 \mp \xi + \xi^2 \mp \xi^3 + \cdots \qquad \textbf{(A-15)}$$

$$(1 \pm \xi)^{-3/2} = 1 \mp \tfrac{3}{2}\xi + \tfrac{15}{8}\xi^2 \mp \tfrac{35}{16}\xi^3 + \cdots \qquad \textbf{(A-16)}$$

$$(1 \pm \xi)^{-2} = 1 \mp 2\xi + 3\xi^2 \mp 4\xi^3 + \cdots \qquad \textbf{(A-17)}$$

In these equations, one takes either the upper or the lower sign throughout.

A-7   BASIC TRIGONOMETRIC DEFINITIONS.
In order to write down the basic trigonometric relationships, consider the right triangle shown in Fig. A-1. Then,

$$\left.\begin{aligned}
&\sin \theta = \frac{b}{c}, \quad \cos \theta = \frac{a}{c}, \quad \tan \theta = \frac{\sin \theta}{\cos \theta} = \frac{b}{a} \\[2mm]
&\csc \theta = \frac{1}{\sin \theta} = \frac{c}{b}, \quad \sec \theta = \frac{1}{\cos \theta} = \frac{c}{a}, \\[2mm]
&\operatorname{ctn} \theta = \frac{\cos \theta}{\sin \theta} = \frac{a}{b}
\end{aligned}\right\} \qquad \textbf{(A-18)}$$

Fig. A-1. A right triangle.

Using the Pythagorean theorem, $a^2 + b^2 = c^2$, we find

$$\left.\begin{aligned}
&\cos^2 \theta + \sin^2 \theta = 1 \\
&1 + \tan^2 \theta = \sec^2 \theta \\
&1 + \operatorname{ctn}^2 \theta = \csc^2 \theta
\end{aligned}\right\} \qquad \textbf{(A-19)}$$

A-8   TRIGONOMETRIC IDENTITIES.
The basic trigonometric functions for the sum and difference of angles are

$$\sin (x + y) = \sin x \cos y + \cos x \sin y \qquad \textbf{(A-20)}$$

$$\sin (x - y) = \sin x \cos y - \cos x \sin y \qquad \textbf{(A-21)}$$

$$\cos (x + y) = \cos x \cos y - \sin x \sin y \qquad \textbf{(A-22)}$$

$$\cos (x - y) = \cos x \cos y + \sin x \sin y \qquad \textbf{(A-23)}$$

$$\tan (x + y) = \frac{\tan x + \tan y}{1 - \tan x \tan y} \qquad \textbf{(A-24)}$$

$$\tan (x - y) = \frac{\tan x - \tan y}{1 + \tan x \tan y} \qquad \textbf{(A-25)}$$

The addition rules are:

$$\sin x + \sin y = 2 \sin \tfrac{1}{2} (x + y) \\ \cdot \cos \tfrac{1}{2} (x - y) \qquad \textbf{(A-26)}$$

$$\sin x - \sin y = 2 \cos \tfrac{1}{2} (x + y) \\ \cdot \sin \tfrac{1}{2} (x - y) \qquad \textbf{(A-27)}$$

$$\cos x + \cos y = 2 \cos \tfrac{1}{2} (x + y) \\ \cdot \cos \tfrac{1}{2} (x - y) \qquad \textbf{(A-28)}$$

$$\cos x - \cos y = -2 \sin \tfrac{1}{2} (x + y) \\ \cdot \sin \tfrac{1}{2} (x - y) \qquad \textbf{(A-29)}$$

$$\tan x + \tan y = \frac{\sin (x + y)}{\cos x \cos y} \qquad \textbf{(A-30)}$$

$$\tan x - \tan y = \frac{\sin (x - y)}{\cos x \cos y} \qquad \textbf{(A-31)}$$

The half-angle formulas are:

$$\sin^2 \tfrac{1}{2}x = \tfrac{1}{2}(1 - \cos x) \qquad \textbf{(A-32)}$$

$$\cos^2 \tfrac{1}{2}x = \tfrac{1}{2}(1 + \cos x) \qquad \textbf{(A-33)}$$

The double-angle formulas are:

$$\sin 2x = 2 \sin x \cos x \qquad \textbf{(A-34)}$$

$$\cos 2x = 2 \cos^2 x - 1 \qquad \textbf{(A-35)}$$

A-9   TRIANGLE RULES.
Consider the general triangle shown in Fig. A-2. We have

Law of sines: $\dfrac{a}{\sin \theta_a} = \dfrac{b}{\sin \theta_b} = \dfrac{c}{\sin \theta_c}$ **(A-36)**

Law of cosines: $a^2 = b^2 + c^2 - 2bc \cos \theta_a$ **(A-37)**

Area of triangle: $A = \tfrac{1}{2}bc \sin \theta_a$ **(A-38)**

Fig. A-2

A-10   RADIAN MEASURE.
The radian measure of an angle θ, such as that between the lines $A'O$ and $B'O$ in Fig. A-3, is defined as the ratio of the length of the intercepted arc $\widehat{AB}$ on any circle centered on O to the radius r of that circle. That is,

$$\theta = \frac{\widehat{AB}}{r} \qquad (\theta \text{ expressed in radians})$$

Fig. A–3

When θ is increased to 360°, the arc length $\overset{\frown}{AB}$ equals the circumference of the circle. Then,

$$360° = 2\pi \text{ rad}$$

from which $1° = \dfrac{2\pi}{360} = 0.01745329\ldots \text{ rad}$

and $\qquad 1 \text{ rad} = \dfrac{360°}{2\pi} = 57.2957795\ldots°$ **(A–39)**

A–11 TRIGONOMETRIC SERIES. Some useful trigonometric series are:

$$\sin\theta = \theta - \frac{\theta^3}{3!} + \frac{\theta^5}{5!} - \frac{\theta^7}{7!} + \cdots$$
$$\theta^2 < \infty \quad \textbf{(A–40)}$$

$$\cos\theta = 1 - \frac{\theta^2}{2!} + \frac{\theta^4}{4!} - \frac{\theta^6}{6!} + \cdots$$
$$\theta^2 < \infty \quad \textbf{(A–41)}$$

$$\tan\theta = \theta + \tfrac{1}{3}\theta^3 + \tfrac{2}{15}\theta^5 + \tfrac{17}{315}\theta^7 + \cdots$$
$$\theta^2 < \frac{\pi^2}{4} \quad \textbf{(A–42)}$$

In these equations the angle θ is expressed in *radians*. From these expressions we see that, when $\theta \ll 1$,

$$\sin\theta \cong \theta \qquad \cos\theta \cong 1 - \tfrac{1}{2}\theta^2 \qquad \tan\theta \cong \theta \quad \textbf{(A–43)}$$

Other useful expansions are

$$\sin^{-1} x = x + \frac{1}{2}\frac{x^3}{3} + \frac{1\cdot 3}{2\cdot 4}\frac{x^5}{5}$$
$$+ \frac{1\cdot 3\cdot 5}{2\cdot 4\cdot 6}\frac{x^7}{7} + \cdots \quad \textbf{(A–44)}$$

$$\cos^{-1} x = \frac{\pi}{2} - \sin^{-1} x \quad \textbf{(A–45)}$$

$$\tan^{-1} x = x - \frac{x^3}{3} + \frac{x^5}{5} - \frac{x^7}{7} + \cdots \quad \textbf{(A–46)}$$

A–12 HYPERBOLIC FUNCTIONS. The most important hyperbolic functions are

$$\sinh x = \tfrac{1}{2}(e^x - e^{-x}) \quad \textbf{(A–47)}$$

$$\cosh x = \tfrac{1}{2}(e^x + e^{-x}) \quad \textbf{(A–48)}$$

$$\tanh x = \frac{e^x - e^{-x}}{e^x + e^{-x}} \quad \textbf{(A–49)}$$

The hyperbolic sine and cosine functions obey the relationship

$$\cosh^2 x - \sinh^2 x \quad \textbf{(A–50)}$$

COMPLEX NUMBERS

Basic Definitions. A *complex number* is an expression of the form $a + ib$, where a and b are ordinary real numbers and where i is the imaginary unit defined by $i^2 = -1$. The *real part* of the complex number $a + ib$ is a, and the *imaginary part* is b.

Graphical Representations of Complex Numbers. A complex number may be plotted as a point in a *complex plane* defined by a horizontal (real) x-axis and a vertical (imaginary) y-axis. A general complex number may then be represented as $z = x + iy$. (In these discussions we reserve the letter z to represent a complex quantity.) Figure B–1, called an *Argand diagram*, shows several points plotted in the complex plane.

Fig. B–1. An Argand diagram representation of several complex numbers $z = x + iy$.

Another convenient way to represent complex numbers is in polar form, namely,

$$z = \rho(\cos\theta + i\sin\theta) \qquad \text{(B–1)}$$

where

$$\left. \begin{array}{l} x = \rho\cos\theta \\ y = \rho\sin\theta \end{array} \right\} \qquad \text{(B–2)}$$

Also,

$$\rho = |z| = \sqrt{x^2 + y^2} \qquad \text{(B–3)}$$

and*

$$\tan\theta = \frac{y}{x} \qquad \text{(B–4)}$$

The various polar quantities are illustrated in Fig. B–2.

The quantity ρ is the *absolute value* (or *amplitude*) of z, $\rho = |z|$. The angle θ, which is measured counterclockwise from the (real) x-axis, is defined in the domain† $0 \leqslant \theta \leqslant 2\pi$, and is referred to as the *argument* (or *phase*) of z.

Fig. B–2. The polar representation of a complex number $z = x + iy$.

An alternative form for writing a complex number (due to Leonhard Euler) is

$$z = \rho e^{i\theta} \qquad \text{(B–5)}$$

This form is obtained by noting that we can *formally* express the exponential function of a complex number,

---

*The possible sign ambiguity that results when calculating $\theta = \tan^{-1}(y/x)$ is resolved by noting the quadrant in which $x + iy$ lies. Recall the similar problem encountered in the discussion of vectors in two dimensions (Section 3–4).

†The domain of θ is sometimes defined to be $-\pi \leqslant \theta \leqslant \pi$.

$\zeta = i\theta$, in terms of the power series

$$e^{\zeta} = 1 + \zeta + \frac{\zeta^2}{2!} + \frac{\zeta^3}{3!} + \cdots + \frac{\zeta^n}{n!} + \cdots$$

$$= 1 + i\theta + \frac{i^2\theta^2}{2!} + \frac{i^3\theta^3}{3!} + \frac{i^4\theta^4}{4!} + \cdots$$

Now, we have $i^2 = -1$, $i^3 = -i$, $i^4 = 1$, and so forth. Thus,

$$e^{i\theta} = \left(1 - \frac{\theta^2}{2!} + \frac{\theta^4}{4!} - \cdots\right) + i\left(\theta - \frac{\theta^3}{3!} + \frac{\theta^5}{5!} - \cdots\right)$$

We also have

$$\cos\theta = 1 - \frac{\theta^2}{2!} + \frac{\theta^4}{4!} - \cdots$$

$$\sin\theta = \theta - \frac{\theta^3}{3!} + \frac{\theta^5}{5!} - \cdots$$

Hence, $e^{i\theta} = \cos\theta + i\sin\theta$ **(B–6)**

so that $z = \rho e^{i\theta} = \rho(\cos\theta + i\sin\theta)$ **(B–7)**

To be more general, we can allow ρ and θ to become complex. For example, suppose that $\theta = \alpha + i\beta$, where α and β are real numbers. Then, with $\rho = 1$, we have

$$e^{i\theta} = e^{i(\alpha+i\beta)} = e^{i\alpha}e^{-\beta}$$

$$= e^{-\beta}(\cos\alpha + i\sin\alpha)$$ **(B–8)**

Notice that

$$|e^{i\theta}|^2 = |\cos\theta + i\sin\theta|^2 = \cos^2\theta + \sin^2\theta = 1$$

Thus, the quantity $e^{i\theta}$, with unit magnitude, is a convenient way to specify the phase of a complex number. Accordingly, $e^{i\theta}$ is sometimes called a *phasor* (or *rotator*). It is particularly useful when $\theta = \omega t$, for then $e^{i\theta} = e^{i\omega t}$ represents a rotation at a rate ω about the origin, as indicated in Fig. B–3.

Some useful results are:

$$\left.\begin{array}{l} e^{i0} = \cos 0 + i\sin 0 = 1 \\[6pt] e^{i(\pi/2)} = \cos\frac{\pi}{2} + i\sin\frac{\pi}{2} = i \\[6pt] e^{i\pi} = \cos\pi + i\sin\pi = -1 \\[6pt] e^{i(3\pi/2)} = \cos\frac{3\pi}{2} + i\sin\frac{3\pi}{2} = -i \end{array}\right\}$$ **(B–9)**

$$\left.\begin{array}{l} e^{i\theta'}\,(\theta' > 2\pi) = e^{i(\theta'-2\pi)} \\[6pt] e^{-i\theta} = e^{i(2\pi-\theta)} \end{array}\right\}$$ **(B–10)**

Thus, $e^{-i(\pi/2)} = e^{i(3\pi/2)} = -i$

Fig. B–3. Use of the phasor $e^{i\theta}$ to represent rotation.

The Euler Formulas.   Using Eq. B–6, namely,

$$e^{i\theta} = \cos\theta + i\sin\theta$$

we can write

$$e^{-i\theta} = \cos(-\theta) + i\sin(-\theta)$$

$$= \cos\theta - i\sin\theta$$

Adding these two expressions and solving for $\cos\theta$, we have

$$\cos\theta = \frac{e^{i\theta} + e^{-i\theta}}{2}$$ **(B–11)**

Similarly, by subtracting we find

$$\sin\theta = \frac{e^{i\theta} - e^{-i\theta}}{2i}$$ **(B–12)**

Equations B–11 and B–12 are known as the *Euler formulas*.

Equality.   Two complex quantities are *equal* if and only if the real parts and the imaginary parts are separately equal. Thus, if we have

$$a + ib = c + id$$

then it is required that $a = c$ and $b = d$. A complex quantity is zero if and only if both the real part and the imaginary part are separately equal to zero.

Addition and Subtraction.   Complex quantities are added and subtracted in accordance with the rule

$$(a + ib) \pm (c + id) = (a \pm c) + i(b \pm d)$$ **(B–13)**

Complex Conjugation.   If a complex quantity is written exclusively in terms of real quantities and the imaginary unit i, then the *complex conjugate* is formed by everywhere replacing i by $-i$. The conjugate is indicated by a superscript asterisk. If $z = x + iy$, the conjugate is

$$z^* = (x + iy)^* = x - iy \qquad \text{(B-14)}$$

Also,
$$z^* = (\rho e^{i\theta})^* = \rho(\cos\theta + i\sin\theta)^*$$
$$= \rho(\cos\theta - i\sin\theta) = \rho e^{-i\theta} \qquad \text{(B-15)}$$

Multiplication. The multiplication of complex quantities follows the rules of ordinary algebra. Thus,

$$(a + ib)(c + id) = ac + iad + ibc + i^2bd$$
$$= (ac - bd) + i(ad + bc) \qquad \text{(B-16)}$$

In Euler form, we have

$$(\rho_1 e^{i\theta_1})(\rho_2 e^{i\theta_2}) = \rho_1\rho_2 e^{i(\theta_1 + \theta_2)} \qquad \text{(B-17)}$$

If a complex quantity, $z = e^{i\theta}$, is multiplied by $e^{i\phi}$ (ϕ real), the magnitude is unchanged but the phasor is rotated counterclockwise by an angle ϕ in the Argand diagram. Thus,

$$z = \rho e^{i\theta} e^{i\phi} = \rho e^{i(\theta + \phi)} \qquad \text{(B-18)}$$

Absolute Values. The square of the absolute value of a complex quantity (that is, $|z|^2$, which is a *real* quantity) is obtained by multiplying the quantity by its complex conjugate. Thus,

$$zz^* = (a + ib)(a - ib) = a^2 - iab + iba - i^2b^2$$
$$= a^2 + b^2 = |z|^2 \qquad \text{(B-19)}$$

In Euler form, we have

$$zz^* = (\rho e^{i\theta})(\rho e^{-i\theta}) = \rho^2 = |z|^2 \qquad \text{(B-20)}$$

Division and Rationalization of a Complex Denominator. The division of complex quantities is most easily given in Euler form, namely,

$$\frac{z_1}{z_2} = \frac{\rho_1 e^{i\theta_1}}{\rho_2 e^{i\theta_2}} = (\rho_1/\rho_2)e^{i(\theta_1 - \theta_2)} \qquad \text{(B-21)}$$

Setting $z_1 = 1$, we have for the reciprocal of a complex quantity,

$$z^{-1} = \frac{1}{\rho e^{i\theta}} = \rho^{-1}e^{-i\theta} \qquad \text{(B-22)}$$

(•Can you use this result to show that $i^{-1} = -i$?)

If a complex quantity is expressed in Cartesian form, the reciprocal is obtained by a process called *rationalization of the denominator*. This consists of multiplying both numerator and denominator by the complex conjugate of the denominator. Thus,

$$\frac{1}{a + ib} = \frac{1}{a + ib} \cdot \frac{a - ib}{a - ib}$$
$$= \frac{a}{a^2 + b^2} - i\frac{b}{a^2 + b^2} \qquad \text{(B-23)}$$

The division of one complex quantity by another expressed in Cartesian form is also accomplished by using the rationalization procedure. Thus,

$$\frac{c + id}{a + ib} = \frac{ac + bd}{a^2 + b^2} + i\frac{ad - bc}{a^2 + b^2} \qquad \text{(B-24)}$$

De Moivre's Theorem. Using the Euler form of a complex quantity, the nth power is expressed as

$$z^n = (\rho e^{i\theta})^n = \rho^n e^{in\theta}$$
$$= \rho^n(\cos n\theta + i\sin n\theta) \qquad \text{(B-25)}$$

when n is a positive integer. If $\rho = 1$, we have de Moivre's theorem, namely,

$$(e^{i\theta})^n = (\cos\theta + i\sin\theta)^n$$
$$= \cos n\theta + i\sin n\theta \qquad \text{(B-26)}$$

Any multiple of 2π can be added to the angle θ without affecting the value of $e^{i\theta}$; that is,

$$e^{i\theta} = e^{i(\theta + 2\pi k)}$$

where k is any integer. Then, the general expression for $z^{1/n}$, the nth root of z, is

$$z^{1/n} = \rho^{1/n}e^{i(\theta + 2\pi k)/n}$$
$$= \rho^{1/n}\left(\cos\frac{\theta + 2\pi k}{n} + i\sin\frac{\theta + 2\pi k}{n}\right) \qquad \text{(B-27)}$$

The n distinct roots of z are obtained by giving k the values $0, 1, 2, \ldots, (n - 1)$, successively. Setting $z = 1$, we see that the n roots of unity are spaced uniformly around the unit circle in an Argand diagram.

The Projection Operators, Re and Im. The operators **Re** and **Im** project the real and imaginary parts, respectively, of the complex quantity on which they act. Thus,

$$\left.\begin{array}{l}\mathbf{Re}(a + ib) = a \\ \mathbf{Im}(a + ib) = b\end{array}\right\} \qquad \text{(B-28)}$$

Notice that

$$z_1 z_2^* + z_1^* z_2 = \rho_1 \rho_2 (e^{i(\theta_1 - \theta_2)} + e^{-i(\theta_1 - \theta_2)})$$

$$= 2\rho_1 \rho_2 \cos(\theta_1 - \theta_2)$$

$$= 2\mathbf{Re}(\rho_1 \rho_2 e^{i(\theta_1 - \theta_2)})$$

$$= 2\mathbf{Re}(z_1 z_2^*) \qquad\qquad \text{(B–29)}$$

Then, we can write

$$|z_1 + z_2|^2 = (z_1 + z_2)(z_1^* + z_2^*)$$

$$= z_1 z_1^* + z_1 z_2^* + z_1^* z_2 + z_2 z_2^*$$

$$= |z_1|^2 + |z_2|^2 + 2\mathbf{Re}(z_1 z_2^*) \qquad \text{(B–30)}$$

ANSWERS TO SELECTED PROBLEMS

CHAPTER 1

1. 17.88 m/s   **3.** 3.15×10^7 s   **5.** 7913 mi   **7.** 10.4 s   **9.** 4045 m^2 **11.** 128.2 y   **13.** $v^2 = Kgh$ where K is a constant of proportionality   **15.** 3.74×10^{-2} m   **17.** 62.4 lb/ft^3

CALCULUS I APPENDIX

3. (a) Domain: $0 \le y \le v_0^2/2g$, Range: $0 \le t \le 2v_0/g$; (c) Domain: $0 \le t \le 2v_0/g$, Range: $0 \le y \le v_0/2g$
13. (a) 1, $-\pi/2$; (b) v_0, a; (c) $t = 0$: $-\alpha, \alpha^2 - \omega^2$; $t \to \infty$: 0, 0   **19.** (a) $x = 1$, maximum; $x = -1$, minimum; points of inflection are $x = 0, \pm\sqrt{3}$; (b) $x = a$, point of inflection; (c) $x = 4n + 1$, maximum; $x = 4n - 1$, minimum
21. (a) $2\,(V/\pi)^{1/3}$; (b) $(4V/\pi)^{1/3}$ **23.** $x = -a/b$ is a minimum   **25.** $1/2\pi$ m   **27.** 120 cm^3

CALCULUS II APPENDIX

3. $\dfrac{(2x + 1)^{5/2}}{10} - \dfrac{(2x + 1)^{3/2}}{6}$ **5.** (a) $x^3/3 + x^2/2 + C_1 x + C_2$; (b) $-4\cos(\pi x/2)/\pi^2 + C_1 x + C_2$; (c) $(x + a)^5/120 + C_1 x^2 + C_2 x + C_3$ **7.** $\frac{1}{2}at^2 + v_0 t + s_0$ **9.** $e^{-t} - 1$ **11.** 1.45×10^{14}; 0.938 m^2

CHAPTER 2

5. $\bar{v} = 2v_1 v_2/(v_1 + v_2)$ **7.** (b) 1 m/s; (c) 2 m/s; (d) 0   **9.** 1.125 m/s   **11.** 2 m/s^2
13. $v_m = 15$ cm/s; $a_m = 45$ cm/s^2 **15.** (a) $-5\pi^2$ m/s^2; (b) 4π m/s^2 **17.** $v = 600$ m/s; $s = 3000$ m   **19.** -4 m/s^2
21. (a) 33.3 s; (b) 925.7 m; (c) 55.6 m/s   **23.** 15.6 m/s   **25.** -16.9 m/s   **27.** 1.79 s   **31.** 1000 m
33. $s(t) = \frac{1}{3}t^3 + t^2 + \frac{1}{2}t + 1$ **35.** 50 km/h   **37.** 3400 m/s^2 **39.** -19.4 g   **41.** -0.20 m/s^2 **43.** 8.72 m/s
45. 0.0384 m   **47.** 914.7 m   **49.** (a) 2 s; (b) 4.4 m; (c) down   **51.** 14 km

CHAPTER 3

1. 58.3 km, 59.0° south of west   **3.** -2.046 units   **11.** 2.668 units   **13.** 63.4° east of north, 7.6° below surface
15. $-10\hat{\mathbf{i}} - 3\hat{\mathbf{j}} - 3\hat{\mathbf{k}}$ **19.** $3\hat{\mathbf{i}} + 6\hat{\mathbf{j}} + 5\hat{\mathbf{k}}$; 90°   **21.** 54.8°   **23.** 3.937; $\hat{\mathbf{u}}_N = \pm(3\hat{\mathbf{i}} - 2\hat{\mathbf{j}} - 7\hat{\mathbf{k}})/\sqrt{62}$
25. $\mathbf{r} = 5\hat{\mathbf{i}} + 7\hat{\mathbf{j}} - 8\hat{\mathbf{k}}$ **29.** 49.1°   **33.** 8.66; $\hat{\mathbf{u}} = (-\hat{\mathbf{i}} + 7\hat{\mathbf{j}} + 5\hat{\mathbf{k}})/8.66$ **37.** 24

CHAPTER 4

1. $y = 2x^2 + 1$ **3.** $\bar{\mathbf{v}} = (-2$ m/s$)\hat{\mathbf{i}} - (9$ m/s$)\hat{\mathbf{j}}$; $|\bar{\mathbf{v}}| = 9.22$ m/s   **7.** $\mathbf{v} = (6t + 2)\hat{\mathbf{i}} + (4t + 4)\hat{\mathbf{j}}$; $\mathbf{a} = 6\hat{\mathbf{i}} + 4\hat{\mathbf{j}}$; $|\mathbf{a}| = 7.21$ m/s^2; $\theta = 33.69°$ **9.** $\mathbf{v} = R(\omega - \omega\cos\omega t)\hat{\mathbf{i}} + R\omega(\sin\omega t)\hat{\mathbf{j}}$; $\mathbf{a} = R\omega^2(\sin\omega t)\hat{\mathbf{i}} + R\omega^2(\cos\omega t)\hat{\mathbf{j}}$; $t = 2n\pi/\omega$;

$\mathbf{r} = 2\pi R n \hat{\mathbf{i}}$; $\mathbf{a} = R\omega^2 \hat{\mathbf{j}}$ **11.** (a) 7.10 m;   (b) 38.57 m;   (c) $v = 20.96$ m/s, $t = 2.22$ s   **13.** 0.112°, 0.195 m
17. 56.6°, 27.7°   **19.** 20.94 m/s; 2194 m/s²   **21.** (a) π rad;   (b) π s⁻¹;   (c) $\pi/10$ m/s;
(d) $a_\phi = \pi/10$ m/s², $a_\rho = \pi^2/10$ m/s²;   (e) 1.036 m/s²   **23.** (a) $\omega = 6t^2 - 6t$, $a_\rho = 0.2(6t^2 - 6t)^2$, $a_\phi = 1.2(2t-1)$;
(b) 0.5 s, 0.45 m/s²;   (c) 0, 1 s; $a_\phi(0) = -1.2$ m/s²; $a_\phi(1) = 1.2$ m/s²   **25.** 12 s   **27.** 3.02 km/h   **29.** 16.6° west of
north; 335 km/h   **33.** $\mathbf{v} = (t^2 + t)\hat{\mathbf{i}} + 7\hat{\mathbf{k}}$; $\mathbf{r} = (t^3/3 + t^2/2)\hat{\mathbf{i}} + 3\hat{\mathbf{j}} + 7t\hat{\mathbf{k}}$ **35.** 3.16 m   **37.** 7.2 m   **39.** $\sqrt{Rg}$
43. (a) 3 km/h;   (b) 7 km/h   **45.** 60 s, 180 m, 240 m

CHAPTER 5

1. 15 m/s, 225 m   **3.** $\sqrt{5}$ N   **5.** 10 m/s   **7.** 0.44 N   **9.** 11.93 N, $\theta = 144°$ **11.** 2.5 m/s², 10 N   **13.** (a) 19.6 N;
(b) 25.6 N;   (c) 15.6 N   **15.** 95.2%   **17.** 7.5°, 589 km   **19.** 1.66 × 10⁶ N   **21.** 50 m   **23.** 20 m/s²   **25.** (a) 6.25 s;
(b) 15.6 m;   (c) $v = 10.13$ m/s, $\theta = 80.9°$ **27.** (a) $g' = 7.32$ m/s², $F_g = 1.46 \times 10^4$ N, $f_g = 659$ N, $v = 7350$ m/s;
(b) 0, still weightless   **29.** 40.2 m   **33.** 175.5 N

CHAPTER 6

1. (a) 1.960 m/s²;   (b) 23.52 N;   (c) 47.04 N   **3.** (a) 3.836 m/s²;   (b) 68.18 N;   (c) 54.55 N;   (d) 61.36 N   **5.** 50.09 N,
76.9° above horizontal   **7.** 862.4 N   **9.** 1.250 m/s²   **11.** (a) 4.876 m/s²;   (b) 24.64 N   **15.** (b) 29.89 rev/min
19. 3.552 s   **21.** 400 N   **23.** (a) $a_1 = 2.2$ m/s², $a_2 = -1.8$ m/s², $a_3 = -0.2$ m/s²;   (b) $T_A = 48$ N, $T_B = 24$ N
25. 0.6003   **27.** 35.71 m/s, 5.601 s   **29.** 0.2166   **35.** 25.7°

CALCULUS III APPENDIX

5. $1/\sqrt{2}$ **7.** $y = [3(1 - e^{-at})/a]^{1/3}$ **9.** (a) $\frac{1}{3}$;   (b) $e - 1$;   (c) 4;   (d) 4;   (e) 1   **11.** 0.40   **13.** $4\pi a^2$

CHAPTER 7

1. 26.0 J, -26.0 J, -26.0 J, 26.0 J   **9.** 18 J   **11.** 15 J, -50.1 J   **15.** 7.91 m/s   **17.** 2.52 m   **19.** -2.28 m/s
23. $v_{max} = 10.3$ m/s, $v_{min} = 9.7$ m/s   **25.** (a) -160 J;   (b) 73.5 J;   (c) 28.8 N;   (d) 0.68   **27.** 2.17 m/s   **29.** 4.91 kW
33. -3.92 J, 0.99 m/s   **35.** 76.3 J   **37.** 789 J, 697 J, 372 J   **39.** 0.27 J   **45.** 14.5 m/s   **47.** 0.49 m
49. 0, 4/3 m; 16/27 J

CALCULUS IV APPENDIX

7. 0.404 m³, 2.9%

CHAPTER 8

3. $(-4 \text{ m}^2)A$; $(4 \text{ m}^2)A$ **5.** $7\frac{7}{36}$ **7.** 60 J; $P = (0, -2a, 0)$, $Q = (2a, 4a, 2a)$ **9.** $-mgR$ **13.** $v_0 \sqrt{m/(\kappa + \mu_k \gamma)}$
15. 12.9 m, 1

CHAPTER 9

1. 15 N   **3.** 3600 N   **7.** (a) $\frac{2}{5}$ m/s;   (b) $48\frac{3}{5}$ J   **9.** 51.6 km/h, 48.4° north of east   **11.** 45°   **15.** $\frac{1}{5}$ h from base along
line perpendicular to base and passing through peak   **17.** $\hat{\mathbf{i}}$ = direction of particle 1, $\hat{\mathbf{j}}$ = direction of particle 2;
$\mathbf{v}_{31} = -\frac{4}{3}v\hat{\mathbf{i}} - \frac{1}{3}v\hat{\mathbf{j}}$, $\mathbf{v}_{32} = -\frac{1}{3}v\hat{\mathbf{i}} - \frac{4}{3}v\hat{\mathbf{j}}$ **19.** (a) 40 m from target;   (b) 5.3 s after first fragment   **21.** 523 kg/s, 1570 kg/s,
1.57 × 10⁶ N, 4.70 × 10⁶ N   **23.** (a) 3.44 m/s²;   (b) 800 kg/s;   (c) 2750 m/s   **25.** (a) 6000 N;   (b) 263 m
27. 291 N   **29.** 331 m/s   **31.** (a) 0.19 m/s;   (b) 3444 m   **33.** 172 m/s   **35.** (a) 5 m/s, 2.5 J, 0.39 N;   (b) 2.3 J, 0.15 m;

(c) 1.9 J, 0.14 m;   (d) 7.1 m   **43.** (0.9, 0), (0.3, 0.1), (-0.3, 0.2), (-0.9, 0.3); none   **47.** (b) 5040 m/s;   (c) 3640 m/s;
(d) 3940 m/s

CHAPTER 10

1. (a) 720 kg · m^2/s;   (b) 360 N · m   **3.** 2.66 × 10^4 kg · m^2/s   **5.** $a = g \sin \alpha$ **7.** (a) $\mathbf{F}_1 = 6\hat{\mathbf{i}} + 12t\hat{\mathbf{j}}$; $\mathbf{F}_2 = 12t\hat{\mathbf{i}} + 4\hat{\mathbf{j}}$;
(b) $\mathbf{a}_{\text{C.M.}} = (6t + 3)\hat{\mathbf{i}} + (6t + 2)\hat{\mathbf{j}}$;   (c) $\mathbf{v}_{12} = (6t^2 - 6t + 5)\hat{\mathbf{i}} + (-6t^2 + 4t)\hat{\mathbf{j}}$ **9.** $\mathbf{r} = 100\hat{\mathbf{j}} - 1690\hat{\mathbf{k}}$ where the origin of
the coordinate system moves with the balloon and the unit vectors are $\hat{\mathbf{i}}$ = east, $\hat{\mathbf{j}}$ = north, $\hat{\mathbf{k}}$ = up.   **13.** $R_2 = 1.914R$
15. (a) v_{12} = 1.6 m/s;   (b) $v_{\text{C.M.}}$ = 2 m/s;   (c) v_1 = 1.5 m/s, v_2 = 2.9 m/s   **17.** (a) X = 0.5 m from top of rod,
V = 4/3 m/s;   (b) ω = 4/3 s^{-1};   (c) 6 J   **19.** (a) $\mathbf{L} = -36\hat{\mathbf{j}} + 12t^2\hat{\mathbf{k}}$;   (b) $\boldsymbol{\tau} = 24t\hat{\mathbf{k}}$;   (c) $\mathbf{P} = (12t + 26)\hat{\mathbf{i}} - 4\hat{\mathbf{j}}$;
(d) $\mathbf{F} = 12\hat{\mathbf{i}}$ **21.** ω_2 = 27.78 s^{-1}

CALCULUS V APPENDIX

3. $6\frac{2}{3}$ **5.** $\sqrt{2} - 1$ **7.** $\frac{3}{2}\pi a$ **9.** $M_x = ab^2/2$ or $a^2b/2$; $I_x = ab^3/3$ or $a^3b/3$ **13.** $2\pi R^2(1 - \cos\beta)$

CHAPTER 11

1. (a) 3.0 N · m;   (b) 5.56 rad/s^2;   (c) 16.67 rad/s   **3.** (a) -3.32×10^{-2} rad/s;   (b) 29.2 rev;   (c) -1.31×10^{-3} N · m
5. -2×10^{-5} **9.** (a) 35 rad/s^2;   (b) ω = 70 rad/s;   (c) T = 0.70 N   **11.** ω = 3.16 rad/s; K_i = 23.8 J; K_f = 36.1 J
13. 11.10 cm from end of rod containing hole   **15.** $x = a^2b/(R^2 - a^2)$ **17.** 5.77 cm from base of triangle along a line
perpendicular to base and passing through center of base   **23.** $\frac{3}{2}MR^2$ **25.** $\frac{1}{4}(R_1^2 + R_2^2)M + \frac{1}{3}M\ell^2$ **29.** 17.6 ms
31. (a) ω = 0.50 s^{-1};   (b) $F = 5 \times 10^4$ N;   (c) 0.90   **33.** (b) W = 144 J;   (c) 2.45 m   **35.** (b) a = 0.82 m/s^2, -0.37%,
-0.41%;   (c) t = 3.5 s;   (d) 4.6%   **37.** ($\frac{3}{5}b$, 0)   **39.** $\frac{3}{20}M\ell^2$ **41.** 3.75×10^{-3} kg · m^2 **43.** $\dfrac{3M}{10}\left(\dfrac{R^2}{2} + \dfrac{h^2}{3}\right)$

CHAPTER 12

1. 66.7 cm   **3.** V = 0.160 m/s; $\omega \cong$ 1.00 rad/s; C.M. located 4.0 mm from center of sphere   **5.** a = 2.439 m/s^2,
$\mu_{s,\text{min}}$ = 0.0062   **7.** a = 7.84 m/s; $\alpha_1 = \alpha_2$ = 49.0 rad/s; T = 15.7 N   **9.** 219.7° clockwise from original position;
C.M. is 10 m to right and 4.9 m below original position   **11.** 3.40 s   **13.** (a) 2.71 m/s;   (b) 2.71 m/s
15. a_c = 0.889 m/s^2; a_b = 1.78 m/s^2 **19.** a_R = 2/5 m/s^2; a_p = 4/5 m/s^2; between rollers and plank, f = 3/5 N;
between rollers and ground, f = 1/5 N

CHAPTER 13

1. 3010 N   **3.** 84.6 g   **5.** Vertical component, 122.5 N; bottom horizontal component 76.6 N directed toward door; top
horizontal, 76.6 N away from door.   **7.** 849 N, 60°   **9.** 0.70   **11.** (a) 73.5 N;   (b) 73.5 N   **13.** 0.789
15. (a) 57,800 N;   (b) F_x = 55,500 N, F_y = 35,800 N   **23.** T = 250 N; F_x = 200 N; F_y = 50 N

CHAPTER 14

1. 2.68×10^{-7} m/s^2 **3.** $[GMm/ar^2]\sin\alpha$ **5.** 2.25×10^{-7} **13.** $F_0 = 0$, $F_R = GMm/R^2$; $U_0 = -3GMm/2R$,
$U_R = -GMm/R$ **17.** 3.04 km/s   **19.** (a) 1.62 m/s^2;   (b) 1.68×10^3 m/s;   (c) 17.2 minutes   **23.** 1.58×10^{10} J
27. (a) 0.076;   (b) 7.35×10^{22} kg;   3.34 g/cm^3 **35.** 6.36×10^{35} J   **37.** 413 kg   **39.** $M_A/M_B \cong 2.1$, $T \cong 50$ y;
(a) $M_A = 4.35 \times 10^{30}$ kg;   (b) 0.60;   (c) a_A = 6.4 A.U., b_A = 5.1. A.U., a_B = 13.6 A.U., b_B = 10.9 A.U.

CHAPTER 15

1. (b) $\xi = 1$, stable   **3.** (b) $r = r_0$, stable   **5.** (b) 0.418 m;   (c) 3.74 m/s;   (d) 0.296 m;   (e) $\omega_0 = 8.94$ rad/s, $T = 0.702$ s; (f) 130.9°   **7.** (a) 2.29 s;   (b) 0.110 m/s   **11.** $\omega = 2\pi/3$; $v_0 = 20\pi$ cm/s   **13.** 1.466 s, 2.932 s **15.** (a) 5.14×10^{-8} N $\cdot$ m;   (b) 0.48 cm   **17.** $r = R/\sqrt{2}$;   1.07 s   **21.** (a) 13.42 N $\cdot$ s/m;   (b) 0.819 s   **25.** (b) $\xi = 0$, unstable; $\xi = 2b/3a$, stable   **29.** 6.62 cm   **31.** 0.357 m   **33.** (a) 0.397 s;   (c) 0.132 J   **35.** 9.75×10^{-4} N **41.** (b) 740 N   **43.** (b) 1.16 s   **45.** (a) 877 N/m;   (b) 10.6 s;   (c) 3/4;   (d) approximately ω_0;   (e) F_0/κ;   (f) 160.5; (g) $0.183 F_0$;   (h) 160.5 F_0

CHAPTER 16

1. 2.80×10^4 N   **3.** 1.25×10^{11} Pa   **5.** 0.422 mm   **7.** 4.64×10^6 N   **9.** 6.92 kg/m³; 0.26% **11.** $(R_2^2 + R_1^2)/(R_2^2 - R_1^2)$ **13.** 0.59 mm   **15.** (a) 0.114°;   (b) 0.44 m from the copper wire   **17.** (a) 0.0415 s; (b) 9.62 kg;   (c) 4.81 kg;   (d) 4.28×10^{-4} m   **19.** 19.3°

CHAPTER 17

1. 2.0×10^6 N   **3.** $p = 4000$ Pa; $\mathbf{F} = (20$ N$)(\hat{\mathbf{i}} + \hat{\mathbf{j}} + \hat{\mathbf{k}})$ **5.** 5.09×10^4 N   **7.** 4.99 atm   **9.** 5.88×10^6 N; 5.88×10^5 N; 1.96×10^5 N   **11.** $f = 218$ N; $\Delta\rho/\rho = 2.78 \times 10^{-4}$ **15.** 10.5 m   **17.** 1.08 cm   **19.** 1.12×10^3 kg **21.** yes   **23.** 15.9 cm   **25.** 3.21 N   **27.** (b) 8.0 Pa   **29.** 96.6 N   **33.** 18.4 km   **37.** 0.424 cm   **39.** 0.032% **43.** $t = 1.25$ mm; $\gamma = 72.0 \times 10^{-3}$ N/m

CHAPTER 18

1. (a) 6.67 m/s;   (b) 0.280 m/s   **3.** 14.8 m/s   **5.** 2.23 m   **9.** 3.91 cm   **11.** (a) 4.24 m/s;   (b) 11.79 m/s; (c) 0.587 atm   **13.** 7.90 W/m; 21.2%   **15.** 0.395   **17.** (a) 14.0 mm/s;   (b) 1.0 mm/s;   (c) 1.18 m/s   **21.** 11.8 cm/s **23.** 849   **25.** (a) 0.0315 mm;   (b) 5.19 mm;   (c) v_∞(fog) = 12.0 cm/s, v_∞(Al) = 5.70 cm/s   **27.** (a) 1.27×10^5; (b) 4.56×10^3 Pa   **29.** $v = 80$ m/s; $p = 0.974 \times 10^5$ Pa   **33.** 0.652×10^{-3} Pa $\cdot$ s;   water   **35.** (a) 100; (b) 7.5×10^{-4} N/m

CHAPTER 19

1. 5.41 km   **5.** (a) 20 m/s;   (b) 8 cm   **7.** 250 m/s   **13.** (a) 2.15 cm;   (b) +21.7°;   (c) 540 cm/s   **15.** 15.1 W; 3.0 J **17.** 2.79×10^3 m/s   **23.** $F_1 = 9F_2/4$ **25.** (a) $F' = 4F$;   (b) 11.1 cm   **27.** 0.596   **33.** $\beta = 0.0033$; $\alpha = 66.7$ **35.** (a) 1.67 m;   (b) 6.11 g;   (c) 0.60 N;   (d) $F_c/F = 1.2\%$ **37.** 587 m/s   **43.** 1.77 m/s; $v_{0.5} = 1.69$ m/s; $v_{0.1} = 0.98$ m/s   **45.** 822 N

CHAPTER 20

1. 1.127   **3.** 2.33 m   **5.** (a) 2.93 N/m²;   (b) 2.63×10^{-6} m   **7.** (a) 17.0 Hz;   (b) 34.0 Hz;   (c) $\nu_{\text{open}} = 17.6$ Hz, $\nu_{\text{closed}} = 35.2$ Hz   **9.** 13.7° C   **11.** (b) closed at both ends;   (c) 80 cm, 414.4 Hz, 207.2 Hz;   (d) 0.257 mm;   (e) 288.8 μJ **13.** 1.18 mm   **15.** 435 Hz, 445 Hz   **17.** (a) $\theta_1 = 12.7°, \theta_2 = 26.2°$;   (b) $\theta_0' = 6.3°, \theta_1' = 19.3°$;   (c) 6.36 mW/m² **19.** 12.7°   **21.** $\lambda_0 = 1.583$ m, $\lambda_{45} = 1.604$ m, $\lambda_{90} = 1.658$ m, $\lambda_{135} = 1.711$ m, $\lambda_{180} = 1.733$ m, $\nu_0 = 209.4$ Hz, $\nu_{45} = 206.7$ Hz, $\nu_{90} = 199.9$ Hz, $\nu_{135} = 193.7$ Hz, $\nu_{180} = 191.3$ Hz   **23.** (a) 205.2 Hz;   (b) 2.5°;   (c) 205.1 Hz   **25.** 50% **27.** (a) 1.1×10^{-11} m;   (b) 1.1×10^{-5} m   **29.** (a) 1900 Hz;   (b) 20 mm;   (c) 400 Hz, 30 mm   **31.** (a) 100 Hz, 200 Hz, 300 Hz, 400 Hz, 500 Hz, 600 Hz, 700 Hz, 800 Hz   **35.** 48 cm × 72 cm × 120 cm   **37.** 33.4°C   **41.** 1.60   **43.** (a) 93.0; (b) 1.30 N/m²;   (c) 28.3 m

CHAPTER 21

3. (a) 4.00×10^{-7} s; (b) 3.20×10^{-7} s   **5.** (a) 2.83×10^8 m/s; (b) 0.577 m   **7.** $1244\ \mu s$ **9.** $0.882c$ **11.** $0.70c$
13. (a) 3.87×10^{26} W; (b) 4.30×10^9 kg/year, 6.81×10^{-14}; (c) 1.47×10^{12} years   **15.** (a) 9.76×10^3;
(b) $1 - 0.53 \times 10^{-8}$; (c) 6.784×10^{-9} m; (d) 9.76×10^3 **19.** (a) 1.09 Å; (b) 26 days
21. $v = 3.22 \times 10^4$ m/s, $\theta = 47.0°$

CHAPTER 22

1. $4.7°$ F   **3.** 13.93 K   **5.** 0.299 m   **7.** 0.14% **11.** 75.7 N   **13.** 3.25 kg   **15.** 0.0931 Cal/kg $\cdot$ deg   **17.** $485.6°$ C
19. 100.4 m^3/h   **21.** (a) 6.03×10^3 Cal/h; (b) 4.0 gal   **25.** 0.0145 deg   **27.** (a) 233.15 K; (b) 433.15 K
29. 67.0 m   **31.** (a) 5.38×10^7 Pa; (b) 10.3% **33.** (b) 0.372×10^{-3} deg^{-1} **35.** (b) 2.895×10^7 Pa; 8.53%

37. $10.6°$ C   **39.** $0.94°$ C   **41.** 1 Cal $= 4118$ J   **43.** $\dfrac{dQ}{dt}$ (double pane)$\Big/ \dfrac{dQ}{dt}$ (single pane) $= 2.36\%$ **45.** 4.15

47. 3.29 cm

CHAPTER 23

1. 1.6736×10^{-27} kg   **3.** $75.757\%\ ^{35}$Cl; $24.243\%\ ^{37}$Cl   **5.** 6.54552×10^{21} molecules/cm^3 **7.** 3.1043 Å
9. (a) 4.002×10^5 Pa (absolute), 43.362 lb/in.^{2} (gauge); (b) 4.488×10^5 Pa (absolute), 50.405 lb/in.^{2} (gauge)
11. 1.8914 atm   **13.** (a) $P_{He} = 0.3153$ atm, $P_{O_2} = 0.6847$ atm; (b) 94.55% **15.** $84.8°$ C   **17.** 53.8% **23.** 1.130
25. 34.4 Å   **27.** (d) $-38.3°$ C, 0.312 atm   **29.** $139.3°$ C   **31.** 0.362 **33.** 46.0% **35.** 21.6 g/m^3

CHAPTER 24

1. (b) 488 m/s; (c) 532 m/s; (d) 1.090 **3.** (a) 131.25 u; (b) 234.0 m/s; (c) $\Delta U_{grav} = 2.136 \times 10^{-25}$ J,
$\bar{K} = 5.967 \times 10^{-21}$ J   **5.** 0.175 Cal/kg $\cdot$ deg   **7.** 647.5 K   **13.** In units of 10^{-5} m^2/s, Ar: 1.39, H$_2$: 10.96, Kr: 0.74,
methane: 1.70, Ne: 3.87, N$_2$: 1.56, Xe: 0.43 **15.** Using D, 3.01 Å; using η, 3.37 Å   **17.** 0.172 Cal/kg $\cdot$ deg
21. 8.381×10^{-2} kg/m^3; 9.5235×10^{-5} m^2/s; 7.982×10^{-6} Pa $\cdot$ s; 1755 m/s; 1.628×10^{-7} m

CHAPTER 25

3. (a) 211.5 J; (b) 0.50 atm   **7.** (a) 4.99×10^3 J; (b) -4.99×10^3 J   **13.** (a) 7.116×10^4 J; (b) 0.702 m^3;
(c) 32.3 K   **17.** (b) 3059 J; (c) 3148 J   **19.** (d) 112.5 J; (e) 562.5 J; (f) 675.0 J   **21.** (a) 0.499 mol; (b) $506.9°$ C,
79.8 atm; (c) 407.8 K; (d) 3830 J; (e) 76.2%

CHAPTER 26

1. 3.44 **5.** (a) 1.833×10^7 J; (b) 1.645×10^6 J, 27.42 min   **7.** $e_{actual}/e_{reversible} = 0.394$ **9.** 0.854 K
11. (a) 1315.8 MW; (b) $dQ_1/dt = 3289.5$ MW, $dQ_2/dt = 1973.7$ MW; (c) 0.551; (d) 2300; (e) 2.62 K, 0.13 K
17. 0.312 Cal/deg   **19.** 1.00 cal/deg   **21.** 1.089×10^2 J/deg; 1.029×10^2 J/deg; 0.942×10^2 J/deg;
0.722×10^2 J/deg   **25.** 6.6% **27.** 7.50×10^5 Cal/h   **29.** 9.10 J/K

CHAPTER 27

1. The force is 0.843 N and attractive. The force will be increased in the case of the conducting spheres. The upper limit
on the new value of the force is 1.78 times the old value.   **3.** (a) The point is 323 cm beyond the charge of $-3.0\ \mu$C.
(b) Same answer as in part (a). (c) The point at which the electric field strength is zero will be a point of zero force on
both positive and negative charges.   **5.** $\mathbf{F} = -4.19\hat{\mathbf{i}} - 1.89\hat{\mathbf{j}}$ N if q_1 is in the upper left corner of the square.

7. $q = 0.204 \, \mu C$ **9.** $F = 1.27 \times 10^5 \, N$ **11.** The force is maximum on a circle of radius $\ell/\sqrt{2}$.   **13.** $\theta_1 = 0.366$; $\theta_2 = 0.353$; $\theta_3 = 0.361$; $\theta_4 = 0.356$; $\theta_5 = 0.359$; $\theta_6 = 0.357$; $\theta_7 = 0.358$; $\theta_8 = 0.358 = 20.5$ °; $2\theta = 41.0°$

CHAPTER 28

1. (a) For $0 < x < 0.20$ m, $\mathbf{E} = 8.99 \times 10^3 \left[\dfrac{3}{x^2} + \dfrac{8}{(0.20 - x)^2} \right] \hat{\mathbf{i}} \, N/C$; for $x > 0.20$ m,

$\mathbf{E} = 8.99 \times 10^3 \left[\dfrac{3}{x^2} - \dfrac{8}{(x - 0.20)^2} \right] \hat{\mathbf{i}} \, N/C$; for $x < 0$, $\mathbf{E} = 8.99 \times 10^3 \left[\dfrac{8}{(0.20 - x)^2} - \dfrac{3}{x^2} \right] \hat{\mathbf{i}} \, N/C$;   (b) $x = -0.316$ m

5. $\mathbf{E} = -\dfrac{24Q\ell^2}{4\pi\epsilon_0 x^4} \hat{\mathbf{i}}$ **7.** 7.08×10^{-12} C   **11.** Outside the sheets $E = 0$; between the sheets $E = \dfrac{\sigma}{\epsilon_0}$.

15. For $r < R$, $\mathbf{E} = \dfrac{r\rho}{2\epsilon_0} \hat{\mathbf{u}}_r$; for $r > R$, $\mathbf{E} = \dfrac{R^2\rho}{2\epsilon_0 r} \hat{\mathbf{u}}_r$ **17.** $\dfrac{\delta Q}{Q} = 4 \times 10^{-3}$ **19.** 33% ionic, 67% covalent

21. $\mathbf{E} = \dfrac{\lambda}{\sqrt{3}\pi\epsilon_0 a} \hat{\mathbf{k}}$

CHAPTER 29

1. For 100 V, $r = 0.72$ m; for 50 V, $r = 1.44$ m; for 25 V, $r = 2.88$ m   **3.** (a) 0.317 J;   (b) 1.44×10^5 V;   (c) $E = 0$

5. $\Phi = \dfrac{\lambda}{2\pi\epsilon_0} \ln\left(1 + \dfrac{1}{2x}\right)$; when $x \gg L$, $\Phi = \dfrac{\lambda}{2\pi\epsilon_0} \cdot \dfrac{L}{2x} = \dfrac{Q}{4\pi\epsilon_0 x}$ where $Q = \lambda L$ **7.** (a) 6 m;   (b) $2 \, \mu C$

9. (a) $\mathbf{E} = -A(y\hat{\mathbf{i}} + x\hat{\mathbf{j}})$;   (b) x/y;   (c) $-y/x$;   (d) the field lines and the equipotentials are mutually perpendicular

11. (a) $\Phi(z) = \dfrac{\sigma}{2\epsilon_0}[(z^2 + R^2)^{1/2} - z - R]$;   (b) $\Phi(z) = \dfrac{-\sigma}{2\epsilon_0} z$ **13.** For $r < R_1$, $\mathbf{E} = Q\hat{\mathbf{r}}/4\pi\epsilon_0 r^2$; for $R_1 < r < R_2$, $\mathbf{E} = 0$; for $r > R_2$, $\mathbf{E} = Q\hat{\mathbf{r}}/4\pi\epsilon_0 r^2$ **15.** $\sigma = 0.0255$ nC/cm^2 **17.** (a) By the superposition principle, adding the charge Q'' at the center merely adds a constant value to the potential on the surface of the sphere;   (b) $Q'' = -Q' = +mQ$; (c) $Q = 2$ nC, $Q' = -2/3$ nC, $Q'' = 2/3$ nC;   (d) 30.0 V   **19.** $a = 5.98 \times 10^{13}$ g   **23.** (a) 2η;   (b) $5\eta/2$;   (c) the field is not conservative since the two values are not equal;   (d) $\nabla \times \mathbf{E} = -\eta x^2 \hat{\mathbf{k}}$, and since this is not zero the field is not conservative;   (e) $-\eta/2$, which confirms the conclusion reached in (d);   (f) $-\eta/2$, which is the same as in (e).

25. $\dfrac{\partial^2\Phi}{\partial x^2} + \dfrac{\partial^2\Phi}{\partial y^2} + \dfrac{\partial^2\Phi}{\partial z^2} = -\dfrac{\rho}{\epsilon_0}$ **27.** (a) 1.798×10^5 V;   (b) 1.272×10^5 V   **29.** (a) $\Phi(x) = \dfrac{\sigma R}{2\epsilon_0 x}[x + R - (x^2 + R^2)^{1/2}]$; (c) $E(x) = \dfrac{\sigma R^2}{2\epsilon_0 x^2}\left[1 - \dfrac{R}{(x^2 + R^2)^{1/2}}\right]$ **31.** (a) On the inner surface of the outer cylinder;   (b) for $\rho < R_1$ and $\rho > R_2$, $E = 0$; for $R_1 < \rho < R_2$, $E = \lambda/2\pi\epsilon_0\rho$;   (c) $V = (\lambda/2\pi\epsilon_0) \ln(R_2/R_1)$ **33.** (a) 2×10^6 eV;   (b) 5.11×10^5 eV; (c) 0.741;   (d) 2.80

CHAPTER 30

1. $S = 4.43$ **3.** $Q = 3.54$ nC   **5.** $Q = 33.4$ nC   **7.** (a) $Q = 1.31 \times 10^{-10}$ C;   (b) $V = 746$ V   **11.** (a) 39.8 J/m^3; (b) 3.98×10^{-2} N   **13.** (a) 0.045 J;   (b) 0.135 J   **21.** (a) 1.95×10^5 V/m;   (b) 2.69×10^{-6} C/m^2; (c) $\sigma = 4.43 \times 10^{-6}$ C/m^2, $\sigma' = 2.69 \times 10^{-6}$ C/m^2 **23.** (a) 45 C;   (b) 41.4 C   **25.** $\dfrac{Q^2}{2A\epsilon_0}\left(t + \dfrac{d}{\kappa}\right)$; this is equal to $Q^2/2C$ **29.** (a) 708 μF;   (b) 708 C   **35.** $F = \dfrac{\epsilon_0 b}{S}(\kappa - 1)V^2\hat{\mathbf{i}}$ **37.** $\alpha = 0.829 \times 10^{-40}$ C · m^2/V; $\ell = 5.7 \times 10^{-16}$ m. This is smaller than a nuclear radius.

CHAPTER 31

1. (a) $I = 1.5 \times 10^5$ A;   (b) $\Delta q = 5.4 \times 10^8$ C   **3.** 1.3×10^{-5} A/m^2 **7.** (a) $\rho = 3.21 \times 10^{-8}$ Ω-m; (b) $J = 6.23 \times 10^6$ A/m;   (c) $I = 4.89 \times 10^{-2}$ A;   (d) $\langle v \rangle = 2.15 \times 10^{-4}$ m/s;   (e) $V = 0.4$ V   **9.** $R = 22.4 \, \Omega$

11. (a) $R = 6.4 \times 10^{17}\,\Omega$;   (b) $I = 1.6 \times 10^{-14}$ A;   (c) $t = 3.1 \times 10^6$ s $= 36$ days   **13.** (a) $I = 0.381$ A;
(b) $E' = 9.80$ V, $r' = 0.75\,\Omega$ **15.** $\Delta I/I = 0.0036$ **17.** (b) $\mathcal{E}_F(\text{Li}) = 4.76$ eV, $\mathcal{E}_F(\text{Cu}) = 7.06$ eV   **21.** 1.05 mA
23. 0.815 μA

CHAPTER 32

1. (a) 27.3 V   (b) The current through the $10\,\Omega$ resistor is 2.73 A. The current through the $20\,\Omega$ resistor is 1.36 A. The current through the $30\,\Omega$ resistor is 0.91 A.   **3.** $3.34\,\Omega$ **5.** The current through the $8\,\Omega$ resistor is $\frac{11}{13}$ A. The current through the $5\,\Omega$ resistor and the $1\,\Omega$ resistor is $\frac{6}{13}$ A. The current through the $3\,\Omega$ resistor and the other $1\,\Omega$ resistor is $\frac{17}{13}$ A.   **7.** $E_1 = \frac{32}{3}$ V   **9.** $R_1 = 2950\,\Omega$, $R_2 = 12{,}000\,\Omega$, $R_3 = 135{,}000\,\Omega$ **13.** $I = \frac{23}{19}$ A directed toward the 3 V battery.   **15.** $I = \frac{14}{9}$ A   **17.** $dI/dT = -1.35$ mA/deg   **19.** $R = \dfrac{R_1^2}{R_1 + R_2}$ **25.** $R = \frac{20}{3}\,\Omega$ **27.** $I = 9.70 \times 10^{-5}$ A

CHAPTER 33

1. The magnitude of the field is 0.0208 T. It is in the negative y direction.   **3.** (a) $v = 3.47 \times 10^7$ m/s;
(b) $t = 2.04 \times 10^{-8}$ s;   (c) The energy is 6.25 MeV.   **5.** The deflection is 5.93 mm. The deflection is non-trivial, but the whole picture moves together.   **7.** $\mathbf{F} = (-2.88\hat{\mathbf{j}})$ N   **13.** (a) $I = 17.82$ A; the current flows in a clockwise direction;
(b) $F = 0.490$ N; the force is down.   **15.** $E = 5.93 \times 10^5$ V/m   **17.** $\Delta t = 14.5$ ns   **19.** (b) $R = 222$ A.U.;
(c) $\omega = 9 \times 10^{-6}$ rad/s   **21.** (a) $T = 0.80$ N;   (b) $\Delta\ell = 3.24 \times 10^{-4}$ cm   **23.** (a) $T = 0.354$ s;   (b) $\Omega = 3.09$ rad/s

CHAPTER 34

1. $B = \mu_0 I/4\pi x$ **3.** $B = \alpha\mu_0 I/4\pi R$ **5.** $B_z = \dfrac{2I\mu_0 a^2}{\pi(a^2 + z^2)(2a^2 + z^2)^{1/2}}$

7. $B = \dfrac{\mu_0 n I}{\sqrt{1 + 4\left(\dfrac{R}{\ell}\right)^2}}$

$\eta = \ell/R$	$B/\mu_0 n I$
0.5	0.243
1	0.447
2	0.707
5	0.928
10	0.981
20	0.995

9. For $r < a$, $B = \mu_0 Ir/2\pi a^2$; for $a < r < b$, $B = \mu_0 I/2\pi r$; for $b < r < c$, $B = \mu_0 I(c^2 - r^2)/2\pi r(c^2 - b^2)$; for $r > c$, $B = 0$ **11.** $\varphi_M = \dfrac{\mu_0 I c}{2\pi} \ln \dfrac{a + b}{a}$ **13.** (a) $a = -b$;   (b) $\mathbf{J} = -(2xb/\mu_0)\hat{\mathbf{k}}$ **15.** $I = 1.96$ A   **17.** $B = \dfrac{\mu_0 I}{4}\left(\dfrac{1}{R_1} - \dfrac{1}{R_2}\right)$; the field is directed into the paper.   **23.** (b) $B_y = \dfrac{a}{2}(x^2 - y^2) + C$

CHAPTER 35

1. $(0.90$ V$)\sin \omega t$ **3.** $(6.81 \times 10^{-2})e^{(1.6\,\text{s}^{-1})t}$ V   **5.** (a) 0.96 V counterclockwise;   (b) 0.96 V counterclockwise
7. 0.90 V; positive polarity in the rotating end   **9.** For $r < a$, $\mathbf{B} = \mu_0 \epsilon_0 E_0 \omega \left(1 - \dfrac{r}{2a}\right)\cos \omega t\, \hat{\mathbf{u}}_\phi$; for $r > a$,

$\mathbf{B} = \dfrac{\mu_0 \epsilon_0 E_0 \omega R}{r} \left(1 - \dfrac{R}{2a}\right) \cos \omega t \, \hat{\mathbf{u}}_\phi$ **13.** (b) 750 μA   **15.** (a) $E_y = \dfrac{E_0}{a^2} y^2$;   (b) $\dfrac{2ZE_0}{a^2}(\hat{\mathbf{i}} - \hat{\mathbf{j}})t$;   (c) $\dfrac{2E_0 t}{\mu_0 a^2}(\hat{\mathbf{i}} + \hat{\mathbf{j}})$

17. 12.6 mV   **19.** $-(3.03 \times 10^{-2} \text{ V}) \cos^2 8\pi t$ **21.** 2.25 V

CHAPTER 36

1. 1.0 V   **3.** $\dfrac{\mu_0 N_c}{2\pi} \ln \dfrac{a+b}{a}$ **7.** -18.8 V cos ωt **11.** (a) 91.1 mH;   (b) 0.182 J   **13.** 1.6×10^{-2} J

15. -3.43×10^{-35} per atom; $\chi_M = -0.92 \times 10^{-9}$ **17.** $B = 0.012$ T; $H = 63.7$ A/m; $M = 9.49 \times 10^3$ A/m

21. 8.05 A in each coil   **23.** $D = 2 \times 10^{-7}$ C/m^2; $E = 5{,}650$ V/m; $P = 1.5 \times 10^{-7}$ C/m^2; $H = 2.5 \times 10^{-7}$ A/m;

$B = 9.42 \times 10^{-12}$ T   **25.** $\dfrac{\mu_0 \pi a^2 b^2}{4r^3}(3 \cos^2 \theta - 1)$ where $\theta = \hat{\mathbf{k}} \cdot \hat{\mathbf{r}}$ **27.** 6.19×10^{16} rad/s   **29.** (a) $H = 54$ A/m;

$B = 190$ G;   (b) $H = 54$ A/m; $B = 0.68$ G

CALCULUS VII APPENDIX

1. (a) $y = C_1 e^{\left(\frac{5+\sqrt{13}}{2}\right)x} + C_2 e^{\left(\frac{5-\sqrt{13}}{2}\right)x}$ **3.** (a) $y = C_1 e^{ax} + C_2 e^{-ax} - x^2 - \dfrac{2}{a^2}$

(b) $y = C_1 + C_2 e^{3x}$

(c) $y = C_1 e^{+2ix} + C_2 e^{-2ix} = A \sin 2x + B \cos 2x$ (b) $y = C_1 e^{-\beta x} + \dfrac{A}{\beta - \alpha} e^{-\alpha x}$

(d) $y = C_1 e^x + C_2 e^{2x}$

CHAPTER 37

1. (a) 9.0×10^{-6} J;   (b) 9.0×10^{-6} J;   (c) 18.0×10^{-6} J;   (d) no change   **3.** (a) 2.25;   (b) 0.105   **5.** (a) 1.31 J/s;
(b) 4.16 J/s;   (c) 5.48 J/s;   (d) the energy is conserved.   **11.** $C = 238$ μF; power absorbed $= 2.97$ kW; power
delivered $= 3.27$ hp   **13.** (a) 705 W   **15.** $L = 26.3$ mH; $R = 17.7\,\Omega$; $C = 22.7$ μF; $\omega_0 = 1294$ rad/s; $\mathcal{Q} = 1.92$
17. $16.6 - i\,29.6$ **19.** Aluminum   **21.** (a) $5.93 + i\,25.5\,\Omega$;   (b) $0.80 - i\,3.44\,\Omega$;   (c) $4.80 - i\,2.94\,\Omega$ **29.** (a) 15.1 Ω;
(b) $I_0 = 5.81$ A, $\delta_L = 31.1°$ **31.** (a) 2.0×10^4 rad/s;   (b) 44.7 Ω;   (c) 141 Ω

CHAPTER 38

1. $\mathbf{H}(\mathbf{r}, t) = (0.53 \text{ A/m})\hat{\mathbf{j}} \cos(kz - \omega t)$ **3.** 1.2×10^7 m/s   **7.** $E_0 = 17.4$ V/m; $H_0 = 0.05$ A/m   **9.** (a) 1.46;
(b) 2.48×10^8 m/s;   (c) 248 m;   (d) 6.46×10^{-12} J/m^3;   (e) 1.60×10^{-3} W/m^2 **11.** $[E(a) - E_0]/E_0 = 3.60 \times 10^{-11}$.
The radiation effects are negligible. Radiation sets in if $R = 10\,\Omega$.   **13.** (a) $E_0 = 27.5$ V/m, $H_0 = 7.28 \times 10^{-2}$ A/m;
(b) 3.33×10^{-9} J/m^3;   (c) 0.719 J

CHAPTER 39

1. 1.76×10^{-25} W   **3.** (a) 8.5×10^{-3} A;   (b) 4.44×10^{-4} W;   (c) $3.08 \times 10^6\,\Omega$ **5.** 276 μs   **9.** (a) 469 Ω;   (b) 93.8 V;
(c) 9.38 W   **13.** (a) 9.5×10^{-8} W;   (b) 2.45×10^{-4} V/m   **15.** (a) $I(x, t) = \dfrac{CV_0 \omega}{k} \cos(kx - \omega t)$;   (b) 74.8 Ω

CHAPTER 40

1. 10 μs   **3.** 3.13×10^8 m/s   **7.** $\theta_2 = 35.2°$; $\theta_3 = 30.3°$ **9.** 1.461   **13.** (a) -0.415;   (b) 0.172   **19.** 59.3%
25. (a) 28.68 μm;   (b) 0.4109 μm;   (c) 4.59 rad   **29.** (a) 1.59

CHAPTER 41

3. $(3, 4, -5)$, $(3, -4, 5)$, $(3, -4, -5)$ **5.** (a) $s' = 33.3$ cm, $m = -\frac{1}{3}$, inverted real;   (b) $s' = 87.5$ cm, $m = -\frac{5}{2}$, inverted real;   (c) $s' = -37.5$ cm, $m = -\frac{5}{2}$, erect virtual;   (d) $s' = -2.17$ cm, $m = -1.09$, erect virtual;   (e) $s' \rightarrow -s$, $m \rightarrow 1$
7. $s = 27$ cm, inverted real; $s = 9$ cm, erect virtual   **9.** $s' = 12$ cm, $m = \frac{1}{3}$ **11.** 2   **15.** 0.050 cm from edge of sphere that is opposite source; real image   **17.** 28.8 cm   **21.** $f = 20/3$ cm, object-to-screen distance $= 30$ cm, $m = 2, \frac{1}{2}$
23. 120 cm   **25.** 8.1°   **27.** 12 cm, 42 cm behind the closer mirror; 15 cm, 39 cm behind the other mirror
29. 2.25 m   **31.** 10.7 cm from point at which axis of lens passes through hemispheric edge of lens   **33.** (a) 20.6 cm;
(b) 11.5 cm   **37.** 0.006, 0.012

CHAPTER 42

3. 2.53 rad   **5.** (b) 0.86 mm from central maximum   **7.** (a) 4.86 cm from top edge of film;   (b) 72.4 nm, 117.2 nm;
(c) 0.497″ of arc   **9.** 113 nm   **11.** 0.443 mm   **13.** 634 nm, orange   **15.** 1.000293; air   **17.** (a) 5 ns;   (b) 1.5 m
19. 1.54 mm   **21.** 325 nm   **25.** 673 nm   **27.** 5.99 Å

CHAPTER 43

1. (a) 0.281 mm;   (b) 6.67 mm, 9.33 mm   **7.** seven   **9.** (a) 412 nm;   (b) 0.183 Å   **11.** 25.8″ of arc   **13.** 0.116 Å
15. 563 nm   **17.** 0.225 mm   **19.** 25.8 cm   **27.** $15.8\sqrt{n}$ cm;   15.8 cm, 22.3 cm, 27.4 cm, 31.6 cm, 35.3 cm, 38.7 cm

CHAPTER 44

1. (a) linearly polarized along x-axis;   (b) linearly polarized along $\hat{\mathbf{i}} \pm \hat{\mathbf{j}}$;   (c) circular polarization; left for $+$, right for $-$;   (d) elliptical polarization; left for $+$, right for $-$ **9.** 31.6°   **11.** 355 nm, 396 nm   **13.** (b) 8.56×10^{-7} m
15. 17   **19.** (a) left circular;   (b) linearly polarized   **21.** 5.82°

CHAPTER 45

1. virtual, erect, 5.5 cm from left edge of lens   **3.** $f = 21.4$ cm; focal points are 14.3 cm to left of left lens, and 17.1 cm to right of right lens; principal points are 7.1 cm to right of left lens, and 4.3 cm to left of right lens.   **5.** (a) 1.11 cm;
(b) -45.0 **9.** $M = 15$ **11.** (a) $M = 5$;   (b) 12 cm   **13.** $\Delta = 0.263$ cm   **15.** $f = 157$ mm, $s = 16.0$ cm
17. $\theta = 17°$ **19.** 3 diopters   **21.** $f = 8.17$ cm; principal point is 0.17 cm to right of lens; focal point is 4.67 cm to right of right edge of glass plate.   **23.** $f_\alpha = f_\rho = 5.002$ cm   **29.** $d = 4.98$ cm, 123 m   **31.** (a) 2 diopters;   (b) 0.091 mm;
(c) yes, 0.23 mm   **33.** (a) -2.72 diopters;   (b) 0.11 mm

CHAPTER 46

1. 2.26 eV   **3.** (a) 5800 K;   (b) 500 nm   **7.** λ_{max} (Cu) $= 279$ nm, λ_{max} (Cs) $= 687$ nm   **9.** λ_{min} (45 kV) $= 0.0276$ nm, λ_{min} (25 kV) $= 0.0498$ nm   **11.** $\lambda' = 0.00743$ nm   **15.** (a) 0.0123 nm;   (b) 9.10×10^{-15} m;   (c) 1.05×10^{-34} m;
(d) 9.95×10^{-40} m;   (e) 3.72×10^{-63} m   **17.** 3.58×10^{-13} m
19. (b) E (min, macro) $= 5.28 \times 10^{-32}$ J $= 3.30 \times 10^{-13}$ eV, E (min, micro) $= 7.72 \times 10^{-27}$ J $= 4.83 \times 10^{-8}$ eV;
(c) $E_{min} = 5.28 \times 10^{-20}$ J $= 0.330$ eV   **21.** $P_{net} = 79.9$ W   **23.** $\sigma = 5.63 \times 10^{-8}$ W/(m²K⁴)
27. (a) $h/e = 4.14 \times 10^{-15}$;   (b) $W_0 = 2.38$ eV   **29.** 0.317 nm   **31.** 0.380 mm

CHAPTER 47

1. $r = 4.55 \times 10^{-14}$ m   **3.** $E = 54.4$ eV   **11.** $\lambda = 1.19$ nm   **13.** $\Delta E = 0.0379$ eV   **15.** (a) $r = 3.49 \times 10^{-15}$ m;
(b) $\Delta E = 14.18$ MeV   **17.** $E_1 = 4.23 \times 10^{-97}$ J, $\rho_1 = 1.20 \times 10^{29}$ m   **19.** (c) $\Delta E = 113$ eV

INDEX

SI UNITS*

QUANTITY	NAME	SYMBOL (Alternate Expression)
SI Base Units		
Length	meter	m
Mass	kilogram	kg
Time	second	s
Electric current	ampere	A
Temperature	kelvin	K
Amount of a substance	mole	mol
SI Derived Units with Special Names		
Frequency	hertz	$Hz \ (s^{-1})$
Force	newton	$N \ (kg \cdot m/s^2)$
Pressure	pascal	$Pa \ (N/m^2)$
Energy, work	joule	$J \ (N \cdot m)$
Power	watt	$W \ (J/s)$
Electric charge	coulomb	$C \ (A \cdot s)$
Electric potential	volt	$V \ (W/A)$
Capacitance	farad	$F \ (C/V)$
Electric resistance	ohm	$\Omega \ (V/A)$
Conductance	siemens	$S \ (A/V)$
Magnetic flux	weber	$Wb \ (V \cdot s)$
Magnetic field strength (flux density)	tesla	$T \ (Wb/m^2)$
Inductance	henry	$H \ (Wb/A)$
SI Derived Units Expressed by Means of Special Names		
Viscosity	pascal·second	$Pa \cdot s$
Surface tension	newton per meter	N/m
Specific heat capacity	joule per kilogram·meter	$J/kg \cdot m$
Entropy	joule per kelvin	J/K
Thermal conductivity	watt per meter·kelvin	$W/m \cdot K$
Electric field strength	volt per meter	V/m
Units to be Used with SI System Temporarily		
(Length)	ångstrom	$Å \ (10^{-10} \ m)$
(Pressure)	atmosphere	$atm \ (1.01 \times 10^5 \ Pa)$
(Radioactive disintegration rate)	curie	$Ci \ (3.7 \times 10^{10} \ s^{-1})$
Units Used with SI System whose Values in SI Units Are Obtained from Experiment		
(Energy)	electronvolt	$eV \ (1.602 \times 10^{-19} \ J)$
(Mass)	unified atomic mass unit	$u \ (1.66 \times 10^{-27} \ kg)$
(Length)	astronomical unit	$A.U. \ (1.50 \times 10^{11} \ m)$

---

*As recommended by the General Conference of Weights and Measures on the International System of Units (1974).